ASTRONOMY AND ASTROPHYSICS ABSTRACTS

A Publication of the Astronomisches Rechen-Institut Heidelberg
Member of the Abstracting Board
of the International Council of Scientific Unions

Volume 21
Literature 1978, Part 1

Edited by
S. Böhme U. Esser W. Fricke I. Heinrich
D. Krahn L. D. Schmadel G. Zech

Springer-Verlag Berlin Heidelberg GmbH 1978

Astronomisches Rechen-Institut Heidelberg
Director: Professor Dr. Walter Fricke

Astronomy and Astrophysics Abstracts
Editors-in-Chief: Ute Esser, Dr. Lutz D. Schmadel

Astronomy and Astrophysics Abstracts is prepared
under the auspices of the International Astronomical Union

ISBN 978-3-662-12315-7   ISBN 978-3-662-12313-3 (eBook)
DOI 10.1007/978-3-662-12313-3

Originally published by Astronomisches Rechen-Institut Heidelberg in 1978.

Softcover reprint of the hardcover 1st edition 1978

2153/3130-543210

Preface

Astronomy and Astrophysics Abstracts, which has appeared in semi-annual volumes since 1969, is devoted to the recording, summarizing and indexing of astronomical publications throughout the world. It is prepared under the auspices of the International Astronomical Union (according to a resolution adopted at the 14th General Assembly in 1970).

Astronomy and Astrophysics Abstracts aims to present a comprehensive documentation of literature in all fields of astronomy and astrophysics. Every effort will be made to ensure that the average time interval between the date of receipt of the original literature and publication of the abstracts will not exceed eight months. This time interval is near to that achieved by monthly abstracting journals, compared to which our system of accumulating abstracts for about six months offers the advantage of greater convenience for the user.

Volume 21 contains literature published in 1978 and received before August 10, 1978; some older literature which was received late and which is not recorded in earlier volumes is also included.

We acknowledge with thanks contributions to this volume by Dr. J. Bouška, Prague, who surveyed journals and publications in Czech and supplied us with abstracts in English, and by Prof. P. Brosche, Bonn, who supplied us with literature concerning some border fields of astronomy.

Dr. Wilfried Hofmann joined our staff in July 1978. His valuable contributions to this volume are acknowledged. We express our warmest thanks again to Ms. Helga Ballmann, Ms. Monika Betz, Ms. Lore Kiefert, Ms. Dimitra Roussou, who typed the text of this volume on IBM 72 Composers and compiled the pages from abstract slips in a perfect form for offset reproduction. We are indebted to Ms. Elisabeth Feigenbutz for punching material for the author index and for the subject index, which finally were printed with a TN chain on a 1403 IBM high-speed printer. Finally, we have to thank Mr. Gernot Klein and Mr. Werner Sanns who supported our task by careful proofreading.

Heidelberg, September 1978

Siegfried Böhme
Ute Esser
Walter Fricke
Inge Heinrich

Dietlinde Krahn
Lutz D. Schmadel
Gert Zech

Contents

Positional Astronomy, Celestial Mechanics

Space Research

Theoretical Astrophysics

Sun

X   Contents

Introduction

Astronomical bibliographies

Astronomy and Astrophysics Abstracts begins documentation and abstracting from the year 1969. For information on astronomical literature before this date consultation of one of the following bibliographies is suggested:
(1) J. J. de Lalande, Bibliographie Astronomique, Paris 1803 (this work covers the time from 480 B. C. to the year 1803, VIII + 966 pages).
(2) J. C. Houzeau, A. Lancaster, Bibliographie générale de l'astronomie, Volume I (in two parts), Bruxelles 1882, 1887, Volume II, Bruxelles 1889. The complete title of Volume II is "Bibliographie générale de l'astronomie ou catalogue méthodique des ouvrages, des mémoires et des observations astronomiques, publiés depuis l'origine de l'imprimerie jusqu'en 1880". A new edition of these volumes was prepared by D. W. Dewhirst in 1964.
(3) Bibliography of Astronomy, 1881 - 1898. The literature of this period was recorded on standard slips by the Observatoire Royal de Belgique. From the material (some 52,000 items) a microfilm version was produced by University Microfilms Limited, Tylers Green, High Wycombe, Buckinghamshire, England, in 1970.
(4) Astronomischer Jahresbericht, 1899 gegründet von Walter Wislicenus, herausgegeben vom Astronomischen Rechen-Institut in Heidelberg (formerly in Berlin), Verlag W. de Gruyter, Berlin. For the period from 1899 to 1968 sixty-eight volumes were published, each of which, in general, covers the literature of one year.
(5) Bulletin Signalétique – Section 120: Astronomie, Physique Spatiale, Géophysique. Published by Centre de Documentation du Centre National de la Recherche Scientifique, Paris. This publication is a continuation of "Bibliographie Mensuelle de l'Astronomie" founded in 1933 by the Société Astronomique de France. The publication is continued.
(6) Referativnyj Zhurnal. Founded in 1953 and published by Vsesoyuznyj Institut Nauchnoj i Tekhnicheskoj Informatsii, Akademiya Nauk, Moskva. The publication is continued.

Concept of Astronomy and Astrophysics Abstracts

This abstracting service aims to present a comprehensive documentation of the literature in all fields of astronomy and astrophysics and their border fields. It appears in semi-annual volumes. Two of these volumes cover the literature of one calendar year. The half-yearly period of issue is regarded as an optimal period for summarizing papers into subject categories and for the presentation of abstracts as quickly as possible after the publication of the original literature.

The recording, summarizing and indexing of astronomical publications of the year 1978 received from January to September 1978 are subjects of Volume 21. It also records a number of papers issued before 1978 but received within this period.
The main characteristics of the concept of Astronomy and Astrophysics Abstracts may be summarized as follows:
(1) The subdivision of astronomy and its border fields into subject categories is facilitated by the fact that the astronomical objects appear to be particularly well suited for the formation of categories. It may be assumed that such subdivisions can be maintained for a long period. Experience shows, however, that progress in research might imply minor changes in the classification scheme.
(2) Each paper has been classified into one of 109 numbered subject categories and given a serial number within the category. In this way each item is numbered by six figures: the first three indicate the number of the category, the following three the serial number within the category. Reference to an abstract in Volume 1 is indicated by "01" before the number of the category; for example: 01.074.028, denotes Volume 1, category 074, abstract 028.
A paper might be classified into more than one category. In this case, its abstract is placed only in one category, whereas in the other categories only cross references are given. These are listed at the end of each category.
(3) Authors' abstracts are used whenever possible. Popular articles are not abstracted.
(4) If possible, titles of papers and abstracts are given in English. A special reference is made to titles which we have not taken in the original language.

Transliteration scheme for the Russian alphabet

The transliteration of the Russian alphabet in use in Astronomy and Astrophysics Abstracts is presented here.

А	а	a	Р	р	r
Б	б	b	С	с	s
В	в	v	Т	т	t
Г	г	g	У	у	u
Д	д	d	Ф	ф	f
Е	е	e	Х	х	kh
Ё	ё	e	Ц	ц	ts
Ж	ж	zh	Ч	ч	ch
З	з	z	Ш	ш	sh
И	и	i	Щ	щ	shch
Й	й	j	Ъ	ъ	"
К	к	k	Ы	ы	y
Л	л	l	Ь	ь	'
М	м	m	Э	э	eh
Н	н	n	Ю	ю	yu
О	о	o	Я	я	ya
П	п	p			

This transliteration was recommended by the Abstracting Board of the International Council of Scientific Unions in 1969. It corresponds essentially to the transliteration proposed by the Academy of Sciences, Moscow, which is used by the Referativnyj Zhurnal. In this case the letters can be read and printed by usual data processing machines.
If the names of Russian authors in the literature are transliterated in a different scheme, we present the names as they are given in the references cited and in addition in brackets according to our transliteration table.

Sources of information

The majority of sources of information for this volume is given in section **001 Periodicals** and in section **008 Observatories, Institutes**. Section 001 records 634 periodicals indicating full titles and publishers. It may be noted that the titles of the periodicals are given in the original languages, and that Russian titles have been transliterated applying the transliteration scheme given above. Section 008 records 100 periodicals; these are publication series of observatories and astronomical institutes. Titles of the periodicals have been given following the recommendations of the "International List of Periodical Title Word Abbreviations" and its additions (see also **Abbreviations**, p. 3). In most cases they permit recognition of the full title without recourse to the key in section 001.

If other abstracting journals have been consulted in order to examine the degree of completeness of our service, we cite these papers and give reference to the abstracting service.

Author index and subject index

The subject category and the serial number have been used as a reference both in the author index and the subject index. These references are more precise than page references. They offer considerable advantages in indexing by means of data processing machines, and are more convenient for the user.

The author index of this volume contains 9245 names. A complete reference comprises six figures, three for the subject category and three for the serial number within the category. In the case of more than one reference to abstracts in one category, the number of the category is given only once and not repeated in the immediately following references. The total number of papers (some do not give names of authors) recorded in this volume amounts to 8300.

We consider the subject index as an approximation to an optimal index covering all fields of astronomy and astrophysics and their border fields. The assigning of one or more key words to a paper is, undoubtedly, a difficult task. Some journals have started giving key words together with the titles of papers. These key words are chosen by the authors themselves. Starting with Volume 18, the subject index was enlarged to a certain extent in order to provide a thesaurus of astronomical and astrophysical terms. This is done not only for the users' convenience, but also with the intention to propose the use of special key words to authors and publishers.

While each volume is scheduled to contain an author index and a subject index, the magnetic tapes containing the index information will be used to produce separate index volumes (authors and subjects) at intervals of five years.

The sorting program for the author and subject indexes is based on the IBM SORT/MERGE Program. This SORT-program sorts blank before hyphen (−) and before letters. Apostrophes are ignored by a special routine.

Examples: a) De Laeter

:

Deacon

:

DeLaeter

b) A Stars

:

Absolute Magnitudes

c) Solar X Rays

:

Solar-Terrestrial Relations

d) Boehm-Vitense

Boehme

The introduction of small and capital letters in the layout caused some difficulties. Special programs had to code the capital letters into small ones. For the layout, a TN chain for a 1403 IBM high-speed printer was used. All the programs were written by Dr. H. Scholl in PL/I. The computations and printing were carried out on an IBM 360/44.

The users are requested to inform us on spelling errors within the author and subject indexes in order to assist us in eliminating mistakes in future cumulative indexes.

Abbreviations

Abbreviations used in Astronomy and Astrophysics Abstracts are primarily based on the 'International List of Periodical Title Word Abbreviations', prepared for the UNISIST/ICSU-AB Working Group on Bibliographic Descriptions (1970).

A.A.B.	Associazione Astrofili Bolognesi	Ber.	Bericht–
Aarg.	Aargang	Bibl.	Bibliot–
AAS	American Astronomical Society	Bibliogr.	Bibliograf–, Bibliograph–
AAVSO	American Association of Variable Star Observers	BIH	Bureau International de l'Heure (Paris)
		Bimest.	Bimestr–
Abh.	Abhandlung–	Bl.	Blatt, Blätter
Abstr.	Abstract–	Bol.	Boletin
Abt.	Abteilung–	Boll.	Bolletino
Acad.	Academi–, Academy	Bul.	Buleten–, Buletin–, Bulten
Accad.	Accademi–	Bull.	Bulletin–, Bullettino
Act.	Active, Activit–	Bur.	Bureau–
Adm.	Administr–	Byul.	Byuleten–, Byuletin–
Adv.	Advanc–	Byull.	Byulleten–
Aehron.	Aehronomi–	C.R.	Comptes Rendus
Aeron.	Aeronom–	Cah.	Cahier–
Aeronaut.	Aeronauti–	Calif.	California
Aerosp.	Aerospace	Cas.	Casopis
AG	Astronomische Gesellschaft	Cent.	Center–, Central, Centrale, Centrally, Centre
AIAA	American Institute of Aeronautics and Astronautics	Cercet.	Cercetary
AJB	Astronomischer Jahresbericht	Chem.	Chemi–
Akad.	Akadem–	Chim.	Chimi–
Ala.	Alabama	Chron.	Chronic–, Chronik, Chronique
Alm.	Almanac–, Almanak–	Chronom.	Chronometr–
Amat.	Amateur–	Cie.	Compagnie
An.	Anais, Anale–, Anali–, Anals	Cienc.	Ciencia–
Anal.	Analis–, Analit–, Analys–, Analyt–	Cient.	Cientific–
Angew.	Angewandt–	Circ.	Circolar–, Circolo, Circolaire–, Circular–, Circulo
Ann.	Annaes, Annal–	Cirk.	Cirkulaer–
Annu.	Annu–	Cl.	Clasa, Classe–
Anst.	Anstalt	Co.	Companies, Company
Anu.	Anual–, Anuar–	Coll.	College
Anz.	Anzeiger	Collect.	Collect–
Appl.	Applied	Colloq.	Colloqui–
Arb.	Arbeit	Colo.	Colorado
Arch.	Archiv–	Comet.	Cometary
Årg.	Årgang	Commentat.	Commentat–
Ariz.	Arizona	Commun.	Communica–
Ark.	Arkiv–	Comput.	Computation, Computer–, Computing
Arkh.	Arkhiv–	Comun.	Comunica–
Artif.	Artifici–	Conf.	Conferen–
ASA	Astronomical Society of Australia	Congr.	Congres–
Asoc.	Asocia–	Conn.	Connecticut
ASP	Astronomical Society of the Pacific	Contract.	Contract–
ASSA	Astronomical Society of Southern Africa	Contrib.	Contribu–
Assem.	Assembl–	Cosm.	Cosmic–
Assoc.	Associ–	Cosmochim.	Cosmochimi–
Assoz.	Assozi–	COSPAR	Committee on Space Research
Astrofis.	Astrofisic–	Crystallogr.	Crystallograph–
Astrofiz.	Astrofizi–	CSIRO	Commonwealth Scientific and Industrial Research Organization
Astrometr.	Astrometr–		
Astron.	Astronom–	Cult.	Cultur–, Cultuur
Astronaut.	Astronauti–, Astronauty–	Curr.	Current
Astrophys.	Astrophys–	D.C.	District of Columbia
ASV	Astronomical Society of Victoria	DDR	Deutsche Demokratische Republik
ASWA	Astronomical Society of Western Australia	Del.	Delaware
		Dep.	Departament, Département, Department
At.	Atom–	Dev.	Development–, Développement–
Atmos.	Atmosf–, Atmosph–	Diss.	Disserta–
BAA	British Astronomical Association	Div.	Divis–
Bayer.	Bayerisch–	Doc.	Document–
Beitr.	Beitrag, Beiträge	Dok.	Dokument–
Beob.	Beobacht–	Dokl.	Doklad–
		Ehksp.	Ehksperiment–
		Eidg.	Eidgenössisch–
		Eksp.	Eksperiment–
		Electron.	Electroni–

Eng.	Engineer–
Environ.	Environment–
Equip.	Equipement, Equipment
Ergeb.	Ergebnis–
ESA	European Space Agency
ESO	European Southern Observatory
ESRO	European Space Research Organization
Eval.	Evaluation–
Exp.	Experiment–
Extraterr.	Extraterrestr–
F. R. Germany	Federal Republic of Germany
Fac.	Facolt–, Faculd–, Facult–
Fak.	Fakult
Fasc.	Fascicul–
Fenn.	Fenni–
Fis.	Fisic–, Fisik–
Fiz.	Fizic–, Fizik–, Fizyk--
Fla.	Florida
Fluid.	Fluidi–
Fond.	Fondation–, Fondazione
Fortschr.	Fortschritt–
Fotogr.	Fotograf–
Found.	Foundation–
Freq.	Frequen–
Fundam.	Fundamenta–
Fys.	Fysik–, Fysisch, Fysisk–
Fyz.	Fyzik–
G.	Giornale
Ga.	Georgia
Gaz.	Gazeta, Gazette
Gazz.	Gazzetta
Gen.	General
Geochem.	Geochem–
Geochim.	Geochim–
Geod.	Geodaes–, Geodaet–, Geodes–, Geodet–, Geodez–
Geofis.	Geofis–
Geofiz.	Geofiz–
Geofys.	Geofys–
Geogr.	Geograf–, Geograph–
Geokhim.	Geokhim–
Geol.	Geolog–, Geolosk–
Geomagn.	Geomagneti–
Geophys.	Geophys–
Ges.	Gesellschaft
Gl.	Glavno–
Glas.	Glasnik
Gos.	Gosudarst–
Gov.	Government–
Grenzgeb.	Grenzgebiet–
GSFC	Goddard Space Flight Center
H. M.	Her Majesty's , His Majesty's
Handb.	Handbook, Handbuch
Her.	Herald–
Hist.	History
Hochsch.	Hochschule
Hoegsk.	Hoegskol–
HR-diagram	Hertzsprung-Russell diagram
Hydrogr.	Hydrograf–, Hydrograph–
IAF	International Astronautical Federation
IAU	International Astronomical Union
IBM	International Business Machines Corporation
ICSU	International Council of Scientific Unions
ICSU-AB	International Council of Scientific Unions – Abstracting Board
IEEE	Institute of Electrical and Electronics Engineers
Ill.	Illinois
Inc.	Incorporated
Ind.	Industr–
Inf.	Informat–, Informaz–, Informe–
Ing.	Ingenieur
INIS	International Nuclear Information System
Inst.	Institut–, Instytut–
Instn.	Institution
Instrum.	Instrument–
Int.	Internationa–, Internazional–
Inter.	Intérieur–, Interior
Interplanet.	Interplanetary
Intez.	Intezet–
Ionos.	Ionosfer–, Ionospher–
Iskusstv.	Iskusstvenn–
Issled.	Issledovan–
Ist.	Istitut–
Izd.	Izdatel–
Izv.	Izvesti–
J.	Joernaal–, Jornal–, Journal–
Jaarb.	Jaarboek–
Jahrb.	Jahrbuch, Jahrbücher
Jahresber.	Jahresbericht–
Jahresschr.	Jahresschrift
Jahrg.	Jahrgang
JPL	Jet Propulsion Laboratory
K.	Königlich–, Koninklijk–, Kunglig–
Kans.	Kansas
Kartogr.	Kartograf–
Kernforsch.	Kernforschung
Kernphys.	Kernphysik–
Khem.	Khemyi–
Khim.	Khimi–
Kim.	Kimija–, Kimya
Kl.	Klass–
Kolloq.	Kolloquium–
Komet.	Kometnyj
Komm.	Kommission–
Konf.	Konfer–
Kongr.	Kongress
Kosm.	Kosmich–
Kosmog.	Kosmogon–
Kozp.	Kozponti
KPNO	Kitt Peak National Observatory
Kut.	Kutato
Ky.	Kentucky
La.	Louisiana
Lab.	Laborato–
Lett.	Letter–, Lettra, Lettre
Libr.	Librair–, Librar–
Mag.	Magasin, Magazin–
Magn.	Magneti–, Magnitn–
Mass.	Massachusetts
Mat.	Matemaat–, Matemat–
Mater.	Material–
Math.	Mathemat–
Md.	Maryland
Meas.	Measur–
Mec.	Mecani–
Mech.	Mechani–
Medd.	Meddelande–, Meddelelse
Meded.	Mededeeling–, Mededeling–
Mekh.	Mekhani–
Mem.	Memento–, Memoir–, Memori–, Memory–, Memuary
Memo.	Memorand–
Mens.	Mensile, Mensual–, Mensuel–
Messtech.	Messtechni–
Meteorol.	Meteorolog–
Mich.	Michigan
Micromec.	Micromecaniq–
Miner.	Mineral, Minerale–, Minerali–
Mineral.	Mineralog–
Minn.	Minnesota
Miss.	Mississippi
MIT	Massachusetts Institute of Technology

Mitt.	Mitteilung–	Prov.	Provinc–, Provints–, Provinz–
Mo.	Missouri	Pubbl.	Pubblicazion–
Mod.	Modern–	Publ.	Publicac–, Publicas–, Publicat–,
Mol.	Molecul–, Molekul–		Publikas–, Publikat–
Mon.	Monat, Monatlich–, Month–	Q.	Quarterly
Monogr.	Monograph–	Quant.	Quantit–
Mont.	Montana	R.	Royal
MPI	Max-Planck-Institut	R. I.	Rhode Island
N. C.	North Carolina	Radiat.	Radiati–
N. D.	North Dakota	Radioact.	Radioactiv–, Radioaktiv–
N. H.	New Hampshire	Radioisot.	Radioisotop–
N. J.	New Jersey	Rap.	Raport–
N. M.	New Mexico	Rapp.	Rapport–
N. Y.	New York	RAS	Royal Astronomical Society
Nablyud.	Nablyudeni–	Rec.	Record–
Nac.	Nacion–	Rech.	Recherche–
Nachr.	Nachricht–	Ref.	Referat–, Reference–, Referieren
NASA	National Aeronautics and Space	Relat.	Related, Relation–
	Administration	Relativ.	Relativit–
Nat.	Natur–	Rend.	Rendicont–
Natl.	National–	Rep.	Report–
Naturforsch.	Naturforsch–	Repr.	Reprint–
Naturwiss.	Naturwissenschaft–	Repub.	Republi–
Nauchn.	Nauchny–	Res.	Research–
Nauk.	Nauka, Naukite, Naukov–, Naukow–	Result.	Resultad–, Resultat–
Naut.	Nautic–	Rev.	Review–, Revisio, Revista, Revue–
Nav.	Naval–	Rezul't.	Rezul'tat–
Navig.	Navigat–	Ric.	Ricerca, Ricerche
Naz.	Nazion–	Riv.	Rivist–
Nebr.	Nebraska	Rundsch.	Rundschau
Nev.	Nevada	S. C.	South Carolina
Newsl.	Newsletter–	S. D.	South Dakota
Not.	Notationes, Notic–, Notise–, Notizi–	SAF	Société Astronomique de France
Nouv.	Nouveau–, Nouvell–	SAI	Società Astronomica Italiana
Nov.	Novoe	Samml.	Sammlung–
Nucl.	Nucléaire–, Nuclear–, Nucl–	SAO	Smithsonian Astrophysical Observatory
Nukl.	Nukle–	SAS	Société Astronomique de Suisse
Numer.	Numeri–	Satell.	Satellite
O-va	Obshchestva	Sb.	Sbornik–
O-vo	Obshchestvo	Schr.	Schrift–
Obs.	Observ–	Schriftenr.	Schriftenreihe
Obz.	Obzor–	Sci.	Scienc–, Scient–, Scienz–
Okla.	Oklahoma	Scr.	Scripta, Scritt–
Opt.	Optic–, Optik–, Optique	Secc.	Seccion–
Oreg.	Oregon	Sect.	Secti–
Oss.	Osserva–	Seer.	Seeria
Pa.	Pennsylvania	Sekc.	Sekci–, Sekcj–
Paleontol.	Paleontolog–	Sekt.	Sektion–, Sektor–
Pap.	Paper–, Papier	Sekts.	Sektsi–
Part.	Particle	Sel.	Seleccion–, Select–, Selek–, Selezione
Perem.	Peremenn–	Selsk.	Selskab–, Selskap–
Period.	Periodi–	Semin.	Séminair–, Seminar–
Petrol.	Petrolog–	Sep.	Separat–
Philos.	Philosoph–	Ser.	Seria–, Serie–, Seriya
Photogr.	Photograf–, Photograph–	Serv.	Servic–, Serviz–
Photogramm.	Photogrammetr–	Sess.	Sessi–
Photom.	Photometr–	Signal.	Signalétique–
Phys.	Physic–, Physik–, Physique–, Physisch–	Simp.	Simpoz–
Pict.	Picture–	Sitzungsber.	Sitzungsbericht–
Planet.	Planetary	Skr.	Skrift–
Pr.	Prac–	Soc.	Sociedad–, Societ–
Prelim.	Prelimin–	Sol.	Solar
Prepr.	Preprint	Soln.	Solnechn–
Prib.	Pribor–	Sonderdr.	Sonderdruck–
Prikl.	Prikladnoj	Soobshch.	Soobshchen–
Prir.	Prirodn–	South .	Southern
Prirodved.	Prirodoved–	Spacecr.	Spacecraft
Probl.	Problem–	Spat.	Spatial–
Proc.	Proceedings	Spec.	Special–
Prod.	Prodott–, Produc–, Produkt,	Spectrosc.	Spectroscop–
Prog.	Progres–	Spectrosk.	Spectroskop–
Propag.	Propagation	Spets.	Spetsial–

Spez.	Spezial–, Speziell–	UK	United Kingdom
SSR	Sovetskaya Sotsialisticheskaya Respublika	Umsch.	Umschau
		UN	United Nations
SSSR	Soyuz Sovetskikh Sotsialisticheskikh Respublik	Univ.	Universidad–, Universit–, Univerzitet–
		US	United States
St.	Saint–, Sankt–, Sant–	USA	United States of America
– St.	– Straße, Street	USSR	Union of Soviet Socialist Republics
Stand.	Standard–, Standart–	Va.	Virginia
Sternw.	Sternwarte–	Var.	Various
Stiint.	Stiintific–	Ver.	Verein–, Verenig–
Stn.	Station, Stazione	Veränderl.	Veränderlich–
Stud.	Studia, Studie–, Studii	Verh.	Verhandl–
Supl.	Suplement–, Supliment–	Vermess.	Vermessung–
Suppl.	Supplement–	Vermessungswes.	Vermessungswesen
Surv.	Survey–	Veröff.	Veröffentlich–
Symp.	Sympos–, Sympoz–	Vesn.	Vesnik
Syst.	System–	Vestn.	Vestnik
Sz.	Szemle	Vetensk.	Vetenskap–
Teach.	Teacher–, Teaching	Vidensk.	Videnskab–, Videnskap
Tec.	Tecni–	Vierteljahresschr.	Vierteljahresschrift–
Tech.	Techni–	Vierteljahrsschr.	Vierteljahrsschrift–
Technol.	Technolog–	VLB	Very Long Baseline
Tecnol.	Tecnolog–	Volcanol.	Volcanolog–
Teh.	Tehnic–, Tehnika, Tehnisk–	Vopr.	Vopros–
Tehnol.	Tehnolog–, Tehnolosk–	Vortr.	Vorträge
Tek.	Tekni–	Vses.	Vsesoyuzn–
Tekh.	Tekhni–	Vt.	Vermont
Tekhnol.	Tekhnolog–	Vyp.	Vypusk–
Teknol.	Teknolog–	Vyssh.	Vyssh–
Telesc.	Telescop–	Vyzk.	Vyzkum–
Telev.	Television–	W. Va.	West Virginia
Tenn.	Tennessee	Wash.	Washington
Teor.	Teoret–, Teori–	West.	Western
Terr.	Terrestr–	Wet.	Wetenschap–, Wetenskap–
Test.	Testing	Wis.	Wisconsin
Tex.	Texas	Wiss.	Wissenschaft–
TH	Technische Hochschule	Wyo.	Wyoming
Theor.	Theoret–, Theori–	Yad.	Yadern–
Tidschr.	Tidschrift–	Z.	Zeitschrift–
Tidskr.	Tidskrift–	ZA	Zero Age
Tidsskr.	Tidsskrift–	ZAED	Zentralstelle für Atomkernenergie-Dokumentation
Top.	Topic–		
Tr.	Trudy	Zap.	Zapisk–, Zapyisk–
Trans.	Transactions, Transazione	Zaved.	Zaveden–
Tsentr.	Tsentral–	Zent.	Zentral
Tsirk.	Tsirkulyar–	Zentralbl.	Zentralblatt
TU	Technical University	Zesz.	Zeszyt
Uch.	Uchen–	Zh.	Zhurnal–
Uchebn.	Uchebn–	Zirk.	Zirkular

Periodicals, Proceedings, Books, Activities

001 Periodicals

AAS Photo-Bull.
AAS (American Astronomical Society) Photo-Bulletin. Published by the Working Group on Photographic Materials. Produced by Eastman Kodak Co., Rochester, N.Y.

AAVSO Bull.
Bulletin of the American Association of Variable Star Observers, 187 Concord Avenue, Cambridge, Mass., 02138, U.S.A.

Acad. R. Belgique, Bull. Cl. Sci.
Académie Royale de Belgique, Bulletin de la Classe des Sciences (Koninklijke Academie van België, Mededelingen van de Klasse der Wetenschappen). 5ᵉ Série, Palais des Académies, Bruxelles.

Acta Astron.
Acta Astronomica. An international quarterly journal. Publisher: Polska Akademia Nauk, Komitet Astronomii (Polish Academy of Sciences, Committee of Astronomy), Warszawa – Wrocław.

Acta Astron. Sinica
Acta Astronomica Sinica. Published by Purple Mountain Observatory, Academia Sinica, Nanking, China.

Acta Astronaut.
Acta Astronautica. Journal of the International Academy of Astronautics. Publisher: Pergamon Press Inc., Elmsford, New York, U.S.A.; Pergamon Press Ltd., Oxford, England.

Acta Cienc. Indica
Acta Ciencia Indica. Pragati Prakashan, Begum Bridge, Post Box No. 62, Meerut-250001, India.

Acta Cosmologica
Acta Cosmologica. Published by Obserwatorium Astronomiczne Universytetu Jagiellońskiego, Kraków, Poland.

Acta Crystallogr. A
Acta Crystallographica, Section A: Crystal Physics, Diffraction, Theoretical and General Crystallography. Munksgaard International Booksellers and Publishers Ltd., 35 Norre Sogade, DK 1370 Kobenhavn K, Denmark.

Acta Electron.
Acta Electronica. 3 Avenue Descartes, BP 15,94 450 Limeil-Brevannes, (Val-de-Marne), France.

Acta Geophys. Polonica
Acta Geophysica Polonica. ARS Polona-Ruch, 00-068 Warszawa, Krakowskie Przedmiescie 7, P.O. Box 1001, Poland.

Acta Geophys. Sinica
Acta Geophysica Sinica. Chinese Academy of Sciences, Department of Geophysical Research. Published by Science Press, Peking, Peoples Republic of China.

Acta Phys. Acad. Sci. Hungaricae
Acta Physica Academiae Scientiarum Hungaricae. Postafiok 24, Budapest 502, Hungary.

Acta Phys. Austriaca
Acta Physica Austriaca. Springer-Verlag, A-1011 Wien, Molkerbastei 5, Postfach 367, Austria.

Acta Phys. Polonica B
Acta Physica Polonica B. ARS Polona-Ruch, Warszawa 1, P.O. Box 154, Poland.

Acta Phys. Sinica
Acta Physica Sinica. Chinese Academy of Sciences, Institute of Physics, Peking, People's Republic of China. [English translation in: Chinese J. Phys. (*USA*)].

Acta Phys. Slovaca
Acta Physica Slovaca. VEDA Publishing House of the Slovak Academy of Sciences, 895 30 Bratislava, Klemensova 27, Czechoslovakia.

Acta Polytech. III
Acta Polytechnica. Series III. Elektrotechnicka fakulta CVUT v Prace, Technicka ul. 2, Praha 6-Dejvice, Czechoslovakia.

Acta Tech. CSAV
Acta Technica Ceskoslovenska Akademie Ved. Academia, Publishing House of the Czechoslovak Academy of Sciences, Vodickova 40, 112 29 Praha 1, Czechoslovakia. John Benjamins N.V., Periodical Trade, Warmoesstraat 54, Amsterdam, Netherlands.

Acta Univ. Carolinae Math. Phys.
Acta Universitatis Carolinae, Mathematica et Physica. Administrace: Matematicko-fyzikálni fakulta University Karlovy, Praha.

Adv. Astron. Astrophys.
Advances in Astronomy and Astrophysics. Publisher: Academic Press, New York—London.

Adv. Phys.
Advances in Physics. Taylor & Francis Ltd., 10—14 Macklin Street London, WC2B 5NF, England.

Aerotec. Missili Spazio
L'Aerotecnica Missili e Spazio. Tamburini Editore S.P.A., Via Pascoli 55, 20133 Milano, Italy.

AIAA J.
AIAA Journal. A Publication of the American Institute of Aeronautics and Astronautics devoted to Aerospace Research and Development. Published by the American Institute of Aeronautics and Astronautics, New York, N.Y.

AIP Conf. Proc.
AIP Conference Proceedings. American Institute of Physics, 335 East 45th Street, New York, N.Y. 10017, USA.

Alta Freq.
Alta Frequenza.Ufficio Centrale AEI-CEI, Viale Monza 259, 20126 Milano, Italy.

American J. Phys.
American Journal of Physics. Published for the American Association of Physics Teachers by the American Institute of Physics, 335 East 47th Street, New York, N.Y. 10017, USA.

American Mineral.
American Mineralogist. Mineralogical Society of America, 1707 L Street, N.W., Washington, DC 20036, USA.

American Sci.
American Scientist. Society of Sigma XI, 345 Whitney Avenue, New Haven, CT 06510, USA.

An. Acad. Brasil. Cienc.
Anais da Academia Brasileira de Ciencias. Caixa Postal 229, ZC-00 Rio de Janeiro gb, Brazil.

An. Fis.
Anales de Física. Real Sociedad Española de Física y Química (Facultad de Ciencias), Ciudad Universitaria, Madrid-3, Spain.

An. Stiint. 'Al. I. Cuza' Iasi (Ser. Noua) I
Analele Stiintifice ale Universitatu 'Al. I. Cuza' din Iasi (Serie Noua), Sectiunea I Fizica. Calea 23 August, Iasi, Rumania.

Ann. Acad. Sci. Fennicae, Ser. A. VI
Annales Academiae Scientiarum Fennicae, Series A VI (Physica). Snellmaninkatu 9–11, 00170 Helsinki-17, Finland.

Ann. Françaises Chronom. Microméc.
Annales Françaises de Chronométrie et de Micromécanique, publication annuelle de l'Observatoire de Besançon, du Centre Technique de l'Industrie Horlogère et de la Société Française de Chronométrie et de Micromécanique. Rédaction et administration: Observatoire de Besançon. Publiées avec le concours du Centre National de la Recherche Scientifique et des organismes corporatifs.

Ann. Geofis.
Annali di Geofisica. Istituto Nazionale di Geofisica, Citta Universitaria, Via Ruggero Bonghi 11/B, 00184 Roma, Italy.

Ann. Géophys.
Annales de Géophysique. Service des Publications du CNRS, 15 Quai Anatole-France, 75700 Paris, France.

Ann. Inst. Henri Poincaré A
Annales de l'Institut Henri Poincaré, Section A (Physique Theorique). 11 Rue Pierre-Curie, Paris 5, France.

Ann. Obs. Astron. Météorol. Toulouse
Annales de l'Observatoire Astronomique et Météorologi-que de Toulouse. Publisher: Gauthier-Villars, Paris.

Ann. Physics
Annals of Physics. Academic Press Inc., 111 Fifth Avenue, New York, NY 10003, USA.

Ann. Physik
Annalen der Physik. 7. Folge. Publisher: Johann Ambrosius Barth, Salomonstr. 18B, Leipzig 701, German Democratic Republic.

Ann. Physique
Annales de Physique. Publisher: Masson et Cie., 120 Boulevard Saint-Germain, Paris 6, France.

Ann. Sci.
Annals of Science. Taylor & Francis Ltd., 10–14 Macklin Street, London, WC2B 5NF, England.

Ann. Soc. Sci. Bruxelles I
Annales de la Société Scientifique de Bruxelles. Série I: Sciences Mathématiques, Astronomiques et Physiques. Rue de Bruxelles 61, B 5000 Namur, Belgium.

Ann. Télécommun.
Annales des Télécommunications. Centre National d'Études des Télécommunications, 38 rue du Général Leclerc, 92 Issy-les-Moulineaux, France.

Ann. Tokyo Astron. Obs.
Annals of the Tokyo Astronomical Observatory. University of Tokyo, Mitaka, Tokyo, Japan.

Ann. Univ.-Sternw. Wien
Annalen der Universitäts-Sternwarte Wien. In Kommission bei Ferd. Dümmlers Verlag, Bonn.

Annu. Rep. Astron. Inst. Greece
Annual Reports of the Astronomical Institutes of Greece. Published by the Greek National Committee for Astronomy. Academy of Athens, Research Center for Astronomy and Applied Mathematics.

Annu. Rev. Astron. Astrophys.
Annual Review of Astronomy and Astrophysics. Publisher: Annual Reviews Inc., Palo Alto, California.

Antenna
L'Antenna. Via Monte Generoso 6/a, 20155 Milano, Italy.

Anz. Österreich. Akad. Wiss. Math.-Naturwiss. Kl.
Anzeiger. Österreichische Akademie der Wissenschaften. Mathematisch-Naturwissenschaftliche Klasse. Publisher: Springer-Verlag, Wien.

APL Tech. Dig.
APL Technical Digest.Applied Physics Laboratory, The John Hopkins University, 8621 Georgia Avenue, Silver Spring, MD 20910, USA.

Appl. Opt.
Applied Optics. A monthly publication of the Optical Society of America. Published for the Optical Society of America by the American Institute of Physics, 335 East 45th Street, New York, NY 10017, USA.

Appl. Phys.
Applied Physics. Springer-Verlag, Heidelberger Platz 3, D-1000 Berlin 33, F. R. Germany.

Appl. Phys. Lett.

Applied Physics Letters. American Institute of Physics, 335 East 45th Street, New York, N.Y. 10017, USA.

Appl. Sci. Res.
Applied Scientific Research. Martinus Nijhoff, Lange Voorhout 9, Den Haag, Netherlands.

Appl. Spectrosc.
Applied Spectroscopy. 428 East Preston Street, Baltimore, MD 21202, USA.

Appl. Spectrosc. Rev.
Applied Spectroscopy Reviews. Marcel Dekker Inc., 95 Madison Avenue, New York, NY 10016, USA.

Arch. Mech.
Archives of Mechanics (Archiwum Mechaniki Stosowanej). Polish Scientific Publishers, Swietokrzyska 21, Warszawa, Poland.

Arch. Ration. Mech. Anal.
Archive for Rational Mechanics and Analysis. Springer-Verlag, Heidelberger Platz 3, D 1000 Berlin 33, F. R. Germany.

Arch. Sci.
Archives des Sciences, éditées par la Société de Physique et d'Histoire Naturelle de Genève. Publisher: Imprimerie Kundig, Genève. Subscription address: Librairie Payot, Genève.

Ark. Astron.
Arkiv för Astronomi. Utgivet av Kungliga Svenska Vetenskapsakademien, Stockholm. Printed by Almqvist & Wiksell, Stockholm.

Ark. Geofys.
Arkiv för Geofysik. Kungliga Svenska Vetenskapsakademien, Stockholm. Printed by Almqvist & Wiksell, Stockholm.

Artif. Satell.
Artificial Satellites. Publication of Polish Scientific Institutions. Polish Academy of Sciences, National Committee of Geophysics and Geodesy, National Committee for Space Research, Warsaw. Space Research Centre, Pałac Kultury i Nauki 2301, 00-901 Warszawa, Poland.

Asoc. Argentina Astron. Bol.
Asociación Argentina de Astronomía. Boletin. Editor: Instituto Argentino de Radioastronomía, Provincia de Buenos Aires, Argentina. Printer: Talleres Gráficos "Renovación", La Plata, República Argentina.

Assoc. Veneta Oss. Stelle Variabili Bull.
Associazione Veneta Osservatori di Stelle Variabili, Bulletin. c/o Gruppo Astrofili di Padova, Corso Garibaldi 41, 35100 Padova, Italy.

Astrofiz. Issled. Izv. Spets. Astrofiz. Obs.
Astrofizicheskie Issledovaniya. Izvestiya Spetsial'noj Astrofizicheskoj Observatorii. Akademiya Nauk SSSR. Publishers: Izdatel'stvo "Nauka", Leningradskoe Otdelenie, Leningrad.

Astrofizika
Astrofizika. Izdatel'stvo Akademii Nauk Armyanskoj SSR, Erevan. [An English translation is published in "Astrophysics"].

Astrometr. Astrofiz.
Astrometriya i Astrofizika. Respublikanskij Mezhvedomstvennyj Sbornik. Akademiya Nauk Ukrainskoj SSR,

Glavnaya Astronomicheskaya Observatoriya. Naukova Dumka, Kiev.

Astron. Astrophys.
Astronomy and Astrophysics. A European Journal. Published by Springer-Verlag, Berlin – Heidelberg– New York.

Astron. Astrophys., Suppl. Ser.
Astronomy and Astrophysics. Supplement Series. A European Journal. Published by the Astronomical Institute Lausanne and Geneva Observatory, CH-1290 Sauverny, Switzerland.

Astron. Circ.
Astronomical Circular. Compiled by the editor section of Acta Astronomica Sinica, Purple Mountain Observatory, Nanking, China.
Edited by the Chinese Astronomical Society.

Astron. Her.
Astronomical Herald. Astronomical Society of Japan, Tokyo Astronomical Observatory, Oosawa Mitaka, Tokyo, Japan.

Astron. J.
The Astronomical Journal. Published for the American Astronomical Society by the American Institute of Physics, New York, N. Y. Editorial Office: Department of Astronomy, Columbia University, New York, N. Y.

Astron. Nachr.
Astronomische Nachrichten. Publisher: Akademie-Verlag, Berlin.

Astron. Rep.
The Astronomical Reports. Polish Amateur Astronomical Society. Polskie Towarzystwo Miłośników Astronomii, Kraków, Poland.

Astron. Schule
Astronomie in der Schule. Zeitschrift für die Hand des Astronomielehrers. Herausgegeben vom Verlag Volk und Wissen, Berlin. Redaktion: Sternwarte Bautzen.

Astron. Soc. Western Australia, Circ.
The Astronomical Society of Western Australia (Inc.), Circular.

Astron. Tidsskr.
Astronomisk Tidsskrift. Edited by Astronomisk Selskab, København; Norsk Astronomisk Selskap, Oslo; Svenska Astronomiska Sällskapet, Stockholm. Printed by John Griegs Boktrykkeri, Bergen.

Astron. Tsirk.
Astronomicheskij Tsirkulyar, izdavaemyj Byuro Astronomicheskikh Soobshchenij Akademii Nauk SSSR. Tipografiya Astrosoveta AN SSSR, Moskva.

Astron. Vestn.
Astronomicheskij Vestnik. Publishers: Izdatel'stvo "Nauka", Moskva.

Astron. Zh. Akad. Nauk SSSR
Astronomicheskij Zhurnal. Akademiya Nauk SSSR. Publishers: Izdatel'stvo "Nauka", Moskva. [An English translation is published in "Soviet Astronomy AJ"].

Astronautik
Astronautik. Organ der Hermann-Oberth-Gesellschaft e.V. Astronautik-Verlag, Druckerei H. Brandt, Delmen-

horst, F.R. Germany.

Astronomia
Astronomia. Periodico trimestrale dell'Unione Astrofili Italiani.

Astronomie
L'Astronomie et Bulletin de la Société Astronomique de France. Société Astronomique de France, Paris.

Astronomy
Astronomy. AstroMedia Corp., 757 North Broadway, Suite 204, Milwaukee, WI 53202, USA.

Astrophys. J.
The Astrophysical Journal. Published for the American Astronomical Society by the University of Chicago Press, Chicago, Illinois.

Astrophys. J., Lett.
The Astrophysical Journal. Letters to the Editors. Published for the American Astronomical Society by the University of Chicago Press, Chicago, Illinois.

Astrophys. J., Suppl. Ser.
The Astrophysical Journal. Supplement Series. Published for the American Astronomical Society by the University of Chicago Press, Chicago, Illinois.

Astrophys. Lett.
Astrophysical Letters. Published by NASA–Goddard Space Flight Center. Gordon and Breach Science Publishers Ltd., New York–London–Paris.

Astrophys. Space Sci.
Astrophysics and Space Science. An International Journal of Cosmic Physics. Published by D. Reidel Publishing Company, Dordrecht, Holland.

Astrophysics
Astrophysics. A cover-to-cover translation of Astrofizika (USSR). Consultants Bureau, New York, N.Y.

Atmos. Environ.
Atmospheric Environment. Pergamon Press Ltd., Headington Hill Hall, Oxford, OX3 0BW, England.

Atomkernenergie
Atomkernenergie. Verlag Karl Thiemig, Pilgersheimerstrasse 38, 8 München 90, Postfach 900740, F.R. Germany.

Atti Accad. Naz. Lincei, Mem. Ser. Ottava
Atti della Accademia Nazionale dei Lincei. Serie Ottava. Memorie. Classe di Scienze fisiche, matematiche e naturali. Sezione I: Matematica, Meccanica, Astronomia, Geodesia e Geofisica. Published by Accademia Nazionale dei Lincei, Roma.

Atti Accad. Naz. Lincei, Rend. Ser. Ottava
Atti della Accademia Nazionale dei Lincei. Serie Ottava. Rendiconti. Classe di Scienze fisiche, matematiche e naturali. Published by Accademia Nazionale dei Lincei, Roma.

Atti Accad. Sci. Torino I
Atti della Accademia delle Scienze di Torino. I. Classe di Scienze Fisiche, Mathematiche e Naturali. Via Accademia delle Scienze 6, Via Maria Vittoria 3, Torino (208), Italy.

Atti Fond. G. Ronchi, Contrib. Ist. Naz. Ottica
Atti della Fondazione Giorgio Ronchi e Contributi dell'

Istituto Nazionale di Ottica. Largo Enrico Fermi 6, 50125 Firenze, Italy.

Atti Soc. Astron. Italiana
Atti della Società Astronomica Italiana. Via Brera 28, Milano. Publisher: Tipografia Baccini & Chiappi, Firenze (Italy).

Australian J. Phys.
Australian Journal of Physics. Published by the Commonwealth Scientific and Industrial Research Organization, 372 Albert Street, East Melbourne, Victoria 3002, Australia.

Australian J. Phys., Astrophys. Suppl.
Australian Journal of Physics, Astrophysical Supplement. Published by Commonwealth Scientific and Industrial Research Organization, 372 Albert Street, East Melbourne, Victoria 3002, Australia.

BAV Rundbrief
BAV Rundbrief. Mitteilungsblatt der Berliner Arbeitsgemeinschaft für Veränderliche Sterne. Editor: BAV Berliner Arbeitsgemeinschaft für Veränderliche Sterne eV., Berlin.

BBSAG Bull.
Bedeckungsveränderlichen Beobachter der Schweizerischen Astronomischen Gesellschaft, [Swiss Astronomical Society's Eclipsing Variable Observers], Bulletin. To be obtained from R. Diethelm, Winterthur, Switzerland.

Bell Syst. Tech. J.
Bell System Technical Journal. American Telephone and Telegraph Co., 195 Broadway, New York, NY 10007, USA.

Ber. Bunsenges. Phys. Chem.
Berichte der Bunsengesellschaft für Physikalische Chemie. Verlag Chemie, 6940 Weinheim/Bergstrasse, Postfach 1260/1280, Germany.

Bild Wiss.
Bild der Wissenschaft. Zeitschrift über die Naturwissenschaften und die Technik in unserer Zeit. Publisher: Deutsche Verlagsanstalt, Stuttgart.

Blick Weltall
Blick in das Weltall. Astronomische Veranstaltungen und Mitteilungen für Sternfreunde. Archenhold-Sternwarte Berlin-Treptow.

Bol. Acad. Cienc. Fis. Mat. Nat.
Boletin de la Academia de Ciencias Fisicas Matematicas y Naturales. Printed by Italgrafica, S.R.L. Republica de Venezuela.

Bol. Astron. Obs. Madrid
Boletín Astronómico del Observatorio de Madrid. Editor: Instituto Geografico y Catastral. General Ibáñez de Ibero, Madrid.

Bol. Inst. Mat., Astron. Fis., Univ. Nac. Córdoba
Boletin del Instituto de Matematica, Astronomia y Fisica, Universidad Nacional de Córdoba (R. A.). Dirección General de Publicaciones, Córdoba (Argentina).

Bol. Inst. Tonantzintla
Boletin del Instituto de Tonantzintla. Instituto Nacional de Astrofisica, Optica y Electronica, Apartados Postales Nos. 216 y 51, Puebla, Pue, Mexico.

Bol. Liga Latinoamericana Astron.
Boletin de la Liga Latinoamericana de Astronomia. Publi-cado por la Asociacion Argentina Amigos de la Astrono-mia, Buenos Aires, Argentina.

Bol. Obs. Ebro
Boletín del Observatorio del Ebro, Tortosa. Printed by Cooperativa Gráfica Dertosense, Tortosa.

Boll. Geod. Sci. Affini
Bolletino di Geodesia e Scienze Affini. Pubblicazione dell'Istituto Geografico Militare, Firenze.

British Astron. Assoc. Circ.
British Astronomical Association, Circular. Editorial Office: 97 Hawkswood Drive, Hailsham, Sussex.

British J. Philos. Sci.
British Journal for the Philosophy of Science. Cambridge University Press, Bentley House, 200 Euston Road, London, NW1 2DB, England.

Bul. Inst. Politeh. Iasi I
Buletinul Institutului Politehnic din Iasi. Sectia I Mate-matica, Mecanica, Teoretica, Fizica. Polytechnic Institute, Iasi, Rumania.

Bull. Acad. Polonaise Sci., Ser. Sci. Tech.
Bulletin de l'Académie Polonaise des Sciences. Série des Sciences Techniques. 00-901 Warszawa, Palac Kultury i Nauki, P. O. Box 20, Poland.

Bull. AFOEV
Bulletin de l'Association Française des Observateurs d'Étoiles Variables. Rédaction et publication: M. Émile Schweitzer, 1, rue Beethoven, 67000 Strasbourg.

Bull. American Astron. Soc.
Bulletin of the American Astronomical Society. Published for the American Astronomical Society by the American Institute of Physics, 335 East 45th Street, New York, N.Y. 10017, USA.

Bull. American Meteorol. Soc.
Bulletin of the American Meteorological Society. 45 Beacon Street, Boston, MA 02108, USA.

Bull. Assoc. Suisse Electr.
Bulletin de l'Association Suisse des Electriciens (Organe Commun de l'Association Suisse des Electriciens (ASE) et de l'Union des Centrales Suisses d'Electricité (UCS)). Seefeldstrasse 301, Case Postale 229, 8008 Zürich, Switzerland.

Bull. Astron. Inst. Czechoslovakia
Bulletin of the Astronomical Institutes of Czechoslovakia. Published under the auspices of the Czechoslovak Acade-my of Sciences by Academia, Praha. Editor: Astronomi-cal Institutes of the Czechoslovak Academy of Sciences, Praha.

Bull. Astron., Obs. R. Belgique
Bulletin Astronomique, Observatoire Royal de Belgique. (Astronomisch Bulletin, Koninklijke Sterrenwacht van België).

Bull. Astron. Soc. India
Bulletin of the Astronomical Society of India. Edited and published by M. S. Vardya, Tata Institute of Fundamental Research, Bombay on behalf of the Astronomical Society of India, Osmania University, Hyderabad.

Bull. Géod.
Bulletin Géodésique, being the Journal of the Interna-tional Association of Geodesy. Nouvelle Série. Publié par le Bureau Central de l'Association Internationale de Géodésie, 39 Rue Gay-Lussac, 75005 Paris, France.

Bull. Geogr. Surv. Inst.
Bulletin of the Geographical Survey Institute. Published by the Geographical Survey Institute, Ministry of Con-struction, Tokyo, Japan.

Bull. Groupe Rech. Géod. Spat.
Groupe de Recherches de Géodésie Spatiale. Bulletin. CNES/Toulouse, France.

Bull. Inst. Space Aeronaut. Sci., Univ. Tokyo, B
Bulletin of the Institute of Space and Aeronautical Science, University of Tokyo, B. Tokyo, Japan.

Bull. Obs. Astron. Belgrade
Bulletin de l'Observatoire Astronomique de Belgrade. Editor: Observatoire Astronomique de Belgrade. Printed by Naucna delo, Belgrade.

Bull. Res. Inst. Sci. Meas. Tohoku Univ.
Bulletin of the Research Institute for Scientific Measure-ments, Tohoku University, Sendai, Japan.

Bull. Sci. Yougoslavie
Bulletin Scientifique. Conseil des Academies des Sciences et des Arts de la RSF de Yougoslavie. Section A: Sciences Naturelles, Techniques et Médicales. Rédaction et Admin-istration: Opatička ul. 18/II, Zagreb, Yougoslavie.

Bull. Seismol. Soc. America
Bulletin of the Seismological Society of America. Seismological Society of America, 2907 Claremont Avenue, Berkeley, CA 94705, USA.

Bull. Signal.
Bulletin Signalétique. Section 120: Astronomie, Physique spatiale, Géophysique. Centre de Documentation du Centre Nationale de la Recherche Scientifique, Paris.

Bull. Signal.
Bulletin Signalétique. Bibliographie des Sciences de la Terre. Section 220, Cahier A: Minéralogie, Géochimie, Géologie extraterrestre. Centre de Documentation du C.N.R.S., Paris; Département Documentation du B.R. G.M., Orléans.

Bull. Soc. Astron. Liège
Bulletin de la Société Astronomique de Liège. Editeur G. Mathys, 3, Avenue Laurent Gilys, 4621 Retinne.

Bull. Soc. Française Minéral. Crystallogr.
Bulletin de la Société Française de Minéralogie et de Cristallographie. Masson et Cie S.A., 120 Boulevard Saint-Germain, F-75280 Paris Cedex 06, France.

Bull. Soc. R. Sci. Liège
Bulletin de la Société Royale des Sciences de Liège. L'Université, 15 Avenue des Tilleurs, Liège, Belgium.

Bull. Tokyo Gakugei Univ. IV
Bulletin of Tokyo Gakugei University. Series IV (Mathe-matics and Natural Sciences). 4-1-1 Nukui-kita-machi, Koganei, Tokyo, Japan.

Byull. Abastuman. Astrofiz. Obs.
Abastumanskaya Astrofizicheskaya Observatoriya, Gora Kanobili. Byulleten'. Akademiya Nauk Gruzinskoj SSR.

Publishers: Izdatel'stvo "Metsniereba", Tbilisi.

Byull. Inst. Astrofiz.
Byulleten' Instituta Astrofiziki, Akademiya Nauk
Tadzhikskoj SSR. Izdatel'stvo Donish, Dushanbe.

Byull. Inst. Teor. Astron.
Byulleten' Instituta Teoreticheskoj Astronomii. Izdatel'-
stvo Nauka, Leningradskoe Otdelenie, Leningrad.

C. R. Acad. Bulgare Sci.
Comptes Rendus de l'Académie Bulgare des Sciences.
(Doklady Bolgarskoj Akademii Nauk). Sofiya, Bulgaria.

C. R. Acad. Sci. Paris
Comptes Rendus hebdomadaires des Séances de l'Aca-
démie des Sciences, publiés par MM. les Secrétaires
Perpétuels. Imprimerie: Gauthier-Villars, Montreuil,
France.

Cah. Fundam. Sci.
Fundamenta Scientiae. Cahiers du Séminaire sur les
Fondements des Sciences. Fundamenta Scientiae,
Groupe CBLL, Centre de Recherche Nucléaires, 67037
Strasbourg Cédex.

Canadian J. Earth Sci.
Canadian Journal of Earth Sciences. National Research
Council of Canada, Ottawa KIA OR6, Canada.

Canadian J. Phys.
Canadian Journal of Physics. Published by the National
Research Council of Canada, Ottawa. Printed in Canada
by the University of Toronto Press, Toronto, Ont.

Carter Obs., Astron. Bull.
Carter Observatory, Astronomical Bulletin. Carter Observatory, P.O. Box 2909, Wellington 1, New Zealand.

Celestial Mech.
Celestial Mechanics. An International Journal of Space
Dynamics. Publishers: D. Reidel Publishing Company,
Dordrecht, Holland.

Cent. Astrophys. Prepr. Ser.
Center for Astrophysics, Preprint Series. Harvard College
Observatory, Smithsonian Astrophysical Observatory.
Center for Astrophysics, 60 Garden St., Cambridge,
Mass. 02138.

Cent. Données Stellaires, Inf. Bull.
Centre de Données Stellaires. Information Bulletin. Compiled at Observatoire de Strasbourg, Strasbourg, France.

Ceskoslovensky Cas. Fis., A
Ceskoslovensky Casopis pro Fisiku. Sekce A. Academia
Publishing House of the Czechoslovak Academy of
Sciences, Vodickova 40, 112 29 Praha 1, Czechoslovakia.

Chem. Phys. Lett.
Chemical Physics Letters. North-Holland Publishing Co.,
P.O. Box 211, Amsterdam-C, Netherlands.

Chinese Astron.
Chinese Astronomy. A selected translation of Acta Astro-
nomica Sinica. Translated by T. Kiang. Publisher:
Pergamon Press, Headington Hill Hall, Oxford, OX3 0BW,
England — Maxwell House, Fairview Park, Elmsford,
New York 10523, USA.

Chinese J. Phys. (*Taiwan*)
Chinese Journal of Physics. Physical Society of the Re-

public of China, Physics Department, National Taiwan
University, Taipei, Taiwan, China.

Ciel Terre
Ciel et Terre. Bulletin de la Société Belge d'Astronomie,
de Météorologie et de Physique du Globe. Administration: Avenue Circulaire, 3, Bruxelles. Printed by Imprimerie R. Louis, Bruxelles.

Circ. Inf.
Circulaire d'Information. Union Astronomique Interna-
tionale. Commission des Etoiles Doubles. Address:
Observatoire de Meudon, Meudon, France.

Circ. Stn. Astron. Int. Latitudine, Carloforte-Cagliari
Circolari della Stazione Astronomica Internazionale di
Latitudine, Carloforte-Cagliari. Serie A printed by Tipo-
Offset "3T", Cagliari. Serie B printed by Multi Copy,
Milano.

Coelum
Coelum. Periodico bimestrale per la Divulgazione dell'
Astronomia. Editor: Osservatorio Astronomico Univer-
sitario di Bologna.

Commentat. Phys.-Math.
Commentationes Physico-Mathematicae. Societas Scientiarum Fennica, Helsinki–Helsingfors. Printed by Kes-
kuskirjapaino–Centraltryckeriet, Helsinki–Helsingfors.

Comments Astrophys.
Comments on Astrophysics. A Journal of Critical Dis-
cussion of the Current Literature. Comments on Modern
Physics: Part C. Publishers: Gordon and Breach, Science
Publishers Ltd., 42 William IV Street, London WC2,
England.

Comments At. Mol. Phys.
Comments on Atomic and Molecular Physics. Gordon
& Breach Science Publishers Ltd., 41 and 42 William IV
Street, London, WC2, England.

Comments Nucl. Part. Phys.
Comments on Nuclear and Particle Physics. Gordon &
Breach Science Publishers Ltd., 41 and 42 William IV
Street, London, WC2, England.

Commun. Math. Phys.
Communications in Mathematical Physics, Springer-
Verlag, Postfach 105280, 6900 Heidelberg 1, F.R.
Germany.

Comput. Math. with Appl.
Computers & Mathematics with Applications. Pergamon
Press Ltd., Headington Hill Hall, Oxford, OX3 OBW,
England.

Comput. Phys. Commun.
Computer Physics Communications. North-Holland
Publishing Co., P.O. Box 211 Amsterdam, Netherlands.

Comun. Obs. Astron. Univ. Coimbra
Comunicações do Observatório Astronomico da Universi-
dade de Coimbra, Portugal.

Comunicaciones
Comunicaciones. Centro de Informacion de Comunicaciones, Ministerio de Comunicaciones, Habana, Cuba.

Contemp. Phys.
Contemporary Physics. Taylor and Francis Ltd., 10 - 14
Macklin Street, London, WC2B 5NF, England.

Contrib. Atmos. Phys.
Contributions to Atmospheric Physics – Beiträge zur Physik der Atmosphäre. Publisher: Friedrich Vieweg & Sohn, Braunschweig.

Contrib. Obs. New Mexico State Univ.
Contributions of the Observatory of New Mexico State University. Published by the Astronomy Department, Box 4500, New Mexico State University, Las Cruces, New Mexico 88003.

COSPAR Inform. Bull.
COSPAR. Information Bulletin. Address: COSPAR Secretariat, Paris.

CQ Radio Amat. J.
CQ Radio Amateur's Journal. 14 Vanderventer Avenue, Port Washington, Long Island, NY 11050, USA.

Cryogenics
Cryogenics. IPC Science and Technology Press Ltd., IPC House 32 High Street, Guildford, Surrey GU1 3EW, England.

CSIO Commun.
CSIO Communications. Central Scientific Instruments Organisation, Sector 30, Chandigarh-160020, India.

Curr. Sci.
Current Science, Current Science Association, Raman Research Institute, Bangalore 6, India.

Czechoslovak J. Phys. B
Czechoslovak Journal of Physics, Section B. Czechoslovak Academy of Science, Akademia, Vodickova 40, 112 29 Praha 1, Czechoslovakia.

Delft Prog. Rep.
Delft Progress Report. Delft University Press, Mijnbouwplein 11, NL-2208 Delft, Netherlands.

Deutsche Geod. Komm. Bayerisch. Akad. Wiss.
Deutsche Geodätische Kommission bei der Bayerischen Akademie der Wissenschaften. Reihe A: Höhere Geodäsie; Reihe B: Angewandte Geodäsie; Reihe C: Dissertationen; Reihe D: Tafelwerke; Reihe E: Geschichte und Entwicklung der Geodäsie. Published by Verlag der Bayerischen Akademie der Wissenschaften, München.

Dokl. Akad. Nauk
Doklady Akademii Nauk SSSR. Seriya Matematika, Fizika. Publishers: Izdatel'stvo "Nauka", Moskva.

Dudley Obs. Rep.
Dudley Observatory Reports. Dudley Observatory, Albany, N.Y., USA.

Dunsink Obs. Publ.
Dunsink Observatory Publications. The Observatory of the School of Cosmic Physics, Dublin Institute for Advanced Studies, Dublin.

Earth Extraterr. Sci.
Earth and Extraterrestrial Sciences. Gordon & Breach Science Publishers Ltd., 41 and 42 William IV Street, London, WC2, England.

Earth Planet. Sci. Lett.
Earth and Planetary Science Letters. A Letter Journal devoted to the Development in Time of the Earth and Planetary System. Publisher: North-Holland Publishing Company, Amsterdam, Netherlands.

El Universo
El Universo. Organo de la Sociedad Astronomica de Mexico, Mexico, D.F.

Electricidade
Electricidade. Empresa Editorial Electrotecnica Edel, Rua de Dona Estefania, 48, 3., Esq., Lisboa 1, Portugal.

Electron. Appl. Ind.
Electronique & Applications Industrielles. Société des Editions Radio, 9 Rue Jacob, 75006 Paris, France.

Electron. Australia
Electronics Australia. 12th Floor, 235-243 Jones Street, Broadway, Sydney 2007, N.S.W., Australia.

Electron. Lett.
Electronics Letters. Institution of Electrical Engineers, Savoy Place, London, WC2R OBL, England.

Electron. Power
Electronics and Power. Journal of the Institution of Electrical Engineers. Savoy Place, London, WC2R OBL, England.

Electron. Prod. Methods Equip.
Electronic Production Methods and Equipment. Kiver-Patterson Publishing, 322 St. John Street, London, EC1 England.

Electronics
Electronics. McGraw-Hill Publishing Co., 1221 Avenue of the Americas, New York, N.Y. 10020, USA.

Elektrotech. Z. B
Elektrotechnische Zeitschrift. Ausgabe B: Der Elektrotechniker. Verband Deutscher Elektrotechniker Publication address: VDE-Verlag, Bismarckstrasse 33, 1000 Berlin 12, F. R. Germany.

Eltek. Aktuell Elektron. A
Elteknik med Aktuell Elektronik Edition A. Ingenjorsforlaget AB, Box 5703, 11487 Stockholm, Sweden.

Endeavour
Endeavour. A review of the progress of science, published in four languages by Imperial Chemical Industries, North Block, Thames House, Millbank, London, SW1P 4QE, England.

EOS Trans. American Geophys. Union
EOS Transactions of the American Geophysical Union. 1707 L Street, N.W., Washington, DC 20036, USA.

ESA Bull.
ESA Bulletin. ESA Space Documentation Service, 8 - 10 rue Mario Nikis, 75738 Paris Cedex 15, France.

ESA J.
ESA Journal. ESA Scientific & Technical Information Branch, ESTEC, Noordwijk, Netherlands.

ESO Bull.
European Southern Observatory, Bulletin. Edited by European Southern Observatory. Office of the Director: Hamburg.

ESO Sci. Prepr.
European Southern Observatory, Scientific Preprints. Available from Preprints Service, ESO-Library c/o CERN, 1211 Geneva 23, Switzerland.

ESO Tech. Rep.
European Southern Observatory, (ESO), Technical
Report. Published by the European Southern Observa-
tory Telescope Project Division, CERN, Geneva,
Switzerland.

Exp. Tech. Phys.
Experimentelle Technik der Physik, VEB Deutscher Ver-
lag der Wissenschaften, Traubenstrasse 10, 108 Berlin 8,
German Democratic Republic.

Feingerätetechnik
Feingerätetechnik. VEB Verlag Technik, Oranienburger
Strasse 13/14, 1020 Berlin, Germany.

Feinwerktech. Messtech.
F & M. Feinwerktechnik und Messtechnik. Fusion of
"Feinwerktechnik" and "Messtechnik" (formerly Zeit-
schrift für Instrumentenkunde) beginning with Jahrgang
82, No. 5 (1974). Publishers: Karl Hanser Verlag,
Kolbergerstr. 22, D-8000 München 80. F.R.Germany.

Finommech. Mikrotech.
Finommechanika-Mikrotechnika.Vadasz utca 16, Budapest 5, Hungary.

Fiz. Szemle
Fizikai Szemle. Kiadja a Lapkiado Vallalat, Budapest VII,
Lenin korut 9–11, Hungary.

Fizika
Fizika. 'Mladost' Export-Import, Zagreb, Ilica 30,
Yugoslavia.

Found. Phys.
Foundations of Physics. Plenum Publishing Co., 8 Scrubs
Lane, Harlesden, London, NW10 6SE, England.

Fra Fys. Verden
Fra Fysikkens Verden. Fysisk Institutt, Universitetet i
Trondheim, Norges Laererhogskole, 7000 Trondheim,
Norway.

Fundam. Cosmic Phys.
Fundamentals of Cosmic Physics. Gordon and Breach
Science Publishers Ltd., New York–London–Paris.

Funkschau
Funkschau. Francis-Verlag, 8 München 37, Postfach
37 01 20, Karlstrasse 37, Germany.

Fys. Tidsskr.
Fysisk Tidsskrift. Subscription address: Jul. Gjellerups
Boghandel, Solvgade 87, 1307 Kobenhavn, Denmark.

G. A.A.B.
Giornale dell'A.A.B. Notiziario trimestrale delle attività
culturali e scientifiche della Associazione Astrofili Bolog-
nesi, Bologna, Italy.

G. Astron.
Giornale di Astronomia, Pubblicazione della Società
Astronomica Italiana. Printed by Tipolitografia Lodigraf
S.p.A. Lodi (MI).

Gen. Relativ. Gravitation
General Relativity and Gravitation. Published under the
auspices of the International Committee on General
Relativity and Gravitation GRG. Publishing Office:
Plenum Publishing Corporation, 227 West 17th Street,
New York, N.Y. 10011, USA.

Geochim. Cosmochim. Acta
Geochimica et Cosmochimica Acta. Journal of the Geo-
chemical Society. Publishing House: Pergamon Press,
Ltd., Oxford.

Geod. Geophys. Veröff., Reihe III
Geodätische und Geophysikalische Veröffentlichungen.
Reihe III: Physik der festen Erde. Herausgegeben vom
Nationalkomitee für Geodäsie und Geophysik bei der
Akademie der Wissenschaften der Deutschen Demokra-
tischen Republik.

Geod. Kartogr.
Geodezja i Kartografia. Komitet Geodezji Polskiej Akademii Nauk. Publisher: Państwowe Wydawnictwo Naukowe, Warszawa.

Geoexploration
Geoexploration. Elsevier Publishing Co., P.O. Box 211,
Amsterdam, Netherlands.

Geomagn. Aehron.
Geomagnetizm i Aehronomiya. Akademiya Nauk SSSR.
Izdatel'stvo "Nauka", Moskva [An English translation
is published in "Geomagnetism and Aeronomy", American Geophysical Union, Washington, D.C.].

Geophys. Astrophys. Fluid Dyn.
Geophysical and Astrophysical Fluid Dynamics. Gordon
and Breach Science Publishers Ltd., 41/42 William IV
Street, London, WC2, England.

Geophys. J. R. Astron. Soc.
The Geophysical Journal of the Royal Astronomical So-
ciety. Published for the Royal Astronomical Society by
Blackwell Scientific Publications, Oxford–Edinburgh.
American Office of the Geophys. J., US Geological
Survey, Stop 967, Box 25046, Federal Center, Denver,
Colorado 80225, USA.

Geophys. Res. Lett.
Geophysical Research Letters. Published monthly by the
American Geophysical Union, Washington, D.C., U.S.A.

Geophysics
Geophysics. Society of Exploration Geophysicists,
P.O. Box 3098, Tulsa, OK 74101, USA.

GEOS
GEOS. Department of Energy, Mines and Resources,
Ottawa, Canada.

Gerlands Beitr. Geophys.
Gerlands Beiträge zur Geophysik. Publisher: Akademische Verlagsgesellschaft Geest & Portig K.-G., Leipzig.

Glasnik Mat.
Glasnik Matematicki. Published by the Society of Math-
ematicians and Physicists of the S. R. of Croatia. Publisher: Drustvo Matematicara i Fizicara S. R. Hrvatske,
Zagreb.

GSFC Doc.
Goddard Space Flight Center, Greenbelt, Maryland.
Available from Technical Information Division, Code
250, Goddard Space Flight Center, Greenbelt, Maryland 20771.

Helvetica Phys. Acta
Helvetica Physica Acta. Schweizerische Physikalische
Gesellschaft. Publisher: E. Birkhäuser, Elisabethenstr. 19,
CH-4000 Basel 10, Switzerland.

Hvar Obs. Bull.
Hvar Observatory Bulletin. Faculty of Geodesy. 41000 Zagreb, Kačićeva 26, Yugoslavia.

I.U.A.A. Bull.
I.U.A.A. Bulletin. International Union of Amateur Astronomers, Contributions. I.U.A.A. c/o Achille Leani, via Bertesi 15, 26100 Cremona, Italy.

IAU Circ.
International Astronomical Union, Circular. Central Bureau for Astronomical Telegrams, Smithsonian Astrophysical Observatory, Cambridge, Mass.

IBM Tech. Disclosure Bull.
IBM Technical Disclosure Bulletin. International Business Machines Co., Armonk, New York, NY 10504, USA.

Icarus
Icarus. International Journal of Solar System Studies. Publisher: Academic Press, New York – London.

ICSU Bull.
ICSU Bulletin. International Council of Scientific Unions. Secretariat: 51, Bd de Montmorency, Paris, France.

IEE J. Microwave Opt. Acoust.
IEE Journal on Microwave, Optics and Acoustics. Institution of Electrical Engineers, Publishing Department, P.O. Box 8, Southgate House, Stevenage, Herts. SG1 1HQ, England.

IEEE Spectrum
IEEE Spectrum. Published monthly by the Institute of Electrical and Electronics Engineers, 345 East 47th Street New York, N.Y. 10017, USA.

IEEE Trans. Aerosp. Electron. Syst.
IEEE Transactions on Aerospace and Electronic Systems. Published by the Institute of Electrical and Electronics Engineers, 345 East 47th Street, New York, N.Y. 10017, USA.

IEEE Trans. Antennas Propag.
IEEE Transactions on Antennas and Propagation. Published by the Institute of Electrical and Electronics Engineers, 345 East 47th Street, New York, N.Y. 10017, USA.

IEEE Trans. Commun.
IEEE Transactions on Communications. Institute of Electrical and Electronics Engineers, 345 East 47th Street, New York, NY 10017, USA.

IEEE Trans. Consum. Electron.
IEEE Transactions on Consumer Electronics. Institute of Electrical and Electronics Engineers, 345 East 47th Street, New York, NY 10017, USA.

IEEE Trans. Electron Devices
IEEE Transactions on Electron Devices. Published by the Institute of Electrical and Electronics Engineers, 345 East 47th Street, New York, N.Y. 10017, USA.

IEEE Trans. Geosci. Electron.
IEEE Transactions on Geoscience Electronics. Published by the Institute of Electrical and Electronics Engineers, 345 East 47th Street, New York, N.Y. 10017, USA.

IEEE Trans. Instrum. Meas.
IEEE Transactions on Instrumentation and Measurement. Published by the Institute of Electrical and Electronics Engineers, 345 East 47th Street, New York, N.Y. 10017, USA.

IEEE Trans. Microwave Theory Tech.
IEEE Transactions on Microwave Theory and Techniques. Published by the Institute of Electrical and Electronics Engineers, 345 East 47th Street, New York, N.Y. 10017, USA.

IEEE Trans. Nucl. Sci.
IEEE Transactions on Nuclear Science. Institute of Electrical and Electronics Engineers, 345 East 47th Street, New York, NY 10017, USA.

IEEE Trans. Plasma Sci.
IEEE Transactions on Plasma Science. Institute of Electrical and Electronics Engineers, 345 East 47th Street, New York, NY 10017, USA.

Ind. Math.
Industrial Mathematics. Industrial Mathematics Society, P.O. Box 159, Roseville, MI 48066, USA.

Indian East. Eng.
Indian and Eastern Engineer. 'Piramal Mansion', 235 Dr. D. Naoroji Road, Bombay 400001, India.

Indian J. Meteorol. Hydrol. Geophys.
Indian Journal of Meteorology, Hydrology & Geophysics. Formerly: Indian J. Meteorol. Geophys. Indian Meteorological Department, Civil Lines, Delhi 110006, India.

Indian J. Phys.
Indian Journal of Physics. 2 - 3 Raja Subodhchandra, Mallik Road, Calcutta 700032, India.

Indian J. Pure Appl. Math.
Indian Journal of Pure and Applied Mathematics. National Institute of Sciences India, Bahadur Shah Zafar Marg, New Delhi 1, India.

Indian J. Pure Appl. Phys.
Indian Journal of Pure and Applied Physics. Council of Scientific and Industrial Research, Hillside Road, New Delhi 110012, India.

Indian J. Radio Space Phys.
Indian Journal of Radio & Space Physics. Council of Scientific & Industrial Research. Editorial address: Publications & Information Directorate, Hillside Road, New Delhi 110012, India.

Inf. Bull. South. Hemisphere
Information Bulletin for the Southern Hemisphere. Editorial Office: Observatorio Astronómico, La Plata, Argentina.

Inf. Bull. Variable Stars
Commission 27 of the I.A.U. Information Bulletin on Variable Stars. Konkoly Observatory, Budapest.

Infrared Phys.
An International Research Journal. Publisher: Pergamon Press Ltd., Oxford, England.

Ingenieur
De Ingenieur. Koninklijk Institut van Ingenieurs. Editorial address; 23 Prinsessegracht, Den Haag, Netherlands.

Inst. Theor. Astrophys., Blindern–Oslo, Rep.
Institute of Theoretical Astrophysics, Blindern–Oslo,

Report. Universitetsforlagets trykningssentral, Oslo.

Int. At. Energy Agency Bull.
International Atomic Energy Agency Bulletin. Kärntner-ring 11, P.O. Box. 590, A-1011 Wien, Austria.

Int. J. Electron.
International Journal of Electronics. Taylor and Francis Ltd., 10—14 Macklin Street, London, WC2B 5BF, England.

Int. J. Theor. Phys.
International Journal of Theoretical Physics. Plenum Publishing Co. Ltd., Davis House, 8 Scrubs Lane, London, NW10 6SE, England.

Irish Astron. J.
The Irish Astronomical Journal. A Quarterly Publication under the auspices of the Observatories of Armagh and Dunsink. Armagh Observatory, Northern Ireland.

Izv. Akad. Nauk Armyan. SSR
Izvestiya Akademii Nauk Armyanskoj SSR. Fizika. Publisher: Izdatel'stvo AN Armyanskoj SSR, Erevan.

Izv. Astron. Ehngel'gardt. Obs.
Izvestiya Astronomicheskoj Ehngel'gardtovskoj Observatorii. Izdatel'stvo Kazanskogo Universiteta, Kazan.

Izv. Glav. Astron. Obs. Pulkovo
Izvestiya Glavnoj Astronomicheskoj Observatorii v Pulkove. Akademiya Nauk SSSR. Izdanie Glavnoj astronomicheskoj observatorii v Pulkove, Leningrad.

Izv. Krymskoj Astrofiz. Obs.
Izvestiya Krymskoj Astrofizicheskoj Observatorii. Akademiya Nauk SSR. Publishers: Izdatel'stvo "Nauka", Moskva.

J. American Assoc. Variable Star Obs.
The Journal of the American Association of Variable Star Observers. Published by The American Association of Variable Star Observers, 187 Concord Avenue, Cambridge, Mass. 02138, USA.

J. Appl. Meteorol.
Journal of Applied Meteorology. American Meteorological Society, 45 Beacon Street, Boston, MA 02108, USA.

J. Appl. Photogr. Eng.
Journal of Applied Photographic Engineering. Society of Photographic Scientists and Engineers, Suite 204, 1330 Massachusetts Avenue, N.W. Washington, D.C. 20005, USA.

J. Appl. Phys.
Journal of Applied Physics. American Institute of Physics, 335 East 45th Street, New York, NY 10017, USA.

J. Astron. Soc. Victoria
The Journal of the Astronomical Society of Victoria. McKinnon—Melbourne.

J. Astron. Soc. Western Australia
The Journal of the Astronomical Society of Western Australia. Edited by the Astronomical Society of Western Australia, Perth, W. A.

J. Astronaut. Sci.
Journal of the Astronautical Sciences. American Astronautical Society, 6060 Duke Street, Alexandria, VA 22304, USA.

J. Atmos. Sci.
Journal of the Atmospheric Sciences.American Meteorological Society, 45 Beacon Street, Boston, MA 02108, USA.

J. Atmos. Terr. Phys.
Journal of Atmospheric and Terrestrial Physics. Pergamon Press Ltd., Oxford, England.

J. Australian Math. Soc., Ser. B
Journal of the Australian Mathematical Society,Series B (Applied Mathematics). Editorial address: Department of Mathematics, University of Western Australia, Nedlands, Western Australia 6009, Australia.

J. British Astron. Assoc.
Journal of the British Astronomical Association. Burlington House, Piccadilly, London, W1V ONL, England.

J. British Interplanet. Soc.
Journal of the British Interplanetary Society. British Interplanetary Society, 12 Bessborough Gardens, London, SW1V 2JJ, England.

J. Colloid Interface Sci.
Journal of Colloid and Interface Science. Academic Press Inc., 111 Fifth Avenue, New York, N.Y. 10003, USA.

J. Comput. Phys.
Journal of Computational Physics. Academic Press Inc., 111 Fifth Avenue, New York, NY 10003, USA.

J. Fluid Mech.
Journal of Fluid Mechanics. Cambridge University Press, Bentley House, 200 Euston Road, London, NW1 2DB, England.

J. Franklin Inst.
Journal of the Franklin Institute. Pergamon Press Ltd., Headington Hill Hall, Oxford, OX3 OBW, England.

J. Geomagn. Geoelectr.
Journal of Geomagnetism and Geoelectricity. Society of Terrestrial Magnetism and Electricity of Japan, Geophysical Institute, Tokyo University, Tokyo 113, Japan.

J. Geophys.
Journal of Geophysics / Zeitschrift für Geophysik. Springer-Verlag, D-6900 Heidelberg, Postfach 105280, F. R. Germany.

J. Geophys. Res.
Journal of Geophysical Research. Published by American Geophysical Union, 1909K Street, N.W. Washington D.C. First section: Space physics; Second section: Physics and chemistry of the solid earth, planetology, geodesy; Third section: Oceans and atmospheres.

J. Hist. Arabic Sci.
Journal for the History of Arabic Science. Institute for the History of Arabic Science, University of Aleppo, Aleppo, Syria.

J. Hist. Astron.
Journal for the History of Astronomy. Published by Science History Publications Ltd., Halfpenny Furze, Mill Lane, Chalfont St Giles, Buckinghamshire,England, and distributed by Neale Watson Academic Publications Inc., 156 Fifth Avenue, New York, NY 10010, USA.

J. Illum. Eng. Inst. Japan
Journal of the Illuminating Engineering Institute of Japan. 3 Yurakucho, 1-chome, Chiyodaku, Tokyo 100, Japan.

J. Illum. Eng. Soc.
Journal of the Illuminating Engineering Society. 345 East 47th Street, New York, NY 10017, USA.

J. Indian Inst. Sci.
Journal of the Indian Institute of Science. Bangalore 560012, India.

J. Inst. Math. Appl.
Journal of the Institute of Mathematics and its Applications. Academic Press Inc. (London) Ltd., 24 - 28 Oval Road, London NW1 7DX, England.

J. Inst. Telev. Eng. Japan
Journal of the Institute of Television Engineers of Japan. Kikai-Shinko Building, 3-5-8 Shiba Park, Minato-ku, Shiba PO, Tokyo, Japan.

J. Instn. Electron. Telecommun. Eng.
Journal of the Institution of Electronics and Telecommunication Engineers. 2 Lodi Road, Institutional Area, New Delhi 110003, India.

J. Low Temp. Phys.
Journal of Low Temperature Physics. Plenum Publishing Corp., 227 West 17th Street, New York, NY 10011, USA.

J. Math. Phys., New York
Journal of Mathematical Physics. American Institute of Physics, 335 East 45th Street, New York, N.Y. 10017, USA.

J. Mech. Eng. Lab.
Journal of Mechanical Engineering Laboratory. Agency of Industrial Science and Technology, Igusa Suginami-ku, Tokyo, Japan.

J. Mol. Spectrosc.
Journal of Molecular Spectroscopy. Academic Press Inc., 111 Fifth Avenue, New York, NY 10003, USA.

J. Navig.
The Journal of Navigation. The Royal Institute of Navigation at the Royal Geographical Society, Kensington Gore, London, SW7 2AT. Scottish Academic Press Ltd., 33 Montgomery Street, Edinburgh EH7 5JX.

J. Opt.
Journal of Optics. Optical Society of India, Department of Applied Physics, University of Calcutta, 92 Acharya Prafulla Chandra Road, Calcutta-9, India.

J. Opt. Soc. America
Journal of the Optical Society of America. American Institute of Physics, 335 East 45th Street, New York, N.Y. 10017, USA.

J. Photogr. Sci.
Journal of Photographic Science. Royal Photographic Society, 14 South Audley Street, London, W1Y 5DP, England.

J. Phys. A
Journal of Physics A, (Mathematical, Nuclear and General). Institute of Physics, 47 Belgrave Square, London, SW1X 8QX, England.

J. Phys. B
Journal of Physics B, (Atomic and Molecular Physics). Institute of Physics, 47 Belgrave Square, London, SW1X 8QX, England.

J. Phys. Chem. Ref. Data
Journal of Physical and Chemical Reference Data. American Chemical Society, 1155 Sixteenth Street, N.W., Washington, DC 20036, USA.

J. Phys. E
Journal of Physics E, (Scientific Instruments). Formerly: J. Sci. Instrum. (GB). Institute of Physics, 47 Belgrave Square, London, SW1X 8QX, England.

J. Phys. G
Journal of Physics G, (Nuclear Physics). Institute of Physics, 47 Belgrave Square, London, SW1X 8QX, England.

J. Phys. Soc. Japan
Journal of the Physical Society of Japan. Room 211, Kikai Shinko Building, Shiba Koen, Minato-ku, Tokyo 105, Japan.

J. Physique
Journal de Physique. Société Française de Physique, 87 bis Avenue du Général Leclerc, 75014 Paris, France.

J. Plasma Phys.
Journal of Plasma Physics. Cambridge University Press, Bentley House, 200 Euston Road, London, NW1 2DB, England.

J. Proc. R. Soc. New South Wales
Journal and Proceedings of the Royal Society of New South Wales. Science House, Gloucester and Essex Streets, Sydney, N.S.W. 2000, Australia.

J. Quant. Spectrosc. Radiat. Transfer
Journal of Quantitative Spectroscopy & Radiative Transfer. Pergamon Press Ltd., Headington Hill Hall, Oxford, OX3 0BW, England.

J. R. Astron. Soc. Canada
The Journal of the Royal Astronomical Society of Canada, devoted to the advancement of astronomy and allied sciences. The Royal Astronomical Society of Canada, 124 Merten Street, Toronto, Ontario, Canada.

J. Radio Res. Lab.
Journal of the Radio Research Laboratories. Chief Planning Section, Radio Research Laboratories, Ministry of Posts & Telecommunications, Nukui-Kitamachi, Konganei-shi, Tokyo 184, Japan.

J. Radioanal. Chem.
Journal of Radioanalytical Chemistry. Elsevier Sequoia S.A., P.O. Box 851, 1001 Lausanne 1, Switzerland. Kultura, H-1389 Budapest, 62, P.O. Box 149, Hungary.

J. Res. Natl. Bur. Stand. B
Journal of Research of the National Bureau of Standards. Section B (Mathematics and Mathematical Physics). US Government Printing Office, Division of Public Documents, Washington, DC 20402, USA.

J. Sci. Res. Banaras Hindu Univ.
Journal of Scientific Research of the Banaras Hindu University. PO-Banaras Hindu University, India.

J. Spacecr. Rockets
Journal of Spacecraft and Rockets. American Institute of Aeronautics and Astronautics, 1290 Avenue of the Americas, New York, N.Y. 10019, USA.

J. Stat. Phys.
Journal of Statistical Physics. Plenum Publishing Corp., 227 West 17th Street, New York, NY 10011, USA.

J. Test. Eval.
Journal of Testing and Evaluation. American Society for Testing and Materials, 1916 Race Street, Philadelphia, PA 19103, USA.

Japanese J. Appl. Phys.
Japanese Journal of Applied Physics. Publication Office, 2nd Toya Kaiji Building, 24-8 Shinbashi, Minato-ku, Tokyo 105, Japan.

Jenaer Rundsch. (Jena Rev.)
Jenaer Rundschau (Jena Review). Publisher: VEB Verlag Technik, Berlin, German Democratic Republic.

JETP Lett.
JETP Letters. A translation of JETP Pis'ma v Redaktsiyu of the Academy of Sciences in the USSR. American Institute of Physics, 335 East 45th Street, New York, NY 10017, USA.

JPL Tech. Mem.
Jet Propulsion Laboratory, California Institute of Technology, Pasadena, California. National Aeronautics and Space Administration. Technical Memorandum.

JPL Tech. Rep.
Jet Propulsion Laboratory, California Institute of Technology, Pasadena, California. National Aeronautics and Space Administration. Technical Report.

K. Danske Vidensk. Selsk. Mat.-Fys. Skr.
Kongelige Danske Videnskabernes Selskab, Matematisk-fysiske Skrifter. Dantes Plads, 5, 1556 Kobenhavn V. Publisher: Munksgaard Publishers, 35 Norre Sogade, DK-1370 Kobenhavn, Denmark.

Kexue Tongbao
Kexue Tongbao. Academia Sinica, Peking, People's Republic of China [English translation in: Kexue Tongbao (Scientia) (USA)].

Kometn. Tsirk. *Kiev*
Kometnyj Tsirkulyar. Gruppa po Issledovaniyu Komet Astrosoveta i Mezhdu·edomstvennyj Geofizicheskij Komitet Akademii Nauk SSSR. Kievskij Universitet im. T. G. Shevchenko.

Komety i Meteory
Komety i Meteory. Akademiya Nauk Tadzhikskoj SSR. Astronomicheskij Sovet Akademii Nauk SSSR. Publishers: Izdatel'stvo "Donish", Dushanbe.

Kosm. Issled.
Kosmicheskie Issledovaniya. Akademiya Nauk SSSR. Publishers: Izdatel'stvo "Nauka", Moskva [An English translation is published as "Cosmic Research", Consultants Bureau, New York, N.Y.].

Kozmos
Kozmos. Popular Astronomical Journal of the Slovak Central Observatory in Hurbanovo. Publisher: Slovenská ústredná hvezdáreň v Hurbanove.

Lab. Equip. Dig.
Laboratory Equipment Digest. Morgan-Grampian Ltd., 30 Calderwood Street, Woolwich, London, SE18 6QH, England.

Lett. Math. Phys.
Letters in Mathematical Physics. D. Reidel Publishing Co., P.O. Box 17, Dordrecht, Netherlands.

L'Universo
L'Universo. Rivista dell'Istituto Geografico Militare. Direzione, Redazione e Amministrazione: Istituto Geografico Militare, Firenze.

Magn. Polya Soln. Pyaten
Magnitnye Polya Solnechnykh Pyaten. (Supplements to Solnechnye Dannye. Byulleten' (*Solar Data*)). Publishers: Izdatel'stvo "Nauka", Leningrad.

Marconi Rev.
Marconi Review. Marconi Co., Marconi House, Chelmsford, Essex, England.

Mater. Pr.
Materiały i Prace. Publications of the Institute of Geophysics, Polish Academy of Sciences. Edited by Państwowe Wydawnictwo Naukowe, Warszawa.

Math. Intelligencer
The Mathematical Intelligencer. Springer-Verlag, Berlin–Heidelberg–New York, 175 Fifth Avenue, New York, NY 10010, USA.

Math. Proc. Cambridge Philos. Soc.
Mathematical Proceedings of the Cambridge Philosophical Society. Formerly: Proceedings of the Cambridge, Philosophical Society (Mathematical and Physical Sciences). Cambridge University Press, Bentley House, 200 Euston Road, London, NW1 2DB, England.

Meccanica
Meccanica, Tamburini Editore, s.p.a., Via G. Pascoli, 55-Milano, Italy.

Mech. Res. Commun.
Mechanics Research Communications. Pergamon Press Ltd., Headington Hill Hall, Oxford, OX3 OBW, England.

Mem. Chubu Inst. Technol.
Memoirs Chubu Institute of Technology. Nagoya, Japan.

Mem. Fac. Eng. Osaka City Univ.
Memoirs of the Faculty of Engineering, Osaka City University. 459 Sugimoto-cho, Sumi Yoshi-kum, Osaka, Japan.

Mem. Fac. Sci. Kyoto Univ.
Memoirs of the Faculty of Science, Kyoto University. Series of Physics, Astrophysics, Geophysics, and Chemistry. Printed by Yamashiro Printing Publishing Co. Ltd., Kamigyo, Kyoto.

Mem. Japan Astron. Study Assoc.
Memoirs of the Japan Astronomical Study Association. c/o National Science Museum, Ueno Park, Taito-ku, Tokyo, Japan.

Mem. R. Astron. Soc.
Memoirs of the Royal Astronomical Society. Published for the Royal Astronomical Society by Blackwell Scientific Publications, Oxford – London – Edinburgh – Melbourne.

Mém. Sci. Rev. Metall.
Mémoires Scientifiques de la Revue de Metallurgie.
5 Rue Paul Cézanne, 75008 Paris, France.

Mem. Soc. Astron. Italiana
Memorie della Società Astronomica Italiana. Presso
Laboratorio di Astrofisica Spaziale, Castella Postale 67,
00044 Frascati, Italy.

Mercury
Mercury. The Journal of the Astronomical Society of the
Pacific. Published by the Astronomical Society of the
Pacific, 1244 Noriega Street, San Francisco, California
94122, USA.

Messenger
The Messenger, El Mensajero. Edited by European
Southern Observatory, Schleißheimer Straße 17, D-8046
Garching bei München, F. R. Germany.

Meteor Sect. Rep.
Netherlands Association for Astronomy and Meteorology.
Meteor Section Report. Meteor Section NVWS, De
Sitterlaan 37, 2313 TK Leiden, Netherlands.

Meteoritics
Meteoritics. The Journal of the Meteoritical Society. Pub-
lished quarterly by The Meteoritical Society and Arizona
State University Bureau of Publications. Editorial address:
Center for Meteorite Studies, The Arizona State Universi-
ty, Tempe, Ariz. 85281, USA.

Meteoritika
Akademiya Nauk SSSR. Komitet po Meteoritam. Publishers: Izdatel'stvo "Nauka", Moskva.

Meteorol. Rundsch.
Meteorologische Rundschau. Springer-Verlag, D-1000
Berlin 33, Heidelberger Platz 3, Germany.

Metrologia
Metrologia. Springer-Verlag, Heidelberger Platz 3,
D-1000 Berlin 33, Germany.

Microwave J.
Microwave Journal. To be obtained from 610 Washington
Street, Dedham Plaza, Dedham, Massachusetts, U.S.A.

Microwaves
Microwaves. Hayden Publishing Co., 50 Essex Street,
Rochelle Park, NJ 07662, USA.

Minor Planet Bull.
The Minor Planet Bulletin. Bulletin of the Minor Planets
Section of the Association of Lunar and Planetary
Observers. Editorial Office: R. G. Hodgson, Dordt
College, Sioux Center, Iowa, U.S.A.

Minor Planet Circ.
Minor Planet Circulars. Published by Cincinnati Observatory. Observatory Place, Cincinnati, Ohio 4528, U.S.A.

Mitt. Astron. Ges.
Mitteilungen der Astronomischen Gesellschaft, Hamburg.
Printed by G. Braun, GmbH, Karlsruhe. To be available
from Astron. Inst. Univ. Bochum.

Mitt. Inst. Theor. Geod. Univ. Bonn
Mitteilungen aus dem Institut für Theoretische Geodäsie
der Universität Bonn, Nußallee 17, 5300 Bonn 1, F.R.
Germany.

Mitt. Veränderl. Sterne (MVS)
Mitteilungen über Veränderliche Sterne. Herausgegeben
von der Sternwarte Sonneberg der Akademie der
Wissenschaften der DDR, Sonneberg, German Democratic Republic.

Mod. Geol.
Modern Geology. Gordon & Breach Science Publishers
Ltd., 41 and 42 William IV Street, London WC2,
England.

Mon. Not. R. Astron. Soc.
Monthly Notices of the Royal Astronomical Society.
Published for the Royal Astronomical Society by Blackwell Scientific Publications, Oxford – London – Edinburgh – Melbourne.

Mon. Notes Astron. Soc. South. Africa
Monthly Notes of the Astronomical Society of Southern
Africa. Published by the Astronomical Society of
Southern Africa, S. A. Astronomical Observatory, Cape
Province, South Africa.

Mon. Notes Int. Polar Motion Serv.
Monthly Notes of the International Polar Motion Service.
Published by the Central Bureau, International Latitude
Observatory of Mizusawa, Mizusawa-shi, Iwate-ken, Japan.

Monitor
Monitor. Formerly: Proc. Instn. Radio Electron. Eng.
Australia. Institution of Radio and Electronics Engineers
Australia. Science House, 157 Gloucester Street, Sydney,
N.S.W. 2000, Australia.

Moon
The Moon. An International Journal of Lunar Studies.
Publisher: D. Reidel Publishing Company, Dordrecht,
Holland.

Nablyud. Iskusstv. Nebesn. Tel
Nablyudeniya Iskusstvennykh Nebesnykh Tel. Published
by Astronomicheskij Sovet Akademii Nauk SSSR,
Moskva.

Nachr. Akad. Wiss. Göttingen II
Nachrichten der Akademie der Wissenschaften in Göttin-
gen. II. Mathematisch-Physikalische Klasse. Vandenhoeck
& Ruprecht, Göttingen.

Nachr. Elektron.
Nachrichten Elektronik. Formerly: Int. Elektron.
Rundsch. Elektro-Welt-Verlag Dr. Hüthig, D-6900
Heidelberg 1, Postfach 102869, Wilckensstr. 3 - 5, F.R.
Germany.

Nachr. Karten-, Vermessungswesen
Nachrichten aus dem Karten- und Vermessungswesen.
Editor: Institut für Angewandte Geodäsie (Abt. II des
Deutschen Geodätischen Forschungsinstituts). Published
by Verlag des Instituts für Angewandte Geodäsie, Frank-
furt a. M.

Nachr. Olbers-Ges. Bremen
Nachrichten der Olbers-Gesellschaft Bremen. Werder-
straße 73, Bremen.

Nachrichtentech. Elektron.
Nachrichtentechnik Elektronik. VEB Verlag Technik,
DDR 102 Berlin, Oranienburger Strasse 13/14, Germany.

NASA Contract. Rep.
NASA Contractor Report. National Aeronautics and

Space Administration, Washington, D.C. For sale by the National Technical Information Service, Springfield, Virginia 22161.

NASA Ref. Publ.
NASA Reference Publication. National Aeronautics and Space Administration. Scientific and Technical Information Office. Washington, D.C. 20546. For sale by the National Technical Information Service, Springfield, Virginia 22161.

NASA Tech. Memo.
NASA Technical Memorandum. National Aeronautics and Space Administration, Washington, D.C. For sale by the National Technical Information Service, Springfield, Virginia 22161.

NASA Tech. Note
NASA Technical Note. National Aeronautics and Space Administration, Washington, D.C. For sale by the National Technical Information Service, Springfield, Virginia 22161.

NASA Tech. Pap.
NASA Technical Paper. National Aeronautics and Space Administration, Washington, D.C. For sale by the National Technical Information Service, Springfield, Virginia 22161.

Natl. Geogr.
National Geographic. Official Journal of the National Geographic Society, Washington, D.C. 17th and M Sts. N.W., Washington, D.C. 20036.

Nature
Nature. Editorial and Publishing Offices: Macmillan Journals Limited, 4 Little Essex Street, London WC2R 3LF, 711 National Press Building, Washington, D.C. 20045.

Naturwiss. Rundsch.
Naturwissenschaftliche Rundschau. Wissenschaftliche Verlagsgesellschaft mbH. Birkenwaldstraße 44, Stuttgart N.

Naturwissenschaften
Die Naturwissenschaften. Publisher:Springer-Verlag, Berlin – Heidelberg – New York.

Nauch. Inf.
Nauchnye Informatsii. Astronomicheskij Sovet Akademii Nauk SSSR, Moskva.

Navigation (*France*)
Navigation. Institut Française de Navigation, 3 avenue Octave-Greard, Paris 7, France.

NBS Monogr.
National Bureau of Standards Monograph. U.S. Government Printing Office, Washington, D.C. 20402.

Nederlands Tijdschr. Natuurkd. A
Nederlands Tijdschrift voor Natuurkunde, Publisher: Martinus Nijhoff, Lange Voorhout 9, Den Haag, Netherlands.

New Phys. (Korean Phys. Soc.)
New Physics (Korean Physical Society). Seoul, Korea.

New Scientist
New Scientist. New Science Publications, 128 Long Acre, London, WC2E 9QH, England.

New Zealand Energy J.
New Zealand Energy Journal. Formerly: New Zealand Electr. J. Technical Publications Ltd., 127 Molesworth Street, P.O.Box 3047, Wellington, New Zealand.

New Zealand J. Sci.
New Zealand Journal of Science. Department of Scientific and Industrial Research, Private Bag, Wellington, New Zealand.

News Lett. Astron. Soc. N.Y.
News Letter of the Astronomical Society of New York. A. G. D. Philip (Editor). Astronomical Society of New York, Dudley Observatory, 100 Fuller Road, Albany, N. Y.

Notas Fis.
Notas de Física. Centro Brasileiro de Pesquisas Físicas, Av. Wenceslau Braz 71, Rio de Janeiro, Brazil.

Nouv. Rev. Opt.
Nouvelle Revue d'Optique. Masson, 120 Boulevard Saint-Germain, F-75280 Paris Cedex 06, France.

Nucl. Instrum. Methods
Nuclear Instruments and Methods. North-Holland Publishing Co., P.O. Box 211, Amsterdam, Netherlands.

Nucl. Phys. A
Nuclear Physics, Volume A. North-Holland Publishing Co., P. O. Box 211, Amsterdam, Netherlands.

Nucl. Track Detect.
Nuclear Track Detection. Pergamon Press Ltd., Headington Hill Hall, Oxford OX3 0BW, England.

Nukleonika
Nukleonika. Polska Akademia Nauk, 00-901 Warszawa, Polac Kultury i Nauki, Poland.

Numer. Math.
Numerische Mathematik. Springer-Verlag, Berlin – Heidelberg – New York.

Nuovo Cimento
Il Nuovo Cimento. Rivista Internazionale e Organo della Società Italiana di Fisica, Series A, B. Via Degli Andalo 2, 40124 Bologna, Italy.

Nuovo Cimento, Lett.
Lettere al Nuovo Cimento, a Cura della Società Italiana di Fisica. Via Degli Andalo 2, 40124 Bologna, Italy.

Nuovo Cimento, Riv.
Rivista del Nuovo Cimento. Società Italiana di Fisica, Via Degli Andalo 2, Bologna 40124, Italy.

Nuovo Cimento, Suppl.
Supplemento al Nuovo Cimento. Società Italiana di Fisica. Via Degli Andalo 2, 40124 Bologna, Italy.

Obs. Artif. Earth Satell.
Observations of Artificial Satellites of the Earth (Nablyudeniya Iskusstvennykh Sputnikov Zemli). Magyar Tudományos Akadémia Csillagvizsgáló Intézete. Budapest.

Obs. Astrophys. Lab., Univ. Helsinki.Rep.
Observatory and Astrophysics Laboratory, University of Helsinki. Report. Helsinki, Finland.

Observatory

The Observatory. A Review of Astronomy. Publishers: The Editors of 'The Observatory', Royal Greenwich Observatory, Herstmonceux Castle, Hailsham, Sussex, England, BN27 1RP.

Occas. Rep. R. Obs. Edinburgh
Occasional Reports of the Royal Observatory, Edinburgh, Blackford Hill, Edinburgh EH9 3HJ, Scotland.

Occultation Newsl.
Occultation Newsletter. Published by the International Occultation Timing Association (I.O.T.A.). 6 N 106 White Oak Lane, St. Charles, Ill. 60174, USA.

Österreich. Z. Vermessungswes. Photogramm.
Österreichische Zeitschrift für Vermessungswesen und Photogrammetrie. Editor and Publisher: Österreichischer Verein für Vermessungswesen und Photogrammetrie, Wien, Austria.

Opt. Acta
Optica Acta. Taylor and Francis Ltd., 10 - 14 Macklin Street, London, WC2B 5NF, England.

Opt. Commun.
Optics Communications. North-Holland Publishing Co., P.O. Box 211, Amsterdam, Netherlands.

Opt. Eng.
Optical Engineering. Society of Photo-Optical Instrumentation Engineers, 337 Tejon Place, Palos Verdes Estates, CA 90274, USA.

Opt. Laser Technol.
Optics and Laser Technology. Formerly: Opt. Technol. IPC Science and Technology Press Ltd., IPC House, 32 High Street, Guildford, Surrey, England.

Opt. Lett.
Optics Letters. A publication of the Optical Society of America. American Institute of Physics, 335 East 45th Street, New York, N.Y. 10017, USA.

Opt. News
Optics News. Publication of the Optical Society, Washington, D.C.

Opt. Pura Apl.
Optica Pura y Aplicada. Serrano 121, Madrid 6, Spain.

Opt. Spectra
Optical Spectra. The Magazine of Optical/Electro-Optical/Laser Technology. Published by The Optical Publishing Co., Inc., 59 Bartlett Ave., P.O. Box 1146, Pittsfield, Mass. 01201, USA.

Optik
Optik. Zeitschrift für das gesamte Gebiet der Licht- und Elektronenoptik. Publishers: Wissenschaftliche Verlagsgesellschaft mbH, Postfach 40, D-7000 Stuttgart, F. R. Germany.

Org. Geochem.
Organic Geochemistry. A publication of the International Association of Geochemistry and Cosmochemistry. Pergamon Press, Headington Hill Hall, Oxford OX3 0BW. Maxwell House, Fairview Park, Elmsford, N.Y. 10523, USA.

Origins of Life
Origins of Life (Formerly Space Life Sciences). An International Journal. Publisher: D. Reidel Publishing Compa-

ny, Dordrecht, Holland.

Orion
Orion. Zeitschrift der Schweizerischen Astronomischen Gesellschaft (SAG). Bulletin de la Société Astronomique de Suisse (SAS). Printed by A. Schudel & Co. AG, 4125 Riehen, Switzerland.

Oss. Astrofis. Catania, Pubbl.
Osservatorio Astrofisico di Catania, Pubblicazione. Printed by Scuola Salesiana del Libro, Catania.

Oss. Mem. Oss. Astrofis. Arcetri
Osservazioni e Memorie dell'Osservatorio Astrofisico di Arcetri. Università Degli Studi di Firenze, Firenze, Italy.

Oyo Buturi
Oyo Buturi. Japan Society of Applied Physics, Room No. 209-2, Kikai-Shinko Building, 21 Shiba-Koen Minato-ku, Tokyo, Japan.

Perem. Zvezdy, Byull.
Peremennye Zvezdy, Byulleten', izdavaemyj Astronomicheskim Sovetom Akademii Nauk SSSR. Published by Astronomicheskij Sovet Akademii Nauk SSSR, Moskva.

Perem. Zvezdy, Prilozhenie
Peremennye Zvezdy, Prilozhenie (The Variable Stars, Supplement). Astronomicheskij Sovet Akademii Nauk SSSR, Moskva.

Philos. Mag.
The Philosophical Magazine. A Journal of Theoretical, Experimental and Applied Physics. Eighth Series. Publisher: Taylor & Francis, Ltd., 10 - 14 Macklin Street, London, WC2B 5NF, England.

Philos. Trans. R. Soc. London, Ser. A
Philosophical Transactions of the Royal Society of London. Series A, Mathematical and Physical Sciences. Carlton House Terrace, London, SW1Y 5AG, England.

Photogr. Sci. Eng.
Photographic Science and Engineering. Society of Photographic Scientists and Engineers, Suite 204, 1330 Massachusetts Avenue N.W., Washington, DC 20005, USA.

Photogramm. Eng. Remote Sensing
Photogrammetric Engineering and Remote Sensing. Formerly: Photogramm. Eng. American Society of Photogrammetry, 105 North Virginia Avenue, Falls Church, VA 22046, USA.

Phys. Abstr.
Physics Abstracts. Science Abstracts, Series A. An INSPEC Publication, published by The Institution of Electrical Engineers in Association with the Institute of Electrical and Electronics Engineers Inc. Printed by Pindar & Son Ltd., Scarborough, N. Yorkshire, England.

Phys. Ber.
Physikalische Berichte. Herausgegeben von der Deutschen Physikalischen Gesellschaft e.V. und von der Deutschen Akademie der Wissenschaften zu Berlin. Physik-Verlag, Weinheim, F.R. Germany.

Phys. Bl.
Physikalische Blätter. Physik-Verlag GmbH, Pappelallee 3, Postfach 1260/1280, D-6940 Weinheim, F. R. Germany.

Phys. Bull.

Physics Bulletin. Published by the Institute of Physics, 47 Belgrave Square, London, SW1X 8QX, England.

Phys. Earth Planet. Inter.
Physics of the Earth and Planetary Interiors. A journal devoted to observational and experimental studies of the Earth and Planetary interiors and their theoretical interpretation by the physical sciences. Publisher: North-Holland Publishing Company, Amsterdam, Netherlands.

Phys. Educ.
Physics Education. Institute of Physics, 47 Belgrave Square, London, SW1X 8QX, England.

Phys. Fluids
The Physics of Fluids. Published by the American Institute of Physics, 335 East 45th Street, New York, NY 10017, USA.

Phys. Lett.
Physics Letters. Volumes A and B. Publisher: North-Holland Publishing Company, Amsterdam.

Phys. Med. Biol.
Physics in Medicine and Biology. Institute of Physics, 47 Belgrave Square, London, SW1X 8QX, England.

Phys. Norvegica
Physica Norvegica. Institute of Physics, University of Oslo, Blindern, per Oslo, Norway.

Phys. Rep. Phys. Lett. C
Physics Reports. Physics Letters Section C. North-Holland Publishing Co., P.O.Box 103, Amsterdam, Netherlands.

Phys. Rev. A
Physical Review A, General Physics. Published for the American Physical Society by the American Institute of Physics, 335 East 45th Street, New York, NY 10017, USA.

Phys. Rev. B
Physical Review B, Solid State. Published for the American Physical Society by the American Institute of Physics 335 East 45th Street, New York, NY 10017, USA.

Phys. Rev. C
Physical Review C, Nuclear Physics. Published for the American Physical Society by the American Institute of Physics, 335 East 45th Street, New York, NY 10017, USA

Phys. Rev. D
Physical Review D, Particles and Fields. Published for the American Physical Society by the American Institute of Physics, 335 East 45th Street, New York, NY 10017, USA.

Phys. Rev. Lett.
Physical Review Letters. Published for the American Physical Society by the American Institute of Physics, 335 East 45th Street, New York, NY 10017, USA.

Phys. Scr.
Physica Scripta. (Formerly Arkiv för Fysik). Published by the Royal Swedish Academy of Sciences, S-104 05 Stockholm 50, Sweden.

Phys. Teach.
Physics Teacher. American Institute of Physics, 335 East 45th Street, New York, NY 10017, USA.

Phys. Today
Physics Today. Published by the American Institute of

Physics, 335 East 45th Street, New York, NY 10017, USA.

Phys. unserer Zeit
Physik in unserer Zeit. Verlag Chemie GmbH, Pappelallee 3, Postfach 1260/1280, D-6940 Weinheim, F. R. Germany.

Physica
Physica. Publishers: North-Holland Publishing Company, Amsterdam, The Netherlands, on request of the Foundation "Physica", Utrecht.

Pis'ma v Astron. Zhurn.
Pis'ma v Astronomicheskij Zhurnal. Akademiya Nauk SSSR. Publishers: Izdatel'stvo 'Nauka', Moskva. Translation in Soviet Astron. Lett.

Planet. Space Sci.
Planetary and Space Science. Pergamon Press Ltd., Headington Hill Hall, Oxford, OX3 OBW, England.

Plasma Phys.
Plasma Physics. Pergamon Press Ltd., Headington Hill Hall, Oxford, OX3 OBW, England.

Pokroky
Pokroky matematiky, fyziky a astronomie. Editor: Jednota čs. matematiků a fyziků. Publisher: Academia, Praha.

Pomiary Autom. Kontrola
Pomiary Automatyka Kontrola. Wydawnictwa Czasopism Technicznych NOT, ul. Czackiego 3/5, Warszawa 1, Poland.

Postępy Astron.
Postępy Astronomii. Czasopismo Poświecone Upowszechnianiu Wiedzy Astronomicznej. Polskie Towarzystwo Astronomiczne, Warszawa. Printed in Poland by Pánstwowe Wydawnictwo Naukowe, Lódź.

Pramāṇa
Pramāṇa. Indian Academy of Sciences, Bangalore 560006, India.

Priroda
Priroda. Publishers: Izdatel'stvo "Nauka", Moskva.

Probl. Kosm. Fiz.
Problemy Kosmichskoj Fiziki. Mezhvedomstvennyj Nauchnoj Sbornik. Izdatel'skoe Obedinenie Vishcha Shkola. Izdatel'stvo pri Kievskom Universitete, Kiev.

Proc. Astron. Soc. Australia
Proceedings of the Astronomical Society of Australia. Published for the Society by Sydney University Press, Sydney.

Proc. IEEE
Proceedings of the IEEE. Published by the Institute of Electrical and Electronics Engineers, 345 East 47th Street, New York, NY 10017, USA.

Proc. Indian Acad. Sci. A
Proceedings of the Indian Academy of Sciences, Section A. Bangalore 560006, India.

Proc. Instn. Electr. Eng.
Proceedings of the Institution of Electrical Engineers. Institution of Electrical Engineers, Savoy Place, London, WC2R OBL, England.

Proc. Int. Latitude Obs. Mizusawa
Proceedings of the International Latitude Observatory of Mizusawa. Published by the International Latitude Observatory of Mizusawa, Japan.

Proc. Japan Acad.
Proceedings of the Japan Academy. Ueno Park, Tokyo, Japan.

Proc. K. Nederlandse Akad. Wet. B
Koninklijke Nederlandse Akademie van Wetenschappen. Proceedings. Series B, Physical Sciences. Publisher: North-Holland Publishing Company, Amsterdam, Netherlands.

Proc. Natl. Acad. Sci. U.S.A.
Proceedings of the National Academy of Sciences of the United States of America. National Academy of Sciences, 2101 Constitution Avenue, Washington, DC 20418, USA.

Proc. R. Instn. Great Britain
Proceedings of the Royal Institution of Great Britain. 21 Albemarle Street, London, W1, England.

Proc. R. Irish Acad., Sect. A
Proceedings of the Royal Irish Academy, Section A (Mathematical, Astronomical and Physical Science). 19 Dawson Street, Dublin 2, Ireland.

Proc. R. Soc. London, Ser. A
Proceedings of the Royal Society of London. Series A: Mathematical and Physical Sciences. Published by the Royal Society, 6 Carlton House Terrace, London, SW1Y 5AG, England.

Proc. Res. Inst. Atmos. Nagoya Univ.
Proceedings of the Research Institute of Atmospherics Nagoya University. Nagoya University, 13 Honohara, 3 Chrome, Toyokawa 442, Japan.

Prog. Aerosp. Sci.
Progress in Aerospace Sciences. Pergamon Press Ltd., Headington Hill Hall, Oxford, OX3 0BW, England.

Prog. Theor. Phys.
Progress of Theoretical Physics. Published for the Research Institute for Fundamental Physics and the Physical Society of Japan. Publication Office: Progress of Theoretical Physics, Yukawa Hall, Kyoto University, 606 Kyoto, Japan.

Prog. Theor. Phys., Suppl.
Supplement of the Progress of Theoretical Physics. Published for the Research Institute for Fundamental Physics and The Physical Society of Japan. Publication Office: Progress of Theoretical Physics. Yukawa Hall, Kyoto University, 606 Kyoto, Japan.

PTB Mitt.
PTB Mitteilungen. Forschen + Prüfen. Fachorgan für Wirtschaft und Wissenschaft. Amts- und Mitteilungsblatt der Physikalisch-Technischen Bundesanstalt. Braunschweig—Berlin. Deutscher Eichverlag, Postfach 3367, D-3300 Braunschweig, F.R. Germany.

Publ. Astron. Soc. Japan
Publications of the Astronomical Society of Japan. Published by the Astronomical Society of Japan. Office of the Society: Tokyo Astronomical Observatory, Mitaka, Tokyo. Agent: Maruzen Co. Ltd. (Export Department), Nihonbashi, Tokyo, Japan.

Publ. Astron. Soc. Pacific
Publications of the Astronomical Society of the Pacific. Published in Provo, Utah, by the Astronomical Society of the Pacific, San Francisco, California. Printed by Brigham Young University Printing Service, Provo, Utah 84602, USA.

Publ. Dominion Astrophys. Obs.
Publications of the Dominion Astrophysical Observatory, Victoria, B.C. National Research Council of Canada.

Publ. Eidg. Sternw. Zürich
Publikationen der Eidgenössischen Sternwarte Zürich. Schulthess Polygraphischer Verlag, Zürich.

Publ. Inst. Geophys., Polish Acad. Sci., F
Publications of the Institute of Geophysics, Polish Academy of Sciences, F: Planetary geodesy. Państwowe Wydawnictwo Naukowe, Warszawa, Poland.

Publ. Inst. R. Meteorol. Belgique A
Publications, Institut Royal Météorologique de Belgique. Serie A. 3 Avenue Circulaire, Uccle-Bruxelles 1180, Belgium.

Publ. Int. Latitude Obs. Mizusawa
Publications of the International Latitude Observatory of Mizusawa. Published by the International Latitude Observatory of Mizusawa, Japan.

Publ. R. Obs. Edinburgh
Publications of the Royal Observatory, Edinburgh. Published by The Royal Observatory, Edinburgh, Scotland.

Publ. Tartu Astrofiz. Obs.
W. Struve nimelise Tartu Astrofüüsika Observatooriumi, Publikatsioonid. Eesti NSV Teaduste Akadeemia, Tartu.

Publ. United States Naval Obs.
Publications of the United States Naval Observatory. Department of the Navy, U.S. Naval Observatory, Washington. U.S. Government Printing Office, Washington, D.C.

Publ. Variable Star Sect. R. Astron. Soc. New Zealand
Publications of the Variable Star Section, Royal Astronomical Society of New Zealand. Director: F. M. Bateson, Greerton, Tauranga, New Zealand.

Pure Appl. Geophys.
Pure and Applied Geophysics. Birkhäuser Verlag, CH-4010 Basel, Elizabethenstrasse 19, Switzerland.

Q. Appl. Math.
Quarterly of Applied Mathematics. American Mathematical Society, P.O. Box 1571, Providence, RI 02901, USA.

Q. Bull. Sol. Act.
International Astronomical Union, Quarterly Bulletin on Solar Activity. Published by the Eidgenössische Sternwarte in Zürich with financial support from UNESCO.

Q. J. R. Astron. Soc.
Quarterly Journal of the Royal Astronomical Society. Burlington House, London, W1V ONL, England.

Q. J. R. Meteorol. Soc.
Quarterly Journal of the Royal Meteorological Society. Cromwell House, High Street, Bracknell, Berks, England.

R. Obs. Ann.
Royal Observatory Annals. Royal Greenwich Observatory, Herstmonceux Castle, Hailsham, East Sussex, BN 27 1RP, England.

Radiat. Eff.
Radiation Effects. Gordon & Breach Science Publishers Ltd., 41 and 42 William IV Street, London, WC2, England.

Radio Commun.
Radio Communication. Radio Society of Great Britain, 35 Doughty Street, London, WC1N 2AE, England.

Radio Sci.
Radio Science. American Geophysical Union, 2901 Byrdhill Road, Richmond, VA 23228, USA.

Radiochem. Radioanal. Lett.
Radiochemical and Radioanalytical Letters. Elsevier Sequoia S.A., P.O. Box 851, CH-1001 Lausanne, Switzerland. Akademiai Kiado, Alkotmany U 21, Budapest 5, Hungary.

Radiotekh. Ehlektron.
Radiotekhnika i Ehlektronika. Moskva TSP-3, Pr. Karl Marx 18, USSR.

Rech. Aerosp.
Recherche Aerospatiale. Office National d'Études et de Recherches Aerospatiales, 29 Avenue de la Division Leclerc, 92320-Chatillon, France.

Recherche
Recherche, 4 Place de l'Odéon, Paris 6, France.

Ref. Zh. 51. Astron.
Referativnyj Zhurnal. 51. Astronomiya. Vsesoyuznyj Institut Nachnoj i Tekhnicheskoj Informatsii. Moskva.

Ref. Zh. 52. Geod. i Aehrosemka
Referativnyj Zhurnal. 52. Geodeziya i Aehrosemka. Vsesoyuznyj Institut Nachnoj i Tekhnicheskoj Informatsii. Moskva.

Ref. Zh. 62. Issled. kosm. prostranstva
Referativnyj Zhurnal. 62. Issledovanie Kosmicheskogo Prostranstva. Vsesoyuznyj Institut Nauchnoj i Tekhnicheskoj Informatsii. Moskva.

Rep. Finnish Geod. Inst.
Reports of the Finnish Geodetic Institute. Suomen Geodeettisen Laitoksen Tiedonantoja. Helsinki, Finland.

Rep. Inst. Phys. Chem. Res.
Reports of the Institute of Physical and Chemical Research. Rikagaku Kenkyushu, Wako-shi, Saitama 351, Japan.

Rep. Ionos. Space Res. Japan
Report of Ionosphere and Space Research in Japan. Institute of Space and Aeronautical Science, University of Tokyo, Komaba, Meguro-ku, Tokyo 153, Japan.

Rep. Math. Phys.
Reports on Mathematical Physics. Pergamon Press Ltd., Headington Hill Hall, Oxford OX3 0BW, England. Subscription address: ARS Polona-Ruch Foreign Trade Enterprise, Krakowskie Przedmiescie 7, 00-068 Warszawa.

Rep. NRL Prog.

Report of NRL Progress. National Technical Information Service, U.S. Department of Commerce, Springfield, VA 22151, USA.

Rep. Obs. Lund
Reports from the Observatory of Lund.

Rep. Prog. Phys.
Reports on Progress in Physics. Published by the Institute of Physics, 47 Belgrave Square, London, SW1X 8QX, England.

Res./Dev.
Research/Development. F.D. Thompson Publications, 1301 South Grove Avenue, Barrington, IL 60010, USA.

Rev. Acad. Cienc. Zaragoza
Revista de la Academia de Ciencias Zaragoza. Academia de Ciencias Exactas, Fisico-Quimicas y Naturales de Zaragoza, Plaza de Paraiso 1, Zaragoza, Spain.

Rev. Astron.
Revista Astronomica. Organo de la Asociación Argentina Amigos de la Astronomia, Avenida Patricias Argentinas 550, Buenos Aires 5, Argentina.

Rev. Brasil. Fis.
Revista Brasileira de Fisica. Sociedade Brasileira de Fisica, Cx. Postal 20553, Sao Paolo SP, Brazil.

Rev. Céthédec
Revue du Céthédec. Centre d'Études Théoriques de la Détection et des Communications, 5 bis, avenue de la Porte de Sèvres, Paris 15, France.

Rev. Fac. Sci. Univ. Istanbul C
Revue de la Faculté des Sciences de l'Université d'Istanbul (Istanbul Universitesi Fen Fakultesi Mecmuasi) Série C (Astronomie, Physique, Chimie). Beyazit, Istanbul, Turkey.

Rev. Geofis.
Revista de Geofisica, Instituto Nacional de Geofisica, Serrano 123, Madrid 2, Spain.

Rev. Geophys. Space Phys.
Reviews of Geophysics and Space Physics (formerly Reviews of Geophysics). Published by the American Geophysical Union, 1909 K Street, N.W., Washington, DC 20006, USA.

Rev. Mexicana Astron. Astrofis.
Revista Mexicana de Astronomia y Astrofisica. Dirección: Instituto de Astronomia, Universidad Nacional Autónoma de México, Apartado Postal 70-264, Mexico 20, D.F., Mexico.

Rev. Mexicana Fis.
Revista Mexicana de Fisica. Sociedad Mexicana de Fisica, Apartado Postal No. 20-364, Mexico 20, D.F, Mexico.

Rev. Mod. Phys.
Reviews of Modern Physics. Published for the American Physical Society by the American Institute of Physics, 335 East 35th Street, New York, NY 10017, USA.

Rev. Phys. Appl.
Revue de Physique Appliquée. Société Française de Physique. 87 bis, Avenue du Général-Leclerc, 75014 Paris, France.

Rev. Polytech.

Revue Polytechnique. Chemin de la Caroline 22, 1213 Petit-Lancy, Genève, Switzerland.

Rev. Radio Res. Lab.
Review of the Radio Research Laboratories. Tokyo, Japan.

Rev. Roumaine Phys.
Revue Roumaine de Physique. Academie Republicii Populare Romine, Boite Postale 134 - 135, Bucuresti, Rumania.

Rev. Sci. Instrum.
Review of Scientific Instruments. American Institute of Physics, 335 East 45th Street, New York, NY 10017, USA.

Rezul't. Nablyud. Iskusstv. Sputnikov Zemli
Rezul'taty Nablyudenij Iskusstvennykh Sputnikov Zemli. Published by Astronomicheskij Sovet Akademii Nauk SSSR, Ryazanskij Gosudarstvennyj Pedagogicheskij Institut, Ryazan'.

Říše hvězd
Říše hvězd. Czechoslovak popular astronomical journal. Publisher: Orbis, Praha.

Rumanian Sci. Abstr.
Rumanian Scientific Abstracts. Natural Sciences. Publishers: The Scientific Documentation Centre of the Academy of the Socialist Republic of Romania, Bucureşti.

Sci. American
Scientific American. 415 Madison Avenue, New York, NY 10017, USA.

Sci. Cult.
Science and Culture. Indian Science News Association, 92 Acharya Prafullachandra Road, Calcutta 9, India.

Sci. Dimension
Science Dimension. National Research Council of Canada, Ottawa K1A OR6, Canada.

Sci. Pap. Coll. Gen. Educ. Univ. Tokyo
Scientific Papers of the College of General Education, University of Tokyo. Komaba, Meguro-ku, Tokyo, Japan.

Sci. Pap. Inst. Phys. Chem. Res.
Scientific Papers of the Institute of Physical and Chemical Research. Rikagaku Kenkyusho, Wako-shi, Saitama 351, Japan.

Sci. Prog.
Science Progress. Blackwell Scientific Publications, Oxford, England.

Sci. Prog. Découverte
Science Progrès Découverte (formerly Science Progrès, La Nature). Revue publiée avec la participation du Palais de la Découverte. Published by Dunod, Editeur, Paris. Imprimerie Bayeusaine, Bayeux.

Sci. Rep. Tôhoku Univ. I.
The Science Reports of the Tôhoku University. First Series (Physics, Chemistry, Astronomy). Published by the Faculty of Science, Tôhoku University, Sendai, Japan.

Sci. Sinica
Scientia Sinica. Edited by Editorial Committee of Scientia Sinica, Peking. Published by Science Press,

Peking, China.

Sci. Sintering
Science of Sintering. Yugoslav Committee for ETAN, P.O. Box 356 (ETAN), 11001 Beograd, Yugoslavia.

Science
Science. American Association for the Advancement of Science, 1515 Massachusetts Avenue, N.W., Washington, D.C. 20005, USA.

Sdelovaci Tech.
Sdelovaci Technika. Publishers of Technical Literature, Spalena 51, 11302 Praha 1, Czechoslovakia.

SIAM J. Appl. Math.
SIAM Journal on Applied Mathematics. Society for Industrial and Applied Mathematics, 33 South 17th Street, Philadelphia, PA 19103, USA.

Siemens Rev.
Siemens Review. Siemens-Aktiengesellschaft, Postfach 325, 8520 Erlangen 2, F. R. Germany.

Simon Stevin
Simon Stevin. De Natuur - en Geneeskundige Vennootschap, Rozier 44, B-9000 Gent, Belgium.

Sitzungsber. Akad. Wiss. DDR
Sitzungsberichte der Akademie der Wissenschaften der DDR. Mathematik-Naturwissenschaften-Technik. Akademie-Verlag, Berlin.

Sitzungsber. Bayerische Akad. Wiss.
Bayerische Akademie der Wissenschaften. Mathematisch-Naturwissenschaftliche Klasse. Sitzungsberichte. Publisher: Verlag der Bayerischen Akademie der Wissenschaften, München.

Sitzungsber. Heidelberger Akad. Wiss.
Sitzungsberichte der Heidelberger Akademie der Wissenschaften. Mathematisch-Naturwissenschaftliche Klasse. Publisher: Springer-Verlag, Heidelberg.

Sitzungsber. Österreich. Akad. Wiss.
Sitzungsberichte. Österreichische Akademie der Wissenschaften. Mathematisch-Naturwissenschaftliche Klasse. Abteilung II: Mathematik, Astronomie, Meteorologie und Technik. Publisher: Springer-Verlag, Wien.

Sky Telesc.
Sky and Telescope. Published by Sky Publishing Corporation, 49-50-51 Bay State Road, Cambridge, Mass. 02138.

Slaboproudy Obz.
Slaboproudy Obzor. Krakovska 8, Praha 1, Czechoslovakia.

Smithsonian Astrophys. Obs., Spec. Rep.
Smithsonian Institution, Astrophysical Observatory, Cambridge, Massachusetts 02138. Research in Space Science. SAO Special Report.

Smithsonian Contrib. Astrophys.
Smithsonian Contributions to Astrophysics. Smithsonian Institution Astrophysical Observatory, Cambridge, Mass. Printed by Smithsonian Institution Press, City of Washington. For sale by the Superintendent of Documents, U.S. Government Printing Office, Washington, D.C.

Smithsonian Year
Smithsonian Year. Annual Report of the Smithsonian

Institution, including the financial report of the Executive Committee of the Boards of Regents. Published by the Smithsonian Institution, Washington, D.C.

Sol. Energy
Solar Energy. Pergamon Press, Maxwell House, Fairview Park, Elmsford, NY 10523, USA

Sol. Phys.
Solar Physics. A Journal for Solar Research and the Study of Solar Terrestrial Physics. Publisher: D. Reidel Publishing Company, Dordrecht, Holland.

Soln. Dannye, Byull.
Solnechnye Dannye. Byulleten'. *(Solar Data).* Publishers: Izdatel'stvo "Nauka", Leningradskoe Otdelenie, Leningrad.

Soobshch. Byurakan. Obs.
Soobshcheniya Byurakanskoj Observatorii. Akademiya Nauk Armyanskoj SSR, Erevan.

Soobshch. Gos. Astron. Inst. Shternberg
Soobshcheniya Gosudarstvennogo Astronomicheskogo Instituta im. P.K. Shternberga. Publishers: Izdatel'stvo Moskovskogo Universiteta, Moskva.

Soobshch. Spets. Astrofiz. Obs.
Soobshcheniya Spetsial'noj Astrofizicheskoj Observatorii. Izdanie Spetsial'noj Astrofizicheskoj Observatorii AN SSSR.

South African Astron. Obs. Circ.
South African Astronomical Observatory, Circulars. S.A. Astronomical Observatory, Observatory, Cape.

South African J. Antarct. Res.
South African Journal of Antarctic Research. Editorial Address: P.O. Box 3718, Johannesburg, South Africa.

South. Stars
Southern Stars. The Journal of the Royal Astronomical Society of New Zealand (Inc.). Address of the Society: P.O. Box 3181, Wellington C1, New Zealand.

Soviet Astron.
Soviet Astronomy. A translation of Astronomicheskij Zhurnal (Astronomical Journal). Published by the American Institute of Physics, New York, N.Y.

Soviet Astron. Lett.
Soviet Astronomy Letters. A translation of "Pis'ma v Astronomicheskij Zhurnal". Published by the American Institute of Physics.

Space Sci. Instrum.
Space Science Instrumentation. An International Journal of Scientific Instruments for Aircraft, Balloons, Sounding Rockets, and Spacecraft. Published by D. Reidel Publishing Company, Dordrecht, Holland.

Space Sci. Rev.
Space Science Reviews. Publishers: D. Reidel Publishing Company, Dordrecht, Holland.

Spaceflight
Spaceflight. Published by the British Interplanetary Society. Printed by Unwin Brothers Ltd., at the Gresham Press, Old Woking, England.

Spaceworld
Spaceworld. Palmer Publications Inc., Amherst, WI

54406, USA.

Spectrosc. Lett.
Spectroscopy Letters. Marcel Dekker Inc., 270 Madison Avenue, New York, NY 10016, USA.

Sterne
Die Sterne. Zeitschrift für alle Gebiete der Himmelskunde. Johann Ambrosius Barth, Leipzig, German Democratic Republic.

Sterne Weltraum
Sterne und Weltraum. Astronomische Monatsschrift. Publisher: Verlag Sterne und Weltraum Dr. Vehrenberg, Düsseldorf, F.R. Germany.

Sternenbote
Sternenbote. Monatsschrift für Österreichs Amateurastronomen. Publisher: Astronomisches Büro, Hermann Mucke, Wien, Austria.

Stockholms Obs. Ann.
Stockholms Observatorium Annaler. Printed by Almquist & Wiksell, Stockholm, Sweden.

Stockholms Obs. Rep.
Stockholms Observatorium, Saltsjöbaden, Sweden, Report.

Strolling Astron.
The Strolling Astronomer. The Journal of The Association of Lunar and Planetary Observers, Publication Office: The Strolling Astronomer, Box 3 AZ, University Park, New Mexico, USA.

Stud. Appl. Math.
Studies in Applied Mathematics. American Elsevier Publishing Co., 52 Vanderbilt Avenue, New York, NY 10017, USA.

Stud. Cercet. Astron.
Studii şi Cercetări de Astronomie. Editura Academiei Republicii Socialiste România. Editorial Office: Observatorul Astronomic, Bucuresti, Rumania.

Stud. Cercet. Fiz.
Studii şi Cercetări de Fizica. Academia Republicii Populare Romine. P.O. Box 134-5, Calca Victoriei 126, Bucureşti, Rumania.

Stud. Geophys. Geod.
Studia geophysica et geodaetica. Published for the Geophysical Institute of the Czechoslovak Academy of Sciences by Academia, Praha.

Stud. Soc. Sci. Torunensis
Studia Societatis Scientiarum Torunensis, Toruń — Polonia. Sectio F (Astronomia).

Stud. Univ. Babeş-Bolyai
Studia Universitatis Babeş-Bolyai. Series Mathematica-Physica. Publishers: Intreprinderea Poligrafica, Cluj.

Tartu Astron. Obs. Teated
Tartu Astronoomia Observatoorium Teated. Eesti NSV Teaduste Akadeemia W. Struve nim. Tartu Astrofüüsika Observatoorium, Tartu.

Technica
Technica. E. Birkhäuser, CH-4010 Basel 10, Switzerland.

Tectonophysics

Tectonophysics. Elsevier Scientific Publishing Co., P.O. Box 211, Amsterdam, Netherlands.

Tegnikon
Tegnikon. Die Suid-Afrikaanse Akademie wir Wetenskap en Kuns, Engelenburghuis, Hamiltonstraat, Postbus 538, Pretoria, S. Africa.

Tehnika
Tehnika. Beograd, Kneza Milosa 7/11, Yugoslavia.

Tek. Ukebl.
Teknisk Ukeblad. Norske Ingeniorforening og den Polytekniske Forening, Kronprinsensgate 17, Oslo 1, Norway.

Telecommun. J.
Telecommunication Journal.(English Edition). International Telecommunications Union,Place des Nations, 1211 Genève 20, Switzerland.

Tellus
Tellus, a bi-monthly Journal of Geophysics. Svenska Geofysiska Foreningen, Arrhenius laboratoriet, Fack, S - 104 05 Stockholm, Sweden.

Tijdschr. Ned. Elektron. Radiogenoot.
Tijdschrift van het Nederlands Elektronica- en Radiogenootschap. Postbus 39, Leidschendam, Netherlands.

Tokyo Astron. Bull.
Tokyo Astronomical Observatory, Japan. Tokyo Astronomical Bulletin.

Tokyo Astron. Obs. Rep.
University of Tokyo, Tokyo Astronomical Observatory, Japan. Report.

Tr. Astrofiz. Inst. Alma-Ata
Trudy Astrofizicheskogo Instituta, Alma-Ata. Akademiya Nauk Kazakhskoj SSR. Publishers: Izdatel'stvo "Nauka" Kazakhskoj SSR, Alma-Ata.

Tr. Astron. Obs., *Leningrad*
Uchenye Zapiski Gosudarstvennogo Universiteta im. A. A. Zhdanova, Seriya matematicheskikh nauk = Trudy Astronomicheskoj Observatorii. Izdatel'stvo Leningradskogo Universiteta, Leningrad.

Tr. Glav. Astron. Obs. Pulkovo
Trudy Glavnoj Astronomicheskoj Observatorii v Pulkove. Akademiya Nauk SSR. Izdanie Glavnoj astronomicheskoj observatorii v Pulkove, Leningrad.

Tr. Inst. Teor. Astron., *Leningrad*
Trudy Instituta Teoreticheskoj Astronomii. Akademiya Nauk SSSR. Publishers: Izdatel'stvo "Nauka", Leningrad.

Tr. Kazan. Gorod. Astron. Obs.
Trudy Kazanskoj Gorodskoj Astronomicheskoj Observatorii. Izdatel'stvo Kazanskogo Universiteta, Kazan.

Tr. Tashkent.Astron. Obs.
Trudy Tashkentskoj Astronomicheskoj Observatorii. Akademiya Nauk Uzbekskoj SSR. Publishers: Izdatel'stvo "FAN" Uzbekskoj SSR, Tashkent.

Trans. Astron. Obs. Yale Univ.
Transactions of the Astronomical Observatory of Yale University. Published by the Observatory, New Haven.

Trans. IAU

Transactions of the International Astronomical Union.
Published and distributed for the IAU (UAI) by D. Reidel Publishing Company, Dordrecht, Holland – Boston, U.S.A.

Trans. Inst. Electron. Commun. Eng. Japan E
Transactions of the Institute of Electronics and Communication Engineers of Japan, Section E. Denshi Tsushin Gakkai, Kikai-Shinko-kaikan Bldg., Shiba Park 21-1-5, Minatoku, Tokyo, Japan. English translation of selected articles appear in Electron & Commun. Jap. (USA) and Syst. Comput. Control (USA).

Trans. R. Soc. Canada
Transactions of the Royal Society of Canada (Mémoires de la Société Royale du Canada). National Library, 395 Wellington Street, Ottawa 4, Canada. University of Toronto Press, Toronto 5, Ontario.

Tsirk. Astron. Inst. Tashkent
Tsirkulyar Astronomicheskogo Instituta. Akademiya Nauk Uzbekskoj SSR. Izdatel'stvo "FAN" Uzbekskoj SSR, Tashkent.

Tsirk. Astron. Obs. L'vov
Tsirkulyar. Astronomicheskaya Observatoriya. L'vovskij Ordena Lenina Gosudarstvennyj Universitet imeni Ivana Franko. Publisher: Izdatel'stvo L'vovskogo Universiteta, L'vov.

Umschau
Umschau in Wissenschaft und Technik. Umschau Verlag, Stuttgarter Str. 18 - 24, D-6000 Frankfurt/M., F.R. Germany.

United States Naval Obs., Circ.
United States Naval Observatory, Circular. U.S. Naval Observatory, Washington, D.C. 20390.

Urania Barcelona
Urania. Revista de Astronomía y Ciencias Afines. Órgano de la Sociedad Astronómica de España y América, Barcelona; Unión Nacional de Astronomía y Ciencias Afines, Madrid, Spain.

Urania Kraków
Urania. Miesięcznik Polskiego Towarzystwa Miłośników Astronomii, Kraków. Publisher: Krakowska Drukarnia Prasowa, Kraków, Poland.

Vasiona
Vasiona. Revue d'Astronomie et d'Astronautique. Bulletin de la Société Astronomique "R. Bosković", Beograd.

Vatican Obs. Publ.
Vatican Observatory Publications, Specola Vaticana, Città del Vaticano, Castel Gandolfo, Italy.

Veröff. Astron. Rechen-Inst. Heidelberg
Veröffentlichungen des Astronomischen Rechen-Instituts Heidelberg. Verlag G. Braun, Karlsruhe, F.R. Germany.

Veröff. Sternw. Sonneberg
Akademie der Wissenschaften der DDR, Zentralinstitut für Astrophysik, Veröffentlichungen der Sternwarte in Sonneberg. Publisher: Akademie-Verlag, Berlin, German Democratic Republic.

Veröff. Zentralinst. Phys. Erde
Akademie der Wissenschaften der DDR, Forschungsbereich Geo- und Kosmowissenschaften. Veröffentlichungen des Zentralinstituts für Physik der Erde, Potsdam,

German Democratic Republic.

Vesmír
Vesmír. Přírodovědecky časopis Čs. akadmie věd. Publisher: Academia, Praha.

Vestn. Khar'kov. Univ.
Vestnik Khar'kovskogo Universiteta. Seriya Astronomicheskaya. Publishers: Izdatel'stvo Khar'kovskogo Universiteta, Khar'kov.

Vestn. Kiev. Univ.
Vestnik Kievskogo Universiteta. Seriya Astronomii. Publishers: Izdatel'stvo Kievskogo Universiteta, Kiev.

Vierteljahrsschr. Naturforsch. Ges. Zürich
Vierteljahrsschrift der Naturforschenden Gesellschaft in Zürich. Printer and Publisher: Leeman AG, Zürich, Switzerland.

Vistas Astron.
Vistas in Astronomy. An international review journal. Pergamon Press, Oxford–New York–Braunschweig.

Weather
Weather. James Glaisher House, Grenville Place, Bracknell, Berks RG12 1BX, England.

Wiad. Telekomun.
Wiadomosci Telekomunikacyjne. Editorial address: Warszawa, Kazimierzowska 52, Poland.

Wireless World
Wireless World. IPC Electrical-Electronic Press Ltd., Dorset House, Stamford Street, London, SE1 9LU, England.

Wiss. Fortschr.
Wissenschaft und Fortschritt. Akademie-Verlag Berlin, Berlin-Treptow.

Wiss. Z. Friedrich-Schiller-Univ. Jena
Wissenschaftliche Zeitschrift der Friedrich-Schiller-Universität. Jena. Mathematisch-Naturwissenschaftliche Reihe; Edited by the Rektor der Friedrich-Schiller-Universität Jena, Am Anger 24, Jena, German Democratic Republic.

Wiss. Z. Humboldt-Univ. Berlin
Wissenschaftliche Zeitschrift der Humboldt-Universität zu Berlin. Mathematisch-Naturwissenschaftliche Reihe. Edited by the Rektor der Humboldt-Universität Berlin, Unter den Linden 6, 108 Berlin, German Democratic Republic.

Yamamoto Circ.
Yamamoto Circular. Published by the Yamamoto

Observatory, Kamitanakami-Kiryutyo, Otu, Siga-Ken, [520-21] Japan.

Z. angew. Math. Mech.
Zeitschrift für angewandte Mathematik und Mechanik. Akademie-Verlag GmbH, 108 Berlin, Leipziger Strasse 3–4, German Democratic Republic.

Z. Elektr. Inf.- Energietech.
Zeitschrift für Elektrische Informations- und Energietechnik. Akademische Verlagsgesellschaft Geest & Portig K.G., DDR 701 Leipzig, Sternwartenstrasse 8, Germany. Formerly: Hochfrequenztech. & Elektroakust. (Germany) and Wiss. Z. Elektrotech. (Germany).

Z. Naturforsch.
Zeitschrift für Naturforschung. Teil A. A Europhysics Journal. Physik–Physikalische Chemie–Kosmophysik. Verlag der Zeitschrift für Naturforschung, Tübingen. P. O. Box 2645, D–7400 Tübingen, F. R. Germany.

Z. Phys. A
Zeitschrift für Physik A. Atoms and Nuclei. Springer-Verlag, Berlin–Heidelberg–New York.

Z. Phys. B
Zeitschrift für Physik B. Condensed Matter and Quanta. Springer-Verlag, Berlin–Heidelberg–New York.

Z. Vermessungswes.
Zeitschrift für Vermessungswesen. Verlag Konrad Wittwer, 7000 Stuttgart 1, Nordbahnhofstrasse 16, Postfach 147, F.R. Germany.

Zemlya i Vselennaya
Zemlya i Vselennaya. Astronomiya, Geofizika, Issledovaniya Kosmicheskogo Prostranstva. Nauchno-Populyarnyj Zhurnal Akademii Nauk SSSR. Publishers: Izdatel'stvo "Nauka", Moskva.

Zenit
Populair wetenschappelijk maandblad over sterrenkunde/weerkunde/ruimtevaart/ruimte-onderzoek/aanverwante wetenschappen en technieken. Bureau: Stichting De Koepel, Utrecht.

Zentralbl. Math. Grenzgeb. – Math. Abstr.
Zentralblatt für Mathematik und ihre Grenzgebiete – Mathematics Abstracts. Publisher: Springer-Verlag, Berlin–Heidelberg–New York.

Zvaigžnota Debess
Latvijas PSR Zinātņu Akadēmijas Radioastrofizikas Observatorijas Populārzinatnisks Gadalaiku Izdevums. Izdevnieciba "Zinātne", Riga.

Journals abstracted completely.

A selected number of periodicals listed in category 001 are central to the subject scope of *Astronomy and Astrophysics Abstracts.* Depending on their relevance, almost all papers of the journals listed below are abstracted in our service.

AAVSO Bull.
Acta Astron.
Acta Astron. Sinica
Acta Cosmologica
Adv. Astron. Astrophys.
AFOEV Bull.
Ann. Tokyo Astron. Obs.
Ann. Univ.-Sternw. Wien
Annu. Rep. Astron. Inst. Greece
Annu. Rev. Astron. Astrophys.
Assoc. Veneta Oss. Stelle Variabili Bull.
Astrofiz. Issled. Izv. Spets. Astrofiz. Obs.
Astrofizika
Astrometr. Astrofiz.
Astron. Astrophys.
Astron. Astrophys., Suppl. Ser.
Astron. J.
Astron. Nachr.
Astron. Schule
Astron. Tidsskr.
Astron. Tsirk.
Astron. Vestn.
Astron. Zh. Akad. Nauk SSSR
Astronomia
Astronomie
Astrophys. J.
Astrophys. J., Lett.
Astrophys. J., Suppl. Ser.
Astrophys. Lett.
Astrophys. Space Sci.
Atti Soc. Astron. Italiana
Australian J. Phys., Astrophys. Suppl.
BAV Rundbrief
BBSAG Bull.
Blick Weltall
Bol. Astron. Obs. Madrid
Bol. Inst. Tonantzintla
British Astron. Assoc. Circ.
Bull. American Astron. Soc.
Bull. Astron. Inst. Czechoslovakia
Bull. Astron., Obs. R. Belgique
Bull. Astron. Soc. India
Bull. Obs. Astron. Belgrade
Bull. Soc. Astron. Liège
Byull. Abastuman. Astrofiz. Obs.
Byull. Inst. Astrofiz.
Byull. Inst. Teor. Astron.
Carter Obs., Astron. Bull.
Celestial Mech.
Cent. Données Stellaires, Inf. Bull.
Chinese Astron.
Circ. Inf.
Circ. Stn. Astron. Int. Latitudine, Carloforte-Cagliari
Coelum
Comments Astrophys.
Comun. Obs. Astron. Univ. Coimbra
Contrib. Obs. New Mexico State Univ.
Dudley Obs. Rep.
Dunsink Obs. Publ.
El Universo
ESO Bull.
ESO Sci. Prepr.
ESO Tech. Rep.
G. A.A.B.
G. Astron.
IAU Circ.

Icarus
Inf. Bull. Variable Stars
Inst. Theor. Astrophys., Blindern—Oslo, Rep.
Irish Astron. J.
Izv. Astron. Ehngel'gardt. Obs.
Izv. Glav. Astron. Obs. Pulkovo
Izv. Krymskoj Astrofiz. Obs.
J. American Assoc. Variable Star Obs.
J. Astron. Soc. Victoria
J. Astron. Soc. Western Australia
J. British Astron. Assoc.
J. Hist. Astron.
J. R. Astron. Soc. Canada
Kometn. Tsirk. *Kiev*
Komety i Meteory
Mem. R. Astron. Soc.
Mem. Soc. Astron. Italiana
Mercury
Messenger
Meteoritics
Meteoritika
Minor Planet Bull.
Minor Planet Circ.
Mitt. Astron. Ges.
Mitt. Veränderl. Sterne (MVS)
Mon. Not. R. Astron. Soc.
Mon. Notes Astron. Soc. South Africa
Mon. Notes Int. Polar Motion Serv.
Moon
Nablyud. Iskusstv. Nebesn. Tel
Nauchn. Inf.
News Lett. Astron. Soc. N.Y.
Obs. Astrophys. Lab., Univ. Helsinki. Rep.
Observatory
Occas. Rep. R. Obs. Edinburgh
Occultation Newsl.
Orion
Oss. Astrofis. Catania, Pubbl.
Oss. Mem. Oss. Astrofis. Arcetri
Perem. Zvezdy, Byull.
Perem. Zvezdy, Prilozhenie
Pis'ma v Astron. Zhurn.
Postępy Astron.
Probl. Kosm. Fiz.
Proc. Astron. Soc. Australia
Proc. Int. Latitude Obs. Mizusawa
Publ. Astron. Soc. Japan
Publ. Astron. Soc. Pacific
Publ. Dominion Astrophys. Obs.
Publ. Eidg. Sternw. Zürich
Publ. Int. Latitude Obs. Mizusawa
Publ. R. Obs. Edinburgh
Publ. Tartu Astrofiz. Obs.
Publ. United States Naval Obs.
Publ. Variable Star Sect. R. Astron. Soc. New Zealand
Q. Bull. Sol. Act.
Q. J. R. Astron. Soc.
R. Obs. Ann.
Rep. Obs. Lund
Rev. Astron.
Rev. Mexicana Astron. Astrofis.
Rezul't. Nablyud. Iskusstv. Sputnikov Zemli
Sky Telesc.
Smithsonian Astrophys. Obs., Spec. Rep.
Smithsonian Contrib. Astrophys.
Smithsonian Year

Sol. Phys.
Soln. Dannye, Byull.
Soobshch. Byurakan. Obs.
Soobshch. Gos. Astron. Inst. Shternberg
Soobshch. Spets. Astrofiz. Obs.
South African Astron. Obs. Circ.
South. Stars
Space Sci. Instrum.
Space Sci. Rev.
Sterne
Sterne Weltraum
Sternenbote
Stockholms Obs. Ann.
Stockholms Obs. Rep.
Strolling Astron.
Tartu Astron. Obs. Teated
Tokyo Astron. Bull.
Tokyo Astron. Obs. Rep.
Tr. Astrofiz. Inst. Alma-Ata

Tr. Astron. Obs., *Leningrad*
Tr. Glav. Astron. Obs. Pulkovo
Tr. Inst. Teor. Astron., *Leningrad*
Tr. Kazan. Gorod. Astron. Obs.
Tr. Tashkent. Astron. Obs.
Trans. IAU
Tsirk. Astron. Inst. Tashkent
Tsirk. Astron. Obs. L'vov
United States Naval Obs., Circ.
Urania Barcelona
Urania Kraków
Vatican Obs. Publ.
Veröff. Astron. Rechen-Inst. Heidelberg
Veröff. Sternw. Sonneberg
Vestn. Khar'kov. Univ.
Vestn. Kiev. Univ.
Vistas Astron.
Yamamoto Circ.

002 Bibliographical Publications, Catalogues, Atlases

002.001 The ESO/Uppsala survey of the ESO (B) Atlas of the Southern Sky. V.
E. B. Holmberg, A. Lauberts, H.-E. Schuster, R. M. West.
Astron. Astrophys., Suppl. Ser., Vol. 31, 15 - 54 (1978).

A systematic search for certain objects (NGC + IC galaxies, all galaxies with a diameter larger than about 1.0, all disturbed galaxies, all star clusters in the Budapest Catalogue, and all listed planetary nebulae) is being carried out by means of the ESO (B) Atlas, covering the southern sky from −90° to −20°. The present paper contains the fifth list of objects in 78 fields. A total of 2395 objects are listed; of those south of −45°, about 75% for the first time.

002.002 Stellar testing catalogue. S. Swierkowska.
Veröff. Zentralinst. Phys. Erde, No. 52, Part 3, (see 012.001), p. 1037 - 1044 (1977).

This paper allocates photographic positions of 126 stars. The photographs were taken with two instruments: a satellite camera PO-2 (F = 996 mm, f = 140 mm), and a Schmidt telescope (1800/900/600). The reference catalogue is the SAO catalogue.

002.003 The stars of low luminosity. W. J. Luyten.
Separate print Univ. Minnesota, Minneapolis, Minn. 75 pp. (1977).

The present catalogue constitutes a continuation of the list of 1055 stars of low luminosity published in 1970. The catalogue gives data for 2907 stars and is arranged in the same way as before except that now positions are given to 1^s and 0.1, red as well as photographic magnitudes are listed, and the proper motion is given to 0.001.

002.004 The Tucson Revised Index of Asteroid Data.
D. Bender, E. Bowell, C. Chapman, M. Gaffey, T. Gehrels, B. Zellner, D. Morrison, E. Tedesco.
Icarus, Vol. 33, 630 - 631 (1978).

Attention is called to the availability of the TRIAD computer file, a compilation of all reliable physical parameters for minor planets.

002.005 Come e che cosa osservare, IX. L. Baldinelli.
G. A. A. B., N. 48, p. 3 - 6 (1977).

Un elenco di atlanti e cataloghi stellari accessibili all'astrofilo e di chiara utilità pratica.

002.006 Katalog über 78 irrtümlich als Impaktstrukturen bezeichnete Objekte. J. Classen.
Orion, 36. Jahrg., 31 - 35 (1978).

002.007 Errors in the S.A.O. Catalog. M. Bischoff.
Bull. Inf. Centre Données Stellaires, No. 14, p. 2 - 10 (1978).

The list in this paper contains errors found in the "Smithsonian Astrophysical Observatory Star Catalog".

002.008 The non-duplicity of 189 ADS stars.
G. G. Douglass.
Bull. Inf. Centre Données Stellaires, No. 14, p. 11 - 14 (1978).

The New General Catalogue of Double Stars published by Aitken (1932) contains 189 double stars that have been omitted from the most recent version of the Index Catalogue of Visual Double Stars. In order to ascertain the duplicity of these stars and to insure the completeness of the Index Catalogue, each of the 189 systems was investigated. The astronomical literature was searched and, in several cases, the objects in question were reobserved.

002.009 Errors in the "Luminous Stars in the Southern
Milky Way".   M. Bischoff.
Bull. Inf. Centre Données Stellaires, No. 14, p. 15 (1978).

During the process of incorporating the LSSMW to the "Catalog of Stellar Identifications" (CSI), a number of errors were detected.

002.010 More errata in Bright Star Catalogue, third edition.
D. Hoffleit.
Bull. Inf. Centre Données Stellaires, No. 14, p. 16 (1978).

002.011 Some errata for bright stars in the "Bibliography of radial velocities" by H. A. Abt and E. S. Biggs.
D. Hoffleit.
Bull. Inf. Centre Données Stellaires, No. 14, 17 - 18 (1978).

002.012 Errata in the "Bibliography of stellar radial velocities" by H. A. Abt and E. S. Biggs. W. Buscombe.
Bull. Inf. Centre Données Stellaires, No. 14, p. 19 (1978).

002.013 Enquiry on radioastronomical data. II.
G. Westerhout, C. Jaschek.
Bull. Inf. Centre Données Stellaires, No. 14, p. 20 - 31 (1978).

002.014 Description of a code-numbering system for identifying stars in magnetic tape catalogues.
J.-C. Mermilliod.
Bull. Inf. Centre Données Stellaires, No. 14, p. 32 - 61 (1978).

The general scheme of the code-numbering system already in use at the Geneva and Lausanne observatories is explained in detail. Complete definitions and bibliographical tables are given.

002.015 Les catalogues de vitesses radiales moyennes.
M. Barbier.
Bull. Inf. Centre Données Stellaires, No. 14, p. 69 - 71 (1978).

At Grenoble, IAU Commission 30 (Radial Velocities) created a Working Group to define a set of rules to obtain the best average radial velocity of a star. The author, president of the group, sent a questionnaire to all members of Commission 30. A summary is given in this note of the answers so far received.

002.016 Catalogue des "Boîtes photometriques d'étoiles".
M. Golay.
Bull. Inf. Centre Données Stellaires, No. 14, p. 72 - 73 (1978).

002.017 A few utilisations of the uvbyβ catalogue.
H. Heck.
Bull. Inf. Centre Données Stellaires, No. 14, p. 77 - 78 (1978).

002.018 The IRAS project and its needs for a catalogue of non-stellar objects. P. R. Wesselius.
Bull. Inf. Centre Données Stellaires, No. 14, p. 108 - 109 (1978).

002.019 IAU Commission 45 Working Group on Spectroscopic and Photometric Data. Catalogs recently published, to be published or in preparation. List VIII.
Bull. Inf. Centre Données Stellaires, No. 14, p. 110 - 112 (1978).

002.020 Catalogs recently published, to be published or in preparation for edition on microfiche. List I.
F. Ochsenbein.
Bull. Inf. Centre Données Stellaires, No. 14, p. 113 - 114 (1978).

002.021 Short description of new catalogues available.

Bull. Inf. Centre Données Stellaires, No. 14, p. 115 - 116 (1978).

Concerning: (1) Polarization Catalog. D. S. Mathewson, V. I. Ford, J. Krautter. (2) Tables of atomic spectral lines for the 10000 Å to 40000 Å region. M. Outred.

002.022   The Directory of Astronomical Data Files.
J. Mead.
Bull. Inf. Centre Données Stellaires, No. 14, p. 117 - 118 (1978).

A Directory of Astronomical Data Files has been obtained by the U.S. Interagency Data Task Force and compiled by the NASA-National Space Science Data Center. The objective is to identify sources of astronomical data, both machine-readable and non-machine-readable, including characteristics of each data set and information on availability.

002.023   Star catalogs available at the C.D.S.: corrections and additions.
Bull. Inf. Centre Données Stellaires, No. 14, p. 120 - 121 (1978).

002.024   Catalogs currently availabe on microfiche.
Bull. Inf. Centre Données Stellaires, No. 14, p. 122 (1978).

002.025   White dwarfs II.   W. J. Luyten.
Univ. Minnesota, Minneapolis, Minnesota. 103 pp. (1977).

The present catalogue forms a continuation of the first List of White Dwarfs published in 1970, and containing 3033 entries. The catalogue gives data for 3513 stars but it should be emphasized that while the large majority of stars listed with colors "b" and "a" are probably genuine white dwarfs, the percentage of those which are high-velocity subdwarfs, or main-sequence red dwarfs with strong H and K emission lines increases rapidly with increasing color. Probably no more than forty percent of the stars with color "k" will prove to be genuine yellow degenerates.

002.026   Viking lander imaging investigation – Picture catalog of primary mission. Experiment data record.
R. B. Tucker.
NASA Ref. Publ., NASA RP-1007. For sale by the National Technical Information Service, Springfield, Virginia 22161. 9 + 558 + 7 pp. Price $ 16.25 (1978).

All the images returned by the two Viking landers during the primary phase of the Viking Mission are presented in this report. Listings of supplemental information which describe the conditions under which the images were acquired are included together with skyline drawings which show where the images are positioned in the field of view of the cameras. Subsets of the images are listed in a variety of sequences to aid in locating images of interest. The format and organization of the digital magnetic tape storage of the images are described. A brief description of the mission and the camera system is also included.

002.027   An astronomically oriented bibliography of atomic autoionization.
S. J. Adelman, S. N. Shore, M. A. Nasson.
Publ. Astron. Soc. Pacific, Vol. 89, 780 - 791 (1977/78).

A critical bibliography of papers dealing with atomic autoionization is presented.

002.028   Catalogue of measurements in the six-colour photometric system (magnetic tape).
C. Nicollier, B. Hauck.
Astron. Astrophys., Suppl. Ser., Vol. 31, 437 - 438 (1978).

This catalogue contains all measurements made in the six-colour system of Stebbins and Whitford, being 1702 measurements for 1297 stars.

002.029   A spectrophotometric survey of stars along the Milky Way. Part IV.   B. A. Rydström.
Astron. Astrophys., Suppl. Ser., Vol. 32, 25 - 56 (1978).

In the present paper a catalogue of spectrophotometric quantities, spectral types, monochromatic magnitudes and colour equivalents is given for all stars brighter than the magnitude $m_{4400} = 10^m 5$ in a region of the Milky Way in Perseus. No absorption is found for stars closer than about $r = 100$ pc. The absorbing clouds are situated at distances closer than 1 kpc and at about 2.5 kpc in the local arm and the Perseus arm, respectively. The space between the two arms is free from absorption. It is also concluded that the Perseus arm continues beyond $l = 140°$, containing not only hydrogen gas but also dust (to at least $l = 150°$), while the content of OB stars decreases abruptly at $l = 140°$.

002.030   A catalog of ultraviolet objective-prism spectra.
K. G. Henize, J. D. Wray, S. B. Parsons, G. F. Benedict.
Bull. American Astron. Soc., Vol. 9, 571 (1978). – Abstract.

002.031   The determination of the proper motions of the AGK3R stars.   T. E. Corbin.
Bull. American Astron. Soc., Vol. 9, 597 (1978). – Abstract.

002.032   Catalog of astronomical image processing facilities.
C. T. Bolton.
Bull. American Astron. Soc., Vol. 9, 700 - 708 (1978).

002.033   A catalogue of absorption lines in QSO spectra.
R. Ellis, S. Phillipps.
Mon. Not. R. Astron. Soc., Vol. 183, 271 - 274 and microfiche MN 183/1 (1978).

002.034   Catalogue of the archives and manuscripts of the Royal Astronomical Society.
J. A. Bennett, with a foreword by D. W. Dewhirst.
Mem. R. Astron. Soc., Vol. 85, 1 - 90 (1978).

002.035   The limiting magnitude of the ESO-B and SRC-J Sky Survey.
G. L. White, with a comment by K. P. Tritton.
Proc. Astron. Soc. Australia, Vol. 3, 128 - 130 (1977).

As part of an identification programme undertaken on the southern section of the Molonglo Deep Sky Survey of Radio Sources, the limiting magnitudes of some of the available ESO-B films and SRC-J films and plates in this area have been determined. The southern section comprises a 45′ wide strip centred about declination $-62°$ and stretching in an irregular fashion from $18^h 25^m$ to $00^h 16^m$.

002.036   Bibliography of interstellar travel and communication – April 1977 update.
E. F. Mallove, R. L. Forward, Z. Paprotny.
J. British Interplanet. Soc., Vol. 31, 225 - 234 (1978).

002.037   Remarks on the catalogue of LMC supergiants published by Stock et al. (1976).
C. Fehrenbach, M. Duflot.
Astron. Astrophys., Suppl. Ser., Vol. 32, 159 - 160 (1978).
In French.

The catalogue of LMC star members published by Stock et al. contains a number of stars known as galactic members. This catalogue is dubious.

002.038   A low latitude 21 cm line survey for the longitude range $20° \leqslant l \leqslant 42°$.
E. Braunsfurth, K. Rohlfs.
Astron. Astrophys., Suppl. Ser., Vol. 32, 177 - 204 (1978).

A high resolution 21 cm line survey for latitudes below $|b| = 3°$ is presented in the form of contour plots. The data with $|b| \geqslant 1°5$ have been measured with the Bonn 100 m telescope, those with $|b| \leqslant 1°5$ have been taken from the new Maryland-Green Bank Galactic 21 cm Line Survey (Westerhout 1972). The survey consists of scans perpendicular to the galactic plane, every $0°3$ in l. The angular and velocity resolutions are about $10'$ and 2 km s^{-1}, respectively.

002.039 The Nançay survey of absorption by galactic neutral hydrogen. I. Absorption towards extragalactic sources.   J. Crovisier, I. Kazès, D. Aubry.
Astron. Astrophys., Suppl. Ser., Vol. 32, 205 - 282 (1978).
A survey of galactic absorption by neutral hydrogen at 21 cm was carried out with the Nançay radio telescope towards 819 extragalactic radio sources. The observational technique as well as the reduction procedures are described. Emission profiles at positions close to those of the radio sources were selected from published surveys. The reliability of the results and comparison with other absorption surveys are discussed. The authors present a catalogue where relevant information on 819 sources is given, as well as emission and absorption profiles towards 386 of these sources where absorption was detected or suspected.

002.040 Decametric survey of discrete sources in the Northern sky. II. Source catalogue in the range of declinations + 10° to+20°.   S. Ya. Braude, A. V. Megn (*Men'*), S. L. Rashkovski, B. P. Ryabov, N. K. Sharykin, K. P. Sokolov, A. P. Tkatchenko (*Tkachenko*), I. N. Zhouck (*Zhuk*).
Astrophys. Space Sci., Vol. 54, 37 - 128 (1978).
Coordinates and absolute flux densities of 329 discrete sources located in the interval of declinations between + 10° and + 20° are presented. The measurements were carried out at 10, 12.6, 14.7, 16.7, 20 and 25 MHz on the UTR-2 radio telescope. The coordinates are compared with the 4C catalogue and other high-frequency surveys. There are ten sources in the Grakovo catalogue which have not been detected elsewhere previously.

002.041 Decametric survey of discrete sources in the Northern sky. III. Low-frequency absolute flux scale of discrete radio sources.   S. Ya. Braude, B. P. Ryabov, N. K. Sharykin, I. N. Zhouck (*Zhuk*).
Astrophys. Space Sci., Vol. 54, 129 - 143 (1978).
A new method was developed for comparing absolute flux scales of the decametric surveys. The absolute scales of the new survey performed with the UTR-2 in the 10 to 25 MHz frequency range are compared with UTR-1 data and surveys at 10.03, 22.25, 26.3 and 38 MHz operating frequencies.

002.042 Decametric survey of discrete sources in the Northern sky. IV. Spectra of 266 discrete sources in the range 10 to 1400 MHz.   S. Ya. Braude, A. V. Megn (*Men'*), B. P. Ryabov, N. K. Sharykin, K. P. Sokolov, I. N. Zhouck (*Zhuk*).
Astrophys. Space Sci., Vol. 54, 145 - 179 (1978).
Spectra of 266 discrete sources within a frequency range of 10 to 1400 MHz have been obtained for a declination interval 10° to 20° on the basis of a new decametric survey carried out at Grakovo on the UTR-2 radio telescope, as well as a number of high-frequency surveys performed at Culgoora, Cambridge, Molonglo, Parkes and Ohio. Spectra of 86% of the sources are linear, those of 10% have positive curvature and, of the remainder, 4% have negative curvature. The spectral index distribution has been investigated and the spectra graphs are plotted.

002.043 Catalogue of Martian craters and statistics of the craters of Mars, Mercury and the Moon.
Yu. N. Lipskij, Zh. F. Rodionova, T. P. Skobeleva, K. I. Dekhtyareva, D. A. Kazimirov, B. D. Sitnikov, G. A. Poroshkova, N. V. Bubnova, V. A. Shuvaeva.
Ministerstvo Vysshego i Srednego Spetsial'nogo Obrazovaniya

SSSR. Moskovskij Gosudarstvennyj Universitet im. M. V. Lomonosova. Gosudarstvennyj Astronomicheskij Institut im. P. K. Shternberga. Akademiya Nauk SSSR. Geologicheskij Institut, Moskva. 69 pp. Price 55 Kop. (1977). In Russian.

002.044 Catalogues of craters of Mercury and the Moon.
Yu. N. Lipskij, Zh. F. Rodionova, T. P. Skobeleva, K. I. Dekhtyareva, D. A. Kazimirov, B. D. Sitnikov, G. A. Poroshkova, N. V. Bubnova, V. A. Shuvaeva.
Ministerstvo Vysshego i Srednego Spetsial'nogo Obrazovaniya SSSR. Moskovskij Gosudarstvennyj Universitet im. M. V. Lomonosova. Gosudarstvennyj Astronomicheskij Institut im. P. K. Shternberga. Akademiya Nauk SSSR. Geologicheskij Institut, Moskva. 55 pp. Price 43 Kop. (1977). In Russian.

002.045 S stars in the Southern Milky Way.
B. E. Westerlund, N. Olander.
Astron. Astrophys., Suppl. Ser., Vol. 32, 401 - 408 (1978).
An objective prism survey in near infrared light has been carried out of the Southern Milky Way from $l = 235°$ to $l = 7°$ and between $b = \pm 5°$. The present catalogue gives positions and estimated visual and infrared magnitudes for 74 S stars found in the survey. Finding charts are given for the identification of the stars.

002.046 A bibliography on the search for extraterrestrial intelligence.   E. F. Mallove, M. M. Connors, R. L. Forward, Z. Paprotny.
NASA Ref. Publ., NASA RP-1021, 132 pp. (1978).
This report presents a uniform compilation of works dealing with the search for extraterrestrial intelligence. Entries are by first author, with cross-reference by topic index and by periodical index. This bibliography updates earlier bibliographies on this general topic while concentrating on research related to listening for signals from extraterrestrial intelligence.

002.047 The 5 GHz strong source surveys. IV. Survey of the area between declination 35 and 70 degrees and summary of source counts, spectra and optical identifications.
I. I. K. Pauliny-Toth, A. Witzel, E. Preuss, H. Kühr, K. I. Kellermann, E. B. Fomalont, M. M. Davis.
Astron. J., Vol. 83, 451 - 474 (1978).
A new zone (S4) has been surveyed at 5 GHz and is essentially complete above 0.5 Jy. A total of 269 sources above this flux density were found in an area of 1.71 sr between declinations of 35° and 70°. Accurate flux densities at 2.7, 5.0, and 10.7 GHz are given, together with source coordinates accurate to between 1 and 5 arcsec, and proposed optical identifications. Sources below the completeness limit are also listed, giving a catalog of 484 sources. The results of the S4 survey are combined with those of the previous 5 GHz surveys.

002.048 Annotations to papers on geomagnetism and aeronomy published in "News of Universities. Radiophysics", Vol. 19, No. 12 (1976); Vol. 20 Nos. 1 - 9 (1977).
Geomagn. Aehron., Vol. 18, 562 - 571 (1978). In Russian.

002.049 Catalogue of 202 stars of the supplementary list of the Catalogue of Faint Stars in the declination zone from −15° to +30°.   V. A. Sinyaev.
Odessk. univ. Odessa, 1977. 19 pp. In Russian. − Abstr. in Ref. zh., 51. Astron., 5.51.121 (1978).

002.050 Catalogue of "strict $0°^{m}01$ photometric star boxes".
M. Golay, N. Mandwewala.
Publ. Obs. Genève, Sér. B, Fasc. 4, 30 pp. (1978).
This catalogue gives the stellar content of all the [0.01 mag] (strict) photometric star boxes made using 7082 stars measured in the UBV $B_1B_2V_1G$ system.

002.051   The ESO/Uppsala survey of the ESO (B) Atlas of
          the Southern Sky – VI.
E. B. Holmberg, A. Lauberts, H.-E. Schuster, R. M. West.
ESO Sci. Prepr. No. 28, 113 pp. (1978). – Submitted to
Astronomy and Astrophysics.
     A systematic search for certain objects (NGC + IC
galaxies, all galaxies with a diameter larger than about 1.'0, all
disturbed galaxies, all star clusters in the Budapest Catalogue,
and all listed planetary nebulae) is being carried out by means
of the ESO (B) Atlas, covering the southern sky from –90° to
–20°. The present paper contains the sixth list of objects in
98 fields. A total of 3162 objects are listed; of those south of
–45°, about 90% for the first time.

002.052   Compiled data of C- and M-type stars in solar
          neighborhood.   T. Mikami.
Ann. Tokyo Astron. Obs., Second Ser., Vol. 17, 1 - 49 (1978).
     The physical and kinematical data for 321 carbon stars
and 1490 M-type stars are compiled from various catalogues
and papers available at present. The stars listed here are
limited to those with the measured radial velocities.

002.053   Declinations of 125 zenith stars from observations
          at the Engelhardt Astronomical Observatory.
A. M. Zulliev.
Izv. Astron. Ehngel'gardt. Obs., Kazan', No. 41 - 42, p. 148 -
153 (1976). In Russian.

002.054   An atlas of stellar spectrophotometry.
          V. V. Smith, J. S. Neff.
Bull. American Astron. Soc., Vol. 10, 413 (1978). – Abstract.

002.055   An atlas of photoelectric spectrophotometry of
          comets.   J. S. Neff, V. V. Smith.
Bull. American Astron. Soc., Vol. 10, 460 (1978). – Abstract.

002.056   The Henry Draper Catalog on microfiche.
          S. Danford, R. Muir.
Bull. American Astron. Soc., Vol. 10, 461 (1978). – Abstract.

002.057   Characteristics of the large stellar data base.
          D. M. Gottlieb.
Bull. American Astron. Soc., Vol. 10, 462 (1978). – Abstract.

002.058   Supplement to the Catalogue of the Crawford
          Library of the Royal Observatory Edinburgh.
M. F. I. Smyth, M. J. Smyth.
Royal Observatory, Edinburgh. 11 + 112 pp. Price £ 5.00
(1977). ISBN 0-902553-18-6.
     This Supplement is a computer-produced index of the
earlier works of the Crawford Collection published before the
year 1800, and of manuscripts, broadsheets and works on
comets published up to the early 1900s. It includes, in addition to the works entered in the original 1890 Catalogue all
those in the same categories, acquired from whatever source,
now belonging to the Royal Observatory Edinburgh, forming
an augmented Crawford Collection. Of the 4376 volumes and
pamphlets indexed, about one-fifth do not appear in the
original Catalogue.

002.059   Astronomy and Astrophysics Abstracts. Vol. 20:
          Literature 1977, Part 2.
S. Böhme, U. Esser, W. Fricke, I. Heinrich, D. Krahn, L. D.
Schmadel, G. Zech (Editors).
Published for Astronomisches Rechen-Institut by Springer-
Verlag, Berlin–Heidelberg–New York. 10 + 786 pp. Price
DM 95.00; ca. US $ 47.50 [Subscription price DM 76.00;
ca. US $ 38.00] (1978).

002.060   List of dissertations defended at sessions of Special-
          ized Councils of the Moscow State University in the
field of astronomy and gravimetry in 1977.

L. N. Bondarenko.
Astron. Tsirk., No. 976, p. 7 - 8 (1977). In Russian.

002.061   Bibliographie des publications ayant rapport avec
          l'astrolabe Danjon ou avec d'autres astrolabes de
types nouveaux (1948–1976).
A. Stoyko, with a preface by S. Débarbat.
Obs. Paris, 28 + 8 pp. (1978).

002.062   An atlas of galactic neutral hydrogen for the region
          $270° \leqslant l \leqslant 310°$; $-7° \leqslant b \leqslant 2°$.   S. L. Garzoli.
Carnegie Inst., Washington, Publ. No. 629, 123 pp. (1972) =
Contrib. Inst. Argentino Radioastron., No. 38. In English and
Spanish.
     This atlas, covering an extensive zone of the Galaxy, is
presented as a complement to existing observations. Published
surveys, some of higher angular resolution in this region, do
not cover the whole zone in systematic fashion. The author
presents diagrams of the distribution of neutral hydrogen
obtained in the region $270° \leqslant l \leqslant 310°$; $-7° \leqslant b \leqslant 2°$, comple-
menting the existing observations. He also shows superposi-
tions of profiles from various surveys for comparison.

002.063   Observations in the 21-cm neutral hydrogen line.
          I. The region $220° \leqslant l \leqslant 294°$; $-29° \leqslant b \leqslant -11°$.
II. The region around the south celestial pole.
E. Bajaja, F. R. Colomb.
Carnegie Inst., Washington, Publ. No. 632, 77 pp. (1973) =
Contrib. Inst. Argentino Radioastron., No. 40. In English and
Spanish.
     The main purpose of this investigation is to study the
distribution of neutral hydrogen away from the galactic plane,
since little information is available for this zone, visible mainly
from the southern hemisphere.

002.064   Dissertations at the Kazan State University.
          N. A. Sakhibullin.
Astron. Tsirk., No. 979, p. 8 (1978). In Russian.

002.065   The Ariel V High-Latitude Catalogue.
          K. A. Pounds.
Ann. New York Acad. Sci., Vol. 302, (see 012.033), 361 - 385
(1977).
     Ariel V has now been operational for more than two
years since its launch on 15 October 1974. One of the payload
of six X-ray experiments is the Sky Survey instrument. Its
principal aims are to carry out a new and more complete sky
survey in the 2–18 keV range. In the present talk the author
reviews the present state of the survey, concentrating on the
observations above 10° galactic latitude, which form the basis
of the Ariel V High-Latitude Catalogue.

002.066   Review of the literature about pulsars. (1968 - 1974).
          V. T. Fedorov.
Nauchn.-issled. radiofiz. inst. Gor'kij. Prepr. 1977, No. 106,
23 pp. In Russian. – Abstr. in Ref. zh., 51. Astron., 6.51.45
(1978).

002.067   Catalogo degli eventi radio solari di tipo IV in pre-
          parazione all'Osservatorio Astronomico di Trieste.
C. Zanelli.
Mem. Soc. Astron. Italiana, Vol. 48, (see 012.037), 237 - 238
(1977).

002.068   New international journal "Physica Solariter-
          restris".   L. I. Miroshnichenko.
Geomagn. Aehron., Vol. 18, 560 - 561 (1978). In Russian.

002.069   Bibliography.
          Z. Kopal, M. Moutsoulas, F. B. Waranius (Editors).
Moon, Vol. 17, 309 - 339, 425 - 470 (1977). – Concerning
literature on lunar topics.

002.070   Union catalogue of printed books of the XV and
          XVI centuries in astronomical European observa-
tories.   Compiled by G. Grassi Conti.
Astronomical Observatory, Rome. 18 + 105 pp. Free on
request. (1977). – Review in J. Hist. Astron., Vol. 9, 152
(1978).

002.071   A guide to the literature of astronomy.
          R. A. Seal.
Libraries Unlimited, Inc., P.O Box 263, Littleton, Colo. 80160.
306 pp. Price $ 25.00 (1977). – Review in Sky Telesc., Vol.
55, 430 (1978).

002.072   Newton and Newtoniana 1672–1975: a biblio-
          graphy.   P. Wallis, R. Wallis.
Dawson, Folkestone, Kent, U.K. 362 pp. Price £ 30.00 (1977).
Review in Nature, Vol. 273, 575 - 576; 1978 (I. B. Cohen).

          Le costellazioni – atlante illustrato.
See Abstr. 003.072.

          General catalogue of stars observed with astrolabes
(1957 - 1975). Individual corrections to places of FK4 stars.
See Abstr. 041.001.

          General catalogue of FK4 stars observed with astro-
labes (1958 - 1975).   See Abstr. 041.002

          The Basle catalogue of blue objects in higher galactic
latitudes in SA51, SA54, SA57, SA82 and in the halo field
around M5 (I).   See Abstr. 113.014.

          An atlas of Copernicus ultraviolet spectra of
Wolf-Rayet stars.   See Abstr. 114.032.

          A statistical analysis of the incompleteness and the

number of optical systems in the IDS.
See Abstr. 118.037.

          Supplement to the Catalogue of RR Lyrae stars
elements.   See Abstr. 122.193.

          Objetos infrarrojos no identificados en el catálogo
IRC.   See Abstr. 133.042.

          Optical identifications from the Molonglo cata-
logues MC2 and MC3. I. Right ascensions $11^h 28^m$ to $17^h 00^m$.
See Abstr. 141.029.

          Culgoora-3 list of radio source measurements.
See Abstr. 141.030.

          The Parkes 2700 MHz survey. (Thirteenth part).
Catalogue for declinations $-15°$ to $-30°$, right ascensions
10^h to 15^h.   See Abstr. 141.031.

          A catalogue of sources found at 610 MHz with the
Westerbork Synthesis Radio Telescope: source parameters and
identifications.   See Abstr. 141.054.

          The Ariel V (SSI) catalogue of high galactic latitude
($|b| > 10°$) X-ray sources.   See Abstr. 142.007.

          Star clusters in the Small Magellanic Cloud.
See Abstr. 159.018.

          A finding list of southern clusters of galaxies – I.
See Abstr. 160.003.

          Superclustering of galaxies.
See Abstr. 160.091.

003 Books

003.001 **La recherche en astrophysique.**
Selected articles from La Recherche, selected and presented by J. Lequeux; with a glossary and a bilbliography. Éditions du Seuil. La Recherche. Société d'Éditions scientifiques, Paris. 255 pp. (1977). ISBN 2-02-004597-4. – The individual contributions are included in their corresponding subject categories – see abstracts 061.010, 066.022, 066.023, 080.007, 093.016, 105.026, 107.015, 125.014, 131.045 - 131.048, 141.508, 142.039, 143.008, 151.010, 156.007, 162. 011 - 162.013.

003.002 **Focus on the stars.**
H. Messel, S. T. Butler (Editors).
Heinemann Educational Books, London. 291 pp. Price DM 24.40 (1977). ISBN 0 435 68282 2. – Review in Observatory, Vol. 98, 72; 1978 (*R. C. Smith*). – The individual contributions are included in their corresponding subject categories see abstracts 015.014, 080.016, 097.050, 115.003, 131.062, 155.010, 162.023.

003.003 **Project Daedalus. The final report on the BIS Starship Study.**
J. British Interplanet. Soc., Suppl., 192 pp. (1978). – The individual papers within the subject scope of Astronomy and Astrophysics Abstracts are included in their corresponding categories – see abstracts 015.025 - 015.030, 117.027, 117.028.

003.004 **Investigations of the sun and red stars. 7.**
A. Balklavs (Editor).
Latvijas PSR Zinātņu akadēmija, Radioastrofizikas observatorija. Akademiya nauk Latvijskoj SSR, Radioastrofizicheskaya observatoriya. Izdatel'stvo "Zinatne", Riga. 60 pp. Price 20 Kop. (1977). In Russian. – The individual contributions are included in their corresponding subject categories – see abstracts 021.014, 114.043, 114.044, 122.112, 122.113, 141.152.

003.005 **Investigations of the sun and red stars. 8.**
A. Balklavs (Editor).
Latvijas PSR Zinātņu akadēmija, Radioastrofizikas observatorija. Akademiya nauk Latvijskoj SSR, Radioastrofizicheskaya observatoriya. Izdatel'stvo "Zinatne", Riga. 92 pp. Price 30 Kop. (1978). In Russian. – The individual contributions are included in their corresponding subject categories – see abstracts 033.020, 072.024, 114.045, 114.046, 122.114.

003.006 **Solar wind and the magnetosphere.**
K. G. Ivanov (Editor).
Inst. zemnogo magnetizma, ionosfery i rasprostraneniya radiovoln. AN SSSR. Moskva. 128 pp. Price 80 Kop. (1976). In Russian. – From Ref. zh., 62. Issled. kosm. prostranstva, 3.62.345 (1978). – See abstracts 062.074, 074.043, 084.019, 084.260, 085.011, 106.038, 106.039.

003.007 **Planetary satellites.**   J. A. Burns (Editor).
The University of Arizona Press, Tucson, Arizon. 23 + 598 pp. Price $ 19.95, £ 7.20, DM 56.20 (1977). ISBN 0-8165-0552-7. – The individual contributions are included in their corresponding subject categories – see abstracts 041.020, 042.050, 042.051, 091.047 - 091.055, 094.012, 097.517, 097.518, 099.525 - 099.527, 100.034, 100.506 - 100.508, 107.032 - 107.035.–Reviews in J. British Astron. Assoc., Vol. 88, 202 - 203; 1978 (*P. Moore*); Sky Telesc., Vol. 55, 255 (1978).
Most of the papers originated as invited review papers presented to the IAU Colloquium No. 28, "Planetary Satellites", Ithaca, N.Y., 18 - 21 August 1974.

003.008 **Problems of solar activity and space system Prognoz.**
S. N. Vernov (Editor).
Nauka, Moskva. 262 pp. Price 1 Rbl. 70 Kop. (1977). In Russian. – Review in Ref. zh., 62. Issled. kosm. prostranstva, 4.62.69 (1978). – See abstracts 073.063 - 073.066, 074.060, 076.022 - 076.025, 078.007 - 078.009, 084.288, 085.014, 143.032.

003.009 **Chinese Astronomy.** A selected translation of Acta Astronomica Sinica. Volumes 14 - 17 (1966 - 1976).
Chinese Astron., Vol. 1, (No.2), 197 - 364 (1977). – Translations of 11.044.020, 17.098.008, 17.102.008, 17.124.101, 19.141.092, 19.066.074, 19.122.111, 19.091.023, 19.098. 015, 19.082.043, 19.045.011, 19.072.054, 19.124.135.

003.010 **Handbook of astronomy, astrophysics and geophysics. Volume I. The Earth: 1. The upper atmosphere, ionosphere and magnetosphere.**
C. W. Gordon, V. Canuto, W. I. Axford (Editors).
Gordon and Breach, Science Publishers, Inc. 5 + 412 pp. Price £ 35.00 (1978). ISBN 0-677-16100-X. – The individual contributions are included in their corresponding subject categories – see abstracts 082.087, 082.088, 083.126, 083. 127, 084.030, 084.410.

003.011 **Annual Review of Earth and Planetary Sciences. Volume 6.**
F. A. Donath, F. G. Stehli, G. W. Wetherill (Editors).
Annual Reviews Inc., 4139 El Camino Way, Palo Alto, California 94306. 9 + 543 pp. Price DM 45.00 (1978). ISBN 8243-2006-9. – The individual contributions within the subject scope of Astronomy and Astrophysics Abstracts are included in their corresponding subject categories – see abstracts 015.038, 099.533, 131.214.

003.012 **Problems of observational and theoretical astronomy.**
O. A. Mel'nikov, V. K. Abalakin (Editors).
Moskva – Leningrad. 271 pp. Price 1 Rbl. 60 Kop. (1977). In Russian. – From Ref. zh., 51. Astron., 5.51.37 (1978). See abstracts 005.014, 005.020, 008.098, 015.041, 031.309, 032.055 - 032.058, 041.030, 042.135, 042.139, 047.015, 051.024, 080.041, 094.023, 104.029, 104.091, 151.091, 151.126.

003.013 **Chinese Astronomy.** A selected translation of Acta Astronomica Sinica. Volumes 15 - 18 (1974 - 1977).
Chinese Astron., Vol. 2, (No. 1), 1 - 170 (1978). – Translations of 13.082.041, 13.099.209, 20.044.017, 20.041.029, 20.072.038, 20.122.088, 20.121.033, 20.122.089, 20.151. 054, 20.141.141, 20.142.096.

003.014 **Reports of planetary geology program, 1977–1978.**
Compiled by R. Strom, J. Boyce.
NASA Tech. Memo. 79729, 14 + 342 pp. (1978). – For sale by the National Technical Information Service, Springfield, NASA Tech. Memo. 79729, 14 + 342 pp. (1978). – For sale by the National Technical Information Service, Springfield, Virginia 22161. – The individual contributions within the subject scope of Astronomy and Astrophysics Abstracts are included in their corresponding categories – see abstracts 011.025, 013.026, 022.107, 022.108, 031.279 - 031.284, 032.541, 032.542, 042.120, 051.025, 074.076, 084.018, 091.064 - 091.068, 092.015 - 092.021, 093.048, 094.145, 094.452 - 094.453, 097.105 - 097.162, 097.520 - 097.524, 098.092, 099.536 - 099.538, 100.040, 100.041, 100.509, 102.033, 102.034, 107.043 - 107.045, 131.220.

003.015 **Mathematical models of the near space.**
V. G. Pivovarov (Editor).
Nauka, Novosibirsk. 288 pp. Price 1 Rbl. 80 Kop. (1977). In
Russian. – Review in Ref. zh., 62. Issled. kosm. prostranstva,
6.62.45 (1978). – See abstracts 062.116, 073.100, 074.089,
074.091 - 074.096, 080.065, 084.329, 084.330, 106.065.

003.016 **Cosmophysical investigations.**
V. I. Utkin, E. Ya. Gidalevich (Editors).
Sverdlovsk. 104 pp. Price 63 Kop. (1977). In Russian.
Review in Ref. zh., 62. Issled. kosm. prostranstva, 6.62.44
(1978). – See abstracts 074.090, 106.066, 106.067, 143.060,
143.070.

003.017 **Drama of the Universe.** G. O. Abell.
Holt, Rinehart and Winston, New York. 456 pp.
Price $ 13.50 (1978). – From Phys. Today, Vol. 31, No. 6,
p. 57 (1978).

003.018 **Initiation à l'astronomie.**
A. Acker, with a preface by C. Fehrenbach.
Masson, Paris–New York–Barcelona–Milan. 8 + 150 pp.
Price F 45.00 (1978). ISBN 2-225-49434-7.
Contents: Rayonnement stellaire. Atmosphère terrestre.
Instruments. Positions et mouvements des astres. Caractér-
istiques physiques des étoiles. Atmosphère et intérieur
stellaires typiques. Etude du soleil. Evolution stellaire. Etoiles
variables. Matière interstellaire. Groupements stellaires.
Notions de cosmologie.

003.019 **Computer methods in image analysis.**
J. K. Aggarwal, R. O. Duda, A. Rosenfeld (Editors).
IEEE Press Selected Reprint Series. IEEE Press, New York,
(U.S. distributor, Wiley, New York). 6 + 466 pp. Price cloth
$ 24.95, paper $ 14.95 (1977). ISBN 0–87 942–089–8.
Review in Proc. IEEE, Vol. 66, 271 (1978).

003.020 **Review questions in astronomy**, preliminary edition.
B. M. Aghassi.
Bentley College Observatory, Waltham, Mass. 115 pp. Price
$ 8.00 (1978). – From Phys. Today, Vol. 31, No. 6, p. 57
(1978).

003.021 **Astrophysical quantities.** K. W. Allen.
Translated from the 3. revised and suppl. English
edition. Mir, Moskva. 448 pp. Price 3 Rbl. 60 Kop. (1977).
In Russian. – Review in Ref. zh., 51. Astron., 4.51.34 (1978).

003.022 **The collapsing universe.** I. Asimov.
Walker and Company, New York. 8 + 204 pp.
Price $ 8.95, £ 4.50 (1977). ISBN 0-8027-0486-7. – Reviews
in Nature, Vol. 272, 784; 1978 (R. Znajek); Sky Telesc., Vol.
56, 62 (1978).
Contents: Particles and forces. The planets. Compressed
matter. White dwarfs. Exploding matter. Neutron stars. Black
holes. Endings and beginnings.

003.023 **Solar heating design by the f-chart method.**
W. A. Beckman, S. A. Klein, J. A. Duffie.
Wiley-Interscience, New York. 18 + 200 pp. Price $ 14.95
(1977). – From Science, Vol. 199, 918 (1978).

003.024 **Skylab, our first space station.** L. F. Belew.
National Aeronautics and Space Administration.
169 pp. Price $ 7.00 (1977). – Reviews in Sky Telesc., Vol.
54, 422 (1977); Spaceflight, Vol. 20, 34 - 35; 1978 (F. I.
Ordway).

003.025 **White dwarfs.** S. I. Blinnikov.
Znanie, Moskva. 64 pp. (1977). In Russian.
Review in Ref. zh., 51. Astron., 5.51.46 (1978).

003.026 **Forschung in der Deutschen Demokratischen
Republik zu Fragen der Physik der Atmosphäre
1971 bis 1975.** W. Böhme (Editor).
Akademie-Verlag, Berlin. 43 pp. Price M 8.00 (1976) = Veröffentlichungen des Meteorologischen Dienstes der DDR,
Nr. 22. – Review in Gerlands Beitr. Geophys., Band 87,
77 - 78; 1978 (H. Pethe).

003.027 **The Milky Way.** B. J. Bok, P. F. Bok.
Translated from the English edition. Mir, Moskva.
296 pp. Price 2 Rbl. 70 Kop. (1978). In Russian. – From
Ref. zh., 51. Astron., 7.51.46 (1978).

003.028 **Closeup: new worlds.** B. Bova, T. E. Bell (Edi-
tors).
St. Martin's Press. 222 pp. Price $ 15.00 (1977). – Review in
Sky Telesc., Vol. 55, 344 - 345 (1978).

003.029 **Space colonies.** S. Brand (Editor).
Penguin, New York. 160 pp. Price $ 5.00 (1977).
From Science, Vol. 200, 809 (1978).

003.030 **Relativistic theories of materials.** A. Bressan.
Springer Tracts in Natural Philosophy, Vol. 29.
Springer-Verlag, Berlin – Heidelberg – New York. 14 + 290
pp. Price DM 98.00 (1978). ISBN 3-540-08177-1, ISBN 0-
387-08177-1.
The purpose of this book is to describe the foundations
of the general relativistic theories that include constitutive
equations, and to present some applications, mainly to elastic
waves, of these theories. This tract is divided into two parts.
In the first part only the Eulerian point of view is considered;
basic equations of general relativity, other than constitutive
equations, are stated in full generality (except for couple stresses which are considered in part 2. Part 1 also thoroughly
covers fluids, including constitutive equations. In the second
part, the Lagrangian point of view is introduced and – primarily following Bressan – some mathematical tools connected
with materials coordinates are elucidated; by means of these
tools, solids and materials of a very general kind – possibly
capable of couple stresses – are dealt with. An introductory
chapter reviews the main ideas underlining the origin and
development of relativity: among others, the mass-energy
equivalence principle, the debated general principle of relativity, its local form, and Fock's privileged harmonic co-ordinates.

003.031 **Atomic processes and applications.** P. G. Burke,
B. L. Moiseiwitsch.
North-Holland, Amsterdam, 533 pp. Price $ 67.50 (about
£ 39) (1976). – Review in Observatory, Vol. 98, 35; 1978
(V. P. Myerscough).

003.032 **Energy and the atmosphere: a physical-chemical
approach.** I. M. Campbell.
John Wiley and Sons, Chichester, Sussex – New York – Sydney. 398 pp. Price hardback £ 14.50, $ 31.00, paperback
£ 5.95, $ 12.50 (1977). – Review in Nature, Vol. 274, 193;
1978 (J. T. Houghton).

003.033 **Group theory and general relativity.** M. Carmeli.
McGraw-Hill, New York – Toronto. 391 pp. Price
£ 16.85 (1977). – Review in Nature, Vol. 271, 690; 1978
(P. Davies).

003.034 **Discovering astronomy.** R. D. Chapman.
W. H. Freeman, San Francisco. 518 pp. Price
$ 20.00 cloth, $ 12.00 paper (1978). – Review in Sky Telesc.,
Vol. 56, 63 (1978).

003.035 **This wild abyss.** G. E. Christianson.
Macmillan. 461 pp. Price $ 12.95 (1978). – Review
in Sky Telesc., Vol. 56, 63 (1978).

003.036 **Isaac Newton and astrology.** T. G. Cowling.
Leeds University Press. 20 pp. Price 65 p. (1977).
Review in Observatory, Vol. 98, 79 (1978).

003.037 **A handbook of quasistellar and BL Lacertae objects.**
E. R. Craine.
Pachart Publishing House, P.O. Box 6721, Tucson, Ariz.
283 pp. Price $ 38.00 (1977). Review in Sky Telesc., Vol. 56,
63 (1978).

003.038 **The runaway universe.** P. Davies.
J. M. Dent and Sons, London. 205 pp. Price £ 5.95
(1978). — Reviews in J. British Astron. Assoc., Vol. 88, 418;
1978 (G. C. McVittie); Nature, Vol. 272, 785 - 786; 1978
(M. Rowan-Robinson).

003.039 **Nuclear cascade process in dense matter.**
A. I. Dem'yanov, V. S. Murzin, L. I. Sarycheva.
Nauka, Moskva. 203 pp. Price 1 Rbl. 20 Kop. (1977).
In Russian. — Review in Ref. zh., 51. Astron., 2.51.45 (1978).

003.040 **Non-stationary phenomena in galaxies.**
Eh. A. Dibaj.
Znanie, Moskva. 64 pp. (1977). In Russian. — Review in Ref.
zh., 51. Astron., 4.51.56 (1978).

003.041 **"Juri Gagarin".** A. Dichtjar.
Verlag Neues Leben, Berlin. 355 pp. Price M 12.50
(1977). — Review in Astron. Schule, 15. Jahrg., 22 - 23;1978
(W. König).

003.042 **Die kleinen Planeten. Planetoide und ihre Entdek-
kungsgeschichte.** J. W. Ekrutt.
Kosmos Bibliothek, Band 296. Kosmos, Gesellschaft der
Naturfreunde, Franckh'sche Verlagshandlung, Stuttgart.
71pp. Price DM 8.80 (1977). ISBN 3-440-00296-9.

003.043 **Time and man.** L. R. B. Elton, H. Messel.
Pergamon Press, Elmsford, N. Y. 114 pp. Price
$ 8.00 cloth, $ 4.00 paper (1978). — From Phys. Today, Vol.
31, No. 5. p. 79 (1978).

003.044 **Two hundred years of flight in America.**
E. M. Emme (Editor).
Univelt Inc., San Diego. 310 pp. Price $ 25.00 (1977).
Review in Sky Telesc., Vol. 56, 62 - 63 (1978).

003.045 **What is the world made of?** G. Feinberg.
Anchor-Doubleday, Garden City, N. Y. 290 pp.
Price $ 10.00 (1977). — Review in Phys. Today, Vol. 31, No. 2,
p. 53 - 54; 1978 (R. A. Carrigan, Jr.).

003.046 **Selected papers. Meteorites and meteor matter.**
V. G. Fesenkov.
Nauka, Moskva. 250 pp. Price 2 Rbl. (1978). In Russian.
Review in Ref. zh., 51. Astron., 7.51.42 (1978).

003.047 **Cosmic evolution: an introduction to astronomy.**
G. B. Field, G. L. Verschuur, C. Ponnamperuma.
Houghton Mifflin, Boston. 450 pp. Price $ 14.95 (1978).
From Phys. Today, Vol. 31, No. 7, p. 53 (1978).

003.048 **Pioneer odyssey.** R. O. Fimmel, W. Swindell,
E. Burgess.
National Aeronautics and Space Administration, Washington,
D.C. [Available from the Superintendent of Documents U.S.
Government Printing Office, Washington, D.C. 20402.]
NASA SP-396. 217 pp. Price $ 9.85 (1977). — Reviews in
Sky Telesc., Vol. 55, 345 (1978); Vol. 56, 56 - 57; 1978
(J. R. Jokipii).

003.049 **The creation of the universe.** D. E. Fisher.
Bobbs-Merrill, Indianapolis—New York. 8 + 120 pp.
Price $ 8.95 cloth, $ 5.95 paper (1977). — From Phys. Today,
Vol. 31, No. 1, p. 83 - 84 (1978).

003.050 **Beyond earth.** R. A. Gallant.
Four Winds Press, New York. 16 + 191 pp. Price
£ 7.95 (1977). — Review in J. British Astron. Assoc., Vol. 88,
311; 1978 (C. A. Ronan).

003.051 **Kvasarer, pulsarer og svarte hull.** F. Golden.
Universitetsforlaget, Oslo. 136 pp. Price N.kr. 45.00
(1977). — Review in Astron. Tidsskr., Årg. 11, 100 (1978).

003.052 **Scientists confront Velikovsky.** D. Goldsmith
(Editor).
Cornell University Press, Ithaca—London. 183 pp. Price
$ 8.95 (1977). — From Sky Telesc., Vol. 55, 170 (1978).

003.053 **Numerical methods of solving inverse problems of
astrophysics.** A. V. Goncharskij, A. M.
Cherepashchuk, A. G. Yagola.
Nauka, Moskva. 335 pp. Price 2 Rbl. 10 Kop. (1978). In
Russian. — Review in Ref. zh., 51. Astron., 7.51.40 (1978).

003.054 **Climatic change.** J. Gribbin (Editor).
Cambridge University Press, Cambridge—London.
292 pp. Price £ 17.50 hardback, £ 6.50 paperback (1978).
Review in Nature, Vol. 273, 477 - 478; 1978 (T. J. Hughes).

003.055 **Our changing planet.** J. Gribbin.
Crowell, New York. 6 + 166 pp. Price $ 7.95
(1977). — From Science, Vol. 200, 349 (1978).

003.056 **Astrolabes of the world.** R. T. Gunther.
The Holland Press, London. 17 + 3 + 609 + 153 pp.
Price £ 40.00 (1976). — Review in J. Hist. Astron., Vol. 9,
69 (1978).

003.057 **Lunar impact: a history of project Ranger.**
R. C. Hall.
NASA History Series. National Aeronautics and Space
Administration, Washington, D. C. [For sale by US Govern-
ment Printing Office]. 17 + 450 pp. Price $ 6.25 (1977).
From Nature, Vol. 274, 300 (1978).

003.058 **Astronomy: the cosmic journey.** W. K. Hartmann.
Wadsworth, Belmont, Calif. 536 pp. (1978). — From
Phys. Today, Vol. 31, No. 5, p. 78 (1978).

003.059 **Daylight and its spectrum.** S. T. Henderson.
Halsted (Wiley), New York. 2nd edition. 349 pp.
Price $ 29.50 (1977). — From Phys. Today, Vol. 31, No. 6,
p. 57 (1978).

003.060 **The cosmonauts.** G. R. Hooper.
Published privately by the author. 80 pp. Price
£ 1.00 U. K.; £ 1.25 overseas. (1977). Duplicated paperback.
Review in Spaceflight, Vol. 20, 37; 1978 (D. J. Shayler).

003.061 **L'exploration du système solaire.** G. Israël, with
a postface de I. Rasool.
Hachette, Paris. 287 pp. (1977). — Review in Astronomie,
Vol. 92, 230; 1978 (B. Morando).

003.062 **Until the sun dies.** R. Jastrow.
W. W. Norton & Co., Inc., New York. 173 pp. Price
$ 8.95 (1977). — Review in Sky Telesc., Vol. 55, 255 (1978).

003.063 **Collected papers of Sir Harold Jeffreys on geophysics and other sciences. Vol. 6, Mathematics,
probability and miscellaneous other sciences.**
H. Jeffreys, B. Swirles (Editors).

Gordon and Breach Science Publishers, London – Paris –
New York. 18 + 622 pp. Price $ 70.00 (1977). – From
Science, Vol. 200, 307 (1978).

003.064 **The night sky book.**   J. Jobb.
Little,Brown, Boston, Mass., 127 pp. Price $ 7.95
(1977). – Review in Sky Telesc., Vol. 55, 344 (1978).

003.065 **Ensamma i rymden?**   K. Johansson, L. O. Lodén,
L. Lundin.
Studieförlaget, Uppsala. 160 pp. Price S.kr. 68.75 (1978).
Review in Astron. Tidsskr., Årg. 11, 100 (1978).

003.066 **Space settlements: a design study.**   R. D. Johnson,
C. Holbrow (Editors).
USGPO, Washington, D. C. 185 pp. Price $ 5.00 (1977).
Review in Phys. Today, Vol. 31, No. 1, p. 69 - 70; 1978
(*P. K. Chapman*).

003.067 **Early history of Planck's radiation law.**
H. Kangro.
Translated from the German edition.
Crane, Russak, New York. 18 + 282 pp. Price $ 39.50 (1977).
From Science, Vol. 199, 170 (1978).

003.068 **Information mechanics.**   F. W. Kantor.
John Wiley & Sons Inc., New York. 397 pp. Price
$ 21.95 (1977). – From Sky Telesc., Vol. 55, 77 (1978).

003.069 **Interstellar medium and the origin of stars.**
S. A. Kaplan.
Znanie, Moskva. 64 pp. (1977). In Russian. – Review in Ref.
zh., 51. Astron., 5.51.47 (1978).

003.070 **Physics of plasma of the solar atmosphere.**
S. A. Kaplan, S. B. Pikel'ner, V. N. Tsytovich.
Nauka, Moskva. 256 pp. Price 1 Rbl. 40 Kop. (1977).
In Russian. – Review in Ref. zh., 51. Astron., 2.51.53 (1978).

003.071 **Exploration of the solar system.**   W. J. Kauf-
mann III.
Macmillan, New York. 575 pp. Price $ 16.95 (1978).
From Phys. Today, Vol. 31, No. 6, p. 57 (1978).

003.072 **Le costellazioni – atlante illustrato.**
J. Klepesta, A. Rükl.
Nicola Teti Editore, Milano, 287 + 4 pp. Price L 3000. – Re-
view in G. Astron., Vol. 3, 259 - 260; 1977 (*E. Proverbio*).

003.073 **Der Mars. Bericht über einen Nachbarplaneten.**
H. W. Köhler.
Friedr. Vieweg & Sohn Verlagsgesellschaft mbH, Braunschweig
6 + 224 pp. (1978). ISBN 3-528-08401-4.
   Contents: Marsbeobachtung mit optischen Geräten. Das
Viking-Programm. Landeplatzsuche und die Landung von
Viking. Phobos und Deimos. Die Marsatmosphäre. Geophysik
des Planeten Mars. Geologie und Topographie. Charakteristi-
sche Oberflächenformen im Bild. Meteorologie auf dem Mars.
An den Landeplätzen von Viking 1 und Viking 2. Physik und
Chemie des Marsbodens (VL1 und VL2). Die Suche nach
Leben auf dem Mars. Zukünftige Marsforschung – Möglich-
keiten und Ausblick.

003.074 **Meteorology of planets.**   K. Ya. Kondrat'ev.
Izdatel'stvo Leningradskogo Universiteta, Leningrad.
294 pp. Price 2 Rbl. 41 Kop. (1977). In Russian.

003.075 **Physics of the atomic nucleus and cosmic rays.**
M. I. Korsunskij (Editor).
Alma-Ata. 212 pp. Price 80 Kop. (1977). In Russian. – From
Ref. zh., 51. Astron., 7.51.44 (1978).

003.076 **Radiophysical investigations of planets.**
N. N. Krupenio.
Nauka, Moskva. 183 pp. Price 64 Kop. (1978). In Russian.
Review in Ref. zh., 51. Astron., 6.51.37 (1978).

003.077 **In search of ancient astronomies.**
E. C. Krupp (Editor).
Doubleday, New York. 300 pp. Price $ 10.00 (1978). – Re-
view in Sky Telesc., Vol. 56, 62 (1978).

003.078 **Poradniki miłośnika astronomii (A handbook of an
amateur astronomer).**
P. G. Kulikowski (*Kulikovskij*).
Jerzy Kreiner, PWN, Warszawa (1976). – Review in Postępy
Astron., Tom 25, 195 - 197; 1977 (*J. Mietelski*). In Polish.

003.079 **Lettres cosmologiques sur l'organisation de l'univers.**
J. H. Lambert.
Reprint of the 1801 translation by A. Darquier, with preface
by J. Merleau-Ponty.
Alain Brieux, 48 rue Jacob, 75006 Paris. 19 + 290 pp. (1977).
Review in J. Hist. Astron., Vol. 9, 134 - 139; 1978 (*M. Hoskin*).

003.080 **Mathematical cosmology, an introduction.**
P. T. Landsberg, D. A. Evans.
Clarendon Press, Oxford; Oxford University Press, Oxford –
London – New York. 10 + 309 pp. Price $ 15.95, DM 48.50
(1977). ISBN 0-19-851136-1.
   Contents: Cosmology. Newtonian gravitation: some
fundamentals. The cosmological differential equation: the
particle model, the continuum model. Some simple Friedmann
models. The classification of the Friedmann models. The
steady-state model. Universes with pressure. Optical effects of
the expansion according to various theories of light. Optical
observations and cosmological models. Appendices: Effects
of a special relativistic theory of light propagation. The mass
of a typical stable particle, the number of such particles in the
universe, and cosmological coincidences. Representation of
cosmological models in the density parameter-deceleration
parameter plane.

003.081 **Methods of determination of astronomical coordinates and problems of terrestrial navigation in
the Antarctic.**   T. E. Lazarev.
Nauka, Moskva. 147 pp. Price 1 Rbl. (1977). In Russian.
Review in Ref. zh., 52. Geod. Aehrosemka, 3.52.98 (1978).

003.082 **Energetic particles in the auroral magnetosphere.**
L. L. Lazutin (Editor).
Apatity. 150 pp. Price 70 Kop. (1977). In Russian. – Review
in Ref. zh., 62. Issled. kosm. prostranstva, 7.62.41 (1978).

003.083 **From Vinland to Mars.**   R. S. Lewis.
Quadrangle/New York Times. 436 pp. Price
$ 15.00 (1976). – Review in Sky Telesc., Vol. 55, 430 (1978).

003.084 **Astronomie selbst erlebt.**   K. Lindner.
Urania Verlag, Leipzig – Jena – Berlin; Aulis Verlag
Deubner & Co KG, Köln. 184 pp. Price M 12.80, DM 24.00
(1977). – Reviews in Sky Telesc., Vol. 55, 526 - 527; 1978
(*J. Ashbrook*); Sterne, 54. Band, 123; 1978 (*S. Marx*).

003.085 **Der Sternhimmel.**   K. Lindner.
Urania-Verlag, Salomonstr. 26/28, Leipzig C1,
DDR. 128 pp. Price M 4.50 (1977). – From Sky Telesc.,
Vol. 55, 170 (1978).

003.086 **Skylab explores the Earth.**
NASA Lyndon B. Johnson Space Center, National
Aeronautics and Space Administration. 517 pp. Price $ 17.50,
£ 29.50 (1977). – Reviews in Nature, Vol. 273, 690; 1978
(*V. H. Anderson*); Sky Telesc., Vol. 55, 430 (1978).

003.087 **Mechanics and motion.**   L. Mackinnon.
Oxford Physics Series. Clarendon Press, Oxford —
Oxford University Press, London. 161 pp. Price £ 6.00 hard-
back, £ 3.25 paperback (1978). ISBN 0-19-851825-0. — From
Nature, Vol. 274, 299 (1978).

003.088 **Worlds within worlds.**
M. Marten, J. Chesterman, J. May, J. Trux.
Martin Secker & Warburg Ltd. 208 pp. Price £ 7.95 hardback,
£ 3.95 paperback (1977). — Review in Spaceflight, Vol. 20,
75 (1978).

003.089 **Astronomy.** Volume 13: **Radioastronomy.**
L. I. Matveenko. Edited by I. S. Shcherbina-
Samojlova.
Itogi nauki i tekhniki. Seriya Astronomiya, tom 13. Moskva.
156 pp. Price 1 Rbl. 20 Kop. (1977). In Russian.

003.090 **The stellar handbook.**
R. J. Maxwell, R. P. Van Zandt.
Astronomical League Book Service, 13 Meadowlark Lane,
Carpentersville, Ill. 60110. 56 pp. Price $ 3.00 (1977).
Review in Sky Telescope, Vol. 55, 170 (1978).

003.091 **Meteorite craters.**   G. J. H. McCall (Editor).
Dowden, Hutchingson & Ross, Inc., Stroudsburg,
Pa. (Distributor, Halsted (Wiley), New York). 16 + 366 pp.
Price $ 44.50 (1977). — Review in Astron. Tidsskr., Årg. 11,
100 (1978).

003.092 **Stellar atmospheres.**   D. Mihalas.
W. H. Freeman, San Francisco. 2nd Edition. 632 pp.
Price $ 24.95 (1978). — From Phys. Today, Vol. 31, No. 5,
p. 78 (1978).

003.093 **The Cambridge encyclopaedia of astronomy.**
S. Mitton (Editor), with a foreword by M. Ryle.
Jonathan Cape, thirty Bedford square, London. 495 pp.
Price £ 15.00, $ 35.00 (1977). ISBN 0-224-01418-8. —
Reviews in Astron. Tidsskr., Årg. 11, 100 (1978); J. British
Astron. Assoc., Vol. 88, 308 - 309; 1978 (*I. Nicolson*); J. Hist.
Astron., Vol. 9, 151 (1978); Nature, Vol. 272, 786 - 787;
1978 (*A. M. Cruise*); Phys. Abstr., Vol. 81, Abstr. 45199
(1978).

003.094 **Exploring the galaxies.**   S. Mitton.
Charles Scribner's Sons, New York. 206 pp. Price
$ 4.95 (1978). — From Sky Telesc., Vol. 55, 527 (1978).

003.095 **Guide to Mars.**   P. Moore.
W. W. Norton & Co., Inc., New York. 214 pp.
Price $ 9.95 (1977). — From Sky Telesc., Vol. 55, 528 (1978).

003.096 **1978 yearbook of astronomy.**
P. Moore, with articles by F. R. Spry, R. Maddison,
J. C. D. Marsh, S. Miyamoto, G. E. Hunt, S. J. Bell Burnell,
S. Mitton, H. Couper, I. Nicolson, P. Moore.
W. W. Norton & Company Inc., New York. 223 pp. Price
$ 10.95, DM 43.80 (1977). ISBN 0-393-06430—1. — Review
in Sky Telesc., Vol. 55, 255 (1978).

003.097 **The story of astronomy.**   P. Moore.
Macdonald & Jane's Publishers, London. 253 pp.
Price £ 6.95 (1977). — Review in Spaceflight, Vol. 20, 75
(1978).

003.098 **The new challenge of the stars: a science fact look**
at science fiction.
P. Moore, D. Hardy. Foreword by A. C. Clarke.
Mitchell Beazley, London. 62 pp. Price £ 4.95 (1978).
Reviews in J. British Astron. Assoc., Vol. 88, 416 - 417; 1978
(*H. Miles*); Nature, Vol. 273, 326; 1978 (*D. W. Hughes*).

003.099 **Maria Mitchell, First Lady of American astronomy.**
H. L. Morgan.
The Westminster Press, Witherspoon, Bldg., Philadelphia,
Pa. 19107. 141 pp. Price $ 7.95 (1977). — Reviews in Sky
Telesc., Vol. 55, 345 (1978); Vol. 55, 523 - 525; 1978 (*D.
Hoffleit*).

003.100 **The universe. Its beginning and end.**   L. Motz.
Scribner, New York. Reprint of the 1975 edition.
14 + 344 pp. Price $ 4.95 (1977). — From Science, Vol. 199,
700 (1978).

003.101 **The new astronomy.**   P. Murdin, L. Murdin.
Reference International, London — New York,
Thomas Y. Crowell Company, New York. 215 pp. Price
£ 5.95, $ 12.95 (1978). ISBN 0-905154-06-1. — From Nature,
Vol. 273, 694 (1978).

003.102 **Problems of detecting life on planets.**
B. G. Murzakov.
Znanie, Moskva. 64 pp. (1977). In Russian. — Reviews in Ref.
zh., 51. Astron., 5.51.40 (1978); 62. Issled. kosm. prostranstva, 3.62.608 (1978).

003.103 **Liv bland miljarder stjärnor.**   N. Mustelin.
Natur och Kultur, Stockholm. 288 pp. Price S.kr.
97.00 (1978). — Review in Astron. Tidsskr., Årg. 11, 100
(1978).

003.104 **Solar flare activity and some indices of the inter-**
planetary and geomagnetic fields, 1957 - 1973.
Eh. I. Nesmyanovich, A. T. Nesmyanovich.
Mezhduved. geofiz. kom. pri Prezidiume AN SSSR. Mater.
Mirovogo tsentr. dannykh B. Moskva. 240 pp. Price 62 Kop.
(1977). In Russian and English. — Review in Ref. zh., 51.
Astron., 5.51.44 (1978).

003.105 **Nikita Ivanovich Popov (1720 — 1782).**
N. I. Nevskaya.
Nauka, Leningrad. 110 pp. Price 35 Kop. (1977). In Russian.
Review in Ref. zh., 51. Astron., 4.51.18 (1978).

003.106 **Deep sky objects — a photographic guide for the**
amateur.   J. Newton.
Gall Publications, Toronto, Canada. 160 pp. Price Canadian
$ 9.10, £ 4.55 paperback; $ 12.15, £ 6.00 hard cover (1977).
ISBN 0-88904-081-8. — Reviews in J. British Astron. Assoc.,
Vol. 88, 200 - 201; 1978 (*H. R. Hatfield*); Sky Telesc., Vol.
55, 170 (1978).

003.107 **Black holes in the universe.**   I. D. Novikov.
Znanie, Moskva. 64 pp. (1977). In Russian. — Review in Ref. zh., 51. Astron., 4.51.57 (1978).

003.108 **Solar activity.**   T. B. Omarov (Editor).
Nauka, Alma-Ata. 109 pp. Price 1 Rbl. (1977). In
Russian. — From Ref. zh., 51. Astron., 3.51.49 (1978).

003.109 **A handbook of radio sources.**   A. G. Pacholczyk.
Pachart, Tucson, Ariz. 234 pp., Price $ 38.00
(1978). — From Phys. Today, Vol. 31, No. 7, p. 53 (1978).

003.110 **Astronomy now.**   J. M. Pasachoff.
W. B. Saunders Company, Philadelphia. 459 pp.
Price $ 11.95 (1978). — Review in Sky Telesc., Vol. 56,
62 (1978).

003.111 **Student guide to contemporary astronomy.**
J. M. Pasachoff, M. L. Kutner, N. Pasachoff.
W. B. Saunders Co., Philadelphia. 156 pp. Price $ 5.10 (1977).
Review in J. R. Astron. Soc. Canada, Vol. 72; 119 - 120;
1978 (*C. Clement*).

003.112 **Astronomische Grundlagen der Ortung und Zeit-bestimmung.** W. Petri.
Methoden der Archäologie. Eine Einführung in ihre natur-wissenschaftlichen Techniken.
C. H. Beck'sche Verlagsbuchhandlung, München. 392 pp. Price DM 35.00 (1978). — Review in Sterne Weltraum, Jahrg. 17, 270; 1978 (*G. D. Roth*).

003.113 **The chamber of physics.** G. Pipping.
Kungl. Vetenskapsakademien (Royal Swedish Academy of Sciences), Stockholm. 250 pp. Price Kr. 70, cloth; Kr. 65, paper (1977). — Review in Sky Telesc., Vol. 55, 429 (1978).

003.114 **The making of geology.** Earth science in Britain 1660 - 1815. R. Porter.
Cambridge University Press, New York. 12 + 288 pp. Price $ 18.95 (1977). — Review in Science, Vol. 199, 166 - 167; 1978 (*K. L. Taylor*).

003.115 **The uncertainty principle and foundations of quantum mechanics. A fifty years' survey.**
W. C. Price, S. S. Chissick (Editors).
Wiley-Interscience, New York. 18 + 572 pp. Price $ 42.00 (1977). — Review in Science, Vol. 199, 168 - 169; 1978 (*A. Shimony*).

003.116 **Astronomy data book.** J. H. Robinson.
John Wiley & Sons, Chichester, England. 270 pp. Price £ 8.25 (1977). — Review in Astron. Tidsskr., Årg.11, 97 - 98; 1978 (*S. Söderhjelm*).

003.117 **Cosmology.** M. Rowan-Robinson.
Oxford Physics Series, No. 15. Clarendon Press, Oxford; Oxford University Press, Oxford — London — New York. 10 + 159 pp. Price $ 13.95, £ 6.00 hardback; £ 2.95 paperback. (1977). ISBN 0-19-851838-2; ISBN 0-19-851839-0 pbk. — Reviews in Nature, Vol. 272, 787 - 788; 1978 (*J. Silk*); Space Sci. Rev., Vol. 21, 703; 1978 (*Y. Ne'eman*); Sky Telesc., Vol. 55, 430 (1978).
Contents: The visible universe. Our Galaxy, and other galaxies. The empirical basis for cosmological theories. The big-bang models. Early stages of the big bang. Observational cosmology. The density of matter in the universe. Other cosmological theories.

003.118 **Precambrian of the northern hemisphere and general features of early geological evolution.**
L. J. Salop, translated from the Russian edition (Leningrad, 1973) by G. M. Young.
Developments in palaeontology and stratigraphy, 3. Ed. Elsevier, New York. 10 + 378 pp. Price $ 59.25 (1977). Review in Science, Vol. 199, 1429; 1978 (*J. Sutton*).

003.119 **Theorien und Ansätze in der griechischen Astronomie — im Kontext benachbarter Wissenschaften betrachtet.** W. G. Saltzer.
Collection des Travaux de l'Académie Internationale d'Histoire des Sciences, No. 23. Franz Steiner Verlag GmbH, Wiesbaden. 8 + 162 pp. Price DM 46.00 (1976). ISBN 3-515-02508-1. Review in J. Hist. Astron., Vol. 9, 149 - 150; 1978(*B. L. van der Waerden*).
Contents: Aristoteles und Aristarch: physikalische Aspekte der heliostatischen Theorie. Die Anfänge der Epizykel-und Exzentertheorie. Aspekte der perspektivischen Optik.

003.120 **Entwicklungsphasen der Erforschung der Leuchten-den Nachtwolken.** E. Schröder.
Akademie-Verlag, Berlin. 6 + 64 pp. (1975). — Review in J. Atmos. Terr. Phys., Vol. 40, 115; 1978 (*K. Rawer*).

003.121 **Rhythmen der Sterne** (Erscheinungen und Bewegung-en von Sonne, Mond und Planeten). J. Schultz.
Philosophisch-Anthroposophischer Verlag, Dornach (Schweiz). 224 pp. Price DM 37.00 (1977). — Review in Sterne Weltraum, Jahrg. 17, 220; 1978 (*T. Schmidt*).

003.122 **The phenomenon book of calendars. 1978 - 1979.**
G. M. Sesti, A. T. Mann IV, M. Flanagan, P. Cowen.
Anchor (Doubleday), Garden City, N.Y. 48 pp. Price $ 6.95 (1977). — From Phys. Today, Vol. 31, No. 4, p. 62 (1978).

003.123 **Motion of the moon relative to the mass centre.**
K. S. Shakirov.
Saratov. univ., Saratov. 84 pp. Price 32 Kop. (1976). In Russian. — Reviews in Ref. zh., 51. Astron., 2.51.42 (1978); 52. Geod. Aehrosemka, 4.52.89 (1978).

003.124 **The restless universe: an introduction to astronomy.**
H. L. Shipman.
Houghton Mifflin, Boston. 464 pp. Price $ 15.95 (1978). From Phys. Today, Vol. 31, No. 7, p. 53 (1978).

003.125 **The moon: steppingstone to outer space.**
D. E. Shuttlesworth, L. A. Williams.
Doubleday, Garden City, N.Y. 10 + 118 pp. Price $ 5.95 (1977). — From Science, Vol. 200, 349 (1978).

003.126 **Amateur radio astronomer's notebook. Vol. 2.**
R. M. Sickels.
Bob's Electronic Service, Ft. Lauderdale, Fla. 33311. 100 pp. Price £ 20.00 (1977). — Review in Sky Telesc., Vol. 55, 343 (1978).

003.127 **The asteroid belt.** A. N. Simonenko.
Znanie, Moskva. 64 pp. (1977). In Russian.
Review in Ref. zh., 51. Astron., 5.51.42 (1978).

003.128 **Elementi di tecnica radioastronomica. L'abc dell' astroradioamatore.** G. Sinigaglia.
Associazione Radiotecnica Italiana, 96 pp. (1977). — Reviews in G. A. A. B., N. 47, p. 14 - 15; 1977 (*L. Baldinelli*); G. Astron., Vol. 4, 53 - 54; 1978 (*S. Ghedini*).

003.129 **Scientific analysis on the pocket calculator.**
J. M. Smith.
John Wiley & Sons, New York. 445 pp. Price $ 13.75 (1977). From Sky Telesc., Vol. 55, 527 (1978).

003.130 **Physical processes in the interstellar medium.**
L. Spitzer, Jr.
A Wiley-Interscience Publication, John Wiley & Sons, New York — Chichester — Brisbane — Toronto. 18 + 318 pp. Price $ 15.95 (1978). ISBN 0-471-02232-2. — Reviews in Phys. Today, Vol. 31, No. 7, p. 48 - 49; 1978 (*M. Jura*); Science, Vol. 200, 676 - 677; 1978 (*G. B. Field*).
Contents: Interstellar matter — an overview. Elastic collisions and kinetic equilibrium. Radiative processes. Excitation. Ionization and dissociation. Kinetic temperature. Optical properties of grains. Polarization and grain alignment. Physical properties of grains. Dynamical principles. Overall equilibrium. Explosive motions. Gravitational motion.

003.131 **Physics of the earth.** F. D. Stacey.
John Wiley & Sons, New York—Chichester, England. 414 pp. (1977). — Review in Geophys. J. R. Astron. Soc., Vol. 52, 366; 1978 (*J. A. Hudson*).

003.132 **Skylab, classroom in space.**
L. B. Summerlin (Editor).
National Aeronautics and Space Administration, NASA SP-401. [Available from the Superintendent of Documents U. S. Government Printing Office, Washington, D.C. 20402]. 182 pp. Price $ 8.25 (1977). — Review in Sky Telesc., Vol. 55, 345

(1978).

003.133   **Journey through the universe: an introduction to astronomy.**   T. L. Swihart.
Houghton Mifflin, Boston. 366 pp. Price $ 15.95 (1978).
From Phys. Today, Vol. 31, No. 7, p. 53 (1978).

003.134   **Acta Geodaetica, Geophysica et Montanistica Academiae Scientiarum Hungaricae.**
A. Tárczy-Hornoch (Editor).
Akadémiai Kiadó, Budapest. 509 pp. (1976). − Review in
Gerlands Beitr. Geophys., Band 87, 247; 1978 (*P. Mauersberger*).

003.135   **Il mistero di Sirio.**   R. K. G. Temple.
Sugar Editore, Milano. 316 pp. Price L 4500.
(1976). − Review in Coelum, Vol. 46, 30 - 33; 1978 (*G. Cavallo*). − See 19.003.166.

003.136   **Photochemical and transport processes in the upper atmosphere.**   L. Thomas, H. Rishbeth.
Pergamon Press, Oxford. 112 pp. Price U.K. £ 5.00, U.S. $
10.00 (1976). − Review in Planet. Space Sci., Vol. 26, 193;
1978 (*P. M. Banks*).

003.137   **Einführung in die Protophysik der Welle. Kymometrie.**   B. Thüring.
Duncker & Humblot/ Berlin. 162 pp. Price DM 76.00 (1978).
ISBN 3-428-04136-4.

003.138   **The invention of the telescope.**   A. Van Helden.
Transactions of the American Philosophical Society,
Philadelphia, Vol. 67, Part 4, 67 pp. Price $ 6.00 (1977).
Reviews in Observatory, Vol. 98, 68 - 69; 1978 (*D. W. Dewhirst*); Sci. American, Vol. 238, No. 6, p. 26; 1978 (*P. Morrison*).

003.139   **Cosmic catastrophes.**   G. L. Verschuur.
Addison-Wesley, Reading, Mass. 214 pp. Price
$ 9.95 cloth, $ 5.95 paper (1978). − Review in Sky Telesc.,
Vol. 56, 63 (1978).

003.140   **Report on active and planned spacecraft and experiments.**
J. I. Vette, R. W. Vostreys (Editors).
National Space Science Data Center, Code 601.4, Goddard
Space Flight Center, Greenbelt, Md. 20771. 279 pp. (1977).
Review in Sky Telesc., Vol. 55, 429 (1978).

003.141   **Evolution of the atmosphere.**   J. C. G. Walker.
Macmillan, New York. 318 pp. Price $ 12.95 (1977).
From Phys. Today, Vol. 31, No. 7, p. 53 (1978).

003.142   **Atmosphere science − an introductory survey.**
J. M. Wallace, P. V. Hobbs.
Academic Press, New York. 467 pp. Price $ 17.95 (1977).
Review in Icarus, Vol. 34, 223 - 224; 1978 (*A. P. Ingersoll*).

003.143   **Origins of NASA names.**
H. T. Wells, S. H. Whiteley, C. E. Karegeannes.
National Aeronautics and Space Administration. 227 pp.
Price $ 3.65 (1976). − Review in Sky Telesc., Vol. 55, 170
(1978).

003.144   **Whitney's star finder. A field guide to the heavens.**
C. A. Whitney.
Knopf, New York. 2nd edition. 20 + 106 pp. Price $ 5.95
(1977). − From Science, Vol. 199, 1239 (1978).

003.145   **The origin of the planets.**   I. P. Williams.
Monographs on Astronomical Subjects. Hilger,

Bristol, England. (U.S. distributor: Crane, Russak, New York).
8 + 108 pp. Price $ 14.00 (1977). − Review in Sky Telesc.,
Vol. 55, 344 (1978).

003.146   **The evolving continents.**   B. F. Windley.
John Wiley & Sons Inc., New York. 18 + 386 pp.
Price cloth $ 29.95, paper $ 11.95 (1977). − Review in
Science, Vol. 199, 60; 1978 (*J. F. Dewey*).

003.147   **Earth's aura.**   L. B. Young.
Alfred A. Knopf, New York. 305 pp. Price $ 12.95
(1977). − Review in Nature, Vol. 271, 691; 1978 (*J. T. Houghton*).

003.148   **Megaliths, myths and men: an introduction to astro-archaeology.**   P. L. Brown.
Taplinger, New York. 324 pp. Price $ 13.95 (1977). − From
Phys. Today, Vol. 30, No. 6, p. 63 (1977).

003.149   **Time and its measurement from billion's parts of a second to milliards of years.**   F. S. Zavel'skij.
4th revised edition. Nauka, Moskva. 288 pp. Price 50 Kop.
(1977). In Russian. − Review in Ref. zh., 51. Astron.,
4.51.33 (1978).

003.150   **R Corona Borealis-type stars.**
B. E. Zhilyaev, M. Ya. Orlov, A. F. Pugach, M. G.
Rodriges, A. G. Totochava. Edited by M. Ya. Orlov.
Naukova dumka, Kiev. 128 pp. Price 1 Rbl. 20 Kop. (1978).
In Russian.

003.151   **Astronomers observe.**   F. Yu. Zigel'.
Nauka, Moskva. 192 pp. Price 40 Kop. (1977). In
Russian. − From Ref. zh., 51. Astron., 4.51.27 (1978).

003.152   **Our incredible universe.** A lecture using 40 colour
slides and tape cassette.
G. P. Slide Co. Inc., Houston, Texas. Running time: 28 minutes
18 seconds. Price £ 10.50. − Review in J. British Astron.
Assoc., Vol. 88, 420 - 421; 1978 (*C. A. Ronan*).

003.153   **Gravitation and relativity. Issue 13.**
Kazan. univ., Kazan. 130 pp. Price 70 Kop. (1976).
In Russian. − Review in Ref. zh., 51. Astron., 4.51.50 (1978).

003.154   **Moon, Mars and meteorites.**
Geological Museum. H. M. Stationery Office,
Atlantic House, Holborn Viaduct. London EC 1P 1BN, England. 36 pp. Price 70p. (1977). − Review in Sky Telesc.,
Vol. 55, 170 (1978).

003.155   **Physical processes in the upper atmosphere.**
Fiz.-tekh. inst. AN TSSR. Ylym, Ashkhabad. 132
pp. Price 90 Kop. (1977). In Russian. − Review in Ref. zh.,
51. Astron., 2.51.40 (1978).

003.156   **La terre, les eaux, l'atmosphère.**
Introduction by J. Coulomb.
Encyclopédie scientifique de l'univers. Bureau des Longitudes.
Gauthier-Villars, Paris. 232 pp. Price DM 62.00 (1977).
ISBN 2-04-006 750-7. − Reviews in Astronomie, Vol. 92,
324 (1978); C. R. Acad. Sci. Paris, Vie Acad., Tome 286, 28
(1978).
Contents: I. Géodésie. Rotation de la terre. Marées
terrestres. Sismologie. Plaques lithosphériques. Tectonique
globale, Le champ magnétique terrestre. Constitution
physique de la terre. II. Les marées. Courants marins. La houle
et les vagues. L'atmosphère. La météorologie. La circulation
des eaux. Neige et glaciers.

003.157   **Investigations on solar activity.**
AN SSSR. Dal'nevost. nauch. tsentr. Ussur. st.

sluzhby Solntsa. Vladivostok. 162 pp. Price 64 Kop. (1977).
In Russian. – Review in Ref. zh., 51. Astron., 2.51.55 (1978).

004 History of Astronomy, Chronology

004.001 Early dead reckoning navigation.
C. H. Cotter.
J. Navig., Vol. 31, 20 - 27 (1978).

004.002 A geologist among astronomers: the rise and fall of
the Chamberlin-Moulton cosmogony, part 1, 2.
S. G. Brush.
J. Hist. Astron., Vol. 9, 1 - 41, 77 - 104 (1978).

004.003 The accuracy of Tycho Brahe's instruments.
W. G. Wesley.
J. Hist. Astron., Vol. 9, 42 - 53 (1978).

004.004 Some megalithic sites in Shetland.
A. Thom, R. L. Merritt.
J. Hist. Astron., Vol. 9, 54 - 60 (1978).

004.005 An almanac ring from the Middle Ages.
S. Moskowitz.
Sky Telesc., Vol. 55, 124 - 125 (1978).

004.006 Notes inédites de Le Verrier.   A. Stoyko.
Astronomie, Vol. 92, 94 - 99 (1978).

004.007 An eighteenth-century telescope.
D. G. Hinds.
J. British Astron. Assoc., Vol. 88, 154 - 157 (1978).

004.008 Tian-Guan guest star (supernova) of 1054 in
history.   D.-c. Wang.
Acta Astron. Sinica, Vol. 18, 248 - 254 (1977). In Chinese.

004.009 The origin of the planetesimal theory.
S. G. Brush.
Origins of Life, Vol. 8, No. 1, p. 3 - 6 (1977). – Abstr. in
Phys. Abstr., Vol. 80, Abstr. 87623 (1977).

004.010 William McKinley's renaissance man – Simon
Newcomb. The life and times of America's greatest
astronomer.   R. A. Schorn.
Bull. American Astron. Soc., Vol. 9, 547 (1977). – Abstract.

004.011 The Abbé de la Caille.   M. W. Feast.
Mon. Notes Astron. Soc. Southern Africa, Vol. 37,
12 - 14 (1978).

004.012 Fulton's orrery.   R. M. Grogans.
J. British Astron. Assoc., Vol. 88, 277 - 280 (1978).

004.013 Le temps de Greenwich.   D. Howse.
Astronomie, Vol. 92, 163 - 172 (1978).

004.014 Reflections concerning Neptune's "ring".
W. G. Hoyt.
Sky Telesc., Vol. 55, 284 - 285 (1978).

004.015 Steps toward the Hertzsprung-Russell diagram.
D. H. DeVorkin.
Phys. Today, Vol. 31, No. 3, p. 32 - 39 (1978).

004.016 Creative collaboration of Near East and Middle East
astronomers in the Xth and early XIth centuries.
Kh. F. Abdulla-Zade.
Redkollegiya zh. Izv. AN TadzhSSR. Otd. fiz.-math. i geol.-
khim. nauk. Dushanbe, 1977. 33 pp. In Russian. – Abstr. in
Ref. zh., 51. Astron., 2.51.2 (1978).

004.017 The quadrants of medieval East.
A. K. Tagi-Zade.
Istor.-astron. issled. Vyp. (No.) 13. Moskva, Nauka, 1977,
p. 183 - 200. In Russian. – Abstr. in Ref. zh., 51. Astron.,
2.51.3 (1978).

004.018 Astronomische Inhalte in bronzezeitlichen Fels-
ritzungen.   W. Brunner-Bosshard.
Orion, 36. Jahrg., 68 - 70 (1978).

004.019 Observations of comets in Greek and Roman sources
before A.D. 410.   A. A. Barrett.
J. R. Astron. Soc. Canada, Vol. 72, 81 - 106 (1978).
Unlike the ancient Chinese, the Greeks and Romans did
not keep careful records of comet appearances; "scientific"
study of comets was more or less restricted to speculation
about their nature. The catalogue of observations presented in
this article is compiled from allusions, often incidental, scattered throughout the corpus of Greek and Roman literature.

004.020 Kalenderbauten. Das Observatorium des Mahārāja
Jai Singh II (1686 - 1743) in Jaipur (Rajasthan).
H. Kern.
Sterne Weltraum, Jahrg. 17, 167 - 170 (1978).

004.021 Copernican revolution as a transition to the classic
type of physical knowledge.   V. P. Hütt.
Tartu Ulikooli toimetised, Uch. zap. Tartus. univ., 1977, vyp.
(No.) 404, p. 72 - 85. In Russian. – Abstr. in Ref. zh., 51.
Astron., 3.51.2 (1978).

004.022 Copernican revolution in the conception of
scientific revolutions. T. Kuna.   N. A. Tarasenko.
Filos. probl. sovrem. estestvozn. Mezhved. nauch. sb., 1977,
vyp. (No.) 43, p. 9 - 15. In Russian. – Abstr. in Ref. zh., 51.
Astron., 3.51.3 (1978).

004.023 Natural sciences in Russia from the sixties of the
XIXth century to 1917. Astronomy.
G. M. Idlis.
Razv. estestvozn. v Rossii (XVIII – nachalo XX veka). Moskva,
1977, p. 319 - 333. In Russian. – Abstr. in Ref. zh., 51.
Astron., 3.51.7 (1978).

004.024 On the mediaeval arabic knowledge of the star
Alpha Eridani.   P. Kunitzsch.
J. Hist. Arabic Sci., Vol. 1, 263 - 267 (1977).

004.025 A reconstruction of the chronology of Mesoamerican calendrical systems.   V. H. Malmström.
J. Hist. Astron., Vol. 9, 105 - 116 (1978).

004.026 Borrowed perceptions: Harriot's maps of the Moon.

T. F. Bloom.
J. Hist. Astron., Vol. 9, 117 - 122 (1978).

004.027   Meteors, meteor showers and meteorites in the
          Middle Ages: from European medieval sources.
U. Dall'Olmo.
J. Hist. Astron., Vol. 9, 123 - 134 (1978).

004.028   Lambert and Herschel.   M. Hoskin.
          J. Hist. Astron., Vol. 9, 140 - 142 (1978).

004.029   Namoratunga: the first archeoastronomical evidence
          in Sub-Saharan Africa.
B. M. Lynch, L. H. Robbins.
Science, Vol. 200, 766 - 768 (1978).
       Namoratunga, a megalithic site in northwestern Kenya,
has an alignment of 19 basalt pillars that are nonrandomly
oriented toward certain stars and constellations. The same
stars and constellations are used by modern eastern Cushitic
peoples to calculate an accurate calendar. The fact that
Namoratunga dates to about 300 B.C. suggests that a pre-
historic calendar based on detailed astronomical knowledge
was in use in eastern Africa.

004.030   Astronomy and civilization in the time of
          Aryabhata I.   N. S. Hetherington.
J. British Astron. Assoc., Vol. 88, 371 - 376 (1978).

004.031   A mystery of the zodiacal light.   C. M. Botley.
          J. British Astron. Assoc., Vol. 88, 380 - 381 (1978).

004.032   Die Geschichte der astronomischen Ortsbestimmung
          auf See. Zur Entwicklung der Navigation nach Ge-
stirnen im 18. und 19. Jahrhundert.   L. Brandt.
Sterne Weltraum, Jahrg. 17, 230 - 235 (1978).

004.033   The great unlucky fellow. Discovery of Neptune.
          A. Marks.
Urania Kraków, Vol. 48, 330 - 335 (1977). In Polish.

004.034   Development of the views upon Galaxy structure.
          P. Rybka.
Urania Kraków, Vol. 48, 354 - 364 (1977). In Polish.

004.035   Isaac Newton and astrology.   T. G. Cowling.
          Leeds University Press, 20 pp. Price 65 p. (1977).

004.036   Kishiyar Dzhili's tables of equatorial coordinates
          of bright stars.   Kh. F. Abdulla-Zade.
Izv. AN TadzhSSR. Otd. fiz.-mat. i geol.-khim. nauk, 1977,
No. 4, p. 25 - 28. In Russian. − Abstr. in Ref. zh., 51. Astron.,
5.51.3 (1978).

004.037   Die schönsten Sternsagen der Griechen. Die Sterne
          des Frühlingshimmels.   D. B. Herrmann.
Archenhold-Sternw., Berlin-Treptow, Vortr. Schr., Nr. 54,
24 pp. (1977).

004.038   Karl Friedrich Zöllner und sein Beitrag zur Rezep-
          tion der naturwissenschaftlichen Schriften Im-
manuel Kants.   D. B. Herrmann.
NTM, Schriftenr. Gesch. Naturwiss., Tech., Med., Band 13,
50 - 53 (1976) = Mitt. Archenhold-Sternw. Berlin-Treptow
Nr. 117.

004.039   Eine Totenmaske Isaac Newtons?
          D. Wattenberg.
NTM, Schriftenr. Gesch. Naturwiss., Tech.,Med., Band 14,
43 - 46 (1977) = Mitt. Archenhold-Sternw. Berlin-Treptow
Nr. 122.

004.040   Die Evolution physikalischer Ideen seit der

Renaissance.   A. Unsöld.
Naturwiss. Rundsch., Band 30, 117 - 122 (1977) = Sonderdr.
Sternw. Kiel, Nr. 224.

004.041   Joseph Everett and the King's College Observatory.
          R. L. Bishop.
J. R. Astron. Soc. Canada, Vol. 72, 138 - 148 (1978).
       Through the efforts of Joseph David Everett, Professor of
Mathematics, Natural Philosophy and Astronomy at King's
College, Windsor, Nova Scotia, an astronomical observatory
was built at that college in 1861. This was apparently the first
observatory to be erected in Nova Scotia for educational
purposes. With Everett's return to Scotland in 1864, interest
in astronomy faded at King's, although during the next decade
the building was used as a meteorological observatory.

004.042   Letters of I. I. and K. L. Littrov to I. M. Simonov
          (1823 - 1846) as a source on history of astronomy.
S. N. Korytnikov.
Tr. Kazan. Gorod. Astron. Obs., No. 41, p. 126 - 133 (1976).
In Russian.

004.043   I. M. Simonov and the Cracow Astronomical
          Observatory (1825).   S. N. Korytnikov.
Tr. Kazan. Gorod. Astron. Obs., No. 41, p. 134 - 136 (1976).
In Russian.

004.044   From the scientific correspondence of I. M.
          Simonov. II. Letters of W. J. Struve to I. M.
Simonov (1839 - 1846).
Introduction, publication and remarks by S. N. Korytnikov.
Tr. Kazan. Gorod. Astron. Obs., No. 41, p. 149 - 159 (1976).
In Russian.

004.045   From the scientific correspondence of I. M. Simonov,
          III. Correspondence I. M. Simonov's and O. W.
Struve's (1847 - 1851).
Introduction, publication and remarks by S. N. Korytnikov.
Tr. Kazan. Gorod. Astron. Obs., No. 41, p. 160 - 167 (1976).
In Russian.

004.046   D. I. Dubyago and the beginning of the Kazan
          astronomical school.   S. N. Korytnikov.
Izv. Astron.Ehngel'gardt. Obs., Kazan', No. 41 - 42, p. 234 -
244 (1976). in Russian.

004.047   Peirce's theory of the geometrical structure of
          physical space.   R. R. Dipert.
ISIS, Vol. 68, 404 - 413 (1977). − Abstr. in Phys. Abstr.,
Vol. 81, Abstr. 28906 (1978).

004.048   The origin of North American astronomy − seven-
          teenth century.   D. K. Yeomans.
ISIS, Vol. 68, 414 - 425 (1977). − Abstr. in Phys. Abstr.,
Vol. 81, Abstr. 28907 (1978).

004.049   Giant mirror of Birr.   P. S. Callahan.
          Appl. Opt., Vol. 17, 678 - 680 (1978).

004.050   Percival Lowell and the Martian civilization.
          H. Yokoo.
Kyorin J. Arts Sci., Vol. 3, 49 - 69 (1976). In Japanese.
       A history of the so-called canal controversy is examined
using annual production rates of astronomical papers on Mars
from 1899 to 1973 (4254 in total). Discussions on the Martian
world are collected, including those of S. Newcomb, E. S.
Morse, S. Arrhenius, A. R. Wallace, H. G. Wells, and Lenin.
Trials of communication with the inhabited Mars are listed.

004.051   Bernard Le Bovier de Fontenelle and his discussions
          on the extraterrestrial civilizations.   H. Yokoo.
Kyorin J. Arts Sci., Vol. 5, 57 - 73 (1977). In Japanese.

The life of the writer of "Entretiens sur la pluralité des mondes", 1686, is described. The profound influence of this book is surveyed. The concept of extraterrestrial inhabitants was the key point of the Copernican revolution, and had an explosive power in the struggle against old thoughts.

004.052 Percival Lowell and his study on the Japanese culture. H. Yokoo.
Kyorin J. Arts Sci., Vol. 4, 67 - 87 (1977). In Japanese.
The life of Lowell is described with special attention to his stay in Japan, 1883 - 1893.

004.053 First investigations of short-wave radiation of the sun. A. V. Yakovleva, L. A. Vedeshin.
Vestn. AN SSSR, 1978, No. 2, p. 105 - 113. In Russian.
From Ref. zh., 62. Issled. kosm. prostranstva, 6.62.2 (1978).

Catalogue of the archives and manuscripts of the Royal Astronomical Society. See Abstr. 002.034.

Supplement to the Catalogue of the Crawford Library of the Royal Observatory Edinburgh. See Abstr. 002.058.

Union catalogue of printed books of the XV and XVI centuries in astronomical European observatories. See Abstr. 002.070.

Newton and Newtoniana 1672–1975: a bibliography. See Abstr. 002.072.

Isaac Newton and astrology. See Abstr. 003.036.

Astrolabes of the world. See Abstr. 003.056.

In search of ancient astronomies. See Abstr. 003.077.

Lettres cosmologiques sur l'organisation de l'univers. See Abstr. 003.079.

The making of geology. See Abstr. 003.114.

Theorien und Ansätze in der griechischen Astronomie – im Kontext benachbarter Wissenschaften betrachtet. See Abstr. 003.119.

The invention of the telescope. See Abstr. 003.138.

Cosmology, history, and theology. See Abstr. 003.148.

Henry Norris Russell and the H-R diagram. See Abstr. 005.004.

XV International Congress of the History of Science. See Abstr. 011.014.

History, astronomy and New Zealand. See Abstr. 013.021.

Télescope et radio-télescope. See Abstr. 032.014.

Report of a Near Eastern sighting of the Crab supernova explosion. See Abstr. 125.061.

005 Biography

005.001 Pavel Petrovich Parenago. B. V. Kukarkin.
Tr. Gos. Astron. Inst. Shternberga, Vol. 47, 5 - 15 (1976). In Russian and English. – Translated from Russian by N. N. Samus'.

005.002 Johann Schöner 1477 - 1547. G. Wolfschmidt.
Sterne Weltraum, 17. Jahrg., 86 - 90 (1978).

005.003 Michelson – great optician (on the occasion of his 125th birthday). Ya. A. Smorodinskij.
Zemlya i Vselennaya, 1978, No. 1, p. 52 - 59. In Russian.

005.004 Henry Norris Russell and the H-R diagram. A. G. D. Philip, L. C. Green.
Sky Telesc., Vol. 55, 306 - 310 (1978).

005.005 P. I. Yashnov (1874 - 1940), a Pulkovo astronomer. G. K. Gorel', M. S. Zverev.
Istor.-astron. issled. Vyp. (No.) 13. Moskva, Nauka, 1977, p. 117 - 146. In Russian. – Abstr. in Ref. zh., 51. Astron., 2.51.6 (1978).

005.006 A forgotten bicentenary: Johann Georg Soldner. S. L. Jaki.
Sky Telesc., Vol. 55, 460 - 461 (1978).

005.007 Commemoration of the birth centenary of Sir James Jeans.
Q. J. R. Astron. Soc., Vol. 19, 158 - 159 (1978).

005.008 Sir James Hopwood Jeans, OM, FRS, 1877 - 1946. W. H. McCrea.
Q. J. R. Astron. Soc., Vol. 19, 160 - 166 (1978).

005.009 Thomas Simpson Evans, 1777 - 1818. N. S. Kiernan.
J. British Astron. Assoc., Vol. 88, 365 - 370 (1978).

005.010 Über das Erstlingswerk Bruno H. Bürgels. Altes und Neues über den Planeten Mars in Mir Boshi (1897). (Aus Anlaß des 30. Todestages Bürgels am 8. Juli 1978).
A. Zenkert.
Sterne, 54. Band, 81 - 90 (1978).

005.011 Joseph Ward – pioneer astronomer and telescope maker. D. Calder.
South. Stars, Vol. 27, 104 - 108 (1978).

005.012 Tycho Brahe (1546 - 1601). M. Gabriel.
Bull. Soc. Astron. Liège, Vol. 40, 17 - 21 (1978).

005.013 P. G. A. Moritz, a graduated student of the Tartu University – former director of the Tbilisi Observatory. V. L. Chenakal.
Rol' Tartus. univ. v razvitii otechestven. nauk. i v podgotovke nauchn.-ped. kadrov. Tartu, 1977, p. 84 - 88. In Russian. Abstr. in Ref. zh., 51. Astron., 5.51.11 (1978).

005.014 N. N. Sytinskaya (1906 - 1974). V. A. Bronshtehn.

Problems of observational and theoretical astronomy, Moscow – Leningrad, 1977, (see 003.012), p. 266 - 268. In Russian. – Abstr. in Ref. zh., 51. Astron., 5.51.15 (1978).

005.015   Julius Scheiner und der erste Lehrstuhl für Astrophysik an der Universität Berlin.
D. B. Herrmann.
NTM, Schriftenr. Gesch. Naturwiss., Tech., Med., Band 14, 33 - 42 (1977) = Mitt. Archenhold-Sternw. Berlin-Treptow Nr. 121.

005.016   "Nicht das Wissen, sondern das Lernen . . . " Carl Friedrich Gauß – Leben und Wirken.
D. B. Herrmann.
Wiss. Fortschr., 17. Jahrg., 148 - 151 (1977) = Mitt. Archenhold-Sternw. Berlin-Treptow Nr. 127.

005.017   C. F. Gauß als Astronom und Physiker.
D. B. Herrmann.
Wiss. Fortschr., 17. Jahrg. (1977) = Mitt. Archenhold-Sternw. Berlin-Treptow Nr. 128.

005.018   Karl Friedrich Zöllner. Versuch einer Analyse seiner philosophischen Position.   J. Hamel.
Mitt. Archenhold-Sternw. Berlin-Treptow Nr. 129, 50 pp. (1977).

005.019   I. A. Littrov – teacher in astronomy at the Kazan University.   S. N. Korytnikov.
Tr. Kazan. Gorod. Astron. Obs., No. 41, p. 111 - 125 (1976). In Russian.

005.020   V. V. Bazykin, (1916 - 1974).
Problems of observational and theoretical astronomy, Moscow–Leningrad, 1977, (see 003.012), p. 269 - 271. In Russian. – Abstr. in Ref. zh., 51. Astron., 7.51.5 (1978).

Maria Mitchell, First Lady of American astronomy.
See Abstr. 003.099.

Nikita Ivanovich Popov (1720 - 1782).
See Abstr. 003.105.

006   Personal Notes

Hannes Alfvén – the first seventy years,
G. Arrhenius, C.-G. Fälthammar, Z. Kopal.
Astrophys. Space Sci., Vol. 55, 3 - 5 (1978).

I. S. Astapovich, on the occasion of his 70th birthday.   V. V. Fedynskij.
Zemlya i Vselennaya, 1978, No. 2, p. 65 - 68. In Russian.

M. K. V. Bappu has been elected an associate of the Royal Astronomical Society (London).
Bull. Astron. Soc. India, Vol. 6, 22 (1978).

D. Bates received the Gold Medal of the Royal Astronomical Society.   A. H. Cook.
Q. J. R. Astron. Soc., Vol. 19, 153 (1978).

J. Bell-Burnell, received the 1978 J. Robert Oppenheimer Memorial Prize.
Phys. Today, Vol. 31, No. 4, p. 68 - 69 (1978).

J. G. Bolton received the Gold Medal of the Royal Astronomical Society.   A. H. Cook.
Q. J. R. Astron. Soc., Vol. 19, 153 - 156 (1978).

B. Buti received the 1977 Sarabhai Research award in Planetary and Space Physics.
Bull. Astron. Soc. India, Vol. 6, 22 (1978).

W. N. Christiansen received la treizième medaille annuelle de l'A.D.I.O.N.   N. R. Labrum.
Assoc. Dév. Int. Obs. Nice, Bull. N. 13 - 14, p. 51 - 61 (1978).

C. H. Cotter received the Gold Medal of the Royal Institute of Navigation.
J. Navig., Vol. 31, 162 - 163 (1978).

J. Delhaye received la quatorzième medaille annuelle de l'A.D.I.O.N.   A. Hayli.
Assoc. Dév. Int. Obs. Nice, Bull. N. 13 - 14, p. 83 - 88 (1978).

J. Elliot, honored by NASA.

Phys. Today, Vol. 31, No. 2, p. 63 - 64 (1978).

Konradin Ferrari d'Occhieppo siebzig Jahre.
H. Haupt.
Sternenbote, 21. Jahrg., 43 - 44 (1978).

W. A. Fowler received the Eddington Medal of the Royal Astronomical Society.
Phys. Today, Vol. 31, No. 6, p. 68 (1978).

H. Friedman, Ehrenpromotion.
Sterne Weltraum, Jahrg. 17, 63 (1978)

S. W. Hawking, received the Albert Einstein Award.
Phys. Today, Vol. 31, No. 4, p. 68 (1978).

T. J. Johnson, received the David Richardson Medal for 1978.
Phys. Today, Vol. 31, No. 4, p. 67 (1978).

A. Mrkos, 60th birthday.
Říše hvězd, Vol. 59, 8 - 9 (1978). In Czech.

R. Müller, 80. Geburtstag.
Sterne Weltraum, Jahrg. 17, 63 (1978).

P. J. E. Peebles received the A. Cressy Morrison Award in Natural Sciences.
Phys. Today, Vol. 31, No. 3, p. 77 (1978).

A. A. Penzias received the Herschel Medal.
A. H. Cook.
Q. J. R. Astron. Soc., Vol. 19, 156 - 157 (1978).

C. Sagan, honored by NASA.
Phys. Today, Vol. 31, No. 2, p. 63 - 64 (1978).

M. Schmidt, director of Hale Observatories.
Phys. Today, Vol. 31, No. 1, p. 91 (1978).

B. V. Sreekantan received the first Homi J. Bhabha

Medal of the Indian National Science Academy.
Bull. Astron. Soc. India, Vol. 5, 115 (1977).

R. W. Wilson received the Herschel Medal.
A. H. Cook.
Q. J. R. Astron. Soc., Vol. 19, 156 - 157 (1978).

N. J. Woolf was appointed the acting director of the new Multi-Mirror Telescope for its final construction and early testing phase.
Phys. Today, Vol. 31, No. 7, p. 56 (1978).

007 Obituaries

A. Armitage, 1902 July 19 - 1976 February 8.
G. J. Whitrow.
Q. J. R. Astron. Soc., Vol. 19, 149 (1978).

R. Baillaud, 1885 November 10 - 1977 July 2.
B. Decaux.
C. R. Acad. Sci. Paris, Vie Acad., Tome 286, 94 - 95 (1978).

A. Brun, 1881 March 28 - 1978 January 6.
E. Schweitzer.
Bull. AFOEV, Tome 11, 71 - 72 (1977).

D. Chalonge, 1895 - 1977.
Astronomie, Vol. 92, 53 (1978).

D. H. Chalonge, 31 January 1895 - 28 November 1977. M. Fracassini, L. E. Pasinetti.
Coelum, Vol. 46, 19 - 20 (1978).

C. Cruz-González, 1941 - 1977 April 8.
P. Pişmiş.
Rev. Mexicana Astron. Astrofis., Vol. 2, 173 (1977).

A. Curtis, 1901 - 1976. S. Dunlop.
J. British Astron. Assoc., Vol. 88, 280 -281 (1978).

W. V. Day, 1921 June - 1977 March.
S. Mitton.
Q. J. R. Astron. Soc., Vol. 19, 246 (1978).

Tauno Honkasalo. Memorial address.
T. J. Kukkamäki.
Acad. Sci. Fennica, Proc. 1975, p. 89 - 92 (1977).

S.-s. Huang, died 1977 September 15.
W. Buscombe.
Phys. Today, Vol. 31, No. 2, p. 64 (1978).

E. L. Johnson died.
Mon. Notes Astron. Soc. Southern Africa, Vol. 37, 2 - 3 (1978).

B. V. Kukarkin, 1909 October 30 - 1977 September 15.
Astron. Tsirk., No. 969, p. 1 - 3 (1977). In Russian.

B. V. Kukarkin, 1909 October 30 - 1977 September 15.
Astron. Zh. Akad. Nauk SSSR, Vol. 55, 202 - 204 (1978). In Russian. English translation in Soviet Astron., Vol. 22, No. 1.

B. V. Kukárkin, 1909 October 30 − 1977 September 15. Translated from Russian by V. Bona.
Coelum, Vol. 46, 72 - 75 (1978).

B. V. Kukarkin, 1909 October 30 - 1977 September

15. P. G. Kulikovskij.
Q. J. R. Astron. Soc., Vol. 19, 247 - 248 (1978).

B. V. Kukarkin died 1977 September 15.
O. Obůrka.
Říše hvězd, Vol. 59, 10 (1978). In Czech.

B. V. Kukarkin, 1909 October 30 - 1977 September 15.
Zemlya i Vselennaya, 1978, No. 1, p. 59 - 60. In Russian.

L. G. Kwasha, 1909 January 30 - 1977 July 28.
E. L. Krinov, A. A. Yavnel'.
Meteoritics, Vol. 13, 161 - 162 (1978).

A. Lallemand, 1904 - 1978 March 24.
Astronomie, Vol. 92, 289 (1978).

G. A. Lange, 1905 - 1977 July 20.
V. P. Tsesevich.
Astron. Tsirk., No. 981, p. 7 - 8 (1978). In Russian.

D. N. Limber, 1928 May - 1977 April 16.
Phys. Today, Vol. 31, No. 1, p. 93 (1978).

S. B. Lorensen, 1942 - 1978. R. M. West.
Messenger, No. 13, p. 12 (1978).

R. A. McIntosh, 1904 January 21 - 1977 May 17.
F. M. Bateson.
South. Stars, Vol. 27, 82 - 87 (1977).

A. A. Nefed'ev, 1910 Nov. 23 - 1976 Sept. 14.
Izv. Astron. Ehngel'gardt. Obs., Kazan', No. 41 - 42, p. 256 (1976). In Russian.

B. P. Ostashchenko-Kudryavtsev, 1877 Jan. 10 - 1956 Oct. 1. K. N. Kuz'menko, V. Kh. Pluzhnikov, V. I. Lats'ko, T. A. Senchuk.
Zemlya i Vselennaya, 1978, No. 3, p. 59 - 61. In Russian.

C. Popovici, 1910 - 1977 January 16.
E. Ţifrea.
Obs. Solaire, Rotation 1637 - 1649. Cent. Astron. Sci. Spatiales, Bucarest. p. 3 - 6 (1977).

G. Ruggieri died. G. Romano.
Coelum, Vol. 45, 244 - 245 (1977).

C. H. Smiley, 1903 September 6 - 1977 July 26.
D. H. Kelley.
J. R. Astron. Soc. Canada, Vol. 72, 46 - 47 (1978).

B. Sticker, 1906 - 1977 August 30.
J. Hist. Astron., Vol. 9, 153 (1978).

J. A. G. Storms, 1910 - 1978.   **W. De Kort.**
Soc. R. Astron. Anvers, 58ᵉ rapport 1977. p. 7 - 8, 10 - 11 (1978). In French and Flemish.

A. D. Thackeray, 1910 June 19 - 1978 February 21.
B. Warner.
Mon. Notes Astron. Soc. South Africa, Vol. 37, 20 - 22 (1978).

Reminiscences of A. D. Thackeray.   D. S. Evans.
Mon. Notes Astron. Soc. South Africa, Vol. 37, 22 - 23 (1978).

A. D. Thackeray, 1910 - 1978 February 21.
M. W. Feast.
Nature, Vol. 274, 100 - 101 (1978).

A. D. Thackeray died 1978 February 21.
Observatory, Vol. 98, 79 (1978).

A. D. Thackeray, died 1978 February 21.
Sky Telesc., Vol. 55, 371 (1978).

P. J. Treanor died 1978 February 18.

Observatory, Vol. 98, 80 (1978).

H. von Klüber, 1901 September 6 - 1978 February 14.   **D. E. Blackwell, D. W. Dewhirst.**
Nature, Vol. 273, 414 (1978).

H. von Klüber died 1978 February 14.
Observatory, Vol. 98, 79 - 80 (1978).

S. Wierzbiński, 1910 August 5 - 1976 July 14.
P. Rybka.
Postępy Astron., Tom 25, 35 (1977). In Polish.

J. Witkowski died 1976 May 26.   **H. Hurnik.**
Postępy Astron., Tom 24, 267 - 268 (1976). In Polish.

F. Zagar, 1900 November 30 - 1976 February 17.
E. Proverbio.
Mem. Soc. Astron. Italiana, Vol. 47, 4 - 10 (1976). In Italian.

E. V. Zvyagina, 1939 - 1978 January 11.
Astron. Tsirk., No. 984, p. 7 - 8 (1978). In Russian.

008 Observatories, Institutes

Reports, communications and publications of observatories and astronomical institutes are recorded in this section; included are numbered series of reprints. Whenever possible, the numbers of the abstracts referring to the publications are given. Observatories and institutes are listed in alphabetical order of their towns. In some cases observatory publications do not give the name of the town; the following list which gives names and towns of some institutions may serve as an aid in such cases.

Aarne Karjalainen Observatory	Oulu, Finland
Algonquin Radio Observatory	Lake Traverse, Ontario, Canada
Allegheny Observatory	Pittsburgh, Pennsylvania, USA
Archenhold-Sternwarte	Berlin-Treptow, German Democratic Republic
Argentine Radioastronomy Institute	Pereyra Iraola, Argentina
Arthur J. Dyer Observatory	Nashville, Tennessee, USA
Astronomical Latitude Station, Polish Academy of Sciences	Borowiec, Poland
Astronomisches Rechen-Institut	Heidelberg, F. R. Germany
Australian National University–Research School of Physical Sciences, Department of Astronomy	Mount Stromlo; Siding Spring, Australia
Bell Laboratories	Murray Hill, New Jersey, USA
Bell Telephone Laboratories	Holmdel, New Jersey, USA
Bosscha Observatory	Lembang, Indonesia
Boyden Observatory	Bloemfontein, South Africa
Bureau International de l'Heure	Paris, France
Cajigal Observatory	Caracas, Venezuela
California Institute of Technology	Pasadena, California, USA
Carter Observatory	Wellington, New Zealand
Catalina Station	Tucson, Arizona, USA
Cavendish Laboratory	Cambridge, England
Cerro Tololo Interamerican Observatory	La Serena, Chile
Ceskoslovenská Akademie Ved Astronomický Ustav	Praha, Czechoslovakia
Chamberlin Observatory, University of Denver	Denver, Colorado, USA
Columbia University, Department of Astronomy	New York, New York, USA
Commonwealth Observatory	Canberra, Australia
Cornell University, Center for Radiophysics and Space Research	Ithaca, New York, USA
Corralitos Observatory	Las Cruces, New Mexico, USA
Crawford Hill Laboratory	Holmdel, New Jersey, USA
David Dunlap Observatory, University of Toronto	Richmond Hill, Ontario, Canada
Dearborn Observatory	Evanston, Illinois, USA
Department of Astronomy and Observatory, Univ. California	Los Angeles, California, USA
Department of Astronomy, University of Texas	Austin, Texas, USA
Deutsches Hydrographisches Institut (DHI)	Hamburg, F. R. Germany
Division Radiophysics, C.S.I.R.O. University Grounds	Sydney, Australia
Dominion Astrophysical Observatory	Victoria, B.C., Canada

Dominion Observatory	Ottawa, Ontario, Canada
Dominion Radio Astrophysical Observatory	Penticton, B. C., Canada,
Dudley Observatory	Albany, New York, USA
Dunsink Observatory	Dublin, Ireland
Dyer Observatory, Vanderbilt University	Nashville, Tennessee, USA
Ege University	Izmir, Turkey
Engelhardt Observatory	Kazan, USSR
Erwin W. Fick Observatory, Iowa State University	Ames, Iowa, USA
European Southern Observatory	La Silla, Chile
Felix Aguilar Observatory	San Juan, Argentina
Fernbank Observatory	Atlanta, Georgia, USA
Five College Observatories	Amherst, Massachusetts, USA
Florida State University Radio Observatory	Tallahassee, Florida, USA
Flower and Cook Observatories, University of Pennsylvania	Philadelphia, Pennsylvania, USA
Fraunhofer Institut	Freiburg, F.R. Germany
George R. Wallace Jr. Astrophysical Observatory	Cambridge, Massachusetts, USA
Georgetown Observatory	Washington, D.C., USA
Glavnaya Astronomicheskaya Observatoriya AN SSSR	Pulkovo, USSR
Goddard Space Flight Center	Greenbelt, Maryland, USA
Goethe Link Observatory, Indiana University	Bloomington, Indiana, USA
"Guido Horn d'Arturo" Observatory	Bologna, Italy
H. M. Nautical Almanac Office, Royal Greenwich Observatory	Greenwich, England
Hale Observatories	Pasadena, California, USA
Harvard College Observatory	Cambridge, Massachusetts, USA
Harvard Radio Astronomy Station	Cambridge, Massachusetts, USA
Haute Provence Observatory	Saint Michel, France
Haystack Observatory	Westford, Massachusetts, USA
Heinrich-Hertz-Institut	Berlin-Adlershof, Germany
Herzberg Institute of Astrophysics	Victoria, B. C.; Ottawa, Canada
High Altitude Observatory, University of Colorado	Boulder, Colorado, USA
Hopkins Observatory	Williamstown, Massachusetts, USA
Horn d'Arturo Observatory	Bologna, Italy
Hvar Observatory	Zagreb, Yugoslavia
IBM Thomas J. Watson Research Center	Yorktown Heights, New York, USA
Indian Institute of Astrophysics	Bangalore, India
Institute for Astronomy, University of Hawaii	Honolulu, Hawaii, USA
Institute for Theoretical Astronomy (Institut Teoreticheskoj Astronomii)	Leningrad, USSR
Institute of Astronomy and Space Science, University of British Columbia	Vancouver, B. C., Canada
Institute of Theoretical Astrophysics, Blindern	Oslo, Norway

Instituto Argentino de Radio-astronomia	Pereyra Iraola, Argentina
Instituto de Astronomía y Física del Espacio (IAFE)	Buenos Aires, Argentina
Instituto Venezolano de Astronomia	Merida, Venezuela
Instituto y Observatorio de Marina	San Fernando (Cádiz), Spain
Inter-American Observatory	Cerro Tololo, La Serena, Chile
International Latitude Observatory	Mizusawa, Japan
Jet Propulsion Laboratory, California Institute of Technology	Pasadena, California
Joint Institute for Laboratory Astrophysics (JILA)	Boulder, Colorado, USA
Kandilli Observatory	Istanbul, Turkey
Kansas University Observatory	Lawrence, Kansas, USA
Kapteyn Astronomical Laboratory	Groningen, Netherlands
Karl-Schwarzschild-Observatorium	Tautenburg, German Democratic Republic
Kenneth Mees Observatory	Rochester, New York, USA
Kitt Peak National Observatory	Tucson, Arizona, USA
Kwasan and Hida Observatories	Kyoto, Japan
Lamont-Hussey Observatory	Bloemfontein, South Africa
Landessternwarte Heidelberg - Königstuhl	Heidelberg, F. R. Germany
Lawrence Livermore Laboratory, University of California	Livermore, California, USA
Leander McCormick Observatory, University of Virginia	Charlottesville, Virginia, USA
Lee Observatory	Beirut, Lebanon
Leopold-Figl-Observatorium	Wien, Austria
Leuschner Observatory	Berkeley, California, USA
Lick Observatory	Santa Cruz, (Mount Hamilton), California, USA
Lindheimer Astronomical Research Center	Evanston, Illinois, USA
Lockheed Palo Alto Research Laboratory	Palo Alto, California, USA
Lockheed Solar Observatory	Saugus, California, USA
Lohrmann-Observatorium der Technischen Universität Dresden	Dresden, German Democratic Republic
Louisiana State University Observatory	Baton Rouge, Louisiana, USA
Lowell Observatory	Flagstaff, Arizona, USA
Lunar and Planetary Laboratory	Tucson, Arizona, USA
Max-Planck-Institut für Astronomie	Heidelberg, F. R. Germany
Max-Planck-Institut für Physik und Astrophysik	München, F. R. Germany
Max-Planck-Institut für Radioastronomie	Bonn, F. R. Germany
McDonald Observatory	Fort Davis, Texas, USA
McGraw-Hill Observatory	Kitt Peak, Arizona, USA
McMath Hulbert Observatory	Pontiac, Michigan, USA
C.E.K. Mees Observatory, University of Rochester	Rochester, New York, USA
Michigan State University Observatory	East Lansing, Michigan, USA
Molonglo Radio Observatory, University of Sydney	Sydney, Australia
Monterey Institute for Research in Astronomy	Carmel Valley, California, USA
Mount Cuba Observatory	Wilmington, Delaware, USA
Mount John Observatory	Lake Tekapo, New Zealand
Mount Palomar Observatory	Pasadena, California, USA
Mount Wilson Observatory	Pasadena, California, USA
Mullard Radio Astronomy Observatory	Cambridge, England
Mullard Space Science Laboratory	London, England
Narrabri Observatory, University of Sydney	Sydney, Australia
National Bureau of Standards	Washington, D.C., USA
National Observatory, USA	Kitt Peak, Arizona, USA
National Radio Astronomy Observatory	Charlottesville, Virginia, USA Green Bank, West Virginia, USA Socorro, New Mexico, USA Tucson, Arizona, USA
National Research Council of Canada	Ottawa, Ontario, Canada
New Mexico State University Observatory	Las Cruces, New Mexico, USA
Nicholas Copernicus Observatory and Planetarium	Brno, Czechoslovakia
Nizamiah & Rangapur Observatories	Hyderabad, India
Nuffield Radio Astronomy Laboratories, Jodrell Bank University of Manchester	Manchester, England
Observatoire Royal de Belgique	Uccle, Belgium
Observatories of the University of Western Ontario	London, Canada
Observatorio Astronomico del Vaticano	Castel Gandolfo, Italy
Observatório Astronômico do Instituto de Física da Universidade Federal do Rio Grande do Sul	Porto Alegre, Rio Grande do Sul, Brazil
Observatorio de Cartuja	Granada, Spain
Observatorio del Ebro	Tortosa, Spain
Observatorio Fabra	Barcelona, Spain
Observatorio Nacional	Rio de Janeiro, Brazil
Observatorio Nacional de Física Cósmica	San Miguel, Argentina
Observatory, University of Michigan	Ann Arbor, Michigan, USA
Ohio State University Radio Observatory	Columbus, Ohio, USA
Ole Roemer-Observatoriet	Aarhus, Denmark
Onsala Space Observatory	Gothenburg, Sweden
Owens Valley Radio Observatory	Big Pine, California, USA
Perkins Observatory, Ohio State and Wesleyan Universities	Delaware, Ohio, USA
Physical Research Laboratory	Ahmedabad, India
Purple Mountain Observatory	Nanking, China
Radcliffe Observatory	Pretoria, South Africa
Raman Research Institute	Bangalore, India
Remeis-Sternwarte	Bamberg, F. R. Germany
Ritter Astrophysical Research Center of the University of Toledo	Toledo, Ohio, USA
Rosemary Hill Observatory	Gainesville, Florida, USA
Royal Radar Establishment, Radio Astronomy Division	Malvern, England
Sacramento Peak Observatory	Sunspot, New Mexico, USA
Sagamore Hill Radio Observatory	Hamilton, Massachusetts, USA
San Fernando Observatory	El Segundo, California, USA
Smithsonian Astrophysical Observatory	Cambridge, Massachusetts, USA
Sonnenobservatorium Kanzelhöhe	Graz, Austria
South African Astronomical	

Observatory	**Cape Town**, South Africa
Specola Astronomica Vaticana	Castel Gandolfo, Italy
Specola di Padova	Asiago, Italy
Sproul Observatory	Swarthmore, Pennsylvania, USA
Stellar Data Center	Strasbourg, France
Sternberg Astronomical Institute	Moskva, USSR
Steward Observatory, University of Arizona	Tucson, Arizona, USA
W. Struve Tartu Astrophysical Observatory	Tartu, USSR
Tata Institute of Fundamental Research	Bombay, India
United States Naval Observatory	Washington, D. C., USA
University of California	Berkeley, California, USA
University of Florida Observatories	Gainesville, Florida, USA
University of Florida, Radio Observatory	Gainesville, Florida, USA
University of Hawaii	Honolulu, Hawaii, USA
University of Illinois Observatory	Urbana, Illinois, USA
University of Kansas Observatory	Lawrence, Kansas, USA
University of Maryland	College Park, Maryland, USA
University of Michigan Observatories	Ann Arbor, Michigan, USA
University of Minnesota	Minneapolis, Minnesota, USA
University of South Florida Observatory	Tampa, Florida, USA

University of Texas, Department of Astronomy	**Austin**, Texas, USA
University of Washington, Astronomy Department	Seattle, Washington, USA
Uttar Pradesh State Observatory	Naini Tal, India
Van Vleck Observatory	Middletown, Connecticut, USA
Vatican Observatory	Castel Gandolfo, Italy
Venezuelan Astronomical Institute	Merida, Venezuela
Wallace Astrophysical Observatory	Cambridge, Massachusetts, USA
Warner and Swasey Observatory	Cleveland, Ohio, USA
Washburn Observatory, University of Wisconsin	Madison, Wisconsin, USA
West Melton Observatory	Christchurch, New Zealand
Wilhelm-Förster Sternwarte	Berlin, Germany
Yale University Observatory	New Haven, Connecticut, USA
Yerkes Observatory	Williams Bay, Wisconsin, USA
Zentralinstitut für Astrophysik, Sternwarte Babelsberg, (Fachbereich Kosmische Physik)	Potsdam-Babelsberg, German Democratic Republic
Zentralinstitut für Astrophysik, Sternwarte in Sonneberg	Sonneberg, German Democratic Republic
Zentralinstitut für solar-terrestrische Physik	Berlin-Adlershof, German Democratic Republic

008.001 Aberdeen

Department of Natural Philosophy, Aberdeen University.   R. V. Jones.
Q. J. R. Astron. Soc., Vol. 19, 198 - 199 (1978). – Report for the year ending 1977 September 30.

008.002 Ames, Iowa

Erwin W. Fick Observatory, Iowa State University, Ames, Iowa. – Report.   W. I. Beavers.
Bull. American Astron. Soc., Vol. 10, 100 - 102 (1978).

008.003 Amherst, Mass.

Five College Astronomy Department: Amherst College, Amherst, Massachusetts; Hampshire College, Amherst,

Massachusetts; Mount Holyoke College, South Hadley, Massachusetts; Smith College, Northampton, Massachusetts; University of Massachusetts, Amherst, Massachusetts. – Reports.   T. Arny.
Bull. American Astron. Soc., Vol. 10, 103 - 106 (1978).

008.004 Ankara

Communications from the Astronomy Department, University of Ankara, Fen Fakültesi, Ankara – Turkey, Nos. 76 (21.075.012), 77 (21.121.095), 78 (21.075.013), 79 (21.114.580), 80 (21.021.030).

008.005 Ann Arbor, Mich.

University of Michigan, Department of Astronomy, Ann Arbor, Michigan. – Report.   W. A. Hiltner.
Bull. American Astron. Soc., Vol. 10, 244 - 247 (1978).

008.006 Arecibo

Arecibo Observatory, Arecibo, Puerto Rico. Report.   H. D. Craft, Jr., T. H. Hankins.
Bull. American Astron. Soc., Vol. 10, 6 - 14 (1978).

008.007 Armagh

Armagh Observatory Contribution, Nos.93 (20.122.043), 95 (20.085.007), 97 (19.122.158).

Armagh Observatory, Leaflet, Nos. 129 (20.008.005), 132 (20.015.010).

008.008 Atlanta, Ga.

Fernbank Observatory, Fernbank Science Center, Atlanta, Georgia. – Report.   R. M. Williamon.
Bull. American Astron. Soc., Vol. 10, 102 - 103 (1978).

008.009 Austin, Tex.

University of Texas at Austin Astronomy Department, Austin, Texas; McDonald, UTRAO, and Millimeter

Observatories. – Reports.   D. S. Evans, H. J. Smith.
Bull. American Astron. Soc., Vol. 10, 325 - 352 (1978).

Outer planet satellite studies. 1976 September -
1978 February 28.
Dep. Astron. McDonald Obs., Austin, Texas, 78712. Semi-
annual status Report, Nos. 7 - 9 (1977/78).

Department of Astronomy and McDonald Observa-
tory of the University of Texas, Austin, Texas, Reprints Nos.
501 (18.122.003), 502, 503 (19.135.016), 504 (19.032.557),
505 (19.131.075), 506 (19.094.161), 507 (19.131.147), 508
(19.122.081), 509 (19.122.060), 510 (19.071.018), 511
(19.008.013), 512 (19.122.085), 513 (19.124.130), 514
(19.114.338), 515 (19.101.018), 516 (19.118.007), 517
(19.066.077), 518 (19.141.527), 519 (19.122.117), 520
(19.119.014), 521 (19.122.113), 522 (19.122.031), 523
(19.131.125), 524 (19.065.055), 525 (19.158.151), 526
(19.093.075), 527 (19.131.125), 528 (19.122.155), 529
(19.114.357), 530 (19.122.156), 531 (20.124.101), 532
(20.099.503), 533 (20.096.006), 534 (20.096.007), 535
(20.154.002), 536 (19.100.023), 537 (20.096.003), 538
(20.158.034), 539 (20.096.004), 540 (19.162.027), 541
(19.097.041), 542 (20.131.039), 543 (20.099.502), 544
(20.134.007), 545 (20.131.045), 546 (18.124.103), 547
(20.114.037), 548 (20.158.044), 549 (19.160.047), 550, 551,
552 (20.131.049), 553 (20.114.533), 554 (20.115.006), 555
(20.096.009), 556 (20.153.012), 557 (20.122.063), 558
(20.131.056), 559 (20.122.029), 560 (20.131.048), 561
(20.122.003), 562 (20.131.093), 563 (20.155.015), 564
(20.131.076), 565 (20.114.018), 566 (20.114.526), 567
(20.099.051), 568 (11.100.211), 569 (19.004.037), 570
(19.124.021), 571 (20.158.506), 572 (20.141.143), 573
(13.141.074), 574 (20.124.005), 575 (20.114.549), 576
(20.101.033), 577 (20.119.016), 578 (20.121.029), 579
(20.117.036), 580 (20.141.147), 581 (20.117.034), 582
(21.114.503), 583 (21.135.004), 584 (21.158.010), 585
(21.142.003), 586 (21.091.005), 587 (21.124.101), 588
(21.131.002), 589, 590 (20.125.002), 591 (20.114.050), 592
(21.114.002), 593 (21.114.507), 594 (21.135.005), 595
(21.124.002), 596 (20.158.073), 597 (20.096.001, 20.096.
008), 598 (21.099.504), 599 (21.114.508).

008.010 Bamberg

Veröffentlichungen der Remeis-Sternwarte Bamberg, Astronomisches Institut der Universität Erlangen–
Nürnberg, Vol. 11, Nr. 121 (21.012.012).

008.011 Bangalore

Kodaikanal Observatory, Bulletins, Series A, Vol. 2,
No. 1 (21.158.270, 21.119.024, 21.080.058, 21.080.059,
21.114.086, 21.073.091, 21.063.036, 21.034.053, 21.031.
423, 21.124.112, 21.114.087, 21.119.025).

008.012 Barcelona

Universidad de Barcelona. Departamento de Fisica
de la Tierra y del Cosmos, Publicacion No. A-26 (21.155.
035), A-29 (21.065.074), A-30 (19.065.038).

008.013 Baton Rouge, La.

Louisiana State University Observatory, Baton

Rouge, Louisiana. – Report.   A. U. Landolt.
Bull. American Astron. Soc., Vol. 10, 218 - 219 (1978).

008.014 Berkeley, Calif.

University of California: Berkeley, Los Angeles,
San Diego, Lick Observatory, and Board of Studies. I. Berkeley
Campus. – Report.
Bull. American Astron. Soc., Vol. 10, 30 - 45 (1978).

008.015 Berlin-Adlershof

Heinrich-Hertz-Institut, Solare Beobachtungsergebnisse. Akademie der Wissenschaften der DDR, Zentralinstitut
für Solar-Terrestrische Physik (Heinrich-Hertz-Institut), Berlin-
Adlershof, HHI Solar Data, Vol. 28, 1977 Oct - Dec., Vol. 29,
1978 Jan.-Feb.

008.016 Berlin-Treptow

Blick in die Sternenwelt 1978 (21.047.018).

Blick in das Weltall, Archenhold-Sternwarte Berlin-
Treptow. Astronomische Veranstaltungen und Mitteilungen
für Sternfreunde, 26. Jahrg., Nos. 1 - 6 (1978).

Mitteilungen der Archenhold-Sternwarte Berlin-
Treptow, Nos. 117 (21.004.038), 121 (21.005.015), 122
(21.004.039), 127 (21.005.016), 128 (21.005.017), 129
(21.005.018).

Archenhold-Sternwarte, Berlin-Treptow, Vorträge
und Schriften, No. 54 (21.004.037).

008.017 Bloemfontein

University of the Orange Free State: Boyden Observatory, Department of Astronomy.   A. H. Jarrett.
Mon. Notes Astron. Soc. South Africa, Vol. 37, 28 - 29
(1978). – Report for 1978.

008.018 Bonn

Max-Planck-Insitut für Radioastronomie, Bonn,
Federal Republic of Germany. – Report for the period 1
October 1976 - 1 October 1977.
V. Pankonin, R. Schwartz, P. G. Mezger.
Bull. American Astron. Soc., Vol. 10, 233 - 237 (1978).

Max-Planck-Gesellschaft zur Förderung der Wissenschaften, M.P.I. f. Radioastronomie, Bonn, Sonderdruck,
Ser. A, Nos. 117 (18.116.026), 171 (19.155.015), 172 (19.
133.010), 173 (19.158.101), 174 (20.022.089), 175 (19.042.
032), 176 (19.141.528), 177 (19.142.051), 178 (19.141.140).

008.019 Borowiec

Polish Academy of Sciences, Astronomical Latitude
Observatory, Borowiec, Poland, Circular Nos. 143 - 144
(21.044.021).

008.020   Boulder, Colo.

Joint Institute for Laboratory Astrophysics of the National Bureau of Standards and the University of Colorado, Boulder, Colorado. – Report.   J. P. Cox.
Bull. American Astron. Soc., Vol. 10, 140 - 151 (1978).

National Center for Atmospheric Research, High Altitude Observatory, Boulder, Colorado. – Report for the period October 1976 - September 1977.   G. Newkirk, Jr.
Bull. American Astron. Soc., Vol. 10, 271 - 277 (1978).

National Oceanic and Atmospheric Administration, Boulder, Colorado. I. Space Environment Laboratory/Space Environment Services Center.   G. R. Heckman.
II. Environmental Data Service, World Data Center A for Solar-Terrestrial Physics.   J. V. Lincoln.
Bull. American Astron. Soc., Vol. 10, 277 - 279 (1978).

008.021   Cambridge, England

University of Cambridge, Institute of Astronomy.
D. Lynden-Bell.
Separate print, The Observatories, Cambridge, 2 + 25 pp. (1977). – Annual report 1976 - 1977.

Mullard Radio Astronomy Observatory, Cavendish Laboratory, University of Cambridge.   M. Ryle.
Q. J. R. Astron. Soc., Vol. 18, 475 - 485 (1977). – Report for the period 1975 October 1 - 1976 September 30.

Mullard Radio Astronomy Observatory, Cavendish Laboratory, University of Cambridge.   M. Ryle.
Q. J. R. Astron. Soc., Vol. 19, 216 - 224 (1978). – Report for the period 1976 October 1 to 1977 September 30.

008.022   Cambridge, Mass.

Center for Astrophysics, Harvard College Observatory and Smithsonian Astrophysical Observatory, Cambridge, Massachusetts. – Reports.
Bull. American Astron. Soc., Vol. 10, 68 - 77 (1978).

008.023   Campinas

Il primo osservatorio astronomico municipale brasiliano: l'«Observatorio do Capricornio» di Campinas.
J. Nicolini, translated by C. Frisoni.
G. A. A. B., N. 47, p. 7 - 9 (1977).

008.024   Cape Town

The South African Astronomical Observatory.
M. W. Feast.
Q. J. R. Astron. Soc., Vol. 19, 137 - 148 (1978). – Report for 10 months ending 1977 January 31.

University of Cape Town: Department of Astronomy. (*B. Warner*). Department of Applied Mathematics. (*G. F. R. Ellis*).
Mon. Notes Astron. Soc. South Africa, Vol. 37, 23 - 28 (1978). Report for 1978.

User guide to telescope facilities – II. South African Astronomical Observatory.   M. Feast.

Q. J. R. Astron. Soc., Vol. 19, 110 - 115 (1978).

008.025   Carloforte

Rapporto annuale per gli anni 1975 e 1976.
E. Proverbio.
Pubbl. Stn. Astron. Int. Latitudine, Carloforte–Cagliari, Nuova Ser., No. 62, 17 pp.

Circolari della Stazione Astronomica Internazionale di Latitudine, Carloforte–Cagliari, Serie A (4), No. 13 (21. 044.033).

Circolari della Stazione Astronomica Internazionale di Latitudine, Carloforte–Cagliari, Serie B (9), No. 12 (21. 021.029).

Pubblicazioni della Stazione Astronomica Internazionale di Latitudine, Carloforte–Cagliari, Nuova Serie, Nos. 47 (21.045.020), 50 (21.046.044), 54 (21.007.000), 56 (21.045.021), 62 (21.008.025), 67 (21.015.044).

008.026   Carmel Valley, Calif.

Monterey Institute for Research in Astronomy, Carmel Valley, California. – Report.   W. B. Weaver.
Bull. American Astron. Soc., Vol. 10, 252 - 253 (1978).

008.027   Castel Gandolfo

Specola Vaticana. Annual report 1977: Report of the Astronomical Observatory.   P. Treanor.
Specola Vaticana, Città del Vaticano. Printed in Vatican City, 14 pp. (1978).

Specola Vaticana, *Castel Gandolfo,* Comunicazione, No. 70 (21.159.003).

Ricerche Astronomiche, Specola Vaticana, Vol. 8, No. 29 (21.121.091).

Ricerche Spettroscopiche, Laboratorio Astrofisico della Specola Vaticana, Vol. 3, Nos 9 (21.105.077), 10 (21.105.078).

008.028   Charlottesville, Va.

National Radio Astronomy Observatory: Charlottesville, Virginia; Green Bank, West Virginia; Socorro, New Mexico; Tucson, Arizona. – Report for the period July 1976 - June 1977.   D. S. Heeschen, D. E. Hogg.
Bull. American Astron. Soc., Vol. 10, 279 - 291 (1978).

008.029   Chicago, Ill.

The University of Chicago, Department of Astronomy and Astrophysics, Chicago, Illinois; The Yerkes Observatory, Williams Bay, Wisconsin. – Reports.
E. N. Parker, L. M. Hobbs.
Bull. American Astron. Soc., Vol. 10, 77 - 83 (1978).

008.030 Cleveland, Ohio

Warner and Swasey Observatory, Case Western Reserve University, Cleveland, Ohio. − Report for the period 1 July 1976 - 30 June 1977.   P. Pesch.
Bull. American Astron. Soc., Vol. 10, 364 - 367 (1978).

008.031 College Park, Md.

University of Maryland, College Park, Maryland. Report for the period 1 September 1976 - 31 August 1977. J. D. Trasco.
Bull. American Astron. Soc., Vol. 10, 223 - 232 (1978).

008.032 Columbus, Ohio

The Observatories of the Ohio State and Ohio Wesleyan Universities, Columbus and Delaware, Ohio. Reports.   A. Slettebak.
Bull. American Astron. Soc., Vol. 10, 294 - 298 (1978).

008.033 Cordoba, Argentina

Observatorio Astronómico (Universidad Nacional de Córdoba, Córdoba, Argentina), Tirada Aparte, Nos. 219 (19.158.139), 220 (20.158.113), 221 (20.158.114), 222 (20.158.095).

008.034 Crimea

Chronicle.
Izv. Krymskoj Astrofiz. Obs., Vol. 58, 120 - 121 (1978). In Russian.

Izvestiya Ordena Trudovogo Krasnogo Znameni Krymskoj Astrofizicheskoj Observatorii, Akademiya Nauk SSSR, Vol. 58 (21.071.029; 21.073.048; 21.080.035; 21.072.032; 21.072.033; 21.077.022; 21.080.036; 21.142.718; 21.142.719; 21.121.036; 21.121.037; 21.121.038; 21.119.009; 21.114.558; 21.158.165; 21.031.008; 21.031.009; 21.008.034).

008.035 Delaware, Ohio

The Observatories of the Ohio State and Ohio Wesleyan Universities, Columbus and Delaware, Ohio. Reports.   A. Slettebak.
Bull. American Astron. Soc., Vol. 10, 294 - 298 (1978).

008.036 Dresden

Technische Universität Dresden, Lohrmann-Observatorium, Zirkular, Nos. 81 - 86 (21.045.009).

008.037 East Lansing, Mich.

Michigan State University, Department of Astronomy and Astrophysics and the Observatory, East Lansing,

Michigan. − Report.   T. P. Stoeckly.
Bull. American Astron. Soc., Vol. 10, 241 - 243 (1978).

008.038 Edinburgh

Occasional Reports of the Royal Observatory, Edinburgh, Nos. 1 (21.159.018), 3 (21.153.038).

008.039 El Segundo, Calif.

The Aerospace Corporation, El Segundo, California. Report.   H. R. Rugge.
Bull. American Astron. Soc., Vol. 10, 1 - 5 (1978).
Research in astronomy at The Aerospace Corporation consists of astrophysics work in the Space Sciences Laboratory, millimeter-wave radio astronomy in the Electronics Research Laboratory, and infrared astronomical studies in the Satellite Systems   Division.

008.040 Evanston, Ill.

Lindheimer Astronomical Research Center and Dearborn Observatory, Northwestern University, Evanston, Illinois. − Report.   J. D. R. Bahng.
Bull. American Astron. Soc., Vol. 10, 206 - 207 (1978).

008.041 Flagstaff, Ariz.

Lowell Observatory, Flagstaff, Arizona. − Report for the period 1 July 1976 - 30 June 1977.   A. A. Hoag.
Bull. American Astron. Soc., Vol. 10, 220 - 223 (1978).

008.042 Fort Davis, Tex.

University of Texas at Austin Astronomy Department, Austin, Texas; McDonald, UTRAO, and Millimeter Observatories. − Reports.   D. S. Evans, H. J. Smith.
Bull. American Astron. Soc., Vol. 10, 325 - 352 (1978).

008.043 Gainesville, Fla.

University of Florida, Tampa, Florida. I. Rosemary Hill Observatory. II. Radio Observatory. − Reports. A. G. Smith.
Bull. American Astron. Soc., Vol. 10, 107 - 111 (1978).

008.044 Genève

Publications de l'Observatoire de Genève, Série B, Fasc. 4 (21.002.050), 5 (21.113.072).

008.045 Graz

Mitteilungen der Universitätssternwarte Graz No. 37 (21.042.108).

008.046 Green Bank, W. Va.

National Radio Astronomy Observatory: Charlottesville, Virginia; Green Bank, West Virginia; Socorro, New Mexico; Tucson, Arizona. – Report for the period July 1976 - June 1977. D. S. Heeschen, D. E. Hogg.
Bull. American Astron. Soc., Vol. 10, 279 - 291 (1978).

National Radio Astronomy Observatory, *Green Bank*, Reprints, Series A, Nos 727 (21.141.092), 728 (21.162.052), 751 (20.141.056), 752 (20.141.057), 753 (20.141.058), 754 (20.141.059), 755 (20.151.010), 756 (20.099.007), 757 (20.160.020), 758 (20.158.038), 759 (20.116.006), 760 (20.131.066), 761 (20.131.077), 762 (20.131.078), 763 (20.131.061), 764 (20.141.514), 765 (20.132.011), 766 (20.158.092), 767 (20.158.090), 768 (20.158.062), 769 (20.131.075), 770 (20.141.049), 771 (20.158.089), 772 (20.155.055), 773 (20.131.076), 774 (20.034.017), 775 (20.141.053), 776 (20.141.066), 777 (20.131.110), 778 (20.122.059), 779 (20.131.151), 780 (20.131.152), 781 (20.141.054), 782 (20.106.013), 783 (20.141.005), 784 (20.141.160), 785 (21.066.239), 786 (20.117.044), 787 (20.141.118), 788 (20.131.149), 789 (20.131.150), 790 (20.131.164), 791 (20.141.134), 792 (20.131.167), 793 (21.122.002), 794 (21.158.004), 795 (21.131.003), 796 (21.131.005), 797 (21.131.006), 798 (21.131.009), 799 (21.141.502), 800 (21.141.019), 801 (21.131.016), 802 (21.131.010), 803 (21.131.011), 804 (21.131.018), 805 (21.131.021), 806 (21.158.005), 807 (20.141.051).

National Radio Astronomy Observatory, *Green Bank*, Reprints, Series B, Nos. 479 (20.162.042), 481 (19.141.070), 482 (20.155.026), 483 (20.131.106), 485 (21.141.006).

008.047 Greenbelt, Md.

Goddard Space Flight Center, Greenbelt, Maryland, GSFC Documents X-661-77-152 (21.142.724), X-921-77-246 (21.081.064), X-921-77-259 (21.081.065).

008.048 Greenwich

The Royal Greenwich Observatory.
A. Hunter, F. G. Smith.
Q. J. R. Astron. Soc., Vol. 19, 116 - 125 (1978). Report for the period 1975 October 1 to 1976 September 30.

H. M. Nautical Almanac Office, Royal Greenwich Observatory, Library Reprint, Nos. 341 (21.010.023), 342 (21.126.002).

008.049 Hamburg

Hamburger Sternwarte. Sonderdruck, Serie A, Nos. 36 (14.158.049), 37 (14.112.001), 38 (14.158.235).

Deutsches Hydrographisches Institut, Hamburg. Zeit- und Breitendienst, 1977 January - September (21.044.022).

008.050 Hamilton, Mass.

Sagamore Hill Radio Observatory, Air Force Geo-

physics Laboratory, Hanscom AFB, Massachusetts. – Report. J. P. Castelli.
Bull. American Astron. Soc., Vol. 10, 311 - 313 (1978).

008.051 Heidelberg

Astronomy and Astrophysics Abstracts, Vol. 19 (20.002.033), Vol. 20 (21.002.059).

Astronomisches Rechen-Institut, Heidelberg, Mitteilungen Serie A, Nos. 110 (21.042.060), 111 (21.151.044), 112 (21.151.082).

Astronomisches Rechen-Institut Heidelberg, Mitteilungen Serie B, Nos. 71 (21.041.008), 73 (21.043.008).

Max-Planck-Institut für Kernphysik, Heidelberg, Separate prints MPI H-1978-V2 (21.032.546), H-1978-V3 (21.032.547), H-1978-V8 (21.104.024).

008.052 Honolulu, Hawaii

University of Hawaii, Institute for Astronomy, Honolulu, Hawaii. – Report. S. C. Wolff.
Bull. American Astron. Soc., Vol. 10, 111 - 121 (1978).

008.053 Houston, Tex.

Rice University, Department of Space Physics and Astronomy, Houston, Texas. – Report. F. C. Michel.
Bull. American Astron. Soc., Vol. 10, 303 - 305 (1978).

008.054 Ithaca, N.Y.

Cornell University, Center for Radiophysics and Space Research, Ithaca, New York. – Report.
Bull. American Astron. Soc., Vol. 10, 87 - 89 (1978).

008.055 Kazan

V. P. Engelhardt and the formation of the new Astronomical Observatory of the Kazan University.
S. N. Korytnikov.
Izv. Astron. Ehngel'gardt. Obs., Kazan', No. 41 - 42, p. 245 - 254 (1976). In Russian.

N. I. Lobachevskij and I. M. Simonov on search of a site for the Kazan Astronomical Observatory (1828 - 1829). Documentary publication with an introduction and remarks by S. N. Korytnikov.
Tr. Kazan. Gorod. Astron. Obs., No. 41, p. 137 - 148 (1976). In Russian.

Izvestiya Astronomicheskoj Ehngel'gardtovskoj Observatorii, Kazan', No. 41 - 42 (21.082.092 , 21.041.037, 21.041.038; 21.045.012; 21.093.047; 21.032.064; 21.032.065; 21.031.274; 21.103.661; 21.103.671. 21.103.711; 21.104.033; 21.082.093; 21.082.094; 21.031.275; 21.002.053; 21.045.013; 21.032.066; 21.032.067; 21.043.009; 21.044.030; 21.085.017; 21.045.014; 21.121.078; 21.034.043; 21.131.218; 21.103.273; 21.104.034; 21.104.035; 21.031.418; 21.004.046; 21.008.055; 21.007.000).

Trudy Kazanskoj Gorodskoj Astronomicheskoj Observatorii, Kazan', No. 41 (21.121.077; 21.064.088; 21. 064.089; 21.064.090; 21.162.157; 21.066.243; 21.104.032; 21.103.621; 21.103.631; 21.102.031; 21.103.641; 21.102. 032; 21.103.651; 21.005.019; 21.004.042; 21.004.043; 21. 008.055; 21.004.044; 21.004.045).

008.056 Kiel

Sonderdrucke der Sternwarte Kiel, Nos. 219 (19. 126.006), 220 (18.007.000), 221 (19.114.011), 222 (19.105. 208), 223 (21.105.071), 224 (21.004.040), 225 (19.116.011), 226 (21.022.104), 227 (19.011.015), 228 (19.064.024), 230 (20.114.502), 231 (20.065.001), 232 (20.064.020), 233 (18. 064.046), 234 (19.063.005), 235 (19.114.343), 236 (19.065. 043).

008.057 Kiev

Astrometriya i Astrofizika, Kiev, Vyp. (Nos.) 34, 35 (1978).

008.058 Kingston, Ontario

Queen's University at Kingston, Ontario. A. H. Bridle. J. R. Astron. Soc. Canada, Vol. 72, 51 - 53 (1978). – Observatory report.

008.059 Kyoto

Contributions from the Department of Astronomy, University of Kyoto, Nos. 69 (18.158.105), 70 (18.162.037), 71 (18.114.336), 72 (18.114.337), 73 (18.151.048), 74 (18. 114.386), 75 (21.032.044), 76 (21.113.058), 77 (18.122. 095), 78 (18.122.179), 79 (19.141.072), 80 (19.114.326), 81 (20.151.006), 82 (20.114.540), 83 (20.151.038).

008.060 La Serena

Kitt Peak National Observatory, Tucson, Arizona; Cerro Tololo Inter-American Observatory, La Serena, Chile. Reports.   L. Goldberg, V. Blanco. Bull. American Astron. Soc., Vol. 10, 152 - 204 (1978).

008.061 La Silla

European Southern Observatory. Annual report 1977.   L. Woltjer. Printed in the Federal Republic of Germany by Fritz König, München. 36 pp. (1978). ISSN 0531-4496.

The Messenger – El Mensajero. Edited by European Southern Observatory (ESO), Garching, Nos. 12, 13 (1978).

ESO Scientific Preprint Nos. 19 (21.042.109), 20 (21.113.074), 21 (21.114.065), 22 (21.125.057), 23 (21.122. 159), 24 (21.114.573), 25 (21.124.110), 26 (21.042.110), 27 (21.041.033), 28 (21.002.051).

008.062 Las Cruces, N. Mex.

New Mexico State University, Department of Astronomy, Las Cruces, New Mexico. – Report.   H. A. Beebe. Bull. American Astron. Soc., Vol. 10, 292 - 294 (1978).

008.063 Lawrence, Kans.

University of Kansas, Lawrence, Kansas. – Report. S. J. Shawl. Bull. American Astron. Soc., Vol. 10, 151 - 152 (1978).

008.064 Lembang

Annals of the Bosscha Observatory, Lembang (Java), Indonesia, Vol. IX, Part 5 (21.118.011).

Contributions from the Bosscha Observatory, Bandung Institute of Technology, Department of Science, No. 63 (20.118.006).

008.065 Liège

Institut d'Astrophysique, Université de Liège, Cointe-Ougrée (Belgique), Collection in 4°, Nos. 307 (18.065. 079), 308 (19.122.027), 309 (19.065.011), 310 (19.065.015), 311 (19.133.019), 312 (19.065.004), 313 (19.066.023), 314 (19.082.029), 315 (19.071.014), 316 (19.114.046), 317 (21.082.091), 318 (20.080.002).

Institut d'Astrophysique, Université de Liège, Cointe-Ougrée (Belgique), Collection in 8°, Nos. 668 (21.066. 240), 669 (18.114.128), 670 (18.065.043), 671 (18.065.106), 672 (21.162.156), 674 (20.065.074).

008.066 Livermore, Calif.

Lawrence Livermore Laboratory, University of California, Livermore, California. – Report.   C. B. Tarter. Bull. American Astron. Soc., Vol. 10, 204 - 206 (1978).

008.067 London, Canada

The Observatories of the University of Western Ontario, London, Canada. – Reports.   W. Wehlau. Bull American Astron. Soc., Vol. 10, 375 - 377 (1978).

008.068 London, England

University College London, Mullard Space Science Laboratory.   R. L. F. Boyd. Q. J. R. Astron. Soc., Vol. 19, 126 - 136 (1978). – Report for the period 1976 October 1 - 1977 September 30.

008.069 Los Alamos, N. Mex.

Los Alamos Scientific Laboratory, University of California, Los Alamos, New Mexico. – Report.   A. N. Cox.

Bull. American Astron. Soc., Vol. 10, 209 - 217 (1978).

008.070   Los Angeles, Calif.

University of California: Berkeley, Los Angeles, San Diego, Lick Observatory, and Board of Studies. II. Los Angeles Campus. – Report.
Bull. American Astron. Soc., Vol. 10, 45 - 54 (1978).

University of California, Los Angeles, Astronomical Papers, Vol. 16, Nos.1(20.159.012), 2 (19.131.010), 3 (19. 141.506), 4 (19.141.004), 5 (19.131.061), 6 (18.132.037), 7 (19.117.015), 8 (19.135.011), 9 (18.131.129), 10 (19.121. 036), 11 (19.160.024), 12 (20.126.006), 13 (19.135.021), 14 (20.114.517), 15 (20.121.009), 16 (20.121.008), 17 (19. 141.050), 18 (20.121.026), 19 (19.142.056), 20 (19.131.163), 21 (20.114.010), 22 (19.142.716), 23 (20.121.005), 24 (19. 126.013), 25 (21.131.215), 26 (20.160.042), 27 (19.122.059), 28 (20.080.035), 29 (20.131.169), 30 (21.121.010), 31 (20.080.050).

008.071   Madison, Wis.

Washburn Observatory, University of Wisconsin, Madison, Wisconsin. – Report.   J. S. Mathis.
Bull. American Astron. Soc., Vol. 10, 368 - 371 (1978).

008.072   Madrid

Boletín Astronómico del Observatorio de Madrid, Instituto Geografico y Catastral, Vol. 10, No. 1 (21.133.041; 21.075.007; 21.133.042).

Universidad Complutense – Facultad de Ciencias, Madrid. Seminario de Astronomia y Geodesia, Publicación, Nos. 90 (20.041.030), 91 (20.031.326), 92 (20.045.016), 93 (21.121.076), 94 (21.080.044), 95 (21.041.032), 96 (21.113.073).

008.073   Manchester

University of Manchester, Nuffield Radio Astronomy Laboratories, Jodrell Bank.   B. Lovell.
Q. J. R. Astron. Soc., Vol. 19, 200 - 207 (1978). – Report for the year ending 1977 September 30.

008.074   Manila

Manila Observatory, Solar Division, Solar maps and activities, 1977 October 1 - 1978 February 28 (21.075.006).

008.075   Mauna Kea

Progress on NASA's 3-meter infrared telescope.
See Abstr. 032.053.

008.076   Menlo Park, Calif.

Stanford Center for Radar Astronomy: Stanford

University, Stanford, California; SRI International, Menlo Park, California. – Report.   V. R. Eshleman, R. L. Leadabrand, A. M. Peterson, J. F. Vesecky.
Bull. American Astron. Soc., Vol. 10, 316 - 320 (1978).

008.077   Middletown, Conn.

Van Vleck Observatory, Wesleyan University, Middletown, Connecticut. – Report.   A. R. Upgren.
Bull. American Astron. Soc., Vol. 10, 358 - 360 (1978).

008.078   Milano-Merate

Contributi dell'Osservatorio Astronomico di Milano-Merate, Nos. 376 (18.114.373), 377 (18.122.101), 379 (20. 114.555), 380 (14.123.031), 381 (18.122.127), 382 (17.114. 357), 383 (13.065.077), 384 (14.122.161), 385 (20.065.054), 386 (19.121.005), 387 (20.113.015), 388 (20.122.036), 389 (20.122.102), 390 (20.123.025).

Pubblicazioni dell'Osservatorio Astronomico di Milano-Merate, Nuova Serie, Nos. 26 (07.122.049), 27 (21. 121.075), 28 (12.121.004).

008.079   Minneapolis, Minn.

University of Minnesota, Minneapolis, Minnesota.
Report.   E. P. Ney.
Bull. American Astron. Soc., Vol. 10, 247 - 252 (1978).

008.080   Mizusawa

Monthly Notes of the International Polar Motion Service, 1977 No. 12, 1978 Nos. 1 - 4 (21.045.010).

Publications of the International Latitude Observatory of Mizusawa, Vol. 11, No. 1 (21.045.011; 21.044.025).

008.081   Moskva

Soobshcheniya Gosudarstvennogo Astronomicheskogo Instituta im. P. K. Shternberga. Izdatel'stvo Moskovskogo Universiteta, Nos. 193 (21.032.007; 21.031. 207; 21.041.003; 21.032.008), 199 (21.155.006; 21.155.007; 21.118.007; 21.031.208; 21.113.020; 21.122.014; 21.034. 005), 200 (21.032.009; 21.032.010; 21.031.209; 21.032.011; 21.041.004; 21.082.010; 21.044.005), 201 (21.042.021), 204 (21.094.123).

Trudy Gosudarstvennogo Astronomicheskogo Instituta im. P. K. Shternberga. Izdatel'stvo Moskovskogo Universiteta, Vol. 47 (21.005.001; 21.158.031; 21.154.008; 21.155.005; 21.152.001; 21.122.013; 21.153.009; 21.118. 006; 21.036.001; 21.112.003; 21.135.001).

008.082   Nashville, Tenn.

Dyer Observatory, Vanderbilt University, Nashville, Tennessee. – Report for the period 1 October 1976 - 30 September 1977.   A. M. Heiser.

Bull. American Astron. Soc., Vol. 10, 98 - 100 (1978).

008.083 Neuchâtel

Observatoire Cantonal de Neuchâtel.
Rapport d'activité pour l'exercice 1977.
J. Bonanomi, G. Fischer.
Obs. Cantonal de Neuchâtel, 18 pp. (1978).

008.084 New Haven, Conn.

Yale University Observatory, New Haven, Connec-
ticut. — Report for the period 1 July 1976 - 30 June 1977.
W. F. van Altena.
Bull. American Astron. Soc., Vol. 10, 377 - 386 (1978).

008.085 New York, N.Y.

Columbia University, Department of Astronomy
and Department of Physics, New York, New York. — Report.
L. B. Lucy.
Bull. American Astron. Soc., Vol. 10, 83 - 86 (1978).

008.086 Nice

Rapport d'activité de l'Observatoire de Nice pour
1975 et 1976.
Assoc. Dév. Int. Obs. Nice, Bull. N. 13 - 14, p. 89 - 127 (1978).
Publications de l'Observatoire de Nice; Services techniques; Seminaires.

La communication et l'Observatoire de Nice.
P. Franck.
Assoc. Dév. Int. Obs. Nice, Bull. N. 13 - 14, p. 129 - 132
(1978).

Les relations extérieures à l'Observatoire de Nice
pendant la période du 1er Juillet 1976 au 31 Decembre 1977.
P. Franck.
Assoc. Dév. Int. Obs. Nice, Bull. N. 13 - 14, p. 133 - 137
(1978).

008.087 Ottawa

National Research Council of Canada, Ottawa,
Ontario, Canada, NRC Nos. 16129 (20.077.040), 16260 (20.
135.015), 16261 (20.132.023), 16262 (20.074.030).

008.088 Oxford

Department of Astrophysics, University of Oxford.
D. E. Blackwell.
Q. J. R. Astron. Soc., Vol. 19, 208 - 215 (1978). — Report for
the period 1976 August 1 - 1977 July 31.

008.089 Palo Alto, Calif.

Lockheed Palo Alto Research Laboratory, Palo Alto,

California. — Report.   H. M. Johnson.
Bull. American Astron. Soc., Vol. 10, 207 - 209 (1978).

008.090 Paris

Rapport annuel pour 1977.
Bureau International de l'Heure, 61 Avenue de l'Observatoire,
75014 Paris. 6 + A12 + B58 + C18 pp. (1978).

Bureau International de l'Heure, (B.I.H.), Circular
D 134 - 140 (21.044.023).

008.091 Pasadena

Hale Observatories, operated by Carnegie Institution
of Washington and California Institute of Technology, Pasadena, California. Annual report of the director, 1976 - 1977.
H. W. Babcock, J. B. Oke.
Reprinted from Carnegie Institution, Washington, Year Book
76, 1976 - 1977, p. 117 - 198 (1977).

008.092 Pereyra Iraola

Posibilidades actuales y futuras de observación de
galaxias en el Instituto Argentino de Radioastronomía.
E. Bajaja.
Bol. Acad. Nac. Cienc. Cordoba, Argentina, Tomo 52, 165 -
170 (1976) = Contrib. Inst. Argentino Radioastron., No. 73.
A short history of IAR with a description of its observations facilities and functioning is given. Two antennae of
30 m diameter with variable distance will be used in aperture
synthesis. A new low noise level receiver will provide new
possibilities for observations requiring high sensitivity.

Instituto Argentino de Radioastronomia, Contri-
buciones Nos. 25 (21.155.044), 29 (21.153.043), 33 (21.131.
248), 38 (21.002.062), 40 (21.002.063), 41 (13.157.010),
71 (21.158.271), 73 (21.008.092).

008.093 Perth

Perth Observatory, Western Australia. Communications No. 3 (21.103.009).

008.094 Philadelphia, Pa.

Flower and Cook Observatory, University of
Pennsylvania, Philadelphia, Pennsylvania 19104. — Report.
B. S. P. Shen.
Bull. American Astron. Soc., Vol. 10, 487 - 488 (1978).

008.095 Potsdam

Astronomische Zeit- und Breitenbestimmungen,
Empfangszeiten von Zeitsignalen, Präzisionszeitvergleiche.
1977 January - April (21.044.024).

Veröffentlichungen des Zentralinstituts Physik der
Erde, *Potsdam*, No. 52, Teil 2+3 (21.012.001).

008.096   Praha

Astronomical Institute of the Czechoslovak Academy of Sciences.   V. Bumba.
Říše hvězd, Vol. 59, 1 - 3 (1978). In Czech.

Fifty years of the Observatory Prague-Petřin.
O. Hlad.
Říše hvězd, Vol. 59, 133 - 137 (1978). In Czech.

Czechoslovak Academy of Sciences, Astronomical Institute, Publication No. 54 (21.012.028).

Académie Tchécoslovaque des Sciences, Institut Astronomique, Temps et Latitude, Janvier - Août 1977 (21.044.028).

008.097   Princeton, N. J.

Princeton University Observatory, Princeton, New Jersey. — Report.   L. Spitzer, Jr.
Bull. American Astron. Soc., Vol. 10, 298 - 303 (1978).

008.098   Pulkovo

A modern astronomical observatory.
N. G. Ponomarev.
Problems of observational and theoretical astronomy, Moscow — Leningrad, 1977, (see 003.012), p. 140 - 147. In Russian. — Abstr. in Ref. zh., 51. Astron., 5.51.14 (1978).
Report on the occasion of the jubilee session of the USSR Academy of Sciences in 1940 dedicated to the 100th anniversary of the Pulkovo Observatory. Published for the first time.

Time Service bulletins.   See Abstr. 044.038.

008.099   Richland, Wash.

Battelle—Northwest Laboratories, Rattlesnake Mountain Observatory, Richland, Washington. — Report.
R. A. Stokes, G. M. Stokes.
Bull. American Astron. Soc., Vol. 10, 23 - 24 (1978).

008.100   Richmond Hill, Ontario

David Dunlap Observatory, University of Toronto, Richmond Hill, Ontario, Canada. — Report for the period 1 July 1976 - 30 June 1977.   D. A. MacRae.
Bull. American Astron. Soc., Vol. 10, 89 - 98 (1978).

008.101   Rochester, N.Y.

C. E. Kenneth Mees Observatory, University of Rochester, Rochester, New York. — Report.   M. P. Savedoff.
Bull. American Astron. Soc., Vol. 10, 237 - 240 (1978).

008.102   Rohnert Park, Calif.

Sonoma State College, Department of Physics and Astronomy, Rohnert Park, California. — Report for the period

September 1976 - August 1977.   G. G. Spear, J. S. Tenn.
Bull. American Astron. Soc., Vol. 10, 313 - 314 (1978).

008.103   Roma

Photographic Journal of the Sun, Osservatorio Astronomico di Roma, Nos. 125 - 130 (21.075.008).

Solar phenomena. Osservatorio Astronomico di Roma, Monthly Bulletin, Nos. 233 - 237 (21.075.009).

008.104   San Diego, Calif.

University of California: Berkeley, Los Angeles, San Diego, Lick Observatory, and Board of Studies. III. San Diego Campus. — Report.   G. Burbidge.
Bull. American Astron. Soc., Vol. 10, 54 - 58 (1978).

Annual report for the Mount Laguna Observatory for 1977. San Diego State University, San Diego, California 92115.   M. S. Snowden.
Bull. American Astron. Soc., Vol. 10, 488 - 490 (1978).

008.105   Santa Cruz, Calif.

University of California: Berkeley, Los Angeles, San Diego, Lick Observatory, and Board of Studies. IV. Lick Observatory. — Report.   D. E. Osterbrock.
Bull. American Astron. Soc., Vol. 10, 58 - 65 (1978).

University of California: Berkeley, Los Angeles, San Diego, Lick Observatory, and Board of Studies. V. Board of Studies in Astronomy and Astrophysics. — Reports.
P. H. Bodenheimer.
Bull. American Astron. Soc., Vol. 10, 65 - 68 (1978).

008.106   Seattle, Wash.

University of Washington Astronomy Department, Seattle, Washington. — Report.   G. Wallerstein.
Bull. American Astron. Soc., Vol. 10, 371 - 375 (1978).

008.107   Sendai

Sendai Astronomiaj Raportoj, Nos. 173 (19.134. 009), 174 (20.122.114), 175 (20.121.064), 176 (20.065.057), 177 (20.151.005), 179 (20.154.039), 181 (20.162.118),182 (20.151.073).

008.108   Shanghai

Time Service Annual Report 1974 (21.044.007), 1975 (21.044.008), 1976 (21.044.018).

008.109   Siding Spring

Anglo-Australian telescope 1975/76. Report of the Anglo-Australian telescope board, 1 July 1975 to 30 June

1976.   J. P. Wild, D. C. Morton.
Australian Government Publishing Service, Canberra, 5 +
34 pp. (1978).

A visit to Siding Spring.   See Abstr. 032.050.

008.110 Socorro, N. Mex.

National Radio Astronomy Observatory: Charlottes-
ville, Virginia; Green Bank, West Virginia; Socorro, New
Mexico; Tucson, Arizona. — Report for the period July 1976 -
June 1977.   D. S. Heeschen, D. E. Hogg.
Bull. American Astron. Soc., Vol. 10, 279 - 291 (1978).

008.111 Sonneberg

Akademie der Wissenschaften der DDR, Zentralinstitut für Astrophysik, Sternwarte Sonneberg. Mitteilungen
der Sternwarte zu Sonneberg, No. 62 (14.064.011; 17.161.
002; 18.121.165; 20.004.089; 19.008.130; 19.031.267).

Zentralinstitut für Astrophysik. Mitteilungen über
Veränderliche Sterne, Sonneberg, Band 8, Heft 2 (21.134.
035; 21.123.047; 21.123.048; 21.123.049; 21.124.111;
21.124.307; 21.123.050; 21.123.051), 3 (21.122.160; 21.113.
077; 21.122.161; 21.123.052; 21.123.053; 21.123.054; 21.
123.055).

008.112 Stanford, Calif.

Stanford Center for Radar Astronomy: Stanford
University, Stanford, California; SRI International, Menlo
Park, California. — Report.   V. R. Eshleman, R. L.
Leadabrand, A. M. Peterson, J. F. Vesecky.
Bull. American Astron. Soc., Vol. 10, 316 - 320 (1978).

008.113 Stony Brook, N.Y.

State University of New York at Stony Brook,
Stony Brook, New York. — Report.   M. Simon.
Bull. American Astron. Soc., Vol. 10, 320 - 325 (1978).

008.114 Swarthmore, Penn.

Sproul Observatory, Swarthmore College, Swarth-
more, Pennsylvania, — Report for the period 1 July 1976 - 30
June 1977.   S. L. Lippincott, W. D. Heintz.
Bull. American Astron. Soc., Vol. 10, 314 - 315 (1978).

Sproul Observatory, Swarthmore, Pennsylvania,
Reprints, Nos. 256 (18.155.092), 257 (17.008.127), 258
(17.117.018), 259 (18.001.002), 260 (18.111.004), 261,
262 (18.118.001), 263 (18.118.015), 264, 265 (19.008.134),
266 (19.118.003), 267 (19.112.004), 268 (19.013.011), 269
(19.117.042), 270 (20.121.015), 271 (20.112.004), 272
(20.117.028), 273 (20.118.002).

008.115 Sunspot, N. Mex.

Sacramento Peak Observatory, Sunspot, New Mexi-

co. — Report.   J. B. Zirker.
Bull. American Astron. Soc., Vol. 10, 309 - 311 (1978).

008.116 Tampa, Fla.

University of Florida, Tampa, Florida. I. Rosemary
Hill Observatory. II. Radio Observatory. — Reports.
A. G. Smith.
Bull. American Astron. Soc., Vol. 10, 107 - 111 (1978).

008.117 Tartu

Tartu Observatory — a center for training astronomers and geodesists.   G. A. Zhelnin.
Rol'Tartus. univ. v razvitii otechestven. nauk. i v podgotovke
nauchn.-ped. kadrov. Tartu, 1977, p. 76 - 80. In Russian.
Abstr. in Ref. zh., 51. Astron., 5.51.9 (1978).

The Astronomical Observatory of the Tartu University in the middle of the nineteenth century (at the time of
professor Mädler's directorship).   H. Eelsalu.
Rol' Tartus. univ. v razvitii otechestven. nauk. i v podgotovke
nauchn.-ped. kadrov. Tartu, 1977, p. 81 - 84. In Russian.
Abstr. in Ref. zh., 51. Astron., 5.51.10 (1978).

Tartu Astronoomia Observatoorium, Teated No. 56
(21.063.022; 21.063.023; 21.114.050).

008.118 Tempe, Ariz.

Arizona State University, Tempe, Arizona. — Report.
S. Starrfield.
Bull. American Astron. Soc., Vol. 10, 14 - 16 (1978).

008.119 Tokyo

Annual report from the Tokyo Astronomical
Observatory, 1977.
Separate print Tokyo Astron. Obs., 96 pp. (1978). In Japanese.

Annals of the Tokyo Astronomical Observatory,
Second Series, Vol. 16, No. 4 (21.104.031). Vol. 17, No. 1
(21.002.052, 21.035.013, 21.066.242, 21.114.066).

Tokyo Astronomical Bulletin, Tokyo Astronomical
Observatory, Second Series, Nos. 251 (21.113.075), 252
(21.041.035), 253 (21.035.012), 254 (21.113.076).

Contributions from the Department of Astronomy,
University of Tokyo, Nos. 228 (18.125.045), 235 (20.074.
068), 236 (20.131.215), 237 (20.133.017), 238 (20.131.216),
239 (20.063.048), 240 (20.114.064), 241 (21.064.026),
242 (21.151.014), 243 (21.151.015), 244 (21.158.054),
245 (21.113.027), 246 (21.113.028).

University of Tokyo, Tokyo Astronomical Observatory, Report (No. 69), Vol. 18, No. 2 (1978).

Tokyo Astronomical Observatory, Reprints
No. 532 (21.122.039).

Time and Latitude Bulletins, Tokyo Astronomical
Observatory, Vol. 51, Nos 3 - 4 (21.044.027).

Data Report of Hydrographic Observations. Series of Astronomy and Geodesy, Maritime Safety Agency, Tokyo, Japan, No. 12 (21.096.012; 21.032.060).

008.120 Toledo, Ohio

Ritter Astrophysical Research Center, The University of Toledo, Toledo, Ohio. – Report.   A. N. Witt.
Bull. American Astron. Soc., Vol. 10, 306 - 308 (1978).

008.121 Torino

Attivita dell'Osservatorio.   M. G. Fracastoro.
Pubbl. Varie Fuori Ser. Oss. Astron. Torino, (Pino Torinese), No. 68, (see 047.019), p. 27 - 30 (1977).

Il telescopio astrometrico da 105 cm dell'Osservatorio Astronomico di Torino.   See Abstr. 032.046.

Contributi dell'Osservatorio Astronomico di Torino (Pino Torinese), Nos 97 (20.098.008), 102 (20.121.010), 104 (20.098.036), 105 (20.098.037), 106 (20.098.038), 107 (21.102.010), 108 (21.118.001), 109 (21.066.241), 110 (21.112.013).

Pubblicazioni Varie Fuori Serie dell'Osservatorio Astronomico di Torino (Pino Torinese), No. 68 (21.047.019).

Osservatorio Astronomico di Torino, Pino Torinese. Time Service, Bulletin No. 17 (21.044.026).

008.122 Treviso

Osservatorio Privato Specola "Ariel", Treviso, Pubblicazione, Nos. 77 (18.141.111), 79 (19.158.128), 80 (19.122.131), 81 (20.141.006), 82 (19.141.185), 83 (20.158.033), 84 (20.158.117).

008.123 Tucson, Ariz.

University of Arizona, Department of Planetary Sciences/Lunar and Planetary Laboratory, Tucson, Arizona. Report from 1 October 1976 to 30 September 1977. W. B. Hubbard.
Bull. American Astron. Soc., Vol. 10, 16 - 23 (1978).

Kitt Peak National Observatory, Tucson, Arizona; Cerro Tololo Inter-American Observatory, La Serena, Chile. Reports.   L. Goldberg, V. Blanco.
Bull. American Astron. Soc., Vol. 10, 152 - 204 (1978).

National Radio Astronomy Observatory: Charlottesville, Virginia; Green Bank, West Virginia; Socorro, New Mexico; Tucson, Arizona. – Report for the period July 1976 - June 1977.   D. S. Heeschen, D. E. Hogg.
Bull. American Astron. Soc., Vol. 10, 279 - 291 (1978).

008.124 Uccle

Bulletin Astronomique. (Astronomisch Bulletin). Observatoire Royal de Belgique (Koninklijke Sterrenwacht van België), Vol. 9, No. 1 (21.098.094; 21.098.095; 21.098.

096; 21.103.306; 21.103.804; 21.103.805; 21.113.086; 21.096.014; 21.077.042; 21.079.203; 21.118.039; 21.118.040).

Observatoire Royal de Belgique (Koninklijke Sterrenwacht van België), Communications (Mededelingen), Série A, Nos. 43, 44 (20.098.046), 45 (20.122.016), 46 (21.114.505), 47 (21.098.033), 48 (21.081.080), 49 (21.117.041).

Observatoire Royal de Belgique (Koninklijke Sterrenwacht van Belgie), Communications (Mededelingen), Série B, Nos 97, 98 (14.071.053), 99 (20.118.018), 100 (20.034.009), 101 (20.118.014), 102 (21.065.047), 103 (14.061.034).

008.125 Urbana, Ill.

University of Illinois at Urbana–Champaign Department of Astronomy, Urbana, Illinois. – Report for the period 1 September 1976 - 31 August 1977.
J. S. Gallagher, J. W. Truran, S. P. Wyatt.
Bull. American Astron. Soc., Vol. 10, 129 - 140 (1978).

008.126 Vancouver, B. C.

University of British Columbia, Institute of Astronomy and Space Science, Vancouver, British Columbia, Canada. – Report from October 1975 to October 1977.
Bull. American Astron. Soc., Vol. 10, 24 - 30 (1978).

008.127 Victoria, B. C.

Dominion Astrophysical Observatory, Victoria, B.C. J. B. Hutchings.
J. R. Astron. Soc. Canada, Vol. 72, 116 (1978).

University of Victoria Observatory, Physics Department, University of Victoria, Victoria, British Columbia, Canada. – Report.   R. M. Pearce.
Bull. American Astron. Soc., Vol. 10, 361 - 362 (1978).

008.128 Villanova, Penn.

Villanova University, Department of Astronomy, Villanova, Pennsylvania. – Report for the period September 1976 - September 1977.   G. P. McCook.
Bull. American Astron. Soc., Vol. 10, 362 - 364 (1978).

008.129 Warszawa

Warsaw University Observatory and Astronomical Institute, Polish Academy of Sciences, Reprint Nos. 383 (21.124.102), 384 (21.158.022), 385 (21.141.028), 386 (21.065.009), 387 (21.117.008).

Politechnika Warszawska, Obserwatorium Astronomiczno-Geodezyjne w Józefosławiu, (Warsaw Technical University, Astronomic-Geodetical Observatory at Józefosław), Latitude Circular, Nos. 61 (21.045.015), 62 - 65 (21.045.016).

Institute of Higher Geodesy and Geodetical Astronomy of the Warsaw Technical University, Publications, New Series, No. 9 (21.045.017; 21.044.029; 21.045.018; 21.045.019).

008.130 Washington, D. C.

U.S. Naval Observatory, Washington, DC. — Report for the period 1 July 1976 - 30 June 1977.
J. C. Smith, G. Westerhout.
Bull. American Astron. Soc., Vol. 10, 352 - 358 (1978).

United States Naval Observatory, *Washington, D.C.,* Circular, No. 157 (21.079.304).

U.S. Naval Observatory, Washington, D.C. Time Service Publications, Series 4, Nos. 570 - 595; Series 7, Nos. 523 - 548; Series 11, No. 224 (21.044.034 - 21.044.036).

National Aeronautics and Space Administration, Washington, DC. — Report.   N. W. Boggess.
Bull. American Astron. Soc., Vol. 10, 253 - 271 (1978).

008.131 Westford, Mass.

Haystack Observatory, Northeast Radio Observatory Corporation (NEROC), Westford, Massachusetts. — Report for the period from 1 July 1976 - 30 June 1977.
P. B. Sebring, M. L. Meeks.
Bull. American Astron. Soc., Vol. 10, 122 - 126 (1978).

008.132 Williams Bay, Wis.

The University of Chicago, Department of Astronomy and Astrophysics, Chicago, Illinois. The Yerkes Observatory, Williams Bay, Wisconsin. — Reports.
E. N. Parker, L. M. Hobbs.
Bull. American Astron. Soc., Vol. 10, 77 - 83 (1978).

008.133 Williamstown, Mass.

Hopkins Observatory—Williams College, Williamstown, Massachusetts. — Report for the 1976 - 1977 academic year.   J. M. Pasachoff.

Bull. American Astron. Soc., Vol. 10, 126 - 128 (1978).

008.134 Wrocław

Wrocław Astronomical Observatory, Reprint, Nos. 103 (21.122.011), 104 (19.122.012).

008.135 Yorktown Heights, N.Y.

IBM Thomas J. Watson Research Center, Yorktown Heights, New York. — Report.   G. Lasher.
Bull. American Astron. Soc., Vol. 10, 128 - 129 (1978).

008.136 Zagreb

Hvar Observatory, Bulletin, No. 1 (21.009.007; 21.032.031; 21.046.022; 21.082.056).

The stellar instrumentation of the Hvar Observatory and its possible development.   See Abstr. 009.007.

Determination of geodetic coordinates of the Hvar Observatory.   See Abstr. 046.022.

Conditions for astronomical observations at the Hvar Observatory.   See Abstr. 082.056.

008.137 Zelenchukskaya

Soobshcheniya Spetsial'noj Astrofizicheskoj Observatorii, Akademiya Nauk SSSR, Zelenchukskaya, Vyp. (Nos.) 16 (21.141.197; 21.031.419; 21.031.420; 21.033.041; 21.034.047) 19 (21.031.278; 21.031.277; 21.031.421) 20 (21.034.054; 21.142.226; 21.113.082; 21.141.223; 21.141.224).

008.138 Zürich

Astronomische Mitteilungen der Eidgenössischen Sternwarte Zürich, Nos. 359 (21.072.057), 361 (21.079.501), 362 (21.072.058), 363 (21.075.011).

009 Notes on Observatories, Planetaria, and Exhibitions

009.001 Budapest finally gets its big planetarium.
G. Schalk.
Sky Telesc., Vol. 55, 9 - 12 (1978).

009.002 Das geplante Sonnenobservatorium auf den Kanari-
schen Inseln.   W. Mattig.
Sterne Weltraum, Jahrg. 17, 51 - 55 (1978).

009.003 A solar observatory in the Canary Islands.
J. B. Zirker.
Sky Telesc., Vol. 55, 216 - 217 (1978).

009.004 Qu'est-ce qu'un observatoire à notre époque.
M. Golay, B. Hauck.
Conférences d'astronomie de l'Observatoire,(see 012.004), p.
205 - 220 (1977).

009.005 Das Sonnen-Observatorium der DFG bei Locarno.
R. Spindler, E. Wiehr.
Sterne Weltraum, Jahrg. 17, 117 - 120 (1978).

009.006 Attrezzatura strumentale dell'Osservatorio di
Arcetri.   A. Cantù, G. Ceppatelli, M. Landini,
A. Righini, G. Tofani.
G. Astron., Vol. 3, 203 - 215 (1977).

009.007 The stellar instrumentation of the Hvar Observatory
and its possible development.   P. Mayer.
Hvar Obs. Bull., No. 1, p. 5 - 13 (1977).

009.008 Posibilidades actuales y futuras de observación de
galaxias en el Instituto Argentino de Radioastronomía.   E. Bajaja.
Astron. extragalactica, (see 012.015), 165 - 170 (1976).
   A short history of IAR with a description of its observation facilities, and functioning is given. Two antennae of 30 m
diameter with variable distance will be used in aperture synthesis. A new low noise level receiver will provide new possibilities for observations requiring high sensitivity.

009.009 Copernicus Astronomical Center opens in Warsaw.
S. Mitton.
Nature, Vol. 273, 92 (1978).

009.010 Rao — a world class observatory.   L. Stenstrom.
Eltek. Aktuell Elektron. A, Vol. 20, No. 17,
p. 20 - 21 (1977). In Swedish. — Abstr. in Phys. Abstr., Vol.
81, Abstr. 28513 (1978).

009.011 Rhodes University: Department of Physics.
E. E. Baart.
Mon. Notes Astron. Soc. South Africa, Vol. 37, 29 - 30 (1978).
Report for 1978.

009.012 University of South Africa: Department of Mathe-
matics, Applied Mathematics and Astronomy.
P. D. Bennewith, J. Wolterbeek.
Mon. Notes Astron. Soc. South Africa, Vol. 37, 30 - 31 (1978).

009.013 Défense du site de l'observatoire.
J. P. Zahn, A. Baglin.
Assoc. Dév. Int. Obs. Nice, Bull. N. 13 - 14, p. 25 - 32 (1978).

009.014 Observatory on mount "Koshka".
Eh. S. Brodskaya, I. I. Dmitrotsa.
Nauchn. Inf., Vyp. (No.) 38, p. 3 - 6 (1976). In Russian.

009.015 Installation of the scientific center for geodynamic

and astronomical investigations.   J. Kovalevsky.
New methods of space geodesy, (see 012.029), p. 18 - 25
(1976). In Russian.
   The installation of CERGA (Center for Studies and Research in Geodynamics and Astronomy), an institute devoted
to the study of the Earth-Moon system, space geodesy and
astrometry continues for over a year. In addition to studies
and installation of instruments (laser ranging systems for
satellites and the Moon, infrared interferometry, Schmidt
telescope, astrolabes, etc...) the activity of this center deals
with the following scientific subjects: Earth rotation and
polar motion, dynamics of higher atmosphere, shape and
dynamics of the Moon, star catalogues, celestial mechanics
and astrometric observations of the Sun.

009.016 The Wyoming Infrared Observatory.
R. D. Gehrz, J. A. Hackwell.
Bull. American Astron. Soc., Vol. 10, 393 - 394 (1978).
Abstract.

009.017 An automated station for the detection of bursts
of cosmic origin at VHF and UHF.   P. Inzani,
G. Sironi, G. Cazzola, S. Cortiglioni, N. Mandolesi, G. Morigi,
G. G. C. Palumbo.
Astrophys. Space Sci., Vol. 56, 239 - 254 (1978).
   An automated system for the detection of bursts of cosmic origin at radio frequencies has been built at Medicina
(Bologna). The station, completely independent, is part of a
network of observatories based in Ireland, United Kingdom,
Switzerland and Italy. The receiving apparatus, operating at
151 and 408 MHz, and the experimental method used
to reject unwanted events is described. Observational programmes include a search for the radio counterpart of X- and γ-
ray bursts, observations of the galactic centre and Cygnus
regions, and a full sky monitoring. Sensitivity limits as well as
preliminary results are discussed.

009.018 Experience of building an astronomical complex
in Smolensk.   A. Ya. Virin.
Problems of astronomy and geodesy, VAGO assembly,
Yerevan, 1975, (see 012.039), p. 173 - 174. In Russian.
Abstr. in Ref. zh., 51. Astron., 6.51.47 (1978).

009.019 On the experience of the planetarium and the
People's Observatory of the Club of Universe Investigators in L'vov.   A. V. Butkevich, G. I. Pashchenko.
Problems of astronomy and geodesy, VAGO assembly Yerevan,
1975, (see 012.039), p. 177 - 184. In Russian. — Abstr. in Ref.
zh., 51. Astron., 7.51.17 (1978).

009.020 Working out an experimental project for a small
planetarium coupled with an observatory (theses of
a report).   I. M. Bezchastnov.
Problems of astronomy and geodesy, VAGO assembly Yerevan,
1975 (see 012.039), p. 169 - 172. In Russian. — Abstr. in Ref.
zh., 51. Astron., 7.51.61 (1978).

009.021 Astronomy Centre, University of Sussex.
L. Mestel.
Q. J. R. Astron. Soc., Vol. 19, 225 - 233 (1978). — Report for
the year ending 1977 September 30.

Enquiry on radioastronomical data. II.
See Abstr. 002.013.

Exploring the infrared universe from Wyoming.
See Abstr. 032.037.

010 Societies, Associations, Organizations

010.001   American Association of Variable Star Observers
(AAVSO)

Meetings and activities of the Society, committee
reports.
J. American Assoc. Variable Star Obs., Vol. 6, 93 - 112 (1977/
78).

Minutes of the general meeting of the AAVSO held
at Nantucket, Massachusetts, October 14 - 15, 1977.
C. B. Ford.
J. American Assoc. Variable Star Obs., Vol. 6, 89 - 92
(1977/78).

The Journal of the American Association of Variable Star Observers, Vol. 6, No. 2 (1977/78).

The American Association of Variable Star Observers,
Bulletin 41, 41 Supplement, 41-A + B.

AAVSO Alert Notice No. 19.

010.002   American Astronomical Society (AAS)

Abstracts of papers presented at the 151st meeting
of the American Astronomical Society, held 9 - 11 January
1978 at Austin, Texas.
Bull. American Astron. Soc., Vol. 9, 553 - 670 (1978).

Abstracts of papers presented at the ninth annual
meeting of the Division for Planetary Sciences of the American
Astronomical Society, held 27 - 30 October 1977 at Boston,
Massachusetts.
Bull. American Astron. Soc., Vol. 9, 491 - 552 (1977).

Abstracts of the papers presented at the Dynamical
Astronomy Division Meeting, held 11 - 13 January 1978 at
Austin, Texas.
Bull. American Astron. Soc., Vol. 10, 481 - 486 (1978).

The 152nd meeting of the American Astronomical
Society held 26 - 28 June 1978 at Madison, Wisconsin.
Abstracts of papers presented.
Bull. American Astron. Soc., Vol. 10, 387 - 480 (1978).

Division on Dynamical Astronomy: 1975 - 1976.
B. G. Marsden.
Bull. American Astron. Soc., Vol. 9, 696 (1978). – Annual
report.

High Energy Astrophysics Division: 1977.
R. Novick.
Bull. American Astron. Soc., Vol. 9, 696 - 697 (1978).
Annual report.

Some historical notes on the DPS/AAS (Division
for Planetary Sciences, American Astronomical Society).
Bull. American Astron. Soc., Vol. 9, 487 - 488 (1977).

Division for Planetary Sciences, American Astro-
nomical Society. Minutes of the annual business meeting,
Honolulu, Hawaii, 20 January 1977.   T. V. Johnson.
Bull. American Astron. Soc., Vol. 9, 489 - 490 (1977).

Division for Planetary Sciences: 1977.
T. V. Johnson.

Bull. American Astron. Soc., Vol. 9, 697 - 698 (1978).
Annual report.

Division of Solar Physics: 1977.   J. D. Bohlin.
Bull. American Astron. Soc., Vol. 9, 698 (1978). – Annual
report.

Task Group on Education in Astronomy (TGEA):
Annual report for 1977.   P. H. Knappenberger, Jr.
Bull. American Astron. Soc., Vol. 9, 699 (1978).

Bulletin of the American Astronomical Society,
Vol. 9, No. 4, Part I, II (1977), Vol. 10, No. 1 (1978).

010.003   Association Française des Observateurs d'Étoiles
Variables (A F O E V)

La vie de l'Association.   E. Schweitzer.
AFOEV Bull., Tome 11, 79 (1977).

Bulletin de l'Association Française des Observateurs
d'Étoiles Variables, Tome 11, No. 3 (1977).

Activité de l'A.F.O.E.V.   E. Schweitzer.
Astronomie, Vol. 92, 100 - 102, 277 - 278 (1978).

010.004   Association of Lunar and Planetary Observers
(ALPO)

Minor planets section news.
Minor Planet. Bull., Vol. 5, 26; Vol. 6, 10 (1978).

General report of position observations by the
A.L.P.O. minor planets section for the year 1977.
See Abstr. 098.055.

The Minor Planet Bulletin, Vol. 5, Nos. 3 - 4; Vol.
6, No. 1, 1 Suppl. (1978).

The Strolling Astronomer. The Journal of the
Association of Lunar and Planetary Observers, Vol. 27, Nos.
3 - 4 (1978).

010.005   Astronomical Society of Australia (ASA)

Society business.
Proc. Astron. Soc. Australia, Vol. 3, 181 - 184 (1977).

Proceedings of the Astronomical Society of Australia, Vol. 3, No. 2 (1977).

Papers presented at the Eleventh Annual General
Meeting held at Monash University, 30, 31 May and 1 June
1977.
Proc. Astron. Soc. Australia, Vol. 3, 90 - 180 (1977). – The
individual contributions are included in their corresponding
subject categories.

010.006   Astronomical Society of Czechoslovakia

No publication received.

010.007 Astronomical Society of the Pacific (ASP)

Mercury. The Journal of the Astronomical Society of the Pacific, Vol. 7, Nos. 1, 2 (1978).

Publications of the Astronomical Society of the Pacific, Vol. 89, No. 532 (1977), Vol. 90, Nos. 523 - 534 (1978).

010.008 Astronomical Society of Southern Africa (ASSA)

Notices.
Mon. Notes Astron. Soc. South. Africa, Vol. 37, 1, 19 (1978).

Monthly Notes of the Astronomical Society of Southern Africa, Vol. 37, Nos. 1 - 2, 3 - 4 (1978).

010.009 Astronomical Society of Victoria (ASV)

No publication received.

010.010 Astronomical Society of Western Australia (ASWA)

Report on proceedings of ordinary meetings.
J. Astron. Soc. Western Australia, Vol. 28, December (1977), Vol. 297 - 301, January - May (1978).

The Astronomical Society of Western Australia, Circular, No. 1 (1978).

The Journal of the Astronomical Society of Western Australia, Vol. 28, 1977 December, Vol. 297 - 301, 1978 January - May.

010.011 Astronomische Gesellschaft (AG)

No publication received.

010.012 Astronomisk Selskab København

No publication received.

010.013 British Astronomical Association (BAA)

Notices.
J. British Astron. Assoc., Vol. 88, 109 - 110, 213 - 216, 321 - 323 (1978).

Meetings and Activities of the Association.
J. British Astron. Assoc., Vol. 88, 111 - 118, 217 - 225, 324 - 333 (1978).

Section reports.
J. British Astron. Assoc., Vol. 88, 158 - 187, 283 - 300, 382 - 406 (1978).

Journal of the British Astronomical Association, Vol. 88, Nos. 2 - 4 (1978).

010.014 British Interplanetary Society (BIS)

Society news.
Spaceflight, Vol. 20, 117 (1978).

Journal of the British Interplanetary Society, Vol. 31, Nos. 1 - 7 (1978); Supplement 1978.

Spaceflight. A publication of the British Interplanetary Society, Vol. 20, Nos. 1 - 7 (1978).

010.015 Committee on Space Research (COSPAR)

No publication received.

010.016 European Space Agency (ESA)

Das Satellitenprogramm der ESA (Stand April 1978).   H. W. Köhler.
Sterne Weltraum, Jahrg. 17, 228 - 229 (1978).

010.017 International Astronautical Federation (IAF)

No publication received.

010.018 International Astronomical Union (IAU)

International Astronomical Union, Information Bulletin, No. 39. 55 pp. (1978).   E. A. Müller.
Printed by D. Reidel, Dordrecht, Holland.
Introduction; The XVIIth General Assembly; Executive Committee; Commissions; International Organizations; IAU Symposia and Colloquia; Meetings co-sponsored by the IAU; Other scientific meetings; IAU publications; Other publications; Membership; Letter to all directors of observatories.

International Astronomical Union, Information Bulletin, No. 40, 27 pp. (1978).   E. A. Müller.
Published by D. Reidel Publishing Company, Dordrecht, Holland.
The XVIIth General Assembly. Executive committee. Commissions: (a) general, (b) commission announcements. International organizations. IAU symposia and colloquia. Meetings cosponsored by the IAU. Other scientific meetings. IAU publications. Other publications. Membership.

010.019 Meteoritical Society

The 40th Annual Meeting of the Meteoritical Society in Cambridge, England, July 24 - 29, 1977.
P. M. Millman.
J. R. Astron. Soc. Canada, Vol. 72, 49 - 50 (1978).

Meteoritics. The Journal of the Meteoritical Society, Vol. 13, No. 1 (1978).

010.020 Nederlandse Vereniging voor Weer- en Sterrenkunde

No publication received.

010.021   Polskie Towarzystwo Astronomiczne (PTA)

Report on the activity of the Executive Council of the Polish Astronomical Society for the period from September 17, 1975 to September 22, 1977.
K. Rudnicki, J. Stodółkiewicz.
Postępy Astron., Tom 26, 63 - 65 (1978). In Polish.

Report on the plenary meeting of the Polish Astronomical Society. Frombork, September 22, 1977.
D. Zaremba.
Postępy Astron., Tom 26, 61 - 63 (1978). In Polish.

010.022   Polskie Towarzystwo Miłośników Astronomii (PTMA)

No publication received.

010.023   Royal Astronomical Society (RAS)

Geophysical Journal of the Royal Astronomical Society, Vol. 52, Nos. 1 - 3; Vol. 53, Nos. 1 - 3; Vol. 54, No. 1 (1978).

Memoirs of the Royal Astronomical Society, Vol. 85 (1978).

Monthly Notices of the Royal Astronomical Society, Vol. 182, Nos. 1 - 3; Vol. 183, Nos. 1 - 3 (1978).

The Quarterly Journal of the Royal Astronomical Society, Vol. 19, Nos. 1 - 2 (1978).

Meetings of the Society.
Observatory, Vol. 98, 37 - 43, 81 - 96 (1978).

RAS specialist discussion on Astronomy with the 3.8-metre UK infrared flux collector, 1977 December 9.
J. Ring, V. C. Reddish.
Observatory, Vol. 98, 96 - 101 (1978).

The Martian satellites – 100 years on. Proceedings of RAS Discussion Meeting, held on 1977 May 13.
G. E. Hunt (Editor).
Q. J. R. Astron. Soc., Vol. 19, 90 - 109 (1978).
This report contains sections provided by contributors who reviewed the current knowledge of the Martian satellites. Historical review (*G. A. Wilkins*). Earth-based observations of the satellites of Mars (*D. Pascu*). Spacecraft observations of Phobos and Deimos (*J. Veverka*). The mass of Phobos (*W. H. Michael, Jr.*). The origin of the Martian satellites (*M. Woolfson*).

010.024   Royal Astronomical Society of Canada (RAS Canada)

The Journal of the Royal Astronomical Society of Canada, Vol. 72, Nos. 1 - 3 (1978).

National Newsletter, Supplement to The Journal of the Royal Astronomical Society of Canada, Vol. 72, Nos. 1 - 3 (1978).

010.025   Royal Astronomical Society of New Zealand (RAS New Zealand)

The Royal Astronomical Society of New Zealand (Inc.). 55th annual report of council for the year ended 1977 September 30.
South. Stars, Vol. 27, 88 - 92 (1977).

Southern Stars, Journal of the Royal Astronomical Society of New Zealand, Vol. 27, Nos. 4 - 5 (1977).

010.026   Schweizerische Astronomische Gesellschaft (SAG)

Orion, Zeitschrift der Schweizerischen Astronomischen Gesellschaft. Bulletin de la Société Astronomique de Suisse, 36. Jahrgang, Nos. 164 - 166 (1978).

010.027   Sociedad Astronómica de México

No publication received.

010.028   Società Astronomica Italiana (SAI)

Notizie della Società.
G. Astron., Vol. 3, 263 - 266; Vol. 4, 59 - 66 (1978).

Note sulla XXI riunione della Società Astronomica Italiana. Tenutasi a Bologna dal 20 al 23 ottobre 1977.
L. Baldinelli, S. Ghedini, A. Maitan.
G. A. A. B., N. 48, p. 7 - 11 (1977).

Giornale di Astronomia, Vol. 3, N. 4 (1977); Vol. 4, N. 1 (1978).

Memorie della Società Astronomica Italiana, Vol. 47, No. 1 (1976), with corrections in Vol. 48, No. 1 (1977); No. 2 (1976), with corrections in Vol. 48, No. 1 (1977); Nos. 3 - 4 (1976); Vol. 48, Nos. 1 - 3 (1977).

010.029   Société Astronomique de France (SAF)

Séances, commissions, activités de la Société.
Astronomie, Vol. 92, 3 - 17, 88 - 91, 108, 134, 162, 177 - 188, 236 - 246 (1978).

Société Astronomique de France, groupe du Languedoc. Activités de l'année scolaire 1976 - 1977.
Astronomie, Vol. 92, 81 (1978).

L'Astronomie et Bulletin de la Société Astronomique de France, Vol. 92, Janvier - Août (1978).

010.030   Société Astronomique "R. Boškovic"

No publication received.

010.031   Société Belge d'Astronomie, de Météorologie et de Physique du Globe

No publication received.

010.032   **Société Chronométrique de France**

No publication received.

010.033   **Svenska Astronomiska Sällskapet**

No publication received.

010.034   **VAGO (Astronomical-Geodetical Society of the USSR)**

Sixth plenary meeting of the Central Council of VAGO of the fifth convocation. Yerevan, 1975, October 21. Tsirk. VAGO, 1977, No. 28, p. 3. In Russian. — Abstr. in Ref. zh., 51. Astron., 7.51.10 (1978).

Sixth assembly of the All-Union Astronomical-Geodetic Society (Yerevan, 1975, October 21 - 24). Tsirk. VAGO, 1977, No. 28, p. 4 - 10. In Russian. — Abstr. in Ref. zh., 51. Astron., 7.51.11 (1978).

First plenary session of the Central Council of VAGO of the sixth convocation. Tsirk. VAGO, 1977, No. 28, p. 10. In Russian. — Abstr. in Ref. zh., 51. Astron., 7.51.12 (1978).

Report on financial activities of VAGO. N. A. Polyakov. Tsirk. VAGO, 1977, No. 28, p. 53 - 56. In Russian. — Abstr. in Ref. zh., 51. Astron., 7.51.13 (1978).

Report of the central revision commission of VAGO. Tsirk. VAGO, 1977, No. 28, p. 57 - 64. In Russian. — Abstr. in Ref. zh., 51. Astron., 7.51.14 (1978).

35 years of the Omsk branch of VAGO. D. N. Fialkov. Tsirk. VAGO, 1977, No. 28, p. 65 - 70. In Russian. — Abstr. in Ref. zh., 51. Astron., 7.51.15 (1978).

Possibilities of VAGO in the organization of amateur sky patrol.   V. I. Koval'. Problems of astronomy and geodesy, VAGO assembly Yerevan, 1975, (see 012.039), p. 126 - 129. In Russian. — Abstr. in Ref. zh., 51. Astron., 7.51.16 (1978).

010.035   **Vereniging voor Sterrenkunde, België**

No publication received.

010.036   **25 Jahre "Vereinigung der Sternfreunde".** G. D. Roth. Sterne Weltraum, Jahrg. 17, 4 - 7 (1978).

010.037   **30 years Crimean society of amateur astronomers.** V. V. Martynenko. Zemlya i Vselennaya, 1978, No. 1, p. 78 - 84. In Russian.

010.038   **The Irish Astronomical Association 1974—1977.** D. E. Beesley. Irish Astron. J., Vol. 13, 40 - 42 (1977).

The Association was founded in 1974 and this report outlines its progress during the 1974/75, 1975/76 and 1976/77 sessions.

010.039   **25 Jahre VdS.   R. Wirz.** Orion, 36. Jahrg., 90 - 91 (1978).

010.040   **Société Astronomique de Liège.** Bulletin de la Société Astronomique de Liège, Vol. 40, 1 - 148, janvier - juin 1978.

010.041   **The Nantucket Maria Mitchell Association.** Seventy-sixth annual report for the year ending December 31, 1977. Edited by the Nantucket Maria Mitchell Assoc., Nantucket, Mass., 56 pp. (1978).

010.042   **Rapport d'activité de l'A.D.I.O.N. par le Secrétaire Général et rapport financier.** P. Delache, E. Schatzman, H. Frisch. Assoc. Dév. Int. Obs. Nice, Bull. N. 13 - 14, p. 33 - 49, 63 - 81 (1978). — Report 1975 and 1976.

010.043   **Société Royale d'Astronomie d'Anvers [Koninklijk Sterrenkundig Genootschap van Antwerpen].** Cinquante-huitième rapport 1977.   L. van Mellaert. Imprimerie: "La Prévoyance", Antwerpen. 11 + 13 pp. (1978). In French and Flemish.

010.044   **Nachrichten der Vereinigung der Sternfreunde e.V.** Sterne Weltraum, Jahrg. 17, 33 - 35, 67 - 70, 106 - 108, 140 - 143, 179 - 183, 214 - 219 (1978).

010.045   **Bulletin of the Astronomical Society of India,** Vol. 5, No. 4 (1977), Vol. 6, No. 1 (1978).

010.046   **Astronomia, Periodico Trimestrale dell'Unione Astrofili Italiani, N. 4 (1977).**

010.047   **Publications of the Astronomical Society of Japan,** Vol. 30, No. 1 (1978).

The Strasbourg Stellar Data Centre. See Abstr. 013.005.

011 Reports on Colloquia, Congresses, Meetings, Symposia, and Expeditions

011.001   Wechselwirkung veränderlicher Sterne mit ihrer
          Umgebung. Kolloquium Nr. 42 der Internationalen
Astronomischen Union (IAU) in Bamberg, 6.- 9. September
1977.   J. Rahe.
Sterne Weltraum, Jahrg. 17, 28 - 29 (1978).

011.002   Telescopes of the future.
          E. Kibblewhite, J. Whelan.
Nature, Vol. 271, 307 - 308 (1978).
        The ESO Conference on Optical Telescopes of the
Future was held at CERN, Geneva on 12 - 15 December,
1977.

011.003   The Eighth Lunar Science Conference.   D. Lal.
          Bull. Astron. Soc. India, Vol. 5, 106 - 110
(1977).

011.004   Report on the IAU Colloquium No. 42.
          S. C. Joshi.
Bull. Astron. Soc. India, Vol. 5, 110 - 112 (1977).

011.005   NATO Advanced Study Institute on energy sources
          and emission mechanisms of quasars, Cambridge,
25 July - 4 August, 1977.   S. Ramadurai.
Bull. Astron. Soc. India, Vol. 5, 112 - 113 (1977).

011.006   Conference on space physics. Havana, 1977, June
          27 - July 4.   M. A. Rimsha.
Zemlya i Vselennaya, 1978, No. 1, p. 61 - 63. In Russian.

011.007   Infrared and submillimetre astronomy. A sym-
          posium in Philadelphia, June 8 - 10, 1976.
J. Hanasz.
Postępy Astron., Tom 25, 37 - 41 (1977). In Polish.

011.008   A breakthrough in cosmology? IAU Colloquium
          No. 37 in Paris, September 6 - 9, 1976.
K. Rudnicki.
Postępy Astron., Tom 25, 43 - 47 (1977). In Polish.

011.009   "Radioastronomy and cosmology", IAU Symposium
          No. 74 in Cambridge, Great Britain, August 16 - 20,
(1976).   J. Krempeć, B. Krygier.
Postępy Astron., Tom 25, 49 - 50 (1977). In Polish.

011.010   Fifth cosmological summer school, Cracow, 1976.
          P. Flin.
Postępy Astron., Tom 25, 51 (1977). In Polish.

011.011   Conference of the directors of the Intercosmos
          program.   E. F. Chugunov.
Zemlya i Vselennaya, 1978, No. 2, p. 75 - 77. In Russian.
Ulan-Bator, 1977, August 9 - 15.

011.012   International symposium on electromagnetic meas-
          urements of distances and influences of atmospheric
refraction.   J. Milewski.
Geod. Kartogr., Vol. 27, 71 - 75 (1978). In Polish.

011.013   Colloquium on physics of supernovae and their
          remnants. Moscow, Astronomical Sternberg Insti-
tute, 1977, May 18 - 20.   N. N. Chugaj.
Astron. Zh. Akad. Nauk SSSR, Vol. 55, 443 - 444 (1978).
In Russian. English translation in Soviet Astron., Vol. 22,
No. 2.

011.014   XV International Congress of the History of Science.
          A. V. Douglas.

J. R. Astron. Soc. Canada, Vol. 72, 108 - 111 (1978).

011.015   Primera reunion regional sobre astronomia extra-
          galactica, 24 - 26 de Abril de 1975, with an address
by E. Holmberg.   T. G. Castellanos.
Astron. extragalactica, (see 012.015), 5 - 7 (1976).

011.016   Icarus Meeting Report: International Symposium
          on Planetary Atmospheres.   B. L. Lutz.
Icarus, Vol. 34, 220 - 222 (1978).
        An International Symposium on Planetary Atmospheres
was held in Ottawa, Canada, on 16–19 August 1977.

011.017   "Neutrino–77" – Bericht über eine Konferenz.
          H. J. Haubold.
Sterne, 54. Band, 41 - 44 (1978).
        Vom 19. bis 24. Juni 1977 fand in Baksan im Nord-
Kaukasus (UdSSR) eine Konferenz über Neutrinophysik und
Neutrino-Astrophysik statt.

011.018   The A.L.P.O. at the fourth N.A.A. convention.
          A. Porter.
Strolling Astron., Vol. 27, 45 (1978).

011.019   Polar nights of Spitzbergen at the disposal of
          astrometry.   V. I. Kiyaev.
Zemlya i Vselennaya, 1978, No. 3, p. 73 - 78. In Russian.

011.020   On the international meeting "Photoelectric obser-
          vations of star transits". Moscow, 1976, July 6 - 8.
Ya. S. Yatskiv.
Astrometriya i Astrofizika, Kiev, Vyp. (No.) 34, p. 59 - 60
(1978). In Russian.

011.021   Report on the 15th International Cosmic Ray
          Conference.   R. R. Daniel.
Bull. Astron. Soc. India, Vol. 6, 18 - 21 (1978).

011.022   Round table conference on 'Training Requirements
          of Astronomers in India'.   K. D. Abhyankar.
Bull. Astron. Soc. India, Vol. 6, 21 (1978).

011.023   The twenty-first Herstmonceux conference. Digital
          methods in astronomy. 1977 September 21 - 22.
Observatory, Vol. 98, 101 - 115 (1978).

011.024   Large-scale characteristics of the Galaxy. 84th
          Symposium of the International Astronomical
Union, held at the University of Maryland.
V. Clube.
Nature, Vol. 274, 112 - 113 (1978).

011.025   Planetary geology field conference on aeolian
          processes.   R. Greeley.
Reports of planetary geology program, 1977–1978,
(see 003.014), p. 337 (1978).

011.026   First conference for co-ordination of synchronous
          X-ray and optical observations. – Moscow, 1977,
May 12 - 13.   A. M. Cherepashchuk.
Astron. Tsirk., No. 980, p. 1 - 3 (1978). In Russian.

011.027   The second European Workshop on White Dwarfs.
          Rome and Monte Porzio Catone, September 10 - 14,
1976.   V. Weidemann.
Mem. Soc. Astron. Italiana, Vol. 48, 5 - 12 (1977).

011.028   Conference on European co-operation in satellite

laser ranging, Katwijk, Netherlands, 18 - 19 January, 1977.   L. Cugusi.
Mem. Soc. Astron. Italiana, Vol. 48, 143 - 147 (1977).

Standard techniques for presentation and analysis

of crater size-frequency data.   See Abstr. 091.046.

Significant achievements in the planetary program— 1976 - 1977.   See Abstr. 107.021.

012 Proceedings of Colloquia, Congresses, Meetings and Symposia

012.001   Proceedings of the 3rd International Symposium Geodesy and Physics of the Earth. Parts 2 and 3.
Weimar, GDR 25 - 31 October 1976.
H. Kautzleben, A. Massevitsch (*A. G. Masevich*), E. Tengström, E. Buschmann (Editors).
Veröff. Zentralinst. Phys. Erde, No. 52, Part 2, 450 pp., Part 3, 392 pp. (1977). For Part 1 see 20.012.035. The individual contributions within the subject scope of Astronomy and Astrophysics Abstracts are included in their corresponding subject categories — see abstracts 002.002, 031.205, 032.006, 034.003, 044.001 - 044.003, 045.001 - 045.005, 046.002 - 046.008, 052.001 - 052.006, 055.001, 055.002, 081.007 - 081.019, 082.007, 082.008, 097.006.

012.002   **The Soviet-American conference on cosmochemistry of the moon and planets.** A conference held in Moscow, U.S.S.R., June 4 - 8, 1974.
J. H. Pomeroy, N. J. Hubbard (Editors), with a preface by N. W. Hinners and a foreword by A. P. Vinogradov.
Natl. Aeronaut. Space Adm., Washington, D.C., NASA SP-370 Parts 1, 2. For sale by the Superintendent of Documents, U.S. Government Printing Office, Washington, D.C. 20402. 10 + 10 + 929 pp. (1977). — The individual contributions are included in their corresponding subject categories — see abstracts 091.009 - 091.012, 092.002, 094.001 - 094.007, 094.106 - 094.122, 094.404 - 094.428, 097.008, 099.007, 102.004, 105.016 - 105.019, 107.005.

012.003   **Summer workshop on near-earth resources.**
A workshop held at La Jolla, California, August 6 - 13, 1977.   J. R. Arnold, M. B. Duke (Editors).
National Aeronautics and Space Administration, Washington, D. C., NASA Conference Publication 2031. For sale by the National Technical Information Service, Springfield, Virginia 22161. 13 + 95 pp. Price $ 5.50 (1978). — Contents: Resources from space; Lunar resources;Asteroids; Lunar bibliography.

012.004   **Conférences d'astronomie de l'Observatoire.** 1976.
Preface by M. Golay.
Observatoire, Sauverny, Genève, 4 + 222 pp. Price Fr. 20.00 (1977). — Review in Orion, 35. Jahrg., 177; 1977 (*P. Gerber*). The individual contributions are included in their corresponding subject categories — see abstracts 009.004, 015.011, 032.014, 080.008, 084.207, 091.032, 115.002, 120.001, 155.009.

012.005   **Problems of stellar convection.** Proceedings of the Colloquium Nr. 38 of the International Astronomical Union, held in Nice, August 16—20, 1976.
E. A. Spiegel, J.-P. Zahn (Editors).
Lecture Notes in Physics, Vol. 71. Springer-Verlag Berlin—Heidelberg—New York. 7 + 363pp. (1977). — ISBN 3-540-08532-7, ISBN 0-387-08532-7. The individual contributions are included in their corresponding subject categories — see abstracts 061.011 - 061.014, 062.018 - 062.021, 065.015 - 065.029, 071.004 - 071.006, 082.015.

012.006   **Abstracts of papers presented at the second U.K. Geophysical Assembly,** held 10—13 April 1978 at Liverpool. With a preface by R. Wilson.
Geophys. J. R. Astron. Soc., Vol. 53, 123 - 189 (1978). — The individual contributions within the subject scope of Astronomy and Astrophysics Abstracts are included in their corresponding subject categories — see abstracts 081.027 - 081.031, 084.220 - 084.222, 097.052, 105.033, 107.018.

012.007   **Quantum gravity. An Oxford Symposium** held 1974, February 15 - 16.
C. J. Isham, R. Penrose, D. W. Sciama (Editors).
Clarendon Press, Oxford University Press, 605 pp. Price $ 15.00 (1975). ISBN 0 19 851943 5. — Review in Sterne Weltraum, Jahrg. 16, 347 - 348; 1977 (*J. Schmid-Burgk*). The individual papers within the subject scope of Astronomy and Astrophysics Abstracts are included in their corresponding categories — see abstracts 066.038 - 066.043, 162.035.

012.008   **Equatorial aeronomy. Fifth International Symposium, Townsville, Australia 25—31 August 1976.**
S. Matsushita, B. B. Balsley, H. Rishbeth (Editors).
J. Atmos. Terr. Phys., Vol. 39, 939 - 1275 (1977). — The individual contributions within the subject scope of Astronomy and Astrophysics Abstracts are included in their corresponding categories — see abstracts 082.023, 083.023 - 083.034, 084.004, 084.230 - 084.233.

012.009   **Décalages vers le rouge et expansion de l'univers.**
Paris, 6 - 7 septembre 1976. (Colloque de l'Union Astronomique Internationale, No. 37). **L'évolution des galaxies et ses implications cosmologiques.** Paris, 8 - 9 septembre 1976. (Colloque International du Centre National de la Recherche Scientifique, No. 263).
C. Balkowski, B. E. Westerlund (Editors), with an preface by S. van den Bergh and concluding remarks by P. Morrison.
Editions du Centre National de la Recherche Scientifique, Paris. 619 pp. Price FF 180.00. ISBN 2-222-02022-0. — The individual contributions are included in their corresponding subject categories — see abstracts 066.052, 112.005, 132.032, 132.033, 141.089 - 141.096, 151.036, 151.037, 158.100 - 158.107, 160.033 - 160.039, 162.039 - 162.056.

012.010   **Space Research, Vol. XVII.**
Proceedings of open meetings of working groups on physical sciences of the nineteenth plenary meeting of COSPAR, Philadelphia, Pennsylvania, USA, 8 - 19 June 1976 and COSPAR/IAGA symposium on minor constituents and excited species, Philadelphia, Pennsylvania, USA, 9 - 10 June 1976.
M. J. Rycroft, A. C. Stickland (Editors), with a foreword by C. de Jager.
Pergamon Press, Oxford—New York—Toronto— Sydney—Paris—Frankfurt. 19 + 860 pp. Price DM 375.00 (1977). ISBN 0-08-021636-6. — The individual contributions within the subject scope of Astronomy and Astrophysics Abstracts are included in their corresponding categories — see abstracts

032.019, 032.020, 032.508, 032.509, 046.014 - 046.017,
051.005, 052.019, 055.004, 072.011, 073.024, 074.028,
076.005, 076.006, 081.032, 082.033 - 082.042, 083.045 -
083.055, 085.005, 085.006, 092.008, 093.022 - 093.025,
094.431 - 094.435, 097.054, 099.033 - 099.037, 102.014 -
102.017, 104.007, 104.008, 106.022 - 106.026, 142.115,
142.709, 159.006.

012.011 Radar probing of the auroral plasma. Proceedings of
the EISCAT Summer School, Tromsø, Norway,
June 5 - 13, 1975.   A. Brekke (Editor).
Universitetsforlaget, Tromsø–Oslo–Bergen. 463 pp. Price
N. Kr. 140.00, $ 28.00, DM 57.55 (1977). ISBN 82-00-02421-0.
Reviews in Astron. Tidsskr., Årg. 10, 139 (1977); Science,
Vol. 199, 1061; 1978 (*D. T. Farley*); Space Sci. Rev., Vol. 21,
630; 1978 (*R. Knuth*). – The individual contributions are included in their corresponding subject categories – see abstracts
013.007, 013.008, 031.225, 031.226, 033.003 - 033.006,
062.045, 062.046, 082.045, 083.058 - 083.063, 084.011,
084.012, 084.249.

**012.012 The interaction of variable stars with their environ-
ment.** Proceedings of the IAU Colloquium No. 42,
Bamberg, September 6 - 9, 1977.
R. Kippenhahn, J. Rahe, W. Strohmeier (Editors).
Veröff. Remeis-Sternw. Bamberg, Astron. Inst. Univ. Erlangen-
Nürnberg, Band XI, Nr. 121. 10 + 649 pp. (1977). – The individual contributions are included in their corresponding
subject categories – see abstracts 064.047 - 064.052, 065.012,
113.039, 114.034, 114.035, 114.544, 114.545, 116.007, 116.
008, 117.021 - 117.023, 121.022, 122.068 - 122.093, 124.
013 - 124.019, 124.104 - 124.106, 124.304, 124.801, 126.
025, 126.026, 131.122 - 131.125, 133.027, 133.028, 155.
018.

012.013 Quasars and active nuclei of galaxies. Proceedings of
the Copenhagen Symposium on 'Active Nuclei'
June 27–July 2 1977.   O. Ulfbeck (Editor).
Phys. Scr., Vol. 17, 135 - 385 (1978). – The individual
contributions are included in their corresponding subject
categories – see abstracts 031.238, 031.239, 032.522,
032.523, 141.120 - 141.143, 142.135, 151.049, 155.024,
158.139 - 158.146, 158.507, 158.508, 160.050.

012.014 Planetary nebulae. Observations and theory.
International Astronomical Union. Symposium No.
76, held at Cornell University, Ithaca, New York, U.S.A.,
June 6–10, 1977.
Y. Terzian (Editor), with symposium conclusions by D. E.
Osterbrock, G. Field.
D. Reidel Publishing Company, Dordrecht, Holland–Boston,
U.S.A. 21 + 376 pp. Price Dfl. 90.00, $ 37.00 cloth, Dfl. 60.00,
$ 24.00 paper (1978). ISBN 90-277-0872-X, ISBN 90-277-
0873-8 paperback. – The individual contributions are included
in their corresponding subject categories – see abstracts
022.070, 064.060, 064.061, 065.054 - 065.059, 113.047,
114.040 - 114.042, 115.012, 117.026, 118.016, 122.106,
126.030, 131.148, 132.047, 132.048, 135.028 - 135.095.

012.015 Astronomia extragalactica. Proceedings of the first
Regional Meeting held in Cordoba, April 24–26,
1975.
Acad. Nac. Cienc. Cordoba, (Rep. Argentina), 214 pp. (1976).
– The individual contributions are included in their corresponding subject categories – see abstracts 009.008, 011.015,
032.032, 032.033, 034.022, 141.147, 142.136, 142.137,
151.051, 151.052, 154.023, 154.024, 158.153 - 158.155,
159.010.

012.016 Weak interaction physics – 1977. Conference held
at Bloomington, USA, 16 - 17 May 1977.
AIP Conf. Proc., No. 37 (1977). – Abstr. in Phys. Abstr.,

Vol. 81, Abstr. 414 (1978).

012.017 27th International Astronautical Congress.
Conference held at Anaheim, Calif., USA, 10 - 16
October 1976.
Acta Astronaut., Vol. 4, No. 5-6 (1977). – See abstracts
032.524, 052.022 - 052.024, 053.008 - 053.010. – See also
20.012.016. – Abstr. in Phys. Abstr., Vol. 81, Abstr. 4054
(1978).

**012.018 Workshop on relation between laboratory and space
plasmas.**   H. Kikuchi, I. Kimura, Y. Higuchi,
Y. Kiwamoto, K. Kawamura, I. Suzuki, S. Nakai, A. Mohri,
T. Tokakura, T. Kawabe, H. Ishizuka, N. Kawashima.
Rep. IPPJ-286, Nagoya Univ., Japan, 70 pp. (1977). – See
abstracts 062.067 - 062.069, 099.045, 106.037, 141.530.
Abstr. in Phys. Abstr., Vol. 81, Abstr. 7821 (1978).

012.019 Transactions of the VIIIth international seminar:
"Active processes on the sun and the problem of
solar neutrinos". Leningrad, 25 - 27 Sept., 1976.
Izv. AN SSSR, Ser. fiz., Vol. 41, 1745 - 1980 (1977).
In Russian. – From Ref. zh., 51. Astron., 3.51.48 (1978).
The individual contributions are included in Astronomy and
Astrophysics Abstracts, Vol. 20 and Vol. 21.

**012.020 Dynamics of planets and satellites and theories of
their motion.** International Astronomical Union,
Colloquium No. 41, held in Cambridge, England, August 17–
19, 1976.   V. G. Szebehely (Editor), with an introduction
by P. J. Message.
Astrophysics and Space Science Library, Vol. 72, Proceedings.
D. Reidel Publishing Company, Dordrecht, Holland – Boston,
U.S.A. 12 + 375 pp. Price Dfl. 90.00, $ 39.00 (1978).
ISBN 90-277-0869-X. – The individual contributions are
included in their corresponding subject categories – see
abstracts 021.018, 042.052 - 042.084, 042.087, 052.034,
052.035, 081.055, 094.013 - 094.015, 102.022, 102.023.

012.021 The Earth's magnetopause regions. Symposium
held at the European Geophysical Society meeting,
7 - 10 September 1976, Amsterdam.
J. Atmos. Terr. Phys., Vol. 40, No. 3, p. 231 - 394 (1978).
The individual contributions are included in their corresponding subject categories – see abstracts 062.075, 084.268 -
084.282.

**012.022 Papers presented during the V. Summer School of
Cosmology, Cracow 1976.**
Acta Cosmologica, Zesz 7, 81 - 159 (1978). – The individual
contributions are included in their corresponding subject
categories – see abstracts 066.120, 162.092 - 162.096.

**012.023 Observational parameters and dynamical evolution
of multiple stars.** Proceedings of the International
Astronomical Union Colloquium No. 33, held at Oaxtepec,
México, 13 - 16 October, 1975.
O. G. Franz, P. Pişmiş (Editors), with concluding remarks
by I. R. King.
Rev. Mexicana Astron. Astrofis., Vol. 3, special issue, 216 pp.
(1977). – The individual contributions are included in their
corresponding subject categories – see abstracts 031.249 -
031.251, 034.021, 042.088 - 042.090, 117.036 - 117.059,
118.024 - 118.028, 119.010, 126.033.

**012.024 Problems of gravitation. Transactions of the 5th
international conference on gravitation and relativity. Tbilisi, 9 - 16 Sept. 1968.**
Tbilisi. 721 pp. Price 2 Rbl. 53 Kop. (1976). In Russian and
English. – From Ref. zh., 51. Astron., 4.51.51 (1978).
See abstracts 061.049, 061.050, 066.122, 066.124,
066.232 - 066.235, 066.237, 066.238, 066.307, 066.308,

126.039, 151.093.

**012.025 Proceedings of the 1st Marcel Grossman meeting
on general relativity.** Trieste, Italy, 7 - 12 July 1975.
R. Ruffini (Editor).
North-Holland Publishing Company, Trieste, Italy. 15 + 671
pp. Price $ 40.95 (1977). – See abstracts 066.128 - 066.134,
066.136 - 066.138, 066.140 - 066.143, 066.147, 066.156 -
066.168, 141.537, 162.099, 162.102 - 162.104.

012.026 Structure and properties of nearby galaxies.
International Astronomical Union Symposium No.
77, held in Bad Münstereifel, F.R.G., August 22 - 26, 1977.
E. M. Berkhuijsen, R. Wielebinski (Editors), with an introduction by J. H. Oort.
D. Reidel Publishing Company, Dordrecht, Holland–Boston,
U.S.A. 18 + 307 pp. Price Dfl. 75,00 (1978). – The individual
contributions are included in their corresponding subject
categories – see abstracts 131.195, 151.067 - 151.069,
155.028, 155.029, 158.177 - 158.199, 161.006.

012.027 The large scale structure of the universe. International Astronomical Union, Symposium No. 79,
held in Tallinn, Estonia, U.S.S.R., September 12 - 16, 1977.
M. S. Longair, J. Einasto (Editors).
D. Reidel Publishing Company, Dordrecht, Holland – Boston,
U.S.A. 20 + 464 pp. Price Dfl. 110.00 (1978). ISBN 90-277-
0895-9, ISBN 90-277-0896-7 pbk. – The individual contributions are included in their corresponding subject categories –
see abstracts 132.063, 141.186 - 141.190, 142.186, 151.078 -
151.087, 158.218 - 158.228, 159.013 - 159.015, 160.067 -
160.092, 161.007, 162.128 - 162.148.

012.028 Magnetic stars.
3rd Conference of the Subcommission No. 4. Committee of the Multilateral Cooperation of the Academies of
Sciences of the Socialist Countries "Physics and the evolution
of stars", held at Praha, February 20 - 24, 1978.
J. Grygar, Z. Mikulášek (Editors).
Publ. Astron. Inst. Czechoslovak Acad. Sci. No. 54, 48 pp.
(1978). – The individual papers are included in their corresponding subject categories – see abstracts 031.417, 034.042,
062.101 - 062.103, 064.085 - 064.087, 065.073, 113.078 -
113.080, 114.067, 114.574, 116.016 - 116.021, 119.019,
119.020, 123.056, 123.057.

**012.029 Transactions of the seminar "New methods of
space geodesy".** Symposium held in Leningrad,
24 - 30 November 1975.
I. D. Zhongolovich (Editor), with an opening address by
T. J. Kukkamaeki.
Nablyudeniya Iskusstvennykh Nebesnykh Tel, No. 15
(chast' 1), 1975. Publikatsiya Nauchnykh Rezul'tatov
Sotrudnichestva "INTERKOSMOS". Astronomicheskij Soviet
Akademii Nauk SSSR, Moskva. 333 pp. (1976). In Russian.
The individual contributions are included in their corresponding subject categories – see abstracts 009.015, 031.276,
033.039, 044.031, 044.032, 046.030 - 046.040, 052.048 -
052.050, 081.068 - 081.070.

012.030 Physics of the hot plasma in the magnetosphere.
Proceedings of the thirtieth Nobel Symposium held
April 2 - 4, 1975 at Kiruna Geophysical Institute, Kiruna,
Sweden.
B. Hultqvist, L. Stenflo (Editors).
Plenum Press, New York – London. 10 + 369 pp. Price DM
87.40, $ 30.00 (1975). ISBN 0-306-33700-2. – Review in
Gerlands Beitr. Geophys., Band 87, 75 - 76; 1978 (*H.-R.
Lehmann*). – The individual contributions are included in
their corresponding subject categories – see abstracts 062.
107, 084.031 - 084.033, 084.311 - 084.322.

012.031 Workshop on cometary missions. ESOC, Darmstadt,
17 - 19 April 1978. Chairmen's summaries and
extended abstracts of invited contributions.
Contributions by C. Arpigny, J. L. Bertaux, J. Bodechtel, B.-K.
Dalmann, H. Fechtig, M. Festou, R. H. Giese, E. Grün, L.
Haser, D. W. Hughes, W.-H. Ip, K. Jockers, H. U. Keller,
E. Keppler, J. Kissel, S. Koutchmy, D. Krankowsky, P. Läm-
merzahl, P. L. Lamy, A. C. Levasseur-Regourd, D. Lukoschus,
D. Malaise, F. Mariani, K. W. Michel, G. Neukum, R. Orfei,
H. Rosenbauer, S. Röser, F. Schlude, H. U. Schmidt, J.
Schubart, G. H. Schwehm, A. Sieber, H. Wänke, M. K.
Wallis, R. H. Zerull.
European Space Agency, SOL (78) 14. Paris, 105 pp. (1978).
This workshop was organised by an ad hoc Panel of
ESA's Solar System Working Group. The purpose of the work-
shop was to involve a cross-section of the interested scientific
community in Europe in providing suggestions for the orienta-
tion of future study and planning work.

012.032 Chemical and dynamical evolution of our Galaxy.
Proceedings of the IAU Colloquium No. 45 held at
the Nicolaus Copernicus University, Torun, Poland, September
7–9, 1977.
E. Basinska-Grzesik, M. Mayor (Editors), with a foreword by
L. Martinet.
Published by Geneva Observatory, 1290 Sauverny, Suisse.
9 + 319 pp. (1977/78). – The individual contributions are
included in their corresponding subject categories – see
abstracts 065.086, 065.087, 114.090, 114.091, 131.249 -
131.252, 151.107 - 151.119, 153.044, 155.045 - 155.063,
156.020, 161.009.

**012.033 Eighth Texas symposium on relativistic astro-
physics.** Conference held at Boston, Mass., USA,
13 - 17 December 1976.
M. D. Papagiannis (Editor).
Annals of the New York Academy of Sciences, Vol. 302, 12 +
689 pp. Price $ 45.00 (1977). ISBN 0-89072-048-7. – The
individual contributions are included in their corresponding
subject categories – see abstracts 002.065, 066.274 - 066.289,
117.083, 117.085, 125.072 - 125.075, 131.253, 134.044,
141.227 - 141.232, 141.555, 142.227 - 142.242, 142.733,
158.276, 162.174 - 162.176.

**012.034 Impact and explosion cratering. Planetary and ter-
restrial implications.** Proceedings of the symposium
on planetary cratering mechanics, Flagstaff, Arizona, September 13 - 17, 1976.
D. J. Roddy, R. O. Pepin, R. B. Merrill (Editors).
Pergamon Press, New York – Oxford – Toronto – Sydney –
Frankfurt. 14 + 1301 pp. Price $ 137.50, £ 70.00, DM 412.50
(1977). ISBN 0-08-022050-9. – The individual contributions
within the subject scope of Astronomy and Astrophysics
Abstracts are included in their corresponding categories – see
abstracts 013.025, 091.073 - 091.088, 093.050, 094.150 -
094.152, 094.456 - 094.461, 097.167 - 097.169, 105.079 -
105.096, 106.063, 106.064.

**012.035 Flare stars. Symposium held in Byurakan after the
inauguration of the 2.6-m telescope, October 5 - 8,
1976.** L. V. Mirzoyan (Editor).
Izdatel'stvo AN ArmSSR, Erevan, 231 pp. Price 2 Rbl. 60 Kop.
(1977). In Russian. – From Ref. zh., 51. Astron., 6.51.43
(1978). – See abstracts 064.112, 112.015, 113.093, 116.025,
122.201, 122.206 - 122.214, 124.114, 125.076, 134.045 -
134.047, 153.046, 153.047.

012.036 Ultraviolet stellar astronomy. Proceedings of the
5th Course of the Advanced School of Astronomy,
held at the «E. Majorana» Centre for Scientific Culture, Erice,
Sicily, May 26 - June 8, 1975.
P. L. Bernacca, A. Renzini (Editors).

Memorie della Società Astronomica Italiana, Vol. 47, (Nos. 3-4), 311 - 632 (1976). − See abstracts 031.312, 031.313, 034.070, 054.024, 064.107, 064.108, 080.067, 114.096 - 114.099, 116.024, 131.258 - 131.261, 135.109, 142.245, 158.278, 159.025.

012.037 Giornate di studio sulla fisica dei brillamenti solari, Firenze, Italy, November 17 - 19, 1975.
With an introduction by G. Noci.
Memorie della Società Astronomica Italiana, Vol. 48, No. 2, p. 159 - 331 (1977). − See abstracts 002.067, 073.094 - 073. 097, 076.048, 077.045, 077.046, 078.017, 085.023, 106.068.

012.038 The sun, a tool for stellar physics. Proceedings of the 6th Course of the Advanced School of Astronomy held at the "E. Majorana" Centre for Scientific Culture, Erice, Italy, August 8 - 21, 1976. A. Zichichi, A. Renzini, M. Rigutti (Editors), with a foreward by B. Caccin, M. Rigutti.
Memorie della Società Astronomica Italiana, Vol. 48, No. 3, p. 335 - 581 (1977). − See abstracts 013.028, 062.117, 063.043, 063.044, 064.110, 071.053, 071.054, 073.098, 073.099. 074.098, 080.068 - 080.070.

012.039 Problems of astronomy and geodesy. Proceedings of the VIth assembly of VAGO, Yerevan, 1975.
Yu. D. Bulanzhe (Editor).
Moskva. 215 pp. Price 1 Rbl. (1976). In Russian. − Abstract in Ref. zh., 51. Astron., 7.51.38 (1978). − See abstracts 009.018 - 009.020, 010.034, 013.032, 013.033, 014.042 - 014.048, 015.036, 032.073, 080.064, 093.051, 094.159, 105.113, 107.049.

012.040 Far infrared technology and applications. Conference held at Reston, Va., USA, 18 - 21 April 1977.
T. S. Hartwick, D. T. Hodges (Editors).
Proceedings of the Society of Photo-optical Instrumentation Engineers, Vol. 105. Society of Photo-optical Instrumentation Engineers, Bellingham, Wash., USA. 7 + 120 pp. (1977). ISBN 0-89252-132-5. − Abstr. in Phys. Abstr., Vol. 81, Abstr. 45711 (1978).

012.041 X-ray imaging. Conference held at Reston, Va., USA, 18 - 21 April 1977.
R. C. Chase, G. Kuswa (Editors).
Proceedings of the Society of Photo-optical Instrumentation Engineers, Vol. 106. Society of Photo-optical Instrumentation Engineers, Bellingham, Wash., USA. 8 + 208 pp. (1977). ISBN 0-89252-133-3. − The individual contributions within the subject scope of Astronomy and Astrophysics Abstracts are included in their corresponding categories − see abstracts 031.316, 031.317, 032.571 - 032.576, 142.247.

012.042 Proceedings of the 15th IUPAP International Conference on cosmic rays. Conference held at Plovdiv, Bulgaria, August 1977.
Central Research Institute for Physics, Budapest, Hungary (1977). − See abstracts 078.018, 078.019, 143.061 - 143.066.

012.043 Satellite communications in the next decade. Proceedings of a symposium, March 12, 1976

Washington, D.C. L. Jaffe (Editor).
Publication of the American Astronomical Society, Science and Technology series Vol. 44. Univelt, Inc., P.O. Box 28130, San Diego, Calif. 177 pp. Price $ 20.00 (1977).− Review in Sky Telesc., Vol. 55, 344 (1978).

012.044 Theoretical principles in astrophysics and relativity. Papers from a symposium, Chicago, May 1975.
N. R. Lebovitz, W. H. Reid, P. O. Vandervoort (Editors).
University of Chicago Press, Chicago. 8 + 258 pp. Price $ 23.00 (1978). − Review in Science, Vol. 200, 675 - 676; 1978 (*D. W. Sciama*).

012.045 Proceedings of the NASA/AVS symposium on the use of the Space Shuttle for science and engineer-ing. Moffett Field, Calif., 9 - 11 May 1977.
J. Vac. Sci. Technol., Vol. 14, No. 6 (1977). − Abstr. in Phys. Abstr., Vol. 81, Abstr. 36426 (1978).

012.046 The search for extraterrestrial intelligence − SETI. P. Morrison, J. Billingham, J. Wolfe (Editors), with contributions by I. S. Rasool, D. L. De Vincenzi, J. Billingham, M. A. Stull, J. L. Greenstein, D. C. Black, B. M. Oliver, C. L. Seeger, J. H. Wolfe, J. N. Cuzzi, S. Gulkis, R. Basler, P. Morrison.
National Aeronautics and Space Administration, Scientific and Technical Information Office. NASA SP-419. [For sale by the Superintendent of Documents, U.S. Government Printing Office Washington, D.C. 20402]. 15 + 276 pp. (1977).

012.047 Coronal holes and high speed wind streams: a monograph from Skylab solar workshop I. J. B. Zirker (Editor).
Colorado Associated U.P., Boulder, Colo. 454 pp. Price $ 10.00 (1977). − From Phys. Today, Vol. 31, No. 6, p. 57 (1978).

012.048 Interaction of radiation with condensed matter. Lectures presented at a Winter College held at Trieste, January - March 1976.
International Atomic Energy Agency, Vienna. Vol. 1: 484 pp. Price $ 32.00, Vol. 2: 492 pp. Price $ 32.00 (1977).
Review in Space Sci. Rev., Vol. 21, 629 - 635; 1978 (*H. G. van Bueren*).

012.049 Essay on the history of rocketry and astronautics. Proceedings of the third through the sixth history symposia of the International Academy of Astronautics.
R. Carghill Hall, NASA Conference Publications No. 2104, U.S. Government Printing Office. 2 Volumes, 714 pp. Price $ 10.00 (1977). − Review in Spaceflight, Vol. 20, 278; 1978 (*R. E. Bilstein*).

012.050 The upper atmosphere and magnetosphere. Proceed-ings of a Symposium at the American Geophysical Union, San Francisco, December 1975.
National Academy of Sciences, Washington. 168 pp. (1977).
From Phys. Today, Vol. 31, No. 5, p. 78 (1978).

013 Reports on Astronomy in Various Countries and Particular Fields, International Cooperation

013.001 **Astronomy in Venezuela today.**
H. Alvarez, D. J. MacConnell.
Sky Telesc., Vol. 55, 103 - 105 (1978).

013.002 **Sociology of astronomical innovation.**
D. W. Hughes.
Nature, Vol. 271, 210 - 211 (1978).

013.003 **Pourquoi les astronomes suisses doivent-ils poursuivre leurs recherches dans le ciel Sud.** M. Golay.
Orion, 36. Jahrg., 16 - 23 (1978).

013.004 **Propositions pour un centre de données galactiques non stellaires.** M. C. Lortet.
Bull. Inf. Centre Données Stellaires, No. 14, p. 79 - 104 (1978).
Some points to be dealt with for starting a Center for Non-Stellar Galactic Data are reviewed.

013.005 **Possibilities for data distribution.**
F. Ochsenbein.
Bull. Inf. Centre Données Stellaires, No. 14, p. 123 - 124 (1978).

013.006 **Photoelectric photometry at Wise Observatory.**
N. V. Vidal, N. Brosch, M. Livio.
Observatory, Vol. 98, 60 - 62 (1978).
The performance for *UBV* photometry of the two-star photometer at Wise Observatory has been assessed. Factors affecting the count rates, and the system spectral response, have been analyzed. Mean extinction coefficients for the site are given. The transformations from the instrumental b and v to the standard system are relatively simple, but the $u-b$ transformation is somewhat more complicated and may induce larger errors.

013.007 **EISCAT – status and updated description.**
K. Folkestad.
Radar probing of the auroral plasma, (see 012.011), p. 215 - 228 (1977).

013.008 **Space Shuttle programme and possible coordination with EISCAT.** K. Folkestad.
Radar probing of the auroral plasma, (see 012.011), p. 451 - 463 (1977).

013.009 **On the feasibility of a southern cooperation.**
Z. Kviz.
Proc. Astron. Soc. Australia, Vol. 3, 180 (1977).

013.010 **The activity and the program of the development of the Department of Planetary Geodesy of the Polish Academy of Sciences.** J. B. Zieliński.
Geod. Kartogr., Vol. 26, 179 - 188 (1977). In Polish.

013.011 **Participation of Polish geodetical community in the work of commission of international cooperation of** the Academy of Sciences of the socialistic countries – "Planetary Geophysical Investigations" (KAPG).
S. Kryński, W. Krzemiński.
Geod. Kartogr., Vol. 26, 189 - 194 (1977). In Polish.

013.012 **Creative cooperation in space.** B. Petrov.
Aviatsiya i kosmonavtika, 1977, No. 10, p. 22 - 23.
In Russian. – Abstr. in Ref. zh., 62. Issled. kosm. prostranstva, 2.62.8 (1978).

013.013 **Wie viele Astronomen arbeiten zur Zeit in der Bundesrepublik Deutschland?**
Sterne Weltraum, Jahrg. 17, 152 (1978).

013.014 **Horizons of astronomy.** Yu. N. Efremov.
Zemlya i Vselennaya, 1978, No. 3, p. 50 - 58.
In Russian.

013.015 **60 years amateur astronomy in Finland.**
O. Vilhu.
Zemlya i Vselennaya, 1978, No. 3, p. 80 - 84. In Russian.

013.016 **Beobachtungen in der Sowjetunion.**
G. V. Schultz.
Sterne Weltraum, Jahrg. 17, 201 - 206 (1978).

013.017 **Scientific astronomical centres in Poland.**
Postepy Astron., Tom 26, 59 (1978). In Polish.

013.018 **Soviet period – a qualitatively new stage of development of astronomy in Estonia.**
G. Zhelnin.
Vopr. istor. nauk. i tekh. Pribaltiki. Tartu, 1977, p. 107 - 112.
In Russian. – Abstr. in Ref. zh., 51. Astron., 4.51.2 (1978).

013.019 **Plans for variable star work at the University of Rhode Island.** W. S. Penhallow.
J. American Assoc. Variable Star Obs., Vol. 6, 59 (1977/78).

013.020 **The revolution in astronomy.**
Astronomy, Vol. 5, No. 11, p. 6 - 15, 18 - 23, 26 - 27 (1977). – Abstr. in Phys. Abstr., Vol. 81, Abstr. 16307 (1978).

013.021 **History, astronomy and New Zealand.** G. Eiby.
South. Stars, Vol. 27, 118 - 124 (1978).

013.022 **Nichtoptische Astronomie.** K.-H. Schmidt.
Astron. Schule, 15. Jahrg., 53 - 56 (1978).

013.023 **Peculiarities of the progress of modern astrophysics.**
V. A. Ambartsumyan.
Oktyabr' i nauk., Moskva, 1977, p. 67 - 89. In Russian. – From Ref. zh., 51. Astron., 5.51.2 (1978).

013.024 **Radio astronomy.**
V. A. Kotel'nikov, A. D. Kuz'min.
Oktyabr' i nauk. Moskva, 1977, p. 236 - 249. In Russian.
Abstr. in Ref. zh., 51. Astron., 5.51.52 (1978).

013.025 **Nuclear cratering experiments: United States and Soviet Union.** M. D. Nordyke.
Impact and explosion cratering, (see 012.034), p. 103 - 124 (1977).

013.026 **Report on the Tharsis workshop.** R. J. Phillips.
Reports of planetary geology program, 1977–1978, (see 003.014), p. 334 - 336 (1978).

013.027 **IUE (*International Ultraviolet Explorer*) Observatory observations.** D. K. West.
Bull. American Astron. Soc., Vol. 10, 442 (1978). – Abstract.

013.028 **Forecast of future in solar physics.**
R. N. Thomas.
Mem. Soc. Astron. Italiana, Vol. 48, (see 012.038), 579 - 581

(1977).

013.029 **Advances in astronomy in the year 1977.**
        J. Grygar.
Říše hvězd, Vol. 59, 46 - 50, 71 - 73, 98 - 103 (1978).
In Czech.

013.030 **Amateur astronomy in Czechoslovakia in the years**
        1948–1978.   O. Obůrka.
Říše hvězd, Vol. 59, 31 - 32, 37 (1978). In Czech.

013.031 **Czechoslovak astronomy in the years 1948–1978.**
        V. Guth.
Říše hvězd, Vol. 59, 25 - 31 (1978). In Czech.

013.032 **Prospective development of astronomy in the USSR**
        during the 10th five-years plan.
Eh. R. Mustel', A. G. Masevich, G. S. Khromov.
Problems of astronomy and geodesy, VAGO assembly Yerevan,
1975, (see 012.039), p. 3 - 11. In Russian. – Abstr. in Ref.
zh., 51. Astron., 7.51.7 (1978).

013.033 **Brief review of development of astrophysics in**
        Soviet Armenia.   L. V. Mirzoyan.
Problems of astronomy and geodesy, VAGO assembly Yerevan,
1975, (see 012.039), p. 30 - 34. In Russian. – Abstr. in Ref.
zh., 51. Astron., 7.51.8 (1978).

      Forschung in der Deutschen Demokratischen

**Republik zu Fragen der Physik der Atmosphäre 1971 bis
1975.**   See Abstr. 003.026.

      Two hundred years of flight in America.
See Abstr. 003.044.

      An automated station for the detection of bursts of
cosmic origin at VHF and UHF.   See Abstr. 009.017.

      Workshop on cometary missions.
See Abstr. 012.031.

      Future photoelectric projects at the Beverly-Begg
Observatory.   See Abstr. 031.261.

      The needs for space astronomy in ground-based data.
See Abstr. 051.004.

      Coordination of the Tromsø partial reflection ex-
periment (PRE) and EISCAT.   See Abstr. 083.063.

      The Innisfree meteorite and the Canadian camera
network.   See Abstr. 105.005.

      History of development of meteoritics in Estonia.
See Abstr. 105.059.

      The pioneering investigations in the field of the
interstellar molecules, 1935 - 1942.   See Abstr. 131.212.

014 Teaching in Astronomy

014.001 **Astronomie auf der Jahrestagung des Deutschen**
        Vereins zur Förderung des mathematischen und
naturwissenschaftlichen Unterrichts.   T. Schmidt-Kaler.
Sterne Weltraum, Jahrg. 17, 30 (1978).

014.002 **Instructive three-dimensional models of cometary**
        orbits from simple materials.   B. F. Shinn.
J. British Astron. Assoc., Vol. 88, 139 - 144 (1978).

014.003 **Synodic period statement of Kepler's third law: a**
        good example of an understandable singularity.
C. S. Zaidins.
American J. Phys., Vol. 45, 875 (1977). – Abstr. in Phys.
Abstr., Vol. 80, Abstr. 87601 (1977).

014.004 **Aufgaben mit gebundener Beantwortung.**
        K. Lindner.
Astron. Schule, 15. Jahrg., 6 - 8 (1978).

014.005 **Zum Einsatz von Arbeitsblättern.**   A. Muster.
        Astron. Schule, 15. Jahrg., 9 - 11 (1978).

014.006 **Die Sonne als Beobachtungsobjekt.**
        W. Küttner.
Astron. Schule, 15. Jahrg., 11 - 13 (1978).

014.007 **Erörterung des Gezeitenproblems in meiner**
        Arbeitsgemeinschaft.   J. Stier.
Astron. Schule, 15. Jahrg., 16 - 18 (1978).

014.008 **Astronomy's right hand rule.**   H. L. Cohen.
        Mercury, Vol. 7, No. 1, p. 18 - 20 (1978).

014.009 **Astronomy and solar energy – an adult education**
        program.   A. C. Meyers III, C. Sumners.
Bull. American Astron. Soc., Vol. 9, 569 - 570 (1978).
Abstract.

014.010 **Astronomy and solar energy – a planetarium show.**
        C. Sumners, A. C. Meyers III.
Bull. American Astron. Soc., Vol. 9, 570 (1978). – Abstract.

014.011 **The integration of solar energy content material into**
        an introductory astronomy curriculum.
C. Sumners, A. C. Meyers, III.
Bull. American Astron. Soc., Vol. 9, 612 - 613 (1978).
Abstract.

014.012 **Astronomische Übung: Die Zeitgleichung.**
        W. Büttner.
Sterne Weltraum, Jahrg. 17, 135 - 138 (1978).

014.013 **Risultati preliminari di una indagine statistica sulla**
        situazione dell'insegnamento delle scienze nella
scuola secondaria in Sardegna.   E. Proverbio, T. Lai.
G. Astron., Vol. 3, 235 - 254 (1977).

014.014 **Methodische Hilfen zur Unterrichtseinheit "Die**
        Sonne" (I).   H. Bienioschek, K. Lindner.
Astron. Schule, 15. Jahrg., 36 - 38 (1978).

014.015 **Photometrische Versuche zum Stoffgebiet "Die**
        Sterne".   H.-W. Klee.
Astron. Schule, 15. Jahrg., 42 - 43 (1978).

014.016 Les jumelles.   R. Bucaille.

Astronomie, Vol. 92, 173 - 176 (1978).

014.017 Laboratory exercises in astronomy — Hubble's law.
A. Evans.
Sky Telesc., Vol. 55, 299 - 301 (1978).

014.018 Wer besucht astronomische Kurse?   A. Kunert.
Sterne Weltraum, Jahrg. 17, 177 - 178 (1978).

014.019 Astronomie und Schule.   A. Kunert.
Sterne Weltraum, Jahrg. 17, 178 (1978).

014.020 The luminosity distance equation in Friedmann
cosmology.   J. Terrell.
American J. Phys., Vol. 45, 869 - 870 (1977). — Abstr. in
Phys. Abstr., Vol. 81, Abstr. 6 (1978).

014.021 Gravitation as Newton first saw it.   B. Gee.
Phys. Educ., Vol. 12, 347 - 350 (1977). — Abstr.
in Phys. Abstr., Vol. 81, Abstr. 4436 (1978).

014.022 Simple gravitation using a programmable pocket
calculator.   R. F. Phillips.
Phys. Educ., Vol. 12, 360 - 363 (1977). — Abstr. in Phys.
Abstr., Vol. 81, Abstr. 4438 (1978).

014.023 Astronomy at PTU.   E. P. Levitan.
Zemlya i Vselennaya, 1978, No. 3, p. 69 - 72.
In Russian.

014.024 Esperienze di astronomia per la scuola dell'obbligo.
M. Rigutti.
G. Astron., Vol. 4, 43 - 51 (1978).

014.025 Newton's telescope.   G. Cantor.
Phys. Educ., Vol. 12, 354 - 356 (1977). — Abstr. in
Phys. Abstr., Vol. 81, Abstr. 8187 (1978).

014.026 Determination of a planet's diameter by occultation
time — an introductory astronomy exercise.
F. X. Hart, J. S. Massey.
American J. Phys., Vol. 45, 914 - 917 (1977). — Abstr. in Phys.
Abstr., Vol. 81, Abstr. 12457 (1978).

014.027 Astrodynamics (for teachers).   A. E. Roy.
Phys. Educ., Vol. 12, 445 - 452 (1977). — Abstr. in
Phys. Abstr., Vol. 81, Abstr. 12509 (1978).

014.028 Methodische Hilfen zur Unterrichtseinheit "Die
Sonne" (II).   H. Bienioschek, K. Lindner.
Astron. Schule, 15. Jahrg., 56 - 58 (1978).

014.029 Implementing a teaching strategies workshop for
college astronomy teachers.   D. B. Hoff.
Bull. American Astron. Soc., Vol. 10, 435 (1978). — Abstract.

014.030 Flexibility, procrastination, and tutor development
in a PSI astronomy course.   J. E. Gaustad.
Bull. American Astron. Soc., Vol. 10, 435 (1978). — Abstract.

014.031 PSI astronomy for astrophysics majors: a long-term
assessment.   M. Zeilik.
Bull. American Astron. Soc., Vol. 10, 435 (1978). — Abstract.

014.032 A sundial project for introductory astronomy
classes.   R. H. Allen.
Bull. American Astron. Soc., Vol. 10, 435 (1978). — Abstract.

014.033 Intermediate astronomy courses for non-science
majors.   D. M. Smith.
Bull. American Astron. Soc., Vol. 10, 435 (1978). — Abstract.

014.034 Further links in the California-Wisconsin axis in
American astronomy.   D. E. Osterbrock.
Bull. American Astron. Soc., Vol. 10, 435 - 436 (1978).
Abstract.

014.035 PROJECT: UNIVERSE astronomy on public television.   D, A. Pierce.
Bull. American Astron. Soc., Vol. 10, 436 (1978). — Abstract.

014.036 Why does the Sun shine?   M. Navenberg, V. F.
Weisskopf.
American J. Phys., Vol. 46, 23 - 31 (1978). — Abstr. in Phys.
Abstr., Vol. 81, Abstr. 36756 (1978).

014.037 Orbits of two-body problem from the Lenz vector.
S. Caplan, H. Fuerstenberg, C. Hayes, D. Kane,
S. Rabay.
American J. Phys., Vol. 46, 68 - 70 (1978). — Abstr. in Phys.
Abstr., Vol. 81, Abstr. 36763 (1978).

014.038 The classical Kepler problem in momentum space.
J. Stickforth.
American J. Phys., Vol. 46, 74 - 75 (1978). — Abstr. in Phys.
Abstr., Vol. 81, Abstr. 36764 (1978).

014.039 Demonstrational appliances for the teaching of
astronomy.   L. S. Toth.
Fiz. Szemle, Vol. 27, 219 - 225 (1977). In Hungarian.
Abstr. in Phys. Abstr., Vol. 81, Abstr. 36772 (1978).

014.040 Demonstration of the Sun's eclipse.   M. Jukka.
Fiz. Szemle, Vol. 27, 219 (1977). In Hungarian.
Abstr. in Phys. Abstr., Vol. 81, Abstr. 36773 (1978).

014.041 Theory and observations in lectures on astronomy.
G. Momchev, A. Nikolov, N. Nikolov, N. Petrov.
Fizika, Vol. 2, No. 6, p. 1 - 6 (1977). In Bulgarian. — Abstr.
in Ref. zh., 51. Astron., 6.51.31 (1978).

014.042 Popularization of astronomical knowledge in the
Primorskij province.   G. I. Kornienko.
Problems of astronomy and geodesy, VAGO assembly Yerevan,
1975, (see 012.039), p. 175 - 176. In Russian. — Abstr. in
Ref. zh., 51. Astron., 7.51.19 (1978).

014.043 Engaging students to review popular lectures.
A. V. Marinbakh.
Problems of astronomy and geodesy, VAGO assembly Yerevan,
1975, (see 012.039), p. 210 - 213. In Russian. — Abstr. in
Ref. zh., 51. Astron., 7.51.20 (1978).

014.044 On programs and manuals for physical-astronomical
sections of pedagogical colleges.
V. V. Radzievskij.
Problems of astronomy and geodesy, VAGO assembly Yerevan,
1975, (see 012.039), p. 154 - 168. In Russian. — Abstr. in
Ref. zh., 51. Astron., 7.51.28 (1978).

014.045 On some current problems of improving teaching in
astronomy at schools in present and future.
E. P. Levitan.
Problems of astronomy and geodesy, VAGO assembly Yerevan,
1975, (see 012.039), p. 130 - 141. In Russian. — Abstr. in
Ref. zh., 51. Astron., 7.51.29 (1978).

014.046 On a connection of astronomy with other subjects
at high schools.   Eh. V. Kononovich.
Problems of astronomy and geodesy, VAGO assembly Yerevan,
1975, (see 012.039), p. 149 - 154. In Russian. — Abstr. in
Ref. zh., 51. Astron., 7.51.30 (1978).

014.047 On a program for a high school course of astronomy.

I. D. Il'evskij.
Problems of astronomy and geodesy, VAGO assembly Yerevan, 1975, (see 012.039), p. 142 - 148. In Russian. — Abstr. in Ref. zh., 51. Astron., 7.51.31 (1978).

014.048 Development of high school lectures on astronomy in some planetaria of the USSR.
I. A. Stamejkina.
Problems of astronomy and geodesy, VAGO assembly Yerevan, 1975, (see 012.039), p. 190 - 201. In Russian. — Abstr. in Ref. zh., 51. Astron., 7.51.32 (1978).

Student Guide to contemporary astronomy.

See Abstr. 003.111.

Our incredible universe.   See Abstr. 003.152.

Round table conference on 'Training Requirements of Astronomers in India'.   See Abstr. 011.022.

Satellite motion in the vicinity of the triangular libration points.   See Abstr. 042.097.

General relativity in two- and three-dimensional space-times.   See Abstr. 066.079.

015 Miscellanea

015.001 Planetary mapping with the airbrush.
R. M. Batson.
Sky Telesc., Vol. 55, 109 - 112 (1978).

015.002 On humanity's role in space.   J. von Puttkamer.
Spaceflight, Vol. 20, 55 - 59 (1978).

015.003 Interstellar colonization.   E. M. Jones.
J. British Interplanet. Soc., Vol. 31, 103 - 107 (1978).

015.004 A signalling strategy for interstellar communication.
I. Ridpath.
J. British Interplanet. Soc., Vol. 31, 108 - 109 (1978).

015.005 Extraterrestrial intelligence: an observational approach.
B. Murray, S. Gulkis, R. E. Edelson.
Science, Vol. 199, 485 - 492 (1978).
Existing antennas can set limits on the presence of radio signals from extraterrestrial intelligent life.

015.006 Eavesdropping: the radio signature of the earth.
W. T. Sullivan III, S. Brown, C. Wetherill.
Science, Vol. 199, 377 - 388 (1978).

015.007 Baikonur.   B. P. Vladimirov.
Zemlya i Vselennaya, 1978, No. 1, p. 64 - 71. In Russian.

015.008 The colonization of Mars.   B. C. Clark.
Bull. American Astron. Soc., Vol. 9, 501 (1977).
Abstract.

015.009 Directed panspermia: a technical and ethical evaluation of seeding the universe.
M. Meot-Ner (Mautner), G. Matloff.
Bull. American Astron. Soc., Vol. 9, 501 (1977). — Abstract.

015.010 Habitable zones about main-sequence stars.
M. H. Hart.
Bull. American Astron. Soc., Vol. 9, 501 - 502 (1977).
Abstract.

015.011 Evolution et vie intelligente dans l'univers.
A. Maeder.
Conférences d'astronomie de l'Observatoire, (see 012.004), p. 187 - 204 (1977).

015.012 Calendrical errors and the bicentennial.
A. A. Hirsch.
Proc. Louisiana Acad. Sci., Vol. 50, 96 - 101 (1977).
A functional diagram shows the relationships between calendar errors, elapsed time, the leap year, the inaccurate 100-year quadricentennially skipped decalation cycle, its error of closure, the 128-year natural decalation period, the Julian and the Gregorian calendars. Gregorian errors have caused 54% of Independence Day anniversaries during its first bicentennium to be observed a day too early.

015.013 An investigation of the astronomical theory of the ice ages using a simple climate-ice sheet model.
D. Pollard.
Nature, Vol. 272, 233 - 235 (1978).

015.014 The radio search for intelligent extraterrestrial life.
F. D. Drake.
Focus on the stars, (see 003.002), p. 257 - 287 (1977).

015.015 The origin of life. Part one: the pre-biotic era.
M. A. Bodin.
J. British Interplanet. Soc., Vol. 31, 129 - 139 (1978).

015.016 The origin of life. Part two: monomers to polymers.
M. A. Bodin.
J. British Interplanet. Soc., Vol. 31, 140 - 146 (1978).

015.017 Exotic life and exobiology.   P. M. Molton.
J. British Interplanet. Soc., Vol. 31, 156 - 160 (1978).

015.018 Astrochemistry, organic molecules, and the origin of life.   F. Hoyle.
Mercury, Vol. 7, No. 1, p. 2 - 7, 17 (1978).

015.019 Some thoughts on Dyson spheres.
K. G. Suffern.
Proc. Astron. Soc. Australia, Vol. 3, 177 - 179 (1977).

015 .020 Comets, ice ages, and ecological catastrophes.
F. Hoyle, C. Wickramasinghe.
Astrophys. Space Sci., Vol. 53, 523 - 526 (1978).
A total mass $\sim 10^{14}$ g added to the Earth's upper atmosphere in the form of small particles of high albedo for visual wavelengths would produce an inverse greenhouse effect, shielding ground level from sunlight but permitting infrared radiation from the ground to escape into space. Such a mass of small particles might be acquired by the Earth in a close

approach to a cometary nucleus. Ice ages and ecodisasters, such as that which occured 6.5×10^7 years ago, could arise from the effects of such an addition of small particles.

015.021 The Christmas Star as a supernova in Aquila.
A. J. Morehouse.
J. R. Astron. Soc. Canada, Vol. 72, 65 - 68 (1978).

Objections to a nova theory for the so-called "Christmas Star" are discussed and shown to be invalid. It is proposed that the Christmas Star was actually a sequence of three events: a series of planetary conjunctions and two novae. The present discussion, however, concentrates on the last of these, the nova of 4 B.C. It is suggested, moreover, that this event was not in fact a nova, but a supernova, and that the binary pulsar PSR 1913+16b is a result of that event.

015.022 On the likelihood of a human interstellar civilization.
P. M. Molton.
J. British Interplanet. Soc., Vol. 31, 203 - 208 (1978).

015.023 On ETI alien probe flux density. E. J. Betinis.
J. British Interplanet. Soc., Vol. 31, 217 - 221 (1978).

An alien probe diffusion equation and corresponding alien probe flux density are developed to see if Earth or the Solar System is being detected by technologically advanced extraterrestrial civilisations. If such is the case, then data in the form of observed probes, anomalous and alien as they may be, may possibly be reconciled with data gathered by Earth-based observers.

015.024 Search for extraterrestrial civilizations.
J. Jugaku, H. Hirabayashi.
J. Japan Soc. Aeronaut. Space Sci., Vol. 26, 29 - 35 (1978). In Japanese.

015.025 Project Daedalus. A. Bond, A. R. Martin.
J. British Interplanet. Soc., Suppl., (see 003.003), p. S5 - S7 (1978).

015.026 Project Daedalus: astronomical data on nearby stellar systems. H. R. Mattinson.
J. British Interplanet. Soc., Suppl., (see 003.003), p. S8 - S18 (1978).

015.027 Project Daedalus: the ranking of nearby stellar systems for exploration. A. R. Martin.
J. British Interplanet. Soc., Suppl., (see 003.003), p. S33 - S36 (1978).

015.028 Project Daedalus: the mission profile.
A. Bond, A. R. Martin.
J. British Interplanet. Soc., Suppl., (see 003.003), p. S37 - S42 (1978).

015.029 Project Daedalus: bombardment by interstellar material and its effects on the vehicle.
A. R. Martin.
J. British Interplanet. Soc., Suppl., (see 003.003), p. S116 - S121 (1978).

015.030 Project Daedalus: the navigation problem.
G. R. Richards.
J. British Interplanet. Soc., Suppl., (see 003.003), p. S143 - S148 (1978).

015.031 US halts search for extra-terrestrial intelligence.
D. Dickson.
Nature, Vol. 273, 92 (1978).

015.032 On the topography of extrasolar earthlike planets.
T. A. Heppenheimer.

Icarus, Vol. 34, 441 - 443 (1978).

The author considers a class of planets which have experienced early, nearly complete differentiation and outgassing, whose mantles are fully convective, and whose crusts are isostatically compensated. The evolutionary model of Hargraves (1976) suggests that in the absence of a runaway greenhouse, such planets may usually possess continent/ocean topographies similar to that of Earth. But if the planet is significantly larger than Earth, and its star of spectral type earlier than G, it may ordinarily be completely water-covered.

015.033 The historicity of the gospels and astronomical events concerning the birth of Christ.
J. Seymour, M. W. Seymour.
Q. J. R. Astron. Soc., Vol. 19, 194 - 197 (1978).

The Gospels of St. Matthew and St. Luke, read in association with known contemporary historical and astronomical events, lead us to the conclusion that Christ was born in 4 BC.

015.034 Problem of communication with civilisations around other stars. D. R. Bates.
Proc. R. Irish Acad., Sect. A, Vol. 77, No. 4, p. 45 - 60 (1977). Abstr. in Phys. Abstr., Vol. 81, Abstr. 16314 (1978).

015.035 On making radio contact with extraterrestrial civilizations. D. R. Bates.
Astrophys. Space Sci., Vol. 55, 7 - 13 (1978).

The problem of attempting to make radio contact with other technological civilizations is examined. It is argued that there are few (perhaps indeed no) omnidirectional radio beacons in the Galaxy primarily because any such beacon, to be effective, would have to be designed for a call-response time of at least some 10^4 years, which is much too long to be likely to be acceptable to governments. A consequence is that any search program by us would have little chance of detecting incoming signals.

015.036 Astronomical objects named in honour of heroes of the Great Patriotic War. O. N. Korottsev.
Problems of astronomy and geodesy, VAGO assembly Yerevan, 1975, (see 012.039), p. 202 - 209. In Russian. – Abstr. in Ref. zh., 51. Astron., 6.51.46 (1978).

015.037 Astronomische Randerscheinungen bei Karl May.
G. Zech.
Sterne Weltraum, Jahrg. 17, 252 - 253 (1978).

015.038 The earth as part of the universe.
F. L. Whipple.
Annu. Rev. Earth Planet. Sci., Vol. 6, (see 003.011), 1 - 8 (1978).

015.039 Astronomy and the problem of outlook on life.
V. A. Ambartsumyan, V. V. Kazyutinskij.
Obshchestv. nauk., 1977, No. 3, p. 104 - 122. In Russian. Abstr. in Ref. zh., 51. Astron., 5.51.1 (1978).

015.040 Motion of the earth in the Galaxy and related important peculiarities of the global distribution of some useful minerals. G. P. Tamrazyan.
Izv. AN ArmSSR. Nauk. o zemle, Vol. 30, No. 6, p. 12 - 17 (1977). In Russian. – Abstr. in Ref. zh., 51. Astron., 5.51. 663 (1978).

015.041 On a visit of inhabitants of other planets to the earth. B. K. Fedyushin.
Problems of observational and theoretical astronomy, Moscow –Leningrad, 1977, (see 003.012), p. 237 - 248. In Russian. Abstr. in Ref. zh., 62. Issled. kosm. prostranstva, 5.62.577 (1978).

015.042 The search for extraterrestrial intelligence: tele-

communications technology.
R. E. Edelson, G. S. Levy.
IEEE Commun. Soc. Mag., Vol. 15, No. 6, p. 20 - 23, 32
(1977). – Abstr. in Phys. Abstr., Vol. 81, Abstr. 32319
(1978).

015.043   The Christmas star.   E. R. Meyer.
          Phys. Teach., Vol. 15, 533 - 537, 549 (1977).
Abstr. in Phys. Abstr., Vol. 81, Abstr. 32364 (1978).

015.044   Fisica e metafisica degli UFO.   E. Proverbio.
          Pubbl. Stn. Astron. Int. Latitudine, Carloforte–
Cagliari, Nuova Ser., No. 67, 27 pp. (1978).

015.045   Chemical evolution and the evolution of the Earth's
          crust.   D. E. Ingmanson, M. J. Dowler.
Origins of Life, Vol. 8, 221 - 224 (1977). – Abstr. in Phys.
Abstr., Vol. 81, Abstr. 40666 (1978).

015.046   Far UV irradiation of model prebiotic atmospheres.
          G. Toupance, A. Bossard, F. Raulin.
Origins of Life, Vol. 8, 259 - 266 (1977). – Abstr. in Phys.
Abstr., Vol. 81, Abstr. 40668 (1978).

015.047   Why does Earth's climate change?   J. Gribbin.
          Astronomy, Vol. 6, No. 2, p. 18 - 23 (1978).
Abstr. in Phys. Abstr., Vol. 81, Abstr. 44975 (1978).

          A bibliography on the search for extraterrestrial
intelligence.   See Abstr. 002.046.

          Problems of detecting life on planets.
See Abstr. 003.102.

          The search for extraterrestrial intelligence – SETI.
See Abstr. 012.046.

Applied Mathematics, Physics

021 Mathematics, Computing

021.001 Least-squares adjustment with probabilistic constraints. H. Eichhorn.
Mon. Not. R. Astron. Soc., Vol. 182, 355 - 360 (1978).

A generalized least-squares algorithm is given which permits one to find estimates of unknown parameters which are samples of a population whose covariance matrix is known. This is required for the realistic rigorous solution of certain problems in astrometry, stellar kinematics and other fields.

021.002 Topographic changes on airless bodies: theoretical considerations. L. B. Ronca, R. B. Furlong.
Moon, Vol. 17, 233 - 257 (1977).

A theoretical model of the evolution of topographic features on airless bodies is based on the erosion caused by the impact of small meteorites and the deposition of material ejected from nearby impacts. Three differing conditions are postulated: (1) the rate of erosion equals the rate of sedimentation, as may be expected on large bodies like the Earth's moon and Mercury, (2) the rate of erosion exceeds the rate of sedimentation, as may be expected in smaller bodies like the asteroids, where portions of the ejecta may reach escape velocity, and (3) the rate of sedimentation exceeds the rate of erosion, as may be expected on bodies during planetary growth.

021.003 All Lyapunov characteristic numbers are effectively computable.
G. Benettin, L. Galgani, A. Giorgilli, J.-M. Strelcyn.
C. R. Acad. Sci. Paris, Tome 286, Sér. A, 431 - 433 (1978). In French.

The authors prove a simple theorem which leads to a numerical method allowing to compute all Lyapunov characteristic numbers. As an example, the authors report numerical results of an application of this method to a non trivial model.

021.004 On estimating gravity anomalies — a comparison of least squares collocation with conventional least squares techniques. P. Argentiero, B. Lowrey.
Bull. Géod., Vol. 51, 119 - 126 (1977). — Abstr. in Phys. Abstr., Vol. 80, Abstr. 90983 (1977).

021.005 Least squares adjustment and collocation. K.-R. Koch.
Bull. Géod., Vol. 51, 127 - 135 (1977). — Abstr. in Phys. Abstr., Vol. 80, Abstr. 90984 (1977).

021.006 A note on the choice of norm when using collocation for the computation of approximations to the anomalous potential. C. C. Tscherning.
Bull. Géod., Vol. 51, 137 - 147 (1977). — Abstr. in Phys. Abstr., Vol. 80, Abstr. 90985 (1977).

021.007 Koordinatentransformationen, sphärische Dreiecke und Taschenrechner. H. Schilt.
Orion, 36. Jahrg., 36 - 38 (1978).

021.008 A set of computer programmes for the adjustment of overlapping plates. K. von der Heide.
Astron. Astrophys., Suppl. Ser., Vol. 32, 141 - 155 (1978). In German.

The author has developed a set of computer programmes which allows one to perform a block adjustment of any set of contemporaneous overlapping plates (plate overlap). The reduction model can be chosen arbitrarily and may contain the individual plate constants as unknowns and other groups of unknowns due to the peculiarities of the measuring machines, the telescopes, the observer etc... The unknowns need not be restricted to the plates. The data input into the programmes is a special arrangement of the coordinate measurements of all stars on all plates including magnitude and effective wavelength (if known), as well as the reference star positions. Beside the solution of the equations, simulations and partial inversion of the normal matrix are possible. The set of programmes is well tested and is being used at present to re - reduce the AGK3 in one single adjustment.

021.009 Programerbar lommekalkulator finner planetposisjonene. O. Trondal.
Astron. Tidsskr., Årg.10, 164 - 168 (1977).

021.010 Planetary positions and lunar elements using an algebraic manipulator.
K. F. Pulkkinen, T. C. Van Flandern.
Bull. American Astron. Soc., Vol. 9, 621 (1978). — Abstract.

021.011 Rechnen mit dem Taschenrechner: Ephemeridenrechnung (Teil I). W. Wepner.
Sterne Weltraum, Jahrg. 17, 132 - 134 (1978).

021.012 Vector space methods of photometric analysis: applications to O stars and interstellar reddening.
D. Massa, C. F. Lillie.
Astrophys. J., Vol. 221, 833 - 850 (1978).

Multivariate statistical and vector space notions are utilized to determine the information content of astronomical photometry. These methods enable a concise formulation of many problems in observational astrophysics. In particular, the questions of atmospheric extension in O stars and interstellar reddening are addressed.

021.013 Rechnen mit dem Taschenrechner: Ephemeridenrechnung (Teil II). W. Wepner.
Sterne Weltraum, Jahrg. 17, 174 - 176 (1978).

021.014 On the control of the method of reciprocal dependence based on the distribution of simultaneous events in a numerical experiment. M. Paupere.
Investigations of the sun and red stars. 7, (see 003.004), p. 33 - 42 (1977). In Russian.

The method of the reciprocal dependence based on the distribution of simultaneous events is controlled by means of a numerical experiment. The coincidence between results obtained by this method and those given by the correlation method is considered.

021.015 **Givens transformation techniques for Kalman filtering.**   C. L. Thornton, G. J. Bierman.
Acta Astronaut., Vol. 4, 847 - 863 (1977). – Abstr. in Phys. Abstr., Vol. 81, Abstr. 4473 (1978).

021.016 **Application of the regularization method to the solution of inverse problems of astrophysics.**
D. Ya. Martynov, A. V. Goncharskij, A. M. Cherepashchuk, A. G. Yagola.
Probl. mat. fiz. i vychisl. mat. Moskva, Nauka, 1977, p. 205 - 218. In Russian. – Abstr. in Ref. zh., 51. Astron., 3.51.785 (1978).

021.017 **Derivatives of the Voigt functions.**   P. Heinzel.
Bull. Astron. Inst. Czechoslovakia, Vol. 29, 159 - 162 (1978).
By means of a suitably introduced complex function $D(w)$ the basic relations for both Voigt functions and their derivatives are derived. The choice of the appropriate variables allows the obtained results to be applied directly to astrophysical problems. For the computations of the partial derivatives of the n-th order of the Voigt functions the simple recurrence relations are given simultaneously with the formulas for these derivatives up to the third order. The method presented assumes the simultaneous knowledge of both Voigt functions. Some astrophysical applications are discussed in the conclusion to this paper.

021.018 **MAC revisited: Mechanised Algebraic Operations on fourth generation computers.**   A. Deprit.
Dynamics of planets and satellites, (see 012.020), p. 109 (1978). – Abstract.

021.019 **Fourier transforms of data sampled at unequal observational intervals.**   D. D. Meisel.
Astron. J., Vol. 83, 538 - 545 (1978).
The problem of obtaining a discrete Fourier transform (DFT) of data sampled at unequal (nonperiodic) intervals often occurs in observational astronomy. Although direct approximations with limited validity can be formulated, optimal numerical solutions can be obtained in procedures which interpolate in the time domain while minimizing distortion in the transform (frequency) domain.

021.020 **Symbolic algebraic computer programs. Part II. Applications and perspectives.**
A. Krasiński, M. Perkowski.
Postępy Astron., Tom 26, 33 - 49 (1978). In Polish.
Examples of definite applications of algebraic programs made in mathematics, physics and astronomy are briefly described. The topics treated in more detail are: symbolic integration, general relativity, quantum electrodynamics and celestial mechanics. The underlying idea and some capabilities of general purpose algebraic manipulation systems are discussed and possibilities of creating new ones in Poland are considered. The impact of algebraic programs on mathematics and the science of artificial intelligence is briefly presented. A comparison of criteria of usefulness and efficiency of various programs and systems is given.

021.021 **Über die Methode der kleinsten Quadrate.**
B. L. van der Waerden.
Nachr. Akad. Wiss. Göttingen II, Jahrg. 1977, 75 - 87 (1977).

021.022 **Die Berechnung der Ephemeriden von Planeten und Kometen mit dem programmierbaren Taschenrechner.**   R. A. Gubser.
Orion, 36. Jahrg., 103 - 110 (1978).

021.023 **Fourier transforms applied to the contrast profile method of measuring vector magnetic fields.**
R. E. Loughhead, R. J. Bray.
Publ. Astron. Soc. Pacific, Vol. 90, 230 - 235 (1978).
The contrast profile method of measuring strong magnetic fields (Loughhead and Bray 1976) offers the advantages of high spatial resolution and freedom from the saturation effects encountered with the conventional magnetograph. This paper explores theoretically the extent to which the application of Fourier transforms to magnetic contrast profile data can facilitate the derivation of the various field and velocity parameters. For a stationary solar feature, Fourier transforms offer no advantage. However, for a moving feature the Fourier sine transform provides a simpler and more sensitive means of inferring the magnitude and direction of the field and the line-of-sight velocity than the original contrast data.

021.024 **A Poisson series processor on a computer – Part I.**   H. Nakai.
Tokyo Astron. Obs. Report No. 69, Vol. 18, 242 - 259 (1978). In Japanese.

021.025 **Spherical Bessel functions j_n and y_n of integer order and real argument.**   R. W. B. Ardill, K. J. M. Moriarty.
Comput. Phys. Commun., Vol. 14, 261 - 265 (1978).

021.026 **A precompiler for Poisson series manipulation.**   R. L. Ricklefs, W. H. Jefferys.
Bull. American Astron. Soc., Vol. 10, 481 - 482 (1978). Abstract.

021.027 **A new treatment of the Hill's differential equation $y'' + (a + b\,f(t))y = 0$.**   J. Meffroy.
Bull. American Astron. Soc., Vol. 10, 483 (1978). – Abstract.

021.028 **Least squares adjustment with stochastic condition equations.**   H. Eichhorn.
Bull. American Astron. Soc., Vol. 10, 485 (1978). – Abstract.

021.029 **Coefficienti e fattori di amplificazione di filtri digitali lineari.**
E. Proverbio, S. Uras, G. Graccione.
Circ. Stn. Astron. Int. Latitudine, Carloforte–Cagliari, Ser. B (9), N. 12, 70 pp. (1976).
A set of tables useful for smoothing time series of equidistant observational data is presented in this paper.

021.030 **Die Bestimmung der Cornu-Hartmann'schen Dispersionsformel durch eine Polynomapproximation.**
I. Yavuz.
Commun. Fac. Sci. Univ. Ankara, Sér. A_3: Astron., Tome 25, 77 - 85 (1977) = Commun. Astron. Dep. Univ. Ankara, No. 80.

021.031 **Instability of an area-preserving polynomial mapping.**   J. H. Bartlett.
Celestial Mech., Vol. 17, 3 - 36 (1978).

021.032 **New methods of pre-calculation of the apparent equatorial coordinates of the sun and planets by means of key-punched electronic computers.**
V. A. Levitskij.
Sudovozhdenie, 1977, No. 21, p. 50 - 57. In Russian. – Abstr. in Ref. zh., 51. Astron., 6.51.127 (1978).

021.033 **Estimates of orbits in the case of incomplete knowledge of the mathematical expectation and matrix of covariance of errors.**   B. Ts. Bakhshiyan, R. R. Nazirov, P. E. Ehl'yasberg.
Inst. kosm. issled. AN SSSR. Prepr., 1977, No. 360, 18 pp. In Russian. – Abstr. in Ref. zh., 62. Issled. kosm.

prostranstva, 6.62.419 (1978).

Description of a code-numbering system for identifying stars in magnetic tape catalogues.
See Abstr. 002.014.

Collected papers of Sir Harold Jeffreys on geophysics and other sciences. Vol. 6, Mathematics, probability and miscellaneous other sciences.   See Abstr. 003.063.

Scientific analysis on the pocket calculator.
See Abstr. 003.129.

Method for finding the periods of variable stars.
See Abstr. 031.203

Mathematical-statistical description of the iterative beam removing technique (method CLEAN).
See Abstr. 031.408.

Spurious results from Fourier analysis of data with closely spaced frequencies.   See Abstr. 031.422.

Determinación de acimutes por observación de la polar.   See Abstr. 041.032.

Stabilization by making use of a generalized Hamiltonian variational formalism.   See Abstr. 042.062.

A special perturbation method: m-fold Runge-Kutta.
See Abstr. 042.063.

Numerical integration of nearly-Hamiltonian systems.   See Abstr. 042.064.

Almanac for computers, 1978.
See Abstr. 047.021.

Determination of orbits from optical and laser observations.   See Abstr. 052.006.

Fast numerical analysis of nuclear electric propulsion orbiter-around-Jupiter missions.   See Abstr. 053.001.

Automatic computation of diffusion rates.
See Abstr.062.037.

Numerical methods for Mie theory of scattering by a sphere.   See Abstr. 063.036.

Consistency of Weyl's geometry as a framework for gravitation.   See Abstr. 066.230.

On the mechanical heating of the external layers of the sun. A methodological discussion.
See Abstr. 080.067.

Multidimensional inequality by Bernstein and evaluations of gravity potential derivatives.
See Abstr. 081.047.

An application of synthetic light curves to the RS CVn-type eclipsing binaries.   See Abstr. 121.017.

Results on the large scale distribution of extragalactic objects obtained by the method of statistical reduction.
See Abstr. 162.131.

022 Physical Papers Related to Astronomy and Astrophysics

022.001   X-ray spectroscopy of multiply-charged ions from laser plasmas.
V. A. Boiko, A. Ya. Faenov, S. A. Pikuz.
J. Quant. Spectrosc. Radiat. Transfer, Vol. 19, 11 - 50 (1978).

022.002   Far-infrared absorption in H_2 and H_2–He mixtures.
G. Birnbaum.
J. Quant. Spectrosc. Radiat. Transfer, Vol. 19, 51 - 62 (1978).

Collision-induced absorption in the translation-rotation band of H_2 and H_2–He mixtures has been measured from 20 to 900 cm^{-1} at 77.4, 195 and 292 K. To establish the accuracy of the results, various sources of error are investigated. The zeroth and first spectral moments are evaluated from experiment and theory for H_2 at the various temperatures.

022.003   Pressure-induced absorption of microwave radiation by the oxygen molecule.   M. Mizushima.
J. Quant. Spectrosc. Radiat. Transfer, Vol. 19, 63 - 68 (1978).

022.004   Ionization equilibrium of some elements of astrophysical importance.   N. K. Jain, U. Narain.
Astron. Astrophys., Suppl. Ser., Vol. 31, 1 - 9 (1978).

Ionization equilibrium for low density plasmas of H through Ca, Cr, Mn, Fe and Ni is investigated in the temperature range 5×10^3 K $- 5 \times 10^8$ K using new collisional ionization and modified dielectronic recombination rate coefficients. The other processes included are autoionization and radiative recombination through continuum and via bound levels. The results are presented in tabular form. Comparison with other results, in a few representative cases, have also been made.

022.005   Photoionization cross sections for some CIII levels.
N. Sakhibullin, A. J. Willis.
Astron. Astrophys., Suppl. Ser., Vol. 31, 11 - 14 (1978).

Photoionization cross sections for a number of levels of CIII computed using the Quantum Defect Method (QDM) are presented. These results are compared to previous calculations, and substantial differences are found. Hydrogenic approximation cross sections approach the QDM values for the higher levels. A simple formula is fitted to the QDM cross sections for some levels.

022.006   Electrical conductivity of high pressure ionized argon.
C. Goldbach, G. Nollez, S. Popović, M. Popović.
Z. Naturforsch., Band 33a, 11 - 17 (1978).

022.007   Radiative recombination of complex ions.
R. J. Gould.
Astrophys. J., Vol. 219, 250 - 261 (1978).

The rates of radiative capture by atomic systems is computed for the first four ionization stages of the abundant elements C, N, O, Ne, Mg, Si, S, and Ar. A simple prescription is given for calculating the rates for systems in higher ionization stages. Results for capture to atomic helium are also given. Computed recombination rates are tabulated for the four ionization stages at an electron temperature of 10^4 K and for the first stage at 10^2 K. A convenient procedure is outlined for evaluating the rates at other temperatures in the general neighborhood of these values.

022.008   Lifetime measurements of some levels of Sm I by use of dye laser excitation.
J. Marek, P. Münster.
Astron. Astrophys., Vol. 62, 245 - 247 (1978).

Using the method of delayed coincidence the lifetimes of 13 levels of neutral samarium in the energy range from

21,000 to 29,000 cm^{-1} have been determined. The levels were selectively populated by the light of a pulsed tunable dye laser, pumped by a nitrogen laser.

022.009   OH main lines masers I : OH/IR stars.
M. Elitzur.
Astron. Astrophys., Vol. 62, 305 - 309 (1978).

Excitation processes that can lead to inversion of the main lines of the OH ground state are discussed. Due to the frequency dependence of the emission coefficient of dust, far-IR emitted by warm enough dust can excite the upper halves of the Λ-doublets of rotational levels more strongly than the lower halves. The cascade back to the ground state will then invert the main lines and it is shown that this mechanism can explain rather well the main lines emission from OH/IR stars.

022.010   Vibrational excitation of H_2 in intense ultraviolet fluxes.   J. M. Shull.
Astrophys. J., Vol. 219, 877 - 885 (1978).

The author studies the vibrational excitation of H_2 in intense ultraviolet radiation fields. The radiative cascade through bound levels of the ground electronic state, following absorption of Lyman and Werner band photons, is modified to include effects of "multiple pumping" and direct photodissociation from the high vibrational levels. Although the column densities in the excited vibrational levels ($v \gtrsim 1$) produced by UV pumping appear too low to explain the recently observed Orion quadrupole emission lines, it is suggested that the far-red lines in the (3−0) and (4−0) bands, produced by UV pumping from O stars near dense molecular clouds, may be detectable.

022.011   Calculated structures and microwave frequencies of HNSi and HSiN.   H. W. Kroto, J. N. Murrell, A. Al-Derzi, M. F. Guest.
Astrophys. J., Vol. 219, 886 - 890 (1978).

Ab initio calculations on the species HNSi and HSiN, including configuration interaction, have shown that HNSi is the more stable isomer by approximately 300 kJ mol^{-1}. The microwave spectra of both species including quadrupole structure have been predicted and are shown to be inconsistent with interstellar emission lines which have been tentatively attributed to HSiN.

022.012   Multiconfiguration Hartree-Fock calculation of magnetic quadrupole transitions of Be isoelectronic sequence.   D. L. Lin, W. Fielder, Jr., L. Armstrong, Jr.
Astrophys. J., Vol. 219, 1093 - 1095 (1978).

The authors have calculated M2 transition rates in the Be isoelectronic sequence up to $Z = 92$ by using both relativistic and nonrelativistic multiconfiguration Hartree-Fock procedures.

022.013   Mean lives in Ne V, Mg VII, and Al VIII.
L. C. McIntyre, Jr., D. J. Donahue, E. M. Bernstein.
Phys. Scr., Vol. 17, 5 - 8 (1978).

022.014   Non-LTE line transfer with diffusion of excited atoms II.
D. F. Düchs, S. Rehker, J. Oxenius.
Z. Naturforsch., Band 33a, 124 - 129 (1978).

Non-LTE radiative transfer in a spectral line due to two-level atoms is studied taking the diffusion of excited atoms into account. Numerical results are presented for the case of a stationary, plane parallel plasma of constant total density and temperature without external radiation and without exchange of matter with the surroundings, assuming pure Doppler broadening of the spectral line.

022.015 Electron impact excitation of highly charged sodium-like ions. M. Blaha, J. Davis.
J. Quant. Spectrosc. Radiat. Transfer, Vol. 19, 227 - 235 (1978).

Oscillator strengths, transition probabilities and collision strengths for transitions between $n = 3$ and $n = 4$ levels in Ca(X), Fe(XVI), Zn(XX), Kr(XXVI) and Mo(XXXII) have been calculated in a non-relativistic approximation. Wave functions of excited states have been obtained using a semi-empirical procedure. Collision strengths for electron-impact excitation have been calculated in a distorted wave approximation without exchange. Relative intensities of certain emission lines in the sodium isoelectronic sequence are density dependent. An example of this dependence is discussed.

022.016 Theoretical study of isocyanoacetylene and the isocyanoethynyl radical. S. Wilson.
Astrophys. J., Vol. 220, 363 - 365 (1978).

Quantum mechanical calculations, using the matrix Hartree-Fock model, have been performed to obtain estimates of the rotation constants of the isocyanoacetylene molecule and the isocyanoethynyl radical which may be detectable in space. A rotation constant, B_e, of 5076 MHz is calculated for HC_2NC while for the radical C_2NC the value 5458 MHz is obtained.

022.017 Proton collisional excitation in the ground configuration of Fe^{+11}. D. A. Landman.
Astrophys. J., Vol. 220, 366 - 369 (1978).

Proton collisional excitation cross sections at solar coronal energies for the Fe^{+11} $(3s^2\,3p^3)$ ground configuration transitions are presented. Semiclassical Coulomb excitation theory is used with direct integration of the resulting Schrödinger equation. The rate constants at coronal temperatures are compared to the corresponding ones for electron impact excitation.

022.018 Low-energy proton reactions on ^{45}Sc of interest in stellar nucleosynthesis. J. S. Schweitzer, Z. E. Switkowski, R. M. Wieland.
Nucl. Phys. A, Vol. A287, 344 - 352 (1977). – Abstr. in Phys. Abstr., Vol. 80, Abstr. 88134 (1977).

022.019 A review of spectroscopic information in the visible and ultraviolet region for diatomic molecules of astrophysical interest. D. K. Hsu, W. H. Smith.
Spectrosc. Lett., Vol. 10, No. 4, p. 181 - 303 (1977). – Abstr. in Phys. Abstr., Vol. 80, Abstr. 88459 (1977).

022.020 Laboratory measurements of the $H_2(4,0)$ quadrupole S(1) line and the 6800 Å band of CH_4.
J. T. Bergstralh, J. W. Brault, J. S. Margolis.
Bull. American Astron. Soc., Vol. 9, 515 (1977). – Abstract.

022.021 Laboratory measurements of the hydrogen (3,0) and (4,0) S1 quadrupole lines.
M. E. Mickelson, L. E. Larson, J. T. Trauger.
Bull. American Astron. Soc., Vol. 9, 515 - 516 (1977). Abstract.

022.022 Photoacoustic measurements of line strengths for HD, CH_4, NH_3 red transitions.
S. Bragg, W. H. Smith.
Bull. American Astron. Soc., Vol. 9, 516 - 517 (1977). Abstract.

022.023 Sputtering of water ice by MeV light ions and its importance for planetary astronomy.
W. L. Brown, L. J. Lanzerotti, J. M. Poate, W. M. Augustyniak.
Bull. American Astron. Soc., Vol. 9, 527 (1977). – Abstract.

022.024 Quantitative plagioclase/pyroxene abundance from reflectance spectra: a calibration.
M. J. Gaffey, L. A. McFadden.
Bull. American Astron. Soc., Vol. 9, 530 (1977). – Abstract.

022.025 The visual spectrum of liquid methane.
J. Caldwell, K. R. Ramaprasad, D. McClure.
Bull. American Astron. Soc., Vol. 9, 537 (1977). – Abstract.

022.026 The red bands of methane.
B. L. Lutz, T. Owen, R. D. Cess.
Bull. American Astron. Soc., Vol. 9, 537 (1977). – Abstract.

022.027 Intensity measurements in the ν_4-fundamental of methane. F. K. Ko, P. Varanasi.
Bull. American Astron. Soc., Vol. 9, 537 (1977). – Abstract.

022.028 Ion-molecule condensation reactions: a mechanism for chemical synthesis in reducing planetary atmospheres. M. Meot-Ner (Mautner).
Bull. American Astron. Soc., Vol. 9, 538 (1977). – Abstract.

022.029 High resolution infrared spectra of germane.
S. J. Daunt, G. W. Halsey, K. Fox, R. J. Lovell, N. M. Gailar.
Bull. American Astron. Soc., Vol. 9, 550 (1977). – Abstract.

022.030 Oscillator strengths of Ti I from hook and emission measurements. M. Kühne, K. Danzmann, M. Kock.
Astron. Astrophys., Vol. 64, 111 -113 (1978).

By a combination of hook and emission measurements it is possible to determine a large set of relative oscillator strengths of Ti I lines without any assumption concerning the plasma state. As plasma light sources, modified cascaded arcs have been used for both hook and emission measurements. The relative f-values have been converted to an absolute scale by means of literature data. The overall uncertainties of the f-values are within 13 - 33%. Comparisons with other experimental data are made indicating excellent to fair agreement.

022.031 The C III transition probabilities.
H. Nussbaumer, P. J. Storey.
Astron. Astrophys., Vol. 64, 139 - 144 (1978).

All the astrophysically interesting transition probabilities for the three lowest configurations $2s^2$, $2s2p$, $2p^2$ in C III have been calculated in intermediate coupling. Particular attention is paid to the electric dipole and magnetic quadrupole intercombination lines $2s^2\,{}^1S_0 - 2s2p^3\,P_1^o$, $^3P_2^o$ at 1909 Å and 1907 Å.

022.032 Oscillator strengths for some Ni II lines in the near-ultraviolet from wall-stabilized arc measurements.
J. Moity.
Astron. Astrophys., Vol. 64, 165 - 172 (1978).

Relative oscillator strengths of some Ni II lines in the near UV have been determined from end-on observations of a modified wall-stabilized arc operated in an argon atmosphere with a small admixture of nickel carbonyl. For most of the Ni II lines, the relative gf-values agree with other experimental and theoretical values to within the experimental uncertainty range of 15%. Using the absolute oscillator strengths of some Ni I lines studied by a number of authors, the Ni II gf-values could be brought to an absolute scale with help of the Saha equation.

022.033 Lifetime measurements of the lowest ^2D-states in K I and Rb I using electric quadrupole excitation.
U. Teppner, P. Zimmermann.
Astron. Astrophys., Vol. 64, 215 - 217 (1978).

Using electric quadrupole excitation by a pulsed dye laser the lifetimes of the $3\,^2D_{3/2,5/2}$-states in potassium and $4\,^2D_{3/2,5/2}$-states in rubidium were investigated by the delayed coincidence method.

022.034 Magnetic quadrupole transitions of the beryllium and magnesium isoelectronic sequences.
C. D. Lin, C. Laughlin, G. A. Victor.
Astrophys. J., Vol. 220, 734 - 736 (1978).

Accurate wave functions obtained from model potential methods are used to calculate the probabilities of the magnetic quadrupole transitions $2\,^3P_2 \rightarrow 2\,^1S_0$ of the beryllium isoelectronic sequence and $3\,^3P_2 \rightarrow 3\,^1S_0$ of the magnesium isoelectronic sequence. The results are compared with previous calculations.

022.035 Theoretical study of HSiO⁺ and HOSi⁺.
S. Wilson.
Astrophys. J., Vol. 220, 737 - 738 (1978).

A quantum-mechanical study of the ions $HSiO^+$ and $HOSi^+$ reveals that the latter is more stable. Rotation constants for both species are predicted.

022.036 Theoretical study of the thioformyl ion.
S. Wilson.
Astrophys. J., Vol. 220, 739 - 741 (1978).

The equilibrium structure of the thioformyl ion has been determined from an ab initio matrix Hartree-Fock calculation and an estimated rotation constant has been derived.

022.037 Experimental determination of the line width parameter $\Delta\nu/p$ for the J = 13, K = 13 inversion line of $^{14}NH_3$.
P. Suzeau, J.-P. Mathelin, B. Lamalle, J. Chanussot.
C. R. Acad. Sci. Paris, Tome 286, Sér. B, 123 - 125 (1978).
In French.

022.038 Theoretical f values for sodium-like ions ($11 \leq Z \leq 26$).   E. Biémont.
Astron. Astrophys., Suppl. Ser., Vol. 31, 285 - 290 (1978).

Theoretical oscillator strengths have been calculated, in both the dipole length and dipole velocity formalisms of the transition operator, for numerous transitions in the sodium iso-electronic sequence. The wavefunctions have been computed for the $nl\,^2L$ ($n \leq 8; l = s, p, d, f; 11 \leq Z \leq 26$) configurations, by use of a fully variational Hartree-Fock method. Trends of oscillator strengths along the sequence and comparisons with previously available results are considered.

022.039 Arbitrarily slow irreversibility.
D. Lynden-Bell.
Observatory, Vol. 98, 64 - 65 (1978).

A simple example demonstrates arbitrarily slow irreversibility in a system with no dissipative forces.

022.040 Radiative lifetimes of resonance levels of Al I and Ag I by dye laser excitation.
K. P. Selter, H.-J. Kunze.
Astrophys. J., Vol. 221, 713 - 716 (1978).

The lifetimes of the $3\,^2D_{3/2,5/2}$ levels in Al I and the $5\,^2P_{3/2}$ level in Ag I have been measured for different vapor densities of an atomic beam by using pulsed dye laser excitation and observing the exponential decay directly. Extrapolation to low densities yields the natural lifetimes of these levels accurately to about 5%.

022.041 Intensity measurements of the CH_4 bands in the region 4350 Å to 10,600 Å.   L. P. Giver.
J. Quant. Spectrosc. Radiat. Transfer, Vol. 19, 311 - 322 (1978).

The CH_4 overtone bands from 4410 to 9990 Å, long known in the spectra of the major planets, were studied at room temperature with a long-path, high-pressure White cell. Band intensities and profiles were measured, and are more complete than other recent laboratory measurements of these bands.

022.042 Polyatomic ion-electron reactions.   E. Herbst.
Bull. American Astron. Soc., Vol. 9, 553 (1978).
Abstract.

022.043 Calculation of the cross sections for the CIV-H, NIV-H, and SIII-H charge-exchange collisions. Significance for the interstellar medium.
R. B. Christensen, W. D. Watson, R. J. Blint.
Bull. American Astron. Soc., Vol. 9, 580 (1978). – Abstract.

022.044 Experimental condensation of magnesium silicate grains.   B. Donn, K. L. Day.
Bull. American Astron. Soc., Vol. 9, 583 (1978). – Abstract.

022.045 High-resolution studies of the 9613-Å band of CH_3D.   B. L. Lutz, R. G. Danehy.
Bull. American Astron. Soc., Vol. 9, 605 (1978). – Abstract.

022.046 Atomic structure calculations involving optimization of radial integrals: energy levels and oscillator strengths for Fe XII and Fe XIII $3p-3d$ and $3s-3p$ transitions.   G. E. Bromage, R. D. Cowan, B. C. Fawcett.
Mon. Not. R. Astron. Soc., Vol. 183, 19 - 28 (1978).

Energy levels and oscillator strengths are calculated for the $3s^2\,3p^q-3s^2\,3p^{q-1}\,3d$ and $3s^2\,3p^q-3s3p^{q+1}$ transition arrays of Fe XII and Fe XIII. Strong configuration interactions are explicitly included in the computations, and the method also involves adjustment of radial energy-integrals F^k, G^k, R^k, in order to minimize differences between observed and calculated energy levels, under the least-squares criterion. The possible use of empirically adjusted radial-integral values in electron–ion collisional–excitation calculations (as a means of improving the accuracy of target wavefunctions) is discussed.

022.047 Si⁻ opacity.   T. L. John, R. J. Williams.
Mon. Not. R. Astron. Soc., Vol. 183, 257 - 264 (1978).

The continuous absorption coefficient of Si^- is calculated from quantum mechanically determined free–free cross-sections and experimental photodetachment data. The results indicate that, apart from hydrogen deficient late-type stars, Si^- is a minor source of opacity in stellar atmospheres.

022.048 Lifetimes of some levels in neutral carbon, nitrogen and oxygen.
J. Bromander, N. Đurić, P. Erman, M. Larsson.
Phys. Scr., Vol. 17, 119 - 121 (1978).

Radiative lifetimes of a number of transitions to excited levels in neutral C, N and O have been directly measured using the High Frequency Deflection technique. With one exception, these lifetimes have not been experimentally studied earlier. The results are compared to Coulomb approximation calculations and to semi-empirical estimates as well as to very recent calculations including configuration interactions. A number of these visible and near infrared transitions have been utilized for estimating the solar C, N, and O abundances, and slight modifications are now suggested in some cases in view of the new f-values.

022.049 The cluster concept in multiple hadron production.   I. M. Dremin, C. Quigg.
Science, Vol. 199, 937 - 941 (1978).

The general features of high-energy collisions of elementary particles are outlined. It is argued that multiple production occurs through the production of hadronic clusters. The history and present status of the cluster concept are surveyed.

022.050 Ultraviolet-photoproduced organic solids synthesized under simulated jovian conditions: molecular analysis.
B. N. Khare, C. Sagan, E. L. Bandurski, B. Nagy.
Science, Vol. 199, 1199 - 1201 (1978).

Khare and Sagan (1973) reported the production of a brownish polymetric material from the near-ultraviolet irradiation of simulated jovian atmospheres with a low hydrogen abundance. Examination of this product indicates that hydrogen sulfide is the initial photon acceptor; the powder resulting after extraction with benzene is 84 percent sulfur, largely S_8. In results reported here, the remaining 16 percent was pyrolyzed and then examined by gas chromatography-mass spectrometry.

022.051 Vibrational populations of the excited states of N_2 under auroral conditions. D. C. Cartwright.
J. Geophys. Res., Vol. 83, 517 - 531 (1978).

The purpose of this paper is to present detailed vibrational populations for all the singlet and triplet excited states of N_2, within 12.5 eV of the ground state, excited under auroral conditions.

022.052 Breit-Pauli approximation for highly ionized boron-like ions, up to Fe XXII. W. Dankwort, E. Trefftz.
Astron. Astrophys., Vol. 65, 93 - 98 (1978).

The authors calculated structure and transition probabilities of very highly ionized boron like ions. Where not yet known the wavelengths of the intercombination lines are estimated. The new calculation uses a simple multi-configuration Hartree Fock (MCHF) ansatz with all Breit-Pauli corrections. It is compared with an earlier calculation (Dankwort and Trefftz, 1977) with elaborate MCHF and spin orbit coupling only.

022.053 Helium-like ion line intensities. I. Stationary plasmas.
R. Mewe, J. Schrijver.
Astron. Astrophys., Vol. 65, 99 - 114 (1978).

The population densities of all levels with principal quantum number $n = 2$ in a number of helium-like ions from C V to Ni XXVII have been evaluated as a function of various parameters.

022.054 Helium-like ion line intensities. II. Non-stationary plasmas. R. Mewe, J. Schrijver.
Astron. Astrophys., Vol. 65, 115 - 121 (1978).

The ratio of the forbidden to intercombination line intensities and the ratio of the forbidden plus intercombination to resonance line intensities of helium-like ions have been evaluated as a function of time for two models of time-varying plasmas representative of solar flares.

022.055 Long-range potentials and Stark broadening of neutral lines.
M. S. Dimitrijević, P. Grujić.
J. Quant. Spectrosc. Radiat. Transfer, Vol. 19, 407 - 416 (1978).

The influence of the dipole polarization and quadrupole potentials on the Stark broadening of isolated neutral lines in a plasma has been investigated within the semiclassical approximation. It has been shown that both the perturbing electron trajectories and the minimum impact parameter may undergo a noticeable change when reaction of the emitter on the impact electron is taken into account. The effects on the half-widths of selected lines have been examined within the impact approximation. Calculated results are compared both with available experimental data and other theoretical results. The importance of including long-range effects is also discussed.

022.056 Unique definitions for the band strength and the
electronic-vibrational dipole moment of diatomic molecular radiative transitions. A. Schadee.
J. Quant. Spectrosc. Radiat. Transfer, Vol. 19, 451 - 453 (1978).

Unique definitions of the band strength and the electronic-vibrational dipole moment are inferred from a recently published fundamental discussion on the line strength of rotational transitions in diatomic molecules. In addition, consistent relations between these quantities and the transition probability or oscillator strength are presented.

022.057 Dissociation energy for the ground state of AlO from true potential energy curve.
N. S. Murthy, S. P. Bagare, B. N. Murthy.
J. Quant. Spectrosc. Radiat. Transfer, Vol. 19, 455 - 459 (1978).

022.058 Determination of the $L\alpha_{1,2}$ and $L\beta$ X-ray emission spectrum of iron using a photoelectron spectrometer.
P. M. Montague, D. S. Urch.
Nature, Vol. 272, 804 - 806 (1978).

022.059 Electron collisional excitation cross sections for Fe III and Fe VI and iron abundances in gaseous nebulae. R. H. Garstang, W. D. Robb, S. P. Rountree.
Astrophys. J., Vol. 222, 384 - 397 (1978).

The authors have calculated electron impact excitation cross sections between most of the low metastable levels of Fe III and Fe VI, using the close-coupling method. In addition, they have recalculated the electric quadrupole and magnetic dipole transition probabilities for these transitions in Fe VI. The statistical-equilibrium equations appropriate to a collisional/radiative model are solved to obtain level populations for each ion. Line emissivities have been calculated for the forbidden lines [Fe III] and [Fe VI]. Comparison is made with intensities in several nebulae and peculiar stars. The iron abundances in the Orion Nebula, NGC 6720, NGC 7009, and NGC 7662 are discussed. NGC 7662 seems to have an iron abundance an order of magnitude smaller than the solar abundance.

022.060 Observed transitions between the levels of the ground configuration in S I. K. B. S. Eriksson.
Astrophys. J., Vol. 222, 398 - 399 (1978).

Accurate wavelengths of three transitions in the S I ground configuration are obtained, and the intensities are discussed in view of the theoretical transition probabilities. The level values are revised. The solar-line identifications are strengthened.

022.061 On the relation of dynamics to statistical mechanics.
I. Prigogine, A. Grecos, C. George.
Celestial Mech., Vol. 16, 489 - 507 (1977).

022.062 On some problems of dynamics of protons trapped by the geomagnetic field.
K. Kudela, Yu. Dubinski.
Izv. AN SSSR. Ser. fiz., Vol. 41, 1864 - 1869 (1977). In Russian. – Abstr. in Ref. zh., 62. Issled. kosm. prostranstva, 2.62.249 (1978).

022.063 The presence of π^--mesons in heavy atomic nuclei.
L. Sh. Grigoryan, G. S. Saakyan.
Astrofizika, Vol. 13, 463 - 471 (1977). In Russian. English translation in Astrophysics, Vol. 13, No. 3.

The possibility of the presence of π^--mesons in heavy atomic nuclei is investigated. The parameters characterizing the state of π^--mesons are found in a semiempirical way. In the nuclei with mass numbers $A \lesssim 200$ there are no mesons, and with $A \gtrsim 200$ there are a few π^--mesons. For decreasing ordinal number Z at given A, the number of pions increases, reaching 5 - 7 particles in isobars with smaller Z.

022.064 On the stable state of cold matter with frozen-in magnetic field at densities lower than nuclear.
V. S. Sekerzhitskij, G. A. Shul'man.
Astrofizika, Vol. 13, 473 - 484 (1977). In Russian. English translation in Astrophysics, Vol. 13, No. 3.

022.065 The broadening of spectral lines by electron scattering. II. Pure absorption in a line.   V. G. Vedmich.
Astrofizika, Vol. 13, 493 - 504 (1977). In Russian. English translation in Astrophysics, Vol. 13, No. 3.
The profiles of a spectral line broadened by electron scattering are calculated. Calculations are made for the Doppler profile of the line absorption coefficient and for the distribution of primary sources depending uniformly and linearly on optical depth. Asymptotic formulae for the emergent intensity are obtained too. These formulae describe the wings of the line.

022.066 On the calculation of the population of the 2S level of the hydrogen atom in a plasma medium.
E. B. Klejman, I. M. Ojringel'.
Astrofizika, Vol. 13, 517 - 522 (1977). In Russian. English translation in Astrophysics, Vol. 13, No. 3.
A calculation of the population of the second level of the hydrogen atom located in a turbulent plasma medium is made. It is shown that the two-quantum $2S-1S$ transition accompanied by plasmon emission may essentially have an influence on the observed intensity of the continuous emission.

022.067 Decrement of radio-line series with high quantum numbers.   S. A. Kaplan, V. V. Kulinich.
Astrofizika, Vol. 13, 523 - 534 (1977). In Russian. English translation in Astrophysics, Vol. 13, No. 3.
A method of a numerical simultaneous solution of the transfer equation and equations of balance for atom levels with high quantum numbers close to each other is proposed. The whole system of levels is divided into separate zones consisting of a small number of levels. Then, the decrement is calculated inside of each zone.

022.068 Superdense degenerate plasma.
G. S. Saakyan, L. Sh. Grigoryan.
Astrofizika, Vol. 13, 669 - 684 (1977). In Russian. English translation in Astrophysics, Vol. 13, No. 4.
It is shown that in degenerate matter at densities of $\rho > 3.4 \times 10^{10}$ g/cm³ the phenomenon of neutronization is replaced by pionization. A qualitative consideration of hadron plasma at densities above nuclear is made. For the whole region of densities the equation of the state of degenerate plasma is obtained.

022.069 Cross-sections and rates for electron excitation of excited positively-charged hydrogen and hydrogenic ions.   I. C. Percival, D. Richards.
Mon. Not. R. Astron. Soc., Vol. 183, 329 - 334 (1978).
Semi-empirical cross-sections and rates are presented for transitions between highly excited levels of positively-charged hydrogenic ions and hydrogen induced by electron impact.

022.070 Atomic and molecular data.   A. Dalgarno.
IAU Symp. No. 76, (see 012.014), p. 139 - 149 (1978).
Excitation, ionization and recombination processes are discussed by Seaton (1977). The author restricts his attention to those processes involving heavy particles that may modify the ionization structure of planetary nebulae and that may affect the molecular composition of condensations and of the transition regions between the ionized nebulae and the surrounding interstellar gas.

022.071 Calculated X-radiation from optically thin plasmas. III. Abundance effects on continuum emission.
E. H. B. M. Gronenschild, R. Mewe.

Astron. Astrophys., Suppl. Ser., Vol. 32, 283 - 305 (1978).
The authors have calculated the continuum spectrum ($\lambda 1-1000$ Å, T 10^5-10^8 K) of an optically thin stationary plasma. Abundance effects on the free-free, free-bound and two-photon emission have been investigated. Individual effective gaunt factors for both low- and high-density plasmas have been computed for the elements H, C, N, O, Ne, Mg, Si, S, Ca, Fe and Ni, respectively.

022.072 Optical properties of silicates in the far ultraviolet.   P. L. Lamy.
Icarus, Vol. 34, 68 - 75 (1978).
Near-normal incidence reflectance measurements in the interval 1026−1640 Å were performed on four silicates already studied in the visible and infrared by Pollack et al. (1973). The author uses a Kramers−Kronig analysis of these data to calculate the complex index of refraction $m = n - ik$. New transmission measurements improve the determination of k in the interval 2500−4500 Å, except for andesite, which is more opaque than found by Pollack et al.

022.073 Pyrolysis of organic compounds in the presence of ammonia. The Viking Mars lander site alteration experiment.   G. Holzer, J. Oró.
Org. Geochem., Vol. 1, 37 - 52 (1977).
The influence of ammonia on the pyrolysis pattern of selected organic substances sorbed on an inorganic phase was investigated. A model for the non-equilibrium production of organic compounds on Jupiter is discussed. The investigation was performed in connection with the Viking lander molecular analysis.

022.074 Absorption oscillator strengths for vibration-rotation transitions of ground-state CO.
K. Kirby-Docken, B. Liu.
Astrophys. J., Suppl. Ser., Vol. 36, 359 - 387 (1978).
Absorption oscillator strengths for the infrared vibration-rotation bands of CO in its ground electronic state are tabulated for values of the rotational and vibrational quantum numbers J and ν up to 110 and 30, respectively.

022.075 Semi-empirical electron-impact ionization cross-sections of some hydrocarbons.
A. Tan, S. T. Wu.
Chinese J. Phys., Vol. 15, No. 1, p. 56 - 59 (1977). − Abstr. in Phys. Abstr., Vol. 81, Abstr. 1089 (1978).

022.076 Heat transfer and heat production in a turbulent moving gas.   K. O. Eschrich.
Astron. Nachr., Band 299, 137 - 144 (1978). In German.
This paper deals with heat transfer and heat production in a gas being in turbulent motion. A general equation shall be derived for the mean temperature which contains coefficients determined by turbulence. This treatment is restricted to weak, homogeneous and stationary turbulence. Especially the heating by density fluctuations is discussed and compared with heating due to internal friction.

022.077 Zeeman splitting of some Fe I spectral lines in a magnetic field. II.
E. N. Zemanek, A. P. Stefanov.
Vestn. Kiev. Univ., Astron., Vyp. (No.) 19, p. 55 - 68 (1977). In Russian.

022.078 Tensors and parameters of polarization of electromagnetic emission.   F. E. Khlystun.
Vestn. Kiev. Univ., Astron., Vyp. (No.) 19, p. 68 - 75 (1977). In Russian.
The article deals with the covariant determination of the polarization tensors and parameters of non-zero electromagnetic emission. Their properties are studied.

022.079 Stark broadening of the Al resonance lines.
T. Bach.
J. Quant. Spectrosc. Radiat. Transfer, Vol. 19, 483 - 491 (1978).

Absolute Stark-broadened profiles of the Al resonance lines ($3p-4s$, $\lambda = 3961.5$ and 3944.0 Å) have been measured under conditions of complete LTE in a conventional shock-tube. Plasma electron densities were determined by using a laser-interferometer. Uncertainties in the measurements (15% for the widths and 13% for the shifts) are discussed. The results are compared with impact-broadening calculations.

022.080 On the Zeeman effect in electronic transitions of diatomic molecules.   A. Schadee.
J. Quant. Spectrosc. Radiat. Transfer, Vol. 19, 517 - 531 (1978).

The Zeeman effect for diatomic molecules is calculated in a representation with Hund's case (a) wave functions as a basis set. The resulting expressions are simpler and easier to handle than those obtained from previous calculations on the basis of case (b) wave functions.

022.081 Line positions and oscillator strengths of rotation-vibration bands of possible interstellar SiH and SiH⁺.
P. D. Singh, F. G. Vanlandingham.
Astron. Astrophys., Vol. 66, 87 - 92 (1978).

Line positions of the $1-0$, $2-1$ and $2-0$ vibrational transitions in the ground electronic state of possible interstellar SiH and SiH⁺ are obtained by a differencing process from laboratory measurements of their electronic bands. In the framework of the realistic Klein-Dunham potential, the transition probabilities and oscillator strengths for these pure vibrational transitions of SiH and SiH⁺ are estimated. The $(J = 1) - (J = 0)$ transitions for SiH⁺ in the vibrational levels $\nu = 0$, 1 and 2 of the $X^1 \Sigma^+$ state are predicted. The possibilities of formation and identification of SiH and SiH⁺ in the interstellar medium are also discussed.

022.082 Non-LTE line formation in a magnetic field. The two-level atom with a frequency independent source function. I. Formulation.   E. Landi Degl'Innocenti.
Astron. Astrophys., Vol. 66, 119 - 127 (1978).

The formation of a Zeeman triplet in a plane-parallel atmosphere with a homogeneous magnetic field is considered on the hypothesis of complete redistribution in frequency. The problem is reduced to the solution of a system of coupled integral equations for the source functions of the various Zeeman sublevels. The relevant properties of the kernels are examined in some detail and some conclusions are obtained concerning the differences in the populations of the Zeeman sublevels. In the limit of complete depolarization of the atom in the excited state, previous results are recovered.

022.083 Fine-structure transitions by electron impact in singly-ionized sulphur.   A. K. Pradhan.
Mon. Not. R. Astron. Soc., Vol. 183, 89P - 92P (1978).

022.084 On the dissociation of nitrogen by electron impact and by e.u.v. photo-absorption.
E. C. Zipf, R. W. McLaughlin.
Planet. Space Sci., Vol. 26, 449 - 462 (1978).

The dissociation of N_2 by electron impact and by e.u.v. photo-absorption is studied, and it is shown that the forbidden predissociation of the numerous $^1\Pi_u$ and $^1\Sigma_u^+$ valence and Rydberg states of N_2 in the 11–24 eV energy range is the dominant mechanism for N atom production.

022.085 Heating and vaporization during hypervelocity particle impact.   G. Eichhorn.
Planet. Space Sci., Vol. 26, 463 - 467 (1978).

022.086 Primary velocity dependence of impact ejecta parameters.   G. Eichhorn.
Planet. Space Sci., Vol. 26, 469 - 471 (1978).

022.087 Spectra of very highly ionized atoms.
E. Ya. Kononov.
Phys. Scr., Vol. 17, 425 - 432 (1978).

Recent laboratory investigations of the spectra of highly ionized atoms are reviewed. These include some interpretations of X-ray and VUV spectra of elements heavier than iron. New theoretical data are compared with experimental ones. The possible accuracy of wavelength measurements for spectra produced with specified light sources is discussed.

022.088 Rotational dependence of Franck-Condon factors for selected band systems of CN, C₂, CO, and CH.
P. H. Dwivedi, D. Branch, J. N. Huffaker, R. A. Bell.
Astrophys. J., Suppl. Ser., Vol. 36, 573 - 586 (1978).

022.089 On dielectronic recombination and resonances in excitation cross sections.   J. C. Raymond.
Astrophys. J., Vol. 222, 1114 - 1116 (1978).

Stabilizing transitions of the recombining electron will enhance the dielectronic recombination rate by way of $\Delta n=0$ transitions of highly charged ions. The enhancement is found to be within the accuracy of the Burgess dielectronic recombination rates for astrophysically important ions. The contribution of resonances to the excitation of several lines in the helium and neon isosequences is estimated from comparison of the dielectronic recombination rates of Jacobs et al. (1977) with the Burgess rate and is found to be comparable to the direct excitation rates.

022.090 Electron impact excitation of Fe XXVI and other one-electron ions.
K. L. Baluja, M. R. C. McDowell.
J. Phys. B, Vol. 10, L673 - L676 (1977). – Abstr. in Phys. Abstr., Vol. 81, Abstr. 9257 (1978).

022.091 An ionization wedge for simulation of the solar flare spectrum.   S. K. Gupta, D. Lal, M. N. Rao.
Nucl. Instrum. Methods, Vol. 146, 497 - 502 (1977). – Abstr. in Phys. Abstr., Vol. 81, Abstr. 12786 (1978).

022.092 On the R_e-function of the A-X system of NO⁺.
P. S. Murty.
Phys. Lett. A, Vol. 63A, 405 - 406 (1977). – Abstr. in Phys. Abstr., Vol. 81, Abstr. 13554 (1978).

022.093 Exploratory experiments of impact craters formed in viscous-liquid targets: analogs for Martian rampart craters?   D. E. Gault, R. Greeley.
Icarus, Vol. 34, 486 - 495 (1978).

Exploratory experimental impact studies have been performed using "soupy" mud as a target material. Although differing in details, the results appear to support the hypothesis that ejecta deposits around a class of Martian craters recently revealed in high-resolution Viking Orbiter images were emplaced as a flow of fluidized materials.

022.094 Water vapor adsorption by sodium montmorillonite at $-5°C$.
D. M. Anderson, M. J. Schwarz, A. R. Tice.
Icarus, Vol. 34, 638 - 644 (1978).

A large amount of interest has recently been expressed pertaining to the quantity of physically adsorbed water by the Martian regolith. Thermodynamic calculations indicate that physical adsorption of more than one or two monomolecular layers is highly unlikely under Martian conditions. Any additional water would find ice to be the state of lowest energy and therefore the most stable form. To test the validity of the thermodynamic calculations the authors have measured adsorption and desorption isotherms of sodium montmorillonite at −5°C. To a first approximation it was found to be valid.

022.095 Stark broadening of Ne II lines.
M. Platiša, M. S. Dimitrijević, N. Konjević.
Astron. Astrophys., Vol. 67, 103 - 105 (1978).
The half-widths of twelve Ne II lines have been measured in a pulsed arc plasma. The electron density of $5.1 \times 10^{16}\,cm^{-3}$ was determined by laser interferometry while the electron temperature 28300 K was measured from a Boltzmann plot of relative intensities of Ne II lines.

022.096 Molecular collision processes. II. Excitation of the fine-structure transition of C^+ in collisions with H_2.   D. R. Flower, J. M. Launay.
J. Phys. B, Vol. 10, 3673 - 3681 (1977). − Abstr. in Phys. Abstr., Vol. 81, Abstr. 25675 (1978).

022.097 Radiative lifetimes for selected astrophysically important resonance transitions of F I, Si II, S I, II, III, P II, and CO.   W. H. Smith.
Phys. Scr., Vol. 17, 513 - 515 (1978).
Radiative lifetimes for selected upper levels for transitions of F I, Si II, S I, II, and III, P II, and the CO molecule, lying between 950 Å and 1350 Å have been measured with the phase shift method. These species have been observed or sought via these transitions with the Copernicus satellite. The measured laboratory lifetimes ranged from 0.5 to 26 ns. Comparison is made with previous measurements where possible and the results of the Copernicus satellite in determining interstellar abundances.

022.098 Lifetimes of excited states in Y I, Y II, and Zr I by beam-foil and beam-sputtering excitation.
T. Andersen, P. S. Ramanujam, K. Bahr.
Astrophys. J., Vol. 223, 344 - 349 (1978).
Radiative lifetimes have been determined by beam-foil and beam-sputtering excitation for some excited states in Y I, Y II, and Zr I to test the reliability of the transition probabilities used so far for solar abundance determinations of yttrium and zirconium. The yttrium abundance in the Sun should be increased by a factor of ~2 to $\log A(Y) = 2.35 \pm 0.20$ by using the new transition probabilities, whereas the zirconium abundance should remain unchanged.

022.099 On the flow of special relativistic fluids through channels.   P. J. Wiita.
Astrophys. Space Sci., Vol. 54, 407 - 415 (1978).
The author derives the equations of motion of a perfect relativistic fluid flowing down a confining channel of varying cross section. The formalism of differential geometry is employed, and he starts from the conservation of energy-momentum. Applications to beam models for extragalactic double radio sources are mentioned.

022.100 On the infrared bands of AlO related to Mira phenomena.   P. S. Murty.
Astrophys. Space Sci., Vol. 54, 509 - 512 (1978).
The calculated band positions and intensity factors for the bands of the infrared $A-X$ transition of AlO, suggest that an investigation for the infrared bands in the 1 to 5 μm region of the spectra of Mira stars and M supergiants, along with the 'blue-green' $(B-X)$ bands, could provide clues to the Mira phenomena. The need for additional laboratory investigation of the infrared electronic transition is emphasized.

022.101 Stellar reaction rates for $^{45}Sc(p, \gamma)^{46}Ti$.
S. B. Solomon, D. G. Sargood.
Astrophys. J., Vol. 223, 697 - 700 (1978).
Stellar rates for the reaction $^{45}Sc(p, \gamma)^{46}Ti$ have been calculated from new experimental data. The problem of properly allowing for the contribution of excited states to the stellar rate is discussed and resolved. This problem arises from the existence of a low-lying isomeric level in ^{45}Sc.

022.102 Bifurcations of straight-line oscillations.
V. V. Markellos.
Astron. Astrophys., Vol. 67, 229 - 240 (1978).
In the Störmer and the spring-pendulum problems perturbed by central force inversely proportional to the square of the distance the author examines the stability of the straight-line (on the symmetry axis of each problem) periodic oscillations under "transversal" perturbations in position and velocity. Critical oscillations are determined (branching points) at which the moving particle can switch to two-dimensional motion without losing periodicity. This provides an outline of the structure of the plane periodic oscillations of each problem. Intervals of the parameter describing the size of the perturbation for which such qualitatively different structures can be expected are determined. The role played by resonances is illustrated.

022.103 On some observations of the thermal emission of iron flakes produced in a grinding wheel.
Y. Öhman.
Astrophys. Space Sci., Vol. 55, 39 - 47 (1978).
A report is given of some laboratory experiments on the thermal emission of glowing iron flakes. Clear effects of polarization are found sometimes in flakes of small size, indicating polarization of a kind similar to that appearing in the thermal emission from narrow metallic filaments. Sudden flashes of light appear in the thermal emission from flakes produced in a grinding wheel. At the same time the flake splits into two parts. It is suggested that the flash is due to tribo-thermo-luminescence. It seems possible that the infrared radiation of the solar corona may contain a faintly polarized component due to thermal emission from dust particles.

022.104 Photoabsorption of alkali and alkaline earth elements calculated by the Scaled Thomas Fermi method.   D. Hofsaess.
Z. Phys. A, Vol. 281, 1 - 13 (1977) = Sonderdr. Sternw. Kiel Nr. 226.

022.105 Autoionization rate coefficients for some ions of astrophysical interest.   H. P. Mital, U. Narain.
Sol. Phys., Vol. 57, 341 - 343 (1978).
Autoionization rate coefficients for some quadruply and quintuply ionized atoms have been computed in the temperature range $10^5 - 10^8$ K. Typical temperature dependence has been found and simple expressions are given for their quick estimation.

022.106 Monochromatic emissive power of black body found quickly by nomograph.   A. Zanker.
Optik, Vol. 49, 409 - 412 (1978).
The behaviour of light, radiated from a black body is excellently described by Max Planck, which obtained his well-known formula from his quantum theory. This formula, however, is awkward to handle, and its solution is tedious and time-consuming. The nomograph presented allows to calculate the desired monochromatic emissive power within a matter of seconds with reasonable accuracy, using three subsequently movements of a ruler only. Besides, this nomograph permits also the quick calculation of a wave length with the maximum radiation intensity, according to the Wien's displacement law.

022.107 Laboratory infrared reflectance studies.
L. A. Lebofsky.
Reports of planetary geology program, 1977−1978, (see 003.014), p. 316 - 318 (1978).

022.108 Diffuse reflectance spectra of azurite, malachite, and chrysocolla.   K. L. Andersen.

Reports of planetary geology program, 1977–1978,
(see 003.014), p. 319 - 321 (1978).

022.109 **Preliminary oscillator strengths in the first spectrum of technetium.** R. H. Garstang.
Bull. American Astron. Soc., Vol. 10, 452 (1978). – Abstract.

022.110 **Neutral bremsstrahlung from molecular hydrogen and nitrogen.** T. L. John.
Astron. Astrophys., Vol. 67, 395 - 398 (1978).

Free-free absorption coefficients of molecular negative ions H_2^-, N_2^- are calculated and compared with other work. Results are fitted to simple analytical expressions suitable for stellar atmosphere computations.

022.111 **Optical constants of liquid methane in the infrared.** L. W. Pinkley, P. P. Sethna, D. Williams.
J. Opt. Soc. America, Vol. 68, 186 - 189 (1978).

022.112 **Absorption spectrum of Ba I between 1770 and 1560 Å.** C. M. Brown, M. L. Ginter.
J. Opt. Soc. America, Vol. 68, 817 - 825 (1978).

022.113 **Reflectivity of gold coated surfaces in the soft X-ray range.**
E. Costa, G. Auriemma, L. Boccaccini, P. Ubertini.
Appl. Opt., Vol. 17, 621 - 623 (1978).

022.114 **Refractive index measurements of a germanium sample.** R. P. Edwin, M. T. Dudermel, M. Lamare.
Appl. Opt., Vol. 17, 1066 - 1068 (1978).

022.115 **High-resolution infrared spectra of ν_3 and $2\nu_3$ of germane.** S. J. Daunt, G. W. Halsey, K. Fox, R. J. Lovell, N. M. Gailar.
J. Chem. Phys., Vol. 68, 1319 - 1321 (1978). – Abstr. in Phys. Abstr., Vol. 81, Abstr. 37636 (1978).

022.116 **Reactions of CH_n^+ ions with molecules at 300K.** N. G. Adams, D. Smith.
Chem. Phys. Lett., Vol. 54, 530 - 534 (1978). – Abstr. in Phys. Abstr., Vol. 81, Abstr. 44357 (1978).

022.117 **Binary and ternary reactions of CH_3^+ ions with several molecules at thermal energies.**
D. Smith, N. G. Adams.
Chem. Phys. Lett., Vol. 54, 535 - 540 (1978). – Abstr. in Phys. Abstr., Vol. 81, 44358 (1978).

022.118 **Laboratory simulation of the lunar magnetosphere.** E. M. (*Eh. M.*) Dubinin, I. M. Podgorny (*Podgornyj*), Yu. M. Potanin, C. P. Sonett.
Geophys. Res. Lett., Vol. 4, 391 - 394 (1977).

The magnetic field disturbance created by the impingement of a supermagnetosonic collision-free plasma upon a hollow, electrically nonconducting body (lunella) has been studied in the laboratory using a plasma accelerator. The results are in qualitative accord with lunar magnetometer data and bear upon the interaction of the solar wind with small non-magnetic bodies.

022.119 **Matrix-isolation applied to high-temperature and interstellar molecules: a review.** W. Weltner, Jr.
Ber. Bunsenges. Phys. Chem., Vol. 82, No. 1, p. 80 - 89 (1978).
Abstr. in Phys. Abstr., Vol. 81, Abstr. 46340 (1978).

022.120 **Microwave spectra of molecules of astrophysical**
interest. XII. Hydroxyl radical. R. A. Beaudet.
J. Phys. Chem. Ref. Data, Vol. 7, No. 1, p. 311 - 362 (1978).
Abstr. in Phys. Abstr., Vol. 81, Abstr. 50542 (1978).

022.121 **Z-dependence of relativistic Hartree-Fock second-order energies for isoelectronic ions with filled L-envelope.** A. V. Shestakov.
Opt. i spektrosk., Vol. 43, 995 - 996 (1977). In Russian.
Abstr. in Ref. zh., 51. Astron., 7.51.150 (1978).

022.122 **Study of the emission spectrum of WR stars. II. Calculation of oscillator strengths for the transitions in C III, N IV and O V ions.** A. F. Kholtygin.
Vestn. LGU, 1977, No. 19, p. 128 - 136. In Russian. – Abstr. in Ref. zh., 51. Astron., 7.51.153 (1978).

022.123 **A heuristic model for spectral-line-profiles.** H. Melcher, E. Gerth.
Exp. Tech. Phys., Vol. 25, 521 - 525 (1977). In German.

By means of the Fourier-transformation for the statistical Poisson-distribution a function $y(x)$ is obtained, which is called in this paper Lorentz-function of the n-th degree: $y = (1 + x^2)^{-n}$. Special cases such as $n = 1$ and $n \to \infty$, resp. are used for representing types of spectral lines. The new general Lorentz-function is suitable for approximating or replacing the complicated Voigt-function.

022.124 **Representation of spectral-line-profiles by means of the Lorentz-function of the n-the degree.**
H. Melcher, E. Gerth.
Exp. Tech. Phys., Vol. 25, 527 - 538 (1977). In German.

It is shown how to fit the Lorentz-function of the n-th degree to profiles of the spectral lines. Some examples are given for analyzing profiles by a new method, called the "cutting-method". Values of the Voigt-function are compared with those of the general Lorentz-function (of the n-th degree). It seems to be impossible to differentiate these two functions by means of experimental methods. The results of the analysis of profiles yielding $n < 1$ may be due to those profiles being composed of two or more components.

An astronomically oriented bibliography of atomic autoionization. See Abstr. 002.027.

Nuclear cascade process in dense matter. See Abstr. 003.039.

What is the world made of? See Abstr. 003.045.

Early history of Planck's radiation law. See Abstr. 003.067.

The chamber of physics. See Abstr. 003.113.

The uncertainty principle and foundations of quantum mechanics. A fifty years' survey. See Abstr. 003.115.

Derivatives of the Voigt functions. See Abstr. 021.017.

New satellite structure of the solar and laser plasma spectra in vicinity of the Lα (Mg XII) line. See Abstr. 071.021.

An experimental investigation of the condensation of silicate grains. See Abstr. 131.131.

Astronomical Instruments and Techniques

031 Astronomical Optics, Methods of Observation and Reduction, Data Processing, Automation

Astronomical Optics

031.001 **On computer simulation in design of astronomical optics.** M. I. Tertitskij.
Astron. Zh. Akad. Nauk SSSR, Vol. 55, 168 - 179 (1978). In Russian. English translation in Soviet Astron., Vol. 22, No. 1.
Interactive methods of computer design work in the field of astronomical optics are considered. This process consists of the following steps: image data calculation for a fixed starting point; shift and turn imitation; free parameters deviation; optimization of the optical system. On every step the aberrations are calculated and the spot diagrams displayed.

031.002 **Quick Hartmann method for the test of astronomical optics.** Eh. A. Vitrichenko, F. K. Katagarov.
Astron. Zh. Akad. Nauk SSSR, Vol. 55, 180 - 185 (1978). In Russian. English translation in Soviet Astron., Vol. 22, No. 1.
An automated Hartmann method is developed. One measures only those points of the Hartmann picture which lie on perpendicular diameters.

031.003 **A self-correcting foil mirror.** H. Krause, G. Muller.
Feinwerktech. Messtech., Vol. 85, 330 - 331 (1977). In German. − Abstr. in Phys. Abstr., Vol. 80, Abstr. 88883 (1977).

031.004 **Neste generasjon teleskoper, I. "Roterende sko". II. "Styrbar tallerken".** R. Brahde.
Astron. Tidsskr., Årg. 11, 1 - 5, 53 - 55 (1978).

031.005 **Influence of measuring mark parameters on the visibility of stars in an optical device.**
V. V. Vasil'ev, V. A. Gavrilov, I. A. Zabelina.
Opt.-mekh. prom-st', 1977, No. 7, p. 53 - 56. In Russian. Abstr. in Ref. zh., 51. Astron., 2.51.242 (1978).

031.006 **Diffuse reflection of light on rough surfaces.** E. M. Koshelyaev, V. P. Borodulin, A. P. Zambrzhitskij, A. A. Puzanov.
Vestn. MGU. Fiz., astron., Vol. 18, No. 5, p. 25 - 34 (1977). In Russian. − Abstr. in Ref. zh., 51. Astron., 3.51.180 (1978).

031.007 **Sharpening stellar images.** A. Buffington, F. S. Crawford, S. M. Pollaine, C. D. Orth, R. A. Muller.
Science, Vol. 200, 489 - 494 (1978).
Atmospherically induced phase perturbations have for years limited the resolution of large optical astronomical telescopes. A prototype telescope system with six movable elements has successfully corrected these phase perturbations. This use of real-time image sharpening has restored stellar images to the diffraction limit (in one dimension) for a 30-centimeter telescope.

031.008 **Method of designing an Offner-type compensating system for testing concave aspheric mirrors.**
G. M. Popov, M. B. Popova.
Izv. Krymskoj Astrofiz. Obs., Vol. 58, 109 - 112 (1978). In Russian.
Some methods of testing concave aspheric mirrors by means of a compensation method are discussed. A comparatively simple method of designing an Offner compensator is described. Several Offner compensators are calculated to match concave mirrors for most wide-spread Ritchey-Chrétien systems.

031.009 **Field aberrations of concentric optical systems.** G. M. Popov.
Izv. Krymskoj Astrofiz. Obs., Vol. 58, 113 - 119 (1978). In Russian.
Off-axis aberrations of concentric systems are discussed when the (non-flat) object is at arbitrary distances from the system. The derived expressions may be useful to design focal reducers of the Meinel type.

031.010 **On the frequency correlation of waves in a medium with large-scale random inhomogeneities.**
S. N. Molodtsov, A. I. Saichev.
Izv. vyssh. uchebn. zaved. Radiofiz., Vol. 20, 1244 - 1246 (1977). In Russian. − Abstr. in Ref. zh., 51. Astron., 4.51.109 (1978).

031.011 **Eyepiece data for the observer.** G. E. Taylor.
J. British Astron. Assoc., Vol. 88, 352 - 355 (1978).

031.012 **Splines application in astronomical optics control by the Hartmann method.** B. P. Artamonov, V. P. Voevudskij.
Astron. Zh. Akad. Nauk SSSR, Vol. 55, 660 - 665 (1978). In Russian. English translation in Soviet Astron., Vol. 22, No. 3.
An algorithm for approximation of optical surfaces with the help of cubic interpolation and smoothing splines is discussed. The advantages of application of smoothing splines as compared to interpolation have been demonstrated. The results of computations of a model mirror with the help of an ALGOL program realizing a smoothing algorithm are presented.

031.013 **Active optics: a new technology for the control of light.** J. W. Hardy.
Proc. IEEE, Vol. 66, 651 - 697 (1978).
In this tutorial-review paper, the basic concepts of active optics systems and their historical evolution are discussed, from

early figure-control systems with a servo bandwidth of less than 1 Hz to the recently developed high-bandwidth systems for atmospheric compensation with bandwidths of several hundred hertz. A critical comparison of the various approaches to wavefront sensing is then made covering both coherent (laser) and incoherent (white-light) systems. Current techniques for wavefront correction including Bragg cells, segmented mirrors, thin-plate deformable mirrors, monolithic mirrors, and membrane mirrors are described. The performance analysis and optimization of closed loop systems is covered using two basic models. The paper concludes with a review of the design and performance of five current experimental active optical systems, with some comments on future applications.

031.014 The Hartmann test for the 105 cm Schmidt tele-scope.   T. Noguchi, T. Soyano, T. Aoki.
Tokyo Astron. Obs. Report No. 69, Vol. 18, 169 - 186 (1978). In Japanese.

031.015 An aspheric objective for a coronagraph.
I. Shimizu, Y. Shimizu, M. Fukatsu, Z. Wakimoto, Y. Nagayama.
Tokyo Astron. Obs. Report No. 69, Vol. 18, 198 - 210 (1978). In Japanese.

031.016 Polishing and Foucault test of the receiving tele-scope mirror for the lunar laser ranging.
H. Sato, K. Tomita, K. Ikeya.
Tokyo Astron. Obs. Report No. 69, Vol. 18, 344 - 359 (1978). In Japanese.

031.017 Optical design of the Viking Lander camera.
W. J. Davis.
Systems integration and optical design. Reston, Va., USA, 18 - 21 April 1977. Proc. Soc. Photo-optical Instrumentation Engineers, Vol. 103. 8 + 128 pp. (1977). S. Refermat (Editor). p. 71 - 75. — Abstr. in Phys. Abstr., Vol. 81, Abstr. 53794 (1978).

031.018 Polarization and interference in optics. V. Lenses, imaging properties of lenses.   J. Ben Uri.
Optik, Vol. 49, 375 - 388 (1978).

031.019 Off-axis aberration of achromatic Schmidt cor-rectors.   A. Greve.
Optik, Vol. 50, 235 - 242 (1978).
For an achromatic Schmidt system (which eliminates the spherical aberration at two wavelengths) expressions of the residual chromatic aberration and the off-axis aberration (distortion) are given. For a specified image quality the ex-pressions allow the specification of the ratio D/N^3 and the form factor A of the achromatic corrector.

031.020 Marechal intensity criteria modified for circular apertures with nonuniform intensity transmission: Dini series approach.   S. Szapiel.
Opt. Lett., Vol. 2, 124 - 126 (1978).
Starting with the Dini expansion of a wave aberration function, a relationship is derived between the Maréchal evaluation of the Strehl definition (MESD) for uniformly il-luminated aperture and the MESD for an aperture with non-uniform intensity transmission. The result is particularly use-ful for rapid evaluation of the far-field peak-intensity degrada-tion due to aberrations and in problems connected with aber-ration balancing techniques, as well as for the estimation of optical system performance when the aperture intensity dis-tribution is varied.

031.021 Achromatic Schmidt corrector plates for the near infrared.   A. Greve.
Infrared Phys., Vol. 18, 7 - 10 (1978).

The paper demonstrates the advantages of near infrared achromatic Schmidt systems, which eliminate spherical aber-ration at two wavelengths, and give a residual spherical aber-ration which fulfils the Strehl/Rayleigh criterion. The theory is illustrated by a few numerical examples.

031.022 On the calculation of deformations of astronomical mirrors with horizontal control.
N. G. Bochkarev, N. A. Sukhova, G. G. Telepneva.
Astron. Tsirk., No. 969, p. 4 - 6 (1977). In Russian.

031.023 Discovery of previously unknown density irregulari-ties of the field of Schmidt cameras.
K. Rudnicki, S. B. Vladimirov, Ya. A. Bellas.
Astron. Tsirk., No. 972, p. 4 - 5 (1977). In Russian.

031.024 Tolerances for centering errors based on wave-optic quality criteria.   C. Hofmann.
Feingeraetetechnik, Vol. 26, 548 - 550 (1977). In German. Abstr. in Phys. Abstr., Vol. 81, Abstr. 33806 (1978).

031.025 Methods for centering optical surfaces.
G. Jaunet, J. P. Marioge, F. Farfal, M. Mullot, B. Bonino.
J. Opt., Vol. 9, 31 - 44 (1978). In French. — Abstr. in Phys. Abstr., Vol. 81, Abstr. 33924 (1978).

031.026 Photoelectric method of determination of a tele-scope's focal distance.
G. A. Terez, Eh. I. Terez.
Astron. Tsirk., No. 993, p. 4 - 5 (1978). In Russian.

031.027 Aberration balancing for grating mountings with large aberrations.   C. H. F. Velzel.
J. Opt. Soc. America, Vol. 68, 38 - 41 (1978).

031.028 On an asymptotic theory of diffraction gratings used in the scalar domain.
E. G. Loewen, M. Nevière, D. Maystre.
J. Opt. Soc. America, Vol. 68, 496 - 502 (1978).

031.029 Testing of aspherical lenses using side band Ronchi test.   D. Malacara, M. Josse.
Appl. Opt., Vol. 17, 17 - 18 (1978).

031.030 Caustic coordinates in Platzeck-Gaviola test of conic mirrors.   A. Cornejo, D. Malacara.
Appl. Opt., Vol. 17, 18 - 19 (1978).

031.031 Testing of flat optical surfaces by the quantitative Foucault method.   M. C. Simon, J. M. Simon.
Appl. Opt., Vol. 17, 132 - 137 (1978).
The complete theory of measurement of optical flat mirrors of circular or elliptical shape using the quantitative Foucault method is described here. It has been used in Córdo-ba since 1939 in a partially intuitive but correct form. The surface, not yet flat and, at times, astigmatic, is assimilated to the sum of a spherical plus a cylindrical dome. The errors of the three possible ways of reckoning are calculated.

031.032 All-reflecting Baker-Schmidt flat-field telescopes.
D. J. Schroeder.
Appl. Opt., Vol. 17, 141 - 144 (1978).
The theory of the Baker-Schmidt flat-field telescope with tilted reflecting corrector and an analysis of the performance of several different all-reflecting Baker-Schmidt systems is presented. A comparison is given between the performance of a flat-field Baker-Schmidt and an all-reflecting Schmidt tele-scope of similar focal ratio.

031.033 Measurement of steep aspheric surfaces.

J. G. Dil, P. F. Greve, W. Mesman.
Appl. Opt., Vol. 17, 553 - 557 (1978).

A method of measuring the shape of high numerical aperture (NA <0.95), convex or concave, aspheric surfaces is described. The aspheric slope may be as large as 1200 waves/rad. The method is applied in two steps. First, a standard measurement is performed to obtain a reference surface. Second, the reproducibility of the fabrication of aspheric surfaces is tested by means of a holographic comparison method. The measuring error is smaller than 0.1 μm.

031.034 Interferometric testing with computer-generated holograms: aberration balancing method and error analysis.   T. Yatagai, H. Saito.
Appl. Opt., Vol. 17, 558 - 565 (1978).

When testing aspheric surfaces with a computer-generated hologram, some problems should be considered. In this paper, first, the authors compare two types of hologram: Lohmann and interference. The phase error in the Lohmann type hologram is estimated, and a method of compensating the error is described. Second, they discuss the relation between the shape of the required wavefront and the number of resolution cells of the hologram. Since testing smaller f number optical elements increases the required number of resolution cells of the hologram, the authors propose the aberration balancing method to reduce the number of resolution cells. The optimum values of the defocus aberration are calculated. Finally, an experimental example for testing an aspheric mirror 150 mm in diameter and 300 mm in focal length is given.

031.035 Soft X-ray imaging with toroidal mirrors.   Y. Sakayanagi, S. Aoki.
Appl. Opt., Vol. 17, 601 - 603 (1978).

The fabrication of small toroidal mirrors in tandem for X-ray imaging is discussed. First, a male mandrel is made by grinding and polishing a molybdenum rod. Then a glass replica is cast and lightly polished. The method of polishing the male mandrel is described. A photograph of a copper mesh taken with a 8.3-Å X-ray is shown.

031.036 Lens design merit functions: rms image spot size and rms optical path difference.   B. Brixner.
Appl. Opt., Vol. 17, 715 - 716 (1978).

The chief lens design problem is to get all the optical paths from object point to image point equal within a small fraction of a wavelength to ensure that spherical wavefronts will be converging on all image points in the field of view. The LASL lens design program minimizes the lateral deviations of the rays from their ideal image points. Results given here show that this procedure also minimizes the optical path difference and that there is a linear relationship between the rms image spot size and the rms optical path difference.

031.037 Four-mirror unobscured anastigmatic telescopes with all-spherical surfaces.   D. R. Shafer.
Appl. Opt., Vol. 17, 1072 - 1074 (1978).

A variety of four-mirror telescopes are described, all corrected for spherical aberration, coma, and astigmatism; all unobscured; and all with spherical surfaces. Performance is discussed with an afocal beam expander and a flat-field telescope as typical design examples.

031.038 Aspheric grating for extreme ultraviolet astronomy.

S. O. Kastner, C. Wade, Jr.
Appl. Opt., Vol. 17, 1252 - 1258 (1978).

A family of plane curves is developed which can diffract incident parallel rays to a point focus. These curves, termed diffractoidal curves, are rotated around an axis to produce surfaces of revolution correspondingly termed diffractoids, whose imaging properties for sources at infinity are studied by ray tracing in a few examples. The paraboloid emerges as a limiting case of the diffractoid. A comparison is made between the stigmatic focusing properties of the diffractoid and the toroidal grating.

031.039 Compound catadioptric telescopes with all spherical surfaces.   R. D. Sigler.
Appl. Opt., Vol. 17, 1519 - 1526 (1978).

Catadioptric, all spherical Cassegrainian and Gregorian telescopes with one and two full aperture corrector lenses are investigated. Appropriate closed form third-order aberration equations are presented, and a variety of aplanatic and anastigmatic solutions are indicated.

031.040 Testing glass reflecting-angles of prisms.   A. S. De Vany.
Appl. Opt., Vol. 17, 1661 - 1662 (1978).

031.041 Imaging properties of concave holographic gratings.   T. Namioka, F. Masuda.
Bull. Res. Inst. Sci. Meas. Tohoku Univ., Vol. 26, No. 1, p. 3 - 17 (1977). In Japanese. — Abstr. in Phys. Abstr., Vol. 81, Abstr. 37971 (1978).

A corrector lens system for Cassegrain telescope.   See Abstr. 032.013.

On external occulting systems for coronographs.   See Abstr. 032.034.

Workshop testing results of the 6-m mirror of the large azimuthal telescope.   See Abstr. 032.048.

STARLAB, a spacelab-based one-meter general purpose telescope.   See Abstr. 032.504.

A survey telescope for space.   See Abstr. 032.505.

High resolution X-ray optics for astronomical and laboratory sources.   See Abstr. 032.571.

Design, fabrication and performance of two grazing incidence telescopes for celestial extreme ultraviolet astronomy.   See Abstr. 032.572.

Design, fabrication and expected performance of the HEAO-B X-ray telescope.   See Abstr. 032.573.

Passive imaging through the turbulent atmosphere: fundamental limits on the spatial frequency resolution of a rotational shearing interferometer.   See Abstr. 082.114.

Method of determination of the optical characteristics of an instrument for measurement of illuminance in the Venus atmosphere.   See Abstr. 093.035.

Methods of Observation and Reduction

031.201 **Analysis of stellar occultation data. Effects of photon noise and initial conditions.**
R. G. French, J. L. Elliot, P. J. Gierasch.
Icarus, Vol. 33, 186 - 202 (1978).

A new inversion technique for obtaining temperature, pressure, and number density profiles of a planetary atmosphere from an occultation light curve is described. This technique employs an improved boundary condition to begin the numerical inversion and permits the computation of errors in the profiles caused by photon noise in the light curve. The authors present their assumptions about the atmosphere, optics, and noise and develop the equations for temperature, pressure, and number density and their associated errors.

031.202 **Phaseless aperture synthesis.**
J. E. Baldwin, P. J. Warner.
Mon. Not. R. Astron. Soc., Vol. 182, 411 - 422 (1978).

The authors present two techniques, of general use in radio astronomical interferometry, for analysing aperture synthesis data without employing phase information. In the first, a trial map is derived from observed amplitudes alone, and in the second, the true distribution of brightness is deduced from the trial map by an iterative procedure. These techniques are similar to those used in solving molecular structures from X-ray diffraction patterns. A practical test, using data from the 5C 7 survey, is presented.

031.203 **Method for finding the periods of variable stars.**
P. Renson.
Astron. Astrophys., Vol. 63, 125 - 129 (1978). In French.

A method is described, which is intended to find the period of a variable phenomenon from a given number of discrete observations. It is a trial method, but more elaborated than Lafler and Kinman's one. A better distribution of the trial periods is proposed. The observational error is taken into account for the criterion used to select the right period. In order to avoid at best to retain a spurious period, in particular an associated period, instead of the right one, trial methods must be applied according to a technique which is also explained. Finally the various possible criteria for the trial method are compared.

031.204 **Calibration of spectrograms using a Lyot filter element.** C. Trefzger, J. Solf.
Astron. Astrophys., Vol. 63, 131 - 136 (1978).

Furenlid (1973) has described a method of calibration for photographic spectrograms which uses the spectral modulation imposed on a continuum light source by means of a Lyot filter element (birefringent quartz crystal plate between two parallel polarizers). The authors present a calibration device based on this principle. Systematic tests show that this calibration method provides higher internal accuracy compared with a conventional step filter method and that systematic errors were not detected in astronomical applications determining equivalent widths of H_γ absorption lines.

031.205 **Transatlantic geodesy by long baseline interferometry.**
W. H. Cannon, R. B. Langley, W. T. Petrachenko.
Veröff. Zentralinst. Phys. Erde, No. 52, Part 2, (see 012.001), p. 281 - 322 (1977).

The authors report geodetic and astrometric results from the analysis of fringe frequency observations from a series of three long baseline interferometry experiments carried out in 1973 between the 46m antenna of the Algonquin Radio Observatory, Lake Traverse, Canada and the 25m antenna at the Chilbolton Field Station, Chilbolton, England. The r.m.s. deviation from the mean of the estimates of the length and orientation of the 5251 km equatorial component of the baseline from all three experiments is 1.05m and 0″.015 respectively. The experiments also yielded positions of five extragalactic radio sources.

031.206 **The performance of telescopes for the observation of meteors.** M. Kresáková.
Bull. Astron. Inst. Czechoslovakia, Vol. 29, 50 - 56 (1978).

The performance of instruments of different aperture, magnification and field of view for the observation of meteors is examined. Formulae for the predicted meteor rates and limiting meteor magnitudes are derived and evaluated.

031.207 **Estimate of the quality of galaxy images taken with the 15″ astrograph.** L. P. Panteleeva.
Soobshch. Gos. Astron. Inst. Shternberga, No. 193, p. 14 - 17 (1977). In Russian.

031.208 **On a method of photometric reduction of slitless spectrograms.** R. I. Noskova, T. A. Birulya.
Soobshch. Gos. Astron. Inst. Shternberga, No. 199, p. 15 - 18 (1977). In Russian.

A method of photometric calibration of slitless spectrograms is suggested. The method was applied to slitless spectrograms of planetary nebulae. The results obtained by the method described agree with the photoelectric observations much better than the results received by the calibration method formerly accepted.

031.209 **On a method of excluding photographic irradiation in photographic positional observations of Venus.**
B. S. Vozdvizhenskij.
Soobshch. Gos. Astron. Inst. Shternberga, No. 200, p. 23 - 32 (1977). In Russian.

A method is proposed for excluding the effect of irradiation from photographic observations of Venus. It is based on an investigation of photographic irradiation for various photographic emulsions on a simulator "artificial planet". As an example the results of the treatment of Venus observations on a wide-angle astrograph are presented.

031.210 **On the phase correction for position observations of planets.** Yu. I. Safronov.
Astron. Zh. Akad. Nauk SSSR, Vol. 55, 138 - 147 (1978). In Russian. English translation in Soviet Astron., Vol. 22, No. 1.

The paper deals with some methods of eliminating phase errors from the results of planetary position observations.

031.211 **Computer image processing – the Viking experience.** W. B. Green.
IEEE Trans. Consum. Electron., Vol. CE-23, 281 - 299 (1977). Abstr. in Phys. Abstr., Vol. 80, Abstr. 91403 (1977).

031.212 **Aperture averaging effects in stellar scintillation.**
R. S. Iyer, J. L. Bufton.
Opt. Commun., Vol. 22, 377 - 381 (1977). – Abstr. in Phys. Abstr., Vol. 80, Abstr. 91519 (1977).

031.213 **The reduction of panoramic photometry. 1. Two search algorithms.** B. Newell, E. J. O'Neil, Jr.
Publ. Astron. Soc. Pacific, Vol. 89, 925 - 928 (1977/78).

The authors describe two algorithms that can be used to search two-dimensional data arrays for image peaks. These algorithms are simple, fast in operation, and require only three lines of the data array in core at any one time. The maximum search provides reliable detection of all image peaks in an array, the DOG search is designed to isolate the individual image peaks in composite images.

031.214 **Correlation analysis of deep galaxy samples – I.**

Techniques with applications to a two-colour sample. S. Phillipps, R. Fong, R. S. Ellis, S. M. Fall, H. T. MacGillivray.
Mon. Not. R. Astron. Soc., Vol. 182, 673 - 685 (1978).
The authors develop techniques for analysing correlations in the distribution of galaxies on very deep photographic plates scanned by automated measuring machines. They include a relativistic generalization of Limber's equation, the basic equation relating spatial and angular pair—correlation functions. The application of the method to the data is internally consistent but gives rise to a discrepancy when comparison is made with the shallow samples. The authors consider that the discrepancy is most likely due to their sample being unrepresentative, but the application of these techniques to a larger sample may allow an empirical measurement of cluster evolution.

031.215   Fourier analysis of Zeeman splitting for integrated stellar profiles.   J. C. Evans.
Bull. American Astron. Soc., Vol. 9, 572 - 573 (1978).
Abstract.

031.216   Sources of errors in trigonometric parallaxes: a Monte Carlo approach.   P. A. Ianna.
Bull. American Astron. Soc., Vol. 9, 599 (1978). − Abstract.

031.217   Fourier techniques for binary star orbit computation and the distance to the Hyades.   D. Monet.
Bull. American Astron. Soc., Vol. 9, 600 (1978). − Abstract.

031.218   A method for detection of faint stellar and planetary companions.   D. A. Drake, M. Kobrick.
Bull. American Astron. Soc., Vol. 9, 625 (1978). − Abstract.

031.219   New techniques in high resolution astrophysics using amplitude interferometry.
D. G. Currie, R. Braunstein, K. M. Liewer.
Bull. American Astron. Soc., Vol. 9, 625 (1978). − Abstract.

031.220   A feasibility study of calibrating stellar photographic equivalent widths against solar photoelectric equivalent widths.   F. N. Edmonds, Jr.
Bull. American Astron. Soc., Vol. 9, 635 (1978). − Abstract.

031.221   Period determination techniques for variable stars.   R. F. Stellingwerf.
Bull. American Astron. Soc., Vol. 9, 643 (1978). − Abstract.

031.222   Time-asymmetry in astrophysical time series.
J. I. Katz, M. C. Weisskopf.
Bull. American Astron. Soc., Vol. 9, 645 (1978). − Abstract.

031.223   A new method of reducing PZT observations.
J. Vondrák.
Bull. Astron. Inst. Czechoslovakia, Vol. 29, 97 - 103 (1978).
A method of reducing PZT plates, which has been used at the Ondřejov Observatory since July 1977, is described. The new method solves, by the method of least squares, simultaneously seven unknown quantities (coordinates of the zenith on a plate, effective focal length, angle of rotation of the rotary, orientation angle of the plate in a measuring machine, clock correction and instantaneous latitude). Each of the four exposed images of a star yields two observation equations.

031.224   Image reconstruction from incomplete and noisy data.   S. F. Gull, G. J. Daniell.
Nature, Vol. 272, 686 - 690 (1978).
Results are presented of a powerful technique for image reconstruction by a maximum entropy method, which is sufficiently fast to be useful for large and complicated images. Although the examples are taken from the fields of radio and X-ray astronomy, the technique is immediately applicabe in spectroscopy, electron microscopy, X-ray crystallography,

geophysics and virtually any type of optical image processing. Applied to radioastronomical data, the algorithm reveals details not seen by conventional analysis, but which are known to exist.

031.225   Incoherent scatter radar observations.
T. Hagfors.
Radar probing of the auroral plasma, (see 012.011), p. 75 - 101 (1977).
The present lecture discusses various observation schemes to obtain results suitable for the estimation of physical parameters of the ionosphere by incoherent scatter observations.

031.226   A method of obtaining the energy distribution of auroral electrons from incoherent scatter radar measurements.   R. R. Vondrak, M. J. Baron.
Radar probing of the auroral plasma, (see 012.011), p. 315 - 330 (1977).

031.227   Infrared measurements at helium-temperature levels.   K.-P. Bartholomä, F. D. Heidt, H. Schwille.
Umschau, 78. Jahrg., 246 - 247 (1978). In German.
Besides the X-rays, ultraviolet, optical and radiowaves regions of the electromagnetic spectrum, the infrared radiation draws presently more and more interest. It is estimated that in the spectral region 0.7 μm−1000 μm more radiation is emitted than in all others combined. Infrared measurements are only effective from the top of the earth atmosphere due to the atmospheric attenuation effects. Another technological problem remains with the infrared detectors which need to be cooled down to the temperature of liquid helium ($\leqslant$ 4.2 K).

031.228   Optical reconstruction of aperture synthesis data.
T. W. Cole.
Proc. Astron. Soc. Australia, Vol. 3, 105 - 107 (1977).

031.229   Esperienze di fotometrica fotoelettrica: l'elaborazione dei dati.   L. Baldinelli, S. Ghedini.
G. Astron., Vol. 3, 217 - 234 (1977).
Lo scopo del presente lavoro è quello di fornire un quadro generale dei metodi di riduzione dei dati osservativi. Ci si riferisce cioè a quella parte di lavoro della fotometria fotoelettrica che viene fatta a tavolino.

031.230   Photographs of deep-sky objects.
H. Vehrenberg.
Sky Telesc., Vol. 55, 295 - 298 (1978).

031.231   Interactive process of computation of equivalent width at the objective-prism astrograph.
M. Azzopardi, A. Bijaoui, J. Marchal, C. Ounnas.
Astron. Astrophys., Vol. 65, 251 - 258 (1978).
In order to determine the absolute magnitude of the Small Magellanic Cloud star members, a method of computation of the Hγ equivalent width on objective-prism plates has been established at the Centre de Dépouillement des Clichés Astronomiques. Problems inherent to this kind of spectrophotometry are pointed out and a description of the instrumentation and the analysis process are presented. A list of Hγ equivalent widths for 50 standard stars of both hemispheres is given.

031.232   A physical parameter method for the design of broad-band X-ray imaging systems to do coronal plasma diagnostics.   S. Kahler, A. S. Krieger.
Sol. Phys., Vol. 56, 351 - 357 (1978).
The technique commonly used for the analysis of data from broad-band X-ray imaging systems for plasma diagnostics is the filter ratio method. This requires the use of two or more broad-band filters to derive temperatures and line-of-sight emission integrals or emission measure distributions as a function of temperature. Here the authors propose an alter-

native analytical approach in which the temperature response of the imaging system is matched to some physical parameter which the experimenter wishes to investigate. They have calculated the temperature response of a system designed to measure the total radiated power along the line of sight of any coronal structure. Other examples are discussed.

031.233 Use of a correlation model for solution of statistical problems in interpretation of observations in radio astronomy.   M. I. Ryabov, N. A. Smirnova.
Odessk. univ. Odessa, 1977. 13 pp. In Russian. – Abstr. in Ref. zh., 51. Astron., 2.51.260 (1978).

031.234 Accuracy and reproducibility of a mass-spectrometer analysis of geological samples and of lunar soil.
G. I. Ramendik.
Zh. analit. khim., Vol. 32, 1990 - 1998 (1977). In Russian. Abstr. in Ref. zh., 62. Issled. kosm. prostranstva, 2.62.103 (1978).

031.235 Die visuelle Beobachtung veränderlicher Sterne.
K.-P. Timm.
Orion, 36. Jahrg., 52 - 58 (1978).

031.236 Sonnenbeobachtung für den Amateur.
W. Lüthi.
Orion, 36. Jahrg., 76 - 79 (1978).

031.237 Coherent signal processing techniques for radio pulses.
J. W. Erkes, I. R. Linscott, J. Irwin, N. R. Powell.
Bull. American Astron. Soc., Vol. 9, 561 (1978). – Abstract.

031.238 Possible use of electronographically recorded low-dispersion slitless spectra in optical quasar surveys.
B. Strömgren.
Phys. Scr., Vol. 17, (see 012.013), 339 - 340 (1978).
    Uses of low-dispersion slitless spectra, recorded electronographically, in spectral classification are discussed, and a method of producing such spectra with a Fehrenbach-type prism placed about 20 cm in front of the focal plane of an f/8.6 reflector is described. A program of stellar classification of faint stars in specially selected areas of high galactic latitude is briefly referred to.

031.239 Computer simulation of low-dispersion spectra obtained using a Fehrenbach-prism in the converging beam of a 1.5 m F/8.6 telescope.   P. Kjaergaard, J. Ravn.
Phys. Scr., Vol. 17, (see 012.013), 341 - 346 (1978).
    A computer simulation of low-dispersion spectra obtained using Fehrenbach-prisms in the converging beam of a 1.5 m F/8.6 telescope has shown that most quasars of redshift smaller than $z = 2.0$ can be distinguished from white dwarfs—only white dwarfs of type F may look like quasars with abnormally weak Mg II emission and with redshift between $z = 0.35$ and $z = 0.95$. With the 1.5 m telescope this technique can with certainty be used down to $B = 18$ magnitude.

031.240 Terrestrial measurement of the performance of high-rejection optical baffling systems.
J. C. Kemp, C. L. Wyatt.
Opt. Eng., Vol. 16, 412 - 416 (1977). – Abstr. in Phys. Abstr., Vol. 81, Abstr. 7802 (1978).

031.241 Interactive image analysis for astronomers.
D. C. Wells.
Computer, Vol. 10, No. 8, p. 30 - 34 (1977). – Abstr. in Phys. Abstr., Vol. 81, Abstr. 7886 (1978).

031.242 Investigation of the method of determination of the azimuth from circumpolar stars with the help of the reconstructed U-5″ astronomical transit with a special circular

net.   B. A. Klassov.
Tr. Novosib. inst. inzh. geod., aehrofotosemki i kartogr., Vol. 37, 165 - 175 (1975). In Russian. – Abstr. in Ref. zh., 52. Geod. Aehrosemka, 3.52.99 (1978).

031.243 Methods of analytical reduction of various aerospace images.   B. A. Novakovskij.
Izv. vyssh. uchebn. zaved. Geod. i aehrofotosemka, 1977, No. 4, p. 87 - 90. In Russian. – Abstr. in Ref. zh., 52. Geod. Aehrosemka, 3.52.211 (1978).

031.244 Photoelectric method of recording star transit moments.   N. N. Pavlov.
Astrometriya i Astrofizika, Kiev, Vyp. (No.) 34, 60 - 63 (1978). In Russian.
    The problems are discussed of increasing sensitivity and wider application of the photoelectric method for recording transits.

031.245 Recording of mean moments of star transits and their errors.
K. Šteins, M. Ogriņš, P. Rozenbergs.
Astrometriya i Astrofizika, Kiev, Vyp. (No.) 34, p. 71 (1978). In Russian.

031.246 Some comments on the method of objectivization of observations with a circumzenithal.
H. Potthoff, K. G. Steinert.
Astrometriya i Astrofizika, Kiev, Vyp. (No.) 34, p. 71 - 73 (1978). In Russian.
    A photoelectric device based on the photon count principle is proposed for objectivization of observations obtained with a circumzenithal. Estimates of some parameters of such a device for observations of stars up to $7^{\text{m}}5$ are given.

031.247 The effect of microclimate on observations of meridian transits of stars and its consideration in photoelectric observations carried out in Dresden.
J. Diettrich.
Astrometriya i Astrofizika, Kiev, Vyp. (No.) 34, p. 73 - 75 (1978). In Russian.
    Brief results are given of studying temperature errors of observations carried out with the photoelectric transit instrument in Dresden. It is concluded that reducing microclimatic effects of their registration is necessary.

031.248 Revealing and eliminating systematic errors in observations with a transit instrument.
L. A. Solov'eva.
Astrometriya i Astrofizika, Kiev, Vyp. (No.) 34, p. 79 - 81 (1978). In Russian.

031.249 Seeing compensation in photoelectric area scanning.
O. G. Franz.
Rev. Mexicana Astron. Astrofis., Vol. 3, (see 012.023), 33 (1977). – Abstract.

031.250 Techniques of perturbation analysis.
P. van de Kamp.
Rev. Mexicana Astron. Astrofis., Vol. 3, (see 012.023), 39 - 41 (1977).
    A historical summary of stellar-perturbation analyses is presented, followed by a discussion of the attainable accuracy and the sources of observational errors. A provisional value for the mass-density of astrometric companions is found to be 0.063 ± 0.018 (p.e.) solar masses per cubic parsec.

031.251 The detection of stellar systems by lunar occultations.   N. M. White.
Rev. Mexicana Astron. Astrofis., Vol. 3, (see 012.023), 43 - 46 (1977).
    A few examples of the resolution of very close multiple

stars by the method of lunar occultation are described. In this way both the power and some of the limitations of this technique may be demonstrated.

031.252 **Digital image centering. II.**
          L. H. Auer, W. F. van Altena.
Astron. J., Vol. 83, 531 - 537 (1978).
   A variety of digital image centering algorithms have been applied to PDS microdensitometer raster scans of a Yerkes Observatory 40-in. refractor parallax series consisting of 22 stars on 26 plates in an attempt to identify the most precise method of locating stellar image positions.

031.253 **Deadtime.**   J. Africano, R. Quigley.
          J. American Assoc. Variable Star Obs., Vol. 6, 53 - 55 (1977/78).
   The concept of deadtime in pulse-counting photometry is discussed, and a method for measuring it is given. It is shown that a large error may result for bright stars if the deadtime correction is not applied.

031.254 **Altitude excursion of the balloon ceiling and occul-
          tation techniques.**   J. P. Naudet.
Ann. Géophys., Vol. 33, 367 - 372 (1977). In French. – Abstr. in Phys. Abstr., Vol. 81, Abstr. 11947 (1978).

031.255 **Solar energy – its measurement.**
          S. Wieder, E. Jaoudi.
American J. Phys., Vol. 45, 981 - 984 (1977). – Abstr. in Phys. Abstr., Vol. 81, Abstr. 12470 (1978).

031.256 **Calibration of two plastic detectors and application
          on study of heavy cosmic rays.**
J. Tripier, M. Debeauvais.
Nucl. Instrum. Methods, Vol. 147, No. 1, p. 221 - 226 (1977). Abstr. in Phys. Abstr., Vol. 81, Abstr. 17500 (1978).

031.257 **Speckle interferometry with a 1 m-telescope.**
          G. P. Weigelt.
Astron. Astrophys., Vol. 67, L11 - L12 (1978).
   A speckle interferometer was constructed and used for binary star measurements. The speckle interferometry measurements were performed with a telescope of relatively small aperture (1 m - telescope of Hoher List Observatory). 30 speckle interferograms per binary star were found to be sufficient to obtain good autocorrelations.

031.258 **Optimizing photographic detection at large tele-
          scopes.**   A. G. Millikan.
J. Photogr. Sci., Vol. 25, 178 - 182 (1977). – Abstr. in Phys. Abstr., Vol. 81, Abstr. 24468 (1978).

031.259 **Two-dimensional maximum entropy reconstruction
          of radio brightness.**   S. J. Wernecke.
Radio Sci., Vol. 12, 831 - 844 (1977). – Abstr. in Phys. Abstr., Vol. 81, Abstr. 28534 (1978).

031.260 **The determination of personal equation.**
          E. G. Moore.
J. British Astron. Assoc., Vol. 88, 377 - 379 (1978).

031.261 **Future photoelectric projects at the Beverly-Begg
          Observatory.**   W. H. Allen.
South. Stars, Vol. 27, 93 - 97 (1978).

031.262 **On numerical methods of correction of observed
          spectral line profiles for the influence of the
          instrument.**   I. Sattarov.
Astron. Zh. Akad. Nauk SSSR, Vol. 55, 649 - 659 (1978). In Russian. English translation in Soviet Astron., Vol. 22, No. 3.
   Numerical methods of correction of observed spectral

line profiles for the influence of the instrumental profile of Burger-van Cittert and of the Fourier transform with regularization are studied. It is shown that in the case of spectral lines with a half-width twice and more times that of the instrumental profile both methods give close results. In the case of lines with a narrower profile the Fourier transform with regularization gives a better solution than the method of Burger-van Cittert.

031.263 **A search for stellar oscillations.**   W. A. Traub,
          J. T. Mariska, N. P. Carleton.
Astrophys. J., Vol. 223, 583 - 588 (1978).
   The authors have used the PEPSIOS spectrometer to search for stellar photospheric oscillations with periods in the range from about 10 to 5000 s, and with velocity sensitivities down to about 1 m s^{-1} (rms) at the 1 σ level. In the nine stars observed, the authors find no evidence for oscillations at the 3 σ level. However, using the same technique on the Sun, they were able to detect the well-known 300 s oscillation with an rms amplitude of 1.6 ± 0.5 m s^{-1} for the integrated disk.

031.264 **Statistically rigorous reduction of optical lunar
          occultation measurements.**
G. Knoechel, K. von der Heide.
Astron. Astrophys., Vol. 67, 209 - 220 (1978).
   If an occultation of a star by the moon is observed through the Earth's atmosphere, the measured occultation pattern is the Fresnel curve which originates at the lunar limb, disturbed by atmospheric scintillation. Two statistically rigorous methods for the reduction of such observations are developed one of which makes use of the fact that the correlation of subsequent data points, introduced into the measurements by the scintillation, can be taken into account by modifying the covariance matrix of the measurements, while in the other method the scintillation is represented in form of additional polynomials weighted in respect to the scintillation power spectrum. The theory is applied to a large number of simulated lunar occultation measurements.

031.265 **Holography at the telescope – an interferometric
          method for recording stellar spectra in thick photo-
          graphic emulsions.**   L. Lindegren, D. Dravins.
Astron. Astrophys., Vol. 67, 241 - 255 (1978).
   Low-resolution spectra ($\lambda/\Delta\lambda \lesssim 100$) are recorded without any dispersive optics by direct focal-plane Lippmann photography using thick holographic emulsions. These record the Fourier transforms of the spectra, enabling spectrum reconstruction by reflected light and analysis with a microspectrophotometer. Spectral resolution is set by emulsion thickness and is independent of seeing and telescope guiding.

031.266 **An echellogram of Saturn.**   D. W. Latham.
          Sky Telesc., Vol. 56, 8 - 9 (1978).

031.267 **Determination of the equatorial coordinates of
          objects by means of an astrotelevision device.**
M. Bagkhanov, Kh. Gul'medov, M. A. Polyakov, Yu. A. Seliverstov.
Fiz. protsessy verkhn. atmos. Ashkhabad, Ylym, 1977, p. 75 - 80. In Russian. – Abstr. in Ref. zh., 51. Astron., 5.51.117 (1978).

031.268 **On a possibility of restoring the image of a faint
          object distorted by the terrestrial atmosphere.**
S. V. Kornienko.
DAN USSR, 1977, Ser. A, No. 10, p. 931 - 933. In Russian. Abstr. in Ref. zh., 51. Astron., 5.51.188 (1978).

031.269 **Bistatic radar measurements of the moon with
          application of a modulated signal.**   A. L. Zajtsev,
V. I. Kaevitser, A. I. Kucheryavenkov, S. S. Matyugov, A. T. Pavel'ev, G. M. Petrov, O. I. Yakovlev.

Radiotekh. i ehlektron., Vol. 23, 2097 - 2104 (1977). In Russian. – Abstr. in Ref. zh., 52. Geod. Aehrosemka, 5.52.76 (1978).

031.270 Measurement of radial velocities from objective prism spectra. I. Use of the A-band as a reference line. H. Maehara, Y. Yamashita.
Tokyo Astron. Obs. Report No. 69, Vol. 18, 153 - 161 (1978). In Japanese.

031.271 Filtering characteristics of smoothing method by parabola fitting. S. Iijima, S. Okazaki.
Tokyo Astron. Obs. Report No. 69, Vol. 18, 162 - 168 (1978). In Japanese.

031.272 Analysis and interpretation of soft X-ray photographs of coronal active regions taken with Fresnel zone plates. I: Image analysis. G. Krämer, H. J. Einighammer, G. Elwert, H. Bräuninger, H. H. Fink, J. Trümper.
Sol. Phys., Vol. 57, 345 - 367 (1978).
Soft X-ray photographs of the Sun taken at O VII 21.6 Å and in a spectral band ranging from 13.2 to 22.1Å have been analysed in order to establish spatially resolved maps of temperature and emission measure for several active regions in the corona. The authors first deal with those aspects of the instrumentation which are important for setting up a suitable image analysis procedure. They discuss the characteristics of the wavelength dependent image formation by zone plates combined with absorption filters. Results of the calibration of the X-ray film are given. Then they describe a specific iterative data reduction procedure and finally present the resulting maps of temperature and emission measure for a selected active region.

031.273 A method of approximation of laser satellite-tracking data. G. S. Kurbasova.
Nauchn. Inf., Vyp. (No.) 38, p. 26 - 29 (1976). In Russian.
A method of approximation of laser satellite-tracking data by means of Chebyshev polynomial interpolation is developed.

031.274 Position observations of the moon at the Engelhardt Astronomical Observatory. S. G. Valeev.
Tr. Astron. Ehngel'gardt. Obs., Kazan', No. 41 - 42, p. 110 - 112 (1976). In Russian.

031.275 Reduction of differential observations of declinations of stars with the NAIRI computer.
L. R. Tokhtas'eva.
Izv. Astron.Ehngel'gardt. Obs., Kazan', No. 41 - 42, p. 144 - 147 (1976). In Russian.

031.276 On the reduction of heterogeneous observations of satellites in a local network. G. Karsky.
New methods of space geodesy, (see 012.029), p. 294 - 304 (1976). In Russian.
Observational equations for a parametrical adjustment of a local satellite network in general form and applied to laser, Doppler and photographic observations are derived. Some remarks as to the procedure of the adjustment and to the transformation of the results to the geocentric datum are given.

031.277 Arrangement of technique of the MANIA experiment. G. S. Tsarevskij, V. F. Shvartsman.
Soobshch. Spets. Astrofiz. Obs., Zelenchukskaya, Vyp. (No.) 19, p. 39 - 51 (1977). In Russian.
Difficulties are examined which arise when attempting to analyze brightness and polarization fluctuations of very faint celestial objects within the time range $10^{-7} - 10^{-2}$ s. The methods of overcoming these difficulties in the MANIA experiment are described.

031.278 The MANIA [Multichannel Analysis of Nanosecond Intensity Alterations] experiment. Astrophysical problems, mathematical methods, instrumentation complex, results of the first observations. V. F. Shvartsman.
Soobshch. Spets. Astrofiz. Obs., Zelenchukskaya, Vyp. (No.) 19, p. 5 - 38 (1977). In Russian.

031.279 Comparative imaging radar planetology. C. Elachi, R. S. Saunders.
Reports of planetary geology program, 1977–1978, (see 003.014), p. 122 - 123 (1978).

031.280 Spectral or time series analysis of surface roughness: correlation with radar backscatter cross-section measurements. G. G. Schaber.
Reports of planetary geology program, 1977–1978, (see 003.014), p. 124 - 126 (1978).

031.281 Radar backscatter from Martian-type geologic surfaces. M. Kobrick.
Reports of planetary geology program, 1977–1978, (see 003.014), p. 127 (1978).

031.282 Radar backscatter from sand dunes. W. E. Brown, Jr., R. S. Saunders.
Reports of planetary geology program, 1977–1978, (see 003.014), p. 137 - 139 (1978).

031.283 Photometric analyses of spacecraft planetary images. B. Hapke.
Reports of planetary geology program, 1977–1978, (see 003.014), p. 309 - 311 (1978).

031.284 Use of the edge enhancer in geologic analysis of planetary surfaces. J. R. Underwood, Jr.
Reports of planetary geology program, 1977–1978, (see 003.014), p. 312 - 313 (1978).

031.285 Atmospheric extinction and photometric reductions near four microns. R. F. Wing, C. P. Rinsland.
Bull. American Astron. Soc., Vol. 10, 408 (1978). – Abstract.

031.286 A technique for improved spatial resolution using the MSFC magnetograph.
M. J. Hagyard, E. A. West.
Bull. American Astron. Soc., Vol. 10, 432 (1978). – Abstract.

031.287 The IUE (*International Ultraviolet Explorer*) spectral image system, IUESIPS. D. A. Klinglesmith.
Bull. American Astron. Soc., Vol. 10, 443 (1978). – Abstract.

031.288 Photometric performance of the IUE. R. C. Bohlin, B. D. Savage.
Bull. American Astron. Soc., Vol. 10, 443 (1978). – Abstract.

031.289 Speckle interferometry at Steward Observatory. E. Hubbard, P. A. Strittmatter, N. J. Woolf, K. Hege, S. P. Worden.
Bull. American Astron. Soc., Vol. 10, 459 (1978). – Abstract.

031.290 Speckle image reconstruction of outer planet satellites. P. Nisenson, R. V. Stachnik.
Bull. American Astron. Soc., Vol. 10, 459 (1978). – Abstract.

031.291 Statistically correlated satellite observations. C. F. Peters.
Bull. American Astron. Soc., Vol. 10, 481 (1978). – Abstract.

031.292 Astrometric reduction of satellite photographic observations. T. C. Duxbury, L. A. Morabito.

Bull. American Astron. Soc., Vol. 10, 481 (1978). – Abstract.

031.293 On the accuracy of the positions of celestial objects determined from a linear model. L. G. Taff.
Bull. American Astron. Soc., Vol. 10, 483 (1978). – Abstract.

031.294 Optimum interpolation of astronomical data sampled at unequal time intervals. D. D. Meisel.
Bull. American Astron. Soc., Vol. 10, 485 (1978). – Abstract.

031.295 Improvement of the methods of obtaining equidensites. S. B. Vladimirov.
Astron. Tsirk., No. 977, p. 5 - 8 (1977). In Russian.

031.296 A new method of pseudo color equidensitometry.
J. Opt., Vol. 9, 1 - 4 (1978). – Abstr. in Phys. Abstr., Vol. 81, Abstr. 32903 (1978).

031.297 On a new principle of measuring photographs of a graduated circle. B. A. Golovko.
Astron. Tsirk., No. 986, p. 3 - 6 (1978). In Russian.

031.298 Restoring with maximum entropy. III. Poisson sources and backgrounds.
B. R. Frieden, D. C. Wells.
J. Opt. Soc. America, Vol. 68, 93 - 103 (1978).
The maximum entropy restoring formalism has previously been derived under the assumptions of (1) zero background and (2) additive noise in the image. However, the noise in the signals from many modern image detectors is actually Poisson, i.e., dominated by single-photon statistics. Hence, the noise is no longer additive. Particularly in astronomy, it is often accurate to model the image as being composed of two fundamental Poisson features: (1) a component due to a smoothly varying background image, such as caused by interstellar dust, plus (2) a superimposed component due to an unknown array of point and line sources (stars, galactic arms, etc.). Given the estimated background, a maximum-likelihood restoring formula was derived for the foreground image. The authors applied this approach to some one-dimensional simulations and to some real astronomical imagery.

031.299 Line shifts due to blending. A. T. Young.
J. Opt. Soc. America, Vol. 68, 246 - 250 (1978).
The effective position of a strong line partially blended with a weaker one is found as a function of their separation for three different line profiles and two line-position criteria. Some practical examples illustrate the use of the results.

031.300 Speckle interferometry at finite spectral bandwidths and exposure times.
D. P. Karo, A. M. Schneiderman.
J. Opt. Soc. America, Vol. 68, 480 - 485 (1978).
The effects of wavelength, spectral bandpass, and exposure time on the speckle lens-atmosphere modulation transfer function have been measured using bright stellar sources.

031.301 High-dispersion astronomical spectroscopy with holographic and ruled diffraction gratings.
D. Dravins.
Appl. Opt., Vol. 17, 404 - 414 (1978).
Holographic gratings cause much less stray light and spectral degradation than classically ruled gratings. Their high groove densities enable high dispersion in first diffraction order and a high spectrograph throughput comparable to the best echelles. Their lower reflective efficiency is compensated by the avoidance of cross dispersers, enabling efficient high-fidelity spectroscopy with single-pass spectrographs. Instrumental profiles of the ESO coudé spectrograph with large holographic and ruled gratings have been studied in detail, and their effects on astronomical spectra are discussed and

compared to those of other instruments.

031.302 Fourier transform spectroscopy in the visible and ultraviolet range. P. Luc, S. Gerstenkorn.
Appl. Opt., Vol. 17, 1327 - 1331 (1978).

031.303 Airborne infrared astronomical observations by Fourier transform spectroscopy. H. P. Larson.
Appl. Opt., Vol. 17, 1352 - 1359 (1978).
Infrared spectroscopic observations from NASA-operated aircraft constitute a rapidly maturing application of FTS methods initially developed for ground-based telescopes. Coupled to airborne telescopes up to 36 in. in diameter, these experiments are now producing new astronomical results as exciting and unexpected as those derived from Connes's first high resolution planetary observations at mountain-top observatories. This review examines the special problems of the ir spectral region that led to aircraft observatories and includes a brief survey of the facilities themselves and their modes of operation.

031.304 Photographing the Sun in H-alpha. E. Hirsch.
Astronomy, Vol. 6, No. 1, p. 36 - 39 (1978).
Abstr. in Phys. Abstr., Vol. 81, Abstr. 40833 (1978).

031.305 Defocusing effects in astronomical speckle interferometry. F. Roddier, G. Ricort, C. Roddier.
Opt. Commun., Vol. 24, 281 - 284 (1978). – Abstr. in Phys. Abstr., Vol. 81, Abstr. 40834 (1978).

031.306 Photograph the Moon! R. Berry.
Astronomy, Vol. 6, No. 2, p. 34 - 39 (1978).
Abstr. in Phys. Abstr., Vol. 81, Abstr. 45238 (1978).

031.307 Infrared upconversion for astronomy.
R. W. Boyd.
Opt. Eng., Vol. 16, 563 - 568 (1977). – Abstr. in Phys. Abstr., Vol. 81, Abstr. 45240 (1978).

031.308 Real-time atmospheric compensation of stellar images by active wavefront control.
A. J. MacGovern.
OSA/IEEE Conference on Laser and Electrooptical Systems, San Diego, CA, USA, 7 - 9 February 1978, IEEE, New York, (1978). p. 4. – Abstr. in Phys. Abstr., Vol. 81, Abstr. 45242 (1978).

031.309 On a technique of visualization of sunspot images observed in the infrared range at $\lambda > 1 \, \mu m$.
N. F. Kuprevich.
Problems of observational and theoretical astronomy, Moscow - Leningrad, (see 003.012), p. 112 - 126. In Russian.
Abstr. in Ref. zh., 51. Astron., 6.51.219 (1978).

031.310 Il "Quantimet" come analizzatore di immagini astronomiche. V. Caloi, V. Castellani.
Mem. Soc. Astron. Italiana, Vol. 47, 85 - 93 (1976).
Viene illustrato il funzionamento dell'analizzatore di immagini "Quantimet" in funzione presso l'Università di Roma. Se ne esaminano in particolare le risposte nello studio delle strutture degli ammassi globulari. Viene in particolare esaminato il caso dell'ammasso ellitico M19.

031.311 Numerical mapping technique applied to the photographic photometry of bright galaxies.
R. Barbon, L. Benacchio, M. Capaccioli.
Mem. Soc. Astron. Italiana, Vol. 47, 263 - 285 (1976).
The numerical mapping technique applied to the reduction of astronomical plates for purpose of surface photometry is thoroughly presented and discussed. The application of the numerical mapping technique to the specific case of galaxy photometry, as presently carried out at Padova Observatory,

is described in the second part.

031.312 Ultraviolet observational techniques – stellar ultraviolet spectrophotometry.   A. D. Code.
Mem. Soc. Astron. Italiana, Vol. 47, (see 012.036), 629 (1976). – Abstract.

031.313 TD1 spectrophotometric observations in the UV.   D. Malaise.
Mem. Soc. Astron. Italiana, Vol. 47, (see 012.036), 630 (1976). Abstract.

031.314 Image restoration by a regularization method: application to seeing integrated profile.
O. Bendinelli, S. Lorenzutta.
Mem. Soc. Astron. Italiana, Vol. 48, 55 - 66 (1977).
   In this paper a particular regularization procedure is developed which yields the solution of Volterra integral equations of the first kind. This technique is used to determine the intensity distribution in Polaris seeing disk starting from its integrated profile.

031.315 Astrometric expectations.   P. Brosche.
   Proceedings of the European Workshop on space oceanography, navigation and geodynamics ('SONG') held at Schloss Elmau, Germany, 16 - 21 January 1978. ESA SP-137, p. 341 - 342 (1978).
   Astrometric aspects of VLBI via satellite are briefly reviewed and some specific needs of the procedures involved are emphasized.

031.316 Analysis of photographic X-ray images.   A. S. Krieger.
X-ray imaging, (see 012.041), p. 24 - 33 (1977). – Abstr. in Phys. Abstr., Vol. 81, Abstr. 49341 (1978).

031.317 Imaging of solar active regions with Fresnel zone plates.   G. Kramer, G. Elwert, H. J. Einighammer, H. Bräuninger, H. H. Fink.
X-ray imaging, (see 012.041), p. 79 - 84 (1977). – Abstr. in Phys. Abstr., Vol. 81, Abstr. 49342 (1978).

031.318 Infrared detection techniques for low-background satellite experiments.   K. Shivanandan.
Far infrared technology and applications, (see 012.040), p. 37 - 39 (1977). – Abstr. in Phys. Abstr., Vol. 81, Abstr. 49343 (1978).

031.319 Infrared and millimeter wave techniques for the Cosmic Background Explorer Satellite.
J. C. Mather.
Far infrared technology and applications, (see 012.040), p. 44 - 50 (1977). – Abstr. in Phys. Abstr., Vol. 81, Abstr. 49344 (1978).

   Viking lander imaging investigation – Picture catalog of primary mission. Experiment data record.
See Abstr. 002.026.

   Decametric survey of discrete sources in the Northern sky. III. Low-frequency absolute flux scale of discrete radio sources.   See Abstr. 002.041.

   RAS specialist discussion on Astronomy with the 3.8-metre UK infrared flux collector, 1977 December 9.
See Abstr. 010.023.

   Photoelectric photometry at Wise Observatory.
See Abstr. 013.006.

   A set of computer programmes for the adjustment of overlapping plates.   See Abstr. 021.008.

   Sharpening stellar images.   See Abstr. 031.007.

   A statistical approach for the determination of relative Zeeman and Doppler shifts in spectrograms.
See Abstr. 031.401

   Initial considerations of minor planet astrometry with the Space Telescope.   See Abstr. 032.544.

   Very-long-baseline radio interferometry.
See Abstr. 033.001.

   Reduction of the baseline ripple on spectra recorded with the Parkes radio telescope.   See Abstr. 033.009.

   Decametric survey of discrete sources in the Northern sky. I. The UTR-2 radio telescope. Experimental techniques and data processing.   See Abstr. 033.021.

   Chromatism in radio telescopes due to blocking and feed scattering.   See Abstr. 033.033.

   Geodetic and astrometric analysis of fringe rate residuals from the Algonquin-Chilbolton 2.8 cm long-baseline interferometer.   See Abstr. 033.039.

   Polarization effects in grating scanners.
See Abstr. 034.007.

   Statistical properties of atmospheric emission in the infrared: I. Instrumentation and observations.
See Abstr. 034.016.

   Meßmethoden in der astronomischen Forschung.
See Abstr. 034.026.

   Self-scanned photodiode array: high performance operation in high dispersion astronomical spectrophotometry.
See Abstr. 034.059.

   Radiointerferometrie als Wegbereiter eines radioastronomischen Bezugssystems.   See Abstr. 041.013.

   Astrometric techniques for the observation of planetary satellites.   See Abstr. 041.020.

   Astrometric accuracy with large reflectors.
See Abstr. 041.033.

   On application of the lunar laser ranging technique for determinations of the earth rotation in view of accuracy of computed topocentric distances of lunar retroreflectors.
See Abstr. 044.003.

   On methods of observations applied in investigation of the earth's rotation.   See Abstr. 044.032.

   Observations of artificial satellites in connection with selected reference stars.   See Abstr. 055.003.

   Fundamental limits to the measurements of the gravitational constant.   See Abstr. 066.188.

   Neutrino bursts and gravitational waves experiments.
See Abstr. 066.244.

   A method for determining laboratory/solar wavelength shifts: application to the C_2 molecule.
See Abstr. 071.017.

   Study of power spectra of quasi-periodic fluctuations of solar radio emission at 3 cm wavelength with periods

shorter than 100 sec.   See Abstr. 077.043.

Apparatus functions of atmospheric oscillations from the solar limb for nonclassical assumptions. See Abstr. 080.037.

Large-scale tropospheric irregularities and their effect on radio astronomical seeing.   See Abstr. 082.001.

Statistic properties of atmospheric emission in the infrared: II. Sky noise correlation. See Abstr. 082.052.

Effects of atmospheric turbulence on the central irradiance in the Fraunhofer diffraction pattern by slit apertures with uni-dimensional quadratic and linear apodisation filters: a generalised expression.   See Abstr. 082.080.

Measurements of atmospheric isoplanatism using speckle interferometry.   See Abstr. 082.116.

Speckle interferometry measurements of atmospheric nonisoplanicity using double stars. See Abstr. 082.117.

Measurement of r_0 with speckle interferometry. See Abstr. 082.118.

Investigation of the comparative effectiveness of the regression analysis and of modifications of gradient methods in connection with the problem of contour finding when reducing photographs of planetary surfaces by machine. See Abstr. 091.044.

Definition and detailed characterization of lunar surface units using remote observations. See Abstr. 094.101.

On using the photometric method for study of the Martian topography.   See Abstr. 097.059.

How to compare the surface of Io to laboratory samples.   See Abstr. 099.522.

Lunar occultation of Saturn. IV. Astrometric results from observations of the satellites.

See Abstr. 100.503.

On the background fog correction in measurements with an iris photometer.   See Abstr. 113.020.

Derivation of brightness distribution across a stellar disk from photoelectric observations of its lunar occultation. See Abstr. 113.062.

Stellar disk diameter measurements by amplitude interferometry.   See Abstr. 115.007.

Computer simulations of speckle interferometry of binary stars in the photon-counting mode. See Abstr. 117.019.

Observations of binary stars by speckle interferometry – I.   See Abstr. 118.023.

Photometry and variability of double stars. See Abstr. 118.028.

Speckle interferometry of spectroscopic binaries. See Abstr. 119.005.

On the application of the Lafler-Kinman method to determination of periods of variable stars by a computer. See Abstr. 122.014.

The large scale structure of extra-galactic radio sources (> 1 arcsecond).   See Abstr. 141.131.

The CTIO surveys for large redshift quasars. See Abstr. 141.141.

UBV photographic photometry in and around the galactic cluster NGC 2527.   See Abstr. 153.038.

The Simkin effect.   See Abstr. 158.043.

Future prospects of infrared observations of active nuclei.   See Abstr. 158.144.

Future prospects of radio observations of active nuclei.   See Abstr. 158.145.

Data Processing, Automation

031.401 A statistical approach for the determination of relative Zeeman and Doppler shifts in spectrograms.
W. W. Weiss, H. Jenkner, H. J. Wood.
Astron. Astrophys., Vol. 63, 247 - 257 (1978).
A statistical technique is described which allows one to find a correlation between two stellar spectrograms digitized with the PDS 1000/PDP 12 microdensitometer system of the Vienna Observatory. The correlation yields a best fit shift in microns of the features in one spectrum with respect to the same features in the other. In the case of classical double Zeeman spectrograms this shift is correlated to the longitudinal component of the magnetic field. Relative radial velocity changes can be determined in the reference frame of the iron comparison spectra. With a method presently under development which is an extension of the same technique, relative radial velocities for objective prism spectrograms taken with

Fehrenbach's double prism can be determined automatically.

031.402 A fast and accurate position measurement and finding chart production system for glass copies of the Palomar Sky Survey.
P. D. Hemenway, G. W. Torrence, C. S. Wolfe.
Bull. American Astron. Soc., Vol. 9, 598 (1978). – Abstract.

031.403 Digitization of astronomical catalogues.
T. A. Nagy, J. M. Mead.
Bull. American Astron. Soc., Vol. 9, 598 - 599 (1978). Abstract.

031.404 Image restoration techniques applied to QSO images.   P. A. Wehinger, S. P. Worden, S. Wyckoff.
Bull. American Astron. Soc., Vol. 9, 608 - 609 (1978). Abstract.

031.405 Computer correction of absolute telescope pointing

errors by use of spherical harmonic functions.
M. A. Powell.
Bull. American Astron. Soc., Vol. 9, 613 (1978). – Abstract.

031.406 **A general system for astronomical image analysis.**
A. Bijaoui, J. Marchal, C. Ounnas.
Astron. Astrophys., Vol. 65, 71 - 75 (1978).
The authors describe the possibilities of a general system for astronomical image analysis. The concepts of the software are exposed. Some astronomical applications are developed: stellar photometry, analysis of bi-dimensional astronomical objects, classical spectrography, bi-dimensional spectrography.

031.407 **Algorithm and program for analysis of information from the "Filin" device.**
L. S. Gurin, V. S. Mokrov, E. I. Moskalenko, K. A. Tsoj.
Inst. kosm. issled. AN SSSR. Pr-316. Moskva, 1977. 18 pp.
In Russian. – Abstr. in Ref. zh., 51. Astron., 2.51.259 (1978).

031.408 **Mathematical-statistical description of the iterative beam removing technique (method CLEAN).**
U. J. Schwarz.
Astron. Astrophys., Vol. 65, 345 - 356 (1978).
The CLEAN method (Högbom, 1974), which is a deconvolution method, is analysed mathematically for the 1-dimensional case. A criterion of convergence is given. In typical applications of the method the solution of the system of equations is not unique and the consequences for the CLEAN solution are discussed. By applying the analysis to maps which are obtained from Fourier transformed data the convergence criterion is shown to be equivalent to the condition that all weights used for the Fourier transform have to be non-negative. The extension of the present analysis to 2-dimensional distributions is briefly discussed.

031.409 **High-speed digital image processing.**
P. Mengers, K. A. Wickersheim.
Res./Dev., Vol. 28, No. 10, p. 42 - 44, 46, 48, 50, 52 (1977).
Abstr. in Phys. Abstr., Vol. 81, Abstr. 12680 (1978).

031.410 **The range measurement system of the laser ranging equipment.** Y. Kozai, A. Tsuchiya, K. Tomita.
Tokyo Astron. Obs. Report No. 69, Vol. 18, 271 - 283 (1978).
In Japanese.

031.411 **The photoelectric circuit of the laser ranging equipment.** A. Tsuchiya.
Tokyo Astron. Obs. Report No. 69, Vol. 18, 284 - 293 (1978).
In Japanese.

031.412 **Range gate of the laser ranging equipment.**
Y. Iizuka.
Tokyo Astron. Obs. Report No. 69, Vol. 18, 294 - 297 (1978).
In Japanese.

031.413 **Timing system of the laser ranging equipment.**
T. Kanda, M. Noguchi.
Tokyo Astron. Obs. Report No. 69, Vol. 18, 298 - 301 (1978).
In Japanese.

031.414 **Telescope drive of the laser ranging equipment.**
Y. Iizuka.
Tokyo Astron. Obs. Report No. 69, Vol. 18, 302 - 309 (1978).
In Japanese.

031.415 **The system controller of the laser ranging equipment.** A. Tsuchiya, N. Kobayashi, H. Sato,
Y. Torii, N. Oshima, M. Noguchi.
Tokyo Astron. Obs. Report No. 69, Vol. 18, 310 - 315 (1978).
In Japanese.

031.416 **Computer system of the laser ranging equipment.**
T. Hirayama, T. Kanda.
Tokyo Astron. Obs. Report No. 69, Vol. 18, 316 - 326 (1978).
In Japanese.

031.417 **On-line running at a telescope.** D. Lange.
Magnetic stars, (see 012.028), p. 31 (1978).

031.418 **Device of data output for printing and punching.**
O. E. Shornikov.
Izv. Astron. Ehngel'gardt. Obs., Kazan', No. 41 - 42, p. 231 - 233 (1976). In Russian.

031.419 **Express processing of spectrograms of extragalactic objects. I. Preliminary processing.**
A. L. Shcherbanovskij, V. L. Afanas'ev.
Soobshch. Spets. Astrofiz. Obs., Zelenchukskaya, Vyp. (No.) 16, p. 25 - 34 (1976). In Russian.
The necessity of rapid processing of spectral observational results for extragalactic objects is discussed. A computer program for rapid preliminary processing of spectrograms obtained with a digital output microphotometer is described. As an example, the spectrum of Markarian 639 processed using this program is presented.

031.420 **A digital system for registration of spectra.**
T. A. Somova, N. N. Somov, V. M. Gurin.
Soobshch. Spets. Astrofiz. Obs., Zelenchukskaya, Vyp. (No.) 16, p. 35 - 41 (1976). In Russian.

031.421 **Program-algorithmic complex of the MANIA experiment (First stage).** V. N. Mansurov, V. F. Shvartsman.
Soobshch. Spets. Astrofiz. Obs., Zelenchukskaya, Vyp. (No.) 19, p. 52 - 71 (1977). In Russian.

031.422 **Spurious results from Fourier analysis of data with closely spaced frequencies.**
G. L. Loumos, T. J. Deeming.
Bull. American Astron. Soc., Vol. 10, 417 (1978). – Abstract.

031.423 **An analog system for fast light curve registration.**
J. C. Bhattacharyya, A. Sundareswaran.
Kodaikanal Obs. Bull., Ser. A, Vol. 2, 69 - 74 (1977).

031.424 **Timing system for comet telescope.**
T. Yamaguchi.
Tokyo Astron. Obs. Report No. 69, Vol. 18, 327 - 333 (1978).
In Japanese.

The twenty-first Herstmonceux conference. Digital methods in astronomy. 1977 September 21 - 22.
See Abstr. 011.023.

Computer image processing – the Viking experience. See Abstr. 031.211.

Interactive process of computation of equivalent width at the objective-prism astrograph.
See Abstr. 031.231.

Automation of a transit instrument.
See Abstr. 032.043.

Decametric survey of discrete sources in the Northern sky. I. The UTR-2 radio telescope. Experimental techniques and data processing. See Abstr. 033.021.

Eventi di tipo III ad alta risoluzione temporale registrati all'Osservatorio Astronomico di Trieste; Descrizione della strumentazione per l'acquisizione e riduzione dei dati polarimetrici. See Abstr. 077.046.

032 Astronomical Instruments, Space Instrumentation

Astronomical Instruments

032.001 The four lives of a 60-inch reflector.
J. Ashbrook.
Sky Telesc., Vol. 55, 20 - 22, 42 (1978).

032.002 The precise adjustment of an equatorial mounting.
B. L. Souther.
Sky Telesc., Vol. 55, 78 - 83 (1978).

032.003 Forefathers of the MMT. L. Jacchia.
Sky Telesc., Vol. 55, 100 - 102 (1978).

032.004 Die transportable Zenitkamera – ein modernes
Instrument zur geographischen Ortsbestimmung.
G. Seeber.
Sterne Weltraum, Jahrg. 17, 45 - 50 (1978).

032.005 The reduction of scattered light in an external oc-
culting disk coronograph.
B. Fort, C. Morel, G. Spaak.
Astron. Astrophys., Vol. 63, 243 - 246 (1978).

The light intensity scattered by the external occulting
disk is the main limitation in the observation of the outer
corona with a space coronograph. In order to estimate this
effect, calculation of the diffracted flux on the entrance coro-
nographic lens is briefly presented for a circular occulting disk.
It is demonstrated experimentally that the apodisation by a
toothed wheel, as suggested by Purcell and Koomen is
effective. To achieve the required degree of optical polish,
electrolytic polishing has been used.

032.006 Digit-visual telescope for satellite observations.
A. Horváth, P. Horváth, I. Péter.
Veröff. Zentralinst. Phys. Erde, No. 52, Part 3, (see 012.001),
p. 953 - 958 (1977). In Russian.

A "digit-visual telescope" is developed in Hungary for
visual observation of satellites. The observational result is
directly recorded on punched tape and/or on digital magnetic
cassette which makes it convenient for further computer processing.

032.007 Just one more photographic zenith instrument for
latitude and time determination. Yu. A. Shokin.
Soobshch. Gos. Astron. Inst. Shternberga, No. 193, p. 3 - 13
(1977). In Russian.

032.008 Distortions of the image of a graduated circle due to
a plane-parallel plate of the lighting system.
V. G. Tauber.
Soobshch. Gos. Astron. Inst. Shternberga, No. 193, p. 22 - 25
(1977). In Russian.

032.009 La méthode de détermination de toutes les erreurs
de division du cercle méridien de l'Institut Sternberg
APM-4. A. G. Oborneva.
Soobshch. Gos. Astron. Inst. Shternberga, No. 200, p. 3 - 20
(1977). In Russian.

L'auteur expose la méthode de détermination de toutes
les erreurs de division des cercles divisés proposée par Lévy;
la technique de son application est démontrée par un exemple.
Il décrit une méthode élaborée d'aprés la méthode de Lévy
pour l'étude de toutes les divisions du cercle méridien de
l'Institut Sternberg APM-4 (division 5').

032.010 Some results of calculations of diameter corrections
with an electronic computer. V. A. Korobova.

Soobshch. Gos. Astron. Inst. Shternberga, No. 200, p. 21 - 22
(1977). In Russian.

The author gives the results of calculation of several different order normal systems of linear equations solved by
three different programmes with an electronic computer.

032.011 Determination of the inclination of reading microscopes. V. G. Tauber, N. N. Kabaeva.
Soobshch. Gos. Astron. Inst. Shternberga, No. 200, p. 33 - 36
(1977). In Russian.

A generalized formula for the determination of the inclination angle of microscopes is derived and an estimate of
its permissible values given.

032.012 Controlled frequency supply for a telescope drive.
J. Hers.
J. British Astron. Assoc., Vol. 88, 149 - 153 (1978).

032.013 A corrector lens system for Cassegrain telescope.
Y. Shibata
Bull. Res. Inst. Sci. Meas. Tôhoku Univ., Vol. 25, No. 2 - 3,
p. 73 - 81 (1976). In Japanese. – Abstr. in Phys. Abstr., Vol.
80, Abstr. 91427 (1977).

032.014 Télescope et radio-télescope. F. Rufener.
Conférences d'astronomie de l'Observatoire, (see
012.004), p. 157 - 185 (1977).

032.015 Il maggior telescopio giapponese realizzato da un
astrofilo.
T. Sato, translated by B. Ellena.
G. A. A. B., N. 47, p. 4 - 6 (1977).

032.016 The 74-inch telescope at Sutherland.
I. S. Glass.
Mon. Notes Astron. Soc. Southern Africa, Vol. 37, 4 - 8 (1978).

A brief history of the 74-inch telescope is given with a
description of its present arrangement at Sutherland. The
recent installation of axial encoders and a computer are described. Mention is made of other improvements already made
or in preparation.

032.017 Progetto e costruzione di un telescopio catadiottrico.
L. Ferioli.
Coelum, Vol. 45, 236 - 243 (1977).

032.018 On a photoelectric control as main component of
automatical guidance of telescopes.
R. Schielicke.
Astron. Nachr., Band 299, 47 - 53 (1978) = Mitt. Univ. Sternw.
Jena, No. 131. In German.

An autoguider based on the time-method is described. It
is characterized by using one multiplier only, taking into account the scintillation, compensating if necessary the back-
lash of the gearing, and using digital techniques. Therefore, it
is possible to use for the guider the special electronics described or a control computer. The accuracy of the guiding is
about ± 0."1.

032.019 Improvements of the Dodaira satellite laser tracker.
Y. Kozai, A. Tsuchiya, K. Tomita, T. Kanda,
H. Sato.
Space Research, Vol. XVII, (see 012.010), 55 - 57 (1977).

This paper reports on recent improvements made to the
satellite laser tracker at Dodaira Station of the Tokyo Astronomical Observatory. Both transmitting and receiving telescopes
of the tracker are on an X-Y mounting to avoid a dead point
at zenith and are controlled by a mini-computer through direct

drive torque motors. The old laser oscillator is still used; however, the pulse width is reduced to 2.5 ns by an electronic chopper and the output energy is compensated by adding an amplifier. The laser beam is transmitted through the coudé focus of a 10 cm transmitting telescope and the returned signal is collected by a 50 cm telescope which is the transmitter for the lunar laser ranging.

032.020 **Lunar laser ranging system for measuring distances with accuracy to ±20 cm.** Yu. L. Kokurin, V. V. Kurbasov, V. F. Lobanov, A. N. Sukhanovsky (*Sukhanovskij*).
Space Research, Vol. XVII, (see 012.010), 77 - 80 (1977).

032.021 **A new method of determining pivot errors of meridian circles.** B. Loibl.
Astron. Astrophys., Vol. 65, 77 - 81 (1978).
It is shown that the methods used so far to determine pivot errors of meridian circles where the angular separations of the bearing points are 90° (Oehler, 1939; Podobed, 1965) are not correct. A solution of this problem is presented which involves the construction of new bearings with angular separations of the bearing points of 45°. This new method was applied to the Hamburg meridian circle at Perth Observatory in 1971 during the observations for the Perth 70 catalogue (Høg and v.d. Heide, 1976). No systematic effects were caused by this new method.

032.022 **El Teide and the flux collector. Infrared astronomy on Tenerife.** T. Jones.
J. British Astron. Assoc., Vol. 88, 257 - 266 (1978).

032.023 **Polar axis alignment of equatorial instruments.** W. R. Vezin.
J. British Astron. Assoc., Vol. 88, 267 - 273 (1978).

032.024 **Calculation of basic systems of large azimuthal telescopes.** L. P. Borodulya, V. V. Bobashov.
Opt.-mekh. prom-st', 1977, No. 8, p. 3 - 5. In Russian.
Abstr. in Ref. zh., 51. Astron., 2.51.102 (1978).

032.025 **Automatized control system for a large azimuthal telescope.** E. M. Neplokhov.
Opt.-mekh. prom-st', 1977, No. 8, p. 41 - 49. In Russian.
Abstr. in Ref. zh., 51. Astron., 2.51.103 (1978).

032.026 **Elements of the polar siderostat theory.** V. G. Banin.
Issled. po geomagn., aehron. i fiz. Solntsa. Moskva, Nauka, 1977, p. 116 - 122. In Russian. − Abstr. in Ref. zh., 51. Astron., 2.51.108 (1978).

032.027 **Sternwarte Kreuzlingen: Aufsehenerregendes Teleskop für hochenergetische Gammastrahlung.** K. Bosshard.
Orion, 36. Jahrg., 79 - 81 (1978).

032.028 **Neues Sonnen-Turmteleskop von Zeiss.**
Orion, 36. Jahrg., 83 - 84 (1978).

032.029 **The zenithal blind spot of a computer-controlled altazimuth telescope.** F. G. Watson.
Mon. Not. R. Astron. Soc., Vol. 183, 277 - 284 (1978).
The zenithal blind spot inherent in the altazimuth mounted telescope is investigated and equations representing it are given. Its size is found to depend mainly on the performance of the azimuth drive in the tracking and fast slewing modes. The blind spot of the 4.2-m telescope for the projected International Observatory on La Palma is discussed.

032.030 **Correcting periodic error in a clock drive.** L. J. Faix.

Sky Telesc., Vol. 55, 439 - 442 (1978).

032.031 **Solar double-telescope at the Hvar Observatory.** P. Ambrož, V. Bumba, K. Havlíček, J. Ptáček, J. Suda.
Hvar Obs. Bull., No. 1, p. 15 - 23 (1977).

032.032 **Búsqueda, observación y análisis espectral de banda ancha de objetos celestes en el rango 50-350 micrones.** E. J. Gandolfi, A. Quaglia, A. Ringuelet.
Astron. extragalactica, (see 012.015), 149 - 153 (1976).
A description is given of the work carried out for the construction of an infrared telescope that enables a systematic search of objects in the 50-350 microwave band, and an observational program for its use is outlined.

032.033 **Current status of CTIO's 4 m telescope.** V. M. Blanco.
Astron. extragalactica, (see 012.015), 155 - 163 (1976).
The history of the construction project for the 4 m telescope at Cerro Tololo is given and its optical and mechanical characteristics are described as well as the observational work presently under way with this instrument.

032.034 **On external occulting systems for coronographs.** A. V. Lenskij.
Soln. Dannye 1977 Byull., No. 11, p. 68 - 74 (1978).
In Russian.
The Rubinowicz representation of Kirchhoff's diffraction integral is suggested to use for diffracted light intensity calculations in the cases of single and multiple disk systems. Approximate expressions are presented for simple and double disks.

032.035 **Building a 6″ reflector. IV. The mounting.**
Astronomy, Vol. 5, No. 9, p. 42 - 45 (1977).
Abstr. in Phys. Abstr., Vol. 81, Abstr. 4102 (1978).

032.036 **Investigation of the limb of the meridian circle of the Astronomical Observatory of the Kiev University.** N. A. Chernega.
Vestn. Kiev. Univ., Astron., Vyp. (No.) 19, p. 91 - 95 (1977).
In Russian.

032.037 **Exploring the infrared universe from Wyoming.** R. D. Gehrz, J. A. Hackwell.
Sky Telesc., Vol. 55, 466 - 473 (1978).

032.038 **Experiment of using graphite-carbide evaporators when making aluminum mirrors.**
V. D. Zhukov, V. F. Novikov, A. N. Shurshakov, G. D. Posov'eva.
Tr. N.−i. tsentra izuch. prirod. resursov, 1977, vyp. (No.) 3, p. 119 - 128. In Russian. − Abstr. in Ref. zh., 51. Astron., 3.51.89 (1978).

032.039 **Photoelectric phase device.** M. I. Malyshev.
Astrometriya i Astrofizika, Kiev, Vyp. (No.) 34, p. 63 - 64 (1978). In Russian.
Some advantages are described of a photoelectric phase device for the transit instrument used by the time service of the Shternberg Astronomical Institute.

032.040 **Photoelectric device with high time resolution.** A. I. Yazev, Eh. P. Medvedkov.
Astrometriya i Astrofizika, Kiev, Vyp. (No.) 34, p. 64 - 66 (1978). In Russian.
Results obtained with an experimental photoelectric device designed by the authors are described.

032.041 **Some features of the Dresden photoelectric phase device attached to the transit instrument and some results of observations.** H. Potthoff.

Astrometriya i Astrofizika, Kiev, Vyp. (No.) 34, p. 66 - 67 (1978). In Russian.

032.042 Photoelectric device for recording moments of star transits in Zagreb.   N. Solarić.
Astrometriya i Astrofizika, Kiev, Vyp. (No.) 34, p. 68 - 69 (1978). In Russian.

032.043 Automation of a transit instrument.
K. Šteins, A. V. Ivanov.
Astrometriya i Astrofizika, Kiev, Vyp. (No.) 34, p. 70 (1978). In Russian.
A description is given of the general plan and some already realized stages of complete automation of the transit instrument APM-10 operated by the time service of the Astronomical Observatory of the Latvian State University.

032.044 The 40/70/120 cm Schmidt telescope.
F. Imagawa, S. Kawai, T. Tsujimura, H. Ohtani, R. Hirata.
Mem. Fac. Sci. Kyoto Univ., Ser. Phys. Astrophys. Geophys. Chem., Vol. 35, 185 - 195 (1977).
A 40/70/120 cm Schmidt telescope was constructed in May 1972. A short description of the optical and mechanical systems is presented with some results of test observations.

032.045 Tecniche e strumentazione per radiometria solare ed atmosferica.   L. Mureddu.
G. Astron., Vol. 4, 5 - 24 (1978).
Una breve rassegna delle tecniche maggiormente usate per misure terrestri di radiometria e spettroradiometria; frequenti riferimenti verranno fatti alle soluzioni adottate dal "World Radiation Center, Davos". Scopo di questo lavoro non è di spiegare in dettaglio la costruzione e l'uso di ogni strumento radiometrico, ma solo di fornire un quadro d'insieme dei problemi, della terminologia e degli sviluppi più recenti in questo campo.

032.046 Il telescopio astrometrico da 105 cm dell'Osservatorio Astronomico di Torino.   W. Ferreri.
G. Astron., Vol. 4, 25 - 42 (1978).

032.047 User guide to telescope equipment. III. Anglo-Australian telescope.   B. L. Webster, J. A. J. Whelan.
Q. J. R. Astron. Soc., Vol. 19, 234 - 239 (1978).

032.048 Workshop testing results of the 6-m mirror of the large azimuthal telescope.   I. M. Kopylov, Yu. P. Korovyakovskij, A. F. Fomenko.
Opt.–mekh. prom-st', 1977, No. 10, p. 3 - 5. In Russian.
Abstr. in Ref. zh., 51. Astron., 4.51.95 (1978).

032.049 A simple scanning automatic device for the horizontal ATsU-5 solar telescope.
V. A. Chugunov.
Soln. akt. Alma-Ata, 1977, p. 95 - 97. In Russian. – Abstr. in Ref. zh., 51. Astron., 4.51.103 (1978).

032.050 A visit to Siding Spring.   N. Williams.
Electron. Australia, Vol. 39, No. 3, p. 12 - 17 (1977). – Abstr. in Phys. Abstr., Vol. 81, Abstr. 16291 (1978).

032.051 Neues Infrarot-Teleskop mit ungeahnten Möglichkeiten.   C. M. Humphries, V. C. Reddish.
Orion, 36. Jahrg., 101 - 103 (1978).

032.052 The United Kingdom's giant infrared reflector.
C. M. Humphries.
Sky Telesc., Vol. 56, 22 - 24 (1978).

032.053 Progress on NASA's 3-meter infrared telescope.
G. M. Smith.
Sky Telesc., Vol. 56, 25 - 27 (1978).

032.054 Ein neues Fernrohr zur H-Alpha-Beobachtung.
B. Wedel.
Sterne Weltraum, Jahrg. 17, 265 (1978).

032.055 Use of azimuthal mounting for large telescopes – reflectors.   N. G. Ponomarev.
Problems of observational and theoretical astronomy, Moscow – Leningrad, 1977, (see 003.012), p. 127 - 135. In Russian. – Abstr. in Ref. zh., 51. Astron., 5.51.12 (1978).

032.056 The work of N. G. Ponomarev "A tracking system as a clock mechanism for a telescope".
N. N. Mikhel'son.
Problems of observational and theoretical astronomy, Moscow – Leningrad, 1977, (see 003.012), p. 136 - 139. In Russian. – Abstr. in Ref. zh., 51. Astron., 5.51.13 (1978).

032.057 New constructions of large optical telescopes.
O. A. Mel'nikov, V. S. Popov.
Problems of observational and theoretical astronomy, Moscow –Leningrad, 1977, (see 003.012), p. 85 - 111. In Russian. Abstr. in Ref. zh., 51. Astron., 5.51.85 (1978).

032.058 Vision of insects and a multi-mirror telescope.
N. N. Mikhel'son.
Problems of observational and theoretical astronomy, Moscow – Leningrad, 1977, (see 003.012), p. 148 - 152. In Russian. – Abstr. in Ref. zh., 51. Astron., 5.51.86 (1978).

032.059 Telescope of the Elbrus spectrograph for cosmic radiation.   L. I. Dorman, I. Ya. Libin, A. A. Chechenov, A. Z. Zhappuev.
Issled. po probl. soln.-zemnoj fiz., Moskva, 1977, p. 79 - 82. In Russian. – Abstr. in Ref. zh., 62. Issled. kosm. prostranstva, 5.62.116 (1978).

032.060 New 60 cm-telescope at Sirahama Hydrographic Observatory.   T. Mori.
Data Rep. Hydrographic Observations, Ser. Astron. Geod., *Tokyo,* No. 12, p. 48 - 54 (1978).

032.061 On the setting of the polar axis of the 105 cm Schmidt telescope.   K. Hamajima, K. Ishida, B. Takase, T. Aoki, T. Soyano.
Tokyo Astron. Obs. Report No. 69, Vol. 18, 226 - 241 (1978). In Japanese.

032.062 On the accuracy of the photochronograph of the AFU-75 camera.   L. S. Shtirberg, S. S. Dzyamko.
Nauchn. Inf., Vyp. (No.) 38, p. 36 - 38 (1976). In Russian.
The accuracy of the time-registering photochronograph of the AFU-75 camera is studied. Some recommendations are proposed to increase the accuracy of fixation of the mean moment of the exposure in the compensational regime of the camera up to 0.0001 s.

032.063 On the beginning of observations with the Danjon astrolabe in Simeiz.   V. I. Sergienko, L. M. Malomyzhev, V. I. Bormotov, D. N. Nagornyuk.
Nauchn. Inf., Vyp. (No.) 38, p. 39 - 49 (1976). In Russian.
Some local difficulties with the mounting and calibration of the Danjon astrolabe in Simeiz are described. The instrumental constants and the principal characteristics of the observational program are presented. The list of the program stars is given in the appendix.

032.064 Investigation of the objective scales of the Engelhardt Astronomical Observatory heliometer.
A. S. Mamakov.
Izv. Astron. Ehngel'gardt. Obs., Kazan', No. 41 - 42, p. 94 -

103 (1976). In Russian.

032.065 The 16″ astrograph of the Engelhardt Astronomical Observatory.
Sh. T. Khabibullin, N. G. Rizvanov.
Izv. Astron. Ehngel'gardt. Obs., Kazan', No. 41 - 42, p. 104 - 109 (1976). In Russian.

032.066 Determination of the value of a revolution of the eyepiece micrometer of the Bamberg zenith telescope of the Engelhardt Astronomical Observatory.
V. V. Lobanova.
Izv. Astron. Ehngel'gardt. Obs., Kazan', No. 41 - 42, p. 175 - 177 (1976). In Russian.

032.067 Determination of the value of a division of Talcott levels and of the value of a revolution of the eyepiece micrometer of the ZTL-180 zenith telescope from observations of latitude pairs.   N. N. Chudinov.
Izv. Astron. Ehngel'gardt. Obs., Kazan', No. 41 - 42, p. 178 - 179 (1976). In Russian.

032.068 Le télescope de Schmidt de l'Observatoire de Tokyo.
B. Takase.
Astronomie, Vol. 92, 296 (1978).

032.069 Les applications effectives du télescope de Schmidt.
A. Fresneau.
Astronomie, Vol. 92, 325 - 327 (1978).

032.070 Wide-field filter camera.
P. G. Johnson, A. L. Kaye, J. Meaburn.
Appl. Opt., Vol. 17, 442 - 446 (1978).
A rugged camera with a 32° field has been constructed and used to detect faint nebulosity through narrow-band interference filters. The optimum bandwidth is determined. Also the merits of using a phosphor output image tube have been examined quantitatively, and a method of cooling the tube has been derived and investigated both experimentally and numerically.

032.071 Compact infrared heat trap field optics.

J. Keene, R. H. Hildebrand, S. E. Whitcomb, R. Winston.
Appl. Opt., Vol. 17, 1107 - 1109 (1978).
A design is presented for achieving compact ir heat trap field optics by incorporation of a lens into the light collector.

032.072 Building a 6″ reflector. V. Two mountings you can build at home.   R. Berry.
Astronomy, Vol. 6, No. 1, p. 50 - 54 (1978). − Abstr. in Phys. Abstr., Vol. 81, Abstr. 40830 (1978).

032.073 Five years of amateur telescope building at VAGO departments (1971 - 1975).   M. M. Shemyakin.
Problems of astronomy and geodesy, VAGO assembly Yerevan, 1975, (see 012.039), p. 185 - 189. In Russian. − Abstr. in Ref. zh., 51. Astron., 7.51.59 (1978).

032.074 Results of investigations of the 6-m azimuthal telescope by the Hartmann method.
Yu. P. Korovyakovskij, M. F. Shabanov.
Opt.-mekh. prom-st', 1977, No. 12, p. 3 - 5. In Russian.
Abstr. in Ref. zh., 51. Astron., 7.51.80 (1978).

Attrezzatura strumentale dell'Osservatorio di Arcetri.   See Abstr. 009.006.

Newton's telescope.   See Abstr. 014.025.

Computer correction of absolute telescope pointing errors by use of spherical harmonic functions.
See Abstr. 031.405.

On-line running at a telescope.   See Abstr. 031.417.

Direct, quasi-telecentric, and telecentric combinations of interference filters with large astronomical Schmidt cameras.   See Abstr. 034.063.

On the height of a telescope dome at a good site.
See Abstr. 082.099.

Space Instrumentation

032.501 Die Erforschung der Sonne mit Ballonteleskopen.
A. Wittmann.
Sterne Weltraum, Jahrg. 17, 8 - 13 (1978).

032.502 On board electronic units for space research.
H. J. Fischer.
Comunicaciones, No. 17, p. 3 - 16 (1975). In Spanish. − Abstr. in Phys. Abstr., Vol. 80, Abstr. 91401 (1977).

032.503 Long-lived landers for Venus.   T. A. Croft.
Bull. American Astron. Soc., Vol. 9, 500 - 501 (1977). − Abstract.

032.504 STARLAB, a spacelab-based one-meter general purpose telescope.   C. Anderson, K. Henize, E. Jenkins, R. O'Connell, A. Smith, B. Smith, T. Stecher.
Bull. American Astron. Soc., Vol. 9, 613 (1978). − Abstract.

032.505 A survey telescope for space.
T. R. Gull, K. G. Henize, D. J. Schroeder.

Bull. American Astron. Soc., Vol. 9, 613 (1978). − Abstract.

032.506 Effects of diffraction in multiple-grid X-ray telescopes.   C. A. Lindsey.
Bull. American Astron. Soc., Vol. 9, 626 (1978). − Abstract.

032.507 Measurements of X-ray source positions by the scanning modulation collimator on HEAO-1.
D. A. Schwartz, H. Gursky, H. Bradt, R. Doxsey, J. Schwarz, R. Dower, G. Fabbiano, R. E. Griffiths, M. Johnston, R. Leach, A. Ramsey, G. Spada.
Bull. American Astron. Soc., Vol. 9, 626 (1978). − Abstract.

032.508 High resolution stellar spectroscopy in the balloon ultraviolet.   C. de Jager, Y. Kondo, K. A. van der Hucht, T. H. Morgan.
Space Research, Vol. XVII, (see 012.010), 741 - 748 (1977).
The JSC/SRL balloon-borne ultraviolet stellar spectrometer (BUSS) comprises a 40 cm telescope, an echelle spectrograph and a SEC Vidicon detector. Operating at an altitude of 40 km this instrument provides spectra of stars with 0.1 Å resolution in the 2000–3400 Å region. Observations made on 19 May 1976 are discussed.

032.509 Preliminary results of the X-ray telescope "Filin" on board Salyut 4.   V. Ya. Berezhnoy, G. M. Grechko, A. A. Gubarev, E. F. Kirillov, P. I. Klimuk, V. G. Kurt, V. I. Sevastyanov (*Sevast'yanov*), L. G. Titarchuk, E. I. Moskalenko, E. K. Sheffer.
Space Research, Vol. XVII, (see 012.010), 757 - 762 (1977).

The paper presents a description of the "Filin" instrumentation aboard Salyut 4 for measuring the X-ray emission of extraterrestrial sources in the range from 0.2 to 10 keV. Preliminary results on some galactic X-ray sources (Sco X-1, Her X-1, Cir X-1, Cyg X-1 and A0620-00) are given. The spectral characteristics are discussed and interpreted in terms of the temperature of the source and the absorption of the radiation by hydrogen between the source and the space station.

032.510 The LPSP instrument on OSO 8. II. In-flight performance and preliminary results.   R. M. Bonnet, P. Lemaire, J. C. Vial, G. Artzner, P. Gouttebroze, A. Jouchoux, J. W. Leibacher, A. Skumanich, A. Vidal-Madjar.
Astrophys. J., Vol. 221, 1032 - 1061 (1978).

The in-flight performance for the first 18 months of operation of the French, pointed instrument on board OSO 8 is described. The angular and spectral resolution, the scattered light level, and various other instrumental parameters are evaluated from the observed data and shown to correspond mostly to nominal design values. The properties of the instrument are discussed, together with their evolution with time. The distribution of the first 8363 orbits between various observing programs is given. Preliminary results are also described. They include studies of the chromospheric network, sunspots and active regions, prominences, oscillations in the chromosphere, chromosphere-corona transition lines, and aeronomy.

032.511 A low energy ion sensor for space measurements with reduced photo-sensitivity.   P. J. L. Wildman.
Space Sci. Instrum., Vol. 3, 363 - 368 (1977).

A low energy ion detector designed to reduce the contribution of photoelectrons to the measured sensor current is described. The ion collector is shielded from direct solar illumination and the internal configuration accelerates the incoming ions onto this collector. Satellite measurements show that the photocurrent to ion current ratio is reduced by a factor of four with this new configuration. Results from the new sensor are compared with simultaneous measurements using a planar electrostatic analyzer on the same satellite.

032.512 The University of Colorado OSO-8 spectrometer experiment. I: Introduction and optical design considerations.   E. C. Bruner, Jr.
Space Sci. Instrum., Vol. 3, 369 - 387 (1977).

The instrument is a conventional 1 m Ebert-Fastie spectrometer fed by a Cassegrainian telescope. The instrument operates in the spectral range 1200 Å to about 2000 Å with spectral resolution of order 0.02 Å. Spatial resolution is about 2.5″ normal to the direction of the slit and is selectable from about 3″ to 15′ along the direction of the slit. Time resolution for the single spectrometer channel is selectable according to the needs of an individual observation and is limited to a maximum sampling rate of 40 ms per data point. The instrument is controlled by an internal general purpose computer.

032.513 Optical design of a stigmatic spectroheliometer for photometric studies of dynamic phenomena at extreme-ultraviolet wavelengths.   M. C. E. Huber, J. G. Timothy.
Space Sci. Instrum., Vol. 3, 389 - 406 (1977).

The design of a stigmatic spectroheliometer for photometric studies of dynamic phenomena in the solar atmosphere of extreme ultraviolet (EUV) wavelengths is described. The normal-incidence spectrometer requires only one reflective surface, and is equipped with a series of exit slits and associated one-dimensional detector arrays that are mounted at the secondary (vertical) foci of the concave diffraction grating. The resulting spectroheliograms would have an average photometric precision of 10% and a spectral purity of 0.1 Å.

032.514 The high energy X-ray detector on the Ariel-5 satellite.   A. R. Engel, M. J. Coe.
Space Sci. Instrum., Vol. 3, 407 - 421 (1977).

The authors describe the Imperial College hard X-ray detector which is used to make spectral measurements in the 26 keV to 1.2 MeV energy range on celestial X-ray sources from the Ariel-5 satellite. Details are given of the design, calibration and in-orbit performance of the detector. A modulation process is used to detect weak signals against a background and the authors give details of the spectrum unfolding techniques used to convert the measured spectra into corrected incident spectra.

032.515 Background sensitivity of a double Compton gamma-ray telescope to atmospheric neutrons.
J. Daugherty, V. Schönfelder.
Space Sci. Instrum., Vol. 3, 423 - 447 (1977).

032.516 A mass spectrometer-beam experiment for investigations of planetary atmospheres.
K. Mauersberger.
Space Sci. Instrum., Vol. 3, 465 - 471 (1977).

An experimental concept for investigation of planetary atmospheres using a mass spectrometer is discussed. High atmospheric pressures are reduced to mass spectrometer operating pressures by means of differential pumping. Sampling is accomplished through the formation of a molecular beam. A flag in front of the mass spectrometer can be used to separate ambient gases from system background. Laboratory tests have confirmed the feasibility of the experiment, particularly in CO_2 dominated atmospheres.

032.517 Hadamard transform X-ray telescope.   S. Miyamoto.
Space Sci. Instrum., Vol. 3, 473 - 481 (1977).

A Hadamard transform X-ray telescope is a type of Dicke's random hole X-ray camera for observing the X-ray sky. Instead of making a random hole pattern mask, a cyclic Hadamard matrix or *PN* sequence is used to make the mask pattern of this telescope. With this mask and a position sensitive X-ray detector, one can get Hadamard transformed image data of the X-ray sky and easily reconstruct the X-ray sky image from the observed data. The Hadamard matrix can be used to make one dimensional X-ray telescopes as well as two - dimensional telescopes.

032.518 Wide field-of-view Hadamard X-ray spectrometer.   S. Miyamoto.
Space Sci. Instrum., Vol. 3, 483 - 489 (1977).

A new type of spectrometer consisting of a flat Bragg crystal and a Hadamard transform X-ray telescope is described. It has a wide field of view and is suitable to observe line spectrum features of widely extended X-ray sources. The signal to noise ratio advantage of this spectrometer is almost the same as that of the curved Bragg crystal spectrometer proposed by Borken and Kraushaar, but its wavelength resolution and spatial resolution are better than for the curved crystal spectrometer.

032.519 Design and performance of an actively collimated phoswich system for X-ray astronomy.
J. L. Matteson, P. L. Nolan, W. S. Paciesas, R. M. Pelling.
Space Sci. Instrum., Vol. 3, 491 - 506 (1977).

The authors discuss the design and performance of a phoswich type scintillation detector system having 34 cm² effective collecting area and a sensitivity of 3×10^{-5} photons

$cm^{-2} s^{-1} keV^{-1}$ for balloon-borne cosmic X-ray source observations.

032.520 Airborne infrared astronomy. J. A. Thomas.
Proc. Astron. Soc. Australia, Vol. 3, 97 - 99 (1977).
Invited paper.

032.521 Spectrometer based on scintillation detectors of different configuration for analysis of spatial distribution of gamma-fields. Eh. M. Iovenko, A. L. Ioffe, A. K. Milovanov, V. N. Nikolaev, N. V. Smirnova, E. I. Yurevich.
Izv. AN SSSR. Ser. fiz., Vol. 41, 1887 - 1898 (1977). In Russian. – Abstr. in Ref. zh., 51. Astron., 2.51.236 (1978).

032.522 The HEAO-B X-ray telescope and its observing program. R. Giacconi.
Phys. Scr., Vol. 17, (see 012.013), 307 - 320 (1978).
A brief description is given of the performance characteristics of the HEAO-B X-ray telescope observatory. Particular emphasis is given to the descriptions of the parameters whose knowledge is required to establish an observational program. The detailed observing program by the Consortium Institutions responsible for the HEAO-B program is summarized.

032.523 Future prospects for spectroscopic and direct work: optical and UV. E. M. Burbidge.
Phys. Scr., Vol. 17, (see 012.013), 321 - 323 (1978).
A description of the main features and proposed instrumentation of the 2.4 m Space Telescope is given. Highlights of work that can be planned on active nuclei of galaxies, QSOs, and BL Lac objects are briefly outlined, involving spectroscopy over wavelengths from 1200 Å to 1 mm, direct imaging with 0˝.1 resolution, and the capability for 0˝.1 resolution along the spectrograph slit. The resolution, the much reduced sky background, and the full wavelength coverage also make possible important observations relevant to cosmology.

032.524 Accuracy of sighting a celestial object by orbiting star-tracking telescope.
B. N. Petrov, E. I. Mitroshin, S. P. Kuz'min, A. I. Zavedeyev, A. G. Chesnokov, K. K. Koloskov.
Acta Astronaut., Vol. 4, (see 012.017), 605 - 616 (1977). Abstr. in Phys. Abstr., Vol. 81, Abstr. 4057 (1978).

032.525 Planetary radio astronomy receiver.
G. J. Lang, R. G. Peltzer.
IEEE Trans. Aerosp. Electron. Syst., Vol. AES-13, 466 - 472 (1977). – Abstr. in Phys. Abstr., Vol. 81, Abstr. 4075 (1978).

032.526 A Large Infrared Telescope on Spacelab (LIRTS).
V. Manno.
ESA J., Vol. 1, No. 1, p. 35 - 42 (1977). – Abstr. in Phys. Abstr., Vol. 81, Abstr. 12070 (1978).

032.527 The ultraviolet stellar spectrograph BUSS.
W. Werner.
Nederlands Tijdschr. Natuurkd. A, Vol. A43, 123 - 128 (1977). In Dutch. – Abstr. in Phys. Abstr., Vol. 81, Abstr. 7880 (1978).

032.528 Bent crystal spectrometer for solar X-ray spectroscopy. C. G. Rapley, J. L. Culhane, L. W. Acton, R. C. Catura, E. G. Joki, J. C. Bakke.
Rev. Sci. Instrum., Vol. 48, 1123 - 1130 (1977). – Abstr. in Phys. Abstr., Vol. 81, Abstr. 7881 (1978).

032.529 Four-channel spectrometer for investigations of corpuscular radiation in the circumterrestrial cosmic space. A. P. Babaev, A. I. Gavrikov, V. I. Lazarev, V. A. Lipovetskij, B. V. Mar'in, M. V. Tel'pov, V. F. Tulinov, V. M. Fejgin.

Tr. N.-i. tsentra izuch. prirod. resursov, 1977, vyp. (No.) 3, p. 27 - 34. In Russian. – Abstr. in Ref. zh., 62. Issled. kosm. prostranstva, 3.62.123 (1978).

032.530 Radiofrequency mass analyzer of increased sensitivity for investigation of the upper atmospheres of the earth and planets. V. A. Kochnev, Yu. P. Barzilovich, K. V. Grechnev, I. A. Kalinina, Yu. A. Shul'chishin.
Inst. kosm. issled. AN SSSR. Pr-332. Moskva, 1977. 26 pp. In Russian. – Abstr. in Ref. zh., 62. Issled. kosm. prostranstva, 3.62.124 (1978).

032.531 An accelerator evaluation of the performance of the Soviet-French wide-gap spark chamber for gamma-ray astronomy. L. Kalinkin, V. D. Kozlov, V. E. Nesterov, V. L. Prokhine, G. V. Veselova, V. Akimov, R. Bazer-Bachi, E. Bonfand, A. Galper (*Gal'per*), M. Gorisse, M. Gros, V. Kirillov-Ugryumov, J. M. Lavigne, J. P. Leray, B. Luchkov, B. Mougin, Y. (*Yu.*) Ozerov, J. A. Paul, A. Popov, A. Raviart, G. Serra, S. Voronov.
Nucl. Instrum. Methods, Vol. 147, 329 - 332 (1977). – Abstr. in Phys. Abstr., Vol. 81, Abstr. 17503 (1978).

032.532 Efficiency of a cosmic ray neutron detector determined by the Monte-Carlo method.
K. Kudela, J. Pavlisko.
Acta Phys. Slovaca, Vol. 27, 282 - 284 (1977). – Abstr. in Phys. Abstr., Vol. 81, Abstr. 21809 (1978).

032.533 A burst length discrimination system for a gas scintillation proportional counter.
E.-A. Leimann, A. van Dordrecht.
Nucl. Instrum. Methods, Vol. 146, 339 - 342 (1977). – Abstr. in Phys. Abstr., Vol. 81, Abstr. 21824 (1978).

032.534 For on-board optical equipment: microchannels. J.-M. Morton, W. Parkes.
Toute Electron., No. 422, p. 35 - 38 (1977). In French. Abstr. in Phys. Abstr., Vol. 81, Abstr. 24436 (1978).

032.535 A narrowband TV system as a viewfinder for a balloon borne solar telescope. I. Sagehashi, Y. Yoshida, N. Niwa.
J. Inst. Telev. Eng., Vol. 31, 378 - 384 (1977). In Japanese. From Phys. Abstr., Vol. 81, Abstr. 24464 (1978).

032.536 A magnetic spectrometer for cosmic ray studies. R. L. Golden, G. D. Badhwar, J. L. Lacy, J. E. Zipse.
Nucl. Instrum. Methods, Vol. 148, 179 - 185 (1978). – Abstr. in Phys. Abstr., Vol. 81, Abstr. 25383 (1978).

032.537 Detector of transient radiation for electron identification in cosmic rays. A. A. Gusev, G. I. Pugacheva, A. F. Titenkov.
Geomagn. Aehron., Vol. 18, 522 - 524 (1978). In Russian.

032.538 KRT-3 antenna of the space radio telescope. V. I. Kostenko, L. I. Matveenko.
Inst. kosm. issled. AN SSSR. Prepr., 1977, No. 340, 22 pp. In Russian. – Abstr. in Ref. zh., 62. Issled. kosm. prostranstva, 5.62.111 (1978).

032.539 On the influence of a Langmuir probe on measurements of electric field fluctuations aboard Intercosmos 10. V. M. Kostin, L. A. Malyavin, G. A. Mikhajlova.
Issled. po probl. soln.-zemnoj fiz., Moskva, 1977, p. 115 - 122. In Russian. – Abstr. in Ref. zh., 62. Issled. kosm. prostranstva, 5.62.120 (1978).

032.540 Device for measurement of the λ 6300 Å line profile and its intensity. V. A. Gladyshev, L. I. Grafova, A. K. Kuz'min, T. M. Mulyarchik, Yu. F. Ermolaev,

Sh. Ya. Mamin, V. A. Notkin, A. V. Skargin, S. R. Tabaldyev.
Konstruir. nauchn. kosm. appar. Moskva, 1977, p. 79 - 88.
In Russian. – Abstr. in Ref. zh., 62. Issled. kosm. prostranstva,
5.62.121 (1978).

032.541 Mars water instrument study.   J. B. Stephens.
Reports of planetary geology program, 1977–1978,
(see 003.014), p. 314 - 315 (1978).

032.542 Tests of a miniaturized chemical analysis system.
T. E. Economou, A. L. Turkevich, E. J. Franzgrote.
Reports of planetary geology program, 1977–1978,
(see 003.014), p. 322 - 323 (1978).

**032.543 Advantages of diamond turned optics for future
missions in X-ray astronomy.**   R. C. Catura.
Bull. American Astron. Soc., Vol. 10, 433 (1978). – Abstract.

**032.544 Initial considerations of minor planet astrometry
with the Space Telescope.**   P. D. Hemenway.
Bull. American Astron. Soc., Vol. 10, 458 - 459 (1978).
Abstract.

**032.545 A Spacelab facility for spectroscopy and polarim-
etry of X-ray sources (EXSPOS).**   V. Manno.
ESA J., Vol. 1, 145 - 153 (1977). – Abstr. in Phys. Abstr.,
Vol. 81, Abstr. 32314 (1978).

**032.546 A capacity type detector for low velocity dust
particles as expected during a slow fly-by cometary
mission.**   N. Pailer, J. Kissel, E. Schneider.
Max-Planck-Inst. Kernphys., Heidelberg, MPI H-1978-V2,
24 pp. (1978).
    Thin foil capacitor detector has been designed which
responds to particle impacts with a decrease in capacitance.
The capacitative decrease is reflected as a frequency increase
in an associated resonance oscillator. Typical resonant frequencies for active sensor areas in the order of several mm^2
were 40 MHz. Frequency changes from impacts could be
measured down to 20 Hz. It is shown that this detector is
capable of recording impacts by particles with velocities as low
as 700 m/sec.

**032.547 An impact-mass-spectrometer for in situ chemical
analysis of cometary particulates to be used on**
board of a flyby-mission.   B.-K. Dalmann, H. Fechtig,
E. Grün, J. Kissel.
Max-Planck-Inst. Kernphys., Heidelberg, MPI H-1978-V3,
20 pp. (1978).
    A time-of-flight mass spectrometer using an impact ioni-
zation source was built to test this concept for measuring the
chemical composition of dust particles on a cometary probe.
The instrument analyzes the positive ions of the plasma produced upon a micrometeoroid impact.

**032.548 A 1 m^2 gamma ray spark chamber telescope for the
energy range 10 - 100 MeV.**   B. Parlier, B.
Agrinier, E. Bonfand, B. Mougin, A. Lecomte, J. Andrejol, J.
C. Courtois, M. Gorisse, J. M. Lavigne, M. Niel, G. Vedrenne,
C. Doulade.
Nucl. Instrum. Methods, Vol. 148, 483 - 488 (1978). – Abstr.
in Phys. Abstr., Vol. 81, Abstr. 36459 (1978).

**032.549 Astro 7 zodiacal light UV polychromator and its
absolute radiometric calibration.**
E. Pitz, C. Leinert, A. Schulz, H. Link.
Appl. Opt., Vol. 17, 730 - 735 (1978).
    The UV polychromator, flown in 1975 on the Astro 7
zodiacal light rocket experiment, and its absolute radiometric
calibration are described. The instrument is essentially a
broadband double polychromator with high stray light suppression. The spectral ranges of 40-nm width are centered

at 180 nm, 220 nm, and 260 nm, respectively.

032.550 Viking infrared thermal mapper.
S. C. Chase, Jr., J. L. Engel, H. W. Eyerly, H. H.
Kieffer, F. D. Palluconi, D. Schofield.
Appl. Opt., Vol. 17, 1243 - 1251 (1978).
    The infrared thermal mapper (IRTM) was designed to
measure the emitted and reflected radiance of Mars. Carried
by the Viking Orbiter, the IRTM contains four small Cassegrainian telescopes which each image the same, seven circular
areas. There is a total of twenty-eight channels in four surface
and one atmospheric thermal bands from 6 μm to 30 μm and a
broad solar reflectance band. All channels are sampled simultaneously, using the spacecraft scanning capability to map the
radiance over small and large areas of the planet. All channels
use thermopile detectors; spectral passbands are determined
by a combination of interference filters, detector lense materials, antireflection coatings, and reststrahlen optics.

**032.551 New monochromators of concave grating suitable
for space telescopes.**   M. Singh.
Appl. Opt., Vol. 17, 1815 - 1823 (1978).

**032.552 An X-ray proportional counter for the Viking
Lander.**   F. L. Glesius, J. C. Kroon, A. J. Castro,
B. C. Clark.
IEEE Trans. Nucl. Sci., Vol. NS-25, 478 - 484 (1978). – Abstr.
in Phys. Abstr., Vol. 81, Abstr. 40792 (1978).

**032.553 Thin window Si(Li) detectors for the ISEE-C
telescope.**   J. T. Walton, H. A. Sommer, D. E.
Greiner, F. S. Bieser.
IEEE Trans. Nucl. Sci., Vol. NS-25, 391 - 394 (1978). – Abstr.
in Phys. Abstr., Vol. 81, Abstr. 40819 (1978).

**032.554 The high energy astronomy observatory X-ray
telescope.**   R. Miller, G. Austin, D. Koch,
N. Jagoda, T. Kirchner, R. Dias.
IEEE Trans. Nucl. Sci., Vol. NS-25, 422 - 429 (1978). – Abstr.
in Phys. Abstr., Vol. 81, Abstr. 40820 (1978).

032.555 The high resolution imaging instrument for HEAO-B.
K. Kubierschky, G. K. Austin, D. C. Harrison, A. G.
Roy.
IEEE Trans. Nucl. Sci., Vol. NS-25, 430 - 436 (1978). – Abstr.
in Phys. Abstr Vol. 81, Abstr. 40821 (1978).

**032.556 The HEAO-B monitor proportional counter
instrument.**   R. Gaillardetz, P. Bjorkholm, R.
Mastronardi, M. Vanderhill, D. Howland.
IEEE Trans. Nucl. Sci., Vol. NS-25, 437 - 444 (1978). – Abstr.
in Phys. Abstr., Vol. 81, Abstr. 40822 (1978).

032.557 Imaging proportional X-ray counter for HEAO.
A. Humphrey, R. Cabral, R. Brissette, R. Carroll,
J. Morris, P. Harvey.
IEEE Trans. Nucl. Sci., Vol. NS-25, 445 - 452 (1978). – Abstr.
in Phys. Abstr., Vol. 81, Abstr. 40823 (1978).

**032.558 The Goddard space flight center solid state spectro-
meter for the HEAO-B mission.**   R. M. Joyce, R.
H. Becker, F. B. Birsa, S. S. Holt, M. P. Noordzy.
IEEE Trans. Nucl. Sci., Vol. NS-25, 453 - 458 (1978). – Abstr.
in Phys. Abstr., Vol. 81, Abstr. 40824 (1978).

**032.559 The focal plane crystal spectrometer for the
HEAO-B satellite.**   J. F. Donaghy, C. R. Canizares.
IEEE Trans. Nucl. Sci., Vol. NS-25, 459 - 463 (1978). – Abstr.
in Phys. Abstr., Vol. 81, Abstr. 40825 (1978).

**032.560 Electronics for a focal plane crystal spectrometer
[on HEAO-B spacecraft].**   R. F. Goeke.

IEEE Trans. Nucl. Sci., Vol. NS-25, 464 - 466 (1978). — Abstr. in Phys. Abstr., Vol. 81, 40826 (1978).

032.561 Interference mirror for soft X-rays. T. J. Jach. IBM Tech. Disclosure Bull., Vol. 20, 1626 - 1627 (1977). — Abstr. in Phys. Abstr., Vol. 81, Abstr. 41385 (1978).

032.562 Proportional counters for the X-ray photometer. M. Hlond. Pomiary Autom. Kontrola, Vol. 24, No. 2, p. 47 - 48 (1978). In Polish. — Abstr. in Phys. Abstr., Vol. 81, Abstr. 41388 (1978).

032.563 A two channel rocket photometer for 5577 Å OI nightglow measurements. A. Drescher. J. Geophys., Vol. 44, No. 1-2, p. 173 - 178 (1977). — Abstr. in Phys. Abstr., Vol. 81, Abstr. 45078 (1978).

032.564 A rocket borne experiment to measure plasma densities in the D-region. M. Friedrich, S. Ulrich, K. Torkar. J. Geophys., Vol. 44, 91 - 98 (1977). — Abstr. in Phys. Abstr., Vol. 81, Abstr. 45126 (1978).

032.565 Infrared instrumentation on NASA airborne observatories. N. W. Boggess. Opt. Eng., Vol. 16, 528 - 531 (1977). — Abstr. in Phys. Abstr., Vol. 81, Abstr. 45224 (1978).

032.566 Rocket-borne infrared astronomical telescopes. K. Shivanandan. Opt. Eng., Vol. 16, 532 - 536 (1977). — Abstr. in Phys. Abstr., Vol. 81, Abstr. 45225 (1978).

032.567 Infrared astronomical satellite. H. H. Aumann, R. G. Walker. Opt. Eng., Vol. 16, 537 - 543 (1977). — Abstr. in Phys. Abstr., Vol. 81, Abstr. 45226 (1978).

032.568 A 102-cm balloon-borne telescope for far-infrared astronomy. G. G. Fazio. Opt. Eng., Vol. 16, 551 - 557 (1977). — Abstr. in Phys. Abstr., Vol. 81, Abstr. 45227 (1978).

032.569 The Imperial College 41-inch telescope for far-infrared balloon astronomy. R. D. Joseph, J. Allen, W. P. S. Meikle, K. C. Sugden, M. F. Kessler, D. L. Rosen, G. Masson. Opt. Eng., Vol. 16, 558 - 562 (1977). — Abstr. in Phys. Abstr., Vol. 81, Abstr. 45228 (1978).

032.570 Design features of the Space Telescope. E. B. Brown. Proceedings of the 12th International Congress on High Speed Photography, Toronto, Canada, 1 - 7 August 1976. Society of the Photo-Optical Instrumentation Engineers, Washington, DC, (1977). p. 519 - 530. — Abstr. in Phys. Abstr., Vol. 81, Abstr. 45229 (1978).

032.571 High resolution X-ray optics for astronomical and laboratory sources. J. K. Silk. X-ray imaging, (see 012.041), p. 113 - 124 (1977). — Abstr. in Phys. Abstr., Vol. 81, Abstr. 45766 (1978).

032.572 Design, fabrication and performance of two grazing incidence telescopes for celestial extreme ultraviolet astronomy. M. Lampton, W. Cash, R. F. Malina, S. Bowyer. X-ray imaging, (see 012.041), p. 93 - 106 (1977). — Abstr. in Phys. Abstr., Vol. 81, Abstr. 49329 (1978).

032.573 Design, fabrication and expected performance of the

HEAO-B X-ray telescope. L. P. Van Speybroeck. X-ray imaging, (see 012.041), p. 136 - 144 (1977). — Abstr. in Phys. Abstr., Vol. 81, Abstr. 49330 (1978).

032.574 The curved crystal X-ray spectrometer for the HEAO-B satellite. C. R. Canizares, G. W. Clark, D. Bardas, T. Markert. X-ray imaging, (see 012.041), p. 154 - 162 (1977). — Abstr. in Phys. Abstr., Vol. 81, Abstr. 49331 (1978).

032.575 A hard X-ray imaging instrument for solar and cosmic sources. G. J. Hurford. X-ray imaging, (see 012.041), p. 163 - 171 (1977). — Abstr. in Phys. Abstr., Vol. 81, Abstr. 49332 (1978).

032.576 High resolution imaging X-ray detector for astronomical measurements. J. P. Henry, E. M. Kellogg, U. G. Briel, S. S. Murray, L. P. Van Speybroeck, P. J. Bjorkholm. X-ray imaging, (see 012.041), p. 196 - 205 (1977). — Abstr. in Phys. Abstr., Vol. 81, Abstr. 49333 (1978).

032.577 The biology instrument for the Viking Mars mission. F. S. Brown, H. E. Adelson, M. C. Chapman, O. W. Clausen, A. J. Cole, J. T. Cragin, R. J. Day, C. H. Debenham, R. E. Fortney, R. I. Gilje, D. W. Harvey, J. L. Kropp, S. J. Loer, J. L. Logan, Jr., W. D. Potter, G. T. Rosiak. Rev. Sci. Instrum., Vol. 49, 139 - 182 (1978). — Abstr. in Phys. Abstr., Vol. 81, Abstr. 49347 (1978).

032.578 Orbiting solar telescope on 'Salyut-4' station. A. V. Bruns, G. M. Grechko, A. A. Gubarjev (*Gubarev*), P. I. Klimuk, V. I. Sevastjanov (*Sevast'yanov*), A. B. Severny (*Severnyj*), N. V. Steshenko. Acta Astronaut., Vol. 4, 1121 - 1125 (1977). — Abstr. in Phys. Abstr., Vol. 81, Abstr. 53783 (1978).

032.579 NASA Space Shuttle laser experiments. R. T. Menzies. OSA/IEEE Conference on Laser and Electrooptical Systems, San Diego, CA, USA, 7 - 9 February 1978. IEEE, New York, USA (1978). p. 78. — Abstr. in Phys. Abstr., Vol. 81, Abstr. 53793 (1978).

Reflectivity of gold coated surfaces in the soft X-ray range. See Abstr. 022.113.

Soft X-ray imaging with toroidal mirrors. See Abstr. 031.035.

Optical design of the Viking Lander camera. See Abstr. 031.042.

Infrared measurements at helium-temperature levels. See Abstr. 031.227.

Airborne infrared astronomical observations by Fourier transform spectroscopy. See Abstr. 031.303.

The reduction of scattered light in an external occulting disk coronograph. See Abstr. 032.005.

Balloon-borne ultraviolet stellar echelle spectrograph. See Abstr. 034.060.

Mylar beam-splitter efficiency in far infrared interferometers: angle of incidence and absorption effects. See Abstr. 034.062.

Early operations with the International Ultraviolet Explorer. See Abstr. 051.026.

The International Ultraviolet Explorer.
See Abstr. 054.022.

The International Ultraviolet Explorer (IUE).
See Abstr. 054.024.

Soft X-ray results from the Wisconsin experiment on OSO-8.   See Abstr. 142.033.

The UCSD/MIT hard X-ray and low energy γ-ray experiment on HEAO-1.   See Abstr. 142.091.

Preliminary hard X-ray results from HEAO-1.
See Abstr. 142.092.

Search for interstellar scattering X-ray halos with HEAO-1 scanning modulation collimator.
See Abstr. 142.095.

Polarization of cosmic X-ray sources.
See Abstr. 142.238.

Blick ins Zentrum der Milchstraße.
See Abstr. 156.006.

033 Radio Telescopes and Equipment

033.001 **Very-long-baseline radio interferometry.**
L. I. Matveenko.
Zemlya i Vselennaya, 1978, No. 1, p. 4 - 11. In Russian.

033.002 **La radioastronomie d'amateur.**   F. Biraud.
Astronomie, Vol. 92, 127 - 133 (1978).

033.003 **The Chatanika radar system.**   M. J. Baron.
Radar probing of the auroral plasma, (see 012.011), p. 103 - 141 (1977).

033.004 **The U. K. multistatic incoherent scatter facility.**
P. H. McPherson.
Radar probing of the auroral plasma, (see 012.011), p. 143 - 157 (1977).
Incoherent scatter measurements have been made in the U.K. since 1962 with three different radar systems. This paper deals mainly with the most recent system, MISCAT, and describes some of the technical aspects of the radar and data processing equipment and the method of obtaining the ionospheric parameters by computer analysis. Some of the factors which affect data quality are discussed.

033.005 **A grid antenna array for the VHF incoherent scatter radar with discrete beam directions.**
M. Tiuri, J. Henriksson.
Radar probing of the auroral plasma, (see 012.011), p. 187 - 206 (1977).

033.006 **Receivers for EISCAT.**   S. Westerlund.
Radar probing of the auroral plasma, (see 012.011), p. 207 - 214 (1977).

033.007 **A possible configuration for the Australian Synthesis Telescope.**   R. N. Manchester.
Proc. Astron. Soc. Australia, Vol. 3, 103 - 105 (1977).
The proposal made to ASTEC for an Australian Synthesis Telescope (AST) is for a high-sensitivity, high-resolution synthesis array to be located at the Australian National Radio Astronomy Observatory, Parkes. A design study group has been investigating the design of such an array. This paper reports on one aspect of this design, the array configuration.

033.008 **An acousto-optical radio spectrograph for spectral integration.**   T. W. Cole, D. K. Milne.
Proc. Astron. Soc. Australia, Vol. 3, 108 - 111 (1977).

033.009 **Reduction of the baseline ripple on spectra recorded with the Parkes radio telescope.**   R. Padman.
Proc. Astron. Soc. Australia, Vol. 3, 111 - 113 (1977).
This paper is a progress report on an experimental in-vestigation of the spectral baseline ripple on the Parkes 64-m radio telescope at frequencies near 5 GHz.

033.010 **Millimetre wavelength performance of the new 16.7m surface of the Parkes radio telescope.**   D. E. Yabsley.
Proc. Astron. Soc. Australia, Vol. 3, 113 - 115 (1977).
A new surface installed recently over the central 16.7-m-diameter zone of the 64-m radio telescope at Parkes has extended the operating range of this instrument to millimetre wavelengths. The overall efficiency of this 16.7-m zone now exceeds 40% at 7-mm wavelength and is ∼ 16% at 3.3 mm.

033.011 **The Tidbinbilla interferometer.**   M. J. Batty, D. L. Jauncey, S. Gulkis, M. J. Yerbury.
Proc. Astron. Soc. Australia, Vol. 3, 115 - 117 (1977).

033.012 **The new radio telescope of the USSR Academy of Sciences, RATAN-600.**
A. B. Berlin, N. A. Esepkina, Yu. K. Zverev, N. L. Kajdanovskij, D. V. Korol'kov, A. I. Kopylov, Eh. I. Korkin, Yu. N. Parijskij, N. F. Ryzhkov, N. S. Soboleva, A. A. Stotskij, O. N. Shivris.
Pribory i tekh. ehksp., 1977, No. 5, p. 8 - 16. In Russian.
Abstr. in Ref. zh., 51. Astron., 2.51.111 (1978).

033.013 **Standard sources for flux calibration and determination of antenna parameters by radio astronomical methods.**   V. I. Altunin, V. P. Ivanov, K. S. Stankevich.
Izv. vyssh. uchebn. zaved. Radiofiz., Vol. 20, 969 - 981 (1977).
In Russian. – Abstr. in Ref. zh., 51. Astron., 2.51.112 (1978).

033.014 **On absolute calibration of a cross-shaped radio interferometer.**
V. G. Zandanov, B. I. Lubyshev, T. A. Treskov.
Issled. po geomagn., aehron. i fiz. Solntsa. Moskva, Nauka, 1977, p. 145 - 148. In Russian. – Abstr. in Ref. zh., 51. Astron., 2.51.113 (1978).

033.015 **On coordinate accuracy of a multi-frequency interferometer.**   P. A. Fridman.
Izv. vyssh. uchebn. zaved. Radiofiz., Vol. 20, 848 - 852 (1977).
In Russian. – Abstr. in Ref. zh., 51. Astron., 2.51.114 (1978).

033.016 **Facilities of an equidistant interferometer for observations of local sources on the sun.**
L. M. Risover, T. A. Treskov.
Issled. po geomagn., aehron. i fiz. Solntsa. Moskva, Nauka, 1977, p. 142 - 144. In Russian. – Abstr. in Ref. zh., 51. Astron., 2.51.116 (1978).

033.017 **An 11.35 cm range radiometer with an ortho-model mixer.**   M. N. Kajdanovskij, Yu. A. Nemlikher,

A. A. Stotskij, I. A. Strukov.
Pribory i tekh. ehksp., 1977, No. 5, p. 145 - 147. In Russian.
Abstr. in Ref. zh., 51. Astron., 2.51.117 (1978).

**033.018 Evaluation of intensity calibration errors of milli-
meter-wave spectrometers.** B. L. Ulich.
Astrophys. Lett., Vol. 19, 93 - 95 (1978).

Millimeter-wave spectrometers currently used for astronomical observations of spectral lines employ superheterodyne
radiometers. The local oscillators are reflex klystrons which
are phase-locked to a harmonic of a low frequency crystal
standard. Phase noise is transferred to the local oscillator
signal, resulting in a broadening of its power density spectrum.
The finite width of the local oscillator signal also broadens the
power density spectrum of the unknown spectral "line" observed in the intermediate-frequency passband of the radiometer. The observed spectral "line" is thus smeared out in
frequency, and the observed intensity is less than its true value.
This intensity calibration error depends on the local oscillator
phase noise, and measurements of this noise may be used to
estimate the magnitude of the intensity calibration error. This
paper presents the results of such measurements on the local
oscillator systems.

033.019 An amateur radio telescope – I.
G. W. Swenson, Jr.
Sky Telesc., Vol. 55, 385 - 390 (1978).

033.020 Short-baseline radiointerferometer for solar patrol.
M. Eliäss.
Investigations of the sun and red stars. 8, (see 003.005), p.
70 - 86 (1978). In Russian.

**033.021 Decametric survey of discrete sources in the
Northern sky. I. The UTR-2 radio telescope. Experi-
mental techniques and data processing.**
S. Ya. Braude, A. V. Megn (*Men'*), B. P. Ryabov, N. K.
Sharykin, I. N. Zhouck (*Zhuk*).
Astrophys. Space Sci., Vol. 54, 3 - 36 (1978).

The paper describes the antenna system, principles of the
electrical beam steering, checks of the pattern parameters and
the receiving facilities of the UTR-2 radio telescope. Principal
characteristics of the instrument are given and possible errors
analysed for the operating frequencies of 10, 12.6, 14.7,
16.7, 20 and 25 MHz. The methods of survey work on discrete
radio sources are described with an analysis of the flux
measurement errors specific for the decametre band. A computer algorithm of data processing is presented.

033.022 Tableau des plus grands radiotélescopes en service.
P. Kohler.
Astronomie, Vol. 92, 226 (1978).

033.023 A new class of millimetre-wave telescope.
N. Fourikis.
Astron. Astrophys., Vol. 65, 385 - 388 (1978).

A telescope is described which consists of a fully steerable
structure supporting an annular reflector and a number of
secondary reflectors which reflect the incoming radiation into
an equal number of receivers. An example is given of such
an annular synthesis telescope utilizing 16 millimetre-wave
receivers to realize 81 independent antenna beams. Its properties are compared to those of an earth-rotation synthesis
telescope and a single-dish multi-beam telescope employing
as many millimetre-wave receivers and having the same resolution. The same structure can be used to support a number
of such annuli each operating over a different waveguide
band.

033.024 An amateur radio telescope – II.
G. W. Swenson, Jr.
Sky Telesc., Vol. 55, 475 - 479 (1978).

**033.025 The classical analysis of the response of a long
baseline radio interferometer.** W. H. Cannon.
Geophys. J. R. Astron. Soc., Vol. 53, 503 - 530 (1978).

A general derivation of the classical response of a long
baseline interferometer is presented. The general theory is then
applied to a pair of antennae attached to the Earth and discussed in the light of a number of geophysical phenomena including the effects of precession, nutation, rotation, polar
motion, annual aberration and Earth tides.

033.026 On the control of the form of the surface.
V. I. Buyakas, Yu. V. Chekulaev.
Kosm. Issled., Vol. 16, 169 - 178 (1978). In Russian.

**033.027 Collecting and computer processing system of radio
astronomical observational data.** S. L. Domin,
V. A. Efanov, E. S. Korsenskaya, V. A. Korsenskij, I. G.
Moiseev, N. S. Nesterov, I. D. Strepko.
Strukt., tekh. sredstva i organ. sistem avtom. nauchn. issled.
Mater. X Vses. shkoly po avtom. Nauchn. issled.,Gelendzhik,
1976. Leningrad, 1977, p. 167 - 170. In Russian. – Abstr. in
Ref. zh., 51. Astron., 4.51.104 (1978).

033.028 Forty years evolution of radioastronomy.
Antenna, Vol. 49, No. 3, p. 110 - 114 (1977).
In Italian. – Abstr. in Phys. Abstr., Vol. 81, Abstr. 12095
(1978).

033.029 Radio interferometer Crimea – Haystack.
L. I. Matveenko, I. G. Moiseev, D. M. Moran,
B. F. Burke, L. R. Kogan, V. A. Efanov.
Pis'ma Astron. Zh., Vol. 4, 51 - 56 (1978). In Russian.

VLBI Crimea – Haystack has operated now at $\lambda = 1.35$
cm. Receivers with masers, hydrogen frequency standard and
Mark II system of registration are used. Sensitivity of the interferometer is equal to 0.1 Jy. The fringe rate stability is better
than 10^{-14}.

033.030 Carbon in space is tracked with a mixer-receiver.
S. Lidholm.
Eltek. Aktuell Elektron. A, Vol. 20, No. 17, p. 22 - 24 (1977).
In Swedish. – Abstr. in Phys. Abstr., Vol. 81, Abstr. 24461
(1978).

033.031 Waveguide system for a very large antenna array.
S. Weinreb, R. Predmore, M. Ogai, A. Parrish.
Microwave J., Vol. 20, No. 3, p. 49 - 52 (1977). – Abstr. in
Phys. Abstr., Vol. 81, Abstr. 24465 (1978).

033.032 A basic radio telescope. J. R. Smith.
Wireless World, Vol. 84, No. 1506, p. 26 - 29 (1978).
Abstr. in Phys. Abstr., Vol. 81, Abstr. 28524 (1978).

**033.033 Chromatism in radio telescopes due to blocking and
feed scattering.** D. Morris.
Astron. Astrophys., Vol. 67, 221 - 228 (1978).

Chromatism in radio telescopes due to feed scattering is
expressed as the product of a feed scattering factor γ, and the
mirror reflection coefficient. Values have been calculated for
γ using the physical optics approximation for single mode
circular TE 11 feeds with plane flanges. Values depend
critically on flange size as energy spilling past the feed aperture is scattered back to the telescope mirror. At the prime
focus, feed scattering will probably dominate over receiver
mismatches as a source of chromatism. Secondary focus telescopes will have larger chromatism in general but in this case
receiver mismatches are likely to dominate.

033.034 An acousto-optical solar radio spectrograph.
T. W. Cole, R. T. Stewart, D. K. Milne.
Astron. Astrophys., Vol. 67, 277 - 279 (1978).

This research note reports on the successful application

of an acousto-optical radio spectrograph (AOS) to the study of rapidly varying spectral features in metre-wavelength solar bursts. In particular, it outlines the use of a digital computer for real-time correction and calibration of the data. This optimizes the recording of the data on photographic film.

033.035   An amateur radio telescope—III.
         G. W. Swenson, Jr.
Sky Telesc., Vol. 56, 28 - 33 (1978).

033.036   Das RATAN-600-Radioteleskop.   R. Schwartz.
Sterne Weltraum, Jahrg. 17, 242 - 245 (1978).

**033.037   Dispersion interferometers for investigation of the
         inhomogeneous structure and dynamics of cosmic
plasma.   M. B. Vasil'ev.**
Radiotekh. i ehlektron., Vol. 23, 411 - 416 (1978). In Russian.
Abstr. in Ref. zh. 62. Issled. kosm. prostranstva, 5.62.263
(1978).

**033.038   Measurement of the aperture efficiency of 6m mm
         wave antenna.   K. Akabane, S. Hata.**
Tokyo Astron. Obs. Report No. 69, Vol. 18, 105 - 115 (1978).
In Japanese.

**033.039   Geodetic and astrometric analysis of fringe rate
         residuals from the Algonquin-Chilbolton 2.8 cm
long-baseline interferometer.   W. H. Cannon, R. B. Langley,
W. T. Petrachenko, N. W. Broten, T. H. Legg, D. N. Fort, J. L.
Yen, P. C. Barber, M. J. S. Quigley.**
New methods of space geodesy, (see 012.029), p. 182 - 227
(1976).
    In March, 1973 a long baseline interferometry experiment was carried out at 2.8 cm using the 46 m radio telescope of the National Research Council at the Algonquin Radio Observatory, Canada and the 25 m radio telescope of the Appleton Laboratory at the Chilbolton Field Station, England. The fringe rates obtained from the six extragalactic radio sources observed were compared with those predicted by a model. The differences between the natural fringe rates and the predicted fringe rates were minimized in a least-squares fashion by adjusting the values of the free parameters of the model.

033.040   VLA: the ultimate in radio astronomy.
Opt. Spectra, Vol. 12, No. 4, p. 45 - 46 (1978).

**033.041   On the question of measurement of circular
         polarization with a variable profile antenna.**
A. N. Korzhavin.
Soobshch. Spets. Astrofiz. Obs., Zelenchukskaya, Vyp. (No.)
16, p. 43 - 62 (1976). In Russian.
    Peculiarities of variable profile antenna (VPA) operation in receiving circularly polarized components of radiation are investigated. The split angle between the right and the

left patterns of the VPA is determined. Possibilities of parasitic circular polarization correction on the basis of matching the right and the left VPA patterns is considered.

**033.042   The continuing search for flare star magnetic fields:
         A Zeeman analyzer system for the KPNO 4-meter
echelle spectrograph.   C. M. Anderson, K. H. Nordsieck.**
Bull. American Astron. Soc., Vol. 10, 452 (1978). – Abstract.

**033.043   A quasioptical feed system for radioastronomical
         observations at millimeter wavelengths.**
P. F. Goldsmith.
Bell Syst. Tech. J., Vol. 56, 1483 - 1501 (1977). – Abstr. in
Phys. Abstr., Vol. 81, Abstr. 32343 (1978).

**033.044   A basic radio telescope. II. Construction, perform-
         ance and testing.   J. R. Smith.**
Wireless World, Vol. 84, No. 1507, p. 83 - 84 (1978). – From
Phys. Abstr., Vol. 81, Abstr. 40828 (1978).

**033.045   Measuring the gravitational astigmatism of a radio
         telescope.   S. von Hoerner.**
IEEE Trans. Antennas Propag., Vol. AP-26, 315 - 318 (1978).
Abstr. in Phys. Abstr., Vol. 81, Abstr. 45222 (1978).

033.046   A convertor for reception in the 70 cm range.
         R. Ambrosini, G. Tomassetti.
Antenna, Vol. 49, 283 - 292 (1977). In Italian. – Abstr. in
Phys. Abstr., Vol. 81, Abstr. 45234 (1978).

033.047   The Soviet RATAN-600 radiotelescope.
Sdelovaci Tech., Vol. 26, No. 2, p. 42 (1978). In
Czech. – Abstr. in Phys. Abstr., Vol. 81, Abstr. 53824 (1978).

**033.048   A submillimeter broad-band superheterodyne
         radiometer.   Yu. V. Lebskij, V. P. Mezentsev,
L. I. Fedoseev, A. A. Shvetsov.**
Izv. vuzov. Radiofiz., Vol. 20, 1676 - 1679 (1977). In Russian. – Abstr. in Ref. zh., 51. Astron., 7.51.93 (1978).

Astronomy. Volume 13: Radioastronomy.
See Abstr. 003.089.

Attrezzatura strumentale dell'Osservatorio di
Arcetri.   See Abstr. 009.006.

Posibilidades actuales y futuras de observación de
galaxias en el Instituto Argentino de Radioastronomía.
See Abstr. 009.008.

Télescope et radio-télescope.   See Abstr. 032.014.

Observaciones de galaxias en la linea de 21 cm del
hidrogeno neutro con el instrumento de sintesis de Westerbork
(Holanda).   See Abstr. 158.153.

034 Astronomical Accessories (Spectrometers, Photometers, etc.)

**034.001   Ein rechnergesteuertes lichtelektrisches Photopola-
rimeter.** Entwicklung und Inbetriebnahme am
123cm-Teleskop auf dem Calar Alto. Messungen im W3-Kom-
plex.   K. Proetel.
Diss. Ruprecht-Karl-Univ., Heidelberg. 5 + 80 pp. (1978).

Das neue lichtelektrische Zweikanal-Photometer wurde
für das 123cm-Teleskop auf dem Calar Alto entwickelt. Es
stellt ein Messystem dar, mit welchem auf Grundlage von
Photonenzählung unter Kontrolle eines Prozessrechners im
Farbbereich von 3000 bis 11000 Å gearbeitet werden kann.
Dabei stand die vielseitige Verwendbarkeit bei weitgehender
Rechnerunterstützung im Vordergrund. Dieses Ziel wurde da-
durch erreicht, daß verschiedene Baugruppen des Gerätes aus-
wechselbar gestaltet wurden.

**034.002   Spectroscopic observations of stars and planetary
nebulae with a multichannel analogue detector
system.**
G. Adrianzyk, J. C. Baietto, J. P. Berger, C. Fehrenbach,
L. Prevot, A. Vin.
Astron. Astrophys., Vol. 63, 279 - 283 (1978).

A new multichannel detector has been developed for
visual stellar and nebular spectroscopy at moderate resolution.
It consists of a cooled Nocticon tube provisionally mounted
at the focus of the Coudé Echelle spectrograph. The video signal is measured and stored in a digital local memory. The on-
line computer provides extended reduction facilities in order
to have a quick look on the collected data. This equipment
has been used for the observation of the star CI Cyg, of some
planetary nebulae (NGC 6210, 6543, 6572) and of some
comparison stars.

034.003   Television pick-up system for satellite location.
E. Maase.
Veröff. Zentralinst. Phys. Erde, No. 52, Part 3, (see 012.001),
p. 959 - 965 (1977). In German.

In the optoelectronic registration of astronomical objects
by means of television systems some special aspects are to be
taken into account, allowing to derive the points of view
necessary for the development of a satellite location device.
The characteristics of the optimal system and of the scanning
beam determine the form and the amplitude of the output
signal and define the conditions for the choice of the pick-up
tube.

034.004   An inexpensive photometer for amateur astronomy.
R. Dick, A. Fraser, F. Lossing, D. Welch.
J. R. Astron. Soc. Canada, Vol. 72, 40 - 45 (1978).

A simple and inexpensive photoelectric photometer,
using the three-colour UBV system and suitable for use by
amateurs, is described.

034.005   Transistor stabilized supply for photomultipliers.
Yu. K. Kolpakov, A. K. Magnitskij.
Soobshch. Gos. Astron. Inst. Shternberga, No. 199, p. 31 - 34
(1977). In Russian.

**034.006   Amplificatore semplificato per fotometro foto-
elettrico.**
G. A. A. B., N. 48, p. 12 - 14 (1977).

034.007   Polarization effects in grating scanners.
D. S. Hayes, K. H. Rex.
Publ. Astron. Soc. Pacific, Vol. 89, 929 - 934 (1977/78).

The authors discuss polarization effects which enter the
measurement of stellar energy distributions with a grating
scanner when the starlight is partially linearly polarized. If
these effects are ignored, the measured energy distributions

may show both broad, shallow spectral features and sharp
discontinuities. A procedure for correcting for these effects is
discussed. The authors limit the discussion to a grating scanner
used at a Cassegrain focus. As an example, they consider
reddened, early-type stars.

**034.008   Spectrophotometry using an intensifier silicon
vidicon.**
W. Weller, W. Herbst, S. Jeffers.
Publ. Astron. Soc. Pacific, Vol. 89, 935 - 938 (1977/78).

A spectrometer has been assembled from a small-grating
spectrograph and a commercially available intensifier silicon
vidicon detector system. The spectrometer exhibits stability of
response and linearity suitable for spectrophotometry and
evidence is presented that spectrophotometric accuracy (within ±0.01 magnitude) is achieved even on nights of low quality.
A spectrum of such quality of the Of star 9 Sge is presented.
The absolute photometry of Kuan and Kuhi (1976) obtained
for several points in the spectral region 420-560 nm has been
used for calibration and allows absolute measurements of line
strengths to be made.

**034.009   High-dispersion spectroscopy with a 4-cm McMullan
electronographic camera.**
D. L. Harmer, R. J. Keyse, L. J. Sieber.
Observatory, Vol. 98, 57 - 59 (1978).

The RGO 30-inch coudé has a simple small-field spectrograph with an accessible focal surface in which a 4-cm
McMullan electronographic camera has been used as the detector. This instrumental combination giving high-dispersion
stellar spectra has been tested and found suitable for equiva-
lent-width measurements.

**034.010   The velocity system of the Radcliffe Cassegrain
d-camera.**   A. D. Thackeray, R. Wood.
Observatory, Vol. 98, 65 - 67 (1978).

**034.011   A high resolution, fast loading electrographic
camera.**   P. J. Griboval.
Bull. American Astron. Soc., Vol. 9, 613 - 614 (1978).
Abstract.

**034.012   Absolute calibration of 2-dimensional detectors in
the rocket ultraviolet.**   R. C. Bohlin.
Bull. American Astron. Soc., Vol. 9, 614 (1978). – Abstract.

034.013   Characteristics of optical multi-element detectors.
R. Rudolph, W. Schlosser, T. Schmidt-Kaler, H. Tüg.
Astron. Astrophys., Vol. 65, L5 - L6 (1978).

The given tabulation summarizes and compares importan
performance characteristics of multi-element electronic detectors available at the present time of development.

**034.014   The Cassegrain échelle spectrograph at Mt. John Ob-
servatory.**   J. B. Hearnshaw.
Proc. Astron. Soc. Australia, Vol. 3, 102 - 103 (1977).

034.015   A small image-intensifier spectrograph.
M. A. Seeds, S. Horan.
Sky Telesc., Vol. 55, 302 (1978).

**034.016   Statistical properties of atmospheric emission in
the infrared: I. Instrumentation and observations**
S. Bensammar, B. de Batz, F. Perrier-Dampierre.
Astron. Astrophys., Vol. 65, 193 - 197 (1978). In French.

The authors describe an infrared photometer used to
study statistical properties of sky noise at the focus of a
telescope. It consists of a beam splitting system which

gives two channels of simultaneous observations. The authors determine their correlation as a function of separation in the focal plane. They search for possible spatial filtering by varying the size of the cold image of the pupil, using an internal mechanism of a liquid helium dewar.

034.017   **Semi-automatic device for spectra processing.**
          V. E. Stepanov, S. G. Bortnik, B. F. Osak, I. B. Maksyutov, N. F. Tyagun, V. D. Trifonov.
Issled. po geomagn., aehron. i fiz. Solntsa. Moskva, Nauka, 1977, p. 130 - 133. In Russian. − Abstr. in Ref. zh., 51. Astron., 2.51.109 (1978).

034.018   **Programmed chronorecorder for astronomical photographic observations under field conditions.**
          V. M. Kolgunov, Yu. Ya. Goncharenko.
Geod. i kartogr., 1977, No. 8, p. 13 - 17. In Russian. − Abstr. in Ref. zh., 51. Astron., 2.51.190; 52. Geod. Aehrosemka, 2.52.87 (1978).

034.019   **Study of the system of a multi-layer interference filter taking into account small absorption in layers.**
          F. A. Korolev, A. Yu. Klement'eva, A. V. Tikhonravov.
Vestn. Mosk. univ. Fiz., Astron., Vol. 18, No. 3, p. 15 - 19 (1977). In Russian. − Abstr. in Ref. zh., 51. Astron., 2.51.258 (1978).

034.020   **A voltage to frequency converter for astronomical photometry.**   E. Dunham, J. L. Elliot.
Publ. Astron. Soc. Pacific, Vol. 90, 119 - 124 (1978).
          A voltage to frequency converter (VFC) for general use with photomultipliers is described. For high light levels, when the dead-time corrections for a photon counter would be excessive, the VFC maintains a linear response and allows the recording of data at high time resolution . Results of laboratory tests are given for the signal-to-noise characteristics, linearity, stability, and transient response of the VFC when used in conjunction with EMI 9658 and RCA C31034 photomultipliers.

034.021   **A new Michelson stellar interferometer.**
          R. Q. Twiss, W. J. Tango.
Rev. Mexicana Astron. Astrofis., Vol. 3, (see 012.023), 35 - 37 (1977).
          A new 1.87 m fixed-baseline interferometer designed for binary star observation is now in operation at the Italian outstation of the Royal Observatory Edinburgh. It has a theoretical resolving power of 0.015 arcsec and a limiting B magnitude of 6.5. Atmospheric seeing is dynamically corrected, allowing a quantitative measurement of fringe visibility.

034.022   **Puesta a punto de un tubo de imágenes para fotografía directa.**   J. H. Calderon.
Astron. extragalactica, (see 012.015), 171 - 192 (1976).

034.023   **A versatile radiometer for infrared emission measurements of the atmosphere and targets.**
          R. J. Huppi.
Opt. Eng., Vol. 16, 485 - 492 (1977). − Abstr. in Phys. Abstr., Vol. 81, Abstr. 4681 (1978).

034.024   **An infrared upconverter for astronomical imaging.**
          R. W. Boyd, C. H. Townes.
Appl. Phys. Lett., Vol. 31, 440 - 442 (1977). − Abstr. in Phys. Abstr., Vol. 81, Abstr. 7883 (1978).

034.025   **A new Michelson spectrophotometer system.**
          H. L. Johnson.
Rev. Mexicana Astron. Astrofis., Vol. 2, 219 - 230 (1977).
          A new and optically very simple Michelson Fourier Transform Spectrophotometer has been developed. The development includes full consideration of the data processing

problems and their solutions. The raw data are recorded in analog form at the telescope, but all succeding work is done by computers; all the necessary software has been written and is in successful operation.

034.026   **Meßmethoden in der astronomischen Forschung.**
          G. Möstl.
Sterne, 54. Band, 34 - 40 (1978).

034.027   **Instrumental profile of the ASP-20 spectrograph of the ATsU-5 telescope.**   V. N. Milovanov, A. S. Kajdash.
Soln. akt. Alma-Ata, 1977, p. 98 - 104. In Russian. − Abstr. in Ref. zh., 51. Astron., 4.51.102 (1978).

034.028   **Luminous material of constant action for photoelectric measurements.**   O. P. Pyl'skaya.
Pribory i tekh. ehksp., 1977, No. 5, p. 207 - 208. In Russian. Abstr. in Ref. zh., 51. Astron., 4.51.201 (1978).

034.029   **Far-infrared photoconductivity of uniaxially stressed germanium.**
          A. G. Kazanskii, P. L. Richards, E. E. Haller.
Appl. Phys. Lett., Vol. 31, 496 - 497 (1977). − Abstr. in Phys. Abstr., Vol. 81, Abstr. 10496 (1978).

034.030   **A spectrophotometric 'scanner' for astronomical applications.**
Nguyen-Doan, D. Dubet, R. Godon, C.-T. Hua, M. Roixel.
Electron. Appl. Ind., No. 239, p. 50 - 55 (1977). In French. Abstr. in Phys. Abstr., Vol. 81, Abstr. 20638 (1978).

034.031   **Digital recorder for an astronomical electrophotometer.**   V. V. Egorov, V. N. Ivanov, I. I. Krishchuk, M. P. Petrov, R. A. Chajchuk.
Astrometriya i Astrofizika, vyp. (No.) 35, p. 102 - 105 (1978). In Russian.
          The design of an automatic punch register for an astronomical electrophotometer is described. Examples of recording and primary reduction of observations of the satellite Pageos are given.

034.032   **An echelle spectrograph.**   J. B. Hearnshaw.
Sky Telesc., Vol. 56, 6 - 8 (1978).

034.033   **Complex muon-neutron detectors for investigations of variations of cosmic rays. I. Block-scheme, electronics, geometry.**   L. I. Dorman, V. Eh. Drejzin, I. Ya. Libin, Kh. M. Khamirzov, E. V. Shinkarenko, A. A. Chechenov, A. Z. Zhappuev.
Issled. po probl. soln-zemnoj fiz., Moskva, 1977, p. 83 - 88. In Russian. − Abstr. in Ref. zh., 62. Issled. kosm. prostranstva, 5.62.118 (1978).

034.034   **Complex muon-neutron detectors for investigations of variations of cosmic rays. II. Calculations of directivity diagrams, meteorological effects and coupling coefficients.**   L. I. Dorman, I. Ya. Libin, A. Z. Zhappuev, A. A. Chechenov, V. G. Yanke.
Issled. po probl. soln.-zemnoj fiz., Moskva, 1977, p. 89 - 94. In Russian. − Abstr. in Ref. zh., 62. Issled. kosm. prostranstva, 5.62.117 (1978).

034.035   **A magic eye for astronomical spectrophotometry.**
          R. Zurbuchen.
Messenger, No. 12, p. 18 - 19 (1978).

034.036   **Catching all the photons: the CCD.**
          W. Wamsteker.
Messenger, No. 13, 10 - 11 (1978).

034.037   **An automated isophotometer for large photo-**

graphic plates.   H. Maehara, K. Ishida.
Tokyo Astron. Obs. Report No. 69, Vol. 18, 132 - 152 (1978).
In Japanese.

034.038 An experiment on sensitivity of the SIT camera.
G. Sasaki, M. Igarashi, K. Chiba.
Tokyo Astron. Obs. Report No. 69, Vol. 18, 219 - 225 (1978).
In Japanese.

034.039 A semi-automated iris photometer for large photographic plates.
K. Ishida, H. Maehara, M. Ohashi.
Tokyo Astron. Obs. Report No. 69, Vol. 18, 260 - 270 (1978).
In Japanese.

034.040 Polarimeter attached to the multi-channel spectrophotometer.   K. Okida, E. Watanabe, M. Shimizu.
Tokyo Astron. Obs. Report No. 69, Vol. 18, 334 - 343 (1978).
In Japanese.

034.041 A semi-automated blink comparator for large photographic plates.   K. Ishida, H. Maehara, K. Hamajima.
Tokyo Astron. Obs. Report No. 69, Vol. 18, 360 - 370 (1978).
In Japanese.

034.042 Photon-counting magnetograph for measurements of circular polarization in narrow spectral ranges.
V. M. Kuvshinov, L. V. Granytsky (*Granitskij*).
Magnetic stars, (see 012.028), p. 30 (1978). — Abstract.

034.043 Photometric systems of the 35-cm meniscus telescope of the Engelhardt Astronomical Observatory.
L. A. Urasin, I. A. Urasina.
Izv. Astron.Ehngel'gardt. Obs., Kazan'; No. 41 - 42, p. 209 - 214 (1976). In Russian.

034.044 Filter glasses with band-pass characteristics for the red region of the visible spectrum.
M. A. Res, J. Bednarik, F. Hengstberger, C. J. Kok.
Optik, Vol. 49, 433 - 444 (1978).
Filters were made with transmission peaks spaced from 1 to 12 nm (about 4 nm on average). This was achieved by colouring seven different base glasses with NiO, by developing "mixed" glasses (e.g. boro-silicate and mixed alkali-alumina-phosphate glasses coloured with NiO) and by introducing supplementary oxides with cut-on and band-reject characteristics into the NiO-coloured glasses.

034.045 Counterdoped extrinsic silicon infrared detectors.
C. T. Elliott, P. Migliorato, A. W. Vere.
Infrared Phys., Vol. 18, 65 - 71 (1978).
A novel type of doping scheme is described which could lead to background-limited operation of extrinsic, photoconductive i.r. detectors in the 3–5 μm and 8–14 μm bands at much higher operating temperatures than is possible with conventional dopants. The scheme involves compensation of a very deep level with a shallow impurity of the opposite type. A theoretical analysis is given which shows that very exact compensation is not necessary for good detector performance.

034.046 Single mesh narrow bandpass filters from the infrared to the submillimeter region.   K. Sakai, T. Yoshida.
Infrared Phys., Vol. 18, 137 - 140 (1978).

034.047 Circular polarization analyzer of the BTA stellar magnetograph.   I. D. Najdenov, G. A. Chuntonov.
Soobshch. Spets. Astrofiz. Obs., Zelenchukskaya, Vyp. (No.) 16, p. 63 - 65 (1976). In Russian.
The design of a circular polarization analyser based on

Fresnel rhombs and operating in a wide wavelength range is presented.

034.048 The timing system for the laser radar of the Simeiz tracking station.   L. S. Shtirberg, V. V. Kurbasov.
Nauchn. Inf., Vyp. (No.) 38, p. 16 - 25 (1976). In Russian.

034.049 An automated photometer.   A. P. Linnell.
Bull. American Astron. Soc., Vol. 10, 413 (1978).
Abstract.

034.050 An economical Cassegrain high dispersion spectrophotometry system.   L. W. Ramsey.
Bull. American Astron. Soc., Vol. 10, 451 (1978). – Abstract.

034.051 A portable Reticon multichannel spectrometer.
K. H. Nordsieck, J. F. McNall, D. E. Michalski.
Bull. American Astron. Soc., Vol. 10, 451 (1978). – Abstract.

034.052 Response characteristics of a Reticon diode array.
J. Percival.
Bull. American Astron. Soc., Vol. 10, 451 (1978). – Abstract.

034.053 An automated spectrum scanner.
M. K. V. Bappu.
Kodaikanal Obs. Bull., Ser. A, Vol. 2, 64 - 68 (1977).
A rapid-scan single channel spectrum in the Ebert-Fastie arrangement has been constructed and successfully used on the Kavalur 102-cm reflector. The spectral region 3000–11000 Å is covered with appropriate dry-ice cooled photomultipliers. Data acquisition is by a 4K computer with suitably developed software.

034.054 Instrumentation complex of the MANIA experiment.   M. I. Demchuk, O. A. Evseev, G. S. Tsarevskij, V. F. Shvartsman, A. K. Yakushev.
Soobshch. Spets. Astrofiz. Obs., Zelenchukskaya, Vyp. (No.) 20, p. 5 - 17 (1977). In Russian.
A description is presented of the engineering side of the MANIA experiment, which is being carried out for the purpose of detecting relativistic objects (neutron stars, black holes) due to super-fast variability of the optical brightness and polarization. The complex includes a telescope, a two-channel photometer-polarimeter, a multichannel amplitude analyzer, and a fast-operating converter "time-to-number", which is on-line with a computer.

034.055 On the possibility of control of the systematic errors connected with the zero drift in electrophotometers.   Eh. I. Terez.
Astron. Tsirk., No. 967, p. 4 - 6 (1977). In Russian.

034.056 On temporal stability of radioluminescent radiators.   Eh. I. Terez.
Astron. Tsirk., No. 991, p. 2 - 4 (1978). In Russian.

034.057 Instrumental profile of a triple Fabry-Perot interferometer for use in solar spectroscopy.
R. E. Loughhead, R. J. Bray, N. Brown.
Appl. Opt., Vol. 17, 415 - 419 (1978).
A universal filter for use in solar spectroscopy is under development, consisting of a computer-controlled triple Fabry-Perot interferometer. This paper explores the extent to which the performance of such a filter is degraded by off-axis effects under the intended conditions of use. Computed instrumental profiles for both on- and off-axis cases are presented. The calculated bandwidths range from 0.021 Å to 0.047 Å at eight wavelengths of interest in the visible region of the spectrum.

034.058 URSULA: a twin-beam multiband computer-con-

trolled photometer for astronomical applications.
G. A. De Biase, L. Paternò, M. Pucillo, G. Sedmak.
Appl. Opt., Vol. 17, 435 - 441 (1978).

**034.059 Self-scanned photodiode array: high performance
operation in high dispersion astronomical spectrophotometry.**   S. S. Vogt, R. G. Tull, P. Kelton.
Appl. Opt., Vol. 17, 574 - 592 (1978).
    The authors have developed a multichannel spectro-
photometric detector system using a 1024 element self-scanned silicon photodiode array, which is now in routine operation with the high dispersion coudé spectrograph of the University of Texas McDonald Observatory 2.7-m telescope. They
discuss operational considerations in the use of such arrays
for high precision and low light level spectrophotometry. A
detailed description of the system is presented. Performance
of the detector as measured in the laboratory and on astronomical program objects is described, and it is shown that
these arrays are highly effective detectors for high dispersion
astronomical spectroscopy.

**034.060 Balloon-borne ultraviolet stellar echelle spectro-
graph.**
R. Hoekstra, T. M. Kamperman, C. W. Wells, W. Werner.
Appl. Opt., Vol. 17, 604 - 613 (1978).

**034.061 Signal-to-noise ratios of multiplexing spectrometers
in high backgrounds.**   R. F. Knacke.
Appl. Opt., Vol. 17, 684 - 685 (1978).

**034.062 Mylar beam-splitter efficiency in far infrared inter-
ferometers: angle of incidence and absorption effects.**   D. A. Naylor, R. T. Boreiko, T. A. Clark.
Appl. Opt., Vol. 17, 1055 - 1058 (1978).

**034.063 Direct, quasi-telecentric, and telecentric combina-
tions of interference filters with large astronomical
Schmidt cameras.**   J. Meaburn.
Appl. Opt., Vol. 17, 1271 - 1274 (1978).
    Several ways of combining narrow bandwidth inter-
ference filters with a Schmidt camera, of 1.2-m aperture, are
described. A completely telecentric method is proposed in
which very narrow filters can be used.

**034.064 Composite bolometers for submillimeter wave-
lengths.**   N. S. Nishioka, P. L. Richards,
D. P. Woody.
Appl. Opt., Vol. 17, 1562 - 1567 (1978).

**034.065 Novel spectrophotometer for the investigation of
short term variability in stellar spectra.**
T. Stiff, S. Jeffers.
Appl. Opt., Vol. 17, 1811 - 1814 (1978).
    A spectrophotometer has been built whose output signal
is a measure of the line strength only. The spectrophotometer
is used to look at the emission feature and the adjacent continuum in rapid succession by means of magnetic modulation
of the electron image of the optical spectrum in an image tube,
thus generating a modulated signal which is detected with a
lockin amplifier. This detection technique essentially subtracts
off an instrumental dark current signal due to sky background
and the signal due to the continuum of the star giving a real
time measure of the line strength only. The design of the
instrument, its laboratory calibration, and some preliminary
observational data are presented.

034.066 The utilization of imaging tubes in astronomy.
M. Kenler.
Acta Electron., Vol. 20, 297 - 301 (1977). In French. – Abstr.
in Phys. Abstr., Vol. 81, Abstr. 40829 (1978).

034.067 Detectors for infrared astronomy.   F. C. Gillett,
E. L. Dereniak, R. R. Joyce.
Opt. Eng., Vol. 16, 544 - 550 (1977). – Abstr. in Phys. Abstr.,
Vol. 81, Abstr. 45231 (1978).

**034.068 Design and operation of an infrared spatial inter-
ferometer.**   D. W. McCarthy, F. J. Low, R.
Howell.
Opt. Eng., Vol. 16, 569 - 574 (1977). – Abstr. in Phys. Abstr.,
Vol. 81, Abstr. 45232 (1978).

034.069 Ultrasoft X-ray source for astrophysical purposes.
C. La Padula, E. Costa, P. Ubertini, S. Ugazio.
Mem. Soc. Astron. Italiana, Vol. 47, 95 - 99 (1976).
    A low power ultrasoft X-ray source stabilized to ± 1% is
described. The instrument was successfully employed to assembly facilities for calibration and test of image multiwire
proportional counters and glancing incidence optics to be
employed in a focussing X-ray astronomy payload.

034.070 Very wide field UV cameras.   M. Viton.
Mem. Soc. Astron. Italiana, Vol. 47, (see 012.036),
377 - 385 (1976).

**034.071 A stable precision astronomical electrometer
amplifier.**   J. P. Oliver.
J. Astron. Soc. Western Australia, Vol. 300, April, p. 2 - 5
(1978).

**034.072 A large area (300 cm²) gas scintillation proportional
counter for X-ray astronomy.**
R. D. Andresen, E.-A. Leimann, A. Peacock, B. G. Taylor.
IEEE Trans. Nucl. Sci., Vol. NS-25, 800 - 807 (1978).
Abstr. in Phys. Abstr., Vol. 81, Abstr. 49321 (1978).

**034.073 A large-area gas scintillation proportional counter
for X-ray astronomy.**
D. F. Anderson, W. H.-M. Ku, R. Novick, M. Scheckman.
IEEE Trans. Nucl. Sci., Vol. NS-25, 813 - 816 (1978). – Abstr.
in Phys. Abstr., Vol. 81, Abstr. 49322 (1978).

**034.074 On long-baseline amplitude interferometers in
astronomical applications.**
A. H. Greenaway, J. C. Dainty.
Opt. Acta, Vol. 25, No. 3, p. 181 - 189 (1978). – Abstr. in
Phys. Abstr., Vol. 81, Abstr. 53832 (1978).

034.075 Astronomical equipment (GDR).
Nauchn. pribory, 1977, No. 14, p. 31 - 34. In Russian. – Abstr. in Ref. zh., 51. Astron., 7.51.145 (1978).

**Refractive index measurements of a germanium
sample.**   See Abstr. 022.114.

An aspheric objective for a coronagraph.
See Abstr. 031.015.

**Il "Quantimet" come analizzatore di immagini
astronomiche.**   See Abstr. 031.310.

**Algorithm and program for analysis of information
from the "Filin" device.**   See Abstr. 031.407.

**Tecniche e strumentazione per radiometria solare
ed atmosferica.**   See Abstr. 032.045.

**Wideband laser-interferometer gravitational-radia-
tion experiment.**   See Abstr. 066.301.

Sternspektroskopie.   See Abstr. 114.011.

035 Clocks and Frequency Standards, Sundials

035.001 Instruments de mesure du temps.
Astronomie, Vol. 92, 60 (1978).

035.002 When sundials appeared in Old Russia.
V. L. Chenakal.
Zemlya i Vselennaya, 1978, No. 1, p. 72 - 75. In Russian.

035.003 Electronic corrector of the digital quartz clock indications. U. Nowak, S. Oszczak.
Postępy Astron., Tom 25, 187 - 190 (1977). In Polish.
The authors present the device used to automatic introduction of corrections for quartz clock indication. This device
was constructed at the Observing Station No. 1151 in Olsztyn.
The device enables (i) connection of the working clock time
to the world time system, (ii) a comparison of quartz generators with an accuracy of half sinusoid (the accuracy of this
comparison is equal to 5×10^{-8} s for generators with standard frequency 5 MHz).

035.004 Rilievo tempi con segnali registrati su nastro mag-
netico. F. Cerchio.
Coelum, Vol. 46, 11 - 18 (1978).

035.005 A latitude-independent sundial.
J. G. Freeman.
J. R. Astron. Soc. Canada, Vol. 72, 69 - 80 (1978).

035.006 Eine weitere Methode zur Theorie der "Homogenen
Sonnenuhr" (Hybrid-Sundial). G. Haslinde.
Sterne, 54. Band, 51 - 59 (1978).

035.007 Sidereal conversion of a Heathkit digital clock.
R. E. Gilbert, R. W. Sinnott.
Sky Telesc., Vol. 55, 433 - 436 (1978).

035.008 Bifilar gnomonics. F. W. Sawyer III.
J. British Astron. Assoc., Vol. 88, 334 - 351 (1978).

035.009 Die Kunstsonnenuhr im Stadtpark zu Dessau.

A. Zenkert.
Sterne, 54. Band, 119 - 120 (1978).

035.010 Cadran solaire a équation. C. Pommier.
Astronomie, Vol. 92, 283 - 287 (1978).

035.011 La construction de cadrans solaires. J. Bosard.
Bull. Soc. Astron. Liège, Vol. 40, 80 - 84 (1978).

035.012 Travel time and accuracy of reception of remote
time signals on short waves, WWV and WWVH, as
received in Tokyo. S. Iijima, G. Shibutani, T. Sakai.
Tokyo Astron. Bull., Second Ser., No. 253, p. 2917 - 2923
(1978).
The effective velocity of the radio time signals, WWV and
WWVH, is obtained for the period from 1968 through 1976
by use of the data of the reception times corrected by UTC
(NBS) minus UTC (TAO). It is found that the mean effective
velocity stays within a limit of 284.5 ± 1 km/ms for both
signals, and that the standard deviation for a single reception
becomes ± 0.23 ms on average.

035.013 Effect of environmental conditions on the rate of a
cesium clock. S. Iijima, K. Fujiwara, H.
Kobayashi, T. Kato.
Ann. Tokyo Astron. Obs., Second Ser., Vol. 17, 50 - 67 (1978).
Effects of environmental conditions on the rate of a
cesium clock is examined experimentally by using two commercial cesium beam oscillators (HP) equipped with super
tubes.

A sundial project for introductory astronomy classes.
See Abstr. 014.032.

Pourquoi l'heure est-elle retardée d'une seconde
de temps à autre? See Abstr. 044.009.

An experiment for the potential blue shift at the
Norikura Corona Station. See Abstr. 066.242.

036 Photographic Auxiliaries

036.001   On the storage of an exposed astronomical film
A-600 U before its photographic development.
O. D. Dokuchaeva.
Tr. Gos. Astron. Inst. Shternberga, Vol. 47, 99 - 104 (1976).
In Russian.

036.002   **Filmkunde für Astroamateure.** Teil I: Bildentstehung,
Schwärzung und Kontrast.   T. Spahni.
Orion, 36. Jahrg., 11 - 16 (1978).

036.003   **Microwave baking of photographic plates – preliminary results.**   S. J. Solomon.
Bull. American Astron. Soc., Vol. 9, 599 (1978). – Abstract.

036.004   **Filmkunde für Astroamateure.** Teil II: Spektrale
Empfindlichkeit und Verarbeitung von Filmen.
T. Spahni.
Orion, 36. Jahrg., 58 - 64 (1978).

036.005   **Orwo-ZU 2 plate with improved properties.**
W. Högner, R. Ziener.
Astron. Nachr., Band 299, 159 - 163 (1978). In German.

036.006   **Recent advances in planetary photography.**
C. F. Capen.
Strolling Astron., Vol. 27, 47 - 51 (1978).

036.007   **Hypersensibilisering av fotografiska emulsioner.**
N. Hansson.
Astron. Tidsskr., Årg. 11, 56 - 64 (1978).

036.008   **A masking technique for astronomical photography.**
J. British Astron. Assoc., Vol. 88, 362 - 364 (1978).

036.009   **A vacuum film holder assembly for astronomical
applications.**   F. P. Andrews.
South. Stars, Vol. 27, 98 - 103 (1978).
A brief evaluation of the suitability of film-based emulsions as opposed to glass plates is made with respect to astronomical applications, and the design of a film holder using a
simple suction system to ensure perfect flatness and registration of sheet film at the focal plane of an astrograph or telescope is outlined.

036.010   **Hypersensitizing processes in astrophotography.**
B. V. Barlow.
Contemp. Phys., Vol. 19, 47 - 64 (1978). – Abstr. in Phys.
Abstr., Vol. 81, Abstr. 32348 (1978).

036.011   **Hypersensitization and testing of astronomical
plates for threshold imagery.**   A. G. Smith.
J. Appl. Photogr. Eng., Vol. 3, 205 - 208 (1977). – Abstr. in
Phys. Abstr., Vol. 81, Abstr. 32351 (1978).

036.012   **On the new astronomical films A-500 H and A-600 H.**
I. I. Brejdo, O. M. Mikhajlova, P. V. Mejklyar,
M. D. Mirmil'shtejn, I. A. Baranova.
Astron. Tsirk., No. 971, p. 4 - 6 (1977). In Russian.

**Optimizing photographic detection at large tele-
scopes.**   See Abstr. 031.258.

Positional Astronomy, Celestial Mechanics

041 Positional Astronomy, Astrometry

041.001 **General catalogue of stars observed with astrolabes (1957 - 1975). Individual corrections to places of FK4 stars.** G. Billaud, G. Guallino, G. Vigouroux.
Astron. Astrophys., Suppl. Ser., Vol. 31, 159 - 165 (1978). In French.
From 1958 to 1975, 20 catalogues of stars have been published. These catalogues spread from $\delta = -63°6$ to $\delta = 79°6$ and give corrections to the positions of the FK4 catalogue. The results are given for 1139 stars with a mean error of 4 ms in α and $0.''06$ in δ. The discussion and the method of computation are described in the main journal.

041.002 **General catalogue of FK4 stars observed with astrolabes (1958 - 1975).**
G. Billaud, G. Guallino, G. Vigouroux.
Astron. Astrophys., Vol. 63, 87 - 95 (1978). In French.
From 1958 up to 1975, 20 catalogues of stars have been published. These catalogues spread from $\delta = -63°6$ to $\delta = 79°6$ and give corrections to the positions given by the fundamental catalogue FK4. When studying the catalogues, a systematic and periodic error of an unknown origin is found. When this is taken into account it allows catalogues to be compared: the observed zones cover one another and it is possible assigning an arbitrary value to one of the constants, to express corrections in a single system. The results obtained for 1139 fundamental stars with a mean error of 4ms in α and $0.''06$ in δ, show systematic and important discrepancies with the FK4 catalogue, the amplitude of which reaches 10ms for $\Delta\alpha$ and $0.''2$ for $\Delta\delta$. The individual corrections will be published in Astronomy and Astrophysics Supplements.

041.003 **On the maximum of weight of a linear function.**
A. P. Gulyaev.
Soobshch. Gos. Astron. Inst. Shternberga, No. 193, p. 18 - 21 (1977). In Russian.
The difficulty of the analytical determination of the best reference stars distribution in series of observations is shown. An investigation of the weights of linear function coefficients as an example is presented.

041.004 **On the system of the Moscow catalogue of faint stars (zone 45°-60°).** L. M. Khommik.
Soobshch. Gos. Astron. Inst. Shternberga, No. 200, p. 37 - 41 (1977). In Russian.
A comparison of the right ascensions and declinations of Moscow KSZ (zone 45-60°) with the AGK3R and Yale Catalogue is made. Using the systematical differences FK4—FK3 and $W2_{50}$—FK3 the author has obtained the differences between the KSZ and the FK3 systems. The right ascension system reproduces the system of the reference catalogue rather well. Judging by the comparison of the differences Δa_a with the analogous differences in the zone 30—45°, one can suppose that there are some systematical errors in the catalogue AGK3R. The declination system of the reference catalogue is reproduced worse.

041.005 **Experiment on Fundamental Catalogue improvement by means of Venus photographic positional observation results.** B. S. Vozdvizhenskij, V. V. Podobed.
Astron. Zh. Akad. Nauk SSSR, Vol. 55, 188 - 190 (1978). In Russian. English translation in Soviet Astron., Vol. 22, No. 1.
On the basis of observations of Venus made at the wide-angle astrograph of the Sternberg Astronomical Institute the FK4 orientation corrections are calculated.

041.006 **Reduction of the Wrocław FKSZ catalogues to the FK4 system.** O. V. Kiyaeva, D. D. Polozhentsev.
Postępy Astron., Tom 24, 215 - 221 (1976). In Polish.

041.007 **Éphémérides et position de Mars en 1975—1976.** S. Débarbat, J. Pham Van, M. Sanchez.
Astron. Astrophys., Vol. 64, 281 - 284 (1978).
This paper contains an analysis of Mars astrolabe observations, details of which have been published elsewhere (Débarbat, 1978; Pham Van et al., 1978; Sanchez, 1978). These observations were obtained at three different stations (Paris, Cerga, San Fernando) and the results have been investigated to study the precision of the observations and to compare them, respectively, with American Ephemeris and with new improved ephemerides (Laubscher, 1971).

041.008 **Accuracy of fundamental positions and proper motions.** T. Lederle.
Bull. Inf. Centre Données Stellaires, No. 14, p. 62 - 68 (1978).
Information on errors in fundamental catalogues. Individual and systematic errors. Mean errors of the FK4. Practical application. Accuracy of the FK5.

041.009 **A simple method of deriving equatorial coordinates from astronomical photographs.**
G. S. Rossano, M. F. A'Hearn.
Publ. Astron. Soc. Pacific, Vol. 89, 922 - 924 (1977/78).
A simple method of deriving equatorial coordinates of moderate accuracy from astronomical photographs is presented. When used with the Palomar Observatory Sky Survey, positions of order 10 arc second accuracy can be derived.

041.010 **Les bases astrométriques de l'astrophysique.** J. Dommanget.
Astronomie, Vol. 92, 109 - 123 (1978).

041.011 **An alternative approach to Brosche's model for catalog comparisons.** H. Yasuda.
Publ. Astron. Soc. Japan, Vol. 30, 173 - 190 (1978).
A method with a new statistical approach to Brosche's (1966) model for catalog comparisons is proposed. The approach is based on the minimum estimate of total errors of catalogued positions. The systematic corrections of the catalogued positions listed in an observational catalog are determined by a statistical criterion. The magnitude equation is represented by linear variations of each significant coefficient

in the adopted model. The method is applied to the catalog comparison Tokyo—FK4 and the Tokyo catalog system is determined with an accuracy of about ±0."01.

041.012 **Real-time astrometry.**   L. G. Taff.
    Bull. American Astron. Soc., Vol. 9, 597 - 598 (1978). – Abstract.

041.013 **Radiointerferometrie als Wegbereiter eines radio-astronomischen Bezugssystems.**   H. G. Walter.
Sterne Weltraum, Jahrg. 17, 157 - 161 (1978).

041.014 **Astrometric accuracy of Palomar Schmidt plates.**   A. Fresneau.
Astron. J., Vol. 83, 406 - 410 (1978).
    It has generally been admitted (Hunstead 1973) that the best accuracy obtainable in position determinations from the Palomar Sky Survey has been nominally set to 0.5 arcsec for the prints and 0.3 arcsec for glass copies. The present study is an attempt to prove that using the original plates, a still better accuracy of 0.15 arcsec can be obtained and to identify the most important phenomena affecting the accuracy of the reduction procedures and to investigate whether their effects can be reduced by least-squares interpolation techniques. A reference grid is used to separate the effects of displacements of the emulsion from those of the metric properties of the curved field. The author concludes that the disadvantages of bending the plates are remarkably small.

041.015 **Observations of Saturn with the astrolabe of the CERGA Observatory during the years 1974, 1975, 1976.**   J. Pham-Van, G. Billaud, H. Choplin, C. Delmas, P. Grudler, G. Guallino, F. Mignard, G. Vigouroux.
Astron. Astrophys., Suppl. Ser., Vol. 32, 323 - 324 (1978). In French.
    This paper contains results for Saturn observed with the Danjon astrolabe. During the 1974-75 winter 26 east and west transits were observed and during the winter 1975-1976, 34. The tables give the date, the zenith distance residuals, the corrections for defective illumination and the indication of the observer.

041.016 **Mars observations with Algiers astrolabe (winter 1973).**   J. Pham-Van.
Astron. Astrophys., Suppl. Ser., Vol. 32, 325 - 326 (1978). In French.
    The results of Mars observations with Danjon astrolabe during winter 1973 at Algiers Observatory are given. Tables contain residuals in zenith distance R, calculated with group corrections, for each transit (east or west), corrections ϕ_1 and ϕ_2 for defective illumination, Julian Dates (UT) and the observer's name.

041.017 **Observations of Saturn with the astrolabe of the Algiers Observatory during the years 1969–1974.**
F. Meyer, J. Pham-Van.
Astron. Astrophys., Suppl. Ser., Vol. 32, 327 - 330 (1978). In French.
    This paper contains the results of the five campaigns of Saturn at the Algiers Observatory. 66 east and west transits were observed. Tables contain residuals (zenith distance) for each transit (east and west) and corrections ϕ_1 and ϕ_2 for defective illumination. Another table contains $\Delta\alpha$ and $\Delta\delta$. 20 double transits have been obtained during the years 1969–1974 to measure Saturn's position.

041.018 **Observations of Mars with the Danjon astrolabe of the San Fernando Observatory.**   M. Sánchez.
Astron. Astrophys., Suppl. Ser., Vol. 32, 331 - 333 (1978). In French.
    This paper contains the results for Mars observed with the Danjon astrolabe through two campaigns (winter 1973-74

and winter 1975-76).

041.019 **Observations of Mars position with the Paris astrolabe.**   S. Débarbat.
Astron. Astrophys., Suppl. Ser., Vol. 32, 335 - 337 (1978). In French.
    This paper contains results for Mars observed with the Paris astrolabe during the 1975–1976 campaign. Twenty transits (Chollet et al. 1977) have been observed between 1975 October 23 and 1976 January 25. Among them, 14 give 7 double transits for which the corrections to American Ephemeris are given in a table (right ascension and declination). Another table gives the same quantities calculated for the ephemerides (Kaplan et al. 1976) based on Clemence's theory and the elements of Mars by Laubscher.

041.020 **Astrometric techniques for the observation of planetary satellites.**   D. Pascu.
Planetary satellites, (see 003.007), p. 63 - 86 (1977).

041.021 **Results of studying astronomical observations of many years with a photoelectric transit instrument.**
M. P. Mishchenko, A. V. Shiryaev.
Astrometriya i Astrofizika, Kiev, Vyp. (No.) 34, p. 75 - 77 (1978). In Russian.
    Results obtained with a photoelectric transit instrument during a period of 30 years are briefly considered.

041.022 **Spectral analysis of a ten-year series of observations obtained with two astrolabes and with one photo-electric transit instrument in Irkutsk.**   V. I. Sergienko, A. I. Yazev, S. A. Sergienko, L. S. Burdejnaya, G. I. Balashova, I. M. Selenchuk, V. S. Kudeeva.
Astrometriya i Astrofizika, Kiev, Vyp. (No.) 34, p. 77 - 79 (1978). In Russian.
    Spectral analysis of the differences of parallel observations with two astrolabes and one photoelectric transit instrument revealed systematic errors of seasonal origin which are peculiar for each instrument. To avoid such errors in studying the irregularity of the earth's rotation the use of different instruments is recommended.

041.023 **Results of concurrent observations with two astrolabes in Irkutsk during the period 1966.0–1975.8.**
V. I. Sergienko, S. A. Sergienko, G. I. Balashova, V. S. Kudeeva.
Astrometriya i Astrofizika, Kiev, Vyp. (No.) 34, p. 82 - 86 (1978). In Russian.

041.024 **Fundamental astrometry and the definition of the pole.**   C. A. Murray.
Q. J. R. Astron. Soc., Vol. 19, 187 - 193 (1978).

041.025 **The orbits of five minor planets and corrections to the FK4 equator and equinox.**
R. L. Branham, Jr.
Astron. J., Vol. 83, 675 - 681 (1978).
    Minor planets 6 (Hebe), 7 (Iris), 8 (Flora), 9 (Metis), and 15 (Eunomia) were used to determine the equator and equinox errors of the FK4 catalog system. Other quantities derived were: the secular variations of the equator and equinox; the elements of the Earth's orbit; the secular variation of the obliquity; and a correction to Newcomb's general precession in longitude. The analysis was based on a total of 14 761 observations in right ascension and 14 515 in declination.

041.026 **Preliminary results of a comparison of catalogues of proper motions of stars with respect to galaxies.**
S. P. Rybka.
Astrometriya i Astrofizika, vyp. (No.) 35, p. 3 - 9 (1978). In Russian.
    Some results are given of comparison of three proper-

motion catalogues with respect to galaxies: Pulkovo-Goloseevo, Pulkovo-Moscow. Systematic errors depending on stellar magnitude were found in proper motions. Precessional corrections Δk and Δn are shown to be affected by these errors.

041.027 On photographic positions of Venus.
A. B. Onegina.
Astrometriya i Astrofizika, vyp. (No.) 35, p. 21 - 30 (1978). In Russian.
Methods are described of measurement and reduction of Venus images on negatives taken with the 400/5500 astrograph. The methods of eliminating atmospheric dispersion and phase effects are also described.

041.028 Photographic position observations of Mars with the 400-mm astrograph at the Main Astronomical Observatory of the Ukrainian Academy of Sciences (1971 - 1972). E. M. Sereda, E. M. Izhakevich.
Astrometriya i Astrofizika, vyp. (No.) 35, p. 31 - 43 (1978). In Russian.
107 positions of Mars are given in the SAO reference system.

041.029 Position observations of Venus at the Main Astronomical Observatory of the Ukrainian Academy of Sciences (1967 - 1971). E. B. Vinnikova, Eh. A. Gerts, A. B. Onegina, E. M. Sereda.
Astrometriya i Astrofizika, vyp. (No.) 35, p. 44 - 50 (1978). In Russian.

041.030 Problems of modern fundamental astrometry.
A. A. Nemiro.
Problems of observational and theoretical astronomy, Moscow – Leningrad, 1977, (see 003.012), p. 10 - 25. In Russian. Abstr. in Ref. zh., 51. Astron., 5.51.116 (1978).

041.031 Treatment of series of fundamental stars by the quasi-absolute method. V. A. Sinyaev.
Odessk. univ. Odessa, 1977. 12 pp. In Russian. – Abstr. in Ref. zh., 51. Astron., 5.51.120 (1978).

041.032 Determinación de acimutes por observación de la polar. Método micrométrico. Programa de cálculo automático. M. J. Sevilla, R. Chueca.
Tec. Topogr., Vol. 5, Núm. 16, 16 pp. (1977) = Univ. Complutense – Fac. Cienc., Madrid, Semin. Astron. Geod., Publ. núm. 95.
El método micrométrico para la determinación del acimut de una dirección por observación de la estrella Polar, consiste, básicamente, en medir el ángulo entre la dirección de la estrella y la de una referencia auxiliar, mediante la utilización del micrómetro del instrumento.

041.033 Astrometric accuracy with large reflectors.
I. R. King.
ESO Sci. Prepr. No. 27, 5 pp. (1978).
Berkeley experience with astrometric measurements on reflector plates is used to estimate the accuracy that ground-based work can achieve in proper motions, and with less certainty, in parallaxes. Space planning should compare with optimal, rather than ordinary, ground-based accuracies; the difference is considerable.

041.034 Declination observations of circumpolar stars with the meridian circle. R. Fukaya.
Tokyo Astron. Obs. Report No. 69, Vol. 18, 211 - 218 (1978). In Japanese.

041.035 Meridian observations of major and minor planets, 1974–1977. H. Yasuda, R. Fukaya, H. Hara, H. Ishii, Y. Adachi, S. Isobe, N. Miyauchi, S. Suzuki.
Tokyo Astron. Bull., Second Ser., No. 252, p. 2895 - 2916 (1977).
This report contains positions of the major planets (Mercury, Venus, Mars, Jupiter, Saturn, Uranus, and Neptune) and of some of minor planets (Ceres, Pallas, Juno, and Vesta) observed with the Tokyo Meridian Circle during the period 1974–1977.

041.036 Possible application of the method of independent calculation of quasi-geocentric directions and distances to artificial satellites. I. I. Dmitrotsa.
Nauchn. Inf., Vyp. (No.) 38, p. 50 - 64 (1976). In Russian.
A method of independent calculation of quasi-geocentric directions and distances of artificial satellites is described together with a geometric interpretation and several practical recommendations. The accuracy of different methods of position determinations of satellites is compared.

041.037 Determination of absolute declinations of stars according to the Sanders-Raimond method.
A. A. Krasikova.
Izv. Astron. Ehngel'gardt. Obs., Kazan', No. 41 - 42, p. 71 - 81 (1976). In Russian.

041.038 Observations of major planets with the meridian circle of the Engelhardt Astronomical Observatory.
Eh. N. Vorob'eva, A. I. Nefed'eva, A. A. Krasikova, A. Yu. Yatsenko.
Izv. Astron. Ehngel'gardt. Obs., Kazan', No. 41 - 42, p. 82 - 85 (1976). In Russian.

041.039 Zero-point determination of the reference system by radio interferometry. H. G. Walter.
Proceedings of the European Workshop on Space Oceanography, Navigation and Geodynamics, Schloss Elmau, Germany, 16 - 21 January 1978, p. 339 - 340.

041.040 Observations of Mars with the photographic zenith tube. A. K. Babcock.
Bull. American Astron. Soc., Vol. 10, 402 (1978). – Abstract.

041.041 Investigation of the influence of the magnitude equation on the position of an object from observations of nova Cygni. N. M. Bronnikova, I. A. Shramko.
Astron. Tsirk., No. 970, p. 7 - 8 (1977). In Russian.

041.042 Azimuth calculations for Polaris by means of special tables. M. I. Kuz'min, I. I. Tishkin.
Geod. i kartogr., 1978, No. 1, p. 18 - 22. In Russian. – Abstr. in Ref. zh., 51. Astron., 7.51.142; 52. Geod. Aehrosemka, 7.52.133 (1978).

041.043 Experimental observations of the sun with the astrolabe at CERGA Observatory during the summer of 1976. F. Chollet, J. Demarq, F. Laclare.
Astron. Astrophys., Suppl. Ser., Vol. 33, 7 - 9 (1978). In French.
This paper reports the experimental results of a second observational campaign of solar positions similar to that which is reported in an earlier paper (Chollet and Laclare 1977). Some changes have been made in the computational procedure, because a new astrolabe is used devoted exclusively to solar observations. The authors have used BIH data to correct the observations. As in the first paper, they have assumed that the instrumental zenith distance remains constant during the whole observing period each day. The results therefore provide no more than a confirmation of the quality of the method. It should be noted that the $\Delta\alpha$ results are very similar to those obtained in 1975. The same is not true for the Y and Δd results, as a consequence of the change of instrument.

Stellar testing catalogue.
See Abstr. 002.002.

The determination of the proper motions of the AGK3R stars.   See Abstr. 002.031.

Catalogue of 202 stars of the supplementary list of the Catalogue of Faint Stars in the declination zone from −15° to +30°.   See Abstr. 002.049.

Declinations of 125 zenith stars from observations at the Engelhardt Astronomical Observatory. See Abstr. 002.053.

Bibliographie des publications ayant rapport avec l'astrolabe Danjon ou avec d'autres astrolabes de types nouveaux (1948–1976).   See Abstr. 002.061.

Polar nights of Spitzbergen at the disposal of astrometry.   See Abstr. 011.019.

A set of computer programmes for the adjustment of overlapping plates.   See Abstr. 021.008.

Transatlantic geodesy by long baseline interferometry.   See Abstr. 031.205.

On the phase correction for position observations of planets.   See Abstr. 031.210.

A new method of reducing PZT observations. See Abstr. 031.223.

Determination of the equatorial coordinates of objects by means of an astrotelevision device. See Abstr. 031.267.

On the accuracy of the positions of celestial objects determined from a linear model.   See Abstr. 031.293.

Astrometric expectations.   See Abstr. 031.315.

A fast and accurate position measurement and finding chart production system for glass copies of the Palomar Sky Survey.   See Abstr. 031.402.

A new method of determining pivot errors of meridian circles.   See Abstr. 032.021.

On the beginning of observations with the Danjon astrolabe in Simeiz.   See Abstr. 032.063.

Planetary theories and observational data. See Abstr. 042.052.

Observations of the light aberration of stars of different spectral types with a vertical circle. See Abstr. 043.005.

Possibilities of application of the radiointerferometric equipment of the Institute of Space Research to geodesy and astrometry.   See Abstr. 046.035.

Perspectives of solving fundamental problems of astrometry with the help of superlong baseline interferometry and special space experiments.   See Abstr. 046.036.

Tests of general relativity using astrometric and radio metric observations of the planets. See Abstr. 066.318.

Observations of Mars with the astrolabe of the CERGA observatory during the winter 1975 - 1976. See Abstr. 097.002.

Observations of Jupiter with the astrolabe of the CERGA observatory during the winter 1976-1977. See Abstr. 099.003.

Observations of Saturn with the astrolabe of the CERGA observatory (January - April 1977). See Abstr. 100.002.

Astrometric studies of 45 Tauri, 97 Tauri, 102 ι Tauri, 31 o^2 Cygni, and selected stars in those regions. See Abstr. 111.007.

Astrometry and photometry of stars in the vicinity of the Hulse-Taylor binary pulsar.   See Abstr. 113.065.

Precise optical positions of radio sources in the FK4-system. Results from a pilot program.   See Abstr. 141.174.

042 Celestial Mechanics, Figure of Celestial Bodies

042.001 Orbital resonance in a dissipative medium.
R. Greenberg.
Icarus, Vol. 33, 62 - 73 (1978).

Orbital resonances tend to force bodies into noncircular orbits. If a body is also under the influence of an eccentricity-reducing medium, it will experience a secular change in semimajor axis which may be positive or negative depending on whether its orbit is exterior or interior to that of the perturbing body. Thus a dissipative medium can promote either a loss or a gain in orbital energy. This process may explain the resonant structure of the asteroid belt and of Saturn's rings. The process would also affect the radial distribution of preplanetary material. Moreover, it provides an explanation for the large amplitude of the Titan-Hyperion libration.

042.002 On the restricted problem of three finite bodies.
M. Šidlichovský.
Bull. Astron. Inst. Czechoslovakia, Vol. 29, 14 - 18 (1978).

The main purpose of this paper is a generalization of the transformation to a rotating and pulsating coordinate system when the motion of the primaries is determined by a general force function $\Phi(r)$. This approach enables one to take into account the oblateness of bodies in the restricted three-body problem. Equations for the determination of libration points are a straightforward consequence of this transformation. For the non-circular orbit of primaries the existence of libration points is possible only for special cases.

042.003 Stabilization of Kepler's problem. A. Stokes.
Celestial Mech., Vol. 16, 27 - 34 (1977).

A regularization of Kepler's problem due to Moser is used to 'stabilize' the equations of motion, that is, imbed a particular solution of Kepler's problem in a Lyapunov stable system.

042.004 Bifurcation points in the planar problem of three bodies. K. Zare.
Celestial Mech., Vol. 16, 35 - 38 (1977).

An essential parameter in the planar problem of three bodies is the product of the square of the angular momentum and of the total energy ($c^2 H$). The role of this parameter, which may be called a bifurcation parameter, in establishing regions of possible motions has been shown by Marchal and Saari (1975) and Zare (1976). There exist critical values of this parameter below which exchange between bodies cannot occur. These critical values may be called bifurcation points. This paper gives an analytical criterion to obtain these bifurcation points for any given masses of the participating bodies.

042.005 Modification of an equation of Szebehely.
F. Morrison.
Celestial Mech., Vol. 16, 39 - 40 (1977).

042.006 Solution of the plane problem of n fixed centres.
G. T. Arazov.
Celestial Mech., Vol. 16, 41 - 44 (1977). In Russian.

For a special choice of parameters the plane problem of the motion of a passively gravitating material point in the gravitation field of n fixed centres is reduced to quadratures.

042.007 On convergence of an asymmetrical body potential expansion in spherical harmonics.
C. (K. V.) Kholshevnikov.
Celestial Mech., Vol. 16, 45 - 60 (1977).

The gravity potential of an arbitrary body T is expanded in a series of spherical harmonics and rigorous evaluations of the general term V_n of the expansion are obtained. It is proved that V_n decreases on the sphere enveloping T according to the power law if the body structure is smooth. For a body

with analytic structure, V_n decreases in geometric progression. The exactness of these evaluations is proved for bodies having irregular and analytic structures. For the terrestrial planets $V_n = 0 (n^{-5/2})$.

042.008 The existence of families of periodic orbits in the N-body problem. J. D. Hadjidemetriou.
Celestial Mech., Vol. 16, 61 - 76 (1977).

It is proved that monoparametric families of periodic orbits of the N-body problem in the plane, for fixed values of all masses, exist in a rotating frame of reference.

042.009 A discussion of long-term numerical solutions of the Jupiter–Saturn–Sun system. P. E. Nacozy.
Celestial Mech., Vol. 16, 77 - 86 (1977).

A detailed discussion of recent numerical studies concerning the stability of the Jupiter–Saturn–Sun system is presented.

042.010 Application of spin-orbit theory to a class of orbit-orbit resonances: the case of Titan-Hyperion.
L. Blitzer.
Celestial Mech., Vol. 16, 87 - 95 (1977).

The orbit-orbit interaction of two satellites of greatly unequal mass is studied under the condition that the more massive satellite moves in a circular (unperturbed) orbit that lies entirely inside the orbit of the smaller (perturbed) body. It is shown that this system is equivalent in every respect to a special case of spin-orbit coupling. On this basis, conditions for resonance are derived, as well as libration periods and bandwidths. Application is made to Saturn's resonant pair of satellites, Titan and Hyperion, which approximate the conditions of this problem.

042.011 On periodic orbits of the second generation and regions of stability. V. V. Markellos, C. G. Zagouras.
Celestial Mech., Vol. 16, 123 - 126 (1977).

A part of the family h of retrograde periodic orbits of the restricted problem ($\mu = 0.5$) is examined in relation to the families of periodic orbits of the second generation branching from it. Many such families are determined and the significance of the configuration of their characteristics for the determination of the boundary of the stability region around this part of h is discussed.

042.012 The effect of oblateness of the bigger primary on collinear libration points in the restricted problem of three bodies. K. B. Bhatnagar, J. M. Chawla.
Celestial Mech., Vol. 16, 129 - 136 (1977).

In the restricted problem of three bodies, the effect of oblateness of the bigger primary appears as an additional term in the potential. As a result, the location of libration points and the roots of the characteristic equation at these points depend not only upon the mass parameter μ but also on the oblateness term I of the bigger primary. Series solutions are developed in terms of μ and I which are used for locating the collinear libration points and for determining the mean motions and characteristic exponents at these points.

042.013 Periodic solutions in Hill's relativistic problem.
V. Singh.
Celestial Mech., Vol. 16, 137 - 142 (1977).

The existence of periodic solutions in Hill's relativistic problem is demonstrated using Poincaré's small parameter method. This method guarantees the convergence of the series representing the periodic solutions.

**042.014 Gravitational harmonics from shallow resonant or-
bits.**   C. A. Wagner, S. M. Klosko.
Celestial Mech., Vol. 16, 143 - 163 (1977).

Until very recently, there has been no identification of
the significant gravitational constraints on the many common
artificial earth satellite orbits in shallow resonance. Without
them it is difficult to compare results derived for different sets
of harmonics from different orbits. With them it is possible to
extend these results to any degree without reintegration of the
orbits. All such constraints are shown to be harmonic in the
argument of perigee with constants determinable from
tracking data. Five such constants (lumped harmonics) have
been derived for the GEOS-2 orbit (order 13, to 30th degree)
whose principal resonant period is 6 days.

**042.015 Periodic elliptic motions in a planar restricted
(N + 1)-body problem.**
R. F. Arenstorf, R. E. Bozeman.
Celestial Mech., Vol. 16, 179 - 189 (1977).

Let $N \geq 2$ mass points (primaries) move on a collinear
solution of relative equilibrium of the N-body problem; i.e.
suitably fixed on a uniformly rotating straight line.
Consider the motion of a massless particle in the gravitational
field of these primaries with arbitrarily given masses. An existence proof for periodic solutions (i.e. closed trajectories in a
rotating coordinate system) will be given, in which the particle
performs nearly keplerian elliptic motions about (and close to)
any one of the primaries.

**042.016 The Kepler problem and geodesic flows in spaces of
constant curvature.**   Yu. S. Osipov.
Celestial Mech., Vol. 16, 191 - 208 (1977).

The main result of this paper is a theorem on the trajectory equivalence of phase flows on isoenergetic surfaces with
a positive energy level in the Kepler problem and perturbed
Kepler problem. The following two facts are crucial for prov-
ing it: firstly, an isomorphism of the phase flow on an isoenergetic surface in the Kepler problem and the geodesic flow in a
constant curvature space. The isomorphism is studied in detail. In particular, all the integrals of the Kepler problem are
obtained proceeding from the group-theory considerations.
The second fact is a generalization of the theorem on structural stability of Anosov flows onto non-compact manifolds.

**042.017 The Titius-Bode law—place for more than ten
planets?**   G. P. Horedt, P. Pop, H. Ruck.
Celestial Mech., Vol. 16, 209 - 213 (1977).

This note presents the results of five numerical examples
concerning the Titius-Bode law. The last section contains the
principal conclusions.

**042.018 On Szebehely's equation for the potential of a
prescribed family of orbits.**   R. Broucke, H. Lass.
Celestial Mech., Vol. 16, 215 - 225 (1977).

In the present note the authors first give a simple proof
of the Dainelli formulas for the force field generating a given
family of orbits. They also show that the Szebehely partial
differential equation for the potential can be derived from the
Dainelli formulas if the energy integral is assumed. The
Szebehely equation can be solved directly with the method of
characteristics or indirectly with the Joukovsky formulas.
Several examples are briefly described in the article. In particular the authors find some rather general potential functions
corresponding to circular motion.

042.019 Modification of an equation of Szebehely.
F. Morrison.
Celestial Mech., Vol. 16, 227 - 228 (1977). – Reprint from
Celestial Mech., Vol. 16, 39 - 40 (1977).

042.020 Keplerian transition matrices for elliptical orbits.
D. L. Hitzl.

Astron. Astrophys., Vol. 63, 429 - 432 (1978).

For the case of an elliptical reference orbit, compact
analytic formulas are given for the four submatrices composing the state transition matrix (matrizant) of the two-body
problem. The transition matrix plays a fundamental role in
determining the effects of perturbations on an orbit and in
analyzing the stability of an orbit. For example, the submatrix
Φ_{rvo} was an essential ingredient in an earlier stability analysis
of second species periodic orbits of the restricted three body
problem.

**042.021 On the expansion of the force function of two finite
bodies.**   G. N. Duboshin.
Soobshch. Gos. Astron. Inst. Shternberga, No. 201, 30 pp.
(1977). In Russian.

The problem of the development of the force function of
the Newtonian attraction of two arbitrary finite bodies having
a definite form and given structure is considered. A general
form of the development is established and some of its
properties are noted. From general formulas the first terms of
the development appropriating in a particular case to the well
known classic approximation are found. The method of finding
the terms of any order is shown. For the case of two axially
symmetric bodies each of which possesses besides a plane of
symmetry perpendicular to the axis of symmetry, the terms of
the fourth order are calculated. The simplest cases are noted.

042.022 On Cassini's laws.   Yu. V. Barkin.
Astron. Zh. Akad. Nauk SSSR, Vol. 55, 113 - 122
(1978). In Russian. English translation in Soviet Astron.,
Vol. 22, No. 1.

Generating periodic Poincaré solutions of the problem
of the rotational motion of a satellite around its own mass
centre on a circular regularly precessing orbit are interpreted
as generalized Cassini's laws. A qualitative analysis of the
analytic existence conditions for motion according to Cassini
is given. The results are used for an improvement of the
Moon's physical libration parameters.

**042.023 On the transformation of the kinematic equation
of Euler motion to a form of Kepler equation.**
K. V. Kholshevnikov, Yu. F. Mishchuk.
Astron. Zh. Akad. Nauk SSSR, Vol. 55, 132 - 137 (1978). In
Russian. English translation in Soviet Astron., Vol. 22, No. 1.

**042.024 On the planar form and symmetry of asymptotic
solutions in the restricted three-body problem.**
L. G. Luk'yanov.
Astron. Zh. Akad. Nauk SSSR, Vol. 55, 156 - 163 (1978). In
Russian. English translation in Soviet Astron., Vol. 22, No. 1.

It is shown that the asymptotic solutions for the libration points in the restricted three-body problem are plane
curves situated in the plane of the orbital motion of the
primaries. Moreover, these solutions are symmetric with
respect to the straight line connecting the primaries. This is
the consequence of the symmetry of solutions which is shown
to exist for the restricted elliptical three-body problem.

**042.025 On a simplified version of the spatial circular restrict-
ed problem of three bodies.**   V. Kh. Karaganchu.
Astron. Zh. Akad. Nauk SSSR, Vol. 55, 164 - 167 (1978). In
Russian. English translation in Soviet Astron., Vol. 22, No. 1.

A simplified version of the spatial circular restricted
problem of three bodies as obtained in terms of the proper
problem by use of the averaging method is stated. This version
admits the complete system of the first integrals at a given
precision of Le Verrier's expansion of the perturbation function.

042.026 Clearing gaps and divisions in belts and rings.
R. Greenberg.
Bull. American Astron. Soc., Vol. 9, 520 (1977). – Abstract.

042.027 A new type of connection between the families of periodic orbits of the restricted problem.
M. Michalodimitrakis.
Astron. Astrophys., Vol. 64, 83 - 86 (1978).

A new type of connection between the families of periodic orbits of the restricted problem is proposed. As an example, the connection between the family c and the family m (for $\mu = 0.5$) is numerically established by continuing a vertical critical orbit of the family c to the third dimension.

042.028 Periodic solutions of an equatorial satellite in the J_2 field. C. Delmas.
Astron. Astrophys., Vol. 64, 267 - 272 (1978). In French.

The problem of an equatorial satellite in the J_2 field is treated by Poincaré's method for finding periodic solutions. Two classes appear: the first of them is periodic in a slightly moving frame and depends on a parameter (energy) for every value of the eccentricity of a generating Keplerian ellipse $(0 < e < 1)$.The second one is periodic in an absolute system of reference, with a fixed energy constant, and for discrete values of the generating eccentricity. Both of these classes are unstable.

042.029 Un étonnant paradoxe sur la chute des corps.
J.-C. Pecker.
Astronomie, Vol. 92, 142 - 145 (1978).

042.030 Stability in dynamical astronomy.
V. Szebehely.
Bull. American Astron. Soc., Vol. 9, 620 (1978). – Abstract.

042.031 Dive's contribution to rotation heterogeneous celestial bodies. J. Meffroy.
Bull. American Astron. Soc., Vol. 9, 621 - 622 (1978). Abstract.

042.032 Second species solutions with an $0(\mu^\nu)$, $1/3 < \nu < 1$, near-moon passage. L. M. Perko.
Celestial Mech., Vol. 16, 275 - 290 (1977).

This paper uses Guillaume's extension of the Breakwell-Perko matching theory for the restricted three-body problem with small mass ratio $\mu > 0$ to establish the existence and asymptotic approximation of one-parameter families of second species periodic solutions with one near-moon passage of $0(\mu^\nu)$, $1/3 < \nu < 1$, during a half-period. The deflection angle at such a near moon passage, $\beta = 0(\mu^{1-\nu})$, approaches zero as the mass ratio μ approaches zero.

042.033 Time regularization of an Adams-Moulton-Cowell algorithm. N. Borderies.
Celestial Mech., Vol. 16, 291 - 308 (1977).

This paper deals with the Adams-Moulton-Cowell multi-step integrator, as described by Oestwinter and Cohen (1972). In order to evaluate the accuracy of the method, the author tests it in the case of the unperturbed two-body motion; numerical instability may arise by integrating first order systems. The accuracy is improved by applying a Sundman transformation of the independent variable. The algorithm is then modified such that the equations of pure Keplerian motion are integrated with respect to the new independent variable without truncation error; numerical experiments show the considerable improvement of accuracy and the reduction of computing time for Keplerian motion.

042.034 The intermediate anomaly. P. Nacozy.
Celestial Mech., Vol. 16, 309 - 313 (1977).

A Sundman time transformation of the form $dt = cr^n ds$, where $n = 3/2$ and c is a constant, is integrated analytically for Keplerian motion. The integral involves an elliptic integral of the first kind. The variable s is called the intermediate anomaly.

042.035 A new kind of 3-body problem.

H. A. G. Robe.
Celestial Mech., Vol. 16, 343 - 351 (1977).

A new kind of restricted 3-body problem is considered. One body, m_1, is a rigid spherical shell filled with an homogeneous incompressible fluid of density ρ_1. The second one, m_2, is a mass point outside the shell and m_3 a small solid sphere of density ρ_3 supposed moving inside the shell and subjected to the attraction of m_2 and the buoyancy force due to the fluid ρ_1. There exists a solution with m_3 at the center of the shell while m_2 describes a Keplerian orbit around it. The linear stability of this configuration is studied assuming the mass of m_3 to be infinitesimal. Explicitly two cases are considered. In the first case, the orbit of m_2 around m_1 is circular. In the second case, this orbit is elliptic but the shell is empty (i.e. no fluid inside it) or the densities ρ_1 and ρ_3 are equal. In each case, the domain of stability is investigated for the whole range of the parameters characterizing the problem.

042.036 The force function of two general bodies.
M. Šidlichovský.
Bull. Astron. Inst. Czechoslovakia, Vol. 29, 90 - 97 (1978).

A method for calculating the force function of two celestial bodies is proposed. It is based only on the knowledge of the Stokes constants of the two bodies, which may be determined, e.g., from observations of the motion of satellites. A general formula for the force function as a function of the two sets of Stokes constants, two sets of Euler angles of the bodies and their position is presented.

042.037 The equilibrium and the stability of the Jeans spheroids with toroidal magnetic fields.
S. K. Trehan, M. Singh.
Astrophys. Space Sci., Vol. 53, 69 - 76 (1978).

The authors examine the effect of a toroidal magnetic field on the equilibrium and stability of homogeneous masses distorted by the tidal effects of a secondary. It is shown that if the toroidal magnetic field is assumed to be axisymmetric about the direction of the line joining the centres of mass of the primary and the secondary, then the equilibrium configuration is a prolate spheroid. Also determined are the characteristic frequencies of the various modes of oscillations belonging to the second harmonics. It is found that the magnetic field shifts these frequencies to higher values than the ones which prevail in the absence of a magnetic field.

042.038 Libration points in the generalised elliptic restricted three body problem. R. K. Choudhry.
Celestial Mech., Vol. 16, 411 - 419 (1977).

The existence of collinear as well as equilateral libration points for the generalised elliptic restricted three body problem has been studied distinct from Kondurar and Shinkarik (1972) where a study has been made for the generalised circular restricted three body problem. Here the coordinates of the libration points have been found to be functions of time t.

042.039 Are higher order resonances really interesting?
J. Sanders.
Celestial Mech., Vol. 16, 421 - 440 (1977).

It is shown that higher order resonances of equilibria of conservative Hamiltonian systems do not give rise to an exchange of energies between the different degrees of freedom for most of the initial values on a certain long time-scale. The topology of these resonances is analysed, using Bott-Morse theory.

042.040 The extended phase space formulation of the Vinti problem. K. T. Alfriend, R. Dasenbrock, H. Pickard, A. Deprit.
Celestial Mech., Vol. 16, 441 - 458 (1977).

The Vinti problem, motion about an oblate spheroid, is formulated using the extended phase space method. The new independent variable, similar to the true anomaly, decouples the radius and latitude equations into two perturbed harmonic oscillators whose solutions to $O(J_2^4)$ are obtained using Lindstedt's method. From these solutions and the solution to the Hamilton-Jacobi equation suitable angle variables, their canonical conjugates and the new Hamiltonian are obtained. The new Hamiltonian, accurate to $O(J_2^4)$ is a function of only the momenta.

042.041   On the perturbations of a close-earth satellite due to lunar inequalities.   D. R. Cok.
Celestial Mech., Vol. 16, 459 - 479 (1977).

The lunar disturbing function for a close-earth satellite is expressed as a sum of products of harmonics of the satellite's position and harmonics of the Moon's position, and the latter are expanded about a rotating and precessing elliptic orbit inclined to the ecliptic. The deviations of the Moon from this approximate orbit are computed from Brown's lunar theory and the perturbations in satellite orbital elements due to these inequalities are derived. Numerical calculations indicate that several perturbations in the position of the satellite's node and perigee have magnitudes on the order of one meter.

042.042   The use of angles and angular rates. I: Initial orbit determination.   L. G. Taff, D. L. Hall.
Celestial Mech., Vol. 16, 481 - 488 (1977).

MIT's Lincoln Laboratory has developed a computer driven, rapidly slewing ($\simeq 4°$ s^{-1}), electro-optical ($\simeq 3$ "resolution) telescope. This enables the rapid measurement of angles and instantaneous angular rates for artificial satellites. The simultaneous acquisition of angles and angular rates constitutes a new initial orbit problem which has been solved. Three different methods of solution are presented including an exact, analytical one. Numerical tests on six widely different satellite orbits indicate that the topocentric distance can be determined to better than 1% (and usually as well as 0.1%) for most satellites after a 5—10 min observation interval.

042.043   Euler solutions in the problem of the translatory-rotary motion of three rigid bodies.
V. V. Vidyakin.
Celestial Mech., Vol. 16, 509 - 526 (1977). In Russian.

The present paper is a continuation of the article (Vidyakin, 1976) in which the author proved the existence of Lagrange (triangle) solutions in the general problem of the translatory-rotary motion of three absolutely rigid bodies. In particular, he has found the conditions for the existence of Lagrange solutions in the case when all the bodies possess a symmetry with respect to three mutually perpendicular planes both in respect to the distribution of matter and in respect to the outward form. The author proves the existence of Euler (rectilinear) solutions in the problem of the translatory-rotary motion of three rigid bodies, assuming that the elementary particles of the rigid bodies are mutually attracted according to the Newtonian law.

042.044   The oscillations and the stability of rotating masses with magnetic fields. V: Existence of the point of bifurcation.   S. K. Trehan, M. Singh.
Astrophys. Space Sci., Vol. 53, 335 - 338 (1978).

Using first variations of the integral properties of equilibrium second-order virial relations, the existence of the point of bifurcation of rotating gaseous masses with magnetic fields is substantiated. With the presence of a magnetic field component along the axis of rotation, it is shown that the point of bifurcation, where the Jacobi ellipsoids branch off from the Maclaurin spheroids, is altered, and in fact shifts to higher values of eccentricity compared to the one (namely, $e = 0.81267$) obtained when there is no magnetic field.

042.045   A separable potential in sphero-conical coordinates satisfying the Laplace equation.   E. I. Burshtejn.
Astron. Zh. Akad. Nauk SSSR, Vol. 55, 426 - 433 (1978).

In Russian. English translation in Soviet Astron., Vol. 22, No. 2.

A general potential function in sphero-conical coordinates is derived which allows separation of the Hamilton-Jacobi equation and which satisfies the Laplace equation. An interpretation of this potential in real space as the potential of single and double layers in which the density distribution is defined by Green's formula is given.

042.046   Solution of the equations of the fifth approximation of the theory of figure for a quadratic density distribution law.   A. V. Kozenko.
Astron. Zh. Akad. Nauk SSSR, Vol. 55, 440 - 442 (1978). In Russian. English translation in Soviet Astron., Vol. 22, No. 2.

An analytical solution of the system of equations of the fifth approximation of the theory of figure is found for a quadratic density distribution in the form of power series with recurrent relations.

042.047   Remark on the polytrope of index 5.
H. A. Buchdahl.
Australian J. Phys., Vol. 31, 115 - 116 (1978).

Contrary to statements in the current literature the gravitational potential energy of the polytrope of index 5 is finite. With ξ_n denoting the least positive zero of the Emden function of index n, the value of the limit as $n \to 5$ of $(5-n)\xi_n$ is determined and made the basis of a simple rational approximation to ξ_n qua function of n.

042.048   Numerical determination of families of three-dimensional periodic orbits bifurcating from plane periodic orbits around both primaries.
C. G. Zagouras, M. Kalogeropoulou.
Astron. Astrophys., Suppl. Ser., Vol. 32, 307 - 321 (1978).

The numerical techniques employed in the determination of families of three-dimensional symmetric periodic orbits of all kinds of symmetry in the restricted three body problem are first outlined. Then seven new families bifurcating from the "vertical critical" orbits ($|a_l| = 1$) of the families l and m of plane symmetric retrograde (in the rotating system) periodic orbits around both primaries for $\mu = 0.5$, are computed on the basis of the above methods. It is found that, in general the families bifurcating from family l have as terminal member an orbit of the family m. Also that the corresponding families originating from family m are terminated with plane periodic orbits going around the collinear equilibrium point L_2.

042.049   Application of averaging methods to the solution of translatory-rotational motion of an axisymmetric body in the attraction field of a sphere.   Yu. V. Barkin.
Vestn. Mosk. univ. Fiz., astron., Vol. 18, No. 4, p. 9 - 16 (1977). In Russian. – Abstr. in Ref. zh., 51. Astron., 3.51.122 (1978).

042.050   Motions of natural satellites.
J. Kovalevsky, J.-L. Sagnier.
Planetary satellites, (see 003.007), p. 43 - 62 (1977).

The equations of motion of planetary satellites are presented and the various types of disturbing forces are reviewed. From a consideration of forces, three classes of problems are defined and described. The motions of all the natural satellites are reviewed and, for each group of them, the latest works on the theory of motion are presented.

042.051   Orbit-orbit resonances among natural satellites.
R. Greenberg.
Planetary satellites, (see 003.007), p. 157 - 168 (1977).

A qualitative, physical description of the orbital resonance phenomenon is stressed. Various examples observed in the satellite systems are discussed, and recent theoretical work on the formation of resonances is summarized.

042.052 Planetary theories and observational data.
R. L. Duncombe, P. K. Seidelmann.
Dynamics of planets and satellites, (see 012.020), p. 3 - 14 (1978).

A brief review is given of planetary theories from Leverrier to Newcomb to the age of computers. The presently used planetary theories are discussed and the process of replacing these theories with new ones is described. Some difficulties in preparing new planetary theories and the observational discrepancies which have been encountered previously are discussed.

042.053 Correspondances entre une théorie générale planétaire en variables elliptiques et la théorie classique de Le Verrier.   L. Duriez.
Dynamics of planets and satellites, (see 012.020), p. 15 - 32 (1978).

In order to improve the determination of the mixed terms in classical theories, the author shows how these terms may be derived from a general theory developed with the same variables (of a Keplerian nature). He finds that the general theory of the first order in the masses already allows to develop the mixed terms which appear at the second order in the classical theory. The short period terms of the classical theory are found the same in the general theory, but without the numerical substitution of the values of the variables.

042.054 Mathematical results of the general planetary theory in rectangular coordinates.
V. A. Brumberg, L. S. Evdokimova, V. I. Skripnichenko.
Dynamics of planets and satellites, (see 012.020), p. 33 - 48 (1978).

Mathematical construction of the general planetary theory has led to the series of two forms for the coordinates of eight major planets (excluding Pluto). The series of the first form are Poisson series where all orbital elements except the semi-major axes occur in literal shape. The series of the second form are polynomial-exponential series with respect to the time and serve to calculate the ephemerides.

042.055 Construction of planetary theory by iterative procedure.   T. V. Ivanova.
Dynamics of planets and satellites, (see 012.020), p. 49 - 52 (1978).

In this paper the method of determination of the planetary perturbations is proposed which is a modification of Dziobek-Brouwer's method. For the sake of simplicity the case of two mutually disturbing planets is considered.

042.056 Qualitative dynamics of the Sun–Jupiter–Saturn system.   V. Szebehely.
Dynamics of planets and satellites, (see 012.020), p. 53 - 55 (1978).

The stability of the three-body problem formed by the Sun, Jupiter and Saturn is investigated using surfaces of zero velocity. The results obtained with the models of the restricted and general problems of three bodies are compared with numerical integration. The system is found to be stable in the sense that Saturn will neither interrupt the (perturbed) binary orbit of Jupiter around the Sun, nor will it escape from the system. It is shown that the known classical triple stellar systems are "more stable" than the solar system, which in turn is "more stable" than the Earth–Moon system.

042.057 A new approach for the construction of long-periodic perturbations.   R. Dvorak.
Dynamics of planets and satellites, (see 012.020), p. 57 - 64 (1978).

The aim of this work is to study perturbations of planets of a period of some thousands of years. The use of an analytical method allows to separate all different influences, e.g. near resonances, and is combined with the very precise method of the numerical integration. A special form of the Lagrange

equations is used where the terms containing the inverse distance from the planet to the perturbing one are separated as it is the most difficult to compute. To develop this a specific formulation has been found where the short periodic terms can precisely be determined. It should be possible to integrate the new form of the Lagrange equations within a reasonable computer-time to determine the long periodic perturbations.

042.058 Construction d'une théorie planétaire au troisième ordre des masses.   J. L. Simon.
Dynamics of planets and satellites, (see 012.020), p. 65 - 75 (1978).

The author recalls briefly the sense which is given to a planetary theory of the Le Verrier type (i.e., with secular variations in the metrical elements). The solutions are developed with elliptical coordinates. The author discusses the results obtained, order by order till the third order of the masses.

042.059 Discussion sur les résultats de théories planétaires.   P. Bretagnon.
Dynamics of planets and satellites, (see 012.020), p. 77 - 85 (1978).

Various aspects of the construction of the developments of the solutions are discussed: problem of convergence in iterative methods, comparisons with numerical integrations and ephemerides, special differences between major and minor planets, and precision of the solutions.

042.060 New results on the commensurability cases of the problem Sun-Jupiter-asteroid.   J. Schubart.
Dynamics of planets and satellites, (see 012.020), p. 137 - 143 (1978).

The short-period terms are removed by averaging from special equations of motion for commensurability cases of the three-dimensional, elliptic, restricted three-body problem. Some examples of retrograde motion corresponding to the 1/1 commensurability, and an application to Hilda-type motion demonstrate the possibilities given by the method.

042.061 A theory of the Trojan asteroids.   B. Garfinkel.
Dynamics of planets and satellites, (see 012.020), p. 145 (1978). – Abstract.

042.062 Stabilization by making use of a generalized Hamiltonian variational formalism.   J. W. Baumgarte.
Dynamics of planets and satellites, (see 012.020), p. 149 - 156 (1978).

A generalized Hamiltonian formalism is established which is invariant not only under canonical transformations but under arbitrary transformations. Moreover the dependent variables, coordinates and momenta, as well as the independent variable are allowed to be transformed. This is to say that instead of the physical time t another independent variable s is used, such that t becomes a dependent variable or, more precisely, an additional coordinate. The formalism under consideration permits also to include nonconservative forces.

042.063 A special perturbation method: m-fold Runge-Kutta.   D. G. Bettis.
Dynamics of planets and satellites, (see 012.020), p. 157 (1978). – Abstract.

042.064 Numerical integration of nearly-Hamiltonian systems.   V. R. Bond.
Dynamics of planets and satellites, (see 012.020), p. 159 - 173 (1978).

042.065 New formulation of de Sitter's theory of motion for Jupiter I–IV. I. Equations of motion and the disturbing function.   K. Aksnes.
Dynamics of planets and satellites, (see 012.020), p. 189 - 206 (1978).

A brief discussion is given of the basic features of de Sitter's theory. The main advantage of his theory is that it contains no small divisors, thanks to the use of elliptic rather than circular intermediate orbits in the first approximation. A 50-year extension of the satellite observations available to de Sitter makes it desirable to rederive the elements of his intermediate orbits, whose perijoves have a common retrograde motion. By means of the generalized Newcomb operators devised by Izsak, the disturbing function is expanded in a form that is very convenient for use with the modified Delaunay variables, G, $L-G$, $H-G$, $l + \omega + \Omega$, l, and Ω and their associated Poincaré variables.

042.066   Theory of motion of Jupiter's Galilean satellites.
            J. H. Lieske.
Dynamics of planets and satellites, (see 012.020), p. 207 (1978). – Abstract.

042.067   A second-order theory of the Galilean satellites of Jupiter.   S. Ferraz-Mello.
Dynamics of planets and satellites, (see 012.020), p. 209 - 236 (1978).
     The theory of the motion of the Galilean satellites of Jupiter is developed up to the second-order terms. The disturbing forces are those due to mutual attractions, to the non-symmetrical internal mass distribution of Jupiter and to the attraction from the Sun. The mean equator of Jupiter is taken as the reference plane and its motion is considered. The integration of the equations is performed. The main characteristic of the theory is that it allows the main frequencies to be kept fixed from the earlier stages, and so, to have a purely trigonometric solution.

042.068   Solar perturbations in Saturnian satellite motions and Iapetus-Titan interactions.   Y. Kozai.
Dynamics of planets and satellites, (see 012.020), p. 237 (1978). – Abstract.

042.069   Improvement of orbits of satellites of Saturn using photographic observations.   A. T. Sinclair.
Dynamics of planets and satellites, (see 012.020), p. 238 (1978). – Abstract.

042.070   New orbits for Enceladus and Dione based on the photographic observations.
W. H. Jefferys, J. D. Mulholland, L. M. Ries.
Dynamics of planets and satellites, (see 012.020), p. 239 (1978). – Abstract.

042.071   Long periodic variation of orbital elements of a satellite perturbed by discrete gravity anomalies.
M. P. Ananda.
Dynamics of planets and satellites, (see 012.020), p. 240 (1978). – Abstract.

042.072   Some considerations on the theoretical determination of the potential by the motion of artificial satellites in the plane case.   J. J. Martinez-Benjamin.
Dynamics of planets and satellites, (see 012.020), p. 259 (1978). – Abstract.

042.073   Families of periodic planetary-type orbits in the N-body problem and their application to the solar system.   J. D. Hadjidemetriou, M. Michalodimitrakis.
Dynamics of planets and satellites, (see 012.020), p. 263 - 281 (1978).
     A new approach to the study of the solar system and planetary systems in general is proposed, through the use of periodic planetary-type orbits of the general N-body problem. In such an orbit, one body has a large mass and the rest N−1 bodies have small but not negligible masses and it can be proved that monoparametric families of periodic orbits of the

N-body problem exist in a rotating frame of reference, all being of the planetary type. Two cases are studied in detail, N = 3 and N = 4.

042.074   Perturbations of critical mass in the restricted three-body problem.   R. K. Sharma, P. V. Subba Rao.
Dynamics of planets and satellites, (see 012.020), p. 283 (1978). – Abstract.

042.075   Gravitational restricted three-body problem: existence of retrograde satellites at large distance.
D. Benest.
Dynamics of planets and satellites, (see 012.020), p. 285 - 303 (1978).
     In the frame of the gravitational restricted three-body problem, the author studies by numerical simulation the retrograde satellites – and their stability – at large distance. In the circular plane case, the stable satellites mostly surround (in the phase-space $X_0 - V_0$) the characteristic of a family of single-periodic orbits where this family is stable, and librates (in the physical space) around a curve which corresponds to the nearest (in the phase-space $X_0 - V_0$) periodic orbit. An analytical analysis of this libration is made for Hill's case. The beginning of the study of the three-dimensional orbits is presented.

042.076   Displacement of the Lagrange equilibrium points in the restricted three body problem with rigid body satellite.   W. J. Robinson.
Dynamics of planets and satellites, (see 012.020), p. 305 - 314 (1978).
     In the restricted problem of three point masses, the positions of the equilibrium points are well known and are tabulated. When the satellite is a rigid body, these values no longer correspond to the equilibrium points. This paper seeks to determine the magnitudes of the discrepancies.

042.077   A new kind of periodic orbit: the three-dimensional asymmetric.   V. V. Markellos.
Dynamics of planets and satellites, (see 012.020), p. 315 - 317 (1978).

042.078   On asymmetric periodic solutions of the plane restricted problem of three bodies, and bifurcations of families.   P. J. Message, D. B. Taylor.
Dynamics of planets and satellites, (see 012.020), p. 319 - 323 (1978).

042.079   Construction de solutions périodiques du problème restreint elliptique par la méthode de Hale.
A. Sergysels-Lamy, R. Sergysels.
Dynamics of planets and satellites, (see 012.020), p. 325 - 331 (1978).
     This paper contains numerical results following the theoretical demonstration already published in Celestial Mech. Vol. 11, 43 (1975).

042.080   Orbital stability in the elliptic restricted three body problem.   C. A. Williams, J. G. Watts.
Dynamics of planets and satellites, (see 012.020), p. 333 - 337 (1978).
     Based on the concept of orbital stability introduced by G. W. Hill, a method is presented to facilitate the determination of the orbital stability of solutions to the planar elliptic restricted problem of three bodies. The invariant relation introduced by Szebehely and Giacaglia (1964) contains an integral which is expanded here about a Keplerian solution to the problem. If the expansion converges, it can be used to determine the conditions for Hill stability. With it one can also define stability in a periodic sense.

042.081   Resonance in the restricted problem of three bodies

with short-period perturbations in the elliptic case.
K. B. Bhatnagar, B. Gupta.
Dynamics of planets and satellites, (see 012.020), p. 339 - 353 (1978).

The motion of an asteroid moving in the gravitational field of Jupiter is considered. The authors consider the orbit to be elliptic. The series occurring in the problem are expanded in powers of a small parameter ϵ, which represents the ratio of the mass of Jupiter to that of the Sun. The perturbations in the osculating elements are obtained up to $O(\epsilon)$.

042.082   Periodic orbits of the first kind in the restricted three-body problem when the more massive primary is an oblate spheroid.   R. K. Sharma.
Dynamics of planets and satellites, (see 012.020), p. 355 (1978). – Abstract.

042.083   Triple collision as an unstable equilibrium. J. Waldvogel.
Dynamics of planets and satellites, (see 012.020), p. 356 (1978). – Abstract.

042.084   Regions of escape on the velocity ellipsoid for the planar three body problem.   G. Bozis.
Dynamics of planets and satellites, (see 012.020), p. 357 - 369 (1978).

The notion of the velocity ellipsoid for the planar three body problem is given. Using the sufficient conditions for escape of one member of a triple system, given by Standish (1971), a region is found on the velocity ellipsoid for which escape is guaranteed.

042.085   Une extension de la méthode de Delaunay. S. Ferraz-Mello.
C. R. Acad. Sci. Paris, Tome 286, Ser. A, 969 - 971 (1978).

A modification of the classical theorem of Jacobi allows the method of Delaunay to be applied to the study of resonance in disturbed hamiltonian dynamic systems which may not be reduced to only one degree of freedom. The extended Delaunay method is more suitable than the usual Bohlin method and does not include operations which change the order of the terms. The determination of a complete solution of the partial differential equation of Delaunay-Poincaré is enough to determine the formal solution to the same order as the square root of the small parameter.

042.086   A model for close encounters in the planetary problem.   L. P. Cox, J. S. Lewis, M. Lecar.
Icarus, Vol. 34, 415 - 427 (1978).

A model is proposed for single close encounters between two small masses, m_1 and m_2, which orbit a much larger mass, M. The main new feature of the model is the assumption of conic motion of the center of mass of m_1 and m_2 in the gravitational field of M. Comparisons of the model with the three-body equations of motion indicate that the model is a useful approximation for $m_1, m_2 \lesssim 10^{-5} M$. The model is therefore applicable for encounters between bodies of the order of an earth mass or smaller in the presence of the sun.

042.087   A note on the development of the reciprocal distance in planetary theory.
G.-i. Hori, M. Yuasa.
Dynamics of planets and satellites, (see 012.020), p. 177 (1978). – Abstract.

042.088   Stable orbits in the circular plane restricted three-body problem.   D. Benest.
Rev. Mexicana Astron. Astrofis., Vol. 3, (see 012.023), 151 - 158 (1977).

The author investigates by numerical exploration the extended regions of stability for retrograde satellites when the mass of the less massive body increases up to $\mu = 0.5$. A pre-

liminary study is made of the characteristic and stability properties of the families of periodic orbits; then the general case of non-periodic orbits is considered. Those numerical explorations show that for stable large non-periodic orbits, the motion can be decomposed into a fast "reference motion" and slow libration; the author studies theoretically this libration in Hill's case for which the "reference motion" is elliptic.

042.089   A complete family of periodic solutions of the planar three-body problem and their stability.
M. Henon.
Rev. Mexicana Astron. Astrofis., Vol. 3, (see 012.023), 159 (1977). – Abstract.

042.090   Three dimensional stability of a rectilinear periodic solution of the three-body problem, for all values of the masses.   M. Hénon.
Rev. Mexicana Astron. Astrofis., Vol. 3, (see 012.023), 161 (1977). – Abstract.

042.091   An invariant relation in the elliptic restricted problem of three bodies. I.
Z. Vrcelj, J. H. Kiewiet de Jonge.
Astron. J., Vol. 83, 514 - 521 (1978).

An invariant relation generalizing the Jacobi integral to the elliptic restricted three-body problem is derived on the basis of the classical perturbation theory and by making use of energy and angular momentum integrals. But the relation contains a nonintegrable term, becoming integrable only in a few special cases.

042.092   Plane periodic motions of two rigid bodies. Yu. V. Barkin.
Vestn. MGU. Fiz., astron., Vol. 18, No. 5, p. 67 - 74 (1977). In Russian. – Abstr. in Ref. zh., 51. Astron., 4.51.118 (1978).

042.093   A class of periodic solutions in the restricted circular three-body problem.   V. P. Evteev.
Dokl. AN TadzhSSR, Vol. 20, No. 7, p. 15 - 18 (1977). In Russian. – Abstr. in Ref. zh., 51. Astron., 4.51.119 (1978).

042.094   On the integration of the Hamilton-Jacobi equation for a class of dynamical systems.
T. B. Omarov, A. A. Bekov.
Vestn. AN KazSSR, 1977, No. 8, p. 65 - 66. In Russian. Abstr. in Ref. zh., 51. Astron., 4.51.130; 62. Issled. kosm. prostranstva, 4.62.563 (1978).

042.095   ALITA system for analytical operations on Poisson series by computer and possibilities of its practical application to celestial mechanics.
A. V. Vasil'eva.
Astron. i geod. Vyp. (No.) 6. Tomsk, 1977, p. 9 - 17. In Russian. – Abstr. in Ref. zh., 51. Astron., 4.51.132 (1978).

042.096   Application of Lie transforms method to Delaunay's problem up to the third order.
T. S. Boronenko.
Astron. i geod. Vyp. (No.) 6. Tomsk, 1977, p. 18 - 25. In Russian. – Abstr. in Ref. zh., 51. Astron., 4.51.135 (1978).

042.097   Satellite motion in the vicinity of the triangular liberation points.
J. A. Blackburn, M. A. H. Nerenberg, Y. Beaudoin.
American J. Phys., Vol. 45, 1077 - 1081 (1977). – Abstr. in Phys. Abstr., Vol. 81, Abstr. 12488 (1978).

042.098   Proof of the stability of Lagrangian solutions at a critical relation of masses.   A. G. Sokol'skij.
Pis'ma Astron. Zh., Vol. 4, 148 - 152 (1978). In Russian.

The stability of Lagrangian solutions of the plane circular

restricted three-body problem at a critical relation of masses of the main bodies is proved. The Lyapunov stability of the equilibrium point of an autonomic Hamilton system with two degrees of freedom, when the frequencies of a linear system are equal (resonance of the second order) and the determining matrix has nonlinear elementary divisors, is proved for the first time.

042.099   The Kepler orbit from initial conditions via the Lenz vector.   S. Caplan, H. Fuerstenberg, C. Hayes, D. Kane, S. Raboy.
American J. Phys., Vol. 45, 1089 - 1090 (1977). − Abstr. in Phys. Abstr., Vol. 81, Abstr. 16602 (1978).

042.100   Sur la classification des orbites périodiques instables d'après Poincaré.   R. Vieira-Martins.
C. R. Acad. Sci. Paris, Tome 286, Sér. A, 1023 - 1025 (1978).
Poincaré classifies linearly unstable periodic orbits for certain problems which have two degrees of freedom. This classification is based on the study of the orbits in the configuration plane. Here the author summarizes Poincaré's reasoning and proposes a simpler form of reasoning which was suggested to the author by Morse's ideas.

042.101   Generating orbits for stable close encounter periodic solutions of the restricted problem.
D. L. Hitzl.
AIAA J., Vol. 15, 1410 - 1418 (1977). − Abstr. in Phys. Abstr., Vol. 81, Abstr. 28489 (1978).

042.102   An invariant relation in the elliptic restricted problem of three bodies. II. Trojan asteroids.
Z. Vrcelj, J. H. Kiewiet de Jonge.
Astron. J., Vol. 83, 664 - 674 (1978).
The invariant relation derived in Part I (042.091) is reduced to an approximate integral valid in the case of the Sun−Jupiter−Trojan system. Nonintegrable terms neglected contain the Jupiter−Sun mass ratio as a factor, and are of the second order in eccentricities or the fourth order in the relative inclination. The approximate integral is then reduced to the ecliptic as a reference plane, allowing the "Jacobi constant" J to be calculated to four decimals for any Trojan.

042.103   On the theory of rotational motion of Mercury and synchronous satellites of planets.   Yu. V. Barkin.
Pis'ma Astron. Zh., Vol. 4, 235 - 239 (1978). In Russian.
For the study of rotational motion of celestial bodies with respect to their own centre of mass an intermediate theory is proposed which is based on the integrable averaging scheme by Delauney−Hill for the problem of the rotational motion of rigid satellites whose centre of mass is moving along an undisturbed elliptic orbit.

042.104   Integrable cases of the Hamilton − Jacobi equation and some non-stationary problems of celestial mechanics.   A. A. Bekov, T. B. Omarov.
Astron. Zh. Akad. Nauk SSSR, Vol. 55, 635 - 644 (1978). In Russian. English translation in Soviet Astron., Vol. 22, No. 3.
New integrable cases for the Hamilton − Jacobi equation have been found. As applications are considered the problem of two fixed centres of variable mass in the presence of an additional force proportional to the velocity of the test particle and the straight-line version of the non-stationary restricted three-body problem.

042.105   The mean radius in Kepler's third law.
J. E. Prussing.
American J. Phys., Vol. 45, 1216 - 1217 (1977). − Abstr. in Phys. Abstr., Vol. 81, Abstr. 24612 (1978).

042.106   Numerical extrapolation of periodic solutions.
V. V. Markellos, P. G. Kazantzis, C. G. Zagouras.
Astrophys. Space Sci., Vol. 54, 379 - 388 (1978).
Numerical procedures are established for the continuation of families of periodic solutions of non-integrable dynamical systems. They are based on the use of the previous known members of a family for non-linear prediction of the next member to be determined. Both symmetric and asymmetric periodic solutions are considered. The procedures are applied and compared in the case of the restricted three-body problem. They are shown to lead to considerable saving of computer time.

042.107   On deviations of the elements of a perturbed elliptic orbit from calculated values owing to the fortuitousness of its initial elements.   Yu. I. Trofimtsev.
Nekotor. vopr. differents. i integral'n. uravnenij i ikh pril., 1977, No. 2, p. 153 - 159. In Russian. − Abstr. in Ref. zh., 51. Astron., 5.51.103 (1978).

042.108   Das Einfangproblem in der Himmelsmechanik.   R. Dvorak.
Sitzungsber. Österreich. Akad. Wiss., Math.-Naturwiss. Kl., Abt. II, 186. Band, 1 - 53 (1977) = Mitt. Univ.-Sternw. Graz Nr. 37.

042.109   Higher order resonances in dynamical systems.   G. Contopoulos.
ESO Sci. Prepr. No. 19, 16 pp. (1978). − Submitted to Celestial Mechanics.
It is shown that the appearance of higher order resonances, (that produce higher order islands on a surface of section), is not an indication of non-integrability. Examples are given and a method is described for constructing integrable Hamiltonians with higher order resonances.

042.110   On the number of isolating integrals in Hamiltonian systems.
G. Contopoulos, L. Galgani, A. Giorgilli.
ESO Sci. Prepr. No. 26, 20 pp. (1978). − Submitted to Phys. Rev. A.

042.111   Gravitational potential of solid bodies in the solar system.   G. Balmino, N. Borderies.
Celestial Mech., Vol. 17, 113 - 119 (1978).
The gravitational potential of a solid body is expanded without approximation in any moving reference frame, in terms of harmonic coefficients relative to fixed axis in the body.

042.112   Periodic three body orbits in the case of small third mass.   P. C. Kammeyer.
Celestial Mech., Vol. 17, 121 - 125 (1978).
Near a symmetric periodic orbit of the plane circular or elliptic restricted problem, the conditions for a symmetric periodic orbit of the plane general three body problem are reduced, under a natural nondegeneracy condition, to the vanishing of a single real valued function. The implicit function theorem and Hörmander's generalized Morse's lemma are then used to analyze the set of zeros of this function.

042.113   A short note on Hamiltonians linear in the momenta.   P. Hagedorn.
Celestial Mech., Vol. 17, 127 - 129 (1978).

042.114   A note on the averaging method.   H. Kinoshita.
Celestial Mech., Vol. 17, 131 - 136 (1978).
When time-averaged equations are used to discuss the secular behavior of dynamical systems, the action-angle variables conjugate to the action variables of the unperturbed motion of the system should be chosen as dependent variables; otherwise, the results are not correct.

042.115 **General time elements for the two body problem.**
A. Stokes.
Celestial Mech., Vol. 17, 137 - 144 (1978).

When integrating a perturbed two-body problem, very often the propagation of the numerical error is reduced by using a new time variable s defined by $dt/ds = |q|^n$, ($|q|$ is the radial distance, t the time). This paper introduces a time element for such transformations, i.e., a new variable t_n is defined so that $dt_n/ds = 1 + $ (perturbing terms) and $t = F_n(t_n)$, where F_n is a known function. The time element equation should be useful in reducing the error in the determination of the time t. F_n is given explicitly for $n = 1$, $\frac{3}{2}$, 2, $\frac{5}{2}$ and 3, and a general expression is given for other values.

042.116 **Disappearance of integrals in systems of more than two degrees of freedom.** G. Contopoulos.
Celestial Mech., Vol. 17, 167 - 172 (1978).

The disappearance of some integrals of motion when two or more resonance conditions are approached at the same time is explained. As an example a Hamiltonian of three degrees of freedom is considered in action-angle variables which in zero order represents three harmonic oscillators, while the perturbation contains two trigonometric terms. One integral disappears if two appropriate resonant conditions are approached sufficiently closely.

042.117 **Keplerian motion and gyration.** M. Vitins.
Celestial Mech., Vol. 17, 173 - 192 (1978).

An equivalence between Keplerian motion and harmonic oscillations has been established by Burdet by using essentially the true anomaly as a new independent variable. In this paper, a relation between these oscillator equations and the motion of a gyroscope is derived. Important numerical and analytical consequences are discussed.

042.118 **A geometric property of Hill's curves.**
V. Matas.
Celestial Mech., Vol. 17, 193 - 194 (1978).

The following is proved in this note: If we construct a circle passing through a given primary in the planar circular restricted three-body problem, center of which is the remaining primary, then the minor arc of this circle with endpoints represented by the triangular libration points represents − when the given primary is excluded − locus of all the points on the Hill's curves that are the least distant points from the given primary.

042.119 **Hill's problem in lunar theory.** J. Henrard.
Celestial Mech., Vol. 17, 195 - 204 (1978).

A power series solution is proposed for Hill's problem considered as an approximation of the main problem of lunar theory. Terms up to degree 28 in the parameter m (the ratio of mean motions) are considered. This solution is intended to be the stepping stone for a solution of the main problem with very high accuracy.

042.120 **Planetary close encounters: changes in a planet-crossing population.** A. Carusi, F. Pozzi, G. Valsecchi.
Reports of planetary geology program, 1977−1978, (see 003.014), p. 22 - 24 (1978).

042.121 **Theory of the Trojan asteroids, Part II.**
B. Garfinkel.
Bull. American Astron. Soc., Vol. 10, 483 (1978). − Abstract.

042.122 **On the isoceles triangle three-body-problem.**
R. Broucke.
Bull. American Astron. Soc., Vol. 10, 484 (1978). − Abstract.

042.123 **On Strömgren's termination principle for families of periodic orbits.** J. M. A. Danby.

Bull. American Astron. Soc., Vol. 10, 485 (1978). − Abstract.

042.124 **Existence and stability of some periodic solutions of rectilinear problems.** P. Kammeyer.
Bull. American Astron. Soc., Vol. 10, 485 (1978). − Abstract.

042.125 **Flexibly coupled motion of rigid bodies.**
D. J. Jezewski, J. D. Donaldson.
Bull. American Astron. Soc., Vol. 10, 485 (1978). − Abstract.

042.126 **Solar influenced orbits of minimum displacement from L_4 and L_5 of the Earth-Moon system.**
B. E. Schutz.
Bull. American Astron. Soc., Vol. 10, 485 - 486 (1978). Abstract.

042.127 **Accuracy of planetary theories, particularly for Mars.** S. E. Babb, Jr.
ISIS, Vol. 68, 426 - 434 (1977). − Abstr. in Phys. Abstr., Vol. 81, Abstr. 32325 (1978).

042.128 **Perturbations of Lagrangian points in the restricted three-body problem.** R. K. Sharma.
Indian J. Pure Appl. Math., Vol. 6, 1099 - 1102 (1975). Abstr. in Phys. Abstr., Vol. 81, Abstr. 32327 (1978).

042.129 **Planetary theory developments, 1973−1976.**
P. K. Seidelmann.
Celestial Mech., Vol. 17, 103 - 112 (1978).

The developments in planetary theories during the period 1973−1976 are reviewed. Emphasis is placed on the efforts to prepare new general theories, on the new techniques of numerical integration and their application, and on the studies of minor planets, particularly of those with periods nearly commensurable with the period of Jupiter.

042.130 **Periodic orbits of distant satellites of stars with inverse motion.** I. N. Latyshev.
Astron. Tsirk., No. 991, p. 7 - 8 (1978). In Russian.

042.131 **On the formulation of Hill's problem for a system star − satellite − Galaxy.** V. V. Ostroumov.
Astron. Tsirk., No. 993, p. 7 - 8 (1978).

042.132 **Periodic orbits around a rotating ellipsoid.**
P. C. Kammeyer.
Celestial Mech., Vol. 17, 37 - 48 (1978).

This paper studies families of symmetric periodic satellite orbits around a rotating triaxial ellipsoid. Existence of the families of orbits is established, and Morse's lemma is used to analyze their bifurcations. Several consequences of the many symmetries of the ellipsoid are discussed.

042.133 **On the classification of motion in the generalized two-dimensional problem of three fixed centers.**
G. T. Arazov, S. A. Gabibov.
Celestial Mech., Vol. 17, 49 - 81 (1978). In Russian.

A qualitative analysis and classification of forms of motion in the problem under consideration have been carried out using a method (applicable to any case of integrability) due to Liouville. All the forms of the two-dimensional motions for any masses (negative and complex as well) at fixed centres corresponding to the real potential have been considered.

042.134 **Time transformations and Cowell's method.**
C. E. Velez, S. Hilinski.
Celestial Mech., Vol. 17, 83 - 99 (1978).

The precise numerical integration of Cowell's equations of satellite motion is frequently performed with an independent variable s defined by an equation of the form $dt = cr^n\, ds$, where t represents time, r the radial distance from the center of attraction, c is a constant, and n is a parameter. This has been primarily motivated by the 'uniformizing' effects of such a transformation resulting in desirable 'analytic' stepsize control for elliptical orbits. This report discusses the 'proper' choice of the parameter n defining the independent variable s for various types of orbits and perturbation models, and develops a criterion for its selection.

042.135 On the problem of integrability of the equations of the three-body problem in finite form.
A. A. Efimov.
Problems of observational and theoretical astronomy, Moscow - Leningrad, 1977, (see 003.012), p. 199 - 229. In Russian. — Abstr. in Ref. zh., 51. Astron., 6.51.100 (1978).

042.136 On the application of the averaging method in the two-centre problem.   A. M. Shul'gin, N. I. Shanchenko.
Dokl. AN UzSSR, 1977, No. 11, p. 23 - 25. In Russian. Abstr. in Ref. zh., 51. Astron., 6.51.103 (1978).

042.137 Qualitative properties of motion of a satellite of a three-axial planet.   Kh. B. Ibragimova, M. Kh. Khasanova.
Dokl. AN TadzhSSR, Vol. 20, No. 12, p. 16 - 19 (1977). In Russian. — Abstr. in Ref. zh., 51. Astron., 6.51.109 (1978).

042.138 Resonance in the restricted problem of three bodies with short-periodic perturbations.
K. B. Bhatnagar, B. Gupta.
Proc. Indian Natl. Sci. Acad., Part A, Vol. 43, No. 2, p. 153 - 168 (1977). — Abstr. in Phys. Abstr., Vol. 81, Abstr. 53803 (1978).

042.139 On the surface of zero radial velocity in the three-body problem.   V. V. Radzievskij.
Problems of observational and theoretical astronomy, Moscow–Leningrad, 1977, (see 003.012), p. 153 - 163. In Russian. Abstr. in Ref. zh., 51. Astron., 7.51.103 (1978).

042.140 Estimate of the value of angular velocity of some equilibrium figures of a rotating liquid.
A. Mardanov.
Vestn. LGU, 1977, No. 19, p. 121 - 127. In Russian. — Abstr. in Ref. zh., 51. Astron., 7.51.118 (1978).

042.141 On the solution of the navigation problem with use of the observed parameters of a system of material points in the gravitational field of a planet.
A. G. Ilyukhin, E. N. Bytsan'.
Slozhn. sistemy upr., Kiev, 1977, p. 47 - 57. In Russian. Abstr. in Ref. zh., 51. Astron., 7.51.121 (1978).

042.142 Determination of phase coordinates of a solid body moving in a Newtonian field of forces.
A. M. Kovalev.
Mekh. tverd. tela. Resp. mezhved. sb., 1978, No. 10, p. 65 - 76. In Russian. — Abstr. in Ref. zh., 62. Issled. kosm. prostranstva, 7.62.373 (1978).

042.143 On periodic motions of an axisymmetric satellite relative to the mass center in a circular orbit. I.
Yu. V. Barkin, A. A. Pankratov.
Vestn. MGU. Fiz., astron., Vol. 19, No. 1, p. 95 - 104 (1978). In Russian. — Abstr. in Ref. zh., 62. Issled. kosm. prostranstva, 7.62.374 (1978).

Astrodynamics (for teachers).
See Abstr. 014.027.

Orbits of two-body problem from the Lenz vec'or.
See Abstr. 014.037.

The classical Kepler problem in momentum space.
See Abstr. 014.038.

Planetary positions and lunar elements using an algebraic manipulator.   See Abstr. 021.010.

Givens transformation techniques for Kalman filtering.   See Abstr. 021.015.

Instability of an area-preserving polynomial mapping.   See Abstr. 021.031.

Bifurcations of straight-line oscillations.
See Abstr. 022.102.

An element formulation for perturbed motion about the center of mass.   See Abstr. 052.018.

Optimal low-thrust, limited power transfers between arbitrary elliptical orbits.   See Abstr. 052.024.

Long term solar radiation effects upon an orbit in the ecliptic.   See Abstr. 052.025.

An application of the stroboscopic method.
See Abstr. 052.034.

Third-order solution of an artificial-satellite theory.
See Abstr. 052.035.

Double-point problems of determination of orbits.
See Abstr. 052.039.

On a method for determination of an orbit from observational results.   See Abstr. 052.040.

The influence of satellite flexibility on orbital motion.   See Abstr. 052.047.

Differing time standards and celestial mechanics.
See Abstr. 066.311.

The equilibrium figure of the earth.
See Abstr. 081.007.

On the solution of the exterior boundary value problem with the aid of series.   See Abstr. 081.055.

Relation of a contracting Earth to the apparent accelerations of the Sun and Moon.   See Abstr. 094.013.

Contribution à l'étude des perturbations planétaires de la Lune.   See Abstr. 094.014.

Kirkwood gaps and stability of conservative periodic systems.   See Abstr. 098.065.

The long period behavior of the orbits of the Galilean satellites of Jupiter.   See Abstr. 099.502.

On regions of possible motion of Jupiter and Saturn satellites.   See Abstr. 099.534.

An observational test for the origin of the Titan-Hyperion orbital resonance.   See Abstr. 100.509.

A dynamical study of escaped satellites of Neptune.
See Abstr. 101.029.

A semianalytical theory for the long-term motion of Pluto.   See Abstr. 101.036.

Distributions of Jovian perturbations on short-period comet orbits. See Abstr. 102.010.

The asteroidal planet as the origin of comets. See Abstr. 102.022.

Comets and the missing planet. See Abstr. 102.023.

Motion of two rigid bodies under the gravitational influence of each other. See Abstr. 117.003.

043 Astronomical Constants

043.001 Numerical values of the constants of the Joint Report of the Working Groups of IAU Commission 4. P. K. Seidelmann.
Celestial Mech., Vol. 16, 165 - 177 (1977).

The numerical values underlying the Joint Report of the Working Groups of IAU Commission 4 on precession, planetary ephemerides, units and time scales are summarized in this report and additional explanation and references are provided.

043.002 Obliquity of the ecliptic in geological time. S.-f. Liu.
Acta Astron. Sinica, Vol. 18, 259 - 263 (1977). In Chinese.

043.003 Precession. A. A. Mikhajlov.
Zemlya i Vselennaya, 1978, No. 2, p. 24 - 29. In Russian.

043.004 Minor planet orbits and the FK4 equator and equinox. R. L. Branham.
Bull. American Astron. Soc., Vol. 9, 622 (1978). – Abstract.

043.005 Observations of the light aberration of stars of different spectral types with a vertical circle. B. Thüring, F. Schmeidler.
Astron. Nachr., Band 299, 55 - 58 (1978). In German.

Measurements of 9 stars of high declination generally confirm W. Becker's relation between the constant of aberration and the spectral type of the stars. The right ascensions and the declinations of the 9 stars are also determined. The relation between aberration and spectral type cannot be explained by atmospheric dispersion. The relation does not affect measurements of trigonometric parallaxes.

043.006 On the new value of the constant of precession and its influence on the proper motions of stars. A. A. Mikhajlov.
Astron. Zh. Akad. Nauk SSSR, Vol. 55, 436 - 437 (1978). In Russian. English translation in Soviet Astron., Vol. 22, No. 2.

The change of Newcomb's value of the general precession is explained and a simple method of deriving corresponding corrections of the proper motions of stars is described.

043.007 The IAU (1976) system of astronomical constants. Ya. S. Yatskiv.
Astrometriya i Astrofizika, Kiev, Vyp. (No.) 34, p. 52 - 58 (1978). In Rusian.

043.008 A rediscussion of determinations of precession and galactic rotation from Lick proper motions referred to galaxies. B. du Mont.
Astron. Astrophys., Vol. 66, 441 - 451 (1978).

A determination of precessional corrections from differences between the published AGK3 proper motions and the Lick proper motions measured with reference to galaxies has been made. The proper motion differences AGK3-Lick$_{corr}$ have yielded the centennial values of the precessional corrections $\Delta p_1 = +0\rlap{.}''95 \pm 0\rlap{.}''15$ (m.e.) and $\Delta\lambda + \Delta e = +1\rlap{.}''21 \pm 0\rlap{.}''16$ (m.e.) per century following from the directly determined values $\Delta n = +0\rlap{.}''38 \pm 0\rlap{.}''06$ (m.e.) and $\Delta k = -0\rlap{.}''34 \pm 0\rlap{.}''11$ (m.e.) per century. An analysis of the proper motions of the faintest Lick stars which, according to their authors, appear to be well measured, yielded Oort's constants $A = +8.5 \pm 2.4$ (m.e.) km s^{-1} kpc^{-1} and $B = -15.2 \pm 1.9$ (m.e.) km s^{-1} kpc^{-1} on the basis of the Oort-Lindblad model of galactic rotation. The μ_δ-solution yielded $A = +13.7 \pm 3.3$ (m.e.) km s^{-1} kpc^{-1}.

043.009 Determination of the aberration constant from latitude observations with the ZTL-180 of the Engelhardt Astronomical Observatory. Yu. G. Yusupov.
Izv. Astron.Ehngel'garth. Obs., Kazan', No. 41 - 42, p. 180 - 181 (1976). In Russian.

043.010 On the effect of the scale value upon the determination of the coefficients of main nutation terms derived from observations of the International Latitude Service. E. A. Yablokov.
Astron. Tsirk., No. 972, p. 1 - 2 (1977). In Russian.

043.011 New method for determination of the earth's precession due to the sun. D. Z. Koenov.
Dokl. AN TadzhSSR, Vol. 20, No. 11, p. 15 - 17 (1977). In Russian. – Abstr. in Ref. zh., 51. Astron., 6.51.124 (1978).

043.012 Possibility of a geophysical determination of the Newtonian gravitational constant. F. D. Stacey.
Geophys. Res. Lett., Vol. 5, 377 - 378 (1978).

The Newtonian gravitational constant G may be determined from a comparison of gravity at the ocean surface and at depth. Uncertainties arising from instrumental limitations and localized gravity anomalies may be reducible to the point of obtaining an improved value for G, but it is in any case of interest to verify that the result coincides with laboratory measurements on a scale 10^7 times smaller.

The orbits of five minor planets and corrections to the FK4 equator and equinox. See Abstr. 041.025.

Planetary theories and observational data. See Abstr. 042.052.

The earth's tides according to the new system of USSR Universal Time for the years 1955 - 1974. See Abstr. 044.013.

On the reduction problem related to the axial rotation of the earth. See Abstr. 044.020.

044 Time, Rotation of the Earth

044.001 Bemerkungen zur Ableitung der Bewegung des Poles und der Weltzeit 1.   W. K. Hristov.
Veröff. Zentralinst. Phys. Erde, No. 52, Part 2, (see 012.001), p. 327 - 331 (1977).

044.002 Dynamic influence of the liquid core on the daily tides of the earth.   B. Bodri, L. Bodri.
Veröff. Zentralinst. Phys. Erde, No. 52, Part 2, (see 012.001), p. 441 - 492 (1977). In Russian.

044.003 On application of the lunar laser ranging technique for determinations of the earth rotation in view of accuracy of computed topocentric distances of lunar retroreflectors.   B. Kołaczek.
Veröff. Zentralinst. Phys. Erde, No. 52, Part 2, (see 012.001), p. 589 - 597 (1977).

044.004 The effect of harmonics in the gravitational potential of the Earth, Moon and Sun on the precession angle of the Earth's rotation axis.   M. Bursa.
Bull. Astron. Inst. Czechoslovakia, Vol. 29, 1 - 10 (1978).
Harmonic terms in the time variations of the precession angle of the Earth's rotation axis, excited by Stokes constants in the gravitational potentials of the Earth, Moon and Sun of the following degrees n and orders k have been derived: Earth (2,2), (3,0), (3,1), (3,3), (4,0), (4,3); Moon (2,0), (2,2), (3,1); Sun (2,0).

044.005 Histoire du service de l'heure de l'Institut Sternberg (1920 - 1940).   M. A. Smirnova.
Soobshch. Gos. Astron. Inst. Shternberga, No. 200, p. 47 - 55 (1977). In Russian.

044.006 Astronomical determinations of time and continental drift.   G. P. Pil'nik.
Izv. AN SSSR. Fiz. Zemli, 1977, No. 7, p. 3 - 8. In Russian.
Abstr. in Ref. zh., 51. Astron., 2.51.162 (1978).

044.007 Time Service Annual Report 1974.
Zi-Ka-Wei Section, Shanghai Obs., Acad. Sinica, Shanghai, China, 105 pp.

044.008 Time Service Annual Report 1975.
Zi-Ka-Wei Section, Shanghai Obs., Acad. Sinica, Shanghai, China, 100 pp.

044.009 Pourquoi l'heure est-elle retardée d'une seconde de temps à autre?   B. Guinot.
Astronomie, Vol. 92, 227 - 229 (1978).

044.010 Lunar methods for 'longitude without time'.   D. H. Sadler.
J. Navig., Vol. 31, 244 - 249 (1978).

044.011 Scientific problems and organization of time observations.   G. P. Pil'nik.
Astrometriya i Astrofizika, Kiev, Vyp. (No.) 34, p. 81 - 82 (1978). In Russian.
Several scientific problems are considered, such as the study of diurnal earth tides, that can be solved using analysis of astronomical time observations. The author gives some suggestions to facilitate the solution of these problems.

044.012 On the precession and nutation of the earth's axis of figure.   C. A. Murray.
Mon. Not. R. Astron. Soc., Vol. 183, 677 - 685 (1978).
The equations of motion of the rigid earth about its centre of mass are developed by the methods of vectorial

mechanics and compared with the classical formulation given by Woolard. It is shown that apparent discrepancies in the coefficients of the Oppolzer terms in the motion of the axis of figure between values listed by Woolard and those recently published by Kinoshita, can largely be accounted for by approximations introduced by Woolard in his integration procedure. In particular the apparent difference between the precessions of the axis of figure and the angular-momentum vector is shown to be spurious.

044.013 The earth's tides according to the new system of USSR Universal Time for the years 1955 - 1974.
V. S. Gubanov, L. I. Yagudin.
Pis'ma Astron. Zh., Vol. 4, 108 - 111 (1978). In Russian.
Short-period variations of Universal Time connected with tides have been investigated according to data of the revision of USSR Universal Time for 20 years. From four waves obtained in the high-frequency region of the spectrum estimates of the Love number and the coefficient of the fortnightly term of the nutation in obliquity were obtained.

044.014 Greenwich Time Report. Royal Greenwich Observatory Time and Latitude Service. 1976 July–September.
Royal Greenwich Obs., Herstmonceux, Sussex, England. 16 pp. (1977). – From Phys. Abstr., Vol. 81, Abstr. 20624 (1978).

044.015 Greenwich Time Report. Royal Greenwich Observatory Time and Latitude Service 1976 October - December.
Royal Greenwich Obs., Herstmonceux Castle, Sussex, England. 17 pp. (1977). – From Phys. Abstr., Vol. 81, Abstr. 24443 (1978).

044.016 Deduction of a new scale of UT for 1955 - 1974.
L. I. Yagudin.
Astrometriya i Astrofizika, vyp. (No.) 35, p. 51 - 61 (1978). In Russian.
A method has been elaborated for redetermining the UT scale of the USSR with the purpose of improving its accuracy. For the interval 1955 - 1974 a new UT scale UT" has been determined in the system of the General Catalogue of the USSR Time Services, in the system of pole coordinates obtained from the BIH with reference to the international conventional origin. A systematic difference of the scales of UT" and BIH of the seasonal character has been obtained.

044.017 Corrections to longitudes of Time Services.
L. I. Yagudin.
Astrometriya i Astrofizika, vyp. (No.) 35, p. 61 - 66 (1978). In Russian.
On the basis of astronomical time determinations made by USSR Time Services during 1955 - 1977 corrections to the longitudes of these observatories were obtained.

044.018 Time Service Annual Report 1976.
Zi-Ka-Wei Section, Shanghai Obs., Acad. Sinica, Shanghai, China, 112 pp.

044.019 Les unités de temps.   P. Paquet.
Bull. Soc. Astron. Liège, Vol. 40, 57 - 62, 85 - 89, 108 - 112 (1978).

044.020 On the reduction problem related to the axial rotation of the earth.
Zh. S. Erzhanov, A. A. Kalybaev, A. K. Egorov.
Vestn. AN KazSSSR, 1977, No. 10, p. 44 - 55. In Russian.
Abstr. in Ref. zh., 51. Astron., 5.51.104 (1978).

044.021 Time and latitude service.
Polish Acad. Sci., Astron. Latitude Obs., Borowiec, Poland, Circ. Nos. 143 - 144 (1977). — 1977 July – December.

044.022 Zeit- und Breitendienst.
Deutsches Hydrogr. Inst., Hamburg (1977). — 1977 January - September.

044.023 Universal time and coordinates of the pole; Emission time of time signals; Coordinated Universal Time; International Atomic Time.
Bureau International de l'Heure, (B.I.H.), Paris, Circ. D134 - D140 (1978). — 1978 January - July.

044.024 Astronomische Zeit- und Breitenbestimmungen, Empfangszeiten von Zeitsignalen, Präzisionszeitvergleiche.
Akad. Wiss. DDR, Zentralinst. Phys. Erde, Abt. Geod. Astron., Jahrg. 1977, Nr. 1 - 2 (1978). — 1977 January - April.

044.025 Nutation and non-polar latitude and longitude variations of the IPMS system.
K. Yokoyama, H. Ishii, I. Sato.
Publ. Int. Latitude Obs. Mizusawa, Vol. 11, 11 - 22 (1977).

Amplitudes of the principal, semi-annual and annual nutation terms are evaluated from the z-term in latitude variation and the τ-term in longitude variation of the IPMS. For the purpose of detecting the non-polar longitude variation due to the improper nutation terms from the time observations, a term $\tau \tan \phi_i$ is added to the equation with which the polar motion and UT1–UTC are determined. The meaning of τ is equivalent to that of z so far as the nutation is concerned. The nutation amplitudes obtained from both z and τ are in good agreement with one another, except for the annual nutation.

044.026 Time Service. C. Moranzino (Editor).
Oss. Astron. Torino (Pino Torinese), Bull. No. 17 (1977). — Results of the time determinations 1977 May - 1977 August.

044.027 International Time and Latitude Service at the Tokyo Astronomical Observatory during 1977.
Z. Suemoto.
Tokyo Astron. Obs., Time and Latitude Bulletins, Vol. 51, Nos. 3 - 4, p. 31 - 56 (1977). — Results of time determinations 1977 July - December.

044.028 Temps et latitude. L. Webrová, V. Ptáček.
Circulaire des observatoires tchécoslovaques.
Acad. Tchécoslovaque Sci., Inst. Astron. (1978). — Janvier - Août 1977.

044.029 Some comments on the 5-th chapter of "The rotation of the earth" by Munk and MacDonald.
W. Opalski.
Prace Naukowe Geod., Nr. 19, p. 35 - 51 (1977).

044.030 Ephemeris time from observations on the 16″ astrograph of the Engelhardt Astronomical Observatory.
S. G. Valeev.
Izv. Astron. Ehngel'gardt. Obs., Kazan', No. 41 - 42, p. 182 - 188 (1976). In Russian.

044.031 Complex studies in planetary dynamics of the earth.
H. Kautzleben, C. Elstner, G. Hemmleb, H. Montag.
New methods of space geodesy, (see 012.029), p. 63 - 84 (1976).

Beim gegenwärtigen Stand der Beobachtungstechnik können Fortschritte im Verständnis der planetaren Dynamik insbesondere durch vergleichende Untersuchungen der Rotationsschwankungen, der Polbewegung und der Gezeiten erzielt werden.

044.032 On methods of observations applied in investigation of the earth's rotation. E. P. Fedorov.
New methods of space geodesy, (see 012.029), p. 253 - 276 (1976). In Russian.

The basic principles of the methods as well as the co-ordinate systems used for the studies of the Earth's rotation are considered; among the methods are those based upon observations of the stars, artificial satellites, moon and radio sources. To analyse the advantages of the various observational methods or their combinations, the principles of the comparison of these methods are developed.

044.033 Time service for 1975. V. Quesada.
Circ. Stn. Astron. Int. Latitudine, Carloforte–Cagliari, Ser. A (4), N. 13, 31 pp. (1976).

044.034 Daily phase values and time differences.
U.S. Naval Obs., Washington, D.C. Time Service
Publ., Ser. 4, Nos. 570 - 595 (1978). — 1978 January - June.

044.035 Preliminary times and coordinates of the pole.
U.S. Naval Obs., Washington, D.C. Time Service
Publ., Ser. 7, Nos. 523 - 548 (1978). — 1978 January - June.

044.036 Time service report.
U.S. Naval Obs., Washington, D.C. Time Service
Publ., Ser. 11, No. 224, 27 pp. (1978). — Times of emission, UTC (USNO, MC) — UTC (Transmitter); Time scales; Observations of time and latitude; Polar coordinates, corrections for seasonal and polar (longitude) variations. — 1975 January - 1976 December.

044.037 Short-period non-uniformity in the earth's rotation.
G. P. Pil'nik.
Astron. Tsirk., No. 988, p. 7 - 8 (1978). In Russian.

044.038 Time Service bulletins.
Glavnaya astronomicheskaya observatoriya Akademiya Nauk SSSR, 1976 (1978), yan.-dek., No. 256 - 267, 28 pp. In Russian. — Abstr. in Ref. zh., 51. Astron., 7.51.134 (1978).

Time and its measurement from billion's parts of a second to milliards of years. See Abstr. 003.149.

La terre, les eaux, l' atmosphère.
See Abstr. 003.156.

The classical analysis of the response of a long baseline radio interferometer. See Abstr. 033.025.

Spectral analysis of a ten-year series of observations obtained with two astrolabes and with one photoelectric transit instrument in Irkutsk. See Abstr. 041.022.

Gezeiten bremsen unseren Planeten. Früher drehte sich die Erde schneller. See Abstr. 081.079.

045 Latitude Determination, Polar Motion

045.001   The determination of the length of the Chandler-
wobble by comparison of the astronomic determined annual period of polar motion with the period calculated from air mass shifts.   H. Jochmann.
Veröff. Zentralinst. Phys. Erde, No. 52, Part 2, (see 012.001),
p. 323 - 325 (1977).

045.002   On some analysis of filtered latitude variations.
R. Galas.
Veröff. Zentralinst. Phys. Erde, No. 52, Part 2, (see 012.001),
p. 569 - 574 (1977).
   The filtered latitude for a few IPMS stations by the use
of harmonic analysis were computed and compared with mean
latitude values obtained by other formula.

045.003   Analyse der Dresdner Breitenbeobachtungen auf
systematische Deklinationskorrektionen.
D. Ponomarev, S. Wächter.
Veröff. Zentralinst. Phys. Erde, No. 52, Part 2, (see 012.001),
p. 575 - 580 (1977) = Mitt. Lohrmann Obs., Dresden, GDR,
Nr. 34.
   Since 1957 latitude determination by the Horrebow-
Talcott method had been performed with the Zeiss-transit
instrument 100/1000 with two short interruptions only. For a
period of 12 years declination corrections were derived from
the observation results of continuous programmes for different
time intervals. The standard error of declination correction
amounts to < ± 0."03. In addition, information is given on
changes in proper motions of pairs of stars from the analysis
of annually computed declination corrections.

045.004   Untersuchungen zur Jahreswelle in den Breitenbestimmungen aus PZT-Beobachtungen in Potsdam.
J. Dittrich, E. Felsmann.
Veröff. Zentralinst. Phys. Erde, No. 52, Part 2, (see 012.001),
p. 581 - 584 (1977).

045.005   Refinement of corrections to the influence of the
atmosphere in studying the motions of the earth's
pole.   M. T. Prilepin, N. S. Zabolotnyj.
Veröff. Zentralinst. Phys. Erde, No. 52, Part 3, (see 012.001),
p. 1029 - 1035 (1977). In Russian.

045.006   Analysis of long-term latitude determinations at
Potsdam Observatory with respect to variations in
their principal components.   J. Höpfner.
Gerlands Beitr. Geophys., Band 86, 449 - 459 (1977).

045.007   An optimal complex AR.MA (autoregressive-
moving average) model of the Chandler wobble.
M. Ooe.
Geophys. J. R. Astron. Soc., Vol. 53, 445 - 457 (1978).
   The character of the Chandler wobble suggests that the
optimal representation of the observed polar motion is a
complex AR.MA model. This paper develops the theory of
such a model and presents a modification of the scalar AR.MA
computer program of Akaike, Arahata & Ozaki. The complex
AR.MA model is applied to the ILS data covering the period
1900–1975. An optimal AR.MA (1,4) model is obtained.

045.008   The spectra of diurnal variations of latitude from
observations in Gorky.   L. D. Kovbasyuk, S. G.
Kulagin, Yu. M. Shur.
Astron. Zh. Akad. Nauk SSSR, Vol. 55, 645 - 648 (1978).
In Russian. English translation in Soviet Astron., Vol. 22,
No. 3.
   The spectra of diurnal variations of latitude transformed
to the low-frequency range have been calculated for a ten-year

interval of observations (1961.5–1972.0) of two bright
zenith stars and four bright Talcott pairs carried out at the
zenith telescope in Gorky. Components with periods of 1.75,
1.17, 1.03, 0.88, 0.55, 0.50, and 0.44 years have been
revealed. Of these the half-year oscillation is the most significant one.

045.009   Breitenbestimmung.
Tech. Univ. Dresden, Lohrmann-Obs., Zirk. Nos.
81 - 86 (1976/77). – 1976 May - 1977 December.

045.010   Monthly notes of the international polar motion
service.
IPMS Mon. Not., Int. Latitude Obs. Mizusawa (Japan). 1977
No. 12, p. 119 - 127; 1978 Nos. 1 - 4, p. 1 - 40 (1978).
Announces the values of latitudes observed at the collaborating
stations during 1977 December - 1978 May. Preliminary values
of the coordinates of the instantaneous pole derived from
them are also given.

045.011   On the statistical model of the Chandler wobble.
M. Ooe, Y. Kaneko, H. Akaike.
Publ. Int. Latitude Obs. Mizusawa, Vol. 11, 1 - 9 (1977).
   The ILS data for the period 1900 to 1969 are analysed
by fitting AR·MA models by the minimum AIC procedure
(Akaike 1974). The best fit AR·MA model chosen for the
Chandler wobble suggests the following conclusions: (1) The
Chandler wobble has only one natural period of 1.182 - 1.187
years and its Q value is 51 - 94. (2) The power of the input
noise which maintains the Chandler wobble is about
$0.3 \times (0."01)^2/\text{cpy}$.

045.012   Results of observations of latitude variation of the
Engelhardt Astronomical Observatory in the years
1966 - 1970.   V. S. Kulagin, D. P. Laas, N. N. Chudinov.
Izv. Astron. Ehngel'gardt. Obs., Kazan', No. 41 - 42, p. 86 -
90 (1976). In Russian.

045.013   Results of observations of the latitude of the Engelhardt Astronomical Observatory at the ZTL-180
in the period 1957 - 1972.   D. P. Laas, V. V. Lobanova,
I. A. Urasina, N. N. Chudinov, Yu. G. Yusupov.
Izv. Astron. Ehngel'gardt. Obs., Kazan', No. 41 - 42, p. 154 -
174 (1976). In Russian.

045.014   The semi-diurnal tidal effect of the moon in observations of the latitude of the Engelhardt Astronomical Observatory.   I. A. Urasina.
Izv. Astron. Ehngel'gardt. Obs. Kazan', No. 41 - 42, p. 194 -
195 (1976). In Russian.

045.015   New system of the Horrebow-Talcott program in
Józefosław.
M. Dukwicz-Łatka, L. Pieczyński,
Warsaw Tech. Univ., Astron.-Geod. Obs. Józefosław. Latitude
Circ. No. 61.

045.016   Results of determination of latitude in Józefosław
by observations of the Horrebow-Talcott pairs.
L. Pieczyński.
Warsaw Tech. Univ., Astron.-Geod. Obs. Józefosław. Latitude
Circ. Nos. 62 - 65 (1977). – 1977 January - December.

045.017   On random excitation and damping of the polar
motion.   W. Opalski.
Prace Naukowe Geod., Nr. 19, p. 7 - 34 (1977).

045.018   Analysis of non polar variations of geographical

latitude at Józefosław. J. B. Rogowski.
Prace Naukowe Geod., Nr. 19, p. 53 - 88 (1977).

045.019 On the mean latitude determination. R. Galas.
Prace Naukowe Geod., Nr. 19, p. 89 - 106 (1977).

045.020 Sur le mouvement relative du pôle d'inertie par rapport au pôle instantané. C. Dramba.
Rend. Semin. Fac. Sci. Univ. Cagliari, Vol. 46, 273 - 280 (1976) = Pubbl. Stn. Astron. Int. Latitudine, Carloforte − Cagliari, Nuova Ser., No. 47.

045.021 On examining the ILS results for secular motion. H. J. M. Abraham.
Rend. Semin. Fac. Sci. Univ. Cagliari, Vol. 47, 25 - 33 (1977)= Pubbl. Stn. Astron. Int. Latitudine, Carloforte−Cagliari, Nuova Ser., No. 56.
The secular wobble and the librations which appear in the motion of the earth's pole are described. Interrelations between these two components and the Chandler wobble are discussed.

The classical analysis of the response of a long baseline radio interferometer. See Abstr. 033.025.

Bemerkungen zur Ableitung der Bewegung des Poles und der Weltzeit 1. See Abstr. 044.001.

On the precession and nutation of the earth's axis of figure. See Abstr. 044.012.

Time and latitude service. See Abstr. 044.021.

Zeit- und Breitendienst. See Abstr. 044.022.

Astronomische Zeit- und Breitenbestimmungen, Empfangszeiten von Zeitsignalen, Präzisionszeitvergleiche. See Abstr. 044.024.

Nutation and non-polar latitude and longitude variations of the IPMS system. See Abstr. 044.025.

International Time and Latitude Service at the Tokyo Astronomical Observatory during 1977. See Abstr. 044.027.

Temps et latitude. See Abstr. 044.028.

Complex studies in planetary dynamics of the earth. See Abstr. 044.031.

Time service report. See Abstr. 044.036.

Experience in studying the temperature fluctuations of the lower atmosphere's layers during latitude determinations. See Abstr. 082.010.

046 Astronomical Geodesy, Satellite Geodesy, Navigation

046.001 A new navigation computer. M. Richey.
J. Navig., Vol. 31, 157 - 161 (1978).

046.002 Alternative determinations of geodetic datum shifts. E. Groten, H. Schaab.
Veröff. Zentralinst. Phys. Erde, No. 52, Part 2, (see 012.001), p. 333 - 368 (1977).
Two previously published computational methods in determining absolute orientation of geodetic datums are applied to the European Datum (1950). Using plumb line deflections, geoid heights and satellite data results for several types of combinations are presented. In comparison to classical least-squares approaches alternatives on applying collocation procedures are discussed and first numerical data are outlined.

046.003 Satellite laser ranging station at Borowiec. S. Schillak, E. Wnuk.
Veröff. Zentralinst. Phys. Erde, No. 52, Part 3, (see 012.001), p. 803 - 811 (1977).

046.004 The new Nd−YAG laser-ranging system for the satellite observation station at Wettzell.
P. Wilson, H. Seeger, K. Nottarp.
Veröff. Zentralinst. Phys. Erde, No. 52, Part 3, (see 012.001), p. 823 - 851 (1977).
The installation of a new laser ranging system for the satellite observation station in Wettzell in early November 1976 will culminate two years of development. This paper describes the system in some detail and reports on the decision making processes influencing the design. A brief summary of tracking to date at the manufacturers plant is given together with some remarks on the results of acceptance testing.

046.005 Satellite laser ranging at Hradec Králové. F. Hovorka, M. Konrád, J. Utěkal.
Veröff. Zentralinst. Phys. Erde, No. 52, Part 3, (see 012.001), p. 853 - 857 (1977).

046.006 Possibilities in improving classical networks by satellite geodesy. S. Mihály.
Veröff. Zentralinst. Phys. Erde, No. 52, Part 3, (see 012.001), p. 859 - 869 (1977).
The accuracies of satellite geodetic,stellar triangulation and classical geodetic networks have been compared for the case of networks extending from 400 to 3000 km. The comparison has been based on a) recent accuracy informations of satellite geodetic measurements and networks, and of stellar triangulation in Finnland, b) the accuracy of the elements measured in the Hungarian astrogeodetic network and c) formulae derived for apriori estimates of the error propagation in stellar triangulation and classical networks. The comparison demonstrates that large differences may occur between horizontal (along and across) and vertical error components at the edge of a given network and between identical error components in various networks on the other hand.

046.007 On optimum conditions for determining the station coordinates by a semi-dynamical method.
Yu. V. Batrakov, T. K. Nikol'skaya.
Veröff. Zentralinst. Phys. Erde, No. 52, Part 3, (see 012.001), p. 871 - 876 (1977). In Russian.

046.008 Determination of station coordinates from laser observations. J. Ádám.
Veröff. Zentralinst. Phys. Erde, No. 52, Part 3, (see 012.001), p. 885 - 911 (1977).
Spatial positions of a station can be computed using its laser measurements with known satellite coordinates. Based on the method of least squares adjustment, algorithms were developed for obtaining the unknown station positions.

Simple mathematical models are also described in detail. Computational results are presented, characterizing the accuracy of the methods under different geometrical, observing and other conditions.

046.009   Azimuth-dependent statistics for interpolating geodetic data.   F. Morrison.
Bull. Géod., Vol. 51, 105 - 118 (1977). – Abstr. in Phys. Abstr., Vol. 80, Abstr. 90982 (1977).

046.010   Infinitesimal transformations of the nonholonomic geodetic frames.   E. Osada.
Boll. Geod. Sci. Affini, Anno 36, 471 - 486 (1977).
In this paper the equations of the infinitesimal transformation of the geodetic frame have been proved as well as the components of the objects of connection and anholonomity in the geodetic system coordinates (B, L, H). Starting from the transformation rules of the infinitesimal transformation for two frames: geodetic and natural (astronomic), some topics between geometric parameters of the gravity field and parameters of the reference surface have been proved.

046.011   Combined use of Doppler devices and photo-astronomical cameras for the institution of vertical deflection points.   G. Birardi.
Boll. Geod. Sci. Affini, Anno 36, 499 - 505 (1977).
An operative proposal is presented for the simultaneous use of Doppler satellite ranging devices, and photo-astronomical zenith cameras to define vertical deflection points. Using this approach, both the geodetic and astronomic terms may be obtained in each location, independent from the local geodetic net, with a sufficient accuracy. Remarkable advantages, both for quickness and homogeneity of results, are envisioned.

046.012   Die astrogeodätischen Lotabweichungsbestimmungen der Abteilung I des Deutschen Geodätischen Forschungsinstituts auf Punkten des Deutschen Hauptdreiecksnetzes in den Jahren 1966 bis 1977.
K. Kaniuth, K. Stuber.
Deutsche Geod. Komm. Bayerische Akad. Wiss., Reihe B: Angew. Geod., Heft Nr. 229, 32 pp. (1978).
This paper deals with astronomic latitude and longitude determinations carried out with Zeiss Ni2 astrolabes on 27 points of the primary triangulation network in Southern Germany, in order to derive deflections of the vertical. The deflections refer to the European Datum ED50; an estimation of the external accuracy of the components ξ and η resulting from observations in two nights gives $\pm 0\overset{\prime\prime}{.}3$.

046.013   Geodesy by radio interferometry: determination of a 1.24-km base line vector with ~5-mm repeatability.   A. E. E. Rogers, C. A. Knight, H. F. Hinteregger, A. R. Whitney, C. C. Counselman III, I. I. Shapiro, S. A. Gourevitch, T. A. Clark.
J. Geophys. Res., Vol. 83, 325 - 334 (1978).
The 1.24-km base line vector between the two antennas of the Haystack Observatory was determined from X band radio interferometric observations of extragalactic sources via a new method that utilizes the precision inherent in fringe phase measurements. This method was employed in 11 separate experiments distributed between October 1974 and January 1976, each being between about 5 and 20 hours in duration. The rms scatters about the means for the vertical and the two horizontal components of the base line obtained from the 11 independent determinations were 7, 5, and 3 mm, respectively. The corresponding scatter for the base line length was 3 mm; the mean differed from the result obtained in a conventional survey by 8 mm, well within the 20-mm uncertainty of the survey.

046.014   Improved gravity field and station coordinate estimates from laser tracking data.
E. M. Gaposchkin.
Space Research, Vol. XVII, (see 012.010), 63 - 71 (1977).

046.015   Intercosmos laser ranging stations.
A. G. Massevitch (*Masevich*), K. Hamal.
Space Research, Vol. XVII, (see 012.010), 73 - 76 (1977).
The laser ranging station built in the framework of the Intercosmos co-operation is described. The distribution of these stations, and the first results of the observations,are given. Plans for further developments are indicated.

046.016   The use of balloons for geodetic research.
J. Kakkuri.
Space Research, Vol. XVII, (see 012.010), 795 - 800 (1977).
Two types of balloon suitable for geodetic research are discussed. Examples are given of how a geodetic network is established using balloons to complement the classical and satellite triangulation methods. Experiments have been conducted from Finland and Libya.

046.017   First experiments on balloon geodesy in the framework of the Intercosmos programme.
A. M. Losinsky (*Losinskij*), A. G. Massevitch (*Masevich*), S. P. Orayevskaya (*Oraevskaya*).
Space Research, Vol. XVII, (see 012.010), 801 - 803 (1977).
A description of the first Soviet-Mongolian experiment on balloon geodesy is presented: the launching of balloons with electronic flashes, the equipment at the ground reception stations, and the reduction methods are described. Plans for further investigations in the framework of the Intercosmos programme are discussed.

046.018   Integrale Auswertung von Satellitenbeobachtungen.
G. Beutler.
Astron.-Geod. Arb. Schweiz, 33. Band, 114 pp. (1977).

046.019   Geodetic space triangulations.   G. Zlatanov.
Geod. Kartogr., Vol. 26, 103 - 115 (1977). In Polish.
The work is an approach to the mathematical adjustment of a three-dimensional geodetic net. Two ways of solution are presented: the first deals with the multi-group adjustments using the conditioned method, and the second describes the parameter method of correlative equations, suited for local, limited triangulation networks.

046.020   Leading programme of satellite works considering scientific and practical aspects of Doppler technique.
J. Śledziński.
Geod. Kartogr., Vol. 27, 3 - 16 (1978). In Polish.
Investigation programme concerning elaboration of technology of satellite Doppler measurements as well as application of satellite observations to establishment of basic geodetic networks are discussed in this paper. Specific errors of satellite Doppler observations and their sources are given.

046.021   On the most advantageous geometric observational schemes with a long-baseline radio interferometer for determination of the length and direction of the chord.
O. S. Razumov, I. A. Basova.
Izv. vyssh. uchebn. zaved. Geod. i aehrofotosemka, 1977, No. 3, p. 3 - 8. In Russian. – Abstr. in Ref. zh., 52. Geod. Aehrosemka, 2.52.88 (1978).

046.022   Determination of geodetic coordinates of the Hvar Observatory.   V. Petković.
Hvar Obs. Bull., No. 1, p. 31 - 40 (1977).

046.023   A method for filtering the observational data in astronomical determinations by the position line method.   B. Buonocore, L. Turturici, A. Vassallo.

Boll. Geod. Sci. Affini, Anno 37, 77 - 104 (1978). In Italian.

A graphical-analytical method for the analysis of random errors and thus the filtering of astronomical observations by the position line method is shown.

046.024 La télémétrie laser: une technique de la géodésie spatiale. J.-L. Hatat, F. Gomez Armario.
Astronomie, Vol. 92, 205 - 210 (1978).

046.025 Present state and courses of the development of satellite geodesy. W. Baran.
Postępy Astron., Tom 26, 3 - 17 (1978). In Polish.

Geometrical and dynamical methods of satellite geodesy are presented. With geometrical methods one can determine the space triangulation net formed of observational stations located on the Earth's surface. Accuracy is of the order of a few metres. Dynamical methods or satellite motion analysis make possible to determine the inhomogeneities of the Earth's gravitational field. Present observations make possible to determine the coefficients of the order of about 30 in the expansion of the gravitational potential into series of spherical functions.

046.026 Dynamical constraints in satellite photogrammetry. J. N. Blanton, J. L. Junkins.
AIAA J., Vol. 15, 488 - 498 (1977). — Abstr. in Phys. Abstr., Vol. 81, Abstr. 15943 (1978).

046.027 Application of modern automatic drafting system to problems of geodesy. H. Wurtzler.
Z. Vermessungswes., Vol. 102, 467 - 474 (1977). In German. Abstr. in Phys. Abstr., Vol. 81, Abstr. 15952 (1978).

046.028 Ein geodätisches Weltdatum aus terrestrischen und Satellitendaten. B. Eitschberger.
Deutsche Geod. Komm. Bayerische Akad. Wiss., Reihe C, Diss., Heft Nr. 245, 188 pp. (1978).

046.029 Complex investigations of geodynamic phenomena with the help of parallel observations of stars and artificial earth satellites. L. V. Rykhlova.
Nauchn. Inf., Vyp. (No.) 38, p. 7 - 11 (1976). In Russian.

046.030 Works in the field of space geodesy carried out by the Astronomical Council of the USSR Academy of Sciences in the years 1970 - 1975. N. P. Erpylev, A. G. Masevich, V. D. Sobolevskij, S. K. Tatevyan.
New methods of space geodesy, (see 012.029), p. 6 - 17 (1976). In Russian.

A short review of the satellite geodesy programs carried out under co-ordination of the Astronomical Council is given. First results of the calculation of chord directions and station positions are presented.

046.031 Use of radiotechnical/Doppler/ methods for observation of AES for the solution of geometrical and dynamical problems of space geodesy. F. Halmos, J. Adam, I. Almar, I. Fejesh.
New methods of space geodesy, (see 012.029), p. 85 - 134 (1976). In Russian.

Die Studie gibt einen Überblick über die geodätische Anwendung der radiotechnischen Satelliten-Beobachtungen. Die mit Radiowellen-Sendern ausgerüsteten künstlichen Erdsatelliten und die Instrumente, die zum Empfang der Radiosignale geeignet sind, werden besprochen. Die geometrischen und dynamisch-geodätischen Bearbeitungsmethoden der Doppler-Beobachtungen werden analysiert. Die annähernden und strengen Lösungen auf dem Gebiete der geometrischen Geodäsie werden geprüft. Die Rolle der Doppler-Beobachtungen bei der Bestimmung des Schwerefeldes wird geschildert und die bisher erzielten Ergebnisse der geometrischen und dynamischen Geodäsie werden zusammengefasst. Anwendungsmöglichkeit der Doppler-Messungen bei der Bestimmung

der Polschwankungen wird dargestellt.

046.032 Analysis of laser ranging observations of satellites. Y. Kozai.
New methods of space geodesy, (see 012.029), p. 144 - 149 (1976).

The author discusses how to analyze laser ranging observations of satellites by an analytical method in order to obtain accurate geodetic informations.

046.033 Aspects of positioning using satellite-borne laser or RF systems. I. I. Mueller, B. H. W. van Gelder, M. Kumar.
New methods of space geodesy, (see 012.029), p. 150 - 154 (1976).

Investigations are being conducted in the United States to determine the feasibility of a satellite-borne laser or RF system which in conjunction with inexpensive reflectors placed on the ground, could repeatedly determine positions with great accuracies. The applications of such a system, such as crustal motion monitoring, subsidence detection, are discussed together with the results of error analyses based on extensive simulations.

046.034 On the AFU–75 network adjustment. G. Vass.
New methods of space geodesy, (see 012.029), p. 155 - 158 (1976).

The paper presents a method for station coordinates determination, using observations from the station to be determined and from a known station. The method can be applied for a local network adjustment.

046.035 Possibilities of application of the radiointerferometric equipment of the Institute of Space Research to geodesy and astrometry. L. R.Kogan, V. I. Kostenko, L. I. Matveenko.
New methods of space geodesy, (see 012.029), p. 171 - 181 (1976). In Russian.

A description of the parameters and usefulness of the interferometric equipment developed at the Institut of Space Research of the USSR Academy of Sciences for the aim of geodesy and astrometry is given. It is shown that this equipment allows to measure the length of a base with an error of 10 cm using radio sources in the water vapour line ($\lambda = 1.35$ cm).

046.036 Perspectives of solving fundamental problems of astrometry with the help of superlong baseline interferometry and special space experiments. V. S. Gubanov, Yu. S. Streletskij, N. D. Umarbaeva, B. A. Firago.
New methods of space geodesy, (see 012.029), p. 228 - 252 (1976). In Russian.

Further development of VLBI technique permits to hope that one shall be able to measure the length of Earth chords with a precision about ±10 cm and angular distances between radio sources with the precision about ±0".01. This method will contribute to geodynamics, geodesy and geophysics immediately. Together with space experiments, such as measuring artificial radio sources on satellites of the Earth and other bodies of the solar system, it may help to study the motion of these objects and to solve the fundamental problem of positional astronomy, i.e. to establish the inertial system of star coordinates.

046.037 Traitement des mesures et résultats de géodésie spatiale par récepteur Doppler. F. Nouel, J. Kovalevsky.
New methods of space geodesy, (see 012.029), p. 277 - 283 (1976).

Après avoir cité les travaux en géodésie spatiale par récepteur Doppler effectués en France depuis 10 ans, on donne les procédures suivies au Groupe de Recherches de Géodésie Spatiale pour dépouiller et exploiter les observations Doppler. Les résultats montrent, pour plusieurs stations, une cohérence interne de 1 à 3 mètres sur les trois coordonnées et présentent seulement quelques mètres d'écart avec les résultats d'Anderle pour le système Europe 50.

046.038 **Variation of latitude and longitude of the station of CERGA from Doppler satellite tracking and precise satellite ephemerides.** N. Capitaine, L. Saint Crit.
New methods of space geodesy, (see 012.029), p. 284 - 289 (1976).

The authors have obtained the variations of latitude and longitude of the station of CERGA from November 1973 to August 1975 from Doppler satellite observations of one TRANSIT satellite for which ephemerides were given by the N.W.L. The two polar coordinates can be obtained from these data. Comparisons with the computed values of variation of latitude and longitude using B.I.H. and D.P.M.S. polar coordinates have been performed.

046.039 **The Finnish stellar triangulation net as a geodetic control for the first-order terrestrial triangulation.** J. Kakkuri.
New methods of space geodesy, (see 012.029), p. 290 - 293 (1976).

046.040 **Some problems of the accuracy of a world geodetic datum out of satellite observations.** B. Eitschberger.
New methods of space geodesy, (see 012.029), p. 305 - 315 (1976).

046.041 **The accuracy of astronomic azimuth determinations.** W. E. Carter, J. E. Pettey, W. E. Strange.
Bull. Géod., Vol. 52, 107 - 113 (1978).

The authors have analyzed repeat determinations, by analysis of variance techniques, to derive realistic estimates of the expected accuracy of typical astronomic azimuths to be used in the readjustment of the North American Datum. They found that the dominant errors are systematic in nature, with a very important source being observer bias, or "personal equation". They found, from an analysis of determinations that were first corrected for observer bias, an increase in the variance of repeat azimuth determinations as a function of latitude that agrees reasonably with theoretical expectations.

046.042 **Earth rotation and network geometry optimization for very long baseline interferometers.** A. Dermanis, I. I. Mueller.
Bull. Géod., Vol. 52, 131 - 158 (1978).

The main objective of the present work is the exploration of the capabilities of VLBI for the recovery of earth rotation and network geometry parameters. For this purpose, a number of characteristic experimental designs based on present and candidate for the near future station locations is chosen. The results from the analysis of simulated observations for each particular design are presented.

046.043 **Astronomical azimuth determinations on triangulation stations in 1962–1970.** M. Ollikainen.
Publ. Finnish Geod. Inst., No. 84, 46 pp. (1977).

046.044 **Determinazione dell'azimut astronomico del punto trigonometrico di Torre S. Pancrazio (Cagliari).** A. Poma.
Rend. Semin. Fac. Sci. Univ. Cagliari, Vol. 46, 89 - 95 (1976)= Pubbl. Stn. Astron. Int. Latitudine Carloforte–Cagliari, Nuova Ser., No. 50.

046.045 **Dependence of the accuracy of determination of the**

position of an observational station on the kind of orbital measurements.   V. V. Bojkov.
Geod. i kartogr., 1978, No. 2, p. 15 - 19. In Russian. – Abstr. in Ref. zh., 62. Issled. kosm. prostranstva, 6.62.346 (1978).

Methods of determination of astronomical coordinates and problems of terrestrial navigation in the Antarctic.   See Abstr. 003.081.

La terre, les eaux, l' atmosphère. See Abstr. 003.156.

Die Geschichte der astronomischen Ortsbestimmung auf See. Zur Entwicklung der Navigation nach Gestirnen im 18. und 19. Jahrhundert.   See Abstr. 004.032.

Transatlantic geodesy by long baseline interferometry.   See Abstr. 031.205.

Investigation of the method of determination of the azimuth from circumpolar stars with the help of the reconstructed U-5″ astronomical transit with a special circular net. See Abstr. 031.242.

On the reduction of heterogeneous observations of satellites in a local network.   See Abstr. 031.276.

Corrections to longitudes of Time Services. See Abstr. 044.017.

Compilation of the ephemerides of star pairs of equal height observed in the plane of a common vertical with the help of a computer.   See Abstr. 047.017.

Iterative methods of determination of station coordinates and orbital satellite elements from Doppler observations.   See Abstr. 052.004.

Lageos orbital acquisition and initial assessment. See Abstr. 055.004.

Determination of the free boundary in potential theory, the boundary problem of physical geodesy. See Abstr. 081.009.

The determination of the earth gravity field by use of satellite gradiometry.   See Abstr. 081.018.

Application of satellite gradiometry measurements to local geoid determination.   See Abstr. 081.035.

Investigation of the Earth gravity field by differential Doppler measurements.   See Abstr. 081.036.

Application of low-low satellite to satellite observations to the investigation of the earth's gravity field by collocation.   See Abstr. 081.037.

Application of low-low satellite observations to local geoid determination.   See Abstr. 081.038.

Evaluation of the GEOS 3 altimetry data by means of spherical functions expansions.   See Abstr. 081.043.

The Clairaut equation and its connection to satellite geodesy and planetology.   See Abstr. 081.057.

Satellite altimetry and the scale factor of the geopotential.   See Abstr. 081.069.

Orientation of the ellipsoid in geodetic networks. See Abstr. 081.083.

047 Ephemerides, Almanacs, Calendars

047.001 Almanacco Astronomico della Rivista Coelum per l'anno 1978.
Coelum, Suppl. al fasc. 11-12, Vol. 45 (Oss. Astron. Univ. Bologna), 28 + 38 pp. Price L. 5000 (1977).

047.002 Efemérides Astronómicas para o ano de 1978.
Edited by Obs. Astron. Univ. Coimbra, printed by Gráfica de Coimbra, Coimbra. 13 + 250 pp. (1977).

047.003 The Air Almanac 1978, July - December.
Her Majesty's Stationery Office, London; United States Naval Observatory, Washington, p. 363 - 732, A 105 + F4 pp. Price £ 7.50 (1978). ISBN 0-11-772129-8.

047.004 Anuarul Observatorului din Bucureşti – 1978.
Editura Academiei Republicii Socialiste România, Bucureşti. 283 pp. Price Lei 11.00 (1977).

047.005 1979 Japanese Ephemeris.
Compiled by A. M. Sinzi, A. Yamazaki, T. Mori, Y. Tano, A. Senda, Y. Kubo, F. Ono, Y. Harada, K. Inoue, T. Jojo.
Astronomical Division, Hydrographic Department, Tokyo, Japan. Pub. No. 684, 6 + 465 pp. (1977).

047.006 The Astronomical Ephemeris for the year 1978.
Issued by Her Majesty's Nautical Almanac Office, London; and Nautical Almanac Office, United States Naval Observatory, Washington.
Her Majesty's Stationery Office, London. 8 + 573 pp. Price £ 8.50 (1976). ISBN 0-11-880627-0.

047.007 The Astronomical Ephemeris for the year 1979.
Issued by Her Majesty's Nautical Almanac Office, London; and Nautical Almanac Office, United States Naval Observatory, Washington.
Her Majesty's Stationery Office, London. 8 + 571 pp. Price £ 10.30 (1978). ISBN 0-11-880680-7.

047.008 Theory of the algorithm for compiling ephemerides of pairs of stars of equal altitude of Pevtsov's method with a computer.   S. S. Uralov, L. V. Neverov.
Izv. vyssh. uchebn. zaved. Geod. i aehrofotosemka, 1977, No. 4, p. 79 - 82. In Russian. – Abstr. in Ref. zh., 51. Astron., 2.51.148 (1978).

047.009 Tables of sunrise, sunset, twilight, moonrise, and moonset 1978.
Prepared under the supervision of W. A. Miñoza.
Republic of the Philippines – Department of National Defense, Philippine Atmospheric, Geophysical and Astronomical Services Administration, Quezon City. National Geophysical and Astronomical Office. 14 + 57 pp. (1977).

047.010 Connaissance des Temps ou des mouvements célestes pour l'an 1979 à l'usage des astronomes et des navigateurs.
Published by Bureau des Longitudes under the supervision of B. Morando.
Gauthier-Villars Editeur, Paris. 42 + 501 + A145 pp. (1978). ISBN 2-04-010199-3.

047.011 The Nautical Almanac for the year 1979.
Issued by Her Majesty's Nautical Almanac Office, London; and Nautical Almanac Office, United States Naval Observatory, Washington. Printed and published by Her Majesty's Stationery Office, London. A 4 + 276 + 35 pp. Price £ 5.50. ISBN 0 11 772180 8.

047.012 Astronomical Ephemeris for 1979.
Prepared by Shikin-zan Observatory, Chinese Academy of Sciences. Academic Publishing House, Peking. 5 + 367 pp. Price $ 5.70 (1978). In Chinese.

047.013 Astronomische Kurz-Kalender +1978 bis +2000.
H. Mucke.
Sternenbote, 21. Jahrg., 2 - 26 (1978).

047.014 Philippine Astronomical Handbook 1978.
Prepared under the supervision of W. A. Miñoza.
Republic of the Philippines – Department of National Defense. Philippine Atmospheric, Geophysical and Astronomical Services Administration, Quezon City. 13 + 61 pp. (1977).

047.015 Permanent calendar.
I. Ya. Golub, L. S. Khrenov.
Problems of observational and theoretical astronomy, Moscow – Leningrad, 1977, (see 003.012), p. 249 - 260. In Russian. Abstr. in Ref. zh., 51. Astron., 5.51.4 (1978).

047.016 A new form of the Nautical Astronomical Almanac.
V. M. Amelin.
Sudovozhdenie, 1977, No. 21, p. 30 - 34. In Russian. – Abstr. in Ref. zh., 51. Astron., 5.51.111 (1978).

047.017 Compilation of the ephemerides of star pairs of equal height observed in the plane of a common vertical with the help of a computer.
S. S. Uralov, L. V. Neverov.
Izv. vyssh. uchebn. zaved. Geod. i aehrofotosemka, 1977, No. 5, p. 76 - 79. In Russian. – Abstr. in Ref. zh., 52. Geod. Aehrosemka, 5.52.80 (1978).

047.018 Blick in die Sternenwelt 1978. Astronomischer Kalender der Archenhold-Sternwarte.
E. Rothenberg.
Archenhold-Sternw., Berlin-Treptow, 48 pp. (1977).

047.019 Annuario 1978.
Edited by Osservatorio Astronomico di Torino.
Pubbl. Varie Fuori Ser., Oss. Astron. Torino, (Pino Torinese), No. 68, 68 pp. (1977).
Cronologia; Coordinate dell'osservatorio; Calendario ed effemeridi del Sole e della Luna; I pianeti nel 1978; Eclissi e occultazioni; Attività dell'osservatorio (M. G. Fracastoro); Misura della rotazione terrestre (C. Moranzino); Gli asteroidi del sistema solare (V. Banfi); Insolazione a Pino Torinese (A. Di Battista, M. G. Fracastoro).

047.020 Astronomical Phenomena for the year 1980.
Issued by the Nautical Almanac Office, United States Naval Observatory, Washington. U.S. Government Printing Office, Washington. 72 pp. (1976).
The material in this pamphlet is partly a preprint of selected pages from 'The American Ephemeris and Nautical Almanac'.

047.021 Almanac for computers, 1978.
L. E. Doggett, G. H. Kaplan, P. K. Seidelmann.
Nautical Almanac Office, United States Naval Obs., Washington, D.C. 20390. A19 + B16 + C27 + D28 + E22 + F8 pp.

047.022 Nautisches Jahrbuch oder Ephemeriden und Tafeln für das Jahr 1979, zur Bestimmung der Zeit, Länge und Breite auf See nach astronomischen Beobachtungen.
Edited by "Deutsches Hydrographisches Institut", Hamburg. 128. Jahrg., 45 + 365 + 30 pp. (1978).

047.023 **Astronomical calendar 1978.**   G. Ottewell.
           Furman University Physics Dep., Greenville,
S.C.29613. 68 pp. Price $ 5.95 (1978). — Review in Sky
Telesc., Vol. 55, 170 (1978).

047.024 **The observer's handbook 1978.**
           J. R. Percy (Editor).
Royal Astronomical Society of Canada, 124 Merton St.,
Toronto, Ont. M4S 2Z2. 128 pp. Price $ 4.00 (1977).
Review in Sky Telesc., Vol. 55, 170 (1978).

047.025 **Army ephemeris, 1978.**
           Headquarters, Department of the Army. 63 pp.
Price $ 2.20 (1977). — Review in Sky Telesc., Vol. 55,
429 - 430 (1978).

A reconstruction of the chronology of Mesoamerican calendrical systems.   See Abstr. 004.025.

Programerbar lommekalkulator finner planetposisjonene.   See Abstr. 021.009.

New methods of pre-calculation of the apparent equatorial coordinates of the sun and planets by means of key-punched electronic computers.   See Abstr. 021.032.

A comparison of observations and ephemerides of Mars, 1950 - 1975.   See Abstr. 097.165.

Space Research

051 Extraterrestrial Research, Spaceflight Related to Astronomy and Astrophysics

051.001 X-ray astronomy: HEAO looks further and sees more.   W. D. Metz.
Science, Vol. 199, 869 - 870 (1978).

051.002 Space Shuttle: high flying Yankee ingenuity.   E. Ulsamer.
Spaceworld, Vol. N-6-162, p. 18 - 23 (1977). – Abstr. in Phys. Abstr., Vol. 80, Abstr. 91406 (1977).

051.003 Asteroidal resources and gravity-assisted low-energy retrieval missions.   B. O'Leary.
Bull. American Astron. Soc., Vol. 9, 500 (1977). – Abstract.

051.004 The needs for space astronomy in ground-based data.   A. Heck.
Bull. Inf. Centre Données Stellaires, No. 14, p. 74 - 76 (1978).
The author briefly presents a few examples where ground-based data are needed in space astronomy. He considers only orbital satellites because interplanetary spacecraft represent a particular kind of space astronomy with its own problems which are not of primary interest here.

051.005 Science on Spacelab.   E. R. Schmerling.
Space Research, Vol. XVII, (see 012.010), 835 - 837 (1977).
This paper summarizes studies which have been conducted to meet the needs of astronomy, solar physics, high energy astrophysics, atmospheric and space physics, and life sciences.

051.006 Neuartige Raumsonden-Missionen zu den Kometen in den achtziger Jahren.   H. Fechtig.
Sterne Weltraum, Jahrg. 17, 121 - 125 (1978).

051.007 Twenty years cosmic era.   B. V. Raushenbakh.
Vestn. AN SSSR, 1977, No. 10, p. 48 - 60. In Russian. – From Ref. zh., 62. Issled. kosm. prostranstva, 2.62.1 (1978).

051.008 Robot probes: exploring hostile environments.   S. A. Kallis, Jr.
Astronomy, Vol. 5, No. 9, p. 18 - 24 (1977). – Abstr. in Phys. Abstr., Vol. 81, Abstr. 4065 (1978).

051.009 GEOS at the centre of a world-wide study of the Earth's magnetosphere.   K. Knott.
ESA Bull., No. 9, p. 12 - 16 (1977). – Abstr. in Phys. Abstr., Vol. 81, Abstr. 4072 (1978).

051.010 Some results of performing the program of scientific investigations during the long spaceflight of Salyut 4.   V. I. Sevast'yanov.
Tr. Desyatykh chtenij, posvyashch. razrab. nauchn. naslediya i razvitiyu idej K. Eh. Tsiolkovskogo, Kaluga, 16 - 19 sent.
1975. Moskva, 1977, p. 9 - 13. In Russian. – Abstr. in Ref. zh., 62. Issled. kosm. prostranstva, 3.62.92 (1978).

051.011 General aspects of the mission Helios 1 and 2. Introduction to a special issue on initial scientific results of the Helios Mission.   H. Porsche.
J. Geophys., Vol. 42, 551 - 559 (1977). – Abstr. in Phys. Abstr., Vol. 81, Abstr. 20615 (1978).

051.012 Studies of cosmic physics by space vehicles.   M. Oda, I. Kondo, T. Takakura, Y. Tanaka.
Bull. Inst. Space Aeronaut. Sci. Univ. Tokyo, B, Vol. 12, 830 - 848 (1976). In Japanese. – From Phys. Abstr., Vol. 81, Abstr. 24417 (1978).

051.013 Future programs in space research.   T. Obayashi.
Bull. Inst. Space Aeronaut. Sci. Univ. Tokyo, B, Vol. 12, 885 - 887 (1976). In Japanese. – From Phys. Abstr., Vol. 81, Abstr. 24418 (1978).

051.014 Observations by balloons.   J. Nishimura.
Bull. Inst. Space Aeronaut. Sci. Univ. Tokyo, B, Vol. 12, 898 - 908 (1976). In Japanese. – From Phys. Abstr., Vol. 81, Abstr. 24419 (1978).

051.015 'Voyager'.
Telecommun. J., Vol. 44, 580 - 586 (1977). – Abstr. in Phys. Abstr., Vol. 81, Abstr. 24435 (1978).

051.016 Problems in the measurement of electron energy distribution function in a space plasma by a probe.
H. Amemiya, K. Shimizu, T. Dote.
Bull. Inst. Space Aeronaut. Sci. Univ. Tokyo, B, Vol. 12, 975 - 988 (1976). In Japanese. – From Phys. Abstr., Vol. 81, Abstr. 28474 (1978).

051.017 Observations by rockets and satellites.   T. Takayanagi.
Bull. Inst. Space Aeronaut. Sci. Univ. Tokyo, B, Vol. 12, 888 - 897 (1976). In Japanese. – From Phys. Abstr., Vol. 81, Abstr. 28481 (1978).

051.018 Antenna concepts for interstellar search systems.   R. P. Basler, G. L. Johnson, R. R. Vondrak.
Radio Sci., Vol. 12, No. 5, p. 845 - 858 (1977). – Abstr. in Phys. Abstr., Vol. 81, Abstr. 28545 (1978).

051.019 Observations with the International Ultraviolet Explorer.
IAU Circ., No. 3173 (1978).

051.020 HEAO-1 observations.
IAU Circ., No. 3206 (1978).

051.021 Voyager's historic view of earth and moon.
Natl. Geogr., Vol. 154, 52 - 53 (1978).

051.022 **Space investigations.**   M. V. Keldysh, M. Ya.
Marov.
Oktyabr' i nauka, Moskva, 1977, p. 90 - 152. In Russian. —
From Ref. zh., 62. Issled. kosm. prostranstva, 5.62.1 (1978).

051.023 **Investigation of space.**   R. Z. Sagdeev.
250 let AN SSSR. Dokl. i mater. yubil. torzhestv.
Moskva, 1977, p. 286 - 293. In Russian. — From Ref. zh.,
62. Issled. kosm. prostranstva, 5.62.2 (1978).

051.024 **On problems of interstellar flights.**
B. K. Fedyushin, S. Ya. Shcherbak.
Problems of observational and theoretical astronomy, Moscow
—Leningrad, 1977, (see 003.012), p. 230 - 236. In Russian.
Abstr. in Ref. zh., 62. Issled. kosm. prostranstva, 5.62.578
(1978).

051.025 **Near-earth asteroids as targets for exploration.**
E. M. Shoemaker, E. F. Helin.
Reports of planetary geology program, 1977—1978,
(see 003.014), p. 20 - 21 (1978).

051.026 **Early operations with the International Ultraviolet
Explorer.**   The IUE Project Staff and Science
Commissioning Team.
Bull. American Astron. Soc., Vol. 10, 460 - 461 (1978).
Abstract.

051.027 **Why send probes to Venus?**   D. Hunter.
Nature, Vol. 274, 306 - 307 (1978).

051.028 **The planetary exploration program.**   D. G. Rea.
IEEE Commun. Soc. Mag., Vol. 15, No. 6, p. 10 -
11, 25 (1977). — Abstr. in Phys. Abstr., Vol. 81, Abstr.
32316 (1978).

051.029 **Super-powerful orbital telescope. What the space
telescope with a 3-m mirror can give to science.**
G. A. Gurzadyan.
Vestn. AN SSSR. 1977, No. 12, p. 80 - 90. In Russian. — Abstr.
in Ref. zh., 51. Astron., 6.51.3 (1978).

051.030 **Optical investigations of Soviet aeronauts
(1961 - 1975).**   A. I. Lazarev, A. G. Nikolaev,
A. E. Mikirov, T. A. Daminova.
Tr. inst. prikl. geofiz., 1977, No. 32, p. 36 - 65. In Russian.
Abstr. in Ref. zh., 62. Issled. kosm. prostranstva, 6.62.28
(1978).

051.031 **Astronautics in the year 1977.**   M. Grün, P.
Koubský.
Říše hvězd, Vol. 59, 113 - 115, 144 - 147 (1978). In Czech.

051.032 **Interstellar flights.**   P. Andrle.

Říše hvězd, Vol. 59, 89 - 95 (1978). In Czech.

051.033 **Journey to the outer planets.**
Spaceworld, Vol. N-11-167, p. 5 - 14 (1977).
Abstr. in Phys. Abstr., Vol. 81, Abstr. 49289 (1978).

051.034 **Mariner 10 expedition to Venus and Mercury.**
A. O. Brunfelt.
Tek. Ukebl., Vol. 124, No. 52a, p. 93 - 94 (1977). In
Norwegian. — Abstr. in Phys. Abstr., Vol. 81, Abstr. 49291
(1978).

051.035 **The Space Shuttle — tomorrow's space transport
system.**   L. Lukacs.
Tek. Ukebl., Vol. 124, No. 52a, p. 162 - 166, 169 (1977). In
Norwegian. — Abstr. in Phys. Abstr., Vol. 81, Abstr. 49292
(1978).

051.036 **Main stages of development of lunar investigations
in the United States by automatic interplanetary
stations.**   A. V. Tkachev.
Iz istor. aviatsii i kosmonavtiki. Vyp. (No.) 30. Moskva, 1976,
p. 168 - 182. In Russian. — From Ref. zh., 52. Geod.
Aehrosemka, 7.52.202 (1978).

051.037 **Space report.**
J. British Interplanet. Soc., Vol. 31, 10, 15, 32, 36,
66, 75 - 78, 146, 160, 175, 180 (1978).

051.038 **Space Report.**
Spaceflight, Vol. 20, 13 - 17, 33, 48 - 54, 105 - 110,
150 - 153, 176 - 186, 217 - 223, 265 - 271 (1978).

From Vinland to Mars.
See Abstr. 003.083.

**Report on active and planned spacecraft and ex-
periments.**   See Abstr. 003.140.

Workshop on cometary missions.
See Abstr. 012.031.

**Proceedings of the NASA/AVS symposium on the
use of the Space Shuttle for science and engineering.**
See Abstr. 012.045.

Pioneer Venus 1978.   See Abstr. 053.016.

Cratering mechanics and future Martian exploration.
See Abstr. 097.169.

**Viking first encounter of Phobos: preliminary
results.**   See Abstr. 097.502.

**The micrometeoroid impact rate on a spacecraft in
close proximity to a cometary nucleus.**   See Abstr. 102.017.

052 Astrodynamics and Navigation of Space Vehicles

052.001 The aerodynamic lift in the satellite dynamics.
L. Sehnal.
Veröff. Zentralinst. Phys. Erde, No. 52, Part 3, (see 012.001), p. 691 - 693 (1977).
The influence of the atmospherical lift on the motion of the oriented artificial satellites is estimated. The changes of the radius-vector of the satellite orbit are computed and compared to those caused by drag.

052.002 Estimation of a possible AES orbit calculation accuracy improvement based on predicted values of geomagnetic disturbance and solar activity indices.
B. V. Kugaenko.
Veröff. Zentralinst. Phys. Erde, No. 52, Part 3, (see 012.001), p. 727 - 738 (1977). In Russian.

052.003 Use of an intermediate AES orbit for solution of geodetic problems. N. Georgiev.
Veröff. Zentralinst. Phys. Erde, No. 52, Part 3, (see 012.001), p. 763 - 784 (1977). In Russian.

052.004 Iterative methods of determination of station co-ordinates and orbital satellite elements from Doppler observations. W. Goral.
Veröff. Zentralinst. Phys. Erde, No. 52, Part 3, (see 012.001), p. 913 - 920 (1977).
Iterative methods of determination of station co-ordinates are given on the basis of minimal number of Doppler observations of a satellite of known orbit. Iterative methods of solution of certain problems in the range of determination of the preliminary orbit are described on the basis of Doppler measurements and known co-ordinate station. Methods are given for determining roughly approximate values of roots sought.

052.005 The application of the intermediate orbits to satellite motion prediction. J. H. Zhagar.
Veröff. Zentralinst. Phys. Erde, No. 52, Part 3, (see 012.001), p. 989 - 1005 (1977). In Russian.
Intermediate orbits, representing special solutions of the generalized two fixed-centre problem and the generalized Garfinkel problem, have been compared in order to find the most suitable one for the ephemeris service. For satellites with small eccentricity the Aksnes intermediate orbit should be preferred as it leads to relatively simple convertion algorithms to the keplerian elements and vice versa.

052.006 Determination of orbits from optical and laser observations.
E. P. Aksenov, S. N. Vashkov'yak, N. V. Emel'yanov.
Veröff. Zentralinst. Phys. Erde, No. 52, Part 3, (see 012.001), p. 1007 - 1017 (1977). In Russian.
A computer programme for orbit improvements by means of optical and laser observations is presented. The orbit derived from the solution of the problem of two fixed centres has been selected as an intermediate orbit. The position of the satellite is determined taking into consideration gravitational and lunisolar perturbations. The programme is written in ALGOL.

052.007 Astronautical generalisation of the effects of relativity due to flyby of a central field of gravitation. K. Vulkov.
J. British Interplanet. Soc., Vol. 31, 111 - 112 (1978).
In space flight in the bounds of our Solar System with flyby of the Sun, and also on interstellar missions and flybys of stars (especially those with very large masses), modern science has to take account of the relativistic effects resulting from the influence of gravitation. Astronautical generalisation of the relations of relativity concerning flyby of a central field of gravitation meets both demands, and promotes the inference of new relativistic effects typical to astronautics.

052.008 Perturbations of a close-Earth satellite due to sunlight diffusely reflected from the Earth. II Variable albedo. D. A. Lautman.
Celestial Mech., Vol. 16, 3 - 25 (1977).
A semianalytical method has been developed to calculate the radiation-pressure perturbations of a close-Earth satellite due to sunlight reflected from the Earth. It is assumed that the satellite is spherically symmetric and that the solar radiation is reflected from the Earth according to Lambert's Law. To account for the increasing reflectivity of the Earth toward the poles, its albedo is assumed to have a latitudinal dependence given by $a = a_0 + a_2 \sin^2 \phi$. The effect of the terminator on the perturbations has been neglected. The perturbations within a particular revolution are given analytically, while the long-range perturbations are obtained by accumulation.

052.009 The motion of a satellite in resonance with the second-degree sectorial harmonic.
S. S. Dallas, R. E. Diehl.
Celestial Mech., Vol. 16, 97 - 121 (1977).
The solution to the motion of a satellite in an eccentric orbit and in resonance with the second-degree sectorial harmonic of the potential field is developed. The method of solution used parallels to the well known von Zeipel method of general perturbations. The solution consists of expressions for the variations of the Delaunay variables. These expressions are composed of the perturbations developed by Brouwer in 1959 for the motion of an artificial satellite plus first-order perturbations due to the second-degree sectorial harmonic (in terms of the Legendre normal elliptic integrals of the first and second kind).

052.010 Determination of the orientation of artificial earth satellites from a two-vectorial system of measurements. A. M. Titov, V. P. Shchukin.
Kosm. Issled., Vol. 16, 3 - 9 (1978). In Russian.

052.011 On the construction of an optimum program of navigation measurements.
Zh. E. Vavulov, M. P. Nevol'ko.
Kosm. Issled., Vol. 16, 10 - 21 (1978). In Russian.

052.012 Flight paths to Jupiter using the Martian gravitational field. K. G. Georgiev, O. V. Papkov.
Kosm. Issled., Vol. 16, 38 - 43 (1978). In Russian.

052.013 On a method for determining the moments of maximum approach of AES.
A. G. Klimenko, V. B. Solomodenko.
Kosm. Issled., Vol. 16, 134 - 136 (1978). In Russian.

052.014 Optimum parameters of a system of gravitational stabilization of a satellite taking into account atmospherical drag. G. M. Belyaeva.
Kosm. Issled., Vol. 16, 139 - 142 (1978). In Russian.

052.015 The effect of orbital eccentricity and nodal regression on spin-stabilized satellites.
V. N. Dvornychenko, R. B. Gerding.
Celestial Mech., Vol. 16, 263 - 274 (1977).
The gyroscopic motion of a spin-stabilized satellite due to gravity gradient torques in a circular orbit has been analyzed to varying degrees in numerous publications. This paper shows

that the restriction to a circular orbit is, in fact, not essential and with a slight increase in complexity, non-circular orbits may be treated. More importantly, a uniform regression of the orbital node can also be accounted for. The general results are expressed in closed form using Jacobian elliptic functions. Finally certain algebraic integrals of the precession are given which can be used to place limits on the excursions of the spin axis without actually solving for the motion.

052.016 Operational requirements and the geometry of a
         station-keeping maneuver.   J. B. Eades, Jr.
Celestial Mech., Vol. 16, 315 - 342 (1977).

052.017 Thrust aided sub-orbits.   J. B. Eades, Jr.
         Celestial Mech., Vol. 16, 353 - 365 (1977).
     A discussion on thrust assisted sub-orbits is presented. This operation may be considered as a candidate means for providing station-keeping, for the interrogation of circulating satellites, and as an operational maneuver for pick-up, launch and/or inspection of experiment modules (and the like) from (say) the SHUTTLE spacecraft.

052.018 An element formulation for perturbed motion about
         the center of mass.
J. D. Donaldson, D. J. Jezewski.
Celestial Mech., Vol. 16, 367 - 387 (1977).
     The perturbed motion of a rigid body about its center of mass is formulated in terms of the six elements: l, the magnitude of the angular momentum vector; h, the total energy; τ and ϵ, two linear functions of the independent variable; and, ψ_1 and θ_1, two Euler angles that orientate the inertial frame with respect to the unperturbed solution. Solutions from the element formulation and the original Euler equations are numerically compared using shuttle-type data.

052.019 Atmospheric lift on an oriented artificial satellite.
         L. Sehnal.
Space Research, Vol. XVII, (see 012.010), 35 - 41 (1977).
     The stable orientation of an artificial earth satellite gives rise to a force due to atmospheric lift perpendicular to the velocity vector. The magnitude of this force is computed taking into account different kinds of molecular interaction with the satellite surface. The orbital perturbations are computed; their magnitudes may be perceptible in the case of an oriented satellite with solar paddles.

052.020 Methods for predicting satellite orbital lifetimes.
         D. G. King-Hele.
J. British Interplanet. Soc., Vol. 31, 181 - 196 (1978).
     In this paper simple graphical methods are presented for estimating the future lifetime of an Earth satellite from its current rate of decay, using theory adapted to an atmosphere with a realistic variation of density with height. The effects of the departure of the Earth and atmosphere from spherical symmetry, and the variations of density with time, are approximated by specifying correction factors. Orbits which experience serious lunisolar perturbations call for numerical-integration methods, and the uses of the computer programs PROD and PTDEC are described.

052.021 Three-dimensional problem of motion of a space
         vehicle with a solar sail in an arbitrarily oriented
geocentric orbit.   E. N. Polyakova, L. K. Grinevitskaya.
Astron. i geod., Vyp. (No.) 6, Tomsk, 1977, p. 76 - 83. In Russian. – Abstr. in Ref. zh., 62. Issled. kosm. prostranstva, 2.62.390 (1978).

052.022 A global analysis of two-terminal trajectories.
         F.-T. Sun.
Acta Astronaut., Vol. 4, (see 012.017), 469 - 493 (1977).
Abstr. in Phys. Abstr., Vol. 81, Abstr. 4085 (1978).

052.023 Optimum low-thrust multiple rendezvous between
         inclined orbits.   K. Uesugi, H. Matsuo.
Acta Astronaut., Vol. 4, (see 012.017), 495 - 509 (1977).
Abstr. in Phys. Abstr., Vol. 81, Abstr. 4086 (1978).

052.024 Optimal low-thrust, limited power transfers between
         arbitrary elliptical orbits.
J. P. Marec, N. X. Vinh.
Acta Astronaut., Vol. 4, (see 012.017), 511 - 540 (1977).
Abstr. in Phys. Abstr., Vol. 81, Abstr. 4087 (1978).

052.025 Long term solar radiation effects upon an orbit in
         the ecliptic.   J. C. Van Der Ha, V. J. Modi.
Acta Astronaut., Vol. 4, 813 - 831 (1977). – Abstr. in Phys. Abstr., Vol. 81, Abstr. 4088 (1978).

052.026 Computation of optimal Earth–Mars and Earth–
         Venus trajectories.   W. Powers, S. Yoshimura.
IEEE Trans. Aerosp. Electron. Syst., Vol. AES-13, 549 - 550 (1977). – Abstr. in Phys. Abstr., Vol. 81, Abstr. 4090 (1978).

052.027 Precision orbit computations for Starlette.
         J. G. Marsh, R. G. Williamson.
Bull. Géod., Vol. 52, 71 - 83 (1978).
     The Starlette satellite was designed to minimize the effects of non-gravitational forces and to obtain the highest possible accuracy for laser range measurements. Analyses of the first four months of laser tracking data from nine stations have confirmed the stability of the orbit and the precision to which the satellite's position can be established.

052.028 Algorithms of control of the descent of a space
         vehicle in a given region of the Martian surface.
N. M. Ivanov, V. D. Belykh, A. I. Martynov.
Kosm. Issled., Vol. 16, 198 - 207 (1978). In Russian.

052.029 Non-gravitational foci and apsides of paths with
         given starting point and semi-latus rectum.
K. Mison.
Acta Tech. CSAV, Vol. 22, 335 - 345 (1977). – Abstr. in Phys. Abstr., Vol. 81, Abstr. 7867 (1978).

052.030 Librational dynamics of satellites with thermally
         flexed appendages.   V. J. Modi, K. Kumar.
J. Astronaut. Sci., Vol. 25, No. 1, p. 3 - 20 (1977). – Abstr. in Phys. Abstr., Vol. 81, Abstr. 7868 (1978).

052.031 Partial differentiations of vis-viva and Lambert
         equations.   I. Sugai.
J. Astronaut. Sci., Vol. 25, No. 1, 63 - 74 (1977). – Abstr. in Phys. Abstr., Vol. 81, Abstr. 7864 (1978).

052.032 On the determination of the integration constants
         in the approximate analytical theory of motion of
a stationary satellite.   E. I. Timoshkova.
Vestn. Leningr. univ., 1977, No. 13, p. 162 - 165. In Russian. – Abstr. in Ref. zh., 51. Astron., 3.51.125 (1978).

052.033 Tidal perturbations on the satellite 1967-92A.
         T. L. Felsentreger, J. G. Marsh.
J. Geophys. Res., Vol. 83, 1837 - 1842 (1978).
     The orbit of the 1967-92A satellite has been studied to ascertain the extent to which tidal forces contribute to orbital perturbations. This study has permitted an estimation of the magnitudes of ocean tide effects on the satellite's inclination.

052.034 An application of the stroboscopic method.
         E. A. Roth.
Dynamics of planets and satellites, (see 012.020), p. 181 - 188 (1978).
     The motion of an orbiter of a satellite of one of the major planets is considered. The orbiter undergoes various

perturbing effects. It is shown that the semi-analytical strobo-scopic method is well suited to take into account all perturbations.

052.035 Third-order solution of an artificial-satellite theory.
H. Kinoshita.
Dynamics of planets and satellites, (see 012.020), p. 241 - 257 (1978).
A third-order solution is developed for the motions of artificial satellites moving in the gravitational field of the Earth, whose potential includes the second-, third-, and fourth-order zonal harmonics. A comparison with the results of numerical integration of the equations of motion indicates that the solution can predict the position of a close-earth satellite with a small eccentricity with an accuracy of better than 1 cm over 1 month.

052.036 Analytical solution of the main problem of the theory of motion of artificial earth satellites by computer.   A. L. Kutuzov.
Astron. i geod. Vyp. (No.) 6. Tomsk, 1977, p. 26 - 31. In Russian. – Abstr. in Ref. zh., 51. Astron., 4.51.128 (1978).

052.037 Globally optimal coplanar orbit transfer.
H. W. Small.
AIAA J., Vol. 15, 1224 - 1230 (1977). – Abstr. in Phys. Abstr., Vol. 81, Abstr. 8236 (1978).

052.038 Orbit determinations from range measurements. II.
C. Botti.
Ann. Geofis., Vol. 29, 89 - 120 (1976). In Italian. – Abstr. in Phys. Abstr., Vol. 81, Abstr. 24439 (1978).

052.039 Double-point problems of determination of orbits.
Yu. S. Savrasov.
Kosm. Issled., Vol. 16, 323 - 330 (1978). In Russian.

052.040 On a method for determination of an orbit from observational results.   E. O. Kotov.
Kosm. Issled., Vol. 16, 331 - 338 (1978). In Russian.

052.041 Some properties of the criterion of accuracy of determination of AES orientation.
A. M. Titov, V. P. Shchukin.
Kosm. Issled., Vol. 16, 339 - 344 (1978). In Russian.

052.042 Approximate analytical solution of the equations of motion at descent in an atmosphere.
V. N. Baranov.
Kosm. Issled., Vol. 16, 361 - 369 (1978). In Russian.

052.043 Algorithm of navigation and control at descent of space vehicles in the phase of aerodynamic braking in the Martian atmosphere.   O. A. Nogov.
Kosm. Issled., Vol. 16, 370 - 377 (1978). In Russian.

052.044 On the motion of bodies in the Jovian atmosphere taking into account their change of mass and form under the action of aerodynamic heating.
Eh. A. Gershbejn, Eh. Ya. Sukhodol'skaya, S. L. Sukhodol'skij, G. A. Tirskij.
Kosm. Issled., Vol. 16, 378 - 387 (1978). In Russian.

052.045 On the determination of the tensor of inertia of a space vehicle on flight.
G. M. Andreev, K. B. Alekseev, M. I. Kiselev.
Kosm. Issled., Vol. 16, 456 - 457 (1978). In Russian.

052.046 General formulae for second order zonal perturbations of artificial satellite orbits.   G. D. Novikov.
Nauchn. Inf., Vyp. (No.) 38, p. 65 - 74 (1976). In Russian.
General analytical formulae for perturbations of the

first and second order of the zonal part of the geopotential (including all terms) are developed. The integration of the disturbed motion is done by the "small parameter" method.

052.047 The influence of satellite flexibility on orbital motion.   A. K. Misra, V. J. Modi.
Celestial Mech., Vol. 17, 145 - 165 (1978).
The orbital perturbations induced by the librational motion and flexural oscillations are studied for satellites having large flexible appendages. Using a Lagrangian procedure, the equations for coupled motion are derived for a satellite having an arbitrary number of appendages in the nominal orbital plane and two flexible members normal to it. The formulation enables one to study the influence of flexibility on both the orbital and attitude motions. Three specific examples, Alouette II, Radio Astronomy Explorer and Tethered Orbiting Interferometer, are considered subsequently and their possible secular drifts estimated.

052.048 Approximation of orbits of artificial earth satellites with the use of series.   N. Georgiev, B. Shustov.
New methods of space geodesy, (see 012.029), p. 159 - 170 (1976). In Russian.
A method for the approximation of orbits with the use of power time series on the basis of a three-dimensional adjustment of laser and photographical satellite observations is developed and an Algol-60 program is written. After an extrapolation of the orbit a correction due to gravitational anomalies is introduced into the calculated satellite positions.

052.049 Improvement of the orbit of a satellite from photographic observations of the lunar surface on a short arc.   V. Ivanova.
New methods of space geodesy, (see 012.029), p. 316 - 320 (1976). In Russian.
Photographic observations of the moon's surface are modeled on a 120° arc of an artifical satellite's orbit. The dependence of the accuracy of determined elements from the position of the arc of the orbit on which the observation is situated has been investigated.

052.050 The main problem of the theory of motion of AES (analytical solution on a computer).
A. L. Kutuzov.
New methods of space geodesy, (see 012.029), p. 329 - 331 (1976). In Russian.
A computer-generated analytical solution of the main problem of satellite theory has been obtained. The solution is given as truncated power series of J_2, the secular terms having fourth-order accuracy with respect to J_2 and the short and long period terms being of third-order accuracy.

052.051 Small eccentricity orbits at the critical inclination.
C. Oesterwinter.
Bull. American Astron. Soc., Vol. 10, 483 (1978). – Abstract.

052.052 Mission analysis for terrestrial satellite and planetary orbiters, with special emphasis on highly eccentric orbits. I. The mathematical background.   E. A. Roth.
ESA J., Vol. 1, 245 - 268 (1977). – Abstr. in Phys. Abstr., Vol. 81, Abstr. 40790 (1978).

052.053 Aspect system for HEAO-B.   D. Koch, R. Hall, H. Tsao, H. Wollman, R. Kilinski.
IEEE Trans. Nucl. Sci., Vol. NS-25, 473 - 477 (1978). – Abstr. in Phys. Abstr., Vol. 81, Abstr. 40791 (1978).

052.054 Luni-solar perturbations in the motion of resonant artificial earth satellites.   K. P. Ivanovskaya.
Nauchn. tr. Omsk. s.-kh. inst., 1977, No. 163, p. 55 - 58. In Russian. – Abstr. in Ref. zh., 51. Astron., 6.51.106; 62. Issled. kosm. prostranstva, 6.62.418 (1978).

052.055 **On isochronous first- and second-order derivatives in problems of space dynamics.**
B. Ts. Bakhshiyan, A. A. Sukhanov.
Inst. kosm. issled. AN SSSR. Prepr., 1977, No. 357, 27 pp. In Russian. – Abstr. in Ref. zh., 62. Issled. kosm. prostranstva, 6.62.416 (1978).

052.056 **Method for estimating the methodical errors of algorithms of forecast of the motion of artificial earth satellites.**   T. I. Bezruchenko, N. A. Golovanova, F. D. Petrovskij.
Radiotekh. inst. AN SSSR. Prepr., 1977, No. 775, 15 pp. In Russian. – Abstr. in Ref. zh., 62. Issled. kosm. prostranstva, 6.62.420 (1978).

052.057 **Qualitative analysis of scattering of nearly circular orbits.**   A. F. Bochkarev, I. V. Belokonov.
Izv. VUZ Aviats. tekh., 1977, No. 4, p. 13 - 19. In Russian. Abstr. in Ref. zh., 62. Issled. kosm. prostranstva, 6.62.427 (1978).

Astrodynamics (for teachers).
See Abstr. 014.027.

Project Daedalus: the navigation problem.
See Abstr. 015.030.

Gravitational harmonics from shallow resonant orbits.   See Abstr. 042.014.

Second species solutions with an $0(\mu^\nu)$, $^1/_3 < \nu < 1$, near-moon passage.   See Abstr. 042.032.

The use of angles and angular rates. I: Initial orbit determination.   See Abstr. 042.042.

Some considerations on the theoretical determination of the potential by the motion of artificial satellites in the plane case.   See Abstr. 042.072.

Flexibly coupled motion of rigid bodies.
See Abstr. 042.125.

Time transformations and Cowell's method.
See Abstr. 042.134.

Maneuver strategies and orbit selection for the ISEE-C halo orbit mission.   See Abstr. 054.023.

Covariance analysis for a relativity mission.
See Abstr. 066.089.

Error analysis of a relativity test with counter-orbiting satellites. See Abstr. 066.320.

Consequences of inaccuracy of gravitational parameter value for orbits in the Earth's gravitational field.
See Abstr. 081.076.

Studiul rotaţiei atmosferei înalte cu ajutorul observaţiilor sateliţilor artificiali ai Pămîntului.
See Abstr. 082.046.

Fragmentation of asteroids and artificial satellites in orbit.   See Abstr. 098.051.

053 Lunar and Planetary Probes and Satellites

053.001 **Fast numerical analysis of nuclear electric propulsion orbiter-around-Jupiter missions.**
G. Vulpetti, R. Melli.
J. British Interplanet. Soc., Vol. 31, 67 - 74 (1978).
    Enlarging the capabilities of a previous computer programme, several hundred of profiles of Earth Jupiter Orbiter Transfer flights can be computed spending a few minutes of machine time. A typical transfer flight is accomplished by means of a Nuclear-Electric Propulsion vehicle inserted into solar gravitational field by chemical units. After jettisoning the powerplant system the spacecraft is braked into a prefixed orbit around Jupiter. Flights are examined maximising net payload on this final orbit when varying kind of booster with or without dynamical and structural constraints.

053.002 **Voyager, die Reise zu den Riesenplaneten.**
H. W. Köhler.
Phys. Bl., Vol. 34, 36 - 37 (1978).

053.003 **Viking to Mars.**
Spaceworld, Vol. N-6-162, p. 24 - 32 (1977).
Abstr. in Phys. Abstr., Vol. 80, Abstr. 91407 (1977).

053.004 **Viking science: tantalizing Viking scientists cautious.**   G. Alexander.
Spaceworld, Vol. N-6-162, p. 9 - 17 (1977). – Abstr. in Phys. Abstr., Vol. 80, Abstr. 91463 (1977).

053.005 **A path with a length of 12 years.**
D. Yu. Gol'dovskij.

Zemlya i Vselennaya, 1978, No. 2, p. 20 - 23. In Russian.

053.006 **Erforschung der Venus – das NASA-Programm Pioneer–Venus.**   H. W. Köhler.
Phys. Bl., 34. Jahrg., 232 - 234 (1978).

053.007 **Pioneer-Venus – die Erkundung des inneren Nachbarplaneten.**   H. W. Köhler.
Sterne Weltraum, Jahrg. 17, 153 - 156 (1978).

053.008 **Mass loss, shape change and real-gas aerodynamic effects on a Jovian atmospheric probe.**
G. D. Walberg, J. J. Jones, W. B. Olstad, K. Sutton, J. N. Moss, R. W. Powell.
Acta Astronaut., Vol. 4, (see 012.017), 555 - 575 (1977). Abstr. in Phys. Abstr., Vol. 81, Abstr. 4055 (1978).

053.009 **The 1981 Jupiter orbiter probe mission.**
J. W. Moore, J. R. Hyde, J. A. Van Allen, R. S. Nunamaker.
Acta Astronaut., Vol. 4, (see 012.017), 577 - 603 (1977). Abstr. in Phys. Abstr., Vol. 81, Abstr. 4056 (1978).

053.010 **Feasibility of study of buoyant Venus station placed by inflated balloon entry.**
R. Akiba, M. Hinada, H. Matsuo.
Acta Astronaut., Vol. 4, (see 012.017), 625 - 639 (1977). Abstr. in Phys. Abstr., Vol. 81, Abstr. 4058 (1978).

053.011 **Lunar polar orbiter: a global survey of the Moon.**

J. D. Burke.
Acta Astronaut., Vol. 4, 907 - 920 (1977). − Abstr. in Phys. Abstr., Vol. 81, Abstr. 4064 (1978).

053.012 Soviet atmospheric and surface Venus probes.
N. L. Johnson.
Spaceflight, Vol. 20, 224 - 228 (1978).

053.013 Maneuver sequence design for the post-Jupiter leg of the Pioneer Saturn mission.
R. B. Frauenholz, W. F. Brady.
J. Spacecr. Rockets, Vol. 14, 395 - 400 (1977). − Abstr. in Phys. Abstr., Vol. 81, Abstr. 7857 (1978).

053.014 The International Sun-Earth Explorer (formerly IME). A. C. Durney.
ESA J., Vol. 1, No. 1, p. 1 - 33 (1977). − Abstr. in Phys. Abstr. Abstr., Vol. 81, Abstr. 12069 (1978).

053.015 Accuracy estimate of the laser rangefinder for Mars rover. T. Ostroski, C. N. Shen.
Modeling and simulation, Vol. 7, Part I, Pittsburgh, Pa., USA, 26 - 27 April 1976. W. G. Vogt, M. H. Mickle (Editors). ISA, Pittsburgh, Pa., USA (1976). p. 521 - 525. − Abstr. in Phys. Abstr., Vol. 81, Abstr. 24437 (1978).

053.016 Pioneer Venus 1978.
Spaceworld, Vol. N-10-166, 28 - 29 (1977). Abstr. in Phys. Abstr., Vol. 81, Abstr. 36436 (1978).

Viking lander imaging investigation − Picture catalog of primary mission. Experiment data record. See Abstr. 002.026.

Pioneer odyssey. See Abstr. 003.048.

054 Artificial Earth Satellites

054.001 Ariel 5: a British triumph. M. J. Coe.
Spaceflight, Vol. 20, 42 - 47 (1978).

054.002 Vertical 5. L. A. Vedeshin.
Priroda, 1978, No. 1, p. 132 - 133. In Russian.

054.003 Prognoz 6. S. A. Nikitin.
Priroda, 1978, No. 1, p. 133. In Russian.

054.004 Intercosmos 17. S. A. Nikitin.
Priroda, 1978, No. 2, p. 137 - 138. In Russian.

054.005 Vertical 6. L. A. Vedeshin.
Priroda, 1978, No. 3, p. 124. In Russian.

054.006 Watching the earth move from space.
C. Wong.
Sky Telesc., Vol. 55, 198 - 202 (1978).

054.007 Kosmos − 1000 up. V. Rich.
Nature, Vol. 272, 574 - 575 (1978).

054.008 Magnetic effects on space vehicles and other celestial bodies. R. H. Wilson, Jr.
Irish Astron. J., Vol. 13, 1 - 13 (1977).

054.009 The orbit of the satellite Ariel 4 (1971−109A).
R. H. Gooding.
J. British Interplanet. Soc., Vol. 31, 168 - 175 (1978).

054.010 Infrared Astronomical Satellite.
W. I. McLaughlin, W. H. de Leeuw.
Spaceflight, Vol. 20, 187 - 191 (1978).

054.011 The special technical features of GEOS.
D. E. Mullinger.
ESA Bull., No. 9, p. 4 - 11 (1977). − Abstr. in Phys. Abstr., Vol. 81, Abstr. 4071 (1978).

054.012 Mission design implications of an inclined elliptical geosynchronous orbit (International Ultraviolet Explorer). O. L. Dial, J. L. Cooley.
J. Spacecr. Rockets, Vol. 14, 401 - 408 (1977). − Abstr. in Phys. Abstr., Vol. 81, Abstr. 4104 (1978).

054.013 Switchboard in the sky.
S. W. Fordyce, L. Jaffe, E. C. Hamilton.
Spaceflight, Vol. 20, 203 - 217 (1978).
A concept is presented for a large satellite platform in geostationary orbit that can support multiple communications satellite systems while providing subsystems support and on-board switching facilities.

054.014 New steps in the development of astronautics: Salyut 6, Soyuz 26, Soyuz 27, Progress.
S. A. Nikitin.
Priroda, 1978, No. 5, p. 122 - 125. In Russian.

054.015 Sneg 3 and gamma astronomy.
S. V. Petrunin, G. I. Kharitonov.
Zemlya i Vselennaya, 1978, No. 3, p. 44 - 47. In Russian.

054.016 Intercosmos 17. M. A. Rimsha.
Zemlya i Vselennaya, 1978, No. 3, p. 48 - 49. In Russian.

054.017 Optimum form of a body from radiative heating along a given trajectory. A. A. Deev, N. N. Pilyugin.
Kosm. Issled., Vol. 16, 163 - 168 (1978). In Russian.

054.018 The Soyuz program. C. Sheldon.
Spaceworld, Vol. N-7-163, 4 - 26 (1977). − Abstr. in Phys. Abstr., Vol. 81, Abstr. 12071 (1978).

054.019 Spacelab.
Spaceworld, Vol. N-8-164, 4 - 8 (1977). − Abstr. in Phys. Abstr., Vol. 81, Abstr. 12072 (1978).

054.020 U.S. reconnaissance satellite programmes.
A. Kenden.
Spaceflight, Vol. 20, 243 - 262 (1978).

054.021 Le satellite OSO 8 depuis 8 mois en orbite.
Assoc. Dév. Int. Obs. Nice, Bull. N. 13 - 14, p. 5 - 23 (1978).

054.022 The International Ultraviolet Explorer.
A. Boggess.

Bull. American Astron. Soc., Vol. 10, 442 (1978). — Abstract.

054.023 **Maneuver strategies and orbit selection for the ISEE-C halo orbit mission.**
D. L. Richardson, D. W. Dunham.
Bull. American Astron. Soc., Vol. 10, 485 (1978). — Abstract.

054.024 **The International Ultraviolet Explorer (IUE).**
F. Macchetto.
Mem. Soc. Astron. Italiana, Vol. 47, (see 012.036), 431 - 451 (1976).
Contents: The IUE spacecraft. The scientific instrument. Instrumental performance. Scientific operations.

054.025 **Space Shuttle and its Spacelab.**   E. Muller.
Bull. Assoc. Suisse Electr., Vol. 69, No. 6, p. 256 - 261 (1978). In French. — Abstr. in Phys. Abstr., Vol. 81, Abstr. 53784 (1978).

054.026 **Spacelab and material processing facilities and experiments.**   G. Seibert.
Proc. R. Soc. London, Ser. A, Vol. 361, 131 - 142 (1978). Abstr. in Phys. Abstr., Vol. 81, Abstr. 53789 (1978).

054.027 **Satellite digest.**
Spaceflight, Vol. 20, 17 - 18, 72 - 73, 116 - 117, 158, 192 - 193, 235 - 236, 276 (1978).

Skylab, classroom in space.
See Abstr. 003.132.

Digit-visual telescope for satellite observations.
See Abstr. 032.006.

Third-order solution of an artificial-satellite theory.
See Abstr. 052.035.

Method for estimating the methodical errors of algorithms of forecast of the motion of artificial earth satellites.   See Abstr. 052.056.

Qualitative analysis of scattering of nearly circular orbits.   See Abstr. 052.057.

Magnetic damping of rotation.
See Abstr. 091.037.

055 Observations of Earth Satellites, Lunar and Planetary Probes

055.001 **The influence of refraction on positional observations of earth's artificial satellites.**
K. Kurzyńska.
Veröff. Zentralinst. Phys. Erde, No. 52, Part 3, (see 012.001), p. 1045 - 1046 (1977).

055.002 **Accuracy investigation of simultaneous photographic and laser observations of artificial satellites.**
J. Adám.
Veröff. Zentralinst. Phys. Erde, No. 52, Part 3, (see 012.001), p. 1059 - 1073 (1977).
Topocentric vector components can be computed using simultaneous photographic observations and laser range measurements carried out at the station. The accuracy of components depends on the observational accuracy and on geometrical conditions. Based on the error propagation principle, an algorithm is developped to determine the mean square errors of the components.

055.003 **Observations of artificial satellites in connection with selected reference stars.**   S. Oszczak.
Postępy Astron., Tom 25, 177 - 186 (1977). In Polish.
The author presents a method of reduction of observations of artificial satellites. This method uses an instrument with an automatic registration of moments of observations and angles towards satellites. This instrument was constructed at the Observing Station No. 1151 in Olsztyn. The methods of star identification and transformation of the system of measured coordinates to the system of actual coordinates of observed stars are given.

055.004 **Lageos orbital acquisition and initial assessment.**
M. R. Pearlman, J. M. Thorp, D. A. Arnold, F. O. Vonbun.
Space Research, Vol. XVII, (see 012.010), 59 - 62 (1977).
The Smithsonian Astrophysical Observatory (SAO) Baker-Nunn camera and laser network provided the orbital acquisition for Lageos, launched on 4 May 1976. Signal-strength and range-noise measurements made by SAO and the National Aeronautics and Space Administration show that the satellite is functioning as anticipated. The standard deviation of measurements of the satellite range is only 7 cm at present.

055.005 **Poljot 1 — 1963-43-1. May - November 1976. Oreol 1 — 1971-119-1. June - October 1976. Visual observations. Horizontal coordinates.**
Rezul'taty Nablyud. Iskusstv. Sputnikov Zemli, vyp. (No.) 66 (206), 77 pp. (1977). In Russian.

055.006 **Poljot 1 — 1963-43-1. October - December 1976. Explorer 19 — 1963-53-1. March - October 1976. Explorer 39 — 1968-66-1. June - December 1976. Oreol 1 — 1971-119-1. August - October 1976. Intercosmos 13 — 1975-22-2. October 1976. Intercosmos 14 — 1975-115-2. October - November 1976. Explorer 19 — 1963-53-1. January-March 1977. Intercosmos 13 — 1975-22-2. January - March 1977. Visual observations. Equatorial coordinates (1950.0).**
Rezul'taty Nablyud. Iskusstv. Sputnikov Zemli, vyp. (No.) 67 (207). 65 pp. (1977). In Russian.

055.007 **Explorer 19 — 1963-53-1. April - May 1977. Intercosmos 14 — 1975-115-2. January - May 1977. Visual observations. Equatorial coordinates (1950.0).**
Rezul'taty Nablyud. Iskusstv. Sputnikov Zemli, vyp. (No.) 68 (208), 68 pp. (1977). In Russian.

055.008 **Poljot 1 — 1963-43-1. January - June 1977. Intercosmos 10 — 1973-82-1. March - April 1977. Intercosmos 13 — 1975-22-2. April - May 1977. Visual observations. Equatorial coordinates (1950.0).**
Rezul'taty Nablyud. Iskusstv. Sputnikov Zemli, vyp. (No.) 69 (209), 62 pp. (1977). In Russian.

055.009 **Explorer 19 — 1963-53-1. October 1976. Explorer 39 — 1968-66-1. October 1976. Intercosmos 14 — 1975-115-2. October 1976. Poljot 1 — 1963-43-1. January - May 1977. Explorer 19 — 1963-53-1. January - May 1977. Intercosmos 14 — 1975-115-2. January - May 1977. Visual**

observations. Horizontal coordinates.
Rezul'taty Nablyud. Iskusstv. Sputnikov Zemli, vyp. (No.)
70 (210), 69 pp. (1977). In Russian.

055.010 Poljot 1 – 1963-43-1. April - July 1977. Explorer
19 – 1963-53-1. April - July 1977. Intercosmos 13–
1975-22-2. February - July 1977. Intercosmos 14 – 1975-115-
2. April - June 1977. Visual observations. Horizontal coordinates.
Rezul'taty Nablyud. Iskusstv. Sputnikov Zemli, vyp. (No.) 71
(211), 67 pp. (1977). In Russian.

055.011 Poljot 1 – 1963-43-1. July - September 1977. Explorer 19 – 1963-53-1. June - September 1977.
Intercosmos 13 – 1975-22-2. July - September 1977. Visual
observations. Equatorial coordinates (1950.0).
Rezul'taty Nablyud. Iskusstv. Sputnikov Zemli, vyp. (No.)
72 (212), 66 pp. (1977). In Russian.

055.012 Some results of laser ranging of artificial satellites
at the Simeiz satellite-tracking station.
I. I. Dmitrotsa, G. S. Kurbasova, L. S. Shtirberg.
Nauchn. Inf., Vyp. (No.) 38, p. 12 - 15 (1976). In Russian.

The apparature for laser ranging of artificial satellites at
the Simeiz satellite-tracking station is described together with
some preliminary observational results.

055.013 Photographic observations of stationary satellites
with the AFU-75 camera. S. S. Dzyamko.
Nauchn. Inf., Vyp. (No.) 38, p. 30 - 35 (1976). In Russian.

A method of photographic observations of stationary
artificial satellites with the AFU-75 camera is discussed. The
coordinates α_0, δ_0 of some stationary satellites observed in
September 1975 are presented.

Television pick-up system for satellite location.
See Abstr. 034.003.

The timing system for the laser radar of the
Simeiz tracking station. See Abstr. 034.048.

Theoretical Astrophysics

061 General Theoretical Problems of Astrophysics, Neutrino Astronomy, Infrared, X-Ray, Gamma-Ray Astronomy, Origin and Abundances of Elements

061.001 Condensation in supernova ejecta and isotopic anomalies in meteorites.   J. M. Lattimer, D. N. Schramm, L. Grossman.
Astrophys. J., Vol. 219, 230 - 249 (1978).

Some of the observed isotopic anomalies in meteorites may be due to presolar grains that originated in supernova explosions. This hypothesis is investigated by performing chemical equilibrium condensation calculations with pressures and compositions that may be appropriate to cooling supernova ejecta. A scenario is described that attempts to explain the observed isotopic anomalies in oxygen, magnesium, neon, and xenon. Four components of the presolar nebula are considered: interstellar gas and dust (enriched by supernovae throughout galactic history) and "last-event" gas and dust produced by a supernova which occurred about 10^6 yr before the meteorites solidified. The four-component scenario may account for the apparent discrepancy in the elapsed time from nucleosynthesis to condensation inferred from magnesium (10^6 yr) and xenon (10^8 yr) isotopic measurements. The lack of anomalies in Ca and Os may indicate that the last-event supernova did not synthesize anything beyond explosive carbon burning, or that matter interior to this zone did not enter the presolar nebula

061.002 s-process studies: the exact solution.   M. J. Newman.
Astrophys. J., Vol. 219, 676 - 689 (1978).

The exact solution to the unbranched s-process network is computed for the range of neutron exposures of astrophysical interest, and the exact solution for the case of an arbitrary degree of degeneracy of the cross-section values is obtained. The validity of the Clayton, Fowler, Hull, and Zimmerman approximate solution is determined by direct comparison with the exact solution. Correction factors for the approximate solution are presented. It is well known that the approximate solution becomes invalid for very small values of the neutron exposure τ. An alternate form of the exact solution developed for small τ becomes easy to evaluate in that regime.

061.003 A full mathematical analysis of the physically thin screen.
S. Bonazzola, L. M. Celnikier, M. Chevreton.
Astrophys. J., Vol. 219, 690 - 699 (1978).

The authors carry out a full mathematical analysis of the physically thin phase-changing screen. The phase can be a function of both frequency and position. A number of analytical limits are presented, and numerical methods are shown for calculating the modulation index for any value of $\phi_0{}^2$.

061.004 On the bulk yields of nucleosynthesis from massive stars.   W. D. Arnett.

Astrophys. J., Vol. 219, 1008 - 1016 (1978).

Preliminary estimates are made of the absolute yields of abundant nuclei synthesized in observed stars. The compositions of nine helium stars of mass $3 \leqslant M_\alpha/M_\odot \leqslant 48$ are presented, taken at or near the instant of instability. These stars of mass M_α are identified with stars of main-sequence mass M. The amount of synthesized matter for each mass $M \geqslant 10\,M_\odot$ is estimated. A variety of choices for the initial mass function are used to calculate the yield per stellar generation. The author finds the present rate of nucleosynthesis in the solar neighborhood to be about 10% of the average rate over galactic history. This result is consistent with many standard models of galactic evolution.

061.005 X-ray and gamma-ray line production by nonthermal ions.   R. W. Bussard, R. Ramaty, K. Omidvar.
Astrophys. J., Vol. 220, 353 - 362 (1978).

The authors have calculated X-ray production at ~ 6.8 keV by the $2p$ to $1s$ transition in fast hydrogen- and helium-like iron ions, following both electron capture to excited levels and collisional excitation. They used a refinement of the Oppenheimer-Brinkman-Kramers approximation to obtain an improved charge-exchange cross-section. The effective X-ray line production cross section was found to be sharply peaked in energy at about 8 to 12 MeV per amu. Since fast ions of similar energies can also excite nuclear levels, the authors have calculated the ratio of selected strong γ-ray line emissivities to the X-ray line emissivity. The authors use these calculations to set limits on the intensity of γ-ray line emission from the galactic center and the radio galaxy Cen A, and they find that these limits are generally lower than those reported in the literature.

061.006 Light elements synthesis in the hot model of the universe.   B. V. Vajner, O. V. Dryzhakova, V. L. Zaguskin, L. S. Marochnik, L. I. Reznitskij.
Astron. Zh. Akad. Nauk SSSR, Vol. 55, 3 - 12 (1978). In Russian. English translation in Soviet Astron., Vol. 22, No. 1.

The results of calculations of light elements nucleosynthesis in the big bang are presented. A new method of solving the equations of nuclear reactions which gives sufficiently high precision and allows to save computation time is described. The results are identical to those of Wagoner (1973). It is shown that large-amplitude temperature and entropy fluctuations lead to an increase of D abundance by 2–3 orders of magnitude, so the D abundance is close to the observed one even if the density of matter is nearly critical. Moreover, entropy fluctuations increase the He^4 abundance by 3–6%.

061.007 Integral circular polarization of spectral lines produced by inhomogeneities of magnetic and velocity

fields.   G. G. Pavlov, Yu. A. Shibanov.
Astron. Zh. Akad. Nauk SSSR, Vol. 55, 79 - 83 (1978). In
Russian. English translation in Soviet Astron., Vol. 22, No. 1.

Integral circular polarization of spectral lines due to a
nonuniform magnetic and velocity field is calculated. The
dependences of integral circular polarization on values of the
magnetic field and velocity gradients as well as on line strength
for different profiles of the absorption coefficient are obtained.

061.008   On particle creation in gravitational collapse.
H. C. Ohanian.
Nuovo Cimento B, Ser. 11, Vol. 41B, 88 - 98 (1977).
Abstr. in Phys. Abstr., Vol. 80, Abstr. 91420 (1977).

061.009   Scalar-particle production near the singularity in
an anisotropic universe. II. Mass creation by gravitational field.   E. Pessa.
Nuovo Cimento B, Ser. 11, Vol. 41B, 99 - 110 (1977).
Abstr. in Phys. Abstr., Vol. 80, Abstr. 91421 (1977).

061.010   L'origine des éléments.   H. Reeves.
La recherche en astrophysique, (see 003.001), p.
70 - 86 (1977).

061.011   Numerical methods in convection theory.
N. O. Weiss.
Problems of stellar convection, (see 012.005), p. 142 - 150
(1977).

Two and three-dimensional computations have enlarged
the understanding of non-linear convection, particularly in
Boussinesq fluids. However, one cannot adequately predict the
relationship between convective heat transport and the super-
adiabatic temperature gradient. Nor is there any indication of
a preferred length scale, other than the depth of the convecting layer, in a compressible fluid.

061.012   Convective dynamos.   S. Childress.
Problems of stellar convection, (see 012.005),
p. 195 - 224 (1977).

Convective dynamo theory can be regarded as combining
two kinds of physical problems, each involving an electrically
conducting fluid medium, but differing in the role of the
magnetic field and in the physical processes described. The
author outlines some of the current work on such systems.

061.013   The boundaries of a convective zone.   A. Maeder.
Problems of stellar convection, (see 012.005),
p. 235 - 236 (1977).

061.014   Thermosolutal convection.   H. E. Huppert.
Problems of stellar convection, (see 012.005),
p. 239 - 254 (1977).

The aim of this contribution is to survey a relatively new
form of convection, which is very easy to investigate in the
laboratory, plays an important role in the oceans and many
chemical engineering situations and is likely to prove essential
in the understanding of some areas of stellar convection.

061.015   Extremal states of matter in astrophysics. Part IV:
The crystalline state of matter: neutron crystals.
M. S. Borczuch, B. Kuchowicz.
Postępy Astron., Tom 24, 235 - 264 (1976). In Polish.

A brief outline of the concept of nuclear matter in
physics is given, and the appearance of this matter in nature is
considered. Current theories of the physical state of high
density nuclear matter (consisting mainly of neutrons) are
reviewed, with an emphasis on the possibility of a transition of
this matter under sufficient pressure from a liquid into a
crystalline state. Some observational evidence from pulsars is
discussed.

061.016   Astrophysical and cosmological information from
the neutral pion production in cosmic space.
B. Kuchowicz, Z. Strugalski.
Postępy Astron., Tom 25, 21 - 26 (1977). In Polish.

Production and decay of neutral pions π^0, relatively well
studied in laboratories, is continuously occurring in cosmic
space. From high energy physics laboratories we obtain
data about cross sections and other characteristics of the
above mentioned processes. This information can be applied
in gamma-ray astronomy – to an analysis of nuclear processes
in space, to a location of γ-ray (and π^0) sources, and to an
estimation of a possible admixture of antimatter in the
neighbourhood of the Galaxy.

061.017   Parametric instabilities III.   M. Luheshi, P.
Stewart.
Astron. Astrophys., Vol. 64, 259 - 265 (1978).

The spectra of parametric instabilities for a strong linearly polarized electric wave are calculated for both transverse
and longitudinal modes. Maximum instability rates are discussed in relation to the Crab nebula.

061.018   Sur la variance relativiste de la section efficace de
collision électron-photon.
S. Kichenassamy, R. Krikorian.
C. R. Acad. Sci. Paris, Tome 286, Sér. B, 127 - 129 (1978).

The relativistic law of transformation of the total cross-
section σ for the collision of two beams of relativistic particles
is established; this permits to avoid the inconsistencies due to
the postulated invariance of σ.

061.019   X-ray radiation emitted by an atom and a diatomic
molecule interacting with ultrarelativistic charged
particles.   A. S. Ambartsumyan, C. Yang.
Izv. Akad. Nauk Armyanskoj SSR, Fizika, Vol. 12, 320 - 328
(1977). In Russian.

061.020   Neutrinos: the ultimate astrophysical probe.
M. S. Turner.
Mercury, Vol. 7, No. 1, p. 9, 11 - 17 (1978).

061.021   A world made of quarks.
Mercury, Vol. 7, No. 1, p. 10 (1978).

061.022   A local proton irradiation model for isotopic
anomalies in the solar system.   T. Lee.
Bull. American Astron. Soc., Vol. 9, 563 (1978). – Abstract.

061.023   Penetration of the solar nebula by recent nucleosynthetic material.   S. H. Margolis.
Bull. American Astron. Soc., Vol. 9, 565 (1978). – Abstract.

061.024   On the magnetic fields of rotating astronomical
objects.   A. Brecher, K. Brecher.
Bull. American Astron. Soc., Vol. 9, 565 - 566 (1978).
Abstract.

061.025   Redistribution of neutrino energy by elastic
scattering.   D. L. Tubbs.
Bull. American Astron. Soc., Vol. 9, 595 (1978). – Abstract.

061.026   The bulk properties and equation of state of hot
dense matter.   J. M. Lattimer.
Bull. American Astron. Soc., Vol. 9, 595 (1978). – Abstract.

061.027   Dynamics of stellar collapse with degenerate
neutrinos.   T. J. Mazurek.
Bull. American Astron. Soc., Vol. 9, 597 (1978). – Abstract.

061.028   Reaction rates for neutrino processes.
D. Shalitin.
Astrophys. Space Sci., Vol. 53, 55 - 67 (1978).

Some integrals involved in neutrino processes are evaluated by transformation to a special system of reference – usually to the center of mass system (CM). Rather simple analytic expressions are obtained for reaction rates and, though less simple, for moments. An interesting result thus obtained is for an isotropic interaction (in CM) of a neutrino with a monoenergetic isotropic gas of extreme relativistic electrons: it is found that the probability of the scattered neutrino to have energy in a certain range is independent of this energy.

061.029 A method for evaluating abundances in long neutron capture and nuclear decay chains.
R. L. Smith.
Astrophys. Space Sci., Vol. 53, 85 - 89 (1978).

The Maclaurin expansion and algebraic addition formula for abundances in long neutron capture reaction chains are given. These permit the calculation of abundances without the numerical difficulties which occur with the use of the well-known Bateman solution.

061.030 Maximum abundant isotopes correlation.
G. S. Anagnostatos.
Astrophys. Space Sci., Vol. 53, 227 - 229 (1978).

The neutron excess of the most abundant isotopes of the elements shows an overall linear dependence upon the neutron number for nuclei between neutron closed shells. This maximum abundant isotopes correlation supports the arguments for a common history of the elements during nucleosynthesis.

061.031 Ultrahigh-energy neutrino astronomy.
S. H. Margolis, D. N. Schramm, R. Silberberg.
Astrophys. J., Vol. 221, 990 - 1002 (1978).

High-energy cosmic ray interactions can produce neutrinos. The authors calculate the neutrino fluxes over a range of energies. The sources considered (and their ranges of importance) are cosmic-ray interactions in the Earth's atmosphere ($E_\nu < 10^4 \, \mathrm{GeV}$), cosmic ray interactions with ambient hydrogen in galaxies ($10^4 \, \mathrm{GeV} < E_\nu < 10^6 \, \mathrm{GeV}$), regions of cosmic ray acceleration—e.g., pulsars—and cosmic ray interactions with the microwave background radiation ($10^8 \, \mathrm{GeV} < E_\nu$). In addition, estimates of the flux from compact sources, such as active galaxies, are made.

061.032 Proton-associated alpha-irradiation in the early solar system: a possible ^{41}K anomaly.   E. Dwek.
Astrophys. J., Vol. 221, 1026 - 1031 (1978).

In this paper the author shows that energetic particle fluences sufficiently high to produce the observed Al-correlated ^{26}Mg excess found in the Allende meteorite are also capable of producing ^{41}Ca by ^{38}Ar$(\alpha, n)^{41}$Ca reactions. The decay of ^{41}Ca in the condensing grains produces a detectable Ca-correlated excess of ^{41}K. A search for a ^{41}K anomaly in an Allende inclusion has thus far yielded negative results. The datum is not in conflict with the irradiation model. A firm negative result will put severe constraints on the model and,considering other objections to this early particle irradiation, will tend to disqualify it as a possible source for the isotopic anomalies found in the solar system.

061.033 Matter–antimatter boundary layers with a magnetic neutral sheet.   B. Lehnert.
Astrophys. Space Sci., Vol. 53, 459 - 465 (1978).

An earlier model of matter–antimatter boundary layers has been extended to include a sheet with a reversed magnetic field. The derived layer thickness is largely unaffected by a magnetic field-reversal, provided that the width of the corresponding magnetic neutral sheet becomes substantially smaller than the layer thickness. This condition is likely to be satisfied within parameter ranges of cosmical interest. The present model represents a crude first approach, and a more rigorous treatment of a quasi-neutral ambiplasma is desirable which also includes the problem of stability.

061.034 Kinetics of neutronization in superdense matter with frozen-in magnetic field.   G. A. Shul'man.
Astrofizika, Vol. 13, 657 - 667 (1977). In Russian. English translation in Astrophysics, Vol. 13, No. 4.

The time of neutronization process of frozen matter is estimated after hydrogen, helium, carbon and iron having achieved the necessary thermodynamical limit density. The limit magnetic fields at a given mass density of the matter are determined.

061.035 Modulation instability in astrophysics.
V. N. Tsytovich.
Issled. po geomagn., aehron i fiz. Solntsa. Vyp. (No.) 42. Moskva, Nauka, 1977, p. 3 - 20. In Russian. – Abstr. in Ref. zh., 51. Astron., 3.51.183 (1978).

061.036 Neutrino astrophysics.   G. E. Kocharov.
Izv. AN SSSR. Ser. fiz., Vol. 41, 1916 - 1948 (1977). In Russian. – Abstr. in Ref. zh., 51. Astron., 3.51.515 (1978).

061.037 Plasma physics and astronomy.
I. S. Shklovskij.
Priroda, 1978, No. 5, p. 26 - 33. In Russian.

061.038 Gamma-ray spectroscopy in astrophysics.
R. Ramaty, T. L. Kline, R. Weaver.
Nature, Vol. 273, 591 - 592 (1978).

061.039 Grain formation through nucleation process in astrophysical environment.
T. Yamamoto, H. Hasegawa.
Prog. Theor. Phys., Vol. 58, 816 - 828 (1977). – Abstr. in Phys. Abstr., Vol. 81, Abstr. 16501 (1978).

061.040 Astrophysical aspects of radiative capture of protons by ^{19}F.   K. M. Subotic.
Fizika, Vol. 9, Suppl. 1, p. 45 - 46 (1977). – Abstr. in Phys. Abstr., Vol. 81, Abstr. 17191 (1978).

061.041 On the diffusion of dust through a gas in a force field.   S. Simons.
Phys. Lett. A, Vol. 64A, 11 - 13 (1977). – Abstr. in Phys. Abstr., Vol. 81, Abstr. 18366 (1978).

061.042 Limits on the mass of the muon neutrino in the absence of muon-lepton-number conservation.
T. Goldman, G. J. Stephenson, Jr.
Phys. Rev. D, Vol. 16, 2256 - 2259 (1977). – Abstr. in Phys. Abstr., Vol. 81, Abstr. 21368 (1978).

061.043 Implications of a nonzero neutrino mass for the process $\gamma\gamma \to \nu\bar{\nu}$.   E. Fischbach, S. P. Rosen, H. Spirack, J. T. Gruenwald, A. Halprin, B. Kayser.
Phys. Rev. D, Vol. 16, 2377 - 2378 (1977). – Abstr. in Phys. Abstr., Vol. 81, Abstr. 21483 (1978).

061.044 Improving the resolution of small energy shifts using ^{67}Zn and ^{181}Ta.   R. V. Pound.
Workshop on new directions in Mössbauer spectroscopy, Argonne, Ill., USA, 1977. AIP Conf. Proc. No. 38, p. 41 - 45 (1977). – Abstr. in Phys. Abstr., Vol. 81, Abstr. 24905 (1978).

061.045 On three approximations for free convection.
V. I. Kovshov.
Astron. Zh. Akad. Nauk SSSR, Vol. 55, 501 - 515 (1978). In Russian. English translation in Soviet Astron., Vol. 22, No. 3.

Substantiation is given for simplifications of the equations of free convection in deep, highly stratified and evolving regions. Simplifications are deduced in three cases: when the temporal and spatial scales of convective disturbances are of the order of, or smaller than, or much smaller than the time scales and the depth of the convective region corresponding to the pressure or the temperature.

061.046   Formation of by-passed isotopes under the influence of neutrinos and possible role of neutrinos in nucleosynthesis.   G. V. Domogatskij, D. K. Nadezhin.
Astron. Zh. Akad. Nauk SSSR, Vol. 55, 516 - 530 (1978). In Russian. English translation in Soviet Astron., Vol. 22, No. 3.

The influence of neutrinos formed during the gravitational collapse of a stellar core on the isotope composition of the material of the stellar envelope is considered. The obtained values of abundance of the major part of by-passed isotopes reproduce quite satisfactorily the shape of the curve of the abundances of the solar system by-passed isotopes. A possible role of neutrinos in the general picture of nucleosynthesis is briefly discussed and it is emphasized that a consistent theory of nucleosynthesis in an ejected stellar envelope cannot be constructed without accounting for the influence of neutrinos.

061.047   On the origin and evolution of isotopes of carbon, nitrogen, and oxygen.   D. Dearborn, B. M. Tinsley, D. N. Schramm.
Astrophys. J., Vol. 223, 557 - 566 (1978).

New calculations of CNO processing in stellar envelopes and other results from nucleosynthesis theory have been used as input to models for chemical evolution in the solar neighborhood. The results are contrasted with those of Audouze, Lequeux, and Vigroux (1975, 1976). A detailed comparison of the authors' predictions with solar system abundances is given, pointing up both successes and problems with the current nucleosynthesis theory. The problems of meteoritic oxygen isotopes are discussed.

061.048   Beta transition rates in hot and dense matter.   K. Takahashi, M. F. El Eid, W. Hillebrandt.
Astron. Astrophys., Vol. 67, 185 - 197 (1978).

Allowed and first-forbidden transition rates of $\beta^\pm$ decays and $e^\pm$ captures under stellar conditions of high temperatures and high densities are reformulated. The paper mainly describes the formalism which is essentially based on the gross theory of nuclear β-decays, but also contains the numerical results of the transition rates of nuclei with the mass number 56. The discussion includes a short but critical review of several different approaches to the astrophysical β-transitions on nuclei as well as on the neutron and proton.

061.049   Collapse of stars and neutrino emission.   O. Kh. Gusejnov.
Probl. gravitatsii. Tbilisi, 1976, p. 110 - 117. In Russian. Abstr. in Ref. zh., 51. Astron., 5.51.755 (1978).

061.050   Photon polarization in the scattering process in the gravitational field of a spherically-symmetric star.
Yu. S. Vladimirov, I. F. Iskhakov.
Probl. gravitatsii. Tbilisi, 1976, p. 81 - 86. In Russian. — Abstr. in Ref. zh., 51. Astron., 5.51.765 (1978).

061.051   Neutrino stability and cosmological helium production.   R. J. Tayler.
Nature, Vol. 274, 232 - 234 (1978).

Heavy leptons with associated neutrinos could lead to an increase in cosmological helium production and their discovery might make the standard theory conflict with observations. If some of the neutrinos are unstable with a short enough half life, and if the properties of the weak interactions allow electron neutrinos to decouple sufficiently long after neutrinos of other types, this problem may be avoided. If other evidence for cosmological helium production which is incompatible with the existence of more stable neutrinos seems sufficiently strong, this will suggest that, if additional neutrinos are dis-

covered, they must be unstable.

061.052   The rapid neutron-capture process and the synthesis of heavy and neutron-rich elements.
W. Hillebrandt.
Space Sci. Rev., Vol. 21, 639 - 702 (1978).

Recent developments in theoretical model-calculations for the synthesis of heavy and neutron-rich chemical elements by means of the rapid neutron-capture process are reviewed. Special emphasis is put on a discussion of possible astrophysical sites, e.g. supernova models, and extrapolated nuclear input data as well. Numerical methods to solve complex nuclear reaction-networks are presented.

061.053   Plasma astrophysics.   V. Canuto, C. K. Chou, L. Fassio-Canuto.
Fundam. Cosmic Phys., Vol. 3, 221 - 339 (1978). Review paper.

061.054   Isotope fractionation under simulated space conditions.   G. Arrhenius, R. Fitzgerald, S. Markus, C. Simpson.
Astrophys. Space Sci., Vol. 55, 285 - 297 (1978).

In order to clarify some of the mechanisms responsible for the isotope fractionation observed in the interstellar cloud medium and in meteorite materials the authors have undertaken a series of laboratory model experiments. Initial studies have been made in the plasma reaction system $(^{14}N^{15}N)$–O.

061.055   X-ray astronomy.   B. Rossi.
Daedalus, Vol. 2, p. 37 - 58 (1977). — Abstr. in Phys. Abstr., Vol. 81, Abstr. 32604 (1978).

061.056   The evolution of the elements.   P. M. Williams.
Contemp. Phys., Vol. 19, No. 1, p. 1 - 24 (1978). Abstr. in Phys. Abstr., Vol. 81, Abstr. 32619 (1978).

061.057   Physical mechanisms affecting the observations of fast transient phenomena in astrophysics.
G. Cavallo.
Mem. Soc. Astron. Italiana, Vol. 48, 109 - 138 (1977).

The role of fast transient phenomena is now increasing in observational astrophysics as, in some energy band at least, searches are not exceedingly expensive, and look comparatively promising. In the present work, the formation and propagation of electromagnetic signals are considered, and mechanisms which affect either the intensity or the temporal characteristics of the signals are examined, with the aim of suggesting limitations and guidelines for future searches.

061.058   The resonating group method for 3-cluster systems based on the use of generator coordinate kernels:
application to the 3α structure of ^{12}C and the 3α reaction in stellar evolution.   Y. Fukushima, M. Kamimura.
J. Phys. Soc. Japan, Vol. 44, Suppl., 225 - 231 (1978). Abstr. in Phys. Abstr., Vol. 81, Abstr. 50108 (1978).

Atomic processes and applications.
See Abstr. 003.031.

Theoretical principles in astrophysics and relativity.
See Abstr. 012.044.

Interaction of radiation with condensed matter.
See Abstr. 012.048.

Exploring the infrared universe from Wyoming.
See Abstr. 032.037.

Sneg 3 and gamma astronomy.
See Abstr. 054.015.

Optically thick X-ray transfer: the shell game.
See Abstr. 063.027.

s-process studies: the effects of a pulsed neutron
flux.   See Abstr. 065.002.

^{26}Al production in explosive carbon burning.
See Abstr. 065.004.

On the surface composition of thermally pulsing
stars of high luminosity and on the contribution of such stars
to the element enrichment of the interstellar medium.
See Abstr. 065.032.

Pure neutral vector boson contribution to neutrino
opacity in stellar matter.   See Abstr. 066.336.

Solar neutrino experiment as a test for cosmology
and cosmogony. I. Davis' experiment as a permanent stimulation for numerous branches of physics and astronomy.
See Abstr. 080.032.

A preliminary determination of the relative abundances of the isotopes of zirconium in R Cygni and V Cancri.
See Abstr. 114.543.

Deuterium in the Galaxy.   See Abstr. 131.126.

Dust, helium abundance.   See Abstr. 131.259.

A momentum distribution of high energy particles
around Crab pulsar.   See Abstr. 141.541.

Time delay between the nucleosynthesis of cosmic
rays and their acceleration to relativistic energies.
See Abstr. 143.005.

Helium production and limits on the anisotropy of
the universe.   See Abstr. 162.003.

Properties of the stars of early generations in the
scale covariant cosmology.   See Abstr. 162.004

Cosmological implications of massive, unstable
neutrinos.   See Abstr. 162.028.

Significance of neutrinos in cosmology.
See Abstr. 162.097.

The universe and deuterium.   See Abstr. 162.185.

062 Hydrodynamics, Magnetohydrodynamics, Plasma

062.001 **Relativistic parametric instabilities in extended extragalactic radio sources.**
A. Ferrari, E. Trussoni, L. Zaninetti.
Mon. Not. R. Astron. Soc., Vol. 182, 49 - 62 (1978).

A general discussion is presented of parametric instabilities of electromagnetic waves in cold plasmas. Previous results for $f \equiv eE/m_e c \omega_0 \gg 1$ and $\ll 1$ are extended and the intermediate range $f \sim 1$, which could be relevant in some astrophysical applications, is analysed by numerical techniques. In the final section a model for particle acceleration and radiation emission by turbulent plasma modes excited in extended radiosources by parametric absorption of strong electromagnetic waves is tentatively discussed.

062.002 **On the dynamic interaction between magnetic fields and convection.**
R. S. Peckover, N. O. Weiss.
Mon. Not. R. Astron. Soc., Vol. 182, 189 - 208 (1978).

A simple two-dimensional model of convection in an electrically conducting Boussinesq fluid with an externally imposed magnetic field has been investigated in a series of numerical experiments. The results are relevant to the interaction between small-scale magnetic fields and photospheric granulation in the Sun.

062.003 **The principle of exchange of stabilities in the atmosphere.** P.-A. Bois.
C. R. Acad. Sci. Paris, Tome 286, Sér. A, 79 - 81 (1978).
In French.

Classical results of hydrodynamic-stability theory in bounded domains are generalized in the case of the atmosphere. In particular, if ζ denotes the altitude normalized by the atmospheric scale, then for a Brunt-Väisälä frequency h (ζ) which is always positive, the atmosphere is stable. If h (ζ) vanishes and changes sign at a point ζ_0, an instability threshold occurs only through a stationary state.

062.004 **Magnetohydrodynamics of atmospheric transients. I. Basic results of two-dimensional plane analyses.**
Y. Nakagawa, S. T. Wu, S. M. Han.
Astrophys. J., Vol. 219, 314 - 323 (1978).

The dynamic behavior of transients in a stratified atmosphere in the presence of a magnetic field is investigated with the two-dimensional plane formulation, i.e., the formulation in which variations of the flow and magnetic field are confined within a plane. It is shown that (1) in the open (mostly radial) magnetic field, the propagation of "bubble-like" density enhancements, the coronal transients, result; while (2) in the closed (mostly azimuthal) field, the density enhancement forms a pair of "horns" near the foot points of field lines similar to that often observed by coronagraphs after flares. Discussions in interpreting the results in terms of the anisotropic propagation of the fast and slow waves (or shocks) are presented together with consideration of possible consequences with the inclusion of the transverse waves.

062.005 **Formation of a rotational accretion column.**
P. Cassen.
Astrophys. J., Vol. 219, 336 - 344 (1978).

The author considers the problem of the flow produced by the accretion of gas from a rotating cloud onto a point gravitational source. It is shown that, for a cloud initially in uniform rotation, the effect of accretion is to produce a Taylor column in the cloud parallel to the rotation axis. Only gas within the column is accreted; the gas outside the column undergoes oscillatory flow, which is probably subject to shear instability. Application of the results to the formation of Jupiter by accretion is discussed.

062.006 **On the effect of gas pressure in the theory of line accretion – I.** S. Yabushita.
Mon. Not. R. Astron. Soc., Vol. 182, 371 - 380 (1978).

The Bondi–Hoyle equation of line accretion is modified. It is shown that for physically interesting values of γ (specific heats ratio) there exists a unique solution such that the gas velocity is subsonic near the stagnation point but supersonic beyond the sonic transition and asymptotically approaches V_∞, the unperturbed relative velocity. In addition, there exists an infinity of solutions such that the velocity is thoroughly subsonic and approaches the value $(\gamma-1) V_\infty/(\gamma + 1)$ at great distance from the star.

062.007 **The exterior source surface for force-free fields.** D. D. Barbosa.
Astron. Astrophys., Vol. 62, 267 - 268 (1978).

The spherical harmonic expansion of force-free fields satisfying $\nabla \times B = \kappa B$ and $\kappa = $ const is presented for boundary values of B_r on two concentric spheres. An upper limit on κ is found which depends on the lowest multipole moment of the prescribed boundary values. For a dipole field this limit is $\kappa A < 4.5$ where A is the outer radius.

062.008 **Normal modes of self-gravitating fluids in perturbed configurations. I. Perturbational-variational procedure.** J. N. Silverman, Y. Sobouti.
Astron. Astrophys., Vol. 62, 355 - 363 (1978) = Contrib. Biruni Obs., Pahlavi Univ., Shiraz, Iran, No. 1.

In connection with a generalized perturbed eigenvalue equation (arising, in particular, from the small oscillations of self-gravitating fluids in perturbed configurations), the authors have developed a perturbational-variational Rayleigh-Ritz (PV-RR) expansion scheme for systematically obtaining the normal modes (eigenvalues and eigenfunctions) in powers of λ; here, λ is an external perturbing parameter characterizing the eigenvalue equation. The perturbation (which enters formally through λ) can originate from a wide variety of sources in astrophysical contexts.

062.009 **Normal modes of self-gravitating fluids in perturbed configurations. II. Perturbational-variational expansion of the g- and p-modes of a non-adiabatic fluid about the adiabatic limit.** Y. Sobouti, J. N. Silverman.
Astron. Astrophys., Vol. 62, 365 - 374 (1978) = Contrib. Biruni Obs., Pahlavi Univ., Shiraz, Iran, No. 2.

The g-eigenfrequencies of an adiabatic fluid are identically zero; their growth rate with departure of the fluid from the adiabatic limit and the subsequent motions, however, are obtainable from an eigenvalue equation. The p-modes of the adiabatic fluid are also solutions of another eigenvalue equation. These two eigenvalue problems are derived and solved. With regard to convection, the information on g-modes should be of particular interest: In a slightly superadiabatic fluid, the first-order g-eigenvalues and their corresponding eigenvectors give the time rate of growth of the convective instabilities and the patterns of convective motions, respectively.

062.010 **An instability of finite amplitude circularly polarized Alfvén waves.** M. L. Goldstein.
Astrophys. J., Vol. 219, 700 - 704 (1978).

A circularly polarized, large-amplitude Alfvén wave in a plasma with finite plasma β is shown to be unstable. The wave decays by coupling to random density and magnetic fluctuations in the ambient medium. Daughter waves are produced at the sidebands which are at the first harmonic of the Alfvén wave frequency and wavenumber. For parameters more typical of the solar corona and solar wind, large decay rates are found for $0.1 \leqslant \beta \leqslant 1$ and $0.1 \leqslant \eta \leqslant 0.9$, where η is the ratio of magnetic energy density of the initial Alfvén wave to that of the background magnetic field. The possible role played by this instability in several astrophysical situations is briefly considered.

062.011 **Stimulated scattering processes in magnetized plasmas.** L. Stenflo.
Phys. Scr., Vol. 17, 9 - 10 (1978).

Stimulated scattering of low frequency modes is considered for the case where the pump wave, which is assumed to be circularly polarized and to propagate parallel to an external magnetic field, is relativistically strong. The results are compared with those of non-relativistic theory in order to extend their range of validity.

062.012 **Threshold of electromagnetic instability in a magnetic neutral sheet.** K. Yamanaka.
Phys. Scr., Vol. 17, 15 - 22 (1978).

062.013 **On the turbulent transport coefficient.** E. Schatzman.
Astron. Astrophys., Vol. 63, L17 - L18 (1978).

A new interpretation of the classical Taylor experiments on rotating fluids seems to be compatible with a turbulent transport coefficient $D = Re^* \nu$, where $Re^* = 200$ and ν is the kinematic viscosity.

062.014 **Magnetic field generation at high magnetic Reynolds number.** E. H. Levy.
Astrophys. J., Vol. 220, 325 - 329 (1978).

The lowest order contribution of finite electrical resistivity to the process of magnetic field regeneration at high magnetic Reynolds number is calculated. It is found that finite resistivity changes the calculated regeneration rate by less than a factor of 2.

062.015 **The root-mean-square magnetic field in turbulent diffusion.** E. Knobloch.
Astrophys. J., Vol. 220, 330 - 334 (1978).

The author uses the exact Eulerian formulation of the problem of diffusion of vector fields by a turbulent velocity field to calculate the mean square magnetic field (the mean energy) to third order in the velocity autocorrelation time. The calculation is carried out for zero ordinary diffusivity and for homogeneous, stationary, isotropic, incompressible, helical turbulence. It is found that the rms magnetic field behaves quite differently from the mean field.

062.016 **Evidence for two-dimensional inertial turbulence in a cosmic-scale low-β plasma.**
M. C. Kelley, P. M. Kintner.
Astrophys. J., Vol. 220, 339 - 345 (1978).

Magnetospheric electric field power spectral densities measured from high-altitude balloons and the Hawkeye 1 satellite are combined in wavenumber space and exhibit a power law spectrum with index -1.6 ± 0.3 below $k = 5 \times 10^{-2}$ km^{-1} and index -2.8 ± 0.3 above $k = 2$ km^{-1}. This spectrum agrees with that predicted for the inertial subrange of an isotropic homogeneous two-dimensional fluid and for a two-dimensional plasma when energy enters the system at the spectral knee and there exists a viscous dissipation mechanism at large wavenumbers. The wavelength of the spectral knee corresponds to that of folds and curls in the aurora and may be due to an instability of the auroral particle beams.

062.017 **A virial theorem for relativistic plasmas.** A. Georgiou.
Phys. Lett. A, Vol. 62A, 484 - 486 (1977). − Abstr. in Phys. Abstr., Vol. 80, Abstr. 89299 (1977).

062.018 **Magnetic fields and convection.** N. O. Weiss.

Problems of stellar convection, (see 012.005), p. 176 - 187 (1977).

In a highly conducting plasma convection is hindered by the imposition of a magnetic field. Convection may set in as direct or overstable modes and behaviour near the onset of instability depends on the ratio of the magnetic to the thermal diffusivity. Vigorous convection produces local flux concentrations with magnetic fields that may be much greater than the equipartition value. The interaction between magnetic fields and convection can be observed in detail on the sun and is essential to any stellar dynamo.

062.019 **Axisymmetric convection with a magnetic field.** D. J. Galloway.
Problems of stellar convection, (see 012.005), p. 188 - 194 (1977).

The non-linear Boussinesq equations describing axisymmetric convection in a cylinder with an initially uniform magnetic field have been integrated forward in time numerically. When the field is weak a strong central fluxrope is formed at the axis. In this case the maximum field strength can be limited either kinematically or by dynamical effects and the equipartition prediction $B^2_{max} \sim 4\pi\mu\rho u^2$ is easily exceeded. If the field is strong oscillations can occur and hysteresis is possible as the field is increased and decreased.

062.020 **Fully developed turbulence, intermittency and magnetic fields.** U. Frisch.
Problems of stellar convection, (see 012.005), p. 325 - 336 (1977).

Turbulence is usually associated with the idea of chaos, i.e. erratic behaviour of some observable quantity. The author demonstrates that there are at least two different kinds of chaos.

062.021 **Turbulence: determinism and chaos.** Y. Pomeau.
Problems of stellar convection, (see 012.005), p. 337 - 348 (1977).

062.022 **The evolution of expanding nonthermal sources. II. Relativistic expansion.**
P. Vitello, F. Pacini.
Astrophys. J., Vol. 220, 756 - 771 (1978).

A hydrodynamical treatment of radio source models expanding relativistically into vacuum is presented. Both a linear and a spherical geometry are considered. The time evolution of the flux density is computed for sources which initially are optically thick and upon expansion become optically thin, and for sources which are initially and at all times optically thin. The spectral evolution is also shown. The results are compared with those for nonrelativistic models, and for relativistic homogeneous models. It was found that apparent superluminous motion is observable only for the linear models, and not for the cases of spherical expansion.

062.023 **Transport of twist in force-free magnetic flux tubes.** K. Jockers.
Astrophys. J., Vol. 220, 1133 - 1136 (1978).

Parker has studied the problem of a force-free magnetic flux tube, part of which is expanded by external forces. For two special magnetic field distributions he has shown that the twist will increase in the expanded section. In this paper Parker's result is generalized. It is demonstrated that expansion of part of a force-free magnetic flux tube will lead to twist transport into the expanded section if the longitudinal component of the magnetic field in the undisturbed flux tube increases toward the center of the flux tube. It is proposed to check the observational evidence as to whether this twist mechanism causes the rotational motions observed in eruptive prominences.

062.024 **Electromagnetic field of an electrically conducting** sphere rotating in a time-independent homogeneous external magnetic field (Unipolar induction). E. Schutzer. Exp. Tech. Phys., Vol. 25, 369 - 383 (1977). In German. Abstr. in Phys. Abstr., Vol. 81, Abstr. 17836 (1978).

062.025 **Unsteady hydrodynamics of spherical gravitational collapse.** A. F. Cheng.
Astrophys. J., Vol. 221, 320 - 326 (1978).

Similarity solutions are obtained which describe the time evolution of spherical, self-gravitating gas clouds with initial density distribution $\rho \sim r^{\alpha}$, where $\alpha > -3$ but otherwise arbitrary, and with pressure $P = \kappa \rho^{\gamma}$, where γ is constant and κ is constant along the trajectory of any fluid element. For all $1 \leqslant \gamma \leqslant 5/3$, appropriate for interstellar clouds, the collapse solutions contain a massive central core near which the gas approaches free fall. If the initial configuration is nearly static, rapid infall begins only behind the front of an expansion wave propagating outward into the nearly static envelope. Several numerical examples are given.

062.026 **Magnetic reconnection:acceleration, heating, and shock formation.** T. Hayashi, T. Sato.
J. Geophys. Res., Vol. 83, 217 - 220 (1978).

062.027 **The role of viscosity and cooling mechanisms in the stability of accretion disks.** T. Piran.
Astrophys. J., Vol. 221, 652 - 660 (1978).

A dispersion relation for linearized perturbation of a geometrically thin accretion disk is derived for an arbitrary viscosity law and cooling mechanism. Using this equation, the author formulates the necessary and sufficient conditions for the stability of a disk. It is shown that the conditions cannot be satisfied by any one of the relevant cooling mechanisms with the standard viscosity law. Small variations in this law lead, however, to a stable configuration.

062.028 **Stability of astrophysical gas flow. I. Isothermal accretion.** R. F. Stellingwerf, J. Buff.
Astrophys. J., Vol. 221, 661 - 671 (1978).

A general numerical scheme is presented that allows a variety of analyses of an arbitrary set of hydrodynamic equations: search for steady states; stability of steady states; nonlinear integration; search for specified nonlinear solutions; stability of nonlinear solutions. As an application, the authors investigate the stability of spherically symmetric, isothermal accretion flows.

062.029 **Wave modes in a magnetic flux sheath.** P. R. Wilson.
Astrophys. J., Vol. 221, 672 - 676 (1978).

The symmetric wave modes of a two-dimensional magnetic flux sheath embedded in a compressible isothermal but nongravitational atmosphere are investigated under more general conditions than those considered earlier by Cram and Wilson (1975). Both two-dimensional "taut wire" and pulsation modes are considered, and explicit solutions are obtained when the thickness of the flux sheath is small compared with the transverse scale size of the perturbations. The solutions for the pulsation mode are compared with those obtained by Defouw (1976) under similar conditions.

062.030 **Ionization equilibrium and radiative cooling of a high temperature plasma.**
C. Breton, C. de Michelis, M. Mattioli.
J. Quant. Spectrosc. Radiat. Transfer, Vol. 19, 367 - 379 (1978).

The results of calculations of ionization equilibrium and of the rate of radiative cooling of a high temperature plasma are presented as a function of the electron temperature from a few eV up to some tens of keV. The most important elements detected in Tokamak plasmas (O, N, C, Fe and Mo) are considered.

062.031 **Rotating protostellar collapse in three space dimensions.** J. E. Tohline.
Bull. American Astron. Soc., Vol. 9, 566 (1978). – Abstract.

062.032 **Multidimensional hydrodynamics.** J. J. Monaghan, R. A. Gingold.
Bull. American Astron. Soc., Vol. 9, 567 (1978). – Abstract.

062.033 **Particle streaming: is the Alfvén velocity the ultimate speed limit?**
G. D. Holman, J. A. Ionson, J. S. Scott.
Bull. American Astron. Soc., Vol. 9, 568 (1978). – Abstract.

062.034 **Cloud confinement by momentum transfer from suprathermal protons.**
D. Rowe, R. E. Stoner, R. L. Ptak.
Bull. American Astron. Soc., Vol. 9, 630 (1978). – Abstract.

062.035 **Thermal convection in compressible and Boussinesq fluids in plane-parallel and spherical geometries.**
P. S. Marcus.
Bull. American Astron. Soc., Vol. 9, 638 (1978). – Abstract.

062.036 **The validity of ionization equilibrium in steady-state flows.** J. A. Joselyn, R. H. Munro, T. E. Holzer.
Bull. American Astron. Soc., Vol. 9, 650 (1978). – Abstract.

062.037 **Automatic computation of diffusion rates.** P. D. Noerdlinger.
Astrophys. J., Suppl. Ser., Vol. 36, 259 - 273 (1978).

The equations for diffusion due to gravity, thermal, and concentration gradients in a multi-component, quiescent plasma are so complicated that the calculation of the diffusion rate(s) should be automated. The author gives an automated calculational scheme that utilizes the inherent reliability of computers in conducting repeated operations, but does not require simultaneous solution of many linear equations for each time step. The problem is defined and the fundamental equations of Burgers are stated. The method for automatically handling coefficients is presented. Numerical results, including the fitting polynomials to the expressions, are given.

062.038 **The spectrum of synchrotron-transient radiation of a charged particle in plasma with random inhomogeneities.** V. V. Tamoykin.
Astrophys. Space Sci., Vol. 53, 3 - 11 (1978).

The radiation of a charge rotating in a circle with constant velocity (in the external magnetic field) in an isotropic plasma with random inhomogeneities of the electron density has been considered. A general expression is obtained for the radiation intensity at the nth harmonic, which is a generalization of the known Shott formula. In the ultra-relativistic case the conditions are clarified under which the inhomogeneity effect on the form of the spectrum of radiation from a particle is essential. An asymptotic formula is derived for the spectral intensity in the region of sufficiently low frequencies. The mechanism of transient radiation in this case is shown to prevail over the synchrotron one.

062.039 **Radio frequency emission in electron beam-plasma and beam-beam interaction.**
V. Krishan, K. P. Sinha, S. Krishan.
Astrophys. Space Sci., Vol. 53, 13 - 19 (1978).

A linear excitation of electromagnetic modes at frequencies $\simeq (n + \frac{1}{2})\omega_{ce}$, in a plasma through which two electron beams are contra-streaming along the magnetic field is investigated. This may be a source of the observed $\frac{3}{2}\omega_{ce}$ emissions at auroral latitudes.

062.040 **Concerning isothermal self-similar blast waves. I: One-dimensional flow and its stability.**
I. Lerche.

Astrophys. Space Sci., Vol. 53, 21 - 37 (1978).

The author investigates the one-dimensional self-similar flow behind a blast wave from a plane explosion in a medium whose density varies with distance as $x^{-\omega}$ with the assumption that the flow is isothermal. The dependence of the solutions on the parameter ω is discussed.

062.041   **Concerning isothermal self-similar blast waves. II: Two-dimensional flow and its stability.**
I. Lerche.
Astrophys. Space Sci., Vol. 53, 39 - 54 (1978).

The author investigates the two-dimensional self-similar flow behind a blast wave from a line explosion in a medium whose density varies with distance as $r^{-\omega}$ with the assumption that the flow is isothermal. The dependence of the solutions on the parameter ω is discussed. It appears that if a two-dimensional isothermal flow pattern is sought with continuous post-shock velocity, it must occur in a non-self-similar combination. The already established invalidity of the adiabatic model (which leads to temperature gradients so large that they violate the assumption of neglect of the heat flux), and the demise of the isothermal model reported here strongly suggest that no self-similar model adequately describes two-dimensional blast flow problems.

062.042   **On the stability of twisted magnetic fields.**
S. K. Trehan, M. Singh.
Astrophys. Space Sci., Vol. 53, 165 - 172 (1978).

The stability of helical magnetic fields is investigated when fluid motions are present along the lines of force. The general dispersion relation is obtained and some limiting cases are examined. It is established that the configuration can be unstable when the velocity field exceeds a certain critical value. This result is to be compared with the case when the helical fields confined by a rigid boundary are stable when the energy density in the velocity field is at least equal to that of the magnetic field.

062.043   **A quasi-linear WKB kinetic theory for nonplanar waves in a nonhomogeneous warm plasma.**
1. Transverse waves propagating along axisymmetric B_0.
J. V. Hollweg.
J. Geophys. Res., Vol. 83, 563 - 574 (1978).

A new set of quasi-linear kinetic equations is presented for transverse waves propagating along an axisymmetric magnetic field configuration. A WKB expansion is used to include effects of nonhomogeneity and nonplanarity of the waves. The equations allow simultaneous calculation of the spatial (and temporal) evolution of wave power spectra and the spatial (and temporal) evolution of particle distribution functions, including for the first time important wave-particle interactions which depend explicitly on the nonplanarity of the waves and the nonhomogeneity of the plasma and fields. The usefulness of the equations is demonstrated for a cold plasma, where a number of new results for wave propagation and acceleration of the plasma have been obtained.

062.044   **A mechanism for current interruption in a collisionless plasma.**   J. R. Kan, S.-I. Akasofu.
J. Geophys. Res., Vol. 83, 735 - 738 (1978).

A mechanism for current interruption in a collisionless plasma is proposed. The fraction of current interrupted by the proposed mechanism depends on the ratio of the ion thermal energy to the electron streaming energy. In a model examined in this paper it is found that for the unstable current to be interrupted and reduced to one half of its original value, the ion to electron energy ratio must be greater than 2. To interrupt the current by a larger fraction, a higher energy ratio is required. Possible application of this current interruption mechanism to magnetospheric substorms is suggested.

062.045   **The scattering of electromagnetic waves from density fluctuations in a plasma.**   T. Hagfors.
Radar probing of the auroral plasma, (see 012.011), p. 15 - 28 (1977).

The scattering of electromagnetic waves by a single electron is developed from first principles. The result is used to derive the relationship of the scattered power spectrum to the spacetime Fourier transform of the electron density fluctuations in a plasma.

062.046   **The origin and properties of thermal fluctuations in a plasma.**   J. Trulsen, N. Bjørna.
Radar probing of the auroral plasma, (see 012.011), p. 29 - 54 (1977).

A general expression is derived for the differential scattering cross-section of radio waves scattered by thermal fluctuations of electron density in a plasma. The derivation is based on the test particle principle.

062.047   **Lagrangian perturbation theory of nonrelativistic fluids.**   J. L. Friedman, B. F. Schutz.
Astrophys. J., Vol. 221, 937 - 957 (1978).

The authors develop a formalism for perturbations of a stationary Newtonian fluid. A description of fluid perturbations in terms of a Lagrangian displacement is introduced. The class of trivial displacements is identified, and an explicit form obtained for the generic trivial. The formal structure of the perturbation equations is presented. The canonical energy and angular momentum are introduced, together with two related dynamically conserved inner products. The authors find that the canonical energy and angular momentum do not vanish on trivial displacements. They therefore cannot be identified with the second-order changes in the system's energy and angular momentum, and following Bardeen (1975), the explicit relation between the canonical and physical conserved quantities is obtained. The inner products are used to characterize a dynamically invariant class of canonical displacements orthogonal to the trivials.

062.048   **Stability of force-free magnetic fields in degenerate stars.**   G. Chanmugam.
Astrophys. J., Vol. 221, 965 - 968 (1978).

The stability of force-free magnetic fields in an axisymmetric, cylindrical, degenerate plasma is considered. It is shown that, under certain conditions, instabilities can arise as the plasma cools, confirming Tayler's results for non-force-free fields. It is suggested that such instabilities may take place in pulsars as they spin down, leading to possible field decay and accounting for the absence of long-period pulsars.

062.049   **Nonlinear astrophysical dynamos: bifurcation of steady dynamos from oscillating dynamos.**
H. Yoshimura.
Astrophys. J., Vol. 221, 1088 - 1099 (1978).

The nonlinear dynamo wave equation, which has been formulated to explore oscillating dynamos, is found also to have steady magnetic field configurations as its stable solutions. The solutions of the nonlinear wave equation, integrated numerically as the initial-boundary-value problem in the rotating spherical geometry, eventually bifurcate into a stationary oscillating state and a stationary steady state, depending on the initial condition adopted in the integration. Both states are stable with respect to small perturbations. The author suggests that the magnetic fields of the earth and planets, and the fields of non-solar-type magnetic stars, especially stars classified as oblique rotators, can be understood as special stationary solutions of the nonlinear dynamo wave equation, which can also have oscillating solutions.

062.050   **Strong waves in a collisionless, homogeneous plasma: waveform, polarization and non-linear effects.**
M. Salvati.
Astron. Astrophys., Vol. 65, 1 - 10 (1978).

A system of protons, electrons and electromagnetic waves is postulated, where the full set of Maxwell and Lorentz equations has to be used, and no linearization is allowed; all the variables depend on time and position through the combination $t - zn/c$. The problem is shown to be equivalent to the classical motion of a point mass in a time-independent potential; the latter is explicitly given in terms of the initial distribution functions. The following limiting cases are investigated by means of analytical methods: very low and very high energy, as compared with the rest energy; very high and very low carriers' density, as measured by the index n; almost transverse and almost longitudinal oscillations. The gaps between these approximate analytical solutions are filled up by means of a numerical study.

062.051   **The effect of a magnetic field on the onset of thermal convection when constant flux boundary conditions apply.**   J. O. Murphy.
Proc. Astron. Soc. Australia, Vol. 3, 164 - 165 (1977).

062.052   **Hydromagnetic waves in magnetic flux tubes.**
P. R. Wilson.
Proc. Astron. Soc. Australia, Vol. 3, 173 - 174 (1977).

062.053   **Entropy and the spontaneous emission of plasma waves.**   R. J. M. Grognard.
Proc. Astron. Soc. Australia, Vol. 3, 174 - 177 (1977).

062.054   **Potential drops above pulsar polar caps: ultrarelativistic particle acceleration along the curved magnetic field.**   E. T. Scharlemann, J. Arons, W. M. Fawley.
Astrophys. J., Vol. 222, 297 - 316 (1978).
   The authors investigate the electrostatic acceleration of relativistic particles by pulsars. The model studied is the steady flow of nonneutral plasma along the narrow tube of curved open field lines, well within the light cylinder of a rotating, magnetized, isolated neutron star. They also investigate the effects of higher multipole moments in the magnetic field. The γ-ray luminosity of the neutron star without pair creation is briefly discussed, as well as the total energy available for particle acceleration and the total energy loss from the star.

062.055   **On the interaction of a weak-R type ionization front and a contact discontinuity.**
J. S. Berry.
Z. Naturforsch., Band 33a, 393 - 397 (1978).

062.056   **Time evolution of the energy spectrum of Alfvén waves due to the non-linear Landau damping.**
N. Tajima, A. Tomimatsu, R. Horiuchi.
Astron. Astrophys., Vol. 65, 239 - 243 (1978).
   By solving exactly the Lifshitz-Tsytovich equations, time evolution of the energy spectrum of Alfvén waves due to the non-linear Landau damping is investigated. It is shown that, irrespective of its initial value (from 1 to infinity), the index of the power law spectrum tends immediately to concentrate on a narrow region between $1.5 \sim 2$, and afterward it approaches slowly the final value 1. The energy decreases monotonously to a certain stationary value. These features are compatible with the observations of Alfvén waves in the solar wind.

062.057   **Acoustic waves in the solar atmosphere. IV. On the efficiency of one-dimensional hydrodynamic codes.**   R. Hammer, P. Ulmschneider.
Astron. Astrophys., Vol. 65, 277 - 277 (1978).
   In this paper the leapfrog type finite difference method and various forms of modified characteristics methods are compared for accuracy and efficiency. Although the choice of the method depends somewhat on the application, the authors found that the most accurate and the most efficient method is a modified characteristics method

with natural cubic spline interpolation.

062.058   **Peculiar lines of the magnetic field in a plasma and reconnection in pinch layers.**   S. I. Syrovatskij.
Izv. AN SSSR. Ser. fiz., Vol. 41, 1782 - 1789 (1977). In Russian. - Abstr. in Ref. zh., 62. Issled. kosm. prostranstva, 2.62.250 (1978).

062.059   **Self-similar flow for accretion with spherical symmetry taking into account the pressure gradient.**
Ya. M. Kazhdan, A. E. Lutskij.
Astrofizika, Vol. 13, 535 - 541 (1977). In Russian. English translation in Astrophysics, Vol. 13, No. 3.
   A self-similar time-dependent solution of the problems on spherical accretion in the gravitating field of a point mass is obtained. The pressure gradient is taken into account in the whole space. The correspondent solution is found for the value of adiabate exponent $\kappa = 5/3$.

062.060   **Thermal radiation from an optically thick plasma with a strong magnetic field.**
G. G. Pavlov, Yu. A. Shibanov.
Astron. Zh. Akad. Nauk SSSR, Vol. 55, 373 - 389 (1978). In Russian. English translation in Soviet Astron., Vol. 22, No. 2.
   Spectrum, polarization and beam pattern of radiation from a photosphere with a strong magnetic field are obtained. The case when the opacity is caused mainly by free—free transitions is considered. Some observational effects connected with a strong magnetic field in the photosphere of radiating objects are discussed: the cyclotron absorption resonances in spectra; the nonmonotonous dependence of radiation directivity on photon frequency; the sign reverse of circular polarization, the jump of the position angle of linear polarization at some wavelengths, etc. The results obtained are valid for interpretation of observations of hot magnetic white dwarfs. It is noted that linear polarization of some white dwarfs can be explained without a hypothesis on the presence of a very strong magnetic field in the coronae of these objects.

062.061   **Bremsstrahlung instability of relativistic electrons in a plasma.**   A. V. Akopyan, V. N. Tsytovich.
Astrofizika, Vol. 13, 717 - 730 (1977). In Russian. English translation in Astrophysics, Vol. 13, No. 4.
   Instability caused by bremsstrahlung of electromagnetic waves by relativistic electrons in a plasma has been investigated. Ultrarelativistic electron isotropic distribution is considered as well as radiation instability occurring by breaking of anisotropic electrons on the ions of a plasma. The electron energy and the radiation frequency which make the reabsorption coefficient negative are determined.

062.062   **The equilibrium statistical mechanics of a one-dimensional self-gravitational system.**
J. J. Monaghan.
Australian J. Phys., Vol. 31, 95 - 103 (1978).
   The configurational integral and the single-particle and pair distributions for a one-dimensional gravitating system are calculated using approximate methods. The Helmholtz free energy and single-particle distributions are in good agreement with the exact analytical results. The pair distribution is new and shows that, in the Vlasov limit, it is approximately the product of two one-particle distributions. The method may be easily extended to more general systems.

062.063   **A similarity model illustrating the effect of rotation on an inflated magnetosphere.**   C. Sozzou.
Planet. Space Sci., Vol. 26, 311 - 317 (1978).
   The effect of rotation on a magnetosphere containing a conducting fluid is investigated by means of a similarity solution. It is shown that rotation increases the current density in the equatorial region at the expense of that in the polar re-

gions. In the case of a compressible fluid it is shown that rotation can confine the plasma in a thin region about the equatorial plane where the centrifugal force is balanced by the Lorentz force. The solution is discussed in terms of the thin equatorial plasma layer observed in the outer magnetosphere of Jupiter.

062.064   Cyclotron instabilities in a hydrogen plasma.
J.-T. Horng.
Chinese J. Phys., Vol. 14, No. 4, p. 85 - 89 (1976). – Abstr. in Phys. Abstr., Vol. 81, Abstr. 1814 (1978).

062.065   Synchro-Compton radiation damping of relativistically strong linearly polarized plasma waves.
E. Asseo, C. F. Kennel, R. Pellat.
Astron. Astrophys., Vol. 65, 401 - 408 (1978).
The authors reconsider the Synchro-Compton effect for a nonlinear self-consistent linearly polarized plane wave of relativistic amplitude in an electron-positron plasma. The radiative damping is necessarily nonlinear and spatially inhomogeneous, since the wave phase velocity depends upon its amplitude. The authors find that radiation reaction would extinguish a pulsar wave in about $2 \times 10^{41}/\dot{W}\omega$ wavelengths, where $\dot{W}$ is the total energy outflow per second, and ω is the frequency.

062.066   Analytical representations of the parameters of a plasma moving adiabatically in a dipole magnetic field in the presence of longitudinal electrostatic fields.
V. G. Il'in, N. K. Osipov.
Kosm. Issled., Vol. 16, 308 - 310 (1978). In Russian.

062.067   Experimental studies on beam-plasma interaction.
Y. Kiwamoto.
Rep. IPPJ-286, (see 012.018), 4 pp. (1977). – Abstr. in Phys. Abstr., Vol. 81, Abstr. 5888 (1978).

062.068   Use of fast ion beams in plasma research.
H. Ishizuka.
Rep. IPPJ-286, (see 012.018), 6 pp. (1977). – Abstr. in Phys. Abstr., Vol. 81, Abstr. 5889 (1978).

062.069   Turbulent plasma phenomena in space and laboratory.   T. Kawabe.
Rep. IPPJ-286, (see 012.018), 4 pp. (1977). – Abstr. in Phys. Abstr., Vol. 81, Abstr. 5945 (1978).

062.070   The role of mean circulation in parity selection by planetary magnetic fields.   M. R. E. Proctor.
Geophys. Astrophys. Fluid Dyn., Vol. 8, 311 - 324 (1977). Abstr. in Phys. Abstr., Vol. 81, Abstr. 7566 (1978).

062.071   Adiabatic collapse of rotating gas clouds.
M. Takahara, K. Nakazawa, S. Narita, C. Hayashi.
Prog. Theor. Phys., Vol. 58, 536 - 548 (1977). – Abstr. in Phys. Abstr., Vol. 81, Abstr. 7875 (1978).

062.072   Current-driven instabilities in a laminar perpendicular shock.   D. S. Lemons, S. P. Gary.
J. Geophys. Res., Vol. 83, 1625 - 1632 (1978).
The linear theory of fully electromagnetic plasma instabilities driven by currents flowing across a magnetic field is investigated, with applications to the laminar perpendicular bow shock.

062.073   Theory of the pulsar atmosphere. II. Arbitrary magnetic and rotational axes; qualitative features.
E. A. Jackson.
Astrophys. J., Vol. 222, 675 - 688 (1978).
An investigation is made of the dominant guiding center motions of particles in the vacuum near-field region of a rotating, magnetized, and charged neutron star, with an arbitrary angle between the rotational and magnetic axes, to determine the qualitative features of the atmosphere formed by such fields.

062.074   Double layers in an infinite plasma.
A. I. Laptukhov.
Solar wind and the magnetosphere, (see 003.006), p. 101 - 112 (1976). In Russian. – Abstr. in Ref. zh., 62. Issled. kosm. prostranstva, 3.62.348 (1978).

062.075   An experiment on the interaction of a magnetized sphere with a supersonic plasma flow.
R. Kist, D. C. Agarwal.
J. Atmos. Terr. Phys., Vol. 40, (see 012.021), 389 - 394 (1978).
The authors started an experimental study on the interaction of a fast plasma beam with a magnetized sphere. The experiment reported in this paper relates to a case where the transition from $\beta > 1$ to $\beta \ll 1$ and $M_A > 1$ to $M_A \ll 1$ is achieved by reducing substantially the initial magnetization of a low density plasma beam. A pulsed magnetic dipole was exposed to an argon plasma beam of velocities ranging from 9.8 to 31 km s^{-1}.

062.076   On the number of unstable modes of an equilibrium.
J. Katz.
Mon. Not. R. Astron. Soc., Vol. 183, 765 - 769 (1978).
The number of unstable modes of an equilibrium may, in a variety of cases of astrophysical interest, be deduced from topological properties of continuous series of equilibria without having to solve an eigenvalue equation.

062.077   The polarization of second harmonic plasma emission.   D. B. Melrose, G. A. Dulk, S. F. Smerd.
Astron. Astrophys., Vol. 66, 315 - 324 (1978).
It is shown that the second harmonic plasma emission is partially polarized in the sense of the ordinary mode of magnetoionic theory only when the Langmuir waves are confined to a small range of angles ($<30°$) to the magnetic field lines. Consequently, Suzuki and Sheridan's observations of the polarization of harmonic Type III emission implies that (at least in the cases reported) the Langmuir waves must be nearly one-dimensional. For the observed polarization of Type III bursts, the implied magnetic field strengths are strong enough for induced scattering to cause the Langmuir waves to become nearly one dimensional, which is consistent with the observed sense of polarization.

062.078   On the theory of modulation instability of a relativistic plasma.   M. Khakimova, F. Kh. Khakimov, V. N. Tsytovich.
Dokl. AN TadzhSSR, Vol. 20, No. 8, p. 20 - 23 (1977). In Russian. – Abstr. in Ref. zh., 51. Astron., 4.51.178 (1978).

062.079   Current sheets and flares in cosmic and laboratory plasma.   S. I. Syrovatskij.
Vestn. AN SSSR. 1977, No. 10, p. 33 - 44. In Russian. Abstr. in Ref. zh., 51. Astron., 4.51.411 (1978).

062.080   "Reconnection" of magnetic force lines in a plasma.
S. I. Syrovatskij.
Priroda, 1978, No. 6, p. 84 - 92. In Russian.

062.081   Stability of aligned magnetoatmospheric flow.
J. A. Adam.
J. Plasma Phys., Vol. 19, Pt. 1, p. 77 - 86 (1978). – Abstr. in Phys. Abstr., Vol. 81, Abstr. 26197 (1978).

062.082   The interaction between a magnetized plasma flow and a magnetized celestial body: a review of magnetospheric studies.   S.-I. Akasofu.
Space Sci. Rev., Vol. 21, 489 - 526 (1978).

Through an intensive study of the magnetospheres of the Earth, Mercury, and Jupiter, the author has begun to understand how a magnetized celestial body interacts with a magnetized plasma flow. Some of the important findings are presented. Some of the findings may have significant implications in interpreting a variety of astrophysical processes associated with the magnetosphere of magnetic stars, pulsars and head-tail galaxies and also with transient processes, such as solar flares and flarings of particular types of variable stars, etc.

062.083 **On the Brinkley-Kirkwood method in magnetic gas dynamics.** S. A. Silich.
Astrometriya i Astrofizika, vyp. (No.) 35, p. 67 - 76 (1978). In Russian.

The Brinkley-Kirkwood method is applied to the magnetic gas dynamics case. The influence of a magnetic field on the kinematics of gas flows is taken into account, but the influence on the energetics of the processes taking place is neglected. Equations are obtained describing the shock wave motion in a flat magnetic and gravity-nonuniform atmosphere without taking into account energy loss due to ionization and radiation. The particular case of strong magnetic gas dynamics shock wave propagation in an isothermal atmosphere is discussed.

062.084 **Collisional contributions to the nonlinear current density of a turbulent plasma.**
R. Jancel, L. Stenflo.
Phys. Scr., Vol. 17, 533 - 534 (1978).

The collisional contributions to the second order nonlinear current density are calculated for a magnetized turbulent plasma in the "weak collisional regime".

062.085 **The stability of an anisotropic plasma jet.**
K. M. Srivastava.
Astrophys. Space Sci., Vol. 54, 263 - 277 (1978).

The modified Chew-Goldenberger-Low equations have been used to study the stability of a resistive anisotropic plasma jet surrounded by a non-conducting compressible gas. The dispersion relation has been obtained and discussed in three limiting situations. The conditions for instability have been obtained. It has been found that the resistivity introduces new modes which make the plasma jet overstable. In the limit of large wavelength disturbances, the jet with finite but high conductivity is found to be unstable. For small wavelength disturbances, the jet is found to be unstable.

062.086 **Double radio sources and the new approach to cosmical plasma physics.** H. Alfvén.
Astrophys. Space Sci., Vol. 54, 279 - 292 (1978).

The methodology of cosmic plasma physics is discussed. A summary is given of laboratory investigations of electric double layers, of the in situ measurements in the magnetosphere by which the importance of electric double layers in the Earth's surrounding is established. The scaling laws between laboratory and magnetospheric double layers are studied. A further extrapolation to galactic phenomena leads to a theory of double radio sources.

062.087 **Plasma-neutral gas interaction in cosmical physics.**
B. Lehnert.
Astrophys. Space Sci., Vol. 55, 25 - 37 (1978).

The influence of plasma-neutral gas interaction on cosmical phenomena is discussed, with applications to steady and non-steady phenomena in cool interstellar and interplanetary clouds, in the photospheres, chromospheres, prominences and filaments of the Sun and other stars, in the planetary ionospheres, as well as in certain cosmogonical theories. Special attention is given to filamentary structures of hot plasmas embedded in cooler regions, and to the internal properties of cool partially ionized clouds.

062.088 **A double layer review.** L. P. Block.
Astrophys. Space Sci., Vol. 55, 59 - 83 (1978).

A review is given of the main results on electrostatic double layers (sometimes called 'space charge layers' or 'sheaths') obtained from theory and laboratory and space experiments up to the spring of 1977. The paper begins with a definition of double layers in terms of potential drop, electric field, and charge separation. Then a review is made of the theoretical results obtained so far. Next, experimental results obtained in the laboratory are compared with the theoretical results. By means of barium jets and satellite probes, double layers have now been found at the altitudes that were previously predicted theoretically. The general potential distribution above the auroral zone, suggested by inverted V-events and electric field reversals, is corroborated.

062.089 **Kinetic envelope solitons in turbulent plasmas.**
B. Buti.
Astrophys. Space Sci., Vol. 55, 85 - 92 (1978).

A non-linear Schrödinger equation which characterizes the non-linear electrostatic waves in collisionless turbulent plasma is derived. Detailed analysis of this equation for the non-linear Langmuir waves is presented to show how the ion dynamics affects the envelope behaviour of these waves. Necessary condition for the existence of Langmuir envelope solitons is found.

062.090 **On Alfvén's critical velocity for the interaction of a neutral gas with a moving magnetized plasma.**
R. K. Varma.
Astrophys. Space Sci., Vol. 55, 113 - 124 (1978).

A theory for the interaction of a neutral gas with a moving magnetized plasma is given. The Alfvén expression for the critical velocity is identified with that for the terminal velocity while another expression for the threshold velocity for interaction is given. The implications of these results to the Alfvén–Arrhenius model for the solar system are discussed.

062.091 **The role of electrostatic instabilities in the critical ionization velocity mechanism.** M. A. Raadu.
Astrophys. Space Sci., Vol. 55, 125 - 138 (1978).

The role of electrostatic instabilities in the critical ionization velocity mechanism is investigated. The analysis is based on the theory developed by Sherman, which interprets Alfvén's critical velocity in terms of a circular process. A general expression for the energy and momentum of ions and electrons associated with an electrostatic mode is derived in terms of the plasma dielectric constant. This is used in the case of the modified two-stream instability to determine the distribution of energy between ions and electrons. An extrapolation from the linear phase then gives an estimate of the energy delivered to the electrons which is compared to that required to ionize the neutral gas.

062.092 **Experimental investigation of the critical ionization velocity in gas mixtures.** I. Axnäs.
Astrophys. Space Sci., Vol. 55, 139 - 146 (1978).

The present report summarizes the results of systematic experiments on the critical ionization velocity as a function of the mixing ratio for binary gas mixtures of H_2, He, N_2, O_2, Ne and Ar.

062.093 **Excited cosmic plasmas.** C. de Jager.
Astrophys. Space Sci., Vol. 55, 147 - 168 (1978).

The author describes some aspects of the energetic radiations of high-energy cosmical plasmas in stellar environments, mainly stellar chromospheres and coronae, and solar and stellar flare-type phenomena. As far as possible he discusses the morphology and physics of these plasmas, and speculates on their origin.

062.094 **Terrestrial and extraterrestrial plasmas.**

R. Lüst.
Astrophys. Space Sci., Vol. 55, 169 - 178 (1978).

Starting point for the discussion are some basic problems which have to be solved in plasma physics before a thermonuclear fusion reactor for the production of useful energy can be built. Phenomena of solar, interplanetary, and planetary plasma can indicate how extraterrestrial plasmas behave.

062.095   Plasma flow in a curved magnetic field.
        L. Lindberg.
Astrophys. Space Sci., Vol. 55, 203 - 225 (1978).

A beam of collisionless plasma is injected along a longitudinal magnetic field into a region of curved magnetic field. Two unpredicted phenomena are observed: The beam becomes deflected in the direction opposite to that in which the field is curved, and it contracts to a flat slab in the plane of curvature of the magnetic field.

062.096   Interaction of energetic particles with a shock wave front in a turbulent medium.
V. N. Vasil'ev, I. N. Toptygin, A. G. Chirkov.
Geomagn. Aehron., Vol. 18, 415 - 422 (1978). In Russian.

062.097   Low-frequency impedance of a flat disc in a magnetoactive plasma.   V. A. Panchenko, E. P. Smirnov.
Geomagn. Aehron., Vol. 18, 529 - 531 (1978). In Russian.

062.098   Geometrical MHD wave coupling.
        J. V. Hollweg, C. G. Lilliequist.
J. Geophys. Res., Vol. 83, 2030 - 2036 (1978).

Refraction and/or magnetic field curvature can lead to 'geometrical mode coupling' between the MHD modes. The effect is linear, but it is a finite-wavelength effect. It is studied here for a simple configuration which is amenable to analysis and which illustrates the basic features of the coupling. In the solar wind the geometric coupling may be operative only in the solar corona or in the interaction regions of high-speed streams. In the latter case the geometrical coupling may provide an explanation for the non-Alfvénic fluctuations but only for long-period waves (greater than several hours) in small interaction regions (< 0.1 AU).

062.099   A fluid description of a hot collisionless plasma and limits to the double-adiabatic approximation associated with the presence of Birkeland currents.
A. Hruška.
J. Geophys. Res., Vol. 83, 2208 - 2210 (1978).

The large-scale structure of a slowly moving plasma is discussed, a modified Chew-Goldberger-Low approximation being used. The double-adiabatic relations for pressure components fail if there is a nonzero current flowing perpendicularly to the binormals to field lines.

062.100   Comment on 'The signature of parallel electric fields in a collisionless plasma' by E. C. Whipple, Jr.
W. Lennartsson, with a reply by E. C. Whipple.
J. Geophys. Res., Vol. 83, 2228 - 2229 (1978).

062.101   Some problems of the nearly symmetric dynamos.
        I. Cupal.
Magnetic stars, (see 012.028), p. 38 - 39 (1978). – Abstract.

062.102   The pumping effect in mean-field magnetohydrodynamics.   F. Krause.
Magnetic stars, (see 012.028), p. 41 - 42 (1978). – Abstract.

062.103   Magnetodynamics mechanism of magnetic stars.
        E. Woyk (Chvojková).
Magnetic stars, (see 012.028), p. 44 - 45 (1978). – Abstract.

062.104   Operator algebra and the stationary states of

stellar magnetic fields.   R. H. Dicke.
Astrophys. Space Sci., Vol. 55, 275 - 283 (1978).

A vector-operator algebra technique for solving magnetic field problems in a toroidal/poloidal representation is illustrated with physical examples. Among the illustrative examples are calculations of necessary and/or sufficient conditions for the existence of stationary magnetic fields in stellar interiors.

062.105   On the stability of helical velocity and magnetic fields.   R. K. Kochhar.
Astrophys. Space Sci., Vol. 55, 395 - 400 (1978).

In this paper the author studies the stability of an infinitely conducting, incompressible, inviscid infinite cylinder with non-parallel helical velocity and magnetic field. It is shown that the system is stable if the energy in the ϕ-component of the velocity field is larger than that in the ϕ-component of the magnetic field.

062.106   On magnetothermoelastic Rayleigh waves with thermal relaxation.   D. S. Chandrasekharaiah.
Gerlands Beitr. Geophys., Band 87, 221 - 228 (1978).

Magnetothermoelastic Rayleigh waves in a perfectly conducting half-space with stress-free boundary are studied by using the equations of electromagnetism and generalized thermoelasticity.

062.107   Double layers.   L. P. Block.
        Physics of the hot plasma in the magnetosphere, (see 012.030), p. 229 - 249 (1975).

The properties of double layers as observed in laboratory gaseous discharges are reviewed. Theories based on macroscopic equations are compared with those based on the coupled Maxwell-Vlasov equations. Two mechanisms for the power supply to double layers are described, and possible electrostatic field configurations in the auroral ionosphere-magnetosphere system are demonstrated.

062.108   A difficult two-dimensional hydrodynamics problem.   E. M. Jones.
Bull. American Astron. Soc., Vol. 10, 434 (1978). -- Abstract.

062.109   Thomson cross sections in a strong magnetic field.
        P. Mészáros, G. Börner.
Bull. American Astron. Soc., Vol. 10, 447 - 448 (1978). Abstract.

062.110   Kelvin-Helmholtz instability for a thin plasma layer in a magnetic field.   V. M. Lipunov.
Astron. Tsirk., No. 993, p. 1 - 2 (1978). In Russian.

062.111   Freezing-in condition for a magnetic field and current sheets in plasma.   S. I. Syrovatskii (*Syrovatskij*).
Astrophys. Space Sci., Vol. 56, 3 - 12 (1978).

In the ideal magnetohydrodynamic approximation it is shown that for physically permissible boundary conditions there may exist some lines on which the freezing-in condition is not valid. Such singular lines are closed magnetic lines of force and lines with both ends on the boundary surface. By analogy with the singular lines of a potential magnetic field the conclusion is made that X-type singular lines are the place where current sheets (sheet pinches) appear in plasma, whereas on O-type singular lines quasi-cylindrical pinches of a usual type appear.

062.112   On the particle number conservation inherent in the diffusion approach of the Fokker-Planck equation.
J. E. Kunstmann.
Astrophys. Space Sci., Vol. 56, 81 - 87 (1978).

Starting from the Fokker-Planck equation the author shows that, in a strict sense, the usual diffusive transport of charged particles in weak magnetic field fluctuations is not valid until the space-integrated number of particles per velocity interval has reached its final overall constant value and is conserved. Large anisotropies are thus not compatible with diffusion. Diffusion becomes exactly valid after an infinite time, but has reached a good approximation to the real transport before that. The particle-number conservation holds for quite general mean magnetic field configurations, and is not limited to the case of a constant mean field.

062.113 The Alfvén-Carlquist double-layer theory of solar flares. S. S. Hasan, D. Ter Haar.
Astrophys. Space Sci., Vol. 56, 89 - 107 (1978).

The authors use the Vlasov equations for ions and electrons to develop a theory of a double layer in which there are both free and trapped electrons and ions. They discuss the applicability of this theory to solar flares, and show that conditions in solar flares may be such that double layers can exist for which the free particles have a power-law energy distribution. These particles will be accelerated in a double layer and may in this way account for the production of high-energy particles during the impulsive phase of solar flares.

062.114 Propagation of plane relativistic shock waves in the presence of a magnetic field. G. Deb Ray, T. K. Chakraborty.
Astrophys. Space Sci., Vol. 56, 119 - 128 (1978).

The authors obtain similarity solutions for the propagation of plane relativistic shock waves in the presence of a transverse magnetic field for the medium, where the nucleon number density obeys a power law of distance from the plane of explosion. The shock surface moves with constant velocity and the total energy of the disturbance is dependent on time. The solutions are applicable only to an isothermal medium or a cold gas.

062.115 Models for convectively driven hydromagnetic dynamos. L. Baker.
Geophys. Astrophys. Fluid Dyn., Vol. 9, 257 - 278 (1978).
Abstr. in Phys. Abstr., Vol. 81, Abstr. 42668 (1978).

062.116 Magnetohydrodynamical instability of a neutral layer. N. V. Erkaev.
Mathematical models of the near space, 1977, (see 003.015), p. 157 - 161. In Russian. – Abstr. in Ref. zh., 62. Issled. kosm. prostranstva, 6.62.265 (1978).

062.117 Short period acoustic heating theory and its application to the construction of model chromospheres. P. Ulmschneider.
Mem. Soc. Astron. Italiana, Vol. 48, (see 012.038), 439 - 474 (1977).

062.118 Behaviour of a rotating electrically conducting sphere in a time-independent homogeneous external magnetic field (cosmic approximation). E. Schmutzer.
Acta Phys. Polonica B, Vol. B9, 303 - 314 (1978). – Abstr. in Phys. Abstr., Vol. 81, Abstr. 49309 (1978).

062.119 On tidal motions of matter of nonconstant density. V. P. Kulesh, V. B. Rzhonsnitskij.
Vestn. LGU, 1977, No. 24, p. 100 - 111. In Russian. – Abstr. in Ref. zh., 51. Astron., 6.51.186 (1978).

Non-LTE line transfer with diffusion of excited atoms II.   See Abstr. 022.014.

Long-range potentials and Stark broadening of neutral lines.   See Abstr. 022.055.

Calculated X-radiation from optically thin plasmas. III. Abundance effects on continuum emission. See Abstr. 022.071.

Convective dynamos.   See Abstr. 061.012.

Plasma astrophysics.   See Abstr. 061.053.

Propagation of low frequency electromagnetic wave in a relativistic electron gas.   See Abstr. 063.008.

Thermodynamic inconsistency of the modified Saha equation at high pressures.   See Abstr. 064.011.

Thickening of axisymmetric accretion disks and flows.   See Abstr. 064.024.

A semitheory for semiconvection. See Abstr. 065.014.

Wave transport in stratified, rotating fluids. See Abstr. 065.027.

Wave generation and pulsation in stars with convective zones.   See Abstr. 065.028.

The hydrodynamics of stellar collapse. See Abstr. 065.033.

Secular instability of rotating Newtonian stars. See Abstr. 065.051.

Exact MHD solutions for rotating, magnetic stars. See Abstr. 065.070.

On the Cauchy problem for the general relativistic perfect magnetofluid.   See Abstr. 066.196.

Tunneling of Alfvén waves in sunspots. See Abstr. 072.024.

Magnetohydrodynamics of atmospheric transients. II. Two-dimensional numerical results for a model solar corona See Abstr. 074.002.

Plasma instabilities of trapped particles in solar magnetic fields.   See Abstr. 077.019.

Nonlinear Langmuir waves during type III solar radio bursts.   See Abstr. 077.031.

Nonlinear astrophysical dynamos: the solar cycle as a nonlinear oscillation of the general magnetic field driven by the nonlinear dynamo and the associated modulation of the differential-rotation–global-convection system. See Abstr. 080.011.

The flux-rope-fibre theory of solar magnetic fields. See Abstr. 080.049.

Composition of the hot plasmas in the magnetosphere.   See Abstr. 084.313.

Hot plasma dynamics within geostationary altitudes. See Abstr. 084.316.

VLF electrostatic waves in the magnetosphere. See Abstr. 084.318.

A hydrodynamic study of a slow nova outburst. See Abstr. 124.008.

Interstellar clouds and the formation of stars. See Abstr. 131.219.

Magneto-parametric instabilities in the Crab

Nebula: II.   See Abstr. 134.026.

Wave production in an ultrarelativistic electron-positron plasma.   See Abstr. 141.502.

Pulse distortion in a shearing plasma.
See Abstr. 141.546.

Errata

062.901   **Errata: "Enhancement of thermonuclear reaction rate due to strong screening"** [Astrophys. J., Vol. 218, 477 - 483 (1977)].
N. Itoh, H. Totsuji, S. Ichimaru.
Astrophys. J., Vol. 220, 742 (1978).

063 Radiative Transfer, Scattering

063.001   **A Monte Carlo study of frequency redistribution in an externally excited medium.**
R. R. Meier, J.-S. Lee.
Astrophys. J., Vol. 219, 262 - 273 (1978).
   The effects of partial frequency redistribution on the transport of resonance radiation through a medium consisting of two-level atoms have been examined for an external continuum source. Models for optical depths up to 10^4 were considered by utilizing new methods of generating the redistribution function and a statistical technique for computing the intensity. The results show significant departures from complete frequency redistribution for source functions and intensities inside the medium. The emission line profiles also exhibit peculiar features as the result of both partial frequency redistribution and excitation from an external source.

063.002   **Solution of the comoving-frame equation of transfer in spherically symmetric flows. V. Multilevel atoms.**
D. Mihalas, P. B. Kunasz.
Astrophys. J., Vol. 219, 635 - 653 (1978).
   Methods that were developed in earlier papers of this series for solving the equation of transfer in the comoving fluid frame in spherically symmetric flows are extended and applied to the calculation of multiline spectra from multilevel, multi-ion model atoms. This direct solution of the coupled transfer and statistical equilibrium equations avoids the approximations inherent in the Sobolev method, and is thus applicable throughout a trans-sonic wind, in which velocities are small in the deeper atmospheric layers but rise to very large terminal velocities at great distances from the star. The results show that subordinate line profiles are sensitive both to the assumed mass-loss rate, and to the assumed structure of the velocity law in the atmosphere.

063.003   **A generalization of the Sobolev method for flows with nonlocal radiative coupling.**
G. B. Rybicki, D. G. Hummer.
Astrophys. J., Vol. 219, 654 - 675 (1978).
   The Sobolev, or escape-probability, method for solving radiative transfer problems in moving atmospheres is generalized to treat flows in which the line-of-sight component of the flow velocity is not monotonic; for these cases the purely local nature of the approximation is lost, and radiative coupling between distant parts of the atmosphere must be taken into account. The method is formulated for a general three-dimensional flow. Extensive numerical results for inverse power-law velocity fields are presented to illustrate the magnitude of the coupling between different parts of the atmosphere.

063.004   **Radiative transfer calculated from a Markov chain formalism.**   L. W. Esposito, L. L. House.
Astrophys. J., Vol. 219, 1058 - 1067 (1978) = Contrib. Five Coll. Astron. Dep., Amherst, Mass., No. 244.
   The theory of Markov chains is used to formulate the radiative transport problem in a general way by modeling the successive interactions of a photon as a stochastic process. Under the minimal requirement that the stochastic process is a Markov chain, the determination of the diffuse reflection or transmission from a scattering atmosphere is equivalent to the solution of a system of linear equations. The authors have verified the speed and accuracy of this formalism for the standard problem of finding the intensity of scattered light from a homogeneous plane-parallel atmosphere with an arbitrary phase function for scattering.

063.005   **Radiative transfer effect in an ionized medium at high temperature.**   S. Miyamoto.
Astron. Astrophys., Vol. 63, 69 - 81 (1978).
   The spectra of radiation emitted from hot plasmas of moderate optical thicknesses are calculated using a numerical calculation method. The X-ray spectra emitted from a spherical hot plasma and from a shell-form plasma are computed. The computed spectra are compared with free-free radiation spectra. The effects of the surrounding plasma of different temperatures on the spectra are computed and discussed.

063.006   **Radiation scattering for a general law of frequency redistribution.**   H. A. Haruthyunian (G. A. Arutyunyan), A. G. Nikoghossian (Nikogosyan).
J. Quant. Spectrosc. Radiat. Transfer, Vol. 19, 135 - 148 (1978).
   Some relatively simple radiative-transfer problems are considered in the general case of non-coherent scattering. The approach proposed in the paper is based on representation of the redistribution function in the form of a bilinear expansion with respect to an appropriately orthonormalized set of functions. Numerical results are given for profiles of spectral lines formed by diffuse reflection from a semi-infinite medium, as well as for those formed in an isothermal atmosphere. These results illustrate the influence of the various approximations for the redistribution law on the solution of the problem.

063.007   **The differential equations of the Voigt function.**   T. Andersen.
J. Quant. Spectrosc. Radiat. Transfer, Vol. 19, 169 - 171 (1978).
   The three second-order partial differential equations of the Voigt function are presented, leading to a powerful and accurate method of determining the Voigt function in the calculation of a line profile in a stellar atmosphere.

063.008   **Propagation of low frequency electromagnetic wave in a relativistic electron gas.**   H.-q. Zhang.
Acta Astron. Sinica, Vol. 18, 221 - 226 (1977). In Chinese.
   In this paper the propagation of low frequency electromagnetic wave with large amplitude in a magnetized plasma is discussed. It is shown that penetration of magnetic-dipole radiation in the medium of the Crab nebula seems to be

possible.

063.009   Two-dimensional radiative transfer. I. Planar geometry.
D. Mihalas, L. H. Auer, B. R. Mihalas.
Astrophys. J., Vol. 220, 1001 - 1023 (1978).

Differential-equation methods for solving the transfer equation in two-dimensional planar geometries are developed. One method, which uses a Hermitian integration formula on ray segments through grid points, proves to be extremely well suited to velocity-dependent problems. An efficient elimination scheme is developed for which the computing time scales linearly with the number of angles and frequencies; the authors are thus able to treat problems with large velocity amplitudes accurately. A very accurate and efficient method for performing a formal solution is also presented. A discussion is given of several examples of periodic media and freestanding slabs, both in static cases and with velocity fields.

063.010   Two-dimensional isotropic scattering in a semi-infinite medium.   A. L. Crosbie, T. L. Linsenbardt.
J. Quant. Spectrosc. Radiat. Transfer, Vol. 19, 257 - 284 (1978).

Exact results are presented for the source function, radiative flux, and intensity at the boundary of a two-dimensional, isotropically scattering, semi-infinite medium subjected to collimated or diffuse radiation. The spatial distributions of incident radiation considered are (1) cosine-varying, (2) semi-infinite step, (3) step at the origin and (4) finite strip. Two-dimensional effects are most pronounced at large albedos.

063.011   Expansions for two-dimensional radiative transfer in an isotropically scattering semi-infinite medium.
A. L. Crosbie.
J. Quant. Spectrosc.Radiat. Transfer, Vol. 19, 285 - 301 (1978).

Small spatial frequency expansions for the source function and radiative flux are obtained for a purely scattering, semi-infinite, two-dimensional medium. Both collimated and diffuse boundary conditions are analyzed. With these expansions, other expansions are obtained which are valid at large optical distances away from the incident radiation. Expansions are presented for a finite strip, circular disk and a Gaussian distribution.

063.012   Optically-thick X-ray transfer: the shell game.
S. H. Langer, R. R. Ross, R. McCray.
Bull. American Astron. Soc., Vol. 9, 626 (1978). – Abstract.

063.013   The effect of negative ions on collision-dominated Thomson scattering.   J. D. Mathews.
J. Geophys. Res., Vol. 83, 505 - 512 (1978).

An expression for the collision-dominated Thomson scatter power spectrum is derived from the fluid or continuum equations. This expression includes the possibility of multiple ion species of arbitrary charge. The total power and power spectral shape are investigated for scattering from a plasma composed of negative ions, positive ions, and electrons. The results indicate that total scattered power is modified by negative ion presence and that a limited amount of information about the negative ion content of the ionospheric D region may be found from Thomson scatter experiments.

063.014   Radiative transfer in atmospheres with spherical symmetry – II. The conservative Milne problem.
E. Simonneau.
J. Quant. Spectrosc. Radiat. Transfer, Vol. 19, 439 - 450 (1978).

A new method of successive approximations is developed to solve transfer problems in spherical coordinates. It is equivalent to methods involving the introduction of a closure rela-

tion for the μ-moment system. From a practical point of view, it is a two-direction method with variable quadrature weights which are easily computable for any depth variation of the opacity.

063.015   Continued fraction expansions in scattering theory and statistical non-equilibrium mechanics.
P. Hänggi, F. Rösel, D. Trautmann.
Z. Naturforsch., Band 33a, 402 - 417 (1978).

063.016   Fermi acceleration and particle pitch angle scattering.   J. S. Scott, W. J. Cocke, R. A. Chevalier, D. G. Wentzel.
Astrophys. Space Sci., Vol. 53, 421 - 427 (1978).

The authors suggest that sharp velocity gradients will exist in fluid-like turbulence in nearly collisionless plasma. This implies effective quenching of Fermi acceleration of thermal particles, but the Fermi acceleration coefficient for relativistic particles remains essentially unchanged.

063.017   A note on the H-functions of transfer problems in multiplying media.   S. R. Das Gupta.
Astrophys. Space Sci., Vol. 53, 517 - 522 (1978).

063.018   On the standard problem of the theory of radiative transfer.   V. V. Ivanov.
Astrofizika, Vol. 13, 505 - 515 (1977). In Russian. English translation in Astrophysics, Vol. 13, No. 3.

The standard problem of the calculation of the radiation field in an isotropically scattering semi-infinite atmosphere with exponentially distributed primary sources is considered.

063.019   Linear Fredholm integral equations for radiative transfer problems in finite plane parallel media. I. Imbedding in an infinite medium.   H. Domke.
Astron. Nachr., Band 299, 87 - 93 (1978).

Linear Fredholm integral equations are derived for the Stokes vector of the radiation emerging from a scattering plane parallel medium of finite optical thickness. The integral equations are obtained by means of imbedding the slab in an infinite medium. They are formulated in terms of Green's function matrices and renormalized for the asymptotic eigenmode. Explicitly, linear integral equations are given for the reflection and transmission matrices. The reciprocity principle is employed to obtain integral equations also for the mean intensity of the inner radiation field in the case of the slab albedo problem.

063.020   Linear Fredholm integral equations for radiative transfer problems in finite plane parallel media. II. Imbedding in a semi-infinite medium.   H. Domke.
Astron. Nachr., Band 299, 95 - 102 (1978).

By means of imbedding in a semi infinite medium linear Fredholm integral equations are derived for transfer problems of polarized radiation in a scattering slab of finite optical thickness. The integral equations are formulated in terms of surface Green's function matrices and renormalized for the zeroth azimuthal Fourier component relative to the asymptotic mode. Integral equations for the Stokes vector of the emergent radiation for any distribution of primary sources as well as integral equations for the inner radiation field of the slab albedo problem are obtained. Asymptotic formulae are given describing the radiation field for slabs of great optical thickness.

063.021   The effects of small-angle scattering on a pulse of radiation with an application of X-ray bursts and interstellar dust.   C. Alcock, S. Hatchett.
Astrophys. J., Vol. 222, 456 - 470 (1978).

The authors have analyzed in detail the general problem of the propagation of a pulse of radiation through a uniform medium containing small-angle scatterers. Simple expressions for the first moment of the arrival time delay and for the second and fourth moments of the arrival angle of the pulse photons are given that are valid for any optical depth. Using the first two of these, the authors find a simple method of distance determination. Joint and individual distribution functions for arrival time and angle are derived in the small and large optical depth limits. An exponential tail in the arrival-time distribution is characteristic in both limits. The analysis has been applied to a burst of X-rays from a source near the galactic center propagating through the dust of the interstellar medium. The authors derive useful constraints on the grain distribution from observations of the burst tails.

063.022   Radiative transfer in a uniformly nongray layer of gas.   A. Heinlo.
Tartu Astron. Obs., Teated No. 56, p. 3 - 36 (1978). In Russian.

Radiative transfer in a nongray plane-parallel layer where the opacity probability distribution function does not change along the spectra and with optical depth is considered. Studying the asymptotic behaviour of the resolvent and X- and Y-functions it turns out that there exists a straight relation between the functional dependence of these quantities and that of the opacity distribution function. It is shown that there exists an essential difference in the temperature structure of a nongray layer compared with that of a gray one.

063.023   Radiative transfer in a finite slab by kernel approximation.   A. Heinlo, T. Viik.
Tartu Astron. Obs., Teated No. 56, p. 37 - 54 (1978).

The integral equation for the resolvent in a finite slab is solved by a kernel approximation method. A Gauss-Christoffel quadrature is used to provide the best fit to the kernel by a finite sum of exponents, resulting in high-accuracy approximate formulas to calculate the X- and Y-functions and their moments which are applicable even to problems of nongray radiative transfer.

063.024   On the equality of the mean escape probability and mean net radiative bracket for line photons.
F. E. Irons.
Mon. Not. R. Astron. Soc., Vol. 182, 705 - 709 (1978).

It is pointed out that while the direct escape probability and net radiative bracket may not be equal at individual points in an isolated source, the mean values of these quantities (obtained by weighting with the source function and volume averaging) are identically equal, irrespective of the state of the source.

063.025   Resonance-line polarization. III. The Hanle effect in a compact non-LTE radiative transfer formulation.   J. O. Stenflo.
Astron. Astrophys., Vol. 66, 241 - 248 (1978).

A non-LTE theory for the Hanle effect (magnetic-field sensitive coherence effects in resonance-line scattering) is developed. The theory allows polarization calculations for multi-level atoms in arbitrary optically thick media, and includes the effect of collisions. The Hanle effect seems to offer a unique possibility of future mappings in three dimensions of the vector magnetic field in the solar chromosphere and corona.

063.026   Scattering of light from particulate surfaces. I. A laboratory assessment of multiple-scattering effects.
J. Veverka, J. Goguen, S. Yang, J. Elliot.
Icarus, Vol. 34, 406 - 414 (1978).

A convenient photometric function for many particulate surfaces is the generalization of the Lommel-Seeliger law derived by Hapke (1963) and Irvine (1966). This generalization accounts for the effects of mutual shadowing among particles, but still assumes that multiple scattering within the surface layer can be neglected – an assumption which is evidently

valid for dark surfaces. The authors describe a series of laboratory measurements which test the range of validity of this basic assumption, and the applicability of the Hapke-Irvine photometric function, for particulate surfaces whose normal reflectances range from 0.04 to 1.04. They find that multiple-scattering effects can be neglected, and that the Hapke-Irvine function can be used, for particulate surfaces whose normal reflectance is about 0.3, or less. The function is definitely inapplicable to surfaces whose normal reflectance exceeds 0.4.

063.027   Optically thick X-ray transfer: the shell game.
S. H. Langer, R. R. Ross, R. McCray.
Astrophys. J., Vol. 222, 959 - 966 (1978).

The authors investigate the radiative transfer of X-rays through a shell that is optically thick to Compton scattering, surrounding a point source of continuum X-rays. The emission and absorption of X-rays due to K-shell transitions of iron are included. The calculations are done in two entirely independent ways: by Monte Carlo simulation, and by solving a Fokker-Planck diffusion equation. The authors also develop a procedure for treating models of compact X-ray sources consisting of incomplete shells.

063.028   On the passage of radiation through moving astrophysical plasmas.   H. C. Ko, C. W. Chuang.
Astrophys. J., Vol. 222, 1012 - 1019 (1978).

The authors discuss the effect of the motion of a refracting medium on the direction of refracted rays, and its implication in astrophysics. It is assumed that, in its own rest frame, the medium is homogeneous, isotropic, lossless, and transparent. A pencil beam of quasi-monochromatic radiation is incident normally upon the surface of a medium moving at uniform velocity in the direction parallel to the surface. For a medium which has only frequency dispersion and satisfies the Landau and Lifshitz transparency condition, it is shown that the refracted beam cannot be bent toward the upstream direction whether the refractive index is greater or smaller than unity. For a nonrelativistic, warm isotropic plasma which has spatial dispersion in addition to the frequency dispersion, the refracted beam is bent toward the upstream direction. These results disagree with the conclusions reached by earlier investigators. The causes for the discrepancy are discussed. Based on the wave kinematical and electrodynamical considerations, the authors have shown that for a plane, quasi-monochromatic wave propagating in moving homogeneous dispersive media of infinite extent, the direction of energy flow coincides with that of the group velocity, but not with that of the Poynting vector.

063.029   On the passage of radiation through inhomogeneous, moving media. XIII. Causality, analyticity, group speed, transparency, and allied phenomena with emphasis on some recent calculations by Ko and Chuang.
I. Lerche.
Astrophys. J., Vol. 222, 1020 - 1031 (1978).

A critical examination is made of some recent calculations by Ko and Chuang (see 063.028).

063.030   Multi-dimensional non-LTE transfer of polarized radiation in magnetic flux tubes.
L. G. Stenholm, J. O. Stenflo.
Astron. Astrophys., Vol. 67, 33 - 45 (1978).

Calculations of multi-dimensional transfer of the Stokes vector in magnetic flux tubes are presented. The Zeeman splitting caused by the magnetic field modifies the multi-dimensional effects in different ways: the line opacities and the probabilities of photon escape in various directions and at various frequencies are changed. These effects are considered in detail for all the Stokes parameters. It is shown how the multi-dimensional effects influence the relation between the apparent field strength observed with a solar magnetograph and the true field strength. The line-ratio method used to derive true field strengths from simultaneous magnetograph recordings in two spectral lines of different Landé factor is tested.

063.031   On coherent scattering in stellar atmospheres.
S. R. Das Gupta.
Phys. Lett. A, Vol. 64A, 342 - 344 (1977). – Abstr. in Phys. Abstr., Vol. 81, Abstr. 28667 (1978).

063.032   Generation of very-high-velocity satellite-features through stimulated Raman scattering of 22.2 GHz H_2O maser-lines in compact H II plasma regions. One-dimensional model.   J. C. Fernandez, G. Reinisch.
Astron. Astrophys., Vol. 67, 163 - 174 (1978).

The authors study a particular kind of stimulated Raman scattering which concerns the plasma of compact H II clouds associated with the masering sources. They examine with particular emphasis the case of extreme high-radial-velocity features (at least 100 km s^{-1}) and obtain a value of the plasma density in the Raman-active part of the H II region which is about 10^6 cm^{-3}.

063.033   Full angle and frequency redistribution effects on the formation of spectral lines.   S. J. McKenna.
Bull. American Astron. Soc., Vol. 10, 457 (1978). – Abstract.

063.034   Monte Carlo simulation of light transfer in circumstellar dust shells.   J.-Y. Daniel.
Astron. Astrophys., Vol. 67, 345 - 353 (1978).

The author presents a numerical analysis of a radiation transfer problem for polarized light, in a context where the scattering source term is the only contributing factor. In this particular case, the scattering is rather anisotropic, and several scatterings occur in the medium: a Monte Carlo simulation turns out to be a convenient way to treat the problem. The aim of the present work is to explore certain phenomena related to multiple scattering, rather than to investigate particular objects, since the simulation can account for the properties of very polarized stars but not in an unambiguous way. This work has also led to a number of interesting results concerning unpolarized light.

063.035   Integral theory of radiative heat transfer with anisotropic scattering and general boundary conditions.   V. C. Boffi, G. Spiga.
J. Math. Phys., Vol. 18, 2448 - 2455 (1977). – Abstr. in Phys. Abstr., Vol. 81, Abstr. 32339 (1978).

063.036   Numerical methods for Mie theory of scattering by a sphere.   G. A. Shah.
Kodaikanal Obs. Bull., Ser. A, Vol. 2, 42 - 63 (1977).

A new procedure for evaluating the basic quantities connected with the Mie theory of scattering by a sphere is presented. Some independent numerical techniques are developed to enable one to achieve the desired accuracy. The method is quite general in the sense that it works for arbitrary but finite values of real, pure imaginary or complex indices of refraction and the size-to-wavelength parameter.

063.037   An exact solution of transfer equations for interlocked multiplets.   S. R. Das Gupta.
Astrophys. Space Sci., Vol. 56, 13 - 20 (1978).

The transfer equations for non-coherent scattering arising from interlocking of principal lines without redistribution are exactly solved in a very simple way by Laplace transform and Wiener-Hopf technique which are easily applied by the use of the new representation of H-functions obtained recently by the author (1977). The emergent intensity in the rth line is expressed in terms of an H-function and a Cauchy type integral admitting of closed form evaluation.

063.038   A new technique for exact and unique solution of transfer equations in finite media.

S. R. Das Gupta.
Astrophys. Space Sci., Vol. 56, 21 - 49 (1978).

The author develops a new method, combined with Laplace transformation and Wiener-Hopf technique, to obtain unique solutions of transport equations in finite media. For this purpose he considers the simple transfer equation for diffuse reflection by a plane-parallel finite atmosphere scattering radiation with moderate anisotropy. It is transformed, by Laplace transformation, into two coupled linear integral equations which are then reduced to two uncoupled Fredholm integral equations admitting of unique solutions by the method of iteration for values of the breadth of the atmosphere greater than that specified, depending on the scattering process.

063.039   The range of validity of the Rayleigh and Thomson limits for Lorenz-Mie scattering.
M. Kerker, P. Scheiner, D. D. Cooke.
J. Opt. Soc. America, Vol. 68, 135 - 137 (1978).

The small-particle limit for scattering by spheres (Rayleigh scattering) has an upper limit on the refractive index. Also, the limiting expression for perfectly reflecting spheres has a lower limit on the refractive index. These limits are delineated here for both absorbing and nonabsorbing spheres.

063.040   Approximate method for radiative transfer in scattering absorbing plane-parallel media.
Y. S. Chou.
Appl. Opt., Vol. 17, 364 - 373 (1978).

063.041   An adjoint Monte Carlo treatment of the equations of radiative transfer for polarized light.
L. L. Carter, H. G. Horak, M. T. Sandford,II.
J. Comput. Phys., Vol. 26, 119 - 128 (1978). – Abstr. in Phys. Abstr., Vol. 81, Abstr. 37799 (1978).

063.042   Use of the method of discrete ordinates for the solution of the transfer equation in the case of an inhomogeneous planetary atmosphere.
L. V. Zasova, E. A. Ustinov.
Inst. kosm. issled. AN SSSR. Prepr., 1977, No. 377, 19 pp.
In Russian. – Abstr. in Ref. zh., 51. Astron., 7.51.166 (1978).

063.043   Non-LTE diagnostics.   A. Skumanich.
Mem. Soc. Astron. Italiana, Vol. 48, (see 012.038), 375 (1977). – Summary.

063.044   Radiative transfer in geometries other than plane-parallel layers.   L. E. Cram.
Mem. Soc. Astron. Italiana, Vol. 48, (see 012.038), 377 - 399 (1977).

Studies of several aspects of radiation transfer in a stellar atmosphere have generally taken place within the confines of plane-parallel or sherically-symmetric geometry. In this paper the author discusses some consequences of introducing more complex and more realistic geometries into the study of astrophysical radiation transfer. He discusses both fine-structure and mean-value analyses. He defines basic concepts and classifies multi-dimensional radiation transfer studies into three classes. Each of the three classes is then considered in some detail.

063.045   Characteristics for scattering of unpolarized light for gamma-distribution of spherical particles. Results of calculations.
V. P. Shari, N. L. Lukashevich.
Inst. prikl. mat. AN SSSR. Prepr., 1977, No. 105, 64 pp. In Russian. – Abstr. in Ref. zh., 51. Astron., 7.51.164 (1978).

Non-LTE line formation in a magnetic field. The two-level atom with a frequency independent source function. I. Formulation.   See Abstr. 022.082.

The scattering of electromagnetic waves from density fluctuations in a plasma.   See Abstr. 062.045.

Radiative shock dynamics. II. Hydrogen continua. See Abstr. 064.021.

Determining velocity gradients from differential Doppler shifts.   See Abstr. 064.064.

Numerical applications to radiative transfer in expanding envelopes: the two-level atom model. See Abstr. 064.065.

Departures from radiative equilibrium in stellar atmospheres. Grey absorption.   See Abstr. 064.102.

Raman scattering in the atmospheres of the major planets.   See Abstr. 091.005.

Atmospheric composition via Raman scattering. See Abstr. 091.021.

Raman scattering in the atmospheres of the major planets.   See Abstr. 091.022.

Radiative transfer in spherical shell atmospheres. See Abstr. 091.023.

Spectral line shapes in spherically symmetric radially moving clouds.   See Abstr. 131.009.

064 Stellar Atmospheres, Stellar Envelopes, Mass Loss

064.001 Iron hydride in stellar atmospheres.
J. R. Mould, S. Wyckoff.
Mon. Not. R. Astron. Soc., Vol. 182, 63 - 68 (1978).

Calculations are presented of the FeH column density in late-type stellar atmospheres. The calculations indicate a strong gravity dependence in the FeH band strength, which is also in accord with observations. Of importance for galaxy population synthesis is the predicted weak dependence of FeH band strength on metal abundance. The blue and green systems of FeH are discussed.

064.002 Mass loss from dynamically unstable stellar envelopes. Y. Tuchman, N. Sach, Z. Barkat.
Astrophys. J., Vol. 219, 183 - 194 (1978).

Results of an extensive study of the dynamical stability and dynamical behavior of envelopes of red giants are given. The region on the luminosity (L)-effective temperature (T_{eff})-plane within which static, thermally equilibrated envelopes are dynamically unstable is reanalyzed and is found to be more complex than has hitherto been thought. The significance of the structure of this region and its dependence on relevant physical parameters is discussed.

064.003 Spectral energy distributions and effective temperature scale of M-giant stars. T. Tsuji.
Astron. Astrophys., Vol. 62, 29 - 50 (1978).

Observed spectral energy distributions of red giant stars are interpreted on the basis of model atmospheres in which molecular line opacities are taken into account by a band-model method. By the comparison of predicted and observed fluxes on a relative absolute scale, it is shown that T_{eff}= 3900, 3600 and 3200 K for M 0 III, M 3 III and M 6 III, respectively. In variables later than M 7, the models based on the classical assumptions may not be fully valid. It can, however, be suggested that the effective temperatures of such variables are below 3000 K and decrease rather rapidly with spectral type. The proposed effective temperature scale of red giant stars which is obtained as the best compromise between model-atmosphere analysis and angular diameter measurements represents an upward revision to the ones previously suggested. Some consequences of this revision are discussed.

064.004 Radiative forces in expanding envelopes.
J. Surdej.
Astron. Astrophys., Vol. 62, 135 - 141 (1978).

The escape probability method introduced by V. V. Sobolev is applied to derive expressions for the radiative forces acting on atoms by the resonance scattering of radiation in envelopes which expand spherically with positive or negative radial velocity gradients. In the latter case, it is shown that the contribution of the whole stellar envelope to the radiative force acting on individual atoms has the same effect as the one due to the stellar core radiation, i.e. a radial outward-acceleration. Numerical applications to a two-level atom model illustrate the behaviour of radiative forces in outward-accelerating and outward-decelerating stellar envelopes.

064.005 On the turbulent energy transport in accretion discs. N. I. Shakura, R. A. Sunyaev (*Syunyaev*), S. S. Zilitinkevich.
Astron. Astrophys., Vol. 62, 179 - 187 (1978).

Convective turbulence may arise in the inner regions of accretion discs where radiative pressure P_r is dominant. This paper presents the predicted vertical structure of the disc in this zone. A parcel of the energy flux transported by turbulence is found, and this does not exceed half the flux transferred by radiation. It is shown that despite the contribution of turbulence to the energy transport over the vertical coordi-

nate z, thermal instability in the inner zone of the disc takes place; the increment of thermal instability decreases due to turbulence by no more than one-half. There is shear turbulence in the outer regions of the disc where $P_g \gg P_r$ (P_g is the gas pressure). In this area, the entropy increasing towards the disc surface interferes with the development of turbulent motion over the z coordinate.

064.006 Observational tests of the shock heating theory for late-type stellar chromospheres.
L. E. Cram, P. Ulmschneider.
Astron. Astrophys., Vol. 62, 239 - 244 (1978) = Mitt. Fraunhofer Inst., Freiburg, No. 149.

The authors combine recent predictions of the positions of stellar temperature minima with a theory of the formation of Ca$^+$ resonance lines, and thus present observable tests of the shock wave heating theory of stellar chromospheres. The quantitative agreement is only satisfactory for a solar-type star with log g = 4 and T_{eff}= 6000 K. The theoretical minima of giant stars are located much deeper and the minima of cool dwarf stars are located much higher than the observations suggest. The authors discuss possible explanations of this disparity.

064.007 Stellar winds and neutral gas in diffuse nebulae.
J. E. Dyson.
Astron. Astrophys., Vol. 62, 269 - 271 (1978).

There is strong evidence that neutral material is associated with density fluctuations existing in the ionized gas of diffuse nebulae. The neutral material may be formed from previously ionized gas by the compressive action of energetic stellar winds from early type stars within the nebulae. Two possibilities are considered here. Firstly, the recombination of the material in the thin shell of swept up ambient gas. Secondly, small density fluctuations in the ionized gas may be left behind by the shell and be compressed by the bubble of hot shocked stellar wind gas during the shell. It is shown that in this latter case, neutral condensations having lifetimes ~10^4 yr may be produced by the compression.

064.008 Ionization conditions in the expanding envelopes of O and B stars.
H. J. G. L. M. Lamers, T. P. Snow, Jr.
Astrophys. J., Vol. 219, 504 - 514 (1978).

The authors study the observational evidence for the presence and absence of warm or hot expanding envelopes and their location in the temperature-luminosity diagram. They use the ionization balance in the envelope as an indicator of temperature. Consequences of hot envelopes for four stellar wind models are discussed.

064.009 Orbital changes of the gaseous ring around Be stars. II. Apsidal motion and slow drift of the orbital elements. S.-S. Huang.
Astrophys. J., Vol. 219, 956 - 962 (1978).

It is shown that the observed data about the apsidal motion of gaseous rings around some Be stars cannot be explained solely by the influence of the equatorial bulge. The author develops a theory of apsidal motion and slow drift of orbital elements that is based on the mixing of slowly escaping gases from the central star with gases in the emission ring around it. Observational implication and physical significance of this theory are discussed.

064.010 The effects of TiO opacity on the atmospheric structure of cool stars.
B. M. Krupp, J. G. Collins, H. R. Johnson.
Astrophys. J., Vol. 219, 963 - 969 (1978) = Publ. Goethe

Link Obs., No. 194.

The authors examine quantitatively the effects of various representations of TiO molecular opacities on the atmospheric structures of cool stars of solar composition. The authors compute model atmospheres which include opacity-sampling (OS) or straight-mean (SM) opacities for three singlet (β, δ, ϕ) and three triplet(α, γ, γ') band systems of TiO. By comparing model atmospheres using OS and SM opacities for TiO with models having no TiO, the authors are able to examine (1) the effects of TiO opacities on the temperature structures of various atmospheres, and (2) the usefulness of straight-mean TiO opacities in model atmosphere calculations.

064.011 Thermodynamic inconsistency of the modified Saha equation at high pressures.   M. A. Sweeney.
Astrophys. J., Vol. 220, 335 - 338 (1978).

The inclusion of a pressure ionization term in the Saha equation violates the thermodynamic Maxwell identities if corresponding changes are not made to the expressions for entropy and pressure. It is demonstrated that the usual application of the Rouse and Stewart-Pyatt models suffers from this limitation. Negative values of the adiabatic gradient in the degenerate dwarf models of Böhm and Straka are explained in terms of this thermodynamic inconsistency.

064.012 Between stars and space.   R. E. Stencel.
Astronomy, Vol. 5, No. 9, p. 34 - 38 (1977).
Abstr. in Phys. Abstr., Vol. 80, Abstr. 91517 (1977).

064.013 A model for stellar microturbulence.
M. G. Edmunds.
Astron. Astrophys., Vol. 64, 103 - 109 (1978).

It is suggested that the microturbulent broadening of spectral lines in late-type stellar atmospheres is caused by the propagation of accoustic waves, generated in the convection zones. It is shown that the wavelength of the waves obeys a condition required for the waves to affect the equivalent widths of strong lines. The variation of the predicted line broadening with temperature and surface gravity is in good agreement with published observations.

064.014 The structure of the winds and coronae of O stars derived from Hα line-profile analyses.
J. P. Cassinelli, G. L. Olson, R. Stalio.
Astrophys. J., Vol. 220, 573 - 581 (1978).

Several recent studies of the emission-line spectra of hot stars have led to the conclusion that there are high-temperature regions or coronae in the stellar winds. To test this idea, the effects of coronal regions on Hα line profiles are studied by using the Sobolev escape-probability method. The analysis is applied to ζ Ori, O9.7 Ib, and it is found that the corona must have a small thickness, less than 10% of the stellar radius. The velocity structure of the wind is deduced from a fit of the theoretical to the observed line profile. The velocity does not increase steadily outward, but reaches a plateau and then at about 2.5 R_* experiences a second rapid acceleration to terminal speeds. It is argued that hard radiation from a small coronal zone can produce the high stages of ionization such as O VI observed in the winds of hot stars. Thus a model which is consistent with the Hα, IR, and UV data of hot stars has a very small coronal zone surrounded by a cool stellar wind.

064.015 Diffusion and line asymmetries in HgMn stars.
G. Michaud.
Astrophys. J., Vol. 220, 592 - 599 (1978).

If abundance anomalies are caused by diffusion in stellar atmospheres, some elements are expected to lie in thin layers above the stellar atmospheres. It is shown in this paper that line asymmetries are then expected to appear, even in stable atmospheres, but only in the lines of some elements. The velocity distribution for those elements is nonisotropic and non-Maxwellian. The line asymmetries can mimic a stellar wind but without destabilization of the whole atmosphere. This may explain the recent observations of Smith and Parsons. This model predicts that He and O lines should not show the asymmetry.

064.016 The ratio of ^{6}Li to ^{7}Li in the atmospheres of carbon stars.   R. Canal, J. Isern, B. Sanahuja.
Astrophys. J., Vol. 220, 606 - 608 (1978).

A recent measurement of the wavelength of the lithium resonance line in the carbon star Y CVn has shown a lithium content of the atmosphere which is about 10% ^{6}Li. Such a result is examined within the framework of a low-energy spallation model.

064.017 Stellar model chromospheres. VI. Empirical estimates of the chromospheric radiative losses of late-type stars.   J. L. Linsky, T. R. Ayres.
Astrophys. J., Vol. 220, 619 - 628 (1978).

The authors develop a method for estimating the non-radiative heating of stellar chromospheres by measuring the net radiative losses in strong Fraunhofer line cores. This method is then applied to observations of the Mg II resonance lines in a sample of 32 stars including the Sun. The authors find at most a small dependence of chromospheric nonradiative heating on stellar surface gravity, contrary to the large effect predicted by recent calculations based on acoustic heating theories.

064.018 H and He II spectra of Of stars.
R. I. Klein, J. I. Castor.
Astrophys. J., Vol. 220, 902 - 923 (1978).

The dynamical models of Of stars developed in a previous publication (Castor, Abbott, Klein 1975) are used as the basis of a set of statistical equilibrium calculations for hydrogen and ionized helium, on the hypothesis of radiative equilibrium in the stellar envelope. The results for the line strengths and for selected line profiles are presented. The Hα strengths are used to derive rates of mass loss from observation, and these are correlated with stellar properties to test whether the stellar wind theory is confirmed. The He II λ4686 strengths, in the form of Hα/λ4686 ratios, are compared with observations to test the validity of the statistical equilibrium theory. Other observable lines of He II are discussed. The Hα and He II λ4686 profiles are compared with those observed.

064.019 A model for γ Cassiopeiae.
R. Poeckert, J. M. Marlborough.
Astrophys. J., Vol. 220, 940 - 961 (1978).

The authors present new linear polarization measurements of the continuum and new line profiles of Hα and Hβ for γ Cas. They use these data, primarily, together with the polarization of the Hα line (Poeckert and Marlborough 1977) and continuum measures in the visible and infrared and upper limit to the radio flux, to obtain a model for the circumstellar envelope of γ Cas which is more realistic than any available.

064.020 Stellar model chromospheres. VII. Capella (G5 III+), Pollux (K0 III), and Aldebaran (K5 III).
W. L. Kelch, J. L. Linsky, G. S. Basri, H.-Y. Chiu, S.-H. Chang, S. P. Maran, I. Furenlid.
Astrophys. J., Vol. 220, 962 - 979 (1978).

Data from high-resolution SEC vidicon spectroscopy with a ground-based telescope (for the Ca II K line) and from spectral scans made with the BUSS ultraviolet balloon spectrograph (for the Mg II h and k lines) are used to derive models of the chromospheres and upper photospheres of three G–K giants. The models are based on partial redistribution analyses of the Ca II K line wings and cores and on the fluxes in the Mg II lines. The photospheres thus computed are hotter than predicted by radiative equilibrium models. T_{min}/T_{eff} is found to decrease with decreasing T_{eff}, while m_0 (the mass column density at the top of the chromosphere) increases with decreasing stellar surface gravity.

064.021 **Radiative shock dynamics. II. Hydrogen continua.**
R. I. Klein, R. F. Stein, W. Kalkofen.
Astrophys. J., Vol. 220, 1024 - 1040 (1978).

The interaction between radiation and a shock wave propagating through a stellar atmosphere is investigated. Departures from local thermodynamic equilibrium (LTE) are permitted in the first two levels of a 10-level hydrogen atom; levels 3−10 are in LTE. All lines are assumed to be in detailed balance, and all collisional bound-bound transitions have been neglected except for those between levels 1 and 2. The angular variation of the radiation field is accounted for by variable Eddington factors.

064.022 **Vibrationally excited carbon monoxide in IRC**
+10216.   N. Z. Scoville, P. M. Solomon.
Astrophys. J., Lett., Vol. 220, L103 - L107 (1978).

The authors have detected emission from the luminous carbon star IRC +10216 at 114.222 GHz, which is the frequency of the $J = 1 \rightarrow 0$ rotational transition from the first excited vibrational state of CO. Assuming that this tentative identification is correct, the transition originates from a level at an energy corresponding to 3100 K above the ground state. The emission is probably a maser situated close to the stellar surface pumped by absorption of 2.3 μm stellar photons in the CO $\nu = 0 \rightarrow 2$ band.

064.023 **Dust particles near stars.**   P. F. Chugajnov.
Zemlya i Vselennaya, 1978, No. 2, p. 39. In Russian.

064.024 **Thickening of axisymmetric accretion disks and**
flows.   W. R. Stoeger.
Mon. Not. R. Astron. Soc., Vol. 182, 647 - 656 (1978).

An absolute upper limit of 5.5 per cent is derived for the fraction of the total energy which can be extracted from accreting plasma inside the marginally stable circular orbit ($r = r_{ms}$) in Schwarzschild geometry without disrupting steady flow. On the basis of this the author presents rough upper limits on the thickening of thin accretion disks and flows under the action of a thermal Pringle−Rees (PR) instability. An estimate of the thickening timescale is also made.

064.025 **A grid of medium cool radiative equilibrium non-**
LTE models: 10 000° K $\leq T_{eff} \leq$ 15 000° K,
$g = 10^4$.   J. Borsenberger, M. Gros.
Astron. Astrophys., Suppl. Ser., Vol. 31, 291 - 296 (1978).

Non-LTE radiative equilibrium model atmospheres are constructed using the method of complete linearization of Auer and Mihalas. Models for different effective temperatures 10 000° K $\leq T_{eff} \leq$ 15 000° K, $g = 10^4$ are tabulated.

064.026 **Model study of the sphericity effects in extended**
atmospheres of late-type stars.
T. Watanabe, K. Kodaira.
Publ. Astron. Soc. Japan, Vol. 30, 21 - 38 (1978).

Spherical static models of nongray LTE atmospheres are constructed to study sphericity effects in the atmospheres of red giants and supergiants. These models differentially demonstrate the sphericity effects, which turn out to be more enhanced than expected through coupling of molecule formation (which strongly depends upon temperature) and atmospheric cooling due to the flux dilution effect. Possible sphericity effects in the atmosphere of Betelgeuse are briefly discussed.

064.027 **Mass loss from P Cygni: the evidence of the Balmer**
lines.   D. Van Blerkom.
Astrophys. J., Vol. 221, 186 - 192 (1978).

Envelopes of stars which are losing mass are divided into two groups according to their velocity distribution. "Rapid

accelerators" are driven to nearly terminal velocity near the photosphere and then coast outward at nearly constant speed. "Gradual accelerators" experience a steady increase in velocity over a sizable fraction of the envelope. The Balmer-line profiles from P Cygni are shown to be consistent with the latter class of envelope model. Very good agreement between theory and observation is shown by a model with $v(r) \propto r$ throughout and with $\dot{M} = 3 \times 10^{-5} M_\odot$ yr^{-1}.

064.028 **Ammonia and OH near T Tauri stars.**
K. R. Lang, R. F. Willson.
Bull. American Astron. Soc., Vol. 9, 554 - 555 (1978).
Abstract.

064.029 **Hydrodynamical instabilities in the envelopes of**
main-sequence stars: constraints implied by the
lithium, beryllium and boron observations.
S. Vauclair, G. Vauclair, E. Schatzman, G. Michaud.
Bull. American Astron. Soc., Vol. 9, 573 (1978). − Abstract.

064.030 **Abundance anomalies in the envelopes of main-**
sequence stars: competition between diffusion
processes and turbulent motions.
G. Vauclair, S. Vauclair, G. Michaud.
Bull. American Astron. Soc., Vol. 9, 574 (1978). − Abstract.

064.031 **Stellar chromospheres for fun and profit.**
O. C. Wilson.
Bull. American Astron. Soc., Vol. 9, 585 (1978). − Abstract.
Russell Lecture.

064.032 **Hydrodynamical mass loss from long period**
variable stars.   S. J. Hill.
Bull. American Astron. Soc., Vol. 9, 594 (1978). − Abstract.

064.033 **Periodic shockwave description of LPV atmospheres.**
L. A. Willson.
Bull. American Astron. Soc., Vol. 9, 594 (1978). − Abstract.

064.034 **A possible explanation for the weak G-band effect.**
J. G. Mengel, A. V. Sweigart.
Bull. American Astron. Soc., Vol. 9, 603 (1978). − Abstract.

064.035 **Supercritical accretion.**
H. L. Burger, J. I. Katz.
Bull. American Astron. Soc., Vol. 9, 632 (1978). − Abstract.

064.036 **Radiative tides in accretion disks.**   S. V. Weber.
Bull. American Astron. Soc., Vol. 9, 632 - 633
(1978). − Abstract.

064.037 **The dependence of Ca II emission widths (log W_0)**
on [Fe/H].   T. E. Lutz, B. E. J. Pagel.
Bull. American Astron. Soc., Vol. 9, 634 (1978). − Abstract.

064.038 **The effective temperatures of G dwarfs.**
J. Hardorp.
Bull. American Astron. Soc., Vol. 9, 634 (1978). − Abstract.

064.039 **Can partial redistribution explain the Ca II H−K**
wing emission lines?   R. E. Stencel.
Bull. American Astron. Soc., Vol. 9, 634 (1978). − Abstract.

064.040 **Cooling by line radiation deep in an atmosphere.**
W. Kalkofen.
Bull. American Astron. Soc., Vol. 9, 635 (1978). − Abstract.

064.041 **Rapid mass loss and supersonic stellar winds.**
D. J. Mullan.
Bull. American Astron. Soc., Vol. 9, 649 - 650 (1978).
Abstract.

064.042 On the circumstellar shell of the Be star γ Cassiopeiae.   J. D. Scargle, E. F. Erickson, F. C. Witteborn, D. W. Strecker.
Bull. American Astron. Soc., Vol. 9, 650 - 651 (1978).
Abstract.

064.043 Nebular observations and stellar coronae. L. Hartmann, J. C. Raymond.
Bull. American Astron. Soc., Vol. 9, 651 (1978). − Abstract.

064.044 Modelling of chromospheric activity in F−M dwarf stars and the sun. W. L. Kelch, J. L. Linsky, S. P. Worden.
Bull. American Astron. Soc., Vol. 9, 651 (1978). − Abstract.

064.045 Constraints on the properties of circumstellar shells from observations of thermal CO and SiO millimeter line emission.   D. L. Lambert, P. A. Vanden Bout.
Astrophys. J., Vol. 221, 854 - 860 (1978).
An attempt to detect CO and SiO microwave emission from Betelgeuse and long-period variables (LPVs) is described. The absence of SiO 86 GHz emission from the Betelgeuse circumstellar shell is shown to require that either all Si is associated into silicate dust grains or SiO molecule formation is inhibited. The former explanation is consistent with published estimates of the column density of silicate grains. The latter explanation is supported by the absence of emission in the CO 115 GHz line. The new detections of circumstellar microwave emission include an observation of the CO 115 GHz line from R Cas (the first LPV for which both SiO 86 GHz emission and CO 115 GHz emission are seen) and of the SiO 86 GHz line from the S star χ Cyg. The absence of the isotopic line ^{30}SiO at 84 GHz shows that the shells around R Cas, R Leo, and χ Cyg have an optical depth in the 86 GHz line of $\tau < 10$.

064.046 Circumstellar shells and mass loss. An investigation of the colours of luminous O−B 3 stars in open clusters.   J. Isserstedt.
Astron. Astrophys., Vol. 65, 57 - 64 (1978). In German.
The colours and reddenings of single O−B 3 stars with $M_V < -4$ in galactic open clusters are investigated. It is shown that the colours of the luminous O−B 3 stars derived from the individual $(B-V)$ and $(U-B)$ measurements and the mean reddenings of the clusters do not agree with any of the intrinsic colours given in the literature. Especially the O-stars are significantly too red. It is shown that this is due to circumstellar shells.

064.047 Spectral line formation in YY Ori and T Tau envelopes.   C. Bertout.
IAU Colloq. No. 42, (see 012.012), p. 72 - 79 (1977).
First results of line profile computations in a collapsing rotating envelope are discussed. As an application, an observed profile of the YY Ori star CoD −35° 10525 is compared to computed profiles. Possible similarities in the formation of YY Ori and T Tau line profiles are briefly mentioned.

064.048 Secular mass loss from cataclysmic variables. J. Smak.
IAU Colloq. No. 42, (see 012.012), p. 365 - 370 (1977).

064.049 Optical information on mass-loss from evolved stars. D. Reimers.
IAU Colloq. No. 42, (see 012.012), p. 559 - 576 (1977).

064.050 Theoretical aspects of mass loss from late-type stars. R. J. Weymann.
IAU Colloq. No. 42, (see 012.012), p. 577 - 590 (1977).

064.051 On mass loss from long period variables. V. G. Gorbatsky (Gorbatskij), O. V. Fedorova.
IAU Colloq. No. 42, (see 012.012), p. 591 - 594 (1977).

064.052 Mass ejection from old stars − an example of a possible mechanism. M. P. FitzGerald, S. Kwok, C. R. Pruton.
IAU Colloq. No. 42, (see 012.012), p. 606 - 608 (1977).

064.053 Mass loss from stars.   D. C. Morton.
Proc. Astron. Soc. Australia, Vol. 3, 96 (1977). −Invited paper.

064.054 Observations of OH and H_2O microwave maser emission from VY Canis Majoris. B. R. Rosen, J. M. Moran, M. J. Reid, R. C. Walker, B. F. Burke, K. J. Johnston, J. H. Spencer.
Astrophys. J., Vol. 222, 132 - 139 (1978).
VLBI observations of the H_2O emission from VY CMa made in 1974 and 1976 show the diameter of the masing region to be about 5×10^{15} cm, and the diameters of the individual components to be about 5×10^{13} cm. The total mass of the gas in the water masing region is estimated to be $10^{-1} M_\odot$, and the density is about 10^9 cm^{-3}. The H_2O and main-line OH maser spectra from single antenna measurements show substantial changes within a few months. Expanding-shell models, with emission from both the limbs and the direction along the line of sight to the star, can explain the observations.

064.055 The combined effects of expansion and rotation on spectral line shapes.   P. Duval, A. H. Karp.
Astrophys. J., Vol. 222, 220 - 225 (1978).
It is common practice to approximate the effects of rotation on spectral line shapes by a convolution. More recently it has been shown that the effect of radial motion without gradient can also be approximated by a convolution. In this paper the authors show that the superposition of these two motions can also be approximated by a convolution, but that the convolving function is not the convolution of the pure radial motion and pure rotation functions. The convolution method is compared to the explicit integration over the stellar disk.

064.056 The circumstellar envelopes of M giants and supergiants.   W. Hagen.
Astrophys. J., Lett., Vol. 222, L37 - L40 (1978).
This Letter summarizes the results of an observational study of the circumstellar gas and dust shells of late-type giants and supergiants. Comparison of the derived amounts of circumstellar gas and dust reveals a large variation of the gas to dust ratio from star to star, with the metals in the stars of spectral class M6 and later appearing almost entirely as grains. Mass loss rates from 10^{-8} to 10^{-6} $M_\odot$ per year are derived, and are shown to be relatively insensitive to the gas to dust ratio, indicating that the mechanism of mass loss may be something other than radiation pressure on grains.

064.057 A line driven Rayleigh-Taylor-type instability in hot stars.   G. D. Nelson, A. G. Hearn.
Astron. Astrophys., Vol. 65, 223 - 227 (1978).
The existence of a Rayleigh-Taylor-type instability in the atmospheres of hot stars, driven by the radiative force associated with impurity ion resonance lines, is demonstrated. In a hot star with an effective temperature of 50 000 K, the instability will grow exponentially with a time scale of approximately 50 s in the layers where the stellar wind velocity is 5% of the thermal velocity of the ion. As a result, radially symmetric stellar winds driven by resonance line radiative forces will break up in small horizontal scale lengths.

064.058 Overstabilization of acoustic modes in a polytropic atmosphere.   H. M. Antia, S. M. Chitre,

D. M. Kale.
Sol. Phys., Vol. 56, 275 - 292 (1978).

The overstability of sound waves in a polytropic atmosphere is examined for disturbances of arbitrary optical thickness. It is concluded that the Cowling-Spiegel mechanism can operate in the solar convective zone, although the κ-mechanism is predominantly responsible for the observed five-minute oscillations.

064.059   On the effect of gas pressure in the theory of line accretion – II.   S. Yabushita.
Mon. Not. R. Astron. Soc., Vol. 183, 459 - 472 (1978).

It is shown that two previously ill-understood features of line accretion originally formulated by Bondi & Hoyle can be removed by including the pressure term in the basic equation. First, the distance of the stagnation point can be determined uniquely if the flow is such that the gas velocity is supersonic in the vicinity of the star and at great distance from it. Secondly, a cut-off distance to the accretion column can also be estimated without ambiguity, when pressure is taken into account. Isothermal accretion is also discussed. The retarding force for various types of solutions is briefly discussed.

064.060   Atmospheres of central stars.   D. G. Hummer.
IAU Symp. No. 76, (see 012.014), p. 171 - 183 (1978).

This review begins with a brief summary of atmospheric models that are of possible relevance to the central stars of planetary nebulae, and then discusses the extent to which these models accord with the observations of both nebulae and central stars. Particular attention is given to the significance of the very high Zanstra temperature implied by the nebular He II λ4686 Å line, and to the discrepancy between the Zanstra He II temperature and the considerably lower temperatures suggested by the appearance of the visual stellar spectrum for some of these objects.

064.061   A shortlived, deep convective envelope for highly evolved, blue stars?   I.-J. Sackmann.
IAU Symp. No. 76, (see 012.014), p. 207 - 208 (1978).
Abstract.

064.062   On the chemical composition and age of β Vir.   T. Gehren.
Astron. Astrophys., Vol. 65, 427 - 433 (1978).

The F8V star β Vir is reanalyzed by means of a differential model atmosphere analysis. The final atmospheric parameters are: T_{eff} = 6170 ± 100 K, log g = 4.25 ± 0.1, [Fe/H] = 0.20 ± 0.1, and ζ = 2.3 ± 0.3 km s^{-1}. Evolutionary isochrones suggest a stellar age of approximately 3 × 10^9 years, in contrast to earlier investigations.

064.063   Nebular observations and stellar coronae.   L. Hartmann, J. C. Raymond.
Astrophys. J., Vol. 222, 541 - 546 (1978).

The authors examine the ways in which nebular observations of He II λ4686 can be used to infer the existence of coronae or hot winds of early-type stars. Discrepancies between the He II Zanstra temperatures and effective temperatures derived from other methods may indicate coronal emission from planetary nebula central stars. Observations of diffuse nebulae about Population I early-type stars may also prove useful in detecting hot winds.

064.064   Determining velocity gradients from differential Doppler shifts.   A. H. Karp.
Astrophys. J., Vol. 222, 578 - 584 (1978).

The problem of spectral line formation in a moving stellar atmosphere is studied. An analytical solution to the transfer equation which is valid when the velocity gradient is large is presented. This solution is then used to show that the difference between the Doppler shifts sometimes observed in lines of different strengths depends on the ionization balance in the atmosphere as well as on the velocity gradient.

064.065   Numerical applications to radiative transfer in expanding envelopes: the two-level atom model.
J. Surdej.
Astron. Astrophys., Vol. 66, 45 - 55 (1978).

The author applies the escape probability method introduced by V. V. Sobolev in order to solve the statistical equilibrium equations relative to a two-level atom model in envelopes which expand spherically with positive or negative radial velocity gradients. It is shown that the level populations behave very differently in outward-accelerating and outward-decelerating envelopes, the degree of excitation appearing much higher in the second case.

064.066   The effect of radiation pressure on spherical accretion.   L. Maraschi, C. Reina, A. Treves.
Astron. Astrophys., Vol. 66, 99 - 101 (1978).

Spherical accretion is studied taking into account the effect of radiation pressure on the flow of the gas. The gas pressure is neglected in the dynamic equations and the opacity is described using an approximate Eddington factor. The paper corrects and extends a previous work on the same subject (Maraschi et al., 1974).

064.067   The onset of compressible convection.   E. Graham, D. R. Moore.
Mon. Not. R. Astron. Soc., Vol. 183, 617 - 632 (1978).

The linear equations for steady marginal convection in a compressible atmosphere with arbitrary physical properties are derived in optimal form for computation. The influence of variation in viscosity upon convection is investigated for two model atmospheres, one polytropic, the other stably stratified in part.

064.068   Stellar model chromospheres. VIII. 70 Ophiuchi A (K0 V) and ε Eridani (K2 V).   W. L. Kelch.
Astrophys. J., Vol. 222, 931 - 940 (1978).

High-resolution Ca II K line profiles and Mg II h and k line fluxes are used to derive photosphere and chromosphere models of two K dwarf active chromosphere stars, ε Eri and 70 Oph A. These models are computed on the basis of partial redistribution diagnostics.

064.069   A search for nebulosity around Sirius.   N. Brosch, I. Nevo.
Observatory, Vol. 98, 136 - 137 (1978).

064.070   Upper limits to X-ray emission from colliding stellar winds.
B. A. Cooke, A. C. Fabian, J. E. Pringle.
Nature, Vol. 273, 645 - 646 (1978).

In a binary system containing two similar mass losing stars, stellar winds can interact and shock to produce X rays. The spectrum and intensity of radiation emitted by the shocked gas gives information on the velocity and rate of mass loss. The authors discuss a range of candidate systems in which such processes may be occurring.

064.071   Stellar wind (a short review).   Eh. Ya. Vil'koviskij, L. V. Tambovtseva.
Soln. akt. Alma-Ata, 1977, p. 66 - 88. In Russian. – Abstr. in Ref. zh., 51. Astron., 4.51.502 (1978).

064.072   Influence of heavy element abundance in stellar atmospheres of spectral classes A–F upon the (B–V, T_e) relation.   L. P. Zajkova, Yu. S. Romanov.
Astrometriya i Astrofizika, vyp. (No.) 35, p. 76 - 84 (1978). In Russian.

A grid of effective temperature scales has been computed with accounting for heavy element content in stellar atmospheres at temperatures ranging from 6000 to 10000 K. Computations are based on Smak's method using specified initial data.

064.073   Estimate of the lower limit of thickness of a layer with anomalous chemical composition in the atmospheres of Ap stars.   V. L. Khokhlova.
Pis'ma Astron. Zh., Vol. 4, 228 - 231 (1978). In Russian.

A method of estimation of the lower limit of the thickness of a layer with anomalous composition in an Ap star atmosphere is proposed, using lines situated shortward and longward of the Balmer jump and are formed at different atmospheric layers.

064.074   Polarization effects in the radiation of an accretion disk.   Yu. N. Gnedin, N. A. Silant'ev.
Astron. Zh. Akad. Nauk SSSR, Vol. 55, 564 - 571 (1978). In Russian. English translation in Soviet Astron., Vol. 22, No. 3.

Radiation from an accretion disk is linearly polarized due to Thomson scattering. In the standard disk model, the polarization vector lies in the orbital plane. A jump in the position angle of the plane of the electric vector oscillations is possible at the transition in the "c" region of the standard disk. The deviations from the standard disk lead to polarization effects described in detail.

064.075   A study of mass loss from the mid-ultraviolet spectrum of α Cygni (A2 Ia), β Orionis (B8 Ia), and η Leonis (A0 Ib).   H. J. G. L. M. Lamers, R. Stalio, Y. Kondo.
Astrophys. J., Vol. 223, 207 - 220 (1978).

The first results on mass loss from high-resolution mid-ultraviolet spectra of α Cyg (A2 Ia), β Ori (B8 Ia), and η Leo (A0 Ib) are given. The spectrum of α Lyr (A0 V) is also included in this study as a comparison spectrum to distinguish photospheric lines from envelope lines. The spectra are compared with one another in selected wavelength regions. The differences in the characteristics of the winds are discussed. The comparison between α Cyg and β Ori shows the variation of mass-loss effects with temperature, whereas the comparison of these two stars with η Leo shows the variation with luminosity.

064.076   The Ca II V/R ratio and mass loss.   R. E. Stencel.
Astrophys. J., Lett., Vol. 223, L37 - L39 (1978).

The V/R ratio of the intensities of the self-reversed emission peaks in the center of the Ca II K line appears to be correlated with spectral type in a way which also correlates with indicators of mass loss from late-type stars and the H-K wing emission lines.

064.077   Energy distributions in main-sequence A and F stars.   E. Böhm-Vitense.
Astrophys. J., Vol. 223, 509 - 525 (1978).

The continuum energy distributions of A and F stars, derived from the scanner observations by Oke, Baschek and Oke, and Böhm-Vitense and Johnson are discussed. It is found that all main-sequence stars with $B - V > 0.22$ have less flux in the UV than predicted by radiative equilibrium models, indicating that convective energy transport is much more important in F stars than hitherto thought. For $0.1 < B - V < 0.2$ an UV flux reduction occurs for all rapidly rotating stars but only for some slow rotators, indicating that rotation does not inhibit convection but may even enhance it. Stars in the young Pleiades and α Per clusters show a somewhat larger UV flux reduction—or a steeper infrared gradient—than the stars in older clusters.

064.078   Extreme subdwarfs. I. Molecular band strengths: a theoretical approach.   P. L. Cottrell.
Astrophys. J., Vol. 223, 544 - 551 (1978).

G-band and MgH band synthetic spectra are computed for a grid of dwarf and subdwarf model atmospheres. These spectra are then used to interpret Greenstein's "CH" and "MgH" classifications. These can either correspond to carbon and magnesium abundance anomalies or can be explained by molecular association effects in the atmospheres of these stars. This is investigated by a detailed analysis of the molecular equilibria in the grid of model atmospheres.

064.079   Hydrodynamical instabilities in the envelopes of main-sequence stars: constraints implied by the lithium, beryllium, and boron observations.   S. Vauclair, G. Vauclair, E. Schatzman, G. Michaud.
Astrophys. J., Vol. 223, 567 - 582 (1978).

The abundances of lithium, beryllium, and boron in main-sequence stars are studied in view of their implications on the hydrodynamics in stellar interiors. Since they are destroyed by nuclear reactions relatively close to the surface, these elements can give a clue to the macroscopic transport processes occurring in stellar envelopes. The determination of the macroscopic transport can also indicate in which stars microscopic diffusion processes can be important. Theoretical variations of the abundances of lithium, beryllium, and boron induced by a combination of either a laminar meridional circulation or a turbulent diffusion with microscopic diffusion and nuclear destruction have been computed. A fitting of the theoretical abundance variations with the observed ones give severe constraints on the allowed macroscopic motions in main-sequence stars. The authors find that overshooting is probably the basic transport process under the outer convection zones in F and G stars, and that it must be responsible for the lithium abundance variations from F to G stars. Mixing due to rotation has probably a smaller effect in these stars. In particular, meridional circulation cannot be the reason for lithium depletion in the Sun, as it would lead to beryllium depletion—contrary to observations. Microscopic diffusion must also be negligible in G stars and in ordinary F stars. The authors conclude that the beryllium-deficient F stars could be a continuation of the Am—Fm sequence toward lower effective temperatures.

064.080   Disk accretion by magnetic neutron stars.   P. Ghosh, F. K. Lamb.
Astrophys. J., Lett., Vol. 223, L83 - L87 (1978).

The authors propose a model for disk accretion by a rotating magnetic neutron star which makes definite predictions regarding the inner radius of the disk, the radial and vertical structure of the transition zone, and the accretion torque. One of the main conclusions is that in disk accretion the transition zone is not thin, and recognition of this is an important step in understanding the period behavior of Her X-1, Cen X-3, and 4U 0900—40.

064.081   Extreme-ultraviolet observations of nearby B stars: constraints on hot circumstellar plasma.   F. Paresce, S. Bowyer, M. Lampton.
Astrophys. J., Lett., Vol. 223, L89 - L92 (1978).

The extreme-ultraviolet telescope on the Apollo-Soyuz Test Project Mission observed several nearby stars of spectral type earlier than B3 in search of thermal emission from possible hot circumstellar plasma. The stars studied in the 45—185 and 114—185 Å passbands of the instrument included λ Sco, ν Sco, β CMa, β Cen, σ Sgr, and α Pav. The experimental data have been used to place upper limits on the emission measure of the plasma and the stellar wind luminosity, using standard models. The data rule out the existence of sizable bubbles of hot plasma around any of the stars surveyed. In all cases the electron density n_e in these regions, if they exist, may not exceed ~ 0.2 electrons cm^{-3}.

064.082   Comparison of TDI-fluxes with those from model-atmosphere including OH-opacity.

S. P. Tarafdar, P. W. Fox.
Astron. Astrophys., Vol. 67, 281 - 285 (1978).

Model atmospheres have been constructed including OH-bound-free opacity for temperatures corresponding to F stars. The emergent fluxes from the models have been compared with their values observed from TDI-satellite. It has been found that OH-bound-free opacity is not very important for F0V stars but starts contributing from F2V with increasing importance towards later spectral type. Comparison of theoretical fluxes with their observed values shows that OH-bound-free opacity can account for ultraviolet missing opacity in F stars for wavelengths were observed flux does not indicate the presence of many lines and alleviates the discrepancy in other parts of the spectral region.

064.083 Massenverlust bei heißen Sternen.
J. Schmid-Burgk.
Sterne Weltraum, Jahrg. 17, 246 - 251 (1978).

064.084 Holes or spots?   J. Mergentaler.
Urania Kraków, Vol. 48, 327 - 330 (1977). In Polish.

064.085 Model atmospheres of magnetic stars.
K. Stepień.
Magnetic stars, (see 012.028), p. 32 - 33 (1978). – Abstract.

064.086 Synthesis of Hγ line profiles in LTE magnetic star atmospheres.   J. Madej.
Magnetic stars, (see 012.028), p. 34 - 35 (1978). – Abstract.

064.087 The study of local characteristics of Ap stars' surfaces.   A. V. Goncharskij, V. V. Stepanov, V. L. Khokhlova, A. G. Yagola.
Magnetic stars, (see 012.028), p. 36 - 37 (1978). – Abstract.

064.088 Temperature distribution in extended hot stellar atmospheres.   V. I. Stebnev.
Tr. Kazan. Gorod. Astron. Obs., No. 41, p. 24 - 30 (1976). In Russian.

064.089 Application of Unsöld's method of iteration of temperature to models of extended atmospheres.
V. G. Grinkevich.
Tr. Kazan. Gorod. Astron. Obs., No. 41, p. 31 - 35 (1976). In Russian.

064.090 Application of Feautrier's method to extended atmospheres.   V. G. Grinkevich.
Tr. Kazan. Gorod. Astron. Obs., No. 41, p. 36 - 40 (1976). In Russian.

064.091 Explanation of anomalous ionization in B supergiants with a corona plus cool wind model.
J. P. Cassinelli.
Bull. American Astron. Soc., Vol. 10, 412 (1978). – Abstract.

064.092 The effects of coronal regions on the ionization structure of stellar winds of O stars.
G. L. Olson.
Bull. American Astron. Soc., Vol. 10, 412 (1978). – Abstract.

064.093 Coronal regions in the winds of early-type stars.
K. B. MacGregor.
Bull. American Astron. Soc., Vol. 10, 412 (1978). – Abstract.

064.094 Spectrophotometry of nebulae around Wolf-Rayet stars.   K. B. Kwitter.
Bull. American Astron. Soc., Vol. 10, 412 - 413 (1978). Abstract.

064.095 Wind models with enhanced equatorial mass loss

rates.   W. M. Rumpl.
Bull. American Astron. Soc., Vol. 10, 413 (1978). – Abstract.

064.096 Composition changes in the envelopes of intermediate mass stars climbing the asymptotic giant branch.   S. A. Becker, I. Iben, Jr.
Bull. American Astron. Soc., Vol. 10, 438 (1978). – Abstract.

064.097 M giant circumstellar envelopes.
W. Hagen, A. M. Boesgaard.
Bull. American Astron. Soc., Vol. 10, 453 - 454 (1978). Abstract.

064.098 Variable Eddington factors for stellar envelope calculations.   J. Eoll.
Bull. American Astron. Soc., Vol. 10, 457 - 458 (1978). Abstract.

064.099 The effects of stellar chromospheric activity on metallicity measurements.
M. S. Giampapa, S. P. Worden.
Bull. American Astron. Soc., Vol. 10, 458 (1978). – Abstract.

064.100 Effects of carbon enhancement on stellar atmospheres.   H. R. Johnson.
Bull. American Astron. Soc., Vol. 10, 458 (1978). – Abstract.

064.101 Discovery of linearly polarized continuum light scattered by the shell around Alpha Orionis.
R. S. McMillan, S. Tapia.
Bull. American Astron. Soc., Vol. 10, 464 (1978). – Abstract.

064.102 Departures from radiative equilibrium in stellar atmospheres. Grey absorption.   L. E. Cram.
Astron. Astrophys., Vol. 67, 301 - 309 (1978) = Mitt. Fraunhofer Inst., Freiburg, No. 155.

The author discusses some of the consequences of departures from radiative equilibrium in stellar atmospheres. Using a discrete ordinates method he solves the radiative transfer equation in a grey atmosphere subjected to a specified distribution of mechanical heating, and determines the resulting temperature changes in LTE and non LTE conditions. He shows how radiative transfer leads to temperature changes in regions that are not directly heated, and how non LTE effects lead to an amplification of the temperature rise produced by a given distribution of heating. An attempt is made to resolve a controversy surrounding the estimation of excess radiative losses in the solar chromosphere.

064.103 A study of the microturbulence in giants in terms of stellar evolution.   R. Foy.
Astron. Astrophys., Vol. 67, 311 - 321 (1978).

Detailed analyses of nine yellow giants are carried out. The results are discussed together with those of a similar analysis of eight other giants. These results provide evidence that the spectroscopic "microturbulence" is actually produced by the atmospheric velocity field and that it is very likely induced in the photosphere by the underlying convective zone, presumably by overshooting.

064.104 Finite Larmor radius effect on thermal-convective instability of a stellar atmosphere.
R. C. Sharma, K. Prakash.
Acta Phys. Acad. Sci. Hungaricae, Vol. 42, No. 2, p. 103 - 109 (1977). – Abstr. in Phys. Abstr., Vol. 81, Abstr. 32438 (1978).

064.105 The application of theoretical models of stellar atmospheres in some astrophysics problems.
E. Tanasescu.
Bul. Inst. Politeh. 'Gheorghe Gheorghiu-Dej' Bucuresti Ser. Mec., Vol. 39, No. 2, p. 3 - 8 (1977). In Rumanian. – Abstr.

in Phys. Abstr., Vol. 81, Abstr. 36557 (1978).

064.106   On the selective absorption of the envelope of
R CrB.   A. F. Pugach.
Astron. Tsirk., No. 987, p. 2 - 5 (1978). In Russian.

064.107   Diagnostic of stellar chromospheres.
F. Praderie.
Mem. Soc. Astron. Italiana, Vol. 47, (see 012.036), 553 - 625 (1976).
   Contents: The solar chromosphere as a methodological tool to study stellar chromospheres. Diagnostic of stellar chromospheres from the continuum. Diagnostic of stellar chromospheres from spectral lines. Physical characteristics of stellar chromospheres.

064.108   Problems in theory of stellar atmospheres.
C. J. Cannon.
Mem. Soc. Astron. Italiana, Vol. 47, (see 012.036), 627 - 628 (1976). − Abstract.

064.109   A grid of model atmospheres and synthetic spectra
for white dwarfs with $5000 < T_{eff} < 7000$ K.
R. Wehrse.
Mem. Soc. Astron. Italiana, Vol. 48, 13 - 26 (1977).

064.110   What is a stellar atmosphere?   R. N. Thomas.
Mem. Soc. Astron. Italiana, Vol. 48, (see 012.038), 339 - 356 (1977).

064.111   Radiative-transfer effects on thermal-convective
instability in a stellar atmosphere with finite gyro-
viscosity and Hall currents.
S. L. Maheshwari, P. K. Bhatia.
Indian J. Phys. A, Vol. 51A, No. 4, p. 237 - 242 (1977).
Abstr. in Phys. Abstr., Vol. 81, Abstr. 53929 (1978).

064.112   Possible radiative effects in the circumstellar dust
related with T Tauri stars.   V. Vanysek.
Flare stars. Symposium 1976, (see 012.035), p. 113 - 115. In Russian. − Abstr. in Ref. zh., 51. Astron., 7.51.518 (1978).

064.113   Determination of the local chemical abundance of
stellar atmospheres from spectral line contours.
V. L. Khokhlova.
Diagnost. plyazmy po konturam spektr. linij. Petrozavodsk, 1977, p. 143 - 146. In Russian. − Abstr. in Ref. zh., 51. Astron., 7.51.524 (1978).

   Stellar atmospheres.   See Abstr. 003.092.

   Si⁻ opacity.   See Abstr. 022.047.

   Non-LTE line formation in a magnetic field. The two-level atom with a frequency independent source function. I. Formulation.   See Abstr. 022.082.

   A heuristic model for spectral-line-profiles.
See Abstr. 022.123.

   Representation of spectral-line-profiles by means of the Lorentz-function of the n-th degree.
See Abstr. 022.124.

   A search for stellar oscillations.
See Abstr. 031.263.

   On the origin and evolution of isotopes of carbon, nitrogen, and oxygen.   See Abstr. 061.047.

   The validity of ionization equilibrium in steady-state flows.   See Abstr. 062.036.

   Short period acoustic heating theory and its application to the construction of model chromospheres.
See Abstr. 062.117.

   Solution of the comoving-frame equation of transfer in spherically symmetric flows. V. Multilevel atoms.
See Abstr. 063.002.

   The differential equations of the Voigt function.
See Abstr. 063.007.

   Two-dimensional radiative transfer. I. Planar geometry.   See Abstr. 063.009.

   On coherent scattering in stellar atmospheres.
See Abstr. 063.031.

   Radiative transfer in geometries other than plane-parallel layers.   See Abstr. 063.044.

   Massive stars evolution with mass-loss.
I. 20 - 100 $M_\odot$ models.   See Abstr. 065.008.

   Convection in stars.   See Abstr. 065.013.

   Correlation of mass loss rates and $^{12}C/^{13}C$ ratios in luminous carbon stars.   See Abstr. 065.037.

   The effect of mass-loss and semiconvection on the evolution of a 15 $M_\odot$ Population I star.   See Abstr. 065.048.

   Effects of mass loss on the evolution of massive stars. I. Main-sequence evolution.   See Abstr. 065.072.

   The evolution of massive stars losing mass and angular momentum, II. The effects of shear due to differential rotation.   See Abstr. 065.078.

   The evolution of O stars and the origin of Wolf-Rayet stars.   See Abstr. 065.079.

   On the bulk yields of nucleosynthesis from massive stars.   See Abstr. 065.086.

   Magnetic structures in photosphere of sun and stars.
See Abstr. 071.054.

   C III density diagnostics in nonequilibrium plasmas.
See Abstr. 073.030.

   Color index and finite extent of the atmosphere.
See Abstr. 113.023.

   The Ca II λ8542 and λ8498 lines as important indicators of stellar chromospheres.   See Abstr. 114.027.

   Infrared observations of late type stars.
See Abstr. 114.035.

   IUE observations of F, G, and K stars and preliminary models for upper chromospheres based on C II, Si II, and Si III.   See Abstr. 114.076.

   Photometric and spectroscopic variability of Ca II H and K in G- and K-type giants.   See Abstr. 114.085.

   Observations of stellar chromospheres.
See Abstr. 114.098.

   Photospheric velocity fields in Tau Scorpii.
See Abstr. 114.507.

The fundamental bands of CO in Arcturus: evidence for an inhomogeneous chromosphere. See Abstr. 114.511.

Direct observations of the circumstellar gas shell of Betelgeuse.   See Abstr. 114.538.

Circumstellar gas and dust shells of luminous M stars.   See Abstr. 114.539.

Absolute energy distribution in Rho Cassiopeiae. See Abstr. 114.540.

Variable stellar wind from the WN-star HD 151932. See Abstr. 114.544.

The expanding envelope of τ Sco (B0 V). See Abstr. 114.560.

Abundances in globular cluster red giants. I. M3 and M13.   See Abstr. 114.565.

Analysis of the subgiant halo star HD 76932. See Abstr. 114.578.

Accretion discs.   See Abstr. 117.060.

Nonstationary accretion of stellar wind. See Abstr. 117.061.

A spectroscopic reinvestigation of the massive binary HD 698.   See Abstr. 119.016.

The infrared eclipse of V444 Cygni and the structure of Wolf-Rayet winds.   See Abstr. 121.012.

On the role of photospheric convection in W Ursae Majoris stars.   See Abstr. 121.021.

Dynamics of the photosphere and shells in the β Cephei star γ Pegasi.   See Abstr. 122.065.

Masses of classical Cepheids and mass loss rate at pre-Cepheid stage.   See Abstr. 122.091.

Mass loss and Cepheid pulsation. See Abstr. 122.092.

Oscillating stars (Miras), mass loss and formation of planetary nebulae (PN).   See Abstr. 122.106.

Optically thick winds in classical novae. See Abstr. 124.001.

The Cassiopeia A progenitor: a consistent evolutionary picture involving supergiant mass loss. See Abstr. 125.012.

Hot white dwarfs as soft X-ray sources. See Abstr. 126.028.

Mass loss from stars and the chemical evolution of the interstellar medium.   See Abstr. 131.125.

Cosmic masers.   See Abstr. 131.199.

Encounters with interstellar clouds: the effect on stellar winds and the possibilities for accretion. See Abstr. 131.253.

UV observations and stellar evolution. See Abstr. 131.258.

Molecular envelopes around evolved stars and the origin of planetary nebulae.   See Abstr. 133.013.

On the origin of planetary nebulae. See Abstr. 135.008.

The anomalous giant branch of NGC 188. See Abstr. 153.013.

Errata

064.901   Errata: "Interstellar bubbles. II. Structure and evolution" [Astrophys. J., Vol. 218, 377 - 395 (1977)]. R. Weaver, R. McCray, J. Castor, P. Shapiro, R. Moore. Astrophys. J., Vol. 220, 742 (1978).

064.902   Erratum: 'On the absolute scale of mass-loss in red giants. I. Circumstellar absorption lines in the spectrum of the visual companion of α Her' [Astron. Astrophys., Vol. 61, 217 - 224 (1977)].   D. Reimers. Astron. Astrophys., Vol. 67, 161 (1978).

065 Stellar Structure and Evolution

065.001 **Ejection of planetary nebulae by helium shell flashes and the planetary distance scale.**
V. Trimble, I.-J. Sackmann.
Mon. Not. R. Astron. Soc., Vol. 182, 97 - 102 (1978).

The authors investigate a consequence of the hypothesis that planetary nebulae are ejected in the nuclear runaway shell flashes that take place in intermediate mass stars when both hydrogen and helium are burning in thin shells. If planetary nebulae with two or more shells are the products of ejection at the same velocity by two or more successive flashes, it should be possible to correlate the separations of the shells with the brightnesses of the central stars. The authors find that the hypothesis is tenable only if the larger of the two distance scales which have been suggested is correct. In this case, the carbon—oxygen cores of the stars apparently have masses of $0.8 \pm 0.15\, M_\odot$, which is interesting in connection with measured white dwarf masses.

065.002 **s-process studies: the effects of a pulsed neutron flux.**   R. A. Ward, M. J. Newman.
Astrophys. J., Vol. 219, 195 - 212 (1978).

Recent investigations of the astrophysical site of the s-process strongly suggest the helium-burning shell of the helium-shell-flashing stars of intermediate mass as the most likely site for the s-process event which produced the solar-system abundances. Previous analyses of branching in the s-process have been in terms of a constant temperature and density environment. The occurrence of periodic thermal instabilities seems to make this assumption inappropriate. The current effort is to reformulate the mathematics of the branched s-process to allow for the influence of thermal pulses. It is shown that the decay of the unstable branch nuclei between neutron irradiations allows the mean s-process neutron flux responsible for the solar-system abundances in such a pulsed environment to be considerably larger than that for a single continuous-exposure event.

065.003 **Urca neutrino-loss rates under conditions found in the carbon-oxygen cores of intermediate-mass stars.**
I. Iben, Jr.
Astrophys. J., Vol. 219, 213 - 225 (1978).

It is shown that in matter which is composed initially of elements in a solar-system distribution and which has undergone first complete hydrogen burning and then complete helium burning, neutrino-loss rates due to 11 Urca pairs either rival or exceed neutrino losses predicted by the charge- and neutral-current theories of weak interactions. The dominant Urca-loss rates are still due to the pairs $^{21}F-^{21}Ne$, $^{23}Ne-^{23}Na$, $^{25}Na-^{25}Mg$, and $^{25}Ne-^{25}Na$, as in matter containing a solar-system distribution of elements that has undergone prior processing during hydrogen- and helium-burning phases. The abundances of these Urca-active pairs are enhanced by one to three orders of magnitude as a consequence of carbon-burning reactions.

065.004 **^{26}Al production in explosive carbon burning.**
J. W. Truran, A. G. W. Cameron.
Astrophys. J., Vol. 219, 226 - 229 (1978).

Nucleosynthesis associated with explosive carbon burning is reexamined, with the aim of providing a realistic prediction of the abundance of ^{26}Al formed under such conditions. The sensitivities of the $^{26}Al/^{27}Al$ production ratio to variations in the temperature, the initial composition, and the degree of neutron competition from heavy nuclei are explored.

065.005 **Neutrino angular momentum loss in rotating stars.**
R. Epstein.
Astrophys. J., Lett., Vol. 219, L39 - L41 (1978).

The author considers how an axisymmetric, uniformly and slowly rotating homogeneous star might be affected by the loss of energy and angular momentum via neutrino emission. For example, these losses can cause a collapsing, rotating star to flatten either faster or slower than would be the case if these losses were not taken fully into account, depending on the efficiency of the angular momentum loss mechanism.

065.006 **Non-radial oscillations of rotating stars and their relevance to the short-period oscillations of cataclysmic variables.**   J. Papaloizou, J. E. Pringle.
Mon. Not. R. Astron. Soc., Vol. 182, 423 - 442 (1978).

The authors consider the usual hypothesis that the short-period coherent oscillations seen in cataclysmic variables are attributable to g modes in a slowly rotating white dwarf. They show that this hypothesis is untenable. The authors investigate the low-frequency spectrum of a rotating pulsating star, taking the effects of rotation fully into account. In this case there are two sets of low-frequency modes, the g modes, and modes similar to Rossby waves in the Earth's atmosphere and oceans, which they designate r modes. They conclude that non-radial oscillations of rotating white dwarfs can account for the properties of the oscillations seen in dwarf novae. Application of these results to other systems is also discussed.

065.007 **Theoretical evolution of extremely metal-poor stars. II.**   R. L. Wagner.
Astron. Astrophys., Vol. 62, 9 - 12 (1978) = Contrib. Louisiana State Univ. Obs., Baton Rouge, No. 128.

The theoretical evolution of stars of mass $3.25\, M_\odot$ and $4.0\, M_\odot$ and initial composition $(X, Z) = (0.739, 10^{-4})$ has been calculated from the main sequence to the blue end of the blueward loop during central helium burning. As expected, these intermediate-mass, very metal-poor stars evolve significantly faster than Population I stars of the same mass and ignite core helium prior to reaching the red giant branch. Implications of the evolutionary calculations for the study of galactic evolution are briefly discussed, and an observational search for pulsation in low-mass subdwarfs is suggested.

065.008 **Massive stars evolution with mass-loss. I. 20 - 100 $M_\odot$ models.**   C. Chiosi, E. Nasi, S. R. Sreenivasan.
Astron. Astrophys., Vol. 63, 103 - 124 (1978).

The evolution of stars with initial masses 20, 30, 40, 60, 80, 100 $M_\odot$ and Population I chemical composition (X = 0.700, Z = 0.02) is calculated, taking into account mass-loss due to stellar winds, from the main sequence up to the early stages of central He-burning. The results are presented in terms of evolutionary tracks, isochrones, loci of constant mass-loss rates and loci of constant mass in the HR diagram. A detailed comparison of the theoretical predictions and observational results is made and possible implications for O, Of, Wolf-Rayet stars and red supergiants are brought out.

065.009 **Stars with shell energy sources. I. Special evolutionary code.**   M. Różyczka.
Acta Astron., Vol. 27, 415 - 427 (1977).

A new version of the Henyey-type stellar evolution code is described and tested. It is shown, as a by-product of the tests, that the thermal time scale of the core of a red giant approaching the helium flash is of the order of the evolutionary time scale. The code itself appears to be a very efficient tool for investigations of the helium flash, carbon flash and the evolution of a white dwarf accreting mass.

065.010 **Evolution of stars with two shell sources: the second loop in the H—R diagram.**

W. Höppner, H. Kähler, M. L. Roth, A. Weigert.
Astron. Astrophys., Vol. 63, 391 - 399 (1978).

A model evolving through the second loop during He-shell source burning can be divided into two basic regions which are treated separately: a contracting C–O core, and an envelope with two shell sources which is close to an equilibrium structure. Sequences of such envelopes for $M = 9\,M_\odot$ are presented, and it is shown how they interact with a contracting core to form a loop in the H–R diagram. Conditions for the occurrence of this evolution are given.

065.011 **The evolution of rotating stars. II. Calculations with time-dependent redistribution of angular momentum for 7 and 10 $M_\odot$ stars.**
A. S. Endal, S. Sofia.
Astrophys. J., Vol. 220, 279 - 290 (1978).

Calculations have been performed for the evolution of rotating stars with realistic, time-dependent redistribution of angular momentum and chemical composition due to gas-dynamical instabilities. Convection and Eddington circulation are found to be the most important mechanisms for changing the angular momentum distribution, while the dynamical shear and Solberg-Hoiland instabilities produce chemical mixing in regions which remain unmixed in non-rotating calculations. The calculations indicate that ignoring the finite time scales associated with angular momentum redistribution is a poor and, often, misleading approximation.

065.012 **Nova models and their problems.** H.-C. Thomas.
IAU Colloq. No. 42, (see 012.012), p. 301 - 310 (1977). – Review paper.

065.013 **Convection in stars.** R. van der Borght.
Proc. Astron. Soc. Australia, Vol. 3, 91 - 95 (1977).

065.014 **A semitheory for semiconvection.**
D. J. Stevenson.
Proc. Astron. Soc. Australia, Vol. 3, 165 - 167 (1977).

065.015 **Historical reminiscences of the origins of stellar convection theory (1930 - 1945).** L. Biermann.
Problems of stellar convection, (see 012.005), p. 4 - 14 (1977).

065.016 **The current state of stellar mixing-length theory.**
D. Gough.
Problems of stellar convection, (see 012.005), p. 15 - 56 (1977).

The basic assumptions of the mixing-length formalism are described, and the theory is developed with a view to representing convection in stars. Directions in which the results might be improved and extended are indicated.

065.017 **On taking mixing-length theory seriously.**
D. O. Gough, E. A. Spiegel.
Problems of stellar convection, (see 012.005), p. 57 - 62 (1977).

The authors interpret the mixing-length theory in terms of the specific model in which a star is composed of a background fluid through which discrete, well-defined parcels of fluid move. The convective model is a two-fluid model loosely resembling the composite of radiation and matter familiar in astrophysics, except that the quasiparticle fluid is more complicated than the photon gas.

065.018 **Observations bearing on the theory of stellar convection II.** E. Böhm-Vitense.
Problems of stellar convection, (see 012.005), p. 63 - 86 (1977).

It is shown that the best way to get information about efficiency of convective energy transport in the hydrogen convection zones in stars other than the sun is presently con-tained in continuous energy distributions of A and F stars. Scans show that convective energy transport must be much more efficient than thought hitherto. The scans indicate that rapid rotation enhances convective energy transport. A figure lists all the observations, which are influenced by the outer convection zone and which can therefore give us information for the author's convection theory.

065.019 **Dynamical instabilities in stars.** P. Ledoux.
Problems of stellar convection, (see 012.005), p. 87 - 102 (1977).

The linear dynamical instability at the origin of convection in stars is reviewed. The case of two or more super-adiabatic regions separated by subadiabatic ones might well deserve more detailed attention.

065.020 **Observations bearing on convection.** K. H. Böhm.
Problems of stellar convection, (see 012.005), p. 103 - 118 (1977).

Solar observations contain a considerable amount of information on the hydrodynamics of stellar convection. The author emphasizes and discusses some special questions. The Li–Be problem and its possible relevance as an indicator of convective overshoot is briefly summarized. Convection may have a stronger influence on the observable properties of He-rich ("non–DA") white dwarfs than of most other stars.

065.021 **Compressible convection.** E. Graham.
Problems of stellar convection, (see 012.005), p. 151 - 155 (1977).

Numerical simulation of compressible convection provides a way of obtaining a detailed picture of stellar convection. At the present time, solutions are still far from the parameter range found in stellar interiors. However, the solutions are well removed from the Boussinesq limit of laboratory convection experiments.

065.022 **Convection in rotating stars.** F. H. Busse.
Problems of stellar convection, (see 012.005), p. 156 - 175 (1977).

It is shown that many features of convection in rotating spheres and spherical shells can be understood on the basis of plane layer models. The phenomenon of differential rotation generated by convection is emphasized. The potential applications and limitations of analytical and numerical models for problems of astrophysical interest are briefly discussed.

065.023 **Penetrative convection in stars.** J.-P. Zahn.
Problems of stellar convection, (see 012.005), p. 225 - 234 (1977).

Penetrative convection occurs in a fluid whenever a convectively unstable region is bounded by a stable domain. This situation is encountered in many stars, and it is also a very common circumstance on Earth: in the oceans and in the atmosphere. The astrophysicists have developed methods to describe stellar convection, some of them are widely inspired by those used by the geophysicists. The same is true for convective penetration, whose study cannot be separated from that of convection itself. The purpose of this review is to recall those methods, and to verify if they are suited to describe the penetration of convective motions into stable surroundings.

065.024 **The URCA convection.** G. Shaviv.
Problems of stellar convection, (see 012.005), p. 255 - 266 (1977).

The possible role that β-decays may play in stellar collapse was first discussed by Gamow and Schoenberg (1940, 1941). The authors proposed this mechanism for extracting quickly the energy content of a star and transporting it outside. In this way they hypothesized that stellar collapse may proceed.

065.025 **Photoconvection.**   E. A. Spiegel.
    Problems of stellar convection, (see 012.005),
p. 267 - 283 (1977).

Convection under the influence of dynamically significant radiation fields occurs routinely in hot stars (Underhill 1949) and probably also in a variety of other objects near the Eddington limit (Joss, Salpeter, and Ostriker 1973). The author considers three aspects: first, he lists a set of approximate equations for plane-parallel photoconvection; then he gives a schematic treatment of the onset of instability; and finally, he outlines some of the arguments for believing that photon bubbles occur in the nonlinear regime.

065.026 **Convection in the helium flash.**   A. J. Wickett.
    Problems of stellar convection, (see 012.005),
p. 284 - 289 (1977).

The evolution of a star through the helium flash depends upon uncertain aspects of convection theory. Observations place some constraints on the theory of convection in stellar cores.

065.027 **Wave transport in stratified, rotating fluids.**
    M. E. McIntyre.
Problems of stellar convection, (see 012.005), p. 290 - 314 (1977).

Momentum and energy transport by buoyancy-Coriolis waves is illustrated by means of a simple model example. The need for careful consideration of a complete problem for mean-flow evolution is emphasised, especially when moving media are involved. Then a recent generalisation of the wave-action and pseudomomentum concepts is introduced, and used to exhibit in a very general way the roles of wave dissipation, forcing, or transience in the mean flow problem, for a certain class of 'nearly-unidirectional' mean flows. This class includes differentially-rotating stellar interiors, which could well be systematically changed by wave transport of angular momentum. Similar results hold for MHD and self-gravitating fluids. Finally the physical distinction between momentum and pseudomomentum is discussed.

065.028 **Wave generation and pulsation in stars with convective zones.**   W. Unno.
Problems of stellar convection, (see 012.005), p. 315 - 324 (1977).

Wave generation processes are classified in (1) strong and (2) weak, and (a) spontaneous and (b) stimulated processes. Then, the case (2b) operating in convective zones is discussed in detail. Both the dynamical and the thermodynamical coupling between pulsation and convection are formulated by use of the diffusion approximation for the turbulent convection. A mixing length variable with time is thereby introduced.

065.029 **Stellar convection.**   D. O. Gough.
    Problems of stellar convection, (see 012.005),
p. 349 - 363 (1977).

065.030 **The shock wave interpretation of individual condensations in Herbig-Haro objects.**   K. H. Böhm.
Astron. Astrophys., Vol. 64, 115 - 118 (1978).

The typical time scales, sizes, radial velocities, and the small filling factors of individual condensations of Herbig-Haro objects can be understood if we interpret the condensations as (approximately) spherical, nonstationary shock waves. Making use of the Sedov-Taylor solution (as modified by Cox) all the typical parameters listed above are shown to have values which are in approximate agreement with observations. The results depend only on the preshock particle density and on the initial energy input into the shock wave.

065.031 **The thermal stability of hot degenerate stars in steady-state accretion.**

E. M. Sion, M. J. Acierno, D. A. Turnshek.
Astrophys. J., Vol. 220, 636 - 639 (1978).

Models of high-luminosity degenerate stars have been constructed in the mass range $1.2 \leqslant M/M_\odot \leqslant 1.38$. They derive all of their energy from hydrogen burning near the surface in a steady state with accretion. The authors have analyzed the thermal stability of these models using both a linear analysis and evolutionary calculations. The models are thermally stable in all cases with the stability increasing with increasing luminosity. The bearing of their results on X-ray sources and pulsations of white dwarfs driven by hydrogen shell burning is briefly discussed.

065.032 **On the surface composition of thermally pulsing stars of high luminosity and on the contribution of such stars to the element enrichment of the interstellar medium.**   I. Iben, Jr., J. W. Truran.
Astrophys. J., Vol. 220, 980 - 995 (1978).

The enhancements of ^{3}He, ^{4}He, ^{12}C, ^{13}C, ^{14}N, ^{22}Ne, and s-process elements at the surfaces of stars that ultimately experience the thermal-pulse phenomenon are studied in some detail. By combining the results of stellar evolution calculations with a standard birthrate function, sources of several elements in the interstellar medium are identified.

065.033 **The hydrodynamics of stellar collapse.**
    K. A. Van Riper.
Astrophys. J., Vol. 221, 304 - 319 (1978).

Adiabatic (no transport of energy) evolutionary calculations are made to study the pure hydrodynamic motions of collapsing cores of massive stars. The results of the evolution, in particular the densities and amplitude of the bounce, and the energy ejected from the surface by a reflected shock wave, are presented for a large number of parameters in the equation of state. Emphasis is placed in finding those parameters most favorable for mass ejection. The ejected energy depends mainly on the adiabatic index in the mantle rather than on the energy in the bounce.

065.034 **Signatures of the ^{22}Ne neutron source in red giants and planetary nebulae.**   J. M. Scalo.
Astrophys. J., Vol. 221, 627 - 634 (1978).

The operation of the ^{22}Ne$(\alpha, n)^{25}$Mg neutron source should affect the relative abundances of the Mg isotopes and the elemental abundances of Mg and Ne in red giants which exhibit s-process enhancements and in planetary nebulae whose progenitors were peculiar red giants. The predicted abundance changes as a function of s-process enhancement are derived using the published s-process calculations for double shell source models of Truran and Iben (1977).

065.035 **Shock structure and neutrino radiation in stellar collapse.**   S. W. Bruenn, J. R. Buchler,
W. R. Yueh.
Astrophys. J., Lett., Vol. 221, L83 - L86 (1978).

Neutrino transport through a shock front in the context of a collapsing stellar core is carefully considered. The authors show (1) that the neutrino transport produces negligible shock broadening and that the shock width is therefore microscopic; and (2) that numerical stellar collapse calculations using a pseudoviscosity can thus seriously underestimate the neutrino flux across the shock unless unusually fine zoning is used in the shock region. This increased neutrino flux should yield the soughtfor stress necessary to produce a supernova explosion.

065.036 **Upper mass limit for stars dying as white dwarfs.**
    W. Romanishin, J. R. P. Angel.
Bull. American Astron. Soc., Vol. 9, 567 (1978). – Abstract.

065.037 **Correlation of mass loss rates and ^{12}C/^{13}C ratios in luminous carbon stars.**   B. Zuckerman.

Bull. American Astron. Soc., Vol. 9, 573 (1978). — Abstract.

065.038 **On neutrino viscosity in collapsing stellar cores.**
D. Kazanas.
Bull. American Astron. Soc., Vol. 9, 595 (1978). — Abstract.

065.039 **Presupernova evolution of massive stars.**
T. A. Weaver, G. B. Zimmerman, S. E. Woosley.
Bull. American Astron. Soc., Vol. 9, 596 (1978). — Abstract.

065.040 **Collapsing stellar cores and supernovae.**
H. Nørgaard, R. I. Epstein, R. Bond.
Bull. American Astron. Soc., Vol. 9, 597 (1978). — Abstract.

065.041 **Off-center He ignition and the subgiant CH star**
phenomenon.   R. L. Wagner.
Bull. American Astron. Soc., Vol. 9, 603 (1978). — Abstract.

065.042 **Convection in pulsating stars.**   M. J. Costello.
Bull. American Astron. Soc., Vol. 9, 637 (1978).
Abstract.

065.043 **Stochastic convection: how constant is the solar**
"constant"?   M. J. Newman, D. Dearborn.
Bull. American Astron. Soc., Vol. 9, 638 (1978). — Abstract.

065.044 **Evolution of the α Centauri system.**
B. P. Flannery, T. R. Ayres.
Bull. American Astron. Soc., Vol. 9, 638 (1978). — Abstract.

065.045 **Photonuclear p-processing in degenerate hydrogen**
burning regions and its relationship to nova out-
bursts.   T. G. Harrison.
Bull. American Astron. Soc., Vol. 9, 638 - 639 (1978).
Abstract.

065.046 **Very hot hydrogen burning.**
S. E. Woosley, R. K. Wallace.
Bull. American Astron. Soc., Vol. 9, 639 (1978). — Abstract.

065.047 **The evolution of massive close binaries. VI: Final**
considerations for the conservative case.
J. P. de Grève, C. de Loore, E. L. van Dessel.
Astrophys. Space Sci., Vol. 53, 105 - 130 (1978).

The results of evolutionary computations for massive
binary systems (initial masses of the primary $\gtrsim 10\,M_\odot$) with
mass ratios between 0.3 and 0.8 are summarized and compared
with observations in order to verify how far one can go with
the conservative assumption of mass exchange. It is found that
conservative mass exchange leads to acceptable first-order
models of W–R and massive X-ray binaries.

065.048 **The effect of mass-loss and semiconvection on the**
evolution of a 15 $M_\odot$ Population I star.
S. R. Sreenivasan, W. J. F. Wilson.
Astrophys. Space Sci., Vol. 53, 193 - 216 (1978).

Evolutionary sequences are computed from the main
sequence to central helium exhaustion for a 15 $M_\odot$ star, with
an initial composition of $X = 0.70$, $Y = 0.27$, $Z = 0.03$. Parallel
sequences are computed to investigate the effects of different
mass loss rates on the evolution of the star. These rates are
chosen to reflect the physical causes of the mass loss, and
occur at all phases of evolution. One sequence without, and
one with, mass loss are recomputed, allowing for semiconvection and full convection in intermediate mass zones, using the
Schwarzschild and Härm criterion for convective neutrality.
Low to moderate rates of mass loss in the early evolutionary
phases shift the evolution to lower luminosities and effective
temperatures, but do not radically alter the form of evolution.
If semiconvective and intermediate fully convective zones are
included, then in a sequence without mass loss these zones
greatly alter the chemical profile of the model. Observations

indicate that the blue supergiant region is wider and bluer than
predicted by previous evolutionary calculations. The present
results show that mass loss widens and reddens this phase.
Hence, the inclusion of other factors will be necessary to reconcile theory and observations.

065.049 **'Primary' elements in H-burning.**
V. Castellani, M. Sacchetti.
Astrophys. Space Sci., Vol. 53, 217 - 221 (1978).

The behaviour of intermediate nuclei taking part in H-
burning is analysed. Comparing time scales for equilibrium
with the time scale of convective mixing, the authors find that
Be_7, C_{13}, N_{15}, O_{17} cannot be assumed everywhere as bona
fide secondary elements in stellar evolutionary computations.
Some consequences of the onset of CNO burning are also discussed.

065.050 **Double core evolution. I. A 16 $M_\odot$ star with a 1 $M_\odot$**
neutron-star companion.
R. E. Taam, P. Bodenheimer, J. P. Ostriker.
Astrophys. J., Vol. 222, 269 - 280 (1978) = Lick Obs. Bull.,
No. 789.

The penetration of an orbiting 1 $M_\odot$ neutron star into a
16 $M_\odot$ supergiant companion is investigated. Primary emphasis
is placed on the structure and evolution of the massive component. It is found that double core evolution leads to either
(1) hydrodynamic ejection of part or all of the envelope or (2)
coalescence of the two cores with little or no mass ejection —
possibly resulting in a Thorne-Żytkow object. Some implications regarding the binary pulsar are discussed in the context
of the double core scenario.

065.051 **Secular instability of rotating Newtonian stars.**
J. L. Friedman, B. F. Schutz.
Astrophys. J., Vol. 222, 281 - 296 (1978).

The effect of gravitational radiation and of viscosity on
the stability of rotating self-gravitating fluids is considered.
Previous criteria governing secular stability to radiation are
shown to fail as a result of the trivial displacements introduced
in a previous paper. The required modification is obtained by
describing perturbations in terms of canonical diplacements
(displacements orthogonal to the trivials). The behavior of
normal modes is discussed and used to elucidate the generic
radiation-induced instability; and certain orthogonality properties are derived.

065.052 **Convection and stellar structure.**
I. W. Roxburgh.
Astron. Astrophys., Vol. 65, 281 - 285 (1978).

The "theory" of convection used to model stellar
convective zones neglects the flux of turbulent kinetic
energy and is therefore only applicable to zones that are
small compared to a scale height, yet the resulting models
have convective zones that are several scale heights thick.
In this paper the author attempts to quantify the error
involved by retaining the kinetic energy flux and assuming
that the viscous dissipation is small. This "theory" leads to
a new criterion to determine the extent of convective
zones which requires the convection to penetrate into
surrounding stable layers.

065.053 **On convection in a periodic gravitational field. II.**
L. N. Ivanov.
Astrofizika, Vol. 13, 703 - 710 (1977). In Russian. English
translation in Astrophysics, Vol. 13, No. 4.

Expressions for the convective energy flux with periodic
changes of the gravitational field are obtained in the mixing-
length approximation. The energy flux is determined by the
ratio of the kinematic τ_c and radiative τ_r time scales of convection. In the case of a red dwarf $\tau_r > \tau_c$, and the phase shift
of the convective flux oscillation is positive; for cepheids
$\tau_c > \tau_r$, and the shift is negative.

065.054   Theory of evolution of the central star.
G. Shaviv.
IAU Symp. No. 76, (see 012.014), p. 195 - 199 (1978).

065.055   Theory of evolution of central stars of planetary
nebulae.   B. Paczyński.
IAU Symp. No. 76, (see 012.014), p. 201 - 205 (1978).

065.056   Hydrogen- and helium-shell flashes and FG Sagittae
phenomenon.
D. Sugimoto, M. Y. Fujimoto, K. Nariai, K. Nomoto.
IAU Symp. No. 76, (see 012.014), p. 208 (1978) — Abstract.

065.057   Proto-planetary nebulae.   B. Zuckerman.
IAU Symp. No. 76, (see 012.014), p. 305 - 313
(1978).
The observation and identification of proto-planetary
nebulae appears reasonably well established. Continued study
of such objects promises to help the understanding of stellar
evolution including, for example, the physical mechanism(s)
responsible for the ejection of planetary nebulae. The elemental composition of these planetaries and their progenitors helps
to regulate the chemical evolution of the Galaxy.

065.058   Radio measurements of possible proto-planetary
nebulae.   C. R. Purton, P. A. Feldman.
IAU Symp. No. 76, (see 012.014), p. 325 (1978). — Abstract.

065.059   A single star model for V 1016 Cygni.
F. J. Ahern, M. P. FitzGerald, K. A. Marsh, C. R.
Purton.
IAU Symp. No. 76, (see 012.014), p. 326 (1978). — Abstract.

065.060   Evolutionary sequences for red giant stars.
A. V. Sweigart, P. G. Gross.
Astrophys. J., Suppl. Ser., Vol. 36, 405 - 437 (1978).
A set of 46 evolutionary sequences has been computed for stars ascending the red giant branch for the first
time. These red giant sequences, extending from the subgiant branch to the onset of helium burning within the
core, cover the following ranges in the helium abundance
Y, the heavy-element abundance Z, and the mass
M: $0.10 \leq Y \leq 0.40$, $0.00001 \leq Z \leq 0.04$, and
$0.70 \leq M \leq 2.20\,M_\odot$. Combinations of these values appropriate to galactic and globular clusters have been
emphasized. The effects of varying the rate of neutrino
emission from zero to twice the normally adopted value
have also been investigated.

065.061   On the rate of pycno-nuclear reactions.
E. Schatzman.
Astron. Astrophys., Vol. 65, L17 - L19 (1978).
Recent improvements in the calculation of the pair correlation function in a dense plasma, and the introduction by
Jancovici (1977) of the technique of the path integral, gives
the possibility of new estimates of the screening correction
factor to thermonuclear reactions. It is important to notice
that the theory gives a relatively low value of the increased
rate factor for carbon burning.

065.062   The low-temperature photonuclear nucleosynthesis
of the bypassed (p-) nuclei in degenerate hydrogen
burning zones and its relationship to nova outbursts.
T. G. Harrison.
Astrophys. J., Suppl. Ser., Vol. 36, 199 - 216 (1978).
The author has investigated the likelihood that the canonical p-nuclei (bypassed nuclei) may have been synthesized, at
least in part, within degenerate hydrogen burning zones via
low temperature photonuclear reactions triggered by photons
released through proton capture by triply heavy hydrogen
(^{3}T). He shows that a likely astrophysical site for the photonuclear synthesis of the nuclei exists in white dwarfs accret-

ing surface layers of hydrogen-rich material, eventually leading to nova outbursts as shown by Starrfield, Truran, and
Sparks. Under the conditions of degenerate hydrogen burning,
the author computes the abundances for several exposure
periods, and finds reasonable agreement with that for observed
solar system p-nuclei, providing the solar s-nuclei distribution
characterizes that for the seed nuclei. He shows that, on a
galactic scale, the observed nova rate leads to essentially the
correct number of heavy p-nuclei relative to hydrogen.

065.063   Nuclear reactions in Ap stars.   M. Koval'skij.
Fiz. ehlementar. chastits i atom. yadra, Vol. 8,
1134 - 1154 (1977). In Russian. — Abstr. in Ref. zh., 51.
Astron., 3.51.576 (1978).

065.064   On the solution of the equilibrium equations for
rapidly rotating stars.   M. J. Clement.
Astrophys. J., Vol. 222, 967 - 975 (1978).
A two-dimensional, finite-difference technique is presented for the solution of the general stellar structure equations in the case of conservative rotation. The relaxation method makes no approximation with respect to the gravitational
potential and is not limited by large central concentration.
Numerical tests on differentially rotating models of upper-
main-sequence stars show that convergence is very rapid and
also that the technique can be applied to a wide range of
stellar masses with rotation rates up to the point of secular
instability.

065.065   On neutrino viscosity in collapsing stellar cores.
D. Kazanas.
Astrophys. J., Lett., Vol. 222, L109 - L111 (1978).
If, as current numerical models show, neutrinos get
trapped and become degenerate in the central region of collapsing stellar cores, then they should significantly contribute
to their shear viscosity. The magnitude of neutrino viscosity
($\sim 10^{11}$ cm^2 s^{-1}) is such that it would tend to make a fast
rotating stellar core stable beyond the point of bifurcation.
The bulk viscosity of such systems is also found to be big, and
of the same order of magnitude. The above effects, along with
thermal conduction, will tend to damp out acoustical waves
generated in a collapsing stellar core after its hydrodynamic
bounce.

065.066   Static stellar models with helium cores. II. Linear
series of a 7 $M_\odot$ star with varying helium core mass.
D. Lauterborn, W. Zeuge.
Astron. Astrophys., Vol. 66, 367 - 376 (1978).
Numerical computations were carried out for a linear
series of static stellar models of a 7 $M_\odot$ star consisting of a
hydrogen rich envelope and a pure helium core of varying mass.
The linear series starts at the hydrogen main sequence and
terminates at the helium main sequence. Discussing core and
envelope structures separately in the $U - V$ diagram, the
physical causes of the various solution branches could be elucidated.

065.067   Theory of red giant variable stars.   R. S. Lipman.
J. American Assoc. Variable Star Obs., Vol. 6, 51 -
52 (1977/78).
The development and structure of red giant variables is
reviewed qualitatively.

065.068   The effect of the heavy element abundance on the
evolution of stars.   C. Alcock, B. Paczyński.
Astrophys. J., Vol. 223, 244 - 251 (1978).
The authors have evolved stars of mass 2, 3, 5, 7, and
10 $M_\odot$ and chemical compositions ($X = 0.7$, $Z = 0.03$),
(0.7, 0.01), (0.7, 0.003), (0.7, 0.001), and (0.7, 0.0004)
past core helium exhaustion (except for the 2 $M_\odot$ stars which
the authors evolved to core helium ignition). The results
demonstrate how the evolution of massive stars is determined

by their initial heavy element abundance and should be useful in the study of galactic evolution.

065.069   Use of Roche coordinates in the problems of small oscillations of rotationally distorted stellar models.
C. Mohan, V. P. Singh.
Astrophys. Space Sci., Vol. 54, 293 - 304 (1978).

The system of Roche coordinates developed by Kopal to study the problems of stars in close binary systems has been used to study the problems of small oscillations of rotationally distorted stars.

065.070   Exact MHD solutions for rotating, magnetic stars.
A. F. Cheng.
Astrophys. Space Sci., Vol. 54, 315 - 321 (1978).

Simple exact solutions of the magnetohydrodynamic equations are found for rotating, magnetic stars. The velocity and magnetic field are axisymmetric and purely toroidal, and the magnetic energy density equals the kinetic energy density. For constant mass density, the solution reduces to that of Chandrasekhar (1956), which is stable even against non-axisymmetric perturbations. For an ideal gas equation of state, the condition for radiative thermal equilibrium is solved to lowest order in the non-spherical perturbation. The velocity, magnetic field and non-spherical pressure and temperature perturbations all vanish within cones centered around the rotation axis, $|\cos \theta| > x_l$ a zero of a Legendre polynomial. Low-order, long-period stellar oscillations may be excited by MHD instabilities near the equatorial region which become damped near the axis.

065.071   The oscillation frequencies of magnetic stellar models.   D. Moss.
Astrophys. Space Sci., Vol. 54, 445 - 452 (1978).

A first-order perturbation theory method developed by Goossens to determine the perturbation to the eigenfrequencies of stellar models caused by the presence of a magnetic field is modified slightly, and applied to models with toroidal and poloidal fields. Some limitations of the analysis are pointed out.

065.072   Effects of mass loss on the evolution of massive stars. I. Main-sequence evolution.   D. S. P. Dearborn, J. B. Blake, K. L. Hainebach, D. N. Schramm.
Astrophys. J., Vol. 223, 552 - 556 (1978).

A systematic study on the effects of mass loss on OB stars is conducted by following the evolution of 15, 30, and $60 M_\odot$ models with varying rates of mass loss. The nucleosynthesis (particularly of the CNO isotopes) is followed in order to allow detailed comparison between the models and actual stars.

065.073   On the slow-rotator-phenomenon of Ap stars.   G. Rüdiger.
Magnetic stars, (see 012.028), p. 43 (1978). – Abstract.

065.074   Convección en estrellas frías.   J. M. Massaguer Navarro.
Summary of a thesis. Univ. Barcelona, Dep. Fis. Tierra Cosmos, Publ. No. A-29, 18 pp. (1977).

065.075   The eddy viscosity approach to stellar convection.   R. G. Deupree.
Bull. American Astron. Soc., Vol. 10, 400 (1978). – Abstract.

065.076   A new look at semiconvection.   L. D. Cloutman.
Bull. American Astron. Soc., Vol. 10, 400 (1978). Abstract.

065.077   Further studies of the helium-driven r-process.   J. J. Cowan, A. G. W. Cameron, J. W. Truran.
Bull. American Astron. Soc., Vol. 10, 437 - 438 (1978).

Abstract.

065.078   The evolution of massive stars losing mass and angular momentum, II. The effects of shear due to differential rotation.   S. R. Sreenivasan, W. J. F. Wilson.
Bull. American Astron. Soc., Vol. 10, 438 (1978). – Abstract.

065.079   The evolution of O stars and the origin of Wolf-Rayet stars.   C. de Loore, J. P. De Grève, D. Vanbeveren.
Astron. Astrophys., Vol. 67, 373 - 379 (1978).

Evolutionary sequences for massive stars (100, 80, 60 and 50 $M_\odot$) were computed, taking mass loss by stellar winds into account. A semi-empirically determined formula for the mass loss rates was used. The computations reveal that if the proportionality constant between $\dot{M}$ (in solar masses per year) and the luminosity (in solar luminosities) has the value 1.95 × $10^{-11} (\dot{M} \sim 10^{-5} M_\odot \text{yr}^{-1})$. These models can explain the overluminosity of the OB-supergiant companions of massive X-ray binaries, and the occurrence of the OBN stars.

065.080   An exact perturbation solution to the equation $\epsilon XY \, dY/dX = X - Y (Y + 1)$.   R. E. Mickens.
Acta Phys. Polonica B, Vol. B9, 75 - 77 (1978). – Abstr. in Phys. Abstr., Vol. 81, Abstr. 32439 (1978).

065.081   On the structure of uniformly rotating polytropes.   B. L. Smith.
Indian J. Pure Appl. Math., Vol. 6, 918 - 930 (1975). – Abstr. in Phys. Abstr., Vol. 81, Abstr. 32441 (1978).

065.082   Approximate analytic solution of a strong shock with radiation near the surface of the star.
S. Ashraf, Z. Ahmad.
Indian J. Pure Appl. Math., Vol. 6, 1090 - 1098 (1975). Abstr. in Phys. Abstr., Vol. 81, Abstr. 32442 (1978).

065.083   Similarity solutions for explosions in radiating stars.   G. Deb Ray, J. B. Bhowmick.
Indian J. Pure Appl. Math., Vol. 7, 96 - 103 (1976). – Abstr. in Phys. Abstr., Vol. 81, Abstr. 32443 (1978).

065.084   Mirror planes in Newtonian stars with stratified flows.   L. Lindblom.
J. Math. Phys., Vol. 18, 2352 - 2355 (1977). – Abstr. in Phys. Abstr., Vol. 81, Abstr. 32444 (1978).

065.085   Pair correlation function in a dense plasma and pycnonuclear reactions in stars.   B. Jancovici.
J. Stat. Phys., Vol. 17, 357 - 370 (1977). – Abstr. in Phys. Abstr., Vol. 81, Abstr. 32445 (1978).

065.086   On the bulk yields of nucleosynthesis from massive stars.   W. D. Arnett.
IAU Colloq. No. 45, (see 012.032), p. 161 - 164 (1977/78).

065.087   s-process elements and galactic evolution.   B. Rocca-Volmerange, J. Audouze.
IAU Colloq. No. 45, (see 012.032), p. 197 - 200 (1977/78).

The evolution of the s-elements is analyzed by taking into account the mass of the stars responsible of their formation and by considering the different possible mechanisms which have been proposed to explain their nucleosynthesis.

065.088   Use of Roche coordinates in the problems of small oscillations of tidally distorted stellar models.
C. Mohan, V. P. Singh.
Astrophys. Space Sci., Vol. 56, 109 - 117 (1978).

The system of Roche coordinates developed by Kopal to study the problems of stars in close binary systems has been used to study the problems of small oscillations of tidally distorted stars.

065.089   On the angular momentum loss of late-type stars.
B. R. Durney, J. Latour.
Geophys. Astrophys. Fluid Dyn., Vol. 9, 241 - 255 (1978).
Abstr. in Phys. Abstr., Vol. 81, Abstr. 45331 (1978).

065.090   Self-similar adiabatic motions of a self-gravitating
gas within stars.   O. I. Bogoyavlenskij.
Pis'ma v ZhEhTF, Vol. 27, 91 - 94 (1978). In Russian. — Abstr.
in Ref. zh., 51. Astron., 6.51.190 (1978).

Low-energy proton reactions on ^{45}Sc of interest in
stellar nucleosynthesis.   See Abstr. 022.018.

Remark on the polytrope of index 5.
See Abstr. 042.047.

On the bulk yields of nucleosynthesis from massive
stars.   See Abstr. 061.004.

The boundaries of a convective zone.
See Abstr. 061.013.

Reaction rates for neutrino processes.
See Abstr. 061.028.

Normal modes of self-gravitating fluids in perturbed configurations. I. Perturbational-variational procedure.
See Abstr. 062.008.

Normal modes of self-gravitating fluids in perturbed
configurations. II. Perturbational-variational expansions of the
g- and p- modes of a non-adiabatic fluid about the adiabatic
limit.   See Abstr. 062.009.

Adiabatic collpase of rotating gas clouds.
See Abstr. 062.071.

On the number of unstable modes of an equilibrium.
See Abstr. 062.076.

Operator algebra and the stationary states of stellar
magnetic fields.   See Abstr. 062.104.

Optical information on mass-loss from evolved stars.
See Abstr. 064.049.

Hydrodynamical instabilities in the envelopes of
main-sequence stars: constraints implied by the lithium,
beryllium, and boron observations.   See Abstr. 064.079.

A study of the microturbulence in giants in 'terms
of stellar evolution.   See Abstr. 064.103.

Nuclear burning in accreting neutron stars and X-ray
bursts.   See Abstr. 066.016.

Ultraviolet observations of nine Wolf−Rayet stars.
See Abstr. 114.015.

Mixing and the strong-cyanogen phenomenon.
See Abstr. 114.038.

Giant problems.   See Abstr. 114.054.

HD 91805 and the nature of the Bidelman-MacConnell weak-G-band stars.   See Abstr. 114.542.

Masses of peculiar red giants.
See Abstr. 115.008.

Evolution in binary systems.   See Abstr. 117.009.

Young W UMa binaries and the interaction with
their environment.   See Abstr. 117.022.

On the vibrational stability of stars in thermal
imbalance.   See Abstr. 122.026.

Double-mode cepheid period ratios from linear and
nonlinear theory.   See Abstr. 122.027.

On estimating the intrinsic properties of Miras from
observational data.   See Abstr. 122.040.

Mass loss and Cepheid pulsation.
See Abstr. 122.092.

An interpretation of the unique triple mode variable
AC Andromedae as an RR Lyrae type star oscillating in the
first three radial over-tones.   See Abstr. 122.115.

On the Rayleigh-Taylor instability in stellar explosions.   See Abstr. 125.005.

Analytic supernova models and black holes.
See Abstr. 125.040.

Immiscibilities in cold, degenerate stars.
See Abstr. 126.027.

Hydrogen shell flashes in a white dwarf with mass
accretion.   See Abstr. 126.032.

Nonaxisymmetric models of collapsing, rotating
protostars.   See Abstr. 131.220.

Models of clouds in adiabatic gravitational contraction.   See Abstr. 131.248.

UV observations and stellar evolution.
See Abstr. 131.258.

Spectra of Cassiopeia A. II. Interpretation.
See Abstr. 134.006.

Origin of planetary nebulae.   See Abstr. 135.084.

PDS photometry of the open cluster NGC 2420.
See Abstr. 153.037.

Post He flash evolution in globular clusters. Observations vs. theory.   See Abstr. 154.016.

Properties of the stars of early generations in the
scale covariant cosmology.   See Abstr. 162.004.

066 Neutron Stars, Relativistic Astrophysics, Background Radiation, Gravitation Theory

066.001 On the existence of ergoregions in rotating stars.
B. F. Schutz, N. Comins.
Mon. Not. R. Astron. Soc., Vol. 182, 69 - 76 (1978).

Very compact rotating stars in general relativity can in principle contain regions, called ergoregions, in which all trajectories must rotate in the direction of the star's rotation. These regions are known to create an instability in the star, and it is therefore of physical interest to know whether realistic stars can contain ergoregions. The authors develop an approximation scheme which enables them to study ergoregions with a minimum of numerical effort. They apply the method to the more realistic degenerate-neutron equation of state (Harrison—Wakano—Wheeler), for which no other ergoregion calculations have been made, and conclude that no realistic neutron star, even with strong differential rotation, can develop an ergoregion. The ergoregion instability is therefore only likely to have astrophysical importance if stars significantly more compact than ordinary neutron stars exist.

066.002 General scalar-tensor theory of gravity with constant G. B. M. Barker.
Astrophys. J., Vol. 219, 5 - 11 (1978).

A special case of the general scalar-tensor theory of gravity is proposed where the Newtonian gravitational constant G does not vary with time. This implies that $\omega = (4 - 3\phi)/(2\phi - 2)$ and the Nordtvedt parameter $\eta = 4\beta - \gamma - 3 = 0$. The resulting cosmological equations are discussed and shown to be consistent with any assumed values for Hubble's constant, the deceleration parameter, and the mean mass density of the universe. The possibility exists that the true theory of gravitation is a scalar-tensor theory, similar to general relativity in its effects at the present epoch, but very different from general relativity in the distant past and in the distant future.

066.003 Compton scattering of microwave background radiation by gas in galaxy clusters.
R. J. Gould, Y. Rephaeli.
Astrophys. J., Vol. 219, 12 - 17 (1978).

Based on data on the X-ray spectrum of the Coma cluster, interpreted as thermal bremsstrahlung, the expected brightness depletion from Compton scattering of the microwave background in the direction of the cluster is computed. The calculated depletion is about one-third that recently observed by Gull and Northover, and the discrepancy is discussed.

066.004 The fate of matter and angular momentum in disk accretion onto a magnetized neutron star.
E. T. Scharlemann.
Astrophys. J., Vol. 219, 617 - 628 (1978).

Observed values for the average rate of decrease of the rotational period for seven binary X-ray sources are used to establish limits on the outer radius R_m of the magnetospheres of the neutron stars. In all cases this radius is found to be significantly less than the "corotation radius", where the Keplerian orbital angular velocity matches the stellar angular velocity. A simple but plausible model for the shape and location of R_m in disk accretion is developed to follow the matter from the disk onto the star; this model permits an estimate of the magnetic dipole moments and radiating areas of the individual neutron stars.

066.005 Asymptotic freedom and dense stellar matter. II. The equation of state for neutron stars.
M. B. Kislinger, P. D. Morley.
Astrophys. J., Vol. 219, 1017 - 1028 (1978).

The behavior of the equation of state for cold, dense hadronic matter in asymptotically free gauge theories is examined as nuclear densities are approached from above. By requiring that the energy per baryon be a nucleon mass in the nuclear density regime the authors determine the on-shell mass for the lightest quarks to be about 355 MeV. They find that hadronic matter is highly incompressible above nuclear densities. Hyperonic matter does not appear at nuclear densities and begins to come in only above $7 \times 10^{14} g\ cm^{-3}$. The softening of the equation of state above $7 \times 10^{14} g\ cm^{-3}$ limits neutron star masses from getting exceptionally high, and the authors find the maximum to be $2.34\ M_\odot$ with central density $10^{15} g\ cm^{-3}$. The ratio of hyperonic matter to normal matter is 10% at $10^{15} g\ cm^{-3}$ for this maximum star mass. The critical density, ρ_c, above which perturbation theory can be reliably used is $6 \times 10^{14} g\ cm^{-3}$ for the currently favored experimental value of the quark fine structure constant.

066.006 Self-similar growth of primordial black holes. I. Stiff equation of state.
G. V. Bicknell, R. N. Henriksen.
Astrophys. J., Vol. 219, 1043 - 1057 (1978).

The authors discuss the possible self-similar growth of primordial black holes in a "stiff" early universe. They conclude that such growth is possible only if the black hole accretes the matter as a null-fluid inside a certain surface, called here the phase-change surface. This has required the authors to match an ingoing self-similar Vaidya solution to the self-similar $p = \epsilon$ matter solution. They also discuss carefully the rarefaction waves that play an essential role in the accretion flow. Numerically they agree with an earlier result by Lin et al. (1976), that the maximum mass reached by primordial black holes in the stiff era is $\sim 1\ M_\odot$. However, the authors' description of the process is quite different from theirs.

066.007 Possible test of the strong principle of equivalence.
K. Brecher.
Astrophys. J., Lett., Vol. 219, L117 - L118 (1978).

The author suggests that redshift determinations of X-ray and γ-ray lines produced near the surface of neutron stars which arise from different physical processes could provide a significant test of the strong principle of equivalence for strong gravitational fields. As a complement to both the high-precision weak-field solar-system experiments and the cosmological time variation searches, such observations could further test the hypothesis that physics is locally the same at all times and in all places.

066.008 A search for exploding black holes.
N. A. Porter, T. C. Weekes.
Sky Telesc., Vol. 55, 113 - 114 (1978).

066.009 The analytic theory of fluid disks orbiting the Kerr black hole.
M. Kozłowski, M. Jaroszyński, M. A. Abramowicz.
Astron. Astrophys., Vol. 63, 209 - 220 (1978).

Abramowicz et al. (1977) have recently shown that a sharp cusp exists on the inner edge of the accretion disk (with constant angular momentum), orbiting the Kerr black hole. The cusp resembles very much a similar cusp located on the Roche lobe in the close binary case (Lagrange L_1 point) and therefore its existence is very important from the physical point of view. In this paper the authors show that the existence of the cusp is a typical phenomenon for any angular momentum distribution. They also discuss the physical importance of the cusps. It is proved that the inner edge of any stable disk cannot be closer to the black hole than the marginally bound circular orbit, $r = r_{mb}$.

066.010 Relativistic, accreting disks.
       M. Abramowicz, M. Jaroszyński, M. Sikora.
Astron. Astrophys., Vol. 63, 221 - 224 (1978).
An analytic theory of the hydrodynamical structure of accreting disks (without self-gravitation but with pressure) orbiting around an axially symmetric, stationary, compact body(e.g. black hole) is presented. The inner edge of the marginally stable accreting disk (i.e. disk with constant angular momentum density) has a sharp cusp located on the equatorial plane between r_{ms} and r_{mb}. The existence of the cusp is also typical for any angular momentum distribution.

066.011 Reflection of X rays by neutron star surfaces.
       R. Lenzen, J. Trümper.
Nature, Vol. 271, 216 - 220 (1978).
A neutron-star surface should act as a good reflector for photons up to hard X-ray energies, due to the high density of the magnetic surface matter. Reflectivities are high, especially in spectral bands above the plasma and cyclotron frequencies depending on the photon polarisation. X-ray reflection may be important for the beaming of the binary neutron star Her X-1 and similar objects.

066.012 Interpretation of observed cosmic microwave background radiation.   S. Pollaine.
Nature, Vol. 271, 426 - 427 (1978).
Alfvén and Mendis (1977) concluded recently that dust grains in galaxies rendered the Universe opaque to the cosmic microwave background at a redshift of $z = 40$, instead of the generally accepted value of $z = 1,500$. The author presents here some arguments to show that their assumptions are unreasonable and that a proper calculation in the standard big bang model shows that dust grains are transparent to the microwave background. The microwave background can give information on an early dense phase of the Universe.

066.013 Anisotropy in blackbody radiation shows Earth's motion.   G. B. Lubkin.
Phys. Today, Vol. 31, No. 1, p. 17 - 19 (1978).

066.014 Supergravity and the unification of the laws of physics.   D. Z. Freedman, P. van Nieuwenhuizen.
Sci. American, Vol. 238, No. 2, p. 126, 128 - 129, 131 - 139, 141 - 143 (1978).
In this new theory the gravitational force arises from a symmetry relating particles with vastly different properties. The ultimate result may be a unified theory of all the basic forces in nature.

066.015 Search for high energy γ-ray bursts from evaporation of primordial black holes.
D. J. Fegan, B. McBreen, D. O'Brien, C. O'Sullivan.
Nature, Vol. 271, 731 - 732 (1978).
The authors report the results of a search for γ rays associated with the evaporation of primordial black holes, and upper limits are presented based on the elementary particle model.

066.016 Nuclear burning in accreting neutron stars and X-ray bursts.   D. Q. Lamb, F. K. Lamb.
Astrophys. J., Vol. 220, 291 - 302 (1978).
The authors consider the general properties of nuclear burning in accreting neutron stars. They discuss the behaviour expected in the pycnonuclear and thermonuclear regimes, the conditions required for thermonuclear runaway, and the energy available from H, He, C, and O burning during an X-ray burst, assuming the burst arises from a thermonuclear flash. By using these results and making conservative assumptions, the authors derive some constraints placed on such models of

X-ray bursts by observation. They find that hydrogen burning appears to be ruled out as the source of the X-ray bursts.

066.017 The deformed figures of the Dedekind ellipsoids in the post-Newtonian approximation to general relativity: corrections and amplifications.
S. Chandrasekhar, D. D. Elbert.
Astrophys. J., Vol. 220, 303 - 313 (1978).
Two errors in the analysis of an earlier paper (Astrophys. J., Vol. 192, 731 (1974)) on the same subject are corrected. It is found that, as a consequence of the corrections, the solution to the post-Newtonian equations (appropriate to determining the deformed figures of the Dedekind ellipsoid) now diverges at a point where the axes of the ellipsoid are in the ratios 1 : 0.3370 : 0.2850. In addition, the fourth-harmonic oscillations of the Dedekind ellipsoid are considered; and it is found that it becomes dynamically unstable when its axes are in the ratios 1 : 0.3121 : 0.2680.

066.018 Relativistic astrophysics – a new branch of science on sky.   I. D. Novikov.
Zemlya i Vselennaya, 1978, No. 1, p. 12 - 15. In Russian.

066.019 Possibility of superlong gravitational waves detection.   M. V. Sazhin.
Astron. Zh. Akad. Nauk SSSR, Vol. 55, 65 - 68 (1978). In Russian. English translation in Soviet Astron., Vol. 22, No. 1.
The influence of superlong gravitational waves on the propagation of electromagnetic pulses is considered. The conditions under which it is possible to detect gravitational waves from binaries are specified. It is shown that it might be possible to detect gravitational radiation from binary superstars with masses $M_1 \approx M_2 \approx 10^{10} M_\odot$

066.020 Pion condensation and abnormal nuclear matter.
       M. Chanowitz, P. J. Siemens.
Phys. Lett. B, Vol. 70B, 175 - 179 (1977). – Abstr. in Phys. Abstr., Vol. 80, Abstr. 88069 (1977).

066.021 Abnormal neutron star matter at ultrahigh densities.
       S. A. Moszkowski, C. G. Kallman.
Nucl. Phys. A, Vol. A287, 495 - 500 (1977). – Abstr. in Phys. Abstr., Vol. 80, Abstr. 91560 (1977).

066.022 Les trous noirs existent-ils?   A. G. W. Cameron.
La recherche en astrophysique, (see 003.001), p. 109 - 121 (1977).

066.023 La relativité générale vérifiée.   J. Lequeux.
La recherche en astrophysique, (see 003.001), p. 235 - 240 (1977).

066.024 Are there force-free fields in neutron star interiors?
       J. Arons, R. G. Spencer.
Astrophys. J., Vol. 220, 640 - 642 (1978).
The force-free character of the magnetic field in the fluid interior of a static neutron star is investigated. It is found that the effects of a finite temperature gradient can upset the force-free configuration which might otherwise exist, but that under some possible conditions of temperature and density the magnetic field in the deep interior may indeed be force free.

066.025 Some problems with the interpretation of recent microwave background observations in the direction of galaxy clusters, or, beware of negative antenna temperatures.
J. C. Tarter.
Astrophys. J., Vol. 220, 749 - 755 (1978).
Recently reported observations of the 3 K microwave background in the direction of rich clusters of galaxies should be viewed as placing stringent limits on the mass of cooler ionized gas within the clusters, rather than as a verification of

thermal bremsstrahlung models for cluster X-ray sources. At the high radio frequencies employed in the observations, there is a positive contribution to the observed source brightness distributions from free-free emission by any cooler gas. This can overwhelm the anticipated inverse Compton diminution of the background radiation, even when the total mass in cooler gas is significantly less than the mass of hot plasma required to explain the X-ray source.

066.026 Radiative effects in supersonic accretion.
R. G. Carlberg.
Astrophys. J., Vol. 220, 1041 - 1050 (1978).

Supersonic gas flow onto a neutron star is investigated. There are two regimes of accretion flow, differentiated by whether the gas can cool significantly before it falls to the magnetosphere. If radiative losses are negligible, the captured gas falls inward adiabatically in a wide accretion column. If the radiative energy-loss time scale is less than the fall time, the gas will cool to some equilibrium temperature which determines the width of the wake. An accreting neutron star generates sufficient luminosity that radiation heating may determine the temperature of the accretion column, provided the accretion column is optically thin. Gas crossing the shock beyond the critical radius forms an extended turbulent wake which gradually merges into the surrounding medium. As a specific example, the flow for the range of parameters suggested for the stellar wind X-ray binaries is considered.

066.027 Post-Newtonian gravitational bremsstrahlung.
M. Turner, C. M. Will.
Astrophys. J., Vol. 220, 1107 - 1124 (1978).

The authors present formulae and numerical results for the gravitational radiation emitted during a low-deflection encounter between two massive bodies ("gravitational bremsstrahlung"). The results are valid through post-Newtonian order within general relativity. They discuss in detail the gravitational waveform (transverse-traceless part of the metric perturbation tensor), the total luminosity and total emitted energy, the angular distribution of emitted energy (antenna pattern), and the frequency spectrum.

066.028 Tidal fields in general relativity: d'Alembert's principle and the test rigid rod.
J. Faulkner, B. P. Flannery.
Astrophys. J., Vol. 220, 1125 - 1132 (1978) = Lick Obs. Bull., No. 773.

To the general relativist, tidal forces are a manifestation of the Riemann tensor; the relativist therefore uses the Riemann tensor to calculate the effects of such forces. In contrast, the authors show that the introduction of gravitational "probes" (or "test rigid rods") and the adoption of a viewpoint closely allied to d'Alembert's principle, give an enormous simplification in cases of interest. No component of the Riemann tensor need be calculated as such. In the corotating orbital case (or Roche problem) the calculation of the relevant distortional field becomes trivial. As a by-product of this investigation, there emerges an illuminating strong field generalization of de Sitter's weak field precession for slowly spinning gyroscopes.

066.029 X-ray spectrum from disk accretion onto massive black holes.
D. G. Payne, D. M. Eardley.
Astrophys. Lett., Vol. 19, 39 - 45 (1977).

The authors discuss a model for X-ray emission from massive black holes, $M \gtrsim 10^3$ M$_\odot$. An optically thin accretion disk gives thermal emission with $kT \sim 10$ keV $(M/10^3$ M$_\odot)^{-1/2}$ $(L/10^{37}$ erg s$^{-1})^{1/2}$. Such a disk could form either as a result of radiative preheating of infalling gas or as a corona on a standard, optically thick disk. Strong emission lines may be visible, and the large blueshifts (and redshifts) due to the black hole will lead to striking effects in the spectrum, notably an abrupt "relativistic edge".

066.030 Charged currrent loops around Kerr holes.
R. L. Znajek.
Mon. Not. R. Astron. Soc., Vol. 182, 639 - 646 (1978).

The author obtains spherical harmonic expansion for the field and potential of a stationary loop of charge and current situated axisymmetrically, but not necessarily equatorially with respect to a Kerr hole. He calculates the potential induced on the hole and shows that it depends only on the charge density of the loop as measured in the Carter frame.

066.031 Superdense neutron matter.
V. Canuto, B. Datta, G. Kalman.
Astrophys. J., Vol. 221, 274 - 283 (1978).

The authors present a relativistic theory of high-density matter which takes into account the short-range interaction due to the exchange of spin-2 mesons. An equation of state is derived and used to compute neutron star properties. The prediction of the theory for the values of maximum mass and moment of inertia for a stable neutron star are 1.75 $M_\odot$ and 1.68×10^{45} g cm^2. The corresponding radius is found to be 10.7 km. The authors find that the inclusion of the spin-2 interaction reduces the disagreement between the relativistic and nonrelativistic theories in their predictions of masses and moments of inertia.

066.032 Bimetric gravitation theory and PSR 1913 + 16.
N. Rosen.
Astrophys. J., Vol. 221, 284 - 285 (1978).

Will and Eardley (1977) found that, for the binary pulsar PSR 1913 + 16, the bimetric theory of gravitation (in contrast to the general relativity theory) predicts the emission of dipole gravitational radiation, which should cause rapid changes in the orbital period, except in certain special cases. It is pointed out that their conclusion that this binary pulsar provides the first feasible test of the viability of the theory may not be valid. This is because there are arguments for believing that both in the general relativity theory and in the bimetric theory a physical system cannot lose (or gain) energy by emitting gravitational radiation.

066.033 Gravitational radiation from stellar collapse: ellipsoidal models.
R. A. Saenz, S. L. Shapiro.
Astrophys. J., Vol. 221, 286 - 303 (1978).

The burst of gravitational radiation emitted during the initial collapse and rebound of a homogeneous ellipsoid with internal pressure is analyzed numerically. A configuration of mass 1.4 $M_\odot$ is assumed to collapse aspherically from rest at an initial density of 10^{10} g cm^{-3}. By following the collapse along two distinct low-entropy adiabats ("hot" and "cold" collapse) on free-fall time scales, the authors estimate the maximum amount of gravitational radiation liberated during the initial implosion of a compact stellar core. They calculate the total gravitational (quadrupole) radiation energy loss, the radiation spectra, and the wave forms for several collapse sequences, parametrized by their initial deviations from spherical symmetry. A minimum estimate of the total neutrino losses during this initial collapse phase is also provided.

066.034 Variable G.
W. H. McCrea.
Observatory, Vol. 98, 52 - 54 (1978).

Newtonian physics requires the gravitational constant to be a universal constant. Any replacement of the feature is a fundamental departure from such physics, and the procedure for effecting it is highly arbitrary. Possibilities in this regard seem more plausible if formulated in terms of variability of mass rather than of the gravitational 'constant'.

066.035 Effect of a changing G on the moment of inertia of the Earth.
R. A. Lyttleton, J. P. Fitch.
Astrophys. J., Vol. 221, 412 - 413 (1978).

A recent value found by Nordtvedt and Will for the rate of increase of the moment of inertia of the Earth resulting from a decreasing G is shown to be based on conflicting physical assumptions. In a realistic treatment, additional purely geophysical causes must also be taken into account, and these imply a decreasing moment of inertia on a scale much outweighing the increase that would occur on any acceptable value of $\dot{G}$. This accords with the intrinsic acceleration of angular velocity established from the ancient-eclipse data.

066.036   Varierer gravitasjonskonstanten med tiden?
P. Wesson.
Astron. Tidsskr., Årg.10, 162 - 163 (1977).

066.037   Neutronstjärnor och svarta hål, I.
G. Larsson-Leander.
Astron. Tidsskr., Årg.11, 13 - 28 (1978).

066.038   An introduction to quantum gravity.
C. J. Isham.
Quantum gravity, (see 012.007), p. 1 - 77 (1975).

066.039   Covariant quantization.   M. J. Duff.
Quantum gravity, (see 012.007), p. 78 - 135 (1975).

066.040   Quantum gravitation: trees, loops and renormalization.   S. Deser.
Quantum gravity, (see 012.007), p. 136 - 173 (1975).

066.041   Particle creation by black holes.   S. W. Hawking.
Quantum gravity, (see 012.007), p. 219 - 267 (1975).

066.042   Impact of quantum gravity theory on particle physics.   A. Salam.
Quantum gravity, (see 012.007), p. 500 - 537 (1975).

066.043   Is physics legislated by cosmogony?
C. M. Patton, J. A. Wheeler.
Quantum gravity, (see 012.007), p. 538 - 605 (1975).

066.044   Gravitational radiation from stellar collapse: ellipsoidal models.   R. A. Saenz, S. L. Shapiro.
Bull. American Astron. Soc., Vol. 9, 595 (1978). − Abstract.

066.045   Maximum mass of neutron stars formed in stellar collapse.   K. A. Van Riper.
Bull. American Astron. Soc., Vol. 9, 597 (1978). − Abstract.

066.046   A new disk source for the Kerr metric.
M. R. Bernstein.
Bull. American Astron. Soc., Vol. 9, 608 (1978). − Abstract.

066.047   Nuclear burning in accreting neutron stars and X-ray bursts.   D. Q. Lamb, F. K. Lamb.
Bull. American Astron. Soc., Vol. 9, 631 (1978). − Abstract.

066.048   A model for disk accretion by magnetic neutron stars.   P. Ghosh, F. K. Lamb.
Bull. American Astron. Soc., Vol. 9, 631 - 632 (1978). Abstract.

066.049   Bursts in accretion disks.
J. D. McKee, W. K. Rose.
Bull. American Astron. Soc., Vol. 9, 633 (1978). − Abstract.

066.050   White holes.   K. Lake.
Nature, Vol. 272, 599 - 601 (1978).
    The author shows that observable white holes ought to be considered as undelayed local inhomogeneities in the big bang, and he distinguishes those white holes which we are most likely to 'see' today.

066.051   Quantum fluctuations in gravitational collapse and cosmology.   J. V. Narlikar.
Mon. Not. R. Astron. Soc., Vol. 183, 159 - 168 (1978).
    It is shown that the conformal degrees of freedom in the metric tensor can be quantized and that this procedure leads to fluctuations around the solutions of the classical Einstein field equations. These fluctuations become progressively more important as the classical solution approaches the space−time singularity. An explicit calculation is given of the quantum mechanical propagator which describes the conformal fluctuations in a collapsing homogeneous ball of dust. The solution can also be applied to the Friedmann models with the conclusion that the Universe need not have originated in a unique classical big bang.

066.052   White holes of types II and III.
K. Lake, R. C. Roeder.
CNRS Colloq. No. 263, (see 012.009), p. 541 - 548 (1977).
    Les auteurs donnent une discussion qualitative des trous blancs dans le contexte d'un espace de Schwarzschild. Ils montrent que les configurations d'expansion (trous blancs) peuvent être classées en types I, II ou III suivant la région de l'espace de Schwarzschild dans lesquelles elles s'étendent à partir de la région IV. Ils montrent que si le trou est de type II ou III, la radiation reçue par un observateur peut avoir un décalage vers le rouge intrinsèque non lié au décalage cosmologique.

066.053   Soft gravitation radiation.
C. Alvegård, K.-E. Eriksson, C. Högfors.
Phys. Scr., Vol. 17, 95 - 102 (1978).
    The methods of soft photon radiation are applied to the linearized theory of gravitation in a radiation gauge similar to the radiation gauge in quantum electrodynamics. The time evolution of soft gravitation emission following any scattering process is studied in detail. Results obtained by Weinberg in a different way are reproduced.

066.054   The gauge in general relativity. II.
L. Stephenson, J. Cohn.
Gen. Relativ. Gravitation, Vol. 9, 21 - 38 (1978).
    The present paper and a companion work are intended as one possible realization of a noncustomary gauge theory of general relativity − as set forth in only broad outline in an earlier work. In this first paper, it is found that both radar echo delay and the perihelion shift differ slightly from their customary expressions. Unfortunately, it is also found that the usual statement of the principle of equivalence does not hold in the present formulation.

066.055   The gauge in general relativity. III.
L. Stephenson, J. Cohn.
Gen. Relativ. Gravitation, Vol. 9, 39 - 51 (1978).
    A noncustomary gauge theory of general relativity, developed in detail in the preceding paper (II), is here applied to cosmology. A universe that is homogeneous and isotropic in the customary gauge, is considered − first generally, and then in more detail for the case where the noncustomary universe is matter dominated and static. With a particular choice of σ equation, this model is solved and a new relation between customary mass density, Hubble constant, and deceleration parameter is found. For a customary deceleration parameter of 1.98, this relation yields a customary mass density of 3.1×10^{-31} g/cm^3 − in good agreement with experiment. Finally, the age of the universe in this model is found to be $> 6.6 \times 10^9$ yr, again in agreement with other estimates.

066.056   Exact one-body solution of the gauge-covariant gravitational field equations.   P. J. Adams.
Gen. Relativ. Gravitation, Vol. 9, 53 - 57 (1978).
    Dirac has recently introduced a new theory of gravity which includes his large numbers hypothesis. The exact one-

body vacuum solution consistent with multiplicative matter creation is obtained and discussed.

066.057 Polarized radiation in relativistic cosmology.
G. Dautcourt, K. Rose.
Astron. Nachr., Band 299, 13 - 23 (1978).

A method to solve the general-relativistic equation of radiative transfer for polarized light incorporating elastic Compton scattering is discussed. The method is based on an expansion in spin-0 and spin-2 spherical harmonics.

066.058 The neutrino radiation for a hot neutron star formation and the envelope outburst problem.
D. K. Nadyozhin (*Nadezhin*).
Astrophys. Space Sci., Vol. 53, 131 - 153 (1978).

With the equations of neutrino heat conductivity being used, the neutrino light curve is calculated for the spherically symmetrical collapse of an iron—oxygen $2\,M_\odot$ star up to the formation of a hot hydrostatically equilibrium neutron star. The maximum neutrino luminosity value is equal to 3×10^{53} erg s^{-1}. For a $10\,M_\odot$ star collapse, the luminosity maximum 3×10^{54} erg s^{-1} takes place just at the moment of the formation of a black hole inside the collapsing star. The total radiated energy in this case is about $0.08\,Mc^2$. The set of calculations, allowing for the deposition of momentum by means of neutrino—nuclear coherent scattering, brings the author to the conclusion that the envelope outburst is only possible if the scattering cross-section is 50 times larger than the value experimentally accepted.

066.059 The angular appearance of white holes.
J. V. Narlikar, R. C. Kapoor.
Astrophys. Space Sci., Vol. 53, 155 - 163 (1978).

It is shown that non-radial light rays emitted from the surface of a white hole can emerge from inside the Schwarzschild barrier. The upper limit on their impact parameter is calculated under the requirement that such rays are blueshifted. The apparent angular size of the white hole determined by blueshifted rays is shown to grow so rapidly in the early stages of its expansion that it produces the appearance of superluminal expansion.

066.060 Static spherical configurations of cold matter in the Einstein—Cartan theory of gravitation.
M. Demiański, M. Prószyński.
Astrophys. Space Sci., Vol. 53, 173 - 179 (1978).

The authors investigate static, spherical configurations of cold catalyzed matter in the Einstein—Cartan theory of gravitation. Assuming that density of spin is proportional to the number density of baryons n and using an equation of state of a degenerate, relativistic Fermi gas, the authors numerically integrated the relativistic equation of equilibrium. They have also studied the stability of those configurations.

066.061 Dodging Heisenberg.   N. MacDonald.
Nature, Vol. 272, 667 (1978).

066.062 Why do collapsed stars rotate so slowly?
K. Brecher, G. Chanmugam.
Astrophys. J., Vol. 221, 969 - 972 (1978).

The case of the puzzlingly slow rotation of neutron stars and white dwarfs is examined. Continuing mass loss during formation can rapidly stop rotation of white dwarfs which have magnetic field strengths of at least 10^6 gauss, for mass loss rates consistent with current theories. If mass loss can occur during formation of neutron stars, magnetic braking may be significant in slowing their rotations as well. The observed slow rotation of the "nonmagnetic" white dwarfs, however, remains a mystery.

066.063 An upper limit to the rate of formation of neutron stars in the Galaxy.   J. G. Hills.
Astrophys. J., Vol. 221, 973 -974 (1978).

Energy conservation and the observed luminosity of the Galaxy place a firm upper limit on the possible rate of formation of neutron stars in the Galaxy. As it is, most of the luminosity is from white-dwarf progenitors. Allowing for this indicates a probable upper limit on the rate of neutron-star formation of one every 53 years. Making the maximum possible allowance for the physical uncertainties establishes a firm upper limit of one every 27 years. The most probable rate of neutron-star formation is one every 75—100 years. The fact that these limits are much smaller than the probable rates inferred recently from pulsar statistics may indicate that these pulsar results are subject to systematic errors.

066.064 Field energies and principles of equivalence.
C. H. McGruder III.
Nature, Vol. 272, 806 - 807 (1978).

The radial free fall motion of a charged body in a gravitational field is calculated here. A simple approach is presented, which includes the effects of the electrostatic and gravitational field energies of the falling body. The result shows that the acceleration is dependent on both mass and charge of the falling body and that both gravitational and electrostatic field energies gravitate repulsively, indicating that bodies of larger mass or larger charge will fall more slowly than bodies of lesser mass or smaller charge. Thus Einstein's principle of equivalence is not valid, as it requires that all physical systems experience the same acceleration of gravity.

066.065 Deflection of polarised radiation: relative phase delay technique.   B. Dennison, G. Melnick,
M. Harwit, T. Sato, C. T. Stelzried, D. Jauncey.
Nature, Vol. 273, 33 - 35 (1978).

Until recently, experiments designed to detect variations in the geodesic motion of test particles of differing internal properties have been confined to particles of non-zero rest mass. The experiment discussed here applies the same question to photons by asking whether oppositely polarised photons fall at the same rate.

066.066 The field equations of metric-affine gravitational theories.   S. J. Aldersley.
Z. Naturforsch., Band 33a, 398 - 401 (1978).

The notions of conservation of charge and dimensional consistency are used to obtain conditions which uniquely characterize the field equations of electromagnetism and gravitation in a metric-affine gravitational framework with a vector potential. Conditions for the uniqueness of the choice of field equations of a metric-affine gravitational theory (in the absence of electromagnetism) follow as a special case. Some consequences are discussed.

066.067 Effect of a gravitational wave on electromagnetic radiation confined in a cavity: I. Boundary coupling.   P. Tourrenc.
Gen. Relativ. Gravitation, Vol. 9, 123 - 140 (1978).

Gravitational radiation is considered within the first-order approximation. A pattern of an electromagnetic cavity is studied: Gravitational waves give rise to a deformation of the planes limiting the cavity. This deformation alters the electromagnetic radiation. Several cases are studied and orders of magnitude are put forward.

066.068 Effect of a gravitational wave on electromagnetic radiation confined in a cavity: II. Intrinsic coupling.
P. Tourrenc.
Gen. Relativ. Gravitation, Vol. 9, 141 - 153 (1978).

Gravitational radiation is considered within the first-order approximation. A classical gravitational background yields transitions between quantum electromagnetic states in a cavity. Orders of magnitude are picked out from special patterns. The interest of second-order effects is emphasized and experimental possibilities are sketched.

066.069   Time machine and geodesic motion in Kerr metric.
      M. Calvani, F. de Felice, B. Muchotrzeb, F.
Salmistraro.
Gen. Relativ. Gravitation, Vol. 9, 155 - 163 (1978).
      The existence of nontrivial causality violation is a peculiar
property of the space-time around a naked singularity. In the
case of Kerr metric with $a > m$ the authors have found that
for a particular class of geodesics that could in principle violate
causality, the conditions for causality violation are never satisfied.

**066.070   Orbital topography and other astrophysical con-
      sequences of Rosen's bimetric theory of gravity.**
W. R. Stoeger.
Gen. Relativ. Gravitation, Vol. 9, 165 - 174 (1978).
      Since Rosen's bimetric theory of gravity provides at
present a worthy devil's advocate for the black hole hypothesis, it is important for eventual observational work to elaborate the astrophysical consequences and possibilities peculiar
to it. The author does this by deriving the orbital topography
of the spherically symmetric solution to Rosen's field equations—which is relevant to the behavior of relativistic axisymmetric accretion flows—and calculating predicted accretion
disk efficiencies, which can be as much as 2.5 times higher
than for a disk in Schwarzschild. Thereafter, a brief treatment
of the shortest kinematic time scale and the time dilations for
in-falling material is given. Finally he shows that Birkhoff's
theorem does not hold in Rosen's theory, and, therefore, that
genuine gravitational monopole radiation is possible. The
energy it carries, however, is not positive definite.

066.071   Self-similar space—times. I: Three solutions.
      R. N. Henriksen, P. S. Wesson.
Astrophys. Space Sci., Vol. 53, 429 - 444 (1978).
      The equations of general relativity are reduced to a set of
simultaneous ordinary differential equations in one dimensionless variable by an appeal to self-similar symmetry. Analytic
solutions are given representing (1) a static, spherically symmetric matter distribution with density $\rho \propto R^{-\theta}$ and $\theta = 2$,
and (2) a non-static dust cloud with θ dependent on epoch.
(3) For a 'stiff' equation of state, the topology of the solutions
has been investigated numerically and the authors identify
a uniquely defined 'critical' solution. They argue that self-similarity implies a canonical space—time cosmologically, and
they also briefly indicate where these space—times may apply
astrophysically.

066.072   Self-similar space—times. II: Perturbation scheme.
      R. N. Henriksen, P. S. Wesson.
Astrophys. Space Sci., Vol. 53, 445 - 457 (1978).
      An exact solution of Einstein's field equations is given
representing an inhomogeneous sphere of matter in static isothermal equilibrium with a density profile of the form
$\rho \propto R^{-2}$. A perturbation analysis shows that initial small density inhomogeneities grow self-similarly in a non-extreme configuration for which general relativistic effects are important.
As a cosmology, the solutions in tandem provide an expanding
cosmology that did not begin as a Big Bang, but evolved from
an inhomogeneous 'primeval atom', tending at late epochs to
a model with homogeneous density. The new static solution
has applications also in other branches of astrophysics.

**066.073   Bimetric gravitation theory on a cosmological
      basis.**   N. Rosen.
Gen. Relativ. Gravitation, Vol. 9, 339 - 351 (1978).
      In the bimetric theory of gravitation the background
metric tensor $\gamma_{\mu\nu}$, previously taken as describing flat space-
time, is now chosen on the basis of a model of the universe. In
accordance with the perfect cosmological principle, it is taken
as describing a space-time of constant curvature. There are

three possible forms, corresponding to $k = 0, 1, -1$. Only for
$k = 1$ (a closed universe) does the model not go through a
singular state; hence this is the appropriate choice. The isotropic solution of the field equations can be chosen to agree
with the present cosmological observations. For small systems
like the solar system the theory gives the same results as before, in agreement with those of general relativity.

066.074   Classical gravity with higher derivatives.
      K. S. Stelle.
Gen. Relativ. Gravitation, Vol. 9, 353 - 371 (1978).

066.075   The motion of spinless particles in general relativity.
      L. J. Gregory, A. H. Klotz.
Australian J. Phys., Vol. 31, 105 - 110 (1978).
      A technique is developed, which uses a general energy—
momentum tensor, to derive the equations of motion in
general relativity. This enables the geodesic equations for
spinless particles to be deduced.

066.076   The irreversible thermodynamics of black holes.
      P. Candelas, D. W. Sciama.
Gen. Relativ. Gravitation, Vol. 9, 183 - 187 (1978).
      The action of quantum fluctuations of the gravitational
field may be regarded as the origin of the dissipative processes
associated with Hawking radiation. In this picture the black
hole possesses internal coherence by virtue of the localization
of its mass. The cumulative effect of the quantum fluctuations
in the geometry is that this coherence is corrupted and the
mass is sapped away.

**066.077   Quantum mechanics of electromagnetically bounded
      spin-$1/2$ particles in an expanding universe: I. Influ-
ence of the expansion.**   J. Audretsch, G. Schäfer.
Gen. Relativ. Gravitation, Vol. 9, 243 - 255 (1978).
      The quantum mechanically described electron in an ex-
ternal electromagnetic field, both embedded in an expanding
universe with shear, is discussed. This is important for the
fundamental question if a quantum mechanically treated
atomic clock in curved space-time (based on a hydrogen atom)
shows proper or gravitational time. Furthermore, contradic-
tory results reported by other authors seem to imply that
quantum mechanics cannot be reconciled with curved space-
time. It is shown that this is not the case for expanding
Robertson—Walker universes.

**066.078   Scattering of gravitational radiation from vacuum
      black holes.**   R. A. Matzner, M. P. Ryan, Jr.
Astrophys. J., Suppl. Ser., Vol. 36, 451 - 481 (1978).
      Scattering and inelastic cross sections are calculated for
gravitational radiation incident along the axis of symmetry of
a vacuum black hole. The authors consider both nonrotating
and rotating black holes and, in the latter case, both corotating
and counterrotating circularly polarized incident radiation.
Graphical results are displayed for the superradiance or absorption, the phase shifts, and the cross section for scattering
of gravitational waves over a range of the parameter values of
interest.

**066.079   General relativity in two- and three-dimensional
      space-times.**   P. Collas.
American J. Phys., Vol. 45, 833 - 837 (1977). — Abstr. in
Phys. Abstr., Vol. 81, Abstr. 5 (1978).

**066.080   Dispersive effects of the electromagnetic polariza-
      tion in post-geometric optics of curved spacetime.**
K. Y. Fu.
Chinese J. Phys., Vol. 14, No. 4, p. 109 - 110 (1976). — Abstr.
in Phys. Abstr., Vol. 81, Abstr. 96 (1978).

**066.081   An approximation method to solve Einstein's field
      equations on a curved background metric.**

G. Lessner.
Int. J. Theor. Phys., Vol. 16, No. 2, p. 99 - 110 (1977).
Abstr. in Phys. Abstr., Vol. 81, Abstr. 109 (1978).

066.082 Conditions for the separation of the Hamilton-
Jacobi equation. C. D. Collinson, J. Fugere.
J. Phys. A, Vol. 10, 1877 - 1885 (1977). − Abstr. in Phys.
Abstr., Vol. 81, Abstr. 110 (1978).

066.083 Initial-value problems and singularities in general
relativity. B. K. Datta.
Pramāṇa, Vol. 9, 229 - 238 (1977). − Abstr. in Phys. Abstr.,
Vol. 81, Abstr. 112 (1978).

066.084 Gravitational energy from a quadratic Lagrangian
with torsion. C. A. Lopez.
Int. J. Theor. Phys., Vol. 16, 167 - 176 (1977). − Abstr. in
Phys. Abstr., Vol. 81, Abstr. 115 (1978).

066.085 Reduced spinor formalism by means of tetrad
calculus with application to Yang's gravitational
field equations. J. Sheng.
Acta Phys. Sinica, Vol. 26, 259 - 273 (1977). In Chinese.
Abstr. in Phys. Abstr., Vol. 81, Abstr. 119 (1978).

066.086 Classical solutions in supergravity.
N. S. Baaklini, S. Ferrara, P. van Nieuwenhuizen.
Nuovo Cimento, Lett., Ser. 2, Vol. 20, 113 - 116 (1977).
Abstr. in Phys. Abstr., Vol. 81, Abstr. 122 (1978).

066.087 On the connection between the Dirac and the Dixon
equation in a weak gravitational field.
R. Catenacci, M. Martellini.
Nuovo Cimento, Lett., Ser. 2, Vol. 20, 282 - 284 (1977).
Abstr. in Phys. Abstr., Vol. 81, Abstr. 124 (1978).

066.088 On the induced cosmological term in quantum
gravity. B. de Wit, R. Gastmans.
Nucl. Phys. B, Vol. B128, 294 - 312 (1977). − Abstr. in Phys.
Abstr., Vol. 81, Abstr. 125 (1978).

066.089 Covariance analysis for a relativity mission.
D. Schaechter, J. V. Breakwell, R. A. Van Patten,
C. W. F. Everitt.
J. Spacecr. Rockets, Vol. 14, 474 - 478 (1977). − Abstr. in
Phys. Abstr., Vol. 81, Abstr. 126 (1978).

066.090 Extension of the York field decomposition to
general gravitationally coupled fields.
J. A. Isenberg, J. M. Nester.
Ann. Physics, Vol. 108, 368 - 386 (1977). − Abstr. in Phys.
Abstr., Vol. 81, Abstr. 127 (1978).

066.091 Gravity, nonintegrable phase factor and motion
associated with a closed curve.
K. Hayashi, T. Shirafuji.
Prog. Theor. Phys., Vol. 58, 353 - 361 (1977). − Abstr. in
Phys. Abstr., Vol. 81, Abstr. 129 (1978).

066.092 Some properties of spin-weighted spheroidal
harmonics.
R. A. Breuer, M. P. Ryan, Jr., S. Waller.
Proc. R. Soc. London, Ser. A, Vol. 358, 71 - 86 (1977).
Abstr. in Phys. Abstr., Vol. 81, Abstr. 4277 (1978).

066.093 Metrical connection in space-time, Newton's and
Hubble's laws. A. Maeder.
Astron. Astrophys., Vol. 65, 337 - 343 (1978).
The theory of gravitation in general relativity is not scale
invariant. The author follows Dirac's proposition of a scale
invariant theory of gravitation (i.e. a theory in which the
equations keep their form when a transformation of scale is

made). He examines some concepts of Weyl's geometry, like
the metrical connection, the scale transformations and invariance, and he discusses their consequences for the equation
of the geodetic motion and for its Newtonian limit. Under
general conditions, the author shows that the only non-vanishing component of the coefficient of metrical connection may
be identified with Hubble's constant. Hubble's law—like
Newton's law—would appear as an intrinsic property of
gravitation, being only the most visible manifestation of a
general effect characterizing the gravitational interaction.

066.094 Spherically symmetric solution of gauge supersym-
metry equations. D. K. Ross.
J. Phys. A, Vol. 10, 2031 - 2035 (1977). − Abstr. in Phys.
Abstr., Vol. 81, Abstr. 4532 (1978).

066.095 Equatorial circular geodesics in the Kerr-Newman
geometry. N. Dadhich, P. P. Kale.
J. Math. Phys., Vol. 18, 1727 - 1728 (1977). − Abstr. in
Phys. Abstr., Vol. 81, Abstr. 4533 (1978).

066.096 Path structures on manifolds.
J. Ehlers, E. Kohler.
J. Math. Phys., Vol. 18, 2014 - 2018 (1977). − Abstr. in Phys.
Abstr., Vol. 81, Abstr. 4535 (1978).

066.097 On the combined Dirac-Einstein-Maxwell field
equations. M. J. Hamilton, A. Das.
J. Math. Phys., Vol. 18, 2026 - 2030 (1977). − Abstr. in Phys.
Abstr., Vol. 81, Abstr. 4536 (1978).

066.098 A non-static source of the Taub solution of Ein-
stein's gravitational equations.
J. Horsky, P. Lorenc, J. Novotny.
Phys. Lett. A, Vol. 63A, 79 - 80 (1977). − Abstr. in Phys.
Abstr., Vol. 81, Abstr. 4537 (1978).

066.099 Einstein-Cartan spheres.
A. K. Raychaudhuri, S. Banerji.
Phys. Rev. D, Vol. 16, 281 - 285 (1977). − Abstr. in Phys.
Abstr., Vol. 81, Abstr. 4538 (1978).

066.100 Gravitational scattering of zero-rest mass plane
waves. W. K. De Logi, S. J. Kovacs, Jr..
Phys. Rev. D, Vol. 16, 237 - 244 (1977). − Abstr. in Phys.
Abstr., Vol. 81, Abstr. 4542 (1978).

066.101 Poincaré gauge theory with Weyl-type Lagrangian
and gravitation. S. Hamamoto.
Prog. Theor. Phys., Vol. 58, 710 - 712 (1977). − Abstr. in
Phys. Abstr., Vol. 81, Abstr. 4545 (1978).

066.102 Properties of the solutions of cold ultradense con-
figurations in the Brans-Dicke theory.
W. F. Bruckman, E. Kazes.
Phys. Rev. D., Vol. 16, 261 - 268 (1977). − Abstr. in Phys.
Abstr., Vol. 81, Abstr. 4546 (1978).

066.103 Generalisation of solutions of the interior Brans-
Dicke equations. W. F. Bruckman, E. Kazes.
Phys. Rev. D, Vol. 16, 269 - 274 (1977). − Abstr. in Phys.
Abstr., Vol. 81, Abstr. 4547 (1978).

066.104 On a quantized scalar field in a Bianchi-type I
universe. H. Nariai.
Prog. Theor. Phys., Vol. 58, 560 - 574 (1977). − Abstr. in
Phys. Abstr., Vol. 81, Abstr. 4549 (1978).

066.105 On a test of Fujii's theory of gravitation.
Y. Yamaguchi.
Prog. Theor. Phys., Vol. 58, 723 - 724 (1978). − Abstr. in
Phys. Abstr., Vol. 81, Abstr. 4550 (1978).

066.106 **The gravitational field of spherically symmetric matter distributions in the Yang-Mills gauge theory of gravity.**   M. Camenzind.
Phys. Lett. A, Vol. 63A, 69 - 72 (1977). — Abstr. in Phys. Abstr., Vol. 81, Abstr. 4773 (1978).

066.107 **Strong gravity with a cosmological constant.**   K. Tennakone.
Indian J. Pure Appl. Phys., Vol. 15, 584 - 585 (1977). — Abstr. in Phys. Abstr., Vol. 81, Abstr. 4820 (1978).

066.108 **A relativistic many-body theory of high density matter.**   S. A. Chin.
Ann. Physics, Vol. 108, 301 - 306 (1977). — Abstr. in Phys. Abstr., Vol. 81, Abstr. 4909 (1978).

066.109 **The role of neutrinos and antineutrinos in the formation and cooling of neutron stars.**
G. E. Brown.
AIP Conf. Proc., No. 37, (see 012.016), p. 63 - 75 (1977). Abstr. in Phys. Abstr., Vol. 81, Abstr. 8038 (1978).

066.110 **Thermodynamic light on black holes.**   P. Davies.
New Scientist, Vol. 75, 238 - 240 (1977). — Abstr. in Phys. Abstr., Vol. 81, Abstr. 8041 (1978).

066.111 **Comments on the optical appearance of white holes.**   D. Dultzin-Hacyan, S. Hacyan.
Rev. Mexicana Astron. Astrofis., Vol. 2, 263 - 268 (1977).
The optical appearance of gas inside and emerging beyond the Schwarzschild sphere is studied. The authors emphasize the importance of taking into account photons moving with nontrivial angular momentum. A semiqualitative analysis shows that no spectral lines could be seen by a distant observer if gas emits from a region near the Schwarzschild sphere. On the basis of their analysis the authors suggest that quasars and BL Lac objects are somewhat different manifestations of the same physical process, namely, a white hole expansion.

066.112 **Sustenance of a black hole in a galactic nucleus.**   G. A. Shields, J. C. Wheeler.
Astrophys. J., Vol. 222, 667 - 674 (1978).
The authors examine the class of models in which a QSO is powered by a massive black hole accreting gas produced in a dense galactic nucleus. The relative importance of tidal disruption, collisions, and stellar evolution as a function of conditions in the nucleus is considered; the possibility of increasing the tidal disruption rate through enhanced relaxation is considered; and arguments are given in favor of models with intermittent periods of high luminosity. The problem is formulated in terms of the required hole mass and accretion rate and the parameters of the nuclear star cluster. Mechanisms for feeding the hole are examined in the context of steady accretion. The alternative of intermittent periods of rapid accretion is discussed.

066.113 **Weak principle of equivalence: corrections due to the earth's field.**   R. M. Nugaev.
Gravitatsiya i teor. otnositel'n. Vyp. (No.) 13, Kazan', Kazan. univ., 1976, p. 98 - 101. In Russian. — Abstr. in Ref. zh., 51. Astron., 3.51.955 (1978).

066.114 **Domains of stationary communications in space-time.**   B. Carter.
Gen. Relativ. Gravitation, Vol. 9, 437 - 450 (1978).
Basic relationships between the causal connectivity properties and the symmetry group invariance properties of space-time manifolds are discussed with a view to applications in the theory of stationary black holes.

066.115 **The dynamics of polarization.**   W. Israel.
Gen. Relativ. Gravitation, Vol. 9, 451 - 468 (1978).
Long-standing problems concerning the structure of the energy-momentum tensor and the thermodynamics of a polarized medium are reexamined in the light of an exact, fully relativistic kinetic model.

066.116 **Classical instability of a naked singularity.**   F. de Felice.
Nature, Vol. 273, 429 - 431 (1978).
The effective potential of the equatorial circular geodesics in Kerr metric shows a sharp discontinuity from positive to negative values, when $a \to m$. It is shown that when a is in the small range $1.088\, m > a > m$, stable circular orbits exist with negative energy (with respect to infinity); this implies that, if accretion takes place, a Kerr naked singularity is slowed down. The efficiency of this process increases as $a \to m$. Therefore, in the above range of a, Kerr naked singularities should decay to a black-hole state.

066.117 **The present appearance of white holes.**   K. Lake, R. C. Roeder.
Nature, Vol. 273, 449 - 450 (1978).
It has been argued (Lake, 1978) that one can (at least in principle) still observe certain types of white hole, most likely those which never expand beyond their 'Schwarzschild spheres' into our space, and which have, in fact, always been visible. Here the authors report the main results of an analysis of the optical properties of such white holes.

066.118 **Distortions of the cosmic radiation spectrum in baryon-symmetric cosmologies.**   B. J. T. Jones, G. Steigman.
Mon. Not. R. Astron. Soc., Vol. 183, 585 - 594 (1978).
In matter-antimatter symmetric cosmologies, annihilation-induced heating distorts the spectrum of the microwave background radiation. For 'violent' models, such as that of Stecker & Puget, the predicted distortions exceed the observational constraints. For 'quiescent' models, such as that of Aldrovandi et al., only small distortions are produced and these are consistent with current observational limits. It is, however, difficult to understand the origin of galaxies in the quiescent models. The retarded recombination in such models vastly increases the photon drag on any annihilation-driven turbulence.

066.119 **Is observation of the gravitational microwave background radiation possible?**   A. Kułak.
Acta Cosmologica, Zesz. 7, 57 - 73 (1978).
The properties of a circular photon beam as the gravitational incoherent radiation detector is considered. Then the possibility of detection of gravitational microwave background radiation is discussed, assuming that its temperature is $T_g \approx 1° K$

066.120 **Rotational motion of matter in general relativity.**   A. Krasiński.
Acta Cosmologica, Zesz. 7, (see 012.022), 119 - 131 (1978).
It is argued that in the further development of cosmology the overidealized assumption leading to Friedmann-Lemaître models should be gradually relaxed, and substituted by more realistic ones taking into account the departures of the real universe from perfect homogeneity and isotropy. The paper presents the mathematical formalism for treating a class of irregular models: those with nonzero vorticity.

066.121 **Superfluidity in neutron stars. III. Relaxation processes between the superfluid and the crust.**
D. Harding, R. A. Guyer, G. Greenstein.
Astrophys. J., Vol. 222, 991 - 1005 (1978). = Contrib. Five-Coll. Astron. Dep., Amherst, Mass., No. 245.
The authors calculate the rates for a variety of processes

whereby the neutron superfluid within a neutron star exchanges momentum with the charged-particle system with which it coexists. They consider the scattering of the normal neutron fluid within vortex lines from phonons (one- and two-phonon processes) impurities and lattice defects in the crust, and momentum transfer brought about by neutron β-decay. They then turn to a discussion of the ways in which the results of previous hydrodynamical calculations are altered by their new relaxation rates. Of particular interest is a major discrepancy between the predicted and the observed spin-down of a pulsar after a period jump, a discrepancy than can be resolved by proposing that neutron stars are bigger than has been previously thought.

066.122 **Observations of the cosmic microwave background .** R. B. Partridge.
Probl. gravitatsii. Tbilisi, 1976, p. 471 - 483. — Abstr. in Ref. zh., 51. Astron., 4.51.737 (1978).

066.123 **Fluctuations of the black-body background radiation in adiabatic and entropy theories of formation of galaxies.** A. G. Doroshkevich, Ya. B. Zel'dovich, R. A. Syunyaev.
Inst. prikl. mat. AN SSSR. Prepr. No. 80. Moskva, 27 pp. In Russian. — Abstr. in Ref. zh., 51. Astron., 4.51.739 (1978).

066.124 **Some experiments on gravitation.** L. I. Schiff. Probl. gravitatsii. Tbilisi, 1976, p. 689 - 701. Abstr. in Ref. zh., 51. Astron., 4.51.771 (1978).

066.125 **Black holes are coloured.** M. J. Perry. Phys. Lett. B, Vol. 71B, 234 - 236 (1977). — Abstr. in Phys. Abstr., Vol. 81, Abstr. 8280 (1978).

066.126 **Wave fronts near a black hole.** R. S. Hanni. Phys. Rev. D, Vol. 16, 933 - 936 (1977). — Abstr. in Phys. Abstr., Vol. 81, Abstr. 8282 (1978).

066.127 **Wave scattering theory and the absorption problem for a black hole.** N. Sanchez.
Phys. Rev. D, Vol. 16, 937 - 945 (1977). — Abstr. in Phys. Abstr., Vol. 81, Abstr. 8283 (1978).

066.128 **Interactions on the extended conformal space-time.** J. Tarski.
Proceedings of the 1st Marcel Grossman meeting on general relativity (see 012.025), p. 165 - 172 (1977). — Abstr. in Phys. Abstr., Vol. 81, Abstr. 8290 (1978).

066.129 **Space-time singularities.** R. Penrose. Proceedings of the 1st Marcel Grossman meeting on general relativity (see 012.025), p. 173 - 181 (1977). Abstr. in Phys. Abstr., Vol. 81, Abstr. 8291 (1978).

066.130 **Maximal surfaces in closed and open spacetimes.** D. Brill.
Proceedings of the 1st Marcel Grossman meeting on general relativity (see 012.025), p. 193 - 206 (1977). — Abstr. in Phys. Abstr., Vol. 81, Abstr. 8294 (1978).

066.131 **A necessary condition for Killing horizon to be event horizon.** P. Hajicek.
Proceedings of the 1st Marcel Grossman meeting on general relativity (see 012.025), p. 207 - 209 (1977). — Abstr. in Phys. Abstr., Vol. 81, Abstr. 8295 (1978).

066.132 **Positive energy in U_4 theory.** E. W. Mielke. Proceedings of the 1st Marcel Grossman meeting on general relativity (see 012.025), p. 213 - 217 (1977). Abstr. in Phys. Abstr., Vol. 81, Abstr. 8296 (1978).

066.133 **Metric solutions of spinning masses.**

A. Tomimatsu, H. Sato.
Proceedings of the 1st Marcel Grossman meeting on general relativity (see 012.025), p. 227 - 228 (1977). — Abstr. in Phys. Abstr., Vol. 81, Abstr. 8297 (1978).

066.134 **Solutions of Einstein-Cartan equations.** A. R. Prasanna.
Proceedings of the 1st Marcel Grossman meeting on general relativity (see 012.025), p. 289 - 295 (1977). — Abstr. in Phys. Abstr., Vol. 81, Abstr. 8299 (1978).

066.135 **Probability distribution for radiation from a black hole in the presence of incoming radiation.**
P. Panangaden, E. M. Wald.
Phys. Rev. D, Vol. 16, 929 - 932 (1977). — Abstr. in Phys. Abstr., Vol. 81, Abstr. 8300 (1978).

066.136 **Inertia and gravity in general relativity.** A. Quale. Proceedings of the 1st Marcel Grossman meeting on general relativity (see 012.025), p. 221 - 226 (1977). Abstr. in Phys. Abstr., Vol. 81, Abstr. 8317 (1978).

066.137 **A new relativistic theory of gravitation.** F. I. Cooperstock, G. J. G. Junevicus.
Proceedings of the 1st Marcel Grossman meeting on general relativity (see 012.025), p. 649 - 652 (1977). — Abstr. in Phys. Abstr., Vol. 81, Abstr. 8319 (1978).

066.138 **Klein paradox and vacuum polarization.** T. Damou.
Proceedings of the 1st Marcel Grossman meeting on general relativity (see 012.025), p. 459 - 482 (1977). — Abstr. in Phys. Abstr., Vol. 81, Abstr. 8329 (1978).

066.139 **Constraints on the gravitational constant at large distances.** D. R. Mikkelsen, M. J. Newman.
Phys. Rev. D, Vol. 16, 919 - 926 (1977). — Abstr. in Phys. Abstr., Vol. 81, Abstr. 8331 (1978).

066.140 **Theoretical significance of present-day gravity experiments.** K. Nordtvedt, Jr.
Proceedings of the 1st Marcel Grossman meeting on general relativity (see 012.025), p. 539 - 544 (1977). — Abstr. in Phys. Abstr., Vol. 81, Abstr. 8332 (1978).

066.141 **Gravitation, relativity and precise experimentation.** C. W. F. Everitt.
Proceedings of the 1st Marcel Grossman meeting on general relativity (see 012.025), p. 545 - 615 (1977). — Abstr. in Phys. Abstr., Vol. 81, Abstr. 8333 (1978).

066.142 **Pair production by photons in a constant electromagnetic field.** L. R. Davis.
Proceedings of the 1st Marcel Grossman meeting on general relativity (see 012.025), p. 511 - 514 (1977). — Abstr. in Phys. Abstr., Vol. 81, Abstr. 8568 (1978).

066.143 **Shock waves in relativistic magnetohydrodynamics.** A. Lichnerowicz.
Proceedings of the 1st Marcel Grossman meeting on general relativity (see 012.025), p. 335 - 339 (1977). — Abstr. in Phys. Abstr., Vol. 81, Abstr. 9770 (1978).

066.144 **Instantaneous Cauchy surfaces, topology change, and exploding black holes.** R. H. Gowdy.
J. Math. Phys., Vol. 18, 1798 - 1801 (1977). — Abstr. in Phys. Abstr., Vol. 81, Abstr. 12087 (1978).

066.145 **A star and a dust shell.** H. Tuohimaa. Nuovo Cimento, Lett., Ser. 2, Vol. 20, 361 - 363 (1977). — Abstr. in Phys. Abstr., Vol. 81, Abstr. 12088 (1978).

066.146 **Approaches to the analysis of relativistic superdense matter.**   R. L. Bowers, A. M. Gleeson, R. D. Pedigo.
Nuovo Cimento B, Ser. 11, Vol. 41B, 441 - 459 (1977).
Abstr. in Phys. Abstr., Vol. 81, Abstr. 12091 (1978).

066.147 **Relativistic elasticity applied to neutron stars.**   H. Quintana.
Proceedings of the 1st Marcel Grossman meeting on general relativity (see 012.025), p. 439 - 448 (1977). — Abstr. in Phys. Abstr., Vol. 81, Abstr. 12267 (1978).

066.148 **'Splendeurs et misères' of Hawking's effect.**   P. Jajicek.
Contacts between high energy physics and other fields of physics. Schladming, Austria, 24 February - 5 March 1977 = Acta Phys. Austriaca, Suppl. 18 (1977). p. 835 - 872. — Abstr. In Phys. Abstr., Vol. 81, Abstr. 12268 (1978).

066.149 **A comment on the symmetries of Kerr black holes.**   C. D. Collinson, P. N. Smith.
Commun. Math. Phys., Vol. 56, No. 3, p. 277 - 279 (1977). Abstr. in Phys. Abstr., Vol. 81, Abstr. 12269 (1978).

066.150 **The Kerr metric and stationary axisymmetric gravitational fields.**   S. Chandrasekhar.
Proc. R. Soc. London, Ser. A, Vol. 358, 405 - 420 (1977). Abstr. in Phys. Abstr., Vol. 81, Abstr. 12272 (1978).

066.151 **The gravitational perturbations of the Kerr black hole. I. The perturbations in the quantities which vanish in the stationary state.**   S. Chandrasekhar.
Proc. R. Soc. London, Ser. A, Vol. 358, 421 - 439 (1977). Abstr. in Phys. Abstr., Vol. 81, Abstr. 12273 (1978).

066.152 **The gravitational perturbations of the Kerr black hole. II. The perturbations in the quantitites which are finite in the stationary state.**   S. Chandrasekhar.
Proc. R. Soc. London, Ser. A, Vol. 358, 441 - 465 (1977). Abstr. in Phys. Abstr., Vol. 81, Abstr. 12274 (1978).

066.153 **Black holes and thermal Green functions.**   G. W. Gibbons, M. J. Perry.
Proc. R. Soc. London, Ser. A, Vol. 358, 467 - 494 (1977). Abstr. in Phys. Abstr., Vol. 81, Abstr. 12275 (1978).

066.154 **Instability of black hole inner horizons.**   J. M. McNamara.
Proc. R. Soc. London, Ser. A, Vol. 358, 499 - 517 (1977). Abstr. in Phys. Abstr., Vol. 81, Abstr. 12276 (1978).

066.155 **Test charge near an extreme charged black hole.**   R. S. Hanni.
Phys. Rev. D, Vol. 16, 1245 - 1246 (1977). — Abstr. in Phys. Abstr., Vol. 81, Abstr. 12277 (1978).

066.156 **The vacuum black hole uniqueness theorem and its conceivable generalisations.**   B. Carter.
Proceedings of the 1st Marcel Grossman meeting on general relativity (see 012.025), p. 243 - 254 (1977). — Abstr. in Phys. Abstr., Vol. 81, Abstr. 12278 (1978).

066.157 **Electromagnetic perturbations of a rotating black hole.**   S. L. Detweiler.
Proceedings of the 1st Marcel Grossman meeting on general relativity (see 012.025), p. 303 - 310 (1977). — Abstr. in Phys. Abstr., Vol. 81, Abstr. 12279 (1978).

066.158 **On relativistic magnetohydrodynamics processes in the fields of black holes.**   R. Ruffini.
Proceedings of the 1st Marcel Grossman meeting on general relativity (see 012.025), p. 349 - 392 (1977). — Abstr. in Phys. Abstr., Vol. 81, Abstr. 12280 (1978).

066.159 **Magnetohydrodynamics near a black hole.**   J. R. Wilson.
Proceedings of the 1st Marcel Grossman meeting on general relativity (see 012.025), p. 393 - 413 (1977). — Abstr. in Phys. Abstr., Vol. 81, Abstr. 12281 (1978).

066.160 **Plasma horizons of a charged black hole.**   R. S. Hanni.
Proceedings of the 1st Marcel Grossman meeting on general relativity (see 012.025), p. 429 - 436 (1977). — Abstr. in Phys. Abstr., Vol. 81, Abstr. 12282 (1978).

066.161 **Magnetic black holes.**   W. Kundt.
Proceedings of the 1st Marcel Grossman meeting on general relativity (see 012.025), p. 437 - 438 (1977). — Abstr. in Phys. Abstr., Vol. 81, Abstr. 12283 (1978).

066.162 **Quantum processes near black holes.**   G. W. Gibbons.
Proceedings of the 1st Marcel Grossman meeting on general relativity (see 012.025), p. 449 - 458 (1977). — Abstr. in Phys. Abstr., Vol. 81, Abstr. 12284 (1978).

066.163 **Classical and quantum states in black hole physics.**   N. Deruelle.
Proceedings of the 1st Marcel Grossman meeting on general relativity (see 012.025), p. 483 - 498 (1977). — Abstr. in Phys. Abstr., Vol. 81, Abstr. 12285 (1978).

066.164 **Effect of zero point fluctuations on the formation of a black hole.**   U. H. Gerlach.
Proceedings of the 1st Marcel Grossman meeting on general relativity (see 012.025), p. 515 - 522 (1977). — Abstr. in Phys. Abstr., Vol. 81, Abstr. 12286 (1978).

066.165 **Particle detectors and black holes.**   W. G. Unruh.
Proceedings of the 1st Marcel Grossman meeting on general relativity (see 012.025), p. 527 - 536 (1977). Abstr. in Phys. Abstr., Vol. 81, Abstr. 12287 (1978).

066.166 **Spontaneous discharge of Kerr-Newman black holes.**   W. T. Zaumen.
Proceedings of the 1st Marcel Grossman meeting on general relativity (see 012.025), p. 537 - 538 (1977). — Abstr. in Phys. Abstr., Vol.81, Abstr. 12288 (1978).

066.167 **Particle creation and geometry.**   P. C. W. Davies.
Proceedings of the 1st Marcel Grossman meeting on general relativity (see 012.025), p. 507 - 510 (1977). — Abstr. in Phys. Abstr., Vol. 81, Abstr. 12442 (1978).

066.168 **Covariant treatment of particle creation in curved space-time.**   H. Rumpf.
Proceedings of the 1st Marcel Grossman meeting on general relativity (see 012.025), p. 523 - 526 (1977). — Abstr. in Phys. Abstr., Vol. 81, Abstr. 12443 (1978).

066.169 **Cosmological solution of Einstein's equations in general relativity.**   B. B. Paul, S. N. Guha Thakurta.
Indian J. Pure Appl. Math., Vol. 8, No. 1, p. 117 - 119 (1977). Abstr. in Phys. Abstr., Vol. 81, Abstr. 12617 (1978).

066.170 **Evolution of time-symmetric gravitational waves: initial data and apparent horizons.**   K. Eppley.
Phys. Rev. D, Vol. 16, 1609 - 1614 (1977). — Abstr. in Phys. Abstr., Vol. 81, Abstr. 12626 (1978).

066.171 The collapse of rotating gas clouds.   Y. Kamiya.
Prog. Theor. Phys., Vol. 58, 802 - 815 (1977).
Abstr. in Phys. Abstr., Vol. 81, Abstr. 12628 (1978).

066.172 Einstein-Maxwell fields in asymptotically null-
spherical coordinates.
S. Persides, I. Ioannides.
Prog. Theor. Phys., Vol. 58, 829 - 841 (1977). — Abstr. in
Phys. Abstr., Vol. 81, Abstr. 12631 (1978).

066.173 Energy-momentum tensor of a massless scalar quan-
tum field in a Robertson-Walker universe.
P. C. W. Davies, S. A. Fulling, S. M. Christensen, T. S. Bunch.
Ann. Physics, Vol. 109, 108 - 142 (1977). — Abstr. in Phys.
Abstr., Vol. 81, Abstr. 12635 (1978).

066.174 Quantised gravitational wave perturbations in
Robertson-Walker universes.
L. H. Ford, L. Parker.
Phys. Rev. D, Vol. 16, 1601 - 1608 (1977). — Abstr. in Phys.
Abstr., Vol. 81, Abstr. 12636 (1978).

066.175 On the spontaneous origin of Newton's constant.
P. Minkowski.
Phys. Lett. B, Vol. 71B, 419 - 421 (1977). — Abstr. in Phys.
Abstr., Vol. 81, Abstr. 12853 (1978).

066.176 Magnetic support against gravitational collapse: a
static axisymmetric interior solution of the Einstein-
Maxwell equations.   H. Ardovan, M. Hossein Partovi.
Phys. Rev. D, Vol. 16, 1664 - 1677 (1977). — Abstr. in Phys.
Abstr., Vol. 81, Abstr. 16288 (1978).

066.177 Hamiltonian formalism for relativistic perfect fluids.
V. Moncrief.
Phys. Rev. D, Vol. 16, 1702 - 1705 (1977). — Abstr. in Phys.
Abstr., Vol. 81, Abstr. 16289 (1978).

066.178 Quantum mechanical approach to small black holes.
E. Spallucci.
Nuovo Cimento, Lett., Ser. 2, Vol. 20, 391 - 394 (1977).
Abstr. in Phys. Abstr., Vol. 81, Abstr. 16453 (1978).

066.179 Upper bounds on collisional Penrose processes near
rotating black-hole horizons.
T. Piran, J. Shaham.
Phys. Rev. D, Vol. 16, 1615 - 1635 (1977). — Abstr. in Phys.
Abstr., Vol. 81, Abstr. 16455 (1978).

066.180 Low-frequency limit of gravitational scattering.
R. A. Matzner, M. P. Ryan, Jr.
Phys. Rev. D, Vol. 16, 1636 - 1642 (1977). — Abstr. in Phys.
Abstr., Vol. 81, Abstr. 16456 (1978).

066.181 Black holes — where physical law ends.   D. Peat.
Sci. Dimension, Vol. 9, No. 3, p. 281, 30 (1977).
Abstr. in Phys. Abstr., Vol. 81, Abstr. 16457 (1978).

066.182 Integral conservation laws for isolated systems by
the method of privileged vector fields in the
general theory of relativity and the external Kerr metric.
S. Manoff.
Acta Phys. Austriaca, Vol. 48, No. 1, p. 41 - 50 (1977). In
German. — Abstr. in Phys. Abstr., Vol. 81, Abstr. 16698
(1978).

066.183 Singularities in conformally flat spacetimes.
F. J. Tipler.
Phys. Lett. A, Vol. 64A, 8 - 10 (1977). — Abstr. in Phys.
Abstr., Vol. 81, Abstr. 16703 (1978).

066.184 Field equations in Treder's tetrad theory that are

deducible from a variational principle.
D.-E. Liebscher.
Ann. Physik, Vol. 34, 402 - 404 (1977). In German. — Abstr.
in Phys. Abstr., Vol. 81, Abstr. 16709 (1978).

066.185 Scale-covariant theory of gravitation and astrophysi-
cal applications.
V. Canuto, P. J. Adams, S. H. Hsieh, E. Tsiang.
Phys. Rev. D, Vol. 16, 1643 - 1663 (1977). — Abstr. in Phys.
Abstr., Vol. 81, Abstr. 16710 (1978).

066.186 Back reaction of a quantized field in the gauge
treatment of gravity.   V. M. Nikolaenko.
Acta Phys. Polonica B, Vol. B8, 911 - 918 (1977). — Abstr. in
Phys. Abstr., Vol. 81, Abstr. 16711 (1978).

066.187 Quantized metric on locally flat space-time and
spontaneous dilatation asymmetry.   H. Saller.
Nuovo Cimento A, Ser. 11, Vol. 42A, 189 - 206 (1977).
Abstr. in Phys. Abstr., Vol. 81, Abstr. 16713 (1978).

066.188 Fundamental limits to the measurements of the
gravitational constant.   F. N. Hooge, J. A. Poulis.
Appl. Sci. Res., Vol. 33, No. 2, p. 191 - 195 (1977). — Abstr.
in Phys. Abstr., Vol. 81, Abstr. 16714 (1978).

066.189 On self-gravitation.   H.-J. Treder, W. Yourgrau.
Phys. Lett. A, Vol. 64A, 25 - 28 (1977). — Abstr.
in Phys. Abstr.; Vol. 81, Abstr. 16971 (1978).

066.190 Strong gravity, black holes, and hadrons.
C. Sivaram, K. P. Sinha.
Phys. Rev. D, Vol. 16, 1975 - 1978 (1977). — Abstr. in Phys.
Abstr., Vol. 81, Abstr. 16973 (1978).

066.191 Abnormal phase in dense neutron star matter.
E. M. Nyman, M. Rho.
Nucl. Phys. A, Vol. A290, 493 - 500 (1977). — Abstr. in Phys.
Abstr., Vol. 81, Abstr. 20849 (1978).

066.192 On the outside of cold black holes.   J. Bicak.
Czechoslovak J. Phys. B, Vol. B28, 1 - 7 (1978).
Abstr. in Phys. Abstr., Vol. 81, Abstr. 20850 (1978).

066.193 Basic properties of a stationary accretion disk
surrounding a black hole.   R. Hoshi.
Prog. Theor. Phys., Vol. 58, 1191 - 1204 (1977). — Abstr. in
Phys. Abstr., Vol. 81, Abstr. 20851 (1978).

066.194 Weak electromagnetic fields around a black hole. I.
Field of a point charge.   R. M. Misra.
Prog. Theor. Phys., Vol. 58, 1205 - 1217 (1977). — Abstr. in
Phys. Abstr., Vol. 81, Abstr. 20852 (1978).

066.195 A method of obtaining the rate of perihelion precession in a system of two charged masses.
P. D. P. Smith.
Gen. Relativ. Gravitation, Vol. 9, 469 - 487 (1978).
    This is the fourth of a series of papers concerned with the
description of motion in combined gravitational and electromagnetic fields by the Klein-Gordon equation. The rate of
perihelion precession of a pair of charged masses, and, for the
electric case, the fine structure, are obtained here in a new way
by using this equation in a coordinate system rotating at a
suitable rate. Applied to the motion of Mercury around the
Sun, the first-order calculation gives a rate of perihelion precession of 42.98 sec arc/century. Consideration of upper limits
on the field strength on the surfaces of the Sun and Mercury
show that the electric part of the rate of precession is far
below 0.1 sec arc/century, and so need not be considered in
the test of general relativity based on Mercury's perihelion precession.

066.196 On the Cauchy problem for the general relativistic perfect magnetofluid.   T. H. Date, N. P. Patil.
Gen. Relativ. Gravitation, Vol. 9, 501 - 509 (1978).
   This paper aims at the study of the Cauchy problem for the general relativistic perfect magnetofluid. The consistency and uniqueness condition to be satisfied on the hypersurface by three unknown quantities in the magnetohydrodynamic field equations is obtained.

066.197 Relativistic equations of motion of extended bodies.   N. Spyrou.
Gen. Relativ. Gravitation, Vol. 9, 519 - 530 (1978).
   The author derives the most general equations of motion in the post-Newtonian approximation of general relativity for a bounded system of extended bodies with arbitrary internal structure and internal motions, and explores in detail the conditions under which the motion of the bodies deviates from the geodesic motion.

066.198 Elasticity and electro-magneto-elasticity of general-relativistic systems.   G. A. Maugin.
Gen. Relativ. Gravitation, Vol. 9, 541 - 549 (1978).
   This essay aims at demonstrating the interest for a genuine continuum mechanics and, more restrictively, elasticity of general-relativistic systems in cases where the Einsteinian gravitational picture cannot be deleted, e.g., in "elastic" gravitational-wave detectors and in the study of elastic oscillations of solid stars.

066.199 Inversion of Limber's relativistic formula.   S. A. Bonometto, F. Lucchin.
Astron. Astrophys., Vol. 67, 153 - 155 (1978).
   Groth and Peebles recently extended Limber's formula to the relativistic case. Fall and Tremaine show the importance of an inversion of Limber's formula yielding $\xi(r)$ when $w(\vartheta)$ is known. In this note the authors extend Fall and Tremaine's inversion to the relativistic case, provided that a number of conditions are fulfilled. The meaning of these conditions is also discussed.

066.200 Anisotropic fluid in a spherical space-time. II. Most probable state of high-frequency null radiation.
D. C. Wilkins.
Ann. Physics, Vol. 109, 401 - 417 (1977). — Abstr. in Phys. Abstr., Vol. 81, Abstr. 21016 (1978).

066.201 A necessary condition for the validity of Huygens' principle on a curved space-time.   R. Goldani.
J. Math. Phys., Vol. 18, 2125 - 2128 (1977). — Abstr. in Phys. Abstr., Vol. 81, Abstr. 21024 (1978).

066.202 A solution-generating theorem with applications in general relativity.   J. R. Ray, Mein Sieng Wei.
Nuovo Cimento B, Ser. 11, Vol. 42B, 151 - 164 (1977).
Abstr. in Phys. Abstr., Vol. 81, Abstr. 21027 (1978).

066.203 On the invariance transformations of the energy-momentum tensor in the Einstein-Maxwell theory.
M. L. Woolley.
J. Phys. A, Vol. 10, 2107 - 2114 (1977). — Abstr. in Phys. Abstr., Vol. 81, Abstr. 21036 (1978).

066.204 The gravitational influence of a beam of light of variable flux.   R. W. Nackoney.
J. Math. Phys., Vol. 18, 2146 - 2152 (1977). — Abstr. in Phys. Abstr., Vol. 81, Abstr. 21037 (1978).

066.205 Tachyons in bi-metric theories of gravitation.   D.-E. Liebscher.
Ann. Physik, Vol. 34, 295 - 297 (1977). — Abstr. in Phys. Abstr., Vol. 81, Abstr. 21038 (1978).

066.206 On regularizations of the dynamical singularities by crossing terms in general relativistic gravitation equations.   D.-E. Liebscher, H.-J. Treder.
Ann. Physik, Vol. 34, 314 - 320 (1977). — Abstr. in Phys. Abstr., Vol. 81, Abstr. 21039 (1978).

066.207 On a generalization of Hilbert's metrical matter tensor in Einstein's field equations.   H.-J. Treder.
Ann. Physik, Vol. 34, 321 - 324 (1977). — Abstr. in Phys. Abstr., Vol. 81, Abstr. 21040 (1978).

066.208 A geometric approach to Dirac's constraints of quantum gravitation.
D. Christodoulou, M. Francaviglia.
Atti Accad. Sci. Torino I, Vol. 110, 379 - 386 (1976). In Italian. — Abstr. in Phys. Abstr., Vol. 81, Abstr. 21044 (1978).

066.209 The need is compelling to perform an experiment capable of determining the velocity of gravitational interaction.   G. Cristea.
Found. Phys., Vol. 7, 871 - 884 (1977). — Abstr. in Phys. Abstr., Vol. 81, Abstr. 21045 (1978).

066.210 (Non)-Abelian gauge field theories of the Fermi gas. Neutron-quark stars.   V. Baluni.
Phys. Lett. B, Vol. 72B, 381 - 384 (1978). — Abstr. in Phys. Abstr., Vol. 81, Abstr. 21341 (1978).

066.211 Strong gravity and Dirac numbers.   V. De Sabbata, P. Rizzati.
Nuovo Cimento, Lett., Ser. 2, Vol. 20, 525 - 528 (1977). Abstr. in Phys. Abstr., Vol. 81, Abstr. 21372 (1978).

066.212 Ray aberration and large-scale anisotropy of the cosmic background radiation.   R. Fabbri.
Nuovo Cimento, Lett., Ser. 2, Vol. 20, 563 - 568 (1977). Abstr. in Phys. Abstr., Vol. 81, Abstr. 24576 (1978).

066.213 Boundary conditions at past null infinity for zero-rest mass fields including gravitation.
G. Leipold, M. Walker.
Ann. Inst. Henri Poincaré, Sect. A, Vol. 27, No. 1, p. 61 - 71 (1977). — Abstr. in Phys. Abstr., Vol. 81, Abstr. 24661 (1978).

066.214 Stress-energy tensor for a two-dimensional evaporating black hole.   W. A. Hiscock.
Phys. Rev. D, Vol. 16, 2673 - 2674 (1977). — Abstr. in Phys. Abstr., Vol. 81, Abstr. 24706 (1978).

066.215 Stationary interacting fields around an extreme Reissner-Nordström black hole.   J. Bicak.
Phys. Lett. A, Vol. 64A, 279 - 281 (1977). — Abstr. in Phys. Abstr., Vol. 81, Abstr. 24710 (1978).

066.216 Electrogravitational conversion cross sections in static electromagnetic fields.   W. K. De Logi,
A. R. Mickelson.
Phys. Rev. D, Vol. 16, 2915 - 2927 (1977). — Abstr. in Phys. Abstr., Vol. 81, Abstr. 24711 (1978).

066.217 Gravitational and acoustic waves in an elastic medium.   B. Carter, H. Quintana.
Phys. Rev. D, Vol. 16, 2928 - 2938 (1977). — Abstr. in Phys. Abstr., Vol. 81, Abstr. 24712 (1978).

066.218 Particle emission rates from a black hole. III. Charged leptons from a nonrotating hole.
D. N. Page.
Phys. Rev. D, Vol. 16, 2402 - 2411 (1977). — Abstr. in Phys. Abstr., Vol. 81, Abstr. 24718 (1978).

066.219 Relativistic Klein-Gordon systems.

K. C. Chung, T. Kodama, A. F. daF. Teixeira.
Phys. Rev. D, Vol. 16, 2412 - 2416 (1977). — Abstr. in Phys. Abstr., Vol. 81, Abstr. 24719 (1978).

066.220 Yang-Mills formulation of gravitational dynamics.
E. E. Fairchild, Jr.
Phys. Rev. D, Vol. 16, 2538 - 2547 (1977). — Abstr. in Phys. Abstr., Vol. 81, Abstr. 24721 (1978).

066.221 Dressing up a Reissner naked singularity.
T. Damour, N. Deruelle.
Phys. Lett. B, Vol. 72B, 471 - 476 (1978). — Abstr. in Phys. Abstr., Vol. 81, Abstr. 24990 (1978).

066.222 Note on the spacetimes of Szekeres.
B. K. Berger, D. M. Eardley, D. W. Olson.
Phys. Rev. D, Vol. 16, 3086 - 3089 (1977). — Abstr. in Phys. Abstr., Vol. 81, Abstr. 28510 (1978).

066.223 Simple tests for proposed interior Kerr metrics.
P. Collas.
Nuovo Cimento, Lett., Ser. 2, Vol. 21, 68 - 72 (1978). Abstr. in Phys. Abstr., Vol. 81, Abstr. 28666 (1978).

066.224 Infrared limit on the fine-scale anisotropy of the cosmic background radiation. N. Caderni,
R. Fabbri, V. De Cosmo, B. Melchiorri, F. Melchiorri,
V. Natale.
Phys. Rev. D, Vol. 16, 2424 - 2429 (1977). — Abstr. in Phys. Abstr., Vol. 81, Abstr. 28863 (1978).

066.225 The cosmological constant, broken gauge theories and 3K black-body radiation. V. Canuto,
J. F. Lee.
Phys. Lett. B, Vol. 72B, 281 - 284 (1977). — Abstr. in Phys. Abstr., Vol. 81, Abstr. 28870 (1978).

066.226 Motion of superlight particles in the gravitational field of a black hole. V. M. Lipunov.
Astrometriya i Astrofizika, vyp. (No.) 35, p. 98 - 101 (1978). In Russian.

The motion of particles with velocities exceeding the velocity of light is analysed in the gravitational field of a non-rotating black hole. It is shown that there are no stable circular orbits and finite motions in the Schwarzschild metric. Particles with $v > c$ are noticed to be able to escape from the space limited by the gravitational radius. The author points out that superlight particles increase their energies from small values to very high ones while approaching the black hole. This fact can lead to observational effects.

066.227 Neuere Ergebnisse der Untersuchung der kosmischen Mikrowellen-Hintergrundstrahlung. Teil I: Beobachtungen. G. Dautcourt.
Sterne, 54. Band, 91 - 95 (1978).

066.228 On tidal resonance. B. Mashhoon.
Astrophys. J., Vol. 223, 285 - 298 (1978).

An approximate theory of the interaction of a weak gravitational wave with a Newtonian self-gravitating two-body system is presented within the framework of Einstein's theory of gravitation. The first-order orbital variation due to the external wave is thus determined. The existence of certain resonances that may possibly lead to the "ionization" of the system is investigated. The polarization dependence of the resonances and some of the astrophysical implications of tidal resonance are explored.

066.229 Neutron star matter at high temperatures and densities. I. Bulk properties of nuclear matter.
J. M. Lattimer, D. G. Ravenhall.
Astrophys. J., Vol. 223, 314 - 323 (1978).

The authors perform calculations of the thermodynamic properties of uniform dense matter at finite temperature using the Skyrme nuclear interaction. The calculations are valid for arbitrary proton concentrations and temperatures. In order to understand the conditions under which one might expect to find nuclei immersed in a sea of nucleons, the authors explore the coexistence of two fluid phases with different proton concentrations. A phase diagram for nuclear matter is presented. The critical temperature, as a function of proton concentration, above which no coexistence is possible and nuclei evaporate, is established.

066.230 Consistency of Weyl's geometry as a framework for gravitation. P. Bouvier, A. Maeder.
Astrophys. Space Sci., Vol. 54, 497 - 508 (1978).

Recent work, quoted in the introduction, concerning a scale covariant theory of gravitation and corresponding astrophysical tests, has called for a revival of the geometrical considerations put forward by Weyl half a century ago, together with a better understanding of the mathematical foundation of Weyl's geometry. The present paper is to be regarded as a contribution in that direction.

066.231 Holes in the microwave background?
M. Birkinshaw, S. F. Gull.
Nature, Vol. 274, 111 (1978).

066.232 Transmutation of photons, electrons and gravitons and the density of gravitons in the universe.
V. De Sabbata.
Probl. gravitatsii. Tbilisi, 1976, p. 526 - 541. — Abstr. in Ref. zh., 51. Astron., 5.51.749 (1978).

066.233 Models of protostars in the generalized theory of gravitation. M. A. Mnatsakanyan, G. S. Saakyan.
Probl. gravitatsii. Tbilisi, 1976, p. 423 - 440. In Russian. Abstr. in Ref. zh., 51. Astron., 5.51.757 (1978).

066.234 Some results on Y-space-time. W. Budich.
Probl. gravitatsii. Tbilisi, 1976, p. 55 - 63. In Russian. — Abstr. in Ref. zh., 51. Astron., 5.51.758 (1978).

066.235 Astrophysical relativity. (Relativistic stars and star clusters: their structure, rotation, pulsation, and emission of gravitational waves). K. S. Thorne.
Probl. gravitatsii. Tbilisi, 1976, p. 571 - 584. — Abstr. in Ref. zh., 51. Astron., 5.51.759 (1978).

066.236 Gamma radiation in the process of spherically-symmetric accretion on black holes in binary stellar systems. P. I. Kolykhalov, R. A. Syunyaev.
Inst. kosm. issled. AN SSSR. Prepr., 1977, No. 374, 18 pp. In Russian. — Abstr. in Ref. zh., 51. Astron., 5.51.762 (1978).

066.237 Prospects for experiments of generation and detection of coherent scalar and tensor gravitational waves in the optical region.
U. Kh. Kopvillem, V. R. Nagibarov.
Probl. gravitatsii. Tbilisi, 1976, p. 333 - 343. In Russian. Abstr. in Ref. zh., 51. Astron., 5.51.769 (1978).

066.238 On the three-body problem in Einstein's theory of relativity. A. P. Ryabushko.
Probl. gravitatsii. Tbilisi, 1976, p. 521 - 525. In Russian. Abstr. in Ref. zh., 51. Astron., 5.51.772 (1978).

066.239 Gravitational and acoustic waves in an elastic medium. B. Carter, H. Quintana.
Phys. Rev. D, Vol. 16, 2928 - 2938 (1977) = Natl. Radio Astron. Obs., Green Bank, Repr. No. 785.

066.240 General relativistic stellar stability. J. Demaret.

Bull. Acad. R. Belgique, Cl. Sci., 5° Sér., Tome 62, 337 - 407 (1976) = Inst. Astrophys., Univ. Liège, Cointe-Ougrée (Belgique), Collect. 8°, No. 668.

The author gives here a derivation of the general relativistic non adiabatic radial pulsation equation for spherically symmetric non rotating stars.

066.241   Isoareal transformations of the Kerr-Newman black holes.   A. Curir, M. Francaviglia.
Acta Phys. Polonica, Vol. B9, No. 1, p. 3 - 10 (1978) = Contrib. Oss. Astron. Torino, (Pino Torinese), No. 109.

066.242   An experiment for the potential blue shift at the Norikura Corona Station.   S. Iijima, K. Fujiwara.
Ann. Tokyo Astron. Obs., Second Ser., Vol. 17, 68 - 78 (1978).

An experiment to detect the potential blue shift resulting from the general relativistic theory has been conducted by carrying a commercial cesium beam clock between the Tokyo Astronomical Observatory and the Norikura Corona Station in turn.

066.243   Equilibrium spherically-symmetric distributions of a relativistic gravitating gas in the general relativity theory.   V. I. Bashkov, Yu. G. Ignat'ev, V. I. Kovtun.
Tr. Kazan. Gorod. Astron. Obs., No. 41, p. 46 - 53 (1976). In Russian.

066.244   Neutrino bursts and gravitational waves experiments.   C. Castagnoli, P. Galeotti, O. Saavedra.
Astrophys. Space Sci., Vol. 55, 511 - 513 (1978).

Several experiments have been performed in many countries to observe gravitational waves or neutrino bursts. Since their simultaneous emission may occur in stellar collapses, the authors evaluate the effect of neutrino bursts on gravitational wave antennas and suggest the usefulness of a time correlation among the different detectors.

066.245   Test of general relativity with data from Viking orbiters and landers.   D. L. Cain, J. D. Anderson, M. S. W. Keesey, T. Komarek, P. A. Laing, E. L. Lau.
Bull. American Astron. Soc., Vol. 10, 396 (1978). – Abstract.

066.246   On the quadratic first integral for the uncharged particle orbits in the uncharged Kerr solution.
A. M. Faridi.
Bull. American Astron. Soc., Vol. 10, 397 (1978). – Abstract.

066.247   Quark-star models with "realistic" equations of state.   W. B. Fechner, P. C. Joss.
Bull. American Astron. Soc., Vol. 10, 397 (1978). – Abstract.

066.248   X-ray bursts from helium-burning flashes on accreting neutron stars.   P. C. Joss.
Bull. American Astron. Soc., Vol. 10, 419 - 420 (1978). Abstract.

066.249   General relativistic hydrodynamic stellar collapse to a hot neutron star or black hole.   K. A. Van Riper.
Bull. American Astron. Soc., Vol. 10, 425 (1978). – Abstract.

066.250   Neutrino mediated shock waves in supernovae.   D. Kazanas.
Bull. American Astron. Soc., Vol. 10, 425 (1978). – Abstract.

066.251   Quark stars with 'realistic' equations of state.   W. B. Fechner, P. C. Joss.
Nature, Vol. 274, 347 - 349 (1978).

The authors investigate the properties of collapsed stars in the context of a class of quark-matter models that has recently been proposed. These models, which are consistent with all experimental nuclear and high-energy physics data, are based on a quantum chromodynamic treatment of quarks

that are assumed to be of low mass ($\lesssim 100$ MeV for the lightest quarks). The authors find that stable quark stars are possible. However, they also conclude that such stars need not have significantly different macroscopic properties from those of neutron-star models based on conventional nuclear-matter equations of state.

066.252   Some solutions of stationary, axially symmetric gravitational field equations.   M. Gurses.
J. Math. Phys., Vol. 18, 2356 - 2359 (1977). – Abstr. in Phys. Abstr., Vol. 81, Abstr. 29010 (1978).

066.253   The canonical formulation of supergravity.   M. Pilati.
Nucl. Phys. B, Vol. B132, 138 - 154 (1978). – Abstr. in Phys. Abstr., Vol. 81, Abstr. 29016 (1978).

066.254   Gravitational and electromagnetic wave flux compared and contrasted.   J. A. Wheeler.
Phys. Rev. D, Vol. 16, 3384 - 3389 (1977). – Abstr. in Phys. Abstr., Vol. 81, Abstr. 29027 (1978).

066.255   Field equations and equations of motion in a completely covariant theory of gravitation. Newtonian approximation.   R. Burghardt.
Acta Phys. Austriaca, Vol. 48, 101 - 110 (1978). In German. Abstr. in Phys. Abstr., Vol. 81, Abstr. 29029 (1978).

066.256   A Lorentz-invariant theory of gravitation.   W. Petry.
Ann. Phys., Vol. 34, 477 - 484 (1977). In German. – Abstr. in Phys. Abstr., Vol. 81, Abstr. 29030 (1978).

066.257   A machian approach to gravitation.   R. Goldoni.
Meccanica, Vol. 11, 236 - 238 (1976). – Abstr. in Phys. Abstr., Vol. 81, Abstr. 29032 (1978).

066.258   The Hamiltonian formalism of the local scale invariant gravitational theory.   M. Omote, M. Kasuya.
Prog. Theor. Phys., Vol. 58, 1627 - 1634 (1977). – Abstr. in Phys. Abstr., Vol. 81, Abstr. 29033 (1978).

066.259   Irreducible mass, unincreasable angular momentum and isoareal transformations for black hole physics.   M. Calvani, M. Francaviglia.
Acta Phys. Polonica B, Vol. B9, 11 - 14 (1978). – Abstr. in Phys. Abstr., Vol. 81, Abstr. 32503 (1978).

066.260   Scalar fields around a charged, rotating black hole.   J. Bicak, Z. Stuchlik, M. Sob.
Czechoslovak J. Phys. B, Vol. B28, 121 - 124 (1978). – Abstr. in Phys. Abstr., Vol. 81, Abstr. 32505 (1978).

066.261   On the nonexistence of white holes.   D. Lohiya, N. Panchapakesan.
Nuovo Cimento, Lett., Ser. 2, Vol. 21, 81 - 82 (1978). Abstr. in Phys. Abstr., Vol. 81, Abstr. 32506 (1978).

066.262   Wave equations from space-time: Galileian limit.   R. Cecchini.
Nuovo Cimento A, Ser. 11, Vol. 43 A, 181 - 192 (1978). Abstr. in Phys. Abstr., Vol. 81, Abstr. 32740 (1978).

066.263   Some static and non-static solutions of Brans-Dicke theory of gravitation.   G. K. Goswami.
J. Math. Phys., Vol. 19, 442 - 445 (1978). – Abstr. in Phys. Abstr., Vol. 81, Abstr. 32744 (1978).

066.264   Gravitation and the Einstein-Sciama force.   M. Gasperini.

Nuovo Cimento, Lett., Ser. 2, Vol. 21, 285 - 287 (1978).
Abstr. in Phys. Abstr., Vol. 81, Abstr. 32745 (1978).

**066.265   The metric-affine gravitational theory as the gauge
theory of the affine group.   E. A. Lord.**
Phys. Lett. A, Vol. 65A, 1 - 4 (1978). – Abstr. in Phys.
Abstr., Vol. 81, Abstr. 32746 (1978).

**066.266   Calculation of the renormalised quantum stress
tensor by adiabatic regularisation in two- and four-
dimensional Robertson-Walker space-times.   T. S. Bunch.**
J. Phys. A, Vol. 11, 603 - 607 (1978). – Abstr. in Phys. Abstr.,
Vol. 81, Abstr. 32747 (1978).

**066.267   Conformal off-mass-shell extension and elimination
of conformal anomalies in quantum gravity.**
E. S. Fradkin, G. A. Vilkovisky.
Phys. Lett. B, Vol. 73B, 209 - 213 (1978). – Abstr. in Phys.
Abstr., Vol. 81, Abstr. 32749 (1978).

**066.268   Integral equations for even parity perturbations of
hot perfect fluid nonrotating neutron stars.**
P. Cazzola, L. Lucaroni, R. Semenzato.
J. Math. Phys., Vol. 19, 237 - 246 (1978). – Abstr. in Phys.
Abstr., Vol. 81, Abstr. 36628 (1978).

**066.269   Isoareal transformations of the Kerr-Newman black
holes.   A. Curir, M. Francaviglia.**
Acta Phys. Polonica B, Vol. B9, 3 - 10 (1978). – Abstr. in
Phys. Abstr., Vol. 81, Abstr. 36630 (1978).

**066.270   Black-holes and tachyons.   V. De Sabbata,
M. Pavsic, E. Recami.**
Univ. Catania, Italy, Report INFN/AE-77/5, 10 pp. (1977).
Abstr. in Phys. Abstr., Vol. 81, Abstr. 36635 (1978).

**066.271   Schwarzschild electrodynamics: black holes, neutron
stars.   M. W. Kearney, L. S. Kegeles, J. M. Cohen.**
Astrophys. Space Sci., Vol. 56, 129 - 190 (1978).
The authors present a number of calculations involving
the production and propagation of electromagnetic waves in
the Schwarzschild metric. They are based on algorithms
developed from the power series solutions of the Schwarzschild radial equation (Regge–Wheeler equation) of Arenstorf,
Cohen and Kegeles. These include the scattering of electromagnetic plane waves from a Schwarzschild black hole. The
authors calculate the power radiated from a radially vibrating
neutron star and find that the radiation can be hyperemissive.
The Schwarzschild radial functions are extensively treated in
the appendices.

**066.272   Radiation hydrodynamics of high-luminosity accre-
tion into black holes.   J. Schmid-Burgk.**
Astrophys. Space Sci., Vol. 56, 191 - 218 (1978).
To extend Shapiro's (1973) calculations of black hole
accretion to the regimes of interstellar gas densities and of
black hole masses for which emergent luminosities are expected to be high, the radiation hydrodynamics of spherically
symmetric gas flows in static isotropic metrics is discussed.
Particular attention has to be paid to radiative transfer through
non-Euclidean spaces, and a method for solving the full
transfer problem is presented. The method is applied to accretion into black holes of mass between $10\,M_\odot$ and $10^5\,M_\odot$.

**066.273   Force-free fields in the vicinity of a Reissner-Nord-
strøm black hole.   E. Evangelidis.**
Astrophys. Space Sci., Vol. 56, 255 - 269 (1978).
The behaviour of a force-free field has been studied in a
Reissner-Nordstrøm metric. An expansion in tensor harmonics
of even-odd parity reduced the radial equations in a differential
equation of the Sturm-Liouville system which was solved
asymptotically in a conveniently defined space coordinate.

Further, it has been possible to regularize the singular behaviour of the Reissner-Nordstrøm metric at the event horizon
and the modified metric to be given explicitly.

**066.274   Quantum field theory in curved space-time: an
overview.   C. J. Isham.**
Ann. New York Acad. Sci., Vol. 302, (see 012.033), 114 - 157
(1977).
Contents: Heuristic quantum field theory aspects.
Bogolubov transformations. Isotropic cosmological models.
Anisotropic cosmological models. The collapsing ball. The
energy-momentum tensor. The falling shell. The eternal black
hole. Accelerated observers in Minkowski space-time. Green's
functions and analyticity. Conclusions and future developments.

**066.275   Black holes and unpredictability.
S. W. Hawking.**
Ann. New York Acad. Sci., Vol. 302, (see 012.033), 158 - 160
(1977).

**066.276   The thermodynamics of black holes.
D. W. Sciama.**
Ann. New York Acad. Sci., Vol. 302, (see 012.033), 161 - 165
(1977).

**066.277   Stress tensor calculations and conformal anomalies.
P. C. W. Davies.**
Ann. New York Acad. Sci., Vol. 302, (see 012.033), 166 - 185
(1977).

**066.278   Particle detectors and black hole evaporation.
W. G. Unruh.**
Ann. New York Acad. Sci., Vol. 302, (see 012.033), 186 - 190
(1977).

**066.279   On the possible existence of quark stars.
G. Chapline, M. Nauenberg.**
Ann. New York Acad. Sci., Vol. 302, (see 012.033), 191 - 196
(1977).

**066.280   Timing effects in rotating neutron stars.
E. J. Schreier.**
Ann. New York Acad. Sci., Vol. 302, (see 012.033), 445 -
459 (1977).
The author discusses observations of rotating neutron
stars, and in particular, what one can learn about neutron stars
from studying various timing effects. He intends to concentrate on the observations and their immediate interpretations.
He finds that X-ray pulsar observations do extend the study
of neutron star structure, since the accretion torques supply
the external stimulus necessary to kick the neutron star; one
can then watch the response and determine magnetic fields
and moments of inertia.

**066.281   The masses of neutron stars: observational con-
straints.   S. A. Rappaport, P. C. Joss.**
Ann. New York Acad. Sci., Vol. 302, (see 012.033), 460 -
470 (1977).
The authors describe the present state of empirical
knowledge of neutron-star masses and discuss the theoretical
implications of this information.

**066.282   "Neutron" stars within the laws of physics.
K. Brecher, G. Caporaso.**
Ann. New York Acad. Sci., Vol. 302, (see 012.033), 471 -
481 (1977).
The main points of this paper are: 1. The upper mass
limit of "neutron" stars crucially depends on the assumed
properties of matter at high densities and on the assumed
nature of strong gravitational fields. 2. Neither the local
properties of matter at supernuclear densities, nor the effects

of gravitation when the fields are large are experimentally known. 3. Using experimentally allowed theories of matter and gravitation, the value of M_{max} is found to lie in the range 1–80 $M_\odot$. 4. A particular value of M_{max}, therefore, cannot be used with certainty as part of a demonstration that a specific object is a black hole.

066.283   Knowledge of neutron stars from X-ray observations.
F. K. Lamb.
Ann. New York Acad. Sci., Vol. 302, (see 012.033), 482 - 513 (1977).
    The author discusses in turn the evidence that can be furnished by X-ray observations regarding neutron star masses, moments of inertia, magnetic moments, surface magnetic fields, and internal structure.

066.284   Neutron stars: general review.   V. Canuto.
    Ann. New York Acad. Sci., Vol. 302, (see 012.033), 514 - 527 (1977).
    The author reviews the work on neutron stars that has been performed in the last five or six years. The general trend of his talk is that of stressing the fact that today one seems to have reached a satisfactory agreement among theorists and observers. What the author is referring to is the apparent clarification of at least one problem, that of the maximum value for the mass of a neutron star.

066.285   The damping of differential rotation in the cores of neutron stars.   D. J. Hegyi.
Ann. New York Acad. Sci., Vol. 302, (see 012.033), 528 - 531 (1977).

066.286   Supergravity.   B. Zumino.
    Ann. New York Acad. Sci., Vol. 302, (see 012.033), 545 - 556 (1977).

066.287   Gravitational radiation damping and energy loss in binary systems.   W. L. Burke.
Ann. New York Acad. Sci., Vol. 302, (see 012.033), 557 - 564 (1977).
    The author discusses the general relativistic two-body problem and its solution using singular perturbation methods.

066.288   Gravitational energy loss in scattering problems.
A. Rosenblum.
Ann. New York Acad. Sci., Vol. 302, (see 012.033), 565 - 568 (1977).
    The author discusses the problem of the energy loss due to gravitational radiation in the scattering of two bodies of equal mass.

066.289   Space-times generated by computers: black holes with gravitational radiation.   L. Smarr.
Ann. New York Acad. Sci., Vol. 302, (see 012.033), 569 - 604 (1977).
    Contents: Space-time kinematics. Space-time dynamics. Numerical methods. Gravitational radiation. Brill waves – a pure radiation space-time. Two black hole collision. Other sources of gravitational radiation.

066.290   Parametrized post-Newtonian approximation and Rastall's gravitational field equations.
L. L. Smalley.
Found. Phys., Vol. 8, 59 - 68 (1978). – Abstr. in Phys. Abstr., Vol. 81, Abstr. 36894 (1978).

066.291   Is gravitational dual charge physical?
E. Lubkin.
Int. J. Theor. Phys., Vol. 16, 551 - 554 (1977). – Abstr. in Phys. Abstr., Vol. 81, Abstr. 36906 (1978).

066.292   Einstein gravitation from scalar nonsingular sources
alone.   J. A. Souza, A. F. daF. Teixeira.
Int. J. Theor. Phys., Vol. 16, 585 - 590 (1977). – Abstr. in Phys. Abstr., Vol. 81, Abstr. 36908 (1978).

066.293   Particle creation near black holes.   R. Wald.
    American Sci., Vol. 65, 585 - 589 (1977). – Abstr. in Phys. Abstr., Vol. 81, Abstr. 40986 (1978).

066.294   On the Christodoulou-Ruffini mass formula.
    M. Yamazaki.
Phys. Lett. A, Vol. 65A, 267 - 268 (1978). – Abstr. in Phys. Abstr., Vol. 81, Abstr. 40990 (1978).

066.295   Occurrence of whimper singularities.
    S. T. C. Siklos.
Commun. Math. Phys., Vol. 58, 255 - 272 (1978). – Abstr. in Phys. Abstr., Vol. 81, Abstr. 41205 (1978).

066.296   Maximal extension of a nonsingular solution in a generalised theory of gravitation.   G. Kunstatter,
J. W. Moffat.
Phys. Rev. D, Vol. 17, 396 - 403 (1978). – Abstr. in Phys. Abstr., Vol. 81, Abstr. 41214 (1978).

066.297   Effects of naked singularities: particle orbits near a two-parameter family of naked singularity solutions.   C. Duncan, F. P. Esposito, S. Lee.
Phys. Rev. D, Vol. 17, 404 - 413 (1978). – Abstr. in Phys. Abstr., Vol. 81, Abstr. 41215 (1978).

066.298   Spinor fields in an Einstein universe. The vacuum-averaged stress-energy tensor.   J. S. Dowker,
B. M. Altaie.
Phys. Rev. D, Vol. 17, 417 - 422 (1978). – Abstr. in Phys. Abstr., Vol. 81, Abstr. 41217 (1978).

066.299   The interaction of the gravitational and electromagnetic fields.   D. Lovelock.
J. Math. Phys., Vol. 19, 586 - 592 (1978). – Abstr. in Phys. Abstr., Vol. 81, Abstr. 41219 (1978).

066.300   Some conformally-flat space-times of a divergence-free Riemann tensor.   L. J. Rose, A. E. G. Stuart,
G. W. Barrett.
J. Phys. A, Vol. 11, 709 - 713 (1978). – Abstr. in Phys. Abstr., Vol. 81, Abstr. 41220 (1978).

066.301   Wideband laser-interferometer gravitational-radiation experiment.   R. L. Forward.
Phys. Rev. D, Vol. 17, 379 - 390 (1978). – Abstr. in Phys. Abstr., Vol. 81, Abstr. 41223 (1978).

066.302   Neutrino cooling of neutron stars in pion-condensed phase.   M. Kiguchi.
Prog. Theor. Phys., Vol. 58, 1766 - 1774 (1977). – Abstr. in Phys. Abstr., Vol. 81, Abstr. 45351 (1978).

066.303   Local calculation of black hole radiance.
    P. Meyer.
Phys. Lett. A, Vol. 65A, 369 - 370 (1978). – Abstr. in Phys. Abstr., Vol. 81, Abstr. 45354 (1978).

066.304   The effect of pressure gradient force on an accretion disk surrounding a black-hole.
R. Hoshi, N. Shibazaki.
Prog. Theor. Phys., Vol. 58, 1759 - 1765 (1977). – Abstr. in Phys. Abstr., Vol. 81, Abstr. 45355 (1978).

066.305   Navier's equation for a viscous fluid in general relativity.   V. I. Bashkov, G. G. Matveev.
Gravitatsii i teor. otnositel'n. Vyp. (No.) 13, Kazan', Kazan. univ., 1976, p. 18 - 25. In Russian. – Abstr. in Ref. zh., 51.

Astron., 6.51.951 (1978).

066.306 Equilibrium distribution of relativistic gas in the gravitational field of a cylindrical beam of plane electromagnetic waves. G. G. Ivanov.
Gravitatsii i teor. otnositel'n. Vyp. (No.) 13, Kazan', Kazan. univ., 1976, p. 57 - 66. In Russian. – Abstr. in Ref. zh., 51. Astron., 6.51.952 (1978).

066.307 Rotational relativistic polytropes and supermassive stars. V. V. Papoyan, D. M. Sedrakyan, Eh. V. Chubaryan.
Probl. gravitatsii, Tbilisi, 1976, p. 460 - 470. In Russian. Abstr. in Ref. zh., 51. Astron., 6.51.954 (1978).

066.308 On radiative spin polarization of an electron moving in a spiral in a magnetic field. A. A. Sokolov, I. M. Ternov, V. G. Bagrov.
Probl. gravitatsii, Tbilisi, 1976, p. 542 - 551. In Russian. Abstr. in Ref. zh., 51. Astron., 6.51.960 (1978).

066.309 Perihelion shift in Minkowski's theory of gravitation taking into account the effect of the Galaxy. R. F. Bilyalov.
Gravitatsii i teor. otnositel'n, Vyp. (No.) 13, Kazan', Kazan. univ., 1976, p. 26 - 29. In Russian. – Abstr. in Ref. zh., 51. Astron., 6.51.965 (1978).

066.310 Le onde gravitazionali: una nuova astronomia? G. Pizzella.
Mem. Soc. Astron. Italiana, Vol. 47, 11 - 22 (1976).

066.311 Differing time standards and celestial mechanics. B. Bertotti, G. Picone.
Mem. Soc. Astron. Italiana, Vol. 48, 77 - 81 (1977).

066.312 Mansouri-Chang gravitation theory. R. Pavelle.
Phys. Rev. Lett., Vol. 40, 267 - 270 (1978).
The author examines the new gauge theory of gravitation theory proposed recently by Mansouri and Chang. It appears that the predictions of the theory are indistinguishable from those of general relativity with regard to the usual tests. The author finds that the theory possesses a remarkable similarity to Einstein's theory with respect to vacuum solutions.

066.313 Unified approach to matter coupling in Weyl and Einstein supergravity.
A. Das, M. Kaku, P. K. Townsend.
Phys. Rev. Lett., Vol. 40, 1215 - 1218 (1978).
The authors present a new unified method for coupling matter to both Poincaré and conformal supergravity. As an example they construct the coupled Maxwell-Weyl and the coupled Maxwell-Einstein supergravity theories.

066.314 Homogeneous and isotropic world models in the Yang-Mills dynamics of gravity. The structure of the adiabats. M. Camenzind.
J. Math. Phys., Vol. 19, 624 - 634 (1978). – Abstr. in Phys. Abstr., Vol. 81, Abstr. 45543 (1978).

066.315 Prospects for detection of gravitational radiation by simultaneous Doppler tracking of several spacecraft.
F. B. Estabrook, H. D. Wahlquist.
Acta Astronaut., Vol. 5, 5 - 7 (1978). – Abstr. in Phys. Abstr., Vol. 81, Abstr. 45549 (1978).

066.316 Measurement of the geodesic deviation of a gyroscope. P. K. Chapman.
Acta Astronaut., Vol. 5, 19 - 25 (1978). – Abstr. in Phys. Abstr., Vol. 81, Abstr. 45550 (1978).

066.317 Equivalence principle tests in Earth orbit.

P. W. Worden, Jr.
Acta Astronaut., Vol. 5, 27 - 42 (1978). – Abstr. in Phys. Abstr., Vol. 81, Abstr. 45551 (1978).

066.318 Tests of general relativity using astrometric and radio metric observations of the planets.
J. D. Anderson, M. S. W. Keesey, E. L. Lau, E. M. Standish, Jr., X X Newhall.
Acta Astronaut., Vol. 5, 43 - 61 (1978). – Abstr. in Phys. Abstr., Vol. 81, Abstr. 45552 (1978).

066.319 Recent developments in the measurement of space time curvature.
J.-P. Richard.
Acta Astronaut., Vol. 5, 63 - 76 (1978). – Abstr. in Phys. Abstr., Vol. 81, Abstr. 45553 (1978).

066.320 Error analysis of a relativity test with counter-orbiting satellites. R. A. Van Patten, J. V. Breakwell, D. Schaechter, C. W. F. Everitt.
Acta Astronaut., Vol. 5, 77 - 86 (1978). – Abstr. in Phys. Abstr., Vol. 81, Abstr. 45554 (1978).

066.321 Superconducting accelerometers for the study of gravitation and gravitational radiation.
W. C. Oelfke, W. O. Hamilton.
Acta Astronaut., Vol. 5, 87 - 96 (1978). – Abstr. in Phys. Abstr., Vol. 81, Abstr. 45555 (1978).

066.322 The reflection of gravitational waves from compact stars. Y. G. (*Yu. G.*) Ignat'ev, A. V. Zakharov.
Phys. Lett. Vol. A, Vol. 66A, 3 - 4 (1978). – Abstr. in Phys. Abstr., Vol. 81, Abstr. 49314 (1978).

066.323 Cosmic betatron. C. D. Ciubotariu.
Stud. Cercet Fiz., Vol. 29, 803 - 809 (1977). In Rumanian. – Abstr. in Phys. Abstr., Vol. 81, Abstr. 49315 (1978).

066.324 Questions about rotating superfluid dynamics: problems of pulsar astrophysics accessible in the laboratory. P. W. Anderson, D. Pines, M. Ruderman, J. Shaham.
J. Low Temp. Phys., Vol. 30, 839 - 847 (1978). – Abstr. in Phys. Abstr., Vol. 81, Abstr. 49469 (1978).

066.325 Study on perturbations in the Reissner-Nordström background geometry. C. H. Lee.
Nuovo Cimento B, Ser. 11, Vol. 44B, 123 - 134 (1978). Abstr. in Phys. Abstr., Vol. 81, Abstr. 49472 (1978).

066.326 Asymptotic eigenfrequency distribution for even-parity perturbations of hot perfect-fluid relativistic stars. P. Cazzola, L. Lucaroni, R. Semenzato.
J. Math. Phys., Vol. 19, 901 - 910 (1978). – Abstr. in Phys. Abstr., Vol. 81, Abstr. 49668 (1978).

066.327 Schwarzschild sphere and classical electron self-energy problem. A. V. Vilenkin, P. I. Fomin.
Nuovo Cimento A, Ser. 11, Vol. 45A, 59 - 77 (1978). – Abstr. in Phys. Abstr., Vol. 81, Abstr. 49681 (1978).

066.328 Covariant decomposition of the equations of motion of the gravitation theory for a rotating system of observers. R. Burghardt.
Acta Phys. Austriaca, Vol. 48, No. 3, p. 181 - 188 (1978). In German. – Abstr. in Phys. Abstr., Vol. 81, Abstr. 49724 (1978).

066.329 A conformal invariant model of localized spinning test particles. C. Duval, H. H. Fliche.
J. Math. Phys., Vol. 19, 749 - 752 (1978). – Abstr. in Phys.

Abstr., Vol. 81, Abstr. 49729 (1978).

066.330   Relativistic spherical stars reformulated.
            E. N. Glass, S. P. Goldman.
J. Math. Phys., Vol. 19, 856 - 859 (1978). — Abstr. in Phys.
Abstr., Vol. 81, Abstr. 49731 (1978).

066.331   Generation of metrics for charged rotating bodies.
            J. R. Meinhardt, E. Leibowitz.
Phys. Lett. A, Vol. 66A, 1 - 2 (1978). — Abstr. in Phys. Abstr.,
Vol. 81, Abstr. 49735 (1978).

066.332   Non-existence stationary, axially symmetric, asymp-
            totically flat solutions of the Einstein equations for
dust.   A. Caporali.
Phys. Lett. A, Vol. 66A, 5 - 7 (1978). — Abstr. in Phys. Abstr.,
Vol. 81, Abstr. 49736 (1978).

066.333   Supergravity as geometry of superspace.
            L. Brink, M. Gell-Mann, P. Ramond, J. H. Schwarz.
Phys. Lett. B, Vol. 74B, 336 - 340 (1978). — Abstr. in Phys.
Abstr., Vol. 81, Abstr. 49744 (1978).

066.334   Local supersymmetry in (2 + 1) dimensions. I.
            Supergravity and differential forms.
P. S. Howe, R. W. Tucker.
J. Math. Phys., Vol. 19, 869 - 873 (1978). — Abstr. in Phys.
Abstr., Vol. 81, Abstr. 49980 (1978).

066.335   A new approach to the physical optics of polarized
            light beams in gravitational fields.   D. Leiter.
Nuovo Cimento B, Ser. 11, Vol. 44B, 275 - 288 (1978).
Abstr. in Phys. Abstr., Vol. 81, Abstr. 50696 (1978).

066.336   Pure neutral vector boson contribution to neutrino
            opacity in stellar matter.
P. K. Deka, S. A. S. Ahmed.
Indian J. Pure Appl. Phys., Vol. 15, 839 - 841 (1977). — Abstr.
in Phys. Abstr., Vol. 81, Abstr. 53928 (1978).

066.337   Zero sound in dense neutron matter.
            P. Haensel.
Nucl. Phys. A, Vol. A298, 139 - 150 (1978). — Abstr. in Phys.
Abstr., Vol. 81, Abstr. 53977 (1978).

066.338   Electromagnetic radiation near black holes and
            neutron stars.
R. F. Arenstorf, J. M. Cohen, L. S. Kegeles.
J. Math. Phys., Vol. 19, No. 4, p. 833 - 837 (1978). — Abstr.
in Phys. Abstr., Vol. 81, Abstr. 53978 (1978).

066.339   Switching off the black-hole evaporation.
            G. Denardo, E. Spallucci.
Nuovo Cimento B, Ser. 11, Vol. 44B, 381 - 393 (1978).
Abstr. in Phys. Abstr., Vol. 81, Abstr. 53979 (1978).

066.340   Killing vectors in plane HH spaces.
            J. D. Finley, III, J. F. Plebanski.
J. Math. Phys., Vol. 19, 760 - 766 (1978). — Abstr. in Phys.
Abstr., Vol. 81, Abstr. 54104 (1978).

066.341   Fluctuations of microwave background radiation.
            R. A. Syunyaev.
Inst. kosm. issled. AN SSSR. Prepr., 1977, No. 371, 21 pp.
In Russian. — Abstr. in Ref. zh., 51. Astron., 7.51.940 (1978).

066.342   Fluctuations of background radiation originating
            during secondary ionization of matter in the uni-
verse.   R. A. Syunyaev.
Inst. kosm. issled. AN SSSR. Prepr., 1977, No. 349, 16 pp.
In Russian. — Abstr. in Ref. zh., 51. Astron., 7.51.941 (1978).

066.343   Tidal disruption of stars and evolution of a massive
            black hole under the conditions of the galactic
center.   V. I. Dokuchaev, L. M. Ozernoj.
Fiz. inst. AN SSSR, Prepr. 1977, No. 137, 16 pp. In Russian.
Abstr. in Ref. zh., 51. Astron., 7.51.980 (1978).

066.344   Black hole in an external magnetic field.
            D. V. Gal'tsov, V. I. Petukhov.
Zh. ehksp. i teor. fiz., Vol. 74, 801 - 818 (1978). In Russian.
Abstr. in Ref. zh., 51. Astron., 7.51.982 (1978).

066.345   Accretion on a rapidly moving gravitating center.
            G. S. Bisnovatyj-Kogan, Ya. M. Kazhdan, A. A.
Klypin, A. E. Lutskij, N. I. Shakura.
Inst. prikl. mat. AN SSSR. Prepr., 1978, No. 20, 32 pp. In
Russian. — Abstr. in Ref. zh., 51. Astron., 7.51.988 (1978).

066.346   Gravitational waves and the limiting stability of
            self-excited oscillators.
V. B. Braginskij, S. P. Vyatchanin.
Zh. ehksp. i teor. fiz., Vol. 74, 828 - 832 (1978). In Russian.
Abstr. in Ref. zh., 51. Astron., 7.51.993 (1978).

The collapsing universe.   See Abstr. 003.022.

Relativistic theories of materials.
See Abstr. 003.030.

Group theory and general relativity.
See Abstr. 003.033.

Nuclear cascade process in dense matter.
See Abstr. 003.039.

Kvasarer, pulsarer og svarte hull.
See Abstr. 003.051.

Black holes in the universe.
See Abstr. 003.107.

Gravitation and relativity. Issue 13.
See Abstr. 003.153.

Theoretical principles in astrophysics and relativity.
See Abstr. 012.044.

Possibility of a geophysical determination of the
Newtonian gravitational constant.   See Abstr. 043.012.

Astronautical generalisation of the effects of
relativity due to flyby of a central field of gravitation.
See Abstr. 052.007.

On particle creation in gravitational collapse.
See Abstr. 061.008.

Dynamics of stellar collapse with degenerate
neutrinos.   See Abstr. 061.027.

The evolution of expanding nonthermal sources. II.
Relativistic expansion.   See Abstr. 062.022.

The spectrum of synchrotron-transient radiation of
a charged particle in plasma with random inhomogeneities.
See Abstr. 062.038.

Propagation of plane relativistic shock waves in the
presence of a magnetic field.   See Abstr. 062.114.

On the turbulent energy transport in accretion
discs.   See Abstr. 064.005.

Thickening of axisymmetric accretion disks and flows. See Abstr. 064.024.

The effect of radiation pressure on spherical accretion. See Abstr. 064.066.

Disk accretion by magnetic neutron stars. See Abstr. 064.080.

The hydrodynamics of stellar collapse. See Abstr. 065.033.

Further studies of the helium-driven r-process. See Abstr. 065.077.

Inner region of accretion discs in binary stellar systems. See Abstr. 117.064.

X-ray emission processes in close binary systems. See Abstr. 117.085.

The unusual object KR Aurigae. See Abstr. 122.077.

Analytic supernova models and black holes. See Abstr. 125.040.

Star formation in shock-compressed layers. See Abstr. 131.057.

Stellar debris clouds in quasars and related objects. See Abstr. 141.011.

Evidence for supermassive accretion disks in QSOs. See Abstr. 141.075.

Gravity power. See Abstr. 141.122.

Accretion and the quasar phenomenon. See Abstr. 141.123.

The black flash model of QSOs. See Abstr. 141.146.

Search for radio emission from objects with continuous optical spectrum and large proper motion. See Abstr. 141.223.

Polar electric fields of aligned, magnetized neutron stars. See Abstr. 141.517.

Pulsar glitches and the thermal/timing instability in neutron stars. See Abstr. 141.549.

Cyclotron emission and beaming mechanisms in magnetized neutron stars: Her X-1. See Abstr. 142.032.

The luminosity enhancement following X-ray bursts. See Abstr. 142.035.

Accreting neutron stars in highly compact binary systems and the nature of 3U 1626–67. See Abstr. 142.053.

A thermonuclear runaway in the hydrogen envelope of a neutron star as a possible model for A0620–00 (V616 Mon). See Abstr. 142.104.

Neutron star and degenerate dwarf models of X-ray bursts. See Abstr. 142.236.

A search for high energy gamma-ray bursts from primordial black holes or other astronomical objects. See Abstr. 142.708.

Gamma-ray lines as probes of the fundamental properties of collapsed stars. See Abstr. 142.726.

The distribution of stars around a massive central black hole in a spherical stellar system. I. Results for test stars with a unique mass and radius. See Abstr. 151.062.

A multi-term trial function stability analysis of isotropic relativistic star clusters. See Abstr. 151.094.

Evolutionary effects of a supermassive body on a large surrounding cloud. See Abstr. 151.123.

Constraints to the mass of a black hole at the Galactic centre. See Abstr. 155.048.

Dynamical evidence for a central mass concentration in the galaxy M87. See Abstr. 158.112.

Emission from the nuclei of nearby galaxies: evidence for massive black holes? See Abstr. 158.195.

Primordial black hole formation in an anisotropic universe. See Abstr. 162.001.

Helium production and limits on the anisotropy of the universe. See Abstr. 162.003.

Cosmology with another theory of gravity. See Abstr. 162.021.

A crucial test of the Dirac cosmologies. See Abstr. 162.034.

Anomalous Hubble expansion and inhomogeneous cosmological models. See Abstr. 162.054.

On the compatibility of paradoxical redshifts, observed in galaxies and quasars, with the theory of gravitation. See Abstr. 162.057.

Possible cosmological consequences of the process of evaporation of primary black holes. See Abstr. 162.066.

Charge in cosmological models. See Abstr. 162.068.

Spatially homogeneous universe. See Abstr. 162.099.

Quantum effects in cosmology and black–and–white hole physics. See Abstr. 162.102.

On a quantized scalar field in some Bianchi-type I universe. II. DeWitt's two vacuum states connected causally. See Abstr. 162.105.

Eigenvalue treatment of cosmological models. See Abstr. 162.109.

Solutions du type Robertson-Walker en théorie unitaire. See Abstr. 162.113.

Fluctuations of the microwave background radiation. See Abstr. 162.137.

The anisotropy in the microwave background radiation in a FIB metric. See Abstr. 162.159.

Errata

066.901   Erratum: "Gravitational radiation instability in
             rotating stars" [Astrophys. J., Lett., Vol. 199,
L157 - L159 (1975)].   J. L. Friedman, B. F. Schutz.

Astrophys. J., Lett., Vol. 221, L99 (1978).

066.902   Erratum: 'On the stability of relativistic systems'
             [Astrophys. J., Vol. 200, 204 - 220 (1975)].
J. L. Friedman, B. F. Schutz.
Astrophys. J., Vol. 222, 1119 (1978).

Sun

071 Photosphere, Spectrum

071.001 **The abundances of the elements in the solar photosphere – VIII. Revised abundances of carbon, nitrogen and oxygen.** D. L. Lambert.
Mon. Not. R. Astron. Soc., Vol. 182, 249 - 271 (1978).

A comprehensive analysis of atomic and molecular lines provides the following recommended solar photospheric abundances for carbon, nitrogen and oxygen log ϵ(C) = 8.67, log ϵ(N) = 7.99 and log ϵ(O) = 8.92 on the standard scale log ϵ(H) = 12.00.

071.002 **Tin in the ultraviolet solar spectrum.** M. S. Allen.
Astrophys. J., Vol. 219, 307 - 313 (1978).

An analysis of the ultraviolet solar tin spectrum has been carried out by using absolute intensity profiles for Sn I $\lambda\lambda$2863, 3034, and 3801. A value for log Sn/H + 12 of 1.85 $\pm$ 0.3 was obtained, in satisfactory agreement with the chondritic abundance.

071.003 **Balloon-borne imagery of the solar granulation. II. The lifetime of solar granulation.**
J. P. Mehltretter.
Astron. Astrophys., Vol. 62, 311 - 316 (1978) = Mitt. Fraunhofer Inst., Freiburg, No. 152.

The phenomenological aspects of the evolution in time of photospheric granulation are described, lifetimes of individual granules determined, and the correlation of the brightness fluctuations at different time-lags measured. The lifetime probability is very nearly given by an exponential law with time constants of the order of 5.9-8.3 min, depending somewhat upon the characteristic processes involved.

071.004 **Evolution pattern of the exploding granules.** O. Namba, R. van Rijsbergen.
Problems of stellar convection, (see 012.005), p. 119 - 125 (1977).

The evolution pattern of the so-called exploding granules has been studied on the basis of a time sequence from Princeton Stratoscope pictures of the solar granulation. Some preliminary results are presented. A new interpretation of the phenomenon is suggested.

071.005 **Granulation observations.** A. Nesis.
Problems of stellar convection, (see 012.005), p. 126 - 127 (1977).

High resolution spectrograms that were obtained in Capri with the coudé refractor are analyzed.

071.006 **Convective overshooting in the solar photosphere; a model granular velocity field.** Å. Nordlund.
Problems of stellar convection, (see 012.005), p. 237 - 238 (1977).

071.007 **Cooling of the Sun's photosphere coincident with increased sunspot activity.** W. C. Livingston.
Nature, Vol. 272, 340 - 341 (1978).

In summer 1975, the author began to monitor the spectroscopic temperature of integrated sunlight. The first 1.5 yr of observations were taken at a time of minimum activity and the full disk temperature was apparently constant at $\pm$1−2K. However, 1977 produced a marked increase of activity with the onset of cycle No. 21. In response to this activity the Sun seems to have cooled by ~6K. To the extent that the Sun is a black body, a 6K change in temperature corresponds to a ~0.5% reduction in luminosity. The inverse relation between activity and temperature is a new discovery, and may prove helpful in explaining certain aspects of the Sun−Earth climate relationship.

071.008 **A north-south asymmetry in the solar brightness.** A. Wittmann.
Astron. Astrophys., Vol. 64, 91 - 95 (1978).

Photoelectric measurements of the continuum centre-to-limb variation of the quiet sun have been made along polar and equatorial diameters. The intensities have been corrected for scattered light, image blurring, and differential extinction. Radiative transfer calculations were used to interpret the measurements. A north-south asymmetry in the solar brightness has been detected: Whereas the equatorial brightness distribution did not depart significantly from the average, the south polar region was hotter (ΔT = + 19 K at τ = 0.15) and north polar region was cooler (ΔT = −19 K) than average.

071.009 **Computation of oscillator strengths by a semi-empirical method for some elements of the iron-group and their solar photospheric abundance. IV: Results for V I, Co I and summary.** E. Biémont.
Sol. Phys., Vol. 56, 79 - 86 (1978).

Oscillator strengths calculated by the STF method for selected transitions of V I and Co I have permitted to deduce, from an LTE study of weak lines, photospheric abundances in agreement with those obtained from carbonaceous chondrites. A summary of results is also presented for all the iron-group elements.

071.010 **One-dimensional speckle interferometry of the solar granulation.**
C. Aime, G. Ricort, J. Harvey.
Astrophys. J., Vol. 221, 362 - 367 (1978).

Rapid photoelectric scans of disk-center solar granulation were made with a telescope with a "one-dimensional" (2 cm X 50 cm) aperture in a direction parallel to the long dimension of the aperture. The theory of speckle interferometry applied to the observations allowed the shape of an axial cut through the two-dimensional spatial power spectrum of the granulation to be determined between 0.3 and 1.4 cycles arcsec^{-1}, in spite of seeing of the order of 3″. The observed exponential decrease of the granulation power spectrum with increasing spatial frequency favors a convective model of solar granulation.

071.011 Hydraulic concentration of magnetic fields in the solar photosphere. VI. Adiabatic cooling and concentration in downdrafts.   E. N. Parker.
Astrophys. J., Vol. 221, 368 - 377 (1978).

The remarkable concentration of the general field of the Sun into isolated intense flux tubes at the visible surface must be a direct consequence of conditions immediately beneath the surface. It is pointed out that the convective heat transport in the magnetic field swept into the downdrafts in the junctions of supergranule boundaries is strongly suppressed by the magnetic field. The net heat transport is reduced to such a degree that the temperature of the downdraft within the field increases nearly adiabatically below the visible surface, and hence is significantly cooler than the surrounding ambient gas. The reduced temperature enhances the downdraft within the field and permits the gravitational field to evacuate the flux tube. The magnetic field is then strongly compressed by the external gas pressure, leading to the extraordinary observed strengths of 1500 gauss or more.

071.012 Molecular hydrogen in the solar atmosphere.
        G. Brueckner, J.-D. F. Bartoe, G. D. Sandlin, M. E. VanHoosier, C. Jordan.
Bull. American Astron. Soc., Vol. 9, 568 (1978). – Abstract.

071.013 Formation of the solar Mg I spectrum.
        A. L. Zachary, E. H. Avrett, R. L. Kurucz, R. K. Loeser, J. E. Vernazza.
Bull. American Astron. Soc., Vol. 9, 568 (1978). – Abstract.

071.014 The solar limb effect.   J. M. Beckers.
        Bull. American Astron. Soc., Vol. 9, 616 (1978).
Abstract.

071.015 The scale of granulation.
        S. A. Musman, G. D. Nelson.
Bull. American Astron. Soc., Vol. 9, 616 - 617 (1978).
Abstract.

071.016 The supergranulation: solar rip currents?
        L. D. Cloutman.
Bull. American Astron. Soc., Vol. 9, 617 (1978). – Abstract.

071.017 A method for determining laboratory/solar wavelength shifts: application to the C_2 molecule.
M. J. Shallis.
Mon. Not. R. Astron. Soc., Vol. 183, 1 - 11 (1978).

A method for measuring wavelength shifts between solar absorption lines at the disc centre and the corresponding lines produced in the Oxford furnace is described. The accuracy of the method is assessed, showing that shifts can be measured to better than 0.1pm (1mÅ). The method is demonstrated by observing the shifts of Fe I lines using, as reference, adjacent telluric lines and an instrumental fiducial. The method is applied to very faint solar lines by observation of lines from the (0,0) Swan band of the C_2 molecule, which give a mean redshift of 0.404 ± 0.086pm (4.04mÅ) at the solar disc centre.

071.018 The abundances of the elements in the solar photosphere – IX: Na to Ca.
D. L. Lambert, R. E. Luck.
Mon. Not. R. Astron. Soc., Vol. 183, 79 - 100 (1978).

Revised photospheric abundances are presented for the elements sodium to calcium. A thorough assessment of current information on the f values is given. Abundance calculations are reported for a selection of recent model solar atmospheres. A comparison of the photospheric and meteoritic abundances shows that the relative abundances are in good agreement.

071.019 Empirical NLTE analyses of solar spectral lines.

II: The formation of the Ba II λ4554 resonance line.
R. J. Rutten.
Sol. Phys., Vol, 56, 237 - 262 (1978).

The center-to-limb behaviour of the Ba II λ4554 resonance line is analyzed together with data from the extreme limb, flash intensities and profiles of other Ba II lines. An empirical NLTE method is employed in which the observed profiles are compared with synthesized profiles based on a standard one-dimensional model atmosphere, with the line source function, the barium abundance, the collisional damping and the atmospheric turbulence as free parameters. Results are given for various parameters (*gf*-values, solar barium abundance and isotope ratios, collisional damping, microturbulence and macroturbulence).

071.020 Observed and theoretical profiles of the Si II lines at λ1814.   G. D. Finn, H. C. McAllister.
Sol. Phys., Vol. 56, 263 - 273 (1978).

Emission line spectra observed at the limb of the solar disc are presented for transitions of singly ionized silicon at λλ1816.93, 1817.45 and 1808.01. The profiles show self-reversals in the line cores. A non-LTE analysis for a simplified model of the silicon ion is presented and calculated line profiles are compared with the observations. The study indicates that some modification of the Vernazza et al. (1973) model of the solar chromosphere in the transition region is needed to reconcile the calculated and observed profiles.

071.021 New satellite structure of the solar and laser plasma spectra in vicinity of the L α (Mg XII) line.
E. V. Aglizki, V. A. Boiko, A. Ya. Faenov, V. V. Korneev, V. V. Krutov, S. L. Mandelstam (*Mandel'shtam*), S. A. Pikuz, U. I. Safronova, J. A. Sylwester, A. M. Urnov, L. A. Vainshtein (*Vajnshtejn*), I. A. Zhitnik.
Sol. Phys., Vol. 56, 375 - 382 (1978).

Spectra in the narrow vicinity of the Mg XII resonance line at λ = 8.42 Å were obtained aboard the satellites 'Intercosmos-4, -7' and the rocket 'Vertical-2', as well as from laser-produced plasma. The high resolution in solar and laboratory spectra made it possible to reveal a new spectral structure close to the Lα (Mg XII) line from both short and long wavelength sides. The main features were observed in all the spectra and were interpreted as a single or group of dielectronic satellite lines due to $2l3l' \rightarrow 1s3l''$ transitions in the He-like ions.

071.022 Preliminary results of a study of short-period oscillations of relative brightness in photospheric facular knots.   A. P. Kramynin.
Issled. po soln. aktivnosti. Vladivostok, 1977, p. 143 - 150. In Russian. – Abstr. in Ref. zh., 51. Astron., 2.51.377 (1978).

071.023 Small-scale magnetic fields and convection in the solar photosphere.   N. O. Weiss.
Mon. Not. R. Astron. Soc., Vol. 183, 63P - 65P (1978).

Theoretical studies of convection in a magnetic field are related to the recently detected intergranular magnetic fields. There is no clear explanation for the reported occurrence of strong fields at the centres of granules.

071.024 On the influence of the granulation structure on a weak absorption line profile.
M. G. Gerbil'skij, V. M. Sobolev.
Soln. Dannye 1977 Byull., No. 11, p. 79 - 85 (1978). In Russian.

A simple model of the structure of granulation is proposed. For this model calculations have been made for a coefficient of asymmetry for a weak photospheric line profile. The dependence of the coefficient on physical properties of the granulation structure has been derived.

071.025 An investigation of the center-limb variation of

lithium Fraunhofer lines and solar lithium abundance. B. T. Babij, R. E. Rykalyuk.
Soln. Dannye 1977 Byull., No. 11, p. 98 - 105 (1978).
In Russian.

Theoretical calculation of the center-to-limb variations of equivalent widths and central depths of solar lithium lines is made assuming LTE with HSRA photospheric models. Agreement with observational data is satisfactory. It is found that for photospheric lithium lg N_L/N_H = 1.11.

071.026 On the profiles of weak absorption lines under flares. N. N. Kondrashova, P. N. Polupan.
Vestn. Kiev. Univ., Astron., Vyp. (No.) 19, p. 14 - 19 (1977).
In Russian.

The profiles of weak Fraunhofer lines under flares are shown to be changed relative to profiles of the neighbouring photosphere. Features of the profiles may be explained by increased temperature and turbulent velocity in the layers of line formation.

071.027 Photospheric models of solar active regions and the network based on the Mg II h and k line wings.
N. D. Morrison, J. L. Linsky.
Astrophys. J., Vol. 222, 723 - 734 (1978).

From a comparison between observed and computed wings of the Mg II resonance lines, the authors derive distributions of temperature versus mass column density for solar photospheric layers in plages and in the chromospheric network. In the active regions, temperatures exceed those in the quiet Sun by up to 200 K near the temperature minimum and up to 400 K in deeper layers. In the observed network structure, the temperature is enhanced by 200 K at the temperature minimum but is the same as that in the quiet Sun at greater depths. The difference in the slope of the temperature distribution between the network and plages is real. Adjacent to the network is a region in which the temperatures are similar to those in the quiet Sun, except immediately below the temperature minimum, where the temperatures are depressed by 150 K.

071.028 The scale of solar granulation. G. D. Nelson, S. Musman.
Astrophys. J., Lett., Vol. 222, L69 - L72 (1978).

The authors derive the observed scale of granulation as a consequence of a physical model. At scales smaller than granulation, horizontal radiative transport reduces temperature fluctuations and the resultant buoyancy forces in the surface layers. At scales only slightly larger than granulation, the dynamical pressure required to drive horizontal motions changes the opacity sufficiently to reduce the apparent contrast. The authors also estimate the maximum possible horizontal extent of a surface driven convective flow as 4.5 times granular scale.

071.029 The lines of rare earths in the spectrum of the sun as a star. N. N. Stepanyan, Z. A. Shcherbakova.
Izv. Krymskoj Astrofiz. Obs., Vol. 58, 3 - 7 (1978). In Russian.

Variability in time of the equivalent widths of seven rare-earth lines was discovered in the solar spectrum. Maximum equivalent widths correspond to high activity of flocculi, to a stronger magnetic field of the sun as a star. Six of the seven lines studied weaken in the sunspot spectrum. No perceptible changes of the equivalent widths are found in the spectrum of flocculi.

071.030 Effect of a travelling sound wave on the profiles of spectral lines. V.
R. I. Kostyk, A. V. Perekhod.
Astrometriya i Astrofizika, Kiev, Vyp. (No.) 34, p. 3 - 5 (1978). In Russian.

Asymmetry of the Fraunhofer lines in a homogeneous medium with plane monochromatic progressive sound waves is derived. The calculations show that asymmetry of Fe I lines with excitation potentials $\chi_{0i} \leqslant$ 2 eV is negative, and with $\chi_{0i} \geqslant$ 3 eV is positive. The asymmetry of Fraunhofer lines was found to increase with height above the photosphere and to decrease with increasing damping constant.

071.031 Was wissen wir von der Granulation?
J. P. Mehltretter.
Sterne Weltraum, Jahrg. 17, 207 - 209 (1978).

071.032 Note on the method of analysis of brightness fluctuations in Fraunhofer lines. V. N. Karpinskij.
Soln. Dannye 1977 Byull., No. 12, p. 79 - 83 (1978).
In Russian.

071.033 Variation of the profiles of medium-strong photospheric lines with heliographic latitude.
B. Caccin, R. Falciani, A. Donati Falchi.
Sol. Phys., Vol. 57, 13 - 17 (1978).

The asymmetric profiles of 11 metallic lines are studied, at μ = 0.3, as functions of φ. Their variations cannot be interpreted as due to temperature effects, but might imply a dependence on φ of the photospheric velocity field.

071.034 A comparison of synthetic and measured solar continuum intensities and limb darkening coefficients.
T. R. Ayres.
Sol. Phys., Vol. 57, 19 - 26 (1978).

Absolute continuum intensities and wavelength-dependent low-order polynomial fits to optical and infrared continuum limb darkening provide useful discriminants among single-component models of the solar photosphere. The thermal structure in best quantitative agreement with the recent center-limb measurements by Pierce and Slaughter (1977) and by Pierce et al. (1977) is the semi-empirical model by Vernazza, Avrett and Loeser (VAL). However, the VAL model M temperatures must be scaled upward by a factor of 1.015±0.005 to be consistent with the Labs and Neckel absolute calibration of continuum high points in the optical region 0.40–0.65 μm.

071.035 Resonance line scattering from optically thin structures located above the solar limb.
L. E. Cram, I. M. Vardavas.
Sol. Phys., Vol. 57, 27 - 36 (1978) = Mitt. Fraunhofer Inst., Freiburg, No. 154.

The authors discuss the formation of emission lines by resonance scattering from optically thin structures located above the solar limb. When the scattered radiation is only partially redistributed in frequency and angle, the resulting coherency is sufficient to affect the interpretation of such lines. In particular the apparent Doppler width of the scattered line may be different from the Doppler width in the scattering structure.

071.036 The solar metallicity and colors from spectral quantification.
D. C. Barry, R. H. Cromwell, S. A. Schoolman.
Astrophys. J., Vol. 222, 1032 - 1042 (1978).

The precise UBV colors of the Sun are extremely sensitive to the solar metals/hydrogen abundance ratio. Quantitative comparison of the spectra of various Hyades and Coma F and G dwarfs with the solar spectrum indicates that the solar metallicity is the same as that of a Hyades dwarf and substantially higher than that of a Coma dwarf. Most of the existing significant discrepancies among previously reported solar UBV colors are removed by correcting those colors for a solar ultraviolet excess of zero. The solar colors from the present spectral quantification procedure are $(B-V)_{Sun}$ = 0.66 and $(U-B)_{Sun}$ = 0.20.

071.037 A morphological interpretation of the spatial power spectrum of the solar granulation. C. Aime.

Astron. Astrophys., Vol. 67, 1 - 6 (1978).

A morphological model of solar granulation is presented as a two-dimensional generalization of a model established by White and Cha (1973) to describe oscillation of the photosphere. It is constructed from a convective hypothesis of the "Bénard cell" type, but by adjustment of parameters it can eventually represent turbulent granulation. Comparison between theoretical and experimental results is satisfactory for the limiting cases of the model which assume that granules are dissimilar or arranged irregularly.

071.038   On the chromium abundance in the solar photo-
**   sphere.**   E. Biémont, N. Grevesse, M. C. E. Huber.
Astron. Astrophys., Vol. 67, 87 - 91 (1978).

The solar abundance of chromium is inferred from high-quality photospheric spectra with the aid of several recent sets of experimental and theoretical oscillator strengths for Cr I. The mean abundance, $A_{Cr} = \log (N_{Cr}/N_H) + 12 = 5.64$, obtained with an LTE-analysis, agrees with the meteoritic value.

071.039   Photospheric faculae. II. Line profiles and magnetic
**   field in the bright network of the quiet sun.**
S. Koutchmy, G. Stellmacher.
Astron. Astrophys., Vol. 67, 93 - 102 (1978).

Spatially high resolution spectra ($\approx 0.''75$) of the three iron lines Fe λ 5576 Å (non split), Fe λ 6301.5 Å and Fe λ 6302.5 Å (triplet) observed in magnetic regions (network) of the quiet sun were analysed. High values of the magnetic field strength (1000–1500 gauss) are found for the quiet network regions.

071.040   On observations of solar emission of the transition
**   zone from photosphere to chromosphere using**
spectra in the Hα region.   Eh. E. Dubov, V. I. Makarov.
Astron. Zh. Akad. Nauk SSSR, Vol. 55, 666 - 669 (1978).
In Russian. English translation in Soviet Astron., Vol. 22, No. 3.

A method of investigation of the solar limb emission with small height steps using the Lyot type coronograph is suggested. In the spectral region λλ 6552.5–6574.5 Å 18 emission lines were observed. Several of these lines are not identified and some of them are not detected in absorption.

071.041   Sur la variation des dimensions des granules photo-
**   sphériques au voisinage des taches solaires.**
C. J. Macris.
C. R. Acad. Sci. Paris, Tome 286, Sér. B, 315 - 316 (1978).

The author studies the variation of the dimensions of the photospherical granules, measured in the vicinity of the sunspot penumbra, with the intensity of the sunspot magnetic fields.

071.042   Measurement of solar disc polarization in a number
**   of Fraunhofer lines and their adjacent continuum.**
II: Improved data, new line measurements.   E. Wiehr.
Astron. Astrophys., Vol. 67, 257 - 260 (1978).

The measuring accuracy of 2×10^{-4} achieved in Paper I (13.071.019) has been increased to 5×10^{-5} by investigating a number of instrumental effects which are described. The observations of Paper I were verified under these conditions.

071.043   The relationship between solar activity and the
**   H and K line cores in integrated sunlight.**
D. E. Jebsen, W. E. Mitchell, Jr.
Sol. Phys., Vol. 57, 309 - 318 (1978).

The authors' investigation confirms that there is a variation with solar activity in the intensities of the cores of the H and K lines of ionized calcium in integrated sunlight. They find the correlations to lie in the range +0.40 to +0.81. A photoelectric determination of size of the Wilson-Bappu emission width gives excellent agreement with Wilson and Bappu's prediction for the Sun.

071.044   The last observable line in hydrogen emission
**   spectrum.**   L. N. Kurochka, L. B. Ribko.
Sol. Phys., Vol. 57, 319 - 328 (1978).

It is shown that outside continuous solar spectrum or diffused solar light in the Earth's atmosphere substantially reduces the contrast and quality of observable emission lines in the hydrogen spectrum. A formula is obtained which enables the electron concentration in hydrogen plasma to be determined by the last observable line with due account for outside (spurious) continuous spectrum.

071.045   Observations of phase delays and the vertical propa-
**   gation of waves in the solar atmosphere.**
B. W. Lites, E. G. Chipman.
Bull. American Astron. Soc., Vol. 10, 415 (1978). – Abstract.

071.046   The height dependence of solar velocity fluctua-
**   tions.**   S. L. Keil.
Bull. American Astron. Soc., Vol. 10, 415 (1978). – Abstract.

071.047   The 300 sec oscillation in Si II 1817: observed
**   impulse response of the solar atmosphere.**
R. Illing.
Bull. American Astron. Soc., Vol. 10, 416 (1978). – Abstract.

071.048   Quiet sun fluctuations of C I 5380 Å.
**   **   C. A. Lindsey, D. A. Landman.
Bull. American Astron. Soc., Vol. 10, 417 (1978). – Abstract.

071.049   High spatial resolution cinematography of the 1600
**   Å solar continuum.**
J.-D. F. Bartoe, G. E. Brueckner.
Bull. American Astron. Soc., Vol. 10, 417 (1978). – Abstract.

071.050   Solar C III line ratios observed from Skylab.
**   **   J. W. Cook, K. R. Nicolas.
Bull. American Astron. Soc., Vol. 10, 439 (1978). – Abstract.

071.051   Investigation of the infrared solar spectrum by the
**   nonlinear optical method.**
M. I. Divlekeev, G. F. Sitnik.
Astron. Tsirk., No. 970, p. 3 - 5 (1977). In Russian.

071.052   Depths of formation of Mg I b lines in the solar
**   atmosphere.**   G. F. Sitnik, A. A. Galal.
Astron. Tsirk., No. 978, p. 5 - 7 (1978). In Russian.

071.053   Modeling the solar photosphere and chromosphere.
**   **   R. G. Athay.
Mem. Soc. Astron. Italiana, Vol. 48, (see 012.038), 401 - 437 (1977).

Contents: Purpose of models. Physical features of the solar atmosphere. Assigning heights to specific features. Reference points and height scales. Photosphere and temperature minimum region. Low and middle chromosphere. High chromosphere. Chromospheric emission lines, Non-thermal line broadening.

071.054   Magnetic structures in photosphere of sun and stars.
**   **   C. Zwaan.
Mem. Soc. Astron. Italiana, Vol. 48, (012.038), 525 - 557 (1977).

Contents: Solar observational constraints on models for magnetic structures in stellar atmospheres. Physics of quasi-stationary magnetic structures. The interplay between magnetic and velocity fields. Magnetic activity in sun - and stars. Observational indications for stellar magnetic structure and activity.

Electron impact excitation of Fe XXVI and other

one-electron ions.   See Abstr. 022.090.

Photographing the Sun in H-alpha.
See Abstr. 031.304.

The LPSP instrument on OSO 8. II. In-flight per-
formance and preliminary results.   See Abstr. 032.510.

On the dynamic interaction between magnetic
fields and convection.   See Abstr. 062.002.

The solar XUV He I and He II emission lines. I.
Intensities and gross center-to-limb behavior.
See Abstr. 073.010.

An observational search for large-scale organization
of five-minute oscillations on the sun.
See Abstr. 074.051.

Coronal heating and its relation to magnetic field
evolution.   See Abstr. 074.083.

High resolution UV solar spectroscopy.
See Abstr. 076.004.

Preliminary results obtained from the solar EUV
experiment on board the D2B Aura satellite.
See Abstr. 076.005.

Relation between circular polarization of moving
type IV bursts and polarity of photospheric magnetic fields.
See Abstr. 077.013.

Radiative equilibrium in the photosphere and glob-
al oscillations of the sun and an upper bound on energy
transport.   See Abstr. 080.002.

Bifurcation of force-free solar magnetic fields: a
numerical approach.   See Abstr. 080.014.

Application of curves of growth to a study of the
solar atmosphere's opacity in the continuum.
See Abstr. 080.028.

Outlooks on the progress of solar physics.
See Abstr. 080.041.

Linear hydrodynamical equations coupled with
radiative transfer in a non-isothermal atmosphere. II. Application to solar photospheric observations.
See Abstr. 080.045.

Very elongated convective cells in the sun.
See Abstr. 080.051.

Spectroscopic differences between solar-type
cluster stars and the Sun.   See Abstr. 114.004.

072 Sunspots, Faculae, Activity Cycles

072.001   Activité solaire.   P. Simon.
Astronomie, Vol. 92, 61 - 81 (1978).
(1) Phénomènes solaires: nombre de taches; Flux 10 cm; Éruptions; Cycles solaires. (2) Phénomènes liés à l'activité solaire: vent solaire et champ magnétique interplanétaire; Rayonnement cosmique; Activité géomagnétique.

072.002   Investigation of some structural and dynamic properties of the large sunspot group of August 1972.   G. S. Minasyans (*Minasyants*).
Bull. Astron. Inst. Czechoslovakia, Vol. 29, 18 - 22 (1978).
Using high resolution photographs of the photosphere (resolution $\gtrsim 1.0''$), obtained at Ondřejov and Alma-Ata, the proper motions and changes of areas of individual nuclei of the large August 1972 sunspot group were investigated. It is shown that the nature of the motion of the nuclei is determined by the structure of the surrounding magnetic field. The mean velocities of the nuclei amount to about 50 m/s, the maxima are between 74 and 134 m/s.

072.003   The role of spiral spots in solar activity in October 1972.   Y.-j. Djng, W.-b. Li, M.-c. Wu, Z.-z. Jiang, Q.-f. Hong, Z.-z. Liu.
Acta Astron. Sinica, Vol. 18, 143 - 157 (1977). In Chinese.
This paper collects the photographic features of 192 knots of 68 flares and some prominences. From correlation analysis of the morphology of these flares and prominences with the fine structures of sunspot groups in this region, the authors verified conclusions previously obtained.

072.004   The origin of broad-band circular polarization in sunspots.   L. H. Auer, J. N. Heasley.
Astron. Astrophys., Vol. 64, 67 - 71 (1978).
The authors show that the net circular polarization in broad-band observations of sunspots reported by Illing, Landman and Mickey (1974, 1975) can be explained if macroscopic motions comparable to the thermal motions are present. Further they prove that in a static LTE atmosphere regardless of the variations of magnetic field and atmospheric structure, it is impossible to produce any net circular polarization in a magnetically split line.

072.005   Die Sonnenflecken 1977.   P. Staiger.
Orion, 36. Jahrg., 27 - 29 (1978).

072.006   Vibration rotation bands of SiO in sunspots.
V. P. Gaur, M. C. Pande, B. M. Tripathi.
Sol. Phys., Vol. 56, 67 - 69 (1978).
The equivalent widths of P(51) and R(43) lines of the 1−0 and P(45) and R(56) lines of the 2−1 vibration rotation bands of SiO near 8 μm region, have been computed for Zwaan's (1974) sunspot model at the centre of the disc. The predicted equivalent widths suggest a possible presence of these SiO bands in the sunspot spectrum.

072.007   Comments on the variation of sunspot-group fragmentation.   M. Kopecký, B. Růžičková-Topolová.
Bull. Astron. Inst. Czechoslovakia, Vol. 29, 65 - 70 (1978).
The paper demonstrates that the variations of the P/R-ratio (P − total area of sunspots, R − Wolf's relative number) are primarily due to the variations of the internal structure of the sunspot groups, i.e. the variations of group fragmentation into individual sunspots. It is shown that equally large sunspot groups as regards area differ in the individual sunspot numbers in various intervals. The dependence of this fragmentation of sunspot groups on their size, the 11-year cycle and on the phase of the 11-year cycle is studied.

072.008   Models of heat flux in the subphotospheric layers of sunspots and the interpretation of umbral granulation.   J. Staude.
Bull. Astron. Inst. Czechoslovakia, Vol. 29, 71 - 79 (1978).
Different models of heat flux in the normal convective zone (deep and shallow reference models) and in an umbral flux tube in magnetostatic equilibrium are investigated by numerical calculations. Recent observations of umbral granulation (Bumba et al., Suda) are explained by residual cellular convection which still transfers a small fraction of energy to the subphotospheric flux tube. Some criteria for the inhibiting effect of the magnetic field on convection derived from linear theories seem to be too strong.

072.009   On the one-year oscillation found by Chistyakov in the drift of sunspots.   J. Tuominen.
Bull. Astron. Inst. Czechoslovakia, Vol. 29, 79 - 81 (1978).
A model of sunspots is proposed which could explain as an aspect effect the one-year oscillation in their drift in latitude and longitude.

072.010   Line profile families of faculae and pores.
E. N. Frazier.
Astron. Astrophys., Vol. 64, 351 - 358 (1978).
It is pointed out that for any given Fraunhofer line, there should exist an entire range of facular line profiles. This paper investigates the consequences of the hypothesis that the structure of faculae and pores is determined to first order by the total magnetic flux, and that therefore, there should be a one parameter family of line profiles, with the magnetic flux being that parameter. Observations from a magnetograph are used to construct this family of line profiles in a statistical manner. Results are given for the two absorption lines, Fe I λ 525.022 nm and Fe I λ 524.706 nm.

072.011   Ultraviolet spectra of solar active regions from Salyut 4.   A. V. Bruns, G. M. Grechko, A. A. Gubarev, P. I. Klimuk, V. I. Sevastyanov (*Sevast'yanov*), A. B. Severny (*Severnyj*), N. V. Steshenko.
Space Research, Vol. XVII, (see 012.010), 509 - 513 (1977).
The results of observations of solar active features with an orbiting solar telescope on Salyut 4 station are discussed. Over 600 spectra of flocculi, brightenings, prominences, sunspots and undisturbed parts of the sun were photographed. The study of the Ly-α profile in the spectrum of the brightening and neighbouring quiet chromosphere shows that in different weak flares the electron density is $\sim 2 \times 10^{13} \mathrm{cm}^{-3}$, the temperature ranges from 15 to $25 \times 10^{3}\,^{\circ}$K and the column densities of hydrogen atoms are $5 \times 10^{19} - 2 \times 10^{20}$ cm^{-2}. This indicates that the ionization of hydrogen in brightenings and flares can exceed the ionization of hydrogen in the quiet chromosphere by a factor of $10^{2} - 10^{3}$.

072.012   A forecast of solar activity for the 21st solar cycle.   J. Xanthakis, C. Poulakos.
Sol. Phys., Vol. 56, 467 - 469 (1978).
Predicted values of some main indices of solar activity for the 21st solar cycle are given. The epoch of maximum of solar activity has been placed in 1980.8 ± 0.1. The predicted peak values of the relative sunspot numbers published by other authors are also given.

072.013   On velocity variations of rotation of the sunspot formation zone.   A. V. Baranov.
Issled. po soln. aktivnosti. Vladivostok, 1977, p. 151 - 156.
In Russian. − Abstr. in Ref. zh., 51. Astron., 2.51.354 (1978).

072.014 Characteristics of the shift of the π-component of the Zeeman triplet during temporal observations of magnetic fields of sunspots.   V. A. Golubev.
Issled. po soln. aktivnosti. Vladivostok, 1977, p. 91 - 103. In Russian. − Abstr. in Ref. zh., 51. Astron., 2.51.372 (1978).

072.015 Selected lines of neutral iron in a sunspot spectrum.   Eh. P. Surkov, L. D. Surkova.
Issled. po soln. aktivnosti. Vladivostok, 1977, p. 104 - 122. In Russian. − Abstr. in Ref. zh., 51. Astron., 2.51.373 (1978).

072.016 Parallel observations of sunspot magnetic fields using Fe I λ5250.2 Å and Cr I λ5247.6 Å lines.
V. A. Golubev, L. F. Lazareva, L. A. Lukashenko, A. V. Mazhuga, F. A. Toropova, V. F. Chistyakov.
Issled. po soln. aktivnosti. Vladivostok, 1977, p. 123 - 142. In Russian. − Abstr. in Ref. zh., 51. Astron., 2.51.374 (1978).

072.017 Temperature and electron density above a sunspot umbra.   R. B. Teplitskaya, S. A. Grigor'eva, V. G. Skochilov.
Issled. po geomagn., aehron i fiz. Solntsa. Vyp. (No.) 42. Moskva, Nauka, 1977, p. 48 - 49. In Russian. − Abstr. in Ref. zh., 51. Astron., 2.51.375 (1978).

072.018 Emergence of new satellite groups as nonstationary and prognostic factor in an active aggregate on the sun.   V. V. Kasinskij.
Issled. po soln. aktivnosti. Vladivostok, 1977, p. 3 - 18. In Russian. − Abstr. in Ref. zh., 51. Astron., 2.51.393 (1978).

072.019 Contrast of spots and flares on the sun.   Eh. P. Surkov.
Issled. po soln. aktivnosti. Vladivostok, 1977, p. 80 - 90. In Russian. − Abstr. in Ref. zh., 51. Astron., 2.51.394 (1978).

072.020 Revealing cyclicity in changes of cosmic ray intensity and in solar activity in the past.
V. A. Dergachev, S. Kh. Tleugaliev.
Izv. AN SSSR. Ser. fiz., Vol. 41, 1899 - 1915 (1977). In Russian. − Abstr. in Ref. zh., 51. Astron., 2.51.410 (1978).

072.021 On the mutual location of sunspots and flares on the sun in 1970.   F. A. Toropova, V. F. Chistyakov.
Issled. po soln. aktivnosti. Vladivostok, 1977, p. 71 - 79. In Russian. − Abstr. in Ref. zh., 51. Astron., 2.51.420 (1978).

072.022 Index of compactness of solar active regions and characteristics of proton events.
V. V. Kasinskij, E. V. Ivanov, V. N. Obridko.
Issled. po geomagn., aehron. i fiz. Solntsa. Vyp. (No.) 42. Moskva, Nauka, 1977, p. 34 - 43. In Russian. − Abstr. in Ref. zh., 51. Astron., 2.51.422 (1978).

072.023 Solar activity display during the minimum of solar cycle No. 20.
M. N. Nazarova, N. K. Pereyaslova, I. E. Petrenko.
Izv. AN SSSR. Ser. fiz., Vol. 41, 1757 - 1764 (1977). In Russian. − Abstr. in Ref. zh., 51. Astron., 2.51.428; 62. Issled. kosm. prostranstva, 2.62.93 (1978).

072.024 Tunneling of Alfvén waves in sunspots.   Yu. D. Zhugzhda, V. Locāns.
Investigations of the sun and red stars, 8, (see 003.005), p. 51 - 69 (1978). In Russian.
    Propagation of Alfvén waves in an inhomogeneous atmosphere is considered. Low-frequency waves transmit energy only due to tunneling. Properties of tunneled Alfvén waves are investigated. The reflection of tunneling waves from a sharp discontinuity is considered. The energy flux of Alfvén waves in a sunspot is estimated, and the reflection of waves from the chromosphere-corona transition layer is calculated.

072.025 On the 11-year cycle of solar activity. I. Detection of the 11-year cycle by methods of linear band filtration.   Yu. R. Rivin.
Soln. Dannye 1977 Byull., No. 11, p. 63 - 68 (1978). In Russian.
    Using a system of linear band filters of somewhat different passband widths the 11-year variation was detected as a time process from a Wolf number series.

072.026 The nature of solar cyclicity.   P. R. Romanchuk.
Vestn. Kiev. Univ., Astron., Vyp. (No.) 19, p. 3 - 14 (1977). In Russian.
    The results of many years' work of the author on the nature of solar activity and solar cycles are presented. An explanation of resonance phenomena of the solar activity caused by the gravitational action of Jupiter and Saturn on the sun is given.

072.027 The long years of the quiet Sun.   K. Hindley.
New Scientist, Vol. 75, 468 (1977). − Abstr. in Phys. Abstr., Vol. 81, Abstr. 7962 (1978).

072.028 Possible variations of solar activity in the past.   N. N. Sazeeva.
Fiz.−tekh. inst. AN SSSR. Prepr. 544. Leningrad, 1977. 31 pp. In Russian. − Abstr. in Ref. zh., 51. Astron., 3.51.489 (1978).

072.029 Relation of flare activity to the approach and separation of sunspots in an active region and to its magnetic properties.   E. Marková.
Bull. Astron. Inst. Czechoslovakia, Vol. 29, 163 - 171 (1978).
    The relation between the flare activity of active regions within the scope of a large complex and the magnetic gradients of these active regions and their daily variations is investigated in the interval of exceptionally high flare activity of June 1970. New indices, characterizing the active region, were defined, e.g., the instantaneous sunspot-area density and the instantaneous sunspot-number density. These indices were determined on the basis of measuring the surface, containing all the sunspots of the complex of active regions enclosed by an envelope. An attempt was made to substitute the surface in the relation for the individual indices by distance. The daily variations of these indices were again compared with the flare activity and some mutual relations were derived.

072.030 On the evolution of magnetic and velocity fields of an originating sunspot group.   G. Bachmann.
Bull. Astron. Inst. Czechoslovakia, Vol. 29, 180 - 184 (1978).
    Magnetographic measurements have been made to derive longitudinal magnetic field strengths, line-of-sight velocities and the brightness distribution in an originating sunspot group. These results and photographs of the group are used to compare the evolution of a relatively simple active region with the present ideas about the evolution of active regions in general.

072.031 The sunspot cycle before the Maunder minimum.   A. Wittmann.
Astron. Astrophys., Vol. 66, 93 - 97 (1978).
    Several hundred observations of giant sunspots have been recorded during the pre-telescopic era, especially in ancient China. The clustering of observations reveals approximately 50 activity maxima during this era, which − in combination with the modern record of sunspot activity − can be used to trace the phase of the 11-year sunspot cycle back to the 5th century B.C. Although long-term modulations exist, it is highly likely that the sunspot cycle persisted without interruption throughout this time span. The mean period is equal to 11.135 yr.

072.032 **Sunspot distribution relative to the sector structure of the interplanetary magnetic field.** M. B. Ogir'.
Izv. Krymskoj Astrofiz. Obs., Vol. 58, 26 - 30 (1978).
In Russian.

All the sunspot groups for the period 1968–1974 were united in 535 complexes of activity by using magnetic field maps of the whole sun, the maps of 9.1 cm radio emission and Hα-spectroheliograms. The geometrical centres of complexes were plotted on the interplanetary magnetic field sector structure cards. Calculation of the magnetic fluxes from the separate sunspot groups and the summary flux from the whole complex do not show any peculiarity in relation to the interplanetary magnetic field sector structure. It may be assumed that there is no connection between sunspots and the interplanetary magnetic field sector structure.

072.033 **On the vertical gradient of magnetic field strength in a sunspot umbra.** M. Dzh. Gusejnov.
Izv. Krymskoj Astrofiz. Obs., Vol. 58, 31 - 34 (1978).
In Russian.

The vertical distribution of the magnetic field in a sunspot umbra is investigated. In deep layers, where the Fe I lines with Rowland intensity $I = 1 - 10$ originate from the magnetic field strength decreases with height; the gradient is $\Delta H/\Delta h \approx 0.8 - 2.3$ gauss/km. In relatively high layers forming the cores of Hα and H_3 Ca II lines, the gradient is $|\Delta H/\Delta h| \approx 0.2 - 0.6$ gauss/km. These values agree well with numerical values for the vertical gradients determined earlier by several authors for transparent models of a sunspot umbra.

072.034 **Forecast of solar and geomagnetic activity for solar cycle No. 21.** A. I. Ol'.
Soln. Dannye 1977 Byull., No. 12, p. 87 - 89 (1978).
In Russian.

Wolf numbers for each year of the 21st cycle of solar activity, sum of the Wolf numbers and the mean Wolf number for the whole cycle, duration of the cycle and its ascending branch have been forecasted. A forecast of indices of geomagnetic perturbations for the year of maximum and the next two years has also been made.

072.035 **On the 11-year cycle of solar activity. II. Analysis of some properties of W_{11}.** Yu. R. Rivin.
Soln. Dannye 1977 Byull., No. 12, p. 89 - 96 (1978).
In Russian.

Some properties of the 11-year cycle variation of the Wolf numbers have been investigated. A dependence of the variation of the leading front duration of a single 11-year cycle on W_{11} as well as the absence of the dependence at the trailing front are affirmed. For the interval considered the 11-year cycle was found to begin on the average a year before it should according to the Wolf numbers. A number of other forecasting properties of W_{11} has been analysed.

072.036 **The Evershed flow in the transition region chromosphere-photosphere.** J. Bønes, P. Maltby.
Sol. Phys., Vol. 57, 65 - 71 (1978).

The Mg b_1 line profile is studied as a function of spatial position in the sunspot region. Comparing the wavelengths of the core and the wing, and in just outside the penumbra, a reversal in the shift is detected. The displacements of the core and the wing are interpreted as horizontal motions directed into the spot in the chromosphere and as a flow directed out of the spot in deeper layers. Systematic wavelength shifts are detected in the line core in some regions outside the penumbra. This is interpreted as a chromospheric velocity field usually directed horizontally away from the spot.

072.037 **On physics of sunspots.** S. O. Obashev, G. S. Minasyants.
Soln. akt. Alma-Ata, 1977, p. 3 - 13. In Russian. – Abstr. in Ref. zh., 51. Astron., 4.51.376 (1978).

072.038 **Photometry of sunspots in the Fe II lines.** V. N. Milovanov.
Soln. akt. Alma-Ata, 1977, p. 14 - 19. In Russian. – Abstr. in Ref. zh., 51. Astron., 4.51.377 (1978).

072.039 **Physical conditions in faculae.** T. M. Minasyants, G. S. Minasyants.
Soln. akt. Alma-Ata, 1977, p. 47 - 56. In Russian. – Abstr. in Ref. zh., 51. Astron., 4.51.379 (1978).

072.040 **Photoelectric measurements of the magnetic field in an active region.** N. N. Morozov.
Soln. akt. Alma-Ata, 1977, p. 30 - 46. In Russian. – Abstr. in Ref. zh., 51. Astron., 4.51.393 (1978).

072.041 **Structural and dynamical properties of a sunspot group in August 1972.** G. S. Minasyants.
Soln. akt. Alma-Ata, 1977, p. 20 - 29. In Russian. – Abstr. in Ref. zh., 51. Astron., 4.51.431 (1978).

072.042 **Some studies on isolated sunspot magnetic configurations and associated Hα-flares and microwave bursts.** T. Chattopadhyay, S. K. Sarkar, M. K. D. Gupta.
Indian J. Radio Space Phys., Vol. 6, No. 2, p. 144 - 147 (1977).
Abstr. in Phys. Abstr., Vol. 81, Abstr. 16387 (1978).

072.043 **The Sun – predictions for future years – look back at the past.** M. Miceli.
Antenna, Vol. 49, No. 4, p. 149 - 157 (1977). In Italian.
Abstr. in Phys. Abstr., Vol. 81, Abstr. 24496 (1978).

072.044 **On sunspot decay.** S. D. Ivanov, V. P. Maksimov.
Pis'ma Astron. Zh., Vol. 4, 232 - 234 (1978). In Russian.

A mechanism of sunspot decay is investigated allowing for turbulent processes in a strong magnetic field. The crucial role of the turbulent pulsation velocity amplitude for the decay time is shown.

072.045 **Der 20. Sonnenfleckenzyklus. Ein zusammenfassender Bericht.** W. Schulze.
Sterne, 54. Band, 96 - 103 (1978).

072.046 **Divergent views about the probable behaviour of sun-spot cycle 21.** B. Egan, M. Shafti.
South. Stars, Vol. 27, 109 - 117 (1978).

072.047 **New light on sunspot darkness and the solar cycle.** F. Albregtsen, P. Maltby.
Nature, Vol. 274, 41 - 42 (1978).

The authors report a variation in the infrared umbra intensity of large sunspots with the phase in the solar cycle.

072.048 **Bright points in sunspot umbrae: morphology, intensities.** A. Adjabshirzadeh, S. Koutchmy.
C. R. Acad. Sci. Paris, Tome 286, Sér. B, 335 - 338 (1978).
In French.

La photométrie photographique précise de 3 clichés (séparés de plusieurs minutes dans le temps) de l'ombre d'une tache solaire a permis de déduire la distribution des points brillants de l'ombre en intensités et en dimensions suivant 2 directions orthogonales. L'anisotropie et la largeur des distributions en dimensions sont discutées.

072.049 **Predicted intensity of the solar maximum.** R. P. Kane.
Nature, Vol. 274, 139 - 140 (1978).

It is desirable to know beforehand, for planning geophysical and other studies, the intensity of a future period of solar activity. The author describes a simpler version of the method of Sargent (1977) correlating R_{max} with the minimum annual aa index in the preceding years.

072.050 **High resolution Hα observations of solar active regions.**
F. Moriyama, H. Morishita, K. Mizugaki.
Tokyo Astron. Obs. Report No. 69, Vol. 18, 187 - 197 (1978).
In Japanese.

072.051 **Comments on the so-called Maunder minimum.**
Yu. I. Vitinsky (*Vitinskij*).
Sol. Phys., Vol. 57, 475 - 478 (1978).
Arguments are presented in favour of the operation of the 11-yr cycle during the Maunder minimum and before it. The laws of differential rotation (1642–1644 and 1899–1901) before the low cycle are shown to differ insignificantly. It is suggested that the Maunder minimum was a result of the 600-yr cycle effect (during its epoch of minimum) on the 80–90 yr cycle.

072.052 **Reflection of Alfvén waves and the cooling of sunspots.** J. H. Thomas.
Bull. American Astron. Soc., Vol. 10, 415 (1978). – Abstract.

072.053 **Thermal models of sunspots.** A. Clark, Jr.
Bull. American Astron. Soc., Vol. 10, 415 (1978). Abstract.

072.054 **Using dynamo theory to predict the sunspot number during solar cycle 21.** P. H. Scherrer,
K. H. Schatten, L. Svalgaard, J. M. Wilcox.
Bull. American Astron. Soc., Vol. 10, 415 (1978). – Abstract.

072.055 **Ephemeral active regions during solar minimum.** K. L. Harvey, S. F. Martin.
Bull. American Astron. Soc., Vol. 10, 417 (1978). – Abstract.

072.056 **Brightness temperature and continuous absorption coefficient in sunspots.** G. F. Sitnik.
Astron. Tsirk., No. 977, p. 1 - 3 (1977). In Russian.

072.057 **How close to the limb can sunspots be seen?**
M. Waldmeier.
Astron. Mitt. Eidg. Sternw. Zürich, Nr. 359, 5 pp. (1978). In German.
A regularly shaped and stable sunspot has been observed at the time of its passage around the sun's western limb. When it was seen for the last time, its distance from the sun's center was $\theta = 88.37°$. This corresponds to a distance of 0.4″ from the limb and to an optical depth of $\overline{\tau} = 0.030$. This level lies about 200 km above $\tau = 1$. The spot extends at least up to this height. No deformation of the sun's limb was observed when the spot was exactly at the limb ($\theta = 90°$).

072.058 **Facules polaires.** S. Cortesi.
Astron. Mitt. Eidg. Sternw. Zürich, Nr. 362, 10 pp. (1978).
Observations of polar faculae carried out in October 1962, at intervals of about 25 min have been used to study their lifetime. The average lifetime amounts to 16 min or less for smaller faculae, 39 min for medium and 143 min or more for large ones. For the study of the latitude distribution additional observations from the sunspot minimum 1975/76 were used. The visibility-function shows a maximum at a distance of about 60° from the center of the sun's disc. The number-density of the faculae increases with latitude, reaching a maximum at 75°. This is in contradiction to earlier studies, where an increase up to the latitude of at least 80° was found.

072.059 **Proper motions of sunspots and the velocity field in active regions.** Dzh. I. Irgashev.
Astron. Tsirk., No. 989, p. 1 - 3 (1978). In Russian.

072.060 **Diverging motions of main sunspots in groups and the magnetic field in active regions.**
Dzh. I. Irgashev.
Astron. Tsirk., No. 989, p. 3 - 5 (1978). In Russian.

072.061 **A study of plasma motion in the penumbral region of the leader of sunspot group No. 240 (1970).**
M. Mamadazimov.
Sb. nauchn. tr. Tashkent Gos. ped. inst., Vol. 138, 60 - 69 (1975). In Russian. – Abstr. in Ref. zh., 51. Astron., 6.51.483 (1978).

072.062 **Characteristic features of the cyclic changes of solar activity after 1610.** A. D. Bonov.
Astrofiz. issled. (NRB), Vol. 2, 3 - 7 (1977). – Abstr. in Ref. zh., 51. Astron., 6.51.543 (1978).

072.063 **Meteoritic evidence for the Maunder minimum in solar activity.** M. A. Forman, O. A. Schaeffer,
G. A. Schaeffer.
Geophys. Res. Lett., Vol. 5, 219 - 222 (1978).
In this paper the authors report some evidence that meteorites in interplanetary space during the Maunder and Spörer minima in solar activity may have experienced a higher level of cosmic ray flux than the current solar-cycle average. If this is interpreted as an enhanced flux only during the last two long minima, these meteorite measurements are consistent with virtually no solar modulation of 400 MeV galactic cosmic rays at that time.

072.064 **Using dynamo theory to predict the sunspot number during solar cycle 21.** K. H. Schatten,
P. H. Scherrer, L. Svalgaard, J. M. Wilcox.
Geophys. Res. Lett., Vol. 5, 411 - 414 (1978).
On physical grounds it is suggested that the sun's polar field strength near a solar minimum is closely related to the following cycle's solar activity. Four methods of estimating the sun's polar magnetic field strength near solar minimum are employed to provide an estimate of cycle 21's yearly mean sunspot number at solar maximum of 140 ± 20.

072.065 **High resolution photography of sunspots on the Ondřejov Observatory.** L. Hejna.
Říše hvězd, Vol. 59, 69 - 71, 77 - 79 (1978). In Czech.

072.066 **Analysis of solar activity to find out the annual variation.** V. P. Antonova, I. A. Pimenov.
Ionos. i soln.-zemnye svyazi. Alma-Ata, 1978, p. 109 - 111. In Russian. – Abstr. in Ref. zh., 51. Astron., 7.51.465 (1978).

Solar activity. See Abstr. 003.108.

Investigations on solar activity.
See Abstr. 003.157.

Dissociation energy for the ground state of AlO from true potential energy curve. See Abstr. 022.057.

Sonnenbeobachtung für den Amateur.
See Abstr. 031.236.

On a technique of visualization of sunspot images observed in the infrared range at $\lambda > 1$ μm.
See Abstr. 031.309.

The LPSP instrument on OSO 8. II. In-flight performance and preliminary results. See Abstr. 032.510.

Cooling of the Sun's photosphere coincident with increased sunspot activity. See Abstr. 071.007.

Photospheric models of solar active regions and the network based on the Mg II h and k line wings.
See Abstr. 071.027.

Sur la variation des dimensions des granules photo-sphériques au voisinage des taches solaires.
See Abstr. 071.041.

The chromosphere above sunspot umbrae. I. Observations of the emission cores in the Ca II H- and K-lines.
See Abstr. 073.027.

The chromosphere above sunspot umbrae. II. The interpretation of the H, K, and IR lines of Ca II.
See Abstr. 073.028.

On physical conditions in the chromosphere above sunspot umbrae.   See Abstr. 073.032.

Flares outside sunspot groups.
See Abstr. 073.035.

Correlation and dispersion fields of the flare index and some other indices of solar activity.
See Abstr. 073.038.

Appearance of large flares in decayed sunspot groups.   See Abstr. 073.039.

On the short-term forecast of chromospheric flares on the basis of characteristics of sunspot groups.
See Abstr. 073.040.

On the dependence of the height of prominences on solar cycle and heliographic latitude.   See Abstr. 073.092.

Provisional sunspot-numbers for December 1977, January - May 1978.   See Abstr. 075.003.

Definitive sunspot numbers for 1977.
See Abstr. 075.004.

Solar phenomena.   See Abstr. 075.009.

Sunspot relative numbers for 1977.
See Abstr. 075.011.

Sunspot observations for the year 1974.
See Abstr. 075.012.

Sunspot observations for the year 1975.
See Abstr. 075.013.

Zürcher Sonnenfleckenrelativzahlen.
See Abstr. 075.014.

L'activité solaire.   See Abstr. 075.015.

Solar activity.   See Abstr. 075.019.

Sunspot numbers.   See Abstr. 075.021.

Irradiation solar flux measurements between 120 and 400 nm. Current position and future needs.
See Abstr. 076.012.

Active region observations of the Mg I and II resonance lines.   See Abstr. 076.036.

The development and structure of bright active regions at 2.8 cm.   See Abstr. 077.010.

Structure of a local source on the sun obtained from eclipse observations with the 22-meter radio telescope in integral and circularly polarized light at 1.35 cm.
See Abstr. 077.022.

Low-energy corpuscular fluxes from non-flare active solar regions.   See Abstr. 078.007.

Phase reversals in the polar magnetic fields of the sun and in the annual and semiannual variations in cosmic ray intensity.   See Abstr. 080.012.

The mutual attraction of magnetic knots.
See Abstr. 080.022.

Emergence of solar magnetic fields and 11-year activity period.   See Abstr. 080.034.

Comparisons of measured and calculated potential magnetic fields.   See Abstr. 080.048.

Variation of E-layer critical frequency with solar activity.   See Abstr. 083.080.

Variations of plasma streams at midlatitudes under conditions of high solar activity.   See Abstr. 083.115.

Planetary climatic effects of long-term solar variability.   See Abstr. 091.016.

A solar cycle variation of the interplanetary magnetic field.   See Abstr. 106.032.

The role of active regions and of the general magnetic field of the sun in the 11-year cycle of cosmic rays.
See Abstr. 143.020.

Eleven-year cycle of cosmic rays in the stratosphere (review).   See Abstr. 143.049.

073 Chromosphere, Flares, Prominences

073.001 Electron densities in solar flare and active region plasmas from a density-sensitive line ratio of Fe IX. U. Feldman, G. A. Doschek, K. G. Widing. Astrophys. J., Vol. 219, 304 - 306 (1978).

The authors demonstrate that the intensity ratio of the two lines of Fe IX at 241.739 and 244.911 Å is sensitive to electron density above about $10^{10}\,cm^{-3}$. They calculate the intensity ratio as a function of density, and apply the result to two spectroheliograms of flares and surrounding plage regions recorded by the Naval Research Laboratory spectroheliograph on Skylab. The authors find that the densities at coronal temperatures of $\sim 10^6\,K$ vary considerably from region to region and can be at least as high as $2 \times 10^{11}\,cm^{-3}$.

073.002 Multifrequency observations of solar filaments at centimeter wavelengths. M. R. Kundu, E. Fürst, W. Hirth, M. Butz. Astron. Astrophys., Vol. 62, 431 - 437 (1978).

On June 15 and 16, 1976 the authors observed solar filaments at five wavelengths in the centimeter range. They found a filament with a much larger size at radio than at optical wavelengths. While the diameter of the radio filament turned out to be about 2', the one of the H_α-filament was only 0.5. This large difference is explained in terms of a filament-cavity model and the observations were used to determine the ratio of the electron densities in the cavity and in the quiet region. The authors found a ratio of about 3:4.

073.003 Influence of rotational motions on the spectral lines of prominence streamers. V. Ruždjak. Bull. Astron. Inst. Czechoslovakia, Vol. 29, 22 - 27 (1978).

Line profiles of rotating streamers resulting from the combined action of rotational, thermal and microturbulent broadenings are calculated for various ratios of the rotational-to-characteristic thermal and microturbulent velocities.

073.004 Parameters of Forbush decreases and their parent flares in the solar cycle 1965 - 1976. L. Křivský, B. Růžičková-Topolová. Bull. Astron. Inst. Czechoslovakia, Vol. 29, 30 - 44 (1978).

The authors determine the flares responsible for Forbush decreases (FDs) established from cosmic ray recordings made at two stations from 1965 to 1976 July (the whole solar cycle No. 20), and examine the main parameters of these FDs in relation to one another and also to the parent flares.

073.005 Solar flares in laboratory. S. I. Syrovatskij. Priroda, 1978, No. 2, p. 143 - 144. In Russian.

073.006 Spectrum synthesis of chromospheric lines of Si II and Si III. D. A. Tripp, R. G. Athay, V. L. Peterson. Astrophys. J., Vol. 220, 314 - 324 (1978).

Profiles of emission lines of Si II and Si III in the EUV spectrum near the solar limb and near disk center are synthesized using a range of temperature and microturbulence models. Reasonably good agreement between computed and observed profiles near disk center is obtained with the Vernazza, Avrett, and Loeser (VAL) temperature model but with a modified microturbulence model. The adopted microturbulent velocity is equal to the mean thermal velocity of hydrogen temperatures above 7000 K and, at lower temperatures, decreases linearly to a value of $1\,km\,s^{-1}$ in the low chromosphere. The 20,000 K temperature plateau in the VAL model is found to be of importance in determining the central intensities in the Si II lines at $\lambda1816$, $\lambda1265$, and $\lambda1533$ and in determining the total intensity of the Si III line at $\lambda1206$.

073.007 EUV models of the chromosphere-corona transition region. N. Raghavan. Bull. Astron. Soc. India, Vol. 5, 100 - 104 (1977).

073.008 On the orientation of magnetic fields in quiescent prominences. J. L. Leroy. Astron. Astrophys., Vol. 64, 247 - 252 (1978).

Polarization measurements (Leroy et al., 1977) have been interpreted with the help of the Hanle effect theory (Sahal-Bréchot et al., 1977) to obtain new data concerning the orientation of magnetic field in quiescent prominences. First, using results of preceding papers, the author is able to show that the prominence magnetic field is not vertical but is probably roughly horizontal. Therefore, in the course of the paper he searches to determine the angle between the line of sight and the lines of force assumed to be horizontal.

073.009 Further measurements of quiescent prominence spectra. D. A. Landman, R. M. E. Illing, M. Mongillo. Astrophys. J., Vol. 220, 666 - 674 (1978).

Measurements of the integrated intensity of Hα, Hγ, and Hδ, He λλ10830 and 4471, Ca^+ H and K, and Sr^+ λ4078 in a number of quiescent prominences are presented. The data are discussed with reference both to other observations and to recent theoretical developments in the field.

073.010 The solar XUV He I and He II emission lines. I. Intensities and gross center-to-limb behavior. S. A. Mango, J. D. Bohlin, D. L. Glackin, J. L. Linsky. Astrophys. J., Vol. 220, 683 - 691 (1978).

The authors derive the center-to-limb variation of the He II λ304 and λ256 lines and He I λ584 and λ537 lines for different solar features, but averaged over the chromospheric supergranulation structure. The general trend is for limb brightening in quiet-Sun regions, limb neutrality in unipolar magnetic regions (UMR), and limb darkening in polar coronal holes. The center-to-limb behavior in these optically thick emission lines indicates collisional excitation and decreasing transition-region temperature gradients with respect to optical depth in the sequence quiet Sun → UMR → coronal hole.

073.011 Evaporative cooling of flare plasma. S. K. Antiochos, P. A. Sturrock. Astrophys. J., Vol. 220, 1137 - 1143 (1978).

The authors investigate a one-dimensional loop model for the evaporative cooling of the coronal flare plasma. The important assumptions are that conductive losses dominate radiative cooling and that the evaporative velocities are small compared with the sound speed. The authors calculate the profile and evolution of the temperature and verify the accuracy of their assumptions for plasma parameters typical of flare regions. The model is in agreement with soft X-ray observations on the evolution of flare temperatures and emission measures. The effect of evaporation is to greatly reduce the conductive heat flux into the chromosphere and to enhance the EUV emission from the coronal flare plasma.

073.012 Delay of the impact of high-energy protons from the solar flare of April 29, 1973. N. N. Volodichev, G. Ya. Kolesov, A. N. Podorol'skij, I. A. Savenko, A. A. Suslov. Kosm. Issled., Vol. 16, 58 - 63 (1978). In Russian.

073.013 Measurement of the relative intensities of Heζ—H8 lines λ 3889 Å in the spectrum of the chromospheric spicules at various heights above the limb. K. Nikolskaya. Sol. Phys., Vol. 56, 71 - 77 (1978).

A series of 31 large scale spectra of the spicules near H8–Heζ lines λ 3889 Å has been photographed at different heights (1500–5000 km) with 53-cm coronagraph. The solar image was 121 mm in diameter, dispersion and spatial resolution were respectively 0.647 Å mm^{-1} and 1–2". Equivalent widths of both lines have been measured and ratio R = EW$_{He\zeta}$/ EW$_{H8}$ was obtained. T_e (h) distribution in the averaged spicule deduced differs from that in the recent spicule model by Beckers (1972): the spicules are colder greatly than it was accepted hitherto ($T_e \leqslant$ 10 000 K). The electron temperature in the quiescent prominence which occurred in the spectra was also estimated: $T_e \leqslant$ 6000 K at h = 4000 and 5000 km.

073.014   **The fine structure of prominences. IV: Spectral observations.**   O. Engvold.
Sol. Phys., Vol. 56, 87 - 106 (1978).

High spatial resolution spectral observations of five hedgerow prominences were made in Hα, He I D$_3$ and Ca II H and K. The observed relations between the lines were not the same in all prominences. The Ca II H and K lines were 2–4 times brighter relative to Hα and D$_3$ than predicted theoretically. The optical thickness of Hα was less than for the H and K lines, the Hα was optically thin in medium faint prominence structures. Faint structures appeared slightly hotter than bright structures.

073.015   **EUV structure of a small flare.**   R. H. Levine.
Sol. Phys., Vol. 56, 185 - 203 (1978).

Observations of a small flare are presented using data from the Harvard spectroheliometer on Skylab. The event is discussed in terms of the magnetic structure of the active region as deduced from the EUV observations and from field line extrapolations. The role of emerging flux in the initial flare brightenings is emphasized. A detailed model of one loop is deduced using the EUV data. This self-consistent model indicates initial heating of the loop modelled near its top, and mass flow into the cool core of the loop, with matter preferentially concentrating in a few distinct knots along the loop. Implications for theories of the flare process are discussed.

073.016   **Analysis of the emission line spectra of a solar flare observed from Skylab.**   C.-C. Cheng.
Sol. Phys., Vol. 56, 205 - 222 (1978).

The EUV emission spectra in the wavelength range 110–1900 Å of the 5 September 1973 flare observed with the NRL slit spectrograph on Skylab are studied. The results are: (1) The chromospheric and transition-zone lines are greatly enhanced during the flare. (2) The chromospheric lines maintain their sharp and gaussian profiles and are not appreciably broadened. The transition zone lines, on the other hand, show a red-shifted component during the initial phase of the flare. (3) The density of the 10^5 K flare plasma, as deduced from density-sensitive lines, is greater than 10^{12} cm^{-3}. The depth of the 10^5 K plasma in the flare transition zone is only of the order of 0.1 km, giving a steep temperature gradient.

073.017   **On the problem of power-law spectrum of particles accelerated in solar flares.**   A. A. Korchak.
Sol. Phys., Vol. 56, 223 - 234 (1978).

The formation of power-law energy spectrum of particles accelerated in solar flares is investigated. The distinct difference between the mechanism and the model of acceleration is pointed out. It is shown that Fermi's model is described by a linear differential equation of the first order and therefore a power-law spectrum is formed only for some special conditions which apparently are not fulfilled for flares. A satisfactory alternative to Fermi's model hasn't yet been found. In conclusion the connection between the mechanism of acceleration and a charge spectrum of accelerated particles is examined.

073.018   **Structure and spectrum of quiescent prominences.**

III. **Application of theoretical models in helium abundance determinations.**   J. N. Heasley, R. W. Milkey.
Astrophys. J., Vol. 221, 677 - 688 (1978).

The solar helium abundance has been investigated by using spectra of quiescent prominences. The physical parameters (T, P, y, ξ, M) describing an isothermal and isobaric model of the prominence can be determined from observations of hydrogen, helium, and ionized calcium lines. Given these parameters, it may be shown that low values of the helium abundance, such as those inferred from solar wind data, are inconsistent with the prominence data. The helium-to-hydrogen ratio is indicated to be 0.10 ± 0.025 by number.

073.019   **Measurements of the Hα and Hβ in quiescent prominences.**   D. A. Landman, M. Mongillo.
Bull. American Astron. Soc., Vol. 9, 565 (1978). – Abstract.

073.020   **The relationships between EUV flares and surges.**   E. J. Schmahl.
Bull. American Astron. Soc., Vol. 9, 568 - 569 (1978). Abstract.

073.021   **Do all flares have impulsive phases?**   D. W. Datlowe, M. J. Elcan, H. S. Hudson, L. E. Peterson.
Bull. American Astron. Soc., Vol. 9, 569 (1978). – Abstract.

073.022   **Active edges of quiescent prominences.**   J. M. Malville, O. Engvold, W. Livingston.
Bull. American Astron. Soc., Vol. 9, 569 (1978). – Abstract.

073.023   **Optical thicknesses and degree of ionization of hydrogen in chromospheric flares of various importances.**   L. N. Kurochka, V. A. Ostapenko,
Bull. Astron. Inst. Czechoslovakia, Vol. 29, 82 - 89 (1978).

A method of determining the optical thicknesses of flares from the intensities of remote lines of the Balmer series is presented. The number of observed lines of the series in flares of different importances is determined from spectral observations of the flares. The optical thicknesses and the degree of ionization were determined by solving the equation of ionization equilibrium. The conclusion was drawn that the optical thicknesses in flares of all importances do not exceed 10–100 units. The degree of ionization in flares is within the interval 10^{-1}–10^4.

073.024   **Evaluation of a prediction technique for low energy solar particle events.**
E. C. Roelof, R. E. Gold, E. P. Keath.
Space Research, Vol. XVII, (see 012.010), 545 - 550 (1977).

A technique has been developed for the prediction of low energy solar charged particle fluxes near 1 AU based on comprehensive analyses of solar and interplanetary plasma, magnetic field and energetic particle observations. The prediction technique has been tested on the period May–November 1967 and the results are satisfactory for the larger particle events.

073.025   **Spectral lines observed in solar flares between 171 and 630 Angstroms.**   K. P. Dere.
Astrophys. J., Vol. 221, 1062 - 1067 (1978).

Several hundred spectral lines emitted in solar flares between 171 and 630 Å have been recorded by the Naval Research Laboratory spectroheliograph aboard Skylab. The wavelengths, identifications, and intensity estimates of these lines are presented, based on measurements of all of the suitable flare plates. Nearly 100 new and unidentified lines have been observed. Identifications of three Fe XXI and two Fe XVII lines are suggested.

073.026   **A phenomenological model of solar flares.**   S. A. Colgate.

Astrophys. J., Vol. 221, 1068 - 1087 (1978).

It is almost universally agreed that the energy of solar flares is derived from the magnetic energy of fields convected to the Sun's surface and subsequently converted to heat and energetic particles within the chromosphere. The circumstances of this conversion in most present models is magnetic flux annihilation at a neutral sheet. The author agrees that the primary energy is magnetic in origin,but he analyzes the constraints of flux annihilation in greater detail and shows how the present evidence of solar cosmic rays, X-rays, γ-rays, and total energy suggests a choice of annihilation not at a neutral point, but by an enhanced dissipation of a field-aligned current.

073.027 The chromosphere above sunspot umbrae. I. Observations of the emission cores in the Ca II H- and K-lines. W. Mattig, F. Kneer.
Astron. Astrophys., Vol. 65, 11 - 16 (1978) = Mitt. Fraunhofer Inst., Freiburg, No. 153.

Photographic spectra of the Ca II H- and K-lines in sunspot umbrae are analysed. The emission features in the line profiles may be classified into minimum emission profiles that also lack self-reversal, and profiles which change rapidly in time and exhibit self-reversals (umbral flashes) that are most often strongly asymmetric. Average intensity profiles from the minimum emission parts of the umbra and characteristic intensities of the more active parts of the umbra are given.

073.028 The chromosphere above sunspot umbrae. II. The interpretation of the H, K, and IR lines of Ca II. F. Kneer, W. Mattig.
Astron. Astrophys., Vol. 65, 17 - 28 (1978) = Mitt. Fraunhofer Inst., Freiburg, No. 150.

The chromospheres above sunspot umbrae are investigated by comparing calculated and observed profiles of the Ca II H, K, and infrared lines. The authors test several model chromospheres in hydrostatic equilibrium. They distinguish between chromospheres which are optically thick and optically thin at the centre of the K line. They are forced to adopt a thick model as reference chromosphere.

073.029 Line formation in the solar chromosphere. I. The C II resonance lines observed with OSO 8. B. W. Lites, R. A. Shine, E. G. Chipman.
Astrophys. J., Vol. 222, 333 - 341 (1978).

The temperature structure of the upper chromosphere is investigated using center-to-limb measurements of the C II resonance lines at 1335 Å from the University of Colorado spectrometer aboard OSO 8. The authors compute hydrostatic equilibrium models of the quiet chromosphere to obtain theoretical spectra of the Lyman lines and continuum as well as the center-to-limb behavior of the C II lines, and find good agreement with observations for a plateau at 16,500 K and do so with about 25% more material than that of Vernazza et al. (1973).

073.030 C III density diagnostics in nonequilibrium plasmas. J. C. Raymond, A. K. Dupree.
Astrophys. J., Vol. 222, 379 - 383 (1978).

Mass flow through a temperature gradient can produce substantial departures of the ionization states from their equilibrium values and can change the character of line emission. The effect of such flows on extreme ultraviolet intensities in the solar atmosphere is evaluated for the beryllium-sequence ion C III. A downflow of material in a transition region can lead to an underestimate of the electron density; outflow of material, as in a stellar or solar wind passing from chromospheric to coronal temperatures, yields a reliable density diagnostic.

073.031 The electron density at 10^5 K in different regions of the solar atmosphere derived from an intersystem line of O IV. U. Feldman, G. A. Doschek.

Astron. Astrophys., Vol. 65, 215 - 222 (1978).

Electron densities are derived for typical solar active regions and flares from EUV spectral lines of O IV, N V, C IV and Si IV. The densities pertain to electron temperatures near 10^5 K. Typical densities found for active regions are $\sim 10^{11}$ cm^{-3}. The density in flares can range from $\sim 10^{11}$ cm^{-3} to above 10^{13} cm^{-3}. The volumes of high density emitting plasma in active regions and flares are quite small, with values ranging from 1.5×10^{23} cm^3 to less than 2.2×10^{20} cm^3. Also, the density in a polar coronal hole is about one-half of the density in a typical quiet Sun region, at a temperature near 6×10^4 K.

073.032 On physical conditions in the chromosphere above sunspot umbrae. R. B. Teplitskaja (*Teplitskaya*), S. A. Grigoryeva (*Efendieva*) (*Grigor'eva (Ehfendieva)*), V. G. Skochilov.
Sol. Phys., Vol. 56, 293 - 303 (1978).

By means of an inversion of H and K Ca II line profiles the temperature and electron density in the chromosphere above the umbrae of two sunspots have been estimated. The temperature gradient 5 K km^{-1} exceeds the corresponding values in both quiet regions and plages. At a height of about 1500 km the umbra becomes hotter than the quiet region. At a temperature of about 10000 K the temperature gradient increases sharply. The electron density at 1500 km is approximately the same as that in the quiet chromosphere at the same height.

073.033 Some results of a two- and three-dimensional spectral analysis of wave processes in the lower chromosphere from filtergrams in the Ba II 4554 + 0.05 Å line. N. V. Larionov, V. E. Merkulenko, V. I. Polyakov, V. I. Skomorovskij.
Issled. po geomagn., aehron. i fiz. Solntsa. Vyp. (No.) 42. Moskva, Nauka, 1977, p. 50 - 59. In Russian. – Abstr. in Ref. zh., 51. Astron., 2.51.362 (1978).

073.034 Characteristics of sympathetic flares on the sun. V. F. Chistyakov.
Issled. po soln. aktivnosti. Vladivostok, 1977, p. 19 - 50. In Russian. – Abstr. in Ref. zh., 51. Astron., 2.51.419 (1978).

073.035 Flares outside sunspot groups. V. F. Chistyakov, K. G. Chistyakova.
Issled. po soln. aktivnosti. Vladivostok, 1977, p. 51 - 70. In Russian. – Abstr. in Ref. zh., 51. Astron., 2.51.421 (1978).

073.036 Unusual phenomena on the sun in 1972 - 1975. V. F. Chistyakov.
Izv. AN SSSR. Ser. fiz., Vol. 41, 1772 - 1775 (1977). In Russian. – Abstr. in Ref. zh., 51. Astron., 2.51.427 (1978).

073.037 On the prediction of recurrent flocculi appearance at the solar limb. V. B. Gumanitskij, V. M. Efimenko, V. V. Tel'nyuk-Adamchuk.
Soln. Dannye 1977 Byull., No. 11, p. 58 - 62 (1978). In Russian.

Using data on flocculi development during their passage across the solar disc the appearance of flocculi for the next solar rotation is predicted. Two methods are used: the method of constructing a prediction function as a function of many variables and the method of potential functions. The first method gave a mean error of 13% for the years 1967 - 1969, and 9% for 1968 - 1969. Both the methods based on the data for 1967 gave an error of 8% for 1968 - 1969. A comparison of the obtained results with those of other authors is made.

073.038 Correlation and dispersion fields of the flare index and some other indices of solar activity. Yu. I. Vitinskij, B. M. Rubashev.

Soln. Dannye 1977 Byull., No. 11, p. 85 - 92 (1978). In Russian.

The results of a study of the longitudinal-latitudinal distribution of correlation coefficients of the flare index with total areas of sunspot groups, total maximum intensities of the sunspot magnetic field, total areas of Ca II plages weighted for brightness, total length of hydrogen filaments, and dispersions of the indices (for 67 revolutions of the sun during 1966-1970) are presented.

073.039 **Appearance of large flares in decayed sunspot groups.**   V. S. Prokudina.
Soln. Dannye 1977 Byull., No. 11, p. 92 - 98 (1978). In Russian.

The appearance of large chromospheric flares in decayed sunspot groups is investigated. It is noted that an abnormally high chromospheric activity remains in the sunspot groups which had large areas at the preceding rotation, but at the given rotation the area of these groups is small or has disappeared at all. The possible cause of the remaining high chromospheric activity in the decayed groups may be a magnetic field, which, being frozen in the matter, is carried out from the spot into the chromosphere, thus creating conditions for acceleration and heating of the electron and proton gas at the level of the chromosphere.

073.040 **On the short-term forecast of chromospheric flares on the basis of characteristics of sunspot groups.**
V. I. Efimenko, V. M. Efimenko, V. V. Tel'nyuk-Adamchuk.
Vestn. Kiev. Univ., Astron., Vyp. (No.) 19, p. 19 - 29 (1977). In Russian.

The parameters which characterize the sunspot group development are used for the short-term forecast of flares. In 1974 - 1975 a forecast is carried out for the following 1–2 days on the basis of the adopted method. The forecast is correct for 75 percent of the flares.

073.041 **Forecast of chromospheric flares from 1 to 7 days ahead.**   P. R. Romanchuk, Yu. I. Izotov, I. Yu. Izotova, V. N. Krivodubskij.
Vestn. Kiev. Univ., Astron., Vyp. (No.) 19, p. 29 - 36 (1977). In Russian.

Two methods of prediction of flare activity of active centers up to 7 days ahead are proposed. The connection of flares of different importance with the Zürich classes of sunspot groups is used in the first method. The dependence of flare activity on the dimensions of the convective elements is taken into account in the second method. The forecast is true for 90 per cent of flares within the period from July 1974 to November 1975.

073.042 **On physical characteristics of a flare of small and average power.**   P. N. Polupan.
Vestn. Kiev. Univ., Astron., Vyp. (No.) 19, p. 36 - 42 (1977). In Russian.

The profiles of the Balmer lines from $H\beta$ to H_{15} in limb and from $H\beta$ to H_{10} in disk flares are presented. The parameters of these profiles and physical characteristics are determined.

073.043 **Determination of the physical parameters of an active prominence from steady state equations for a hydrogen atom by means of the least squares method.**
E. G. Rudnikova, M. N. Pasechnik, S. V. Pasechnik.
Vestn. Kiev. Univ., Astron., Vyp. (No.) 19, p. 42 - 49 (1977). In Russian.

The relation between the main parameters of prominence matter from the statistical equilibrium equations for the hydrogen atom by means of the least squares method is determined.

073.044 **Statistical analysis of 6600 chromospheric flares in**

1965 - 1966.   V. M. Rossada.
Vestn. Kiev. Univ., Astron., Vyp. (No.) 19, p. 49 - 55 (1977). In Russian.

073.045 **Forbidden lines of Fe XIX, Fe XX, and Fe XXI in solar flares.**   K. G. Widing.
Astrophys. J., Vol. 222, 735 - 739 (1978).

A critical compilation of wavelengths and intensities for 33 highly ionized lines observed in solar flare spectra between 320 and 600 Å is given. Seven of the lines are identified as magnetic dipole transitions within the ground configurations of Fe XIX, Fe XX, and Fe XXI. One line is a forbidden transition of Ni XXI. Absolute intensities of the forbidden lines are computed for the flare of 1973 December 17 and compared with observation.

073.046 **On the recognition of proton and non-proton flares.**
B. G. Dolgoarshinnykh, G. V. Kuklin.
Issled. po geomagn., aehron. i fiz. Solntsa. Vyp. (No.) 42. Moskva, Nauka, 1977, p. 44 - 47. In Russian. — Abstr. in Ref. zh., 51. Astron., 3.51.498 (1978).

073.047 **The form and energy of the shock waves from the solar flares of August 2, 4, and 7, 1972.**
G. N. Zastenker, V. V. Temny (*Temnyj*), C. d'Uston, J. M. Bosqued.
J. Geophys. Res., Vol. 83, 1035 - 1041 (1978).

The shape of the shock waves associated with the August 1972 solar flares was determined by comparing the plasma measurement data from the Prognoz, Prognoz 2, Pioneer 9, and Pioneer 10 space probes with observations of radio source scintillations and comet brightness. A nearly spherical shock wave form was obtained for two flares on August 2, and essentially nonspherical shock wave forms were obtained for flares on August 4 and 7. The energy and ejected mass determined from the Prognoz 2 measurements of the additional energy and mass flux behind each shock wave are 10^{32}–10^{33} ergs and 10^{16}–10^{17} g, respectively.

073.048 **On the size and brightness of moustaches.**   A. N. Babin.
Izv. Krymskoj Astrofiz. Obs., Vol. 58, 8 - 12 (1978). In Russian.

From cinematographic ($H\alpha \pm 1$ Å)-filter observations of the occultation of bright compact moustaches by the moon during the partial solar eclipse on 29 April 1976 has been found: (1) the true size of moustaches is equal to 0.62 ± 0.05 seconds of arc; (2) the brightness distribution across moustaches is close to rectangular; (3) the true excessive brightness of moustaches is equal to ~150% in units of the undisturbed continuum of the solar disc center.

073.049 **On the influence of survival probability of $H\alpha$-quanta on the variation of the source function.**
V. V. Zharkova.
Astrometriya i Astrofizika, Kiev, Vyp. (No.) 34, p. 5 - 13 (1978). In Russian.

The $H\alpha$-source functions obtained for $\lambda_{32} = 1$ and $\lambda_{32} \neq 1$ (λ_{32} is the $H\alpha$-output coefficient) in prominences and flares with optical thickness $\tau_{23}^0 = 1, 10, 50, 100$ are compared. In flares the difference of λ_{32} from unity is considerable; it decreases with growing optical thickness, the difference of the source function increases. For prominences λ_{32} is very close to unity.

073.050 **Some properties of the development of chromospheric surges.**   M. M. Makhmudov, Yu. V. Platov.
Soln.Dannye 1977 Byull., No. 12, p. 65 - 73 (1978). In Russian.

The variation of various parameters of chromospheric surges is studied from spectral observations in the H_α line.

The surge formation is accompanied by an intensive increase of nonthermal velocities and appearance of rotational motions.

073.051   On peculiarities of motion in an emission surge.
M. N. Stoyanova.
Soln. Dannye 1977 Byull., No. 12, p. 73 - 79 (1978).
In Russian.
Profiles of spectral lines are studied for an emission surge on the solar disc. A comparison is made between the observed profiles and theoretical ones for different physical factors. None of the considered theoretical profiles gives satisfactory agreement with observation. It is concluded that a model of medium consisting of many small elements and moving with variable velocity gradient inside it is most suitable.

073.052   Simultaneous observations of the solar chromosphere in the Hα hydrogen and Ca II K lines.
R. T. Sotnikova.
Soln. Dannye 1977 Byull., No. 12, p. 97 - 100 (1978).
In Russian.
Simultaneous observations of the solar chromosphere in two lines (Hα and Ca II K) have been made. The results of a preliminary comparative analysis of the undisturbed chromospheric fine structure are given. The most brightest features are identical in both pictures. The faintest features on the Hα filtergrams have no corresponding details on the Ca II filtergrams. As a result the mean diameter of the Hα network is smaller than that of Ca II.

073.053   On the correspondence of optical and radio events in an eruptive (post-maximum) phase of the proton flare of August 4, 1972.
V. N. Ishkov, Eh. I. Mogilevskij, V. P. Nefed'ev.
Soln. Dannye 1978 Byull., No. 1, p. 72 - 76 (1978).
In Russian.
Evidence is given in favour of a repeated energy release not only in the beginning but also during the solar flare.

073.054   Some circumstances of flare appearance outside sunspot groups.
K. G. Chistyakova, V. F. Chistyakov.
Soln. Dannye 1978 Byull., No. 1, p. 76 - 83 (1978).
In Russian.
The number of flares outside sunspot groups is equal to 5 per cent of the total number of solar flares. Forty per cent of them precede the appearance of sunspot groups by about 10 days. The decay of the flare activity continues for about 10 days after the vanishing of sunspot groups. The flares appear on the longitudes where the sunspot number is twice less and the number of flares is twice larger than that on longitudes ±20° eastward and westward. A new phenomenon was discovered: the locations of occurrences of flares as well as those of sunspots gradually move from West to East. A suggestion is made that deep layers of the sun rotate in opposite directions. The rotation period is about 7—12 days.

073.055   Oscillations of a magnetic rope and solar flare mechanism.   A. A. Solov'ev.
Soln. Dannye 1978 Byull., No. 1, p. 84 - 87 (1978).
In Russian.
It is shown that with an oscillation amplitude of a magnetic rope large enough turbulence of plasma begins which can lock the current channel and result in an electric discharge of solar flare type. An approximate mathematical description of this process is given. The theoretical evaluation of the released energy agrees with observational data.

073.056   Peculiarities of the solar flare of August 2, 1972.
I. F. Nikulin, G. F. Sitnik.
Soln. Dannye 1978 Byull., No. 1, p. 96 - 98 (1978).
In Russian.

Structure and motion of the flare surges of August 2, 1972 were studied. On the whole the motion was directed to the photosphere with radial velocity up to 150 km/sec and was greatly influenced by the magnetic field.

073.057 D_3 spicules and the lower chromosphere.
K. A. Marsh.
Sol. Phys., Vol. 57, 37 - 48 (1978).
High resolution filtergrams of the solar limb in D_3 and off-band Hα have been used to investigate the spatial structure of the D_3 chromosphere. It was found that spicules provide the major contribution to the intensity of the D_3 emission band observed above the limb, with the remainder of the emission coming from a semi-homogeneous background component at low heights. In coronal holes, the D_3 emission is confined to isolated emission patches, and these patches contain a fine structure resembling normal chromospheric spicules.

073.058   Downflow in the supergranulation network and its implications for transition region models.
G. W. Pneuman, R. A. Kopp.
Sol. Phys., Vol. 57, 49 - 64 (1978).
The authors wish to explore the theoretical implications of some recent observations from space of velocity fields in the solar atmosphere – namely, the discovery of the presence of quasi-steady downflows in the transition region network overlying the supergranulation boundaries. The coincidence of the estimated particle-mass flux associated with this flow with that observed moving upward in the form of spicules in the same region leads to suggest that this downflow may represent spicular material returning to the chromosphere after being heated to coronal temperatures.

073.059   A theory of the onset of solar eruptive processes.
J. Birn, H. Goldstein, K. Schindler.
Sol. Phys., Vol. 57, 81 - 101 (1978).
The present theory assumes that solar eruptive processes – such as flares or eruptive prominences – occur at a critical stage when a configuration evolves in a quasi-static way. The onset criterion is not based on standard linear stability considerations but on the fact that under suitable conditions static equilibrium configurations cease to exist. For the major part of the paper the configurations are characterized by a single parameter. In a particularly simple (one-dimensional) case the authors discuss solutions depending on two parameters. The results can be discussed in terms of catastrophe theory. The theory is valid for two space dimensions and contains topological changes of the magnetic field, although the latter feature is not necessary for the theory to apply. The theory of two-dimensional force-free fields is contained as a special case.

073.060   Threshold effect in second-stage acceleration.
H. S. Hudson.
Sol. Phys., Vol. 57, 237 - 240 (1978).
Proton fluxes observed at Earth have a much flatter size distribution than do other parameters of solar flares more representative of total energy. Peak proton flux varies at least as rapidly as the fourth power of total flare energy. An absolute threshold may exist, in view of the flatness of the proton distribution.

073.061   The Lα/Hα ratio in solar flares, quasars, and the chromosphere.   H. Zirin.
Astrophys. J., Lett., Vol. 222, L105 - L107 (1978).
The ratio of Lα to Hα is around unity in flares, quasars, and the solar chromosphere and prominences. The weakness of Lα is shown to be essentially due to photon trapping and de-excitation, but it is argued that the surprising stability of this ratio is due to the role of these lines in cooling the plasma rather than the accidental parameters used in various models of these widely different phenomena.

073.062 Dynamics of an atmosphere irradiated by soft X-rays in flares.   J. C. Hénoux, Y. Nakagawa.
Astron. Astrophys., Vol. 66, 385 - 393 (1978).

In a previous paper it was shown that the lower chromosphere can be heated substantially by the soft X-ray radiation at the wavelengths 1–300 Å arising from a high temperature plasma (10^7 K) in the corona during a flare. The analysis was, however, based on the approximation of steady state. In this paper, therefore, a more realistic dynamic response is examined utilizing numerical calculations. The results indicate that with continuous and constant irradiation by the soft X-ray flux, (1) the atmosphere undergoes an oscillating expansion characterized by the period of oscillation approximately 270 s before reaching a new steady state in about 30 min, (2) in the oscillatory motions, increase of temperature coincides with downward motions and decrease of temperature with upward motions, and (3) the observed time variation of Hα line profile in flares appears to be in agreement with the initial part of the oscillatory behavior.

073.063 Some characteristics of pre-flare increases.
G. M. Blokh, G. Ya. Kolesov, B. M. Kuzhevskij,
Problems of solar activity and space system Prognoz, (see 003.008), p. 125 - 131 (1977). In Russian. – Abstr. in Ref. zh., 51. Astron., 4.51.406; 62. Issled. kosm. prostranstva, 4.62.131 (1978).

073.064 On the problem of forecast of solar flares.
S. I. Syrovatskij.
Problems of solar activity and space system Prognoz, (see 003.008), p. 5 - 22 (1977). In Russian. – Abstr. in Ref. zh., 51. Astron., 4.51.410 (1978).

073.065 Motion of a shock wave due to the solar flare on 4 August 1972 according to observations of type II bursts.   V. P. Grigor'eva.
Problems of solar activity and space system Prognoz, (see 003.008), p. 175 - 179 (1977). In Russian. – Abstr. in Ref. zh., 51. Astron., 4.51.416; 62. Issled. kosm. prostranstva, 4.62.135 (1978).

073.066 A study of non-relativistic electrons generated during solar flares.   N. L. Grigorov, V. G. Kurt,
Yu. I. Logachev, V. N. Lutsenko, T. N. Markelova, N. F. Pisarenko, I. A. Savenko, I. P. Shestopalov.
Problems of solar activity and space system Prognoz, (see 003.008), p. 64 - 94 (1977). In Russian. – Abstr. in Ref. zh., 51. Astron., 4.51.419 (1978).

073.067 Peculiarities of interaction of solar cosmic radiation with the earth's magnetosphere and the interplanetary space in the initial anisotropic phase of flares.
N. A. Mikirova.
Tr. IPG, 1977, vyp. (No.) 35, p. 19 - 23. In Russian. – Abstr. in Ref. zh., 62. Issled. kosm. prostranstva, 4.62.406 (1978).

073.068 Observations of a solar flare in monochromatic light, and at 136 and 90 megahertz.
H. R. Hatfield.
J. British Astron. Assoc., Vol. 88, 356 - 359 (1978).

073.069 Particle acceleration in the current sheet of solar flares.   L. A. Pustil'nik.
Astron. Zh. Akad. Nauk SSSR, Vol. 55, 607 - 616 (1978). In Russian. English translation in Soviet Astron., Vol. 22, No. 3.

The applicability and effectiveness of the main mechanisms of particle acceleration under real conditions of solar flares are analysed. It has been shown that the realization of the mechanisms of pure regular acceleration by the electric field and pure stochastic acceleration in the turbulence field faces the difficulties of principle in both the theoretical treatment and the comparison with observations. A mechanism of "quasi-diffusive" acceleration including the processes of both types is proposed.

073.070 Eine außergewöhnlich aktive Protuberanzen-Erscheinung.   W. Paech.
Sterne Weltraum, Jahrg. 17, 266 (1978).

073.071 Formation of shock waves from chromospheric flares and their evolution in the solar wind.
T. V. Stepanova, A. M. Uralov.
Issled. po geomagn., aehron. i fiz. Solntsa. Vyp. (No.) 42. Moskva, Nauka, 1977, p. 83 - 91. In Russian. – Abstr. in Ref. zh., 51. Astron., 5.51.353 (1978).

073.072 Hα, hard X-ray, and microwave emissions in the impulsive phase of solar flares.   D. F. Neidig, Jr.
Sol. Phys., Vol. 57, 385 - 398 (1978).

The author has studied the observational relationship between the location of flare sites in active regions and three other observables, viz., Hα line width, hard X-ray burst parameters, and peak microwave fluxes. Qualitative relationships are derived on the assumption of proportionality between the spectral maximum frequency of the associated microwave burst and the field strength in the microwave source. The relationship inferred between the power in thick target electrons (derived from the hard X-ray burst) and the column density of second-level hydrogen atoms (derived from the Hα line widths) is compared with calculations by Brown (1973) and Canfield (1974).

073.073 Hα profiles from electron-heated solar flares.
J. C. Brown, R. C. Canfield, M. N. Robertson.
Sol. Phys., Vol. 57, 399 - 408 (1978).

The authors briefly review the status of models of optical flare heating by electron bombardment. They recompute Brown's (1973) flare model atmospheres using considerably revised radiative loss rates, based on Canfield's (1974) method applied to Hα, Lα, and H⁻. Profiles of Hα are computed and compared with observation. The computed profiles agree satisfactorily with those observed during the large 1972 August 7 flare, if spatial and velocity inhomogeneities are assumed. The electron injection rate inferred from Hα is one order of magnitude less than that inferred from hard X-rays, for this event. This may be due to either (1) the neglect of a mechanism that reduces the thick-target electron injection rate or (2) failure to incorporate important radiative loss terms.

073.074 Scattering of fast flare electrons in solar atmosphere and their X-ray spectrum.   G. Elwert,
R. R. Rausaria.
Sol. Phys., Vol. 57, 409 - 413 (1978).

The evolution of the energy distributions of fast flare electrons injected towards the chromosphere are computed by the Monte Carlo method for different depths. Using these distributions, power law bremsstrahlung spectra having spectral indices increasing with photon energies are obtained.

073.075 The radio spectrum of chromospheric structure.
G. J. Hurford, K. A. Marsh, W. Graf, M. A. Janssen,
E. T. Olsen.
Bull. American Astron. Soc., Vol. 10, 431 (1978). – Abstract.

073.076 Thermally unstable current sheets in prominences.
J. M. Malville.
Bull. American Astron. Soc., Vol. 10, 439 (1978). – Abstract.

073.077 Doppler shifts measured in the O VI line from OSO-8 observations above and in the vicinity of plage McMath 13738.   P. Lemaire, A. Skumanich, G. Artzner,
P. Gouttebroze, J. C. Vial, R. M. Bonnet, P. McWhirter.

Bull. American Astron. Soc., Vol. 10, 440 (1978). – Abstract.

073.078 Observations of solar flare X-ray continua.
J. R. Lemen, W. H.-M. Ku, R. Novick, J. H.
Parkinson, N. J. Veck.
Bull. American Astron. Soc., Vol. 10, 440 - 441 (1978).
Abstract.

073.079 OSO-8, radio and X-ray observations of the 19 April
1977 flare.   A. Skumanich, A. Jouchoux,
J. Castelli, P. Lemaire, G. Artzner, P. Gouttebroze, J. C. Vial,
R. M. Bonnet.
Bull. American Astron. Soc., Vol. 10, 441 (1978). – Abstract.

073.080 ATM evidence for a low non-thermal proton/electron
energy flux ratio in solar flares.
R. C. Canfield, J. W. Cook.
Bull. American Astron. Soc., Vol. 10, 441 (1978). – Abstract.

073.081 Evidence that the mass of the thermal X-ray plasma
in solar flares is supplied by conduction-driven evap-
oration.   R. L. Moore.
Bull. American Astron. Soc., Vol. 10, 442 (1978). – Abstract.

073.082 Numerical simulations of the decay phase of com-
pact flares.   K. R. Krall, S. K. Antiochos.
Bull. American Astron. Soc., Vol. 10, 442 (1978). – Abstract.

073.083 VLA observations of solar flares.
K. A. Marsh, H. Zirin, G. J. Hurford.
Bull. American Astron. Soc., Vol. 10, 454 (1978). – Abstract.

073.084 Soft X-ray observations during the preflare phase of
the solar flare phenomenon.
C. J. Wolfson, L. W. Acton, J. W. Leibacher.
Bull. American Astron. Soc., Vol. 10, 456 - 457 (1978).
Abstract.

073.085 X-ray analysis of the 29 July 1973 flare.
J. Nolte, M. Gerassimenko, A. Krieger, R. Krogstad,
R. Petrasso, F. Seguin, Z. Švestka.
Bull. American Astron. Soc., Vol. 10, 457 (1978). – Abstract.

073.086 Torsional oscillations in an active prominence.
M. Schindler, J. M. Malville.
Bull. American Astron. Soc., Vol. 10, 461 (1978). – Abstract.

073.087 Some new measurements of line emission in quies-
cent prominences.   D. A. Landman, M. Mongillo.
Bull. American Astron. Soc., Vol. 10, 462 (1978). – Abstract.

073.088 Radiative heating in chromospheric flares.
M. E. Machado.
Bull. American Astron. Soc., Vol. 10, 462 (1978). – Abstract.

073.089 Interpretations of the moving emission front ob-
served with the flare of 5 Sep. 1973.   S. F. Martin.
Bull. American Astron. Soc., Vol. 10, 462 (1978). – Abstract.

073.090 Solar flares – what will the 1970s provide?
W. J. Weber.
Nederlands Tijdschr. Natuurkd. A, Vol. A43, No. 4, p. 188 -
191 (1977). In Dutch. – Abstr. in Phys. Abstr., Vol. 81,
Abstr. 32432 (1978).

073.091 Intensity fluctuations in the solar chromosphere.
K. R. Sivaraman, P. P. Venkitachalam.
Kodaikanal Obs. Bull., Ser. A, Vol. 2, 34 - 41 (1977).
High resolution spectrograms of the Balmer lines obtained at five positions from the centre to the limb have been
used to study the r.m.s. intensity fluctuations and their dependence on the heliocentric angle. The intensity fluctuations
at the line centre as well as at several wavelength positions
within the lines have been derived and the variations of these
fluctuations within the lines have been studied.

073.092 On the dependence of the height of prominences on
solar cycle and heliographic latitude.
V. Dermendzhiev.
Astrofiz. issled. (NRB), Vol. 2, 8 - 19 (1977). In Russian.
Abstr. in Ref. zh., 51. Astron., 6.51.547 (1978).

073.093 On the spatial distribution of solar prominences.
V. Dermendzhiev.
Astrofiz. issled. (NRB), Vol. 2, 20 - 24 (1977). In Russian.
Abstr. in Ref. zh., 51. Astron., 6.51.548 (1978).

073.094 Aspetti osservativi del fenomeno brillamento.
R. Pallavicini.
Mem. Soc. Astron. Italiana, Vol. 48, (see 012.037), 161 - 196
(1977).
The paper reviews observations of electromagnetic and
particle emission from solar flares with special emphasis on the
relationships between characteristic features of flares in different spectral bands. Three main phases in the development
of a typical event are identified: precursor, impulsive phase
and gradual phase. High spatial resolution observations of
soft X-ray flares are discussed in detail and a comparison with
simultaneous EUV and Hα images is performed. Considerable
attention is paid to magnetic fields associated with flares and
to the location of the sources of different emission with respect to the magnetic field structure.

073.095 Emissione dei flares in luce di sodio e posizione
relativa ai campi magnetici.   A. Cacciani.
Mem. Soc. Astron. Italiana, Vol. 48, (see 012.037), 201 - 211
(1977).

073.096 Meccanismi di emissione nei brillamenti solari.
G. Noci.
Mem. Soc. Astron. Italiana, Vol. 48, (see 012.037), 297 -
319 (1977).

073.097 Meccanismi dei brillamenti.   C. Chiuderi.
Mem. Soc. Astron. Italiana, Vol. 48, (see 012.037),
321 - 331 (1977).

073.098 Chromospheric motions observed by the University
of Colorado experiment on OSO-8.   R. G. Athay.
Mem. Soc. Astron. Italiana, Vol. 48, (see 012.038), 509 - 523
(1977).
Contents: Review of previous observations. The University of Colorado experiment on OSO-8. Line widths. Systematic flows. Oscillations. Vertical phase relations.

073.099 The upper chromosphere and corona. Observations
and interpretation.   A. H. Gabriel.
Mem. Soc. Astron. Italiana, Vol. 48, (see 012.038), 559
(1977). – Summary.

073.100 On a model of chromospheric flares on the sun.
V. S. Sokolov, A. G. Kosovichev, V. S. Slavin.
Mathematical models of the near space, 1977, (see 003.015),
p. 216 - 234. In Russian. – Abstr. in Ref. zh., 51. Astron.,
7.51.440 (1978).

073.101 Problems of the theory of solar flares.
S. B. Pikel'ner, S. A. Kaplan.
Izv. VUZ..Radiofiz., Vol. 20, 1310 - 1317 (1977). In Russian.
Abstr. in Ref. zh., 51. Astron., 7.51.442 (1978).

Solar flare activity and some indices of the inter-
planetary and geomagnetic fields, 1957 - 1973.
See Abstr. 003.104.

An ionisation wedge for simulation of the solar flare spectrum.   See Abstr. 022.091.

The LPSP instrument on OSO 8. II. In-flight performance and preliminary results.   See Abstr. 032.510.

Transport of twist in force-free magnetic flux tubes. See Abstr. 062.023.

Current sheets and flares in cosmic and laboratory plasma.   See Abstr. 062.079.

The Alfvén-Carlquist double-layer theory of solar flares.   See Abstr. 062.113.

Short period acoustic heating theory and its application to the construction of model chromospheres. See Abstr. 062.117.

Diagnostic of stellar chromospheres. See Abstr. 064.107.

Observed and theoretical profiles of the Si II lines at λ1814.   See Abstr. 071.020.

On the profiles of weak absorption lines under flares.   See Abstr. 071.026.

On observations of solar emission of the transition zone from photosphere to chromosphere using spectra in the Hα region.   See Abstr. 071.040.

The height dependence of solar velocity fluctuations.   See Abstr. 071.046.

Modeling the solar photosphere and chromosphere. See Abstr. 071.053.

Activité solaire.   See Abstr. 072.001.

The role of spiral spots in solar activity in October 1972.   See Abstr. 072.003.

Ultraviolet spectra of solar active regions from Salyut 4.   See Abstr. 072.011.

Emergence of new satellite groups as nonstationary and prognostic factor in an active aggregate on the sun. See Abstr. 072.018.

Contrast of spots and flares on the sun. See Abstr. 072.019.

On the mutual location of sunspots and flares on the sun in 1970.   See Abstr. 072.021.

Relation of flare activity to the approach and separation of sunspots in an active region and to its magnetic properties.   See Abstr. 072.029.

The Evershed flow in the transition region chromosphere-photosphere.   See Abstr. 072.036.

Some studies on isolated sunspot magnetic configurations and associated Hα-flares and microwave bursts. See Abstr. 072.042.

On the generation of coronal and interplanetary shock waves during solar flares of 2 - 13 VIII, 1972. See Abstr. 074.036.

Polar coronal holes and the variation with latitude of the height of the Hα chromosphere. See Abstr. 074.079.

L'activité solaire.   See Abstr. 075.015.

Time-varying oscillations in the solar soft X-ray flux as observed from Skylab.   See Abstr. 076.008.

Analysis of X-ray observations of the 15 June 1973 flare in active region NOAA 131.   See Abstr. 076.009.

Generation of gamma radiation and neutrons in solar flares.   See Abstr. 076.010.

Analysis of the profiles of the solar S I lines at 1807 and 1900 Å.   See Abstr. 076.014.

Hard X-ray flares from the sun. See Abstr. 076.017.

The inter-relationship of hard X-ray and EUV bursts during solar flares.   See Abstr. 076.019.

Impulsive extreme-ultraviolet and hard X-ray emission during solar flares.   See Abstr. 076.021.

Solar X-ray flares in July – December 1972. See Abstr. 076.022.

On the anisotropy of X-ray emission of solar flares. See Abstr. 076.023.

Gamma-ray bursts during solar flares on 2, 4, and 7 August, 1972 according to observations on the Prognoz 2 cosmic station.   See Abstr. 076.025.

Impulsive solar X-ray bursts.   See Abstr. 076.029.

Active region observations of the Mg I and II resonance lines.   See Abstr. 076.036.

Solar flare X-ray line studies. See Abstr. 076.037.

Hard X-ray time profiles and acceleration processes in large solar flares.   See Abstr. 076.040.

Pre-onset visible, EUV, and X-ray flare structures. See Abstr. 076.043.

Observations of a moving X-ray source associated with an eruptive prominence at the limb. See Abstr. 076.045.

Tecniche osservative e discussione critica dell'emissione XUV.   See Abstr. 076.048.

Manifestation of pulsation instability in solar radio emission preceding proton flares.   See Abstr. 077.011.

A non-homogeneous model of the chromosphere and of the lower corona for interpretation of data on the quiet sun's radio emission in the millimeter wavelength range. See Abstr. 077.030.

Radio data and theoretical models for shock waves generated by intense solar flares.   See Abstr. 077.037.

Emissioni radio solari associate ai brillamenti cromosferici.   See Abstr. 077.045.

Investigation of propagation of solar cosmic radiation from eastern flares on May 28, June 15 and August 2, 1972 aboard Prognoz. I. Experimental results of measurements of solar particles in flares and analysis of integral proton spectra.   See Abstr. 078.002.

Super-adiabatic acceleration of charged particles and flares on the sun and stars.   See Abstr. 078.005.

Direct measurements of proton influx from the solar flare of August 7, 1972 on the Prognoz 2 station. See Abstr. 078.008.

Registration of plasma from solar flares of August 1972 in the earth's orbit.   See Abstr. 078.009.

On the peculiarity of emergence of solar flare accelerated electrons into the interplanetary medium. See Abstr. 078.011.

The composition of corotating energetic particle streams.   See Abstr. 078.015.

Raggi cosmici durante i brillamenti solari. See Abstr. 078.017.

Study of coronal and interplanetary propagation of solar particles following the E45° solar flare on July 29, 1973. See Abstr. 078.019.

Bifurcation of force-free solar magnetic fields: a numerical approach.   See Abstr. 080.014.

Response of the solar atmosphere to infalling surge material.   See Abstr. 080.056.

Numerical modelling of propagation of shock waves in the solar atmosphere and in the solar wind. See Abstr. 080.065.

Shock waves on Earth and in space. See Abstr. 082.059.

VLF sudden phase anomalies during flares of August 1972.   See Abstr. 083.135.

Solar flare effects on Earth's circumpolar currents. See Abstr. 084.001.

Pulsations of the earth's magnetic field before solar proton flares.   See Abstr. 084.291.

Effect of solar flares on lower tropospheric temperature and pressure.   See Abstr. 085.019.

Possible correlation of Io's posteclipse brightening with major solar flares.   See Abstr. 099.501.

Correlations between the typical hydromagnetic structure of interplanetary streams of a strong flare and variations of energetic particles.   See Abstr. 106.039.

The interplanetary scattering mean free path from 1 to 3×10^3 MV.   See Abstr. 106.040.

On the observation of a flare-generated shock wave at 9.7 AU by Pioneer 10.   See Abstr. 106.041.

Onde d'urto interplanetarie generate da brillamenti solari.   See Abstr. 106.068.

074 Corona, Solar Wind

074.001 Physical parameters defining the changing structure of a coronal hole.
J. A. Vorpahl, R. M. Broussard.
Astrophys. J., Vol. 219, 300 - 303 (1978).
The authors have used the S-056 X-ray data to determine significant quantitative values for physical parameters defining the changing structure of a coronal hole. The study suggests that changes in a coronal hole may be explained by the loss of material along weakened, less-confining magnetic field lines in the corona.

074.002 Magnetohydrodynamics of atmospheric transients. II. Two-dimensional numerical results for a model solar corona.   S. T. Wu, M. Dryer, Y. Nakagawa, S. M. Han.
Astrophys. J., Vol. 219, 324 - 335 (1978).
A systematic study of dynamic response of the inner solar corona is made within the context of two-dimensional, time-dependent plane hydromagnetics. Effects of the magnetic field configuration are illustrated for initially open (radial) and closed (azimuthal) magnetic fields by comparison with the nonmagnetic response. The channeling or blocking effects by the magnetic fields on the mass motion of solar plasma as the consequence of the evolution of fast and slow mode MHD shock waves are demonstrated. Some physically significant applications of the results, useful for the interpretation of observations, are discussed.

074.003 Polar enhancements of interplanetary Lα through solar wind asymmetries.
P. A. Isenberg, E. H. Levy.
Astrophys. J., Lett., Vol. 219, L59 - L62 (1978).
A solar wind velocity which increases toward the solar poles, or a solar wind flux which decreases toward the poles, or both, can produce polar enhancements in observed, resonantly scattered Lα from the interstellar hydrogen gas. Recently Broadfoot and Kumar have reported observations of polar enhancements in Lα. Through the use of a simple model, the authors estimate the magnitude of the latitudinal variations in the solar wind which are necessary to account for such enhancements.

074.004 Radial dependence of solar wind properties deduced from Helios 1/2 and Pioneer 10/11 radio scattering observations.   R. Woo.
Astrophys. J., Vol. 219, 727 - 739, with a correction in Vol. 223, 704 - 705 (1978).
The author presents the results of radio scattering measurements conducted at 2.3 GHz over an extensive heliocentric distance range $(1.7 - 180\,R_\odot)$ of the solar wind. Solar wind velocity of 24 km s^{-1} at $1.7\,R_\odot$ is obtained. Phase or Doppler scintillations, which are shown to be proportional to $\sigma_{ne}\,v^{5/6}$ (v is the solar wind velocity), have been measured out to 180 $R_\odot$. Beyond $10\,R_\odot$ the radial dependence of the phase scintillations is roughly $R^{-1.3}$, and, within the assumptions that $\sigma_{ne} \propto n_e$ and $v^{5/6} \sim v$, suggest that the solar wind is slightly converging in the equatorial region between approximately 20 and $180\,R_\odot$.

074.005 Magnetic fields in a coronal condensation.
T.-j. Cao.
Acta Astron. Sinica, Vol. 18, 158 - 165 (1977). In Chinese.
Using the potential field model the author has calculated the magnetic structure of the coronal condensation which appeared on the east limb of the solar disk at the 22 September 1968 eclipse. The comparison between calculated magnetic field geometry and observed contour of the coronal condensation shows, in general, an agreement. From this analysis, one can see that the magnetic field extending to the lower

coronal region is approximately the potential field during the steady period of an active region.

074.006 The triggering of plasma turbulence during fast flux emergence in the solar corona.
J. Heyvaerts, M. Kuperus.
Astron. Astrophys., Vol. 64, 219 - 234 (1978).
During flux emergence from the solar photosphere into the corona, current sheets are likely to be formed in the regions of contact between old and new flux. The processes of importance for the physics of these regions are discussed. Particular attention is paid to the case of fast emergence, in which no reconnection can take place at first. It is shown that in the one-dimensional phase of the development of the sheet, it is almost impossible to trigger microturbulence. The two-dimensional phase of the flow induced by this compression is then studied analytically and it is shown that this flow evacuates the region of the sheet where the pinching is strongest. The behavior of the current sheet after turbulence sets in is crudely discussed.

074.007 Dynamics of the quiescent solar corona.
R. Rosner, W. H. Tucker, G. S. Vaiana.
Astrophys. J., Vol. 220, 643 - 665 (1978).
A model for the quiescent, inhomogeneous solar corona is developed, based upon the concept of loop structures as the basic structural element of the corona. The results show that (1) hydrostatic solutions are stable only if the temperature maximum is located at the top of loop structures, and the deposition scale length of the coronal heating mechanism is comparable with (or larger than) the loop scale size; (2) the loop temperature ($\sim T_{max}$), pressure p, and size L are related by the expression $T_{max} \sim 1.4 \times 10^3 (pL)^{1/3}$, a relation which contains no free parameters; (3) coronal heating models based upon the coronal magnetic field (e.g., Alfvén mode dissipation and coronal current heating) are consistent with the loop model, while acoustic mode damping is not.

074.008 Expansion and broadening of coronal loop transients: a theoretical explanation.
T. C. Mouschovias, A. I. Poland.
Astrophys. J., Vol. 220, 675 - 682 (1978).
The authors explore the consequences of the assumption that a coronal loop transient (observed by the white-light coronagraph aboard Skylab) is a twisted rope of magnetic field lines expanding and broadening in the background coronal plasma and magnetic field. They show that the expansion (i.e., the outward motion of the loop top) can be accounted for by the azimuthal component of the field, B_{az}; the observed broadening of the loop as it moves outward can be accounted for by the longitudinal component of the field, B_l. The two components of the field must satisfy the inequality $1.41\,B_l > B_{az} > B_l$. The authors predict that, as the loop rises, the width of its top portion should vary proportionally with the distance from the Sun's center.

074.009 On the evolution of strong irregularities in the solar wind plasma.   I. S. Veselovskij.
Geomagn. Aehron., Vol. 18, 3 - 8 (1978). In Russian.

074.010 A model of interaction of the solar wind with the magnetosphere. Statement of the problem.
N. V. Erkaev, V. G. Pivovarov.
Geomagn. Aehron., Vol. 18, 106 - 112 (1978). In Russian.

074.011 Stable solar wind and geomagnetic activity.
A. Prigancová.
Geomagn. Aehron., Vol. 18, 113 - 116 (1978). In Russian.

074.012 **The decay of coronal loops brightened by flares and transients.**   A. S. Krieger.
Sol. Phys., Vol. 56, 107 - 120 (1978).

The sizes and shapes of X-ray emitting loops brightened by flares and other coronal transients have been derived from the Skylab S-054 photographs. This information has been combined with estimates of temperature and emission measure derived from the photographs and from Solrad data to compute brightness decay times attributable to various coronal energy loss mechanisms. The computed decay.times are compared to those actually observed. Examples are presented of the brightness decay of soft X-ray flare kernels, post-flare loops, and the coronal X-ray enhancement associated with an $H\alpha$ filament disappearance.

074.013 **On the temperature distribution in the solar corona.**   H. P. Mital, U. Narain.
Sol. Phys., Vol. 56, 121 - 123 (1978).

An improved value of coronal temperature is obtained by the degree of ionization method taking various processes into consideration. Comparison with some of the existing results has also been made.

074.014 **Extreme ultraviolet observations of coronal holes. II. Association of holes with solar magnetic fields and a model for their formation during the solar cycle.**
J. D. Bohlin, N. R. Sheeley, Jr.
Sol. Phys., Vol. 56, 125 - 151 (1978).

Extreme-ultraviolet Skylab and ground-based solar magnetic field data have been combined to study the origin and evolution of coronal holes. It is shown that holes exist only within the large-scale unipolar magnetic cells into which the solar surface is divided at any given time. A well-defined boundary zone usually exists between the edge of a hole and the neutral line which marks the edge of its magnetic cell. Three pieces of observational evidence are offered to support the hypothesis that the magnetic lines of force from a hole are open. Kitt Peak magnetograms are used to show that, at least on a relative scale, the average field strengths within holes are quite variable, but indistinguishable from the field strengths in other quiet parts of the Sun's surface. Finally it is shown that the large, equatorial holes characteristic of the declining phase of the last solar cycle during Skylab (1973−74) were all formed as a result of the mergence of bipolar magnetic regions, confirming an earlier hypothesis by Timothy et al. (1975). Systematic application of this model to the different aspects of the solar cycle correctly predicts the occurrence of both large, equatorial coronal holes and the polar cap holes.

074.015 **Coronal hole evolution by sudden large scale changes.**   J. T. Nolte, M. Gerassimenko, A. S. Krieger, C. V. Solodyna.
Sol. Phys., Vol. 56, 153 - 159 (1978).

The authors have compared sudden shifts in coronal hole boundaries observed by the S-054 X-ray telescope on Skylab between May and November, 1973, within 1 day of CMP of the holes, at latitudes $\leq 40°$, with the long-term evolution of coronal hole area. They find that large-scale shifts in boundary locations can account for most if not all of the evolution of coronal holes. The temporal and spatial scales of these large-scale changes imply that they are the results of a physical process occurring in the corona. The authors conclude that coronal holes evolve by magnetic field lines opening when the holes are growing, and by fields closing as the holes shrink.

074.016 **A survey of coronal holes and their solar wind associations throughout sunspot cycle 20.**
R. M. Broussard, N. R. Sheeley, Jr., R. Tousey, J. H. Underwood.
Sol. Phys., Vol. 56, 161 - 183 (1978).

To gain insight into the relationships between solar activity, the occurrence and variability of coronal holes, and the association of such holes with solar wind features such as high-velocity streams, a study of the period 1963−1974 was made. This period corresponds approximately with sunspot cycle 20. The primary data used for this work consisted of X-ray and XUV solar images obtained from rockets. The investigation revealed that: (1) The polar coronal holes prominent at solar minimum, decreased in area as solar activity increased and were small or absent at maximum phase. (2) During maximum, coronal holes occurred poleward of the sunspot belts and in the equatorial region between them. The observed equatorial holes were small and persisted for one or two solar rotations only; some high latitude holes had lifetimes exceeding two solar rotations. (3) During 1963−74 whenever XUV or X-ray images were available, nearly all recurrent solar wind streams of speed > 500 km s^{-1} were found associated with coronal holes at less than 40° latitude.

074.017 **The quiet coronal X-ray spectrum of highly ionized oxygen and nitrogen.**   D. L. McKenzie, H. R. Rugge, J. H. Underwood, R. M. Young.
Astrophys. J., Vol. 221, 342 - 349 (1978).

An X-ray emission-line spectrum of a region of the quiet solar corona in the wavelength range 17 to 25 Å was obtained on 1976 May 11. A differential emission measure function describing the observed corona is derived from the spectral data, and the uncertainties in this function are analyzed by using a new technique.

074.018 **Expansion of the solar wind in high-speed streams.**   Y. C. Whang, T. H. Chien.
Astrophys. J., Vol. 221, 350 - 361 (1978).

The purpose of this paper is to show that the expansion of the solar wind in streamtubes with rapid divergence in their supersonic section can accelerate the wind to a speed of 600 km s^{-1} or more at 1 AU. The authors also study acceleration of the solar wind due to Alfvén waves which propagate according to the WKB solution, whose effect is one order of magnitude less than the effect of rapid divergence in the supersonic section of streamtubes.

074.019 **Ion acoustic waves in the solar wind.**   D. A. Gurnett, L. A. Frank.
J. Geophys. Res., Vol. 83, 58 - 74 (1978).

Plasma wave measurements on the Helios 1 and 2 spacecraft have revealed the occurrence of electric field turbulence in the solar wind at frequencies between the electron and ion plasma frequencies. Wavelength measurements with the Imp 6 spacecraft now provide strong evidence that these waves are short-wavelength ion acoustic waves which are Doppler-shifted upward in frequency by the motion of the solar wind. Comparison of the Helios results with measurements from the earth-orbiting Imp 6 and 8 spacecraft shows that the ion acoustic wave turbulence detected in interplanetary space has characteristics essentially identical to those of bursts of electrostatic turbulence generated by protons streaming into the solar wind from the earth's bow shock.

074.020 **Acceleration of solar wind He^{++}. 3. Effects of resonant and nonresonant interactions with transverse waves.**   J. V. Hollweg, J. M. Turner.
J. Geophys. Res., Vol. 83, 97 - 113 (1978).

In two previous papers (Hollweg, 1974 and Chang and Hollweg, 1976) a fundamentally new physical process which may be relevant to solar wind heavy ion dynamics was described, and some features were developed. In the present paper, resonant wave-particle interactions, which were not considered in papers 1 and 2, will be examined. The basic physical effect to be considered is the acceleration of the α particle flow by resonant pitch angle scattering. It will be found that including the resonant acceleration in a simple model for the α particle flow can lead to improved agreement between theory and observation.

**074.021 The cooling of solar wind protons from Mariner 2
           data.   M. Eyni, R. Steinitz.**
J. Geophys. Res., Vol. 83, 215 - 216 (1978).

The authors examine the possibility that cooling of solar
wind protons has escaped detection in the Mariner 2 data. After
dividing the data into two velocity ranges and introducing an
intermediate quantity r to decrease the strong dependence of
temperature on velocity, the authors find that the low-velocity
data exhibit cooling.

**074.022 Coronal mass increases prior to surface Hα eruptions
           from the sun.   B. V. Jackson.**
Bull. American Astron. Soc., Vol. 9, 569 (1978). − Abstract.

**074.023 The shape of network elements in the solar
           transition zone.   E. J. Eadon, D. E. Billings.**
Bull. American Astron. Soc., Vol. 9, 569 (1978). − Abstract.

**074.024 Coronal hole observations at 1420 and 2600 MHz
           with the Arecibo radiotelescope.**
M. D. Papagiannis, F. L. Wefer.
Bull. American Astron. Soc., Vol. 9, 617 (1978). − Abstract.

**074.025 Observations of enhanced radio emission at 15 GHz
           from a coronal hole region.**
F. L. Wefer, M. D. Papagiannis.
Bull. American Astron. Soc., Vol. 9, 617 - 618 (1978).
Abstract.

**074.026 Least solar distance of interplanetary dust and the
           origin of the solar wind.   D. B. Beard, J. Reimer.**
Bull. American Astron. Soc., Vol. 9, 618 (1978). − Abstract.

**074.027 Simulation of the solar wind interaction with non-
           magnetic celestial bodies.**
I. M. Podgorny (*Podgornyj*), Yu. V. Andrijanov (*Andriyanov*).
Planet. Space Sci., Vol.26, 99 - 109 (1978).

Laboratory experiments are described to simulate the
solar wind flowing around non-magnetic planets for three
cases: non-conducting and ideally conducting planets, and a
planet with a gaseous shell. A glass sphere was used as a model
of a non-conducting planet (the Moon). Spatial distributions of
plasma density and magnetic field strength that have been obtained agree with the data from measurements in space.

**074.028 The modern and ancient flux of solar wind particles,
           solar flare particles and micrometeoroids.**
G. Poupeau, R. S. Rajan, R. M. Walker, E. Zinner.
Space Research, Vol. XVII, (see 012.010), 599 - 604 (1977).

Measurements of microcraters, tracks and implanted ions
allow us to determine relative fluxes of micrometeoroids, solar
flare and solar wind particles. Lunar rocks span the period up
to 10^6 years. Lunar drill cores provide a record of up to 10^9
years. However, for a direct comparison of the solar activity
and the micrometeoroid flux ion probe solar wind measurements are required. Gas-rich meteorites take us even further
into the past ($>4 \times 10^9$ years) as well as to different solar
distances. Preliminary results show that also in these objects
microcraters are present although they occur in only a small
fraction of solar flare irradiated grains. Quantitative analysis
of the microcrater-solar flare track relationship is in progress.

**074.029 Dynamic MHD modeling of the solar wind disturb-
           ances during the August 1972 events.**
M. Dryer, C. Candelaria, Z. K. Smith, R. S. Steinolfson, E. J.
Smith, J. H. Wolfe, J. D. Mihalov, P. Rosenau.
J. Geophys. Res., Vol. 83, 532 - 540 (1978).

A time-dependent one-dimensional MHD theoretical
model (Steinolfson et al., 1975) is tested by using plasma and
magnetic field observations of Pioneer 9 and Pioneer 10
during the August 1972 events on the sun and in the interplanetary medium. These spacecraft were nearly aligned along a

common heliocentric radius during these events, considered
now to be the most spectacular and best-documented events
during solar cycle 20. The observations of Pioneer 9 at 0.78
AU were used as 'input' for the theoretical model. The plasma
and magnetic field forcing functions were superimposed upon
a preexisting ambient solar wind at this inner boundary, and
the response was simulated as far as 8 AU. The simulated
'output' at 2.2 AU is compared directly with the Pioneer 10
observations at 2.2 AU.

**074.030 Additional evidence consistent with solar cycle
           variations in the solar wind.   D. S. Intriligator.**
Astrophys. J., Vol. 221, 1009 - 1013 (1978).

The activity of the solar wind varies in phase with sunspot activity from 1964 through 1972. The purpose of this
paper is to confirm and extend the first report of this result
in Intriligator (1974), to point out that other researchers'
results confirm this observation (although the authors did not
indicate this), and to suggest that the rise in stream activity in
1973 implies a shift in the primary source for streams near
solar minimum.

**074.031 The structure of the X-ray bright corona above ac-
           tive region McMath 12628 and derived implications
           for the description of equilibria in the solar atmosphere.**
J. P. Pye, K. D. Evans, R. J. Hutcheon, M. Gerassimenko, J. M.
Davis, A. S. Krieger, J. F. Vesecky.
Astron. Astrophys., Vol. 65, 123 - 138 (1978).

X-ray filtergrams, obtained with the SKYLAB/ATM
S-054 telescope, and X-ray line spectra, simultaneously obtained with a rocket-borne spectrometer, have been jointly
used in an effort to understand the structure of the corona
above the active region McMath 12628. Temperature and density maps of the region's high temperature loop system are
given, discussed and used to reduce a simplified model of the
region which is suitable for further calculation of the energetics
of the loop system.

**074.032 Heating of coronal plasma by anomalous current
           dissipation.**
R. Rosner, L. Golub, B. Coppi, G. S. Vaiana.
Astrophys. J., Vol. 222, 317 - 332 (1978).

The authors show that the observed high temperature and
inhomogeneous structure of the solar corona, as well as the
long-term spatial and temporal evolution of coronal features,
is economically explained by in situ heating of the coronal
plasma via anomalous current dissipation. The basic geometrical structure is a loop configuration heated by nearly field-
aligned currents occupying a small fraction of the total loop
volume.

**074.033 On the collisional theory of the anisotropic solar
           wind plasma.   D. Summers.**
Sol. Phys., Vol. 56, 429 - 438 (1978).

The collisional equations for the solar wind assuming
steady, spherically symmetric flow and including thermal
conduction and an anisotropic proton temperature are analysed in the absence of a heat source function. It is found that
the model based purely on Coulomb collisions predicts values
for the proton temperature anisotropy in the vicinity of the
Earth that are much smaller than those observed, and that increasing the coronal base temperature serves to decrease the
predicted anisotropy still further.

**074.034 Topological structure of coronal-interplanetary
           magnetic field.   T. Yeh.**
Sol. Phys., Vol. 56, 439 - 447 (1978).

The topological structure of the coronal-interplanetary
magnetic field is determined by the arrangement of the neutral
lines at the coronal base. It is characterized by nested separatrices, which are interfaces between closed and open field lines
or between oppositely directed open field lines, in the coronal-

interplanetary space. In the neighborhoods of these separatrices there are important electric currents. These currents form the basis for the sheet-current model.

074.035 Electron densities, temperatures and nonthermal velocities in the corona at different heliographic latitudes.   V. E. Stepanova, N. F. Tyagun.
Issled. po geomagn., aehron. i fiz. Solntsa. Vyp. (No.) 42. Moskva, Nauka, 1977, p. 60 - 72. In Russian. – Abstr. in Ref. zh., 51. Astron., 2.51.367 (1978).

074.036 On the generation of coronal and interplanetary shock waves during solar flares of 2 - 13 VIII, 1972.
S. Pinter, M. Dryer.
Izv. AN SSSR. Ser. fiz., Vol. 41, 1849 - 1860 (1977). In Russian. – Abstr. in Ref. zh., 62. Issled. kosm. prostranstva, 2.62.135 (1978).

074.037 Asymmetry of the region of interaction between the solar wind and Venus from data of the
automatic interplanetary stations Venera 9 and Venera 10.
S. A. Romanov.
Kosm. Issled., Vol. 16, 318 - 319 (1978). In Russian.

074.038 Solar wind stream interfaces.   J. T. Gosling, J. R. Asbridge, S. J. Bame, W. C. Feldman.
J. Geophys. Res., Vol. 83, 1401 - 1412 (1978).
Measurements with Los Alamos Scientific Laboratory instrumentation aboard Imp 6, 7 and 8 reveal that approximately one third of all high-speed solar wind streams observed at 1 AU contain a sharp boundary (of thickness less than $\sim 4 \times 10^4$ km) near their leading edge, called a stream interface, which separates plasma of distinctly different properties and origins. A superposed epoch analysis of plasma data has been performed for 23 discontinuous stream interfaces observed during the interval March 1971 through August 1974.

074.039 Interplanetary scintillation measurements of the electron density power spectrum in the solar wind.
W. A. Coles, J. K. Harmon.
J. Geophys. Res., Vol. 83, 1413 - 1420 (1978).
The spatial spectrum of electron density in the solar wind is estimated over the frequency range 2×10^{-3} to 3×10^{-2} km^{-1} by using 74 MHz observations of interplanetary scintillation. The inversion technique by which the spatial spectrum is estimated from the observed temporal spectrum requires weak scattering. The shape of the spectrum varies significantly from day to day, but no mean variation with solar distance has been observed.

074.040 Numerical simulation of MHD (*magnetohydrodynamic*) shock waves in the solar wind.
R. S. Steinolfson, M. Dryer.
J. Geophys. Res., Vol. 83, 1576 - 1582 (1978).
The effects of the interplanetary magnetic field on the propagation speed of shock waves through an ambient solar wind are examined by numerical solutions of the time-dependent nonlinear equations of motion. The numerical method is used to simulate (starting at 0.3 AU) the large deceleration of a shock observed in the lower corona by ground-based radio instrumentation and the more gradual deceleration of the shock in the solar wind observed by the Pioneer 9 and Pioneer 10 spacecraft.

074.041 A self-consistent solution of the three-fluid model for the solar wind.   N. Metzler, M. Dryer.
Astrophys. J., Vol. 222, 689 - 695 (1978).
The three-fluid model equations for the solar wind are solved simultaneously in a self-consistent manner. The model consists of the individual continuity, momentum, and energy equations for electrons, protons, and α-particles without considering the α-particle as a minor species. The importance and

the meaning of the "critical points" are discussed along with the proper manner of choosing them. A possible explanation for the observed streaming velocities of the α-particles being larger than the proton streaming velocities is given.

074.042 The emission of Langmuir waves in the solar corona.   R. D. Robinson.
Astrophys. J., Vol. 222, 696 - 706 (1978).
The author discusses the Langmuir wavenumber spectrum emitted from several representative electron distributions. For electrons having isotropic pitch angle distributions the spectrum produced is well defined and determined primarily by the electron energy distribution. For anisotropic pitch angle distributions induced emission may result, with consequent interaction between particles and waves. Both loss-cone and streaming distributions may cause Langmuir waves to be amplified. For the streaming distribution, this generation is primarily in the direction of streaming; for the loss-cone distribution, wave generation perpendicular to the field lines is favored.

074.043 Variation of solar wind parameters at the front of an anisotropic shock wave.
S. A. Grib, A. N. Solunin.
Solar wind and the magnetosphere, (see 003.006), p. 19 - 27 (1976). In Russian. – Abstr. in Ref. zh., 62. Issled. kosm. prostranstva, 3.62.353 (1978).

074.044 Direct observation of the latitudinal extent of a high-speed stream in the solar wind.
R. Schwenn, M. D. Montgomery, H. Rosenbauer, H. Miggenrieder, K. H. Mühlhäuser, S. J. Bame, W. C. Feldman, R. T. Hansen.
J. Geophys. Res., Vol. 83, 1011 - 1017 (1978).
Solar wind speeds measured from the Helios 1 solar probe between 0.31 and 1 AU and the earth-orbiting Imp 7/8 satellites have been correlated with coronal holes as determined from K coronal brightness measurements. During perihelion passage the space probe crossed the northern boundary of the high-speed stream associated with an equatorward extension of the south polar coronal hole. While this same stream continued to be observed by Imp satellites at −5° latitude, it was no longer observable from Helios 1 at +5° latitude. The conclusion is that sharp boundaries separate high-speed flows from the surrounding solar wind. The thickness of the boundary in latitude appears to be narrower than about 10°.

074.045 A steady three-fluid coronal expansion for nonspherical geometries.   J. Joselyn, T. E. Holzer.
J. Geophys. Res., Vol. 83, 1019 - 1026 (1978).
A steady three-fluid model of the solar coronal expansion, in which ^{4}He^{++} ions (alphas) are treated as a nonminor species, is developed for nonspherically symmetric flow geometries of the general sort thought to be characteristic of coronal holes. It is found that the very high mass fluxes in the low corona, which are associated with rapidly diverging flow geometries, lead to a locally enhanced frictional coupling between protons and alphas and consequently to a significant reduction of the He/H abundance ratio in the lower corona from that normally predicted by multifluid models. In the models considered, the frictional drag on the protons by the alphas (a process neglected in most studies) is found to play an important role near the sun.

074.046 Plasma fluctuations in the solar wind.
M. Neugebauer, C. S. Wu, J. D. Huba.
J. Geophys. Res., Vol. 83, 1027 - 1034 (1978).
Ogo 5 plasma and magnetic field data are used to compute power spectra of solar wind fluctuations over the frequency interval $10^{-3}-10^{-1}$ Hz. The authors confirm the validity of the assumption made in earlier papers that the power spectra calculated from total flux measurements are

approximately equal to the power spectra of density fluctuations times the square of the average solar wind speed. All cases studied show evidence of the presence of Alfvén waves in this frequency range.

074.047   Coronal holes at 11.5 and 21 cm observed with the Arecibo radio telescope.
M. D. Papagiannis, F. L. Wefer.
Nature, Vol. 273, 520 - 522 (1978).

The observations reported here made at 1,420 and 2,600 MHz with the 1,000 ft Arecibo radio telescope show the presence of coronal holes very clearly and provide for the first time an excellent matching of coronal hole images obtained from decimetre solar radio maps and 10,830 Å spectroheliograms.

074.048   Colour of the solar corona on March 7, 1970.
A. K. Ajmanov.
Soln. Dannye 1977 Byull., No. 12, p. 83 - 87 (1978).
In Russian.

The solar corona of March 7, 1970 has been investigated. Three-colour photometry shows colour excess of coronal radiation in red and green regions of the spectrum at distances near (3–4) $R_\odot$ from the sun's centre.

074.049   Pulsations in the solar wind and on the ground.
T. A. Plyasova-Bakounina (*Plyasova-Bakunina*),
Yu. V. Golikov, V. A. Troitskaya, P. C. Hedgecock.
Planet. Space Sci., Vol. 26, 547 - 553 (1978).

Measurements of the pulsation activity recorded by the HEOS-1 satellite in the solar wind upstream from the Earth's bow shock are compared with records of Pc 3–4 activity at the Soviet Borok Observatory. The authors obtained eight events with closely similar periods at the satellite and at Borok, while another event showed similar onset times but had rather different dominant periodicities at the two locations.

074.050   A search for the origin of very low electron temperatures.
A. Geranios.
Planet. Space Sci., Vol. 26, 571 - 579 (1978).

VLET's (Very Low Electron Temperature) are regions in the solar wind (lasting 12–30 h) in which the electron temperature is abnormally low. A statistical analysis of many of these events is made with respect to the dependence of this phenomenon on interplanetary plasma and field parameters.

074.051   An observational search for large-scale organization of five-minute oscillations on the sun.
P. H. Dittmer, P. H. Scherrer, J. M. Wilcox.
Sol. Phys., Vol. 57, 3 - 11 (1978).

The large-scale solar velocity field has been measured over an aperture of radius 0.8 $R_\odot$ on 121 days between April and September, 1976. Measurements are made in the line Fe I 5123.730 Å, employing a velocity subtraction technique similar to that of Severny et al. (1976). Comparisons of the amplitude and frequency of the five-minute resonant oscillation with the geomagnetic C9 index and magnetic sector boundaries show no evidence of any relationship between the oscillations and coronal holes or sector structure.

074.052   Observational evidence of continual heating in X-ray emitting coronal loops.
M. Gerassimenko, C. V. Solodyna, J. T. Nolte.
Sol. Phys., Vol. 57, 103 - 110 (1978).

A 90 s time resolution study of the soft X-ray emission from three active region loops shows the emission to be constant to about two percent over the half hour period of observation. Soft X-ray observations in two wavebands are used to deduce the temperature and density of these loops. The data unambiguously demonstrate that energy is supplied to each loop during the observations. If heating is due to discrete events, the time interval between events is shown to be less

than 10 min, which is short relative to the radiative cooling time of the loops.

074.053   Can coronal loop transients be driven magnetically?
U. Anzer.
Sol. Phys., Vol. 57, 111 - 118 (1978).

The author investigates the possibility of magnetic driving of loop transients. The action of local magnetic forces to balance gravity in a coasting loop and to confine the loop has been proposed by Mouschovias and Poland (1977). In this paper the author uses similar configurations but deals with the global field structure and presents models which show both the initial phase of large acceleration and the later phase of almost constant velocity. He uses very simple one-dimensional models consisting of a ring current which is subjected to gravitational attraction. The velocity curves calculated for these models are in good agreement with the observations.

074.054   λ5303 Fe XIV density models of the inner solar corona.   R. R. Fisher.
Sol. Phys., Vol. 57, 119 - 128 (1978).

The Sacramento Peak Observatory's 40 cm coronagraph was used with an emission line photometer to observe the distribution of λ5303 Fe XIV brightness as a function of position angle, height above the limb, and time. These data were used to construct models of the volume emissivity as a function of solar latitude and longitude. These models in turn yield estimates of the distribution of electron density in the lower solar corona as a function of latitude and longitude for several specific periods in 1973 and 1975.

074.055   Short term evolution of coronal hole boundaries.
J. T. Nolte, A. S. Krieger, C. V. Solodyna.
Sol. Phys., Vol. 57, 129 - 139 (1978).

In an examination of the evolution of coronal hole boundaries on a time scale of ~1 day, the authors find that 38% of all the boundaries of coronal holes observed near central meridian passage during the Skylab period shifted in location by >1° heliocentric in ~1 day. Of these boundary changes, 70% were on a scale <3 times the average supergranulation cell size. However, large-scale shifts in the boundary locations also occurred, which involved changes in the X-ray emission from these areas of the Sun. X-ray emitting structures on the borders of isolated and evolving holes were less clearly defined than those on the boundaries of well-established, elongated holes. There were generally more changes in the boundaries of the most rapidly evolving holes, but no simple relationship between the amount of change and the rate of hole growth or decay.

074.056   Motions and mass changes of a persistent coronal streamer.   A. I. Poland.
Sol. Phys., Vol. 57, 141 - 153 (1978).

A coronal streamer was observed by the white light coronagraph on Skylab during 5 successive limb passages between 1 June, 1973 and 6 August, 1973. The Skylab data give independent measures of coronal brightness and polarization, as functions of time. These permit the distinction between changes in the coronal streamer's appearance due to solar rotation and actual structural changes. The streamer's visual appearance changed slightly between successive limb passages indicating that it was not a steady state feature. Comparison of the outer coronal observations with observations from lower in the solar atmosphere indicate that the streamer was associated with a complex of solar activity consisting of active regions and filaments. This complex of activity shifted southward by the same amount as the streamer.

074.057   The association of nonthermal electrons with non-flaring coronal transients.
D. F. Webb, M. R. Kundu.
Sol. Phys., Vol. 57, 155 - 173 (1978).

A close temporal and spatial association has been found between erupting filaments/coronal transients and radio noise storm continua. The three transients studied occurred away from active regions and are members of a class not usually accompanied by chromospheric emission. Calculations confirmed that observed microwave radiation from the transients is due to thermal bremsstrahlung. The results are consistent with an interpretation of heating of an increased amount of coronal plasma by nonthermal, 10–100 keV electrons. Three possibilities for the source of the material are described.

074.058 **Height gradient of the intensity of the Fe X**
λ6374 Å line.   A. S. Zubtsov.
Soln. akt. Alma-Ata, 1977, p. 89 - 94. In Russian. – Abstr. in Ref. zh., 51. Astron., 4.51.367 (1978).

074.059 **Evolution of the large-scale structure of the solar**
wind at large heliocentric distances.
V. N. Malyshkin.
Issled. po geomagn., aehron. i fiz. Solntsa. Vyp. (No.) 42. Moskva, Nauka, 1977, p. 92 - 98. In Russian. – Abstr. in Ref. zh., 51. Astron., 4.51.421 (1978).

074.060 **Characteristics of the proton and α-components of**
the solar wind after the passage of interplanetary
shock waves from measurements aboard the Prognoz station.
A. A. Zertsalov, O. L. Vajsberg, G. N. Zastenker, V. V. Temnyj, M. Z. Khokhlov.
Problems of solar activity and space system Prognoz, (see 003.008), p. 179 - 184 (1977). In Russian. – Abstr. in Ref. zh., 62. Issled. kosm. prostranstva, 4.62.400 (1978).

074.061 **A hypothesis of the repulsion on the space charged**
particles by light and the origin of the solar wind.
T.-w. Wei.
Kexue Tongbao, Vol. 22, No. 7, p. 291 - 299 (1977). In Chinese. – From Phys. Abstr., Vol. 81, Abstr. 16388 (1978).

074.062 **On the possiblity of measuring the solar wind**
velocity from observations of scintillations of H_2O
and OH maser sources.   R. L. Sorochenko, V. I. Shishov.
Pis'ma Astron. Zh., Vol. 4, 140 - 142 (1978). In Russian.
A method is proposed to measure the solar wind velocity near the sun from measurements of the relative shift of diffraction patterns of separate components of H_2O and OH maser sources.

074.063 **Helios-1 Faraday rotation experiment: results and**
interpretations of the solar occultations in 1975.
H. Volland, M. K. Bird, G. S. Levy, C. T. Stelzried, B. L. Seidel.
J. Geophys., Vol. 42, 659 - 672 (1977). – Abstr. in Phys. Abstr., Vol. 81, Abstr. 20790 (1978).

074.064 **Time delay occultation data of the Helios space-**
craft and preliminary analysis for probing the solar
corona.   P. Edenhofer, P. B. Esposito, R. T. Hansen, S. F. Hansen, E. Luneburg, W. L. Martin, A. I. Zygielbaum.
J. Geophys., Vol. 42, 673 - 698 (1977). – Abstr. in Phys. Abstr., Vol. 81, Abstr. 20791 (1978).

074.065 **Limb-brightening observations from the OSO-7**
satellite. III. Comparison of EUV line intensities
of Fe XII, Fe XI, Fe XV, Si X and S XII, Si IX and S XI with predictions.   S. O. Kastner, H. E. Mason.
Astron. Astrophys., Vol. 67, 119 - 127 (1978).
Continuing a study of heliocentric dependence of EUV emission line intensities observed by the Goddard OSO-7 spectroheliograph in 1972, the variation of lines of the ions Fe XII, Fe XI, Fe XV, Si X and S XII, Si IX and S XI is compared with the results of individual calculations for these ions, including theoretical intensities presented here for Fe XII and

Fe XI. Agreement is found to be good for Fe XII and reasonable for some of the lines of the other ions which in general are weaker in intensity. Several apparent anomalies are found however which may be due to unknown line components near the wavelengths observed.

074.066 **Occultation of polarized radio emission by the**
solar corona in the period of weak solar activity.
A. B. Berlin, D. V. Korol'kov, Yu. N. Parijskij, N. S. Soboleva, G. M. Timofeeva.
Pis'ma Astron. Zh., Vol. 4, 191 - 192 (1978). In Russian.
An attempt was made to find the small-scale structure of the magnetic field of plasma near the sun from variation of the position angle of the electric vector of radio emission of the Crab nebula during its occultation by the solar corona. The change of the position angle of the polarization vector doesn't exceed 0°.5 at 4 cm wavelength and corresponds to a sectorial structure of the interplanetary magnetic field.

074.067 **Long-term correlation between latitude-dependent**
solar activity and solar wind streams.
S. Cuperman, M. Dryer.
Astrophys. J., Vol. 223, 601 - 604 (1978).
The long-term correlation between the latitude-dependent 5303 Å coronal line intensity and the solar wind plasma streams measured at 1 AU heliocentric distance during the period 1962–1970 is investigated. It is found that (1) the intensity of the light radiation is anticorrelated with the high-speed solar wind streams, and (2) the magnitude of the anticorrelation coefficient increases with heliographic latitude to a value ~ 0.6 at about 30°, after which it changes only a little. The possible identification of coronal regions of very low green line emission with coronal holes and the resulting relationship between the latitude-dependent holes and high-speed solar wind flows are also discussed.

074.068 **MHD solitons in the solar wind.**
A. V. Gul'el'mi, N. M. Bondarenko.
Geomagn. Aehron., Vol. 18, 423 - 426 (1978). In Russian.

074.069 **Shock-associated energetic proton events at large**
heliocentric distances.
I. D. Palmer, J. T. Gosling.
J. Geophys. Res., Vol. 83, 2037 - 2046 (1978).
Enhancements of energetic protons ($\gtrsim 0.5$ MeV) in association with forward-reverse shock pairs have been observed at large heliocentric distances. An interpretation of the time profiles of these events is offered in terms of a model of solar wind stream structure.

074.070 **Long-term variations of selected solar wind proper-**
ties: Imp 6, 7, and 8 results.
W. C. Feldman, J. R. Asbridge, S. J. Bame, J. T. Gosling.
J. Geophys. Res., Vol. 83, 2177 - 2189 (1978).
Variations in solar wind ion characteristics observed between March 18, 1971, and January 6, 1977, are presented in order to study long-term trends in large-scale interplanetary structures. Past investigations of long-term solar wind variability are extended to include the minimum of solar cycle 20. In addition to the He abundance, relative He-H velocity difference, and He-H temperature ratio the authors consider the solar wind proton density, speed, rms velocity variation, temperature, and number flux, as well as the kinetic and total energy fluxes.

074.071 **Coronal magnetic fields.**   G. A. Dulk, D. J.
McLean.
Sol. Phys., Vol. 57, 279 - 295 (1978).
The observational evidence on the strength of the coronal magnetic field above active regions is reviewed. Recent advances in observations and plasma theory are used to determine which data are the more reliable and to revise some

earlier estimates of field strength. The results from the different techniques are found to be in general agreement, and the relation $B = 0.5((R/R_\odot)-1)^{-1.5}$ G, $1.02 \lesssim R/R_\odot \lesssim 10$ is consistent with all the data to within a factor of about 3.

074.072   Physical conditions in the corona for a bipolar magnetic region.   J. A. Vorpahl.
Sol. Phys., Vol. 57, 297 - 308 (1978).
    The author has used the S-056 X-ray data from Skylab to determine quantitative values for the coronal conditions characterizing a new bipolar magnetic region (BMR). In particular, he includes: (a) the time variation of the total soft X-ray flux from the BMR as a function of time; (b) the temporal and spatial variation of the temperature and emission measure; (c) the variation with time of thermal energy density; (d) the (calculated) magnetic field configuration and magnetic flux density in the corona; and (e) the temporal variation of the magnetic field energy in the corona. Detailed comparisons are made between the configuration of X-ray features and the magnetic field topology.

074.073   The forbidden transitions within $3s^2\,3p^5\,3d$ of Fe IX and Ni XI and $3s^2\,3p^4\,3d$ of Fe X and Ni XII.
B. Edlén, R. Smitt.
Sol. Phys., Vol. 57, 329 - 339 (1978).
    Improved level values for $3s^2\,3p^5\,3d$ and $3s^2\,3p^4\,3d$ of Fe IX and Fe X are derived from the observed forbidden lines with special regard to recent accurate measurements in the vacuum-ultraviolet part of the coronal spectrum. A procedure for estimating expected relative intensities is proposed and used for an additional check on the consistency of the identifications, which now comprise 12 lines of Fe IX and 11 lines of Fe X. Finally, by use of a suitable extrapolation technique the wavelengths of corresponding lines in nickel are predicted, which leads to some new identifications and a total of 6 identified lines of Ni XI and 4 of Ni XII.

074.074   Local instabilities of Alfvén waves in high speed streams.   B. Bavassano, M. Dobrowolny, G. Moreno.
Sol. Phys., Vol. 57, 445 - 465 (1978).
    A two fluid stability analysis of an inhomogeneous solar wind plasma leads to the prediction of possible instabilities of both Alfvénic and magnetoacoustic waves driven by local velocity gradients. The authors have performed a detailed study, based on Pioneer 6 magnetic and plasma data relative to several high speed streams in the solar wind, on the direction of propagation of the transverse waves which are found within the streams and on their association with velocity gradients within the stream structure. The analysis leads to the conclusion that the observed Alfvén waves may be consistent with the hypothesis of local generation through one of the above mentioned instabilities.

074.075   Ion cyclotron instability in the solar wind.
G. S. Lakhina.
Sol. Phys., Vol. 57, 467 - 473 (1978).
    An ion cyclotron instability, arising because of the relative drift between the beam and the main components of the proton distribution function in the solar wind at 1 AU, is studied.

074.076   Solar wind sputtering: an ineffective weathering process of airless bodies.   R. E. Scott.
Reports of planetary geology program, 1977—1978, (see 003.014), p. 305 - 307 (1978).

074.077   Long-wavelength p mode oscillations as an energy source for the solar wind.
E. J. Rhodes, Jr., R. K. Ulrich.
Bull. American Astron. Soc., Vol. 10, 416 (1978). – Abstract.

074.078   High velocity jets in the "quiet" sun as a possible source of the solar wind and the heating of the corona.   G. E. Brueckner, J.-D. F. Bartoe.
Bull. American Astron. Soc., Vol. 10, 416 (1978). – Abstract.

074.079   Polar coronal holes and the variation with latitude of the height of the Hα chromosphere.
D. M. Rabin, R. L. Moore.
Bull. American Astron. Soc., Vol. 10, 430 - 431 (1978). Abstract.

074.080   Experimental search for coronal Alfvén waves.   C. W. Querfeld, J. V. Hollweg.
Bull. American Astron. Soc., Vol. 10, 431 (1978). – Abstract.

074.081   Boundary-layer convection in the quiescent solar corona.   J. A. Ionson.
Bull. American Astron. Soc., Vol. 10, 432 - 433 (1978). Abstract.

074.082   The dynamical properties of the solar corona from intensities and line widths of EUV forbidden lines of Si VIII, Fe XI, and Fe XII.
C. C. Cheng, G. A. Doschek, U. Feldman.
Bull. American Astron. Soc., Vol. 10, 439 - 440 (1978). Abstract.

074.083   Coronal heating and its relation to magnetic field evolution.
L. Golub, R. Rosner, G. S. Vaiana.
Bull. American Astron. Soc., Vol. 10, 440 (1978). – Abstract.

074.084   Radio and white light observations of the 21 August 1973 coronal transient.   T. E. Gergely, M. R. Kundu, R. H. Munro, A. I. Poland.
Bull. American Astron. Soc., Vol. 10, 456 (1978). – Abstract.

074.085   Prompt solar proton events and coronal mass ejections.   S. W. Kahler, E. Hildner, M. A. I. Van Hollebeke.
Bull. American Astron. Soc., Vol. 10, 456 (1978). – Abstract.

074.086   Coronal holes, solar wind streams and geomagnetic activity during the new solar cycle.
N. R. Sheeley, J. W. Harvey.
Bull. American Astron. Soc., Vol. 10, 461 (1978). – Abstract.

074.087   Anomalous acceleration of minor ions in the solar wind.
J. F. McKenzie, W.-H. Ip, W. I. Axford.
Nature, Vol. 274, 350 - 351 (1978).
    Observations of the solar wind plasma have shown that the bulk speed of helium ions often exceeds that of the protons, provided conditions are such that Coulomb collisions are ineffective in coupling the two species together. The authors describe here a mechanism which could produce this unexpected effect for any minor ion species, based on the assumption that the sun is a copious emitter of Alfvén waves, as is indeed observed.

074.088   Three-colour photometry of the solar corona of July 10, 1972.   A. K. Ajmanov.
Astron. Tsirk., No. 980, p. 3 - 4 (1978). In Russian.

074.089   Laboratory modelling of collective processes in the solar wind plasma.   V. A. Derevyanko, Yu. P. Zakharov, A. V. Makukha, G. A. Tarnavskij.
Mathematical models of the near space, 1977, (see 003.015), p. 204 - 215. In Russian. – Abstr. in Ref. zh., 62. Issled. kosm. prostranstva, 6.62.230 (1978).

074.090   Model of shockless deceleration of the solar wind.

E. Ya. Gidalevich.
Cosmophysical investigations, Sverdlovsk, 1977, (see 003.016),
p. 3 - 11. In Russian. — Abstr. in Ref. zh., 62. Issled. kosm.
prostranstva, 6.62.232 (1978).

074.091 Numerical investigation of some models of inter-
         action of the solar wind with cosmic objects.
O. M. Belotserkovskij, V. Ya. Mitnitskij.
Mathematical models of the near space, 1977, (see 003.015),
p. 3 - 13. In Russian. — Abstr. in Ref. zh., 62. Issled. kosm.
prostranstva, 6.62.236 (1978).

074.092 Numerical modelling of interaction between the
         solar wind plasma and the Moon, Mars and Venus.
A. S. Lipatov.
Mathematical models of the near space, 1977, (see 003.015),
p. 89 - 96. In Russian. — Abstr. in Ref. zh., 62. Issled. kosm.
prostranstva, 6.62.235 (1978).

074.093 Experimental results of an investigation of inter-
         action between the solar wind and the earth's
magnetosphere.   T. K. Breus, K. I. Gringauz.
Mathematical models of the near space, 1977, (see 003.015),
p. 100 - 140. In Russian. — Abstr. in Ref. zh., 62. Issled. kosm.
prostranstva, 6.62.256 (1978).

074.094 Interaction between the solar wind stream and the
         geomagnetic field taking into account the inter-
planetary magnetic field.   A. P. Kropotkin.
Mathematical models of the near space, 1977, (see 003.015),
p. 42 - 49. In Russian. — Abstr. in Ref. zh., 62. Issled. kosm.
prostranstva, 6.62.257 (1978).

074.095 Modelling the physical processes in the solar wind.
         I. S. Veselovskij.
Mathematical models of the near space, 1977, (see 003.015),
p. 50 - 66. In Russian. — Abstr. in Ref. zh., 62. Issled. kosm.
prostranstva, 6.62.258 (1978).

074.096 Some results of investigation of interaction between
         the solar wind and the earth's magnetosphere.
V. G. Pivovarov.
Mathematical models of the near space, 1977, (see 003.015),
p. 14 - 23. In Russian. — Abstr. in Ref. zh., 62. Issled. kosm.
prostranstva, 6.62.260 (1978).

074.097 Il vento solare lontano.
         A. Egidi, G. Moreno.
Mem. Soc. Astron. Italiana, Vol. 47, 23 - 37 (1976).
     Contents: Modelli del vento solare "quieto" a grandi
distanze eliocentriche. Interazione del vento solare con il
mezzo interstellare. Osservazioni dirette del vento solare oltre
1 UA.

074.098 The solar corona and the origin of the solar wind.
         C. Chiuderi.
Mem. Soc. Astron. Italiana, Vol. 48, (see 012.038), 561 - 577
(1977).

074.099 Radio and EUV studies of coronal holes.
         K. V. Sheridan, S. F. Smerd.
Izv. VUZ. Radiofiz., Vol. 20, 1331 - 1337 (1977). In Russian.
Abstr. in Ref. zh., 51. Astron., 7.51.418 (1978).

074.100 On radio wave propagation in an inhomogeneous
         magnetic field of the solar corona.
V. V. Zheleznyakov, E. Ya. Zlotnik.
Izv. VUZ. Radiofiz., Vol. 20, 1444 - 1461 (1977). In Russian.
Abstr. in Ref. zh., 51. Astron., 7.51.424 (1978).

         Coronal holes and high speed wind streams: a
monograph from Skylab solar workshop I.

See Abstr. 012.047.

         Proton collisional excitation in the ground configu-
ration of Fe^{+11}.   See Abstr. 022.017.

         On some observations of the thermal emission of
iron flakes produced in a grinding wheel.
See Abstr. 022.103.

         A physical parameter method for the design of
broad-band X-ray imaging systems to do coronal plasma
diagnostics.   See Abstr. 031.232.

         Analysis and interpretation of soft X-ray photo-
graphs of coronal active regions taken with Fresnel zone
plates. I: Image analysis.   See Abstr. 031.272.

         Analysis of photographic X-ray images.
See Abstr. 031.316.

         Magnetohydrodynamics of atmospheric transients.
I. Basic results of two-dimensional plane analyses.
See Abstr. 062.004.

         The exterior source surface for force-free fields.
See Abstr. 062.007.

         An instability of finite amplitude circularly polarized
Alfvén waves.   See Abstr. 062.010.

         Time evolution of the energy spectrum of Alfvén
waves due to the non-linear Landau damping.
See Abstr. 062.056.

         Experimental studies on beam-plasma interaction.
See Abstr. 062.067.

         Geometrical MHD wave coupling.
See Abstr. 062.098.

         EUV models of the chromosphere-corona transition
region.   See Abstr. 073.007.

         Evaporative cooling of flare plasma.
See Abstr. 073.011.

         Downflow in the supergranulation network and its
implications for transition region models.
See Abstr. 073.058.

         Formation of shock waves from chromospheric
flares and their evolution in the solar wind.
See Abstr. 073.071.

         The upper chromosphere and corona. Observations
and interpretation.   See Abstr. 073.099.

         Observation and analysis of Fe XVIII solar X-ray
emission.   See Abstr. 076.002.

         Die Röntgenabbildung der Sonne und ihr Beitrag zur
Koronaforschung. Teil I.   See Abstr. 076.018.

         Die Röntgenabbildung der Sonne und ihr Beitrag
zur Koronaforschung. Teil II.   See Abstr. 076.030.

         OSO-7 observations of coronal hard X-ray sources.
See Abstr. 076.039.

         Extremely high resolution photographic X-ray
images.   See Abstr. 076.044.

A reinterpretation of the centimetric quiet sun and the density in the long distance arches.
See Abstr. 077.001.

An extremely narrow-band solar type IV_{dm} burst with fine structure.   See Abstr. 077.003

Coronal magnetic structure near the source region of a type II radio burst.   See Abstr. 077.007.

A non-homogeneous model of the chromosphere and of the lower corona for interpretation of data on the quiet sun's radio emission in the millimeter wavelength range.
See Abstr. 077.030.

On a possible relation of the S-component of radio emission with solar wind velocity.   See Abstr. 077.057.

Rigidity-independent propagation of cosmic rays in the solar corona.   See Abstr. 078.013.

Rigidity-independent coronal propagation and escape of solar protons and α particles.
See Abstr. 078.014.

Prompt solar proton events and coronal mass ejections.   See Abstr. 078.016.

Linear force-free fields in the lower corona.
See Abstr. 080.015.

Alfvén waves in the solar atmosphere.
See Abstr. 080.023.

Numerical modelling of propagation of shock waves in the solar atmosphere and in the solar wind.
See Abstr. 080.065.

On the mechanical heating of the external layers of the sun. A methodological discussion.
See Abstr. 080.067.

Peculiarities of solar-wind flow around the magnetosphere and the magnetopause position.
See Abstr. 084.237.

Energy flow and closure of current systems in the magnetosphere.   See Abstr. 084.263.

Entry of solar wind plasma into the magnetosphere.
See Abstr. 084.312.

Solar wind sputtering: an ineffective weathering process on airless bodies.   See Abstr. 091.020.

Mercury's helium exosphere after Mariner 10's third encounter.   See Abstr. 092.012.

Plasma electron observations in the wake of Venus.
See Abstr. 093.033.

Limitations on the parameters of the solar wind in modelling lunar electromagnetic induction.
See Abstr. 094.128.

Solar-wind sputtering of the martian atmosphere.
See Abstr. 097.056.

Solar wind effect on Jupiter's non-Io-related radio emission.   See Abstr. 099.039.

The flute instability as the trigger mechanism for disruption of cometary plasma tails.   See Abstr. 102.030.

Are plasma tail disconnection events in comets caused by high speed solar-wind streams or by reversals in the interplanetary magnetic field?   See Abstr. 102.035.

Interplanetary gas. XXII. Interaction of comet Kohoutek's ion tail with the compression region of a solar-wind corotating stream.   See Abstr. 103.402.

Meteor radar rates, geomagnetic activity and solar wind sector structure.   See Abstr. 104.027.

Interplanetary scintillation.
See Abstr. 106.021.

Connection of proton flux increase in the interplanetary space in August - September 1973 with variations of solar wind velocity.   See Abstr. 106.033.

Plasma disturbances caused by the Helios spacecraft in the solar wind.   See Abstr. 106.049.

Connection of increases of proton streams in September - November 1973 with changes in solar wind velocity according to data of the AIS (*Automatic Interplanetary Station*) Mars 7.   See Abstr. 106.053.

The radial variation of corotating energetic particle streams in the inner and outer solar system.
See Abstr. 106.057.

Long wavelength minimum source diameters: severe solar wind scattering contamination?
See Abstr. 141.209.

Cosmic ray diffusion in the solar wind with non-spherical boundary.   See Abstr. 143.060.

Errata

074.901   Erratum: 'Development of a complex of activity in the solar corona' [Sol. Phys., Vol. 54, 65 - 105 (1977)].   R. Howard, Z. Švestka.
Sol. Phys., Vol. 56, 471 (1978).

075 Solar Patrol

075.001 A survey of the 127 MHz Toruń solar radio data.
K. M. Borkowski.
Postępy Astron., Tom 25, 135 - 159 (1977). In Polish.

The paper presents (i) a complete survey of mean monthly and mean yearly values of the flux density from the Sun, (ii) a review of outstanding occurrences observed at 127 MHz since 1958 at the Toruń Observatory.

075.002 Observations solaires. Rotations 1637 - 1649 (13 janvier - 31 décembre 1976). E. Ţifrea,
V. Dinulescu, A. Dimitriu, S. Dinulescu, G. Mariş, I. D. Niţă.
Cent. Astron. Sci. Spatiales. Editura Acad. Repub. Socialiste România, Bucarest. 48 pp. Price Lei 4.00 (1977).

075.003 Provisional sunspot-numbers for December 1977, January - May 1978.
Yamamoto Circ., No. 1873 (1978). In Japanese.

075.004 Definitive sunspot numbers for 1977.
M. Waldmeier.
J. Geophys. Res., Vol. 83, 2232 (1978).

075.005 Solare Beobachtungsergebnisse (Solar data).
Akad. Wiss. DDR, Zentralinst. Solar-Terr. Phys. (Heinrich-Hertz-Inst.), DDR-1199 Berlin-Adlershof. HHI Solar Data, Vol. 28, 75 - 103 (1977); Vol. 29, A-F, 1 - 24 (1978). 1977 October - 1978 February. Solar radio emission.

075.006 Solar maps and activity. F. J. Heyden,
A. G. Ambion, E. R. Gallardo, V. L. Badillo.
Manila Observatory, Solar Division (1977/78). — 1977 October 1 - 1978 February 28.

075.007 Actividad solar en 1976. I. Números relativos de Wolf. II. Estadística de manchas y superficie de las mismas. M. Lopez Arroyo.
Bol. Astron. Obs. Madrid, Vol. 10, No. 1, p. 23 - 58 (1977).

075.008 Daily Hα chromosphere pictures, daily K_{232} chromosphere pictures, daily white light photosphere pictures. M. Cimino (Editor).
Photographic Journal of the Sun, Oss. Astron. Roma, Nos. 125 - 130. — 1977 March 23 - 1977 September 3. Solar rotation 1653 - 1658.

075.009 Solar phenomena.
M. Cimino, M. Torelli, V. Croce, F. Casamassima.
Oss. Astron. Roma, Mon. Bull. N. 233 - 236 (1977). — 1977 September - December, 1978 January: Daily total areas of sunspot-groups; Heliographic position, classification and area of sunspot-groups; Longitudinal sunspot magnetic fields; Hours of K-line cinematographic patrol; Hours of Hα cinematographic patrol; Explanations.

075.010 Nombres relatifs moyens de l'activité solaire pour l'année 1977. M. Waldmeier.
Astronomie, Vol. 92, 304 (1978).

075.011 Sunspot relative numbers for 1977.
M. Waldmeier.
Astron. Mitt. Eidg. Sternw. Zürich, Nr. 363, 10 pp. (1978).

075.012 Sunspot observations for the year 1974.
N. Doğan.
Commun. Fac. Sci. Univ. Ankara, Sér. A_3 : Astron., Tome 25, 1 - 24 (1977) = Commun. Astron. Dep. Univ. Ankara, No. 76.

In this paper the sunspot groups seen between the solar rotation number 1610 and 1622, inclusive, and their development and the distribution of these sunspot groups in the northern and southern latitudes are given.

075.013 Sunspot observations for the year 1975.
N. Doğan.
Commun. Fac. Sci. Univ. Ankara, Sér. A_3 : Astron., Tome 25, 35 - 54 (1977) = Commun. Astron. Dep. Univ. Ankara, No. 78.

In this paper the sunspot groups seen between the solar rotation number 1623 and 1636, inclusive, and their development and the distribution of these sunspot groups in the northern and southern latitudes are given.

075.014 Zürcher Sonnenfleckenrelativzahlen.
Sterne Weltraum, Jahrg. 17, 38, 74, 146, 185, 220, 274 (1978).

075.015 L'activité solaire. M.-J. Martres, G. Zlicaric.
Astronomie, Vol. 92, 54 - 55, 103, 151, 199, 247, 291, 331 (1978). — Rotations 1656 - 1662.

075.016 Daily maps of the sun and geophysical graphs.
Soln. Dannye 1977 Byull., No. 11, p. 1 - 50; No. 12, p. 1 - 64; 1978, No. 1, p. 1 - 71. In Russian.

075.017 Magnetic fields of sunspots.
Prilozhenie k Byulletenyu "Soln. Dannye", 1977, Nos. 11, 12; 1978, No. 1. In Russian.

075.018 Solar and solar system activity. R. J. J. Langton.
J. British Astron. Assoc., Vol. 88, 164 - 167, 289 - 291, 389 - 392 (1978). — Report of the Radio Astronomy Section.

075.019 Solar activity. K. J. Medway, H. Hill.
J. British Astron. Assoc., Vol. 88, 158 - 163, 283 - 288, 382 - 388 (1978). — Report of the Solar Section.

075.020 Geomagnetic and solar data. J. V. Lincoln.
J. Geophys. Res., Vol. 83, 232, 744, 1191, 1679, 2230 - 2231 (1978).

075.021 Sunspot numbers.
Sky Telesc., Vol. 55, 92, 187, 272, 364, 453, 548; Vol. 56, 84 (1978).

Short-baseline radiointerferometer for solar patrol.
See Abstr. 033.020.

076 UV, X, Gamma Radiation

076.001 Backscatter, anisotropy, and polarization of solar hard X-rays.   T. Bai, R. Ramaty.
Astrophys. J., Vol. 219, 705 - 726 (1978).

Hard X-rays incident upon the photosphere with energies $\gtrsim 15\,\mathrm{keV}$ have high probabilities of backscatter due to Compton collisions with electrons. This effect has a strong influence on the spectrum, intensity, and polarization of solar hard X-rays—especially for anisotropic models in which the primary X-rays are emitted predominantly toward the photosphere. The authors have carried out a detailed study of X-ray backscatter, and have investigated the interrelated problems of anisotropy, polarization, center-to-limb variation of the X-ray spectrum, and Compton backscatter in a coherent fashion. The results of this study are compared with observational data. The authors also discuss the characteristics of the brightness distribution of the X-ray albedo patch created by the Compton backscatter.

076.002 Observation and analysis of Fe XVIII solar X-ray emission.   H. R. Rugge, A. B. C. Walker, Jr.
Astrophys. J., Vol. 219, 1068 - 1078 (1978).

X-ray spectra from the solar corona, obtained with a crystal spectrometer on the OV1-17 satellite, are used to analyze Fe XVIII emission lines between ~14 and ~16 Å. The first comprehensive and accurate determinations of the coronal Fe XVIII wavelengths and relative intensities are made for the $2p-3d$ and $2p-3s$ transitions. Eighteen emission lines or line blends of Fe XVIII were observed and analyzed, including all X-ray lines previously observed in hot laboratory plasmas in this wavelength region. The relative intensities for the Fe XVIII lines are used to deduce relative effective collision strengths for a number of transitions. The measured energy flux emitted in the X-ray region by the corona in the form of Fe XVII and Fe XVIII emission lines is compared for a variety of coronal conditions.

076.003 Polarization of the solar UV light at the limb.   F. Goutail.
Astron. Astrophys., Vol. 64, 73 - 82 (1978).

This paper describes a new method to measure the polarization of the solar UV light at the limb. The instrumentation utilized is presented and then the results of the observations. The rate of polarization and the angle of vibration is reported at four spectral bands centered around 2160 Å, 2260 Å, 2940 Å and 3500 Å for points distributed along a solar perimeter at 50" from the limb ($\mu = 0.3$).

076.004 High resolution UV solar spectroscopy.   R. M. Bonnet.
Space Sci. Rev., Vol. 21, 379 - 409 (1978). − Review paper presented at the Vth Conference on UV and X-ray Spectroscopy of Astrophysical and Laboratory Plasmas, London, July 4−7, 1977.

The advantages of high resolution UV spectroscopy for the investigation of the solar atmosphere are stressed while the limitations in the areas of instrumentation and diagnosis are discussed. The recent achievements (made essentially by Skylab, OSO-8 and rocket instruments) are reviewed and discussed.

076.005 Preliminary results obtained from the solar EUV experiment on board the D2B Aura satellite.
J. P. Delaboudinière, F. Millier.
Space Research, Vol. XVII, (see 012.010), 519 - 524 (1977).

The solar EUV experiment on board D2B has been operating successfully since 1 November 1975. Three main sets of data are being gathered: (l) Spectroheliograms at a resolution of one arc minute, within prominent solar lines

formed at temperatures between 2×10^4 and $3 \times 10^{6\,\circ}\mathrm{K}$, providing detailed observations of individual active regions. (2) Disc integrated values of the solar flux in the range 17−127 nm at a resolution of 0.8 to 0.6 nm suitable to define the spectral energy input to the upper atmosphere. (3) Simultaneous measurements of solar extinction at sunset and sunrise at eight different wavelengths, providing a partially redundant set of information for remote sounding of atmospheric composition and temperature between 100 and 400 km.

076.006 EUV flux variations with solar rotation observed during 1974−1976 from the AE-C satellite.
H. E. Hinteregger, D. E. Bedo, J. E. Manson, D. R. Skillman.
Space Research, Vol. XVII, (see 012.010), 533 - 544 (1977).

Solar EUV fluxes in the spectral range 140−1850 Å have been observed by spectrophotometers on the satellites AE-C, D, and E. Variations over the long period of one or two years cannot be verified quantitatively, as the observed small variations are of the same magnitude as possible variations of instrumental sensitivities and estimated uncertainties of absolute values from the rocket experiment which established the calibration of the AE-C instrument. Fortunately, no similar difficulty exists for the interpretation of observed EUV variations within smaller time periods up to that of a full solar rotation. Results from AE-C observations for many different wavelength groups for the year 1974 are shown in some detail.

076.007 Solar X-ray emission and metre-wave radio bursts.   R. A. Duncan.
Proc. Astron. Soc. Australia, Vol. 3, 154 - 157 (1977).

The author compares scatter plots of the position of type I and type III radio bursts, recorded by the Culgoora radioheliograph, with Skylab X-ray maps. He briefly discusses the 27-day recurrence pattern of the radio sources.

076.008 Time-varying oscillations in the solar soft X-ray flux as observed from Skylab.
D. L. Teuber, R. M. Wilson, W. Henze, Jr.
Astron. Astrophys., Vol. 65, 229 - 231 (1978).

Evidence is presented for a 256-second oscillation in the solar soft X-ray flux during the flare of 15 June 1973. A power spectrum analysis was used with proportional counter data from the NASA-MSFC/The Aerospace Corporation S-056 experiment on Skylab. No oscillations have been observed in the X-ray flux during quiet periods.

076.009 Analysis of X-ray observations of the 15 June 1973 flare in active region NOAA 131.   K. R. Krall, E. J. Reichmann, R. M. Wilson, W. Henze, Jr., J. B. Smith, Jr.
Sol. Phys., Vol. 56, 383 - 404, with corrections in Vol. 57, 485, 487 (1978).

Observations and analyses of the 1B/M3 flare of 15 June, 1973 in active region NOAA 131 (McMath 12379) are presented. The X-ray observations, consisting of broadband photographs and proportional counter data are used to infer temperatures, emission measures, and densities for the flaring plasma. Also discussed are Hα and magnetic field observations of the flare and the active region. Finally, results of numerical calculations, including thermal conduction, radiative loss and chromospheric evaporation, are in qualitative agreement with the decay phase observations.

076.010 Generation of gamma radiation and neutrons in solar flares.   I. A. Ibragimov, G. E. Kocharov.
Izv. AN SSSR. Ser. fiz., Vol. 41, 1832 - 1841 (1977). In Russian. − Abstr. in Ref. zh., 51. Astron., 2.51.406 (1978).

076.011 Todays knowledge of the solar EUV output and the

future needs for more accurate measurements for aeronomy.   G. Schmidtke.
Planet. Space Sci., Vol. 26, 347 - 353 (1978).

In recent years discrepancies arose in the determination of the solar EUV output. It is difficult and often impossible to remove these discrepancies from the observational data reported so far. However, the EUV data show evidence for a strong variability during the Solar Cycle 20. The measuring methods applied so far create uncertainties of the order of ± 30% or less. Therefore, new methods have to be developed for more accurate measurements. Two approaches offer the possibility of overcoming todays shortcomings.

076.012   Irradiation solar flux measurements between 120 and 400 nm. Current position and future needs.
P. C. Simon.
Planet. Space Sci., Vol. 26, 355 - 365 (1978). Paper presented at the XXth Cospar Meeting, Tel Aviv, 7 - 18 June 1977.

The irradiation solar fluxes between 120 and 400 nm are reviewed and discussed. The disagreements between the recent observations are pointed out, emphasizing the future needs in this wavelength range for aeronomic purposes. Interpretation of the available data as function of the solar activity cannot explain their discrepancies, showing that the solar variability during the eleven-year cycle is still unknown.

076.013   The solar XUV He I and He II emission lines. II. Intensity ratios and distribution functions.
D. L. Glackin, J. L. Linsky, S. A. Mango, J. D. Bohlin.
Astrophys. J., Vol. 222, 707 - 715 (1978).

From high-resolution solar images the authors show that the He II λ256 line intensity is very nearly uncorrelated with He II λ304 at the same location on the Sun and that the λ256 line is formed mainly by the photoionization-recombination process. They also derive center-to-limb variations of He II λ304 and λ256 and He I λ584 and λ537 for network and cell regions separately. The authors conclude that the appearance of dark coronal holes in the helium lines is not a geometrical effect involving the chromospheric network, but is rather an intrinsic property of the atmosphere in both network and cell regions. They suggest that the network and cells can be treated as isolated atmospheres in the solution of the transfer equation in the helium lines.

076.014   Analysis of the profiles of the solar S I lines at 1807 and 1900 Å.
L. R. Doherty, H. C. McAllister.
Astrophys. J., Vol. 222, 716 - 722 (1978).

High-resolution profiles of five lines of S I have been obtained from rocket echelle spectra of a quiet region near the solar limb. Lines in the 1807 Å multiplet show strong central reversals; the 1900 Å lines are nearly Gaussian in shape. Coherent scattering provides the best fit to the wings of the reversed lines. Thus the S I 1807 Å multiplet provides further evidence for the importance of partial redistribution in the formation of strong chromospheric lines. Computed absolute intensities are less than those observed by more than a factor of 2. Changes in the model which reduce this discrepancy are discussed.

076.015   Investigation of the X and gamma radiation of the sun by means of the RGS-1 apparatus.
A. V. Baskakov, Yu. G. Derevitskij, A. G. Enikeev, G. E. Kocharov, G. A. Matveev, A. S. Melioranskij, V. O. Najdenov, A. A. Sementsov, Yu. N. Starbunov, Yu. E. Charikov.
Izv. AN SSSR. Ser. fiz., Vol. 41, 1808 - 1818 (1977). In Russian. – Abstr. in Ref. zh., 62. Issled. kosm. prostranstva, 3.62.122 (1978).

076.016   The sun in X-rays.   I. P. Tindo.
Zemlya i Vselennaya, 1978, No. 3, p. 18 - 27. In Russian.

076.017   Hard X-ray flares from the sun.   V. B. Bhatia.
Bull. Astron. Soc. India, Vol. 6, 4 - 8 (1978).

A review of solar flare hard X-rays is presented. Various features of these flares have been considered to pinpoint the location of the source of emission and the nature of the emission of these flares. It is shown that the arguments for the location of the source in the dense chromosphere and against it are finely balanced, so that nothing conclusive may be said about the location of the source. Similarly, arguments for and against the non-thermal nature of the emission do not allow any definite conclusion.

076.018   Die Röntgenabbildung der Sonne und ihr Beitrag zur Koronaforschung. Teil I.   G. Elwert.
Sterne Weltraum, Jahrg. 17, 193 - 200 (1978).

076.019   The inter-relationship of hard X-ray and EUV bursts during solar flares.
A. G. Emslie, J. C. Brown, R. F. Donnelly.
Sol. Phys., Vol. 57, 175 - 190 (1978).

A comparison is made between the flux-versus-time profile in the EUV band and the thick target electron flux profile as inferred from hard X-rays for a number of moderately large solar flares.

076.020   Pulsations in solar hard X-ray bursts.
B. Lipa.
Sol. Phys., Vol. 57, 191 - 204 (1978).

The author has analyzed the hard X-ray emission from 28 large solar events, searching for pulsations in intensity profiles. Periodicity occurred in 26 events, usually soon after the onset, with periods in the range 10–100 s. Pulsations occurring at common frequencies in different energy bands are observed to be closely in phase. Periodic behavior in hard X-ray emission is related to that at microwave and decametric wavelength. The observations are discussed briefly in terms of two models.

076.021   Impulsive extreme-ultraviolet and hard X-ray emission during solar flares.
R. F. Donnelly, S. R. Kane.
Astrophys. J., Vol. 222, 1043 - 1053 (1978).

Impulsive solar bursts of hard X-rays observed with moderate spectral resolution and high time resolution by the OGO 5 satellite were compared with concurrent enhancements of extreme-ultraviolet (EUV) radiation observed by means of sudden frequency deviations. The impulsive EUV flux rises and decays, with a time structure most like that of the lowest-energy hard X-rays observed – the 10–20 keV X-rays – but slower than hard X-rays with photon energy greater than 32 keV. These results are interpreted in terms of the dominance of 10–25 keV electrons in exciting the high-density impulsive EUV source.

076.022   Solar X-ray flares in July – December 1972.
N. V. Illarionova, A. S. Melioranskij, I. N. Rozantsev, I. A. Savenko.
Problems of solar activity and space system Prognoz, (see 003.008), p. 58 - 63 (1977). In Russian. – Abstr. in Ref. zh., 51. Astron., 4.51.399; 62. Issled. kosm. prostranstva, 4.62.127 (1978).

076.023   On the anisotropy of X-ray emission of solar flares.
O. M. Kovrizhnykh, I. A. Savenko, L. M. Chupova.
Problems of solar activity and space system Prognoz, (see 003.008), p. 95 - 110 (1977). In Russian. – Abstr. in Ref. zh., 51. Astron., 4.51.400; 62. Issled. kosm. prostranstva, 4.62.129 (1978).

076.024   Temporal associations of hard solar X-ray bursts according to data from the Prognoz automatic stations.   M. I. Kudryavtsev, A. S. Melioranskij, I. A. Savenko, V. M. Shamolin.

Problems of solar activity and space system Prognoz, (see 003.008), p. 23 - 30 (1977). In Russian. – Abstr. in Ref. zh., 51. Astron., 4.51.404; 62. Issled. kosm. prostranstva, 4.62.125 (1978).

076.025   Gamma-ray bursts during solar flares on 2, 4, and 7 August, 1972 according to observations on the Prognoz 2 cosmic station.   F. Albernhe, G. Vedrenne, F. Cambou, M. I. Kudryavtsev, O. B. Likin, A. S. Melioranskij, N. I. Nazarova, V. M. Pankov, N. F. Pisarenko, I. A. Savenko, R. Talon, V. M. Shamolin.
Problems of solar activity and space system Prognoz, (see 003.008), p. 30 - 58 (1977). In Russian. – Abstr. in Ref. zh., 51. Astron., 4.51.405; 62. Issled. kosm. prostranstva, 4.62.126 (1978).

076.026   Observation of the centre-to-limb variations of the Sun in the vacuum ultraviolet region.
K. Nishi, K. Higashi, A. Yamaguchi, Z. Suemoto.
Bull. Inst. Space Aeronaut. Sci.,Univ. Tokyo, B, Vol. 12, 959 - 973 (1976). In Japanese. – Abstr. in Phys. Abstr., Vol. 81, Abstr. 28604 (1978).

076.027   CO fluorescence in the extreme-ultraviolet solar spectrum.   J.-D. F. Bartoe, G. E. Brueckner, G. D. Sandlin, M. E. VanHoosier, C. Jordan.
Astrophys. J., Lett., Vol. 223, L51 - L53 (1978).
Emission lines in the fourth positive system of CO have been identified in the extreme-ultraviolet solar spectrum 1540–1660 Å. These lines are excited by the C IV transition-zone lines at 1548 and 1551 Å. They are strong in the spectrum of a sunspot and in parts of the adjacent active region. Some of them appear as weak, broad emission features in the quiet sun.

076.028   Calibrated full disk solar H I Lyman-α and Lyman-β profiles.   P. Lemaire, J. Charra, A. Jouchoux, A. Vidal-Madjar, G. E. Artzner, J. C. Vial, R. M. Bonnet, A. Skumanich.
Astrophys. J., Lett., Vol. 223, L55 - L58 (1978).
Resolved solar H I Lα and Lβ profiles have been recorded by the French LPSP experiment on OSO 8. Intensity observations at the center and at the limb have been combined to obtain flux-equivalent profiles. Comparison of the flux profiles with unresolved calibration rocket profiles allows one to obtain an absolute calibration.

076.029   Impulsive solar X-ray bursts.   C. J. Crannell, K. J. Frost, C. Mätzler, K. Ohki, J. L. Saba.
Astrophys. J., Vol. 223, 620 - 637 (1978).
Selection criteria for a set of 22 impulsive spike bursts are described and the observational data for these events are presented. Physical parameters derived from these observations and the relationships between the various parameters are presented. The significance of the flare parameters is discussed in light of instrumental and analytic effects and characteristics of the impulsive flare process. The conclusions are summarized.

076.030   Die Röntgenabbildung der Sonne und ihr Beitrag zur Koronaforschung. Teil II.   G. Elwert.
Sterne Weltraum, Jahrg. 17, 236 - 241   (1978).

076.031   The characteristics of impulsive solar EUV bursts.   A. G. Emslie, R. W. Noyes.
Sol. Phys., Vol. 57, 373 - 383 (1978).
The authors examine a number of high time resolution intensity-time profiles of EUV impulsive bursts. These bursts are found to be synchronous (to within the instrumental time resolution of 5.5 s) in all wavelengths observed, corresponding to emissions from temperatures ranging from upper chromospheric to coronal. The distribution with temperature of a suitably defined emission measure parameter is also

examined as a function of time throughout the bursts and a marked similarity in the shape of this distribution, both between different events and throughout the time history of any particular event, is noted.

076.032   OSO-8 observations of the Lyman-α profile at the limb and in a prominence.   E. G. Chipman.
Bull. American Astron. Soc., Vol. 10, 418 (1978). – Abstract.

076.033   Nonthermal broadening of extreme ultraviolet emission lines near the solar limb.
J. T. Mariska, U. Feldman, G. A. Doschek.
Bull. American Astron. Soc., Vol. 10, 432 (1978). – Abstract.

076.034   The growth of filaments by the condensation of coronal arches.   J. M. Davis, A. S. Krieger.
Bull. American Astron. Soc., Vol. 10, 439 (1978). – Abstract.

076.035   EUV observations of a solar active region.
K. P. Dere.
Bull. American Astron. Soc., Vol. 10, 440 (1978). – Abstract.

076.036   Active region observations of the Mg I and II resonance lines.   M. S. Allen.
Bull. American Astron. Soc., Vol. 10, 440 (1978). – Abstract.

076.037   Solar flare X-ray line studies.   N. J. Veck, W. H.-M. Ku, J. R. Lemen, R. Novick, J. H. Parkinson.
Bull. American Astron. Soc., Vol. 10, 441 (1978). – Abstract.

076.038   OSO-8 measurements of rapid downflow in the transition zone during the impulsive phase of solar flares.   B. W. Lites, E. C. Bruner, E. R. Hansen.
Bull. American Astron. Soc., Vol. 10, 441 (1978). – Abstract.

076.039   OSO-7 observations for coronal hard X-ray sources.   H. S. Hudson.
Bull. American Astron. Soc., Vol. 10, 454 (1978). – Abstract.

076.040   Hard X-ray time profiles and acceleration processes in large solar flares.   T. Bai.
Bull. American Astron. Soc., Vol. 10, 454 (1978). – Abstract.

076.041   Rapid time variations in solar hard X-ray emission.   D. W. Datlowe.
Bull. American Astron. Soc., Vol. 10, 454 - 455 (1978). Abstract.

076.042   Spectral evolution of multiply-impulsive solar bursts.   J. T. Karpen, C. J. Crannell, K. J. Frost.
Bull. American Astron. Soc., Vol. 10, 455 (1978). – Abstract.

076.043   Pre-onset visible, EUV, and X-ray flare structures.   E. J. Schmahl, C. Solodyna, J. B. Smith, C. C. Cheng.
Bull. American Astron. Soc., Vol. 10, 456 (1978). – Abstract.

076.044   Extremely high resolution photographic X-ray images.   A. S. Krieger, J. M. Davis, R. Haggerty.
Bull. American Astron. Soc., Vol. 10, 460 (1978). – Abstract.

076.045   Observations of moving X-ray source associated with an eruptive prominence at the limb.
J. M. Mosher, C. J. Wolfson.
Bull. American Astron. Soc., Vol. 10, 460 (1978). – Abstract.

076.046   The solar flux in the far ultraviolet: origin, relative variations.   R. M. Bonnet.
Ann. Géophys., Vol. 33, 409 - 422 (1977). In French.
Abstr. in Phys. Abstr., Vol. 81, Abstr. 40884 (1978).

076.047   Intensity of solar X-rays in the E region measured

by proportional counters carried on rockets.
A. Loidl, H. Schwentek.
J. Geophys., Vol. 44, 117 - 123 (1977). — Abstr. in Phys.
Abstr., Vol. 81, Abstr. 45072 (1978).

076.048   Tecniche osservative e discussione critica dell'emissione XUV.   M. Landini.
Mem Soc. Astron. Italiana, Vol. 48, (see 012.037), 239 - 252 (1977).

Dopo un breve riepilogo delle caratterisťiche più evidenti dell'emissione X nel caso di un brillamento si analizzano i metodi seguiti per ricavare informazioni sulle temperature e la densità di un brillamento. Si describe un metodo per la costruzione di un modello di temperatura e si discutono le implicazioni che si possono ottenere sullo studio del bilancio energetico.

A physical parameter method for the design of broad-band X-ray imaging systems to do coronal plasma diagnostics.   See Abstr. 031.232.

Analysis and interpretation of soft X-ray photographs of coronal active regions taken with Fresnel zone plates. I: Image analysis.   See Abstr. 031.272.

Imaging of solar active regions with Fresnel zone plates.   See Abstr. 031.317.

A hard X-ray imaging instrument for solar and cosmic sources.   See Abstr. 032.575.

Electron densities in solar flare and active region plasmas from a density-sensitive line ratio of Fe IX.
See Abstr. 073.001.

The solar XUV He I and He II emission lines. I. Intensities and gross center-to-limb behavior.
See Abstr. 073.010.

EUV structure of a small flare.
See Abstr. 073.015.

Analysis of the emission line spectra of a solar flare observed from Skylab.   See Abstr. 073.016.

Spectral lines observed in solar flares between 171 and 630 Angstroms.   See Abstr. 073.025.

C III density diagnostics in nonequilibrium plasmas.
See Abstr. 073.030.

Dynamics of an atmosphere irradiated by soft X-rays in flares.   See Abstr. 073.062.

Hα, hard X-ray, and microwave emission in the impulsive phase of solar flares.   See Abstr. 073.072.

Scattering of fast flare electrons in solar atmosphere and their X-ray spectrum.   See Abstr. 073.074.

Doppler shifts measured in the O VI line from OSO-8 observations above and in the vicinity of plage McMath 13738.
See Abstr. 073.077.

Observations of solar flare X-ray continua.
See Abstr. 073.078.

OSO-8, radio and X-ray observations of the 19 April 1977 flare.   See Abstr. 073.079.

ATM evidence for a low non-thermal proton/electron energy flux ratio in solar flares.   See Abstr. 073.080.

Evidence that the mass of the thermal X-ray plasma in solar flares is supplied by conduction-driven evaporation.
See Abstr. 073.081.

Soft X-ray observations during the preflare phase of the solar flare phenomenon.   See Abstr. 073.084.

X-ray analysis of the 29 July 1973 flare.
See Abstr. 073.085.

Radiative heating in chromospheric flares.
See Abstr. 073.088.

Aspetti osservativi del fenomeno brillamento.
See Abstr. 073.094.

Meccanismi di emissione nei brillamenti solari.
See Abstr. 073.096.

Physical parameters defining the changing structure of a coronal hole.   See Abstr. 074.001.

The quiet coronal X-ray spectrum of highly ionized oxygen and nitrogen.   See Abstr. 074.017.

The structure of the X-ray bright corona above active region McMath 12628 and derived implications for the description of equilibria in the solar atmosphere.
See Abstr. 074.031.

Observational evidence of continual heating in X-ray emitting coronal loops.   See Abstr. 074.052.

Short term evolution of coronal hole boundaries.
See Abstr. 074.055.

Physical conditions in the corona for a bipolar magnetic region.   See Abstr. 074.072.

On the quasi-periodic structure of the impulsive burst of March 1, 1970.   See Abstr. 077.038.

The energy balance and pressure in the solar transition zone.   See Abstr. 080.054.

Bowen fluorescence in the solar transition region.
See Abstr. 080.055.

EUV absorption analysis of thermospheric structure from AE-satellite observations of 1974—1976.
See Abstr. 082.042.

Effect of solar X-ray and Lyman-α radiation on ionospheric absorption.   See Abstr. 083.134.

Cosmic flare transients: constraints upon models for energy storage and release derived from the event frequency distribution.   See Abstr. 142.150.

Errata

076.901   Errata: 'The analysis and interpretation of solar X-ray photographs' [Sol. Phys., Vol. 53, 417 - 433 (1977)].   J. H. Underwood, D. L. McKenzie.
Sol. Phys., Vol. 57, 485 (1978).

077 Radio, Infrared Radiation

077.001 A reinterpretation of the centimetric quiet sun and the density in the long distance arches.
P. Lantos.
Astron. Astrophys., Vol. 62, 69 - 74 (1978).

A reinterpretation of the classical characteristics of the centimetric quiet sun is proposed in terms of a coronal model including coronal holes and the long distance arches seen in X ray photographs. The irregular presence of coronal holes is invoked in order to explain both the center-to-limb effects and the cycle variation of the central brightness temperature of the radio quiet sun.

077.002 Correlation between drift rate and polarization in solar type III radio bursts.
A. O. Benz, P. Zlobec.
Astron. Astrophys., Vol. 63, 137 - 145 (1978).

The authors have tested a sample of 463 type III bursts at 237 MHz and have found a statistically significant correlation between polarization and frequency drift rate by several independent tests. The correlation increases with higher polarization, smaller total bandwidth and a polarization spike before maximum intensity. It is shown that the groups of "constant", resp. "spike" polarization bursts are drawn from statistically different populations.

077.003 An extremely narrow-band solar type IV_{dm} burst with fine structure.   G. L. Tarnstrom, A. O. Benz.
Astron. Astrophys., Vol. 63, 147 - 154 (1978).

On 1970 January 30 a type IV_{dm} solar radio burst occurred with a halfpower bandwidth of only 40 MHz. Since Benz and Tarnstrom (1976) have shown that this burst cannot be caused by synchrotron emission, it is concluded that its concurrent fine structure is a signature for plasma wave emission. In addition, the various fine structure observed in this event has been used to determine the magnetic field assuming plasma wave emission and the most widely spread interpretations of fine structures for this case. From the separation of fibre absorption and emission ridges magnetic field strengths in the range 5.7 ± 1.5 Gauss have been found.

077.004 Analysis of data of the total solar flux observations at 3.2 cm obtained at the Peking Observatory during the years 1965–1975.   Q.-j. Fu, X.-c. Li, Y.-y. Liu.
Acta Astron. Sinica, Vol. 18, 166 - 181 (1977). In Chinese.

In this paper data of the total flux observations of the sun at λ 3.2 cm obtained at the Peking Observatory during the years 1965–1975 have been analysed.

077.005 Observations of type III bursts in the interplanetary space aboard the stations Prognoz and Prognoz 2.
V. P. Grigor'eva, S. A. Kaplan.
Kosm. Issled., Vol. 16, 73 - 77 (1978). In Russian.

077.006 Stereoscopic direction finding analysis of a type III solar radio burst: evidence for emission at $2f_p^-$.
D. A. Gurnett, M. M. Baumback, H. Rosenbauer.
J. Geophys. Res., Vol. 83, 616 - 622 (1978).

Stereoscopic direction finding measurements from the Imp 8, Hawkeye 1, and Helios 2 spacecraft over base line distances of a substantial fraction of an astronomical unit are used to directly determine the three-dimensional trajectory of a type III solar radio burst. By comparing the observed source positions with the direct in situ solar wind plasma density measurements obtained by Helios 1 and 2 near the sun the relationship of the emission frequency to the local plasma frequency can be determined directly without any modeling assumptions. These comparisons show that the type III radio emission occurs near the second harmonic, $2f_p^-$, of the local

electron plasma frequency. Other characteristics of the type III radio emission, such as the source size, which can be obtained from this type of analysis are also discussed.

077.007 Coronal magnetic structure near the source region of a type II radio burst.   R. T. Stewart.
Proc. Astron. Soc. Australia, Vol. 3, 157 - 159 (1977).

This paper describes the first radioheliograph observations of a type II burst near coronal magnetic structures depicted by soft X-ray pictures. It is shown that three separate lanes in a type II burst are associated with three separate source regions each located almost radially above a soft X-ray loop. To explain these observations a model is derived in which a wide-fronted m.h.d. disturbance travels from the flare explosion along magnetic field lines and then intersects successively with three coronal streamer structures each located above a soft X-ray loop.

077.008 Three frequency observations of a moving type IV burst at the limb.   G. J. Nelson.
Proc. Astron. Soc. Australia, Vol. 3, 159 - 162 (1977).

077.009 6 centimeter observations of solar bursts with 6" resolution.
C. E. Alissandrakis, M. R. Kundu.
Astrophys. J., Vol. 222, 342 - 356 (1978).

Nine 6 cm solar bursts were observed with the Westerbork Synthesis Radio telescope on 1974 May 8–10. The bursts were of weak to moderate intensity. One-dimensional fan-beam scans with resolution of up to 6" were obtained every 30 s. The spatial structure and the time evolution of the bursts are discussed.

077.010 The development and structure of bright active regions at 2.8 cm.   A. Donati Falchi, M. Felli, P. Pampaloni, G. Tofani.
Sol. Phys., Vol. 56, 335 - 350 (1978).

The development of three intense active centers during their appearance on the solar disk is examined using high resolution observations at 2.8 cm. Each region shows a very bright component with brightness temperature $>10^6$ K and size smaller than 20". The development of the bright components has been investigated on different time scales. The problem of the bright cores height is discussed.

077.011 Manifestation of pulsation instability in solar radio emission preceding proton flares.   M. M. Kobrin, A. I. Korshunov, S. I. Arbuzov, V. V. Pakhomov, V. M. Fridman, Yu. V. Tikhomirov.
Sol. Phys., Vol. 56, 359 - 373 (1978).

The investigation of solar radio emission fluctuations at the wavelength λ ≅ 3 cm led to the discovery of a visible increase in pulsations with periods of about 30–120 min prior to proton flares. These pulsations were observed before all (seven) proton flares included in the cycle of observations from 1969 to 1974. The phenomenon was not found to occur before non-proton flares. The assumption is made that the observed pulsations are a manifestation of pre-flare instability in coronal structures. Estimates have been made for fluctuations of the gyro-resonance radiation from the regions above spots associated with the magnetic field variations when a groove instability of a coronal condensation is developed. They are in good agreement with the observational data.

077.012 A study of type V solar radio bursts. II. A theoretical model.   R. D. Robinson.
Sol. Phys., Vol. 56, 405 - 416 (1978).

A model for the solar type V event is developed. This

model assumes that the basic difference between type III and type V bursts is the evolution of the electron beam. For a type V this beam rapidly elongates, so that it takes progressively longer times to pass higher plasma levels. Physical process influencing the beam development, including Coulomb collisions, non-linear interactions with Langmuir waves and wave-particle scattering from various hydromagnetic wave modes is discussed. The model is compared with previously derived models and with observations.

077.013 Relation between circular polarization of moving type IV bursts and polarity of photospheric magnetic fields.   K. Kai.
Sol. Phys., Vol. 56, 417 - 427 (1978).

Two-dimensional, high-resolution observations of about 30 moving type IV bursts allow us to compare the polarization structure of the radio sources high in the corona with the distribution of magnetic fields measured at the photospheric level. Left- and right-handed circularly polarized moving type IV bursts are associated with active regions dominated by magnetic fields of plus and minus polarity respectively. The above relation between the sense of polarization and the polarity of magnetic field is contrary to what would be expected from the generally accepted synchroton hypothesis.

077.014 Occurrence of type II radio emission during the development of a solar radio flare.
L. Křivský, S. Pinter.
Izv. AN SSSR. Ser. fiz., Vol. 41, 1861 - 1863 (1977). In Russian. – Abstr. in Ref. zh., 51. Astron., 2.51.400 (1978).

077.015 On a mechanism of generation of "drift pair" solar radio bursts.   V. V. Zajtsev, B. N. Levin.
Astron. Zh. Akad. Nauk SSSR, Vol. 55, 390 - 398 (1978). In Russian. English translation in Soviet Astron., Vol. 22, No. 2.

A mechanism of "drift pair" bursts generation is suggested, based on cyclotron instability of a fast electron beam on plasma waves near a double plasma resonance region and their subsequent scattering into electromagnetic radiation. The main characteristics of "drift pair" bursts are explained within the framework of this mechanism.

077.016 On the intensity ratio of the second and third harmonics in type III solar radio bursts.
E. Ya. Zlotnik.
Astron. Zh. Akad. Nauk SSSR, Vol. 55, 399 - 403 (1978). In Russian. English translation in Soviet Astron., Vol. 22, No. 2.

It is shown that the intensity ratio of the third and the second harmonics can be much greater than the one predicted by Zheleznyakov and Zlotnik (1974), if the angular spectrum of the plasma waves induced by an electron beam and responsible for generation of type III radio bursts is quasi-onedimensional. Numerical estimates for the February 23, 1973 event, described by Takakura and Yousef (1974), are given.

077.017 On circular polarization of solar microwave bursts.   V. A. Kovalev.
Soln. Dannye 1977 Byull., No. 11, p. 74 - 79 (1978). In Russian.

The results of polarization observations of some microwave bursts at 3 GHz in July 1974 are discussed. Attention is drawn to the fact that intensity increase is accompanied by circular polarization decrease. This effect is explained by a change of the optical depth of the emitting plasma in the magneto-bremsstrahlung model of the burst.

077.018 First results of the STEREO-5 experiment: evidence of ionospheric intensity scintillation of solar radio bursts at decameter wavelengths?
M. Poquérusse, J. L. Steinberg.

Astron. Astrophys., Vol. 65, L23 - L26 (1978).

Simultaneous observations of solar radio bursts at 30 and 60 MHz from the Earth (Nançay) and from a planetary probe (Mars-7) show unexpectedly large differences (up to 3db) between the peak intensities of the two records, even when the angle probe–Sun–Earth θ is small (<3°). These intensity fluctuations are stronger at 30 MHz and their time scale ranges between 1 s and 1 min. The most likely interpretation is ionospheric intensity scintillation, a phenomenon generally considered as negligible in solar bursts observations.

077.019 Plasma instabilities of trapped particles in solar magnetic fields.   M. Berney, A. O. Benz.
Astron. Astrophys., Vol. 65, 369 - 384 (1978).

This work has been motivated by suggestions in the literature that type IV_{dm} radio bursts are the result of an electrostatic loss-cone instability. In order to investigate the trapping conditions of electrons in solar magnetic fields the authors have studied both the electromagnetic (whistler) and the electrostatic (hydrodynamic) instability in a magnetized plasma. It is argued that whistler growth could be limited by field curvature in a twisted magnetic field or by refraction. The electrostatic instability, on the other hand, will in any case reach a state of saturation, which is in a quasi-equilibrium with the losses of untrapped particles.

077.020 On the radio properties of the proton region (June-July, 1974).   Le Bach Yen.
Bull. Astron. Inst. Czechoslovakia, Vol. 29, 172 - 179 (1978).

The active region McMath 13043 from 28 June to 10 July, 1974 was the source of intensive radio bursts. Several bursts, recorded at Ondřejov, are mentioned and evaluated. Since the activity of the region close to the minimum of the solar cycle (1976) was rare and was not perturbed by other active sources on the Sun, the slowly varying component (S-component) was studied on certain selected frequency (its spectral pattern) and compared with the minimum flux of the quiet Sun.

077.021 Polarization and location of metric radiobursts in relationship with the emergence of a new magnetic field.   J. Heyvaerts, A. Kerdraon, A. Mangeney, M. Pick, C. Slottje.
Astron. Astrophys., Vol. 66, 81 - 86 (1978).

Spike bursts have been observed in association with some type III burst groups. In this event, the spikes appear according to bandwidth and duration to be mini-type I bursts. Their circular polarization cannot be explained by an emission of the ordinary mode in the magnetic field of the dominant photospheric polarity. This observation is interpreted by the expansion at $0.3 R_\odot$ of magnetic loops related to the emergence of a satellite polarity. A model of coronal field is proposed for the whole active center. Application of the theory of Mangeney and Veltri (1976) for type I bursts leads to a consistent picture of the generation of these spikes.

077.022 Structure of a local source on the sun obtained from eclipse observations with the 22-meter radio telescope in integral and circularly polarized light at 1.35 cm.
S. L. Domnin, V. A. Efanov, V. A. Korsenskij, I. G. Moiseev, N. S. Nesterov.
Izv. Krymskoj Astrofiz. Obs., Vol. 58, 35 - 39 (1978). In Russian.

Parameters of the radio source connected with the sunspot group N 24 were determined. The source in both circularly polarized (V) and integral (I) light consisted of three components with dimensions of 8, 12 and 30 seconds of arc, whose positions coincided in I and V. The brightness temperatures of the components were (0.6–3.1) × 10^5 K in I and (3.5–19) × 10^3 K in V.

077.023 On periodic fluctuations in the continuous noise

storm for the period of November 16–23, 1975.
G. F. Eliseev.
Soln. Dannye 1978 Byull., No. 1, p. 88 - 96 (1978).
In Russian.

The continuous radio emission was considered at 208 MHz frequency during November 16–23, 1975. Quasi-periodic fluctuations of the radio flux density with periods of 20–200 sec were found. They might be connected with rapid processes in the chromosphere. The fluctuations may either enhance or subdue the continuum. The appearance of oscillations was accompanied by a simultaneous change in the spectrum of the noise storm source.

077.024 Structure and evolution of solar radio bursts at 26.4 MHz.
H. S.-L. Chen, S. D. Shawhan.
Sol. Phys., Vol. 57, 205 - 227 (1978).

Two dimensional source brightness distributions at 26.4 MHz for solar bursts of spectral type II, III, IV, and V are derived from observations with a multiple-baseline, time-sharing interferometer system. It was designed explicitly to study the large angle (~40′ halo) component of low frequency solar bursts first reported by Weiss and Sheridan (1962). Thirty-two bursts occurring in the interval of June–August, 1975, were fit with a circular gaussian 'core' and an elliptical gaussian 'halo' component. All burst types were found to have some large angle structure. Two processes for producing the core-halo structure of type III bursts are compared.

077.025 Angular sizes of stria-burst sources in the range 24–26 MHz. E. P. (Eh. P.) Abranin, L. L.
Bazelyan, N. Yu. Goncharov, V. V. Zaitsev (Zajtsev), V. A. Zinichev, V. O. Rapoport, Ya. G. Tsybko.
Sol. Phys., Vol. 57, 229 - 235 (1978).

The UTR-2 antenna has been used to measure angular sizes of sources of narrow-band short-lived solar stria-bursts at frequencies 24–26 MHz. The majority of these sources have apparent diameters between 20 and 40′. According to this parameter they do not differ noticeably from that of type III bursts at the same frequency. The short duration of the stria-bursts prevents explanation of the large diameter by scattering in the solar corona.

077.026 On the directivity diagram of solar microwave radio bursts. M. N. Belovskij, Yu. P. Ochelkov.
Dokl. AN SSSR, Vol. 236, 1331 - 1333 (1977). In Russian. Abstr. in Ref. zh., 51. Astron., 4.51.414 (1978).

077.027 Electromagnetic loss-cone instability and type IV solar radio emission. A. V. Stepanov.
Pis'ma Astron. Zh., Vol. 4, 193 - 196 (1978). In Russian.

The electromagnetic loss-cone instability at electron cyclotron harmonics in coronal magnetic traps (sources of type IV burst) is examined. It is shown that in sufficiently strong magnetic fields the linear mechanism of radio emission from 50 keV electrons may predominate over the nonlinear plasma mechanism.

077.028 On the generation of solar radio spike bursts by electron streams. G. P. Chernov.
Astron. Zh. Akad. Nauk SSSR, Vol. 55, 572 - 585 (1978). In Russian. English translation in Soviet Astron., Vol. 22, No. 3.

Besides known spectral data investigations of time profiles of spikes indicate also spike generation by electron streams. For explanation of the spikes' narrow band and short duration it is supposed that the electron streams which excite spikes are born instantaneously near the emission region, have small sizes and small velocity dispersion. A calculation model of time profiles of spikes is suggested. Satisfactory coincidence of theoretical and observed profiles allows to determine the linear dimension of streams, velocity dispersion

and decay constant. Strong polarization of spikes is explained by the exit conditions of radiation from sources where the magnetic field strength is equal to ~ 3 - 4 gauss.

077.029 Origin of the components of the complex solar radio emission event of January 14, 1971.
Yu. B. Vedeneev, V. F. Mel'nikov.
Astron. Zh. Akad. Nauk SSSR, Vol. 55, 586 - 597 (1978). In Russian. English translation in Soviet Astron., Vol. 22, No. 3.

The following main features of the intense complex event associated to a proton flare have been revealed: 1) the moments of the beginning and of the end for the microwave and for the type V continua almost coincided, 2) short time-scale (~ 10 sec) microwave bursts observed in a broad frequency range (500–9100 MHz) on the microwave continuum background were accompanied by type III bursts on the type V background, 3) all the type III bursts had a high-frequency cut-off not higher than 250 MHz. A model for the motion of electrons ejected form the acceleration region has been suggested, as well as a mechanism for generation of the main components of the event by these electrons.

077.030 A non-homogeneous model of the chromosphere and of the lower corona for interpretation of data on the quiet sun's radio emission in the millimeter wavelength range. N. A. Kuznetsova.
Astron. Zh. Akad. Nauk SSSR, Vol. 55, 598 - 606 (1978). In Russian. English translation in Soviet Astron., Vol. 22, No. 3.

A method of calculation of the spectrum and of the radio brightness distribution over the quiet sun's disk for a non-homogeneous model of the chromosphere and of the lower corona, including spiculae, is suggested. A number of non-homogeneous models of the solar atmosphere, allowing to account for the observed change of the spectral index of the sun's radio emission in the millimeter range at wavelengths 5–9 mm, is considered.

077.031 Nonlinear Langmuir waves during type III solar radio bursts. D. R. Nicholson, M. V. Goldman,
P. Hoyng, J. C. Weatherall.
Astrophys. J., Vol. 223, 605 - 619 (1978).

Type III solar radio bursts are thought to be associated with intense levels of electron beam excited Langmuir waves. The purpose of this paper is to present a two-dimensional treatment of these waves, including the relevant wave-wave nonlinearities, while ignoring plasma inhomogeneity, wave-particle interactions, and effects of the background magnetic field.

077.032 The height of the 9 cm solar emission.
M. Waldmeier.
Sol. Phys., Vol. 57, 369 - 371 (1978).

The large values of the heights, their dependence on latitude and time as found by Graf and Bracewell can satisfactorily be explained by assuming a constant height of around 18 000 km and by taking into account that the microwave sources do not lay above the spots but above the plages.

077.033 On the coronal source regions of U bursts. From observations with the three-frequency radioheliograph and the spectropolarimeter at Culgoora.
S. Suzuki.
Sol. Phys., Vol. 57, 415 - 422 (1978).

The projected source positions at 43, 80, and 160 MHz and the sense and degree of circular polarization in the range 24 to 220 MHz, are used: (1) To substantiate the hypothesis that metric U bursts originate in high coronal, magnetic loops. (2) To strengthen the hypothesis that U-burst radiation is in the ordinary magneto-ionic mode.

077.034 Reply to the paper: 'Solar radio type III bursts and coronal density structures' by Y. Leblanc and J. de la Noë.   C. Mercier.
Sol. Phys., Vol. 57, 423 - 428 (1978).

Leblanc and de la Noë (see 20.077.030) used the set of data published by Mercier and Rosenberg (1974) on the type III burst at 169 MHz. They conclude that type III bursts are associated with low density coronal structures and occur in low density regions. The author shows that their methods cannot lead to firm conclusions; he points out some inconsistencies in their results.

077.035 On the relation of SPA (sudden phase anomalies) measured at VLF to solar microwave burst energies.
P. Kaufmann, L. R. Piazza.
Sol. Phys., Vol. 57, 479 - 481 (1978).

A surprisingly good correlation has been found for SPA measured at VLF propagation ($\Delta\phi$) and 7 GHz solar microwave burst energies (E_μ). The data are correlated in the form $\Delta\phi = a \log E_\mu + b$ and include all kind of solar events, irrespectively from type, complexity or duration. Soft X-ray peak fluxes (I_x) have a known similar correlation to SPA, and a functional relationship of the form $E_\mu^a \sim I_x^k$ can be established. As one practical application, the energies from solar events can be reasonably well inferred from SPA data, which are quite reliable and easily obtainable.

077.036 Fine structure and time variation of the quiet sun at 1.3 cm.   A. P. Rao, M. R. Kundu.
Bull. American Astron. Soc., Vol. 10, 431 - 432 (1978). Abstract.

077.037 Radio data and theoretical models for shock waves generated by intense solar flares.   A. Maxwell.
Bull. American Astron. Soc., Vol. 10, 454 (1978). – Abstract.

077.038 On the quasi-periodic structure of the impulsive burst of March 1, 1970.   H. Wiehl, A. Magun, J. A. Ionson, C. Crannell, K. Frost, C. Mätzler.
Bull. American Astron. Soc., Vol. 10, 455 (1978). – Abstract.

077.039 Possible explanations for the turnovers in two solar microwave burst spectra.
I. A. Ahmad, M. R. Kundu.
Bull. American Astron. Soc., Vol. 10, 455 (1978). – Abstract.

077.040 Spectral development of solar cm-λ bursts.
G. L. Tarnstrom.
Bull. American Astron. Soc., Vol. 10, 461 (1978). – Abstract.

077.041 Longitude dependence of the duration of solar radio type III bursts at decameter wavelengths.
J. de la Noë, M. G. Aubier, S. M. Youssef.
Astron. Astrophys., Vol. 67, 333 - 338 (1978).

Measurements of a large sample of type III bursts at 29, 37 and 60 MHz for a half solar rotation display a longitude dependence of the burst duration, of exciter duration and of decay time. Each of these quantities increases toward low longitudes. Tentative explanations are proposed which are based either on the combination of fundamental and harmonic modes being observed near central meridian or on a variation of the intrinsic directivity of the radiation in the course of the emission.

077.042 Observations radioélectriques solaires faites sur 600 MHz en 1976 au Laboratoire de Radioastronomie de Humain-Rochefort.   C. Gonze, R. Gonze.
Bull. Astron. Obs. R. Belgique, Vol. 9, 33 - 42 (1978).

077.043 Study of power spectra of quasi-periodic fluctuations of solar radio emission at 3 cm wavelength with periods shorter than 100 sec.   A. A. Bezotosnyj,

O. G. Gontarev.
Ionos. i soln.-zemnye svyazi. Alma-Ata, Nauka, KazSSR, 1978, p. 112 - 116. In Russian. – Abstr. in Ref. zh., 51. Astron., 6.51.497 (1978).

077.044 On the influence of atmospheric fluctuations on the investigation of quasi-periodic fluctuations of the solar radio radiation flux at 3.2 cm wavelength.
A. A. Bezotosnyj.
Ionos. i soln.-zemnye svyazi. Alma-Ata, Nauka, KazSSR, 1978, p. 116 - 121. In Russian. – Abstr. in Ref. zh., 62. Issled. kosm. prostranstva, 6.62.325 (1978).

077.045 Emissioni radio solari associate ai brillamenti cromosferici.   A. Abrami.
Mem. Soc. Astron. Italiana, Vol. 48, (see 012.037), 213 - 229 (1977).

077.046 Eventi di tipo III ad alta risoluzione temporale registrati all'Osservatorio Astronomico di Trieste; descrizione della strumentazione per l'acquisizione e riduzione dei dati polarimetrici.
L. Lampi, P. Santin, P. Zlobec.
Mem. Soc. Astron. Italiana, Vol. 48, (see 012.037), 231 - 236 (1977).

077.047 The fine structure of solar radio bursts at meter wavelengths.   V. V. Fomichev, I. M. Chertok.
Izv. VUZ. Radiofiz., Vol. 20, 1255 - 1301 (1977). In Russian. Abstr. in Ref. zh., 51. Astron., 7.51.447 (1978).

077.048 Some comments on the observed relationship between type I and type III solar radio bursts.
R. T. Stewart.
Izv. VUZ. Radiofiz., Vol. 20, 1338 - 1358 (1977). In Russian. Abstr. in Ref. zh., 51. Astron., 7.51.448 (1978).

077.049 Recent data on type II solar radio bursts.
D. J. McLean, G. J. Nelson.
Izv. VUZ. Radiofiz., Vol. 20, 1359 - 1368 (1977). In Russian. Abstr. in Ref. zh., 51. Astron., 7.51.449 (1978).

077.050 Remarks on plasma emission and its application to solar radio bursts.   D. V. Melrose.
Izv. VUZ. Radiofiz., Vol. 20, 1369 - 1378 (1977). In Russian. Abstr. in Ref. zh., 51. Astron., 7.51.450 (1978).

077.051 On the theory of type II and III solar radio bursts.
V. V. Zajtsev.
Izv. VUZ. Radiofiz., Vol. 20, 1379 - 1398 (1977). In Russian. Abstr. in Ref. zh., 51. Astron., 7.51.451 (1978).

077.052 Type III bursts with fine structure in the spectrum in the decameter wavelength range.
L. L. Bazelyan, V. A. Zinichev, V. O. Rapoport.
Izv. VUZ. Radiofiz., Vol. 20, 1399 - 1412 (1977). In Russian. Abstr. in Ref. zh., 51. Astron., 7.51.452 (1978).

077.053 Fine structure in solar radio emission.
P. M. McCulloch, G. R. A. Ellis.
Izv. VUZ. Radiofiz., Vol. 20, 1413 - 1431 (1977). In Russian. Abstr. in Ref. zh., 51. Astron., 7.51.453 (1978).

077.054 Observations of the circular polarization in type III bursts in the first and second harmonics in the frequency range 24 to 220 MHz.
S. Suzuki, K. V. Sheridan.
Izv. VUZ. Radiofiz., Vol. 20, 1432 - 1443 (1977). In Russian. Abstr. in Ref. zh., 51. Astron., 7.51.454 (1978).

077.055 Observations of the radio burst on July 4, 1974 with radar stations (ECRS).   S. I. Avdyushin,

M. M. Alibegov, Yu. F. Barabanshchikov.
Tr. Inst. prikl. geofiz. Gl. upr. gidrometeorol. sluzhby pri Sov.
Min. SSSR, 1977, No. 35, p. 34 - 36. In Russian. – Abstr. in
Ref. zh. 51. Astron., 7.51.455 (1978).

**077.056 Some parameters of radio bursts associated with
solar cosmic ray flares.**
M. N. Belovskij, N. K. Pereyaslova.
Tr. Inst. prikl. geofiz. Gl. upr. gidrometeorol. sluzhby pri Sov.
Min. SSSR, 1977, No. 35, p. 51 - 55. In Russian. – Abstr. in
Ref. zh., 51. Astron., 7.51.456 (1978).

**077.057 On a possible relation of the S-component of radio
emission with solar wind velocity.**
M. N. Belovskij, L. G. Kuz'micheva.
Tr. Inst. prikl. geofiz. Gl. upr. gidrometeorol. sluzhby pri Sov.
Min. SSSR, 1977, No. 35, p. 56 - 58. In Russian. – Abstr. in
Ref. zh., 51. Astron., 7.51.457 (1978).

Catalogo degli eventi radio solari di tipo IV in pre-
parazione all'Osservatorio Astronomico di   Trieste.
See Abstr. 002.067.

An acousto-optical solar radio spectrograph.
See Abstr. 033.034.

The polarization of second harmonic plasma emission.   See Abstr. 062.077.

Some studies on isolated sunspot magnetic configurations and associated Hα-flares and microwave bursts.
See Abstr. 072.042.

Multifrequency observations of solar filaments at
centimeter wavelengths.   See Abstr. 073.002.

Motion of a shock wave due to the solar flare on
4 August 1972 according to observations of type II bursts.
See Abstr. 073.065.

Hα, hard X-ray, and microwave emissions in the
impulsive phase of solar flares.   See Abstr. 073.072.

The radio spectrum of chromospheric structure.
See Abstr. 073.075.

OSO-8, radio and X-ray observations of the 19 April
1977 flare.   See Abstr. 073.079.

VLA observations of solar flares.
See Abstr. 073.083.

Meccanismi di emissione nei brillamenti solari.
See Abstr. 073.096.

The emission of Langmuir waves in the solar corona.
See Abstr. 074.042.

Coronal holes at 11.5 and 21 cm observed with the
Arecibo radio telescope.   See Abstr. 074.047.

Radio and white light observations of the 21 August
1973 coronal transient.   See Abstr. 074.084.

A survey of the 127 MHz Toruń solar radio data.
See Abstr. 075.001.

Solare Beobachtungsergebnisse (Solar data).
See Abstr. 075.005.

Solar X-ray emission and metre-wave radio bursts.
See Abstr. 076.007.

On quantitative diagnostics of proton bursts from
characteristics of microwave radio bursts at $\sim$9 GHz frequency.
See Abstr. 078.012.

On the change of the diameter of the radio image
of the sun at short centimeter wavelengths.
See Abstr. 080.036.

Radioastronomy and amateurs.
See Abstr. 141.154.

078 Cosmic Radiation

078.001 Solar cosmic rays travel to the sun.
Yu. V. Mineev.
Zemlya i Vselennaya, 1978, No. 1, p. 32 - 36. In Russian.

**078.002 Investigation of propagation of solar cosmic radia-
tion from eastern flares on May 28, June 15 and
August 2, 1972 aboard Prognoz. I. Experimental results of
measurements of solar particles in flares and analysis of integral proton spectra.**   B. M. Kuzhevskij, V. L. Maduev,
N. F. Pisarenko, I. A. Savenko, I. P. Shestopalov.
Kosm. Issled.,Vol. 16, 64 - 72 (1978). In Russian.

**078.003 Bidirectional anisotropies in solar cosmic ray events:
evidence for magnetic bottles.**
I. D. Palmer, F. R. Allum, S. Singer.
J. Geophys. Res., Vol. 83, 75 - 90 (1978).
The occurrence of bidirectional anisotropies in low
energy solar proton and electron events has been investigated
in the period 1967–1973 by using data obtained from five
satellites, viz., Explorer 34 and 41 and Vela 5B, 6A, and 6B.

Some 16 observations of bidirectional distribution were
detected in this time period. The characteristics of two of
these electron events are discussed in detail. The average duration over all events is 9 hours, implying a spatial scale of
$\approx$0.13 AU, and the events show a strong correlation with the
minimum of a Forbush decrease. The two maxima in the
particle distributions are aligned with the interplanetary magnetic field.

**078.004 Solar cosmic radiation and interplanetary shock
waves on April 29 - 30, 1973.**
N. N. Volodichev, N. L. Grigorov, G. Ya. Kolesov, O. M.
Kovrizhnykh, M. I. Kudryavtsev, B. M. Kuzhevskij, V. G. Kurt,
Yu. I. Logachev, N. F. Pisarenko, I. A. Savenko, A. A. Suslov,
L. M. Chupova, V. F. Shesterikov, I. P. Shestopalov.
Izv. AN SSSR. Ser. fiz., Vol. 41, 1794 - 1807 (1977). In Rus-
sian. – Abstr. in Ref. zh., 62. Issled. kosm. prostranstva,
2.62.137 (1978).

078.005 Super-adiabatic acceleration of charged particles and

flares on the sun and stars. A. A. Rumyantsev.
Izv. AN SSSR. Ser. fiz., Vol. 41, 1790 - 1793 (1977). In Russian. − Abstr. in Ref. zh., 62. Issled. kosm. prostranstva, 2.62.138 (1978).

078.006 Investigation of propagation of solar cosmic radiation from the eastern flares on May 28, June 15 and August 2 in 1972 aboard the Prognoz satellite. II. Model of propagation of solar particles in the interplanetary medium.
B. M. Kuzhevskij, V. L. Maduev, N. F. Pisarenko, I. A. Savenko, I. P. Shestopalov.
Kosm. Issled., Vol. 16, 250 - 256 (1978). In Russian.

078.007 Low-energy corpuscular fluxes from non-flare active solar regions. M. A. Zel'dovich, B. M. Kuzhevskij, Yu. I. Logachev.
Problems of solar activity and space system Prognoz, (see 003.008), p. 110 - 124 (1977). In Russian. − Abstr. in Ref. zh., 51. Astron., 4.51.408; 62. Issled. kosm. prostranstva, 4.62.130 (1978).

078.008 Direct measurements of proton influx from the solar flare of August 7, 1972 on the Prognoz 2 station. N. N. Volodichev, N. L. Grigorov, G. Ya. Kolesov, E. I. Morozova, A. N. Podorol'skij, I. A. Savenko, A. A. Suslov.
Problems of solar activity and space system Prognoz, (see 003.008), p. 131 - 136 (1977). In Russian. − Abstr. in Ref. zh., 51. Astron., 4.51.409; 62. Issled. kosm. prostranstva, 4.62.132 (1978).

078.009 Registration of plasma from solar flares of August 1972 in the earth's orbit. O. L. Vajsberg, F. Cambou, G. N. Zastenker, A. A. Zertsalov, V. V. Temnyj, M. Z. Khokhlov, H. Espange, C. d'Uston.
Problems of solar activity and space system Prognoz, (see 003.008), p. 155 - 174 (1977). In Russian. − Abstr. in Ref. zh., 51. Astron., 4.51.423; 62. Issled. kosm. prostranstva, 4.62.134 (1978).

078.010 Peculiarities of spatial and temporal characteristics of solar cosmic radiation in the events of 1972, April 19 and 1974, July 5. N. A. Mikirova.
Tr. IPG, 1977, vyp. (No.) 35, p. 24 - 27. In Russian. − Abstr. in Ref. zh., 62. Issled. kosm. prostranstva, 4.62.405 (1978).

078.011 On the peculiarity of emergence of solar flare accelerated electrons into the interplanetary medium. O. M. Kovrizhnykh, B. M. Kuzhevskij, L. M. Chupova.
Pis'ma Astron. Zh., Vol. 4, 197 - 200 (1978). In Russian.
From analysis of data on solar X-ray fluxes at 0.5−3 Å and solar electron fluxes of energy greater than 30 keV follows that the larger is the flare and the more particles are accelerated in it, the larger fraction of particles leaves the sun.

078.012 On quantitative diagnostics of proton bursts from characteristics of microwave radio bursts at ∼9 GHz frequency. S. T. Akin'yan, M. M. Alibegov, V. D. Kozlovskij, I. M. Chertok.
Geomagn. Aehron., Vol. 18, 410 - 414 (1978). In Russian.

078.013 Rigidity-independent propagation of cosmic rays in the solar corona.
G. Newkirk, Jr., D. G. Wentzel.
J. Geophys. Res., Vol. 83, 2009 - 2015 (1978).
The authors propose a mechanism for the coronal transport of solar cosmic rays: the bird cage model, in which particles are normally transferred between adjacent flux tubes by field line reconnection produced by the rearrangement of the field in the supergranulation network. The observed size of coronal flux loops and the rate of reorganization of flux in the supergranulation network lead to 'diffusion' rates which

are both independent of rigidity and energy and consistent with the observed propagation rates.

078.014 Rigidity-independent coronal propagation and escape of solar protons and α particles.
C. Perron, V. Domingo, R. Reinhard, K.-P. Wenzel.
J. Geophys. Res., Vol. 83, 2017 - 2029 (1978).
A statistical study of 42 solar proton and α particle events has been performed in the energy range of 9−36 MeV/nucleon (nuc) as measured by the Space Science Department/European Space Agency experiment on Heos 2 during 1972−1974. From one event to another there is a high variability of the p/α ratio at equal energy per nucleon ranging from about 4 to 1000 at 10 MeV/nuc. It is found that the lower value of the p/α ratios increases with azimuthal distance from the flare site.

078.015 The composition of corotating energetic particle streams.
R. E. McGuire, T. T. von Rosenvinge, F. B. McDonald.
NASA Tech. Memo., NASA TM 78105, 22 pp. (1978).
The relative abundances of 1.5 - 23 MeV/nucleon ions in corotating nucleon streams are compared with ion abundances in particle events associated with solar flares and with solar and solar wind abundances. He/O and C/O ratios are found to be a factor of the order 2-3 greater in corotating streams than in flare-associated events. The distribution of H/He ratios in corotating streams is found to be much narrower and of lower average value than in flare-associated events.

078.016 Prompt solar proton events and coronal mass ejections. S. W. Kahler, E. Hildner, M. A. I. van Hollebeke.
Sol. Phys., Vol. 57, 429 - 443 (1978).
The authors have used data from the HAO white light coronograph and AS&E X-ray telescope on Skylab to investigate the coronal manifestations of 18 prompt solar proton events observed with the GSFC detectors on the IMP-7 spacecraft during the Skylab period. They find evidence that a mass ejection is a necessary condition for the occurrence of a prompt proton event. They suggest that: (1) the occurrence of mass ejection events facilitates the escape of protons − whether accelerated at low or high altitudes − to the interplanetary medium; and (2) there may exist a proton acceleration region above or around the outward moving ejecta far above the flare site.

078.017 Raggi cosmici durante i brillamenti solari.
E. Antonucci.
Mem. Soc. Astron. Italiana, Vol. 48, (see 012.037), 257 - 268 (1977).
Contents: Fenomenologia degli eventi di raggi cosmici solari. Eventi impulsivi di particelle solari. Classificazione dei brillamenti in base all'emissione corpuscolare. Composizione delle particelle solari. Raggi cosmici solari prodotti da razioni nucleari.

078.018 Correlation of propagation characteristics of solar cosmic rays detected on board the spatially separated space probes Mars-7 and Prognoz-3. T. Gombosi, A. J. Somogyi, G. Ya. Kolesov, V. G. Kurt, B. M. Kuzhevskii (*Kuzhevskij*), Yu. I. Logachev, I. A. Savenko.
15th IUPAP International Conference on cosmic rays, (see 012.042), p. 85 - 92 (1977). − Abstr. in Phys. Abstr., Vol. 81, Abstr. 49247 (1978).

078.019 Study of coronal and interplanetary propagation of solar particles following the E45° solar flare on July 29, 1973. T. Gombosi, A. J. Somogyi, G. Ya. Kolesov, V. G. Kurt, B. M. Kuzhevskii (*Kuzhevskij*), Yu. I. Logachev, I. A. Savenko, I. P. Shestopalov.
15th IUPAP International Conference on cosmic rays, (see 012.042), p. 93 - 101 (1977). − Abstr. in Phys. Abstr., Vol. 81,

Abstr. 49248 (1978).

Index of compactness of solar active regions and characteristics of proton events.   See Abstr. 072.022.

Evaluation of a prediction technique for low energy solar particle events.   See Abstr. 073.024.

A study of non-relativistic electrons generated during solar flares.   See Abstr. 073.066.

Shock-associated energetic proton events at large heliocentric distances.   See Abstr. 074.069.

Prompt solar proton events and coronal mass ejections.   See Abstr. 074.085.

Some parameters of radio bursts associated with solar cosmic ray flares.   See Abstr. 077.056.

Polar-cap absorption — observations and theory. See Abstr. 083.127.

Magnetosheath distortion of pitch angle distributions of solar protons.   See Abstr. 084.223.

Interplanetary propagation of flare-associated energetic particles.   See Abstr. 106.044.

The radial variation of corotating energetic particle streams in the inner and outer solar system. See Abstr. 106.057.

079 Solar Eclipses

Point discharge currents during an eclipse of the Sun.
See Abstr. 082.077.

Solar eclipse 1977 October 12

079.101 North America's partial eclipse.
Sky Telesc., Vol. 55, 6 - 8 (1978).

079.102 October's partly cloudy solar eclipse.
Sky Telesc., Vol. 55, 15 - 18 (1978).

079.103 More about October's solar eclipse.
Sky Telesc., Vol. 55, 181 - 187 (1978).

079.104 Infrared observations of the 1977 total solar eclipse.
J. M. Pasachoff, M. T. Sandford, II, C. F. Keller, Jr.
Bull. American Astron. Soc., Vol. 10, 431 (1978). – Abstract.

Solar eclipse 1976 April 29

079.201 L'eclisse parziale di sole del 29/IV/1976.
E. Sassone Corsi, P. Sassone Corsi.
Astronomia, N. 4, p. 22 - 27 (1977).

079.202 Results of observations of the annular solar eclipse on 29 April, 1976 at 3.2 and 5.2 cm wavelengths.
G. Ya. Smol'kov, A. A. Dutov, V. G. Zandanov, B. I. Lubyshev, V. P. Nefed'ev, V. S. Polukhin, N. N. Potapov, A. Ya. Smol'kov, A. V. Storozhko, Yu. V. Tarbeev, T. A. Treskov, I. Yu. Ul'danov, M. A. Khaitov.
Issled. po geomagn., aehron. i fiz. Solntsa. Vyp. (No.) 42.
Moskva, Nauka, 1977, p. 76 - 82. In Russian. – Abstr. in Ref. zh., 51. Astron., 4.51.392 (1978).

079.203 Observation visuelle de l'éclipse de soleil du 29 avril 1976 à l'astrographe double de 40 cm.
H. Debehogne, G. Roland.
Bull. Astron. Obs. R. Belgique, Vol. 9, 43 - 44 (1978).

079.204 Results of observations of the partial solar eclipse on April 29, 1976 at frequencies of 755 MHz, 610 MHz and 326 MHz. D. Blums, G. Ozoliņš, M. Eliass.
Astron. Tsirk., No. 988, p. 2 - 3 (1978). In Russian.

Quasi-periodic fluctuations in electron content during a partial solar eclipse. See Abstr. 083.003.

Ionospheric observations of the ring-shaped solar eclipse on April 29, 1976 in Ashkhabad.
See Abstr. 083.066.

Solar eclipse 1979 February 26

079.301 The total eclipse of February 26, 1979.
Sky Telesc., Vol. 55, 486 - 488 (1978).

079.302 Weather prospects for February's eclipse.
E. M. Brooks.
Sky Telesc., Vol. 56, 4 - 5 (1978).

079.303 Predictions for the 1979 solar eclipse.
F. Espenak.
J. R. Astron. Soc. Canada, Vol. 72, 149 - 161 (1978).
Predictions are presented which delineate the limits of the path of totality and describe the central-line circumstances for the 1979 solar eclipse. Local circumstances for 70 cities in Canada and the United States are included and weather prospects are investigated. The general characteristics of this eclipse are discussed and an examination of the history of the saros family puts this eclipse into perspective.

079.304 Total solar eclipse of 26 February 1979.
A. D. Fiala, M. R. Lukac.
United States Naval Obs. Circ. No. 157, 47 pp. (1977).

Solar eclipse 1980 February 16

079.401 Total solar eclipse of 1980. J. C. Bhattacharyya.
Bull. Astron. Soc. India, Vol. 6, 16 - 17 (1978).

079.402 Total solförmörkelse i Kenya den 16 februari 1980.
P.-Å. Björklund.
Astron. Tidsskr., Årg. 11, 88 - 91 (1978).

Solar eclipse 1952 February 25

079.501 Terrestrial observations at the solar eclipse of February 25, 1952. M. Waldmeier.
Astron. Mitt. Eidg. Sternw. Zürich, Nr. 361, 23 pp. (1978).
In German.

Solar eclipse 1975 May 11

079.601 Results of observations of the partial solar eclipse on May 11, 1975 at frequencies of 755 MHz and 326 MHz. D. Blums, G. Ozoliņš, M. Eliass.
Astron. Tsirk., No. 988, p. 1 - 2 (1978). In Russian.

080 Atmosphere, Figure, Internal Constitution, Neutrinos, Magnetic Fields, Rotation, Miscellanea

080.001 Differential rotation rates for short-lived regions of emerging magnetic flux.
L. Golub, G. S. Vaiana.
Astrophys. J., Lett., Vol. 219, L55 - L57 (1978).

The authors have measured the synodic rotation rates of a sample of compact X-ray emission features lasting from 1 day to 7 days, thus bridging the transition between X-ray bright points and active regions. The rotation rate is found to be a function of the lifetime, or size, of the feature; shorter-lived smaller features rotate more slowly than long-lived ones. The rotation rate for features lasting 2 days or less is consistent with that of the photospheric gas. The longest-lived features rotate at a rate ~5% higher, consistent with the sunspot rotation rate.

080.002 Radiative equilibrium in the photosphere and global oscillations of the sun and an upper bound on energy transport. R. H. Dicke.
Mon. Not. R. Astron. Soc., Vol. 182, 303 - 314 (1978).

The effect of non-local radiation equilibrium in the photosphere acting as a constraint on global oscillations ($\omega < 0.02$, $l \leqslant 3$) is determined. It is shown that for a correctly chosen effective specific heat ratio $\tilde{\gamma}$ the wave equation is very nearly adiabatic in form. The theory is used to compute upper bounds on energy transport to the upper atmosphere using the Grec-Fossat observations. These upper bounds are 3–5 orders of magnitude too small to be significant for the 1966 oblateness observations.

080.003 The interpretation of C III and O V emission line ratios in the sun. P. L. Dufton, K. A. Berrington, P. G. Burke, A. E. Kingston.
Astron. Astrophys., Vol. 62, 111 - 120 (1978).

Using new atomic data for electron collision rates and Einstein A-coefficients, the observed C III and O V emission line ratios in the solar transition region have been reinterpreted. From the C III ratio $R = I(^3P^e \rightarrow ^3P^o)/I(^1P^o \rightarrow ^1S^e)$, a quiet sun electron density (N_e) of $3 \times 10^9\,\mathrm{cm}^{-3}$ is deduced. For O V, the ratio, R, observed in the quiet sun cannot be reproduced with $N_e > 10^8\,\mathrm{cm}^{-3}$, but this discrepancy can be removed by 10% charges in selected collision rates.

080.004 On the solar rotation elements i and Ω as determined by Doppler velocity measurements of the solar plasma. H. Wöhl.
Astron. Astrophys., Vol. 62, 165 - 177 (1978) = Habilitationsschr. Univ. Göttingen.

The basic assumptions, experimental techniques and the results of two attempts to derive the solar rotation elements i (the angle between the ecliptic and the solar equator) and Ω (the angle between the crosspoint of the solar equator with the ecliptic and the equinox point) by Doppler velocity measurements of the solar plasma are given. Both methods are mainly based on the solar differential rotation. The first method depends on the determination of the maximum of the differential rotation velocity as well as on recently detected humps in the shape of the differential rotation velocity. The second method makes use of the full shape of the differential rotation velocity. The derived solar rotation elements are $i = (6.77 \pm 0.31)°$ and $\Omega = (76.31 \pm 0.65)°$ for 1976.

080.005 Gab es Änderungen der Solarkonstanten in historischer Zeit? H. Wöhl.
Sterne Weltraum, Jahrg. 17, 25 - 27 (1978).

080.006 On the theoretical spectrum of the free oscillations of the sun. S. V. Vorontsov, V. N. Zharkov.
Astron. Zh. Akad. Nauk SSSR, Vol. 55, 84 - 95 (1978). In Russian. English translation in Soviet Astron., Vol. 22, No. 1.

The spectrum of quadrupole oscillations of solar polytropes is calculated and the usual scheme of classification of the oscillations on p, f, g modes is discussed. On the example of the oscillations of the standard solar model it is shown that such a scheme may lead to misunderstandings in the identification of the observed periods with the theoretical normal-mode spectrum. An additional scheme of classification of the oscillations of the outer regions is proposed.

080.007 L'énigme des neutrinos du soleil. R. Omnès.
La recherche en astrophysique, (see 003.001), p. 55 - 69 (1977).

080.008 Le soleil. Y. Chmielewski.
Conférences d'astronomie de l'Observatoire, (see 012.004), p. 1 - 28 (1977).

Le soleil calme; structure interne, l'atmosphère solaire. Le soleil actif. Conclusions et relations soleil-terre.

080.009 What further with the solar neutrinos? B. Kuchowicz.
Postępy Astron., Tom 25, 15 - 19 (1977). In Polish.

The discrepancy between Davis' observational results and theoretical predictions forms the basis for a high influx of hypotheses in which new physical processes are introduced, standard assumptions of stellar evolution theory are modified etc. At the same time, the reaction $^{37}\mathrm{Cl} + \nu_e \rightarrow ^{37}\mathrm{Ar} + e^-$ applied in the above mentioned experiment still has not been directly verified in any laboratory neutrino source. An application of another nuclid besides $^{37}\mathrm{Cl}$ to solar neutrino detection may be equally necessary for solving the neutrino puzzle.

080.010 A solar abundance study using recent Ti I oscillator strengths. M. Gehlsen, H. Holweger, K. Danzmann, M. Kock, M. Kühne.
Astron. Astrophys., Vol. 64, 285 - 288 (1978).

The f-values measured by Kühne et al. (1977) lead to a solar abundance of titanium, $\log \epsilon_{Ti} = 4.94 \pm 0.12$ on the scale $\log \epsilon_H = 12$. Comparison with abundances following from two independent sets of f-values (Ellis, 1976; Whaling et al., 1977) shows basic consistency of absolute scales. The solar ratio $Ti/Ca = 3.8 \times 10^{-2}$ ($\pm 25\%$) matches that found in chondritic meteorites of types C1 or C2, but contrasts with other types. Further subjects of abundance analysis are the solar velocity field, constraints on departures from LTE, and the accuracy of the STFD oscillator strengths of Kurucz and Peytremann (1975).

080.011 Nonlinear astrophysical dynamos: the solar cycle as a nonlinear oscillation of the general magnetic field driven by the nonlinear dynamo and the associated modulation of the differential-rotation–global-convection system. H. Yoshimura.
Astrophys. J., Vol. 220, 692 - 711 (1978).

The nonlinear reaction of the magnetic field on the field-generating fluid motions in the dynamo process is formulated within the framework of the dynamo equation of the mean magnetohydrodynamics. If this nonlinear reaction mechanism is working in the process of the solar cycle, the differential-rotation–global-convection system operating the solar dynamo must be undergoing an oscillatory modulation associated with the solar cycle. As observable evidences of the operation of the dynamo process, the extended period of near absence of the surface activity of the Sun in the 17th to 18th centuries, the rapid resumption of the magnetic activity to the normal level of the present solar cycle, and the observed variation of the differential rotation at the surface are discussed. Numerical solutions of the dynamo equation as an initial-boundary-value problem show that this nonlinear reaction mechanism works so efficiently that steady oscillations can be achieved quickly from arbitrary initial conditions of negligible field level even without the other possible nonlinear mechanism of the magnetic flux eruption from the system. This shows that the nonlinear dynamo equation governing the solar cycle can be applied to other general astrophysical dynamos.

080.012 **Phase reversals in the polar magnetic fields of the sun and in the annual and semiannual variations in cosmic ray intensity.**
E. Antonucci, D. Marocchi, G. E. Perona.
Astrophys. J., Vol. 220, 712 - 718 (1978).

Annual and semiannual variations in galactic cosmic ray intensity have been computed for the period 1954–1973. These variations show a fairly constant phase over several years. However, an abrupt reversal of the direction of both vectors, representing the annual and the semiannual variations, is associated with the reversal of the polarities of the solar polar magnetic fields, which occurs near sunspot maximum. Furthermore, it is well known that the annual and semiannual variations may be related to the first- and second-order derivatives, with respect to the solar latitude, of the cosmic ray density. Consequently, the cosmic ray gradient, perpendicular to the solar equatorial plane, regularly changes its direction at the solar activity maximum, displaying a 22 year periodicity.

080.013 **Vertical motions in an intense magnetic flux tube.**
B. Roberts, A. R. Webb.
Sol. Phys., Vol. 56, 5 - 35 (1978).

The recent discovery of localised intense magnetic fields in the solar photosphere is one of the major surprises of the past few years. The authors consider the theoretical nature of small amplitude motions in such an intense magnetic flux tube, within which the field strength may reach 2 kG. They give a systematic derivation of the governing 'expansion' equations for a vertical, slender tube, taking into account the dependence upon height of the buoyancy, compressibility and magnetic forces. Several special cases (e.g., the isothermal atmosphere) are considered as well as a more realistic, non-isothermal, solar atmosphere. The expansion procedure is shown to give good results in the special case of a uniform basic-state, in which gravity is negligible and for which a more exact treatment is possible.

080.014 **Bifurcation of force-free solar magnetic fields: a numerical approach.** K. Jockers.
Sol. Phys., Vol. 56, 37 - 53 (1978).

Numerical calculations of two-dimensional force-free fields as models of solar active regions are presented. For a given 'toroidal' component of the photospheric magnetic field two branches of solutions are numerically obtained which merge at the critical point of maximum allowed toroidal magnetic field. Depending on boundary conditions magnetic islands may or may not form. The results are discussed with respect to their relevance to the flare process.

080.015 **Linear force-free fields in the lower corona.**
D. D. Barbosa.
Sol. Phys., Vol. 56, 55 - 66 (1978).

The author has computed the surface Green's function for linear force-free magnetic fields, where $\nabla \times \mathbf{B} = \kappa \mathbf{B}$ and κ is a constant, for application to low coronal levels of the solar atmosphere. Boundary conditions are imposed on the normal component of $\mathbf{B}$ on two parallel planes which delineate the force-free volume. This procedure ensures that the magnetic field energy remains bounded, and that the field lines have a smooth behavior. A simple bipolar source distribution is treated and representative field line tracings are shown.

080.016 **The sun.** J. P. Wild.
Focus on the stars, (see 003.002), p. 71 - 110 (1977).

080.017 **Observations of long period solar oscillations.**
R. Stebbins.
Bull. American Astron. Soc., Vol. 9, 568 (1978). – Abstract.

080.018 **The spectral type of the sun.** H. J. Reitsema.
Bull. American Astron. Soc., Vol. 9, 635 (1978).
Abstract.

080.019 **Has the sun a companion star?**
E. L. Wright, with a reply by E. R. Harrison.
Nature, Vol. 272, 649 (1978).

080.020 **Observations of oscillations of the Sun.**
V. A. Kotov, A. B. Severny (*Severnyj*), T. T. Tsap.
Mon. Not. R. Astron. Soc., Vol. 183, 61 - 78 (1978).

Doppler shifts of a solar line of Fe I at λ 5123.7, averaged over a large central zone of the Sun, relative to the mean position of this line for the outer rim, were measured during 1974–76. The power spectra of the difference (central zone) – (rim) show the strongest concentration of power in the range of periods P = 120–180 min. High-resolution analysis of fluctuations in this range for 6001 values of periods P has shown the highest significance of five oscillations with P = 134.498, 148.359, 160.004, 171.099, 175.061 min. The 160-min oscillation is most significant; it lasts four years and shows a coherence in phase.

080.021 **Bestimmungen der Rotationselemente i und Ω der Sonne.** H. Wöhl.
Sterne Weltraum, Jahrg. 17, 126 - 128 (1978).

080.022 **The mutual attraction of magnetic knots.**
E. N. Parker.
Astrophys. J., Vol. 222, 357 - 364 (1978).

It is observed that the magnetic knots associated with active regions on the Sun have an attraction for each other during the formative period of the active regions, when new magnetic flux is coming to the surface. The purpose of this paper is to show that the mutual attraction of knots during the formative stages of a sunspot region may be understood as the mutual hydrodynamic attraction of the rising flux tubes. Two rising tubes attract each other, as a consequence of the wake of the leading tube when one is moving behind the other, and as a consequence of the Bernoulli effect when rising side by side.

080.023 **Alfvén waves in the solar atmosphere.**
J. V. Hollweg.
Sol. Phys., Vol. 56, 305 - 333 (1978).

The author examines the propagation of Alfvén waves in the solar atmosphere. The principal theoretical virtues of this work are: (1) The full wave equation is solved without recourse to the small-wavelength eikonal approximation. (2) The background solar atmosphere is realistic, consisting of an HSRA/VAL representation of the photosphere and chromosphere, a 200 km thick transition region, a model for the upper transition region below a coronal hole (provided by R. Munro) and the Munro-Jackson model of a polar coronal hole.

080.024 **On the supposed anticorrelation of solar polar and equatorial rotation rates.**
T. L. Duvall, Jr., L. Svalgaard.
Sol. Phys., Vol. 56, 463 - 466 (1978).

Howard and Harvey (1970) analyzed Mt. Wilson Doppler shifts to obtain a daily measure of the Sun's differential rotation. The data were fitted to give an angular velocity of the form $\omega = a + b \sin^2 B + c \sin^4 B$ (B = heliographic latitude). Changes in a, b, c were found to be correlated. In this paper, the anticorrelation of b and c is shown to be due to numerical coupling.

080.025   On the discreteness of the spectrum of scales of motion in the solar convection zone.
V. P. Savchenko.
Astron. Zh. Akad. Nauk SSSR, Vol. 55, 404 - 410 (1978).
In Russian. English translation in Soviet Astron., Vol. 22, No. 2.

A new heuristic approach to the calculation of characteristic scales of motion in the solar convection zone is suggested. It is shown that cell convection is inevitable and more beneficial energetically. As a consequence, the meridional circulation and differential rotation must be functions of longitude, latitude and time. The largest and the smallest scales of motion are discussed briefly.

080.026   Magnetfeld-Feinstrukturen auf der Sonne.
E. Wiehr.
Sterne Weltraum, Jahrg. 17, 162 - 166 (1978).

080.027   Has the Sun really got a companion star?
H. F. Henrichs, R. F. A. Staller.
Nature, Vol. 273, 132 - 134 (1978).

Harrison (1977) has pointed out that the six radio pulsars with the smallest observed spin-down rates ($\dot{P}$) are grouped together in a relatively small region of the sky, roughly in the direction of the galactic centre. To explain this particular effect he suggested that the intrinsic values of $\dot{P}$ are actually larger than the observed ones, because the barycentre of the solar system would be accelerated in that direction. Harrison then suggested that this acceleration would be due to an hitherto undetected companion star of the Sun. The authors show that the apparent clustering of these radio pulsars is a consequence of a strong observational selection effect and conclude that the distribution of all radio pulsars seems to be symmetric around the galactic centre.

080.028   Application of curves of growth to a study of the solar atmosphere's opacity in the continuum.
A. K. Kirillov.
Soln. Dannye 1977 Byull., No. 11, p. 51 - 57 (1978). In Russian.

The curve-of-growth method for a short wavelength region of the visible solar spectrum was used to determine the general members of Fe I atoms through the atmosphere. A difference between the Fe I atom members in $\lambda \approx 3000$ Å and $\lambda \approx 4000$ Å, which is due to a change in opacity of the solar atmosphere continuum in the λ 3000 - 4000 Å region, has been found using semi-empirical oscillation strengths.

080.029   The low solar neutrino flux: a search for a resonance in ^{6}Be.
A. B. McDonald, T. K. Alexander, J. R. Beene, H. B. Mak.
Nucl. Phys. A, Vol. A288, 529 - 532 (1977). – Abstr. in Phys. Abstr., Vol. 81, Abstr. 619 (1978).

080.030   Solar neutrinos.   J. N. Bahcall.
AIP Conf. Proc., No. 37, (see 012.016), p. 53 - 62 (1977). – Abstr. in Phys. Abstr., Vol. 81, Abstr. 4171 (1978).

080.031   Analysis of time variations of the magnetic field using observations made on a solar magnetograph.
A. A. Golovko.
Issled. po geomagn., aehron. i fiz. Solntsa. Vyp. (No.) 42. Moskva, Nauka, 1977, p. 26 - 33. In Russian. – Abstr. in Ref. zh., 51. Astron., 3.51.409 (1978).

080.032   Solar neutrino experiment as a test for cosmology and cosmogony. I. Davis' experiment as a permanent stimulation for numerous branches of physics and astronomy.
B. Kukhovich.
Izv. AN SSSR. Ser. fiz., Vol. 41, 1949 - 1959 (1977). In Russian. – Abstr. in Ref. zh., 51. Astron., 3.51.516 (1978).

080.033   Estimate of the rate of ^{37}Ar production by cosmic ray neutrinos evaluated from solar neutrino detection experiment.   G. V. Domogatskij, R. A. Ehramzhyan.
Izv. AN SSSR. Ser. fiz., Vol. 41, 1969 - 1971 (1977). In Russian. – Abstr. in Ref. zh., 51. Astron., 3.51.517 (1978).

080.034   Emergence of solar magnetic fields and 11-year activity period.   V. L. Kuznetsov, S. I. Syrovatskij.
Kratkie soobshch. po fiz., 1977, No. 8, p. 36 - 40. In Russian. Abstr. in Ref. zh., 51. Astron., 4.51.342 (1978).

080.035   Magnetic fields, velocity fields and brightness in the central region of the solar disk.   T. T. Tsap.
Izv. Krymskoj Astrofiz. Obs., Vol. 58, 13 - 25 (1978). In Russian.

The longitudinal magnetic fields, velocity fields and brightness at the center of the solar disk are studied. It is found that the average magnetic field strength recorded in the iron line λ 5233 Å is 18 gauss for the elements of N-polarity and 23 gauss for the elements of S-polarity. Comparison of the magnetic field with the radial velocities recorded in the λ5250 and 5233 Å lines has shown that radial velocities are close to zero in the regions of maximal longitudinal magnetic field.

080.036   On the change of the diameter of the radio image of the sun at short centimeter wavelengths.
A. F. Bachurin, A. S. Dvoryashin, N. N. Eryushev, L. I. Tsvetkov.
Izv. Krymskoj Astrofiz. Obs., Vol. 58, 40 - 43 (1978). In Russian.

It is found that the radio image at the limb of the solar disk at the 1.9–3.5 cm wavelength band does not have circular symmetry. The ratio of the diameter of the radio image of the sun along the equator to the diameter along the central meridian increases with wavelength and the level of solar activity.

080.037   Apparatus functions of atmospheric oscillations from the solar limb for nonclassical assumptions.
Yu. A. Nagovitsyn.
Soln. Dannye 1977 Byull., No. 12, p. 101 - 106 (1978). In Russian.

The problem of calculating the apparatus function of atmospheric oscillations of the solar limb is considered for unclassical assumptions. The solution is used to determine the location of the solar limb and apparatus function profile.

080.038   Nonlinear stage of instability due to local Joule-overheating in the solar active regions.
V. S. Sokolov, A. G. Kosovichev.
Sol. Phys., Vol. 57, 73 - 79 (1978).

The numerical solution by a computer of the system of magnetohydrodynamics equations in the one-dimensional approximation serves as the basis for studying the non-linear stage of the instability due to local Joule-overheating of zones with large values of magnetic field gradients in the active regions of the sun. The authors have demonstrated the formation of a system of current layers responsible for efficient transformation of magnetic energy into Joule heat and kinetic energy of the macroscopic motion. The specific features of quasi-stationary skinning of magnetic field with gravitation have been noted.

080.039   Lepton mixing and the 'solar neutrino puzzle'.
S. M. Bilenky, B. Pontecorvo.

Comments Nucl. Part. Phys., Vol. 7, No. 5, p. 149 - 152
(1977). – Abstr. in Phys. Abstr., Vol. 81, Abstr. 16373 (1978).

**080.040   Long-period oscillations of the apparent solar
diameter: observations.** T. M. Brown, R. T.
Stebbins, H. A. Hill.
Astrophys. J., Vol. 223, 324 - 338 (1978).

New observations of the time-varying component of the
apparent solar diameter are reported. Power spectra derived
from these observations reveal narrow-band oscillations at
frequencies consistent with the normal mode frequencies of a
standard solar model. The amplitudes of these oscillations are
discussed and related to the observations of other investigators.
A detailed analysis of the experiment and its associated sources
of error is presented and used to show that there is a very
small likelihood that the power spectrum peaks are due to
nonsolar causes.

080.041   Outlooks on the progress of solar physics.
V. A. Krat.
Problems of observational and theoretical astronomy, Mos-
cow – Leningrad, 1977, (see 003.012), p. 3 - 9. In Russian.
Abstr. in Ref. zh., 51. Astron., 5.51.302. (1978).

**080.042   Background from internal radioactivity in some
radiochemical detectors of solar neutrinos.**
Yu. I. Zakharov.
Izv. AN SSSR. Ser. fiz., Vol. 41, 1972 - 1974 (1977). In
Russian. – Abstr. in Ref. zh., 62. Issled. kosm. prostranstva,
5.62.274 (1978).

080.043   Is there anything new on the sun?
R. Pallavicini, G. Poletto.
Mercury, Vol. 7, 23 - 26, 33 (1978).

**080.044   Contraste y determinacion por métodos astrofisicos
de fuerzas de oscilador.**
M. Rego, M. J. Fernandez Figueroa.
Rev. Real Acad. Cienc., Madrid, Tomo 71, 141 - 145 (1977) =
Univ. Complutense – Fac. Cienc., Madrid, Semin. Astron.
Geod., Publ. núm. 94.

A test of the gf values of Fe I computed by Kurucz has
been carried out by spectral synthesis from a solar model. As
a consequence of the test results, new values have been computed.

**080.045   Linear hydrodynamical equations coupled with
radiative transfer in a non-isothermal atmosphere.
II: Application to solar photospheric observations.**
B. Schmieder.
Sol. Phys., Vol. 57, 245 - 253 (1978).

In a previous paper (Schmieder, 1977), the author solved
simultaneously the hydrodynamical and radiative transfer
equations, so he does not have to assume any relaxation time
of the atmosphere. In this paper, he uses that theory to
interpret photospheric observations of the Mg I line at 5172 Å.
For periods between 400 and 140 s, the phase-shifts observed
between velocities and the phase shifts between intensity and
velocity fluctuations are explained by the existence of
radiative dissipation coupled with evanescent waves or upward propagating waves, according to the frequency. For
smaller periods partial or total reflections must be considered.
The results relative to radiative dissipation are expressed in
terms of the variation of a relaxation time with frequency
through the atmosphere ($10^{-3} < \tau_{5000} < 1$).

080.046   Motions in solar magnetic tubes. I: The downflow.
R. G. Giovanelli, C. Slaughter.
Sol. Phys., Vol. 57, 255 - 260 (1978).

The line-centre-magnetogram technique has been used to
measure the average velocity in magnetic elements in plages
and isolated magnetic elements (including dipoles) in Ca II

8542, Mg I 5183, Fe I 8688 and C I 9111. The velocities vary
from 0.6 km s⁻¹ downflow in the line of deepest origin to
zero in the highest. The smooth curve obtained by combining
these with the results of other investigators is in conformity
with Giovanelli's (1977) theory of inflow in the neighborhood of the temperature minimum.

**080.047   Hydromagnetic-gravity wave critical levels in the
solar atmosphere.** O. El Mekki, I. A. Eltayeb,
J. F. McKenzie.
Sol. Phys., Vol. 57, 261 - 266 (1978).

It is shown that the discontinuous jump in the vertical
wave energy flux of slow hydromagnetic-gravity waves, occur-
ing at a critical level, which is accompanied by wave absorption, and the existence of a reflection point imply that slow
waves are trapped in the solar atmosphere. Thus such a system
behaves as a leaky wave guide.

**080.048   Comparisons of measured and calculated potential
magnetic fields.** M. J. Hagyard, D. Teuber.
Sol. Phys., Vol 57, 267 - 278 (1978).

Photospheric line-of-sight and transverse magnetic field
data for an isolated sunspot are described. A study of the
linear polarization patterns and of the calculated transverse
field lines indicates that the magnetic field of the region is
very nearly potential. The Hα fibril structures of this region
corroborate this conclusion. A potential field calculation is
described using the measured line-of-sight fields together with
assumed Neumann boundary conditions; both are necessary
and sufficient for a unique solution. The computed transverse
fields are then compared with the measured transverse fields
to verify the potential field model and assumed boundary
values. The implications of these comparisons on the validity
of magnetic field extrapolations using potential theory are
discussed.

080.049   The flux-rope-fibre theory of solar magnetic fields.
J. H. Piddington.
Astrophys. Space Sci., Vol. 55, 401 - 425 (1978).

The flux-rope theory of solar magnetic fields is reviewed
briefly and, together with the dynamo theory, compared with
various observational results. Dynamo and related theories
are based on fields controlled by the plasma, and it is shown
that such fields cannot account for the strong surface fields or
even emerge without becoming tangled. Observations which
appear uniquely explicable in terms of powerful (⪞ 4000 G),
helically twisted flux ropes and their many twisted flux
fibres (≈ 3 X 10¹⁸ Mx) are listed. The results provide a strong
case for the flux-rope theory against the entrenched dynamo
theory.

**080.050   Polar deceleration and convection zone depth for
the sun.** P. A. Gilman.
Bull. American Astron. Soc., Vol. 10, 400 (1978). – Abstract.

080.051   Very elongated convective cells in the sun.
E. A. Geronicholas.
Bull. American Astron. Soc., Vol. 10, 400 (1978). – Abstract.

080.052   Secular changes in solar rotation, 1888 - 1964.
J. A. Eddy, R. W. Noyes, J. G. Wolbach, A. A.
Boornazian.
Bull. American Astron. Soc., Vol. 10, 400 - 401 (1978).
Abstract.

080.053   Large scale flows near the solar poles.
J. M. Beckers.
Bull. American Astron. Soc., Vol. 10, 430 (1978). – Abstract.

**080.054   The energy balance and pressure in the solar transi-
tion zone.** K. R. Nicolas, J.-D. F. Bartoe,
G. E. Brueckner, M. E. VanHoosier.

Bull. American Astron. Soc., Vol. 10, 432 (1978). — Abstract.

080.055   Bowen fluorescence in the solar transition region.
J. C. Raymond.
Bull. American Astron. Soc., Vol. 10, 432 (1978). — Abstract.

080.056   Response of the solar atmosphere to infalling surge
material.   D. Webb, E. Schmahl.
Bull. American Astron. Soc., Vol. 10, 455 - 456 (1978).
Abstract.

080.057   Neutrons, protons, and solar neutrinos.
M. J. Newman.
Bull. American Astron. Soc., Vol. 10, 457 (1978). — Abstract.

080.058   Fundamental flux tubes in the solar magnetic fields.
1. Significance of the flux-amount; possibility of
rising and durability.   M. H. Gokhale.
Kodaikanal Obs. Bull., Ser. A, Vol. 2, 10 - 18 (1977).

For a satisfactory understanding of the solar magnetic
phenomena it may be advantageous first to interrelate pheno-
menologically observations involving different scales of length
and time and in different layers of the solar atmosphere. As
a first step towards the construction of such a phenomenological model it is tentatively concluded that a large number
of the solar magnetic phenomena may be due to the rise of
'fundamental flux tubes'of magnetic fluxes $\sim 10^{17} - 10^{18}$ Mx
(FFT's) presumably originating in large($>10^5$ km) depths in the
convection zone and rising across the observable layers of the
solar atmosphere through bands of successively larger sizes.

080.059   Fundamental flux tubes in the solar magnetic fields.
2. Generation of the 'FFT's and the periodic field-
reversal.   M. H Gokhale.
Kodaikanal Obs. Bull., Ser. A, 19 - 27 (1977).

It is conceivable that the periodic reversal of the Sun's
large-scale magnetic field and the concentration of the field
in strong,thin flux tubes on small scales could both result
from the following likely processes near the base of the conveo
tion zone. It is likely that in a "base layer" of thickness $\sim 10^5$ km
near the base of the convection zone a small part ($\sim 0.4\%$) of
the energy flux from below goes into an azimuthal magneto-
accoustic oscillation and once in ~ 11 y or so such an oscillation reaches a critical amplitude $\sim 10^4$ cm s^{-1} when it undergoes a 'shock transition' to a cellular (azimuthal) convective
mode. The shock transition will involve creation of electric
current sheaths and magnetic flux sheaths of field intensity
$\sim 10^4$ G and thickness $\sim 10^2$ km each near the 'nodal' meridian
planes. The electric current sheaths would provide the 'reversed' magnetic field on one side.

080.060   Resonance oscillations on the sun taking into account friction.   A. I. Khlystov.
Astron. Tsirk., No. 978, p. 1 - 3 (1978). In Russian.

080.061   Estimate of the contribution of various planets to
the amplitude of resonance oscillations on the sun.
A. I. Khlystov.
Astron. Tsirk., No. 978, p. 3 - 5 (1978). In Russian.

080.062   The amplitudes of resonance oscillations on the
sun under stationary conditions.   A. I. Khlystov.
Astron. Tsirk., No. 981, p. 2 - 3 (1978). In Russian.

080.063   The non-stationary oscillation regime of the solar
convective zone.   A. I. Khlystov.
Astron. Tsirk., No. 985, p. 3 - 5 (1978). In Russian.

080.064   Fast oscillations of the rotation of the sun.
V. F. Chistyakov.
Problems of astronomy and geodesy, VAGO assembly Yerevan,
1975, (see 012.039), p. 107 - 108. In Russian. — Abstr. in

Ref. zh., 51. Astron., 6.51.451 (1978).

080.065   Numerical modelling of propagation of shock waves
in the solar atmosphere and in the solar wind.
V. V. Zakajdakov, V. S. Synakh.
Mathematical models of the near space, 1977, (see 003.015),
p. 33 - 41. In Russian. — Abstr. in Ref. zh., 51. Astron.,
6.51.504 (1978).

080.066   On the statistical description of solar and solar-
geophysical phenomena.   Eh. I. Mogilevskij.
Inst. zemn. magn., ionos. i rasprostr. radiovoln AN SSSR.
Prepr., 1977, No. 12, 18 pp. In Russian. — Abstr. in Ref. zh.,
62. Issled. kosm. prostranstva, 6.62.273 (1978).

080.067   On the mechanical heating of the external layers of
the sun. A methodological discussion.
P. Souffrin.
Mem. Soc. Astron. Italiana, Vol. 47, (see 012.036), 365 - 376
(1976).

Contents: Statement of the problem. The mathematical
formulation and its degradation. Mixing length theory for
convective zones. The heating of the solar corona. I. Schwarzschild (1948). II. Schatzman (1949). The theory of Brinkley
and Kirkwood.

080.068   The solar interior and solar neutrinos.
R. T. Rood.
Mem. Soc. Astron. Italiana, Vol. 48, (see 012.038), 357 - 374
(1977).

080.069   Velocity fields in the solar atmosphere: theory.
J. W. Leibacher.
Mem. Soc. Astron. Italiana, Vol. 48, (see 012.038), 475 - 497
(1977).

080.070   Velocity fields in the solar atmosphere (observations).   F. L. Deubner.
Mem. Soc. Astron. Italiana, Vol. 48, (see 012.038), 499 - 507
(1977).

In these lectures oscillatory and quasi oscillatory motions
as observed in the photosphere and lower chromosphere are
discussed as well as convective motions on different horizontal
scales.

080.071   Proposed solar-neutrino experiment using ^{71}Ga.
J. N. Bahcall, B. T. Cleveland, R. Davis, Jr.,
I. Dostrovsky, J. C. Evans Jr., W. Frati, G. Friedlander, K.
Lande, J. K. Rowley, R. W. Stoenner, J. Weneser.
Phys. Rev. Lett., Vol. 40, 1351 - 1354 (1978).

A solar-neutrino experiment that uses ^{71}Ga as a detector
can distinguish between broad classes of explanations for the
discrepancy between prediction and observation in the ^{37}Cl
experiment. A radiochemical experiment with the required
amount of ^{71}Ga is feasible.

080.072   Magnetic field transfer in the convective zone of the
sun.   S. I. Vajnshtejn, A. A. Ruzmajkin.
Inst. prikl. mat. AN SSSR. Prepr., 1978, No. 15, 16 pp. In
Russian. — Abstr. in Ref. zh., 51. Astron., 7.51.399 (1978).

080.073   Characteristic scales on the solar surface and the
structure of the interior of the convective zone.
V. P. Savchenko.
Tr. Inst. prikl. geofiz. Gl. upr. gidrometeorol. sluzhby pri Sov.
Min. SSSR, 1977, No. 35, p. 59 - 71. In Russian. — Abstr. in
Ref. zh., 51. Astron., 7.51.400 (1978).

080.074   Motions and magnetic fields on the sun.
V. A. Krat.
Izv. VUZ. Radiofiz., Vol. 20, 1302 - 1309 (1977). In Russian.
Abstr. in Ref. zh., 51. Astron., 7.51.408 (1978).

Physics of plasma of the solar atmosphere.
See Abstr. 003.070.

Why does the Sun shine?   See Abstr. 014.036.

Fourier transforms applied to the contrast profile
method of measuring vector magnetic fields.
See Abstr. 021.023.

Lifetimes of some levels in neutral carbon, nitrogen
and oxygen.   See Abstr. 022.048.

A search for stellar oscillations.
See Abstr. 031.263.

Tecniche e strumentazione per radiometria solare
ed atmosferica.   See Abstr. 032.045.

Die Erforschung der Sonne mit Ballonteleskopen.
See Abstr. 032.501.

Neutrinos: the ultimate astrophysical probe.
See Abstr. 061.020.

On the dynamic interaction between magnetic
fields and convection.   See Abstr. 062.002.

Magnetohydrodynamics of atmospheric transients.
I. Basic results of two-dimensional plane analyses.
See Abstr. 062.004.

Magnetic fields and convection.
See Abstr. 062.018.

On the stability of twisted magnetic fields.
See Abstr. 062.042.

Nonlinear astrophysical dynamos: bifurcation of
steady dynamos from oscillating dynamos.
See Abstr. 062.049.

Acoustic waves in the solar atmosphere. IV. On
the efficiency of one-dimensional hydrodynamic codes.
See Abstr. 062.057.

The interaction between a magnetized plasma flow
and a magnetized celestial body: a review of magnetospheric
studies.   See Abstr. 062.082.

Overstabilization of acoustic modes in a polytropic
atmosphere.   See Abstr. 064.058.

Hydrodynamical instabilities in the envelopes of
main-sequence stars: constraints implied by the lithium,
beryllium, and boron observations.   See Abstr. 064.079.

Stochastic convection: how constant is the solar
"constant"?   See Abstr. 065.043.

Observations of phase delays and the vertical propagation of waves in the solar atmosphere.
See Abstr. 071.045.

The 300 sec oscillation in Si II 1817: observed
impulse response of the solar atmosphere.
See Abstr. 071.047.

Magnetic fields in a coronal condensation.
See Abstr. 074.005.

High resolution UV solar spectroscopy.
See Abstr. 076.004.

Plasma instabilities of trapped particles in solar
magnetic fields.   See Abstr. 077.019.

Compilation of solar and meteoritic abundances.
See Abstr. 105.071.

Dynamics of a cloud of fast electrons travelling
through the plasma (*solar electrons in interplanetary space*).
See Abstr. 106.037.

The sun among the stars. I. A search for solar spectral analogs.   See Abstr. 114.006.

The role of active regions and of the general magnetic
field of the sun in the 11-year cycle of cosmic rays.
See Abstr. 143.020.

Influence of the total magnetic field of the sun on
modulation of cosmic rays in the stratosphere at mean latitudes.   See Abstr. 143.022.

Evolutionary abundances – the sun and the Hyades
stars.   See Abstr. 153.041.

Mass changing cosmology and solar evolution.
See Abstr. 162.180.

Earth

081 Figure, Composition, and Gravity

081.001 **Pangaea − fact or fiction?** A. Hoffer.
Gerlands Beitr. Geophys., Band 86, 460 - 462 (1977).
The gradual decrease in the geometrical flattening of the planet earth in hydrostatic response to its decreasing spin velocity, together with the simultaneous process of outgassing, led to a significant degree of shrinkage of surface area, which must have been accompanied by continuing differentiation of more granitic crustal material from the planet's interior. Under such conditions, the only manner in which crustal fragments could have moved at any time is so as to approach, one another from all directions simultaneously.

081.002 **A model of time-periodic mantle flow.**
F. H. Busse.
Geophys. J. R. Astron. Soc., Vol. 52, 1 - 12 (1978).
The instability of a layer consisting of a lighter viscous fluid on top of a heavier less viscous fluid is considered in the case when the heavy fluid is adiabatically stratified and the light fluid contains heat sources and possesses a lower heat conductivity. A perturbation in the thickness of the upper fluid layer causes horizontal temperature variations in the lower fluid. The motions induced by thermal buoyancy can interact with the distortion of the interface in such a way that the initial perturbation is reinforced in the form of an overstable oscillation. It is proposed that this mechanism is relevant to the problem of time-dependent flow in the Earth's mantle.

081.003 **Detailed gravity-tide spectrum between one and four cycles per day.**
R. J. Warburton, J. M. Goodkind.
Geophys. J. R. Astron. Soc., Vol. 52, 117 - 136 (1978).

081.004 **The heating of the Earth during its formation.**
V. S. Safronov.
Icarus, Vol. 33, 3 - 12 (1978).
The thermal state of the Earth accumulating from solid bodies is investigated. The conductivity equation is deduced for a growing spherically symmetrical planet which takes into account heating by impacts of bodies, by radioactivity, and by compression of its material. The cooling is produced mainly by impact mixing, which is approximated by extrapolating the parameters from known impact craters to larger sizes. The solution of a more simple conductivity equation for a uniformly heated plane parallel layer with moving boundaries is found.

081.005 **The end of the expanding earth hypothesis?**
P. J. Smith.
Nature, Vol. 271, 301 (1978).

081.006 **Limits to the expansion of Earth, Moon, Mars and Mercury and to changes in the gravitational constant.**
M. W. McElhinny, S. R. Taylor, D. J. Stevenson.
Nature, Vol. 271, 316 - 321 (1978).
New estimates of the palaeoradius of the Earth for the past 400 Myr from palaeomagnetic data limit possible expansion to less than 0.8%, sufficient to exclude any current theory of Earth expansion. The lunar surface has remained static for 4,000 Myr with possible expansion limited to 0.06%, the martian surface suggests a small possible expansion of 0.6% while the surface of Mercury supports a small contraction. Observations of Mercury, together with reasonable assumptions about its internal structure, indicate that G decreases at a rate of less than 8×10^{-12} yr^{-1}, in constant mass cosmologies, and 2.5×10^{-11} yr^{-1} in Dirac's multiplicative creation cosmology.

081.007 **The equilibrium figure of the earth.**
E. Grafarend, K. Hauer.
Veröff. Zentralinst. Phys. Erde, No. 52, Part 2, (see 012.001), p. 239 - 255 (1977).
Based on the method of virial equations, the integro-differential equations governing the physical figure of the earth are derived. Volume forces of gravitational type and surface forces of stress type are assumed. Initial stress is represented by the H. Takeuchi (1951) model. Equilibrium figures for different earth models − elastic, elastic under initial stress, liquid − are studied. It is proved that non-liquid earth models allow ellipsoidal structures of Maclaurin type.

081.008 **Models for the auto- and cross covariances between mass density anomalies and first and second order derivatives of the anomalous potential of the earth.**
C. C. Tscherning.
Veröff. Zentralinst. Phys. Erde, No. 52, Part 2, (see 012.001), p. 261 - 268 (1977).

081.009 **Determination of the free boundary in potential theory, the boundary problem of physical geodesy.**
P. Holota.
Veröff. Zentralinst. Phys. Erde, No. 52, Part 2, (see 012.001), p. 269 - 279 (1977).

081.010 **The surface mass distribution of the earth and the geoid figure.** G. Barta.
Veröff. Zentralinst. Phys. Erde, No. 52, Part 2, (see 012.001), p. 369 - 378 (1977).

081.011 **The geoid problem in nowadays works.**
W. Dobaczewska.
Veröff. Zentralinst. Phys. Erde, No. 52, Part 2, (see 012.001), p. 397 - 403 (1977).

081.012 **Formule de Whittaker pour le calcul du potentiel terrestre.** H. M. Dufour.
Veröff. Zentralinst. Phys. Erde, No. 52, Part 2, (see 012.001), p. 405 - 417 (1977).

081.013 On the definition of the Listing-geoid taking into consideration different height systems.
D. Lelgemann.
Veröff. Zentralinst. Phys. Erde, No. 52, Part 2, (see 012.001), p. 419 - 439 (1977).

081.014 Influence of sea tides on gravimetric observations of the tides of the earth in Eastern Europe.
B. P. Pertsev.
Veröff. Zentralinst. Phys. Erde, No. 52, Part 2, (see 012.001), p. 493 - 497 (1977). In Russian.

081.015 Interpolation of deflection of the vertical from horizontal gradients of gravity.   L. Völgyesi.
Veröff. Zentralinst. Phys. Erde, No. 52, Part 2, (see 012.001), p. 561 - 567 (1977).

081.016 Lateral inhomogeneities of the density in the upper mantle of the earth.   K. Arnold, D. Schoeps.
Veröff. Zentralinst. Phys. Erde, No. 52, Part 2, (see 012.001), p. 667 - 672 (1977).

081.017 Taking into account the inertia forces in analysing observations of tides of the earth.
N. N. Parijskij, B. P. Pertsev.
Veröff. Zentralinst. Phys. Erde, No. 52, Part 2, (see 012.001), p. 683 - 685 (1977). In Russian.

081.018 The determination of the earth gravity field by use of satellite gradiometry.   J. Łatka.
Veröff. Zentralinst. Phys. Erde, No. 52, Part 3, (see 012.001), p. 785 - 789 (1977).

081.019 Determination of the harmonic coefficients of 14th-order from the inclination changes of the Interkosmos 9 and 10 satellites.   J. Klokočnik.
Veröff. Zentralinst. Phys. Erde, No. 52, Part 3, (see 012.001), p. 1027 (1977). In Russian. – Abstract.

081.020 Evaluation of the 14th-order harmonics from resonant inclination variations.
J. Klokočnik, J. Kostelecký.
Bull. Astron. Inst. Czechoslovakia, Vol. 29, 10 - 14 (1978).

The orbital inclination changes due to the 14th-order resonance of 9 satellites at inclinations between 32° and 115° were already analysed by several authors (Walker, 1975; Wagner, 1976; Klokočnik, 1976, 1977) and the lumped co-efficients were obtained. Here, these lumped coefficients are used to reveal the individual harmonic coefficients of the 14th-order and odd degree. The results from several methods of adjustment (including Kaula's rule in collocation) are obtained, discussed and compared with the resonant Reigber and Balmino (1976) solution and with several comprehensive Earth gravitational field models.

081.021 Untersuchungen zur Anwendung von Sampling-Funktionen in der Geodäsie.   H. F. Schmidt.
Deutsche Geod. Komm. Bayerische Akad. Wiss., *München,* Reihe C: Dissertationen, Heft Nr. 241, 148 pp. (1978).

081.022 The Precambrian temperatures of the earth.
S. Epstein.
Bull. American Astron. Soc., Vol. 9, 497 (1977). – Abstract.

081.023 A model for the evolution of large terrestrial impact basins.   J. B. Hartung.
Bull. American Astron. Soc., Vol. 9, 530 - 531 (1977). Abstract.

081.024 Discussion on the existence and uniqueness of the solution of Molodensky's problem in gravity space.
F. Sansò.

Atti Accad. Naz. Lincei, Ser. Ottava, Rend., Vol. 61, 260 - 268 (1977).

081.025 New determination of the absolute value of gravity in Potsdam.   Yu. D. Bulanzhe, G. P. Arnautov, E. N. Kalish, Yu. F. Stus', V. G. Tarasyuk, G. Harnisch.
Gerlands Beitr. Geophys., Band 87, 9 - 18 (1978). In Russian.

In July 1976 new absolute gravity measurements were performed at the International Gravity Stations Ledovo (near Moscow) and Potsdam. These measurements were initiated by the Academy of Sciences of the Soviet Union. The instrument used was a laser gravity meter with a free-falling corner reflector. A deviation of −13.960 ± 0.017 mGal between the resulting gravity value and the level of the Potsdam gravity system was obtained. This value is in good agreement with other measurements of the gravity difference between both stations using relative instruments. The disagreement amounts to 0.011 ± 0.024 mGal.

081.026 The bodily tide and the yielding of the Earth due to tidal loading.   C. L. Pekeris.
Geophys. J. R. Astron. Soc., Vol. 52, 471 - 478 (1978).

The theory of the bodily tide and of the yielding of the Earth to tidal loading is re-examined, with the purpose of checking the standard formula for the gravity tide which is used in the interpretation of tidal gravity measurements. Some remarks are made concerning the Green's function occurring in the theory of the gravity tide.

081.027 The earth's topography: What does it reveal about internal processes?   H. N. Pollack.
Geophys. J. R. Astron. Soc., Vol. 53, (see 012.006), 138 (1978). – Abstract.

081.028 A new method for locating the electrically conducting fluid of a planet from external magnetic observations: application to the Earth.   R. Hide, S. R. C. Malin.
Geophys. J. R. Astron. Soc., Vol. 53, (see 012.006), 145 (1978). – Abstract.

081.029 Hydrodynamic modeling of flows occurring in the earth's core.   C. R. Carrigan.
Geophys. J. R. Astron. Soc., Vol. 53, (see 012.006), 145 (1978). – Abstract.

081.030 Numerical simulations of convection in the core.
D. Gubbins.
Geophys. J. R. Astron. Soc., Vol. 53, (see 012.006), 145 (1978). – Abstract.

081.031 Properties of the Earth's deep interior revealed by the secular variation.   D. R. Barraclough, S. R. C. Malin.
Geophys. J. R. Astron. Soc., Vol. 53, (see 012.006), 150 (1978). – Abstract.

081.032 Detailed gravimetric geoid computations in North America.   J. G. Marsh, E. S. Chang.
Space Research, Vol. XVII, (see 012.010), 43 - 48 (1977).

081.033 The effect of a general aspherical perturbation on the free oscillations of the earth.
J. H. Woodhouse, F. A. Dahlen.
Geophys. J. R. Astron. Soc., Vol. 53, 335 - 354 (1978).

The Lagrangian governing the infinitesimal elastic-gravitational oscillations of a completely general earth model with interior fluid–solid boundaries is given. Rayleigh's principle is then used to derive a formula for calculating the first-order perturbations in the eigenfrequencies due to an arbitrary slight perturbation of a spherically symmetric, non-rotating, isotropic starting model. The perturbations considered include rotation, asphericity, elastic anisotropy, and a deviatoric initial stress, as well as a change in the positions of both welded and fluid—solid boundaries.

081.034 Das Geoid in der Schweiz.   W. Gurtner.
Astron.-Geod. Arb. Schweiz, 32. Band, 8 + 103 + 4 pp. (1978).

081.035 Application of satellite gradiometry measurements to local geoid determination.
J. Kryński, K. P. Schwarz.
Geod. Kartogr., Vol. 26, 87 - 101 (1977). In Polish.

Satellite gradiometry is studied as a means to improve the geoid in local areas from a limited data coverage. Least-squares collocation is used for this purpose because it allows to combine heterogeneous data in a consistent way and to estimate the integrated effect of the attenuated spectrum. In this way accuracy studies can be performed in a general and reliable manner.

081.036 Investigation of the Earth gravity field by differential Doppler measurements.   J. B. Zieliński.
Geod. Kartogr., Vol. 26, 235 - 239 (1977). In Polish.

One of the new proposed methods of the Earth gravity field investigation is the satellite-to-satellite-tracking method. The possibility of measurement of the relative velocity of two objects situated on the same low orbit (below 300 km) and about 200 km away one from another is considered in the paper. If this relative velocity could be measured with ca. 0.05 mm/sec precision it would be possible to determine the geoid with the accuracy about ± 1 m and with 3° resolution.

081.037 Application of low-low satellite to satellite observations to the investigation of the earth's gravity field by collocation.   J. Kryński.
Geod. Kartogr., Vol. 26, 241 - 253 (1977). In Polish.

081.038 Application of low-low satellite observations to local geoid determination.   J. Kryński.
Geod. Kartogr., Vol. 27, 17 - 27 (1978). In Polish.

Low-low satellite observations are studied as a means to improve the geoid in local areas from a limited data coverage. Least squares collocation is used for this purpose because it allows to combine heterogeneous data in a consistent way and estimate the integrated effect of the attenuated spectrum. In this way accuracy studies can be performed in a general and reliable manner. The influence of measuring errors is discussed and it is shown that only low-low satellite observations with accuracies better than 0.5 mm/sec will give an improvement of the geoid.

081.039 A systematic approach to modeling the geopotential with point mass anomalies.
J. P. Reilly, E. H. Herbrechtsmeier.
J. Geophys. Res., Vol. 83, 841 - 844 (1978).

A point mass model is used to predict oceanic gravity anomalies from simulated altimeter data. Gravity anomalies from land areas are also combined with altimetry as hybrid data inputs to the mathematical model. The geocentric radii of the point mass anomaly sets are determined.

081.040 Determination of the gravitational field of the earth by way of an ordinary layer on its surface from measurements of the parameters of satellite motion and terrestrial gravimetric data.   V. F. Eremeev, M. I. Yurkina.
Tr. TsNII geod., aehrosemki i kartogr. Moskva, 1977, vyp. (No.) 205, 164 pp. In Russian. − Abstr. in Ref. zh., 52. Geod. Aehrosemka, 2.52.72 (1978).

081.041 Accuracy of deflections of vertical determined from discrete solution of the geodetic boundary value problem.   J. Nedoma.
Pure Appl. Geophys., Vol. 115, 639 - 645 (1977). − Abstr. in

Phys. Abstr., Vol. 81, Abstr. 3702 (1978).

081.042 Interpretation of the gravity effect due to two-dimensional asymmetrical triangular prism by Fourier transforms.   S. Sengupta.
Pure Appl. Geophys., Vol. 115, 647 - 654 (1977). − Abstr. in Phys. Abstr., Vol. 81, Abstr. 3703 (1978).

081.043 Evaluation of the GEOS 3 altimetry data by means of spherical functions expansions.
K.-R. Koch, E.-U. Fischer.
Z. Vermessungswes., Vol. 102, 349 - 354 (1977). In German. Abstr. in Phys. Abstr., Vol. 81, Abstr. 3708 (1978).

081.044 Accurate formula expressing the difference between the normal gravity and its radial component.
G. Blaha.
Bull. Géod., Vol. 52, 19 - 23 (1978).

A simple formula is presented giving the value of $\gamma - \gamma_r$ to better than 0.001 mgal associated with an arbitrary reference ellipsoid, where γ is the normal gravity and γ_r is its radial component. Further simplifications of this formula are possible, depending on the desired accuracy. Since in the actual field $g - g_r$ equals $\gamma - \gamma_r$ to a good approximation, this formula makes it possible to work in terms of g_r rather than in terms of the measured quantity g. Such a choice is attractive mainly because the spherical harmonic expansion of g_r is very simple.

081.045 The definition of the telluroid.   E. W. Grafarend.
Bull. Géod., Vol. 52, 25 - 37 (1978).

The threedimensional mapping of the earth surface onto the best approximative figure, the telluroid, is analyzed by computing the threedimensional mapping equations and the distortion tensor of Lagrangian and Eulerian type. The angular distortion is given in terms of the distortion tensor.

081.046 Molodensky's problem in gravity space: a review of the first results.   F. Sanso.
Bull. Géod., Vol. 52, 59 - 70 (1978).

In the last year a new formulation of Molodensky's problem has been given, in which the gravity vector $\vec{g}$ has been considered as the independent variable of the problem, while the position vector $\vec{x}$ is the dependent. This new approach has the great advantage to transform the problem of Molodensky which is of free boundary type, into a fixed boundary problem for a non linear differential equation. In this paper the first results of the study of the new approach are summarized, without going into many mathematical details. The problem of Molodensky for the rotating earth is also discussed.

081.047 Multidimensional inequality by Bernstein and evaluations of gravity potential derivatives.
V. A. Antonov, K. V. Kholchevnikov.
Astron. Nachr., Band 299, 131 - 135 (1978). In German.

In different problems of celestial mechanics it is often necessary to estimate the effect of the truncated higher harmonics of the gravity potential on the motion of a test particle. As a rule the magnitude of this effect is strictly connected with the gravitational acceleration, i.e. partials of the potential. But the mathematical theory of attraction gives the estimations for the potential itself. However, as far as the general term of development (spherical harmonic) possesses a definite reserve of the smoothness, the authors have succeeded in passing from potential estimations to its partials estimations. The mathematical method based on a multidimensional generalization (obtained here) of an inequality by Bernstein is used. By the way several inequalities connecting different norms of spherical harmonics are proved.

081.048 The spherical-harmonics expansion of the gravitational potential of the Earth in the external space and its convergence.   K. Arnold.

Gerlands Beitr. Geophys., Band 87, 81 - 90 (1978).

The paper considers the expansion in spherical harmonics of the gravitational potential of the Earth in the external space and shows that this series expansion uniformly converges in the external space of the Earth and on the Earth's surface.

081.049 Modern standards for gravity surveys.
C. Morelli.
Indian J. Meteorol. Hydrol. Geophys., Vol. 27, 445 (1976).
Abstr. in Phys. Abstr., Vol. 81, Abstr. 7558 (1978).

081.050 On the argument of the variation method of the theory of the figure of the earth.   Yu. M. Nejman.
Izv. vyssh. uchebn. zaved. Geod. i aehrofotosemka, 1977, No. 4, p. 21 - 25. In Russian. – Abstr. in Ref. zh., 52. Geod. Aehrosemka, 3.52.62 (1978).

081.051 Normal gravity in space and on the surface of a reference ellipsoid.   K. M. Kartvelishvili.
Soobshch. AN GruzSSR, Vol. 87, 605 - 608 (1977). In Russian. – Abstr. in Ref. zh., 52. Geod. Aehrosemka, 3.52. 63 (1978).

081.052 Numerical analysis and prediction of the geopotential by using remote sensing data of the atmosphere with satellites.   P. N. Belov, A. I. Burtsev.
Primenenie dannykh nablyud. so sputnikov dlya vosstanovl. i chisl. analiza meteorol. polej. Leningrad, Gidrometeoizdat, 1977, p. 3 - 20. In Russian. – Abstr. in Ref. zh., 62. Issled. kosm. prostranstva, 3.62.118 (1978).

081.053 Statistical model for gravity, topography, and density contrasts in the earth.   S. K. Jordan.
J. Geophys. Res., Vol. 83, 1816 - 1824 (1978).

A statistical model of topographic heights and density contrasts within the earth is inferred from the empirical worldwide, free air gravity anomaly covariance function and the Bullen and Haddon earth model HB_1. Models for relief and loading at the crust-mantle and mantle-core boundaries are also inferred. A worldwide Bouguer anomaly model is derived by suppressing terrain-related portions of the free air anomaly model.

081.054 Simultanbestimmungen der Lotabweichungskomponenten ξ und η mit dem Prismenastrolabium. Vb. Beobachtungen im 48. Breitengrad im Jahre 1972.
A. Rödde.
Deutsche Geod. Komm. Bayerisch. Akad. Wiss., Reihe B: Angewandte Geod., Heft Nr. 199, 102 pp. (1977).

The author reports astronomic determinations of longitudes and latitudes performed simultaneously on 19 stations on the profile of the geoid at 48° of latitude, as well as the calculation of the components ξ and η of the deviation of the vertical.

081.055 On the solution of the exterior boundary value problem with the aid of series.
M. S. Petrovskaya.
Dynamics of planets and satellites, (see 012.020), p. 175 - 176 (1978). – Abstract.

081.056 Thermal history of the earth and luminosity of the sun.   L. I. Miroshnichenko.
Priroda, 1978, No. 5, p. 132 - 133. In Russian.

081.057 The Clairaut equation and its connection to satellite geodesy and planetology.
P. Lanzano, J. C. Daley.
AIAA J., Vol. 15, 1231 - 1237 (1977). – Abstr. in Phys. Abstr., Vol. 81, Abstr. 11795 (1978).

081.058 The physical background of the shape of the Earth.

G. Barta, A. Hajosy.
Fiz. Szemle, Vol. 27, No. 5, p. 182 - 189 (1977). In Hungarian.
Abstr. in Phys. Abstr., Vol. 81, Abstr. 15944 (1978).

081.059 Statistical geodesy – an engineering perspective.
R. A. Nash, Jr., S. K. Jordan.
Proc. IEEE, Vol. 66, 532 - 550 (1978).

081.060 Marées terrestres.
P. Melchior (Editor).
Bull. Inf., (Obs. R. Belgique, Bruxelles), Nos. 77, 78, p. 4504 - 4725 (1978).

081.061 Geodesia. – Connection in the actual gravity field of the Earth in terms of the disturbing potential.
E. Livieratos.
Atti Accad. Naz. Lincei, Ser. Ottava, Rend., Cl. Sci. fis. mat. nat., Vol. 62, 523 - 528 (1977).

The author derives the Christoffel's symbols of the second kind referred to the actual gravity field of the Earth, as a sum of the corresponding Christoffel's symbols of the second kind referred to a model gravity field (e.g. a Somigliana-Pizzetti type of model field or a model field derived, for instance, from a satellite solution), and of the "connection disturbancies" caused by the disturbing potential of the Earth.

081.062 Solution of M.S. Molodenskij's boundary problem.
N. L. Lekishvili.
Tr. Vychisl. tsentr. AN GruzSSR, Vol. 17, 128 - 137 (1977).
In Russian. – Abstr. in Ref. zh., 52. Geod. Aehrosemka, 5.52. 47 (1978).

081.063 On an optimum stabilizer in the variation method of the solution of Molodenskij's problem.
Yu. M. Nejman, M. V. Kuznetsov.
Izv. VUZ. Geod. i aehrofotosemka, 1977, No. 5, p. 36 - 42. In Russian. – Abstr. in Ref. zh., 52. Geod. Aehrosemka, 5.51.48 (1978).

081.064 Gravity model improvement using GEOS-3 (GEM *(Goddard Earth Model)* **9 and 10).**
F. J. Lerch, S. M. Klosko, R. E. Laubscher, C. A. Wagner.
GSFC Doc. X-921-77-246, Prepr., 12 + 121 pp. (1977).

The mass constant of the Earth, GM, has been estimated from the laser data as $398600.64 \pm .02$ km^3/sec^2, a value which is principally determined from LAGEOS. The speed of light used was 299792.5 km/sec. Geocentric station positions were determined for approximately 150 stations in GEM 10. These station coordinates, their mean sea level heights and altimetry data provide an estimate for the mean radius of the earth of $a_e = 6378140 \pm 1$ m. Accuracy estimates derived for the potential coefficients have been verified with independent data sets.

081.065 Ocean gravity and geoid determination.
W. D. Kahn, J. W. Siry, R. D. Brown, W. T. Wells.
GSFC Doc. X-921-77-259, Prepr., 6 + 20 pp. (1977).

Gravity anomalies have been recovered in the North Atlantic and the Indian Ocean regions. Geoids derived from the altimeter solutions are consistent with altimetric sea surface height data to within the precision of the data, about ±2 meters.

081.066 Gravity anomalies determined from tracking the Apollo-Soyuz.
F. O. Vonbun, W. D. Kahn, W. T. Wells, T. D. Conrad.
NASA Tech. Memo., NASA TM 78031, 33 pp. (1977).

In July of 1975 a Geodynamics Experiment was performed during the Apollo-Soyuz Mission to assess the feasibility of tracking and recovering high frequency components of the Earth's gravity field. For this experiment, a synchronous orbiting tracking station namely the Applications Technology Satellite 6 (ATS-6) was utilized to track a low orbiting spacecraft namely the Apollo. Gravity anomaly blocks of magnitude of 5 milligals or larger with wavelengths of 500 to 1000 km have been recovered within the region bounded by latitudes 52°S to 52°N and longitudes 0° to 115°E.

081.067 **An information theory approach to the density of the earth.**   M. A. Graber.
NASA Tech. Memo., NASA TM 78034, 34 pp. (1977).
    Information theory can develop a technique which takes experimentally determined numbers and produces a uniquely specified "best" density model satisfying those numbers. This technique does not depend on previous density models for starting conditions; it is self-starting. A model was generated using five numerical parameters: the mass of the earth, its moment of inertia, three zero-node torsional normal modes (L = 2, 8, 26).

081.068 **Questions of combined use of satellite, gravimetric and astronomical-geodetical data for determination of the figure and gravitational field of the earth.**
L. P. Pellinen, O. M. Ostach.
New methods of space geodesy, (see 012.029), p. 26 - 62 (1976). In Russian.
    The existing (gravimetric, astrogeodetic and satellite) methods for determination of the gravitational field and the figure of the earth are discussed from the point of view of increasing the accuracy of the combined solutions.

081.069 **Satellite altimetry and the scale factor of the geopotential.**   M. Burša.
New methods of space geodesy, (see 012.029), p. 135 - 143 (1976). In Russian.

081.070 **Corrections to the coefficients of the gravitational potential of a planet due to the nonspherical form of its surface.**   Eh. M. Yagudina.
New methods of space geodesy, (see 012.029), p. 321 - 328 (1976). In Russian.
    Numerical values of the corrections to the coefficients of the geopotential due to deviation of the nonspherical surface from a sphere were obtained supposing that the surface is the geoid and the development of the radius-vector of the geoid in spherical functions contains only zonal harmonics.

081.071 **Variational formulation of the geodetic boundary value problem.**   S. M. Nakiboglu.
Bull. Géod., Vol. 52, 93 - 100 (1978).
    A variational principle for the Stokes boundary value problem is derived using the Euler–Lagrange theory. The resulting variational principle is then transformed into an equation determining the semi-major axis of the best fitting ellipsoid which fulfills the condition $U_0 = W_0$. The computations using three different geopotential models yield the semi-major axis of the earth ellipsoid as $a = 6378145.4$ metres for the flattening f = 1/298.2564. The corresponding equatorial gravity and the geopotential number are computed as $\gamma_a = 978029.59$ mgals and $U_0 = W_0 = 6.26367371 \cdot 10^6$ kgalmeters respectively.

081.072 **Modified equations for the determination of odd zonal harmonics.**   X. Berger, G. Ligier.
Bull. Géod., Vol. 52, 101 - 105 (1978).
    The development of the analytical theory of the motion of an artificial satellite (Berger, 1972–1975) points out the great importance of the second order, third order and coupling terms between zonal harmonics. These terms have been added to the equations established by King–Hele and Cook (1968–1973) for the determination of odd zonal harmonics. This solution is compared with theirs. It satisfies all the equations much better and especially the equations

relating to near-critical inclination satellites.

081.073 **Direct gravity formula for the geodetic reference system 1967.**   D. Nagy.
Bull. Géod., Vol. 52, 159 - 164 (1978) = Contrib. Earth Phys. Branch, Dep. Energy Mines Resources, Ottawa, Canada, No. 720.
    A direct gravity formula, polynomial in latitude, has been obtained from the series expansion of the closed gravity formula of Somigliana by a telescoping procedure. The use of the 7 coefficient formula gives a result as accurate as the closed expression.

081.074 **On the tide-generating forces.**   M. Heikkinen.
Publ. Finnish Geod. Inst., No. 85, 150 pp. (1978).

081.075 **Applications of thermodynamics to fundamental earth physics.**   F. D. Stacey.
Geophys. Surv., Vol. 3, 175 - 204 (1977). – Abstr. in Phys. Abstr., Vol. 81, Abstr. 36125 (1978).

081.076 **Consequences of inaccuracy of gravitational parameter value for orbits in the Earth's gravitational field.**   K. Mison.
Acta Tech. CSAV, Vol. 22, No. 6, p. 718 - 736 (1977). Abstr. in Phys. Abstr., Vol. 81, Abstr. 40460 (1978).

081.077 **Hydrodynamic model of the earth's evolution.**   V. P. Myasnikov, E. G. Markaryan.
Dokl. AN SSSR, Vol. 237, 1055 - 1058 (1977). In Russian. Abstr. in Ref. zh., 51. Astron., 6.51.244 (1978).

081.078 **Gravity of the normal earth given in the form of a non-levelled ellipsoid of rotation.**
K. M. Kartvelishvili, A. S. Gabuniya.
Soobshch. AN GruzSSR, Vol. 88, 65 - 68 (1977). In Russian. Abstr. in Ref. zh., 52. Geod. Aehrosemka, 6.52.53 (1978).

081.079 **Gezeiten bremsen unseren Planeten. Früher drehte sich die Erde schneller.**
P. Brosche, J. Sündermann.
Deutsche Forschungsgem. Mitt., No. 2, p. 14 - 15 (1978).

081.080 **A trans-world tidal gravity profile.**
B. Ducarme, P. Melchior.
Phys. Earth Planet. Interiors, Vol. 16, 257 - 276 (1978) = Obs. R. Belgique, Commun., Sér. A, No. 48.

081.081 **Attenuation models of the earth.**
D. L. Anderson, R. S. Hart.
Phys. Earth Planet. Interiors, Vol. 16, 289 - 306 (1978).

081.082 **Effects on core formation.**   G. H. Shaw.
Phys. Earth Planet. Interiors, Vol. 16, 361 - 369 (1978).

081.083 **Orientation of the ellipsoid in geodetic networks.**
G. L. Strang Van Hees.
Delft Prog. Rep., Vol. 3, No. 1, p. 35 - 38 (1977). – Abstr. in Phys. Abstr., Vol. 81, Abstr. 48955 (1978).

    Climatic change.   See Abstr. 003.054.

    Precambrian of the northern hemisphere and general features of early geological evolution.
See Abstr. 003.118.

    Physics of the earth.   See Abstr. 003.131.

    La terre, les eaux, l' atmosphère.
See Abstr. 003.156.

On estimating gravity anomalies – a comparison of least squares collocation with conventional least squares techniques. See Abstr. 021.004.

A note on the choice of norm when using collocation for the computation of approximations to the anomalous potential. See Abstr. 021.006.

Gravitational harmonics from shallow resonant orbits. See Abstr. 042.014.

The effect of harmonics in the gravitational potential of the Earth, Moon and Sun on the precession angle of the Earth's rotation axis. See Abstr. 044.004.

On the precession and nutation of the earth's axis of figure. See Abstr. 044.012.

Improved gravity field and station coordinate estimates from laser tracking data. See Abstr. 046.014.

Watching the earth move from space. See Abstr. 054.006.

Shock waves on Earth and in space. See Abstr. 082.059.

Investigations of the diurnal variation of inclinations of the earth's surface. See Abstr. 085.017.

082 Atmosphere Including Refraction, Scintillation, Extinction, Airglow, Site Testing

082.001 **Large-scale tropospheric irregularities and their effect on radio astronomical seeing.**
P. J. Hargrave, L. J. Shaw.
Mon. Not. R. Astron. Soc., Vol. 182, 233 - 239 (1978).
Observations with the Cambridge 5-km Radio Telescope have revealed the existence of tropospheric water-vapour irregularities having scale sizes greater than the maximum baseline of the telescope. Unlike the smaller scale irregularities (0.3−1.2 km) studied previously, they show no marked seasonal or diurnal variations, although there is some evidence that they occur more often and with greater magnitude in the summer months. The effect of these irregularities on aperture synthesis telescopes is to produce radial sidelobes, especially near the main response. They are also a hindrance to the determination of accurate source positions in astrometric programmes.

082.002 **The evolution of the atmosphere of the Earth.**
M. H. Hart.
Icarus, Vol. 33, 23 - 39 (1978).
Computer simulations of the evolution of the Earth's atmospheric composition and surface temperature have been carried out. The program took into account changes in the solar luminosity, variations in the Earth's albedo, the greenhouse effect, variation in the biomass, and a variety of geochemical processes. Results indicate that prior to two billion years ago the Earth had a partially reduced atmosphere, which included N_2, CO_2, reduced carbon compounds, some NH_3, but no free H_2. Surface temperatures were higher than now, due to a large greenhouse effect.

082.003 **Paleoatmospheric temperature structure.**
D. A. Morss, W. R. Kuhn.
Icarus, Vol. 33, 40 - 49 (1978).
This study presents calculated radiative-convective, mean global, vertical temperature profiles of the Earth's evolving atmosphere extending back to about 4.5 billion years. The authors assume the atmosphere evolved from one rich in carbon dioxide and deficient in oxygen, and thus ozone. The resultant temperature profiles are obviously dependent upon the assumed concentrations and distributions of the radiatively active gases.

082.004 **A bright future for the night sky.**
R. Pike, R. Berry.
Sky Telesc., Vol. 55, 126 - 129 (1978).

082.005 **Possible climatic and biological impact of nearby supernovae.**   G. E. Hunt.
Nature, Vol. 271, 430 - 431 (1978).
Clark et al. (1977) were unclear whether the supernovae removal of ozone from the atmosphere of the Earth would heat or cool the surface of the planet. There is evidence, reported here, which enables a more precise statement to be made of the effect upon the global Earth. The reduction in the concentration of ozone will cool the stratosphere, troposphere and surface layers of the Earth.

082.006 **Mass spectrometric measurement of the positive ion composition in the stratosphere.**
E. Arijs, J. Ingels, D. Nevejans.
Nature, Vol. 271, 642 - 644 (1978).

082.007 **Analysis of the 27-days fluctuation of the thermospheric density.**   E. Illés-Almár.
Veröff. Zentralinst. Phys. Erde, No. 52, Part 3, (see 012.001),

p. 719 - 725 (1977). In Russian.

082.008 **New results concerning the geomagnetic effect in the upper atmosphere.**
I. Almár, A. Horváth, E. Illés-Almár.
Veröff. Zentralinst. Phys. Erde, No. 52, Part 3, (see 012.001), p. 739 - 751 (1977).
89 equivalent duration values based on more than 30000 satellite observations were used to derive relations between different geomagnetic parameters and the correlated changes in upper atmospheric density during magnetospheric storms. The results demonstrate how the storm-time relative density increase depends on the intensity of the magnetospheric storm, on the altitude of the perigee and on its geocentric distance from the center of the diurnal bulge.

082.009 **Synthetic atmospheric transmittance spectra near 15 and 4.3 μm.**   J. Susskind, J. E. Searl.
J. Quant. Spectrosc. Radiat. Transfer, Vol. 19, 195 - 215 (1978).
Synthetic monochromatic atmospheric transmittance spectra are presented for infrared intervals in the vicinity of the 15 and 4.3 μm CO_2 bands. The intervals are nominally 20 cm^{-1} in width, and the spectra partition the intervals [560, 780] cm^{-1} and [2180, 2400] cm^{-1}, respectively. The spectra are for a vertical atmospheric path. The transmittance model for the calculations is described.

082.010 **Experience in studying the temperature fluctuations of the lower atmosphere's layers during latitude determinations.**   A. O. Mogilin.
Soobshch. Gos. Astron. Inst. Shternberga, No. 200, p. 42 - 46 (1977). In Russian.
Preliminary results of the temperature field distribution of the lower atmosphere's layers are given. The influence of the temperature field on the observed latitudes is estimated.

082.011 **Leuchtende Nachtwolken.**   W. Schröder.
Phys. Bl., Vol. 34, 133 - 135 (1978).

082.012 **Comparison of the Solar Optical Observatory testing sites at Shahe, Xinglong and Huairou. II. Observation of atmospheric temperature fluctuation close to ground.**
J.-m. Wang, Z.-y. Qian, G.-x. Ai, Z.-x. Shi, Y.-f. Hu.
Acta Astron. Sinica, Vol. 18, 182 - 191 (1977). In Chinese.
Measurements and comparison of temperature fluctuation very close to ground for three different terrains which are plain (Shahe), mountainous terrain (Xinglong) and reservoir (Huairou) have been investigated. Heights of optical solar observatory towers have been proposed: fifteen - twenty meters for the water surface, thirty - forty meters for mountainous terrain, forty - fifty meters for the plain.

082.013 **Estimation of vertical temperature profiles as a system identification problem.**
H. H. Kagiwada, R. E. Kalaba.
Joint automatic control conference, Part II. San Francisco, Calif., USA, 22 - 24 June 1977. IEEE, New York, USA (1977), p. 809 - 814. − Abstr. in Phys. Abstr., Vol. 80, Abstr. 91342 (1977).

082.014 **Escape of hydrogen from the primitive earth's atmosphere.**   Y. L. Yung, M. B. McElroy.
Bull. American Astron. Soc., Vol. 9, 497 (1977). − Abstract.

082.015 **Some aspects of convection in meteorology.**
R. S. Lindzen.

Problems of stellar convection, (see 012.005), p. 128 - 141 (1977).

Various aspects of convection in meteorology which may have some relevance for astrophysics are discussed. In particular the role of convection in determining the gross thermal structure of the atmosphere, the treatment of convective turbulence in the boundary layer, and the larger scale organization of convection are dealt with.

082.016 **Preliminary results of an experimental investigation of the upper atmosphere's infrared radiation aboard the artificial earth satellite Meteor.** Yu. M. Kondrat'ev, Yu. A. Martynov, E. A. Pryakhin, V. G. Sochnev, V. F. Tulinov, S. G. Yakovlev.
Kosm. Issled., Vol. 16, 142 - 144 (1978). In Russian.

082.017 **Calculation of ionization losses of albedo particles in the upper layers of the atmosphere with the earth's magnetic field taken into account.** R. N. Basilova, E. I. Kogan-Laskina, G. I. Pugacheva.
Geomagn. Aehron., Vol. 18, 14 - 18 (1978). In Russian.

082.018 **On a possible interconnection of semi-annual and annual variations of oxygen in the upper atmosphere.** M. K. Ivel'skaya, V. V. Katyushina, N. N. Klimov.
Geomagn. Aehron., Vol. 18, 91 - 95 (1978). In Russian.

082.019 **Variations in atmospheric water vapor: baseline results from Smithsonian observations.**
R. G. Roosen, R. J. Angione.
Publ. Astron. Soc. Pacific, Vol. 89, 814 - 822 (1977/78).

Analysis is made of midmorning measurements of absolute humidity at ground level, and spectroscopic determinations of total atmospheric precipitable water vapor above the field stations of the Astrophysical Observatory of the Smithsonian Institution (APO). Large seasonal variations are apparent. Substantial longer-term variations are also present. The observed mean amounts of precipitable water vapor for various APO sites show reasonable agreement with Kuiper's generalized predictions. Study of the relation between precipitable water vapor and surface humidity shows a strong positive correlation, but the variance is so large that surface humidity is not a reliable indicator of precipitable water vapor for any particular day.

082.020 **Simultaneous nighttime lidar measurements of atmospheric sodium and potassium.**
G. Megie, F. Bos, J. E. Blamont, M. L. Chanin.
Planet. Space Sci., Vol. 26, 27 - 35 (1978).

082.021 **Interferometric studies of the twilight and night-glow sodium D-line profiles.**
D. P. Sipler, M. A. Biondi.
Planet. Space Sci., Vol. 26, 65 - 73 (1978).

A 150 mm aperture, pressure scanned Fabry-Perot interferometer has been used to study the midlatitude twilight and nightglow sodium D-line profiles. Nightglow intensity measurements obtained with the interferometer indicate a highly variable behaviour, ranging from near-constant intensities, to monotonically falling, and to rising and falling intensities during the night.

082.022 **On the relationship of 6300 Å airglow to parameters of the F-region of the ionosphere.**
P. A. Hopgood, P. L. Dyson.
Planet. Space Sci., Vol. 26, 93 - 98 (1978).

082.023 **Mesospheric winds at equatorial latitudes: a review on observational aspects.** R. F. Woodman.
J. Atmos. Terr. Phys., Vol. 39, 941 - 958 (1977).

082.024 **Movements of airglow structures within the inter-**
tropical arc observed from southern Kenya.
N. J. Skinner, E. H. Carman, M. P. Heeran.
J. Atmos. Terr. Phys., Vol. 39, 1395 - 1398 (1977).

082.025 **The thermalization of meteoric ionization.**
W. J. Baggaley, T. H. Webb.
J. Atmos. Terr. Phys., Vol. 39, 1399 - 1403 (1977).

The thermal properties of electrons produced by collisional ionization in meteors are discussed and the processes by which thermalization is achieved are examined. It is shown that the electron cooling time varies from about 10^{-3} s at a height of 80 km to about 10^{-1} s at 115 km with faster cooling being operative in bright meteors where electron deexcitation by positive ion collisions is important.

082.026 **Baseline atmospheric extinction.**
R. J. Angione, R. G. Roosen, G. Fouts.
Bull. American Astron. Soc., Vol. 9, 614 (1978). – Abstract.

082.027 **On the relation between diffusion coefficients and height from radar meteor echoes.** G. Forti.
J. Atmos. Terr. Phys., Vol. 40, 89 - 93 (1978).

Diffusion coefficients derived from radar meteor echoes and corrected for atmospheric wind shears have been used to investigate a linear relation between diffusion and height in the atmosphere. The results of a least square fit and of a method in which both variables are considered subject to errors are presented and discussed. The present results seem to indicate that the relation between diffusion coefficient and height is not as simple as supposed. Scale height values derived from these data are not consistent with accepted values in the atmosphere.

082.028 **Photoelectrons in the upper atmosphere: a formulation incorporating effects of transport.**
M. J. Prather, M. B. McElroy, J. Rodriguez.
Planet. Space Sci., Vol. 26, 131 - 138 (1978).

An efficient scheme is described for computation of photoelectron energy spectra. The method, similar to one used in radiative transfer theory, incorporates effects due to spatial and angular redistribution of electrons. These effects play a dominant role at heights above 400 km.

082.029 **Un modèle bidimensionnel du comportement de l'ozone dans la stratosphère.** G. Brasseur.
Planet. Space Sci., Vol. 26, 139 - 159 (1978).

082.030 **Fluctuations in mid-latitude 6300 Å airglow and their relationship to F-region irregularities.**
P. L. Dyson, P. A. Hopgood.
Planet. Space Sci., Vol. 26, 161 - 169 (1978).

Observations of fluctuations in the 6300 Å airglow emission at night have been studied using a tilting filter photometer. Spectral analysis of the fluctuations shows that at times the fluctuations are primarily due to wavelike disturbances with relatively well defined periods. At other times the spectrum of the fluctuations contains significant power only at long periods and there is a relatively sharp cut-off at the short period end.

082.031 **The near infrared nightglow continuum.**
J. F. Noxon.
Planet. Space Sci., Vol. 26, 191 - 192 (1978).

New measurements of the nightglow continuum near 8500 Å show the brightness to be comparable with that in the visible and to be below 1 R/A; there does not appear to be the marked increase of intensity with wavelength assumed in some recent theoretical treatments.

082.032 **Direct measurement of conjugate photoelectrons and predawn 630 nm airglow.**
G. G. Shepherd, J. F. Pieau, T. Ogawa, T. Tohmatsu,

K. Oyama, Y. Watanabe.
Planet. Space Sci., Vol. 26, 211 - 217 (1978).

082.033 **Satellite temperature measurements in the 40–90 km region by the pressure modulator radiometer.**
M. D. Austen, J. J. Barnett, P. D. Curtis, J. T. Houghton,
C. G. Morgan, C. D. Rodgers, E. J. Williamson.
Space Research, Vol. XVII, (see 012.010), 111 - 115 (1977).

082.034 **The distribution of nitric oxide and its variations near the mesopause derived from ionospheric observations.**   J. Taubenheim.
Space Research, Vol. XVII, (see 012.010), 271 - 278 (1977).

082.035 **Optical observation of enhanced nitric oxide abundance in the lower ionosphere following extended auroral activity.**   G. Witt, J. Stegman.
Space Research, Vol. XVII, (see 012.010), 285 (1977).
    A rocket payload measuring the atmospheric density profile, the density of atomic oxygen by the silver-film method, and optical emissions at several wavelengths is discussed. Preliminary results indicate that the 540-nm airglow continuum volume emission rate has a maximum value >1 rayleigh/Å at 102 km altitude, leading to a value of $1.5 \times 10^9 \, cm^{-3}$ for the density of nitric oxide molecules there. This high value is presumed to be due to earlier auroral activity.

082.036 **Rocket-borne sodium nightglow measurement during post-auroral conditions.**
J. Stegman, G. Witt.
Space Research, Vol. XVII, (see 012.010), 287 - 290 (1977).

082.037 **Density and composition of the neutral atmosphere at 140 km from Atmosphere Explorer C satellite data.**   F. A. Marcos, K. S. W. Champion, W. E. Potter,
D. C. Kayser.
Space Research, Vol. XVII, (see 012.010), 321 - 327 (1977).

082.038 **Correlative satellite measurements of atmospheric mass density by accelerometers, mass spectrometers and ionization gauges.**
F. A. Marcos, C. R. Philbrick, C. J. Rice.
Space Research, Vol. XVII, (see 012.010), 329 - 334 (1977).

082.039 **Preliminary results obtained from the low-g accelerometer CACTUS.**   F. Barlier, Y. Boudon,
J. F. Falin, R. Futaully, J. P. Villain, J. J. Walch, A. M.
Mainguy, J. P. Bordet.
Space Research, Vol. XVII, (see 012.010), 341 - 347 (1977).

082.040 **Variations in atmospheric composition and density during a geomagnetic storm.**
C. R. Philbrick, J. P. McIsaac, G. A. Faucher.
Space Research, Vol. XVII, (see 012.010), 349 - 353 (1977).

082.041 **Comparison of drag and mass spectrometer measurements during small geomagnetic disturbances.**
G. M. Keating, E. J. Prior, K. Chang, J. Y. Nicholson III,
U. von Zahn.
Space Research, Vol. XVII, (see 012.010), 355 - 361 (1977).

082.042 **EUV absorption analysis of thermospheric structure from AE-satellite observations of 1974–1976.**
H. E. Hinteregger, L. M. Chaikin.
Space Research, Vol. XVII, (see 012.010), 525 - 532 (1977).

082.043 **A high-altitude rocket measurement of nitric oxide.**   R. J. Thomas.
J. Geophys. Res., Vol. 83, 513 - 516 (1978).
    The nitric oxide density was measured from 110 to 300 km by a rocket photometer during the day. The small measured peak density, about $6.2 \times 10^6 \, cm^{-3}$ at 111 km, can probably be attributed to the period of very low solar magnetic activity preceding the rocket flight. This experiment was coordinated with a similar measurement made by the Ultraviolet Nitric Oxide Experiment aboard the Atmosphere Explorer C satellite; the measurements are in good agreement.

082.044 **North-south aligned equatorial airglow depletions.**   E. J. Weber, J. Buchau, R. H. Eather, S. B. Mende.
J. Geophys. Res., Vol. 83, 712 - 716 (1978).
    Initial observations of equatorial and near-equatorial 6300-Å O I airglow show the existence of north-south aligned regions of airglow depletion. Airglow fine structure associated with the boundaries of the dark bands has been observed down to the 2.5-km resolution limit of the instrument. Simultaneous airborne ionospheric soundings indicate that these regions of airglow depletion are characterized by an increase in the virtual height of the F layer. A simple model of field-aligned electron density depletion in the bottomside of the F layer explains both the airglow observations and the ionospheric soundings.

082.045 **Thermospheric structure and the F region.**   H. Rishbeth.
Radar probing of the auroral plasma, (see 012.011), p. 231 - 256 (1977).
    This article reviews in a general way the structure and dynamics of the thermosphere, particularly in relation to the properties of the ionospheric F region.

082.046 **Studiul rotaţiei atmosferei înalte cu ajutorul observaţiilor sateliţilor artificiali ai Pămîntului.**
M. M. Cîrşmaru.
Anu. Obs. Bucuresti – 1978, (see 047.004), p. 241 - 280 (1977).

082.047 **The effect of increased carbon dioxide concentrations on stratospheric ozone.**   R. E. Boughner.
J. Geophys. Res., Vol. 83, 1326 - 1332 (1978).
    This paper presents one-dimensional model results for the steady state ozone behavior when the CO_2 concentration is twice its ambient level which account for coupling between chemistry and temperature. When the CO_2 level doubled, the total ozone burden increased in relation to the ambient burden by $1.2–2.5\%$, depending on the vertical diffusion coefficient used. Above 30 km, ozone concentrations were larger than the ambient values, a maximum increase of 16% being reached at 43 km.

082.048 **Upper-atmosphere winds from the orbit of 1971–37B, Cosmos 408 rocket.**   C. J. Brookes.
J. British Interplanet. Soc., Vol. 31, 176 - 180 (1978).

082.049 **Upper atmosphere winds from the orbit of 1973–101A.**   D. M. C. Walker.
J. British Interplanet. Soc., Vol. 31, 197 - 199 (1978).

082.050 **Atmospheric halos.**   D. K. Lynch.
    Sci. American, Vol. 238, No. 4, p. 144 - 152 (1978).
    Rings around the sun and moon and related apparitions in the sky are caused by crystals of ice. Precisely how they are formed is still a challenge to modern physics.

082.051 **Night-time variation of short period fluctuations (2–15 min) in the oxygen green line.**
H. Teitelbaum, M. Petitdidier.
J. Atmos. Terr. Phys., Vol. 40, 223 - 227 (1978).
    Measurements of short period fluctuations in the oxygen green line intensity at the 100 km level show a systematic variation during the night with a minimum energy around midnight. This result when compared with other results obtained by different methods leads to a possible interpretation linked to the effect of the terminator.

082.052 **Statistical properties of atmospheric emission in the infrared: II. Sky noise correlation.**
S. Bensammar.
Astron. Astrophys., Vol. 65, 199 - 205 (1978). In French.

It was found that the fluctuations of the infrared thermal atmospheric emission (sky noise) are strongly correlated for a separation of 2' on the sky and a fortiori for smaller separations. Correlations practically indistinguishable from 99 or 100% were frequently observed. The width of the temporal correlation function of the fluctuations of the thermal emission varied from about 100 ms for a good night down to 5 ms for a night with agitation. It is shown that it is more effective to compensate sky noise by simultaneous observations in two points of the field than by the modulation of a single beam on the sky, unless such modulation is performed at high frequency, at least 100 Hz.

082.053 **On the tensor of viscous tensions of the neutral component of the upper atmosphere.**
A. V. Pavlov.
Geomagn. Aehron., Vol. 18, 281 - 285 (1978).

082.054 **Zur numerischen Berechnung der Normalrefraktion.**
H. Beuchat.
Orion, 36. Jahrg., 64 - 67 (1978).

082.055 **Application of a two-layer atmospheric model to the calibration of millimeter observations.**
M. L. Kutner.
Astrophys. Lett., Vol. 19, 81 - 87 (1978).
The combined effect of atmospheric water vapor and oxygen on the calibration of millimeter observations is considered. For most purposes, one can adequately describe the atmosphere as having an upper, oxygen, layer and a lower, water, layer. An important feature is that the two layers radiate with different effective temperatures. In the application to calibration the opacity of the oxygen is calculated, and that of the water is measured. The uncertainties introduced are analyzed, and it is found that, of the atmospheric parameters, the water opacity probably causes the largest error.

082.056 **Conditions for astronomical observations at the Hvar Observatory.** V. Vujnović.
Hvar Obs. Bull., No. 1, p. 41 - 48 (1977).

082.057 **Variations in air density from January 1972 to April 1975 at heights near 200 km.**
D. M. C. Walker.
Planet. Space Sci., Vol. 26, 291 - 309 (1978).

082.058 **Magnetic ordering of the polar airglow.**
J. E. Frederick, P. B. Hays.
Planet. Space Sci., Vol. 26, 339 - 345 (1978).

082.059 **Shock waves on Earth and in space.** I. I. Glass.
Prog. Aerosp. Sci., Vol. 17, 269 - 286 (1977).
Abstr. in Phys. Abstr., Vol. 81, Abstr. 3897 (1978).

082.060 **Atmospheric turbidity at Athens Observatory.**
B. Katsoulis.
Pure Appl. Geophys., Vol. 115, 583 - 591 (1977). – Abstr. in Phys. Abstr., Vol. 81, Abstr. 7768 (1978).

082.061 **Rocket investigations of corpuscular radiation in the upper atmosphere at polar latitudes of the northern and southern hemispheres.** V. F. Tulinov, V. M. Fejgin.
Tr. N.-i. tsentra izuch. prirod. resursov, 1977, vyp. (No.) 3, p. 35 - 40. In Russian. – Abstr. in Ref. zh., 62. Issled. kosm. prostranstva, 3.62.392 (1978).

082.062 **Corpuscular radiation in the polar mesosphere of the southern hemisphere.** V. F. Tulinov, V. V.

Tulyakov, V. M. Fejgin, V. A. Lipovetskij, Yu. M. Zhuchenko.
Tr. N.-i. tsentra izuch. resursov, 1977, vyp. (No.) 3, p. 45 - 48. In Russian. – Abstr. in Ref. zh., 62. Issled. kosm. prostranstva, 3.62.393 (1978).

082.063 **The sizes of particles in noctilucent clouds: implications for mesospheric water vapor.** M. Gadsden.
J. Geophys. Res., Vol. 83, 1155 - 1156 (1978).
The amount of water vapor involved in the formation of noctilucent clouds may be a considerable fraction of the water vapor available in the mesosphere. If the cloud particles consist principally of ice, estimates of the amount of water in the clouds will correspond critically with the size of the particles assumed to be involved.

082.064 **Note on 'Studies of the sensitivity of the components of the earth's radiation balance to changes in cloud properties using a zonally averaged model'** by G. E. Hunt. [J. Quant. Spectrosc. Radiat. Transfer, Vol. 18, 295 - 307 (1977)]. J. Walker.
J. Quant. Spectrosc. Radiat. Transfer, Vol. 19, 553 - 554 (1978).

082.065 **Empirical model for general circulation of the atmosphere at ionospheric levels above 100 km.**
G. V. Vergasova, E. I. Zhovty, E. S. Kazimirovsky (*Eh. S. Kazimirovskij*).
Planet. Space Sci., Vol. 26, 387 - 398 (1978).

082.066 **Interpretation of OGO-5 line shape measurements of Lyman-α emission from terrestrial exospheric hydrogen.** J. L. Bertaux.
Planet. Space Sci., Vol. 26, 431 - 447 (1978).
The velocity distribution of hydrogen atoms in the terrestrial exosphere was measured as a function of radial distance (up to 7 Earth radii, E_R) with the help of a Lyman-α hydrogen absorption cell, flown in 1968 on board the OGO-5 satellite. This paper contains the final analysis of the measurements.

082.067 **Correlations among $O_2(0-1)$ atmospheric band, OH(8$-$3) band and [OI] 5577 Å line and among P_1 (2), $P_1(3)$ and $P_1(4)$ lines of OH(8$-$3) band.**
K. Misawa, I. Takeuchi.
J. Atmos. Terr. Phys., Vol. 40, 421 - 428 (1978).

082.068 **Rocket altitude atmospheric X-rays from magnetospheric electrons at low and middle latitudes.**
J. G. Luhmann, J. B. Blake.
J. Atmos. Terr. Phys., Vol. 40, 465 - 469 (1978).
Like auroral electrons, quasitrapped magnetospheric electrons mirroring in the upper atmosphere at low and middle latitudes will generate X-rays by the bremsstrahlung and K_a line excitation processes. These atmospheric X-rays may contribute a diffuse background to rocket-borne astronomy experiments.

082.069 **K-shell X-ray fluxes at satellite altitude from proton precipitation.** J. G. Luhmann, J. B. Blake.
J. Atmos. Terr. Phys., Vol. 40, 471 - 473 (1978).
A sample calculation is presented of K-shell X-ray emission resulting from energetic proton precipitation into the upper atmosphere. It is found that intense fluxes of K-shell X-rays at low satellite altitudes (450 nautical miles) can result from observed fluxes of precipitating protons.

082.070 **The quenching of OH* in the atmosphere.**
E. J. Llewellyn, B. H. Long, B. H. Solheim.
Planet. Space Sci., Vol. 26, 525 - 531 (1978).

082.071 **The excitation of the infrared atmospheric oxygen bands in the nightglow.**

E. J. Llewellyn, B. H. Solheim.
Planet. Space Sci., Vol. 26, 533 - 538 (1978).

082.072 On the dispersal of artificially-injected gases in the night-time atmosphere.   R. W. Schunk.
Planet. Space Sci., Vol. 26, 605 - 610 (1978).

082.073 The scattering of sunlight from noctilucent cloud particles.   M. Gadsden.
Ann. Géophys., Vol. 33, 357 - 362 (1977). − Abstr. in Phys. Abstr., Vol. 81, Abstr. 11945 (1978).

082.074 The polarization of noctilucent clouds.   M. Gadsden.
Ann. Géophys., Vol. 33, 363 - 366 (1977). − Abstr. in Phys. Abstr., Vol. 81, Abstr. 11946 (1978).

082.075 Analytic infrared transmissivities of the atmosphere.   H. L. Kuo.
Contrib. Atmos. Phys., Vol. 50, 331 - 349 (1977). − Abstr. in Phys. Abstr., Vol. 81, Abstr. 11998 (1978).

082.076 Model construction for atomic oxygen concentrations observed by incoherent scattering.
D. Alcayde, P. Bauer.
Ann. Géophys., Vol. 33, 305 - 320 (1977). In French. Abstr. in Phys. Abstr., Vol. 81, Abstr. 12030 (1978).

082.077 Point discharge currents during an eclipse of the Sun.   A. Madhusudhana Rao.
Indian J. Radio Space Phys., Vol. 6, No. 2, p. 148 - 149 (1977). Abstr. in Phys. Abstr., Vol. 81, Abstr. 16094 (1978).

082.078 Extinction coefficients for multimodal atmospheric particle size distributions.
K. Willeke, J. E. Brockmann.
Atmos. Environ., Vol. 11, 995 - 999 (1977). − Abstr. in Phys. Abstr., Vol. 81, Abstr. 24370 (1978).

082.079 Neutral winds in the upper atmosphere.   A. Brekke.
Fra Fys. Verden, Vol. 39, 66 - 71 (1977). In Norwegian. Abstr. in Phys. Abstr., Vol. 81, Abstr. 24399 (1978).

082.080 Effects of atmospheric turbulence on the central irradiance in the Fraunhofer diffraction pattern by slit apertures with uni-dimensional quadratic and linear apodisation filters: a generalised expression.
P. K. Mondal, S. Subramanyam.
Atti Fond. Giorgio Ronchi, Vol. 32, 897 - 900 (1977). Abstr. in Phys. Abstr., Vol. 81, Abstr. 25796 (1978).

082.081 Increased atmospheric carbon dioxide and stratospheric ozone.
K. S. Groves, S. R. Mattingly, A. F. Tuck.
Nature, Vol. 273, 711 - 715 (1978).
    A column model in which atmospheric photochemistry and radiative transfer have been coupled together, and with vertical transport represented by eddy exchange coefficients, has been used to calculate the effect of increasing carbon dioxide amounts upon stratospheric ozone. Cooling in the upper stratosphere results in enhanced quantities of ozone.

082.082 The O I (6300 Å) airglow.   P. B. Hays, D. W. Rusch, R. G. Roble, J. C. G. Walker.
Rev. Geophys. Space Phys., Vol. 16, 225 - 232 (1978).
    The measurement capability of the Atmosphere Explorer (AE) satellite series has been used to investigate the detailed chemistry of $O(^1D)$ atoms in the earth's thermosphere. Measurements of the O I $(^3P-^1D)$ transition at 6300 Å on AE have been used to infer the strength of the O_2^+ dissociative recombination source of $O(^1D)$, (β_{1D}), the quenching rate of $O(^1D)$ by molecular nitrogen (K_{N_2}), and the value of the photodissociation rate of O_2 (J_{O_2}).

082.083 Modelling of O and O_2 distribution in the lower thermosphere taking into account turbulent mixing and vertical motions.   R. A. Akmaev, G. M. Shved.
Geomagn. Aehron., Vol. 18, 487 - 490 (1978). In Russian.

082.084 Experience in modelling of O and O_2 distribution in the lower thermosphere taking into account a rising tide.   R. A. Akmaev, G. M. Shved.
Geomagn. Aehron., Vol. 18, 546 - 548 (1978). In Russian.

082.085 On oscillations of intensity of night emission of the upper atmosphere in periods of stratospheric warming.   L. M. Fishkova.
Geomagn. Aehron., Vol. 18, 549 - 550 (1978). In Russian.

082.086 Neutral composition measurements between 90- and 220-km altitude by rocket-borne mass spectrometer.
H. Trinks, D. Offermann, U. von Zahn, C. Steinhauer.
J. Geophys. Res., Vol. 83, 2169 - 2176 (1978).
    On June 29, 1974, at 1514 LST a neutral gas mass spectrometer with a cryogenically cooled ion source was flown above Wallops Island as part of the Aladdin 74 program. This paper presents the obtained density height profiles of N_2, O_2, O, and Ar.

082.087 Hydrogen in the upper atmosphere.   B. A. Tinsley.
The Earth: 1, (see 003.010), p. 1 - 100 (1978).

082.088 The equatorial electrojet.   D. M. Cunnold.
The Earth: 1, (see 003.010), p. 101 - 127 (1978).
    The theory of the electrojet is presented. Recent calculations which stress the importance of meridional currents in electrojet theory are reviewed. A limited discussion is also given of the variability of the electrojet and of techniques used in its measurement.

082.089 Energetic spectrum of the vibration of the image of the sun.   U. I. Il'yasov.
Fiz. protsessy verkhn. atmos. Ashkhabad, Ylym, 1977, p. 81 - 85. In Russian. − Abstr. in Ref. zh., 51. Astron., 5.51.94 (1978).

082.090 Regionalization of the Turkmen Soviet Socialist Republic according to clear weather.
A. P. Savrukhin, V. G. Khamidulina.
Fiz. protsessy verkhn. atmos. Ashkhabad, Ylym, 1977, p. 59 - 64. In Russian. − Abstr. in Ref. zh., 51. Astron., 5.51.95 (1978).

082.091 Confirming the presence of hydrofluoric acid in the upper stratosphere.
R. Zander, G. Roland, L. Delbouille.
Geophys. Res. Lett., Vol. 4, 117 - 120 (1977) = Inst. Astrophys., Univ. Liège, Cointe-Ougrée (Belgique), Collect. 4°, No. 317.

082.092 Astronomical refraction. Part III. Refraction anomalies.   A. I. Nefed'eva.
Izv. Astron. Ehngel'gardt. Obs., Kazan', No. 41 - 42, p. 3 - 70 (1976). In Russian.

082.093 Investigation of the aerosol component of the atmosphere from twilight observations.
L. B. Gusakovskaya, D. I. Stepanov.
Izv. Astron. Ehngel'gardt. Obs., Kazan', No. 41 - 42, p. 129 - 140 (1976). In Russian.

082.094 Daily variations of the astronomical refraction.

A. I. Nefed'eva.
Izv. Astron.Ehngel'gardt. Obs., Kazan', No. 41 - 42, p. 141 - 143 (1976). In Russian.

082.095 Experimental measurements of atmospheric aerosol inhomogeneities.   M. J. Post.
Opt. Lett., Vol. 2, 166 - 168 (1978).

082.096 Temperature effect on the atmospheric transmission function in the 15-μm region.   J. Y. Wang.
Opt. Lett., Vol. 2, 169 - 171 (1978).

082.097 The water vapour content in the western European atmosphere obtained from infrared measurements.
A. Greve.
Infrared Phys., Vol. 18, 127 - 132 (1978).
    The author reports on measurements of the water vapour content in the western European atmosphere obtained from infrared attenuation measurements, made during a site survey for a millimetre-wave radio observatory.

082.098 On night-time measurements of temperature fluctua-tions in the ground layer on the eastern summit near Mt. Majdanak in summer 1975 by means of the F-1 in-strument.   A. Eh. Gur'yanov.
Astron. Tsirk., No. 975, p. 3 - 5 (1977). In Russian.

082.099 On the height of a telescope dome at a good site.
P. V. Shcheglov.
Astron. Tsirk., No. 975, p. 6 - 8 (1977). In Russian.

082.100 First astronomical observations at Shorbulak.
I. I. Kanaev.
Astron. Tsirk., No. 976, p. 2 - 4 (1977). In Russian.

082.101 Site testing results, telescopic image quality and atmospheric structure for some observatories.
P. V. Shcheglov.
Astron. Tsirk., No. 976, p. 4 - 6 (1977). In Russian.

082.102 On the size distribution of atmospheric aerosol particles of different composition.   A. Meszaros.
Atmos. Environ., Vol. 11, 1075 - 1081 (1977). – Abstr. in Phys. Abstr., Vol. 81, Abstr. 36218 (1978).

082.103 Direct solar radiation for Rayleigh atmosphere.
M. Chandra.
Indian J. Pure Appl. Phys., Vol. 15, 804 - 805 (1977).
Abstr. in Phys. Abstr., Vol. 81, Abstr. 36274 (1978).

082.104 Statistical analysis of wavefront random deforma-tions produced by atmospheric turbulence near the ground. II. Correlation function estimation by numerical processing.   F. Martin, J. Borgnino.
J. Opt., Vol. 9, 15 - 24 (1978). In French. – Abstr. in Phys. Abstr., Vol. 81, Abstr. 36280 (1978).

082.105 Radiative transfer of atomic and molecular resonant emissions in the upper atmosphere. II. The 9.6 micrometer emission of atmospheric ozone.
H. Yamamoto.
J. Geomagn. Geoelectr., Vol. 29, No. 3, p. 153 - 179 (1977).
Abstr. in Phys. Abstr., Vol. 81, Abstr. 36350 (1978).

082.106 Estimate of day seeing at Shorbulak.
I. I. Kanaev.
Astron. Tsirk., No. 980, p. 7 - 8 (1978). In Russian.

082.107 Local night seeing at Shorbulak.   I. I. Kanaev.
    Astron. Tsirk., No. 981, p. 6 - 7 (1978). In Russian.

082.108 On the structural function of night-time tempera-

ture in the atmospheric ground layer at Mt. Majda-nak.   A. Eh. Gur'yanov.
Astron. Tsirk., No. 986, p. 6 - 8 (1978). In Russian.

082.109 Conditions for photoelectric photometry on Mt. Majdanak.
K. Zdanavičius, J. Sūdžius.
Astron. Tsirk., No. 988, p. 4 - 5 (1978). In Russian.

082.110 Integral luminosity investigations of the night sky on the Assy-Turgen Plateau, the region of the Alpine expedition of the Astrophysical Institute of the Academy of Sciences of the Kazakh SSR.   A. V. Didenko.
Astron. Tsirk., No. 988, p. 5 - 7 (1978). In Russian.

082.111 Measurement of the atmospheric correlation scale.
M. G. Miller, P. L. Zieske.
J. Opt. Soc. America, Vol. 67, 1680 - 1685 (1977).

082.112 Worldwide data on reduced visibility due to air-borne dust.
G. B. Hoidale, B. D. Hinds, R. B. Gomez.
J. Opt. Soc. America, Vol. 67, 1688 - 1690 (1977).

082.113 Determination of observational conditions at Dalanzadgad.
N. Tugjisurun, D. Khaltar, D. Shagdarsurun.
Astron. Tsirk., No. 993, p. 6 - 7 (1978). In Russian.

082.114 Passive imaging through the turbulent atmosphere: fundamental limits on the spatial frequency resolu-tion of a rotational shearing interferometer.
J. J. Burke, J. B. Breckinridge.
J. Opt. Soc. America, Vol. 68, 67 - 77 (1978).
    The authors compute the signal-to-noise (S/N) ratio to be expected when their 180° rotationally shearing interferometer is used for image recovery at the diffraction limit of a large telescope. The variance and covariance of the irradiance fluc-tuations at the detector array are shown to yield measures of the high-frequency spatial spectrum of the source. The authors consider four fundamental sources of noise: temporal fluctua-tions of the source, space-time fluctuations of the atmosphere, shot noise in the detected photocurrents, and the effects of finite sampling. S/N is found to be directly proportional to the angular resolution of the telescope, the single-frame integra-tion time, the square root of the number of frames, the cube of the operating wavelength, the quantum efficiency of the detector, and the average spectral irradiance from the source on the pupil. It is inversely proportional to the cube of the field angle subtended by the source (or part thereof) under study.

082.115 Modal compensation of atmospheric turbulence phase distortion.   J. Y. Wang, J. K. Markey.
J. Opt. Soc. America, Vol. 68, 78 - 87 (1978).

082.116 Measurements of atmospheric isoplanatism using speckle interferometry.
P. Nisenson, R. V. Stachnik.
J. Opt. Soc. America, Vol. 68, 169 - 175 (1978).
    Measurements of isoplanatism for speckle interferometry and speckle imaging applications have been made at a 1.57 meter aperture telescope in Hawaii. The measurements were obtained from optically produced spatial power spectra of short-exposure images showing pairs of stars with different angular separations. The result of this process is a sequence of plots of correlation versus spatial frequency in the image for 0.25, 0.5, 1.9, and 4.7 arc sec separation binary stars. Substan-tial correlation is found to at least 0.6 of the diffraction limit cutoff for the 4.7 arc sec pair.

082.117 Speckle interferometry measurements of atmo-

spheric nonisoplanicity using double stars.
A. M. Schneiderman, D. P. Karo.
J. Opt. Soc. America, Vol. 68, 338 - 347 (1978).
 The parameter characterizing nonisoplanatic effects in speckle interferometry can be measured by observing double stars. An experimental method is developed which permits estimation of this parameter in the presence of inherently large photon and detector noises as well as unequal double-star component brightnesses. Data are presented on five double stars ranging from 0.22 to 3.48 arc sec in separation. The results are compared to theory and the implications to both speckle interferometry and compensated imaging are discussed.

082.118   Measurement of r_0 with speckle interferometry.
          A. M. Schneiderman, D. P. Karo.
J. Opt. Soc. America, Vol. 68, 348 - 351 (1978).

082.119   Conditional fading statistics of scintillation.
          P. A. Pincus, R. A. Elliott, J. R. Kerr.
J. Opt. Soc. America, Vol. 68, 756 - 760 (1978).

082.120   Sky radiance during a total solar eclipse: a theoreti-
          cal model.   G. E. Shaw.
Appl. Opt., Vol. 17, 272 - 276 (1978).

082.121   Ultraviolet limit of solar radiation at the earth's
          surface with a photon counting monochromator.
L. M. Garrison, L. E. Murray, A. E. S. Green.
Appl. Opt., Vol. 17, 683 - 684 (1978).

082.122   Estimation of reflected light from the Earth's sur-
          face in a Rayleigh atmosphere.   S. Sekine.
J. Illum. Eng. Inst. Japan, Vol. 61, 588 - 597 (1977). In
Japanese. – Abstr. in Phys. Abstr., Vol. 81, Abstr. 40663
(1978).

082.123   Observations of the oxygen green line airglow during
          the Winter Anomaly Campaign 1975/76.
H. Lauche.
J. Geophys., Vol. 44, 35 - 38 (1977). – Abstr. in Phys. Abstr.,
Vol. 81, Abstr. 45075 (1978).

082.124   Measurement of the O_2($^1\Delta_g$) height profile during
          a winter anomaly event.   W. Bangert, V. Amann.
J. Geophys., Vol. 44, 165 - 172 (1977). – Abstr. in Phys.
Abstr., Vol. 81, Abstr. 45077 (1978).

082.125   Preliminary results of measurements of atmospheric
          emission and absorption at infrared wavelength at
the Gornergrat.   B. Carli, G. Dall'oglio, C. Dilworth, F.
Melchiorri, F. Mencaraglia, V. Natale, E. Puplett, L. Rossi,
P. Saraceno, E. G. Tanzi.
Mem. Soc. Astron. Italiana, Vol. 47, 101 - 108 (1976).
 Measurements of atmospheric emission and absorption in the infrared have been performed during September and October 1974 and April 1975 at the Gornergrat. A preliminary analysis of data is presented.

082.126   Site selection.   K. Menzel.
          Astron. Soc. Western Australia, Circ., No. 1, 14 pp.
(1978).

082.127   Twilight ozone measurement by solar occultation
          from AE 5.   B. Guenther, R. Dasgupta, D. Heath.
Geophys. Res. Lett., Vol. 4, 434 - 436 (1977).
 The BUV on AE 5 was used for a solar occultation measurement of atmospheric ozone. The number densities determined by this measurement were 3.7×10^{10} cm^{-3} at 50 km, 5.1×10^9 at 60 km, 3.9×10^8 at 70 km and 3.0×10^7 at 80 km.

082.128   The earth hydrogen exobase near a solar minimum.

A. Vidal-Madjar.
Geophys. Res. Lett., Vol. 5, 29 - 32 (1978).
 Measurements of the solar Lyman alpha flux and of the exobase hydrogen density distribution were made by the University of Paris experiment aboard the Orbiting Solar Observatory (OSO-5) during the period August 1974 to August 1975. The results are presented individually and the average exobase hydrogen density is observed to reach 3.5×10^5 atoms cm^{-3} (at 500 km) for an average exospheric temperature as low as 725 K.

082.129   Consequences of a past encounter of the earth with
          an interstellar cloud.   C. P. McKay, G. E. Thomas.
Geophys. Res. Lett., Vol. 5, 215 - 218 (1978).
 The authors consider some direct consequences of a past encounter on the earth's upper atmosphere, namely those involving the altered chemistry produced by an inflow of neutral interstellar hydrogen.

082.130   Atmospheric light pollution.   D. M. Finch.
          J. Illum. Eng. Soc., Vol. 7, 105 - 117 (1978).
Abstr. in Phys. Abstr., Vol. 81, Abstr. 49139 (1978).

082.131   The twinkling of stars.
          E. Jakeman, G. Parry, E. R. Pike, P. N. Pusey.
Contemp. Phys., Vol. 19, No. 2, p. 127 - 145 (1978). – Abstr.
in Phys. Abstr., Vol. 81, Abstr. 53726 (1978).

082.132   Observational testing of new tables of astronomical
          refraction.   A. Yu. Yatsenko.
Astron. obs. pri Kazan. univ. Kazan', 1978. 14 pp. In Russian.
Abstr. in Ref. zh., 51. Astron., 7.51.139 (1978).

     Forschung in der Deutschen Demokratischen
Republik zu Fragen der Physik der Atmosphäre 1971 bis
1975.   See Abstr. 003.026.

     Entwicklungsphasen der Erforschung der Leuchten-
den Nachtwolken.   See Abstr. 003.120.

     Photochemical and transport processes in the upper
atmosphere.   See Abstr. 003.136.

     Evolution of the atmosphere.
See Abstr. 003.141.

     Earth's aura.   See Abstr. 003.147.

     Physical processes in the upper atmosphere.
See Abstr. 003.155.

     La terre, les eaux, l' atmosphère.
See Abstr. 003.156.

     The upper atmosphere and magnetosphere.
See Abstr. 012.050.

     Terrestrial measurement of the performance of
high-rejection optical baffling systems.   See Abstr. 031.240.

     Atmospheric extinction and photometric reductions
near four microns.   See Abstr. 031.285.

     Speckle interferometry at finite spectral bandwidths
and exposure times.   See Abstr. 031.300.

     A two channel rocket photometer for 5577 Å OI
nightglow measurements.   See Abstr. 032.563.

     A versatile radiometer for infrared emission meas-
urements of the atmosphere and targets.
See Abstr. 034.023.

The influence of refraction on positional observations of earth's artificial satellites.   See Abstr. 055.001.

The principle of exchange of stabilities in the atmosphere.   See Abstr. 062.003.

The range of validity of the Rayleigh and Thomson limits for Lorenz-Mie scattering.   See Abstr. 063.039.

Substorms and transmission of oscillations of the ionospheric pressure into the neutral atmosphere. See Abstr. 083.015.

Global effect of auroral particle and Joule heating in the undisturbed thermosphere.   See Abstr. 084.009.

An analytic study of impact ejecta trajectories in the atmospheres of Venus, Mars, and Earth. See Abstr. 093.002.

A comparison of Martian and terrestrial stratospheric thermal structures.   See Abstr. 097.021.

Hydrogen-implanted interplanetary dust grains: their role in hydrogen chemistry of the upper atmosphere. See Abstr. 106.055.

Stjernefotometri i nordlyssonen, I. Feltstasjon i Skibotn.   See Abstr. 113.029.

Stjernefotometri i nordlyssonen, II. Observasjoner og betraktninger.   See Abstr. 113.030.

083 Ionosphere

083.001   Plasma fluxes between ionosphere and protono-
           sphere.
L. Kersley, H. Hajeb-Hosseinieh, K. J. Edwards.
Nature, Vol. 271, 427 - 429 (1978).

083.002   Post-geomagnetic storm protonospheric replenish-
           ment.
L. Kersley, H. Hajeb-Hosseinieh, K. J. Edwards.
Nature, Vol. 271, 429 - 430 (1978).

083.003   Quasi-periodic fluctuations in electron content
       .   during a partial solar eclipse.
S. Vaidyanathan, C. R. Reddi, B. V. Krishnamurthy.
Nature, Vol. 271, 40 - 41 (1978).
      The authors report the observation of quasi-periodic
fluctuations in Faraday rotation angle Ω and the 1 MHz
modulation phase delay ϕ of 40 MHz transmissions from
ATS-6 geosynchronous satellite recorded at Trivandrum (dip
$0°57'S$, geographical longitude $76°57'E$) during the partial
solar eclipse on 29 April 1976.

083.004   Ray tracing study of effects of ionospheric irregu-
           larities on HF and VHF radio waves.
N. C. Mathur, R. P. Pandey.
J. Instn. Electron. Telecommun. Eng., Vol. 23, 121 - 123
(1977). − Abstr. in Phys. Abstr., Vol. 80, Abstr. 91362
(1977).

083.005   VLF phase shifts and amplitude variations during
           SIDs.   K. Baba.
Mem. Chubu Inst. Technol. Vol. 12A, 77 - 82 (1976).
Abstr. in Phys. Abstr., Vol. 80, Abstr. 91363 (1977).

083.006   Investigations of the sporadic radio radiation of the
           sun and of the parameters of the earth's ionosphere
aboard Intercosmos − Copernicus 500. 3. Natural radio noise
of ionospheric plasma.   V. I. Aksenov, T. V. Efimova,
G. P. Komrakov, V. V. Pisareva, R. Schreiber, H. Hanasz.
Kosm. Issled., Vol. 16, 78 - 86 (1978). In Russian.

083.007   Twice-charged ions of atomic oxygen in the dis-
           turbed ionosphere.   S. V. Avakyan.
Kosm. Issled., Vol. 16, 144 - 148 (1978). In Russian.

083.008   Velocity of ionospheric plasma heating by solar
           radiation. I. Heating by photoelectrons.
A. G. Kolesnik, V. I. Chernyshev.
Geomagn. Aehron., Vol. 18, 24 - 32 (1978). In Russian.

083.009   Ionization profiles of low-energetic solar and
           magnetospheric particles in the ionosphere.
P. Velinov.
Geomagn. Aehron., Vol. 18, 50 - 56 (1978). In Russian.

083.010   On a possibility of estimating the E_s-parameters
           from reflection coefficients.   V. L. Pozigun.
Geomagn. Aehron., Vol. 18, 57 - 61 (1978). In Russian.

083.011   A comparison of day and night effects of internal
           gravitational waves in the F2-region of the midlati-
tude ionosphere.   V. M. Smertin, A. A. Namgaladze.
Geomagn. Aehron., Vol. 18, 146 - 148 (1978). In Russian.

083.012   NO, H_2O dissociation and winter anomaly of the
           ionospheric absorption.   V. I. Krasovskij.
Geomagn. Aehron., Vol. 18, 151 - 153 (1978). In Russian.

083.013   Effective frequency of electron collisions in the

E-region of the ionosphere.   N. P. Danilkin,
P. F. Denisenko, V. I. Vodolazkin, S. M. Sushchij, Yu. N.
Faer.
Geomagn. Aehron., Vol. 18, 154 - 155 (1978). In Russian.

083.014   Pulsation spectra of low-frequency radiation in the
           outer ionosphere and short-period oscillations of
the geomagnetic field on the earth's surface.
Ya. I. Likhter, V. I. Larkina.
Geomagn. Aehron., Vol. 18, 162 - 163 (1978). In Russian.

083.015   Substorms and transmission of oscillations of the
           ionospheric pressure into the neutral atmosphere.
L. M. Alekseeva, L. G. Tashkinova.
Geomagn. Aehron., Vol. 18, 169 - 171 (1978). In Russian.

083.016   Joule dissipation in F-region and equatorial spread-F
           events.   T. N. Rajaraman, J. N. Desai,
S. S. Degaonkar, K. D. Cole.
Nature, Vol. 272, 516 (1978).
      The authors describe recent experimental measurements
of neutral atmosphere temperatures at F-region heights at Mt
Abu which agree with the theoretical prediction of very great
atmospheric heating in equatorial spread-F events.

083.017   Photoelectron flux buildup in the plasmasphere.
           G. P. Mantas, H. C. Carlson, V. B. Wickwar.
J. Geophys. Res., Vol. 83, 1 - 15 (1978).
      Processes which confine photoelectrons to the plasma-
sphere (e.g., collisional backscattering from the thermosphere
and magnetic trapping due to pitch angle redistribution
through Coulomb collisions in the plasmasphere) tend to increase the steady state photoelectron flux in the plasmasphere
above the amplitude level that would otherwise have been
attained. Theoretical calculations are presented of steady state
photoelectron fluxes in the plasmasphere, for specified atmospheric and ionospheric conditions. The transparency of the
plasmasphere and the backscattering properties of the thermosphere are investigated. The buildup effect due to collisional
backscatter alone is calculated, and the combined buildup
effect of pitch angle diffusion and backscatter is estimated. It
is found that the inclusion of these effects increases the steady
state photoelectron flux amplitude in the plasmasphere by
about 50% over the value obtained when the buildup effects
are neglected.

083.018   Heating of the ambient ionosphere by an artificially
           injected electron beam.
D. G. Cartwright, S. J. Monson, P. J. Kellogg.
J. Geophys. Res., Vol. 83, 16 - 24 (1978).
      An electrostatic analyzer on the electron accelerator of
the Electron Echo 2 experiment showed that the electrons of
the background plasma were heated to $10,000°K$ or more
within 8 ms of the start of gun pulses. The degree of heating
was dependent on the orientation of the rocket with respect to
the magnetic field but was not measurably dependent on
ambient electron density or neutral atmosphere density or on
the pitch angle at which the 40-keV electron beam was injected. This heating was also accompanied by an increase of
plasma density.

083.019   Diffusion-thermal effects in the topside of the
           nocturnal equatorial ionosphere.
G. J. Bailey, R. J. Moffett.
J. Geophys. Res., Vol. 83, 145 - 150 (1978).

083.020   Ogo 6 observations of small-scale irregularity struc-
           tures associated with subtrough density gradients.

S. Basu.
J. Geophys. Res., Vol. 83, 182 - 190 (1978).
In this paper the author presents high-resolution Ogo 6 ionospheric irregularity data which suggest that plasma instabilities are associated with mid-latitude subtrough density gradients in the region $L \sim 2$–4.

083.021 **Thermal structure of the primitive ionosphere.**
O. Ashihara, M. Shimizu, T. Shimazaki.
J. Geophys. Res., Vol. 83, 191 - 194 (1978).
Exospheric neutral and electron temperatures have been estimated for the primitive upper atmosphere and ionosphere with various oxygen content in the scheme of the authors' previous model (Shimizu and Shimazaki, 1976). The exospheric neutral temperature has been shown to be rather insensitive to the change of oxygen content, justifying the previous assumption for the temperature variation, while the exospheric electron temperature has been found to be quite sensitive to the compositional change, mainly owing to the strong dependence of electron density on the oxygen concentration.

083.022 **The effect of HF-induced plasma instabilities on ionospheric electron temperatures.**
R. L. Showen, R. A. Behnke.
J. Geophys. Res., Vol. 83, 207 - 209 (1978).
A connection is shown between the rise in electron temperature and the presence of plasma instabilities during an ionospheric heating experiment.

083.023 **Atmospheric waves and the equatorial ionosphere.**
T. Beer.
J. Atmos. Terr. Phys., Vol. 39, 971 - 979 (1977).

083.024 **D-region processes at equatorial latitudes.**
P. A. J. Ratnasiri.
J. Atmos. Terr. Phys., Vol. 39, 999 - 1009 (1977).
The current understanding of the ionization production and loss processes, and electron and ion density distributions in the undisturbed D-region are reviewed with special reference to the equatorial latitudes. The importance of combining ground-based measurements with rocket-borne measurements in the development of D-region electron density profile models is stressed.

083.025 **Ionospheric drifts.** B. H. Briggs.
J. Atmos. Terr. Phys., Vol. 39, 1023 - 1033 (1977).
Work on ionospheric drifts carried out since 1968 is reviewed under the following headings: introduction, drifts in the D-region, drifts in the E-region (non-electrojet and electrojet), drifts in the F-region, and miscellaneous points.

083.026 **E-region dynamics.** B. B. Balsley.
J. Atmos. Terr. Phys., Vol. 39, 1087 - 1096 (1977).
This review outlines the basic types of dynamic processes observed in the equatorial E-region. Time scales of these dynamic motions range from a few seconds to a few hours. Observed spatial scales include turbulent-like three-meter irregularity motions as well as macroscopic dynamic motions having dimensions of a few hundred kilometers. Evidence showing large day-to-day variations in E-region currents, electric fields, and neutral winds will be presented and possible causal processes are discussed.

083.027 **Structure of the equatorial F-region, topside and bottomside – a review.** G. Rajaram.
J. Atmos. Terr. Phys., Vol. 39, 1125 - 1144 (1977).

083.028 **The early morning development and the evening decay of electron content—latitude profiles at low latitudes and their dependence upon solar declination.**
G. O. Walker, C. B. Poon.
J. Atmos. Terr. Phys., Vol. 39, 1145 - 1154 (1977).

083.029 **High to low latitude variations in the evening summer total electron content and F-region electron density.** E. A. Essex.
J. Atmos. Terr. Phys., Vol. 39, 1155 - 1158 (1977).

083.030 **Dynamics of the equatorial F-region.**
H. Rishbeth.
J. Atmos. Terr. Phys., Vol. 39, 1159 - 1168 (1977).
This paper reviews the large-scale dynamics of the equatorial F-region. Its main topic is electric fields: section 1 deals with their origin and effects on the F2-layer, in particular the 'equatorial fountain' and the post-sunset rise of the F2-layer. Section 2 deals with neutral air winds and their F2-layer effects, with discussion of what may happen if the horizontal winds are convergent or divergent. Both these sections conclude with a list of current problems. Section 3 deals with aspects of the equatorial F2-layer that are not simply classifiable as 'electric fields' or 'winds', while section 4 briefly recapitulates some major unsolved problems.

083.031 **Low-latitude ionospheric height changes associated with geomagnetic substorms.** G. G. Bowman.
J. Atmos. Terr. Phys., Vol. 39, 1169 - 1173 (1977).

083.032 **Equatorial electron densities—seasonal and solar cycle changes.** G. Rajaram, R. G. Rastogi.
J. Atmos. Terr. Phys., Vol. 39, 1175 - 1182 (1977).

083.033 **Preliminary comparisons of VHF radar maps of F-region irregularities with scintillations in the equatorial region.** S. Basu, J. Aarons, J. P. McClure, C. LaHoz, A. Bushby, R. F. Woodman.
J. Atmos. Terr. Phys., Vol. 39, 1251 - 1261 (1977).

083.034 **On the relationship between the plasma density profile measured in the equatorial E- and F-regions and simultaneous energetic particle and spread-F observations.**
M. C. Kelley, W. E. Swartz, Y. Tayan, R. Torbert.
J. Atmos. Terr. Phys., Vol. 39, 1263 - 1268 (1977).

083.035 **Ionospheric electrostatic fields.** D. Möhlmann.
J. Atmos. Terr. Phys., Vol. 39, 1325 - 1332 (1977).
A perturbation-theoretical method has been used to solve the ionospheric potential equation of the dynamo-theory. The ionospheric large-scale electrostatic potential and force fields are given in analytical expression depending on the coefficients of an appropriate expansion of the wind fields. Numerical and graphical examples are given.

083.036 **Ionospheric model effects on thermospheric calculations.** B. K. Ching, J. M. Straus.
J. Atmos. Terr. Phys., Vol. 39, 1389 - 1393 (1977).

083.037 **Effects of a magnetic storm on the plasmaspheric electron content.**
W. Degenhardt, G. K. Hartmann, R. Leitinger.
J. Atmos. Terr. Phys., Vol. 39, 1435 - 1440 (1977).

083.038 **F-region temperatures measured simultaneously by satellite and incoherent scatter radar.**
P. H. McPherson.
J. Atmos. Terr. Phys., Vol. 39, 1459 - 1464 (1977).

083.039 **The calculated and observed ionospheric properties during Atmospheric Explorer-C satellite crossings over Millstone Hill.** R. G. Roble, A. I. Stewart, M. R. Torr, D. W. Rusch, R. H. Wand.
J. Atmos. Terr. Phys., Vol. 40, 21 - 33 (1978).

083.040 **Multi-satellite scintillations. Spread-F and sporadic-E over Brisbane—2.** L. A. Hajkowicz.
J. Atmos. Terr. Phys., Vol. 40, 99 - 104 (1978).

083.041 **Total electron content observations during the 23 October 1976 solar eclipse over south-eastern Australia.** F. W. Morton, E. A. Essex.
J. Atmos. Terr. Phys., Vol. 40, 111 - 114 (1978).

Total electron content observations were taken on a triangular array to investigate travelling ionospheric disturbances (TIDs), induced by the solar eclipse of 23 October, 1976 over south—eastern Australia. The observations were taken in the path of totality. No TIDs were detected which could be attributed to the eclipse.

083.042 **Effect of diffusion—thermal processes on the high-latitude topside ionosphere.**
R. W. Schunk, W. J. Raitt, A. F. Nagy.
Planet. Space Sci., Vol. 26, 189 - 191 (1978).

The authors studied the extent to which diffusion—thermal heat flow affects H^+ temperatures in the high-latitude topside ionosphere. Such a heat flow occurs whenever there are H^+—O^+ relative drifts. From their study they found that at high-latitudes, where H^+ flows up and out of the topside ionosphere, diffusion—thermal heat flow acts to reduce H^+ temperatures by 500—600 K at altitudes above about 900 km.

083.043 **Helium ion outflow from the terrestrial ionosphere.** W. J. Raitt, R. W. Schunk, P. M. Banks.
Planet. Space Sci., Vol. 26, 255 - 268 (1978).

083.044 **Induced proton flow in a collapsing post-sunset ionosphere and protonosphere.**
J. A. Murphy, R. J. Moffett.
Planet. Space Sci., Vol. 26, 281 - 288 (1978).

Results are calculated for a dipole magnetic field tube situated at $L = 4$ and acceleration terms are included in the momentum equations. Proton flow into the ionosphere results from decay of the F2-layer. Changes in temperatures and temperature gradients following sunset may not enhance the H^+ flow. Under extreme conditions the H^+ flow remains subsonic. It seems unlikely that an interhemispheric flux of protons can directly maintain the nighttime F2-layer.

083.045 **Ionization ledges and the counter-electrojets in the equatorial ionosphere.**
R. Raghavarao, P. Sharma, A. R. Jain.
Space Research, Vol. XVII, (see 012.010), 417 - 421 (1977).

083.046 **Ionization ledges in the equatorial ionosphere: a method of detecting cross-equatorial winds.**
R. Raghavarao, P. Sharma, M. R. Sivaraman.
Space Research, Vol. XVII, (see 012.010), 423 - 425 (1977).

083.047 **Ionization by energetic electrons in the midlatitude night-time E region.** L. G. Smith, H. D. Voss.
Space Research, Vol. XVII, (see 012.010), 427 - 432 (1977).

Observations of the electron density profile and flux of energetic electrons obtained in two rocket flights at Wallops Island near midnight are presented. The ionization rates of the upper E region deduced from the electron density profiles are found to support the dependence on Kp established in previous observations. Calculations of the ionization rates using the observed electron fluxes show agreement with the values derived from the electron density profiles.

083.048 **ESRO 4: an electron density model of the F2 layer for quiet solar conditions.**
W. Köhnlein, W. J. Raitt.
Space Research, Vol. XVII, (see 012.010), 439 - 444 (1977).

083.049 **Relationship between plasma density, temperature and suprathermal electron flux in the topside ionosphere.** K. Spenner, K. Rawer, H. Wolf.
Space Research, Vol. XVII, (see 012.010), 445 - 449 (1977).

083.050 **Ground-based and rocket experiments to determine electron collision frequency and electron density profiles at heights 80–500 km.** A. V. Biryukov, N. P. Danilkin, P. F. Denisenko, G. M. Kucherenko, I. A. Knorin, V. A. Rudakov, V. V. Sotsky (*Sotskij*), S. M. Sustchy (*Sushchij*), Yu. N. Faer, L. A. Shnyreva.
Space Research, Vol. XVII, (see 012.010), 451 - 455 (1977).

083.051 **Electron temperature and density measured aboard the Vertical 3 rocket.** S. Chapkunov, G. Gdalevich, M. Petrounova, T. Ivanova, L. Bankov.
Space Research, Vol. XVII, (see 012.010), 457 - 459 (1977).

083.052 **Vertical distribution of ion concentration measured aboard the Vertical 3 rocket.** K. B. Serafimov, S. K. Chapkunov, M. C. Petrounova, T. N. Ivanova.
Space Research, Vol. XVII, (see 012.010), 461 - 463 (1977).

083.053 **In situ measurements of ion composition and the principal ion—molecule reactions in the F2 region.** A. D. Danilov, Yu. A. Romanovsky (*Romanovskij*), V. K. Semenov, Yu. A. Agafonov.
Space Research, Vol. XVII, (see 012.010), 465 - 470 (1977).

083.054 **Mass spectrometer measurements of the F2 region neutral and ion composition aboard the orbital station Salyut 4.** A. S. Loevsky, L. I. Pogulyaevsky, (*Pogulyaevskij*), Yu. A. Romanovsky (*Romanovskij*), E. G. Ul'yanov.
Space Research, Vol. XVII, (see 012.010), 471 - 476 (1977).

083.055 **Molecular ions in the region of the midlatitude trough at heights of 210–250 km from Intercosmos 8 observations.** Ts. P. Dachev, G. A. Stanev.
Space Research, Vol. XVII, (see 012.010), 477 - 481 (1977).

083.056 **Time variations of HF-induced plasma waves.** R. L. Showen, D. M. Kim.
J. Geophys. Res., Vol. 83, 623 - 628 (1978).

083.057 **Effect of the day night ionospheric conductivity gradient on polar cap convective flow.** G. Atkinson, D. Hutchison.
J. Geophys. Res., Vol. 83, 725 - 729 (1978).

083.058 **Electrostatic waves in the ionosphere.** E. Leer.
Radar probing of the auroral plasma, (see 012.011), p. 55 - 72 (1977).

The author studies electrostatic waves that can propagate in the ionosphere. These may be excited under certain conditions and can scatter a high frequency radar signal.

083.059 **Electric fields, neutral winds and currents derived from Chatanika.** A. Brekke.
Radar probing of the auroral plasma, (see 012.011), p. 285 - 314 (1977).

083.060 **Field-aligned plasma flow and the Polar Wind: a review.** M. Blanc.
Radar probing of the auroral plasma, (see 012.011), p. 331 - 383 (1977).

083.061 **Characteristics of the polar D-region determined by radar and satellite measurements.** T. M. Watt.
Radar probing of the auroral plasma, (see 012.011), p. 385 - 405 (1977).

083.062 **On the coordination of EISCAT measurements with rocket and satellite observations.** B. Hultqvist.
Radar probing of the auroral plasma, (see 012.011), p. 419 - 428 (1977).

The scientific interest of combining EISCAT measurements of the thermal ionospheric plasma with sounding rocket and/or satellite measurements of the hot plasma distribution function and other variables is discussed briefly. Some examples are presented where such coordinated measurements are of great interest.

083.063 Coordination of the Tromsø partial reflection experiment (PRE) and EISCAT. A. Haug.
Radar probing of the auroral plasma, (see 012.011), p. 429 - 449 (1977).

A redesigned partial reflection experiment has been installed and put in operation at Ramfjord, Tromsø. In this paper the theory for the experiment is briefly outlined and it is shown that the partial reflection experiment is suitable for a number of different studies of the lower ionosphere.

083.064 An improved model of the variation of electron concentration with height in the ionosphere.
J. R. Dudeney.
J. Atmos. Terr. Phys., Vol. 40, 195 - 203 (1978).

An empirical model of the variation of electron concentration with height is described which overcomes some limitations found in practice with a previous widely used model (Bradley and Dudeney, 1973). In particular, the new model will generate more realistic variations of electron concentration with real height and virtual height, both including and excluding an $F1$-ledge. The model has no gradient discontinuities and will reproduce cases in which the $F1$-ledge does not have a true turning point. The model should prove very valuable for a wide range of propagation problems and for certain aeronomical applications.

083.065 Some characteristics of the quiet polar D-region and mesosphere obtained with the partial reflection method. K. Schlegel, A. Brekke, A. Haug.
J. Atmos. Terr. Phys., Vol. 40, 205 - 213 (1978).

It is shown that from a detailed analysis of the amplitude behaviour of electromagnetic waves partially reflected in the D-region valuable information about the reflection process and the dynamics of the mesosphere can be obtained. The results indicate that the process which is responsible for the partial reflections is different for different heights in the D-region and also changes with the season. From the temporal fluctuations of the amplitudes of the reflected waves, estimates of various turbulence parameters of the mesosphere have been deduced. Their mean values are in satisfactory agreement with results obtained by other methods.

083.066 Ionospheric observations of the ring-shaped solar eclipse on April 29, 1976 in Ashkhabad.
O. Ovezgel'dyev, M. Berkeliev, L. P. Korsunova, E. V. Nepomnyashchaya.
Izv. AN TurkmSSR. Ser. fiz.-tekh., khim. i geol. nauk, 1977, No. 4, p. 48 - 54. In Russian. – Abstr. in Ref. zh., 62. Issled. kosm. prostranstva, 2.62.242 (1978).

083.067 Calculation of the parameters of the mid-latitude upper ionosphere for high solar activity conditions.
M. A. Nikitin, B. E. Serebryakov.
Geomagn. Aehron., Vol. 18, 218 - 223 (1978). In Russian.

083.068 Calculation of the night-time magneto-quiet F2-region of the mid-latitude ionosphere.
A. V. Mikhajlov, G. I. Ostrovskij.
Geomagn. Aehron., Vol. 18, 224 - 228 (1978). In Russian.

083.069 Model of the effective recombination coefficient for the mid-latitude ionosphere.
Yu. K. Chasovitin, V. B. Shushkova.
Geomagn. Aehron., Vol. 18, 235 - 242 (1978). In Russian.

083.070 Heating velocities of ionospheric plasma by solar

radiation. II. Heating by radiation in the Schumann-Runge continuum and on account of chemical reactions.
V. I. Chernyshev, A. G. Kolesnik, M. N. Vlasov.
Geomagn. Aehron., Vol. 18, 243 - 250 (1978). In Russian.

083.071 Reaction of the ionosphere on rapid changes of large-scale electric fields. II. Molecular ions.
M. G. Deminov, V. P. Kim.
Geomagn. Aehron., Vol. 18, 251 - 256 (1978). In Russian.

083.072 Modeling of electric fields of magnetospheric origin in the ionosphere. I.
V. E. Zakharov, M. A. Nikitin.
Geomagn. Aehron., Vol. 18, 257 - 262 (1978). In Russian.

083.073 Ionospheric effects of magnetospheric substorms at night time. G. N. Pushkova, L. A. Yudovich.
Geomagn. Aehron., Vol. 18, 263 - 266 (1978).

083.074 Observations of hydrogen emissions and sporadic ionization of the E-region.
K. I. Gorelyj, O. M. Pirog.
Geomagn. Aehron., Vol. 18, 354 - 355 (1978).

083.075 Ion structure and the phenomenon of the winter anomaly in the D-region of the ionosphere.
V. V. Koshelev, S. G. Fedchenko.
Geomagn. Aehron., Vol. 18, 356 - 358 (1978).

083.076 Ionosphere research by the K.N.M.I.
H. J. A. Vesseur.
Nederlands Tijdschr. Natuurkd. A, Vol. A43, No. 3, p. 128 - 133 (1977). In Dutch. – Abstr. in Phys. Abstr., Vol. 81, Abstr. 4029 (1978).

083.077 Measurements of the electric field and conductivity of the auroral ionosphere during substorms.
G. A. Vnuchkov, I. A. Zhulin, I. M. Kopaev, N. K. Osipov, Ya. E. Pidgirnyak, V. S. Tsybul'skij.
Kosm. Issled., Vol. 16, 310 - 312 (1978). In Russian.

083.078 Concentration and frequency of electron collisions in the ionosphere from measurements at rocket starts of Vertical type in 1975. A. V. Biryukov, N. P. Danilkin, P. F. Denisenko, I. A. Knorin, G. M. Kucherenko, V. A. Rudakov, V. V. Sotskij, S. M. Sushchij, Yu. N. Faer, L. A. Shnyreva.
Kosm. Issled., Vol. 16, 315 - 317 (1978). In Russian.

083.079 Ionospheric scintillation to 1 GHz.
M. L. Heron.
Monitor, Vol. 38, 101 - 103 (1977). – Abstr. in Phys. Abstr., Vol. 81, Abstr. 7827 (1978).

083.080 Variation of E-layer critical frequency with solar activity. S. S. Kouris.
Proc. Instn. Electr. Eng ., Vol. 124, 1007 - 1008 (1977). Abstr. in Phys. Abstr., Vol. 81, Abstr. 7828 (1978).

083.081 Simultaneous observations of equatorial ionospheric scintillation on four frequencies.
R. G. Rastogi, M. R. Deshpande, N. M. Vadher, K. Davies, P. B. Parikh.
Nature, Vol. 273, 285 - 287 (1978).

This note describes recordings of ionospheric scintillations at equatorial latitudes occurring simultaneously on 40, 140, 360 and 860 MHz.

083.082 A study of electron density spectra of traveling ionospheric disturbances. A. L. Hearn, K. C. Yeh.
J. Geophys. Res., Vol. 83, 1442 - 1446 (1978).

Special runs of the Arecibo incoherent scatter radar were

made to provide high spatial resolution (8.7 km) and high temporal resolution (about 1 min) electron density data at discrete heights in the altitude range 200–500 km. The observed electron density fluctuations were spectrally analyzed and interpreted in terms of gravity wave induced traveling ionospheric disturbances.

083.083   Ionospheric heating by radio waves: predictions for Arecibo and the satellite power station.
F. W. Perkins, R. G. Roble.
J. Geophys. Res., Vol. 83, 1611 - 1624 (1978).
   The effect of resistive heating by radio waves on ionospheric temperatures, electron densities, and airglow emissions is examined by using numerical ionospheric structure and heat balance codes. Two cases are studied: (1) a 3-GHz, 10-GW microwave beam from a proposed satellite power station and (2) 1-MW and 3-MW beams of 15-MHz radio waves launched by the Arecibo antenna. By intent, these two cases have similar intensities and geometries of resistive heating. The most dramatic heating effects are predicted to occur in the E region, where a thermal runaway will take place.

083.084   A note on lunar tides in the ionosphere.
J. V. Evans.
J. Geophys. Res., Vol. 83, 1647 - 1652 (1978).
   In this paper the author reviews the information available concerning the influence of the weaker lunar (gravitational) tide on the ionosphere and suggests that mode coupling may play an important role in the behavior of this tide also.

083.085   Composition and variations of average-energy corpuscular streams.   V. F. Tulinov, V. M. Fejgin, Yu. M. Zhuchenko, V. A. Lipovetskij, M. A. Savel'ev, V. V. Tulyakov, T. A. Zhuchenko, L. S. Novikov, G. F. Tulinov.
Tr. N.-i. tsentra izuch. prirod. resursov, 1977, vyp. (No.) 3, p. 49 - 55. In Russian. – Abstr. in Ref. zh., 62. Issled. kosm. prostranstva, 3.62.364 (1978).

083.086   Energetic characteristics of low-energy electron streams.   V. F. Tulinov, V. M. Fejgin, Yu. M. Zhuchenko, V. A. Lipovetskij, T. A. Zhuchenko, M. A. Savel'ev, V. A. D'yachenko, A. P. Babaev, L. S. Novikov, G. F. Tulinov.
Tr. N.-i. tsentra izuch. prirod. resursov, 1977, vyp. (No.) 3, p. 41 - 44. In Russian. – Abstr. in Ref. zh., 62. Issled. kosm. prostranstva, 3.62.365 (1978).

083.087   A determination of F region convective electric fields from rocket measurements of ionospheric thermal ion spectra.   B. G. Morgan, R. L. Arnoldy.
J. Geophys. Res., Vol. 83, 1055 - 1061 (1978).
   A detector package designed to measure the bulk flow of heavy ions was developed for sounding rocket studies of the lower ionosphere. From least squares fits of a shifted, drifting Maxwellian distribution to direct measurements of the ionospheric particles' thermal energy spectra, average values of rocket payload potential, ion temperature, and bulk plasma flow were obtained.

083.088   A numerical study of the lunar tide in the mid-latitude $F2$-region of the ionosphere – I. Oscillations of the electron density.   S. Handa, H. Maeda.
J. Atmos. Terr. Phys., Vol. 40, 395 - 404 (1978).
   The drift due to the lunar electric field which is communicated from the ionospheric E-region causes a re-distribution of the electron density and a neutral air motion in the $F2$-region. The lunar tide in the electron density of the $F2$-region due to the lunar electric field has been obtained and discussed.

083.089   A numerical study of the lunar tide in the mid-

latitude $F2$-region of the ionosphere – II. Oscillations of the ion and neutral gas velocities.
S. Handa, H. Maeda.
J. Atmos. Terr. Phys., Vol. 40, 405 - 408 (1978).
   The lunar tides in the ion and neutral velocities are dealt with in this paper. It is found that the lunar tidal oscillations in the drifts and neutral winds markedly depend on the solar time.

083.090   A search for E-region disturbances induced by ion cloud releases at F-region heights.
T. B. Jones, C. T. Spracklen.
J. Atmos. Terr. Phys., Vol. 40, 409 - 420 (1978).

083.091   HF Doppler observations of 23 October, 1976 total solar eclipse over south–eastern Australia.
D. W. Cornelius, E. A. Essex.
J. Atmos. Terr. Phys., Vol. 40, 497 - 502 (1978).

083.092   Mid-latitude ionospheric scintillations of VHF radio signals associated with peculiar fluctuations of Faraday rotation.   K. Sinno, M. Kan.
J. Atmos. Terr. Phys., Vol. 40, 503 - 506 (1978).

083.093   A numerical computer investigation of the polar F-region ionosphere.   B. J. Watkins.
Planet. Space Sci., Vol. 26, 559 - 569 (1978).
   Results of a numerical computer investigation of the geomagnetically quiet, high latitude F-region ionosphere are presented. A mathematical model of the steady state polar convective electric field pattern is used in conjunction with production and loss processes to solve the continuity equation for the ionization density in a unit volume as it moves across the polar cap and through the auroral zones.

083.094   Resonance region of a PC5 micropulsation examined by a dual auroral radar system.
A. D. M. Walker, R. A. Greenwald, W. F. Stuart, C. A. Green.
Nature, Vol. 273, 646 - 649 (1978).
   Since January 1977 a new radar auroral backscatter facility, STARE – Scandinavian Twin Auroral Radar Experiment – has been in operation in northern Scandinavia. Because of the spatial and temporal resolution of the STARE system, it is well suited for the study of ionospheric electric fields associated with PC5 geomagnetic pulsations (period 150–600 s). Here the authors present the first results of the analysis of such an event which occurred between 1230 and 1300 UT on 2 February 1977.

083.095   Electric fields in the ionosphere produced by polar field-aligned currents.
K. Maekawa, H. Maeda.
Nature, Vol. 273, 649 - 650 (1978).
   Two regions of field-aligned currents of magnetospheric origin have recently been found along the auroral oval and region 1 currents flow at a slightly higher latitude than those in region 2. The authors discuss the effects that these currents seem to have on the distribution of electric fields in the ionosphere, in addition to the S_q electric field of ionospheric origin.

083.096   Transmission of polar electric fields to the equator.
T. Kikuchi, T. Araki, H. Maeda, K. Maekawa.
Nature, Vol. 273, 650 - 651 (1978).
   To clarify the mechanism of the instantaneous horizontal transmission of the polar electric field, the authors have analysed the response of the ionosphere to a suddenly impressed electric field.

083.097   The relative effects of electric fields and neutral winds on the formation of the equatorial sporadic E layer.   R. Dagar, P. Verma, O. P. Nagpal, C. S. G. K. Setty,

Ann. Géophys., Vol. 33, 333 - 340 (1977). – Abstr. in Phys. Abstr., Vol. 81, Abstr. 12045 (1978).

083.098 A simplified model of polar cap electric fields.
N. D'Angelo.
Ann. Géophys., Vol. 33, 341 - 345 (1977). – Abstr. in Phys. Abstr., Vol. 81, Abstr. 12046 (1978).

083.099 Ionospheric observations by space vehicles.
S. Miyazaki, T. Itoh, T. Mukai, I. Aoyama,
T. Ogawa, S. Kato, T. Tohmatsu, N. Matsuura.
Bull. Inst. Space Aeronaut. Sci., Univ. Tokyo, B, Vol. 12, 866 - 884 (1976). In Japanese. – From Phys. Abstr., Vol. 81, Abstr. 24402 (1978).

083.100 Analysis of the electron density in the lower and upper ionosphere from measurements of VLF propagation modes by means of K-9M-53 rocket.
I. Nagano, M. Mambo, I. Kimura.
Bull. Inst. Space Aeronaut. Sci., Univ. Tokyo, B, Vol. 12, 947 - 958 (1976). In Japanese. – Abstr. in Phys. Abstr., Vol. 81, Abstr. 24403 (1978).

083.101 Phase and amplitude scintillation frequencies produced by propagation through turbulent layers.
J. C. Baker.
IEEE Trans. Commun., Vol. COM-26, No. 1, p. 174 - 177 (1978). – Abstr. in Phys. Abstr., Vol. 81, Abstr. 24405 (1978).

083.102 Deducing an ionospheric current using geomagnetic data on Earth's surface. Y. Suzuki.
Mem. Fac. Eng. Osaka City Univ., Vol. 17, 35 - 45 (1976). In Japanese. – Abstr. in Phys. Abstr., Vol. 81, Abstr. 24406 (1978).

083.103 Combined effects of terrestrial diffraction and ionospheric reflection on medium-frequency sky wave propagation. A. Kinase.
Proc. Instn. Electr. Eng., Vol. 124, 1125 - 1126 (1977). Abstr. in Phys. Abstr., Vol. 81, Abstr. 28410 (1978).

083.104 A model of F_2 peak electron densities in the main trough region of the ionosphere.
B. W. Halcrow, J. S. Nisbet.
Radio Sci., Vol. 12, 815 - 820 (1977). – Abstr. in Phys. Abstr., Vol. 81, Abstr. 28417 (1978).

083.105 Goals and status of the international reference ionosphere. K. Rawer, D. Bilitza, S. Ramakrishnan.
Rev. Geophys. Space Phys., Vol. 16, 177 - 181 (1978).
Under the supervision of an international steering committee, basic data for establishing ionospheric profiles were gathered, critically reviewed, and used for establishing vertical profiles of the most important parameters of the ionosphere, namely, electron density, electron and ion temperature, and (positive) ion composition. The data sources used are described in detail, their reliability and coverage is discussed, and remaining problems are reviewed.

083.106 Results of measurement of electron and ion concentrations and of the electron temperature with the Vertical 4 rocket. G. L. Gdalevich, S. K. Chapkanov, L. B. Bankov, V. F. Gubskij, Ts. P. Dachev, Kh. M. Petrunova.
Kosm. Issled., Vol. 16, 394 - 397 (1978). In Russian.

083.107 Diurnal variations in the D-region during a storm after-effect. P. H. G. Dickinson, F. D. G. Bennett.
J. Atmos. Terr. Phys., Vol. 40, 549 - 558 (1978).
Measurements of electron concentration in the D- and lower E-regions of the ionosphere are reported for seven rocket flights from South Uist, Scotland, in April 1973. The principal feature of the results is that the electron concentra-tions below 85 km varied between 4 and 10 times the concentrations found on normal days. The variability was correlated with changes in radio wave absorption.

083.108 The effect of D-region absorption on the forward scatter radiometeor amplitude distribution index.
B. Ramachandra Rao, E. Bhagiratha Rao, M. Srirama Rao, S. Raja Ratnam, D. A. V. Krishna Rao.
J. Atmos. Terr. Phys., Vol. 40, 571 - 576 (1978).

083.109 Measurements of the electron temperature at satellites and peculiarities of its behaviour in the region of the main ionospheric hole. V. V. Afonin, O. P. Kolomijtsev, Yu. G. Mizun.
Geomagn. Aehron., Vol. 18, 432 - 435 (1978). In Russian.

083.110 Large-scale irregularities in the E_s-layer of the auroral ionosphere. N. N. Volkov, R. S. Kukushkina.
Geomagn. Aehron., Vol. 18, 436 - 439 (1978). In Russian.

083.111 On ionospheric generation of VLF-radiation.
M. S. Kovner, V. A. Kuznetsova, Ya. I. Likhter.
Geomagn. Aehron., Vol. 18, 466 - 472 (1978). In Russian.

083.112 Modelling of electric fields of magnetospheric origin in the ionosphere. II.
V. E. Zakharov, M. A. Nikitin.
Geomagn. Aehron., Vol. 18, 495 - 502 (1978). In Russian.

083.113 A possibility of using an electrostatic approach in the analysis of electric field penetration in a near-earth plasma along the force lines of the geomagnetic fields.
B. N. Gershman, A. V. Samsonov.
Geomagn. Aehron., Vol. 18, 503 - 506 (1978). In Russian.

083.114 Diffusion of the function of distribution of an electron beam in the ionosphere depending on the distance from the injection point. N. I. Izhovkina.
Geomagn. Aehron., Vol. 18, 525 - 526 (1978). In Russian.

083.115 Variations of plasma streams at midlatitudes under conditions of high solar activity.
M. A. Nikitin, B. E. Serebryakov.
Geomagn. Aehron., Vol. 18, 526 - 528 (1978). In Russian.

083.116 Semi-empirical model of the ionosphere in stationary and non-stationary approximations.
V. M. Polyakov, M. K. Ivel'skaya, V. E. Sukhodol'skaya, G. E. Sutyrina, G. V. Shapranova.
Geomagn. Aehron., Vol. 18, 531 - 533 (1978). In Russian.

083.117 Charts of prognoses of planetary distribution of median values of geometrical parameters of the F2-layer and empirical models. T. A. Anufrieva.
Geomagn. Aehron., Vol. 18, 533 - 535 (1978). In Russian.

083.118 Decrease of electron concentration under the influence of neutral waves in the ionosphere.
Eh. G. Mirmovich, P. T. Prudnikov, Z. B. Rojkhvarger.
Geomagn. Aehron., Vol. 18, 535 - 536 (1978). In Russian.

083.119 On dynamic properties of the ionospheric plasma in periods of magnetospheric substorms.
O. P. Kolomijtsev, N. P. Sergeenko, L. A. Yudovich.
Geomagn. Aehron., Vol. 18, 537 - 539 (1978). In Russian.

083.120 Ionospheric effects of magnetospheric substorms.
E. E. Goncharova, N. N. Klimov, V. M. Shashun'-kina, L. A. Yudovich.
Geomagn. Aehron., Vol. 18, 539 - 542 (1978). In Russian.

083.121 Influence of the temperature gradient in the ionosphere on the formation of disturbances by the
solar terminator.   V. M. Somsikov.
Geomagn. Aehron., Vol. 18, 542 - 544 (1978). In Russian.

083.122 On limits of application of the electrostatic approximation to diffusion spreading of weak irregularities
of ionospheric plasma.   L. E. Zhmur.
Geomagn. Aehron., Vol. 18, 544 - 546 (1978). In Russian.

083.123 Interplanetary magnetic field variations and the
electromagnetic state of the equatorial ionosphere.
V. L. Patel.
J. Geophys. Res., Vol. 83, 2137 - 2144 (1978).
   The effect of interplanetary magnetic field variations on
the low-latitude ionosphere has been studied by using the
simultaneous occurrence of three phenomena—the disappearance of equatorial E_{sq}, the reversal or decrease in intensity of
the equatorial electrojet as implied from magnetogram signatures, and a northward turning of the z component of the
interplanetary magnetic field B_z observed by the Explorer 43
satellite.

083.124 Low-altitude plasma line anisotropy.
E. S. Oran, P. J. Palmadesso.
J. Geophys. Res., Vol. 83, 2190 - 2194 (1978).
   Plasma line observations obtained from incoherent radar
backscatter have been used as a ground-based method for
deriving information about the size and anisotropy of the
ionospheric photoelectron fluxes.

083.125 Low-latitude E region ionization by energetic ring
current particles.   L. R. Lyons, A. D. Richmond.
J. Geophys. Res., Vol. 83, 2201 - 2204 (1978).
   Energetic neutral particles resulting from the charge
exchange of ring current ions with geocoronal hydrogen are an
important quiet time source of E region ionization at night
and can account for the observed increases in nighttime
ionization that correlate with geomagnetic activity.

083.126 Electron plasma resonances in the topside ionosphere.   J. R. McAfee.
The Earth: 1, (see 003.010), p. 129 - 175 (1978).
   The well-developed theory for the resonances at the
plasma and upper hybrid frequencies. presented in detail, may
point to explanations at the cyclotron harmonics and also at
the maximum frequencies of the Bernstein modes. The theory
not only provides a check on the dispersion relations for electrostatic waves, but raises the possibility of using the observations to measure electron temperatures. Some of the other
resonant-like behavior, observed more sporadically, can be explained in principle by considering possible nonlinear effects.
These include the low-frequency subsidiary resonances and
the diffuse resonances, as well as the resonance at twice the
upper hybrid frequency.

083.127 Polar-cap absorption — observations and theory.
G. C. Reid.
The Earth: 1, (see 003.010), p. 269 - 302 (1978).
   The term 'polar-cap absorption' is now generally employ-
ed to describe the entire range of ionospheric effects caused
by energetic solar particles.

083.128 Ionosphere and the interplanetary magnetic field.
N. F. Solonitsyna.
Ionos. i soln.-zemnye svyazi. Alma-Ata, Nauka KazSSR, 1978,
p. 57 - 66. In Russian. — Abstr. in Ref. zh., 62. Issled. kosm.
prostranstva, 5.62.372 (1978).

083.129 Estimation of energy release from precipitating
proton flux in equatorial ionosphere.
A. Z. Bochev, B. C. N. Rao, V. C. Jain.

C. R. Acad. Bulgare Sci., Vol. 30, 1415 - 1418 (1977).
Abstr. in Phys. Abstr., Vol. 81, Abstr. 32268 (1978).

083.130 High-frequency reflection and transmission coefficients for a semi-infinite plasma medium with
gradual boundary.   A. Hizal.
J. Phys. D, Vol. 11, 261 - 269 (1978). — Abstr. in Phys. Abstr.,
Vol. 81, Abstr. 32290 (1978).

083.131 Coupling between equatorial and auroral ionospheres
during polar substorms.   R. G. Rastogi.
Proc. Indian Acad. Sci., Sect. A, Vol. 86A, 409 - 416 (1977).
Abstr. in Phys. Abstr., Vol. 81, Abstr. 32291 (1978).

083.132 Features of electron density and temperature in the
500 - 3500 km region of the plasmasphere. I. Dawn
and dusk sectors.   G. Rjaaram, A. C. Das.
Proc. Indian Acad. Sci., Sect. A, Vol. 86A, 423 - 434 (1977).
Abstr. in Phys. Abstr., Vol. 81, Abstr. 32293 (1978).

083.133 Vertical lifting of ionization during geomagnetic
storms from satellite measurements of ion composition.   M. K. Goel, B. C. N. Rao, S. Chandra, E. J. Maier.
J. Geomagn. Geoelectr., Vol. 29, No. 3, p. 143 - 151 (1977).
Abstr. in Phys. Abstr., Vol. 81, Abstr. 36395 (1978).

083.134 Effect of solar X-ray and Lyman-α radiation on
ionospheric absorption.   D. N. Madhusudhana
Rao, K. V. V. Ramana.
Indian J. Radio Space Phys., Vol. 6, 32 - 34 (1977). — Abstr.
in Phys. Abstr., Vol. 81, Abstr. 40737 (1978).

083.135 VLF sudden phase anomalies during flares of
August 1972.   R. C. Dubey, S. K. Gupta, M. C.
Pande.
Indian J. Radio Space Phys., Vol. 6, 67 - 68 (1977). — Abstr.
in Phys. Abstr., Vol. 81, Abstr. 40739 (1978).

083.136 Stimulated plasma waves in the ionosphere.
R. F. Benson.
Radio Sci., Vol. 12, 861 - 878 (1977). — Abstr. in Phys. Abstr.,
Vol. 81, Abstr. 45128 (1978).

083.137 Experimental and theoretical first approach to f_h
plasma resonance from a relaxation sounding rocket
experiment.   B. Higel, H. de Feraudy.
Radio Sci., Vol. 12, 879 - 889 (1977). — Abstr. in Phys. Abstr.,
Vol. 81, Abstr. 45129 (1978).

083.138 HF excited instabilities in space plasmas.
H. C. Carlson, Jr., L. M. Duncan.
Radio Sci., Vol. 12, 1001 - 1013 (1977). — Abstr. in Phys.
Abstr., Vol. 81, Abstr. 45130 (1978).

083.139 Numerical modelling of ionospheric chemistry and
transport processes.   E. S. Oran, T. R. Young.
J. Phys. Chem., Vol. 81, 2463 - 2467 (1977). — Abstr. in
Phys. Abstr., Vol. 81, Abstr. 49202 (1978).

083.140 Dynamical model of the midlatitude ionosphere for
a height range from 100 to 1000 km.
A. A. Namgaladze, K. S. Latishev (*Latyshev*), Yu. N. Korenkov,
L. P. Zakharov.
Acta Geophys. Polonica, Vol. 25, No. 3, p. 173 - 182 (1977).
Abstr. in Phys. Abstr., Vol. 81, Abstr. 53768 (1978).

083.141 Electromagnetic wave propagation in the iono-
sphere: a general treatment.
C. L. Roy, B. Chakrabarty, B. Mishra, S. Sen Gupta.
Indian J. Phys. A, Vol. 51A, No. 4, p. 243 - 248 (1977).
Abstr. in Phys. Abstr., Vol. 81, Abstr. 53770 (1978).

Incoherent scatter radar observations.
See Abstr. 031.225.

A rocket borne experiment to measure plasma
densities in the D-region.   See Abstr. 032.564.

The U.K. multistatic incoherent scatter facility.
See Abstr. 033.004.

Experimental studies on beam-plasma interaction.
See Abstr. 062.067.

Use of fast ion beams in plasma research.
See Abstr. 062.068.

The effect of negative ions on collision-dominated
Thomson scattering.   See Abstr. 063.013.

First results of the STEREO-5 experiment: evidence
of ionospheric intensity scintillation of solar radio bursts at
decameter wavelengths?   See Abstr. 077.018.

On the relation of SPA (*sudden phase anomalies*)
measured at VLF to solar microwave burst energies.
See Abstr. 077.035.

North-south aligned equatorial airglow depletions.
See Abstr. 082.044.

Thermospheric structure and the F region.
See Abstr. 082.045.

Empirical model for general circulation of the
atmosphere at ionospheric levels above 100 km.

See Abstr. 082.065.

Frequency analysis of 4- to 6-keV electrons associated with an auroral arc.   See Abstr. 084.010.

IMFP effects on the equatorial geomagnetic field
and ionosphere – a review.   See Abstr. 084.232.

Damping of geomagnetic pulsations by the ionosphere.   See Abstr. 084.236.

Self-consistent particle and parallel electrostatic
field distributions in the magnetospheric-ionospheric auroral
region.   See Abstr. 084.243.

Current systems in the magnetosphere and ionosphere.   See Abstr. 084.249.

Controlled experiment in the ionosphere with a
rocket-borne plasma gun.   See Abstr. 084.257.

Field aligned distribution of plasma mantle and
ionospheric plasmas.   See Abstr. 084.277.

Dynamics and interconnection of structures of
magnetospheric convection and the polar ionosphere.
See Abstr. 084.301.

Mechanisms for driving Birkeland currents.
See Abstr. 084.322.

Influence of the interplanetary magnetic field on
the night ionosphere at mean latitudes.
See Abstr. 106.015.

084 Aurorae, Magnetic Field, Radiation Belts

Aurorae

084.001 **Solar flare effects on Earth's circumpolar currents.**
L. Křivský.
Bull. Astron. Inst. Czechoslovakia, Vol. 29, 27 - 29 (1978).
The effects of solar proton flares and enhanced solar
activity, indicated by intensive radio bursts in the decimetre
range, on the ionospheric circumpolar currents (Auroral
Electrojet Index) are investigated. A significant increase in the
level of the Auroral Electrojet was observed after the occurrence of these events.

084.002 **The altitude profile of the N_2^+ first negative rota-
tional temperature in an auroral arc.**
G. J. Romick, V. Degen, W. J. Stringer, K. Henriksen.
J. Geophys. Res., Vol. 83, 91 - 96 (1978).

084.003 **Electron acceleration in an array of auroral arcs.**
D. A. Bryant, D. S. Hall, D. R. Lepine.
Planet. Space Sci., Vol. 26, 81 - 92 (1978).
Energy spectra of electrons encountered on a rocket
flight across an array of auroral arcs are employed to test three
related models of electron acceleration. All three are based on
a potential difference existing between the source plasma in
the magnetosphere and the observation point in the ionosphere. One of the models provides a satisfactory fit to the
observed spectra. Two alternative mechanisms are suggested to

explain this model.

084.004 **Opportunities for analysis of ring current composition change through observation of the equatorial
aurora.**   B. A. Tinsley.
J. Atmos. Terr. Phys., Vol. 39, 1203 - 1205 (1977).

084.005 **Aurorae and closed magnetic field lines.**
D. D. Duthie, M. W. J. Scourfield.
J. Atmos. Terr. Phys., Vol. 39, 1429 - 1434 (1977).
The positions of auroral forms imaged by a TV system
are compared with the positions of field lines along which
whistlers propagate. Diffuse aurora occurs on closed field
lines and indirect evidence shows that this is also the case for
pulsating aurora.

084.006 **Observations of auroral westward traveling surges
and electron precipitations.**
C.-I. Meng, A. L. Snyder, Jr., H. W. Kroehl.
J. Geophys. Res., Vol. 83, 575 - 585 (1978).
Precipitations of low-energy electrons above the auroral
westward traveling surge were examined from a number of
surges observed by the DMSP 32 satellite in the fall of 1974.
From the limited surges observed, differences in precipitations were detected, associated with different auroral forms
in the vicinity of a surge.

084.007 **Auroral electron distribution function.**

R. L. Kaufmann, P. B. Dusenbery, B. J. Thomas, R. L. Arnoldy.
J. Geophys. Res., Vol. 83, 586 - 598 (1978).

The electron velocity distribution function is presented in the energy range 25 eV < E < 15 keV during a rocket flight over an active aurora. The electron measurements are compared with optical observations by all–sky cameras and a television system. The most striking feature of the velocity distribution function is a broad plateau produced by down-coming electrons.

084.008   On the polarization and origin of auroral kilometric radiation.   D. A. Gurnett, J. L. Green.
J. Geophys. Res., Vol. 83, 689 - 696 (1978).

The purpose of this paper is to present the initial results obtained from observations of auroral kilometric radiation at radial distances of about $2.0 R_E$ over the auroral zone. These measurements provide important new evidence on the mode of propagation and origin of the auroral kilometric radiation.

084.009   Global effect of auroral particle and Joule heating in the undisturbed thermosphere.   B. B. Hinton.
J. Geophys. Res., Vol. 83, 707 - 711 (1978).

From the compositional variations observed with the neutral atmosphere composition experiment on Ogo 6 and a simplified model of thermospheric dynamics, global average values of non–EUV heating are deduced.

084.010   Frequency analysis of 4- to 6-keV electrons associated with an auroral arc.   J. S. Murphree, H. R. Anderson.
J. Geophys. Res., Vol. 83, 730 - 734 (1978).

Theoretical considerations indicate that auroral electrons may be modulated by an electrostatic wave generated by instabilities in the ionosphere. A high time resolution experiment has been designed to detect such variations and covers the frequency ranges of 0–2.5 kHz and 230 kHz to 9.9 MHz. The energy range sampled is 4–6 keV, and the pitch angle coverage is designed to be from 0° to 180°. The results of the experiment set an upper limit of approximately 5% on the electron modulations in the appropriate frequency ranges and establish the necessity for high-intensity fluxes before this upper limit can be lowered.

084.011   Expected influence of auroral clutter on the EISCAT experiments.   A. Egeland.
Radar probing of the auroral plasma, (see 012.011), p. 161 - 185 (1977).

084.012   On the coordination of EISCAT measurements with ground based auroral observations.
G. Gustafsson.
Radar probing of the auroral plasma, (see 012.011), p. 409 - 417 (1977).

Optical measurements in combination with those of EISCAT are discussed. The optical aurora is useful in mapping the precipitation of energetic particles over a large area and can therefore be used to put the radar measurements in a larger frame of reference.

084.013   Geheimnisvolles Polarlicht.   W. Borst.
Phys. unserer Zeit, Vol. 9, 2 - 20 (1978).

084.014   Rocket measurements of oxygen and nitrogen emissions in the aurora.   A. J. Deans, G. G. Shepherd.
Planet. Space Sci., Vol. 26, 319 - 333 (1978).

084.015   O I (7774 Å) and O I (8446 Å) emissions in aurora.   A. B. Christensen, M. H. Rees, G. J. Romick, G. G. Sivjee.
J. Geophys. Res., Vol. 83, 1421 - 1425 (1978).

Simultaneous ground-based photometric measurements of auroral O I (6300 Å), O I (7774 Å), O I (8446 Å), and N_2^+ (4278 Å) emissions and model calculations of their intensities are presented.

084.016   Chatanika radar observations of the latitudinal distributions of auroral zone electric fields, conductivities, and currents.   J. L. Horwitz, J. R. Doupnik, P. M. Banks.
J. Geophys. Res., Vol. 83, 1463 - 1481 (1978).

084.017   Rapid scan Doppler velocity maps of the UHF diffuse radar aurora.   D. R. Moorcroft, R. T. Tsunoda.
J. Geophys. Res., Vol. 83, 1482 - 1492 (1978).

This paper presents the first results of rapid scan Doppler spectra of the diffuse radar aurora obtained with the Homer 398-MHz phased array radar. Each spectrum was obtained from three consecutive 1–ms radar pulses spanning an interval of 0.04 s. The spectral resolution obtainable with these short samples was improved by using a maximum entropy method of spectral analysis. Most of the results were characterized by relatively narrow spectra having Doppler velocities which were nearly constant over an extended azimuth sector (referred to as a Doppler velocity plateau). Typically, there is a positive plateau to the west and a negative plateau to the east. This situation exactly parallels the observations in the equatorial electrojet of 'Type 1' irregularities, and it is likely that these velocity plateaus arise from the same mechanism.

084.018   Comparison of incoherent scatter radar and photometric measurements of the energy distribution of auroral electrons.   R. R. Vondrak, R. D. Sears.
J. Geophys. Res., Vol. 83, 1665 - 1667 (1978).

Simultaneous radar and photometric measurements of the mean energy of precipitating electrons have been made at Chatanika, Alaska, during a variety of auroral conditions. The radar measurements of the altitude profile of electron density were used to obtain the differential energy spectrum of the precipitating flux. Photometric measurements of 4278- and 6300-Å emission intensities over the same region and time span were analyzed in terms of the mean energy parameter. The mean energy parameter values determined by the two techniques show good agreement, which tends to confirm quantitatively the theoretical predictions of Rees and Luckey (1974).

084.019   Index of aurorae activity.   Ya. I. Fel'dshtejn, G. V. Starkov, V. G. Vorob'ev.
Solar wind and the magnetosphere, (see 003.006), p. 113 - 118 (1976). In Russian. – Abstr. in Ref. zh., 62. Issled. kosm. prostranstva, 3.62.352 (1978).

084.020   A theory of quiet auroral arcs.   T. Sato.
J. Geophys. Res., Vol. 83, 1042 - 1048 (1978).

Being fed by the large-scale magnetospheric convection, the coupled ionosphere-magnetosphere system is subject to a feedback instability for a conjugate perturbation elongating in the east-west direction, having a width of several tens of kilometers at the ionospheric level, standing along the magnetic field line. The growth time is as short as several minutes even when no hot electrons are involved in the field-aligned current. Nonlinear development of the feedback instability is numerically investigated, taking hot electrons into account.

084.021   Geometrical considerations of 136 MHz amplitude scintillation in the auroral oval.
I. S. Mikkelsen, J. Aarons, E. Martin.
J. Atmos. Terr. Phys., Vol. 40, 479 - 483 (1978).

084.022   Electric field oscillations measured near an auroral arc.   P. M. Kintner, L. J. Cahill, Jr.
Planet. Space Sci., Vol. 26, 555 - 558 (1978).

A sounding rocket, launched into the expansive phase of an auroral substorm, measured bursts of electric field oscillations with a typical period of one second and a magnitude exceeding 20 mV/m. The oscillations appear to be due to an MHD wave propagating along the magnetic field. The bursts were observed as the sounding rocket passed over the southern border of an auroral arc. The southern border coincided with an increase in 1−5 keV electron flux and an increase in field-aligned current.

084.023 Precipitation of charged particles by a parallel electric field. A. D. Johnstone.
Planet. Space Sci., Vol. 26, 581 - 594 (1978).

A time-dependent model of the effect of a parallel electric field on particle precipitation from a closed field-line has been constructed and the results are presented.

084.024 On the intensification of λ 6300 Å [O I] radiation in rays of type A aurorae. L. V. Moiseeva.
Geomagn. Aehron., Vol. 18, 491 - 494 (1978). In Russian.

084.025 Major auroras in early May.
Sky Telesc., Vol. 56, 77, 79 - 82 (1978).

084.026 The latitudinal distributions of auroral zone electric fields and ground magnetic perturbations and their response to variations in the interplanetary magnetic field.
J. L. Horwitz, J. R. Doupnik, P. M. Banks, Y. Kamide, S.-I. Akasofu.
J. Geophys. Res., Vol. 83, 2071 - 2084 (1978).

Chatanika observations of latitudinal distributions of convection electric fields ($E_\perp$) are compared with isointensity ΔH contours in latitude and time from the Alaskan magnetometer chain and with the north-south component of the interplanetary magnetic field (IMF B_{zm}) from Imp-J. The most important findings are that (1) southward (northward) IMF B_{zm} transitions caused rapid equatorward (poleward) shifts of the electric field and ΔH patterns and (2) southward IMF B_z transitions, magnetospheric substorms, and local time transitions of the Harang discontinuity can all lead to northward-to-southward transitions of the electric field in the midnight sector.

084.027 Shear instability: auroral arc deformation and anomalous momentum transport.
A. Miura, T. Sato.
J. Geophys. Res., Vol. 83, 2109 - 2117 (1978).

Temporal evolution of a two-dimensional electrostatic shear instability is numerically studied with special attention to rotational deformation of auroral arcs and momentum transport across the magnetospheric boundary.

084.028 Rocket-borne measurements of particles and ion convection in dayside aurora.
P. W. Daly, B. A. Whalen.
J. Geophys. Res., Vol. 83, 2195 - 2200 (1978).

084.029 Pulsating aurora. A. D. Johnstone.
Nature, Vol. 274, 119 - 126 (1978).

084.030 Auroral particle precipitation-observations.
B. A. Whalen, I. B. McDiarmid.
The Earth: 1, (see 003.010), p. 177 - 268 (1978).

To set the framework for a discussion of auroral particle precipitation observations, a summary of the current understanding of the geomagnetic field topology is presented along with a brief description of the terms most commonly used in the presentation of these data. Auroral optical observations are then discussed and the gross characteristics of particle precipitation inferred from these measurements are outlined. The significant direct observations of auroral precipitation are summarized and the data are interpreted in terms of energiza-

tion and precipitation mechanisms and source regions.

084.031 Auroral electron beams near the magnetic equator. C. E. McIlwain.
Physics of the hot plasma in the magnetosphere, (see 012.030), p. 91 - 112 (1975).

Intense beams of electrons travelling parallel to the local magnetic field have been observed at a magnetic latitude of 11° and a radial distance of 6.6 R_e. Although the origin and destiny of these electrons is as yet unknown, considerations of the total energy and the number of particles transported guarantee that they must play a dominant role in many key magnetospheric processes.

084.032 A study of auroral displays photographed from the DMSP-2 and ISIS-2 satellites. S.-I. Akasofu.
Physics of the hot plasma in the magnetosphere, (see 012.030), p. 113 - 136 (1975).

Recent studies of auroral photographs taken from the DMSP and ISIS-2 satellites are briefly summarized. A particular emphasis is made in describing: (1) New large-scale morphological features, such as the distinction between discrete and diffuse auroras. (2) The roles of the north-south component of the interplanetary magnetic field on large-scale auroral dynamics. (3) The relationships between the auroral electrojet and large-scale auroral features.

084.033 Evidence for the low altitude acceleration of auroral particles. D. S. Evans.
Physics of the hot plasma in the magnetosphere, (see 012.030), p. 319 - 340 (1975).

There have been an ever increasing number of observations which strongly suggest that particle acceleration may occur at low altitude; less than 6000 km above the atmosphere along the high latitude magnetic line of force. This paper is intended to review these observational data and to briefly describe the nature of the low altitude acceleration mechanisms.

084.034 Peculiarities of the planetary distribution of pulsating aurorae. S. A. Chernous.
Geomagn. issled., 1977, No. 21, p. 16 - 24. In Russian.
Abstr. in Ref. zh., 62. Issled. kosm. prostranstva, 7.62.282 (1978).

Energetic particles in the auroral magnetosphere. See Abstr. 003.082.

Vibrational populations of the excited states of N_2 under auroral conditions. See Abstr. 022.051.

A method of obtaining the energy distribution of auroral electrons from incoherent scatter radar measurements. See Abstr. 031.226.

Evidence for two-dimensional inertial turbulence in a cosmic-scale low-β plasma. See Abstr. 062.016.

Double layers. See Abstr. 062.107.

Optical observation of enhanced nitric oxide abundance in the lower ionosphere following extended auroral activity. See Abstr. 082.035.

Rocket-borne sodium nightglow measurement during post-auroral conditions. See Abstr. 082.036.

Neutral winds in the upper atmosphere. See Abstr. 082.079.

On the nature of large auroral zone electric fields at 1-R_E altitude. See Abstr. 084.261.

Magnetic Field

084.201   ARAKS – controlled or puzzling experiment?
F. Cambou, J. Lavergnat, V. V. Migulin, A. I.
Morozov, B. E. Paton, R. Pellat, A. Kh. Pyatsi, H. Rème,
R. Z. Sagdeev, W. R. Sheldon, I. A. Zhulin.
Nature, Vol. 271, 723 - 726 (1978).
The injection of an electron beam into the magnetosphere
has been considered a straightforward technique for studying
the large-scale structure of the Earth's environment. The large
current released in the magnetosphere by the ARAKS experiment has produced many results which are not yet well understood.

084.202   Possible solar eclipse effect 23 October 1976.
G. L. M. Scheepers, with a reply by F. E. M. Lilley,
D. V. Woods.
Nature, Vol. 271, 91 - 92 (1978).

**084.203   Palaeomagnetic directions and pole positions – XV.
Pole numbers 15/1 to 15/232.**
M. W. McElhinny, J. A. Cowley.
Geophys. J. R. Astron. Soc., Vol. 52, 259 - 276 (1978).

084.204   Energetic oxygen ions in the magnetosphere.
D. T. Young.
Nature, Vol. 271, 303 (1978).

084.205   New views of the nature of terrestrial magnetism.
L. I. Miroshnichenko.
Priroda, 1978, No. 2, p. 140 - 141. In Russian.

**084.206   Characteristic peculiarities of the earth's magnetic
field.**   G. N. Petrova.
Zemlya i Vselennaya, 1978, No. 1, p. 44 - 47. In Russian.

084.207   L'espace interplanétaire.   G. Goy.
Conférences d'astronomie de l'Observatoire, (see
012.004), p. 53 - 72 (1977).
Quelques chiffres et comparaisons; La vie et l'espace
interplanétaire; Une première particularité: le champ magnétique terrestre; Les ceintures de van Allen; Les comètes; Les
météorites.

**084.208   Experimental evidence of the existence of an open
and closed model of the magnetosphere. II.**
Eh. M. Dubinin, I. M. Podgornyj, Yu. N. Potanin.
Kosm. Issled., Vol. 16, 87 - 95 (1978). In Russian.

**084.209   α-particle acceleration near the earth's magneto-
sphere.**   G. Ya. Kolesov, B. M. Kuzhevskij,
K. N. Sharvina.
Kosm. Issled., Vol. 16, 148 - 152 (1978). In Russian.

**084.210   On a possibility of investigating the plasma turbu-
lence in the magnetosphere.**   V. A. Gudkova.
Kosm. Issled., Vol. 16, 152 - 155 (1978). In Russian.

**084.211   On eigenvalues of a two-dimensional model of the
magnetosphere.**   Yu. A. Baurov, V. P. Shabanskij.
Geomagn. Aehron., Vol. 18, 101 - 105 (1978). In Russian.

**084.212   Isolated bursts of Pc1b-type geomagnetic pulsations
at high latitudes.**   Eh. T. Matveeva, V. A.
Troitskaya, F. Z. Fejgin.
Geomagn. Aehron., Vol. 18, 122 - 128 (1978). In Russian.

**084.213   Electromechanical effects of short-period variations
of the geomagnetic field.**
S. I. Braginskij, V. M. Fishman.

Geomagn. Aehron., Vol. 18, 135 - 143 (1978). In Russian.

**084.214   Diagnosis of plasma condensations from magnetic
field records at a geostationary satellite.**
N. A. Barkhatov, P. A. Bespalov.
Geomagn. Aehron., Vol. 18, 144 - 146 (1978). In Russian.

**084.215   On the origin of global fine structure formations
in the magnetosphere during substorms.**
M. I. Panasyuk, Eh. N. Sosnovets, L. V. Tverskaya,
O. V. Khorosheva.
Geomagn. Aehron., Vol. 18, 172 - 175 (1978). In Russian.

**084.216   On the dynamics of the Caspian centre of secular
variations of the earth's magnetic field.**
V. I. Pochtarev.
Geomagn. Aehron., Vol. 18, 183 - 185 (1978). In Russian.

**084.217   Secular variations of the magnetic field in San José
Las Lajas (Cuba).**
V. I. Afanas'eva, B. Enriques, J. Reis.
Geomagn. Aehron., Vol. 18, 185 - 187 (1978). In Russian.

**084.218   A study of two polar magnetic substorms with a
two-dimensional magnetometer array.**
J. R. Bannister, D. I. Gough.
Geophys. J. R. Astron. Soc., Vol. 53, 1 - 26 (1978).
The paper reports studies of the three-dimensional magnetospheric–ionospheric current systems which produced polar magnetic substorms on 1974 September 7 and September
18. The authors discuss these two substorms which involve
westward current in the ionosphere above the array, but which
cannot be fitted with uniform current density.

**084.219   Observational constraints on the generation process
of the Earth's magnetic field – correction.**
D. Gubbins.
Geophys. J. R. Astron. Soc., Vol. 53, 113 - 115 (1978). – See
18.084.220.

**084.220   Synthetic results obtained from modified Alldredge-
Hurwitz radial dipole models.**
K. M. Creer, T. E. Hogg.
Geophys. J. R. Astron. Soc., Vol. 53, (see 012.006), 150
(1978). – Abstract.

**084.221   Does the geomagnetic annual variation depend on
local time?**   S. R. C. Malin, A. M. Isikara.
Geophys. J. R. Astron. Soc., Vol. 53, (see 012.006), 150 - 151
(1978). – Abstract.

**084.222   Applications of linear inverse theory to geomagnetic
field analysis.**   D. Gubbins, K. A. Whaler.
Geophys. J. R. Astron. Soc., Vol. 53, (see 012.006), 151
(1978). – Abstract.

**084.223   Magnetosheath distortion of pitch angle distribu-
tions of solar protons.**   I.D. Palmer, P.R. Higbie.
J. Geophys. Res., Vol. 83, 30 - 38 (1978).
The propagation of energetic solar protons of $\lesssim 1$ MeV into
the magnetosheath is investigated through three-dimensional
pitch angle distributions measured on Vela satellites. Distortions are observed in the magnetosheath as compared with isotropic or unidirectional distributions normally expected in
interplanetary space. Two types of distortions are observed
which are characterized by breaks in the distributions at
$\mu_0 < 0$ or $\mu_0 > 0$, where μ is the cosine of the pitch angle. The
distributions in the magnetosheath are explained by a
Liouville transformation, if particle motion across the bow
shock and through the magnetosheath is assumed to be adiabatic. The results show that one must be cautious in inferring

the true interplanetary anisotropy from that measured in the magnetosheath.

084.224 Evidence of drift waves at the plasmapause.
P. M. Kintner, D. A. Gurnett.
J. Geophys. Res., Vol. 83, 39 - 44 (1978).

As the Hawkeye 1 spacecraft crosses the plasmapause at high altitudes ($R > 3 R_E$), a band of electric field noise is often detected in the frequency channels from 1.7 to 178 Hz. No corresponding magnetic field noise is detected, indicating that the noise is electrostatic (or at least quasi-electrostatic), and the electric field is polarized perpendicular to the plasma density gradient. The noise is only detected when the scale length of the plasmapause is $0.1 R_E$ or less, indicating that a large density gradient is required to produce the noise.

084.225 Correlated electric field and low-energy electron measurements in the low-altitude polar cusp.
P. M. Kintner, K. L. Ackerson, D. A. Gurnett, L. A. Frank.
J. Geophys. Res., Vol. 83, 163 - 168 (1978).

This paper presents simultaneous electric field and plasma measurements from the Hawkeye 1 satellite at low altitudes over the southern polar cap. The primary objective of this study is to further investigate the relationship of the plasma flow to the spatial structure of the polar cusp plasma distribution by using the improved spatial resolution of the Hawkeye 1 measurements compared to the earlier Injun 5 results.

084.226 Longitudinal extension of the substorm-associated long-period hydromagnetic waves.
C. S. Wang, T. Lee, J. S. Kim.
J. Geophys. Res., Vol. 83, 210 - 214 (1978).

The suggestion of a resonant geomagnetic cavity rather than resonances of a single field line was based on the longitudinal extension (~115°) of the north-south chain. In this paper the size of the resonant geomagnetic cavity will be discussed in terms of magnetograms recorded at stations in an east-west chain situated along a geomagnetic latitudinal circle of about 49° in the northern hemisphere.

084.227 Comment on 'Solar wind plasma injection at the dayside magnetospheric cusp' by P. H. Reiff, T. W. Hill, and J. L. Burch.   W. J. Heikkila, with a reply by P. H. Reiff, T. W. Hill, J. L. Burch.
J. Geophys. Res., Vol. 83, 227, 229 - 231 (1978).

084.228 Some features of Pc5 pulsations during a solar cycle.   D. R. K. Rao, J. C. Gupta.
Planet. Space Sci., Vol. 26, 1 - 20 (1978) = Div. Geomagn., Earth Phys. Branch, Dep. Energy, Mines, Resources, Ottawa, Canada, Contrib. No. 671.

084.229 Low energy electrons and protons in the magnetosphere.   M. B. Bavassano-Cattaneo, V. Formisano.
Planet. Space Sci., Vol. 26, 51 - 63 (1978).

Observations made by HEOS-2 of low energy electrons and protons in the high latitude magnetosphere are presented. Plasma in the magnetosphere is observed in the cusp (which extends down to low altitudes) and over large areas adjacent to the high latitude magnetopause both on the dayside and on the nightside (the entry layer and the plasma mantle respectively). A comparative study of the plasma properties in the various parts of the magnetosphere is performed.

084.230 The equatorial counter-electrojet − a review of its geomagnetic aspects.   P. N. Mayaud.
J. Atmos. Terr. Phys., Vol. 39, 1055 - 1070 (1977).

084.231 On the course of the geomagnetic daily variation in low latitudes.
C. A. Onwumechili, P. O. Ezema.
J. Atmos. Terr. Phys., Vol. 39, 1079 - 1086 (1977).

084.232 IMFP effects on the equatorial geomagnetic field and ionosphere − a review.   S. Matsushita.
J. Atmos. Terr. Phys., Vol. 39, 1207 - 1215 (1977).

It has recently been suggested that the interplanetary magnetic field polarity (IMFP) has a remarkable influence on geomagnetic and ionospheric phenomena over the equatorial region, in addition to the effect over the polar region. All recent reports with regard to those equatorial effects are critically examined, and a new result of geomagnetic field effects with their seasonal variations is presented. After a discussion of possible mechanisms of the effects, suggestions for future investigations are provided.

084.233 Spectral characteristics of geomagnetic field variations at low and equatorial latitudes.
W. H. Campbell.
J. Atmos. Terr. Phys., Vol. 39, 1217 - 1227 (1977).

084.234 Lunar geomagnetic tides and the ocean dynamo.
D. M. Schlapp.
J. Atmos. Terr. Phys., Vol. 39, 1453 - 1457 (1977).

A modified fixed solar hour method has been used to analyse data from geomagnetic observatories for which there exist long time series. The results show that the lunar tide at night is not due entirely to the oceanic dynamo but contains a component from causes above the earth's surface.

084.235 Criticism of reconnection models of the magnetosphere.   W. J. Heikkila.
Planet. Space Sci., Vol. 26, 121 - 129 (1978).

084.236 Damping of geomagnetic pulsations by the ionosphere.   R. S. Newton, D. J. Southwood, W. J. Hughes.
Planet. Space Sci., Vol. 26, 201 - 209 (1978).

084.237 Peculiarities of solar-wind flow around the magnetosphere and the magnetopause position.
T. V. Kuznetsova, M. I. Pudovkin.
Planet. Space Sci., Vol. 26, 229 - 236 (1978).

MHD problems of solar wind interaction with the Earth's magnetosphere on the day-side are investigated. It is shown that the observed regularities may be adequately explained within the bounds of MHD-flow theory which includes the stagnation line at the nose of the magnetosphere. The ratio k of the magnetic field pressure to the plasma pressure in the vicinity of the subsolar point of the magnetosphere, which determines the magnitude of the interplanetary magnetic field penetrating into the magnetosphere, was estimated.

084.238 The statistical magnetic signature of magnetospheric substorms.   M. N. Caan, R. L. McPherron, C. T. Russell.
Planet. Space Sci., Vol. 26, 269 - 279 (1978).

The characteristic magnetic signatures of magnetospheric substorms both on the ground and in space have been determined from the analysis of ~ 1800 substorm events. The timing and properties of these events were objectively determined according to explicit mathematical criteria by a computer pattern-recognition program. This program processed daily magnetograms from a mid-latitude network of geomagnetic observatories.

084.239 Long-period irregular magnetic pulsation, Pi3.   T. Saito.
Space Sci. Rev., Vol. 21, 427 - 467 (1978).

In the 1973 Scientific Assembly, the International Association of Geomagnetism and Aeronomy proposed with Resolution No. 11 to settle two new classes of magnetic pulsations; Pc6 having a sinusoidal waveform with periods longer than 600 s and Pi3 having an irregular waveform with periods longer than 150 s. The present paper reviews the studies on these pulsations putting a stress on Pi3.

084.240 Geos I: identification of natural magnetospheric emissions.   P. Christiansen, P. Gough, G. Martelli, J.-J. Bloch, N. Cornilleau, J. Etcheto, R. Gendrin, D. Jones, C. Béghin, P. Décréau.
Nature, Vol. 272, 682 - 686 (1978).
Natural noise bands, a persistent feature of the Earth's magnetosphere, have been identified for the first time, using novel techniques on the recently launched Geos I satellite. It is argued that these waves, in the electron cyclotron harmonic mode, are radiated incoherently by suprathermal electrons.

084.241 High-latitude, ground-based observations of ULF plasma waves in the magnetosphere.   J. C. Samson.
Geophys. J. R. Astron. Soc., Vol. 53, 319 - 334 (1978).

084.242 Large-scale characteristics of field-aligned currents associated with substorms.   T. Iijima, T. A. Potemra.
J. Geophys. Res., Vol. 83, 599 - 615 (1978).
The purpose of this paper is to present the statistical characteristics of large-scale field-aligned currents during various phases of substorms that were determined from the magnetometer data acquired with the Triad satellite during a large number of substorms. These data were obtained throughout almost all magnetic local time sectors, and representative examples of Triad data are presented.

084.243 Self-consistent particle and parallel electrostatic field distributions in the magnetospheric-ionospheric auroral region.   Y. T. Chiu, M. Schulz.
J. Geophys. Res., Vol. 83, 629 - 642 (1978).
The variation of the self-consistent electrostatic potential along the magnetic field is calculated by application of the principle of quasi-neutrality to the plasma components distributed along an auroral field line. The equilibrium plasma consists of hot anisotropic magnetospheric plasma, ionospheric plasma evaporated or extracted upward by the parallel electrostatic field, and backscattered electrons. It is shown that the above charged particle populations can support a potential difference of up to several kilovolts between the equator and the ionosphere along an auroral field line.

084.244 Transmission of Alfvén waves through the Earth's bow shock: theory and observation.   A. B. Hassam.
J. Geophys. Res., Vol. 83, 643 - 653 (1978).
The object of this paper is to investigate the transmission of Alfvén waves through the bow shock from both a theoretical and an observational standpoint. The paper is divided into two main parts: the first part extends the theory of Alfvén wave transmission through strong fast MHD shocks; the second part presents the results of a data analysis performed on Explorer 35 data taken in the vicinity of the bow shock.

084.245 Multiple—satellite studies of magnetospheric substorms: distinction between polar magnetic substorms and convection—driven negative bays.
T. Pytte, R. L. McPherron, E. W. Hones, Jr., H. I. West, Jr.
J. Geophys. Res., Vol. 83, 663 - 679 (1978).

084.246 Comparison of magnetic field perturbations at high latitudes with charged particle and IMF measurements.   I. B. McDiarmid, J. R. Burrows, M. D. Wilson.
J. Geophys. Res., Vol. 83, 681 - 688 (1978).
The magnetometer aligned along the spin axis of the Isis 2 satellite has been used to study magnetic field perturbations in the dawn-dusk sectors of the auroral oval and in the polar cap. The field perturbations have been compared with simultaneous particle measurements and with measurements of the interplanetary magnetic field.

084.247 Measurements of the Poynting vector of standing hydromagnetic waves at geosynchronous orbit.
W. D. Cummings, S. E. DeForest, R. L. McPherron.
J. Geophys. Res., Vol. 83, 697 - 706 (1978).

084.248 Stormtime Pc 5 magnetic pulsations observed at synchronous orbit and their correlation with the partial ring current.   J. N. Barfield, R. L. McPherron.
J. Geophys. Res., Vol. 83, 739 - 743 (1978).

084.249 Current systems in the magnetosphere and ionosphere.   R. Boström.
Radar probing of the auroral plasma, (see 012.011), p. 257 - 284 (1977).
A discussion is given of the currents that flow in the high latitude ionosphere and in the magnetosphere and problems that can be investigated using the EISCAT facility are identified. Topics discussed are the magnetic effects of the combined ionospheric-magnetospheric current systems, the relation between the currents and electric fields, and mechanisms for driving the large-scale and small-scale currents.

084.250 Satellite observations of polar, magnetotail lobe, and interplanetary electrons at low energies.
P. F. Mizera, J. F. Fennell.
Rev. Geophys. Space Phys., Vol. 16, 147 - 153 (1978).
Low-altitude satellite observations of the low-energy electron fluxes that populate the polar regions are summarized and classified into two groups: class 1, the very low intensity distributions, and class 2, the more intense, often structured distributions observed during magnetically disturbed conditions. High-altitude observations of electron fluxes, including the solar wind and the tail lobes, are presented to suggest that class 1 observations are the result of direct access of interplanetary electrons through the lobes into the polar regions. Class 2 observations may be due in part to magnetospheric processes.

084.251 Geomagnetic bays and *Pc*5 pulsation substorms at high latitudes.   J. C. Gupta.
J. Atmos. Terr. Phys., Vol. 40, 169 - 181 (1978).

084.252 Numerical modeling of the convection of magnetospheric plasma.
V. N. Senatorov, O. I. Shumilov, G. V. Popov.
Geomagn. Aehron., Vol. 18, 300 - 306 (1978).

084.253 Excitation of hydromagnetic waves of increasing frequency in the earth's magnetosphere.
A. V. Gul'el'mi, N. A. Zolotukhina.
Geomagn. Aehron., Vol. 18, 307 - 311 (1978).

084.254 Anomaly of the variable geomagnetic field in Central Asia.   A. A. Avagimov, O. N. Zhdanova.
Geomagn. Aehron., Vol. 18, 375 - 376 (1978).

084.255 A nearly axisymmetric model of an hydromagnetic dynamo of the earth.   S. I. Braginskij.
Geomagn. Aehron., Vol. 18, 340 - 351 (1978).

084.256 A one-dimensional gasdynamical model of magnetospheric convection.   M. V. Samokhin.
Moon, Vol. 17, 409 - 423 (1977).
The plasma flow in the equatorial plane of the magnetosphere is examined within the framework of a one-dimensional model in which all quantities are supposed to depend only on the distance along the Sun—Earth axis. The following models are considered: (1) the gasdynamical model in which the Ampère force is ignored, (2) the magnetohydrodynamical model in which the normal component of the Ampère force on the magnetopause is taken into account.

084.257 **Controlled experiment in the ionosphere with a rocket-borne plasma gun.** N. Kawashima, S. Sasaki, O. Kaneko, A. Yamori, R. Okamura, S. Okamura.
Planet. Space Sci., Vol. 26, 367 - 373 (1978).

084.258 **Small-scale transverse magnetic disturbances in the polar regions observed by Triad.**
N. A. Saflekos, T. A. Potemra, T. Iijima.
J. Geophys. Res., Vol. 26, 1493 - 1502 (1978).
Small-scale transverse magnetic field disturbances are often observed with the magnetometer experiment on board the Triad satellite at latitudes poleward of the large-scale field-aligned current regions associated with auroral phenomena (Iijima and Potemra, 1976). In contrast with the magnetic variations associated with the large-scale auroral field-aligned currents the polar cap magnetic variations are smaller in amplitude ($\sim 100 \, \gamma$) and show variations over smaller latitude ranges ($\sim 0.2°$). The small-scale variations are transverse to the main geomagnetic field and therefore explainable by field-aligned currents, but these currents are not necessarily sheets aligned in the geomagnetic east-west direction as the auroral field-aligned currents often are.

084.259 **Energization of charged particles to high energies by an induced substorm electric field within the magnetotail.** R. J. Pellinen, W. J. Heikkila.
J. Geophys. Res., Vol. 83, 1544 - 1550 (1978).

084.260 **Dynamics of magnetospheric disturbances on March 27, 1968.** A. D. Shevkin, Ya. I. Fel'dshtejn, L. Z. Sizova, N. F. Shevnina.
Solar wind and the magnetosphere, (see 003.006), p. 58 - 73 (1976). In Russian. − Abstr. in Ref. zh., 62. Issled. kosm. prostranstva, 3.62.357 (1978).

084.261 **On the nature of large auroral zone electric fields at 1-R_E altitude.**
S. D. Shawhan, C.-G. Fälthammar, L. P. Block.
J. Geophys. Res., Vol. 83, 1049 - 1054 (1978).
Mechanisms that may support magnetic-field-aligned electric fields in collisionless plasma are discussed in the light of recent magnetospheric observations, which for the first time allow a quantitative test of the theoretical models. Data from barium ion releases which indicate large field-aligned potential drops and direct electric field probe measurements at high altitude which reveal electric fields of several hundred millivolts per meter are discussed.

084.262 **Electric field topology near the dayside magnetopause.** W. J. Heikkila.
J. Geophys. Res., Vol. 83, 1071 - 1078 (1978).
A direct test for the occurrence of the reconnection electric field E_R tangential to the dayside magnetopause is possible by the use of in situ spacecraft observations of magnetic field and plasma parameters. One particular pass of Heos 2 on June 18, 1973, is discussed. The expected energization is about 1 keV/particle, but none is observed, nor is there any evidence of an enhanced outflow as predicted by reconnection theories. An alternate appraisal, from a global viewpoint, shows that the effect of a reconnection electric field should be one of major proportions and could not be missed. The apparent conclusion is that no reconnection electric field tangential to the dayside magnetopause exists; thus the magnetopause must be very nearly an equipotential surface.

084.263 **Energy flow and closure of current systems in the magnetosphere.** G. Atkinson.
J. Geophys. Res., Vol. 83, 1089 - 1103 (1978).
The result of an attempt to express the net energy input to the magnetosphere−ionosphere system in terms of upstream solar wind parameters is presented. Seven types of electric current system within the magnetosphere are defined and discussed, and it is found that only three of these systematically transfer energy into the magnetosphere from the solar wind. One of these is the magnetotail current system, and the other two form the field-aligned current observed at the northern edge of the auroral oval.

084.264 **Geomagnetic pulsations observed simultaneously on three geostationary satellites.**
W. J. Hughes, R. L. McPherron, J. N. Barfield.
J. Geophys. Res., Vol. 83, 1109 - 1116 (1978).

084.265 **Anomalous transport produced by kinetic Alfvén wave turbulence.** A. Hasegawa, K. Mima.
J. Geophys. Res., Vol. 83, 1117 - 1123 (1978).
Alfvén wave turbulence is shown to produce anomalous transport phenomena if the wavelength in the direction perpendicular to the magnetic field is comparable to the ion gyroradius. Applications to diffusion of radiation particles and ring current, heating of plasmapause electrons, and formation of the stable auroral red arcs as well as viscous interaction between the solar wind and the magnetosphere are discussed.

084.266 **Comment on 'On hot tenuous plasmas, fireballs, and boundary layers in the Earth's magnetotail' by L. A. Frank, K. L. Ackerson, and R. P. Lepping.**
E. W. Hones, Jr., with a reply by L. A. Frank, R. J. DeCoster, K. L. Ackerson.
J. Geophys. Res., Vol. 83, 1183 - 1190 (1978).

084.267 **Investigation of geomagnetic-field secular variation around the Sverdlovsk magnetic observatory.**
V. A. Shapiro, N. A. Ivanov, L. N. Ivanova.
Geophys. J. R. Astron. Soc., Vol. 53, 497 - 502 (1978).

084.268 **The history of the magnetopause regions.**
J. W. Dungey.
J. Atmos. Terr. Phys., Vol. 40, (see 012.021), 231 - 234 (1978).

084.269 **Recent techniques of observations and results from the magnetopause regions.** A. Bahnsen.
J. Atmos. Terr. Phys., Vol. 40, (see 012.021), 235 - 256 (1978).

084.270 **The entry layer.** G. Paschmann, G. Haerendel, N. Sckopke, H. Rosenbauer.
J. Atmos. Terr. Phys., Vol. 40, (see 012.021), 257 - 259 (1978).
This paper summarizes briefly the plasma observations in the magnetospheric entry layer and its environment made by the Max-Planck Institute (MPI) instrument aboard the Heos 2 satellite.

084.271 **The plasma mantle. A survey of magnetotail boundary layer observations.**
N. Sckopke, G. Paschmann.
J. Atmos. Terr. Phys., Vol. 40, (see 012.021), 261 - 278 (1978).

084.272 **Where do charged particles enter in and exit from the magnetosphere? Some HEOS-2 measurements.**
V. Domingo, D. E. Page, K.-P. Wenzel.
J. Atmos. Terr. Phys., Vol. 40, (see 012.021), 279 - 291 (1978).
The review concentrates on HEOS-2 spacecraft results in two main areas: (1) Solar protons penetrating the magnetosphere; and (2) the magnetopause electron layer.

084.273 **On the mechanism of plasma penetration and energetic electron escape across the dayside magnetopause.** V. Formisano, V. Domingo, K.-P. Wenzel.

J. Atmos. Terr. Phys., Vol. 40, (see 012.021), 293 - 300 (1978).

084.274 The magnetopause: microstructure and interaction with magnetospheric plasma.   D. M. Willis.
J. Atmos. Terr. Phys., Vol. 40, (see 012.021), 301 - 322 (1978).

084.275 Structure of tangential discontinuities at the magnetopause: the nose of the magnetopause.
M. Roth.
J. Atmos. Terr. Phys., Vol. 40, (see 012.021), 323 - 329 (1978).

084.276 Penetration of solar wind plasma elements into the magnetosphere.   J. Lemaire, M. Roth.
J. Atmos. Terr. Phys., Vol. 40, (see 012.021), 331 - 335 (1978).

084.277 Field aligned distribution of plasma mantle and ionospheric plasmas.
J. Lemaire, M. Scherer.
J. Atmos. Terr. Phys., Vol. 40, (see 012.021), 337 - 342 (1978).
The density and bulk velocity distributions of warm magnetosheath particles and cold ionospheric O^+ and H^+ ions are calculated along a polar cusp (cleft) magnetic field line.

084.278 Microscopic plasma processes related to reconnection.   G. Haerendel.
J. Atmos. Terr. Phys., Vol. 40, (see 012.021), 343 - 353 (1978).
Microscopic plasma processes that have been discussed in the literature in relation to reconnection are reviewed. Evidence for eddy convection in the polar cusp region and its possible consequences for mass transport into the magnetosphere and for merging are discussed.

084.279 A static-state field-line reconnection model for the Earth's magnetosphere.   G.-H. Voigt.
J. Atmos. Terr. Phys., Vol. 40, (see 012.021), 355 - 365 (1978).
A magnetostatic boundary value problem has been solved within the magnetospheric cavity and in the interplanetary space in order to incorporate the interplanetary magnetic field in a quantitative magnetospheric vacuum-B-field model. The model is valid for the field-line topology on the dayside including the polar cusp region.

084.280 High-energy electron drift echoes at the geostationary orbit.
G. Chanteur, R. Gendrin, S. Perraut.
J. Atmos. Terr. Phys., Vol. 40, (see 012.021), 367 - 371 (1978).
It is demonstrated that the origin of drift echoes is less likely to be an injection of high-energy particles, but more likely to be the consequence of a redistribution of particles among different L-shells after a sudden compression or expansion of the magnetosphere, as originally proposed by Brewer et al. (1969).

084.281 Convection and wave-particle interactions.
J. Solomon, R. Pellat.
J. Atmos. Terr. Phys., Vol. 40, (see 012.021), 373 - 378 (1978).
The authors study, by means of the adiabatic invariants and Liouville's theorem, the deformation of the distribution functions of substorm injected particles in the magnetosphere.

084.282 Movement and distribution of the energetic particles which generate Pc 1.   F. Glangeaud, A. Knob,
J. L. Lacoume, V. Troitskaya, S. Krilov, W. Gorbine.

J. Atmos. Terr. Phys., Vol. 40, (see 012.021), 379 - 387 (1978).

084.283 Evidence from Iceland on geomagnetic reversal during the Wisconsinan Ice Age.
J. W. Peirce, M. J. Clark.
Nature, Vol. 273, 456 - 458 (1978).
The authors report that a recent geomagnetic event is the only mechanism which can explain both the Icelandic and axial valley reversed outcrops.

084.284 An important statistical consideration, and the effect of the interplanetary magnetic field on the quiet time variation of the geomagnetic field.
C. L. Carter, K. D. Cole.
Planet. Space Sci., Vol. 26, 403 - 412 (1978).

084.285 The earth's palaeomagnetosphere as the third type of planetary magnetosphere.
T. Saito, T. Sakurai, K. Yumoto.
Planet. Space Sci., Vol. 26, 413 - 422 (1978).
From the viewpoint of dynamical topology, planetary magnetospheres are classified into three types. Type 1 is a planetary magnetosphere with $\vec{\Omega} \parallel \vec{M} \perp \vec{V}$ and includes the magnetospheres of Mercury, present Earth, Jupiter, and Uranus near the equinoctial points on its orbit. Type 2 is $\vec{\Omega} \parallel \vec{M} \parallel \vec{V}$ and includes the magnetosphere of Uranus near the solstitial points (Siscoe, 1971, 1975), while type 3 is $\vec{\Omega} \perp \vec{M}$ and $\vec{\Omega} \perp \vec{V}$. The purpose of this paper is to propose that the Earth's magnetosphere frequently changed to type 3 in the past, and to show various peculiar characteristics of the type 3 magnetosphere.

084.286 Geomagnetic Sq field at successive Universal Times.
A. Suzuki.
J. Atmos. Terr. Phys., Vol. 40, 449 - 463 (1978).
A method for analyzing the geomagnetic variation field at a fixed Universal Time (UT) is presented and applied to six quiet days in September 1964.

084.287 What happened to the high-latitude palaeomagnetic poles?   P. L. Lapointe, J. L. Roy, W. A. Morris.
Nature, Vol. 273, 655 - 657 (1978).

084.288 Energetic electrons of magnetospheric origin.
V. N. Lutsenko, N. S. Nikolaeva, N. F. Pisarenko, I. P. Shestopalov.
Problems of solar activity and space system Prognoz, (see 003.008), p. 231 - 242 (1977). In Russian. – Abstr. in Ref. zh., 62. Issled. kosm. prostranstva, 4.62.411 (1978).

084.289 Geomagnetic variations with periods from 13 to 30 years.   L. R. Alldredge.
J. Geomagn. Geoelectr., Vol. 29, No. 2, p. 123 - 135 (1977). Abstr. in Phys. Abstr., Vol. 81, Abstr. 15962 (1978).

084.290 Modulation of geomagnetic lunar variation by the sector polarity of IMF.   D. R. K. Rao, B. R. Arora
J. Geomagn. Geoelectr., Vol. 29, No. 2, p. 137 - 142 (1977). Abstr. in Phys. Abstr., Vol. 81, Abstr. 15963 (1978).

084.291 Pulsations of the earth's magnetic field before solar proton flares.
M. V. Bystrov, M. M. Kobrin, S. D. Snegirev.
Pis'ma Astron. Zh., Vol. 4, 143 - 144 (1978). In Russian.
An analysis of records of the earth's magnetic field is given for periods preceding solar proton flares early in August 1972 and early in July 1974. Quasi-periodic pulsations have been found with periods 20–200 min. These pulsations increase sharply on days preceding flares. Components of the fluctuation spectrum with periods 65, 90, 180 min correlate with similar pulsations in the solar radio emission.

084.292 Correction of pole positions for palaeomagnetic
         data from the Mesozoic of Australia.   R. Green.
Tectonophysics, Vol. 43, No. 3 - 4, p. T11 - T13 (1977).
Abstr. in Phys. Abstr., Vol. 81, Abstr. 20356 (1978).

084.293 On the upstream wave boundary outside the Earth's
         bow shock.   L. Diodato.
Ann. Geofis., Vol. 29, 121 - 131 (1976). In Italian. — Abstr.
in Phys. Abstr., Vol. 81, Abstr. 24409 (1978).

084.294 Research in magnetospheric and space plasma
         physics.   T. Obayashi, H. Oya, H. Takeuchi,
B. Makino, I. Kimura, A. Nishida, N. Kawashima.
Bull. Inst. Space Aeronaut. Sci., Univ. Tokyo, B, Vol. 12,
849 - 865 (1976). In Japanese. — From Phys. Abstr., Vol. 81,
Abstr. 24410 (1978).

084.295 Saturation of the substorm mode by magnetic
         trapping.   M. Sinha.
IEEE Trans. Plasma Sci., Vol. PS-5, 213 - 215 (1977). — Abstr.
in Phys. Abstr., Vol. 81, Abstr. 24412 (1978).

084.296 Transitional field configurations and geomagnetic
         reversal.   K. A. Hoffman, M. Fuller.
Nature, Vol. 273, 715 - 718 (1978).
   Records of various geomagnetic reversals which occurred
over the past 25 Myr suggest that transitional field configurations are largely controlled by low-order zonal components.
Given sufficient data further clarification is possible as
octupole and quadrupole dominated fields are palaeomag-
netically distinguishable. Moreover, because each geometry
may reflect a reversal process which starts in a particular
region of the core, configurational characteristics of the geomagnetic dynamo during polarity transitions are, in principle,
identifiable.

084.297 The driving force for magnetospheric convection.
         F. S. Johnson.
Rev. Geophys. Space Phys., Vol. 16, 161 - 167 (1978).
   The most popular concept for the driving force for magnetospheric convection has been that merging of the interplanetary magnetic field with the geomagnetic field at the
front of the magnetosphere establishes a good electrically
conducting path between the solar wind and the polar ionosphere; the flow of electrical energy then provides the driving
force. This model has had impressive success in predicting some
phenomena, but it also has some serious defects. The alternative view for the driving force has been that a viscous interaction exists between the solar wind and the outer layer of the
magnetosphere. A third possibility suggested by recent observations is that solar wind plasma enters the magnetosphere rather
readily around the neutral points, where the flow is turbulent
and the geomagnetic field is weak, and then spreads throughout a boundary layer covering virtually the entire inner surface
of the magnetopause. This plasma can expand and drive magnetospheric convection.

084.298 On the orientation of hydromagnetic waves in the
         magnetosphere.   A. Hasegawa, L. J. Lanzerotti.
Rev. Geophys. Space Phys., Vol. 16, 263 - 266 (1978).
   The authors review evidence gathered from theoretical
considerations and from measurements on the ground and in
space for a $90°$ rotation of the major axis of the polarization
ellipse of geomagnetic pulsation signals between the geomagnetic equator and the ground. The evidence at present supports
the existence of the rotation phenomenon.

084.299 Problems related to macroscopic electric fields in the
         magnetosphere.   C.-G. Fälthammar.
Astrophys. Space Sci., Vol. 55, 179 - 201 (1978).
   A particularly interesting feature of magnetospheric elec-
tric fields is the fact that they can have substantial components

along the geomagnetic field. Several physical mechanisms have
been identified by which such electric fields can be supported
even when collisions between particles are negligible. Comments are made on the magnetic-mirror effect, anomalous
resistivity, collisionless thermoelectric effect and electric
double layers, emphasizing key features and differences and
their significance in the light of recent observational data.

084.300 Effects of electric fields in periods of magneto-
         spheric substorms from data of Doppler sounding.
L. A. Lobachevskij, N. P. Sergeenko, O. S. Sergeenko,
L. A. Yudovich.
Geomagn. Aehron., Vol. 18, 427 - 431 (1978). In Russian.

084.301 Dynamics and interconnection of structures of
         magnetospheric convection and the polar iono-
sphere.   N. K. Osipov, A. M. Mozhaev.
Geomagn. Aehron., Vol. 18, 480 - 486 (1978). In Russian.

084.302 Influence of bounce effects on the generation of
         Pc 4- and 5-type geomagnetic pulsations.
A. M. Buloshnikov, E. A. Gerasimovich, M. B. Gokhberg,
Yu. G. Khabazin.
Geomagn. Aehron., Vol. 18, 507 - 510 (1978). In Russian.

084.303 Division of variations of the geomagnetic field into
         quiet and disturbed components by the method of
natural orthogonal components.   V. P. Golovkov, N. E.
Papitashvili, Yu. S. Tyupkin, E. P. Kharin.
Geomagn. Aehron., Vol. 18, 511 - 515 (1978).

084.304 On the possibility of artificial localization of a
         magnetospheric substorm.   I. A. Zhulin, V. M.
Mishin, E. V. Mishin, V. M. Chmyrev.
Geomagn. Aehron., Vol. 18, 551 - 553 (1978). In Russian.

084.305 Geophysical peculiarities of the substorm on De-
         cember 8, 1976.   N. A. Ivanov, A. S. Lakin,
B. A. Undzenkov, B. M. Chistoserdov, B. L. Shirman.
Geomagn. Aehron., Vol. 18, 554 - 557 (1978). In Russian.

084.306 A unified view of substorm sequences.
         Y. Kamide, S. Matsushita.
J. Geophys. Res., Vol. 83, 2103 - 2108 (1978).
   In an attempt to eliminate the 'growth phase' controversy
a unified phenomenological model of substorm sequences is
presented by taking recent findings into account. The following three fundamental requirements are important in synthesizing the vast amount of ground-based and satellite observations:
(1) substorms can occur regardless of the direction of the
IMF (interplanetary magnetic field) except during a large
prolonged northward IMF which yields a 'ground' state of the
magnetosphere, (2) substorm occurrence probability increases
with an increase of available energy in the magnetotail which
exceeds the ground state energy, and (3) substorms are, in
general, intense when the available energy is large.

084.307 Energization of polar cusp electrons at the noon
         meridian.   J. D. Craven, L. A. Frank.
J. Geophys. Res., Vol. 83, 2127 - 2132 (1978).
   Observations gained with an electrostatic analyzer on
board the low-altitude satellite Ariel 4 demonstrate that the
directional, differential spectra of polar cusp electron intensities are regulated by the sign of the elevation angle of the
interplanetary magnetic field.

084.308 Plasma flow pulsations in earth's magnetic tail.
         F. V. Coroniti, L. A. Frank, R. P. Lepping, F. L.
Scarf, K. L. Ackerson.
J. Geophys. Res., Vol. 83, 2162 - 2168 (1978).

084.309 Processes at the magnetotail boundary: comments

on 'On hot tenuous plasmas, fireballs, and boundary layers in the earth's magnetotail' by L. A. Frank, K. L. Ackerson, and R. P. Lepping.   E. W. Hones, Jr.
J. Geophys. Res., Vol. 83, 2216 - 2227 (1978).

084.310   Ergebnisse des Langmuir-Sonden-Experiments auf Interkosmos-10. (I).   A. Best, H.-D. Bettac, K. Flemming, H.-R. Lehmann, R. Treumann, C.-U. Wagner.
Gerlands Beitr. Geophys., Band 87, 161 - 176 (1978).

This paper describes the instruments and analysis of so far obtained Langmuir-probe measurements on board the Intercosmos-10 satellite. Electron density and temperature profiles including their peculiarities are presented.

084.311   Electric current structure of the magnetosphere.   H. Alfvén.
Physics of the hot plasma in the magnetosphere, (see 012. 030), p. 1 - 22 (1975).

084.312   Entry of solar wind plasma into the magnetosphere.   G. Haerendel, G. Paschmann.
Physics of the hot plasma in the magnetosphere, (see 012. 030), p. 23 - 43 (1975).

It is not the authors' intention to review data and theoretical ideas bearing on the general subject of entry of solar wind plasma into the magnetosphere. They want to restrict themselves to the entry through the polar cusp or cleft, on which recently available HEOS2 data contribute some new information.

084.313   Composition of the hot plasmas in the magnetosphere.
R. G. Johnson, R. D. Sharp, E. G. Shelley.
Physics of the hot plasma in the magnetosphere, (see 012. 030), p. 45 - 68 (1975).

The results now available on the He^+, He^{++}, and O^+ ions in the hot magnetospheric plasmas provide increasing evidence of the importance of mass and charge composition measurements for investigating the complex electrodynamic processes within and at the boundaries of the magnetosphere. The present measurements on these ions are still greatly limited in energy range, in mass range and resolution, and in detection sensitivities. The data have thus far been acquired only at low altitudes. Although still limited, the results from the energetic oxygen and helium measurements are beginning to provide sufficient definition of the characteristics of these particles that current theoretical models of the processes which act on them can be more realistically evaluated.

084.314   Magnetospheric plasma regions and boundaries.   W. J. Heikkila.
Physics of the hot plasma in the magnetosphere, (see 012. 030), p. 69 - 90 (1975).

Contents: Bow shock and magnetosheath. Magnetopause. Plasma sheet and its outer boundary. High latitude lobes. Lower latitude boundaries.

084.315   Recent observations relating to the dynamics and origin of the magnetotail plasma sheet.
E. W. Hones, Jr.
Physics of the hot plasma in the magnetosphere, (see 012. 030), p. 137 - 158 (1975).

This paper reviews the substorm behavior of the plasma sheet and the evidence that a substorm expansive phase onset signals the sudden formation of a magnetic neutral line across a near region of the magnetotail. Several types of observations are presented as evidence that solar wind plasma enters the plasma sheet via convection along the cleft, through the magnetotail surface layer, through the magnetotail lobes and finally earthward from a magnetotail neutral line. These observations were made, however, during periods of southward interplanetary magnetic field.

084.316   Hot plasma dynamics within geostationary altitudes.   D. J. Williams.
Physics of the hot plasma in the magnetosphere, (see 012. 030), p. 159 - 185 (1975).

The author presents observations and results pertaining to hot plasma dynamics as observed in the earth's magnetosphere. He begins with a discussion of protons during geomagnetic storms and presents briefly considerations of energy densities, hot-cold plasma interactions, stable and turbulent regions, the distribution function, charge exchange, SAR ARC generation, and adiabatic and non-adiabatic effects. He then presents the behavior of the energetic electron population during quiet and storm times along with a discussion of the mechanisms responsible for the quiet time and post injection recovery behavior.

084.317   Acceleration processes in the plasma sheet.   J. W. Dungey.
Physics of the hot plasma in the magnetosphere, (see 012. 030), p. 187 - 199 (1975).

The environment of the plasma sheet is described and the possible importance of noise and parallel electric fields near the outer boundary noted, but attention is concentrated on the neutral sheet. A significant feature in the plasma sheet is that the parallel pressure must exceed the perpendicular pressure.

084.318   VLF electrostatic waves in the magnetosphere.   M. Ashour-Abdalla, C. F. Kennel.
Physics of the hot plasma in the magnetosphere, (see 012. 030), p. 201 - 227 (1975).

The stability of electrostatic waves above the electron cyclotron frequency in an electron plasma consisting of a cold component and a hot component with a weak "loss-cone" perpendicular velocity distribution is investigated.

084.319   Plasma turbulence in the magnetosphere with special regard to plasma heating.   A. A. Galeev.
Physics of the hot plasma in the magnetosphere, (see 012. 030), p. 251 - 270 (1975).

The author restricts his attention to the problem of cold magnetospheric plasma heating. He considers: 1. Cold plasma heating due to linear and nonlinear absorption of electromagnetic waves, excited due to instabilities in the hot magnetospheric plasma. 2. Heating of the ionosphere by electron plasma waves, excited due to beam instability of the electrons precipitating into the ionosphere.

084.320   Characteristics of instabilities in the magnetosphere deduced from wave observations.   F. L. Scarf.
Physics of the hot plasma in the magnetosphere, (see 012. 030), p. 271 - 289 (1975).

This report contains a very general summary of the types of unstable plasma distributions encountered in the magnetosphere, along with an assessment of the present state of knowledge concerning instability characteristics based on wave observations.

084.321   Some experimentally determined characteristics of the turbulence in the magnetosphere.
B. Hultqvist.
Physics of the hot plasma in the magnetosphere, (see 012. 030), p. 291 - 318 (1975).

The magnetospheric turbulence is discussed on the basis of hot plasma measurements in and near the loss cone, mainly by means of the ESRO 1A and 1B satellites. The observational results are given.

084.322   Mechanisms for driving Birkeland currents.   R. Boström.
Physics of the hot plasma in the magnetosphere, (see 012. 030), p. 341 - 362 (1975).

The author has considered the basic mechanisms for generating Birkeland currents in the magnetosphere, that is mechanisms for redirecting transverse currents to form Birkeland currents. Using MHD equations he finds that both the plasma convection and pressure can account for currents of the observed magnitude. The author has also considered secondary Birkeland currents of ionospheric origin which are currents tending to discharge ionospheric polarization electric fields.

084.323 **A critical study of Gliddon's model of the protonosphere.**   B. Singh, V. B. Johri.
Acta Phys. Acad. Sci. Hungaricae, Vol. 42, 277 - 281 (1977). Abstr. in Phys. Abstr., Vol. 81, Abstr. 32294 (1978).

084.324 **Bounds on the fluid velocity and the magnetic field in the Earth's core imposed by hydromagnetic consideration of an $\alpha\omega$-dynamo.**   H. Watanabe.
J. Geomagn. Geoelectr., Vol. 29, 191 - 209 (1977). Abstr. in Phys. Abstr., Vol. 81, Abstr. 36101 (1978).

084.325 **Magnetospheric modulation of geomagnetic activity. I. Harmonic analysis of quasi-logarithmic indices Km, Kn, and Ks.**   D. Damaske.
Ann. Géophys., Vol. 33, 461 - 478 (1977). – Abstr. in Phys. Abstr., Vol. 81, Abstr. 40468 (1978).

084.326 **Analysis of a hundred year series of magnetic activity indices. IV. Various components of the annual wave at subauroral latitudes.**   P. N. Mayaud.
Ann. Géophys., Vol. 33, 479 - 501 (1977). In French. Abstr. in Phys. Abstr., Vol. 81, Abstr. 40469 (1978).

084.327 **On the annual line in geomagnetic spectrum.**   B. N. Bhargava, G. K. Rangarajan.
Ann. Géophys., Vol. 33, 513 - 517 (1977). – Abstr. in Phys. Abstr., Vol. 81, Abstr. 40470 (1978).

084.328 **Peculiarities of the MHD-flow by the magnetopause and generation of the electric field in the magnetosphere.**   M. I. Pudovkin, V. S. Semenov.
Ann. Géophys., Vol. 33, 423 - 427 (1977). – Abstr. in Phys. Abstr., Vol. 81, Abstr. 40760 (1978).

084.329 **On modelling the magnetic field and distribution of charged particles in the earth's magnetosphere.**
A. E. Antonova.
Mathematical models of the near space, 1977, (see 003.015), p. 141 - 156. In Russian. – Abstr. in Ref. zh., 62. Issled. kosm. prostranstva, 6.62.255 (1978).

084.330 **Origin of the electric field in the earth's magnetosphere.**   M. V. Samokhin.
Mathematical models of the near space, 1977, (see 003.015), p. 162 - 173. In Russian. – Abstr. in Ref. zh., 62. Issled. kosm. prostranstva, 6.62.264 (1978).

084.331 **Climatic change and geomagnetic field reversals: a statistical correlation.**   C. S. M. Doake.
Earth Planet. Sci. Lett., Vol. 38, 313 - 318 (1978).

084.332 **IGRF (*International Geomagnetic Reference Field*) comparisons.**   E. Dawson, L. R. Newitt.
Phys. Earth Planet. Interiors, Vol. 16, P1 - P6 (1978).

Solar flare activity and some indices of the interplanetary and geomagnetic fields, 1957 - 1973.
See Abstr. 003.104.

The upper atmosphere and magnetosphere.
See Abstr. 012.050.

On some problems of dynamics of protons trapped by the geomagntic field.   See Abstr. 022.062.

GEOS at the centre of a world-wide study of the Earth's magnetosphere.   See Abstr. 051.009.

A mechanism for current interruption in a collisionless plasma.   See Abstr. 062.044.

Experimental studies on beam-plasma interaction. See Abstr. 062.067.

Turbulent plasma phenomena in space and laboratory.   See Abstr. 062.069.

Double layers in an infinite plasma. See Abstr. 062.074.

An experiment on the interaction of a magnetized sphere with a supersonic plasma flow. See Abstr. 062.075.

The interaction between a magnetized plasma flow and a magnetized celestial body: a review of magnetospheric studies.   See Abstr. 062.082.

A fluid description of a hot collisionless plasma and limits to the double-adiabatic approximation associated with the presence of Birkeland currents.   See Abstr. 062.099.

Forecast of solar and geomagnetic activity for solar cycle No. 21.   See Abstr. 072.034.

Peculiarities of interaction of solar cosmic radiation with the earth's magnetosphere and the interplanetary space in the initial anisotropic phase of flares. See Abstr. 073.067.

A model of interaction of the solar wind with the magnetosphere. Statement of the problem. See Abstr. 074.010.

Stable solar wind and geomagnetic activity. See Abstr. 074.011.

Experimental results of an investigation of interaction between the solar wind and the earth's magnetosphere. See Abstr. 074.093.

Interaction between the solar wind stream and the geomagnetic field taking into account the interplanetary magnetic field.   See Abstr. 074.094.

Some results of investigation of interaction between the solar wind and the earth's magnetosphere. See Abstr. 074.096.

Properties of the Earth's deep interior revealed by the secular variation.   See Abstr. 081.031.

New results concerning the geomagnetic effect in the upper atmosphere.   See Abstr. 082.008.

Variations in atmospheric composition and density during a geomagnetic storm.   See Abstr. 082.040.

Comparison of drag and mass spectrometer measurements during small geomagnetic disturbances. See Abstr. 082.041.

Rocket altitude atmospheric X-rays from magnetospheric electrons at low and middle latitudes.

See Abstr. 082.068.

The equatorial electrojet.   See Abstr. 082.088.

Plasma fluxes between ionosphere and protono-sphere.   See Abstr. 083.001.

Post-geomagnetic storm protonospheric replenish-ment.   See Abstr. 083.002.

Pulsation spectra of low-frequency radiation in the outer ionosphere and short-period oscillations of the geo-magnetic field on the earth's surface.   See Abstr. 083.014.

Low-latitude ionospheric height changes associated with geomagnetic substorms.   See Abstr. 083.031.

Effects of a magnetic storm on the plasmaspheric electron content.   See Abstr. 083.037.

Effect of the day night ionospheric conductivity gradient on polar cap convective flow.   See Abstr. 083.057.

Field-aligned plasma flow and the Polar Wind: a review.   See Abstr. 083.060.

Ionospheric effects of magnetospheric substorms at night time.   See Abstr. 083.073.

Composition and variations of average-energy corpuscular streams.   See Abstr. 083.085.

Energetic characteristics of low-energy electron streams.   See Abstr. 083.086.

Resonance region of a PC5 micropulsation examined by a dual auroral radar system.   See Abstr. 083.094.

Modelling of electric fields of magnetospheric

origin in the ionosphere. II.   See Abstr. 083.112.

Ionospheric effects of magnetospheric substorms. See Abstr. 083.120.

Coupling between equatorial and auroral iono-spheres during polar substorms.   See Abstr. 083.131.

The latitudinal distributions of auroral zone electric fields and ground magnetic perturbations and their response to variations in the interplanetary magnetic field. See Abstr. 084.026.

Auroral electron beams near the magnetic equator. See Abstr. 084.031.

Dynamics of ring current protons during the storm on 25. 1. 1974.   See Abstr. 084.406.

On a comparison of anomalous magnetic fields of the moon and earth.   See Abstr. 094.137.

Meteor radar rates, geomagnetic activity and solar wind sector structure.   See Abstr. 104.027.

Interplanetary magnetic field and geomagnetic activity.   See Abstr. 106.045.

Variations of the Z-component of the interplanetary magnetic field stimulating an intensive main phase of a flare-type magnetic storm.   See Abstr. 106.056.

On the asymmetry effect of cosmic radiation in high-latitude zones of the earth's magnetosphere. See Abstr. 143.035.

Galactic cosmic radiation in the interior regions of the magnetosphere.   See Abstr. 143.053.

Radiation Belts

084.401   Stimulated excitation of Alfvén waves in the near-earth plasma by impulse radio radiation.
A. V. Gul'el'mi, B. V. Dovbnya, B. I. Klajn, V. A. Parkhomov.
Geomagn. Aehron., Vol. 18, 179 - 181 (1978). In Russian.

084.402   Energetic ionized helium in the quiet time radiation belts: theory and comparison with observation.
W. N. Spjeldvik, T. A. Fritz.
J. Geophys. Res., Vol. 83, 654 - 662 (1978).
   The nonadiabatic behavior of energetic α particles and singly ionized helium ions in the earth's inner magnetosphere is modeled for average quiet time conditions. The theoretical quiet time predictions are compared with direct observation of energetic helium ions in the lower MeV range obtained from the satellites ATS 6 and Explorer 45 at and below the geostat-ionary orbit, respectively. To the extent of this data it is found that the theory simulates the most important charac-teristics of the radiation belt helium ion population.

084.403   Investigation of electrons with energies of 0.3–3 MeV at the satellite Prognoz-4.   Yu. V. Mineev,
I. A. Savenko, V. G. Savel'ev, E. S. Spir'kova.
Geomagn. Aehron., Vol. 18, 203 - 206 (1978). In Russian.

084.404   The outer radiation belt during a strong magnetic storm.
S. N. Emel'yanenko, S. N. Kuznetsov, V. G. Stolpovskij.
Geomagn. Aehron., Vol. 18, 207 - 213 (1978). In Russian.

084.405   Life-time of electrons of the radiation belts of the earth and VLF-radiation.
A. V. Zakharov, S. N. Kuznetsov.
Geomagn. Aehron., Vol. 18, 352 - 354 (1978).

084.406   Dynamics of ring current protons during the storm on 25. 1. 1974.
A. S. Kovtyukh, M. I. Panasyuk, Eh. N. Sosnovets.
Kosm. Issled., Vol. 16, 226 - 237 (1978). In Russian.

084.407   Theory for charge states of energetic oxygen ions in the Earth's radiation belts.   W. N. Spjeldvik,
T. A. Fritz.
J. Geophys. Res., Vol. 83, 1583 - 1594 (1978).
   Fluxes of geomagnetically trapped energetic oxygen ions have been studied in detail. Ion distributions in radial loca-tions below the geostationary orbit, energy spectra between 1 keV and 100 MeV, and the distribution over charge states have been computed for equatorially mirroring ions. Both ionospheric and solar wind oxygen ion sources have been considered, and it is found that the charge state distributions

in the interior of the radiation belts are largely independent of the charge state characteristics of the sources.

084.408 Relative abundance and spectrum of α-particles in the inner radiation belt of the earth from measurements aboard AES Prognoz 5.   V. N. Lutsenko, N. S. Nikolaeva.
Kosm. Issled., Vol. 16, 459 - 462 (1978). In Russian.

084.409 Rapid populations of electron intensity in the capture region.   S. N. Emel'yanenko, S. N. Kuznetsov, G. B. Lopatina, V. G. Stolpovskij.
Geomagn. Aehron., Vol. 18, 520 - 522 (1978). In Russian.

084.410 The physical mechanisms of the inner Van Allen belt.   M. Walt, T. A. Farley.
The Earth: 1, (see 003.010), p. 303 - 412 (1978).

This paper is neither a review nor a presentation of new research results. It rests in large measure on the published work of others, often rewritten here with the benefit of hindsight to simplify the presentation, to emphasize what is currently thought to be important, and to standardize notation and terminology. The brief historical reviews are limited to one specific aspect of inner zone research — the attempt to describe quantitatively the effect of various mechanisms on the distribution function of energetic electrons and protons in the inner radiation zone.

085 Solar-Terrestrial Relations

085.001 Analysis of a possible Sun—weather correlation.   E. J. Gerety, R. H. Olson, W. O. Roberts.
Nature, Vol. 272, 231 - 232 (1978).

085.002 On the transmission of energy of the solar corpuscular stream into the lower layers of the earth's atmosphere by infrared radiation.   B. A. Mirtov.
Geomagn. Aehron., Vol. 18, 19 - 23 (1978). In Russian.

085.003 Large-amplitude standing planetary waves induced in the troposphere by the sun.   J. W. King, A. J. Slater, A. D. Stevens, P. A. Smith, D. M. Willis.
J. Atmos. Terr. Phys., Vol. 39, 1357 - 1367 (1977).

Investigations have been made of the global morphologies of two different sun-weather relationships: the effect of the 27.5-day solar rotations and the effect of the 11-year sunspot cycles on tropospheric pressure. In both cases solar-induced perturbations occur in the form of large-amplitude standing planetary waves.

085.004 Influence of the solar cycle on the abundance of carbon dioxide in the earth's atmosphere.
B. W. Rust.
Bull. American Astron. Soc., Vol. 9, 565 (1978). − Abstract.

085.005 Effect of solar activity on neutral temperatures in the 50—90 km altitude region.
S. Ramakrishna, R. Seshamani.
Space Research, Vol. XVII, (see 012.010), 135 - 139 (1977).

085.006 Twilight measurement of scattered Lyman-α intensities at the equator.
B. H. Subbaraya, P. N. Pareek, S. Prakash, V. Kumar.
Space Research, Vol. XVII, (see 012.010), 515 - 518 (1977).

085.007 Initiation of non-tropical thunderstorms by solar activity.   J. R. Herman, R. A. Goldberg.
J. Atmos. Terr. Phys., Vol. 40, 121 - 134 (1978).

085.008 On the geographic conditionality of solar-atmospherical relations.
N. B. Mulyukova, Eh. R. Mustel', V. E. Chertoprud.
Astron. Zh. Akad. Nauk SSSR, Vol. 55, 437 - 440 (1978). In Russian. English translation in Soviet Astron., Vol. 22, No. 2.

085.009 Solar modulation of atmospheric electrification and possible implications for the Sun—weather relationship.   R. Markson.
Nature, Vol. 273, 103 - 109 (1978).

Atmospheric electrical processes may help explain statistical findings indicating that variable solar activity, such as solar flares or the Earth's position in the extended solar magnetic field, affects the weather. An atmospheric electrical mechanism bypasses difficulties associated with solar heating mechanisms. Understanding Sun—weather relationships offers a basis for long range weather forecasting.

085.010 Dynamics of the upper atmosphere's layers, influence of solar activity and anthropogenic factors on the climate.   L. R. Rakipova.
Sovrem. fundament. i prikl. issled. Glav. geofiz. obs. Leningrad, Gidrometeoizdat, 1977, p. 53 - 57. In Russian. − Abstr. in Ref. zh., 51. Astron., 3.51.509 (1978).

085.011 Comparison of solar wind parameters and terrestrial geomagnetic variations with recordings of the geostationary ATS-1 satellite in periods of 20 - 300 seconds.
N. A. Barkhanov, A. E. Levitin.
Solar wind and the magnetosphere, (see 003.006), p. 83 - 92 (1976). In Russian. − Abstr. in Ref. zh., 62. Issled. kosm. prostranstva, 3.62.356 (1978).

085.012 Time extrapolation of solar and geomagnetic indices.   E. K. Mol'kentin.
Tr. Glav. geofiz. obs., 1977, vyp. (No.) 386, p. 96 - 100. In Russian. − Abstr. in Ref. zh., 51. Astron., 4.51.448 (1978).

085.013 Observation of short-periodic pulsations of electron fluxes and gamma-quanta in the upper layers of the atmosphere and their relations to periodic oscillations of the solar diameter.   A. M. Gal'per, V. M. Grachev, V. V. Dmitrienko, V. G. Kirillov-Ugryumov, A. V. Kurochkin, V. I. Luchkov, S. E. Ulin, Eh. M. Shermanzon, Yu. T. Yurkin.
Izv. AN SSSR. Ser. fiz., Vol. 41, 1765 - 1771 (1977). In Russian. − Abstr. in Ref. zh., 51. Astron., 4.51.456 (1978).

085.014 Dependence of the position of the front of a circumterrestrial shock wave and of the magnetopause on the parameters of the solar wind and the plasma structure of the magnetopause from data of traps of charged particles aboard the Prognoz and Prognoz 2 stations.
V. V. Bezrukikh, T. K. Breus, M. I. Verigin, P. A. Majsuradze, A. P. Remizov, Eh. K. Solomatina.
Problems of solar activity and space system Prognoz, (see 003.008), p. 242 - 255 (1977). In Russian. − Abstr. in Ref. zh., 62. Issled. kosm. prostranstva, 4.62.416 (1978).

085.015 **Atmospheric water vapour of extraterrestrial origin: a discussion of its possible role in Sun—weather relationships.**   D. M. Willis.
J. Atmos. Terr. Phys., Vol. 40, 513 - 528 (1978).

This paper discusses in detail the speculative suggestion that the influx of extraterrestrial hydrogen to the Earth's atmosphere constitutes a significant variable source of atmospheric water vapour (sometimes called 'solar rain' or 'extraterrestrial rain') which can be invoked to explain Sun—weather relationships.

085.016 **Solar plages and the vorticity of the earth's atmosphere.**   R. H. Olson, W. O. Roberts, H. D. Prince, E. R. Hedeman.
Nature, Vol. 274, 140 - 142 (1978).

Three superimposed epoch analyses of the vorticity area index at 500 mbar were carried out. The results lead the authors to conclude that the location on the solar disk of very active plages plays an important part in determining their meteorological influence.

085.017 **Investigations of the diurnal variation of inclinations of the earth's surface.**   I. A. Urasina.
Izv. Astron. Ehngel'gardt. Obs., Kazan', No. 41 - 42, p. 189 - 193 (1976). In Russian.

085.018 **Influence of the sun on coherent quantum systems.**   O. B. Khavroshkin, V. V. Tsyplakov.
Astron. Tsirk., No. 968, p. 3 - 4 (1977). In Russian.

085.019 **Effect of solar flares on lower tropospheric temperature and pressure.**   S. Jeevananda Reddy, K. Rama Rao.
Indian J. Radio Space Phys., Vol. 6, No. 1, p. 44 - 50 (1977). Abstr. in Phys. Abstr., Vol. 81, Abstr. 40615 (1978).

085.020 **The shaky machine (*solar variations and the Earth's climate*).**   T. E. Bell.
Astronomy., Vol. 6, No. 2, p. 6 - 17 (1978). − Abstr. in Phys. Abstr., Vol. 81, Abstr. 45307 (1978).

085.021 **Seasonal change of S_q-variations from data of the global network of stations in the year of maximum solar activity.**   V. N. Pogrebnoj, V. V. Kazakov, B. T. Zhumabaev.
Ionos. i soln.-zemnye svyazi, Alma-Ata, Nauka, KazSSR, 1978, p. 103 - 108. In Russian. − Abstr. in Ref. zh., 51. Astron., 6.51.577 (1978).

085.022 **Influence of particle streams accelerated in solar flares on the radiation situation in the stratosphere.**
O. A. Barsukov, O. A. Bogdanova, E. V. Kolomeets, T. I. Sysoeva, V. L. Shmonin.
Prikl. i teor. fiz., 1977, No. 9, p. 125 - 129. In Russian. Abstr. in Ref. zh., 51. Astron., 6.51.580 (1978).

085.023 **Effetti dei brillamenti sull'atmosfera terrestre.**   G. Tagliaferri.
Mem. Soc. Astron. Italiana, Vol. 48, (see 012.037), 253 - 255 (1977).

085.024 **Response of the atmospheric circulation to conditions in space.**   V. F. Loginov.
Tr. VNII gidrometeorol. inf. − Mirov. tsentr dannykh, 1977, No. 37, p. 117 - 130. In Russian. − Abstr. in Ref. zh., 62. Issled. kosm. prostranstva, 7.62.279 (1978).

085.025 **Response of atmospheric circulation to proton flares on the sun.**   L. T. Trofimenko.
Tr. VNII gidrometeorol. inf. − Mirov. tsentr dannykh, 1977, No. 37, p. 131 - 134. In Russian. − Abstr. in Ref. zh., 62. Issled. kosm. prostranstva, 7.62.280 (1978).

New international journal "Physica Solariterrestris".   See Abstr. 002.068.

Activité solaire.   See Abstr. 072.001.

On the statistical description of solar and solar-geophysical phenomena.   See Abstr. 080.066.

Velocity of ionospheric plasma heating by solar radiation. I. Heating by photoelectrons.   See Abstr. 083.008.

The early morning development and the evening decay of electron content—latitude profiles at low latitudes and their dependence upon solar declination.   See Abstr. 083.028.

Equatorial electron densities—seasonal and solar cycle changes.   See Abstr. 083.032.

Calculation of the parameters of the mid-latitude upper ionosphere for high solar activity conditions.   See Abstr. 083.067.

Heating velocities of ionospheric plasma by solar radiation. II. Heating by radiation in the Schumann-Runge continuum and on account of chemical reactions.   See Abstr. 083.070.

Planetary System

091 Physics of the Planetary System (Planetary Atmospheres, Figure, Interiors, Magnetic Fields, Rotation, etc.)

091.001 Ice clathrate as a possible source of the atmospheres of the terrestrial planets.
G. T. Sill, L. L. Wilkening.
Icarus, Vol. 33, 13 - 22 (1978).

The presence and compositions of atmospheres on the terrestrial planets do not follow directly from condensation models which would have Earth accreting near 500° K. No single mechanism yet proposed adequately accounts for the abundances of noble gases and carbon and nitrogen in the atmospheres. The authors show that the composition of clathrates forming at low temperatures in cold regions of the nebula can be predicted. Addition of about 1 ppm clathrate material to the Earth can explain observed abundances of Ar, Kr, and Xe. Condensation and adsorption processes occurring at 400 - 500° K are necessary to explain the observed abundances of Ne, H_2O, C, and N. Possible sources of clathrates could be cometary bodies formed in the outer solar system.

091.002 On the ortho-para equilibrium of H_2 in the atmospheres of the Jovian planets.
W. H. Smith.
Icarus, Vol. 33, 210 -216 (1978).

The ratio for the equivalent widths for the unsaturated H_2 quadrupole transitions observed in the Jovian planets is calculated and compared with a large number of observations. The comparison indicates that equilibrium hydrogen may be present in Jupiter and Saturn, while Uranus and Neptune exhibit ratios not in accord with equilibrium hydrogen. Observations which can differentiate among the possible states of H_2 are proposed.

091.003 Circular polarization of sunlight reflected by planetary atmospheres. Y. Kawata.
Icarus, Vol. 33, 217 - 232 (1978).

Multiple scattering calculations are performed in order to investigate the nature of the circular polarization of sunlight reflected by planetary atmospheres. Contour diagrams as a function of size parameter and phase angle are made for the integrated light from a spherical but locally plane-parallel atmosphere of spherical particles. To investigate the origin of the circular polarization, results are also computed for second-order scattering and for a simpler semiquantitative model of scattering by two particles. Observations of the circular polarization of the planets are presently too meager for accurate deduction of cloud particle properties.

091.004 Millimetre observations of planets, galactic and extra-galactic sources. M. Rowan-Robinson, P. A. R. Ade, E. I. Robson, P. E. Clegg.
Astron. Astrophys., Vol. 62, 249 - 254 (1978).

The authors have improved measurements of millimetre temperatures for Mercury, Venus, Mars, Saturn and Uranus,

relative to Jupiter. They have also detected millimetre emission from the galactic dust cloud M 17 S and from two CO positions in W43. Improved fluxes are given for positions centred on Sag B2 and W 51 A, and upper limits for 15 other sources. The authors have detected the quasar 3C 345 and possibly the QSO BL Lac and the radio galaxy 3C 274 (Vir A). 3C 120 appears to be variable over a time-scale of years.

091.005 Raman scattering in the atmospheres of the major planets. W. D. Cochran, L. M. Trafton.
Astrophys. J., Vol. 219, 756 - 762 (1978).

A method is developed for calculating the rate at which photons are Raman scattered as a function of frequency and depth in an inhomogeneous anisotropically scattering atmosphere. This method is used to determine the effects of Raman scattering by H_2 in the atmospheres of the major planets. Raman scattering causes an insufficient decrease in the blue and ultraviolet to explain the albedos of all of the planets; an additional source of extinction is necessary in this spectral region.

091.006 Planetary occultations. D. W. Dunham.
Occultation Newsl., Vol. 1, 134 - 135 (1978).

091.007 Planetary occultation update. D. W. Dunham.
Occultation Newsl., Vol. 1, 136 - 137 (1978).

091.008 Near-opposition limb darkening of solids of planetary interest.
J. Veverka, J. Goguen, S. Yang, J. L. Elliot.
Icarus, Vol. 33, 368 - 379 (1978).

This paper presents a laboratory study of the limb darkening near opposition, of particulate materials of planetary interest and concentrates on the wavelength dependence of this limb darkening. The authors find that near zero phase the scattering properties of most particulate materials can be described adequately by Minnaert's law. However, there are materials for which such a representation is totally inadequate. Examples are bronzite and graphite, materials that tend to fracture into flakes having mirrorlike surfaces. In addition, there are materials, such as olivine, whose scattering properties within deep absorption bands show definite departures from Minnaert's law at large angles of incidence or emission.

091.009 The chemical composition of the cores of the terrestrial planets and the Moon.
O. L. Kuskov, N. I. Khitarov.
Cosmochemistry of the moon and planets, (see 012.002), p. 231 - 242 (1977).

Using models of the quasi-chemical theory of solutions, the activity coefficients of silicon are calculated in the melts Fe-Si, Ni-Si, and Fe-Ni-Si. The oxidation-reduction conditions are studied, and the fugacity of oxygen in the mantles of the planets and at the core-mantle boundary are calculated. It is concluded that silicon can enter into the composition of the outer core of the Earth and Venus, but probably does not enter into the composition of the cores of Mercury, Mars, and the Moon, if in fact the latter possesses one.

091.010 On the mechanism of the magnetic dynamo of the planets.   Sh. Sh. Dolginov.
Cosmochemistry of the moon and planets, (see 012.002), p. 493 - 498 (1977).
    Results of testing the effectiveness of the theory of precessional dynamos in the generation of the magnetic fields of the planets are presented. A formula is given with which the magnetic state of Earth and of the planets Mars, Jupiter, and Venus can be satisfactorily described.

091.011 Studies of chemical abundances in the outer solar system.   T. Owen.
Cosmochemistry of the moon and planets, (see 012.002), p. 887 - 892 (1977).

091.012 Use of ground-based telescopes in determining the composition of the surfaces of solar system objects.
T. B. McCord, J. B. Adams.
Cosmochemistry of the moon and planets, (see 012.002), p. 893 - 922 (1977) = Planet. Astron. Lab., Dep. Earth Planet. Sci., Mass. Inst. Technol., Cambridge, Mass., Publ. No. 102.
    This article reviews efforts to use ground-based optical telescope measurements to determine the composition of the surfaces of the solar system objects. Spectral reflectance is measured at the telescope using several instruments: mainly a dual-beam 24-filter photoelectric photometer, a two-dimensional integrating television-like imaging system, and a Michaelson interferometer spectrometer.

091.013 Re-evaluating Bode's law of planetary magnetism.
        C. T. Russell.
Nature, Vol. 272, 147 - 148 (1978).
    The author brings up to date the magnetic Bode's law based on recent analysis of the moments of the terrestrial planets.

091.014 Tectonic asymmetry of the earth and other planets.
        Yu. M. Pushcharovskij, V. V. Kozlov, E. D. Sulidi-Kondrat'ev.
Priroda, 1978, No. 3, p. 32 - 41. In Russian.

091.015 Use of equivalency in estimating the historical size distribution of impact craters.   W. W. Mullins.
Icarus, Vol. 33, 624 - 629 (1978).
    A previous analysis of a stochastic model of lunar-type impact cratering is extended to utilize geological age data by defining a more general statistic $\Omega_i (t)$ to be the number of equivalent whole craters of original diameter d_i and age $\leq t$ in an observational area A; each crater is taken to be equivalent to the fraction of its rim (or area) that is in A and not occluded by later craters. By integration of a new gain—loss differential equation, a generalization of the previous basic equation is obtained. It is shown that use of the Ω_i permits, in principle, a reconstruction of the historical values of the process functions and correctly compensates for the effect of overlap by removing the false bias favoring large craters that results from the usual method of crater counting. Possible generalizations of the gain—loss equation are indicated.

091.016 Planetary climatic effects of long-term solar variability.   J. A. Eddy.
Bull. American Astron. Soc., Vol. 9, 497 (1977). — Abstract.

091.017 An atmospheric instability mechanism on a non-rotating planet and its application to Venus.
L. Elson.

Bull. American Astron. Soc., Vol. 9, 508 - 509 (1977). Abstract.

091.018 Tides in the atmospheres of the giant planets.
        H. Houben, P. Gierasch.
Bull. American Astron. Soc., Vol. 9, 510 - 511 (1977). Abstract.

091.019 Chemistry of the outer planets.   J. S. Lewis.
        Bull. American Astron. Soc., Vol. 9, 514 - 515 (1977). — Abstract.

091.020 Solar wind sputtering: an ineffective weathering process on airless bodies.
R. E. Scott, R. L. Huguenin.
Bull. American Astron. Soc., Vol. 9, 532 - 533 (1977). Abstract.

091.021 Atmospheric composition via Raman scattering.
        L. Trafton, J. H. Woodman.
Bull. American Astron. Soc., Vol. 9, 533 (1977). — Abstract.

091.022 Raman scattering in the atmospheres of the major planets.   W. D. Cochran, L. M. Trafton.
Bull. American Astron. Soc., Vol. 9, 533 (1977). — Abstract.

091.023 Radiative transfer in spherical shell atmospheres.
        C. N. Adams, G. W. Kattawar.
Bull. American Astron. Soc., Vol. 9, 534 (1977). — Abstract.

091.024 Tectonic patterns: implications for planetary history.   H. J. Melosh.
Bull. American Astron. Soc., Vol. 9, 541 (1977). — Abstract.

091.025 Interactions of volcanism and ice — earth, Mars and icy satellites.   C. C. Allen.
Bull. American Astron. Soc., Vol. 9, 541 (1977). — Abstract.

091.026 Volcanism on the inner planets.   D. Walker.
        Bull. American Astron. Soc., Vol. 9, 542 (1977). Abstract.

091.027 Put up or shut up: What do we really know about terrestrial planetary interiors?   S. C. Solomon.
Bull. American Astron. Soc., Vol. 9, 542 (1977). — Abstract.

091.028 Dynamo generation of planetary magnetic fields.
        E. H. Levy.
Bull. American Astron. Soc., Vol. 9, 542 (1977). — Abstract.

091.029 Planetary magnetospheres: an overview.
        N. F. Ness.
Bull. American Astron. Soc., Vol. 9, 543 (1977). — Abstract.

091.030 Determination of the bulk composition of terrestrial planetary bodies utilizing geochemical and cosmochemical methods: a selective review.   M. J. Drake.
Bull. American Astron. Soc., Vol. 9, 545 (1977). — Abstract.

091.031 Gravitating systems of colliding particles in planetary discs.   A. Brahic.
Bull. American Astron. Soc., Vol. 9, 550 (1977). — Abstract.

091.032 L'exploration du système solaire.   P. Bartholdi.
        Conférences d'astronomie de l'Observatoire, (see 012.004), p. 29 - 51 (1977).

091.033 Method for calculating the brightness characteristics of planetary surfaces.   A. S. Panfilov.
Kosm. Issled., Vol. 16, 96 - 106 (1978). In Russian.

091.034 Narrowband submillimeter photometry of the

planets.
D. K. Lynch, J. R. Smith, D. D. Cudaback, M. W. Werner.
Bull. American Astron. Soc., Vol. 9, 622 (1978). – Abstract.

091.035   Stellar occultations by turbulent, non-uniform planetary atmospheres.
R. V. E. Lovelace, J. A. Scannell.
Bull. American Astron. Soc., Vol. 9, 625 - 626 (1978).
Abstract.

091.036   Crustal evolution in the silicate planets.
P. D. Lowman, Jr.
Naturwissenschaften, 65. Jahrg., 117 - 124 (1978).
This paper reviews recent discoveries from space missions to the Moon, Mercury, Mars, and Venus, and their implications for planetary development. It appears that all these planets have undergone similar sequences of crustal evolution, the chief difference among them being how far each has evolved.

091.037   Magnetic damping of rotation.   E. J. Öpik.
Irish Astron. J., Vol. 13, 14 - 21 (1977).
It is concluded that, because of shielding effects, there is not much of magnetic braking or rotation of celestial bodies by external fields. On the other hand, damping of the rotation of artificial satellites by the geomagnetic field is very prominent.

091.038   Regularities in the interior structure of planets.
V. V. Piotrovskij.
Izv. vyssh. uchebn. zaved. Geod. i aehrofotosemka, 1977, No. 3, p. 123 - 128. In Russian. – Abstr. in Ref. zh., 51. Astron., 2.51.264 (1978).

091.039   On early stages of evolution of atmospheres and climate on terrestrial planets.
V. I. Moroz, L. M. Mukhin.
Inst. kosm. issled. AN SSSR. Pr-337. Moskva, 1977. 51 pp.
In Russian. – Abstr. in Ref. zh., 51. Astron., 2.51.266 (1978).

091.040   Theoretical investigations of possibilities of radio spectroscopy of planetary atmospheres by automatic interplanetary stations.   A. P. Naumov.
Izv. AN SSSR. Fiz. atmos. i okeana, Vol. 13, 943 - 948 (1977).
In Russian. – Abstr. in Ref. zh., 51. Astron., 2.51.267; 62. Issled. kosm. prostranstva, 2.62.112 (1978).

091.041   Creep in geodynamic processes.
U. R. Vetter, R. O. Meissner.
Tectonophysics, Vol. 42, No. 1, p. 37 - 54 (1977). – Abstr. in Phys. Abstr., Vol. 81, Abstr. 3744 (1978).

091.042   Flight characteristics of probes in the atmospheres of Mars, Venus and the outer planets.
P. F. Intrieri, C. E. De Rose, D. B. Kirk,
Acta Astronaut., Vol. 4, 789 - 799 (1977). – Abstr. in Phys. Abstr., Vol. 81, Abstr. 7854 (1978).

091.043   Chemistry of the planets of our solar system.
A. P. Vinogradov.
XI Mendeleevsk. sezd po obshch. i prikl. khim. Plenar. dokl. Alma-Ata, 1975. Moskva, Nauka, 1977, p. 139 - 169.
In Russian. – Abstr. in Ref. zh., 51. Astron., 3.51.246; 62. Issled. kosm. prostranstva, 3.62.174 (1978).

091.044   Investigation of the comparative effectiveness of the regression analysis and of modifications of gradient methods in connection with the problem of contour finding when reducing photographs of planetary surfaces by machine.   A. V. Agapov, B. I. Kolosov, S. T. Mosin, A. G. Eremeev.
Inst. kosm. issled. AN SSSR. Pr-329. Moskva, 1977. 23 pp.
In Russian. – Abstr. in Ref. zh., 52. Geod. Aehrosemka,

3.52.207 (1978).

091.045   Laser heterodyne spectroscopy measures planetary winds.
Phys. Today, Vol. 31, No. 5, p. 17 - 19 (1978).

091.046   Standard techniques for presentation and analysis of crater size-frequency data.
R. Arvidson, J. Boyce, C. Chapman, M. Cintala, M. Fulchignoni, H. Moore, G. Neukum, P. Schultz, L. Soderblom, R. Strom, A. Woronow, R. Young.
NASA Tech. Memo., NASA TM–79730, 3 + 20 pp.
Price $ 4.00 (1978).
In September 1977, a crater studies workshop was held for the purpose of developing standard data analysis and presentation techniques. This report contains the unanimous recommendations of the participants. This first meeting considered primarily crater size-frequency data. Future meetings will treat other aspects of crater studies such as morphologies.

091.047   Introducing the satellites.
D. Morrison, D. P. Cruikshank, J. A. Burns.
Planetary satellites, (see 003.007), p. 3 - 17 (1977).
The physical properties and orbits of the known satellites in the solar system are briefly described and tabulated. The evaluation of satellite sizes and masses is presented to calculate mean densities.

091.048   Properties of satellite orbits: ephemerides, dynamical constants, and satellite phenomena.
K. Aksnes.
Planetary satellites, (see 003.007), p. 27 - 42 (1977).
The general character of satellite motion is described from a geometric and physical point of view without reference to mathematics. This is followed by a similar discussion of ordinary and mutual satellite phenomena. The accuracies of published ephemerides are discussed on the basis of the observations and theories available for the various satellites. The determination of physical parameters from positional observations of satellites and photometric observations of ordinary and mutual satellite phenomena is discussed.

091.049   Rotation histories of the natural satellites.
S. J. Peale.
Planetary satellites, (see 003.007), p. 87 - 112 (1977).
Several aspects of the rotational history and current rotation states of the natural satellites are considered. Only the regular satellites as well as Iapetus, Hyperion, Triton, and the Moon are tidally evolved. The remaining irregular satellites essentially retain their primordial spins except for a relaxation to principal axis rotation shared by all the satellites.

091.050   Orbital evolution.   J. A. Burns.
Planetary satellites, (see 003.007), p. 113 - 156 (1977).
This review is concerned with the orbital evolution of particles in circum-planetary space. The discussion is largely qualitative, emphasizing simple physical principles. It attempts to give an understanding of the important phenomena and to provide a guide to the current literature. Processes which govern the orbital evolution of large bodies, that is, the bodies seen today, will be distinguished from those which principally affect small bodies – dust particles.

091.051   Photometry of satellite surfaces.   J. Veverka.
Planetary satellites, (see 003.007), p. 171 - 209 (1977).
This paper reviews some basic photometric concepts as they apply to satellites; then the photometric data for each satellite are summarized. A few topics of special interest are

discussed in some detail, and a number of important observations are listed which should be carried out in the immediate future.

091.052   Polarimetry of satellite surfaces.   J. Veverka.
Planetary satellites, (see 003.007), p. 210 - 231 (1977).

Available polarization observations of the satellites are reviewed, and the question of what information about satellite surfaces such observations contain is discussed critically. For satellites with negligible atmospheres, polarization measurements should provide useful information on surface texture and on the opacity of the surface materials. However, they are unlikely to yield specific compositional information. For Titan, the only satellite known to have an extensive atmosphere, polarimetry indicates the presence of optically thick clouds; it should be possible to deduce the cloud particle characteristics from such data, especially if more extensive observations, covering the full range of available phase angles and wavelengths, are obtained.

091.053   Satellite spectrophotometry and surface compositions.   T. V. Johnson, C. B. Pilcher.
Planetary satellites, (see 003.007), p. 232 - 268 (1977).

A review of currently available spectrophotometric data for planetary satellites, excluding Earth's Moon, is presented. These data are interpreted in terms of surface composition with consideration given to the uniqueness of such identifications.

091.054   Radiometry of satellites and of the rings of Saturn.   D. Morrison.
Planetary satellites, (see 003.007), p. 269 - 301 (1977).

Brightness temperatures at wavelengths between 8 μ and 3.7 cm are tabulated for satellites of known size; except for Titan, they are consistent with blackbody emission values; For Titan a more complex thermal radiation spectrum with three separate regimes is identified. Radiometric brightness measurements, when used in conjunction with photometry, are shown to determine satellite sizes and albedos; this technique is calibrated by using satellites and asteroids with radii evaluated from other methods. Eclipse radiometry indicates the thermophysical properties of surface layers; such observations show that the uppermost coverings of Phobos, Callisto, and Ganymede have remarkably low thermal conductivities. Infrared and radar detections of the rings of Saturn are surveyed; the important constraints that they provide on ring particle sizes and bulk composition are outlined.

091.055   Stellar occultations by planetary satellites.   B. O'Leary.
Planetary satellites, (see 003.007), p. 302 - 315 (1977).

Prediction accuracies have improved, and recent success with two of Jupiter's Galilean satellites shows that it is possible to determine atmospheric and geometric properties of solar system bodies with great precision by using simple photoelectric equipment at small and moderate aperture telescopes. This paper reviews recent work in the field, observational techniques, and future opportunities.

091.056   The evolution of icy satellite interiors and surfaces.   G. J. Consolmagno, J. S. Lewis.
Icarus, Vol. 34, 280 - 293 (1978).

The satellites of the outer solar system planets are thought to be mixtures of ices and rocky material, in which decay of radioactive nuclides can lead to internal melting and solid-state convection. Time-dependent models indicate that melting will reach its maximum extent approximately 2.0 GYr after formation; bodies of radius <500 km will never melt, and those <750 km in radius will be totally refrozen by present. Surface water flows are not expected for bodies of

<1500-km radius. However, even small (100 km) bodies may be unstable against solid-state convection, and their surfaces may show signs of tectonism. Other processes altering the surfaces include sublimation and photolysis of ices.

091.057   The expected frequency of doublet craters.   A. Woronow.
Icarus, Vol. 34, 324 - 330 (1978).

The observed abundances of doublet craters do not necessarily indicate a nonrandom process. Previous studies maintained that planetary surfaces have more doublet craters than would result from randomly located impacts. A new probabilistic calculation and more realistic Monte Carlo simulations demonstrate that the expected and observed number of doublet craters do not differ significantly.

091.058   Stellar occultations by turbulent planetary atmospheres: a wave-optical theory including a finite scale height.   W. B. Hubbard, J. R. Jokipii, B. A. Wilking.
Icarus, Vol. 34, 374 - 395 (1978).

A generalized wave-optical theory of stellar occultations by a turbulent planetary atmosphere is developed. The finite scale height of the atmosphere is retained for the first time. It is found that the finite scale height of the atmosphere affects the scintillations observed during the occultation in a number of ways which are most easily understood in terms of an effective Fresnel scale. The authors demonstrate the validity of a phase-changing screen approximation for occultation by a turbulent atmosphere in parameter ranges of general interest. Using this approximation various statistical properties of the fluctuating intensity are calculated explicitly.

091.059   Effects of turbulence on average refraction angles in occultations by planetary atmospheres.   R. Eshleman, B. S. Haugstad.
Icarus, Vol. 34, 396 - 405 (1978).

Four separable effects of atmospheric turbulence on average refraction angles in occultation experiments are derived from a simplified analysis, and related to more general formulations by B. S. Haugstad. The major contributors are shown to be due to gradients in height of the strength of the turbulence, and the sense of the resulting changes in refraction angles is explained in terms of Fermat's principle. Because the results of analyses of such gradient effects by W. B. Hubbard and J. R. Jokipii are expressed in other ways, a special effort is made to compare all of the predictions on a common basis. The authors conclude that there are fundamental differences, and use arguments based on energy conservation and Fermat's principle to help characterize the discrepancies.

091.060   Evolution of terrestrial planets and giant circular structures.   V. L. Avdeev, A. M. Nikishin.
Izv. vyssh. uchebn. zaved., Geol. i razvedka, 1977, No. 10, p. 33 - 37. In Russian. — Abstr. in Ref. zh., 51. Astron., 4.51.209 (1978).

091.061   On the influence of turbulence on thermal conditions of the thermospheres of planets.
M. N. Izakov.
Kosm. Issled., Vol. 16, 403 - 411 (1978). In Russian.

091.062   Spectral line profiles in a planetary corona: a collisional model.
R. A. Prisco, J. W. Chamberlain.
J. Geophys. Res., Vol. 83, 2157 - 2161 (1978).

The theory of spectral line profiles for a planetary corona (Chamberlain, 1976) is extended to include the effects of H-H+ resonance charge exchange in which satellite particles are omitted (Chamberlain, 1977). A computational model is detailed, and sample profiles are presented. Observations of Lyman alpha or Balmer alpha profiles would have an important application to an understanding of the distribution of satellite particles and the nonthermal component of escaping hydrogen. The problem of satellites is seen to merit further investigation.

091.063 Conjonctions planétaires serrées.   E. Goffin.
Astronomie, Vol. 92, 305 - 308 (1978).

091.064 **Cooling histories of terrestrial planets.**
G. Schubert, P. Cassen, R. E. Young.
Reports of planetary geology program, 1977–1978, (see 003.014), p. 60 - 62 (1978).

091.065 **The tectonics of filled basins on the terrestrial planets.**   S. C. Solomon, J. W. Head.
Reports of planetary geology program, 1977–1978, (see 003.014), p. 66 - 68 (1978).

091.066 **Application of anisotropy of magnetic susceptibility measurements to transport mechanisms on planetary surfaces.**   J. H. Howard III, B. B. Ellwood.
Reports of planetary geology program, 1977–1978, (see 003.014), p. 106 - 108 (1978).

091.067 **On predicting densities of doublet craters.**
A. Woronow.
Reports of planetary geology program, 1977–1978, (see 003.014), p. 143 (1978).

091.068 **Subglacial volcanism – terrestrial landforms and implications for planetary geology.**   C. C. Allen.
Reports of planetary geology program, 1977–1978, (see 003.014), p. 194 - 195 (1978).

091.069 **Stellar and planetary observations with the Voyager photopolarimeter experiment.**   C. F. Lillie.
Bull. American Astron. Soc., Vol. 10, 458 (1978). – Abstract.

091.070 **Why do planets have rings?**   L. A. Willson.
Astronomy, Vol. 5, No. 12, p. 6 - 16 (1977).
Abstr. in Phys. Abstr., Vol. 81, Abstr. 32373 (1978).

091.071 **Some problems of the physics of planets.**
Z. Pokorny.
Ceskoslovensky Cas. Fis., A, Vol. 27, 562 - 578 (1977). In Czech. – Abstr. in Phys. Abstr., Vol. 81, Abstr. 32374 (1978).

091.072 **Planetary dynamos.**   M. Stix.
J. Geophys., Vol. 43, No. 5-6, p. 695 - 717 (1977).
In German. – Abstr. in Phys. Abstr., Vol. 81, Abstr. 40852 (1978).

091.073 **Why study impact craters?**   E. M. Shoemaker.
Impact and explosion cratering, (see 012.034), p. 1 - 10 (1977).
The thesis of this paper is that impact of solid bodies is the most fundamental process that has taken place on the terrestrial planets. The terrestrial planets were formed by this process; the last stage of accretion is still proceeding at a very slow rate. It is suggested that the moon is a sample of the protoearth's mantle that was spun off when the earth had grown to about one-half its present mass. Much more research on impact craters is needed to solve nearly every aspect of the accretion process.

091.074 **A summary of explosion cratering phenomena relevant to meteor impact events.**
H. F. Cooper, Jr.
Impact and explosion cratering, (see 012.034), p. 11 - 44 (1977).
Theoretical and experimental studies of cratering and related phenomena from buried explosions are reviewed with emphasis on information applicable to the study of impact craters. Phenomena considered include crater volume and shape, ejecta, airblast, and ground shock. Based on a number of theoretical studies and experiments, a fairly consistent picture of the evolution of a crater (and of the mechanisms important to the cratering process) emerges.

091.075 **Application of high explosion cratering data to planetary problems.**   V. R. Oberbeck.
Impact and explosion cratering, (see 012.034), p. 45 - 65 (1977).
A review is presented of the conditions of explosion or nuclear cratering required to simulate impact crater formation. Some planetary problems associated with three different aspects of crater formation are illustrated and solutions based on high-explosion data are suggested. Structures of impact craters and properly selected explosion craters formed in layered media are discussed and related to the structure of lunar basins.

091.076 **Cratering mechanisms observed in laboratory-scale high-explosive experiments.**
A. J. Piekutowski.
Impact and explosion cratering, (see 012.034), p. 67 - 102 (1977).

091.077 **Crater morphometry from bistatic radar.**
R. A. Simpson, G. L. Tyler, H. T. Howard.
Impact and explosion cratering, (see 012.034), p. 481 - 487 (1977).
Bistatic radar data can be used to identify and measure the dimensions of anomalously scattering regions on planetary surfaces. A simple case in which spectral features in Apollo 14 echoes were correlated with parts of the lunar crater Lansberg has been investigated quantitatively. Generalization of this technique will permit detection and measurement of surface regions which scatter anomalously (as a result of centimeter to meter scale roughness) on Mars, where high altitude photography does not have sufficient resolution, on Venus, where orbital photography is impossible, and on other targets.

091.078 **Historical variations in the density and distribution of impacting debris in the inner solar system: evidence from planetary imaging.**   L. A. Soderblom.
Impact and explosion cratering, (see 012.034), p. 629 - 633 (1977).

091.079 **Shock wave, a possible source of magnetic fields?**
B. A. Ivanov, B. A. Okulessky (*Okulesskij*), A. T. Basilevsky (*Bazilevskij*).
Impact and explosion cratering, (see 012.034), p. 861 - 867 (1977).
One possible effect due to impact cratering is the transformation of part of the kinetic energy of the projectile to energy in the electromagnetic field. Shock-induced polarization of piezo- and dielectric material could be the mechanism by which this is accomplished. The mechanism suggested is attractive because it gives a possible explanation for the high values of remanent magnetization displayed by some lunar samples, without calling for the hypothesis of a strong global magnetism for the moon in the past epochs.

091.080 **The theory of cratering phenomena, an overview.**
C. P. Knowles, H. L. Brode.
Impact and explosion cratering, (see 012.034), p. 869 - 895 (1977).
The essential features of detailed continuum mechanics solutions of cratering action associated with explosive or hypervelocity impact sources are explained. The initial energy coupling factors include radiation transport for nuclear explosive sources. The physical factors which must be accounted for in such calculations include material properties from

ionization, dissociation, vaporization, melting, and crystal phase changes. The hydrology and geologic details of a crater site are equally important. The mathematical models and the numerical techniques are reviewed.

091.081   Ejecta formation: calculated motion from a shallow-buried nuclear burst, and its significance for high velocity impact cratering.   J. G. Trulio.
Impact and explosion cratering, (see 012.034), p. 919 - 957 (1977).
    Several years ago a calculation was made at ATI of ground motion during the ejecta-forming stages of a 1 MT nuclear explosion at a depth of 15 feet in granite. That calculation, and its relation to craters formed by high velocity impact events of a kind entered into by meteoroids, are the subject of this paper.

091.082   Numerical simulations of a 20-ton TNT detonation on the earth's surface and implications concerning the mechanics of central uplift formation.
G. W. Ullrich, D. J. Roddy, G. Simmons.
Impact and explosion cratering, (see 012.034), p. 959 - 982 (1977).
    Central uplifts are common features observed in craters on the Earth, the Moon, Mars, and Mercury. Since these uplifts do not occur in all craters, they should be useful in providing strong constraints on both planetary evolution and numerical cratering simulations.

091.083   Review and comparison of hypervelocity impact and explosion cratering calculations.
K. N. Kreyenhagen, S. H. Schuster.
Impact and explosion cratering, (see 012.034), p. 983 - 1002 (1977).
    Two-dimensional (axisymmetric) finite-difference numerical code calculations of several hypervelocity impacts, near-surface nuclear bursts, and near-surface high explosive cratering events are reviewed and compared.

091.084   Crater-related ground motions and implications for crater scaling.   H. F. Cooper, Jr., F. M. Sauer.
Impact and explosion cratering, (see 012.034), p. 1133 - 1163 (1977).

091.085   Scaling of cratering experiments – an analytical and heuristic approach to the phenomenology.
B. G. Killian, L. S. Germain.
Impact and explosion cratering, (see 012.034), p. 1165 - 1190 (1977).
    The phenomenology of cratering can be thought of as consisting of two phases. The first phase, where the effects of gravity are negligible, consists of the energy source dynamically imparting its energy to the surroundings, rock and air. The second phase consists of the rock, with its already developed velocity field, being "thrown out".

091.086   Influence of gravitational fields and atmospheric pressures on scaling of explosion craters.
A. J. Chabai.
Impact and explosion cratering, (see 012.034), p. 1191 - 1214 (1977).
    Methods of obtaining scaling rules from dimensional analysis are reviewed. Scaling rules to describe the phenomena of cratering by explosions or by hypervelocity impacts are presented and discussed in terms of the requirements of similitude. Some sources of similarity violation are outlined and some experiments are suggested which may eliminate a suspected major source of similarity violation present in all explosion cratering experiments conducted to date in accelerated coordinate systems.

091.087   Craters from surface explosions and energy depen-

dence. – A retrospective view.   L. J. Vortman.
Impact and explosion cratering, (see 012.034), p. 1215 - 1229 (1977).

091.088   Energies of formation for ejecta blankets of giant impacts.   S. K. Croft.
Impact and explosion cratering, (see 012.034), p. 1279 - 1296 (1977).
    Ejection energy-diameter scaling relations for impact craters have been investigated by a simple ballistic ejection model. Scaling relations of the form $E = AD^B$, derived from least-squares fits to calculated ejection energies over a range of crater diameters, vary according to ejection angle, ejecta blanket distribution, planetary radius, and the depth-diameter ratio. Possible values of the above parameters are drawn from laboratory and field geologic crater data.

091.089   Correlation between X-ray K-lines of some elements and planetary masses.   N. A. Shirokov.
VNII yader. geofiz. i geokhim. Moskva, 1978, 12 pp. In Russian. – Abstr. in Ref. zh., 51. Astron., 6.51.225 (1978).

091.090   On the relative locations of the bow shocks of the terrestrial planets.   C. T. Russell.
Geophys. Res. Lett., Vol. 4, 387 - 390 (1977).
    The observed bow shock encounters at Mercury, Venus and Mars are least square fit using the same technique so that their sizes and shapes can be intercompared. The shock front of Mercury most resembles the terrestrial shock in shape, and the shock stand off distance is consistent with the observed moment. The shapes of the Venus and Mars shock fronts more resemble each other than the earth's and the stand off distances are consistent with direct interaction of the solar wind with the ionosphere on the dayside.

091.091   The diameter-frequency relations for craters on planets and satellites.   J. Bican, J. Bouška.
Acta Univ. Carolinae Math. Phys., Vol. 19, No, 2, p. 7 - 11 (1978).
The diameter-frequency relations for craters on Mars and on Phobos have been derived and compared with those relations for craters on the Moon, Mercury and the Earth.

Closeup: new worlds.   See Abstr. 003.028.

L'exploration du système solaire.
See Abstr. 003.061.

Exploration of the solar system.
See Abstr. 003.071.

Meteorology of planets.   See Abstr. 003.074.

Radiophysical investigations of planets.
See Abstr. 003.076.

Icarus Meeting Report: International Symposium on Planetary Atmospheres.   See Abstr. 011.016.

Nuclear cratering experiments: United States and Soviet Union.   See Abstr. 013.025.

Topographic changes on airless bodies: theoretical considerations.   See Abstr. 021.002.

Far-infrared absorption in H_2 and H_2-He mixtures.
See Abstr. 022.002.

Sputtering of water ice by MeV light ions and its importance for planetary astronomy.   See Abstr. 022.023.

The red bands of methane.

See Abstr. 022.026.

Ion-molecule condensation reactions: a mechanism for chemical synthesis in reducing planetary atmospheres. See Abstr. 022.028.

High-resolution studies of the 9613-Å band of CH_3D. See Abstr. 022.045.

Analysis of stellar occultation data. Effects of photon noise and initial conditions.   See Abstr. 031.201.

Comparative imaging radar planetology. See Abstr. 031.279.

Spectral or time series analysis of surface roughness: correlation with radar backscatter cross-section measurements.   See Abstr. 031.280.

Speckle image reconstruction of outer planet satellites.   See Abstr. 031.290.

Statistically correlated satellite observations. See Abstr. 031.291.

A mass spectrometer-beam experiment for investigations of planetary atmospheres.   See Abstr. 032.516.

Meridian observations of major and minor planets, 1974–1977.   See Abstr. 041.035.

Motions of natural satellites. See Abstr. 042.050.

Radiative transfer calculated from a Markov chain formalism.   See Abstr. 063.004.

Scattering of light from particulate surfaces. I. A laboratory assessment of multiple-scattering effects. See Abstr. 063.026.

Numerical investigation of some models of interaction of the solar wind with cosmic objects. See Abstr. 074.091.

Numerical modelling of interaction between the solar wind plasma and the Moon, Mars and Venus. See Abstr. 074.092.

The evolution of the Moon and the terrestrial planets.   See Abstr. 094.001.

Significant achievements in the planetary program— 1976 - 1977.   See Abstr. 107.021.

White dwarfs and giant planets. See Abstr. 126.040.

092 Mercury

092.001 Tectonic implications for the gravity structure of Caloris Basin, Mercury.
H. J. Melosh, D. Dzurisin.
Icarus, Vol. 33, 141 - 144 (1978).
    Studies of tectonic landforms associated with Caloris Basin on Mercury suggest that isostatic adjustment has occurred in response to basin excavation, and that the smooth plains inside Caloris were emplaced significantly before isostatic equilibrium was attained. Combined with dynamical considerations, this leads the authors to propose that the Caloris region is characterized by a circular negative or zero free air gravity anomaly centered inside Caloris, and an annular positive anomaly which coincides with extensive tracts of young smooth plains outside the basin. This proposed gravity pattern differs markedly from that associated with mare-filled basins on the Moon.

092.002 Television observations of Mercury by Mariner 10.
B. C. Murray, M. J. S. Belton, G. E. Danielson, M. E. Davies, D. E. Gault, B. Hapke, B. O'Leary, R. G. Strom, V. Suomi, N. Trask.
Cosmochemistry of the moon and planets, (see 012.002), p. 865 - 885 (1977).
    The morphology and optical properties of the surface of Mercury resemble those of the Moon in remarkable detail, recording a very similar sequence of events; chemical and mineralogical similarity of the outer layers is implied. Mercury is probably a differentiated planet with an iron-rich core. Differentiation is inferred to have occurred very early. No evidence of atmospheric modification of any landform is found. Large-scale scarps and ridges unlike lunar or Martian features may reflect a unique period of planetary compression near the end of heavy bombardment, perhaps related to contraction of the core.

092.003 Impact melts on Mercury and the Moon.
B. R. Hawke, M. J. Cintala.
Bull. American Astron. Soc., Vol. 9, 531 (1977). – Abstract.

092.004 Mercury: detection of a 1000nm ferrous band.
L. Tepper, B. Hapke, J. Woodman, M. Perry.
Bull. American Astron. Soc., Vol. 9, 532 (1977). – Abstract.

092.005 Mercury: interferometry at 1.35 cm wavelength and determination of thermophysical parameters.
L. M. Golden.
Bull. American Astron. Soc., Vol. 9, 532 (1977). – Abstract.

092.006 Derivation of the temperature and emissivity of Mercury as a result of the removal of the thermal component in a Mercury/Sun albedo spectrum.   R. N. Clark.
Bull. American Astron. Soc., Vol. 9, 532 (1977). – Abstract.

092.007 Fe/silicate fractionation and the origin of Mercury.
S. J. Weidenschilling.
Bull. American Astron. Soc., Vol. 9, 546 (1977). – Abstract.

092.008 Experimental determination of Mercury's mass and oblateness.   P. B. Esposito, J. D. Anderson, A. T. Y. Ng.
Space Research, Vol. XVII, (see 012.010), 639 - 644 (1977).
    The ratio of the mass of the sun to that of Mercury deduced from the first and third flybys of Mercury by Mariner 10 is 6023600 ± 600 and 6023700 ± 300 respectively. Additional analysis should reduce these uncertainties by a factor of 2–5. Assuming an equatorial radius of 2439 km, the mean density of the planet is 5.44 g cm^{-3}. The gravitational oblateness deduced from the first encounter is discussed. Data received from the third encounter are better suited for this determination and yield $J_2 = (8 \pm 6) \times 10^{-5}$. In addition, the

Mercury 3 data are shown to be sensitive to local gravity effects or anomalies.

092.009 **Mercury's craters from earth.**   A. T. Young.
Icarus, Vol. 34, 208 - 209 (1978).
Craters on Mercury were discovered by See in 1901, using the 26-in. Naval Observatory refractor.

092.010 **IAU nomenclature for albedo features on the planet Mercury.**   A. Dollfus, C. R. Chapman, M. E. Davies, O. Gingerich, R. Goldstein, J. Guest, D. Morrison, B. A. Smith.
Icarus, Vol. 34, 210 - 214 (1978).
The International Astronomical Union has endorsed a nomenclature for the albedo features on Mercury. Designations are based upon the mythological names related to the god Hermes; they are expressed in Latin form. The dark-hued albedo features are associated with the generic term Solitudo. The light-hued areas are designated by a single name without generic term. The 32 names adopted are allocated on the Mercury map.

092.011 **Remarks on the nomenclature of Mercury.**   L. E. Krumenaker.
Icarus, Vol. 34, 215 - 219 (1978).
Longitudes of surface markings observed on Mercury by E. M. Antoniadi are determined. The most frequent of the locations are used to determine the most probable positions on Mercury's surface for the names defined on Antoniadi's 88-day planisphere. It is suggested that these locations can serve as a valid and accurate basis for Mercurian nomenclature.

092.012 **Mercury's helium exosphere after Mariner 10's third encounter.**   S. A. Curtis, R. E. Hartle.
J. Geophys. Res., Vol. 26, 1551 - 1557 (1978).
From a comparison of the Mariner 10 third-encounter UV spectrometer data with intensities generated from a newly constructed model exosphere, the authors have derived a new value of 4.5×10^{-4} for the fraction of the solar wind He^{++} flux to be intercepted and captured by Mercury's magnetosphere if the observed He atmosphere is maintained by the solar wind. If an internal source for He prevails, the corresponding upper bound for the global outgassing rate is estimated to be 4.5×10^{22} s^{-1}.

092.013 **New names of craters on Mercury.**   G. A. Burba.
Priroda, 1978, No. 5, p. 33. In Russian.

092.014 **Mercury: magnetic field and interior.**   N. F. Ness.
Space Sci. Rev., Vol. 21, 527 - 553 (1978).
Between 1965 and 1975, our knowledge of Mercury and its physical characteristics improved dramatically. Radar studies of the planetary orbit and rotation rate and Mariner 10 spacecraft studies of its surface, atmosphere, magnetic field and plasma environment provided startling new results on what had been the least understood member of the terrestrial planets. With a highly cratered surface and a modest magnetic field, Mercury is a differentiated planet with fractionally the largest iron core of all.

092.015 **Review of Mercurian geology.**   R. G. Strom.
Reports of planetary geology program, 1977−1978, (see 003.014), p. 72 - 74 (1978).

092.016 **The stratigraphy of the Caloris basin.**
J. F. McCauley, J. E. Guest, N. J. Trask, G. G. Schaber, R. Greeley, D. E. Gault, H. E. Holt.
Reports of planetary geology program, 1977−1978, (see 003.014), p. 75 - 77 (1978).

092.017 **Geologic history of the Victoria Quadrangle (H-2) Mercury.**   E. A. King, G. E. McGill.
Reports of planetary geology program, 1977−1978, (see 003.014), p. 78 (1978).

092.018 **Structural lineament pattern analysis of the Caloris surroundings: a pre-Caloris pattern on Mercury.**
P. Thomas.
Reports of planetary geology program, 1977−1978, (see 003.014), p. 79 - 82 (1978).

092.019 **Crater production and erosion on Mercury.**   W. P. O'Donnell.
Reports of planetary geology program, 1977−1978, (see 003.014), p. 144 - 146 (1978).

092.020 **In search of ancient astroblemes: Mercury.**   R. A. de Hon.
Reports of planetary geology program, 1977−1978, (see 003.014), p. 150 - 152 (1978).

092.021 **Mercury geologic mapping.**   H. E. Holt.
Reports of planetary geology program, 1977−1978, (see 003.014), p. 327 (1978).

092.022 **Mercurian bright patches: evidence for physio-chemical alteration of surface material?**
D. Dzurisin.
Geophys. Res. Lett., Vol. 4, 383 - 386 (1977).
Morphologically and photometrically anomalous patches of highly-reflective material exist inside several large mercurian craters. Local physio-chemical alteration along impact-induced fractures may have been involved in production of these uniquely mercurian features.

092.023 **On the apparent source depth of planetary magnetic fields.**   R. C. Elphic, C. T. Russell.
Geophys. Res. Lett., Vol. 5, 211 - 214 (1978).
Two simple assumptions regarding the ratios of the strengths of the field contributions of the multipole moments of the terrestrial magnetic field at its effective source depth are used to examine the consistency between the apparent source depths for the magnetic fields of Mercury and Jupiter and the present understanding of their interior structure.

Catalogue of Martian craters and statistics of the craters of Mars, Mercury and the Moon.
See Abstr. 002.043.

Catalogues of craters of Mercury and the Moon.
See Abstr. 002.044.

On the theory of rotational motion of Mercury and synchronous satellites of planets.   See Abstr. 042.103.

The interaction between a magnetized plasma flow and a magnetized celestial body: a review of magnetospheric studies.   See Abstr. 062.082.

Limits to the expansion of Earth, Moon, Mars and Mercury and to changes in the gravitational constant.
See Abstr. 081.006.

Application of anisotropy of magnetic susceptibility measurements to transport mechanisms on planetary surfaces.
See Abstr. 091.066.

Modification of fresh crater landforms: evidence from the Moon and Mercury.   See Abstr. 094.129.

Crater shape-size profiles for fresh craters on the Moon and Mercury.   See Abstr. 094.145.

Problems of geology of the moon, Mercury and
Mars.   See Abstr. 094.159.

Planetary cartography.   See Abstr. 097.159.

093 Venus

**093.001   Independent radio-occultation studies of Venus'
      atmosphere.   P. D. Nicholson, D. O. Muhleman.**
Icarus, Vol. 33, 89 - 101 (1978).

Two independent analyses of the dual-frequency radio-
occultation experiment performed by Mariner 10 at Venus
are presented. Using closed-loop frequency data obtained at
NASA's Goldstone facility, the authors have computed S-
and X-band pressure-temperature profiles for Venus' neutral
atmosphere, and an S-band profile of the nightside ionosphere. These two Mariner 10 profiles are compared with the
Mariner 5 occultation profile and in situ measurements by
Veneras 8, 9, and 10.

**093.002   An analytic study of impact ejecta trajectories in the
      atmospheres of Venus, Mars, and Earth.**
M. E. Tauber, D. B. Kirk, D. E. Gault.
Icarus, Vol. 33, 529 - 536 (1978).

Calculations have been made to determine the effects of
atmospheric drag and gravity on impact ejecta trajectories on
Venus, Mars, and the Earth. The equations of motion were
numerically integrated for a broad range of sizes, initial velocities, and initial elevation angles. A dimensionless parameter
was found from approximate analytic solutions which correlated the ejecta range, final impact angle, and final impact
velocity for all three planets.

093.003   A geochemical model of the Venus troposphere.
      C. P. Florensky (*K. P. Florenskij*), V. P. Volkov,
O. V. Nikolaeva.
Icarus, Vol. 33, 537 - 553 (1978).

A geochemical model of the Venus troposphere
(0−65 km) based on a theoretical generalization of the available instrumental data and equilibrium calculations in the
O−H−C−N−S−Cl system is proposed. The model is based on
the concept of geochemical zoning of the troposphere.

093.004   Venus: the 17- to 38-micron spectrum.
      R. A. Reed, W. J. Forrest, J. R. Houck, J. B.
Pollack.
Icarus, Vol. 33, 554 - 557 (1978).

The Venus emission spectrum was measured from the
NASA Lear Jet on five nights in June 1975. A cooled
grating spectrometer with a resolution of $\lambda/\Delta\lambda \simeq 25$ over
the spectral interval 17 to 38 μm was used. The main
features in the observed spectrum are consistent with the
theoretical emission spectrum of a haze of aqueous sulfuric
acid droplets suspended in a CO_2 atmosphere.

**093.005   A forced barotropic model as an analog of the
      stratosphere of Venus.**
W. B. Rossow, G. P. Williams.
Bull. American Astron. Soc., Vol. 9, 508 (1977). − Abstract.

**093.006   Possible detection of an atmospheric planetary
      wave on Venus.   S. S. Limaye, V. E. Suomi.**
Bull. American Astron. Soc., Vol. 9, 509 (1977). − Abstract.

093.007   Persistence and size of thermal anomalies on Venus.
      R. A. Brown, R. M. Goody, J. Apt.
Bull. American Astron. Soc., Vol. 9, 509 (1977). − Abstract.

093.008   Venus: still no rapid retrograde rotation.
      R. A. J. Schorn, A. T. Young, L. D. G. Young,
D. Crisp, M. J. S. Belton.
Bull. American Astron. Soc., Vol. 9, 509 (1977). − Abstract.

093.009   Ionosphere of Venus during Mariner 10 encounter.
      R. E. Hartle, S. J. Bauer, H. G. Mayr.
Bull. American Astron. Soc., Vol. 9, 512 (1977). − Abstract.

**093.010   Airborne spectroscopic observations of Venus be-
      tween 12 and 20 microns.**
G. S. Orton, H. H. Aumann.
Bull. American Astron. Soc., Vol. 9, 512 (1977). − Abstract.

**093.011   Spacecraft isophotes of Venus support acid aerosol,
      not isotropic scattering.**
A. T. Young, G. W. Kattawar.
Bull. American Astron. Soc., Vol. 9, 512 (1977). − Abstract.

**093.012   Liquid content of the lower clouds of Venus from
      Mariner 10 radio occultation.**
A. Kliore, C. Elachi, J. B. Cimino.
Bull. American Astron. Soc., Vol. 9, 513 (1977). − Abstract.

093.013   Is the inverse phase effect on Venus real?
      L. D. G. Young.
Bull. American Astron. Soc., Vol. 9, 513 (1977). − Abstract.

093.014   Rotation of Venus.
      W. M. de Campli, I. I. Shapiro.
Bull. American Astron. Soc., Vol. 9, 520 (1977). − Abstract.

093.015   Radar images of the surface of Venus.
      D. B. Campbell.
Bull. American Astron. Soc., Vol. 9, 532 (1977). − Abstract.

093.016   Vénus.   M. Ya. Marov.
      La recherche en astrophysique, (see 003.001), p.
39 - 54 (1977).

**093.017   Nocturnal atmosphere of Venus from results of a
      radio transillumination with Venera 9 and Venera
10.   O. I. Yakovlev, A. I. Efimov, S. S. Matyugov, T. S.**
Timofeeva, E. V. Chub, G. D. Yakovleva.
Kosm. Issled.,Vol. 16, 113 - 119 (1978). In Russian.

**093.018   Preliminary results of simultaneous use of the two
      satellites Venera 9 and Venera 10 for a two-fre-
quency radio transillumination of the nocturnal Venus iono-
sphere.   S. L. Azarkh, M. B. Vasil'ev, V. A. Vinogradov,**
A. S. Vyshlov, M. A. Kolosov, A. P. Mestehrton, N. A.
Savich, V. A. Samovol, L. N. Samoznaev, A. I. Sidorenko,
Yu. N. Shnygin.
Kosm. Issled., Vol. 16, 120 - 126 (1978). In Russian.

**093.019   Attempt of improving H_2O abundance determina-
      tion in the Venus atmosphere according to data of a
narrow-band photometry aboard Venera 9 and Venera 10.**
E. A. Ustinov, V. I. Moroz.
Kosm. Issled., Vol. 16, 127 - 133 (1978). In Russian.

093.020 Estimates of the turbulence conditions of the
         atmosphere near the Venus surface from data of the
automatic interplanetary stations Venera 9 and Venera 10.
G. S. Golitsyn.
Kosm. Issled., Vol. 16, 156 - 158 (1978). In Russian.

093.021 **Venus in motion.**   J. L. Anderson, M. J. S. Belton,
         G. E. Danielson, N. Evans, J. M. Soha.
Astrophys. J., Suppl. Ser., Vol. 36, 275 - 284, plates 1–60
(1978).
    A comprehensive set of television pictures of Venus
taken by the Mariner 10 spacecraft is presented. Included is a
chronological sequence of television images illustrating the
development variety, and circulation of the Venus upper atmospheric phenomena as viewed in the near-ultraviolet. The
higher resolution images have been assembled into global mo-
saics to facilitate comparison. Figures and tables describing
the imaging sequences have been included to provide a guide
to the more complete set of 3400 Venus images on file at
the National Space Science Data Center.

093.022 **The first panoramas of the Venusian surface:**
         geological-morphological analysis of pictures.
C. P. Florensky (*K. P. Florenskij*), A. T. Basilevsky
(*Bazilevskij*), A. A. Pronin.
Space Research, Vol. XVII, (see 012.010), 645 - 649 (1977).
    The results of geological-morphological analysis of tele-
vision pictures of the Venusian surface taken by Venera
9 and Venera 10 probes are described. The Venera 9
picture shows the surface of a rather steep (20–30°) slope
covered by plate-like sharp-edged fragments with sizes some
tens of centimetres. Between the fragments the surface has
the texture of gravel. The Venera 10 picture shows a plain-
like landscape having hard-rock outcrops protruding through
the regolith.

093.023 **Investigations of the density of the Venusian sur-**
         face rocks by Venera 10.   Yu. A. Surkov.
F. F. Kirnozov, V. K. Khristianov, B. N. Korchuganov,
V. N. Glazov, V. F. Ivanov.
Space Research, Vol. XVII, (see 012.010), 651 - 657 (1977).
    In October 1975 Venera 10 made a soft landing on the
sunlit side of the planet Venus. The results of densitometric
measurements have shown that the rock in the landing site
has a density of 2.8 ± 0.1 g cm^{-3}, in agreement with the concept of the basaltic composition of the planet's crust. The
relatively high density of the uppermost layer is indicative of
the low intensity of erosion processes taking place on the
planet.

093.024 **Investigations of Venusian gamma-radiation by**
         Venera 9 and Venera 10.   Yu. A. Surkov.
F. F. Kirnozov, V. N. Glazov, G. A. Fedoseyev (Fedoseev).
Space Research, Vol. XVII, (see 012.010), 659 - 662 (1977).
    The results of determining the content of natural radio-
active elements in Venusian rocks are reviewed. Investigations
were made with the aid of gamma-ray spectrometers carried
by Venera 9 and Venera 10 probes. It is concluded that there
is a crust on Venus which contains a large amount of basic
rocks.

093.025 **Some results of dual-frequency radio occultation**
         exploration of the night-time ionosphere of Venus
with satellites Venera 9 and 10.   Yu. N. Alexandrov
(*Aleksandrov*), M. B. Vasilyev (*Vasil'ev*), A. S. Vyshlov,
V. M. Dubrovin, A. L. Zaitsev (*Zajtsev*), M. A. Kolosov,
N. A. Savich, V. A. Samovol, L. N. Samoznaev, A. I.
Sidorenko, D. Ya. Shtern, A. F. Hasyanov (*Khasyanov*).
Space Research, Vol. XVII, (see 012.010), 663 - 666 (1977).
    Examples of the recordings of the reduced phase dif-
ference and electron density height profiles in the night-time
ionosphere of Venus are presented for the most characteristic

cases. The Venera 9 and 10 radio occultation experiments
have shown that the parameters of the night-time Venusian
ionosphere suffer great changes with time.

093.026 **Radiative fluxes and equilibrium temperature**
         profile in the Venus atmosphere.
A. S. Ginzburg, A. S. Safraj.
Izv. AN SSSR. Fiz. atmos. i okeana, Vol. 13, 936 - 942 (1977).
In Russian. – Abstr. in Ref. zh., 51. Astron., 2.51.272 (1978).

093.027 **Properties of the clouds of Venus, as inferred from**
         airborne observations of its near-infrared reflectiv-
ity spectrum.   J. B. Pollack, D. W. Strecker, F. C.
Witteborn, E. F. Erickson, B. J. Baldwin.
Icarus, Vol. 34, 28 - 45 (1978).
    The authors have measured the shape and absolute value
of Venus' reflectivity spectrum in the 1.2- to 4.0-μm spectral region. Comparing these spectra with synthetic spectra
generated with a multiple-scattering computer code, the
authors infer a number of properties of the Venus clouds.
The comparison enables the authors to test the sulfuric acid
hypothesis, to obtain a good determination of acid concentration, and to estimate a globally averaged value for the cloud
optical depth.

093.028 **High-dispersion spectroscopic observations of**
         Venus near superior conjunction. II. The carbon
dioxide band at 7883 Å.   A. T. Young, L. D. G. Young,
R. A. J. Schorn.
Icarus, Vol. 34, 46 - 51 (1978).
    Nine plates of the 7883-Å CO_2 band were taken be-
tween phase angles 7.2 and 10.7° in 1971. A curve-of-growth
analysis of 28 rotational lines in the band indicates an
average rotational temperature of $236 \pm 8°K$; the average
slope of the curve of growth was 0.63 ± 0.06. The results
for this band are compared to those for the 7820-Å band.

093.029 **The clouds of Venus: an anisotropic scattering**
         model.   J. P. Lestrade, J. W. Chamberlain.
Icarus, Vol. 34, 52 - 62 (1978).
    The authors use a homogeneous, semi-infinite, atmo-
spheric model with an analytical phase function to derive
values for the total optical thickness and single-scattering
albedo for the atmosphere of Venus. The principal source
of data for this analysis is the equivalent widths of the
7820-Å CO_2 band. Estimates of the "true" values as would
be derived from a Mie-scattering analysis are obtained
through an application of approximate similarity relations
to the isotropic values. If the authors assume an average
anisotropy factor for the aerosols of 0.7, they find an optical
thickness of 50 to 63 and a local albedo of 0.9992 to
0.9995.

093.030 **Venus daytime atmosphere from results of a**
         radio transillumination with the Venera 9 and
Venera 10 satellites.   M. A. Kolosov, O. I. Yakovlev,
S. S. Matyugov, T. S. Timofeeva, E. V. Chub, G. D.
Yakovleva. I. Eh. Kalashnikov.
Kosm. Issled., Vol. 16, 278 - 284 (1978). In Russian.

093.031 **Radio wave fluctuations and turbulence of the**
         Venus night atmosphere from radio transillumina-
tion data with the Venera 9 satellite.
T. S. Timofeeva, O. I. Yakovlev, A. I. Efimov.
Kosm. Issled., Vol. 16, 285 - 293 (1978). In Russian.

093.032 **Interpretation of the CO_2 absorption bands observ-**
         ed in the Venus infrared spectrum between 1 and
2.5 μm.   J.-Y. Mandin.
J. Mol. Spectrosc. Vol. 67, 304 - 321 (1977). – Abstr. in
Phys. Abstr., Vol. 81, Abstr. 7896 (1978).

093.033   **Plasma electron observations in the wake of Venus.**
C. M. Yeates, K. W. Ogilvie, G. L. Siscoe.
J. Geophys. Res., Vol. 83, 1524 - 1530 (1978).

Two plasma regimes were observed in the Venus wake, one characterized by negative-going anisotropies accompanied by high bulk speeds and low magnetic variance and the other characterized by positive-going anisotropies accompanied by low bulk speeds and high magnetic variance. The two modes alternate in time in a stochastic bimodal fashion as interplanetary magnetic field discontinuities propagate past Venus and switch on the positive anisotropies. The positive mode probably represents enhanced planetary ion pick up which results in instabilities, the damping of which energizes the electrons in the distant wake.

093.034   **Electroconductivity of the Venus ionosphere.**
L. P. Mishina, Eh. B. Fajnberg, L. L. Van'yan.
Issled. prostranstv. -vremennoj strukt. geomagn. polya. Moskva, Nauka, 1977, p. 178 - 183. In Russian. – Abstr. in Ref. zh., 51. Astron., 3.51.255; 62. Issled. kosm. prostranstva, 3.62.176 (1978).

093.035   **Method of determination of the optical characteristics of an instrument for measurement of illuminance in the Venus atmosphere.**   Yu. M. Golovin, B. E. Moshkin, N. I. Protasov, A. P. Ehkonomov.
Inst. kosm. issled. AN SSSR. Pr-341. Moskva, 1977. 22 pp. In Russian. – Abstr. in Ref. zh., 62. Issled. kosm. prostranstva, 3.62.119 (1978).

093.036   **A comprehensive model of the Venus ionosphere.**
R. H. Chen, A. F. Nagy.
J. Geophys. Res., Vol. 83, 1133 - 1140 (1978).

The coupled time-dependent continuity-momentum and energy balance equations were simultaneously solved for CO_2^+, O_2^+, O^+, He^+, and H^+ densities and electron and ion temperatures for an altitude range of 120–500 km. The solar zenith angle was varied from $0°$ at the subsolar point to $90°$ at the terminator. The two-stream photoelectron transport method was used to find the heating rates for the ambient electrons. The effects of different boundary conditions were investigated, and a nightside ionosphere was calculated with electron precipitation as the source of ionization. The results of these model calculations agree well with measurements in the region of maximum electron density.

093.037   **Radiophysical investigations of Venus from space vehicles.**   M. A. Kolosov, O. I. Yakovlev.
Zemlya i Vselennaya, 1978, No. 3, p. 33 - 37. In Russian.

093.038   **The elongation of Venus: 1977 June.**
J. H. Robinson.
J. British Astron. Assoc., Vol. 88, 393 - 396 (1978). – Report of the Mercury and Venus Section.

093.039   **Spectral composition of solar radiation in the Venus atmosphere from results of illuminance measurements aboard Venera 9 and Venera 10.**
B. E. Moshkin, A. P. Ehkonomov, Yu. M. Golovin.
Kosm. Issled., Vol. 16, 412 - 418 (1978). In Russian.

093.040   **On the induced magnetosphere of Venus.**
A. S. Lipatov.
Kosm. Issled., Vol. 16, 429 - 433 (1978). In Russian.

093.041   **A propos de la périodicité des passages de Vénus.**
J. Bouzon.
Astronomie, Vol. 92, 264 (1978).

093.042   **Fabry-Perot measurement of retrograde rotation of Venus' atmosphere.**
IAU Circ., No. 3232 (1978).

093.043   **Photoelectrons and electron temperatures in the Venus ionosphere.**   D. M. Butler, R. S. Stolarski.
J. Geophys. Res., Vol. 83, 2057 - 2065 (1978).

Solar wind induced magnetic fields may be present in the dayside ionosphere of Venus. A scale analysis and detailed calculations of the effects of different field line orientations on photoelectron transport and electron temperature show that horizontal magnetic fields considerably alter the profiles of electron heating and temperature and photoelectron flux. Direct injection of solar wind electrons has been considered, and the effects are minor. No evidence is found to support the suggestion that a sharp gradient in electron temperature is the cause of certain features in the Mariner 10 electron density profile.

093.044   **On some characteristics of the Venus cloud layer from results of density and temperature measurements aboard AIS Venera 4.**
A. I. Lifshits, V. V. Mikhnevich.
Tr. Inst. prikl. geofiz. Glav. upr. gidrometeorol. sluzhby pri Sov. Min. SSSR, 1977, No. 35, p. 72 - 76. In Russian. – Abstr. in Ref. zh., 62. Issled. kosm. prostranstva, 5.62.244 (1978).

093.045   **Distribution of water vapour in the Venus upper thermosphere from data of the automatic stations Venera 4, 5, 6.**   B. M. Andrejchikov.
Geokhimiya, 1978, No. 1, p. 11 - 15. In Russian. – Abstr. in Ref. zh., 62. Issled. kosm. prostranstva, 5.62.245 (1978).

093.046   **Phase effect of Venus at 3.4 mm wavelength radiation.**   S. Hata, K. Akabane.
Tokyo Astron. Obs. Report No. 69, Vol. 18, 116 - 131 (1978). In Japanese.

093.047   **The phase effect of Venus.**   A. A. Nefed'ev.
Izv. Astron.Ehngel'gardt. Obs., Kazan', No. 41 - 42, p. 91 - 93 (1976). In Russian.

093.048   **Venus tristatic radar status.**   R. Jurgens.
Reports of planetary geology program, 1977–1978, (see 003.014), p. 140 - 141 (1978).

093.049   **A three-dimensional model of dynamical processes in the Venus atmosphere.**
R. E. Young, J. B. Pollack.
J. Atmos. Sci., Vol. 34, 1315 - 1351 (1977). – Abstr. in Phys. Abstr., Vol. 81, Abstr. 32376 (1978).

093.050   **Probable distribution of large impact basins on Venus: comparison with Mercury and the Moon.**
G. G. Schaber, J. M. Boyce.
Impact and explosion cratering, (see 012.034), p. 603 - 612 (1977).

Low resolution (80 km) 12.5 cm wavelength radar maps have been used to identify probable large impact basins on Venus. The radar data indicate that the cratering record preserved on Venus may be very similar to that of the Moon, Mars, and Mercury.

093.051   **On the structure of individual regions of the Venus surface and on the period of the Venus rotation based on 1972 radar observations at 39 cm wavelength.**
V. K. Golovkov, B. I. Kuznetsov, G. M. Petrov, A. F. Khasyanov.
Problems of astronomy and geodesy, VAGO assembly Yerevan, 1975, (see 012.039), p. 61 - 76. In Russian. – Abstr. in Ref. zh., 51. Astron., 6.51.239; 52. Geod. Aehrosemka, 6.52.119 (1978).

093.052   **Water vapor distribution in the upper troposphere of Venus based on automatic stations Venera 4, 5, 6 data.**   B. M. Andrejchikov.

Geokhimiya, 1978, No. 1, p. 11 - 15. In Russian. – Abstr. in Ref. zh., 51. Astron., 6.51.240 (1978).

093.053   **Estimates of wind velocity on Venus.**
        V. G. Vasin, M. Ya. Marov.
Inst. prikl. mat. AN SSSR. Prepr., 1977, No. 92, 14 pp. In Russian. – Abstr. in Ref. zh., 51. Astron., 6.51.241; 62. Issled. kosm. prostranstva, 6.62.205 (1978).

093.054   **Results of a radio transillumination of the neutral atmosphere of Venus and bistatic sounding of its surface with the satellites Venera 9, 10.**   M. A. Kolosov, O. I. Yakovlev, T. S. Timofeeva, E. V. Chub, A. I. Efimov, S. S. Matyugov, A. G. Pavel'ev, A. I. Kucheryavenkov, I. Eh. Kalashnikov.
Usp. fiz. nauk, Vol. 123, 697 - 698 (1977). In Russian. Abstr. in Ref. zh., 62. Issled. kosm. prostranstva, 6.62.204 (1978).

093.055   **Geologic interpretation of new observations of the surface of Venus.**   R. S. Saunders, M. C. Malin.
Geophys. Res. Lett., Vol. 4, 547 - 550 (1977).
    New radar observations of the surface of Venus provide further evidence of a diverse and complex geologic evolution. The radar bright feature "Beta" is interpreted to be a large volcanic construct, analogous to terrestrial and martian shield volcanoes. Two large, quasi-circular areas of low reflectivity, examples of a class of features interpreted to be impact basins are shown in altimetry maps to be depressions.

093.056   **The planet Venus.**   Z. Pokorný.
        Říše hvězd, Vol. 59, 52 - 56 (1978). In Czech.

093.057   **Mass-spectral investigations of the chemical composition of the Venus atmosphere aboard the automatic interplanetary stations Venera 9 and Venera 10.**
Yu. A. Surkov, V. F. Ivanova, A. N. Pudov, B. I. Verkin, N. N. Bagrov, A. P. Pilipenko.
Geokhimiya, 1978, No. 4, p. 506 - 513. In Russian. – Abstr. in Ref. zh., 62. Issled. kosm. prostranstva, 7.62.160 (1978).

Determination of a planet's diameter by occultation time – an introductory astronomy exercise.
See Abstr. 014.026.

Long-lived landers for Venus.
See Abstr. 032.503.

On photographic positions of Venus.
See Abstr. 041.027.

Position observations of Venus at the Main Astronomical Observatory of the Ukrainian Academy of Sciences (1967 - 1971).   See Abstr. 041.029.

Observations of major planets with the meridian circle of the Engelhardt Astronomical Observatory.
See Abstr. 041.038.

Erforschung der Venus – das NASA-Programm Pioneer–Venus.   See Abstr. 053.006.

Pioneer-Venus – die Erkundung des inneren Nachbarplaneten.   See Abstr. 053.007.

Feasibility of study of buoyant Venus station placed by inflated balloon entry.   See Abstr. 053.010.

Asymmetry of the region of interaction between the solar wind and Venus from data of the automatic interplanetary stations Venera 9 and Venera 10.
See Abstr. 074.037.

An atmospheric instability mechanism on a nonrotating planet and its application to Venus.
See Abstr. 091.017.

Electron cooling by excitation of carbon dioxide.
See Abstr. 097.065.

094 Moon: Dynamics, Global Properties, Local Properties

Moon, Dynamics

094.001 The evolution of the Moon and the terrestrial planets. M. N. Toksöz, D. H. Johnston.
Cosmochemistry of the moon and planets, (see 012.002), p. 295 - 327 (1977).

The thermal evolutions of the Moon, Mars, Venus, and Mercury are calculated theoretically starting from cosmochemical condensation models. An assortment of geological, geochemical, and geophysical data are used to constrain both the present-day temperatures and the thermal histories of the planets' interiors. Such data imply that the planets were heated during or shortly after formation and that all the terrestrial planets started their differentiations early in their history.

094.002 Heat flow and thermal history of the Moon. Ye. A. (*E. A.*) Lyubimova.
Cosmochemistry of the moon and planets, (see 012.002), p. 329 - 346 (1977).

An analysis is made of current heat flow data and thermal models of lunar evolution which satisfy the diverse information that has accumulated on internal processes.

094.003 Calculations of the Moon's thermal history at different concentrations of radioactive elements, taking into account differentiation on melting.
O. I. Ornatskaya, Ya. I. Al'ber, I. L. Ryazantseva.
Cosmochemistry of the moon and planets, (see 012.002), p. 347 - 366 (1977).

094.004 Riddles about the origin and thermal history of the moon. B. Yu. Levin, S. V. Mayeva (*Maeva*).
Cosmochemistry of the moon and planets, (see 012.002), p. 367 - 385 (1977).

094.005 Mechanical processes affecting differentiation of protolunar material. W. M. Kaula.
Cosmochemistry of the moon and planets, (see 012.002), p. 805 - 813 (1977) = Univ. California, Los Angeles, Publ. Inst. Geophys. Planet. Phys. No. 1342.

Mechanisms prior to lunar formation are sought to account for the loss of volatiles, the depletion of iron, and the enrichment of plagioclase. Some of the same mechanisms are necessary to account for achondritic, stony-iron, and iron meteorites.

094.006 The origin of the Moon. Ye. L. (*E. L.*) Ruskol.
Cosmochemistry of the moon and planets, (see 012.002), p. 815 - 822 (1977).

This paper discusses fractionation of the chemical compositions of the Moon and the Earth and the thermal history of the Moon for formation of the Moon from an Earth-orbiting swarm of bodies, during the accumulation of the Earth.

094.007 Formation and composition of the Moon. D. L. Anderson.
Cosmochemistry of the moon and planets, (see 012.002), p. 823 - 845 (1977) = Div. Geol. Planet. Sci., Calif. Inst. Technol., Pasadena, Calif., Contrib. No. 2482.

Many of the properties of the Moon, including the "enrichment" in Ca, Al, Ti, U, Th, Ba, Sr, and the REE and the "depletion" in Fe, Rb, K, Na, and other volatiles can be understood if the Moon represents a high-temperature condensate from the solar nebula.

094.008 Secularly unstable Maclaurin spheroids and the

origin of the moon. J. A. O'Keefe, E. C. Sullivan.
Bull. American Astron. Soc., Vol. 9, 546 (1977). – Abstract.

094.009 Nebular condensation models and the place of origin of the Moon. R. J. Malcuit, R. R. Winters.
Bull. American Astron. Soc., Vol. 9, 546 (1977). – Abstract.

094.010 Lunar tidal acceleration determined from laser range measures. O. Calame, J. D. Mulholland.
Science, Vol. 199, 977 - 978 (1978).

Lunar laser range measures covering the period 1969 to 1976 have been used to determine the anomalous secular acceleration in the mean longitude of the moon, commonly attributed to the effect of tidal friction in the earth. The acceleration determined is -24.6 ± 1.6 arc seconds per century squared, against an atomic time scale, where the uncertainty is the formal standard deviation of the solution. The realistic uncertainty is surely larger, as evidenced by the ensemble of solutions performed with various models and observation sets. The determined value is in good agreement with the conventional value and with several recent determinations by other methods.

094.011 Model of the gravitational field of the moon from observations of motion of its artificial satellites
Luna 10, 12, 14, 19 and 22. Eh. L. Akim, Z. P. Vlasova.
Dokl. AN SSSR, Vol. 235, 38 - 41 (1977). In Russian.
Abstr. in Ref. zh., 52. Geod. Aehrosemka, 2.52.84 (1978).

094.012 Origin of Earth's Moon. J. A. Wood.
Planetary satellites, (see 003.007), p. 513 - 529 (1977).

A brief review is given of the three most common hypotheses of the origin of the Moon: fission from the Earth, capture by the Earth and binary accretion. The last, which has the most adherents today, is considered most extensively.

094.013 Relation of a contracting Earth to the apparent accelerations of the Sun and Moon.
R. A. Lyttleton.
Dynamics of planets and satellites, (see 012.020), p. 87 (1978). Abstract.

094.014 Contribution à l'étude des perturbations planétaires de la Lune. N. Abu el Ata.
Dynamics of planets and satellites, (see 012.020), p. 113 - 124 (1978).

The need for accurate ephemerides describing the motion of the Moon warrants a new determination of the inequalities in the Moon's coordinates due to the action of planets (direct and indirect). Here the author exposes the different aspects of the problem and the methods to treat them. The difficulties that arose during the work and the elimination of their effects are explained.

094.015 Hamiltonian theory of the libration of the Moon. J. Henrard, M. Moons.
Dynamics of planets and satellites, (see 012.020), p. 125 - 135 (1978).

The feasibility of applying the Lie transform method to the problem of the physical libration of the Moon is investigated. By a succession of canonical transformations, the Hamiltonian of the problem is brought under a form suitable for perturbation technique. The mean value of the inclination of the angular momentum upon the ecliptic and the frequencies of the free libration are computed.

094.016 Lunar tidal acceleration obtained from satellite-

derived ocean tide parameters.
C. C. Goad, B. C. Douglas.
J. Geophys. Res., Vol. 83, 2306 - 2310 (1978).

Analysis of 100 sets of mean elements of Geos 3 computed at 2-day intervals has yielded observation equations for the M_2 ocean tide from the long periodic variations of the inclination and node of the orbit. The authors obtain for the tidal acceleration in lunar longitude the value $\dot{n} = -27.4 \pm 3$ arc sec/$(100 \text{ yr})^2$ in excellent agreement with the most recent determinations from ancient and modern astronomical data. The mean elements of Geos 3 are also presented in tabular form.

094.017 **On the history of the lunar orbit.**
         V. N. Zharkov, A. P. Trubitsyn.
Izv. AN SSSR. Fiz. Zemli, 1976, No. 8, p. 3 - 6. In Russian.
Abstr. in Ref. zh., 62. Issled. kosm. prostranstva, 4.62.171 (1978).

094.018 **Contribution of tidal dissipation to lunar thermal history.**   S. J. Peale, P. Cassen.
Reports of planetary geology program, 1977–1978, (see 003.014), p. 63 - 65 (1978).

094.019 **A note on the dynamical evolution of the lunar orbit.**   T. C. Van Flandern, R. S. Harrington.
Bull. American Astron. Soc., Vol. 10, 482 (1978). – Abstract.

094.020 **Of the effects of lunar dissipation on the lunar physical librations.**
C. F. Yoder, W. S. Sinclair, J. G. Williams.
Bull. American Astron. Soc., Vol. 10, 482 (1978). – Abstract.

094.021 **Lunar tidal acceleration determined from laser range measures.**   O. Calame, J. D. Mulholland.
Bull. American Astron. Soc., Vol. 10, 482 (1978). – Abstract.

094.022 **Laser ranging to the Moon: eight years of scientific progress.**   E. Faller.
Laser 77 Opto-Electronics. Conference, Munich, FRG, 20 - 24 June 1977. W. Waidelich (Editor). IPC Sci. Technol. Press. 772 pp. (1977). p. 1 - 11. – Abstr. in Phys. Abstr., Vol. 81,

Abstr. 36476 (1978).

094.023 **The surfaces of perigee and apogee of the moon.**
         V. V. Radzievskij.
Problems of observational and theoretical astronomy, Moscow-Leningrad, 1977, (see 003.012), p. 164 - 180. In Russian.
Abstr. in Ref. zh., 51. Astron., 7.51.116 (1978).

Bibliography.   See Abstr. 002.069.

Motion of the moon relative to the mass centre.
See Abstr. 003.123.

Position observations of the moon at the Engelhardt Astronomical Observatory.   See Abstr. 031.274.

Lunar laser ranging system for measuring distances with accuracy to ±20 cm.   See Abstr. 032.020.

On Cassini's laws.   See Abstr. 042.022.

Application of Lie transforms method to Delaunay's problem up to the third order.   See Abstr. 042.096.

Hill's problem in lunar theory.
See Abstr. 042.119.

The effect of harmonics in the gravitational potential of the Earth, Moon and Sun on the precession angle of the Earth's rotation axis.   See Abstr. 044.004.

The semi-diurnal tidal effect of the moon in observations of the latitude of the Engelhardt Astronomical Observatory.   See Abstr. 045.014.

Limits to the expansion of Earth, Moon, Mars and Mercury and to changes in the gravitational constant.
See Abstr. 081.006.

The lunar capture hypothesis revisited.
See Abstr. 107.028.

Moon, Global Properties

094.101 Definition and detailed characterization of lunar surface units using remote observations.
J. W. Head, C. Pieters, T. McCord, J. Adams, S. Zisk.
Icarus, Vol. 33, 145 - 172 (1978).

Remote sensing techniques and data may be subdivided into three principal types according to how they are used: (1) defining techniques help to define unit boundaries and extent; (2) characterizing techniques allow classification and characterization of physical features, lithology, or chemical composition; (3) supporting techniques provide additional useful information but are not fundamental to the definition or characterization of units. Defined units represent a fundamental subdivision of the rocks in a planetary crust and thus represent processes and sequences of events. Detailed consideration of unit definition and characterization is presented using the mare deposits of the Imbrium basin as an example.

094.102 Electromagnetic sounding of the Moon using Apollo 16 and Lunokhod 2 surface magnetometer observations. (Preliminary results).
L. L. Vanyan (*Van'yan*), T. A. Vnutchokova, E. B. Fainberg (*Eh. B. Fajnberg*), E. A. Eroschenko (*Eroshenko*), P. Dyal, C. W. Parkin, W. D. Daily.
Moon, Vol. 17, 259 - 262 (1977).

A new technique of deep electromagnetic sounding of the Moon using simultaneous magnetic field measurements at two lunar surface sites is described. The method, used with the assumption that deep electrical conductivity is a function only of lunar radius, has the advantage of allowing calculation of the external driving field from two surface site measurements only, and therefore does not require data from a lunar orbiting satellite. A transient response calculation is presented for the example of a magnetic field discontinuity of February 13, 1973, measured simultaneously by Apollo 16 and Lunokhod 2 surface magnetometers.

094.103 Events interpreted as volcanic, associated with the impact craters Gagarin and Aitken on the lunar farside. C. S. Beals, R. W. Tanner.
J. R. Astron. Soc. Canada, Vol. 72, 3 - 14 (1978).

Lunar nearside maria show signs of recurrent lava flows over considerable periods of time. An examination of photographs of the two farside craters Gagarin and Aitken suggests that a series of volcanic events occurred there also. Many lunar features are thought to be due to the combined effects of impact and volcanism, and may have terrestrial counterparts.

094.104 The ancient lunar core dynamo. S. K. Runcorn.
Science, Vol. 199, 771 - 773 (1978).

Lunar paleomagnetism provides evidence for the existence of an ancient lunar magnetic field generated in an iron core. Paleointensity experiments give a surface field of 1.3 gauss, 4.0×10^9 years ago, subsequently decreasing exponentially. Thermodynamic arguments give a minimum value of the heat source in the core at that time: known sources, radioactive and other, are quantitatively implausible, and it is suggested that superheavy elements were present in the early moon.

094.105 Lunar crater Giordano Bruno: A.D. 1178 impact observations consistent with laser ranging results.
O. Calame, J. D. Mulholland.
Science, Vol. 199, 875 - 877 (1978).

The hypothesis of Hartung, that the impact formation of lunar crater Giordano Bruno (103° east, 36° north) was observed and recorded 800 years ago, is considered in the context of data from the Luna 24 mission and laser range observations. It is concluded that (1) the event would certainly have been visible, and (2) current determinations of the free libration in

longitude in the moon's rotation are consistent with the hypothesis. Such a study cannot prove Hartung's interpretation, but it is nonetheless supportive of it.

094.106 Geochemical zoning and early differentiation in the moon. S. R. Taylor, P. Jakeš.
Cosmochemistry of the moon and planets, (see 012.002), p. 55 - 61 (1977) = Lunar Sci. Inst., Houston, Tex., Contrib. No. 229.

094.107 Evolution of the Moon: the 1974 model.
H. H. Schmitt.
Cosmochemistry of the moon and planets, (see 012.002), p. 63 - 80 (1977).

The interpretive evolution of the Moon can be divided into seven major stages: (1) The Beginning: 4.6 billion years ago. (2) The Melted Shell: 4.6 to 4.4 billion years ago. (3) The Cratered Highlands: 4.4 to 4.1 billion years ago. (4) The Large Basins: 4.1 to 3.9 billion years ago. (5) The Light-colored Plains: 3.9 to 3.8 billion years ago. (6) The Basaltic Maria: 3.8 to 3.0 (?) billion years ago. (7) The Quiet Crust: 3.0 (?) billion years ago to the present.

094.108 Lunar elemental analysis obtained from the Apollo gamma-ray and X-ray remote sensing experiment.
J. I. Trombka, J. R. Arnold, I. Adler, A. E. Metzger, R. C. Reedy.
Cosmochemistry of the moon and planets, (see 012.002), p. 153 - 182 (1977).

Gamma-ray and X-ray spectrometers carried in the Service Modules of the Apollo 15 and Apollo 16 spacecraft were employed for compositional mapping of the lunar surface. A large-scale compositional map of over 10 percent of the lunar surface was obtained from an analysis of the observed spectra.

094.109 To the problem about the origin of lunar maria and continents (Mössbauer investigations).
T. V. Malysheva.
Cosmochemistry of the moon and planets, (see 012.002), p. 243 - 251 (1977).

A comparative study of Mössbauer spectra of regolith returned by the Luna 16 and Luna 20 spacecraft has shown that the distribution of iron among the mineral phases is quite characteristic of the landing sites.

094.110 The surface abundance and stratigraphy of lunar rocks from data about their albedo.
V. V. Shevchenko.
Cosmochemistry of the moon and planets, (see 012.002), p. 253 - 258 (1977).

094.111 In-situ measurements of lunar heat flow.
M. G. Langseth, S. J. Keihm.
Cosmochemistry of the moon and planets, (see 012.002), p. 283 - 293 (1977) = Lamont-Doherty Geol. Obs., Columbia Univ., Palisades, N.Y., Contrib. No. 2141.

During the Apollo program two successful heat-flow measurements were made in situ on the lunar surface. Comparison with earlier determinations of heat flow, using the microwave emission spectrum from the Moon, gives support to the high gradients and heat flows observed in situ.

094.112 Results from the Apollo passive seismic experiment.
G. Latham, Y. Nakamura, J. Dorman, F. Duennebier, M. Ewing, D. Lammlein.
Cosmochemistry of the moon and planets, (see 012.002), p. 389 - 401 (1977).

094.113 Lunar mascons.
V. N. Zharkov, A. P. Trubitsyn.
Cosmochemistry of the moon and planets, (see 012.002),

p. 403 - 405 (1977).

In 1968, Muller and Sjogren discovered large positive anomalies in the gravitational field of the Moon and introduced the concept of mascons as the sources of these anomalies. In this review, the authors attempt to summarize ideas set forth in the literature and related in some manner to lunar mascons.

094.114   Some geologic observations concerning lunar geophysical models.   J. W. Head.
Cosmochemistry of the moon and planets, (see 012.002), p. 407 - 416 (1977).

Observations and problems discussed here deal with the characterization of the upper 25 km of the lunar crust; the tectonic style of the crust; the formation of mascons within major basins; analysis of lunar magnetic anomalies; and the history of the lunar crust.

094.115   Deep electromagnetic sounding of the moon with Lunokhod 2 data.   L. L. Van'yan, I. V. Yegorov (*Egorov*), E. B. Faynberg (*Eh. B. Fajnberg*).
Cosmochemistry of the moon and planets, (see 012.002), p. 443 - 446 (1977).

094.116   Lunar electrical conductivity, permeability, and temperature from Apollo magnetometer experiments.   P. Dyal, C. W. Parkin, W. D. Daily.
Cosmochemistry of the moon and planets, (see 012.002), p. 447 - 491 (1977).

Magnetometers have been deployed at four Apollo sites on the Moon to measure remanent and induced lunar magnetic fields. Measurements from this network of instruments have been used to calculate the electrical conductivity, temperature, magnetic permeability, and iron abundance of the lunar interior.

094.117   The intensity of the ancient lunar field from magnetic studies on lunar samples.
A. Stephenson, D. W. Collinson, S. K. Runcorn.
Cosmochemistry of the moon and planets, (see 012.002), p. 499 - 503 (1977).

Palaeointensity determinations on Apollo 11, 16, and 17 rocks have indicated that from 3.9 to 4.0 AE ago the strength of the surface lunar magnetic field was about 1.3 Oe, while there is evidence from younger rocks that a field of about one quarter of this value was present at a later time (3.6 AE).

094.118   Pre-mare cratering and early solar system history.   G. W. Wetherill.
Cosmochemistry of the moon and planets, (see 012.002), p. 553 - 567 (1977).

An evaluation of the application of the high extralunar flux in pre-mare times to more general problems of early solar system history is attempted by combining the results of dynamic studies with lunar chronological data.

094.119   The role of exogenic factors in the formation of the lunar surface.   K. P. Florenskiy (*Florenskij*), A. T. Bazilevskiy (*Bazilevskij*), A. V. Ivanov.
Cosmochemistry of the moon and planets, (see 012.002), p. 571 - 584 (1977).

094.120   Meteoritic material on the Moon.
J. W. Morgan, R. Ganapathy, H. Higuchi, E. Anders.
Cosmochemistry of the moon and planets, (see 012.002), p. 659 - 689 (1977).

Three types of meteoritic material are found on the Moon: micrometeorites, ancient planetesimal debris from the "early intense bombardment", and debris of recent, crater-forming projectiles. Their amounts and compositions have been determined from trace element studies.

094.121   Primary cosmic rays on the lunar surface.
S. N. Vernov, A. K. Lavrukhina.
Cosmochemistry of the moon and planets, (see 012.002), p. 691 - 696 (1977).

The present article reports some results from studies of the temporal variations of cosmic rays using data on the radioactivity of cosmogenic Na^{22} and Al^{26} in lunar samples returned by the Luna 16 and Luna 20 automatic spacecraft, as well as data from the instruments of Luna 9.

094.122   Antipodes on the Moon.
Yu. N. Lipskiy (*Lipskij*), Zh. F. Rodionova.
Cosmochemistry of the moon and planets, (see 012.002), p. 755 - 761 (1977).

In the department of physics of the Moon and planets of the Shternberg State Astronomical Institute, cartometric work was done in order to determine the areas of the lunar maria and large craters.

094.123   Data for compiling the third edition of the complete lunar map, scale 1 : 5 000 000.
Yu. N. Lipskij, V. A. Nikonov, Zh. F. Rodionova, V. I. Chikmachev, V. V. Shevchenko, N. B. Lavrova, L. I. Volchkova, Yu. P. Pskovskij.
Soobshch. Gos. Astron. Inst. Shternberga, No. 204, 59 pp. (1977). In Russian.

094.124   Gardening process of the lunar surface region.
J. Iriyama, M. Honda.
Mem. Chubu Inst. Technol., Vol. 12A, 121 - 124 (1976). In Japanese. – Abstr. in Phys. Abstr., Vol. 80, Abstr. 91452 (1977).

094.125   Distribution of radioactive elements in the lunar crust.   E. L. Haines, R. E. Parker, A. E. Metzger.
Bull. American Astron. Soc., Vol. 9, 529 - 530 (1977). Abstract.

094.126   Photometric function of the Moon.   A. F. Cook.
Bull. American Astron. Soc., Vol. 9, 531 (1977). Abstract.

094.127   Lunar radar map at 7.5 m wavelength.
T. W. Thompson.
Bull. American Astron. Soc., Vol. 9, 531 - 532 (1977). Abstract.

094.128   Limitations on the parameters of the solar wind in modelling lunar electromagnetic induction.
B. A. Hobbs, R. L. Parker.
Geophys. J. R. Astron. Soc., Vol. 52, 433 - 439 (1978).

A striking feature of the day-side response of the Moon to periodic fluctuations in the solar wind is the rapid rise, and subsequent fall, in the amplitude of the transfer function as the inducing field frequency increases. Before the response of a conductivity model representing the Moon can be calculated at a given frequency, the parameters (v, θ) (where v is the solar wind speed and θ is the angle between the solar wind velocity and the magnetic field propagation direction) have to be specified. The authors have determined constraints on the parameter space (v, θ). In particular, they determine the region of the (v, θ) space in which conductivity models may be found that satisfy their data pair.

094.129   Modification of fresh crater landforms: evidence from the Moon and Mercury.
M. C. Malin, D. Dzurisin.
J. Geophys. Res., Vol. 83, 233 - 243 (1978).

The morphology of fresh lunar and mercurian craters provides insight into processes of crater formation and modification. Measurements determined for mercurian craters and compared to previously presented lunar data are depth/

diameter, central peak and wall-related mass movement frequencies as functions of diameter, crater rim wall width/rim diameter, rim diameter/floor diameter, and central peak height/rim diameter. Two important results are as follows: (1) there is no evidence for direct gravity scaling of crater morphology, although some slight ($\sim g^{1/4}$ to $g^{1/8}$) scaling relationship may be indicated, and (2) mass movements are responsible for the change in depth/diameter relationship observed near 2-km depth and 10-km diameter. The latter result is helpful in explaining gravitational and topographic data which suggest low-density regions beneath large, fresh craters.

094.130 **Imbrian-age highland volcanism on the moon: the Gruithuisen and Mairan domes.**
J. W. Head III, T. B. McCord.
Science, Vol. 199, 1433 - 1436 (1978).

The Gruithuisen and Mairan domes on the moon represent morphologically and spectrally distinct nonmare extrusive volcanic features of Imbrian age. The composition, morphology, and age relationships of the domes indicate that nonmare extrusive volcanism in the northern Procellarum region of the moon continued until about 3.3×10^9 to 3.6×10^9 years ago and was partially contemporaneous with the emplacement of the main sequence of mare deposits.

094.131 **Tidal stresses in the moon.**
C. H. Cheng, M. N. Toksöz.
J. Geophys. Res., Vol. 83, 845 - 853 (1978).

The tidal stresses in a radially heterogeneous moon are calculated numerically as a function of time and location by using the latest moon model obtained from seismic data. Theoretical results are compared with the observed features of deep moonquakes and in particular with those of the most active A_1 hypocenter.

094.132 **A general cratering-history model and its implications for the lunar highlands.** A. Woronow.
Icarus, Vol. 34, 76 - 88 (1978).

Through analysis of a large number of Monte Carlo and Markov Chain simulations, a model for determining crater accumulation and crater obliteration histories has been derived. The model generally applies to populations of large craters. Application of the model to the lunar highlands is made.

094.133 **The centre-to-limb variation of the Moon's brightness at 2 and 6 cm wavelength.**
W. Hirth, M. Butz, L. Velden, E. Fürst.
Moon, Vol. 17, 395 - 400 (1977).

A map of the Moon at 2 cm wavelength is presented. The angular ($\simeq 1'$arc) and temperature resolution (< 0.1 K) is sufficient to study systematic details of the brightness distribution. In particular, the centre-to-limb variation is considered. An estimate of the dielectric constant ϵ is possible ($1.4 \leqslant \epsilon \leqslant 2.5$). The existence of a temperature gradient in the lunar surface layers is used to derive the depth of penetration of electromagnetic waves (L_e), which is $L_e \simeq 8$ m for 2 cm wavelength. The parameters derived from the 2 cm map are found to be compatible with those obtained from a former observation at 6 cm wavelength.

094.134 **Periodicity in lunar seismic activity and earthquakes.**
D. Sadeh, K. Wood.
J. Geophys. Res., Vol. 83, 1245 - 1249 (1978).

Lunar seismic activity detected by Apollo seismographs clearly shows a monthly periodicity. Using 36 months of data, Lammlein et al. (1974) found that the period is 27.2 days, the lunar nodical period. A search for such periodicity in earthquakes, based on the 8 years covered in the NOAA catalog, shows an analogous effect at half the lunar sidereal period, 13.65 days. The probability that the effect occurs randomly is assessed. The highest earthquake rate occurs at the two times

each month when the moon crosses declination zero.

094.135 **Calculation of the dependence of the stream of gamma-quanta from the moon and Mars on the relief and distance to the surface.** Yu. A. Surkov, L. P. Moskaleva, O. S. Manvelyan.
Kosm. Issled., Vol. 16, 301 - 306 (1978). In Russian.

094.136 **Zum gegenwärtigen Stand der Erforschung von Struktur und Ursachen des magnetischen Feldes des Mondes.** W. Mundt, K. Rother.
Sterne, 54. Band, 11 - 24 (1978).

094.137 **On a comparison of anomalous magnetic fields of the moon and earth.** V. N. Lugovenko, A. G. Popov, A. L. Kharitonov.
Issled. prostranstv. –vremennoj strukt. geomagn. polya. Moskva, Nauka, 1977, p. 184 - 189. In Russian. – Abstr. in Ref. zh., 51. Astron., 3.51.312; 62. Issled. kosm. prostranstva, 3.62.168 (1978).

094.138 **On the propagation of radio waves along the lunar surface.** V. A. Bader, O. I. Yakovlev, E. P. Novichikhin.
Radiotekh. i ehlektron., Vol. 23, 2091 - 2096 (1977). In Russian. – Abstr. in Ref. zh., 51. Astron., 3.51.318 (1978).

094.139 **Neotectonics of the moon.** I. N. Galkin.
Priroda, 1978, No. 5, p. 131 - 132. In Russian.

094.140 **Lunar atmosphere past and present.**
D. W. Hughes.
Nature, Vol. 273, 489 - 490 (1978).

094.141 **On recent lunar atmosphere.** Yu. B. Chernyak.
Nature, Vol. 273, 497 - 501 (1978).

Two independent kinds of lunar expedition data are considered which put forward some strong theoretical evidence of the existence of a considerable atmosphere on the Moon during the greater part of its history. The transformation of a meteoritic flux when passing through an arbitrary atmosphere has been analysed and tentative estimates of the lunar atmospheric power presented.

094.142 **Formation of lunar basin rings.**
C. A. Hodges, D. E. Wilhelms.
Icarus, Vol. 34, 294 - 323 (1978).

The origin of the multiple concentric rings that characterize lunar impact basins, and the probable depth and diameter of the transient crater have been widely debated. As an alternative to prevailing "megaterrace" hypotheses, the authors propose that the outer scarps or mountain rings that delineate the topographic rims of basins define the transient cavities, enlarged relatively little by slumping, and thus are analogous to the rim crests of craters like Copernicus; inner rings are uplifted rims of craters nested within the transient cavity. The magnitude of slumping that occurs on all scarps is insufficient to produce major inner rings from the outer. Terrestrial analogs suggest two possible mechanisms for producing rings.

094.143 **On the validity of the representation of the moon's figure by spherical and sampling-function expansions.**
N. A. Chujkova.
Astron. Zh. Akad. Nauk SSSR, Vol. 55, 617 - 627 (1978). In Russian. English translation in Soviet Astron., Vol. 22, No. 3.

An analysis of all known lunar topographic altitudes has been carried out. Estimates of the global properties of the moon's figure are derived.

094.144 **The moon in Heiligenschein.** R. L. Wildey.
Science, Vol. 200, 1265 - 1267 (1978).

An analysis of 25 photometric digital images of the moon has been carried out to obtain a single image in a new mapping parameter, the Heiligenschein exponent. The data necessarily represent a range of lunar phases, but all are within 10 hours of full moon. The new parameter characterizes the rate at which lunar features brighten as their local phase angles approach zero. Although considerable contrast is present in this parameter, there is only a small correlation with normal albedo.

094.145 **Crater shape-size profiles for fresh craters on the Moon and Mercury.**   E. I. Smith, J. A. Hartnell.
Reports of planetary geology program, 1977–1978, (see 003.014), p. 147 - 149 (1978).

094.146 **The position of the moon's mass center in the plane of the sky.**   I. G. Chugunov.
Astron. Tsirk., No. 973, p. 2 - 3 (1977). In Russian.

094.147 **The figure of the marginal zone of the moon.**   I. G. Chugunov.
Astron. Tsirk., No. 973, p. 3 - 4 (1977). In Russian.

094.148 **The lunar ellipsoid.**
A. S. Dubrovskij, Yu. A. Chikanov.
Astron. Tsirk., No. 985, p. 7 - 8 (1978). In Russian.

094.149 **The systems of heights on the maps of the marginal zone of the moon.**   I. G. Chugunov.
Astron. Tsirk., No. 992, p. 7 - 8 (1978). In Russian.

094.150 **Fourier analysis of planimetric lunar crater shape — possible guide to impact history and lunar geology.**
D. T. Eppler, D. Nummedal, R. Ehrlich.
Impact and explosion cratering, (see 012.034), p. 511 - 526 (1977).

094.151 **Equations of state and impact-induced shock-wave attenuation on the moon.**
T. J. Ahrens, J. D. O'Keefe.
Impact and explosion cratering, (see 012.034), p. 639 - 656 (1977) = Div. Geol. Planet. Sci., California Inst. Technol., Pasadena, Calif., Contrib. No. 2844.
The major new result described here is that different impact velocities and meteorite shock impedances give rise to significantly different spatial peak shock attenuation rates in the far-field.

094.152 **Crater modification by gravity: a mechanical analysis of slumping.**   H. J. Melosh.
Impact and explosion cratering, (see 012.034), p. 1245 - 1260 (1977).

094.153 **The distortion of the moon due to convection.**
P. Cassen, R. E. Young, G. Schubert.
Geophys. Res. Lett., Vol. 5, 294 - 296 (1978).
Numerical calculations of the dynamical ellipticity of the moon due to finite amplitude solid state convection indicate that convection could be the cause of the non-hydrostatic gravitational figure, but only if the lunar lithosphere is capable of resisting global scale deformation. Thus lithospheric inhomogeneities and surface loads could also contribute substantially to the disequilibrium of the gravity figure. The calculations also show that it is unlikely that the geometrical distortion is due to convection.

094.154 **Is the moon really as smooth as a billiard ball? Remarks concerning recent models of sputter-fractionation on the lunar surface.**   B. Hapke, W. Cassidy.
Geophys. Res. Lett., Vol. 5, 297 - 300 (1978).
Two recent discussions of chemical and isotopic fractionation by solar wind sputtering (Pillinger et al., 1976; Switkowski et al., 1977) are not applicable to the moon because the authors have failed to include major effects caused by the real geometry of the lunar surface.

094.155 **Comments on: "Is the moon really as smooth as a billiard ball? ...."**.   C. T. Pillinger, A. J. T. Jull.
Geophys. Res. Lett., Vol. 5, 301 - 303 (1978).
Hapke and Cassidy (1978), in describing their ideas concerning sputter deposition, have criticised a number of aspects of a model for chemical fractionation and metallic iron production which is based on preferential sputtering. Questions referring to the fate of sputtered atoms, the surface roughness of the moon and the existence of metal rich layers may all be answered. Doubts may be expressed concerning the validity of some arguments used by Hapke and Cassidy since the regolith in their model is predominantly static with only episodic turnover.

094.156 **Evidence for a non-random magnetization of the moon.**   L. Hood, C. T. Russell, P. J. Coleman, Jr.
Geophys. Res. Lett., Vol. 5, 305 - 308 (1978).
An approximate inversion, using a specific model, of Apollo subsatellite vector magnetometer data has provided estimates for the locations, scale sizes, and magnetization characteristics of strongly magnetized lunar crustal materials across a portion of the lunar far side.

094.157 **On the possible existence of superheavy elements in the primeval moon.**   S. K. Runcorn.
Earth Planet. Sci. Lett., Vol. 39, 193 - 198 (1978).
The remarkably complete and well-dated record of the primeval history of the moon poses two problems concerned with early heat sources. The early melting and differentiation, especially the formation of a lunar core, requires a source with a half-life of a few times 10^8 years. Convection and magnetic field generation in the iron core requires a heat source soluble in iron with a half-life of the order of 10^8 years. It is shown that both requirements are quantitatively met by supposing that superheavy elements existed in the early moon and the relationship of this result with the search for their existence today is discussed.

094.158 **A lunar terminator configuration.**   O. E. Berg.
Earth Planet. Sci. Lett., Vol. 39, 377 - 381 (1978).
Although introduced more than two decades ago, electrostatic soil transport on the lunar surface remains a controversial subject due, primarily, to insufficiently high surface fields and particle charges to initiate and sustain the phenomenon. This paper introduces a realistic geometrical configuration for the terminators which plausibly provides adequate fields and charges. Myriads of miniature sunlit islands and dark valleys are formed in the vicinity of a zig-zag terminator line as it moves. The islands and valleys are surrounded by areas of opposite electrical polarity, providing local field regions up to thousands of volts per centimeter. The result is random directional movement of electrostatically charged lunar fines. It is shown that the mechanism can readily account for the denuding of certain asteroids while others exhibit a layer of dust on their surface.

094.159 **Problems of geology of the moon, Mercury and Mars.**   Yu. A. Khodak.
Problems of astronomy and geodesy, VAGO assembly Yerevan, 1975, (see 012.039), p. 77 - 87. In Russian. — Abstr. in Ref. zh., 51. Astron., 7.51.211 (1978).

094.160 **Investigation of the thermal history of the moon with the most probable concentrations of radioactive elements.**   O. I. Ornatskaya, N. M. Tsejtlin, Ya. I. Al'ber, I. P. Ryazantseva.
Nauchn.-issled. radiofiz. inst. Gor'kij, Prepr., 1977, No. 104, 26 pp. In Russian. — Abstr. in Ref. zh., 62. Issled. kosm. prostranstva, 7.62.132 (1978).

Catalogue of Martian craters and statistics of the craters of Mars, Mercury and the Moon. See Abstr. 002.043.

Catalogues of craters of Mercury and the Moon. See Abstr. 002.044.

Bibliography. See Abstr. 002.069.

Lunar impact: a history of project Ranger. See Abstr. 003.057.

Moon, Mars and meteorites. See Abstr. 003.154.

Borrowed perceptions: Harriot's maps of the Moon. See Abstr. 004.026.

Quantitative plagioclase/pyroxene abundance from reflectance spectra: a calibration. See Abstr. 022.024.

Laboratory simulation of the lunar magnetosphere. See Abstr. 022.118.

Bistatic radar measurements of the moon with application of a modulated signal. See Abstr. 031.269.

Main stages of development of lunar investigations in the United States by automatic interplanetary stations. See Abstr. 051.036.

Lunar polar orbiter: a global survey of the Moon. See Abstr. 053.011.

Simulation of the solar wind interaction with non-magnetic celestial bodies. See Abstr. 074.027.

Impact melts on Mercury and the Moon. See Abstr. 092.003.

Heat flow and thermal history of the Moon. See Abstr. 094.002.

Formation and composition of the Moon. See Abstr. 094.007.

Geometry of lunar grabens: implications for shallow crustal structure. See Abstr. 094.452.

Moon, Local Properties

094.401 The thermal and deformational history of Apollo 15418, a partly shock-melted lunar breccia.
G. L. Nord, Jr., J. M. Christie, J. S. Lally, A. H. Heuer.
Moon, Vol. 17, 217 - 231 (1977).
A thermal and mechanical history of lunar gabbroic anorthosite 15418 (1140g) has been deduced from petrographic examination of both exterior and interior thin sections and electron microprobe analysis and transmission electron microscopy of interior thin sections. The authors suggest that the rock underwent two major shock events – an early brecciation and annealing that produced a recrystallized breccia, followed by a second shock event that melted the surface of the rock, vitrified the interior plagioclase and heavily deformed the mafic phases.

094.402 Basalts from Mare Crisium.
G. Ryder, H. Y. McSween, Jr., U. B. Marvin.
Moon, Vol. 17, 263 - 287 (1977).
The stratified core sample returned from Mare Crisium by the Luna 24 unmanned space probe is composed primarily of a new variety of subophitic to ophitic basalt with very low contents of TiO_2 and MgO. Granular metabasalts have the same bulk composition, but mineral phases exhibit less compositional variation. Fine-grained impact melts have similar compositions and are apparently derived from these basalts. The authors conclude that the basalts, which are chemically distinct from the very-low-titanium basalts found elsewhere on the Moon, represent the local surface flows of Mare Crisium.

094.403 Tectonic pattern of the Crüger/Rima Sirsalis – Rima de Gasparis region of the Moon. J. Raitala.
Moon, Vol. 17, 289 - 308 (1977).
Tectonic lunar units were studied in an area of about 540000 km² in the southwestern part of the Moon's visible disk. The area is situated in the vicinity of Mare Humorum, Oceanus Procellarum, and Mare Orientale.

094.404 Differentiation of the matter of the Moon.
A. P. Vinogradov.
Cosmochemistry of the moon and planets, (see 012.002), p. 5 - 33 (1977).
There is a great general similarity between the surface rocks of the Moon and the basic rocks of the Earth's crust, such as tholeiitic basalts, although certain differences have been found between lunar and terrestrial rocks. It is the differences between lunar and terrestrial rocks to which the current report essentially addresses itself.

094.405 A survey of lunar rock types and comparison of the crusts of earth and moon. J. A. Wood.
Cosmochemistry of the moon and planets, (see 012.002), p. 35 - 53 (1977).
The principal known types of lunar rocks are briefly reviewed, and their chemical relationships discussed.

094.406 Petrogenesis of lunar rocks: Rb-Sr constraints and lack of H_2O. A. L. Albee, A. J. Gancarz.
Cosmochemistry of the moon and planets, (see 012.002), p. 81 - 90 (1977) = Div. Geol. Planet. Sci., California Inst. Technol., Pasadena, Calif., Contrib. No. 2474.

094.407 Lunar highland rock types: their implications for impact-induced fractionation.
W. C. Phinney, J. L. Warner, C. H. Simonds.
Cosmochemistry of the moon and planets, (see 012.002), p. 91 - 126 (1977) = Lunar Sci. Inst., Houston, Tex., Contrib. No. 230.

094.408 Lunar igneous rocks and the nature of the lunar interior. J. F. Hays, D. Walker.
Cosmochemistry of the moon and planets, (see 012.002), p. 127 - 135 (1977).

094.409 A chemical model for lunar non-mare rocks.
N. J. Hubbard, J. M. Rhodes.
Cosmochemistry of the moon and planets, (see 012.002), p. 137 - 151 (1977).

094.410 Comparison of lunar rocks and meteorites: implications to histories of the Moon and parent meteorite bodies.   M. Prinz, R. V. Fodor, K. Keil.
Cosmochemistry of the moon and planets, (see 012.002), p. 183 - 199 (1977).

094.411 Radioactivity of the Moon, planets, and meteorites.   Yu. A. Surkov, G. A. Fedoseyev (Fedoseev).
Cosmochemistry of the moon and planets, (see 012.002), p. 201 - 218 (1977).
This paper reviews and summarizes analytical data for the content of natural radioactive elements in meteorites, eruptive terrestrial rocks, and also in lunar samples returned by Apollo missions and the Luna series of automatic stations.

094.412 New data for the Luna 20 core and a survey of published chemical data.   N. J. Hubbard, A. P. Vinogradov, G. I. Ramendik, M. S. Chupakhin.
Cosmochemistry of the moon and planets, (see 012.002), p. 259 - 262 (1977).

094.413 Chemical composition of crystalline rock fragments from Luna 16 and Luna 20 fines.
A. Cimbálníková, M. Palivcová, J. Frána, A. Maštalka.
Cosmochemistry of the moon and planets, (see 012.002), p. 263 - 275 (1977).

094.414 Investigation of the composition of the Luna 16 lunar sample.   L. Bakos, M. Chayka (Chajka), L. Cher, A. Cheke, N. N. Dogadkin, A. Elek, K. Kulchar, A. Nagy, D. L. Nagy, E. Szabo, B. Forzats, E. Zemplen.
Cosmochemistry of the moon and planets, (see 012.002), p. 277 - 280 (1977).

094.415 Magnetic and dielectric properties of lunar samples.   D. W. Strangway, G. W. Pearce, G. R. Olhoeft.
Cosmochemistry of the moon and planets, (see 012.002), p. 417 - 431 (1977).

094.416 Magnetic field in Le Monnier Bay according to data of Lunokhod 2.   Sh. Sh. Dolginov, Ye. G. Yeroshenko (E. G. Eroshenko), L. N. Zhuzgov, V. A. Sharova, G. A. Vnuchkov, B. A. Okulessky (Okulesskij), A. T. Bazilevsky (Bazilevskij).
Cosmochemistry of the moon and planets, (see 012.002), p. 433 - 441 (1977).
The results of the first traverse measurement of the magnetic field on the surface of the Moon are analyzed. The mean value of the magnetic field in the portion of Le Monnier Bay investigated is estimated at 20–30 gammas. An anomaly of the field (10–15 gammas) was disclosed, which is confined to craters that exceed 50 meters in size.

094.417 Early history of the moon: implications of U-Th-Pb and Rb-Sr systematics.
M. Tatsumoto, P. D. Nunes, D. M. Unruh.
Cosmochemistry of the moon and planets, (see 012.002), p. 507 - 523 (1977).

094.418 ^{39}Ar-^{40}Ar dating of basalts and rock breccias from Apollo 17 and the Malvern achondrite.
T. Kirsten, P. Horn.
Cosmochemistry of the moon and planets, (see 012.002), p. 525 - 540 (1977).

094.419 The exposure history of the Apollo 16 site: an assessment based on methane and hydrolysable carbon.   C. T. Pillinger, G. Eglinton, A. P. Gowar, A. J. T. Jull, J. R. Maxwell.
Cosmochemistry of the moon and planets, (see 012.002), p. 541 - 551 (1977).

094.420 Microcraters on lunar samples.
H. Fechtig, W. Gentner, J. B. Hartung, K. Nagel, G. Neukum, E. Schneider, D. Storzer.
Cosmochemistry of the moon and planets, (see 012.002), p. 585 - 603 (1977).

094.421 The micrometeoroid complex and evolution of the lunar regolith.
F. Horz, D. A. Morrison, D. E. Gault, V. R. Oberbeck, W. L. Quaide, J. F. Vedder, D. E. Brownlee, J. B. Hartung.
Cosmochemistry of the moon and planets, (see 012.002), p. 605 - 635 (1977).

094.422 Lunar highlands breccias generated by major impacts.   O. B. James.
Cosmochemistry of the moon and planets, (see 012.002), p. 637 - 658 (1977).

094.423 Results of special mechanical analyses of Luna 16 material.   H. Stiller, H. Vollstädt, R. Wasch, P. Bankwitz, E. Bankwitz, F. C. Wagner, J. Schön.
Cosmochemistry of the moon and planets, (see 012.002), p. 697 - 702 (1977) = Akad. Wiss. DDR, Zentralinst. Phys. Erde, Potsdam, Rep. No. 409.

094.424 The analysis of various size, visually selected and density and magnetically separated fractions of Luna-16 and -20 samples.
G. Eglinton, A. P. Gowar, A. J. T. Jull, C. T. Pillinger, S. O. Agrell, J. E. Agrell, J. V. P. Long, S. H. U. Bowie, P. R. Simpson, R. D. Beckinsale, J. J. Durham, G. Turner, P. H. Cadogan, T. C. Gibb, R. Greatrex, N. N. Greenwood, D. W. Collinson, S. K. Runcorn, A. Stephenson, S. A. Durrani, J. H. Fremlin, F. S. W. Hwang, H. A. Kahn.
Cosmochemistry of the moon and planets, (see 012.002), p. 703 - 727 (1977).

094.425 Mössbauer spectroscopy of iron in the Luna 20 regolith.   T. Zemčík, K. Raclavský.
Cosmochemistry of the moon and planets, (see 012.002), p. 729 - 734 (1977).

094.426 The main peculiarities of the processes of the deformation and destruction of lunar soil.
A. K. Leonovich, V. V. Gromov, A. D. Dmitriyev (Dmitriev), V. N. Penetrigov, P. S. Semenov, V. V. Shvarev.
Cosmochemistry of the moon and planets, (see 012.002), p. 735 - 743 (1977).

094.427 The radiation history of material returned by the Soviet automatic stations Luna 16 and Luna 20, according to track studies.   L. L. Kashkarov, L. I. Genayeva (Genaeva), A. K. Lavrukhina.
Cosmochemistry of the moon and planets, (see 012.002), p. 745 - 754 (1977).

094.428 Measurement of the optical properties of lunar rocks in the transition zone, resulting from observations made by Lunokhod 2.
Yu. N. Lipskiy (Lipskij), V. V. Shevchenko.
Cosmochemistry of the moon and planets, (see 012.002), p. 763 - 767 (1977).

094.429 Unsampled Flamsteed basalts: soil spectra (0.6 to 2.5 µm).
C. Pieters, J. Adams, J. Head, T. McCord, S. Zisk.
Bull. American Astron. Soc., Vol. 9, 530 (1977). – Abstract.

094.430 Chemical composition of regolith from Mare Crisium.   Yu. A. Surkov, G. M. Kolesov, I. N. Ivanov, A. P. Shpanov.
Kosm. Issled., Vol. 16, 107 - 112 (1978). In Russian.

094.431 The cumulative flux of interplanetary particles related to production and equilibrium distributions of lunar craters. D. G. Ashworth.
Space Research, Vol. XVII, (see 012.010), 605 - 610 (1977).

Current lunar crater production and equilibrium distribution curves are reviewed. Equilibrium distribution curves, lunar crater erosion rates and rock lifetimes on the lunar surface are derived from the production curves by using proven analytical techniques. It is shown that the equilibrium curves are in good agreement with equilibrium distributions measured on lunar rocks and with other statistically derived distributions. The predicted erosion rates on the lunar surface are also in very good agreement with erosion measurements made on typical lunar rocks. It is shown that ambiguity still exists between the curves derived from lunar rock crater counts and those derived from satellite measurements.

094.432 Lunar surface microscale transportation phenomena: I. J. A. M. McDonnell, R. P. Flavill.
Space Research, Vol. XVII, (see 012.010), 611 - 616 (1977).

Experimental measurements of the hypervelocity impacts of microparticles on lunar rock have established the comminution distribution of spallation type debris generated on lunar surface rocks; most of this matter is expelled at low velocities in the region of <0.5 km s^{-1}. Using this size distribution the debris size-frequency spectrum has been evaluated from observed lunar rock crater statistics to derive the expected debris flux.

094.433 Lunar surface microscale transportation phenomena: II.
R. P. Flavill, W. C. Carey, J. A. M. McDonnell.
Space Research, Vol. XVII, (see 012.010), 617 - 622 (1977).

Results of scanning electron microscope studies of submicrometre debris grains on lunar samples 15205, 52, 2 and 60015, 6 are presented. From particle orientation and size-frequency distribution, a sputter rate of 0.02 Å/yr is measured for debris grains on 15205. The solar wind erosion rate for 60015 is shown to be much lower.

094.434 Groups of micrometeoroids in the Earth–Moon system. J. S. Dohnanyi.
Space Research, Vol. XVII, (see 012.010), 623 - 625 (1977).

It is shown that the groups of micrometeoroids observed aboard HEOS 2 could well have been produced by the impact of a meteorite (of $\gtrsim 1$ kg) on the lunar surface.

094.435 Electrostatic dust transport and Apollo 17 LEAM experiment.
J. W. Rhee, O. E. Berg, H. Wolf.
Space Research, Vol. XVII, (see 012.010), 627 - 629 (1977).

The Lunar Ejecta and Meteorite (LEAM) experiment has been in operation since December 1973 when it was deployed in the Taurus-Littrow region of the moon by the Apollo 17 crew. A specialized analysis based on more than twenty-two lunations of the impact data shows that all of the events recorded by the sensors during the terminator passages are essentially lunar surface microparticles carrying a high electrostatic charge.

094.436 Light element geochemistry of the Apollo 12 site.
J. F. Kerridge, I. R. Kaplan, C. C. Kung, D. A. Winter, D. L. Friedman, D. J. DesMarais.
Geochim. Cosmochim. Acta, Vol. 42, 391 - 402 (1978) = Publ. No. 1740, Inst. Geophys. Planet. Phys., Univ. Calif., Los Angeles.

Analytical techniques of improved sensitivity have revealed details of the concentrations and isotopic compositions of light elements for a comprehensive suite of samples from the Apollo 12 regolith. These samples show a wide spread in maturity, although maximum contents observed for solar wind elements are less than observed at other sites, possibly reflecting relative recency of craters at the Apollo 12 site. Isotopic composition of nitrogen is consistent with the idea that $^{15}N/^{14}N$ in the solar wind has increased with time, at least a major part of this increase having occurred in the past 3.1 Gyr. Sulfur isotope systematics support a model in which sulfur is both added to the regolith, by meteoritic influx, and lost, by an isotopically selective process. Most soils from this site are heavily contaminated with terrestrial carbon.

094.437 Colorimetric scheme of the lunar disk in the IR-region of the spectrum.
Vu Thy Jen, N. N. Evsyukov, D. I. Shestopalov.
Astron. Zh. Akad. Nauk SSSR, Vol. 55, 434 - 435 (1978). In Russian. English translation in Soviet Astron., Vol. 22, No. 2.

094.438 The nature and possible origin of lava-like material within Cheniér Crater. C. J. Villella.
Moon, Vol. 17, 343 - 352 (1977).

This paper considers the origin of certain tongues of lava-like material in Cheniér Crater, a meteorite crater located about 63 km northeast of the major crater, Tsiolkovsky, on the lunar far side. The author contends that the tongues originated from subsurface movement of magma generated as a result of the meteorite impact which created Tsiolkovsky Crater.

094.439 Different ages of lunar light plains.
G. Neukum.
Moon, Vol. 17, 383 - 393 (1977).

The crater populations of 18 lunar light plains (Cayley plains) show a variation in relative ages by a factor of about 4 in crater frequency of regions in the surroundings of the Orientale resp. Imbrium basin, and by a factor of greater than 25 for more distant sites. Thus the idea of a Moon-wide synchronism in the emplacement of the lunar light plains with the formation of the basins Imbrium or Orientale cannot be supported.

094.440 X-ray diffraction studies of shocked lunar analogs. R. E. Hanss, B. R. Montague, C. P. Galindo.
Proceedings of the 25th Annual Conference on Applications of X-ray Analysis, Denver, Colo., USA, 4 - 6 August 1976. Advances in X-ray Analysis, Vol. 20. Plenum, New York (1977). p. 337 - 344. – Abstr. in Phys. Abstr., Vol. 81, Abstr. 7002 (1978).

094.441 Instrumental neutron activation analysis of lunar samples and the identification of primary matter in the lunar highlands. H. Wänke, H. Kruse, H. Palme, B. Spettel.
J. Radioanal. Chem., Vol. 38, 363 - 378 (1977). – Abstr. in Phys. Abstr., Vol. 81, Abstr. 7890 (1978).

094.442 Analysis of the results of a measurement of electric parameters of lunar soil and its analogues.
A. R. Golovkin.
Opt.–fiz. izmer. Moskva, Izd. standartov, 1977, p. 141 - 144. In Russian. – Abstr. in Ref., zh. 51. Astron., 3.51.314; 62. Issled. kosm. prostranstva, 3.62.163 (1978).

094.443 Investigation of electric parameters of lunar soil and its analogues at different temperatures.
A. R. Golovkin.
Opt.–fiz. izmer. Moskva, Izd. standartov, 1977, p. 145 - 149. In Russian. – Abstr. in Ref. zh., 51. Astron., 3.51.315; 62. Issled. kosm. prostranstva, 3.62.164 (1978).

094.444 Accuracy and precision of a mass-spectrometric analysis of geological and lunar samples.
G. I. Ramendik.
Zh. analit. khim., Vol. 32, 1990 - 1998 (1977). In Russian.

Abstr. in Ref. zh., 51. Astron., 3.51.316 (1978).

094.445 **Bistatic lunar radio location with a modulated signal.**   A. L. Zajtsev, V. I. Kaevitser, A. I. Kucheryavenkov, S. S. Matyugov, A. G. Pavel 'ev, G. M. Petrov, O. I. Yakovlev.
Radiotekh. i ehlektron., Vol. 23, 2097 - 2104 (1977). In Russian. — Abstr. in Ref. zh., 51. Astron., 3.51.319; 62. Issled. kosm. prostranstva, 3.62.170 (1978).

094.446 **Determination of some zonal statistical parameters from colour space photographs of the moon.**
Yu. B. Petrov, A. E. Altynov.
Izv. vyssh. uchebn. zaved. Geod. i aehrofotosemka, 1977, No. 3, p. 79 - 83. In Russian. — Abstr. in Ref. zh., 52. Geod. Aehrosemka, 3.52.206 (1978).

094.447 **Electromagnetic sounding of the moon with the help of synchronous observations of magnetic variations aboard Apollo 16 and Lunokhod 2 (preliminary results).**
L. L. Van'yan, T. A. Vnuchkova, Eh. B. Fajnberg, E. G. Eroshenko, P. Dyal, C. W. Parkin, W. D. Daily.
Issled. prostranstv.-vremennoj strukt. geomagn. polya. Moskva, Nauka, 1977, p. 174 - 177. In Russian. — Abstr. in Ref. zh., 62. Issled. kosm. prostranstva, 3.62.166 (1978).

094.448 **Smallest iron particles in lunar regolith.**
V. S. Urusov.
Priroda, 1978, No. 5, p. 95 - 99. In Russian.

094.449 **Distribution of TiO$_2$ over the lunar surface (from an analysis of the photometric catalogue of J. S. Mikhail).**   L. R. Lisina.
Astrometriya i Astrofizika, Kiev, Vyp. (No.)34, p. 44 - 51 (1978). In Russian.
The slope of the spectral reflectivity curve between 0.402 and 0.564 μ was obtained for 90 regions on the lunar surface as a result of analyzing Mikhail's photometric catalogue. Using an empirical relation between the slope of the reflectance curve and the TiO$_2$ content, the author has obtained this content for all the regions. For mare basalt the value of TiO$_2$ content varies from 1.3 to 9%, for non-mare basalt it does not exceed 2%, for mare crater it is 1.3 to 3%.

094.450 **Zr and Nb partition coefficients: implications for the genesis of mare basalts, KREEP, and sea floor basalts.**   I. S. McCallum, M. P. Charette.
Geochim. Cosmochim. Acta, Vol. 42, 859 - 869 (1978).

094.451 **Electronic microanalysis of lunar iron.**
R. I. Mints, V. I. Grokhovskij, T. M. Petukhova, V. K. Rudenko.
Izv. vuz. Chernaya metall., 1977, No. 12, p. 123 - 126. In Russian. — Abstr. in Ref. zh., 62. Issled. kosm. prostranstva, 5.62.238 (1978).

094.452 **Geometry of lunar grabens: implications for shallow crustal structure.**   M. P. Golombek.
Reports of planetary geology program, 1977—1978, (see 003.014), p. 103 - 105 (1978).

094.453 **The significance of buried craters associated with basins on the Moon and Mars.**   J. S. King, D. H. Scott.
Reports of planetary geology program, 1977—1978, (see 003.014), p. 153 - 156 (1978).

094.454 **Enhanced brightness of crater Aristarchus on 3 March 1977.**
A. G. Kolonin, K. S. Khachaturyan, V. I. Kirichenko.
Astron. Tsirk., No. 983, p. 8 (1978). In Russian.

094.455 **A list of spatial coordinates of points on the lunar far side in the region of Mare Orientale.**
V. I. Chikmachev.
Astron. Tsirk., No. 986, p. 1 - 3 (1978). In Russian.

094.456 **Size-dependence in the shape of fresh impact craters on the moon.**   R. J. Pike.
Impact and explosion cratering, (see 012.034), p. 489 - 509 (1977).

094.457 **A stratigraphic model for Bessel Crater and southern Mare Serenitatis.**   R. A. Young.
Impact and explosion cratering, (see 012.034), p. 527 - 538 (1977).

094.458 **Nested-crater model of lunar ringed basins.**
D. E. Wilhelms, C. A. Hodges, R. J. Pike.
Impact and explosion cratering, (see 012.034), p. 539 - 562 (1977).
The authors propose a model for the origin of impact-basin rings whereby the main topographic rim of a basin approximates the limit of excavation and inner rings approximate the rims of craters formed inside the transient crater by some perturbation in the cratering process.

094.459 **Origin of outer rings in lunar multi-ringed basins: evidence from morphology and ring spacing.**
J. W. Head.
Impact and explosion cratering, (see 012.034), p. 563 - 573 (1977).

094.460 **Impact melt on lunar crater rims.**
B. R. Hawke, J. W. Head.
Impact and explosion cratering, (see 012.034), p. 815 - 841 (1977).
Deposits of lava-like material around relatively fresh lunar craters are interpreted as impact melt on the basis of deposit distribution, lack of volcanic sources, morphology of the material, and time of emplacement. Evidence is presented indicating that the melt which is concentrated on the rim was emplaced during and slightly after the modification stage of the impact cratering event. A model is suggested for the emplacement of large amounts of impact melt on crater rims.

094.461 **A model for wind-extension of the Copernicus ejecta blanket.**
D. E. Rehfuss, D. Michael, J. C. Anselmo, N. K. Kincheloe.
Impact and explosion cratering, (see 012.034), p. 1123 - 1132 (1977).

094.462 **Water in regolith from Mare Crisium (Luna 24)?**
M. V. Akhmanova, B. V. Dement'ev, M. N. Markov.
Geokhimiya, 1978, No. 2, p. 285 - 288. In Russian. — Abstr. in Ref. zh., 51. Astron., 6.51.342 (1978).

094.463 **Metallic phases in the Luna 24 soil samples.**
J. J. Friel, J. I. Goldstein.
Geophys. Res. Lett., Vol. 4, 481 - 483 (1977).
The metal and sulfide phases in the Luna 24 soil samples were studied with the optical microscope and the electron microprobe. The compositions of the metal particles fall into three groups based on their Ni and Co contents: (1) Samples of meteoritic composition. (2) Samples of submeteoritic, low Ni and low Co contents. (3) Samples of high Co content.

094.464 **High-silica glass inclusions in olivine of Luna-24 samples.**   E. Roedder, P. W. Weiblen.
Geophys. Res. Lett., Vol. 4, 485 - 488 (1977).
Optical examination of nine polished grain mounts of Luna-24 drillcore material (0.09—0.50 mm size) revealed melt inclusions in olivine crystals.

094.465 **Luna 24: opaque mineral chemistry of gabbroic and basaltic fragments from Mare Crisium.**
S. E. Haggerty.
Geophys. Res. Lett., Vol. 4, 489 - 492 (1977).

This report summarizes the results to date on spinels and ilmenites in fragments from the gabbroic horizon at 170 cm, and of the opaque minerals in a suite of basalts from between the 77 cm and 174 cm levels.

094.466 **Gabbros from Mare Crisium: an analysis of the Luna 24 soil.** A. E. Bence, T. L. Grove, T. Scambos.
Geophys. Res. Lett., Vol. 4, 493 - 496 (1977).

Monomineralic and lithic fragments from the Luna 24 drill core are samples of a mare ferrogabbro and a rarer gabbro of possible highlands origin. Mineral phase chemistries of the mare ferrogabbro suggest the crystallization near the lunar surface of an evolved Fe-rich multiply saturated liquid.

094.467 **Ferrobasalts from Mare Crisium: Luna 24.**
D. T. Vaniman, J. J. Papike.
Geophys. Res. Lett., Vol. 4, 497 - 500 (1977).

Soils from the Soviet Luna 24 mission to Mare Crisium contain fragments of ferrobasalt and ferrogabbro. The ferrobasalt is very low in Ti content, similar to very low Ti "VLT" basalts from Apollo 17. Brown glasses and micropoikilitic fragments similar to the ferrobasalt occur in the Crisium soils, and green glasses similar to the Apollo 17 VLT basalt have been found. There is growing evidence that very low Ti basalts may be an abundant mare basalt type; there is a possibility that this suite may be derived by crystal fractionation from magma similar to green glass in composition.

094.468 **Mare Crisium: regional stratigraphy and geologic history.** J. B. Adams, J. W. Head, III, T. B. McCord, C. Pieters, S. Zisk.
Geophys. Res. Lett., Vol. 5, 313 - 316 (1978).

Spectral reflectance measurements of five Luna 24 samples and new telescopic reflectance spectra of 10-20 km areas of seven sites in Mare Crisium have been used to calibrate multispectral images of mare units. Based on these data, three major mare units are defined in the Crisium basin and their stratigraphy is interpreted. It is concluded that subsidence occurred throughout the emplacement of mare units, including extensive warping and downfaulting of the inner part of the Crisium basin.

094.469 **Origin of magnetization in lunar breccias: an example of thermal overprinting.**
W. A. Gose, D. W. Strangway, G. W. Pearce.
Earth Planet. Sci. Lett., Vol. 38, 373 - 384, with a correction in Vol. 39, 444 (1978) = Univ. Texas Marine Sci. Inst. Contrib. No. 85.

Twenty six samples from seven hand specimens, collected from the station 6 at the Apollo 17 landing site, were studied magnetically. The authors propose that the natural remanent magnetization in these breccias is the vector sum of two magnetizations, a pre-impact magnetization and a partial thermoremanence acquired during breccia formation.

094.470 **^{39}Ar–^{40}Ar systematics of two millimeter-sized rock fragments from Mare Crisium.**
A. Stettler, F. Albarède.
Earth Planet. Sci. Lett., Vol. 38, 401 - 406 (1978).

094.471 **Early lunar differentiation: 4.42-AE-old plagioclase clasts in Apollo 16 breccia 67435.**
B. Dominik, E. K. Jessberger.
Earth Planet. Sci. Lett., Vol. 38, 407 - 415 (1978).

094.472 **Lunar initial ^{143}Nd/^{144}Nd: differential evolution of the lunar crust and mantle.**
G. W. Lugmair, K. Marti.
Earth Planet. Sci. Lett., Vol. 39, 349 - 357 (1978).

The Sm–Nd evolution of Apollo 15 green glass is discussed. The ICE age (intercept with chondritic evolution) of 3.8 ± 0.4 AE overlaps the range of reported ^{39}Ar–^{40}Ar ages (T_2) and implies a distinct source region for green glass, characterized by very low and unfractionated REE abundances. The authors studied the Sm-Nd system of various lunar rock types. The results obtained from a limited number of rocks clearly indicate differential Sm–Nd evolution for the lunar crust and mantle.

094.473 **^{87}Rb–^{87}Sr age of soils and rock from Mare Crisium Luna 24.** J. L. Birck, C. J. Allègre.
Phys. Earth Planet. Interiors, Vol. 16, P10 - P14 (1978) = Contrib. Inst. Phys. Globe NS No. 269.

The soil and one gabbroic fragment from Mare Crisium, brought back by the Luna-24 mission, have been measured through the ^{87}Rb/^{87}Sr method. The soil is one of the least radiogenic from the moon, comparable with Luna-16 soils. The gabbroic rock, which has a very low Rb/Sr ratio, gives an internal isochron age of 3.74 ± 0.58 Ga and an initial Sr ratio of 0.699085 ± 0.000070.

094.474 **Impact flows and crater scaling on the moon.**
J. D. O'Keefe, T. J. Ahrens.
Phys. Earth Planet. Interiors, Vol. 16, 341 - 351 (1978) = Contrib. Div. Geol. Planet. Sci., Cal. Inst. Technol. No. 2682.

094.475 **Investigation of regolith returned by the automatic lunar station Luna 20 with the method of X-ray and photoelectric spectroscopy.** V. V. Nemoshkalenko, V. G. Aleshin, Yu. P. Dikov, O. A. Bogatikov, V. L. Barsukov, A. V. Ivanov, E. P. Moiseenko.
Dokl. AN SSSR, Vol. 238, 1079 - 1082 (1978). In Russian. Abstr. in Ref. zh., 62. Issled. kosm. prostranstva, 7.62.129 (1978).

094.476 **Roentgenographic investigation of lunar soil particles returned by the automatic station Luna 24.**
N. R. Khisina, V. I. Mokeeva, A. Ya. Volkova, T. V. Sletova, V. I. Bukin, K. I. Tobelko, E. S. Makarov.
Geokhimiya, 1978, No. 3, p. 323 - 332. In Russian. – Abstr. in Ref. zh., 62. Issled. kosm. prostranstva, 7.62.130 (1978).

Main stages of development of lunar investigations in the United States by automatic interplanetary stations.
See Abstr. 051.036.

Solar wind sputtering: an ineffective weathering process of airless bodies. See Abstr. 074.076.

Crater morphometry from bistatic radar.
See Abstr. 091.077.

Lunar occultation of Saturn. IV. Astrometric results from observations of the satellites.
See Abstr. 100.503.

Determination of lithium and halogens and the significance of lithium to the understanding of cosmochemical processes. See Abstr. 105.060.

Phosphor in meteorites and lunar samples.
See Abstr. 105.070.

Impact conditions required for formation of melt by jetting in silicates. See Abstr. 105.091.

095 Lunar Eclipses

095.001   The partial lunar eclipse of April 4, 1977.
W. H. Haas.
Strolling Astron., Vol. 27, 77 - 79 (1978).

095.002   Totale Mondfinsternis 16. September 1978.

L. D. Schmadel.
Sterne Weltraum, Jahrg. 17, 221 - 222 (1978).

Occultations during the lunar eclipse of 1978
March 24.   See Abstr. 096.001.

096 Lunar Occultations

096.001   Occultations during the lunar eclipse of 1978
March 24.   D. W. Dunham.
Occultation Newsl., Vol. 1, 131 - 132 (1978).

096.002   Erroneous star positions from occultations.
D. Herald.
Occultation Newsl., Vol. 1, 132 - 133 (1978).

096.003   Lunar occultations of planets.   M. D. Reynolds.
Occultation Newsl., Vol. 1, 138 (1978).

096.004   Extended-coverage U.S.N.O. total occultation
predictions.   D. W. Dunham.
Occultation Newsl., Vol. 1, 138 - 140 (1978).

096.005   Tre strykande ockultationer av Aldebaran i
Skandinavien 1979.   J. Meeus.
Astron. Tidsskr., Årg.10, 168 - 170 (1977).

096.006   Occultations during lunar eclipses and cluster
passages.   D. W. Dunham.
Occultation Newsl., Vol. 1, 145 - 148 (1978).

096.007   Grazing occultations.   D. W. Dunham.
Occultation Newsl., Vol. 1, 148 - 149 (1978).

096.008   The occultation of υ Leonis.
G. H. Jacoby, C. M. Price.
Publ. Astron. Soc. Pacific, Vol. 90, 113 - 116 (1978).
A lunar occultation of υ Leo (V = 4.3, G9 III) has been
observed with a 16-inch (41-cm) telescope from UCLA. Analysis of the occultation curve yields an angular diameter of
$0.0021 \pm 0\overset{.}{.}0007$ if the disk is assumed to be uniformly
illuminated.

096.009   Occultations of Jupiter satellites.   J. A. Buarque.
Publ. Astron. Soc. Pacific, Vol. 90, 117 - 118
(1978).
Observations of lunar occultations are presented for
Callisto, Europa, and Ganymede. The resulting linear diameters are 4912 km, 3141 km, and 5062 km, respectively. The
date of the events was 1977 January 1.

096.010   Observations of occultations of stars by the moon
at the Astronomical Observatory of the Kiev University in 1975.   A. K. Osipov, A. A. Zhitetskij.
Vestn. Kiev. Univ., Astron., Vyp. (No.) 19, p. 95 - 100 (1977).
In Russian.

096.011   Occultations d'étoiles brillantes par la lune.
J. Meeus.
Astronomie, Vol. 92, 279 - 282 (1978).

096.012   Occultation observation in 1976.
T. Mori, Y. Harada, M. Kawada.
Data Rep. Hydrographic Observations, Ser. Astron. Geod.,
Tokyo, No. 12, p. 1 - 47 (1978).

096.013   Lunar occultation observations at the University of
Illinois Prairie Observatory.   R. Radick.
Bull. American Astron. Soc., Vol. 10, 458 (1978). — Abstract.

096.014   Occultations d'étoiles par la lune, observées à
l'équatorial de 45 cm de 1974 à 1977.
J. Dommanget, E. Van Dessel.
Bull. Astron. Obs. R. Belgique, Vol. 9, 32 (1978).

The detection of stellar systems by lunar occultations.   See Abstr. 031.251.

Statistically rigorous reduction of optical lunar
occultation measurements.   See Abstr. 031.264.

Lunar occultations of minor planets in 1978.
See Abstr. 098.038.

Lunar occultation of (15) Eunomia.
See Abstr. 098.039.

Lunar occultation of Saturn. III. How big is Iapetus?
See Abstr. 100.502.

Lunar occultation of Saturn. IV. Astrometric results
from observations of the satellites.   See Abstr. 100.503.

Derivation of brightness distribution across a stellar
disk from photoelectric observations of its lunar occultation.
See Abstr. 113.062.

New double stars.   See Abstr. 118.015.

Occultation astrometry of the Beta Scorpii system.
See Abstr. 118.022.

Occultation of the S255 molecular cloud in the
CO line.   See Abstr. 131.104.

097 Mars, Mars Satellites

Mars

097.001 Arecibo radar observations of Martian surface characteristics near the equator.
R. A. Simpson, G. L. Tyler, D. B. Campbell.
Icarus, Vol. 33, 102 - 115 (1978).

Mars radar observations at 12.6-cm wavelength indicate that many of what were potential Viking landing sites along the planet's equator are rougher than interpretations of Mariner 9 images suggested. Study of spectral shapes indicates the Hagfors scattering law remains the best descriptor of quasi-specular surface scattering properties in an average sense; widespread variations in the surface argue against its indiscriminate use, however. Backscattering at moderate (25 - 40°) incidence angles was studied qualitatively and was found to be significantly above the level predicted by a strictly quasi-specular (e.g., Hagfors) process; it also is variable over the surface.

097.002 Observations of Mars with the astrolabe of the CERGA observatory during the winter 1975 - 1976.
J. Pham-Van, H. Choplin, C. Delmas, P. Grudler, G. Guallino, F. Mignard, G. Vigouroux.
Astron. Astrophys., Suppl. Ser., Vol. 31, 171 - 173 (1978).
In French.

This paper contains results of the first campaign at the CERGA observatory. During the 1975 - 1976 winter, 41 east and west transits were observed. The computed positions with which the observed positions are compared were derived from the tabular values of the American Ephemeris for the right ascension and the declination.

097.003 La planète Mars en 1975–1976.   J. Dragesco.
Astronomie, Vol. 92, 3 - 17 (1978).

097.004 Viking et la planète Mars.   A. Dollfus.
Astronomie, Vol. 92, 19 - 49 (1978).

097.005 Chemical interpretation of Viking Lander 1 life detection experiment.   E. V. Ballou, P. C. Wood, T. Wydeven, M. E. Lehwalt, R. E. Mack.
Nature, Vol. 271, 644 - 645 (1978).

097.006 Gravity field of Mars.   M. Burša.
Veröff. Zentralinst. Phys. Erde, No. 52, Part 2, (see 012.001), p. 257 - 259 (1977).

097.007 Mars: topographic control of clouds, 1907–1973.
W. K. Hartmann.
Icarus, Vol. 33, 380 - 387 (1978).

Mariner 9 high-resolution photos and topographic information were used to make a topographic analysis of "blue" and "red" cloud positions reported over a 66-year period in Lowell Observatory records. A sample of 77 "blue" cloud sites lay preferentially at the highest Martian elevations; 60% centered precisely on the seven major volcanic mountain peaks (unknown when the clouds were observed); another 16% lay on substantial slopes or contacts between cratered terrain and lower plains. The median altitude of blue cloud sites was 2.1 km above the global topographic median. Of 88 "possible dust clouds" (chosen by additional criteria), about two-thirds occur at borders between light and dark areas, in the light regions. These sites may have thin veneers of dust, and current depositional or denudational activity. Median altitude of "possible dust cloud" sites was 0.5 km below the global topographic median.

097.008 The magnetic field of Mars estimated from the data of plasma measurements by Soviet artificial satellites of Mars.   K. I. Gringauz, V. V. Bezrukikh, T. K. Breus, M. I. Verigin, A. P. Remizov.
Cosmochemistry of the moon and planets, (see 012.002), p. 859 - 863 (1977).

The dimensions of the obstacle forming the shock wave of Mars are estimated by use of electron trap data from Mars 2, 3, and 5. The mean altitude of the obstacle at the subsolar point can be convincingly explained if the obstacle is the magnetosphere of Mars. On the assumption that Mars has its own dipole magnetic field, the magnetic moment of Mars is estimated, $M_m \cong 2 \times 10^{22}$ gs cm^3.

097.009 Martian dust storms – a mechanism for transportation of life?   G. Day.
Spaceflight, Vol. 20, 83 - 88 (1978).

097.010 Neue Karten und Globen des Mars und Phobos.
J. Blunck.
Sterne Weltraum, 17. Jahrg., 81 - 85 (1978).

097.011 Mars synthetic topographic mapping.
S. S. C. Wu.
Icarus, Vol. 33, 417 - 440 (1978).

Topographic contour maps of Mars are compiled by the synthesis of data acquired from various scientific experiments of the Mariner 9 mission, including S-band radio-occultation, the ultraviolet spectrometer, the infrared radiometer, the infrared interferometer spectrometer and television imagery, as well as Earth-based radar information collected at Goldstone, Haystack, and Arecibo Observatories. The entire planet is mapped at scales of 1 : 25,000,000 and 1 : 5,000,000 using Mercator, Lambert, and polar stereographic map projections.

097.012 Radar studies of the Martian surface at centimeter wavelengths: the 1975 opposition.
G. S. Downs, R. R. Green, P. E. Reichley.
Icarus, Vol. 33, 441 - 453 (1978).

The Goldstone radar system was operated at wavelengths of 3.5 and 12.6 cm to probe the Martian surface during the 1975 opposition. Regions studied in detail by range—Doppler techniques are Syrtis Major, Sinus Meridiani, and the crater Schiaparelli.

097.013 The dynamics of and the heat transfer by baroclinic eddies and large-scale stationary topographically forced long waves in the Martian atmosphere.
A. M. Gadian.
Icarus, Vol. 33, 454 - 465 (1978).

The large-scale dynamics of the Martian winter atmosphere are analyzed and discussed. Results obtained from a baroclinic instability analysis, including a Newtonian radiational damping term, are compared with the Mariner 9 data. The beta plane baroclinic poleward heat fluxes are contrasted with those produced using spherical geometry, from the stationary orographically forced long waves.

097.014 A statistical study of crater-associated wind streaks in the north equatorial zone of Mars.
J. Veverka, K. Cook, J. Goguen.
Icarus, Vol. 33, 466 - 482 (1978).

The authors study the physical characteristics of all crater-related streaks within the north equatorial zone of Mars (between 0 and 30°N). The survey includes 25% of the surface area of Mars and covers the latitude band of major interest to the 1976 Viking missions.

097.015 **Chemical weathering on Mars. Thermodynamic stabilities of primary minerals (and their alteration products) from mafic igneous rocks.**   J. L. Gooding. Icarus, Vol. 33, 483 - 513 (1978).

Chemical weathering on Mars is examined theoretically from the standpoint of heterogeneous equilibrium between solid mineral phases and gaseous O_2, H_2O, and CO_2 in the Martian atmosphere. Thermochemical calculations are performed in order to identify important gas-solid decomposition reactions involving the major mineral constituents of mafic igneous rocks.

097.016 **Theoretical models for Mars and their seismic properties.**   E. A. Okal, D. L. Anderson. Icarus, Vol. 33, 514 - 528 (1978).

Theoretical seismic properties of the planet Mars are investigated on the basis of the various models which have been proposed for the internal composition of the planet. A detailed discussion is given of the seismic properties which could—in principle—help answer the following two questions: Is Mars' core liquid or solid? Does Mars have a partially molten asthenosphere in its upper mantle?

097.017 **Paleozoic era on Mars.**   J. Boynton. Bull. American Astron. Soc., Vol. 9, 497 (1977). Abstract.

097.018 **A polar symmetric model of the Martian atmosphere with horizontal eddy transport.** J. A. Pirraglia. Bull. American Astron. Soc., Vol. 9, 509 (1977). – Abstract.

097.019 **Occultation of ϵ Gem by Mars: further evidence for atmospheric waves.**   R. G. French, J. L. Elliot. Bull. American Astron. Soc., Vol. 9, 509 - 510 (1977). Abstract.

097.020 **Infrared observations of local Martian dust storms.** A. R. Peterfreund, T. Z. Martin. Bull. American Astron. Soc., Vol. 9, 510 (1977). – Abstract.

097.021 **A comparison of Martian and terrestrial stratospheric thermal structures.**   B. J. Conrath. Bull. American Astron. Soc., Vol. 9, 510 (1977). – Abstract.

097.022 **Abundance of volatiles on Mars. I. The model.** E. Anders, T. Owen. Bull. American Astron. Soc., Vol. 9, 513 (1977). – Abstract.

097.023 **Abundance of volatiles on Mars. II. Predictions and tests.**   T. Owen, E. Anders. Bull. American Astron. Soc., Vol. 9, 513 (1977). – Abstract.

097.024 **Latitude variation of O_2 dayglow and inferred O_3 abundance on Mars.** W. A. Traub, N. P. Carleton, P. Connes, J. F. Noxon. Bull. American Astron. Soc., Vol. 9, 513 - 514 (1977). Abstract.

097.025 **High resolution Martian atmosphere modeling between 2800 and 5600 cm^{-1}.** W. G. Egan, W. L. Fischbein, T. Hilgeman, L. L. Smith. Bull. American Astron. Soc., Vol. 9, 514 (1977). – Abstract.

097.026 **The vertical distribution of scattering particles in the Martian atmosphere.**   W. A. Baum, L. J. Martin, L. H. Wasserman, J. B. Wellman, S. Eberle. Bull. American Astron. Soc., Vol. 9, 514 (1977). – Abstract.

097.027 **The vertical distribution of scatterers in the Martian atmosphere.**   R. A. Kahn, J. B. Pollack. Bull. American Astron. Soc., Vol. 9, 514 (1977). – Abstract.

097.028 **The pole direction and precession of Mars.** R. D. Reasenberg, R. B. Goldstein, P. E. MacNeil, I. I. Shapiro, R. W. King. Bull. American Astron. Soc., Vol. 9, 520 (1977). – Abstract.

097.029 **Chemical heterogeneity among soil and crust samples at each of the Viking landing sites.** B. C. Clark, A. K. Baird, P. Toulmin III, H. J. Rose, Jr., K. Keil. Bull. American Astron. Soc., Vol. 9, 527 (1977). – Abstract.

097.030 **Petrologic interpretation of Viking XRF analysis based on reflectance spectra and the photochemical weathering model.**   M. Maderazzo, R. Huguenin. Bull. American Astron. Soc., Vol. 9, 527 - 528 (1977). Abstract.

097.031 **Mars: near-infrared spectral reflectance and compositional implication.** T. B. McCord, R. N. Clark, R. L. Huguenin. Bull. American Astron. Soc., Vol. 9, 528 (1977). – Abstract.

097.032 **Mars: surface mineralogy from reflectance spectra.** R. L. Huguenin, J. B. Adams, T. B. McCord. Bull. American Astron. Soc., Vol. 9, 528 (1977). – Abstract.

097.033 **Lag magnetite in Martian dust.** R. L. Huguenin, K. L. Andersen. Bull. American Astron. Soc., Vol. 9, 528 (1977). – Abstract.

097.034 **How does the Martian regolith control of the Martian atmospheric pressure?** F. P. Fanale, W. A. Cannon. Bull. American Astron. Soc., Vol. 9, 528 - 529 (1977). Abstract.

097.035 **Photoemission measurements during photochemical weathering.**   C. Poole, R. Huguenin. Bull. American Astron. Soc., Vol. 9, 529 (1977). – Abstract.

097.036 **Surface source of O_2 on Mars.**   R. Huguenin, K. Miller, M. Williams, D. Stutman. Bull. American Astron. Soc., Vol. 9, 529 (1977). – Abstract.

097.037 **Particle motion on Mars inferred from the Viking Lander cameras.**   C. Sagan, D. Pieri, P. Fox, R. E. Arvidson, E. A. Guinness. Bull. American Astron. Soc., Vol. 9, 538 (1977). – Abstract.

097.038 **Topographic mapping of Mars from orbit.** K. R. Blasius. Bull. American Astron. Soc., Vol. 9, 538 (1977). – Abstract.

097.039 **Evaporation of ice-choked rivers: application to Martian channels.**   D. Wallace, C. Sagan. Bull. American Astron. Soc., Vol. 9, 539 (1977). – Abstract.

097.040 **Small channels in the Margaritifer Sinus Region: implications for climatic change.**   D. C. Pieri. Bull. American Astron. Soc., Vol. 9, 539 (1977). – Abstract.

097.041 **Martian rift valleys: evidence for episodic crustal uplift.**   H. Frey. Bull. American Astron. Soc., Vol. 9, 539 (1977). – Abstract.

097.042 **Retreat of the Martian south polar cap.** H. H. Kieffer. Bull. American Astron. Soc., Vol. 9, 540 (1977). – Abstract.

097.043 **North polar mapping of Mars with the Viking thermal mappers.**   F. D. Palluconi. Bull. American Astron. Soc., Vol. 9, 540 (1977). – Abstract.

097.044 Electrical mapping of Martian permafrost.
G. R. Olhoeft.
Bull. American Astron. Soc., Vol. 9, 540 (1977). – Abstract.

097.045 Lava erosional channels on Mars.
J. A. Cutts, K. R. Blasius.
Bull. American Astron. Soc., Vol. 9, 541 - 542 (1977).
Abstract.

097.046 Vertical distribution of water vapor on Mars.
D. W. Davies.
Bull. American Astron. Soc., Vol. 9, 550 (1977). – Abstract.

097.047 Photometric mapping of Mars with the Viking
infrared thermal mappers. E. D. Miner.
Bull. American Astron. Soc., Vol. 9, 551 (1977). – Abstract.

097.048 Does Mars have a magnetosphere?
A. Johnstone.
Nature, Vol. 272, 399 (1978).

097.049 Chemical composition of the Martian surface.
V. I. Chesnokov.
Zemlya i Vselennaya, 1978, No. 2, p. 56 - 58. In Russian.

097.050 Mars – the view from Mariner 9. C. Sagan.
Focus on the stars, (see 003.002), p. 111 - 143
(1977).

097.051 The surface of Mars.
R. E. Arvidson, A. B. Binder, K. L. Jones.
Sci. American, Vol. 238, No. 3, p. 76 - 77, 80 - 89 (1978).
The Viking spacecraft have provided an unparalleled view
of it from orbit and from the ground, adding much evidence
on how it has been shaped by volcano, meteorite impact, water
and wind.

097.052 A history of the Martian atmosphere – a geophysi-
cal approach. A. Henderson-Sellers.
Geophys. J. R. Astron. Soc., Vol. 53, (see 012.006), 186
(1978). – Abstract.

097.053 Mars-Programm Viking im Rückblick.
H. W. Köhler.
Phys. Bl., 34. Jahrg., 175 - 186 (1978).

097.054 Simultaneous polarimetry of Mars from the Mars 5
spacecraft and ground-based telescopes.
A. Dollfus, L. V. Ksanfomaliti, V. I. Moroz.
Space Research, Vol. XVII, (see 012.010), 667 - 671 (1977).
The polarization of the Martian light has been observed
both from the ground and from the Mars 5 spacecraft. The
behaviour of the polarization during the dust storm of July
and August 1973 is discussed. At the time of the Mars 5
observations made in February 1974 the dust had settled; the
observations are unambiguously interpreted as showing that
the Martian surface is covered by a finely divided soil. Also,
the occurrence of small (~100–200 km) dust storms has been
discovered.

097.055 Morphology of Martian rampart craters.
P. J. Mouginis-Mark.
Nature, Vol. 272, 691 - 694 (1978).
Preliminary results from a morphological analysis of the
lobate flows arount Martian rampart craters are presented.
It is proposed that lobe formation is not entirely associated
with ballistic deposition of the crater's ejecta because the
proximal ends of the lobes are found closer to the primary
crater rim than to the region of secondary cratering associated
with either lunar or Mercurian craters. Lobe surface area is
shown to be proportional to crater diameter and an estimate
of 30–60 m is deduced for the lobe thickness for craters between 10 and 35 km diameter.

097.056 Solar-wind sputtering of the martian atmosphere.
P. K. Haff, Z. E. Switkowski, T. A. Tombrello.
Nature, Vol. 272, 803 - 804 (1978).
Beside mass loss, sputtering may lead to preferential loss
of light elements and isotopes. This phenomenon occurs during
ion bombardment of many alloys and compounds and has been
proposed as the source of the isotopic fractionation observed
on the surface of fine lunar dust grains subjected to solar wind
bombardment. It may also occur in the martian atmosphere
where certain isotopic anomalies have been reported. Within
a collision cascade, lighter isotopes recoil on average with
higher velocities so that they will be preferentially lost from
the top of the atmosphere.

097.057 Principles for naming features of the Martian relief.
G. A. Burba.
Izv. vyssh. uchebn. zaved. Geol. i razvedka, 1977, No. 10,
p. 180 - 191. In Russian. – Abstr. in Ref. zh., 51. Astron.,
2.51.273 (1978).

097.058 On the structure of the cryolithosphere of Mars.
R. O. Kuz'min.
Probl. kriolitologii. Moskva, 1977, p. 7 - 25. In Russian.
Abstr. in Ref. zh., 51. Astron., 2.51.275 (1978).

097.059 On using the photometric method for study of the
Martian topography. V. V. Botvinova.
Astron. Zh. Akad. Nauk SSSR, Vol. 55, 419 - 425 (1978).
In Russian. English translation in Soviet Astron., Vol. 22,
No. 2.
The conditions providing a most effective use of the
photometric method for determining inclinations on the Martian surface are examined. It is shown that the method works
best near the wavelength $0.4~\mu$. Estimates of the inclinations
of exterior and interior walls made for three craters on the
surface of Mars gave the following values: $3°-4°$ for external
and $7°- 8°$ for internal walls.

097.060 Martian occultation of ϵ Gem as observed from the
C. E. Kenneth Mees Observatory.
R. G. French, J. D. Goguen, J. G. Duthie.
Icarus, Vol. 34, 182 - 187 (1978).
Ground-based observations of the occultation of ϵ Gem
by Mars on April 8, 1976 have been reduced in the manner
of French et al. to yield the scale height and temperature
profiles of the Martian atmosphere for number densities
between 10^{13} and 10^{15} cm^{-3}. The deduced variations in
temperature are remarkably similar to those obtained by
Elliot et al. and to the in situ measurements from the
Viking landers.

097.061 The occultation of ϵ Geminorum by Mars as ob-
served at Princeton. E. J. Groth, J. B. Klopfen-
stein, W. C. Wickes, J. Caldwell.
Astron. J., Vol. 83, 442 - 446 (1978).
The 8 April 1976 occultation of ϵ Gem by Mars was observed in two colors with the Princeton high-speed photometry
system. Data were observed continuously for about 20 min
with a resolution of 1 ms. Assuming an isothermal atmosphere,
the scale heights are 7.75 ± 0.85 km (immersion) and
9.56 ± 0.94 km (emersion).

097.062 On the magnetic field of Mars.
Sh. Sh. Dolginov.
Kosm. Issled., Vol. 16, 257 - 268 (1978). In Russian.

097.063 The polar regions of Mars. J. Cutts.
Astronomy, Vol. 5, No. 10, p. 10 - 17 (1977).
Abstr. in Phys. Abstr., Vol. 81, Abstr. 7899 (1978).

097.064 Martian climate: past, present and future.
J. Gribbin.
Astronomy, Vol. 5, No. 10, p. 18 - 24 (1977). — Abstr. in
Phys. Abstr., Vol. 81, Abstr. 7900 (1978).

097.065 Electron cooling by excitation of carbon dioxide.
M. A. Morrison, A. E. Greene.
J. Geophys. Res., Vol. 83, 1172 - 1174 (1978).
   The cooling of a heated electron gas by electron impact
excitation of CO_2 is believed to be an important process in the
neutral atmospheres of Mars and Venus. Electron energy loss
rates are calculated for a variety of excitation processes as a
function of electron temperature at a CO_2 gas temperature of
200 K. Vibrational excitation is found to be the most efficient
cooling mechanism for the range of electron temperatures
studied. The contributions of various vibrational energy loss
processes to electron cooling are discussed.

097.066 Carbonate formation in Marslike environments.
M. C. Booth, H. H. Kieffer.
J. Geophys. Res., Vol. 83, 1809 - 1815 (1978).
   Carbonate growth was examined within rock powders
subjected to Marslike environmental simulations. Rates of
growth under experimental conditions were $10^{12} - 10^{13}$ molecules $cm^{-3} s^{-1}$ with or without an aqueous phase of H_2O
present and were found to be proportional to CO_2 pressure,
H_2O abundance, and particle surface area. Direct ultraviolet
illumination of powders was found not to affect carbonate
growth significantly, but photochemistry of absorbed H_2O
was thought to play an important role in chemical alteration
activity.

097.067 Mars 1975 - 76 aphelic apparition — A.L.P.O.
       report I.   C. F. Capen, R. B. Rhoads.
Strolling Astron., Vol. 27, 63 - 76 (1978).

097.068 Zur Chemie der Marsoberfläche. Mineralogische und
       petrologische Implikationen.
K. Keil, B. C. Clark, A. K. Baird, P. Toulmin III, H. J. Rose, Jr.
Naturwissenschaften, 65. Jahrg., 231 - 238 (1978).
   Analyses of 13 samples of Martian surface materials with
the Viking X-ray fluorescence spectrometers show SiO_2
similar to that of terrestrial mafic rocks, whereas Fe_2O_3, Cl,
and S are higher and Al_2O_3, K_2O, Rb, Sr, Y, and Zr are lower.
Low totals suggest presence of CO_2, H_2O, and Na_2O. Duricrust
fragments are higher in S than fines, but samples from both
landing sites are surprisingly similar. The authors suggest that
Martian surface materials are aeolian deposits of complex
mixtures of weathering products of mafic-ultramafic rocks,
possibly consisting of iron-rich clays, sulfates, iron oxides,
carbonates, and chlorides.

097.069 The control net of Mars: May 1977.   M. E. Davies.
J. Geophys. Res., Vol. 83, 2311 - 2312 (1978).
   In May 1977 a planet-wide control net of Mars was computed by means of a large single-block analytical triangulation
using 17,224 measurements of 3037 control points on 928
Mariner 9 pictures. Areographic coordinates of the control
points are available on microfiche. The direction of the spin
axis and the rotation rate as determined by Viking were used.
The angle V, measured from Mars' vernal equinox along the
equator to the prime meridian (Airy-0), was found to be $V =$
$148°37 + 350°891986$ (J.D. 2,433,282.5).

097.070 Mars: the role of the regolith in determining atmo-
       spheric pressure and the atmosphere's response to
insolation changes.   F. P. Fanale, W. A. Cannon.
J. Geophys. Res., Vol. 83, 2321 - 2325 (1978).
   The authors present a quantitative model for atmosphere-
regolith exchange of CO_2 on Mars based on new laboratory
measurements of CO_2 adsorption on ground rock at temperatures of 158°, 175°, 196°, and 231° K and CO_2 pressures from

1.0 to 80 mbar. The model is consistent with Viking observations, whereas models involving a massive residual CO_2 cap
and no long-term atmosphere-regolith CO_2 exchange are not.
The model describes the role of the regolith as a CO_2 store-
house, as a long-term buffer of the atmospheric pressure, and
as a major factor in determining the response of the atmosphere to postulated changes in surface insolation.

097.071 Main tectonic elements of Mars and their comparison
       with the structures of other terrestrial planets and
the moon.   Ya. G. Kats, V. V. Kozlov, E. D. Sulidi-Kondrat'ev,
V. E. Khain.
Izv. vyssh. uchebn. zaved. Geol. i razvedka, 1977, No. 10, p.
3 - 8. In Russian. — Abstr. in Ref. zh., 62. Issled. kosm.
prostranstva, 4.62.179 (1978).

097.072 Principles and methods of compiling a tectonic map
       of Mars.   V. A. Avdeev, Ya. G. Kats, V. V.
Kozlov, E. D. Sulidi-Kondrat'ev.
Izv. vyssh. uchebn. zaved. Geol. i razvedka, 1977, No. 10, p.
9 - 15. In Russian. — Abstr. in Ref. zh., 62. Issled. kosm.
prostranstva, 4.62.180 (1978).

097.073 Fault systems on Mars.
       V. V. Kozlov, E. D. Sulidi-Kondrat'ev.
Izv. vyssh. uchebn. zaved. Geol. i razvedka, 1977, No. 10, p.
16 - 23. In Russian. — Abstr. in Ref. zh., 62. Issled. kosm.
prostranstva, 4.62.181 (1978).

097.074 Volcanism on Mars.
       E. D. Sulidi-Kondrat'ev, V. V. Kozlov.
Izv. vyssh. uchebn. zaved. Geol. i razvedka, 1977, No. 10, p.
24 - 32. In Russian. — Abstr. in Ref. zh., 62. Issled. kosm.
prostranstva, 4.62.182 (1978).

097.075 On some exogenous processes on Mars.
       N. V. Makarova.
Izv. vyssh. uchebn. zaved. Geol. i razvedka, 1977, No. 10, p.
38 - 45. In Russian. — Abstr. in Ref. zh., 62. Issled. kosm.
prostranstva, 4.62.183 (1978).

097.076 On the geological importance of details of the
       albedo of Mars (for example the region Thaumasia—
Erythraeum Mare).   G. A. Burba.
Izv. vyssh. uchebn. zaved. Geol. i razvedka, 1977, No. 10, p.
53 - 59. In Russian. — Abstr. in Ref. zh., 62. Issled. kosm.
prostranstva, 4.62.184 (1978).

097.077 Experience of elaboration of the geographic founda-
       tion for maps of Mars.
G. A. Burba, N. N. Bobina.
Izv. vyssh. uchebn. zaved. Geol. i razvedka, 1977, No. 10, p.
60 - 64. In Russian. — Abstr. in Ref. zh., 62. Issled. kosm.
prostranstva, 4.62.185 (1978).

097.078 On the classification of ring structures of Mars.
       V. L. Avdeev, Ya. G. Kats, A. M. Nikishin.
Izv. vyssh. uchebn. zaved. Geol. i razvedka, 1977, No. 10, p.
171 - 173. In Russian. — Abstr. in Ref. zh., 62. Issled. kosm.
prostranstva, 4.62.186 (1978).

097.079 Exploring Mars using photographs obtained with
       the help of automatic interplanetary probes.
M. Oluic.
Tehnika, Vol. 32, 1517 - 1519 (1977). In Croatian. — Abstr.
in Phys. Abstr., Vol. 81, Abstr. 12132 (1978).

097.080 Global color variations on the Martian surface.
       L. A. Soderblom, K. Edwards, E. M. Eliason, E. M.
Sanchez, M. P. Charette.
Icarus, Vol. 34, 446 - 464 (1978).
   Surface materials exposed throughout the equatorial

region of Mars have been classified and mapped on the basis of spectral reflectance properties determined by the Viking II Orbiter vidicon cameras. Frames acquired at each of three wavelengths (0.45 ± 0.03 μm, 0.53 ± 0.05 μm, and 0.59 ± 0.05 μm) were mosaicked by computer. The classical dark region between the equator and $\sim 30°$ S in the Martian highlands is composed of two units: (1) an ancient unit consisting of topographic highs which is among the reddest on the planet ($0.59/0.45$ $\mu m \simeq 3$); and (2) intermediate age, smooth, intercrater volcanic plains displaying numerous mare ridges which are among the least red on Mars ($0.59/0.45$ $\mu m \simeq 2$). The relatively young shield volcanoes are, like the oldest unit, dark and very red. Two probable eolian deposits are recognized in the intermediate and high albedo regions.

097.081 Distribution and relations of 4- to 10-km-diameter craters to global geologic units of Mars.
C. D. Condit.
Icarus, Vol. 34, 465 - 478 (1978).

By correlating the 1:25,000,000 geologic map of Mars of Scott and Carr (1977) with 4- to 10-km-diameter crater density data from Mariner 9 images, the average crater density for 23 of the equatorial geologic–geomorphic units on Mars was computed. Condit believes the average crater density values are accurate indicators of the relative age of the geologic units considered. The statistical validity of these average values is strongest for the geologic units of the largest areal extent. The relative ages as obtained from the average crater density values for the seven largest geologic units, from youngest to oldest, are presented.

097.082 Mars, highlands–lowlands: Viking contributions to Mariner relative age studies.   D. H. Scott.
Icarus, Vol. 34, 479 - 485 (1978).

Stratigraphic relations between lowland plains and highlands, two major types of Martian geologic-terrain units, were not directly distinguishable on Mariner-9 images. Morphologic characteristics and crater densities suggested that the lava plains beneath their eolian cover were younger than adjacent highland rocks. Alternatively, the lowland plains could be the older unit. Viking photos across five areas of the highland–lowland boundary, however, tend to confirm the younger age of the plains-forming lava flows. A time interval of several hundred million years probably occurred between the retreat of the highland scarp and its latest embayment by lava extrusions in the lowlands.

097.083 Structural evolution of Arsia Mons, Pavonis Mons, and Ascreus Mons: Tharsis region of Mars.
L. S. Crumpler, J. C. Aubele.
Icarus, Vol. 34, 496 - 511 (1978).

Analysis of Viking Orbiter data suggests that Arsia Mons, Pavonis Mons, and Ascreus Mons, three large shield volcanoes of the Tharsis volcanoes of Mars, have had similar evolutionary trends. Arsia Mons appears to have developed in the following sequence: (1) construction of a main shield volcano, (2) outbreak of parasitic eruption centers on the northeast and southwest flanks, (3) volcano-tectonic subsidence of the summit and formation of concentric fractures and grabens, and (4) continued volcanism along a fissure or rift bisecting the main shield. In terms of this sequence Pavonis Mons has developed to stage (3) and Ascreus Mons has evolved to stage (2). This interpretation is supported by crater frequency–diameter distributions in the 0.1–to 3.0 km-diameter range.

097.084 The mantle of Mars: some possible geological implications of its high density.
T. R. McGetchin, J. R. Smyth.
Icarus, Vol. 34, 512 - 536 (1978).

The authors explore high pressure ($\sim$ 30 kbar) mineralogies which they believe to be petrologically plausible and consistent with the Mars mantle density. They find that, indeed, the model Martian mantle has a different phase assemblage than the Earth and that partial melting of this mantle would be likely to produce ultrabasic (ferrobasaltic) melts of very low viscosity. Implications, which are consistent with geology and geophysics, are presented.

097.085 Thermal history and evolution of Mars.
M. N. Toksöz, A. T. Hsui.
Icarus, Vol. 34, 537 - 547 (1978).

A theoretical thermal evolution model of Mars is constructed, utilizing as constraints the available geophysical and geological data, including those provided by the Viking missions. The calculation includes conduction and subsolidus mantle convection. Calculated models indicate that Martian evolution can be roughly characterized by four different stages. (1) Core formation and crust differentiation. (2) Heating, expansion, and mantle differentiation. (3) Mature phase. (4) Cooling period. The models suggest that the core is molten, and the calculated surface heat flux is 35 erg cm^{-2} sec^{-1}.

097.086 Location of Viking 1 Lander on the surface of Mars.
E. C. Morris, K. L. Jones, J. P. Berger.
Icarus, Vol. 34, 548 - 555 (1978).

A location of the Viking 1 Lander on the surface of Mars has been determined by correlating topographic features in the lander pictures with similar features in the Viking orbiter pictures. Radio tracking data narrowed the area of search for correlating orbiter and lander features and an area was found on the orbiter pictures in which there is good agreement with topographic features on the lander pictures. This location, when plotted on the 1:250,000 scale photomosaic of the Yorktown Region of Mars (U.S. Geological Survey, 1977) is at 22.487°N latitude and 48.041°W longitude.

097.087 Crater streaks in the Chryse Planitia region of Mars: early Viking results.
R. Greeley, R. Papson, J. Veverka.
Icarus, Vol. 34, 556 - 567 (1978).

High-resolution images of Chryse Planitia and eastern Lunae Planum from the early revolutions of Viking Orbiter I permit detailed analyses of crater-associated streaks and interpretation of related eolian processes. A total of 614 light and dark streaks were studied and treated statistically in relation to: (1) morphology, morphometry, and orientation, (2) "parent" crater size and morphology, (3) terrain type in which they occurred, (4) topographic elevation, and (5) meteorological data currently being acquired by Viking Lander I.

097.088 CO$_2$ permafrost and Martian topography.
R. St J. Lambert, V. E. Chamberlain.
Icarus, Vol. 34, 568 - 580 (1978).

The role of CO_2 permafrost as an erosive agent on Mars is considered. It is argued that conditions may exist in which CO_2 permafrost is extensive on Mars, provided that adequate CO_2 is available: the maximum ratio of $H_2O:CO_2$ required in the subsurface pore space system is 17:3. Erosional processes likely to result from such permafrost are block slumping, leading to canyon development; pit chains along faults; chaotic terrain where massive permafrost destruction has occurred; large-scale flows of slurry; and perhaps even the flash floods which create channels.

097.089 Origin of the stepped topography of the Martian poles.   A. D. Howard.
Icarus, Vol. 34, 581 - 599 (1978).

The circumpolar stepped topography observed within the Martian polar regions can have originated from one of a limited number of processes, including (1) erosion of resistant layers, (2) erosion rates inversely proportional to slope gradient, (3) basal sapping, and (4) bistable rates of erosion and deposition.

097.090   Martian fretted terrain: flow of erosional debris.
    S. W. Squyres.
Icarus, Vol. 34, 600 - 613 (1978).

Viking orbital photographs of two regions of Martian fretted terrain have revealed a number of landforms which appear to possess distinct flow lineations. The features are attributed to the deformation and flow of a mass consisting of erosional particles and ice incorporated from the atmosphere. A plastic deformation model is presented which is consistent with the known mechanical properties of rock glaciers and with the observed features of the landforms.

097.091   Some consequences of a liquid water saturated
    regolith in early Martian history.
A. O. Fuller, R. B. Hargraves.
Icarus, Vol. 34, 614 - 621 (1978).

Flooding of low-lying areas of the Martian regolith may have occurred early in the planet's history when a comparatively dense primitive atmosphere existed. If this model is valid, the following are some pedogenic and mineralogical consequences to be expected. Fluctuation of the water table in response to any seasonal or longer term causes would have resulted in precipitation of ferric oxyhydroxides with the development of a vesicular duricrust (or hardpan). Disruption of such a crust by scarp undercutting or frost heaving accompanied by wind deflation of fines could account for the boulders visible on Utopia Planitia in the vicinity of the second Viking lander site.

097.092   Possible fossil H_2O liquid-ice interfaces in the
    Martian crust.
L. A. Soderblom, D. B. Wenner.
Icarus, Vol. 34, 622 - 637 (1978).

The purpose of this paper is to present morphological evidence which suggests that below the permafrost zone containing H_2O ice, a second zone exists in which liquid water is stable and at least temporally resided. It is suggested that this stratification implanted a discontinuity in the Martian crust. This discontinuity is reflected in both the chemical and the physical state of materials above and below the boundary.

097.093   Implications of abundant hygroscopic minerals in
    the Martian regolith.   B. C. Clark.
Icarus, Vol. 34, 645 - 665 (1978).

Converging lines of evidence suggest that a significant portion of the Martian surface fines may consist of salts and smectite clays. Salts can form stoichiometric hydrates as well as eutectic solutions with depressed freezing points; clays contain bound water of constitution and adsorb significant quantities of water from the vapor phase. The formation of ice may be suppressed by these minerals in some regions on Mars, and their presence in abundance would imply important consequences for atmospheric and geologic processes and the prospects for exobiology.

097.094   The Viking biological experiments on Mars.
    H. P. Klein.
Icarus, Vol. 34, 666 - 674 (1978).

The essential findings of the three biological experiments aboard the two Viking Mars landers are reviewed and compared. All three of the experiments yielded significant data in repeated tests of Martian surface samples. Most of the findings are inconsistent with a biological basis. The combined data suggest the presence of several classes of oxidants on Mars and these would account for most of the observations. An explanation for the apparent small synthesis of organic matter in the pyrolytic release experiment remains obscure.

097.095   Atomic hydrogen on Mars: measurements at solar
    minimum.

J. S. Levine, D. S. McDougal, D. E. Anderson, Jr., E. S. Barker.
Science, Vol. 200, 1048 - 1051 (1978).

The Copernicus Orbiting Astronomical Observatory was used to obtain measurements of Mars Lyman-α (1215.671-angstrom) emission at the solar minimum. The Copernicus measurements, coupled with the Viking in situ measurements of the temperature ($170° \pm 30°K$) of the upper atmosphere of Mars, indicate that the atomic hydrogen number density at the exobase of Mars (250 kilometers) is about 60 times greater than that deduced from Mariner 6 and 7 Lyman-α measurements obtained during a period of high solar activity.

097.096   Primitive atmosphere and implications for the for-
    mation of channels on Mars.
Y. L. Yung, J. P. Pinto.
Nature, Vol. 273, 730 - 732 (1978).

It has been suggested that primitive Mars had a reducing atmosphere, composed mainly of methane. Such an atmosphere, as the authors show here, could be polymerised by solar ultraviolet radiation to produce higher hydrocarbons. These compounds are low viscosity liquids at today's temperature on Mars, and could contribute to the formation of channels.

097.097   Model of daily variations in the composition of the
    Martian atmosphere.   O. P. Krasitskij.
Kosm. Issled., Vol. 16, 434 - 442 (1978). In Russian.

097.098   Estimate of the density of Martian soil from radio-
    physical measurements in the 3-cm range.
N. N. Krupenio.
Kosm. Issled., Vol. 16, 443 - 452 (1978). In Russian.

097.099   Method and results of computing optical properties
    of the Martian atmosphere in the period of the
dust storm in 1971.   Zh. M. Dlugach.
Astrometriya i Astrofizika, vyp. (No.) 35, p. 85 - 97 (1978). In Russian.

A numerical method is expounded for computing the intensity of radiation diffusely reflected from a semi-infinite homogeneous atmosphere with anisotropic scattering. The case of phase function with a sharp forward peak is considered separately. The model of a homogeneous semi-infinite layer is adopted for the Martian atmosphere at the period of the maximum of the dust storm in 1971. This layer is assumed to consist of dust particles with refractive index 1.59 and mean geometrical radius 10 μ. In the limits of this model such optical characteristics as scattering indicatrix, single scattering albedo, phase function, spherical and geometrical albedos of the planet are calculated. The calculated brightness distributions over the intensity equator are in satisfactory agreement with observational data.

097.100   Die Tage des Planeten Mars.   N. Giesinger.
Sternenbote, 21. Jahrg., 34 - 38 (1978).

097.101   Mars.
IAU Circ., Nos. 3164, 3169 (1978).

097.102   Mars.
Yamamoto Circ., No. 1876 (1978). In Japanese.

097.103   Excess ^{15}N in the Martian atmosphere and cosmic
    rays in the early solar system.
S. Yanagita, M. Imamura.
Nature, Vol. 274, 234 - 235 (1978).

Viking 1 and 2 have collected various data on the amount and composition of the Martian atmosphere. The isotopic ratios $^{15}N/^{14}N$, $^{40}Ar/^{36}Ar$ and $^{129}Xe/^{132}Xe$ have been found to be distinctly different from the terrestrial values. The authors propose that the enrichment of ^{15}N amounting to 62% may be caused by nuclear reactions with intense cosmic rays in the early solar system.

097.104 Requiem pour des canaux.
P. Moore, translated by P. Wallach.
Astronomie, Vol. 92, 297 - 303 (1978).

097.105 Topical problems in Martian geophysics.
R. J. Phillips.
Reports of planetary geology program, 1977—1978,
(see 003.014), p. 69 - 70 (1978).

097.106 The nature and origin of the Martian planetary
dichotomy (is still a problem). M. C. Malin,
R. J. Phillips, R. S. Saunders.
Reports of planetary geology program, 1977—1978,
(see 003.014), p. 83 - 85 (1978).

097.107 Geological implications of regional color variations
in the Martian equatorial regions.
L. A. Soderblom.
Reports of planetary geology program, 1977—1978,
(see 003.014), p. 86 - 88 (1978).

097.108 Viking contributions: Martian highland-lowland
stratigraphy. D. H. Scott.
Reports of planetary geology program, 1977—1978,
(see 003.014), p. 89 - 90 (1978).

097.109 Stratigraphy of Amenthes quadrangle and vicinity.
K. Hiller, G. Neukum.
Reports of planetary geology program, 1977—1978,
(see 003.014), p. 91 - 93 (1978).

097.110 Structural evolution of the Claritas-Fossae area of
Mars. P. Masson.
Reports of planetary geology program, 1977—1978,
(see 003.014), p. 94 - 96 (1978).

097.111 Polygonal fractures of the Martian plains.
E. C. Morris, J. R. Underwood.
Reports of planetary geology program, 1977—1978,
(see 003.014), p. 97 - 99 (1978).

097.112 Mars: near infrared spectral reflectance and
compositional implication. T. B. McCord, R. N.
Clark, R. L. Huguenin.
Reports of planetary geology program, 1977—1978,
(see 003.014), p. 109 - 111 (1978).

097.113 Characterization of Mars surface units.
T. B. McCord, R. B. Singer.
Reports of planetary geology program, 1977—1978,
(see 003.014), p. 112 - 114 (1978).

097.114 Martian surface composition: comparison of remote
spectral studies and in situ X-ray fluorescence
analysis. O. B. Toon, B. N. Khare, J. B. Pollack, C. Sagan.
Reports of planetary geology program, 1977—1978,
(see 003.014), p. 115 (1978).

097.115 The composition of Mars. K. A. Goettel.
Reports of planetary geology program, 1977—1978,
(see 003.014), p. 116 - 117 b (1978).

097.116 Mars: petrologic units in the Margaritifer Sinus and
Coprates Quadrangle. R. L. Huguenin, J. W.
Head, T. R. McGetchin.
Reports of planetary geology program, 1977—1978,
(see 003.014), p. 118 - 120 (1978).

097.117 Radar topography of the Arsia Mons — Claritas
Fossae region of Mars. R. S. Saunders, L. E. Roth,

C. Elachi, G. Schubert.
Reports of planetary geology program, 1977—1978,
(see 003.014), p. 128 - 131 (1978).

097.118 Radar profiles and the depth/diameter ratios of
large Martian craters. L. E. Roth, C. Elachi,
R. S. Saunders, G. Schubert.
Reports of planetary geology program, 1977—1978,
(see 003.014), p. 132 - 134 (1978).

097.119 Mars bistatic radar. R. A. Simpson, G. L. Tyler,
H. T. Howard.
Reports of planetary geology program, 1977—1978,
(see 003.014), p. 135 - 136 (1978).

097.120 Topographic confirmation of 500 km degraded
crater north of Ladon Valles, Mars.
R. S. Saunders, L. E. Roth, C. Elachi, G. Schubert.
Reports of planetary geology program, 1977—1978,
(see 003.014), p. 157 - 159 (1978).

097.121 Areal distribution of rampart craters on Mars.
C. C. Allen.
Reports of planetary geology program, 1977—1978,
(see 003.014), p. 160 - 161 (1978).

097.122 Martian rampart craters: crater processes that may
affect diameter-frequency distributions.
J. M. Boyce, D. J. Roddy.
Reports of planetary geology program, 1977—1978,
(see 003.014), p. 162 - 165 (1978).

097.123 Interior morphometry of fresh Martian craters:
preliminary Viking results. C. A. Wood, M. J.
Cintala, J. W. Head.
Reports of planetary geology program, 1977—1978,
(see 003.014), p. 166 - 168 (1978).

097.124 Central pit craters, peak rings, and the Argyre
basin. C. A. Hodges.
Reports of planetary geology program, 1977—1978,
(see 003.014), p. 169 - 171 (1978).

097.125 Mars chronology. G. Neukum, K. Hiller, J.
Henkel, J. Bodechtel.
Reports of planetary geology program, 1977—1978,
(see 003.014), p. 172 - 174 (1978).

097.126 Intermediate diameter (2–10 km) crater density
maps of Mars. C. D. Condit, D. A. Johnson.
Reports of planetary geology program, 1977—1978,
(see 003.014), p. 175 - 177 (1978).

097.127 Crater density determination in the Tharsis region
of Mars. R. S. Saunders, T. Gregory.
Reports of planetary geology program, 1977—1978,
(see 003.014), p. 178 (1978).

097.128 The Viking I landing site crater diameter-frequency
distribution. A. L. Dial, Jr.
Reports of planetary geology program, 1977—1978,
(see 003.014), p. 179 - 181 (1978).

097.129 Multi-stage evolution of the Tharsis shield volcanoes
of Mars from Viking orbiter imagery.
L. S. Crumpler, J. C. Aubele, W. E. Elston.
Reports of planetary geology program, 1977—1978,
(see 003.014), p. 196 - 197 (1978).

097.130 Rheologic properties of Arsia Mons lavas.
J. B. Plescia, R. S. Saunders.
Reports of planetary geology program, 1977—1978,

(see 003.014), p. 198 - 201 (1978).

097.131   Small volcanic constructs in the Chryse Planitia
region of Mars.   R. Greeley, E. Theilig.
Reports of planetary geology program, 1977–1978,
(see 003.014), p. 202 (1978).

097.132   Tharsis Province of Mars: deformational history and
fault sequence.   D. U. Wise, M. P. Golombek,
G. E. McGill.
Reports of planetary geology program, 1977–1978,
(see 003.014), p. 203 - 204 (1978).

097.133   Possible intrusive activity on Mars.   P. Schultz.
Reports of planetary geology program, 1977–1978,
(see 003.014), p. 205 - 207 (1978).

097.134   Surface textures of sand sized particles abraded
under Earth and Martian conditions.
D. H. Krinsley, R. Greeley.
Reports of planetary geology program, 1977–1978,
(see 003.014), p. 225 - 227 (1978).

097.135   Windform patterns on Earth and on Mars: implications for similarities of eolian processes on two
planets.   C. S. Breed, W. A. Ward, J. F. McCauley.
Reports of planetary geology program, 1977–1978,
(see 003.014), p. 228 - 229 (1978).

097.136   Contrast reversal on Mars.   J. Veverka, P. Thomas,
J. Burt, T. Thorpe.
Reports of planetary geology program, 1977–1978,
(see 003.014), p. 230 (1978).

097.137   Wind-deposited sand atop terrestrial boulders:
possible significance for the Viking Lander 1 site on
Mars.   R. S. U. Smith.
Reports of planetary geology program, 1977–1978,
(see 003.014), p. 234 - 235 (1978).

097.138   Viking implications for Martian aeolian dynamics.
R. Arvidson.
Reports of planetary geology program, 1977–1978,
(see 003.014), p. 238 - 240 (1978).

097.139   Wind transport rates on Mars.   B. White, R.
Greeley.
Reports of planetary geology program, 1977–1978,
(see 003.014), p. 241 - 243 (1978).

097.140   Mars: a model for the formation of dunes and
related structures.   R. Greeley.
Reports of planetary geology program, 1977–1978,
(see 003.014), p. 244 - 245 (1978).

097.141   Eolian activity in the Tharsis/Syria Planum region
of Mars: a summary of Viking and Mariner 9 results.
J. Veverka, P. Thomas, S. Lee, R. Greeley.
Reports of planetary geology program, 1977–1978,
(see 003.014), p. 246 (1978).

097.142   Morphometry of streamlined erosional forms in
terrestrial and Martian channels.   V. R. Baker,
R. C. Kochel.
Reports of planetary geology program, 1977–1978,
(see 003.014), p. 251 - 253 (1978).

097.143   The role of liquefaction in channel development on
Mars.   D. Nummedal.
Reports of planetary geology program, 1977–1978,
(see 003.014), p. 257 - 259 (1978).

097.144   Formation of Martian flood features by release of
water from confined aquifers.   M. H. Carr.
Reports of planetary geology program, 1977–1978,
(see 003.014), p. 260 - 262 (1978).

097.145   Analysis of shape-frequency histograms of terrestrial and Martian remnant landforms.
D. T. Eppler, P. J. Brown, R. Ehrlich, D. Nummedal.
Reports of planetary geology program, 1977–1978,
(see 003.014), p. 263 - 266 (1978).

097.146   Small channels on Mars from Viking Orbiter.
D. Pieri.
Reports of planetary geology program, 1977–1978,
(see 003.014), p. 267 (1978).

097.147   Junction angles of Martian channels.
D. Pieri, C. Sagan.
Reports of planetary geology program, 1977–1978,
(see 003.014), p. 268 (1978).

097.148   Channeling history of Maja Vallis.   E. Theilig,
R. Greeley.
Reports of planetary geology program, 1977–1978,
(see 003.014), p. 269 - 271 (1978).

097.149   Stability analysis for the origin of Martian fluvial
features.   D. E. Thompson.
Reports of planetary geology program, 1977–1978,
(see 003.014), p. 272 - 274 (1978).

097.150   Large scale erosive flows associated with Chryse
Planitia, Mars: source and sink relationships.
K. R. Blasius, J. A. Cutts, W. J. Roberts.
Reports of planetary geology program, 1977–1978,
(see 003.014), p. 275 - 276 (1978).

097.151   Chaotic terrain and channels associated with Chryse
Planitia, Mars: an alternative erosional model.
J. A. Cutts, K. R. Blasius, W. J. Roberts.
Reports of planetary geology program, 1977–1978,
(see 003.014), p. 277 - 279 (1978).

097.152   The flow mechanics and resulting erosional and
depositional features of explosive volcanic density
currents on Earth and Mars.   C. E. Reimers, P. D. Komar.
Reports of planetary geology program, 1977–1978,
(see 003.014), p. 280 - 281 (1978).

097.153   Martian landslides – classification and genesis.
E. H. Christiansen, J. W. Head.
Reports of planetary geology program, 1977–1978,
(see 003.014), p. 285 - 287 (1978).

097.154   Landslides in the Valles Marineris, Mars.
B. K. Lucchitta.
Reports of planetary geology program, 1977–1978,
(see 003.014), p. 288 - 290 (1978).

097.155   Volatile evolution.   F. P. Fanale.
Reports of planetary geology program, 1977–1978,
(see 003.014), p. 292 - 294 (1978).

097.156   Viking soil: chemical activity from surface frosts?
R. L. Huguenin.
Reports of planetary geology program, 1977–1978,
(see 003.014), p. 298 - 300 (1978).

097.157   Surface materials of the Viking landing sites-extend-
ed mission.   H. J. Moore, C. R. Spitzer, P. Cates,
K. Bradford, R. F. Scott, R. E. Hutton, R. W. Shorthill.

Reports of planetary geology program, 1977–1978, (see 003.014), p. 301 - 302 (1978).

097.158 **Geochemical studies within simulated Martian environments.** M. C. Booth, H. H. Kieffer.
Reports of planetary geology program, 1977–1978, (see 003.014), p. 303 - 304 (1978).

097.159 **Planetary cartography.** R. Batson.
Reports of planetary geology program, 1977–1978, (see 003.014), p. 325 - 326 (1978).

097.160 **The control net of Mars.** M. E. Davies.
Reports of planetary geology program, 1977–1978, (see 003.014), p. 328 - 329 (1978).

097.161 **Mars geologic mapping.** D. H. Scott.
Reports of planetary geology program, 1977–1978, (see 003.014), p. 330 - 332 (1978).

097.162 **Standard techniques for presentation and analysis of crater size-frequency data.** – A report of the crater analysis techniques working group. R. Arvidson, J. Boyce, C. Chapman, M. Cintala, M. Fulchignoni, H. Moore, G. Neukum, P. Schultz, L. Soderblom, R. Strom, A. Woronow, R. Young.
Reports of planetary geology program, 1977–1978, (see 003.014), p. 338 - 339 (1978).

097.163 **The ephemeris of Mars.** E. M. Standish, Jr.
Bull. American Astron. Soc., Vol. 10, 482 (1978). Abstract.

097.164 **The use of Viking range data in the development of Mars'planetary and physical ephemerides.**
M. S. W. Keesey, D. L. Cain.
Bull. American Astron. Soc., Vol. 10, 482 (1978). – Abstract.

097.165 **A comparison of observations and ephemerides of Mars, 1950 - 1975.**
Y. Kubo, R. L. Duncombe, P. K. Seidelmann.
Bull. American Astron. Soc., Vol. 10, 482 (1978). – Abstract.

097.166 **The dark and bright streaks of Mars.** R. O. Kuzmin.
Mod. Geol., Vol. 6, No. 3, p. 139 - 146 (1978). – Abstr. in Phys. Abstr., Vol. 81, Abstr. 32381 (1978).

097.167 **Martian fresh crater morphology and morphometry – a pre-Viking review.** M. J. Cintala.
Impact and explosion cratering, (see 012.034), p. 575 - 591 (1977).

097.168 **Distribution and emplacement of ejecta around martian impact craters.** M. H. Carr.
Impact and explosion cratering, (see 012.034), p. 593 - 602 (1977).

097.169 **Cratering mechanics and future Martian exploration.** H. Masursky.
Impact and explosion cratering, (see 012.034), p. 635 - 637 (1977).

097.170 **Principles and methods of construction of a tectonic map of Mars.** V. A. Avdeev, Ya. G. Kats, V. V. Kozlov, E. D. Sulidi-Kondrat'ev.
Izv. vyssh. uchebn. zaved. Geol. i razvedka, 1977, No. 10, p. 9 - 15. In Russian. – Abstr. in Ref. zh., 51. Astron., 6.51.248 (1978).

097.171 **Great dust storms on Mars and cap recession rate, from 1877 to 1971.** G. de Mottoni y Palacios.

Mem. Soc. Astron. Italiana, Vol. 47, 211 - 216 (1976).
An examination of the observations of the martian surface features collected during the perihelion oppositions from 1877 to 1971 shows that an exceptionally high rate of dissolution of the south polar cap and the occurrence of great dust storms are related phenomena.

097.172 **Mars: regolith adsorption and the relative concentrations of atmospheric rare gases.** F. P. Fanale, W. A. Cannon, T. Owen.
Geophys. Res. Lett., Vol. 5, 77 - 80 (1978).
Laboratory measurements of Kr and Xe adsorption on samples of ground montmorillonite, limonite and basalt at $-77°$ C have been performed. Results suggest that 1) most degassed martian Xe could be adsorbed on a regolith with a large effective surface area, and 2) if this is the case the elemental composition of martian nonradiogenic rare gas (counting that in the regolith) may be virtually identical to that of ordinary chondrites.

097.173 **The magnetic field of Mars: Mars 3 evidence re-examined.** C. T. Russell.
Geophys. Res. Lett., Vol. 5, 81 - 84 (1978).
Published reports permit the reconstruction of the Mars 3 trajectory on which the discovery of the Martian magnetic field by Dolginov and co-workers is based. It is concluded that the observed magnetic field was draped over the Martian obstacle as expected if the field were simply shocked and compressed solar wind magnetic field. There is no conclusive evidence that Mars 3 ever entered a Martian magnetosphere.

097.174 **The magnetic field of Mars: Mars 5 evidence re-examined.** C. T. Russell.
Geophys. Res. Lett., Vol. 5, 85 - 88 (1978).
It is the purpose of this article to examine the magnetic field data from the Mars 5 spacecraft to see if there is any evidence for penetrations of a Martian magnetotail by Mars 5.

097.175 **On the magnetic field of Mars: Mars 2 and 3 evidence.** Sh. Sh. Dolginov.
Geophys. Res. Lett., Vol. 5, 89 - 92 (1978).
Recently the existence of an intrinsic field of Mars has been questioned. In this note the author reviews the evidence from Mars 2 and 3 from which his original conclusion of a Martian magnetic field was deduced and reaffirms his original conclusions.

097.176 **On the magnetic field of Mars: Mars 5 evidence.** Sh. Sh. Dolginov.
Geophys. Res. Lett., Vol. 5, 93 - 95 (1978).
The measurements obtained by Mars 5 reaffirm the conclusion of an intrinsic Martian magnetic field based on the Mars 2 and 3 data. The bow shock positions are equivalent to those observed on the earlier missions. Furthermore, a magnetotail is observed whose direction is independent of interplanetary sector structure.

097.177 **On search for life on Mars.**
A. A. Imshenetskij, B. G. Murzakov.
Mikrobiol., Vol. 46, 1103 - 1113 (1977). In Russian. – Abstr. in Ref. zh., 62. Issled. kosm. prostranstva, 7.62.165 (1978).

Viking lander imaging investigation – Picture catalog of primary mission. Experiment data record.
See Abstr. 002.026.

Catalogue of Martian craters and statistics of the craters of Mars, Mercury and the Moon.
See Abstr. 002.043.

Der Mars. Bericht über einen Nachbarplaneten.
See Abstr. 003.073.

Guide to Mars.   See Abstr. 003.095.

Moon, Mars and meteorites.   See Abstr. 003.154.

Report on the Tharsis workshop.
See Abstr. 013.026.

Pyrolysis of organic compounds in the presence of ammonia. The Viking Mars lander site alteration experiment.
See Abstr. 022.073.

Exploratory experiments of impact craters formed in viscous-liquid targets: analogs for Martian rampart craters?
See Abstr. 022.093.

Water vapor adsorption by sodium montmorillonite at −5°C.   See Abstr. 022.094.

Diffuse reflectance spectra of azurite, malachite, and chrysocolla.   See Abstr. 022.108.

Mars water instrument study.   See Abstr. 032.541.

The biology instrument for the Viking Mars mission.
See Abstr. 032.577.

Éphémérides et positions de Mars en 1975–1976.
See Abstr. 041.007.

Mars observations with Algiers astrolabe (winter 1973).   See Abstr. 041.016.

Observations of Mars with the Danjon astrolabe on the San Fernando Observatory.   See Abstr. 041.018.

Observations of Mars position with the Paris astrolabe.   See Abstr. 041.019.

Photographic position observations of Mars with the 400-mm astrograph at the Main Astronomical Observatory of the Ukrainian Academy of Sciences (1971 - 1972).
See Abstr. 041.028.

Observations of major planets with the meridian circle of the Engelhardt Astronomical Observatory.
See Abstr. 041.038.

Observations of Mars with the photographic zenith tube.   See Abstr. 041.040.

Accuracy of planetary theories, particularly for Mars.   See Abstr. 042.127.

Viking to Mars.   See Abstr. 053.003.

Viking science: tantalizing Viking scientists cautious.
See Abstr. 053.004.

Limits to the expansion of Earth, Moon, Mars and Mercury and to changes in the gravitational constant.
See Abstr. 081.006.

Interactions of volcanism and ice − earth, Mars and icy satellites.   See Abstr. 091.025.

Application of anisotropy of magnetic susceptibility measurements to transport mechanisms on planetary surfaces.
See Abstr. 091.066.

An analytic study of impact ejecta trajectories in the atmospheres of Venus, Mars, and Earth.
See Abstr. 093.002.

Calculation of the dependence of the stream of gamma-quanta from the moon and Mars on the relief and distance to the surface.   See Abstr. 094.135.

Problems of geology of the moon, Mercury and Mars.   See Abstr. 094.159.

The significance of buried craters associated with basins on the Moon and Mars.   See Abstr. 094.453.

Phobos transit of Mars as viewed by the Viking cameras.   See Abstr. 097.515.

Possible effects of Mars on the symmetry plane of interplanetary dust.   See Abstr. 106.019.

Mars Satellites

097.501 A model of Phobos.   R. J. Turner.
   Icarus, Vol. 33, 116 - 140 (1978).
   A model of the Martian satellite Phobos was constructed at a scale of 1 : 60 000 using 25 Mariner 9 photorecords and a solar-simulation technique. Measurements of the crater diameters D, depths d, ratios d/D, longitude and latitude locations of the centers, IAU designations, crater shapes, and rim class are given in a catalog of 260 depressions. A 100-m contour-interval topographic map has been drawn from measurements of the model. This is rendered on an elliptical form of a Lambert equal-area polar projection.

097.502 Viking first encounter of Phobos: preliminary results.   R. H. Tolson, T. C. Duxbury, G. H. Born, E. J. Christensen, R. E. Diehl, D. Farless, C. E. Hildebrand, R. T. Mitchell, P. M. Molko, L. A. Morabito, F. D. Palluconi, R. J. Reichert, H. Taraji, J. Veverka, G. Neugebauer, J. T. Findlay.
Science, Vol. 199, 61 - 64 (1978).
   During the last 2 weeks of February 1977, an intensive scientific investigation of the martian satellite Phobos was conducted by the Viking Orbiter-1 (VO-1) spacecraft. More than 125 television pictures were obtained during this period and infrared observations were made. An estimate of the mass of Phobos was obtained by observing the effect of Phobos's gravity on the orbit of VO-1 as sensed by Earth-based radiometric tracking. The low density, together with the low albedo and the recently determined spectral reflectance, suggest that Phobos is compositionally similar to type I carbonaceous chondrites.

097.503 The composition of Phobos: evidence for carbonaceous chondrite surface from spectral analysis.
K. D. Pang, J. B. Pollack, J. Veverka, A. L. Lane, J. M. Ajello.
Science, Vol. 199, 64 - 66 (1978).
   A reflectance spectrum of Phobos (from 200 to 1100 nanometers) has been compiled from the Mariner 9 ultraviolet spectrometer, Viking lander imaging, and ground-based photometric data. The spectral albedo of Phobos bears a

striking resemblance to that of asteroids (1) Ceres and (2) Pallas. Comparison of the reflectance spectra of asteroids with those of meteorites has shown that the spectral signature of Ceres is indicative of a carbonaceous chondritic composition. A physical explanation of how the compositional information is imposed on the reflectance spectrum is given. It seems reasonable to believe that the surface composition of Phobos is similar to that of carbonaceous chondrites.

097.504 Multicolor observations of Phobos with the Viking lander cameras: evidence for a carbonaceous chondritic composition. J. B. Pollack, J. Veverka, K. Pang, D. Colburn, A. L. Lane, J. M. Ajello. Science, Vol. 199, 66 - 69 (1978).

The reflectivity of Phobos has been determined in the spectral region from 0.4 to 1.1 micrometers from images taken with a Viking lander camera. The reflectivity curve is flat in this spectral interval and the geometric albedo equals 0.05 ± 0.01. These results, together with Phobos's reflectivity spectrum in the ultraviolet, are compared with laboratory spectra of carbonaceous chondrites and basalts. The spectra of carbonaceous chondrites are consistent with the observations, whereas the basalt spectra are not. These findings raise the possibility that Phobos may be a captured object rather than a natural satellite of Mars.

097.505 New light on Phobos, the inner satellite of Mars. G. Fielder. Nature, Vol. 271, 613 - 614 (1978).

097.506 Photometry and distribution of dark material on Phobos. J. Goguen, J. Veverka, T. Duxbury. Bull. American Astron. Soc., Vol. 9, 517 (1977). – Abstract.

097.507 Grooves on Phobos: morphology, distribution, age, and possible origin. P. Thomas, J. Veverka, T. Duxbury. Bull. American Astron. Soc., Vol. 9, 517 (1977). – Abstract.

097.508 The surface of Phobos: summary of the latest Viking Orbiter results. J. Veverka, P. Thomas, T. Duxbury. Bull. American Astron. Soc., Vol. 9, 517 - 518 (1977). Abstract.

097.509 Deimos encounter by Viking Orbiter-2: preliminary imaging results. T. C. Duxbury, J. Veverka. Bull. American Astron. Soc., Vol. 9, 518 (1977). – Abstract.

097.510 Spectral albedo and composition of Deimos: inferences on the origin of Martian satellites. K. D. Pang, J. W. Rhoads, A. L. Lane, J. M. Ajello. Bull. American Astron. Soc., Vol. 9, 518 (1977). – Abstract.

097.511 Are Phobos and Deimos made of the same material? L. M. French, J. Veverka. Bull. American Astron. Soc., Vol. 9, 518 (1977). – Abstract.

097.512 An origin by capture for the Martian satellites? J. B. Pollack, J. A. Burns. Bull. American Astron. Soc., Vol. 9, 518 - 519 (1977). Abstract.

097.513 On the origin of the satellites of Mars. A. Harris. Bull. American Astron. Soc., Vol. 9, 519 (1977). – Abstract.

097.514 Mars satellites. E. L. Ruskol. Zemlya i Vselennaya, 1978, No. 2, p. 52 - 56. In Russian.

097.515 Phobos transit of Mars as viewed by the Viking cameras. T. C. Duxbury. Science, Vol. 199, 1201 - 1202 (1978).

A Viking orbiting spacecraft successfully obtained pictures of the martian satellite Phobos with Mars in the background. This is the first time that a single picture was obtained from a spacecraft which contained both a planet and a moon and had significant surface detail visible on both. These pictures showed Phobos to be smaller than previously thought. The image of Phobos can be used as a control point to determine the map coordinates of surface features on Mars.

097.516 Origin of the grooves on Phobos. P. Thomas, J. Veverka, T. Duxbury. Nature, Vol. 273, 282 - 284 (1978).

The authors present new data on the grooves which tend to support the cratering hypothesis and seem to rule out the tidal mechanism.

097.517 Phobos and Deimos. J. B. Pollack. Planetary satellites, (see 003.007), p. 319 - 345 (1977).

097.518 Phobos and Deimos: geodesy. T. C. Duxbury. Planetary satellites, (see 003.007), p. 346 - 362 (1977).

Results of geodesy studies of Phobos and Deimos based on Mariner 9 imaging data are presented. This analysis includes a review of the surface coverage and high resolution pictures obtained and the determined sizes, shapes, topographies, and librations of the two Martian satellites. Also, exploration of Phobos and Deimos by missions such as Viking is discussed.

097.519 On some seismic effects on small bodies of the solar system. G. A. Lejkin. Pis'ma Astron. Zh., Vol. 4, 145 - 147 (1978). In Russian.

An impact of a large meteoroid may create a system of stationary waves in a small body of the solar system. This wave system can manifest itself as a regular system of grooves if the body is covered by a regolith layer. This mechanism could be responsible for the system of grooves and the lack of small craters on Phobos.

097.520 Phobos globe/Mars topography. T. C. Duxbury. Reports of planetary geology program, 1977–1978, (see 003.014), p. 35 (1978). – Abstract.

097.521 Phobos: photometry and origin of dark markings on crater floors. J. Goguen, J. Veverka, T. C. Duxbury. Reports of planetary geology program, 1977–1978, (see 003.014), p. 36 - 37 (1978).

097.522 A comparative study of surface features on Phobos and Deimos. P. Thomas, J. Veverka, A. Bloom, T. C. Duxbury. Reports of planetary geology program, 1977–1978, (see 003.014), p. 38 - 40 (1978).

097.523 Characteristics of the cratering process on small bodies: Phobos and Deimos. M. J. Cintala, J. W. Head, J. Veverka. Reports of planetary geology program, 1977–1978, (see 003.014), p. 41 - 43 (1978).

097.524 Crater densities on Phobos and Deimos: summary of Viking results. P. Thomas, J. Veverka, T. C. Duxbury. Reports of planetary geology program, 1977–1978, (see 003.014), p. 44 - 45 (1978).

097.525 Tidal evolution of the Martian satellites' orbits.

R. H. Tolson, H. Smith, Jr.
Bull. American Astron. Soc., Vol. 10, 481 (1978). – Abstract.

097.526 **The mass of Phobos.**   R. H. Tolson, W. T. Black-
shear, M. L. Mason, G. M. Kelly.
Geophys. Res. Lett., Vol. 4, 551 - 554 (1977).

Radio tracking data associated with the February 1977
encounters between the Martian satellite Phobos and the
Viking Orbiter I spacecraft have been analyzed to determine
the gravitational constant (GM) of Phobos. The result of the
data analysis was an estimate of $(7.3 \pm 0.7) \times 10^{-4}$ km^3/sec^2
for GM of Phobos.

097.527 **The mass of Phobos from Viking flybys.**
E. J. Christensen, G. H. Born, C. E. Hildebrand,
B. G. Williams.
Geophys. Res. Lett., Vol. 4, 555 - 557 (1977).

The mass of the Martian satellite Phobos has been
determined by processing radiometric tracking data obtained
from the Viking-1 spacecraft. The authors' best estimate for
the gravitational constant of Phobos is $(6.6 \pm 0.8) \times 10^{-4}$
km^3/s^2. The corresponding density of Phobos based on a
volume estimate of 4800 ± 960 km^3 from Mariner 9 imaging
is 2.0 ± 0.5 gm/cm^3.

Der Mars. Bericht über einen Nachbarplaneten.
See Abstr. 003.073.

The Martian satellites – 100 years on.
See Abstr. 010.023.

Neue Karten und Globen des Mars und Phobos.
See Abstr. 097.010.

Mars-Programm Viking im Rückblick.
See Abstr. 097.053.

Errata

097.901 **Erratum: "Rotational variations in the radio bright-
ness of Mars"** [Astrophys. J., Lett., Vol. 213,
L131 - L134 (1977)].
B. H. Andrew, G. A. Harvey, F. H. Briggs.
Astrophys. J., Lett., Vol. 220, L61 (1978).

098 Minor Planets

098.001 Kowal's strange slow-moving object.
J. Ashbrook.
Sky Telesc., Vol. 55, 4 - 5 (1978).

098.002 Asteroid 1 Ceres : evidence for water of hydration.
L. A. Lebofsky.
Mon. Not. R. Astron. Soc., Vol. 182, 17P - 21P (1978).
An absorption feature centred near 3.0 μm has been discovered in the infrared spectrum of asteroid 1 Ceres. This spectrum has been compared with laboratory spectra of meteorites and shows great similarity to the spectra of type II carbonaceous chondrites. By analogy this suggests the presence of about 10 - 15 per cent water in the form of water of hydration on the surface of Ceres. This is the first evidence of water in the surface material of an asteroid.

098.003 Photoelectric lightcurve and the period of rotation of the asteroid 200 Dynamene: a further object with low spin rate. H. J. Schober.
Astron. Astrophys., Suppl. Ser., Vol. 31, 175 - 178 (1978).
The minor planet 200 Dynamene is an object with a diameter of 123 km, belonging to the C type asteroids with low albedo. From photoelectric observations in December 1975 with the 60 cm telescope of the OHP, France, a rotation period of $P_{syn} = 19^h$ or $0\overset{d}{.}79$ was found. The lightcurve shows double wave characteristics, typical for rotating asteroids, with a total amplitude of only $0\overset{m}{.}095$. 200 Dynamene belongs to the group of asteroids with low spin rate. An updated histogram of the frequency distribution of asteroids' rotation periods is given, including now 116 objects.

098.004 The discovery of Chiron: some reflections.
R. G. Hodgson.
Minor Planet Bull., Vol. 5, 21 - 22 (1978).

098.005 The considerations of porosity in minor planet densities. W. W. Watson.
Minor Planet Bull., Vol. 5, 22 (1978).

098.006 Minor planets at unusually favorable opposition in 1978. F. Pilcher.
Minor Planet Bull., Vol. 5, 23 - 25 (1978).

098.007 Selected ephemerides for 1977/78.
Minor Planet Bull., Vol. 5, 25 - 26 (1978).

098.008 Minor planet news.
Minor Planet Bull., Vol. 5, 26 (1978).

098.009 Minor planet occultations.
Minor Planet Bull., Vol. 5, 26 (1978).

098.010 Astéroïdes. Additif-rectificatif à l'article de Froeschlé et Scholl (l'Astronomie, février 1977).
P. Kohler.
Astronomie, Vol. 92, 102 (1978). – See 19.098.003.

098.011 Minor planets and related objects. XXIV. Photometric observations for (5) Astraea.
R. C. Taylor.
Astron. J., Vol. 83, 201 - 204 (1978).
Astraea has a sidereal period of 16 h 48 min 42.61 s (± 0.03 s) with a retrograde sense of rotation. The North Pole is at $9°$ latitude and 148° ecliptic longitude. The mean absolute magnitude is $V(1, 0) = 7.50$ and the colors are $B-V = +0.77$ and $U-B = +0.38$. The phase coefficients are +0.014, +0.004, and +0.001 mag/deg in V, $B-V$, and $U-B$, respectively.

098.012 Der Kleine Planet (85) Io in der Opposition 1977.
L. D. Schmadel.
Sterne Weltraum, 17. Jahrg., 98 - 102 (1978).

098.013 Close-approach minor planets. R. Buckley.
J. British Astron. Assoc., Vol. 88, 119 - 138 (1978).

098.014 An investigation of the identity of orbits of a number of new asteroids.
Planetary Laboratory of Purple Mountain Observatory, Academia Sinica.
Acta Astron. Sinica, Vol. 18, 203 - 210 (1977). In Chinese.
Starting from over one hundred preliminary orbits of minor planets discovered at the Purple Mountain Observatory and collecting over 7000 positions, the authors employed a computer to make a rather systematic study of their identity.

098.015 Radar detection of the asteroid 1 Ceres.
G. H. Pettengill, S. J. Ostro, I. I. Shapiro, D. B. Campbell, R. R. Green.
Bull. American Astron. Soc., Vol. 9, 502 (1977). – Abstract.

098.016 Asteroid mass determinations: a search for further encounter opportunities.
D. R. Davis, D. F. Bender.
Bull. American Astron. Soc., Vol. 9, 502 - 503 (1977). Abstract.

098.017 Remote spectroscopic identification of carbonaceous chondrite mineralogies: applications to Ceres and Pallas. M. Feierberg, H. Larson, U. Fink.
Bull. American Astron. Soc., Vol. 9, 503 (1977). – Abstract.

098.018 UBV photometry of distant asteroids and faint satellites. J. Degewij, J. Gradie, B. Zellner.
Bull. American Astron. Soc., Vol. 9, 503 (1977). – Abstract.

098.019 UBV and 10μ photometric observations of Eos and Koronis family members.
J. Gradie, G. Rieke, B. Zellner.
Bull. American Astron. Soc., Vol. 9, 503 - 504 (1977). Abstract.

098.020 The phase function of an asteroid at very small phase angles.
E. Bowell, K. Lumme, L. J. Martin.
Bull. American Astron. Soc., Vol. 9, 504 (1977). – Abstract.

098.021 Regolith evolution on small bodies.
K. Housen, L. L. Wilkening, R. Greenberg, C. R. Chapman.
Bull. American Astron. Soc., Vol. 9, 504 (1977). – Abstract.

098.022 The brittle fragmentation of asteroids.
J. L. Remo, A. A. Johnson.
Bull. American Astron. Soc., Vol. 9, 504 (1977). – Abstract.

098.023 Primordial metamorphism of asteroids via electrical induction in a T-Tauri-like solar wind.
F. Herbert, C. P. Sonett.
Bull. American Astron. Soc., Vol. 9, 505 (1977). – Abstract.

098.024 Asteroid systematics. C. R. Chapman.
Bull. American Astron. Soc., Vol. 9, 505 (1977). Abstract.

098.025 1 to 4 micron studies of asteroids and satellites.
L. A. Lebofsky.

Bull. American Astron. Soc., Vol. 9, 505 (1977). – Abstract.

098.026 Visual and radiometric photometry of 1580 Betulia.
L. A. Lebofsky, G. J. Veeder, M. J. Lebofsky,
D. L. Matson.
Bull. American Astron. Soc., Vol. 9, 551 (1977). – Abstract.

098.027 Physical properties of asteroids. Part II.
Z. Musielak.
Postępy Astron., Tom 25, 161 - 168 (1977). In Polish.
A review of results of polarimetric, radiometric and
spectrophotometric observations is presented.

098.028 Origin of slow moving object Kowal.
R. C. Smith.
Nature, Vol. 272, 229 - 230 (1978).
The possibility is examined that slow moving object
Kowal has been flung into its present orbit by a sort of gravi-
tational sling-shot, having been successively perturbed, or even
temporarily captured, by Jupiter and Saturn.

**098.029 Minor Planet Circulars, (MPC), Nos. 4265 - 4358
(1977/1978).**
Edited by Cincinnati Observatory under the supervision of
P. Herget.
A repository of nearly all new data for numbered and un-
numbered minor planets: Observations, elements and ephemerides, identifications, newly assigned numbers and names,
occultations.

098.030 Photographic photometry of 110 main-belt asteroids.
C.-I. Lagerkvist.
Astron. Astrophys., Suppl. Ser., Vol. 31, 361 - 381 (1978).
Photographic photometry has been carried out for 101
numbered and 9 newly discovered asteroids. Rotation periods
have been determined for about two dozens of the observed
asteroids. For 37 of the observed asteroids lightcurves are presented. The present study shows that there is a correlation
between the percentage of variable asteroids and the absolute
magnitudes. Family asteroids seem to have shorter rotation
periods than ordinary field asteroids. There are indications
that small asteroids have shorter rotation periods than large
asteroids. Indications are presented that the physical parameters of the asteroids depend upon the mean distance and the
orbital inclination. No correlation appears to exist between
the physical parameters and the orbital eccentricity.

**098.031 Nomina de los asteroides descubiertos hasta la
fecha (Orden alfabético).   P. L. Blanco Peñalver.**
Bol. Acad. Cienc. Fis. Mat. Nat., Año 37, Numero Extra-
ordinario, p. 163 - 197 (1977).

**098.032 Photoelectric lightcurves of the minor planets 29
Amphitrite, 121 Hermione and 185 Eunike.**
H. Debehogne, A. Surdej, J. Surdej.
Astron. Astrophys., Suppl. Ser., Vol. 32, 127 - 133 (1978).
Asteroids 29 Amphitrite, 121 Hermione and 185 Eunike
were observed with a photoelectric photometer attached to
the 50 cm telescope at the European Southern Observatory.
The lightcurve of 29 Amphitrite presents a triple maximum
and minimum as remarkable features and a total amplitude of
0.11 mag. The synodic rotation period derived for this asteroid
is found to be $5^h 23^m 24^s \pm 4^s$. The minor planet 121 Hermione
only displayed very smooth light variations (~ 0.03 mag)
during 16 hours of observation. No rotation period could be
deduced from the measurements. Finally, asteroid 185 Eunike
appeared with light variations greater than 0.12 mag and with
a very probable synodic rotation period of $10^h 50^m \pm 2^m$.

**098.033 Observations of Minor Planets at ESO La Silla by
means of the GPO (f = 400 cm, $\varnothing$ = 40 cm) in
August 1976.   H. Debehogne, A. Surdej, J. Surdej.**

Astron. Astrophys., Suppl. Ser., Vol. 32, 135 - 140 (1978). In
French.
In August 1976, the authors have observed Minor Planets
at ESO La Silla, most of them indicated as having to be observed. Because of the sky quality, magnitude 17 was easily
attained. Three new asteroids were discovered. Measures and
reductions were performed at Uccle with the Ascorecord
measuring machine (0.1 μ) and by means of five reference stars
on the Siemens 4004 ORB computer (SAO catalogue registered on magnetic tape by the Strasbourg Observatory, mean
least squares method).

098.034 Das Objekt zwischen Saturn- und Uranusbahn.
W. Kranzer.
Phys. Bl., 34. Jahrg., 192 (1978).

098.035 Far-infrared observation of Ceres.
E. L. Wright, G. G. Fazio, M. T. Stier, F. J. Low.
Bull. American Astron. Soc., Vol. 9, 563 (1978). – Abstract.

098.036 Occultations of stars by minor planets.
D. W. Dunham.
Bull. American Astron. Soc., Vol. 9, 621 (1978). – Abstract.

**098.037 Expected shape distribution of asteroids obtained
from laboratory impact experiments.**
A. Fujiwara, G. Kamimoto, A. Tsukamoto.
Nature, Vol. 272, 602 - 603 (1978).
The authors examine the shape of the fragments produced
in laboratory high-velocity impact experiments between two
solid bodies, and they find that the obtained shape distribution may be applied to asteroids of radius less than a few km.

098.038 Lunar occultations of minor planets in 1978.
D. W. Dunham.
Occultation Newsl., Vol. 1, 141 (1978).

098.039 Lunar occultation of (15) Eunomia.
D. W. Dunham.
Occultation Newsl., Vol. 1, 141 - 142 (1978).

098.040 Predictions of planetary occultations.
D. W. Dunham
Occultation Newsl., Vol. 1, 142 - 143 (1978).

098.041 Passages of comets and asteroids near the Earth.
L'. Kresák.
Bull. Astron. Inst. Czechoslovakia, Vol. 29, 103 - 114 (1978).
A list of 82 approaches of different comets and asteroids
to within 0.20 A.U. of the Earth is compiled. Parameters relevant to encounter geometry and brightness are computed for
each event, and the geocentric trajectories are presented. The
effects of observational selection depending on absolute brightness, heliocentric distance, geocentric distance, phase angle,
and atmospheric extinction at small solar elongations are evaluated, and the problem of conversion of the absolute magnitudes into body diameters is discussed.

**098.042 The comet and asteroid population of the earth's
environment.   L'. Kresák.**
Bull. Astron. Inst. Czechoslovakia, Vol. 29, 114 - 125 (1978).
The average flux and number density in circumterrestrial
space is determined for different types and sizes of interplanetary objects. For body diameters exceeding 1 km, the average
rate of approaches to within 0.1 A.U. per century is estimated
at 2.5 for long-period comets, 0.2 for short-period comets of
the Halley type, 1.0 for short-period comets of the Jupiter
family, 20 to 30 for asteroids of the Amor type, and 120 to
170 for asteroids of the Apollo type. The earth would collide
with a body of this size once in 1.5–2 million years. The
directional patterns of streaming with respect to a fixed point
on the earth's orbit, and with respect to the moving earth are

derived. For the long-period comets this is entirely consistent with a random distribution of orbital planes. The directional pattern of the short-period comets, with direct motions prevailing even in the geocentric reference frame, is practically indistinguishable from that of the earth-approaching asteroids.

098.043 Kleine Planeten – II. Quartal 1978.
L. D. Schmadel.
Sterne Weltraum, Jahrg. 17, 144 - 145 (1978).

098.044 Conjonctions de petites planètes avec des étoiles brillantes.   J. Meeus.
Astronomie, Vol. 92, 189 - 191 (1978).

098.045 Minor planet observing programs for smaller observatories.   R. G. Hodgson.
Minor Planet. Bull., Vol. 5, 27 - 30 (1978).

098.046 On the possible diameter of 1977 UB ("Chiron").
D. Wallentinsen.
Minor Planet. Bull., Vol. 5, 30 (1978).

098.047 Discovery and observations of two new planets: 1978 CA and 1978 DA.   R. G. Hodgson.
Minor Planet. Bull., Vol. 5, 31 - 32 (1978).

098.048 On the spin rate of the S-type asteroids.
C. P. Renschen.
Astron. Nachr., Band 299, 103 - 106 (1978).
A relation between the radii and rotation rates of S-type asteroids is discussed. According to this relation these objects do not rotate with the same spin rate.

098.049 The size of minor planet 6 Hebe.
G. E. Taylor, D. W. Dunham.
Icarus, Vol. 34, 89 - 92 (1978).
The size of the minor planet Hebe is determined from timed visual observations of an occultation of a star. For an oblate planet the equatorial diameter is found to be 195 ± 6 km, with a corresponding polar diameter of 170 km. With an assumed circular cross section its diameter is found to be 186 ± 9 km.

098.050 Lightcurves, phase function and pole of the asteroid 22 Kalliope.   F. Scaltriti, V. Zappalà, R. Stanzel.
Icarus, Vol. 34, 93 - 98 (1978).
Photoelectric observations of 22 Kalliope give a synodic rotation period of $4^h 08^m 52^s.1 \pm 0^s.4$. The magnitude–phase relation shows an opposition effect, starting near a phase angle of 7° to 8°, as usual. From the V magnitude of the maximum light, an effort was made to find the coordinates of the pole of 22 Kalliope using the method of Sather (1976) for 39 Laetitia, assuming a triaxial ellipsoid. The authors found $\lambda_{pole} = 215 \pm 10°$ and $\beta_{pole} = 45 \pm 15°$. From the adopted geometrical albedo $p_V = 0.123$ (Chapman et al., 1975) the resulting dimensions were roughly $215 \times 160 \times 130$ km^3.

098.051 Fragmentation of asteroids and artificial satellites in orbit.   W. Wiesel.
Icarus, Vol. 34, 99 - 116 (1978).
The author develops expressions for the probability density in the orbital elements resulting from a spherical explosion in orbit. This theory is then applied to both the asteroid families and the satellite breakups. The observation of similarities between satellite breakups and asteroid families leads to the conclusion that the asteroid families are also the result of catastrophic fragmentation events.

098.052 Astrometric observations of the minor planet 433 Eros during the opposition of 1975.
K. S. Rumstay.
Astron. J., Vol. 83, 447 - 448 (1978).

A series of 31 astrometric observations of the minor planet 433 Eros, made during its 1975 opposition, is presented. The observations were made at Van Vleck Observatory and are reduced to topocentric positions.

098.053 Un nouvel astéroïde exceptionnel: 1977 UB.
M.-A. Combes, J. Meeus.
Astronomie, Vol. 92, 231 - 235 (1978).

098.054 186 Celuta: a slowly spinning asteroid.
C.-I. Lagerkvist, B. Pettersson.
Astron. Astrophys., Suppl. Ser., Vol. 32, 339 - 342 (1978).
The asteroid 186 Celuta was observed photoelectrically and photographically during August and September, 1977. The synodic period was determined to $0^d.816$ and the lightcurve amplitude was found to be about $0^m.4$. From the colour $B-V = 0^m.86$ and $U-B = 0^m.38$, 186 Celuta is of compositional type S. The absolute magnitude was found to be $V_0 (1,0) = 9^m.31$.

098.055 General report of position observations by the A.L.P.O. minor planets section for the year 1977.
F. Pilcher.
Minor Planet. Bull., Vol. 6, 1 - 10 (1978).

098.056 Revised guide to "New names of minor planets".
J. LoGuirato.
Minor Planet Bull., Vol. 6, 11 - 12 (1978).

098.057 Unusual minor planet.   D. Ya. Martynov.
Zemlya i Vselennaya, 1978, No. 3, p. 37. In Russian.

098.058 The total number of the Apollo asteroids and their chance rediscoveries.   L'. Kresák.
Bull. Astron. Inst. Czechoslovakia, Vol. 29, 149 - 154 (1978).
The estimates of the total number of the Apollo asteroids based on the observation that none of them has been rediscovered by chance, are discussed. It is shown that the adopted sampling-with-replacement analogy does not apply to the problem. The point is that for a steadily increasing number of new objects, reliable orbits and ephemerides tend to preclude a chance rediscovery. Curiously enough, strict probability tests making use of all available data do neither reject the unsound hypothesis that all Apollo asteroids of $D > 1$ km have already been detected, nor do they fix any upper limit to the number of the unknown objects.

098.059 Asteroid lightcurves simulated by the rotation of a three-axes ellipsoid model.   A. Surdej, J. Surdej.
Astron. Astrophys., Vol. 66, 31 - 36 (1978).
Simulating the rotation of an asteroid by a three-axes ellipsoid model, asteroid lightcurves are presented under different general assumptions. It is shown that the good agreement between laboratory work and the mathematical model presented here is in favour of the Hapke-Irvine relation for describing the scattering properties of a small surface element. The authors confirm by further arguments that typical asteroid phase coefficients cannot be interpreted unambiguously.

098.060 Photoelectric photometry of asteroids 37, 80, 97, 216, 270, 313, and 471.
F. Scaltriti, V. Zappalá.
Icarus, Vol. 34, 428 - 435 (1978).
Photoelectric observations of several minor planets were made from 1975 to 1977 at the Astronomical Observatory of Torino. Lightcurves and period information are given for 7 objects: 37 Fides, 80 Sappho, 97 Klotho, 216 Kleopatra, 270 Anahita, 313 Chaldaea, and 471 Papagena.

098.061 The mini-planet.   I. Ridpath.
New Scientist, Vol. 76, 406 - 407 (1977). – Abstr.

in Phys. Abstr., Vol. 81, Abstr. 20721 (1978).

**098.062   Minor planets and related objects. XXV. *UBV*
photometry of 145 faint asteroids.**
J. Degewij, J. Gradie, B. Zellner.
Astron. J., Vol. 83, 643 - 650 (1978).

Magnitudes and colors on the *UBV* system are presented
for 145 minor planets, including 31 objects in the Eos
family, 14 in the Koronis family, 6 in the Nysa family, 11 in
the Themis family, 4 Hungarias, 7 Hildas, 8 Trojans, and
several objects in unusual orbits.

098.063   Visual and infrared photometry of asteroids.
G. J. Veeder, D. L. Matson, J. C. Smith.
Astron. J., Vol. 83, 651 - 663 (1978).

The authors report the results of a survey of asteroids at
0.56, 1.6, and 2.2 μ. The observations of thirty asteroids are
reduced to relative reflectances R_λ at 1.6 μ and 2.2 μ [such
that R_λ (0.56 μ) is scaled to unity]. These relative reflectances
have important implications for the classification of asteroids.
Low albedo asteroids show a significant range in their infrared
relative reflectances, but carbonaceous chondritic-type
material remains a good candidate for their surface composition. Many S-type asteroids are found to have significantly
brighter infrared relative reflectances than M-type asteroids.
Lunar-like, dark glass on the surfaces of these objects is
probably ruled out. A metallic phase is the most plausible
candidate which implies a multiple component surface
composition perhaps similar to certain stony-iron meteorites.
The data suggest that such a metallic phase may be common in
the inner asteroid belt.

**098.064   Asteroid surface materials: mineralogical characteri-
zations from reflectance spectra.**
M. J. Gaffey, T. B. McCord.
Space Sci. Rev., Vol. 21, 555 - 628 (1978).

The authors discuss the utilization of general and specific
diagnostic spectral features and parameters to interpret most
of the published high quality reflectance spectra of asteroids.
This provides the most complete and sophisticated mineralo-
gically- and petrologically-based interpretation of the surface
materials of these asteroids. A range of mineral assemblages
similar to certain meteorite classes have been identified.

**098.065   Kirkwood gaps and stability of conservative
periodic systems.**   T. Kiang.
Nature, Vol. 273, 734 - 736 (1978).

The author shows that the Hilda region is stable in a well-
defined sense, and the Hecuba region is unstable.

098.066   1977 VA.
IAU Circ., Nos. 3158, 3213 (1978).

098.067   Occultations by minor planets.
IAU Circ., No. 3158 (1978).

098.068   1977 HB.
IAU Circ., Nos. 3159, 3223 (1978).

098.069   1977 YA (formerly comet Lovas 1977t).
IAU Circ., Nos. 3162, 3165, 3169, 3183 (1978).

098.070   Observations of minor planets.
IAU Circ., No. 3166 (1978). – Concerning 1960 UA,
1977 RA, 1977 VA.

098.071   1978 CA (object Schuster).
IAU Circ., Nos. 3174, 3179, 3185, 3189, 3191,
3193, 3200, 3212 (1978).

098.072   1978 DA.
IAU Circ., Nos. 3183, 3185, 3189, 3192, 3193,

3203, 3212, 3229 (1978).

**098.073   Improvement of the elements of minor planet
orbits in the Institute for Theoretical Astronomy
of the USSR Academy of Sciences.**   A. S. Sufiyanova.
Tr. Samarkand. univ., 1975, No. 256, p. 128 - 134. In Russian. – Abstr. in Ref. zh., 51. Astron., 7.51.114 (1978).

098.074   1977 UB.
IAU Circ., No. 3215 (1978).

098.075   1977 RA.
IAU Circ., No. 3219 (1978).

**098.076   Occultation of SAO 85009 by (2) Pallas on 1978
May 29.**
IAU Circ., Nos. 3221, 3224 (1978).

098.077   1975 YA.
IAU Circ., No. 3231 (1978).

098.078   1977 UB (Kowal).
Yamamoto Circ., No. 1872 (1977), No. 1873 (1978).
In Japanese.

098.079   1977 YA (formerly comet Lovas (1977t)).
Yamamoto Circ., Nos. 1874, 1876, 1879 (1978).
In Japanese.

098.080   1978 CA (object Schuster).
Yamamoto Circ., Nos. 1879, 1880 (1978). In
Japanese.

098.081   1978 DA.
Yamamoto Circ., No. 1879 (1978). In Japanese.

098.082   Object Yakiimo 2.
Yamamoto Circ., No. 1883 (1978). In Japanese.

**098.083   Occultation of SAO 165132 by (9) Metis on 1978
July 12.**
Yamamoto Circ., No. 1888 (1978). In Japanese.

**098.084   Photoelectric lightcurves and rotation period of the
minor planet 59 Elpis.**
H. Debehogne, A. Surdej, J. Surdej.
Astron. Astrophys., Suppl. Ser., Vol. 33, 1 - 5 (1978).

The minor planet 59 Elpis was observed photoelectrically
with the ESO 50 cm telescope at La Silla (Chile) during the
1976 opposition. The lightcurve appears fairly continuous
with two nearly symmetric maxima and minima. The synodic
rotation period is found to be $13^h 41^m 24^s \pm 18^s$ and the maximum amplitude about 0.1 mag.

098.085   Positions of selected minor planets.
S. Vaghi, V. Zappalà.
Astron. Astrophys., Suppl. Ser., Vol. 33, 11 - 13 (1978).

The precise positions are given of 18 minor planets observed during the period March 1976 - June 1977 at the
Observatory of Torino.

**098.086   Primordial metamorphism of asteroids via electrical
induction in a T Tauri-like solar wind.**
F. Herbert, C. P. Sonett.
Astrophys. Space Sci., Vol. 55, 227 - 239 (1978).

Simple evolutionary models of asteroids of various sizes
and solar distances have been constructed assuming unipolar
electrical induction heating due to passage of the Sun through
a T Tauri phase with an increased magnetic field. Typical T
Tauri conditions and an elementary solar wind model were
used to calculate induced currents in models assuming electrical conductivities appropriate for carbonaceous material. It

is found that maximum heating, in some cases sufficient for melting, occurs for model asteroids at the inner edge of the belt and with (model-dependent) radii from 25 to 250 km. This effect, if operant, would have produced a primordial distribution of metamorphosed asteroids primarily occurring at small solar distance and intermediate size. The model seems relevant to the dichotomy between Ceres and Vesta.

098.087 **Ephemerides of minor planets for 1979.**
Editor: Institut Teoreticheskoj Astronomii Akademii Nauk SSSR, under the editorship of N.S. Yakhontova. Izdatel'stvo Nauka, Leningradskoe Otdelenie, Leningrad. 222 pp. Price 3 Rbl. 20 Kop. (1978). In Russian and English.
Contents: Introduction, p. 3 - 8; Information on new elements, p. 9 - 13; New elements, p. 14 - 18; Elements, p. 19 - 56; Opposition dates, p. 57 - 69; Ephemerides, p. 70 - 199; Ephemerides of bright planets, p. 200 - 213; Ephemerides of some unusual planets, p. 214 - 220; Critical list, p. 221.

098.088 **Chiron: a new planet in the solar system.**
Messenger, No. 12, p. 6 (1978).

098.089 **The close encounter with 1978 CA and 1978 DA.**
Messenger, No. 13, p. 2 - 3 (1978).

098.090 **Photometric observations of 1978 CA and 1978 DA.**
J. Surdej, A. Surdej.
Messenger, No. 13, p. 3 - 4 (1978).

098.091 **The sizes of 1978 CA and 1978 DA.**
J. Degewij.
Messenger, No. 13, p. 5 (1978).

098.092 **Asteroids as geologic materials: collisions and fragmentation.** C. R. Chapman, D. R. Davis, R. Greenberg, J. Wacker.
Reports of planetary geology program, 1977–1978, (see 003.014), p. 25 - 27 (1978).

098.093 **The discovery and orbit of 1977 UB.**
C. T. Kowal, W. Liller, B. G. Marsden.
Bull. American Astron. Soc., Vol. 10, 481 (1978). – Abstract.

098.094 **Observations photographiques de petites planètes effectuées à l'astrographe double de 40 cm au cours de l'année 1976.**
H. Debehogne.
Bull. Astron. Obs. R. Belgique, Vol. 9, 2 - 11 (1978).

098.095 **Observations photographiques de petites planètes effectuées en 1976 à l'équatorial GPO de 40 cm de l'Observatoire Austral Européen (ESO) à La Silla (Chili).**
H. Debehogne.
Bull. Astron. Obs. R. Belgique, Vol. 9, 12 - 18 (1978).

098.096 **Observations photographiques de petites planetes effectuées en 1976 à la caméra astrographique de 25 cm de l'Observatoire National de Rio de Janeiro.**
H. Debehogne, R. R. de Freitas Mourão.
Bull. Astron. Obs. R. Belgique, Vol. 9, 18 - 19 (1978).

098.097 **Object Kowal 1977 UB.** J. Bouška.
Říše hvězd, Vol. 59, 37 - 40 (1978). In Czech.

098.098 **1977 UB.**
British Astron. Assoc. Circ., No. 583 (1978).

098.099 **1977 YA (formerly comet Lovas 1977t).**
British Astron. Assoc. Circ., Nos. 585, 587 (1978).

098.100 **Chiron – the celestial centaur.** K. Hindley.
New Scientist, Vol. 77, 300 - 301 (1978). – Abstr.

in Phys. Abstr., Vol. 81, Abstr. 53863 (1978).

The Tucson Revised Index of Asteroid Data.
See Abstr. 002.004.

Die kleinen Planeten. Planetoide und ihre Entdeckungsgeschichte. See Abstr. 003.042.

The asteroid belt. See Abstr. 003.127.

The orbits of five minor planets and corrections to the FK4 equator and equinox. See Abstr. 041.025.

Orbital resonance in a dissipative medium.
See Abstr. 042.001.

New results on the commensurability cases of the problem Sun-Jupiter-asteroid. See Abstr. 042.060.

A theory of the Trojan asteroids.
See Abstr. 042.061.

An invariant relation in the elliptic restricted problem of three bodies. II. Trojan asteroids.
See Abstr. 042.102.

Theory of the Trojan asteroids, Part II.
See Abstr. 042.121.

Planetary theory developments, 1973–1976.
See Abstr. 042.129.

Minor planet orbits and the FK4 equator and equinox. See Abstr. 043.004.

Asteroidal resources and gravity-assisted low-energy retrieval missions. See Abstr. 051.003.

Near-earth asteroids as targets for exploration.
See Abstr. 051.025.

Planetary occultations. See Abstr. 091.006.

Planetary occultation update.
See Abstr. 091.007.

Photographic positions of comets, minor planets and Pluto observed during 1970 - 1974.
See Abstr. 103.009.

Observations of comets and asteroids at the Kleť Observatory in the year 1976. See Abstr. 103.010.

Infrared (JHK) photometry of meteorites and asteroids. See Abstr. 105.050.

Astronomically observable crater-forming projectiles.
See Abstr. 106.064.

Origin of asteroids and the missing planet.
See Abstr. 107.020.

Model consideration of the bombardment event of the asteroidal belt by the planetesimals scattered from the Jupiter zone. See Abstr. 107.024.

Errata

098.901 **Errata: "Mass content and mass distribution of the asteroid system"** [Bull. Astron. Inst. Czechoslovakia, Vol. 28, 65 - 82 (1977)]. L'. Kresák.
Bull. Astron. Inst. Czechoslovakia, Vol. 29, 192 (1978).

099 Jupiter, Jupiter Satellites

Jupiter

099.001 The D/H and C/H ratios in Jupiter from the CH₃D
phase.   R. Beer, F. W. Taylor.
Astrophys. J., Vol. 219, 763 - 767 (1978).
    Previous determinations of the D/H ratio in Jupiter have
given inconsistent results. New, higher resolution observations
of CH_3D in Jupiter suggest that the apparent CH_3D column
abundance is variable, and that both the ratios D/H and C/H
are greater than the "cosmic" values by substantial factors.

099.002 Evidence for CO in Jupiter's atmosphere from air-
borne spectroscopic observations at 5 microns.
H. P. Larson, U. Fink, R. R. Treffers.
Astrophys. J., Vol. 219, 1084 - 1092 (1978).
    High-altitude (12.4 km) spectra of Jupiter are analyzed
for the presence of CO absorption lines. A line-by-line comparison of Jupiter's spectrum with that of carbon monoxide
is presented, as well as a correlation analysis that includes the
influence of other gases present in Jupiter's atmosphere. The
resulting evidence points strongly to the presence of carbon
monoxide in Jupiter's atmosphere, thus strengthening Beer's
evidence for it. Possible explanations for the existence and
observability of Jovian CO, including convection from hotter,
deeper layers or decomposition of organic molecules, are
explored.

099.003 Observations of Jupiter with the astrolabe of the
CERGA observatory during the winter 1976-1977.
G. Vigouroux, P. Grudler, G. Guallino, F. Mignard, J. Pham-
Van.
Astron. Astrophys., Suppl. Ser., Vol. 31, 169 - 170 (1978).
In French.
    This paper contains the results of the first campaign at
the CERGA observatory. During the 1976-1977 winter 36
east and west transits were observed. A table gives the date,
the zenith distance residuals, the corrections for defective
illumination as usual, the indications of the observer and of
the reference groups.

099.004 Drift rates of Jovian S-bursts.   J. J. Riihimaa.
Astron. Astrophys., Vol. 63, L27 - L28 (1978).
    The average drift rates of Jupiter's S-bursts are related
to the Jovicentric declination of the Earth. The study is made
with positive values of the declination. The rates do not seem
to depend markedly on the CML or Io's longitude in region B.

099.005 On the production and interaction of planetary
solitary waves: applications to the Jovian atmo-
sphere.   T. Maxworthy, L. G. Redekopp, P. D. Weidman.
Icarus, Vol. 33, 388 - 409 (1978).
    The authors present further evidence that strengthens the
case for the interpretation of many features in the Jovian
atmosphere as solitary Rossby waves (solitons). These include:
a mechanism whereby such waves can evolve from the
instability of the basic shear flows; further interpretation of the
interaction between observed features, and comparison
with calculations of the interaction between planetary solitons
of a restricted class; and calculations of soliton morphology for
a type of shear flow other than the type considered originally
by Maxworthy and Redekopp.

099.006 Détermination quantitative de l'effect de phase dans
la mesure des longitudes sur Jupiter.   S. Cortesi.
Icarus, Vol. 33, 410 - 413 (1978).
    The author has quantitatively determined the phase exag-
geration effect (Phillips effect) as a function of the planet's

phase angle for the correction of the longitude of spots on the
Jupiter disk. This was done on the basis of over 1000 visual
observations of the longitude of permanent details of Jupiter's
surface compared with photographic observations.

099.007 The interior structure of Jupiter (Consequences of
Pioneer 10 data).   R. Smoluchowski.
Cosmochemistry of the moon and planets, (see 012.002),
p. 849 - 858 (1977). — Paper also published in Icarus, see
13.099.075.

099.008 Jovian S-burst observations at 32 MHz.
M. D. Desch, R. S. Flagg, J. May.
Nature, Vol. 272, 38 - 40 (1978).

099.009 Jupiter's S bursts and Io.   P. S. Whitham.
Nature, Vol. 272, 40 - 41 (1978).
    It is concluded that the S bursts are emitted by electron
streams that are accelerated outwards from the feet of the Io
flux tube in Jupiter's ionosphere.

099.010 The influence of differential rotation of planets
on their gravitational fields.
P. P. Vasil'ev, A. B. Efimov, V. P. Trubitsyn.
Astron. Zh. Akad. Nauk SSSR, Vol. 55, 148 - 155 (1978). In
Russian. English translation in Soviet Astron., Vol. 22, No. 1.
    The influence of differential rotation on figures and
gravitational fields is investigated for simple models of Jupiter
and Saturn. The square of angular velocity is taken to be a
power function of distance from the axis, with power indices
$2i = 2, 4, 6$. The coefficients of the expansion of the gravitational potential are calculated up to $k = 12$. The formula connecting the small parameter m of the theory and the mean
radius s_1 of the differentially rotating planet with the observational data is given.

099.011 Photometry and polarimetry of Jupiter at large
phase angles. I. Analysis of imaging data of a promi-
nent belt and a zone from Pioneer 10.
M. G. Tomasko, R. A. West, N. D. Castillo.
Icarus, Vol. 33, 558 - 592 (1978).
    Limb-darkening curves are derived from Pioneer 10 imag-
ing data for Jupiter's South Tropical Zone (-18 to $-21°$ latitude) and north component of the South Equatorial Belt (-5
to $-8°$ latitude) in red and blue light at phase angles of 12, 23,
34, 109, 120, 127, and 150°. Inhomogeneous scattering models are computed and compared with the data to constrain the
vertical structure and the single-scattering phase functions of
the belt and the zone in each color.

099.012 Chemical structure of the deep atmosphere of Jupi-
ter.   S. S. Barshay, J. S. Lewis.
Icarus, Vol. 33, 593 - 611 (1978).
    The authors report here the equilibrium abundances
calculated for a system of over 500 compounds of 27 selected
elements along a nominal Jupiter adiabat. Several species predicted to be of negligible abundance in the visible upper
troposphere if chemical equilibrium is exactly attained are
found to be potential tracers of rapid vertical motions. Vertical mixing of certain species, especially CO, PH_3, AsH_3, GeS,
and GeH_4, may provide detectable quantities of these species
near the visible cloudtops due to quenching and incomplete
equilibration of the rapidly rising, rapidly cooling gas. Observational prospects for detecting such tracers of deep circulation
are discussed.

099.013 A detailed analysis of Jovian cloud features in the
north tropical current.   R. Beebe.

Bull. American Astron. Soc., Vol. 9, 510 (1977). – Abstract.

099.014 A search for minor molecules in the Jovian spectrum
between 7 and 13 microns.
T. Encrenaz, M. Combes.
Bull. American Astron. Soc., Vol. 9, 515 (1977). – Abstract.

099.015 The spectrum of Jupiter at 1.9 and 2.7 microns.
H. P. Larson, U. Fink.
Bull. American Astron. Soc., Vol. 9, 515 (1977). – Abstract.

099.016 Detection of H_2 on Jupiter and the orthopara ratio.
W. H. Smith.
Bull. American Astron. Soc., Vol. 9, 515 (1977). – Abstract.

099.017 D/H in Jupiter's atmosphere.
M. Combes, T. Encrenaz, T. Owen.
Bull. American Astron. Soc., Vol. 9, 516 (1977). – Abstract.

099.018 The D/H ratios on Jupiter, Saturn, and Uranus based
on new HD and H_2 data.
J. T. Trauger, F. L. Roesler, M. E. Mickelson.
Bull. American Astron. Soc., Vol. 9, 516 (1977). – Abstract.

099.019 Tidal dissipation in the solid cores of the major
planets.   S. F. Dermott.
Bull. American Astron. Soc., Vol. 9, 520 (1977). – Abstract.

099.020 Observing Jupiter's optical emission nebula.
R. A. Brown.
Bull. American Astron. Soc., Vol. 9, 525 (1977). – Abstract.

099.021 The extended sodium and sulfur clouds of Jupiter.
C. B. Pilcher, W. V. Schempp.
Bull. American Astron. Soc., Vol. 9, 526 (1977). – Abstract.

099.022 Jovian magnetospheric plasma in the optical emission region.   F. C. Michel.
Bull. American Astron. Soc., Vol. 9, 527 (1977). – Abstract.

099.023 The phase function of the Jovian clouds from
Pioneer 11.   M. G. Tomasko.
Bull. American Astron. Soc., Vol. 9, 533 - 534 (1977).
Abstract.

099.024 Infrared limb-darkening on Jupiter: a theoretical
study.   W. I. Newman, C. Sagan.
Bull. American Astron. Soc., Vol. 9, 534 (1977). – Abstract.

099.025 Spatially resolved photometry of Jupiter in the
6190, 7250, and 8900 Å bands of methane.
R. West.
Bull. American Astron. Soc., Vol. 9, 534 (1977). – Abstract.

099.026 Models for the reflective spectrum of Jupiter's
North Equatorial Belt.   K. Rages, C. Sagan.
Bull. American Astron. Soc., Vol. 9, 534 - 535 (1977).
Abstract.

099.027 Jupiter's ultraviolet ammonia absorption.
W. Macy, E. Barker.
Bull. American Astron. Soc., Vol. 9, 535 (1977). – Abstract.

099.028 Le osservazioni visuali di Giove.   G. Favero.
Astronomia, N. 4, p. 12 - 21 (1977).

099.029 Observations of Jupiter around 80 μm at high
spectral resolution.   J. P. Baluteau, A. Marten,
E. Bussoletti, M. Anderegg, A. F. M. Moorwood, J. E.
Beckman, N. Coron.
Astron. Astrophys., Vol. 64, 61 - 65 (1978).
    New, high-resolution air-borne, spectral observations of

Jupiter in the 105–140 cm^{-1} range are presented. The procedures used to remove the effects of telluric opacity and for
the brightness temperature calibration are described. The resulting spectrum is compared with recent calculations for
different Jovian models. Some constraints on the NH_3
column density above the temperature inversion level are derived.

099.030 Zeitliche Veränderungen auf Jupiter 1971–1976/77.
J. Böing.
Orion, 36. Jahrg., 24 - 26 (1978).

099.031 On the abundance of deuterium in Jupiter's atmosphere.   M. Combes, T. Encrenaz, T. Owen.
Astrophys. J., Vol. 221, 378 - 381 (1978).
    The authors have calculated the ratio of deuterium to
hydrogen in the Jovian atmosphere using a new approach.
They obtain D/C from weak lines of HD and CH_4 in the visible
region of the spectrum and then derive C/H from stronger
methane and hydrogen absorptions near 1 μm. The result is
D/H = 2.3 ± 1.1 × 10^{-5}, within the range of local interstellar
values.

099.032 The abundance and distribution of carbon monoxide
in Jupiter.   R. Beer, F. W. Taylor.
Bull. American Astron. Soc., Vol. 9, 563 (1978). – Abstract.

099.033 The post-Pioneer Jupiter.   R. Smoluchowski.
Space Research, Vol. XVII, (see 012.010), 673 -
686 (1977). – Review paper.

099.034 The Pioneer 11 imaging experiment of Jupiter.
W. Swindell, J. Fountain.
Space Research, Vol. XVII, (see 012.010), 687 - 701 (1977).
    Pioneer 11 flew by Jupiter in December 1974, and
obtained several hundred images of the planet. Imaging data
were displayed to maintain the proper shape of the planet.
Color images were made by synthesizing green data from
red and blue data. Pictures created from Pioneer 11 imaging
data show complex detail within the Red Spot as well as indications of flow around it. Bright spots with trailing plumes
are seen in the Equatorial Zone. The North Polar Region is
devoid of belt structure, but numerous irregular cells are seen
in red light. The Galilean satellites were imaged with a resolution of several hundred kilometers.

099.035 Pioneer 10 and 11 radio occultations by Jupiter.
A. J. Kliore, P. M. Woiceshyn, W. B. Hubbard.
Space Research, Vol. XVII, (see 012.010), 703 - 710 (1977).
    Radio occultation data received from Pioneers 10 and 11
during their encounters with Jupiter have been analyzed using
an inversion technique which accounts for the oblateness of
Jupiter's atmosphere. The temperature profiles derived from
the different measurements are consistent, showing a temperature inversion between the 10 and 100 millibar levels, with
temperatures between 130 and 170°K at the 10 millibar level,
and 80–120°K at 100 millibars.

099.036 The atmosphere of Jupiter from earth-based and
spacecraft observations in the thermal infrared.
G. S. Orton.
Space Research, Vol. XVII, (see 012.010), 711 - 717 (1977).
    The temperature and cloud structure, relative abundances
of H_2 and He, and the global climatology of Jupiter's atmosphere have been deduced from Pioneer 10 and 11 infrared
radiometer data, along with earth-based observations of the
spectrum at 8–14 μm and 12–24 μm. The H_2 and He abundances are near those expected from "solar" composition. The
effective planetary temperature is 125 ± 3 K. Temperatures
at 1.0 bar are near 165 K and drop to 100–110 K at 0.1 bar,
for an overlying thermal inversion which reaches 133–145 K
near 0.03 bar.

099.037   Distribution and dynamics of energetic particles in the Jovian magnetosphere.   J. A. Van Allen.
Space Research, Vol. XVII, (see 012.010), 719 - 731 (1977).

Following a brief comparative review of the magnetospheres of Earth and Jupiter, some current work on analysis and interpretation of Pioneer 10 and 11 observations at Jupiter is sketched. The topics selected are (1) the distribution of high energy electrons and inferences therefrom, (2) diffusion coefficients, (3) the magnetodisc, (4) recirculation of energetic particles, and (5) the magnetotail.

099.038   The abundance of carbon monoxide in Jupiter.
R. Beer, F. W. Taylor.
Astrophys. J., Vol. 221, 1100 - 1109 (1978).

New spectra of Jupiter in the 5 μm window region have been acquired, from which the authors deduce that (a) the presence of CO in Jupiter is verified, (b) the column abundance is approximately 1.6×10^{-2} cm amagat ($\equiv 4.3 \times 10^{17}$ mol cm^{-2}), and (c) the CO is probably nonuniformly mixed in the atmosphere, being concentrated into the stratosphere.

099.039   Solar wind effect on Jupiter's non-Io-related radio emission.   T. Terasawa, K. Maezawa, S. Machida.
Nature, Vol. 273, 131 - 132 (1978).

The authors show that the non-Io-related decametric radio activities are affected by solar wind conditions around Jupiter.

099.040   Jupiter-Io electrodynamic interaction.
J. H. Piddington.
Moon, Vol. 17, 373 - 382 (1977).

Strong interaction between Jupiter and its satellite Io is revealed by the control of the decametric radiation, by the distributions of energetic particles, and perhaps by the location of the boundary of Jupiter's plasmasphere near Io's magnetic flux tube. Two opposed theories of this interaction depend on different relative motions of Io and its flux tube. In one case the flux tube is frozen into Io and moves with Io, while in the plasma-sheath model Io moves freely across magnetic field lines. It is shown that the plasma-sheath model is unacceptable, and that Io must drive its flux tube through the magnetosphere.

099.041   Light and dark spots in the equatorial regions of Jupiter.   G. C. Browne, A. J. Meadows.
Planet. Space Sci., Vol. 26, 335 - 338 (1978).

The lifetimes and motions of dark and light spots in the equatorial regions of Jupiter have been studied for the 1972 apparition. The difference in lifetimes of the two types of spot is demonstrated, as are their differing modes of formation and disappearance. The way in which the spots interact is examined in detail, and it is pointed out that this may have implications for the fluid dynamics of the Jovian atmosphere.

099.042   A self-consistent model of the Jovian plasma sheet.
G. T. Vickers.
Planet. Space Sci., Vol. 26, 381 - 385 (1978).

An axially-symmetric, rapidly-rotating magnetosphere containing low-energy plasma is considered. The resulting plasma sheet is presumed isothermal and thin compared with the radius of the sheet. Solutions of the model equations are found which include the effects of centrifugal, pressure and electro-magnetic forces. These solutions show that the sheet has a constant thickness and that the pressure decays exponentially with distance from the equatorial plane. The calculated curves for the magnetic induction field are compared with the observed field of Jupiter.

099.043   Actualité jovienne.   J. Dragesco.
Astronomie, Vol. 92, 236 - 237 (1978).

099.044   The thermal structure of Jupiter from infrared spectral measurements by means of a filtered iterative inversion method.   D. Gautier, A. Lacombe, I. Revah.
J. Atmos. Sci., Vol. 34, 1130 - 1137 (1977). – Abstr. in Phys. Abstr., Vol. 81, Abstr. 7918 (1978).

099.045   A model of Jovian short-duration bursts.
K. Kawamura, I. Suzuki.
Rep. IPPJ-286, (see 012.018), 5 pp. (1977). – Abstr. in Phys. Abstr., Vol. 81, Abstr. 7919 (1978).

099.046   Acceleration of protons at 32 Jovian radii in the outer magnetosphere of Jupiter.
A. W. Schardt, F. B. McDonald, J. H. Trainor.
J. Geophys. Res., Vol. 83, 1104 - 1108 (1978).

During the inbound pass of Pioneer 10 a rapid tenfold increase of the 0.2- to 5-MeV proton flux was observed at 32 R_J (Jovian radii). The total event lasted for 30 min. Before and after the event the proton flux was characteristic of the low flux level normally encountered between crossings of the magnetic equator.

099.047   Transport properties in the Jovian atmosphere.
L. Biolsi.
J. Geophys. Res., Vol. 83, 1125 - 1131 (1978).

Transport properties in a Jupiter-like atmosphere (89 mol % hydrogen and 11 mol % helium) are obtained by using the method of the kinetic theory of gases. The transport collision integrals are calculated by fitting various two-body semiempirical interaction potentials. The collision integrals are used to calculate the binary diffusion coefficients, viscosity, and 'total' thermal conductivity of the pure gases and the gas mixtures at 1-atm pressure from 1000°K to 25,000°K.

099.048   Measured photographic latitudes on Jupiter in 1976 - 77.   R. L. Hull.
Strolling Astron., Vol. 27, 76 - 77 (1978).

099.049   Upper limits to trace constituents in Jupiter's atmosphere from an analysis of its 5-μm spectrum.
R. R. Treffers, H. P. Larson, U. Fink, T. N. Gautier.
Icarus, Vol. 34, 331 - 343 (1978).

A high-resolution (0.6 cm^{-1}) spectrum of Jupiter at 5 μm recorded at the Kuiper Airborne Observatory is used to determine upper limits to the column density of 19 molecules. The upper limits to the mixing ratios of SiH$_4$, H$_2$S, HCN, and simple hydrocarbons are discussed with respect to current models of Jupiter's atmosphere. These upper limits are compared to expectations based upon the solar abundance of the elements. This analysis permits upper limit measurements (SiH$_4$), or actual detections (GeH$_4$), of molecules with mixing ratios with hydrogen as low as 10^{-9}.

099.050   Germane in the atmosphere of Jupiter.
U. Fink, H. P. Larson, R. R. Treffers.
Icarus, Vol. 34, 344 - 354 (1978).

High-altitude spectra of Jupiter obtained from the Kuiper Airborne Observatory are analyzed for the presence of germane (GeH$_4$) in Jupiter's atmosphere. Comparison with laboratory spectra shows that the strong Q branch of the ν_3 band of germane at 2111 cm^{-1} is prominent in the Jovian spectra. The abundance of germane in Jupiter's atmosphere is 0.006 ($\pm$0.003) cm-am corresponding to a mixing ratio of 0.6 ppb. This trace amount of germane is consistent with chemical equilibrium calculations if the germane present at ~1000°K is carried up by convection to the spectroscopically observable region at ~300°K.

099.051   Equation of state for hydrogen and the structure of Jupiter.   V. N. Zharkov, V. P. Trubitsyn, I. A. Tsarevskij.

Fazovye perekhody met.-diehlektr. Kratk. soderzh. dokl., predstavl. na II Vses. konf. po fazovym perekhodam met.-diehlektr., L'vov, 1977. Moskva – L'vov, 1977, p. 21 - 23. In Russian. – Abstr. in Ref. zh., 51. Astron., 4.51.242 (1978).

099.052 **Jupiter: présentation 1976.** F. Jetzer.
Orion, 36. Jahrg., 119 - 123 (1978).

099.053 **Jovian plasmaspheres.** G. L. Siscoe.
J. Geophys. Res., Vol. 83, 2118 - 2126 (1978).
The author obtains a set of model density profiles for three types of plasmaspheres. He finds that if the photoelectron flux from the planetary atmosphere is strongly trapped, the associated plasmasphere should dominate in number density. However, if the photoelectron flux is in diffusive equilibrium with the atmosphere, the Io-derived plasmasphere should dominate in number density. In either case the Io-derived plasmasphere should dominate in mass density because of the heavy ions. The heavy ion plasmasphere can be expected to have important dynamical effects on the magnetosphere in the region beyond about $30\,R_J$. The interstellar plasmasphere can reach densities approaching $10^2\,\mathrm{cm}^{-3}$ when Jupiter is in the high-interstellar-density portion of its orbit.

099.054 **Reflectivity of the Jovian disk center at $0.3-1.1\,\mu m$.**
V. D. Vdovichenko, S. M. Gajsin.
Astron. Tsirk., No. 971, p. 6 - 8 (1977). In Russian.

099.055 **Ultraviolet emission from Jovian atmosphere.**
O. N. Singh, R. N. Singh.
Indian J. Radio Space Phys., Vol. 6, No. 1, p. 77 - 79 (1977). Abstr. in Phys. Abstr., Vol. 81, Abstr. 40865 (1978).

099.056 **On the determination of the mass of Jupiter from observations of its VIIth and VIIIth satellites.**
L. E. Bykova, V. A. Yurga.
Astron. i geod. Vyp. (No.) 6. Tomsk, 1977, p. 42 - 46. In Russian. – Abstr. in Ref. zh., 52. Geod. Aehrosemka, 6.52.130 (1978).

099.057 **Longitudinal control of Jovian magnetopause motion.** A. J. Dessler.
Geophys. Res. Lett., Vol. 5, 65 - 68 (1978).
The magnetopause crossings of the Pioneer 10 and 11 spacecraft in Jovian magnetic coordinates (system III) are largely restricted in longitude to one hemisphere of Jupiter. This hemisphere is the one that has been identified by Vasyliunas (1975) as the "active hemisphere". This finding is interpreted as indicating that the magnetopause of the active hemisphere moves inward and outward with a radial speed that is typically faster than that of the inactive hemisphere.

099.058 **On the contribution of water products from Galilean satellites to the Jovian magnetosphere.**
L. J. Lanzerotti, W. L. Brown, J. M. Poate, W. M. Augustyniak.
Geophys. Res. Lett., Vol. 5, 155 - 158 (1978).
The authors demonstrate, by using the recent experimental results on the "sputtering" rates of water ice by energetic protons, that significant fluxes of heavy atoms can be eroded from Europa, Ganymede, and Callisto. These atoms may play an important role in contributing to an atmosphere around Ganymede. If they can escape from the satellites and be ionized, they can be important contributors to the heavy ions that may dominate the plasma centrifugal distortion of the Jovian magnetosphere.

099.059 **Possible origins of time variability in Jupiter's outer magnetosphere. 3. Variations in the heavy ion plasma.** A. Eviatar, C. F. Kennel, M. Neugebauer.
Geophys. Res. Lett., Vol. 5, 287 - 290 (1978).
The authors discuss the implications of a heavy ion plasma in the Jovian magnetosphere. The plasma electron density varies on time scales comparable with that of radial diffusion. This plasma can enhance a super-Alfvénic planetary wind by increasing the radial mass flux. A mechanism by which the heavy ion plasma can regenerate itself via self-sputtering from the surface of Io followed by ionization first by solar ultra-violet and later by electron impact is proposed.

Intensity measurements of the CH_4 bands in the region 4350 Å to 10,600 Å. See Abstr. 022.041.

Ultraviolet-photoproduced organic solids synthesized under simulated jovian conditions: molecular analysis. See Abstr. 022.050.

Pyrolysis of organic compounds in the presence of ammonia. The Viking Mars lander site alteration experiment. See Abstr. 022.073.

Semi-empirical electron-impact ionization cross-sections of some hydrocarbons. See Abstr. 022.075.

Optical constants of liquid methane in the infrared. See Abstr. 022.111.

High-resolution infrared spectra of ν_3 and $2\nu_3$ of germane. See Abstr. 022.115.

Mass loss, shape change and real-gas aerodynamic effects on a Jovian atmospheric probe. See Abstr. 053.008.

The 1981 Jupiter orbiter probe mission. See Abstr. 053.009.

Formation of a rotational accretion column. See Abstr. 062.005.

A similarity model illustrating the effect of rotation on an inflated magnetosphere. See Abstr. 062.063.

The interaction between a magnetized plasma flow and a magnetized celestial body: a review of magnetospheric studies. See Abstr. 062.082.

Studies of chemical abundances in the outer solar system. See Abstr. 091.011.

On the apparent source depth of planetary magnetic fields. See Abstr. 092.023.

Io-phase motion and jovian decametre source locations. See Abstr. 099.520.

The interplanetary modulation and transport of Jovian electrons. See Abstr. 106.027.

Formation of Jupiter, Saturn, Uranus and their regular satellites. See Abstr. 107.006.

On the origin and evolution of Jupiter and Saturn. See Abstr. 107.019.

Jupiter Satellites

099.501 Possible correlation of Io's posteclipse brightening with major solar flares.   R. M. Nelson, B. W. Hapke.
Icarus, Vol. 33, 203 - 209 (1978).

In 6 of the 7 instances where posteclipse brightening of Io·has been reported a major solar flare occurred. In none of the 18 instances where no posteclipse brightening was observed did such a flare occur. It is proposed that a phenomenon associated with a major solar flare causes an increase in the trapped particle flux at Io's orbit by an order of magnitude. The posteclipse brightening may be caused by thermoluminescence of Io's surface material upon emergence. Alternatively, it is possible that the increase in trapped particle flux would warm the surface, creating a temporary atmosphere which would precipitate during eclipse cooling and vaporize in the period of warming after reemergence.

099.502 The long period behavior of the orbits of the Galilean satellites of Jupiter.   B. Brown.
Celestial Mech., Vol. 16, 229 - 259 (1977).

Some of the results of an investigation into the long period behavior of the orbits of the Galilean satellites of Jupiter are presented. The disturbing function was expanded as a Poisson series in the modified Keplerian elements referred to a Jovicentric coordinate system. The differential equations for the modified Keplerian elements were then formed, and all short period perturbations were removed using Kamel's perturbation method. Approximate analytical solutions for these differential equations are derived, and the general form of the solutions are given.

099.503 Photoelectric photometry of JV.   R. L. Millis.
Icarus, Vol. 33, 319 - 321 (1978).

UBV photometry of JV near greatest western elongation shows this satellite to be about one magnitude fainter than previously believed. It is observed to be very red, having a B−V color index near + 1.5 mag. Unlike the Galilean satellites, JV appears to be a low-albedo object.

099.504 On the distribution of sodium in the vicinity of Io.   L. Trafton, W. Macy, Jr.
Icarus, Vol. 33, 322 - 335 (1978).

The authors investigate the contribution of scattering in the telescope to their measurements of the size of Io's sodium cloud and to the distribution of emission intensity in the cloud. The brightest regions, within 30 "of Io near opposition and along the equatorial plane, are relatively undistorted but regions further than 45 "away and not close to the equatorial plane are very likely to consist of mainly scattered light. Portions of the cloud in the vicinity of the magnetic equator are also mostly scattered light when Io is near extreme magnetic latitude. The equatorial torus, however, extends up to 20 arcmin from Jupiter. The large size of the cloud is thus confirmed.

099.505 Possible flyby measurements of Galilean satellite interior structure.   W. B. Hubbard, J. D. Anderson.
Icarus, Vol. 33, 336 - 341 (1978).

Flyby encounters of the Galilean satellites from a Jupiter orbiter spacecraft could yield information about the second-degree gravity harmonics of these satellites. The authors have calculated the expected values of these harmonics for a range of plausible interior models in hydrostatic equilibrium.

099.506 Images of Io's sodium cloud.
D. L. Matson, B. A. Goldberg, T. V. Johnson, R. W. Carlson.
Science, Vol. 199, 531 - 533 (1978).

The first direct images of Io's sodium cloud are reported and analyzed. The observed cloud extends for more than 10^5 kilometers along Io's orbit and is a somewhat "banana-shaped" partial toroid. More sodium atoms precede Io than follow it. A model based on the escape of sodium from a specific localized area on Io provides a reasonable fit to the observed intensity distribution whereas   isotropic escape does not.

099.507 Icy craters on Europa, Ganymede, and Callisto?
S. J. Ostro, G. H. Pettengill.
Bull. American Astron. Soc., Vol. 9, 523 - 524 (1977). Abstract.

099.508 Eclipse determinations of Galilean satellite limb darkening and Callisto's orbit.
D. W. Smith, T. F. Greene.
Bull. American Astron. Soc., Vol. 9, 524 (1977). − Abstract.

099.509 Spectral reflectivities of the Galilean satellites and Titan.   R. M. Nelson, B. W. Hapke.
Bull. American Astron. Soc., Vol. 9, 524 (1977). − Abstract.

099.510 Observations of the near infrared spectra of the Galilean satellites.   J. Pollack, F. Witteborn, D. Strecker, E. Erickson, B. Baldwin, T. Bunch.
Bull. American Astron. Soc., Vol. 9, 524 (1977). − Abstract.

099.511 Reflectance spectra for sodium and potassium doped ammonia and water frosts: implications for Io's surface.   M. D. A. Rosen, F. M. Pipkin.
Bull. American Astron. Soc., Vol. 9, 524 - 525 (1977). Abstract.

099.512 Io's surface: a sulfur sublimate model.
B. Hapke, R. Nelson.
Bull. American Astron. Soc., Vol. 9, 525 (1977). − Abstract.

099.513 Images of Io's sodium cloud.
F. J. Murcray.
Bull. American Astron. Soc., Vol. 9, 525 (1977). − Abstract.

099.514 Modeling Io's sodium cloud images.
W. H. Smyth, M. B. McElroy.
Bull. American Astron. Soc., Vol. 9, 525 (1977). − Abstract.

099.515 Images of Io's sodium cloud.   D. L. Matson, B. A. Goldberg, T. V. Johnson, R. W. Carlson.
Bull. American Astron. Soc., Vol. 9, 525 - 526 (1977). Abstract.

099.516 Images of Io's sodium cloud: constraints on cloud geometry.
B. A. Goldberg, D. L. Matson, T. V. Johnson, R. W. Carlson.
Bull. American Astron. Soc., Vol. 9, 526 (1977). − Abstract.

099.517 Io: models for neutral cloud dynamics.
R. W. Carlson, D. L. Matson, T. V. Johnson.
Bull. American Astron. Soc., Vol. 9, 526 (1977). − Abstract.

099.518 A search for new features in the spectrum of Io.
R. A. Brown.
Bull. American Astron. Soc., Vol. 9, 526 (1977). − Abstract.

099.519 Escape of sodium from Io.
G. V. Ramanathan, S. H. Gross.
Bull. American Astron. Soc., Vol. 9, 527 (1977). − Abstract.

099.520 Io-phase motion and jovian decametre source locations.   M. D. Desch.
Nature, Vol. 272, 339 - 340 (1978).

The author concludes (1) that the observed variation in Io phase is due solely to the latitudinal changes in Earth-Jupiter viewing geometry described by the jovicentric declination of the Earth and (2), that the Io—B (Io—C) source emission is beamed from northern (southern) magnetic latitudes.

099.521 Galilean satellites: analysis of photometric eclipses.
J. H. Lieske.
Astron. Astrophys., Vol. 65, 83 - 92 (1978).

The series of photometric eclipses of the Galilean satellites observed at Harvard from 1878 to 1903 have been combined with the relatively few available post-1903 photometric eclipse observations in order to evaluate the satellite parameters which occur in the new theory of motion of the satellites. The results indicate that the new theory satisfactorily can predict ephemeris positions of the Galilean satellites over spans of nearly a century and that it will be a useful tool for analyzing future observations.

099.522 How to compare the surface of Io to laboratory samples. J. Veverka, J. Goguen, S. Yang, J. L. Elliot.
Icarus, Vol. 34, 63 - 67 (1978).

Since one does not know the photometric functions of various parts of Io, one cannot convert the observed geometric albedo of the satellite to a parameter more directly measurable in the laboratory. One must therefore convert laboratory reflectances to geometric albedos before quantitative comparisons between Io's surface and a laboratory sample are made. This procedure involves determining the wavelength dependence of the sample's photometric function. For substances such as sulfur, whose reflectance varies strongly with wavelength, it is incorrect to assume that the photometric function, and hence the ratio (laboratory reflectance/ geometric albedo) is independent of wavelength. In general, unless the laboratory reflectance is precisely the geometric albedo, a wavelength-dependent correction factor must be determined before the laboratory sample can be compared quantitatively with Io's surface.

099.523 Mutual phenomena of Jupiter's five inner satellites in 1979. K. Aksnes, F. A. Franklin.
Icarus, Vol. 34, 188 - 193 (1978).

The authors first predict eclipses and occultations of the Galilean satellites in 1979 and find that, although circumstances are generally poor, about 75 events are observable. They have then made a special point of including 40 eclipses of J5 (Amalthea) by the Galilean satellites in the hope that both visual and far-infrared light curves can be obtained—the former giving accurate astrometric information for J5, and the latter possibly bearing on its surface structure or composition.

099.524 Die Erscheinungen der Jupitermonde in den Beobachtungsperioden 1975/76 und 1976/77.
P. Ahnert.
Sterne, 54. Band, 45 - 50 (1978).

099.525 Photometry of the Galilean satellites.
D. Morrison, N. D. Morrison.
Planetary satellites, (see 003.007), p. 363 - 378 (1977).

The Galilean satellites have been studied photometrically by Stebbins(1927), Stebbins and Jacobsen (1928), Harris (1961), Johnson (1971), Blanco and Catalano (1974), and Morrison et al. (1974). From these observations, the authors derive the dependence of the magnitudes and colors on both solar and orbital phase angles.

099.526 Io's surface and the histories of the Galilean satellites. F. P. Fanale, T. V. Johnson, D. L. Matson.
Planetary satellites, (see 003.007), p. 379 - 405 (1977).

The authors consider the problem of the chemical evolution of Io's surface in the context of the general problem of the histories of the Galilean satellites.

099.527 Picture of Ganymede. T. Gehrels.
Planetary satellites, (see 003.007), p. 406 - 411 (1977).

099.528 Galilean satellites: 1976 radar results.
D. B. Campbell, J. F. Chandler, S. J. Ostro, G. H. Pettengill, I. I. Shapiro.
Icarus, Vol. 34, 254 - 267 (1978).

Radar observations of the Galilean satellites, made in late 1976 using the 12.6-cm radar system of the Arecibo Observatory, have yielded mean geometric albedos of 0.04 ± 0.01, 0.69 ± 0.17, 0.37 ± 0.09, and 0.15 ± 0.04, for Io, Europa, Ganymede, and Callisto, respectively. The albedo for Io is about 40% smaller than that obtained approximately a year earlier, while the albedos for the outer three satellites average about 70% larger than the values previously reported for late 1975. For Europa, Ganymede, and Callisto, the ratios of the echo received in one mode of circular polarization to that received in the other were: 1.61 ±0.20, 1.48 ± 0.27, and 1.24 ± 0.19, respectively, with the dominant component having the same sense of circularity as that transmitted. This behavior has not previously been encountered in radar studies of solar system objects, whereas the corresponding observations with linear polarization are "normal".

099.529 Icy craters on the Galilean satellites?
S. J. Ostro, G. H. Pettengill.
Icarus, Vol. 34, 268 - 279 (1978).

A model which possibly accounts for the unusual radar scattering behavior observed for Europa, Ganymede, and Callisto postulates a thick surface layer of ice saturated with nearly hemispherical craters. In the development of this model it is noted that a single reflection at normal incidence reverses the rotational sense of circularly polarized incident radiation, in conflict with the radar observations which show an echo predominantly not reversed. Randomly oriented reflecting facets, either of ice on the surface or of rocks in the interior, cannot yield the observed behavior since too few of the total possible backscattering configurations meet the above requirement. Hemispherical surface craters, on the other hand, favor 45° dual reflection. A model consisting of such craters in ice is investigated and found capable of explaining the observed results, not only in respect to polarization, but in respect to albedo and angular scattering law as well.

099.530 On the application of a generalized iterative method when evaluating the parameters of celestial bodies.
A. M. Chernitsov.
Astron. i geod. Vyp. (No.) 6. Tomsk, 1977, p. 47 - 55. In Russian. — Abstr. in Ref. zh., 51. Astron., 4.51.137 (1978).

099.531 Jupiter: double and triple satellite phenomena.
W. E. Fox.
J. British Astron. Assoc., Vol. 88, 360 - 361 (1978).

099.532 A cometary ionosphere model for Io.
P. A. Cloutier, R. E. Daniell, Jr., A. J. Dessler, T. W. Hill.
Astrophys. Space Sci., Vol. 55, 93 - 112 (1978).

The authors propose a model in which the material for Io's atmosphere and ionosphere is drawn from the ionosphere of Jupiter through a Birkeland current system that is driven by the potential induced across Io by the Jovian corotation electric field. They argue that the ionization near Io is caused by a comet-like interaction between the corotating plasma and Io's atmosphere.

099.533 The Galilean satellites of Jupiter: four worlds.
T. V. Johnson.
Annu. Rev. Earth Planet. Sci., Vol. 6, (see 003.011), 93 - 125 (1978).

099.534 On regions of possible motion of Jupiter and Saturn
satellites.   M. Kh. Khasanova.
Dokl. AN TadzhSSR, Vol. 20, No. 8, p. 16 - 19 (1977). In
Russian. − Abstr. in Ref. zh., 51. Astron., 5.51.99 (1978).

099.535 Phénomènes remarquables des satellites de Jupiter.
J. Meeus.
Astronomie, Vol. 92, 309 - 312 (1978).

099.536 Processes effecting the surfaces of the Galilean
satellites.   J. B. Pollack.
Reports of planetary geology program, 1977−1978,
(see 003.014), p. 46 - 48 (1978).

099.537 Physical processes affecting water on the Galilean
satellites.   C. B. Pilcher, N. G. Purves.
Reports of planetary geology program, 1977−1978,
(see 003.014), p. 49 - 50 (1978).

099.538 X-ray spectroscopy of the Galilean satellites.
R. H. Parker, A. E. Metzger, J. L. Luthey, R. Pehl,
H. W. Schnopper.
Reports of planetary geology program, 1977−1978,
(see 003.014), p. 51 - 53 (1978).

099.539 Long term light variations of Io and Titan.
C. Blanco, S. Catalano.
Mem. Soc. Astron. Italiana, Vol. 48, 103 - 108 (1977).
    A long-term variation of the V magnitude at mean
opposition of Io and Titan is shown to be correlated with the
orbital motion of Jupiter and Saturn respectively. The amplitude of the variation is about 0^m2 for both satellites. The
maximum of brightness occurs near the perihelion of the
respective planet.

099.540 The inner satellites of the planet Jupiter.
K. Beneš.
Říše hvězd, Vol. 59, 115 - 117 (1978). In Czech.

New formulation of de Sitter's theory of motion
for Jupiter I−IV. I. Equations of motion and the disturbing
function.   See Abstr. 042.065.

Theory of motion of Jupiter's Galilean satellites.
See Abstr. 042.066.

A second-order theory of the Galilean satellites of
Jupiter.   See Abstr. 042.067.

Occultations of Jupiter satellites.
See Abstr. 096.009.

Volatile evolution.   See Abstr. 097.155.

Jupiter's S bursts and Io.   See Abstr. 099.009.

Observing Jupiter's optical emission nebula.
See Abstr. 099.020.

The extended sodium and sulfur clouds of Jupiter.
See Abstr. 099.021.

The Pioneer 11 imaging experiment of Jupiter.
See Abstr. 099.034.

Solar wind effect on Jupiter's non-Io-related
radio emission.   See Abstr. 099.039.

Jupiter-Io electrodynamic interaction.
See Abstr. 099.040.

On the contribution of water products from
Galilean satellites to the Jovian magnetosphere.
See Abstr. 099.058.

Observations photographiques de la comète P/d'Arrest et des satellites de Jupiter effectuées en 1976 à la caméra
astrographique de 25 cm de l'Observatoire National de Rio de
Janeiro.   See Abstr. 103.805.

100 Saturn, Saturn Satellites

Saturn

100.001   The 22.7 micron brightness of Saturn's rings versus declination of the sun.
I. G. Nolt, A. Tokunaga, F. C. Gillett, J. Caldwell.
Astrophys. J., Lett., Vol. 219, L63 - L66 (1978).

Equatorial scans of Saturn's rings were obtained with the KPNO 4 m telescope in 1977 March to provide ring-brightness data at 22.7 μm for $|B'| = 16°.3$. A comparison of these new results with an earlier measurement by Rieke shows that a specific intensity decrease of ~30% has occurred for the A and B rings since 1973, when $|B'|$ was ~26°. The results also confirm the existence of a bright infrared C ring component, which has increased in specific intensity by ~35% over the same period.

100.002   Observations of Saturn with the astrolabe of the CERGA observatory (January - April 1977).
G. Guallino, P. Grudler, F. Mignard, J. Pham-Van, G. Vigouroux.
Astron. Astrophys., Suppl. Ser., Vol. 31, 167 - 168 (1978).
In French.

This paper gives results for Saturn observed with the astrolabe at the CERGA observatory. 44 east and west transits were observed.

100.003   Azimuthal brightness variation of Saturn's ring-A and size of particles.   I. Ferrin.
Nature, Vol. 271, 528 - 529 (1978).

Observations of the brightness of ring A of Saturn, indicate that it is not constant in azimuth. The A ring is fainter in quadrants following conjunction of the particles with the Earth—Saturn line, and brighter in quadrants preceding conjunction. Because observations do not show any time dependence, the effect is stationary. A possible explanation for this fact has been given by Colombo et al. (1976). Here the author gives an alternative explanation for this fact, which would be valid if these particles in ring A are large, which the author shows is indeed the case, using a recent measurement of the total mass of the rings.

100.004   Saturn's rings: particle composition and size distribution as constrained by microwave observations.
I. Radar observations.   J. N. Cuzzi, J. B. Pollack.
Icarus, Vol. 33, 233 -262 (1978).

The authors have calculated the radar backscattering characteristics of a variety of compositional and structural models of Saturn's rings and compared them with observations of the absolute value, wavelength dependence, and degree of depolarization of the rings' radar cross section (reflectivity). If the rings are many particles thick, irregular centimeter- to meter-sized particles composed primarily of water ice attain sufficiently high albedos and scattering efficiencies to explain the radar observations. In that case, the wavelength independence of radar reflectivity implies the existence of a broad particle size distribution. A narrower size distribution with $\bar{a}$ ~ 6 cm is also a possibility. A monolayer of very large ice "particles" that exhibit multiple internal scattering may not yet be ruled out. Observations of the variation of radar reflectivity with the opening angle of the rings will permit further discrimination between ring models that are many particles thick and ring models that are one "particle" thick.

100.005   A measurement of the brightness temperature of Saturn's rings at 8-mm wavelength.
M. A. Janssen, E. T. Olsen.
Icarus, Vol. 33, 263 - 278 (1978).

The authors have measured the brightness temperature of Saturn's rings at 8-mm wavelength. They obtain for the ring brightness temperature $T_R = 12.7 \pm 2°$K, with the assumption that the rings are of uniform brightness and the region of emission coincides with the visible A and B rings. This result is higher than comparable results obtained at centimeter wavelengths and may indicate a small increase in the thermal emission from the rings at 8mm. The low brightness temperature places significant constraints on the nature of the ring particles, and implies that they must be either highly metallic or of limited size and composed of a low-loss dielectric material such as water ice.

100.006   On the azimuthal brightness variations of Saturn's rings.   F. A. Franklin, G. Colombo.
Icarus, Vol. 33, 279 - 287 (1978).

The authors present a simple, semiquantitative explanation that accounts both for the presence of the azimuthal brightness variations in Saturn's ring A and for their absence in ring B. Their explanation avoids any ad hoc reliance on albedo variations and/ or synchronous rotation of ring particles. Instead, it requires only some degree of self-gravitation between nearby orbiting bodies. A bias in the particle distribution and corresponding photometric effects are thereby produced—the latter corresponding very closely to the variations observed in ring A. Their absence in ring B is primarily a consequence of the higher optical thickness and decreasing importance of self-gravitation in that ring.

100.007   Five-color photometry of Saturn and its rings.
K. Lumme, H. J. Reitsema.
Icarus, Vol. 33, 288 - 300 (1978) = Contrib. Five Coll. Obs., Amherst, Mass., No. 243.

Analysis of 206 high-quality plates from three recent apparitions taken in five colors has yielded several photometric parameters for Saturn and its A and B rings. Phase curves and geometric albedos are derived for two regions on Saturn and for each ring. The phase coefficients of the rings are found to be independent of the ring-plane inclination angle. The observations of the ring tilt effect indicate that the particles of ring A may also have lower single-scattering and geometric albedos. The color dependence of the geometric albedo of the particles in ring B is shown to be very similar to that of Europa (J II). The authors find for ring A an optical thickness of 0.50 ($0.45 \le \tau_A \le 0.57$) and for the Cassini division, 0.018 ± 0.004.

100.008   Models of Saturn's interior: Evidence for phase separation.   M. Podolak.
Icarus, Vol. 33, 342 - 348 (1978).

Models of Saturn's interior have been constructed based on an accumulation picture of planet formation. It was found that central pressures were ~ 90 Mb, and central temperatures ~ 10^4 K. In sharp contrast to Jupiter, which requires large amounts of heavy material in the envelope to match the observed gravitational quadrupole moment, Saturn requires an almost solar envelope. Indeed, the ratio of enhanced material in the envelope to material in the core is less than ~ 0.1, while the corresponding value for Jupiter is ~ 2.

100.009   Saturn 1976—1977.   A. W. Heath.
J. British Astron. Assoc., Vol. 88, 168 - 179 (1978).
Report of the Saturn section.

100.010   The temperature enhancement at the South polar region of Saturn.
A. Tokunaga, J. Caldwell, F. Gillett, I. Nolt.
Bull. American Astron. Soc., Vol. 9, 511 (1977). – Abstract.

100.011 Saturn: predicted seasonal variation of thermal flux.
W. M. Sinton, J. Good.
Bull. American Astron. Soc., Vol. 9, 511 (1977). – Abstract.

100.012 Observation of HD on Saturn and Uranus.
W. Smith, W. Macy.
Bull. American Astron. Soc., Vol. 9, 516 (1977). – Abstract.

100.013 Some observations of the Saturn D-ring.
S. M. Larson.
Bull. American Astron. Soc., Vol. 9, 520 - 521 (1977).
Abstract.

100.014 Phase curves for Saturn's rings.   L. W. Esposito.
Bull. American Astron. Soc., Vol. 9, 521 (1977).
Abstract.

100.015 Extensions to the classical calculation of the effect
of mutual shadowing in diffuse reflection.
L. W. Esposito.
Bull. American Astron. Soc., Vol. 9, 521 (1977). – Abstract.

100.016 Low tilt angle photometry and the thickness of
Saturn's ring.   K. Lumme, W. M. Irvine.
Bull. American Astron. Soc., Vol. 9, 521 (1977). – Abstract.

100.017 Azimuthal brightness variations of Saturn's rings.
L. W. Esposito, W. M. Irvine, K. Lumme, W. A.
Baum.
Bull. American Astron. Soc., Vol. 9, 521 (1977). – Abstract.

100.018 A possible explanation for the azimuthal brightness
variations in ring A.   F. A. Franklin.
Bull. American Astron. Soc., Vol. 9, 522 (1977). – Abstract.

100.019 Structure of the Saturn rings from reflectance
polarimetry.   A. Dollfus.
Bull. American Astron. Soc., Vol. 9, 522 (1977). – Abstract.

100.020 Saturn's rings: the variation in the near-infrared
albedo spectrum with phase angle.   R. N. Clark.
Bull. American Astron. Soc., Vol. 9, 522 (1977). – Abstract.

100.021 Interferometric observation of Saturn and the rings
at 3.71 cm wavelength.
F. P. Schloerb, D. O. Muhleman, G. L. Berge.
Bull. American Astron. Soc., Vol. 9, 522 - 523 (1977).
Abstract.

100.022 Visible and near-infrared spectral photometry of the
rings of Saturn from silicon diode vidicon images.
R. B. Singer.
Bull. American Astron. Soc., Vol. 9, 523 (1977). – Abstract.

100.023 Ice on Saturn's rings and satellites.
R. Smoluchowski.
Bull. American Astron. Soc., Vol. 9, 523 (1977). – Abstract.

100.024 UBV pinhole scans of Saturn's disk.
O. G. Franz, M. J. Price.
Bull. American Astron. Soc., Vol. 9, 535 (1977). – Abstract.

100.025 The 5μ spectrum of Saturn.
U. Fink, H. P. Larson.
Bull. American Astron. Soc., Vol. 9, 535 (1977). – Abstract.

100.026 Magnetic field of Saturn.   R. Smoluchowski.
Bull. American Astron. Soc., Vol. 9, 542 - 543
(1977). – Abstract.

100.027 Water ice sputtering and the atmospheres of the rings
of Saturn and Uranus.

A. F. Cheng, L. J. Lanzerotti.
Bull. American Astron. Soc., Vol. 9, 550 (1977). – Abstract.

100.028 The 22.7 μm brightness of the rings of Saturn versus
inclination to the sun.
I. G. Nolt, A. Tokunaga, F. C. Gillett, J. Caldwell.
Bull. American Astron. Soc., Vol. 9, 551 (1977). – Abstract.

100.029 Recent studies of Saturn's rings.
K. Lumme, L. W. Esposito, W. M. Irvine, W. A.
Baum, H. M. Ferguson, S. E. Jones, L. S. Martin, D. T. Thompson, W. Benton, A. Harris, P. V. Birch, J. B. Stuart, G. Wren.
Bull. American Astron. Soc., Vol. 9, 620 (1978). – Abstract.

100.030 Photometric confirmation of the Encke division in
Saturn's ring A.   H. J. Reitsema.
Nature, Vol. 272, 601 - 602 (1978).
The observed brightness of Iapetus during eclipse represents a spatial convolution of light from the limb-darkened disk of the Sun with the transmission of the rings and the non-uniform reflectivity over the disk of Iapetus. The author has examined this physical model through a computer simulation of the eclipse and uses this to give photometric confirmation of the Encke division in ring A.

100.031 A rocket observation of the far-ultraviolet spectrum
of Saturn.   H. Weiser, H. W. Moos.
Astrophys. J., Vol. 222, 365 - 369 (1978).
Far-ultraviolet (1160–1750 Å) spectra of the Saturnian disk and the ring system have been obtained by using a very sensitive rocket-borne spectrograph with a microchannel plate detector. H I λ1216 was the only atomic spectral line emission detected in the planet and the rings. A weak signal from the disk between 1300 Å and 1500 Å was observed. Geometric disk albedos, averaged over 50 Å, were determined from 1500 Å to 1700 Å. Measurements of the ring reflectivity longward of 1650 Å are compatible with H_2O frost but not NH_3 frost.

100.032 Major satellites cause wavy deformation of Saturn's
rings.   M. S. Bobrov.
Nature, Vol. 273, 284 - 285 (1978).
The author shows that allowing for gravitational perturbations of particle orbits by satellites and the oblateness of Saturn one may obtain a model satisfying the observational data. The main features of the model are: (1) specific wavy deformation of the rings which thus appear twisted; (2) particle collisions are rare (and there is therefore only slow energy dissipation) because the neighbouring particles move in nearly the same phase.

100.033 Detection of HD on Saturn and Uranus, and the
D/H ratio.   W. Macy, Jr., W. H. Smith.
Astrophys. J., Lett., Vol. 222, L73 - L75 (1978).
The authors have observed the HD 5–0 $R(0)$ line on Saturn and Uranus. Values for the D/H abundance ratio computed from models for the atmospheres of these two planets are $5.5 \pm 2.9 \times 10^{-5}$ for Saturn and $3.0 \pm 1.2 \times 10^{-5}$ for Uranus. The D/H ratio for Jupiter of $5.1 \pm 0.7 \times 10^{-5}$ found from a measurement of the HD 4–0 $P(1)$ line is bracketed by the values they find for Saturn and Uranus.

100.034 Saturn's rings: a new survey.
A. F. Cook, F. A. Franklin.
Planetary satellites, (see 003.007), p. 412 - 419 (1977).
Observations of radar returns from Saturn's rings, together with radio interferometry of their absorption of radiation from the disk, combine to require an effective radius of ring particles of about 6 cm or larger. The authors suggest that the ring particles may also include, in addition to the known constituent ice, a mixture of the clathrated hydrate of methane and ammonia hydrate.

100.035 The velocity dispersion in Saturn's rings.
        P. Goldreich, S. Tremaine.
Icarus, Vol. 34, 227 - 239 (1978).
        The velocity dispersion in a differentially rotating disk
of particles such as Saturn's rings is determined by the details
of the collision process. Collisions give rise to a viscous stress
that converts orbital energy into random motions. Since the
collisions are not perfectly elastic, the energy in random motions is dissipated as heat. With increasing velocity dispersion
the latter process becomes more important relative to the
former because the collisions become less elastic. The velocity
dispersion adjusts so that the effects of these two processes
balance. If the rings are about as old as the solar system then
their radial width implies that the velocity dispersion of the
ring particles is less than 0.2 cm sec^{-1}. The corresponding
vertical thickness is then less than 10 m. The authors discuss
the effects of collisions on the particles in Saturn's rings. If the
particles are made of ice they are eroded by collisions and
accrete the collisional debris. The time scale for erosion and
accretion is probably shorter than the age of the solar system.

**100.036 The formation of the Cassini division in Saturn's
        rings.**   P. Goldreich, S. Tremaine.
Icarus, Vol. 34, 240 - 253 (1978).
        The satellite Mimas excites a trailing spiral density wave
in Saturn's rings at the position of the 2 : 1 resonance. The
density wave carries negative angular momentum and propagates outward. The wave is damped by a combination of
nonlinear and viscous effects, and its negative angular momentum is transferred to the ring particles. Consequently, the
particles just outside the 2 : 1 resonance spiral inward, opening a gap. The inner edge of the gap is close to the resonance
position in agreement with the location of the inner edge of
the Cassini division. Despite its tiny mass, Mimas is able to
clear a gap as wide as the Cassini division.

100.037 Iapetus probes the rings of Saturn.   K. Hindley.
        New Scientist, Vol. 76, 140 - 141 (1977). – Abstr.
in Phys. Abstr., Vol. 81, Abstr. 16342 (1978).

100.038 Saturne: présentation 1976/77.   F. Jetzer.
        Orion, 36. Jahrg., 123 - 124 (1978).

**100.039 The external gravitational field of Saturn and of its
        rings.**   V. P. Trubitsyn, A. M. Bobrov, P. P.
Vasil'ev, Ya. A. Lev.
Astron. Zh. Akad. Nauk SSSR, Vol. 55, 628 - 634 (1978).
In Russian. English translation in Soviet Astron., Vol. 22,
No. 3.
        The contributions to the external gravitational field of
Saturn from its rings and from the tides in the planet caused
by the rings are calculated. The gravitational momenta of the
rings and the tidal corrections to Saturn's momenta up to the
tenth harmonic are determined.

**100.040 Comments on the magnetic field and internal heat
        of Saturn, Uranus and Neptune.**
R. Smoluchowski.
Reports of planetary geology program, 1977–1978,
(see 003.014), p. 58 - 59 (1978).

100.041 Amorphous ice on Saturn's rings.
        R. Smoluchowski.
Reports of planetary geology program, 1977–1978,
(see 003.014), p. 295 - 297 (1978).

100.042 Polarization of the central region of Saturn's disk.
        L. A. Sigua.
Astron. Tsirk., No. 974, p. 2 - 3 (1977). In Russian.

100.043 Electropolarimetric study of Saturn and its rings.
        L. A. Sigua.
Astron. Tsirk., No. 985, p. 5 - 7 (1978). In Russian.

    An echellogram of Saturn.   See Abstr. 031.266.

    Planetary radio astronomy receiver.
See Abstr. 032.525.

    **Observations of Saturn with the astrolabe of the
CERGA Observatory during the years 1974, 1975, 1976.**
See Abstr. 041.015.

    **Observations of Saturn with the astrolabe of the
Algiers Observatory during the years 1969–1974.**
See Abstr. 041.017.

    Orbital resonance in a dissipative medium.
See Abstr. 042.001.

    **Studies of chemical abundances in the outer solar
system.**   See Abstr. 091.011.

    Radiometry of satellites and of the rings of Saturn.
See Abstr. 091.054.

    **The influence of differential rotation of planets
on their gravitational fields.**   See Abstr. 099.010.

    **The D/H ratios on Jupiter, Saturn, and Uranus based
on new HD and H$_2$ data.**   See Abstr. 099.018.

    **Tidal dissipation in the solid cores of the major
planets.**   See Abstr. 099.019.

    **Formation of Jupiter, Saturn, Uranus and their
regular satellites.**   See Abstr. 107.006.

    On the origin and evolution of Jupiter and Saturn.
See Abstr. 107.019.

Saturn Satellites

100.501 **On the theory of the motion of Iapetus.**
M. Rapaport.
Astron. Astrophys., Vol. 62, 235 - 238 (1978).

The author has compared Struve's and Sinclair's theories
of Iapetus with observations obtained from 1878 to 1973.
He shows that the addition of some disturbing terms and a
correction of the mean motion and of the longitude of epoch
of Iapetus yield a satisfactory representation of the observations. For the ecliptic and equinox of 1950.0 as reference
system, the author gets $n = 4°5379503 \pm 2 \times 10^{-7}/\text{day}$
$\lambda_0 = 76°4203 \pm 0°005$.

100.502 **Lunar occultation of Saturn. III. How big is Iapetus?**
J. Veverka, J. Burt, J. L. Elliot, J. Goguen.
Icarus, Vol. 33, 301 - 310 (1978).

By considering both the orbital lightcurve of Iapetus and
data obtained during the March 30, 1974, occultation of the
satellite by the Moon, the authors obtain information about
the brightness distribution on the bright face of Iapetus and
derive an accurate value for the satellite's radius. From the observed orbital lightcurve the authors find that the trailing face
of Iapetus must consist predominantly of a single bright material with an effective limb-darkening parameter of $k = 0.62^{+0.10}_{-0.12}$.
Given this result the occultation observations imply a radius of
718^{+87}_{-78} km. If the patchy albedo model proposed by Morrison
et al. represents the surface of Iapetus accurately (as far as the
relative albedo distribution is concerned) then the radius of
Iapetus is 724 ± 60 km.

100.503 **Lunar occultation of Saturn. IV. Astrometric re-
sults from observations of the satellites.**
D. W. Dunham, J. L. Elliot.
Icarus, Vol. 33, 311 - 318 (1978).

The method of determining local lunar limb slopes, and
the consequent time scale needed for diameter studies, from
accurate occultation timings at two nearby telescopes is described. The results for the photoelectric observations made at
Mauna Kea Observatory during the occultation of Saturn's
satellites on March 30, 1974, are discussed. Analysis of all
observations of occultations of Saturn's satellites during 1974
indicates possible errors in the ephemerides of Saturn and its
satellites.

100.504 **Detection of the Kuiper bands in the spectrum of
Titan.**
R. Danehy, T. Owen, B. L. Lutz, J. H. Woodman.
Bull. American Astron. Soc., Vol. 9, 537 (1977). – Abstract.

100.505 **Mutual phenomena of Saturn's satellites in
1979–1980.**   K. Aksnes, F. A. Franklin.
Icarus, Vol. 34, 194 - 207 (1978).

Using two sets of orbital elements and the radii of the
Saturnian satellites 1 (Mimas) through 7 (Hyperion), the
authors find that from October 1979 until August 1980
nearly 300 mutual eclipses and occultations involving these
bodies will occur. To allow for the expected errors in the
satellite ephemerides, the authors repeat these calculations
in order to obtain the additional events that occur when all
satellite radii (save Titan's) are increased by 1000 km. A

third listing predicts eclipses of satellites by (the shadow of)
the ring.

100.506 **Titan's atmosphere and surface.**   D. M. Hunten.
Planetary satellites, (see 003.007), p. 420 - 437
(1977).

100.507 **Thermal radiation from Titan's atmosphere.**
J. Caldwell.
Planetary satellites, (see 003.007), p. 438 - 450 (1977).

The temperature inversion model of Titan's atmosphere
is modified to include the thermal emission from a plausible
amount of C_2H_2. Infrared photometry of Titan from 8 to
14 μ agrees with model calculations for 0.5 cm-atm C_2H_6 and
1.0 cm-atm C_2H_2 and an optically thin dust layer in a temperature inversion region at $160°K$. Measurements of the C_2H_2
emission at 13.7 μm may provide an indirect check on the
presence of H_2 on Titan.

100.508 **Variability of Titan: 1896–1974.**
L. E. Andersson.
Planetary satellites, (see 003.007), p. 451 - 459 (1977).

Visual and photoelectric photometry of Titan for 1896–
1900, 1922, 1951–56, and 1967–74 is discussed and the
absolute magnitude is derived for each apparition covered. The
coverage is too spotty to determine whether the variation is
periodic or irregular, but a strict periodic variation with P =
29.5 yr (the orbital period of Saturn) is excluded. No attempt
is made here to explain Titan's behavior.

100.509 **An observational test for the origin of the
Titan-Hyperion orbital resonance.**   S. J. Peale.
Reports of planetary geology program, 1977–1978,
(see 003.014), p. 54 - 56 (1978).

Orbital resonance in a dissipative medium.
See Abstr. 042.001.

**Application of spin-orbit theory to a class of orbit-
orbit resonances: the case of Titan-Hyperion.**
See Abstr. 042.010.

**Solar perturbations in Saturnian satellite motions
and Iapetus-Titan interactions.**   See Abstr. 042.068.

**Improvement of orbits of satellites of Saturn using
photographic observations.**   See Abstr. 042.069.

**New orbits for Enceladus and Dione based on the
photographic observations.**   See Abstr. 042.070.

**Spectral reflectivities of the Galilean satellites and
Titan.**   See Abstr. 099.509.

**On regions of possible motion of Jupiter and Saturn
satellites.**   See Abstr. 099.534.

Long term light variations of Io and Titan.
See Abstr. 099.539.

Ice on Saturn's rings and satellites.
See Abstr. 100.023.

101 Uranus, Neptune, Pluto, Transplutonian Planets

101.001 A redetermination of the Uranus rotation period.
J. T. Trauger, F. L. Roesler, G. Münch.
Astrophys. J., Vol. 219, 1079 - 1083 (1978).

The rotation velocity of Uranus has been measured by a comparison of spectroscopic profiles for the 5281.8 Å Fraunhofer line reflected from Uranus and the Moon. This method yields a rotation velocity which is insensitive to atmospheric seeing conditions. The authors' value for the equatorial velocity is 3.5 ± 0.4 km s^{-1}.

101.002 On the thermal structure of Uranus from infrared measurements.
R. Courtin, D. Gautier, A. Lacombe.
Astron. Astrophys., Vol. 63, 97 - 101 (1978).

Thermal profiles of Uranus are inferred from presently available infrared measurements of this planet, and compared with the radiative-convective equilibrium models of Wallace. A supersaturation of CH_4 in the Uranian stratosphere is not required to explain the infrared features of this planet. The effective temperature of the retrieved profiles is $T_e = 57°K \pm 2.5°K$ for a relative abundance of hydrogen of 85%. There is no evidence of an internal heat source and a significant enhancement of helium compared with solar abundance is unlikely.

101.003 Occultations by Neptune and by the rings of Uranus.   D. W. Dunham.
Occultation Newsl., Vol. 1, 131 (1978).

101.004 Predicted occultations by Neptune, 1978 - 1980.
A. R. Klemola, W. Liller, B. G. Marsden, J. L. Elliot.
Astron. J., Vol. 83, 205 - 207 (1978) = Lick Obs. Bull. No. 796.

Predictions and signal-to-noise ratios are supplied for 26 occultations by and appulses to Neptune during 1978 - 1980.

101.005 Uranus and the shape of elliptical rings.
R. L. Lucke.
Nature, Vol. 272, 148 (1978).

The author accounts for the variable width of the ϵ ring by differences in the orbital eccentricities of the individual particles composing the ring.

101.006 Photoelectrical observation and analysis of occultation of SAO 158687 by the Uranian ring.
Planetary laboratory, Purple Mountain Observatory, stellar division of the Peking Observatory.
Acta Astron. Sinica, Vol. 18, 255 - 258 (1977). In Chinese.

The computational results show that the main ring is an elliptic one with a small inclination to the equatorial plane of Uranus, the semi-major axis a and the eccentricity e are obtained. The radius of Uranus is assumed to be 26450 km in the computations.

101.007 Stellar occultations and the rings of Uranus.
J. L. Elliot.
Bull. American Astron. Soc., Vol. 9, 498 (1977). – Abstract.

101.008 Airborne observations of the occultation of SAO 158687 by Uranus and its rings.
J. L. Elliot, E. Dunham, D. Mink.
Bull. American Astron. Soc., Vol. 9, 498 (1977). – Abstract.

101.009 The rings of Uranus: their widths and optical thicknesses.
R. L. Millis, L. H. Wasserman, J. L. Elliot, E. Dunham.
Bull. American Astron. Soc., Vol. 9, 498 (1977). – Abstract.

101.010 The radii of the rings of Uranus.   L. H. Wasserman,
R. L. Millis, J. L. Elliot, E. Dunham.
Bull. American Astron. Soc., Vol. 9, 498 - 499 (1977).
Abstract.

101.011 The albedo of the rings around Uranus.
W. A. Baum, B. Thomsen, B. L. Morgan.
Bull. American Astron. Soc., Vol. 9, 499 (1977). – Abstract.

101.012 Methane-band detection of the Uranian ring system.
B. A. Smith, H. J. Reitsema, D. E. Weistrop.
Bull. American Astron. Soc., Vol. 9, 499 (1977). – Abstract.

101.013 Evolution of the Uranus rings.   J. Boynton.
Bull. American Astron. Soc., Vol. 9, 499 - 500 (1977). – Abstract.

101.014 Observation of an internal heat source on Neptune.
M. T. Stier, W. A. Traub, G. G. Fazio, F. J. Low.
Bull. American Astron. Soc., Vol. 9, 511 (1977). – Abstract.

101.015 Rotation of Neptune.   D. P. Cruikshank.
Bull. American Astron. Soc., Vol. 9, 511 (1977).
Abstract.

101.016 Rotation period of Neptune.
D. Slavsky, H. Smith, B. Smith, J. Africano.
Bull. American Astron. Soc., Vol. 9, 512 (1977). – Abstract.

101.017 The (5-0) H_2 quadrupole lines in Uranus' spectrum.
L. Trafton.
Bull. American Astron. Soc., Vol. 9, 535 - 536 (1977). Abstract.

101.018 Photometric variations of Uranus in the methane bands.   G. W. Lockwood, D. T. Thompson.
Bull. American Astron. Soc., Vol. 9, 536 (1977). – Abstract.

101.019 Evidence of an increase in the microwave brightness temperature of Uranus.
M. J. Klein, J. A. Turegano.
Bull. American Astron. Soc., Vol. 9, 536 (1977). – Abstract.

101.020 Limb-brightening on Uranus in the λ7300Å CH_4 band.   M. J. Price, O. G. Franz.
Bull. American Astron. Soc., Vol. 9, 536 (1977). – Abstract.

101.021 Image tube spectra of Uranus, Neptune, Triton and Pluto from 6800–9000 Å.
D. C. Benner, U. Fink, R. H. Cromwell.
Bull. American Astron. Soc., Vol. 9, 536 (1977). – Abstract.

101.022 Ethane and methane emission by Neptune.
W. Macy, W. Sinton.
Bull. American Astron. Soc., Vol. 9, 537 (1977). – Abstract.

101.023 Observations of Uranus ring occultation events.
W. B. Hubbard, B. A. Smith, B. Zellner.
Bull. American Astron. Soc., Vol. 9, 551 (1977). – Abstract.

101.024 Structure of the Uranian upper atmosphere from airborne observations of the occultation of SAO 158687.   E. Dunham, J. L. Elliot, D. Mink.
Bull. American Astron. Soc., Vol. 9, 552 (1977). – Abstract.

101.025 Atmospheres of Uranus and Neptune.
G. E. Hunt.
Nature, Vol. 272, 403 - 404 (1978).

101.026 **On the rotation period of Neptune.**
    D. P. Cruikshank.
Astrophys. J., Lett., Vol. 220, L57 - L59 (1978).
    Two plausible values for the rotation period of Neptune
have been deduced from photometric light curves at 1.25 and
2.2 μm. They are 0.7572 ± 0.002 and 0.8160 ± 0.002 days;
additional photometric data will be required to ascertain which
period is correct. Evidence is presented for changes in the
atmosphere of Neptune with a time scale of several days.

101.027 **A photometric test of rotation periods for Uranus
    and time variations of methane-band strengths.**
G. W. Lockwood, D. T. Thompson.
Astrophys. J., Vol. 221, 689 - 693 (1978).
    Intermediate-band photometry of Uranus in the reflected
solar continuum at 5508 and 7529 Å and in the methane bands
at 6190 and 7261 Å has been analyzed for variations related to
the rotation of the planet. No periodic variations were detected
at a level of 0.003 mag, but a 1% increase in the strengths of
the methane bands was observed over a 2 month interval.

101.028 **Opdagelsen af Uranus' ringe.**   H. Nielsen.
    Astron. Tidsskr., Årg.10, 146 - 153 (1977).

101.029 **A dynamical study of escaped satellites of Neptune.**
    R. S. Harrington, T. C. Van Flandern.
Bull. American Astron. Soc., Vol. 9, 621 (1978). – Abstract.

101.030 **On the origin of Uranus' rings.**   W.-H. Ip.
    Nature, Vol. 272, 802 - 803 (1978).
    The model presented here is basically an application of
the concept of jet stream formation in the early history of the
Solar System, but also offers the opportunity of observational
test. Possible confirmation of the inner satellite predicted here
will be of great interest in cosmogony.

101.031 **La découverte des anneaux d'Uranus: une aventure
    astronomique vécue.**   R. L. Millis.
Astronomie, Vol. 92, 157 - 161 (1978).

101.032 **Evidence for the depletion of ammonia in the
    Uranus atmosphere.**   S. Gulkis, M. A. Janssen,
E. T. Olsen.
Icarus, Vol. 34, 10 - 19 (1978).
    The theoretical disk brightness temperature spectra for
Uranus are computed and compared with the observed
microwave spectrum. It is shown that the emission observed
at short centimeter wavelengths originates deep below the
region where ammonia would ordinarily begin to condense.
The authors demonstrate that this result is inconsistent with
a wide range of atmospheric models in which the partial
pressure of NH_3 is given by the vapor-pressure equation in
the upper atmosphere.

101.033 **Interpretation of the 6818.9-Å methane line in
    terms of inhomogeneous scattering models for
Uranus and Neptune.**
W. Macy, Jr., J. Gelfand, W. H. Smith.
Icarus, Vol. 34, 20 - 27 (1978).
    The authors have obtained high-resolution spectra of
Uranus and Neptune in the methane transition near 6800 Å,
and in particular, the 6818.9-Å feature. Calculated equivalent
widths for this line using recently proposed models of the
atmospheres of these two planets indicate that the C/H ratio
is greater than or equal to 5×10^{-3} below the CH_4 saturation level. This value is 12 times the solar mixing ratio. The
half-widths of the computed line profiles are in agreement
with the observed half-widths.

101.034 **On the deuterium-to-hydrogen ratio in the atmo-
    sphere of Uranus.**   L. Trafton.
Astrophys. J., Vol. 222, 740 - 743 (1978).

    Photoelectric observations and recently published labora-
tory measurements of HD imply an upper limit on the
deuterium-to-hydrogen ratio in Uranus's atmosphere of
$D/H < 9.6 \times 10^{-5}$.

101.035 **Limb brightening on Uranus: the visible spectrum,
    II.**   M. J. Price, O. G. Franz.
Icarus, Vol. 34, 355 - 373 (1978).
    New narrow-band (100 Å) photoelectric area-scanning
photometry of the Uranus disk is reported. Observations were
concentrated on the two strong CH_4 bands at λ 6190 and
7300 Å. Coarse quantitative determinations of the true intensity distribution over the Uranus disk were made. For the
λ 6190-Å CH_4 band, Uranus exhibits a disk of essentially
uniform intensity except for a hint of polar brightening. For
the λ 7300-Å CH_4 band, moderate limb brightening is apparent. Specifically, the true intensities at the center and limb
of the planetary disk are approximately in the proportion $1:2$.

101.036 **A semianalytical theory for the long-term motion of
    Pluto.**   P. E. Nacozy, R. E. Diehl.
Astron. J., Vol. 83, 522 - 530 (1978).
    This paper presents a semianalytical solution for the long-
term motion of Pluto. First, the theory is discussed and the
dynamical model presented. The model consists of a modified
circular restricted problem of three bodies having a hypothetical oblate Sun and Neptune as primaries and Pluto with negligible mass. The model admits solutions that are modified
periodic orbits of the third kind. It is shown that Pluto oscil-
lates (librates) about one of these periodic solutions. The solution gives the long-term motion of the eccentricity, inclination, and perihelion of Pluto. The solution includes both the
Pluto-Neptune and Pluto-Uranus resonant effects, as well as
the nonresonant effects due to Jupiter and Saturn.

101.037 **Pluto – Ein Überblick des heutigen Erkenntnisstan-
    des.**   C. P. Renschen.
Sterne, 54. Band, 114 - 118 (1978).

101.038 **Occultations by Uranian rings.**
    IAU Circ., No. 3215 (1978).

101.039 **Neptune II (Nereid).**
    IAU Circ., No. 3219 (1978).

101.040 **Detection of H_2 quadrupole lines belonging to the
    (5–0) overtone band in the spectrum of Uranus.**
L. Trafton.
Astrophys. J., Vol. 223, 339 - 343 (1978).
    The author has detected the $S_5(0)$ and $S_5(1)$ quadrupole
lines of H_2 in the atmosphere of Uranus. They lie at the
wavelengths predicted by the energy levels derived from ultraviolet emission spectra of H_2. Their equivalent widths are in
approximate agreement with both a radiative-convective
atmosphere and a deep radiative atmosphere. The validity of
each model and the theoretical line strengths published for
the (5–0) band are discussed.

101.041 **Predicted occultations by Neptune in 1978.**
    Yamamoto Circ., No. 1875 (1978). In Japanese.

101.042 **Observations of the rings of Uranus from their
    occultation of SAO 158687.**   J. L. Elliot.
Bull. American Astron. Soc., Vol. 10, 439 (1978). – Abstract.

101.043 **The occultation of BD–15°3969: a second detec-
    tion of the rings of Uranus.**
R. L. Millis, L. H. Wasserman.
Bull. American Astron. Soc., Vol. 10, 458 (1978). – Abstract.

101.044 **The rings of Uranus: invisible and impossible?**
    T. C. Van Flandern.

Bull. American Astron. Soc., Vol. 10, 481 (1978). − Abstract.

101.045 **Pluto's satellite.** D. W. Hughes.
Nature, Vol. 274, 309 (1978).

101.046 **Electropolarimetric observations of Uranus.**
R. A. Chigladze.
Astron. Tsirk., No. 978, p. 7 - 8 (1978). In Russian.

101.047 **New discovery on rings of Uranus.**
Indian East. Eng., Vol. 119, No. 5, p. 235 (1977).
Abstr. in Phys. Abstr., Vol. 81, Abstr. 45274 (1978).

Reflections concerning Neptune's "ring".
See Abstr. 004.014.

The great unlucky fellow. Discovery of Neptune.
See Abstr. 004.033.

Correlation between X-ray K-lines of some elements and planetary masses. See Abstr. 091.089.

The D/H ratios on Jupiter, Saturn, and Uranus based on new HD and H_2 data. See Abstr. 099.018.

Observation of HD on Saturn and Uranus.
See Abstr. 100.012.

Water ice sputtering and the atmospheres of the rings of Saturn and Uranus. See Abstr. 100.027.

Detection of HD on Saturn and Uranus, and the D/H ratio. See Abstr. 100.033.

Comments on the magnetic field and internal heat of Saturn, Uranus and Neptune. See Abstr. 100.040.

Photographic positions of comets, minor planets and Pluto observed during 1970 - 1974.
See Abstr. 103.009.

Formation of Jupiter, Saturn, Uranus and their regular satellites. See Abstr. 107.006.

102 Comets (Origin, Structure, Atmospheres, Dynamics)

102.001   New osculating orbits for 110 comets and analysis of original orbits for 200 comets.
B. G. Marsden, Z. Sekanina, E. Everhart.
Astron. J., Vol. 83, 64 - 71 (1978).

New osculating orbits are presented for 110 nearly parabolic comets. Combining these with selected orbit determinations from other sources, the authors consider a total of 200 orbits where the available observations yield a result of very good (first class) or good (second class) quality. For each of these, the original and future orbits (referred to the barycenter of the solar system) are calculated. The Oort effect (a tendency for original $1/a$ values to range from 0 to $+100 \times 10^{-6}$ AU^{-1}) is clearly seen among the first-class orbits but not among the second-class orbits. Modifications in original $1/a$ values due to the effects of nongravitational forces are considered.

102.002   Relative motions of fragments of the split comets. II. Separation velocities and differential decelerations for extensively observed comets.   Z. Sekanina.
Icarus, Vol. 33, 173 -185 (1978).

A multiparameter model has been devised to discriminate the effects of the velocity of separation of two fragments of a split comet on their observed motions from the effects of the postsplit differential nongravitational forces. The iterative differential-correction procedure is applied to five extensively observed split comets: West 1975n, Ikeya-Seki 1965 VIII, Wirtanen 1957 VI, the Southern Comet 1947 XII, and Periodic Brooks 2 1889 V.

102.003   Synthetic spectra of C_2 in comets.
M. F. A'Hearn.
Astrophys. J., Vol. 219, 768 - 772 (1978).

With the key result of Krishna Swamy and O'Dell, a complete fluorescence equilibrium calculation has been used to predict the detailed Swan-band spectrum in comets. A rotational Boltzmann temperature of approximately 3000 K and small changes in published molecular constants lead to a theoretical spectrum in excellent agreement with that observed. The Fox-Herzberg bands are predicted to be quite weak compared with previously observed ultraviolet features in cometary spectra. The Ballik-Ramsay bands should be observable in the near-infrared.

102.004   Comets: data, problems, and objectives.
F. L. Whipple.
Cosmochemistry of the moon and planets, (see 012.002), p. 923 - 929 (1977).

A highly abridged review of new relevant results from the observations of Comet Kohoutek (1973f) is followed by an outline summary of our basic knowledge concerning comets, both subjects being confined to data related to the nature and origin of comets rather than the phenomena (for example, plasma phenomena are omitted).

102.005   Airglow from the inner comas of comets.
T. E. Cravens, A. E. S. Green.
Icarus, Vol. 33, 612 - 623 (1978).

The intensities of radiation from the inner comas of comets which are composed primarily of water and carbon monoxide have been calculated. Only "airglow" emissions initiated by the absorption of extreme ultraviolet radiation have been considered. The photoionizations of H_2O, CO, CO_2, and N_2 are the most important emission sources, although photoelectron excitation is also considered. In the inner coma (collision region) the airglow mechanisms are shown to be possible competitors with the usually assumed resonance scattering and fluorescence mechanisms which are appropriate for the outer coma and tail.

102.006   Comets are more primitive than C 1 chondrites.
A. H. Delsemme.
Bull. American Astron. Soc., Vol. 9, 506 - 507 (1977). Abstract.

102.007   Physical and dynamical evolution of long-period comets.   P. R. Weissman.
Bull. American Astron. Soc., Vol. 9, 507 (1977). – Abstract.

102.008   On the brightness variations and surface structures of cometary nuclei.   F. L. Whipple.
Bull. American Astron. Soc., Vol. 9, 507 (1977). – Abstract.

102.009   Composition variation among comets.
M. F. A'Hearn, R. L. Millis.
Bull. American Astron. Soc., Vol. 9, 507 (1977). – Abstract.

102.010   Distributions of Jovian perturbations on short-period comet orbits.   H. Rickman, S. Vaghi.
Astron. Astrophys., Suppl. Ser., Vol. 31, 389 - 400 (1978).

The authors present statistical distributions of perturbations by Jupiter on the orbital elements of short-period comets. Two groups of hypothetical comets are considered, each containing 1000 members. The cometary orbits all have low inclinations and are randomly oriented. One group is characterized by perihelia near Jupiter's orbit and aphelia near the orbit of Saturn, while the comets of the other group have aphelia near Jupiter's orbit and perihelia at "observable" distances from the Sun. All perturbation distributions are observed to be extremely peaked. The random probability per orbital period for a close encounter with Jupiter, which may lead to considerable perturbations, is very small for the orbits in question ($\approx$ 2% for a passage within 0.41 a.u.).

102.011   Monochromatic brightness variations in comets.
D. A. Mendis, G. D. Brin.
Bull. American Astron. Soc., Vol. 9, 564 (1978). – Abstract.

102.012   Plasma tail disconnection events in comets and magnetic reconnection at sector boundaries.
M. B. Niedner, J. C. Brandt.
Bull. American Astron. Soc., Vol. 9, 618 (1978). – Abstract.

102.013   Cometary ray closing rate.   E. P. Moore.
Bull. American Astron. Soc., Vol. 9, 618 (1978).

102.014   Dust evolution from comets.   Z. Sekanina.
Space Research, Vol. XVII, (see 012.010), 573 - 584 (1977).

The studies of the evolution of cometary debris are reviewed. The subject is divided into three major sections: (1) the developments in the immediate vicinity of the cometary nucleus, which is the source of the dust; (2) the formation of the dust tail; and (3) the blending of the debris with the dust component of interplanetary matter. The importance of the physical theory of comets is emphasized for the understanding of the early phase of the evolution of cometary dust. A physico-dynamical model designed to analyze the particle-emission mechanism from the distribution of light in the dust tails is described and the results are presented. Increased attention is paid to large particles because of their importance for the evolution of the zodiacal cloud. Finally, implications are discussed for the future in situ investigations of comets.

102.015   Significance of dust particle measurements with a simple impact plasma detector on a cometary mission.   E. Grün, H. Fechtig.
Space Research, Vol. XVII, (see 012.010), 585 - 586 (1977).

The use of an impact plasma detector to measure the mass distribution, rate and velocity of dust particles emitted by a comet is investigated briefly. The mass threshold is $\sim 10^{-15}$ g, at an impact speed of 15 km s^{-1}. By comparison with theoretical model calculations, some information on the composition of the particles is obtainable.

102.016 Experiment for compositional analysis of cometary particulates. J. F. Friichtenicht, E. Grün.
Space Research, Vol. XVII, (see 012.010), 587 - 592 (1977).

An impact ionization time of flight mass spectrometer has been successfully used aboard Helios A and B for studying micrometeoroids. It is suggested that an instrument operating on the same principle could be used to determine the relative abundance of elements in the particulate matter near comets.

102.017 The micrometeoroid impact rate on a spacecraft in close proximity to a cometary nucleus.
D. W. Hughes.
Space Research, Vol. XVII, (see 012.010), 593 - 598 (1977).

Using recent estimates of the mass loss per revolution of comet P/Encke calculations are made of the total mass of micrometeoroid dust which is expected to impact with unit area of a spacecraft as it passes close by the nucleus of the comet. Assuming that the spacecraft has a collision cross-section of 10 m^2 there is a high probability of the spacecraft being struck by a 7×10^{-3} g particle at fly-by. At a 13 km s^{-1} impact velocity this particle would penetrate aluminium of thickness 0.5 cm. The probability of encountering larger particles decreases drastically with particle size. It is concluded that a close fly-by of the comet P/Encke will not unduly endanger the spacecraft.

102.018 Comets. H. B. Ridley.
J. British Astron. Assoc., Vol. 88, 226 - 247 (1978).

102.019 Investigation of the interaction of solar radiation with an artificial comet.
E. A. Kajmakov, Yu. I. Svetov.
Izv. AN SSSR. Ser. fiz., Vol. 41, 1884 - 1886 (1977). In Russian. – Abstr. in Ref. zh., 62. Issled. kosm. prostranstva, 2.62.116 (1978).

102.020 On the origin of the Oort cloud.
S. F. Dermott, T. Gold.
Astron. J., Vol. 83, 449 - 450 (1978).

If the Sun lost an appreciable fraction of its mass in two or more mass-loss events during the early stages of planet accretion, then a small fraction of bodies in the original solar system disk would have had a high probability of being ejected with near-zero total energies.

102.021 Monochromatic brightness variations of comets. II. Core-mantle model. D. A. Mendis, G. D. Brin.
Moon, Vol. 17, 359 - 372 (1977).

A class of comets, of which Comet Kohoutek (1973f) is typical, show total as well as monochromatic brightness asymmetries about perihelion. They are fainter after perihelion than before at the same heliocentric distance. A model of the cometary nucleus consisting of a growing non-volatile dust mantle surrounding a volatile icy core is used to discuss this phenomenon. Numerical results are obtained for Comet Kohoutek (1973f). It is found that dust mantles of thickness in the range of 10–75 cm can be grown by perihelion passage for various values of the thermal conductivity of the dust if there is no substantial dust blow-off by the sublimating volatiles.

102.022 The asteroidal planet as the origin of comets.
T. C. van Flandern.
Dynamics of planets and satellites, (see 012.020), p. 89 - 99 (1978).

Ovenden has raised seemingly plausible dynamical arguments which suggest that a 90-earth-mass planet existed in the present location of the asteroid belt until 16×10^6 years ago, and then rapidly disintegrated. If the long-period comets originated from the recent disintegration of such a planet, several improbable characteristics of their orbits would be predicted, including a tendency for those orbits which are least perturbed to return to the site of the original break-up. The author compares observed characteristics of long-period comet orbits with expected characteristics, based on the missing planet hypothesis. The conclusion is that long-period comet orbits are wholly consistent with the hypothesis.

102.023 Comets and the missing planet.
M. W. Ovenden, J. Byl.
Dynamics of planets and satellites, (see 012.020), p. 101 - 107 (1978).

Integrating backwards in time in the circular restricted three-body problem Galaxy-Sun-Comet, for both the real long-period comets and fictitious random sets of orbital elements, the authors have confirmed van Flandern's conclusion that there is a statistically-significant clustering of the orbits of real long-period comets, in heliocentric direction, some 5×10^6 years ago. The authors suggest that the "event" discovered by van Flandern is not the explosive disruption of a planet formerly in the asteroid belt, but the latest in a series of minor catastrophies, such as the collisional break-up of a pair of large asteroids.

102.024 Are comets dirty snowballs or dust swarms?
W. I. Axford, H. U. Keller.
Nature, Vol. 273, 427 - 428 (1978).

102.025 On the detection of newly created CN radicals in comets. B. Donn, R. J. Cody.
Icarus, Vol. 34, 436 - 440 (1978).

Laboratory investigations of CN radical formation by photodissociation of parent molecules have suggested the possibility of observing emission lines in cometary spectra from newly formed CN radicals. These laboratory studies have shown that high initial internal excitation of CN is the rule with excitation of rotational levels N up to 70. In the collisionless environment of the cometary atmosphere this initial excitation would yield a corresponding distribution for the lowest vibrational level of the ground $X^2\Sigma^+$ state. The calculations show that it is feasible with present observational techniques to detect photochemically excited lines with $N \sim 30$ in the 0–0 band of the violet system.

102.026 On orbital dynamics of hypothetical comets with retrograde motion. M. A. Mamedov.
Dokl. AN AzSSR, Vol. 33, No. 4, p. 10 - 13 (1977). In Russian. – Abstr. in Ref. zh., 51. Astron., 4.51.134 (1978).

102.027 Determination of the density of glassy water ice.
E. A. Kajmakov, S. Aliev.
Dokl. AN TadzhSSR, Vol. 20, No. 9, p. 18 - 20 (1977). In Russian. – Abstr. in Ref. zh., 51. Astron., 4.51.273 (1978).

102.028 A model of comet comae. I. Gas-phase chemistry in one dimension. P. T. Giguere, W. F. Huebner.
Astrophys. J., Vol. 223, 638 - 654 (1978).

A model of gas-phase chemical kinetics in comet comae has been developed. Its purpose is to allow characterization of comet nuclei by predicting abundances of observable coma species. The one-dimensional calculation follows an isothermal shell of outstreaming molecules, positive ions, and atoms that starts at the nucleus with a simple set of mother molecules. Rate equations for 441 photo and chemical reactions are solved simultaneously at many points in the expansion of the shell. The calculation includes effects of the ultraviolet opacity of the mother molecules. Results are presented for four combinations of hypothetical mother molecules: H_2O, CO_2, NH_3, and CH_4.

102.029 Interplanetary gas. XXIII. Plasma tail disconnection events in comets: evidence for magnetic field line reconnection at interplanetary sector boundaries?
M. B. Niedner, Jr., J. C. Brandt.
Astrophys. J., Vol. 223, 655 - 670 (1978).

Attention is focused on a form of cometary activity which has been known for some time but is poorly understood: the discarding of a plasma tail by a comet. The authors find a link between plasma tail rejections and conditions in the solar wind. A model is presented in which a disconnected tail is the end result of magnetic field line reconnection in the cometary ionosphere caused by the traversal of a magnetic sector boundary. Observations of plasma tails appear to be the best and only method at present of mapping the interplanetary sector structure out of the ecliptic plane.

102.030 The flute instability as the trigger mechanism for disruption of cometary plasma tails.
W.-H. Ip, D. A. Mendis.
Astrophys. J., Vol. 223, 671 - 675 (1978).

The sporadic disruption of the plasma tails of some comets has recently been explained by Niedner and Brandt, see 102.029. An alternative model is proposed here in which the tail disruption is the end result of compression of the cometary ionosphere by high-velocity solar wind streams, triggering the flute instability in the marginally stable tangential discontinuity surface which separates the cometary ionosphere from the solar wind plasma. According to the authors' model, the tail disruption events could occur at all heliographic latitudes, whereas the Niedner-Brandt model should predict that these events are restricted to low heliographic latitudes.

102.031 Some remarks on the selection of observations in determination of orbits of comets.
E. D. Kondrat'eva, E. A. Reznikov.
Tr. Kazan. Gorod. Astron. Obs., No. 41, p. 71 - 76 (1976).
In Russian.

102.032 Selection of observations of comets when improving their orbits.   I. Yu. Evdokimov.
Tr. Kazan. Gorod. Astron. Obs., No. 41, p. 84 - 94 (1976).
In Russian.

102.033 Gravitational interaction between fragments of a split comet.   Z. Sekanina.
Reports of planetary geology program, 1977–1978, (see 003.014), p. 28 - 31 (1978).

102.034 The rotation of comet nuclei.   F. L. Whipple.
Reports of planetary geology program, 1977–1978, (see 003.014), p. 32 - 34 (1978).

102.035 Are plasma tail disconnection events in comets caused by high speed solar-wind streams or by reversals in the interplanetary magnetic field?
J. C. Brandt, M. B. Niedner.
Bull. American Astron. Soc., Vol. 10, 416 (1978). – Abstract.

102.036 The convolution of cometary brightness profiles by circular diaphragms.   M. R. Combi.
Bull. American Astron. Soc., Vol. 10, 460 (1978). – Abstract.

102.037 Cross-sections for a capture of a comet by the Sun-Jupiter binary system.
M. J. Valtonen, K. A. Innanen.
Bull. American Astron. Soc., Vol. 10, 484 (1978). – Abstract.

102.038 Calculation of series of orbits of some comets observed in one apparition.   I. N. Murav'eva.
Kazan. univ., Kazan', 1977. 14 pp. In Russian. – Abstr. in Ref. zh., 51. Astron., 6.51.117 (1978).

102.039 Distribution of nodes of cometary orbits and meteor matter in the ecliptic plane.
O. V. Dobrovol'skij.
Dokl. AN TadzhSSR, Vol. 20, No. 10, p. 13 - 17 (1977). In Russian. – Abstr. in Ref. zh., 51. Astron., 6.51.119 (1978).

102.040 On a possibility of modelling cometary phenomena in the near space.   E. A. Kajmakov, Yu. I. Svetov, V. I. Sharkov.
Fiz.-tekh. inst. AN SSSR. (Prepr.), 1977, 12 pp. In Russian. Abstr. in Ref. zh., 51. Astron., 6.61.415 (1978).

102.041 Dispersion of liquid mixtures for an experiment on modelling cometary phenomena in space.
E. A. Kajmakov, Yu. I. Svetov.
Fiz.-tekh. inst. AN SSSR. (Prepr.), 1977, 8 pp. In Russian. Abstr. in Ref. zh., 51. Astron., 6.51.416 (1978).

102.042 Comets and meteor streams.   M. Buhagiar.
J. Astron. Soc. Western Australia, Vol. 300, April, p. 6 - 9 (1978).

Workshop on cometary missions.
See Abstr. 012.031.

A capacity type detector for low velocity dust particles as expected during a slow fly-by cometary mission. See Abstr. 032.546.

An impact-mass-spectrometer for in situ chemical analysis of cometary particulates to be used on board of a flyby-mission.   See Abstr. 032.547.

Neuartige Raumsonden-Missionen zu den Kometen in den achtziger Jahren.   See Abstr. 051.006.

L'espace interplanétaire.   See Abstr. 084.207.

Passages of comets and asteroids near the Earth. See Abstr. 098.041.

The comet and asteroid population of the earth's environment.   See Abstr. 098.042.

Meteor stream formation after cometary decay. See Abstr. 104.007.

Ordinary chondrites and the origin of comets. See Abstr. 105.025.

Observations of interplanetary sector structure to high latitudes using disconnection events in cometary plasma tails.   See Abstr. 106.058.

Comparison of size distributions of meteoroids at injection by short period comets, near the earth and elsewhere. See Abstr. 106.061.

Molecole interstellari e cometarie.
See Abstr. 131.255.

Errata

102.901 Erratum: 'What is a cometary nucleus?'[Q. J. R. Astron. Soc., Vol. 18, 213 - 233 (1977)].
R. A. Lyttleton.
Q. J. R. Astron. Soc., Vol. 19, 150 (1978).

103 Comets: Listed Objects

103.001 **Les comètes de l'année 1976.** C. Bertaud.
Astronomie, Vol. 92, 135 - 142 (1978).

103.002 **Comets in 1974.** B. G. Marsden, E. Roemer.
Q. J. R. Astron. Soc., Vol. 19, 38 - 58 (1978).

103.003 **Comets in 1975.** B. G. Marsden, E. Roemer.
Q. J. R. Astron. Soc., Vol. 19, 59 - 89 (1978).

103.004 **Kometer 1977.** H. Q. Rasmusen.
Astron. Tidsskr., Årg. 11, 93 - 94 (1978).

103.005 **Roman numeral designations of comets in 1976.**
IAU Circ., No. 3187 (1978).

103.006 **Observations of comets.**
Yamamoto Circ., No. 1872 (1977). In Japanese.
Concerning P/Schwassmann-Wachmann 1, P/Arend-Rigaux (1977k), P/Chernykh (1977*l*), comet Kohler (1977m), comet Tsuchinshan (1977q).

103.007 **Possible comet Lovas.**
Yamamoto Circ., No. 1873 (1978). In Japanese.

103.008 **Observations of comets.**
Yamamoto Circ., No. 1885 (1978). In Japanese.
Concerning P/Gehrels 3 (1975o), P/Arend-Rigaux (1977k), P/Chernykh (1977*l*), comet Kohler (1977m), comet Tsuchinshan (1977q), P/Wild 2 (1978b), comet Lovas (1976 XII).

103.009 **Photographic positions of comets, minor planets and Pluto observed during 1970 - 1974.**
M. P. Candy, D. N. Harwood, D. J. Gans.
Perth Obs. Commun. No. 3, p. 29 - 65 (1977).

103.010 **Observations of comets and asteroids at the Kleť Observatory in the year 1976.** A. Mrkos.
Acta Univ. Carolinae Math. Phys., Vol. 19, No. 2, p. 13 - 24 (1978).
Concerning comets: West (1975n), Bradfield (1975p), Harlan (1976g), P/Faye, P/d'Arrest, P/Klemola, P/Schwassmann - Wachmann 1; and asteroids: 661 Cloelia, 1079 Mimosa, 29 Amhitrite, 84 Klio, 183 Istria, 516 Amherstia, 684 Hilburg, 172 Baucis, 221 Eos, 211 Izolda, 944 Hidalgo.

103.011 **Recoveries of periodic comets.**
British Astron. Assoc. Circ., No. 583 (1978).
Concerning: 1977r P/Kojima, 1977s P/van Biesbroeck.

103.012 **Other comets.**
British Astron. Assoc. Circ., No. 583 (1978).
Concerning: P/Faye 1976i, P/Klemola 1976j, P/Grigg-Skjellerup 1977b, Kohler 1977m, P/Chernykh 1977*l*.

103.013 **Recoveries of periodic comets.**
British Astron. Assoc. Circ., No. 587 (1978).
Concerning: 1978e P/Tsuchinshan (1), 1978g P/Clark.

103.014 **Comet digest.** J. E. Bortle.
Sky Telesc., Vol. 55, 13 - 14, 206, 215, 292 - 293, 381, 483, 496; Vol. 56, 19 (1978).

An atlas of photoelectric spectrophotometry of comets. See Abstr. 002.055.

Observations of comets in Greek and Roman sources before A.D. 410. See Abstr. 004.019.

New osculating orbits for 110 comets and analysis of original orbits for 200 comets. See Abstr. 102.001.

Periodic comet Arend-Rigaux

103.101 **Comet Arend-Rigaux: not dead yet.**
J. Degewij.
Sky Telesc., Vol. 55, 14 (1978).

103.102 **Periodic comet Arend-Rigaux (1977k).**
IAU Circ., Nos. 3181, 3195 (1978).

103.103 **P/Arend-Rigaux (1977k).**
Yamamoto Circ., Nos. 1879, 1882 (1978). In Japanese.

Comet 1977m Kohler

103.121 **Comète Kohler (1977m).** C. Bertaud.
Astronomie, Vol. 92, 243 - 244 (1978).

103.122 **Comet Kohler (1977m).**
IAU Circ., Nos. 3157, 3159, 3168, 3181, 3205, 3229 (1978).

103.123 **Comet Kohler (1977m).**
Yamamoto Circ., Nos. 1873, 1875, 1877, 1879, 1883, 1887 (1978). In Japanese.

103.124 **Photoelectric UBV observations of comet Kohler 1977m.**
V. M. Lyutyj, V. P. Tarashchuk.
Astron. Tsirk., No. 985, p. 1 - 3 (1978). In Russian.

Periodic comet Smirnova-Chernykh

103.131 **Periodic comet Smirnova-Chernykh (1975 VII).**
IAU Circ., No. 3157 (1978).

Periodic comet Sanguin

103.141 **Periodic comet Sanguin (1977p).**
IAU Circ., No. 3158 (1978).

103.142 **P/Sanguin (1977p).**
Yamamoto Circ., No. 1872 (1977). In Japanese.

103.143 **Comet P/Sanguin 1977p.**
British Astron. Assoc. Circ., No. 583 (1978).

Periodic comet Schuster

103.151 **Periodic comet Schuster (1977o).**
IAU Circ., Nos. 3160 with a corrigendum 3166, 3187, 3233 (1978).

103.152 **Periodic comet Schuster (1977o).**
Yamamoto Circ., No. 1873 (1978). In Japanese.

Periodic comet Chernykh

103.161   **Periodic comet Chernykh (1977 l).**
IAU Circ., Nos. 3160, 3163, 3171, 3205, 3223, 3236 (1978).

103.162   **Periodic comet Chernykh (1977l).**
Yamamoto Circ., No. 1873 (1978). In Japanese.

Comet 1969 IX Tago-Sato-Kosaka

103.201   **Hydrogen and hydroxyl production rates of Comet Tago-Sato-Kosaka (1969 IX).**
H. U. Keller, C. F. Lillie.
Astron. Astrophys., Vol. 62, 143 - 147 (1978).
    Comet Tago-Sato-Kosaka (1969 IX) was observed with the ultraviolet photometers on OAO-2 from January 16.41 to January 29.89, 1970, while its heliocentric distance increased from 0.78 to 1.03 a.u. The production rates of hydrogen and hydroxyl are derived from Lyman alpha (1216 Å) and the OH (3090 Å) emission. The variation of the hydrogen and hydroxyl production ran parallel to one another while their ratio was about 3 : 1. These results are consistent with the assumption that vaporization of water ice controlled the production rate of gas during this interval. The hydrogen production rates of four non-periodic comets are compared.

103.202   **Forms of tail rays – comet 1969 IX.**
F. D. Miller.
Bull. American Astron. Soc., Vol. 9, 508 (1977). – Abstract.

Periodic comet Ashbrook-Jackson

103.211   **Periodic comet Ashbrook-Jackson (1977g).**
IAU Circ., Nos. 3161, 3223 (1978).

Periodic comet Whipple

103.221   **Periodic comet Whipple (1977h).**
IAU Circ., Nos. 3161, 3219 (1978).

Comet 1976 IX Lovas

103.231   **Comet Lovas (1976k).**
IAU Circ., No. 3163 (1978).

Comet 1977q Tsuchinshan

103.241   **Comet Tsuchinshan (1977q).**
IAU Circ., Nos. 3165, 3203, 3228 (1978).

103.242   **Comet Tsuchinshan (1977q).**
Yamamoto Circ., Nos. 1873, 1874 (1978). In Japanese.

103.243   **New comet Tsuchinshan 1977q.**
British Astron. Assoc. Circ., No. 583 (1978).

Periodic comet Wild 2

103.251   **Periodic comet Wild 2 (1978b).**

IAU Circ., Nos. 3166, 3167, 3170, 3173, 3177, 3193, 3207, 3217, 3221, 3228 (1978).

103.252   **Periodic comet Wild 2 (1978b).**
Yamamoto Circ., Nos. 1876 - 1880, 1882, 1887, 1888 (1978). In Japanese.

103.253   **New periodic comet Wild (2) 1978 b.**
British Astron. Assoc. Circ., Nos. 585, 587 (1978).

Periodic comet Kopff

103.261   **Periodic comet Kopff (1976b).**
IAU Circ., No. 3166 (1978).

Periodic comet Giacobini-Zinner

103.271   **Periodic comet Giacobini-Zinner (1978h).**
IAU Circ., Nos. 3167, 3216 (1978).

103.272   **Periodic comet Giacobini-Zinner (1978h).**
Yamamoto Circ., No. 1885 (1978). In Japanese.

103.273   **Observations of the comets 1972d and 1972h.**
L. T. Markova.
Izv. Astron.Ehngel'gardt. Obs., Kazan', No. 41 - 42, p. 223 - 224 (1976). In Russian.

Comet 1976 VI West

103.301   **Splitting of Comet West 1975n: photography and narrow-band photometry.**
C. B. Cosmovici, F. Strafella, L. Dimagli, A. D'Innocenzo, G. Leggieri, C. Nesta, A. Perrone.
Astron. Astrophys., Vol. 63, 83 - 86 (1978).
    Photographic and photometric observations of Comet West 1975n have been carried out on March 12 and 13, 1976. The splitting of the parent nucleus in four nuclei could be resolved. Narrow-band photometry with interference filters at nine wavelengths in the range 4000 - 7000 Å has shown strong C_3, C_2 and NH_2 emission with fluxes of 29.8×10^{-14}, 12.5×10^{-14} and $7.6 \times 10^{-14} W/m^2$ Å resp.

103.302   **Infrared observations of Comet West (1975n). I. Observational results.**   M. Oishi, K. Kawara, Y. Kobayashi, T. Maihara, K. Noguchi, H. Okuda, S. Sato, T. Iijima, T. Ono.
Publ. Astron. Soc. Japan, Vol. 30, 149 - 159 (1978).
    For Comet West (1975n) the authors have made extensive observations including photometry between 1 and 20 μm, polarimetry at 1.00, 1.25, 1.65, and 2.25 μm, spectrophotometry in the 2.8—3.6 μm region and a measurement of brightness distribution of thermal radiation at 10.7 μm. The observational results are presented.

103.303   **Infrared observations of Comet West (1975n). II. A model of the cometary dust.**   M. Oishi, H. Okuda, N. C. Wickramasinghe.
Publ. Astron. Soc. Japan, Vol. 30, 161 - 171 (1978).
    On the basis of the infrared photometric and polarimetric observations of Comet West (1975n) (see 103.302), a cometary dust model is constructed. The features of the energy spectra and the scattering angle dependences of the polarization can be reproduced by a mixture of metallic dust particles (graphite or iron) and dielectric dust particles (silicate).

103.304 **Spectrophotometry of comet West (1975n) after the perihelion passage.** U. S. Chaubey.
Astrophys. Space Sci., Vol. 54, 233 - 237 (1978).

In the wavelength range 350 to 650 nm, the flux distribution of comet West (1975n) is presented for various dates following perihelion passage. The variations with heliocentric distance in the flux of the emission features of the CN band at 388 nm, the C_2 bands at 474 nm, 516 nm and 563 nm and Na at 589 nm have been discussed. It is concluded that the comet was dust rich.

103.305 **Propagating inhomogeneities in the dust tail of comet West 1975.**
S. Koutchmy, P. Lamy.
Nature, Vol. 273, 522 - 524 (1978).

To study the striae (striated tail) of comet West 1975 the authors have examined numerous original and duplicate pictures, and selected four observations which reveal that the morphology was relatively well conserved during the evolution of the dust tail over a period of more than 4 days. The authors have identified three striae which seem to propagate in space while retaining their basic form. These striae are made of dust grains.

103.306 **Observations photographiques de la comète West (1975n) effectuées en 1976 à l'astrographe double de 40 cm (f = 2m) d'Uccle.** H. Debehogne.
Bull. Astron. Obs. R. Belgique, Vol. 9, 20 - 22 (1978).

Comet 1978a West

103.311 **Comet West (1978a).**
IAU Circ., Nos. 3162, 3164, 3165, 3178, 3193, 3199, 3213, 3225, 3230 (1978).

103.312 **Comet West (1978a).**
Yamamoto Circ., Nos. 1874, 1875, 1876, 1879, 1881, 1882, 1884 (1978). In Japanese.

103.313 **New comet West 1978a.**
British Astron. Assoc. Circ., No. 585 (1978).

Comet 1978c Bradfield

103.321 **Comet Bradfield (1978c).**
IAU Circ., Nos. 3170, 3172, 3175, 3176, 3177, 3185, 3189, 3195, 3196, 3199, 3202, 3207, 3233 (1978).

103.322 **Comet Bradfield (1978c).**
Yamamoto Circ., Nos. 1877 - 1882 (1978). In Japanese.

103.323 **Radio observations of comet Bradfield (1978c).**
D. Despois, E. Gerard, J. Crovisier, I. Kazès.
Messenger, No. 13, p. 8 - 10 (1978).

103.324 **New comet Bradfield 1978c.**
British Astron. Assoc. Circ., Nos. 584, 585 (1978).

Periodic comet Tritton

103.331 **Periodic comet Tritton (1978d).**
IAU Circ., Nos. 3175, 3186, 3194, 3198 (1978).

103.332 **Periodic comet Tritton (1978d).**
Yamamoto Circ., Nos. 1879, 1882 (1978). In Japanese.

103.333 **New periodic comet Tritton 1978d.**
British Astron. Assoc. Circ., No. 587 (1978).

Periodic comet Kojima

103.341 **Periodic comet Kojima (1977r).**
IAU Circ., Nos. 3181, 3232 (1978).

103.342 **Periodic comet Kojima (1977r).**
Yamamoto Circ., No. 1872 (1977), No. 1874 (1978). In Japanese.

Periodic comet Holmes

103.351 **Periodic comet Holmes.**
IAU Circ., No. 3187 (1978).

Comet 1973 XII Kohoutek

103.401 **Comet Kohoutek (1973f) observations attempted at decametric radio waves.** Eh. P. Abranin, L. L. Bazelyan, N. Yu. Goncharov, V. A. Zinichev, L. M. Shul'man.
Astron. Zh. Akad. Nauk SSSR, Vol. 55, 123 - 131 (1978). In Russian. English translation in Soviet Astron., Vol. 22, No. 1.

This paper presents information on comet Kohoutek observations attempted in the period from 25.12.73 to 16.1.74.

103.402 **Interplanetary gas. XXII. Interaction of comet Kohoutek's ion tail with the compression region of a solar-wind corotating stream.** M. B. Niedner, Jr., E. D. Rothe, J. C. Brandt.
Astrophys. J., Vol. 221, 1014 - 1025 (1978).

A large-scale disturbance in the ion tail of comet Kohoutek 1973f is apparent on photographs taken on late January 19 and early January 20 of 1974. Solar-wind properties obtained by the IMP 8 satellite in Earth orbit were corotated to the comet to ascertain whether an event in the interplanetary medium was responsible for the disturbance. The results indicate that, at the time of formation of the perturbed tail structure, the comet was entering the compression region of the strong high-speed solar-wind stream. The wind sock theory of ionic comet-tail orientations was used in conjunction with the corotated satellite observations to generate a theoretical time history of the position angle made by the comet tail on the sky plane. Comparison of the observed and predicted position angles showed that the complex tail morphology on January 20 is the signature of a large, rapid change in the polar component of the solar-wind bulk velocity on the forward edge of the density buildup.

Monochromatic brightness variations of comets. II. Core-mantle model. See Abstr. 102.021.

Periodic comet Tsuchinshan 1

103.421 **Periodic comet Tsuchinshan 1 (1978e).**
IAU Circ., Nos. 3188, 3204 (1978).

103.422 **P/Tsuchinshan 1 (1978e).**
Yamamoto Circ., No. 1881 (1978). In Japanese.

Periodic comet Wolf-Harrington

103.431   Periodic comet Wolf-Harrington (1977j).
IAU Circ., No. 3188 (1978).

Periodic comet Gehrels 3

103.441   Periodic comet Gehrels 3 (1975 o)
IAU Circ., Nos. 3196, 3214 (1978).

103.442   P/Gehrels 3 (1975o).
Yamamoto Circ., No. 1882 (1978). In Japanese.

Comet 1975 II Schuster

103.451   Comet Schuster (1975 II).
IAU Circ., No. 3200 (1978).

Periodic comet van Biesbroeck

103.461   Periodic comet van Biesbroeck (1977s).
IAU Circ., Nos. 3207, 3227 (1978).

103.462   P/van Biesbroeck (1977s).
Yamamoto Circ., No. 1872 (1977), Nos. 1882,
1888 (1978). In Japanese.

Periodic comet Encke

103.501   Post-perihelion interference filter photometry of
Comet Encke in 1974.
R. L. Newburn, Jr., T. V. Johnson.
Bull. American Astron. Soc., Vol. 9, 508 (1977). – Abstract.

103.502   Observations of radio OH in periodic comet Encke.
J. C. Webber, L. E. Snyder, J. Ensinger.
Bull. American Astron. Soc., Vol. 9, 564 (1978). – Abstract.

103.503   Periodic comet Encke.
IAU Circ., No. 3210 (1978).

Periodic comet Shajn-Schaldach

103.521   Periodic comet Shajn-Schaldach.
IAU Circ., No. 3217 (1978).

Periodic comet Daniel

103.531   Periodic comet Daniel.
IAU Circ., No. 3219 (1978).

Periodic comet Gunn

103.541   Periodic comet Gunn.
IAU Circ., No. 3229 (1978).

Periodic comet Tempel 2

103.561   Periodic comet Tempel 2 (1977d).
IAU Circ., No. 3212 (1978).

Comet 1978f Meier

103.571   Comet Meier (1978f).
IAU Circ., Nos. 3214, 3216, 3218, 3220, 3222,
3224, 3226, 3227, 3237 (1978).

103.572   Comet Meier (1978f).
Yamamoto Circ., Nos. 1884 - 1888 (1978). In
Japanese.

103.573   New comet Meier 1978f.
British Astron. Assoc. Circ., Nos. 586, 587 (1978).

Comet 1976 XII Lovas

103.581   Comet Lovas (1976 XII).
IAU Circ., No. 3214 (1978).

Periodic comet Jackson-Neujmin

103.591   Periodic comet Jackson-Neujmin.
IAU Circ., No. 3215 (1978).

Comet 1977e Helin

103.601   Comet Helin 1977 e.   E. F. Helin, J. W.
Sulentic, B. A. Goldberg, J. A. Howell, H. L. Giclas.
Bull. American Astron. Soc., Vol. 9, 550 (1977). – Abstract.

Periodic comet Clark

103.611   Periodic comet Clark (1978g).
IAU Circ., No. 3216 (1978).

103.612   Possible P/Clark (1973 V), periodic comet Clark
(1978g).
Yamamoto Circ., Nos. 1884, 1885 (1978). In Japanese.

Periodic comet Pons-Winnecke

103.621   Investigation of the motion of comet Pons-Winnecke
from 1909 - 1964.   E. A. Reznikov.
Tr. Kazan. Gorod. Astron. Obs., No. 41, p. 60 - 65 (1976).
In Russian.

Periodic comet Tempel-Tuttle

103.631   Evolution of the orbit of comet Tempel-Tuttle over
400 years (1660 - 2060).   E. D. Kondrat'eva.
Tr. Kazan. Gorod. Astron. Obs., No. 41, p. 66 - 70 (1976).
In Russian.

Periodic comet Schaumasse

103.641   **Evolution of the orbit of comet Schaumasse over 400 years (1660 - 2060).**   K. P. Matsukov.
Tr. Kazan. Gorod. Astron. Obs., No. 41, p. 77 - 83 (1976). In Russian.

Comet 1968 IV Tago-Honda-Yamamoto

103.651   **New elements of comet 1968 IV Tago-Honda-Yamamoto.**   L. A. Markelov.
Tr. Kazan. Gorod. Astron. Obs., No. 41, p. 95 - 101 (1976). In Russian.

Comet 1957 III Arend-Roland

103.661   **Photometry of comet Arend-Roland.**   L. A. Urasin, A. S. Shotina.
Izv. Astron. Ehngel'gardt. Obs., Kazan', No. 41 - 42, p. 113 - 114 (1976). In Russian.

Comet 1968 VI Honda

103.671   **Equidensitometry of comet Honda (1968 VI).**   L. A. Urasin, S. K. Fomin, A. N. Chernysheva.
Izv. Astron. Ehngel'gardt. Obs., Kazan', No. 41 - 42, p. 115 - 117 (1976). In Russian.

Periodic comet Schwassmann-Wachmann 1

103.701   **Rotation and outbursts of comet P/Schwassmann-Wachmann 1.**   F. L. Whipple.
Bull. American Astron. Soc., Vol. 9, 563 - 564 (1978). Abstract.

103.702   **Periodic comet Schwassmann-Wachmann 1.**
IAU Circ., Nos. 3165, 3171, 3182, 3188, 3210, 3236 (1978).

103.703   **Periodic comet Schwassmann-Wachmann 1.**
Yamamoto Circ., Nos.1873, 1879, 1881, 1884 (1978). In Japanese.

Comet 1970 II Bennett

103.711   **Equidensitometry of comet Bennett (1969i).**   L. A. Urasin, V. G. Efremov, V. T. Popova.
Izv. Astron. Ehngel'gardt. Obs., Kazan', No. 41 - 42, p. 118 - 121 (1976). In Russian.

Comet 1972 IX Sandage

Observations of the comets 1972d and 1972h.
See Abstr. 103.273.

Periodic comet d'Arrest

103.801   **Meteoroids from periodic Comet d'Arrest.**   Z. Sekanina, H. E. Schuster.
Astron. Astrophys., Vol. 65, 29 - 35 (1978).
The anomalous tail of periodic Comet d'Arrest was photographed and photometrically studied. The derived particle-size distribution function varies inversely according to slightly higher than the 4-th power of the size. Particle dimensions span a few orders of magnitude, the mean size being in the general range of 0.01 to 0.1 cm depending on the assumed bulk density. The product of the particle albedo and of the production rate of meteoroids that remain gravitationally bound to the solar system is found to be about 3000 g s^{-1} at the time of maximum activity and some 200 g s^{-1} when averaged over the comet's revolution period.

103.802   **The light curve of the nucleus of comet d'Arrest.**   T. D. Fay, Jr., W. Wisniewski.
Icarus, Vol. 34, 1 - 9 (1978).
This paper reports new photoelectric photometry of the nucleus of P/d'Arrest (1976e) during 4, 5, and 6 August 1976 (UT), just before perihelion passage. The comet appeared as a bright stellar nucleus with a faint surrounding coma. The average visual magnitude from the nuclear region (16-arcsec diaphragm) was 11.60. The visual brightness varies with an amplitude of 0.15 ± 0.02 mag. There is evidence for a period of 5.17 ± 0.01 hr. The authors interpret such a periodic variation as due to the rotation of the nucleus. The colors of the nuclear region are very similar to those of a solar-type star, and they do not vary significantly with aspect angle.

103.803   **Periodic comet d'Arrest (1976e): observations during the unusually favorable apparition of 1976.**
D. Wallentinsen.
Strolling Astron., Vol. 27, 51 - 63 (1978).

103.804   **Observations photographiques de la comète P/d'Arrest effectuées en 1976 à l'équatorial GPO de 40 cm de l'Observatoire Austral Européen (ESO) à La Silla (Chili).**
H. Debehogne, J. Surdej, A. Surdej.
Bull. Astron. Obs. R. Belgique, Vol. 9, 23 - 25 (1978).

103.805   **Observations photographiques de la comète P/d'Arrest et des satellites de Jupiter effectuées en 1976 à la caméra astrographique de 25 cm de l'Observatoire National de Rio de Janeiro.**   H. Debehogne, R. R. de Freitas Mourão.
Bull. Astron. Obs. R. Belgique, Vol. 9, 25 - 26 (1978).

Comet 1858 VI Donati

103.901   **Rotation period of comet Donati.**   F. L. Whipple.
Nature, Vol. 273, 134 - 135 (1978).
The period is 4.621 ± 0.004 h.

104 Meteors, Meteor Streams

104.001   Meteor magnitudes and enduring trains.
    W. J. Baggaley.
Observatory, Vol.98, 8 - 11 (1978).

104.002   The size distribution of large meteor bodies.
    W. J. Baggaley.
Bull. Astron. Inst. Czechoslovakia, Vol. 29, 57 - 59 (1978).

    Because the influx on the earth of very large meteoroids (mass $\sim 10^6$ g) is small, information on their mass distribution function is sparse. Such large bodies occasionally produce meteors visible in full daylight. The occurrence frequency of such meteors determined from reports extending over many years indicates an incremental mass distribution index, $s \simeq 1.6$.

104.003   Observational data concerning the meteoric 5577 Å emission.   W. J. Baggaley.
Bull. Astron. Inst. Czechoslovakia, Vol. 29, 59 - 61 (1978).

    It is shown that the examination of observational data on visual trains and radar echoes provides information relevant to the mechanism responsible for the meteoric 5577 Å radiation.

104.004   Some results of an investigation of time and height dependences of the initial radius of ionized meteor traces.   I. A. Delov.
Geomagn. Aehron., Vol. 18, 166 - 169 (1978). In Russian.

104.005   Photographic data of: 24 double station meteors. 2 double station meteor trains. 3 meteor spectra.
1965 - 1975.
Meteor Sect. Rep. 1, 29 pp.

104.006   Anomalous LF radiowave records associated with meteoric ionisation.
W. J. Baggaley, R. G. T. Bennett.
Nature, Vol. 272, 650 (1978).

104.007   Meteor stream formation after cometary decay.
    D. W. Hughes.
Space Research, Vol. XVII, (see 012.010), 565 - 570 (1977).

    Theories put forward to explain the formation of a meteor stream from the dust debris emitted by an icy conglomerate cometary nucleus as it decays near perihelion are reviewed. These theories are then applied, using the orbital parameters of the comets responsible for the Quadrantids, Perseids and Geminids, to predict the form of the meteor streams produced and the way in which the parameters of the streams should vary with time and meteoroid mass. Recent measurements of the width of meteor streams, mass of dust in streams, across stream flux profiles, radiant distributions and   annual flux variations are used to test the veracity of the above theories.

104.008   Micrometeoroid swarms.
    J. S. Dohnanyi, H. Fechtig.
Space Research, Vol. XVII, (see 012.010), 571 - 572 (1977).

    The hypothesis that micrometeoroid swarms are fragments produced by meteoroids that enter the earth's atmosphere tangentially and break up is examined. Whilst it is shown that all but two of the swarms observed aboard HEOS 2 may be explained in this way, it is concluded that this process is unlikely to be the main process producing micrometeoroid swarms.

104.009   Influence of the earth's magnetic field on the distribution of duration of meteor radio echoes.
R. A. Aminova, O. I. Bel'kovich, A. M. Nasyrov.
Meteorn. rasprostr. radiovoln. Vyp. (No.) 12 - 13. Kazan', Kazan. univ., 1976, p. 181 - 194. In Russian. — Abstr. in Ref.

zh., 51. Astron., 2.51.330 (1978).

104.010   Variation of some statistical parameters of a meteor complex.   Yu. A. Pupyshev, V. V. Sidorov, A. M. Stepanov, K. K. Kostylev, V. A. Makarov, T. K. Filimonova.
Meteorn. rasprostr. radiovoln. Vyp. (No.) 12 - 13. Kazan', Kazan. univ., 1976, p. 223 - 230. In Russian. — Abstr. in Ref. zh., 51. Astron., 2.51.334 (1978).

104.011   On the choice of wavelength for radio measuring of orbits of small meteor particles.
N. S. Andrianov.
Meteorn. rasprostr. radiovoln. Vyp. (No.) 12 - 13, Kazan', Kazan. univ., 1976, p. 27 - 32. In Russian. — Abstr. in Ref. zh., 51. Astron., 2.51.336 (1978).

104.012   On the period of the Geminid meteor stream.
    J. Jones.
Mon. Not. R. Astron. Soc., Vol. 183, 539 - 546 (1978).

    The author   devised a new method for analysing the periodicities from year to year in meteor-shower activity. This appears to give more reproducible results than have been previously obtained and the orbital period of Geminid radio meteors is found to be close to 1.49 yr. When interpreted in terms of the Poynting–Robertson effect, the data indicate a stream age of 4.7×10^3 yr which is sufficiently long to explain the lack of very large concentrations of particles in the stream.

104.013   Form of the light curve on the initial section of disruption of a meteor body.
V. G. Kruchinenko.
Vestn. Kiev. Univ., Astron., Vyp. (No.) 19, p. 75 - 81 (1977). In Russian.

104.014   Study of the effect of electron distribution on meteor radio echoes along trails.
G. I. Kolomiets.
Vestn. Kiev. Univ., Astron., Vyp. (No.) 19, p. 81 - 83 (1977). In Russian.

    Some questions are discussed concerning the problem of electron distribution influence on the diffraction pattern along trails. Nonuniformity of the ionization profile and wind gradient affect the diffraction pattern.

104.015   Change of the form of a meteor stream at different distances from the perihelion under the action of planetary perturbations and of the Poynting-Robertson effect.   L. M. Sherbaum.
Vestn. Kiev. Univ., Astron., Vyp. (No.) 19, p. 83 - 91 (1977). In Russian.

    The change of meteor stream form on the basis of a calculation of the disturbed motion in the course of 230 years is investigated. The displacement of the same particles under the action of the Poynting-Robertson effect in 10^3–10^4 years is obtained. The action of both effects is compared.

104.016   Drift and luminosity decay of the Perseid meteor stream.   L. I. Nasyrova, G. A. Nasyrov, V. T. Khrapach.
Fiz. protsessy verkhn. atmos. Ashkhabad, Ylym, 1977, p. 86 - 91. In Russian. — Abstr. in Ref. zh., 51. Astron., 3.51. 366 (1978).

104.017   Modelling of meteor matter influx based on radar observations. Communication 3. Adjustment of the transformation operator model.   Yu. I. Voloshchuk, I. P. Khopova.

Radiotekh. Resp. mezhved. nauch.-tekh. sb., 1977, vyp. (No.) 43, p. 3 - 7. In Russian. – Abstr. in Ref. zh., 51. Astron., 3.51.374 (1978).

104.018 Pressure of the interplanetary plasma on charged meteor bodies. A. A. Dmitrievskij.
Meteor. rasprostr. radiovoln. Vyp. (No.) 12 - 13. Kazan', Kazan. univ., 1976, p. 98 - 108. In Russian. – Abstr. in Ref. zh., 51. Astron., 3.51.375 (1978).

104.019 The mass distribution and sources of interplanetary boulders. L'. Kresák.
Bull. Astron. Inst. Czechoslovakia, Vol. 29, 135 - 149 (1978).
Combining the rate of passages of the comets and asteroids near the Earth with the terrestrial influx of meteorites and fireballs, an attempt is made at reconstructing the mass distribution of interplanetary bodies in the intermediate, essentially unobservable size range. The two compositionally different classes, suggested by meteor observations, are found to constitute separate populations of substantially different distribution characteristics.

104.020 The height dependence of the error in ambipolar diffusion coefficient measurement.
V. Novotný.
Bull. Astron. Inst. Czechoslovakia, Vol. 29, 155 - 158 (1978).
The scatter of measured values in the ambipolar diffusion coefficient D of meteor trails under influence of the height gradient of the wind is discussed. It was found that the error in D has a height dependence with a minimum in the height interval of 94–96 km. A method of selecting the data on the basis of high range-resolution meteor radar station observations is suggested.

104.021 Catalogue of bands and spectral lines identified in meteor spectra. Kh. Gul'medov, S. Mukhamednazarov, V. Smirnov, N. Smirnova.
Fiz. protsessy verkhn. atmos. Ashkhabad, Ylym, 1977, p. 92 - 131. In Russian. – Abstr. in Ref. zh., 51. Astron., 4.51.308 (1978).

104.022 Application of the method of statistical modelling of meteor phenomena for studying meteor radio wave propagation. I. V. Kostylev, N. T. Svetashkova.
Astron. i geod. Vyp. (No.) 6. Tomsk, 1977, p. 63 - 75. In Russian. – Abstr. in Ref. zh., 51. Astron., 4.51.312 (1978).

104.023 On statistical revealing of meteor streams, associations and background based on the theory of brightness fluctuations. G. V. Andreev, R. G. Lazarev.
Astron. i geod. Vyp. (No.) 6. Tomsk, 1977, p. 56 - 62. In Russian. – Abstr. in Ref. zh., 51. Astron., 4.51.314 (1978).

104.024 Micrometeoroids within ten earth radii.
H. Fechtig, E. Grün, G. Morfill.
Max-Planck-Inst. Kernphys., Heidelberg, MPI H-1978-V8, 68 pp. (1978). – Submitted to Planetary and Space Science.

104.025 Investigation of the RLS parameters of the Gissar Astronomical Observatory for faint meteor observations. L. N. Rubtsov, R. G. Lazarev, G. V. Andreev, O. Alimov.
Izv. AN TadzhSSR. Otd. fiz.-mat. i geol.-khim. nauk, 1977, No. 3, p. 44 - 52. In Russian. – Abstr. in Ref. zh., 51. Astron., 4.51.318 (1978).

104.026 The stars fell like rain. K. Hindley.
New Scientist, Vol. 76, 78 - 80 (1977). – Abstr. in Phys. Abstr., Vol. 81, Abstr. 16358 (1978).

104.027 Meteor radar rates, geomagnetic activity and solar wind sector structure. B. A. Lindblad.

Nature, Vol. 273, 732 - 734 (1978).
The study of solar-influenced meteor rate changes on a time scale of days has not received adequate attention. Evidence for a short-term dependence of meteor radar rates on geomagnetic activity and solar corpuscular radiation is presented here.

104.028 The spectrum of a bright meteor obtained with the Wampler scanner. J. Stauffer, H. Spinrad.
Publ. Astron. Soc. Pacific, Vol. 90, 222 - 225 (1978).
The authors present here the spectrum of a bolide obtained (accidentally) with the Lick 120-inch (3.05-m) Cassegrain scanner system. The spectrum is notable because of the well-calibrated intensity scale available with the Lick system and because few high-quality spectra of members of the sporadic meteor population have been previously obtained.

104.029 Some unresolved problems of the physics of meteor phenomena. V. A. Bronshtehn.
Problems of observational and theoretical astronomy, Moscow–Leningrad, 1977, (see 003.012), p. 44 - 84. In Russian. – Abstr. in Ref. zh., 51. Astron., 5.51.281 (1978).

104.030 Density of meteoroids.
P. B. Babadzhanov.
Dokl. AN TadzhSSR, Vol. 20, No. 10, p. 10 - 12 (1977). In Russian. – Abstr. in Ref. zh., 51. Astron., 5.51.286 (1978).

104.031 Analysis of the spectra of Leonid meteors.
K. Nagasawa.
Ann. Tokyo Astron. Obs., Second Ser., Vol. 16, 157 - 187 (1978).
Physical and chemical properties of meteor plasma are investigated in an attempt to obtain the chemical composition of meteoroids from their spectra. In this analysis, the effective thermal equilibrium state is assumed for the meteor plasma by defining an effective excitation temperature. For calculating the diffusion of the meteor column, two kinds of models, (1) an elastic sphere model and (2) a pressure-balance model, are considered.

104.032 Optimum number of major planets for allowance of perturbations when studying the orbital evolution of condensations within meteor streams. Yu. V. Evdokimov.
Tr. Kazan. Gorod. Astron. Obs., No. 41, p. 54 - 59 (1976). In Russian.

104.033 Formation of the initial radius of a meteor train. V. S. Tokhtas'ev.
Izv. Astron. Ehngel'gardt. Obs., Kazan', No. 41 - 42, p. 122 - 128 (1976). In Russian.

104.034 Influence of recombination with electron stabilization on the limitation of duration of meteor radio echoes. V. S. Tokhtas'ev.
Izv. Astron. Ehngel'gardt. Obs., Kazan', No. 41 - 42, p. 225 - 227 (1976). In Russian.

104.035 Dependence of the ionization coefficient on velocity. V. S. Tokhtas'ev.
Izv. Astron. Ehngel'gardt. Obs., Kazan', No. 41 - 42, p. 228 - 230 (1976). In Russian.

104.036 On the inadequacy of application of effective magnitudes of meteors to the determination of meteor masses and atmospheric parameters. V. A. Smirnov.
Astron. Tsirk., No. 973, p. 4 - 6 (1977). In Russian.

104.037 True distribution of sporadic meteor radiants from radar observations. I. P. Khopova.
Astron. Tsirk., No. 973, p. 6 - 7 (1977). In Russian.

104.038   Influence of secular perturbations upon the structure of a meteor swarm.   E. N. Kramer, E. A. Timchenko-Ostroverkhova, I. S. Shestaka.
Astron. Tsirk., No. 987, p. 5 - 7 (1978). In Russian.

104.039   Simultaneous spectral observations of meteors and meteor trains.   O. O. Ovezgel'dyev, Kh. D. Gul'medov, R. I. Shafiev, S. Mukhamednazarov.
Astron. Tsirk., No. 989, p. 5 - 8 (1978). In Russian.

104.040   Degree of non-uniformity of meteor emission in spectral lines of Na I, Fe I, Ca II.
V. A. Smirnov, I. N. Kovshun.
Astron. Tsirk., No. 990, p. 3 - 5 (1978). In Russian.

104.041   On the estimate of the frequency of inelastic type II collisions in a meteor coma.   I. N. Kovshun.
Astron. Tsirk., No. 991, p. 4 - 6 (1978). In Russian.

104.042   On the scale of meteor masses.   I. N. Kovshun.
Astron. Tsirk., No. 992, p. 5 - 7 (1978). In Russian.

104.043   Attenuation of LF radio signals due to meteor trains.
B. Saha, A. K. Sen.
Ann. Géophys., Vol. 33, 503 - 511 (1977). – Abstr. in Phys. Abstr., Vol. 81, Abstr. 40726 (1978).

104.044   LF radio absorption events during major meteor showers in relation to ionization of different meteoric constituents.   A. K. Sen, B. Saha.
Indian J. Radio Space Phys., Vol. 6, 25 - 28 (1977). – Abstr. in Phys. Abstr., Vol. 81, Abstr. 40736 (1978).

104.045   Geminid meteor shower observed over Waltair during 1961-66.   M. Srirama Rao, P. V. Srirama Rao, B. Lokanadham.
Indian J. Radio Space Phys., Vol. 6, No. 1, p. 74 - 75 (1977). Abstr. in Phys. Abstr., Vol. 81, Abstr. 40881 (1978).

104.046   Report of observations of the 1977 Geminid meteor shower.
J. Astron. Soc. Western Australia, Vol. 297, January, p. 5 - 7 (1978).

104.047   Fireballs.   J. Bouška.
Vesmír, Vol. 57, 133 - 136 (1978). In Czech.

104.048   Twenty years of the satellite research of micro-

meteoroids.   J. Štohl, I. Kapišinský.
Kozmos, Vol. 9, 2 - 4 (1978). In Slovak.

104.049   Fireball of 1978 April 16.
British Astron. Assoc. Circ., No. 587 (1978).

Selected papers. Meteorites and meteor matter.
See Abstr. 003.046.

Meteors, meteor showers and meteorites in the Middle Ages: from European medieval sources.
See Abstr. 004.027.

The performance of telescopes for the observation of meteors.   See Abstr. 031.206.

The thermalization of meteoric ionization.
See Abstr. 082.025.

On the relation between diffusion coefficients and height from radar meteor echoes.   See Abstr. 082.027.

The micrometeoroid impact rate on a spacecraft in close proximity to a cometary nucleus.   See Abstr. 102.017.

Distribution of nodes of cometary orbits and meteor matter in the ecliptic plane.   See Abstr. 102.039.

Comets and meteor streams.
See Abstr. 102.042.

A test of the comet hypothesis of the Tunguska Meteor Fall: nature of the meteor "thermal" explosion paradox.   See Abstr. 105.099.

Albedos and size distribution of meteoroids from 0.3 to 4.8 AU.   See Abstr. 106.003.

Comparison of size distributions of meteoroids at injection by short period comets, near the earth and elsewhere.
See Abstr. 106.061.

The nature of the present interplanetary crater-forming projectiles.   See Abstr. 106.063.

Astronomically observable crater-forming projectiles.
See Abstr. 106.064.

105 Meteorites, Meteorite Craters

105.001 A "chondritic" eucrite parent body : inference from trace elements.
J. W. Morgan, H. Higuchi, H. Takahashi, J. Hertogen.
Geochim. Cosmochim. Acta, Vol. 42, 27 - 38 (1978).

Four eucrites were analyzed by instrumental and radiochemical neutron activation for a total of 33 major, minor and trace elements.

105.002 Further studies of trace elements in C3 chondrites.
H. Takahashi, M.-J. Janssens, J. W. Morgan, E. Anders.
Geochim. Cosmochim. Acta, Vol. 42, 97 - 106 (1978).

Five carbonaceous chondrites were analyzed by radiochemical neutron activation analysis for 20 elements. These data, together with earlier measurements on seven additional C3s, are interpreted in the light of petrographic studies by McSween (1977) and revised condensation temperatures (Wai and Wasson, 1977).

105.003 Noble gases in the Allende and Abee meteorites and a gas-rich mineral fraction: investigation by stepwise heating.
B. Srinivasan, R. S. Lewis, E. Anders.
Geochim. Cosmochim Acta, Vol. 42, 183 - 198 (1978).

Noble gases in three meteoritic samples were examined by stepwise heating, in an attempt to relate peaks in the outgassing curves to specific minerals: NeKrXe in Allende (C3V) and an Allende residue insoluble in HF-HCl, and Xe in Abee (E4).

105.004 The magnetic effects of brecciation and shock in meteorites: I. The LL-chondrites.
A. Brecher, J. Stein, M. Fuhrman.
Moon, Vol. 17, 205 - 216 (1977).

The authors explore in depth and analyze in detail the magnetic behavior of 8LL-chondrites and examine some implications for their mode of formation and evolutionary history, in the broader context of other pertinent types of data (petrology, magnetic mineralogy, radiochronology, cooling rates, etc.).

105.005 The Innisfree meteorite and the Canadian camera network.
I. Halliday, A. T. Blackwell, A. A. Griffin.
J. R. Astron. Soc. Canada, Vol. 72, 15 - 39 (1978).

The events which led to the establishment of the camera network in western Canada known as the Meteorite Observation and Recovery Project (MORP) are described. The network consists of 12 small observatories, each equipped with five cameras, a meteor detector and exposure control systems. A bright fireball was observed. Two MORP stations in Alberta photographed the event from which a predicted impact point was calculated, leading to the recovery of a 2-kg meteorite near Innisfree, Alberta. Measurements of the weak radioactivity produced by cosmic-ray activity were made. The results from this meteorite are of special interest because the meteor photographs provide a reliable orbit for the object before impact. Innisfree is only the third meteorite for which such a well-defined orbit is available.

105.006 A technique for preparing a polished thin section from a diamond-containing meteorite.
G. Moreland, R. Johnson.
Meteoritics, Vol. 12, 397 - 398 (1977).

105.007 Where did the Twin City, Georgia, meteorite come from?
R. Knox, Jr.
Meteoritics, Vol. 12, 399 - 400 (1977).

105.008 Experimental vaporization of the Holbrook chondrite.
J. L. Gooding, D. W. Muenow.
Meteoritics, Vol. 12, 401 - 408 (1977).

Knudsen cell-quadrupole mass spectrometry has been used to determine quantitatively the composition of the vapor phase produced by heating samples of the Holbrook chondrite to $1300°C$.

105.009 Analyses of some meteorites from the British Museum (Natural History) collection.
A. J. Easton, C. J. Elliott.
Meteoritics, Vol. 12, 409 - 416 (1977).

105.010 Noble gases in the unique chondrite, Kakangari.
B. Srinivasan, E. Anders.
Meteoritics, Vol. 12, 417 - 424 (1977).

The authors have measured the five noble gases in Kakangari to obtain its cosmic-ray exposure age and to characterize its primordial noble gases.

105.011 Four new iron meteorite finds.
E. R. D. Scott, J. T. Wasson, R. W. Bild.
Meteoritics, Vol. 12, 425 - 435 (1977).

Four new irons are described: Buenaventura (IIIB) from Chihuahua, Mexico: mass 114 kg; Denver City (anomalous) from Texas, USA: mass 26.1 kg; Kinsella (IIIB) from Alberta, Canada: mass 3.7 kg; and Tacoma (IA) from Washington, USA: mass 17 g. Denver City is unique, i.e. not related to any other known iron. Tacoma is the smallest iron meteorite recorded.

105.012 Habit planes of platelike schreibersite in hexahedrites.
E. Randich, K. H. Eckelmeyer.
Meteoritics, Vol. 12, 437 - 442 (1977).

The authors have measured the habit planes of plate rhabdites with respect to the parent kamacite phase in eight hexahedrites. This was accomplished using X-ray diffraction plus two-surface trace analysis. Results show that plate rhabdites form on either (001) or (122) planes in the parent kamacite. The meteorite Uwet exhibits a (112) habit plane which is possibly an anomaly due to the occurrence of a large shock event during rhabdite nucleation.

105.013 Adelaide and Bench Crater—members of a new subgroup of the carbonaceous chondrites.
M. J. Fitzgerald, J. B. Jones.
Meteoritics, Vol. 12, 443 - 458 (1977).

The composition and mineralogy of Bench Crater and a new carbonaceous chondrite (Adelaide) are compared. They, Kakangari and possibly certain carbonaceous xenoliths from other meteorites constitute a distinct chemical subgroup of the carbonaceous chondrites characterized primarily by a calcium-to-aluminium ratio (atomic) of about 0.5. It is proposed to call this group the Kakangari (CK) group.

105.014 Abstracts of papers published in Meteoritika, Vol. 36, 190 - 195 (1977).
Translated by J. E. Agrell, E. R. D. Scott.
Meteoritics, Vol. 12, 467 - 475 (1977).

105.015 The orbit of the Dhajala meteorite.
G. M. Ballabh, A. Bhatnagar, N. Bhandari.
Icarus, Vol. 33, 361 - 367 (1978).

Observations of the trail caused by the meteorite which fell around Dhajala, Gujarat (India), on 28 January 1976 have been used to compute the probable orbit of the meteoroid in space. Taking $V_\infty = 21.5$ km/sec as indicated by the measured mass ablation of the meteorite, the orbital elements are deduced to be $a = 1.8$ AU, $e = 0.59$, $i = 27°6$, $\omega = 109°1$, $\Omega = 307°8$, and $q = 0.74$.

105.016   The effect of temperature and pressure on the distribution of iron group elements between metal and olivine phases in the process of differentiation of protoplanetary material.   A. P. Vinogradov, N. P. Il'yin (*Il'in*), L. N. Kolomeytseva (*Kolomejtseva*).
Cosmochemistry of the moon and planets, (see 012.002), p. 219 - 229 (1977).
The distribution patterns of Ni, Co, Mn, and Cr were studied in olivines of various origins: from meteorites (chondrites, achondrites, pallasites), which are likely analogs of the protoplanetary material, to peridotite inclusions in kimberlite pipes, which are analogs of mantle material.

105.017   Clues in the rare gas isotopes to early solar system history.   J. H. Reynolds.
Cosmochemistry of the moon and planets, (see 012.002), p. 771 - 780 (1977).

105.018   Heterogeneities in the solar nebula.
R. N. Clayton, L. Grossman, T. K. Mayeda, N. Onuma.
Cosmochemistry of the moon and planets, (see 012.002), p. 781 - 785 (1977).

105.019   Chemical fractionation in the solar nebula.
L. Grossman.
Cosmochemistry of the moon and planets, (see 012.002), p. 787 - 796 (1977).

105.020   Barium and neodymium isotopic anomalies in the Allende meteorite.
M. T. McCulloch, G. J. Wasserburg.
Astrophys. J., Lett., Vol. 220, L15 - L19 (1978) = Div. Geol. Planet. Sci., Calif. Inst. Technol., Pasadena, Contrib. No. 2979 (252).
Isotopic anomalies have been found for Ba and Nd in two inclusions from the Allende meteorite. These inclusions are typical Ca-Al-rich objects associated with early condensates from the solar nebula but have distinctive O and Mg isotopic anomalies of the FUN type. The observations are interpreted as the result of a nearby supernova explosion which produced elements over a wide mass range and injected them into the early solar nebula shortly before condensation.

105.021   Calcium isotopic anomalies in the Allende meteorite.
T. Lee, D. A. Papanastassiou, G. J. Wasserburg.
Astrophys. J., Lett., Vol. 220, L21 - L25 (1978) = Div. Geol. Planet. Sci., Calif. Inst. Technol., Pasadena, Contrib. No. 2978 (251).
The authors report isotopic anomalies in Ca which were found in two Ca-Al-rich inclusions of the Allende meteorite. The Ca data, when corrected for mass fractionation by using $^{40}Ca/^{44}Ca$ as a standard, show nonlinear isotopic effects in ^{48}Ca of +13.5 per mil and in ^{42}Ca of +1.7 per mil for one sample. The second sample shows a ^{48}Ca depletion of −2.9 per mil, but all other isotopes are normal.

105.022   Optical and spectral properties of the low albedo meteorites.   M. Gaffey.
Bull. American Astron. Soc., Vol. 9, 505 (1977). − Abstract.

105.023   Some genetic constraints from meteoritic paleomagnetism.   A. Brecher.
Bull. American Astron. Soc., Vol. 9, 505 - 506 (1977). Abstract.

105.024   Louisville meteorite: ordinary chondrite L6 with

shock metamorphic textures.   G. Hunt.
Bull. American Astron. Soc., Vol. 9, 506 (1977). − Abstract.

105.025   Ordinary chondrites and the origin of comets.
C. A. Wood.
Bull. American Astron. Soc., Vol. 9, 506 (1977). − Abstract.

105.026   Des grains de poussière interstellaire dans les météorites.   J. Audouze, G. Poupeau.
La recherche en astrophysique, (see 003.001), p. 32 - 38 (1977).

105.027   Thermonuclear origin of Ne-E.
M. Arnould, H. Nørgaard.
Astron. Astrophys., Vol. 64, 195 - 213 (1978).
The possibility of explaining the special meteoritic neon component Ne-E in the framework of an extra-solar origin model, and more specifically in the context of the canonical explosive nucleosynthesis model, has been examined in detail.

105.028   Cooling rates of seven hexahedrites.
E. Randich, J. I. Goldstein.
Geochim. Cosmochim. Acta, Vol. 42, 221 - 233 (1978).

105.029   The cosmic ray record in the San Juan Capistrano meteorite.   R. C. Finkel, C. P. Kohl, K. Marti, B. Martinek, L. Rancitelli.
Geochim. Cosmochim. Acta, Vol. 42, 241 - 250 (1978).

105.030   Meteoritic material at five large impact craters.
H. Palme, M.-J. Janssens, H. Takahashi, E. Anders, J. Hertogen.
Geochim. Cosmochim. Acta, Vol. 42, 313 - 323 (1978).
Sixteen crater samples were analyzed by radiochemical neutron activation analysis for Ge, Ir, Ni, Os, Pd and Re.

105.031   Multiple parent bodies of polymict brecciated meteorites.   R. N. Clayton, T. K. Mayeda.
Geochim. Cosmochim. Acta, Vol. 42, 325 - 327 (1978).
In three brecciated meteorites, Bencubbin, Cumberland Falls and Plainview, the oxygen isotopic compositions of different rock types within each meteorite were determined to seek genetic relationships between them. In all cases the isotopic compositions are not consistent with derivation from a single parent body. There is no evidence that chondrites and achondrites could be derived from a common parent body. Lithic fragments in brecciated meteorites provide a wider range of rock types than is represented by known macroscopic meteorites. Collisions between some meteorite parent bodies were of sufficiently low velocity that fragments of both are preserved in breccias.

105.032   High temperature heating of the Allende meteorite.
K. Notsu, N. Onuma, N. Nishida, H. Nagasawa.
Geochim. Cosmochim. Acta, Vol. 42, 903 - 907 (1978).

105.033   Interpretation of $^{40}Ar-^{39}Ar$ release patterns and ages of chondritic meteorites.   M. C. Enright, G. Turner.
Geophys. J. R. Astron. Soc., Vol. 53, (see 012.006), 177 (1978). − Abstract.

105.034   Ion microprobe evidence for the presence of excess ^{26}Mg in an Allende anorthite crystal.
J. G. Bradley, J. C. Huneke, G. J. Wasserburg.
J. Geophys. Res., Vol. 83, 244 - 254 (1978).

105.035   The beryllium and boron abundance in the carbonaceous chondrites.   E. Dwek.
Bull. American Astron. Soc., Vol. 9, 574 (1978). − Abstract.

105.036 Fe^{57} Mössbauer spectroscopy of the Leoville meteorite.   F. W. Oliver.
Planet. Space Sci., Vol. 26, 289 - 290 (1978).

The Mössbauer spectra has been measured for the Leoville meteorite. Experimental results identify it as a type III
carbonaceous meteorite. The spectrum shows the principle
iron components to be olivine, magnetite, metallic iron, troilite
and phyllosilicate material.

105.037 Meteoritic component and impact melt composition
at the Lac à l'Eau Claire (Clearwater) impact structures, Quebec.   R. A. F. Grieve.
Geochim. Cosmochim. Acta, Vol. 42, 429 - 431 (1978).

105.038 Formation of the Bencubbin polymict meteoritic
breccia.   G. W. Kallemeyn, W. V. Boynton,
J. Willis, J. T. Wasson.
Geochim. Cosmochim. Acta, Vol. 42, 507 - 515 (1978).

105.039 Quench temperatures of Moore County and other
eucrites: residence time on eucrite parent body.
C. J. Hostetler, M. J. Drake.
Geochim. Cosmochim. Acta, Vol. 42, 517 - 522 (1978).

105.040 Tunguska: the final answer.   I. Ridpath.
New Scientist, Vol. 75, 346 - 347 (1977). – Abstr.
in Phys. Abstr., Vol. 81, Abstr. 7642 (1978).

105.041 Meteoritenkrater in Streuellipsen auf Erde, Mond
und Planeten. Teil I: Irdische Meteoritenkrater in
Streuellipsen.   J. Classen.
Sterne, 54. Band, 25 - 33 (1978).

105.042 The Jiling meteorite.   T. Tsung.
Sky Telesc., Vol. 55, 464 - 465 (1978).

105.043 Analysis of ablation debris from natural and artificial iron meteorites.   M. B. Blanchard, A. S. Davis.
J. Geophys. Res., Vol. 83, 1793 - 1808 (1978).

Artificial ablation studies have been performed on iron
and nickel-iron samples by using an arc-heated plasma of
ionized air. Experiment conditions simulated a meteoroid
traveling about 12 km/s at an altitude of 70 km. The artificially produced fusion crusts and ablation debris show features
very similar to natural fusion crusts of the iron meteorites
Boguslavka, Norfork, and N'Kandhla and to magnetic
spherules recovered from deep-sea Mn nodules.

105.044 Tysnes Island: an unusual clast composed of solidi-
fied, immiscible, Fe-FeS and silicate melts.
L. L. Wilkening.
Meteoritics, Vol. 13, 1 - 9 (1978).

105.045 The Inman, McPherson County, Kansas meteorite.
K. Keil, G. Lux, D. G. Brookins, E. A. King,
T. V. V. King, E. Jarosewich.
Meteoritics, Vol. 13, 11 - 22 (1978).

105.046 Petrology of the Yamato meteorites (j), (k), (l),
and (m) from Antarctica.
K. Yagi, J. F. Lovering, M. Shima, A. Okada.
Meteoritics, Vol. 13, 23 - 45 (1978).

The Yamato (j), (k), (l), and (m) meteorites collected
from near the Yamato Mountains in December, 1973, are
respectively an H-4 and L-5 chondrite, a howardite, and an
L-5 chondrite. Yamato (l), the howardite, is a polymict breccia of diogenite and eucrite clasts. Olivine in the chondrites
ranges in composition from Fo$_{75}$ to Fo$_{80}$, whereas in the
howardite, where it is rare, the composition is about Fo$_{60}$.

105.047 Grain size and petrography of C2 and C3 carbonace-

ous chondrites.   T. V. V. King, E. A. King.
Meteoritics, Vol. 13, 47 - 72 (1978).

105.048 The mineralogy of a rhönite-bearing calcium alumi-
num rich inclusion in the Allende meteorite.
L. H. Fuchs.
Meteoritics, Vol. 13, 73 - 88 (1978).

The meteoritic occurrence of rhönite-bearing calcium
and aluminum rich inclusions (CAIs) appears to be restricted
to the Allende meteorite. These inclusions, however, are very
scarce as only four are known to the author. This report is
primarily concerned with the descriptive mineralogy of one
of these rare CAIs.

105.049 An analytical scheme for chrondritic meteorites.
T. C. Hughes, P. Hannaker.
Meteoritics, Vol. 13, 89 - 100 (1978).

A chemical scheme is reported for the analysis of chondritic materials and results are presented for the Cocklebiddy
meteorite. Analytical problems associated with chondrite
inhomogeneity including phase leaching selectivity and the
completeness of individual phase attack are considered. Data
obtained by this method can be used to classify chondrites
according to the accepted weight and molar ratio methods.

105.050 Infrared (JHK) photometry of meteorites and
asteroids.
M. Leake, J. Gradie, D. Morrison.
Meteoritics, Vol. 13, 101 - 120 (1978).

The authors present JHK colors observed for ten asteroids and synthesized JHK colors for seven meteorite groups,
samples of iron and nickel metal, pyroxene, olivine, feldspar,
a lunar anorthite and some terrestrial mineral samples. Pronounced differences are apparent between the chondritic and
achondritic meteorite classes. The authors find small but
significant differences between the JHK colors of the pre-
dominant C and S classes of asteroids. All JHK colors of
asteroids observed here fall within the limited domain defined
by the various chondritic and iron-rich meteorites but are
strikingly different from those of most achondritic meteorites.

105.051 The Adelaide meteorite.
R. Davy, S. G. Whitehead, G. Pitt.
Meteoritics, Vol. 13, 121 - 140 (1978).

A preliminary description and chemical analysis of the
newly-discovered Adelaide meteorite is given. This meteorite
is a C2 chondrite with C3 affinities. However, the meteorite
is unlike both C2 and C3 chondrites in having a Ca/Al
ratio of 0.47.

105.052 Vein formation in the C1 carbonaceous chrondrites.
S. M. Richardson.
Meteoritics, Vol. 13, 141 - 159 (1978).

Veins in the C1 chondrites Orgueil, Alais, and Ivuna have
been deposited during an extended period of impact brecciation and leaching. At least three generations of mineralization,
dominated successively by carbonates, calcium sulfate, and
magnesium sulfate, can be recognized.

105.053 Shunak – a meteoritic crater.   A. I. Dabizha.
Priroda, 1978, No. 5, p. 140. In Russian.

105.054 Meteoritic crater under the layer of ice of the
Antarctic?   B. Yu. Levin.
Zemlya i Vselennaya, 1978, No. 3, p. 28 - 32. In Russian.

105.055 The Tunguska object: a fragment of comet Encke?
L'. Kresák.
Bull. Astron. Inst. Czechoslovakia, Vol. 29, 129 - 134 (1978).

It is suggested that the Tunguska object was an extinct
fragment separated from the nucleus of the periodic comet

Encke. Evidence in support of this conclusion is presented.

105.056 **Le meteoriti.**   P. Tempesti.
Coelum, Vol. 46, 45 - 62 (1978).

105.057 **Structure of taenite in two iron meteorites.**
J. F. Albertsen, G. B. Jensen, J. M. Knudsen.
Nature, Vol. 273, 453 - 454 (1978).
The authors have studied taenite (f.c.c. iron-nickel alloy) from iron meteorite Toluca, using Mössbauer and X-ray techniques. Here they report the main results of these investigations and compare the data with those reported previously.

105.058 **Mössbauer effect studies of taenite lamellae of an iron meteorite Cape York (III. A).**
J. F. Albertsen, M. Aydin, J. M. Knudsen.
Phys. Scr., Vol. 17, 467 - 472 (1978).

105.059 **History of development of meteoritics in Estonia.**
A. O. Aaloé.
Vopr. istor. nauk. i tekh. Pribaltiki, Tartu, 1977, p. 180 - 182. In Russian. – Abstr. in Ref. zh., 51. Astron., 4.51.13 (1978).

105.060 **Determination of lithium and halogens and the significance of lithium to the understanding of cosmochemical processes.**
G. Dreibus, B. Spettel, H. Wänke.
J. Radioanal. Chem., Vol. 38, 391 - 403 (1977). – Abstr. in Phys. Abstr., Vol. 81, Abstr. 12111 (1978).

105.061 **Chemical studies of the Mundrabilla iron meteorite by neutron activation.**   O. Müller.
J. Radioanal. Chem., Vol. 38, 499 - 511 (1977). – Abstr. in Phys. Abstr., Vol. 81, Abstr. 12171 (1978).

105.062 **Some noble metals and copper content of Italian meteorites as determined by destructive neutron activation analysis.**   M. T. Ganzerli-Valentini, V. Caramella-Crespi, S. Meloni, L. Maggi, P. Borroni.
J. Radioanal. Chem., Vol. 38, 513 - 521 (1977). – Abstr. in Phys. Abstr., Vol. 81, Abstr. 12172 (1978).

105.063 **The origin of black magnetic spherules through a study of their chemical, physical and mineralogical characteristics.**   M. Del Monte, T. Nanni, M. Tagliazucca.
Ann. Geofis., Vol. 29, 9 - 25 (1976). – Abstr. in Phys. Abstr., Vol. 81, Abstr. 24494 (1978).

105.064 **A preliminary survey on the Kirin meteorite shower.**
Joint Investigation Group of the Kirin Meteorite Shower, Acad. Sinica, Peking, China.
Sci. Sinica, Vol. 20, 502 - 512 (1977). In Chinese. – Abstr. in Phys. Abstr., Vol. 81, Abstr. 24495 (1978).

105.065 **Meteoritenkrater in Streuellipsen auf Erde, Mond und Planeten. Teil II: Irdische Meteoritenkrater in Streuellipsen.**   J. Classen.
Sterne, 54. Band, 104 - 113 (1978).

105.066 **Boron concentrations in carbonaceous chondrites.**
M. R. Weller, M. Furst, T. A. Tombrello, D. S. Burnett.
Geochim. Cosmochim. Acta, Vol. 42, 999 - 1009 (1978) = California Inst. Technol., Pasadena, Contrib. No. 2983.

105.067 **A microstructural study of the Tishomingo meteorite.**   L. K. Ives, M. B. Kasen, R. E. Schramm, A. W. Ruff, R. P. Reed.
Geochim. Cosmochim. Acta, Vol. 42, 1051 - 1066 (1978).

105.068 **Manifestation of neotectonics in a meteorite crater.**

G. P. Vdovykin, Yu. L. Kisarev.
Geotektonika, 1978, No. 1, p. 122 - 125. In Russian. – Abstr. in Ref. zh., 51. Astron., 5.51.299 (1978).

105.069 **Planar elements in a biotite from shock-metamorphosed rocks of explosion meteorite craters of the Ukrainian shield.**   E. P. Gurov.
Zap. Vses. mineral. o-va, Vol. 106, 715 - 719 (1977). In Russian. – Abstr. in Ref. zh., 51. Astron., 5.51.300 (1978).

105.070 **Phosphor in meteorites and lunar samples.**
K. B. Mur.
Fosfor v okruzh. srede. Moskva, 1977, p. 13 - 23. In Russian. Abstr. in Ref. zh., 62. Issled. kosm. prostranstva, 5.62.239 (1978).

105.071 **Compilation of solar and meteoritic abundances.**
Appendix to "The solar Na/Ca and S/Ca ratios: a close comparison with carbonaceous chondrites".
H. Holweger.
Sonderdr. Sternw. Kiel, No. 223, 5 pp. (1977).

105.072 **The earth's accretion rate of cosmic dust derived from neon isotopes in terrestrial formations.**
G. S. Anufriev, A. Ya. Krylov, V. P. Pavlov, T. I. Mazina.
Dokl. AN SSSR, Vol. 237, 284 - 287 (1977). In Russian. Abstr. in Ref. zh., 51. Astron., 4.51.317 (1978).

105.073 **Noble gases in the Murchison meteorite: possible relics of s-process nucleosynthesis.**
B. Srinivasan, E. Anders.
Science, Vol. 201, 51 - 56 (1978).
The Murchison carbonaceous chondrite contains a new type of xenon component, enriched by up to 50 percent in five of the nine stable xenon isotopes, mass numbers 128 to 132. Krypton shows a similar but smaller enrichment in the isotopes 80 and 82. Neon and helium are highly enriched in the isotopes 22 and 3. These patterns are strongly suggestive of three nuclear processes believed to take place in red giants: the s process (neutron capture on a slow time scale), helium burning, and hydrogen shell burning. If this interpretation is correct, then primitive meteorites contain yet another kind of alien, presolar material: dust grains ejected from red giants.

105.074 **Tunguska: collision with a comet.**   J. E. Oberg.
Astronomy, Vol. 5, No. 12, p. 18 - 24 (1977).
Abstr. in Phys. Abstr., Vol. 81, Abstr. 32396 (1978).

105.075 **Occurrence of chromiferous sulphides and oxides in the Allende chondrite.**
M. Christophe-Michel-Levy, B. Cervelle, C. Desnoyers.
Bull. Soc. Française Minéral Crystallogr., Vol. 100, No. 5, p. 258 - 262 (1977). In French. – Abstr. in Phys. Abstr., Vol. 81, Abstr. 32411 (1978).

105.076 **Observations on the spatial distribution of Dhajala meteorite fragments in the strewnfield.**
D. Lal, J. R. Trivedi.
Proc. Indian Acad. Sci., Sect. A, Vol. 86A, 393 - 407 (1977). Abstr. in Phys. Abstr., Vol. 81, Abstr. 32423 (1978).

105.077 **Der Mundrabilla Meteorit.**   H. H. Weinke.
Ric. Spettrosc., Lab. Astrofis. Specola Vaticana, Castel Gandolfo, Vol. 3, (N. 9), 531 - 545 (1977).

105.078 **The Angra dos Reis iron meteorite.**
H. Malissa, Jr., M. Wichtl, W. Kiesl, E. W. Salpeter.
Ric Spettrosc., Lab. Astrofis. Specola Vaticana, Castel Gandolfo, Vol. 3, (N. 10), 547 - 574 (1978).

105.079 **Tabular comparisons of the Flynn Creek impact**

crater, United States, Steinheim impact crater, Germany and Snowball explosion crater, Canada. D. J. Roddy.
Impact and explosion cratering, (see 012.034), p. 125 - 162 (1977).

105.080 Large-scale impact and explosion craters: comparisons of morphological and structural analogs. D. J. Roddy.
Impact and explosion cratering, (see 012.034), p. 185 - 246 (1977).
Morphological and structural comparisons between natural impact and explosion craters show that a well-developed set of analogs exist.

105.081 Terrestrial impact structures: principal characteristics and energy considerations.
M. R. Dence, R. A. F. Grieve, P. B. Robertson.
Impact and explosion cratering, (see 012.034), p. 247 - 275 (1977) = Contrib. Earth Phys. Branch, Dep. Energy Mines Resources, Ottawa, Canada, No. 661.
A general model has been developed for the mechanics of formation of terrestrial craters in crystalline rock. Absolute energy values are estimated and energy scales are given. The subdivision of crater structure into simple and complex types in terrestrial craters parallels classifications adopted for other planetary bodies. Application of the results of the present analysis to craters on the other planets requires allowing for gravity differences and possible differences in physical properties, particularly those affected by water content.

105.082 Pre-impact conditions and cratering processes at the Flynn Creek Crater, Tennessee. D. J. Roddy.
Impact and explosion cratering, (see 012.034), p. 277 - 308 (1977).

105.083 The Steinheim Basin – an impact structure. W. Reiff.
Impact and explosion cratering, (see 012.034), p. 309 - 320 (1977).

105.084 Deformation at the Decaturville impact structure, Missouri. T. W. Offield, H. A. Pohn.
Impact and explosion cratering, (see 012.034), p. 321 - 341 (1977).

105.085 The Ries impact crater.
J. Pohl, D. Stöffler, H. Gall, K. Ernstson.
Impact and explosion cratering, (see 012.034), p. 343 - 404 (1977).

105.086 Impact cratering phenomenon for the Ries multiring structure based on constraints of geological, geophysical, and petrological studies and the nature of the impacting body. E. C. T. Chao, J. A. Minkin.
Impact and explosion cratering, (see 012.034), p. 405 - 424 (1977).

105.087 Shallow drilling in the "Bunte Breccia" impact deposits, Ries Crater, Germany.
F. Hörz, H. Gall, R. Hüttner, V. R. Oberbeck.
Impact and explosion cratering, (see 012.034), p. 425 - 448 (1977).

105.088 Rochechouart impact crater: statistical geochemical investigations and meteoritic contamination.
P. Lambert.
Impact and explosion cratering, (see 012.034), p. 449 - 460 (1977).

105.089 Buried impact craters in the Williston Basin and adjacent area. H. B. Sawatzky.

Impact and explosion cratering, (see 012.034), p. 461 - 480 (1977).

105.090 Shock attenuation at terrestrial impact structures. P. B. Robertson, R. A. F. Grieve.
Impact and explosion cratering, (see 012.034), p. 687 - 702 (1977) = Contrib. Earth Phys. Branch, Dep. Energy Mines Resources, Ottawa, Canada, No. 688.

105.091 Impact conditions required for formation of melt by jetting in silicates. S. W. Kieffer.
Impact and explosion cratering, (see 012.034), p. 751 - 769 (1977).

105.092 Dynamical implications of the petrology and distribution of impact melt rocks.
W. C. Phinney, C. H. Simonds.
Impact and explosion cratering, (see 012.034), p. 771 - 790 (1977).

105.093 Cratering processes: as interpreted from the occurrence of impact melts.
R. A. F. Grieve, M. R. Dence, P. B. Robertson.
Impact and explosion cratering, (see 012.034), p. 791 - 814 (1977) = Contrib. Earth Phys. Branch, Dep. Energy Mines Resources, Ottawa, Canada, No. 660.

105.094 Some peculiarities of selective evaporation in target rocks after meteoritic impact.
O. V. Parfenova, O. I. Yakovlev.
Impact and explosion cratering, (see 012.034), p. 843 - 859 (1977).

105.095 Simple Z model of cratering, ejection, and the overturned flap. D. E. Maxwell.
Impact and explosion cratering, (see 012.034), p. 1003 - 1008 (1977).

105.096 Numerical simulation of a very large explosion at the earth's surface with possible application to tektites. E. M. Jones, M. T. Sandford II.
Impact and explosion cratering, (see 012.034), p. 1009 - 1024 (1977).

105.097 The primordial magnetic field preserved in chondrules of the Allende meteorite.
M. Lanoix, D. W. Strangway, G. W. Pearce.
Geophys. Res. Lett., Vol. 5, 73 - 76 (1978).
The authors have studied individual chondrules from the Allende meteorite in an effort to determine the ancient magnetic field present when the chondrules formed.

105.098 Evidence for a primordial magnetic field during the meteorite parent body era. C. P. Sonett.
Geophys. Res. Lett., Vol. 5, 151 - 154 (1978).
A total of 247 measurements of natural magnetic remanence and effective susceptibility of meteorites tabulated by Herndon et al. (1972) from Russian data is investigated. An estimate of the primordial field strength cannot be made with certainty because the field strength $H = 1.8/\rho\,^2N$, where ρ is density and N is the average carrier demagnetization factor; N can vary from zero to $4\,\pi$. Thus an extreme lower limit for H is 0.02 oersted for $N = 4\,\pi$.

105.099 A test of the comet hypothesis of the Tunguska Meteor Fall: nature of the meteor "thermal" explosion paradox. V. C. Liu.
Geophys. Res. Lett., Vol. 5, 309 - 312 (1978).
The hypothesis of an impinging comet used to construct the theory of the Great Tunguska Meteor Fall of 1908 is shown to be incompatible with the apparent existence of the intense terminal explosion of the meteor. A critical examina-

tion of and proposal for the mechanism of meteor explosions are made to shed light on the "thermal" explosion paradox of the Tunguska Meteoroid.

105.100   Size and shape of near-spherical Allegan chondrules.
P. M. Martin, A. A. Mills.
Earth Planet. Sci. Lett., Vol. 38, 385 - 390 (1978).

105.101   A disaggregation and thin section analysis of the size and mass distribution of the chondrules in the Bjurböle and Chainpur meteorites.   D. W. Hughes.
Earth Planet. Sci. Lett., Vol. 38, 391 - 400 (1978).

105.102   The abundance of barium in stony meteorites.
J. R. De Laeter, D. J. Hosie.
Earth Planet. Sci. Lett., Vol. 38, 416 - 420 (1978).
The concentration of Ba in 7 carbonaceous chondrites, 18 ordinary chondrites, 3 achondrites and 1 stony-iron meteorite has been determined by the stable isotope dilution technique using solid source mass spectrometry. Analysis of the C1 chondrite Orgueil indicates a small adjustment of the "cosmic" abundance of Ba to 4.2 on the Si = 10^6 abundance scale. The present work provides a more complete coverage of a number of meteorite classes than has so far been available for the abundance of Ba in stony meteorites.

105.103   Nitrogen abundances and isotopic compositions in stony meteorites.   C.-C. Kung, R. N. Clayton.
Earth Planet. Sci. Lett., Vol. 38, 421 - 435 (1978).
Nitrogen contents range from a few parts per million in ordinary chondrites and achondrites to several hundred parts per million in enstatite chondrites and carbonaceous chondrites. Four major isotopic groups are recognized: (1) C1 and C2 carbonaceous chondrites have δ ^{15}N of +30 to +50‰; (2) enstatite chondrites have δ ^{15}N of −30 to −40‰; (3) C3 chondrites have low δ ^{15}N with large internal variations; (4) ordinary chondrites have δ ^{15}N of −10 to +20‰. The major variations are primary, representing isotopic abundances established at the time of condensation and accretion.

105.104   Neon in gas-rich samples of the carbonaceous chondrites Mokoia, Murchison, and Cold Bokkeveld.
S. P. Smith, J. C. Huneke, G. J. Wasserburg.
Earth Planet. Sci. Lett., Vol. 39, 1 - 13 (1978) = Div. Geol. Planet. Sci., Cal. Inst. Technol., Contrib. No. 2893 (239).

105.105   Comparative uranium-thorium-lead and rubidium-strontium study of the Saint Sèverin amphoterite: consequences for early solar system chronology.
G. Manhes, J. F. Minster, C. J. Allègre.
Earth Planet. Sci. Lett., Vol. 39, 14 - 24 (1978) = Contrib. Inst. Phys. Globe NS No. 271.

105.106   Investigations on cosmic-ray-produced nuclides in iron meteorites. 1. The measurement and interpretation of rare gas concentrations.   H. Voshage, H. Feldmann.
Earth Planet. Sci. Lett., Vol. 39, 25 - 36 (1978).

105.107   Chronology and chemical history of the parent body of basaltic achondrites studied by the ^{87}Rb−^{87}Sr method.   J. L. Birck, C. J. Allègre.
Earth Planet. Sci. Lett., Vol. 39, 37 - 51 (1978) = Contrib. Inst. Phys. Globe NS No. 270.

105.108   The composition and origin of large microporphyritic chondrules in the Manych (L−3) chondrite.
R. T. Dodd.
Earth Planet. Sci. Lett., Vol. 39, 52 - 66 (1978).

105.109   A new approach to decipher the origin of the carbonaceous chondrite fission krypton and xenon.
H. W. Müller, O. A. Schaeffer.

Earth Planet. Sci. Lett., Vol. 39, 358 - 362 (1978).
It is suggested that the carbonaceous chondrite fission krypton and xenon, as measured in the primitive meteorites, may have been produced by nuclear fission induced by CNO flare particles in the few−MeV/nucleon energy range on very heavy target elements such as Au, Hg, Tl, Pb, and Bi. It is speculatively proposed that the locale of this process has been the T-Tauri phase of the sun.

105.110   Tungsten in iron meteorites.   E. R. D. Scott.
Earth Planet. Sci. Lett., Vol. 39, 363 - 370 (1978).

105.111   Chondrule mass distribution and the Rosin and Weibull statistical functions.   D. W. Hughes.
Earth Planet. Sci. Lett., Vol. 39, 371 - 376 (1978).

105.112   Compositions of droplet chondrules in the Manych (L−3) chondrite and the origin of chondrules.
R. T. Dodd.
Earth Planet. Sci. Lett., Vol. 40, 71 - 82 (1978).

105.113   Investigations on cosmic-ray-produced nuclides in iron meteorites. 2. New results on ^{41}K/^{40}K−^{4}He/^{21}Ne exposure ages and the interpretation of age distributions.
H. Voshage.
Earth Planet. Sci. Lett., Vol. 40, 83 - 90 (1978).
The cosmic ray exposure ages of 16 iron meteorites were determined. The ages measured in the present and in previous experiments are summarized and presented in form of various histograms characterizing the age distributions of the different chemical groups separately.

105.114   Preatmospheric size of the Barwell meteorite: cosmic-ray track, fusion crust and thermoluminescence studies.   C. Bagolia, N. Doshi, D. Lal.
Nucl. Track Detect., Vol. 2, No. 1, p. 29 - 35 (1978). − Abstr. in Phys. Abstr., Vol. 81, Abstr. 53893 (1978).

105.115   On estimation of mass ablation of meteorites based on studies of cosmic-ray tracks.
S. K. Gupta, D. Lal.
Nucl. Track Detect., Vol. 2, No. 1, p. 37 - 49 (1978). − Abstr. in Phys. Abstr., Vol. 81, Abstr. 53894 (1978).

105.116   Outbursts of the Boltyshian meteorite crater in the Ukrainian shield.   E. P. Gurov, A. A. Val'ter.
Geol. zh., Vol. 37, vyp. (No.) 6, p. 79 - 84 (1977). In Russian. Abstr. in Ref. zh., 51. Astron., 7.51.385 (1978).

105.117   Archaeological research of the Kaali meteorite crater.
V. Lougas.
Izv. AN EhstSSR. Obshchestv. nauk, Vol. 27, No. 1, p. 64 - 66 (1978). − Abstr. in Ref. zh., 51. Astron., 7.51.386 (1978).

105.118   Present state of the Tunguska meteorite problem.
N. V. Vasil'ev.
Problems of astronomy and geodesy, VAGO assembly Yerevan, 1975, (see 012.039), p. 88 - 106. In Russian. − Abstr. in Ref. zh., 51. Astron., 7.51.390 (1978).

Katalog über 78 irrtümlich als Impaktstrukturen bezeichnete Objekte.   See Abstr. 002.006.

Selected papers. Meteorites and meteor matter.
See Abstr. 003.046.

Meteorite craters.   See Abstr. 003.091.

Moon, Mars and meteorites.   See Abstr. 003.154.

Meteors, meteor showers and meteorites in the Middle Ages: from European medieval sources.

See Abstr. 004.027.

Condensation in supernova ejecta and isotopic anomalies in meteorites.   See Abstr. 061.001.

Proton-associated alpha-irradiation in the early solar system: a possible ^{41}K anomaly.   See Abstr. 061.032.

Isotope fractionation under simulated space conditions.   See Abstr. 061.054.

Meteoritic evidence for the Maunder minimum in solar activity.   See Abstr. 072.063.

L'espace interplanétaire.   See Abstr. 084.207.

Meteoritic material on the Moon. See Abstr. 094.120.

Comparison of lunar rocks and meteorites: implications to histories of the Moon and parent meteorite bodies. See Abstr. 094.410.

Radioactivity of the Moon, planets, and meteorites. See Abstr. 094.411.

^{39}Ar-^{40}Ar dating of basalts and rock breccias from Apollo 17 and the Malvern achondrite. See Abstr. 094.418.

The brittle fragmentation of asteroids. See Abstr. 098.022.

Comets are more primitive than C 1 chondrites. See Abstr. 102.006.

Aluminum-26 as a planetoid heat source in the early solar system.   See Abstr. 107.004.

Fractionation in the solar nebula. II. Condensation of Th, U, Pu and Cm.   See Abstr. 107.048.

The helium-driven r-process in supernovae. See Abstr. 125.043.

Average abundances of galactic cosmic rays with $Z > 50$ from studies of meteoritic olivines. See Abstr. 143.026.

Long-term averaged abundances of VVH cosmic ray nuclei from studies of olivines from Marjalahti meteorite. See Abstr. 143.057.

Errata

105.901   Erratum: 'Composition, mineralogy and origin of
               group IC iron meteorites [Earth Planet. Sci. Lett.,
Vol. 37, 273 - 284 (1977)].   E. R. D. Scott.
Earth Planet. Sci. Lett., Vol. 39, 308 (1978).

106 Interplanetary Matter, Interplanetary Magnetic Field, Zodiacal Light

106.001   On the isotropy of the phase function in scattering by interplanetary particles.
A.-C. Levasseur-Regourd, R. Dumont.
C. R. Acad. Sci. Paris, Tome 286, Sér. B, 61 - 64 (1978). In French.

The method previously suggested to inverse the zodiacal light brightness integral, and the information already inferred about the volume phase function of the interplanetary space, recently received important support from the results of Helios 1-2 space probes. The same method applied to various photometric data from 5 to 45° elongation shows an isotropic trend of the phase function at similar scattering angles, or at most a moderate rise. All models implying an intense forward scattering by interplanetary grains no longer seem tenable.

106.002   On the spatial distribution of interplanetary dust near 1 AU.   C. Leinert, M. Hanner, E. Pitz.
Astron. Astrophys., Vol. 63, 183 - 187 (1978).

Zodiacal light observations obtained with the space probes Helios 1 and Helios 2 between 0.98 and 0.85 AU were analyzed for angular distribution and radial gradient of brightness. The radial brightness gradient can best be explained by a spatial distribution of the dust particles in the size range $10-100$ μm of $n(r) \sim r^{-1.3\pm0.2}$. No change in the angular brightness distribution was found, which supports the inherent assumption that the scattering properties of interplanetary dust depend little on heliocentric distance. A comparison is made with spatial distributions derived from space observations of individual particles and from radio meteors.

106.003   Albedos and size distribution of meteoroids from 0.3 to 4.8 AU.   A. F. Cook.
Icarus, Vol. 33, 349 - 360 (1978).

Comparison is made between the run of number density of meteoroids from penetration detectors aboard Helios A (masses below 10^{-8} g) and Pioneer 10 (masses near and above 3×10^{-9} g), the source function of the zodiacal light deduced from photometric observations aboard Helios A and Pioneer 10, counts versus brightness of objects passing by Pioneer 10 from the Sisyphus experiment and the distribution of meteoroids deduced from radar and optical meteors at the Earth. The penetration detectors appear to find some increase in number density toward the Sun and a flat distribution outward to 5.2 AU. The overall behavior of the zodiacal light is that the relative distribution over direction is unchanged while the source scattering function diminishes as the inverse 1.4 power of distance from the Sun. The author deduces a mean geometric albedo of meteoroids of 0.006 at 1 AU from the Sun.

106.004   Interplanetary scintillations of PSR 0531 + 21 at 74 MHz.   J. W. Armstrong, W. A. Coles.
Astrophys. J., Vol. 220, 346 - 352 (1978).

The flux density of the 74 MHz compact source in the Crab Nebula is estimated from interplanetary scintillation observations using a new method of analysis. The results removes the earlier disagreement with very-long-baseline interferometry determinations. It also shows a 45% secular decrease in the flux from 1971.4 to 1975.4, in good agreement with observations of the 74 MHz pulsing flux. The differences between this new analysis and the method which had been used to date are described. A procedure for the reinterpretation of other meterwavelength scintillation observations is discussed.

106.005   Carbonaceous chrondritic material in the solar system.   L. L. Wilkening.
Naturwissenschaften, 65. Jahrg., 73 - 79 (1978).

106.006   The three-dimensional structure of the interplanetary magnetic field.   N. P. Korzhov.
Astron. Zh. Akad. Nauk SSSR, Vol. 55, 96 - 106 (1978). In Russian. English translation in Soviet Astron., Vol. 22, No. 1.

The analysis of observations of the white-light corona performed aboard OSO-7 gives evidence for the existence of peculiar coronal ribbon structures which may be observed at the limb as coronal streams. It is found that penetration through the surface, formed by the prolongation of these structures into the interplanetary space, is accompanied by a change of polarity of the interplanetary magnetic field. The sector structure observed is the intersection of this three-dimensional shape with the ecliptic plane. A large-scale three-dimensional structure of the interplanetary magnetic field which existed in May - July 1973 has been found and its connection with several phenomena in the solar atmosphere is established.

106.007   Dynamics of small particles in the solar system.   R. S. Rajan, S. J. Weidenschilling.
Bull. American Astron. Soc., Vol. 9, 519 (1977). – Abstract.

106.008   Lorentz scattering of interplanetary dust.   G. Consolmagno, J. R. Jokipii.
Bull. American Astron. Soc., Vol. 9, 519 - 520 (1977). Abstract.

106.009   Low energy cosmic ray erosion of ice grains in interplanetary and interstellar media.
L. J. Lanzerotti, W. L. Brown, J. M. Poate, W. M. Augustyniak.
Nature, Vol. 272, 431 - 433 (1978).

The authors show that the erosion of water ice from interplanetary particles by solar energetic particles is likely to be the dominant process for determining a grain lifetime in the solar system. Furthermore, they point out that energetic particle erosion of grain surfaces may be an important process in interstellar clouds.

106.010   Zodiacal light photopolarimetry. IV. Annual variations of brightness and the symmetry plane of the zodiacal cloud. Absence of solar-cycle variations.
R. Dumont, A. C. Levasseur-Regourd.
Astron. Astrophys., Vol. 64, 9 - 16 (1978).

A reanalysis of zodiacal light observations from the D2A spacecraft (1971–1973) and from the Tenerife station (1964–1975) is presented. After the short period enhancements have been eliminated, faint annual variations of brightness are observed at high and medium ecliptic latitudes. The authors treat the geometrical problem of deducing from the amplitude and phase of the oscillation observed the most probable inclination ($1°5 \pm 0°4$) and ascending node ($96° \pm 15°$) of the symmetry plane of the zodiacal cloud. No variation related to solar activity is found.

106.011   Search for a dust free zone around the sun from the Helios 1 solar probe.
C. Leinert, M. Hanner, H. Link, E. Pitz.
Astron. Astrophys., Vol. 64, 119 - 122 (1978).

Observations of the zodiacal light made from the Helios 1 solar probe near perihelion have been analyzed for the presence of a dust-free zone. No change in the relative angular brightness distribution of zodiacal light was found while the spacecraft approached the sun from 0.5 to 0.31 AU. This is consistent with a continuation of the dust distribution observed at larger heliocentric distances, $n(r) \sim r^{-1.3}$, to at least 0.09 AU, the limiting distance observable from Helios 1.

106.012   The Doppler shift from zodiacal light.

G. López Rodriguez, C. Sánchez Magro.
Astron. Astrophys., Vol. 64, 161 - 163 (1978).

The Doppler shift in and out of the ecliptic plane for elliptical and circular orbits of the dust is calculated for several differing density distribution functions. The theoretical calculations have been applied to some models with prograde orbits.

106.013   Intensity increase of protons and α-particles at the front of an interplanetary shock wave.
G. M. Blokh, B. M. Kuzhevskij, I. A. Savenko.
Geomagn. Aehron., Vol. 18, 9 - 13 (1978). In Russian.

106.014   Influence of fluctuations of the B_z-component of the interplanetary magnetic field on the equatorial electrojet in different intervals of local time.
B. V. Rezhenov, Ya. I. Fel'dshtejn.
Geomagn. Aehron., Vol. 18, 117 - 121 (1978). In Russian.

106.015   Influence of the interplanetary magnetic field on the night ionosphere at mean latitudes.
B. S. Shapiro.
Geomagn. Aehron., Vol. 18, 148 - 151 (1978). In Russian.

106.016   IMF effects on short-period fluctuations at low latitudes.   A. K. Agarwal, R. G. Rastogi, N. Nityananda, B. P. Singh.
Nature, Vol. 272, 517 - 518 (1978).

The authors conclude that there is a strong correlation between B_z variations and the fluctuations in magnetic fields in all regions and at all periods. The influence of B_z, now established for short periods at low-latitudes, could be a good method of studying variations in the IMF.

106.017   Brightness and polarization of the zodiacal light: multicolor results from Skylab.
J. L. Weinberg, R. C. Hahn.
Bull. American Astron. Soc., Vol. 9, 564 (1978). – Abstract.

106.018   The general inversion of the zodiacal light brightness integral.   D. W. Schuerman.
Bull. American Astron. Soc., Vol. 9, 564 - 565 (1978). Abstract.

106.019   Possible effects of Mars on the symmetry plane of interplanetary dust.
N. Y. Misconi, J. L. Weinberg, R. C. Hahn, D. E. Beeson.
Bull. American Astron. Soc., Vol. 9, 620 - 621 (1978). Abstract.

106.020   Stochastic and dynamic temperature changes in the interplanetary gas.   M. K. Wallis, M. H. A. Hassan.
Planet. Space Sci., Vol. 26, 111 - 120 (1978).

For interplanetary helium gas reaching 1 a.u., the collisional heating ranges from effective transverse increase of 200 K and radial increase of 1500 K in the downstream wake, to several thousand K increase in radial temperature of the secondary component transverse to the initial gas stream. In interpreting 584 Å sky background radiation observations, the dynamical changes in the velocity spread have to be taken into account for helium gas that is initially hot, when Doppler shifts relative to the solar emission line are significant; the present solutions being the thermal approximations to the distribution function reveal the appropriate radial temperature as a function of space.

106.021   Interplanetary scintillation.   W. A. Coles.
Space Sci. Rev., Vol. 21, 411 - 425 (1978).

The use of interplanetary scintillations for probing otherwise inaccessible regions of the solar wind is reviewed. A comparison with space-craft observations in the ecliptic is used as a calibration for the scintillation observations. Recent observations at high latitudes and near the Sun are discussed from this viewpoint. A new analysis which uses both scintillation and angular scattering observations to estimate the electron density spectrum is introduced. The spectrum appears to have a high frequency 'cutoff' which varies slowly with solar distances and may also have a relatively flat region just below the cutoff frequency.

106.022   Limits to the extent of a dust free zone around the sun derived from Helios 1 zodiacal light experiment.
C. Leinert, H. Link, E. Pitz, M. Hanner.
Space Research, Vol. XVII, (see 012.010), 553 - 554 (1977).

Variations of the intensity of the zodiacal light with angular distance from the sun have been measured aboard Helios 1, the sun–spacecraft distance varying from 0.5 to 0.3 AU. The measurements show no evidence for the existence of a "dust-free" zone around the sun at >0.09 AU.

106.023   Radial distribution of beta meteoroids from the Pioneers 8 and 9 cosmic dust experiments.
J. W. Rhee, O. E. Berg, H. Wolf.
Space Research, Vol. XVII, (see 012.010), 555 - 557 (1977).

The heliocentric radial distribution of the flux of hyperbolic cosmic dust has been derived on the basis of more than seven years of impact data from the Pioneer 8 and Pioneer 9 microparticle experiments. Analysis indicates that the distribution of hyperbolic cosmic dust particles of masses 10^{-12} $\sim 10^{-15}$ g decreases between 0.75 AU and 0.94 AU. Near 0.95 AU it starts increasing, reaches a maximum at 1.05 AU and then starts decreasing again. Assuming a particle speed of 50 km s^{-1} the average spatial concentration is derived to be $10^{-8} \sim 10^{-9}$ particle m^{-3}.

106.024   Results of the asteroid–meteoroid particle experiment on Pioneer 11.
R. K. Soberman, S. L. Neste, K. Lichtenfeld.
Space Research, Vol. XVII, (see 012.010), 559 - 564 (1977).

The Asteroid–Meteoroid Detector (AMD), an electro-optical instrument that detects and measures particles in space by sensing the sunlight reflected from them, is part of the payload of both the Pioneer 10 and 11 spacecraft. Detailed analysis yielded 51 events between 1.0 and 3.5 AU, which were used to determine the particle concentration distribution in that region of space. As with the Pioneer 10 results, the exponent β of the size dependence ($N \propto a^{\beta}$ where a is the radius of the particle) varies from about -1.7 for 100 μm radius particles to about -3.2 for 10 cm bodies.

106.025   An upper limit for the flux of interstellar dust grains in the solar system.
D. A. Tomandl, O. E. Berg.
Space Research, Vol. XVII, (see 012.010), 631 - 635 (1977).

Data from the LEAM (a micro-meteorite detector at the Apollo 17 landing site) have been examined for evidence of interstellar (IS) dust grains traversing the solar system. The analysis technique considers IS grains approaching the solar system from the local solar apex. A model calculates the grains' hyperbolic orbits into the solar system and predicts the impact directions on the moon. The observations are then compared with the predicted impact directions to measure the IS dust flux. No evidence has been found (at the $97\frac{1}{2}$% confidence level) for a flux greater than $\sim 6 \times 10^{-5}$ m^{-2} s^{-1} for particles $\geqslant 2 \times 10^{-14}$ g.

106.026   Neutral interstellar gas in the vicinity of the planets.
P. W. Blum, H. J. Fahr.
Space Research, Vol. XVII, (see 012.010), 733 - 738 (1977).

The neutral interplanetary gases of interstellar origin that stream into the solar system are focused by the sun as well as by the planets. The focusing effect of the Earth, Jupiter,

Saturn, and Uranus is investigated here, and is found to be most effective for Jupiter, where the maximum density increase of interplanetary hydrogen may be up to 80 times the undisturbed density. For helium the effect is twice as strong. Under the assumption of a low interstellar gas temperature, this increase amounts to the order of $10^{12} - 10^{13}$ atoms cm^{-2} for Jupiter for both hydrogen and helium. For Saturn the column density increase is one order of magnitude less and for Uranus two orders of magnitude less. For the Earth the helium column density increase at low interstellar temperatures is $10^8 - 10^{10}$ atoms cm^{-2}.

106.027 The interplanetary modulation and transport of Jovian electrons.   T. F. Conlon.
J. Geophys. Res., Vol. 83, 541 - 552 (1978).

Simultaneous measurements by Pioneer 11 of the 3- to 6-MeV Jovian electron flux, interplanetary magnetic field magnitude, and solar wind speed during the pre—Jupiter encounter period reveal that electron transport across the average field direction was greatly inhibited in corotating interaction regions (CIR's) and enhanced in rarefaction regions. Since CIR's are regions of compressed solar wind plasma, these results suggest that cross-field transport is affected by the degree of compression of the solar wind. The solution of a convection-diffusion equation for a model that includes the effects of CIR's and assumes that Jupiter is a time-independent point source of electrons describes the spatial dependence of the peak electron fluxes and the time-intensity profiles of individual electron flux increases.

106.028 Energetic protons associated with interplanetary active regions 1–5 AU from the sun.
M. E. Pesses, J. A. Van Allen, C. K. Goertz.
J. Geophys. Res., Vol. 83, 553 - 562 (1978).

The purposes of this paper are to describe the intensity, anisotropy, and spectra of energetic protons ($0.61 \leq E_p \leq 3.41$ MeV) in events occurring in the range 1–5 AU and to do an extended and detailed study of the association between such events and interplanetary active regions.

106.029 Observations of the interplanetary sector structure up to heliographic latitudes of 16°: Pioneer 11.
E. J. Smith, B. T. Tsurutani.
J. Geophys. Res., Vol. 83, 717 - 724 (1978).

A study of the interplanetary sector structure at heliographic latitudes up to 16° N is reported. The study is based on magnetic field measurements made on board Pioneer 11 as the spacecraft along the post-Jupiter-encounter trajectory. Preliminary measurements are used to determine the dominant polarity of the interplanetary magnetic field during 43 successive solar rotations including Pioneer's ascent to its maximum latitude and motion inward from 5 to 3.7 AU. The Pioneer 11 observations imply that the boundary between adjacent sectors corresponds physically to a current sheet surrounding the sun and lying nearly parallel to the solar equatorial plane.

106.030 Heliocentric distance dependence of the interplanetary magnetic field.   K. W. Behannon.
Rev. Geophys. Space Phys., Vol. 16, 125 - 145 (1978).

Recent and ongoing planetary missions have provided and are continuing to provide extensive observations of the variations of the interplanetary magnetic field (IMF) both in time and with heliocentric distance from the sun. Large time variations in both the IMF and its fluctuations are observed. These are produced predominantly by dynamical processes in the interplanetary medium associated with stream interactions. Magnetic field variations near the sun are propagated to greater heliocentric distances, a process also contributing to the observed variability of the IMF.

106.031 Large fluffy particles: a possible explanation of the optical properties of interplanetary dust.
R. H. Giese, K. Weiss, R. H. Zerull, T. Ono.
Astron. Astrophys., Vol. 65, 265 - 272 (1978).

It is shown that models based on Mie-theory and small ($\lesssim 1~\mu m$) particles are no longer consistent with observational results. The main contribution to zodiacal light seems to be based on larger ($> 10~\mu m$) particles. It is shown by model calculations and by results of microwave analog measurements that, if polarization is taken into account, absorbing particles of fluffy structure are possibly an important component of the interplanetary dust cloud.

106.032 A solar cycle variation of the interplanetary magnetic field.   G. L. Siscoe, N. U. Crooker, L. Christopher.
Sol. Phys., Vol. 56, 449 - 461 (1978).

A systematic and simple solar cycle variation of the magnetic field is shown to exist in the behavior of occurrence-frequency histograms of the magnitude of the solar ecliptic, north—south component of the field.

106.033 Connection of proton flux increase in the interplanetary space in August - September 1973 with variations of solar wind velocity.   N. V. Alekseev, P. V. Vakulov, Yu. I. Logachev, I. A. Savenko, B. Ya. Shcherbovskij.
Izv. AN SSSR. Ser. fiz., Vol. 41, 1827 - 1831 (1977). In Russian. – Abstr. in Ref. zh., 62. Issled. kosm. prostranstva, 2.62.136 (1978).

106.034 Acceleration of energetic particles to relativistic energies in the interplanetary medium.
N. P. Chirkov, A. T. Filippov.
Izv. AN SSSR. Ser. fiz., Vol. 41, 1776 - 1781 (1977). In Russian. – Abstr. in Ref. zh., 62. Issled. kosm. prostranstva, 2.62.139 (1978).

106.035 The influence of the source sizes on interplanetary scintillation spectra. Theory.
V. I. Shishov, T. D. Shishova.
Astron. Zh. Akad. Nauk SSSR, Vol. 55, 411 - 418 (1978). In Russian. English translation in Soviet Astron., Vol. 22, No. 2.

Analytical expressions for the spatial and temporal interplanetary scintillation spectra of radio sources with finite angular size are obtained. A qualitative analysis has shown that the extension of the interplanetary medium and velocity dispersion on the line of sight is essential for the investigation of the temporal interplanetary scintillation spectra.

106.036 Sectorial structure of the interplanetary magnetic field and energetic particles in the interplanetary space.   N. M. Gordeeva, M. A. Zel'dovich, B. M. Kuzhevskij.
Kosm. Issled., Vol. 16, 294 - 300 (1978). In Russian.

106.037 Dynamics of a cloud of fast electrons travelling through the plasma (*solar electrons in interplanetary space*).   T. Takakura.
Rep. IPPJ-286, (see 012.018), 6 pp. (1977). – Abstr. in Phys. Abstr., Vol. 81, Abstr. 7853 (1978).

106.038 Diagnostics of the azimuthal component of the interplanetary magnetic field from observations of the geomagnetic field on the earth's surface.
Ya. I. Fel'dshtejn, P. V. Sumaruk, E. P. Kharin.
Solar wind and the magnetosphere, (see 003.006), p. 39 - 46 (1976). In Russian. – Abstr. in Ref. zh., 62. Issled. kosm. prostranstva, 3.62.346 (1978).

106.039 Correlations between the typical hydromagnetic

structure of interplanetary streams of a strong flare and variations of energetic particles.
L. V. Evdokimova, K. G. Ivanov.
Solar wind and the magnetosphere, (see 003.006), p. 28 - 32 (1976). In Russian. – Abstr. in Ref. zh., 62. Issled. kosm. prostranstva, 3.62.347 (1978).

106.040   The interplanetary scattering mean free path from 1 to 3 × 10³ MV.
R. D. Zwickl, W. R. Webber.
J. Geophys. Res., Vol. 83, 1157 - 1161 (1978).

The authors' intent is to unify all of the previous diffusion coefficient studies in a single picture that is both self-consistent and up to date in its interpretation. They have re-analyzed the published electron and proton solar particle event data from 1967 to 1974. These data, which cover a broad energy range, are presented in a radial mean free path (λ_r) versus rigidity plot. From this plot the authors can then ask and attempt to answer several basic physics questions concerning the interplanetary propagation of charged particles.

106.041   On the observation of a flare-generated shock wave at 9.7 AU by Pioneer 10.
M. Dryer, M. A. Shea, D. F. Smart, H. R. Collard, J. D. Mihalov, J. H. Wolfe, J. W. Warwick.
J. Geophys. Res., Vol. 83, 1165 - 1168 (1978).

The authors discuss the circumstantial evidence for the 'tracking' of a solar-flare-generated shock wave from the sun to and beyond Jupiter to Pioneer 10 at 9.7 AU. They are now in a position to infer the shock's velocity as a function of heliocentric distance. In the present case the shock velocity V_{shock} appears to be proportional to $R^{-0.08}$, where R is the heliocentric radius in astronomical units. The ambient convecting solar wind is, then, responsible for the nearly constant shock velocity.

106.042   The interstellar wind: Mariner 10 measurements of hydrogen (1216 Å) and helium (584 Å) interplanetary emission.   A. L. Broadfoot, S. Kumar.
Astrophys. J., Vol. 222, 1054 - 1067 (1978).

Measurements of diffuse 1216 and 584 Å emissions from interplanetary space were made from the Mariner 10 spacecraft during 1973 December and 1974 January. The data from four roll calibration maneuvers, in which the UV spectrometer field of view scanned the sky as the spacecraft rolled about the spacecraft-Sun axis, are presented. The Mariner 10 sky maps obtained show the general features of the interstellar wind.

106.043   An interpretation of Mariner 10 helium (584 Å) and hydrogen (1216 Å) interplanetary emission observations.   J. M. Ajello.
Astrophys. J., Vol. 222, 1068 - 1079 (1978).

The Mariner 10 ultraviolet spectrometer measured the interplanetary emissions of both He (584 Å) and H(1216 Å) on 1974 January 28 at time of solar minimum. The heliocentric distance was 0.75 AU. A first analysis of the observations by a simple model, which employs the Copernicus measurement for the velocity of the interstellar wind of 22 ± 3 km s⁻¹, shows that a simultaneous measurement of both emissions results in a self-consistent determination of the physical properties of the interstellar wind.

106.044   Interplanetary propagation of flare-associated energetic particles.
L. S. Ma Sung, J. A. Earl.
Astrophys. J., Vol. 222, 1080 - 1096 (1978).

A propagation model which combines a Gaussian profile for particle release from the Sun, with interplanetary particle densities predicted by focused diffusion, is proposed to explain the propagation history of flare-associated energetic particles. This model, which depends on only two parameters,

successfully describes the time-intensity profiles of 30 proton and electron events originating from the western hemisphere of the Sun.

106.045   Interplanetary magnetic field and geomagnetic activity.   P. V. Sumarik, Ya. I. Fel'dshtejn.
Geofiz. sb. AN USSR, 1977, vyp. (No.) 79, p. 86 - 89. In Russian. – Abstr. in Ref. zh., 51. Astron., 4.51.458 (1978).

106.046   Radial variation of the interplanetary magnetic field between 0.3 AU and 1.0 AU. Observations by the Helios-1 spacecraft.
G. Musmann, F. M. Neubauer, E. Lammers.
J. Geophys., Vol. 42, 591 - 598 (1977). – Abstr. in Phys. Abstr., Vol. 81, Abstr. 20729 (1978).

106.047   Dynamical effects on circumsolar dust grains.
G. Schwehm, M. Rohde.
J. Geophys., Vol. 42, 727 - 735 (1977). – Abstr. in Phys. Abstr., Vol. 81, Abstr. 20730 (1978).

106.048   Micrometeoroid orbits observable by the Helios micrometeoroid detector (E10).   K. D. Schmidt.
J. Geophys., Vol. 42, 737 - 741 (1977). – Abstr. in Phys. Abstr., Vol. 81, Abstr. 20731 (1978).

106.049   Plasma disturbances caused by the Helios spacecraft in the solar wind.   U. Isensee.
J. Geophys., Vol. 42, 581 - 589 (1977). – Abstr. in Phys. Abstr., Vol. 81, Abstr. 24416 (1978).

106.050   Observations of zodiacal light from Helios 1 and 2.
C. Leinert, E. Pitz, M. Hanner, H. Link.
J. Geophys., Vol. 42, 699 - 704 (1977). – Abstr. in Phys. Abstr., Vol. 81, Abstr. 24483 (1978).

106.051   Interpretation of the optical properties of interplanetary dust.   R. H. Giese.
J. Geophys., Vol. 42, 705 - 716 (1977). – Abstr. in Phys. Abstr., Vol. 81, Abstr. 24484 (1978).

106.052   Micrometeoroid data from the first two orbits of Helios 1.   E. Grün, H. Fechtig, J. Kissel, P. Gammelin.
J. Geophys., Vol. 42, 717 - 726 (1977). – Abstr. in Phys. Abstr., Vol. 81, Abstr. 24485 (1978).

106.053   Connection of increases of proton streams in September - November 1973 with changes in solar wind velocity according to data of the AIS (Automatic Interplanetary Station) Mars 7.
Yu. V. Mineev, I. A. Savenko, E. S. Spir'kova.
Kosm. Issled., Vol. 16, 398 - 402 (1978). In Russian.

106.054   Energy spectrum of radio waves emitted by Venera 10 during the radio transillumination of the circumsolar medium and determination of the characteristics of plasma turbulence.   A. I. Efimov, O. I. Yakovlev, V. M. Razmanov, V. I. Rogal'skij, V. K. Shtrykov.
Kosm. Issled., Vol. 16, 419 - 428 (1978). In Russian.

106.055   Hydrogen-implanted interplanetary dust grains: their role in hydrogen chemistry of the upper atmosphere.   R. Parthasarathy.
J. Atmos. Terr. Phys., Vol. 40, 565 - 570 (1978).

An evaluation of past measurements of submicron dust grains in the interplanetary medium, by means of zodiacal light observations, satellite-borne detectors and rocket-borne collectors, has been made to specify the earthward flux of the grains. Using laboratory data on ion-implantation, an estimate of the solar wind implanted hydrogen in the grains is then derived. A hypothesis is offered that the hydrogen is released in the upper atmosphere, probably as H_2O.

106.056 Variations of the Z-component of the interplanetary magnetic field stimulating an intensive main phase of a flare-type magnetic storm.
M. S. Bobrov, E. S. Davydova.
Geomagn. Aehron., Vol. 18, 553 - 554 (1978). In Russian.

106.057 The radial variation of corotating energetic particle streams in the inner and outer solar system.
M. A. I. Van Hollebeke, F. B. McDonald, J. H. Trainor,
T. T. von Rosenvinge.
NASA Tech. Memo., NASA TM 78040, 40 pp. (1977).
The radial gradient of long-lived, corotating energetic particle streams is measured using observations of 0.9–2.2 MeV protons from Helios 1 and 2 between 0.3 AU and 1 AU, IMP 7 at 1 AU, Pioneer 11 at 3.8 AU and Pioneer 10 between 9 AU and 10 AU. A positive gradient of ~350% per AU is found between 0.3 AU and 1 AU. Between 1 AU and some 3–5 AU, the gradient is variable with an average value of 100% per AU which is consistent with the earlier statistical results obtained from the IMP 7 and Pioneer 11 data.

106.058 Observations of interplanetary sector structure to high latitudes using disconnection events in cometary plasma tails.   M. B. Niedner.
Bull. American Astron. Soc., Vol. 10, 416 - 417 (1978).
Abstract.

106.059 Infrared observations of the zodiacal dust cloud.
S. D. Price, L. P. Marcotte.
Bull. American Astron. Soc., Vol. 10, 459 (1978). – Abstract.

106.060 Infrared emission of the interplanetary dust cloud.
T. L. Murdock, S. D. Price.
Bull. American Astron. Soc., Vol. 10, 459 (1978). – Abstract.

106.061 Comparison of size distributions of meteoroids at injection by short period comets, near the earth and elsewhere.   A. F. Cook.
Bull. American Astron. Soc., Vol. 10, 459 - 460 (1978).
Abstract.

106.062 The zodiacal light from 1500 Å to 60 micron. Mie scattering and thermal emission.
S. Röser, H. J. Staude.
Astron. Astrophys., Vol. 67, 381 - 394 (1978).
Using the grain size spectrum derived from measurements in situ, the spatial dust distribution deduced from Helios and Pioneer results, and Mie theory with measured refractive indices of obsidian, andesite, olivine, magnetite and graphite, a wavelength dependent model of the zodiacal light (ZL) is discussed. A good absolute fit of the observed visual brightness and polarization over the whole sky is obtained. There are difficulties in explaining the features observed in the near IR radiation from the F Corona.

106.063 The nature of the present interplanetary crater-forming projectiles.   G. W. Wetherill.
Impact and explosion cratering, (see 012.034), p. 613 - 615 (1977).

106.064 Astronomically observable crater-forming projectiles.
E. M. Shoemaker.
Impact and explosion cratering, (see 012.034), p. 617 - 628 (1977).

106.065 Numerical experiments on the investigation of physical processes in the near space by means of the method of "large particles".   Yu. M. Davydov, L. V. Shidlovskaya.
Mathematical models of the near space, 1977, (see 003.015),

p. 67 - 88. In Russian. – Abstr. in Ref. zh., 62. Issled. kosm. prostranstva, 6.62.229 (1978).

106.066 Geometric form of the front of interplanetary shock waves in the plane of the ecliptic.
V. A. Belyaev.
Cosmophysical investigations, Sverdlovsk, 1977, (see 003.016), p. 68 - 77. In Russian. – Abstr. in Ref. zh., 62. Issled. kosm. prostranstva, 6.62.231 (1978).

106.067 On the propagation of interplanetary shock waves.
S. F. Nosov, E. Ya. Gidalevich, V. I. Utkin.
Cosmophysical investigations, Sverdlovsk, 1977, (see 003.016), p. 20 - 22. In Russian. – Abstr. in Ref. zh., 62. Issled. kosm. prostranstva, 6.62.233 (1978).

106.068 Onde d'urto interplanetarie generate da brillamenti solari.   G. Moreno.
Mem. Soc. Astron. Italiana, Vol. 48, (see 012.037), 269 - 293 (1977).

106.069 A note on the complications of the Compton-Getting effect for low energy charged particle measurements in interplanetary space.   E. Keppler.
Geophys. Res. Lett., Vol. 5, 69 - 72 (1978).
The response of thin window semi-conductor detectors to various energetic ions is discussed. Different ions produce response at a given energy loss at different energies per nucleon. The Compton-Getting effect causes further shifts to lower and higher energy ranges dependent upon direction, hence a particular differential channel may count particles coming from quite different energy ranges in the solar wind frame. The implications of these effects are discussed.

Solar flare activity and some indices of the interplanetary and geomagnetic fields, 1957 - 1973.
See Abstr. 003.104.

Astro 7 zodiacal light UV polychromator and its absolute radiometric calibration.   See Abstr. 032.549.

Cyclotron instabilities in a hydrogen plasma.
See Abstr. 062.064.

Sunspot distribution relative to the sector structure of the interplanetary magnetic field.   See Abstr. 072.032.

Evaluation of a prediction technique for low energy solar particle events.   See Abstr. 073.024.

Peculiarities of interaction of solar cosmic radiation with the earth's magnetosphere and the interplanetary space in the initial anisotropic phase of flares.
See Abstr. 073.067.

Polar enhancements of interplanetary Lα through solar wind asymmetries.   See Abstr. 074.003.

Least solar distance of interplanetary dust and the origin of the solar wind.   See Abstr. 074.026.

The modern and ancient flux of solar wind particles, solar flare particles and micrometeoroids.
See Abstr. 074.028.

Dynamic MHD modeling of the solar wind disturbances during the August 1972 events.   See Abstr. 074.029.

Topological structure of coronal-interplanetary magnetic field.   See Abstr. 074.034.

On the generation of coronal and interplanetary

shock waves during solar flares of 2 - 13 VIII, 1972.
See Abstr. 074.036.

Numerical simulation of MHD (*magnetohydrodynamic*) shock waves in the solar wind.   See Abstr. 074.040.

Solar cosmic radiation and interplanetary shock waves on April 29 - 30, 1973.   See Abstr. 078.004.

Effect of the day night ionospheric conductivity gradient on polar cap convective flow.   See Abstr. 083.057.

Interplanetary magnetic field variations and the electromagnetic state of the equatorial ionosphere.
See Abstr. 083.123.

Ionosphere and the interplanetary magnetic field.
See Abstr. 083.128.

The latitudinal distributions of auroral zone electric fields and ground magnetic perturbations and their response to variations in the interplanetary magnetic field.
See Abstr. 084.026.

IMFP effects on the equatorial geomagnetic field and ionosphere – a review.   See Abstr. 084.232.

Comparison of magnetic field perturbations at high latitudes with charged particle and IMF measurements.
See Abstr. 084.246.

Satellite observations of polar, magnetotail lobe, and interplanetary electrons at low energies.
See Abstr. 084.250.

An important statistical consideration, and the effect of the interplanetary magnetic field on the quiet time variation of the geomagnetic field.   See Abstr. 084.284.

Groups of micrometeoroids in the Earth–Moon system.   See Abstr. 094.434.

Dust evolution from comets.
See Abstr. 102.014.

Significance of dust particle measurements with a simple impact plasma detector on a cometary mission.
See Abstr. 102.015.

Experiment for compositional analysis of cometary particulates.   See Abstr. 102.016.

Interplanetary gas. XXIII. Plasma tail disconnection events in comets: evidence for magnetic field line reconnection at interplanetary sector boundaries?
See Abstr. 102.029.

Are plasma tail disconnection events in comets caused by high speed solar-wind streams or by reversals in the interplanetary magnetic field?   See Abstr. 102.035.

Interplanetary gas. XXII. Interaction of comet Kohoutek's ion tail with the compression region of a solar-wind corotating stream.   See Abstr. 103.402.

Meteoroids from periodic Comet d'Arrest.
See Abstr. 103.801.

Pressure of the interplanetary plasma on charged meteor bodies.   See Abstr. 104.018.

The mass distribution and sources of interplanetary boulders.   See Abstr. 104.019.

Micrometeoroids within ten earth radii.
See Abstr. 104.024.

Change of interstellar gas parameters in stellar-wind-dominated astrospheres: solar case.
See Abstr. 131.173.

The interstellar wind and its influence on the interplanetary environment.   See Abstr. 131.214.

Quiet time interplanetary cosmic ray anisotropies observed from Pioneer 10 and 11.   See Abstr. 143.024.

Influence of an electric field of unipolar induction on the distribution function of cosmic radiation in the interplanetary space.   See Abstr. 143.034.

On particle streams with $Z \geqslant 2$ as measured aboard AES Cosmos 490.   See Abstr. 143.046.

Anisotropic diffusion of cosmic rays in the interplanetary space. 5. Solar-daily variations.
See Abstr. 143.050.

Determination of the diffusion coefficient of cosmic rays in the interplanetary medium on the basis of Forbush decreases 1967 - 1972.   See Abstr. 143.070.

107 Cosmogony of the Planetary System

107.001 Planet formation: mechanism of early growth.
W. K. Hartmann.
Icarus, Vol. 33, 50 - 61 (1978).

Experiments in vacuum and in air quantify mechanics of collisions, rebound, and fragmentation at low velocities, under the conditions usually postulated for the preplanetary environment in the primitive solar nebula. Contrary to widespread assumptions, accretionary growth of the largest meteoroid- and asteroid-sized bodies in a given swarm results spontaneously from the simple mechanics of these collisions, without other ad hoc sticking mechanisms. Granular surfaces form, either by gravitational collapse of dust swarms or by rapid formation of regolith surfaces on solid planetesimals; these surfaces strongly promote further growth by retarding rebound. Growth of large bodies increases modal collision velocities, causing fragmentation of smaller bodies and eventual production of interstellar dust as a by-product of planetesimal interactions.

107.002 The formation of iron, stone, and mixed planetesimals in the early solar system.
P. S. Wesson, A. Lermann.
Icarus, Vol. 33, 74 - 88 (1978).

The interactions of dust grains with each other in a finite-temperature solar nebula are examined, taking into account the important fact that such grains would carry net steady-state charges like those of grains in interstellar clouds. This charge is given by the well-known Spitzer relation. It provides a screening mechanism that operates during accretion and results in bodies of differing compositions depending on the local temperature in the nebula. The planetesimals are of iron and stone and mixed composition in the inner solar system, but of mixed composition only in the outer solar system.

107.003 Solar influence on planetary evolution.
M. J. Handbury, I. P. Williams.
Observatory, Vol. 98, 19 - 22 (1978).

107.004 Aluminum-26 as a planetoid heat source in the early solar system.   J. M. Herndon, M. A. Herndon.
Meteoritics, Vol. 12, 459 - 465 (1977).

The results of thermal model calculations which assume ^{26}Al as a heat source are presented. The relation between ^{26}Al content, the maximum central temperature obtained, and the time interval after formation until central cooling commences is elucidated. Because of the heating times required, these results constrain maximum permissible planetoid dimensions more severely than do previous calculations which assume a high initial temperature.

107.005 Time scale for the formation of the Earth and planets and its role in their geochemical evolution.
V. S. Safronov.
Cosmochemistry of the moon and planets, (see 012.002), p. 797 - 803 (1977).

The duration of the process of formation of the Earth and planets is discussed. A short time scale for formation of the Earth ($10^4 - 10^5$ years) has been proposed, not from a consideration of the rate of its growth, but from geochemical and geophysical considerations. The initial mass of the solar nebula is discussed. The important role of large bodies in the process of formation of the planets is noted.

107.006 Formation of Jupiter, Saturn, Uranus and their regular satellites.   W.-s. Dai.
Acta Astron. Sinica, Vol. 18, 192 - 202 (1977). In Chinese.

In the Jupiter-Saturn region of the nebular disk, grains accumulated to form planetary cores. When the masses of planetary cores increased to the order of 10^{25} grams, they started to accrete gas, and around the cores were gradually formed flat rotating gaseous envelopes. When grains and small ice particles entered the gaseous envelopes, the bulk fell into the cores, but a small part had the e, i values of their orbits decreased because of gas drag, and formed around the cores very flat planetesimal disks, from which regular satellites were later formed. The gaseous envelopes finally collapsed to become the intermediate layers and the outer layers of Jupiter and Saturn. No gaseous envelope was formed around Uranus.

107.007 Limits for masses, distances and accumulation times of the terrestrial planets.
A. V. Vityazev, G. V. Pechernikova, V. S. Safronov.
Astron. Zh. Akad. Nauk SSSR, Vol. 55, 107 - 112 (1978). In Russian. English translation in Soviet Astron., Vol. 22, No. 1.

The growth of a planetary embryo is considered, its supply zone increasing due to the increase of its mass. The upper limit of the mass of the planet and the size of its supply zone are estimated for the model when its growth is not limited by the neighbouring planets. For the terrestrial planets the calculated values of these quantities do not differ much from the observed ones.

107.008 A capture hypothesis for the satellites and rings of the major planets.   S. F. Singer.
Bull. American Astron. Soc., Vol. 9, 519 (1977). – Abstract.

107.009 Isotopic evidence for a supernova origin of the solar system.   D. N. Schramm.
Bull. American Astron. Soc., Vol. 9, 544 (1977). – Abstract.

107.010 Accretion as part of the equilibrium condensation model.   S. S. Barshay, J. S. Lewis.
Bull. American Astron. Soc., Vol. 9, 544 (1977). – Abstract.

107.011 Accretion of the terrestrial planets from solid planetesimals.   L. P. Cox.
Bull. American Astron. Soc., Vol. 9, 544 (1977). – Abstract.

107.012 Planetesimals to planets: a simulation of collisional evolution.
J. F. Wacker, R. Greenberg, W. K. Hartmann, C. R. Chapman.
Bull. American Astron. Soc., Vol. 9, 544 - 545 (1977). Abstract.

107.013 Review of accretion processes and solar system chronology.   G. W. Wetherill.
Bull. American Astron. Soc., Vol. 9, 545 (1977). – Abstract.

107.014 A former major planet of the solar system.
T. C. Van Flandern.
Bull. American Astron. Soc., Vol. 9, 546 - 547 (1977). Abstract.

107.015 L'origine du système solaire.   H. Reeves.
La recherche en astrophysique, (see 003.001), p. 12 - 31 (1977).

107.016 Supernovae and lunar melting.   C. E. Singer.
Nature, Vol. 272, 239 (1978).

The view that a supernova explosion preceded the formation of the solar system by at most a few million years has gained support from the interpretation of meteoritic isotope anomalies by Latimer et al. (1977). The author suggests here another significant piece of evidence for this viewpoint, not mentioned in their report – that supernova-produced ^{26}Al (half life, 0.7 Myr) is also the only known physical heat source for melting of the lunar interior.

107.017 Blow-off of the protoplanetary cloud by a T-Tauri like solar wind. G. P. Horedt.
Astron. Astrophys., Vol. 64, 173 - 178 (1978).

The interaction of a T-Tauri like solar wind with the protoplanetary cloud is studied. A simple analytical model for the hydrostatic structure of the protoplanetary cloud is used. The T-Tauri like solar wind is able to transport to infinity several times its initial mass, as shown by a simplified analytical and numerical treatment.

107.018 Isotopes and the early solar system. G. Turner.
Geophys. J. R. Astron. Soc., Vol. 53, (see 012.006), 138 (1978). – Abstract.

107.019 On the origin and evolution of Jupiter and Saturn. J. R. Donnison, I. P. Williams.
Astrophys. Space Sci., Vol. 53, 241 - 243 (1978).

Arguments are presented which show that Kumar's (1977) contention that Jupiter and Saturn are unlikely to have formed from an initially extended state by contraction is not valid.

107.020 Origin of asteroids and the missing planet. E. J. Öpik.
Irish Astron. J., Vol. 13, 22 - 39 (1977).

Mention is made of a recently raised possibility that, in the place of the present asteroid belt, there has been a planet of 90 earth masses which recently (16 million years ago) vanished, leaving the asteroids behind. The author points out that there is very little reason to accept the reality of this suggestion: according to the mode of removal of the hypothetical planet, either we, or the asteroids should not be there. Yet both exist. One can assume that the asteroids were in the belt for all the time of the "regular" existence of the Solar System, or for all the 4.5-odd billion years. Any "unusual" happenings must be relegated to the time of rapid evolution in the beginning, without subsequent substantial changes in the structure of the Solar System.

107.021 Significant achievements in the planetary program— 1976 - 1977. J. W. Head (Editor).
NASA Contract. Rep., NASA CR-2956. For sale by the National Technical Information Service, Springfield, Virginia 22161. 4 + 48 pp. Price $ 5.25 (1978).

Recent developments in planetology research as reported at the NASA Planetology Program Principal Investigators meeting, held May 23 - 25, 1977 in St. Louis, Missouri, are summarized. Important developments are summarized in topics ranging from solar system evolution, comparative planetology, and geologic processes, to techniques and instrument development for future exploration.

107.022 Numerical experiments on planetesimal aggregation during the formation of the solar system. K. Hourigan.
Proc. Astron. Soc. Australia, Vol. 3, 169 - 171 (1977).

Prentice (1974, 1978) has shown how a discrete system of gaseous rings may be detached from the collapsing solar nebula at the present orbital radii of the planets. In the following, the early stages of planetesimal aggregation within such a gaseous ring are investigated.

107.023 Formation of the satellite systems of the major planets. A. J. R. Prentice.
Proc. Astron. Soc. Australia, Vol. 3, 172 - 173 (1977).

In this paper an outline is presented of a theory for the formation of the regular satellite systems which uses the theory of supersonic turbulent convection developed for explaining the formation of the planetary system.

107.024 Model consideration of the bombardment event of the asteroidal belt by the planetesimals scattered from the Jupiter zone. W.-H. Ip.
Icarus, Vol. 34, 117 - 127 (1978).

The temporal evolutions of the planetesimals scattered from the Jupiter zone for different masses of the proto-Jupiter [(a) 0.1 and (b) 1.0 of the present mass] are investigated. Due to the combined effects of the orbital evolution of the planetesimals and the elimination of these projectiles either via impact capture or injection into escape velocity by the outer planets, the whole scattering process lasts about 10^8 yr for case (a) and about 10^7 yr for case (b). Consideration of the collisional interaction of these projectiles with the asteroids indicates that the corresponding bombardment effect could be rather appreciable. Also, the asteroids on the inner edge of the main asteroid belt would have been bombarded more severely than those on the outer edge. From this point of view, the structure of the asteroid belt could be affected significantly not only by Jupiter's gravitational perturbation effect but also by its early scattering process.

107.025 Satellite formation, II. A. W. Harris.
Icarus, Vol. 34, 128 - 145 (1978).

Harris and Kaula (1975) derived equations of growth and orbital evolution for a planet and satellite as both bodies accumulate mass from the protoplanetary cloud. In this paper, those results are extended to include a more complete model of satellite mass gain, the evolution of the orbit in the absence of tidal friction, and the evolution of orbital eccentricity, including the resulting heating of the satellite by tidal energy dissipation.

107.026 Gravitational N-body problem on the accretion process of terrestrial planets.
T. Matsui, H. Mizutani.
Icarus, Vol. 34, 146 - 172 (1978).

Numerical integration of the gravitational N-body problem has been carried out for a variety of protoplanetary clusters in the range $N = 100$ to 200. It is shown graphically how the dispersed N bodies accumulate to a single planet through mutual collisions. The velocity distribution and size distribution of bodies are also investigated as functions of time in the accretion process. The root mean square velocity of bodies in a cluster increases with time in an early stage of accretion but decreases with time in a late stage of accretion. Accretion rates of planets are found to be dependent strongly on the initial number density distribution, the initial size distribution, and the initial velocity distribution of bodies. A substantial mass of bodies escapes from the cluster.

107.027 Mass removed by the outer planets in the early solar system. J. A. Fernández.
Icarus, Vol. 34, 173 - 181 (1978).

The changes in the heliocentric energies of particles due to close encounters with the outer planets are analyzed. Two stages in the evolution of the planets are proposed. The first phase was characterized by the rapid growth of the proto-planets. The second phase was characterized by the "spreading" of the residual solid matter toward the inner and outer regions of the solar system. For the outer planets the probabilities that a particle is ejected out of the solar system or achieves a near-parabolic orbit, after a close encounter, are calculated. The particles in near-parabolic orbits might have originated the cometary cloud. It is found that the ratio between these particles and the total number of ejected particles increases with the planet—Sun distance. It is estimated that the solid material from the region of the outer planets ejected from the solar system ranged from tens to hundreds of terrestrial masses.

107.028 **The lunar capture hypothesis revisited.**
　　　R. R. Winters, R. J. Malcuit.
Moon, Vol. 17, 353 - 358 (1977).

　　Recent work on planetary formation processes have suggested that ancient planetary bodies could have been warmer and, therefore, more easily deformable soon after formation than at present. By use of the estimates for the elastic parameters believed to be appropriate for a warm ancient Moon and Earth, it is shown that the energy of deformation of the planetary bodies during a close gravitational encounter was sufficient to effect capture.

107.029 **Conservation laws and mass distribution in the planet formation process.**
P. Farinella, P. Paolicchi.
Moon, Vol. 17, 401 - 408 (1977).

　　Within the framework of the nebular theory of the origin of the solar system, conservation laws are applied to the condensation of a ring shaped cloud of orbiting particles. The final configuration is assumed to be a point-like planet in a circular orbit around the Sun. For this reason, it is possible to relate the masses of the planets with the interplanetary distances. This relation is confirmed satisfactorily by the observed masses and orbital radii of several planets and satellites of the solar system.

107.030 **Planetary system and crystal system.**
　　　H. Havemann.
Astron. Nachr., Band 299, 145 - 158 (1978). In German.

　　The present hypothesis concerns a possible causal connexion between a first phase of evolution of the planetary system and a crystal system. It revives a paper the author published 50 years ago in this journal, and attempts to explain the then shown numerical relations through a process based upon crystallo-genetical principles. The known Titius-Bode series $4 + 2^n\ 3$ approaching the planetary system coincides with the half numbers of planes of a series of crystal forms beginning with the octahedron and continuing with the octahedron corners being blunted first by cubic planes and then in a rooflike manner by secondary, tertiary etc. planes whose number doubles step by step. As the author has shown, the satellite systems of Jupiter and Saturn coincide with similar series $[3 + 2^n\ 3$ and $2^n\ (3; 4; 5)]$ which also can be coordinated with crystal series.

107.031 **More puzzles about the early solar system.**
　　　M. Edmunds.
Nature, Vol. 273, 337 - 338 (1978).

107.032 **Formation of the outer planets and satellites.**
　　　A. G. W. Cameron.
Planetary satellites, (see 003.007), p. 463 - 471 (1977).

　　Recent models of Jupiter, Saturn, Uranus and Neptune suggest that all of these outer planets contain rocky and icy materials relative to hydrogen and helium in excess of solar composition. This points to a formation process in which condensed cores of the planets first collect together within the primitive solar nebula; when these cores become massive enough substantial gravitational concentrations of gas in the solar nebula are formed in the vicinity of the planetary cores.

107.033 **The critical velocity phenomenon and the origin of the regular satellites.**   B. R. De, H. Alfvén,
G. Arrhenius.
Planetary satellites, (see 003.007), p. 472 - 491 (1977).

　　The authors discuss a physical concept leading to a theory of origin of the regular satellites, that is, the planetary satellites that are likely to have formed in situ. In the course of their discussion they recognize that because of the striking similarity between the satellite systems and the planetary system, it is both logically proper and aesthetically appealing to try to examine the premise that these two systems have an identical genesis.

107.034 **Preliminary thermal history models of icy satellites.**
　　　G. J. Consolmagno, J. S. Lewis.
Planetary satellites, (see 003.007), p. 492 - 500 (1977).

　　The authors present here preliminary results of thermal history calculations for solar system bodies formed out of low-temperature condensates. Several diverse compositional and accretional models are examined, including integral and differential accretion modes of both chemically equilibrated and unequilibrated condensates. Critical sizes for the onset of melting and differentiation, and ages at the time of initial melting, are presented for several specific models.

107.035 **The accumulation of satellites.**
　　　V. S. Safronov, E. L. Ruskol.
Planetary satellites, (see 003.007), p. 501 - 512 (1977).

　　An analysis of the formation processes of the regular satellites is given. The natural satellites are considered to coalesce from a circumplanetary swarm which itself is fed by heliocentric particles which are captured following inelastic collisions. Characteristic times are calculated for the important physical processes. Reasonable values for the mass of the satellite swarm can be found from this model.

107.036 **Dynamic measurement of matter creation and Earth expansion.**   P. Wesson.
Nature, Vol. 273, 572 (1978).

107.037 **A new tidal theory for the origin of the solar system.**
　　　M. M. Woolfson.
Q. J. R. Astron. Soc., Vol. 19, 167 - 174 (1978).

107.038 **Some problems of accumulation of planets.**
　　　B. Yu. Levin.
Pis'ma Astron. Zh., Vol. 4, 102 - 107 (1978). In Russian.

　　It is shown that chaotic velocities of planetesimals in the protoplanetary swarm were much smaller than previously adopted (Safronov, 1969). This leads to some modification of O. Yu. Shmidt's scheme of growth of planets from material of their feeding zones and to removal of the author's arguments (1972) in favour of a protoplanetary nebula with massive and extended periphery.

107.039 **Experiment med "Titius-Bodes lag".**
　　　Å. Wallenquist.
Astron. Tidsskr., Årg. 11, 65 - 68 (1978).

107.040 **Is the solar system stable?**   J. Moser.
　　　Math. Intelligencer, Vol. 1, 65 - 71 (1978).

107.041 **A note on the problem of jet stream formation.**
　　　W.-H. Ip.
Astrophys. Space Sci., Vol. 55, 267 - 269 (1978).

107.042 **Formation of the terrestrial planets and 'rocky comets'.**   M. Shimizu.
Inst. Space Aeronaut. Sci., Univ. Tokyo, Meguro, Tokyo, ISAS Res. Note 58, 45 pp. (1978).

107.043 **From dark clouds to planets and satellites.**
　　　H. Alfvén.
Reports of planetary geology program, 1977−1978, (see 003.014), p. 5 - 12 (1978).

107.044 **Structure and evolution of giant gaseous protoplanets.**   A. G. W. Cameron.
Reports of planetary geology program, 1977−1978, (see 003.014), p. 13 - 15 (1978).

107.045 **Evidence for a primordial magnetic field during the meteorite parent body era.**   C. P. Sonett.

Reports of planetary geology program, 1977–1978, (see 003.014), p. 16 - 18 (1978).

107.046 Numerical modelling of the process of formation of planets from the protoplanetary cloud.
N. N. Kozlov, T. M. Ehneev.
Inst. prikl. mat. AN SSSR. Prepr., 1977, No. 134, 80 pp. In Russian. – Abstr. in Ref. zh., 51. Astron., 6.51.222 (1978).

107.047 Origin and evolution of the giant planets.
Z. Pokorný.
Říše hvězd, Vol. 59, 139 - 144 (1978). In Czech.

107.048 Fractionation in the solar nebula. II. Condensation of Th, U, Pu and Cm.   W. V. Boynton.
Earth Planet. Sci. Lett., Vol. 40, 63 - 70 (1978).

107.049 Cosmogony of the solar system.
S. K. Vsekhsvyatskij.
Problems of astronomy and geodesy, VAGO assembly Yerevan, 1975, (see 012.039), p. 41 - 60. In Russian. – Abstr. in Ref. zh., 51. Astron., 7.51.206 (1978).

The origin of the planets.
See Abstr. 003.145.

A geologist among astronomers: the rise and fall of the Chamberlin-Moulton Cosmogony, part 1.
See Abstr. 004.002

The origin of the planetesimal theory.
See Abstr. 004.009.

A discussion of long-term numerical solutions of the Jupiter–Saturn–Sun system.   See Abstr. 042.009.

The Titius-Bode law–place for more than ten planets?   See Abstr. 042.017.

A local proton irradiation model for isotopic anomalies in the solar system.   See Abstr. 061.022.

Penetration of the solar nebula by recent nucleosynthetic material.   See Abstr. 061.023.

Proton-associated alpha-irradiation in the early solar system: a possible ^{41}K anomaly.   See Abstr. 061.032.

Grain formation through nucleation process in astrophysical environment.   See Abstr. 061.039.

The heating of the Earth during its formation.
See Abstr. 081.004.

Thermal history of the earth and luminosity of the sun.   See Abstr. 081.056.

Pre-mare cratering and early solar system history.
See Abstr. 094.118.

The asteroidal planet as the origin of comets.
See Abstr. 102.022.

Comets and the missing planet.
See Abstr. 102.023.

Cross-sections for a capture of a comet by the Sun-Jupiter binary system.   See Abstr. 102.037.

Chemical fractionation in the solar nebula.
See Abstr. 105.019.

Some genetic constraints from meteoritic paleo-magnetism.   See Abstr. 105.023.

The primordial magnetic field preserved in chondrules of the Allende meteorite.   See Abstr. 105.097.

Evidence for a primordial magnetic field during the meteorite parent body era.   See Abstr. 105.098.

Is the solar system entering a nearby interstellar cloud?   See Abstr. 131.209.

From the origin of the universe to the earliest geological times.   See Abstr. 162.025.

The cosmogonical separation of matter and anti-matter.   See Abstr. 162.153.

Stars

111 Parallaxes, Distances

111.001 **A spectrophotometric parallax for U Geminorum.**
R. A. Wade.
Bull. American Astron. Soc., Vol. 9, 557 (1978). — Abstract.

111.002 **Trigonometric parallaxes determined with the KPNO 4-m telescope.**
W. F. van Altena, C. L. Mora.
Bull. American Astron. Soc., Vol. 9, 599 (1978). — Abstract.

111.003 **Systematic errors of Naval Observatory trigonometric parallaxes.**
R. S. Harrington, V. V. Kallarakal.
Bull. American Astron. Soc., Vol. 9, 599 (1978). — Abstract.

111.004 **Astrometric constants for 18 Puppis, BD + 44°1847, Lalande 26325, G137–8 and selected reference stars.** J. L. Russell, S. L. Kipp.
Astron. J., Vol. 83, 305 - 307 (1978).

18 Puppis, BD + 44°1847, Lalande 26325, G137–8, and selected reference stars were studied as part of the Allegheny Observatory's continuing parallax program. The measurements of the stars on the plates for each region were reduced by the central overlap technique for the astronomic constants position, proper motion, and parallax. The final Allegheny absolute parallaxes for the central stars are 18 Puppis, 0.047 ± 0.005 arcsec; BD + 44°1847, 0.034 ± 007 arcsec; Lalande 26325, 0.015 ± 0.008 arcsec; and G137–8, 0.011 ± 0.009 arcsec.

111.005 **Parallax and astrometric orbit of G24—16 from the Sproul plate series.** J. L. Hershey.
Astron. J., Vol. 83, 308 - 309 (1978).

The red dwarf G24—16, at a distance of 8.7 pc, was discovered to be an astrometric binary by Harrington. It has now been studied from the Sproul plate series, which also clearly shows a perturbation of 1.5 μ half-amplitude. The parallax and photocentric semimajor axis from each of the two plate series are in excellent agreement indicating, as before, a companion of 0.07–0.11 $M_\odot$. The masses would be determined by one observation of separation and Δm, which may be possible with a space telescope.

111.006 **Stellar distances by parallax and relative magnitude: a laboratory model.** A. Spero.
American J. Phys., Vol. 45, 1124 - 1125 (1977). — Abstr. in Phys. Abstr., Vol. 81, Abstr. 16610 (1978).

111.007 **Astrometric studies of 45 Tauri, 97 Tauri, 102 ι Tauri, 31 o² Cygni, and selected stars in those regions.** S. L. Kipp, L. A. Breakiron, J. L. Russell, J. W. Stein.
Astron. J., Vol. 83, 636 - 639 (1978).

Parallaxes, proper motions, and positions were determined for 45 Tauri, 97 Tauri, 102 Tauri, 31 Cygni and selected stars in those regions. The absolute parallaxes of the central stars were found to be 0".032 ± 0".008, 0".020 ± 0".007, 0".020 ± 0".004, and 0".005 ± 0".003, respectively. A reference star in the field of 102 Tauri showed a parallax of 0".036 ± 0".005. The astrometric results for 31 Cygni, combined with spectroscopic and photometric data, yield a parallax of 0".0028 ± 0".0006 and a photocentric semimajor axis of + 0".0055 ± 0".0034.

111.008 **Trigonometric parallaxes for southern hemisphere stars.** L. H. Auer, K. Auer, E. D. Hoffleit, W. F. van Altena.
Astron. J., Vol. 83, 640 - 642 (1978).

New Yale parallaxes are reported for 11 stars and two companions. Of the 13 stars 12 lie in the southern hemisphere. Photoelectric photometry is available for only three of the stars and observers are urged to obtain data for the rest.

111.009 **The significance of individual errors of trigonometric parallaxes.** L. A. Breakiron, A. R. Upgren.
Bull. American Astron. Soc., Vol. 10, 402 (1978). — Abstract.

Sources of errors in trigonometric parallaxes: a Monte Carlo approach. See Abstr. 031.216.

Kinematical, orbital, photometric, and spectroscopic properties of the southern high-velocity stars.
See Abstr. 112.004.

Stars in reflection nebulae near the Herbig-Haro objects in the Gum nebula. See Abstr. 113.022.

A check of the absolute luminosity calibrations of the $uvby\beta$ photometric system by means of small trigonometric parallaxes. See Abstr. 113.045.

The distances of nearby cool carbon stars.
See Abstr. 115.010.

On the statistical parallax of O type runaway stars. See Abstr. 115.011.

The Stein 2051 system. See Abstr. 118.026.

Parallax and orbit of ADS 8887.
See Abstr. 118.036.

112 Proper Motions, Radial Velocities, Space Motions

112.001 Formal values for constants of solar motion and galactic rotation from proper motions.
W. Dieckvoss.
Astron. Astrophys., Vol. 62, 445 - 446 (1978).
Results of a statistical study of 228170 stars from AGK 3 and the southern part of the Smithsonian Catalogue of positions for the spectral types A, F, G, K and photographic magnitudes from $8.^m0$ to $10.^m9$ are presented.

112.002 Astrometric study of the Sproul plate series on van Maanen's star, including gravitational redshift.
J. L. Hershey.
Astron. J., Vol. 83, 197 - 200 (1978).
The Sproul plate series on van Maanen's star, ranging from 1937 to 1976, has been measured on the two-coordinate Grant machine with a 13-star reference field. An astrometrically determined radial velocity of +25 ± 18 km/s has been found. The value of +14 ± 18 km/s for the apparent gravitational redshift, as determined simply from the difference of spectroscopic and astrometric radial velocities, is consistent with statistical values for non-DA-type white dwarfs.

112.003 Proper motions and positions of stars in wide surroundings of the Hyades cluster. II.
N. M. Artyukhina, P. N. Kholopov.
Tr. Gos. Astron. Inst. Shternberga, Vol. 47, 105 - 221 (1976).
In Russian.
Photographic proper motions for 6728 stars brighter than $12.^m5-13.^m0$ pg are derived in six regions of the sky near the Hyades cluster. The proper motions were derived using Astrographic Catalogues (Carte du Ciel) and six plates taken with the wide-angle astrograph of the Sternberg Astronomical Institute in Moscow. A catalogue of positions and relative proper motions of the stars considered is also given. Probable members of the cluster are indicated in tables.

112.004 Kinematical, orbital, photometric, and spectroscopic properties of the southern high-velocity stars.
W. Buscombe, H. J. Augensen.
Bull. American Astron. Soc., Vol. 9, 614 - 615 (1978).
Abstract.

112.005 Solar motion with respect to the nearest galaxies.
G. de Vaucouleurs, W. L. Peters, H. G. Corwin, Jr.
IAU Colloq. No. 37, (see 012.009), p. 149 - 153 (1977).
Solutions for solar motion with respect to 12 classical members of the restricted local group and 9 additional proposed members of an extended local group, all within 2 Mpc, suggest that the local group – both restricted and extended – is expanding at a rate, H* = 47 ± 12 to 57 ± 11 km s^{-1} Mpc^{-1}, which is significantly lower than its free-space (asymptotic) value $H_0 \cong 90$ km s^{-1} Mpc^{-1}, as required by the rotating-expanding model of the local supercluster.

112.006 Mean secular parallax at low galactic latitude.
R. C. Stone.
Astron. J., Vol. 83, 393 - 405 (1978).
Relative proper motions with an average standard error of ± 0.11 arcsec/ century in each coordinate have been determined for 1332 stars in 62 regions at very low galactic latitude. Photoelectric photometry has been done for 295 of these stars. An analysis of these motions has given a solar apex for faint stars near that defined by the basic solar motion and an accurate calibration of the mean secular parallax (h/ρ) at very low latitude. Application of the calibration is used to establish the velocity ellipsoid of O stars and to identify 25 runaway stars. Twenty percent of the O stars are runaways, and almost all O stars with height $z > 80$ pc above the galactic plane are runaways.

112.007 Space velocities and ages of nearby early-type stars.
P. J. Grosbøl.
Astron. Astrophys., Suppl. Ser., Vol. 32, 409 - 421 (1978).
Photometric distances and space velocities have been calculated for 458 B0–A0 stars with apparent magnitudes $m_v \leq 6.^m5$. Using Strömgren's $ubvy\beta$ photometry the effective temperature T_e and the position dM_b in bolometric magnitude over the zero-age-main-sequence of the stars were derived. These quantities were used to obtain age and mass for 423 of the stars by interpolation in the models of stellar evolution (Hejlesen 1975) for the chemical composition $(X, Z) = (0.7, 0.03)$. Finally a relation to derive interstellar reddening for normal stars in the intermediate group is given.

112.008 La mesure par corrélation des vitesses radiales au prisme objectif dans un champ stellaire situé à la latitude galactique $b_{II} = -30°$. C. Fehrenbach, R. Burnage.
C. R. Acad. Sci. Paris, Tome 286, Sér. B, 289 - 292 (1978).
Les mesures des vitesses radiales de 250 étoiles on été obtenues avec un prisme objectif à champ normal. Les vitesses sont connues avec une erreur quadratique moyenne de 5 km s^{-1} (deux spectres à 200 Å mm^{-1} retournés). Un histogramme est publié. Trois étoiles à grande vitesse radiale sont mises en évidence. L'une a aussi des mouvements propres connus, c'est une étoile à très grande vitesse spatiale.

112.009 The motions of K and M dwarf stars of different ages. A. R. Upgren.
Astron. J., Vol. 83, 626 - 635 (1978).
Space motions are calculated for a sample of 145 dwarf stars which covers the spectral classes K2-M2 and which is unbiased toward high velocity. The stars are divided into young disk, old disk, and halo groups in three different ways based on their U-, V-, and W-motion components and also on the calcium emission intensities of the H- and K-spectral lines. Rigorous solutions for the solar motion, velocity ellipsoid and vertex deviation have been calculated for each group using each method of population discrimination. Comparisons are made with similar groups of stars defined in former investigations, in which the high-velocity selection effect is present.

112.010 Relative proper motions of Trapezium-type multiple systems. I. A. I. Yatsenko.
Astrometriya i Astrofizika, vyp. (No.) 35, p. 9 - 21 (1978).
In Russian.
The paper gives relative proper motions for 385 reference stars as well as for the components of the following Trapezium systems: ADS 1209, ADS 11421, ADS 14885, ADS 14969, ADS 15679, ADS 16474, ADS 16953.

112.011 On the dispersion effect of peculiar motions of reference stars when computing photographic proper motions. A. A. Latypov.
Izv. AN UzSSR. Ser. fiz.-mat. nauk, 1977, No. 5, p. 22 - 25. In Russian. – Abstr. in Ref. zh., 51. Astron., 5.51.118 (1978).

112.012 Proper motion survey with the 48-inch Schmidt telescope. LI. Hertzsprung diagrams for one hundred and fifteen thousand proper motion stars. W. J. Luyten.
Sep. print Univ. Minnesota, Minneapolis, Minnesota, 10 pp. (1977).

112.013 Considerazioni sulle stelle vicine e il moto spaziale del Sole. M. G. Fracastoro.

Atti Accad. Naz. Lincei, Ser. Ottava, Rend., Cl. Sci. fis. mat. nat., Vol. 62, 804 - 815 (1977) = Contrib. Oss. Astron. Torino, (Pino Torinese), No. 110.

Using Gliese's (1969) and Woolley's (1970) catalogues, and applying criteria which are described in the present paper, the u, v and w components of the basic solar motion have been deduced, obtaining $w_\odot$ = +6.5, $u_\odot$ = +10.6, $v_\odot$ = +12.4 namely $V_\odot$ = 17.3 km/s. The apex coordinates are consequently $l = 48°6$ and $b = +22°2$. The asymmetry of the v-components is tentatively accounted for. The $V_\odot$ does not fit with the average radial velocities of B-type stars in the direction of apex and antapex.

112.014   **Proper motions of 84 Mira Ceti-type stars.**
          D. K. Karimova, E. D. Pavlovskaya.
Peremennye Zvezdy, Vol. 20, 369 - 374 (1977). In Russian.

112.015   **Motions of flare stars in the solar vicinity.**
          V. Ivanovska.
Flare stars. Symposium 1976, (see 012.035), p. 56 - 59. In Russian. — Abstr. in Ref. zh., 51. Astron., 7.51.796 (1978).

The stars of low luminosity.   See Abstr. 002.003.

Some errata for bright stars in the "Bibliography of radial velocities" by H. A. Abt and E. S. Biggs. See Abstr. 002.011.

Les catalogues de vitesses radiales moyennes. See Abstr. 002.015.

The determination of the proper motions of the AGK3R stars.   See Abstr. 002.031.

Die Bestimmung der Cornu-Hartmann'schen Dispersionsformel durch eine Polynomapproximation. See Abstr. 021.030.

Measurement of radial velocities from objective prism spectra. I. Use of the A-band as a reference line. See Abstr. 031.270.

Accuracy of fundamental positions and proper motions.   See Abstr. 041.008.

On the new value of the constant of precession and its influence on the proper motions of stars. See Abstr. 043.006.

A rediscussion of determinations of precession and galactic rotation from Lick proper motions referred to galaxies. See Abstr. 043.008.

Astrometric constants for 18 Puppis, BD + 44° 1847, Lalande 26325, G137−8, and selected reference stars. See Abstr. 111.004.

Astrometric studies of 45 Tauri, 97 Tauri, 102 ι Tauri, 31 o^2 Cygni, and selected stars in those regions. See Abstr. 111.007.

Runaway stars as witnesses to supernova explosions. See Abstr. 117.007.

A comment on the solution of binary star orbits: the question of unseen companions. See Abstr. 117.067.

The Stein 2051 system.   See Abstr. 118.026.

On some peculiarities of kinematics of RR Lyrae variables.   See Abstr. 122.013.

Long-period variables: stellar and expansion velocities.   See Abstr. 122.028.

LP 182−44.   See Abstr. 126.036.

A systematic comparison of four methods to derive stellar space densities.   See Abstr. 151.053.

Proper motions of bright red giants in globular clusters. I. The globular cluster M13.   See Abstr. 154.040.

Proper motions of bright red giants in globular clusters. II. The globular clusters M92 and M3. See Abstr. 154.041.

Common-proper-motion pairs in the South Galactic Cap.   See Abstr. 155.012.

On the origin of intermediate-latitude OB stars. See Abstr. 155.038.

The absolute solar motion and the discrete redshift. See Abstr. 158.113.

113 Magnitudes, Colors, Photometry

113.001 Photoelectric observations of 14 Ap stars.
K. D. Rakosch, W. Fiedler.
Astron. Astrophys., Suppl. Ser., Vol. 31, 83 - 98 (1978).
In German.

$U'B'V'$ observations of 14 Ap stars have been carried out between July 1963 and July 1964 at Lowell Observatory. The best elements of the light variation for the observed stars are given.

**113.002 Empirical calibrations of the uvby, β systems. II.
The B-type stars.** D. L. Crawford.
Astron. J., Vol. 83, 48 - 63 (1978).

The author has used data for B-type stars brighter than $V = 6.5$ in the north and south and for numerous open clusters to calibrate the $uvby$, β indices in terms of intrinsic color and absolute magnitude. The accuracy of predicting intrinsic colors, $(b-y)_0$, and hence color excesses, is estimated to be better than 0.01 mag. The index β is the primary measure of luminosity, and the calibration of β in terms of absolute magnitude M_v is given. Average values of the indices for each MK type are given. The agreement between the two techniques of classification is very good.

113.003 Observations of faint red stars at intermediate galactic latitude. J. E. Penfold.
Mon. Not. R. Astron. Soc., Vol. 182, 283 - 291 (1978).

$UBVR$ observations of 45 red stars in an area at intermediate galactic latitude ($l = 354°$, $b = -31°$) have been obtained in order to examine the statistics of faint red stars. The results provide direct observational evidence that the large number of very red faint stars found by Weistrop does not exist. This confirms Weistrop's recent re-evaluation of her earlier data. Further, it appears that there is no contamination of the red dwarf population by a significant number of faint, red, halo giants.

113.004 Optical photometry of Cygnus X-1 : 1972–1976.
E. N. Walkner, A. Rolland Quintanilla.
Mon. Not. R. Astron. Soc., Vol. 182, 315 - 329 (1978).

The magnitude of HDE 226868, the optical counterpart of Cygnus X-1, has been measured on 349 nights between 1972 April and 1976 December. The best-fit period to these data is 5.6015 ± 0.0006 days. It is found that the mean brightness of the star changes by 0.02 mag on a timescale of ~150 day. The observations show that there was no change much greater than 0.001 mag in either the 5.6 or 150-day light-curves associated with the X-ray high states. There is some evidence for an overall brightness change during the five years of ~ 0.01 mag.

113.005 Anomalous light curve of Cyg X-1 during the X-ray increase of April–May 1975.
G. Natali, R. Fabrianesi, R. Messi.
Astron. Astrophys., Vol. 62, L1 - L3 (1978).

Optical measurements in UBVRI broad bands of HD 226868 (=Cyg X-1) were made during the April–May 1975 increase of X-ray source. Violent oscillations in the V band were present during the high state of Cyg X-1. Following observations showed only sporadic flickering in the V and B bands.

113.006 Analysis of the absolute magnitudes of stars in the same seven colour photometric "star box".
M. Golay.
Astron. Astrophys., Vol. 62, 189 - 197 (1978).

Arguments are given to justify the assumption that the single star members of the same 0.01 "photometric box" have the same absolute magnitude within a range of about $0\overset{m}{.}1$.

The special case of binary stars in the same photometric box is examined and 6 known stars in Praesepe are suspected to be binaries. The hypothesis that the absolute magnitudes of stars in the same box are equal is applied to compare the individual distances of Hyades stars given by two authors using the convergent point method.

113.007 HD 72968 (3 Hya)– another low amplitude photometric double wave Ap star.
H. M. Maitzen, R. Albrecht, A. Heck.
Astron. Astrophys., Vol. 62, 199 - 204 (1978).

Extensive photoelectric intermediate band photometry of the low amplitude light variations of HD 72968 at ESO– La Silla yields evidence for a double wave variation in all filters with only slightly different maxima. Thus, the real period, in terms of the Oblique Rotator Theory the rotational period, is twice as large as the previously published.

113.008 A systematic investigation of multicolor photometric systems. I. The UBV, RGU and uvby systems.
R. Buser.
Astron. Astrophys., Vol. 62, 411 - 424 (1978).

Detailed spectral energy distributions for stars of various spectral types and luminosity classes have been used along with the response functions of the systems, to compute synthetic colors. A standard law of interstellar extinction has been applied to obtain the reddening parameters for the UBV and RGU systems. The response function of the U-passband of the UBV system has been revised in order to reduce the non-linearities existing in the transformations of the synthetic $U-B$ colors to the mean observational system. For the broad-band systems, the results are in good agreement with observations.

113.009 A systematic investigation of multicolor photometric systems. II. The transformations between the UBV and RGU systems. R. Buser.
Astron. Astrophys., Vol. 62, 425 - 430 (1978).

Improved transformation equations for the UBV and RGU systems have been derived from synthetic colors computed from detailed spectral energy distributions of main-sequence stars. The effects of interstellar reddening have been accounted for, including the variations of the reddening parameters of the two systems due to (1) different degrees of extinction, and (2) the intrinsic colors of the stars.

113.010 The variable shell star HR 5999 and its environment.
P. S. Thé, H. R. E. Tjin A Djie.
Astron. Astrophys., Vol. 62, 439 - 443 (1978).

The variable shell star HR 5999 with hydrogen lines in emission, was observed on the Walraven $VBLUW$ photometric system, together with several stars in its environment. Physical characteristics of HR 5999, which have been derived from these observations, are presented.

113.011 Variable stars of small amplitude. II. Light variations of 7 supergiants of types B2 to G0 in the Geneva observatory photometric system.
F. Rufener, A. Maeder, G. Burki.
Astron. Astrophys., Suppl. Ser., Vol. 31, 179 - 187 (1978).

A long series of observations of 7 Ia supergiants in the region of h and χ Persei were made, totalling 583 measurements in the magnitude and colour indices of the Geneva system. The observed stars were especially chosen so as to constitute an upper envelope of the supergiant branch between B2 and G0. All these stars are found to be variable with a peak-to-peak amplitude between $0\overset{m}{.}035$ and $0\overset{m}{.}065$ in m_V. The semi-periodic character of this variation can be detected in almost all cases.

113.012   Study of the lightcurve of the Of star HD 153919.
J. A. van Paradijs, G. Hammerschlag-Hensberge,
E. J. Zuiderwijk.
Astron. Astrophys., Suppl. Ser., Vol. 31, 189 - 197 (1978).

Photoelectric five-colour observations are presented for
the O6f star HD 153919, optical counterpart of the X-ray
binary 3U1700−37. Combining these data with those obtained
by other observers, the authors derive an orbital period of
$P = 3.41117 \pm 0.00041$ days. The lightcurve is variable: remarkable differences occur between the average results of
different observers. The observed mean lightcurve of HD
153919 cannot be reproduced by light variations predicted
from a model of a tidally distorted corotating primary. Possible reasons for this discrepancy are given.

**113.013   Photoelectric observations of the Ap stars HD
125248, HD 134793 and HD 184905.**
C. Blanco, F. A. Catalano, G. Strazzulla.
Astron. Astrophys., Suppl. Ser., Vol. 31, 205 - 208 (1978).

Lightcurves of the Ap stars HD 125248, HD 134793 and
HD 184905 are presented. The periods $9\overset{d}{.}29477$, $2\overset{d}{.}78$ and
$1\overset{d}{.}845251$ are found, respectively, for the stars HD 125248,
HD 134793 and HD 184905 to be in accordance with the
values previously known.

**113.014   The Basle catalogue of blue objects in higher galactic
latitudes in SA51, SA54, SA57, SA82 and in the
halo field around M5 (I).**   H. Steppe.
Astron. Astrophys., Suppl. Ser., Vol. 31, 209 - 241 (1978).

A systematic search for blue objects in fields of 4.7 × 4.7
square degrees on all 48″ Schmidt plates of the Basle halo
programme is in process at the Astronomical Institute of the
Basle University. The present catalogue contains the positions
and the photographic data, together with the finding charts
for a total of 1906 objects on the following fields: SA51,
SA54, SA57, SA82 and M5.

**113.015   The relation between the blanketing parameter
Δm_2 and [Fe/H] abundances.**   B. Hauck.
Astron. Astrophys., Vol. 63, 273 - 274 (1978).

The author has derived a new relation between Δm_2 values of the Geneva system and the [Fe/H] values taken from
the catalogue of Morel et al. The result is [Fe/H] = $5.14\Delta m_2 +$
0.09. He has also examined the role of the choice of sources
for the [Fe/H] determinations and the linearity of the relation.

**113.016   Photoelectric photometry at the Hvar Observatory.
III. The Ap star CQ UMa.**
Z. Mikulášek, P. Harmanec, J. Grygar, F. Žďárský.
Bull. Astron. Inst. Czechoslovakia, Vol. 29, 44 - 49 (1978).

The UBV observations of the Ap star CQ UMa, obtained
at the Hvar Observatory in 1973 and 1975, are combined with
all other available photometric data to get an improved value
of the period $P = (2.449967 \pm 0.000025)$ days. The authors
discuss the position of the variable in the colour diagram.

**113.017   On the photometric variations of the red giant
HD 65750 and of the surrounding reflection nebula
IC 2220.**   J. Dachs, J. Isserstedt, J. Rahe.
Astron. Astrophys., Vol. 63, 353 - 362 (1978).

For the southern variable M giant HD 65750, a lightcurve between 1963 and 1977 has been constructed from 77
photographic magnitudes derived from plates taken between
1963 and 1972, and from 83 photoelectric UBV measurements, 49 of which are new observations obtained between
1971 and 1977. It is concluded that the star is an irregular
variable. The reddening of the star increases with increasing
light loss. It is suggested that the light variations of the star
are caused by varying dust extinction in the surrounding
nebula. The implications of the maximum observed rate of the
light variations of the star on models of the circumstellar
nebula are discussed. The evolutionary state of HD 65750 and

the possibility that the reflection nebula IC 2220 has been
shed by its central star are discussed.

**113.018 UBV photometry of faint blue stars near the
galactic anticenter.**   F. R. Chromey.
Astron. J., Vol. 83, 162 - 166 (1978).

The photoelectric UBV colors of 96 very blue stars from
the lists of Rubin et al. (1971, 1974) have been observed. At
least half the program stars appear to be young, luminous
objects at large distances from the galactic center. Fifteen stars
have apparent distances of over 30 kpc from the center and
would be significant probes of galactic structure if their distances could be confirmed from spectra. On the other hand,
these 15 stars may all be relatively nearby subluminous objects.
In this case, the distribution of young stars seems to end at
about 22 kpc from the galactic center.

**113.019   Airborne photometric observations between 1.25
and 3.25 μ of late-type stars.**
H. L. Nordh, S. G. Olofsson, G. C. Augason.
Astron. J., Vol. 83, 188 - 193 (1978).

The stars α Aur (G5 III + G0 III), α Boo (K2 IIIp),
α Ori (M1−M2 Ia−Ib), α Sco (M1.5 Iab), μ Gem (M3 III), and
α Her (M5 Ib−II) have been observed using interference filters
in five photometric bands between 1.25 and 3.25 μ. By
calibrating the photometer in the laboratory against a stabilized blackbody source, relative flux curves have been derived.
The energy distributions and the strength of molecular features are discussed. The most interesting result obtained is that
the flux of μ Gem and α Her in the filter centered at 3.25 μ
seems to be depressed by at least some tenths of a magnitude.
Tentatively this depression is proposed to be due to the wings
of the two vibration−rotation bands ($\sim 2.7\ \mu$) of hot water
vapor. Since water vapor is an important opacity source and
since its abundance is a sensitive C/O indicator, the proposed
interpretation makes renewed efforts to detect water bands in
early M stars highly desirable.

**113.020   On the background fog correction in measurements
with an iris photometer.**   G. A. Ponomareva.
Soobshch. Gos. Astron. Inst. Shternberga, No. 199, p. 19 - 21
(1977). In Russian.

The question of the applicability of Weaver's formula
(1962) for fog corrections in measurements with an iris photometer is considered. It is shown that this formula should be
used for small background fog only.

113.021   VRI photometry of E and F region stars.
A. W. J. Cousins.
Mon. Notes Astron. Soc. Southern Africa, Vol. 37, 8 - 10
(1978).

**113.022   Stars in reflection nebulae near the Herbig-Haro
objects in the Gum nebula.**   W. Herbst.
Publ. Astron. Soc. Pacific, Vol. 89, 795 - 796 (1977/78).

$UBVRI$ data are presented for three stars in reflection
nebulae near the Herbig-Haro objects recently discovered in
the Gum nebula. The likely distance to all three stars as well
as the Herbig-Haro objects, and a portion of the Gum nebula
is 4000 pc. One of the Hα-emission stars in this region appears
to be variable on a short time scale.

113.023   Color index and finite extent of the atmosphere.
M. S. Vardya.
Publ. Astron. Soc. Pacific, Vol. 89, 811 - 813 (1977/78).

It has been found that the color index, in general, is not
independent of radius, but is proportional approximately to
five times the logarithm of the ratio of radii at the two wavelength bands. Two examples have been given, where this effect
may be significant: (1)Large differences in effective temperatures, derived from two different color indices, may reflect the
wavelength dependence of the radius. (2) Failure of and incon-

sistent results derived from the Wesselink method when applied to cepheid and other variable stars may be due to significant contribution from the change of the wavelength-dependent radius during the cycle.

113.024 On the nature of the infrared excesses in the pre-main-sequence A and F stars in NGC 2264.
A. E. Rydgren.
Publ. Astron. Soc. Pacific, Vol. 89, 823 - 828 (1977/78).

BVRI photometry is presented for seven early-type stars in NGC 2264 which have IR excesses; these new observations provide additional information on the spectral-energy distributions of these stars. In the unusual star W90 the presence of a circumstellar dust shell is confirmed. In the other six stars the nature of the IR excess is not obvious from inspection of the observed energy distributions. Models for the energy distributions of the six stars are constructed in two ways: (1) from the appropriate photosphere and the continuous emission from a hot gaseous envelope, and (b) from the photosphere plus the IR excess from W90. The latter models provide the better description of the observations. The author also finds no correlation between IR excess and either Balmer line or Balmer continuum emission. These results suggest that the IR excesses in all seven stars are due to circumstellar dust shells rather than hot gaseous envelopes.

113.025 Observations of variable M stars in the DDO photo-electric system.   D. W. Dawson.
Publ. Astron. Soc. Pacific, Vol. 89, 919 - 921 (1977/78).

Effects of spectral features of late-type stars on the color indices of the intermediate-band DDO system are discussed in the framework of the potential of this system for identifying M giants and dwarfs. Spectral classification is given for the M variables EI Peg, DW Peg, Z Sex, EG Lyr, and o Cet.

113.026 Four-colour photometry of southern early-type stars.   D. Kilkenny.
Mon. Not. R. Astron. Soc., Vol. 182, 629 - 637 (1978).

Strömgren four-colour photometry is given for 209 southern hemisphere early-type stars at intermediate galactic latitudes. MK types are available for about half of the stars and spectral types for all of them have been estimated from the colours.

113.027 Absolute magnitudes of M-type stars in the solar neighborhood.   T. Mikami.
Publ. Astron. Soc. Japan, Vol. 30, 191 - 206 (1978).

Absolute magnitudes are studied for 1490 M-type stars in the solar neighborhood by a statistical analysis of radial velocities presently available and of proper motions listed in the AGK3 catalogue. The kinematical parameters such as solar motion and velocity ellipsoid necessary for the investigation are also examined. The mean visual absolute magnitude of M-type supergiants is -5.8 ± 0.5 (m.e.), that of M-type giants varies with the spectral subtype from -1.0 ± 0.2 at M0–1 to $+0.5\pm0.6$ at M7–8, and that of early M-type dwarfs is $+8.6\pm0.3$.

113.028 Absolute magnitudes of F-, G-, and K-type stars in the solar neighborhood.   T. Mikami.
Publ. Astron. Soc. Japan, Vol. 30, 207 - 217 (1978).

Absolute magnitudes of F-, G- and K-type stars in the solar neighborhood are studied statistically. In the analysis, radial velocities in the Lick system and proper motions listed in the AGK3 catalogue are used. The author confirms the previous results, except that the absolute magnitudes of K-type giants are about 0.7 mag brighter than Blaauw's (1963) values.

113.029 Stjernefotometri i nordlyssonen, I. Feltstasjon i Skibotn.   H. K. Myrabø.
Astron. Tidsskr., Årg.10, 154 - 157 (1977).

113.030 Stjernefotometri i nordlyssonen, II. Observasjoner og betraktninger.   H. K. Myrabø.
Astron. Tidsskr., Årg.10, 158 - 161 (1977).

113.031 Polarimetry of Be stars at 1.25 and 2.2 microns.
T. J. Jones.
Bull. American Astron. Soc., Vol. 9, 572 (1978). – Abstract.

113.032 Applications of stellar surface brightness to eclipsing binaries and nearby stars.   C. H. Lacy.
Bull. American Astron. Soc., Vol. 9, 624 (1978). – Abstract.

113.033 AG Draconis, 1975 - 1977.
J. Lutz, T. Lutz, J. Kaler.
Bull. American Astron. Soc., Vol. 9, 642 - 643 (1978). Abstract.

113.034 Narrow-band photoelectric photometry of V1357 Cyg (Cyg X-1).
E. F. Guinan, R. H. Koch, J. D. Dorren, J. Siah.
Bull. American Astron. Soc., Vol. 9, 645 (1978). – Abstract.

113.035 Five-color photometry of Am stars.
E. E. Mendoza, T. Gomez, S. Gonzalez.
Bull. American Astron. Soc., Vol. 9, 652 (1978). – Abstract.

113.036 Infrared photometry of early-type stars – I.
M. J. Smyth, K. Nandy.
Mon. Not. R. Astron. Soc., Vol. 183, 215 - 222 (1978).

K (2.2 µm) magnitudes are given for reddened early-type stars in the galactic anticentre region $160° < l < 230°$, in conjunction with ultraviolet photometry from the TD1 satellite. A value of 3.12 ± 0.05 for the extinction ratio R is derived for the general field and for the Mon O B2 region. The strength of the interstellar absorption feature at 2200 Å is calibrated in terms of A_V. Many of the emission-line stars show infrared excess, and there are indications that in some of these the 2200 Å absorption is depleted.

113.037 A method for reddening determination in the *uvby* photometric system.   H. G. Reimann.
Astron. Nachr., Band 299, 43 - 46 (1978) = Mitt. Univ. Sternw. Jena, No. 133.

A new method for reddening determination in the *uvby* system is described. Using a Q_{vby} parameter calibrated in terms of $b - y$ it is possible to determine simultaneously luminosity, spectral type, reddening, and metallicity of a star. The colour excess ratios for the *uvby* photometric system are given and also analytical functions for the ZAMS are presented.

113.038 Intermediate-band photometry of late-type stars. V. Calibration of indices.   O. J. Eggen.
Astrophys. J., Vol. 221, 881 - 892 (1978).

Additional data for several groups and clusters have been used to improve the provisional $(M_1,[Fe/H]_\odot)$-relation in paper II (Eggen, 1977) and to test the usefulness of C_1 for luminosity determination. In general, C_1 is more useful for isolation of peculiar stars than for accurate luminosity determination of disk stars, but more success in the latter application appears possible for halo objects. An interpolation table for $[Fe/H]_\odot$, on the basis of $(R-I, M_1)$, is constructed for the range $[Fe/H]_\odot = +0.1$ to -1.8.

113.039 Simultaneous photometry and spectroscopy of the peculiar B[e] star with infrared excess HD 45677.
M. Klutz, A. Surdej, J. Surdej, J. P. Swings.
IAU Colloq. No. 42, (see 012.012), p. 126 - 127 (1977). Abstract.

113.040 Search for light variations of HD 101065.
A. Przybylski.

Proc. Astron. Soc. Australia, Vol. 3, 143 - 144 (1977).

HD 101065 can be considered as a single star with no spectral variations but with possible small long period brightness variations of the order of 0.02 to 0.03 magnitudes.

113.041 Elementi di fotometria stellare ad uso dei dilettanti.
P. Tempesti.
G. Astron., Vol. 3, 267 - 353 (1977).

113.042 Infrared photometry, bolometric magnitudes, and
effective temperatures for giants in M3, M13, M92,
and M67.   J. G. Cohen, J. A. Frogel, S. E. Persson.
Astrophys. J., Vol. 222, 165 - 180 (1978).

Broad-band infrared J-, H-, and K-magnitudes and narrow-band CO and H_2O indices are presented for a selection of giants reaching 3 mag below the red-giant tip in the globular clusters M3, M13, and M92, and in the old open cluster M67. A new set of model atmosphere calculations is described, and the predicted colors are compared with the broad-band data. A basic result of the calculations is that reliable effective temperatures can be derived from $V - K$ colors independent of metal abundance or surface gravity. The infrared data further allow empirical bolometric magnitudes to be derived, and thus the theoretical H-R diagram, log $L/L_\odot$ versus log T_{eff}, can be constructed from observational parameters. This H-R diagram is compared to a set of evolutionary tracks for metal-poor stars computed by Rood (1972). The narrow-band CO indices are shown to be metal-sensitive for metal-poor giants. The relative metallicities of M3 and M13, as derived from the CO indices of their stars, are reversed from those derived from lines of the heavier elements.

113.043 Intermediate-band photometry of late-type stars. VI.
Main-sequence stars near the sun.   O. J. Eggen.
Astrophys. J., Vol. 222, 191 - 202 (1978).

Photometry for a complete sample of main-sequence stars in the Bright Star Catalog and south of declination +30°, with $R - I \geqslant +0.19$ mag, is discussed. Twelve percent of the stars are overabundant compared with the Hyades stars, and only 5% are underabundant compared with the Arcturus group members. The largest percentage, 39%, have about one-half the solar abundance, but only 17% are comparable with the sun. The use of C_1 to isolate peculiar stars is discussed, and attention is called to the multiple system Herschel 2621.

113.044 Intermediate-band photometry of late-type stars. VII.
The HR 1614 group of overabundant stars.
O. J. Eggen.
Astrophys. J., Vol. 222, 203 - 208 (1978).

In a sample of 44 stars with (U, V) near $(0, -60)$ km s^{-1}, it is found that 60% are overabundant with respect to the sun, show strong blanketing effects in $(b-y)$, and form a color-luminosity array with many of the characteristics of a very old cluster.

113.045 A check of the absolute luminosity calibrations
of the $uvby\beta$ photometric system by means of
small trigonometric parallaxes.   H. U. Nørgaard-Nielsen.
Astron. Astrophys., Vol. 65, 287 - 294 (1978).

The author has checked the absolute luminosity calibrations of the $uvby\beta$-photometric system by means of small trigonometric parallaxes. The photometric calibrations are confirmed within the obtained accuracy (for F stars $\sim 0^m\!.1$, for A stars and "Intermediate Group" $\sim 0^m\!.2$ and for B stars $\sim 0^m\!.3$). There remains the uncertainty of the zeropoint correction $+0^{''}\!.003$ to the trigonometric parallaxes.

113.046 Stellar angular diameters and visual surface brightness – III. An improved definition of the relationship.   T. G. Barnes, D. S. Evans, T. J. Moffett.
Mon. Not. R. Astron. Soc., Vol. 183, 285 - 304 (1978).

The relationship between visual surface brightness param-

eter F_v and the $(V-R)_0$ colour index is improved by adding 40 new stars to the 52 stars used in the previous calibration. Explicit corrections for limb-darkening effects on the observed angular diameters are introduced. New $BVRI$ photometry is reported for 39 stars used in the calibration. The new calibration strengthens the previous assertion that the $F_v-(V-R)_0$ relation is independent of luminosity class and applicable to all spectral types O4–M8, S, and C.

113.047 Ultraviolet photometric variations in the central
star of IC 418.   D. P. Gilra, S. R. Pottasch,
P. R. Wesselius, R. J. Van Duinen.
IAU Symp. No. 76, (see 012.014), p. 210 (1978). – Abstract.

113.048 Photometry of the "ultraviolet" stellar group in
Auriga.   A. M. Heiser, C. L. Uckotter, D. G.
Uckotter.
Publ. Astron. Soc. Pacific, Vol. 90, 105 - 107 (1978).

Ultraviolet stellar spectra obtained from a space observatory have suggested the probable existence of a small cluster of early-type stars in the constellation of Auriga. UBV and $uvby$ photometry of these stars indicate that this small group apparently contains two subgroups at somewhat different distances and with distinctly different amounts of interstellar reddening. Two or three luminous stars may be part of the more distant subgroup.

113.049 A new stellar standard sequence in the Coma cluster
of galaxies. (Preliminary report).
F. Börngen, N. Richter.
Astron. Nachr., Band 299, 117 - 120 (1978).

Photographic magnitudes in the UBV system of 39 stars in the Coma cluster of galaxies are given.

113.050 Variable stars of small amplitude. III. Semi-period
of variation for seven B2 to G0 supergiant stars.
G. Burki, A. Maeder, F. Rufener.
Astron. Astrophys., Vol. 65, 363 - 367 (1978).

The photometric observations of seven B2 to G0 Ia-type supergiants in the Geneva system (Paper II) are analysed. All these stars are found to be semi-periodic variables. A new method of auto-correlation is used to determine the semi-period of variation P and the accuracy of these determinations is in most cases very good. The P-value increases from 15–20 days for the B supergiants to more than 80 days for the G0 Ia star. This fact confirms the dependence on the spectral type of the semi-period–luminosity relation for the supergiant stars, as previously found by Maeder and Rufener (1972; Paper I).

113.051 Note on the Am stars in the α, λ photometry.
B. Hauck.
Rev. Mexicana Astron. Astrofis., Vol. 2, 231 - 234 (1977).

The α, λ data published by Mendoza concerning the Am stars have been analysed using the Geneva system. The results confirm Mendoza's conclusions and the blanketing and luminosity vectors in the α, λ plane are shown.

113.052 Photoelectric UBV observations of RR Lyrae vari-
able stars. Second list.
B. B. Bookmeyer, W. S. Fitch, T. A. Lee, W. Z. Wiśniewski,
H. L. Johnson.
Rev. Mexicana Astron. Astrofis. Vol. 2, 235 - 258 (1977).

Photoelectric UBV observations are given for 70 RR Lyrae stars selected from van Herk's list of field variables with known proper motion or radial velocity. Magnitudes and colors at maximum and minimum light were obtained for 64 stars.

113.053 Hα and OI photometry of Ap stars.
E. E. Mendoza V.
Rev. Mexicana Astron. Astrofis., Vol. 2, 259 - 261 (1977).

$\alpha(16)$, $\Lambda(9)$ photometry of 23 Ap stars is presented. The results indicate that Ap and normal A-stars are neatly separated in the $\alpha(16)$, $\Lambda(9)$-plane. The Am and Ap form a continuous sequence in $\alpha(16)$ with strongest absorption for the Ap stars. A similar sequence is defined by the normal main sequence A-stars. The normal A, Am and Ap-stars also form a continuous sequence in $\Lambda(9)$. The strongest absorption corresponds to the normal A-stars, the weakest to the Ap stars.

113.054 Photometry of the Be star HD 88661.
M. J. Stift.
Astron. Astrophys., Suppl. Ser., Vol. 32, 343 - 345 (1978). In French.

Photoelectric observations of the Be star HD 88661 obtained in January 1977 show a variability of $0^m.1$ in V within a few nights. The variations appear to be non-periodic and do not affect the colour indices. Comparison with previous observations reveals the presence of long-term brightness variations.

113.055 Extremely red stars at $l^{II} \approx 113°$, $-10° \leq b^{II} \leq +90°$.   C. Poulakos.
Astron. Astrophys., Suppl. Ser., Vol. 32, 395 - 399 (1978).

Finding charts, coordinates, magnitude and colour estimates and spectra are given for 19 very red stars in the region $l^{II} \approx 113°$, $-10° < b^{II} < +90°$. These stars have been detected on the Palomar Observatory Sky Survey prints and from direct and spectral plates taken with the Haute Provence 90 cm Schmidt telescope. There is indication that the stars are intrinsically red rather than being reddened.

113.056 Ultraviolet photometric observations of Ap and Am stars.   W. van Dijk, A. Kerssies, G.
Hammerschlag-Hensberge, P. R. Wesselius.
Astron. Astrophys., Vol. 66, 187 - 195 (1978).

ANS five colour ultraviolet photometric observations of 79 hot and cool Ap stars and 26 Am stars are presented. The positions of Ap and Am stars in ultraviolet colour-colour diagrams differ from those of normal main-sequence stars of the same spectral type. The deviation is largest for the cooler Ap stars. HR 5857 shows ultraviolet light variations of which the amplitude gradually decreases towards longer wavelengths.

113.057 Photometry and polarimetry of AM Herculis.
J. A. Bailey, D. H. P. Jones, G. E. Parkes, K. O. Mason.
Mon. Not. R. Astron. Soc., Vol. 183, 73P - 77P (1978).

Some new high-speed photometric and polarimetric observations of AM Herculis are presented. Several properties are noted to have changed, in particular the size and structure of the periodic linear-polarization events. There is no evidence for periodic variation of the polarization apart from the 3-hr orbital period.

113.058 A survey of extremely red stars in the Orion region with 40/70/120 cm Schmidt telescope.
H. Ohtani, T. Ichikawa.
Mem. Fac. Sci. Kyoto Univ., Ser. Phys. Astrophys. Geophys. Chem., Vol. 35; 197 - 201 (1977).

The Orion region was surveyed in red and near infrared region with the 40/70/120 cm Schmidt telescope. By means of the composite photograph, fifty-five extremely red stars redder than $V-R = 1.85$ were detected. One of them was identified to an infrared source IRC00085 whose optical counterpart had not been known precisely.

113.059 On the numbers of yellow stars in the Large Magellanic Cloud. (Paper II).
P. R. Warren, R. A. Bywater.
Observatory, Vol. 98, 120 - 121 (1978).

113.060 Problems of standards in wide-band UBV photom-

etry.   E. Rybka.
Postępy Astron., Tom 26, 19 - 30 (1978). In Polish.

The generation of photometric standards is described, especially for the case of the UBV system. Bibliographic data of the most important papers concerning UBV photometric standards are given.

113.061 UBVRI photometry of 225 Am stars.
E. E. Mendoza V., T. Gómez, S. González.
Astron. J., Vol. 83, 606 - 614 (1978).

The UBVRI photometry of 225 Am stars is presented. The data are used to obtain mean color deficiencies.

113.062 Derivation of brightness distribution across a stellar disk from photoelectric observations of its lunar occultation.   M. B. Bogdanov.
Astron. Zh. Akad. Nauk SSSR, Vol. 55, 490 - 495 (1978). In Russian. English translation in Soviet Astron., Vol. 22, No. 3.

A method of analysis of photoelectric observations of lunar occultations allowing to find the strip distribution of brightness across a stellar disk is described. The method is based on the digital solution of an integral equation which connects the observed light curve with the brightness distribution across the source. The method can also be used for the analysis of lunar occultations of radio sources.

113.063 HDE 245770.
IAU Circ., Nos. 3167, 3184 (1978).

113.064 Possible optical counterpart (LSI +61°303) for GT 0236+610.
IAU Circ., Nos. 3170, 3180, 3210 (1978).

113.065 Astrometry and photometry of stars in the vicinity of the Hulse-Taylor binary pulsar.
C.-Y. Shao, W. Liller.
Astrophys. J., Vol. 223, 266 - 267 (1978).

Deep blue and near-infrared (0.7–0.9 μm) plates are used to derive coordinates, magnitudes, and colors for faint stars in the vicinity of the Hulse-Taylor binary pulsar. No convincing candidates are found, and new limits are set on the optical visibility of the system.

113.066 HDE 245770.
Yamamoto Circ., No. 1877 (1978). In Japanese.

113.067 A photoelectric UBV sequence in the region of the wing of the Small Magellanic Cloud.
K. L. V. Johansson.
Astron. Astrophys., Suppl. Ser., Vol. 33, 107 - 114 (1978).

A catalogue is presented giving photoelectric UBV observations ($7-V-13$) of 44 stars in the region of the wing of the Small Magellanic Cloud (SMC). For stars with HD numbers the MK classes as found in the Michigan Catalogue (1975) have been used to determine a mean colour excess of $\bar{E}_{B-V} = 0^m.06 \pm 0.05$. One of the stars observed, no. 42, is most likely a supergiant binary and a member of the SMC.

113.068 Photometry of HD 193793.   J. D. Fernie.
Inf. Bull. Variable Stars, No. 1377 (1978).

113.069 A photometric sequence for OI 090.4 and additional information on CSV 1180.   J. H. Baumert.
Inf. Bull. Variable Stars, No. 1410, 3 pp. (1978).

113.070 uvby photometry of the Ap star HD 145102.
D. Vanbeveren, H. Hensberge.
Inf. Bull. Variable Stars, No. 1423 (1978).

113.071 Light variation of CoD $-37°$ 9248.   O. J. Eggen.
Inf. Bull. Variable Stars, No. 1425 (1978).

113.072   Propriétés photométriques des étoiles G, K, M en
relation avec la structure et l'évolution galactique.
M. Grenon.
Publ. Obs. Genève, Sér. B, Fasc. 5, 3 + 299 pp. (1978).

This work contains a description of the properties of G, K and M stars necessary to interpret their behaviour in the Geneva photometry. The properties of the Geneva system are discussed and a multidimensional classification is proposed for the G and K stars, dwarfs and giants. Calibrations in absolute magnitude, effective temperature and metal content are provided as well as methods for the detection of peculiarities and multiplicity.

113.073   Fotometría R G U en un campo del anticentro
galáctico, cerca del NGC 581.
J. M. García-Pelayo.
Rev. Real Acad. Cienc., Madrid, Tomo 71, 537 - 584 (1977) =
Univ. Complutense – Fac. Cienc., Madrid, Semin. Astron.
Geod., Publ. Núm. 96.

113.074   uvbyR surface photometry of the 30 Doradus
region II.   A. C. Danks, J. W. Hartsuiker.
ESO Sci. Prepr. No. 20, 9 pp. (1978). – Submitted to
Astronomy and Astrophysics, Suppl. Series.

113.075   Near-infrared photometry of late-type stars with
balloon-borne telescope.   K. Kodaira, W. Tanaka,
T. Onaka, T. Watanabe, H. Yoshida.
Tokyo Astron. Bull., Second Ser., No. 251, p. 2889 - 2893
(1977).

113.076   Rapid light variation of the companion of Mira Ceti
in the 1977 light minimum.   Y. Yamashita,
K. Ichimura, Y. Shimizu, M. Nakagiri.
Tokyo Astron. Bull., Second Ser., No. 254, p. 2925 - 2928
(1978).

Photoelectric monitoring in the UBV system was carried out for Mira Ceti in the faint phase. Records are presented for light variations which are attributable to the companion. It is concluded that time scales of light variations are almost independent on the 14-year period established spectroscopically for the activity of the companion.

113.077   V 377 Cassiopeiae = BD + 58° 28 = HD 1479 zeigt
keine Veränderlichkeit.   S. Rößiger.
Mitt. Veränderl. Sterne (MVS), Band 8, 45 - 46 (1978).

The results of photoelectric UBV observations of the star during more than two years are presented. These measurements and additional UBV values from two other papers lead to the conclusion that the star is not variable.

113.078   The UBV photometry of the Ap star CQ UMa in
the years 1973 and 1975.
J. Grygar, Z. Mikulášek, P. Harmanec, F. Žďárský.
Magnetic stars, (see 012.028), p. 21 (1978). – Abstract.

113.079   The nature of light variations of the Ap star CQ
UMa.   Z. Mikulášek.
Magnetic stars, (see 012.028), p. 22 (1978). – Abstract.

113.080   On the separation of Ap stars in the Vilnius photometric system.   A. S. Nikolov, I. K. Illiev.
Magnetic stars, (see 012.028), p. 24 (1978). – Abstract.

113.081   Narrow band photometry of the Pleiades.
E. E. Mendoza.
Bull. American Astron. Soc., Vol. 10, 413 (1978). – Abstract.

113.082   Search for super-fast variability of objects with
continuous spectrum according to the MANIA
experiment program in 1973.   O. A. Evseev, V. N.
Mansurov, N. M. Nesterenko, G. S. Tsarevskij, V. F. Shvartsman.
Soobshch. Spets. Astrofiz. Obs., Zelenchukskaya, Vyp. (No.)
20, p. 30 - 38 (1977). In Russian.

Results are presented of a search for optical variability in the range 2×10^{-4} to 10 sec in a number of objects which are "candidates" for black holes: stellar-like radio source OJ 287, quasar 3C 273 and six stars with large proper motions and without lines in the spectra (DC-type white dwarfs). No variability was found in these objects at the level $SA \geqslant 0.03 - 0.15$ for different time ranges (S is the averaged normalized power of flares, A is the averaged normalized amplitude).

113.083   Brightness distribution across the disk of the carbon
star TX Piscium from observations of its occultation
by the moon.   M. B. Bogdanov.
Astron. Tsirk., No. 967, p. 2 - 4 (1977). In Russian.

113.084   On the photometry of faint stars with an I-Isocon.
A. N. Abramenko, E. S. Agapov, V. F. Anisimov,
N. D. Galinskij, M. A. Polyakov, V. V. Prokof'eva.
Astron. Tsirk., No. 968, p. 1 - 2 (1977). In Russian.

113.085   The possibility of using discrete methods for stellar
photographic photometry.
P. D. Chobanov, S. B. Vladimirov.
Astron. Tsirk., No. 974, p. 6 - 8 (1977). In Russian.

113.086   Fotometrische waarneming van de RRs-ster RS
Gru aan de sterrenwacht Serra da Piedade (Brazilië).
E. W. Elst.
Bull. Astron. Obs. R. Belgique, Vol. 9, 27 - 31 (1978).

The short periodical RRs-star Gru, was observed during several nights in the period 1974–1975 at the observatory of Serra da Piedade (Belo Horizonte, Minas Gerais, Brasil). The photometrical V-magnitude observations are tabulated.

113.087   Poly-parametric stellar photographic photometry.
S. B. Vladimirov.
Astron. Tsirk., No. 980 p. 4 - 7 (1978). In Russian.

113.088   Photometric standard +35° 35.
M. S. Kazanasmas, N. A. Mis'kin, L. F. Tomak.
Astron. Tsirk., No. 982, p. 5 - 7 (1978). In Russian.

113.089   UBV observations of FG Sagittae in 1977.
V. P. Arkhipova, G. V. Zajtseva, R. I. Noskova,
M. V. Savel'eva.
Astron. Tsirk., No. 987, p. 1 - 2 (1978). In Russian.

113.090   Brightness distribution across the disk of λ Aquarii
from an analysis of observations of its lunar occultation.   M. B. Bogdanov.
Astron. Tsirk., No. 993, p. 2 - 4 (1978). In Russian.

113.091   On the separation of Ap stars in the Vilnius photometric system.   A. S. Nikolov, I. Kh. Iliev.
Astron. Tsirk., No. 994, p. 1 - 4 (1978). In Russian.

113.092   Statistical investigation of light and colour changes
of Ap stars.   R. Radkov.
Astrofiz. issled. (NRB), Vol. 2, 25 - 29 (1977). In Russian.
Abstr. in Ref. zh., 51. Astron., 6.51.635 (1978).

113.093   Search for photometric flares of Ap stars.
W. Schöneich.
Flare stars. Symposium 1976, (see 012.035), p. 161 - 163.
In Russian. – Abstr. in Ref. zh., 51. Astron., 6.51.636 (1978).

The stars of low luminosity.   See Abstr. 002.003.

Catalogue of measurements in the six-colour photometric system (magnetic tape).   See Abstr. 002.028.

A spectrophotometric survey of stars along the Milky Way. Part IV.   See Abstr. 002.029.

The limiting magnitude of the ESO-B and SRC-J Sky Survey.   See Abstr. 002.035.

S stars in the Southern Milky Way. See Abstr. 002.045.

Catalogue of "strict $0.^m01$ photometric star boxes". See Abstr. 002.050.

Vector space methods of photometric analysis: applications to O stars and interstellar reddening. See Abstr. 021.012.

Esperienze di fotometrica fotoelettrica: l'elaborazione dei dati.   See Abstr. 031.229.

Ultraviolet observational techniques – stellar ultraviolet spectrophotometry.   See Abstr. 031.312.

Energy distributions in main-sequence A and F stars. See Abstr. 064.077.

A catalogue of carbon stars in the Large Magellanic Cloud.   See Abstr. 114.001.

Equivalent width of O I λ 7774 and absolute magnitude relation for F and G supergiants. See Abstr. 114.031.

HD 91805 and the nature of the Bidelman-MacConnell weak-G-band stars.   See Abstr. 114.542.

Spectroscopic observations of 27 CMa from 0.14 to 4.7 microns.   See Abstr. 114.573.

Four-colour photometry of eclipsing binaries. XIB: Lightcurves of RS Chamaeleontis.   See Abstr. 121.051.

Four-colour photometry of eclipsing binaries. IXa: AI Hya, photometric observations.   See Abstr. 121.053.

Osservazioni delle binarie ad eclisse RZ Cancri ed Y Camelopardalis.   See Abstr. 121.075.

Determinación fotométrica del tipo espectral de la componente desconocida de una estrella binaria eclipsante. Aplicación a la variable eclipsante V 453 Scorpii. See Abstr. 121.076.

Photoelectric observations and apsidal motion of Y Cygni.   See Abstr. 121.091.

Intrinsic colours and physical properties of Cepheids.

See Abstr. 122.003.

The photometric variability of 1 Persei. See Abstr. 122.033.

HR 2557: a luminous δ Scuti star. See Abstr. 122.034.

UBV photometric observations of SX Phe. See Abstr. 122.111.

Light variations of the H and K emission star HD 206860.   See Abstr. 122.148.

Reddening lines of classical cepheids on the UBV two-colour diagram.   See Abstr. 122.198.

Nouvelle recherche de périodes d'étoiles Ap observées à l'ESO-I.   See Abstr. 123.024.

Ultraviolet observations of planetary nebulae. III. Variability of the central star. See Abstr. 135.009

Photometry of AM Herculis: a slow optical pulsar? See Abstr. 142.017.

Optical pulses in HD 153919 = 3U 1700–37. See Abstr. 142.193.

Evidence of a periodicity in the optical light curve of HD 226868 (Cyg X-1) during the X-flare of April - May 1975.   See Abstr. 142.225.

A photometric study of the Orion OB 1 association. III. Subgroup analyses.   See Abstr. 152.002.

Optical and infrared properties of the newly formed stars in Canis Major R1.   See Abstr. 152.004.

A four-color and Hβ study of the galactic cluster IC 4756.   See Abstr. 153.036.

UBV photographic photometry in and around the galactic cluster NGC 2527.   See Abstr. 153.038.

Luminous variable stars in M31 and M33. See Abstr. 158.012.

Photoelectric photometry and summary of photoelectric observations of star clusters in the Andromeda Nebula. See Abstr. 158.031.

Recent photometry of OJ287. See Abstr. 158.514.

Studies of the Large Magellanic Cloud stellar content. III. Spectral types and *V* magnitudes of 1822 members. See Abstr. 159.002.

114 Spectra, Temperatures, Chemical Composition, Spectra of Individual Stars

Spectra, Temperatures, Chemical Composition

114.001 A catalogue of carbon stars in the Large Magellanic Cloud.
B. E. Westerlund, N. Olander, H. B. Richer, D. R. Crabtree.
Astron. Astrophys., Suppl. Ser., Vol. 31, 61 - 82 (1978).
      The authors present coordinates, finding charts and near infrared I magnitudes for 302 probable carbon stars in the Large Magellanic Cloud. Photoelectrically observed colours in $R-I$ are also given for 77 of them. Their surface distribution is shown.

114.002 The chemical composition of late-type supergiants. III. Carbon, nitrogen, and oxygen abundances for 19 G and K Ib stars.   R. E. Luck.
Astrophys. J., Vol. 219, 148 - 164 (1978).
      Through an analysis of high-resolution, high signal-to-noise data, carbon, nitrogen, and oxygen abundances are investigated in 19 G and K Ib stars. The final analysis yields abundances for 16 of these objects. The derived abundances show that, relative to solar values, these stars are heavily deficient in carbon and enhanced in nitrogen, and that there is also a moderate oxygen deficiency. There is a tendency for lower carbon abundances to be associated with low $^{12}C/^{13}C$ ratios.

114.003 The helium triplet-to-singlet ratio in T Tauri stars.   T. J. Schneeberger, J. L. Linsky, S. P. Worden.
Astron. Astrophys., Vol. 62, 447 - 448 (1978).
      The He I lines at λ5876 and λ6678 are used to determine the triplet-to-singlet ratio for T Tauri stars DF Tau and BP Tau. The ratio is approximately 3. Line profiles and absolute intensities are presented.

114.004 Spectroscopic differences between solar-type cluster stars and the Sun.   D. C. Barry.
Astrophys. J., Vol. 219, 942 - 946 (1978).
      2 Å resolution spectrograms reveal the strength of 3871 Å (CN) relative to 3859 Å (Fe I) and 3883 Å (CN) relative to 3878 Å (Fe I) to be systematically greater in late F and G dwarfs in the Hyades cluster than in the Coma cluster. This difference is explained as a normal consequence of the general enrichment of the metals in the Hyades and the dependence of the CN abundance on the product of the C and N abundances. The 3859/3871 and 3878/3883 ratios in the spectrum of the Ursa Major cluster dwarf HD 115043 indicate that its surface gravity and metallicity are similar to members of the Coma cluster. The simplest interpretation of the Sun's spectrum is that its surface gravity and metallicity are similar to members of the Hyades cluster.

114.005 Absolute ultraviolet spectrophotometry from the TD1 satellite. X. The ultraviolet spectrum of the Ap stars.
C. Jamar, D. Macau-Hercot, F. Praderie.
Astron. Astrophys., Vol 63, 155 - 174 (1978).
      77 Ap stars are compared to 344 normal stars in the ultraviolet from 1350 to 2550 Å by means of the spectra of the S 2/68 Ultraviolet Bright Star Spectrophotometric Catalogue. The various Ap subclasses have been compared to the normal stars by means of a long basis (λ 2100 vs. λ 5500 Å) index and a typical UV (λ 1400 vs. λ 2100 Å) index. The Hg Mn stars are quasi-normal in the whole range. The Si and SrCrEu stars present a normal flux level around 2100 Å but are very flux deficient at shorter wavelengths. The flux deficiency at λ 1400 Å proves to be a powerful classification criterion both to discriminate Ap stars from normal ones and to detect new members of the Ap group (7 new such stars have been discovered).

114.006 The sun among the stars. I. A search for solar spectral analogs.   J. Hardorp.
Astron. Astrophys., Vol. 63, 383 - 390 (1978).
      77 solar type stars in parts of the northern and southern skies have been photoelectrically scanned with 20 Å resolution to find stars whose ultraviolet spectra (3640–4100 Å) match that of the sun. Down to $V = 6^m6$ there seem to be only two: HR 7504 in the northern and HR 2290 in the southern hemisphere. No G2V star matches the sun, they all have weaker ultraviolet absorptions, most of them much weaker. Stars that do match have a $B-V$ of 0^m66. The search shows that G type dwarfs are hotter than was thought before.

114.007 Spectral types for early-type stars observed by Skylab.   N. G. Roman.
Astron. J., Vol. 83, 172 - 175 (1978).
      Spectral types on the MK system are given for 246 stars, primarily between 6.5 and 9.0 visual magnitudes, observed in the ultraviolet by Skylab.

114.008 Manganese to iron ratios in F and G dwarf stars.   T. G. R. Beynon.
Astron. Astrophys., Vol. 64, 145 - 152 (1978).
      Manganese to iron abundance ratios in F and G dwarf stars are surveyed paying particular attention to accurate correction for the effects of hyperfine structure.

114.009 The OB stars in the vicinity of the long period cepheid VZ Puppis.   R. J. Havlen.
Astron. Astrophys., Vol. 64, 295 - 298 (1978).
      UVB, β photoelectric of catalogued OB stars surrounding VZ Puppis reveals only a few stars at the cepheid's distance of 5.6 kpc. Good evidence is found, however, for the existence of an OB association at 4 kpc in agreement with other young spiral tracers in and below the plane. A comparison of the distribution of these tracers with that of H I confirms that the Schmidt model rotation curve is not valid below the plane and suggests that there are distinct kinematic groups among the Population I objects in this direction.

114.010 Spectra of giant stars in distant satellites of the Galaxy.
A. P. Cowley, F. D. A. Hartwick, W. L. W. Sargent.
Astrophys. J., Vol. 220, 453 - 457 (1978) = Contrib. Dominion Astrophys. Obs. No. 349 = NRC 16108.
      The spectra of individual giant stars in 14 distant low-surface-brightness satellites of the Galaxy are compared with those in M92, M15, and M71. The main results of the comparison are that (1) no gross differences in spectral appearance exist among stars in the dwarf spheroidal galaxies, the outlying globular clusters, and the inner halo clusters, and (2) a large range in metal-line strengths exists at large distances from the galactic center.

114.011 Sternspektroskopie.   E. Pollmann.
Orion, 36. Jahrg., 4 - 9 (1978).

114.012 Angewandte Spektralklassifizierung.

E. Pollmann.
Orion, 36. Jahrg., 9 - 11 (1978).

114.013 Spectral types in the Orion OB1 association.
H. A. Abt, H. Levato.
Publ. Astron. Soc. Pacific, Vol. 89, 797 - 802 (1977/78).

Spectral types are derived for 152 stars in the Northwest, Belt, and Outer Sword regions. The classifications show: (1) stars above the ZAMS in the O9—B3 and the B7—A3 regions; most of these have greater luminosities than class V or are double-lined binaries or peculiar stars; (2) eleven Ap stars; (3) four moderate Be stars; (4) only two stars with broad hydrogen lines; (5) 14 "sn" stars that have both sharp and broad He I lines, perhaps due to tenuous shells; (6) one classical shell star; (7) several Sirius-type Am stars.

114.014 Observations and analysis of carbon monoxide in cool stars at five microns.
T. R. Geballe, E. R. Wollman, J. H. Lacy, D. M. Rank.
Publ. Astron. Soc. Pacific, Vol. 89, 840 - 850 (1977/78).

Portions of the 5-μm fundamental vibration-rotation band of CO have been observed and analyzed for a number of late-type giant and supergiant stars. Analysis using synthetic spectra computed from model atmospheres yields values for $^{12}C/^{13}C$ of approximately 30 in α Her, 15 in α Sco, and 10 in α Ori. $^{16}O/^{17}O$ is ~ 500 in α Her; similar values are indicated for α Ori and α Sco. No evidence for values of $^{16}O/^{18}O$ significantly different from the terrestrial value is found for these three stars. The elemental abundance ratio [C/H] is found to be lower than the cosmic value for all three stars; however, the ratios are dependent on the assumed continuum temperatures at 5 μm, which are somewhat uncertain. The present results imply the mixing of CN-cycle products of main-sequence burning into the envelopes of these stars.

114.015 Ultraviolet observations of nine Wolf—Rayet stars.
A. J. Willis, R. Wilson.
Mon. Not. R. Astron. Soc., Vol. 182, 559 - 594 (1978).

The most extensive set of observations of the ultraviolet spectra of Wolf—Rayet stars yet obtained are presented. These were made with the sky survey experiment (S2/68) on the ESRO satellite *TD-1* and cover nine objects — three WC, three WN and three WC + O binaries. These data have been combined with ground-based observations and analysed for both the continuum energy distributions and line strengths. The strengths of helium, carbon and nitrogen lines in four WR stars (one WC and three WN) were analysed using a non-LTE treatment to determine the abundances of these species. The evolutionary status of the WR stars is discussed in the light of these results and the products of thermonuclear burning. It is concluded that they are in the helium burning phase, with the WN stars being less evolved than the WC. By considering the location of the WR stars in the H—R diagram in the light of recent evolutionary models of massive stars with heavy mass loss, it appears that the recent suggestion by Conti that the Of stars are the progenitors of the WR class is very plausible.

114.016 Line blocking in the near ultraviolet spectrum of early type stars. I. Observed line blocking factors for
132 stars. H. J. G. L. M. Lamers, E. A. Müller, F. Llorente de Andrés.
Astron. Astrophys., Suppl. Ser., Vol. 32, 1 - 16 (1978).

The line blocking factors in the ultraviolet spectrum of 132 stars of types O4 to A7 observed by the S59 experiment are presented. The blocking factors are given for intervals of 10 Å wide in the wavelength ranges of 2070–2150 Å, 2505–2585 Å and 2780–2860 Å as well as for these whole intervals. The accuracy of the blocking factors is given in terms of a 2σ error. The intrinsic continuum colors $m^c_{2100} - m^c_{2500}$, $m^c_{2500} - m^c_{2800}$, and $m^c_{2500} - V_0$ are also given.

114.017 Observational studies of the Herbig Ae/Be stars. II.

Polarimetry.
L. M. Garrison, Jr., C. M. Anderson.
Astrophys. J., Vol. 221, 601 - 607 (1978).

Multicolor polarimetry is presented for a large sample of the pre—main-sequence Herbig Ae/Be stars. Several methods are employed for the removal of the interstellar component of polarization, and the resulting intrinsic polarizations as a function of wavelength resemble those shown by the more classical emission-line B stars.

114.018 A rocket ultraviolet spectrographic survey of Orion stars.
G. R. Carruthers, C. B. Opal, H. M. Heckathorn.
Bull. American Astron. Soc., Vol. 9, 571 (1978). – Abstract.

114.019 The measurement of rotation and macroturbulence in O-type stars using Fourier transform analysis.
D. Ebbets.
Bull. American Astron. Soc., Vol. 9, 572 (1978). – Abstract.

114.020 The abundance of boron in B- and A-type stars.
A. M. Boesgaard, W. D. Heacox.
Bull. American Astron. Soc., Vol. 9, 573 - 574 (1978). Abstract.

114.021 Support from stellar spectroscopy for a dependence of the relative abundance of sodium on overall
metallicity. R. C. Peterson.
Bull. American Astron. Soc., Vol. 9, 604 (1978). – Abstract.

114.022 Interpretation of He I 10830 line profiles in early type stars of high luminosity. D. D. Meisel.
Bull. American Astron. Soc., Vol. 9, 605 - 606 (1978). Abstract.

114.023 Relative isotopic abundances of Zr in some S stars.
A. C. Zook.
Bull. American Astron. Soc., Vol. 9, 636 (1978). – Abstract.

114.024 Molecular constituents identified in the 3—4 μm spectral region of cool stars.
S. T. Ridgway, D. N. B. Hall, D. F. Carbon.
Bull. American Astron. Soc., Vol. 9, 636 (1978). – Abstract.

114.025 A search for TcO in S stars. D. N. Davis.
Bull. American Astron. Soc., Vol. 9, 636 - 637 (1978). – Abstract.

114.026 Molecular spectra of pure S stars.
S. Wyckoff, R. E. S. Clegg.
Bull. American Astron. Soc., Vol. 9, 637 (1978). – Abstract.

114.027 The Ca II λ8542 and λ8498 lines as important indicators of stellar chromospheres.
J. L. Linsky, D. M. Hunten, R. Sowell, D. L. Glackin.
Bull. American Astron. Soc., Vol. 9, 651 (1978). – Abstract.

114.028 Atmospheric compositions of HgMn stars.
W. D. Heacox.
Bull. American Astron. Soc., Vol. 9, 651 - 652 (1978). Abstract.

114.029 The ^{3}He stars. M. R. Hartoog.
Bull. American Astron. Soc., Vol. 9, 652 (1978). Abstract.

114.030 Spectrophotometry of normal and peculiar B, A, and F stars. S. J. Adelman.
Bull. American Astron. Soc., Vol. 9, 652 (1978). – Abstract.

114.031 Equivalent width of O I λ 7774 and absolute magnitude relation for F and G supergiants.

N. Kameswara Rao, S. G. V. Mallik.
Mon. Not. R. Astron. Soc., Vol. 183, 211 - 214 (1978).

Luminosity calibration is established for F5–G2 supergiants using the equivalent width measures of O I λ 7774 and Fe I λ 7748.

114.032   **An atlas of Copernicus ultraviolet spectra of Wolf-Rayet stars.**   H. M. Johnson.
Astrophys. J., Suppl. Ser., Vol. 36, 217 - 240 (1978).

This atlas reproduces spectral scans of γ^2 Vel, HD 50896, and HD 92740, and tabulates line identifications. Spectral features are classified as interstellar, photospheric absorption, UV-displaced P Cygni absorption, P Cygni emission, and simple emission. Ultraviolet-edge velocities of P Cygni profiles are tabulated. P Cygni profile types are summarized, and some specific intercomparisons are made between the Wolf-Rayets and between them and OB stars that were previously known to have ultraviolet P Cygni profiles.

114.033   **Metallicism and pulsation: the marginal metallic line stars.**   D. W. Kurtz.
Astrophys. J., Vol. 221, 869 - 880 (1978).

Evidence is presented that HR 4594 and HR 8210 are pulsating marginal Am stars. It is suggested that (1) classical Am stars do not pulsate, (2) evolved Am stars may pulsate, and (3) marginal Am stars may pulsate. It is further suggested that, within the Am domain, temperature, age, rotation, and pulsation are sufficient to determine whether a star will be Am, marginal Am, or spectrally normal.

114.034   **Spectrophotometry of B and A-type supergiants.**   L. E. Campusano.
IAU Colloq. No. 42, (see 012.012), p. 128 - 136 (1977).

Continuum fluxes measured with a two-channel spectrum scanner and a 40 Å slit are presented, in a differential manner, for six supergiants over the wavelength range 3390–8090 Å. Five of these stars presented variability detectable on a time scale of a few days.

114.035   **Infrared observations of late type stars.**   K. M. Merrill.
IAU Colloq. No. 42, (see 012.012), p. 446 - 494 (1977).

Substantiative mass-loss resulting in appreciable circumstellar dust envelopes is common in late-type stars. The evolutionary history and physical state of a cool star determine the chemistry within the outer stellar atmosphere mirrored by the molecular and particulate material present in the envelope. The observational consequences of this debris determined by moderate spectral resolution ($\lambda/\Delta\lambda \sim 50$–$100$) infrared spectrophotometry are reviewed.

114.036   **Abundance analyses of metal-poor stars. I. Blue spectra of 15 high-velocity dwarfs.**   R. Peterson.
Astrophys. J., Vol. 222, 181 - 190 (1978).

The abundance analysis of a variety of metal-poor stars is undertaken in this series, in an effort to extract accurate relative abundances of the heavy elements over a wide range of stellar metallicities and temperatures. This paper presents the results derived from photometric Cassegrain echelle spectra obtained for 15 dwarfs and subgiants selected from the high-velocity catalog of Eggen.

114.037   **Photoelectric two-dimensional spectral classification of M supergiants.**   N. M. White, R. F. Wing.
Astrophys. J., Vol. 222, 209 - 219 (1978).

A photoelectric system defined by eight narrow bands between 0.7 and 1.1 μm has been used to measure nearly all M supergiants that have been classified on the MK system. The photometric TiO and CN indices reproduce the two-dimensional MK classifications to the accuracy of the MK types themselves. Mean fluxes and spectral classifications are presented for 128 stars.

114.038   **Mixing and the strong-cyanogen phenomenon.**   D. Deming.
Astrophys. J., Vol. 222, 246 - 262 (1978).

A model atmosphere and spectrum synthesis calibration has been obtained for the cyanogen band strength index of DDO photometry. Synthetic 41 − 42 indices have been computed in the DDO system for a variety of CNO compositions chosen to schematically represent varying degrees of mixing from a hydrogen-burning shell. Only a slight strengthening of the CN index, δCN, is found for the degree of mixing from a hydrogen-burning shell which is predicted by standard stellar evolution calculations before and during core helium burning of 1–$3 M_\odot$ stars. Greater amounts of mixing result in a greater weakening of the CN band strength than is observed. Existing stellar evolution calculations for the first ascent of the giant branch are consistent with the CN band strengths of metal-rich stars. A theoretical calibration of δCN in terms of [metals/H] has been obtained and is in reasonable agreement with the existing empirical calibration. It is concluded that the majority of strong-CN Population I stars are more likely to be super-metal-rich than nitrogen-rich.

114.039   **Absolute spectral energy distributions and [Fe/H] values of metal-poor stars and globular clusters.**
C. G. Christensen.
Astron. J., Vol. 83, 244 - 265 (1978).

Absolute spectral energy distributions for 65 metal-poor stars, spanning a wide range in temperature, luminosity, and metal content, are given. Five local globular clusters and five M31 globulars are also described. Empirical relationships between [Fe/H], color, and certain feature indices have been calibrated and applied to 19 objects with previously undetermined [Fe/H] values.

114.040   **Observations of central stars.**   J. H. Lutz.
IAU Symp. No. 76, (see 012.014), p. 185 - 193 (1978).

Contents: Classifications and population types. Temperatures. Binaries. Peculiar central stars.

114.041   **The effective temperature of the central star, and a criterion for complete absorption of hydrogen ionizing photons by the nebula.**
S. R. Pottasch, P. R. Wesselius, C.-C. Wu, R. J. Van Duinen.
IAU Symp. No. 76, (see 012.014), p. 210 - 211 (1978). Abstract.

114.042   **Peculiar central stars and related objects.**   J. Lutz.
IAU Symp. No. 76, (see 012.014), p. 212 (1978). − Abstract.

114.043   **Carbon stars near open clusters at galactic latitude $4°\!.5 < |b| < 9°\!.5$ and between longitudes 68° and 184°.**   A. Alksnis, Z. Alksne, I. Platais.
Investigations of the sun and red stars. 7, (see 003.004), p. 11 - 16 (1977). In Russian.

By visual inspection of objective prism photographs of 58 partly overlapping fields of 5 degrees diameter, 302 carbon stars and about 50 clusters have been identified and the cases of possible cluster membership examined. Nine of the carbon stars occur less than three cluster radii from seven open clusters.

114.044   **Continuum spectra of late-type stars.**   J. I. Straume.
Investigations of the sun and red stars. 7, (see 003.004), p. 28 - 32 (1977). In Russian.

Based on previously calculated stellar model atmospheres (1976, 1977) the energy distribution in the continuum spectra of late-type stars was calculated for effective temperatures $T_e = 3000, 3500, 4000$ and 4500 K. The source of continuous opacity was absorption by H, H^-, H_2^-.

114.045 New carbon stars in the zone centered at galactic
latitude b = +7°.
A. Alksnis, Z. Alksne, V. Ozoliņa, I. Platais.
Investigations of the sun and red stars. 8, (see 003.005), p.
5 - 25 (1978). In Russian.

In a nearly 5° wide zone oriented parallel to the galactic
equator and centered at galactic latitude $b = +7°$, 75 new
carbon stars have been found between longitudes $l = 68°$ and
$l = 184°$. Equatorial and galactic coordinates, spectral types
and identification charts are given for these stars.

114.046 Zirconium stars in open clusters and in the spiral
arms of the Galaxy. U. Dzērvītis.
Investigations of the sun and red stars. 8, (see 003.005), p.
32 - 50 (1978). In Russian.

By combining all published positions of zirconium stars
and open clusters, a list has been compiled of all cases where
zirconium stars could be possible members of open clusters.
Supposing such a membership, in cases having corresponding
photometric data absolute magnitudes of zirconium stars are
evaluated. Maxima and minima in the visible distribution of
zirconium stars in galactic longitude are shown to correspond
to directions along and across nearby spiral arms and its bifurcations.

114.047 Spectrophotometry of B, A, and F stars.
S. J. Adelman.
Astrophys. J., Vol. 222, 547 - 555 (1978).

Energy distributions of 33 normal B, A, and F stars for
$\lambda\lambda 3300-7100$ are given that are consistent with the Hayes-
Latham calibration of Vega. The derived upper limit on the
photometric accuracy using a comparison of observed and
synthetic $u-b$ and $b-y$ indices is 0.010 mag and 0.008 mag,
respectively. Effective temperatures are found for selected
spectral types.

114.048 Relative abundances in metal-poor stars. I. The
aluminum-to-iron ratio. R. Peterson.
Astrophys. J., Vol. 222, 595 - 599 (1978).

Motivated by specific predictions of explosive nucleosynthesis that indicate increasing over-deficiencies of aluminum at low overall metal abundances and armed with new
high-quality, high-resolution echelle-Kron spectra, the author
has determined the aluminum-to-iron ratio in nearly a dozen
high-velocity metal-poor dwarfs and subgiants with metal
abundances 1/4 to 1/100 solar.

114.049 On the theory of silicon spectra in O and B stars.
L. W. Kamp.
Astrophys. J., Suppl. Ser., Vol. 36, 143 - 171 (1978).

Tables of equivalent widths are presented for 59 spectral
lines of silicon II, III, and IV for a grid of models corresponding to spectral types O6−B5, luminosity classes III−V. The
computations were done in non-LTE and in LTE for an abundance $N(Si) = 3 \times 10^{-5} N(H)$ (approximately solar abundance)
and no microturbulence, using the non-LTE unblanketed
model atmospheres of Mihalas. The diagnostic capabilities of
the grid are investigated in terms of the parameters: effective
temperature, gravity, silicon abundance, and microturbulence.
Results are obtained for 11 well-observed stars, with spectral
types from B0.2 to B5, luminosity classes IV and V. The
computed equivalent widths are also compared with mean
observations of many stars (taken mainly from the literature)
versus spectral type.

114.050 Standard approximations to line formation in
stellar spectra. A. Sapar, A. Heinlo.
Tartu Astron. Obs., Teated No. 56, p. 55 - 77 (1978).

114.051 Ultraviolet fluxes of Be stars. D. Briot.
Astron. Astrophys., Vol. 66, 197 - 203 (1978).
A general comparison of ultraviolet fluxes between

1400 Å and 2740 Å of Be stars to those of B stars has been
made using data from the "Ultraviolet Bright Star Spectrophotometric Catalogue" (Jamar et al., 1967). A total of
117 Be stars and 167 B stars in the range B 0−B 9 was
studied. For spectral types later than B 3, no excess or
deficiency in the mean ultraviolet flux of the Be stars
compared to the B stars was detected. For B 1, B 2 and
B 3 stars a very faint mean excess is detected longwards of
$\lambda = 1900$ Å.

114.052 Classification of stars according to unbroadened
low-dispersion spectra and some results of its
application to the study of galactic structure.
V. I. Kuznetsov.
Astrometriya i Astrofizika, Kiev, Vyp. (No.) 34, p. 30 - 43
(1978). In Russian.

Advantages are shown of using faint stars 13^{m}5−15^{m}0
(B) for stellar statistics, the spectral classification of these
stars being carried out according to broadened spectrograms.
A description is given of some special features of the methods
used for the study of interstellar extinction in the directions
of the clusters NGC 2129, NGC 6834, NGC 6913 and NGC
7854.

114.053 New observations of ultraviolet variability in Wolf-
Rayet stars.
W. M. Burton, R. G. Evans, B. Patchett, C.-C. Wu.
Mon. Not. R. Astron. Soc., Vol. 183, 605 - 615 (1978).

Observations of ultraviolet variability in Wolf-Rayet stars
have been made with the ANS satellite Ultraviolet Photometer
Experiment. Significant variations are detected in several of
the observed stars, the timescale of the variability ranging from
a few minutes to several months.

114.054 Giant problems. M. G. Edmunds.
Nature, Vol. 273, 487 - 488 (1978).

114.055 Spectral classification of ultraviolet objects.
R. Barbier, F. Dossin, C. Jaschek, M. Jaschek,
M. Klutz, J. P. Swings, J. M. Vreux.
Astron. Astrophys., Vol. 66, L9 - L10 (1978).

A sample of 53 ultraviolet objects from the catalogs of
Carnochan and Wilson has been studied with an objective
prism in order to determine the origin of their UV excess.

114.056 Lithium line formation in carbon stars.
R. de la Reza, F. Querci.
Astron. Astrophys., Vol. 67, 7 - 14 (1978).

Kinetic equilibrium (KE) calculations of neutral lithium
lines in carbon stars have been performed. The influence of
possible chromospheric radiation into the photosphere is examined. The authors show that the results can be used to
distinguish a hot chromosphere from a cold one. Independent
LTE calculations have been performed for comparison with
results of the KE calculations. Appreciable non-LTE effects
are found only for the resonance transition $\lambda 6708$ Å.

114.057 A new investigation of the energy distribution in
the spectra of eight stars being used as spectro-
photometric standards. A. V. Kharitonov, I. N. Glushneva.
Astron. Zh. Akad. Nauk SSSR, Vol. 55, 496 - 500 (1978).
In Russian. English translation in Soviet Astron., Vol. 22,
No. 3.

Data on the energy distribution in the spectra of stars
used as spectrophotometric standards are obtained on the
basis of observations made independently at the Astrophysical
Institute at Alma-Ata and the Sternberg Astronomical
Institute, as well as by using published data.

114.058 CN strengths and $^{12}C/^{13}C$ ratios in Population I K
giants. D. Deming.
Astrophys. J., Lett., Vol. 223, L31 - L36 (1978).

An analysis of existing data on CN strengths in Population I K giants shows that giants with very low $^{12}C/^{13}C$ ratios ($\lesssim 15$) tend to have weak CN bands. It is also pointed out that several ^{13}C-rich giants are weak-CN stars because they are metal-poor. Considerations suggest that the initial CNO group abundance may be an important factor in producing large ^{13}C enhancements in K giants.

114.059 **Identification of the 3.9 micron absorption band in carbon stars.** J. D. Bregman, J. H. Goebel, D. W. Strecker.
Astrophys. J., Lett., Vol. 223, L45 - L46 (1978).

With a 1.7% spectral resolution filter-wheel spectrometer, 1.2—5.2 micron spectra have been obtained of a sample of carbon stars. The previously unidentified 3.9 micron band is attributed to a combination of CS and C_2H_2. The authors also report the first observation of SiO in a carbon star.

114.060 **Spectral classification from the ultraviolet line features of S2/68 spectra. III. Early A-type stars.**
A. Cucchiaro, D. Macau-Hercot, M. Jaschek, C. Jaschek.
Astron. Astrophys., Suppl. Ser., Vol. 33, 15 - 26 (1978).

This paper concerns the classification of more than 600 A0–(A3–A4) spectra obtained with the sky survey telescope (S2/68) mounted on the TD1 satellite. As in the two previous papers of this series, a statistical study of the spectra is made to establish ultraviolet classification criteria. The scheme thus established permits separation of 14% of the stars showing an abnormal ultraviolet spectrum.

114.061 **The lithium abundance in weak-G band stars.**
M. R. Hartoog.
Publ. Astron. Soc. Pacific, Vol. 90, 167 - 169 (1978) = Lick Obs. Bull., No. 790.

Lithium abundances are derived for ten weak-G band giants. The stars are found to have at least normal lithium abundances, and may have enhanced lithium relative to normal giants of similar temperature. Also, broad lines found in the G5 giant HR 1023 can be interpreted as rotation with a $V \sin i$ of 17 ± 5 km s^{-1}. Possible interpretations of these results are discussed.

114.062 **Spectral types in the Pleiades.**
H. A. Abt, H. Levato.
Publ. Astron. Soc. Pacific, Vol. 90, 201 - 203 (1978).

New spectral classification of 49 of the brightest stars in the Pleiades shows the following peculiarities: (1) three Ap(Si) stars, of which one is a member of a $2\overset{d}{.}46111$ double-lined binary with an Am secondary, (2) an Ap(Hg, Mn) star that may be a member of the cluster, (3) four Am stars, of which two are only marginally abnormal, and (4) three shell spectra that include the pronounced case of Pleione and two mild shell spectra called "sn". The Ap + Am system should be checked for eclipses.

114.063 **Spectrophotometry of low-luminosity stars with the Carnegie image tube.** J. W. Christy.
Publ. Astron. Soc. Pacific., Vol. 90, 207 - 215 (1978).

An observational technique with the Carnegie image tube is described which results in relative spectrophotometry for very faint stars. Relative intensity curves from 4500 Å to 6700 Å with 25 Å resolution were obtained for 15 low-luminosity stars ranging from V = 9.5 to V = 15.5. The large Mg b and Na D absorption for Barnard's Star (Spinrad 1973) is confirmed. Spectral indices for the M dwarfs at $M_V = 13$ are discussed and it is found that these can be measured with internal errors of $\pm 0\overset{m}{.}02$ per 25 Å bandwidth. Among these stars are both components of the common proper-motion pair, G107-69/70 (sdM5/DC).

114.064 **Peculiarities of peculiar stars. XII, XIII.**
B. Kuchowicz.

Urania Kraków, Vol. 48, 322 - 327, 364 - 371 (1977). In Polish.

114.065 **Wolf-Rayet stars in the Small Magellanic Cloud.**
J. Breysacher, B. E. Westerlund.
ESO Sci. Prepr. No. 21, 20 pp. (1978).– Submitted to Astronomy and Astrophysics.

Preliminary results from a spectroscopic study of the four known Wolf-Rayet stars in the Small Magellanic Cloud are presented. The spectral classification of the Wolf-Rayet stars and their components is considered as well as possible spectral variations.

114.066 **Differential curve-of-growth analyses of Mn-Hg stars II. Analyses of 53 Aur, HR 6997, and 112 Her, and final summary.** K. Kodaira, M. Takada.
Ann. Tokyo Astron. Obs., Second Ser., Vol. 17, 79 - 92 (1978).

114.067 **A spectroscopic study of Bp stars.**
O. Vilhu, I. Tuominen.
Magnetic stars, (see 012.028), p. 13 - 15 (1978). – Abstract.

114.068 **A successful search for OH emission in early-type emission-line stars with IR excesses.**
L. E. Davis, E. R. Seaquist, C. R. Purton.
Bull. American Astron. Soc., Vol. 10, 392 (1978). – Abstract.

114.069 **Identification of the 3.9 micron absorption band in carbon stars.**
J. D. Bregman, J. H. Goebel, D. W. Strecker.
Bull. American Astron. Soc., Vol. 10, 407 (1978). – Abstract.

114.070 **Infrared spectra of Wolf-Rayet stars.**
K. M. Merrill, J. H. Black.
Bull. American Astron. Soc., Vol. 10, 407 (1978). – Abstract.

114.071 **Scans of the 4-μ SiO bands in late-type stars.**
C. P. Rinsland, R. F. Wing.
Bull. American Astron. Soc., Vol. 10, 408 (1978). – Abstract.

114.072 **Spectrophotometry of Ap stars.**
D. M. Pyper, S. J. Adelman.
Bull. American Astron. Soc., Vol. 10, 414 (1978). – Abstract.

114.073 **A sequence of objective-prism spectra of the cooler stars, G5–M.** R. C. McNeil.
Bull. American Astron. Soc., Vol. 10, 414 (1978). – Abstract.

114.074 **Observations of low-luminosity optical counterparts of X-ray sources.**
J. R. Thorstensen, P. A. Charles, S. Bowyer.
Bull. American Astron. Soc., Vol. 10, 434 (1978). – Abstract.

114.075 **Ultraviolet spectroscopy of Hg-Mn stars with IUE (*International Ultraviolet Explorer*).**
D. S. Leckrone, W. D. Heacox.
Bull. American Astron. Soc., Vol. 10, 443 (1978). – Abstract.

114.076 **IUE observations of F, G, and K stars and preliminary models for upper chromospheres based on C II, Si II, and Si III.** W. L. Kelch, J. L. Linsky.
Bull. American Astron. Soc., Vol. 10, 443 (1978), – Abstract.

114.077 **A first look at IUE far ultraviolet spectra of K and M stars — α Ori, α Boo, and ϵ Eri.**
G. S. Basri, J. L. Linsky.
Bull. American Astron. Soc., Vol. 10, 443 (1978). – Abstract.

114.078 **Ultraviolet spectra of M stars.**
R. F. Wing, K. G. Carpenter.
Bull. American Astron. Soc., Vol. 10, 444 (1978). – Abstract.

114.079 **Ultraviolet spectroscopy of hot, subluminous stars with the IUE.**   S. R. Heap.
Bull. American Astron. Soc., Vol. 10, 444 (1978). – Abstract.

114.080 **Scanner studies of composite spectra. I. Dwarfs.**   W. I. Beavers, D. B. Cook, E. W. Fick.
Bull. American Astron. Soc., Vol. 10, 451 - 452 (1978). Abstract.

114.081 **Spectroscopy of distant blue stars near the galactic anticenter.**   F. R. Chromey.
Bull. American Astron. Soc., Vol. 10, 452 (1978). – Abstract.

114.082 **Technetium in late-type stars.**   S. J. Little, I. R. Little-Marenin, C. A. Leslie.
Bull. American Astron. Soc., Vol. 10, 453 (1978). – Abstract.

114.083 **Mid-UV emission lines in late-type giants and supergiants.**   R. E. Stencel, K. A. van der Hucht, Y. Kondo.
Bull. American Astron. Soc., Vol. 10, 453 (1978). – Abstract.

114.084 **Profiles of Hα in B-type stars.**   S. C. Wolff, J. H. Heasley, R. J. Wolff, W. H. Smith.
Bull. American Astron. Soc., Vol. 10, 453 (1978). – Abstract.

114.085 **Photometric and spectroscopic variability of Ca II H and K in G- and K-type giants.**
S. L. Baliunas, L. W. Hartmann, W. Liller, A. H. Vaughan, Jr., E. H. Avrett, A. K. Dupree.
Bull. American Astron. Soc., Vol. 10, 461 - 462 (1978). Abstract.

114.086 **The near-infrared spectra of Wolf-Rayet stars.**   M. K. V. Bappu, K. S. Ganesh, K. K. Scaria.
Kodaikanal Obs. Bull., Ser. A, Vol. 2, 28 - 33 (1977).

The near-infrared spectra of three Wolf-Rayet stars of the carbon sequence and five of the nitrogen sequence have been studied. Wavelength identifications and intensity scans are presented to show the emission line characteristics of these objects in the 6800 Å to 8200 Å domain of the spectrum.

114.087 **A survey of red stars in the direction of the Large Magellanic Cloud. 1. The 30 Doradus region.**
M. K. V. Bappu, M. Parthasarathy, K. K. Scaria.
Kodaikanal Obs. Bull., Ser. A, Vol. 2, 85 - 88 (1977).

The authors have commenced a survey of red stars in the direction of the Large Magellanic Cloud using the technique of ultra-low dispersion spectroscopy. This first paper is a listing of the red stars in the 30 Doradus region for which finding charts and coordinates on the Hodge-Wright Atlas charts are provided.

114.088 **Observations of polarization of some R CrB stars.**   M. Ya. Orlov, M. H. Rodriguez.
Astron. Tsirk., No. 969, p. 3 - 4 (1977). In Russian.

114.089 **Correlation between [Fe/H] metal abundance and ultraviolet excess $\delta_{0.6}$ (U−B) for metal-deficient stars.**   V. A. Marsakov, A. A. Suchkov.
Astron. Tsirk., No. 973, p. 1 - 2 (1977). In Russian.

114.090 **Some regularities in chemical composition of unevolved main sequence stars and synthesis of the elements.**   V. L. Khokhlova.
IAU Colloq. No. 45, (see 012.032), p. 177 - 179 (1977/78).

The published data on chemical abundances derived from fine analysis for about 80 G, F and A main sequence stars are analysed.

114.091 **Carbon and nitrogen in halo stars.**   C. Sneden, R. C. Peterson.
IAU Colloq. No. 45, (see 012.032), p. 201 - 204 (1977/78).

A new study of carbon and nitrogen abundances has been carried out for a sample of halo dwarf stars. This sample was limited to relatively bright high velocity stars with substantial metal deficiencies ($-0.65 \geq$ [Fe/H] ≥ -2.00). Synthesis of CH molecular bands for nine program stars indicates that [C/Fe] = -0.1 ± 0.2 in halo stars.

114.092 **New peculiar stars.**   Ts. Radoslavova.
Astron. Tsirk., No. 979, p. 5 - 7 (1978). In Russian.

114.093 **Solar-type stars.**   N. E. Kurochkin.
Astron. Tsirk., No. 983, p. 5 - 7 (1978). In Russian.

114.094 **Concerning the differences between the hydrogen lines in low-dispersion spectra of normal and peculiar stars.**   Ts. S. Radoslavova.
Soobshch. AN GruzSSR, Vol. 88, 341 - 343 (1977). In Russian. – Abstr. in Ref. zh., 51. Astron., 6.51.612 (1978).

114.095 **On the particular chemical composition of Ap stars.**   R. Radkov, S. Milcheva.
Astrofiz. issled. (NRB), Vol. 2, 30 - 33 (1977). – Abstr. in Ref. zh., 51. Astron., 6.51.634 (1978).

114.096 **Ultraviolet observations of early-type supergiants.**   M. Hack.
Mem. Soc. Astron. Italiana, Vol. 47, (see 012.036), 417 - 426 (1976).

The spectrum variability characterizing early-type supergiants is described and the quality of information which can be obtained by means of ultraviolet observations is emphasized. A review of the existing ultraviolet observations is given.

114.097 **Ultraviolet observations of peculiar A-type stars.**   M. Hack.
Mem. Soc. Astron. Italiana, Vol. 47, (see 012.036), 427 - 430 (1976). – Abstract.

114.098 **Observations of stellar chromospheres.**   S. Catalano.
Mem. Soc. Astron. Italiana, Vol. 47, (see 012.036), 631 (1976). Summary.

114.099 **The complementarity between TD1 S2/68 and Copernicus observations.**   J. P. Swings.
Mem. Soc. Astron. Italiana, Vol. 47, (see 012.036), 632 (1976). Summary.

A catalog of ultraviolet objective-prism spectra.
See Abstr. 002.030.

An atlas of stellar spectrophotometry.
See Abstr. 002.054.

Study of the emission spectrum of WR stars. II. Calculation of oscillator strengths for the transitions in C III, N IV and O V ions.   See Abstr. 022.122.

Fourier analysis of Zeeman splitting for integrated stellar profiles.   See Abstr. 031.215.

A feasibility study of calibrating stellar photographic equivalent widths against solar photoelectric equivalent widths.   See Abstr. 031.220.

Interactive process of computation of equivalent width at the objective-prism astrograph.
See Abstr. 031.231.

On numerical methods of correction of observed spectral line profiles for the influence of the instrument.

See Abstr. 031.262.

High resolution stellar spectroscopy in the balloon ultraviolet.   See Abstr. 032.508.

Spectrophotometry using an intensifier silicon vidicon.   See Abstr. 034.008.

Iron hydride in stellar atmospheres.
See Abstr. 064.001.

Diffusion and line asymmetries in HgMn stars.
See Abstr. 064.015.

H and He II spectra of Of stars.
See Abstr. 064.018.

The combined effects of expansion and rotation on spectral line shapes.   See Abstr. 064.055.

Extreme-ultraviolet observations of nearby B stars: constraints on hot circumstellar plasma.
See Abstr. 064.081.

Comparison of TDI-fluxes with those from model-atmosphere including OH-opacity.
See Abstr. 064.082.

Explanation of anomalous ionization in B super-giants with a corona plus cool wind model.
See Abstr. 064.091.

M giant circumstellar envelopes.
See Abstr. 064.097.

Determination of the local chemical abundance of stellar atmospheres from spectral line contours.
See Abstr. 064.113.

Stellar and planetary observations with the Voyager photopolarimeter experiment.   See Abstr. 091.069.

Infrared photometry, bolometric magnitudes, and effective temperatures for giants in M3, M13, M92, and M67.
See Abstr. 113.042.

Luminosity and carbon enhancement in N-type carbon stars.   See Abstr. 115.017.

Duplicity of late B-type stars.
See Abstr. 117.032.

Red and near infrared spectra of pre-main sequence stars. I. A preliminary investigation of T Tauri stars.
See Abstr. 122.073.

A spectrophotometric comparison of the physical properties of classical cepheids and nonvariable supergiants.
See Abstr. 122.184.

Spectrophotometric observations of symbiotic stars.
See Abstr. 122.190.

On the influence of magnetic fields on the polariza-tion of stars connected with some cometary nebulae.
See Abstr. 134.043.

Spectroscopic studies of open clusters — a search for Ap stars.   See Abstr. 153.012.

BVRI photometry of late-type Hα-emission stars in NGC 2264.   See Abstr. 153.019.

The red giants of Trumpler 5.
See Abstr. 153.020.

Nucleosynthesis in the Galaxy and the chemical composition of old halo stars.   See Abstr. 155.031.

Metal enrichment in the first ages of the Galaxy.
See Abstr. 155.054.

Automated faint object detection.
See Abstr. 158.266.

Ultra-low dispersion spectroscopy of stars and galaxies.   See Abstr. 158.270.

Studies of the Large Magellanic Cloud stellar con-tent. III. Spectral types and V magnitudes of 1822 members.
See Abstr. 159.002.

Late type giants in the Large Magellanic Cloud.
See Abstr. 159.010.

Spectra of Individual Stars

114.501　**Emission line variations in RS Canum Venaticorum-type binaries.**　E. J. Weiler.
Mon. Not. R. Astron. Soc., Vol. 182, 77 - 92 (1978).

Photoelectric spectrophotometry of the H and K and Hα emission lines in six close binaries is reported. UX Ari, RS CVn, Z Her, AR Lac, LX Per and SZ Psc were observed. Analysis of these data revealed a marginal relation between emission intensity and phase in three systems: UX Ari, RS CVn and Z Her. Extensive chromospheric activity on the later-type component in each system can account for both the current spectrophotometric results and the photometric peculiarities.

114.502　**Atmospheric analyses of southern peculiar A stars.**　A. J. Kearsley, G. Wegner.
Mon. Not. R. Astron. Soc., Vol. 182, 117 - 125 (1978).

Spectroscopic observations of five southern Ap stars (HD 136347, 166469, 183806, 187474 and 189832) are reported and new data are given on their spectra and radial velocities. A curve of growth analysis is carried out for four of the stars using their Hγ line profiles and spectral energy distributions in the vicinity of the Balmer jump to estimate T_{eff} and log g from the atmospheric models of Carbon and Gingerich. The Ap nature of all these stars is confirmed.

114.503　**A hot subluminous star: HDE 283048.**　M. Laget, A. Vuillemin, S. B. Parsons, K. G. Henize, J. D. Wray.
Astrophys. J., Vol. 219, 165 - 167 (1978).

The star HDE 283048 located at $\alpha = 3^h 50^m 3$, $\delta = +25°36'$ shows a strong ultraviolet continuum. Ground-based observations indicate a hot-dominated composite spectrum. Several lines of evidence suggest that the hot component is a hot subdwarf.

114.504　**Infrared Fe II lines in Eta Carinae and a possible interpretation of infrared excesses.**
A. D. Thackeray.
Mon. Not. R. Astron. Soc., Vol. 182, 11P - 15P (1978).

The identification of very strong emission lines in the near infrared spectrum of Eta Carinae with newly recognized high-level transitions of Fe II raises the possibility that the infrared excesses of hot emission-line stars may be due to dielectronic recombination of Fe II. Johansson's Fe II lines also need to be considered in the interpretation of the infrared spectra of supernovae.

114.505　**Predicted presence and tentative identification of new molecules in the pure S star R Cyg.**
A. J. Sauval.
Astron. Astrophys., Vol. 62, 295 - 298 (1978).

A new approach to the problem of the identification of molecular features observed in the spectra of pure S stars has been attempted. From calculations of the chemical equilibrium in cool stellar models, including a very large number of compounds, a first selection has been made among molecules which ought to be present in S stars. On the basis of the author's prediction, he has searched for bands of these new compounds in spectrograms of R Cyg recently obtained by Wyckoff and Wehinger. The presence of some new molecules, such as HfO and TaO, in this pure S star is proposed.

114.506　**Heavy element abundances in Ap stars from ultraviolet data. I. The bright reference stars α Lyrae and α Canis Majoris A.**　A. A. Boyarchuk, T. P. Snow, Jr.
Astrophys. J., Vol. 219, 515 - 521 (1978).

Curve-of-growth analysis is used to derive chemical abundances in α Lyr and α CMa, based on ultraviolet spectra obtained with Copernicus. This analysis is the first step in a program to study the abundances of the heavy elements mercury and platinum and the short-lived element technetium in the atmospheres of Ap and Am stars. Ultraviolet Fe II lines are used to establish the curves of growth for α Lyr and α CMa A, and abundances of a variety of elements, along with upper limits on Hg, Pt, and Tc, are derived. One new element, cadmium, is observed.

114.507　**Photospheric velocity fields in Tau Scorpii.**　M. A. Smith, A. H. Karp.
Astrophys. J., Vol. 219, 522 - 531 (1978).

The recent discovery of a stellar wind in τ Sco (B0 V) has drawn attention to the photospheric velocity fields in hot stars. Fourier analysis of several high-resolution photoelectric profiles observed in this star is employed to infer a 5 km s^{-1} rotational velocity as well as 6 km s^{-1} radial-tangential macroturbulence. A microturbulence of 3 km s^{-1} is also determined from strong and weak line profiles. This value is consistent with that determined from a recent non−LTE curve-of-growth analysis. The most interesting aspect of this study is the discovery of slightly depressed blue wings in virtually all the lines observed.

114.508　**The remarkable extent of the circumstellar gas shell surrounding Betelgeuse.**　A. P. Bernat, R. K. Honeycutt, J. E. Kephart, C. E. Gow, M. T. Sandford II, D. L. Lambert.
Astrophys. J., Vol. 219, 532 - 537 (1978). = Publ. Goethe Link Obs., Indiana Univ., Bloomington, No. 193.

A two-dimensional television system has been used to study the spatial extent of the K I λ7699 resonance-line emission from the gas shell surrounding the M supergiant Betelgeuse. This emission has been traced out to a minimum radius of 29'' (~ 600 stellar radii).

114.509　**Observations of technetium stars.**　F. Sanner.
Astrophys. J., Vol. 219, 538 - 542 (1978).

Two new technetium stars have been found: the MS irregular variable OP Her and the MS semiregular variable T Cet. An intercomparison of the spectra of these and 13 other red giants, five of which are technetium stars, indicates that the strength of Tc I is closely correlated with the strengths of other s-process element lines, but not necessarily with spectroscopic criteria of temperature, luminosity, and carbon abundance. This correlation, together with the otherwise heterogeneous nature of the sample, suggests that radioactive decay of technetium has not yet occurred to an appreciable extent in cool S and C variables. The seven technetium stars observed here are surrounded by massive circumstellar gas envelopes, as indicated by strong optical circumstellar lines.

114.510　**Linear polarization variability in the emission line O star Lambda Cephei.**　D. P. Hayes.
Astrophys. J., Vol. 219, 952 - 955 (1978).

Systematic and intensive monitoring of the emission line O star, λ Cep, shows that statistically significant polarization changes occur over a 1 day interval. No shorter term variability (hours) has been detected.

114.511　**The fundamental bands of CO in Arcturus: evidence for an inhomogeneous chromosphere.**
J. N. Heasley, S. T. Ridgway, D. F. Carbon, R. W. Milkey, D. N. B. Hall.
Astrophys. J., Vol. 219, 970 - 978 (1978).

The authors compare new high-spectral-resolution observations of the fundamental vibration-rotation bands of CO in the Arcturus spectrum with synthetic spectra for a representative set of existing model atmospheres of this star. They conclude that there is not a homogeneous chromosphere typical of that model on Arcturus.

114.512　**The helium abundance in a globular-cluster star.**　E. A. Mallia.
Observatory, Vol. 98, 11 - 13 (1978).

114.513 Sanduleak's puzzling emission-line object.
W. L. Martin.
Observatory, Vol. 98, 22 - 23 (1978).

114.514 The spectrum of the bright YY Orionis star CoD − 35° 10525.
I. Appenzeller, R. Mundt, B. Wolf.
Astron. Astrophys., Vol. 63, 289 - 295 (1978).

Cassegrain and coudé spectrograms and simultaneous photoelectric UBV-observations of the bright southern T Tauri star CoD − 35° 10525 are described. The spectra show the red-shifted absorption components and the complex structure of the Balmer emission lines which are characteristic of the YY Orionis stars. No Fe II or Ti II emission or absorption lines could be detected . The He I and He II emission lines appear in unusual strength. From the photometric observations the authors find a strong ultraviolet excess and brightness variations of about 0.5 mag within a few days. The observations can be explained by the assumption that CoD − 35° 10525 is a protostar.

114.515 The Hα line of HR 9076.   H. Pedersen.
Astron. Astrophys., Vol. 63, 305 - 306 (1978).
A photoelectrically measured Hα shell profile of HR 9076 (ε Tuc) is reported.

114.516 Search for stellar TcO.   F. Sanner.
Astron. J., Vol. 83, 194 - 196 (1978).
The spectra of three oxygen-type technetium stars have been searched for bands of TcO, based on recent laboratory identifications by Cherniawsky (1977). No evidence for this molecule has been found. Its absence in stellar spectra can probably be attributed to a low degree of molecular association and heavy blending in the regions of its bandheads.

114.517 Optical and infrared observations of the new emission-line object HM Sagittae.
K. Davidson, R. M. Humphreys, K. M. Merrill.
Astrophys. J., Vol. 220, 239 - 244 (1978).
The blue spectrum and photometry from 0.4 to 12 μm of HM Sge in 1977 June are discussed. These data permit the estimation of various physical parameters of interest including size scales, gas densities, a few ionic abundances, and the total energy flux. The object may be a symbiotic star; the possible relationship with planetary nebulae is not clear.

114.518 The manganese abundance in HD 94028.
T. G. R. Beynon.
Astron. Astrophys., Vol. 64, 299 - 301 (1978).
Peterson (1976) has reported an order of magnitude overdeficiency of manganese compared to iron in the subdwarf HD 94028. In this paper some uncertainties in Peterson's analysis are pointed out. An alternative analysis procedure, when applied to the same observational data, yields a much more moderate manganese overdeficiency.

114.519 The spectral and spatial distribution of radiation from Eta Carinae. I. A spherical dust shell model approach.   R. M. Mitchell, G. Robinson.
Astrophys. J., Vol. 220, 841 - 852 (1978).
An extended spherical circumstellar dust shell model in which the radiative transfer problem is treated accurately is developed for η Car. It is found that both the spectral energy distribution, including the 10 μm silicate feature, and the spatial distribution of radiation can be well represented by the model assuming that the shell consists of a mixture of grains with realistic opacity laws. In particular it is concluded that the satisfactory representation of the 10 μm feature requires, in addition to silicates, grains which possess an emissivity maximum in the 12−14 μm region. The possibility that these grains consist of corundum (Al_2O_3) is investigated. It is concluded that there is some evidence for there being an enrichment of corundum grains, relative to the silicate material, in the outer regions.

114.520 High excitation emission lines of HBV475.
S. Tamura.
Astrophys. Lett., Vol. 19, 57 - 60 (1977).
High excitation emission lines are found by spectroscopic observations of HBV475 which were made with the Cassegrain spectrograph of the 74-inch reflector of the Okayama Astrophysical Observatory in August 1974, October 1975, and August 1976. The line intensity ratio of I([Fe VII], 6087)/ I([Fe VII], 5159) suggests that the electron density of the Fe^{+6} zone is of the order of 10^6 cm^{-3} ~ 10^7 cm^{-3} if an electron temperature of 20,000 °K is assumed for this zone. From a comparison of the intensity of He II 4686 with that observed by Mammano and Righini (1973), it is suggested that the temperature of the exciting star has increased abruptly from around 60,000 K to 100,000 K.

114.521 Line identification list for the Ap star HD 215038.
J. B. Rice.
Publ. Astron. Soc. Pacific, Vol. 89, 770 - 779 (1977/78).
A line identification list covering the spectral region λλ3585−4920 is presented for the relatively unknown Ap star HD 215038. The most abundant lines can be attributed to Fe II, Cr II, Ti II, and Si II. No lines of Zr and Mn were identified. The spectrum of HD 215038 is shown to be similar to the spectra of three other Si-Ap stars.

114.522 Metal abundances and microturbulence in seven solar-type stars: spectroscopic data.
T. Gehren, D. Reimers, L. Berthold, J. Berthold, R. Hennig.
Astron. Astrophys., Suppl. Ser., Vol. 31, 297 - 306 (1978).
The authors present the basic spectroscopic data for the analyses of seven solar-type stars including the computational results for individual spectral lines in the wavelength region λ 4400 to 6800 Å. Line strengths are compared with previous observations of other authors.

114.523 Observations of the highly evolved carbon star CRL 3099.   R. D. Gehrz, J. A. Hackwell, D. Briotta.
Astrophys. J., Lett., Vol. 221, L23 - L27 (1978).
Large-amplitude infrared light variations are reported for the peculiar infrared star CRL 3099. High-resolution 2.8−4.1 μm and 7.8−13 μm spectra suggest that CRL 3099 is a carbon star which is highly reddened by a dense circumstellar shell. Small graphite grains apparently are the primary opacity source for the circumstellar shell over the 4−20 μm spectral region. The authors suggest that CRL 3099 represents an advanced evolutionary stage which is characterized by the formation of large amounts of cool circumstellar dust and by large-amplitude bolometric variations in the central star.

114.524 The ultraviolet spectrum of Alpha Cygni.
R. Barbier, J. P. Swings, A. Delcroix, P. Hornack, J. B. Rogerson.
Astron. Astrophys., Suppl. Ser., Vol. 32, 69 - 81 (1978).
Absorption features in the ultraviolet spectrum of α Cyg were identified on the basis of Copernicus 0.4 Å resolution tracings in the region 1800−2550 Å. A comparison between observations and synthetic spectra (L. T. E. approximation) leads to T_{eff} ~ 8500°K, log g = 2.5. The line identifications indicate that Al II, Al III, Si II, Si III, Ca II, Cr II, Mn II, Fe II, Fe III, Co II, Ni II, Cu II and Zn II are either definitely or probably present. C I, N II, Si I and S II are possibly present. Of the 568 absorption features listed, secure identifications have not been found for 67. Ionized iron atoms in an expanding shell may provide the principal contribution to four percent of the observed features.

114.525 Carbon stars possibly associated with clusters in the Large Magellanic Cloud.   N. Sanduleak.

Astrophys. J., Vol. 221, 586 - 587 (1978).

A list is given of nine carbon stars which fall in or very near clusters in the Large Magellanic Cloud (LMC). Statistical arguments indicate that at least several of these may be physical members. The low frequency of occurrence of carbon stars in LMC open clusters conforms with the apparent rarity of this phenomenon in our own Galaxy.

114.526 The Bowen mechanism in HZ Herculis.
B. Margon, J. Cohen.
Bull. American Astron. Soc., Vol. 9, 560 (1978). – Abstract.

114.527 Contents of the near UV spectra of Arcturus and Procyon.   K. van der Hucht, R. E. Stencel, R. Faraggiana, Y. Kondo.
Bull. American Astron. Soc., Vol. 9, 570 (1978). – Abstract.

114.528 Continuum modeling of ultraviolet rocket spectra (900–3100 Å) for five early-type stars.
G. H. Mount, W. H. Brune, P. D. Feldman.
Bull. American Astron. Soc., Vol. 9, 570 - 571 (1978). Abstract.

114.529 Turbulence in α Boo.   D. F. Gray.
Bull. American Astron. Soc., Vol. 9, 572 (1978). Abstract.

114.530 The ^{12}C/^{13}C ratio of V460 Cygni.
J. F. Dominy, D. L. Lambert, K. H. Hinkle, D. N. B. Hall, S. T. Ridgway.
Bull. American Astron. Soc., Vol. 9, 604 (1978). – Abstract.

114.531 Supermetallicity and the strong-lined binary systems.   D. Deming, D. Butler.
Bull. American Astron. Soc., Vol. 9, 604 (1978). – Abstract.

114.532 The abundance of iron in Mu Leonis.
D. Branch, J. Bonnell, J. Tomkin.
Bull. American Astron. Soc., Vol. 9, 607 (1978). – Abstract.

114.533 Spectrophotometry of the Hα region in the spectrum of HR 1099.   D. A. Fraquelli.
Bull. American Astron. Soc., Vol. 9, 624 (1978). – Abstract.

114.534 Low resolution airborne spectra of Alpha Lyr, Beta Gem, Alpha Boo, Alpha Tau and Alpha Ori in the 1.25 to 4.0 micron spectral region.   G. C. Augason, D. W. Strecker, F. C. Witteborn, E. F. Erickson, W. L. Bailey.
Bull. American Astron. Soc., Vol. 9, 635 (1978). – Abstract.

114.535 Lines in the ultraviolet spectrum of θ Aurigae.
W. P. Bidelman.
Bull. American Astron. Soc., Vol. 9, 635 (1978). – Abstract.

114.536 Spectrum variations in the superluminous star HR 8752.   R. E. Luck, D. L. Lambert.
Bull. American Astron. Soc., Vol. 9, 643 (1978). – Abstract.

114.537 Precision digicon spectroscopy of HDE 226868: the nature of Cygnus X-1.   R. Harms, B. Margon.
Bull. American Astron. Soc., Vol. 9, 645 (1978). – Abstract.

114.538 Direct observations of the circumstellar gas shell of Betelgeuse.   A. P. Bernat, R. K. Honeycutt, J. E. Kephart, D. L. Lambert, C. E. Gow, M. T. Sandford.
Bull. American Astron. Soc., Vol. 9, 650 (1978). – Abstract.

114.539 Circumstellar gas and dust shells of luminous M stars.   W. Hagen.
Bull. American Astron. Soc., Vol. 9, 650 (1978). – Abstract.

114.540 Absolute energy distribution in Rho Cassiopeiae.

S. C. Joshi, B. S. Rautela.
Mon. Not. R. Astron. Soc., Vol. 183, 55 - 59 (1978).

The continuous energy distribution of ρ Cas obtained in the wavelength range 340–710nm has been compared with the energy distribution of δ CMa in the same wavelength range. The circumstellar shell associated with ρ Cas seems to modify the continuum of the star appreciably. Variations in the Balmer emission of ρ Cas taking place within a day are noticed.

114.541 A detailed study of the spectrum of the binary X-ray source HD 153919 [3 U 1700-37]. II. Analysis of the radial velocities in the blue spectral region.
G. Hammerschlag-Hensberge.
Astron. Astrophys., Vol. 64, 399 - 405 (1978).

The results of an analysis of radial velocities of 76 spectrograms in the blue spectral region are presented for the O6f star HD 153919, the optical counterpart of the X-ray source 3 U 1700-37. From these new data, obtained over three years, the author derived a new value for the orbital period: P = 3$\overset{d}{.}$4111 ± 0$\overset{d}{.}$0002. The results for the orbital analysis of all lines together yield: e = 0.16 ± 0.08, K = 19.04 ± 1.27 km s^{-1}, $\bar{\omega}$ = 5° ± 22°. This yields a mass function $f(M)$ = 0.0023 ± 0.0005 $M_\odot$. Apart from velocity variations due to the orbital motion, the velocities of some spectral lines show a strong extra phase dependence. A plausible reason for this observed phase dependent Balmer progression is a longitudinal variation in the density and/or velocity profile of the strong stellar wind of the Of star.

114.542 HD 91805 and the nature of the Bidelman-MacConnell weak-G-band stars.   P. L. Cottrell, J. Norris.
Astrophys. J., Vol. 221, 893 - 906 (1978).

High-dispersion spectra of the weak-G-band star HD 91805 and DDO intermediate-band photometry of a number of the weak-G-band stars of Bidelman and MacConnell have been analyzed, and lead to the following results: (1) For HD 91805, [Fe/H] ~ 0.0, [C/H] ~ −1.4, [N/H] ~ +0.6, and [O/H] ~ +0.1. The simplest explanation of these abundance ratios is that material processed in the CN cycle (but not the NO cycle) has been introduced into the surface layers. (2) The majority of the Bidelman-MacConnell objects are in the same phase of evolution as the Hyades clump stars. They are probably core helium-burning objects, subsequent to the helium flash. (3) Of the mechanisms considered, carbon depletion due to meridional circulation (Paczyński 1973) requires the least radical departure from now well understood astrophysical processes.

114.543 A preliminary determination of the relative abundances of the isotopes of zirconium in R Cygni and V Cancri.   A. C. Zook.
Astrophys. J., Lett., Vol. 221, L113 - L116 (1978).

A measurement of the relative abundances of the isotopes of zirconium in two S-type stars, R Cyg and V Cnc, using the isotopic splitting of the (0, 1) band head of the recently studied $^1\Pi$–$^1\Sigma$ system of ZrO, gives the following preliminary result: ^{90}Zr:^{91}Zr:^{92}Zr:^{93}Zr:^{94}Zr:^{96}Zr = 0.50:0.15:0.20:0.05:0.10:<0.01 in R Cyg, and 0.49:0.12: 0.17:0.05:0.16:<0.01 in V Cnc. These results tend to support the usual belief that s-process nucleosynthesis is currently taking place in the S stars.

114.544 Variable stellar wind from the WN-star HD 151932.   W. Seggewiss.
IAU Colloq. No. 42, (see 012.012), p. 633 - 639 (1977).

The effects of the interaction of strong Wolf-Rayet stellar wind with the surrounding interstellar medium in ring-type nebulae are briefly discussed. The outstanding variability of the envelope of the WN 7 star HD 151932 is explained by small density fluctuations. Variable WR wind may be the reason for the filamentary structure observed in ring-type nebulae.

114.545　Absolute energy distribution in Rho Cassiopeiae.
　　　　S. C. Joshi, B. S. Rautela.
IAU Colloq. No. 42, (see 012.012), p. 640 (1977). – Abstract.

114.546　Two of the most metal-deficient stars.
　　　　M. S. Bessell.
Proc. Astron. Soc. Australia, Vol. 3, 144 - 145 (1977).

114.547　The violet opacity of carbon stars.
　　　　J. D. Bregman, J. N. Bregman.
Astrophys. J., Lett., Vol. 222, L41 - L43 (1978) = Lick Obs. Bull., No. 792.

　　　　Eight carbon stars have been observed with a photoelectric spectrum scanner from 3200 to 6800 Å, and compared with M and K giants. The spectra of six carbon stars increase in brightness shortward of 3900 Å, indicating that the violet opacity in these stars is dominated by C_3, not SiC. Maximum absorber temperatures are computed and are consistent with this conclusion. The data for the two coolest stars are less complete, and the authors are unable to choose between SiC and C_3 as the violet opacity source.

114.548　The generalized Compton effect in the spectrum of Canopus.　　M. Missana.
Astrophys. Space Sci., Vol. 53, 339 - 343 (1978).

　　　　The redshift Δ_c caused by the scattering of photons in the chromosphere of Canopus and in the interstellar matter is obtained from the measurements of wavelength, intensity and equivalent width of 191 spectral lines published in 1942. The result is $\Delta_c = 50 \pm 35$ mÅ with a new radial velocity $V_r = -3.3 \pm 2.4$ km s^{-1}. The reliability of the results is briefly discussed.

114.549　Spectroscopic study of the eclipsing binary V 367 Cygni.　　C. Aydin, M. Hack, N. Yilmaz.
Astrophys. Space Sci., Vol. 53, 345 - 370 (1978).

　　　　Seventeen coudé spectrograms (dispersion 20 Å mm^{-1}) of the β Lyrae-type eclipsing binary V 367 Cygni ($P = 18.6$ d) have been studied. An anomalous behavior for the radial velocities of the spectrograms taken during one cycle (406) was observed; it is suggested that gas eruption under form of prominences may explain it. The spectrum is dominated by shell lines very similar to those present in the spectrum of the supergiant A9 Ia ϵ Aurigae. The underlying stellar spectrum is classified as A5 I.

114.550　The emisssion-line spectrum of MWC 342 in 1974–1975.　　N. Brosch, E. M. Leibowitz, N. Spector.
Astron. Astrophys., Vol. 65, 259 - 263 (1978).

　　　　Identifications and relative intensities of emission lines in the spectrum of MWC 342 are given. The spectrograms of the star in the visible region are dominated by hydrogen and permitted Fe II emission lines. With the exception of [O I] and two [Fe II] lines, no forbidden lines are present in the spectrum. The observed line intensities and the continuum distribution are consistent with a model of a B 3 III star surrounded by a thin envelope of gas and dust, with a gas density of the order of 5×10^{10} cm^{-3}.

114.551　Mg II h and k emission from luminous M stars.
　　　　A. P. Bernat, D. L. Lambert.
Mon. Not. R. Astron. Soc., Vol. 183, 17P– 20P (1978).

　　　　Copernicus observations of the M2 giant β Peg and the M5 supergiant α Her show Mg II h and k emission lines. In α Her, the profiles are similar to those previously reported for α Ori and α Sco; the k line is asymmetric and the h line is symmetric. The possibility is discussed that the k line asymmetry is produced by overlying absorption. In β Peg, both h and k lines are asymmetric. Line formation in an expanding chromosphere is indicated. The new observations deviate from the most recent Wilson–Bappu type relation for Mg II h and k line widths.

114.552　Intermediate and heavy lanthanides in the silicon star HD 192913, and the problem of the lanthanide abundance distributions in peculiar A stars.
C. R. Cowley, H. M. Crosswhite.
Publ. Astron. Soc. Pacific, Vol. 90, 108 - 112 (1978) = Dominion Astrophys. Obs., Contrib. No. 351 = NRC No. 16290.

　　　　The spectra of several intermediate and heavy lanthanides, including Ho III, and Er III have been positively identified in the spectrum of the silicon star HD 192913. Both the identification of these ions and their occurrence in a silicon star are unusual. These results are discussed in terms of an empirical two-component model of the distribution of lanthanide abundances in Ap stars.

114.553　Spectroscopic investigations of the magnetic Ap star HD 9996.　　G. Scholz.
Astron. Nachr., Band 299, 81 - 85 (1978).

　　　　Spectroscopic measurements of the Ap star HD 9996 yielded the radial velocity and the strength of the effective magnetic field. From the reversal of the magnetic field periods smaller than 4.92 years can be excluded. The author cannot state an exact period, but a value of 20.60 years seems to be the most probable one. Further he finds hints at a short period of about 1.8 days for the variations in the radial velocities as well as in the magnetic field strengths. The elements of the binary motion were derived.

114.554　Variations rapides du profil de la raie H$_a$ dans l'étoile T Tauri.　　C. Fehrenbach, Y. Andrillat.
C. R. Acad. Sci. Paris, Tome 286, Sér. B, 187 - 190 (1978).

　　　　The authors have found important variations of the H$_a$ line profile in the star T Tau, on 14 spectra obtained at intervals of 6 or 7 min. This extraordinary fast changes which they show for the first time are difficult to explain with the current theoretical models.

114.555　Objective prism study of the 1977 spectrum of HDE 245770 = A0535 + 26.
G. B. Baratta, R. Viotti, A. Altamore.
Astron. Astrophys., Vol. 65, L21 - L22 (1978).

　　　　Objective prism plates of the optical counterpart of the transient X-ray source A0535 + 26 obtained during 1977 show strong H$_a$ in emission and marked Balmer excess. Many other emission lines are probably present in the red-infrared. No large spectral time variation was found.

114.556　Light-element abundances in the weak G-band star HR 6766.　　C. Sneden, D. L. Lambert, J. Tomkin, R. C. Peterson.
Astrophys. J., Vol. 585 - 594 (1978).

　　　　The authors report a model atmosphere analysis of iron-peak elements, the CNO group, and lithium in HR 6766. Synthetic spectra have been generated to derive the carbon, nitrogen, oxygen, and lithium abundances. Additionally, they have determined the $^{12}C/^{13}C$ ratio from a curve-of-growth analysis. The observations are discussed, and the model atmosphere and abundance analyses are presented. The luminosity and mass of HR 6766 are discussed. Some preliminary comments on other stars of this class are presented, and the implications of this work for stellar evolution discussed.

114.557　A detailed study of the spectrum of the binary X-ray source HD 153919 (3U 1700-37). I. Radial velocity data in the blue spectral region.
G. Hammerschlag-Hensberge, C. de Loore, E. P. J. van den Heuvel.
Astron. Astrophys., Suppl. Ser., Vol. 32, 375 - 377 (1978).

　　　　Radial velocity measurements of selected lines in the

spectrum of the Of star HD 153919 (3U 1700-37) are presented.

114.558 On possible energy stellar standards in the spectral UV region λ1000–3500 Å. A. M. Zvereva.
Izv. Krymskoj Astrofiz. Obs., Vol. 58, 89 - 103 (1978). In Russian.

On the basis of available published data of space observations an analysis of the absolute ultraviolet fluxes from γ Ori, α CMa, α Leo and η UMa within the spectral region λ1000–3500 Å is made. In the spectral region λ > 1500 Å these stars can be considered as available standards of absolute fluxes of energy in the ultraviolet. The stars cannot be used as standards in the spectral region λ < 1500 Å because of synchronous temporal variations in the fluxes of all stars. These variations are correlated with solar activity (Wolf numbers).

114.559 C₃ and infrared spectrophotometry of Y Canum Venaticorum. J. H. Goebel, J. D. Bregman, D. W. Strecker, F. C. Witteborn, E. F. Erickson.
Astrophys. J., Lett., Vol. 222, L129 - L132 (1978).

The 1.2 to 5.6 μm spectrum of the carbon star Y CVn is presented and discussed. Comparison of Y CVn near 5 μm with laboratory spectra provides possible evidence for the presence of the linear triatomic molecule C_3. The authors observe, for the first time in a carbon star, the clearly formed band heads of the CN red system between 1.2 and 2.3 μm. They present corroborative evidence for the presence of the molecules HCN and C_2H_2, and a discussion of the relative contributions of C_3, HCN, and C_2H_2 to the 3.1 μm absorption band. Spectra of two other carbon stars, TX Psc and S Cep, are presented for comparison.

114.560 The expanding envelope of τ Sco (B0 V).
H. J. G. L. M. Lamers, J. B. Rogerson, Jr.
Astron. Astrophys., Vol. 66, 417 - 430 (1978).

The ultraviolet resonance lines of C III, C IV, N III, N V, O VI, Si IV and P V in the Copernicus spectrum of τ Sco (B0 V) show extended violet absorption wings, indicating mass ejection. The observed profiles are compared with predictions by means of a line-fitting method. The mass loss rate is $7.0 \pm 1.6 \times 10^{-9} M_\odot$ yr^{-1}. The expansion velocity increases rapidly outwards to at least about 2000 km s^{-1}. The acceleration of the envelope is due to UV-resonance lines. The observations of τ Sco do not agree with the isothermal coronal wind model, nor with the model for cold radiation-driven stellar winds. They agree qualitatively with the imperfect-flow model.

114.561 On the variability of the Hα hydrogen line in the spectrum of the magnetic variable star β CrB.
Yu. V. Glagolevskij, K. I. Kozlova, N. S. Polosukhina.
Pis'ma Astron. Zh., Vol. 4, 138 - 139 (1978). In Russian.

114.562 Possible optical counterpart of 2S 1702–363.
IAU Circ., Nos. 3221, 3226 (1978).

114.563 LS II +33°5, a new hydrogen-deficient B-type star.
J. S. Drilling.
Astrophys. J., Lett., Vol. 223, L29 (1978) = Contrib. Louisiana State Univ. Obs., Baton Rouge, No. 134.

The star +33°5 in Volume 2 of the Case-Hamburg Luminous Stars catalog has been identified as a hydrogen-deficient star. The spectrum shows strong lines due to He I, O II, C II, N II, Mg II, Si II, and Si III; the Balmer lines are weak or absent. A spectral class near B2 is suggested both by the line spectrum and by the B−V and U−B colors.

114.564 Analysis of spectra of α CMi and α Cen A observed with the orbiting stellar ultraviolet spectrophotometer S59 in ESRO's TD1A satellite.
C. de Jager, T. M. Kamperman, L. Neven.
Astrophys. Space Sci., Vol. 54, 343 - 353 (1978).

Ultraviolet spectra of α CMi and α Cen A taken with moderate spectral resolution (approx. 1.8 Å) are used to analyse whether a determination of stellar chemical abundances of Fe and Cr and of the photospheric parameters is possible.

114.565 Abundances in globular cluster red giants. I. M3 and M13. J. G. Cohen.
Astrophys. J., Vol. 223, 487 - 508 (1978).

A detailed abundance analysis of five red giants in M13 and three in M3 has been performed using line-blanketed model atmospheres. Abundances have been determined for over 20 elements in each star. The mean Fe deficiency is − 1.6 dex for M13 and − 1.8 dex for M3, while M3 is marginally more metal-rich in Mg and Si than M13. Although all elements (with the probable exception of Na) have the same abundance in each of the five M13 stars, considerable scatter exists for Na and Ca and perhaps other elements in the M3 stars. Fe and elements heavier than Fe have the same abundance in each of the three M3 red giants.

114.566 Spectroscopy and possible orbital periods for HDE 245770 (= A0535 + 26). J. B. Hutchings, J. E. Bernard, D. Crampton, A. P. Cowley.
Astrophys. J., Vol. 223, 530 - 536 (1978) = Dominion Astrophys. Obs., Victoria, B.C., Canada, Contrib. No. 356 = NRC No. 16295.

The candidate star for the X-ray transient pulsing source 0535 + 26 is a B0 Ve star, at a probable distance of 1.3 kpc. This implies observed X-ray luminosities of less than or equal to $10^{36.7}$ ergs s^{-1}. Because the optical spectrum contains only weak and possibly variable line features, radial velocity determination is difficult. Radial velocities, derived from a cross-correlation technique, and line intensities were examined for periodicity. Together with X-ray data, these suggest possible periods near 28, 48, or 94 days, with an amplitude ~ 20 km s^{-1}. None of the orbits implied is consistent with an X-ray pulse period change that is due entirely to orbital motion. The implications of these results are discussed, and a probable range of $1.2 < M_X/M_\odot < 2.0$ is derived.

114.567 Infrared spectra of HM Sagittae and V1016 Cygni.
R. C. Puetter, R. W. Russell, B. T. Soifer, S. P. Willner.
Astrophys. J., Lett., Vol. 223, L93 - L95 (1978).

Spectrophotometry of HM Sge from 2 to 13 μm is presented along with 2 to 4 μm spectrophotometry of V1016 Cyg. From 2.5 to 8 μm, the spectrum of HM Sge can be represented by a 950 K blackbody, and a strong silicate emission feature is seen from 8 to 13 μm. Both HM Sge and V1016 Cyg show evidence for CO absorption at 2.3 μm. It is suggested that the infrared radiation from these objects arises from a combination of emission by optically thin dust and by the reddened photosphere of a cool star.

114.568 The ultraviolet spectrum of the manganese stars α And and β Tau. C. Aydin, M. Hack.
Astron. Astrophys., Suppl. Ser., Vol. 33, 27 - 61 (1978).

An identification list of the lines observed in two U2, V2 scans of the Mn star α And, together with central depths, is given for the regions 1030–1500 and 2000–3000 Å. Selected portions of the spectrum of the mild Mn star β Tau are compared with α And. The ultraviolet spectrum of α And is compared with published Copernicus data for ζ Dra (B6 III), α Lyr (A0 V) and with the Ap Cr star ε UMa. Comparison of the observed Ly α and continuous flux of α And and β Tau with the theoretical models by Kurucz et al. (1975) show that the observed spectra cannot be fitted by a single LTE model.

114.569 Spectrophotometric study of the Large Magellanic Cloud emission-line star Hen S 22.
G. Muratorio.
Astron. Astrophys., Suppl. Ser., Vol. 33, 125 - 139 (1978).

Line identifications have been made from 20 and 74 Å mm^{-1} spectra of the Large Magellanic Cloud star HD 34664 (Hen S22). Most of Fehrenbach's identifications (1971) are confirmed and about 160 new lines were measured, 35 of which remain unidentified. The Balmer lines from H8 to H31 have a P Cyg type profile with a mean velocity difference of about 70 km s^{-1} between emission and absorption. [Fe III] is present but He II and Ti II have not been found. The equivalent width, the intensity of the peak above the continuum and the width at half intensity are given for all the lines as well as the radial velocity in the case of those identified. Equivalent width variations are analysed.

114.570   **Wolf-Rayet stars in the Small Magellanic Cloud.**
          J. Breysacher, B. E. Westerlund.
Astron. Astrophys., Vol. 67, 261 - 265 (1978).
          Preliminary results from a spectroscopic study of the four known Wolf-Rayet stars in the Small Magellanic Cloud are presented. The spectral classification of the Wolf-Rayet stars and their components is considered as well as possible spectral variations.

114.571   **Emission lines in the spectrum of Vega.**
          H. L. Johnson, W. Z. Wisniewski.
Publ. Astron. Soc. Pacific, Vol. 90, 139 - 143 (1978).
          The authors have found that the infrared lines of Ca II and O I in the spectrum of Vega have small violet-shifted emission satellites. They suggest, therefore, that Vega has a relatively thin, extended atmosphere, which may be expanding. The measured equivalent widths of the emission satellites range from 0.02 Å to 0.07 Å, averaging about 0.05 Å. It is the very high photometric precision of the spectra produced by the new Michelson spectrophotometer system which makes possible the detection and measurement of such minute details.

114.572   **The helium variable HD 64740 – an X-ray binary?**
          K. Hunger.
Messenger, No. 12, p. 12 - 13 (1978).

114.573   **Spectroscopic observations of 27 CMa from 0.14 to 4.7 microns.**   A. C. Danks, L. Houziaux.
ESO Sci. Prepr. No. 24, 16 pp. (1978). – Submitted to Astronomy and Astrophysics.
          Observations of 27 CMa are presented extending from 0.14 to 4.7 microns. The UV spectrum (0.14 μ to 0.33 μ) was obtained by satellites ANS and TD 1. Stellar atmosphere model fits to these observations give an electron temperature of T_e = 20,000 and a gravity g = 6.300 cm s^{-2}. The visible spectrum exhibits complex line profiles in the Balmer series and the spectral type is determined to be B3 IV. From IR photometry the star is seen to have an IR excess.

114.574   **Spectroscopic investigation of the Ap star 53 Cam.**
          G. Scholz.
Magnetic stars, (see 012.028), p. 10 (1978). – Abstract.

114.575   **Infrared spectrophotometry of carbon stars.**
          J. H. Goebel, D. W. Strecker, F. C. Witteborn,
J. D. Bregman, E. F. Erickson.
Bull. American Astron. Soc., Vol. 10, 407 (1978). – Abstract.

114.576   **IUE (*International Ultraviolet Explorer*) observations of the RS CVn stars HR 1099 and λ And.**
T. R. Ayres, J. L. Linsky, N. D. Morrison.
Bull. American Astron. Soc., Vol. 10, 444 (1978). – Abstract.

114.577   **High resolution IUE observations of α Aur: Is the outer atmosphere of Capella similar to a sunspot?**
T. R. Ayres, J. L. Linsky.
Bull. American Astron. Soc., Vol. 10, 444 - 445 (1978). Abstract.

114.578   **Analysis of the subgiant halo star HD 76932.**
          B. Barbuy.
Astron. Astrophys., Vol. 67, 339 - 344 (1978).
          The chemical composition of the southern high velocity star HD 76932 has been determined by means of a detailed analysis. The atmospheric model with θ_{eff} = 0.86 and log g = 3.5 was calculated by an interpolation in the grid of line-blanketed model atmospheres computed by Peytremann (1974). All the elements are found to be deficient by about the same factor of −1.1 dex with respect to the sun.

114.579   **The near-ultraviolet spectrum of the shell star ς Tau (HD 37202).**   A. M. Hubert-Delplace,
K. A. van der Hucht.
Astron. Astrophys., Vol. 67, 399 - 407 (1978).
          In the 2100–2800 Å wavelength range, the absorption features and the flux values of the shell star ς Tau have been determined and compared with different early type stars. Strong shell features are identified, the composite spectrum of ς Tau looks like the supergiant η CMa (B5 Ia) but the line blocking is higher than that of η CMa. When it is compared to the B2 stars, ς Tau shows a deficiency of about 0.3 mag at λλ2100, 2550, 2800 Å. The flux values of ς Tau and the comparison stars are compared to the predictions of the non-LTE models of Mihalas (1972) and to the predicted fluxes from the blanketing models of Kurucz et al. (1974).

114.580   **Études spectrales de quelques étoiles O–B.**
          N. Yilmaz.
Commun. Fac. Sci. Univ. Ankara, Sér. A$_3$: Astron., Tome 25, 55 - 76 (1977) = Commun. Astron. Dep. Univ. Ankara, No. 79.
          Les spectres de deux étoiles naines HD 202214,HD 19374 et deux étoiles supergéantes HD 152234, HD 152236 des types O et B ont été étudiés et déterminés leurs paramètres atmosphériques et leurs abondances à l'aide des modèles d'atmosphère en état ETL et NETL. Les vitesses radiales ont également été mesurées. Outre que l'atmosphère de l'étoile HD 152236 il est probable que l'atmosphère de la supergéante HD 152234 et la partie extérieure de la naine HD 202214 soient en expansion.

114.581   **The spectrum of CH Cygni in July 1977.**
          G. N. Dzhimshelejshvili, V. A. Oshchepkov,
V. V. Natriashvili.
Astron. Tsirk., No. 979, p. 1 - 2 (1978). In Russian.

114.582   **Discovery of a new star with Hα emission in Sge.**
          O. D. Dokuchaeva.
Astron. Tsirk., No. 983, p. 3 - 4 (1978). In Russian.

114.583   **Possible Hα-emission in the HM Sge pre-outburst spectrum.**   V. P. Arkhipova, O. D. Dokuchaeva,
Yu. A. Shokin.
Astron. Tsirk., No. 983, p. 4 - 5 (1978). In Russian.

          A model for γ Cassiopeiae.
See Abstr. 064.019.

          Stellar model chromospheres. VII. Capella (G5 III+), Pollux (K0 III), and Aldebaran (K5 III).
See Abstr. 064.020.

          Vibrationally excited carbon monoxide in IRC +10216.   See Abstr. 064.022.

          On the chemical composition and age of β Vir.
See Abstr. 064.062.

          Stellar model chromospheres. VIII. 70 Ophiuchi A (K0 V) and ε Eridani (K2 V).   See Abstr. 064.068.

A study of mass loss from the mid-ultraviolet spectrum of α Cygni (A2 Ia), β Orionis (B8 Ia), and η Leonis (A0 Ib). See Abstr. 064.075.

The variable shell star HR 5999 and its environment. See Abstr. 113.010.

AG Draconis, 1975 - 1977. See Abstr. 113.033.

Simultaneous photometry and spectroscopy of the peculiar B[e] star with infrared excess HD 45677. See Abstr. 113.039.

The nature of light variations of the Ap star CQ UMa. See Abstr. 113.079.

Determinacion fotometrica del tipo espectral de la componente desconocida de una estrella binaria eclipsante. Aplicación a la variable eclipsante V 453 Scorpii. See Abstr. 121.076.

Steam in RR Telescopii and Henize 2-38. See Abstr. 122.037.

Is there mass loss in β Canis Majoris stars? See Abstr. 122.093.

An apparent eruptive variable star in Scorpius. See Abstr. 123.021.

The spectral evolution of nova HR Del (1967) during its decline. See Abstr. 124.203.

Spectroscopy of the extreme-ultraviolet source Feige 24: the binary orbit and the mass of the white dwarf. See Abstr. 126.037.

The Herbig-Haro objects in the Gum Nebula. See Abstr. 131.027

High-resolution 2-μ spectroscopy of Cyg OB II No. 12. See Abstr. 131.153.

Circumstellar methane in the infrared spectrum of IRC +10°216. See Abstr. 133.033.

On the nature of the peculiar emission-line object RX Puppis. See Abstr. 133.035.

The spectrum of the Antares nebula. See Abstr. 134.011.

The spectrum of HM Sagittae: a planetary nebula excited by a Wolf-Rayet star. See Abstr. 135.098.

An analysis of the EUV and X-ray spectrum in Capella. See Abstr. 142.130.

The soft X-ray spectrum of Capella: discovery of intense line emission. See Abstr. 142.185.

Ultraviolet spectroscopy of HZ Herculis with the IUE satellite. See Abstr. 142.217.

Ultraviolet observations of X-ray sources with IUE. See Abstr. 142.218.

Near-infrared luminosity-sensitive features in M dwarfs and giants, and in M31 and M32. See Abstr. 158.115.

Errata

114.901 Erratum: 'Absolute energy distributions of α Lyrae and 109 Virginis from 3295 Å to 9040 Å' [Astron. Astrophys., Vol. 61, 679 - 684 (1977)]. H. Tüg, N. M. White, G. W. Lockwood. Astron. Astrophys., Vol. 66, 469 (1978).

115 Luminosities, Masses, Diameters, HR-Diagrams and Others

115.001   Masses of hot main-sequence stars.
D. M. Popper.
Astrophys. J., Lett., Vol. 220, L11 - L14 (1978).
Masses of some hot main-sequence stars are reexamined on the basis of preliminary reduction of new spectrographic material. Downward revisions are required of the published values for the most massive known main-sequence system, V382 Cyg, while published values for V453 Cyg and V478 Cyg require little revision. The large masses published for EO Aur are illusory, since the lines of the components are un-resolved. The more massive, hotter component of U Oph is also the more luminous, despite difficulties in interpreting some photometric observations.

115.002   Des milliards de réacteurs nucléaires: les étoiles.
M. Golay.
Conférences d'astronomie de l'Observatoire, (see 012.004), p. 73 - 101 (1977).
Les apparences; Distance des étoiles proches; Couleur des étoiles et température; Composition chimique des étoiles; Les masses des étoiles et les étoiles doubles; Les diamètres stellaires; Les densités moyennes de la matière stellaire; Un diagramme fondamental de l'astrophysique.

115.003   A new look at the stars.
R. Hanbury Brown.
Focus on the stars, (see 003.002), p. 145 - 185 (1977).

115.004   Intrinsic lines in the $(B-V)$, $(V-I_{KC})$ diagram.
A. W. J. Cousins.
Observatory, Vol. 98, 54 - 56 (1978).
The relationship between $(B-V)$ and $(V-I_{KC})$ for different luminosity classes is discussed, with special reference to supergiants.

115.005   Masses of red giants. I. Mean initial mass from visual binary data.
J. M. Scalo, J. F. Dominy, W. A. Pumphrey.
Astrophys. J., Vol. 221, 616 - 626 (1978).
The authors present an independent method for the determination of the frequency distribution of the initial masses of red giants based on an analysis of visual binary systems containing a red giant and a main-sequence star. The apparent frequency distribution of the spectral types of the secondaries is converted to a true distribution of secondary masses by accounting for certain selection effects. This distribution is then transformed into a distribution of initial masses of red giant components by assuming a form for the distribution of mass ratios. The result is $0.8\,M_\odot < \langle M_{RG} \rangle < 1.2\,M_\odot$, in satisfactory agreement with the theoretical prediction. No evidence for a significant variation of initial mass with red giant spectral type was found. Implications of this result for the interpretation of lithium abundances and $^{12}C/^{13}C$ ratios in red giants are discussed.

115.006   The stellar birthrate and initial mass function of stars.   G. E. Miller, J. M. Scalo.
Bull. American Astron. Soc., Vol. 9, 566 (1978). — Abstract.

115.007   Stellar disk diameter measurements by amplitude interferometry.
K. M. Liewer, R. H. Braunstein, D. G. Currie, S. L. Knapp.
Bull. American Astron. Soc., Vol. 9, 598 (1978). — Abstract.

115.008   Masses of peculiar red giants.
J. Scalo, G. Miller.
Bull. American Astron. Soc., Vol. 9, 603 (1978). — Abstract.

115.009   Potentialities and problems of the visual surface brightness relation.   D. S. Evans.
Bull. American Astron. Soc., Vol. 9, 643 - 644 (1978).
Abstract.

115.010   The distances of nearby cool carbon stars.
J. Bergeat, F. Sibille, M. Lunel.
Astron. Astrophys., Vol. 64, 423 - 431 (1978).
Distance ratios are provided for 38 cool carbon stars on the basis of a previous study (Bergeat et al., 1976). The validation of this distance scale is obtained through an analysis of stellar velocities. A relationship is established between proper motions and the distance scale. Luminosities and radii are derived for cool carbon stars which permit a discussion of their evolutionary status. Finally, evaluations are given for the rate of mass ejection corresponding to large graphite grains.

115.011   On the statistical parallax of O type runaway stars.   L. Carrasco, M. Crézé.
Astron. Astrophys., Vol. 65, 279 - 280 (1978).
The hypothesis about the possible underluminosity of the O-type runaway stars is tested by applying the statistical parallax method to a sample of 19 stars. It is found that indeed these stars are underluminous on the mean by 2.7 ± 0.5 magnitudes with respect to low velocity normal O stars.

115.012   The position of central stars in the Hertzsprung-Russell diagram.
S. R. Pottasch, P. R. Wesselius, C.-C. Wu, R. J. Van Duinen.
IAU Symp. No. 76, (see 012.014), p. 211 (1978). — Abstract.

115.013   The H-R diagram as an astronomical tool.
A. G. D. Philip, L. C. Green.
Sky Telesc., Vol. 18, 395 - 398 (1978).

115.014   The semi-period—luminosity—colour relation for supergiant stars.   G. Burki.
Astron. Astrophys., Vol. 65, 357 - 362 (1978).
The semi-period—luminosity—colour relation for late B- to G-type supergiants, first described by Maeder and Rufener (1972), is determined by using all the supergiants with known semi-period of variation. The parallelism between the period vs. M_v relation of the classical cepheids and the semi-period vs. M_v relation of the supergiants of the same spectral type is confirmed.

115.015   A test of absolute magnitude calibrations for early-type stars.   D. Egret.
Astron. Astrophys., Vol. 66, 275 - 281 (1978).
Absolute magnitudes have been computed for stars with spectral type A 0 and earlier using $H\beta$ and MK calibrations separately. The results are compared and standard deviations are given in each case.

115.016   Absolute luminosity calibration of F stars.
A. Heck.
Astron. Astrophys., Vol. 66, 335 - 342 (1978).
The author deals with luminosity calibrations of F-type stars by a statistical parallax method based on the principle of maximum likelihood.

115.017   Luminosity and carbon enhancement in N-type carbon stars.   H. R. Johnson.
Astrophys. J., Vol. 223, 238 - 243 (1978) = Publ. Goethe Link Obs., Indiana Univ., Bloomington, No. 192.
Recent observational data indicate the likelihood of a

relation between luminosity and carbon enhancement in
N-type, irregularly variable carbon stars in the sense that the
more luminous carbon stars appear to have smaller enhancements of carbon. No relation appears between carbon
enhancement and red colors. The observational data are the
luminosities of Peery, the colors and CN band indices of
Baumert, the CO indices of Faÿ and Ridgway, and the C_2
indices of Gow.

115.018 Notes on the H-R diagram.   A. G. D. Philip.
Sky Telesc., Vol. 56, 20 - 21 (1978).

115.019 Spectrophotometric determination of the luminosity
of peculiar stars identified on objective prism plates.
Ts. S. Radoslavova.
Soobshch. AN GruzSSR, Vol. 88, No. 1, p. 61 - 64 (1977). In
Russian. — Abstr. in Ref. zh. 51. Astron., 7.51.532 (1978).

Steps toward the Hertzsprung-Russell diagram.
See Abstr. 004.015.

Influence of heavy element abundance in stellar
atmospheres of spectral classes A—F upon the $(B-V, T_e)$
relation.   See Abstr. 064.072.

The occultation of v Leonis.   See Abstr. 096.008.

Proper motion survey with the 48-inch Schmidt
telescope. LI. Hertzsprung diagrams for one hundred and
fifteen thousand proper motion stars.   See Abstr. 112.012.

A check of the absolute luminosity calibrations
of the $uvby\beta$ photometric system by means of small
trigonometric parallaxes.   See Abstr. 113.045.

Stellar angular diameters and visual surface brightness — III. An improved definition of the relationship.
See Abstr. 113.046.

A determination of some fundamental parameters of
α Leo.   See Abstr. 116.022.

A spectroscopic reinvestigation of the massive
binary HD 698.   See Abstr. 119.016.

Old disk subdwarfs.   See Abstr. 126.010.

The progenitor masses and the luminosity function
of white dwarfs.   See Abstr. 126.035.

The effect of stellar mass on kinematical age
determination.   See Abstr. 151.116.

The blue horizontal-branch stars of NGC 6752.
See Abstr. 154.019.

Chemical properties and age of the components of
the galactic halo.   See Abstr. 155.050.

Luminous red stars as extragalactic distance
indicators.   See Abstr. 158.069.

The extragalactic distance scale. I. A review of
distance indicators: zero points and errors of primary indicators.   See Abstr. 162.150.

Errata

115.901 Addendum:"Radii of nearby stars: an application
of the Barnes-Evans relation" [Astrophys. J., Suppl.
Ser., Vol. 34, 479 - 492 (1977)].   C. H. Lacy.
Astrophys. J., Suppl. Ser., Vol. 36, 621 (1978).

116 Magnetic Fields, Figure, Rotation, Radio Radiation

116.001 **The nature of Beta Lyrae and its radio emission.**
  R. F. Jameson, A. R. King.
Astron. Astrophys., Vol. 63, 285 - 287 (1978).
  The authors have modelled the H II region surrounding
β Lyr which is responsible for its radio emission. The region
has radius 50 AU, temperature $10^{4\,\circ}$ K and electron density
$10^7\,cm^{-3}$. 237 $L_\odot$ of Lyman continuum radiation are needed
to excite the region. To provide this they conclude the secondary is a B2.5V or a B5II star; it cannot be a black hole.

116.002 **High-resolution polarization observations inside the**
  spectral lines of magnetic Ap stars. II. Observations
of α^2 Canum Venaticorum.   E. F. Borra, A. H. Vaughan.
Astrophys. J., Vol. 220, 924 - 930 (1978).
  The authors present observations, obtained with a photon-
counting polarimeter and a Fabry-Perot interferometer, of the
circular polarization and intensity profiles across the line Fe II
$\lambda4520.2$ in the magnetic Ap star α^2 Canum Venaticorum. They
discuss briefly the H_e curve obtained from their data.

116.003 **The radio source associated with the G-type super-**
  giant HR 8752.
L. A. Higgs, P. A. Feldman, J. Smoliński.
Astrophys. J., Lett., Vol. 220, L109 - L112 (1978).
  Radio observations are reported for a source detected in
an earlier survey of late-type supergiants which confirm the
radio-star identification of HR 8752 (G0 Ia) and yield a spectrum which is characteristic of thermal emission from a
circumstellar envelope.

116.004 **On the intrinsic rotation of magnetic variables.**
  R. Rajamohan, G. S. D. Babu.
Mon. Not. R. Astron. Soc., Vol. 182, 773 - 776 (1978).
  The intrinsic rotational velocities of magnetic variables
with periods greater than a day are found to be related to their
colour index or equivalently to their masses. It is also found
that the components of spectroscopic binaries rotate slower
than the single stars of similar colours.

116.005 **Radio emission from ζ Puppis and γ² Velorum at**
  two centimetres.   D. C. Morton, A. E. Wright.
Mon. Not. R. Astron. Soc., Vol. 182, 47P - 51P (1978).
  Radio fluxes of 6.7 ± 1.5 and 67 ± 5 mJy at 14.7 GHz
have been detected from the stars ζ Pup and γ²Vel respectively
with the 65-m telescope at Parkes. This is the first detection of
ζ Pup, while that of γ²Vel is consistent with previous measurements at lower frequencies.

116.006 **A sensitive 5 GHz survey of single G and K sub-**
  giants.   D. M. Gibson.
Bull. American Astron. Soc., Vol. 9, 600 (1978). − Abstract.

116.007 **Radio emission from close binary systems and novae.**
  R. M. Hjellming.
IAU Colloq. No. 42, (see 012.012), p. 279 - 291 (1977).
  This review of the radio emission properties of close
binary systems and novae is partly concerned with surveying
the star systems that exhibit continuum radio emission, and
partly concerned with discussing the implications of the observed radio emission. The phenomena encountered range from
purely thermal continuum radio emission for the case of
ionized nova shells to strong, non-thermal continuum radio
emission produced by relativistic particles in both 'normal'
and X-ray emitting close binary systems.

116.008 **Is the close binary XY Ursae Majoris a radio star?**
  E. H. Geyer.
IAU Colloq. No. 42, (see 012.012), p. 292 - 300 (1977).

  The optical position of the short period eclipsing binary
XY UMa, which shows large scale star spot activities, coincides
within 2 minutes of arc with a radio source, listed in the 4C-,
4CP-, OK-catalogues, and the position obtained with the
Effelsberg radio telescope. If the radio source is identical with
the binary, it would imply important consequences about the
extragalactic nature of high latitude 4C-sources, the radio
background and the high radio luminosity of the binary.

116.009 **The magnetic field geometry of HD 215441.**
  E. F. Borra, J. D. Landstreet.
Astrophys. J., Vol. 222, 226 - 233 (1978).
  The effective field of Babcock's star, HD 215441, has
been measured as a function of phase using a photoelectric
Balmer-line Zeeman analyzer. The effective and mean surface
field curves can be simultaneously fitted in an approximate
manner with a decentered dipole magnetic field distribution
having the field axis at $30°-35°$ to the rotation axis. It is
shown that the "symmetric rotator" model (a combination of
dipole and quadrupole fields having reflection symmetry or
antisymmetry through the stellar equatorial plane) proposed
by Krause and Oetken as an alternative to the "oblique
rotator" does not give an acceptable description of the field
of HD 215441.

116.010 **Angular velocity distribution in a star which is a**
  component of a close binary system and has a fast
rotating core.   V. G. Gorbatskij.
Astrofizika, Vol. 13, 485 - 492 (1977). In Russian. English
translation in Astrophysics, Vol. 13, No. 3.
  There are reasons to suppose the presence of a fast
rotating core inside some stars, in particular on the red giant
stage. In case the star is the component of a close binary with
nonsynchronous rotation and has an outer convective zone
there must be a sink of angular momentum due to dynamical
tides. A simplified model is considered having a source of
angular momentum in the centre and a sink on the outer
border. Stationary distribution of angular velocity is found
for this model on the assumption that the dynamical viscosity
coefficient is a function of temperature. The calculations
have shown that the temperature distribution is only slightly
affected by rotation. The motion is stable only in cases when
the viscosity coefficient diminishes with the distance from
centre.

116.011 **The magnetic deformation of β CrB.**   M. J. Stift.
  Mon. Not. R. Astron. Soc., Vol. 183, 443 - 444
(1978).
  Arguments are advanced in favour of the precession model for βCrB. It is shown that the apparent contradiction between this model and current theories of magnetic main-sequence stars might possibly be overcome.

116.012 **A survey of nine Beta Cephei stars for magnetic**
  fields.   R. J. Rudy, J. C. Kemp.
Mon. Not. R. Astron. Soc., Vol. 183, 595 - 603 (1978).
  Nine β Cephei variables have been observed for the presence of magnetic fields by measuring the circular polarization in the wings of the Balmer lines. Of the 80 nightly measurements on the nine stars, only one, on β Cep itself, exceeded
the 3 σ level of significance.

116.013 **Southern search for OH from M supergiants.**
  P. F. Bowers, F. J. Kerr.
Astron. J., Vol. 83, 487 - 491 (1978).
  Results are reported of a search for OH emission from
about 50 southern M supergiants, and it is concluded that
most optically identified M supergiants do not have substantial

emission at 1612 MHz. From examination of the published OH profiles of unidentified Type II OH/IR stars, it is also concluded that only a small percentage (~10%) of these stars have spectra similar to those of known OH supergiants, suggesting that Type II OH supergiants are an extremely rare class of object. A list of unidentified sources which may be supergiants is given.

116.014   Interferometric observations of weak radio flares from a red dwarf star.   R. J. Davis,
B. Lovell, H. P. Palmer, R. E. Spencer.
Nature, Vol. 273, 644 - 645 (1978).

Measurement of the radio emission associated with flares on M-type red dwarf stars has been difficult because of the sporadic and transient nature of the phenomena, and the danger of confusion with terrestrial sources of emission. The authors have used a long baseline interferometer for such measurements, and report here observations which yielded clear records of weak radio flares from the star YZ CMi.

116.015   Is HD26676 an unusual radio star?
D. M. Popper, B. Margon.
Nature, Vol. 274, 137 (1978).

Strom and Harris (1977) have detected a faint radio source whose position is coincident, to within an error region of 3 X 18 arc s, with HD26676. As they stress, this association would be remarkable, as all previously identified radio stars are radio and optical variables and/or binaries. The authors report here that no such evidence exists for HD26676 or the radio source.

116.016   Interpretation of magnetic field measurements of Ap stars.   L. Oetken.
Magnetic stars, (see 012.028), p. 8 (1978). – Abstract.

116.017   Measurements of the effective magnetic field strength of 52 Her.   E. Gerth.
Magnetic stars, (see 012.028), p. 9 (1978). –Abstract.

116.018   Spectroscopic investigation of the magnetic star Epsilon UMa.   S. Hubrig.
Magnetic stars, (see 012.028), p. 11 (1978). – Abstract.

116.019   The short time variability of magnetic stars.
W. Schöneich.
Magnetic stars, (see 012.028), p. 16 - 18 (1978). – Abstract.

116.020   Intrinsic polarization of the magnetic variable HD 71866.   I. Tuominen, V. Piirola.
Magnetic stars, (see 012.028), p. 25 - 27 (1978). – Abstract.

116.021   Remarks on non-axisymmetric magnetic fields in presence of a differential rotation.   K.-H. Rädler.
Magnetic stars, (see 012.028), p. 40 (1978). – Abstract.

116.022   A determination of some fundamental parameters of α Leo.   G. Sonneborn, G. W. Collins II.
Bull. American Astron. Soc., Vol. 10, 410 (1978). – Abstract.

116.023   A possible alternative method for determining rotational velocities in early type stars.
T. J. Kuzma.
Bull. American Astron. Soc., Vol. 10, 413 (1978). – Abstract.

116.024   Ap stars.   C. Megessier.
Mem. Soc. Astron. Italiana, Vol. 47, (see 012.036), 632 (1976). – Summary.

116.025   Stellar magnetism as a decisive factor of flare activity and of evolution of UV Ceti- and T Tauri-type variable stars.   R. E. Gershberg.

Flare stars. Symposium 1976, (see 012.035), p. 181 - 197. In Russian. – Abstr. in Ref. zh.,51. Astron., 7.51.544 (1978).

Fourier analysis of Zeeman splitting for integrated stellar profiles.   See Abstr. 031.215.

Nonlinear astrophysical dynamos: bifurcation of steady dynamos from oscillating dynamos.
See Abstr. 062.049.

Magnetodynamics mechanism of magnetic stars.
See Abstr. 062.103.

Secular instability of rotating Newtonian stars.
See Abstr. 065.051.

On the existence of ergoregions in rotating stars.
See Abstr. 066.001.

Why do collapsed stars rotate so slowly?
See Abstr. 066.062.

Photometry and polarimetry of AM Herculis.
See Abstr. 113.057.

The measurement of rotation and macroturbulence in O-type stars using Fourier transform analysis.
See Abstr. 114.019.

The lithium abundance in weak-G band stars.
See Abstr. 114.061.

Peculiarities of peculiar stars. XII, XIII.
See Abstr. 114.064.

Spectroscopic investigations of the magnetic Ap star HD 9996.   See Abstr. 114.553.

On the variability of the Hα hydrogen line in the spectrum of the magnetic variable star β CrB.
See Abstr. 114.561.

Large radio flares in RS CVn binaries.
See Abstr. 117.079.

A search for centimetric wavelength emission from UV Ceti stars.   See Abstr. 122.109.

On the adiabatic pulsations of tidally perturbed rotating stars.   See Abstr. 122.120.

Rotating variable stars.   See Abstr. 122.162.

Stellar magnetism as a decisive factor of flare activity and of the evolution of UV Ceti- and T Tauri-type variable stars.   See Abstr. 122.166.

Photoelectric observations of the Ap star HD 34452 (IQ Aur) in ten spectral regions.   See Abstr. 122.170.

Some remarks on the influence of rotation of stars on flare activity.   See Abstr. 122.208.

On the light variability of the magnetic star HD 192678.   See Abstr. 123.025.

On the rotating magnetic white dwarfs.
See Abstr. 126.024.

The Cygnus X region. X: The riddle of the Gamma Cygni radio source resolved.   See Abstr. 141.012.

117 Binary and Multiple Stars, Planetary Companions, Theory

117.001 On a basic objection to the contact binary model of Shu, Lubow and Anderson.
J. Hazlehurst, S. Refsdal.
Astron. Astrophys., Vol. 62, L9 - L11 (1978).

A basic objection is raised against the contact binary model of Shu, Lubow and Anderson (1976). The temperature (entropy) discontinuity across the Roche lobe of the secondary will in a rather short time (~thermal time scale of the secondary's outer convective zone) disappear, due to the heating up of the underlying regions.

117.002 The influence of the nonsphericity and of the radiation pressure force on the moment of inertia of a rigidly rotating star and on the stability criterion for binaries.
D. Vanbeveren.
Astron. Astrophys., Vol. 62, 59 - 63 (1978).

It is shown that the moment of inertia of a star deformed by rotation and/or the attraction of the other star in a binary system differs no more than 65% from that of a spherical star of similar mass distribution. Taking into account the influence of the radiation pressure force, this difference decreases with increasing force. The influence of the deformation and of the radiation pressure force on the stability criterion for binaries without mass loss is considered.

117.003 Motion of two rigid bodies under the gravitational influence of each other.   K. B. Bhatnagar.
Astron. Astrophys., Vol. 62, 217 - 221 (1978).

The motion of two rigid bodies under the mutual gravitational attraction has been studied under the conditions: (1) The inertia ellipsoids of A and B for their respective mass centres $A*$ and $B*$ are general ellipsoids; (2) either the distance between $A*$ and $B*$ is considerably greater than the greatest dimension of either body, or the ellipticities of the inertia ellipsoids of A and B are small; (3) the angular velocities of A and B in an inertial reference frame R, are initially parallel to one of the principal axes for both the bodies and (4) the mass-centres $A*$ and $B*$ move in a plane whose orientation is fixed in R. These results can be applied to binary star systems under the given restrictions.

117.004 Roche-lobe overflow in X-ray binaries.
G. J. Savonije.
Astron. Astrophys., Vol. 62, 317 - 338 (1978).

It is examined whether Roche-lobe overflow can be the main mechanism of mass transfer that powers the low-mass as well as the massive X-ray binaries. Detailed numerical computations of the initial phase of Roche-lobe overflow were performed in order to determine the precise time development of mass transfer from normal stars with masses ranging from $1.5 \, M_\odot$ up to $20 \, M_\odot$ to compact companions with masses of $1 \, M_\odot$ and $1.5 \, M_\odot$. It appears that if in massive X-ray binaries $(M_1 \sim 16 \, M_\odot)$ the mass transfer starts when the primary star is still burning hydrogen in its core, it may take some 10^4 yr before $\dot{M}$ exceeds the critical accretion rate of the neutron star. For Her X-1 and Cen X-3 X-ray lifetimes of the order of 10^5 and 10^4 yr are predicted, respectively.

117.005 Orbital angular momentum loss via gravitational radiation and mass-transfer rates in close binary systems.   W. Y. Chau.
Astrophys. J., Vol. 219, 1038 - 1042 (1978).

The relationship between period changes and mass-transfer rates for close binary systems is discussed, including orbital angular momentum losses. In systems where gravitational radiation is the dominant loss mechanism, a simple criterion for evaluating the effects as nonnegligible can be formulated, and limits on the mass-transfer rate can also be set based just on the sign of the period change. The system DQ Herculis is used as an example for these considerations.

117.006 Stellar parameters of five early type companions of X-ray sources.   P. S. Conti.
Astron. Astrophys., Vol. 63, 225 - 235 (1978).

Using the spectral type to estimate the effective temperatures, and the eclipse duration to estimate the stellar radii, the author derives the positions of the five companions of these X-ray sources in an HR diagram. From masses derived from the orbital parameters it then appears that all the companions are overluminous. All five systems are at or close to their Roche or tidal lobes, consistent with their corotation or noncorotation. Under the assumption that the X-ray source is powered by accretion from a stellar wind, the necessary velocity is derived. The accretion velocity could be brought into agreement with the predicted values if the efficiency factor ρ, for converting potential energy into X-radiation, were of order 0.01. Both a stellar wind and an overflow of material from a Roche or tidal lobe are involved in the present evolutionary state of all five systems.

117.007 Runaway stars as witnesses to supernova explosions.   R. K. Kochhar.
Nature, Vol. 271, 527 - 528 (1978).

The hypothesis that single runaway stars are released from close binaries in which the companion star explodes as a carbon detonation supernova and is totally disrupted is proposed.

117.008 Theoretical *UBV* colours of accretion discs in cataclysmic variables.
A. Schwarzenberg-Czerny, M. Różyczka.
Acta Astron., Vol. 27, 429 - 436 (1977).

The radiation of model stellar atmospheres with appropriate temperatures and gravities is integrated over the surface of the stationary and optically thick accretion disc. The *UBV* colours of the disc are obtained and compared with observations of some close binary systems containing luminous discs. A rough estimate of the amounts of matter flowing through the disc appears to be possible. However, the agreement between the observed and computed colours is not very good. The necessity of taking into account the emission from the optically thin parts of the disc seems to be inevitable.

117.009 Evolution in binary systems.   W. K. Hartmann.
Astronomy, Vol. 5, No. 9, p. 10 - 11 (1977).
Abstr. in Phys. Abstr., Vol. 80, Abstr. 91571 (1977).

117.010 Stars with companions: more than meets the eye.   W. K. Hartmann.
Astronomy, Vol. 5, No. 9, p. 6 - 9, 12 - 14 (1977). – Abstr. in Phys. Abstr., Vol. 80, Abstr. 91572 (1977).

117.011 Apsidal motion of rings in binary systems.   K. G. Castle.
Publ. Astron. Soc. Pacific, Vol. 89, 862 - 873 (1977/78).

The rates of change of semimajor axis, eccentricity, and orientation have been computed for a variety of elliptic rings in binary systems. The rings were allowed to surround either one or both stars. Thin-ring and thick-disk configurations were found with rates of apsidal rotation appropriate to the rate of change of the relative strengths of the wings of the Balmer lines (V/R) in Be stars. These disks can produce a radial-velocity-excitation gradient for the shell lines of either sign. A very narrow range of thin rings was found with rates of apsidal rotation which may be appropriate for the distortion wave in RS CVn systems.

117.012 **W92: a possible gravitationally contracting binary system in NGC 2264.** M. F. Walker.
Publ. Astron. Soc. Pacific, Vol. 89, 874 - 881 (1977/78) = Lick Obs. Bull., No. 780.

Periodic light variations have been detected in the star W92 in NGC 2264. The form of the light curve is unusual, being asymmetric with a slow rise from minimum to a maximum at $0.7P$, followed by a rapid decline to the succeeding minimum. The amplitude of the variation ranges from $0^{m}04$ to $0^{m}10$ with little or no change in color from yellow to ultraviolet. While present data are inadequate to determine whether W92 is a member of NGC 2264, they suggest that the system could be an extremely young, gravitationally contracting binary.

117.013 **Evolution of the α Centauri system.**
B. P. Flannery, T. R. Ayres.
Astrophys. J., Vol. 221, 175 - 185 (1978).

The authors discuss observational material available in the literature relevant to the evolutionary status of α Cen; they describe theoretical evolutionary sequences appropriate for the A and B components for two values of metallicity $Z = 0.02$ and 0.04, and a reference solar model ($Z = 0.02$); they present a partial reanalysis of the observed abundances, based on the equivalent measurements of French and Powell (1971), which provide a consistency check on the evolutionary models; finally they discuss and apply their results in a more general context.

117.014 **A polarimetric determination of binary inclinations: results for five systems.**
R. J. Rudy, J. C. Kemp.
Astrophys. J., Vol. 221, 200 - 210 (1978).

A method is presented for estimating the orbital inclinations of binary systems from phase-locked polarization variations when those variations arise from certain single-scattering processes, principally scattering by optically thin regions of extrastellar material. In the lowest-order approximation the variable polarization, over an orbital period, twice traces out an ellipse in polarization space. The eccentricity of this ellipse, is related to the orbital inclination by $\epsilon = (\sin^2 i)/(1 + \cos^2 i)$. The method is applied to five binaries with known inclinations: AO Cassiopeiae and u Herculis, for which observations have already been published; and Algol, U Sagittae, and V444 Cygni, for which the authors present their observations for the first time.

117.015 **The unseen companion to the red dwarf CC 20,986.**
S. L. Lippincott, E. Borgman.
Bull. American Astron. Soc., Vol. 9, 598 (1978). − Abstract.

117.016 **IW Persei, an Am: ellipsoidal variable with a possible white dwarf companion.** D. W. Kurtz.
Bull. American Astron. Soc., Vol. 9, 623 (1978). − Abstract.

117.017 **Changing light curves of CG Cygni.** K. G. Castle, E. F. Milone, R. M. Robb, D. Swadron.
Bull. American Astron. Soc., Vol. 9, 623 - 624 (1978). Abstract.

117.018 **Non-stellar material in U Cephei.**
R. C. Crawford.
Bull. American Astron. Soc., Vol. 9, 624 - 625 (1978). Abstract.

117.019 **Computer simulations of speckle interferometry of binary stars in the photon-counting mode.**
J. C. Dainty.
Mon. Not. R. Astron. Soc., Vol. 183, 223 - 236 (1978).

Vector autocorrelation calculations have been carried out on photon-limited computer-simulated images of binary stars. The variation of signal, background and noise with the number of pictures, the average number of detected photons per picture, the average number of additive noise photons per picture and the magnitude difference of the binary has been studied and shown to agree with a simple theory. The calculations indicate that binary stars as faint as approximately 18 mag may be resolved to the diffraction limit in 35 min of observing time on a large telescope.

117.020 **Binaries among B2−B5 IV, V absorption and emission stars.** H. A. Abt, S. G. Levy.
Astrophys. J., Suppl. Ser., Vol. 36, 241 - 258 (1978).

Samples of 42 B2−B5 IV, V stars and 21 Be stars with similar types have been studied for duplicity on the basis of 20 coudé spectra of each. Orbital elements are given for 10 newly discovered SB1's; published or improved elements are quoted for seven other spectroscopic binaries, three visual doubles, and 11 common-proper-motion pairs. It is apparent that there are two types of binaries represented among the B stars. Those with periods shorter than about 10 years have secondaries that do not fit the van Rhijn function. The authors conclude that two-thirds of the primaries have stellar-type secondaries and perhaps the remainder have black-dwarf or planetary companions. Among the systems with periods greater than 10 years, the distant companions fit the van Rhijn function. At least one-quarter of the primaries have distant companions, and those companions may themselves be bifurcation doubles.

117.021 **Observational evidence for mass exchange in close binary systems.** H. Drechsel, J. Rahe, G. Wolfschmidt, Y. Kondo, G. E. McCluskey, Jr.
IAU Colloq. No. 42, (see 012.012), p. 371 - 382 (1977).

117.022 **Young W UMa binaries and the interaction with their environment.** F. Van't Veer.
IAU Colloq. No. 42, (see 012.012), p. 388 - 392 (1977).

It is argued that late type W UMa binaries undergo mass ejection, mainly through the second Lagrangian point, which causes a steady decrease of their mass ratio m_2/m_1. This makes it reasonable to believe that the W UMa stage is a state of transition between T Tau and normal main sequence stars.

117.023 **Period changes of W Ursae Majoris stars.**
J. M. Kreiner.
IAU Colloq. No. 42, (see 012.012), p. 393 - 397 (1977).

117.024 **The binary frequency of the OBN and OBC stars.**
C. T. Bolton, G. L. Rogers.
Astrophys. J., Vol. 222, 234 - 245 (1978).

New radial velocity measurements for the OBCN stars have been combined with published data in order to determine the binary frequency of the OBN and OBC stars. At least 50%, and possibly 100%, of the OBN stars are members of short-period binaries. None of the OBC stars are members of short-period binaries. The data indicate that the OBCN stars have approximately normal masses and luminosities for their spectral types. There is weak evidence that the OBN and OBC stars are kinematically distinct groups. The implications of these results for suggested mechanisms for producing OBN and OBC stars are discussed.

117.025 **Improved mean time of disintegration of non-rotating triple systems.** Zh. P. Anosova.
Vestn. Leningr. univ., 1977, No. 13, p. 158 - 161. In Russian. Abstr. in Ref. zh., 51. Astron., 2.51.644 (1978).

117.026 **Binary nuclei of planetary nebulae.** A. Acker.
IAU Symp. No. 76, (see 012.014), p. 209 (1978).
Abstract.

117.027 **Project Daedalus: the evidence for planetary companions of Barnard's star.** A. R. Martin.
J. British Interplanet. Soc., Suppl., (see 003.003), p. S19 - S23

(1978).

117.028 Project Daedalus: an analysis of the photometric data on Barnard's star, and its implications.
A. R. Martin, A. Bond.
J. British Interplanet. Soc., Suppl., (see 003.003), p. S24 - S32 (1978).

117.029 The structure of close binaries.
I. W. Roxburgh, P. S. Williams.
Astrophys. Space Sci., Vol. 54, 199 - 209.(1978).

A method for calculating the structure of a close binary component is presented. It is seen that the effect of binary distortion is to shift the zero age main sequence to the right. Attempts to construct contact systems with these models confirm the results of earlier workers that this is not possible.

117.030 Compagnons invisibles dans les systèmes d'étoiles doubles.   P. Baize.
Astronomie, Vol. 92, 211 - 225 (1978).

117.031 On the formation of planets in binary star systems.
T. A. Heppenheimer.
Astron. Astrophys., Vol. 65, 421 - 426 (1978).

This paper presents a theory for the existence of regions wherein planets may grow in binary star systems. The numerical results are insensitive to assumed model parameters, particularly with respect to nebular density. The criterion for growth by accretion is that planetesimal collision velocities do not exceed a critical value, under the joint influences of nebular drag and of companion-star secular perturbations. It is found that even large orbital semimajor axes ($\sim$ 50 AU) do not ordinarily permit planet growth. Planet growth appears possible only when the companion is very small or its orbit nearly circular, as in the binary system Sun-Jupiter.

117.032 Duplicity of late B-type stars.   S. C. Wolff.
Astrophys. J., Vol. 222, 556 - 569 (1978).

High-dispersion spectroscopic observations of 83 stars are used to determine the binary characteristics of late B-type stars. The paper describes the observations and discusses the distribution of mass ratios, the extent to which rotation is synchronized with orbital revolution, and the frequency of close binaries.

117.033 The young close binary HD 47732 in NGC 2264.
R. H. Koch, C. S. Sutton, K. H. Choi, D. E. Kjer, R. Arquilla.
Astrophys. J., Vol. 222, 574 - 577 (1978).

HD 47732 is a short-period ellipsoidal variable with some complications evident in its yellow light curve. The interpretation of the light curve is entirely dependent upon spectroscopic information. The seeming best spectral classifications lead to a contact configuration. Spectral types that are only slightly later lead to a semidetached description. Either interpretation of HD 47732 implies a distance of at least 1.2 kpc, considerably greater than the commonly accepted distance for the cluster NGC 2264 of which it is a member.

117.034 The remarkable system AM Herculis/3U 1809 + 50. II. A single accretion column model.
G. Chanmugam, R. L. Wagner.
Astrophys. J., Vol. 222, 641 - 646 (1978) = Contrib. Louisiana State Univ. Obs., Baton Rouge, No. 131.

A model for AM Herculis in which the X-radiation, optical radiation, and polarized optical radiation arise in a single accretion column is presented. The coincidence of the primary optical minimum with the standstill in the fractional circular polarization and the X-ray maximum is explained as caused by cyclotron self-absorption. The relationship of AM Her to VV Puppis and AN Ursae Majoris is briefly discussed.

117.035 Effects of libration on the light curve of a close binary system.
R. S. Becker, T. K. Bolland, S. Y. Shieh.
Astrophys. J., Vol. 222, 647 - 651 (1978).

The authors have calculated the effects of libration on the light curve of a close binary system, using a simple model. The model indicates that the libration can produce significant effects, of the order of 10% of the modulation in magnitude due to corotation. This has implications for systems of current interest such as X-ray binaries.

117.036 Some thoughts on multiple stars.   D. S. Evans.
Rev. Mexicana Astron. Astrofis.,Vol. 3, (see 012.023), 13 - 20 (1977).

The occultation and spectroscopic approaches to the discovery of multiple stars are reviewed, and the technique of spectroscopic measurement recapitulated. A notation for multiple stars is proposed. An examination of the gravitational interaction levels suggests an average value of Δ log P for one hierarchical step close to 3.0 and Δ log a near 2.0 consistent with a binding ratio near 1 : 10000.

117.037 The discovery and observation of multiple stellar systems.   A. H. Batten, C. D. Scarfe.
Rev. Mexicana Astron. Astrofis., Vol. 3, (see 012.023), 21 - 31 (1977).

The difficulties of discovering and observing multiple stellar systems are reviewed and discussed with especial reference to those systems which are best observed by a combination of visual and spectroscopic or visual and photometric means. Attention is drawn to the difficulties that often exist in reconciling the results of two different methods of observation. A brief discussion of the possibility of detecting spectroscopically invisible companions of low mass is included in the paper.

117.038 Multiplicity of solar-type stars.   H. A. Abt.
Rev. Mexicana Astron. Astrofis., Vol. 3, (see 012.023), 47 - 51 (1977).

A systematic search for binaries in a sample of 123 bright field stars of type F3—G2 IV or V is described. Combination of the results for 25 newly discovered spectroscopic binaries with those of 21 spectroscopic, 23 visual, and 25 common-proper-motion pairs previously known brings to 88 the total number of companions identified in the sample. The distribution of the 88 periods shows a single maximum; the median period is 14 years. The frequencies of singles:doubles: triples: quadruples are found to be 42: 46: 9: 2. Less than half of the stars are thus observed to be single.

117.039 Note on the multiplicity of dMe stars.
S. L. Lippincott.
Rev. Mexicana Astron. Astrofis., Vol. 3, (see 012.023), 53 - 55 (1977).

There is much current interest in flare stars of the type dMe with particular concern as to their ages and mechanism responsible for flaring. Some fifty Me stars are on the Sproul Astrometric program. They have been discussed as flare stars or potential flare stars (Lippincott 1971). At least half are known to be members of binary or multiple systems.

117.040 Effets de sélection dans la découverte des systèmes multiples.   P. Muller.
Rev. Mexicana Astron. Astrofis., Vol. 3, (see 012.023), 63 - 65 (1977).

En étudiant les systèmes multiples issus d'un couple de découverte ancienne grâce à l'emploi ultérieur d'un instrument plus puissant, on constate qu'un grand nombre d'étoiles doubles n'ont jamais été examinées avec un tel instrument; il en résulte que parmi ces couples plusieurs centaines sont des systèmes au moins triples que nous ignorons.

117.041 Étoiles multiples et amas ouverts.   J. Dommanget.
Rev. Mexicana Astron. Astrofis., Vol. 3, (see 012.
023), 67 - 72 (1977).

It is generally suspected that multiple systems and galactic clusters are objects of different kinds. Their frequency
distribution as a function of the number of components, n,
as well as their sizes and their masses, show important differences between both categories and also the probable existence of a limit lying between n = 10 and n = 20 components.
The basic difference between these objects seems to be that
the multiple systems finally appear to be components of the
stellar medium in the same way as single and double stars,
whereas the galactic clusters are groups of components of a
particularly dense part of this medium.

117.042 On the spectrophotometric identification of late
type binaries in open clusters.
C. Bettis, D. Branch.
Rev. Mexicana Astron. Astrofis., Vol. 3, (see 012.023), 73 -
76 (1977).

Using both measured values and model atmospheres, the
equivalent widths of the Na D lines of hypothetical late type
main sequence binaries are calculated. Enhancements in the
D lines of 90% over the value for a single star whose B−V
color is the same as the composite B−V are possible. It is suggested that this effect may be used to identify binaries in
galactic clusters, in particular those binaries whose components differ by more than 3 mag in brightness.

117.043 Kappa Pegasi – a quadruple system of possible low
total mass.   W. R. Beardsley, M. W. King.
Rev. Mexicana Astron. Astrofis., Vol. 3, (see 012.023), 85 -
87 (1977).

Analyses of early Lick spectra and recent high dispersion
Kitt Peak National Obs. spectra indicate that κ Peg is a quadruple system having all four spectra visible under favorable conditions. Preliminary spectroscopic orbital elements are presented for the AB system, and for Bb. The Bab system has been
found to be the previously known 5.97-day spectroscopic
system. The total mass is found to be between 2 and 3 solar
masses only. Thus the brightest of the four stars Ba (F5IV) −
or possibly all four of the stars − must have an abnormally
small mass for its luminosity.

117.044 Psi Sagittarii, a system with three evolved components.   F. C. Fekel, Jr.
Rev. Mexicana Astron. Astrofis., Vol. 3, (see 012.023), 89 -
92 (1977).

The visual binary Psi Sagittarii (ADS 12214, HR 7292)
is a spectroscopic triple system with orbital periods of 10.78
days and 20.0 years. Periastron passage in the long period
orbit occurred on JD 2442410 ± 30 days. Masses derived from
the parameters of the spectroscopic and visual orbits are 2.7
$M_\odot$ for the visual primary, 2.1 $M_\odot$ for the short period pair
primary, and 1.7 $M_\odot$ for the short period pair secondary. The
distance to the system is 100 parsecs. The orbital inclination
of the close pair is nearly the same numerically as that of the
visual orbit. Suggested spectral classifications indicate that all
three components have evolved off the main sequence. Mass
exchange in the close pair has not yet occurred.

117.045 Sur les systèmes stellaires multiples Hu 1399 et
Cor 197.   R. R. de Freitas Mourão.
Rev. Mexicana Astron. Astrofis., Vol. 3, (see 012.023), 95 -
97 (1977).

The multiple systems Hu 1399 and Cor 197 have been
studied. A deviation of the residuals has been established. The
relative rectilinear motion of the components AC of the system Hussey 1399 can be used to obtain information about the
invisible companion.

117.046 Statistical properties of trapezium systems.

C. Allen, M. Tapia, L. Parrao.
Rev. Mexicana Astron. Astrofis., Vol. 3, (see 012.023), 119 -
126 (1977).

A catalogue of more than 900 multiple systems of trapezium type has recently been completed. In constructing the
catalogue an attempt was made to exclude optical systems.
The authors study the galactic distribution of trapezia of different spectral types, and they compare it with that of ordinary multiple systems and of visual binaries from the IDS. An
estimate is made of the expected number of pseudo-trapezia
among each spectral type. This study shows that most of
our trapezia are true trapezia and should therefore be young.
This conclusion holds for trapezia of all spectral types.

117.047 Trapezia and infrared sources.
A. Poveda, C. Allen, J. Warman.
Rev. Mexicana Astron. Astrofis., Vol. 3, (see 012.023), 127 -
132 (1977).

From dynamical considerations, a significant fraction of
multiple systems of trapezium type is expected to be quite
young. A search for positional coincidences between trapezia
from various catalogues and infrared sources at 2.2 microns
was undertaken. It was found that between 12% and 20% of
the trapezia can be associated with infrared sources. Various
statistical tests indicate that these coincidences are not likely
to be accidental. The nature of the infrared sources associated
with trapezia is briefly discussed.

117.048 UBVRI photometry of stars in trapezium type
systems.   J. Warman, J. Echevarría.
Rev. Mexicana Astron. Astrofis., Vol. 3, (see 012.023), 133 -
138 (1977).

The authors present UBVRI photometric data for 50
stars forming 14 systems of the trapezium type. Color-magnitude and color-color diagrams indicate the presence of ultraviolet excess and anomalous behavior in the infrared part of
the spectrum. This is consistent with the dynamical ages for
stars in trapezia.

117.049 A review of the dynamics of classical triple stars.
R. S. Harrington.
Rev. Mexicana Astron. Astrofis., Vol. 3, (see 012.023), 139 -
143 (1977).

Four ways of studying the dynamics of triple systems
are outlined, namely the analytical, the qualitative, the numerical, and the observational approach. A discussion of the stability of classical (revolutional, hierarchical) triples summarizes the results of analytical developments, based on the application of von Zeipel's method to the general three-body
problem, and of numerical experimentation. Finally, a summary is given of recent investigations concerning the dynamics
of triples other than the revolutional systems.

117.050 Review of the dynamical aspects of triple systems.
V. Szebehely.
Rev. Mexicana Astron. Astrofis., Vol. 3, (see 012.023), 145 -
150 (1977).

A classification of possible motions of triple systems is
presented emphasizing the transient phenomena occurring in
addition to the final (asymptotic) outcome and clarifying the
discrepancies between the astronomical and mathematical
formulations. A conjectured possible instability is described
and it is shown that systems with negative total energy and
low angular momentum may lead to instability and to the
formation of binaries. The ejected or escaping star may have
high velocity if the triple close approach preceding the escape
is sufficiently close. The computational results of several systematic series of such escapes are applied to various stellar configurations.

117.051 Numerical experiments on the decay of three-body
systems.   M. J. Valtonen, S. J. Aarseth.

Rev. Mexicana Astron. Astrofis., Vol. 3, (see 012.023), 163 - 166 (1977).

Numerical results are presented for calculations of stellar three-body systems. Among the triple systems with similar initial separations between each of the stars, the equal-mass systems are most stable with typical half-lives of about 40 crossing times. In the other extreme, systems with a very small total angular momentum or a large mass difference have half-lives of only about 10 crossing times. The escaping star is usually the lightest one, while the escape probability for the heaviest star approaches zero when its mass exceeds the combined mass of the other two stars by more than a factor of about 5. The final binary becomes very eccentric ($e_f \gtrsim 0.8$) when the three stars are restricted to a plane and do not have a very high total angular momentum, or when the total angular momentum is very small.

117.052   The statistical effect of encounters on wide binaries.
         I. R. King.
Rev. Mexicana Astron. Astrofis., Vol. 3, (see 012.023), 167 - 168 (1977).

Previous treatments by Ambartsumian and by Cruz-González are combined into a Fokker-Planck equation that describes the time evolution of the distribution of orbit sizes of a set of wide binaries, as encounters with field stars statistically change the orbits. A solution can be found that predicts the distribution of separations greater than about 3000 AU.

117.053   The lifetime of binary stars.   D. C. Heggie.
         Rev. Mexicana Astron. Astrofis., Vol. 3, (see 012. 023), 169 - 174 (1977).

The concept of the "lifetime" of a population of binaries is applicable only to those of low binding energy, i.e. wide pairs. Of several possible methods of estimating a lifetime, that based on the average change in binding energy due to non-disruptive encounters appears to be the best.

117.054   Planetary systems and stellar multiplicity.
         S.-S. Huang.
Rev. Mexicana Astron. Astrofis., Vol. 3, (see 012.023), 175 - 187 (1977).

The author discusses the origin of planetary systems on one hand and binary and multiple stars on the other. First he shows that phenomenological differences between these two kinds of celestial objects are due to their genetic difference. The basic point is that formation of a planetary system around a star has to be a minor event in the life history of the star while formation of a binary or multiple system has to be an event that is important equally to all components of the system. The dominant mechanism of binary and multiple system formation is that they were formed naturally as they are, each from perhaps a single condensation in the interstellar medium. However such a single mechanism of formation cannot satisfactorily explain the observed spread of binaries in mean separations between two components (or equivalently orbital periods). That several stars could be formed from a single condensation requires the existence of pre-stellar nuclei which are briefly discussed.

117.055   The origin of multiple stars by condensation in dense nebulae.   A. Poveda.
Rev. Mexicana Astron. Astrofis., Vol. 3, (see 012.023), 189 - 195 (1977).

The classical theories of double star formation are briefly reviewed in relation to the origin of multiple stars. Modern observations in the infrared are discussed and their relevance as clues to the process of multiple star formation is stressed. These observations, together with dynamical considerations and data on the statistics of trapezia and hierarchical multiple systems, support the view that multiple stars, in general, are born as trapezia, that is, as independent condensations.

117.056   A new trapezium system associated with a compact H II region.   S. Sharpless, J. L. Pipher, M. P. Savedoff, S. Schurmann.
Rev. Mexicana Astron. Astrofis., Vol. 3, (see 012.023), 197 (1977).

117.057   The formation of multiple systems by dynamical interaction in clusters.   S. J. Aarseth.
Rev. Mexicana Astron. Astrofis., Vol. 3, (see 012.023), 199 - 206 (1977).

This review is mainly devoted to a discussion of binary formation and evolution in stellar systems, as described by N-body techniques. The simplest formation mechanisms consist of ejection from bound triple systems and capture arising from hyperbolic three-body encounters. Numerical calculations show that the evolution of clusters containing several hundred members is invariably dominated by one central binary which absorbs a large fraction of the total energy. It is suggested that a visual binary in the core of the Hyades cluster may have been associated with this process.

117.058   Multiple stars as a result of the dynamical evolution of small stellar systems.   A. Hayli.
Rev. Mexicana Astron. Astrofis., Vol. 3, (see 012.023), 207 - 208 (1977).

117.059   Multiple star formation from N-body system decay.
         R. S. Harrington.
Rev. Mexicana Astron. Astrofis., Vol. 3, (see 012.023), 209 - 210 (1977).

The dynamical decays of random, unstable, negative-energy, N-body systems have been followed numerically, to investigate the statistics of stable decay products. From initial quadruple systems, about 20% of the products were stable triples while 30% resulted from initial quintuple systems. These numbers are consistent with the observed ratio of triples to binaries, especially for the higher-multiplicity initial systems.

117.060   Accretion discs.   B. Paczyński.
         Postępy Astron., Tom 26, 31 - 32 (1978).
Observational data concerning the existence of accretion discs in close binary systems are presented. The various types of astronomical objects in which one supposes the existence of such discs are described.

117.061   Nonstationary accretion of stellar wind.
         R. A. Syunyaev.
Pis'ma Astron. Zh., Vol. 4, 75 - 80 (1978). In Russian.
Accretion by neutron stars of low velocity ($\upsilon \sim 300$ km/s) stellar wind may be nonstationary and may lead to the phenomena of transient sources and bursters.

117.062   A new active period and possible model of CH Cygni.   L. Luud, J. Vennik, M. Pehk.
Pis'ma Astron. Zh., Vol. 4, 87 - 90 (1978). In Russian.
The beginning of a new active period of CH Cygni found using Tartu *UBVIJHK* and spectral observations is described. A model for CH Cygni as a double system with a transient accretion disk is proposed.

117.063   New data on unseen companions of 61 Cygni.
         A. N. Deutsch.
Pis'ma Astron. Zh., Vol. 4, 95 - 98 (1978). In Russian.
The maximum entropy method of spectral analysis has been applied for calculation of periods of unseen companions of 61 Cyg from observations at Pulkovo and Sproul Observatories. Two periods of 6 and 12 years are found with certainty. The orbital elements and mass (4 masses of Jupiter) of the 6-year period companion are estimated.

117.064   Inner region of accretion discs in binary stellar

systems.   V. Yu. Shulikovskij.
Pis'ma Astron. Zh., Vol. 4, 134 - 137 (1978). In Russian.

A model of the inner region of an accretion disc with account of vertical gas velocity is built up. The temperature in the central plane of the disk appears to exceed that in standard theory and depends on the accretion rate. Ejection of part of the gas from the inner regions of the disk leads to formation of a corona around the disk.

117.065   Evolved contact systems of spectral type O.
          II. AO Cassiopeiae.   D. P. Schneider,
K.-C. Leung.
Astrophys. J., Vol. 223, 202 - 206 (1978).

UBV observations of AO Cas by Koch were analyzed with Wilson and Devinney's approach. The system is found to be in contact. A 3% overcontact configuration fitted the light curves better than any previous model. The inclination is very low, 51°; the light variation is due entirely to tidal distortion. As expected from their location on the period–spectral-type diagram, both components are evolved.

117.066   The magnetic-binaries problem: motion of a charged
          particle in the region of a magnetic-binary system.
A. G. Mavraganis.
Astrophys. Space Sci., Vol. 54, 305 - 313 (1978).

The combination of the two-body problem with that of the motion of a charged particle in a magnetic-dipole field (Störmer's problem) is treated in this article. Thus, in a long series of articles, a first step is taken to study the behavior of a charged particle in a magnetic-binary system. This first step includes a description of a simple model of such a system, the derivation of the equations of motion of the particle in the system, and a discussion on this foundation of the problem.

117.067   A comment on the solution of binary star orbits:
          the question of unseen companions.   D. G. Monet.
Astrophys. J., Lett., Vol. 223, L101 - L103 (1978).

A solution is presented for the problem of interpreting measured coordinates for proper motion, parallax, and orbital motion around a suspected but unseen companion. This method has the advantage of an analytic solution with complete error analysis from a single, weighted, least-squares fit to the data in each coordinate.

117.068 α Sculptoris – a long period binary.   O. Vilhu.
          Inf. Bull. Variable Stars, No. 1378, 3 pp. (1978).

117.069   F, G and K type dwarf and subgiant, close binaries
          with H and K emission.   O. J. Eggen.
Inf. Bull. Variable Stars, No. 1426, 31 pp. (1978).

117.070   The variable linear polarization of the O-star binary
          HD 47129 (Plaskett's star).
R. J. Rudy, L. C. Herman.
Publ. Astron. Soc. Pacific, Vol. 90, 163 - 166 (1978).

The linear polarization of the binary HD 47129 has been observed in the *B* filter on 60 nights. The polarization has a variable component which is synchronous with the 14.4-day binary period. This intrinsic component reaches a maximum at quadratures and may be explained by scattering from a corotating region of extrastellar material lying between the stars. A best value of 71° is found for the orbital inclination.

117.071   CC 20,986, a new unresolved astrometric binary.
          S. L. Lippincott, E. R. Borgman.
Publ. Astron. Soc. Pacific, Vol. 90, 226 - 229 (1978).

The nearby M3 V star CC 20,986 shows a perturbation from astrometric plates taken with the 61-cm Sproul refractor. The interpretation indicates an unseen companion revolving about the visible component in a period of 3ʸ72 with a likely range in minimum mass of 0.06 to 0.08 $M_\odot$.

117.072   Can we find a place to live near a multiple star?
          R. S. Harrington, B. J. Harrington.
Mercury, Vol. 7, 34 - 37 (1978).

117.073   The enigmatic structure of the symbiotic binary
          AG Peg.   C. M. Anderson, J. S. Gallagher.
Bull. American Astron. Soc., Vol. 10, 410 (1978). – Abstract.

117.074   New photoelectric photometry of AW UMa.
          R. H. Koch, P. R. Eisenhardt, E. J. Woodward.
Bull. American Astron. Soc., Vol. 10, 410 (1978). – Abstract.

117.075   Calculation of binary star orbits for systems with
          high eccentricity.   D. G. Monet.
Bull. American Astron. Soc., Vol. 10, 411 (1978). – Abstract.

117.076   Coplanarity, masses, and stability of close multiple
          stars.   F. C. Fekel.
Bull. American Astron. Soc., Vol. 10, 411 (1978). – Abstract.

117.077   Three most massive contact binary systems.
          K. C. Leung, D. P. Schneider.
Bull. American Astron. Soc., Vol. 10, 412 (1978). – Abstract.

117.078   A theory for the RS CVn binaries.
          D. S. Hall, S. N. Shore.
Bull. American Astron. Soc., Vol. 10, 418 (1978). – Abstract.

117.079   Large radio flares in RS CVn binaries.
          P. A. Feldman.
Bull. American Astron. Soc., Vol. 10, 418 (1978). – Abstract.

117.080   Variations in the distortion wave of RS CVn.
          E. W. Ludington.
Bull. American Astron. Soc., Vol. 10, 418 - 419 (1978). Abstract.

117.081   Serendipitous UV observations of α Virginis.
          G. W. Collins II.
Bull. American Astron. Soc., Vol. 10, 452 (1978). – Abstract.

117.082   X- and γ-ray bursts from magnetic reconnection in
          very short period binary systems.   A. Treves.
Astron. Astrophys., Vol. 67, 441 - 442 (1978).

Magnetic interaction between the components of a very close binary system is considered. Its importance with regard to X- and γ-ray bursts is discussed.

117.083   Mass-transfer effects in binary star evolution.
          B. P. Flannery.
Ann. New York Acad. Sci., Vol. 302, (see 012.033), 36 - 46 (1977).

The article reviews the processes that can be expected to occur in the first phase of mass-exchange in binary systems, particularly those likely to produce binary X-ray sources. Recent studies, following the pioneering investigation of Benson, indicate that the system will evolve into contact followed by a phase of rapid and substantial loss of mass and angular momentum. Thus, earlier evolutionary schemes proposed for X-ray systems will require alteration in the sense of demanding progenitors of higher mass and greater initial periods.

117.084   Kinematics of binaries.   N. E. Kurochkin.
          Astron. Tsirk., No. 996, p. 4 - 7 (1978). In Russian.

117.085   X-ray emission processes in close binary systems.
          J. E. Pringle.
Ann. New York Acad. Sci., Vol. 302, (see 012.033), 6 - 13 (1977).

It now seems generally agreed that the bright binary X-ray sources discovered and studied by the UHURU and subsequent X-ray satellites are systems in which mass transfer is taking place onto a compact object. For this reason most of this paper is devoted to a review of the state of the art as regards accretion onto compact objects.

On the topography of extrasolar earthlike planets. See Abstr. 015.032.

Techniques of perturbation analysis. See Abstr. 031.250.

The detection of stellar systems by lunar occultations.   See Abstr. 031.251.

Speckle interferometry with a 1 m-telescope. See Abstr. 031.257.

The effect of oblateness of the bigger primary on collinear libration points in the restricted problem of three bodies.   See Abstr. 042.012.

A complete family of periodic solutions of the planar three-body problem and their stability. See Abstr. 042.089.

Secular mass loss from cataclysmic variables. See Abstr. 064.048.

The evolution of massive close binaries. VI: Final considerations for the conservative case.   See Abstr. 065.047.

Use of Roche coordinates in the problems of small oscillations of rotationally distorted stellar models. See Abstr. 065.069.

Use of Roche coordinates in the problems of small oscillations of tidally distorted stellar models. See Abstr. 065.088.

The analytic theory of fluid disks orbiting the Kerr black hole.   See Abstr. 066.009.

Relativistic, accreting disks. See Abstr. 066.010

Possibility of superlong gravitational waves detection.   See Abstr. 066.019.

Gamma radiation in the process of spherically-symmetric accretion on black holes in binary stellar systems. See Abstr. 066.236.

Gravitational radiation damping and energy loss in binary systems.   See Abstr. 066.287.

Parallax and astrometric orbit of G24—16 from the Sproul plate series.   See Abstr. 111.005.

Relative proper motions of Trapezium-type multiple systems. I.   See Abstr. 112.010.

Wolf-Rayet stars in the Small Magellanic Cloud. See Abstr. 114.065.

Emission line variations in RS Canum Venaticorum-type binaries.   See Abstr. 114.501.

A detailed study of the spectrum of the binary X-ray source HD 153919 (3U 1700-37). I. Radial velocity data in the blue spectral region.   See Abstr. 114.557.

Spectroscopy and possible orbital periods for HDE 245770 (= A0535 + 26).   See Abstr. 114.566.

A new look at the stars.   See Abstr. 115.003.

Angular velocity distribution in a star which is a component of a close binary system and has a fast rotating core.   See Abstr. 116.010.

Multiplicity among visual binaries. See Abstr. 118.024.

Contact binaries. III. Early-type systems. See Abstr. 121.023.

BV 845: a contact system with a large temperature difference between components.   See Abstr. 121.025.

Period—mass ratio relations for eclipsing binaries with periods not exceeding 5 days.   See Abstr. 121.027.

The variable lightcurve of 44 i Boo observations and implications.   See Abstr. 121.033.

Flare active binary systems. EQ Peg and its brighter companion BD + 19 5116 A. See Abstr. 122.123.

X Persei (= 4U 0352+30).   See Abstr. 122.139.

A photometric study of the recurrent nova WZ Sagittae at minimum light.   See Abstr. 124.002.

Evolution of nova binaries: the inevitability of helium runaways.   See Abstr. 124.028.

The expanding shell of nova DQ Herculis 1934. See Abstr. 124.602.

Asymmetric supernova explosions and the origin of binary pulsars.   See Abstr. 125.054.

Hard X-ray observations of white dwarf binary systems.   See Abstr. 126.006.

Stability of accreting white dwarfs in close binary systems.   See Abstr. 126.025.

Spectroscopy of the extreme-ultraviolet source Feige 24: the binary orbit and the mass of the white dwarf. See Abstr. 126.037.

Magnetic braking during star formation. See Abstr. 131.110.

The Hulse-Taylor pulsar and gravitational spin precession.   See Abstr. 141.537.

Neutrinos from binary pulsars. See Abstr. 141.552.

Accreting neutron stars in highly compact binary systems and the nature of 3U 1626—67. See Abstr. 142.053.

The Bowen mechanism in HZ Herculis. See Abstr. 142.128.

Theoretical calculations of the optical pulsation from HZ Herculis.   See Abstr. 142.139.

Evolutionary model computations for globular

cluster X-ray sources.   See Abstr. 142.146.

On the turn-off of the reflection effect in the HZ Her system.   See Abstr. 142.153.

On the nature of the galactic bulge X-ray sources. See Abstr. 142.223.

Low-mass X-ray binaries and their relation to the non-X-ray sources.   See Abstr. 142.227.

Evolutionary processes in X-ray binaries and their progenitor systems.   See Abstr. 142.228.

Origin and present evolution of massive X-ray binaries.   See Abstr. 142.229.

X-ray binaries.   See Abstr. 142.245.

Binaries in open clusters: the effects of rotation on determinations of frequency and mass ratio. See Abstr. 153.008.

Errata

117.901   Erratum: 'Tidal friction in close binary stars' [Astron. Astrophys., Vol. 57, 383 - 394 (1977)].
J.-P. Zahn.
Astron. Astrophys., Vol. 67, 162 (1978).

118 Visual Binaries and Multiple Stars

118.001   **Photographic measures of double stars.**
L. Armanelli, A. Delgrosso, R. Pannunzio.
Astron. Astrophys., Suppl. Ser., Vol. 31, 121 - 126 (1978).
204 measures of double stars are given, obtained by photographic observations made with the astrometric reflector of the Astronomical Observatory of Torino.

118.002   **The orbit of LDS 720BC = CD−32°16135.**
R. H. Wilson, Jr.
Mon. Not. R. Astron. Soc., Vol. 182, 345 - 347 (1978).
A new orbit with period 675 yr is shown to represent satisfactorily all available observations of the binary star LDS 720BC. Also the resulting dynamical parallax agrees well with the trigonometric.

118.003   **Period-eccentricity relation of visual double stars.**
M. Fulconis, A. Bücher.
Astron. Astrophys., Vol. 62, 121 - 125 (1978). In French.
The problem of relation between period and eccentricity has been treated from the point of view of the correlation coefficient. 378 binaries with the best determined orbital elements have been retained. The mean value for the periods is 40.44 years with a standard deviation of 2.7 years. For an eccentricity with a standard deviation of 0.705 it equals 0.494.

118.004   **New double stars.**   D. W. Dunham.
Occultation Newsl., Vol. 1, 140 (1978).

118.005   **New zodiacal double stars, 1977 December 23.**
Occultation Newsl., Vol. 1, 140 (1978).

118.006   **The orbits of seventeen visual binary stars.**
G. A. Starikova.
Tr. Gos. Astron. Inst. Shternberga, Vol. 47, 96 - 98 (1976). In Russian.
The orbital elements of the visual binary stars ADS 520, 3064, 4299, 6354, 6828, 6851, 7284, φ 47, φ 309, ADS 9831, 9909, I 1336, ADS 10786, 13944, 14073, 14787, 16173 are given.

118.007   **The orbits of twenty-four visual binary stars.**
G. A. Starikova.
Soobshch. Gos. Astron. Inst. Shternberga, No. 199, p. 12 - 14 (1977). In Russian.
The orbital elements of visual binary stars ADS 61, ADS

713, ADS 784, ADS 1183, ADS 1729, ADS 2028, JC 8, ADS 2612, ADS 3159, ADS 4971, ADS 6664, ADS 6825, ADS 6851, ADS 7054, ADS 7555, ADS 9247, ADS 9757, ADS 10374, ADS 13723, ADS 14761, ADS 14775, ADS 16708, ADS 16819, ADS 17149 are given.

118.008   **Reexamination of suspected unresolved binaries.**
W. D. Heintz.
Astrophys. J., Vol. 220, 931 - 934 (1978).
An update is given on 11 objects in which some evidence for the presence of unseen companions had been reported. Indication of submotions in some micrometrically observed binaries continues. There is no evidence for any companion of substellar mass.

118.009   **HD 165590 – a spectroscopic-visual triple system.**
C. L. Morbey, A. H. Batten, D. H. Andrews,
W. A. Fisher.
Publ. Astron. Soc. Pacific, Vol. 89, 851 - 856, with a correction in Vol. 90, 236 (1977/78) = Dominion Astrophys. Obs., Victoria, B.C., Canada, Contrib. No. 346 = NRC No. 16105.
HD 165590 is known to be a visual binary with a period of 20 years and extremely high orbital eccentricity. New instrumental and computer techniques are used to show that one member of the visual pair is a spectroscopic binary with a period of 0.879511 day.

118.010   **κ Pegasi, a triple system.**
D. J. Barlow, C. D. Scarfe.
Publ. Astron. Soc. Pacific, Vol. 89, 857 - 861 (1977/78) = Dominion Astrophys. Obs., Victoria, B.C., Canada, Contrib. No. 345 = NRC No. 16104.
Spectra of both visual components of κ Peg have been observed on high-dispersion coudé spectrograms of the combined light, obtained at the Dominion Astrophysical Observatory from 1969 to 1976. Radial-velocity measurements fail to confirm the claim of Beardsley and King that the system is quadruple, and an alternative interpretation of their result is presented.

118.011   **Photographic observations of visual double stars.**
E. van Albada-van Dien.
Astron. Astrophys., Suppl. Ser., Vol. 31, 353 - 360 (1978) = Ann. Bosscha Obs., Lembang, Indonesia, Vol. IX, part 5.
Relative distances and position angles are given for 95 pairs of mostly southern visual double stars, measured on

photographic plates taken with the 60 cm visual refractor of the Bosscha Observatory between 1953 and 1969.

118.012 Helium abundance of metal-poor binaries.
H. Saio.
Publ. Astron. Soc. Japan, Vol. 30, 139 - 148 (1978).

The luminosity, radius, and age of evolutionary models for metal-poor stars are expressed as functions of chemical composition, mass and a parameter which indicates the phase of evolution. Introducing into these formulae the mass, luminosity, effective temperature, and spectroscopic metal-to-hydrogen ratio for each of the primary components of visual binaries ADS 520, η Cas, 99 Her, μ Cas, and 85 Peg, the helium abundance and age are estimated. The helium contents of these objects are found to be similar to those of Population I stars.

118.013 64 Ori – a mid-B type triple system.
F. C. Fekel.
Bull. American Astron. Soc., Vol. 9, 623 (1978). – Abstract.

118.014 The apparent orbit of VY CMa AB: the rotating holey dust cloud hypothesis. G. Wallerstein.
Bull. American Astron. Soc., Vol. 9, 623 (1978). – Abstract.

118.015 New double stars. D. W. Dunham.
Occultation Newsl., Vol. 1, 150 (1978).

118.016 Central star of NGC 3132: a visual binary.
L. Kohoutek, S. Laustsen.
IAU Symp. No. 76, (see 012.014), p. 207 (1978). – Abstract.

118.017 Improved orbits of the visual binary δ Equulei.
D. M. Popper, M. M. Dworetsky.
Publ. Astron. Soc. Pacific, Vol. 90, 71 - 72 (1978).

Additional coudé spectrograms of δ Equ combined with analysis of the available visual double star observations allow the authors to determine the properties of this F7 V system with high precision. The average values of the mass and visual absolute magnitude are $1.19\,M_\odot$, and 3^m95, respectively.

118.018 Relative motions in 17 visual double stars.
P. Muller.
Astron. Astrophys., Suppl. Ser., Vol. 32, 165 - 172 (1978). In French.

Equations of the relative motion in α and δ for 17 pairs are presented.

118.019 Orbits of four double stars. P. Muller.
Astron. Astrophys., Suppl. Ser., Vol. 32, 173 - 176 (1978). In French.

Orbital elements, O–C and ephemerides for the visual binaries ADS 3017, 6276, 6776 and 8311 are given.

118.020 Photographic measures of double stars. II.
H. M. Jeffers, S. Vasilevskis.
Astron. J., Vol. 83, 411 - 435 (1978) = Lick Obs. Bull. No. 800.

Measures are given of 302 stars photographed on 657 plates with the 36-in. refractor, mostly from 1948 to 1960. A total of 773 plates, originals and their copies, have been measured, most of them (73%) by Ejnar Hertzsprung and his collaborators. Unmeasured plates in Lick files are listed in an appendix.

118.021 Micrometer observations of double stars. I.
D. J. MacConnell.
Astron. J., Vol. 83, 436 - 437 (1978).

This paper reports visual micrometer measures of 83 stars made during the period 1976.98–1977.91.

118.022 Occultation astrometry of the Beta Scorpii system.
D. S. Evans, J. L. Elliot, D. M. Peterson.
Astron. J., Vol. 83, 438 - 441 (1978).

Results of occultation observations of β Scorpii made at several observatories are combined. Occultation observations of binary stars with separations less than about 0.5 arcsec where both components are occulted by the same lunar limb feature can yield high-quality astrometric results. For greater separations, although reasonable mean values are deducible, the fact that the components are occulted by different and imperfectly known structures introduces errors larger than those affecting astrometry by conventional methods. The parameters for AB are $\Delta m = 3.31 \pm 0.09$ (s.e. of mean), $\rho = 0.463 \pm 0.017$ arcsec, $\theta = 116°.5 \pm 2°.6$. For AC, $\Delta m = 2.41 \pm 0.09$ (s.e. of mean), $\rho = 13.5 \pm 4.2$ arcsec, $\theta = 20°.6 \pm 10°.5$.

118.023 Observations of binary stars by speckle interferometry – I.
B. L. Morgan, D. R. Beddoes, R. J. Scaddan, J. C. Dainty.
Mon. Not. R. Astron. Soc., Vol. 183, 701 - 710 (1978).

Speckle interferometry has been used to study a large number of binary-star systems and variable stars. Thirty-five observations on 30 objects are presented. These objects include six radial-velocity or photometric variable stars. Of these, new components are reported for the systems ϵ UMa, γ Boo and W Sgr and the system σ Sco is resolved into three components, confirming the results of Nather, Churms & Wild. The technique provides a very accurate means of measuring the separations of these systems.

118.024 Multiplicity among visual binaries.
C. E. Worley.
Rev. Mexicana Astron. Astrofis., Vol. 3, (see 012.023), 57 - 61 (1977).

A brief review of the history of multiple-star discovery is followed by a discussion of selection effects, discovery probability, and detection techniques. The incidence of multiplicity and the properties of multiples in various samples of binaries are investigated.

118.025 Sur la masse d'une binaire spectroscopique membre d'une étoile double visuelle. P. Couteau.
Rev. Mexicana Astron. Astrofis., Vol. 3, (see 012.023), 77 - 83 (1977).

The author describes a method which interposes a relation mass-luminosity in order to calculate the dynamic parallax and the masses of the components in a system of known orbit which involves a spectroscopic binary. He applies the method to a few cases.

118.026 The Stein 2051 system. K. Aa. Strand.
Rev. Mexicana Astron. Astrofis., Vol. 3, (see 012.023), 93 (1977). – Abstract.

118.027 Recherche sur l'existence possible d'un compagnon astrométrique dans le couple visuel ADS 9476 – Aitken 14. R. R. de Freitas Mourão.
Rev. Mexicana Astron. Astrofis., Vol. 3, (see 012.023), 99 - 101 (1977).

Observations made during the last 70 years of the visual binary star Aitken 14 reveal systematic deviations in its motion, suggesting the existence of an astrometric companion.

118.028 Photometry and variability of double stars.
R. L. Walker.
Rev. Mexicana Astron. Astrofis., Vol. 3, (see 012.023), 103 - 107 (1977).

Selected topics of double star photometry are discussed together with current observational programs and new observing techniques. A revised catalog of visual binaries with one or more variable components is described.

118.029 **Orbits of 12 visual binary stars.**
G. A. Starikova.
Pis'ma Astron. Zh., Vol. 4, 99 - 101 (1978). In Russian.

118.030 **Orbites nouvelles.**
Circ. Inf., Nos. 74, 75 (1978).

118.031 **Étoiles doubles nouvelles. Étoiles doubles découvertes à Nice.Lunette de 50 cm.   P. Couteau.**
Circ. Inf., No. 74 (1978).

118.032 **Étoiles doubles nouvelles: Lunette de 50 cm (Nice); Lunette de 61 cm (Sproul).**
P. Couteau, W.-D. Heintz.
Circ. Inf., No. 75 (1978).

118.033 **Index catalogue.   W.-D. Heintz.**
Circ. Inf., No. 75 (1978).

118.034 **Radial-velocity variations of OΣ 341.**
A. H. Batten, C. L. Morbey.
Circ. Inf., No. 75 (1978).

118.035 **HD 165590.**
IAU Circ., No. 3208 (1978).

118.036 **Parallax and orbit of ADS 8887.**
C. Ambruster.
Publ. Astron. Soc. Pacific, Vol. 90, 219 - 221 (1978).
    The author presents a new parallax and orbit for the late-type 9th magnitude system ADS 8887. The Sproul absolute parallax is +0$\overset{''}{.}$051 ± 0$\overset{''}{.}$005 (p.e.); a combined absolute parallax of 0$\overset{''}{.}$053 ± 0$\overset{''}{.}$004 (p.e.) is adopted. This absolute parallax, combined with newly derived orbital elements (P = 229$\overset{y}{.}$0, a = 2$\overset{''}{.}$14), yields a total mass for the system of 1.2 $M_\odot$. This appears to resolve a long-standing discrepancy inherent in earlier orbits. No evidence fo an unseen third body was found.

118.037 **A statistical analysis of the incompleteness and the number of optical systems in the IDS.**
A. Poveda, C. Allen.
Bull. American Astron. Soc., Vol. 10, 409 (1978). − Abstract.

118.038 **Elimination of orbit ambiguity for visual binaries of equal magnitude.   R. L. Walker.**
Bull. American Astron. Soc., Vol. 10, 410 (1978). − Abstract.

118.039 **Orbit of the visual binary ADS 3588 = IDS 04546 S 1632.   E. Van Dessel.**
Bull. Astron. Obs. R. Belgique, Vol. 9, 44 - 46 (1978).

118.040 **Orbites nouvelles pour six étoiles doubles visuelles.**
J. Dommanget.
Bull. Astron. Obs. R. Belgique, Vol. 9, 47 - 59 (1978).
    New orbits are given for the binaries IDS No. 03524 S 0127; 06116 N 0959 Aa; 08140 N 4743 Cc; 12592 S 1243; 14448 S 3641 and 19545 S 3852. The dynamical parallaxes are computed for each of them as well as physical elements of their components.

118.041 **Wide physical pairs among bright A and B stars.**
I. N. Latyshev.
Astron. Tsirk., No. 966, p. 7 - 8 (1977). In Russian.

The non-duplicity of 189 ADS stars.
See Abstr. 002.008.

Evolution of the α Centauri system.
See Abstr. 065.044.

Supermetallicity and the strong-lined binary systems.   See Abstr. 114.531.

Masses of red giants. I. Mean initial mass from visual binary data.   See Abstr. 115.005.

Psi Sagittarii, a system with three evolved components.   See Abstr. 117.044.

Coplanarity, masses, and stability of close multiple stars.   See Abstr. 117.076.

White dwarf formation in two-body systems: Sirius AB, Procyon AB, 40 Eri BC and Stein 2051 AB.
See Abstr. 126.033.

119 Spectroscopic Binaries

119.001 **Spectroscopic binary orbits from photoelectric radial velocities. Paper 18 : HR 6388.**
R. F. Griffin.
Observatory, Vol. 98, 14 - 18 (1978).

119.002 **1976 - 1977 photometry of UX Ari, HR 1099 (V711 Tau) and λ And.   H. J. Landis, L. P.** Lovell, D. S. Hall, G. W. Henry, T. R. Renner.
Astron. J., Vol. 83, 176 - 182 (1978).
    The authors present differential observations in V of the UBV system. The light curve of UX Ari showed a wave. The amplitude in V was 0.08 mag and the phase of minimum was 0$\overset{p}{.}$8. The wave appears to be migrating retrograde and may be accelerating. The light curve of HR 1099 showed a wave very similar to the one seen in 1975 - 1976. The amplitude in V was unchanged from earlier years and the phase of minimum had migrated retrograde 0$\overset{p}{.}$074 since 1976.0, suggesting an average migration period of 13 ± 1 yr. The light curve of λ And showed three maxima and two minima. Epochs of these extrema suggested a photometric period of 54.2 ± 0.5 days.

119.003 **Spectroscopic binary orbits from photoelectric radial velocities. Paper 19: HD 147508.**
R. F. Griffin.
Observatory, Vol. 98, 47 - 49 (1978).

119.004 **Study of the spectroscopic binaries in NGC 2516.**
F. Gieseking.
Astron. Astrophys., Suppl. Ser., Vol. 32, 17 - 23 (1978). In German.
    For each of 14 selected stars in the region of the open cluster NGC 2516 twelve new radial velocity determinations are presented. The radial velocities, whose epochs are distributed over a time interval of 88 days, were measured on objective-prism plates taken with the radial velocity astrograph (GPO) of the European Southern Observatory. For the reduction of the measurements the same special method was applied, which was described earlier (Gieseking 1976), Among all of the 14 stars investigated, only the stars Cox 29, 91 and d definitively turned out to be spectroscopic binaries. By combining the GPO measurements and those of Abt and Levy, for the first

time the calculation of orbital elements of these systems succeeded. Whereas the double star nature of Cox 11 and Cox 20, which was supposed by Abt and Levy, turned out to be doubtful, the GPO measurements indicate that Cox 15 and Cox 19 may display long-time radial velocity changes. A possible double star nature of Cox 15 would be especially interesting, because in that case a single unexplained decrease in brightness, which was observed by Cox (1955), may be interpreted as eclipsing light variation. This star would then be the first eclipsing binary among Ap stars.

119.005   Speckle interferometry of spectroscopic binaries.
H. A. McAlister.
Bull. American Astron. Soc., Vol. 9, 599 - 600 (1978). Abstract.

119.006   HD 49798: a strange composite sdO star.
G. Goy.
Astron. Astrophys., Vol. 64, 445 - 446 (1978).
HD 49798 is a "single line" sdO binary star. It is shown that the existence of an F or G companion is to be excluded (first hypothesis). 2 other solutions remain possible: a white dwarf (second hypothesis) or a collapsed object (third hypothesis) associated with the sdO star.

119.007   41 Sextantis – a new metallic line binary.
T. F. Worek, W. R. Beardsley, M. W. King.
Astron. J., Vol. 83, 303 - 304 (1978).
The metallic line A star 41 Sextantis is observed to be a single-line spectroscopic binary with a period of 6.1670 days and a nearly circular orbit.

119.008   The spectroscopic orbit of HD 47732.
W. R. Beardsley, T. S. Jacobsen.
Astrophys. J., Vol. 222, 570 - 573 (1978) = Contrib. Dominion Astrophys. Obs., Victoria, B.C., No. 352 = NRC No. 16291.
The star HD 47732, a member of the open cluster NGC 2264, has been found to be a double-lined spectroscopic binary with component spectral types estimated to be B1.5 IV and B2 V, and to possess a very short period of 1.3041 days. The spectroscopic orbit is derived; by utilizing these elements and the parameters of normal B stars, an inclination close to 39° is estimated. The stars are nearly in contact; thus eclipses are a possibility.

119.009   Spectral classification and search method for cool companions of spectroscopic binaries in the near infrared.
V. D. Bychkov, Eh. A. Vitrichenko, A. G. Shcherbakov.
Izv. Krymskoj Astrofiz. Obs., Vol. 58, 81 - 88 (1978). In Russian.
Spectroscopic investigations of several standard stars and early-type single lined spectroscopic binaries in the region from 8300 to 8900 Å are presented. A quantitative excitation criterion of spectral classification of stars in the near infrared is found from observational data on main-sequence stars. The criterion is effective for the spectral classes A0 to K0. A method of search for cool companions of spectroscopic binaries is developed on the principles of spectral observations in the near infrared.

119.010   Does the system Theta Virginis represent a new class of pulsating star?
W. R. Beardsley, E. R. Zizka.
Rev. Mexicana Astron. Astrofis., Vol. 3, (see 012.023), 109 - 112 (1977).
Analysis of the radial velocity variation of θ Virginis A has revealed that, superposed on an orbital variation with a period of 17 years, there is present an ultrashort periodic variation of 0.15-day. This secondary cycle is attributed to pulsation. The question of what type of pulsating star θ Virginis A may represent is discussed.

119.011   Spectroscopic binary orbits from photoelectric radial velocities. Paper 20: HD 181330.
R. F. Griffin.
Observatory, Vol. 98, 118 - 119 (1978).

119.012   On the circularity of orbit of certain spectroscopic binaries.   E. E. Bassett.
Observatory, Vol. 98, 122 - 124 (1978).

119.013   Bright new spectroscopic binaries discovered from composite spectra, I. HR 6902, 9 Cygni, and HD 135774–5.   E. M. Hendry.
Astron. J., Vol. 83, 615 - 617 (1978).
The initial results of a search for spectroscopic binaries among known composite spectra are discussed. The radial velocities and orbital elements for three of these stars, HR 6902, 9 Cygni, and HD 135774–5 are presented.

119.014   Masses and luminosities for the spectroscopic/speckle interferometric binary 12 Persei.
H. A. McAlister.
Astrophys. J., Vol. 223, 526 - 529 (1978).
Speckle interferometric observations of the double-lined 331 day spectroscopic binary 12 Per are combined with the spectroscopic elements for a preliminary determination of the masses of both components and the distance to the system. Five speckle observations of angular separation and position angle were used to determine the additional elements: $a = 0.''057 \pm 0.''003$, $i = 123.°0 \pm 2.°0$, $\Omega = 227.°9 \pm 1.°5$. The values lead to masses of $M_1 = 1.25 \pm 0.20$, $M_2 = 1.08 \pm 0.17$ and a parallax of $0.''046 \pm 0.''002$. Adopting Colacevich's estimate of $\Delta m \approx 0.3$ then gives $M_{v1} = 3.8 \pm 0.1$ and $M_{v2} = 4.1 \pm 0.1$. The two stars are found to have their expected locations on the empirical mass-luminosity relation.

119.015   Photoelectric photometry of Sigma Geminorum and HR 4665.   C. R. Chambliss.
Inf. Bull. Variable Stars, No. 1432, 4 pp. (1978).

119.016   A spectroscopic reinvestigation of the massive binary HD 698.   J. B. Hutchings, J. E. Bernard.
Publ. Astron. Soc. Pacific, Vol. 90, 179 - 183 (1978) = Dominion Astrophys. Obs., Victoria, B.C., Canada, Contrib. No. 354 = NRC No. 16293.
The B7 (+ B?) spectroscopic binary HD 698 has been investigated using coudé spectrograms. The mass ratio is considered to be close to unity, indicating masses near 15 $M_\odot$ for the stars, rather than the much higher values proposed by Pearce (1932). The primary is in a post-main-sequence evolutionary stage and appears to have a stellar wind, and may be losing mass in excess of $10^{-6}\ M_\odot$ year^{-1}. It may have had an initial mass some 10% higher. The secondary is probably an ~B1 star near the end of its main-sequence lifetime. No tidal interactions are detectable in the system.

119.017   Observations of a second maximum for the very long period spectroscopic binary 5 Lacertae.
E. M. Hendry.
Publ. Astron. Soc. Pacific, Vol. 90, 184 - 187 (1978).
A study of 195 spectra of the bright composite star 5 Lac (K7 Ib + A8: V) has revealed it to be a long-period spectroscopic binary of high eccentricity (period = 41.95 years, $e = 0.49$). Orbital elements and a mass ratio for the system are presented as are comments on the variable H and K emission features.

119.018   Note on two new southern spectroscopic binaries.
F. Gieseking.
Publ. Astron. Soc. Pacific, Vol. 90, 204 - 206 (1978).
By use of the Fehrenbach objective-prism method two

new southern spectroscopic binaries (HD 67084 and HD 67198) have been discovered. The measured radial velocities of both stars are presented. Preliminary orbital elements are given for HD 67084.

119.019   Remarks on the radial velocity of 73 Draconis.
   E. Bartl.
Magnetic stars, (see 012.028), p. 12 (1978). − Abstract.

**119.020   A note on the eccentricity of the orbits of spectro-
   scopic binaries with Ap-components.**
E. Zhelwanowa, W. Schöneich.
Magnetic stars, (see 012.028), p. 28 - 29 (1978). − Abstract.

**119.021   Investigation of the double emission-line star
   V450 Cassiopeiae.**   G. A. Ponomareva.
Peremennye Zvezdy, Vol. 20, 435 - 440 (1977). In Russian.
   V450 Cas is the emission-line variable component of a double star. Variations of the combined light of the two components were investigated on 127 plates taken during 1949−76. The star shows irregular brightness fluctuations with rapid decreases and increases by $0\overset{m}{.}5 - 1^m$ in $1 - 2^d$. The colour indices of the variable component ($B-V = +1\overset{m}{.}7$, $U-B = +0\overset{m}{.}5 - +0\overset{m}{.}3$) give evidence for the presence of an ultraviolet excess after correction for interstellar reddening. The emission-line variable V450 Cas is similar to T Tauri stars by the character of its light variations and the position on the two-colour diagram.

**119.022   Polarimetric evidence for circumstellar material in
   RS CVn.**   R. J. Pfeiffer.
Bull. American Astron. Soc., Vol. 10, 418 (1978). − Abstract.

**119.023   HR 4665: an unusual new long period RS CVn
   binary.**   B. W. Bopp, F. Fekel.
Bull. American Astron. Soc., Vol. 10, 419 (1978). − Abstract.

119.024   Nu Centauri: a new Beta Cephei star.
   R. Rajamohan.
Kodaikanal Obs. Bull., Ser. A, Vol. 2, 6 - 9 (1977).
   Spectrographic observations of the single-lined binary ν Centauri during the years 1968 - 1972 were utilised to derive fresh orbital elements of this system. The measured radial velocities indicate that the primary component is a β Cephei type with a period of 0.1750 days.

119.025   The Wolf-Rayet spectroscopic binary HD 214419.
   M. K. V. Bappu, P. Visvanadham.
Kodaikanal Obs. Bull., Ser. A, Vol. 2, 89 - 94 (1977).
   Velocity curves of the system HD 214419 are obtained for the N IV 4058 and He II 4686 emission lines from prismatic spectra taken in 1951 and 1952. The emission and absorption features experience several phase dependent changes that are described in detail. A common envelope model ex-

plains well the He II 4686 light and velocity changes. The system parameters are best derived from N IV 4058.

**The Ca II λ8542 and λ8498 lines as important
indicators of stellar chromospheres.**   See Abstr. 114.027.

**Spectroscopic investigations of the magnetic Ap
star HD 9996.**   See Abstr. 114.553.

On the intrinsic rotation of magnetic variables.
See Abstr. 116.004.

**Kappa Pegasi − a quadruple system of possible low
total mass.**   See Abstr. 117.043.

**Coplanarity, masses, and stability of close multiple
stars.**   See Abstr. 117.076.

New double stars.   See Abstr. 118.004.

HD 165590 − a spectroscopic-visual triple system.
See Abstr. 118.009.

κ Pegasi, a triple system.
See Abstr. 118.010.

64 Ori − a mid-B type triple system.
See Abstr. 118.013.

**Sur la masse d'une binaire spectroscopique membre
d'une étoile double visuelle.**   See Abstr. 118.025.

**The masses of cool supergiants: the interacting
eclipsing system AZ Cassiopeiae.**   See Abstr. 121.010.

**YY Geminorum. I. Photometry and absolute
dimensions.**   See Abstr. 121.046.

AR and HK Lacertae.   See Abstr. 121.047.

**The period distribution of eclipsing and spectro-
scopic binary systems, I.**   See Abstr. 121.049.

The orbital period variation of SZ Psc.
See Abstr. 121.071.

**Evidence for the duplicity of the anomalous cepheid
Y Ophiuchi.**   See Abstr. 122.154.

Photoelectric observations of V 711 Tauri.
See Abstr. 123.026.

**UBV observations of HR 1099 on February 22nd,
1977.**   See Abstr. 123.032.

120   Variable Stars: Ephemerides, Miscellanea

120.001   Les étoiles extraordinaires.   G. Burki.
   Conférences d'astronomie de l'Observatoire, (see 012.004), p. 103 - 119 (1977).
   Étoiles jeunes; Étoiles doubles; Étoiles variables; Étoiles

superdenses; La nébuleuse du Crabe.

120.002   63rd name-list of variable stars.
   P. N. Kholopov, N. P. Kukarkina, N. B. Perova.
Inf. Bull. Variable Stars, No. 1414, 10 pp. (1978).

121 Eclipsing Binaries

121.001   Optical and infrared observations of β Lyrae.
R. Viotti, M. Ferrari-Toniolo, M. Marcocci,
G. Natali, P. Persi, G. Spada, P. Saraceno.
Astron. Astrophys., Vol. 62, 287 - 289 (1978).

Coordinated optical and infrared photometric observations of β Lyr at phase 0.14 have been used to study the energy distribution of the system. The analysis of the observational data shows that β Lyr has a large infrared excess. This can be explained by a free-free emission from an optically thick gaseous disk with an electron density distribution of the type $\rho \propto r^{-2.7}$, surrounding the secondary.

121.002   II. Revised photometric elements of the eclipsing binaries LY Aur, AS Eri, AW Peg, λ Tau.
B. Cester, B. Fedel, G. Giuricin, F. Mardirossian, M. Mezzetti.
Astron. Astrophys., Vol. 62, 291 - 294 (1978).

Photometric light curves of the double-lined spectrum close binary systems LY Aur, AS Eri, AW Peg, and λ Tau have been analyzed by Wood's program in order to obtain homogeneous photometric elements. Absolute masses, radii and luminosities are given. AS Eri and AW Peg are confirmed respectively as sd and sd-d systems; λ Tau might be an sd-system and LY Aur appears to be a contact system.

121.003   A detailed photometric study of the eclipsing binary AC Boo.   S. Mancuso, L. Milano, G. Russo.
Astron. Astrophys., Vol. 63, 193 - 198 (1978).

Four epoch lightcurves of AC Bootis, a W UMa system with W-type lightcurves, have been analysed by means of two different direct methods of solution, viz by Wilson and Wood's model respectively. The analysis of the gravity-brightening coefficients indicates that the components have convective envelopes. Moreover, the authors find that the hotter component is the one with lower mass, and that the degree of overcontact is small (0.06) for Wilson's model. Some discrepancies between the two methods are pointed out.

121.004   Infrared light curves of AM Herculis.
R. F. Jameson, R. Akinci, D. J. Adams, A. B. Giles, A. McCall.
Nature, Vol. 271, 334 - 335 (1978).

121.005   The linear polarization of VV Cephei into the 1977 eclipse totality.   R. H. Koch, R. J. Pfeiffer.
Astron. J., Vol. 83, 183 - 187 (1978).

Numerous filtered measures of the linear polarization of the supergiant eclipsing system VV Cep are presented. The observed polarization is shown to be variable on a long time scale. Limiting interstellar polarizations are evaluated. A modulation in the intrinsic red polarization is shown to be 180° out of phase with an independently observed light modulation in a cycle of about 150 days. Very small polarization decreases begin during the ingress phase of the body eclipse, but observations are few during this interval. It is shown that the red, green, and blue polarizations decrease after totality. Possible causes of this decrease are discussed.

121.006   Continuing mass transfer in U Cephei: observations and analysis outside primary eclipse.
E. C. Olson.
Astrophys. J., Vol. 220, 251 - 260 (1978).

Large, rapid changes in the light curves of U Cep outside eclipse were observed during transient states of high mass transfer. The most prominent feature was a broad, deep dip which appeared near orbital phase 0.6, and removed up to 46% of the ultraviolet light of the B star. Circumstellar-cloud and cool-spot models are explored as possible explanations; the latter is shown to be in substantial agreement with observations. Spots can evidently appear or disappear on a time scale of about one week, in response to an increase or decrease of the contemporary rate of mass transfer.

121.007   Light elements of TT Hydrae.
A. G. Kulkarni, K. D. Abhyankar.
Bull. Astron. Soc. India, Vol. 5, 105 (1977).

121.008   Spectroscopic observations of VV Cep during the ingress phase of the 1976/1977 eclipse.
C. Möllenhoff, K. Schaifers.
Astron. Astrophys., Vol. 64, 253 - 258 (1978).

The authors report about spectroscopic observations of the phase when the hot companion of VV Cep gradually disappeared behind the primary star. The hot companion is surrounded by an extended Balmer line emitting envelope. The variations of the Hβ and Hγ lines are followed by a densely spaced time series of spectrograms. The derived variations can be explained by a rotating ring-shaped envelope which gradually disappears behind the primary star.

121.009   Ultraviolet photometry from the Orbiting Astronomical Observatory. XXXI. The shape and atmospheric structure of the primary component of UW Canis Majoris.   J. A. Eaton.
Astrophys. J., Vol. 220, 582 - 591 (1978).

Modern, multicolor photoelectric light curves of the early-type eclipsing system UW CMa are presented and discussed. The shapes of the eclipses indicate that the Of primary has a sharp limb. Thus the atmosphere appears to consist of a photospheric region of low geometrical extension beneath an expanding region of low continuum optical depth. This expanding region is nonuniform; a stable condensation in the flow produces a complication in the light curve which is centered at phase 0.58 of secondary eclipse. At some wavelengths an analogous feature is detected during primary eclipse as well.

121.010   The masses of cool supergiants: the interacting eclipsing system AZ Cassiopeiae.
A. P. Cowley, J. B. Hutchings, D. M. Popper.
Publ. Astron. Soc. Pacific, Vol. 89, 882 - 895 (1977/78) = Dominion Astrophys. Obs., Victoria, B.C., Canada, Contrib. No. 347 = NRC No. 16106.

Data are presented from 63 spectrograms covering two cycles of the nine-year eclipsing system AZ Cas. The orbital elements are derived and it is seen that mass is lost from the M supergiant primary at periastron passage in an eccentric ($e = 0.55$) orbit. Several arguments suggest masses of ~13 and $18 M_\odot$ for the B and M stars, respectively. Spectroscopic phenomena are discussed which indicate the complexity of the mass flows in the system. A short discussion is given of the observed masses of cool supergiants.

121.011   Four-colour photometry of eclipsing binaries, XB: lightcurves of χ² Hydrae.
J. V. Clausen, B. Nordström.
Astron. Astrophys., Suppl. Ser., Vol. 31, 307 - 312 (1978).

uvby lightcurves of the bright double-lined B type eclipsing binary χ² Hydrae are presented. A detailed analysis of the lightcurves is published separately.

121.012   The infrared eclipse of V444 Cygni and the structure of Wolf-Rayet winds.   L. Hartmann.
Astrophys. J., Vol. 221, 193 - 199 (1978).

The density structure of the wind of the WN5 component of V444 Cyg is modeled by combining infrared eclipse observations of the free-free continuum with previous optical work. The results show the wind expands at a nearly constant velocity at large continuum optical depths, in contrast with present theories of O star winds driven by radiation pressure in lines.

121.013 **A photometric study of the Algol system TW Draconis.**   K. Walter.
Astron. Astrophys., Suppl. Ser., Vol. 32, 57 - 67 (1978) = Mitt. Astron. Inst. Univ. Tübingen Nr. 139.

Photoelectric observations in yellow and blue with the 40 cm reflector of the Astronomical Institute of the University of Tübingen obtained in Sicily in 1964–1965 and at Tübingen in 1969–1972 are presented and analysed. By the extension of the observations over a large number of nights mean lightcurves in yellow and blue were obtained which cover the phases, also outside of minima, rather uniformly. They were evaluated with regard to the presence of gas stream effects. A photometric solution was derived which was based on identical geometric properties of the system for the lightcurves of both colours and gives a satisfying representation of the observations according to the gas stream model of Algol systems.

121.014 **Secondaries of eclipsing binaries. I. Detection of the secondary of Delta Librae.**   J. Tomkin.
Astrophys. J., Vol. 221, 608 - 615 (1978).

The secondary Ca II infrared triplet lines of the Algol-type eclipsing binary δ Lib have been detected both during and outside of primary eclipse. Well-defined velocity curves for the primary and secondary are obtained. A mass ratio M_1/M_2 of 2.9 ± 0.1 and primary and secondary masses of 4.9 ± 0.2 and 1.7 ± 0.2 $M_\odot$ are derived. The evidence for a third member of the system are discussed and found to be unconvincing.

121.015 **Anomalous period change of UX Ursae Majoris in 1977.**   R. Quigley, J. Africano.
Bull. American Astron. Soc., Vol. 9, 557 - 558 (1978). Abstract.

121.016 **Detection of the secondary of Delta Librae.**   J. Tomkin.
Bull. American Astron. Soc., Vol. 9, 623 (1978). – Abstract.

121.017 **An application of synthetic light curves to the RS CVn-type eclipsing binaries.**   J. A. Eaton.
Bull. American Astron. Soc., Vol. 9, 624 (1978). – Abstract.

121.018 **Linearized limb-darkening coefficients for use in analysis of eclipsing binary light curves.**
H. M. Al-Naimiy.
Astrophys. Space Sci., Vol. 53, 181 - 192 (1978).

Linear limb-darkening coefficients u required in the analysis of eclipsing binary curves, are tabulated for a wide range of effective temperature (50 000° to 4000°), wavelength λ (0.2 $\mu \leqslant \lambda \leqslant$ 2.2 μ), and gravities g (2.0 $\leqslant$ log $g \leqslant$ 5.0). The computation is based on the comprehensive range of model atmospheres of Carbon and Gingerich (1969). The results are compared with the theoretical values of Hosokawa (1957), Kopal (1959) and Grygar et al. (1972), and examined in relation to empirically determined values of u from analyses of eclipsing binary light curves. An improved agreement between theory and observation for the calculated limb-darkening coefficients of the present work is noted.

121.019 **The evaluation of certain integrals encountered in the Fourier analysis of light curves of eclipsing variables.**   O. Demircan.
Astrophys. Space Sci., Vol. 53, 257 - 260 (1978).

121.020 **A model of V356 Sagittarii.**
R. E. Wilson, C. N. Caldwell.
Astrophys. J., Vol. 221, 917 - 925 (1978).

A model is advanced to explain the peculiar features of the unusual eclipsing binary V356 Sgr, for which previous analysis has not been very successful. The most obvious peculiarities were pointed out by Popper and consist of the absence of a significant reflection effect and anomalous depths for the eclipses. In the authors' model, a thick, opaque ring or disk of recently transferred matter surrounds the present primary star. The ring is rather substantial and is not similar to the emission-line rings in systems such as RW Tau and U Cep. They have produced a first-order, but quantitative, computer model of the disk (along with the rest of the system) and have obtained its approximate parameters by the method of differential corrections. The authors account for the presence of the disk as a consequence of the known fast rotation of its central star. Numerous observational arguments indicate that V356 Sgr corresponds to a later evolutionary stage of a system like β Lyr.

121.021 **On the role of photospheric convection in W Ursae Majoris stars.**   L. Anderson, F. H. Shu.
Astrophys. J., Vol. 221, 926 - 928 (1978).

The authors correct the derivation of the source function in the atmospheres of contact binaries given in an earlier communication by Anderson and Shu. This correction affects the cases when convection is present in the photosphere. In their new treatment photospheric convection is more efficient for reducing limb darkening. This result does not, however, modify the numerical examples considered in the earlier paper.

121.022 **The BQ[] evolved massive eclipsing binary RY Scuti.**   F. Ciatti, A. Mammano, R. Margoni, A. Vittone, G. Strazzulla.
IAU Colloq. No. 42, (see 012.012), p. 386 - 387 (1977).

Photometric and spectroscopic observations of RY Scuti are presented. These data indicate an advanced stage of this β Lyrae-type massive system, whose further evolution could lead to a SN-II event with the formation of a system like Cyg X-1.

121.023 **Contact binaries. III. Early-type systems.**   O. J. Eggen.
Astron. J., Vol. 83, 288 - 302 (1978).

About 1000 (u, v, b, y, β) observations of some 60 early-type contact systems are discussed. The photometric indices for most of the stars show little variation and are given in summary form. Loci in the ([$u-b$], M_V) plane for 0.75-, 1.5-, and 3.0-day contact binaries are constructed from evolutionary models of 4-, 6-, 8-, 12-, 15-, and 20-$M_\odot$ stars. Most of the field variables are reasonably well reproduced by these models and estimates of their age are therefore available. Several systems of special interest are discussed.

121.024 **LB3459, a short-period eclipsing binary system containing two O-type subdwarfs.**
D. Kilkenny, R. W. Hilditch, J. E. Penfold.
Mon. Not. R. Astron. Soc., Vol. 183, 523 - 531 (1978).

LB 3459 (=CPD −69°389 = HDE 269696) has been found to be an eclipsing binary system with a period of 0.261539 day. A preliminary analysis of the available photometric data, high (time) resolution observations during eclipse as well as conventional $uvby$ photometry, demonstrates that both components of the system are O-type subdwarfs. The orbital inclination is 90°. The importance of this system together with problems concerning the effective temperatures appropriate for each component are emphasized.

121.025 **BV 845: a contact system with a large temperature difference between components.**
K.-C. Leung, J. J. Darland.

Publ. Astron. Soc. Pacific, Vol. 90, 97 - 100 (1978).

The photoelectric observations of Chambliss are re-analyzed with the Wilson and Devinney approach. BV 845 is found to be a marginal contact system, having a mass ratio of 0.32. There is a very large difference (3800 K) in polar temperature between the components. Its location in the period–spectral-type diagram suggests that this is an evolved system. The authors' improved photometric solution is different from that of Chambliss.

121.026 The eclipsing Small Magellanic Cloud Wolf-Rayet binary HD 5980.   M. Hoffmann, M. J. Stift, A. F. J. Moffat.
Publ. Astron. Soc. Pacific, Vol. 90, 101 - 104 (1978).

A photoelectric light curve showing eclipses is presented for the first time for an extragalactic W-R binary HD 5980. The period is 25.56 ± 0.08 days, rendering the separation of the two stars possibly large enough to minimize interaction effects so that a detailed study of the stratification of a W-R envelope is feasible.

121.027 Period–mass ratio relations for eclipsing binaries with periods not exceeding 5 days.
S. D. Sinvhal, J. B. Srivastava.
Astrophys. Space Sci., Vol. 54, 239 - 245 (1978).

The dependence of the mass ratio of eclipsing binary systems on the period has been studied. It has been found that semi-detached systems obey the relation $m_2/m_1 = 0.64 - 0.079P$ or $m_2/m_1 = 0.29 - 0.028P$ depending upon whether the mass ratio is > 0.3 or < 0.3. The mass ratio of detached systems generally appears to be independent of the period.

121.028 Lichtkurven- und Periodenschwankungen bei AH Virginis.   K. Porzel.
BAV Rundbrief,27. Jahrg., 8 - 12 (1978).

121.029 Interpretation of Beta Lyrae. IV. Hydrogen emission profiles and the pattern of gaseous flow in the system.   S.-S. Huang.
Astrophys. J., Vol. 222, 627 - 634 (1978).

The author presents an interpretation of hydrogen emission profiles found in β Lyrae by attributing the asymmetry of the two emission peaks of both Hα and Hγ at different phases to the absorption of an expanding-rotating gaseous stream. He proposes that different relative shifts of the absorption line with respect to the emission line exist at different phases resulting in an apparent change in the profile with phase. Based on this idea, the expected profiles at four different phases have been computed for the Hα and Hγ lines. They show a general trend close to what has been observed. Based on this same idea an apparent discrepancy between the observed results of the Hγ profile obtained at two different epochs about 15 years apart becomes understandable. Finally a physical picture that can produce this periodic shift of the emission line relative to the absorption line is proposed and discussed. A general pattern of the gaseous flow in the system is derived from this study.

121.030 Hot/cool spots observed in the ANS ultraviolet light curves of U Cephei.
Y. Kondo, G. E. McCluskey, C.-C. Wu.
Astrophys. J., Vol. 222, 635 - 640 (1978).

Ultraviolet light curves of the interacting eclipsing binary U Cephei, reported to have been undergoing major outbursts in 1974 through 1975, were obtained in 1975 with the Astronomical Netherlands Satellite (ANS) ultraviolet photometers. A 21 data-point light curve was obtained at each of the 1550, 1800, 2200, 2500, and 3300 Å wavelength regions. Two very unusual features have been noted in the light curves. (a) The second maximum light at phase 0.75 is decidedly brighter than the first maximum at phase 0.25 by a factor of about 1.3 ∼

1.8. (b) The secondary minimum is displaced, from phase 0.5, to 0.6 in all of the light curves and is deeper at shorter wavelengths, contrary to what might be expected in an eclipsing binary consisting of two normal stars radiating approximately in a Planckian fashion. This anomalous behavior in the light curves may be explained in terms of a hot spot on the trailing side of the B component and a "lesser" hot spot on the trailing side of the G component, both of which result from the impacting gas streams in this binary.

121.031 Revised spectrographic and photometric elements of ZZ Boo.   B. Cester, G. Giuricin, F. Mardirossian, M. Mezzetti.
Astron. Astrophys., Suppl. Ser., Vol. 32, 347 - 350 (1978).

Some inconsistencies between spectrographic and photoelectric elements of the double spectrum eclipsing binary ZZ Boo revealed that a set of spectrographic observations should be interchanged. With the newly deduced elements the star appears to be a normal detached system on the upper border of the main sequence. Both components agree well with an isochrone of age 1×10^9 yrs and chemical composition $Y = 0.30$, $Z = 0.004$.

121.032 Revised photometric elements of ten detached systems.   B. Cester, B. Fedel, G. Giuricin, F. Mardirossian, M. Mezzetti.
Astron. Astrophys., Suppl. Ser., Vol. 32, 351 - 359 (1978).

Photometric lightcurves of 10 detached double-lined eclipsing binary systems with spectra later than A0, have been analyzed by means of the Wood programme in order to obtain homogeneous photometric elements. Masses, radii and luminosities are given. RT And, V805 Aql, ZZ Boo, EI Cep, VZ Hya, TX Her, CM Lac and WZ Oph are easily interpreted in the frame of current evolutionary views. The anomalous behaviour of WX Cep and UV Leo and some misunderstandings concerning the role of the components of ZZ Boo and EI Cep are discussed.

121.033 The variable lightcurve of 44 i Boo observations and implications.   H. W. Duerbeck.
Astron. Astrophys., Suppl. Ser., Vol. 32, 361 - 374 (1978).

B and V lightcurves of the W UMa system i Boo, obtained between 1975 and 1977, are presented and discussed. One U lightcurve is also available. The lightcurves are strongly variable with time. An analysis indicates that the intensity ratio of the components is also variable, while the combined brightness of the stars has remained constant to ± 0^m003 over three years. Minor time-dependent irregularities of the lightcurves are interpreted as resulting from a stream of matter concentrated near the inner Lagrangian point of the binary system, which seems to be in marginal contact only.

121.034 Bestimmung der Systemkonstanten von Beta Persei.   B. Kunzmann.
BAV Rundbrief, 27. Jahrg., 12 - 14 (1978).

121.035 MR Cyg and u Her: two similar eclipsing binaries.   S. Söderhjelm.
Astron. Astrophys., Vol. 66, 161 - 168 (1978).

Most of the available photoelectric observations of u Her and MR Cyg are analyzed, and photometric elements are determined. Several sets of elements are found which give almost equal theoretical lightcurves. Both MR Cyg and u Her are semi-detached, post mass-transfer objects. For these, and for a number of similar massive systems, the period-changes are found to be smaller than for more "typical" Algols of smaller mass.

121.036 Spectrophotometric study of the eclipsing-variable system V599 Aql.   T. M. Rachkovskaya.
Izv. Krymskoj Astrofiz. Obs., Vol. 58, 56 - 63 (1978).
In Russian.

The spectrum of the eclipsing-variable system V599 Aql has been studied and radial velocity measurements have been obtained. It is shown that the system V599 Aql consists of two main-sequence stars of spectral type B3.5; $M_{V_1} = -1.75$; $M_{V_2} = -0.51$. Electron pressure and relative content of hydrogen and helium in the atmosphere of the primary are the same as in the atmospheres of normal stars. The following parameters have been obtained: axial rotation, radii.

121.037 Absolute spectrophotometry of β Lyr.
V. I. Burnashev, M. Yu. Skul'skij.
Izv. Krymskoj Astrofiz. Obs., Vol. 58, 64 - 74 (1978).
In Russian.

A model of the system β Lyr is proposed on the basis of energy distribution measurements in its spectra. The model consists of B8 III and A5 III stars and a gaseous envelope having $T_e = 2 \times 10^4$ K.

121.038 Investigation of the emission line of He I λ 10830 Å in the spectrum of Beta Lyrae.
M. B. Girnyak, M. Yu. Skul'skij, G. I. Shanin, A. G. Shcherbakov.
Izv. Krymskoj Astrofiz. Obs., Vol. 58, 75 - 80 (1978).
In Russian.

Observations of the He I λ10830 Å line in the spectrum of β Lyr show that about 70 per cent of the radiation from this line are emitted by an external envelope surrounding the spectroscopic binary. The outer regions of the envelope expand with a velocity of about 130 km/s. The remaining 30 per cent of radiation may be formed by gas streams and a ring which envelops the faint component.

121.039 RS Canum Venaticorum binaries. P. P. Eggleton.
Bull. Astron. Soc. India, Vol. 6, 9 - 11 (1978).

121.040 Evolved contact systems of spectral type A: AU Puppis, V535 Arae, and V1073 Cygni.
K.-C. Leung, D. P. Schneider.
Astrophys. J., Vol. 222, 917 - 923 (1978).

The photoelectric observations of AU Pup, V535 Ara, and V1073 Cyg were reanalyzed with Wilson and Devinney's approach. All three systems are found to be contact configurations. AU Pup has an unusually large degree of overcontact, 72%. V1073 Cyg is found to be an evolved contact system since the radius of the primary is twice that of a zero-age main-sequence star of corresponding mass. All three systems are found to be evolved according to their locations in the period-spectral-type diagram. Improved photometric parameters are reported for all three systems.

121.041 Evolved contact systems of spectral type O. I. UW Canis Majoris.
K.-C. Leung, D. P. Schneider.
Astrophys. J., Vol. 222, 924 - 930 (1978).

The light curves of UW CMa by Seyfert and by Ogata and Hukusaku were analyzed using the Wilson and Devinney approach. The system is found to be an evolved contact system. The unusual variations in the light curves may be caused by a highly nonuniform surface brightness of the common envelope. The photometric solution favors a mass ratio m_2/m_1 less than unity. It is suggested that UW CMa has components of 46 and 34 $M_\odot$, making it one of the most massive contact systems known so far.

121.042 Detection of the secondary of Algol.
J. Tomkin, D. L. Lambert.
Astrophys. J., Lett., Vol. 222, L119 - L122 (1978).

The authors have detected the Na D lines of the secondary of Algol (β Persei). The velocity curve obtained from 10 measurements of the velocity of the secondary has a semi-amplitude of 201 ± 6 km s^{-1}. A mass ratio, m_A/m_B, of 4.6 ±0.1 and masses m_A, m_B, and m_C of 3.7 ± 0.3, 0.81 ± 0.05, and 1.7 ± 0.2 $m_\odot$, respectively, are derived. The secondary fills its Roche lobe.

121.043 Four-colour photometry of eclipsing binaries. IX b. AI Hya, light curves and photometric elements.
H. E. Jørgensen, B. Grønbech.
Astron. Astrophys., Vol. 66, 377 - 383 (1978).

Photometric elements are derived for the eclipsing binary AI Hya. The elements are very well determined since the primary minimum is a totality. The bigger and cooler star seems to be a δ Scuti variable with a pulsation constant $Q = 0.0259 \pm 0.0004$ corresponding to a radial first overtone oscillation.

121.044 A period for RY Lyncis.
G. Samolyk, G. Wedemayer.
J. American Assoc. Variable Star Obs., Vol. 6, 49 - 50 (1977/78).

Newly determined times of minima, combined with published values, permit a determination of the period, 1.43497 days, for this eclipsing binary.

121.045 Four-colour photometry of eclipsing binaries, XA: photometric elements, absolute dimensions and
helium abundance of χ² Hydrae. J. V. Clausen, B. Nordström.
Astron. Astrophys., Vol. 67, 15 - 22 (1978).

The light curves of the double-lined B-type eclipsing binary χ² Hydrae have been observed in the Strömgren four-colour *uvby* system. The model of an eclipsing binary and the light curve simulation programme WINK described by Wood (1971, 1972) have been used to determine the photometric elements. Combination of these with the spectroscopic data by Andersen (1975) yields the absolute dimensions.

121.046 YY Geminorum. I. Photometry and absolute dimensions. K.-C. Leung, D. P. Schneider.
Astron. J., Vol. 83, 618 - 625 (1978).

Over 2000 photoelectric observations in V, R, and I of YY Geminorum (dMle + dMle) from six nights in 1971 were analyzed. There is evidence of a period increase since 1950. Secondary variations were detected between phases 0.52–0.90. It is suggested that the secondary variations are the result of a starspot on the secondary rather than on the primary as was reported by Kron (1952). A photometric solution was derived, and the absolute dimensions for both components were determined by combining the photometric solution and the spectroscopic solution of Bopp (1974). The effective temperature calculated from the luminosities derived from parallax and radii were higher by about 200 K than the value indicated by Johnson's infrared color temperature calibration for lower main sequence stars.

121.047 AR and HK Lacertae.
IAU Circ., No. 3208 (1978).

121.048 UU Sagittae: eclipsing nucleus of the planetary nebula Abell 63. H. E. Bond, W. Liller, E. J. Mannery.
Astrophys. J., Vol. 223, 252 - 259 (1978) = Contrib. Louisiana State Univ. Obs., Baton Rouge, No. 133.

UU Sge, the central star of the low-surface-brightness planetary nebula Abell 63, is shown to be an eclipsing binary with an orbital period of 0.465 days. Primary minimum is due to total eclipse of the very hot primary star by a cool secondary. The observational material is consistent with a system containing a sdO primary of ~ 0.9 $M_\odot$ and 0.4 $R_\odot$ and a dK secondary of ~ 0.7 $M_\odot$ and 0.7 $R_\odot$, separated by ~ 3 $R_\odot$. The hot star heats the hemisphere of dK star facing it to ~ 10,000 K. Evidence is assembled indicating that nebula ejection by binary-star processes may not be uncommon.

121.049 The period distribution of eclipsing and spectro-

scopic binary systems, I.
P. Farinella, P. Paolicchi.
Astrophys. Space Sci., Vol. 54, 389 - 406 (1978).

The authors analyse the period distribution of eclipsing and spectroscopic binary systems, using various methods to take into account selection effects on observational data, with the purpose of deriving the actual distribution curve. The discussion of results shows the presence of some secondary maxima in the distribution which are probably of a non-statistical character. They could be regarded as an indirect clue to the discrete nature of the star formation process, according to the spiral density wave theory of galactic evolution by Lin et al. (1969).

121.050   The W Ursae Majoris system TZ Bootis.
          M. Hoffmann.
Astron. Astrophys., Suppl. Ser., Vol. 33, 63 - 85 (1978).

Spectroscopic and photoelectric observations of TZ Boo are presented, earlier material of this system is rediscussed. Indications for a third body in the system have been found. The radial velocity curve seems to be disturbed by the influence of circumstellar matter. The period probably varies. Some geometrical and physical elements could be determined. The mass ratio has been found to be about 0.22. Strong lightcurve changes have been detected and are explained by the occurrence of both solar-like starspots and circumstellar matter. The large UV-excess of TZ Boo is discussed briefly.

121.051   Four-colour photometry of eclipsing binaries. XI B:
          Lightcurves of RS Chamaeleontis.
J. V. Clausen, B. Nordström.
Astron. Astrophys., Suppl. Ser., Vol. 33, 87 - 90 (1978).

uvby lightcurves of the bright double-lined A type eclipsing binary RS Chamaeleontis are presented.

121.052   Revised photometric elements of 14 detached sys-
          tems.   B. Cester, B. Fedel, G. Giuricin,
F. Mardirossian, M. Mezzetti.
Astron. Astrophys., Suppl. Ser., Vol. 33, 91 - 102 (1978).

Photometric lightcurves of 14 double-lined eclipsing binaries, known as detached systems, have been analyzed by means of the Wood programme in order to obtain homogeneous photometric elements. Masses, radii, and luminosities are given. V382 Cyg appears to be a contact binary. σ Aql, WW Aur, AR Aur, CW Cep, V453 Cyg, RX Her, and U Oph are normal detached systems easily interpreted in the context of current evolutionary views. The detached binaries ZZ Cep and IM Mon are shown to be anomalous. The detached nature of AH Cep, TT Aur, MR Cyg, and VV Ori turns out to be doubtful. Some incoherencies concerning the three latter binaries are also discussed.

121.053   Four-colour photometry of eclipsing binaries. IXa:
          AI Hya, photometric observations.
H. E. Jørgensen, B. Grønbech.
Astron. Astrophys., Suppl. Ser., Vol. 33, 103 - 105 (1978).

Photometric observations in the uvby system are carried out for the eclipsing binary AI Hya.

121.054   Fourier analysis of the light curves of eclipsing
          variables, XIV.   Z. Kopal, O. Demircan.
Astrophys. Space Sci., Vol. 55, 241 - 261 (1978).

The aim of the paper is to utilize the results obtained in the preceding papers of this series for the development of practical procedures for obtaining the elements of any eclipsing system from the observed photometric data by their analysis in the frequency-domain, for any type of eclipses, any proximity of the two components, and any degree of the law of limb-darkening of the eclipsed star.

121.055   Times of minima of eclipsing variables.
          C. D. Scarfe, D. J. Barlow.

Inf. Bull. Variable Stars, No. 1379, 4 pp. (1978).

121.056   UV Psc – a short period RS CVn star.
          A. R. Sadik.
Inf. Bull. Variable Stars, No. 1381, 4 pp. (1978).

121.057   Some comments on AB Cas.
          B. N. Irkaev, O. A. Chekanikhina, M. S. Frolov.
Inf. Bull. Variable Stars, No. 1382 (1978).

121.058   Photoelectric minima of eclipsing variable VW
          Cephei.   C. Cristescu.
Inf. Bull. Variable Stars, No. 1383 (1978).

121.059   118 day optical variations in VV Cep.
          G. P. McCook, E. F. Guinan.
Inf. Bull. Variable Stars, No. 1385, 3 pp. (1978).

121.060   KO Aur.   T. Berthold.
          Inf. Bull. Variable Stars, No. 1396 (1978).

121.061   Elements of CSV 260.   H. Busch.
          Inf. Bull. Variable Stars, No. 1397 (1978).

121.062   Photoelectric photometry of the multiple star
          system Eta Orionis.   C. R. Chambliss.
Inf. Bull. Variable Stars, No. 1398, 4 pp. (1978).

121.063   Period changes of the binary system RX
          Cassiopeiae.   J. M. Kreiner.
Inf. Bull. Variable Stars, No. 1403, 3 pp. (1978).

121.064   The distortion wave in SS Bootis – direct migration.
          D. S. Hall, G. W. Henry.
Inf. Bull. Variable Stars, No. 1404, 3 pp. (1978).

121.065   BO And, a large amplitude eclipsing binary.
          E. H. Geyer, A. Hänel.
Inf. Bull. Variable Stars, No. 1405, 4 pp. (1978).

121.066   Photoelectric V light curve and minima of RT And.
          S. Mancuso, L. Milano, G. Russo, C. Sollazzo.
Inf. Bull. Variable Stars, No. 1409, 3 pp. (1978).

121.067   Photoelectric minima and light curves of the
          eclipsing variable DO Cassiopeiae.   O. Tümer.
Inf. Bull. Variable Stars, No. 1413 (1978).

121.068   Observations of UV Psc and XY Cet.
          H. Al-Naimiy, E. Budding, D. Jassur, A. R. Sadik.
Inf. Bull. Variable Stars, No. 1415 (1978).

121.069   New photoelectric measures of Eta Orionis.
          R. H. Koch.
Inf. Bull. Variable Stars, No. 1418, 4 pp. (1978).

121.070   Photometry of AI Phe.   B. Reipurth.
          Inf. Bull. Variable Stars, No. 1419 (1978).

121.071   The orbital period variation of SZ Psc.
          S. Catalano, A. Frisina, M. Rodono, F. Scaltriti.
Inf. Bull. Variable Stars, No. 1427 (1978).

121.072   The far ultraviolet spectrum of the binary system
          Epsilon Aurigae.   M. Hack, P. L. Selvelli.
Inf. Bull. Variable Stars, No. 1439 (1978).

121.073   New eclipsing binary system HR 3872A.
          S. M. Jakate.
Inf. Bull. Variable Stars, No. 1440, 3 pp. (1978).

121.074   A discussion of period changes in the white dwarf

eclipsing binary system V471 Tauri.
J. P. Oliver, S. M. Rucinski.
Inf. Bull. Variable Stars, No. 1444, 4 pp. (1978).

121.075 **Osservazioni delle binarie ad eclisse RZ Cancri ed Y Camelopardalis.** P. Broglia, P. Conconi.
Pubbl. Oss. Astron. Milano-Merate, Nuova Ser., No. 27, 23 pp. (1973).

The photoelectric observations of the two eclipsing systems RZ Cnc and Y Cam, obtained at the Merate Observatory, are given.

121.076 **Determinación fotométrica del tipo espectral de la componente desconocida de una estrella binaria eclipsante. Aplicación a la variable eclipsante V 453 Scorpii.** T. J. Vives.
Rev. Real Acad. Cienc., Madrid, Tomo 70, 773 - 818 (1976) = Univ. Complutense – Fac. Cienc., Madrid, Semin. Astron. Geod., Publ. núm. 93.

121.077 **Program for an analysis of light curves of eclipsing binary systems with reciprocal methods with the help of the NAIRI computer.** M. I. Lavrov, N. V. Lavrova.
Tr. Kazan. Gorod. Astron. Obs., No. 41, p. 3 - 23 (1976). In Russian.

121.078 **Sphere – ellipsoid model for semi-detached eclipsing binary stars and its application to the RW Tauri system.** M. I. Lavrov, N. V. Lavrova.
Izv. Astron. Ehngel'gardt. Obs., Kazan', No. 41 - 42, p. 196 - 208 (1976). In Russian.

121.079 **Express method for determination of the orbital elements of eclipsing binaries taking into account the limb darkening coefficient.** E. I. Lenderman.
Peremennye Zvezdy, Prilozhenie, Vol. 3, 149 - 195 (1977). In Russian.

Tables of functions for Tsesevich's express method for a partial occultation eclipse and the limb darkening coefficients from 0.0 to 1.0 are given. They were used for the solution of the light curve of RS Lep.

121.080 **Variable stars in a region of intensive star formation.** N. E. Kurochkin.
Peremennye Zvezdy, Vol. 20, 325 - 330 (1977). In Russian.

Three variable stars in the region of the T-association Cyg T-2 are investigated. One star "w" is probably related to the FU Ori-type (Wenzel, Geßner, 1975). Two others, V 502 Cyg and "c" (202142), are eclipsing variables. Probably, they are β Lyrae stars at an early evolutionary stage. The star "c" is connected with a nebulosity or globule.

121.081 **CX Serpentis.** V. P. Tsesevich.
Peremennye Zvezdy, Vol. 20, 331 - 343 (1977). In Russian.

It is shown that the light curve of this eclipsing star is similar to that of an "old" nova.

121.082 **A preliminary study of the spectrum of BM Orionis.** V. S. Shevchenko, M. M. Zakirov.
Peremennye Zvezdy, Vol. 20, 361 - 368 (1977). In Russian.

BM Ori was observed spectroscopically near primary minimum (phases 0.0066 and 0.0154) and out of eclipse (phases 0.3268 and 0.7933). The star has no own emission in the hydrogen lines, which does not confirm Huang's hypothesis. The spectrum has no indication of lines of a secondary component at primary minimum and out of eclipse. The authors' interpretation of the observations of BM Ori is that its secondary component is a star of small luminosity and small mass ($< 0.3\ M_\odot$) and has an extensive dust envelope.

121.083 **The close binary system V337 Aql. I. Photoelectric observations.** V. Ya. Alduseva.
Peremennye Zvezdy, Vol. 20, 375 - 379 (1977). In Russian.

In 1974–75 UBV photometry of V337 Aql was carried out. An increase of the light amplitude (compared with the results of photoelectric observations in 1965–66) of about 0^m05 was found. The masses of both components are found to be $M_1 \sim 16\ M_\odot$ and $M_2 \sim 10\ M_\odot$. For a mass ratio ~ 0.6 both components are nearly filling their critical Roche lobes. Variations of the light curve are supposed to be connected with mass exchange between the components.

121.084 **Representation of the light curve of an eclipsing binary with a single system of elements (partial eclipse).** A. M. Shul'berg.
Peremennye Zvezdy, Vol. 20, 461 - 466 (1977). In Russian.

The analysis of a synthetic light curve shows that the problem of determination of elements of an eclipsing system in which partial eclipses occur has several solutions. The method of representation of a light curve of an eclipsing binary star by a single system of elements (in case of partial eclipse) is suggested.

121.085 **Photoelectric observations of SZ Camelopardalis.** T. S. Polushina.
Peremennye Zvezdy, Vol. 20, 473 - 480 (1977). In Russian.

245 photoelectric UBV–observations of SZ Cam made in 1971 are presented. Epoch of primary minimum is $JD_\odot = 2441252.3810$.

121.086 **Investigations of eclipsing binary systems in the USSR during the period 1972 - 1975.** A. M. Cherepashchuk.
Peremennye Zvezdy, Vol. 20, 481 - 492 (1977). In Russian.

121.087 **Current observations of the white dwarf eclipsing binary V471 Tauri.** T. D. Oswalt.
Bull. American Astron. Soc., Vol. 10, 410 (1978). – Abstract.

121.088 **Differential spectrophotometry of massive eclipsing binary stars.** G. L. Clements, J. S. Neff.
Bull. American Astron. Soc., Vol. 10, 410 - 411 (1978). Abstract.

121.089 **Synthetic light curve solution for the binary star system AK Her in R and I.** T. A. Nagy.
Bull. American Astron. Soc., Vol. 10, 411 (1978). – Abstract.

121.090 **The eclipsing binary V78 Omega Centauri.** M. H. Liller, S. M. Lichten.
Bull. American Astron. Soc., Vol. 10, 411 (1978). – Abstract.

121.091 **Photoelectric observations and apsidal motion of Y Cygni.** D. J. K. O'Connell.
Ric. Astron., Specola Vaticana, Castel Gandolfo, Vol. 8, (No. 29), 543 - 554 (1977).

Photoelectric observations of Y Cygni are presented. The apsidal motion of the system is rediscussed.

121.092 **On emission sources observed in the spectrum of β Lyrae.** A. N. Dadaev.
Astron. Tsirk., No. 966, p. 3 - 4 (1977). In Russian.

121.093 **The coefficients of linear limb darkening for the components of selected eclipsing binary systems.** M. I. Lavrov.
Astron. Tsirk., No. 971, p. 1 - 2 (1977). In Russian.

121.094 **Revision of the photometric elements of EK Cep.** M. I. Lavrov, N. V. Lavrova.
Astron. Tsirk., No. 971, p. 3 - 4 (1977). In Russian.

121.095 The period and photoelectric light curves of AB Andromedae.   Z. Tüfekçioğlu.
Commun. Fac. Sci. Univ. Ankara, Sér. A₃: Astron., Tome 25, 25 - 33 (1977) = Commun. Astron. Dep. Univ. Ankara, No. 77.
    Photoelectric observations of AB Andromedae made during the fall and winter of 1976 for seven nights are presented. Three primary and three secondary minima are found. The period change is investigated. New light curves are given.

121.096 The period increase of BX Andromedae.
    L.-z. Fang, Y.-y. Zhou, F.-z. Cheng, Y.-q. Chu, L.-t. Shen, S.-y. Jiang.
Kexue Tongbao, Vol. 22, 431 - 432 (1977). In Chinese. From Phys. Abstr., Vol. 81, Abstr. 36642 (1978).

121.097 Non-linear limb darkening coefficients for the components of four eclipsing binary systems.
M. I. Lavrov.
Astron. Tsirk., No. 990, p. 6 - 7 (1978). In Russian.

121.098 69th and 70th list of minima of eclipsing binaries.
    C. Agnesoni, P. Albert, M. Benucci, R. Boninsegna, J. Bourgeois, A. Buzzoni, T. Carradori, J.-P. Clovin, P. Danthine, R. Diethelm, J.-L. Duquesne, F. Ferraro, M. Franchini, M. Frangeul, A. Frère, R. Germann, Z. Hevesi, R. Leyman, K. Locher, E. Lucentini, E. Nezry, C. Pampaloni, A. del Parigi, H. Peter, R. Piazza, C. Plasmati, E. Poretti, P. Ralincourt, J. Remis, A. Royer, G. Troispoux, F. Vespe, N. Zaccaria, A. Livi, M. Penna.
BBSAG Bull., No. 36, p. 1 - 5, No. 37, p. 1 - 5 (1978).

121.099 A reinterpretation of BL Eridani.   K. Locher.
BBSAG Bull., No. 36, p. 6 (1978).

121.100 A finding chart for NO Vulpeculae.   K. Locher.
BBSAG Bull., No. 37, p. 5 (1978).

121.101 Synthetic light curves of two eclipsing binary systems, U Sge and AW UMa.   H. M. K. Al-Naimiy.
Astrophys. Space Sci., Vol. 56, 219 - 238 (1978).
    Photometric curve fits have been investigated by means of numerical quadratures to develop theoretical light curves appropriate to stars built up in accordance with the Roche model. The method has been applied to two systems with contact components. (1) U Sge, spherical primary totally eclipsed by a contact component secondary. Improved photometric elements of the system have been found, and compared with those obtained by Kopal's method in the frequency domain. The outcome of the curve fitting corresponds well with the results of an analysis in the frequency domain. (2) AW UMa, exhibiting the shallowest minima known for totally eclipsing W UMa systems. The physical and geometrical elements of the system have been found, and the contact nature of the two components confirmed.

121.102 Nuove variazioni di periodo e di curva di luce del sistema binario W UMa.
L. Baldinelli, S. Ghedini.
Mem. Soc. Astron. Italiana, Vol. 48, 83 - 90 (1977).
    Photoelectric observations of the eclipsing binary W UMa are reported. They confirm a new change in the period and light curve of this variable star.

121.103 The long period eclipsing binary EE Cephei.
    L. Baldinelli, S. Ghedini.
Mem. Soc. Astron. Italiana, Vol. 48, 91 - 102 (1977).
    Photoelectric and photographic observations of EE Cephei minima are reported. Brief discussions about the spectral class, light curve variations and the probable presence of a variable star in the comparison sequence follow.

    Application of the regularization method to the solution of inverse problems of astrophysics.
See Abstr. 021.016.

    Applications of stellar surface brightness to eclipsing binaries and nearby stars.   See Abstr. 113.032.

    Spectroscopic study of the eclipsing binary V 367 Cygni.   See Abstr. 114.549.

    Masses of hot main-sequence stars.
See Abstr. 115.001.

    The nature of Beta Lyrae and its radio emission.
See Abstr. 116.001

    Is the close binary XY Ursae Majoris a radio star?
See Abstr. 116.008.

    A polarimetric determination of binary inclinations: results for five systems.   See Abstr. 117.014.

    Observational evidence for mass exchange in close binary systems.   See Abstr. 117.021.

    Young W UMa binaries and the interaction with their environment.   See Abstr. 117.022.

    Period changes of W Ursae Majoris stars.
See Abstr. 117.023.

    The spectroscopic orbit of HD 47732.
See Abstr. 119.008.

    Progressi nell'interpretazione e studio delle stelle variabili.   See Abstr. 122.029.

    The photometric variability of 1 Persei.
See Abstr. 122.033.

    Spectrophotometric observations of the nova-like variable RW Trianguli.   See Abstr. 122.044.

    A new nova-like variable LS 55°-8.
See Abstr. 122.046.

    Z Chamaeleontis – Schlüssel zum Verständnis der U-Gem-Sterne.   See Abstr. 122.117.

    The U Geminorum star, EY Cygni, and a probable eclipsing star, V839 Cygni.   See Abstr. 122.128.

    On the membership of the RR$_S$ star V 65 and the EA binary V 78 in Omega Centauri (NGC 5139).
See Abstr. 122.145.

    Rotating variable stars.   See Abstr. 122.162.

    Researches with the Schmidt telescopes. VIII. The stars in a field around Beta Persei.   See Abstr. 122.204.

    Seven variable stars in Sagittarius.
See Abstr. 123.006.

    The period of SX Cassiopeiae.
See Abstr. 123.039.

    Photometry of AM Herculis: a slow optical pulsar?
See Abstr. 142.017.

    HR 1099 (= 4U 0336+01?)
See Abstr. 142.160.

V861 Scorpii = OAO 1653–40.
See Abstr. 142.178.

Optical properties of the accretion disk in the system HZ Her (Her X-1).   See Abstr. 142.230.

A search for new variables in the globular cluster Omega Centauri.   See Abstr. 154.029.

Errata

121.901 Errata: "New coefficients of polynomials approximating eclipsing binary solutions" [Bull. Astron. Inst. Czechoslovakia, Vol. 27, 301 - 308 (1976)].
M. I. Lavrov, N. G. Sokolova.
Bull. Astron. Inst. Czechoslovakia, Vol. 29, 64 (1978).

122 Intrinsic Variables, Spectrum Variables, Flare Stars, Pulsation Theory

122.001 **Airborne infrared spectrophotometry of Mira variables.**   D. W. Strecker, E. F. Erickson, F. C. Witteborn.
Astron. J., Vol. 83, 26 - 31 (1978).
    Infrared (1.2–4.0 μ, 2% resolution) spectra of the Mira variables R Cas, R Leo, and NML Tau (IK Tau) which have been measured using the Kuiper Airborne Observatory are reported.

122.002 **New silicon monoxide masers.**   D. F. Dickinson, L. E. Snyder, L. W. Brown, D. Buhl.
Astron. J., Vol. 83, 36 - 40 (1978).
    Seven new Mira variable stars with silicon monoxide maser emission have been found. All are later than M4 and heavily reddened. Two were observed near maximum light. One, Y Cas, seems to bear out the correlation of 43-GHz SiO maser luminosity and mean spectral type at maximum light. RT Aql, however, falls about 1 mag below the expected brightness. A catalog of SiO maser stars is given.

122.003 **Intrinsic colours and physical properties of Cepheids.**   J. W. Pel.
Astron. Astrophys., Vol. 62, 75 - 94 (1978).
    This paper presents a discussion of extensive photometry in the Walraven *VBLUW* system for 170 Cepheids in the southern Milky Way. It is shown that there exists a narrow intrinsic Cepheid locus in the $(V−B)−(B−L)$ diagram. Using this locus, colour excesses are determined for 156 Cepheids. It is suggested that at least 25% of all Cepheids are members of a binary system. A comparison with the colour excesses of other studies reveals systematic differences: on the average, the *VBLUW* excesses are about $0\overset{m}{.}05−0\overset{m}{.}10$ smaller in $E(B−V)$. Using model atmosphere spectra from Kurucz, temperatures, gravities, bolometric lightcurves, radius variations, and the equilibrium values of these quantities are derived from the intrinsic colours of 98 Cepheids with $P < 11^d$.

122.004 **Masses and radii of multimode Cepheid variables based upon observed periods.**   J. O. Petersen.
Astron. Astrophys., Vol. 62, 205 - 215 (1978).
    Periods for radial oscillations of stellar envelope models, calculated by linear adiabatic theory, are used to construct curves of constant mass in four different types of period ratio diagrams for six compositions. By means of these calibration curves the accuracy in determinations of masses and radii for different types of multimode Cepheids is investigated. The period ratio method seems to yield valuable new information on basic stellar parameters. While the period ratio method for mass determination seems well established from pulsation theory, its observational verification is questionable. Some observational tests of the validity of the method are discussed.

122.005 **Stellar pulsation and the abundance of helium.**   E. G. Schmidt.
Astrophys. J., Vol. 219, 543 - 549 (1978).
    Cox, King, and Tabor suggested that the appearance of nonvariable stars within the Cepheid-strip could be explained by a range in the helium abundance of Population I stars. In order to study this possibility, the author has obtained spectra of the main-sequence B stars in the galactic cluster NGC 129, which contains a nonvariable Cepheid-strip star, and M25, which contains a relatively hot Cepheid. No definite results are obtained.

122.006 **Ultraviolet photometry from the Orbiting Astronomical Observatory. XXVIII. Ultraviolet light curves for α Lupi and BW Vulpeculae.**   J. R. Lesh.
Astrophys. J., Vol. 219, 947 - 951 (1978).
    Photometric data from the Wisconsin Experiment Package on OAO-2 have been used to construct light curves at three ultraviolet wavelengths for α Lup and at seven wavelengths for BW Vul. Both stars are well-known variables of the β Cephei (β Canis Majoris) type. The light curves for α Lup are in good agreement with the radial-velocity period. A temperature variation of 400–500 K is derived. The BW Vul light curves confirm recent ephemerides based on a secularly varying period and show a stillstand near light maximum at some wavelengths. Both stars exhibit increasing light amplitude at the shortest ultraviolet wavelengths.

122.007 **The anomalous cepheid masses.**   A. N. Cox.
Sky Telesc., Vol. 55, 115 - 118 (1978).

122.008 **Kataklysmische Veränderliche.**   J. Krautter.
Sterne Weltraum, Jahrg. 17, 56 - 62 (1978).

122.009 **Photometric variations and period of the spectrum variable HR 3413.**   P. Renson, A. Heck, J. Manfroid.
Astron. Astrophys., Suppl. Ser., Vol. 31, 199 - 203 (1978).
In French.
    Series of *uvby* photometric measures of the Ap star HD 73340 have been made. The analysis of the results gives for the period: $P = 2\overset{d}{.}6679$. The range is nearly 0.04 mag in b and in v, but less in y and somewhat more in u, in which there is a secondary minimum. The light maximum and that of the intensity of the Si II lines seem to coincide more or less, but it is not so for the minima because of the different anharmonicisms of the variations.

122.010 **Intrinsic amplitude and molecular absorption of Mira stars.**   L. Celis S.
Astron. Astrophys., Vol. 63, 53 - 61 (1978).
    The visual amplitude of the Mira stars is determined by the maximum magnitude variation and by the variation of the atmospheric molecular condensation of TiO, in the time interval taken by the star to change the stellar magnitude from minimum to maximum. This conclusion is based on the interpretation of an empirical relation, which the author found among the characteristic elements of each luminous curve. The amplitude relation that he finds for the Mira stars is applicable to the semi-regular stars, too, if the luminous curve is SRa type. It is probable that this may be a general relation for the long-period variables that is, the variation phenomenon is the same for both types of stars but there is only one difference: the quantity of the molecular condensation in their atmospheres.

122.011 **Search for β Cephei stars south of declination −20°.**
I. Incidence of light variability among early B giants and subgiants—summer objects.
M. Jerzykiewicz, C. Sterken.
Acta Astron., Vol. 27, 365 - 387 (1977).
    A list has been compiled of all Bright Star Catalogue objects spectroscopically similar to β Cephei variables, and located south of declination −20°. Nearly one thousand differential *uvby* observations have been obtained of sixty-eight of these stars.

122.012 **Ultraviolet photometry with the Astronomical Netherlands Satellite ANS: the helium-spectrum variables CU Vir and SX Ari.**   M. R. Molnar, C.-C. Wu.
Astron. Astrophys., Vol. 63, 335 - 340 (1978).
    The ANS ultraviolet photometric experiment observed the helium-spectrum variables CU Vir and SX Ari. A single observation of SX Ari was obtained showing ultraviolet colors similar to those for a late B star. For CU Vir 18 high quality

observations resulted in light curves showing a possible "null region" near 2000Å where the adjacent continua vary in opposition with respect to each other. An oblique-rotator model of the silicon distribution determined from spectral variations is used in a new technique to predict light curves induced by silicon backwarming. There is sufficient qualitative agreement to state that silicon is a major source for the photometric variations of CU Vir, although there are other secondary sources.

122.013 On some peculiarities of kinematics of RR Lyrae variables.

D. K. Karimova, B. V. Kukarkin, E. D. Pavlovskaya.
Tr. Gos. Astron. Inst. Shternberga, Vol. 47, 76 - 88 (1976). In Russian.

The kinematic properties of two groups of RR Lyrae (RRab) stars selected on the basis of their metal index $(k-b)_2$ were studied. The elements of the osculating orbits and the velocity ellipsoid for each group are derived. The stars of higher metal content $(k-b)_2 \geqslant 0.17$ have $e < 0.5, i < 17°$. The stars with $(k-b)_2 < 0.17$ have $0 < e < 1$ and $0 < i < 180°$; 21% of these stars have retrograde orbits. The parameters of the velocity ellipsoid and the angular velocity of galactic rotation are also different for the two groups.

122.014 On the application of the Lafler-Kinman method to determination of periods of variable stars by a computer.   G. P. Pustovojt, Yu. A. Fadeev.

Soobshch. Gos. Astron. Inst. Shternberga, No. 199, p. 22 - 30 (1977). In Russian.

The analytical dependence of the dispersion parameter on the current period and a formula for the error estimation are obtained. On the basis of the results of numerical modelling of light curves the influence of the number of the observations, as well as errors and duration of the observations on the accuracy of the period determination is investigated. A possible modification of the Lafler-Kinman method is proposed.

122.015 Hydrodynamic model of the activity of T Tauri-type stars.   R. E. Gershberg, P. P. Petrov.

Priroda, 1978, No. 1, p. 118 - 121. In Russian.

122.016 Key to the evolution of cepheids.

Yu. N. Efremov.
Priroda, 1978, No. 3, p. 125 - 127. In Russian.

122.017 Infrared spectroscopy of Mira variables. I. R Leonis: the CO and OH vibration-rotation overtone bands.

K. H. Hinkle.
Astrophys. J., Vol. 220, 210 - 228 (1978).

Nine high-resolution ($\Delta\sigma \approx 0.15$ cm^{-1}) spectra of the 4000 to 6700 cm^{-1} region have been obtained of the typical Mira variable R Leonis at a variety of phases. Radial velocities, central depths, and equivalent widths of the carbon monoxide and hydroxyl lines have been measured. It is shown that two separate atmospheric regions contribute to the spectrum. A ~ 3500 K region identified with the stellar photosphere is the source of most CO and OH lines. A ~ 1000 K region, which is probably the inner portion of the circumstellar shell, contributes low-excitation absorption lines.

122.018 The photometric radius and pulsation mass of FG Sagittae.   C. A. Whitney.

Astrophys. J., Vol. 220, 245 - 250 (1978).

Photoelectric data for 1960 - 1977 are used to determine angular diameters and absolute radii of FG Sge. Oscillations of light, when interpreted as radial pulsations, give $M = 0.4\ M_\odot$.

122.019 The light-curve of RY Sagittarii.   I. D. Howarth.

J. British Astron. Assoc., Vol. 88, 145 - 148 (1978).

122.020 Two new rapid variables in the Ophiuchus-Scorpius region and one variable star of the RRc type near the boundary of M4.

First group of stellar division, Peking Observatory, third group of stellar division, Purple Mountain Observatory.
Acta Astron. Sinica, Vol. 18, 216 - 220 (1977). In Chinese.

Continuing the search for flare stars in Ophiuchus and Scorpius, from May to July 1976, two new rapid variables were found. According to their light curves, these variable stars are of the flare type. In the inspection of the photographic material for flare stars, one variable star of the RRc type was found near the boundary of M4.

122.021 General properties of hot ultrashort-period Cepheids.   Y.-h. Chu.

Acta Astron. Sinica, Vol. 18, 264 - 272 (1977). In Chinese.

Hot ultrashort-period Cepheids are a type of variables with period $\leqslant 0\overset{d}{.}1$, amplitude $\geqslant 0\overset{m}{.}2$, and spectral type A. They form an extended spherical distribution with little concentration to the galactic plane. They are not only found in globular clusters, but also in moving stellar groups of old disk population. On the colour-magnitude diagram they are situated near and more left to the hotter edge of the Cepheid instability strip, so their temperature is higher than that of other types of pulsating variables.

122.022 Is the star Tr 27−102 (= HD 159378) a long-period cepheid?   A. M. van Genderen, P. S. Thé.

Astron. Astrophys., Vol. 64, L1 - L2 (1978).

A discussion is presented on photometric observations of the G0Ia supergiant Tr 27−102 (= HD 159378) in the open cluster Trumpler 27. This supergiant, situated at the blue edge of the cepheid instability strip, shows a light variation with a period of 70 to 90 days and a maximum amplitude of $0\overset{m}{.}12$. Whether the star can be considered as a cepheid or as another type of variable supergiant is discussed.

122.023 Flare stars as candidates for a new class of transient soft X-ray sources.   H. Lategan.

Astron. Astrophys., Vol. 64, L5 - L7 (1978).

A recently discovered flare star, EQ Vir, is proposed as a possible optical counterpart to a transient soft X-ray source discovered by the Nagoya-Leiden group, and the possibility that all the members of the new class of transient X-ray sources proposed by Hayakawa et. al. (1976) are flare stars is discussed.

122.024 Flare activity in BY Draconis stars.   I. C. Busko, C. A. O. Torres.

Astron. Astrophys., Vol. 64, 153 - 160 (1978).

The flare activity of some BY Dra variables was studied, together with their quiescent variability. TW PsA showed to be a special case: although it displays quiescent variability (period = 10.3 days), its flare incidence level is at least 10 times lower than expected for a star of its luminosity. The flare incidence level observed for the star AU Mic in 1974 was significantly lower than that obtained in 1970 by Kunkel.

122.025 Long-term light variations of four supergiants in the Large Magellanic Cloud.   J. V. Feitzinger.

Astron. Astrophys., Vol. 64, 243 - 246 (1978).

A differential B and V photometry was carried out for four supergiants in the Large Magellanic Cloud. Light variations on time scales of days are firmly established, but light variations on time scales of hours must be ruled out for the observed stars. There is no indication of any regular or of a systematic increase or decrease in luminosity on a time scale of seven years. The origin of these light variations, especially regular or semi-regular pulsations and/or unsteady mass outflow are discussed.

122.026 On the vibrational stability of stars in thermal

imbalance.   J. R. Buchler.
Astrophys. J., Vol. 220, 629 - 635 (1978).

Pulsational stability conditions are derived for stars in thermal imbalance as a simple application of a two-time asymptotic method, previously developed as a practical tool to evolve stars through pulsational phases. In the case of soft oscillations the known results are recovered, but in a more straightforward and transparent way. In the case of hard oscillations, not accessible to the usual linear analysis, a simple method is exposed for studying the nature of the stability and, in particular, the range of stability.

122.027   Double-mode cepheid period ratios from linear and nonlinear theory.
A. N. Cox, S. W. Hodson, D. S. King.
Astrophys. J., Vol. 220, 996 - 1000 (1978).

In order to learn whether the periods of equilibrium stellar models, determined from linear nonadiabatic theory, are applicable to real stars, a nonlinear hydrodynamic calculation near its limit cycle has been analyzed for its periods. The model considered, with mass of $6.0\,M_\odot$, $T_e = 5888$ K, and $2820\,L_\odot$, and the composition $X = 0.602$, $Z = 0.044$, is slowly switching its pulsation mode from the fundamental (6.03 d) to the first overtone (4.39 d). Therefore the ratios of periods can be obtained at two epochs with differing mode-amplitude ratios and nonlinear coupling between the two principal modes. The first overtone to fundamental period ratio is found to agree with the linear theory value regardless of the stage of mode-switching. It is concluded that the period ratios from linear theory are applicable to real radially finite-amplitude pulsating stars.

122.028   Long-period variables: stellar and expansion velocities.
D. F. Dickinson, M. J. Reid, M. Morris, R. Redman.
Astrophys. J., Lett., Vol. 220, L113 - L116 (1978).

The authors have detected emission from the $J = 2-1$ transition in the ground-vibrational state of SiO in 10 long-period variable stars. The line profiles are best understood as the result of thermal emission from an expanding circumstellar envelope. These lines clearly define the stellar radial velocity with uncertainties of about 1 km s^{-1}. Where comparisons are possible, the stellar radial velocity always falls near the mid-point of the double-peaked OH maser emission, and displaced toward the blue from the optical absorption lines. The circumstellar expansion velocities deduced from the SiO line widths are correlated with stellar period in much the same way as are the velocity separations of the two OH maser peaks.

122.029   Progressi nell'interpretazione e studio delle stelle variabili.   L. Rosino.
Coelum, Vol. 45, 221 - 235 (1977).

Le variabili del tipo R Coronae Borealis. Binarie ad eclisse. Oltre i confini della Galassia.

122.030   Blue variables at high galactic latitude.
J. L. Greenstein, H. C. Arp, S. Shectman.
Publ. Astron. Soc. Pacific, Vol. 89, 741 - 745 (1977/78).

Eruptive blue variables have been observed distant from the galactic plane, from 0.5 to possibly 5 kpc. The objects surveyed are RU LMi, PG 2337 + 12, AM Her, AN UMa, AT Cnc, of which the first is the most distant. RU LMi is a U Gem star, AT Cnc is nova-like, AM Her and AN UMa are magnetic degenerate stars. Line breadths are greatest in PG 2337 + 12, which was observed at a deep minimum where it appeared to be a K dwarf with weak emission lines. The accretion emission apparently maintains the brightness of these stars far above their normal luminosity. A new phenomenon was discovered in AM Her, in which coudé spectra display broad emission wings in λ4686 of He II, with changing velocity and strength compared to a relatively sharp peak.

122.031   Technetium in T Cephei.   F. Sanner.
Publ. Astron. Soc. Pacific, Vol. 89, 768 - 769 (1977/78).

Lines of Tc I have been identified in the Me variable T Cep. The presence of technetium in other Me variables is discussed in view of a correlation between the degree of heavy-metal line enhancement in long-period variables and the velocity displacement between emission and absorption lines.

122.032   The period of δ Ceti.   M. C. Lane.
Publ. Astron. Soc. Pacific, Vol. 89, 905 - 911 (1977/78).

A study of the period of the β Cep star δ Cet shows that the star has undergone an increase in its pulsation period since the analysis by van Hoof (1968). From the results of van Hoof's study, and of the present one, it is most likely that the period was constant until about 1965. The period then began to increase at about 0.7 sec century^{-1}. Various alternative interpretations are considered: one being that the behavior of the phase-shift diagram from 1900 to 1965 reflects binary motion.

122.033   The photometric variability of 1 Persei.
D. W. Kurtz.
Publ. Astron. Soc. Pacific, Vol. 89, 939 - 940 (1977/78).

1 Per is announced to be photometrically variable and a possible eclipsing binary. It is suggested that a determination of the period and the nature of the light variability of this star may help solve the problem of this star's unexplained radial-velocity behavior.

122.034   HR 2557: a luminous δ Scuti star.   D. W. Kurtz.
Publ. Astron. Soc. Pacific, Vol. 89, 941 - 944 (1977/78).

HR 2557 is shown to be light variable. It is suggested that this star is one of the most luminous δ Scuti stars known.

122.035   A search for oscillations in 11 dwarf novae during their outbursts.   I. Nevo, D. Sadeh.
Mon. Not. R. Astron. Soc., Vol. 182, 595 - 606 (1978).

The authors have observed 11 dwarf novae during their outbursts, and analysed their light curves for the presence of oscillations. They have found oscillations in four objects: KT Per (22.5−29.5 s), SY Cnc (33.5−20.3 s), AH Her (24.1−25.0 s) and EM Cyg (16.6 s). Oscillations were found to occur during both the ascending and descending parts of the outbursts.

122.036   Observations of rapid blue variables – XVIII. Rapid oscillations in dwarf novae.   B. Warner, A. J. Brickhill.
Mon. Not. R. Astron. Soc., Vol. 182, 777 - 792 (1978).

Observations of rapid oscillations in brightness during outbursts of the dwarf novae CN Ori, AH Her, Z Cha, VW Hyi and V436 Cen are described. Correlations of oscillation period with orbital phase are found. Evidence is given for the existence of a period−luminosity relationship. These observations put new restrictions on possible models for the rapid oscillations in dwarf novae.

122.037   Steam in RR Telescopii and Henize 2-38.
D. A. Allen, D. H. Beattie, T. J. Lee, J. M. Stewart, P. M. Williams.
Mon. Not. R. Astron. Soc., Vol. 182, 57P - 60P (1978).

Low-resolution scans in the 1.9−2.6 μm atmospheric window reveal steam (H_2O) and CO absorption bands in the spectra of the symbiotic stars RR Tel and He 2-38. The steam absorption in RR Tel is particularly intense while the CO is weak, implying the presence in the system of a Mira variable seen near minimum light. In He 2-38 the steam band is weaker while the CO is stronger, as expected for a Mira seen near maximum.

122.038 **Infrared variable stars in the Milky Way field at R.A. 19^{h}16^m — Dec. + 18°28'(1950).**
L. Rosino, L. Guzzi.
Astron. Astrophys., Suppl. Ser., Vol. 31, 313 - 335 (1978).

An infrared survey of the Milky Way field around the globular cluster Palomar 10 in Sagitta has led to the discovery of 123 Mira variables. Positions, finding charts, elements and lightcurves of these variables are given.

122.039 **The companion of Mira Ceti in the 1976 light minimum.** Y. Yamashita, H. Maehara, Y. Norimoto.
Publ. Astron. Soc. Japan, Vol. 30, 219 - 222 (1978).

Several spectrograms were obtained for Mira Ceti near the light minimum in 1976. The companion was found to be in an enhanced stage. Radial velocities were measured for several emission and absorption components. The velocity data available thus far indicate that the velocity of each component varies systematically with a 14-yr period. An implication of the velocity variations is briefly discussed.

122.040 **On estimating the intrinsic properties of Miras from observational data.** J. H. Cahn, S. P. Wyatt.
Astrophys. J., Vol. 221, 163 - 174 (1978).

The authors explore several of the consequences of the model of late stellar evolution recently investigated by Wood and Cahn (1977). The significance of data on spectral classification of Miras as they relate to evolutionary patterns is explored. The authors apply the findings to several specific objects: Mira, IK Tau, and the planetary nebula NGC 7027. They estimate the integrated rate of ejection of matter by aging stars into interstellar space.

122.041 **Pulsational stabilities of a star in thermal imbalance: comparison between the methods.**
Sastri K. Vemury (*V. K. Sastri*).
Astrophys. J., Vol. 221, 258 - 267 (1978).

The stability coefficients for quasi-adiabatic pulsations for a model in thermal imbalance are evaluated using the dynamical energy (DE) approach, the total (kinetic plus potential) energy (TE) approach, and the small amplitude (SA) approaches. From a comparison among the methods, it is found that there can exist two distinct stability coefficients under conditions of thermal imbalance as pointed out by Demaret. For a prenova model with a thin hydrogen-burning shell in thermal imbalance, several radial modes are found to be unstable both for radial displacements and for velocity amplitudes. However, a new kind of pulsational instability also appears, viz., while the radial displacements are unstable, the velocity amplitudes may be stabilized through the thermal imbalance terms.

122.042 **Photometry of cepheid variables in the Large Magellanic Cloud.** C. J. Butler.
Astron. Astrophys., Suppl. Ser., Vol. 32, 83 - 126 (1978).

Photographic photometry in B and V is presented for 98 variables in the Dunsink LMC II region, of which 83 are classical cepheids.

122.043 **The radii and temperatures of classical Cepheids.** B. C. Cogan.
Astrophys. J., Vol. 221, 635 - 644 (1978).

Radius and temperature determinations of classical Cepheids have been assembled from the literature and plotted along with theoretical relationships in ($\log P$, $\log R$)- and ($\log P$, $\log T_e$)-diagrams. Different methods of radius determination give significantly different results.

122.044 **Spectrophotometric observations of the nova-like variable RW Trianguli.** K. D. Borne.
Bull. American Astron. Soc., Vol. 9, 556 - 557 (1978). Abstract.

122.045 **The wavelength dependence of 27 s pulsations in WZ Sge.**
J. Middleditch, J. E. Nelson, G. A. Chanan.
Bull. American Astron. Soc., Vol. 9, 557 (1978). – Abstract.

122.046 **A new nova-like variable LS 55°-8.**
J. Africano, R. Quigley.
Bull. American Astron. Soc., Vol. 9, 558 (1978). – Abstract.

122.047 **Water in the 2μ infrared spectrum of Miras.**
K. H. Hinkle.
Bull. American Astron. Soc., Vol. 9, 571 - 572 (1978). Abstract.

122.048 **Simultaneous photometry and time resolved spectra of flare star AD Leo.**
T. J. Schneeberger, S. P. Worden, J. L. Linsky, W. McClintock.
Bull. American Astron. Soc., Vol. 9, 593 (1978). – Abstract.

122.049 **Chromospheric emission lines in the quiescent spectrum of the flare star AD Leo.**
M. S. Giampapa, J. L. Linsky, T. J. Schneeberger, S. P. Worden.
Bull. American Astron. Soc., Vol. 9, 593 (1978). – Abstract.

122.050 **Magnetic observations of the spotted flare star BY Draconis.** S. S. Vogt.
Bull. American Astron. Soc., Vol. 9, 593 (1978). – Abstract.

122.051 **Detection of a quasi-thermal shell associated with the SiO maser.**
L. E. Snyder, D. F. Dickinson, L. W. Brown, D. Buhl.
Bull. American Astron. Soc., Vol. 9, 593 (1978). – Abstract.

122.052 **Correlation of SiO maser luminosity with bolometric magnitude of Mira variables.** J. H. Cahn.
Bull. American Astron. Soc., Vol. 9, 593 - 594 (1978). Abstract.

122.053 **High resolution spectra of OH maser emission from long period variable stars.** J. D. Fix.
Bull. American Astron. Soc., Vol. 9, 594 (1978). – Abstract.

122.054 **Radial velocity monitoring of χ Cyg, T Cep, and o Ceti.**
J. N. Pierce, L. A. Willson, W. I. Beavers.
Bull. American Astron. Soc., Vol. 9, 594 (1978). – Abstract.

122.055 **The calcium abundance of the RR Lyraes in Omega Centauri.** R. A. Bell, A. Manduca.
Bull. American Astron. Soc., Vol. 9, 602 - 603 (1978). Abstract.

122.056 **A detailed accretion disc model for SS Cygni.**
A. L. Kiplinger.
Bull. American Astron. Soc., Vol. 9, 633 (1978). – Abstract.

122.057 **An explanation of the anomalies of the Magellanic Cloud Cepheids.** J. J. De Yoreo, A. H. Karp.
Bull. American Astron. Soc., Vol. 9, 642 (1978). – Abstract.

122.058 **Visual surface brightnesses and the distances of Cepheid variables.**
T. G. Barnes, B. J. Beardsley.
Bull. American Astron. Soc., Vol. 9, 642 (1978). – Abstract.

122.059 **δ Scuti variables in the α Persei cluster.**
M. H. Slovak.
Bull. American Astron. Soc., Vol. 9, 643 (1978). – Abstract.

122.060 **Further observations of the double-mode Cepheid V367 Scuti in the open cluster NGC 6649.**
B. F. Madore, R. S. Stobie, S. van den Bergh.

Mon. Not. R. Astron. Soc., Vol. 183, 13 - 18 (1978).

New photoelectric observations of the Cepheid V367 Sct are presented. An independent least-squares Fourier analysis of all available data for this Cepheid confirms in detail the presence of two modes with periods of 6.29307 and 4.38466 day reported by Efremov & Kholopov.

122.061 **Time variations of the 22-GHz H₂O maser emission spectrum of the star S Persei.**
G. G. Cox, E. A. Parker.
Mon. Not. R. Astron. Soc., Vol. 183, 111 - 118 (1978).

Time variations of the 22-GHz water vapour maser spectrum of the supergiant star S Per are presented for the period 1974 September—1977 March, together with the light curve for the same period. Some infrared data for $1.2-10$ μm are also given. Individual spectral components vary with different phases, and with phase delays between themselves and the optical emission of up to a few hundred days. The spectrum is consistent with models in which the H_2O molecules are situated in an expanding and/or rotating circumstellar envelope.

122.062 **Temperature variations in ρ Phe.**
D. J. Sullivan, H. J. Trodahl.
Mon. Not. R. Astron. Soc., Vol. 183, 201 - 204 (1978).

Multicolour observations of the delta Scuti Star ρ Phe are reported and the data are used to estimate the temperature variations of this star. The resulting estimate of the amplitude of the temperature variation is anomalously small in comparison to the magnitude variation, supporting a suggestion that this star is not a simple radial pulsator.

122.063 **A large-scale OH sky survey at 1612 MHz. Part II. The galactic distribution and kinematics of the unidentified type II OH/IR stars.**   P. F. Bowers.
Astron. Astrophys., Vol. 64, 307 - 318 (1978).

The results of a large-scale OH sky survey at 1612 MHz are discussed. The survey has been designed to detect OH maser emission of the type commonly associated with late-type stars (Mira variables and M supergiants) in order to study the galactic distribution and kinematics of these objects at large distances from the sun. The spatial distribution of the OH/IR stars is similar to that of young objects such as giant H II regions, supernova remnants, and massive CO clouds, but the OH/IR stars have a larger z-extent ($\sigma_z \sim 110$ pc) and velocity dispersion ($\gtrsim 30$ km s^{-1}) than these other galactic constituents. It is suggested that the sources may be a rare class of Mira variables ($\sim 1\%$) with periods longer than 500 days, ages between 10^8 and 5×10^8 yr, and masses $\gtrsim 2.5$ $M_\odot$.

122.064 **Simultaneous observations of variable stars. IV. Additional measurements of SX Phe.**
R. Haefner, K. Metz, R. Schoembs.
Astron. Astrophys., Vol. 64, 319 - 322 (1978) = Veröff. Sternw. München, Band 7, No. 23.

Extensive polarimetric and photometric observations of SX Phe have been performed. The star has been found to be nearly unpolarized. No periodic or non-periodic fast variations in the light curves could be detected. There is a phaseshifted component in the H_β-index curve, which can be described by a shockwave.

122.065 **Dynamics of the photosphere and shells in the β Cephei star γ Pegasi.**   M. A. Smith, M. L. McCall.
Astrophys. J., Vol. 221, 861 - 868 (1978).

High spectral and time resolution observations of Si III $\lambda4567$ have been made through a cycle of γ Peg, a small-amplitude β Cephei star. Various lines of evidence, including the lack of large profile variations, the variation in line strength, and the presence of a weak shell component at most phases, suggest that the spectral variations in this star are pro-duced by radial pulsation. Analysis of a satellite spectral feature shows that two separate rising and falling shells are present during the pulsational cycle. Several new and interesting astrophysical aspects arise from consideration of the dynamics of this shell ejection.

122.066 **Theoretical models of 10 day Cepheids.**
C. G. Davis, D. K. Davison.
Astrophys. J., Vol. 221, 929 - 936 (1978).

New hydrodynamic techniques coupled with improved opacities have produced light curves that can now be compared directly with the observations. A new dynamic zoning method due to J. Castor has resolved the front of the hydrogen ionization region, allowing effects due to zoning to be separated from the real effects occurring in the limited-amplitude nonlinear pulsation models. New low-temperature opacities have been corrected for blanketing in the atmosphere to produce color-T_{eff} relationships which are in general agreement with the observations (Kraft-Oke) and a color-color loop for a 10 day Cepheid that is in good agreement with the observation of the Cepheid S Nor (Fernie).

122.067 **A progression in the characteristics of the period-amplitude diagrams for Cepheids in different galaxies.**   A. M. van Genderen.
Astron. Astrophys., Vol. 65, 147 - 149 (1978).

It is shown that the enveloping lines in the period-amplitude diagrams for Population I Cepheids in the SMC and LMC differ significantly from each other. The LMC line is intermediate between that of the SMC and that of the Galaxy and M 31. A correlation between this progression and the average metal abundancy in a stellar system is indicated. The Cepheids in NGC 6822 are likely of the SMC/LMC type.

122.068 **Recent observations of the highly variable star DR Tau.**
C. Bertout, J. Krautter, C. Möllenhoff, B. Wolf.
IAU Colloq. No. 42, (see 012.012), p. 65 (1977). – Abstract.

122.069 **Quasicyclic light changes in three low mass T Tauri stars.**   H. Mauder.
IAU Colloq. No. 42, (see 012.012), p. 66 - 71 (1977).

122.070 **The frequency of YY Orionis objects among the T Tauri stars.**   I. Appenzeller.
IAU Colloq. No. 42, (see 012.012), p. 80 - 87 (1977).

A list of 24 T Tauri stars belonging to the YY Orionis subclass is presented. From a statistical analysis it is estimated that at least 75% (and possibly all) UV-excess T Tauri stars are YY Orionis stars. Since about 50% of all known T Tauri stars show a strong UV-excess, the percentage of YY Orionis stars among the T Tauri stars is estimated to be 40 - 50%. This relative high percentage is in good agreement with the present theory of the formation and early evolution of low mass stars.

122.071 **Simultaneous spectroscopic and photometric observations of the YY Orionis star S CrA.**
R. Mundt, B. Wolf.
IAU Colloq. No. 42, (see 012.012), p. 88 - 91 (1977).

122.072 **Spectroscopic and photometric observations of YY Orionis.**   M. F. Walker.
IAU Colloq. No. 42, (see 012.012), p. 92 - 99 (1977).

122.073 **Red and near infrared spectra of pre-main sequence stars. I. A preliminary investigation of T Tauri stars.**
Y. Andrillat, J. P. Swings.
IAU Colloq. No. 42, (see 012.012), p. 100 - 105 (1977).

230 Å mm^{-1} spectra of 30 stars of T Tauri or related type were obtained in the 8000—11000 Å region. A preliminary qualitative analysis of the observations leads to possible correlations between the intensities of emissions of the Ca II

triplet, Paschen series and He I 10830 Å, and the spectral type, K−L color index of the star, or the [O I] and Fe II intensities in the visible spectrum.

122.074 Flare stars in stellar aggregates.   L. V. Mirzoyan.
IAU Colloq. No. 42, (see 012.012), p. 106 - 119 (1977).

122.075 Flare star observations in the Praesepe region.
I. Jankovics.
IAU Colloq. No. 42, (see 012.012), p. 120 - 125 (1977).

122.076 The variable shell star HR 5999.
P. S. Thé, H. R. E. Tjin A Djie.
IAU Colloq. No. 42, (see 012.012), p. 137 - 143 (1977).

122.077 The unusual object KR Aurigae.
M. D. Popova, E. A. (*Eh. A.*) Vitrichenko.
IAU Colloq. No. 42, (see 012.012), p. 144 - 147 (1977).
The photometric and spectrographic data of the peculiar variable KR Aur have been analyzed. The Hβ line contour shows an accretion with velocity of 3200 km/s. The assumption is made that KR Aur may be a black hole.

122.078 Aspects of VW Hyi.   R. Schoembs.
IAU Colloq. No. 42, (see 012.012), p. 218 - 226 (1977).
A coherent sinusoidal oscillation of approximately 88 s and $0.^m005$ amplitude has been detected at the end of the long eruption of VW Hyi in December 1975. This result favours models in which oscillations are caused by rotation of inhomogeneities in the inner disc. A model is proposed to explain the periodic light variations with a 3% longer period than the orbital period of VW Hyi during long eruptions by a sudden and short reverse transport of matter from the inner parts of the accretion disc.

122.079 A beat phenomenon in dwarf nova eruptions.
N. Vogt.
IAU Colloq. No. 42, (see 012.012), p. 227 - 233 (1977).
Photoelectric observations of the dwarf nova VW Hyi, obtained at the end of the December 1975 supermaximum, are presented. After decline from the outburst, the superhump period ($0.^d07622$) combines with the orbital period ($0.^d07427$) to a beat phenomenon: the O−C's and the light curves of the orbital hump vary systematically with the phase of the beat period for at least one week after recovery from the supermaximum.

122.080 HM Sagittae: an eruptive variable resembling V1016 Cygni.   F. Ciatti, A. Mammano, A. Vittone.
IAU Colloq. No. 42, (see 012.012), p. 234 - 235 (1977). Abstract.

122.081 Current status of HM Sge measurements.
H. E. Matthews, W. A. Sherwood.
IAU Colloq. No. 42, (see 012.012), p. 236 - 237 (1977).
A summary of radio and infrared observations of HM Sge made between late May and September 1st, 1977, is given.

122.082 AM Herculis type stars − magnetic binaries.
W. Krzemiński.
IAU Colloq. No. 42, (see 012.012), p. 238 - 241 (1977).
A new class of close binary stars, proposed to be called polars, has been discovered during recent year; its prototype is AM Her. The three known polars, AM Her, AN UMa, and VV Pup, are polarization, spectrum, radial velocity, optical, and infrared variables of short periods.

122.083 Disk structure in U Geminorum.
J. Madej, B. Paczyński.
IAU Colloq. No. 42, (see 012.012), p. 313 - 321 (1977).

A hot spot is a very prominent feature in dwarf novae at minimum light. It is believed to be formed at the outer rim of the accretion disk by the stream of gas flowing from the inner Lagrangian point (Smak 1871). No model of the hot spot exists in the literature. One of the problems is lack of knowledge of the structure of outer rim of the disk. In this paper the authors present a simple model of the rim. In future they intend to use this model for a study of the hot spot structure.

122.084 Nuclei of planetary nebulae in the pulsation instability strip.   W. Dziembowski.
IAU Colloq. No. 42, (see 012.012), p. 342 - 345 (1977).
Pulsation properties of relevant stellar models are investigated with the linear nonadiabatic method. Short-period variability of FG Sge is interpreted as a pulsation phenomenon.

122.085 Possible binary nature of the emission-line variable object V 1329 Cygni (= HBV 475).
J. Grygar, L. Hric, D. Chochol.
IAU Colloq. No. 42, (see 012.012), p. 383 - 385 (1977).

122.086 Radio molecular line observations of late type stars.
A. Winnberg.
IAU Colloq. No. 42, (see 012.012), p. 495 - 520 (1977).
In this paper the author reviews observational work on OH, H_2O and SiO maser lines and thermal emission from CO and SiO in Mira-type variables, late-type supergiants and carbon-stars. He also deals with some important theoretical work and model building for these objects.

122.087 The metal-rich RR Lyraes: their origin and bearing on the age and helium abundance of the galactic disk.
R. P. Kraft.
IAU Colloq. No. 42, (see 012.012), p. 521 - 558 (1977).
The author considers further the question of the origin and evolution of metal-rich RR Lyraes (MR-RR's) in the solar vicinity. If an old disk giant (ODG) is similar to a giant in a galactic cluster like NGC 188, then it must lose $\sim 0.5\,M_\odot$ if it is to pump the MR-RR domain; thus the probability that an ODG will metamorphose into an RR Lyrae is about 1/200th the probability that its metal-poor (halo) counterpart will do so.

122.088 Further evolution of V 1016 Cygni, 1974 - 77.
F. Ciatti, A. Mammano, A. Vittone.
IAU Colloq. No. 42, (see 012.012), p. 595 (1977). − Abstract.

122.089 Maser line profiles and mass outflow of long-period variable stars.   F. M. Olnon.
IAU Colloq. No. 42, (see 012.012), p. 596 - 600 (1977).

122.090 On the photometric variations of the irregular variable red giant HD 65750 in the reflection nebula IC 2220.   J. Dachs, J. Isserstedt, J. Rahe.
IAU Colloq. No. 42, (see 012.012), p. 601 - 605 (1977).
The light-curve between 1964 and 1977 for the variable M2II giant HD 65750 = V341 Car is derived from 77 photographic and 83 photoelectric UBV measurements and analyzed. It is concluded that the light variations of the star are irregular and due to variable extinction in the circumstellar nebula. The appearance of the visible reflection nebula IC 2220 into which HD 65750 is embedded, has been found to vary on a time scale of four years.

122.091 Masses of classical Cepheids and mass loss rate at pre-Cepheid stage.   M. Takeuti.
IAU Colloq. No. 42, (see 012.012), p. 609 - 612 (1977).
The mass-luminosity relation for single-mode Cepheids obtained by Saio, Kobayashi, and Takeuti (1977) indicates that the masses of classical Cepheids seem to be less than those that result from evolutionary calculations without substantial mass loss. The mass loss rate at the red giant phase

should be 10^{-5} solar masses per year. Some of the beat Cepheids seem to form a group among the variable stars.

122.092 Mass loss and Cepheid pulsation.   C. G. Davis, Jr.
IAU Colloq. No. 42, (see 012.012), p. 613 - 624 (1977).

This paper serves two purposes: (1) to discuss the latest improvements in nonlinear pulsation theory indicating the ability to resolve features such as the "Christy bump" on the light curves and (2) to show from the results of a bump model and recent observations that mass loss is one of the possible explanations for the mass discrepancy problem between evolutionary and pulsation theories.

122.093 Is there mass loss in β Canis Majoris stars?
J. Rountree Lesh, A. H. Karp.
IAU Colloq. No. 42, (see 012.012), p. 625 - 632 (1977).

122.094 Fourier analysis of beat cepheids.   D. J. Faulkner.
Proc. Astron. Soc. Australia, Vol. 3, 124 - 128 (1977).

122.095 A possible new variable star in Perseus.
G. M. Hurst.
J. British Astron. Assoc., Vol. 88, 274 - 276 (1978).

122.096 Two Z Camelopardalis variables: AB Draconis 1969–1975, CN Orionis 1963–1975.
I. D. Howarth.
J. British Astron. Assoc., Vol. 88, 292 - 300 (1978). – A report of the Variable Star Section.

122.097 Variable stars in the Leo I dwarf galaxy.
P. W. Hodge, F. W. Wright.
Astron. J., Vol. 83, 228 - 233 (1978).

This preliminary report on variable stars in the Leo I dwarf galaxy identifies 23 variables, including many intrinsically relatively bright objects. Periods and light curves have been found for 12 of them, all of which are examples of those abnormally bright variables found also, but in smaller numbers, in other dwarf ellipticals and commonly called "anomalous cepheids." Periods range from 0.8 to 2.4 days and absolute magnitudes range from $M_B \sim + 0.1$ to ~ -1.6.

122.098 Periodicity of the spectrum variable ω Oph.
P. Renson, H. M. Maitzen.
Astron. Astrophys., Vol. 65, 299 - 300 (1978). In French.

UBV photometric measures of ω Oph show that the luminosity varies periodically. The period is probably a little less than $0\overset{d}{.}75$ or $3\overset{d}{.}0$. Other observations are necessary to remove the ambiguity between these associated periods.

122.099 Spectrophotometric studies of nonstable stars. III. The spectrum of FG Sagittae in 1968 - 1973.
D. Chalonge, L. Divan, L. V. Mirzoyan.
Astrofizika, Vol. 13, 437 - 447 (1977).

The results of a spectrophotometric study of FG Sagittae based on observations made in 1968 - 1973 and covering the region 3100–6100 Å are presented. In this period the spectrum of FG Sagittae changed continuously and according to the BCD-classification transformed from type A3Ib to F7Ia. Its variation was especially sharp between June and September 1970.

122.100 Spectral and photometric observations of fast irregular variables. III. VX Cas, UX Ori, BN Ori and WW Vul – results of U, B, V, J, H, K, L photometry.
E. A. Kolotilov, G. V. Zajtseva, V. I. Shenavrin.
Astrofizika, Vol. 13, 449 - 461 (1977). In Russian. English translation in Astrophysics, Vol. 13, No. 3.

In 1975 - 76 the authors have carried out optical (UBV) and infrared (JHKL) photometric observations of the irregular variables VX Cas, UX Ori, BN Ori, and WW Vul. For each star optical and infrared measurements in most cases were made simultaneously. The photometric behaviour of the variables during the period of observations is described and, as far as possible, the observed variability of the fluxes in the optical and spectral regions is compared.

122.101 Flare activity of UV Cet-type stars.
R. E. Gershberg.
Astrofizika, Vol. 13, 553 - 589 (1977). In Russian. English translation in Astrophysics, Vol. 13, No. 3.

A review of different aspects of the flare activity of red dwarf stars is given. It contains 3 sections: flares, activity out of flares, and interpretation of flare observations and a physical model of flare activity.

122.102 The period-age relation for cepheids.
Yu. N. Efremov.
Astron. Zh. Akad. Nauk SSSR, Vol. 55, 272 - 285 (1978). In Russian. English translation in Soviet Astron., Vol. 22, No. 2.

A list of 119 cepheid members of 55 clusters and associations of the Magellanic Clouds, the Galaxy, and M31 is given. The period-age relation is found from the data on 64 cepheids in 29 clusters for which age determinations are available, the ages of extragalactic clusters were determined mainly from their integral colours. The $U-B$ colours are found to be much better age parameters than the $B-V$ values. The composite period-age relation is found as follows: $\lg t = 8.157 \, (\pm 0.037) - 0.677 \, (\pm 0.047) \lg P, (2^d < P < 50^d)$. This relation agrees well with the theoretical one.

122.103 Observations of the Hα emission line in the spectrum of χ Cyg.   T. E. Derviz, I. S. Savanov.
Astron. Zh. Akad. Nauk SSSR, Vol. 55, 358 - 362 (1978). In Russian. English translation in Soviet Astron., Vol. 22, No. 2.

Variations of the Hα emission profile have been obtained for the long-period variable χ Cyg during one cycle. Intensities, equivalent widths and displacements of the emission line were measured. Before maximum the Hα profile consisted of two components, which existed no longer than 10 days. Some explanations of that fact are discussed.

122.104 On possible cyclic recurrence of flare activity of flare stars in the Pleiades.
L. V. Mirzoyan, G. B. Oganyan.
Astrofizika, Vol. 13, 561 - 568 (1977). In Russian. English translation in Astrophysics, Vol. 13, No. 4.

It is shown that according to statistics of flares only half of the probable members of the Pleiades aggregate of low luminosities possessed flare activity during all observations. For an explanation of this discrepancy with the idea on the evolutionary meaning of the phase of flare stars, across which all dwarf stars pass, two assumptions are suggested: cyclic recurrence of flare activity and large dispersion in the duration of the flare activity phase in the life of stars of the same luminosity. Some evidence in favour of the possible cyclic recurrence of flare activity is presented.

122.105 On the theory of flare stars.
V. P. Grinin, V. V. Sobolev.
Astrofizika, Vol. 13, 587 - 603 (1977). In Russian. English translation in Astrophysics, Vol. 13, No. 4.

It is supposed that an optical flare takes place in a transition layer between chromosphere and photosphere, where the atomic concentration is of the order of $10^{15}–10^{17}$ cm^{-3}. Hydrogen gas of this concentration is considered for temperatures 5000–20000°. Several parameters of the emission of the gas are calculated, namely, colour indices U–B and B–V, Balmer jump, and spectral energy distribution. The results of the calculations are in general agreement with observational data. It is concluded that an optical flare is mainly a secondary phenomenon caused by an "explosion" in the upper atmospheric layers. During the initial phase of the flare there may be present radiation caused by the "explosion" itself. Brief discussion is given of the problem of flare energetics as well as of the similarities between stellar and solar flares.

122.106   Oscillating stars (Miras), mass loss and formation of planetary nebulae (PN).
I. Tuchman, N. Sack, Z. Barkat.
IAU Symp. No. 76, (see 012.014), p. 323 (1978). – Abstract.

122.107   Probable identification of the recurrent nova U Scorpii at minimum.   R. F. Webbink.
Publ. Astron. Soc. Pacific, Vol. 90, 57 - 60 (1978).

A faint star, $B \approx 19.2$, $(B-R) \approx + 0.^m5$, is found $1.''5 \pm 1.''5$ from the corrected position of U Sco obtained in 1863 at Madras Observatory. According to the absolute magnitude-rate of decline relationship for classical novae, this recurrent nova should be 22 to 95 kpc distant and have a giant companion.

122.108   Changes in AM Herculis during maximum and minimum states.   P. Szkody.
Publ. Astron. Soc. Pacific, Vol. 90, 61 - 70 (1978).

Visual light curves of AM Her in U, B, V, and R were obtained over a time span of seven months which included the minimum and maximum states. Fourier analysis of the data on four nights shows repetition of periods on minute time scales. The changes from cycle to cycle and over a period of months are discussed in relation to a binary model with regions that must be blocked to account for the data. A comparison made with cataclysmic variables reveals major differences between the two types of systems.

122.109   A search for centimetric wavelength emission from UV Ceti stars.   T. J. Moffett, H. F. Helmken, S. R. Spangler.
Publ. Astron. Soc. Pacific, Vol. 90, 93 - 95 (1978).

Simultaneous radio and optical observations were obtained on three UV Cet flare stars: YZ CMi, AD Leo, and Wolf 424. The total simultaneous monitoring time was 33.9 hours during which time 41 optical flares were detected. Two enhancements were detected at 1420 MHz in close temporal association with optical flares, but are of uncertain validity. The results presented indicate that 1420 MHz emission during a typical stellar flare is less than 0.10 Jy.

122.110   Note on the radial-velocity variations of HR 6684.
D. H. McNamara.
Publ. Astron. Soc. Pacific, Vol. 90, 96 (1978).

Radial-velocity measurements of HR 6684 exhibit a periodic variation in the same period as the light variations. The velocity amplitude and mean velocity are $2k = 12.8$ km s^{-1} and $\gamma = -16.6$ km s^{-1}, respectively.

122.111 UBV photometric observations of SX Phe.
E. W. Elst.
Astron. Astrophys., Suppl. Ser., Vol. 32, 161 - 164 (1978).

UBV photometric observations of SX Phe, obtained during September 1973 at the European Southern Observatory, are presented.

122.112   Observations of the carbon star RW LMi (CIT 6) in 1975/76.   A. Alksnis, I. Eglītis.
Investigations of the sun and red stars. 7, (see 003.004), p. 5 - 10 (1977). In Russian.

U, B, V and R(0.63) magnitudes of the carbon star RW LMi measured on photographs during the observation period 1975/76 are given. At the latest minimum phase the star was fainter than ever observed. Light fluctuations as previously

recorded were also present.

122.113   Photographic observations of RU Aurigae and SVS 2184.   I. Daube.
Investigations of the sun and red stars. 7, (see 003.004), p. 17 - 27 (1977). In Russian.

The results of photographic photometry in R, V, B are presented. R, V, B magnitudes of comparison stars, as well as identification charts for the variables and comparison stars are given.

122.114   Light variations of the carbon stars RW LMi = CIT 6 during the 1976/77 observing season.
A. Alksnis, I. Eglītis.
Investigations of the sun and red stars. 8, (see 003.005), p. 26 - 31 (1978). In Russian.

The results of photographic photometric observations of the carbon star RW LMi = CIT 6 obtained during 1976/77 in R(0.63), V, B, and U are presented. The radiation of the star has passed through the maximum phase of long-period variations, and it was lower than that at the four previous maxima. The colour indices B–V and B–R (0.63) reached unusually high values before the latest maximum phase and afterwards continually returned to the usual values.

122.115   An interpretation of the unique triple mode variable AC Andromedae as an RR Lyrae type star oscillating in the first three radial over-tones.   J. O. Petersen.
Astron. Astrophys., Vol. 65, 451 - 453 (1978).

A new interpretation of the observed periods in AC Andromedae as the first three radial over-tones is investigated. It is shown that this mode identification agrees well with a stellar model of about 0.5 solar masses belonging to the suprahorizontal branch. In this picture AC Andromedae is a c-type RR Lyrae variable. Further observational tests of this possibility are considered.

122.116   Cepheid masses and Cepheid winds.
A. N. Cox, G. Michaud, S. W. Hodson.
Astrophys. J., Vol. 222, 621 - 626 (1978).

Masses of Cepheids with bumps in their light and velocity curves (bump Cepheids) and those with two pulsation modes (beat or double-mode Cepheids) can be made equal to masses obtained from the evolution theory mass-luminosity relation if they have helium-enriched convection zones. The probable cause of the composition inhomogeneity seems to be a Cepheid wind which depletes hydrogen more than helium as apparently occurs in the solar wind. A layer with $Y = 0.75$ and $Z = 0.02$ external to the 70,000 K level allows 7 $M_\odot$ Cepheids to give the proper periods and light and velocity curve bumps at their evolution theory luminosity.

122.117   Z Chamaeleontis – Schlüssel zum Verständnis der U-Gem-Sterne.   M. Fernandes.
BAV Rundbrief, 27. Jahrg., 1 - 5 (1978).

122.118   Neuere Untersuchungen über SZ Lyn.
W. Braune.
BAV Rundbrief, 27. Jahrg., 15 - 17 (1978).

122.119   Neue Elemente von sechs helleren Cepheiden.
R. Lukas.
BAV Rundbrief, 27. Jahrg., 18 - 19 (1978).

122.120   On the adiabatic pulsations of tidally perturbed rotating stars.   V. Ureche, N. Lungu.
Bull. Astron. Inst. Czechoslovakia, Vol. 29, 185 - 188 (1978).

The adiabatic pulsations of a rotating star in the presence of a perturbing mass are studied. The main results of a previous paper (Ureche and Lungu, 1975) remain valid but the instability of the star appears at lower angular velocities. The tidal effect diminishes the thickness of the pulsating envelope in the substellar point and at the antipode, where the maximum instability appears. A condition of stability, similar to that obtained by Ledoux (1965), is obtained.

122.121   A photographic photometry of 4 new flare stars.   C. Poulakos.
Bull. Astron. Inst. Czechoslovakia, Vol. 29, 189 - 190 (1978).

Results of a photographic survey of flare stars in the region $\alpha = 1^h 57^m$; $\delta = +61°12'$ (1950) are presented. The survey was carried out with the help of multiple exposure plates, down to a limiting magnitude $B = 16.0^m$, as well as with the help of objective prism plates. Four new flare stars were discovered and their photometric and spectroscopic results are given.

122.122   Spectrographic observations of three suspected Delta Scuti variables.   E. Antonello, F. Arienti, M. Fracassini, L. E. Pasinetti.
Astron. Astrophys., Vol. 66, 37 - 43 (1978).

The authors give a short review of some theoretical and observational problems regarding Delta Scuti variables. Three suspected A 2 Delta Scuti stars (σ And, θ Peg and 2 Lyn), of the survey begun in 1970 at the Milano-Merate Observatory are analysed.

122.123   Flare active binary systems. EQ Peg and its brighter companion BD + 19 5116 A.   M. Rodonò.
Astron. Astrophys., Vol. 66, 175 - 185 (1978).

The UV Cet-type flare variability of both components of the visual binary BD + 19 5116 AB (EQ Peg AB) is confirmed and the individual components activity level is determined by means of area scanner observations. A list of binary or multiple systems located within 25 pc of the sun and having at least one member of UV Cet- or BY Dra-type is given. It includes about 60% of the known flare stars in the solar neighborhood.

122.124   Reddenings of Cepheids using *BVI* photometry.   J. F. Dean, P. R. Warren, A. W. J. Cousins.
Mon. Not. R. Astron. Soc., Vol. 183, 569 - 583 (1978).

The reddening free locus $(B-V)$, $(V-I)_{KC}$ is set up for Cepheids by piecing together the loci for individual Cepheids and then fixing the zero point from stars in clusters and associations whose reddenings are believed to be known.

122.125   The composite spectra of FU Orionis stars.   J. R. Mould, D. N. B. Hall, S. T. Ridgway, P. Hintzen, M. Aaronson.
Astrophys. J., Lett., Vol. 222, L123 - L126 (1978).

Fourier transform spectra (4000–7000 cm^{-1}) of FU Ori and V1057 Cyg show strong water vapor and weak CO in absorption. Their optical spectra remain those of F–G supergiants, however. The new infrared spectra demand reappraisal of the FU Orionis phenomenon. Three models are discussed: (1) a single star with a circumstellar dust shell; (2) a binary in a common circumstellar envelope; and (3) a rapidly rotating star with a flattened, possibly disklike, shell. The rapid rotator appears the more promising.

122.126   Another look at the RR Lyrae stars in the Palomar-Groningen fields.   S. V. M. Clube, F. G. Watson.
Observatory, Vol. 98, 124 - 132 (1978).

122.127   A new look at SW Bootis.   P. O. Taylor.
J. American Assoc. Variable Star Obs., Vol. 6, 56 - 58 (1977/78).

Visual observations since 1965 indicate that the primary period of SW Bootis increased abruptly circa 1960. It is also suggested that fluctuations in amplitude may be due to a secondary period of 12.997.

122.128   The U Geminorum star, EY Cygni, and a probable eclipsing star, V839 Cygni.   A. T. Piening.
J. American Assoc. Variable Star Obs., Vol. 6, 60 - 61 (1977/78).

Observational data are collected and discussed for EY Cygni, a U Gem variable. A partial analysis of the short period variable, V839 Cygni, suggests that it is an eclipsing binary.

122.129   Beat phenomena in the light curve of VX Sgr.   H. A. Smith.
J. American Assoc. Variable Star Obs., Vol. 6, 63 (1977/78). Abstract.

122.130   The light curve of the dwarf nova, VW Hydri.   F. M. Bateson.
New Zealand J. Sci., Vol. 20, No. 1, p. 73 - 122 (1977). Abstr. in Phys. Abstr., Vol. 81, Abstr. 12245 (1978).

122.131   The continuum of HM Sagittae.   V. P. Arkhipova, O. D. Dokuchaeva.
Pis'ma Astron. Zh., Vol. 4, 91 - 94 (1978). In Russian.

The continuum of HM Sge in the region $\lambda\lambda$ 3600–7000 is investigated on spectrograms taken in December 1976. The energy distribution corrected for interstellar extinction is found to be anomalous and cannot be due to any single source. The emission Balmer jump is noticed. A model of HM Sge consisting of a dense nebula surrounding a hot star is proposed. The relative energy contributions of the stellar and nebular continua at $\lambda \sim 5000$ Å are found to be 2.5 : 1.

122.132   Periods for nineteen RR Lyrae variables in NGC 1851.   A. Wehlau, M. H. Liller, S. Demers, C. Coutts Clement.
Astron. J., Vol. 83, 598 - 605 (1978).

Periods and light curves for 12 previously known and seven newly discovered RR Lyrae stars in the globular cluster NGC 1851 are presented. The variables are probably cluster members. With $\langle P_{ab} \rangle = 0.573$ days and $N_c/N_{ab} = 0.36$, NGC 1851 is classified as Oosterhoff type I.

122.133   Rho Cassiopeiae, 1964 - 1975.   J. Bailey.
J. British Astron. Assoc., Vol. 88, 397 - 401 (1978). Report of the Variable Star Section.

122.134   SU Tauri, 1962 - 1975.   J. Bailey.
J. British Astron. Assoc., Vol. 88, 402 - 406 (1978). Report of the Variable Star Section.

122.135   Narrow-band photoelectric observations of AM Her.   A. A. Aslanov, V. G. Kornilov, A. M. Cherepashchuk.
Pis'ma Astron. Zh., Vol. 4, 183 - 186 (1978). In Russian.

Narrow-band photoelectric observations of AM Her in the bands $\lambda\lambda$ 4795, 6320 (continuum) and $\lambda\lambda$ 4686, 6563 (emission lines of He II and Hα) are presented. These observations have shown the existence of accretion structure around the degenerated star of the binary system.

122.136   Spectrum of HM Sge in the near infrared.   G. I. Shanin.
Pis'ma Astron. Zh., Vol. 4, 187 - 190 (1978). In Russian.

Twenty-one spectrograms of the emission-variable object HM Sge in the near infrared have been obtained during August 3 - 13, 1977. The lines of O I, H, Ca II, N I, Mg I, He I, He II, [S II], [S III], [Fe II] at $\lambda\lambda$ 8100–11400 have been identified. Similarity of the HM Sge spectrum with spectra of peculiar Be stars MWC 434, MWC 17, MWC 84, V1016 Cyg is noticed.

122.137   Infrared amplitudes of flares of UV Ceti stars.   N. N. Kilyachkov, V. S. Shevchenko.
Pis'ma Astron. Zh., Vol. 4, 224 - 227 (1978). In Russian.

100 hours of patrol observations of EV Lac and 30 hours of UV Cet were made in August - September 1976 at Mt.

Majdanak. In the high-speed photometry regime, 32 flares of EV Lac and 89 flares of UV Cet have been registered.

122.138 HD129041: a new δ Scuti star. D. W. Kurtz.
Mon. Notes Astron. Soc. South Africa, Vol. 37, 32 - 34 (1978).

The A3V star HD 129041 (V = 7.4) is announced to be a δ Scuti variable with P = 0.035 days and a peak to peak light variation of about 0.01 mag.

122.139 X Persei (= 4U 0352+30).
IAU Circ., Nos. 3182, 3201, 3205 (1977).

122.140 X-ray pulsations from SS Cygnis.
IAU Circ., No. 3235 (1978).

122.141 Undulations of a B-type star: 53 Persei.
M. A. Smith, M. L. McCall.
Astrophys. J., Vol. 223, 221 - 233 (1978).

Profile variations of Si II λλ4128, 4130 have been monitored photoelectrically during six epochs in the large-amplitude line profile variable 53 Persei (B4.5 V). For 33 of 34 profiles it is possible to match the observations in detail with a periodic nonradial (traveling-wave) velocity field by keeping constant the pulsational amplitude and stellar rotational velocities and by changing the phase in accordance with the time of observation. The simplest interpretation of the results is that 53 Per is a complex nonradial oscillating star.

122.142 The structure and saturation of OH maser lines from long-period variable stars. J. D. Fix.
Astrophys. J., Lett., Vol. 223, L25 - L27 (1978).

Detailed spectra of the 18 cm OH maser lines from about 20 long-period variable stars have been measured at the Arecibo Observatory. Examination of the main-line emission spectra from these stars reveals that most, if not all, of the sources have emission throughout the velocity interval between the two principal velocity groups. The relatively small values of maximum to minimum flux within the lines argue that the lines are produced in saturated or partially saturated masers.

122.143 Molecular hydrogen emission from T Tauri stars.
S. Beckwith, I. Gatley, K. Matthews, G. Neugebauer.
Astrophys. J., Lett., Vol. 223, L41 - L43 (1978).

Emission in the $v = 1 \to 0\, S(1)$ line of molecular hydrogen has been detected in T Tauri; it has been searched for and is not seen in four other T Tauri stars. For T Tau a measurement of the atomic hydrogen Bγ line strength and an upper limit on the strength of the $v = 2 \to 1\, S(1)$ line of H_2 are also given.

122.144 A new variable star in M 33. G. Romano.
Astron. Astrophys., Vol. 67, 291 - 292 (1978).

A variable star of Hubble-Sandage type has been discovered close to an external spiral arm of M 33 about 17′ from the centre. Its light curve has been determined by examining photographs obtained at Asiago during the last 17 years. The star varies between 16.5 and 17.8 pg. Median absolute magnitude: −7.55.

122.145 On the membership of the RRₛ star V 65 and the EA binary V 78 in Omega Centauri (NGC 5139).
E. H. Geyer, N. Vogt.
Astron. Astrophys., Vol. 67, 297 - 299 (1978).

Radial velocities obtained from image tube spectrograms of the 0.063-day RRₛ-variable V 65 and the 1.168-day EA-binary in the globular cluster NGC 5139 show that both stars are not members of this cluster. Their spectral types and colour index indicate that interstellar reddening to a distance of 2.4 kpc in the cluster's direction is less than $0^m.1$.

122.146 On the period of the Ap variable HR 8861.

C. Sezer.
Inf. Bull. Variable Stars, No. 1384 (1978).

122.147 On the period change of the β CMa variable BW Vul. Z. Tunca.
Inf. Bull. Variable Stars, No. 1386 (1978).

122.148 Light variations of the H and K emission star HD 206860.
C. Blanco, S. Catalano, E. Marilli.
Inf. Bull. Variable Stars, No. 1387, 3 pp. (1978).

122.149 The beat period of CY Aquarii. A. Figer.
Inf. Bull. Variable Stars, No. 1388, 3 pp. (1978).

122.150 Variations de période des céphéides.
M. Petit.
Inf. Bull. Variable Stars, No. 1402, 3 pp. (1978).

122.151 Photoelectric maximum of SZ Lyncis.
J. Africano.
Inf. Bull. Variable Stars, No. 1408 (1978).

122.152 The shift-period-amplitude relation for pulsation variables. E. W. Elst.
Inf. Bull. Variable Stars, No. 1442 (1978).

122.153 Bemerkungen zu harmonisch analysierten Licht-und Farbkurven von 102 Delta-Cephei-Sternen.
Eine Arbeit aus dem Philips Europa Wettbewerb.
J. Adam.
Sterne Weltraum, 17. Jahrg., 258 - 264 (1978).

122.154 Evidence for the duplicity of the anomalous cepheid Y Ophiuchi. H. A. Abt, S. G. Levy.
Publ. Astron. Soc. Pacific, Vol. 90, 188 - 190 (1978).

The mean velocity of this 17-day cepheid seems to show a variation due to binary motion with a period of 2612 days. Orbital elements are given. It is unlikely that the light of the companion is responsible for the small ranges in light, color, and radial velocity of the cepheid.

122.155 A new nova-like variable − LS55° −8.
J. Africano, R. Quigley.
Publ. Astron. Soc. Pacific, Vol. 90, 191 - 193 (1978).

LS55° −8 a suspected cataclysmic variable, was observed using a high-speed photometer. The light curve shows non-periodic rapid light variations amounting to about 0.4 magnitude. The mean light level for each night was constant to a few percent. No eclipses, recurrent features, or coherent oscillations were found in the 20 hours of observations.

122.156 uvby analysis of HR 8102.
G. C. Kilambi, D. L. DuPuy, C. A. Koegler.
Publ. Astron. Soc. Pacific, Vol. 90, 194 - 200 (1978).

A uvby analysis of HR 8102 is presented. From recent data, a mean period of variation is found to be $0^d.097$. A mean effective temperature of 7640 K and a mean surface gravity log g_{eff} = 3.91 are derived over a pulsational cycle. The absolute visual magnitude is found to be $1^m.45$. These values suggest that HR 8102 is a normal Population I pulsating star and a member of the δ Scuti group.

122.157 CD −42°14462: a dwarf nova in permanent out-burst. H. E. Bond.
Publ. Astron. Soc. Pacific, Vol. 90, 216 - 218 (1978) = Contrib. Louisiana State Univ. Obs., No. 132.

The historial light curve of the nova-like variable CD −42°14462 = V3885 Sgr has been determined on 270 plates from the archival collection at Harvard. The object has remained at about $B = 10^m.3$, with random excursions of a few tenths of a magnitude up or down, since 1899. The bulk of the observational evidence supports the conclusion that CD −42°14462 is a dwarf nova in a state of permanent outburst.

122.158   RR Lyrae stars.   J. Lub.
Messenger, No. 13, p. 15 - 17 (1978).

122.159   Ultraviolet photometry of the cepheid β Doradus from the A.N.S. satellite.
J. Lub, J. van Paradijs, J. W. Pel, P. R. Wesselius.
ESO Sci. Prepr. No. 23, 22 pp. (1978). − Submitted to Astronomy and Astrophysics.

The authors present extensive photometric observations of β Dor, covering the wavelength range from 1800 to 5500 Å. The ultraviolet spectrum is probably a sensitive indicator of shock wave phenomena in Cepheid atmospheres. The authors discuss the general properties of the lightcurves of β Dor.

122.160   Photoelectric observations and extinction of the irregular variables BH and BO Cephei.
W. Wenzel, V. Brückner.
Mitt. Veränderl. Sterne (MVS), Band 8, 35 - 44 (1978).

Photoelectric UBV observations of the light curves of the two F stars BH and BO Cep are discussed.

122.161   Photographische Lichtkurve von BH Cephei für die Jahre 1964 bis 1972.   I. Bradl.
Mitt. Veränderl. Sterne (MVS), Band 8, 46 - 47 (1978).

Estimates on roughly 350 Sonneberg Sky Patrol plates yielded a relatively complete light curve of BH Cep for the years 1964 to 1972, which might serve for comparison and completion of photoelectric observations.

122.162   Rotating variable stars.   J. R. Percy.
J. R. Astron. Soc. Canada, Vol. 72, 162 - 166 (1978).

This paper introduces the rotating variable stars, and gives a table and figure which show the relationship, at various positions on the H−R diagram, between their rotation period and their equatorial rotational velocity.

122.163   Observations of three symbiotic stars: Z Andromedae, AX Persei, and CI Cygni.
M. I. Myalkovskij.
Peremennye Zvezdy, Prilozhenie, Vol. 3, 71 - 91 (1977). In Russian.

122.164   AR Serpentis.   V. P. Tsesevich.
Peremennye Zvezdy, Prilozhenie, Vol. 3, 93 - 124 (1977). In Russian.

Over 1800 observations distributed within the interval of 2415148−2442240 JD have been studied. It is shown that the period P has considerable fluctuations which result in significant period variability of very large Blažko effect. The latter has changed from 110.7 to 114 days.

122.165   Investigation of the Blažko effect in XZ Cygni.
V. P. Bezdenezhnyj.
Peremennye Zvezdy, Prilozhenie, Vol. 3, 125 - 148 (1977). In Russian.

On the basis of over 2000 photographic, photovisual and visual observations made by the author, variations of the light curve of XZ Cygni with Blažko effect have been studied. The original observations of XZ Cygni are given in a table. The dependence of different parameters (amplitude of light variations, magnitude and colour index at maximum light and O−C residuals) on the phase of the Blažko effect has been considered. These results have been compared with those obtained by other authors. The primary period of XZ Cyg is found to decrease in 1970. New osculating elements are given.

122.166   Stellar magnetism as a decisive factor of flare activity and of the evolution of UV Ceti- and T Tauri-type variable stars.   R. E. Gershberg.
Peremennye Zvezdy, Vol. 20, 283 - 298 (1977). In Russian.

Considerations of the main stellar statistical and astrophysical features of UV Cet-type flare stars lead to the conclusion that the evolution of these variables as well as other stars is determined by their masses and chemical composition; the third evolutionary factor is a magnetic field that is responsible, eventually, for all manifestations of flare activity. The main physical features of the T Tau-type stars may be understood in the framework of the hypothesis that these stars have large scale magnetic fields with strengths close to a critical level at which the field and subphotospheric convection begin to interact strongly, which results in a significant decrease of the photospheric brightness, as in a sunspot. This hypothesis indicates a way for understanding the FU Ori-type flares and the Herbig-Haro objects. Within the framework of the idea of the essential role of stellar magnetism in UV Cet- and T Tau-type stars, possibilities of their physical and genetic commonality are discussed.

122.167   Radial velocities of RZ Lyrae.   Yu. S. Romanov.
Peremennye Zvezdy, Vol. 20, 299 - 311 (1977). In Russian.

Radial velocities in RZ Lyrae have been determined from absorption lines of hydrogen, ionized calcium and other metals for different phases of the main light variation and Blažko effect. The radial velocity amplitude from hydrogen lines is varying with the phase of Blažko effect from 130 km/sec to nearly 245 km/sec. Relations of amplitudes of radial velocities which have been determined from hydrogen absorption lines, from those of ionized calcium and other metals in variable stars of RR Lyrae-type with $\Delta S \geqslant 5$ are considered.

122.168   New variable stars in the globular cluster M3.
P. N. Kholopov.
Peremennye Zvezdy, Vol. 20, 313 - 324 (1977). In Russian.

B, V magnitudes of 40 comparison stars for RR Lyrae-type variables in the central region of M3 are determined. Ten new variables are discovered. New maps of the central region of the cluster are given.

122.169   Photoelectric UBV observations of V1329 Cyg in 1973 - 1976 and some remarks on its variability.
V. P. Arkhipova.
Peremennye Zvezdy, Vol. 20, 345 - 354 (1977). In Russian.

Photoelectric UBV observations of V1329 Cyg in 1973−76 are given. The star has strong brightness oscillations with amplitude of 1 mag and period of 960^d, observed long before outburst. The colour indices B−V, U−B show a composite nature of the spectrum, consisting of several components. The U−B colour varies significantly with phase of the 960^d period and has a maximum near minimum brightness. The B−V colour does not show noticeable variations. The shape of the light curve in 1971−76 differs strongly from that of preoutburst.

122.170   Photoelectric observations of the Ap star HD 34452 (IQ Aur) in ten spectral regions.   A. S. Nikolov.
Peremennye Zvezdy, Vol. 20, 355 - 359 (1977). In Russian.

On the basis of photoelectrical observations made in 1973 the light curves of the magnetic variable HD 34452 in ten regions of the spectrum are obtained. The observations are well presented with the period by Deutsch (1947) of 2.4660 days. A double wave with maxima near phases ∼ 0.60 and 0.20 is observed on all the light curves. The amplitude changes of the light curve with wavelength are investigated.

122.171   Observations of the T Tauri-type star DI Cephei. I.
G. F. Gahm, R. E. Gershberg, P. P. Petrov, A. G.

Shcherbakov, E. A. Kolotilov, G. V. Zajtseva, G. I. Shanin.
Peremennye Zvezdy, Vol. 20, 381 - 389 (1977). In Russian.

The T Tauri star DI Cephei was observed simultaneously
with 5 telescopes during 4 consecutive nights from September
2 to 5, 1974. The purpose of this investigation was to study
the interplay between time variations in the stellar continuum
and possible changes in the profiles and intensities of different
emission and absorption lines. As it turned out, however, the
star showed only moderate changes in brightness over the observational period. Little evidence of time variations in equivalent widths and profiles of the emission lines of Hα and the
calcium triplet is shown.

122.172 On some photometric features of R Corona
Borealis-type stars.   A. F. Pugach.
Peremennye Zvezdy, Vol. 20, 391 - 401 (1977). In Russian.

From an analysis of visual and photoelectric observations
of R CrB stars the following data were obtained. 1. A coefficient K of photometric activity in per cent. 2. Interstellar extinction and photometric parallaxes have been calculated by
ascribing to all the stars a luminosity $M_V = -4^m5$. Space distribution between spiral arms gives evidence for old age of the
R CrB stars. 3. There is a strong relation between unreddened
colours $(B-V)_0$ and brightness change rate on the descending
branch.

122.173 Infrared observations of variable stars. I. R, I, J,
K, L, M, N photometry of 8 objects.
E. A. Kolotilov, A. A. Liberman, O. G. Taranova.
Peremennye Zvezdy, Vol. 20, 413 - 417 (1977). In Russian.

For 8 late type variables IR observations including photoelectric scans in the $0.6 - 1.1 \mu$ spectral range and multicolor photometry in the $1.25 - 10 \mu$ region were carried out.
From these IR data the energy distributions in the $0.6 - 10 \mu$
range were constructed.

122.174 RR Lyrae-type variables V146 and V222 in the
globular cluster M3.   P. N. Kholopov.
Peremennye Zvezdy, Vol. 20, 419 - 428 (1977). In Russian.

New estimates and light variation elements of M3
cluster variables V146 (Max hel = $2442190.368 + 0^d502193 \cdot E$)
and V222 = vZ 1198 (Max hel = $2441767.366 + 0^d5967455 \cdot E$)
are obtained. The variability of V127 = vZ 1194 is not confirmed. The observations of V146 published by Greenstein
(1935) are really V222 observations.

122.175 On the change of the period of the variable V3 in
the globular cluster M3.   P. N. Kholopov.
Peremennye Zvezdy, Vol. 20, 429 - 434 (1977). In Russian.

Using all the published and new light estimates of the
variable V3 it is possible to suggest that the period of the
variable has diminished in the interval between 1900 and 1921.
In the interval J.D.2413372 − 15161 Max hel =
$2413386.535 + 0^d559055 \cdot E$, in the interval J.D. 2422730 −
42851 Max hel = $2442137.412 + 0^d5582045 \cdot E$.

122.176 High-sensitivity OH observations of Mira variables.
P. F. Bowers, R. P. Sinha.
Bull. American Astron. Soc., Vol. 10, 392 (1978). − Abstract.

122.177 Use of the statistical theory of overshoots for
representing unobserved parts of light curves of
variable stars.   I. A. Klyus.
Peremennye Zvezdy, Vol. 20, 441 - 445 (1977). In Russian.

The purpose of this work is to investigate the representation of unobserved parts of light curves of variable stars with
the aid of the theory of overshoots of statistic characteristics
on light curves of nonperiodic and semiregular variable stars.

122.178 Statistical analysis of the light curves of the variable
stars μ Cep, T Ori, T Cha, RU Lup and AK Sco.
I. A. Klyus.

Peremennye Zvezdy, Vol. 20, 446 - 460 (1977). In Russian.

The light curves of variables μ Cep, T Ori, and T Cha have
been tested for stationarity and normality of functions of
light distribution. Calculations have shown that the hypothesis on stationarity of light variations in these stars can be
adopted, whereas the hypothesis on normality of functions of
light distribution in these stars is rejected. Eight peaks have
been found in the statistic spectrum of μ Cep at 4410^d, 720^d,
364^d, 182^d, 122^d, 73^d, 42^d and 20^d. A peak at 3^d3 has been
found in the statistic spectrum of T Cha. The correlation
function of RU Lup gives evidence for the existence of a
period equal to 3^d8, which is also a mean RU Lup light curve
cycle. Periodic components for the variables T Ori and AK Sco
have not been found.

122.179 Photoelectric observations of two classical cepheids.
O. P. Vasil'yanovskaya.
Peremennye Zvezdy, Vol. 20, 467 - 471 (1977). In Russian.

The results obtained from 1972 - 1975 observations of
FN Aql and SV Per are given. It is shown that the UBV lightcurve forms of SV Per changed in time.

122.180 High resolution observations of maser stars.
A. St. C. Dinger, D. F. Dickinson, L. E. Snyder.
Bull. American Astron. Soc., Vol. 10, 392 (1978). − Abstract.

122.181 Are cepheid light and velocity curves the same?
T. F. Adams, C. G. Davis,
Bull. American Astron. Soc., Vol. 10, 401 (1978). − Abstract.

122.182 Are Beta Cephei stars non-radial pulsators?
A. P. Odell.
Bull. American Astron. Soc., Vol. 10, 401 (1978). − Abstract.

122.183 A search for northern hemisphere double mode
cepheids.   A. A. Henden.
Bull. American Astron. Soc., Vol. 10, 401 (1978). − Abstract.

122.184 A spectrophotometric comparison of the physical
properties of classical cepheids and nonvariable
supergiants.   R. S. Patterson, J. S. Neff.
Bull. American Astron. Soc., Vol. 10, 401 (1978). − Abstract.

122.185 A search for Beta Cephei stars. II: NGC 4755.
S. M. Jakate.
Bull. American Astron. Soc., Vol. 10, 401 - 402 (1978).
Abstract.

122.186 Infrared spatial interferometry of Mira variables.
D. W. McCarthy, F. J. Low, R. Howell.
Bull. American Astron. Soc., Vol. 10, 406 - 407 (1978).
Abstract.

122.187 Backwarming sources in Ap stars.   M. R. Molnar.
Bull. American Astron. Soc., Vol. 10, 413 - 414
(1978). − Abstract.

122.188 Broadband photometry of AM CVn.
R. J. Panek.
Bull. American Astron. Soc., Vol. 10, 419 (1978). − Abstract.

122.189 The spectrum of T Tauri star BM And.
T. J. Schneeberger.
Bull. American Astron. Soc., Vol. 10, 453 (1978). − Abstract.

122.190 Spectrophotometric observations of symbiotic stars.
C. D. Keyes.
Bull. American Astron. Soc., Vol. 10, 453 (1978). − Abstract.

122.191 Group of galactic RR Lyrae variables with periods
of about 0^d4.   N. E. Bukhantsova, N. S. Nikolov.
Astron. Tsirk., No. 966, p. 5 - 7 (1977). In Russian.

122.192 **Outbursts of EY Cygni.** I. D. Howarth.
Astron. Tsirk., No. 970, p. 2 - 3 (1977). In Russian
and English.

122.193 **Supplement to the Catalogue of RR Lyrae stars**
elements. B. N. Firmanyuk.
Astron. Tsirk., No. 979, p. 7 (1978). In Russian.

122.194 **Peculiarities of light variation in DR And.**
Yu. S. Romanov, V. P. Tsesevich, L. I. Shakun,
P. S. Zgonyajko.
Astron. Tsirk., No. 981, p. 4 - 6 (1978). In Russian.

122.195 **Elements of the Blažhko effects in SW And.**
L. E. Lysova, Yu. S. Romanov, B. N. Firmanyuk.
Astron. Tsirk., No. 982, p. 3 - 4 (1978). In Russian.

122.196 **Rapid variability and flare emission of T Tauri.**
G. V. Zajtseva.
Astron. Tsirk., No. 984, p. 3 - 6 (1978). In Russian.

122.197 **Possible periodic light variations of V1057 Cyg from**
photometric observations 1976—77.
E. A. Kolotilov.
Astron. Tsirk., No. 984, p. 6 - 7 (1978). In Russian.

122.198 **Reddening lines of classical cepheids on the UBV**
two-colour diagram. G. R. Ivanov, M. S. Mitev.
Astron. Tsirk., No. 992, p. 1 - 2 (1978). In Russian.

122.199 **On period changes of long-period cepheids in**
globular clusters. A. S. Rastorguev.
Astron. Tsirk., No. 994, p. 5 - 6 (1978). In Russian.

122.200 **Search for flare stars in the vicinity of FU Ori.**
G. A. Ponomareva.
Astron. Tsirk., No. 996, p. 7 - 8 (1978). In Russian.

122.201 **Flare stars in the region of the nebulae NGC 7000**
and IC 5068–70. M. K. Tsvetkov.
Flare stars. Symposium 1976, (see 012.035), p. 75 - 86.
In Russian. – Abstr. in Ref. zh., 51. Astron., 6.51.661 (1978).

122.202 **On the nature of dwarf novae.**
I. G. Mitrofanov.
Pis'ma Astron. Zh., Vol. 4, 219 - 223 (1978). In Russian.
The phenomena of dwarf novae are shown to be connected
with hydrogen accumulation on a local region of the surface
of a degenerate dwarf near the magnetic poles. The increase
of luminosity results from thermonuclear burning in this
hydrogen layer. A magnetic field of $3 \times 10^6 - 3 \times 10^7$ gauss
appears to be necessary. The origin of optical and X radiation
from dwarf novae is discussed.

122.203 **The variability of RZ Cephei.**
B. Cester, I. Todoran.
Mem. Soc. Astron. Italiana, Vol. 47, 217 - 228 (1976).
New observations of the RRc-type star RZ Cep are reported. The period is clearly variable and a new interpretation
of its behaviour is given. The main feature of the light curve is
a pre-maximum hump, as bright as the broad maximum, which
may show a variability, perhaps connected with variations in
the colour index; further observations with a high speed
photometry are therefore required.

122.204 **Researches with the Schmidt telescopes. VIII. The**
stars in a field around Beta Persei.
G. Pinto, G. Romano.
Mem. Soc. Astron. Italiana, Vol. 47, 229 - 244 (1976).
Twenty-nine variable stars of a field (6° in diameter)
around Beta Persei have been studied on the photographs taken
with the Schmidt telescope (40/50/100 cm) of Asiago Obser-

vatory. During the research two new variable stars have been
discovered (GR 271 and GR 272).

122.205 **Flare stars.** K. Menzel.
J. Astron. Soc. Western Australia, Vol. 297,
January, p. 2 - 3 (1978).

122.206 **On the electrodynamical implications of flare stars.**
Ch. G. Papaz.
Flare stars. Symposium 1976, (see 012.035), p. 175 - 180. In
Russian. – Abstr. in Ref. zh., 51. Astron., 7.51.548 (1978).

122.207 **On the optical emission of stellar flares.**
V. P. Grinin, V. V. Sobolev.
Flare stars. Symposium 1976, (see 012.035), p. 169 - 174. In
Russian. – Abstr. in Ref. zh., 51. Astron., 7.51.549 (1978).

122.208 **Some remarks on the influence of rotation of stars**
on flare activity. V. G. Gorbatskij.
Flare stars. Symposium 1976, (see 012.035), p. 198 - 201. In
Russian. – Abstr. in Ref. zh., 51. Astron., 7.51.550 (1978).

122.209 **Phenomenological picture of flare phenomena on**
stars. V. S. Oskanyan.
Flare stars. Symposium 1976, (see 012.035), p. 11 - 22. In
Russian. – Abstr. in Ref. zh., 51. Astron., 7.51.552 (1978).

122.210 **Statistical properties of flare activity of UV Ceti-**
type stars. N. I. Shakhovskaya.
Flare stars. Symposium 1976, (see 012.035), p. 46 - 55. In
Russian. – Abstr. in Ref. zh., 51. Astron,, 7.51.553 (1978).

122.211 **On rapid variations of flare stars.** D. S. Evans.
Flare stars. Symposium 1976, (see 012.035), p. 40 -
45. In Russian. – Abstr. in Ref. zh., 51. Astron., 7.51.554
(1978).

122.212 **On slow variations in flare stars.** D. S. Evans.
Flare stars. Symposium 1976, (see 012.035), p. 35 -
39. In Russian. – Abstr. in Ref. zh., 51. Astron., 7.51.555
(1978).

122.213 **On a flare of an unusual variable.** M. D. Popova.
Flare stars. Symposium 1976, (see 012.035), p.
164 - 165. In Russian. – Abstr. in Ref. zh., 51. Astron.,
7.51.559 (1978).

122.214 **Spectrophotometric observations of FG Sagittae.**
L. Divan, L. V. Mirzoyan.
Flare stars. Symposium 1976, (see 012.035), p. 133 - 140. In
Russian. – Abstr. in Ref. zh., 51. Astron., 7.51.564 (1978).

R Corona Borealis-type stars.
See Abstr. 003.150.

On the infrared bands of AlO related to Mira
phenomena. See Abstr. 022.100.

Method for finding the periods of variable stars.
See Abstr. 031.203

Period determination techniques for variable stars.
See Abstr. 031.221.

The continuing search for flare star magnetic fields:
A Zeeman analyzer system for the KPNO 4-meter echelle
spectrograph. See Abstr. 033.042.

Hydrodynamical mass loss from long period
variable stars. See Abstr. 064.032.

Periodic shockwave description of LPV atmospheres.

See Abstr. 064.033.

Constraints on the properties of circumstellar shells from observations of thermal CO and SiO millimeter line emission.   See Abstr. 064.045.

Secular mass loss from cataclysmic variables. See Abstr. 064.048.

On mass loss from long period variables. See Abstr. 064.051.

Observations of OH and H_2O microwave maser emission from VY Canis Majoris.   See Abstr. 064.054.

On the selective absorption of the envelope of R CrB.   See Abstr. 064.106.

Convection in pulsating stars. See Abstr. 065.042.

On convection in a periodic gravitational field. II. See Abstr. 065.053.

Theory of red giant variable stars. See Abstr. 065.067.

Super-adiabatic acceleration of charged particles and flares on the sun and stars.   See Abstr. 078.005.

A spectrophotometric parallax for U Geminorum. See Abstr. 111.001.

Proper motions of 84 Mira Ceti-type stars. See Abstr. 112.014.

Motions of flare stars in the solar vicinity. See Abstr. 112.015.

Photoelectric photometry at the Hvar Observatory. III. The Ap star CQ UMa.   See Abstr. 113.016.

Color index and finite extent of the atmosphere. See Abstr. 113.023.

Variable stars of small amplitude. III. Semi-period of variation for seven B2 to G0 supergiant stars. See Abstr. 113.050.

Photoelectric UBV observations of RR Lyrae variable stars. Second list.   See Abstr. 113.052.

Rapid light variation of the companion of Mira Ceti in the 1977 light minimum.   See Abstr. 113.076.

Fotometrische waarneming van de RRs-ster RS Gru aan de sterrenwacht Serra da Piedade (Brazilië). See Abstr. 113.086.

UBV observations of FG Sagittae in 1977. See Abstr. 113.089.

Search for photometric flares of Ap stars. See Abstr. 113.093.

The helium triplet-to-singlet ratio in T Tauri stars. See Abstr. 114.003.

Metallicism and pulsation: the marginal metallic line stars.   See Abstr. 114.033.

New observations of ultraviolet variability in Wolf-Rayet stars.   See Abstr. 114.053.

Observations of polarization of some R CrB stars. See Abstr. 114.088.

Observations of technetium stars. See Abstr. 114.509.

Variations rapides du profil de la raie H_α dans l'étoile T Tauri.   See Abstr. 114.554.

Infrared spectra of HM Sagittae and V1016 Cygni. See Abstr. 114.567.

The helium variable HD 64740 — an X-ray binary? See Abstr. 114.572.

The spectrum of CH Cygni in July 1977. See Abstr. 114.581.

Possible $H\alpha$-emission in the HM Sge pre-outburst spectrum.   See Abstr. 114.583.

The semi-period—luminosity—colour relation for supergiant stars.   See Abstr. 115.014.

A survey of nine Beta Cephei stars for magnetic fields.   See Abstr. 116.012.

Stellar magnetism as a decisive factor of flare activity and of evolution of UV Ceti- and T Tauri-type variable stars.   See Abstr. 116.025.

Theoretical *UBV* colours of accretion discs in cataclysmic variables.   See Abstr. 117.008.

W92: a possible gravitationally contracting binary system in NGC 2264.   See Abstr. 117.012.

Note on the multiplicity of dMe stars. See Abstr. 117.039.

Does the system Theta Virginis represent a new class of pulsating star?   See Abstr. 119.010.

Nu Centauri: a new Beta Cephei star. See Abstr. 119.024.

Anomalous period change of UX Ursae Majoris in 1977.   See Abstr. 121.015.

Four-colour photometry of eclipsing binaries. IX b. AI Hya, light curves and photometric elements. See Abstr. 121.043.

YY Geminorum. I. Photometry and absolute dimensions.   See Abstr. 121.046.

Variable stars in a region of intensive star formation.   See Abstr. 121.080.

Recent nova and variable-star research with the Harvard Photographic Collection.   See Abstr. 124.010.

New observations of the pulsating white dwarf R548.   See Abstr. 126.012.

The interaction of accretion flux with stellar pulsation.   See Abstr. 126.026.

Polarization properties of T Tauri stars. See Abstr. 131.236.

A low-detection limit search for OH emission from infrared stars.   See Abstr. 133.015.

Detection of both soft and hard X-ray emission from SS Cygni with ANS.   See Abst. 142.025.

The optical counterpart of Aquila X-1 (3U 1908+00). See Abstr. 142.044.

HEAO-1 observations of soft X-ray emission from U Geminorum.   See Abstr. 142.079.

Soft X-radiation from cataclysmic variable stars. See Abstr. 142.080.

The optical variability of the X-ray binary AM Herculis.   See Abstr. 142.126.

Cosmic flare transients: constraints upon models for energy storage and release derived from the event frequency distribution.   See Abstr. 142.150.

Further observations of nonperiodic optical flickering in HZ Herculis.   See Abstr. 142.194.

Search for super-fast optical variability of X-ray sources and T Tau-type stars according to the MANIA experiment program in 1973–74.   See Abstr. 142.226.

Veränderliche Sterne in Sternassoziationen. See Abstr. 152.005.

A photometric survey of the α Persei cluster for δ Scuti variables.   See Abstr. 153.033.

The age of the galactic cluster Lyngä 6. See Abstr. 153.048.

Variable stars in the globular cluster NGC 5286. See Abstr. 154.001.

Variable stars in the globular cluster Messier 19. See Abstr. 154.007.

Three luminous carbon stars in the Large Magellanic Cloud.   See Abstr. 159.001.

Red variable stars in the Magellanic Clouds – II. The field of NGC 371 in the SMC.   See Abstr. 159.008.

Red variable stars in the Magellanic Clouds – III. Carbon stars in the field of NGC 419 (SMC). See Abstr. 159.009.

The extragalactic distance scale. I. A review of distance indicators: zero points and errors of primary indicators.   See Abstr. 162.150.

123   Variable Stars: Lists of Observations

123.001   Variabilità del sistema 88 Her.
       L. Baldinelli, S. Ghedini.
G. A. A. B., N. 48, p. 11 (1977).

123.002   Visual observations of the recent minimum of RY
       Sagittarii.   J. V. Vincent.
Mon. Notes Astron. Soc. Southern Africa, Vol. 37, 11 (1978).

123.003   La page de l'observateur.   M. Duruy.
       Bull. AFOEV, Tome 11, 76 (1977).

123.004   Tableaux des observations faites par l'AFOEV de
       septembre à décembre 1977.
Bull. AFOEV, Tome 11, 80 - 108 (1977).

123.005   Observations of V1901 Sgr and two new variables
       in Cygnus.   G. L. Harns.
J. American Assoc. Variable Star Obs., Vol. 6, 62 - 63 (1977/78).

123.006   Seven variable stars in Sagittarius.   R. M. Foster.
       J. American Assoc. Variable Star Obs., Vol. 6, 64 -
66 (1977/78).
       New or updated periods are given for six long period
variables and one eclipsing variable. Finder charts are supplied.

123.007   A suspected new variable in Coma Berenices.
       J. W. Briggs.
J. American Assoc. Variable Star Obs., Vol. 6, 67 (1977/78).

123.008   Six variable stars in Sagittarius.
       S. M. Vardamis.
J. American Assoc. Variable Star Obs., Vol. 6, 68 - 69 (1977/78).
       Finder charts and updated periods are given for six long
period variables.

123.009   Periods for five Sagittarius variables.
       M. J. Wolfson.
J. American Assoc. Variable Star Obs., Vol. 6, 70 - 71 (1977/78).
       Updated periods are given for V1652, 1659, and 1660
Sgr and new periods for two previously unpublished variables
in Sagittarius. Finder charts are provided.

123.010   A note on DL Cassiopeiae.   J. W. Briggs.
       J. American Assoc. Variable Star Obs., Vol. 6, 72
(1977/78).
       The nature of the period changes of DL Cas is still
ambiguous.

123.011   A call for photoelectric observations of V389 Cygni.
       D. Hoffleit.
J. American Assoc. Variable Star Obs., Vol. 6, 73 - 75 (1977/78).

123.012   AX Persei.
       IAU Circ., No. 3178 (1978).

123.013   U Geminorum.
       IAU Circ., Nos. 3180, 3232 (1978).

123.014   4U 1249–28.
       IAU Circ., No. 3184 (1978).

123.015   Novalike object in Serpens.
       IAU Circ., No. 3186 (1978).

123.016   VV Puppis.
       IAU Circ., Nos. 3187, 3197 (1978).

123.017   T Tauri.
       IAU Circ., No. 3190 (1978).

123.018 UW Centauri.
IAU Circ., No. 3198 (1978).

123.019 AM Herculis.
IAU Circ., No. 3224 (1978).

123.020 U Geminorum.
Yamamoto Circ., No. 1880 (1978). In Japanese.

123.021 An apparent eruptive variable star in Scorpius.
N. Sanduleak, C. B. Stephenson, D. J. MacConnell.
Inf. Bull. Variable Stars, No. 1376 (1978).

123.022 A new variable star in NGC 5139.
C. R. Fourcade, J. R. Laborde, E. Yurquina.
Inf. Bull. Variable Stars, No. 1380 (1978).

123.023 A suspected symbiotic star in the region of the
Small Magellanic Cloud.   N. Sanduleak.
Inf. Bull. Variable Stars, No. 1389 (1978).

123.024 Nouvelle recherche de périodes d'étoiles Ap
observées à l'ESO-I.   P. Renson.
Inf. Bull. Variable Stars, No. 1391 (1978).

123.025 On the light variability of the magnetic star
HD 192678.   K. Panov.
Inf. Bull. Variable Stars, No. 1392 (1978).

123.026 Photoelectric observations of V 711 Tauri.
P. Flin, M. Sztajno.
Inf. Bull. Variable Stars, No. 1394, 3 pp. (1978).

123.027 Observations of HDE 245770 during the recent
X-ray flare-up of its counterpart A0535+26.
S. Rössiger.
Inf. Bull. Variable Stars, No. 1395 (1978).

123.028 HD 11285 and HD 14940: two new pulsating
variables.   W. W. Weiss.
Inf. Bull. Variable Stars, No. 1400, 4 pp. (1978).

123.029 A new semiregular variable in Cancer.
M. Huruhata.
Inf. Bull. Variable Stars, No. 1401 (1978).

123.030 A red variable in Circinus.
A. M. Orsatti, J. C. Muzzio.
Inf. Bull. Variable Stars, No. 1406, 3 pp. (1978).

123.031 Gliese 867 − a new flare-star system.
P. B. Byrne.
Inf. Bull. Variable Stars, No. 1407, 4 pp. (1978).

123.032 UBV observations of HR 1099 on February 22nd,
1977.   E. F. Milone, R. M. Robb.
Inf. Bull. Variable Stars, No. 1411 (1978).

123.033 Confirmation of the δ Sct-type variability of θ^2
Tauri.   H. W. Duerbeck.
Inf. Bull. Variable Stars, No. 1412 (1978).

123.034 Photoelectric monitoring of the flare star YZ CMi.
K. Ichimura, Y. Shimizu.
Inf. Bull. Variable Stars, No. 1416 (1978).

123.035 Is 20 Cep a variable star?   M. Saito.
Inf. Bull. Variable Stars, No. 1420 (1978).

123.036 New variable stars in Triangulum.   G. Romano.
Inf. Bull. Variable Stars, No. 1421 = 1433 (1978).

123.037 Variability of HD 47396 (CSV 798).
M. Huruhata.
Inf. Bull. Variable Stars, No. 1422 (1978).

123.038 Ultraviolet flare in FK Ser.   J. Darius.
Inf. Bull. Variable Stars, No. 1429 (1978).

123.039 The period of SX Cassiopeiae.   B. S. Whitney.
Inf. Bull. Variable Stars, No. 1430, 4 pp. (1978).

123.040 Finder charts for fifteen Sagittarius variables.
D. Hoffleit.
Inf. Bull. Variable Stars, No. 1431 (1978).

123.041 Long period variables.   C. B. Stephenson.
Inf. Bull. Variable Stars, No. 1434 (1978).

123.042 A new flare star in Cassiopeia?   R. Hudec.
Inf. Bull. Variable Stars, No. 1435 (1978).

123.043 12.15 minute light variations in Przybylski's star,
HD 101065.   D. W. Kurtz.
Inf. Bull. Variable Stars, No. 1436 (1978).

123.044 Visual observations of Mira type variable stars.
U. Hopp.
Inf. Bull. Variable Stars, No. 1437 (1978).

123.045 V 1057 Cygni.   P. Moore.
Inf. Bull. Variable Stars, No. 1438 (1978).

123.046 Observations of the BY Draconis variable Gliese 182.
B. W. Bopp, C. A. O. Torres, I. C. Busko, G. R.
Quast.
Inf. Bull. Variable Stars, No. 1443, 4 pp. (1978).

123.047 18 veränderliche Sterne im Feld η Arae.
H. Geßner.
Mitt. Veränderl. Sterne (MVS), Band 8, 19 - 23 (1977).

123.048 Beobachtungsergebnisse des Arbeitskreises "Veränderliche Sterne" im Kulturbund der DDR.
Mitt. Veränderl. Sterne (MVS), Band 8, 24 - 28 (1977).

123.049 Visuelle Beobachtungen von kurzperiodischen
Veränderlichen.   H. Busch.
Mitt. Veränderl. Sterne (MVS), Band 8, 28 - 30 (1977).

123.050 Visuelle Beobachtungen von NQ Vul, Nova Sge
1977, HM Sge und R CrB.   D. Böhme.
Mitt. Veränderl. Sterne (MVS), Band 8, 33 (1977).

123.051 Ein neuer veränderlicher Stern (= S 10803).
W. Götz.
Mitt. Veränderl. Sterne (MVS), Band 8, 33 - 34 (1977).

123.052 Periodenänderung von RY Scuti.
E. Splittgerber.
Mitt. Veränderl. Sterne (MVS), Band 8, 48 - 49 (1978).

123.053 Photographische Beobachtungen von Mirasternen
auf Platten der Sonneberger Himmelsüberwachung.
E. Splittgerber.
Mitt. Veränderl. Sterne (MVS), Band 8, 50 - 51 (1978).

123.054 Bearbeitung von 25 Veränderlichen am Südhimmel
(Feld β Gruis, Teil II).   I. Meinunger.
Mitt. Veränderl. Sterne (MVS), Band 8, 51 (1978).

123.055 Photographische Beobachtungen von V 393 Cygni
(BD +42°3548).   K. Reichenbächer.
Mitt. Veränderl. Sterne (MVS), Band 8, 52 (1978).

123.056 **Statistical investigation of the short-term light variations of the star HD 219749 observed with the twin telescope.**   K. Panov.
Magnetic stars, (see 012.028), p. 19 - 20 (1978). — Abstract.

123.057 **Spectroscopic investigation of α^2 CVn type stars at the Toruń Observatory.**
A. Burnicki, A. Woszczyk.
Magnetic stars, (see 012.028), p. 23 (1978). — Abstract.

123.058 **Infrared observations of variable Wolf-Rayet stars.**
J. A. Hackwell, R. D. Gehrz, G. L. Grasdalen.
Bull. American Astron. Soc., Vol. 10, 407 - 408 (1978). Abstract.

123.059 **Observations of variable stars, January - December 1977. Rep. No. 32.**   H. Betlem, G. Comello, H. Feijth, R. Jansen, N. A. v. d. Meij, F. Nieuwenhout, I. P. W. Oosterveld, J. P. A. Veerkamp, S. v. d. Wai.
Nederlandse Ver. Weer- Sterrenkunde. Kapteyn Astron. Lab., Groningen, Netherlands. 2 + 5 pp. (1978).

123.060 **Photographic observations of the object A 0620–00.**
Yu. V. Borisov, Yu. Derevyashkin, S. Dejch.
Astron. Tsirk., No. 969, p. 7 (1977). In Russian.

123.061 **New variable SVS 2246 near the NGC 7129 nebula.**
A. L. Gyul'budagyan, R. A. Sarkisyan.
Astron. Tsirk., No. 972, p. 7 - 8 (1977). In Russian.

123.062 **Preliminary designations of SVS stars.**
Astron. Tsirk., No. 973, p. 7 - 8 (1977). In Russian.

123.063 **Two new U Gem-type stars.**   N. E. Kurochkin.
Astron. Tsirk., No. 974, p. 4 - 5 (1977). In Russian.

123.064 **New carbon variable stars.**   I. Daube.
Astron. Tsirk., No. 974, p. 5 - 6 (1977). In Russian.

123.065 **On two variable stars in Cepheus.**
A. L. Gyul'budagyan, R. A. Sarkisyan.
Astron. Tsirk., No. 982, p. 4 - 5, with a correction in No. 986, p. 8 (1978). In Russian.

123.066 **New variable star SVS 2261.**   V. P. Goranskij.
Astron. Tsirk., No. 990, p. 8 (1978). In Russian.

123.067 **Observations of two R CrB-type stars.**
A. Eh. Rozenbush.
Astron. Tsirk., No. 991, p. 1 - 2 (1978). In Russian.

123.068 **Photoelectric photometry of RU Cam.**
G. V. Zajtseva, V. M. Lyutyj.
Astron. Tsirk., No. 994, p. 4 - 5 (1978). In Russian.

123.069 **Information on the photoelectric observations of variable stars deposited at the Odessa Astronomical Observatory.**   V. P. Tsesevich, E. N. Makarenko.
Astron. Tsirk., No. 995, p. 8 (1978). In Russian.

123.070 **AX Persei.**
British Astron. Assoc. Circ., No. 585 (1978).

123.071 **Maxima and minima of long period variables.**
J. A. Mattei.
AAVSO Bull. 41, 8 pp. (1978). — 1978 annual predictions.

123.072 **Predicted intervals brighter than magnitude 11.0 and fainter than magnitude 13.5, during the period January, 1978 through February, 1979.**
Compiled by P. O. Taylor, M. J. Taylor, C. B. Ford.
AAVSO Bull. 41-A + B, 15 pp. (1978).

123.073 **Variable Star Notes.**   J. A. Mattei.
J. R. Astron. Soc. Canada, Vol. 72, 61 - 64, 178 - 180 (1978).

123.074 **List of stars.**
Peremennye Zvezdy, Prilozhenie, Vol. 2, 463 - 464 (1976). In Russian.

Quasars PKS 1217 + 023 and LB 2136 on plates of a supernovae survey.   See Abstr. 141.196.

124 Novae

124.001 Optically thick winds in classical novae.
G. T. Bath.
Mon. Not. R. Astron. Soc., Vol. 182, 35 - 48 (1978).

The photospheric properties of continuously outflowing optically thick winds are derived using Christy's fit to the Cox–Stewart opacities and an approximate treatment of scattering in translucent conditions. The decline behaviour of classical novae has been synthesized by photospheric sequences in which the mass-loss rates slowly decrease and the wind progressively thins. The conclusions of earlier models based on pure scattering opacities are generally confirmed.

124.002 A photometric study of the recurrent nova WZ Sagittae at minimum light. E. L. Robinson, R. E. Nather, J. Patterson.
Astrophys. J., Vol. 219, 168 - 182 (1978).

The authors have acquired a new set of high speed photometric observations of the recurrent nova WZ Sge; using these observations, they develop a quantitative model for the system. WZ Sge is a binary system with an orbital period of 81^m38^s and a separation of 3.2×10^{10} cm consisting of an unevolved star with a mass of $0.019\,M_\odot$ orbiting a white dwarf with a mass of $0.38\,M_\odot$. Mass is being transferred from the late-type star to the white dwarf in a stream through the inner Lagrangian point. The transferred material forms an accretion disk with a radius of 9.7×10^9 cm around the white dwarf. The accretion disk is optically thick and is the sole source of the broad hydrogen absorption lines seen in the spectrum of WZ Sge.

124.003 On the magnitude-frequency relation for recurrent novae. J. E. Steiner.
Astron. Astrophys., Vol. 62, 273 - 274 (1978).

It is shown that there is a relation between the eruption magnitude and the time interval between two eruptions for individual recurrent novae. With this relation it is possible to predict the next eruption epoch and to infer some conclusions about the nature of the eruptions of recurrent novae.

124.004 The evolution of a slow nova model with a $Z = 0.03$ envelope from pre-explosion to extinction.
D. Prialnik, M. M. Shara, G. Shaviv.
Astron. Astrophys., Vol. 62, 339 - 348 (1978).

A model for slow nova explosions is presented. The authors find: (1) CNO equilibrium burning on a timescale of $10^{5.5}$ s produces enough energy for mass ejection. (2) The rise in luminosity stops close to the Eddington limit and the outer envelope layers accelerate via the continuous action of radiation pressure. (3) The mass outflow has two phases: a gentle outflow at the beginning and then a rapid outflow. In both phases the authors find $\dot{m} \simeq$ constant, or equivalently, steady state outflow. (4) The nova's "shut-off" mechanism is the exhaustion of the envelope's mass. In this "slow nova" model it took about 200 days for 95% of the envelope to be ejected and to leave behind a hot white dwarf. (5) The isotope ratios C^{12}/C^{13}, N^{14}/N^{15} and O^{16}/O^{17} are in good agreement with observations. (6) The behaviour of $L_{BOL}(t)$ agrees well with observations. Several additional consequences are discussed.

124.005 The number of outbursts of a classical nova.
H. C. Ford.
Astrophys. J., Vol. 219, 595 - 596 (1978).

The minimum number of outbursts in classical novae is determined by comparing the nova frequency and stellar death rate in the nuclear bulge of M31. The minimum number of outbursts, which is derived by assuming that all dying stars become novae, is shown to lie between 160 and 660. Because only a fraction of the dying stars will be in binary systems that lead to novae, the true number of outbursts will exceed the minimum number by a factor which may be as large as 100.

124.006 Four novae: V 1229 Aquilae (1970), IV Cephei (1971), V 368 Scuti (1970) and FH Serpentis (1970).
I. D. Howarth.
J. British Astron. Assoc., Vol. 88. 180 - 187 (1978). – A report of the Variable Star Section.

124.007 Grain formation in nova envelopes.
T. Yamamoto, S. Nishida.
Prog. Theor. Phys., Vol. 57, 1939 - 1950 (1977). – Abstr. in Phys. Abstr., Vol. 80, Abstr. 91548 (1977).

124.008 A hydrodynamic study of a slow nova outburst.
W. M. Sparks, S. Starrfield, J. W. Truran.
Astrophys. J., Vol. 220, 1063 - 1075 (1978).

The authors have used a Lagrangian, implicit, hydrodynamics computer code incorporating a full nuclear-reaction network to follow a thermonuclear runaway in the hydrogen-rich envelope of a $1.25\,M_\odot$ white dwarf. In this evolutionary sequence the envelope was assumed to be of normal (solar) composition and the resulting outburst closely resembles that of the slow nova HR Del. In contrast, the previous CNO-enhanced models resemble fast nova outbursts. The slow nova model ejects material by radiation pressure when the high luminosity of the rekindled hydrogen shell source exceeds the local Eddington luminosity of the outer layers. This is in contrast to the fast nova outburst where ejection is caused by the decay of the β^+-unstable nuclei.

124.009 An alternative mechanism for the nova phenomenon.
G. W. Collins II, C. B. Foltz.
Publ. Astron. Soc. Pacific, Vol. 89, 896 - 900 (1977/78).

A new mechanism for the origin of the nova outburst is presented. A magnetic white dwarf in a binary system undergoes collapse. This collapse induces pressure disturbance which propagates along the field lines, expelling the polar regions of the star. It is found that the energy of the outburst depends on the magnetic field of the configuration.

124.010 Recent nova and variable-star research with the Harvard Photographic Collection. L. J. Chaisson.
Publ. Astron. Soc. Pacific, Vol. 89, 901 - 904 (1977/78).

This report contains a summary of research conducted in 1976 and early 1977 with the Harvard plate stacks on novae, possible novae, and an emission variable. Light curves for He 3-1481 and HM Sge are included, as well as the results of searches for earlier outbursts of recent novae.

124.011 The temperatures of circumstellar shells.
E. P. Ney, K. M. Merrill.
Bull. American Astron. Soc., Vol. 9, 556 (1978). – Abstract.

124.012 Helium abundances of six old novae.
G. J. Ferland.
Bull. American Astron. Soc., Vol. 9, 557 (1978). – Abstract.

124.013 Optical observations of recent novae. B. Wolf.
IAU Colloq. No. 42, (see 012.012), p. 151 - 181 (1977).

The observations of two rather different classical novae, V 1500 Cyg (= Nova Cyg 1975) and NQ Vul (= Nova Vul 1976), are presented and compared. The light curve of Nova Cyg 1975 is very smooth. Nova Vul 1976 has a completely different lightcurve with extremely strong rapid irregular variations of considerably amplitude. There is a remarkable second maximum about 14 days after discovery. The spectroscopic evolution of Nova Cyg 1975 was very fast. The nebular stage was already reached 9 days after maximum. Whereas the high velocity absorption systems were almost negligible in Nova Cyg 1975, these systems were prominent in Nova Vul 1976.

124.014   Spectroscopic observations of old novae.
A. Bianchini.
IAU Colloq. No. 42, (see 012.012), p. 192 - 193 (1977).

124.015   Post-outburst spectroscopy of classical novae.
S. Wyckoff, P. A. Wehinger.
IAU Colloq. No. 42, (see 012.012), p. 201 - 213 (1977).

SIT Vidicon spectra (3500–5100 Å) with time resolutions of several minutes and spectral resolutions of 6 Å have been obtained of a sample of old novae at minimum light. For comparison, spectra of Sco X-1 (V818 Sco) are also presented. Velocity-resolved image-tube spectra of RR Pic 1925 indicate Doppler variations in the He II λ4686 emission line with a period ~3 hours, thus confirming the binary nature of RR Pic.

124.016   The interaction of a nova with its circumstellar envelope.   I. Malakpour.
IAU Colloq. No. 42, (see 012.012), p. 214 - 217 (1977).

124.017   Element abundance analyses of novae.
R. E. Williams.
IAU Colloq. No. 42, (see 012.012), p. 242 - 273 (1977).

The different methods by which element abundances in novae have been determined are reviewed. Curve of growth studies of novae at maximum light have indicated CNO nuclei to be greatly enhanced with respect to hydrogen in certain objects. These results are questionable because they depend upon an assumed temperature distribution in the photosphere which is probably too steep to be realistic. Emission line analyses of novae, generally obtained in the period of early decline, also indicate possible heavy element enhancement, however these results are tentative because of uncertainties in the parameters of the emitting gas. It is suggested that useful abundance determinations of nova ejecta might be obtained from studies of old, extended nova shells.

124.018   Non uniformity expected in nova ejection.
M. Friedjung.
IAU Colloq. No. 42, (see 012.012), p. 322 - 326 (1977).

It is shown that according to the model of continued ejection of nova envelopes by radiation pressure, discrete clouds are probably formed. Their properties are discussed.

124.019   The evolution of a slow nova model with a Z = .03 envelope from pre-explosion to extinction.
D. Prialnik, M. M. Shara, G. Shaviv.
IAU Colloq. No. 42, (see 012.012), p. 346 - 364 (1977).

A model for slow nova explosions is presented. The model consists of a 0.8 $M_\odot$ C/O core and an envelope of 10^{-4} $M_\odot$ with solar composition. The envelope is assumed to have been accreted from a companion. The nuclear runaway produces luminosity close to the Eddington luminosity; this ejects 95% of the envelope.

124.020   The space density, recurrence rate and classification of novae.   G. T. Bath, G. Shaviv.
Mon. Not. R. Astron. Soc., Vol. 183, 515 - 522 (1978).

A value for the recurrence period of classical nova outbursts ~10^5 yr is obtained from the space densities of white dwarfs and novae in our Galaxy. The authors compare this with the recurrence period of ~30 yr of known recurrent novae and conclude that two classes of galactic novae exist. Those with main-sequence companions and long recurrence periods and those with red giants and short recurrence periods. The properties of the two classes are examined and compared with those of dwarf novae. The basis of the Kukarkin–Parene-

go relation is examined.

124.021   On ^{7}Li production in nova explosions.
S. Starrfield, J. W. Truran, W. M. Sparks,
M. Arnould.
Astrophys. J., Vol. 222, 600 - 603 (1978).

Calculations of ^{7}Li production occurring as a concomitant of thermonuclear runaways in hydrogen envelopes of white dwarfs are reported. It is found that sufficient ^{7}Li can be produced in models displaying fast-nova-like features to suggest that the corresponding objects represent significant contributors to the ^{7}Li enrichment of galactic matter. The sensitivities of these results to various assumptions and uncertainties are discussed.

124.022   The pre-maximum spectrum of a Magellanic Cloud nova.   J. A. Graham.
Bull. American Astron. Soc., Vol. 10, 452 (1978). – Abstract.

124.023   Nova in Large Magellanic Cloud.
IAU Circ., Nos. 3204, 3206 (1978).

124.024   A comment on the slow nova in Scorpius.
I. Lundström, B. Stenholm.
Inf. Bull. Variable Stars, No. 1393 (1978).

124.025   AS 239: the declining phase of a slow nova?
D. A. Allen.
Inf. Bull. Variable Stars, No. 1399, 5 pp. (1978).

124.026   Probable nova Bootis 1962.   H. Gessner.
Inf. Bull. Variable Stars, No. 1428 (1978).

124.027   Novae observed at La Silla.   H. W. Duerbeck.
Messenger, No. 13, p. 6 - 7 (1978).

124.028   Evolution of nova binaries: the inevitability of helium runaways.
R. F. Webbink, J. W. Truran, J. S. Gallagher.
Bull. American Astron. Soc., Vol. 10, 438 (1978). – Abstract.

Nova models and their problems.
See Abstr. 065.012.

Photonuclear p-processing in degenerate hydrogen burning regions and its relationship to nova outbursts.
See Abstr. 065.045.

The low-temperature photonuclear nucleosynthesis of the bypassed (p-) nuclei in degenerate hydrogen burning zones and its relationship to nova outbursts.
See Abstr. 065.062.

Radio emission from close binary systems and novae.   See Abstr. 116.007.

Orbital angular momentum loss via gravitational radiation and mass-transfer rates in close binary systems.
See Abstr. 117.005.

Pulsational stabilities of a star in thermal imbalance: comparison between the methods.
See Abstr. 122.041.

The wavelength dependence of 27 s pulsations in WZ Sge.   See Abstr. 122.045.

Spectra of the Crab Nebula, AM Her, and Nova Ophiuchi from HEAO-1.   See Abstr. 142.089.

Continuum and emission line variations in Sco X-1 and old novae.   See Abstr. 142.100.

Nova-driven winds in globular clusters.
See Abstr. 154.027.

The extragalactic distance scale. I. A review of distance indicators: zero points and errors of primary indicators.   See Abstr. 162.150.

Nova Cygni 1975 = V1500 Cygni

124.101   The helium abundance of ejecta from V1500 Cygni
(Nova Cygni 1975).   G. J. Ferland.
Astrophys. J., Vol. 219, 589 - 594 (1978).

The author has studied the He emission spectrum of V1500 Cygni (Nova Cygni 1975) in the nebular phase of the outburst. The steep Balmer decrement is interpreted as evidence for optical depth in the Balmer series. The principal result is that the He abundance is identical to that found in planetary nebulae, $He/H = 0.11 \pm 0.01$. The color temperature of the ionizing radiation field increased from below 5×10^4 K on day 10 to about 1.7×10^5 K by day 300. High temperatures during the late decline are supported by the presence of strong lines of [Ne V] and [Fe VII].

124.102   Photometric observations of Nova V1500 Cygni.
I. Semeniuk, A. Kruszewski, A. Schwarzenberg-Czerny, T. Chlebowski, M. Mikołajewski, J. Wołczyk.
Acta Astron., Vol. 27, 301 - 318 (1977).

UBV photometric observations of Nova Cygni 1975 = V1500 Cygni were obtained in the time interval from September 1975 to October 1976. The short period variability discovered by Tempesti (1975) and Koch and Ambruster (1975) has always been present but with greatly varying amplitude in a range from 0.03 to 0.7 mag. The period decreased from $0^d.1410$ in September 1975 when the short term variations were first discovered, to $0^d.1380$ in autumn 1976. The roughly sinusoidal light curve changes its shape from cycle to cycle and there are irregular fluctuations superposed.

124.103   Electron temperatures and densities and other
properties of the shell of Nova Cygni 1975.
J. S. Neff, V. V. Smith.
Bull. American Astron. Soc., Vol. 9, 555 - 556 (1978).
Abstract.

124.104   Nova V1500 Cyg — photometry and period variations study.
I. Semeniuk, A. Kruszewski, A. Schwarzenberg-Czerny, T. Chlebowski.
IAU Colloq. No. 42, (see 012.012), p. 194 - 200 (1977).

This study aims to present photometric observations of Nova Cygni V1500 1975, and to discuss the period changes of brightness variation on the basis of all available data.

124.105   Chemical composition of Nova Cygni 1975.
A. A. Boyarchuk.
IAU Colloq. No. 42, (see 012.012), p. 274 - 278 (1977).

The chemical composition of Nova Cygni 1975 was determined by a curve-of-growth analysis. A strong excess of helium and the CNO -group abundances and a normal Si abundance was found.

124.106   Nova Cygni 1975: a model for the variability of the
3 hour period.   A. C. Fabian, J. E. Pringle.
IAU Colloq. No. 42, (see 012.012), p. 311 - 312 (1977).
Abstract.

124.107   Der Helligkeitsabfall der Novae im späten Stadium
am Beispiel der Nova V1500 Cyg.
A. Hänel.
BAV Rundbrief, 27. Jahrg., 5 - 8 (1978).

124.108   Results of observations of nova Cygni 1975.
O. Eh. Aab, N. F. Vojkhanskaya, T. A. Kartasheva, S. M. Morozova.
Astron. Zh. Akad. Nauk SSSR, Vol. 55, 553 - 558 (1978).
In Russian. English translation in Soviet Astron., Vol. 22, No. 3.

The time evolution of hydrogen emission line contours has been traced and it is shown that at the beginning of the phase of decline of brightness the physical conditions were different in different parts of the envelope. Radial velocities have been measured. The time evolution of the velocities was found to be different for different parts of the envelope.

124.109   V1500 Cygni.
IAU Circ., Nos. 3170, 3184 (1978).

124.110   The continuous energy distribution of Nova Cygni
1975.   W. Wamsteker.
ESO Sci. Prepr. No. 25, 14 pp. (1978). — Submitted to Astronomy and Astrophysics.

Infrared observations of the early stage of Nova Cygni 1975 are presented. The results show that at the longer wavelengths the maximum is reached considerably later than in the visual. A model for the continuum radiation is proposed that explains the complete radiative evolution of this nova in terms of thermal bremsstrahlung. The change of the optical thickness in a plasma shell and the wavelength dependence of the optical thickness under expansion gives good agreement with the observed times of maximum brightness. The time of onset of radiative energy loss from the nova is determined to be 29.9 August 1975.

124.111   Weitere Beobachtungen der Nova Cygni 1975
(V 1500 Cyg).
Mitt. Veränderl. Sterne (MVS), Band 8, 30 - 31 (1977).

124.112   Early nebular stage of nova Cygni, 1975.
T. P. Prabhu.
Kodaikanal Obs. Bull.,Ser. A, Vol. 2, 75 - 84 (1977).

The emission line profiles of the spectrum of the envelope of nova V1500 Cyg, 1975 are studied. The spectral region covers $\lambda\lambda 3700-4500$, $H\beta$ and $\lambda\lambda 6500-8800$. The velocity structure of the profiles and the time variation of both the velocities and intensities of the emission peaks of $H\alpha$ are interpreted as due to an expanding system of rings at different latitudes with respect to a common polar axis. The envelope was optically thick to Balmer line radiation and was thinning out rapidly during the period of observation.

124.113   Photographic photometry of nova Cygni 1975
(V1500 Cyg).
N. M. Bronnikova, V. A. Sokolova, I. A. Shramko.
Astron. Tsirk., No. 970, p. 1 - 2 (1977). In Russian.

124.114   A study of nova Cygni 1975.   A. A. Boyarchuk.
Flare stars. Symposium 1976, (see 012.035),
p. 155 - 160. In Russian. — Abstr. in Ref. zh., 51. Astron., 6.51.688 (1978).

124.115   X-ray and UV observations of Nova Cygni 1975.
Z. Urban.
Říše hvězd, Vol. 59, 5 - 8 (1978). In Czech.

Nova Delphini 1967 = HR Delphini

124.201   Nova Delphini 1967. II. Physical properties and
element abundances in the nebular state.
R. R. Robbins, A. Sanyal.
Astrophys. J., Vol. 219, 985 - 993 (1978).

Physical properties of the nebulosity of Nova Delphini

1967 during its "nebular" stages are derived from high dispersion spectra taken during the period 1967 July to 1969 August. The electron temperature and electron density were determined from the ratio of auroral to nebular line intensities of [O III] and [N II]. The authors compute the oxygen and helium abundances and note the existence of optical depth effects in the helium line intensities. The abundance results are discussed. From Hβ photometry, the mass of the nebula is determined to be 2.9×10^{30} g, or less if there is self-absorption in the Balmer spectrum. Various categories of emission lines are established, and the variation with time of the high-and low-velocity components of the nebular lines of [O III] is considered.

124.202 Photoionization models of the envelope of Nova Delphini 1967 in the nebular stage. I. Observational data.   R. Tylenda.
Acta Astron., Vol. 27, 389 - 413 (1977).

The paper contains the observational material and a preliminary discussion of the observations made from a point of view of construction of model envelopes. Intensities of emission lines have been derived from spectroscopic observations of the nova taken in three observational seasons, i.e. June-July 1969, 1971-72, and August 1975. Discussion of possible mechanisms of ionization and heating of a cosmic gas indicates that the nova envelope is ionized mainly by extreme-ultraviolet radiation emitted by a central source. The observed Balmer decrement and a critical analysis of results of other authors show that circumstellar extinction is negligible for Nova Del 1967. The continuum of the nova is produced by the central star.

124.203 The spectral evolution of nova HR Del (1967) during its decline.   P. Rafanelli, L. Rosino.
Astron. Astrophys., Suppl. Ser., Vol. 31, 337 - 352 (1978).

As a result of examining spectra obtained at Asiago from 1967 to 1976, a description is given of the spectral evolution of the slow nova HR Del, particularly during its decline. The nova remained near maximum brightness for about one year, displaying large brightness fluctuations. During this period not less than nine different absorption systems were observed. The final decline began in May 1968 and the nova entered the nebular stage at the end of July. The nebular spectrum in 1968–1969 was characterized by the presence of wide emission bands of H, HeI, NII and forbidden lines of atoms in different stages of ionization. In the last spectra taken in 1974–1976, when the nova was near minimum brightness, the nebular lines of [OIII] $\lambda\lambda$4959–5007 were still present.

124.204 On the circumstellar envelope around nova Delphini 1967.   L. I. Antipova.
Pis'ma Astron. Zh., Vol. 4, 177 - 182 (1978). In Russian.

On the basis of an analysis of published data the expansion velocity of the envelope ejected by nova Delphini 1967 (HR Del) is found to decrease with time. The change of widths of narrow emission lines is discussed. Both observational phenomena may be explained assuming the existence of an extensive circumstellar envelope. It is supposed that explosions of ultraslow novae may result in the formation of at least some planetary nebulae.

124.205 On the pre-maximum and principal absorption line systems in the spectrum of nova Delphini 1967 (HR Del).   L. I. Antipova.
Astron. Zh. Akad. Nauk SSSR, Vol. 55, 531 - 539 (1978). In Russian. English translation in Soviet Astron., Vol. 22, No. 3.

The evolution of the pre-maximum and principal absorption spectra of HR Del is discussed on the basis of an analysis of existing data and the author's measurements. It is found that the principal spectrum appears a few days before maximum brightness of the star. The only regular spectral

line components observed during the pre-maximum stage are those formed by continuous injection of the matter from the nova surface. At the same time the HR Del spectrum showed rather intensive short-lived components due to temporal eruptive injections from the star.

124.206 Peculiarities of the emission lines in the spectrum of HR Delphini.   D. V. Rajkova.
Astron. Zh. Akad. Nauk SSSR, Vol. 55, 540 - 552 (1978). In Russian. English translation in Soviet Astron., Vol. 22, No. 3.

The profiles of emission lines in the spectrum of HR Del have been studied from spectra obtained with dispersions of 14 and 37 Å/mm. It is shown that in the pre-maximum stage the emission lines consist of a usual P Cyg-type component and a sharp component. The behaviour of the Ti II lines shows some peculiarities.

Nova Vulpeculae 1976 = NQ Vulpeculae

124.301 The isothermal dust condensation of Nova Vulpeculae 1976.   E. P. Ney, B. F. Hatfield.
Astrophys. J., Lett., Vol. 219, L111 - L115 (1978).

Nova Vulpeculae 1976 has been followed with broadband photometry at wavelengths of 0.5 to 12.5 microns from day 3 to day 235. Three definite phases are identified. They are expanding pseudophotosphere, free-free expansion, and dust condensation. A very long (day 80 to day 220) period of isothermal infrared emission ($T \approx 900$ K) is accompanied by an exponential decrease in infrared luminosity and is interpreted in terms of continuous mass ejection in the nova wind.

124.302 The distance and reddening of Nova Vulpeculae 1976.   B. W. Carney.
Bull. American Astron. Soc., Vol. 9, 556 (1978). – Abstract.

124.303 Spectroscopic observations of Nova Vulpeculae 1976.   J. W. Younger.
Bull. American Astron. Soc., Vol. 9, 556 (1978). – Abstract.

124.304 Spectral evolution of Nova Vulpeculae 1976.   P. Rafanelli, A. Vittone.
IAU Colloq. No. 42, (see 012.012), p. 182 - 184 (1977).

124.305 NQ Vulpeculae.
IAU Circ., No. 3231 (1978).

124.306 NQ Vulpeculae.
Yamamoto Circ., No. 1888 (1978). In Japanese.

124.307 Visuelle Beobachtungen der Nova Vulpeculae 1976 (NQ Vul).
Mitt. Veränderl. Sterne (MVS), Band 8, 32 (1977).

Nova Scuti 1975 = V373 Scuti

124.401 Comments on the spectrum of Nova Scuti 1975.   J. S. Gallagher.
Astrophys. J., Vol. 221, 211 - 219 (1978).

Optical spectrophotometric observations of Nova Scuti 1975 obtained on 1975 August 2 and 31 (at 3 and 4 mag, respectively, below maximum) and a few near-infrared spectra provide a basis for a discussion of the emission spectrum. That lines from O I, [O I], [O II], [O III], N I, N II, [N II], N III, H, He I, and He II are most prominent indicates a wide range in density and degree of ionization within the nova. The problem of coronal lines in declining novae is reviewed.

Spectroscopic observations of recent novae (N Per 1974, N Sct 1975).   See Abstr. 124.801.

Spectral evolution of nova V400 Per (1974) and nova V373 Sct (1975).   See Abstr. 124.802.

Nova Sagittarii 1975 No. 2 = V3964 Sagittarii

124.411   Nova Sagittarii 1975 No. 2 (V3964 Sgr) on plates of the collection of the Sternberg Institute.
V. P. Goranskij.
Astron. Tsirk., No. 983, p. 1 - 3 (1978). In Russian.

Nova Ophiuchi 1977 = H1705−25

124.501   Nova Ophiuchi 1977: an X-ray nova.
R. E. Griffiths, H. Bradt, R. Doxsey, H. Friedman, H. Gursky, M. Johnston, R. Leach, A. Longmore, D. A. Malin, P. Murdin, M. J. Ricketts, D. A. Schwartz, J. Schwarz, M. G. Watson.
Bull. American Astron. Soc., Vol. 9, 627 (1978). − Abstract.

124.502   Nova Ophiuchi 1977.
IAU Circ., No. 3197 (1978).

Nova Ophiuchi 1977: an X-ray nova.
See Abstr. 142.059.

The X-ray light curve of Nova Ophiuchi 1977 (H1705−25).   See Abstr. 142.060.

X-ray observations of the transient sources H1743−32 and H1705−25 (Nova Ophiuchi 1977).
See Abstr. 142.081.

Nova Herculis 1934 = DQ Herculis

124.601   The 71 second oscillation of DQ Herculis.
J. Patterson, E. L. Robinson, R. E. Nather.
Bull. American Astron. Soc., Vol. 9, 557 (1978). − Abstract.

124.602   The expanding shell of nova DQ Herculis 1934.
D. C. Jenner.
Bull. American Astron. Soc., Vol. 10, 427 (1978). − Abstract.

Nova Sagittarii 1977

124.701   Nova Sgr 1977.   B. Hidayat, S. D. Wiramihardja.
Astron. Astrophys., Vol. 65, 143 - 145 (1978).
The absolute magnitude of Nova Sgr 1977 has been determined to be $M_v = -6.5$ at maximum on April 2, 1977. Using $m_{max} = 8^m8$, and minimum estimate of interstellar absorption of 0^m9, the distance of the nova was derived to be 7 kpc.

124.702   The coordinates of nova Sagittarii 1977.
A. Sh. Khatisashvili.
Astron. Tsirk., No. 972, p. 3 (1977). In Russian.

Nova Sagittarii 1978 = V3876 Sagittarii = IRC−20494

124.751   Nova Sagittarii 1978 = IRC −20494 = V3876

Sagittarii.
IAU Circ., Nos. 3209, 3211, 3213 (1978).

124.752   Nova Sagittarii 1978 = V3876 Sgr = IRC −20494.
Yamamoto Circ., Nos. 1883, 1884 (1978). In Japanese.

Nova Persei 1974 = V400 Persei

124.801   Spectroscopic observations of recent novae (N Per 1974, N Sct 1975).   L. Rosino.
IAU Colloq. No. 42, (see 012.012), p. 185 - 191 (1977).
Continuing the systematic study of novae, carried out at Asiago since 1958, two recent galactic novae, V400 Per (1974) and V373 Sct (1975) have been observed, after the announcement of the discovery, during their decline towards minimum, with the spectrographs applied to the telescopes of 122 cm and 182 cm of the Asiago Astrophysical Observatory, equipped with image-tubes.

124.802   Spectral evolution of nova V400 Per (1974) and nova V373 Sct (1975).   L. Rosino.
Astrophys. Space Sci., Vol. 55, 383 - 394 (1978).
Photographic and spectroscopic observations of the two galactic novae, V400 Per and V373 Sct, which appeared in 1974 and 1975, have been carried out at Asiago. The light curves of the two novae were characterized by the presence of brightness oscillations during the early decline. The spectral evolution was quite normal.

Nova Sagittae 1977 = HS Sagittae

124.851   Nova Sagittae 1977 (HS Sagittae).
IAU Circ., Nos. 3184, 3225 (1978).

124.852   HS Sagittae (nova Sagittae 1977).
Yamamoto Circ., No. 1888 (1978). In Japanese.

Nova Serpentis 1978

124.901   Nova Serpentis 1978.   R. Lukas.
BAV Rundbrief, 27. Jahrg., 19 (1978).

124.902   Nova Serpentis 1978.
IAU Circ., Nos. 3188, 3195, 3198, 3201, 3206, 3222, 3226, 3235 (1978).

124.903   Nova Serpentis 1978.
Yamamoto Circ., Nos. 1880, 1881, 1883, 1888 (1978). In Japanese.

Nova Monocerotis 1975 = V616 Monocerotis

124.951   A provisional optical light curve of the X-ray recurrent nova V616 Monocerotis = A0620−00.
R. F. Webbink.
Dep. Astron. Univ. Illinois, Urbana, Illinois 61801. 73 pp. (1978).
This report is circulated in its present form in response to numerous requests for such a compilation. It is not intended to be definitive, but every effort has been made to make it comprehensive, and as such it may be regarded as a working document.

125 Supernovae, Supernova Remnants

125.001 **Supernovae in the Coma cluster of galaxies.**
R. Barbon.
Astron. J., Vol. 83, 13 - 19 (1978).

All the supernovae so far discovered in the Coma cluster have been identified and their light curves reanalyzed. Unpublished observations are reported for SN 1963c, SN 1963m, and SN 1962i, which is found to belong to a background galaxy. For type I supernovae, comparison with the average light curve provided a mean magnitude at maximum m_0 = 15.5 ± 0.2, corresponding to an absolute magnitude M_p = −18.7 if a distance modulus of 34.2 for the cluster is used.

125.002 **The X-ray spectrum and structure of the Puppis A supernova remnant.** J. C. Zarnecki, J. L. Culhane, A. Toor, F. D. Seward, P. A. Charles.
Astrophys. J., Lett., Vol. 219, L17 - L21 (1978).

New observations have been made by the Ariel 5 satellite of the supernova remnant Puppis A in the energy range 1.5–6 keV. These findings are combined with previous soft X-ray data and crystal spectrometer results to show that the X-ray spectrum of Puppis A is characterized by a two-component model. The new data show that these two components have temperatures of 12.6 and 2.5×10^6 K with a column density of 8×10^{20} cm^{-2}, a value which favors a lower source distance than previously assumed. Reanalysis of the Copernicus data shows that this cooler region, which is responsible for the O VIII line emission, is spatially coincident with the soft X-ray region recently delineated by others.

125.003 **The acceleration of high-velocity clouds in supernova remnants.**
C. F. McKee, L. L. Cowie, J. P. Ostriker.
Astrophys. J., Lett., Vol. 219, L23 - L28 (1978).

Interstellar clouds passed by blast waves emanating from supernova explosions will be accelerated by the ram pressure of the expanding interior shocked gas. The authors present numerical and analytical solutions for cloud acceleration in this environment, comparing the results with recent observations of faint, high-velocity (greater than 100 km s^{-1}) filaments observed in Cygnus and Vela. Several predictions are made, the most important of which is that fast clouds of neutral hydrogen with column densities of $\sim 10^{19}$ cm^{-2} should be observable in 21 cm studies of SNRs.

125.004 **An accurate optical and radio comparison for the Puppis A supernova remnant.**
C. Goudis, J. Meaburn.
Astron. Astrophys., Vol. 62, 283 - 285 (1978).

A very deep Hα+ [N II] photograph of the optical filaments of the Puppis A supernova remnant is compared accurately to the best map of the non-thermal radio emission. Some of the brightest filaments are closely correlated with the peak brightnesses along the radio ridge. Others are not.

125.005 **On the Rayleigh-Taylor instability in stellar explosions.** R. A. Chevalier, R. I. Klein.
Astrophys. J., Vol. 219, 994 - 1007 (1978).

After a shock wave has accelerated the outer layers of a star during a stellar explosion, a rarefaction wave moves back into the stellar material, resulting in the conversion of internal energy into kinetic energy. This additional acceleration can be Rayleigh-Taylor unstable. Two-dimensional hydrodynamic calculations are presented for the case of Type II supernovae, showing that the instability results in the ejection of a clumpy shell. It is unlikely that radiation transport can damp the instability for normal Type II supernovae. The authors further conjecture that the structure of nova shells is a consequence of this instability.

125.006 **Narrow band photography of the Monoceros Loop.**
R. P. Kirshner, T. R. Gull, R. A. R. Parker.
Astron. Astrophys., Suppl. Ser., Vol. 31, 261 - 270 (1978).

Narrow band photographs of the Monoceros Loop show a ring-like structure of diameter 4° in [S II] and in Hα+[N II] that lies near, but outside the remnant's non-thermal radio emission. Sharp filaments in Hα+[N II], and the outer extent of the [O III] are more nearly coincident with the radio extent. The distribution of [O III] is inhomogeneous with a marked concentration toward the eastern edge of the remnant. Comparison with models of remnant evolution favors a distance of 780 pc and a relation with the Mon OB1 association rather than with the Rosette.

125.007 **The Monoceros supernova remnant.**
R. D. Davies, K. H. Elliott, C. Goudis, J. Meaburn, N. J. Tebbutt.
Astron. Astrophys., Suppl. Ser., Vol. 31, 271 - 284 (1978).

Deep Hα+[N II] photographs are compared to a variety of radio maps of the Monoceros supernova remnant. Two distinct components of the ionized gas associated with this object are clearly identified. One is diffuse, whereas the other has a fine filamentary structure. New observations of the velocities of this diffuse component are compared to previous measurements and to velocity features found in the interstellar absorption line profiles in several stars. A complicated picture of the velocity field emerges.

125.008 **Supernova remnants in M 33.**
S. D'Odorico, P. Benvenuti, F. Sabbadin.
Astron. Astrophys., Vol. 63, 63 - 68 (1978).

Three supernova remnants (SNR) have been identified in a field 8′ in diameter centered on the southern arm of M 33. These objects are the three strongest radio sources in that field in the high resolution 21 cm radio map of M 33 obtained by Israel and van der Kruit (1974). From this result it appears that SNR may seriously contaminate H II region population counts based on radio fluxes in M 33.

125.009 **One more young supernova remnant in our Galaxy.**
Yu. P. Pskovskij.
Priroda, 1978, No. 2, p. 138 - 140. In Russian.

125.010 **Optical identification of the supernova remnant G206.9 + 2.3 = PKS 0646 + 06.**
S. van den Bergh.
Astrophys. J., Vol. 220, 171 (1978).

125.011 **A search for a fossil Strömgren sphere surrounding the Tycho supernova remnant.**
R. J. Reynolds, P. M. Ogden.
Astrophys. J., Vol. 220, 172 - 176 (1978).

The authors have used a 15 cm diameter Fabry-Perot spectrometer to obtain scans of [O III] λ5007 in 13 directions within an 8° diameter region centered on Tycho's supernova remnant. The data set an upper limit of 1.1×10^{-7} ergs cm^{-2} s^{-1} sr^{-1} for the intensity of any [O III] λ5007 emission from the interstellar gas surrounding this remnant, which is significantly less than the intensity predicted by Kafatos and Morrison for a "fossil Strömgren sphere" surrounding the remnant. The authors conclude that the total flux of any burst of ultraviolet ($\sim$ 40 eV) photons from the Tycho supernova was less than 1.5×10^{49} ergs. However, these observations do not exclude the possibility that a large burst of radiation was emitted primarily in the form of hard (greater than $\sim$ 100 eV) UV radiation.

125.012 **The Cassiopeia A progenitor: a consistent evolution-**

ary picture involving supergiant mass loss.
S. A. Lamb.
Astrophys. J., Vol. 220, 186 - 192 (1978).

The observed abundances in the Cas A supernova remnant are used together with results of evolutionary calculations of massive stars to construct an evolutionary picture for the Cas A progenitor star. A consistent picture is presented which involves two periods of mass loss, one while the star was an early-type supergiant, the other in a later evolutionary phase. A review of current observational evidence for mass loss from early-type supergiants and the proposed evolutionary picture for the Cas A progenitor suggest that some massive stars lose over half of their hydrogen envelopes prior to crossing the Hertzsprung gap.

125.013 Optical remnant of the supernova of 1181.
S. van den Bergh.
Sky Telesc., Vol. 55, 196 - 197 (1978).

125.014 Pourquoi les supernovae explosent-elles?
J. Audouze.
La recherche en astrophysique, (see 003.001), p. 103 - 108 (1977).

125.015 Coincidence of compact supernova remnants with three COS-B γ-ray sources.   R. C. Lamb.
Nature, Vol. 272, 429 - 430 (1978).

Positional coincidences and a 'near miss' of four small-diameter supernova remnants (SNR) with three of the 11 unidentified COS-B γ-ray sources are reported.

125.016 A maximum likelihood estimate of the supernova rate in Sc galaxies.
P. G. Craven, G. Cavallo, G. G. C. Palumbo.
Astron. Astrophys., Vol. 64, 87 - 90 (1978).

The authors show how the maximum likelihood principle may be used to provide a rigorous statistical analysis of supernova explosions in external galaxies. The method allows for the simultaneous evaluation of several parameters in a multi-parameter model, and can be used to analyse a list of observed supernovae when the number of galaxies which have not produced events is unknown. An analysis is presented of the supernovae discovered by the Palomar supernova search in bright ($m_{pg} \leq 13.0$) Sc and ScI galaxies.

125.017 An observational approach to a problem of filament formation in old supernova remnants.
T. A. Lozinskaya.
Astron. Astrophys., Vol. 64, 123 - 126 (1978).

The interaction between the blast wave and inhomogeneities in the interstellar medium is a possible source of the formation of optical filaments in old supernova remnants. In order to check this theoretical model the correlation between integral brightness and the halfwidth of H_a-line in filaments of Simeiz 147 and VRO 42.05.01 were investigated.

125.018 Are supernovae radio sources? A search for radio emission from young supernova remnants.
R. L. Brown, A. P. Marscher.
Astrophys. J., Vol. 220, 467 - 473 (1978).

The authors conducted a search for radio emission at 11 and 3.7 cm from 46 recent supernovae. None of these supernovae was detected at a flux density greater than 5–10 mJy. These negative results cannot be explained by internal absorption and are thus due to intrinsically weak synchrotron emission in young supernova remnants.

125.019 Origin and evolution of the radio emission from immediate postoutburst supernovae.
A. P. Marscher, R. L. Brown.
Astrophys. J., Vol. 220, 474 - 483 (1978).

The authors explore several models for the radio emission from immediate postoutburst supernovae under the assumption that the expanding remnant consists of a homogeneously mixed distribution of relativistic particles, magnetic field, and thermal plasma. The evolutionary models they investigate are the following: (1) an adiabatic expansion model; (2) a model incorporating the existence of a central pulsar; and (3) variations on the first two models. The authors compare the theoretical evolution of the radio emission for each model with observations of SN 1970g. The only model capable of accounting for the radio evolution of SN 1970g is one in which (1) a pulsar similar in nature to the early Crab Nebula pulsar determines the evolution of the magnetic field, and (2) relativistic electrons are continuously accelerated up to some maximum energy that is inversely proportional to time owing to the expansion of the remnant.

125.020 Type II supernovae: analysis of the observed spectra from 8 to 15 months following the explosion.
C. Gordon.
Astrophys. J., Vol. 220, 484 - 489 (1978).

A study of the spectra of two Type II supernovae, showing a slow decrease in their light curve after $t \approx 100$ days, has been made. For these supernovae, at $t > 100$ days, the envelope is shown to be ionized by radiation from a central source, probably a pulsar. A tentative model of such an envelope is derived.

125.021 Supernova radio pulse searches and possible improvements in sensitivity.
W. P. S. Meikle, S. A. Colgate.
Astrophys. J., Vol. 220, 1076 - 1086 (1978).

A successful search for radio pulses from extragalactic supernovae could result in measurement of the intergalactic ion density, direct measurement of the deceleration parameter q_0, improved supernova statistics, and early detection of supernovae that would give more information about their early evolution and also improve their usefulness as "standard candles". The authors discuss sensitivity limitations involved in attempts to detect radio pulses from extragalactic supernovae and other possible sources, making the assumption that the intergalactic plasma density does not exceed the closure density of the universe.

125.022 The light of the supernova outburst. III. External excitation of the expanding gas shell.
M. Abdulwahab, P. Morrison.
Astrophys. J., Vol. 220, 1087 - 1100 (1978).

The fluorescence-reverberation model for late spectra of Type I supernovae (SN I) has been extended to identify and describe the time evolution of all the major lines in the spectra. Forbidden-line emission is shown to be a highly significant component of late SN I spectra. The model includes three related sources of optical light: (i) the expanding gas shell, which dominates the early spectra by its blackbody continuum emission (with a few absorption lines) but rapidly decays, (ii) the fluorescent circumstellar medium that gives rise to the strong long-lasting lines of He II, (iii) the protracted re-excitation of the cooled outermost layer of the gas shell by the distant fluorescent decay, to produce forbidden lines.

125.023 The early cooling and escape of SN shock accelerated cosmic rays.
S. A. Colgate, A. G. Petschek.
Bull. American Astron. Soc., Vol. 9, 567 - 568 (1978).
Abstract.

125.024 Enhanced molecular abundances and ionization levels in molecular clouds near supernova remnants.
A. Wootten.
Bull. American Astron. Soc., Vol. 9, 589 - 590 (1978).
Abstract.

125.025 Nuclear weak interaction rates of ^{56}Ni in the supernova environment.
G. Fuller, M. J. Newman, W. A. Fowler.
Bull. American Astron. Soc., Vol. 9, 596 (1978). – Abstract.

125.026 Supernovae in molecular clouds.
A. Sivaramakrishnan, J. C. Wheeler.
Bull. American Astron. Soc., Vol. 9, 596 (1978). – Abstract.

125.027 The Cas A progenitor: a consistent evolutionary picture involving supergiant mass loss.
S. A. Lamb.
Bull. American Astron. Soc., Vol. 9, 596 (1978). – Abstract.

125.028 Type I supernovae, R CrB stars, and the Crab Nebula.   J. C. Wheeler.
Bull. American Astron. Soc., Vol. 9, 596 (1978). – Abstract.

125.029 X-ray bursts from type II supernova explosions.
R. I. Klein, R. A. Chevalier.
Bull. American Astron. Soc., Vol. 9, 610 (1978). – Abstract.

125.030 Prompt thermal emission from type II supernovae – limits on the ultraviolet and soft X-ray burst.
S. W. Falk.
Bull. American Astron. Soc., Vol. 9, 610 - 611 (1978). Abstract.

125.031 Detection of the Fe X 6374 Å coronal line in the supernova remnants Cygnus Loop and IC443.
B. E. Woodgate, R. L. Lucke, D. G. Socker, R. P. Kirshner.
Bull. American Astron. Soc., Vol. 9, 646 (1978). – Abstract.

125.032 Soft X-ray emission from the new supernova remnant in Cygnus.   W. A. Snyder, R. C. Henry,
A. F. Davidsen, S. Shulman, G. Fritz, D. Yentis, H. Friedman.
Bull. American Astron. Soc., Vol. 9, 646 (1978). – Abstract.

125.033 The *p*-process in supernovae.
S. E. Woosley, W. M. Howard.
Astrophys. J., Suppl. Ser., Vol. 36, 285 - 304 (1978).
The nucleosynthetic origin of the rare proton-rich isotopes, usually called "*p*-process" isotopes, is examined. A particularly interesting context for this synthesis is found to be explosive events characterized by peak temperatures in the range 2.0 to 3.0 $\times$ 10^9 K. At these temperatures a series of photodisintegration reactions operating upon a distribution of *r*- and *s*-process seeds produces an abundance pattern that displays striking similarities to that of the *p*-process nuclei in the solar system. Requisite conditions for this model are expected to occur naturally in those zones of supernovae that have experienced helium and perhaps carbon burning prior to explosion. Implications for supernova structure, presupernova evolution, and cosmochronology are discussed and a critical discussion of other current *p*-process models is presented.

125.034 On the interpretation of the spectra of type I supernovae.   R. E. Meyerott.
Astrophys. J., Vol. 221, 975 - 989 (1978).
Predictions of the spectra expected from a radioactive excitation source of the late-time luminosity of type I supernovae are presented. It is shown that such a source will give rise to two electron energy distributions: one is a nonequilibrium distribution of high-energy primary and secondary electrons with energies from 20 eV to ~1 MeV, and the other is a thermal equilibrium distribution that can be characterized by an electron temperature whose value is not predicted, but could be consistent with the observed color temperature. For a star deficient in hydrogen but having normal abundances of He and other elements, much of the electron energy is deposited in He and is then transferred to the minor species by energy-

and charge-transfer collisions from the He ions and metastable atoms.

125.035 Asymmetries in the shells of supernova remnants – evidence of large-scale gradients in the surrounding medium.   J. L. Caswell.
Proc. Astron. Soc. Australia, Vol. 3, 130 - 131 (1977).

125.036 Soft X-ray emission from a newly discovered supernova remnant in Cygnus.   W. A. Snyder, A. F.
Davidsen, R. C. Henry, S. Shulman, G. Fritz, H. Friedman.
Astrophys. J., Lett., Vol. 222, L13 - L15 (1978).
The authors have detected a significant soft X-ray flux consistent with emission from the newly discovered supernova remnant in the constellation Cygnus. Emission is observed in the 0.5–2.0 keV band, with a possible detection in the 0.1–0.4 keV region.

125.037 The Hubble diagram for supernovae.
D. Branch, C. Bettis.
Astron. J., Vol. 83, 224 - 227 (1978).
The Hubble diagram for type I and type II supernovae is plotted and discussed. Least-squares fits to several subsets of the data are determined and compared to the results of previous investigations. If the properties of type I supernovae do not depend on parent galaxy type, then the best estimate of the mean absolute magnitude at maximum brightness is obtained from supernovae in elliptical galaxies, $M_{pg} = -18.70 + 5$ log(H_0/100), with $\sigma < 0.5$ mag. At maximum brightness type II supernovae may be intrinsically fainter than those of type I by 1.25 mag or less.

125.038 Evidence of X-ray emission from W44.
E. H. B. M. Gronenschild, R. Mewe, J. Heise,
A. J. F. den Boggende, J. Schrijver, A. C. Brinkman.
Astron. Astrophys., Vol. 65, L9 - L12 (1978).
With the X-ray instruments aboard the ANS satellite the authors have detected X-ray emission with an observed intensity of about 8 $\times$ 10^{-11} erg cm^{-2} s^{-1} (1.0-3.5 keV) from a region of 1.5° $\times$ 0.5° (R.A. $\times$ Decl.) overlapping the supernova remnant W44.

125.039 Photometric classification and observed characteristics of type II supernovae.   Yu. P. Pskovskij.
Astron. Zh. Akad. Nauk SSSR, Vol. 55, 350 - 357 (1978).
In Russian. English translation in Soviet Astron., Vol. 22, No. 2.
The relations between the elements of light curves, color curves, and expansion velocities of the envelopes of type II supernovae and their photometric classes, introduced by the author, are considered.

125.040 Analytic supernova models and black holes.
R. C. Adams, J. M. Cohen.
Int. J. Theor. Phys., Vol. 16, No. 1, p. 35 - 52 (1977).
Abstr. in Phys. Abstr., Vol. 81, Abstr. 4258 (1978).

125.041 Further studies of particle acceleration in Cassiopeia A.   R. A. Chevalier, W. R. Oegerle,
J. S. Scott.
Astrophys. J., Vol. 222, 527 - 536 (1978).
The authors have further investigated models for statistical particle acceleration in the supernova remnant Cas A. Simple (three-parameter) models involving continuous second order Fermi acceleration and variable relativistic particle injection can reproduce the observed radio properties of Cas A. Models dominated by adiabatic expansion losses are preferable to those dominated by particle escape. A model predicting the high-frequency nonthermal spectrum of Cas A indicates that the spectrum turns down in the optical regime due to synchrotron losses. The authors also comment on other models for the relativistic electron content of Cas A.

125.042 **Search for X-ray line emission from Cassiopeia A.**
H. L. Kestenbaum, K. S. Long, R. Novick, M. C. Weisskopf, R. S. Wolff.
Astrophys. J., Vol. 222, 537 - 540 (1978).

The large-area graphite crystal spectrometer aboard OSO 8 has been used to search for X-ray line emission from Cas A from ions of Si, S, and Fe over the energy band 1.865 to 8 keV. No evidence was obtained for line emission. The upper limits for the Si and S line fluxes are used to establish upper limits on the abundances of these elements. The limit on the Fe line flux is consistent with the strength of the emission feature reported in proportional counter experiments. Questions of nonequilibrium in ionization in Cas A and the effects on X-ray line emission are considered.

125.043 **The helium-driven r-process in supernovae.**
J. W. Truran, J. J. Cowan, A. G. W. Cameron.
Astrophys. J., Lett., Vol. 222, L63 - L67 (1978).

The discovery of r-process anomalies in two inclusions in the Allende meteorite, together with their associated oxygen and magnesium anomalies, has caused the authors to examine the consequences of supernova shocks in the helium zones of massive stars. They find that powerful r-processes can operate under such conditions. The details of these processes will vary in different stellar masses. The studied Allende inclusions apparently did not receive material which had been very extensively r-processed.

125.044 **Continuum observations of supernova remnant G 127 + 0.5 at 2695 MHz.** C. J. Salter, T. Pauls, C. G. T. Haslam.
Astron. Astrophys., Vol. 66, 77 - 80 (1978).

The supernova remnant G 127 + 0.5 has been mapped at 2695 MHz with a resolution of 4'.4. The structure and spectrum of the source are discussed. The spectral index α is 0.6. The point-like source lying near the direction to the center of the remnant (Caswell, 1977) is shown to have a flat spectrum above ~ 2000 MHz. No evidence for variability is found in this source.

125.045 **Coronal lines in supernova remnant spectra.**
P. Murdin D. H. Clark, J. L. Culhane.
Mon. Not. R. Astron. Soc., Vol. 183, 79P - 84P (1978).

The detection of coronal line emission offers the possibility of studying the shock-heated plasma in supernova remnants (SNRs). The authors have searched for the λ5303 [Fe XIV] coronal line in the SNRs N49 and Vela, and detected it in the former at an intensity suggesting a plasma temperature in the region of 2×10^6 K. However, they failed to confirm a previously proposed detection for the Vela SNR.

125.046 **A Near-Eastern sighting of the supernova explosion of 1054.** K. Brecher, E. Lieber, A. E. Lieber.
Nature, Vol. 273, 728 - 730 (1978).

Astronomers have long been puzzled by the apparent failure of observers in Europe and the Near East to see and report the Crab supernova of 1054. However, the authors now present evidence of such a sighting.

125.047 **Supernova in anonymous galaxy.**
IAU Circ., No. 3158, with a corrigendum 3166 (1978).

125.048 **Ultraviolet spectrum of supernova remnant.**
IAU Circ., No. 3209 (1978).

125.049 **Possible supernova in MCG −4−32−23.**
IAU Circ., Nos. 3221, 3224 (1978).

125.050 **Studies of N132D, a supernova remnant similar to Cassiopeia A in the Large Magellanic Cloud.**
B. M. Lasker.

Astrophys. J., Vol. 223, 109 - 121 (1978).

The results of a photographic and spectroscopic study of a young supernova remnant in the Large Magellanic Cloud, N132D, are presented. The remnant consists of a number of stationary and rapidly moving knots located almost centrally in a larger disk and ring. In the disk and in certain of the knots the O/H abundance ratios are high by factors in the order of 10 to 100, which supports the suggestion that processed stellar material from the supernova explosion is being dissipated into the interstellar medium.

125.051 **Supernova in anonymous galaxy.**
Yamamoto Circ., No. 1875 (1978). In Japanese.

125.052 **Possible supernova in MCG-4-32-23.**
Yamamoto Circ., No. 1886 (1978). In Japanese.

125.053 **Upper limits for the microwave pulsed emission from supernova explosions in clusters of galaxies.**
G. G. C. Palumbo, N. Mandolesi, G. Morigi, G. A. Baird, T. Delaney, W. P. S. Meikle, R. W. P. Drever, J. V. Jelley, J. H. Fruin, R. B. Partridge.
Astrophys. Space Sci., Vol. 54, 355 - 363 (1978).

Between 1972 and 1975 an international collaborative search was carried out for prompt 10 GHz emission at the onset of supernovae. The motivations and techniques involved in this effort are described, and the results of the three years' work are summarized. No pulses from supernovae were detected, the best upper limit being 4×10^{43} ergs in a 40 MHz band at 10 GHz for a pulse time-scale $\lesssim 0.5$ s. Methods for improving this limit are briefly described.

125.054 **Asymmetric supernova explosions and the origin of binary pulsars.** W. Sutantyo.
Astrophys. Space Sci., Vol. 54, 479 - 488 (1978).

The author investigates the effect of asymmetric supernova explosions on the orbital parameters of binary systems with a compact component. He relates such explosions to the origin of binary pulsars. The degree of asymmetry of the explosion is represented by the kick velocity gained by the exploding star due to the asymmetric mass ejection. He examines the mean survival probability of the binary system ($\langle f \rangle$) for various degrees of asymmetry in the explosion. He derives that $\langle f \rangle$ would be as high as 0.25. Such values of $\langle f \rangle$ can be obtained if the mass of the exploding stars is, in general, not large ($\lesssim 10 \, M_\odot$).

125.055 **The spectrum of Tycho's supernova remnant.**
R. P. Kirshner, R. A. Chevalier.
Astron. Astrophys., Vol. 67, 267 - 271 (1978).

The optical spectrum of filaments in Tycho's supernova remnant shows only Hα and Hβ emission over the wavelength range 4700–8000 Å. The authors suggest that the optical filaments producing this emission may be density enhancements at a temperature near 10^5 K. Even though they consider non-equilibrium processes which can lead to much lower electron temperatures than equilibrium models, it does not seem possible to reconcile the 21 cm absorption data, the optical proper motions, and the optical surface brightness.

125.056 **Supernova remnants: what do they tell us about massive star evolution and the interstellar medium?**
M. Dennefeld.
Messenger, No. 13, p. 20 - 23 (1978).

125.057 **Spectra of supernova remnants in M33.**
I. J. Danziger, P. G. Murdin, D. H. Clark, S. D'Odorico.
ESO Sci. Prepr. No. 22, 18 pp. (1978). – Submitted to Monthly Notices of the Royal Astronomical Society.

Spectral data are presented confirming the classification of three nebulae in M33 as supernova remnants. A comparison with other supernova remnants and model calculations has been made. Several lines of evidence indicate a nitrogen abundance slightly lower than that in the Galaxy, and consistent with the abundance trends in HII regions at a similar radial distance in M 33 reported by Searle (1971) and Smith (1976).

125.058 **Supernova models with slow energy pumping and galactic supernova remnants.**   V. P. Utrobin.
Astrophys. Space Sci., Vol. 55, 441 - 457 (1978).

The study of supernova (SN) models with slow energy pumping is continued. A simple relationship connecting the important SN parameters is obtained. A comparison of the extragalactic type I SN observations with the results of calculations is performed and an investigation of the galactic type I SN remnants is carried out. It was established that the Crab nebula resulted from the outburst of a peculiar SN. The unique properties of such SNs, including SN 1054, are due to the low intensity of energy pumping ($L_e \sim 10^{42}$ erg s^{-1}). The mass of the envelope of the Crab nebula is evaluated to be $M_e \approx 0.7\ M_\odot$.

125.059 **Low energy X-ray line emission from Cas A.**
P. Charles, K. Mason, W. Cash, M. Lampton, S. Bowyer.
Bull. American Astron. Soc., Vol. 10, 420 - 421 (1978). Abstract.

125.060 **Extragalactic distance determinations from supernova light curve models.**
S. Schurmann, S. W. Falk.
Bull. American Astron. Soc., Vol. 10, 421 - 422 (1978). Abstract.

125.061 **Report of a Near Eastern sighting of the Crab supernova explosion.**   K. Brecher, E. Lieber, A. E. Lieber.
Bull. American Astron. Soc., Vol. 10, 424 - 425 (1978). Abstract.

125.062 **The interaction of mantle and envelope in type II supernova models.**   S. W. Falk.
Bull. American Astron. Soc., Vol. 10, 425 (1978). − Abstract.

125.063 **Excitation mechanism for the late time spectra of type I supernovae.**   R. E. Meyerott.
Bull. American Astron. Soc., Vol. 10, 425 - 426 (1978). Abstract.

125.064 **A new study of N132D, a Magellanic Cloud supernova remnant similar to Cas A.**   B. M. Lasker.
Bull. American Astron. Soc., Vol. 10, 426 (1978).

125.065 **Radial velocities of the filaments in the SNR S147.**
C. N. Arnold, R. P. Kirshner.
Bull. American Astron. Soc., Vol. 10, 426 (1978). − Abstract.

125.066 **Optical emission from the SNR G290.1 - 0.8.**
R. P. Kirshner, P. F. Winkler.
Bull. American Astron. Soc., Vol. 10, 426 (1978). − Abstract.

125.067 **Electrographic imagery of the Cygnus Loop at [Ne V] λ3426.**
H. M. Heckathorn, S. Shulman, R. T. Giuli.
Bull. American Astron. Soc., Vol. 10, 427 (1978). − Abstract.

125.068 **Spectra of the SNR 3C58.**
R. A. Fesen, R. P. Kirshner.
Bull. American Astron. Soc., Vol. 10, 427 (1978). − Abstract.

125.069 **Survival of supernova ejecta.**
W. C. Straka, E. M. Jones.
Bull. American Astron. Soc., Vol. 10, 427 - 428 (1978).

Abstract.

125.070 **Coronal line emission from Puppis A.**
R. L. Lucke, J. C. Zarnecki, B. E. Woodgate, D. G. Socker.
Bull. American Astron. Soc., Vol. 10, 433 (1978). − Abstract.

125.071 **Old supernova remnants as magnetohydrodynamic wave fronts. Formation of filamentary structure.**
Y. Sofue.
Astron. Astrophys., Vol. 67, 409 - 420 (1978).

Expanding shells of old supernova remnants (SNR) are simulated by magnetohydrodynamic (MHD) wave fronts. The author examines the three-dimensional behavior of the shells interacting with the inhomogeneous interstellar medium and clouds. The wave front undergoes strong refraction and reflection by the clouds. Subsequent focusing of the waves produces filamentary structures and bright patches, when the cloud size is much smaller than the shell radius. If the cloud size is comparable with or greater than the shell radius, the SNR becomes greatly deformed from a sphere. Morphological features of a typical old SNR, S 147, are fit reasonably well by a model which postulates interaction with several interstellar clouds of radii 8−12 pc.

125.072 **Statistics of supernovae in external galaxies.**
G. A. Tammann.
Ann. New York Acad. Sci., Vol. 302, (see 012.033), 61 - 80 (1977).

Contents: The SN frequency in external galaxies. The SN frequency in our Galaxy. Some observable properties of SNe. Supernova progenitors.

125.073 **Continuum light from supernovae.**
R. P. Kirshner.
Ann. New York Acad. Sci., Vol. 302, (see 012.033), 81 - 89 (1977).

125.074 **Supernovae as phenomena of high-energy astrophysics.**   W. D. Arnett.
Ann. New York Acad. Sci., Vol. 302, (see 012.033), 90 - 100 (1977).

The concept "supernova" includes a number of different theoretical ideas and observationally distinct phenomena. The list includes extragalactic events, historical galactic events, supernova remnants, pulsars, some black holes, and compact X-ray sources. Furthermore, supernovae are thought to be exploding stars, the accelerators of cosmic rays, the site of nucleosynthesis, and an important contributor to the heating and dynamics of the interstellar medium. This paper discusses some observational consequences of theoretical ideas, emphasizing the newest developments and what currently seem to be the most fundamental aspects of the problem.

125.075 **The supernova remnant Cassiopeia A.**
R. A. Chevalier.
Ann. New York Acad. Sci., Vol. 302, (see 012.033), 106 - 113 (1977).

To summarize the results for Cas A, in the optical region one has the opportunity to observe the heavy element abundances in the core and the envelope of a massive star. In the X-ray range, one has the opportunity of observing to what extent these elements are mixed in with the surrounding medium. Searches for the Si XIV and S XVI lines will be of special interest. Time-dependent radio changes in the remnant shed light on the acceleration process of relativistic electrons. In addition, Cas A appears to be an ideal site for the acceleration of cosmic rays observed at earth.

125.076 **Results of studies on physics of supernovae.**
Eh. R. Mustel'.
Flare stars. Symposium 1976, (see 012.035), p. 141 - 154.

125.077   Modello numerico per l'evoluzione di un resto di supernova.
R. Fusco-Femiano, F. Matteucci, A. Preite-Martinez.
Mem. Soc. Astron. Italiana, Vol. 48, 41 - 53 (1977).
In Russian. – Abstr. in Ref. zh., 51. Astron., 6.51.696 (1978).

Tian-Guan guest star (supernova) of 1054 in history.   See Abstr. 004.008.

The Christmas Star as a supernova in Aquila.
See Abstr. 015.021.

Calculated X-radiation from optically thin plasmas. III. Abundance effects on continuum emission.
See Abstr. 022.071.

Condensation in supernova ejecta and isotopic anomalies in meteorites.   See Abstr. 061.001.

The bulk properties and equation of state of hot dense matter.   See Abstr. 061.026.

Reaction rates for neutrino processes.
See Abstr. 061.028.

The rapid neutron-capture process and the synthesis of heavy and neutron-rich elements.   See Abstr. 061.052.

Presupernova evolution of massive stars.
See Abstr. 065.039.

Collapsing stellar cores and supernovae.
See Abstr. 065.040.

Neutrino mediated shock waves in supernovae.
See Abstr. 066.250.

Runaway stars as witnesses to supernova explosions.
See Abstr. 117.007.

Ionization of interstellar H_2 clouds by supernovae.
See Abstr. 131.196.

The role of a supernova in star formation.
See Abstr. 131.232.

Observations of negative velocity hydrogen in IC 443.   See Abstr. 132.029.

Spectra of Cassiopeia A. II. Interpretation.
See Abstr. 134.006.

Some remarkable spectra in the outskirts of the Crab Nebula.   See Abstr. 134.009.

Observations of the optical remnant of SN 1181 = 3C 58.   See Abstr. 141.037.

Radio interferometric observations of the bright core of CTB 80.   See Abstr. 141.207.

Some implications of off-center explosions in stars.
See Abstr. 141.551.

The relationship of pulsars to supernovae.
See Abstr. 141.555.

X-ray emission from MSH 14–63 – probable remnant of the A.D. 185 supernova.   See Abstr. 142.047.

Low energy X-ray observations of supernova remnant MSH 14-63.   See Abstr. 142.084.

Pulsars and cosmic rays in the dense supernova shells.
See Abstr. 143.031.

Optical evidence for a very large, expanding cavity associated with the I Orion OB association and Barnard's loop.
See Abstr. 152.006.

Supernova remnants in open clusters.
See Abstr. 153.002.

Supernova in NGC 4340 = Kulikovskij's supernova

125.101   Kulikovskij's supernova.   Eh. A. Dibaj.
Priroda, 1978, No. 1, p. 134 - 135. In Russian.

126 Low-Luminosity Stars, Subdwarfs, White Dwarfs, Degenerate Stars

126.001 The gravitational redshift of the white dwarf CoD −38°10980.   G. Wegner.
Mon. Not. R. Astron. Soc., Vol. 182, 111 - 116 (1978).
A gravitational redshift of $V_{RS} = +44 \pm 7$ (p.e.) km/s is measured for the 11 mag DA white dwarf CoD −38°10980. The values of V_{RS} and log $R/R_\odot$, using the photometric distance, are consistent with the theoretical Hamada and Salpeter (1961) mass–radius relations for zero-temperature interiors composed of the lighter elements He, C and Mg, and a pure Fe interior is probably excluded.

126.002 Strömgren photometry of southern white dwarfs.   M. S. Bessell, D. T. Wickramasinghe.
Mon. Not. R. Astron. Soc., Vol. 182, 275 - 281 (1978).
Colours of southern white dwarfs in the *uvby* (Ström-gren four-colour) system have been obtained. The results are compared with those of Graham. The extensive absolute photometry of white dwarfs published by Greenstein has also been transferred into the four-colour system and both sets of results are compared with model atmosphere calculations. The scatter in log (g) is higher than previously supposed, and the evidence for an increase in $\langle$log $(g)\rangle$ at the cooler ($T_e < 10\,000$K) end of the DA sequence is discussed.

126.003 The rate of formation of white dwarfs in stellar systems.   J. G. Hills.
Astrophys. J., Vol. 219, 550 - 558 (1978).
The author shows that it is possible to determine the white-dwarf formation rate in a stellar system to high accuracy from the integrated stellar luminosity of the system. The rate of formation and the accumulated numbers of neutron stars in stellar systems are also determined.

126.004 LP 701−29: a new "lowest luminosity" degenerate star with a heavily blanketed spectrum.
C. C. Dahn, P. M. Hintzen, J. W. Liebert, H. S. Stockman, H. Spinrad.
Astrophys. J., Vol. 219, 979 - 984 (1978).
The authors report the discovery of a new, very cool white dwarf with a parallax corresponding to $M_v \sim 16$ and a heavily line-blanketed spectrum. They present astrometry, *BVI* photometry, narrow-band scanner spectrophotometry, and image-tube spectrophotometry for this star. Atmospheric parameters are estimated, and a comparison with the properties of the low-luminosity degenerate star VB 11 is presented.

126.005 Accretion belts on white dwarfs.   R. Kippenhahn, H. - C. Thomas.
Astron. Astrophys., Vol. 63, 265 - 272 (1978).
Matter which falls onto a slowly rotating white dwarf from an accreting disk forms a rapidly rotating belt, whose chemical composition and angular momentum are mixed with the surrounding white dwarf material by shear instabilities. The structure and evolution of accretion belts is computed in a simplified model. After a sufficient amount of material is accreted, hydrogen is ignited at the bottom of the belt. The model presented here differs drastically from the models of spherically symmetric accretion, on which current nova theories are based.

126.006 Hard X-ray observations of white dwarf binary systems.   M. J. Coe, A. R. Engel, J. J. Quenby.
Nature, Vol. 272, 37 - 38 (1978).
Using the Imperial College hard X-ray scintillation telescope on Ariel 5, several short period (up to 12 h) binary star systems have been studied and are described here. This study was carried out after a very hard X-ray flux was detected from AM Herculis leading to a search for similar signals from optically similar star systems.

126.007 Three new hot subdwarfs: AGK 2 + 81°266, BD + 39°3226, BD + 34°1543.
J. Berger, A.-M. Fringant.
Astron. Astrophys., Vol. 64, L9 - L11 (1978).
Spectra of AGK 2 + 81°266, BD + 39°3226, BD + 34°1543 are described and the following types are respectively proposed: sdO, sdOp (He rich) and composite F + sdB.

126.008 Virial theorem, energy content, and mass-radius-relation for white dwarfs.   D. Koester.
Astron. Astrophys., Vol. 64, 289 - 294 (1978).
A simple model for the gravitational contraction of white dwarfs is discussed using the virial theorem. In the second part of the paper numerical calculations are reported which lead to a new mass-radius-relation for finite temperature white dwarfs and to quantitative results concerning the contribution of different forms of energy to the luminosity.

126.009 A new pulsating white dwarf: GD 154.   E. L. Robinson, R. J. Stover, R. E. Nather, J. T. McGraw.
Astrophys. J., Vol. 220, 614 - 618 (1978).
The authors report additional results of a survey of luminosity variations in white dwarfs. GD 154, a DA white dwarf with $B-V = +0.18$, is found to vary at either of two dominant periods, 1186 or 780 s. The variations are probably caused by nonradial *g*-mode pulsations, but the extreme length of the 1186 s pulsation period forces the authors to suggest that the white dwarf is pulsating in a very high overtone ($k \approx 10-30$).

126.010 Old disk subdwarfs.   J. R. Mould, D. B. McElroy.
Astrophys. J., Vol. 220, 935 - 939 (1978).
Measurements of CaH and TiO band strengths are presented for M dwarfs of the old disk population which lie below the main sequence. Their band strengths imply a composition intermediate between that of halo subdwarfs and that of disk dwarfs.

126.011 Thermonuclear runaways in white dwarfs.   R. E. Taam.
Astrophys. Lett., Vol. 19, 47 - 51 (1977).
The thermal evolution of white dwarf interiors has been investigated for white dwarfs of 0.456 $M_\odot$ and 1 $M_\odot$ which accrete hydrogen rich material at $\sim 10^{-13} M_\odot$/year. Energy transport by electron conduction heated the white dwarf cores to central temperatures $\sim 7 \times 10^6$ °K. Comparison of the subsequent thermonuclear runaway with earlier work indicated that compressional heating is most effective in initiating the runaway at densities corresponding to the partially degenerate regions of the white dwarf.

126.012 New observations of the pulsating white dwarf R548.   R. J. Stover, E. L. Robinson, R. E. Nather.
Publ. Astron. Soc. Pacific, Vol. 89, 912 - 918 (1977/78).
R548 is a luminosity variable DA white dwarf. Two close pairs of pulsation frequencies are simultaneously present in its light curve, one at 213 s and one at 274 s. An improved ephemeris obtained by directly modeling the pulsations is presented. The authors demonstrate that the amplitudes and periods of all four pulsations are highly stable.

126.013 On the spectrum of LDS 678B (EG 131).   D. T. Wickramasinghe, M. S. Bessell, J. A. J. Whelan.
Mon. Not. R. Astron. Soc., Vol. 182, 53P - 56P (1978).

IDS data obtained on the AAT of the 'intermediate' DA white dwarf LDS 678B (EG 131) are presented. The spectrum does not show any evidence of hydrogen-line absorption. From the absence of Hα the authors deduce a lower limit of ~ 10^4 for the atmospheric helium to hydrogen ratio by number.

126.014 X radiation from accreting non-magnetic degenerate dwarfs: the regime at large electron scattering optical depths.   N. D. Kylafis, D. Q. Lamb.
Bull. American Astron. Soc., Vol. 9, 632 (1978). – Abstract.

126.015 X and UV radiation from accreting magnetic degenerate dwarfs.   A. R. Masters, D. Q. Lamb.
Bull. American Astron. Soc., Vol. 9, 632 (1978). – Abstract.

126.016 Effects of electron conduction on accretion by white-dwarf stars.   J. Imamura, R. H. Durisen.
Bull. American Astron. Soc., Vol. 9, 633 (1978). – Abstract.

126.017 Evolutionary models of accreting massive white dwarfs.
E. M. Sion, M. Acierno, S. Tomczyk.
Bull. American Astron. Soc., Vol. 9, 634 (1978). – Abstract.

126.018 LP790–29: a cool, magnetic degenerate with a deep absorption trough.
J. Liebert, J. R. P. Angel, H. S. Stockman.
Bull. American Astron. Soc., Vol. 9, 636 (1978). – Abstract.

126.019 Spectral features in strongly magnetic white dwarfs.   J. R. P. Angel.
Bull. American Astron. Soc., Vol. 9, 636 (1978). – Abstract.

126.020 A study of white dwarfs using the visual surface brightness relation.
T. J. Moffett, T. G. Barnes, D. S. Evans.
Bull. American Astron. Soc., Vol. 9, 636 (1978). – Abstract.

126.021 A new pulsating white dwarf: GD 154.   E. L. Robinson, R. J. Stover, R. E. Nather, J. T. McGraw.
Bull. American Astron. Soc., Vol. 9, 642 (1978). – Abstract.

126.022 New observations of the pulsating white dwarf R548.
R. J. Stover, E. L. Robinson, R. E. Nather.
Bull. American Astron. Soc., Vol. 9, 642 (1978). – Abstract.

126.023 UV 1758+36, a hot subluminous B star.   J. R. Giddings, M. M. Dworetsky.
Mon. Not. R. Astron. Soc., Vol. 183, 265 - 269 (1978).
A new subdwarf B star has been discovered by the ultra-violet sky survey telescope (S2/68). Its ultraviolet photometry corresponds to an almost unreddened early B star with V = 11.5 mag. The visible spectrum is dominated by strong broad Balmer lines and is characteristic of the subdwarf B stars. A preliminary analysis of available data yields atmospheric parameters T_{eff} = 32 500 K, log (g) = 5.25 and N_{He}/N_{H} = 0.017. The radial velocity may be variable.

126.024 On the rotating magnetic white dwarfs.   M. K. Das, J. N. Tandon.
Astrophys. J., Vol. 221, 958 - 964 (1978).
The equilibrium structure of a rotating magnetic white dwarf has been obtained for different values of the central condensation parameter. The effect of rotation and general magnetic field vanishing at the equilibrium surface on the critical mass of a white dwarf is examined. It is found that the presence of rotation and a magnetic field does not affect the critical mass value significantly.

126.025 Stability of accreting white dwarfs in close binary systems.   R. Sienkiewicz, W. Dziembowski.
IAU Colloq. No. 42, (see 012.012), p. 327 - 339 (1977).
The authors present results of thermal and vibrational stability analysis for 1 $M_\odot$ white dwarf models corresponding to various accretion rates $\gtrsim 10^{-11}$ $M_\odot$/y. Accretion is assumed to be spherically symmetric and stationary. Thermal instability due to nuclear burning of hydrogen (at lower accretion rates) and helium (at higher rates) was found. At medium rates two growing thermal modes are simultaneously present. Vibrational instability was found for all models except those corresponding to highest accretion rates. Among objects in which these instabilities may be important are symbiotic stars and nuclei of planetary nebulae.

126.026 The interaction of accretion flux with stellar pulsation.   J. C. Papaloizou, J. E. Pringle.
IAU Colloq. No. 42, (see 012.012), p. 340 - 341 (1977).

126.027 Immiscibilities in cold, degenerate stars.   D. J. Stevenson.
Proc. Astron. Soc. Australia, Vol. 3, 167 - 168 (1977).

126.028 Hot white dwarfs as soft X-ray sources.   F. Wesemael.
Astron. Astrophys., Vol. 65, 301 - 304 (1978).
The suggestion that hot, helium- and metal-poor white dwarfs may be significant thermal soft X-ray emitters is investigated using explicit model atmosphere calculations. These calculations, together with the soft X-ray survey of Vanderhill et al., place an improved upper limit on the space density n of white dwarfs with $T_e \gtrsim 10^5$ K. The author finds $n \lesssim 4 \times 10^{-6}$ pc^{-3} for an assumed interstellar neutral hydrogen density of n_H = 0.1 cm^{-3}. This limit confirms, and in some cases extends, previous extreme-ultraviolet and soft X-ray limits and suggests that neutrino emission processes may dominate the pre-white-dwarf evolution of these stars.

126.029 A dipole model for magnetic white dwarf BPM 25114.   B. Martin, D. T. Wickramasinghe.
Mon. Not. R. Astron. Soc., Vol. 183, 533 - 538 (1978).
Theoretical spectra have been computed for the magnetic DA white dwarf BPM 25114 assuming a centred-dipole field distribution. The polar field strength is deduced to be (3.6 ± 0.2) × 10^7 G. The observations suggest that the angle i between the line of sight and the magnetic axis varies by perhaps 30° during a rotation period. The authors predict strong variable wavelength-dependent circular (max ~8 per cent) and linear (max ~4 per cent) polarization for this star.

126.030 Are light DA white dwarfs the progenitors of planetaries?   A. Finzi.
IAU Symp. No. 76, (see 012.014), p. 323 - 324 (1978). Abstract.

126.031 The space density of hot white dwarfs.   D. Koester.
Astron. Astrophys., Vol. 65, 449 - 450 (1978).
The expected space density of hot white dwarfs that might be observed as sources of EUV and soft X-ray radiation is calculated.

126.032 Hydrogen shell flashes in a white dwarf with mass accretion.   B. Paczyński, A. N. Żytkow.
Astrophys. J., Vol. 222, 604 - 611 (1978).
The evolution of a 0.8 $M_\odot$ white dwarf accreting hydrogen-rich matter (X = 0.7, Z = 0.03) was studied through a number of hydrogen shell flashes in order to discover those properties of the evolutionary models which are independent of the initial conditions.

126.033 White dwarf formation in two-body systems: Sirius

AB, Procyon AB, 40 Eri BC and Stein 2051 AB.
T. P. Roark.
Rev. Mexicana Astron. Astrofis., Vol. 3, (see 012.023), 113 - 118 (1977).

The orbital histories of the systems Sirius AB, Procyon AB, 40 Eri BC and Stein 2051 AB are examined. It is assumed that the changes in the orbital elements are a direct result of mass loss from the white dwarf precursor component. This loss is considered to be isotropic and no exchange between the components is allowed. These histories, coupled with observed atmospheric abundances, theoretical ages and cooling times lead to the conclusion that stars as massive as $10\,M_\odot$ can be white dwarf progenitors. Indeed, in the case of Sirius B, there is evidence that the progenitor mass was not less than $8\,M_\odot$ and that the star did not lose mass in an explosive manner.

126.034 The nature of CD −31°622.
M. S. Bessell, P. A. Ianna.
Astrophys. J., Lett., Vol. 222, L127 - L128 (1978).

CD −31°622, suggested by Eggen as possibly being a degenerate star at a distance of about 1 pc, is shown to be a subdwarf on the basis of its essential zero parallax and spectrum. It is found to be deficient in metals by at least a factor of 100 with respect to the Sun.

126.035 The progenitor masses and the luminosity function of white dwarfs.   F. D' Antona, I. Mazzitelli.
Astron. Astrophys., Vol. 66, 453 - 461 (1978).

The dependence of the luminosity function (LF) of white dwarfs on the progenitor mass function is investigated. The authors find, at low luminosity, a departure of the LF of white dwarfs with respect to the LF obtained if constant birthrate and deathrate are assumed, due to the fact that low progenitor masses cannot give rise, at present, to very low luminosity white dwarfs. A tentative explanation of hot white dwarfs different spectral types in terms of different progenitor masses is given.

126.036 LP 182−44.
IAU Circ., No. 3210 (1978).

126.037 Spectroscopy of the extreme-ultraviolet source Feige 24: the binary orbit and the mass of the white dwarf.   J. R. Thorstensen, P. A. Charles, B. Margon, S. Bowyer.
Astrophys. J., Vol. 223, 260 - 265 (1978).

The authors report results of coudé spectroscopy of the extreme-ultraviolet white dwarf Feige 24. Radial velocities of the Hα, He I λ5876, and He I λ6678 emission lines, and the underlying M dwarf absorption features, were determined. The velocities show a binary period of $4^d2319 \pm 0^d0015$. The emission-line and absorption-line velocities agree in phase, which indicates that the emission lines originate in the atmosphere of the M dwarf secondary as a result of reprocessing of the EUV radiation. After modeling this effect, the authors used the observed amplitude of the emission-line variability to place a lower limit on the orbital inclination. From these data and other data they show that the mass of the white dwarf lies between 0.46 and $1.24\,M_\odot$. The authors briefly discuss some possible implications for the evolution of binary stars.

126.038 On the kinematics of the variable DA stars.
E. M. Sion, O. L. Lupie, K. N. Young.
Publ. Astron. Soc. Pacific, Vol. 90, 154 - 156 (1978).

The authors examine the kinematics of the known variable DA white dwarfs relative to a sample of observed nonvariable DA white dwarfs in the same color range. The variable DA stars no not form a kinematically distinct group relative to the nonvariable stars. Possible implications based upon the kinematics are briefly discussed.

126.039 Models of quasi-degenerate white dwarfs.
Yu. L. Vartanyan.
Probl. gravitatsii. Tbilisi, 1976, p. 64 - 74. In Russian. − Abstr. in Ref. zh., 51. Astron., 5.51.756 (1978).

126.040 White dwarfs and giant planets.   W. B. Hubbard.
Fundam. Cosmic Phys., Vol. 3, 167 - 219 (1978).

It is possible to regard a giant planet, such as Jupiter, as a limiting case of a hydrogen-rich white dwarf. The author reviews the classical (and largely analytic) theory for the structure of white dwarfs. A corresponding analytic theory for giant planets, which contains the essential features of their interior physics, is then constructed. He presents analytical calculations for zero-temperature interior structure, thermal structure and cooling rate, and effects of rotation.

126.041 Circular polarization and magnetic fields of white dwarfs.   C. Chiuderi, M. Silvi.
Mem. Soc. Astron. Italiana, Vol. 47, 65 - 72 (1976).

A simple method to estimate the strength of the surface magnetic field of a white dwarf from the degree of circular polarization in its continuum spectrum is proposed. The method is applied to the observed spectra of G99-37 and G227-35 giving a fairly good fit in the optical range.

126.042 The peculiar white dwarf BPM 25114.
G. Wegner.
Mem. Soc. Astron. Italiana, Vol. 48, 27 - 40 (1977).

Observations of the peculiar white dwarf BPM 25114 are reported. Photometric VBU observations indicate that it is light variable on a time scale near 2.84 days. Spectroscopic observations made both in white light and with a Zeeman analyzer suggest that changes in the profile and polarization of the Hβ line are in phase with the light variations. Simple theoretical calculations suggest that these observed variations are consistent with the oblique rotator model for this star and a dipole magnetic field 4×10^7 Gauss.

The stars of low luminosity.   See Abstr. 002.003.

White dwarfs II.   See Abstr. 002.025.

White dwarfs.   See Abstr. 003.025.

The second European Workshop on White Dwarfs. See Abstr. 011.027.

Stability of force-free magnetic fields in degenerate stars.   See Abstr. 062.048.

Extreme subdwarfs. I. Molecular band strengths: a theoretical approach.   See Abstr. 064.078.

A grid of model atmospheres and synthetic spectra for white dwarfs with $5000 < T_{eff} < 7000$ K. See Abstr. 064.109.

Non-radial oscillations of rotating stars and their relevance to the short-period oscillations of cataclysmic variables.   See Abstr. 065.006.

The thermal stability of hot degenerate stars in steady-state accretion.   See Abstr. 065.031.

Upper mass limit for stars dying as white dwarfs. See Abstr. 065.036.

Why do collapsed stars rotate so slowly? See Abstr. 066.062.

Astrometric study of the Sproul plate series of van Maanen's star, including gravitational redshift.

See Abstr. 112.002.

Spectrophotometry of low-luminosity stars with the Carnegie image tube.   See Abstr. 114.063.

Ultraviolet spectroscopy of hot, subluminous stars with the IUE.   See Abstr. 114.079.

A hot subluminous star: HDE 283048.
See Abstr. 114.503.

The manganese abundance in HD 94028.
See Abstr. 114.518.

A discussion of period changes in the white dwarf eclipsing binary system V471 Tauri.   See Abstr. 121.074.

Current observations of the white dwarf eclipsing binary V471 Tauri.   See Abstr. 121.087.

Blue variables at high galactic latitude.
See Abstr. 122.030.

An alternative mechanism for the nova phenomenon.
See Abstr. 124.009.

On ^{7}Li production in nova explosions.
See Abstr. 124.021.

Planetary nebulae and white dwarfs.
See Abstr. 135.090.

Search for radio emission from objects with continuous optical spectrum and large proper motion.
See Abstr. 141.223.

Extreme ultraviolet observation of Sirius: evidence against a photospheric origin of the 0.25 keV flux.
See Abstr. 142.061.

Hydrodynamic simulations of white dwarf — main-sequence star collisions.   See Abstr. 151.029.

Hydrodynamic simulations of white-dwarf — main-sequence star collisions.   See Abstr. 151.058.

Interstellar Matter, Infrared Sources, Gaseous Nebulae, Planetary Nebulae

131 Interstellar Matter, Star Formation

131.001 Detection of hyperfine structure of interstellar Na I in the α Cygni sight-line.
R. C. Wayte, I. Wynne-Jones, J. C. Blades.
Mon. Not. R. Astron. Soc., Vol. 182, 5P - 10P (1978).

A Michelson interferometer, used in a study of interstellar Na I in α Cygni, has resolved the hyperfine structure in the $^2S_{1/2}$ ground state of Na I for the first time. Analysis of one cloud which shows the hyperfine splitting yields $N = 3.0 \times 10^{11}$ cm^{-2} and $b = 0.38$ km/s. This latter value corresponds to $T_k \sim 200$ K, assuming that only thermal motion contributes to the observed line width. The small b value indicates that the internal cloud velocity is subsonic.

131.002 New stellar water masers. D. F. Dickinson, G. N. Blair, J. H. Davis, N. L. Cohen.
Astron. J., Vol. 83, 32 - 35 (1978).

Eight new water masers have been detected, at least six being associated with late type stars. Two of the eight have multiple features with substantial velocity spread suggesting they may be supergiants. Two others may be supergiants, but their spectra cannot rule out origin in an H II region. One object, OH 43.8−0.1, is exceptionally strong; if stellar, it is the most intense water maser star yet found (with the exception of VY CMa).

131.003 A comparison of neutral hydrogen 21 cm observations with UV and optical absorption-line measurements. R. Giovanelli, M. P. Haynes, D. G. York, J. M. Shull.
Astrophys. J., Vol. 219, 60 - 71 (1978).

Several absorption components detected in visible or UV lines have been identified with emission features in new high-resolution, high signal-to-noise 21 cm observations. Stars for which direct overlap is obtained are HD 28497, λ Ori, μ Col, HD 50896, ρ Leo, HD 93521, and HD 219881. With the use of the inferred H I column densities from 21 cm profiles, rather than the integrated column densities obtained from Lα, more reliable densities can be derived from the existence of molecular hydrogen. Hence the cloud thicknesses are better determined.

131.004 Observations of the interstellar K I λ4044 line. R. M. Crutcher.
Astrophys. J., Vol. 219, 72 - 73 (1978).

Equivalent widths of interstellar K I λ4044 lines have been measured toward three stars in order to resolve uncertainties in the interstellar column densities of K I due to possible saturation of the stronger K I λ7699 line. Toward ρ Oph, the author finds $N_{KI} = 1.7 \times 10^{12}$ cm^{-2}; while toward HD 195592, $N_{KI} < 8.8 \times 10^{12}$ cm^{-2}. The latter limit is an order of magnitude less than the published result, which was based on the λ7699 line.

131.005 A radio search for interstellar CO⁺, HCN⁺, HNC⁺, and CN⁺ ions. J. M. Hollis, B. L. Ulich, L. E. Snyder, D. Buhl, F. J. Lovas.
Astrophys. J., Vol. 219, 74 - 76 (1978).

The $N = 1-0$ transition of CO⁺, with components $J = 1/2-1/2$ at 117.692 GHz and $J = 3/2-1/2$ at 118.102 GHz, has been searched for but not found in molecular clouds showing strong emission from the $J = 1-0$ transition of carbon monoxide. Rotational transition rest frequency calculations were made for HCN⁺, HNC⁺, and CN⁺; subsequent interstellar searches were conducted for these ions, but none was detected.

131.006 Observations of interstellar sulfur monoxide. C. A. Gottlieb, E. W. Gottlieb, M. M. Litvak, J. A. Ball, H. Penfield.
Astrophys. J., Vol. 219, 77 - 94 (1978).

Millimeter-wave emission in transitions of ^{32}SO and of ^{34}SO is reported. A survey toward 18 sources shows that the galactic distribution of SO is similar to that of CS and H$_2$S. The ^{34}SO lines are about as strong as predicted from the terrestrial abundance ratio ^{34}S/^{32}S. Predictions of line intensities are used to derive SO abundances in Orion A, NGC 2264, W49, and eight other sources. The SO column density is $\lesssim 2 \times 10^{14}$ cm^{-2}, except toward the Kleinmann-Low nebula, where the value is $\sim 10^{16}$ cm^{-2}. The H$_2$S to SO ratio agrees well with predictions based on an ion-molecule reaction scheme, but the CS to SO ratio may be higher than predicted by an order of magnitude.

131.007 CO observations of a molecular cloud complex associated with the bright rim near VY Canis Majoris. C. J. Lada, M. J. Reid.
Astrophys. J., Vol. 219, 95 - 104 (1978).

The authors present extensive CO observations of a large molecular cloud complex (∼15 pc) associated with a bright rim near the peculiar star VY CMa. The molecular complex is found to consist of two clouds which have different radial velocities and physical properties. The possibility that these two clouds may be in near-collision is discussed. The physical association of the bright rim with the cloud complex indicates that the clouds are at the same distance, 1.5 kpc, as the stars which excite the rim. The authors conclude that VY CMa is at the same distance. The resulting luminosity of VY CMa ($5 \times 10^5 L_\odot$) indicates that the star is very massive and places constraints on interpretation of its evolutionary state.

131.008 Spectrophotometry of OH 26.5 + 0.6 from 2 to 40 microns. W. J. Forrest, F. C. Gillett, J. R. Houck, J. F. McCarthy, K. M. Merrill, J. L. Pipher, R. C. Puetter, R. W. Russell, B. T. Soifer, S. P. Willner.
Astrophys. J., Vol. 219, 114 - 120 (1978).

Airborne and ground-based observations show that OH

26.5 + 0.6 has strong 10 μm and weak 18 μm silicate absorptions superposed on an overall energy distribution much like a blackbody. The flux level, color temperature, and depth of the 10 μm absorption have varied during 2 years of observations. A model of the source as a late-type variable star that has ejected an optically thick dust shell is suggested; the mass-loss rate implied is greater than $\sim 10^{-5}$ $M_\odot$ year^{-1}. The fact that significant flux from the source is observed between 4 and 7 μm is evidence that oxygen-rich dust has significant opacity in that wavelength range.

131.009 Spectral line shapes in spherically symmetric radially moving clouds.
T. B. H. Kuiper, E. N. Rodriguez Kuiper, B. Zuckerman.
Astrophys. J., Vol. 219, 129 - 140 (1978).

The authors present a method for the analysis of spectral line shapes arising in homogeneous, moving gas clouds in which velocity and molecular line excitation have power-law dependences on radial distance from the center. Analytical expressions are obtained for radial flows, for both optically thick and optically thin lines. The additional case of an optically thick line from a differentially rotating cloud is considered qualitatively. The method is applied to the interpretation of the ^{12}CO line observed in the direction of the Kleinmann-Low infrared nebula in Orion.

131.010 Microwave detection of interstellar deuterated ammonia.
B. E. Turner, B. Zuckerman, M. Morris, P. Palmer.
Astrophys. J., Lett., Vol. 219, L43 - L47 (1978).

Singly deuterated ammonia (NH$_2$D) has been detected in the molecular cloud Sgr B2 at 85926.2 MHz. The apparent abundance ratio NH$_2$D/NH$_3$ is ~ 0.017 in Sgr B2, higher than the DCN/HCN ratio by a factor of ~ 20, but lower than the NH$_2$D/NH$_3$ ratio in Ori A by a factor of ~ 3. The NH$_2$D results indicate a highly specific deuteration process, the possible nature of which is discussed.

131.011 Deuterated ammonia toward the Orion Nebula.
E. N. Rodriguez Kuiper, B. Zuckerman, T. B. H. Kuiper.
Astrophys. J., Lett., Vol. 219, L49 - L53 (1978).

The authors observed the two 1_{11}–1_{01} rotation-inversion transitions of NH$_2$D toward the Kleinmann-Low infrared nebula (KL) in Orion. The column density for NH$_2$D is $\sim 5 \times 10^{13}$ cm^{-2}, and the abundance ratio [NH$_2$D]/[NH$_3$] in this direction is ~ 0.05. Nondetection of two E-type transitions in CH$_3$OD suggest [CH$_3$OD]/[CH$_3$OH] < 1/10 toward KL.

131.012 H I observations of Shostak's high-velocity cloud.
R. J. Cohen, I. F. Mirabel.
Mon. Not. R. Astron. Soc., Vol. 182, 395 - 399 (1978).

The high-velocity cloud at l = 39°.4, b =+4°.7 reported by Shostak (1977) has been surveyed in the 21-cm hydrogen line at high sensitivity, with an angular resolution of 12 arcmin and a velocity resolution of 1.8 km/s. The cloud consists of a small core with a tail elongated perpendicular to the galactic plane. Possible interpretations of the object are discussed, and it is suggested that the object may be an intergalactic gas cloud associated with the Local Group of galaxies.

131.013 A small neutral hydrogen velocity bridge.
P. L. Baker.
Astron. Astrophys., Vol. 62, 65 - 68 (1978).

A disturbance in the local neutral interstellar medium has been reobserved in 21 cm radiation with the high spatial resolution of the Arecibo 1000' telescope. Previous low resolution observations showed that the anomalous perturbed gas of the feature appears spatially coincident with an absence of a nearly equal amount of normal velocity gas indicating the disturbance is essentially kinematic. With higher spatial resolu-

tion, the absence of gas shows no new structure but the anomalous velocity gas is resolved into three distinct structural entities. A bright velocity-bridge connects the normal velocity components to a small intermediate velocity cloud. The angular size and intensity of the bridge imply an H I density of 270 cm^{-3}.

131.014 The temperature probability function and infrared emissivity of dust heated by randomly distributed stars.
B. Baschek, G. Gampper, G. Traving.
Astron. Astrophys., Vol. 62, 223 - 228 (1978).

The temperature probability function of interstellar dust grains which are heated by randomly distributed stars is calculated upon the assumption of (1) uniformly distributed dust of uniform size and optical properties, and (2) stars of uniform luminosity. The absorption of the heating stellar radiation by the interstellar medium is taken into account by introducing a shielding distance r_D. The probability distribution depends essentially on μ, the number of stars within r_D. The infrared emission of dust is calculated assuming different power laws for the dust absorption coefficient, and the applicability to astronomical objects is briefly discussed.

131.015 The structure and spectrum of a colliding-cloud system and its possible relationship to QSOs.
E. Daltabuit, G. M. MacAlpine, D. P. Cox.
Astrophys. J., Vol. 219, 372 - 380 (1978).

The detailed structure and spectrum to be expected from a very high-velocity cloud-cloud collision are determined, and the results are compared with observed characteristics of QSOs. It is found that shock waves with $v \lesssim 1000$ km s^{-1} may provide an important source of ionizing radiation in QSOs and related objects.

131.016 Upper limits on the abundance of the sulfur dimer in molecular clouds.   H. S. Liszt.
Astrophys. J., Vol. 219, 454 - 457 (1978).

The author has searched for magnetic-dipole transitions among spin-split rotational levels in ^{32}S$_2$. The absence of such emission shows that less than a few percent of the presumably available sulfur nuclei resides in this species. Several other lines observed during the search are briefly discussed.

131.017 Candidate interstellar molecules formed from ion-molecule reactions of NO.
G. H. Loew, D. S. Berkowitz, S. Chang.
Astrophys. J., Vol. 219, 458 - 466 (1978).

Molecular orbital calculations using the MINDO/3 method were performed on a series of small molecules to explore reactions of NO which could lead to formation of interstellar molecules. Specifically, equilibrium geometries, isomeric energies, heats of reactions, and reaction pathways were calculated for plausible ion-molecule reactions involving NO and NO$^+$. The results of these calculations suggest that NHO$^+$ and HON$^+$ could be present in observable abundance in interstellar clouds. These species can be formed by the reaction of NO with H$_3^+$ but not by NO$^+$ with H$_2$. Moreover, they are stable to dissociation and do not react with H$_2$ to re-form NO or to form isomers of HNOH$^+$.

131.018 Atomic hydrogen in galactic molecular clouds.
W. B. Burton, H. S. Liszt, P. L. Baker.
Astrophys. J., Lett., Vol. 219, L67 - L72 (1978).

Comparing observations of galactic carbon monoxide clouds with high-resolution observations of atomic hydrogen, we see that the molecular emission is strongly correlated with the appearance of narrow minima in the H I profiles. The steep wings of these minima suggest that they are due to self-absorption. Similarities of the spatial and kinematic behavior in the molecular and atomic observations lead to attribute these features to a residual amount of very cold (T_s < 20 K) H I in molecular clouds.

131.019 Coronal gas in the Galaxy. I. A new survey of interstellar O VI.   E. B. Jenkins.
Astrophys. J., Vol. 219, 845 - 860 (1978).

Ultraviolet spectra of 40 O and B type stars have been intensively scanned in the vicinity of the O VI transitions at 1032 and 1038 Å with the Copernicus satellite. This survey of the high-temperature (log T ~5.5) phase of interstellar gas is an extension of the earlier observations of interstellar O VI by Jenkins and Meloy (1974). Plots of absorption optical depths versus radial velocity are presented for both transitions in each star. Total column densities (or their upper limits), velocity centroids, and velocity widths have been extracted from the profiles and tabulated, with the results of Jenkins and Meloy included. Qualitatively, the highly variable O VI densities and velocities seem to exhibit no systematic patterns or regional trends; there is no recognizable correlation in the behavior of coronal gas with pronounced features of galactic structure.

131.020 Observations of interstellar chlorine and phosphorus.   M. Jura, D. G. York.
Astrophys. J., Vol. 219, 861 - 869 (1978).

Observations of interstellar Cl I and Cl II and upper limits for HCl have been obtained toward 10 stars with the Copernicus satellite. The authors can show that the average chlorine depletion is not greater than a factor of 3. Even for clouds of widely different densities, they find that the chlorine abundance is uniform to within a factor of 3, and the results are consistent with the view that the interstellar medium within 1000 pc of the Sun is well mixed. Qualitatively, the observations support models of interstellar chlorine chemistry, and the results can be used to constrain models of interstellar clouds. The authors have also obtained estimates of the phosphorus gas-phase abundance.

131.021 The energetics of molecular clouds. II. The S140 molecular cloud.   G. N. Blair, N. J. Evans II, P. A. Vanden Bout, W. L. Peters III.
Astrophys. J., Vol. 219, 896 - 913 (1978).

This paper is the second in a series dealing with the energetics of molecular clouds. The techniques presented in an earlier paper are here applied to the S140 molecular cloud. This cloud contains a single, strong near-infrared source. The conclusions of the energetics analysis of this source are similar to those from the S255 cloud, except that here the near-infrared source appears to be the dominant heat source for the dust in the cloud. Predictions of the expected far-infrared flux agree well with recent observations. The molecular line data for this source are fitted by a radiative transport model for a collapsing cloud. By constraining the model to fit several lines of several molecules, the temperature and density distributions in the cloud are appraised.

131.022 Magnetic alignment of interstellar grains.   W. W. Duley.
Astrophys. J., Lett., Vol. 219, L129 - L132 (1978).

It is shown that mixed MgO FeO SiO grains should be superparamagnetic or weakly ferromagnetic. The resulting enhancement in magnetic susceptibility implies that 0.1–0.3 μm radius grains with icy mantles on MgO FeO SiO cores can be aligned in magnetic fields $H \leqslant 1 \times 10^{-6}$ gauss. In both cases alignment occurs through the Davis and Greenstein mechanism with an enhancement in χ'' owing to spin-lattice relaxation for ferromagnetic cores and to superparamagnetic relaxation for superparamagnetic cores.

131.023 The detection of cyanohexatriyne, $H(C \equiv C)_3 CN$, in Heiles's Cloud 2.   H. W. Kroto, C. Kirby, D. R. M. Walton, L. W. Avery, N. W. Broten, J. M. MacLeod, T. Oka.
Astrophys. J., Lett., Vol. 219, L133 - L137 (1978).

The $J = 9 \rightarrow 8$ and $21 \rightarrow 20$ rotational transitions of the heavy linear molecule cyanohexatriyne, $H(C \equiv C)_3 CN$, have been detected in Heiles's Cloud 2, a dark cloud at a distance of 115 pc in Taurus. This molecule, which contains eight heavy atoms and has a molecular weight of 99 amu, was synthesized specifically for the purpose of interstellar detection. Statistical-equilibrium calculations indicate a total column density of 2.7×10^{13} molecules cm^{-2}. This abundance is down by a factor of only 2 from that of $H(C \equiv C)_2 CN$.

131.024 A large-scale OH sky survey at 1612 MHz. Part I. The observations.   P. F. Bowers.
Astron. Astrophys., Suppl. Ser., Vol. 31, 127 - 145 (1978).

Results are presented for a large-scale 1612 MHz OH sky survey. A large range in galactic longitude ($l = 12°$ to 234°), galactic latitude ($b = -2°.5$ to 2°.5), and velocity ($V_{LSR} = -200$ km s^{-1} to 200 km s^{-1}) has been observed. A substantial portion of sky at higher latitudes has also been observed. Approximately 40 previously unknown sources have been discovered and the majority of these have OH characteristics similar to those for OH/IR stars, but few of the sources have known infrared or optical counterparts. Positions and profiles of the new sources are presented.

131.025 The structure of the interstellar medium.   C. Heiles.
Sci. American, Vol. 238, No. 1, p. 74 - 84 (1978).

131.026 Interstellar heating by photoelectrons from negative oxygen.   M. Rudkjøbing.
Astron. Astrophys., Vol. 63, 189 - 192 (1978).

Photoelectrons produced by the ionization of interstellar negative oxygen are a very important source of kinetic energy, if the λ2200 absorption band is an O$^-$ auto-ionization resonance. Hot H I regions will then extend as far into the clouds as the λ2200 optical thickness permits. The clear division observed to exist between H I clouds with and without a high H_2 content, which is also the division between CH and CH$^+$ regions, is interpreted as that between cold and hot regions. An effective heating of reddening clouds by the O$^-$ dissociation mechanism seems to remove one difficulty from the interpretation of the diffuse λ4430 absorption band as due to H$^-$.

131.027 The Herbig-Haro objects in the Gum Nebula.   M. A. Dopita.
Astron. Astrophys., Vol. 63, 237 - 241 (1978).

A detailed set of spectrophotometric observations has been obtained of the two newly discovered (Schwartz, 1977) Herbig-Haro objects associated with a dark cloud embedded in the Gum Nebula. The objects appear to be physically connected and are associated with a dust-embedded emission line star. They are most likely to be circumstellar radiating shock waves driven by some multiple eruptive phenomena occurring at the central star, perhaps associated with an FU Orionis type event. Their age is only 2 - 300 years and from the shock models, the density of the dark cloud is between 45 and 95 cm^{-3}.

131.028 Radioastronomical frequency for interstellar NH$^+$.   I. D. L. Wilson, W. G. Richards.
Nature, Vol. 271, 137 (1978).

The authors consider the as yet unobserved but very likely interstellar species NH$^+$. The results of the calculations predict Λ-doubling in the lowest rotational level of ^{14}NH$^+$ to be 13,625 MHz. The predicted frequency is within the range of available radiotelescopes and the calculations could provide the first example of a case where theory has predicted a radiosource rather than merely confirmed its identification.

131.029 Identification of interstellar polysaccharides and related hydrocarbons.   F. Hoyle, A. H. Olavesen, N. C. Wickramasinghe.
Nature, Vol. 271, 229 - 231 (1978).

Infrared transmittance spectra of several polysaccharides which are suitably calibrated for comparison with astronomical data have been obtained. The authors show that a mean 2.5–15 μm spectrum computed from the measurements is remarkably close to that required to explain a wide range of astronomical data. In the 9.5–12 μm waveband an additional source of opacity seems to be necessary. The close agreement between the spectrum of this excess opacity and the absorption spectrum of propene C_3H_6 points strongly to the identification of hydrocarbons of this type which may be associated with polysaccharide grains in interstellar space.

131.030 **Interstellar nitroxyl.**
            J. B. Pickles, D. A. Williams.
Nature, Vol. 271, 335 - 336 (1978).

The authors have considered a large network of reactions involving the chemistry of interstellar nitrogen, and are able to estimate the densities of some species which take part in reactions affecting HNO formation and loss and which are reported here.

131.031 **Preplanetary disk?**   A. Fabian.
Nature, Vol. 271, 503 - 504 (1978).

131.032 **Infra-red emission lines from molecules in grain mantles.**   L. J. Allamandola, C. A. Norman.
Astron. Astrophys., Vol. 63, L23 - L26 (1978).

The hitherto unidentified infra-red emission features in NGC 7027 and other objects are shown to be due to fluorescence from vibrationally excited molecules in grain mantles. Using laboratory data to match the observed spectra, positive identification is made of CH_4, H_2O, NH_3, C_2H_2, H_2CO, CO and NO. This provides the most direct evidence for the composition and existence of complex molecular mantles on grains, the first evidence for such luminescence, and positive identification of methane in the interstellar medium.

131.033 **Intensity of interstellar Lyman alpha.**
            B. T. Draine, E. E. Salpeter.
Nature, Vol. 271, 730 - 731 (1978).

The following conclusions are reached. First, the mean Lα intensity in galactic H I is at most about 3×10^{-5} erg cm^{-2} s^{-1} sr^{-1}. Second, the Lα energy density due to H II regions and supernovae is very patchy, varying over length scales of $\sim$ 10 pc or so. Third, within zones of enhanced Lα surrounding supernovae and H II regions the Lα intensity may be large compared to the continuum background in the 10.2–13.6 eV band. Finally, in most galactic H I the energy density of Lα is small compared to that of continuum ultraviolet.

131.034 **Coronal gas in the Galaxy. II. A statistical analysis of O VI absorptions.**   E. B. Jenkins.
Astrophys. J., Vol. 220, 107 - 123 (1978).

Results from the survey of interstellar O VI by Jenkins and Meloy are analyzed to synthesize a global description of the properties of the coronal gas. Tests for correlations of column densities or velocities with properties of the target stars showed no evidence for a circumstellar origin for the absorption lines. An overall average density n(O VI) = 2.8 $\times$ 10^{-8} cm^{-3} was found in the galactic plane, with a decrease which approximately follows exp $(- z/300$pc$)$ away from the plane.

131.035 **Elemental depletions in the interstellar medium.**
            W. W. Duley, T. J. Millar.
Astrophys. J., Vol. 220, 124 - 128 (1978).

The authors propose a model involving the chemical reaction of ionized elements with small oxide grain surfaces that results in selective depletion. This model leads to a quantitative prediction of depletions which agrees with observed depletions in the spectra of ζ Oph and o Per.

131.036 **Intermediate-velocity gas in the local interstellar medium.**   L. L. Cowie, D. G. York.
Astrophys. J., Vol. 220, 129 - 137 (1978).

Scans of interstellar ultraviolet absorption lines of N I, N II, and Si III for 17 stars are combined with previously published data for 30 stars. The authors tabulate the extremal velocities at which detectable absorption occurs and show that these are correlated for the three species. The data suggest that intermediate-velocity gas (20 km s^{-1} < $|v|$ < 60 km s^{-1}), best known from Na I and Ca II absorption, contains both neutral and ionized hydrogen. Features characteristic of intermediate-velocity isothermal shocks ($|v|$> 60 km s^{-1}) are conspicuously rare. The intermediate-velocity gas may be in the form of clouds containing both H I and H II regions or of radiative shocks propagating in the interstellar medium; in the latter case the gas should be detectable in Hα emission.

131.037 **An interstellar cloud density from Copernicus observations of CO in the spectrum of ζ Ophiuchi.**
A. M. Smith, K. S. Krishna Swamy, T. P. Stecher.
Astrophys. J., Vol. 220, 138 - 148 (1978).

Interstellar CO absorption bands in Copernicus spectra of ζ Oph have been studied. Absorption profiles, computed under the assumption that excitation is due to collisions with H_2 molecules and interaction with the 3 K background radiation field, were fitted to the reduced data of nine bands. When a gas kinetic temperature of 56 K is assumed, the best-fit condition implies a hydrogen nuclei density of 120 cm^{-3}, a CO column density of 1.2×10^{15} cm^{-2}, and a radial-velocity dispersion b of 0.9 km s^{-1}. The relevance of these results to existing ideas concerning the ζ Oph interstellar clouds is discussed. It is suggested that the strongest interstellar component is not circumstellar in origin but is instead part of a supernova remnant. Simple calculations are made to establish the plausibility of the supernova remnant identification. This suggestion is also supported by Heiles's 21 cm pictures.

131.038 **Very long baseline interferometric observations of the hydroxyl masers in VY Canis Majoris.**
M. J. Reid, D. O. Muhleman.
Astrophys. J., Vol. 220, 229 - 238 (1978) = Owens Valley Radio Obs., Div. Geol. Planet. Sci., Calif. Inst. Technol., Contrib. No. 2911.

This paper presents the results of spectral-line very long baseline interferometric (VLBI) observations of the OH emission from VY CMa. The main-line (1665 and 1667 MHz) emission was mapped with an angular resolution of 0 ".02 by analyzing interferometer phase data. The main-line emission comes from many maser components of apparent size less than 0 ".03 which are separated by up to 0 ".05. New maser features near the center of the OH spectra were detected and found to lie within the region encompassed by the low-velocity OH emission. The emitting regions at all three OH transitions appear elongated, suggestive of an expanding disk of circumstellar gas and dust. However, no strong evidence of rotation is found in the maps; the maser spectra and the relative locations of maser components can be explained by a radial outflow of the circumstellar material.

131.039 **Staubscheiben als Geburtsstätten der Sterne.**
            R. Gerwin.
Phys. Bl., Vol. 34, 130 - 133 (1978).

131.040 **On the physical connection between the maser sources W75S and N.**   E. E. Lekht.
Astron. Zh. Akad. Nauk SSSR, Vol. 55, 76 - 78 (1978). In Russian. English translation in Soviet Astron., Vol. 22, No. 1.

The detection of new features in the 1667 MHz line profile of the maser source W75N is reported. At the radial velocity of −3.5 km/sec an absorption line originating in a cold hydroxyl cloud has been observed towards a compact H II region close to W75N. It is shown that the two compact H II regions (H II W75N and DR21) as well as the maser sources W75N and S are immersed in the same cold hydroxyl cloud. Some OH emission features of the source W75N are found to be variable.

131.041 **On the possibility of registration of Dirac monopole by nonlinear effects in the emission of interstellar masers.** S. G. Rautian, V.-P. Safonov, G. I. Smirnov.
Phys. Lett. B, Vol. 70B, 278 - 280 (1977). – Abstr. in Phys. Abstr., Vol. 80, Abstr. 91609 (1977).

131.042 **Observational evidence for supernova-induced star formation.** W. Herbst, G. E. Assousa.
Bull. American Astron. Soc., Vol. 9, 543 (1977). – Abstract.

131.043 **Four UV observations of the interstellar wind by Mariner 10.**
J. M. Ajello, N. Witt, P. W. Blum, A. L. Broadfoot.
Bull. American Astron. Soc., Vol. 9, 544 (1977). – Abstract.

131.044 **Astrophysical implications of numerical simulations of planet-formation.** W. K. Hartmann, R. Greenberg, J. F. Wacker, C. R. Chapman.
Bull. American Astron. Soc., Vol. 9, 545 (1977). – Abstract.

131.045 **Les molécules de l'espace.**
A. Penzias, P. Encrenaz.
La recherche en astrophysique, (see 003.001), p. 156 - 167 (1977).

131.046 **Molécules interstellaires: la liste n'est pas close.**
J. Lequeux.
La recherche en astrophysique, (see 003.001), p. 168 - 174 (1977).

131.047 **Molécules interstellaires: physique et chimie de H_2 et CO.** J. Lequeux.
La recherche en astrophysique, (see 003.001), p. 175 - 182 (1977).

131.048 **Le milieu interstellaire vu de OAO-3.** J. Lequeux.
La recherche en astrophysique, (see 003.001), p. 183 - 189 (1977).

131.049 **The absorption of OH and H_2CO in the direction of Cassiopeia A.** G. de Jager, D. A. Graham, R. Wielebinski, R. S. Booth, G. M. Gruber.
Astron. Astrophys., Vol. 64, 17 - 21 (1978).
Spectra showing the distribution of 6 cm H_2CO and 18 cm OH absorption in the direction of Cas A have been obtained, with a resolution of 2.'6 and 8' respectively. Large variations are seen across the source. No discrepancies appear between the OH and H_2CO spectra that cannot be ascribed to the difference in telescope resolution. Contour maps of formaldehyde absorption are similar to high resolution maps of H I absorption at small negative velocities in the Perseus arm region but differ from H I maps at large negative velocities.

131.050 **Destruction of interstellar grains by sputtering.**
T. Onaka, F. Kamijo.
Astron. Astrophys., Vol. 64, 53 - 60 (1978).
The present paper deals with the sputtering process as a mechanism for the destruction of interstellar grains. A Monte Carlo simulation based on a single scattering model has been applied to this process under the assumptions of random collisions and no inelastic collisions. The simulation has been performed for high-energy (5 keV–5 MeV) proton bombardments on spherical graphite grains and dirty ice grains. A new formula for a spherical target is proposed, in which the boundary effect is taken into account.

131.051 **Small-scale structure in the distribution of neutral hydrogen.** U. J. Schwarz, P. R. Wesselius.
Astron. Astrophys., Vol. 64, 97 - 101 (1978).
Two continuum sources 3C10 and 3C58 were studied with the Westerbork synthesis telescope for 21-cm absorption across the sources and for 21-cm emission in their surroundings on a scale of 0.5 to a few arc minutes. The authors find that there are H I features of diameter 0.5 to 3 pc.

131.052 **Rotation and velocity structure in the core of the optical condensation B213 NW and its relation to the parent gas.** F. O. Clark, D. R. Johnson.
Astrophys. J., Vol. 220, 500 - 509 (1978).
A small region around the high-density core of the optical condensation referred to as B213 NW has been studied in some detail. The 2 cm formaldehyde spectral line was used as a probe for the study. The gas in this region is observed to consist of several fragments, the most massive of which appears to be rotating. The other fragment observed with the 2 cm line appears to be unbound and will escape from B213 NW. The relationship between the gas in this dense core and the parent cloud is discussed. A new method of determining distance to dense gas clouds which exhibit rotation is outlined.

131.053 **CO observations of a high-latitude molecular cloud associated with NGC 7023.**
D. M. Elmegreen, B. G. Elmegreen.
Astrophys. J., Vol. 220, 510 - 515 (1978).
The authors present a ^{12}CO map of a molecular cloud associated with NGC 7023. Measurements of ^{13}CO and CS were made to obtain column densities and densities. Optical evidence for recent star formation in this cloud is supported by the detection of a ^{12}CO peak at the position of a weak 6 cm continuum source. No optical counterpart is present. The ^{12}CO line widths and line-center velocities suggest, furthermore, that an additional velocity feature is in this same area of the cloud.

131.054 **H_2 cooling, dissociation, and infrared emission in shocked molecular clouds.**
J. M. Shull, D. J. Hollenbach.
Astrophys. J., Vol. 220, 525 - 537 (1978).
The authors present models of interstellar shocks in molecular clouds over ranges of ambient molecular density $10^3 \leqslant n_0 \leqslant 10^7$ cm^{-3} and shock velocity $6 \leqslant v_s \leqslant 14$ km s^{-1}. Using estimates of H_2–H_2 collisional-excitation rates, they derive the H_2 radiative cooling rates from vibrational-rotational quadrupole transitions as a function of $n(H_2)$ and temperature T, and tabulate the emissivities integrated through the shock of the strongest infrared lines in the $v = 1 \to 0$, $2 \to 0$, and $2 \to 1$ bands of H_2.

131.055 **Heat sources for bright-rimmed molecular clouds: CO observations of NGC 7822.**
B. G. Elmegreen, D. F. Dickinson, C. J. Lada.
Astrophys. J., Vol. 220, 853 - 863 (1978).
Observations of the 2.6 mm carbon monoxide line in the bright rim NGC 7822 reveal that the peak excitation and column density of the molecule lie in a ridge ahead of the ionization front. Several possible heat sources for NGC 7822 and for other bright-rimmed molecular clouds (BRMC) are considered, after a summary of relevant observations. Cosmic rays are discussed and are shown to provide an excellent heat source for low-temperature Bok globules, but not for BRMC, which are significantly warmer. The authors examine the proposal of Lada and Black (1976) that energy deposition by a shock wave may be adequate to account for the observed CO enhancement in some BRMC. Other macroscopic heat sources are then investigated, including the viscous heating of gas by Alfvén waves or turbulence. Finally, the possibility that radiation from the Cepheus IV OB association heats the gas in NGC 7822 by grain-related processes is discussed. The con-

clusions stress the importance of internal macroscopic energy like Alfvén waves or turbulence as a possible heat source for BRMC.

131.056 Atomic and molecular observations of the ρ Ophiuchi dark cloud. P. C. Myers, P. T. P. Ho, M. H. Schneps, G. Chin, V. Pankonin, A. Winnberg. Astrophys. J., Vol. 220, 864 - 882 (1978).

The ρ Ophiuchi dark cloud has been observed in microwave spectral lines of H, OH, H_2CO, ^{12}CO, ^{13}CO, and NH_3. Comparison with maps of SO, HCN, and HCO^+ indicates that each map has a different spatial extent radius ranging from ~ 0.1 to 6 pc. On a scale less than ~ 1 pc, the cloud is organized into two main fragments. Gas density varies within the fragments as $n \propto r^{-1.3}$. The line width of a given species increases with its spatial extent, which suggests radial contraction of the fragments with a velocity that is about half that expected in free fall. Abundances of H_2CO, OH, and CO are similar to those in other large dark clouds, and are strongly correlated with extinction.

131.057 Star formation in shock-compressed layers. B. G. Elmegreen, D. M. Elmegreen. Astrophys. J., Vol. 220, 1051 - 1062 (1978).

The purpose of this paper is to determine the growth rates and masses of perturbations in compressed, isothermal, plane-parallel layers with no magnetic fields. This is done by numerically calculating the dispersion relation for gravitational instabilities in a pressure-bounded layer. Simple expressions for the maximum growth rates and for the wavelength of the mode at which this maximum growth occurs are given as functions of layer size and external pressure. The results are applied to estimate the behavior of the compressed layers that occur behind isothermal shocks in astrophysical situations. The conditions for gravitational instabilities and star formation behind shocks that are derived here are a refinement of those given by Elmegreen and Lada (1977).

131.058 On the microwave detection of interstellar CH_4, SiH_4, and GeH_4. A. H. Barrett. Astrophys. J., Lett., Vol. 220, L81 - L85 (1978).

Centrifugal distortion effects in spherical top molecules give rise to a large number of infrared ($\Delta J = \pm 1$) and microwave ($\Delta J = 0$) transitions of electric dipole character. Examination of the energy levels and transitions reveals two lines in every XH_4 spherical top molecule which appear to be excellent candidates for being interstellar masers.

131.059 Molecular synthesis in interstellar clouds: radiative association reactions of CH_3^+ ions. D. Smith, N. G. Adams. Astrophys. J., Lett., Vol. 220, L87 - L92 (1978).

A survey of laboratory measurements of several binary and collision-stabilized association reactions of CH_3^+ ions is presented. It is shown that the association reactions can proceed via radiative stabilization at gas kinetic rates in interstellar clouds and that the products of the reactions can result in observed interstellar molecules.

131.060 The kinetic temperature in the interior of the ζ Ophiuchi cloud from Copernicus observations of interstellar C_2. T. P. Snow, Jr. Astrophys. J., Lett., Vol. 220, L93 - L96 (1978).

Copernicus data on the stronger Mulliken band at 2313 Å are described which provide limits on the rotational temperature as well as the total abundance of C_2 toward ζ Oph.

131.061 On the column density of the interstellar Mg II to Sirius and other nearby stars. Y. Kondo, D. L. Talent, E. S. Barker, R. J. Dufour, J. L. Modisette. Astrophys. J., Lett., Vol. 220, L97 - L102 (1978).

The Mg II resonance doublet features at 2795.5 and 2802.7 Å have been observed with the Copernicus V1 spectrometer by using a special high signal-to-noise ratio mode in stars α CMa, α Lyr, α Gru, α Leo, α PsA, and α Aql. The column density of the interstellar Mg II has been determined to α CMa (Sirius) for the first time; it is 2.44×10^{12} cm^{-2}. A reliable determination of this quantity has been made for α Lyr, α Gru, and α Leo. The determinations for α Aql and α PsA are somewhat uncertain.

131.062 Molecules in space – galactochemistry. R. D. Brown. Focus on the stars, (see 003.002), p. 187 - 213 (1977).

131.063 A search for diffuse interstellar bands in far-ultraviolet wavelengths. T. P. Snow, Jr., D. G. York, M. Resnick. Publ. Astron. Soc. Pacific, Vol. 89, 758 - 764 (1977/78).

Far-ultraviolet data obtained at 0.2 Å resolution with Copernicus are used to search for diffuse interstellar bands between 1114 Å($8.98\mu^{-1}$) and 1450 Å($6.90\mu^{-1}$). Extinction curves are derived using unreddened comparison stars for the five program stars, and residuals are calculated from quadratic fits to these curves; the residual curves themselves are then averaged together to suppress noise introduced by guidance drifts and mismatching stellar lines. In the final average curve for the four stars with similar UV extinction, only one feature, centered near 1416 Å($7.06\mu^{-1}$) survives the statistical tests and is considered possibly real. The upper limit on the central depth of other features is 10% for $E(B-V) \simeq 0.3$.

131.064 Interstellar $\lambda 4430$ in the Carina nebula, NGC 3603, and VI Cygni. N. R. Walborn. Publ. Astron. Soc. Pacific, Vol. 89, 765 - 767 (1977/78).

Central depths of the $\lambda 4430$ interstellar band have been measured in the spectra of stars in the Carina nebula (NGC 3372), NGC 3603, and VI Cyg (Cygnus OB2). The central depth of nearly 30% in the NGC 3603 central object (HD 97950) is probably the greatest yet observed for this feature. In the $\lambda 4430, E_{B-V}$ plane, the NGC 3603 point lies far above the Carina nebula and VI Cyg relations.

131.065 Heating of the interstellar medium due to dissipation of weak shock waves. Y. Sabano, M. Tosa. Publ. Astron. Soc. Japan, Vol. 30, 67 - 75 (1978).

The authors study the effect of dissipative heating by weak shock waves with special emphasis on the two-phase model of the interstellar medium on the assumption that the interstellar gas is heated by successive passage of weak shock waves in addition to the background heating due to cosmic rays or soft X-rays. It is shown that dissipative heating has a significant effect on the thermal balance state of the interstellar gas and that the equilibrium pressure of the interstellar phase-change is fairly raised.

131.066 Thermal instabilities due to formation of carbon monoxide in contracting interstellar clouds. Y. Sabano, Y. Kannari. Publ. Astron. Soc. Japan, Vol. 30, 77 - 89 (1978).

The thermal stability of contracting interstellar clouds is investigated taking the formation and destruction processes for CO molecules into account. An instability criterion is derived for the condensation mode by linear perturbation analysis. It is found that CO clouds become unstable in the course of gravitational contraction and subcondensations with ordinary stellar masses can be formed.

131.067 On the interstellar abundance of H_2O^+. D. C. B. Whittet, R. C. Wayte, I. Wynne Jones. Observatory, Vol. 98, 44 - 46 (1978).

An attempt to detect interstellar H_2O^+ by means of a spectral feature at $\lambda 6199$ Å, observed in comets, was unsuccessful. An upper limit of $7 \times 10^{-10}/\text{cm}^3$ was estimated for the space density.

131.068 Energetic secondary electrons and the nonthermal galactic radio background: a probe of the magnetic field in interstellar clouds.   A. P. Marscher, R. L. Brown. Astrophys. J., Vol. 221, 588 - 597 (1978).

The authors find that $B < 80$ microgauss (μG) is characteristic of dense clouds ($n \gtrsim 300$ cm^{-3}). However, even if the mean cloud field is as low as 35 μG, the integrated synchrotron radiation from secondary electrons in interstellar clouds contributes a significant fraction of the nonthermal brightness along the galactic equator. The authors further conclude that along the galactic plane the contribution of bremsstrahlung γ-radiation from secondary electrons in dense clouds is comparable with that from neutral pion decays. Out of the plane, this fraction is smaller.

131.069 Observations of interstellar lithium toward ζ Persei and ϵ Aurigae.
P. A. Vanden Bout, R. L. Snell, S. S. Vogt, R. G. Tull. Astrophys. J., Vol. 221, 598 - 600 (1978).

The $\lambda6708$ line of interstellar ^{7}Li I has been detected toward the stars ζ Per and ϵ Aur, with equivalent widths of 0.87 ± 0.07 mÅ and 0.77 ± 0.08 mÅ, respectively, which yield column densities of $N(^7\text{Li I}) = (2.9 \pm 0.2) \times 10^9$ cm^{-2} and $(2.6 \pm 0.3) \times 10^9$ cm^{-2}, respectively.

131.070 The detection of interstellar diatomic carbon toward ζ Ophiuchi.
F. H. Chaffee, Jr., B. L. Lutz. Astrophys. J., Lett., Vol. 221, L91 - L93 (1978).

The $2-0$, $Q(2)$ line of the Phillips system of C_2 has been detected toward ζ Oph by using a PEPSIOS with the Mount Hopkins 1.5 m telescope. This detection provides indirect evidence for formation of such molecules through ion-molecule reactions in the gas phase of the interstellar medium. It further provides a possible optical "thermometer" for interstellar clouds.

131.071 Neutral hydrogen absorption by high-velocity clouds.   H. E. Payne, J. M. Dickey, E. E. Salpeter, Y. Terzian.
Astrophys. J., Lett., Vol. 221, L95 - L98 (1978).

Neutral hydrogen emission-absorption observations are reported for some high-velocity clouds at moderately low galactic latitudes. Absorption features have been detected for the first time, in the direction of the background source 3C 123 at -57 and -73 km s^{-1}, and possibly in the direction of 4C 33.48 at -128 km s^{-1}. From these observations a temperature range from $\sim$75 K to a few hundred K is suggested. Some H I spatial variations on a small scale (at most $\sim$4') have been found.

131.072 Sizes of dense clouds.
M. A. Gordon, W. B. Burton.
Bull. American Astron. Soc., Vol. 9, 553 (1978). – Abstract.

131.073 On mass estimates from CO emission.
P. L. Baker.
Bull. American Astron. Soc., Vol. 9, 553 - 554 (1978). Abstract.

131.074 Giant molecular clouds in the Galaxy: a survey of the distribution and physical properties of GMC's.
P. M. Solomon, D. B. Sanders, N. Z. Scoville.
Bull. American Astron. Soc., Vol. 9, 554 (1978). – Abstract.

131.075 CO observations of the NGC 2264 molecular cloud.
R. M. Crutcher.
Bull. American Astron. Soc., Vol. 9, 555 (1978). – Abstract.

131.076 The helium problem in Sagittarius B2.
E. J. Chaisson, S. M. Lichten, L. F. Rodriguez.
Bull. American Astron. Soc., Vol. 9, 555 (1978). – Abstract.

131.077 Radiation transport in molecular clouds.
J. Kwan.
Bull. American Astron. Soc., Vol. 9, 555 (1978). – Abstract.

131.078 Fragment interactions and the stellar mass spectrum.
W. A. Pumphrey, J. M. Scalo.
Bull. American Astron. Soc., Vol. 9, 566 (1978). – Abstract.

131.079 The fragmentation of isothermal rings and star formation.   M. L. Norman, J. R. Wilson.
Bull. American Astron. Soc., Vol. 9, 567 (1978). – Abstract.

131.080 Calculations of 3-dimensional collapse and fragmentation.   R. B. Larson.
Bull. American Astron. Soc., Vol. 9, 567 (1978). – Abstract.

131.081 Abundance variations in interstellar molecules.
H. A. Wootten, N. J. Evans II, R. L. Snell, P. A. Vanden Bout.
Bull. American Astron. Soc., Vol. 9, 574 (1978). – Abstract.

131.082 Observations of interstellar lithium.
P. A. Vanden Bout, R. Snell, R. Tull, S. Vogt.
Bull. American Astron. Soc., Vol. 9, 575 (1978). – Abstract.

131.083 A new dynamical model for the Orion molecular cloud.   P. T. P. Ho, A. H. Barrett.
Bull. American Astron. Soc., Vol. 9, 575 (1978). – Abstract.

131.084 Observations of ammonia in the Orion molecular cloud.   T. L. Wilson, D. Downes, J. Bieging.
Bull. American Astron. Soc., Vol. 9, 575 (1978). – Abstract.

131.085 Detection of the J = 3–2 lines of HCN, HNC and HCO$^+$ in the Orion molecular cloud.
P. J. Huggins, T. G. Phillips, G. Neugebauer, M. W. Werner, P. G. Wannier, D. Ennis.
Bull. American Astron. Soc., Vol. 9, 576 (1978). – Abstract.

131.086 Rotational temperature of SO$_2$ in OMC-1.
H. M. Pickett, J. H. Davis.
Bull. American Astron. Soc., Vol. 9, 576 (1978). – Abstract.

131.087 Detection of the $3_{13}-2_{20}$ transition of interstellar H$_2$^{18}O.   T. G. Phillips, P. J. Huggins, J. Y. Kwan, N. Z. Scoville, P. G. Wannier.
Bull. American Astron. Soc., Vol. 9, 576 (1978). – Abstract.

131.088 VLBI studies of the W3 OH water maser.
T. S. Giuffrida, P. E. Greenfield, B. F. Burke, A. D. Haschick, J. M. Moran, O. E. H. Rydbeck, B. O. Rönnäng, L. Barth, K. S. Yngvesson, L. I. Matveyenko (*Matveenko*), V. I. Kostenko, L. R. Kogan, I. G. Moiseev.
Bull. American Astron. Soc., Vol. 9, 576 (1978). – Abstract.

131.089 Maser time variations.
B. F. Burke, T. S. Giuffrida, A. D. Haschick.
Bull. American Astron. Soc., Vol. 9, 576 - 577 (1978). Abstract.

131.090 A compilation of interstellar gas properties.
P. C. Myers.
Bull. American Astron. Soc., Vol. 9, 580 (1978). – Abstract.

131.091 Observed properties of three rotating gas clouds.
F. O. Clark, D. R. Johnson.
Bull. American Astron. Soc., Vol. 9, 580 - 581 (1978). Abstract.

131.092 **B stars and the structure of the interstellar medium.**
T. M. Bania, J. G. Lyon.
Bull. American Astron. Soc., Vol. 9, 581 (1978). – Abstract.

131.093 **Herbig-Haro objects as shocked clouds.**
R. D. Schwartz.
Bull. American Astron. Soc., Vol. 9, 581 (1978). – Abstract.

131.094 **Distribution of reddening.**   G. Kron.
Bull. American Astron. Soc., Vol. 9, 581 (1978).
Abstract.

131.095 **Ice mantles and abnormal extinction in the ρ Oph
cloud.**   D. H. Harris.
Bull. American Astron. Soc., Vol. 9, 582 (1978). – Abstract.

131.096 **Optical properties of interstellar oxide grains.**
T. J. Millar, S. MacLean, W. W. Duley.
Bull. American Astron. Soc., Vol. 9, 582 (1978). – Abstract.

131.097 **Magnetic properties of interstellar oxide grains.**
W. W. Duley.
Bull. American Astron. Soc., Vol. 9, 582 (1978). – Abstract.

131.098 **Ionic neutralization in collisions with dust grains.**
J. C. Weisheit, R. J. Upham, Jr.
Bull. American Astron. Soc., Vol. 9, 582 - 583 (1978).
Abstract.

131.099 **New constraints on the composition of interstellar
grain mantles.**   R. S. McMillan.
Bull. American Astron. Soc., Vol. 9, 583 (1978). – Abstract.

131.100 **A possible solution to the CH^+ abundance problem
in the interstellar gas.**
M. Elitzur, W. D. Watson.
Bull. American Astron. Soc., Vol. 9, 588 (1978). – Abstract.

131.101 **Advancing H_2 dissociation fronts.**   R. London.
Bull. American Astron. Soc., Vol. 9, 588 (1978).
Abstract.

131.102 **Study of the extensive CO emission region south-
west of M17.**
B. G. Elmegreen, C. J. Lada, D. F. Dickinson.
Bull. American Astron. Soc., Vol. 9, 588 - 589 (1978).
Abstract.

131.103 **CO observations of a high latitude molecular cloud
associated with NGC 7023.**
D. M. Elmegreen, B. G. Elmegreen.
Bull. American Astron. Soc., Vol. 9, 589 (1978). – Abstract.

131.104 **Occultation of the S255 molecular cloud in the CO
line.**   W. L. Peters, N. J. Evans II.
Bull. American Astron. Soc., Vol. 9, 589 (1978). – Abstract.

131.105 **Millimeter and infrared observations of the S88
molecular cloud.**
G. N. Blair, M. Felli, N. J. Evans II, P. Vanden Bout, F. Israel.
Bull. American Astron. Soc., Vol. 9, 589 (1978). – Abstract.

131.106 **Radiative transfer effects and the interpretation of
interstellar molecular line observations.**
C. M. Leung.
Bull. American Astron. Soc., Vol. 9, 590 (1978). – Abstract.

131.107 **Observations of dark clouds.**   R. L. Snell.
Bull. American Astron. Soc., Vol. 9, 590 - 591
(1978). – Abstract.

131.108 **Deuterated interstellar molecules and a study of**

molecules in the Taurus dark cloud.
B. E. Turner, B. Zuckerman.
Bull. American Astron. Soc., Vol. 9, 591 (1978). – Abstract.

131.109 **A high resolution study of 6-cm formaldehyde ab-
sorption in L 134.**
J. M. Pasachoff, R. L. Dickman, M. L. Kutner, K. D. Tucker.
Bull. American Astron. Soc., Vol. 9, 591 (1978). – Abstract.

131.110 **Magnetic braking during star formation.**
R. C. Fleck.
Bull. American Astron. Soc., Vol. 9, 601 (1978). – Abstract.

131.111 **6-cm recombination lines from the molecular cloud/
H II region interface in NGC 1977.**
M. L. Kutner, N. J. Evans II, M. Guelin, K. D. Tucker.
Bull. American Astron. Soc., Vol. 9, 611 - 612 (1978).
Abstract.

131.112 **Recombination lines from H II and C II regions.**
P. Silverglate, Y. Terzian.
Bull. American Astron. Soc., Vol. 9, 612 (1978). – Abstract.

131.113 **Observations of λ21 cm absorption by high velocity
clouds.**
H. Payne, J. M. Dickey, E. E. Salpeter, Y. Terzian.
Bull. American Astron. Soc., Vol. 9, 615 (1978). – Abstract.

131.114 **Diatomic oxide interstellar grains.**
T. J. Millar, W. W. Duley.
Mon. Not. R. Astron. Soc., Vol. 183, 177 - 185 (1978).
    The formation of diatomic oxide grains in the inter-
stellar medium is briefly discussed and the infrared properties of several oxide materials are presented. It is shown
that SiO has a broad, featureless absorption profile centered
at 9.6 μm while FeO has an absorption peak at 19 μm. A
discussion of the wavelength dependence of polarization at
10 μm is also given.

131.115 **Is water-ice the carrier of the 3 μm-absorption in infrared objects?**   T. Mukai, S. Mukai, K. Noguchi.
Astrophys. Space Sci., Vol. 53, 77 - 84 (1978).
    It is shown that Mie theory predictions of extinction for
pure water-ice with the optical constant measured at 100 K do
not fit in detail the observed 'ice' absorption feature in infrared objects, although the authors attempt to explain the observations by considering size distribution and shape of the
grains.

131.116 **Study of the region $348° < l < 12°, +3° < b < +17°$
in the 21 cm line.**   M. L. Franco, W. G. L. Pöppel.
Astrophys. Space Sci., Vol. 53, 91 - 103 (1978).
    The study of the region $348° < l < 12°, +3° < b < +17°$
reveals strong kinematic asymmetries of the interstellar gas,
with a notorious predominance of positive radial velocities,
referred to the LSR. The principal cause of this is a very
intense source, a ridge-like object which presents very special
characteristics and seems to be identical with Lindblad's feature A. It should, therefore, be related to Gould's Belt. The
object's velocity dependence on l is in good agreement with
Lindblad's model. It seems highly probable that feature A
presents both the cold as well as the hot component of the
interstellar medium.

131.117 **Silicon monoxide and the 10 μm interstellar feature.**
W. W. Duley, S. MacLean, T. J. Millar.
Astrophys. Space Sci., Vol. 53, 223 - 225 (1978).
    Laboratory spectra of SiO particles of 1 μm radius show
a broad structureless extinction peak at 9.6 μm. The wavelength dependence of extinction from SiO, an amorphous
silicon oxide, provides a good match to that of interstellar
dust.

131.118   The C II and S II regions in the Rho Ophiuchi dark cloud.   V. Pankonin, C. M. Walmsley.
Astron. Astrophys., Vol. 64, 333 - 340 (1978).

The distribution of ionized carbon and sulfur in the Rho Ophiuchi dark cloud was mapped. Recombination line observations were made at 1.6 GHz, 5 GHz, and 9 GHz. The ionized carbon is centered on a near infrared point source which is also coincident with a radio continuum source and an extended far infrared source, but the maximum in the molecular distribution is displaced by $\sim 5'$. There are two components to the C II. There is a high density ($n_e \sim 15$ cm^{-3}), compact "core" ($d \sim 0.1$ pc) which dominates the high frequency line measurements, and there is a lower density ($n_e \sim 1$ cm^{-3}), extended "halo" ($d \sim 0.5$ pc) which dominates the lower frequency line observations. The sulfur line may not originate in exactly the same volume as the carbon line.

131.119   The distribution of color excesses and interstellar reddening material in the solar neighborhood.
P. B. Lucke.
Astron. Astrophys., Vol. 64, 367 - 377 (1978).

Color excesses and photometric distances are calculated in the UBV system for some 4000 O and B stars using data from recent photometric and spectroscopic catalogs. These color excesses and distances are used to construct contour plots of equal mean color excess in the galactic plane. Additional contour plots show the distribution of the mean excesses perpendicular to the galactic plane. The distribution of reddening material is characterized by a strong correlation with the Gould belt, a region of low reddening out to 500 pc between 210° and 255° longitude extending to 2 kpc near 240°, and very heavy obscuration between 10° and 80°.

131.120   Ammonia survey of small dark clouds.
P. T. P. Ho, R. N. Martin, A. H. Barrett.
Astrophys. J., Lett., Vol. 221, L117 - L120 (1978).

Of 29 regions surveyed, NH$_3$ has been detected toward 12 sources, with five additional marginal detections. The authors derive $n(H_2) \gtrsim 5 \times 10^3$ cm^{-3}, and $\Delta V \sim 0.35$ km s^{-1} after corrections for hyperfine overlapping. By comparing with $\Delta V(^{13}CO)$, they support the suggestion that systematic motion and not microturbulence is probably the line-broadening mechanism in dark clouds.

131.121   Detection of a new transition of SiO in OH/IR stars.
E. Scalise, Jr., J. R. D. Lépine.
Astron. Astrophys., Vol. 65, L7 - L8 (1978).

SiO maser emission from the $J = 1-0$ rotational transition has been detected in the third vibrational state ($\nu = 3$) in OH/IR stars.

131.122   Herbig-Haro objects.   K. H. Böhm.
IAU Colloq. No. 42, (see 012.012), p. 3 - 24 (1977).

The available observational information on the geometrical structure, emission line and continuous spectra, line profiles, radial velocities and linear polarization of Herbig-Haro objects is briefly reviewed. The author emphasizes the inhomogeneous structure of the "classical" Herbig-Haro objects and the appearance of small "condensations" with radii of $\sim 300 - 900$ a.u. The apparent paradox of the presence of "gaseous nebula type" as well as "reflection type" Herbig-Haro objects is discussed. A purely empirical model of the regions of line formation as well as different theoretical models for the line forming regions are discussed.

131.123   Observational evidence on the interaction of Orion population stars and the interstellar medium.
G. Grasdalen.
IAU Colloq. No. 42, (see 012.012), p. 25 - 37 (1977).

131.124   Structure and appearance of envelopes around protostars.   H. W. Yorke.

IAU Colloq. No. 42, (see 012.012), p. 38 - 64 (1977).

131.125   Mass loss from stars and the chemical evolution of the interstellar medium.   P. Biermann.
IAU Colloq. No. 42, (see 012.012), p. 641 - 649 (1977).

Mass loss from stars returns processed as well as unprocessed material to the interstellar medium, and thus enriches it in helium and heavy elements. In this brief review the author outlines the theory of the chemical evolution of the interstellar medium.

131.126   Deuterium in the Galaxy.   R. D. Brown.
Proc. Astron. Soc. Australia, Vol. 3, 100 - 101 (1977). – Invited paper.

131.127   H$_2$CO and OH observations of a molecular cloud near RCW 36.   J. B. Whiteoak, F. F. Gardner.
Proc. Astron. Soc. Australia, Vol. 3, 147 - 150 (1977).

Many cold molecular clouds have been detected as a result of their association with dense optical obscuration. The H II region RCW 36 is contained within a prominent dust lane. By means of H$_2$CO and OH observations the authors have found that this lane is associated with a dense elongated molecular cloud.

131.128   The role of grain motions in star formation.
M. J. Krautschneider.
Proc. Astron. Soc. Australia, Vol. 3, 168 - 169 (1977).

131.129   Infrared radiation from dark globules.
R. G. Spencer, C. M. Leung.
Astrophys. J., Vol. 222, 140 - 152 (1978).

Theoretical models are constructed by which to study the infrared emission from dark globules heated by the interstellar radiation field (ISRF). The effects of cloud parameters (grain type, optical depth, and density inhomogeneity) on the emergent spectrum and infrared surface brightness are studied. A physical interpretation of the results is presented. To help remove ambiguities from interpretations of future observations, the observable effects of a grain mixture, variation of the ISRF, as well as beam dilution are examined in detail. The expected flux densities in the infrared of a typical globule are presented for different beam sizes.

131.130   The far-infrared emission of interstellar matter between galactic longitudes $l = 36°$ and $l = 55°$.
G. Serra, J. L. Puget, C. E. Ryter, J. J. Wijnbergen.
Astrophys. J., Lett., Vol. 222, L21 - L25 (1978).

The authors report far-infrared ($115-196$ μm) observations to the diffuse emission from the galactic plane between $l = 36°$ and $l = 55°$. From the flux measurements and CO data, they obtain a far-infrared luminosity normalized per hydrogen atom, $L_{IR}^{H} = 1.8 \times 10^{-30}$ W per H atom between $l = 41°$ and $l = 45°$. This value is significantly greater than that expected from the dust heated by the average stellar radiation field in the solar neighborhood, and is indicative of a larger star-formation rate per unit mass of interstellar gas in the inner region of the Galaxy.

131.131   An experimental investigation of the condensation of silicate grains.   K. L. Day, B. Donn.
Astrophys. J., Lett., Vol. 222, L45 - L48 (1978).

Amorphous magnesium silicate smoke particles were condensed from hydrogen and argon atmospheres containing Mg and SiO. A wide range of initial compositions were observed, but all particles could be recrystallized into forsterite (Mg$_2$SiO$_4$) by heating to 1000° C in vacuum. The amount of smoke formed decreased rapidly with temperatures between 300 and 800 K at reactant partial pressures of about 1 torr.

131.132   The birth of massive stars.   M. Zeilik.
Sci. American, Vol. 238, No. 4, p. 110 - 118 (1978).

The birth of stars many times hotter and more massive than the sun may be triggered by shock waves traveling through large cool clouds of interstellar gas and dust.

131.133 Distribution of extinction in the Corona Australis dark cloud complex. G. S. Rossano.
Astron. J., Vol. 83, 234 - 240 (1978).

General star counts have been used over a region of 92 sq deg to derive the photographic extinction due to the dark cloud complex in Corona Australis. The constituent clouds in this complex are found to be systematically distributed in an extensive region of low obscuration. The mass of this extended background material and lower limits on the masses of the embedded clouds have been derived.

131.134 Distribution of extinction in several dark cloud complexes in Scorpio and Ophiuchus.
G. S. Rossano.
Astron. J., Vol. 83, 241 - 243 (1978).

General star counts have been used to derive the distribution of extinction in several dark cloud complexes in Scorpio and Ophiuchus. The morphology of these complexes suggests that a hierarchy of fragmentation is responsible for the distribution of material in this region.

131.135 Interstellar catalysis. I: The theory of H_2 formation. R. G. Tabak.
Astrophys. Space Sci., Vol. 53, 279 - 294 (1978).

Although it is generally accepted that most, if not all, of the molecular hydrogen in interstellar space is formed through recombination reactions on grains, the exact mechanism by which this is accomplished is far from certain. It is suggested that the chemisorption of hydrogen on transition metal grains may be just that formation mechanism. After separating the adsorption rate equations from those of desorption and using experimentally determined parameters, it is shown that transition metal grains can successfully catalyze as much H_2 as the theoretical maximum predicted for cold ice grains, even though metal grains are probably less than 10% as abundant (by mass) than dielectrics.

131.136 Gas, dust and molecules in the Galaxy. R. J. Quiroga.
Astrophys. Space Sci., Vol. 53, 295 - 333 (1978).

The relative abundances of cool neutral hydrogen, carbon monoxide and formaldehyde are studied using all the available observational data in the literature. The obtained mean values $N_{H I}/N_{H_2CO}$, $N_{H I}/N_{CO}$, N_{CO}/N_{H_2CO} are approximately constant in the dark clouds of the solar neighbourhood and in the distant molecular clouds. From a synthesis based on the results obtained, a cycle is postulated for the neutral hydrogen in the Galaxy: condensation and cooling of gas → molecular formation → gravitational collapse and star formation → gas dissipation and heating by cosmic rays and UV radiation.

131.137 New ionized regions in the Rho Ophiuchi dark cloud. E. Falgarone, D. A. Cesarsky, P. J. Encrenaz, R. Lucas.
Astron. Astrophys., Vol. 65, L13 - L16 (1978).

The authors present an extended set of radio observations of the Rho Oph complex. New recombination line observations confirm the presence of a strongly peaked emission of carbon and sulphur, and reveal an extended plateau of mostly carbon emission. The region of peak emission (Oph Center) is characterized by a SII region two times more extended than the CII region. Continuum observations indicate the presence of several point and extended sources in the direction of the complex. CO observations show a very suggestive correlation between double-peaked CO profiles and the distribution of line and continuum emission. This is the only clear correlation found between CO

and the other radio observations.

131.138 On a mechanism for formaldehyde polymer formation in the interstellar space. V. I. Gol'danskij.
Dokl. AN SSSR, Vol. 235, 1053 - 1055 (1977). In Russian. Abstr. in Ref. zh., 51. Astron., 2.52.601 (1978).

131.139 Thermal instabilities in a nonstationary medium. Yu. A. Shchekinov.
Astron. Zh. Akad. Nauk SSSR, Vol. 55, 311 - 317 (1978). In Russian. English translation in Soviet Astron., Vol. 22, No. 2.

The problem of thermal instabilities under nonstationary conditions is considered. Instability criteria for perturbations of different wavelengths are obtained. Thermal instabilities as a possible mechanism of star formation are discussed.

131.140 On a mechanism of creation of population inversion of linear molecular levels.
D. A. Varshalovich, V. K. Khersonskij.
Astron. Zh. Akad. Nauk SSSR, Vol. 55, 328 - 333 (1978). In Russian. English translation in Soviet Astron., Vol. 22, No. 2.

Nonequilibrium populations of rotational levels of CO, SiO, and CS molecules are calculated for interstellar medium conditions. Necessary and sufficient conditions under which nonequilibrium collisional pumping leads to level population inversion have been elucidated. Inversion occurs only in a narrow range of gas density values under sufficiently high kinetic temperature.

131.141 Mechanism of "surviving" of cosmic relativistic specks of dust.
Ya. S. Elenskij, A. L. Suvorov.
Astrofizika, Vol. 13, 731 - 735 (1977). In Russian. English translation in Astrophysics, Vol. 13, No. 4.

It is shown that the accumulation of electrical charge on relativistic specks of dust as a result of their interaction with the atoms of interstellar gas and photons can be sometimes compensated by evaporation of specks of dust material in strong electric fields.

131.142 The destruction and growth of dust grains in interstellar space – I. Destruction by sputtering.
M. J. Barlow.
Mon. Not. R. Astron. Soc., Vol. 183, 367 - 395 (1978).

The processes governing the destruction and growth of dust grains in interstellar space are investigated with a view to establishing the conditions required for the existence of ice mantles. In this paper sputtering by particles with energies in the eV to GeV range is considered. The sputtering of grains in H II regions, in the inter-cloud medium, and in shock waves produced by cloud–cloud collisions and by supernova remnants, is investigated. Of these, supernova remnants are shown to be the most important. Destruction rates are estimated for grains bombarded by MeV and GeV cosmic rays.

131.143 The destruction and growth of dust grains in interstellar space – II. Destruction by grain surface reactions, grain–grain collisions and photodesorption.
M. J. Barlow.
Mon. Not. R. Astron. Soc., Vol. 183, 397 - 415 (1978).

Chemical reactions on the surfaces of ice grains are shown to be unimportant as destruction agents. The classical Oort–van de Hulst destruction mechanism of grain–grain collisions during cloud collisions is shown to be ineffective. It is argued that photodesorption is the dominant destruction mechanism for ice grains held together by weak van der Waals dispersion forces. The timescale for destruction of an ice grain of radius 10^{-5} cm by the interstellar ultraviolet radiation field is derived to be $\sim 5 \times 10^4$ yr, much shorter than for other destruction mechanisms.

131.144   The destruction and growth of dust grains in interstellar space — III. Surface recombination, heavy element depletion and mantle growth.   M. J. Barlow.
Mon. Not. R. Astron. Soc., Vol. 183, 417 - 434 (1978).

It is shown that metallic elements in interstellar clouds will be trapped on the surfaces of graphite and iron grains, leading to their differential gas-phase depletion, whereas nonmetal elements will be ejected as saturated hydrides from such grains (and as monohydrides from other grains), with no consequent depletion. In diffuse clouds the growth of mantles is prevented by photodesorption and by the ejection of nonmetal atoms during surface recombination, but in dense clouds which have sufficient ultraviolet shielding, and where the nonmetals are in saturated molecules, mantle growth will take place.

131.145   Time variability of astrophysical masers.   M. Salem, M. S. Middleton.
Mon. Not. R. Astron. Soc., Vol. 183, 491 - 500 (1978).

The time-dependent equations of radiative transfer are solved for a maser source much longer than it is wide. Transitions are assumed to occur between two energy levels only. It is found that the intensity may vary rapidly although the physical conditions in the masing region, including the rates of pumping and relaxation, are constant. Under certain circumstances the observed intensity of a saturated maser may vary quasi-periodically with large amplitude and a period (in years) equal to the maser length (in light years); the maser may seem to appear and disappear.

131.146   Observations of interstellar HC$_5$N and HC$_7$N in dark dust clouds.
L. T. Little, G. H. Macdonald, P. W. Riley, D. N. Matheson.
Mon. Not. R. Astron. Soc., Vol. 183, 45P - 50P (1978).

The $J = 9 \rightarrow 8$ transition of cyanodiacetylene has been mapped in emission in the source TMC2, with a 2.3-arcmin beam. The source has an angular extent 3×4 arcmin, similar to that of the NH$_3$ source. The $J = 22 \rightarrow 21$ transition of interstellar HC$_7$N has been detected in Heiles 2 and TMC2. The relative abundances of HC$_5$N and HC$_7$N appear to be similar in both Heiles 2 and TMC2.

131.147   Observations of the interstellar [^{16}OH]/[^{18}OH] abundance ratio.   J. B. Whiteoak, F. F. Gardner.
Mon. Not. R. Astron. Soc., Vol. 183, 67P - 71P (1978).

The 1639-MHz line of ^{18}OH has been detected towards three H II regions well away from the centre of the Galaxy. In two cases (RCW 38 and RCW 57) the [^{16}OH]/[^{18}OH] abundance ratios (310, 220) are below the terrestrial [^{16}O]/[^{18}O] ratio (489), while in the third (NGC 2024) the ratio is about twice terrestrial. The result for NGC 2024 is the first instance of a low ^{18}O content in the interstellar medium.

131.148   The Herbig-Haro objects — breakup of protostellar cocoons.   M. A. Dopita.
IAU Symp. No. 76, (see 012.014), p. 324 (1978). — Abstract.

131.149   Density of interstellar magnesium along the line of sight to two nearby stars.   T. H. Morgan, J. N. McDonnold, J. L. Modisette, Y. Kondo.
Publ. Astron. Soc. Pacific, Vol. 90, 89 - 92 (1978).

The equivalent widths of the interstellar Mg II resonance doublet absorptions along the line of sight to the stars β Lib and α Leo have been measured and the Mg II column density and b value calculated. The best values are $5.7 \times 10^{+12}$ ions cm^{-2} and 5 km s^{-1} for α Lib and $1.86 \times 10^{+12}$ and 4 km s^{-1} for α Leo. These suggest that (at least along the lines of sight to these two stars) the nearby interstellar medium is quite tenuous (N_H in the range 0.08 to 0.015 atoms cm^{-3}).

131.150   A new method for determining the reddenings of extragalactic objects.   D. Burstein, C. Heiles.
Astrophys. Lett., Vol. 19, 69 - 73 (1978) = Lick Obs. Bull. No. 791.

The different values of the coefficient $A = \Delta E(\text{B}-\text{V})/\Delta \csc|b|$ for galactic objects and external galaxies derived by previous investigators can be understood in terms of the distribution of the various objects on the sky and the criteria used in their selection. The irregular variation of galactic extinction with both latitude and longitude is again emphasized. From a re-examination of the relations among galaxy counts, hydrogen column density, and the reddening of halo objects and external galaxies, the authors have determined that reddening can be predicted with reasonable accuracy from the other two sets of data.

131.151   An upper limit to the interstellar abundance of the HCN dimer.   J. W. V. Storey, A. C. Cheung.
Astrophys. Lett., Vol. 19, 89 - 91 (1978).

The authors have searched without success for the $J = 7-6$ transition of HCN...HCN in several interstellar clouds known to be rich in HCN. An upper limit to the abundance of the dimer relative to the monomer is typically 1 percent. It is suggested that this implies a low abundance for similar weakly bound species.

131.152   Star counts and visual extinctions in dark nebulae.   R. L. Dickman.
Astron. J., Vol. 83, 363 - 372 (1978).

Application of star count techniques to the determination of visual extinctions in compact, fairly high-extinction dark nebulae is discussed. Particular attention is devoted to the determination of visual extinctions for a cloud having a possibly anomalous ratio of total to selective extinction.

131.153   High-resolution 2-μ spectroscopy of Cyg OB II No. 12.   S. G. Kleinmann, D. N. B. Hall, S. T. Ridgway, E. L. Wright.
Astron. J., Vol. 83, 373 - 375 (1978).

A high-resolution 2-μ spectrum of the heavily obscured ($E_{B-V} = 3.25$) star Cyg OB 2 No. 12 was obtained. From it the authors infer upper limits to the column densities of intervening interstellar gas, $N(\text{H}_2) < 2.8 \times 10^{23}$ cm^{-2} and $N(^{12}\text{CO}) \leq 10^{18}$ cm^{-2}. Emission features seen in the 2-μ spectrum of No. 12 correspond to radial velocities V_{LSR} near 10 km/s.

131.154   Interstellar catalysis. II. Comparison of the theory of H$_2$ formation with observations.
R. G. Tabak.
Astrophys. Space Sci., Vol. 54, 211 - 232 (1978).

Since gas-phase reactions alone cannot account for the observed abundances of H$_2$ in the typical interstellar cloud, one or more surface reactions are probably involved. Of the three possible candidates, only the catalytic production of H$_2$ on transition metal grains is supported by laboratory evidence. It is found that metal grains can produce as much interstellar H$_2$ as the best physical adsorption mechanism under optimum conditions if the extinction in the visible is less than 5$\overset{m}{.}$0. The three critical parameters for efficient catalysis (activation energy of desorption, grain temperature, and the number density of available sites) are examined, and it is shown that catalytic reactions are efficient producers of H$_2$ under all but the most unfavorable conditions.

131.155   Radio and infrared observations of the OH/H$_2$O source G12.2–0.1.   P. A. Shaver, A. C. Danks.
Astron. Astrophys., Vol. 65, 323 - 327 (1978).

High-resolution $\lambda 6$ and 21-cm radio observations and $\lambda 1-5$ μm infrared observations are presented of the OH/H$_2$O maser source G12.2–0.1. The strongest H$_2$O emission originates within 4"(0.07 pc) of a compact radio source and intense near-infrared source; it is suggested that the infrared source may be the hot (> 1000 K) cocoon of a newly-formed

O star. The OH and weak H_2O sources are located at the edge of a ridge of radio emission, possibly an edge-on ionization front.

131.156 Structure of rotating clouds and quasi-static evolution under external pressure.
Y. Viala, N. Bel, S. Bonazzola.
Astron. Astrophys., Vol. 65, 393 - 400 (1978).

The authors study the influence of a uniform rotation upon the equilibrium conditions of a self-gravitating interstellar cloud; the equation of state is assumed to be that of polytrope with negative index and the cloud is acted upon by an external pressure. For a given angular velocity, equilibrium is possible only when the external pressure falls between two critical values: these limits approach each other with increase of mass and no equilibrium is possible beyond a critical mass. A quasi-static analysis of the evolution of a cloud having constant angular momentum shows that the centrifugal forces in the equatorial plane are such that the pressure sometimes does not reach its critical value for collapse.

131.157 Chemical and thermal equilibrium in dark clouds.
J. Clavel, Y. P. Viala, N. Bel.
Astron. Astrophys., Vol. 65, 435 - 448 (1978).

Chemical and thermal balance equations together with the transfer equation for UV radiation have been solved in a self-consistent way for dark clouds of uniform density $n(H) = 10^4$ cm^{-3} and visual extinction $A_V \cong 20$. In the absence of embedded stars, gas-grain collisions play a negligible role in the heating of the cloud which is mainly due to the cosmic ray ionization of H_2 and the H_2 formation of grains. Chemical reactions could contribute significantly to the heating if the heat is entirely released as thermal kinetic energy. Gas temperature of $\sim 10°$ K is achieved in the inner parts of the cloud which is consistent with the temperature deduced from CO line observations in most dark clouds with no embedded sources.

131.158 Parameters of the interstellar medium from data of measurements of scattered Lα-radiation with Venera 9.
J.-L. Bertaux, J. E. Blamont, M. C. Bourgin, V. G. Kurt, A. S. Smirnov, E. Yu. Shafer.
Kosm. Issled., Vol. 16, 269 - 277 (1978). In Russian.

131.159 The law of interstellar extinction.
H. L. Johnson.
Rev. Mexicana Astron. Astrofis., Vol. 2, 175 - 180 (1977).

The new 13-color medium-narrow band photometry published by Johnson and Mitchell has been used to investigate the law of interstellar extinction in several regions of the sky. The extinction laws in these regions are different, and must indicate that there are physical differences of the interstellar material which produces the extinction in the several regions studied.

131.160 Time-dependent CO formation and fractionation.
H. S. Liszt.
Astrophys. J., Vol. 222, 484 - 490 (1978).

Chemical fractionation effects are discussed in the context of a detailed numerical model which follows the spatial and temporal behavior of 23 species. Results are presented for purely gas-phase processes and for gas-grain coupled schemes, in and out of equilibrium. The author discusses systematic effects relevant to interpretation of the available data, and concludes that both astration and chemical fractionation effects are required to explain the observations.

131.161 Optical interstellar lines toward 18 stars of low reddening. III.
L. M. Hobbs.
Astrophys. J., Vol. 222, 491 - 507 (1978).

New high-resolution, photoelectric scans of one or both of the interstellar D lines of NA I are reported for 16 stars with $E(B-V) \leqslant 0.05$ and for two other stars with $E(B-V) \leqslant 0.10$, along with either new or previous scans of the interstellar K line of Ca II for nine of these stars. In each direction, the resulting column densities of Na I and Ca II are compared with those of H I and, where available, H_2, as deduced previously from ultraviolet interstellar lines measured from OAO 2 and OAO 3. Three principal conclusions follow. (1) In 14 of the 18 cases, essentially normal Na I/H ratios are found. (2) Toward γ Peg and μ^1 Sco, unusually low Na I/H ratios are observed. (3) Toward υ Sco and σ Sgr, the absence of detectable Na I, H I, and H_2 lines shows that unusually little neutral gas of any kind is present along these two light paths.

131.162 What are the products of polyatomic ion–electron dissociative recombination reactions?
E. Herbst.
Astrophys. J., Vol. 222, 508 - 516 (1978).

A statistical theory of polyatomic ion–electron dissociative recombination reactions is presented and applied to the calculation of branching ratios for the various sets of neutral products of the reactions $HCNH^+ + e$, $H_3O^+ + e$, $CH_3^+ + e$, and $NH_4^+ + e$, all thought to be important in interstellar cloud chemistry.

131.163 Carbon recombination lines and the neutral hydrogen clouds near the Orion nebula.
W. L. Boughton.
Astrophys. J., Vol. 222, 517 - 526 (1978).

Both new and previously published observations of carbon recombination lines toward Orion A have been used to construct two-cloud (foreground and background) models for the carbon-line emission.

131.164 Detection of $H_2^{18}O$ and an abundance estimate for interstellar water.
T. G. Phillips, N. Z. Scoville, J. Kwan, P. J. Huggins, P. G. Wannier.
Astrophys. J., Lett., Vol. 222, L59 - L62 (1978).

Detection is reported of the $3_{13} - 2_{20}$ (203.4075 GHz) transition of interstellar $H_2^{18}O$ in the molecular clouds of Orion and DR 21 (OH). An excitation calculation shows that far-infrared radiation from dust provides the primary mechanism for populating the 3_{13} state. The authors estimate the density of para-$H_2^{18}O$ in the Orion cloud as 4×10^{-3} cm^{-3} and that of water (ortho- and para-) as 8 cm^{-3}. The estimated [H_2O/H_2] abundance ratio is $\sim 10^{-5}$.

131.165 The dissociative recombination of $CH^+ X^1\Sigma^+ (\nu = 0)$.
J. B. A. Mitchell, J. W. McGowan.
Astrophys. J., Lett., Vol. 222, L77 - L79 (1978).

Knowledge of the rate coefficient for the dissociative recombination of $CH^+ X^1\Sigma^+(\nu = 0)$ is of vital importance to the theory of molecule formation in diffuse interstellar clouds. The authors report the first measured values of the recombination cross section for this process.

131.166 Ultraviolet spectra of organic molecules and the interstellar medium.
W. G. Egan, T. Hilgeman.
Nature, Vol. 273, 369 - 370 (1978).

The authors feel that although cellulose (or other organics) might exist in the interstellar medium, they are not yet prepared to reject the presence of H_2O ice and silicates.

131.167 Microwave spectral lines in galactic dust globules.
R. N. Martin, A. H. Barrett.
Astrophys. J., Suppl. Ser., Vol. 36, 1 - 51 (1978).

In order to better understand galactic dust globules, a program of mapping several molecular transitions in these clouds has been undertaken. The results of observations of the $J = 1 \rightarrow 0$ rotational transitions of CO, ^{13}CO, $C^{18}O$, and CS, the $J = 2 \rightarrow 1$ rotational transitions of CS and $C^{34}S$, the $J, K = 1,1$ and $J, K = 2,2$ inversion transitions of NH_3, the $J_{KK'} = 1_{11} \rightarrow 1_{10}$ and $J_{KK'} = 2_{12} \rightarrow 2_{11}$ transitions of H_2CO, and the OH $^2\Pi_{3/2} F = 2 \rightarrow 2$ and $F = 1 \rightarrow 1$ transitions are reported. Twelve globules have been selected for observation; seven of these were studied in detail and the remainder observed only sparsely. The simultaneous observation of many molecular transitions has proven useful in obtaining reliable physical parameters for the dust globules. The results are consistent with the notion that globules are small, collapsing, and/or rotating dark nebulae which could be locations for future star formation.

131.168 Galactic neutral hydrogen emission-absorption observations from Arecibo.
J. M. Dickey, E. E. Salpeter, Y. Terzian.
Astrophys. J., Suppl. Ser., Vol. 36, 77 - 114 (1978).

The authors have observed the interstellar medium in the direction of 27 extragalactic sources at high and intermediate galactic latitudes ($|b^{II}| > 5°$) at 21 cm wavelength. Each region has been observed for approximately 2 hours to detect optical depths as low as 10^{-2} in most cases. The data allow the derivation of optical depths, harmonic mean spin temperatures, and column densities for the neutral hydrogen as a function of velocity in the range $-200 < v < 200$ km s^{-1} (LSR) with resolution as fine as 0.25 km s^{-1}.

131.169 Interstellar titanium.   G. M. Stokes.
Astrophys. J., Suppl. Ser., Vol. 36, 115 - 141 (1978).

High-dispersion photoelectric scans of the interstellar $\lambda 3384$ line of Ti II have been obtained for 68 stars. Column densities of Ti II have been derived for the various individual kinematic components in the scans. These column densities, and those integrated over all components in the respective directions, have been compared to corresponding results for Na I, Ca II, K I, H I, and H_2.

131.170 Structure and kinematics of H_2O sources in clusters of newly-formed OB stars.   R. Genzel, D. Downes,
J. M. Moran, K. J. Johnston, J. H. Spencer, R. C. Walker, A. Haschick, L. I. Matveyenko (Matveeenko), L. R. Kogan, V. I. Kostenko, B. Rönnäng, O. E. H. Rydbeck, I. G. Moiseev.
Astron. Astrophys., Vol. 66, 13 - 29 (1978).

The authors report VLBI observations at 22 GHz of 12 galactic water vapor sources in regions of star formation. In all of the sources the H_2O lines are concentrated in "centers of activity" of size $\sim 10^{16}$ cm. The authors think that each "center of H_2O activity" should be identified with the envelope of a young, massive star. The examples of Orion-KL, W51 S, and possibly W49 N, show that there is H_2O emission at the same position and at the same velocity as OH maser radiation at 1665 MHz.

131.171 The correlation of the interstellar extinction law with the wavelength of maximum polarization.
D. C. B. Whittet, I. G. van Breda.
Astron. Astrophys., Vol. 66, 57 - 63 (1978).

Infrared photometry is presented for 56 southern early-type stars which have interstellar polarization data available in the literature. New and previously published data are combined to investigate the relationship between the colour excess ratio E_{V-K}/E_{B-V} and the wavelength of maximum polarization (λ_{max}).

131.172 Star formation rates in the Galaxy.
L. F. Smith, P. Biermann, P. G. Mezger.
Astron. Astrophys., Vol. 66, 65 - 76 (1978).

Data relevant to giant H II regions in the Galaxy are collected. The production rate for Lyman continuum photons by O stars in giant H II regions is 4.7×10^{52} s^{-1} in the whole Galaxy. The corresponding present rate of star formation is $5\, M_\odot$/yr, of which 74% occurs in main spiral arms, 13% in the interarm region and 13% in the galactic center. The star formation rates, the observed heavy element and deuterium

abundances in the solar neighbourhood are compared to model predictions based on star formation proportional to a power of the gas surface density. The mass function is terminated at $M_u = 100\, M_\odot$ above and M_1 below. $M_u = 50\, M_\odot$ is also considered.

131.173 Change of interstellar gas parameters in stellar-wind-dominated astrospheres: solar case.
H. J. Fahr.
Astron. Astrophys., Vol. 66, 103 - 117 (1978).

The general problem of the change of the distribution function of neutral interstellar gas that penetrates into a stellar-wind-dominated star environment is investigated in detail. For the case of the solar system the author determines the temperatures of interstellar helium and hydrogen and shows that in the inner part of the heliosphere they deviate strongly from their undisturbed interstellar values, which are assumed to be about 10^3 K.

131.174 Infra-red molecular line emission from grain surfaces in dense clouds.   L. J. Allamandola,
C. A. Norman.
Astron. Astrophys., Vol. 66, 129 - 135 (1978).

The authors examine the possibility of observing the radiative decay of the lowest vibrational state of a number of interstellar molecules embedded in grains that have been vibrationally excited by either photon absorption in the visible to ultraviolet range and subsequent relaxation, radical reactions, or molecule-grain collisions. They list a number of molecules which may be observable in future infra-red experiments.

131.175 Observations of the $J = 1-0$ transition of CS at 49 GHz in southern molecular clouds.
F. F. Gardner, J. B. Whiteoak.
Mon. Not. R. Astron. Soc., Vol. 183, 711 - 725 (1978).

The $J = 1-0$ transition of CS at 49 GHz has been detected at Parkes in 37 out of a total of 57 molecular concentrations, most of which are associated with H II regions. The CS column density exceeds 10^{14} cm^{-2}. It was found that the CS, CO, H_2CO and H II velocities agreed within about 2 km/s on average. The CS distribution was partly mapped over the nearby H II regions Ori A, NGC 6334 and M17. The results for the Orion cloud do not support the suggestion that the whole cloud rotates.

131.176 Association of cyanodiacetylene emission in Heiles 2 with a cloud collision?   L. T. Little, P. W. Riley,
G. H. Macdonald, D. N. Matheson.
Mon. Not. R. Astron. Soc., Vol. 183, 805 - 811 (1978).

A map of the $J = 9 \to 8$ transition of cyanodiacetylene emission in Heiles 2 dust cloud shows the source to be highly extended with dimensions 16 by $\sim 2 \times 10^{17}$ cm. Although it may have stopped collapsing along its width, it appears to be unstable to contraction along its length. The position, elongation and LSR velocity of the source suggest that it may result from a cloud collision between the two halves of Heiles 2.

131.177 First detection of a non-metastable ammonia line in absorption.   P. T. P. Ho, A. H. Barrett.
Mon. Not. R. Astron. Soc., Vol. 183, 93P - 96P (1978).

The authors report the first detection of the $(J, K) = (2, 1)$ non-metastable line of NH_3 in absorption toward the compact H II region W3 (OH). They also compare metastable observations of W3(OH) with different angular resolutions.

131.178 Effect of the spin of an interplanetary dust on its motion.   H. Hasegawa, A. Fujiwara, C. Koike,
T. Mukai.
Mem. Fac. Sci. Kyoto Univ., Ser. Phys. Astrophys. Geophys. Chem., Vol. 35, 131 - 139 (1977).

Effect of the spin of an interplanetary dust on its motion was studied. The spin effect is larger for larger dust and for smaller heliocentric distance. The equation of heat conduction of a rotating sphere under the irradiation of sun light was solved. The numerical result was also given.

131.179 Collisional excitation of interstellar molecules: H_2.
S. Green, R. Ramaswamy, H. Rabitz.
Astrophys. J., Suppl. Ser., Vol. 36, 483 - 496 (1978).

Cross sections for the important rotational transitions in molecular H_2 collisions are presented for rotational levels to $j = 11$ and energies to 20,000 K. These have been obtained from extensive theoretical calculations. The resulting cross sections are fitted to a simple polynomial in the energy, and the coefficients are tabulated for easy utilization.

131.180 Photoelectric heating of interstellar gas.
B. T. Draine.
Astrophys. J., Suppl. Ser., Vol. 36, 595 - 619 (1978).

Photoelectric emission from interstellar grains is reexamined, and it is argued that some of the assumptions made by other authors lead to an overestimate of the heating rate associated with this process, particularly at temperatures $T \gtrsim 3000$ K. Steady-state solutions for the temperature of diffuse gas are found. A steady-state model with intercloud H I heated by soft X-rays and clouds heated by grain photoemission is in accord with some observations but lacks intermediate-temperature H I. The time-dependent cooling of a fossil H II region is calculated; grain photoelectric heating significantly prolongs the time required for the gas to cool.

131.181 Giant molecular clouds: size distribution.
D. B. Sanders, P. M. Solomon.
Bull. American Astron. Soc., Vol. 9, 554 (1978). – Abstract.

131.182 The abundance and distribution of interstellar C_2H.
K. D. Tucker, M. L. Kutner.
Astrophys. J., Vol. 222, 859 - 862 (1978).

The previously reported microwave discovery of interstellar C_2H has been confirmed by the detection of two hitherto unobserved hyperfine components of the $N = 1-0$ transition. Slightly refined values of the fine-structure and hyperfine-structure coupling constants based on the new observations are presented. Maps of C_2H emission in Orion and in DR 21 show integrated intensities which vary slowly with position. Failure to detect C_2H emission from dark clouds has yielded upper limits to the C_2H column density in those objects. Negative results of searches for ^{13}CCH and for propane (C_3H_8) are presented.

131.183 Interstellar absorption lines in the spectrum of Zeta Puppis.   D. C. Morton.
Astrophys. J., Vol. 222, 863 - 880 (1978).

High-resolution scans of the far-ultraviolet spectrum of ζ Pup, combined with the available ground-based data, have yielded equivalent widths for 147 interstellar absorption lines identified with 35 ions and isotopes of 19 elements, as well as 136 lines of H_2. In addition, 52 unidentified UV lines have been listed. The profiles have shown that at least two velocity components are present in the interstellar gas, separated by about 10 km s^{-1}. This paper also lists new results on wavelengths and f-values for many interstellar lines.

131.184 Molecular cooling and thermal balance of dense interstellar clouds.
P. F. Goldsmith, W. D. Langer.
Astrophys. J., Vol. 222, 881 - 895 (1978).

The authors analyze in detail the cooling produced by line emission from a variety of molecular and atomic species, including those observed as well as theoretically expected in dense interstellar clouds.

131.185 Far-ultraviolet studies.III. A search for light scattered at large angles by dust.
R. C. Henry, R. Anderson, P. D. Feldman, W. G. Fastie.
Astrophys. J., Vol. 222, 902 - 908 (1978).

The Apollo 17 far-ultraviolet spectrometer was used during trans-Earth coast to measure the brightness of two regions at moderate galactic latitudes. The signal obtained can be entirely accounted for by light from stars in the field of view, with no contribution from galactic-plane starlight scattered off interstellar dust. The authors conclude that the interstellar grains are extremely poor large-angle scatterers in the far-ultraviolet.

131.186 Formation of molecular CH^+ in interstellar shocks.
M. Elitzur, W. D. Watson.
Astrophys. J., Lett., Vol. 222, L141 - L144 (1978).

Formation of CH^+ is considered in the hot gas immediately behind shock fronts that may be present in the neutral gas clouds of the interstellar medium. Special attention is given to shocks which can result from expansion of the H II region around the stars toward which the observations are made. Column densities near 10^{13} cm^{-2} in agreement with observations of diffuse clouds are obtained for likely values of uncertain physical parameters. Column densities for other species including CH, OH, and H_2O, are also computed and are compatible with observations of diffuse clouds. Formation by shocks is thus proposed as a solution to the long-standing CH^+ abundance problem.

131.187 The DCO^+/HCO^+ abundance ratio and the electron density in cool interstellar clouds.
W. D. Watson, L. E. Snyder, J. M. Hollis.
Astrophys. J., Lett., Vol. 222, L145 - L147 (1978).

The $J = 1-0$ transition of $H^{13}CO^+$ has been detected in L134 at one-third the level (integrated antenna temperatures) of $D^{12}CO^+$. Scaling of the DCO^+-to-$H^{13}CO^+$ abundance ratio and high-resolution studies of line profiles also presented here then provide more reliable information than previously available about the abundances of DCO^+ and $H^{12}CO^+$ in the representative dark cloud L134.

131.188 Observational studies of star formation: a personal overview.   M. Cohen.
Q. J. R. Astron. Soc., Vol. 19, 177 - 186 (1978).

131.189 OH excitation in interstellar clouds.
J. Guibert, M. Elitzur, Nguyen-Q-Rieu.
Astron. Astrophys., Vol. 66, 395 - 405 (1978).

The authors have developed a detailed model which can account for the excitation temperatures derived from observations of interstellar clouds in the four OH ground state transitions. Their model involves collisions with neutral and charged particles, and far-infrared radiation from dust. It explains anomalies observed in satellite lines, and suggests that weak main line inversions may occur in the presence of hot dust grains ($T_g > 100$ K). It has been used to determine physical conditions in typical clouds exhibiting satellite line anomalies.

131.190 Detection of the $^2\Pi_{3/2}$, $J = 9/2$ Λ-doublet line of OH.
A. Winnberg, C. M. Walmsley, E. Churchwell.
Astron. Astrophys., Vol. 66, 431 - 435 (1978).

The authors have searched for the OH $J = 9/2$ Λ-doublet transitions at 23.8 GHz in six galactic 18-cm maser emission sources. They have detected the $F = 5-5$ and probably the $F = 4-4$ transitions in W3(OH). The lines observed in W3(OH) are in absorption and hence there is no evidence for inversion of the $J = 9/2$ Λ-doublet. However, the $F = 5$ levels of the doublet seem to be overpopulated relative to $F = 4$ if one compares with the ratio expected in L.T.E. The authors estimate on the basis of their measurement a column density of $10^{14}-10^{15}$ cm^{-2} of OH molecules in the $J = 9/2$ state. They estimate a total column density for OH of $10^{16}-10^{17}$ cm^{-2}.

131.191   Observations of CH in the direction of Sgr B2.
B. H. Andrew, L. W. Avery, N. W. Broten.
Astron. Astrophys., Vol. 66, 437 - 439 (1978).

Observations of CH in the direction of Sgr B2 show the presence of all three hyperfine components of the ground state Λ-doubling. There are seven or more velocity components, some arising in the Sgr B2 cloud itself, others in the intervening medium. Column densities are calculated, and confirm that in dense clouds the abundance of CH decreases relative to other molecules.

131.192   On the temperatures of diffuse interstellar clouds.
S. P. Tarafdar.
Observatory, Vol. 98, 115 - 117 (1978).

131.193   Interstellar molecules, galactochemistry and the origin of life.   R. D. Brown.
Interdisciplinary Sci. Rev., Vol. 2, No. 2, p. 124 - 139 (1977). Abstr. in Phys. Abstr., Vol. 81, Abstr. 12338 (1978).

131.194   Interstellar clouds and molecular hydrogen.
M. Jura.
American Sci., Vol. 65, 446 - 454 (1977). – Abstr. in Phys. Abstr., Vol. 81, Abstr. 16491 (1978).

131.195   Global dynamics of the interstellar gas, magnetic field, and cosmic rays.   E. H. Levy.
IAU Symp. No. 77, (see 012.026), p. 57 - 65 (1978).

The author reviews the physical principles which underlie the dynamical equilibrium and stability of a composite system of gas, magnetic field, and cosmic rays. What is of particular concern here are those aspects which control the distribution of magnetic field and cosmic rays, and thus influence the morphology of galaxies as seen in nonthermal radio emission.

131.196   Ionization of interstellar H_2 clouds by supernovae.
T. P. Stecher, D. A. Williams.
Astron. Astrophys., Vol. 67, 115 - 118 (1978).

Two destruction mechanisms for molecular hydrogen in the H I regions of the interstellar medium were given by Stecher and Williams (1967). The second destruction mechanism is dominant in the strong photon fields that exist when a supernova explosion occurs. The authors show that ionization of molecular hydrogen can occur at considerable distances from such supernovae events while the atomic hydrogen remains neutral. The effect should be observable in the dispersion measure of new pulsars.

131.197   Molecular observations of a possible proto-solar nebula in a dark cloud in Taurus.
E. Churchwell, G. Winnewisser, C. M. Walmsley.
Astron. Astrophys., Vol. 67, 139 - 147 (1978).

The authors have detected in TMC 1 (Taurus molecular cloud 1), a small molecular condensation located near the south–eastern edge of Heiles' cloud 2, strong emission from the $J = 9-8$ transition of HC_5N and the $J = 1-0$ transition of HC_3N. In addition they report measurements of the $2_{11}-2_{12}$ transition of H_2CO in absorption, and the $J, K = 1,1$ transition of NH_3 in emission as well as OH observations. The rotation period of TMC 1 is at least 2×10^6 yr and it is probably an example of star formation on a small scale.

131.198   Near-infrared observations of interstellar polarization.   H. M. Dyck, T. J. Jones.
Astron. J., Vol. 83, 594 - 597 (1978).

The authors have observed interstellar linear polarization at three wavelengths between 1 and 2.2 μ for seven stars. They find that the polarization tends to decrease much less rapidly to wavelengths longward of that at which maximum polarization occurs than predicted by Serkowski's empirical relation. The observations appear to be consistent with a model in which the polarization is produced by a mixture of graphite and silicates rather than by pure silicates.

131.199   Cosmic masers.   D. F. Dickinson.
Sci. American, Vol. 238, No. 6, p. 68 - 70, 73 - 79 (1978).

Intense radiation at microwave frequencies is emitted by certain nebular regions and stellar atmospheres. It is generated by maser action, which does for microwaves what laser action does for light.

131.200   Effect of interstellar gas turbulence on coagulation of interstellar dust.   V. S. Kessel'man.
Astron. Zh. Akad. Nauk SSSR, Vol. 55, 482 - 486 (1978). In Russian. English translation in Soviet Astron., Vol. 22, No. 3.

A solution of the coagulation equation for interstellar dust particles in turbulent gas is obtained. The effect of turbulence on the growth of interstellar dust is investigated.

131.201   Cooling of primordial matter by molecular hydrogen.   V. K. Khersonskij, D. A. Varshalovich.
Astron. Zh. Akad. Nauk SSSR, Vol. 55, 487 - 489 (1978). In Russian. English translation in Soviet Astron., Vol. 22, No. 3.

The cooling function of molecular hydrogen, Λ_{H_2}, under high temperatures, $T \sim 6000-10\,000$ K, is calculated. It is shown that Λ_{H_2} is influenced considerably by nonequilibrium effects caused by the difference in the mechanisms of excitation (collisions) and deactivation (mainly radiative) under conditions when the kinetic temperature is much greater than the radiative one. The vibrational excitation of H_2 was found to give the main contribution to the energy outflow and can therefore play an important role in the formation of the first generation of stars from the matter of relict chemical composition.

131.202   A search for emission from vibrationally excited H_2.   W. A. Traub, N. P. Carleton, J. H. Black.
Astrophys. J., Vol. 223, 140 - 146 (1978).

The authors have searched for the 8150 Å emission line of interstellar H_2 in Orion, in NGC 7027, and in the ρ Oph region, with negative results. In order that the results be consistent with the recently reported observations of H_2 emission lines near 2 μm in Orion, the visual extinction must be greater than 10 mag for a 2500 K thermal excitation model, or greater than 6–11 mag for a shock excitation model. Finally, a new ultraviolet excitation model is introduced.

131.203   Radiation transport and the kinematics of molecular clouds.   J. Kwan.
Astrophys. J., Vol. 223, 147 - 160 (1978).

The analysis in this paper hinges on the assumption that most molecular regions observed are characterized by density and temperature distributions which decrease with distance from the center. If a molecular cloud consists of several individual regions (e.g., KL and OMC-2 in the Orion Molecular Cloud, M17 SW, and another CO bright spot in the M17 cloud), the author's analysis is useful only if, around each individual region, the density and temperature decrease with radius. The author gives arguments for the plausibility of this situation. The techniques for calculating the radiative transfer and emergent line profiles are outlined. Characteristic properties of these profiles due to either the systematic motion assumption or the turbulent motion assumption are discussed. Molecular line observations at the M17 SW region and at the Kleinmann-Low nebula are analyzed.

131.204   Infrared colors and the diffuse interstellar bands.
C. Sneden, R. D. Gehrz, J. A. Hackwell, D. G. York, T. P. Snow.
Astrophys. J., Vol. 223, 168 - 179 (1978).

Broad-band infrared photometric measurements have

been gathered for 105 stars which exhibit diffuse interstellar bands in their spectra. All normal stars obey a single reddening law, and a value of $R = 3.08 \pm 0.15$ is derived. No single-valued relation between the color excesses and the diffuse band strengths appears to exist. This casts doubt on whether dust grains which produce the visual and infrared extinctions are the carriers for the diffuse interstellar features.

131.205 Reddening and polarization of young stars: intra-cluster or circumstellar? M. Breger.
Astrophys. J., Vol. 223, 180 - 184 (1978).

Intracluster dust and circumstellar shells have both been used in the recent literature to explain most of the observed polarization of the highly reddened early-type stars in Orion. Because of the apparent contradiction, the physical arguments used are examined. The author finds that the main arguments against the intracluster hypothesis (and in favor of the circumstellar hypothesis) are afflicted by numerical error, are inappropriate, or are inconclusive. While both effects cause polarization and reddening of the early-type stars, intracluster dust effects are considerably more important.

131.206 On the angular momentum in star formation. G. P. Horedt.
Astrophys. Space Sci., Vol. 54, 253 - 261 (1978).

The author discusses the rotation of interstellar clouds which are in a stage immediately before star formation. Cloud collisions seem to be the principal cause of the observed rotation of interstellar clouds. The rotational motion of the clouds is strongly influenced by turbulence. It is suggested that a rotating gas cloud contracts into a ring-like structure which fragments into self-gravitating subcondensations. By collisions and gas accretion these subcondensations accrete into binary systems surrounded by circumstellar clouds. The obtained values are well within the observational limits.

131.207 The sedimentation of grains in interstellar clouds. B. P. Flannery, M. Krook.
Astrophys. J., Vol. 223, 447 - 457 (1978).

Gravitational sedimentation of dust grains is analyzed for two idealized, spherically symmetric models of interstellar clouds: (1) a uniform-density, uniform-temperature model, which permits exact analytic solution, and (2) an isothermal, hydrostatic equilibrium model. An analytic solution is found for sedimentation in a uniform magnetic field; this leads to pancake-shaped dust distribution. The authors present a scenario for the formation of massive clouds, which would imply that interstellar clouds have highly inhomogeneous dust distribution.

131.208 The diffuse interstellar band $\lambda 5780$. E. G. Schmidt.
Astrophys. J., Vol. 223, 458 - 463 (1978).

Photoelectric measurements have been made of the strength of the $\lambda 5780$ interstellar band in the spectra of 50 early-type stars. The purpose of this work is to provide accurate strengths which have reliable error estimates. A comparison with previous work shows serious systematic differences between various sets of data. The strength of the band is correlated with the visible extinction, the strength of the 2200 Å absorption maximum, and the extinction at 1550 Å, but considerable intrinsic scatter is present in all three cases. This seems to indicate that the agent responsible for the continuous diffuse band is different from that responsible for the continuous interstellar absorption.

131.209 Is the solar system entering a nearby interstellar cloud? A. Vidal-Madjar, C. Laurent, P. Bruston, J. Audouze.
Astrophys. J., Vol. 223, 589 - 600 (1978).

A model, based on different observations of the local interstellar medium, indicates the presence of a very close interstellar cloud in front of the Scorpius-Ophiuchus association (almost in the direction of the galactic center) approaching the solar system from a distance of about 0.03 pc at a velocity of about $15-20$ km s^{-1}. These observations are: (1) a H density gradient, (2) the anisotropy of the far-UV flux (around 950 Å), and (3) a large variation of the D/H ratio in two different lines of sight. A mechanism based on a selective radiation pressure effect that acts on deuterium atoms and not on hydrogen atoms explains satisfactorily the large spread in the deuterium abundance in the local interstellar medium. The operation of this mechanism requires that the geometrical configuration remain stable for approximately 10^7 years. This requirement implies the existence of a nearby interstellar cloud.

131.210 The detection of HC$_9$N in interstellar space. N. W. Broten, T. Oka, L. W. Avery, J. M. MacLeod, H. W. Kroto.
Astrophys. J., Lett., Vol. 223, L105 - L107 (1978).

With a molecular weight of 123 amu, and 11 atoms, HC$_9$N (cyano-octatetra-yne) is the heaviest and largest molecule yet detected in interstellar space. The $J = 18 \to 17$ and $J = 25 \to 24$ transitions have been observed in Heiles's Cloud 2 by using a molecular constant obtained by extrapolation from the lighter cyanopolyyne molecules. The column density is estimated to be 3.2×10^{12} cm^{-2}, down by a factor of 4 from that of HC$_9$N in the same source.

131.211 Observations of the $J = 13\text{-}12$ transition of HC$_7$N at 14.7 GHz. F. F. Gardner, J. B. Whiteoak, G. Winnewisser.
Astron. Astrophys., Vol. 67, L23 (1978).

The $J = 13\text{-}12$ rotational transition of cyanohexatriyne (HC$_7$N) at 14.7 GHz was detected towards Heiles Cloud 2 and possibly Sgr B2.

131.212 The pioneering investigations in the field of the interstellar molecules, 1935 - 1942. P. Swings.
Astrophys. Space Sci., Vol. 55, 263 - 265 (1978).

131.213 Extragalactic molecular hydrogen. C. G. Wynn-Williams.
Nature, Vol. 274, 114 - 115 (1978).

131.214 The interstellar wind and its influence on the interplanetary environment. G. E. Thomas.
Annu. Rev. Earth Planet. Sci., Vol. 6, (see 003.011), 173 - 204 (1978).

The author examines the details of the interaction of the Sun's emissions with the interstellar gas. First he discusses the general astronomical setting within which the solar system is imbedded. He next discusses the physical processes by which the interstellar gas and the Sun's emissions mutually interact. Within this framework he then describes a theoretical model of the neutral gas distribution and predicts the backscattered intensity, which will then be compared to the measurements. Finally, he describes the measurements of the ultraviolet emission lines scattered from the two observed interstellar species, hydrogen and helium, and summarizes the more recent observations in terms of what has been learned about the gas density, temperature, and flow velocity.

131.215 Interstellar clouds and molecular hydrogen. M. Jura.
American Sci., Vol. 215, 517 - 520 (1977).

131.216 How stars are born. A. C. Danks, P. A. Shaver.
Messenger, No. 12, p. 17 - 18 (1978).

131.217 Sign-posts of star formation in interstellar clouds south of declination -30 degrees. G. F. Gahm.
Messenger, No. 13, p. 13 - 15 (1978).

131.218 **Investigation of the interstellar absorption in the region of NGC 6866.**   I. A. Dubyago.
Izv. Astron. Ehngel'gardt. Obs., Kazan', No. 41 - 42, p. 215 - 222 (1976). In Russian.

131.219 **Interstellar clouds and the formation of stars.**   H. Alfvén, P. Carlqvist.
Astrophys. Space Sci., Vol. 55, 487 - 509 (1978).
    Contents: I. The new approach to cosmical plasma physics. II. Hydromagnetics of cosmic plasma clouds. III. Star formation in a dusty plasma cloud.

131.220 **Nonaxisymmetric models of collapsing, rotating protostars.**   A. P. Boss, S. J. Peale.
Reports of planetary geology program, 1977−1978, (see 003.014), p. 2 - 4 (1978).

131.221 **Stellar water masers.**   S. G. Kleinmann, D. F. Dickinson, D. G. Sargent.
Bull. American Astron. Soc., Vol. 10, 391 (1978). − Abstract.

131.222 **Five new H_2O masers located near Herbig-Haro objects.**   L. F. Rodriguez, J. M. Moran, D. F. Dickinson, A. L. Gyulbudaghian (*Gyul'budagyan*).
Bull. American Astron. Soc., Vol. 10, 391 (1978). − Abstract.

131.223 **Observations of the hyperfine components of deuterated ammonia.**   D. H. Blake, P. Palmer.
Bull. American Astron. Soc., Vol. 10, 393 (1978). − Abstract.

131.224 **Absorption of nonmetastable NH_3 levels toward W3(OH).**   T. Pauls, T. L. Wilson.
Bull. American Astron. Soc., Vol. 10, 404 (1978). − Abstract.

131.225 **Observations of a narrow HC_5N line in Taurus.**   P. C. Myers, P. T. P. Ho.
Bull. American Astron. Soc., Vol. 10, 404 (1978). − Abstract.

131.226 **Observations of self-reversal in W3IRS5.**   E. J. Brackmann, N. Z. Scoville.
Bull. American Astron. Soc., Vol. 10, 405 (1978). − Abstract.

131.227 **CO emission in the direction of Cas A.**   J. M. Dickey.
Bull. American Astron. Soc., Vol. 10, 405 (1978). − Abstract.

131.228 **Observations of the J = 3→2 transition of $^{12}C^{16}O$ in molecular clouds.**   N. R. Erickson, A. L. Betz.
Bull. American Astron. Soc., Vol. 10, 405 (1978). − Abstract.

131.229 **A search for CaO at mm-wavelengths in stars and molecular clouds.**   E. Churchwell, W. H. Hocking, G. Winnewisser, J. Percival.
Bull. American Astron. Soc., Vol. 10, 405 (1978). − Abstract.

131.230 **Observational consequences of scattering clouds near galaxies.**   M. Jura.
Bull. American Astron. Soc., Vol. 10, 421 (1978). − Abstract.

131.231 **H_2O sources and the formation of OB stars.**   R. Genzel, D. Downes, J. Moran, K. Johnston, J. Spencer, C. Walker, A. Haschick, L. Matveyenko (*Matveenko*), L. Kogan, V. Kostenko, B. Rönnäng, O. Rydbeck, I. Moiseev.
Bull. American Astron. Soc., Vol. 10, 436 (1978). − Abstract.

131.232 **The role of a supernova in star formation.**   S. H. Margolis.
Bull. American Astron. Soc., Vol. 10, 437 (1978). − Abstract.

131.233 **Measurements of interstellar extinction in the Orion nebula from the International Ultraviolet Ex-** plorer.   B. D. Savage, R. C. Bohlin.
Bull. American Astron. Soc., Vol. 10, 445 (1978). − Abstract.

131.234 **Interstellar scattering, the north polar spur, and scintars.**   J. J. Rickard, W. M. Cronyn.
Bull. American Astron. Soc., Vol. 10, 462 - 463 (1978). Abstract.

131.235 **Tidal interactions between interstellar clouds.**   F. O. Clark.
Bull. American Astron. Soc., Vol. 10, 463 (1978). − Abstract.

131.236 **Polarization properties of T Tauri stars.**   P. Bastien, J. D. Landstreet.
Bull. American Astron. Soc., Vol. 10, 463 (1978). − Abstract.

131.237 **Polarization from multiple scattering by dust.**   R. L. White.
Bull. American Astron. Soc., Vol. 10, 463 (1978). − Abstract.

131.238 **Evolution of interstellar dust in the solar neighborhood.**   E. Dwek, J. M. Scalo.
Bull. American Astron. Soc., Vol. 10, 463 - 464 (1978). Abstract.

131.239 **Isotopic abundances in interstellar carbon monosulfide.**   M. A. Frerking, R. W. Wilson, R. A. Linke, P. G. Wannier.
Bull. American Astron. Soc., Vol. 10, 464 (1978). − Abstract.

131.240 **New interstellar diffuse features in the wavelength region 6500-8900 Å.**   F. Sanner, R. Snell, P. Vanden Bout.
Bull. American Astron. Soc., Vol. 10, 464 (1978). − Abstract.

131.241 **A new determination of the interstellar Faraday rotation.**   M. Simard-Normandin, P. P. Kronberg.
Bull. American Astron. Soc., Vol. 10, 465 (1978). − Abstract.

131.242 **A survey of interstellar C I.**   E. B. Jenkins, E. J. Shaya.
Bull. American Astron. Soc., Vol. 10, 465 (1978). − Abstract.

131.243 **Optical interstellar lines in the spectra of the Pleiades.**   R. E. White.
Bull. American Astron. Soc., Vol. 10, 465 (1978). − Abstract.

131.244 **B stars and the structure of the interstellar medium.**   T. M. Bania, J. G. Lyon.
Bull. American Astron. Soc., Vol. 10, 465 (1978). − Abstract.

131.245 **Interstellar absorption lines toward Cygnus OB2.**   S. P. Souza, B. L. Lutz.
Bull. American Astron. Soc., Vol. 10, 465 (1978). − Abstract.

131.246 **A model for the maser sources associated with H II regions.**   M. Elitzur, T. de Jong.
Astron. Astrophys., Vol. 67, 323 - 332 (1978).
    The authors suggest that the OH maser spots associated with compact H II regions are located at the compressed shell in between the shock and ionization fronts. The authors describe the relevant OH observations and the general phenomenological framework and the model − both the physical conditions in the shell and the details of the chemistry. The results of the calculations are presented − both for a young and a relatively old maser source. The authors compare various pumping rates and conclude that the OH masers have to be collisionally pumped. They also discuss the uncertainties in the model.

131.247 **High-resolution mapping of the H I absorption lines in the direction of NGC 2024, Orion A, M 17 and**

W 49.   I. A. Lockhart, W. M. Goss.
Astron. Astrophys., Vol. 67, 355 - 372 (1978).

A cloud fitting procedure has been developed; this procedure as applied to Orion A and W 12 leads to clouds with typical sizes of $\sim$1.0 pc, H I densities $\sim$125 cm^{-3} and H I masses of $\sim$13 $M_{\odot}$. For Orion A at least two clouds at -3 km s^{-1} have been detected, and these objects may be physically associated with the H II region. Near the dark lane and bay in Orion there is an enhanced area of H I absorption. In the direction of W 49 the distribution of H I is quite complex. The evidence suggests that the distance of W 49 B is in the range 12.5 to 14 kpc. A physical association between the two sources W 49 A and W 49 B cannot be ruled out.

131.248 Models of clouds in adiabatic gravitational contraction.   W. G. L. Pöppel.
Rev. Math. Fis. Teor., Vol. 21, 173 - 183 (1971) = Contrib. Inst. Argentino Radioastron., No. 33.

Models of spherical clouds of atomic hydrogen in adiabatic gravitational contraction, with initially uniform distributions of temperature and density, were computed. The models could be appliable to the first (and also most extended) stages of the contraction of a primordial cloud.

131.249 Evolution of maser sources.   V. V. Burdjuzha (*Burdyuzha*), T. V. Ruzmaikina (*Ruzmajkina*).
IAU Colloq. No. 45, (see 012.032), p. 53 - 56 (1977/78).

131.250 Interstellar and stellar abundances across the galactic disk.   M. Peimbert.
IAU Colloq. No. 45, (see 012.032), p. 149 - 159 (1977/78).

Observational evidence related to the chemical composition across the disk of the Galaxy is reviewed. The H$_2$ density distribution derived for the Galaxy is poorly known, consequently it is still not possible to compare theoretical models of the chemical evolution of the Galaxy with the gaseous density distribution. The H$_2$ density distribution is particularly sensitive to the fraction of carbon atoms embedded in CO molecules and to the possible presence of a C/H abundance gradient.

131.251 Carbon stars and interstellar dust.   S. Krawczyk, J. Krempeć, J. Gertner.
IAU Colloq. No. 45, (see 012.032), p. 181 - 185 (1977/78).

131.252 Planetary nebulae and nitrogen enrichment in the Galaxy.   D. C. V. Mallik.
IAU Colloq. No. 45, (see 012.032), p. 187 - 191 (1977/78).

131.253 Encounters with interstellar clouds: the effect on stellar winds and the possibilities for accretion.
M. J. Newman, R. J. Talbot, Jr.
Ann. New York Acad. Sci., Vol. 302, (see 012.033), 665 - 668 (1977).

131.254 HC$_7$N a winner – interstellar heavyweight crowned.   W. Cherwinski.
Sci. Dimension, Vol. 9, No. 5, p. 24 - 26 (1977). – Abstr. in Phys. Abstr., Vol. 81, Abstr. 41029 (1978).

131.255 Molecole interstellari e cometarie.
C. B. Cosmovici, E. D'Anna.
Mem. Soc. Astron. Italiana, Vol. 47, 39 - 63 (1976).

Contents: Scoperta delle molecole interstellari e loro importanza. Reazioni chimiche nella fase gassosa. Formazione delle molecole interstellari. Fotochimica, campo di radiazione interstellare e durata di vita delle molecole. Molecole cometarie. Prospettive per il futuro.

131.256 Nubi fredde e grani interstellari.
E. Bussoletti, A. Borghesi.
Mem. Soc. Astron. Italiana, Vol. 47, 125 - 158 (1976).

131.257 Problemi di osservazione della materia interstellare nel visible e nell'uv.   P. J. Treanor.
Mem. Soc. Astron. Italiana, Vol. 47, 167 - 176 (1976).

131.258 UV observations and stellar evolution.
R. Kippenhahn.
Mem. Soc. Astron. Italiana, Vol. 47, (see 012.036), 337 - 364 (1976).

Contents: Star formation and T Tauri stars. Early-type stars. Binary evolution. Chromospheric activity.

131.259 Dust, helium abundance.   L. F. Smith.
Mem. Soc. Astron. Italiana, Vol. 47, (see 012.036), 407 - 415 (1976).

131.260 A UV picture of the gas in the interstellar medium.   D. G. York.
Mem. Soc. Astron. Italiana, Vol. 47, (see 012.036), 493 - 551 (1976).

The nature of diffuse interstellar regions, having volume densities from 10^5 cm^{-3} to less than 0.01 cm^{-3} is discussed. The author concludes that 1) there is no diffuse intercloud medium of any significance in terms of mass or pressure, with $T_e < 10^{5\circ}$K; 2) relative abundances, while uncertain, show similar qualitative patterns in both reddened and unreddened lines of sight; 3) rather large pressure differences between various types of interstellar regions are inferred formally, so that it certainly cannot be demonstrated that the interstellar medium is in pressure equilibrium. Brief discussion of molecular processes and heating processes in dense clouds is included. Important problems requiring future UV work are mentioned.

131.261 The interstellar extinction law.   R. J. Davis.
Mem. Soc. Astron. Italiana, Vol. 47, (see 012.036), 631 (1976). – Summary.

The Nançay survey of absorption by galactic neutral hydrogen. I. Absorption towards extragalactic sources.
See Abstr. 002.039.

An atlas of galactic neutral hydrogen for the region $270^\circ \leqslant l \leqslant 310^\circ$; $-7^\circ \leqslant b \leqslant 2^\circ$.   See Abstr. 002.062.

Observations in the 21-cm neutral hydrogen line. I. The region $220^\circ \leqslant l \leqslant 294^\circ$; $-29^\circ \leqslant b \leqslant -11^\circ$. II. The region around the south celestial pole.   See Abstr. 002.063.

Interstellar medium and the origin of stars.
See Abstr. 003.069.

Physical processes in the interstellar medium.
See Abstr. 003.130.

Vector space methods of photometric analysis: applications to O stars and interstellar reddening.
See Abstr. 021.012.

OH main lines masers I : OH/IR stars.
See Abstr. 022.009.

Vibrational excitation of H$_2$ in intense ultraviolet fluxes.   See Abstr. 022.010.

Calculated structures and microwave frequencies of HNSi and HSiN.   See Abstr. 022.011.

Theoretical study of isocyanoacetylene and the isocyanoethynyl radical.   See Abstr. 022.016.

A review of spectroscopic information in the visible and ultraviolet region for diatomic molecules of astrophysical interest.   See Abstr. 022.019.

Theoretical study of HSiO$^+$ and HOSi$^+$.
See Abstr. 022.035.

Theoretical study of the thioformyl ion.
See Abstr. 022.036.

Polyatomic ion-electron reactions.
See Abstr. 022.042.

Calculation of the cross sections for the CIV-H, NIV-H, and SIII-H charge-exchange collisions. Significance for the interstellar medium.   See Abstr. 022.043.

Experimental condensation of magnesium silicate grains.   See Abstr. 022.044.

Line positions and oscillator strengths of rotation-vibration bands of possible interstellar SiH and SiH$^+$.
See Abstr. 022.081.

Molecular collision processes. II. Excitation of the fine-structure transition of C$^+$ in collisions with H$_2$.
See Abstr. 022.096.

Reactions of CH$_n$$^+$ ions with molecules at 300K.
See Abstr. 022.116.

Binary and ternary reactions of CH$_3$$^+$ ions with several molecules at thermal energies.   See Abstr. 022.117.

Matrix-isolation applied to high-temperature and interstellar molecules: a review.
See Abstr. 022.119.

Carbon in space is tracked with a mixer-receiver.
See Abstr. 033.030.

Polarization effects in grating scanners.
See Abstr. 034.007.

Isotope fractionation under simulated space conditions.   See Abstr. 061.054.

The effects of small-angle scattering on a pulse of radiation with an application of X-ray bursts and interstellar dust.   See Abstr. 063.021.

Ammonia and OH near T Tauri stars.
See Abstr. 064.028.

Discovery of linearly polarized continuum light scattered by the shell around Alpha Orionis.
See Abstr. 064.101.

The shock wave interpretation of individual condensations in Herbig-Haro objects.   See Abstr. 065.030.

On the surface composition of thermally pulsing stars of high luminosity and on the contribution of such stars to the element enrichment of the interstellar medium.
See Abstr. 065.032.

Consequences of a past encounter of the earth with an interstellar cloud.   See Abstr. 082.129.

Low energy cosmic ray erosion of ice grains in interplanetary and interstellar media.
See Abstr. 106.009.

An upper limit for the flux of interstellar dust

grains in the solar system.   See Abstr. 106.025.

Neutral interstellar gas in the vicinity of the planets.   See Abstr. 106.026.

From dark clouds to planets and satellites.
See Abstr. 107.043.

Structure and evolution of giant gaseous proto-planets.   See Abstr. 107.044.

Classification of stars according to unbroadened low-dispersion spectra and some results of its application to the study of galactic structure.   See Abstr. 114.052.

The spectrum of the bright YY Orionis star CoD $-35°10525$.   See Abstr. 114.514.

The spectral and spatial distribution of radiation from Eta Carinae. I. A spherical dust shell model approach. See Abstr. 114.519.

The origin of multiple stars by condensation in dense nebulae.   See Abstr. 117.055.

The variable linear polarization of the O-star binary HD 47129 (Plaskett's star).   See Abstr. 117.070.

A large-scale OH sky survey at 1612 MHz. Part II. The galactic distribution and kinematics of the unidentified type II OH/IR stars.   See Abstr. 122.063.

Radio molecular line observations of late type stars. See Abstr. 122.086.

Enhanced molecular abundances and ionization levels in molecular clouds near supernova remnants.
See Abstr. 125.024.

Supernovae in molecular clouds.
See Abstr. 125.026.

Supernova remnants: what do they tell us about massive star evolution and the interstellar medium?
See Abstr. 125.056.

Old supernova remnants as magnetohydrodynamic wave fronts. Formation of filamentary structure.
See Abstr. 125.071.

CO and H$_2$O observations of the H II region NGC 281.   See Abstr. 132.004.

Ammonia emission and absorption toward W3 (OH).
See Abstr. 132.006

Dust clouds in H II regions. The Dragon in M8.
See Abstr. 132.010.

H$_2$ dissociation fronts outside expanding compact H II regions.   See Abstr. 132.018.

Far infrared lamellar grating observations of H II regions/molecular clouds.   See Abstr. 132.020.

Recombination lines from W33. Evidence for a cool H II region with mass flow?   See Abstr. 132.034.

Observations of NH$_3$ in southern sources.
See Abstr. 132.037 .

Carbon monoxide observations of molecular gas at

the boundary of H II regions.   See Abstr. 132.067.

Radial diameters of Type II OH/IR sources.
See Abstr. 133.007

Models of infrared emission from dusty and diffuse
H II regions.   See Abstr. 133.010.

Near-infrared polarimetry of compact infrared
sources associated with H II regions and molecular clouds.
See Abstr. 133.012.

Infrared observations of the galactic center. IV.
The interstellar extinction.   See Abstr. 133.014.

Accurate interferometer positions of H_2O masers.
See Abstr. 133.018.

Discovery of $20\,\mu m$ sources associated with
molecular clouds.   See Abstr. 133.024.

The possibility of detecting DC.
See Abstr. 133.029.

Calculations of infrared fluxes from galactic sources
for a polysaccharide grain model.   See Abstr. 133.031.

Radio identification of an infrared object associated
with the passage of a shock front.   See Abstr. 133.039.

Star formation at a front: far-infrared observations
of AFGL 333.   See Abstr. 133.050.

CO observations along an ionization front of the
California Nebula (NGC 1499).   See Abstr. 134.001.

2.1 micron H_2 emission: high-spectral-resolution
observations of the Orion Nebula.   See Abstr. 134.003.

Planetary nebulae and the interstellar medium.
See Abstr. 135.088.

21-cm observations of NGC 7822.
See Abstr. 141.210.

18-cm OH observations of W33A.
See Abstr. 141.225.

Interstellar scattering of pulsars by a turbulent
interstellar plasma.   See Abstr. 141.507.

The Cygnus X region. XI. Maps of visual extinction A_V.   See Abstr. 142.147.

Gamma rays from dense interstellar clouds.
See Abstr. 142.713.

Star formation rates in normal and peculiar galaxies.
See Abstr. 151.001.

A model for winds from galactic disks.
See Abstr. 151.018.

On star formation and galactic evolution.
See Abstr. 151.023.

Stochastic star formation and spiral structure of
galaxies.   See Abstr. 151.075.

Dynamics of the gas in spiral galaxies.
See Abstr. 151.108.

Chemical inhomogeneities in galaxies.
See Abstr. 151.113.

Observational tests on star formation IV: birthplaces of massive stars in open star clusters.
See Abstr. 153.004.

Differential reddening in the young cluster NGC
2264.   See Abstr. 153.035.

Search for neutral hydrogen in the galactic cluster
NGC 2287.   See Abstr. 153.043.

An unsuccessful search for carbon monoxide in
globular clusters.   See Abstr. 154.004.

Nova-driven winds in globular clusters.
See Abstr. 154.027.

Carbon monoxide in the Galaxy. III. The overall
nature of its distribution in the equatorial plane.
See Abstr. 155.004

Probing our Galaxy.   See Abstr. 155.010.

UBV surface brightness photometry of the Milky
Way in Scorpius from the space probe Helios 1.
See Abstr. 155.022.

Eruptive phenomena near the galactic centre.
See Abstr. 155.024.

Spatial distribution of stars and interstellar dust in
the direction of the galactic anticenter.   See Abstr. 155.027.

Structure galactique et formation stellaire dans le
bras spiral de la Carène.   See Abstr. 155.030.

Abundances and chemical evolution of the galactic
center.   See Abstr. 155.045.

Vertical and radial distributions of gas and young
objects in the Galaxy and Schmidt's law of star formation.
See Abstr. 155.051.

Scattering of radio emission from the compact object in Sagittarius A.   See Abstr. 156.014.

Gas distribution, motions and dynamics for some
dwarf irregular galaxies.   See Abstr. 158.017

Further detections of H_2O emission from external
galaxies.   See Abstr. 158.040.

Polarization in elliptical galaxies?
See Abstr. 158.235.

Observaciones de galaxias en la línea de 21 cm del
hidrógeno neutro con el instrumento de síntesis de Westerbork (Holanda).   See Abstr. 158.271.

The extragalactic background light and slow star
formation in galaxies.   See Abstr. 162.022.

High redshift 21-cm lines.   See Abstr. 162.052.

131.901 Erratum: "The energetics of molecular clouds. I.
Methods of analysis and application to the S255
molecular cloud" [Astrophys. J., Vol. 217, 448 - 463 (1977)].
N. J. Evans II, G. N. Blair, S. Beckwith.
Astrophys. J., Vol. 221, 382 (1978).

132 H I, H II Regions

132.001 Infrared studies of 30 Doradus. I. The 2-μ sources of the inner region. A. R. Hyland, J. A. Thomas, G. Robinson.
Astron. J., Vol. 83, 20 - 25 (1978).

A 2-μ survey of the central region of 30 Doradus has revealed a significant population of M supergiants intermingled with the well-known bright WN and OB stars. The results suggest that the 30 Doradus region has undergone two major bursts of star formation in the last 10^7 yr.

132.002 High-resolution radio maps of three H II regions in the Perseus arm. M. Birkinshaw.
Mon. Not. R. Astron. Soc., Vol. 182, 401 - 410 (1978).

The Cambridge 5-km telescope has been used to make high-resolution radio maps of Sh 156 (IC 1470) and Sh 159 at 5 and 15 GHz, and Sh 157 at 5 GHz. Bright compact regions are found inside optical nebulae, asymmetrically placed about the exciting stars. Radiation pressure from these stars dominates the dynamics of the nebulae, and may produce the radio emission minima near their centres.

132.003 Selective absorption of Lyman continuum photons by dust in H II regions.
N. Panagia, L. F. Smith.
Astron. Astrophys., Vol. 62, 277 - 282 (1978).

The correlation between the ionized He abundance and the infrared excess from H II regions is compared with that expected if the effect is due to selective absorption of He-ionizing photons by dust in the H II regions. Comparison with the IR observations of Emerson and Jennings (1977) yields a value $\alpha_0 = \sigma_{He}/\sigma_H = 4 \pm 1$ for the ratio of absorption cross sections for He- and H-ionizing photons. This value can be reduced to 2.5 or less if the number ratio of He- to H-ionizing photons from the ionizing stars is significantly less than predicted by Auer and Mihalas (1972) non-LTE model atmospheres.

132.004 CO and H_2O observations of the H II region NGC 281. B. G. Elmegreen, C. J. Lada.
Astrophys. J., Vol. 219, 467 - 473 (1978).

CO observations of the region around NGC 281 show two molecular cloud fragments that appear to be separated by a channel of ionized gas at the southern edge of the H II region. An H_2O maser has been discovered at the northern edge of one cloud, near the interface between the cloud and the H II region. Near the maser position, ^{12}CO temperatures and line widths reach a maximum value. A ridge of enhanced continuum emission from the H II region is also found to be coincident with the maser position, suggesting that the maser source is adjacent to an ionization front which is progressing into the back side of the molecular cloud. The dynamical history of the region is considered.

132.005 Detection of the impact broadening of a recombination line in W51. K. R. Lang, R. F. Willson.
Astrophys. J., Vol. 219, 474 - 476 (1978).

Observations of the 166α and 239γ recombination line profiles of W51 show that the γ line is twice as wide as the α line, although the two lines are at nearly the same frequency. This extra broadening can be explained in terms of the theory of impact broadening by inelastic electron collisions, if the electron density is 10^3 cm^{-3}.

132.006 Ammonia emission and absorption toward W3(OH). T. L. Wilson, J. Bieging, D. Downes.
Astron. Astrophys., Vol. 63, 1 - 6 (1978).

The authors have observed both absorption and emission from the (1,1), (2,2) and (3,3) metastable levels of NH_3 to-ward the compact H II region W3(OH). The intensity ratios of the hyperfine satellite components in the absorption lines are consistent with an optically thick gas in local thermodynamic equilibrium, whereas the same intensity ratios in the emission lines imply that the emitting gas toward W3(OH) is optically thin. The true optical depth of the NH_3 absorption, obtained from the ratio of the hyperfine components is > 3. This value is ~ 10 times the apparent optical depth, and implies that the continuum source is not uniformly covered by the absorbing gas. The total column density of NH_3 emission toward W3(OH) is $4 \cdot 10^{14}$ cm^{-2} and the column density of the NH_3 seen in absorption is 10^2 times larger.

132.007 The obscuration and excitation of the North American (NGC 7000) and Pelican (IC 5070) Nebular Complex. C. Goudis, P. G. Johnson.
Astron. Astrophys., Vol. 63, 259 - 263 (1978).

A contour map of the obscuration of light over the North American and Pelican Nebular Complex has been produced by comparing the thermal radio brightness with the Hα brightness of the complex. An overall correlation of the distinct features of the obscuring cloud, the various Hα/[N II] maxima and the similar maxima of the thermal radio emission is also shown and the implications of this correlation are discussed.

132.008 Observations of H II regions near OH 43.79−0.13 and OH 48.61 + 0.02.
H. E. Matthews, P. A. Shaver, W. M. Goss, H. J. Habing.
Astron. Astrophys., Vol. 63, 307 - 311 (1978).

Observations of two H II region complexes near the Type I OH sources OH 43.79−0.13 and OH 48.61 + 0.02 have been made using the Westerbork Synthesis Telescope at 21 and 6 cm. The resolution is 25" and 7" in right ascension at the two wavelengths, respectively. Compact continuum sources have been detected in the direction of both OH sources.

132.009 Aperture synthesis observations of galactic H II regions. VIII. S106 and S235: regions of star formation. F. P. Israel, M. Felli.
Astron. Astrophys., Vol. 63, 325 - 334 (1978).

The authors observed the peculiar emission nebulae S106 (M1−19) and S235 (M1−82) with the Westerbork Synthesis Radio Telescope at frequencies of 5.0 and 1.4 GHz. Their analysis indicates that S106 consists of two different ionization front structures, and two (possibly three) obscured compact components. The area around S235, which contains also the nebulosities S231, S232 and S233, is studied. The identification of the radio sources is obtained taking also into account lower resolution observations. Values for the parameters of the identified H II regions are given.

132.010 Dust clouds in H II regions. The Dragon in M8. P. W. J. L. Brand, W. J. Zealey.
Astron. Astrophys., Vol. 63, 345 - 352 (1978).

A small area of M8 including an elongated dust cloud has been mapped at 8 μm spacings on a UK 48" Schmidt plate in the blue, using the COSMOS measuring machine. The map has been calibrated to relative intensity, and scans of intensity made across the cloud boundary. The apparently sharp edge is resolved, and after photographic effects are dealt with the structure of the edge is used to provide a best estimate of dust density in the cloud.

132.011 Optical polarization measurements in W 3 and M 17. A. Schulz, K. Proetel, T. Schmidt.
Astron. Astrophys., Vol. 64, L13 - L15 (1978).

The authors present observations of the polarization and brightness of stars inside the dusty H II regions W 3 and M 17, and of the interstellar foreground polarization. The intrinsic polarization of W 3 and M 17 objects rises to 16% and 29%, respectively, a regular alignment of polarization angles being obvious in both cases.

132.012 The structure and helium abundance of G 0.7–0.0 (Sgr B2).   C. Thum, P. G. Mezger, V. Pankonin, J. Schraml.
Astron. Astrophys., Vol. 64, L17 - L19 (1978).

Continuum scans of G 0.7–0.0 (Sgr B2) at 22 GHz reveal that one of the compact thermal components has a turnover frequency greater than 10 GHz. Hydrogen and helium recombination lines at 14.7 and 22.4 GHz were observed, and the ratio of ionized helium to ionized hydrogen was measured to be 0.050 and 0.062 at these frequencies. The data indicate that the fractional ionization of helium is different for the various components of G0.7–0.0.

132.013 Abundance gradients in the Galaxy derived from H II regions.
M. Peimbert, S. Torres-Peimbert, J. F. Rayo.
Astrophys. J., Vol. 220, 516 - 524 (1978).

Photoelectric spectrophotometry of emission lines in the 3700–7400 Å range is presented for the η Carina nebula, NGC 2359, and NGC 2467. From these observations the chemical abundances of He, N, O, Ne, S, Ar, and Cl relative to hydrogen are derived. These abundances are used, together with those of the Orion Nebula and M8, to obtain abundance gradients across the disk of the Galaxy.

132.014 Far-infrared observations of H II regions near the galactic center.
I. Gatley, E. E. Becklin, M. W. Werner, D. A. Harper.
Astrophys. J., Vol. 220, 822 - 830 (1978).

This paper presents and discusses far-infrared observations of nine H II regions within 1° of the galactic center, including Sgr A, Sgr B2, Sgr C, and G0.5–0.0. The far-infrared luminosity, color temperature, and optical depth of these regions and the ratio of infrared flux to radio-continuum flux lie in the range characteristic of spiral-arm H II regions. The far-infrared results are therefore consistent with the idea that the galactic center H II regions are ionized by luminous, early-type stars. Steep, systematic gradients in far-infrared color temperature and optical depth are seen along the galactic plane between Sgr B2 and G0.5–0.0; the appearance of this area is similar to that of regions of star formation in the spiral arms.

132.015 The structure of the emission nebula M 1–78 at 15.4 GHz.   P. F. Scott, S. Harris.
Mon. Not. R. Astron. Soc., Vol. 182, 657 - 660 (1978).

It is not known whether M 1–78 is a planetary nebula or a young H II region. In an attempt to settle this issue, it has been observed at a frequency of 15.4 GHz with an angular resolution 0.65 × 0.83 arcsec and the optical position of the nebula has been re-measured. It is concluded that the evidence is equivocal and the nature of M 1–78 remains in doubt.

132.016 A survey of HII regions in M 31.
A. Pellet, N. Astier, A. Viale, G. Courtès, A. Maucherat, G. Monnet, F. Simien.
Astron. Astrophys., Suppl. Ser., Vol. 31, 439 - 461 (1978).

An Hα survey, using a focal reducer and narrow-band interference filters, has allowed the authors to detect nearly a thousand HII regions in M 31. The catalogue gives their positions, dimensions, and surface brightness. The general distribution of diameters is highly analogous to that of the Galaxy, but differs from that of M 33. Some HII regions of M 31 and M 33, chosen either on account of their large sizes or high brightness, have been analysed in detail; their emission measures and excitation parameters have been calculated. The dimensions of the largest HII regions show that the Galaxy is intermediate between M 31 and M 33, the intensities of the

HII regions of M 31 being 3.5 times lower than those of M 33.

132.017 Observations of [O III] 88 micron line emission from H II regions and the galactic center.
F. W. Dain, G. E. Gull, G. Melnick, M. Harwit, D. B. Ward.
Astrophys. J., Lett., Vol. 221, L17 - L21 (1978).

The authors have detected the [O III] 88 μm line in seven H II regions, M17, W51, M8, M42, and W3, and two components of NGC 6357, as well as in the galactic center region Sgr A. They also have obtained an upper limit for the line intensity in Sgr B2. For the H II regions they obtain reasonable agreement between their observations and predictions based on a theory of line intensities developed by Petrosian and Simpson. This theory has been applied to H II region models based on radio observations. The line intensities for Sgr A and Sgr B2 are far below those expected from the radio models, but may be consistent with the low ionized helium content of these regions if the sources lack energetic photons. The results indicate that in some H II regions the emission from far-infrared fine-structure lines should amount to several percent of the total infrared flux.

132.018 H$_2$ dissociation fronts outside expanding compact H II regions.   D. J. Hollenbach, J. K. Hill.
Bull. American Astron. Soc., Vol. 9, 555 (1978). – Abstract.

132.019 A search for ^{3}He in H II regions and planetary nebulae.   R. T. Rood, T. L. Wilson, G. Steigman.
Bull. American Astron. Soc., Vol. 9, 575 (1978). – Abstract.

132.020 Far infrared lamellar grating observations of H II regions/molecular clouds.
J. G. Duthie, T. L. Herter, M. P. Savedoff, J. L. Pipher.
Bull. American Astron. Soc., Vol. 9, 581 - 582 (1978). Abstract.

132.021 High resolution observations of the H I cloud associated with IC5146 (S125).
J. A. Irwin, R. S. Roger.
Bull. American Astron. Soc., Vol. 9, 600 (1978). – Abstract.

132.022 Near-infrared and CO observations of W40 and W48.   M. Zeilik, C. J. Lada.
Bull. American Astron. Soc., Vol. 9, 606 (1978). – Abstract.

132.023 Comparison of far-infrared luminosities of H II regions with luminosity of exciting stars.
V. A. Hughes, J. P. Vallee, M. R. Viner.
Bull. American Astron. Soc., Vol. 9, 606 - 607 (1978). Abstract.

132.024 Observations of [O III] 88μ line emission from H II regions and the galactic center.
F. W. Dain, G. E. Gull, G. Melnick, M. Harwit, D. B. Ward.
Bull. American Astron. Soc., Vol. 9, 612 (1978). – Abstract.

132.025 Model H II regions derived from ATLAS 5 and ATLAS 6 stellar atmospheres.
S. Torres-Peimbert, I. Cruz-Gonzales, M. Peimbert.
Bull. American Astron. Soc., Vol. 9, 612 (1978). – Abstract.

132.026 H I absorption spectra towards galactic H II regions.   Γ. J. Lockman, E. W. Greisen.
Bull. American Astron. Soc., Vol. 9, 615 (1978). – Abstract.

132.027 Far infrared observations of extragalactic H II regions.
P. M. Harvey, I. Gatley, H. A. Thronson.
Bull. American Astron. Soc., Vol. 9, 629 (1978). – Abstract.

132.028 Observations of M17 and NGC 2024 at 21 cm.
W. Löbert, W. M. Goss.

Mon. Not. R. Astron. Soc., Vol. 183, 119 - 127 (1978).

The H II regions M17 and NGC 2024 (W12) have been mapped at 21 cm with a resolution of ~ 50 arcsec using the Fleurs synthesis telescope. The northern source in M17 is resolved into three components and at the northeastern edge shows good correspondence with the structure of the nebula as observed in the lines of Hα and [N II]. The sharp optical and radio boundaries in this region are suggestive of an ionization front. The source NGC 2024 is shown to be a double. The two individual components have rms electron densities of ~ $2 \times 10^3 \, cm^{-3}$ and masses of ionized hydrogen of ~ 0.2 $M_\odot$.

132.029 Observations of negative velocity hydrogen in IC 443. L. K. DeNoyer.
Mon. Not. R. Astron. Soc., Vol. 183, 187 - 193 (1978).

Negative velocity hydrogen associated with the supernova remnant IC 443 was observed with the Cambridge Half-Mile telescope. There are several H I condensations with size < 1 × 3 pc, density > 200 cm^{-3}, and velocity width > 40 km/s. The condensations are coincident with those radio continuum features which do not have associated optical filaments. It is suggested that the condensations are within the remnant.

132.030 Two-micron line emission from the H II region G333.6−0.2. C. G. Wynn-Williams, E. E. Becklin, K. Matthews, G. Neugebauer.
Mon. Not. R. Astron. Soc., Vol. 183, 237 - 244 (1978).

Spectrophotometry of the H II region G333.6−0.2 with $\lambda/\Delta\lambda \simeq 100$ shows strong emission lines of hydrogen and helium super-imposed on a continuum. Spatial variations in the equivalent widths of the lines suggest the presence of very hot (≳ 600 K) dust grains within the region. The shape of this very powerful H II region indicates that its dense ionized core is being continuously replenished from a reservoir of neutral gas.

132.031 On the abundance of sulphur in H II regions. B. E. J. Pagel.
Mon. Not. R. Astron. Soc., Vol. 183, 1P - 4P (1978).

Sulphur abundances determined from [S III] λ 6312 using the currently favoured ionization correction procedure in Orion and in the H II regions of the Magellanic Clouds show a marked trend with the degree of ionization measured by O^{++}/O. An explanation is suggested and used to propose an improved ionization correction scheme. The abundance ratio of sulphur to oxygen in both LMC and SMC H II regions is about the same as in Orion.

132.032 Westerbork 21-cm observations of the low and high velocity H I in Stephan's Quintet. R. J. Allen, W. T. Sullivan III.
CNRS Colloq. No. 263, (see 012.009), p. 445 - 450 (1977).

Des cartes à haute résolution d'hydrogène neutre à basse et haute vitesses dans le quintet de Stephan sont présentées. L'émission à basse vitesse a lieu comme attendue, dans la galaxie NGC 7320, et n'apparait pas inhabituelle dans ses caractéristiques. L'hydrogène neutre à haute vitesse, s'il est situé à 120 Mpc, a une masse de 1.6 × $10^{10} M_\odot$ et provient d'un grand nuage d'une taille de ~150 kpc décalé de ~100 kpc (3′ E) par rapport à la position du quintet. Les galaxies spirales à haute vitesse du quintet étant extrêmement pauvres en hydrogène, les auteurs suggèrent que le nuage H I est le résultat d'une collision de type Spitzer-Baade de deux galaxies auparavant riches en gaz.

132.033 H I observations of interacting groups of galaxies. G. S. Shostak.
CNRS Colloq. No. 263, (see 012.009), p. 489 - 496 (1977).

L'auteur teste l'idée d'Arp à savoir que des groupes multiples de galaxies en intéraction sont éjectés hors de systèmes brillants proches de nous. Les observations à 21 centimètres indiquent que, si ceci est le cas, les membres du groupe en intéraction ont pour la plupart des masses H I comparables ou inférieures aux galaxies naines, bien que la largeur des raies et la morphologie de ces galaxies suggèrent des masses totales normales. D'autres conséquences de l'hypothèse d'éjection sont brièvement discutées.

132.034 Recombination lines from W33. Evidence for a cool H II region with mass flow?
J. H. Bieging, V. Pankonin, L. F. Smith.
Astron. Astrophys., Vol. 64, 341 - 349 (1978).

Radio recombination line spectra have been observed at several positions in W33 at 1.7, 5, and 8.6 GHz, and the continuum emission was mapped at 8.6 GHz. In the lines, two velocity components, which are blended in the observed profiles, are extended across most of the H II region complex. A third velocity component is more localized. The velocity components of the recombination lines may be caused by ordered flows of ionized gas surrounding individual O-stars and interacting with the molecular clouds known to be associated with W33.

132.035 The helium problem in Sagittarius B2.
E. J. Chaisson, S. M. Lichten, L. F. Rodriguez.
Astrophys. J., Vol. 221, 810 - 815 (1978).

New observations of hydrogen and helium radio recombination lines emitted from the Sgr B2 nebula near the galactic center have yielded values for an ionized abundance ratio, $N(He^+)/N(H^+)$, considerably less than 10% by number. These measurements are consistent, however, with a normal total helium-to-hydrogen ratio of 8%–10%, provided the nebular dust preferentially absorbs helium-ionizing photons, giving rise to a region of He II with a radius about 80% that of H II. The authors' model is consistent with a dust absorptivity which is a smoothly increasing function of frequency shortward of 912 Å; an absorptivity that increases dramatically near 500 Å, as has been previously suggested, is not required by the observations.

132.036 A comparative study of high-radiofrequency and far-infrared observations of galactic H II regions.
L. F. Rodriguez, E. J. Chaisson.
Astrophys. J., Vol. 221, 816 - 824 (1978).

Nine galactic H II regions were mapped at 23.4 GHz (13 mm) with an angular resolution of 80″ and a positional accuracy of 15″. The integrated flux densities agree with the expected values extrapolated from lower-frequency observations. The positions and deconvolved Gaussian angular sizes permit a derivation and discussion of the physical parameters of several heretofore unstudied source components. A comparison with similarly resolved far-infrared maps suggests that the far-infrared emission arises generally from both inside and outside the ionized volume and that strong dust depletion, if present at all, is confined to an inner fraction of the Strömgren radius.

132.037 Observations of NH_3 in southern sources.
R. A. Batchelor, F. F. Gardner, S. H. Knowles, U. Mebold.
Proc. Astron. Soc. Australia, Vol. 3, 152 - 154 (1977).

The authors report the results of a preliminary southern hemisphere investigation of ammonia in the direction of H II regions using the inner 17-m surface of the Parkes telescope.

132.038 The analysis of radio emission from H II regions: consequences of improper analytic methods.
F. J. Lockman, R. L. Brown.
Astrophys. J., Vol. 222, 153 - 164 (1978).

The authors discuss the validity of some common analytic methods and approximations used to derive physical conditions in emission nebulae from observations of their radio recombination-line and continuum emission, and show that most either are dominated by systematic effects or involve invalid assumptions. They conclude that many phenomena apparently discovered through recombination-line observations (e.g., a hardening of the radiation field at the edge of H II regions, or a gradient in nebular temperatures across the Galaxy) are not unambiguously established.

132.039   W3(OH): a "runaway" compact object?
V. A. Hughes, M. R. Viner.
Astrophys. J., Lett., Vol. 222, L27 - L31 (1978).

The radio nebula W3(OH) has a number of unique properties that could be explained if it were produced by a compact object of $8-11\,M_\odot$ moving through a dense interstellar cloud.

132.040   Oblique refraction of an ionisation front.
W. B. Fu.
Nature, Vol. 273, 36 - 37 (1978).

A problem, of interest to both the astronomer and theoretical astrophysicist, concerns the interaction between an ionisation front and a cloud of high density molecular gas, as strong observational evidence indicates that dense pockets of molecular gas exist in the vicinity of some O and B stars. This paper describes an investigation of the interaction between a planar ionisation front and a planar cloud of high density molecular gas.

132.041   On the effects of inhomogeneities in high excitation H II regions.   J. Manfroid.
Astrophys. Space Sci., Vol. 53, 479 - 487 (1978).

The author discusses the effects of inhomogeneities in H II regions. It is shown that some emission lines are strongly affected when the filling factor is substantially smaller than 1.

132.042   M43—an emission nebula in Orion.
C. Thum, D. Lemke, U. Fahrbach, A. Frey.
Astron. Astrophys., Vol. 65 207 - 213 (1978).

New radio continuum, radio recombination line and infrared data on M43 are presented. The infrared luminosity of the nebula and the luminosity of the single ionizing star, NU Ori, are found to be nearly equal. The dust in front of NU Ori appears to have a lower ultraviolet extinction cross section than dust in the general interstellar medium, and may be concentrated in a neutral layer close to the ionized region. The abundance of ionized helium is compatible with line blanketed stellar atmosphere models.

132.043   609 MHz aperture synthesis observations of S 104.   K. W. Weiler, P. A. Shaver.
Astron. Astrophys., Vol. 65, 305 - 306 (1978).

The flat spectrum of an optically thin H II region is confirmed for S 104 at 609 MHz. Two nearby sources are also found to be thermal and to have flat spectra between 609 and 1415 MHz.

132.044   Neutral hydrogen in the region between the spiral arms of the Galaxy. Cloud structure.
I. V. Gosachinskij, I. A. Rakhimov.
Astron. Zh. Akad. Nauk SSSR, Vol. 55, 292 - 298 (1978). In Russian. English translation in Soviet Astron., Vol. 22, No. 2.

In the region $l = 60-74°$, $b = \pm15°$, $r = 10-12$ kpc between the Orion and Perseus arms, 81 H I clouds with masses from 3×10^3 to $30 \times 10^3 M_\odot$ are found. The half-thickness of the clouds' z-distribution is about ~2 kpc. The mass distribution of the clouds has the form $M^{-3/2}$. A continuation of this dependence to lower masses gives a mean cloud mass of $190\,M_\odot$ and a mean gas density in the clouds of 0.15 cm^{-3}.

132.045   On the origin of high-velocity H I clouds.
Yu. A. Shchekinov.
Astrofizika, Vol. 13, 711 - 715 (1977). In Russian. English translation in Astrophysics, Vol. 13, No. 4.

An interpretation of the correlation $N_H(-v_r)$ for high-velocity H I clouds is proposed. It is assumed that the negative velocity of clouds is a result of motion towards the galactic plane. Plausible formation mechanisms of high-velocity clouds are discussed. Supernova explosion is apparently unsatisfactory. It seems that cloud formation due to collision between galactic and intergalactic gas (e.g. intergalactic wind) is more favourable.

132.046   High resolution radio observations of G49.5−0.4 (W51).   P. F. Scott.
Mon. Not. R. Astron. Soc., Vol. 183, 435 - 442 (1978).

The H II region G49.5−0.4 (W51) has been mapped at frequencies of 5 and 15 GHz with respective angular resolutions of 2 × 8 and 0.65 × 2.6 arcsec. The brighter of the two main components is optically-thin and resolved by the beam at the higher frequency; its radio brightness distribution is consistent with that of the corresponding 20 µm infrared source. The second component comprises a compact core with ridges of emission extending a distance of some 1.5 arcsec (2 pc). The data are consistent with the second component being adjacent to a dense molecular cloud, on the edge of which more OB stars are forming.

132.047   Planetary nebulae and related objects.
N. Panagia.
IAU Symp. No. 76, (see 012.014), p. 315 - 322 (1978).

The author discusses some aspects of H II regions with special emphasis on the problem of how to recognize them.

132.048   Recombination lines from compact H II regions.
P. Silverglate, Y. Terzian.
IAU Symp. No. 76, (see 012.014), p. 327 (1978). − Abstract.

132.049   High-resolution observations of the W33 complex at 2.8, 6, 18 and 21 cm.   W. M. Goss, H. E. Matthews, A. Winnberg.
Astron. Astrophys., Vol. 65, 307 - 312 (1978).

The H II-region complex W33 has been mapped in the continuum at three wavelengths: at 2.8 cm (HPBW 80″), 6 cm (5″.6 × 30″) and 21 cm (25″ × 140″). The region consists of a number of weak discrete features and one strong component (G12.80−0.20). These features are superimposed on a broad background. A new determination of the positions of the OH Type I emission sources W33A and W33B has been made with an accuracy of 2″.

132.050   Internal motions in H II regions. IV. The ring nebula NGC 2359.
P. Pişmiş, E. Recillas-Cruz, I. Hasse.
Rev. Mexicana Astron. Astrofis., Vol. 2, 209 - 217 (1977).

On five Fabry-Pérot interferograms radial velocities from the Hα line at 338 points on and around the ring nebula NGC 2359 have yielded +71.0 ± 7.5 km s^{-1} for the mean overall velocity of the H II region. This value is taken to represent the radial velocity of the WN5 star HD 56925, the source of the excitation of the H II region. Using the Schmidt model the kinematic distance of the object is 4.0 kpc. Attention is concentrated on two well defined elliptical filaments around the exciting star. The mean velocities of outer and inner filaments are 84 ± 5 and 56 ± 7 km s^{-1}, respectively, showing clearly that with respect to the star the outer ring is receding while the inner one approaching the observer.

132.051   A high resolution search for small scale structure in Sharpless HII regions at 4.995 GHz. The catalogue.
M. Felli, R. H. Harten, H. J. Habing, F. P. Israel.
Astron. Astrophys., Suppl. Ser., Vol. 32, 423 - 428 (1978).

To increase the sample of known small scale radio emission features in HII regions, a 6 arcsec resolution survey of 77

Sharpless HII regions was made with the Westerbork Synthesis Radio Telescope. Strip scan distributions were obtained for 103 fields, with a limiting flux density of 30 mJy. Of this sample, 51 fields were reobserved more extensively and two-dimensional maps were obtained. From these maps, positions and flux densities of several new sources were determined. A list of all observed fields and the peak flux density detected in each field is given. The position, maximum size and flux density of all sources with $S_\nu > 25$ mJy are given.

132.052 Ionization and temperature structure of HII regions: the influence of the star, the gas density and its chemical composition.   G. Stasinska.
Astron. Astrophys., Suppl. Ser., Vol. 32, 429 - 438 (1978).

Model HII regions have been constructed by solving the radiative transfer, ionization equilibrium and thermal balance equations. The varying parameters were the characteristics of the star, the gas density distribution and its chemical composition. Results are given for the ionization and temperature structure, the emission line spectrum and the variation of optical depth inside the nebula. The influence of the diffuse radiation is discussed. The author emphasizes in particular those results which do not fit into the classical picture of an HII region.

132.053 First optical detection of W 51 and observations of new H II regions and exciting stars.
D. Crampton, Y. M. Georgelin, Y. P. Georgelin.
Astron. Astrophys., Vol. 66, 1 - 11 (1978) = Dominion Astrophys. Obs., Victoria, Canada, Contrib. No. 340 = NRC No. 15909.

The authors have obtained the first optical detection of three Hα sources near the W 51 complex. Radial velocities of these regions obtained by Fabry-Perot interferograms show the connection of these Hα regions with the radio complexes although their positions are not identical. Optical detection of some other H 109α sources such as W 48 and G 69.9 + 1.5 is discussed. Photometric and spectroscopic observations of recently identified stars in H II regions, and radial velocities of 23 new H II regions, have been obtained for $20° < l < 115°$. From these optical results the authors confirm the existence of another arm between 90 and 112° beyond the Perseus arm.

132.054 Molecular observations of Sharpless 106.
R. Lucas, A. M. Le Squéren, I. Kazès, P. J. Encrenaz.
Astron. Astrophys., Vol. 66, 155 - 160 (1978).

The authors have detected and studied a molecular cloud associated with the Sharpless H II region S 106. The detection of CO, ^{13}CO, HCO$^+$, HCN, H$_2$CO and OH lines in emission and H$_2$CO and OH lines in absorption are discussed. The cloud is massive ($2.6 \times 10^4 M_\odot$) and shows evidence of slow rotation.

132.055 Empirical methods for determining elemental abundances tested on model H II regions.
G. Stasinska.
Astron. Astrophys., Vol. 66, 257 - 267 (1978).

On the basis of model H II regions of different chemical composition, exciting stars and density distribution, it is shown that the temperature fluctuations inside nebulae with single exciting stars usually have less influence on the empirical determination of the elemental abundances than observational errors. However, for high overabundances ($Z > 2Z_\odot$) they may lead to underestimates by factors 5 - 10.

132.056 Near-infrared and CO observations of W40 and W48.
M. Zeilik II, C. J. Lada.
Astrophys. J., Vol. 222, 896 - 901 (1978).

Coordinated infrared observations from 1.2 to 20 µm and ^{12}CO/^{13}CO observations of W40 (G28.8 + 3.5) and W48

(G35.2−1.7) reveal that the H II regions, consisting of compact infrared components, lie on the edge of neighboring molecular clouds. Dynamical evidence supports a blister model for the development of the H II regions from their associated molecular clouds.

132.057 Far-infrared observations of H II regions in M33.
I. Gatley, P. M. Harvey, H. A. Thronson, Jr.
Astrophys. J., Lett., Vol. 222, L133 - L136 (1978).

Three H II regions in M33, namely IC 133, NGC 595, and NGC 604, have been observed at far-infrared wavelengths. Only NGC 604, the strongest radio source of the three, was detected. The far-infrared radiation appears to be thermal emission from dust. The total power radiated in the far-infrared is comparable to the total luminosity of the exciting stars of the H II region.

132.058 8−13µm spectrophotometry of the compact H II region G45.1+0.1.
H. Hefele, J. Schulte in den Bäumen.
Astron. Astrophys., Vol. 66, 465 - 467 (1978).

Observations of the 8−13 µm spectrum with spectral resolution 2% of an infrared source associated with the compact galactic H II region G45.1+0.1 are described. Three emission lines have been detected. These lines are emitted from ionized gas in a high-excitation nebula, thereby lending support to the conclusion from earlier work that the H II region must be excited by a luminous young star of very early spectral type. The silicate absorption feature is only marginally present.

132.059 Observations of the H$_2$O source W51 with the Crimea − Haystack radio interferometer.
L. I. Matveenko, D. M. Moran, B. F. Burke, I. G. Moiseev, V. I. Kostenko, L. R. Kogan.
Pis'ma Astron. Zh., Vol. 4, 57 - 63 (1978). In Russian.

VLBI Crimea − Haystack observations of W51 had been made in February 1976 at 1.35 cm. Some features of W51 were partially resolved. W51 has three centers of activity.

132.060 A double Hnα line galactic source G37.7 + 0.1.
P. Silverglate, Y. Terzian.
Astron. J., Vol. 83, 583 - 585 (1978).

Radio observations of the galactic thermal source G37.7 + 0.1 are reported. These indicate double recombination lines with a velocity separation of ~ 40 km/s. Various models to explain the observations are investigated.

132.061 The properties of weak shocks in H II regions.
P. W. J. L. Brand, J. S. Mathis.
Astrophys. J., Vol. 223, 161 - 167 (1978).

The structure of the cooling region behind weak shocks (speeds ≤ 30 km s^{-1}, isothermal Mach number, $M_0 \lesssim 3$) in H II regions is determined by numerical integration. The observable quantities are the various line strengths in the equilibrium regions ahead of and behind the shock. The line strengths can be approximately understood in terms of the effects of the compression upon their parent ionic abundances. Qualitative criteria are given for distinguishing shocks from changes in path length or discontinuities in extinction from point to point in the nebula. The determination of M_0 from line ratios is discussed; the problem is the separation of the contribution of unshocked material superposed upon the shock.

132.062 The far-ultraviolet flux from λ Orionis A.
R. J. Harms, P. A. Strittmatter, R. E. Williams.
Astrophys. J., Vol. 223, 234 - 237 (1978).

Digicon observations of emission-line intensities in the H II region around λ Ori are used to determine the effective UV ionizing flux from λ Ori A. It appears that the flux shortward of 353 Å (and perhaps shortward of 504 Å) is significantly less than predicted from calculations of static stellar atmospheres. The discrepancy between theory and observation may be a consequence of the existence of an extended atmosphere due to radiation-pressure—driven mass loss.

132.063 **Observations of high velocity H I clouds in the Local Group.** W. K. Huchtmeier.
IAU Symp. No. 79, (see 012.027), p. 49 (1978).

132.064 **Element abundances in gaseous nebulae: a model analysis of ionization-correction formulae.**
S. A. Grandi, S. A. Hawley.
Publ. Astron. Soc. Pacific, Vol. 90, 125 - 131 (1978) = Lick. Obs. Bull., No. 795.
From model calculations of H II regions the authors conclude that the standard ionization-correction formulae adequately account for unobserved ionization stages of N and S, and to a lesser extent, Ne and He. However, the model calculations do not explain the apparent overabundance of Ne and S calculated with the standard formulae for certain nebulae.

132.065 **A giant H II ring and energetic stellar winds in the Large Magellanic Cloud.** K. H. Elliott, C. Goudis, J. Meaburn, J. Pilkington.
Astrophys. Space Sci., Vol. 55, 475 - 486 (1978).
A large ring, $\simeq$ 50 pc diameter, of $H\alpha$ + [N II] emission in the LMC has been investigated using several instruments. It is suggested that stellar winds from O and B stars could play a vital role in the formation of many of the structures revealed.

132.066 **The shape of recombination lines from expanding nebulae.** F. J. Lockman.
Bull. American Astron. Soc., Vol. 10, 399 (1978). – Abstract.

132.067 **Carbon monoxide observations of molecular gas at the boundary of H II regions.**
J. Bally, N. Z. Scoville.
Bull. American Astron. Soc., Vol. 10, 405 (1978). – Abstract.

132.068 **Radio, infrared, and optical observations of the compact H II region S235A.**
J. Krassner, J. L. Pipher, S. Sharpless.
Bull. American Astron. Soc., Vol. 10, 406 (1978). – Abstract.

132.069 **The galactic extended low-density H II region and its relation to star formation and diffuse IR emission.** P. G. Mezger.
Bull. American Astron. Soc., Vol. 10, 406 (1978). – Abstract.

132.070 **Observations of the [O III] 51.8μ line in the H II regions M42, M17 and W51.**
G. Melnick, G. E. Gull, M. Harwit.
Bull. American Astron. Soc., Vol. 10, 406 (1978). – Abstract.

132.071 **Reddening of H II regions in M31.** C. Krishna Kumar.
Bull. American Astron. Soc., Vol. 10, 406 (1978). – Abstract.

132.072 **Abundance determinations from spectrophotometric observations of four Perseus arm H II regions.**
D. L. Talent, R. J. Dufour.
Bull. American Astron. Soc., Vol. 10, 406 (1978). – Abstract.

132.073 **Search for radio emission in the 18-cm lines of the OH molecule from high-latitude clouds of neutral hydrogen.** G. M. Rudnitskij, A. Sh. Ataev.
Astron. Tsirk., No. 996, p. 1 - 2 (1978). In Russian.

Plasma astrophysics. See Abstr. 061.053.

On the interaction of a weak-R type ionization

front and a contact discontinuity. See Abstr. 062.055.

Generation of very-high-velocity satellite-features through stimulated Raman scattering of 22.2 GHz H_2O maser-lines in compact H II plasma regions. One-dimensional model. See Abstr. 063.032.

A new trapezium system associated with a compact H II region. See Abstr. 117.056.

CO observations of a molecular cloud complex associated with the bright rim near VY Canis Majoris. See Abstr. 131.007.

The temperature probability function and infrared emissivity of dust heated by randomly distributed stars. See Abstr. 131.014.

Interstellar heating by photoelectrons from negative oxygen. See Abstr. 131.026.

On the physical connection between the maser sources W75S and N. See Abstr. 131.040.

Small-scale structure in the distribution of neutral hydrogen. See Abstr. 131.051.

6-cm recombination lines from the molecular cloud/ H II region interface in NGC 1977. See Abstr. 131.111.

Recombination lines from H II and C II regions. See Abstr. 131.112.

H_2CO and OH observations of a molecular cloud near RCW 36. See Abstr. 131.127.

Observations of the interstellar [^{16}OH]/[^{18}OH] abundance ratio. See Abstr. 131.147.

Radio and infrared observations of the OH/H_2O source G12.2–0.1. See Abstr. 131.155.

Star formation rates in the Galaxy. See Abstr. 131.172.

Observations of the $J = 1−0$ transition of CS at 49 GHz in southern molecular clouds. See Abstr. 131.175.

First detection of a non-metastable ammonia line in absorption. See Abstr. 131.177.

Photoelectric heating of interstellar gas. See Abstr. 131.180.

Absorption of nonmetastable NH_3 levels toward W3(OH). See Abstr. 131.224.

A model for the maser sources associated with H II regions. See Abstr. 131.246.

High-resolution mapping of the H I absorption lines in the direction of NGC 2024, Orion A, M 17 and W49. See Abstr. 131.247.

Dust, helium abundance. See Abstr. 131.259.

Infrared observations of the galactic center. I. Nature of the compact sources. See Abstr. 133.002.

Models of infrared emission from dusty and diffuse H II regions. See Abstr. 133.010.

Near-infrared polarimetry of compact infrared sources associated with H II regions and molecular clouds. See Abstr. 133.012.

Carbon recombination-line mapping of the Orion Nebula.   See Abstr. 134.002.

Observations of the Rosette nebula NGC 2237 at decametric wavelengths.   See Abstr. 134.024.

The Cygnus X region. X: The riddle of the Gamma Cygni radio source resolved.   See Abstr. 141.012.

H I clouds and the theory of formation of galaxies. See Abstr. 151.092.

UBV photometry of the luminous young cluster NGC 3603.   See Abstr. 153.007

Interferometer measurements of H I absorption in the direction of 16 extragalactic radio sources. See Abstr. 155.002.

Studies of low-frequency recombination lines from the direction of the Galactic Centre and other galactic sources. See Abstr. 156.001.

Detection of H II regions and ionized interarm gas in the Sombrero Galaxy (NGC 4594)   See Abstr. 158.035.

M83. III. Excitation and chemical abundances of gaseous nebulae.   See Abstr. 158.093.

The composition gradient across M101. See Abstr. 158.173.

On the anomaly of the far UV extinction in the 30 Doradus region.   See Abstr. 159.004.

The N119 complex in the Large Magellanic Cloud. See Abstr. 159.021.

21-cm observations of redshifts and their implications for the distance scale.   See Abstr. 162.042.

Errata

132.901 Erratum: 'On an H II condensation in NGC 2175' [Rev. Mexicana Astron. Astrofis., Vol. 2, 59 - 63 (1977)].   P. Pişmiş.
Rev. Mexicana Astron. Astrofis., Vol. 2, 269 (1977).

132.902 Erratum: 'Radio observations of small galactic nebulae' [Astron. Astrophys., Vol. 42, 9 - 16 (1975)].
I. Kazès, A. M. Le Squeren, F. Gadéa.
Astron. Astrophys., Vol. 66, 470 (1978).

133 Infrared Sources

133.001 An infrared search for OH/IR stars.   I. S. Glass.
Mon. Not. R. Astron. Soc., Vol. 182, 93 - 96 (1978).
Possible infrared identifications are given for 13 of 15 OH/IR-type sources discovered during a recent 1612-MHz radio survey. The plausibility of these identifications is discussed and ways of making them more certain are considered.

133.002 Infrared observations of the galactic center. I. Nature of the compact sources.   E. E. Becklin, K. Matthews, G. Neugebauer, S. P. Willner.
Astrophys. J., Vol. 219, 121 - 128 (1978).
Photometry from 1.25 to 12 μm and spectrophotometry from 8 to 13 μm of the compact sources found in the galactic-center region are reported. In addition, revised 10 and new 20 μm maps with 2".3 resolution are given. The nature of the compact sources is discussed. Some are best identified as stars or star clusters; the brightest source at 2 μm is probably a supergiant, and the infrared source near the nonthermal radio source is probably a stellar cluster. One of the sources is observationally similar to extremely red OH/infrared stars. Other sources have luminosities and linear sizes similar to those of compact H II regions; emission from optically thin silicate dust is seen in these.

133.003 A high spatial resolution map of the Orion Nebula at 33 μ.   C. A. Beichman, H. M. Dyck, T. Simon.
Astron. Astrophys., Vol. 62, 261 - 263 (1978).
The authors present a 33 μ map of the Orion Nebula showing the Kleinmann-Low Nebula and the Trapezium region at high spatial resolution. The emission from the Trapezium can be explained by radiation from a single temperature grain population with long wavelength characteristics similar to known silicates. The Kleinmann-Low Nebula appears larger

at 33 μ than at 20 μ, in accordance with predictions of a model for KL as a centrally heated, optically thick cloud. The authors set an upper limit to the emission from the Becklin-Neugebauer point source.

133.004 An observational study of the AFCRL infrared sky survey. III. Further searches for AFCRL/AFGL sources and an evaluation of the contents of the mid-infrared sky.   M. J. Lebofsky, D. G. Sargent, S. G. Kleinmann, G. H. Rieke.
Astrophys. J., Vol. 219, 487 - 493, with a correction in Vol. 223, 704 (1978).
New observations and a reexamination of the results from earlier searches for AFGL/AFCRL 11 μm sources indicate that the AFGL catalog is a reliable source of information of the mid-infrared sky. From these analyses, an evaluation is made of the contents of the mid-infrared sky at the flux limit of the AFGL catalog and below. Methods to distinguish late-type stars from other classes of sources will become increasingly important as infrared sky surveys achieve lower flux limits.

133.005 Lamellar grating observations of the Orion Nebula from 100 to 500 microns.
J. L. Pipher, J. G. Duthie, M. P. Savedoff.
Astrophys. J., Vol. 219, 494 - 497 (1978).
The authors present the results of spectroscopic observations of the Kleinmann-Low nebula in the 100–500 μm region taken from the Gerard P. Kuiper Airborne Observatory. These results, in conjunction with previous work at shorter wavelengths, are consistent with a dilute blackbody spectrum modified by grain emissivity and spatial variations in the grain density.

133.006 Strong far-infrared emission from a compact source in Sharpless 140.
P. M. Harvey, M. F. Campbell, W. F. Hoffmann.
Astrophys. J., Vol. 219, 891 - 895 (1978).

High-spatial-resolution observations of S140 are presented at wavelengths from 35 to 175 μm. A single strong far-infrared source has been found coincident with a compact near-infrared and H_2O maser source and the center of a molecular cloud. The far-infrared spectrum suggests either significant absorption in the 20 μm region or a two-temperature dust structure. The infrared luminosity of $\sim 2 \times 10^4 L_\odot$ is comparable to that of a late O or early B star, although no radio continuum flux has been seen from this object. The source is quite likely a dust-embedded protostar.

133.007 Radial diameters of Type II OH/IR sources.
G. V. Schultz, W. A. Sherwood, A. Winnberg.
Astron. Astrophys., Vol. 63, L5 - L7 (1978).

Twelve Type II OH/IR sources have been monitored at approximately half-yearly intervals during the last two years in the 1612-MHz OH line. Intensity variations of nine of the sources show a phase-lag of the high-velocity component flux relative to the low-velocity component flux. If the time-lags are interpreted as differences in path length the mean OH shell diameter is $(4\pm2) \times 10^{16}$ cm in agreement with values obtained through indirect arguments.

133.008 Observations of Brackett-alpha emission in the region of the BN object. R. R. Joyce, M. Simon, T. Simon.
Astrophys. J., Vol. 220, 156 - 158 (1978).

The authors have mapped the hydrogen Bα emission around the region of the Becklin-Neugebauer object. An extended distribution of Bα emission around the BN object has been found. This emission appears to arise in the foreground nebula M42. They have observed a local maximum of Bα emission at the BN object, whose strength is in quantitative agreement with previous observations by Grasdalen. A marginal detection of the Bγ line toward BN suggests the visual extinction may be approximately 35 mag to the object.

133.009 The effect of multiple grain components on infrared radiation transfer and the 10 micron silicate feature.
C. L. Sarazin.
Astrophys. J., Vol. 220, 165 - 170 (1978).

Interstellar dust is composed of several types of grains which have different size and chemical compositions. The presence of different grain componenets affects the radiative transfer in infrared sources, both for the general continuum emission and for the narrower spectral features. The 10 μm silicate feature is particularly affected; the author shows that a silicate absorption feature can occur in an emitting region in which all grain temperatures are constant along the line of sight. A "multigrain" model for the silicate feature is successfully compared with observations of the Orion ionization-front source and a recently observed set of southern H II regions.

133.010 Models of infrared emission from dusty and diffuse H II regions. P. A. Aannestad.
Astrophys. J., Vol. 220, 538 - 555 (1978).

Models for the infrared emission from amorphous core-mantle dust within diffuse ($n_e \leqslant 10^3$ cm^{-3}) H II regions with neutral shells that are optically thin in the infrared have been calculated. Calculations of the total infrared luminosity, the relative contribution by Lα photons, the infrared spectral distribution, and the size of the dust-depleted regions are presented as functions of the ultraviolet optical depths in the ionized and neutral regions and for stellar temperatures of 35,000 and 48,000 K. For core-mantle grains most of the infrared energy is emitted between 30 and 70 μm, relatively independent of whether the dust is within or outside the H II region.

133.011 The infrared spectra of CRL 618 and HD 44179 (CRL 915).
R. W. Russell, B. T. Soifer, S. P. Willner.
Astrophys. J., Vol. 220, 568 - 572 (1978).

Spectrophotometry from 4 to 8 μm is reported for the infrared sources CRL 618 and HD 44179 (CRL 915). In addition, 2–4 μm spectrophotometry of CRL 618 is reported. Except for marginal detection of emission lines at 2.1 and 2.45 μm, the spectrum of CRL 618 is featureless, consistent with graphite being the major constituent of the circumstellar dust cloud. Strong emission bands at 6.2 μm and 7.7 μm are found in the spectrum of HD 44179.

133.012 Near-infrared polarimetry of compact infrared sources associated with H II regions and molecular clouds. H. M. Dyck, R. W. Capps.
Astrophys. J., Lett., Vol. 220, L49 - L51 (1978).

The authors present new 1.6–3.8 μm linear polarization observations of 15 compact infrared sources associated with H II regions and molecular clouds. Seven of the sources show more than 10% polarization in this range of wavelengths, while only three are unpolarized. The authors find a strong correlation between the degree of infrared compactness and the polarization in the sense that the more compact sources show the largest polarization.

133.013 Molecular envelopes around evolved stars and the origin of planetary nebulae.
B. Zuckerman, P. Palmer, D. P. Gilra, B. E. Turner, M. Morris.
Astrophys. J:, Lett., Vol. 220, L53 - L56 (1978).

The authors surveyed a total of 37 late-type giant stars and IRC and CRL infrared sources in the $J = 1 \to 0$ rotational transition of carbon monoxide. Broad CO emission lines were detected toward six stars. Combination of these results with other CO observations of similar objects suggests that a majority of the red-giant stars with the largest mass-loss rates are carbon-rich (C/O $>$ 1) or S-type (C/O $\sim$ 1). Infrared surveys tend to support this assertion. The authors suggest that these stars will evolve into planetary nebulae.

133.014 Infrared observations of the galactic center. IV. The interstellar extinction.
E. E. Becklin, K. Matthews, G. Neugebauer, S. P. Willner.
Astrophys. J., Vol. 220, 831 - 835 (1978).

Infrared observations of the compact sources and background in the galactic center are used to derive the 1.25–12.5 μm interstellar extinction law. The depth of the interstellar 10 μm silicate absorption feature is derived from observations of the M supergiant IRS 7. The ratio of visual to silicate extinction is found to be 8 ± 3. From the colors of individual sources and the background, it is concluded that there are no more than 6 mag of visual extinction within the central 3 pc of the galactic center.

133.015 A low-detection limit search for OH emission from infrared stars. J. D. Fix, J. M. Weisberg.
Astrophys. J., Vol. 220, 836 - 840 (1978).

The authors have used the 300 m telescope of the Arecibo Observatory to examine 154 cool luminous stars for 18 cm OH emission. Six of the stars (RU Ari, R Com, T Com, RX Oph, UU Peg, and RT Vir) were found to show OH emission. For the stars without OH emission, the authors have established detection limits several times smaller than those of previous surveys.

133.016 Spectroscopic observations of optically identified AFGL/CRL infrared sources. II.
M. Cohen, L. V. Kuhi.
Publ. Astron. Soc. Pacific, Vol. 89, 829 - 839 (1977/78).

Spectra have been obtained of 34 optical candidates for AFGL/CRL sources. Nineteen are classifiable as late-type K or

M stars, eleven as carbon stars, and four as heavily obscured or nebulous objects. Definite indications of circumstellar extinction are found for a few late M giants.

133.017 The effects of rotation on microwave spectral line profiles: a study of CRL 437.
M. H. Schneps, R. N. Martin, P. T. P. Ho, A. H. Barrett.
Astrophys. J., Vol. 221, 124 - 136 (1978).

Microwave observations of the peculiar IR source CRL 437 are reported and discussed in terms of a dynamic model; in particular, the effects of cloud rotation on spectral line profiles are considered. The line shapes are asymmetric and undergo a shape mirror reversal about an axis projected onto the cloud. The line shapes and mirror-symmetry reversal are simulated by a model which includes rotational and radial mass motions as well as an excitation temperature gradient. The uniqueness of this model, which is used to derive kinetic parameters for the source, is discussed.

133.018 Accurate interferometer positions of H_2O masers.
J. R. Forster, W. J. Welch, M. C. H. Wright, A. Baudry.
Astrophys. J., Vol. 221, 137 - 144 (1978).

The authors report accurate interferometer positions for strong water-vapor masers in the regions ON 1, W51, VY CMa, Sgr B2, W75, and NGC 7538. The absolute position uncertainties are mostly less than ~0''.5. Among the 13 H_2O maser sources or groups of spots for which accurate interferometer positions are available, no infrared point sources are coincident in eight cases, OH masers may be coincident in six cases, and H_2O masers lie at the edge of compact H II regions in five cases. In four out of the latter five cases, no infrared emission is seen toward the compact H II regions. Thus the maser emission does not arise in a dense shell of dust, a cocoon that completely surrounds the O star and its compact H II region, because such a shell would have a large infrared luminosity. It appears, rather, that the maser emission arises in separate fragments or condensations.

133.019 Infrared detection of luminous stars in M31 and M33.   R. M. Humphreys, J. W. Warner.
Astrophys. J., Lett., Vol. 221, L73 - L75 (1978).

Four intrinsically very luminous blue variables in M31 and M33 were recently detected at 2.2 μm. These are the most distant single stars to be observed in the infrared. These stars are in many respects similar to the η Carinae-type and S-Doradus type variables in the Milky Way and LMC.

133.020 A new, compact far-infrared source in the W31 region.   G. G. Fazio, C. J. Lada, D. E. Kleinmann, E. L. Wright, P. T. P. Ho, F. J. Low.
Astrophys. J., Lett., Vol. 221, L77 - L81 (1978).

In a survey of the W31 region, a compact far-infrared source was detected at the position of the radio continuum source G10.6−0.4. Associated with this infrared source and in close coincidence with its position is an H II region with a compact core and extended halo as well as OH and H_2O masers. An extended molecular cloud, centered on these sources, has also been detected. The source probably represents an early stage of stellar evolution.

133.021 Infrared spectra of protostars.   R. C. Puetter, R. W. Russell, B. T. Soifer, S. P. Willner.
Bull. American Astron. Soc., Vol. 9, 571 (1978). − Abstract.

133.022 What high resolution infrared spectroscopy is telling us about sources such as the Becklin Neugebauer object and IRC +10°216.   D. N. B. Hall.
Bull. American Astron. Soc., Vol. 9, 604 - 605 (1978). Abstract.

133.023 A near IR photographic sky survey.   E. R. Craine.
Bull. American Astron. Soc., Vol. 9, 606 (1978). − Abstract.

133.024 Discovery of 20 μm sources associated with molecular clouds.
C. A. Beichman, E. E. Becklin, R. W. Capps, H. M. Dyck.
Bull. American Astron. Soc., Vol. 9, 606 (1978). − Abstract.

133.025 Infrared maps of W3 from 4.9 microns to 20 microns.
J. A. Hackwell, R. D. Gehrz, J. R. Smith, D. A. Briotta.
Astrophys. J., Vol. 221, 797 - 809 (1978).

Infrared maps have been made of the region around W3 IRS1, IRS3, and IRS5 at eight wavelengths between 4.9 μm and 20 μm. The infrared maps of IRS1 and IRS3 are very similar in appearance to radio maps made at 5 GHz, implying that the ionized gas is well mixed with the emitting dust in these regions. The 20 μm map shows a previously undetected complex of infrared sources to the west of IRS3. The 8.0 μm to 12.7 μm spectra of IRS1 show a silicate absorption feature. A simple model of IRS1 is proposed in which the 4.9 μm to 20 μm emission comes from silicate dust inside the H II region which is heated principally by trapped Lα photons. The observations of IRS3 are consistent with its being similar to, but less massive than, IRS1. The source IRS3 has a central dust-depleted H II region surrounded by a cool cloud.

133.026 Far-infrared observations of sources associated with double-lobed reflection nebulae.
S. G. Kleinmann, D. G. Sargent, H. Moseley, D. A. Harper, R. F. Loewenstein, C. M. Telesco, H. A. Thronson, Jr.
Astron. Astrophys., Vol. 65, 139 - 141 (1978).

The authors present far-infrared observations of GL 915, 618, 2688 and OH 0739-14. The data provide information on the total flux arising in each of these sources, and on the wavelength-dependence of the emissivity of the associated dust. Two sources, GL 618 and GL 2688, have far-infrared continua similar to those recently observed in planetary nebulae.

133.027 IR energy distribution of OH/IR sources.
W. A. Sherwood, E. Kreysa, G. V. Schultz.
IAU Colloq. No. 42, (see 012.012), p. 434 - 437 (1977).

Some infrared counterparts of Type II OH/IR sources have been observed between 1 μ and 30 μ. In spite of their having very similar OH emission spectra, their energy distributions differ considerably among themselves from "flat" (1.25 μ to 4.9 μ) to "steep" (2.2 μ to 20 μ). The energy distributions are compared with more familiar objects.

133.028 OH/IR stars near the galactic centre.   B. Baud.
IAU Colloq. No. 42, (see 012.012), p. 438 - 445 (1977).

Observations at 1612 MHz around the galactic centre have revealed the existence of 43 OH/IR stars. Sources within 6° longitude from the galactic centre appear to be associated with the central parts of the Galaxy. They have a large velocity dispersion and a mean radial velocity of 0 km s^{-1}, indicating that there is no net rotation around the centre. Within 2° from the centre there is a strong relative increase in the number of sources with a velocity separation between the two emission peaks of less than 30 km s^{-1}. This is discussed in terms of a possible interaction between the circumstellar shell and the ambient medium.

133.029 The possibility of detecting DC.
N. S. Witte, R. D. Brown.
Proc. Astron. Soc. Australia, Vol. 3, 146 - 147 (1977).

133.030 Infrared sources in the vicinity of 2S1728−337.
I. S. Glass.
Nature, Vol. 273, 35 - 36 (1978).

The X-ray burst source MXB1728−34, which is believed

to be associated with the steady source 2S1728−337, lies in a heavily obscured portion of the Milky Way. The infrared photometer described previously (1973), attached to the 1.88-m telescope at Sutherland, was used to search for optically invisible sources at K (2.2 μm) in the region of the error circle of 2S1728−337. The results are presented here: five infrared sources were located.

133.031 Calculations of infrared fluxes from galactic sources for a polysaccharide grain model.
F. Hoyle, N. C. Wickramasinghe.
Astrophys. Space Sci., Vol. 53, 489 - 505 (1978).

Using transmittance data appropriate for grain material which is predominantly comprised of polysaccharides the authors have computed infrared fluxes from several types of galactic infrared sources. The model used in these computations involves polysaccharide condensation in material flowing out from O-type stars. With the exception of rather minor discrepancies they show that it is possible to match the 2.1−13 μ observations of a wide range of galactic infrared sources.

133.032 Equivalent widths of the 10-μm interstellar silicate feature and dust temperatures in infrared sources.
J. Dorschner, C. Friedemann, J. Gürtler.
Astrophys. Space Sci., Vol. 54, 181 - 185 (1978).

For 22 infrared sources showing the 10-μm silicate absorption band, the optical depths, dust temperatures and equivalent widths have been calculated for two assumed simple models.

133.033 Circumstellar methane in the infrared spectrum of IRC +10°216.   D. N. B. Hall, S. T. Ridgway.
Nature, Vol. 273, 281 - 282 (1978).

The authors reported previously the detection of lines due to circumstellar CO, C_2H_2 and HCN in the infrared spectrum of the cool carbon-rich object IRC +10°216. In this letter they announce the further detection of lines due to circumstellar CH_4 in 3.3-μm spectra of this source.

133.034 Platt particles in M 17?   C. D. Andriesse.
Astron. Astrophys., Vol. 66, 169 - 173 (1978).

The temperature differences in the extended infrared source of M 17 cannot be ascribed to a spatial separation of warm and cool dust. They are interpreted in terms of grains with sizes of 10 Å exhibiting temperature fluctuations up to about 150 K, whereas the larger, common grains have an equilibrium temperature of about 36 K. An analytical expression is derived for the temperature distribution of the particles. The absence of spatial temperature variations and the presence of 10 Å particles is discussed in the context of a shock front.

133.035 On the nature of the peculiar emission-line object RX Puppis.   M. Klutz, O. Simonetto, J. P. Swings.
Astron. Astrophys., Vol. 66, 283 - 288 (1978).

In addition to exhibiting complex Balmer and He I lines as well as low excitation permitted and forbidden emission lines, the spectrum of RX Puppis is rich in sharp absorptions due essentially to neutral metals. Variations in the IR flux, a brightening in the visible, and new radio data are also taken into account in the attempt to construct a physical model of RX Puppis with a central star (late B giant), an expanding envelope with high velocities, an almost stationary shell, and a dust shell (or ring).

133.036 Observations of five moderate-luminosity far-infrared sources in Orion and Monoceros.
H. A. Thronson, Jr., D. A. Harper, J. Keene, R. F. Loewenstein, H. Moseley, C. M. Telesco.
Astron. J., Vol. 83, 492 - 499 (1978).

Five regions in Orion and Monoceros have been studied at wavelengths between 40 and 400 μ. The observations include broadband photometry of OMC-1 (the KL nebula), OMC-2, OMC-3, NGC 2024, and Mon R2 (VdB 67) with beam sizes of 1−3.5 arcmin, submillimeter measurements of OMC-1, -2, and -3 with a 9-arcmin beam, and photometry and spatial scans of OMC-2 at an angular resolution of ~20 arcsec. The observed areas include far-infrared sources with luminosities between 100 $L_\odot$ and 2.5 × 10^5 $L_\odot$. The infrared data are used to estimate the masses and radial density gradients of the radiating dust clouds and the total mass of the stars associated with the clouds.

133.037 A 5 GHz survey of infrared sources.
J. W. Warner, J. H. Black.
Astron. J., Vol. 83, 586 - 593 (1978).

A search has been made for radio emission at 5 GHz in the directions of 104 unidentified or nonstellar infrared sources. Of the 48 confirmed infrared sources in this sample, 42% can be identified with radio sources, many of which are related to visible H II regions. Several of these radio and infrared sources warrant further study.

133.038 High-resolution 1.5−5 micron spectroscopy of the Becklin-Neugebauer source in Orion.
D. N. B. Hall, S. G. Kleinmann, S. T. Ridgway, F. C. Gillett.
Astrophys. J., Lett., Vol. 223, L47 - L50 (1978).

High-resolution 1.5 to 5 μm spectra of the Becklin-Neugebauer source in Orion exhibit both Bα and Bγ emission localized within 1.″4 of the continuum source, and CO absorption arising in the obscuring molecular cloud. The data support the hypothesis that BN is a young B0 or B1 star surrounded by a compact H II region and an expanding dust envelope. The object is apparently driving massive, supersonic expansion of the surrounding cloud material into OMC-1.

133.039 Radio identification of an infrared object associated with the passage of a shock front.
J. P. Vallée, V. A. Hughes.
Astrophys. J., Lett., Vol. 223, L97 - L100 (1978).

A small-scale infrared object, previously detected near the large-scale optical shell in the H II region IC 1805, has now been detected at radio wavelengths with a supersynthesis instrument. The radio identification of this infrared object provides evidence for star formation associated with the passage of a shock front, seen optically as a large-scale shell and possibly created by stellar winds.

133.040 The infrared source associated with Sh 2-149.
R. W. Russell.
Astron. Astrophys., Vol. 67, 273 - 275 (1978).

Observations at optical and infrared wavelengths of the bright infrared point source in the direction of Sh 2-149 are used to determine a spectral type (M1) and luminosity class (Ib−II) for the object. These observations contradict the suggestion of Bergeat et al. (1975) that the object is a highly reddened O star. The estimated visual extinction ($A_v \sim 5-7^m$) and luminosity class derived from the present observations are used to estimate the distance ($d \sim 1-3$ kpc) to the infrared source.

133.041 Fuentes de emisión infrarroja.   J. F. Lahulla.
Bol. Astron. Obs. Madrid, Vol.10, No. 1, p. 3 - 21 (1977).

This article reviews the results from IR-sky surveys. The author discusses the infrared stars, diffuse infrared galactic objects, H II regions, OH microwave emission sources in the center of the Galaxy and extragalactic sources.

133.042 Objetos infrarrojos no identificados en el catálogo IRC.   J. F. Lahulla.
Bol. Astron. Obs. Madrid, Vol. 10, No. 1, p. 59 - 72 (1977).

The present paper summarizes the techniques for the identification and the classification of unidentified infrared objects listed in the two-micron Survey Catalog.

133.043 High resolution OH maser observations of IRC +10420.
J. C. Webber, R. L. Mutel, J. D. Fix, J. M. Benson.
Bull. American Astron. Soc., Vol. 10, 392 (1978). – Abstract.

133.044 Monitoring of OH maser emission from late-type stars.   P. R. Jewell, M. Elitzur, J. C. Webber, L. E. Snyder.
Bull. American Astron. Soc., Vol. 10, 392 - 393 (1978). Abstract.

133.045 Progress on a near IR photographic sky survey.
E. R. Craine.
Bull. American Astron. Soc., Vol. 10, 394 (1978). – Abstract.

133.046 Airborne observations of IRC +10216 in the region 3000-4400 cm^{-1}.   L. L. Smith, J. Krassner, T. Hilgeman.
Bull. American Astron. Soc., Vol. 10, 394 (1978). – Abstract.

133.047 Far infrared maps of the OMC-1, OMC-2 region.
J. Keene, D. A. Harper, R. F. Loewenstein.
Bull. American Astron. Soc., Vol. 10, 394 (1978). – Abstract.

133.048 Submillimeter observations of NGC 2024 (W12, Orion B).
D. K. Lynch, J. R. Smith, D. D. Cudaback.
Bull. American Astron. Soc., Vol. 10, 395 (1978). – Abstract.

133.049 Changes in the 0.36μm-3.5μm spectral-flux distribution of extragalactic nonthermal sources.
J. J. Puschell, W. A. Stein, S. L. O'Dell.
Bull. American Astron. Soc., Vol. 10, 403 (1978). – Abstract.

133.050 Star formation at a front: far-infrared observations of AFGL 333.
H. A. Thronson, P. M. Harvey, O. Steward, I. Gatley.
Bull. American Astron. Soc., Vol. 10, 437 (1978). – Abstract.

OH main lines masers I : OH/IR stars.
See Abstr. 022.009.

Airborne infrared astronomy.
See Abstr. 032.520.

Millimetre observations of planets, galactic and extra-galactic sources.   See Abstr. 091.004.

A successful search for OH emission in early-type emission-line stars with IR excesses.   See Abstr. 114.068.

Infrared Fe II lines in Eta Carinae and a possible interpretation of infrared excesses.
See Abstr. 114.504.

Southern search for OH from M supergiants.
See Abstr. 116.013.

Trapezia and infrared sources.   See Abstr. 117.047.

A large-scale OH sky survey at 1612 MHz. Part II. The galactic distribution and kinematics of the unidentified type II OH/IR stars.   See Abstr. 122.063.

Spectrophotometry of OH 26.5 + 0.6 from 2 to 40 microns.   See Abstr. 131.008.

Spectral line shapes in spherically symmetric radial- ly moving clouds.   See Abstr. 131.009.

Deuterated ammonia toward the Orion Nebula.
See Abstr. 131.011.

A large-scale OH sky survey at 1612 MHz. Part I. The observations.   See Abstr. 131.024.

Is water-ice the carrier of the 3 μm-absorption in infrared objects?   See Abstr. 131.115.

The C II and S II regions in the Rho Ophiuchi dark cloud.   See Abstr. 131.118.

Infared radiation from dark globules.
See Abstr. 131.129.

The far-infrared emission of interstellar matter between galactic longitudes $l = 36°$ and $l = 55°$.
See Abstr. 131.130.

Radio and infrared observations of the OH/H$_2$O source G12.2–0.1.   See Abstr. 131.155.

Stellar water masers.   See Abstr. 131.221.

Absorption of nonmetastable NH$_3$ levels toward W3 (OH).   See Abstr. 131.224.

Observations of self-reversal in W3 IRS5.
See Abstr. 131.226.

Infrared studies of 30 Doradus. I. The 2-μ sources of the inner region.   See Abstr. 132.001.

Near-infrared and CO observations of W40 and W48.
See Abstr. 132.022.

Comparison of far-infrared luminosities of H II regions with luminosity of exciting stars.
See Abstr. 132.023.

Observations of [O III] 88μ line emission from H II regions and the Galactic Center.   See Abstr. 132.024.

M43—an emission nebula in Orion.
See Abstr. 132.042.

High resolution radio observations of G49.5–0.4 (W51).   See Abstr. 132.046.

Near-infrared and CO observations of W40 and W48.
See Abstr. 132.056.

8–13 μm spectrophotometry of the compact H II region G45.1+0.1.   See Abstr. 132.058.

Observations of the 51.8 micron [O III] emission line in Orion.   See Abstr. 134.028.

Observations of the molecular hydrogen emission from the Orion Nebula.   See Abstr. 134.034.

High resolution Fabry-Perot observations of 2.1 micron H$_2$ emission from the Orion nebula.
See Abstr. 134.037.

The galactic distribution of OH/IR stars.
See Abstr. 155.018.

The infrared emission of the Galactic center.
See Abstr. 156.008.

The spectral-flux distribution (0.36–3.5 μm) of nonstellar light from the broad-line radio galaxies 3C 227 and 3C 382.   See Abstr. 158.013.

10 micron observations of bright galaxies.

See Abstr. 158.047.

Star formation rates and infrared radiation. See Abstr. 158.065.

134 Emission Nebulae, Reflection Nebulae

134.001 CO observations along an ionization front of the California Nebula (NGC 1499).
D. M. Elmegreen, B. G. Elmegreen.
Astrophys. J., Vol. 219, 105 - 113 (1978).

A dark cloud on the northern edge of the California Nebula has been mapped at 3′ resolution in ^{12}CO and sampled in ^{13}CO and CS. The conversion of ^{13}CO to H_2 column densities is discussed. Irregularities of the southern cloud boundary (at the ionization front) are attributed to an instability that may result from the time-dependent flux of the runaway star ξ Persei, which ionizes the H II region. An unresolved globule or filament close to the ionization front may have become separated from the main cloud as the result of a Kelvin-Helmholtz instability.

134.002 Carbon recombination-line mapping of the Orion Nebula.   T. B. H. Kuiper, N. J. Evans II.
Astrophys. J., Vol. 219, 141 - 147 (1978).

The authors have mapped the distribution of the C75α line in the Orion Nebula with a spatial resolution of 1′.3 over a field of view of $α × δ = 4′ × 8′$. They find that ionized carbon is more extended than the radio continuum distribution. The emission is generally consistent with a model in which the line is formed in a thin sheet in the molecular cloud immediately adjacent to the H II region. However, the spatial extent of the emission does suggest that some of the emitting gas may be cooler and less dense than has previously been assumed.

134.003 2.1 micron H_2 emission: high-spectral-resolution observations of the Orion Nebula.
R. R. Joyce, D. Y. Gezari, N. Z. Scoville, I. Furenlid.
Astrophys. J., Lett., Vol. 219, L29 - L31 (1978).

High-resolution (45 km s^{-1}) observations of the 2.12 μm molecular hydrogen emission line in the Orion Nebula reveal a single feature with $ΔV$ (FWHP) ≤ 30 km s^{-1} at a LSR velocity of +9.5 ± 4 km s^{-1}. The results support the suggestion that the H_2 emission is collisionally excited in a thin shock-heated layer inside the cool molecular cloud. For a spherical geometry the shock velocity is, therefore, constrained to be at most 15 km s^{-1}.

134.004 Photoelectric polarization maps of two bipolar reflection nebulae.
G. D. Schmidt, J. R. P. Angel, E. A. Beaver.
Astrophys. J., Vol. 219, 477 - 486 (1978).

The authors present maps of linear polarization and surface brightness for the scattering lobes of CRL 2688 and M 1-92. These are used to infer the characteristics of illumination of the dust, the scattering optical depth and grain mass of the lobes, and certain properties of the scattering grains. The relation between bipolar reflection nebulae and planetary nebulae is discussed, and a possible mechanism for the production of bipolar nebulae is suggested.

134.005 Proper motions in the outer shell of Eta Carinae.
N. R. Walborn, B. M. Blanco, A. D. Thackeray.

Astrophys. J., Vol. 219, 498 - 503 (1978).

Proper motions have been determined for discrete features in the complex outer shell of η Carinae, from plates taken at Radcliffe Observatory during 1949–1950 and at Cerro Tololo Inter-American Observatory during 1975. Two condensations north of the central object have tangential velocities in excess of 1000 km s^{-1}, the highest yet observed in the η Car nebulosity. On the other hand, the outermost features move more slowly, which implies ejection possibly during the fifteenth century, on the assumption of uniform motions.

134.006 Spectra of Cassiopeia A. II. Interpretation.
R. A. Chevalier, R. P. Kirshner.
Astrophys. J., Vol. 219, 931 - 941 (1978).

The observations of Cas A presented in Paper I are discussed in the context of shock wave emission models computed by Raymond. The models for the brightest quasi-stationary flocculi indicate an overabundance of He and N relative to H by an order of magnitude. This implies that the gas was originally part of the H-burning core and that presupernova mass loss penetrated this region. The abundances of the fast-moving material support the hypothesis that this is uncontaminated material from the core of a massive star. Relative to O, the authors have abundance estimates for S, Ar, Ca, Fe, and Ne and have upper limits for H, He, N, Mg, and C. These results are compared to the stellar evolution calculations of Arnett and Schramm.

134.007 A trip to the Crab Nebula.   F. D. Seward.
J. British Interplanet. Soc., Vol. 31, 83 - 92 (1978).

Because so many observations of the Crab Nebula have been made, physical conditions inside it can be calculated. In this paper, the author assumes that the Earth is transported to a point inside the Crab Nebula, and he compares the environment there with the environment here.

134.008 Infrared polarimetry of three bipolar nebulae.
T. J. Jones, H. M. Dyck.
Astrophys. J., Vol. 220, 159 - 164 (1978).

The authors present measurements of the linear polarization of three bipolar nebulae–CRL 2688 (the Egg Nebula), R Mon, and HD 44179 (the Red Rectangle)–at 1.2 and 2.2 μm and in the visible. All three sources show a decrease in polarization longward of 1 μm which is interpreted in terms of scattering by dust grains. There is no evidence for grains larger than 0.3 μm in radius in any of the sources. For the Egg Nebula there is enough information to extract the intrinsic spectrum of the central source.

134.009 Some remarkable spectra in the outskirts of the Crab Nebula.   K. Davidson.
Astrophys. J., Vol. 220, 177 - 185 (1978).

Spectra have been obtained of regions near the edges of the detectable supernova remnant. Certain characteristics of these are almost unprecedented among nebular spectra.

134.010 Spectrophotometry of filaments in the Crab Nebula.

J. S. Miller.
Astrophys. J., Vol. 220, 490 - 499 (1978) = Lick Obs. Bull., No. 776.

Photoelectric spectrophotometry of filaments in the Crab Nebula has been obtained and intensities for a number of emission lines between 3400 and 7400 Å are presented. Temperatures determined from [O III] lines range from 11,400 to 16,100 K, while the low-ionization species N^+, O^+, and S^+ indicate lower temperatures near 8000 K. Abundance calculations suggest that oxygen, nitrogen, and neon, relative to hydrogen, are within a factor of 10 of what is considered normal. There is little doubt that helium is considerably overabundant, and it is likely that the Crab Nebula contains over 1 $M_\odot$ of helium.

134.011   The spectrum of the Antares nebula.
        J. P. Swings, G. W. Preston.
Astrophys. J., Vol. 220, 883 - 886 (1978).

The [Fe II]-rich nebula around the B companion of Antares is shown to be asymmetrical, with a region of diameter about 3.''5 exhibiting strong lines, surrounded by a zone of weaker lines whose NW—SE extent may reach 15'' (i.e., ~2000 AU). The electron density and temperature cannot be determined from the plates covering the wavelength region 3200—4500 Å because of important deviations from the statistical distribution of the populations of the metastable levels of Fe II.

134.012   A precision measurement of the X-ray polarization of the Crab Nebula without pulsar contamination.
M. C. Weisskopf, E. H. Silver, H. L. Kestenbaum, K. S. Long, R. Novick.
Astrophys. J., Lett., Vol. 220, L117 - L121 (1978).

The linear X-ray polarization of the Crab Nebula has been precisely measured at 2.6 keV and 5.2 keV with the OSO 8 graphite crystal polarimeters. The 1.4 ms time resolution of these instruments permitted the removal of any contribution to the polarization from the pulsar. The nebular polarization is 19.2% ± 1.0% at a position angle of 156.°4 ± 1.°4 at 2.6 keV. At 5.2 keV the corresponding results are 19.5% ± 2.8% at 152.°6 ± 4.°0.

134.013   OH observations of four bipolar nebulae.
        J. D. Fix, R. L. Mutel.
Astrophys. J., Lett., Vol. 19, 37 - 38 (1977).

A high sensitivity search for OH emission from the four bipolar nebulae M3-99, Parsamyan 21, CRL 2688, and M1-92 has been made. Positive results were obtained only for M1-92. The absence of significant circular polarization at the main lines makes M1-92 untypical either of other bipolar nebulae or stellar OH masers with associated nebulosities.

134.014   The Merope nebula revisited: the phase function of dust grains in the ultraviolet.   A. N. Witt.
Publ. Astron. Soc. Pacific, Vol. 89, 750 - 757 (1977/78).

New photoelectric surface brightness measurements of the Merope reflection nebula in 14 positions are reported. The change of the surface brightness distribution with wavelength south of and immediately surrounding 23 Tau has been analyzed with the aid of suitable multiple scattering models. It is concluded that the far ultraviolet scattering and the steep increase in interstellar extinction in the far ultraviolet are caused primarily by very small (10^{-6} cm or less) grains with an isotropic scattering phase function.

134.015   Studies of bipolar nebulae — V. The general phenomenon.   N. Calvet, M. Cohen.
Mon. Not. R. Astron. Soc., Vol. 182, 687 - 704 (1978).

Optical spectrophotometry, infrared photometry and radio continuum observations are combined to investigate the class of bipolar nebulae defined purely morphologically. The evolutionary status of these objects is considered. A few re-

present pre-main sequence objects; most seem to be protoplanetary nebulae; and a small number are evolved planetary nebulae.

134.016   Studies of bipolar nebulae. IV. Mz 3 (= PK 331 − 1°1).
        M. Cohen, M. P. FitzGerald, W. Kunkel, B. M. Lasker, P. S. Osmer.
Astrophys. J., Vol. 221, 151 - 162 (1978) = Contrib. Univ. Waterloo Obs., Waterloo, Ontario, Canada, No. 64.

Optical photography and spectroscopy of the bipolar nebula Mz 3 and its exciting star are combined with infrared observations and preexisting radio data. Elemental abundances are obtained for the bright lobes of this nebula. The nebula is characterized by T_e = 17,000 K and N_e = 18,000 cm^{-3}, from the forbidden emission lines, and an overall mean density of 2300 cm^{-3}. The spectral type of the exciting star is O9.5 or B0. Multiaperture infrared observations and star counts around Mz 3 strongly suggest the existence of a circumstellar dust and gas disk. It is suggested that Mz 3 may represent a protoplanetary nebula.

134.017   A search for the ground state S(2) line of molecular hydrogen in the Orion nebula.
E. T. Young, R. F. Knacke.
Bull. American Astron. Soc., Vol. 9, 576 (1978). — Abstract.

134.018   Structure of the Orion Nebula: the ionized gas.
        V. Pankonin, M. Walmsley, M. Harwit.
Bull. American Astron. Soc., Vol. 9, 611 (1978). — Abstract.

134.019   The kinematics of the Rosette nebula.
        G. Rossano.
Bull. American Astron. Soc., Vol. 9, 612 (1978). — Abstract.

134.020   Impact-broadening theory confirmed by observations of radiofrequency recombination lines W51.
K. R. Lang, R. F. Willson.
Mon. Not. R. Astron. Soc., Vol. 183, 5P - 9P (1978).

Impact broadening has been detected in the radiofrequency recombination lines emitted by the emission nebula W51 (G 49.5−0.4). A comparison of the impact broadening of the 166α, 239γ and 263δ lines indicates that the average electron density in the 3 arcmin beamwidth is N_e = $10^{3.0 \pm 0.1}$ cm^{-3}. Continuum measurements of W51 suggest that this electron density refers to the integrated contribution of both self-absorbed, dense (N_e = $10^3 - 10^4$ cm^{-3}) compact sources and the tenuous regions between them. The observations are consistent with theoretical predictions of a broadening which increases with the 4.4th power of the principal quantum number.

134.021   Light scattering in an inhomogeneous reflection nebula.   N. V. Voshchinnikov.
Astron. Zh. Akad. Nauk SSSR, Vol. 55, 322 - 327 (1978). In Russian. English translation in Soviet Astron., Vol. 22, No. 2.

A spherical nebula with deficiency of smaller size particles in a region surrounding a central star has been considered. Single scattering was assumed. Mie's theory has been used to calculate the scattering functions for spherical silicate and dirty ice particles. It has been shown that for an inhomogeneous nebula the degree of polarization can decrease considerably, especially in V and R colour bands.

134.022   Polarimetric study of the nebula NGC 6618 and neighbouring stars.
V. A. Hagen-Thorn, N. V. Voshchinnikov.
Astrofizika, Vol. 13, 569 - 586 (1977). In Russian. English translation in Astrophysics, Vol. 13, No. 4.

Results of polarimetric observations for six areas of the nebula and for twenty-six neighbouring stars are presented. For all areas of the nebula the degree of polarization in the continuum is found to be higher than that in the emission lines. Thus the existence of intrinsic polarization in the nebula is evident. The determination of the parameters of interstellar polarization by the observations of neighbouring stars may be done with some confidence only for western areas of the nebula. These parameters are found to be equal to those observed in emission lines. The intrinsic polarization in the continuum is shown to arise from scattering of the radiation of stars settled in the dark lane on the dust located in the nebula.

134.023 Spectroscopy of the Crab nebula.
R. A. Fesen, R. P. Kirshner, R. A. Chevalier.
Publ. Astron. Soc. Pacific, Vol. 90, 32 - 35 (1978).

Detailed inspection of a spectrogram of the Crab nebula covering $\lambda\lambda 4450-6100$ reveals the presence of faint forbidden lines of iron and nitrogen, most of which have not been observed before in the Crab. Only some filaments showed these emission features which are of low ionization and are emitted near the high density cores of filaments.

134.024 Observations of the Rosette nebula NGC 2237 at decametric wavelengths.   V. V. Krymkin.
Astrophys. Space Sci., Vol. 54, 187 - 197 (1978).

The results of observations of the Rosette emission nebula NGC 2237 with the radio telescope UTR-2 at frequencies 12.6, 14.7, 16.7, 20.0 and 25.0 MHz are given in the form of contours of constant brightness temperature. Density weighted mean values for the non-thermal radio emissivity between the Sun and the source (7.9×10^{-41} W m^{-3} Hz^{-1} $ster^{-1}$ at 25.0 MHz) and the ratio of the intensity of emissivity generated before the area and the intensity of galactic radio emissivity appearing beyond the area equal to 1.3 have been obtained. The electron temperature T_e = 3600 K, the optical depth (about unit at 25 MHz), the measure of emission (ME = 3500 cm^{-6} pc), the electron density N_e = 8 cm^{-3} and the nebular mass $16.6 \times 10^{+3}$ $M_\odot$ have been determined.

134.025 On Herbig—Haro objects.   A. L. Gyul'budagyan.
Dokl. AN ArmSSR, Vol. 65, 35 - 38 (1977). In Russian. – Abstr. in Ref. zh., 51. Astron., 3.51.608 (1978).

134.026 Magneto-parametric instabilities in the Crab Nebula: II.   G. S. S. Sweeney, P. Stewart.
Astron. Astrophys., Vol. 66, 139 - 153 (1978).

The authors analyse cold relativistic ion-electron and electron-positron models of the plasma instabilities associated with the propagation of a strong low frequency electromagnetic wave in a magnetised plasma in both the linear and non-linear regimes. Results are found for a large range of parameters including those which may be applicable to the Crab Nebula.

134.027 Ultraviolet photometry from the Orbiting Astronomical Observatory. XXX. The Orion reflection nebulosity.   A. N. Witt, C. F. Lillie.
Astrophys. J., Vol. 222, 909 - 916 (1978).

New surface-brightness measurements are presented that cover the region of Orion in nine intermediate-width band-passes ranging from 4250 to 1550 Å. The authors confirm the existence of an extended ultraviolet reflection nebulosity in this area, and derive the characteristics of its spectrum and its spatial distribution. The observations are consistent with a model in which the dense molecular cloud complex in Orion is illuminated by the foreground Orion aggregate of early-type stars.

134.028 Observations of the 51.8 micron [O III] emission line in Orion.   G. Melnick, G. E. Gull, M. Harwit, D. B. Ward.
Astrophys. J., Lett., Vol. 222, L137 - L140 (1978).
The authors report the first observations of the 51.8 μm fine-structure transition $p^2 : {}^3P_2 \to {}^3P_1$ for doubly ionized oxygen.

The observed line strength in the Orion Nebula is $5 \pm 3 \times 10^{-15}$ W cm^{-2}, in good agreement with the theoretical predictions of Simpson.

134.029 Optical colours and polarization of a model reflection nebula. III. Composite and mixture of grains in the nebula with the star in the rear.   G. A. Shah.
Pramāna, Vol. 9, 461 - 470 (1977). – Abstr. in Phys. Abstr., Vol. 81, Abstr. 16502 (1978).

134.030 Radio spectroscopy of C II regions associated with reflection nebulae: NGC 2023, M78, and others.
V. Pankonin, C. M. Walmsley.
Astron. Astrophys., Vol. 67, 129 - 137 (1978).

Observations were made of carbon recombination line spectra toward several reflection nebulae which are illuminated by B type stars. OH and H_2CO spectra were observed in order to sample some of the characteristics of the molecular clouds in the vicinity of these nebulae.

134.031 The Gum nebula.   R. J. Reynolds.
Phys. Teach., Vol. 15, 497 - 499 (1977). – Abstr. in Phys. Abstr., Vol. 81, Abstr. 24545 (1978).

134.032 Cometary nebula near NGC 7023.
Yamamoto Circ., No. 1874 (1978). In Japanese.

134.033 Is gas in the Orion nebula depleted?
S. Aiello, I. Guidi, M. Perinotto.
Astrophys. Space Sci., Vol. 54, 417 - 423 (1978).

Depletion of heavy elements has been recognized to be important in the understanding of the chemical composition of the interstellar medium. This problem is also relevant to the study of H II regions. This paper investigates the gaseous depletion in the physical conditions of the Orion nebula. The authors conclude that very probably no depletion of heavy elements, due to sticking on dust grains, took place during the lifetime of the Orion nebula.

134.034 Observations of the molecular hydrogen emission from the Orion Nebula.   S. Beckwith, S. E. Persson, G. Neugebauer, E. E. Becklin.
Astrophys. J., Vol. 223, 464 - 470 (1978).

Observations of the molecular hydrogen emission at 2 μm in Orion are presented. The data consist of maps at both 13"and 5"resolution in the $v = 1 \to 0$ $S(1)$ emission line; intensity measurements of the $v = 1 \to 0$ $S(0)$, $S(1)$, $S(2)$, and $Q(3)$ transitions and of the $v = 2 \to 1$ $S(1)$ transition are used to derive a vibrational temperature of 2000 ± 300 K and the associated column densities at the two peaks. The data presented here improve upon previous data of the Orion H_2 emission and support the suggestion that the hydrogen is excited in a thin sheet.

134.035 Note on the variability of the cometary nebula at 20^h45^m, $+67°8$.   W. Wenzel.
Mitt. Veränderl. Sterne (MVS), Band 8, 19 (1977).
Some scattered observations on Sonneberg plates from 1939 to 1962 confirm the strange variability of the object.

134.036 Observations of inversion lines of nonmetastable states of $^{14}NH_3$ in Orion.
H. J. Nystrom, P. Palmer, B. Zuckerman.
Bull. American Astron. Soc., Vol. 10, 393 (1978). – Abstract.

134.037 High resolution Fabry-Perot observations of 2.1 micron H_2 emission from the Orion nebula.
P. M. Ogden, F. L. Roesler, R. J. Reynolds, F. Scherb, H. P. Larson, M. Daehler.
Bull. American Astron. Soc., Vol. 10, 395 (1978). – Abstract.

134.038 An echelle spectrograph study of the Rosette nebula.

W. F. Fountain, G. A. Gary, C. R. O'Dell.
Bull. American Astron. Soc., Vol. 10, 397 - 398 (1978).
Abstract.

134.039 Nitrogen and helium abundances in NGC 6888.
R. A. R. Parker.
Bull. American Astron. Soc., Vol. 10, 399 (1978). – Abstract.

134.040 Emission line intensities in the Crab nebula.
K. Davidson.
Bull. American Astron. Soc., Vol. 10, 426 - 427 (1978).
Abstract.

134.041 Photographs of the Orion Nebula in Hα, Hβ and He I λ 10830.
C. E. Gow, M. T. Sandford, R. K. Honeycutt.
Astron. Astrophys., Vol. 67, 435 - 436 (1978).
A reddening map of the region constructed by taking the ratio of the Hβ and Hα photographs is compared to the intensity distribution in Hα and He I.

134.042 On cometary nebulae.   A. L. Gyul'budagyan.
Astron. Tsirk., No. 967, p. 6 - 8 (1977). In Russian.

134.043 On the influence of magnetic fields on the polarization of stars connected with some cometary nebulae.   A. L. Gyul'budagyan.
Astron. Tsirk., No. 968, p. 6 - 8 (1977). In Russian.

134.044 Possible gamma-ray line from the Crab Nebula.
M. Leventhal, C. J. MacCallum, A. C. Watts.
Ann. New York Acad. Sci., Vol. 302, (see 012.033), 532 - 537 (1977).

134.045 Spectrophotometric study of the cometary nebula NGC 2261 and its nucleus R Mon.
J. L. Greenstein, M. A. Kazaryan, T. Yu. Magakyan, Eh. E. Khachikyan.
Flare stars. Symposium 1976, (see 012.035), p. 125 - 126. In Russian. – Abstr. in Ref. zh., 51. Astron., 7.51.653 (1978).

134.046 On the formation of a new cometary nebula.
A. L. Gyul'budagyan, A. S. Amirkhanyan.
Flare stars. Symposium 1976, (see 012.035), p. 127 - 128. In Russian. – Abstr. in Ref. zh., 51. Astron., 7.51.654 (1978).

134.047 On the variability of a cometary nebula.
T. Yu. Magakyan, A. S. Amirkhanyan.
Flare stars. Symposium 1976, (see 012.035), p. 129 - 130. In Russian. – Abstr. in Ref. zh., 51. Astron., 7.51.655 (1978).

Electron collisional excitation cross sections for Fe III and Fe VI and iron abundances in gaseous nebulae. See Abstr. 022.059.

Parametric instabilities III.   See Abstr. 061.017.

Stellar winds and neutral gas in diffuse nebulae. See Abstr. 064.007.

Spectrophotometry of nebulae around Wolf-Rayet stars.   See Abstr. 064.094.

Interplanetary scintillations of PSR 0531 + 21 at 74 MHz.   See Abstr. 106.004.

On the photometric variations of the red giant HD 65750 and of the surrounding reflection nebula IC 2220. See Abstr. 113.017.

Stars in reflection nebulae near the Herbig-Haro objects in the Gum nebula.   See Abstr. 113.022.

A survey of extremely red stars in the Orion region with 40/70/120 cm Schmidt telescope.   See Abstr. 113.058.

Infrared Fe II lines in Eta Carinae and a possible interpretation of infrared excesses. See Abstr. 114.504.

On the photometric variations of the irregular variable red giant HD 65750 in the reflection nebula IC 2220. See Abstr. 122.090.

Deuterated ammonia toward the Orion Nebula. See Abstr. 131.011.

CO observations of a high-latitude molecular cloud associated with NGC 7023.   See Abstr. 131.053.

CO observations of a high latitude molecular cloud associated with NGC 7023.   See Abstr. 131.103.

Detection of $H_2^{18}O$ and an abundance estimate for interstellar water.   See Abstr. 131.164.

A search for emission from vibrationally excited H_2.   See Abstr. 131.202.

Measurements of interstellar extinction in the Orion nebula from the International Ultraviolet Explorer. See Abstr. 131.233.

The structure of the emission nebula M 1–78 at 15.4 GHz.   See Abstr. 132.015.

M43–an emission nebula in Orion. See Abstr. 132.042.

A high spatial resolution map of the Orion Nebula at 33 μ.   See Abstr. 133.003.

Lamellar grating observations of the Orion Nebula from 100 to 500 microns.   See Abstr. 133.005.

Far-infrared observations of sources associated with double-lobed reflection nebulae.   See Abstr. 133.026.

The [O III] lines as a quantitative indicator of nebular central-star temperature.   See Abstr. 135.016.

ANS: X-rays from the direction of the Orion Nebula (M 42).   See Abstr. 142.011.

Very-low-excitation compact nebulae in the Magellanic Clouds.   See Abstr. 159.005.

135 Planetary Nebulae

135.001 The complex motions of the neutral and ionized gas within the Dumb-bell Nebula (NGC 6853) — II.
C. Goudis, D. McMullan, J. Meaburn, N. J. Tebbutt, D. L. Terrett.
Mon. Not. R. Astron. Soc., Vol. 182, 13 - 25 (1978).

An electronographic insect-eye Fabry-Perot spectrograph has been used to obtain profiles of the [O III], [N II] and [O I] emission lines at many positions over the Dumb-bell Nebula. A cylindrical (or ellipsoidal) shell model of the nebula, expanding radially with respect to its axis, and not the central star, explains many of the observations. Very unusual motions are observed near the exciting star.

135.002 Detection of molecular hydrogen emission from five planetary nebulae.
S. Beckwith, S. E. Persson, I. Gatley.
Astrophys. J., Lett., Vol. 219, L33 - L38 (1978).

The $v = 1 \rightarrow 0$ $S(1)$ line of molecular hydrogen has been detected in five planetary nebulae. They are the Ring Nebula (M57, NGC 6720), BD+30°3639, Hb 12, CRL 618, and CRL 2688. A region in the northeast of the Ring Nebula has been mapped in both the $v = 1 \rightarrow 0$ $S(1)$ molecular hydrogen line and the Brackett γ line of atomic hydrogen. The H_2 emission is not spatially correlated with the $B\gamma$, but is correlated with the [O I] emission as determined from interference filter photographs.

135.003 Ultraviolet radiation from planetary nebulae. II. Radiation from the central stars.
S. R. Pottasch, P. R. Wesselius, C.-C. Wu, H. Fieten, R. J. van Duinen.
Astron. Astrophys., Vol. 62, 95 - 110 (1978).

Intermediate band measurements of planetary nebulae have been made in five spectral regions between λ 1500 Å and λ 3300 Å. The authors discuss the separation of the nebular light from that of the central star for 19 nebulae and present the flux originating from the central star in this spectral region. The flux is compared with blackbody radiation and simple model atmospheres. The effective temperatures of the central stars determined with the help of the UV data are sometimes different from values used in the literature. This has consequences for the luminosity. On the basis of distances taken from the literature, the authors also discuss the radii of the central stars and their position in the Hertzsprung-Russell diagram.

135.004 Gas-phase abundances of iron and carbon in planetary nebulae. G. A. Shields.
Astrophys. J., Vol. 219, 559 - 564 (1978).

Five high-excitation planetary nebulae are found to have order-of-magnitude deficiencies of iron in the gas phase, in agreement with an earlier result for NGC 7027. Most of the nebulae cluster in the range [Fe/O] = −1.2±0.4. These results indicate that efficient condensation of iron into grains is characteristic of planetary nebulae. Carbon abundances are determined from recombination lines of C II, III, and IV. The nebulae all have carbon abundances larger than solar. Most of the total carbon may be in the gas phase in contrast to the case for iron. Empirical collision strengths for [Fe III], [Fe V], and [Fe VI] are derived from line intensities of NGC 7027 and a detailed model of its ionization structure.

135.005 Ionization structure and composition of the planetary nebula NGC 7027. G. A. Shields.
Astrophys. J., Vol. 219, 565 - 574 (1978).

The emission-line spectrum of NGC 7027 is analyzed with the aid of detailed models based on photoionization equilibrium theory. Chemical abundances are derived for He, C, N, O,

Ne, Ar, and Fe. A C/O ratio ~6 times larger than solar is indicated by optical recombination lines, forbidden lines, and ultraviolet lines. N/O is more than twice the solar value. Fe/O is a factor 25 lower than solar; the author shows that condensation of iron into grains does not conflict with the absence of internal reddening in NGC 7027.

135.006 The rocket-ultraviolet spectrum and models of the planetary nebula NGC 7662.
R. C. Bohlin, J. P. Harrington, T. P. Stecher.
Astrophys. J., Vol. 219, 575 - 584 (1978).

An ultraviolet spectrum of NGC 7662 was obtained with a rocket-borne telescope in a 130 s exposure by using a microchannel plate detector and film. The observed fluxes are given for both the lines and the continuum on an absolute basis. The correction for interstellar extinction with $E(B-V) = 0.22$ has been determined on the basis of the observed and calculated line ratios for the hydrogenic recombination line of He II at 1640 Å to Hβ. A detailed set of models of the thermal and ionization structure of NGC 7662 have been constructed for the purpose of interpreting the rocket observations. Predictions of three models are presented and compared with the ultraviolet as well as the existing visible, infrared, and radio observations. The spectral energy distribution of the central star for the models was specified so that the observed emission lines could be fitted.

135.007 Spectrophotometry of planetary nebulae. I. Physical conditions. T. Barker.
Astrophys. J., Vol. 219, 914 - 930 (1978) = Lick Obs. Bull., No. 769.

Emission-line intensities have been measured photoelectrically for 35 planetary nebulae in the spectral region 3400–7400 Å. Reddening-corrected observations of the lower Balmer decrement agree well with theoretical calculations. There is also generally good agreement between the electron temperatures and densities determined from different forbidden-line intensity ratios, although there are some cases of systematic disagreement, especially in the case of [O II]. Electron temperature based on the strength of the Balmer continuum are consistent with those found from forbidden lines. Bowen fluorescent efficiencies are generally close to the predicted value of 0.5.

135.008 On the origin of planetary nebulae.
S. Kwok, C. R. Purton, M. P. FitzGerald.
Astrophys. J., Lett., Vol. 219, L125 - L127 (1978) = Contrib. Univ. Waterloo Obs., Ontario, Canada, No. 66.

The stellar wind from a red giant produces an extensive circumstellar cool nebula of appreciable mass. The authors suggest that in some cases mass loss continues until the hot core of the star is exposed, and that the stellar wind from the remnant star collides with the circumstellar nebula, producing a relatively dense shell of gas which increases in mass and radius at a constant rate. It is shown that such a shell, when ionized by radiation from the central star, has the characteristics of a planetary nebula. V1016 Cygni is used as an example of a star which has recently undergone the transition from a red giant to a young low-mass planetary nebula.

135.009 Ultraviolet observations of planetary nebulae. III. Variability of the central star.
D. P. Gilra, S. R. Pottasch, P. R. Wesselius, R. J. van Duinen.
Astron. Astrophys., Vol. 63, 297 - 301 (1978).

Extensive UV observations of planetary nebulae observed with the Netherlands Astronomical Satellite (ANS) were searched for photometric variations of the central stars. Only five cases of variations were found : IC 418, A 78, He 2-131,

NGC 6543 and V-V 1-7. In no case do the variations exceed about 15%.

135.010   Charge transfer reactions: a consistent model of the planetary nebulae NGC 7027.
D. Péquignot, S. M. V. Aldrovandi, G. Stasinska.
Astron. Astrophys., Vol. 63, 313 - 324 (1978).

It is demonstrated that the ionization structure of NGC 7027 cannot be explained by static models based on the standard photoionization theory unless some complementary physical process is included. The main discrepancies are successfully eliminated by including charge transfer reactions with hydrogen for the following ions: C^+, C^{+3}, N^{+2}, O^{+2}, Ne^{+2}, S^+ and S^{+3}. An excellent fit to the observations is obtained with a model consisting of a high excitation component at $N_H = 2.5 \times 10^5 \, cm^{-3}$ and a lower excitation one at $N_H = 6 \times 10^4 \, cm^{-3}$. The elemental abundances by number are H, He, C, N, O, Ne, Mg, S = 1.0, 0.100, 3.0(−3), 2.0(−4), 7.3(−4), 2.2(−4), 3.5(−5), 1.7(−5), the nitrogen value depending critically on the assumed charge transfer rate for N^{+2}.

135.011   Analysis of the catalogue of spectral line intensities of planetary nebulae.   E. B. Kostyakova.
Tr. Gos. Astron. Inst. Shternberga, Vol. 47, 222 - 231 (1976). In Russian.

Revised absolute intensities of spectral lines for 54 planetary nebulae are given.

135.012   Spectrophotometry of planetary nebulae. II. Chemical abundances.   T. Barker.
Astrophys. J., Vol. 220, 193 - 209 (1978) = Lick Obs. Bull., No. 770.

The relative abundances of H, He, Ne, O, and, less reliably, N and S have been determined for 37 planetary nebulae. Helium abundances are found to be affected by incomplete ionization of helium, an effect that is strongly correlated with the abundance of S^+. Helium abundances in planetaries are also apparently affected by enrichment of nebular material by CNO−processed material. The most important result is that planetaries of widely varying kinematical properties have nearly identical (approximately solar) abundances. Apparently most planetaries belong to a metal-rich population, or oxygen and neon abundances are not representative of heavy element abundances, perhaps because oxygen and neon were synthesized at a faster rate than heavier elements in the early history of the Galaxy.

135.013   Kinematics of planetary nebulae from radial velocities.   J. O. Peralta.
Astron. Astrophys., Vol. 64, 127 - 129 (1978).

Radial velocities of planetary nebulae with $|b| \leq 12°$ are analyzed by means of the theoretical permitted areas in the radial velocity versus galactic longitude diagram. It is found that a maximum eccentricity of 0.2 for the whole sample seems to be an appropriate value, whereas for type B nebulae a maximum eccentricity of 0.1 describes quite well the velocity patterns of these objects at all distances. A new distance scale is adopted.

135.014   The planetary nebula in the Fornax dwarf galaxy.
I. J. Danziger, M. A. Dopita, T. G. Hawarden, B. L. Webster.
Astrophys. J., Vol. 220, 458 - 466 (1978).

A planetary nebula has been identified near the center of the Fornax dwarf elliptical galaxy. Quantitative spectrophotometry reveals that it is slightly reddened, and belongs to an intermediate-excitation class, with a normal helium abundance. Since the abundances of nitrogen, oxygen, and argon relative to hydrogen are all 2 to 3 times lower than in the Orion Nebula, the material in the nebula, which has a mass of at least 0.1 $M_\odot$, probably represents at least a second generation of processed matter. The implications of this conclusion are

briefly discussed.

135.015   Improved abundances in three halo planetary nebulae.   S. A. Hawley, J. S. Miller.
Astrophys. J., Vol. 220, 609 - 613 (1978) = Lick Obs. Bull., No. 777.

The authors have obtained new spectrophotometric observations of three halo planetary nebulae: 49+88°1, 108−76°1, and K648. Using the observed line intensities, they are able to compute temperatures, densities, and abundances of helium, oxygen, nitrogen, and neon with respect to hydrogen. Their data show that, for all three nebulae, the helium abundance is normal but oxygen is underabundant with respect to disk planetaries. That the abundance of nitrogen varies from object to object suggests that mixing in differing amounts may have taken place. In 49+88°1, the only halo planetary with detectable [S II] or [S III] emission, sulfur is underabundant by a factor of about 10.

135.016   The [O III] lines as a quantitative indicator of nebular central-star temperature.   J. B. Kaler.
Astrophys. J., Vol. 220, 887 - 889 (1978).

New and revised blackbody temperatures (T_s) for exciting stars of nebulae are calculated by Stoy's method. In low-excitation optically thick nebulae the $I(\lambda 5007)[O \, III]/I(H\beta)$ ratio correlates well enough with T_s to be used alone as an accurate indicator of the calculated blackbody temperature. For nebulae for which $\lambda 4686$ He II is weak [less than 0.05 $I(H\beta)$] or absent, for which $T_s \lesssim 68,000$ K, it is found that log $T_s = 4.426 + 4.827 \times 10^{-4} I(\lambda 5007) − 1.374 \times 10^{-7} I^2(\lambda 5007)$, where $I(H\beta) = 100$. A similar correlation exists for $I(\lambda 3868)[Ne \, III]$.

135.017   Electron temperatures of four planetary nebulae from radio continuum observations.
Gopal-Krishna.
Mon. Not. R. Astron. Soc., Vol. 182, 723 - 726 (1978).

Lunar occultations of four planetary nebulae, namely NGC 2392, 6445, 6537 and 7009, have been observed at 327 MHz. In each case, the estimated flux density fits a thermal spectrum and the derived angular size at 327 MHz is consistent with the optical size. For NGC 7009 an electron temperature $T_e = 5000^{+1500}_{-1000}$ K is estimated. The above radio estimate of T_e is only about half of that deduced from intensities of the [O III] lines, indicating the possibility of substantial temperature fluctuations within this nebula. For the remaining three nebulae, limits to T_e are derived.

135.018   Spectrophotometry of planetary nebulae. III. Sulfur abundances.   T. Barker.
Astrophys. J., Vol. 221, 145 - 150 (1978).

Photoelectric scanner observations have been made to determine the [S III] $\lambda 9069$ and $\lambda 9532$ line intensities in 20 of the galactic planetary nebulae studied in Papers I and II (1978, in press). All 20 planetaries have essentially solar S abundances, and there is no significant correlation between S abundances and the kinematical properties of the nebulae. Since this result is identical to that found previously for the O and Ne abundances, it is felt that it is likely that planetaries belong to a predominantly metal-rich population.

135.019   The infrared spectrum of BD +30°3639 and possible celestial grain constituents.   R. W. Russell,
R. C. Puetter, B. T. Soifer, S. P. Willner.
Bull. American Astron. Soc., Vol. 9, 582 (1978). − Abstract.

135.020   The radio-continuum properties of the new planetary nebula HM Sagittae.
P. A. Feldman, C. R. Purton, M. Ryle, E. R. Seaquist.
Bull. American Astron. Soc., Vol. 9, 600 - 601 (1978). Abstract.

135.021   VLA radio maps of four planetary nebulae.

B. Balick, R. Hjellming, C. Bignell.
Bull. American Astron. Soc., Vol. 9, 601 (1978). – Abstract.

135.022 **Internal motions in the planetary nebula M1-67.**
P. Pişmiş, E. Recillas-Cruz.
Bull. American Astron. Soc., Vol. 9, 601 (1978). – Abstract.

135.023 **On the origin of planetary nebulae.**
S. Kwok, C. R. Purton, P. M. FitzGerald.
Bull. American Astron. Soc., Vol. 9, 601 (1978). – Abstract.

135.024 **Spectrophotometry of nebulae around Wolf-Rayet stars.** K. B. Kwitter.
Bull. American Astron. Soc., Vol. 9, 644 (1978). – Abstract.

135.025 **The spectrum of the nebula associated with FG Sagittae.** S. A. Hawley, J. S. Miller.
Astrophys. J. Vol. 221, 851 - 853 (1978) = Lick Obs. Bull. No. 778.

The authors present new spectrophotometric observations of the nebulosity associated with FG Sge. The emission-line spectrum appears to be that of a low-ionization planetary nebula. Observations of temperature and density indicators allow chemical abundances to be computed for He, N, O, Ne, and S^+. The derived total abundances are completely normal for planetary nebulae. The lack of observable He^+ $\lambda 4686$ suggests that the central star must have been relatively cool when the nebula was excited.

135.026 **Molecules in planetary nebulae.** J. H. Black.
Astrophys. J., Vol. 222, 125 - 131 (1978).

It is shown that significant abundances of simple molecules, like H_2, H_2^+, HeH^+, OH, and CH^+, can exist in the transition zones of ionized nebulae. The dominant molecular processes in such regions, where neutral atoms and free electrons mingle at moderately high temperatures, are discussed. Equilibrium abundances of molecules are computed for a model transition zone. It is possible that the observed rotation-vibration lines of H_2 in NGC 7027 arise in such a region.

135.027 **Spectral studies of compact planetary nebulae. Objects of moderate excitation.**
L. N. Kondrat'eva.
Astron. Zh. Akad. Nauk SSSR, Vol. 55, 334 - 344 (1978). In Russian. English translation in Soviet Astron., Vol. 22, No. 2.

Emission line intensities of twenty-three planetary nebulae are presented in the 3700–7300 Å region. Corrections for interstellar absorption $C(H\beta)$, electron temperatures and densities are determined. The nebulae studied are classical planetaries, but of lower excitation.

135.028 **Statistics – spatial and velocity distribution of planetary nebulae.** J. H. Cahn, S. P. Wyatt.
IAU Symp. No. 76, (see 012.014), p. 3 - 9 (1978).

The local spatial distribution of planetaries in the galactic plane and perpendicular to the plane is presented. The authors show the results that can be obtained on the kinematical properties of the planetaries in our general part of the galactic system. They conclude by moving farther afield and exploring another part of the Galaxy and the Galaxy as a whole. The distribution of planetaries is presented and recent estimates of the total planetary population of our Galaxy are summarized. The last section is devoted to what little one can conclude about the overall kinematical properties of the system of planetaries.

135.029 **Planetary nebulae in the Magellanic Clouds.**
B. L. Webster.
IAU Symp. No. 76, (see 012.014), p. 11 - 18 (1978).

Selection effects in the search for planetary nebulae in the Magellanic Clouds have not seriously distorted the observed space distribution, but have affected the apparent excitation classes. A difference between the properties of the SMC and LMC planetaries remains. A group of high-excitation nebulae with strong lines of He and N has been isolated. If these are excluded, the helium abundance appears to be normal in each Cloud. The oxygen abundance in LMC planetaries is about half the Orion value and may be even lower in the SMC objects.

135.030 **Planetary nebulae in the Andromeda galaxy and its companions.** H. C. Ford.
IAU Symp. No. 76, (see 012.014), p. 19 - 34 (1978).

135.031 **The distance scale of planetary nebulae.**
W. Liller.
IAU Symp. No. 76, (see 012.014), p. 35 - 45 (1978).

In conclusion, it seems obvious that at least for optically thin nebulae, it is potentially possible to derive accurate distance ($\pm 10\%$). All that is needed is a careful calibration of the scale. At the present time we do not have such a calibration but several should be possible.

135.032 **New and misclassified planetary nebulae.**
L. Kohoutek.
IAU Symp. No. 76, (see 012.014), p. 47 - 62 (1978).

Since the "Catalogue of Galactic Planetary Nebulae" 226 new objects have been classified as planetary nebulae. They are summarized in a table which gives the designations, names, coordinates and the references to the discovery. Further 9 new objects have been added and called 'proto-planetary nebulae', but their status is still uncertain. Only 34 objects have been included in the present list of misclassified planetary nebulae although the number of doubtful cases is much larger.

135.033 **The planetary nebula in the Fornax dwarf galaxy.**
I. J. Danziger, M. Dopita, T. G. Hawarden, B. L. Webster.
IAU Symp. No. 76, (see 012.014), p. 63 (1978). – Abstract.

135.034 **Lists of confirmed planetary nebulae in the Magellanic Clouds.**
N. Sanduleak, D. J. MacConnell, A. G. D. Philip.
IAU Symp. No. 76, (see 012.014), p. 64 (1978). – Abstract.

135.035 **A new synthetic distance scale for planetary nebulae.**
A. Acker.
IAU Symp. No. 76, (see 012.014), p. 65 - 66 (1978). Abstract.

135.036 **The planetary nebula in the galactic cluster NGC 2818.** R. J. Dufour, E. C. Hack.
IAU Symp. No. 76, (see 012.014), p. 66 - 67 (1978). Abstract.

135.037 **Advances in optical studies of planetary nebulae.**
J. S. Miller.
IAU Symp. No. 76, (see 012.014), p. 71 - 77 (1978) = Contrib. Lick Obs., No. 411.

This review contains a brief survey of a few of the issues that have been the concern of optical studies of planetary nebulae since the last IAU Symposium on this subject in 1967. Also included are a short discussion of advances in optical instrumentation and remarks on some selected important problems which should be the concern of future optical investigations of planetaries.

135.038 **Ultraviolet observations of planetary nebulae.**
G. A. Gurzadyan.
IAU Symp. No. 76, (see 012.014), p. 79 - 91 (1978).

Contents: Shortwave spectra of the planetary nebulae IC 2149 and NGC 7027. Wide-band short-wave photometry of planetary nebulae. Continuous spectrum of the central star.

Two-photon emission.

135.039   Advances in ultraviolet observations.
S. R. Pottasch.
IAU Symp. No. 76, (see 012.014), p. 93 - 102 (1978).
Several groups have been involved in recent years in making ultraviolet observations of planetary nebulae from rockets and satellites. About 30 or 40 nebulae have been measured up to the present, some with reasonably high spectral resolution, more with band pass filters of 100 Å or more in width. The author discusses: The accuracy of the measurements. The possibility of using the measurements to determine extinction. The separation of the nebular component from the stellar component. The interpretation of the stellar flux in terms of a stellar temperature. The interpretation of the nebular flux. The interpretation of the C IV emission lines. Central stars with A type spectra.

135.040   Advances in infrared observations of planetary nebulae.   D. M. Rank.
IAU Symp. No. 76, (see 012.014), p. 103 - 110 (1978).
The author concentrates his discussion of emission lines to a relatively small number of recent observations at both shorter and longer wavelengths which have produced new data for line emission. The observations include the emission of permitted lines of atomic hydrogen and helium, forbidden lines of less abundant ions, and the more exotic forbidden transitions of molecular hydrogen.

135.041   Planetary nebulae: advances in radio observations.   Y. Terzian.
IAU Symp. No. 76, (see 012.014), p. 111 - 120 (1978).
The most recent review on planetary nebulae at radio wavelengths appeared in 1974 by Thompson. Major advances in this subject include extensive continuum observations at 5 GHz of southern planetary nebulae; the highly successful aperture synthesis observations providing brightness temperature distributions with a resolution of a few arc seconds; the high sensitivity observations of radio recombination lines which have allowed the unambiguous detection of collisional broadening in NGC 7027; the very recent detection of molecular CO emission from NGC 7027, IC 418 and NGC 6543, and CO and OH emission from possible proto-planetary nebulae.

135.042   Observations of NGC 7662 from 1300 to 2850 Å.
R. C. Bohlin, T. P. Stecher.
IAU Symp. No. 76, (see 012.014), p. 121 (1978). – Abstract

135.043   Interpretation of the ultraviolet spectrum of NGC 7662.
R. C. Bohlin, J. P. Harrington, T. P. Stecher.
IAU Symp. No. 76, (see 012.014), p. 121 - 122 (1978).
Abstract.

135.044   The detailed spectrophotometry of 8 planetary nebulae in the spectral region λλ6,000–11,000.
R. I. Noskova.
IAU Symp. No. 76, (see 012.014), p. 122 (1978). – Abstract.

135.045   New absolute intensities of the emission lines of 15 planetary nebulae. The members of the galactic centre group.
B. A. Vorontsov-Velyaminov (*Vorontsov-Vel'yaminov*), E. G. (*E. B.*) Kostyakova, O. D. Dukuchaeva (*Dokuchaeva*), V. P. Arhipova (*Arkhipova*).
IAU Symp. No. 76, (see 012.014), p. 123 (1978). – Abstract.

135.046   Emission lines in the near infrared spectra of faint planetary nebulae.   Y. Andrillat, L. Houziaux.
IAU Symp. No. 76, (see 012.014), p. 123 - 124 (1978).
Abstract.

135.047   Observations of cool dust in planetary nebulae.
H. Moseley, D. A. Harper.
IAU Symp. No. 76, (see 012.014), p. 124 - 125 (1978).
Abstract.

135.048   Medium resolution spectroscopy of NGC 7027 from 16 to 38 microns.
J. F. McCarthy, W. J. Forrest, J. R. Houck.
IAU Symp. No. 76, (see 012.014), p. 125 - 126 (1978).
Abstract.

135.049   Observations of infrared fine-structure lines: [S III].
L. T. Greenberg.
IAU Symp. No. 76, (see 012.014), p. 126 (1978). – Abstract.

135.050   Observations of infrared fine structure emission.
J. H. Lacy, F. Baas, S. Beck, C. H. Townes.
IAU Symp. No. 76, (see 012.014), p. 127 (1978).
Abstract.

135.051   Radio spectroscopy of planetary nebulae.
E. J. Chaisson.
IAU Symp. No. 76, (see 012.014), p. 127 - 128 (1978).
Abstract.

135.052   Advances in studies of physical processes.
M. J. Seaton.
IAU Symp. No. 76, (see 012.014), p. 131 - 137 (1978).
Contents: Excitation of forbidden lines. Transition probabilities for np^3 $^2D_{5/2} \rightarrow {}^4S$, $^2D_{3/2} \rightarrow {}^4S$. Hydrogenic récombination spectra. Permitted lines other than those of H I, He I and He II.

135.053   Ionization models of planetary nebulae.
J. P. Harrington.
IAU Symp. No. 76, (see 012.014), p. 151 - 157 (1978).

135.054   Emissivities of Fe III and Fe VI and iron abundances in planetary nebulae.
R. H. Garstang, W. D. Robb.
IAU Symp. No. 76, (see 012.014), p. 159 - 160 (1978).
Abstract.

135.055   The Bowen fluorescence mechanism for spherically symmetric planetary nebulae.
R. Wehrse, A. Peraiah.
IAU Symp. No. 76, (see 012.014), p. 160 - 161 (1978).
Abstract.

135.056   Models of planetary nebulae.
M. Perinotto.
IAU Symp. No. 76, (see 012.014), p. 161 (1978). – Abstract.

135.057   A theoretical model of NGC 7027.
D. Péquignot.
IAU Symp. No. 76, (see 012.014), p. 162 (1978). – Abstract.

135.058   The ionisation structure of IC 418.
P. D. Atherton, T. R. Hicks, N. K. Reay, S. P. Worswick.
IAU Symp. No. 76, (see 012.014), p. 163 (1978). – Abstract.

135.059   Electron temperature and density mapping in planetary nebulae.   N. K. Reay, S. P. Worswick.
IAU Symp. No. 76, (see 012.014), p. 163 - 164 (1978).
Abstract.

135.060   The electron temperature in the He^{++} regions in planetary nebulae.   R. Tylenda.
IAU Symp. No. 76, (see 012.014), p. 164 - 165 (1978).
Abstract.

135.061 Spectrophotometry of six planetary nebulae.
C. T. Hua.
IAU Symp. No. 76, (see 012.014), p. 165 (1978). – Abstract.

135.062 The absorption of He II Lyα photons in planetary
nebulae.   D. R. Flower, M. Perinotto.
IAU Symp. No. 76, (see 012.014), p. 166 (1978). – Abstract.

135.063 Echelle studies on the symmetrical stratified plane-
tary nebula NGC 3918.   M. A. Dopita.
IAU Symp. No. 76, (see 012.014), p. 166 (1978).
Abstract.

135.064 On the physical differences between the planetary
nebulae of the galactic centre group and the planetaries of the general galactic field.   E. B. Kostyakova.
IAU Symp. No. 76, (see 012.014), p. 167 (1978). – Abstract.

135.065 On the variability of the planetary nebulae.
E. B. Kostyakova.
IAU Symp. No. 76, (see 012.014), p. 209 - 210 (1978).
Abstract.

135.066 Time-dependent effects in planetary nebulae caused
by thermal pulses in central stars. R. Tylenda.
IAU Symp. No. 76, (see 012.014), p. 211 (1978). – Abstract.

135.067 Chemical abundances in planetary nebulae.
M. Peimbert.
IAU Symp. No. 76, (see 012.014), p. 215 - 224 (1978).
Planetary nebulae (PN) can be divided into four types
depending on their chemical composition. In order of decreasing heavy element abundances the types are: 1) He and N rich,
2) intermediate population, 3) high velocity, and 4) halo population. Well defined gradients across the galactic disk of He, N
and O are derived from type 2)PN. By comparing the O, Ne
and S abundances of PN of types 3) and 4) with the Fe abundances of stars of similar population it is found that the O, Ne
and S enrichment in the Galaxy probably took place before
the Fe enrichment.

135.068 Some aspects of chemical abundance determinations
in planetary nebulae.   L. H. Aller.
IAU Symp. No. 76, (see 012.014), p. 225 - 233 (1978).

135.069 The abundances of He, N, Ne, Ar and Cl.
J. B. Kaler.
IAU Symp. No. 76, (see 012.014), p. 235 - 244 (1978).
It is the purpose of this paper to examine various aspects
of the abundances of the elements He, N, Ne, Ar and Cl. Results from forthcoming papers will be summarized. Broadly,
there are two parts to this work. First is the examination of
N/O and He/H ratios which show variations among nebulae.
Second is the study of the element ratios Ne/O, Ar/O, Cl/O
for which no, or infrequent, variation would be expected. The
unifying theme in this discussion is the examination of various
ion ratios in planetaries as a function of nebular excitation or
central star temperature.

135.070 The uniformity of chemical composition of galactic
planetary nebulae.   T. Barker.
IAU Symp. No. 76, (see 012.014), p. 245 (1978). – Abstract.

135.071 Improved abundances in three halo planetary
nebulae.   S. A. Hawley, J. S. Miller.
IAU Symp. No. 76, (see 012.014), p. 245 - 246 (1978).
Abstract.

135.072 Chemical abundances of planetary nebulae in NGC
185, 205, and 221.   D. C. Jenner, H. C. Ford.
IAU Symp. No. 76, (see 012.014), p. 246 (1978). – Abstract.

135.073 Theoretical abundances in planetaries.
A. Finzi, R. Yahel.
IAU Symp. No. 76, (see 012.014), p. 247 (1978). – Abstract.

135.074 Aluminum, titanium and magnesium abundances in
planetary nebula IC 2149.   G. A. Gurzadyan.
IAU Symp. No. 76, (see 012.014), p. 247 - 248 (1978).
Abstract.

135.075 Evolution and gas dynamics of planetary nebulae.
W. G. Mathews.
IAU Symp. No. 76, (see 012.014), p. 251 - 261 (1978).
Since the last IAU Symposium on Planetary Nebulae a
number of significant theoretical and observational advances
have been made in the study of nebular dynamics. This review
emphasizes the evolution of nebulae after they have been
ejected from the central star and have become optically thin
to most radiation.

135.076 Morphology of planetary nebulae.
E. R. Capriotti.
IAU Symp. No. 76, (see 012.014), p. 263 - 273 (1978).

135.077 Dust in planetary nebulae: observational considerations.   B. Balick.
IAU Symp. No. 76, (see 012.014), p. 275 - 279 (1978).
The author reviews the recent observational literature
which relates to questions concerning the dust in planetary
nebulae. He explores the amount of dust, its temperature, and
briefly reopens questions of its chemical composition.

135.078 Dust in planetary nebulae.   J. S. Mathis.
IAU Symp. No. 76, (see 012.014), p. 281 - 287
(1978).
The author concentrates on theoretical aspects of dust
in planetary nebulae (PN). He considers the questions, how
much dust is there in PN? What is its composition? What
effects does it have on the ionization structure? On the dynamics of the nebula?

135.079 The filamentary structure of the Ring Nebula.
L. E. Goad.
IAU Symp. No. 76, (see 012.014), p. 289 (1978). – Abstract.

135.080 The structure of NGC 6720: the Ring Nebula in
Lyra.   P. D. Atherton, T. R. Hicks, N. K. Reay,
S. P. Worswick, W. H. Smith.
IAU Symp. No. 76, (see 012.014), p. 290 - 291 (1978).
Abstract.

135.081 Examples of multiple shell structures in planetary
nebulae and certain peculiar emission nebulae.
T. R. Gull.
IAU Symp. No. 76, (see 012.014), p. 291 - 292 (1978).
Abstract.

135.082 Dynamical effects of radiation pressure in evolved
planetary nebulae. C. B. Tarter, J. C. Weisheit.
IAU Symp. No. 76, (see 012.014), p. 292 (1978). – Abstract.

135.083 Some properties of dust in planetary nebulae.
N. Panagia.
IAU Symp. No. 76, (see 012.014), p. 292 (1978). – Abstract.

135.084 Origin of planetary nebulae.   I. W. Roxburgh.
IAU Symp. No. 76, (see 012.014), p. 295 - 303
(1978).
The first task is to see what the theory of stellar evolution has to say about the internal structure and evolution of
stars from the main sequence to the giant phase. The second
task is to highlight the uncertainties in these calculations. The
author considers various proposed theories and comments on

them. He considers the problem of binary stars.

135.085   Optical histories of some possible embryonic plane-
            tary nebulae.   P. A. Feldman, C. R. Purton.
IAU Symp. No. 76, (see 012.014), p. 326 - 327 (1978).
Abstract.

135.086   Millimeter-wave observations of planetary nebulae.
            H. M. Johnson.
IAU Symp. No. 76, (see 012.014), p. 328 (1978). − Abstract.

135.087   The "Butterfly" nebula M2−9: its possible relation
            to B[e] stars and proto-planetaries.
Y. Andrillat, J. P. Swings.
IAU Symp. No. 76, (see 012.014), p. 328 - 329 (1978).
Abstract.

135.088   Planetary nebulae and the interstellar medium.
            E. E. Salpeter.
IAU Symp. No. 76, (see 012.014), p. 333 - 339 (1978).
    The requirements of the interstellar medium for ioniza-
tion, heating and "stirring" are reviewed. The role of planetary nebulae is compared with that of supernova remnants and
of OB-stars.

135.089   Planetary nebulae and chemical evolution of the
            Galaxy.   B. M. Tinsley.
IAU Symp. No. 76, (see 012.014), p. 341 - 352 (1978).
    The aim of this paper is to illustrate methods and con-
cepts that can be developed in the light of growing knowledge
about PN and their precursors.

135.090   Planetary nebulae and white dwarfs.
            V. Weidemann.
IAU Symp. No. 76, (see 012.014), p. 353 (1978).
Abstract.

135.091   Gas-phase abundance anomalies and the origin of
            planetary nebulae.   G. A. Shields, J. M. Scalo.
IAU Symp. No. 76, (see 012.014), p. 354 (1978). − Abstract.

135.092   Model analysis of 108−76°1.
            B. Balick, G. O. Boeshaar.
IAU Symp. No. 76, (see 012.014), p. 354 - 355 (1978).
Abstract.

135.093   A study of the planetary nebulae Abell 30 and
            Abell 78.   M. Cohen, H. S. Hudson,
S. L. O'Dell, W. A. Stein.
IAU Symp. No. 76, (see 012.014), p. 356 (1978). − Abstract.

135.094   Changes in emission line intensities of HBV475.
            S. Tamura.
IAU Symp. No. 76, (see 012.014), p. 357 (1978).
Abstract.

135.095   Density distribution and chemical abundances in
            NGC 7027.   A. Preite-Martinez, N. Panagia.
IAU Symp. No. 76, (see 012.014), p. 357 (1978). − Abstract.

135.096   Emission-line widths in the spectrum of the possible
            proto-planetary nebula HM Sagittae.
G. Wallerstein.
Publ. Astron. Soc. Pacific, Vol. 90, 36 - 38 (1978).
    The author has observed the new emission-line object
HM Sge in the region of λ6200 to λ6700 at 6 Å mm^{-1}. The
line profiles show a huge disparity in width ranging from
35 km sec^{-1} for [N II] to 1700 km sec^{-1} for Hα. He suggests
the interpretation that the great width of Hα is due to electron
scattering in the central source while most of the other lines
are broadened kinematically as part of the expanding
planetary nebula. The very narrow [N II] lines may be formed

in a larger region that was formerly the expanding envelope
of a cool star.

135.097   Ionization and abundances in the Dumbbell
            nebula.   S. A. Hawley, J. S. Miller.
Publ. Astron. Soc. Pacific, Vol. 90, 39 - 44 (1978) = Lick
Obs. Bull. No. 785.
    The authors present line intensities measured at six
positions in the Dumbbell nebula exhibiting a wide range of
ionization. At each position the electron density, temperatures,
and ionic abundances of helium, oxygen, nitrogen, neon, and
sulfur are computed. The authors estimate the total abundances of nitrogen, neon, and sulfur by assuming that the
oxygen ionization distribution is a good guide to the ionization distribution of these other elements.

135.098   The spectrum of HM Sagittae: a planetary nebula
            excited by a Wolf-Rayet star.   L. W. Brown,
W. A. Feibelman, R. W. Hobbs, C. W. McCracken.
Astrophys. Lett., Vol. 19, 75 - 79 (1978).
    Eight image tube spectrograms of HM Sagittae were ob-
tained during July and August, 1977. More than 70 emission
lines, including several broad emission features, have been
identified. An analysis of the spectra indicates that HM Sge is
a planetary nebula excited by a Wolf-Rayet star. The most
conspicuous Wolf-Rayet feature is that attributed to a blend
of C III at 4650 Å and He II at 4686 Å.

135.099   Photoelectric photometry and physical conditions
            of planetary nebulae.
S. Torres-Peimbert, M. Peimbert.
Rev. Mexicana Astron. Astrofis., Vol. 2, 181 - 207 (1977).
    Photoelectric spectrophotometry of emission lines in the
3400−7400 Å range is presented for thirty-three planetary
nebulae, covering a galactocentric radial range of about 10 kpc.
These observations are analyzed to obtain information about
individual objects and our galaxy. The logarithmic abundances
by number of a typical planetary nebula of the solar neighborhood are H = 12.00, He = 11.04, C = 9.5, N = 8.33, O = 8.87
and Ne = 8.28. The He/H, O/H and Ne/H abundance ratios
are very similar to those of the Orion Nebula while the C/H
and N/H abundance ratios are larger by factors of nine and
four, respectively.

135.100   The structure of NGC 6720.
            P. D. Atherton, T. R. Hicks, N. K. Reay, S. P.
Worswick, W. Hayden Smith.
Astron. Astrophys., Vol. 66, 297 - 305 (1978).
    A Fabry-Perot interferometer has been used to measure
velocity structure, at the [O III] λ 5007 Å wavelength, in
the envelope of NGC 6720, and an electronographic image
converter has been used to obtain monochromatic exposures
in emission lines covering a wide range of ionisation potential. These data are shown to imply that the envelope is
shell-like in form and not toroidal, as previously suggested.

135.101   Recent infrared spectroscopy of planetary nebulae.
            B. Jones, S. P. Willner.
Bull. American Astron. Soc., Vol. 10, 398 (1978). − Abstract.

135.102   90 GHz and 151 GHz observations of planetary
            nebulae.   P. A. Marionni, R. P. Sinha, R. W. Hobbs.
Bull. American Astron. Soc., Vol. 10, 398 (1978). − Abstract.

135.103   Charge transfer involving C^+ and N^+ in planetary
            nebulae.   S. E. Butler.
Bull. American Astron. Soc., Vol. 10, 398 (1978). − Abstract.

135.104   Planetary nebulae in the nuclei of M31 and M32.
            D. G. Lawrie, H. C. Ford.
Bull. American Astron. Soc., Vol. 10, 398 (1978). − Abstract.

135.105 The enrichment of helium and nitrogen in planetary nebulae. J. B. Kaler.
Bull. American Astron. Soc., Vol. 10, 399 (1978). – Abstract.

135.106 HM Sge as a young planetary nebula.
S. Kwok, C. R. Purton.
Bull. American Astron. Soc., Vol. 10, 399 (1978). – Abstract.

135.107 Absolute spectrophotometry of 3 planetary nebulae.
E. B. Kostyakova, N. K. Mal'shakova, N. A. Gorynya.
Astron. Tsirk., No. 976, p. 1 - 2 (1977). In Russian.

135.108 Le nebulose planetarie. M. Perinotto.
Mem. Soc. Astron. Italiana, Vol. 47, 177 - 209(1976).
A summary is given of the present knowledge of planetary nebulae, focusing on the interpretation of the optical, radio and infrared spectrum.

135.109 Theoretical predictions of the ultra-violet spectra of planetary nebulae. D. R. Flower.
Mem. Soc. Astron. Italiana, Vol. 47, (see 012.036), 313 - 335 (1976).
The author discusses those aspects of the physics of planetary nebulae which are believed to be relevant to the prediction of their ultra-violet emission line spectra. Possible excitation mechanisms (collisional, radiative recombination, resonant absorption, and fluorescence) are reviewed. The principles underlying calculations of the ionisation and thermal structure of high excitation planetary nebulae are presented.

The ESO/Uppsala survey of the ESO (B) Atlas of the Southern Sky. V. See Abstr. 002.001.

The ESO/Uppsala survey of the ESO (B) Atlas of the Southern Sky – VI. See Abstr. 002.051.

Atomic and molecular data. See Abstr. 022.070.

Spectroscopic observations of stars and planetary nebulae with a multichannel analogue detector system. See Abstr. 034.002

Mass ejection from old stars – an example of a possible mechanism. See Abstr. 064.052.

Atmospheres of central stars. See Abstr. 064.060.

A shortlived, deep convective envelope for highly evolved, blue stars? See Abstr. 064.061.

Nebular observations and stellar coronae. See Abstr. 064.063.

Ejection of planetary nebulae by helium shell flashes and the planetary distance scale. See Abstr. 065.001.

Signatures of the ^{22}Ne neutron source in red giants and planetary nebulae. See Abstr. 065.034.

Theory of evolution of the central star. See Abstr. 065.054.

Theory of evolution of central stars of planetary nebulae. See Abstr. 065.055.

Hydrogen- and helium-shell flashes and FG Sagittae phenomenon. See Abstr. 065.056.

Proto-planetary nebulae. See Abstr. 065.057.

Radio measurements of possible proto-planetary nebulae. See Abstr. 065.058.

A single star model for V 1016 Cygni. See Abstr. 065.059.

Ultraviolet photometric variations in the central star of IC 418. See Abstr. 113.047.

Observations of central stars. See Abstr. 114.040.

The effective temperature of the central star, and a criterion for complete absorption of hydrogen ionizing photons by the nebula. See Abstr. 114.041.

Peculiar central stars and related objects. See Abstr. 114.042.

The position of central stars in the Hertzsprung-Russell diagram. See Abstr. 115.012.

Binary nuclei of planetary nebulae. See Abstr. 117.026.

Central star of NGC 3132: a visual binary. See Abstr. 118.016.

UU Sagittae: eclipsing nucleus of the planetary nebula Abell 63. See Abstr. 121.048.

Nuclei of planetary nebulae in the pulsation instability strip. See Abstr. 122.084.

Spectrophotometric studies of nonstable stars. III. The spectrum of FG Sagittae in 1968 - 1973. See Abstr. 122.099.

Oscillating stars (Miras), mass loss and formation of planetary nebulae (PN). See Abstr. 122.106.

Are light DA white dwarfs the progenitors of planetaries? See Abstr. 126.030.

Planetary nebulae and nitrogen enrichment in the Galaxy. See Abstr. 131.252.

The structure of the emission nebula M 1−78 at 15.4 GHz. See Abstr. 132.015.

A search for ^{3}He in H II regions and planetary nebulae. See Abstr. 132.019.

Planetary nebulae and related objects. See Abstr. 132.047.

The shape of recombination lines from expanding nebulae. See Abstr. 132.066.

Molecular envelopes around evolved stars and the origin of planetary nebulae. See Abstr. 133.013.

Studies of bipolar nebulae – V. The general phenomenon. See Abstr. 134.015.

A search for [O III] emission in the central part of the X-ray globular cluster M15. See Abstr. 154.006.

Planetary nebulae in Local Group galaxies. V. The Andromeda galaxy. See Abstr. 158.011.

Identifications of faint planetary nebulae in the Magellanic Clouds. See Abstr. 159.023.

Radio Sources, Quasars, Pulsars, Extreme UV, X-Ray, Gamma-Ray Sources, Cosmic Radiation

141 Radio Sources, Quasars, Pulsars

Radio Sources, Quasars

141.001 Quasars in ultraviolet – comparing oranges and
oranges.   D. Overbye.
Sky Telesc., Vol. 55, 31 - 33 (1978).

141.002 Discovery of an X-ray QSO.
        G. R. Ricker, G. W. Clarke, R. E. Doxsey, R. G.
Dower, J. G. Jernigan, J. P. Delvaille, G. M. MacAlpine,
R. M. Hjellming.
Nature, Vol. 271, 35 - 37 (1978).

The authors report the discovery of an X-ray emitting
QSO, the first to be initially identified from X-ray observations. Following the convention for optical QSOs, they have
designated the optical object MR2251−178. The X-ray
luminosity (2−11 keV) of this object, presently $\sim 5 \times 10^{44}$
erg s^{-1}, has been as large as $\sim 1.6 \times 10^{45}$ erg s^{-1} (in 1975) and
exceeds the optical luminosity by a factor of ~ 10. Among
known compact X-ray sources, only 3C 273 has a greater
luminosity. Also, in radio observations at 4,885 MHz a point-
like radio source coincident with MR 2251−178 has been
discovered.

141.003 The angular diameter—redshift test for quasi-stellar
        radio sources with large redshifts.
A. Hooley, M. S. Longair, J. M. Riley.
Mon. Not. R. Astron. Soc., Vol. 182, 127 - 145 (1978).

The overall angular sizes of the radio structures of a
complete sample of 40 quasars in the 3CR catalogue and of
19 quasars which have redshifts greater than 1.5 in two complete samples of quasars in the 4C catalogue have been determined. The results are consistent with the hypothesis that the
overall physical sizes of the radio structures of quasars with
the highest radio luminosity do not change with cosmological
epoch. However, because of the small statistical sample, the
results are also consistent with weak evolution of the physical
sizes with cosmological epoch.

141.004 Quasars near bright galaxies – results from the
        Jodrell Bank 966-MHz survey.
I. W. A. Browne, A. M. Cohen.
Mon. Not. R. Astron. Soc., Vol. 182, 181 - 187 (1978).

The authors have investigated the distribution of bright
galaxies around quasars from the Jodrell Bank 966-MHz
survey. No statistical evidence is found to support the hypothesis that nearby bright galaxies and quasars are physically
associated. Some selection effects which lead to a correlation
between pair separation and galaxy redshift are discussed.

141.005 The 5C 6 and 5C 7 surveys of radio sources.
        T. J. Pearson, A. J. Kus.

Mon. Not. R. Astron. Soc., Vol. 182, 273 - 274 and Microfiche MN 182/1 (1978).

The sources detected in the two surveys are listed in
tables, which include 535 sources stronger than 10 mJy at 408
MHz and 121 stronger than 1.5 mJy at 1407 MHz. Right
ascensions and declinations are given with an uncertainty
varying between about 0.5 and 10 arcsec rms, depending on
the intensity and the frequency of observation. The tables
also include descriptions of optical objects visible on the
Palomar Sky Survey within about 20 arcsec of the radio
sources. All numerical data are given on microfiche only.

141.006 The radio source near NGC 2823.
        E. B. Fomalont, A. H. Bridle.
Mon. Not. R. Astron. Soc., Vol. 182, 1P - 3P (1978).

Observations of the radio source near NGC 2823 at 6 cm
with four antennas of the VLA (*very large array*) have shown
the source to have a double structure with no obvious relation
to the optical galaxy.

141.007 New observations of the radio absorption line in
        3C 286, with potential application to the direct
measurement of cosmological deceleration.
M. M. Davis, L. S. May.
Astrophys. J., Vol. 219, 1 - 4 (1978).

Observations at Arecibo of the 839.4 MHz absorption
line in the spectrum of 3C 286 have failed to detect any structure within the profile. A simple intervening-galaxy model is
consistent with both the present data and recent VLBI results.
The redshift measurement is inherently of very high accuracy;
application of this type of observation to the direct measurement of cosmological deceleration appears feasible on a time
scale of decades.

141.008 Redshifts of southern radio sources. III.
        D. L. Jauncey, A. E. Wright, B. A. Peterson,
J. J. Condon.
Astrophys. J., Lett., Vol. 219, L1 - L5 (1978).

The authors report further observations of 35 objects
associated with Parkes radio sources. Redshifts are given for
25 QSOs and two radio galaxies. Three QSOs which have only
a weak ultraviolet excess or appear neutral in color have redshifts greater than 2.5. The neutral stellar object PKS 2126−15
has a redshift of 3.27.

141.009 Optical structure in some low-redshift quasars.
        M. R. S. Hawkins.
Mon. Not. R. Astron. Soc., Vol. 182, 361 - 369 (1978).

Electronographic observations of seven quasars in the
redshift range z = 0.2–0.4 have been obtained in up to three
colours. Five of the quasars are shown to have non-stellar
images on R plates but in no case is a quasar image distin-

guishable from a star image on a B plate. These measurements are shown to be consistent with the presence of underlying galaxies.

141.010 An estimate of the counts of faint radio sources at 5 GHz. J. V. Wall.
Mon. Not. R. Astron. Soc., Vol. 182, 381 - 393 (1978).

A recent search for 5-GHz emission from nearby E and S0 galaxies produced a frequency distribution of flux densities dominated by background fluctuations. The results indicate a very slow convergence for the 5-GHz count in comparison with $N(S)$ relations at lower frequencies. This suggests that a flat-spectrum population dominates the 5-GHz count at levels below ~ 20 mJy, and calculations have been carried out to determine whether the recognized populations of objects with appropriate spectra, i.e. 'flat-spectrum' quasars, or 'active' galaxies, could be responsible. If it is the former, the cosmological evolution of the objects must be very strong and there can be no redshift cut-off for $z < 10$; if the latter, this low-luminosity population also must undergo strong evolution.

141.011 Stellar debris clouds in quasars and related objects. J. G. Hills.
Mon. Not. R. Astron. Soc., Vol. 182, 517 - 536 (1978).

In a previous paper (see 13.158.021) the author proposed that the power of a QSO or of a Seyfert nucleus may be provided by a process in which a large black hole in a dense galactic nucleus tidally breaks apart stars and eventually accretes their debris. In this paper he gives a more detailed analysis of this mechanism.

141.012 The Cygnus X region. X: The riddle of the Gamma Cygni radio source resolved.
J. W. M. Baars, H. R. Dickel, H. J. Wendker.
Astron. Astrophys., Vol. 62, 13 - 20 (1978).

Direct evidence is presented which shows that the so-called γ Cyg radio source contains several unrelated components. The nebula nearest to the star is a typical H II-region whose probable exciting star is discussed. The non-thermal part contains several long, stretched-out filaments with very steep intensity gradients. These filaments are part of a large supernova remnant. There is no physical relation between the radio components and the star, γ Cyg. Upper limits to the radio emission of the star are given.

141.013 M33: radiocontinuum emission at 4850 MHz.
A. von Kap-herr, E. M. Berkhuijsen, R. Wielebinski.
Astron. Astrophys., Vol. 62, 51 - 57 (1978).

A map of M33 at 6.2 cm with a resolution of $2\overset{.}{.}6$ is presented. The extended radio emission generally coincides with the optically bright parts of the nebula. The integrated radio flux is discussed; the spectral index is 0.52 ± 0.08. Optically seen H II regions and diffuse H_α emission probably contribute more than 58% to the total flux at 6.2 cm within 40' from the centre. Data on point sources in the area is collected in a table. Some individual sources are discussed.

141.014 On the correlation between the absolute magnitudes of quasi-stellar objects and the velocities of ejected gas. J. J. Perry, S. L. O'Dell.
Astron. Astrophys., Vol. 62, 229 - 234 (1978).

For all QSOs which show absorption-line systems, $\beta = [(1+z_e)^2 - (1+z_a)^2]/[(1+z_e)^2 + (1+z_a)^2]$ is plotted against the absolute visual magnitude (evaluated under the assumption that the emission-line redshifts are of cosmological origin, and for $q_0 = 0$ and 1). The form of the correlation is approximately $\beta \propto L^{1/2}$, where L is the cosmological luminosity. The trend of the data lends support to the hypothesis that the QSOs are at their redshift distances, and that many of the absorption systems arise in gas which is intrinsic to the QSOs and which is ejected from the central objects at velocities $c\beta$,

correlated with the luminosity. The form of this correlation is consistent with radiation-driven gas.

141.015 Multifrequency observations of very large radio galaxies: I. 3C 326. A. G. Willis, R. G. Strom.
Astron. Astrophys., Vol. 62, 375 - 392 (1978).

Observations of the radio source 3C 326 at wavelengths of 49, 21 and 6 cm with the Westerbork radio telescope are discussed. The data show conclusively that 3C 326 is a giant radio galaxy. A compact, flat spectrum nuclear source is associated with the optical galaxy. 3C 326 has considerable linear polarization, percentage polarizations of up to 50% and 35% being found at 21 and 49 cm, respectively.

141.016 Radio continuum observations of the edge-on disc galaxy NGC 891. R. J. Allen, J. E. Baldwin, R. Sancisi.
Astron. Astrophys., Vol. 62, 397 - 409 (1978).

The authors present observations of the radio continuum emission of the edge-on disc galaxy NGC 891 at $\lambda\lambda 49.2$, 21.2 and 6.0 cm. The results provide information on the spatial distribution of radio emissivity and spectral index. The radio continuum emission is nonthermal and can roughly be described as consisting of two components with different spatial distributions; a highly flattened component coinciding with the equatorial plane of the galaxy, and a "thick disc" component with an axial ratio of about 3.5/1. Radio emission at 21 cm has been detected at z distances of 6kpc above the plane of the galaxy. The spectral index of the radiation in the range 49–6 cm is approximately uniform at −0.65 within a distance of 12kpc from the centre along the major axis. In the z-direction the spectral index is about −0.65 from $z = 0$ to $z \approx 2.5$kpc. There is a strong indication that the spectrum steepens at higher z.

141.017 Two quasars near a peculiar pair of galaxies. J. Stocke, H. Arp.
Astrophys. J., Vol. 219, 367 - 371 (1978).

Two quasars ($z = 1.87, 2.19$) have been found near the peculiar pair of interacting galaxies NGC 7714–7715 (Arp 284, VV 51) with $z = 0.01$. The two quasars, the two interacting galaxies, and a third, smaller but disturbed, galaxy all fall closely along a line.

141.018 The spectrum of the halo surrounding the quasar PHL 1070. D. C. Morton, T. B. Williams, R. F. Green.
Astrophys. J., Vol. 219, 381 - 386 (1978).

The quasar PHL 1070 and its surrounding halo of light have been observed with the Palomar multichannel spectrophotometer using circular and annular apertures. The central quasar has a magnitude $m_V = 16.6$ and an emission line redshift of z_{em} 0.076 ± 0.002. The halo light has a magnitude $m_V = 19.0$ and a possible absorption line system with redshift $z_{abs} = 0.067 \pm 0.008$. The magnitude, possible absorption lines, and probable spectral energy distribution of the halo are consistent with those of a normal galaxy at the redshift of the quasar.

141.019 Relativistic blast-wave model for superlight motion in compact double radio sources.
A. P. Marscher.
Astrophys. J., Vol. 219, 392 - 399 (1978).

The author explores the observational characteristics of a relativistic blast wave which is propagating through an external medium which is confined to a thin disk and whose density and magnetic field decrease with radius. He finds that the blast wave appears as a double or triple radio source to an observer viewing the system within 20° of edge-on. The components of the source appear to separate with a velocity exceeding $2c$. The dependence of flux density on the observer's inclination angle provides a strong selection effect against sources which do not exhibit the well-separated double or triple structure. The characteristics of the model are favorably compared with observations of the sources 3C 120 and 3C 345.

141.020 High-velocity gas in the jet in the radio galaxy DA 240.
E. M. Burbidge, H. E. Smith, G. R. Burbidge.
Astrophys. J., Vol. 219, 400 - 403 (1978).

Spectrograms of the radio galaxy identified with the source DA 240, obtained with the Kitt Peak 4 m telescope and Ritchey-Chrétien spectrograph, reveal the presence of ionized gas in the jet discovered earlier. This gas has a large negative velocity displacement of 3400 km s^{-1} relative to the center of the galaxy. This is further evidence for the ejection of gas in a violent event in the nucleus of a radio galaxy. The velocity gradient in this gas is less than ~50 km s^{-1} over 14 kpc.

141.021 On the systematic optical identification of the remaining 3C radio sources. II. New data for 50 fields. J. Kristian, A. Sandage, B. Katem.
Astrophys. J., Vol. 219, 803 - 817 (1978).

Data obtained with the Palomar 5 m and 1.2 m telescopes provide proposed candidates or new likely identifications in 14 fields. Previous identifications by others are confirmed, or new data on redshifts, SIT photometry, or optical positions are given for 24 sources. Fifteen fields are still empty to the limit of the present material. Accurate ($\lesssim 1''$) optical positions are listed for a number of field stars near each of 40 sources. Identification charts are given for most fields.

141.022 He I lines in the spectra of QSOs and Seyfert galaxies. H. Netzer.
Astrophys. J., Vol. 219, 822 - 828 (1978).

Detailed calculations of He I line strengths for several models of QSOs and Seyfert galaxy nuclei are given. It is shown that an enhancement of the He I $\lambda 5876$ and He I $\lambda 7065$ lines by a factor of ~2–10, is caused near quasars in emission regions 1–3 pc in diameter (0.1–0.3 pc for Seyfert galaxies). Line enhancement is practically independent of the gas density as long as $N_H \gtrsim 2 \times 10^9$ cm^{-3}, and is a function of distance from the ionizing continuum source. The broad and strong He I lines observed in the spectrum of some active objects are discussed.

141.023 On the use of line shapes in the analysis of QSO absorption spectra. J. C. Weisheit.
Astrophys. J., Vol. 219, 829 - 835 (1978).

Profiles fitted by Morton and Morton (1972) to high-resolution measurements of the rich absorption spectrum of PHL 957 reveal many more asymmetrical features than one expects due to the overlapping of randomly positioned, symmetrical lines. The author argues that the asymmetry is intrinsic if these lines arise in extended rotating envelopes of intervening galaxies. He then describes how the shapes of several lines in a single redshift system can be used to discriminate between this explanation and the alternative explanation that the asymmetrical lines are due to blends of overlapping symmetrical features which arise in adjacent clouds radiatively accelerated by the QSO.

141.024 Historical lightcurves of three QSOs.
B. Q. McGimsey, H. R. Miller.
Astron. Astrophys., Suppl. Ser., Vol. 31, 147 - 149 (1978).

The archive plate collection of the Harvard College Observatory was searched to investigate the long-term optical behavior of the QSOs 1548+056, 1721+343 and 1722+33. Both 1721+343 and 1722+33 were found to be moderately variable while the data concerning 1548+056 are inconclusive.

141.025 *UBV* photometry of southern quasars and quasars candidates. G. Adam.
Astron. Astrophys., Suppl. Ser., Vol. 31, 151 - 157 (1978).

UBV photoelectric measurements are presented, up to $V = 17.3$, for a set of proposed identifications of southern radio sources. 68% of the observed objects show quasar-like colours (i.e. $U-B < -0.4$) and have, roughly, a 95% probability of being really quasars. Several objects show variability, with either a long or a short (a few days) time scale.

141.026 Farbbeobachtungen von NGC 5128 = Centaurus A.
S. van den Bergh, R. J. Dufour.
Sterne Weltraum, Jahrg. 17, 19 - 20 (1978).

141.027 Statistical investigation of compact emitting regions in extragalactic radio sources.
B. Anderson, H. P. Palmer, P. J. Richards, B. Speed, R. S. Warwick.
Nature, Vol. 271, 636 - 638 (1978).

Extragalactic radio sources of high total luminosity usually contain compact emitting regions. Radio interferometers sensitive to structures with an angular scale $\leqslant 1$ arc s have been used at Jodrell Bank to investigate the frequency of occurrence of such compact regions. Large samples of sources, covering a wide range of flux densities, were studied. This work extends to lower flux densities than previous investigations. The frequency of occurrence of compact regions varies as a function of flux density. This is best explained as a cosmological effect involving the strong evolution of source properties with redshift.

141.028 New complete sample of identified radio sources. I. Selection of data. A. Sołtan.
Acta Astron., Vol. 27, 339 - 364 (1977).

A numerous sample of identified extragalactic radio sources is constructed. The sample, consisting of both galaxies and quasars, is 95% complete. Optical and radio selection limits are replaced by their linear combination. For each object in the sample flux densities are determined in comoving reference frame at optical and radio domains.

141.029 Optical identifications from the Molonglo catalogues MC2 and MC3. I. Right ascensions 11^h 28^m to 17^h 00^m. C. Hazard, H. S. Murdoch.
Australian J., Phys., Astrophys. Suppl., No. 42, p. 1 - 44 (1977).

This paper presents the results of an examination of the fields around 350 sources in the Molonglo radio source catalogues MC2 and MC3 for the R.A. interval 11^h 28^m to 17^h 00^m. The 300 sources with a flux density $\geqslant 0.45$ Jy form an essentially complete radio sample and of these 21% are identified with BSOs (most of which have now been spectroscopically confirmed as QSOs) and 22% with galaxies of 18^m or brighter.

141.030 Culgoora-3 list of radio source measurements.
O. B. Slee.
Australian J. Phys., Astrophys. Suppl., No. 43, p. 1 - 123 (1977).

The positions, flux densities and beam broadening of 1946 sources detected at 80 MHz with the Culgoora radioheliograph (the Culgoora-1 and Culgoora-2 lists) have been re-examined with the same instrument at 160 MHz. A further 99 sources of special interest have been observed at 80 and 160 MHz. Contour maps of 163 sources resolved by the 1.'9 arc beam at 160 MHz are presented. For sources common to the Culgoora-3 and 4C catalogues, a comparison is made between the Culgoora 160 MHz positions and flux densities and the corresponding 178 MHz results in the 4C catalogue.

141.031 The Parkes 2700 MHz survey. (Thirteenth part). Catalogue for declinations −15° to −30°, right ascensions 10^h to 15^h.
A. Savage, A. E. Wright, J. G. Bolton.
Australian J. Phys., Astrophys. Suppl., No. 44, p. 1 - 20

(1977).

A catalogue of 404 radio sources is presented from a 2700 MHz survey of 0.316 sr. The catalogue is essentially complete for sources with $S_{2700} \geqslant 0.25$ Jy, corresponding to a source density in this area of 900 sources per steradian. The catalogue includes the results of examination of the Palomar Sky Survey prints at the positions of the sources.

141.032   Investigation of 410-MHz fine structure in candidate low-frequency variable sources.
S. R. Spangler, K. A. Meyers.
Astron. J., Vol. 83, 147 - 152 (1978).

Interplanetary scintillation observations at 410 MHz have been made of 19 steep-spectrum sources identified as variable or probably variable at 365 MHz. Scintillations were detected in 15 of the sources with derived fractions of the source flux in a sub-second-of-arc component ranging from 5% to 100%. Power spectra of the scintillation signals revealed evidence of finite source structure at the level of 0.1–0.2 arcsec in three sources. These sources and those which did not show scintillations are unlikely to be variables at the present time, but in the remaining cases the detected compact components could be the loci of variability activity.

141.033   An 8085-MHz survey of 0.181 steradian of sky with $|b| > 10°$.   G. A. Seielstad, M. S. Ewing.
Astron. J., Vol. 83, 157 - 161 (1978).

A survey at 8085 MHz of 0.181 sr of sky having $|b| > 10°$ and $|\delta| \leqslant 1°\!.5$ has revealed 64 sources, 35 of which constitute a sample 85% complete to $S_{8085} \leqslant 0.40$ Jy. More than two-thirds of the sources can be optically identified, and two-thirds of these appear stellar.

141.034   1 millimeter continuum observations of extragalactic objects.   J. H. Elias, D. J. Ennis, D. Y. Gezari, M. G. Hauser, J. R. Houck, K. Y. Lo, K. Matthews, D. Nadeau, G. Neugebauer, M. W. Werner, W. E. Westbrook.
Astrophys. J., Vol. 220, 25 - 41 (1978).

Continuum observations of 23 extragalactic objects have been made at a wavelength of 1 mm; nine of the 23 have been definitely detected at this wavelength. The sources detected include seven from which the 1 mm emission appears to be nonthermal and two from which the emission appears to be thermal radiation from dust. Observations at 3.4 mm and 9.5 mm were also made of the brighter objects in order to obtain reliable energy distributions.

141.035   The existence of Ca II absorption lines in the spectrum of the quasar 3C 232 due to the galaxy NGC 3067.   A. Boksenberg, W. L. W. Sargent.
Astrophys. J., Vol. 220, 42 - 46 (1978).

The authors have discovered Ca II H and K absorption lines in the spectrum of the quasar 3C 232 (z_{em} = 0.5303) at the same redshift as that of the galaxy NGC 3067, which lies 1°.9 away on the plane of the sky. The H and K lines have a velocity of (1406 ± 11) km s^{-1}, in excellent agreement with the 21 cm neutral hydrogen absorption line found in the system by Haschick and Burke. The 21 cm line is much narrower than are the Ca II lines. The authors present arguments that the H and K lines and the 21 cm line are produced in the outer halo rather than the outer disk of NGC 3067. The implications of the discovery concerning the nature of QSO absorption lines in general are discussed.

141.036   Evidence for nonuniform radial distribution of quasars, regardless of the nature of their redshifts.
R. F. Green, M. Schmidt.
Astrophys. J., Lett., Vol. 220, L1 - L4 (1978).

Preliminary results from the Palomar Bright Quasar Survey provide compelling evidence for strong cosmological evolution of optically selected quasars, if their redshifts are cosmological. The results are incompatible with the local hypothesis of quasars (i.e. noncosmological redshifts) unless the space density increases strongly with distance.

141.037   Observations of the optical remnant of SN 1181 = 3C 58.   S. van den Bergh.
Astrophys. J., Lett., Vol. 220, L9 - L10 (1978).

This letter reports the discovery of the optical counterpart to the Crab-like radio source 3C 58. This supernova remnant consists of short diffuse filaments and knots.

141.038   The separation velocity of radio components in QSO's.   J.-h. You, F.-z. Cheng, L.-z. Fang.
Acta Astron. Sinica, Vol. 18, 239 - 247 (1977).   In Chinese.

For QSO with double radio components, the separation velocity of radio components with respect to the optical parent body can be derived from the radio flux ratio of the two components and the evolution law of radio luminosity. The observational data of 26 samples have been analyzed, and the following results obtained: the separation velocity is within the range 0.2–0.14 c, the average velocity ~0.09 c and the upper limit of the initial eruption velocity ~0.35 c.

141.039   Radiogalaxy Virgo A.   B. V. Komberg.
Zemlya i Vselennaya, 1978, No. 1, p. 16 - 20.   In Russian.

141.040   On the nature of the extragalactic radio recombination lines.   P. A. Shaver, E. Churchwell, C. M. Walmsley.
Astron. Astrophys., Vol. 64, 1 - 8 (1978).

New observations of the H166α, H110α, and H76α recombination lines from M82 are used to show that the observed line emission is predominantly stimulated. It is argued that the recently detected H102α line from NGC 253 may also be due primarily to stimulation by nonthermal background radiation.

141.041   Relativistic Kelvin-Helmholtz instabilities in extragalactic radio sources.
A. Ferrari, E. Trussoni, L. Zaninetti.
Astron. Astrophys., Vol. 64, 43 - 52 (1978).

Kelvin-Helmholtz instability of fluids streaming at relativistic velocity is studied in the framework of continuous energy supply models for extended extragalactic radio sources. The general dispersion relation is derived for beams with cylindrical symmetry and is solved for a range of combinations of the main astrophysical parameters. The final discussion shows that relativistic supersonic flows tend to be stable and allow a long-term flux of energy from the parent galaxy to radio regions.

141.042   On the possible identification of Mg II λ2934 emission in QSOs.
S. A. Grandi, M. M. Phillips.

A broad, weak emission feature at a rest wavelength of approximately 2950 Å is found to be nearly always present in QSO spectra. A correlation of the strength of λ2950 with the Mg II λ2798 resonance doublet suggests that the emission may be Mg II λ2934. Two excitation mechanisms for Mg II λ2934 are suggested and discussed: fluorescence by H I Lβ and fluorescence by N V λ1240. Both mechanisms are shown to lead to appreciable excitation of Mg II λ2934 with respect to Mg II λ2798. It is suggested that Mg II λ2934 may be consistent with a redshift determined from the centroid of the broad components of the Balmer lines. Evidence in support of the proposed emission mechanisms is outlined. Problems such as the apparent weakness of Mg II λλ1737, 1750 emission in the spectra of high-redshift QSOs are also discussed.

141.043   VLBI observations of 3C 345 and NRAO 512 in right and left circular polarization.

C. R. Menyuk, I. I. Shapiro, J. J. Wittels, H. F. Hinteregger,
C. A. Knight, A. E. E. Rogers, A. R. Whitney, T. A. Clark,
L. K. Hutton.
Astrophys. J., Lett., Vol. 220, L27 - L29 (1978).

Using data for each polarization separately, the authors
estimated several parameters describing a model of the fine
structure of radio brightness of 3C 345, and, subsequently, the
angular separation between 3C 345 and NRAO 512. The results
for the two polarizations are in approximate agreement, indicating that, to the limit of resolution ($\sim$ 0.5 milli-arcsec), the
fourth Stokes parameter, V, is not significantly different from
zero within the compact components of these radio sources.
The corresponding quantitative limits on the degree of circular
polarization are 0.08 ± 0.07 for 3C 345 and 0.01 ± 0.08 for
NRAO 512.

141.044   Imaging polarimetry of the jets of M87 and 3C 273.
G. D. Schmidt, B. M. Peterson, E. A. Beaver.
Astrophys. J., Lett., Vol. 220, L31 - L36 (1978).

Photoelectric polarization maps of the jets associated with
M87 and 3C 273 are presented. For the M87 jet, polarization
of individual knots reaches 20%. For the 3C 273 jet as a whole,
the observations yield P = 3.7% ± 4.1%. This measurement is
consistent with inverse Compton models of the optical emission; however, a synchrotron origin cannot be excluded if the
small-scale structure of the 3C 273 jet is similar to that of the
jet of M87.

141.045   New observations of the absorption spectrum of
PKS 0237−23 and their implications for the origin
of quasar absorption lines.
T. Boroson, W. L. W. Sargent, A. Boksenberg, R. F. Carswell.
Astrophys. J., Vol. 220, 772 - 782 (1978).

The QSO PKS 0237−23 has been observed with the
University College London image photon-counting system at
the coudé focus of the Hale telescope. Six observations made
over 3 years have been combined to give high-resolution (0.7 Å)
spectral coverage from λ3700 to λ4300. Wavelengths and
equivalent widths of 193 absorption lines in this spectral
region have been tabulated. Most of the lines are identified in
45 redshift systems based on C IV doublets. Eleven of these
have been labeled "certain", 12 have been labeled "probable",
and 22 have been called "possible". A large statistical excess
of these systems shows the Si II λ1533 excited fine-structure
transition.

141.046   A linear polarization survey of bright QSOs.
H. S. Stockman, J. R. P. Angel.
Astrophys. J., Lett., Vol. 220, L67 - L71 (1978).

With 44 QSOs observed, the authors have investigated the
frequency distribution of polarization for all QSOs as a class
and for the observational subclasses of radio-selected and
optically selected QSOs. The preliminary results indicate that
the large majority of QSOs, represented by those selected
optically, have little intrinsic optical polarization ($P \lesssim$ 0.5%).
Radio emitters, however, tend to have higher polarizations
($P \lesssim$ 1.5%). The authors have obtained several interesting
results for individual QSOs.

141.047   The Lα/Hα intensity ratio in PKS 0237−23.
A. R. Hyland, E. E. Becklin, G. Neugebauer.
Astrophys. J., Lett., Vol. 220, L73 - L75 (1978).

The ratio $I(L\alpha)/I(H\alpha)$ has been measured to be 1.7 ± 0.6
in the quasar PKS 0237−23. The measured ratio is similar to
that measured in 3C 273 by Davidsen et al. and to that inferred
by Baldwin from a study of a large number of quasars; it is a
factor of 5 to 14 smaller than expected from recombination
theory. If the Lα radiation is resonantly scattered and there
are modest amounts of dust within the ionized region, the
measured ratio may be interpreted in terms of recombination
theory, with Lα being depleted by absorption by the dust.

141.048   Cygnus A at 99 GHz: observations of the three
principal components and interpretation of the
central source.
R. W. Hobbs, S. P. Maran, M. Kafatos, L. W. Brown.
Astrophys. J., Lett., Vol. 220, L77 - L80 (1978).

The three principal emission components of Cygnus A
have been observed at 99 GHz, the highest frequency at which
radio measurements of this source have been accomplished.
The observations show no definite indication of a high-
frequency cutoff in the spectrum of the compact central component, which perhaps may be attributed to an optically thin
synchrotron source that peaks at a frequency of several
hundred GHz.

141.049   Le point sur les quasars.   D. Proust.
Bull. AFOEV, Tome 11, 73 - 75 (1977).

141.050   Free-form analysis of the cosmological evolution of
radio sources.   J. G. Robertson.
Mon. Not. R. Astron. Soc., Vol. 182, 617 - 628 (1978).

This paper presents a simple new iterative method of
solving for the evolution function required to reconcile observed radio source counts with the standard general relativistic
cosmological models. A luminosity function is assumed together with a turn-on luminosity above which the source density evolves. It is then possible to solve for the redshift dependence without any constraints on its functional form (i.e. a free-
form solution). The new method has been applied to the
counts at 408 MHz from the Molonglo Cross telescope, using
the Einstein−de Sitter cosmology and a piecewise continuous
form for the radio luminosity function. It is found that the
evolution function is not very sensitive to moderate changes in
the assumed radio luminosity function, but that the predicted
redshifts do show considerable sensitivity to this function.

141.051   Detection of X-ray emission from the giant radio
lobes of Cen A.   B. A. Cooke, A. Lawrence, G. C.
Perola.
Mon. Not. R. Astron. Soc., Vol. 182, 661 - 671 (1978).

The University of Leicester Sky Survey Instrument on
the Ariel V X-ray astronomy satellite has detected, at high
statistical significance, spatially extended X-ray emission from
the giant radio lobes of Cen A. Interpreting the X-rays as
Inverse Compton Scattered photons of the microwave background by the radio-emitting electrons, the authors estimate
the average strength of the magnetic field to be 0.7 μG, a
value almost equal to the equipartition value computed with
protons having the same energy content as electrons. There
is strong evidence that the peaks of the X-ray emission do not
coincide with the radio peaks but are further from the galaxy.

141.052   The statistical properties of compact features in
radio sources as a function of flux density.
B. Speed, R. S. Warwick.
Mon. Not. R. Astron. Soc., Vol. 182, 761 - 772 (1978).

A statistical study has been made of the properties of
compact components in radio sources based on observations
with a long-baseline interferometer over a wide flux density
range. The observations are compared with the expected properties of radio sources calculated on the basis of a complete
sample of 3CR sources. The best agreement with the observations is achieved when strong density evolution and linear-size
evolution are included in the calculations. Some aspects of the
results are explained by a correlation of component compactness (in linear size) with radio luminosity for the high-luminosity sources in the 3CR sample.

141.053   Survey of the optical variability of compact extra-
galactic objects. II. Objects from 17^h to 22^h.
C. Barbieri, G. Romano, M. Zambon.
Astron. Astrophys., Suppl. Ser., Vol. 31, 401 - 407 (1978).

This paper reports observations made at Asiago Observatory of the optical variability of quasi stellar objects located from 17^h to 22^h right ascension. The discovery of a new variable object at 5 arcmin from PKS 2216−03 is also reported. A possible correlation between the optical and the radio behaviour of 3C 446 is strengthened by the data.

141.054 **A catalogue of sources found at 610 MHz with the Westerbork Synthesis Radio Telescope: source parameters and identifications.** J. K. Katgert.
Astron. Astrophys., Suppl. Ser., Vol. 31, 409 - 426 (1978).

A description is given of a catalogue of 347 sources detected in eight fields observed at 610 MHz with the Westerbork Synthesis Radio Telescope. An identification programme has been carried out for all sources; the results of this are also included. Spectral index data are given for 182 sources, and two complete samples are defined.

141.055 **Twin-beam models for double radio sources. I. Steady-state configurations.** P. J. Wiita.
Astrophys. J., Vol. 221, 41 - 50 (1978).

Continuous ejection models for extragalactic symmetric double radio sources are discussed, with emphasis on the twin-exhaust proposal of Blandford and Rees, since it accounts for the data reasonably well. The author constructs models for Cygnus A that have considerably more compact inner regions than those of Blandford and Rees, and he argues that his version may provide a more natural explanation of the alignment between inner and outer radio emission regions. His models require a hotter, denser, confining gas housed within a galactic nucleus; the nucleus is assumed to supply the energy and plasma for the entire double source. Suggestions for further relevant observations are made.

141.056 **Twin beam models for double radio sources. II. Dynamical calculations.** P. J. Wiita.
Astrophys. J., Vol. 221, 436 - 448 (1978).

Numerical experiments of continuous ejection models for symmetric double radio sources are performed. The equations for relativistic spherical flow and for nonrelativistic aspherical expansion are derived after a discussion of the physical conditions in both the internal plasma and the confining gas cloud is given. Models for different values of internal luminosity, and for variations in the parameters of the external gas are computed, and the scaling law is verified. The most important variables turn out to be the dimensionless luminosity and the eccentricity of the confining cloud. Low-power models tend to form bubbles: the expanding blob of plasma bifurcates. Stronger sources lead to oppositely directed jets that sometimes attain relativistic expansion velocities. It is found that the flatter the cloud, the sooner bubbles form.

141.057 **Compact radio sources in and near bright galaxies.** J. J. Condon, L. L. Dressel.
Astrophys. J., Vol. 221, 456 - 467 (1978).

Compact radio sources in galaxies stronger than 35 mJy at 2380 MHz from the Arecibo survey of galaxies brighter than $m_p = +14.5$ have been detected and observed at 2695 and 8085 MHz. Accurate radio and optical positions show that all compact radio sources identifiable with these galaxies are located in their nuclei. Five new BSO (blue stellar object)/galaxy pairs were discovered. The observations are interpreted in terms of accretion by massive black holes in galactic nuclei.

141.058 **An optical and radio study of quasars.** R. A. Sramek, D. W. Weedman.
Astrophys. J., Vol. 221, 468 - 480 (1978).

A low-dispersion spectroscopic search for optical quasars has produced 91 candidates for which probable redshifts could be measured. Coordinates and finding charts are given. These and 156 other optically discovered quasars were observed at Arecibo to a 5 σ flux limit of about 14 mJy at 2380 MHz.

Eighteen were detected and flux limits are given for the others. A local luminosity function is derived for quasars and class 1 Seyfert galaxies. This is compared with the space density of high-redshift quasars from the optical survey.

141.059 **A linear polarization survey of bright QSO's.** H. S. Stockman, J. R. P. Angel.
Bull. American Astron. Soc., Vol. 9, 577 (1978). – Abstract.

141.060 **VLA observations of optical QSO's.** R. Sramek, D. Weedman.
Bull. American Astron. Soc., Vol. 9, 577 (1978). – Abstract.

141.061 **An attempt to detect faint objects near QSO's with low redshift absorption systems.** T. A. Boroson, B. M. Peterson, R. J. Weymann, H. R. Butcher.
Bull. American Astron. Soc., Vol. 9, 577 (1978). – Abstract.

141.062 **Absorption lines in low redshift QSOs.** B. M. Peterson, P. A. Strittmatter.
Bull. American Astron. Soc., Vol. 9, 577 - 578 (1978). Abstract.

141.063 **Attempts to identify the "unidentified" absorption lines in the spectra of quasi-stellar objects.** D. H. Roberts, E. M. Burbidge, G. R. Burbidge, H. E. Smith, V. Junkkarinen.
Bull. American Astron. Soc., Vol. 9, 578 (1978). – Abstract.

141.064 **Angular diameters of quasars.** R. I. Potash, J. F. C. Wardle.
Bull. American Astron. Soc., Vol. 9, 578 (1978). – Abstract.

141.065 **Lα destruction in dusty QSOs.** G. J. Ferland, H. Netzer.
Bull. American Astron. Soc., Vol. 9, 578 (1978). – Abstract.

141.066 **Discovery of an X-ray QSO.** G. R. Ricker, G. W. Clark, R. E. Doxsey, R. G. Dower, J. G. Jernigan, J. P. Delvaille, G. M. MacAlpine, R. M. Hjellming.
Bull. American Astron. Soc., Vol. 9, 578 (1978). – Abstract.

141.067 **Proper motions and distances of quasars.** Y. P. Varshni.
Bull. American Astron. Soc., Vol. 9, 578 - 579 (1978). Abstract.

141.068 **VLA observations of extended sources in rich clusters of galaxies.** F. N. Owen, J. O. Burns, L. Rudnick.
Bull. American Astron. Soc., Vol. 9, 584 (1978). – Abstract.

141.069 **Thermal synchrotron emission.** T. W. Jones, P. E. Hardee.
Bull. American Astron. Soc., Vol. 9, 584 (1978). – Abstract.

141.070 **850 MHz interplanetary scintillation observations of steep spectrum extragalactic radio sources.** S. R. Spangler, K. A. Meyers.
Bull. American Astron. Soc., Vol. 9, 584 - 585 (1978). Abstract.

141.071 **Radio observations of the head-tail radio galaxy NGC 1265.** F. N. Owen, J. O. Burns, L. Rudnick, T. Jones.
Bull. American Astron. Soc., Vol. 9, 585 (1978). – Abstract.

141.072 **A model for superluminal radio sources.** J. N. Bahcall, M. Milgrom.
Bull. American Astron. Soc., Vol. 9, 586 (1978). – Abstract.

141.073 **The angular scale of background fluctuations at**

1400 MHz.
S. J. Goldstein, K. C. Turner, R. T. Rood.
Bull. American Astron. Soc., Vol. 9, 607 (1978). − Abstract.

**141.074	Hubble plot and space density of quasars in a FIB metric.**	J. M. Barnothy.
Bull. American Astron. Soc., Vol. 9, 607 - 608 (1978). Abstract.

**141.075	Evidence for supermassive accretion disks in QSOs.**
G. A. Shields.
Bull. American Astron. Soc., Vol. 9, 608 (1978). − Abstract.

**141.076	Dynamically self-consistent ionization and line profile calculations for QSO emission lines.**
W. R. Oegerle, R. J. Weymann.
Bull. American Astron. Soc., Vol. 9, 608 (1978). − Abstract.

**141.077	Direct imaging of quasars and BL Lac objects.**
P. A. Wehinger, S. Wyckoff.
Bull. American Astron. Soc., Vol. 9, 608 (1978). − Abstract.

**141.078	Recombination line mapping of partially ionized regions toward Orion A.**
D. Jaffe, V. Pankonin.
Bull. American Astron. Soc., Vol. 9, 611 (1978). − Abstract.

**141.079	Detection of the Zeeman effect in Orion A.**
T. Troland, C. Heiles.
Bull. American Astron. Soc., Vol. 9, 611 (1978). − Abstract.

**141.080	Coordinated photometric observations of strong extragalactic 90 GHz sources.**
J. J. Puschell, W. A. Stein.
Bull. American Astron. Soc., Vol. 9, 618 - 619 (1978). Abstract.

**141.081	Superrelativistic expansion in 1633+38?**
W. D. Cotton, J. J. Wittels, I. I. Shapiro, C. Angulo.
Bull. American Astron. Soc., Vol. 9, 619 (1978). − Abstract.

**141.082	A relativistic blast wave model for compact variable radio sources.**	E. B. Hessel, J. F. C. Wardle.
Bull. American Astron. Soc., Vol. 9, 619 (1978). − Abstract.

**141.083	Orbital dynamics of the radio galaxy 3C 129.**
G. G. Byrd, M. J. Valtonen.
Bull. American Astron. Soc., Vol. 9, 628 (1978). − Abstract.

**141.084	Regarding the possibility that PHL 957 has a Lyman-alpha halo.**	K. Davidson.
Bull. American Astron. Soc., Vol. 9, 630 - 631 (1978). Abstract.

**141.085	Do compact radio sources contain non-relativistic plasma?**	J. S. Scott, W. A. Christiansen.
Bull. American Astron. Soc., Vol. 9, 631 (1978). − Abstract.

**141.086	Particle acceleration and radio source dynamics.**
G. Rolison, W. A. Christiansen, J. S. Scott, A. G. Pacholczyk.
Bull. American Astron. Soc., Vol. 9, 647 (1978). − Abstract.

**141.087	Current-carrying beams in astrophysics: models for double radio sources and jets.**	G. Benford.
Mon. Not. R. Astron. Soc., Vol. 183, 29 - 48 (1978).

Beams with both azimuthal and axial currents can have very low self-forces acting on the particles, i.e. $J \times B$ is minimized, and thus enjoy long synchrotron lifetimes. The author shows that these spinning beams may be stable if the long-wavelength modes are suppressed by the inertia of a surrounding plasma sheath. He explores some applications to double radio sources and jets.

**141.088	The radio structure of 3C 123 at 2.7 and 15 GHz.**
J. M. Riley, G. G. Pooley.
Mon. Not. R. Astron. Soc., Vol. 183, 245 - 255 (1978).

3C 123 has been mapped at 2.7 and 15 GHz with resolutions 3.6 × 7.2 and 0.65 × 1.3 arcsec, in total power and polarization. A compact central source, coincident with the optical galaxy, has been discovered. The polarization data suggest the presence of large amounts of 'cold' matter within the components.

**141.089	The 3C 303 system.**	P. P. Kronberg.
IAU Colloq. No. 37, (see 012.009), p. 219 - 222 (1977).

La combinaison des cartes radio en synthèse d'ouverture à haute résolution de 3C 303 et de plaques optiques montre que les principales composantes de la radio source peuvent être identifiées respectivement avec une galaxie de magnitude 17 et un objet de magnitude 22. L'objet de magnitude 22 a un compagnon plus brillant de $m_v \sim 20$ et de décalage vers le rouge 1.57. Ce système montre le meilleur cas (non prouvé à 100%) d'une association physique d'objets possédant des décalages vers le rouge différents.

**141.090	Spectroscopy and the Hubble diagram of faint radio galaxies.**	H. Spinrad, H. E. Smith.
IAU Colloq. No. 37, (see 012.009), p. 251 - 272 (1977).

Les auteurs discutent les identifications récentes et les observations des radiogalaxies faibles, au décalage spectrale $z = 0.75$, en les présentant dans des diagrammes Hubble.

**141.091	Hubble diagram of quasars.**
V. Petrosian, M. A. Soldate.
IAU Colloq. No. 37, (see 012.009), p. 273 - 282 (1977).

Les auteurs montrent que le diagramme de Hubble (densité de flux - décalage spectral) des quasars ne peut-être utilisé pour tester l'hypothèse que leur décalage vers le rouge est cosmologique. Toutefois, ils montrent que cette relation peut-être utilisée pour tester l'hypothèse locale dans laquelle le décalage cosmologique des quasars est supposé être petit et la majeure partie du décalage être due à quelque cause non cosmologique. Une telle hypothèse générale et locale peut-être rejetée avec une certitude supérieure à 95%. Les auteurs montrent que ce pourcentage peut augmenter jusqu'à 99% en faisant des suppositions moins générales mais très plausibles.

**141.092	Super-light motion in radio sources and its implications for the distance scale problem.**
K. I. Kellermann, D. B. Shaffer.
CNRS Colloq. No. 263, (see 012.009), p. 347 - 363 (1977).

Although some of the earlier VLBI data permitted interpretations of the apparent super-light motion with multi-component stationary models, the more detailed data now available, at least for 3C 120 and 3C 345, appear to require an actual change in component separation by up to a factor of ten. This corresponds to apparent linear velocities of up to 10 times the speed of light if these sources are located at the distance corresponding to their redshift. Thus, these observations provide some evidence against the cosmological distance of these objects.

**141.093	21 cm observations of 3C 286.**	A. M. Wolfe.
CNRS Colloq. No. 263, (see 012.009), p. 511 - 515 (1977).

Les observations VLB de la raie 21 cm en absorption décalée vers le rouge dans le QSO 3C 286 montre que le gaz s'étend dans une zone $\geqslant 0.01''$ avec une variation de vitesse de 3 kms^{-1}. Ces observations rejettent l'hypothèse d'une enveloppe de gaz poussée par les photons émis, mais ne sont pas inconsistantes avec une absorption par une galaxie sur la ligne de visée.

141.094 Size limits for expanding light sources.
J. Terrell.
CNRS Colloq. No. 263, (see 012.009), p. 517 - 520 (1977).

Les fluctuations rapides des brillances optiques et radio observées dans les quasars, ainsi que dans les objets BL Lac, imposent des limites supérieures sévères sur les dimensions possibles de ces sources; contrairement à l'opinion répandue, ces limites ne sont pas notablement modifiées par l'hypothèse d'une source lumineuse en expansion rapide. Le conflit entre les limites de variation des dimensions des quasars et la surface minimale pouvant émettre une énergie suffisante, en supposant des distances cosmologiques, n'est pas résolu per une expansion relativiste.

141.095 Radio-quiet QSOs in the region of Hercules cluster of galaxies.
A. A. Hoag, E. M. Burbidge, H. E. Smith.
CNRS Colloq. No. 263, (see 012.009), p. 521 - 524 (1977).

141.096 The CTIO objective-prism survey and SIT vidicon observations of southern quasars.
P. S. Osmer, M. G. Smith.
CNRS Colloq. No. 263, (see 012.009), p. 525 - 528 (1977).

The authors describe the application of two new techniques to the study of quasars. The first makes use of low-dispersion objective-prism plates to discover high-redshift quasars (Smith, 1975). The $L\alpha$ emission can be seen directly on the spectrograms. The second involves a two-dimensional SIT vidicon tube for spectrophotometry at medium resolution of the emission lines (Ingerson, Lasker, and Osmer, 1976).

141.097 Search for optical identifications in the 5C4 radio survey. H.-E. Fröhlich, H. Tiersch.
Astron. Nachr., Band 299, 1 - 12 (1978).

With the aid of a statistical approach regarding only the nearest optical object to a given radio position the authors have found identification rates of nearly 30 per cent up to the blue and red limits of the Palomar Sky Survey prints covering the region of the 5C4 radio survey. Roughly 60 per cent of these identifications are blue starlike objects and therefore possible quasar candidates. The remaining identifications are galaxies. Probabilities for the first neighbours to be an optical counterpart and finding charts for highly reliable objects are given.

141.098 Redshift and radio magnitude of quasars.
D. Basu.
Astrophys. Space Sci., Vol. 53, 245 - 249 (1978).

Significant correlation has been obtained between redshift and radio magnitude of quasars at 3 GHz emitted frequency.

141.099 A trend in the gaps in redshift distribution of QSOs.
D. Basu.
Astrophys. Space Sci., Vol. 53, 251 - 256 (1978).

Analysis of gaps in the emission line redshift distribution of 537 QSOs shows a single definite trend in the distribution of the gaps. The trend is similar for radio-quiet objects, radio sources and all QSO's taken together. However, the trend does not suggest any periodicity in redshifts in any of the three cases.

141.100 The structure of the Orion region in the radiocontinuum. W. Reich.
Astron. Astrophys., Vol. 64, 407 - 421 (1978).

A 1420 MHz map of a region of $25° \times 28°$ in Orion is presented, which has the high sensitivity of 40 mK ($\geqslant$3 times r.m.s. noise in T_B) and is absolutely calibrated. This map is compared with maps at 408 MHz and at 820 MHz to look for

radio spurs and to derive the distribution of the spectral index. A detailed discussion is given of the region of Barnard's Loop which surrounds the Orion association I and of the λ Orionis region north of Barnard's Loop.

141.101 New highly variable radio source, possible counterpart of γ-ray source CG135 + 1.
P. C. Gregory, A. R. Taylor.
Nature, Vol. 272, 704 - 706 (1978).

The discovery of a highly variable radio source which coincides in position with cosmic γ-ray source CG135 + 1 is reported.

141.102 Thermal continuum from accretion disks in quasars.
G. A. Shields.
Nature, Vol. 272, 706 - 708 (1978).

The author argues that supermassive accretion in disks may explain the flat optical continuum observed from some QSOs and Seyfert galaxies.

141.103 PKS 0528−250: a neutral stellar object at z = 2.812, with no emission lines and a rich absorption-line spectrum.
D. L. Jauncey, A. E. Wright, B. A. Peterson, J. J. Condon.
Astrophys. J., Lett., Vol. 221, L109 - L112 (1978).

Image-tube dissector scanner observations of the 19th mag neutral stellar object identified with the radio source PKS 0528−250 show that it possesses a rich absorption-line spectrum similar to that expected from an intervening galaxy at $z = 2.812$. The object is unusual in having no emission lines but strong $L\alpha$ absorption at 4635 Å.

141.104 Interpretation of quasar colours in the near IR.
A. R. Hyland, M. P. Schwarz.
Proc. Astron. Soc. Australia, Vol. 3, 137 - 140 (1977).

The authors investigate the effects of typical quasar emission lines and continua on the predicted infrared colours of quasars as a function of redshift. It is shown that, although $(V−K)$ is affected by such lines, by virtue of its long wavelength baseline, it is still a good indicator of the continuum slope of the source up to $z = 3.0$.

141.105 Radio and optical spectra of sources from the Molonglo Deep Survey. H. S. Murdoch, G. L. White.
Proc. Astron. Soc. Australia, Vol. 3, 142 - 143 (1977).

The authors present preliminary results of further radio and optical work being carried out on the Molonglo Deep Survey (Robertson 1977) which reaches a density of $\sim$ 20,000 sources per sterad at a limiting flux density of 84 mJy at 408 MHz in three small areas of sky at declinations $-20°$ and $-62°$.

141.106 Variable quasi-stellar objects. I. Photometry and completeness of faint blue objects in the Sandage-Luyten survey field 8^h48^m, +18°. P. D. Usher.
Astrophys. J., Vol. 222, 40 - 53 (1978).

Optical variability is investigated for 820 LB objects brighter than $B \sim 19$ mag in the Sandage-Luyten survey field at $\alpha(1855) = 8^h48^m$, $\delta(1855) = +18°$. Photometric statistics of variability and mean magnitudes are given for over 170 objects of color class $I(U − V \leqslant -0.4)$ and for some redder objects. Estimates are given of the efficacy of the criterion of variability as a means to select quasi-stellar object (QSO) candidates. About 70% of the objects selected for both blueness and variability are found to be QSOs. The degree of completeness of cataloged blue objects in the field has been investigated with the help of the original three-color survey plate, resulting in 49 additional blue objects, most of which have color indices $U − V \lesssim 0.0$. Positions, magnitudes, estimated colors, and finding charts are given for these sources. Selection effects for faint blue objects in the field are examined, and incremental correction factors of ~20% to 40% to surface densities of LB QSOs are derived. The variability statistics B_{min} and B_{max} are suggested as a means for separating extreme physical conditions in QSOs, and for combating selection effects.

141.107 A model for Cygnus A: a radio source powered by M87 type knots.   M. J. Valtonen.
Astrophys. J., Vol. 222, 78 - 83 (1978).

The author points out that (1) the magnitude of the optical object, which coincides with component B of radio source Cygnus A, is as expected if the object is physically associated with the radio component and if it is a scaled-up version of a knot in the jet of M87; that (2) the existence of such a knot in component B was previously predicted; and that (3) if two such knots were to be ejected from a galaxy to a distance of about 100 kpc, the resulting radio source would probably look much like Cygnus A. Under the assumption that this ejection took place, the author derives the main physical parameters of the radio source. The masses of the supermassive black holes required to power the knots probably exceed 10^{10} $M_\odot$.

141.108 The historical light curve of PKS 2134 + 004, a highly luminous QSO.   E. W. Gottlieb, W. Liller.
Astrophys. J., Lett., Vol. 222, L1 - L2 (1978).

The historical light curve of PKS 2134 + 004 reveals that during outbursts in 1937 and 1949, it reached an apparent B magnitude of 14.8. The redshift of this QSO, $z = 1.93$, implies that if at a cosmological distance, its absolute magnitude at peak brightness was $M_B = -30.7$, assuming $q_0 = 0$ and $H_0 = 100$ km s^{-1} Mpc^{-1}. At peak brightness, PKS 2134 + 004 was one of the most luminous objects in the universe.

141.109 Optical identifications of radio sources in the NRAO 5-GHz survey: the "S2" and "intermediate" surveys. D. B. Shaffer.
Astron. J., Vol. 83, 209 - 218 (1978).

Optical identifications are given for radio sources in two of the NRAO 5-GHz surveys: S2 (complete to 0.6 f.u.) and I (complete to 0.25 f.u.). Most (88%) of the sources in the S2 survey are identified, whereas only 69% of the I survey sources are identified. Down to a limit of 0.1 f.u., the fraction of identified sources in the 5-GHz surveys that are quasars is constant, indicating that some of the empty fields are probably quasars that are below the Sky Survey limit.

141.110 Are quasar redshifts randomly distributed?
R. J. Weymann, T. Boroson, J. D. Scargle.
Astrophys. Space Sci., Vol. 53, 265 - 278 (1978).

A statistical analysis of possible clumping (not periodicity) of emission line redshifts of QSO's shows the available data to be compatible with random fluctuations of a smooth, non-clumped distribution. This result is demonstrated with Monte Carlo simulations as well as with the Kolmogorov–Smirnov test.

141.111 Are quasi-stellar objects gravitational masers?
P. K. Biswas.
Astrophys. Space Sci., Vol. 53, 371 - 375 (1978).

The present paper constitutes an attempt to interpret a quasar as a gravitationally pumped maser.

141.112 The differential radio source count at 1400 MHz from the GB2 sky survey.   J. Machalski.
Astron. Astrophys., Vol. 65, 157 - 164 (1978).

The corrected differential count of radio sources at 1400 MHz from the GB2 catalogue is presented after a detailed analysis of flux density errors. The complete samples of GB2 flat- ($\alpha_{1.4}^{0.4} < 0.5$) and very steep- ($\alpha_{1.4}^{0.4} \gtrsim 1.0$) spectrum sources give an evident discrepancy in the source counts with the statistical significance of about 3σ. There

is a relative deficiency of intense sources with steep spectra. The slope of their count (about 1.9 ± 0.1) remains steady in the whole range of flux density investigated, while for the flat-spectrum sources the slope decreases with decreasing of intensity.

141.113 Evolution of radio radiation of non-stationary sources in the "Hedgehog" model.
Yu. A. Kovalev, V. P. Mikhajlutsa.
Inst. kosm. issled. AN SSSR. Moskva, 1977, 27 pp. In Russian. – Abstr. in Ref. zh., 51. Astron., 2.51.763 (1978).

141.114 On the distribution of compact groups of compact galaxies and radio sources.
V. A. Sanamyan, Eh. A. Arutyunyan.
Astrofizika, Vol. 13, 543 - 545 (1977). In Russian. English translation in Astrophysics, Vol. 13, No. 3.

It is shown that the density of radio sources in the regions of compact groups of compact galaxies is evidently higher in comparison with the mean density of the surrounding region.

141.115 On the nature of the quasistationary component of variable radio sources emission at centimeter wavelengths.   V. N. Kuril'chik.
Astron. Zh. Akad. Nauk SSSR, Vol. 55, 286 - 291 (1978). In Russian. English translation in Soviet Astron., Vol. 22, No. 2.

It is shown that the observed quasistationary radio emission of variable extragalactic radio sources at centimeter wavelengths can be explained in a model of quasistationary outflow of relativistic electrons from the nucleus into the outer stationary tube of the magnetic field. This tube is situated in the pole of the magnetic field of the nucleus and expands with the distance from it. The spectrum of such an inhomogeneous radio source is calculated.

141.116 A possible correlation between discrete radio sources and the fine structure of the galactic background radio radiation.   M. G. Larionov, V. N. Sidorenkov.
Astron. Zh. Akad. Nauk SSSR, Vol. 55, 299 - 306 (1978). In Russian. English translation in Soviet Astron., Vol. 22, No. 2.

An apparent correlation between the radio sources of the Ohio survey of the sky and extended features of fine structure of the continuum radiation of the Galaxy was obtained as a result of a statistical investigation. The probability of projection of radio sources in loop I and loop II regions on extended "ridges" of the continuum galactic radio emission has been evaluated under the assumption of a random distribution of these sources.

141.117 Redshifts for 24 optically selected QSOs.
A. Savage, J. G. Bolton, K. P. Tritton, B. A. Peterson.
Mon. Not. R. Astron. Soc., Vol. 183, 473 - 477 (1978).

The authors report the spectroscopic observations on 52 ultraviolet-excess objects in two southern fields. Twenty-six are probably QSOs and redshifts have been determined for 24.

141.118 Investigations of the optical fields of 3CR radio sources to faint limiting magnitudes – II.
R. A. Laing, M. S. Longair, J. M. Riley, E. J. Kibblewhite, J. E. Gunn.
Mon. Not. R. Astron. Soc., Vol. 183, 547 - 548 (1978). Microfiche MN 183/2.

A deep optical survey of the fields of 30 3CR radio sources has been carried out with the Hale 200-in. and the Mayall 4-m telescopes. These sources were either previously unidentified or associated with very faint objects for which confirmation was required. High resolution observations ($\theta \sim$

2 arcsec) are available for all 30 sources. For most sources the radio and optical frames of reference have been related with a precision of 0.3–0.5 arcsec.

141.119 Some evidence supporting the cosmological evolution of flat-spectrum quasars. G. M. Blake.
Mon. Not. R. Astron. Soc., Vol. 183, 21P - 27 P (1978).

Application of the luminosity-volume test to a complete sample of quasi-stellar radio sources selected at 1400 MHz gives $\langle V/V_m \rangle$ values of about 0.70 ± 0.05 for both the steep ($\alpha > 0.5$) and flat ($\alpha < 0.5$) spectrum sources. In view of this finding, it may be premature to conclude that flat-spectrum quasars do not take part in the cosmological evolution of powerful radio sources.

141.120 The local space density of quasars and active nuclei. M. Schmidt.
Phys. Scr., Vol. 17, (see 012.013), 135 - 136 (1978).

Local space densities are given for quasars as a function of their optical and radio luminosities, radio galaxies, clusters of galaxies, Seyfert galaxies, giant elliptical galaxies, and field galaxies. The local space density of dead quasars is estimated. A substantial fraction of all galaxies may have a dead quasar as nucleus.

141.121 Emission-line spectra of QSOs and their redshifts. E. M. Burbidge.
Phys. Scr., Vol. 17, (see 012.013), 165 - 169 (1978).

A brief review of work on the emission-line spectra of QSOs is given. Data on radio-emitting and radio-quiet objects are discussed, with sections dealing with the lines that are observed, the distributions of redshifts obtained from the emission lines, and a selection of particular problems posed by current observations. These include a discussion of QSOs with atypical spectra, the Lyman/Balmer/Paschen relative line strengths, the abundances of the elements, correlations of observed properties, and gas flow within the emission-line region.

141.122 Gravity power. D. Lynden-Bell.
Phys. Scr., Vol. 17, (see 012.013), 185 - 191 (1978).

Double whirlpools above and below a black hole cause the beaming and double structure of radio sources. The evidence for the widely held view that gravitational energy powers quasars is reviewed. After a brief discussion of other models an account is given of the standard accretion disk and ways in which such disks may power galactic nuclei and quasars. Finally the problem of creating large black holes cosmogonically is considered.

141.123 Accretion and the quasar phenomenon. M. J. Rees.
Phys. Scr., Vol. 17, (see 012.013), 193 - 200 (1978).

Quasars are interpreted as massive black holes in galactic nuclei which are fuelled by capturing gas or stars from their surroundings. This paper discusses the supply of gas, the nature of the inflow (disclike or quasispherical) and the radiation processes.

141.124 Absorption in the spectra of quasi-stellar objects. E. M. Burbidge.
Phys. Scr., Vol. 17, (see 012.013), 201 - 204 (1978).

A review of the key discoveries concerning absorption lines in QSOs is given. A number of recent studies bearing on the nature and origin of the absorption are discussed. It is concluded that it is most likely that the bulk of the absorption is intrinsic to the QSOs.

141.125 The physics of QSO absorption line regions. A. Boksenberg.
Phys. Scr., Vol. 17, (see 012.013), 205 - 214 (1978).

The physical properties of the absorption systems observed in high redshift QSOs are described in some detail. Apart from the very broad absorption lines seen in a few objects, which almost certainly are due to intrinsic mass outflow, the great majority of the narrow absorption lines commonly present in QSO spectra, ranging in relative velocity from a large fraction of c outward to a few thousand km s^{-1} inward, all of which are characteristically similar and resemble interstellar lines, can be most naturally explained as being produced in cosmologically distributed intervening material consisting of both enriched and primordial gas.

141.126 Statistic of absorption lines and line-locking. R. F. Carswell.
Phys. Scr., Vol. 17, (see 012.013), 215 - 216 (1978).

The location and nature of the clouds which cause absorption lines in the spectra of many high redshift QSOs remains controversial. Most statistical tests performed so far have been inconclusive, though there have been indications that the distribution of numbers of lines is not consistent with that expected from intervening galaxies. There also is a significant peak in numbers corresponding to a velocity difference between emission and absorption regions of about $0.1\ c$. This type of behavior is consistent with radiation pressure driven outflow and 'line-locking', though it is difficult to understand theoretically how the latter mechanism can operate in QSOs.

141.127 A summary of the properties of $Z_{abs} > Z_{em}$ systems and some comments on the origins of QSO absorption line systems. R. J. Weymann, R. E. Williams.
Phys. Scr., Vol. 17, (see 012.013), 217 - 223 (1978).

141.128 On the radiative acceleration of quasar absorption line clouds. P. Goldreich.
Phys. Scr., Vol. 17, (see 012.013), 225 - 228 (1978).

Radiation may expel dense gas clouds from quasar emission line regions. Small clouds can be accelerated to velocities approaching c if they are confined so as to maintain high densities. Models for absorption clouds based on radiatively driven instabilities in expanding quasar envelopes predict column densities that are much larger than those deduced from observation.

141.129 On the origin of quasar absorption lines. J. N. Bahcall.
Phys. Scr., Vol. 17, (see 012.013), 229 - 235 (1978).

Many predictions have been made on the basis of the cosmological hypothesis for the origin of quasar absorption lines. The observational evidence relating to these predictions is summarized. It appears likely that many of the observed absorption lines originate in material that is separated from the emitting quasar by cosmological distances. Two special problems are briefly discussed: the interpretation of redshift splittings and the origin of absorption redshifts larger than their associated emission redshifts.

141.130 The line-locking hypothesis, absorption by intervening galaxies, and the $z = 1.95$ peak in redshifts. G. Burbidge.
Phys. Scr., Vol. 17, (see 012.013), 237 - 241 (1978).

141.131 The large scale structure of extra-galactic radio sources (> 1 arcsecond). A. G. Willis.
Phys. Scr., Vol. 17, (see 012.013), 243 - 255 (1978).

Current data on the structure of extended extragalactic radio sources are reviewed. The major technique used to study such sources is aperture synthesis. Thus the review begins with a brief description of the techniques used in aperture synthesis and the advantages and disadvantages of the method. Section 2 contains a discussion of the source properties. The following Sections 3 and 4 describe the general morphological characteristics of extended extragalactic radio sources. Finally, Section 5 attempts to relate both morphological characteristics and integrated source parameters directly to activity within the nuclei of radio galaxies and quasars.

141.132 Structure and time variations of compact radio sources in galaxies and quasars.   K. I. Kellermann.
Phys. Scr., Vol. 17, (see 012.013), 257 - 263 (1978).

141.133 Extended and compact extragalactic radio sources: interpretation and theory.
R. D. Blandford, M. J. Rees.
Phys. Scr., Vol. 17, (see 012.013), 265 - 274 (1978).
The interpretation of strong double radio sources in terms of the "beam" model is reviewed. Implications of this model for source evolution and the nature of radio trails are briefly discussed. A final section is concerned with deductions about the properties of compact and variable components in quasars and galactic nuclei.

141.134 Energetics of compact radio sources.   L. Woltjer.
Phys. Scr., Vol. 17, (see 012.013), 275 (1978).
The model sensitivity of the energetics of compact radio sources is discussed, and it is shown that very reasonable total energies may be inferred.

141.135 Self consistent models for the emissive regions of quasars and Seyfert nuclei.   S. Collin-Souffrin.
Phys. Scr., Vol. 17, (see 012.013), 293 - 300 (1978).
Models for the region emitting the forbidden and the permitted lines of quasars and Seyfert nuclei are reviewed. First, the mode of energy supply is discussed, and it is concluded that, while the supply is predominantly radiative in the region emitting the forbidden lines, as generally assumed, it is probably partly kinetic in the region emitting the permitted lines. The models include stationary photoionized models, where the thermal equilibrium and the transfer are solved, models in which the transfer and the thermal equilibrium are not taken into account, and time dependent photoionized models. A brief discussion of the other kinds of possible models is given.

141.136 Models of radiation-driven outflow.
R. Kippenhahn.
Phys. Scr., Vol. 17, (see 012.013), 301 - 305 (1978).
The physics of radiative heating and of radiative acceleration in the neighbourhood of a nonthermal energy source is discussed. As an application radiatively driven winds flowing out from the source can be investigated. One can show that very high terminal velocities can be reached. Instabilities of the wind solutions produce shock waves moving outwards supersonically relative to the outflowing wind. These shock waves have such properties that they produce absorption lines which resemble the observed absorption line systems in QSOs.

141.137 The cosmological evolution of quasars.
M. Schmidt.
Phys. Scr., Vol. 17, (see 012.013), 329 - 332 (1978).
There is compelling evidence that quasars, both optically selected and radio selected ones, exhibit strong cosmological evolution. If the space density is an exponential of cosmic time, then the characteristic time of decline is one-tenth of the age of the Universe. Quasars with flat radio spectrum appear to show less cosmological evolution.

141.138 Clustering, pairing and anisotropy of QSOs.
D. Wills.
Phys. Scr., Vol. 17, (see 012.013), 333 - 337 (1978).
Evidence for non-uniformity of the distribution of QSOs over the sky is reviewed. Topics discussed include associations of QSOs with galaxies, pairing of QSOs and the distribution of radio-detected QSOs over the sky. Periodic structure in the distribution of QSO emission-line redshifts is also discussed, and recent work on the surface density of QSOs at various

apparent magnitudes is summarised. Very little compelling evidence exists at present against the view that QSOs are distributed uniformly across the sky and are unassociated with bright galaxies or with other QSOs.

141.139 On the periodicity in the distribution of quasar redshifts.   P. Kjaergaard.
Phys. Scr., Vol. 17, (see 012.013), 347 - 351 (1978).
The periodicity in the distribution of quasar redshifts is explained in terms of selection effects. Special attention is drawn to a selection effect caused by the redshift dependent influence of the strong emission lines on the limiting magnitude for detecting quasars. It is shown that the redshift distribution of quasars selected by a combination of UV-excess information and agreement between radio and optical position is intermediate between the redshift distribution of the two groups of quasars selected by one of the two criteria. It is also shown that the distribution of redshifts for UV-excess selected quasars is very similar to the variation of the ultra-violet excess as a function of redshift. This evidence indicates that strong selection effects are in play.

141.140 The Hubble diagrams for quasars.
J. N. Bahcall, E. L. Turner.
Phys. Scr., Vol. 17, (see 012.013), 353 - 356 (1978).
The optical Hubble diagram for a complete sample of 112 quasars is constructed for $q_0 = 1$. With no correction for selection effects, the Hubble diagram shows a strong correlation of magnitude with redshift (slope 4.43 ± 0.44; correlation coefficient of 0.956). These uncorrected results place an interesting limit on luminosity evolution. The optical and radio Hubble diagrams are also discussed after applying the proper corrections for selection effects.

141.141 The CTIO surveys for large redshift quasars.
P. S. Osmer.
Phys. Scr., Vol. 17, (see 012.013), 357 - 359 (1978).
Lyman α emission in large redshift quasars is readily detectable on slitless spectrograms taken with an objective prism on the Curtis Schmidt and with a transmission grating-prism combination on the 4 m telescope. This provides a new survey method, independent of color, for finding radio-quiet quasars in large numbers. Spectroscopic observations with the CTIO SIT vidicon system have been carried out for more than 50 of the candidates, with the result that the basic properties of the surveys are known. A preliminary analysis of the data indicates that the space density of quasars is at least constant, if not increasing, over the interval $1.9 < z < 3.25$.

141.142 The nature of the unidentified high latitude radio sources.   M. S. Longair.
Phys. Scr., Vol. 17, (see 012.013), 361 - 365 (1978).
A new optical survey of the fields of 3CR radio sources indicates that about 90—95 per cent of the high latitude 3CR sources can be identified with radio galaxies and quasars with apparent magnitudes less than about 23. At the faintest magnitudes, $20 < m < 23$, all the new identifications are galaxies. These results suggest that the unidentified high latitude radio sources are radio galaxies and that like quasars they exhibit strong cosmological evolution.

141.143 Quasars and galactic evolution.   L. Woltjer.
Phys. Scr., Vol. 17, (see 012.013), 367 (1978).
The evolution of quasars is discussed. It is noted that substantial clustering may be present at faint magnitudes. The relationship between quasar evolution and galactic evolution is considered.

141.144 Centimetric radio emission from bright optical quasars.   D. B. Shaffer, R. F. Green.
Publ. Astron. Soc. Pacific, Vol. 90, 22 - 23 (1978).
A search for radio emission from six bright quasars, which

had been found on the basis of their ultraviolet color excess, has been conducted at 2.7 and 8.1 GHz. One quasar was detected as a weak, flat-spectrum radio source. The others were undetected at strengths ≤ 0.01 flux unit.

141.145 Variability of the optical counterpart of OA 33 .
F. F. Donivan, Jr., J. T. Pollock, A. G. Smith, R. J. Leacock, R. L. Scott, P. L. Edwards, M. R. Gearhart.
Publ. Astron. Soc. Pacific, Vol. 90, 24 - 27 (1978) = Contrib. Rosemary Hill Obs., Gainesville, Fla., No. 76.

The flat-spectrum radio source OA 33 (5C3.50) has been identified with a faint blue stellar object that shows strong evidence of optical variability of about one magnitude in the approximate range 19^m to 20^m . Its proximity to M31 and NGC 205 makes it a candidate for searches for superimposed spectral lines, either at optical or radio wavelengths.

141.146 The black flash model of QSOs.
G. A. Shields, J. C. Wheeler.
Gen. Relativ. Gravitation, Vol. 9, 189 - 193 (1978).

A model for QSOs and the nuclei of Seyfert galaxies is proposed in which mass lost from stars in a galactic nucleus repeatedly builds to a critical density and then collapses to the center where it accretes onto a massive black hole ($\sim 10^8 M_\odot$), emitting great luminosity. This model describes a means of starting with an ordinary nucleus and developing conditions found in QSOs. By invoking intermittent flashes the authors overcome a difficulty previously encountered in similar models in which plausible sources of mass in reasonable galactic nuclei fail by a factor of $\sim 10^{-2}$ in fueling a black hole at QSO luminosities.

141.147 Fluxo de radio fontes discretas em 22.2 GHz.
E. Scalise, Jr., M. A. Bräz.
Astron, extragalactica, (see 012.015), 41 - 44 (1976).

Results of observations of discrete sources in 22 GHz obtained at Itapetinga are given.

141.148 Classical double sources in the directions of rich clusters of galaxies.   J. O. Burns, F. N. Owen, L. Rudnick.
Astron. J., Vol. 83, 312 - 321 (1978).

Interferometric observations are reported of two classical double sources (0816 + 526 and 1232 + 414), both of which are in the directions of rich clusters of galaxies. Total intensity and polarization maps are presented at 2695 and 8085 MHz.

141.149 A radio continuum survey of isolated pairs of galaxies.   J. T. Stocke, W. G. Tifft, M. A. Kaftan-Kassim.
Astron. J., Vol. 83, 322 - 347 (1978).

The observational results of a radio continuum survey of isolated pairs of galaxies from the catalog of Karachentsev (1972) is presented. Detailed radio and optical data exist for the 61 detections above the all-catalog limit of 40 mJy at $\lambda = 11$ cm. Three interesting classes of radio detections are discussed including seven pairs with radio halos larger than and/or not coincident with the optical extend of the galaxies.

141.150 Excess radio emission from close pairs of galaxies.
J. T. Stocke.
Astron. J., Vol. 83, 348 - 359 (1978).

Statistics from a radio continuum survey of approximately 600 galaxy pairs [Stocke, Tifft, and Kaftan—Kassim (1978)] show that close pairs are radio sources at more than twice the number than more widely spaced pairs. This correlation holds both for ellipticals and for spirals. A detailed consideration of possible selection effects shows that these do not bias the results. Further investigation of the galaxy pairs and other radio samples generalize the above correlation to one between radio sources and the local galaxy density. Such a relationship suggests that radio sources are triggered by mass transferral

between galaxies during close encounters.

141.151 Selection effects in redshift distribution of QSOs.
D. Basu.
Nature, Vol. 273, 130 - 131 (1978).

Two selection effects have been applied (Basu 1973, 1975) to the distribution of emission line redshifts of QSOs. However, since this hypothesis was proposed, emission-line redshift values have almost trebled. The author reports here a new analysis of 639 redshifts to discover whether they exhibit these selection effects for various values of peak width.

141.152 On a method for determination of the distribution of radio brightness of the sun and other sources of cosmic radio emission.   A. Balklavs, V. Locāns.
Investigations of the sun and red stars. 7, (see 003.004), p. 43 - 55 (1977). In Russian.

A method for calculation of the possible radio brightness distribution for a cosmic radio emission source is considered in the case if reliable information is given only on the source complex visibility module. Examples of calculation made by this method are given. A good accordance is shown between the calculated and real distribution of radio brightness of cosmic radio emission sources. Especially good results are obtained in cases when an additional information about the cosmic radio emission source is available.

141.153 Studies of new complete samples of quasi-stellar radio sources from the 4C and Parkes catalogs.
D. Wills, R. Lynds.
Astrophys. J., Suppl. Ser., Vol. 36, 317 - 358 (1978).

New spectroscopic observations of nearly 100 QSOs are reported, and are combined with earlier data to compile observationally complete samples of QSOs selected from the 4C (178 MHz) and Parkes (2700 MHz) radio source surveys. Adopting the hypothesis that the QSO redshifts are cosmological, the earlier evidence for strong evolution in the QSO properties is amply confirmed. The V/V_m test gives a mean value of 0.683 ± 0.024 rms for $q_0 = 1$, compared with the value (0.5) that is appropriate for a uniform distribution of objects in comoving space. However, if the redshifts are not cosmological in origin, and the QSOs are nearby, the V/V_m test indicates that they are distributed uniformly in space. The data are used to derive the bivariate (radio, optical) luminosity functions at $z = 1$. The three 178 MHz samples agree well with each other, and differ from the 2700 MHz samples in that the slope of the luminosity function is steeper in the radio direction, and flatter in the optical direction, for the 178 MHz data.

141.154 Radioastronomy and amateurs.   M. Duse.
Antenna, Vol. 49, No. 1, p. 21 - 25 (1977). In Italian. – Abstr. in Phys. Abstr., Vol. 81, Abstr. 7887 (1978).

141.155 Optical emission in the SW radio lobe of 3C 33.
S. M. Simkin.
Astrophys. J., Lett., Vol. 222, L55 - L58 (1978).

A diffuse patch of optical emission has been detected in the SW radio lobe of 3C 33. The brightest feature in this patch is coincident with the peak radio emission of the lobe at 5 GHz and with a scintillating radio source known to be contained in the lobe. An optical spectrum of this object shows emission lines from He II, Hβ, and [O III]. These lines have a redshift which is 3300 ± 600 km s^{-1} less than that of the central optical galaxy associated with the source.

141.156 Cosmological evolution of quasars and energy source models.   A. S. Zentsova.
Fiz.–tekh. inst. AN SSSR. Prepr. 543. Leningrad, 1977, 17 pp. In Russian. – Abstr. in Ref. zh., 51. Astron., 3.51.918 (1978).

141.157   UV radiation of quasar 3C 273.
A. G. Doroshkevich.
Priroda, 1978, No. 5, p. 127 - 128. In Russian.

141.158   Polarization in the very large radio galaxy NGC 6251 at 610 MHz.   A. G. Willis, A. S. Wilson, R. G. Strom.
Astron. Astrophys., Vol. 66, L1 - L4 (1978).

Maps of the extended radio galaxy NGC 6251 have been made with the Westerbork telescope at 610 MHz. The north-west lobe is very highly polarized, with degrees of linear polarization up to 70%. The lack of Faraday depolarization implies a very low internal thermal electron density. The radio jet is also polarized, but more weakly.

141.159   On the Hubble diagrams of quasars and the shapes of their optical luminosity functions derived by a new method.   G. Setti, G. Zamorani.
Astron. Astrophys., Vol. 66, 249 - 255 (1978).

The authors show that the study of the Hubble diagram for the brightest quasars can hardly lead, at present, to any significant conclusion on the cosmological evolution of these objects and on the nature of their redshifts. This is because it requires a precise knowledge of the luminosity function, which can only be obtained after making assumptions about the model of the universe and the strength of the cosmological evolution itself.

141.160   Quasar luminosity.   F. G. Smith.
Nature, Vol. 273, 428 (1978).

141.161   Relative quasar luminosities determined from emission line strengths.   J. A. Baldwin, W. L. Burke, C. M. Gaskell, E. J. Wampler.
Nature, Vol. 273, 431 - 435 (1978).

Observations of flat radio spectrum QSOs selected from complete samples confirm the strong correlation between continuum luminosity and emission line equivalent width. The r.m.s. scatter about a mean relationship is about 0.6 mag. The data indicate that the luminosity of QSO emission lines increase as the 1/3 power of the continuum luminosity. Unless the zero point of the relationship between emission line equivalent width and continuum luminosity depends on redshift, both the local hypothesis and zero pressure models of the Universe in which $q_0 \approx 0$ are ruled out by the data at about the 99% confidence level.

141.162   4U0241 + 61: a luminous low-redshift QSO.
K. M. V. Apparao, G. F. Bignami, L. Maraschi, H. Helmken, B. Margon, R. Hjellming, H. V. Bradt, R. G. Dower.
Nature, Vol. 273, 450 - 453 (1978).

A precise (30") celestial position for the faint X-ray source 4U0241 + 61 and its identification at optical and radio wavelengths are reported. Despite its low galactic latitude, the source is extragalactic: a luminous ($M_v = -25.8 \pm 1.8$), low-redshift ($z = 0.0438$) QSO with relatively intense radio emission ($0.2-0.5$ Jy).

141.163   Do compact radio sources contain non-relativistic plasma?   J. S. Scott, W. A. Christiansen, with a reply by J. F. C. Wardle.
Nature, Vol. 273, 572 - 573 (1978).

141.164   Statistical counts of faint sources at 2380 MHz.
J. J. Condon, L. L. Dressel.
Astrophys. J., Vol. 222, 745 - 751 (1978).

The areal density of extragalactic radio sources in the flux density range 1 mJy $< S <$ 30 mJy at 2380 MHz is determined by a statistical analysis of the confusion $P(D)$ distribution observed with a 2.8 pencil beam at Arecibo. Number counts obtained by the $P(D)$ method and by aperture-synthe-

sis surveys are compared. The effective spectral index between 408 and 2380 MHz is found to depend only weakly on 2380 MHz flux density. Observations required for further interpretation of the 2380 MHz number counts are discussed.

141.165   26.3 MHz radio source survey. III. Correlation with extragalactic X-ray sources.
W. C. Erickson, T. A. Matthews, M. R. Viner.
Astrophys. J., Vol. 222, 761 - 778 (1978).

The low-frequency radio characteristics of radio sources associated with 22 well identified extragalactic X-ray sources are discussed. 86% of the X-ray sources having error boxes less than 1 deg^2 are thought to be well identified with clusters of galaxies and isolated active galaxies. The authors confirm that the probability of a cluster being an X-ray source is dependent on whether or not the cluster has a dominating galaxy. Clusters having dominating galaxies have X-ray luminosities and 26 MHz luminosities about 10 times greater than clusters having no dominating galaxies. At 1400 MHz the difference is small or nonexistent. In addition, there is a correlation of the X-ray luminosity with the radio power at both 26 and 1400 MHz. In an effort to encourage further observations which could elucidate these correlations the authors have included a list of radio sources which might be expected to have associated X-ray sources along with rough predictions of the expected X-ray fluxes.

141.166   PKS 1402+044: a red object with a redshift of 3.20.
B. A. Peterson, D. L. Jauncey, A. E. Wright, J. J. Condon.
Astrophys. J., Lett., Vol. 222, L81 - L83 (1978).

Scanner observations of the 18.5 mag red object identified with the radio source PKS 1402+044 show it to have an emission redshift of 3.20. Although the object appears non-stellar on the Palomar Sky Survey, deep plates taken with the 1.2 m UK Schmidt telescope show it to be quite stellar in appearance, with several faint galaxies nearby.

141.167   9.5-mm flux density measurements of a sample of sources from the 5-GHz S4 survey.
A. Witzel, I. I. K. Pauliny-Toth, B. J. Geldzahler, K. I. Kellermann.
Astron. J., Vol. 83, 475 - 477 (1978).

Measurements at 9.5 mm are presented of a complete sample of 56 sources with flux densities at 5 GHz larger than 1 Jy and with flat spectra. Most of the sources in the sample show a steepening in the spectra in the frequency range $5-20$ GHz, indicating that few of these sources contain components which are opaque at short centimeter wavelengths.

141.168   The distribution of QSO redshift $(1 + z_i)/(1 + z_j)$.
J. E. Drew.
Astron. Astrophys., Vol. 66, 343 - 354 (1978).

Following a general survey conducted by Burbidge and Burbidge (1975) to consider the case for line-locking in QSO spectra and a later statistical study by Sargent and Boroson (1977), the method of the latter and the QSO sample of the former are investigated further. A peak is found to occur in the distribution of $(1 + z_{em})/(1 + z_{abs})$ at 1.11 with a chance probability of 3×10^{-5}. It is concluded that the present data suggests the peak is real and that a physical interaction between the QSO emission line regions and absorption systems concerned is required to explain its occurrence.

141.169   A determination of the bivariate size-luminosity function of radio galaxies.
G. Gavazzi, G. C. Perola.
Astron. Astrophys., Vol. 66, 407 - 416 (1978).

Using the maximum radio size of the galaxies in two radio-optically complete samples of sources from the 3C and the B2 catalogues, the size-luminosity function at $z < 0.1$ of radio galaxies $\rho(D, P)$ has been obtained over the range

22.6 < log *P* < 26.2 (at 1415 MHz). It is found that the function $\rho(D, P)$ is not separable, and that the average size increases with the power.

141.170 Radio and optical observations of the galaxy 3C 293.
A. N. Argue, J. M. Riley, G. G. Pooley.
Observatory, Vol. 98, 132 - 135 (1978).

The radio galaxy 3C 293 has unusual radio and optical structures. The authors present radio maps with the Cambridge 5-km telescope at 2.7 and 15 GHz, and a photograph taken with the Isaac Newton Telescope.

141.171 Gigantic energy source is discovered in a galaxy by improved interferometry.   M. Fritz.
Schweiz. Tech. Z., No. 37 - 38, p. 933 - 934 (1977). In German. – Abstr. in Phys. Abstr., Vol. 81, Abstr. 16558 (1978).

141.172 High-resolution observations of compact radio sources at 1.35 cm wavelength.
I. I. K. Pauliny-Toth, E. Preuss, A. Witzel, R. Genzel, K. I. Kellermann, D. B. Shaffer, L. I. Matveenko, I. G. Moiseev, L. R. Kogan, V. I. Kostenko, B. Rönnäng.
Pis'ma Astron. Zh., Vol. 4, 64 - 69 (1978). In Russian.

Observations of four compact radio sources 3C 84, 3C 273, 3C 345 and NRAO 150 have been made at a wavelength of 1.35 cm with interferometers having baselines up to 550 million wavelength. Models of brightness temperature distributions have been obtained and show that three of the sources contain very compact components with brightness temperatures of about 10^{12} K.

141.173 Molecular hydrogen lines in the spectra of quasars.
D. A. Varshalovich, S. A. Levshakov.
Pis'ma Astron. Zh., Vol. 4, 115 - 117 (1978). In Russian.

Identification of some absorption lines in the spectra of 4C 25.05 and OQ 172 with the redshifted lines of the Lyman and Werner series of molecular hydrogen H_2 is proposed.

141.174 Precise optical positions of radio sources in the FK 4-system. Results from a pilot program.
C. de Vegt, U. K. Gehlich.
Astron. Astrophys., Vol. 67, 65 - 71 (1978).

An observational program for the determination of precise positions of the optical counterparts of extragalactic radio sources in the FK4 system has been started at the Hamburg Observatory in 1975. Results from a pilot program concerning 17 sources on the northern hemisphere are reported.

141.175 Hot spots in the outer lobes of extragalactic radio sources.   V. K. Kapahi.
Astron. Astrophys., Vol. 67, 157 - 160 (1978).

A plot of the angular sizes of individual lobes of double radio sources against redshift suggests that lobe sizes may undergo similar evolutionary effects with epoch as the lobe separations. An angular resolution of better than 0."5 may be necessary in order to see "hot spots", similar to those in Cyg-A, in the lobes of sources at high redshifts. The evolutionary effects in lobe sizes together with the limited resolution of the existing observations could account for the recently reported correlations between source luminosity and the intensity and size of "hot spots" in the 3 CR source sample.

141.176 What are radio astronomers looking for?   B. Hoglund.
Eltek. Aktuell Elektron A, Vol. 20, No. 17, p. 18 - 19 (1977). In Swedish. – Abstr. in Phys. Abstr., Vol. 81, Abstr. 24589 (1978).

141.177 Frequency dependence of compact structure in extended extragalactic radio sources.
S. R. Spangler, K. A. Meyers.

Astron. J., Vol. 83, 547 - 559 (1978).

The authors have conducted 850 MHz interplanetary scintillation observations of 31 radio sources, 26 of which have had previous IPS measurements at 430 and/or 81.5 MHz. Most of the sources have steep radio spectra and therefore are probably double radio sources with compact components at the extremities of the lobes. A comparison of scintillation characteristics at three frequencies leads to the following results: (a) The observed scintillation indices tend to decrease with increasing frequency. (b) The angular diameters decrease with increasing frequency. This frequency dependence of the scintillation characteristics may be accounted for if the scintillating components consist of several compact subcomponents. Finally, the data indicate that the typical spectral index difference between the scintillating and nonscintillating components ($\equiv \zeta$) is small, upper limits being of the order of 0.2 to 0.4.

141.178 A high declination search at 8 GHz for compact radio sources.   J. J. Wittels, I. I. Shapiro,
D. S. Robertson, C. C. Counselman, H. F. Hinteregger, C. A. Knight, A. E. E. Rogers, A. R. Whitney, T. A. Clark, L. K. Hutton, C. Ma, A. E. Niell, G. M. Resch, B. O. Rönnäng, O. E. H. Rydbeck.
Astron. J., Vol. 83, 560 - 565 (1978).

With the Haystack-NRAO interferometer the authors observed 37 sources whose declinations were above 50°. Seven of these sources have compact cores with diameters smaller than 5 milliarcsec and with correlated flux densities greater than about 0.5 Jy; the remaining sources have no cores with flux densities above about 0.3 Jy, the sensitivity limit of the interferometer. All diameter estimates were based on an assumed circular Gaussian distribution of radio brightness. Positions for the detected sources were also obtained from the interferometric data, the uncertainty in these coordinate estimates ranging from 0.04 to 0.6 arcsec.

141.179 Formation of the relativistic electron spectrum in nonstationary sources of synchrotron radio emission.   V. I. Altunin.
Astron. Zh. Akad. Nauk SSSR, Vol. 55, 559 - 563 (1978). In Russian. English translation in Soviet Astron., Vol. 22, No. 3.

It is shown that at the beginning of the development of nonstationary sources of nonthermal radio emission conditions of the "synchrotron turbulent boiler" are realized in them. As a result, a power spectrum of relativistic electrons with $\gamma \sim 1$, sharply bounded at high energies is formed. For a number of cases the magnetic field magnitude and the plasma electron concentration are determined in the region of formation of the spectrum of emitting particles.

141.180 Radio counterpart for γ-ray source CG 135+1.
IAU Circ., No. 3164 (1978).

141.181 Variable quasi-stellar objects. II. Photometry and completeness of faint blue objects in the Sandage-Luyten survey field $15^h 10^m$, +24°.   P. D. Usher, K. J. Mitchell.
Astrophys. J., Vol. 223, 1 - 12 (1978).

Optical variability is investigated for 381 faint blue LB objects in the Sandage-Luyten survey field at $\alpha(1855) = 15^h 10^m$, $\delta(1855) = +24°$. Mean magnitudes are derived for all but a few of the bluest (color class I) objects, and for a complete sample of 82 redder objects of color classes II and III. The morphology of all sources has been investigated, resulting in three color class I and two color class II LB sources that appear to be resolved. Five other cases of contiguous nebulosities are noted. Positions, estimated colors, mean magnitudes, and finding charts for additional blue objects are given. Seven of these are judged to be variable.

141.182 The dynamics of relativistic plasmoids and rapidly varying radio sources.   W. A. Christiansen, J. S. Scott, W. T. Vestrand.
Astrophys. J., Vol. 223, 13 - 24 (1978).

Variability and superluminal expansion of compact radio sources are investigated in terms of a theoretical model for the dynamics of radio sources involving relativistic motion of ram-pressure confined plasmoids (defined to be aggregates of magnetized plasma and relativistic particles). It is shown that relativistic spherical blast waves are subject to fragmentation leading to the formation of a number of discrete, relativistic plasmoids. It is shown that the initial temperature of the thermal component of the plasma fragments is likely to be on the order of 10^{12} K.

141.183 The mapping of compact radio sources from VLBI data.   A. C. S. Readhead, P. N. Wilkinson.
Astrophys. J., Vol. 223, 25 - 36 (1978).

The authors describe a method for recovering the brightness distributions of compact radio sources from VLBI observations of closure phase and fringe amplitude. Their approach is to deduce the visibility phase from these data. The authors show that reliable "hybrid" maps of complexity comparable with that of early synthesis results on extended sources can be obtained with networks of four or more existing telescopes.

141.184 PKS 2126—15: a bright quasi-stellar object with neutral color and a redshift of 3.27.
D. L. Jauncey, A. E. Wright, B. A. Peterson, J. J. Condon.
Astrophys. J., Lett., Vol. 223, L1 - L3 (1978).

Scanner observations of the bright 17.3 mag neutral stellar object identified with the radio source PKS 2126—15 show it to be a QSO with an emission-line redshift of 3.27.

141.185 The radio source 0915+320: a "wide-angle-tail" source in a group of galaxies.
E. B. Fomalont, A. H. Bridle.
Astrophys. J., Lett., Vol. 223, L9 - L12 (1978).

The radio source 0915+320 has been mapped with the VLA at a frequency of 4.9 GHz. The observations provide the optical identification by detection of a small-diameter "core" component coincident with an elliptical galaxy which may be associated with a poor group of fainter spirals. The extended radio structure has an almost-collinear, bifurcated distribution with the highest-brightness regions at the inner part of the radio lobes. This morphology is strikingly similar to the "wide-angle-tail" structures hitherto reported only in rich clusters of galaxies.

141.186 Quasars and cosmological evolution.
M. Schmidt.
IAU Symp. No. 79, (see 012.027), p. 289 - 293 (1978).

The first evidence that quasars exhibit a non-uniform distribution in the universe was obtained ten years ago from a sample of 33 quasars in the 3CR catalogue. Much larger samples of radio quasars are now available and the author discusses the results in terms of an exponential decrease of quasar numbers with cosmic time. Only a few samples of optically selected quasars exist. It is shown that their counts indicate a non-uniform space distribution, even if the redshifts are non-cosmological.

141.187 The surface density of quasars.
J. G. Bolton, A. Savage.
IAU Symp. No. 79, (see 012.027), p. 295 - 303 (1978).

The authors have carried out what they believe to be currently the most comprehensive investigation into the surface density of quasars. Three techniques have been used in this investigation. These are: (1) A search for ultra-violet excess stellar objects and their subsequent classification. (2) A search for emission-line stellar objects on objective prism plates. (3) Identification of radio sources from deep radio surveys.

141.188 A survey programme for QSO and related objects.
H. Lorenz.
IAU Symp. No. 79, (see 012.027), p. 303 - 304 (1978).
Short communication.

141.189 The large scale distribution of radio sources.
M. S. Longair.
IAU Symp. No. 79, (see 012.027), p. 305 - 314 (1978).

Three topics are described: (1) the large scale distribution of extragalactic radio sources on the celestial sphere; (2) the problems of identifying optically quasars and radio galaxies in that region of the source counts where they diverge most significantly from the predictions of all uniform world models; (3) the problems of interpretation of the source counts, some models for the spatial distribution of sources and the most important observations for defining more precisely the evolution of the radio source population.

141.190 QSO absorption lines and intergalactic hydrogen clouds.   B. A. Peterson.
IAU Symp. No. 79, (see 012.027), p. 389 - 392 (1978).

In order to study the properties of the clouds in the intergalactic medium, spectra were obtained of QSOs with redshifts in the range $2 < z < 3.6$. The QSOs were selected to have similar luminosities. The preliminary results from a study of these spectra are reported here.

141.191 The spectrum of the quasar B2 1225 + 31.
M. S. Wilkerson, G. Coleman, G. Gilbert, P. A. Strittmatter, R. E. Williams, J. A. Baldwin, R. F. Carswell, S. A. Grandi.
Astrophys. J., Vol. 223, 364 - 372 (1978).

The spectrum of the intrinsically bright QSO B2 1225 + 31 ($z_{em} \sim 2.2$) has been analyzed in the range $3140\ \text{Å} < \lambda < 8000\ \text{Å}$, using image-tube scanner and echelette data. The emission lines are abnormally weak and broad, although their relative intensities are comparable to those of other QSOs. The absorption-line spectrum is exceedingly rich shortward of emission $L\alpha$. One absorption system ($z_A \sim 1.7941$) is very well established, with 25 resonance-line identifications. Several other systems are suggested but the evidence is in no case compelling. No strong evidence for line locking or abnormal abundances was found in this object.

141.192 Preliminary results of linear polarization measurements at 99 GHz.   R. W. Hobbs, S. P. Maran, L. W. Brown.
Astrophys. J., Vol. 223, 373 - 377 (1978).

The results of the first linear polarization measurements of extrasolar radio sources at a frequency above 35 GHz are presented. For the quasars 3C 273 and 3C 345, the observed position angles are in accord with the predictions of Inoue, and the measurements are consistent with the suggestion that the compact components responsible for the high-frequency emission are affected by high degrees of internal Faraday depolarization. For the radio galaxy 3C 274 (Virgo A) the observed position angle and degree of polarization are consistent with the model of Hobbs and Waak.

141.193 Isotropy of radio sources in the B2 catalogue.   C. Fanti, C. Lari, M. C. Olori.
Astron. Astrophys., Vol. 67, 175 - 184 (1978).

The three major sections of the B2 catalogue are analysed for isotropy. All instrumental effects possibly present in the catalogue are accurately described and eliminated. Both power spectrum analysis (PSA) and binning analysis (BA) are performed. The $\log N/\log S$ relation is analysed in areas selected according to the value of their own source density in the flux interval $0.5 \div 1.0$ Jy and no significative discrepancy is found among the various regions. The authors estimate which volume source density contrast may be consistent with results of PSA and BA, and they obtain a value not greater than about 10 for clustering at least 60 Mpc in size.

141.194 Radio sources around 3C 465, in the cluster of galaxies Abell 2634.   R. Wielebinski, C. G. T. Haslam, J. R. Baker, P. P. Kronberg.
Astron. Astrophys., Vol. 67, 293 - 296 (1978).

Continuum maps of an area centered on the cluster of galaxies A2634 have been made with the 100-m radio telescope at both 2650 and 4800 MHz. Of the sources in the field, only the two extended sources 3C465 and 2337+268 are known to be cluster members. Both show a central condensation surrounded on either side by linearly extended features along which the spectral index of the radio emission gradually steepens. Spectra of the point sources in the field have also been determined. No low level emission which could be attributed to the intracluster medium has been detected.

141.195 Quasi-stellar objects.   H. E. Smith.
Mercury, Vol. 7, 27 - 33 (1978).

141.196 Quasars PKS 1217 + 023 and LB 2136 on plates of a supernovae survey.   V. P. Goranskij.
Peremennye Zvezdy, Vol. 20, 403 - 411 (1977). In Russian.

The light curves of two quasars are obtained using the photographic material of a supernovae survey. The observations were carried out during 17 years (1960–1977). PKS 1217 + 023 varies slowly in the time scale of some years. The limits of B magnitude are $15.2 - 16.5$. LB 2136 = 4C 49.22 varies in the limits of B magnitude $16.0 - 17.5$. Bursts of some hundred days duration were observed. A table presents the original observations. Possible errors of photographic observations of such objects are discussed.

141.197 A list of small-diameter extragalactic radio sources for radio astrometry. I. Very-long-baseline inter-ferometer measurements.   N. D. Umarbaeva.
Soobshch. Spets. Astrofiz. Obs., Zelenchukskaya, Vyp. (No.) 16, p. 5 - 23 (1976). In Russian.

Some possibilities to use observations of artificial and natural radio sources with very-long-baseline interferometers (VLBI) for radio astrometry are briefly discussed. On the basis of VLBI observations a complete list of extragalactic radio sources which can be used for such investigations has been compiled.

141.198 The 1977 outburst of the quasar 3C 446.
J. S. Miller, H. B. French.
Bull. American Astron. Soc., Vol. 10, 387 (1978). – Abstract.

141.199 Variability of the central source in Centaurus A (NGC 5128).   K. M. Price, W. Graf.
Bull. American Astron. Soc., Vol. 10, 388 (1978). – Abstract.

141.200 Short-lived clouds in atmospheres of QSO's and active galaxies.   L. J. Caroff, J. Eilek.
Bull. American Astron. Soc., Vol. 10, 388 (1978). – Abstract.

141.201 Flux density variations in flat spectrum sources at 8 GHz: a search for new variable sources.
T. J. Rugare, H. D. Aller.
Bull. American Astron. Soc., Vol. 10, 403 (1978). – Abstract.

141.202 Particle reacceleration in radio galaxies.
J. A. Eilek.
Bull. American Astron. Soc., Vol. 10, 403 (1978). – Abstract.

141.203 Surface brightness and shape of radio jets.
M. J. Valtonen.
Bull. American Astron. Soc., Vol. 10, 403 (1978). – Abstract.

141.204 Positrons in compact radio sources.
P. D. Noerdlinger.
Bull. American Astron. Soc., Vol. 10, 404 (1978). – Abstract.

141.205 Variability in extragalactic low frequency radio sources and "ripple radiation".
W. J. Cocke, A. G. Pacholczyk, F. A. Hopf.
Bull. American Astron. Soc., Vol. 10, 404 (1978). – Abstract.

141.206 Application of the small-pitch angle synchrotron theory to a wide class of variable compact extra-galactic objects.
W. J. Cocke, M. S. Giampapa, A. G. Pacholczyk.
Bull. American Astron. Soc., Vol. 10, 404 (1978). – Abstract.

141.207 Radio interferometric observations of the bright core of CTB 80.
P. E. Angerhofer, M. R. Kundu.
Bull. American Astron. Soc., Vol. 10, 427 (1978). – Abstract.

141.208 A comparison of radio and optical features in sixteen radio sources.   F. D. Ghigo.
Bull. American Astron. Soc., Vol. 10, 428 (1978). – Abstract.

141.209 Long wavelength minimum source diameters: severe solar wind scattering contamination?
W. M. Cronyn, J. J. Rickard.
Bull. American Astron. Soc., Vol. 10, 428 (1978). – Abstract.

141.210 21-cm observations of NGC 7822.   E. J. Grayzeck.
Bull. American Astron. Soc., Vol. 10, 436 (1978). Abstract.

141.211 The hydrogen line intensity ratios in quasi-stellar objects.
H. E. Smith, R. C. Puetter, S. P. Willner, J. L. Pipher.
Bull. American Astron. Soc., Vol. 10, 448 (1978). – Abstract.

141.212 Internal absorption by dust as a possible explanation for the observed correlation between C IV $\lambda 1549$ equivalent width and continuum luminosity in QSOs.
J. M. Shuder, G. M. MacAlpine.
Bull. American Astron. Soc., Vol. 10, 448 (1978). – Abstract.

141.213 Low resolution spectrophotometry of eight optical quasars with $3.0 < z < 3.5$.   P. S. Osmer.
Bull. American Astron. Soc., Vol. 10, 449 (1978). – Abstract.

141.214 Coordinated centimeter, millimeter, infrared, and visual polarimetry of compact nonthermal sources.
T. W. Jones, J. J. Puschell, W. A. Stein, J. W. Warner, L. Rudnick, F. N. Owen.
Bull. American Astron. Soc., Vol. 10, 449 (1978). – Abstract.

141.215 Optical pumping and fine structure absorption in quasars.
C. L. Sarazin, B. P. Flannery, G. B. Rybicki.
Bull. American Astron. Soc., Vol. 10, 449 - 450 (1978). Abstract.

141.216 Video camera photographs of the low redshift QSO 0241+622. Evidence that the QSO is in the center of a spiral galaxy.   H. C. Ford.
Bull. American Astron. Soc., Vol. 10, 450 (1978). – Abstract.

141.217 Further evidence that some quasars are not at their redshift distances.   H. C. Arp.
Bull. American Astron. Soc., Vol. 10, 450 (1978). – Abstract.

141.218 Ultraviolet spectra of quasars observed with IUE.   M. Schmidt, R. F. Green, A. L. Lane, F. B. Estabrook, H. D. Wahlquist.

Bull. American Astron. Soc., Vol. 10, 450 (1978). — Abstract.

141.219 The distance to the quasar PKS 0837-12.
     P. A. Wehinger, S. Wyckoff, A. Boksenberg.
Bull. American Astron. Soc., Vol. 10, 450 - 451 (1978).
Abstract.

141.220 A galaxy associated with 3C 273.   A. Stockton.
     Nature, Vol. 274, 342 - 343 (1978).
     The author reports here the discovery of a galaxy
associated with 3C 273. Within the measurement errors, the
redshift is identical to the accepted redshift $z = 0.158$ of
3C 273.

**141.221 Apparent superluminal expansion velocities in the
          dipole magnetic field model.**
M. Milgrom, J. N. Bahcall.
Nature, Vol. 274, 349 - 350 (1978).
     Very long baseline interferometry observations of com-
pact radio sources have shown that the components of some
radio sources seem to move apart with relative velocities
more than twice the speed of light. The authors present the
valid geometrical consequences of the dipole field model of
extragalactic radio sources, show that they agree with available
observations, and make some additional predictions.

**141.222 Characteristics of the radio source near the super-
          giant star HD 18391.**
L. A. Higgs, P. A. Feldman, J. Smoliński.
Astron. Astrophys., Vol. 67, 431 - 434 (1978).
     Further radio observations of a source detected in an
earlier survey of late-type supergiant stars confirm its chance
positional coincidence with HD 18391 (G0 Ia). This radio
source now appears to be either extragalactic or a thermal
galactic object with an unusually high electron temperature.

**141.223 Search for radio emission from objects with
          continuous optical spectrum and large proper
motion.**   V. M. Malofeev, V. F. Shvartsman.
Soobshch. Spets. Astrofiz. Obs., Zelenchukskaya, Vyp. (No.)
20, p. 39 - 53 (1977). In Russian.
     Theoretical estimates concerning intensity and spectrum
of radio emission from isolated black holes of stellar masses
are presented. A program is described of the search for radio
emission from objects with continuous optical spectrum and
large proper motion that are usually reckoned among DC-type
white dwarfs. The results of observations of 20 DC-objects on
wavelengths $\lambda = 3.5$ m and $\lambda = 2.8$ m are presented. Results
of additional observations of eight DC-objects are presented.
The results of the observations incline to two conclusions:
(1) the density of black holes in the vicinity of the Sun is much
smaller than that of ordinary stars; (2) the rotational periods
of DC-type white dwarfs, as a rule, exceed 100 s.

**141.224 Observations of quasi-simultaneous radio spectra
          and fast radio variability of objects with continuous
optical spectrum.**   S. A. Pustil'nik.
Soobshch. Spets. Astrofiz. Obs., Zelenchukskaya, Vyp. (No.)
20, p. 54 - 68 (1977). In Russian.
     For 5 radio objects with continuous optical spectrum
the results of observations with the RATAN-600 radiotelescope at wavelengths 2.08, 3.9, 6.52 cm are presented. Quasi-
simultaneous radio spectra are constructed. For some objects
rapid variability with characteristic time of 1 day to 1 week
is discovered or suspected.

141.225 18-cm OH observations of W33A.
     M. I. Pashchenko.
Astron. Tsirk., No. 981, p. 1 (1978). In Russian.

**141.226 On the decimetric radio variability of extragalactic
          sources.   V. N. Kuril'chik.**

Astron. Tsirk., No. 982, p. 1 - 3 (1978). In Russian.

141.227 The angular size of quasars.
     J. F. C. Wardle, R. Potash.
Ann. New York Acad. Sci., Vol. 302, (see 012.033), 605 -
610 (1977).

141.228 Spectral properties of quasars.
     D. O. Richstone.
Ann. New York Acad. Sci., Vol. 302, (see 012.033), 611 -
612 (1977).

141.229 Quasar theories.   M. J. Rees.
     Ann. New York Acad. Sci., Vol. 302, (see 012.033),
613 - 636 (1977).
     Contents: The emission lines. The absorption lines. Com-
pact radio components: "superlight" expansion. Accretion
onto massive black holes: the supply of gas. Modes of accretion. The continuum radiation. The evolutionary context:
"dead" quasars.

**141.230 The physics of radio galaxies — constraints and
          hints provided by observational data.**
H. van der Laan.
Ann. New York Acad. Sci., Vol. 302, (see 012.033), 637 -
655 (1977).
     This paper attempts to present the author's main con-
clusions drawn from a large body of data acquired with the
Westerbork Synthesis Radio Telescope.

141.231 The cosmic evolution of radio source populations.
     J. V. Wall.
Ann. New York Acad. Sci., Vol. 302, (see 012.033), 656 -
664 (1977).
     It is the aim of this contribution to show that progress is
now possible in determining the details of the evolution of
different source populations. Progress in the observations, in
particular the definition of source counts at different frequencies, is first discussed, followed by the results from recent
analyses of these and associated data. The concluding section
presents a summary together with a non-exhaustive list of observations essential for further advances.

**141.232 The interaction of radio galaxies with the inter-
          galactic medium.   D. S. De Young.**
Ann. New York Acad. Sci., Vol. 302, (see 012.033), 669 -
680 (1977).
     The existence or non-existence of a general, all pervasive
intergalactic medium has been a question of controversy for
many years. What are considered is the nature and consequences of any interaction that exists between the extended
extragalactic radio sources and any ambient gaseous material
in the neighborhood of these sources.

**141.233 Radiogalaxies and their interaction with the ex-
          ternal medium.   G. C. Perola.**
Mem. Soc. Astron. Italiana, Vol. 47, 287 - 301 (1976).
     Contents: Morphology of extended radio sources. Head-
tail radio sources. "Normal" doubles.

     A catalogue of absorption lines in QSO spectra.
See Abstr. 002.033.

     The Nançay survey of absorption by galactic neutral
hydrogen. I. Absorption towards extragalactic sources.
See Abstr. 002.039.

     Decametric survey of discrete sources in the
Northern sky. II. Source catalogue in the range of declinations + 10° to +20°.   See Abstr. 002.040.

     Decametric survey of discrete sources in the

Northern sky. III. Low-frequency absolute flux scale of discrete radio sources. See Abstr. 002.041.

Decametric survey of discrete sources in the Northern sky. IV. Spectra of 266 discrete sources in the range 10 to 1400 MHz. See Abstr. 002.042.

The 5 GHz strong source surveys. IV. Survey of the area between declination 35 and 70 degrees and summary of source counts, spectra and optical identifications. See Abstr. 002.047.

A handbook of quasistellar and BL Lacertae objects. See Abstr. 003.037.

Kvasarer, pulsarer og svarte hull. See Abstr. 003.051.

Astronomy. Volume 13: Radioastronomy. See Abstr. 003.089.

A handbook of radio sources. See Abstr. 003.109.

An automated station for the detection of bursts of cosmic origin at VHF and UHF. See Abstr. 009.017.

On the flow of special relativistic fluids through channels. See Abstr. 022.099.

Phaseless aperture synthesis. See Abstr. 031.202.

Possible use of electronographically recorded low-dispersion slitless spectra in optical quasar surveys. See Abstr. 031.238.

Geodetic and astrometric analysis of fringe rate residuals from the Algonquin-Chilbolton 2.8 cm long-baseline interferometer. See Abstr. 033.039.

Perspectives of solving fundamental problems of astrometry with the help of superlong baseline interferometry and special space experiments. See Abstr. 046.036.

Relativistic parametric instabilities in extended extragalactic radio sources. See Abstr. 062.001.

The evolution of expanding nonthermal sources. II. Relativistic expansion. See Abstr. 062.022.

Double radio sources and the new approach to cosmical plasma physics. See Abstr. 062.086.

The $L\alpha/H\alpha$ ratio in solar flares, quasars, and the chromosphere. See Abstr. 073.061.

Millimetre observations of planets, galactic and extra-galactic sources. See Abstr. 091.004.

Possible optical counterpart (LSI +61°303) for GT 0236+610. See Abstr. 113.064.

Are supernovae radio sources? A search for radio emission from young supernova remnants. See Abstr. 125.018.

Origin and evolution of the radio emission from immediate postoutburst supernovae. See Abstr. 125.019.

Supernova radio pulse searches and possible im-

provements in sensitivity. See Abstr. 125.021.

Continuum observations of supernova remnant G 127 + 0.5 at 2695 MHz. See Abstr. 125.044.

Optical emission from the SNR G290.1 - 0.8. See Abstr. 125.066.

The structure and spectrum of a colliding-cloud system and its possible relationship to QSOs. See Abstr. 131.015.

A large-scale OH sky survey at 1612 MHz. Part I. The observations. See Abstr. 131.024.

The helium problem in Sagittarius B2. See Abstr. 131.076.

Interstellar scattering, the north polar spur, and scintars. See Abstr. 131.234.

The analysis of radio emission from H II regions: consequences of improper analytic methods. See Abstr. 132.038.

609 MHz aperture synthesis observations of S 104. See Abstr. 132.043.

A double $H n \alpha$ line galactic source G37.7 + 0.1. See Abstr. 132.060.

A 5 GHz survey of infrared sources. See Abstr. 133.037.

Radio identification of an infrared object associated with the passage of a shock front. See Abstr. 133.039.

Carbon recombination-line mapping of the Orion Nebula. See Abstr. 134.002.

The kinematics of the Rosette nebula. See Abstr. 134.019.

Radio and optical identification of the X-ray source GX 17 + 2. See Abstr. 142.048.

X-ray active galaxies and QSOs. See Abstr. 142.187.

Radio observations of the X-ray galaxy NGC 2110. See Abstr. 142.211.

Interferometer measurements of H I absorption in the direction of 16 extragalactic radio sources. See Abstr. 155.002.

ESO 113-IG45: galaxy and/or quasar? See Abstr. 158.007.

The spectral-flux distribution (0.36–3.5 μm) of nonstellar light from the broad-line radio galaxies 3C 227 and 3C 382. See Abstr. 158.013.

Radio and X-ray variability of the nucleus of Centaurus A (NGC 5128). See Abstr. 158.014.

Spiral and irregular galaxies at 2.7 and 8.1 GHz IV: NGC 3079 and NGC 4490. See Abstr. 158.019.

Optical spectra of radio galaxies. See Abstr. 158.048.

Virgo A at 1.3 centimeters.   See Abstr. 158.058.

The discrete redshift and asymmetry in H I profiles. See Abstr. 158.060.

Orbital dynamics of the radio galaxy 3C 129. See Abstr. 158.061.

An optical-radio jet in NGC 7385. See Abstr. 158.071.

Do head−tail radio galaxies have large velocities with respect to the intracluster medium?   See Abstr. 158.077.

Spectroscopic observations of NGC 5128. See Abstr. 158.086.

An observational model of the ionized gas in Seyfert and radio-galaxy nuclei.   See Abstr. 158.088.

Anomalous redshifts in galaxies and quasars. See Abstr. 158.106.

The Hubble diagram for Seyfert galaxies and related objects.   See Abstr. 158.107.

The radio properties of Seyfert galaxies. See Abstr. 158.110.

Detection of neutral hydrogen absorption at $z = 0.045$ in the peculiar galaxy 1506 + 34. See Abstr. 158.123.

The Andromeda nebula radio emission at 102.5 MHz.   See Abstr. 158.135.

Spectrophotometry of three radio galaxies. See Abstr. 158.138.

Optical emission-line spectra of Seyfert galaxies and radio galaxies.   See Abstr. 158.139.

The continuous optical spectra of active nuclei. See Abstr. 158.140.

Infrared observations of Seyfert galaxies and quasars.   See Abstr. 158.141.

Radio observations of active and normal nuclei. See Abstr. 158.142.

Physical state of the emission-line region. See Abstr. 158.143.

Progressos nas investigações de objetos extragalaticos austrais em ondas milimetricas.   See Abstr. 158.154.

Doppler-shift distributions. II. The relativistic gas.   See Abstr. 158.157.

Optical emission in the radio lobes of radio galaxies. See Abstr. 158.160.

The Arecibo 2380 MHz survey of bright galaxies. See Abstr. 158.163.

The radio halo of NGC 891. See Abstr. 158.164.

Curtis Schmidt-thin prism survey for extragalactic emission-line objects. University of Michigan list IV. See Abstr. 158.170.

Radio properties of active nearby spiral galaxies. See Abstr. 158.192.

NGC 4472: a very weak radio galaxy? See Abstr. 158.202.

High frequency radio observations of Seyfert galaxies.   See Abstr. 158.204.

Forbidden lines in quasars and active galaxies. See Abstr. 158.217.

Radio recombination line studies of M82 and other galaxies.   See Abstr. 158.232.

Radio continuum emission from the barred spiral galaxy NGC 5383.   See Abstr. 158.238.

Seyfert galaxies and quasars. See Abstr. 158.243.

Extragalactic observations by IUE (*International Ultraviolet Explorer*).   See Abstr. 158.267.

Nuclei galattici e radio-sorgenti. See Abstr. 158.277.

On active nuclei of galaxies and QSOs. See Abstr. 158.278.

Physical problems associated with BL Lac objects and QSOs.   See Abstr. 158.508.

A comprehensive radio study of the $z = 0.524$ absorption system in AO 0235 + 164.   See Abstr. 158.509.

On the extragalactic nature of Lacertids. See Abstr. 158.510.

Quasars and BL Lac objects as active nuclei of giant galaxies.   See Abstr. 158.515.

Multifrequency observations of the variable quasi-stellar radio source 0735+178.   See Abstr. 158.516.

Nonthermal properties of BL Lac objects and the relation to normal QSOs.   See Abstr. 158.517.

Large Magellanic Cloud sources at 3.4-cm wavelength.   See Abstr. 159.011.

A radio survey of clusters of galaxies. I. 11.1 cm observations of A591, A754, A1066, A1314, A1517, A2094, A2142, A2255, A2256, A2319 and A2462. See Abstr. 160.001.

Velocities of head-tail radio galaxies with respect to the intracluster medium.   See Abstr. 160.025.

Galaxy counts in the regions surrounding head−tail radio sources.   See Abstr. 160.032.

X-ray and radio emission from clusters of galaxies: the heating of intracluster gas by relativistic electrons. See Abstr. 160.043.

Observations of clusters containing radio tail galaxies.   See Abstr. 160.061.

Radio properties of Abell clusters. See Abstr. 160.077.

Radio observations of clusters of galaxies: the tail sources.   See Abstr. 160.078.

Extended radio emission in clusters of galaxies: recent Westerbork observations.   See Abstr. 160.079.

Clusters with extended radio emission at high frequencies.   See Abstr. 160.080.

Radio emission of Abell clusters in the GB and GB2 regions.   See Abstr. 160.084.

Rotation measures as a cosmological test. See Abstr. 162.037.

The redshift distribution of quasi-stellar objects as an indicator of clustering on gigaparsec scale. See Abstr. 162.161.

Hubble's constant from Sc-galaxies and quasars. See Abstr. 162.170.

Pulsars

141.501 On the role of finite inertia and resistivity in axisymmetric pulsar magnetospheres.   Y.-M. Wang.
Mon. Not. R. Astron. Soc., Vol. 182, 157 - 177 (1978).
    Inertial and resistive effects in axisymmetric pulsar magnetospheres are examined as a function of the plasma density.

141.502 Wave production in an ultrarelativistic electron-positron plasma.
P. E. Hardee, W. K. Rose.
Astrophys. J., Vol. 219, 274 - 287 (1978).
    Theoretical models describing the generation and acceleration of charged particles at the polar caps of rotating magnetized neutron stars predict the production of electron-positron pairs in the strong magnetic field, $\sim 10^{12}$ gauss, associated with neutron stars. The pair production mechanism generates an ultrarelativistic electron-positron plasma that streams out along the magnetic field lines which extend from the polar caps to the light cylinder. The problem of generation and propagation of electromagnetic waves in an ultrarelativistic electron-positron plasma has not been adequately considered in previous models and is necessary to the understanding of pulsar emission. The authors consider the wave propagation properties of such a medium.

141.503 Direct observation of pulsar microstructure at timescales down to 6 μs.   N. Bartel.
Astron. Astrophys., Vol. 62, 393 - 396 (1978).
    Radio pulse emission of PSR 1133 + 16 was sampled in two channels of different frequencies at data rates up to 1 MHz. The occurrence of individual pulses delayed in time corresponding to the dispersion delay of the interstellar medium demonstrates the existence of microstructure at $\sim$ 2700 MHz on time scales down to 6 μs and fluxes of up to 500 Jy, correlated over 20 MHz bandwidth. This is the fastest modulation yet published. The results are discussed in terms of the relativistic beaming model.

141.504 Revised estimate of gravitational radiation from Crab and Vela pulsars.   M. Zimmermann.
Nature, Vol. 271, 524 - 525 (1978).
    The author reports that the amplitude of the gravitational waves from the Crab pulsar (PSR0531+21) is likely to be within 2 orders of magnitude of 10^{-27}, but that the Vela pulsar (PSR0833—45) is likely to produce waves of amplitude a factor 10 to 100 larger.

141.505 PSR 1919+21: a temporal variability of the P_3 period.   A. Wolszczan.
Astron. Astrophys., Vol. 63, 425 - 427 (1978).
    The results of a power spectrum analysis of a train of 3968 pulses received from PSR 1919+21 at 1420 MHz are presented. It is shown that the P_3 period of this pulsar may vary in a quasi-periodic manner on the timescale of about 30 min, within a narrow range of values corresponding to the width of the P_3 feature in the average power spectrum. Some theoretical implications of this new effect are discussed.

141.506 Little green men, white dwarfs, or what?
S. J. Bell Burnell.
Sky Telesc., Vol. 55, 218 - 221 (1978).

141.507 Interstellar scattering of pulsars by a turbulent interstellar plasma.
D. S. Bugnolo.
Astron. Zh. Akad. Nauk SSSR, Vol. 55, 69 - 75 (1978). In Russian. English translation in Soviet Astron., Vol. 22, No. 1.
    The stochastic transport theory, originally developed by the author for the study of electromagnetic propagation in media characterized by a random phase velocity, has been applied to the problem of the propagation of a pulsar signal in a turbulent interstellar plasma. The theoretical results have been used to predict the amplitude fluctuations of the pulsar and the correlation of the signal according to the frequency. The theoretical results have been compared to the results of a few experiments and used to determine the mean scale size of the turbulence.

141.508 Les pulsars, radiophares de l'espace.
H.-Y. Chiu.
La recherche en astrophysique, (see 003.001), p. 87 - 102 (1977).

141.509 A detailed, high resolution study of high frequency radio emission from PSR 1133 + 16.
D. C. Ferguson, J. H. Seiradakis.
Astron. Astrophys., Vol. 64, 27 - 42 (1978).
    Observations of PSR 1133 + 16 at wavelengths of 21 cm, 18 cm, and 11 cm, and with time resolutions of 52 microseconds or better are analyzed. The total power measurements of integrated and individual pulses are discussed in terms of pulse widths, fluxes, microstructure peculiarities, and the variation of such quantities with frequency. The polarization measurements are discussed along with possible model interpretations. A summary is given of what the authors believe the measurements imply about the nature of the emission from PSR 1133 + 16 and from pulsars in general.

141.510 A phenomenological pulsar model.
F. C. Michel.
Astrophys. J., Vol. 220, 1101 - 1106 (1978).
    The author adopts particle injection energies and rates previously calculated for the stellar wind generation by rotating magnetized neutron stars. He assumes that the ambient space-charge density being emitted to form this wind is

bunched. These considerations immediately place the coherent radio frequency luminosity from such bunches at around 10^{28} ergs s^{-1} for typical pulsar parameters. A comparable amount of incoherent radiation is emitted for typical (1 second) pulsars. For very rapid pulsars, however, the latter component grows more rapidly than the available energy sources. The author attributes the comparatively low radio luminosity of the Crab and Vela pulsars to both components being limited in the same ratio.

141.511 **Phase stability in the drifting subpulse pattern of PSR 0809 + 74.** S. C. Unwin, A. C. S. Readhead, P. N. Wilkinson, M. S. Ewing.
Mon. Not. R. Astron. Soc., Vol. 182, 711 - 721 (1978).
Observations of PSR 0809 + 74 have been made at 225 MHz to study the stability of its drifting subpulse pattern, and in particular the effect of pulse nulls on the phase of subpulse bands. It is found that a jump of phase occurs at each null. The magnitude of the jump implies that the subpulse bands cease drifting for the duration of the null. If allowance is made for these phase jumps, the periodicity of the subpulse bands varies in the long term by less than one part in 250. The relevance of these results for pulsar models is discussed.

141.512 **The properties of charge-separated pulsar magneto-spheres.** G. A. E. Wright.
Mon. Not. R. Astron. Soc., Vol. 182, 735 - 749 (1978).
The general properties of a cold charge-separated plasma are investigated, including both electromagnetic and inertial effects. There is found to be a 'magneto-vortex' vector which is frozen into the plasma and an equation similar in form to the 'force-free' condition is established. An aligned rotator model is considered and general expressions for its energy, angular momentum and particle loss are derived. General wave propagation is briefly discussed.

141.513 **High-energy pulsèd gamma rays from pulsars.** S. K. Gupta, P. V. Ramana Murthy, B. V. Sreekan-tan, S. C. Tonwar.
Astrophys. J., Vol. 221, 268 - 273 (1978).
The authors looked at 10 pulsars for pulsed emission of γ-rays at energies greater than 500 GeV, using an atmospheric Cerenkov emission technique. They have detected 3.85 σ and 3.56 σ signals from PSR 0950 + 08 and PSR 0531 + 21 (Crab pulsar), respectively. If they attribute the signal from the Crab pulsar to the main pulse, they find a further 2.2 σ signal in the same data, precisely at the place predicted for interpulse by the observations on low-energy γ-rays from Cos-B satellite. The statistical significance of the results is discussed.

141.514 **Radio observations of the Crab pulsar: timing and scattering variations.** D. J. Helfand.
Bull. American Astron. Soc., Vol. 9, 561 (1978). – Abstract.

141.515 **Observational constraints on physical models of radio pulsar emission.** L. Oster, W. Sieber.
Bull. American Astron. Soc., Vol. 9, 562 (1978). – Abstract.

141.516 **Stability of pulsar magnetic fields.** G. Chanmugam.
Bull. American Astron. Soc., Vol. 9, 562 (1978). – Abstract.

141.517 **Polar electric fields of aligned, magnetized neutron stars.** K. B. Baker, P. A. Sturrock.
Bull. American Astron. Soc., Vol. 9, 562 (1978). – Abstract.

141.518 **Second derivatives of pulsar rotation frequencies.** G. E. Gullahorn, J. M. Rankin.
Bull. American Astron. Soc., Vol. 9, 562 (1978). – Abstract.

141.519 **Observations of pulsar dispersion spectra.** I. R. Linscott, J. W. Erkes, N. R. Powell.
Bull. American Astron. Soc., Vol. 9, 562 (1978). – Abstract.

141.520 **Neutral hydrogen absorption in the spectra of four pulsars.** J. M. Weisberg, J. M. Rankin, V. Boriakoff.
Bull. American Astron. Soc., Vol. 9, 580 (1978). – Abstract.

141.521 **Stability of force-free pulsar wind zone.** S. Hinata.
Astrophys. J., Vol. 221, 1003 - 1008 (1978).
The effect of the two-stream instability which may develop in the pulsar wind zone is studied. The excited waves produce an effective friction and the momentum diffusion on the plasma particles. It is therefore impossible for the two components of the plasma particles to maintain the relative velocity which Cheng and Ruderman showed is required by the force-free condition. In other words, the force-free condition will not be satisfied in the presence of the turbulence. Because of the diffusion in the momentum space caused by the turbulence, the same instability will not develop in the downstream after the initial turbulent energy is dissipated. It is argued that the moderate momentum spread does not quench the instability near the star. The resulting coherent radiation peaks at much higher frequency than the radio band.

141.522 **A new pulsar survey.** A. G. Lyne.
Proc. Astron. Soc. Australia, Vol. 3, 118 - 120 (1977).
A new survey of the southern sky for pulsars is being carried out jointly by the University of Sydney and the CSIRO Division of Radiophysics. This paper provides a brief description of the experiment and an account of progress to date.

141.523 **The natural wave modes in a pulsar magnetosphere.** D. B. Melrose, R. J. Stoneham.
Proc. Astron. Soc. Australia, Vol. 3, 120 - 122 (1977).
The purpose of this paper is to explore the properties of the natural wave modes of a relativistically streaming electron-positron gas and to apply the results to the interpretation of the polarization characteristics of pulsar radio emission.

141.524 **Pulsars—radio and X-ray.** J. Bell Burnell.
J. British Astron. Assoc., Vol. 88, 248 - 256 (1978).

141.525 **On a klystron mechanism of generation of vibrations in a pulsar magnetosphere.** Yu. A. Rylov.
Astrophys. Space Sci., Vol. 53, 377 - 402 (1978).
It has been shown that vibrations can be generated in the electron cap of the neutron star (Rylov, 1976, 1977; Jackson, 1976) under certain conditions. The mechanism of generation is like that in a klystron. The electron gas of the cap plays the role of the klystron resonant circuit. The electron beam penetrating the electron cap and returning to the star's surface plays the role of the klystron electron beam. The bunching electron stream along the magnetic axis acts like a strongly directed antenna. The conditions in which it is possible to generate these vibrations were also investigated. The energy of the accelerated primary electrons, the frequency of radiated radiowaves and the degree of the radiation directivity are evaluated.

141.526 **Pulsar radiation patterns. IV. Four-frequency study of the single-pulse structure of PSR 0329 + 54.** L. Oster, W. Sieber.
Astron. Astrophys., Vol. 65, 173 - 177 (1978).
Data on PSR 0329 + 54 at four frequencies between 1400 and 5000 MHz were analyzed in order to study the frequency dependence of some basic parameters entering the polar-cap sparking model for radio pulsars. The authors find that all features which presumably map the sparking region are practically identical at the four frequencies, as

required by the physical model.

141.527 Pulsar radiation patterns. V. Synopsis of results and conclusions concerning the physical mechanism.   L. Oster, W. Sieber.
Astron. Astrophys., Vol. 65, 179 - 185 (1978).

The authors present a synopsis of their observational results, attempting to relate the highlights to tangible physical problems in the framework of the polar-cap sparking model. In particular, they discuss questions such as the relevance of single-pulse observations for the mapping of the sparking area as opposed to the radiation amplification levels, the mutual distance and the overlapping of the sparking areas, the precession motion in the electric and magnetic fields and its as yet unexplained variability, and the relevance of the by now established types of frequency variations.

141.528 Responses to perturbations of the force-free aligned pulsar atmosphere.   E. A. Jackson.
Mon. Not. R. Astron. Soc., Vol. 183, 445 - 457 (1978).

To clarify the likely structure of the pulsar atmosphere, the response of various plasma configurations near a rotating neutron star with aligned rotational and dipole magnetic axes is investigated. The response of various plasma shell atmospheres is examined and confirms the likelihood of the plasma atmosphere previously predicted from a near-vacuum analysis, provided the density is not too large. Larger density shells are found to break up into 'striated' configurations, containing two electron and one or two ion groups, which however may not develop into stable configurations. Criteria on the plasma density related to this and other responses of the plasma shells are discussed.

141.529 Pulsars covered by the dense envelopes as high-energy neutrino sources.   H. Sato.
Prog. Theor. Phys., Vol. 58, 549 - 559 (1977). − Abstr. in Phys. Abstr., Vol. 81, Abstr. 8034 (1978).

141.530 A model of the radio emission mechanism in pulsars.   K. Kawamura, I. Suzuki.
Rep. IPPJ-286, (see 012.018), 5 pp. (1977). − Abstr. in Phys. Abstr., Vol. 81, Abstr. 8036 (1978).

141.531 Polarization characteristics of southern pulsars − II. 640-MHz observations.   P. M. McCulloch, P. A. Hamilton, R. N. Manchester, J. G. Ables.
Mon. Not. R. Astron. Soc., Vol. 183, 645 - 676 (1978).

Observations of integrated pulse profiles and polarization characteristics at frequencies near 640 MHz are presented for 43 pulsars. Significant polarization is observed for most objects; the mean linear polarization for the whole sample is 20 per cent and the mean circular polarization is 13 per cent. Circular polarization is generally of one sign across the integrated profile, but in two pulsars sign changes are observed near the centre of the profile.

141.532 Pulsar magnetospheres with parallel electric fields−I.   R. Buckley.
Mon. Not. R. Astron. Soc., Vol. 183, 771 - 778 (1978).

It is demonstrated that the conventional force-free approach to the physics of pulsar magnetospheres contains a fundamental inconsistency, when applied to the region beyond the light cylinder. The component of electric current parallel to the magnetic field that follows from Ampere's law is incompatible with the corresponding expression derived from the appropriate particle equations of motion, for systems radiating finite electromagnetic energy.

141.533 Observational limits on the location of pulsar emission regions.   J. M. Cordes.
Astrophys. J., Vol. 222, 1006 - 1011 (1978).

The author assumes that a pulsar's beam is associated with the polar region of a magnetic dipole. Under the assumption that high radio frequencies originate closer to the star than do low frequencies, he places limits on the location of the emission region from the apparent absence of differential retardation and aberration in pulse timing measurements.

141.534 Spectra and microstructure of pulsar signals in a relativistic electron beam model.
J. G. Kirk, D. ter Haar.
Astron. Astrophys., Vol. 66, 359 - 365 (1978).

The radio emission mechanism introduced in a previous paper is extended to provide an explanation for some of the features of the spectra of pulsars. Observations are used to fix the γ-value of the beam particles at around 500. The position of the radiating particles in the magnetosphere is also determined. A more detailed investigation of the polarization and microstructure is presented. Various criticisms of beam-plasma models are discussed.

141.535 Deriving pulsar sub-pulse widths from equivalent widths of integrated profiles.   D. C. Ferguson.
Astron. Astrophys., Vol. 66, 463 - 464 (1978).

The sub-pulse widths of pulsars are found to be closely related to the equivalent widths of integrated profiles for pulsars of types S and C. Conversion factors are derived which may enable sub-pulse widths to be found for pulsars which are too weak to allow measurement of individual pulses.

141.536 Electromagnetic processes in a pulsar's magnetosphere and acceleration of cosmic rays.
V. G. Krivdik, A. K. Yukhimuk.
Geofiz. sb.AN USSR, 1977, vyp. (No.) 79, p. 78 - 82. In Russian. − Abstr. in Ref. zh., 51. Astron., 4.51.551 (1978).

141.537 The Hulse-Taylor pulsar and gravitational spin precession.   N. D. Hari Dass, V. Radhakrishnan.
Proceedings of the 1st Marcel Grossman meeting on general relativity (see 012.025), p. 279 - 280 (1977). − Abstr. in Phys. Abstr., Vol. 81, Abstr. 12264 (1978).

141.538 Pulsars.   A. Hewish.
Proc. R. Inst. Great Britain, No. 49, p. 201 - 207 (1976). − Abstr. in Phys. Abstr., Vol. 81, Abstr. 16450 (1978).

141.539 Gamma-ray emission from pulsars.   M. Salvati, E. Massaro.
Astron. Astrophys., Vol. 67, 55 - 63 (1978).

A model for the production of gamma rays in a pulsar environment is presented, together with numerical computations fitted to the observations of PSR 0833−45. It is assumed that the primary particles are accelerated close to the star surface and then injected along the open field lines, which cause them to emit curvature radiation. The equation describing the particles' braking is integrated exactly up to the first order in the pulsar rotational frequency, and the transfer problem for the curvature photons is solved with aberration, Doppler shift, and pair production absorption being taken into account. It is found that the observed gamma rays originate in the innermost regions of the magnetosphere, where the open lines' bundle is narrow and the geometrical beaming is effective.

141.540 On the klystron mechanism of generation of oscillations in a pulsar's magnetosphere.
Yu. A. Rylov.
Inst. kosm. issled. AN SSSR. Prepr., 1977, No. 363, 46 pp. In Russian. − Abstr. in Ref. zh., 51. Astron., 6.51.698 (1978).

141.541 A momentum distribution of high energy particles around Crab pulsar.   S. Hinata.
Astrophys. Space Sci., Vol. 55, 427 - 439 (1978).

The author determines the momentum distribution of the relativistic particles near the Crab pulsar from the observed X- and γ-ray spectra ($10^3 \sim 10^9$ eV), provided that the curvature radiation is responsible for it. The author also discusses pulse shape and polarization of high-energy photons.

141.542　Secular decrease in the visible intensity of the Crab pulsar.　J. Kristian.
Bull. American Astron. Soc., Vol. 10, 425 (1978). – Abstract.

141.543　Electromagnetics of pulsar polar gap sparking.　L. Oster.
Bull. American Astron. Soc., Vol. 10, 445 - 446 (1978). Abstract.

141.544　Microstructure polarization measurements of PSR2016+28.　P. R. Backus, J. M. Cordes.
Bull. American Astron. Soc., Vol. 10, 446 (1978). – Abstract.

141.545　Observations of high-energy X-radiation from PSR 0531+21.　G. S. Maurer, C. J. Crannell,
B. R. Dennis, J. F. Dolan, K. J. Frost, L. E. Orwig.
Bull. American Astron. Soc., Vol. 10, 446 (1978). – Abstract.

141.546　Pulse distortion in a shearing plasma.　A. K. Harding, E. Tademaru.
Bull. American Astron. Soc., Vol. 10, 447 (1978). – Abstract.

141.547　Progress report on the binary pulsar PSR1913+16.　L. A. Fowler, J. H. Taylor.
Bull. American Astron. Soc., Vol. 10, 447 (1978). – Abstract.

141.548　Model for pulsar integrated polarization.　D. C. Ferguson.
Bull. American Astron. Soc., Vol. 10, 447 (1978). – Abstract.

141.549　Pulsar glitches and the thermal/timing instability in neutron stars.　G. Greenstein.
Bull. American Astron. Soc., Vol. 10, 447 (1978). – Abstract.

141.550　Curvature effects of a pulsar's magnetic field lines.　E. Asseo, R. Pellat, M. Rosado.
Bull. American Astron. Soc., Vol. 10, 447 (1978). – Abstract.

141.551　Some implications of off-center explosions in stars.　B. A. Fryxell, W. D. Arnett.
Bull. American Astron. Soc., Vol. 10, 448 (1978). – Abstract.

141.552　Neutrinos from binary pulsars.　D. Eichler.
Bull. American Astron. Soc., Vol. 10, 448 (1978). Abstract.

141.553　On a possibility of realizing Joesephson's nonstationary effect in pulsars.　V. A. Chernobaj.
Astron. Tsirk., No. 966, p. 1 - 3 (1977). In Russian.

141.554　Correction of the dispersion measure for pulsar PSR 2223+65.　V. A. Izvekova.
Astron. Tsirk., No. 974, p. 1 - 2 (1977). In Russian.

141.555　The relationship of pulsars to supernovae.　J. H. Taylor.
Ann. New York Acad. Sci., Vol. 302, (see 012.033), 101 - 105 (1977).
Direct evidence regarding the relationship of pulsars to supernovae is rather slim. How else can one reach conclusions regarding the evolutionary histories of pulsars? The author reviews one such method, in which the pulsars are treated as a galactic population and such properties as the spatial density, luminosity function, and average lifetime are inferred from the statistics of the population and the known observational biases. The results are not inconsistent with the hypo-

thesis that pulsars are born in supernova explosions, but the implied pulsar birth rate is high enough to prove that the majority of galactic supernovae must be unobservable from Earth and must not produce long-lived remnants.

141.556　On the geometrical configurations of the pulsar models.　H. Yokoo.
Kyorin J. Arts Sci., Vol. 2, 25 - 30 (1975). In Japanese.
The behaviors of pulsed emission of oblique rotator models of pulsars are studied. Special attention is paid to the double pulse structure. Applying the fan beam model to Crab pulsar's double pulse, the author finds that the magnetic axis must be inclined at 40° to the rotational axis. In the precessing oblique rotator model of the 35 days periodicity of Her X-1, the pencil beam model produces the inversion of intensity ratio of two pulses during the ON stage of X-ray activity and the fan beam model produces the change of the time interval of two pulses during the ON stage. In the nonoblique rotator models, the observed shortening of pulsation periods of Her X-1 is explained by the growth of the white dwarf mass by accreted matter.

Review of the literature about pulsars. (1968 - 1974). See Abstr. 002.066.

Kvasarer, pulsarer og svarte hull. See Abstr. 003.051.

Coherent signal processing techniques for radio pulses.　See Abstr. 031.237.

A full mathematical analysis of the physically thin screen.　See Abstr. 061.003.

Extremal states of matter in astrophysics. Part IV: The crystalline state of matter: neutron crystals. See Abstr. 061.015.

Potential drops above pulsar polar caps: ultrarelativistic particle acceleration along the curved magnetic field. See Abstr. 062.054.

Synchro-Compton radiation damping of relativistically strong linearly polarized plasma waves. See Abstr. 062.065.

Theory of the pulsar atmosphere. II. Arbitrary magnetic and rotational axes; qualitative features. See Abstr. 062.073.

Thomson cross sections in a strong magnetic field. See Abstr. 062.109.

Bimetric gravitation theory and PSR 1913+16. See Abstr. 066.032.

Superfluidity in neutron stars. III. Relaxation processes between the superfluid and the crust. See Abstr. 066.121.

Questions about rotating superfluid dynamics: problems of pulsar astrophysics accessible in the laboratory. See Abstr. 066.324.

Has the Sun really got a companion star? See Abstr. 080.027.

Interplanetary scintillations of PSR 0531 + 21 at 74 MHz.　See Abstr. 106.004.

Asymmetric supernova explosions and the origin of binary pulsars.　See Abstr. 125.054.

Ground-based observations of the X-ray pulses from PSR 0531+21.   See Abstr. 142.038.

An extended X-ray source centred on the Vela pulsar.   See Abstr. 142.046.

Pulsars and cosmic rays in the dense supernova shells. See Abstr. 143.031.

Neutron beams in active galactic nuclei.

See Abstr. 158.231.

Errata

141.901 Erratum: "3C 123: a distant first-ranked cluster galaxy at z = 0.637" [Astrophys. J., Lett., Vol. 199, L3 - L4 (1975)].   H. Spinrad.
Astrophys. J., Lett., Vol. 220, L135 (1978).

142 Extreme UV, X-Ray, Gamma-Ray Sources

Extreme UV, X-Ray Sources

142.001   Observation of X-ray eclipses from LMC X-4.
F. Li, S. Rappaport, A. Epstein.
Nature, Vol. 271, 37 - 38 (1978).

The authors report here the detection of several eclipses of LMC X-4, which confirms it as the second known extragalactic X-ray binary. They also discuss briefly the inferred value of the mass of the compact X-ray star.

142.002   Discovery of eclipsing nature of LMC X-4.
N. E. White.
Nature, Vol. 271, 38 - 40 (1978).

Six days of data taken from an extended Ariel V observation of the LMC are shown and X-ray eclipses 5 h long can be clearly discerned every 1.4 d. The predicted optical phase zero times are, within the uncertainties, in good agreement with the observed mid-eclipse times, thus confirming the identification and making this the first X-ray eclipsing binary system to be found in the LMC.

142.003   A search for rapid optical periodicities from Cyg X-1.
E. L. Robinson, R. E. Nather, J. Africano, B. Smith.
Nature, Vol. 271, 40 (1978).

142.004   Observation of Cygnus X-1 in the energy range 100 keV–3 MeV.   P. Mandrou, M. Niel,
G. Vedrenne, A. Dupont, K. Hurley.
Astrophys. J., Vol. 219, 288 - 291 (1978) = Rep. Centre d'Étude Spatiale des Rayonnements, No. 77-693.

Cyg X-1 was observed 1976 June 5–6, in the energy range 100 keV–3 MeV by means of an 80 cm^2 balloon-borne γ-ray telescope, while the source was in the "low" state. The energy spectrum is given and compared with previous results at lower and higher energies.

142.005   The Comptonization of iron X-ray features in compact X-ray sources.   R. R. Ross, R. Weaver,
R. McCray.
Astrophys. J., Vol. 219, 292 - 299 (1978).

The authors describe the formation of X-ray spectral features due to iron in a relatively cool cloud of gas with Thomson depth $\tau_T > 1$ surrounding a compact source of continuum X-rays. Coupled equations are solved for the ionization structure of the cloud and for the radiative transfer of the X-rays. Photoionization suppresses the strength of emission lines and absorption edges. Comptonization of the radiation broadens emission lines, fills in absorption edges, and produces a high-energy cutoff. In order to describe multiple scattering, the authors derive a Fokker-Planck equation which includes an important modification of the Kompaneets equation. Narrow resonance lines are treated with an escape probability formalism.

142.006   The case for a burst from 3U0614 + 09.
J. H. Swank, R. H. Becker, E. A. Boldt, S. S. Holt, P. J. Serlemitsos.
Mon. Not. R. Astron. Soc., Vol. 182, 349 - 353 (1978).

A single flare event with a rise time $\sim$ 7s and a decay time $\sim$ 40s was observed by Goddard Cosmic X-ray Spectroscopy Experiment on OSO-8 during observations of 3U0614 + 09 in 1975 September. The possibility is discussed that the source was 3U0614 + 09, which has been compared to the low-mass binary Sco X-1. The spectra during the event could be fitted with blackbody ($kT =$

0.8–1.1 keV) or thermal bremsstrahlung ($kT = 1.1$–2.5 keV) models with absorption by cool material. In the blackbody model the source was $\sim 10^6$ cm in size.

142.007   The Ariel V (SSI) catalogue of high galactic latitude ($|b| > 10°$) X-ray sources.
B. A. Cooke, M. J. Ricketts, T. Maccacaro, J. P. Pye, M. Elvis, M. G. Watson, R. E. Griffiths, K. A. Pounds, I. McHardy, D. Maccagni, F. D. Seward, C. G. Page, M. J. L. Turner.
Mon. Not. R. Astron. Soc., Vol. 182, 489 - 515 (1978).

The 2A catalogue is the result of 10 000 orbits of observation by the Leicester University Sky Survey Instrument on the Ariel V satellite and it contains 105 X-ray sources with $|b| > 10°$. The procedures and criteria used in establishing these sources and measuring their intensities and positions are described. New optical identifications are suggested.

142.008   X-ray spectra and variability of some Seyfert galaxies and other high latitude sources.
J. P. Stark, J. Bell Burnell, J. L. Culhane.
Mon. Not. R. Astron. Soc., Vol. 182, 23P - 29P (1978).

Five extragalactic X-ray sources have been observed by the MSSL spectrometer on Ariel V. The data are consistent with the hypothesis that a significant amount of material lies in the line of sight to 2A1348+700, 1410–029 and 4U1916–79. There is also some evidence for variability in the sources 2A1410–029, 1415+255 and 4U1916–79.

142.009   Gas stripping from spirals within X-ray clusters of galaxies.   D. Tytler, N. V. Vidal.
Mon. Not. R. Astron. Soc., Vol. 182, 33P - 37P (1978).

Empirical evidence is presented supporting the process of gas stripping from spirals by a hot intergalactic medium in clusters. Galaxies within five southern X-ray clusters were morphologically classified. The new data are combined with pre-existing data to show a strong correlation between the percentage of (E + S0) galaxies and the cluster X-ray luminosity. On the basis of this correlation and some other new criteria, the optical identification of 3U1252–28 with an individual galaxy (rather than the cluster core) is suggested.

142.010   The distribution of X-ray sources in the Milky Way.
E. J. A. Meurs.
Astron. Astrophys., Vol. 62, L5 - L8 (1978).

The fourth Uhuru catalogue is analysed for the spatial and intensity distributions of the galactic X-ray sources. It appears that (1) there is no convincing evidence for an excess of bright sources around the galactic centre, (2) the observations of the weak sources are heavily confused, (3) almost all sources may have a population I origin.

142.011   ANS: X-rays from the direction of the Orion Nebula (M 42).   A. J. F. den Boggende, R. Mewe,
E. H. B. M. Gronenschild, J. Heise, J. E. Grindlay.
Astron. Astrophys., Vol. 62, 1 - 7 (1978).

The authors present measurements of X-rays from the direction of the Orion Nebula M 42 (= NGC 1976) region obtained with the instruments aboard the Astronomical Netherlands Satellite ANS. It is concluded that the observed X-ray emission comes from within a rectangular region in the centre of the nebula. Good fits to the observed data are obtained with power law spectra with $\alpha = 1.6^{+1.0}_{-0.8}$ and $N_H = 1.8(<10) \times 10^{20}$ atoms/cm^2 when the soft X-ray result is included. Various interpretations of the results are discussed. None of the candidate objects, proposed earlier as a possible X-ray source (viz. θ^2 Ori A, θ^1 Ori B = BM Ori and θ^1 Ori C) show clear evidence for emitting X-rays. However, the observed X-ray signal may be explained as emission from coronae (with $T \cong 4 \times 10^6$ K and $\epsilon \sim 10^{57}-10^{58}$ cm^{-3}) around T Tauri stars within the nebula. Finally, the present observations cannot exclude the possibility of X-ray emission from an extensive part in the direction of the centre of the nebula.

142.012 Formation of X-ray sources by collisions in globular clusters and in the bulge.   A. Finzi.
Astron. Astrophys., Vol. 62, 149 - 157 (1978).

The author suggests that the X-ray sources in globular clusters and in the galactic bulge are formed in collisions of compact objects (neutron stars or, perhaps, black holes) and normal stars. After the collision the two stars coalesce and the normal star forms a thick envelope around the compact object. Absorption of X-rays from the accreting compact object causes the envelope to expand gradually until it evolves into a transparent envelope of very low density and radius $\sim 10^{17.5}$ cm around the compact object. The author has calculated the rate of formation of X-ray sources by collisions. In globular clusters, collisions seem to produce many more sources than are actually observed. However, the lifetimes of sources in the dense cores of globular clusters are probably much shorter than $10^{8.5}$ years, since stellar collisions cause evaporation of matter from the envelopes.

142.013 X-ray observations of the variability of Cygnus X-1 in the high state.   G. Manzo, D. Molteni, N. R. Robba, J. M. Bonnet-Bidaud, S. Ilovaisky.
Astron. Astrophys., Vol. 62, 265 - 266 (1978).

The ESA COS-B satellite has observed the Cygnus region in the energy range (2–12) keV, between November 28 and December 24, 1975. The Cygnus X-1 X-ray light curve for this period, during which the source appears to be in the "high" state, is reported; a search for a possible attenuation of the X-ray emission at the superior conjunction brings to the conclusion that the evidence for the existence of such a feature is dubious.

142.014 COS-B X-ray observations of Cygnus X-3.
J. M. Bonnet-Bidaud, S. Ilovaisky, G. Manzo, D. Molteni, N. R. Robba.
Astron. Astrophys., Vol. 62, 275 - 276 (1978).

The ESA COS-B satellite has observed the Cygnus region in the energy range (2–12 keV), between 1975 November 28.42 and December 24.26. The Cygnus X-3 4.8 h modulation is detected. The source was in a low state with positive evidence for an increase in flux during the month of observation.

142.015 A model for compact X-ray sources in galactic nuclei: application to the giant elliptical galaxy M87.   W. G. Mathews.
Astrophys. J., Vol. 219, 408 - 412 (1978) = Lick Obs. Bull., No. 767.

A compact source of X-rays has recently been observed from the elliptical galaxy M87 (NGC 4486). The unresolved source contributes $\sim 12\%$ of the total X-ray flux at ~ 1 keV, but has not been detected at 0.25 keV. This source can be understood in terms of emission from hot gas within a blast wave in the nuclear gas cloud (radius ~ 80 pc). Soft X-rays are absorbed by the unshocked gas (external to the shock), the same gas responsible for optical emission lines observed in the nucleus of M87. Quite apart from the specific case of M87, blast waves generated by expanding compact radio sources in galactic nuclei may account for an appreciable fraction of the total flux in extragalactic X-ray sources.

142.016 Possible detection of an extreme-ultraviolet source at 500 Å.   W. Cash, S. Bowyer, J. Freeman, M. Lampton, F. Paresce.
Astrophys. J., Vol. 219, 585 - 588 (1978).

Evidence for a new extreme-ultraviolet source in the constellation Pavo is presented. The potential source, which was observed in the 500–780 Å band, has an intensity of 35 ± 9 mJy. Positional overlap suggests possible identification with the ultraviolet-intense star HD 192273 and/or the high-galactic-latitude X-ray source 3U 1959–69.

142.017 Photometry of AM Herculis: a slow optical pulsar?   W. C. Priedhorsky, W. Krzeminski.
Astrophys. J., Vol. 219, 597 - 604 (1978).

Multicolor photometry of the X-ray binary AM Her suggests that the red component of the optical flux is closely related to the source of optical circular polarization in the system. The authors conclude from the periodic modulation of flux in the U through R bands, which is particularly well defined when plotted as color curves, that the primary and secondary minima are neither eclipses by a secondary star nor eclipses by a hot spot. They suggest that the primary minimum in the visual light curve is the eclipse of a region of intense optical emission in the magnetic field near the surface of a degenerate dwarf by that dwarf itself.

142.018 Hercules X-1 = HZ Herculis: a multiperiod variable?   C. L. Wolff, Y. Kondo.
Astrophys. J., Vol. 219, 605 - 612 (1978).

The authors propose a theory which accounts for various periodicities in the X-ray binary HZ Her = Her X-1. It is based on fundamental rotation properties of the free modes of oscillation of a star. The proposed model, containing only one free parameter, gives a physical basis for five observed periods including 1.7 day, 35 day, and 10–12 year periods.

142.019 Photometry of slow X-ray pulsars. III. The GX 17 + 2 candidate.   B. Margon.
Astrophys. J., Vol. 219, 613 - 616 (1978).

The author presents the first photometry of the faint candidate star for the optical counterpart of the intense X-ray source GX 17 + 2 (= 3U 1813 – 14). This source is of particular interest because of its associated radio emission, and the existence of a 1913 s periodic modulation of the X-ray flux, which makes GX 17 + 2 the slowest X-ray pulsar yet observed. Photoelectric photometry on three nights yields $V = 17.51$, $(B-V) = 1.26$, $(U-B) = 1.03$, with no evidence for variability. The results indicate that any binary period in GX 17 + 2 is very likely longer than 2 days, exceeding that of all the known low-mass X-ray systems.

142.020 Centaurus X-3: the periodicity of the extended lows.   T. J. Chester.
Astrophys. J., Lett., Vol. 219, L77 - L79 (1978).

All published X-ray data are consistent with the hypothesis that Cen X-3 is in a low state every 27 days. The probability that this effect could be due to random sampling is only 0.01. If, as suggested by several other arguments, an accretion disk is present near Cen X-3, the disk may be responsible for this periodic behavior.

142.021 X-ray line spectroscopy for clusters of galaxies. I.   J. N. Bahcall, C. L. Sarazin.
Astrophys. J., Vol. 219, 781 - 794 (1978).

Theoretical models for the intracluster gas in X-ray clusters of galaxies are studied. The limits of validity of all the published models are derived from self-consistent arguments and existing data. New results are presented for the well model (gas trapped in the potential well of a single galaxy) and the wind model. A scaling law is proved that allows one to apply the results calculated for spherical systems to the nonspherical groupings that are often observed. The particular line intensities that can best be used to discriminate among theoretical models or to diagnose the gas (i.e., determine abundances, temperatures, ionization states, and other properties) are listed in tables along with the required experimental sensitivity and resolution.

142.022   **Short-term time variability of Cygnus X-1. II.**
   P. G. Sutherland, M. C. Weisskopf, S. M. Kahn.
Astrophys. J., Vol. 219, 1029 - 1037 (1978).

   A complete theory for shot-noise analysis is presented. The theory takes into account the effects of counting statistics, time resolution, and finite observing times on the autocorrelation function and on the moments of data assumed to consist of shot noise plus a steady component. When applied to X-ray observations of Cyg X-1 in the high and low states, the results provide evidence that both the typical correlation time (~0.5 s) and the fraction of flux in shots (~30%) are constant, independent of the state.

142.023   **Evidence for X-ray emission from superclusters of galaxies determined from Uhuru.**
S. S. Murray, W. Forman, C. Jones, R. Giacconi.
Astrophys. J., Lett., Vol. 219, L89 - L93 (1978) = Cent. Astrophys., Harvard Coll. Obs., Smithsonian Astrophys. Obs., Cambridge, Mass., Prepr. Ser., No. 787.

   The authors detect X-ray emission from three class 2 clusters of rich clusters of galaxies. They find 12 candidate superclusters of distance class 5 and six clusters within the area of sky covered by the 4U catalog. The probability that these three X-ray sources accidentally coincide with the superclusters is less than 0.003. They find equally low probabilities that the X-ray emission is due to either a single luminous cluster or to the combined emission of all members of the supercluster. A possible explanation of these sources is thermal bremsstrahlung emission from a hot tenuous gas pervading the supercluster. Should such regions of enhanced gas density be found to be associated with all groups of clusters (multiplicity ⩾ 2), then this gas may provide a significant fraction of the mass required to close the universe.

142.024   **Evidence for strong cyclotron line emission in the hard X-ray spectrum of Hercules X-1.**
J. Trümper, W. Pietsch, C. Reppin, W. Voges, R. Staubert, E. Kendziorra.
Astrophys. J., Lett., Vol. 219, L105 - L110 (1978).

   The authors present further results of the Hercules X-1 balloon observation on 1976 May 3 which confirm the existence of a strong line feature at ~58 keV in the pulsed (1.24 s) X-ray spectrum. The most likely interpretation of this line is electron cyclotron emission at the basic frequency from the hot polar plasma of the rotating neutron star. The corresponding magnetic field strength is 5.3×10^{12} gauss. They have searched for the second-harmonic cyclotron line and provide evidence of its detection.

142.025   **Detection of both soft and hard X-ray emission from SS Cygni with ANS.**
J. Heise, R. Mewe, A. C. Brinkman, E. H. B. M. Gronenschild, A. J. F. den Boggende, J. Schrijver, D. R. Parsignault, J. E. Grindlay.
Astron. Astrophys., Vol. 63, L1 - L3 (1978).

   A soft (0.16 - 0.284 keV) X-ray signal was detected while SS Cygni was in its quiescent optical state. The discovery of hard X-ray emission suggests that SS Cygni is a source similar to the AM Her/3U1809 + 50 system.

142.026   **X-ray lines and magnetic field of Her X-1.**
   K. Brecher, M. P. Ulmer.
Nature, Vol. 271, 135 - 137 (1978).

142.027   **Positions of galactic X-ray sources: $320° < l^{II} < 340°$.**
   K. M. V. Apparao, H. V. Bradt, R. G. Dower, R. E. Doxsey, J. G. Jernigan, F. Li.
Nature, Vol. 271, 225 - 228 (1978).

   Precise (20–40'') positions of seven X-ray sources in the celestial region $320° < l^{II} < 340°$ are reported here. These include a recurrent transient X-ray source (MX1608–52) and

a source 2S1553–542 coincident with a γ-ray source within the given errors of position. The positions reported here reduce the previously reported areas of the error region for six of the sources by factors ranging between 10 and 100. In the case of MX1608–52, a preliminary report of this work led to the identification of an optical candidate. The present results add confidence to the proposed radio candidates for 4U1624–49 and 4U1642–45. But they eliminate proposed possible optical candidates for the sources 4U1543–62, 2S1553–542, 4U1624–49 and 4U1642–45.

142.028   **Dual character of the rapid burster and a classification of X-ray bursts.**
J. A. Hoffman, H. L. Marshall, W. H. G. Lewin.
Nature, Vol. 271, 630 - 633 (1978).

   Different kinds of X-ray bursts, as observed from the rapid burster (MXB1730–335) and from other sources can be classified into two types which may have different origins and production mechanisms. Type I bursts occur at intervals of hours, days or longer. Their spectra almost always soften during burst decay, and their average spectrum during the first few seconds of a burst is generally harder than the spectrum of the associated persistent X-ray emission (if present). Type II bursts occur at intervals of several seconds to minutes, and their spectra do not soften during burst decay. The rapid burster can produce up to several thousand type II bursts per day. It is shown that the rapid burster also produces type I bursts.

142.029   **The nature of Aquila X-1.**
   B. Margon, J. I. Katz, L. D. Petro.
Nature, Vol. 271, 633 - 634 (1978).

   A recent very accurate X-ray position for Aql X-1 has permitted the optical identification of the source. The authors report here photometry and spectroscopy of the counterpart and the stars in the immediately surrounding field, and use these data to arrive at some inferences regarding the nature of this unusual X-ray system.

142.030   **The helium variable HD 64740: a candidate for the X-ray source 4 U 0750-49?**
D. Groote, J. P. Kaufmann, K. Hunger.
Astron. Astrophys., Vol. 63, L9 - L11 (1978).

   The helium variable HD 64740 is identified tentatively as the X-ray source 4 U 0750-49, on the basis of a faint emission at HeII λ4686 Å.

142.031   **On the contribution of active galactic nuclei to the diffuse X-ray background.**   Y. Avni.
Astron. Astrophys., Vol. 63, L13 - L16 (1978).

   The author shows that cosmological evolution has a pronounced effect on the contributions of active galactic nuclei to the diffuse X-ray background. He finds the dependence of such contributions on the form and amount of density evolution, on the deceleration parameter, and on the formation epoch. The author finds in particular that X-ray Seyferts can account for all of the observed 2–10 keV background when the effects of evolution are considered; the required amount of evolution is intermediate between the evolution of quasars and no evolution.

142.032   **Cyclotron emission and beaming mechanisms in magnetized neutron stars: Her X-1.**
P. Mészáros.
Astron. Astrophys., Vol. 63, L19 - L22 (1978).

   The recently reported feature at 53 keV in Her X-1, if interpreted as cyclotron emission, imposes restrictions on the possible beaming mechanisms. Pencil beam schemes seem ruled out, while occultation schemes may work, but only in conjunction with a fan-beam scheme. Pure fan-beam schemes probably offer the most satisfactory and simple mechanism.

142.033 Soft X-ray results from the Wisconsin experiment on OSO-8. A. N. Bunner.
Astrophys. J., Vol. 220, 261 - 271 (1978).

The Wisconsin Soft X-Ray Astronomy instrument on the Orbiting Solar Observatory (OSO) No. 8 is described and its capabilities are discussed. Results are presented for three point sources which have composite spectra with soft X-ray components: AM Herculis, Hercules X-1, and η Carinae.

142.034 Cygnus X-3: A cocooned X-ray binary pulsar? M. Milgrom, D. Pines.
Astrophys. J., Vol. 220, 272 - 278 (1978).

It is shown that observations of both the hard X-ray and γ-ray spectrum and the γ-ray light curves of Cyg X-3 find a natural explanation in a model for the system of a fast binary pulsar (with a period between 10 and 30 ms) surrounded by a cocoon (of scattering optical depth unity) located $\sim 10^{12}$ cm from the pulsar. The possibility of direct observation of pulsations is discussed, and attention is called to an anomalous behaviour of γ-rays in the vicinity of 0.5 MeV predicted by the model.

142.035 The luminosity enhancement following X-ray bursts. W. A. Baan.
Astrophys. J., Lett., Vol. 220, L5 - L7 (1978).

It is suggested that the luminosity enhancements observed after the termination of X-ray bursts from several sources are due to rapid thermonuclear burning in the material which accreted onto the star during the main burst. The conditions in the burst material are such that helium burning is initiated during the burst and continues after the accretion flow stops. Several other burning processes are considered. Owing to an approximate temperature-independent energy-production rate, the ratio of the burst energy and the enhancement luminosity results in an estimate for the surface potential of the neutron star. The resulting neutron-star masses are of the order of 1 $M_\odot$ for solar abundances.

142.036 X Persei and the X-ray source 3U0352+30. H.-r. Hang, J.-e. Mo.
Acta Astron. Sinica, Vol. 18, 211 - 215 (1977). In Chinese.

According to the authors' spectrophotometric gradient measurements of X Per, they determined its interstellar visual total absorption as $A_v = 1^m 50 \pm 0^m 30$. As compared with the hydrogen column density N_H value given by X-ray observations of 3U0352+30 (using the empirical transformation formula connecting N_H and A_v), it is shown that these two objects are equidistant as observed from the earth. This supports the opinion that the star X Per might be the optical counterpart of the X-ray source 3U0352+30.

142.037 Flares of invisible stars. P. R. Amnuehl'.
Zemlya i Vselennaya, 1978, No. 1, p. 21 - 26. In Russian.

142.038 Ground-based observations of the X-ray pulses from PSR 0531+21.
G. C. Rumi, S. Leschiutta, A. Sabatino.
Nuovo Cimento B, Ser. 11, Vol. 41B, 29 - 45 (1977). – Abstr. in Phys. Abstr., Vol. 80, Abstr. 91444 (1977).

142.039 Les étoiles à rayons X. C. Ryter.
La recherche en astrophysique, (see 003.001), p. 122 - 128 (1977).

142.040 X-ray and radio observations of GX 17 + 2 and GX 13 + 1. N. E. White, K. O. Mason, P. W. Sanford, H. M. Johnson, R. C. Catura.
Astrophys. J., Vol. 220, 600 - 605 (1978).

X-ray data on the sources 3U 1813−14 (GX 17 + 2) and 3U 1811−17 (GX 13 + 1), obtained by Copernicus between 1975 and 1977, are examined. These reveal correlated intensity and spectral slope variations in GX 17 + 2. In the case of GX 13 + 1 no significant variability was seen in either intensity or spectrum. A simultaneous search made from the NRAO simultaneously with some of the X-ray observations revealed no significant radio emission from either source.

142.041 X-ray sources in nearby superclusters. E. M. Kellogg.
Astrophys. J., Lett., Vol. 220, L63 - L65 (1978).

A higher-order clustering of the X-ray sources in nearby clusters of galaxies (distance class 4 or less) is reported. Out of 48 objects, at least 15 are probably clustered into seven superclusters. The superclustered X-ray sources consist of X-ray bright spots centered on rich clusters, with no detectable emission in between. This implies that any hot gas pervading these superclusters is highly clumped.

142.042 Discovery of synchronous binary polarization in Cygnus X-1.
J. C. Kemp, M. S. Barbour, L. C. Herman, R. J. Rudy.
Astrophys. J., Lett., Vol. 220, L123 - L126 (1978).

Using 180 nights of data from 1975 and 1977, the authors have found a stable polarization variation in the V band which is periodic on the orbital cycle of $5^d 600$. The peak-to-peak amplitude is 0.25% and is largely in the position-angle Stokes parameter. The amplitude in the ultraviolet is much smaller. The V-band polarization curves suggest a surprisingly large inclination; the single-scattering model gives $i = (76 \pm 8)°$. The authors also summarize their recent photometry.

142.043 Optical identification of 4U 1608−52. J. E. Grindlay, W. Liller.
Astrophys. J., Lett., Vol. 220, L127 - L130 (1978).

The probable optical counterpart of the recurrent flaring (or transient) X-ray source 4U 1608−52 was discovered as a $m_I \approx 18.2$ object on a deep IV-N plate. Comparison with a nearly identical 60 minute exposure IV-N plate reveals that this object must have brightened by more than 1.8 mag between 1976 August and 1977 August. The object is within the $\sim 10'' \times 20''$ combined error box obtained by SAS 3 and HEAO 1 for 4U 1608−52, which underwent a transient source outburst in 1977 July. Constraints on source models are discussed.

142.044 The optical counterpart of Aquila X-1 (3U 1908+00). J. Thorstensen, P. Charles, S. Bowyer.
Astrophys. J., Lett., Vol. 220, L131 - 134 (1978).

The authors have discovered a highly variable star very close to the center of the SAS 3 error circle for the flaring X-ray source Aquila X-1 (3U 1908+00). Photographic photometry over the last 4 years shows covariability; the optical counterpart brightens from $B \gtrsim 20$ to $B \sim 17$ mag during times of X-ray outburst, thus establishing the identification. Spectrophotometry of this star during its quiescent state shows spectral features of a K0 star with no obvious peculiarities; its distance must be at least 1.7 kpc.

142.045 A new measurement of the spin-up rate of SMC X-1. D. Yentis, S. Shulman, J. D. McKee, W. K. Rose.
Astrophys. Lett., Vol. 19, 53 - 56 (1977).

Observations of the X-ray pulsar, SMC X-1, made from an Aerobee rocket and an Apollo spacecraft have been re-analyzed. The rotation periods found have been combined with observations made from UHURU, SAS-3, and Ariel 5 to obtain a new measurement of the spin-up rate of SMC X-1.

142.046 An extended X-ray source centred on the Vela pulsar. A. Smith.
Mon. Not. R. Astron. Soc., Vol. 182, 39P - 46P (1978).

Presented here are observations of Vela X taken with the Sky Survey Instrument on board Ariel V. The medium energy X-ray source 2U 0832−45 is found to be extended by 2° and may have a relationship to the Vela pulsar that is similar to the situation in the Crab Nebula. A relative deficiency of X-ray emission is indicated from 2U 0832−45 compared with the Crab suggesting a lower magnetic field in Vela ($\sim 2 \times 10^{-5}$ G).

142.047   X-ray emission from MSH 14−63 − probable remnant of the A.D. 185 supernova.

P. F. Winkler, Jr.
Astrophys. J., Vol. 221, 220 - 224 (1978).

Observations during 1972 from the MIT experiment on OSO 7 detected a 2−10 keV X-ray flux of 1.7×10^{-10} ergs cm^{-2} s^{-1} from a source coincident with MSH 14−63, the most probable candidate for association with the A.D. 185 supernova. Model calculations indicate that a mass in excess of 5 $M_\odot$ must have been ejected in the A.D. 185 supernova event.

142.048   Radio and optical identification of the X-ray source GX 17 + 2.   R. M. Hjellming.

Astrophys. J., Vol. 221, 225 - 227 (1978).

VLA 6 cm radio observations of the compact X-ray source GX 17 + 2 (3U 1813−14) provide an improved position for a highly variable radio source, placing it within an arcsecond of a 17.5 mag dwarf G star.

142.049   On the physical reality of the millisecond bursts in Cygnus X-1: bursts and shot noise.

M. C. Weisskopf, P. G. Sutherland.
Astrophys. J., Vol. 221, 228 - 233 (1978).

The method of data analysis used to interpret the millisecond temporal structure of Cyg X-1 is discussed. In particular, the authors examine the effects produced by the shot-noise variability of this source which occurs on time scales of ~0.5 s. Taking into account the recent discovery that only ~30% of the flux may be in the shots, they find that spurious "millisecond bursts" will be detected. A comparison of the properties of these bursts with currently published experimental data is performed.

142.050   Thermal limit for spherical accretion and X-ray bursts.   J. E. Grindlay.

Astrophys. J., Vol. 221, 234 - 257 (1978).

This paper describes a simple analytic calculation of the effects of X-ray heating of the accreting gas in a spherical accretion X-ray source. High-temperature sources are considered, and thus only continuum (bremsstrahlung and Compton) cooling processes are included to calculate the effects of X-ray preheating on the accretion flow. The major result of this work is that X-ray heating can limit the steady source luminosity for accretion rates and source parameters within a range and that unstable flows and repeated bursts can result for accretion rates in a range exceeding this thermal limit.

142.051   Hard X-ray spectrum of NGC 4151.

G. Auriemma, L. Angeloni, B. M. Belli, A. Bernardi, D. Cardini, E. Costa, A. Emanuele, F. Giovannelli, P. Ubertini.
Astrophys. J., Lett., Vol. 221, L7 - L11 (1978).

The X-ray source 4U 1206 + 39 associated with the Seyfert galaxy NGC 4151 was observed with the high-pressure proportional chamber of the HXR76 experiment, on board the Transatlantic Balloon Facility. The spectrum in the 25−190 keV range may be fitted with a power law with spectral slope $\alpha = 0.9 \pm 0.2$. Data are consistent with previous observations in the same range, and no long-term variability is indicated. Comparison with the low-energy spectrum of Ariel 5 shows a break at about 40 keV where the spectral index becomes harder. Data are also consistent with the Miso γ experiment in the range 0.15−20 MeV.

142.052   An upper limit to the linear X-ray polarization of

Scorpius X-1.   M. C. Weisskopf, H. L. Kestenbaum, K. S. Long, R. Novick, E. H. Silver.
Astrophys. J., Lett., Vol. 221, L13 - L16 (1978).

A measurement of Sco X-1 with the graphite crystal polarimeters on the OSO 8 satellite yields upper limits (3 σ) to the linear polarization of 2.7% and 4.9% at 2.6 keV and 5.2 keV, respectively. The results are discussed in terms of an accretion-disk model.

142.053   Accreting neutron stars in highly compact binary systems and the nature of 3U 1626−67.

P. C. Joss, Y. Avni, S. Rappaport.
Astrophys. J., Vol. 221, 645 - 651 (1978).

The authors discuss the existence of pulsing X-ray sources that consist of neutron stars in highly compact binary systems (orbital periods $\lesssim$ 0.3), undergoing accretion from low-mass late-type dwarf or degenerate-dwarf companions. An appropriate mass transfer rate can be driven by the decay of the orbit due to gravitational radiation, a self-excited wind, and/or the evolution of the companion. Such a system may result from the evolution of a cataclysmic variable, wherein a degenerate dwarf collapses to form a neutron star after accreting sufficient mass to exceed the Chandrasekhar limit. The authors apply this model to the 7.7 X-ray pulsar 3U 1626−67, and demonstrate that it can explain the apparent lack of Doppler shifts in the X-ray pulsations from this source. The model may also account for other observed properties of the source.

142.054   On two new X-ray sources in the SMC and the high luminosities of the Magellanic X-ray sources.

G. Clark, R. Doxsey, F. Li, J. G. Jernigan, J. van Paradijs.
Astrophys. J., Lett., Vol. 221, L37 - L41 (1978).

The discovery of two new X-ray sources, SMC X-2 and SMC X-3, in the Small Magellanic Cloud is reported. It is shown that the luminosity distribution of the known Magellanic X-ray sources, which are now nine in number, is shifted toward higher luminosities with respect to that of similar sources in the Galaxy, and that the cause of the shift is probably an underabundance of heavy elements in the material accreted by the X-ray sources.

142.055   Compact and diffuse X-ray sources in the core of the Perseus cluster (Abell 426).   H. Helmken,

J. P. Delvaille, A. Epstein, M. J. Geller, H. W. Schnopper, J. G. Jernigan.
Astrophys. J., Lett., Vol. 221, L43 - L47 (1978).

The X-ray emission from the core of the Perseus cluster (Abell 426) has been observed in the 2−10 keV energy band. The X-ray data imply a diffuse source and are consistent with but do not require the presence of a coincident, compact source. The centroid of the X-ray emission is also coincident with the radio nucleus of NGC 1275 (3C 84A).

142.056   Location of the Norma transient with the HEAO 1 scanning modulation collimator.   G. Fabbiano,

H. V. Bradt, R. E. Doxsey, H. Gursky, D. A. Schwartz, J. Schwarz.
Astrophys. J., Lett., Vol. 221, L49 - L52 (1978).

The authors have obtained a precise position for an X-ray transient source in Norma. The location uncertainty includes a variable star previously suggested to be the optical counterpart. This transient is associated with the steady X-ray source MX 1608−52 and probably with an X-ray burst source. A binary system containing a low-mass primary and a neutron-star or black-hole secondary of a few solar masses is consistent with the observations.

142.057   Position for the rapid burster MXB 1730−335 determined with the scanning modulation collimator

on HEAO 1.   R. Doxsey, H. Bradt, H. Gursky, M. Johnston, D. A. Schwartz, J. Schwarz.
Astrophys. J., Lett., Vol. 221, L53 - L55 (1978).

The authors report a precise position for the X-ray rapid burster MXB 1730−335 determined with the scanning modulation collimator experiment on HEAO 1. The position, accurate to ~20″, includes the core of the globular cluster discovered by Liller.

142.058 **SAS 3 observations of two X-ray transient events with precursors.** J. A. Hoffman, W. H. G. Lewin, J. Doty, J. G. Jernigan, M. Haney, J. A. Richardson.
Astrophys. J.,Lett., Vol. 221, L57 - L62 (1978).

SAS 3 has observed two unusual fast transient X-ray events from different sources, one lasting ~150s and one ~1500s. Both events were preceded by precursor pulses which lasted a few seconds and which rose and fell in less than 0.4 s. The precursors were separated from the "main" events by several seconds, during which no X-rays were detected. There are similarities between the two main events and X-ray bursts in both their temporal and spectral evolution.

142.059 **Nova Ophiuchi 1977: an X-ray nova.** R. E. Griffiths, H. Bradt, R. Doxsey, H. Friedman, H. Gursky, M. Johnston, A. Longmore, D. F. Malin, P. Murdin, D. A. Schwartz, J. Schwarz.
Astrophys. J., Lett., Vol. 221, L63 - L67 (1978).

The authors report the observation by the HEAO 1 satellite experiments of the X-ray nova in Ophiuchus discovered by Kaluzienski and Holt which the present authors designate H1705−25. They report the discovery of a 16.5 mag optical nova (Nova Ophiuchi 1977) in one of the error boxes, and they identify the X-ray source with the optical nova. They classify the object as being of the A0620−00 and A1524−61 type, i.e., a low-mass binary system.

142.060 **The X-ray light curve of Nova Ophiuchi 1977 (H1705−25).** M. G. Watson, M. J. Ricketts, R. E. Griffiths.
Astrophys. J., Lett., Vol. 221, L69 - L71 (1978).

The authors present the X-ray light curve of Nova Ophiuchi 1977 (H1705−25), which shows unambiguously the fast rise and slow decline characteristic of "classical" X-ray transients. Comparison with the light curves of previous transient X-ray sources indicates that Nova Ophiuchi 1977 may be in the same class as A0620−00.

142.061 **Extreme ultraviolet observation of Sirius: evidence against a photospheric origin of the 0.25 keV flux.**
W. Cash, S. Bowyer, M. Lampton.
Astrophys. J., Lett., Vol. 221, L87 - L89 (1978).

New upper limits on the extreme ultraviolet flux from Sirius have been obtained with a rocket-borne grazing incidence imaging telescope. These limits are a factor of 10 below previous limits and rule out the existing photospheric model for the generation of 0.25 keV radiation from Sirius B. An upper limit of 30,000 K is placed on the effective temperature of Sirius B. Constraints on coronal models for the system are discussed.

142.062 **Recent extragalactic results from the Ariel V SSI.** J. P. Pye, M. Elvis, A. Lawrence, R. S. Warwick.
Bull. American Astron. Soc., Vol. 9, 558 (1978). − Abstract.

142.063 **New active galaxy identifications of 2A X-ray sources.** M. J. Ward, A. S. Wilson, M. V. Penston, K. Tritton, M. Elvis, T. Maccacaro.
Bull. American Astron. Soc., Vol. 9, 558 (1978). − Abstract.

142.064 **High spatial resolution X-ray observations of the Perseus cluster of galaxies with the scanning modulation collimator on HEAO-1.** J. Schwarz, M. Johnston, R. E. Doxsey, D. A. Schwartz, H. Gursky, H. Bradt.
Bull. American Astron. Soc., Vol. 9, 558 (1978). − Abstract.

142.065 **X-ray spectra of clusters of galaxies from OSO-8.** B. W. Smith, R. F. Mushotzky, P. J. Serlemitsos, E. A. Boldt, S. S. Holt.
Bull. American Astron. Soc., Vol. 9, 559 (1978). − Abstract.

142.066 **Theoretical X-ray maps of rich clusters of galaxies.** O. B. R. Strimpel, J. J. Binney.
Bull. American Astron. Soc., Vol. 9, 559 (1978). − Abstract.

142.067 **Soft X-ray structure of the Coma cluster of galaxies.** P. Gorenstein, D. Fabricant, K. Topka, F. R. Harnden, Jr.
Bull. American Astron. Soc., Vol. 9, 560 (1978). − Abstract.

142.068 **Short observations of bright stars with a rocket borne imaging X-ray telescope.**
K. Topka, D. Fabricant, F. R. Harnden, Jr., P. Gorenstein.
Bull. American Astron. Soc., Vol. 9, 560 (1978). − Abstract.

142.069 **Spectral features in the X-ray spectrum of Vela X-1.** R. H. Becker, R. E. Rothschild, E. A. Boldt, S. S. Holt, S. H. Pravdo, P. J. Serlemitsos, J. H. Swank.
Bull. American Astron. Soc., Vol. 9, 560 (1978). − Abstract.

142.070 **X-ray polarization measurements of binary pulsars and their possible impact on theories for pulsar beaming.** E. H. Silver, H. L. Kestenbaum, K. S. Long, R. Novick, M. C. Weisskopf.
Bull. American Astron. Soc., Vol. 9, 561 (1978). − Abstract.

142.071 **Optical observations of the pulsar 3U 1626−67 and other X-ray sources.** J. E. Grindlay.
Bull. American Astron. Soc., Vol. 9, 561 (1978). − Abstract.

142.072 **SAS-3 positions of galactic X-ray sources and their optical counterparts.** H. Bradt, K. Apparao, R. Dower, R. Doxsey, J. G. Jernigan.
Bull. American Astron. Soc., Vol. 9, 579 (1978). − Abstract.

142.073 **Precise location of X-ray sources associated with the globular clusters NGC 1851, NGC 6441, NGC 6624, NGC 6712, and NGC 7078.** J. G. Jernigan.
Bull. American Astron. Soc., Vol. 9, 579 (1978). − Abstract.

142.074 **Observation of new X-ray sources in the Small Magellanic Cloud.**
F. Li, G. Clark, G. Jernigan, J. van Paradijs.
Bull. American Astron. Soc., Vol. 9, 579 (1978). − Abstract.

142.075 **A measurement of the time sense of the shot noise pulses from Cygnus X-1.**
M. C. Weisskopf, P. G. Sutherland, C. R. Canizares.
Bull. American Astron. Soc., Vol. 9, 579 - 580 (1978). Abstract.

142.076 **The HEAO A-2 diffuse background experiment.** G. Garmire, P. C. Agrawal, I. R. Tuohy, G. R. Riegler, P. Charles, C. S. Bowyer, M. Lampton, E. Boldt, S. Holt, F. Marshall, R. Rothschild, P. Serlemitsos.
Bull. American Astron. Soc., Vol. 9, 591 (1978). − Abstract.

142.077 **Soft X-ray observations of AM Herculis from HEAO-1.** I. R. Tuohy, F. K. Lamb, G. P. Garmire, K. O. Mason.
Bull. American Astron. Soc., Vol. 9, 591 (1978). − Abstract.

142.078 **The 2−60 keV view of AM Her.** J. H. Swank, E. A. Boldt, S. S. Holt, M. Lampton, R. E. Rothschild, P. J. Serlemitsos.
Bull. American Astron. Soc., Vol. 9, 591 - 592 (1978). Abstract.

142.079 HEAO-1 observations of soft X-ray emission from
U Geminorum.   K. O. Mason, P. A. Charles,
S. M. Kahn, J. R. Thorstensen, F. M. Walter.
Bull. American Astron. Soc., Vol. 9, 592 (1978). — Abstract.

142.080 Soft X-radiation from cataclysmic variable stars.
F. Cordova, I. R. Tuohy, J. Nugent, G. Garmire.
Bull. American Astron. Soc., Vol. 9, 592 (1978). — Abstract.

142.081 X-ray observations of the transient sources
H1743−32 and H1705−25 (Nova Ophiuchi 1977).
L. J. Kaluzienski, S. S. Holt, J. H. Swank, E. A. Boldt, P. J.
Serlemitsos.
Bull. American Astron. Soc., Vol. 9, 592 (1978). — Abstract.

142.082 Soft X-ray observations of Circinus X-1.
S. M. Kahn, P. A. Charles, K. O. Mason, F. M. Walter.
Bull. American Astron. Soc., Vol. 9, 592 (1978). — Abstract.

142.083 Soft X-ray spectra of the Crab Nebula and Sco X-1
observed from HEAO-1.
P. A. Charles, S. M. Kahn, F. M. Walter, S. Bowyer.
Bull. American Astron. Soc., Vol. 9, 592 (1978). — Abstract.

142.084 Low energy X-ray observations of supernova
remnant MSH 14-63.
J. Nugent, I. R. Tuohy, G. P. Garmire.
Bull. American Astron. Soc., Vol. 9, 592 (1978). — Abstract.

142.085 HEAO-1 observations of new soft X-ray sources.
F. M. Walter, P. A. Charles, S. M. Khan, S. Bowyer.
Bull. American Astron. Soc., Vol. 9, 592 (1978). — Abstract.

142.086 HEAO-1 observations of soft X-ray emission from
clusters of galaxies.
G. R. Riegler, S. M. Lea, G. Reichert.
Bull. American Astron. Soc., Vol. 9, 592 (1978). — Abstract.

142.087 Probable balloon observation of the variable X-ray
source 3U0042+32.   J. G. Laros, W. A. Wheaton.
Bull. American Astron. Soc., Vol. 9, 594 - 595 (1978).
Abstract.

142.088 X-ray observations of NGC 4151 and A2319 from
OSO-8 and HEAO-1.
R. F. Mushotzky, E. A. Boldt, S. S. Holt, P. J. Serlemitsos.
Bull. American Astron. Soc., Vol. 9, 609 (1978). — Abstract.

142.089 Spectra of the Crab Nebula, AM Her, and Nova
Ophiuchi from HEAO-1.   R. Rothschild, D.
Gruber, F. Knight, J. Matteson, P. Nolan, F. Marshall, J.
Swank, G. Reigler, I. Touhey, P. Charles, M. Lampton.
Bull. American Astron. Soc., Vol. 9, 609 (1978). — Abstract.

142.090 High energy X-ray observations by HEAO-1 of the
Crab Nebula, Cyg X-1, and the galactic center.
D. Gruber, J. Matteson, F. Knight, P. Nolan, W. Baity, L.
Peterson, J. Hoffman, W. Lewin, W. Wheaton, F. Primini.
Bull. American Astron. Soc., Vol. 9, 609 (1978). — Abstract.

142.091 The UCSD/MIT hard X-ray and low energy γ-ray
experiment on HEAO-1.
J. Matteson, W. Baity, D. Gruber, F. Knight, P. Nolan,
L. Peterson, J. Hoffman, W. Lewin.
Bull. American Astron. Soc., Vol. 9, 610 (1978). — Abstract.

142.092 Preliminary hard X-ray results from HEAO-1.
W. A. Wheaton, J. A. Hoffman, F. A. Primini,
A. Scheepmaker, W. H. G. Lewin, J. L. Matteson, W. A. Baity,
D. Gruber, M. Pelling, L. E. Peterson.
Bull. American Astron. Soc., Vol. 9, 610 (1978). — Abstract.

142.093 Discovery of iron line emission in galactic bulge
X-ray sources.
D. R. Parsignault, J. E. Grindlay.
Bull. American Astron. Soc., Vol. 9, 615 - 616 (1978).
Abstract.

142.094 Fluctuations in the high galactic latitude X-ray
background.
R. A. Shafer, E. A. Boldt, S. S. Holt, P. J. Serlemitsos.
Bull. American Astron. Soc., Vol. 9, 616 (1978). — Abstract.

142.095 Search for interstellar scattering X-ray halos with
HEAO-1 scanning modulation collimator.
G. Spada, H. Bradt, R. Doxsey, R. Griffiths, H. Gursky,
D. Schwartz, J. Schwarz.
Bull. American Astron. Soc., Vol. 9, 626 - 627 (1978).
Abstract.

142.096 Observation of the extended X-ray source in the
Crab Nebula by the SMC experiment on HEAO-1.
U. G. Briel, H. Bradt, R. Doxsey, H. Gursky, M. Oda, A.
Ramsey, J. Schwarz, D. Schwartz.
Bull. American Astron. Soc., Vol. 9, 627 (1978). — Abstract.

142.097 Positions of X-ray burst sources determined with
the HEAO-1 scanning modulation collimator.
R. Doxsey, H. Bradt, H. Gursky, M. Johnston, D. Schwartz,
J. Schwarz.
Bull. American Astron. Soc., Vol. 9, 627 (1978). — Abstract.

142.098 Location of the Norma transient/burster with
HEAO-1 scanning modulation collimator.
G. Fabbiano, H. Gursky, D. A. Schwartz, J. Schwarz, H. Bradt,
R. E. Doxsey.
Bull. American Astron. Soc., Vol. 9, 627 (1978). — Abstract.

142.099 Large population of low luminosity stellar X-ray
sources – implications for observers.
F. R. Harnden Jr., A. Lightman, P. Gorenstein.
Bull. American Astron. Soc., Vol. 9, 627 (1978). — Abstract.

142.100 Continuum and emission line variations in Sco
X-1 and old novae.   S. Wyckoff, P. A. Wehinger.
Bull. American Astron. Soc., Vol. 9, 627 - 628 (1978).
Abstract.

142.101 The linear X-ray polarization of Scorpius X-1.
K. S. Long, H. L. Kestenbaum, R. Novick, E. H.
Silver, M. C. Weisskopf.
Bull. American Astron. Soc., Vol. 9, 628 (1978). — Abstract.

142.102 Optical identification and Sco-like appearance of
the X-ray burster 4U1735−44 = MXB1735−44.
J. E. McClintock, C. R. Canizares, D. E. Backman.
Bull. American Astron. Soc., Vol. 9, 628 (1978). — Abstract.

142.103 Rocket observations of the far-ultraviolet radiation
field at high galactic latitudes.   R. C. Anderson,
W. Brune, R. C. Henry, P. D. Feldman, W. G. Fastie.
Bull. American Astron. Soc., Vol. 9, 629 - 630 (1978).
Abstract.

142.104 A thermonuclear runaway in the hydrogen envelope
of a neutron star as a possible model for A0620−00
(V616 Mon).   S. Starrfield, J. W. Truran, W. M. Sparks.
Bull. American Astron. Soc., Vol. 9, 631 (1978). — Abstract.

142.105 Hydrodynamic studies of spherical accretion onto
a compact X-ray source.
J. Buff, R. F. Stellingwerf.
Bull. American Astron. Soc., Vol. 9, 633 (1978). — Abstract.

142.106 **Discovery of the optical counterpart of Aquila X-1 (3U 1908+00).**
J. R. Thorstensen, P. A. Charles, S. Bowyer.
Bull. American Astron. Soc., Vol. 9, 644 (1978). – Abstract.

142.107 **Circinus X-1: new phenomena near the 16^{d}6 transition.** R. Dower, H. Bradt, C. Canizares.
Bull. American Astron. Soc., Vol. 9, 644 (1978). – Abstract.

142.108 **Spectrum of Cygnus X-1 from ~50 keV to ~3 MeV.**
C. A. Meegan, G. J. Fishman, R. C. Haymes.
Bull. American Astron. Soc., Vol. 9, 644 (1978). – Abstract.

142.109 **Radiative transfer theory for small angle scattering: a possible model for X-ray burst tails.**
S. P. Hatchett, C. Alcock.
Bull. American Astron. Soc., Vol. 9, 645 - 646 (1978).
Abstract.

142.110 **Dual character of the rapid burster and a classification of X-ray bursts.**
J. A. Hoffman, H. L. Marshall, W. H. G. Lewin.
Bull. American Astron. Soc., Vol. 9, 646 (1978). – Abstract.

142.111 **Further analysis of SAS-3 observations of the rapid burster (MXB1730–335).**
H. Marshall, M. P. Ulmer, W. H. G. Lewin, J. A. Hoffman.
Bull. American Astron. Soc., Vol. 9, 646 (1978). – Abstract.

142.112 **X-ray sources and induced surrounding clouds.**
K. L. Chan, W. Y. Chau.
Bull. American Astron. Soc., Vol. 9, 646 (1978). – Abstract.

142.113 **X-ray source counts at high galactic latitude.**
R. S. Warwick, J. P. Pye.
Mon. Not. R. Astron. Soc., Vol. 183, 169 - 175 (1978).

The Ariel V Catalogue is used to investigate the number–intensity distribution of X-ray sources at high galactic latitude. The results are consistent with a uniform distribution of sources in Euclidean space.

142.114 **The recurrent X-ray transient A 0538–66.**
N. E. White, G. F. Carpenter.
Mon. Not. R. Astron. Soc., Vol. 183, 11P - 15P (1978).

Ariel V observed two X-ray outbursts from the LMC in 1977 June and July: the events were separated by 17 days and each lasted several hours. The data are consistent with both outbursts being due to a previously unreported transient source that the authors designate A 0538–66. No evidence was found for regular pulsation.

142.115 **Her X-1 pulsar in the 12.5–40 keV energy range.**
D. Boclet, J. Claisse, P. Durouchoux, P. Pagnier, R. Rocchia.
Space Research, Vol. XVII, (see 012.010), 763 - 767 (1977).

A balloon flight performed from Aire-sur-l'Adour, France, in May 1975 permitted monitoring of Her X-1 for about 3500 s with a relative exposure larger than 0.75. The light curve of the pulsar, between 15 and 25 keV, exhibits a double peak structure with a 99% confidence for the main pulse. The spectrum and pulsar profile are compared with other observational results and discussed.

142.116 **Faint X-ray sources detected near COS B γ-ray positions.** P. F. Julien, H. F. Helmken.
Nature, Vol. 272, 699 - 701, with a correction in Vol. 273, 170 (1978).

The authors describe first, the reanalysis by the point summation technique of three weak 4U sources which lie within COS B error circles; second, a survey of the remaining COS B positions which has uncovered a new X-ray source within the CG075 + 0 error circle; and third, the results of an investigation of the region surrounding CG195 + 4 and CG189 + 1.

142.117 **Positions of galactic X-ray sources:**
$-20° < l^{II} < +6°$. J. G. Jernigan, K. M. V. Apparao, H. V. Bradt, R. E. Doxsey, R. G. Dower, J. E. McClintock.
Nature, Vol. 272, 701 - 704 (1978).

The precise (20–60") positions of nine X-ray sources in the vicinity of the galactic centre are reported.

142.118 **Extragalactic X-ray astronomy.** J. L. Culhane.
Q. J. R. Astron. Soc., Vol. 19, 1 - 37 (1978).

The author describes the classification and identification of extragalactic X-ray sources. The current status of the X-ray log N–log S relationship is discussed and its connection with the diffuse background X-ray flux examined. A number of unidentified X-ray sources have been found at high galactic latitudes. The extended sources associated with clusters of galaxies and the compact sources located in the nuclei of active galaxies together make up the bulk of the presently known and identified extragalactic X-ray sources. These two source categories are treated in some detail.

142.119 **High-galactic-latitude X-ray sources and southern clusters of galaxies.** P. M. Lugger.
Astrophys. J., Vol. 221, 745 - 755 (1978).

A complete optical survey has been made of 31 unidentified high-galactic-latitude ($|b| \geqslant 10°$) X-ray sources from the Uhuru and MX catalogs south of declination $-27°$. Thirteen clusters of galaxies (Abell distance classes $D \leqslant 4$) are proposed as optical identifications. Using all presently available data, the distribution according to Rood and Sastry morphological type of X-ray clusters is found to be consistent with that of the overall Abell population.

142.120 **The optical counterpart of 3U 1956+11.**
B. Margon, J. R. Thorstensen, S. Bowyer.
Astrophys. J., Vol. 221, 907 - 911 (1978).

The authors have identified a strong candidate for the optical counterpart of the X-ray source 3U 1956+11 (= 2S 1957 + 115, = 4U 1957 + 11). This source has an accurate X-ray position ($\lesssim 25"$) available from SAS-3 observations, and the candidate is located within 10" of this position. Photometry and spectrophotometry of the object and the surrounding field have been obtained at the Lick and Kitt Peak Observatories. For the candidate the authors find $V = 18.7$, $(B-V) = 0.3$, $(U-B) = -0.6$; these colors are similar to Sco X-1 and 3U 0614+09.

142.121 **Extended observations of Vela X-1 by OSO 8.**
R. H. Becker, R. E. Rothschild, E. A. Boldt, S. S. Holt, S. H. Pravdo, P. J. Serlemitsos, J. H. Swank.
Astrophys. J., Vol. 221, 912 - 916 (1978).

The Goddard Space Flight Center Cosmic X-ray experiment aboard OSO 8 viewed the X-ray binary pulsar, Vela X-1, on three occasions from late 1975 through late 1976. The X-ray spectrum is well represented by a power law modified by photoelectric absorption, a high-energy cutoff, and a line feature at ~6.8 keV. When combined with other observations, the measurements show that the pulse period is not decreasing monotonically. The three eclipses observed all indicate a significant eclipse flux.

142.122 **On the orbital variations of the optical polarization in Cygnus X-1.** M. Milgrom.
Astron. Astrophys., Vol. 65, L1 - L4 (1978).

The recent observations of the orbital variations of the optical polarization in Cyg X-1 are examined. It is shown that the value of the inclination angle cannot be determined from the data without assuming a detailed model for the mechanism which produces the polarization. The data is consistent with a geometry which is mirror-symmetric about the plane which is perpendicular to the orbital plane and contains the centers of the stars. The suppression of the U-band variations is interpreted as a photospheric effect.

142.123   Balloon observations of the X-ray pulsar Vela XR-1.
M. L. Duldig, M. W. Emery, A. G. Fenton, K. B. Fenton, J. G. Greenhill, R. M. Thomas.
Proc. Astron. Soc. Australia, Vol. 3, 117 - 118 (1977).
   The initial flight of the University of Tasmania balloonborne X-ray telescope was made from Parkes on Dec. 2, 1976. During the flight, enhanced X-ray emission was observed from the directions of 3U0900-40 (Vela XR-1), GX301-2 and the Galactic Centre. In this paper the authors report on the performance of the payload during the 11 hour flight and describe the preliminary results thus far obtained.

142.124   Observations of 4U 1608−52.   R. M. Thomas.
Proc. Astron. Soc. Australia, Vol. 3, 122 - 124 (1977).
   There have been a number of X-ray observations of sources very close to 4U 1608−52 and in this paper the author proposes the identification of these sources with 4U 1608−52. He thereby obtains additional spectral information and can now rule out a globular cluster as the optical counterpart of the object.

142.125   The 20keV break in the diffuse X-ray background spectrum.   R. M. Hudson, R. M. Thomas, M. L. Duldig.
Proc. Astron. Soc. Australia, Vol. 3, 131 - 133 (1977).
   The authors report an independent determination of the location of the break (change in spectral index) in the spectrum of the diffuse X-ray background by applying a simple analysis technique to data already in the literature.

142.126   The optical variability of the X-ray binary AM Herculis.   E. Feigelson, L. Dexter, W. Liller.
Astrophys. J., Vol. 222, 263 - 268 (1978).
   The authors present an optical light curve of AM Herculis from 1890 to 1976 obtained from 338 blue photographic plates of the Harvard College Observatory. Variations up to 3 mag on long time scales (greater than 10^2 days) and 1 mag on short time scales (less than 10^{-1} days) are seen. The observations were analyzed for periodic variability using a least-squares technique; no true periodicities were found with periods between 2 and 10,000 days or around the binary period of 0.13 days.

142.127   A balloon observation of the diffuse cosmic X-radiation above 20 keV.
R. L. Kinzer, W. N. Johnson, J. D. Kurfess.
Astrophys. J., Vol. 222, 370 - 378 (1978).
   A measurement of the diffuse cosmic X-ray background between 20 and 165 keV has been made with a balloon-borne experiment. The cosmic diffuse spectrum obtained by extrapolation of the measured growth curves to the top of the atmosphere can be represented by a single power law of $[67(+21, -18)] E^{(-2.17 \pm 0.07)}$; the best fit for a single-temperature thermal model gives a temperature of 1.1×10^9 K.

142.128   The Bowen mechanism in HZ Herculis.
B. Margon, J. G. Cohen.
Astrophys. J., Lett., Vol. 222, L33 - L36 (1978).
   The authors have obtained a series of high-dispersion blue and ultraviolet spectra of HZ Her. In addition to the N III λ4634−4641 emission reported by previous observers, the spectra reveal for the first time the presence of O III λ3444.1 emission. This provides conclusive evidence that the Bowen fluorescence mechanism is active in HZ Her/Her X-1, in agreement with theoretical predictions by McClintock et al. and Hatchett et al., and suggests that this is also the correct expla-

nation for the λ4640 emission in many such X-ray systems.

142.129   Eine neue Periode im Röntgendoppelstern Cyg X-1?
W. Eichendorf, M. Reinhardt.
Naturwissenschaften, 65. Jahrg., 205 (1978).

142.130   An analysis of the EUV and X-ray spectrum in Capella.   M. M. Katsova, M. A. Livshits.
Astron. Zh. Akad. Nauk SSSR, Vol. 55, 363 - 372 (1978). In Russian. English translation in Soviet Astron., Vol. 22, No. 2.
   From the line intensities at λ ~ 1500 Å the differential emission measure dY/dT as a function of temperature was calculated. Two difficulties of the previously proposed interpretation of these observations based on analogy with the solar chromosphere−corona transition region are outlined. In contrast to the model of a thin transition layer without motion, a "dynamical" model is proposed.

142.131   Variations of Circinus X-1 in the infrared.
I. S. Glass.
Mon. Not. R. Astron. Soc., Vol. 183, 335 - 340 (1978).
   JHKLM observations of the Mayo et al. optical candidate for Cir X-1 show that it varies with the same period as the X-ray source, thus securing the identification. The shape of the light curve is unique, a sudden rise near the time of the X-ray drop being followed by a gradual decay. The colour of the source remains constant.

142.132   An optical candidate for 2A 0042 + 323 (= 3U 0042 + 32).
P. Charles, J. Thorstensen, S. Bowyer.
Mon. Not. R. Astron. Soc., Vol. 183, 29P - 33P (1978).
   The authors have obtained spectrophotometric observations of all stars brighter than 20 mag in the 1 arcmin radius *SAS*-3 error circle for the high-latitude X-ray source 2A 0042 + 323 and most of the stars brighter than 18 mag within 2 arcmin of the *SAS*-3 best position. None of these stars seemed unusual except one star (V ~19.2 mag) within the error circle which displayed weak He II λ4686 emission. There is evidence for this star being variable. Details of this star are provided and spectral types are given for the other stars observed.

142.133   The periodic behaviour of 2A 0042 + 323 (= 3U 0042 + 32).   M. G. Watson, M. J. Ricketts.
Mon. Not. R. Astron. Soc., Vol. 183, 35P - 38P (1978).
   The high galactic latitude 'transient' X-ray source 2A 0042 + 323 has been extensively observed by the Ariel V Sky Survey Instrument. These new observations indicate a possible binary nature. This possibility is discussed with particular reference to the recently proposed optical identification.

142.134   The structure of the Virgo cluster X-ray source.
P. J. N. Davison.
Mon. Not. R. Astron. Soc., Vol. 183, 39P - 43P (1978).
   Evidence is presented for the presence of a second extended X-ray emission region in the Virgo cluster in addition to that which is known to surround M87. This second region, which dominates at energies above ~7 keV, is located in the vicinity of M84 and 86. The presence of two distinct emission regions, at different temperatures, provides a natural explanation for the observation by several experimental groups of a power-law energy spectrum.

142.135   Extragalactic X-ray sources.   R. Giacconi.
Phys. Scr., Vol. 17, (see 012.013), 159 - 164 (1978).
   Observations of X-ray emission from the nuclei of active galaxies are briefly summarized and their relevance to models for galactic nuclei discussed. Particular emphasis is given to the recent observations of X-ray emission from Seyfert gal-

axies, their luminosity function, and the time-scale for variability.

142.136 Fuentes de rayos X extragalácticas.
   H. Quintana.
Astron. extragalactica, (see 012.015), 45 - 77 (1976).

A review of X-ray observations of extragalactic objects is presented and the latest results are analysed with respect to optical identification of sources. The different types of physical processes causing the radiation observed in each case are discussed. Observations of clusters of galaxies are examined deductions on their relation with the existence of intergalactic gas are made and their possible cosmological implications are discussed.

142.137 A model to determine an upper limit to the distance of variable radio sources.
W. G. Fogarty, N. J. Schuch.
Astron. extragalactica, (see 012.015), 107 - 112 (1976).

A model of a variable radio source is constructed that enables the authors to establish an upper limit of its distance. Its application to concrete cases permits them to conclude that, if the model is correct, one has to reject the cosmological interpretation of the redshift of some sources.

142.138 The wide band X-ray energy spectrum of Sco X-1.
   S. Miyamoto, M. Matsuoka, M. Oda, Y. Ogawara.
Astron. Astrophys., Vol. 65, 329 - 335 (1978).

Two rocket observations of Sco X-1 covering wide energy range are reported. The observed continuum spectra are compared with the radiation spectrum from a spherical optically thick hot plasma, and the probable ranges of the values of the radius τ_{es} in electron scattering length and the temperature kT of the X-ray emitting hot plasma are derived. It is concluded that the simultaneous measurements of the continuum X-ray spectrum over wide energy range and the X-ray line emissions are important to investigate the structure of the hot plasma of Sco X-1.

142.139 Theoretical calculations of the optical pulsation from HZ Herculis.   T. J. Chester.
Astrophys. J., Vol. 222, 652 - 663 (1978).

A formalism is presented that enables one to calculate the amount of pulsed optical radiation emergent from a stellar atmosphere exposed to pulsed X-rays. The formalism takes account of the time evolution of the X-ray energy deposited in a time-independent atmosphere. For HZ Her, the calculated optical pulsation agrees with the observed amplitude.

142.140 Searches for correlated X-ray and radio emission from X-ray burst sources.   H. M. Johnson,
R. C. Catura, P. A. Lamb, N. E. White, P. W. Sanford,
J. A. Hoffman, W. H. G. Lewin, J. G. Jernigan.
Astrophys. J., Vol. 222, 664 - 666 (1978).

The NRAO Green Bank interferometer has been used to monitor MXB 1730–335 and MXB 1837+05 during periods when 68 X-ray bursts were detected by X-ray observations. No significant radio emission was detected from these objects, or from MXB 1820–30 and MXB 1906+00, which emitted no bursts throughout the simultaneous observations. The data place upper limits on radio emission from these objects in the 2695 and 8085 MHz bands.

142.141 Some recent results in X-ray astronomy.
   J. J. Quenby.
Nature, Vol. 273, 338 - 339 (1978).

142.142 Positions of galactic X-ray sources: $55° < l^{II} < 320°$.
   R. G. Dower, K. M. V. Apparao, H. V. Bradt, R. E.
Doxsey, J. G. Jernigan, J. Kulik.
Nature, Vol. 273, 364 - 367 (1978).

Precise (20–40'') celestial positions of 10 galactic X-ray sources in the region $55° < l^{II} < 320°$ are reported.

142.143 Discovery of 3.6-s X-ray pulsations from 4U 0115+63.
   L. Cominsky, G. W. Clark, F. Li, W. Mayer, S.
Rappaport.
Nature, Vol. 273, 367 - 369 (1978).

The authors have discovered a 3.61-s pulsation in the hard, recurrent transient X-ray source 4U 0115+63 and have measured its position to ~ 30 arc s.

142.144 Algorithms for localization of celestial sources of X-ray and gamma-ray bursts with the help of some space instruments.   G. A. Mersov.
Inst. kosm. issled. AN SSSR. Pr-344. Moskva, 1977. 15 pp.
In Russian. – Abstr. in Ref. zh., 62. Issled. kosm. prostranstva, 3.62.125 (1978).

142.145 Erste Röntgenquelle in der Großen Magellanschen Wolke identifiziert.
W. Eichendorf, M. Reinhardt.
Naturwissenschaften, 65. Jahrg., 253 - 255 (1978).

142.146 Evolutionary model computations for globular cluster X-ray sources.   D. Lauterborn.
Astron. Astrophys., Vol. 66, 269 - 274 (1978).

The evolution of a binary system consisting initially of a 1 $M_\odot$ main sequence star and a 100 $M_\odot$ black hole is followed numerically, both with and without the inclusion of angular momentum losses due to gravitational radiation emitted from the system. The evolution of the system is followed till the primary mass is reduced to 0.022 $M_\odot$. The period of the system decreases from initially 0.63 d to 0.12 d. For the production of X-rays by the matter falling onto the black hole a specific conversion rate of 10% of the rest mass is assumed. Binary systems of this type are proposed as a model for the globular cluster X-ray sources and for the γ-ray bursts.

142.147 The Cygnus X region. XI. Map of visual extinction A_V.   H. R. Dickel, H. J. Wendker.
Astron. Astrophys., Vol. 66, 289 - 295 (1978).

Optical observations of the Hα emission and observations of the continuum emission at 2695 MHz have been used to derive the distribution of visual extinction A_V across the Cygnus X complex. Maps of convolved $S_{H\alpha}$, T_b and A_V are presented and discussed.

142.148 Parameter estimation for X-ray sources: positions.
   Y. Avni.
Astron. Astrophys., Vol. 66, 307 - 309 (1978).

The author shows that the sizes of the positional error boxes for X-ray sources can be determined by using an estimation method which he has previously formulated generally and applied in spectral analyses. He explains how this method can be used by scanning X-ray telescopes, by rotating modulation collimators and by HEAO-A.

142.149 X-ray and optical observations of 3U 0900–40 (Vela X-1).   P. A. Charles, K. O. Mason,
N. E. White, J. L. Culhane, P. W. Sanford, A. F. J. Moffat.
Mon. Not. R. Astron. Soc., Vol. 183, 813 - 820 (1978).

Copernicus and Ariel V observations of 3U 0900–40 are presented. During 1975 February a region of low intensity was observed between binary phase 0.5 and eclipse which was caused by an increase in photoelectric absorption. The 283-s period measured by Copernicus in 1975 February was 0.017 s longer than that observed by SAS-3 in 1975 July. This is consistent with the secular decrease in period expected for an accreting neutron star. An upper limit of 0.004 mag is placed on the amplitude of broadband 283-s optical pulsations.

142.150 Cosmic flare transients: constraints upon models

for energy storage and release derived from the event frequency distribution.   R. Rosner, G. S. Vaiana. Astrophys. J., Vol. 222, 1104 - 1108 (1978).

Flare behavior for a variety of transient sources, including the Sun, flare stars, and a transient cosmic X-ray source, is examined. It is found that, although these sources span an energy release rate of over 10 decades, the flare frequency as a function of energy released follows a similar power law at large energies for all these sources: the flare frequency distributions at low energies, however, differ substantially. This result is used to develop a model for the general flaring phenomenon which allows a unified description of the flaring process for these diverse sources and which permits one to infer information concerning the modes of energy storage and release.

142.151 High-resolution observations of X-ray sources at the galactic center.   R. G. Cruddace, G. Fritz, S. Shulman, H. Friedman, J. McKee, M. Johnson. Astrophys. J., Lett., Vol. 222, L95 - L98 (1978).

A region about $1°$ in radius and centered on the galactic nucleus has been surveyed during a rocket flight, using X-ray detectors equipped with $10' \times 4°$ (FWHM) collimators. The extended source 4U 1743–29 has been resolved into at least four discrete sources, having luminosities (2–10 keV) between 4×10^{36} and 3.5×10^{37} ergs s^{-1}.

142.152 Periodic timing residuals in pulsating binary X-ray sources and orbital precession in Hercules X-1. Y. Avni, D. Q. Lamb, F. K. Lamb, M. Milgrom. Astrophys. J., Lett., Vol. 222, L113 - L117 (1978).

Intrinsic variations of pulse frequency and reflection of pulses produce systematic residuals in pulse arrival times from pulsating binary X-ray sources. The authors use an iterative technique to calculate higher-order terms in the arrival time curve of a variable-frequency binary source. They also formulate a general framework for studying time delays caused by reflection of pulses. By applying these methods to Her X-1, the authors find that reflection effects and variations in pulse frequency due to accretion torques, stellar wobble, and changes in pulse shape can mimic the effects of orbital precession with the 35 d period. They suggest observational tests to resolve this ambiguity in the interpretation of the timing data.

142.153 On the turn-off of the reflection effect in the HZ Her system.   G. S. Bisnovatyj-Kogan, N. G. Bochkarev, E. A. Karitskaya, A. M. Cherepashchuk, N. I. Shakura. Pis'ma Astron. Zh., Vol. 4, 81 - 86 (1978). In Russian.

For the system HZ Her = Her X-1 possible photometric effects of ellipsoidal form of the optical component and of eclipse by the disc surrounding the neutron star are calculated and discussed for the period of absence of the reflection effect.

142.154 On the possibility of detecting the duplicity of X-ray sources by using the Doppler effect. B. L. Novak. Pis'ma Astron. Zh., Vol. 4, 173 - 176 (1978). In Russian.

A procedure to detect the duplicity of X-ray sources based on spectral analysis of time succession of counts in using a fixed energetic range is suggested. An estimate of duration of observation necessary for detection of duplicity is obtained.

142.155 X-ray flare. IAU Circ., No. 3161 (1978).

142.156 4U 0115+63. IAU Circ., Nos. 3163 with a corrigendum 3169, 3171, 3238 (1978).

142.157 GX 301–2. IAU Circ., No. 3165 (1978).

142.158 X-ray sources. IAU Circ., No. 3169 (1978).

142.159 Circinus X-1. IAU Circ., Nos. 3171, 3217 (1978).

142.160 HR 1099 (= 4U 0336+01?) IAU Circ., Nos. 3173, 3176, 3180, 3184, 3197, 3217 (1978).

142.161 Brief X-ray transient. IAU Circ., No. 3174 (1978).

142.162 2A 1052+606 = SAO 015338? IAU Circ., No. 3176 (1978).

142.163 4U 1538–52. IAU Circ., Nos. 3184, 3197, 3201 (1978).

142.164 Hercules X-1. IAU Circ., No. 3184 (1978).

142.165 Persistent X-ray emission from MXB1659–29. IAU Circ., Nos. 3190, 3229 (1978).

142.166 Other X-ray burst sources. IAU Circ., No. 3190 (1978).

142.167 Periodicity in 4U 1700–37. IAU Circ., No. 3193 (1978).

142.168 X-ray burst. IAU Circ., No. 3193 (1978).

142.169 4U 1630–47. IAU Circ., Nos. 3197, 3227 (1978).

142.170 4U 1837+04 and 4U 1538–52. IAU Circ., No. 3197 (1978).

142.171 H 1743–32. IAU Circ., No. 3203 (1978).

142.172 MXB1730–335. IAU Circ., Nos. 3204, 3208, 3211 (1978).

142.173 A0535+26. IAU Circ., Nos. 3208, 3219 (1978).

142.174 Mk 421 = 2A 1102+384. IAU Circ., Nos. 3212, 3221, 3224 (1978).

142.175 4U 1145–61. IAU Circ., No. 3225 (1978).

142.176 Optical bursts from MXB1735–44. IAU Circ., No. 3230 (1978).

142.177 4U 1626–67. IAU Circ., No. 3233 (1978).

142.178 V861 Scorpii = OAO 1653–40. IAU Circ., No. 3234 (1978).

142.179 Aquila X-1. IAU Circ., No. 3235 (1978).

142.180 4U 1658–48. IAU Circ., No. 3238 (1978).

142.181 A0327+43. IAU Circ., No. 3238 (1978).

142.182   **K-fluorescence lines in spectra of X-ray binaries.**
M. M. Basko.
Astrophys. J., Vol. 223, 268 - 281 (1978).

The reflection of the X-rays in K lines of heavy elements from a cold surface is calculated. The radiative transfer is treated rigorously, with the albedo in K lines expressed in terms of the H-function for isotropic scattering. This approach is also applied to evaluate the spectral shape of the low-energy wing of the line, formed by the singly scattered K photons. A detailed discussion is given of the main characteristics of K-emission from the X-ray binaries. The prospects of observations of the narrow $K\alpha$ lines in X-ray binaries are analyzed, with special attention paid to the new information that can be obtained from such observations. The minimum equivalent width which one could expect in the case of the iron $K\alpha$ line is evaluated for a number of well-known X-ray binaries.

142.183   **The proper-motion approach to counterparts of gamma-ray burst sources and high-latitude X-ray transient sources.**   H. M. Johnson.
Astrophys. J., Vol. 223, 282 - 284 (1978).

Reasons to expect association of high-energy activity with nearby stars are outlined. The need to resort to proper-motion data to identify these stars is explained. Results of a search for proper-motion stars in, or very near, the error boxes of four high-latitude X-ray transient sources are given. The transients are 4U 0336 + 01, 2A 1102 + 384, Cen XR-4, and MX 2346 − 65.

142.184   **Short-term time variability of Cygnus X-1. III.**
M. C. Weisskopf, P. G. Sutherland, J. I. Katz,
C. R. Canizares.
Astrophys. J.,Lett., Vol. 223, L17 - L20 (1978).

Data from a SAS 3 observation of Cygnus X-1 have been analyzed, assuming a simple shot-noise model for the time variability. The results confirm the suggestion of Sutherland, Weisskopf, and Kahn that the shot parameters are constant when the source is in its low state. The first significant measurement of the time sense of the shot-noise pulses was performed, with the result that the average pulse shape is time-symmetric, not asymmetric as has been previously assumed.

142.185   **The soft X-ray spectrum of Capella: discovery of intense line emission.**   W. Cash, S. Bowyer,
P. A. Charles, M. Lampton, G. Garmire, G. Riegler.
Astrophys. J.,Lett., Vol. 223, L21 - L23 (1978).

A soft X-ray spectrum of Capella has been obtained with the HEAO 1 satellite. Only models which include intense line emission at 0.85 keV can adequately explain the data. The spectrum is consistent with emission from an isothermal solar-abundance plasma of temperature $(10.5 \pm 2.8) \times 10^6$ K. It is also consistent with a blend of temperatures between 4×10^6 and 2×10^7 K. The intensity of X-ray emission is constant over 2 days.

142.186   **Cosmological information from X-ray observations.**
H. Gursky.
IAU Symp. No. 79, (see 012.027), p. 327 - 337 (1978).

The author reviews the present state of understanding of extragalactic X-ray astronomy. He presents what little cosmological information one has obtained to date and he discusses how and possibly where new information of this kind may emerge. He is going on the presumption that the key areas are the nature of the diffuse X-ray background and of the discrete X-ray sources.

142.187   **X-ray active galaxies and QSOs.**   B. A. Cooke.
Nature, Vol. 274, 16 - 17 (1978).

142.188   **Observations of low-luminosity X-ray sources in Vela-Puppis.**   S. H. Pravdo, R. H. Becker, E. A.
Boldt, S. S. Holt, P. J. Serlemitsos, J. H. Swank.

Astrophys. J., Vol. 223, 537 - 543 (1978).

The authors present results of a study of the X-ray emission from a small portion of the galactic plane near galactic longitude 260°. This region contains at least six low-luminosity X-ray sources within ~ 10° of PSR 0833 − 45, which is near the center of the Gum nebula. The X-ray source 4U 0833 − 45, associated with the Vela pulsar, is observed at twice its 4U catalog intensity. The lack of X-ray pulsations at the pulsar period (greater than 99% nonpulsed), the nonthermal power law spectrum, and models of the X-ray source distribution in this region suggest that a large fraction of the X-rays come from an extended source ~ 1° in radius. The observation of a high-temperature ($T_{eff} \geq 10^8$ K) spectrum in a field of view containing only Puppis A among known sources has led to the discovery of a new *OSO 8* source, OS 0752 − 39. Other spectra from this region are discussed.

142.189   **Position and pulse profile of the X-ray transient 4U0115+63.**   M. Johnston, H. Bradt, R. Doxsey,
H. Gursky, D. Schwartz, J. Schwarz.
Astrophys. J., Lett., Vol. 223, L71 - L73 (1978).

The celestial position of the 3.6 s X-ray pulsar, 4U 0115+ 63, has been measured to a precision of $12''$ in one dimension with the scanning modulation collimator experiment on HEAO I. This result, together with a position obtained with SAS 3, has led to the identification of an early-type star as the optical counterpart. The authors infer an X-ray luminosity of ~3×10^{37} ergs s^{-1} (3–13 keV) from the observed flux of about 30% of that of the Crab and an estimated distance of 5 kpc. The pulse profile over the energy range 3–13 keV is similar to that of Cen X-3.

142.190   **Optical identification of 4U 1735−44 (=MXB 1735−44) and its similarity to Scorpius X-1.**
J. E. McClintock, C. R. Canizares, D. E. Backman.
Astrophys. J., Lett., Vol. 223, L75 - L78 (1978).

The galactic bulge X-ray source and burst source 4U 1735−44 (=MXB 1735−44) is firmly identified with a $B = 17.6$ blue star on the basis of the presence of $\lambda\lambda 4640$–4650 and He II $\lambda 4686$ emission in the spectrum of the star. A spectrum of Sco X-1 taken an hour earlier is remarkably similar to the spectrum of the 4U 1735−44 counterpart with respect to type and strength of emission features, absence of absorption features, and shape of the continua. Other similarities between 4U 1735−44 and Sco X-1 are discussed.

142.191   **The optical counterparts of SMC X-2 and SMC X-3.**   D. Crampton, J. B. Hutchings, A. P. Cowley.
Astrophys. J., Lett., Vol. 223, L79 - L81 (1978) = Dominion Astrophys. Obs., Victoria, B.C., Contrib. No. 358 = NRC No. 16297.

The spectra of the Sanduleak-Philip optical candidates of SMC X-2 and SMC X-3 are remarkably similar to each other and to the optical X-ray counterparts X Per, γ Cas, and HDE 245770. Although it is probable that they are the optical counterparts of SMC X-2 and X-3, their variable X-ray luminosity, when detected, was considerably higher than that of optically similar variable sources in the Galaxy. He II $\lambda 4686$ emission is present, although very weak, in both spectra, and appears to have large velocity and intensity changes on a time scale of days.

142.192   **Variable optical flickering in Cygnus X-2.**
S. A. Ilovaisky, C. Chevalier, M. Chevreton,
S. Bonazzola.
Astron. Astrophys., Vol. 67, 287 - 289 (1978).

High-speed photometry of the Cygnus X-2 optical candidate has revealed the existence of low-level (0.1–0.2%) flickering activity on one night when the object was faint; this activity was not present on the following night, the object being at the same flux level. A peak at $P = 10.9$ min in the power spectrum of the first night is significant.

142.193   Optical pulses in HD 153919 = 3U 1700−37.
A. Kruszewski.
Inf. Bull. Variable Stars, No. 1424, 3 pp. (1978).

142.194   Further observations of nonperiodic optical flickering in HZ Herculis.
P. A. Vanden Bout, T. J. Moffett.
Publ. Astron. Soc. Pacific, Vol. 90, 149 - 153 (1978).
High-speed photometric observations of the HZ Her/
Her X-1 system show that nonperiodic optical flickering is present during both the ON and OFF portions of the 35-day cycle. The optical flickering is broad band and not confined to emission lines. There is a weak correlation between the amplitude of the flickering and orbital phase but efforts to identify the source of the flickering were unsuccessful.

142.195   Optical observations of galactic X-ray transients.
M. Pakull.
Messenger, No. 13, p. 17 - 18 (1978).

142.196   Observation of soft X-ray emission from SMC X-1.
A. N. Bunner, W. T. Sanders, J. A. Nousek, W. L. Kraushaar.
Bull. American Astron. Soc., Vol. 10, 389 (1978). − Abstract.

142.197   Precise positions of LMC X-1, X-2, and X-3 with the HEAO-1 Modulation Collimator.   M. Johnston,
H. Bradt, R. Doxsey, H. Gursky, D. A. Schwartz, J. Schwarz.
Bull. American Astron. Soc., Vol. 10, 390 (1978). − Abstract.

142.198   Constraints on thermal models for the soft X-ray background from observations between 100 and 1200 eV.   J. A. Nousek, P. M. Fried.
Bull. American Astron. Soc., Vol. 10, 390 (1978). − Abstract.

142.199   HEAO-1 measurement of the energy dependence of the fluctuations in the cosmic X-ray background.
R. A. Shafer, E. Boldt, A. E. Szymkowiak, S. Holt, R. Rothschild, P. Serlemitsos.
Bull. American Astron. Soc., Vol. 10, 390 (1978). − Abstract.

142.200   Position and structure of the X-ray source in Cen A measured with the Modulation Collimator on HEAO-1.   R. Doxsey, H. Bradt, H. Gursky, D. A. Schwartz, J. Schwarz.
Bull. American Astron Soc., Vol. 10, 390 - 391 (1978). Abstract.

142.201   High energy X-ray observations of extragalactic objects.   W. A. Baity, D. E. Gruber, F. K. Knight, J. L. Matteson, L. E. Peterson, F. A. Primini, E. Y. Tsiang, B. A. Cooke.
Bull. American Astron. Soc., Vol. 10, 391 (1978). − Abstract.

142.202   Hard X-ray spectral and temporal analysis of Cygnus X-1 by HEAO A-4.   P. L. Nolan, F. K. Knight, J. L. Matteson, L. E. Peterson, R. E. Rothschild, J. A. Hoffman, F. A. Primini, S. K. Howe.
Bull. American Astron. Soc., Vol. 10, 391 (1978). − Abstract.

142.203   NGC 5506 and NGC 7582: counterparts to X-ray sources.   R. G. Dower, G. R. Ricker, J. G. Jernigan, R. E. Doxsey, H. V. Bradt, H. W. Schnopper.
Bull. American Astron. Soc., Vol. 10, 403 (1978). − Abstract.

142.204   Limits on the optical detection of bursting X-ray source 2S1916-053.
P. M. Rybski, J. O. Patterson.
Bull. American Astron. Soc., Vol. 10, 419 (1978). − Abstract.

142.205   A 4.8 periodicity in the spectra of Cyg X-3.
R. H. Becker, J. L. Robinson-Saba, E. A. Boldt,

S. S. Holt, S. H. Pravdo, P. J. Serlemitsos, J. H. Swank.
Bull. American Astron. Soc., Vol. 10, 420 (1978). − Abstract.

142.206   A search for spectral variations with binary phase in a high state of Cyg X-3.   J. L. Robinson-Saba, R. H. Becker, S. H. Pravdo, E. A. Boldt, S. S. Holt, P. J. Serlemitsos, J. H. Swank.
Bull. American Astron. Soc., Vol. 10, 420 (1978). − Abstract.

142.207   Shot noise in Cygnus X-1: evidence for "growing" shots.
S. Kahn, F. Marshall, W. Predhorsky.
Bull. American Astron. Soc., Vol. 10, 420 (1978). − Abstract.

142.208   Recent observational results on X-ray burst sources.
H. Marshall, L. Cominsky, J. A. van Paradijs, J. A. Hoffman, G. Jernigan, W. Wheaton, W. H. G. Lewin.
Bull. American Astron. Soc., Vol. 10, 420 (1978). − Abstract.

142.209   High resolution crystal spectroscopy of galactic X-ray sources with OSO-8.
W. H.-M. Ku, G. A. Chanan, D. J. Helfand, K. S. Long, R. Novick.
Bull. American Astron. Soc., Vol. 10, 420 (1978). − Abstract.

142.210   X-ray mapping with HEAO-1.
A. E. Szymkowiak, R. A. Shafer, E. A. Boldt, S. S. Holt, P. J. Serlemitsos, R. E. Rothschild.
Bull. American Astron. Soc., Vol. 10, 421 (1978). − Abstract.

142.211   Radio observations of the X-ray galaxy NGC 2110.
P. E. Greenfield, B. F. Burke.
Bull. American Astron. Soc., Vol. 10, 433 (1978). − Abstract.

142.212   Hercules X-1; spectral hard X-ray observations from HEAO-1.   D. E. Gruber, J. L. Matteson, W. A. Baity, L. E. Peterson, W. A. Wheaton, C. Dobson, P. Campbell, W. Lewin.
Bull. American Astron. Soc., Vol. 10, 433 (1978). − Abstract.

142.213   New hard X-ray sources seen by HEAO-A2.
F. E. Marshall, E. A. Boldt, S. S. Holt, R. F. Mushotzky, R. E. Rothschild, P. J. Serlemitsos.
Bull. American Astron. Soc., Vol. 10, 433 - 434 (1978). Abstract.

142.214   X rays from the region of the Orion trapezium.
R. L. Kelley, H. V. Bradt, R. G. Dower, R. E. Doxsey.
Bull. American Astron. Soc., Vol. 10, 434 (1978). − Abstract.

142.215   SAS-3 observations of 4U0115+63.
S. A. Rappaport, G. W. Clark, L. Cominsky, P. C. Joss, F. Li.
Bull. American Astron. Soc., Vol. 10, 434 (1978). − Abstract.

142.216   The high-energy X-ray spectrum of Her X-1 observed from OSO-8.   B. R. Dennis, G. S. Maurer, E. P. Cutler, C. J. Crannell, J. F. Dolan, K. J. Frost, L. E. Orwig.
Bull. American Astron. Soc., Vol. 10, 434 (1978). − Abstract.

142.217   Ultraviolet spectroscopy of HZ Herculis with the IUE satellite.   R. J. Davis, J. Black, A. K. Dupree, H. Gursky, L. W. Hartmann, T. A. Matilsky, J. C. Raymond.
Bull. American Astron. Soc., Vol. 10, 444 (1978). − Abstract.

142.218   Ultraviolet observations of X-ray sources with IUE.
A. K. Dupree.
Bull. American Astron. Soc., Vol. 10, 445 (1978). − Abstract.

142.219   The effect of vacuum birefringence on the polariza-

tion of X-ray pulsars.
G. A. Chanan, R. Novick, E. H. Silver.
Bull. American Astron. Soc., Vol. 10, 445 (1978). − Abstract.

142.220 HEAO-1 observations of 4U1626-67.
S. H. Pravdo, E. A. Boldt, S. S. Holt, R. E. Roth-
schild, P. J. Serlemitsos, J. H. Swank.
Bull. American Astron. Soc., Vol. 10, 446 (1978). − Abstract.

**142.221 Pulse profiles and spectra of the 3.6 sec transient
X-ray pulsar A0115+63, observed by the HEAO A4
hard X-ray experiment.** W. A. Wheaton, F. A. Primini,
E. Y. Tsiang, S. K. Howe, P. Campbell, J. A. Hoffman,
W. H. G. Lewin, D. A. Gruber, P. L. Noland, R. E. Rothschild,
R. M. Pelling, L. E. Peterson.
Bull. American Astron. Soc., Vol. 10, 446 (1978). − Abstract.

**142.222 HEAO A4 hard X-ray observations of Cygnus X-3,
the gamma ray source CG195+4, and a burst from
MXB1728-34.** B. A. Cooke, F. A. Primini, W. A. Wheaton,
C. A. Dobson, A. Scheepmaker, J. A. Hoffman, W. H. G.
Lewin, W. A. Baity, J. L. Matteson, F. K. Knight, A. R. Good.
Bull. American Astron. Soc., Vol. 10, 446 (1978). − Abstract.

142.223 On the nature of the galactic bulge X-ray sources.
M. Milgrom.
Astron. Astrophys., Vol. 67, L25 - L28 (1978).
A model is proposed for the class of galactic X-ray
sources which include: the galactic bulge sources, the globular cluster sources, Sco X-1, Cyg X-2 and possibly the steady
components of X-ray bursters. The model reproduces the
main observable properties of the sources under consideration:
(a) No X-ray pulsations. (b) Rare occurrence of eclipses. (c)
Soft X-ray spectrum. (d) Lower than expected ratios of optical to X-ray luminosities.

**142.224 A soft X-ray source in the vicinity of the Am star
HR976.** A. J. F. den Boggende, R. Mewe,
J. Heise, A. C. Brinkman, E. H. B. M. Gronenschild, J. Schrijver.
Astron. Astrophys., Vol. 67, L29 - L31 (1978).
A region of 34′ in diameter around the position of
HR976 (= HD20210) was observed with the soft and medium
energy X-ray instrument on-board ANS. A 3.8 σ signal in the
0.16−0.284 keV interval and an upper limit in the 1−3.5 keV
interval was obtained. Tentatively identifying this X-ray
source with HR976 (at a distance of about 53 pc) yields an
X-ray luminosity of 2.0 (± 1) $\times$ 10^{30} erg/s (0.16−0.284 keV)
and a 3 σ-upper limit $\leqslant$ 1.3 $\times$ 10^{31} erg/s (1−3.5 keV).

**142.225 Evidence of a periodicity in the optical light curve
of HD 226868 (Cyg X-1) during the X-flare of
April − May 1975.** G. Natali, R. Messi.
Astron. Astrophys., Vol. 67, L33 - L34 (1978).
The photoelectrical data taken on HD 226868, optical
counterpart of Cyg X-1, during the "high state" of the X-ray
source in April - May 1975 show an oscillation in the light
curve with a period of $\approx$ 62 minutes. Some hypothesis are
considered about the nature of the phenomenon.

**142.226 Search for super-fast optical variability of X-ray
sources and T Tau-type stars according to the
MANIA experiment program in 1973−74.**
G. M. Beskin, O. A. Evseev, V. N. Mansurov, L. A. Pustil'nik,
G. S. Tsarevskij, V. F. Shvartsman.
Soobshch. Spets. Astrofiz. Obs., Zelenchukskaya, Vyp. (No.)
20, p. 18 - 29 (1977). In Russian.
Results are presented of a search for optical variability
in the objects V818 Sco, V1357 Cyg, HZ Her, identified
with the X-ray sources Sco X-1, Cyg X-1, Her X-1. In the
course of 1973 observations no variability was registered in
the time range 10^{-4} to 3 s at the level SA $\geqslant$ 0.02−0.05 (S is

the averaged normalized power of flares, A is their averaged
normalized amplitude). In 1974 the search for variability was
carried out in the range 10^{-6} to 10 s. There are grounds to
believe that on June 18, 1974, fast fluctuations of brightness
with characteristic times τ_1 and τ_2 (10^{-3} s $< \tau_1 < 10^{-2}$ s;
10^{-1} s $< \tau_2 < 10$ s) were registered at levels $(SA)_{\tau_1} \sim 1$ and
$(SA)_{\tau_2} \sim 0.3$, respectively. The search for variability in the
T Tau-type stars (SU Aur, RY Tau, NU Ori, AB Aur, T Tau)
was carried out in the range 10^{-4} to 1 s. No variability was
found; for all objects SA $<$ 0.02 - 0.05.

**142.227 Low-mass X-ray binaries and their relation to the
non-X-ray sources.** A. P. Cowley.
Ann. New York Acad. Sci., Vol. 302, (see 012.033), 1 - 5
(1977).

**142.228 Evolutionary processes in X-ray binaries and their
progenitor systems.** E. P. J. van den Heuvel.
Ann. New York Acad. Sci., Vol. 302, (see 012.033), 14 - 35
(1977).
Some recent refinements in the evolutionary scenario
for massive X-ray binaries are discussed, in particular the
effects of mass loss, and the braking of the rotation of a
neutron star in a stellar wind. The possible origins of low-mass
X-ray binaries are discussed, and some recent advancements in
the ideas about mass transfer by Roche-lobe overflow are reviewed, notably the effects of nonsynchronous rotation.

**142.229 Origin and present evolution of massive X-ray
binaries.** J. Ziolkowski.
Ann. New York Acad. Sci., Vol. 302, (see 012.033), 47 - 54
(1977).

**142.230 Optical properties of the accretion disk in the sys-
tem HZ Her (Her X-1).** G. S. Bisnovatyi-Kogan
(*Bisnovatyj-Kogan*), A. W. Goncharskyi (*A. V. Goncharskij*),
B. V. Komberg, A. M. Cherepashchuk, A. G. Jagola (*Yagola*).
Ann. New York Acad. Sci., Vol. 302, (see 012.033), 55 - 60
(1977).

142.231 Nature of X-ray bursters. H. Gursky.
Ann. New York Acad. Sci., Vol. 302, (see 012.
033), 197 - 209 (1977).
Contents: Discovery of X-ray bursters. Associations with
other objects. The relation to γ-ray bursts. Bursts from Cyg
X-1? Summary − What are the X-ray bursters?

142.232 X-ray burst sources. W. H. G. Lewin.
Ann. New York Acad. Sci., Vol. 302, (see 012.
033), 210 - 228 (1977).

**142.233 The astronomy of X-ray sources and the physics of
accretion.** J. P. Ostriker.
Ann. New York Acad. Sci., Vol. 302, (see 012.033), 229 - 243
(1977).
Compact galactic X-ray sources appear to be divided into
relatively few types, with galactic center, globular cluster, and
bursting sources having various properties like each other and
unlike both the systems thought to contain neutron stars (e.g.,
Her X-1, SMC X-1) and those hypothesized to contain $\sim 10^1$
$M_\odot$ black holes (Cyg X-1, Cir X-1). Most of the mechanisms
proposed for accreting gas onto a compact object are intrinsically unstable, and one should not be surprised at the
variety of irregular bursting, flaring, etc. behavior found in the
majority of sources.

142.234 X-ray bursts and their extended tails.
W. A. Baan.
Ann. New York Acad. Sci., Vol. 302, (see 012.033), 244 - 247
(1977).
The author briefly summarizes a possible mechanism for
the production of X-ray bursts. He discusses an accretion

model for bursts from a magnetized rotating neutron star. He mentions a mechanism for the observed enhancement in the steady luminosity of a burst source following an X-ray burst.

142.235   Optical observations of X-ray globular clusters.
        W. Liller.
Ann. New York Acad. Sci., Vol. 302, (see 012.033), 248 - 260 (1977).
        The author presents a class of objects called X-ray globular clusters and asks if there is any optical evidence for the existence of such phenomena as centrally located massive black holes, close binary star systems, or interstellar gas. He divides his paper into three parts: The first deals with star distributions and the observed dynamical properties of globulars; the second with photometry, broad and narrow band; and the last with spectroscopy.

142.236   Neutron star and degenerate dwarf models of X-ray
        bursts.   D. Q. Lamb, F. K. Lamb.
Ann. New York Acad. Sci., Vol. 302, (see 012.033), 261 - 299 (1977).
        The authors survey compact star (either a neutron star or a degenerate dwarf) models of X-ray bursts. These models fall into tow distinct categories. Under "Accretion models", the authors discuss bursts that are produced by instabilities that involve the magnetosphere of an accreting neutron star or degenerate dwarf. Under "Nuclear burning models", they discuss bursts that are produced by thermonuclear flashes in the freshly deposited outermost layers of an accreting neutron star. The principal alternative to compact star models of the X-ray bursts is a model in which such bursts are produced by accretion onto a massive black hole. Under "Are burst sources X-ray binaries or massive black holes? ", they therefore discuss observations that appear potentially able to distinguish between the compact star and massive black hole models of the bursts.

142.237   Extragalactic X-ray sources.
        H. W. Schnopper, J. P. Delvaille.
Ann. New York Acad. Sci., Vol. 302, (see 012.033), 300 - 311 (1977).
        Since the launch of SAS-3 on May 7, 1975, approximately 25 percent of the observing time has been devoted to extragalactic X-ray astronomy. The authors discuss here a sample from the variety of high-latitude sources encountered. A brief description of the properties of the rotating modulation collimator on SAS-3 is given.

142.238   Polarization of cosmic X-ray sources.
        R. Novick, M. C. Weisskopf, E. H. Silver, H. L. Kestenbaum, K. S. Long, R. S. Wolff.
Ann. New York Acad. Sci., Vol. 302, (see 012.033), 312 - 328 (1977).
        The authors review the theory and design of the OSO-8 polarimeters. They give new precise values for the polarization of the Crab Nebula at 2.6 and 5.2 keV, and they give preliminary values for the polarization of the Crab pulsar, Cyg X-1, and Cyg X-2 at 2.6 keV.

142.239   X-ray signatures: new time scales and spectral
        features.   E. Boldt.
Ann. New York Acad. Sci., Vol. 302, (see 012.033), 329 - 348 (1977).
        In searching for new X-ray signatures as well as for fine structure in established ones, the author's group at the Goddard Space Flight Center has made quite extensive observations outside of the usual temporal-spectral regime. During this talk the author describes a variety of results that have come from this program. This involves a brief review of the observations of Sco X-1, a status report on the study of the rapid variations in Cygnus X-1, including preliminary results on a new effect, and finally, a summary of the recent observa-

tions of various spectral features attributable to iron in several objects, ranging from clusters of galaxies to the X-ray binary Her X-1.

142.240   Recent X-ray results from Ariel V and Copernicus.
        P. W. Sanford.
Ann. New York Acad. Sci., Vol. 302, (see 012.033), 386 - 402 (1977).
        This paper presents recent results from the Mullard Space Science Laboratory instruments on Ariel V and Copernicus. The author discusses the nearby clusters of galaxies, source A0430−62 (3 U0426−63), source 3C 273, Centaurus A, source 3C 390.3 (3 U1825+18), and the slow rotator, X Persei (3 U0352+30).

142.241   Search for periodicity in optical light variations of
        Cyg X-2.   M. M. Basko.
Ann. New York Acad. Sci., Vol. 302, (see 012.033), 403 - 407 (1977).
        The principal conclusion of the author's optical study of the X-ray source Cyg X-2 is as follows: He has found a regular component in brightness variation with period $P_1 = 5\overset{d}{.}92 \pm 0.05$ (or $P_2 = 1\overset{d}{.}203 \pm 0.002$, or $P_3 = 0\overset{d}{.}8528 \pm 0.001$, or twice any of these three values) and with full amplitude $\sim 0\overset{m}{.}25$ (B-filter), which persists only when the X-ray luminosity of Cyg X-2 is high enough.

142.242   Evidence for strong cyclotron emission in the hard
        X-ray spectrum of Her X-1.
J. Trümper, W. Pietsch, C. Reppin, B. Sacco, E. Kendziorra, R. Staubert.
Ann. New York Acad. Sci., Vol. 302, (see 012.033), 538 - 544 (1977).
        The authors report on the discovery of a strong line feature in the Her X-1 spectrum at 53 keV, which they interpret as cyclotron emission. The corresponding magnetic field is 4.6×10^{12} gauss.

142.243   Finding out X-ray double systems using the Doppler
        effect.   B. L. Novak.
Inst. kosm. issled. AN SSSR. Prepr., 1977, No. 361, 15 pp. In Russian. − Abstr. in Ref. zh., 51. Astron., 6.51.843 (1978).

142.244   The effect of vacuum birefringence in a magnetic
        field on polarization and directivity of radiation of
X-ray pulsars.   Yu. N. Gnedin, G. G. Pavlov, Yu. A. Shibanov.
Pis'ma Astron. Zh., Vol. 4, 214 - 218 (1978). In Russian.
        The polarization of vacuum in a magnetic field is shown to cause no significant depolarization of X-ray radiation for the models of X-ray pulsars which are generally accepted now. The vacuum birefringence can change the polarization and angular distribution of radiation of an X-ray pulsar. This change depends on the density and magnetic field distributions in a radiating region and its neighbourhood.

142.245   X-ray binaries.   E. P. J. van den Heuvel.
        Mem. Soc. Astron. Italiana, Vol. 47, (see 012.036), 453 - 492 (1976).
        Contents: The observational evidence: generation of X rays by accretion. Evolution of close binaries with possible applications to massive X-ray binaries. Types of accretion in X-ray binaries. Predictions made about the UV behaviour of X-ray binaries.

142.246   New X-ray pulsars.   Z. Urban.
        Říše hvězd, Vol. 59, 117 - 119 (1978). In Czech.

142.247   The X-ray generation − an overview of X-ray
        astronomy 1960 - 1980.   S. S. Murray.
X-ray imaging, (see 012.041), p. 8 - 18 (1977). − Abstr. in Phys. Abstr., Vol. 81, Abstr. 49567 (1978).

The Ariel V High-Latitude Catalogue.
See Abstr. 002.065.

Time-asymmetry in astrophysical time series.
See Abstr. 031.222.

Measurements of X-ray source positions by the scanning modulation collimator on HEAO-1.
See Abstr. 032.507.

Preliminary results of the X-ray telescope "Filin" on board Salyut 4.   See Abstr. 032.509.

X-ray astronomy: HEAO looks further and sees more.   See Abstr. 051.001.

On the effect of gas pressure in the theory of line accretion − I.   See Abstr. 062.006.

Radiative transfer effect in an ionized medium at high temperature.   See Abstr. 063.005

Optically-thick X-ray transfer: the shell game.
See Abstr. 063.012.

The effects of small-angle scattering on a pulse of radiation with an application of X-ray bursts and interstellar dust.   See Abstr. 063.021.

On the turbulent energy transport in accretion discs.   See Abstr. 064.005.

Upper limits to X-ray emission from colliding stellar winds.   See Abstr. 064.070.

Disk accretion by magnetic neutron stars.
See Abstr. 064.080.

The thermal stability of hot degenerate stars in steady-state accretion.   See Abstr. 065.031.

The evolution of massive close binaries. VI: Final considerations for the conservative case.   See Abstr. 065.047.

The fate of matter and angular momentum in disk accretion onto a magnetized neutron star.
See Abstr. 066.004.

Reflection of X rays by neutron star surfaces.
See Abstr. 066.011.

Nuclear burning in accreting neutron stars and X-ray bursts.   See Abstr. 066.016.

Radiative effects in supersonic accretion.
See Abstr. 066.026.

X-ray spectrum from disk accretion onto massive black holes.   See Abstr. 066.029.

Nuclear burning in accreting neutron stars and X-ray bursts.   See Abstr. 066.047.

X-ray bursts from helium-burning flashes on accreting neutron stars.   See Abstr. 066.248.

Timing effects in rotating neutron stars.
See Abstr. 066.280.

Knowledge of neutron stars from X-ray observations.
See Abstr. 066.283.

Optical photometry of Cygnus X-1: 1972−1976.
See Abstr. 113.004.

Anomalous light curve of Cyg X-1 during the X-ray increase of April−May 1975.   See Abstr. 113.005.

Study of the lightcurve of the Of star HD 153919.
See Abstr. 113.012.

Narrow-band photoelectric photometry of V1357 Cyg (Cyg X-1).   See Abstr. 113.034.

HDE 245770.   See Abstr. 113.063.

HDE 245770.   See Abstr. 113.066.

Spectral classification of ultraviolet objects.
See Abstr. 114.055.

Observations of low-luminosity optical counterparts of X-ray sources.   See Abstr. 114.074.

The Bowen mechanism in HZ Herculis.
See Abstr. 114.526.

Precision digicon spectroscopy of HDE 226868: the nature of Cygnus X-1.   See Abstr. 114.537.

A detailed study of the spectrum of the binary X-ray source HD 153919 [3 U 1700-37]. II. Analysis of the radial velocities in the blue spectral region.
See Abstr. 114.541.

Objective prism study of the 1977 spectrum of HDE 245770 = A0535 + 26.   See Abstr. 114.555.

A detailed study of the spectrum of the binary X-ray source HD 153919 (3U 1700-37). I. Radial velocity data in the blue spectral region.   See Abstr. 114.557.

Possible optical counterpart of 2S 1702−363.
See Abstr. 114.562.

Spectroscopy and possible orbital periods for HDE 245770 (= A0535 + 26).   See Abstr. 114.566.

The helium variable HD 64740 − an X-ray binary?
See Abstr. 114.572.

Roche-lobe overflow in X-ray binaries.
See Abstr. 117.004.

Stellar parameters of five early type companions of X-ray sources.   See Abstr. 117.006.

The remarkable system AM Herculis/3U 1809 + 50. II. A single accretion column model.   See Abstr. 117.034.

Nonstationary accretion of stellar wind.
See Abstr. 117.061.

X- and γ-ray bursts from magnetic reconnection in very short period binary systems.   See Abstr. 117.082.

Mass transfer effects in binary star evolution.
See Abstr. 117.083.

X-ray emission processes in close binary systems.
See Abstr. 117.085.

Flare stars as candidates for a new class of transient soft X-ray sources.   See Abstr. 122.023.

Changes in AM Herculis during maximum and minimum states.   See Abstr. 122.108.

Narrow-band photoelectric observations of AM Her. See Abstr. 122.135.

X-ray pulsations from SS Cygni. See Abstr. 122.140.

Observations of HDE 245770 during the recent X-ray flare-up of its counterpart A0535+26. See Abstr. 123.027.

Nova Ophiuchi 1977: an X-ray nova. See Abstr. 124.501.

A provisional optical light curve of the X-ray recurrent nova V616 Monocerotis = A0620–00. See Abstr. 124.951.

X-ray bursts from type II supernova explosions. See Abstr. 125.029.

Prompt thermal emission from type II supernovae – limits on the ultraviolet and soft X-ray burst. See Abstr. 125.030.

Soft X-ray emission from the new supernova remnant in Cygnus.   See Abstr. 125.032.

Soft X-ray emission from a newly discovered supernova remnant in Cygnus.   See Abstr. 125.036.

Evidence of X-ray emission from W44. See Abstr. 125.038.

Search for X-ray line emission from Cassiopeia A. See Abstr. 125.042.

Low energy X-ray line emission from Cas A. See Abstr. 125.059.

Coronal line emission from Puppis A. See Abstr. 125.070.

Hard X-ray observations of white dwarf binary systems.   See Abstr. 126.006.

X radiation from accreting non-magnetic degenerate dwarfs: the regime at large electron scattering optical depths. See Abstr. 126.014.

X and UV radiation from accreting magnetic degenerate dwarfs.   See Abstr. 126.015.

Effects of electron conduction on accretion by white-dwarf stars.   See Abstr. 126.016.

Hot white dwarfs as soft X-ray sources. See Abstr. 126.028.

Discovery of an X-ray QSO. See Abstr. 141.002.

Detection of X-ray emission from the giant radio lobes of Cen A.   See Abstr. 141.051.

Discovery of an X-ray QSO. See Abstr. 141.066.

26.3 MHz radio source survey. III. Correlation with extragalactic X-ray sources.   See Abstr. 141.165.

Pulsars—radio and X-ray.   See Abstr. 141.524.

Neutrinos from binary pulsars. See Abstr. 141.552.

Possible X-ray counterparts of γ-ray sources. See Abstr. 142.710.

Results of observations of the gamma-ray flux from Cyg X-3.   See Abstr. 142.718.

On a model of gamma-quanta generation by the X-ray source Cyg X-3.   See Abstr. 142.719.

The proper-motion approach to counterparts of gamma-ray burst sources and high-latitude X-ray transient sources.   See Abstr. 142.728.

Hard X-ray observations of cosmic γ-ray sources. See Abstr. 142.730.

A search for [O III] emission in the central part of the X-ray globular cluster M15.   See Abstr. 154.006.

Optical studies of the X-ray globular cluster NGC 6624.   See Abstr. 154.017.

Distribution and energy spectrum of diffuse soft X-rays.   See Abstr. 157.002.

Far-ultraviolet studies. IV. Spectroscopy of north and south galactic pole regions observed from Apollo 17. See Abstr. 157.007.

ESO 113-IG45: galaxy and/or quasar? See Abstr. 158.007.

The enormous mass of the elliptical galaxy M87: a model for the extended X-ray source. See Abstr. 158.009.

Radio and X-ray variability of the nucleus of Centaurus A (NGC 5128).   See Abstr. 158.014.

The X-ray structure of NGC 5128. See Abstr. 158.029.

The X-ray emitting galaxy Centaurus A. See Abstr. 158.049.

Concurrent radio, infrared, optical, and X-ray observations of the nucleus of NGC 4151. See Abstr. 158.067.

Seyfert galaxies as X-ray sources. See Abstr. 158.099.

A new model for X-ray emission from NGC 4151. See Abstr. 158.108.

NGC 5506: an almost Seyfert galaxy. See Abstr. 158.168.

X-ray observations of galaxies. See Abstr. 158.208.

MCG–5–23–16.   See Abstr. 158.210.

MCG–6–30–15.   See Abstr. 158.213.

Uhuru observations of X-ray emission from Seyfert galaxies.   See Abstr. 158.215.

X-rays from Markarian 541. See Abstr. 158.253.

An upper limit to the absorption of 1 keV diffuse X-rays by the Small Magellanic Cloud. See Abstr. 159.022.

An X-ray and optical study of seven clusters of galaxies.   See Abstr. 160.005.

The origin and distribution of gas within rich clusters of galaxies: the evolution of cluster X-ray sources over cosmological time scales.   See Abstr. 160.009.

Soft X-ray spectra of the Coma and Perseus clusters of galaxies: constraints on the models.   See Abstr. 160.010.

The X-ray luminosity function of Abell clusters. See Abstr. 160.015.

The X-ray cluster of galaxies Klemola 44. See Abstr. 160.023.

The evolution of cluster X-ray sources. See Abstr. 160.026.

X-ray clusters of galaxies – radio limits to the ionized gas.   See Abstr. 160.027.

Comparison of X-ray structure of clusters of galaxies with thermal models of the intracluster gas. See Abstr. 160.028.

X-ray and radio emission from clusters of galaxies: the heating of intracluster gas by relativistic electrons. See Abstr. 160.043.

A search for X-ray emission from Abell clusters and superclusters.   See Abstr. 160.048.

Origin of X-ray emission from clusters. See Abstr. 160.050.

An energy-dependent map of the X-ray emission of the Perseus cluster.   See Abstr. 160.060.

X-ray observations of clusters of galaxies. See Abstr. 160.081.

Primeval clusters of galaxies and the X-ray background.   See Abstr. 162.145.

Gamma-Ray Sources

142.701 **Distance limit for a class of model γ-ray burst sources.**
W. K. H. Schmidt.
Nature, Vol. 271, 525 - 527 (1978).
   The author points out that MeV photons have actually been observed in γ-ray bursts, and that this means that non-relativistic sources cannot be further away than a few kpc from the Sun and therefore must be galactic.

142.702 **γ-rays from γ Geminorum?**
R. E. Davies, A. C. Fabian, J. E. Pringle.
Nature, Vol. 271, 634 - 635 (1978).

142.703 **Evidence that cosmic γ-ray bursts are galactic.**
   R. S. White, J. M. Ryan, R. B. Wilson, A. D. Zych, W. B. Dayton.
Nature, Vol. 271, 635 - 636 (1978).
   The five bursts observed by detectors on balloons along with the observations from satellites at larger burst sizes strongly suggest that the γ-ray bursts are galactic. The larger bursts, seen by satellites, are probably near, within a few hundred pc, and the smaller ones seen by balloons are probably much further away, perhaps thousands of pc.

142.704 **High-energy γ-rays and their bursts in air shower cores.**
S. Dake, M. Hazama, K. Jitsuno, Y. Nakanishi, K. Nishikawa, M. Sakata, Y. Yamamoto, Y. Hatano.
Nuovo Cimento B, Ser. 11, Vol. 41B, 55 - 87 (1977). – Abstr. in Phys. Abstr., Vol. 80, Abstr. 91393 (1977).

142.705 **γ-ray burst observed at balloon altitude.**
   J. Nishimura, M. Fujii, Y. Tawara, M. Oda, Y. Ogawara, T. Yamagami, S. Miyamoto, M. Kajiwara, H. Murakami, M. Yoshimori, M. Nakagawa, T. Sakurai.
Nature, Vol. 272, 337 - 338 (1978).

A small γ-ray burst was found during ~150 h of observations and its celestial position was determined with a precision of ~0.3°.

142.706 **A model for clusters of discrete gamma-ray sources.**
   M. Abdulwahab, P. Morrison.
Astrophys. J., Lett., Vol. 221, L33 - L36 (1978).
   The authors propose a model for certain discrete sources of ~100 MeV γ-rays. The γ-rays originate, as for the usual diffuse source, from the interaction of cosmic-ray protons with the gas of the interstellar medium (mostly in π^0 decay). But the cosmic rays are enhanced, because they have not spread far from their localized origin in strong type I supernova explosions, and the gas is locally dense within the volumes of small random interstellar clouds. Such sources have distinct properties. The model is fitted quantitatively to three γ-sources reported near the Crab Nebula which tentatively show those features.

142.707 **Position of the 16 August 1976 gamma burst from Solrad 11 and Vela 5 observations.**
J. G. Laros, W. D. Evans, R. W. Klebesadel.
Bull. American Astron. Soc., Vol. 9, 594 (1978). – Abstract.

142.708 **A search for high energy gamma-ray bursts from primordial black holes or other astronomical objects.**   N. A. Porter, T. C. Weekes.
Mon. Not. R. Astron. Soc., Vol. 183, 205 - 210 (1978).
   Two systems of separated atmospheric Cerenkov detectors have been used to search for bursts of high energy γ-rays which have been predicted theoretically from primordial black holes of about 10^{15} g in mass. Using the Hagedorn nuclear model an upper limit of 0.04 events/pc³ yr is set for the rate of black hole explosions in the Galaxy.

142.709 **Recent high energy gamma-ray results from SAS-2.**
   D. J. Thompson, C. E. Fichtel, R. C. Hartman, D. A. Kniffen, R. C. Lamb, G. F. Bignami, H. B. Ögelman, M. E. Özel, T. Tümer.

Space Research, Vol. XVII, (see 012.010), 769 - 774 (1977).

Recent developments in gamma-ray astronomy due to the results from SAS-2 have focused on two areas. First, the emission from the plane of the Galaxy is the dominant feature in the gamma-ray sky. Second, searches of the SAS-2 data for emission from localized sources have shown three strong discrete gamma-ray sources: the Crab nebula and PSR 0531 + 21, the Vela supernova remnant and PSR 0833 − 45, and a source near galactic coordinates 193°+ 3° which does not appear to be associated with other known celestial objects. Evidence has also been found for pulsed gamma-ray emission from two other radio pulsars PSR 1818 − 04 and PSR 1747 −46. A localized source near longitudes 76°−80° may be associated with the X-ray source Cyg X-3.

142.710   Possible X-ray counterparts of γ-ray sources.

L. Maraschi, T. Markert, K. M. V. Apparao, H. Bradt, H. Helmken, W. Wheaton, W. A. Baity, L. E. Peterson.
Nature, Vol. 272, 679 - 681 (1978).

The error circles for the 10 unidentified COS-B γ-ray sources have been examined for X-ray emission. Three of these sources were found to contain known X-ray sources within their error circles: the positional uncertainty of CG135 + 1 contains 4U0241 + 61, CG312−1 contains MX1406−61 and 4U1416−62, and CG327−0 contains 2S1553−542. Spectra and variability, as well as positional coincidence, are used to argue that 4U0241 + 61, MX1406−61 and 2S1553−542 are possible X-ray counterparts to the γ-ray sources.

142.711   Gamma-ray lines: a new window to the universe.

R. E. Lingenfelter, R. Ramaty.
Phys. Today, Vol. 31, No. 3, p. 40 - 47 (1978).

Line emission in the gamma-ray band, an emerging branch of astronomy based on balloon and satellite data, probes the physics of nucleosynthesis, the interstellar medium, solar flares, supernovae and neutron stars.

142.712   Energy spectrum, time structure, and arrival direction of the 1976 August 16 cosmic gamma-ray burst: an observation at balloon altitude.

M. Sommer, D. Müller.
Astrophys. J., Lett., Vol. 222, L17 - L20 (1978).

A strong cosmic γ-ray burst was observed by balloon-borne instrumentation in coincidence with several satellite detectors on 1976 August 16. The onset of the burst was at 16:15:28 UT, and the duration was 32 s with considerable fluctuations during this time. The energy spectrum of the burst was measured over the range 100 keV to 2 MeV, and was found to fit approximately a power law $\sim E^{-2.2}$. The total intensity was derived to be $\sim 10^{-4}$ ergs cm^{-2} above 100 keV energy. The multielement detector used for this experiment allowed a rough determination of the arrival direction of the burst. The resulting error box appears to exclude a common origin for this cosmic γ-ray burst and a radio event which has been observed at approximately the same time.

142.713   Gamma rays from dense interstellar clouds.

F. Lebrun, J. A. Paul.
Astron. Astrophys., Vol. 65, 187 - 191 (1978).

The recent COS-B observations of the high energy (> 300 MeV) gamma-ray emission of the galactic disc have related at $10° < l^{II} < 40°$ a component at low positive latitudes, prominent in the range $25° < l^{II}.< 35°$. The position of this excess coincides with large dark clouds in the Lynds survey. In this paper, the authors investigate whether a diffuse production of gamma rays in the close-by interstellar medium can account for the observed excess emission.

142.714   Ausserirdische Gammastrahlung lässt Koinzidenz-Gebirge entstehen.   K. Bosshard.

Orion, 36. Jahrg., 81 - 82 (1978).

142.715   A qualitative study of cosmic fireballs and γ-ray bursts.   G. Cavallo, M. J. Rees.

Mon. Not. R. Astron. Soc., Vol. 183, 359 - 365 (1978).

If a large amount of energy is suddenly converted into a concentrated burst of (MeV) γ rays, the prolific creation of electron–positron pairs will inhibit the escape of photons until they have been degraded below the pair-production threshold. This sets general constraints on the possible luminosities of rapidly varying γ-ray sources and suggests why the observed γ-ray bursts have an approximately standardized and 'soft' spectrum.

142.716   Diffuse γ-ray background from Seyfert galaxies.

J. E. Grindlay.
Nature, Vol. 273, 211 (1978).

It is suggested that the diffuse background spectrum observed at hard X-ray and γ-ray energies can be accounted for by Seyferts, given their individual (hard) spectra.

142.717   Cosmic gamma rays.   M. Hillas.

New Scientist, Vol. 74, 194 - 196 (1977). − Abstr. in Phys. Abstr., Vol. 81, Abstr. 7843 (1978).

142.718   Results of observations of the gamma-ray flux from Cyg X-3.   B. M. Vladimirskij, Yu. I. Neshpor,

A. A. Stepanyan, V. P. Fomin.
Izv. Krymskoj Astrofiz. Obs., Vol. 58, 44 - 50 (1978).
In Russian.

From an analysis of observations of gamma-ray quanta with energy $> 2 \times 10^{12}$ eV from Cyg X-3 in the years 1972–75 it is shown that the gamma-ray emission is periodical. The best value of the period which was derived from all measurements is equal to $0.199684 \pm 2 \times 10^{-6}$ day.

142.719   On a model of gamma-quanta generation by the X-ray source Cyg X-3.   A. A. Stepanyan.

Izv. Krymskoj Astrofiz. Obs., Vol. 58, 51 - 55 (1978).
In Russian.

It is pointed out that the spectra of the periodic component of three gamma-ray sources: the pulsar PSR 0833–45 in the Vela nebula, the pulsar NP 0532 in the Crab nebula and the X-ray source Cyg X-3 are similar. Since the gamma-ray spectrum of the Vela pulsar agrees with the pion decay, the assumption that all three spectra are of same origin seems justified. It is pointed out that the width of the 10^8 eV gamma-ray pulse corresponds to the angular distribution generated by low-energy neutral pions.

142.720   Diffuse gamma radiation.

C. E. Fichtel, G. A. Simpson, D. J. Thompson.
Astrophys. J., Vol. 222, 833 - 849 (1978).

An examination of the intensity, energy spectrum, and spatial distribution of the diffuse γ-radiation observed by SAS 2 away from the galactic plane in the energy range above 35 MeV has shown that it consists of two components. One component is generally correlated with galactic latitudes, the atomic hydrogen column density as deduced from 21 cm measurements, and the continuum radio emission, believed to be synchrotron emission. It has an energy spectrum similar to that in the plane and joins smoothly to the intense radiation from the plane. It is therefore presumed to be of galactic origin. The other component is apparently isotropic, at least on a coarse scale, and has a steep energy spectrum. The galactic component is interpreted in terms of its implications for both the local and more distant regions of the Galaxy. The apparently isotropic γ-radiation is discussed particularly with regard to the constraints placed on possible models by the steep energy spectrum, the intensity, and the upper limit on the anisotropy.

142.721   Cosmic gamma radiation.   G. Börner.

Contacts between high energy physics and other

fields of physics. Schladming, Austria, 24 February - 5 March 1977 = Acta Phys. Austriaca, Suppl. 18 (1977). p. 789 - 834. Abstr. in Phys. Abstr., Vol. 81, Abstr. 12059 (1978).

142.722 **No evidence of pulsed radio emission at 325 MHz from CG195+4.**
N. Mandolesi, G. Morigi, G. Sironi.
Astron. Astrophys., Vol. 67, L5 - L6 (1978).

CG195+4 is a γ-ray source pulsating with a period of 59 sec. No counterpart of the source in other bands of the electromagnetic spectrum has been so far detected. In December 1977 the source region has been monitored at frequency 325 MHz; no evidence of pulsed emission was found.

142.723 **New limits on gamma-ray bursts.**
G. J. Fishman, C. A. Meegan, J. W. Watts, Jr., J. H. Derrickson.
Astrophys. J., Lett., Vol. 223, L13 - L15 (1978).

Two balloon flights of large-area scintillation crystal detector arrays indicate that the rate of weak γ-ray bursts is significantly below that expected from a uniform distribution of burst sources. This result, combined with the data from stronger bursts, gives strong evidence for a galactic confinement of burst sources.

142.724 **A distant limit for a class of model gamma-ray burst sources.** W. K. H. Schmidt.
GSFC Doc. X-661-77-152, Prepr., 2 + 10 pp. (1977).

Gamma ray burst sources are presumably not larger than 10^9 cm as inferred from observed flux variations. If they are homogeneous and isotropically radiating, then from photon density considerations, they would have to be optically thick due to gamma-gamma pair production when assumed to be too far away. Deviations of observed photon spectra from an exponential shape around 1 MeV lead to an upper limit of the possible distance of such sources of only 2 kpc from the sun. Thus the sources must be galactic unless the radiation is highly beamed or emerges from a relativistically moving shell.

142.725 **Final results on the diffuse gamma ray emission observed by SAS-2.**
C. E. Fichtel, G. S. Simpson, D. J. Thompson.
Bull. American Astron. Soc., Vol. 10, 395 (1978). – Abstract.

142.726 **Gamma-ray lines as probes of the fundamental properties of collapsed stars.**
K. Brecher, A. Burrows.
Bull. American Astron. Soc., Vol. 10, 395 - 396 (1978). Abstract.

142.727 **HEAO-1 A-4 observations of two cosmic gamma ray bursts.** F. K. Knight, J. L. Matteson, A. R. Good, L. E. Peterson.
Bull. American Astron. Soc., Vol. 10, 396 (1978). – Abstract.

142.728 **The proper-motion approach to counterparts of gamma-ray burst sources and high-latitude X-ray transient sources.** H. M. Johnson.
Bull. American Astron. Soc., Vol. 10, 396 (1978). – Abstract.

142.729 **Origin of the diffuse cosmic gamma rays.**
A. W. Wolfendale.
Nature, Vol. 274, 314 - 315 (1978).

142.730 **Hard X-ray observations of cosmic γ-ray sources.** M. J. Coe, J. J. Quenby, A. R. Engel.
Nature, Vol. 274, 343 - 344 (1978).

The authors report on hard X-ray studies of the COS B γ-ray error boxes and suggest that unresolved sources of this type could account for the galactic disk γ-ray emission.

142.731 **Origin of MeV-excess in the energy spectrum of diffuse cosmic γ rays.** V. Schönfelder.
Nature, Vol. 274, 344 - 346 (1978).

The origin of the diffuse cosmic γ radiation has been widely discussed. It is either interpreted as the result of interactions in intergalactic space or as the result of the superposition of many unresolved external galaxies. Strong support for the latter hypothesis is given by the author's measurements of the energy spectrum of the Seyfert galaxy NGC4151 up to 10 MeV with the MPI Compton telescope which are reported here.

142.732 **On the identification of celestial γ-ray sources.** E. Massaro, L. Scarsi.
Nature, Vol. 274, 346 - 347 (1978).

The observations from COS B have provided a new and more detailed picture of the high energy γ-ray emission from the Galaxy. The authors discuss here the first catalogue with 13 localised sources which has been compiled (Hermsen et al., 1977).

142.733 **COS-B observations of localized high-energy gamma-ray emission.** L. Koch.
Ann. New York Acad. Sci., Vol. 302, (see 012.033), 349 - 360 (1977).

It appears that pointlike sources are now being resolved in the γ-ray sky above the diffuse background. The size of the error boxes makes it difficult to establish an identification on a purely positional basis. However, since two of them are pulsars, it is suggested that a search is made at other wavelengths (e.g., radio) for unseen pulsars in the γ error boxes.

Search for high energy γ-ray bursts from evaporation of primordial black holes. See Abstr. 066.015.

X- and γ-ray bursts from magnetic reconnection in very short period binary systems. See Abstr. 117.082.

Coincidence of compact supernova remnants with three COS-B γ-ray sources. See Abstr. 125.015.

Possible gamma-ray line from the Crab Nebula. See Abstr. 134.044.

New highly variable radio source, possible counterpart of γ-ray source CG135 + 1. See Abstr. 141.101.

Radio counterpart for γ-ray source CG 135+1. See Abstr. 141.180.

High-energy pulsed gamma rays from pulsars. See Abstr. 141.513.

Gamma-ray emission from pulsars. See Abstr. 141.539.

Algorithms for localization of celestial sources of X-ray and gamma-ray bursts with the help of some space instruments. See Abstr. 142.144.

The proper-motion approach to counterparts of gamma-ray burst sources and high-latitude X-ray transient sources. See Abstr. 142.183.

HEAO A4 hard X-ray observations of Cygnus X-3, the gamma ray source CG195+4, and a burst from MXB1728-34. See Abstr. 142.222.

Non-thermal electron bremsstrahlung in the Galactic Disk. See Abstr. 157.001.

Spectral characteristics of the galactic gamma radiation observed by COS-B. See Abstr. 157.003.

143 Cosmic Radiation

143.001 The acceleration of cosmic rays in shock fronts – I.
A. R. Bell.
Mon. Not. R. Astron. Soc., Vol. 182, 147 - 156 (1978).

It is shown that charged particles can be accelerated to high energies in astrophysical shock fronts. Fast particles are prevented from streaming away upstream of a shock front by scattering off Alfvén waves which they themselves generate. This scattering confines the particles to the region around the shock and results in first-order Fermi acceleration due to the particles crossing the shock many times. The consequent energy spectrum is a power law with an index close to that observed for galactic cosmic rays. The discussion relates to particles which are already relativistic, and their initial acceleration from thermal energies is not considered.

143.002 The acceleration of cosmic rays in shock
fronts – II.   A. R. Bell.
Mon. Not. R. Astron. Soc., Vol. 182, 443 - 455 (1978).

The acceleration to relativistic energies of the high-energy tail of the particle distribution produced by a shock front is discussed. In order to be accelerated, particles need to be able to pass through the shock without being strongly deflected and it is argued, using the Earth's bow shock as an example, that a shock front produces large numbers of protons, and probably electrons also, which satisfy this condition. The resulting energy spectrum of these initially non-relativistic particles is calculated. The synchrotron radio emission from the energetic electrons in a shocked gas is calculated, and the theoretical and observed flux densities of two supernova remnants (Tycho's SNR and Cas A) are compared and found to agree satisfactorily.

143.003 Effects of particle drift on cosmic ray transport. II.
Analytical solution to the modulation problem with
no latitudinal diffusion.   P. A. Isenberg, J. R. Jokipii.
Astrophys. J., Vol. 219, 740 - 749 (1978).

The authors present an analytical solution of the modulation equations for a particular form of the diffusion tensor and drift velocities. The functional form of the drift velocities is qualitatively similar to that derived for the solar wind in Paper I, so the solutions presented here will provide further insight into the effects of drifts on the solar modulation of galactic cosmic rays.

143.004 Ionization, dissociation, and heating efficiencies of
cosmic rays in a gas of molecular hydrogen.
T. E. Cravens, A. Dalgarno.
Astrophys. J., Vol. 219, 750 - 752 (1978).

Detailed calculations are carried out of the efficiencies with which cosmic rays with energies between 1 MeV and 100 MeV traversing a gas of molecular hydrogen produce H_2^+ ions, H^+ ions, H atoms, and H^- ions, and estimates are given of the heat deposited in the neutral gas.

143.005 Time delay between the nucleosynthesis of cosmic
rays and their acceleration to relativistic energies.
A. Soutoul, M. Cassé, E. Juliusson.
Astrophys. J., Vol. 219, 753 - 755 (1978).

If cosmic rays originate in a supernova explosion, they might be accelerated immediately after their synthesis. This would prevent electron-capture decay of unstable isotopes, and could result in large differences between cosmic-ray and solar-system abundances. Existing data on the abundances of iron, cobalt, and nickel in cosmic rays show great similarity to the solar system abundances, and the authors therefore conclude that more than a year elapses between the synthesis of the cosmic-ray nuclei and their acceleration to relativistic

energies.

143.006 Cosmic ray source abundance of calcium.
C. Perron.
Nature, Vol. 271, 425 - 426 (1978).

The author concludes that it is very unlikely that the observed cosmic ray calcium abundance can be entirely due to interstellar spallation.

143.007 An estimation of the primary proton spectrum
between 10^{12} and 10^{14} eV.
T. K. Gaisser, F. Siohan, G. B. Yodh.
J. Phys. G, Vol. 3, L241 - L244 (1977). – Abstr. in Phys. Abstr., Vol. 80, Abstr. 91389 (1977).

143.008 L'origine du rayonnement cosmique.
J. Audouze, M. Meneguzzi.
La recherche en astrophysique, (see 003.001), p. 190 - 203 (1977).

143.009 Magnetic bottles for cosmic rays.
Nature, Vol. 272, 312 (1978).

143.010 Charge and energy spectra of cosmic rays with
$Z \gtrsim 60$: the Skylab experiment.
E. K. Shirk, P. B. Price.
Astrophys. J., Vol. 220, 719 - 733 (1978).

A 1.2 m² array of Lexan track detectors, 1 g cm⁻² thick, inside a 1 g cm⁻² aluminum wall of the Skylab workshop, detected 104 cosmic-ray nuclei with $Z \gtrsim 65$, seven nuclei with $Z \gtrsim 88$, three nuclei with $Z \gtrsim 94$, and no superheavy nuclei ($Z \gtrsim 110$). The shape of the energy spectrum for events with $Z \geqslant 65$ is consistent with that for Fe. The charge distribution is inconsistent with a source of solar-system composition (even with preferential acceleration of elements with low ionization potential) and is quite consistent with a source of predominantly r-process material.

143.011 Particle acceleration by astrophysical shocks.
R. D. Blandford, J. P. Ostriker.
Astrophys. J., Lett., Vol. 221, L29 - L32 (1978).

A new mechanism is proposed for acceleration of a power-law distribution of cosmic rays with approximately the observed slope. High-energy particles in the vicinity of a shock are scattered by Alfvén waves carried by the converging fluid flow leading to a first-order acceleration process in which the escape time is automatically comparable to the acceleration time. Shocks from supernova explosions propagating through the interstellar medium can account for the acceleration of galactic cosmic rays. Similar processes occurring in extragalactic radio sources can lead to efficient in situ acceleration of relativistic electrons.

143.012 Ultraheavy cosmic rays: theoretical implications of
recent observations.
J. B. Blake, K. L. Hainebach, D. N. Schramm, J. D. Anglin.
Astrophys. J., Vol. 221, 694 - 702 (1978).

The recent extreme ultraheavy cosmic-ray observations ($Z \geqslant 70$) are compared with r-process models. A detailed cosmic ray propagation calculation is used to transform the calculated source distributions to those observed at the Earth. The r-process production abundances are calculated using different mass formulae and β-rate formulae; an empirical estimate based on the observed solar-system abundances is used also. There is the continued strong indication of an r-process dominance in the extreme ultraheavy cosmic rays. However, it is shown that the observed high actinide/Pt ratio in the cosmic rays cannot be fitted with the same r-process calculation which also fits the solar-system material. An estimate is also made of the expected relative abundance of superheavy elements in the cosmic rays if the anomalous heavy xenon in carbonaceous chondrites is due to a fissioning superheavy element.

143.013 Ionization models of cosmic ray sources.
M. Cassé, P. Goret.
Astrophys. J., Vol. 221, 703 - 712 (1978).

The authors briefly recall their coronal-type ionization model and discuss the relevant physical conditions. These conditions are likely to be encountered in common astrophysical sites. The authors tentatively consider the extent to which powerful eruptive stars (flare stars) could contribute to the supply of the galactic cosmic rays.

143.014 Studier af den galaktiske partikelstråling.
I. L. Rasmussen.
Astron. Tidsskr., Årg.11, 6 - 12 (1978).

143.015 The cosmic-ray isotopes.   P. Meyer.
Nature, Vol. 272, 675 - 679, with a correction in Vol. 273, 170 (1978).

Recent measurements suggest that the isotopic composition of the cosmic-ray sources is similar to that of the solar system and that cosmic-ray acceleration follows nucleosynthesis after a considerable delay. Cosmic rays may therefore be a sample of the present interstellar medium rather than fresh supernova ejecta. Observations of the abundance of radioactive isotopes lead to a containment time of cosmic rays in the Galaxy an order of magnitude larger than had been suspected.

143.016 The isotopic composition of cosmic-ray helium from 123 to 279 MeV per nucleon: a new measurement and analysis.   H. W. Leech, J. J. O'Gallagher.
Astrophys. J., Vol. 221, 1110 - 1123 (1978).

Differential energy spectra of primary cosmic-ray helium and hydrogen nuclei in the energy range 123–279 MeV per nucleon were measured on a balloon flight from Thompson, Manitoba, in 1973 August during a time of relatively quiet interplanetary conditions. This represents the first reported ^{3}He–^{4}He measurement in this energy range since 1966. The observed average ^{3}He/(^{3}He + ^{4}He) ratio in the modulated spectra was found to be 0.105 ± 0.012, which is larger than the average value of ~0.05 measured by satellites below 100 MeV per nucleon during the same time period, indicating that the so-called anomalous ^{4}He component does not contribute significantly to the spectrum in the energy range covered by this experiment. The interstellar ^{3}He/(^{3}He + ^{4}He) ratio, when corrected for solar modulation by using up-to-date models, lies between 0.11 and 0.29.

143.017 Galactic cosmic rays in a finite solar cavity.
G. M. Webb, L. J. Gleeson.
Proc. Astron. Soc. Australia, Vol. 3, 162 - 164 (1977).

The authors model mathematically the propagation of galactic cosmic-rays in the solar cavity and study the effects of changing the physical parameters; in particular the radius of the cavity.

143.018 Astronomia con particelle di alta energia: i raggi cosmici.   N. Mandolesi, G. G. C. Palumbo.
Coelum, Vol. 46, 1 - 10 (1978).

143.019 Cosmic-ray streaming perpendicular to the mean magnetic field. II: The gyrophase distribution function.   M. A. Forman, J. R. Jokipii.
Astrophys. Space Sci., Vol. 53, 507 - 513 (1978).

143.020 The role of active regions and of the general magnetic field of the sun in the 11-year cycle of cosmic rays.
T. N. Charakhch'yan.

Izv. AN SSSR. Ser. fiz., Vol. 41, 1746 - 1756 (1977). In Russian. – Abstr. in Ref. zh., 51. Astron., 2.51.423 (1978).

143.021 Nature and origin of cosmic radiation: history and presence.   V. L. Ginzburg, I. V. Dorman.
Priroda, 1978, No. 4, p. 10 - 29. In Russian.

143.022 Influence of the total magnetic field of the sun on modulation of cosmic rays in the stratosphere at mean latitudes.   T. N. Charakhch'yan.
Geomagn. Aehron., Vol. 18, 193 - 202 (1978). In Russian.

143.023 On the sources of ultra-high energy cosmic rays.
G. Cavallo.
Astron. Astrophys., Vol. 65, 415 - 419 (1978).

The acceleration mechanisms for ultra-high energy cosmic rays ($E \geq 10^{20}$ eV) are discussed in relation to the problem of the sources. No one-step mechanism among those which have been studied in detail seems to be capable of accelerating particles to these energies. On the other hand, multistep mechanisms are subjected to size constraints which make it unlikely that these sources are located within the Galaxy. Photomeson production losses, however, confine the sources to the neighborhood of our Galaxy, and the most likely candidate which satisfies the size constraints and the energy requirements appears to be Cen A.

143.024 Quiet time interplanetary cosmic ray anisotropies observed from Pioneer 10 and 11.
W.-H. Ip, W. Fillius, A. Mogro-Campero, L. J. Gleeson, W. I. Axford.
J. Geophys. Res., Vol. 83, 1633 - 1640 (1978).

The authors have made measurements of cosmic ray anisotropies in interplanetary space, using the Pioneer 11 and the Pioneer 10 detector. The east-west and the north-south anisotropy have been determined. A comparison of the anisotropy and magnetic field data suggests that a north-south anisotropy could be due at least in part to the gradient drift effect and perhaps in part to an additional streaming independent of the magnetic field polarity.

143.025 On the propagation of cosmic rays in the Galaxy.
J. Ormes, P. Freier.
Astrophys. J., Vol. 222, 471 - 483 (1978).

The details of the leaky-box model for cosmic-ray propagation are presented. The nuclear composition data indicate that the propagation is rigidity dependent. Consequences of the model at energies above those currently observed and for the rarer components of the radiation are examined. The model predicts that the spectra of $Z > 30$ nuclei will most closely reflect the source spectrum.

143.026 Average abundances of galactic cosmic rays with $Z > 50$ from studies of meteoritic olivines.
V. P. Perelygin, S. G. Stetsenko, D. Lhagvasuren, O. Otgonsuren, P. Pellas, B. Jakupi.
Obedin. inst. yader. issled. Prepr. E7–10667. Dubna, 1977. 11 pp. In Russian. – Abstr. in Ref. zh., 51. Astron., 3.51.388 (1978).

143.027 Construction and application of the ASK precision apparatus for automatic continuous recording of the μ-meson component of cosmic rays, organization of an All-Union network of cosmic ray stations, experimental investigations, application of the method of calculating meteorological effects of cosmic rays and obtaining first characteristics of cosmic ray variations of extraterrestrial origin.
S. N. Vernov, Yu. G. Shafer, G. V. Shafer, A. I. Kuz'min.
Fundament. issled. Fiz.-mat. i tekh. nauk. Novosibirsk, Nauka, 1977, p. 129 - 133. In Russian. – Abstr. in Ref. zh., 62. Issled. kosm. prostranstva, 3.62.213 (1978).

143.028 Astronomia con particelle di alta energia: i raggi
cosmici.   N. Mandolesi, G. G. C. Palumbo.
Coelum, Vol. 46, 63 - 71 (1978).

143.029 The history of cosmic rays in a dynamical halo:
a retrodictive probability approach.   F. C. Jones.
Astrophys. J., Vol. 222, 1097 - 1103 (1978).
    The author examines a particular model of cosmic-ray
propagation in the Galaxy that contains convective outflow
in the halo. When this convection is important, the solutions
for the cosmic-ray density are quite different from those obtained by considering diffusion alone. The author derives an
equation for propagating, backward in time, the particle population that is in the disk at the present time and finds that,
unlike the bulk of the cosmic rays, these particles have never
been very far from the disk.

143.030 Ultrahigh-energy neutrino emission from cosmic
ray sources.   D. Eichler.
Astrophys. J., Vol. 222, 1109 - 1113 (1978).
    It is shown that sources of cosmic rays may emit ultra-
high-energy ($E_\nu > 10^2$ GeV) neutrinos at levels detectable by
DUMAND, depending on how much target material is near the
source. In particular, if cosmic ray sources are in dense clouds,
they can be discrete neutrino sources. The time dependence
of any positive signal could provide information about the
acceleration mechanism.

143.031 Pulsars and cosmic rays in the dense supernova shells.
V. S. Berezinsky (*Berezinskij*), O. F. Prilutsky.
Astron. Astrophys., Vol. 66, 325 - 334 (1978).
    Cosmic rays injected by a young pulsar in a dense super-
nova shell are considered. The fluxes of gamma and neutrino
radiation generated through decays of pions in an expanding
shell are calculated.

143.032 Cosmic ray effects on 4 - 5 August 1972 according to
Prognoz 2 measurements.   N. N. Volodichev,
N. L. Grigorov, G. Ya. Kolesov, E. I. Morozova, A. N.
Podorol'skij, I. A. Savenko, A. A. Suslov.
Problems of solar activity and space system Prognoz,
(see 003.008), p. 136 - 143 (1977). In Russian. − Abstr. in
Ref. zh., 51. Astron., 4.51.425; 62. Issled. kosm. prostranstva,
4.62.133 (1978).

143.033 On the theory of the 27-day modulation of galactic
cosmic rays.   B. D. Naskidashvili, L. Kh.
Shatashvili.
Geofiz. sb. AN USSR, 1977, vyp. (No.) 79, p. 90 - 93. In
Russian. − Abstr. in Ref. zh., 51. Astron., 4.51.429 (1978).

143.034 Influence of an electric field of unipolar induction
on the distribution function of cosmic radiation in
the interplanetary space.
V. M. Dvornikov, Yu. G. Matyukhin.
Issled. po geomagn., aehron. i fiz. Solntsa. Moskva, Nauka,
1977, p. 99 - 103. In Russian. − Abstr. in Ref. zh., 62.
Issled. kosm. prostranstva, 4.62.408 (1978).

143.035 On the asymmetry effect of cosmic radiation in high-
latitude zones of the earth's magnetosphere.
N. K. Pereyaslova, M. N. Nazarova, I. E. Petrenko, S. I.
Avdyushin, Yu. M. Kulagin.
Dokl. AN SSSR, Vol. 237, 288 - 290 (1977). In Russian.
Abstr. in Ref. zh., 62. Issled. kosm. prostranstva, 4.62.412
(1978).

143.036 Similarity solutions of the steady state cosmic-ray
equation of transport.   G. M. Webb.
J. Australian Math. Soc., Ser. B, Vol. 19, 432 - 451 (1976).
Abstr. in Phys. Abstr., Vol. 81, Abstr. 16254 (1978).

143.037 Relative abundances of low energy cosmic ray nuclei
of $Z \geqslant 7$ using plastic detectors.
G. S. Kainth, V. S. Bhatia, S. Paruthi.
Indian J. Radio Space Phys., Vol. 6, No. 2, p. 132 - 135 (1977).
Abstr. in Phys. Abstr., Vol. 81, Abstr. 16260 (1978).

143.038 Dosimetric significance of cosmic radiation in the
altitude of SST and in free space.
O. C. Allkofer.
Radiat. Eff., Vol. 34, No. 1 - 3, p. 113 - 122 (1977). − Abstr.
in Phys. Abstr., Vol. 81, Abstr. 20295 (1978).

143.039 Spectral shape of cosmic rays over the Galaxy.
M. Giler, J. Wdowczyk, A. W. Wolfendale.
J. Phys. A, Vol. 11, 199 - 208 (1978). − Abstr. in Phys. Abstr.,
Vol. 81, Abstr. 20589 (1978).

143.040 Propagation of cosmic rays of energy 1–100 GeV in
the Galaxy.
M. Giler, J. Wdowczyk, A. W. Wolfendale.
J. Phys. A, Vol. 11, 209 - 219 (1978). − Abstr. in Phys. Abstr.,
Vol. 81, Abstr. 20590 (1978).

143.041 Ultra heavy cosmic ray nuclei—analysis and results.
P. H. Fowler.
Nucl. Instrum. Methods, Vol. 147, No. 1, p. 183 - 194 (1977).
Abstr. in Phys. Abstr., Vol. 81, Abstr. 20596 (1978).

143.042 High resolution study of nucleonic cosmic rays with
$Z > 34$.   P. H. Fowler, C. Alexandre, V. M.
Clapham, D. L. Henshaw, C. O'Ceallaigh, D. O'Sullivan,
A. Thompson.
Nucl. Instrum. Methods, Vol. 147, No. 1, p. 195 - 199 (1977).
Abstr. in Phys. Abstr., Vol. 81, Abstr. 20597 (1978).

143.043 Measurement of the cosmic ray element abundances
between $\simeq 300$ and $\simeq 750$ MeV/N in the region from
nickel to krypton using Lexan track detectors.
P. H. Fowler, D. L. Henshaw, C. O'Ceallaigh, D. O'Sullivan,
A. Thompson.
Nucl. Instrum. Methods, Vol. 147, No. 1, p. 201 - 203 (1977).
Abstr. in Phys. Abstr., Vol. 81, Abstr. 20598 (1978).

143.044 Isotope composition of cosmic ray nuclei.
W. Enge.
Nucl. Instrum. Methods, Vol. 147, No. 1, p. 211 - 220 (1977).
Abstr. in Phys. Abstr., Vol. 81, Abstr. 20599 (1978).

143.045 Lunar and meteoritic mineral track detectors and
the composition of the galactic cosmic radiation.
W. Krätschmer.
Nucl. Instrum. Methods, Vol. 147, No. 1, p. 205 - 209 (1977).
Abstr. in Phys. Abstr., Vol. 81, Abstr. 20603 (1978).

143.046 On particle streams with $Z \geqslant 2$ as measured aboard
AES Cosmos 490.   R. N. Basilova, L. F. Kalinkin,
E. I. Kogan-Laskina, G. I. Pugacheva, I. A. Savenko.
Kosm. Issled., Vol. 16, 457 - 459 (1978). In Russian.

143.047 Calculation of LET-spectra of heavy cosmic ray
nuclei at various absorber depths.   W. Heinrich.
Radiat. Eff., Vol. 34, No. 1 - 3, p. 143 - 148 (1977). − Abstr.
in Phys. Abstr., Vol. 81, Abstr. 20296 (1978).

143.048 The charge composition and energy spectra of
cosmic-ray nuclei from 3000 MeV per nucleon to
50 GeV per nucleon.   J. A. Lezniak, W. R. Webber.
Astrophys. J., Vol. 223, 676 - 696 (1978).
    The authors present new data on the energy spectrum
and charge composition of cosmic-ray nuclei with
$3 \leq Z \leq 28$.

143.049 Eleven-year cycle of cosmic rays in the stratosphere (review). T. N. Charakhch'yan.
Geomagn. Aehron., Vol. 18, 385 - 405 (1978). In Russian.

143.050 Anisotropic diffusion of cosmic rays in the interplanetary space. 5. Solar-daily variations.
L. I. Dorman, N. P. Milovidova.
Geomagn. Aehron., Vol. 18, 406 - 409 (1978). In Russian.

143.051 Les rayons cosmiques. J. Vandermeulen.
Bull. Soc. Astron. Liège, Vol. 40, 11 - 16, 40 - 42 (1978).

143.052 Relative abundance of antiprotons and antihelium in the primary cosmic radiation.
G. D. Badhwar, R. L. Golden, J. L. Lacy, J. E. Zipse, R. R. Daniel, S. A. Stephens.
Nature, Vol. 274, 137 - 139 (1978).
The importance of determining the abundance of antiparticles in the primary cosmic rays has already been discussed (Steigmann, 1977). The authors report here on the measurement of $\bar{p}/p$ and $\overline{He}/He$ in the 4–100 GeV/c range made from a balloon flight in May 1976 of a superconducting magnet spectrometer from Palestine, Texas, under 5.8 g cm^{-2} of residual atmosphere. The upper limits derived are the most stringent to date. The upper limit on $\overline{He}/He$ suggests that the local group of galaxies and possibly the Virgo supercluster contain predominantly normal matter.

143.053 Galactic cosmic radiation in the interior regions of the magnetosphere.
V. V. Suvorov, M. V. Tel'tsov.
Vestn. Mosk. univ. Fiz., astron., Vol. 18, No. 3, p. 41 - 46 (1977). In Russian. – Abstr. in Ref. zh., 62. Issled. kosm. prostranstva, 5.62.266 (1978).

143.054 On the choice of boundary conditions in the approximate solution of boundary value problems describing the diffusion of galactic cosmic rays.
M. V. Alaniya, M. A. Aleksidze, L. I. Dorman.
Soobshch. AN GruzSSR, Vol. 88, No. 1, p. 69 - 72 (1977). In Russian. – Abstr. in Ref. zh., 62. Issled. kosm. prostranstva, 5.62.272 (1978).

143.055 The highest-energy cosmic rays. J. Linsley.
Sci. American, Vol. 239, No. 1, p. 48 - 58 (1978).

143.056 Energy loss in the solar system and modulation of cosmic radiation. J. Kota.
Hungarian Acad. Sci., Budapest. Report KFKI-1977-52, 9 pp. (1977). – Abstr. in Phys. Abstr., Vol. 81, Abstr. 32308 (1978).

143.057 Long-term averaged abundances of VVH cosmic ray nuclei from studies of olivines from Marjalahti meteorite. V. P. Perelygin, S. G. Stetsenko, P. Pellas, D. Lhagvasuren, O. Otgonsuren, B. Jakupi.
Nucl. Track Detect., Vol. 1, No. 3 - 4, p. 199 - 205 (1977). Abstr. in Phys. Abstr., Vol. 81, Abstr. 36414 (1978).

143.058 Measurement of the Li abundance in cosmic rays. B. Byrnak.
Nuovo Cimento A, Ser. 11, Vol. 44A, No. 1, p. 1 - 12 (1978). Abstr. in Phys. Abstr., Vol. 81, Abstr. 40785 (1978).

143.059 Study of the north-south asymmetry of cosmic rays. A. Kh. Bychkovskaya, A. G. Zusmanovich, E. V. Kolomeets.
Prikl. i teor. fiz., 1977, No. 9, p. 148 - 153. In Russian. Abstr. in Ref. zh., 51. Astron., 6.51.515 (1978).

143.060 Cosmic ray diffusion in the solar wind with nonspherical boundary. S. F. Nosov.
Cosmophysical investigations, Sverdlovsk, 1977, (see 003.016), p. 31 - 34. In Russian. – Abstr. in Ref. zh., 62. Issled. kosm. prostranstva, 6.62.251 (1978).

143.061 On the origin and propagation of ultra high energy cosmic rays. P. Kiraly.
15th IUPAP International Conference on cosmic rays, (see 012.042), p. 53 - 60 (1977). – Abstr. in Phys. Abstr., Vol. 81, Abstr. 49242 (1978).

143.062 Energy loss in the solar system and modulation of cosmic radiation. J. Kota.
15th IUPAP International Conference on cosmic rays, (see 012.042), p. 73 - 83 (1977). – Abstr. in Phys. Abstr., Vol. 81, Abstr. 49245 (1978).

143.063 Further evidences of the anisotropy observed at Musala station. T. Gombosi, J. Kota, A. J. Somogyi, A. Varga, B. Betev, L. Katsarski (Katsarskij), S. (Sh.) Kavlakov, I. Khirov.
15th IUPAP International Conference on cosmic rays, (see 012.042), p. 9 - 15 (1977). – Abstr. in Phys. Abstr., Vol. 81, Abstr. 49253 (1978).

143.064 Fluctuations of ~10^{14} eV cosmic rays.
G. Erdos, T. Gombosi, J. Kota, A. J. Owens, A. J. Somogyi, A. Varga.
15th IUPAP International Conference on cosmic rays, (see 012.042), p. 17 - 27 (1977). – Abstr. in Phys. Abstr., Vol. 81, Abstr. 49254 (1978).

143.065 Some thoughts on the Musala anisotropy: pitch angle distribution or what else?
J. Kota, A. J. Somogyi.
15th IUPAP International Conference on cosmic rays, (see 012.042), p. 29 - 38 (1977). – Abstr. in Phys. Abstr., Vol. 81, Abstr. 49255 (1978).

143.066 Analysis of data obtained with fast-timing angular distribution measurements.
G. Erdos, A. J. Somogyi, A. Varga.
15th IUPAP International Conference on cosmic rays, (see 012.042), p. 39 - 52 (1977). – Abstr. in Phys. Abstr., Vol. 81, Abstr. 49256 (1978).

143.067 Partially averaged field approach to cosmic ray diffusion.
F. C. Jones, T. J. Birmingham, T. B. Kaiser.
Phys. Fluids, Vol. 21, 347 - 360 (1978). – Abstr. in Phys. Abstr., Vol. 81, Abstr. 53776 (1978).

143.068 Computer simulation of the velocity diffusion of cosmic rays.
T. B. Kaiser, T. J. Birmingham, F. C. Jones.
Phys. Fluids, Vol. 21, 361 - 373 (1978). – Abstr. in Phys. Abstr., Vol. 81, Abstr. 53777 (1978).

143.069 Cosmic pion spectrum at the top of the atmosphere. K. Sarkar, D. P. Bhattacharyya, D. Basu.
Indian J. Phys. A, Vol. 51A, No. 4, p. 231 - 236 (1977). Abstr. in Phys. Abstr., Vol. 81, Abstr. 53780 (1978).

143.070 Determination of the diffusion coefficient of cosmic rays in the interplanetary medium on the basis of Forbush decreases 1967 - 1972.
S. I. Ayubasheva, E. Ya. Gidalevich.
Cosmophysical investigations, Sverdlovsk, 1977, (see 003.016). In Russian. – Abstr. in Ref. zh., 62. Issled. kosm. prostranstva, 7.62.186 (1978).

143.071 Physics of high energies and of cosmic rays.

R. U. Bejsembaev, Yu. N. Vavilov, A. G. Dubovyj.
Fiz. inst. AN SSSR, Prepr., 1977, No. 144. In Russian.
Abstr. in Ref. zh., 62. Issled. kosm. prostranstva, 7.62.187
(1978).

143.072   On the nature of sidereal-daily variations of cosmic
          rays in the energy region below 100 BeV.
E. V. Kolomeets, I. P. Leongard, L. A. Mirkin, R. A. Chumba-
lova.
Fiz. atom. yadra i kosm. luchej, 1977, p. 24 - 26. In Russian.
Abstr. in Ref. zh., 62. Issled. kosm. prostranstva, 7.62.191
(1978).

Calibration of two plastic detectors and application
on study of heavy cosmic rays.   See Abstr. 031.256.

Ultrahigh-energy neutrino astronomy.
See Abstr. 061.031.

Revealing cyclicity in changes of cosmic ray in -
tensity and in solar activity in the past.
See Abstr. 072.020.

Parameters of Forbush decreases and their parent
flares in the solar cycle 1965 - 1976.

See Abstr. 073.004.

Phase reversals in the polar magnetic fields of the
sun and in the annual and semiannual variations in cosmic ray
intensity.   See Abstr. 080.012.

Estimate of the rate of ^{37}Ar production by cosmic
rays neutrinos evaluated from the solar neutrino detection
experiment.   See Abstr. 080.033.

Investigations on cosmic-ray-produced nuclides in
iron meteorites. 1. The measurement and interpretation of
rare gas concentrations.   See Abstr. 105.106.

The early cooling and escape of SN shock accelerated cosmic rays.   See Abstr. 125.023.

Electromagnetic processes in a pulsar's magnetosphere and acceleration of cosmic rays.   See Abstr. 141.536.

Cosmic gamma rays.   See Abstr. 142.717.

Galactic γ-ray spectra, the flux of cosmic ray electrons and cosmic ray gradients.   See Abstr. 157.004.

Stellar Systems

151 Kinematics and Dynamics of Stellar Systems, Evolution of Galaxies

151.001 Star formation rates in normal and peculiar galaxies.
R. B. Larson, B. M. Tinsley.
Astrophys. J., Vol. 219, 46 - 59 (1978).

The authors have constructed an extensive grid of galaxy models with decreasing star formation rates (SFRs) and with bursts on various time scales. Normal galaxies have colors that are consistent with a monotonically decreasing SFR, and very few can have experienced large variations in SFR with time scales $\lesssim 5 \times 10^8$ yr. In contrast, the peculiar galaxies have a large scatter in colors that is consistent with bursts as short as 2×10^7 yr involving up to ~5% of the total mass. Nearly all of this scatter is associated with galaxies showing evidence of tidal interaction. The results provide evidence for a "burst" mode of star formation associated with violent dynamical phenomena.

151.002 Galactic models with variable spiral structure.
R. A. James, J. A. Sellwood.
Mon. Not. R. Astron. Soc., Vol. 182, 331 - 344 (1978).

The authors have run a series of three-dimensional computer simulations of disc galaxies, in which the self-consistent potential of the disc stars is supplemented by that arising from a small uniform Population II sphere. The models show variable spiral structure, which is more pronounced for thin discs. In addition, the thin discs form weak bars. In one case there is a variable spiral structure associated with this bar. The relaxed discs are cool outside resonance regions.

151.003 A non-linear theory of spiral density waves.
C. A. Norman.
Mon. Not. R. Astron. Soc., Vol. 182, 457 - 472 (1978).

A non-linear theory of spiral density waves is developed using the tight winding approximation. When the amplitude of the wave exceeds a certain critical value the wave satisfies a non-linear dispersion relation. An exact solution to this system is found using the technique of inverse scattering. Contrary to the case for linear waves, the non-linear waves are shown to be spirals which do not tend to wind up.

151.004 Dynamical evolution of elliptical protogalaxies.
A. Di Fazio, F. Occhionero.
Astron. Astrophys., Vol. 62, 349 - 354 (1978).

Generalizing a suggestion by Gott and Thuan (1976), the authors idealize elliptical protogalaxies as pressure supported Maclaurin spheroids of pure stars and they study their nondissipative evolution by the numerical integration of the appropriate differential equations for the two semiaxes. The authors show that the finite amplitude oscillations of the structure damp out generally in a few collapse time scales, leaving a relaxed configuration. The present considerations indicate that the oscillations in the equatorial plane damp out faster, in all cases, than the oscillations along the polar axis, which ought to be important in the chemical mixing of the

protogalaxy. Secondly, since the final ellipticity is not a monotonically increasing function of angular momentum, protogalaxies of high angular momentum assume formally, after relaxation, a low ellipticity configuration.

151.005 Chemische Entwicklung von Populationen in Galaxien.
C. Trefzger.
Sterne Weltraum, Jahrg. 17, 14 - 18 (1978).

151.006 Close binaries and evolution of star clusters.
V. I. Dokuchaev, L. M. Ozernoj.
Astron. Zh. Akad. Nauk SSSR, Vol. 55, 27 - 36 (1978). In Russian. English translation in Soviet Astron., Vol. 22, No. 1.

Formation of "hard" pairs (i.e. pairs which have binding energy greater than the mean kinetic energy of a cluster star) inside a star cluster may drastically influence the evolution of that cluster. The authors show that, if at final stages of secular contraction of the cluster's core (i.e. approximately during 40 relaxation times) the number of hard pairs will be comparable with the total number of stars, then exchange of energy between these pairs and the remainder of the cluster is able to provide the transition from contraction of the core to its expansion. The formation of the necessary number of pairs in systems with large number of stars occurs in the most effective way not at triple encounters but by means of double encounters of stars accompanied by tidal effects. The picture developed is discussed in application to the nucleus of our Galaxy and to globular clusters.

151.007 Rotationally symmetric oscillations of a gravitating ellipsoid of rotation having a corona.
S. N. Nuritdinov.
Astron. Zh. Akad. Nauk SSSR, Vol. 55, 37 - 40 (1978). In Russian. English translation in Soviet Astron., Vol. 22, No. 1.

The effect of a spherical corona on rotationally symmetric oscillations of an inner ellipsoidal subsystem is studied. The marginal dependence of the ratio of the corona density to that of the ellipsoid upon the eccentricity of the latter is found.

151.008 On the hydrodynamics of steady-state encounterless spherical systems. L. P. Osipkov.
Astron. Zh. Akad. Nauk SSSR, Vol. 55, 41 - 46 (1978). In Russian. English translation in Soviet Astron., Vol. 22, No. 1.

Applicability of hydrodynamic equations of arbitrary order to stellar dynamics is discussed. The chain of momental equations for steady-state spherical systems is investigated. It is proved that the absence of an excess in the velocity distribution leads to well-known relations between velocity dispersions obtained in ellipsoidal dynamics.

151.009 The structure of the field of motion directions and

the forms of the potential.
T. A. Agekyan, M. G. Saginashvili.
Astron. Zh. Akad. Nauk SSSR, Vol. 55, 47 - 55 (1978). In
Russian. English translation in Soviet Astron., Vol. 22, No. 1.

A new particular solution is found for the system of
partial differential equations connecting the potential with the
functions which determine the gradient of the field of motion
directions normal to the trajectory. The equations of orbital
trajectories and equations for stable periodic orbits are obtained for the family of potentials found.

151.010   La dynamique des galaxies spirales.
L. Weliachew.
La recherche en astrophysique, (see 003.001), p. 129 - 149
(1977).

**151.011   Star escape from isolated clusters. Part I: The
so-called classical theory.   T. Kwast.**
Postępy Astron., Tom 25, 3 - 13 (1977). In Polish.

Some theories of stellar escape from clusters are presented. These theories base on assumptions related to cluster
structure, stellar velocity distribution and escape mechanism.
It turns out that the rate of star loss from clusters is mostly
affected by the mechanism of escape. This mechanism may
consist in a cumulative effect of many distant encounters or
in few close encounters. This second eventuality seems more
realistic.

**151.012   Star escape from isolated clusters. Part III: Generali-
zation of the Fokker-Planck equation.**
T. Kwast.
Postępy Astron., Tom 25, 169 - 175 (1977). In Polish.

Methods of a generalization of the Fokker-Planck equation are presented. The generalizations do not affect the order
of magnitude of the star escape rate.

**151.013   Stellar orbits and the persistence of spiral structure
in galaxies.   R. Frahm, K. O. Thielheim.**
Astrophys. J., Lett., Vol. 220, L43 - L47 (1978).

The persistence of the spiral structure of galaxies is discussed in the context of the position probability of individual
stars. The influence of the spiral-shaped perturbation to the
axially symmetric background potential on this position
probability is considered for our Galaxy in some special cases
as a function of the assumed rotational velocity of the spiral
pattern. The authors find that, for given background and spiral
potential, such a mechanism supporting the persistence of the
spiral arms exists only in certain regions of rotational velocity
of the spiral pattern.

**151.014   Thermodynamic stability of rotating gaseous
cylinders. I. Stability analysis by linear series.**
S. Inagaki, I. Hachisu.
Publ. Astron. Soc. Japan, Vol. 30, 39 - 55 (1978).

The effects of rotation are incorporated in the thermodynamic stability of self-gravitating systems. Unperturbed
states are assumed to be isothermal and uniformly rotating
gaseous cylinders. The cylinders are immersed in an external
medium. Eight kinds of boundary conditions are considered
for the interaction between the cylinder and the external
medium. It is shown that there exists a linear series appropriate to study stability for each case. The stability analyses
show that certain equilibrium configurations are secularly unstable.

151.015   On the gravothermal catastrophe.   Y. Nakada.
Publ. Astron. Soc. Japan, Vol. 30, 57 - 66 (1978).

Gravothermal instability is examined for a collision-
frequent system. The computation of the linearized perturbation equation shows that "gravothermal" instability takes
place when the central concentration exceeds Antonov's
(1962) critical value. The time scale and other features of

instability are also investigated.

151.016   Negative mass instability of flat galaxies.
R. V. E. Lovelace, R. G. Hohlfeld.
Astrophys. J., Vol. 221, 51 - 61 (1978).

Although considerable progress has been made in understanding the wave modes and instabilities of rotationally
flattened galaxies, many important questions remain unan-
swered. The present work discusses a new resonant instability
(Lovelace and Hohlfeld 1976) of a fluid disk model of flat
galaxies. The authors propose that this instability may act to
bring about a marginally stable state. For this marginal state
there is a unique self-similar solution which has a flat rotation
curve.

**151.017   Resonant stellar orbits in spiral galaxies. V. Numeri-
cal studies of orbits in the region of the inner
Lindblad resonance.   D. G. Monet, P. O. Vandervoort.**
Astrophys. J., Vol. 221, 87 - 104 (1978).

The epicyclic approximation for the orbits of stars in the
region of the inner Lindblad resonance of a spiral galaxy is
tested by comparing surfaces of section constructed from
numerical solutions of the exact equations of motion with
those constructed from numerical solutions of the epicyclic
equations. The second part of the paper concerns the population of the two families of tube orbits in the region of the
inner Lindblad resonance with stars. It is explained that special
interest attaches to two particular prescriptions for populating
the resonant orbits; in each prescription, one family of tube
orbits is populated with a relative excess of stars and the other
family with a relative deficiency of stars. The distribution of
stars in the four-dimensional phase space consistent with these
prescriptions is described with the aid of the surfaces of section constructed from the numerical solutions of the exact
equations of motion. Finally, it is explained how, under certain
conditions, considerations of self-consistency in the equilibrium of a spiral galaxy favor the population of the family of
tube orbits called $R2$ by Contopoulos with a relative excess
of stars and the family of tube orbits called $R1$ with a relative
deficiency of stars.

151.018   A model for winds from galactic disks.
J. M. Bardeen, B. K. Berger.
Astrophys. J., Vol. 221, 105 - 113 (1978).

A hydrodynamic model is constructed to represent a
wind originating in the disk of a spiral galaxy. The gas flow is
assumed to be steady, nondissipative, axisymmetric, and self-
similar, scaling in radius. Various heating processes are represented by a pressure-density relation which is isothermal near
the disk and adiabatic far from the disk. Wind solutions are
found for a range of values of initial parameters in the disk. In
a regime where the self-similarity might be expected to be valid,
scaling parameters can be chosen which are reasonable for a
large spiral galaxy with a hot, tenuous corona.

151.019   Bar-driven spiral waves in disk galaxies.
J. M. Huntley, R. H. Sanders, W. W. Roberts, Jr.
Astrophys. J., Vol. 221, 521 - 538 (1978).

The authors investigate the response of rotating disks of
gas to barlike perturbations in galactic gravitational fields. In
particular, two-dimensional, time-dependent, numerical hydrodynamical calculations have been performed in order to determine the steady-state response of disks of gas to rotating, barlike perturbations. Two types of barlike perturbations are considered here: oval distortions in the axisymmetric gravitational
field of the disk, and heterogeneous prolate spheroids.

**151.020   Resonant stellar orbits in spiral galaxies. VI. The
equilibrium of a stellar disk in the regions of the
Lindblad resonances.   P. O. Vandervoort.**
Astrophys. J., Vol. 221, 539 - 553 (1978).

The epicyclic theory of resonant stellar orbits developed

in earlier papers in this series is incorporated into a representation of the equilibrium of a spiral galaxy in the regions of the Lindblad resonances. This is accomplished by generalizing the epicyclic theory of the equilibrium of an axisymmetric stellar disk. In the axisymmetric case, the distribution function giving the density of stars in the four-dimensional phase space is a function of the well-known epicyclic integrals: the radius of the circular orbit to which the epicyclic motion is referred and the energy of the epicyclic motion. The present generalization consists of writing the distribution function as the same function of perturbed epicyclic integrals. Thus the paper deals mainly with the construction of suitable forms of the perturbed integrals.

151.021 **Chemical evolution and the formation of galactic disks.** B. M. Tinsley, R. B. Larson.
Astrophys. J., Vol. 221, 554 - 561 (1978).
The chemical evolution of two collapse models for the formation of disk galaxies is calculated in detail, and the results are compared with properties of the authors' and other spiral galaxies. The models show at least qualitative agreement with empirical stellar and interstellar abundance gradients and with color gradients in spiral galaxies. The outer parts of the model disks are also in general agreement with the metallicity and age distributions of stars in the solar neighborhood, and with correlations between metallicity and kinematics for nearby stars.

151.022 **An approximate theory for the core collapse of two-component gravitating systems.**
A. P. Lightman, S. M. Fall.
Astrophys. J., Vol. 221, 567 - 579 (1978).
An approximate theory is proposed for the dynamical evolution of the central regions of an isolated, spherical star cluster containing stars of two different masses. A characteristic result is that the late stages of core collapse are driven by self-interactions of heavy particles within an inner, self-gravitating subsystem, with the early stages driven by conductive heat transfer to the light particles. Approximate formulae for different regimes of evolution are derived as functions of the initial mass ratios. Also considered is the endpoint evolution of the dense inner core of heavy stars, when binary formation and dissipative processes associated with the finite size of stars (tidal interactions and stellar coalescence) become important. For parameters appropriate to globular clusters with solar type stars, the analysis suggests that the core evaporates down to $\lesssim 100$ stars before the above processes dominate the final evolution.

151.023 **On star formation and galactic evolution.**
M. Kaufman.
Bull. American Astron. Soc., Vol. 9, 566 - 567 (1978).
Abstract.

151.024 **Cluster winds with central accretion.**
R. H. Durisen, J. O. Burns.
Bull. American Astron. Soc., Vol. 9, 616 (1978). – Abstract.

151.025 **Galaxy collisions by three-dimensional N-body integrations.** B. F. Smith, R. H. Miller.
Bull. American Astron. Soc., Vol. 9, 622 (1978). – Abstract.

151.026 **Tidal interactions in close galactic encounters.**
R. C. Reynolds.
Bull. American Astron. Soc., Vol. 9, 622 (1978). – Abstract.

151.027 **Two dimensional calculation of gas flow in barred spiral galaxies.** L. S. Liebovitch, C. C. Lin.
Bull. American Astron. Soc., Vol. 9, 639 (1978). – Abstract.

151.028 **Bar-driven spiral waves in disk galaxies.**
J. M. Huntley, R. H. Sanders, W. W. Roberts, Jr.

Bull. American Astron. Soc., Vol. 9, 639 - 640 (1978).
Abstract.

151.029 **Hydrodynamic simulations of white dwarf – main-sequence star collisions.** M. M. Shara, G. Shaviv.
Bull. American Astron. Soc., Vol. 9, 640 (1978). – Abstract.

151.030 **Computer experiments on the effect of retrograde stars.** T. A. Zang, F. Hohl.
Bull. American Astron. Soc., Vol. 9, 640 (1978). – Abstract.

151.031 **The equilibrium of a spiral galaxy in the region of the inner Lindblad resonance.** P. O. Vandervoort.
Bull. American Astron. Soc., Vol. 9, 640 (1978). – Abstract.

151.032 **Instability of galactic boundary layer flows.**
A. M. Waxman.
Bull. American Astron. Soc., Vol. 9, 640 - 641 (1978).
Abstract.

151.033 **A study of planar stellar orbits in a spiral potential.**
C. L. Berry, D. J. de Smet.
Bull. American Astron. Soc., Vol. 9, 641 (1978). – Abstract.

151.034 **Stochastic star formation and spiral structure of galaxies.** P. E. Seiden, H. Gerola.
Bull. American Astron. Soc., Vol. 9, 641 (1978). – Abstract.

151.035 **Collapse and relaxation of rotating stellar systems.**
F. Hohl, C. M. Costner.
Bull. American Astron. Soc., Vol. 9, 641 (1978). – Abstract.

151.036 **Comment on photometric evolutionary corrections.**
J. P. Huchra.
IAU Colloq. No. 37, (see 012.009), p. 179 - 181 (1977).
De récentes études de galaxies très bleues montrent que la formation des étoiles a lieu avec une fonction initiale de masse plus riche en étoiles de masse élevée que la fonction de Salpeter.

151.037 **Galactic evolution and the interpretation of cosmological tests.** B. M. Tinsley.
IAU Colloq. No. 37, (see 012.009), p. 223 - 242 (1977).
Most of the cosmological tests involving distant galaxies seem to be so sensitive to evolution that they cannot provide useful estimates of the parameters defining the cosmological model. This conclusion is not altogether disappointing, for it opens alternative prospects: the data on distant galaxies will help to answer some key questions related to the evolution of stellar populations, the epoch of first star formation in galaxies, and interactions among galaxies in clusters.

151.038 **Possibilities of observing effects due to Lin's density-wave theory.** J. Reiche.
Astron. Nachr., Band 299, 35 - 42 (1978) = Mitt. Univ. Sternw. Jena, No. 130. In German.
After summarizing the most important effects predicted by the density-wave theory the possibilities of observing these effects are discussed. The observation of structural effects in our own Galaxy is very difficult, mainly because of the inaccurate distances available. Only rough structures can be obtained. That is also true for observations near such points in which the line of sight is tangential to the spiral arms. Observations of kinematical effects in our own Galaxy and of structural effects (kinematical ones are not discussed) in extragalactic spiral systems should yield some useful results.

151.039 **Periodic orbits near the particle resonance in galaxies.** G. Contopoulos.
Astron. Astrophys., Vol. 64, 323 - 332 (1978).
Near the particle resonance of a spiral galaxy the almost circular periodic orbits that exist inside the resonance (direct)

or outside it (retrograde) are replaced by elongated trapped orbits around the maxima of the potential L_4 and L_5. These are the long-period trapped periodic orbits. The long-period orbits shrink to the points L_4, L_5 for a critical value of the Hamiltonian h. The evolution of the periodic orbits with h is followed, theoretically and numerically, from the untrapped orbits to the long-periodic orbits and then to the short-periodic orbits, mainly in the case of a bar. Another family of periodic orbits reaching corotation is trapped at the inner Lindblad resonance. A numerical study of this family is given in the case of a bar. In a spiral model of our Galaxy the author founds an orbit of this type reaching the distance of the sun.

151.040 Density wave theory for spiral galaxies: the regime of finite spiral arm inclination in stellar dynamics.
G. Bertin, J. W -K. Mark.
Astron. Astrophys., Vol. 64, 389 - 397 (1978).

The authors have studied the behaviour of spiral density waves in the regime where effects of finite inclination of spiral arms are important. Provided that the disk stars have moderately small epicycles, an appropriate asymptotic solution can still be obtained for the stellar dynamic equations governing wave phenomena in thin disk galaxies. The major new astrophysical effect of this regime is an enhanced amplification of spiral density waves near their region of corotation with the disk stars. The basic mechanism for amplification is still related to an outward transport of angular momentum.

151.041 Boundary layer circulation in disk-halo galaxies.
A. M. Waxman.
Astrophys. J., Vol. 222, 61 - 77 (1978).

This paper concerns some fluid dynamical interactions between the gaseous components of a galactic disk and a surrounding halo. An idealized model of a disk-halo system is constructed which exhibits the main qualitative features of the resulting flow. The author argues that these interactions, when taken in concert, give rise to a meridional circulation extending to several scale heights above the galactic plane. He gives two examples of circulation patterns that are representative of the general phenomenon. He points out some interesting features of the theoretical flow patterns and their implications for systems undergoing disk formation as well as for observed features in our own Galaxy.

151.042 The equilibrium of cool stellar disks.
E. Zweibel.
Astrophys. J., Vol. 222, 103 - 109 (1978).

A new sequence of self-gravitating thin disks of finite radius is constructed. Phase-space moments of the modified Schwarzschild stellar distribution function are shown to have an expansion in powers of the ratio of the star's random kinetic energy to gravitational potential energy.

151.043 Radial modes of oscillation of cool stellar disks.
E. Zweibel.
Astrophys. J., Vol. 222, 110 - 124 (1978).

Equations governing long wavelength radial modes of oscillation for stellar disks are derived. The formulation is exact for disks with zero velocity dispersion and becomes inaccurate for perturbations with wavenumber less than the epicycle size. The effects of differential rotation, halos, and velocity dispersion on axisymmetric stability are studied for some of the disk models constructed in Paper I (see abstract 151.042).

151.044 Entweichgeschwindigkeiten und mittlere Geschwindigkeiten in Stern- und Galaxienhaufen.
W. Lohmann.
Astrophys. Space Sci., Vol. 53, 411 - 413 (1978) = Astron. Rechen–Inst., Heidelberg, Mitt. Ser. A.

The ratio of the escape velocity at the centre to the mean velocity amounts to ≈ 2.5 for open star clusters, ≈ 2.7 for globular clusters, and 2.8 for the Coma cluster of galaxies.

151.045 Density oscillations in a uniformly rotating disk galaxy.   E. Athanassoula.
Astron. Astrophys., Vol. 65, 295 - 297 (1978).

The author studies oscillations of certain models of uniformly rotating stellar disks. The distribution function is assumed to be a function of the Jacobi Integral in a frame of reference in which the perturbation is stationary. A normal-modes analysis is carried out and the bar-modes of the system are calculated explicitly.

151.046 Late stages of evolution of open star clusters.
V. M. Danilov.
Astrofizika, Vol. 13, 685 - 696 (1977). In Russian. English translation in Astrophysics, Vol. 13, No. 4.

Results of J. G. Hills (1975) of numerical experiments for double stars with single stars in open clusters are used to solve the equations of the dynamical evolution of star clusters. Star clusters evolve in an unclassical way at the later stage of evolution. There are several variants of cluster development: contraction, stabilisation and expansion of the cluster. The presence of galactic gas-dust clouds and double stars increases the value of the least critical stellar concentration in open clusters essentially.

151.047 Non-linear travelling waves of stellar density in a model of a homogeneous medium.
S. N. Nuritdinov.
Astrofizika, Vol. 13, 697 - 702 (1977). In Russian. English translation in Astrophysics, Vol. 13, No. 4.

The existence of non-linear travelling waves of stellar density in a model with constant phase density is proved. The dependence of the wavelength from the amplitude in a coordinate system with respect to which the wave is at rest is found. It has been established that with the increase of the phase velocity the region of existence of the non-linear travelling waves becomes narrow. The travelling waves contrary to the stationary ones do not disintegrate into a number of parts.

151.048 On the rotation of elliptical galaxies.
J. Binney.
Mon. Not. R. Astron. Soc., Vol. 183, 501 - 514 (1978).

The tensor virial theorem is applied to elliptical galaxies having similar isophotes. In the absence of residual velocity anisotropy the shape of such a galaxy turns out to determine uniquely the ratio of the rotational and random kinetic energies independently of the galaxy's radial density-profile or the speed with which its form rotates. Curves are displayed of the expected projected 'rotation' velocities of elliptical galaxies under a variety of assumptions about their forms and degrees of residual anisotropy. If ellipticals are all prolate there should be only a weak correlation between apparent ellipticity and rotation velocity. This possibility is consistent with Illingworth's data. A better fit to the observations is provided by anisotropic oblate models. The limitations of these models as well as the implications of galactic rotation for cosmogony are discussed.

151.049 Dissipative processes, galaxy formation and "early" star formation.   M. J. Rees.
Phys. Scr., Vol. 17, (see 012.013), 371 - 376 (1978).

Scenarios are described for the process whereby the primordial material is transformed from gas at $z \simeq 1000$ into bound systems and stars at the present epoch. The alleged "missing mass" could consist of stars which condensed at $z \gtrsim 100$ before galaxies had formed as gravitationally bound units.

151.050 On the origin of proto-galactic eddies.
A. D. Chernin.
Astrophys. Lett., Vol. 19, 97 - 100 (1978).

Irrotational entropy perturbations seem to be a preferable form of proto-galactic structure. Eddies adequate for galactic rotation appear in turbulent layers produced by violent collisions of supersonic matter currents at epoch $Z = 2-7$.

151.051 Photometric and chemical evolution of galaxies.
R. E. Souza, L. Arakaki.
Astron. extragalactica, (see 012.015), 97 - 106 (1976).

Preliminary results on evolutionary models for galaxies are presented taking into account both chemical and photometric properties, and an assumed slow variation of the lower limit-mass in the IMF (initial mass function).

151.052 La formación de galaxias como proceso termodinámico. J. L. Sérsic.
Astron. extragalactica, (see 012.015), 113 - 148 (1976).

A thermodynamic description of the process of collapse and fragmentation of a protogalactic cloud is given. It is shown that the process will be stationary provided the medium is transparent. At that stage dissipative structures are continuously created in order to keep the system in the non-equilibrium stationary state.

151.053 A systematic comparison of four methods to derive stellar space densities. A. M. Spaenhauer.
Astron. Astrophys., Vol. 65, 313 - 321 (1978).

The problem of deriving stellar space densities for stars with a gaussian luminosity function has put forth several numerical methods for solving the fundamental equation of stellar statistics. In this paper four different methods of deriving stellar space densities are statistically tested and compared by means of realistic computer generated space distributions.

151.054 Orbits in a slowly growing spiral field near the corotation resonance in a galaxy.
M. O. Mennessier, L. Martinet.
Astron. Astrophys., Vol. 65, 409 - 414 (1978).

The authors discuss the conditions for the trapping of orbits at the corotation resonance in a spiral galaxy in the case of a slow introduction of the spiral perturbation. The set of initial conditions for phase coordinates leading to trapped orbits does not qualitatively differ very much from those obtained when the perturbation is abruptly imposed. The considerations are useful for the study of the stellar density distribution near the corotation circle in both slow and abrupt cases.

151.055 Spatial structure of protoclusters and formation of galaxies. A. G. Doroshkevich, Eh. M. Saar, S. F. Shandarin.
Inst. prikl. mat. AN SSSR. Prepr. No. 72. Moskva, 1977.
46 pp. Price 15 Kop. In Russian. — Abstr. in Ref. zh., 51. Astron., 3.51.936 (1978).

151.056 Statistical problems of the theory of galaxy formation. A. G. Doroshkevich, S. F. Shandarin.
Inst. prikl. mat. AN SSSR. Prepr. No. 73. Moskva, 1977.
36 pp. Price 12 Kop. In Russian. — Abstr. in Ref. zh., 51. Astron., 3.51.937 (1978).

151.057 On a possibility of stabilization of chains of galaxies by a continuous background.
A. F. Steklov.
Astrometriya i Astrofizika, Kiev, Vyp. (No.) 34, p. 14 - 18 (1978). In Russian.

If physically real chains of galaxies have no dense background, they should be certainly young. Their lifetime is 1.4×10^9 years for model conditions and $(3 - 5) \times 10^8$ years for the observed radial velocities. In fact the spherical non-evolving homogeneous background stabilizes a chain when the ratio of the background mass to the total mass of galaxies

in the chain is > 10.

151.058 Hydrodynamic simulations of white-dwarf — main-sequence star collisions.
M. M. Shara, G. Shaviv.
Mon. Not. R. Astron. Soc., Vol. 183, 687 - 700 (1978).

The authors have performed two-dimensional hydrodynamic simulations of head-on collisions between a 1 $M_\odot$ white dwarf and a 0.13 $M_\odot$ main-sequence star. Temperatures of $3-5 \times 10^8$ K are achieved by most of the main-sequence star mass, and nuclear reactions release over 10^{49} erg in approximately 1 hr. Such collisions should sharply modify the luminosity function, the luminosity, the energy balance and the evolution of dense galactic nuclei.

151.059 Twisted and warped disks as consequences of heavy halos. J. Binney.
Mon. Not. R. Astron. Soc., Vol. 183, 779 - 797 (1978).

Observationally testable consequences of the hypothesis that galactic disks are confined less by their self-gravity than that of some hot stellar component are sought. Theoretical and observational studies are reviewed which suggest that elliptical galaxies are typically tri-axial in form. It is argued that any hot stellar component is likely to be similarly aspherical. The motion of a galactic disk about such a population is calculated. Alternative explanations of galactic warps are reviewed and their relation to the present proposal discussed.

151.060 The excitation and evolution of density waves.
P. Goldreich, S. Tremaine.
Astrophys. J., Vol. 222, 850 - 858 (1978) = Div. Geol. Planet. Sci., California Inst. Technol., Pasadena, Contrib. No. 2958.

The authors study the linear oscillations of a thin self-gravitating gas sheet. The unperturbed velocity field of the sheet is a parallel shear flow. A Coriolis acceleration is included to simulate the effects of rotation. The sheet exhibits Lindblad resonances, and it can sustain both short and long wavelength density waves. The authors derive equations which govern the excitation and evolution of density waves in all regions of space, including the Lindblad resonances and the forbidden region around corotation.

151.061 Steepest descent technique and stellar equilibrium statistical mechanics. III. Stability of various ensembles. G. Horwitz, J. Katz.
Astrophys. J., Vol. 222, 941 - 958 (1978).

The stability of mean fields in stellar systems under various constraints has been analyzed by a steepest descent technique applied to the relevant thermodynamic potential. Necessary and sufficient conditions of stability with respect to arbitrary perturbations about mean field solutions have been obtained. Instabilities with respect to spherically symmetric perturbations are associated with a change of sign of thermodynamic quantities. With a fixed center of mass, the clusters are stable with respect to arbitrary nonspherical perturbations. Numerical results are given.

151.062 The distribution of stars around a massive central black hole in a spherical stellar system. I. Results for test stars with a unique mass and radius. J. R. Ipser.
Astrophys. J., Vol. 222, 976 - 990 (1978).

A massive black hole that forms in a stellar system should quickly seek out the center of the system, due to dynamical friction. This leads to the question of how stars should distribute themselves around a central black hole in a stellar system. A correct prediction of this star distribution is crucial to (1) the interpretation of observational attempts to detect the hole's gravitational influence on the surrounding stars; (2) estimates of the rate at which the hole consumes stars; and (3) an understanding of the hole's influence on the overall dynamics of the stellar system. This paper presents the details of the author's calculation of the steady-state star distribution around a massive black hole lying at the center of a spherical stellar system.

151.063 Studies of planar stellar orbits. I. Motion in the spiral galactic potential of Barbanis and Woltjer.
C. L. Berry, D. J. De Smet.
Astron. J., Vol. 83, 500 - 513 (1978).

Results of a numerical study of orbits in a planar model of a spiral galaxy are presented. The study is divided into two parts. In the first part the galactic model is fixed; it has tightly wound, rather pronounced spiral arms and a pattern speed of 12.5 km s^{-1} kpc^{-1}. Well-defined invariant curves in the surface of section were found for nearly all orbits calculated. Several strong resonances in addition to the inner Lindblad resonance were discovered in this region (many weak resonances were also found). In the second part of the study two values of the energy (corresponding to resonance regions) were selected and changes in the surfaces of section at fixed energy as the amplitude of the spiral pattern was varied were studied. At both energies resonant periodic orbits were found to appear at amplitudes of the spiral pattern 20%–40% below those used in part one of the study.

151.064 Models of gravitating discs in phase description.
L. P. Osipkov.
Pis'ma Astron. Zh., Vol. 4, 70 - 74 (1978). In Russian.

The integral equation which allows to find the symmetric part of the phase density when the surface density is known, is investigated. Several special cases admitting a unique solution are considered. Some new phase models of Maclaurin and Kuzmin–Toomre discs are constructed as examples. Radial velocity dispersions are found for these models. A method to determine the anti-symmetric part of the phase density is suggested.

151.065 A method for determination of a complete class of potentials forming the velocity field of given multiplicity.   T. A. Agekyan, N. P. Pit'ev, M. G. Saginashvili, S. P. Yakimov.
Pis'ma Astron. Zh., Vol. 4, 121 - 124 (1978). In Russian.

A method that permits to get equations for a complete class of potentials forming the velocity field of given multiplicity is proposed. Solutions of these equations also permit to determine the gradient of the direction field with respect to the normal to a trajectory and to study the structure of the direction field.

151.066 On the evolution of dense star clusters.
G. S. Bisnovatyj-Kogan.
Pis'ma Astron. Zh., Vol. 4, 130 - 133 (1978). In Russian.

Evolutional tracks of different dense star clusters are investigated qualitatively, taking into account evaporation of stars and stellar collisions, up to the moment of beginning of relativistic collapse or disruption of the cluster.

151.067 The evolution of disk galaxies.
S. E. Strom, K. M. Strom.
IAU Symp. No. 77, (see 012.026), p. 69 - 95 (1978).

The authors review recent optical wavelength studies of spiral and S0 galaxies which appear to influence our understanding of disk-system evolution. Particular emphasis is placed on the effects of environment on evolutionary processes, since it appears likely that the addition or removal of gas during the lifetime of a disk system may often be dominant in controlling its appearance.

151.068 The dynamics of the spiral galaxy M81.
H. C. D. Visser.
IAU Symp. No. 77, (see 012.026), p. 105 - 112 (1978).

The first step is the construction of an axisymmetric mass model on the basis of a rotation curve. The observed rotation curve derived from the HI measurements is distorted by

density-wave motions. After correction for this effect a mass model can be constructed consisting of two spheroids in the central regions (R < 3 kpc), representing the nucleus and the bulge, and a Toomre disk.

151.069 A confrontation of density wave theories with observations.   A. J. Kalnajs.
IAU Symp. No. 77, (see 012.026), p. 113 - 130 (1978).

It would be a mistake to think that the density wave theories of spiral structure have reached the maturity where they can make unconditional predictions which can be tested. On the contrary, they are still very dependent on observations for help and guidance.

151.070 The age dependence of stellar velocity dispersion in a scale-covariant (*theory of*) gravitation.
P. Magnenat, L. Martinet, A. Maeder.
Astron. Astrophys., Vol. 67, 51 - 53 (1978).

The equation of motion of a one-dimensional oscillator is solved in the equivalent Newtonian approximation of a scale-invariant theory of gravitation and adapted to the stellar motions perpendicular to the galactic plane. In this framework, the problem of the age dependence of the stellar velocity dispersions in our Galaxy has to be considered in a different light.

151.071 On the criterion of applicability of numerical models of interacting galaxies.   A. M. Fridman.
Pis'ma Astron. Zh., Vol. 4, 207 - 209 (1978). In Russian.

On the basis of the well-known criterion of stabilization of disk instability by a large central mass the incorrectness of the numerical models used by many authors and based on the hypothesis on the decisive role of gravitational interaction between galaxies during the spiral structure formation is shown.

151.072 Self-suppression of Jeans instability in a rotating gravitating disk.   A. G. Morozov.
Pis'ma Astron. Zh., Vol. 4, 210 - 213 (1978). In Russian.

Equations describing the "warming-up" of a gravitating disk of stars resulting from the oscillating of Jeans instability are derived. This "warming-up" leads to suppression of instability. The stabilization time of the disk is calculated.

151.073 Entwicklung von Sternsystemen.
K.-H. Schmidt.
Sterne, 54. Band, 65 - 80 (1978).

151.074 Free collapse of a rotating sphere of stars.
R. H. Miller.
Astrophys. J., Vol. 223, 122 - 128 (1978).

The free-fall collapse of a system of 115,000 stars was studied by means of a three-dimensional simulation on the ILLIAC IV computer. The system started from a spherical shape with uniform density and rigid rotation which balanced the gravitational force in the equatorial plane. The system settled down into a "hot" prolate "bar" in about two initial rotation periods. This bar rotates about a short axis and is a long-lived form. Detailed discussion of the development of this system leads to several important dynamical inferences presented in detail.

151.075 Stochastic star formation and spiral structure of galaxies.   H. Gerola, P. E. Seiden.
Astrophys. J., Vol. 223, 129 - 139 (1978).

Self-propagating star formation in a differentially rotating disk is capable of producing persistent large-scale spiral features. The authors have extended the original deterministic model in two crucial ways. First, they implement a stochastic description of the star formation process; second, they increase the size of the array to avoid the dominance of boundary effects. The results show that this model can generate spiral features that appear very similar to those of real galaxies, in particular with respect to the density and pitch angle of these features.

151.076 Steepest descent technique and stellar equilibrium statistical mechanics. IV. Gravitating systems with an energy cutoff. J. Katz, G. Horwitz, A. Dekel.
Astrophys. J., Vol. 223, 299 - 310 (1978).

A statistical thermodynamic analysis of stability criteria for star cluster models with an energy cutoff is developed. Both the Woolley and the King models are analyzed for different thermodynamic constraints, grand canonical, canonical, and microcanonical ensembles being treated. Necessary and sufficient conditions for thermodynamic stability about a mean field approximation are obtained as conditions on a small number of eigenvalues of a Lynden-Bell type operator. Onsets of instabilities with respect to spherically symmetric perturbations of the mean field are essentially indicated by zeros and infinities of standard thermodynamic quantities. Numerical results are given for stability limits of both models with an energy cutoff.

151.077 Steepest descent technique and stellar equilibrium statistical mechanics. V. Relativistic systems with an energy cutoff. G. Horwitz, J. Katz.
Astrophys. J., Vol. 223, 311 - 313 (1978).

Relativistic star clusters with an energy cutoff, at fixed energy and number of particles, are thermodynamically unstable for central redshifts greater than 0.55. The conclusions are derived from statistical mechanics analysis, giving necessary and sufficient conditions of stability of mean field configurations. It is proved that isolated clusters become unstable when the heat capacity becomes equal to zero. It is also shown that the binding energy, as a function of the redshift at the center, passes through a minimum at the onset of instability.

151.078 Stabilization of systems of galaxies by subclustering. L. M. Ozernoy (*Ozernoj*), M. Reinhardt.
IAU Symp. No. 79, (see 012.027), p. 98 - 100 (1978). – Short communication.

151.079 Interacting systems. A. Toomre.
IAU Symp. No. 79, (see 012.027), p. 109 - 116 (1978).

The author presents two examples of galaxies that seem to have interacted recently with impressive consequences: (1) the well-known Messier 51 system NGC 5194/5 = VV 1, and (2) the so-called Cartwheel, a ring within a ring, separated by vaguely spokelike features and accompanied by two smaller galaxies.

151.080 On the tidal origin of M51-type systems. B. Vorontsov-Velyaminov (*B. A. Vorontsov-Vel'yaminov*).
IAU Symp. No. 79, (see 012.027), p. 117 (1978). – Short communication.

151.081 The frequency of ring galaxies and the probability of their formation by collisions.
V. Dostal (*Dostal'*), V. Metlov.
IAU Symp. No. 79, (see 012.027), p. 117 - 120 (1978). Short communication.

151.082 Encounters of spherical galaxies: N-body simulations and comparison with theoretical predictions.
P. Biermann, R. Wielen.
IAU Symp. No. 79, (see 012.027), p. 121 - 122 (1978).

151.083 Computer simulations of galaxy clustering. S. J. Aarseth.
IAU Symp. No. 79, (see 012.027), p. 189 - 196 (1978).

The aim of the present work is to account for the observed distribution of galaxies in terms of the gravitational instability picture. Specifically the author assumes that all the matter is contained in galaxies. The evolution of such a system can then be studied by N-body simulations once the initial conditions are specified. The approach is essentially experimental; a variety of models are computed and the results are compared with observations.

151.084 Mathematical approach to the problem of clustering. J. Burczyk, A. Zieba.
IAU Symp. No. 79, (see 012.027), p. 199 - 200 (1978). Short communication.

151.085 The evolution of galaxies: evidence from optical observations. B. M. Tinsley.
IAU Symp. No. 79, (see 012.027), p. 343 - 355 (1978).

Lookback observations confirm that the stellar populations of galaxies change with time. Color changes in ellipticals correspond to the predicted evolution of the main-sequence turnoff point, and excess blue galaxies in distant clusters and faint field samples indicate that many S0s were actively forming stars in the visible past. Information on the primeval stages of galaxy evolution is still in the form of upper limits to the numbers, formation times, and luminosities (or surface brightnesses) of galaxies that were extremely bright. Theoretical "predictions" for such stages will surely prove to be simply idealizations with which the real primeval chaos can be contrasted.

151.086 On the dynamical evolution of clusters of galaxies. J. P. Ostriker.
IAU Symp. No. 79, (see 012.027), p. 357 - 375 (1978).

At the relatively recent epoch of $3 < z < 10$, galaxies formed from fluctuations of unknown origin in the expanding universe. Large scale fluctuations took longer to develop, and clusters of galaxies separated out at the very recent epoch of $1 < z < 3$. The inner parts of the clusters collapsed, violently relaxed and adjusted to a nearly isothermal state. The outer parts continue to expand but at a decelerating rate. In the inner parts interactions with cluster gas strips gas from galaxies and inhibits the formation of secondary discs. Relatively fast tidal interactions between galaxies strips off the dark material from the outer parts of individual galaxies leaving systems with conventional sizes and masses and distributing the dark matter throughout the inner parts of the cluster. Then on a longer time scale the giant systems tend to accumulate at the center as supergiant low surface brightness cD systems.

151.087 Non-linear effects in flat gravitating systems. A. M. Fridman.
IAU Symp. No. 79, (see 012.027), p. 450 (1978). – Short communication.

151.088 The warping of the galactic plane and changes of ellipticity. B. Abramenko.
Astrophys. Space Sci., Vol. 54, 323 - 342 (1978).

Observational evidence on the widespread occurrence of warping of the outer part of the galactic plane in many galaxies is presented and various hypotheses for its explanation are reviewed. None is found to be able to account for all the cases reported. A theory has been developed which explains the phenomena mentioned as natural consequences of the non-steady nature of galactic systems due to time-dependent metric. This manifests itself in the appearance of tangential acceleration which leads necessarily to variability of the orbital plane and orbital eccentricity in dependence on the radius vector of the orbits.

151.089 A model of the formation of elliptical galaxies. W. K. Brown.
Astrophys. Space Sci., Vol. 54, 365 - 378 (1978).

An analytical model of elliptical galaxy formation is formulated that yields predicted mass distributions. A good fit is obtained when the model is tested against the luminosity profile of NGC 3379. The present theory is identical to a parallel theory of solar system formation which suggests that the mass distribution of the solar system should be similar to that of NGC 3379. Comparison supports this conjecture.

151.090 **Density wave-star interaction of a differentially rotating spiral system in a scalar field.**
E. Evangelidis.
Astrophys. Space Sci., Vol. 54, 467 - 478 (1978).

The effect of the differential rotation on the statistical behaviour of a dynamical system has been studied in the presence of a scalar field. The existence of neutrally stable regimes in a disk-like system has been shown which formally correspond to the Landau levels of a charged particle whose motion is quantized in a magnetic field. The existence also of the so-called anomalous Doppler effect has been pointed out, and a modified form is given to include the differential rotation. Finally, quantitative results are given concerning the instability of the system.

151.091 **On the statement of the ergodic problem in stellar dynamics. (Comments on the classification of integrals of motion in stellar systems).**   L. P. Osipkov.
Problems of observational and theoretical astronomy, Moscow — Leningrad, 1977, (see 003.012), p. 181 - 189. In Russian. — Abstr. in Ref. zh., 51. Astron., 5.51.659 (1978).

151.092 **H I clouds and the theory of formation of galaxies.**
A. G. Doroshkevich, S. F. Shandarin.
Inst. prikl. mat. AN SSSR. Prepr. No. 84. Moskva, 1977. 25 pp. Price 10 Kop. In Russian. — Abstr. in Ref. zh., 51. Astron., 5.51.680 (1978).

151.093 **The problem of formation of galaxies and the "photon whirlwind" hypothesis.**
L. M. Ozernoj, A. D. Chernin.
Probl. gravitatsii. Tbilisi, 1976, p. 441 - 446. In Russian. Abstr. in Ref. zh., 51. Astron., 5.51.748 (1978).

151.094 **A multi-term trial function stability analysis of isotropic relativistic star clusters.**   K. G. Suffern.
Astrophys. Space Sci., Vol. 55, 351 - 382 (1978).

A multi-term trial function technique is developed for studying the dynamic stability of isotropic relativistic star clusters by using the variational principle originated by Ipser and Thorne (1968). The technique is applied to $n = 4$ polytropic clusters, and low-temperature isothermal clusters. The $n = 4$ polytropic clusters are proved to be dynamically unstable if their central redshifts are greater than $z_c = 0.412$.

151.095 **Accretion-induced overstability of density waves in a self-gravitating disk.**   A. Ambastha, R. K. Varma.
Astrophys. Space Sci., Vol. 55, 459 - 473 (1978).

A two-component differentially rotating disk of self-gravitating particles is considered in the hydrodynamical framework. This system is shown to sustain two pairs of density waves, corresponding to the familiar Jeans modes and an acoustic type of modes. As a result of mass and momentum transfer from the gaseous to the stellar component, the acoustic modes suffer a strong damping, whereas the Jeans modes which were oscillatory, now become overstable provided the thermal velocity of stars is larger than that of gaseous component. The waves with frequencies near the corotation have a rather large growth rate. This amplification can explain the maintenance of spiral structure and a 'selective' amplification could even determine the wave-frequency.

151.096 **Surface brightness distributions for galaxies with periodic waves of star formation.**
J. C. Davis, Jr., R. J. Talbot, Jr., E. B. Jensen.

Bull. American Astron. Soc., Vol. 10, 423 (1978). — Abstract.

151.097 **Effective integrals in a triaxial stellar system.**
M. Schwarzschild.
Bull. American Astron. Soc., Vol. 10, 430 (1978). — Abstract.

151.098 **Galactic stellar orbits in a time-dependent potential.**
C. Allen, E. Moreno.
Bull. American Astron. Soc., Vol. 10, 430 (1978). — Abstract.

151.099 **Global stability conditions for flat galaxies by the Nyquist method.**
R. V. E. Lovelace, R. G. Hohlfeld.
Bull. American Astron. Soc., Vol. 10, 430 (1978). — Abstract.

151.100 **Instability in bounded self-gravitating spherical clouds.**   D. L. Book, I. B. Bernstein.
Bull. American Astron. Soc., Vol. 10, 437 (1978). — Abstract.

151.101 **On instabilities of collapsing isothermal clouds.**
H. Gerola, J. Buff.
Bull. American Astron. Soc., Vol. 10, 437 (1978). — Abstract.

151.102 **Stellar orbits in galactic bars.**   P. O. Vandervoort.
Bull. American Astron. Soc., Vol. 10, 483 (1978). Abstract.

151.103 **Particle motions in a self-consistent prolate bar.**
B. F. Smith, R. H. Miller.
Bull. American Astron. Soc., Vol. 10, 483 (1978). — Abstract.

151.104 **The Liouville equation, equations of motion and the rotation curve of galaxies.**   P. Pişmiş.
Bull. American Astron. Soc., Vol. 10, 484 (1978). — Abstract.

151.105 **The dynamical importance of binaries in star clusters.**   I. R. King.
Bull. American Astron. Soc., Vol. 10, 484 (1978). — Abstract.

151.106 **On the origin of star complexes and globular clusters.**
Yu. N. Efremov.
Astron. Tsirk., No. 975, p. 1 - 3 (1977). In Russian.

151.107 **Formation of spiral galaxies.**   R. B. Larson.
IAU Colloq. No. 45, (see 012.032), p. 3 - 10 (1977/78).

Gravitational collapse of protogalaxies. Formation of spheroidal systems. Formation of discs. The timescale for disc formation.

151.108 **Dynamics of the gas in spiral galaxies.**
W. W. Roberts, Jr.
IAU Colloq. No. 45, (see 012.032), p. 11 - 30 (1977/78).

The dynamics of the gaseous component in disk-shaped galaxies is thought to play an important governing role in star formation, molecule formation, and the degree of development of spiral structure. The prospect that density waves and galactic shock waves are present on the large-scale has received support in recent years from a variety of observational studies. This is particularly apparent in M81 and other external spirals, and for the tracers of HI and CO in our own Galaxy.

151.109 **Evolution of galaxies governed by disk dynamics and spiral structure.**   R. J. Talbot, Jr.
IAU Colloq. No. 45, (see 012.032), p. 31 - 45 (1977/78).

The evolutionary changes in disk galaxies caused by the continuing process of star formation are discussed. The author mentions briefly some of the basic aspects of models which have been used to discuss the past evolution.

151.110 **Gravitational separation of elements during certain**

stages of galactic evolution. E. Basinska-Grzesik.
IAU Colloq. No. 45, (see 012.032), p. 47 - 52 (1977/78).

After some preliminary estimations of the diffusion effects in a simple model of our Galaxy in the vicinity of the Sun (Basinska and Iwanowska 1974, Basinska-Grzesik 1974) the author started to examine the conditions for the gravitational separation of elements in: a) some different regions and evolutionary phases of our Galaxy, b) in other galaxies. The results are presented for the simplified models of the protoglobulars, of the spherical protogalaxies (including the spherical component of our own Galaxy) and for the column of gas perpendicular to the galactic plane in the solar neighbourhood.

151.111 A model for a two component galaxy.
R. Caimmi, N. Dallaporta, L. Secco.
IAU Colloq. No. 45, (see 012.032), p. 57 - 60 (1977/78).

151.112 A model for the chemical evolution of galactic disks. C. Chiosi.
IAU Colloq. No. 45, (see 012.032), p. 61 - 66 (1977/78).

151.113 Chemical inhomogeneities in galaxies.
M. G. Edmunds.
IAU Colloq. No. 45, (see 012.032), p. 67 - 71 (1977/78).

151.114 Early evolution of galaxies. Preliminary results.
J. P. Chieze, B. Lazareff, L. Vigroux.
IAU Colloq. No. 45, (see 012.032), p. 115 - 118 (1977/78).

This paper is a progress report on studies of coupled dynamical and chemical evolution of galaxies. The authors present a preliminary version of a model of galactic hot winds and the results obtained for the early evolution of a $1.2 \times 10^{12}\ M_\odot$ galaxy.

151.115 The non-linear theory of spiral structure.
G. Contopoulos.
IAU Colloq. No. 45, (see 012.032), p. 229 - 240 (1977/78).

The main steps of the non-linear theory of spiral structure are described. Near each of the main resonances the basic periodic orbits are calculated, and the sets of non-periodic orbits that follow them are found.

151.116 The effect of stellar mass on kinematical age determination. W. Iwanowska.
IAU Colloq. No. 45, (see 012.032), p. 283 - 287 (1977/78).

It is believed that kinematical characteristics of a group of stars, as the velocity dispersion or the lag in rotation, change systematically with time. This kinematical evolution is widely used for the determination of stellar ages.

151.117 Halo mass in spiral galaxies. C. K. Terzides.
IAU Colloq. No. 45, (see 012.032), p. 297 - 300 (1977/78).

151.118 On boundary layer flows in disk-halo galaxies.
A. M. Waxman.
IAU Colloq. No. 45, (see 012.032), p. 303 - 306 (1977/78).

151.119 Connections between chemical and dynamical evolution. B. M. Tinsley.
IAU Colloq. No. 45, (see 012.032), p. 309 - 319 (1977/78).

Dynamical processes strongly affect the chemical enrichment of gas in galaxies, so abundances in stars and the interstellar medium can be used as probes of the dynamical history of the Galaxy. The author discusses some examples of connections between chemical and dynamical evolution.

151.120 Space density of stars in King's model of star clusters. A. S. Rastorguev.
Astron. Tsirk., No. 994, p. 7 - 8 (1978). In Russian.

151.121 On an inaccuracy in King's model of star clusters.
A. S. Rastorguev.
Astron. Tsirk., No. 995, p. 2 - 4 (1978). In Russian.

151.122 Distribution of stars in the vicinity of a massive compact body. V. I. Dokuchaev, L. M. Ozernoj.
Zh. ehksperim. i teor. fiz., Vol. 73, 1587 - 1598 (1977). In Russian. – Abstr. in Ref. zh., 51. Astron., 6.51.863 (1978).

151.123 Evolutionary effects of a supermassive body on a large surrounding cloud.
C. Maceroni, R. Fusco Femiano.
Mem. Soc. Astron. Italiana, Vol. 47, 73 - 84 (1976).

The authors study the dynamical evolution of a spherically symmetric cloud containing a 'spinar' in its centre. It is assumed that the 'spinar' loses energy through the magnetic dipole mechanism. This energy, transformed in magnetic field and particles energy, produces a pressure force pushing outwards the surrounding matter. The process leads to the formation of a cavity around the 'spinar', filled with relativistic electrons and magnetic field. The dynamical evolution of the cavity and the spectral evolution of its radiation is studied.

151.124 Numerical developments on the dynamical evolution of clusters with more stellar populations.
L. Angeletti, P. Giannone.
Mem. Soc. Astron. Italiana, Vol. 47, 245 - 254 (1976).

The fluid-dynamical method was used by Larson (1970) to study the dynamical evolution of large star clusters consisting of one stellar population (each star has the same mass). In this paper the authors extend the method to more star populations under the same hypotheses of symmetry and collisional relaxation valid for one population. The method has been applied to the study of the dynamical evolution of some clusters with two stellar populations.

151.125 Discrete spiral modes in disk galaxies: some numerical examples based on density wave theory.
G. Bertin, Y. Y. Lau, C. C. Lin, J. W -K. Mark, L. Sugiyama.
Proc. Natl. Acad. Sci. USA, Vol. 74, 4726 - 4729 (1977). Abstr. in Phys. Abstr., Vol. 81, Abstr. 54053 (1978).

151.126 Evolution of galaxies. J. Einasto.
Problems of observational and theoretical astronomy, Moscow – Leningrad, 1977, (see 003.012), p. 26 - 43. In Russian. – Abstr. in Ref. zh., 51. Astron., 7.51.856 (1978).

151.127 Evolution of a flat ring of gravitating mass points.
S. I. Ipatov.
Inst. prikl. mat. AN SSSR. Prepr., 1978, No. 2, 61 pp. In Russian. – Abstr. in Ref. zh., 51. Astron., 7.51.989 (1978).

Non-stationary phenomena in galaxies.
See Abstr. 003.040.

All Lyapunov characteristic numbers are effectively computable. See Abstr. 021.003.

On the formulation of Hill's problem for a system star – satellite – Galaxy. See Abstr. 042.131.

Gravitational N-body problem on the accretion process of terrestrial planets. See Abstr. 107.026.

The formation of multiple systems by dynamical interaction in clusters. See Abstr. 117.057.

Multiple stars as a result of the dynamical evolution of small stellar systems. See Abstr. 117.058.

Multiple star formation from N-body system decay. See Abstr. 117.059.

On the angular momentum in star formation.
See Abstr. 131.206.

Evolution of maser sources.   See Abstr. 131.249.

Kinematics and structure of the Ursa Major star cluster.   See Abstr. 153.039.

Present and past death rates for globular clusters.
See Abstr. 154.003.

Chemical evolution of the Galaxy at the initial rapid-collapse phase.   See Abstr. 155.034.

Dynamical effects in the central region of the Galaxy.   See Abstr. 155.046.

Radial motions near the Galactic centre and the origin of Galactic spiral structure.   See Abstr. 155.049.

Dynamical history of the stars in relation with age, kinematics, abundances, birthrate.   See Abstr. 155.056.

The cosmogonical significance of the z distribution of stars.   See Abstr. 155.058.

Possible pattern speeds of a density wave in our Galaxy.   See Abstr. 155.060.

A new value for the rotation speed of the spiral structure.   See Abstr. 155.062.

Velocity dispersion in the bulge of M 31; dynamical model.   See Abstr. 158.185.

The spiral structure of M 31.
See Abstr. 158.186.

Interacting galaxies: the kinematics of NGC 4038/39 and the HI bridge between M 81 and NGC 3077.
See Abstr. 158.198.

Angular momentum in the Local Group.
See Abstr. 158.236.

Protogalaxy interactions in newly formed clusters: galaxy luminosities, colors, and intergalactic gas.
See Abstr. 160.018.

The evolution of mass and tidal radius of cluster galaxies.   See Abstr. 160.058.

Optical studies of small groups of galaxies.
See Abstr. 160.069.

Interaction between intergalactic medium and galaxies.   See Abstr. 161.009.

The evolution of the universe.
See Abstr. 162.023.

The internal structure of protoclusters and the formation of galaxies.   See Abstr. 162.141.

152 Stellar Associations

152.001 On the dynamical evolution of the subsystem of bright stars of the ζ Persei association.
G. N. Duboshin, V. P. Dolgachev, E. P. Kalinina, A. I. Rybakov, P. N. Kholopov.
Tr. Gos. Astron. Inst. Shternberga, Vol. 47, 64 - 75 (1976).
In Russian.
The possible dynamical evolution of the subsystem of 17 bright members of the ζ Persei association is studied by numerical solution of the equations of motion for the members of the subsystem. It is supposed that this subsystem is part of the stationary star cluster. For the stationarity of the subsystem the mass of the association must reach 500–600 thousand solar masses. It seems probable that at least some of the stars considered leave the association with velocities exceeding the escape velocity.

152.002 A photometric study of the Orion OB 1 association. III. Subgroup analyses.
W. H. Warren, Jr., J. E. Hesser.
Astrophys. J., Suppl. Ser., Vol. 36, 497 - 572 (1978).
The four principal subgroups of the association are examined in detail using individual distances and reddening values determined for their B-type members from $uvby\beta$ data presented in the previous papers of this series.

152.003 Star complexes. Yu. N. Efremov.
Pis'ma Astron. Zh., Vol. 4, 125 - 129 (1978).
In Russian.
A number of star complexes are picked out in the Galaxy on the basis of data on cepheids. They are vast stellar groups with mean diameter of about 600 pc and ages of several tens of million years. A star complex includes stars having originated in the same gas-dust complex. The local system is one of such star complexes. The cepheid period dispersion indicates duration of star formation in a complex of about $(2-5) \times 10^7$ years. There is a possibility that almost all young stars and clusters are connected with some complex.

152.004 Optical and infrared properties of the newly formed stars in Canis Major R1. W. Herbst, R. Racine, J. W. Warner.
Astrophys. J., Vol. 223, 471 - 482 (1978).
$UBVRIJHKL$ photometry and MK spectral types have been obtained for stars illuminating nebulae in the CMa R1 association. The color-magnitude diagram for the association is unusual; many stars over the entire spectral range of the association (B0 to A1) lie more than 1 mag above the zero-age main sequence (ZAMS) in V. An age of ~ 3×10^5 years is suggested for most of the stars, compatible with the suggestion that a supernova explosion triggered star formation in this region. A comparison is made between the observations and theoretical models of pre–main-sequence stars. Some of the latest-type stars (A0–A1) lie close to the ZAMS, suggesting that either (1) star formation in the association was not coeval, or (2) isochrones derived from Iben's pre–main-sequence models do not represent real stellar groups.

152.005 Veränderliche Sterne in Sternassoziationen.
H. Schmidt, F. Gieseking.
Forschungsbericht des Landes Nordrhein-Westfalen, Nr. 2654/ Fachgruppe Physik/Chemie/Biologie. Westdeutscher Verlag, Opladen. 56 pp. Price DM 16.00 (1977). ISBN 3-531-02654-2.
Review in Sterne, 54. Band, 123; 1978 (*W. Wenzel*).
Die vorliegenden Untersuchungen befassen sich mit der Durchmusterung der Sternassoziationen Cygnus T 1, Cepheus OB 2 und Perseus OB 2 nach RW Aurigae- und T Tauri-Veränderlichen und deren Überwachung. Eine Reihe neuer Objekte dieser Art wurde gefunden. Lichtkurven, Farbenindizes und Spektren werden, soweit sie erfaßt werden konnten, diskutiert.

152.006 Optical evidence for a very large, expanding cavity associated with the I Orion OB association and Barnard's loop. R. J. Reynolds, P. M. Ogden.
Bull. American Astron. Soc., Vol. 10, 426 (1978). – Abstract.

Photometry of the "ultraviolet" stellar group in Auriga. See Abstr. 113.048.

Narrow band photometry of the Pleiades.
See Abstr. 113.081.

Spectral types in the Orion OB1 association.
See Abstr. 114.013.

Variable stars in a region of intensive star formation. See Abstr. 121.080.

Observations of young star clusters in the 18-cm OH lines. See Abstr. 153.024.

153 Galactic Clusters

153.001   **A reappraisal of the gap in the HR diagram of M67.**
J. G. Morgan, P. P. Eggleton.
Mon. Not. R. Astron. Soc., Vol. 182, 219 - 231 (1978).

Consideration of the effects of unresolved binaries and observational errors, and careful application of the Schwarzschild criterion at the boundary of the convective core, show that the observed HR diagram of M67 can be modelled with stars computed conventionally, i.e. without using convective overshooting. The computed isochrone is used to show that M67 may have a constant mass function.

153.002   **Supernova remnants in open clusters.**
C. Krishna Kumar.
Astrophys. J., Lett., Vol. 219, L13 - L15 (1978).

Evidence is offered suggesting that the SNR G291.0 − 0.1 is located in the open cluster Tr 18. The turnoff mass of the cluster is 6 $M_\odot$.

153.003   **The broad giant branch of Melotte 66.**
T. G. Hawarden.
Mon. Not. R. Astron. Soc., Vol. 182, 31P - 32P (1978).

New photoelectric photometry confirms the apparent width of the giant branch seen on an earlier, photographic colour−magnitude diagram of the very old open cluster Melotte 66. It is shown that this width is unlikely to be caused by differential reddening.

153.004   **Observational tests on star formation IV: birthplaces of massive stars in open star clusters.**
G. Burki.
Astron. Astrophys., Vol. 62, 159 - 164 (1978).

The distribution of massive stars ($M \gtrsim 4\,M_\odot$) resulting from proto-cluster cloud fragmentation is studied with the help of eleven very young clusters. In the six youngest of these clusters (age $\lesssim 5 \times 10^6$ y), the proportion of stars more massive than about 20 $M_\odot$ (with respect to the number more massive than about 4 $M_\odot$) is, on the average, more than 2.5 times larger in the outer regions ($r \geqslant 0.58 r_d$) of the apparent cluster surface than in the central regions. The fact that many very massive stars are formed in the outer regions of open clusters may be accounted for by theoretical conjectures showing that the critical Jeans mass M_J for the fragmentation of a proto-cluster interstellar cloud ought to vary with the distance R to the cloud centre according to the qualitative relation $M_J \sim R$.

153.005   **An upper limit to the mass and velocity dispersion of M67.**   B. J. McNamara, W. L. Sanders.
Astron. Astrophys., Vol. 62, 259 - 260 (1978).

A best value of $v = 0.95$ km/s and a one sigma upper limit of $v < 1.48$ km/s for the one dimensional velocity dispersion in M67 is determined from an analysis of the cluster star proper motions. Using the best value and the virial theorem, a probable cluster mass of $M \cong 1.6 \times 10^3 M_\odot$ (and a 1 σ upper limit of $4 \times 10^3 M/M_\odot$) is found. This is to be compared with a direct cluster mass of $1.1 \times 10^3 M_\odot$ based on star counts, cluster membership information and the normal M/L main sequence relation.

153.006   **Problems in the interpretation of cluster membership probabilities.**   B. J. McNamara, T. J. Schneeberger.
Astron. Astrophys., Vol. 62, 449 - 451 (1978).

The degree to which the combination of proper motions of various accuracies and deleting from the vector point diagram influences computed cluster membership probabilities using the maximum likelihood technique as employed by Sanders (1971) is investigated. Results obtained from the recently published proper motion data for M 11 suggest that both of these effects can have a significant influence on the calculated membership probabilities.

153.007   ***UBV* photometry of the luminous young cluster NGC 3603.**   S. van den Bergh.
Astron. Astrophys., Vol. 63, 275 - 277 (1978).

NGC 3603 is located at a kinematical distance of $7.2^{+0.8}_{-1.0}$ kpc. This result is consistent with the photometric distance modulus of the cluster that is obtained from new photometry. At a distance of ∼ 7 kpc the H II region associated with NGC 3603 is among the most massive optically-visible emission regions in the Galaxy. The integrated colours of NGC 3603 are similar to those of the central object in the 30 Doradus Nebula. The latter object is, however, 1.0 mag brighter than NGC 3603.

153.008   **Binaries in open clusters: the effects of rotation on determinations of frequency and mass ratio.**
V. L. Trimble, J. P. Ostriker.
Astron. Astrophys., Vol. 63, 433 - 438 (1978).

A re-examination of the HR diagrams for 175 A and F stars in four young clusters (Coma, Alpha Persei, Praesepe and the Pleiades) shows that the number of stars that appear to be binaries because they fall significantly above the main sequency is a very sensitive function of where the MS is drawn, as are the mass ratios deduced for the binaries. Various plausible main sequences yield apparent binary frequencies of 7−87%, with a most probable value of 45−50% as an upper limit to the percentage of the stars that are binaries with $M_2/M_1 \geqslant 0.5$. Thus, rotating stars and binaries cannot be distinguished without data in addition to the HR diagram, and it is not possible to say on the basis of available information that the open cluster stars and the field stars differ in binary frequency or distribution of mass ratios.

153.009   **The open cluster NGC 7127.**
T. A. Uranova, G. S. Tsarevskij.
Tr. Gos. Astron. Inst. Shternberga, Vol. 47, 89 - 95 (1976).
In Russian.

The open cluster NGC 7127 is examined on the basis of photographic UBV magnitudes measured in a region of 12' and published 1974.

153.010   **Investigation of the open star cluster NGC 6811.**
K. A. Barkhatova, P. E. Zakharova, L. P. Shashkina.
Astron. Zh. Akad. Nauk SSSR, Vol. 55, 56 - 61 (1978). In Russian. English translation in Soviet Astron., Vol. 22, No. 1.

On the basis of photographic UBV photometry of 2000 stars in the region of NGC 6811 the luminosity function of the cluster was determined. This function has a deficiency of faint stars which may be explained as a result of dynamical evolution of the cluster. The times of the beginning and the end of star formation in NGC 6811 are deduced by an analysis of the total mass of main sequence stars per unit interval in $\log M/M_\odot$. Star formation in this cluster occupied a period of about 7×10^8 years. Some cluster characteristics, such as colour-excess, distance modulus, distribution of yellow giants on the H−R diagram, were redetermined.

153.011   **Investigation of the galactic cluster NGC 1245.**
S. S. Peruanskij, V. P. Ryadchenko.
Astron. Zh. Akad. Nauk SSSR, Vol. 55, 62 - 64 (1978). In Russian. English translation in Soviet Astron., Vol. 22, No. 1.

Luminosity functions are found for the whole cluster, its nucleus and the corona.

153.012 Spectroscopic studies of open clusters – a search for Ap stars. W. van Rensbergen, G. Hammerschlag-Hensberge, E. P. J. van den Heuvel.
Astron. Astrophys., Vol. 64, 131 - 137 (1978).

New MK spectral types and radial velocities were determined for 49 B and A stars in the open clusters NGC 2169, 2251, 2287, 2301 and 2422. Of these stars 6 turned out to be Ap, Bp stars and 9 Am stars; none of these belong to NGC 2169. Two Bp and eight of the Am stars had not been recognized before. The ages of the clusters range from $(5 \pm 2) \times 10^7$ yr for NGC 2169 to $(3 \pm 0.5) \times 10^8$ yr for NGC 2251.

153.013 The anomalous giant branch of NGC 188.
B. A. Twarog.
Astrophys. J., Vol. 220, 890 - 901 (1978).

The giant branch of NGC 188 has been analyzed relative to that of M67, and it is shown that a significant difference exists between the two clusters in that the giant branch of NGC 188 is intrinsically redder than the giant branch of M67. It is concluded that the most probable cause of the difference and of the observed scatter in the more luminous region of the giant branch in NGC 188 is the combined effect of mass loss, both early on in post–main-sequence evolution and at the tip of the giant branch, and a difference in chemical composition between the two clusters; NGC 188 has lower helium and/or higher CNO abundance than M67. The effect of this on the ages and distance moduli of the clusters is discussed.

153.014 Photometric study of open clusters. Composite HR diagrams. J. C. Mermilliod.
Bull. Inf. Centre Données Stellaires, No. 14, p. 106 - 107 (1978).

The data collected in the "Catalogue of UBV photometry and MK types in open clusters" (Mermilliod 1976) have been used to study the HR diagrams of open clusters.

153.015 Two sparse open clusters in the region of Collinder 132. J. J. Clariá.
Publ. Astron. Soc. Pacific, Vol. 89, 803 - 810 (1977/78).

Photoelectric measurements in the UBV system are presented for 35 stars in the region of Cr 132. The intensity of the Hβ line was measured for 18 stars, most of them B stars. These UBV-Hβ data are interpreted as demonstrating the existence of two separate physical groups, called Cr 132a and Cr 132b, located at 560 pc and 330 pc from the sun. The clusters have nuclear ages of 6.0×10^7 and 1.6×10^8 years in Sandage's scale and are slightly reddened by $E(B-V) = 0^m02$ and 0^m03, respectively. The eclipsing binary FF CMa (HD 55173) appears to be most probably a member of Cr 132a.

153.016 NGC 6383 – I. The central core.
M. P. FitzGerald, P. D. Jackson, M. Luiken, E. J. Grayzeck, A. F. J. Moffat.
Mon. Not. R. Astron. Soc., Vol. 182, 607 - 616 (1978).

NGC 6383 is a young open cluster with a strong central core and a possible extended halo. From photoelectric UBV photometry and MK spectroscopy of stars in the core, the authors find the following: reddening, $E_{B-V} = 0.33 \pm 0.02$ (sd), apparent distance modulus $V - M_V = 11.9 \pm 0.25$, distance $d = 1.5 \pm 0.2$ kpc, radius $R = 1.25$ arcmin (which corresponds to 0.54 pc at the derived distance), and an age of 1.7 ± 0.4 Myr. The central star, HD 159176 (a spectroscopic binary consisting of two O7 V stars, each of mass $35 \pm 5\,M_\odot$, radius $15 \pm 2\,R_\odot$, and age 2.8 ± 0.5 Myr) is probably older than the rest of the cluster core and might have initiated star formation in the core and beyond.

153.017 Study of the intermediate-age galactic cluster NGC 2281. I. UBV photoelectric observations, binary frequency, and the luminosity function of bright members. M. Yoshizawa.
Publ. Astron. Soc. Japan, Vol. 30, 123 - 138 (1978).

Results of UBV photoelectric photometry in NGC 2281 are presented for 105 stars brighter than 15 mag. By using both color-magnitude and two-color diagrams as well as proper motion data, 59 stars brighter than 13 mag are found to be cluster members and four stars to be possible members. The cluster seems to possess a rather short main sequence. The luminosity function of the cluster reaches its maximum at $M_v \sim 3.0$ mag and shows a decreasing trend beyond that value. The existence of binary stars in the cluster is implied and a binary frequency of 25% or more is estimated through the analysis of the color-magnitude diagram of the cluster. The binary members concentrate towards the cluster center more intensely than single ones. Binary and single members resemble each other in luminosity function.

153.018 A new determination of the Hyades distance modulus. R. B. Hanson.
Bull. American Astron. Soc., Vol. 9, 585 - 586 (1978).
Abstract.

153.019 BVRI photometry of late-type Hα-emission stars in NGC 2264. A. E. Rydgren.
Bull. American Astron. Soc., Vol. 9, 637 (1978). – Abstract.

153.020 The red giants of Trumpler 5.
J. Piccirillo, J. K. Kalinowski, R. F. Wing.
Bull. American Astron. Soc., Vol. 9, 637 (1978). – Abstract.

153.021 The open cluster Cr 140 revisited.
P. M. Williams.
Mon. Not. R. Astron. Soc., Vol. 183, 49 - 53 (1978).

In a study based on photoelectric UBV photometry, the cluster Cr 140 is found to be about 40 Myr old and to be situated 420 ± 20 pc away.

153.022 UBV **photometry of the open clusters NGC 6604 and NGC 6704.** D. Forbes, D. L. DuPuy.
Astron. J., Vol. 83, 266 - 277 (1978).

The open clusters NGC 6604 and NGC 6704 have been studied with UBV photoelectric and photographic photometry, resulting in improved distance estimates and color–magnitude diagrams.

153.023 The nearby open cluster Collinder 140.
J. J. Clariá, P. Rosenzweig.
Astron. J., Vol. 83, 278 - 287 (1978).

Photoelectric UBV data of 77 stars in the region of Cr 140, as well as β indices of 31 stars and DDO data of one red star, are reported and discussed. In addition, MK spectral types of the four brightest B stars in the region are also given. These data, together with existing radial velocity and proper motion measurements, show that Cr 140 is a real, dynamically stable, open cluster. The true distance modulus was found to be $V_0 - M_v = 7.78$. A nuclear age of 2.1×10^7 yr and a lower limit to the contraction age of 2.3×10^7 yr were also obtained for the cluster.

153.024 Observations of young star clusters in the 18-cm OH lines. G. M. Rudnitskij.
Astron. Zh. Akad. Nauk SSSR, Vol. 55, 345 - 349 (1978).
In Russian. English translation in Soviet Astron., Vol. 22, No. 2.

In 1974–1975 observations of 12 young galactic stellar clusters and associations have been conducted at the large radio telescope of the Nançay Radio Astronomy Station (France) in the 1667 and 1665 MHz lines of the OH molecule ground state. The 1667 MHz emission has been detected in the Cep OB IV association. Besides that, the 1667 MHz emission was observed in some points of the NGC 2264 region. For the remaining clusters, upper limits on the intensity of OH emission have been determined, the antenna temperature being less than 0.15–0.25°.

153.025 A photometric and spectroscopic study of the open cluster NGC 2169.   C. L. Perry, P. D. Lee, J. V. Barnes.
Publ. Astron. Soc. Pacific, Vol. 90, 73 - 80 (1978) = Contrib. Louisiana State Univ. Obs., No. 129.

Intermediate- and narrow-band photometry, MK spectral types, and radial velocities are presented for 18 stars in the vicinity of the northern open cluster NGC 2169. Analyses yield a mean cluster color excess of $E(B-V) = 0^m20$ and a distance modulus of 10^m25. An upper limit to the nuclear age of 23×10^6 yr is obtained for the aggregate. The radial velocities confirm the variability of the two proposed β Cephei stars discovered photometrically. Six of the 18 stars are probable nonmembers, including the two β Cephei variables and the three late-type stars, according to various criteria.

153.026 Multicolor photometry of the ζ Sculptoris open cluster.   C. L. Perry, D. K. Walter, D. L. Crawford.
Publ. Astron. Soc. Pacific, Vol. 90, 81 - 88 (1978) = Contrib. Louisiana State Univ. Obs., No. 130.

New multicolor photometry is presented for 23 stars in the vicinity of the ζ Scl open cluster. A number of methods for the determination of intrinsic colors and absolute magnitudes were used to compute a mean cluster color excess of $E(B-V) = 0^m00$ and a distance modulus of $V_0 - M_v = 6^m9$. Several criteria were employed in an examination of cluster membership; 16 of the 23 stars are probable cluster members. The metal deficiency found by Eggen is not confirmed. An age of approximately 50×10^6 yr was obtained for the aggregate.

153.027 When do star clusters have unphysical distribution functions?   K. G. Suffern.
J. Phys. A, Vol. 10, 1897 - 1903 (1977). – Abstr. in Phys. Abstr., Vol. 81, Abstr. 8066 (1978).

153.028 The differential blanketing of the main-sequence and near-main-sequence M67 stars relative to the Hyades and Coma.   B. J. Taylor.
Astrophys. J., Suppl. Ser., Vol. 36, 173 - 197 (1978).

The principal aim of this paper is to compare the blanketing of the Hyades, Coma, and M67 main-sequence stars and the M67 F subgiants through the use of red photometry. The necessary reddening corrections have been derived by using several techniques which are probably or certainly insensitive to blanketing. From the reddening-corrected photometry, one finds that the differential Hyades–M67 blocking is greater in absolute magnitude for early G than for late F stars; that the effects of evolution on the F IV M67 stars are clearly discernible in the ultraviolet and marginally discernible in $B-V$; and that the F stars on and near the M67 main sequence have approximately the same blocking as the Coma cluster and less blocking than the Hyades.

153.029 Be 94, a young distant cluster in Cepheus.   S. Wramdemark.
Astron. Astrophys., Vol. 66, 137 - 138 (1978).

Photoelectric UBV photometry was made for 13 stars in Be 94. Four of the stars were also measured in Hβ. The distance of the cluster is 5 kpc and the interstellar extinction in front of the cluster amounts to $A_V = 2.0$ mag. The cluster is situated in a probable spiral arm beyond the Perseus arm.

153.030 Age determinations of open clusters.   M. Patenaude.
Astron. Astrophys., Vol. 66, 225 - 239 (1978).

The purpose of this work was to construct a set of evolutionary tracks and the corresponding isochrones for a few chemical compositions typical of the Population I in order to calibrate the HR Diagram in terms of mass and age in the spectral range late B to G. The chemical compositions

chosen were $(X, Y, Z) = (0.70, 0.27, 0.03)$ and $(X, Y, Z) = (0.72, 0.27, 0.01)$. The sets of isochrones were compared with photoelectric UBV observations of some open clusters to determine their ages and the masses of the stars near the turnup.

153.031 Study of the galactic cluster NGC 581.   R. Sagar, U. C. Joshi.
Bull. Astron. Soc. India, Vol. 6, 12 - 16 (1978).

Photoelectric magnitudes and colours in the UBV system have been determined for 73 stars in NGC 581. The value of the interstellar reddening is 0^m38. The distance modulus of the cluster is estimated to be $12^m16 \pm 0^m1$ and an age of 4×10^7 years is ascribed.

153.032 OAO 2 observations of the Alpha Persei cluster.   M. R. Molnar, T. C. Stephens, A. D. Mallama.
Astrophys. J., Vol. 223, 185 - 191 (1978).

The authors present an analysis of ultraviolet photometry of the Alpha Persei cluster obtained with the OAO 2 satellite. With these data, supplemented by KPNO coudé spectra, they studied 20 stars ranging in spectral type from B3 to A2. These stars were examined for rotation effects as predicted by the models of Collins (1974) and Collins and Sonneborn (1977).

153.033 A photometric survey of the α Persei cluster for δ Scuti variables.   M. H. Slovak.
Astrophys. J., Vol. 223, 192 - 201 (1978).

To investigate the incidence of δ Scuti variables in the α Persei cluster, a program of differential photometry was undertaken which included 24 A- and F-type stars ($V \leq 9.48$) in or near the δ Scuti instability strip. The new differential B and V photometry for the cluster is presented and is compared with the multicolor photometry published by Mitchell (1960). A discussion of the individual variables and nonvariable stars, as well as the assignment of the variables to the class of δ Scuti variables, is given. A summary of the results of the survey and a comparison to other cluster surveys conclude the paper.

153.034 The metal abundance of the old open cluster NGC 2243.   J. Norris, T. G. Hawarden.
Astrophys. J., Vol. 223, 483 - 486 (1978).

DDO intermediate band photometry has been obtained for six stars in the field of the old open cluster NGC 2243. Both the cyanogen index $(\langle \delta CN \rangle = -0.10)$ and the ultraviolet colors indicate that this cluster is the most metal-deficient open cluster presently known. The authors estimate $[A/H] = -0.7$. This result, together with an age of 5×10^9 years and the position of NGC 2243 some 1.3 kpc below the Galactic plane, is consistent with Larson's two-phase models of the formation of disk galaxies.

153.035 Differential reddening in the young cluster NGC 2264.   A. Young.
Publ. Astron. Soc. Pacific, Vol. 90, 144 - 148 (1978).

A study of intermediate dispersion spectra, and broadband photometry of 69 stars in the very young cluster NGC 2264 reveals the presence of extensive differential reddening. Statistical inquiries lead to the inference that the source of the reddening is intracluster dust clouds, but that these clouds may not be randomly distributed within the cluster.

153.036 A four-color and Hβ study of the galactic cluster IC 4756.   E. G. Schmidt.
Publ. Astron. Soc. Pacific, Vol. 90, 157 - 162 (1978).

Four-color and Hβ photometry has been obtained of 27 stars which are proper-motion members of the galactic cluster IC 4756. Using the earliest main-sequence stars, which are A stars, the true distance modulus was found to be 8.05 and the color excess was found to vary across the area of the cluster from $E(b-y) = 0\overset{m}{.}127$ to $E(b-y) = 0\overset{m}{.}227$. The calibration of the four-colour photometry in terms of absolute magnitude does not seem to apply to this cluster so the distance modulus is somewhat uncertain. Ten stars which appear in the red-giant region were observed and it is concluded that seven are members, two are doubtful members, and one is a foreground star.

153.037 PDS photometry of the open cluster NGC 2420.
R. D. McClure, B. Newell, J. V. Barnes.
Publ. Astron. Soc. Pacific, Vol. 90, 170 - 178 (1978).

A PDS microdensitometer has been used to remeasure the plates of NGC 2420 used by McClure, Forrester, and Gibson (1974) as the basis for their iris photometry. The authors find that their PDS data are characterized by internal errors that are slightly less than those of the iris measurements, and that the PDS data have external errors that are up to a factor of 2 smaller than those of the iris data; a further improvement in external accuracy can be made by averaging the two sets of measurements. A comparison between the authors' new C-M diagram for NGC 2420 and isochrones by Ciardullo and Demarque (1977) emphasizes the difficulty involved in identifying the main-sequence gap that corresponds to the rapid hydrogen-core-exhaustion phase of evolution.

153.038 UBV photographic photometry in and around the galactic cluster NGC 2527. R. J. Dodd.
Occas. Rep. R. Obs. Edinburgh, No. 3, 12 pp. (1977). ISSN 0309-099X.

A major multi-colour photometric investigation of the region centred on $1 = 250°$ in the galactic plane was recently started. This report outlines the procedures adopted for the reduction of iris diaphragm photometer readings to photographic magnitudes and colours and lists the results of the reduction.

153.039 Kinematics and structure of the Ursa Major star cluster. R. Wielen.
Bull. American Astron. Soc., Vol. 10, 408 - 409 (1978). Abstract.

153.040 The old open cluster Melotte 66.
B. J. Anthony-Twarog, R. D. McClure, B. A. Twarog.
Bull. American Astron. Soc., Vol. 10, 409 (1978). − Abstract.

153.041 Evolutionary abundances − the sun and the Hyades stars. P. J. Flower.
Bull. American Astron. Soc., Vol. 10, 437 (1978). − Abstract.

153.042 Possible near-by open clusters. I. N. Latyshev.
Astron. Tsirk., No. 969, p. 7 - 8 (1977). In Russian.

153.043 Search for neutral hydrogen in the galactic cluster NGC 2287. W. G. L. Pöppel, E. R. Vieira.
Rev. Brasileira Fis., Vol. 2, 59 - 66 (1972) = Contrib. Inst. Argentino Radioastron., No. 29.

Observations were made in the direction of the galactic cluster NGC 2287 and its neighbourhood with the radiotelescope at Parque Pereyra Iraola. The results do not give a definite conclusion about the presence of neutral hydrogen in the cluster because of the complexity of the region. An upper limit to neutral hydrogen associated with this cluster and within 1° of it is found to be 600 $M_\odot$.

153.044 Abundances and ages of disk clusters.
R. D. McClure, B. A. Twarog.
IAU Colloq. No. 45, (see 012.032), p. 193 - 196 (1977/78).

153.045 Linear diameters and distances of open clusters.
K. A. Barkhatova, O. P. Pyl'skaya.
Astron. Tsirk., No. 995, p. 4 - 7 (1978). In Russian.

153.046 On the possibility of determining the age of aggregates by flare star observations.
Eh. S. Parsamyan.
Flare stars. Symposium 1976, (see 012.035), p. 87 - 93.
In Russian. − From Ref. zh., 51. Astron., 6.51.852 (1978).

153.047 Results of flare observations in stellar aggregates.
V. A. Ambartsumyan, L. V. Mirzoyan.
Flare stars. Symposium 1976, (see 012.035), p. 63 - 73.
In Russian. − Abstr. in Ref. zh., 51. Astron., 6.51.853 (1978).

153.048 The age of the galactic cluster Lyngä 6.
L. Di Prospero.
Mem. Soc. Astron. Italiana, Vol. 47, 255 - 262 (1976).

The age of the galactic open cluster Lyngä 6 has been determined according to two evolutionary methods. An analogous age was obtained for TW Normae, a Cepheid in the line of sight of the cluster, according to the period-age relationship. The present results strengthen the hypothesis that TW Normae is an effective member of the cluster. The cluster age is estimated to be 4×10^7 years.

The ESO/Uppsala survey of the ESO (B) Atlas of the Southern Sky − VI. See Abstr. 002.051.

Fourier techniques for binary star orbit computation and the distance to the Hyades. See Abstr. 031.217.

Energy distributions in main-sequence A and F stars. See Abstr. 064.077.

Astrometric studies of 45 Tauri, 97 Tauri, 102 ι Tauri, 31 o² Cygni, and selected stars in those regions. See Abstr. 111.007.

Proper motions and positions of stars in wide surroundings of the Hyades cluster. II. See Abstr. 112.003.

On the nature of the infrared excesses in the pre-main-sequence A and F stars in NGC 2264. See Abstr. 113.024.

Intermediate-band photometry of late-type stars. V. Calibration of indices. See Abstr. 113.038.

Infrared photometry, bolometric magnitudes, and effective temperatures for giants in M3, M13, M92, and M67. See Abstr. 113.042.

Spectroscopic differences between solar-type cluster stars and the Sun. See Abstr. 114.004.

Carbon stars near open clusters at galactic latitude $4°5 < |b| < 9°5$ **and between longitudes 68° and 184°.** See Abstr. 114.043.

Zirconium stars in open clusters and in the spiral arms of the Galaxy. See Abstr. 114.046.

Spectral types in the Pleiades. See Abstr. 114.062.

W92: a possible gravitationally contracting binary system in NGC 2264. See Abstr. 117.012.

Etoiles multiples et amas ouverts. See Abstr. 117.041.

On the spectrophotometric identification of late type binaries in open clusters. See Abstr. 117.042.

Study of the spectroscopic binaries in NGC 2516.

See Abstr. 119.004.

Stellar pulsation and the abundance of helium.
See Abstr. 122.005.

δ Scuti variables in the α Persei cluster.
See Abstr. 122.059.

Further observations of the double-mode Cepheid V367 Scuti in the open cluster NGC 6649.
See Abstr. 122.060.

Flare stars in stellar aggregates.
See Abstr. 122.074.

Flare star observations in the Praesepe region.
See Abstr. 122.075.

On possible cyclic recurrence of flare activity of flare stars in the Pleiades.   See Abstr. 122.104.

A search for Beta Cephei stars. II: NGC 4755.
See Abstr. 122.185.

Investigation of the interstellar absorption in the region of NGC 6866.   See Abstr. 131.218.

The planetary nebula in the galactic cluster NGC 2818.   See Abstr. 135.036.

Entweichgeschwindigkeiten und mittlere Geschwindigkeiten in Stern- und Galaxienhaufen.   See Abstr. 151.044.

Errata

153.901   **Erratum: "A study of some loose clusterings in the southern Milky Way" [Astron. Astrophys., Suppl. Ser., Vol. 29, 31 - 50 (1977)].   L. O. Lodén.**
Astron. Astrophys., Suppl. Ser., Vol. 32, 157 - 158 (1978).

154   Globular Clusters

154.001   **Variable stars in the globular cluster NGC 5286.**
     M. H. Liller, S. M. Lichten.
Astron. J., Vol. 83, 41 - 47 (1978).
     A photometric study with B plates of 14 previously discovered and two new variable stars in NGC 5286 allows determination of the characteristics of the light variation. The 16 variables are probably members of the cluster.

154.002   **Membership and other problems of the globular cluster M22.   T. Lloyd Evans.**
Mon. Not. R. Astron. Soc., Vol. 182, 293 - 302 (1978).
     Radial velocity observations of stars in M22 show that all the strong-lined stars in the field are non-members. Many of these stars are distinguishable from cluster members by their position in the $B-V$, $V-I_K$ diagram. There are no stars to the blue of the giant branch and brighter than $V = 12.4$ within 7 arcmin of the cluster centre. The presence of three hot stars fainter than this is typical of clusters with a very blue horizontal branch.

154.003   **Present and past death rates for globular clusters.**
     A. P. Lightman, W. H. Press, S. F. Odenwald.
Astrophys. J., Vol. 219, 629 - 634 (1978).
     The observed distribution of globular cluster central relaxation times is fitted very well by a death rate function which is a power law in cosmological time. The authors fit for this function and infer that (i) existing clusters are collapsing smoothly to some state in which they are no longer recognizable as globular clusters; there is no excess of clusters "hanging up at the edge" of collapse; (ii) the number of clusters already "dead" is between ~2 to ~1000 times the number now present, depending on the (uncertain) time of cluster formation; (iii) the remaining lifetime of a cluster is in the range ~50−120 central relaxation time scales, in good agreement with previous theoretical investigations of core collapse.

154.004   **An unsuccessful search for carbon monoxide in globular clusters.**
T. H. Troland, J. E. Hesser, C. Heiles.
Astrophys. J., Vol. 219, 873 - 876 (1978).
     The authors have searched for 2.6 mm $^{12}C^{16}O$ emission from globular clusters with suspected dark clouds and from

globular clusters which are X-ray sources. No lines are observed, but the sensitivity of the search is insufficient to rule out the possibility that the apparent dark clouds are real features with characteristics similar to galactic-plane dust clouds.

154.005   **The chemical abundances of extreme population II systems.   R. Canterna, R. A. Schommer.**
Astrophys. J., Lett., Vol. 219, L119 - L123 (1978).
     The average [Fe/H] of eight extremely distant globular clusters and the dwarf galaxies Draco and Ursa Minor is found to be −2.2 ± 0.3, similar to that of M 92 and M 15. The one exception to this is Pal 11, with [Fe/H] = −0.7; however, a new distance modulus places its galactocentric distance less than 7 kpc. Several stars in NGC 7006, Pal 11, Pal 12, and Ursa Minor show evidence of enhanced [CNO]. Beyond 20 kpc, no convincing evidence for a [Fe/H] gradient with galactocentric distance is found.

154.006   **A search for [O III] emission in the central part of the X-ray globular cluster M15.**
M. Aurière, P. Laques, J. L. Leroy.
Astron. Astrophys., Vol. 63, 341 - 344 (1978).
     To detect unambiguously gas in globular cluster centers the authors have looked for emission in the 5007 Å line of [O III]. A photometric study has been performed for the X-ray globular cluster M15. Outside of the well known K 648, no [O III] emission has been detected.

154.007   **Variable stars in the globular cluster Messier 19.**
     C. Coutts Clement, H. Sawyer Hogg.
Astron. J., Vol. 83, 167 - 171 (1978).
     Periods have been determined for seven of the eight variables in and around the globular cluster M19. Four of these are Population II cepheids with periods from 2.4 to 16.9 days and three are RR Lyrae variables. All of the cepheids and two of the RR Lyrae stars may be members. The only galactic globular clusters known to have more cepheids within their boundaries are ω Centauri and Messier 14.

154.008   **Direct determination of the integral blanketing effect of globular clusters.**
B. V. Kukarkin, N. N. Kireeva.

Tr. Gos. Astron. Inst. Shternberga, Vol. 47, 34 - 36 (1976). In Russian.

Comparison of unblanketed intrinsic colours of globular clusters according to Kukarkin and Kireeva (1974) with apparent colours derived from interstellar reddening by data of Crawford and Barnes (1974) gives the possibility to determine the integrated blanketing effect. The results are given in a table and by an equation.

154.009 **Radial velocities of stars in the fields of the globular clusters M10, M12, M71 and M92.**
R. G. Gratton, R. Nesci.
Mon. Not. R. Astron. Soc., Vol. 182, 61P - 63P (1978).

Radial velocities for three globular clusters are derived from measurements of individual stars. The results are $v_r = + 67$ km/s for M10, $v_r = -44$ km/s for M12 and $v_r = -19$ km/s for M71.

154.010 **TiO band strengths in metal-rich globular clusters.**
J. R. Mould, D. B. McElroy.
Astrophys. J., Vol. 221, 580 - 585 (1978).

Observations are reported of the run of TiO band strength (7120 Å) with effective temperature for stars at the tip of the giant branch in globular clusters. Relative to field stars of the same de-reddened 1 micron gradient, the cluster giants are seen to be weak in TiO. A parameter is suggested which measures the onset of TiO absorption and provides a metallicity ranking of the clusters in agreement with existing determinations.

154.011 **A search for CO toward globular clusters.**
M. H. Schneps, P. T. P. Ho, A. H. Barrett, R. B. Buxton, P. C. Myers.
Bull. American Astron. Soc., Vol. 9, 590 (1978). – Abstract.

154.012 **A slitless-spectroscopic survey of the nearer globular clusters.**   H. E. Bond.
Bull. American Astron. Soc., Vol. 9, 601 - 602 (1978). Abstract.

154.013 **A systematic reinvestigation of the galactic globular clusters: a progress report.**
J. E. Hesser, S. J. Shawl.
Bull. American Astron. Soc., Vol. 9, 602 (1978). – Abstract.

154.014 **Photometry of the anomalous halo cluster Pal 12 to the main sequence.**   R. Canterna, W. E. Harris.
Bull. American Astron. Soc., Vol. 9, 602 (1978). – Abstract.

154.015 **Horizontal-branch stars in the log T_{eff}, log g diagram.**   A. G. D. Philip.
Bull. American Astron. Soc., Vol. 9, 602 (1978). – Abstract.

154.016 **Post He flash evolution in globular clusters. Observations vs. theory.**   E. Green.
Bull. American Astron. Soc., Vol. 9, 603 (1978). – Abstract.

154.017 **Optical studies of the X-ray globular cluster NGC 6624.**   C. R. Canizares, J. E. McClintock, J. E. Grindlay, W. Liller, W. A. Hiltner.
Bull. American Astron. Soc., Vol. 9, 616 (1978). – Abstract.

154.018 **Intermediate band electronography of globular clusters – I. The red giants in M5 (NGC 5904).**
C. D. Pike.
Mon. Not. R. Astron. Soc., Vol. 183, 101 - 110 (1978).

Electronographic observations of the globular cluster M5 have been obtained in the DDO intermediate band photometric system using a 4-cm McMullan camera. An automatic Gaussian profile-fitting reduction procedure has given DDO colours with estimated errors of ≤ 0.03 mag. Stars on the upper asymptotic giant branch appear anomalous in the DDO temperature–gravity diagram.

154.019 **The blue horizontal-branch stars of NGC 6752.**
B. Newell, E. M. Sadler.
Astrophys. J., Vol. 221, 825 - 832 (1978).

The authors show that the wide horizontal-branch gap, discovered in NGC 6752 by Cannon and Lee, lies at the same value of effective temperature (log T_{eff} = 4.33) as gap 2 of the high-latitude faint blue star samples discussed by Newell and Newell and Graham. The authors demonstrate that the bluest stars in NGC 6752 have $\langle T_{eff} \rangle \approx 27{,}000$ K and log $\langle L/L_\odot \rangle \approx$ 1.25, consistent with membership in an extended horizontal branch, and suggesting that these objects are the cluster analogs of the spectroscopically defined field subdwarf B stars. The large number of possible sdB stars in NGC 6752 (there are at least 15 stars below the horizontal-branch gap) is difficult to explain by using the hypothesis that the sdB stars are binaries.

154.020 **The LMC cluster Hodge 11.**
K. C. Freeman, S. C. B. Gascoigne.
Proc. Astron. Soc. Australia, Vol. 3, 136 - 137 (1977).

154.021 **Photoelectric UBV and DDO photometry in the anomalous globular cluster NGC 2808.**
W. E. Harris.
Publ. Astron. Soc. Pacific, Vol. 90, 45 - 52 (1978).

Results from a new program of photoelectric UBV and DDO photometry in the uniquely anomalous globular cluster NGC 2808 are reported. The combined new data confirm that NGC 2808 is moderately metal-rich ($\delta(U-B) \simeq 0.14$, [Fe/H] (DDO) $\simeq -0.8$), with a predominantly red horizontal branch and essentially no RR Lyrae stars. The existence of the isolated group of extreme blue stars ($(B-V)_0 < 0.0$) on the horizontal branch is also confirmed, with H-B stars of intermediate color ($0.0 \lesssim (B-V)_0 \lesssim 0.4$) almost completely absent. The most plausible interpretation seems to remain the one in which the red and blue groups of H-B stars are segregated in mass by $0.05-0.10 M_\odot$.

154.022 **The helium abundance in galactic globular clusters.**
R. G. Deupree, J. G. Eoll, S. W. Hodson, R. W. Whitaker.
Publ. Astron. Soc. Pacific, Vol. 90, 53 - 56 (1978).

The problem of the helium abundance in galactic globular clusters is discussed in light of two recent theoretical developments—the inclusion of semiconvection in horizontal-branch models and the explanation of the red edge of the RR Lyrae gap. These two results predict substantially different helium abundances. The authors propose alterations in red-giant models which can reduce, but by no means remove, the discrepancy.

154.023 **La naturaleza del sistema estelar en Reticulum.**
S. Demers, W. E. Kunkel.
Astron. extragalactica, (see 012.015), 89 - 91 (1976).

Discussion of the H-R diagrams of the system in Reticulum leads to the conclusion that it is a globular cluster of our Galaxy, although quite far away in the direction of the Magellanic Clouds.

154.024 **Observaciones del sistema estelar en Retículo, hechas en Córdoba.**   J. L. Sérsic, J. C. Arias.
Astron. extragalactica, (see 012.015), 93 - 96 (1976).

154.025 **On the metallicity function of globular clusters.**
A. M. Ehjgenson.
Astrofizika, Vol. 13, 545 - 547 (1977). In Russian. English translation in Astrophysics, Vol. 13, No. 3.

The question on the statistical significance of secondary peaks in the metallicity function of globular clusters which Marsakov and Suchkov (1976) have noted is raised. The hypothesis of the existence of only one top of this function is plausible.

**154.026 Color—magnitude diagrams for the inner parts of
47 Tuc.   M. S. Chun, K. C. Freeman.**
Astron. J., Vol. 83, 376 - 392 (1978).

To understand the origin of the radial color changes observed in 47 Tuc, one needs detailed color—magnitude diagrams for the brighter stars in the inner parts of this cluster. These data are presented here for stars down to the horizontal branch. The authors discuss also their procedure for minimizing errors in iris photometry resulting from the variable cluster background.

**154.027 Nova-driven winds in globular clusters.
E. H. Scott, R. H. Durisen.**
Astrophys. J., Vol. 222, 612 - 620 (1978).

Recent sensitive searches for Hα emission from ionized intracluster gas in globular clusters have set upper limits that conflict with theoretical predictions. The authors suggest that nova outbursts heat the gas, producing winds that resolve this discrepancy. The incidence of novae in globular clusters, the conversion of kinetic energy of the nova shell to thermal energy of the intracluster gas, and the characteristics of the resultant winds are discussed. Calculated emission from the nova-driven models does not conflict with any observations to date. Some suggestions are made concerning the most promising approaches for future detection of intracluster gas on the basis of these models. The possible relationship of nova-driven winds to globular cluster X-ray sources is also considered.

154.028 The globular cluster NGC 6266.   G. Alcaino.
Astron. Astrophys., Suppl. Ser., Vol. 32, 379 - 386 (1978).

A colour magnitude diagram (CMD) for the RR Lyraes rich globular cluster NGC 6266 has been obtained from 290 stars measured photographically, calibrated from a photoelectric sequence of 23 stars to a limiting magnitude of $V = 16^m7$. The SE sector of the cluster is obscured by a dust cloud that absorbs about $A_V \sim 0^m75$ relative to the NW. For the less reddened zone the deduction of $E_{(B-V)} = 0^m50$ and $\langle V \rangle_{RR} = 15^m86$ for the mean visual light for the RR Lyraes, places the object at 5.7 kpc from the Sun, 0.7 kpc from the galactic plane and 3.6 kpc from the galactic centre.

**154.029 A search for new variables in the globular cluster
Omega Centauri.   B. Niss, H. E. Jørgensen,
S. Laustsen.**
Astron. Astrophys., Suppl. Ser., Vol. 32, 387 - 393 (1978).

A systematic search for eclipsing binaries brighter than the turn-off point in the globular cluster Omega Centauri, using a photographic method and blinking, has resulted in 83 new candidates. Of these, 29 are with a reasonable certainty variables. Half a dozen are possible eclipsing, and one is almost certainly an eclipsing binary. It is shown that only few eclipsing binaries can be expected to be found if the frequency of binaries in the cluster is comparable to the frequency among the field stars.

**154.030 Nuclear brightness profiles of the X-ray globular
cluster M15 compared to M3, M5, and M13.
W. A. Feibelman.**
Astron. J., Vol. 83, 482 - 486 (1978).

Microdensitometer traces of the nucleus of M15 indicate that its central peak is narrower in the red than in the blue region. This suggests á concentration of red giants at the nucleus.

**154.031 Pre-horizontal branch evolution and the Oosterhoff
dichotomy in galactic globular clusters.**
F. Caputo, V. Castellani, A. Tornambè.
Astron. Astrophys., Vol. 67, 107 - 113 (1978).

The approach of Population II stars to their ZAHB (zero-age horizontal branch) locations is studied. One finds evidence

that, after the helium flash, pre-HB evolution runs with gravitational time scale until ZAHB effective temperatures are overtaken. The final approach to the ZAHB location is realized by a decrease of the star effective temperature with (O_{16}, p) burning time scale. Under the hypothesis that a hysteresis mechanism is efficient for RR Lyrae pulsators the authors find that pre-HB evolutionary history can control the actual mode of pulsation in ZAHB pulsators. On this basis a discontinuity in the properties of fundamental pulsators is expected, possibly to be connected with the well known Oosterhoff dichotomy observed in galactic globular clusters.

154.032 NGC 1851.
IAU Circ., Nos. 3209, 3226 (1978).

**154.033 The metallicity of globular clusters in the halo of
M87.   R. Racine, J. B. Oke, L. Searle.**
Astrophys. J., Vol. 223, 82 - 87 (1978).

The energy distributions of one 20th and two 21st magnitude globular clusters in the halo of the giant elliptical galaxy M87 (NGC 4486) of the Virgo cluster have been obtained between 0.32 and 1.1 μm. Comparisons with globular clusters in M31 show line strength indices L ranging from 4 ± 2 to 9.5 ± 2.0, with a mean for the three clusters of 7.5 ± 1.5. On average they thus have metal deficiencies relative to the Sun by a factor of about 5. The distribution over abundance in the halo globular clusters of M87 is the same as that in M31, but different from that in the halo of the Galaxy.

**154.034 The space distribution of globular clusters in M49.
W. E. Harris, P. L. Petrie.**
Astrophys. J., Vol. 223, 88 - 93 (1978).

The radial distribution and total population of globular clusters around the giant Virgo elliptical galaxy M49 (NGC 4472) have been derived by counts of faint $(V > 19)$ stellar objects around the galaxy. Down to a counting limit of $V \approx 23.3$ and out to a radius $r = 20'$, the authors estimate M49 to contain 2000 ± 600 clusters. Unlike M87, the M49 cluster system appears to be almost spherically symmetric in structure, and its outer boundary may be more sharply defined, lacking the extended outer-halo envelope of M87. Possible causes for the striking difference in total cluster populations (2000 in M49 versus more than ~ 4500 in M87 down to comparable limiting magnitudes) are briefly discussed.

**154.035 The observed star distributions in the globular
clusters M10 and M12.**
R. Buonanno, C. E. Corsi, V. Castellani, F. Smriglio.
Astron. Astrophys., Suppl. Ser., Vol. 33, 141 - 143 (1978).

The apparent star distribution in the globular clusters M10 and M12 has been studied. Large irregularities in the distribution of the more luminous stars are found, maybe supporting the hypothesis of a recent gravitational disturbance of both clusters.

**154.036 The age of the globular cluster NGC 6752.
B. W. Carney.**
Bull. American Astron. Soc., Vol. 10, 409 (1978). – Abstract.

**154.037 Population synthesis of globular clusters.
W. Tobin.**
Bull. American Astron. Soc., Vol. 10, 409 (1978). – Abstract.

**154.038 Membership in the globular cluster M13.
K. M. Cudworth.**
Bull. American Astron. Soc., Vol. 10, 409 (1978). – Abstract.

**154.039 The two components of the core of the globular
cluster M53.   S. B. Vladimirov, D. P.
Kapusnyakova, P. G. Augustin.**
Astron. Tsirk., No. 982, p. 7 - 8 (1978). In Russian.

154.040 **Proper motions of bright red giants in globular clusters. I. The globular cluster M13.**
N. Spasova.
Astrofiz. issled. (NRB), Vol. 2, 34 - 45 (1977). In Russian. Abstr. in Ref. zh., 51. Astron., 6.51.858 (1978).

154.041 **Proper motions of bright red giants in globular clusters. II. The globular clusters M92 and M3.**
N. Spasova.
Astrofiz. issled. (NRB), Vol. 2, 46 - 58 (1977). In Russian. Abstr. in Ref. zh., 51. Astron., 6.51.856 (1978).

154.042 **Photoelectric scanning of globular clusters. I. The inner structure of M3.**
O. Bendinelli, G. Parmeggiani, S. Lorenzutta.
Mem. Soc. Astron. Italiana, Vol. 48, 67 - 76 (1977).
The photometric structure of the inner part of M3, here obtained, which is confirmed by direct, independent, photoelectric and electrographic measurements of the surface brightness, agrees with Kholopov's hypothesis, based upon star counts, that density waves exist in globular clusters.

The ESO/Uppsala survey of the ESO (B) Atlas of the Southern Sky – VI. See Abstr. 002.051.

Il "Quantimet" come analizzatore di immagini astronomiche. See Abstr. 031.310.

A possible explanation for the weak G-band effect. See Abstr. 064.034.

Infrared photometry, bolometric magnitudes, and effective temperatures for giants in M3, M13, M92, and M67. See Abstr. 113.042.

A new stellar standard sequence in the Coma cluster of galaxies. (Preliminary report). See Abstr. 113.049.

Spectra of giant stars in distant satellites of the Galaxy. See Abstr. 114.010.

Absolute spectral energy distributions and [Fe/H] values of metal-poor stars and globular clusters. See Abstr. 114.039.

Abundances in globular cluster red giants. I. M3 and M13. See Abstr. 114.565.

The calcium abundance of the RR Lyraes in Omega Centauri. See Abstr. 122.055.

Periods for nineteen RR Lyrae variables in NGC 1851. See Abstr. 122.132.

On the membership of the RR$_S$ star V 65 and the

EA binary V 78 in Omega Centauri (NGC 5139). See Abstr. 122.145.

New variable stars in the globular cluster M3. See Abstr. 122.168.

RR Lyrae-type variables V146 and V222 in the globular cluster M3. See Abstr. 122.174.

On the change of the period of the variable V3 in the globular cluster M3. See Abstr. 122.175.

On period changes of long-period cepheids in globular clusters. See Abstr. 122.199.

The rate of formation of white dwarfs in stellar systems. See Abstr. 126.003.

Formation of X-ray sources by collisions in globular clusters and in the bulge. See Abstr. 142.012.

Precise location of X-ray sources associated with the globular clusters NGC 1851, NGC 6441, NGC 6624, NGC 6712, and NGC 7078. See Abstr. 142.073.

Evolutionary model computations for globular cluster X-ray sources. See Abstr. 142.146.

Optical observations of X-ray globular clusters. See Abstr. 142.235.

Close binaries and evolution of star clusters. See Abstr. 151.006.

An approximate theory for the core collapse of two-component gravitating systems. See Abstr. 151.022.

Entweichgeschwindigkeiten und mittlere Geschwindigkeiten in Stern- und Galaxienhaufen. See Abstr. 151.044.

On the origin of star complexes and globular clusters. See Abstr. 151.106.

Photometric studies of composite stellar systems. I. CO and *JHK* observations of E and S0 galaxies. See Abstr. 158.034.

Photometric studies of composite stellar systems. II. Observations of H$_2$O absorption and the coolest stellar component of E and S0 galaxies. See Abstr. 158.045.

Distances of nearby groups and clusters, and the local value of the Hubble ratio. See Abstr. 160.037.

155 Structure and Evolution of the Galaxy

155.001   Galactic structure around $l = 140°$, $b = 0°$.
L. S. Sparke, R. J. Dodd.
Mon. Not. R. Astron. Soc., Vol. 182, 1 - 12 (1978).
Photographic photometry of 15.75 square degrees of sky near $l = 140°$, $b = 0°$ is analysed to investigate galactic structure there. That and data from the literature indicate no discontinuity in the Local arm of the galaxy, but that the Perseus arm as traced optically is partially interrupted for tens of degrees. An explanation in terms of the density wave theory of spiral structure is suggested.

155.002   Interferometer measurements of H I absorption in the direction of 16 extragalactic radio sources.
R. S. Roger, J. L. Caswell, J. D. Murray, D. J. Cole, D. J. Cooke.
Mon. Not. R. Astron. Soc., Vol. 182, 209 - 218 (1978).
The Parkes interferometer has been used to measure the galactic H I absorption spectrum for 16 extragalactic sources. Absorption was detected for 12 sources and values of H I column density and spin temperature are derived in these directions. In addition, observations of 30 Doradus were made at two interferometer baselines over a complete range of projected position angles; these measurements show that the absorption by foreground H I in the Large Magellanic Cloud has two main velocity components, and reveal previously unrecognized spatial structure in the H I.

155.003   Milky Way photographs from Queensland.
A. A. Page.
Sky Telesc., Vol. 55, 120 - 123 (1978).

155.004   Carbon monoxide in the Galaxy. III. The overall nature of its distribution in the equatorial plane.
W. B. Burton, M. A. Gordon.
Astron. Astrophys., Vol. 63, 7 - 27 (1978).
The authors compare the CO characteristics with those derived from observations of the λ 21-cm hyperfine transition of atomic hydrogen. The combined H I and CO terminal velocity data give a new galactic rotation curve applicable in the first longitude quadrant of the galactic equator. The CO clouds follow the same kinematics as the H I gas. Sixty-five percent of the CO intensities emanate from the annulus $4 < R < 8$ kpc, whereas only 36% of the H I distribution lie there. Gas in any form is depleted at $0.4 < R < 4$ kpc. Within the molecular annulus the CO distribution is very clumpy.

155.005   Study of the integral spectrum of the Milky Way.
E. B. Kostyakova.
Tr. Gos. Astron. Inst. Shternberga, Vol. 47, 37 - 63 (1976). In Russian.
A survey and the results of the investigation of the Milky Way integrated spectrum carried out by the author in both hemispheres during several years are given. The spectrophotometric temperatures, T_c, of 20 bright Milky Way regions were obtained. The colour distribution of the Milky Way with galactic longitude shows reddening toward the galactic centre; the effect is real, not caused by interstellar absorption. A method is given for estimating the upper limit of distances to individual "hot" Milky Way star clouds. The revealed colour distribution in the Milky Way is explained by different stellar composition of the Galaxy in various directions, according to current ideas on the structure and stellar composition of galaxies. The energy flux of 17 bright clouds was measured in the visual region: it was calculated for the infrared ($\lambda \mp 9750$ Å), too, using a black-body model with the observed temperature T_c.

155.006   Comparison of optical indicators of the spiral structure of the Galaxy.   E. B. Kostyakova.
Soobshch. Gos. Astron. Inst. Shternberga, No. 199, p. 3 - 6 (1977). In Russian.
The known optical indicators of the spiral structure of the Galaxy are put together. The joint picture shows several spatial condensations with considerable scattering of individual objects; therefore, it is rather difficult to trace some clearly localized spiral arms in the solar neighbourhood. The locations of the revealed condensations agree well with the results got by the author earlier in the study of the Milky-Way integrated spectrum.

155.007   Estimate of the energy flux from bright Milky-Way regions.   E. B. Kostyakova.
Soobshch. Gos. Astron. Inst. Shternberga, No. 199, p. 7 - 11 (1977). In Russian.
By studying the Milky Way integrated spectrum, the energy flux of 17 bright Milky Way star clouds was calculated for the visual spectral region λ 4000–6400 Å and for the infrared one ($\lambda = 9750$ Å), using for the latter a black-body model with the observed temperature T_c.

155.008   Neutral hydrogen in the region between spiral arms of the Galaxy. Parameters of the gas layer.
I. V. Gosachinskij, I. A. Rakhimov.
Astron. Zh. Akad. Nauk SSSR, Vol. 55, 22 - 26 (1978). In Russian. English translation in Soviet Astron., Vol. 22, No. 1.
The results of 21 cm observations of H I in the region $l = 60°-74°$, $b = \pm 15°$, $r = 8-14$ kpc between the Orion and Perseus spiral arms are reported.

155.009   Voie Lactée et galaxies.   L. Martinet.
Conférences d'astronomie de l'Observatoire, (see 012.004), p. 121 - 155 (1977).
La distribution des étoiles dans notre galaxie; Les mouvements des étoiles; Structure spirale de notre galaxie; Evolution de la galaxie; Les galaxies extérieures.

155.010   Probing our Galaxy.   B. J. Bok.
Focus on the stars, (see 003.002), p. 7 - 59 (1977).

155.011   2.4-micron observation of the Galaxy and the galactic structure.   T. Maihara, N. Oda,
T. Sugiyama, H. Okuda.
Publ. Astron. Soc. Japan, Vol. 30, 1 - 19 (1978).
A near-infrared (2.4 μm) map of the galactic center region has been obtained by a balloon-borne telescope with a 1° resolution. The brightness distribution of the scanned area from $l = 350°$ to 27° within $|b| \lesssim 10°$ contains the galactic nuclear bulge, a very narrow ridge of the disk component, and some additional fine structures. A tentative model of the Galaxy presented here defines an explicit form of the 2.4-μm emissivity distribution as well as that of the interstellar dust distribution across the Galaxy. The model gives the total luminosity of the nuclear bulge to be $2 \times 10^{10} L_\odot$, which is comparable to that of M 31.

155.012   Common-proper-motion pairs in the South Galactic Cap.   J. Spencer Jones, J. B. Alexander.
Observatory, Vol. 98, 49 - 51 (1978).
In this note, the authors report a search for common-proper-motion companions to stars near the Sun in the direction of the South Galactic Cap. The source of nearby stars is R.O. Annals 5. The possible companions are selected from the catalogue of Luyten & La Bonte.

155.013   Radial velocities for outlying satellites and their implications for the mass of the Galaxy.

F. D. A. Hartwick, W. L. W. Sargent.
Astrophys. J., Vol. 221, 512 - 520 (1978).

The first radial velocities have been measured for stars in outlying satellites of the Galaxy — the globular clusters NGC 7492, Palomar 1, 2, 3, 4, 11, 12, 13, and 14, and the dwarf spheroidal galaxies Draco, Sculptor, and Ursa Minor. The velocities of the globular clusters M71 and NGC 7006 were remeasured. These data have been used to estimate the mass of the Galaxy out to about 60 kpc from the center. A new method of analysis has been used to derive the mass. The detailed results depend critically on the assumed form of the velocity ellipsoid for the outer clusters. The velocities do not support Lynden-Bell's suggestion that certain outlying systems are associated with high-velocity neutral hydrogen clouds. An instrumental color-magnitude diagram for the distant cluster Pal 14 is given in the appendix.

155.014   Om Vintergatans spiralstruktur.
S. Wramdemark.
Astron. Tidsskr., Årg.11, 29 - 35 (1978).

155.015   The galactic density wave, molecular clouds and star formation – II.   F. N. Bash.
Bull. American Astron. Soc., Vol. 9, 640 (1978). – Abstract.

155.016   A study of galactic absorption as revealed by the Rubin et al. sample of Sc galaxies.   P. Teerikorpi.
Astron. Astrophys., Vol. 64, 379 - 388 (1978).

The Rubin et al. (1976) sample of Sc galaxies has been used to study the galactic absorption. A method employing the observed Hubble moduli HM and colours, and which takes into account the suggested dependence of colour on luminosity, leads to the result that there is a slope of about $0.^m4$ in the cosecant diagram towards the Orion arm, while the opposite (centre) direction is characterized by a smaller absorption, on the average, in front of the sample galaxies. It is concluded that the Rubin et al. anisotropy of the Hubble moduli, the so-called Rubin-Ford effect, is more probably caused by galactic absorption than by galactic motion.

155.017 UBV photometry in the nuclear bulge of the Galaxy.
B. Loibl, R. Schröder, U. Haug.
Astron. Astrophys., Vol. 65, 65 - 70 (1978).

UBV photometry in three low absorption fields centered at $l \approx 0°$ and $b = -7°$, $-10.^o5$ and $-16°$ shows a strong decrease of the areal density of red giants with $(U–B)_0 > 1.^m2$ in the nuclear bulge with increasing negative galactic latitude.

155.018   The galactic distribution of OH/IR stars.
H. J. Habing.
IAU Colloq. No. 42, (see 012.012), p. 401 - 433 (1977).
Review paper.

155.019   Die Spiralstruktur des Milchstraßensystems und der Aufbau seines Kerngebietes.   J. Reiche.
Astron. Schule, 15. Jahrg., 27 - 31 (1978).

155.020   Determination of the spiral pattern speed of the Galaxy.   M. A. Gordon.
Astrophys. J., Vol. 222, 100 - 102 (1978).

The carbon monoxide abundance as a function of galactocentric radius, in conjunction with a newly determined galactic rotation curve, gives a spiral pattern speed of 11.5 ± 1.5 km s^{-1} kpc^{-1}. This value is based upon the assumption that the inner Lindblad resonance occurs where the CO abundance exhibits a sharp discontinuity, at a radius of 4 kpc.

155.021   Neutral hydrogen in the Galaxy and the galactic shocks.   T. Sawa.
Astrophys. Space Sci., Vol. 53, 467 - 478 (1978).

To discriminate the galactic shock theory from the linear density-wave theory in comparison with neutral hydrogen data in the Galaxy, the author calculates model-line profiles and $T_b(l,v)$ (brightness temperature) diagrams of 21-cm line both for the two theories in the longitude range $15° < l < 100°$. It is shown that major differences between the two models appear in the tangential directions of spiral arms and of inter-arm regions. The inter-arm region appears as a trough of the brightness temperature in the shock model. An observed trough on a $T_b(l,v)$ diagram at $l = 80°-100°$, $v = -20$ km s^{-1} is reproduced reasonably well by the shock model, while the linear model fails to reproduce it. Effects of the galactic shocks on the terminal velocity is also discussed.

155.022 UBV surface brightness photometry of the Milky Way in Scorpius from the space probe Helios 1.
M. Hanner, C. Leinert, E. Pitz.
Astron. Astrophys., Vol. 65, 245 - 249 (1978).

The zodiacal light experiment on Helios measures the brightness of the Milky Way along four strips of constant ecliptical latitude. UBV brightnesses for the Milky Way in Scorpius were derived from Helios 1 measurements and compared to previous photometries and predictions from star counts. The recent photometric survey by Classen shows the best agreement with the Helios data.

155.023   On the spatial density of stars in the surroundings of the sun.   I. N. Latyshev.
Astron. Zh. Akad. Nauk SSSR, Vol. 55, 318 - 321 (1978).
In Russian. English translation in Soviet Astron., Vol. 22, No. 2.

The number of stars with large proper motions and the distributions of nearby stars according to spatial and tangential velocities were considered. It is found that the spatial density of stars in the solar vicinity is not less than 0.15 stars per cubic parsec.

155.024   Eruptive phenomena near the galactic centre.
J. H. Oort.
Phys. Scr., Vol. 17, (see 012.013), 175 - 184 (1978).

Four categories of expanding features are described: (1) Features observed in the 21 cm radiation of atomic hydrogen, (2) agglomerations of dense molecular clouds, (3) an extended ionized-hydrogen feature, (4) high-velocity "clouds" of ionized gas within $\sim 1/2$ pc from the centre.

155.025   Das Zentralgebiet unseres Sternsystems.
W. Pfau.
Sterne, 54. Band, 1 - 10 (1978).

155.026   Fluctuations in the brightness of the diffuse galactic light and in the brightness of the Milky Way.   K. Mattila, H. Scheffler.
Astron. Astrophys., Vol. 66, 211 - 224 (1978).

It is the purpose of this paper to investigate the diffuse galactic light (DGL) fluctuations both as a separate component and in connection with fluctuations of the total brightness of the Milky Way. The authors expect that the fluctuations of the DGL will be valuable in studies concerning the albedo and the scattering function of the grains. They also discuss the possibility of using the DGL fluctuations for deriving information on the statistical parameters of the interstellar clouds.

155.027   Spatial distribution of stars and interstellar dust in the direction of the galactic anticenter.
N. B. Kalandadze, L. N. Kolesnik.
Astrometriya i Astrofizika, Kiev, Vyp. (No.) 34, p. 19 - 29 (1978). In Russian.

A 18 sq. deg. region around the galactic cluster NGC 2129 has been selected to study the space distribution of stars and dust. Spectral types and, where possible, luminosity classes were observed. Photometric data in the BV colour system have been obtained for 2047 stars to the limit magnitude of

$V = 13^m$. The run of the interstellar absorption with distance is shown in a figure. Distances for stars of all spectral classes have been calculated. The resulting stellar space densities as functions of the distance from the sun are derived.

155.028 Hot gas in the Galaxy: how extensive is it?
F. H. Shu.
IAU Symp. No. 77, (see 012.026), p. 139 - 145 (1978).

Recent satellite detections of O VI absorption and soft X-ray emission leave little doubt that, locally, a substantial fraction of the volume of space between interstellar clouds must be filled with rarified and highly ionized gas at temperatures ranging from 2×10^5 to $> 10^6$ K. The physical state of this gas contrasts sharply with the theoretical picture of a largely neutral, warm, intercloud medium. The author reviews the evidence, observational and theoretical, concerning how extensive hot gas at $\sim 10^6$ K might be.

155.029 High velocity clouds: galactic or extragalactic?
R. Giovanelli.
IAU Symp. No. 77, (see 012.026), p. 293 - 298 (1978).

155.030 Structure galactique et formation stellaire dans le bras spiral de la Carène.
A. Ardeberg, E. Maurice.
C. R. Acad. Sci. Paris, Tome 286, Sér. B, 293 - 296 (1978).

From results of spectrographic and photometric observations the authors are able to resolve a number of aggregates of very young stars and interstellar matter situated along the Carina spiral arm. There is a positive residual between stellar and interstellar-medium (calcium and ionized hydrogen gas) radial velocities. This residual peaks on the inner side of the spiral feature. The authors suggest that this may be due to recent passage of a gas shock front providing the conditions favourable for massive star formation.

155.031 Nucleosynthesis in the Galaxy and the chemical composition of old halo stars.
M. Spite, F. Spite.
Astron. Astrophys., Vol. 67, 23 - 31 (1978).

Three very metal-deficient stars (HD 128279, HD 84903 and HD 184711) were observed and analysed together with HD 122563 (a standard for very metal-deficient stars). These three stars are extremely metal deficient and they are now with HD 122563 and HD 140283 the most underabundant known stars. To overcome the relatively poor accuracy of abundance analysis, equivalent widths of other very metal-deficient stars have been gathered in the literature and processed in the same way as the measurements of the authors' spectra. Thus, a sample of eleven stars is built, as homogeneously as possible. From this sample a few trends can be ascertained.

155.032 A local structure–kinematic relation in the Galaxy.
R. B. Shatsova, T. G. Malysheva.
Pis'ma Astron. Zh., Vol. 4, 167 - 172 (1978). In Russian.

A method of investigating local structure of the Galaxy by the parameters of stellar velocity distribution is suggested. The efficiency of this method is examined on OB-stars, Cepheids and Mira Ceti stars with $P = 300 - 350^d$ in the region of the Local system and outside of it.

155.033 Orbital elements and kinematics of the halo stars and of the old disk population: evidence for active phases in the evolution of the Galaxy.   V. A. Marsakov, A. A. Suchkov.
Astron. Zh. Akad. Nauk SSSR, Vol. 55, 472 - 481 (1978). In Russian. English translation in Soviet Astron., Vol. 22, No. 3.

The distributions of orbital eccentricities and of angular momenta for the halo stars and for the old disk population

are considered. The distributions have gaps separating the halo from the disk and dividing the halo population into three groups. From the point of view of star formation during the collapse at the early stages of evolution, the gaps give evidence that there were in the Galaxy long periods of delay of star formation. A comparison of the distributions concerned with those of heavy elements for different populations of the Galaxy leads to the conclusion that this delay is associated with phases of enrichment of the Galaxy with metals: in the enrichment periods star formation was probably suppressed. The kinematics and the orbital elements of the halo stars and of the old disk population allow to conclude that there was no significant relaxation in the halo; the halo subsystems are not stationary, they perform radial oscillations with respect to the galactic centre.

155.034 Chemical evolution of the Galaxy at the initial rapid-collapse phase.   R. Caimmi.
Astrophys. Space Sci., Vol. 54, 453 - 466 (1978).

Equations for the chemical evolution of the Galaxy are derived, accounting for (i) the dynamical evolution of the Galaxy (i.e. the collapse of the proto-galaxy), and (ii) either a variable mass-spectrum in the birth-rate stellar function of the type $B(m, t) = \psi(t)\varphi(m, t)$, or a constant mass-spectrum with variable lower mass limit for star birth: $m_{mf} = m_{mf}(Z)$. Simple equations are adopted for the collapse of the proto-galaxy, accounting for the experimental data (i.e. axial ratio and major semi-axis) relative to the halo and to the disk, and best fitted for a rapid collapse; gas density is assumed to be always uniform. The theoretical results are not in complete agreement with the observed data.

155.035 Distribución de las velocidades residuales de las estrellas en el entorno del Sol.
J. J. de Orus Navarro.
I Asamblea Nacional de Astronomia y Astrofisica, Tenerife, Vol. I, 121 - 124 (1976) = Univ. Barcelona, Dep. Fis. Tierra Cosmos, Pubbl. No. A-26.

155.036 A survey for faint M stars in the direction of the South Galactic Pole.
P. Pesch, N. Sanduleak.
Bull. American Astron. Soc., Vol. 10, 414 (1978). – Abstract.

155.037 Distribution and population characteristics of S stars.
S. J. Yorka, R. F. Wing.
Bull. American Astron. Soc., Vol. 10, 414 (1978). – Abstract.

155.038 On the origin of intermediate-latitude OB stars.
F. House, D. Kilkenny.
Astron. Astrophys., Vol. 67, 421 - 429 (1978).

An attempt is made to trace the origin of early-type stars observed at appreciable distances from the galactic plane. Because uncertainties in the proper motions make space motions and hence dynamical lifetimes rather inaccurate, a theory of oscillations normal to the plane has been used to compute radial velocities for 138 intermediate-latitude OB stars. These theoretical values are then compared with the observed radial velocities, and it is found that the low velocity stars were probably ejected from the plane some time after formation whilst the high velocity stars were ejected very soon after formation. Velocities of ejection perpendicular to the plane are computed and show a narrow distribution with mean $|Z_0| = 7$ km s^{-1} together with a spread of velocities from about 40 to over 200 km s^{-1}. The data are in reasonable agreement with a "sling" effect and "runaway" origin for the stars of our sample.

155.039 Energy balance in the centre of our Galaxy.
A. Tutukov, E. Krügel.
Astron. Astrophys., Vol. 67, 437 - 440 (1978).

The authors qualitatively discuss the balance of energy for the central part of our galaxy ($r < 100$ pc). In order to reconcile the high thermal radio flux with the low X-ray flux they propose that star formation occurs in bursts.

155.040 **Determination of the parameters of the spiral structure of the Galaxy from stellar motions.**
Yu. N. Mishurov, E. D. Pavlovskaya, A. A. Suchkov.
Astron. Tsirk., No. 967,p. 1 - 2 (1977). In Russian.

155.041 **Galactocentric distance of the sun.**
V. T. Belikov, V. V. Syrovoj.
Astron. Tsirk., No. 968, p. 5 - 6 (1977). In Russian.

155.042 **An estimate of the velocity dispersion and surface mass density of the galactic disk based upon a determination of the parameters of the spiral structure derived from stellar kinematics.** A. A. Suchkov.
Astron. Tsirk., No. 972, p. 6 - 7 (1977). In Russian.

155.043 **On the chemical evolution of the galactic disk.**
A. A. Suchkov.
Astron. Tsirk., No. 977, p. 4 - 5 (1977). In Russian.

155.044 **Estudio de la estructura del hemisferio occidental galáctico en la zona de Carina.** S. Garzoli.
An. Soc. Cient. Argentina, Tomo 188, 66 pp. (1969) = An. Com. Invest. Cient., Vol. 1, No. 1, (Nueva Ser.) = Contrib. Inst. Argentino Radioastron., No. 25.

155.045 **Abundances and chemical evolution of the galactic center.** J. Audouze.
IAU Colloq. No. 45, (see 012.032), p. 79 - 101 (1977/78).
From observations of the galactic center using various techniques radioastronomy, millimeter waves (molecules) — infrared and gamma rays, the interstellar matter of this region appears to have been strongly processed into stars: the gas density is much lower than in the solar neighbourhood. From CO measurements one knows that there are many molecular clouds such as SgrB2 where stars are forming now. From IR measurements, there are some indication that low mass stars are relatively more numerous in such regions than in the external regions of the Galaxy. Simple models of chemical evolution are reviewed here.

155.046 **Dynamical effects in the central region of the Galaxy.** R. H. Sanders.
IAU Colloq. No. 45, (see 012.032), p. 103 - 113 (1977/78).
The author discusses the rotation curve, the gas density, and the gas velocities within the inner 4 kpc of the Galaxy.

155.047 **Spectral line observations of the galactic centre region.** R. D. Davies, R. J. Cohen.
IAU Colloq. No. 45, (see 012.032), p. 119 - 120 (1977/78).

155.048 **Constraints to the mass of a black hole at the Galactic centre.** L. M. Ozernoy (*Ozernoj*).
IAU Colloq. No. 45, (see 012.032), p. 121 - 124 (1977/78).
The estimates for the black hole mass range from 10^7 - 10^{11} $M_\odot$ (Novikov and Thorne, 1973) to 10^4 $M_\odot$ (Shklovskii, 1976). The author proposes a method to obtain an upper limit to the black hole mass based on the calculations of the inevitable growth of the mass in the course of tidal disruption of stars surrounding the assumed hole at the Galactic centre.

155.049 **Radial motions near the Galactic centre and the origin of Galactic spiral structure.**
T. Schmidt-Kaler.
IAU Colloq. No. 45, (see 012.032), p. 125 - 129 (1977/78).

155.050 **Chemical properties and age of the components of the galactic halo.** V. Castellani.
IAU Colloq. No. 45, (see 012.032), p. 133 - 147 (1977/78).

155.051 **Vertical and radial distributions of gas and young objects in the Galaxy and Schmidt's law of star formation.** J. Guibert, J. Lequeux, F. Viallefond.
IAU Colloq. No. 45, (see 012.032), p. 165 - 168 (1977/78).
The thickness of the disc distribution of young objects is shown to be significantly smaller than that of the interstellar gas over the whole Galaxy, as predicted by Schmidt's law of star formation. The radial distribution of the density of young objects at $z = 0$, compared with that of the gas, is also consistent with Schmidt's law. The exponent in this law is found to be between 1.5 and 2. A brief discussion of simple models of galactic evolution using Schmidt's law is made.

155.052 **Observational constraints for the chemical evolution of the solar neighbourhood.** M. Grenon.
IAU Colloq. No. 45, (see 012.032), p. 169 - 172 (1977/78).
The basic material for this preliminary investigation is formed of the Northern stars of Gliese's catalogue, completely measured down to the K5 spectral type in the Geneva system. The physical properties and new spatial velocities are derived and, after removal of non members of the local sphere, 230 stars can be used here.

155.053 **A composition gradient in the galactic disk.**
K. A. Janes.
IAU Colloq. No. 45, (see 012.032), p. 173 - 175 (1977/78).

155.054 **Metal enrichment in the first ages of the Galaxy.**
F. Spite, M. Spite.
IAU Colloq. No. 45, (see 012.032), p. 205 - 206 (1977/78).
It is obviously interesting to study in detail the chemical composition of extreme Population II stars since they may give indications about the processes which took place in the very first ages of the Galaxy.

155.055 **Spatial distribution of metallicity in a local part of the Galactic disk.** A. Strobel, J. Strobel.
IAU Colloq. No. 45, (see 012.032), p. 207 - 209 (1977/78).

155.056 **Dynamical history of the stars in relation with age, kinematics, abundances, birthrate.**
M. Mayor.
IAU Colloq. No. 45, (see 012.032), p. 213 - 228 (1977/78).
Review paper.

155.057 **Interaction between the Galactic disk and halo components.** J. P. Ostriker.
IAU Colloq. No. 45, (see 012.032), p. 241 - 246 (1977/78).

155.058 **The cosmogonical significance of the z distribution of stars.** L. G. Balazs.
IAU Colloq. No. 45, (see 012.032), p. 271 - 274 (1977/78).

155.059 **What happened to stellar drifts?** S. V. M. Clube.
IAU Colloq. No. 45, (see 012.032), p. 275 - 278 (1977/78).

155.060 **Possible pattern speeds of a density wave in our Galaxy.** P. J. Grosbøl.
IAU Colloq. No. 45, (see 012.032), p. 279 - 282 (1977/78).

155.061 **Vertical force law K_z and chemical gradient in the Galaxy.** L. Martinet, M. Grenon.
IAU Colloq. No. 45, (see 012.032), p. 289 - 291 (1977/78).

155.062 **A new value for the rotation speed of the spiral structure.** J. Palouš.
IAU Colloq. No. 45, (see 012.032), p. 293 - 296 (1977/78).

155.063 The effect of crossing times through the solar
neighbourhood on the observed stellar age and
metallicity distributions.   C. Turon Lacarrieu, M. Mayor,
L. Martinet.
IAU Colloq. No. 45, (see 012.032), p. 301 - 302 (1977/78).

155.064 Statistics of close stellar approaches to the sun.
I. N. Latyshev.
Astron. Tsirk., No. 987, p. 7 - 8 (1978). In Russian.

155.065 A determination of solar motion from radial veloc-
ities of nearby stars.
K. F. Ogorodnikov, L. P. Osipkov.
Astron. Tsirk., No. 995, p. 1 - 2 (1978). In Russian.

155.066 Evidence in favour of the chemical evolution of the
Galaxy during the halo formation stage.
V. A. Marsakov, A. A. Suchkov.
Astron. Tsirk., No. 996, p. 2 - 4 (1978). In Russian.

A spectrophotometric survey of stars along the
Milky Way. Part IV.   See Abstr. 002.029.

S stars in the Southern Milky Way.
See Abstr. 002.045.

An atlas of galactic neutral hydrogen for the region
$270° \leqslant l \leqslant 310°; -7° \leqslant b \leqslant 2°$.   See Abstr. 002.062.

Observations in the 21-cm neutral hydrogen line.
I. The region $220° \leqslant l \leqslant 294°; -29° \leqslant b \leqslant -11°$. II. The
region around the south celestial pole.   See Abstr. 002.063.

The Milky Way.   See Abstr. 003.027.

Development of the views upon Galaxy structure.
See Abstr. 004.034.

Pourquoi les astronomes suisses doivent-ils poursuivre
leurs recherches dans le ciel Sud.   See Abstr. 013.003.

A rediscussion of determinations of precession and
galactic rotation from Lick proper motions referred to galaxies.
See Abstr. 043.008.

On the bulk yields of nucleosynthesis from massive
stars.   See Abstr. 061.004.

On the origin and evolution of isotopes of carbon,
nitrogen, and oxygen.   See Abstr. 061.047.

Theoretical evolution of extremely metal-poor
stars. II.   See Abstr. 065.007.

s-process elements and galactic evolution.
See Abstr. 065.087.

Tidal disruption of stars and evolution of a massive
black hole under the conditions of the galactic center.
See Abstr. 066.343.

Formal values for constants of solar motion and
galactic rotation from proper motions.   See Abstr.112.001.

Solar motion with respect to the nearest galaxies.
See Abstr. 112.005.

Mean secular parallax at low galactic latitude.
See Abstr. 112.006.

Observations of faint red stars at intermediate galactic latitude.   See Abstr. 113.003.

The Basle catalogue of blue objects in higher galactic
latitudes in SA51, SA54, SA57, SA82 and in the halo field
around M5 (I).   See Abstr. 113.014.

UBV photometry of faint blue stars near the galactic
anticenter.   See Abstr. 113.018.

Absolute magnitudes of M-type stars in the solar
neighborhood.   See Abstr. 113.027.

Absolute magnitudes of F-, G-, and K-type stars in
the solar neighborhood.   See Abstr. 113.028.

Fotometría R G U en un campo del anticentro
galáctico, cerca del NGC 581.   See Abstr. 113.073.

Spectra of giant stars in distant satellites of the
Galaxy.   See Abstr. 114.010.

Zirconium stars in open clusters and in the spiral
arms of the Galaxy.   See Abstr. 114.046.

Spectroscopy of distant blue stars near the galactic
anticenter.   See Abstr. 114.081.

Carbon and nitrogen in halo stars.
See Abstr. 114.091.

On the statistical parallax of O type runaway stars.
See Abstr. 115.011.

Infrared variable stars in the Milky Way field at R.A.
$19^h 16^m$ – Dec. + $18°28'$ (1950).   See Abstr. 122.038.

A large-scale OH sky survey at 1612 MHz. Part II.
The galactic distribution and kinematics of the unidentified
type II OH/IR stars.   See Abstr. 122.063.

Coronal gas in the Galaxy. I. A new survey of interstellar O VI.   See Abstr. 131.019.

Study of the region $348° \leqslant l \leqslant 12°, +3° \leqslant b \leqslant +17°$
in the 21 cm line.   See Abstr. 131.116.

The distribution and color excesses and interstellar
reddening material in the solar neighborhood.
See Abstr. 131.119.

Gas, dust and molecules in the Galaxy.
See Abstr. 131.136.

Star formation rates in the Galaxy.
See Abstr. 131.172.

Interstellar and stellar abundances across the
galactic disk.   See Abstr. 131.250.

Planetary nebulae and nitrogen enrichment in the
Galaxy.   See Abstr. 131.252.

High-resolution radio maps of three H II regions in
the Perseus arm.   See Abstr. 132.002.

Abundance gradients in the Galaxy derived from
H II regions.   See Abstr. 132.013.

Far-infrared observations of H II regions near the
galactic center.   See Abstr. 132.014.

Observations of [O III] 88 micron line emission from
H II regions and the galactic center.   See Abstr. 132.017.

Neutral hydrogen in the region between the spiral arms of the Galaxy. Cloud structure.   See Abstr. 132.044.

First optical detection of W 51 and observations of new H II regions and exciting stars.   See Abstr. 132.053.

OH/IR stars near the galactic centre.
See Abstr. 133.028.

Planetary nebulae and chemical evolution of the Galaxy.   See Abstr. 135.089.

Quasars and galactic evolution.
See Abstr. 141.143.

Optical observations of galactic X-ray transients.
See Abstr. 142.195.

The cosmic-ray isotopes.   See Abstr. 143.015.

Close binaries and evolution of star clusters.
See Abstr. 151.006.

Negative mass instability of flat galaxies.
See Abstr. 151.016.

Possibilities of observing effects due to Lin's density-wave theory.   See Abstr. 151.038.

A systematic comparison of four methods to derive stellar space densities.   See Abstr. 151.053.

The age dependence of stellar velocity dispersion in a scale-covariant (*theory of*) gravitation.   See Abstr. 151.070.

Gravitational separation of elements during certain stages of galactic evolution.   See Abstr. 151.110.

Connections between chemical and dynamical evolution.   See Abstr. 151.119.

Abundances and ages of disk clusters.
See Abstr. 153.044.

Angular momentum in the Local Group.
See Abstr. 158.236.

Automated faint object detection.
See Abstr. 158.266.

The Magellanic Stream: the turbulent wake of the Magellanic Clouds in the halo of the Galaxy.
See Abstr. 159.007.

On the Magellanic Stream, the mass of the Galaxy and the age of the Universe.   See Abstr. 159.013.

156 Galactic Magnetic Field, Galactic Radio, Infrared Radiation

156.001 Studies of low-frequency recombination lines from the direction of the Galactic Centre and other galactic sources.   A. Pedlar, R. D. Davies, L. Hart, P. A. Shaver.
Mon. Not. R. Astron. Soc., Vol. 182, 473 - 488 (1978).

Recombination lines at 242, 328 and 408 MHz are reported from the direction of the Galactic Centre and the H II regions W35 and W43. The lines from the direction of the Galactic Centre can only be explained in terms of stimulated emission which appears to originate in an extended, low-density, low-emission measure H II region in the line-of-sight to the Galactic Centre.

156.002 Infrared observations of the galactic center. II. [Ne II] emission.   S. P. Willner.
Astrophys. J., Vol. 219, 870 - 872 (1978).

The [Ne II] line flux has been measured with a 5″ diaphragm at eight positions in the galactic center (Sgr A—West) region. The line fluxes— and hence the density of ionized gas— are not correlated with the presence of discrete, compact sources of 10 μm continuum radiation. The discrete sources probably contain sources of energy that heat nearby dust instead of being regions of enhanced dust density.

156.003 Submillimeter observations of the galactic center.   R. H. Hildebrand, S. E. Whitcomb, R. Winston, R. F. Stiening, D. A. Harper, S. H. Moseley.
Astrophys. J., Lett., Vol. 219, L101 - L104 (1978).

The authors have mapped a 15′ × 15′ region surrounding Sgr A at a mean wavelength of 540 μm. The principal feature is a ridge about 10′ long running parallel to the galactic equator and approximately centered on Sgr A but with no peak at that point. The ridge coincides with the 25 and 55 km s^{-1} clouds seen in molecular line observations. The authors estimate the

mass of the clouds and discuss their positions with respect to the galactic center.

156.004 A model for the radiation from the galactic centre.   E. Krügel, A. V. Tutukov.
Astron. Astrophys., Vol. 63, 375 - 382 (1978).

The equations of radiative transport are solved numerically for the nucleus of our galaxy. The authors assume that the dust is heated by O stars and K giants. The distributions of gas, grains, and stars are chosen in accordance with radio and infrared data. Computed spectra are compared with observations.

156.005 Infrared observations of the galactic center. III. 2.2 micron spectroscopy.   G. Neugebauer, E. E. Becklin, K. Matthews, C. G. Wynn-Williams.
Astrophys. J., Vol. 220, 149 - 155 (1978).

Observations of the Bγ line at 2.17 μm are presented for the central few parsecs of the galactic-center region. The surface-brightness distribution of the line with 3.″3 and 6.″6 resolution has been sampled in 25 locations. The results show that the ionized gas is concentrated within the central parsec of the galactic-center region; the fine-scale structure seen at 10 μm does not appear to be present in the ionized gas. A comparison of the Bγ line strength with radio observations shows that the 2.2 μm extinction is consistent with the 2.7 mag determined from the reddening of the late-type stars. Observations of the CO band at 2.3 μm in the central 30″ are also presented; they strengthen the interpretation that the 2 μm flux arises from late-type giant stars.

156.006 Blick ins Zentrum der Milchstraße.   W. Hofmann.
Umschau, 78. Jahrg., 150 - 151 (1978).

A dry ice cooled infrared telescope was used to study the central region of the Milky Way in the $1-3.5 \mu m$ wavelength region from board the balloon-borne gondola THISBE.

156.007 Le centre de notre galaxie.   J. Lequeux.
La recherche en astrophysique, (see 003.001), p. 150 - 155 (1977).

156.008 The infrared emission of the Galactic center.
G. H. Rieke, C. M. Telesco, D. A. Harper.
Astrophys. J., Vol. 220, 556 - 567 (1978).
The authors present ground-based and airborne infrared observations of the Galactic center near Sgr A. The principal results include: (1) A map at 56 μm (28" beam FWHM) which shows a ridge of emission in which are immersed two peaks, one at the infrared cluster and another 1' to the southwest. Early-type stars may produce much of the luminosity seen as thermally reradiated far-infrared flux. (2) A map of the infrared cluster at 34 μm (8".5 beam) which indicates that the distribution at this wavelength is similar to that at 21 μm, with most of the compact sources which emit at $\lambda < 20 \mu m$ emitting relatively weakly at 34 μm. (3) A map at 10 μm which combines observations obtained with 1".5, 2".3, 3", 5".5, and 20" beams and shows for the first time the extended 10 μm structure.

156.009 Nonthermal galactic emission below 10 megahertz.
J. C. Novaco, L. W. Brown.
Astrophys. J., Vol. 221, 114 - 123 (1978).
The Radio Astronomy Explorer 2 (RAE 2) lunar orbiting satellite has provided new measurements of the nonthermal galactic radio emission at frequencies below 10 MHz. Measurements of the emission spectra are presented for the center, anticenter, north polar, and south polar directions at 22 frequencies between 0.25 and 9.18 MHz. Survey maps of the spatial distribution of the observed low-frequency galactic emission at 1.31, 2.20, 3.93, 4.70, 6.55, and 9.18 MHz are presented.

156.010 Infrared Brackett α and γ observations of the galactic center.
J. Bally, D. Y. Gezari, R. R. Joyce, N. Z. Scoville.
Bull. American Astron. Soc., Vol. 9, 606 (1978). – Abstract.

156.011 Near infrared photographic study of the galactic center region.   J. E. Grindlay, W. Liller.
Bull. American Astron. Soc., Vol. 9, 615 (1978). – Abstract.

156.012 A thin disc model of the Galactic dynamo.
A. M. Soward.
Astron. Nachr., Band 299, 25 - 33 (1978).
An axisymmetric $\alpha\omega$-dynamo model of the Galactic disc is investigated. The disc is thin and its width, $2h$, varies slowly with distance, $\tilde{\omega}$, from the axis of symmetry. The strength of the α-effect, α, varies linearly with axial distance, z, while differential rotation, $d\Omega/d\tilde{\omega}$, remains constant. Otherwise α and $d\Omega/d\tilde{\omega}$ are arbitrary functions of $\tilde{\omega}$. The results are applied to the special case of the oblate spheroid first investigated by Stix (1976) and later by White (1977). In the limit of small axis ratio the new results agree with those obtained by White and confirm that the dynamo numbers computed by Stix are too small.

156.013 The velocities of 6cm recombination line emission originating near the galactic nucleus.
F. F. Gardner, J. B. Whiteoak.
Proc. Astron. Soc. Australia, Vol. 3, 150 - 152 (1977).

156.014 Scattering of radio emission from the compact object in Sagittarius A.   D. C. Backer.
Astrophys. J., Lett., Vol. 222, L9 - L12 (1978).
The recent observation of anomalous scattering of the compact object in Sgr A at 1 GHz indicates that the true brightness distribution of the object has not been revealed by present interferometer experiments. The scattering apparently occurs in a thermal plasma within 100 pc of the Sgr A object containing weak density fluctuations on scale sizes of 10^5 cm or less.

156.015 On frequency and angular variations of the spectral index of the galactic radio emission in the wavelength range of 0.7–2 m.   E. N. Vinyajkin.
Astron. Zh. Akad. Nauk SSSR, Vol. 55, 307 - 310 (1978). In Russian. English translation in Soviet Astron., Vol. 22, No. 2.
Values of the spectral index of the Galaxy radio emission in the ranges of 151.5–240 MHz and 240–408 MHz have been obtained. In the region with the coordinates $\delta = 35°$, $10^h \leqslant \alpha \leqslant 17^h 30^m$, the spectral index increases essentially near the frequency 240 MHz and in the region with the coordinates $\delta = 52°$, $11^h 30^m \leqslant \alpha \leqslant 19^h 30^m$ near the frequency 408 MHz.

156.016 Stellar polarization and the structure of the magnetic field of our Galaxy.   R. S. Ellis, D. J. Axon.
Astrophys. Space Sci., Vol. 54, 425 - 443 (1978).
The stellar polarization data have been examined using a new catalogue containing accurate stellar distances. On the assumption of a magnetic alignment hypothesis, correlations on the larger distance scale indicate the existence of a dominant regular magnetic field, although its characteristics are difficult to determine. The distribution of the polarization vectors away from the galactic plane has been examined and it is proposed that the two largest loop structures, previously identified as Supernova remnants, are linked by the regular field. Incremental polarization maps have been produced but they show little correlation with the spiral structure. The polarization appears to be saturated at about 1 kpc from the Sun, which is explained as the result of an observational selection effect.

156.017 Far infrared survey of the galactic plane in Cygnus.
M. F. Campbell, H. Thronson, W. F. Hoffmann.
Bull. American Astron. Soc., Vol. 10, 394 (1978). – Abstract.

156.018 A near infrared photographic survey of the northern Milky Way.   J. G. Hoessel, J. H. Elias, R. A. Wade, J. P. Huchra.
Bull. American Astron. Soc., Vol. 10, 394 (1978). – Abstract.

156.019 Balloon borne observation of 2.4-μm brightness distribution of the galactic central region. II.
T. Maihara, N. Oda, T. Sugiyama, H. Okuda.
Bull. American Astron. Soc., Vol. 10, 394 (1978). – Abstract.

156.020 The Galaxy magnetic field.   A. A. Ruzmaikin (Ruzmajkin).
IAU Colloq. No. 45, (see 012.032), p. 73 - 75 (1977/78).
The origin and dynamics of magnetic fields depend crucially on the dynamics of a gas in the Galaxy. In turn the magnetic field is needed to isotropize the cosmic rays (the influence on chemistry) and is of importance for the formation of stars (local dynamics).

A low latitude 21 cm line survey for the longitude range $20° \leqslant l \leqslant 42°$.   See Abstr. 002.038.

The Nançay survey of absorption by galactic neutral hydrogen. I. Absorption towards extragalactic sources. See Abstr. 002.039.

Galactic neutral hydrogen emission-absorption observations from Arecibo.   See Abstr. 131.168.

The galactic extended low-density H II region and

its relation to star formation and diffuse IR emission.
See Abstr. 132.069.

Infrared observations of the galactic center. I.
Nature of the compact sources.   See Abstr. 133.002.

Infrared observations of the galactic center. IV.
The interstellar extinction.   See Abstr. 133.014.

A possible correlation between discrete radio sources
and the fine structure of the galactic background radio radiation.   See Abstr. 141.116.

2.4-micron observation of the Galaxy and the
galactic structure.   See Abstr. 155.011.

Eruptive phenomena near the galactic centre.
See Abstr. 155.024.

157 Galactic Extreme UV, X, Gamma Radiation

157.001   Non-thermal electron bremsstrahlung in the Galactic Disk.   R. Schlickeiser, K. O. Thielheim.
Mon. Not. R. Astron. Soc., Vol. 182, 103 - 109 (1978).

Cosmic gamma-ray production by bremsstrahlung in the interstellar gas involves cosmic ray electrons with energies below 1 GeV. In this energy range the electron energy spectrum is rather poorly known and quite uncertain. The authors estimate the influence of this uncertainty to the gamma-ray bremsstrahlung production rate. Especially it is shown that the integral production rate above 100 MeV is uncertain to a factor 8. The consequences for galactic and extragalactic gamma-ray astronomy are discussed.

157.002   Distribution and energy spectrum of diffuse soft X-rays.   S. Hayakawa, T. Kato, F. Nagase, K. Yamashita, Y. Tanaka.
Astron. Astrophys., Vol. 62, 21 - 28 (1978).

The diffuse component of soft X-rays was observed along the galactic longitudes $150°$ and $330°$. The energy spectrum obtained after subtraction of the general extragalactic component is essentially the same except in the regions inside Loop I and crossing Loop II. The spectrum is fitted with thermal emission spectra of temperatures 0.13 keV for the cosmic abundances and 0.1 keV for the abundance with depletion of metallic elements, both without appreciable interstellar absorption. The X-ray emission measure varies over the ranges $(0.6-1.2)\ 10^{-2}\ cm^{-6}\ pc$ and $(1.6-3.2)\ 10^{-2}\ cm^{-6}\ pc$ in respective cases. The result suggests that a majority of galactic diffuse soft X-rays are emitted from a hot plasma surrounding the solar system.

157.003   Spectral characteristics of the galactic gamma radiation observed by COS-B.   J. A. Paul, K. Bennett, G. F. Bignami, R. Buccheri, P. Caraveo, W. Hermsen, G. Kanbach, H. A. Mayer-Hasselwander, L. Scarsi, B. N. Swanenburg, R. D. Wills.
Astron. Astrophys., Vol. 63, L31 - L33 (1978).

The ESA satellite COS-B has measured the high energy gamma-ray emission from several regions of the galactic disc. The energy spectra of this emission have been measured. The energy spectra of the gamma radiation in the range 50 MeV to a few GeV derived from six different longitude intervals have essentially the same shape, which resembles neither a pure $\pi°$ nor a power law form. These results are discussed in the light of the recent discovery by COS-B of several point sources along the galactic disc.

157.004   Galactic γ-ray spectra, the flux of cosmic ray electrons and cosmic ray gradients.
A. W. Strong, A. W. Wolfendale, K. Bennett, R. D. Wills.
Mon. Not. R. Astron. Soc., Vol. 182, 751 - 760 (1978).

Recent measurements of cosmic γ rays above 35 MeV give spectral shapes which indicate unexpectedly high fluxes of primary electrons in the cosmic ray flux in the producing regions. The γ-ray data indicate that the electron/proton ratio in cosmic rays is probably much higher on average in the Galaxy than the usual estimate for the solar vicinity. Consistency with the absolute gamma-ray intensities requires a simultaneous increase in electron flux and decrease in proton flux relative to solar values in directions away from the Galactic Centre, together with spatial variability.

157.005   A search for cosmic spectral lines from HEAO-1.   J. Matteson, D. Gruber, P. Nolan, L. Peterson.
Bull. American Astron. Soc., Vol. 9, 609 (1978). – Abstract.

157.006   The contribution of discrete sources to the γ-ray emission of the Galaxy.
G. F. Bignami, P. Caraveo, L. Maraschi.
Astron. Astrophys., Vol. 67, 149 - 152 (1978).

The contribution of discrete sources to the galactic γ-ray emission is discussed on the basis of the recent observations of γ-ray sources by the COS B satellite. It is shown that, if the sources are uniformly distributed throughout the Galaxy, they can account for at least 40% of the galactic γ-ray luminosity. The whole galactic luminosity can be due to discrete sources if their density within 5 kpc of the galactic centre is substantially higher than in the outer regions of the Galaxy.

157.007   Far-ultraviolet studies. IV. Spectroscopy of north and south galactic pole regions observed from Apollo 17.   R. C. Henry, P. D. Feldman, W. G. Fastie, A. Weinstein.
Astrophys. J., Vol. 223, 437 - 446 (1978).

A scanning $(1180-1680\ Å)$ spectrometer has been used on trans-Earth coast to observe dark high-galactic-latitude regions of the sky. Difficulties with instrumentally scattered solar-system $L\alpha$ are overcome, and the intensity and spectrum of the cosmic far-ultraviolet background have been obtained. The result is consistent with models for the radiation field expected from a hot, dense intergalactic plasma. Other possible origins for the radiation field are discussed.

157.008   Features in the diffuse X-ray background between 0.5 and 1.2 keV within $40°$ of the galactic center.
P. M. Fried, J. A. Nousek.
Bull. American Astron. Soc., Vol. 10, 390 (1978). – Abstract.

157.009   Gamma-ray emission from the galactic plane.   D. J. Thompson, C. E. Fichtel, R. C. Hartman, D. A. Kniffen, G. A. Simpson.
Bull. American Astron. Soc., Vol. 10, 395 (1978). – Abstract.

157.010   X-ray observations of the North Polar Spur.   D. C. Iwan.
Bull. American Astron. Soc., Vol. 10, 434 - 435 (1978).

Abstract.

157.011   **X-ray outbursts in our Galaxy.**   W. H. G. Lewin.
American Sci., Vol. 65, 605 - 613 (1977). — Abstr.
in Phys. Abstr., Vol. 81, 41084 (1978).

Far-ultraviolet studies. III. A search for light scattered at large angles by dust.   See Abstr. 131.185.

High energy X-ray observations by HEAO-1 of the Crab Nebula, Cyg X-1, and the galactic center.
See Abstr. 142.090.

High-resolution observations of X-ray sources at the galactic center.   See Abstr. 142.151.

Recent high energy gamma-ray results from SAS-2.
See Abstr. 142.709.

Diffuse gamma radiation.   See Abstr. 142.720.

Final results on the diffuse gamma ray emission observed by SAS-2.   See Abstr. 142.725.

On the identification of celestial γ-ray sources.
See Abstr. 142.732.

COS-B observations of localized high-energy gamma-ray emission.   See Abstr. 142.733.

X-ray characteristics of normal galaxies.
See Abstr. 158.263.

158 Single and Multiple Galaxies, Peculiar Objects

Single and Multiple Galaxies

158.001 Ten new southern galaxies with broad emission lines.
R. M. West, T. M. Borchkhadze, J. Breysacher,
S. Laustsen, H.-E. Schuster.
Astron. Astrophys., Suppl. Ser., Vol. 31, 55 - 60 (1978).

Spectroscopic and photometric observations have revealed two new Seyfert galaxies of class 2 (IC 4870 [105–IG11], 185–IG13), and eight other galaxies with strong emission lines (054–IG15, 079–G16, 148–IG02, 153–IG16, 153–IG21, 184–IG65, 286–IG19, 290–G01).

158.002 Near-infrared polarization of the nucleus of M31.
R. F. Jameson, J. Hough.
Mon. Not. R. Astron. Soc., Vol. 182, 179 - 180 (1978).

The nucleus of M31 is found to have zero polarization at wavelengths of 1.2, 1.6 and 2.2 μm. This suggests the absence of non-thermal infrared radiation and dust close to the nucleus, if it is assumed that the magnetic field has a constant direction in the observed region.

158.003 Survey of bright galactic nuclei.
W. C. Keel, D. W. Weedman.
Astron. J., Vol. 83, 1 - 10 (1978).

Photographic and partial photoelectric UBV observations are given for the 448 galaxies thought to have the apparently brightest nuclei. The galaxies are ranked and quantitatively calibrated in order of their nuclear brightness. Objects of special interest because of their similarity to Seyfert galaxies and their unusually high surface brightness are pointed out specifically.

158.004 Neutral hydrogen in the elliptical galaxy NGC 4636.
G. R. Knapp, S. M. Faber, J. S. Gallagher.
Astron. J., Vol. 83, 11 - 12 (1978).

H I observations of the elliptical galaxy NGC 4636 show that it contains $\sim 8 \times 10^8 M_\odot$ of neutral hydrogen. The relative H I content is $M_{H I}/L_{pg} = 0.04\ M_\odot/L_\odot$, comparable with the values for two other detected ellipticals, NGC 1052 and NGC 4278.

158.005 The velocity field of the barred spiral galaxy NGC 5383. C. J. Peterson, V. C. Rubin, W. K. Ford, Jr., N. Thonnard.
Astrophys. J., Vol. 219, 31 - 45 (1978).

The velocity field of the barred spiral galaxy NGC 5383 has large-scale deviations from the pattern expected for only circular motions. Velocity gradients across the nucleus show a simple sinusoidal dependence on position angle, but the interpretation in terms of rotational and radial motions is dependent upon the adopted geometry of the galaxy. Exterior to the nucleus, the kinematics are more complex and are discussed in terms of two models. A comparison with the barred spiral NGC 3351 shows NGC 5383 to be more luminous and bluer, and to have a higher hydrogen mass and higher angular momentum.

158.006 The stellar population and dust lane in Centaurus A (= NGC 5128). A. W. Rodgers.
Astrophys. J., Lett., Vol. 219, L7 - L11 (1978).

Multichannel scans of the energy distribution of three regions 1′ from the nucleus of NGC 5128 are described. Two regions lie outside the central dust lane and a third is in it. The stellar energy distribution corresponds very closely to that of a "standard" giant elliptical galaxy. A reddening law is derived for the matter in the dust lane and found to be similar to that in Perseus. Characteristics of the dust lane are discussed.

158.007 ESO 113-IG45: galaxy and/or quasar?
R. M. West, A. C. Danks, G. Alcaino.
Astron. Astrophys., Vol. 62, L13 - L16 (1978).

Spectroscopy, UBV photometry and photography have been obtained of the extraordinary 13th magnitude object ESO 113-IG45 identified as a Seyfert galaxy by Fairall (1977). The nucleus is stellar-like and several times more luminous than the surrounding envelope which has a well-developed lane-structure. It is the intrinsically most luminous Seyfert nucleus yet known, and may be described as a "quasar in the center of a (spiral) galaxy". It is probably associated with the X-ray source 2A0120-591.

158.008 Dynamics of early-type galaxies. III. The rotation curve of the S0 galaxy NGC 4762.
F. Bertola, M. Capaccioli.
Astrophys. J., Vol. 219, 404 - 407 (1978).

The authors present the rotation curve of the flat S0 galaxy NGC 4762, derived from measurements of the absorption lines of Ca II H and K out to a distance of 1′.1 (5 kpc) from the nucleus. This curve shows a moderate central gradient and a steady increase up to a value of 165 km s^{-1}, with no definite indication that the turnover velocity has been reached. Using a simple model, the authors estimate that the mass up to the last observed point is $3.5 \times 10^{10} M_\odot$. The corresponding mean mass-to-light ratio has the small value $M/L_B = 4$. It is pointed out that S0's and spiral galaxies appear to possess similar amounts of angular momentum.

158.009 The enormous mass of the elliptical galaxy M87: a model for the extended X-ray source.
W. G. Mathews.
Astrophys. J., Vol. 219, 413 - 423 (1978) = Lick Obs. Bull., No. 766.

An analysis of the X-ray data from the Virgo cluster indicates that the mass of the giant elliptical galaxy M87 may be $\sim 10^{14} M_\odot$ or greater. This large mass is required in order to confine the extended thermal X-ray source to its observed projected size – provided the gas which radiates X-rays is essentially isothermal ($T = 3 \times 10^7$ K) and in hydrostatic equilibrium. Observations of polarized radio emission from the core source in M87 provide further indirect support for the existence of a massive, low-luminosity halo. The gas at $T \approx 3 \times 10^7$ K which surrounds M87 cools at its center in less than a Hubble time, and produces the H II region which is observed there. The total mass of hot gas in M87 is, very approximately, $5 \times 10^{12} M_\odot$.

158.010 Observations of the outflow of gas from the nucleus of NGC 253 and its implications for the stellar population of the nucleus. M.-H. Ulrich.
Astrophys. J., Vol. 219, 424 - 436 (1978).

The ionized gas in the central region of the nearby spiral galaxy NGC 253 has been observed spectrographically with the following three objectives: (i) to obtain data on the velocity field as accurately as possible in position and velocities; (ii) to examine critically a model proposed earlier where gas is flowing out of the nucleus; (iii) and in the event that the presence of this outflow can be established unambiguously, to explore the implications of such a phenomenon for the stellar population of the nucleus.

158.011 Planetary nebulae in Local Group galaxies. V. The Andromeda galaxy. H. C. Ford, G. H. Jacoby.
Astrophys. J., Vol. 219, 437 - 444 (1978).

The authors present identifications of 315 planetary nebulae in seven fields of M31. The nebulae are isolated by comparing pairs of λ5007 on-line and off-line image intensifier photographs of fields along the major axis and minor axis of M31. Equatorial coordinates are derived for the nebulae. A strong concentration to the center is shown. A derived planetary count-to-luminosity ratio is roughly constant across M31 and indicates a total of 2700 planetary nebulae in M31 within 3 mag of the brightest nebula. A simple model for the luminosity evolution of planetary nebulae is used to predict a lower limit of 5400 planetary nebulae in M31 with radii less than 0.6 pc. Comparison of the number of bright planetary nebulae in M31 with the luminosity function of solar-neighborhood planetary nebulae results in an estimate of 7,000–27,000 planetary nebulae in M31 with radii less than 0.6 pc.

158.012   **Luminous variable stars in M31 and M33.**
            R. M. Humphreys.
Astrophys. J., Vol. 219, 445 - 451 (1978).
    Spectra and *UBVRI* photometry of eight luminous blue variables in M31 and M33 reveal the presence of both ultraviolet and infrared excess radiation. These supergiants, including four Hubble-Sandage variables, have spectra characterized by a strong ultraviolet continuum, no apparent Balmer discontinuity, and emission lines of H, He I, Fe II, and [Fe II]. The hot, very luminous supergiants are spectroscopically and photometrically similar to η Car and S Dor and related objects in our Galaxy and the LMC. Together they form a recognizable group of luminous blue stars in four separate galaxies in the Local Group.

158.013   **The spectral-flux distribution (0.36–3.5 μm) of
            nonstellar light from the broad-line radio galaxies**
3C 227 and 3C 382.   S. L. O'Dell, J. J. Puschell, W. A. Stein, J. W. Warner, M.-H. Ulrich.
Astrophys. J., Vol. 219, 818 - 821 (1978).
    Optical-infrared (0.36–3.5 μm) photometric data of the radio galaxy 3C 382, obtained in 1976, show a distinct excess at the longest wavelengths, an ultraviolet excess, and a significant increase in brightness since previous photometry by Sandage. A flux excess at infrared wavelengths has also been observed from 3C 227. These properties demonstrate the presence of emission not attributable to stars, probably originating in the nucleus, extending into the infrared portion of the spectrum.

158.014   **Radio and X-ray variability of the nucleus of
            Centaurus A (NGC 5128).**   J. H. Beall, W. K.
Rose, W. Graf, K. M. Price, W. A. Dent, R. W. Hobbs, E. K. Conklin, B. L. Ulich, B. R. Dennis, C. J. Crannell, J. F. Dolan, K. J. Frost, L. E. Orwig.
Astrophys. J., Vol. 219, 836 - 844 (1978).
    Centaurus A (NGC 5128) has been observed by the authors at radio frequencies of 10.7, 31.4, 85.2, and 89 GHz and at X-ray energies greater than 20 keV. These observations, together with the results which have been reported by other workers, are interpreted in terms of models of the nucleus of this radio galaxy. The source exhibits significant variability in all the observed radio frequencies. The observed radio and X-ray intensities show some concurrent variations but do not track one another throughout the observations. A model of the source in which X-rays are produced by inverse Compton scattering of blackbody photons by relativistic electrons is proposed to explain these observations. The observed variations in the electromagnetic spectrum are shown to be consistent with adiabatic expansion of a trapped plasma in conjunction with turbulent accelerations of the relativistic electrons.

158.015   **The discontinuity between elliptical and disk
            galaxies.**   F. Bertola, M. Capaccioli.
Astrophys. J., Lett., Vol. 219, L95 - L96 (1978).
    An interpretation of the Hubble morphological classification is discussed. Through the presentation of a number of lines of observational evidence, it is shown that two main classes of galaxies are recognizable: one contains spheroidal

systems (elliptical galaxies); the other contains those galaxies showing both a spheroidal component and a disk (S0 and spiral galaxies). These two classes exhibit a strong discontinuity in angular momentum per mass unit.

158.016   **What are Zwicky's compact galaxies?**
            A. P. Fairall.
Observatory, Vol. 98, 1 - 7 (1978).
    Properties of Zwicky's compact galaxies are reviewed and samples of them are compared with ordinary galaxies in terms of appearances on various photographs, central surface brightnesses, spectroscopy and spatial distributions. Compact galaxies do not appear to form a separate class of galaxy. At best they are elliptical and spiral galaxies deficient in their outer regions; in part they are a property of the Palomar Sky Survey. Spatial distributions suggest that tidal stripping may be responsible for the deficiency in the outer regions.

158.017   **Gas distribution, motions and dynamics for some
            dwarf irregular galaxies.**   R. B. Tully, L.
Bottinelli, J. R. Fisher, L. Gouguenheim, R. Sancisi, H. van Woerden.
Astron. Astrophys., Vol. 63, 37 - 47 (1978).
    The magellanic irregular galaxies DDO 125 (M_{pg} = $-15^m.9$) and Ho I ($M_{pg} = -14^m.4$) have been observed with the Westerbork Synthesis Radio Telescope in the 21-cm H I line. Both galaxies are found to have well-ordered velocity fields which can be described as solid-body rotation over most of the disk, with a hint of a velocity turnover toward the extremities. For both, mass models have been fitted. The masses derived are of order $5 \cdot 10^8 \, M_\odot$, and mass/luminosity ratios are of order 3 in both systems.

158.018   **A 2.7 GHz continuum survey of the M 81 - M 82
            system.**
D. T. Emerson, P. P. Kronberg, R. Wielebinski.
Astron. Astrophys., Vol. 63, 49 - 52 (1978).
    The authors have conducted a search for low-level 11 cm continuum emission from the intergalactic gas stream between the spiral galaxy M 81 and the irregular galaxy M 82. No such emission was detected. The improved 2.7 GHz integrated flux density for M 81 combined with a revised 4.8 GHz flux density defines the spectrum of this galaxy more accurately. A test for faint disc emission in the plane of M 82 gives a negative result.

158.019   **Spiral and irregular galaxies at 2.7 and 8.1 GHz IV:
            NGC 3079 and NGC 4490.**
E. R. Seaquist, L. Davis, R. C. Bignell.
Astron. Astrophys., Vol. 63, 199 - 208 (1978).
    The authors present aperture synthesis observations of the peculiar edge-on spiral NGC 3079 and the interacting pair NGC 4490/4485 at 2695 MHz (2.7 GHz) and 8085 MHz (8.1 GHz). NGC 3079 is resolved into a compact nuclear component and an extended component which takes the form of a ridge parallel to the plane of the galaxy. The most outstanding feature is a triple source aligned roughly perpendicular to the galactic ridge. NGC 4490 shows a complex distribution of radio emitting regions some of which coincide with H II complexes in this object. The authors discuss the distribution of radio emission and conclude that the nonthermal emission is strongest in regions where star formation is most active.

158.020   **Upper limits to CO emission in five external galaxies.**
            F. Combes, P. J. Encrenaz, R. Lucas, L. Weliachew.
Astron. Astrophys., Vol. 63, 303 (1978).
    Upper limits on the CO line intensity are presented for the five following galaxies M81, NGC6822, II Zw40, II Zw70, II Zw71.

158.021   **Is Fairall 9 an X-ray Seyfert galaxy?**   G.R. Ricker.
            Nature, Vol. 271, 334 (1978).

F-9 seems to be the brightest Seyfert galaxy known in the Southern Hemisphere. If the identification of F-9 with 2AO120−591 (4UO105−59) suggested here is correct, then its X-ray luminosity (2−10 keV) is approximately $3 \cdot 10^{44}$ erg s^{-1}, which is comparable to that of the brighter X-ray emitting Seyfert galaxies.

158.022 Optical polarimetry of the Seyfert galaxy NGC 4151.
A. Kruszewski.
Acta Astron., Vol. 27, 319 - 337 (1977).
New polarimetric and photometric observations of the Seyfert galaxy NGC 4151 are presented. The observed features, namely the curved polarized flux spectrum, the characteristic for the interstellar polarization dependence of the nuclear polarization on wavelength, the relative constancy in time and large rotation of the position angles with wavelength, all indicate that the observed polarization is not produced by the synchrotron radiation or other nonthermal mechanism. It seems that the variable nuclear component is originally unpolarized and the observed polarization is introduced by the circum-nuclear dust envelope.

158.023 Joint optical and 21-cm line study of the peculiar galaxy NGC 3448.
L. Bottinelli, R. Duflot, L. Gougenheim.
Astron. Astrophys., Vol. 63, 363 - 374, with a correction in Vol. 67, 443 (1978).
The peculiar galaxy NGC 3448 has been investigated from both H I 21-cm line observations and photographs and spectra. From its H I properties, its relatively high excitation, its patchy distribution of H II regions and its low oxygen abundance, NGC 3448 does not fit an S0 type. On the contrary, the authors suggest that NGC 3448 is a late-type galaxy (Sd or Sm), with a very peculiar nucleus. NGC 3448 exhibits a very unusual velocity field in its central parts and the authors present two possible interpretations: (1) it is a single galaxy exhibiting a strong absorption in its central regions, (2) it is a double interacting galaxy with two different rotating systems.

158.024 Rotation and mass of NGC 4088.
N.Carozzi-Meyssonnier.
Astron. Astrophys., Vol. 63, 415 - 424 (1978). In French.
Two photographs of NGC 4088 show the presence of an H II region near the nucleus. Numerous lanes of dust explain the relatively low and constant ratios: Hα/[N II] and [N II]/[S II]. Image-tube spectra confirm the important asymmetry of the galaxy visible on the photographs. The rotation curve for the north-east region is 100 km s^{-1} lower than for the south-west region. The author explains this difference either by ejection of matter from a nuclear hot-spot, or by attraction and tidal effect from the nearby galaxy NGC 4085. Assuming a distance of 11 Mpc, the mass out to 115″ (north-east) and 95″ (south-west) is about $2.5 \times 10^{10} M_\odot$ and the mass to luminosity ratio is about 1.

158.025 Surface brightness and color distributions of elliptical and S0 galaxies. I. The Coma cluster elliptical galaxies.
K. M. Strom, S. E. Strom.
Astron. J., Vol. 83, 73 - 134 (1978).
U and R surface brightness profiles are presented for 96 elliptical and S0 galaxies in the Coma cluster. Analysis of the surface brightness structure and color distributions within these systems leads to the following basic conclusions: (1) Lower-luminosity galaxies have sharper surface brightness falloffs than those characteristic of higher-luminosity galaxies. (2) Galaxies that have the same luminosity can differ widely in their structural properties. (3) The distribution of structural properties for a given luminosity depends on the location of a galaxy within the cluster. (4) The frequency distribution of derived ellipticities is independent of the location of a galaxy within the cluster and of galaxy luminosity. (5) Most

systems that exhibit color changes become bluer with increasing distance from the nucleus. (6) The frequency of systems that show large changes in halo color (on scales of tens of kiloparsecs) is greater among luminous ellipticals. (7) Systems that exhibit halo color gradients have more extended halos than other galaxies of comparable luminosity. (8) The contours of constant color (and presumably constant mean chemical composition) are nearly identical to, but slightly flatter than, the contours of constant surface brightness.

158.026 Apparent associations between bright galaxies and compact galaxies.
M. J. Valtonen, P. Teerikorpi, A. N. Argue.
Astron. J., Vol. 83, 135 - 138 (1978).
The areas around seven highly inclined spiral galaxies have been searched for possible companions, which were classified according to their compactness and shape. Compact spherical galaxies were found preferentially along the major-axis direction of the central spirals, which may indicate real physical associations. A study of the relative orientations of the compact galaxies cataloged by Zwicky and bright nearby galaxies indicates a similar effect.

158.027 H I in the elliptical galaxy NGC 1052.
G. R. Knapp, J. S. Gallagher, S. M. Faber.
Astron. J., Vol. 83, 139 - 146 (1978) = Lick Obs. Bull. No. 784.
The galaxy contains $8 \times 10^8 M_\odot$ of H I; the ratio of H I mass to photographic luminosity is $0.06 M_\odot/L_\odot$. This value is high for an elliptical galaxy but is typical of ellipticals which have detectable 21-cm emission. The properties of NGC 1052 are compared with those of NGC 4278, a second elliptical having large amounts of neutral gas. The 21-cm radio lines and the optical absorption lines are narrower in NGC 1052 than in NGC 4278, despite the fact that NGC 1052 is intrinsically more luminous. This result suggests that at least part of the scatter seen in the velocity dispersion−absolute magnitude relation for elliptical galaxies is intrinsic to the objects themselves.

158.028 87.2 GHz continuum observations of M82, NGC 253, and NGC 1068. M. Jura, R. W. Hobbs, S. P. Maran.
Astron. J., Vol. 83, 153 - 156 (1978).
Observations in the radio continuum at 87.2 GHz have resulted in measurements of M82 and NGC 253 and in a probable detection (2.8-σ level) of the type 2 Seyfert galaxy NGC 1068. By comparing their measurements with those of the Ne II line at 12.8 μ and assuming that the radio emission is produced mainly by thermal bremsstrahlung, the authors find that neon has at least a solar abundance in NGC 253, while it is likely that neon is at least slightly overabundant within NGC 1068 and appreciably overabundant within M82.

158.029 The X-ray structure of NGC 5128.
J. P. Delvaille, A. Epstein, H. W. Schnopper.
Astrophys. J., Lett., Vol. 219, L81 - L83 (1978).
During an observation of NGC 5128 (Cen A) by the rotating modulation collimator on board the SAS-3 X-ray observatory in 1975 early June, data were obtained which suggest the presence of an extended source with ~3′ FWHM surrounding an extremely compact object located at the radio/IR nucleus. The point component exhibited a 25% increase in intensity over a period of 2 to 5 hours.

158.030 A jet in the nucleus of NGC 6251.
A. C. S. Readhead, M. H. Cohen, R. D. Blandford.
Nature, Vol. 272, 131 - 134 (1978).
A core and a jet of length ~1.7 pc have been found in the nucleus of NGC 6251. The nuclear jet is aligned with both the 200-kpc jet and the outer lobes, which have a separation of ~ 3 Mpc; the alignment persists over a range of scales $\gtrsim 5 \times 10^6 : 1$. The observations provide strong evidence that energy and momentum are transferred continuously from the

central core to the outer lobes by a beam.

**158.031   Photoelectric photometry and summary of photo-
electric observations of star clusters in the
Andromeda Nebula.**
A. S. Sharov, V. M. Lyutyj, V. F. Esipov.
Tr. Gos. Astron. Inst. Shternberga, Vol. 47, 16 - 33 (1976).
In Russian.

Results of UBV photoelectric observations of 44 star
clusters in the Andromeda Nebula M 31 are presented. The
total number of photoelectrically measured M 31 clusters has
increased from 99 to 131. A summary of photoelectric measures of clusters in M 31 and its satellites, transformed to a
homogeneous system as far as possible, is obtained. The
authors note a number of objects which may be background
compact galaxies according to their localization on the two-
colour $(U-B)-(B-V)$ diagram. Some clusters may be similar
to the galactic globular cluster NGC 7492. It is possible that
the line of normal colours and the lines of increasing absorption in the two-colour diagrams of the clusters in M 31 and
our Galaxy do not coincide.

**158.032   Interacting nonequilibrium systems near bright gal-
axies. I.**
J. W. Sulentic, H. C. Arp, G. A. di Tullio.
Astrophys. J., Vol. 220, 47 - 61 (1978).

A survey was conducted to identify interacting double
and multiple galaxies within 1° to 99 bright ($10.0 \leq m_{pg} \leq$
12.0) spiral galaxies. An analogous survey was made in 99
"control" fields chosen to be free of bright galaxies. A total
of 96 interacting systems were found in the galaxy fields compared with 49 in control fields. A catalog of interacting
systems and discussion of the sample properties are given. The
interacting systems are found out as far as 100 to 200 kpc
from the central spiral, approaching the region of the recently
postulated " halos" of spiral galaxies. Surprisingly, however,
these physical companions are concentrated along the minor
axis of the average edge-on spiral, confirming an original result
of Holmberg. A possible interpretation of this result would be
an ejection origin for the companion galaxies.

**158.033   An optical and infrared study of NGC 2768 and
NGC 3115.**   K. M. Strom, S. E. Strom, D. C.
Wells, W. Romanishin.
Astrophys. J., Vol. 220, 62 - 74 (1978).

Wide-band optical and infrared photometric measurements of the galaxies NGC 2768 (E5) and NGC 3115 (E7/S0)
are used to test the efficacy of the color indices ($U - V$) and
($U - R$) as metal-abundance indicators. The data permit a
detailed study of the surface brightness and color distribution
in NGC 2768; this study provides additional support for
gas-dynamical models of E galaxy formation.

**158.034   Photometric studies of composite stellar systems.
I. CO and JHK observations of E and S0 galaxies.**
J. A. Frogel, S. E. Persson, M. Aaronson, K. Matthews.
Astrophys. J., Vol. 220, 75 - 97 (1978).

New multiaperture infrared photometric observations of
the central regions of 51 early-type galaxies and of the integrated light of five globular clusters are presented. These data
are compared with selected optical observations and with various model predictions. The main results of the work are: (1)
the observed parameters for the brighter galaxies, particularly
the CO index and the $V - K$ color, agree with the predictions
of stellar synthesis models characterized by giant-dominated
populations with $M/L_v < 10$; (2) the galaxian broad-band colors tend to redden with increasing luminosity and decreasing
aperture size; (3) for the globular clusters, there is evidence
that the integrated colors become redder with increasing metallicity; and (4) in bright galaxies the relative changes of
$U - V$, $V - J$, and $J - K$ as functions of radius may differ
from the relative changes as functions of luminosity at a fixed

radius.

**158.035   Detection of H II regions and ionized interarm gas in
the Sombrero Galaxy (NGC 4594).**
F. Schweizer.
Astrophys. J., Vol. 220, 98 - 106 (1978).

New spectroscopic observations of NGC 4594 unambiguously show the presence of H II regions and ionized interarm
gas in that galaxy. The brightest H II regions appear to be excited by early O-type stars, giving evidence for the formation
of high-mass stars. Ionized gas of low density pervades the
visible disk of the galaxy and apparently corotates with the
stars. The unusually high rotational velocities reveal a very
massive galaxy. When combined with available photometric
data, these velocities also indicate a large radial increase of the
mass-to-light ratio similar to that observed in M31.

158.036   On the distribution of CO and H I in M31.
D. T. Emerson.
Astron. Astrophys., Vol. 63, L29 - L30 (1978).

A point-by-point comparison between the CO observations of M31 by Combes et al. (1977) and aperture synthesis
H I data (Emerson 1974, 1976) is made. No significant difference is found in the mean H I/CO ratio between the NE
and SW halves of M31, nor is there clear evidence that CO is
found preferentially on the inside of H I spiral arms.

158.037   Statistical research of the Seyfert galaxies.
R.-l. Liu.
Acta Astron. Sinica, Vol. 18, 227 - 238 (1977).   In Chinese.

On classifying the 70 Seyfert galaxies (SyG) by the criterion of the existence and or the non-existence of the jet and
spiral, the author discovers the fact that the jet and spiral are
incompatible in the morphological scheme of the SyG's. The
jet is the main characteristic property of most of the 2nd class
SyG while the spiral is that of the 1st class. On further classifying the 43 SyG's possessing spectroscopic data, the author
could divide the 1st class SyG into three subclasses: non-active,
moderately active and active type; while that of the 2nd class
SyG could be subdivided into two subclasses: non-active and
active type.

**158.038   The galaxy M 82. Map of the surface brightness and
the distribution of colour indices (U–B) and (B–V)
on the disc of the galaxy.**   B. P. Artamonov.
Astron. Zh. Akad. Nauk SSSR, Vol. 55, 13 - 21 (1978). In
Russian. English translation in Soviet Astron., Vol. 22, No. 1.

A map of the surface brightness $B/^{\square''}$ of the galaxy is
constructed and standardization of brightness in the U, B, V
system of the photographic measurements by comparison
with photoelectric ones is performed. The photometric centre
of brightness of M 82 is close to the dynamic centre. The
discovered phenomenon of the asymmetry of the distribution
of colours in the southern and northern parts of M 82 is discussed on the basis of a rough three-component model of the
galaxy. Values of the optical thickness of dust as function of
the distance from the centre along the major axis of the galaxy
have been obtained.

**158.039   The large-scale distribution of neutral hydrogen in
the elliptical galaxy NGC 4636.**
L. Bottinelli, L. Gouguenheim.
Astron. Astrophys., Vol. 64, L3 - L4 (1978).

The elliptical galaxy NGC 4636 is studied at three adjacent points along the east-west diameter of the galaxy with a
distance between successive points of 4 'using the Nançay
radiotelescope. The H I has a very wide extent; the H I diameter is 1.3 times larger than the photometric diameter. The
relative extent of the gas is much wider than in other early
galaxies, about 3 times larger for example than in the lenticular
galaxy NGC 5102.

158.040 Further detections of H_2O emission from external galaxies. W. K. Huchtmeier, A. Witzel, H. Kühr, I. I. K. Pauliny-Toth, J. Roland.
Astron. Astrophys., Vol. 64, L21 - L24 (1978).

The detection of several sources showing 22 GHz water vapour maser emission in extragalactic systems is reported. The 6_{16} - 5_{23} transition of H_2O was detected in three H II regions in M 33 (including IC 133, which contains the first detected extragalactic H_2O-maser source) and in two H II regions in the Sc galaxy IC 342.

158.041 Observations of neutral hydrogen in early-type galaxies. J. H. Bieging.
Astron. Astrophys., Vol. 64, 23 - 26 (1978).

Measurements of 21-cm H I line emission from eight early-type galaxies are reported. NGC 1023, 2685, 4278, and 4546 were detected. Limits are given for NGC 2880, 3610, 4459, and 4762.

158.042 A compact, high-redshift object silhouetted in front of the E galaxy NGC 1199. H. Arp.
Astrophys. J., Vol. 220, 401 - 417 (1978).

A high-surface-brightness emission line object of redshift $cz = 13,300$ km s^{-1} appears silhouetted against an E galaxy of $cz = 2600$ km s^{-1}. Photographs with different telescopes show that there is a circular area around the compact object that is darker than the background light of the E galaxy. Spectrophotometric observations show that this obscuration reddens as it absorbs the light of the E galaxy behind it, as would be expected of dust associated with the compact object. It is concluded that the compact object is slightly in front of, but at approximately the same distance as, the E galaxy, and that most of its redshift is of origin other than Doppler motion of recession.

158.043 The Simkin effect. W. G. Tifft.
Astrophys. J., Vol. 220, 418 - 425 (1978).

The Simkin effect is discussed for Steward Observatory spectrogram measurements and shown by K – H residuals and comparisons with other redshift sources, including sky-subtracted values, probably not to exceed 30 km s^{-1}. Large effects are, however, shown to exist in Kintner and other redshift sources – up to ±400 km s^{-1} – while the Humason, Mayall, and Sandage redshifts appear to be unaffected. The effect that is seen appears to be explained entirely by 4047 Å mercury interference and to be virtually independent of galaxy magnitude. Such interference cannot explain redshift-magnitude bands in either form or amount of shift.

158.044 The Mg index in old galaxy populations: a theoretical approach. J. R. Mould.
Astrophys. J., Vol. 220, 434 - 441 (1978).

The Mg index at 5175 Å is studied by using the techniques of spectrum synthesis. Population models based on theoretical luminosity functions permit an abundance calibration of the index in integrated light. The index is shown to be strongly sensitive to metallicity, but only weakly sensitive to the slope of the initial mass function and to the Mg isotope ratio. A metallicity difference of a factor of 2.5 is derived between M31 and M32 from Spinrad and Taylor's scanner observations.

158.045 Photometric studies of composite stellar systems. II. Observations of H_2O absorption and the coolest stellar component of E and S0 galaxies.
M. Aaronson, J. A. Frogel, S. E. Persson.
Astrophys. J., Vol. 220, 442 - 448 (1978).

Multiaperture observations of the H_2O absorption feature near 2.0 microns are presented for the nuclei of 37 early-type galaxies, for five globular clusters, and for a selection of stars. The large H_2O absorption found in galaxies indicates that at 2 microns a significant contribution from a stellar giant component at least as late as M5 in spectral type is present. The observations are compared with recently published synthesis models of Tinsley and Gunn and of O'Connell; best agreement is found with O'Connell's models. It does not appear that a significant contribution of carbon stars can account for the discrepancy between the infrared data and the models of Tinsley and Gunn.

158.046 On the corona of M87.
G. de Vaucouleurs, J.-L. Nieto.
Astrophys. J., Vol. 220, 449 - 452 (1978).

The east-west luminosity profile of M87 at ~1' resolution is derived. The law $\mu_1 = 14.615 + 2.658r^{1/4}$ (B mag arcsec^{-2}, r in arcsec) holds out to $r \sim 270''$, beyond which a luminosity excess of up to 0.5 mag is observed. This second component begins suddenly at $r \approx 4.5$ and reaches a maximum at $5' < r < 6'$. The corona is not certainly detected beyond 20'. The new data confirm Oemler's suspicion that the brighter outer corona indicated by earlier McDonald data was due to systematic error in the extrapolated sky level.

158.047 10 micron observations of bright galaxies.
G. H. Rieke, M. J. Lebofsky.
Astrophys. J., Lett., Vol. 220, L37 - L41 (1978).

Fifty-two galaxies with $m_{pg} < 11.0$ have been measured at 10 μm by using small beam diameters, and 22 of this group have been detected. This unbiased galaxy sample demonstrates that infrared-bright galactic nuclei are common, and that these nuclei have very small mass-to-luminosity ratios, with an average $M/L \sim 0.2$ for the nuclei of large spiral galaxies.

158.048 Optical spectra of radio galaxies.
S. A. Grandi, D. E. Osterbrock.
Astrophys. J., Vol. 220, 783 - 789 (1978) = Lick Obs. Bull., No. 783.

Based on Lick image-tube–image-dissector scans of 35 radio galaxies known to have emission lines, the authors find a high degree of association between the spectroscopic classification "broad-line" and the morphological classification "N". Relative line intensities are presented for nine broad-line radio galaxies (BLRG). These data, combined with previously published results, illustrate three spectroscopic differences between BLRG and Seyfert 1 galaxies: BLRG have much weaker Fe II emission, steeper Hα/Hβ ratios, and larger [O III] λ5007/Hβ ratios than Seyfert 1 galaxies.

158.049 The X-ray emitting galaxy Centaurus A.
R. F. Mushotzky, P. J. Serlemitsos, R. H. Becker, E. A. Boldt, S. S. Holt.
Astrophys. J., Vol. 220, 790 - 797 (1978).

The authors report X-ray observations of Cen A in 1975 and 1976. The moderate-resolution X-ray spectral data reported allow to determine for the first time the column density of Fe in an extragalactic object. In addition, the broad bandwidth and high sensitivity of these observations, together with simultaneous radio observations (Beall et al. 1978) and IR observations (Hildebrand et al. 1977), severely constrain models of the X-ray emitting and absorbing regions and allow to develop a unified model of this X-ray source.

158.050 High resolution observations of neutral hydrogen in M31 – III. Spectra: major axis – velocity plots.
D. T. Emerson.
Mon. Not. R. Astron. Soc., Vol. 182, 793 - 795 (1978), Microfiche MN 182/2.

The data from high-resolution observations of the neutral hydrogen in M31 (Emerson 1974, 1976) are presented in a form more suitable for direct comparison with future spectral line observations of M31.

158.051 The structure of the isophotes of elliptical galaxies.
D. Carter.

Mon. Not. R. Astron. Soc., Vol. 182, 797 - 799 (1978), Microfiche MN 182/2.

Models of elliptical galaxies make predictions of the shapes of the isophotes of elliptical galaxies. This paper describes procedures for measurement of parameters representing the shapes of the isophotes of elliptical galaxies, and presents results for a sample of 19 elliptical and S0 galaxies in the region of the cluster Abell 2197 and Abell 2199. The full text of the paper appeared in Microfiche MN 182/2.

158.052   **Periodicity of the Seyfert galaxy NGC 4151 in the optical.**   E. T. Belokon', M. K. Babadzanjanz (*Babadzhanyants*), V. M. Lyuty (*Lyutyj*).
Astron. Astrophys., Suppl. Ser., Vol. 31, 383 - 388 (1978).

Results of an analysis of the optical variability for the Seyfert galaxy NGC 4151 are presented. A homogeneous set of photoelectrical *UBV* observations obtained for about eight years (1968–1975) has been analysed. The light variations have been divided into slowly and rapidly varying components. A 130 day period was found for the "rapid" component using a modification of the trial period method introduced by Jurkevich. Mean lightcurves were obtained in all three colours. A pulse of about 70 days length with amplitude of 0.8 mag in the *U* band was found to be the characteristic feature of the 130 day period.

158.053   **Eight compact and interacting galaxies with emission-line spectra.**   B. E. Westerlund, N. Å. S. Bergvall, A. B. G. Ekman, A. Lauberts.
Astron. Astrophys., Suppl. Ser., Vol. 31, 427 - 436 (1978).

Radial velocities, projected dimensions and morphological descriptions are given for eight southern galaxies or interacting systems together with estimated blue magnitudes of the well-defined objects. Equivalent widths have been determined for the strongest emission lines and used, whenever possible, to estimate electron temperatures, densities and abundances.

158.054   **Surface photometry of barred spiral galaxies. I. NGC 7479 and NGC 7743.**   S. Okamura.
Publ. Astron. Soc. Japan, Vol. 30, 91 - 121 (1978).

Detailed surface photometry of the two barred spiral galaxies NGC 7479 and NGC 7743 is carried out by means of a computerized digital reduction method. Luminosity distributions are first analyzed in the standard manner of surface photometry to yield isophotes, integrated luminosity distributions, and photometric parameters. Both galaxies are decomposed into central bulge, bar, spiral arms, and underlying disk. Characteristics of each component are investigated through analyses of photometric properties and profiles along the bar and the spiral arms.

158.055   **The central regions of M82.**   R. W. O'Connell, J. J. Mangano.
Astrophys. J., Vol. 221, 62 - 79 (1978).

The authors discuss new observations of the structure, kinematics, and emission-line spectra of the high-surface-brightness star clusters in the center of M82 and the inner 600 pc of the adjacent filamentary network. They find evidence of intense bursts of star formation continuing to the present in the central parts of the galaxy but also occurring in the disk $\sim 10^8$ yr ago. They also find that the inner regions of the filaments are ionized and that their line radiation is intrinsic. They believe that the best interpretation at present is a modified version of the tidal model of Gottesman and Weliachew (1977).

158.056   **Rotation of the nuclear region of M31.**   C. J. Peterson.
Astrophys. J., Vol. 221, 80 - 86 (1978).

The author reports on velocity measurements made with a third set of spectra from Mount Hamilton. The details of the observational data and the velocity measurements of this study are given. Comparison with other studies is made. The ob-

served rotation along the major axis with the expected rotational velocities of two-component dynamical models, one due to the author and a more sophisticated model published by Ruiz (1976), is compared.

158.057   **The Seyfert galaxy NGC 4235.**   G. O. Abell, T. S. Eastmond, D. C. Jenner.
Astrophys. J., Lett., Vol. 221, L1 - L2 (1978).

The spiral galaxy NGC 4235 was discovered to have an emission-line spectrum characteristic of type 1 Seyfert galaxies. Although its apparent spatial position might suggest that it is a member of the Virgo cluster, the radial velocity of the galaxy suggests that it is more distant than the Virgo cluster.

158.058   **Virgo A at 1.3 centimeters.**   J. R. Forster, J. Dreher, M. C. H. Wright, W. J. Welch.
Astrophys. J., Lett., Vol. 221, L3 - L6 (1978).

A new 1.3 cm polarization map of Virgo A with 6 '' by 10 ''.5 resolution is presented. The morphology and polarization of Virgo A are briefly discussed, and the spectral indices of the two extended components and the source at the galactic nucleus are derived from a comparison with the 6 cm Cambridge map. M 87 appears to be a wide-tail radio source.

158.059   **The double nucleus of Markarian 374.**   R. J. Terlevich.
Observatory, Vol. 98, 63 (1978).

The optical spectrum of the two components of the nucleus of Markarian 374 has been observed. The eastern component shows the Seyfert spectrum, the western appears to be an F–G foreground star.

158.060   **The discrete redshift and asymmetry in H I profiles.**   W. G. Tifft.
Astrophys. J., Vol. 221, 449 - 455 (1978).

A classification system for galaxy H I profiles is presented which takes note of the asymmetrical distribution of intensity. The distribution of asymmetry is shown to relate to line width in a distinctly nonrandom manner. The asymmetry can be related directly to properties of the redshift when interpreted as a discrete variable.

158.061   **Orbital dynamics of the radio galaxy 3C 129.**   G. G. Byrd, M. J. Valtonen.
Astrophys. J., Vol. 221, 481 - 485 (1978).

The radio galaxy 3C 129 possesses a long radio tail which is shaped in the first approximation like a segment of an ellipse. The authors use this segment to reconstruct the whole orbit of 3C 129 around a nearby supergiant galaxy and calculate the complete set of orbital parameters under two interpretations of the tail. They discuss the possible perturbations by the nearby radio galaxy 3C 129.1 and the effects of buoyancy on the shape of the tail. They also compare the dynamical ages with the properties of radiating electron populations in the tail.

158.062   **A theoretical and observational analysis of He I triplet lines in class 1 Seyfert galaxy spectra.**   F. R. Feldman, G. M. MacAlpine.
Astrophys. J., Vol. 221, 486 - 500 (1978).

The authors have performed a detailed calculation involving the 11 most relevant levels of the helium atom. The equations of statistical equilibrium were solved for several densities and temperatures over a run of optical depth in the λ10830 line. The results for emission-line strengths are presented in graphical form. Atomic parameters and their determinations are discussed in the Appendix.

158.063   **[Fe XI] λ7892 emission in Seyfert galaxies.**   S. A. Grandi.
Astrophys. J., Vol. 221, 501 - 506 (1978) = Lick Obs. Bull.

No. 781.

The coronal line [Fe XI] λ7892 has been found in the spectra of the Seyfert galaxies NGC 1068, NGC 4051, NGC 4151, NGC 5548, Mrk 9, Mrk 335, Mrk 486, Arakelian (Akn) 202, and Akn 564. Upper limits for [Fe XI] λ7892 have been measured for 13 other Seyfert galaxies. There is no correlation between X-ray emission and [Fe XI] λ7892 emission in Seyfert galaxies.

158.064 **Near-infrared velocity dispersions for the nuclear bulges of M31 and M81.** C. Pritchet.
Astrophys. J., Vol. 221, 507 - 511 (1978).

Internal velocity dispersions have been derived for the bulges of the galaxies M31 (NGC 224) and M81 (NGC 3031) by using observations of the near-infrared Ca II triplet. A power spectral technique devised by Illingworth has been employed to obtain σ = 145 km s^{-1} for M31 and σ = 165 km s^{-1} for M81. The distribution function of projected velocity in the bulges of M31 and M81 is consistent with a Gaussian distribution.

158.065 **Star formation rates and infrared radiation.**
C. Struck-Marcell, B. M. Tinsley.
Astrophys. J., Vol. 221, 562 - 566 (1978).

Infrared colors and bolometric luminosities of galaxies are related to their ages and star formation rates. Models for old galaxies show that stellar age gradients could contribute to the $V-K$ color gradients that have been observed in S0 galaxies. Under certain simplifying assumptions the following results hold within a factor of 2: If a region of a galaxy has $M/L_{bol} < 0.5$ solar units, then (1) its star formation rate in $M_{\odot}$ Gyr^{-1} is roughly 0.2 $L_{bol}/L_{\odot}$ and (2) the time scale for star formation to build the total mass is roughly $6(M/L_{bol})$ Gyr.

158.066 **En galax i ett vintergatsfönster.** G. Lyngå.
Astron. Tidsskr., Årg.11, 36 - 43 (1978).

158.067 **Concurrent radio, infrared, optical, and X-ray observations of the nucleus of NGC 4151.**
J. H. Beall, W. K. Rose, B. R. Dennis, C. J. Crannell, J. F. Dolan, K. J. Frost, L. E. Orwig.
Bull. American Astron. Soc., Vol. 9, 558 - 559 (1978). Abstract.

158.068 **Surface density of galaxies around radio galaxies.**
R. Buta, M-H. Ulrich.
Bull. American Astron. Soc., Vol. 9, 583 (1978). – Abstract.

158.069 **Luminous red stars as extragalactic distance indicators.** R. M. Humphreys.
Bull. American Astron. Soc., Vol. 9, 586 (1978). – Abstract.

158.070 **H II regions as extragalactic distance indicators.**
R. C. Kennicutt.
Bull. American Astron. Soc., Vol. 9, 586 (1978). – Abstract.

158.071 **An optical-radio jet in NGC 7385.**
R. W. Ekers, S. M. Simkin.
Bull. American Astron. Soc., Vol. 9, 586 (1978). – Abstract.

158.072 **Surface brightness, ellipticity and color distributions for elliptical galaxies. I. Spiral-rich clusters.**
K. M. Strom, S. E. Strom.
Bull. American Astron. Soc., Vol. 9, 587 (1978). – Abstract.

158.073 **Surface brightness, ellipticity and color distributions for elliptical galaxies. II. cD and spiral-poor clusters.**
S. E. Strom, K. M. Strom.
Bull. American Astron. Soc., Vol. 9, 587 (1978). – Abstract.

158.074 **Dynamical properties along the Hubble sequence.**
V. C. Rubin, N. Thonnard, W. K. Ford, Jr.
Bull. American Astron. Soc., Vol. 9, 587 (1978). – Abstract.

158.075 **Kron-Shane corrections and the faint-end anomaly in Zwicky galaxy counts.** J. E. Felten.
Bull. American Astron. Soc., Vol. 9, 587 (1978). – Abstract.

158.076 **Imaging polarimetry of the jets of M87 and 3C 273.**
G. D. Schmidt, B. M. Peterson.
Bull. American Astron. Soc., Vol. 9, 608 (1978). – Abstract.

158.077 **Do head—tail radio galaxies have large velocities with respect to the intracluster medium?**
M-H. Ulrich.
Bull. American Astron. Soc., Vol. 9, 618 (1978). – Abstract.

158.078 **Detection of an optical halo surrounding the spiral galaxy NGC 4565.** D. J. Hegyi, G. L. Gerber.
Bull. American Astron. Soc., Vol. 9, 619 (1978). – Abstract.

158.079 **The dwarf galaxy 1346—35: optical and neutral hydrogen properties.** N. Thonnard, V. C. Rubin.
Bull. American Astron. Soc., Vol. 9, 619 (1978). – Abstract.

158.080 **An H I survey of Seyfert galaxies.**
T. M. Heckman, B. Balick, W. T. Sullivan III.
Bull. American Astron. Soc., Vol. 9, 619 (1978). – Abstract.

158.081 **[Fe XI] λ7892 emission in Seyfert galaxies.**
S. A. Grandi.
Bull. American Astron. Soc., Vol. 9, 619 - 620 (1978). Abstract.

158.082 **Photometry of ring galaxies.** L. A. Thompson.
Bull. American Astron. Soc., Vol. 9, 623 (1978). Abstract.

158.083 **A study of binary galaxies: the dependence of M and M/L on pair separation.** S. D. Peterson.
Bull. American Astron. Soc., Vol. 9, 628 - 629 (1978). Abstract.

158.084 **Nuclear ring structures in N4314 and N4535.**
G. F. Benedict, G. W. van Citters, J. T. McGraw, P. M. Rybski.
Bull. American Astron. Soc., Vol. 9, 629 (1978). – Abstract.

158.085 **Near infrared spectroscopic observations of galaxies with active nuclei.**
B. Atwood, J. A. Frogel, B. M. Lasker.
Bull. American Astron. Soc., Vol. 9, 629 (1978). – Abstract.

158.086 **Spectroscopic observations of NGC 5128.**
J. A. Graham.
Bull. American Astron. Soc., Vol. 9, 630 (1978). – Abstract.

158.087 **Dynamics of the nucleus of M31.**
P. L. Schechter, J. E. Gunn.
Bull. American Astron. Soc., Vol. 9, 630 (1978). – Abstract.

158.088 **An observational model of the ionized gas in Seyfert and radio-galaxy nuclei.** D. E. Osterbrock.
Bull. American Astron. Soc., Vol. 9, 647 (1978). – Abstract.

158.089 **Observations of an unusual Seyfert galaxy.**
B. J. Wills, D. Wills, A. K. Uomoto, S. Vogt, R. G. Tull, P. Rybski, T. Montemayor, P. Kelton, F. Ghigo, J. N. Douglas, F. Bash.
Bull. American Astron. Soc., Vol. 9, 647 (1978). – Abstract.

158.090 **The nature of Tololo emission line galaxies.**
T. J. Bohuski, A. P. Fairall, D. W. Weedman.
Bull. American Astron. Soc., Vol. 9, 647 - 648 (1978). Abstract.

158.091 M83. I. Deconvolution of stellar population distributions and extinction.
E. B. Jensen, R. J. Talbot, Jr., J. C. Davis, Jr., R. J. Dufour.
Bull. American Astron. Soc., Vol. 9, 648 (1978). — Abstract.

158.092 M83. II. Stellar spiral structure and star formation history.
R. J. Talbot, Jr., E. B. Jensen, R. J. Dufour.
Bull. American Astron. Soc., Vol. 9, 648 (1978). — Abstract.

158.093 M83. III. Excitation and chemical abundances of gaseous nebulae.
R. J. Dufour, R. J. Talbot, Jr., E. B. Jensen.
Bull. American Astron. Soc., Vol. 9, 648 (1978). — Abstract.

158.094 On relations between masses, radii, and metallicities of elliptical galaxies.   B. M. Tinsley.
Bull. American Astron. Soc., Vol. 9, 649 (1978). — Abstract.

158.095 Polarization in elliptical galaxies?   M. Jura.
Bull. American Astron. Soc., Vol. 9, 649 (1978). — Abstract.

158.096 Properties of smooth-arm spiral galaxies.
M. S. Wilkerson, S. E. Strom, K. M. Strom.
Bull. American Astron. Soc., Vol. 9, 649 (1978). — Abstract.

158.097 Hα photometry and Fabry-Perot interferometry of late-type galaxies.   G. de Vaucouleurs.
Bull. American Astron. Soc., Vol. 9, 649 (1978). — Abstract.

158.098 Photometry and Fabry-Perot interferometry of the weak-barred spiral galaxy NGC 253.   W. D. Pence.
Bull. American Astron. Soc., Vol. 9, 649 (1978). — Abstract.

158.099 Seyfert galaxies as X-ray sources.
M. Elvis, T. Maccacaro, A. S. Wilson, M. J. Ward, M. V. Penston, R. A. E. Fosbury, G. C. Perola.
Mon. Not. R. Astron. Soc., Vol. 183, 129 - 157 (1978).
The authors report a specific search of the Ariel V data for X-ray emission from a large sample of Seyfert galaxies. The total number of Seyfert galaxies which probably emit X-rays now stands at 15. The authors review the X-ray observations and discuss the identifications from a statistical point of view, as well as giving notes on individual sources and possible alternative identifications. Correlations of the X-ray luminosity with properties in other wavebands are investigated and the physical association between the X-ray emission and the other components of the Seyfert nucleus discussed. The X-ray luminosity function of Seyfert galaxies and their contribution to the isotropic X-ray background radiation are derived.

158.100 Some remarks concerning the velocity and spatial distributions of galaxies.   S. M. Fall, B. J. T. Jones.
IAU Colloq. No. 37, (see 012.009), p. 141 - 148 (1977).
Les auteurs discutent quelques aspects des rapports entre la distribution spatiale et la distribution des vitesses des galaxies en insistant surtout sur la recherche d'anisotropie. Ils incluent à la fois les relations dynamiques et les effets de sélection qui pourraient être importants dans de telles études.

158.101 The influence of the night sky on velocity determinations from optical absorption lines.
S. M. Simkin.
IAU Colloq. No. 37, (see 012.009), p. 167 - 172 (1977).
It has been known for quite some time that measurements of galaxy radial velocities derived from absorption line spectra can have systematic errors of 50–150 km/s (de Vaucouleurs and de Vaucouleurs, 1967; Roberts, 1972). Even for fairly bright galaxies, errors of this magnitude can be attrib-

uted to distortion of the galaxy spectrum by that of the night sky (Simkin, 1972). This contribution reports recent computer experiments which show that even larger systematic errors (500–600 km/sec) can arise from night sky distortions.

158.102 Systematic errors in galaxy redshifts.
B. M. Lewis.
IAU Colloq. No. 37, (see 012.009), p. 173 - 178 (1977).
The doubling of the sample of data has generally weakened all of the systematic errors that appeared to depend on type, colour and inclination. It has allowed the scope of the "Roberts' effect" to be delineated. Some sources of optical velocities require a small systematic correction to bring them into agreement with the 21 cm-line velocities.

158.103 The mass to light ratio of Markarian galaxies.
J. P. Huchra.
IAU Colloq. No. 37, (see 012.009), p. 289 - 299 (1977).
Une estimation du rapport moyen masse sur luminosité des galaxies de Markarian est déterminée à partir de bonnes mesures de vitesses pour des paires serrées de galaxies de Markarian. Une comparaison avec les données rassemblées par Turner pour des galaxies de champ montre que les galaxies de Markarian et les galaxies de champ ont le même rapport moyen masse sur luminosité. Le degré de regroupement en amas ou en paire des galaxies de Markarian est comparé avec celui trouvé pour les galaxies de champ. Les galaxies de Markarian ne montrent pas une tendance plus grande que les autres galaxies à se regrouper en paires.

158.104 Dynamical state of pairs of galaxies.
I. D. Karachentsev.
IAU Colloq. No. 37, (see 012.009), p. 321 - 335 (1977).
In this report the author confines himself to a discussion of the data on the dynamics and morphological features of double galaxies from catalogue "Catalogue of isolated pairs of galaxies in northern hemisphere" (Karachentsev, 1972) which contains 603 northern sky pairs whose apparent magnitudes of components are brighter than $15^m.7$.

158.105 Mass-to-light ratios of binary galaxies.
E. L. Turner.
IAU Colloq. No. 37, (see 012.009), p. 337 - 345 (1977).
The occurrence of apparently bound pairs of galaxies permits a statistical estimation of galaxian masses and mass-to-light ratios. Such estimates are superior in some ways to those obtained from rotation curve and velocity dispersion studies of individual galaxies and virial theorem analyses of groups and clusters. Unlike the former, binary galaxy mass determinations can measure material well outside the optical object (e.g., a dark halo). And, unlike groups and clusters, binary systems are simple enough to permit the use of an explicit model of the individual galaxy orbits. Because the method is fundamentally a statistical one, it is critically important that the binary systems used in the analysis represent a well-defined statistical sample and that the analysis take into account any biases in the selection of the sample.

158.106 Anomalous redshifts in galaxies and quasars.
H. Arp.
CNRS Colloq. No. 263, (see 012.009), p. 377 - 409 (1977).
Les associations apparentes entre galaxies et quasars sont récapitulées. Les probabilités d'associations accidentelles sont discutées et les relations entre la séparation des quasars et la nature de la galaxie associée sont présentées. De nouveaux cas d'associations apparentes entre galaxies de faible décalage vers le rouge et de grand décalage sont présentés et discutés. Le phénomène des décalages systématiquement plus grands pour les galaxies non elliptiques dans les amas et les groupes de galaxies est abordé.

158.107 The Hubble diagram for Seyfert galaxies and related

objects. E. Ye.Khachikian (*Eh. E. Khachikyan*),
D. W. Weedman.
CNRS Colloq. No. 263, (see 012.009), p. 411 - 426 (1977).

Les auteurs présentent les diagrammes magnitude-dé-
calage spectral et diamètre angulaire-décalage spectral pour
les galaxies de Seyfert. Tous deux montrent des corrélations
indiquant que les décalages vers le rouge sont cosmologiques.
Les luminosités des noyaux des Seyfert de classe I sont com-
parées à celles des QSO en utilisant des nouvelles mesures de
flux de raies en émission.

158.108 A new model for X-ray emission from NGC 4151.
R. K. Manchanda, E. Costa, B. M. Belli, P. Ubertini.
Astrophys. Space Sci., Vol. 53, 231 - 239 (1978).

The authors present an analysis of the reported spectral
features of NGC 4151 in X-rays. It is shown that the origin of
X-rays from the source is inconsistent with a single production
mechanism. The authors suggest a new two-component model
in which soft X-rays arise from the black-body emission of a
tiny hot nucleus with $T \sim 2 \times 10^7$ K and the hard X-ray
photons are generated in an extended region by inverse
Compton scattering of electrons with the infrared photons.

158.109 The giant spiral galaxy M 101. III. Integral prop-
erties of several companion galaxies obtained from
neutral hydrogen measurements. R. J. Allen, J. M. van der
Hulst, W. M. Goss, W. Huchtmeier.
Astron. Astrophys., Vol. 64, 359 - 366 (1978).

The authors present new observations of the total profiles of the neutral hydrogen emission in several galaxies which
are considered to be companions of M 101. Estimates of the
systemic velocity, hydrogen mass, and total mass are obtained
for each galaxy and a discussion of the inaccuracies is given.
Finally, new evidence is presented in favour of a physical
association between M 101 and the luminosity class IV–V
galaxy N5477; the velocity of the latter galaxy lies within
15 km s^{-1} of the nearby gas associated with M101 itself. In addition an extension of a faint H I arm in the outer regions of
M101 is found near N5477.

158.110 The radio properties of Seyfert galaxies.
A. G. de Bruyn, A. S. Wilson.
Astron. Astrophys., Vol. 64, 433 - 444 (1978).

The properties of Seyfert galaxies at radio wavelengths
are discussed from a statistical point of view. In general, the
radio emission is associated with the nucleus of the galaxy
and has a steep, non-thermal spectrum. However, at least 10%
of optically selected Seyferts exhibit a flat spectrum, very
compact, core source. Double radio sources outside the
optical boundaries of the galaxies are not found associated
with optically selected Seyferts. Relations between the radio,
optical, and infrared properties are explored. Correlations
appear to exist between the continuum luminosity at λ 21 cm
and the luminosity in the [O III] λ 5007 Å line and between
the continuum luminosities at λ 21 cm and λ 10μ.

158.111 Evidence for a supermassive object in the nucleus
of the galaxy M87 from SIT and CCD area photometry. P. J. Young, J. A. Westphal, J. Kristian, C. P. Wilson,
F. P. Landauer.
Astrophys. J., Vol. 221, 721 - 730 (1978).

Two-dimensional SIT and CCD detectors have been used
to measure the surface brightness of the peculiar elliptical
radio galaxy M87. The data are given in some detail and are
compared with earlier photographic results. A model-independent dynamical analysis, using the photometric data combined
with spectrographic results by Sargent et al., shows that the
nucleus of M87 contains a compact mass of low luminosity,
with $M = 5 \times 10^9 M_\odot$.

158.112 Dynamical evidence for a central mass concentra-
tion in the galaxy M87. W. L. W. Sargent,

P. J. Young, A. Boksenberg, K. Shortridge, C. R. Lynds,
F. D. A. Hartwick.
Astrophys. J., Vol. 221, 731 - 744 (1978).

The elliptical galaxies NGC 3379 (E1) and M87 (E0) have
been observed spectroscopically with the University College
London Image Photon Counting System. Analysis of the redshifts and velocity dispersions as a function of radius by a
Fourier method is presented. The authors compare the measurements made in the galaxies with models. A model-independent analysis is described. They describe a procedure (also
model independent) by which to obtain the mass-to-light ratio
as a function of radius. They conclude that the observations
of M87 are entirely consistent with the presence of a central
black hole of $\sim 5 \times 10^9 M_\odot$.

158.113 The absolute solar motion and the discrete redshift.
W. G. Tifft.
Astrophys. J., Vol. 221, 756 - 775 (1978).

The concept of the discrete redshift is applied to dwarf
galaxy H I redshifts and line profiles. After removing the solar
motion, derived from the data, it is shown that strong periodicities predicted by the discrete-redshift model are present.
Strong related correlations exist in line profile shape. A model of the redshift based upon ultimate discrete levels uniformly
spaced near 12 km s^{-1} is developed. Each cycle of 72 km s^{-1}
contains a pattern of six sublevels. A model of redshift levels
in terms of energy changes is also introduced.

158.114 The nature of Tololo emission-line galaxies.
T. J. Bohuski, A. P. Fairall, D. W. Weedman.
Astrophys. J., Vol. 221, 776 - 787 (1978).

Photometric and spectroscopic observations are presented
for some existing Tololo emission-line galaxies along with a
new list of such objects. The emission lines from more than
80% of Tololo galaxies seem to be produced by hot stars.
About 2% are class 1 Seyferts, and less than 10% are class 2
Seyferts. Virtually all Tololo galaxies would be discovered
with the criteria of a Markarian-type survey, but only about
one-third of the Markarian galaxies have sufficiently strong
emission lines to be included in the Tololo listings.

158.115 Near-infrared luminosity-sensitive features in M
dwarfs and giants, and in M31 and M32.
J. G. Cohen.
Astrophys. J., Vol. 221, 788 - 796 (1978).

Observations are presented of prominent near-infrared
spectral features in M dwarfs and M giants which elucidate the
behavior of these features (8183–8195 Å doublet of Na I,
the Ca II triplet, 9910 Å FeH band, and TiO bands) as a function of luminosity and effective temperature. These spectral
features have been measured from near-infrared spectra of the
nuclei of M31 and M32. A luminosity function similar to that
of the solar neighborhood is supported by the observations.
The measurements indicate an enhancement of metallicity in
the nucleus of M31 as compared with that of M32.

158.116 The morphological distribution of bright galaxies in
the *UVK* color plane. M. Aaronson.
Astrophys. J., Lett., Vol. 221, L103 - L107 (1978).

Multiaperture K magnitudes have been obtained for a
large number of galaxies ranging in Hubble type from early
spiral to Magellanic irregular. The data allow formation of
accurate integrated $V-K$ colors. Galaxies in which dust effects
are probably minimal lie along a well-defined morphological
sequence in the UVK color plane that is analogous to the well-
known UBV color-morphology relation. The general nature of
the UVK color-morphology relation can be accounted for by
the composite effect of mixing red and blue stars. Because the
$V-K$ color is considerably more sensitive than $B-V$ to age
effects, the UVK color-morphology relation should provide
strong constraints on future synthesis studies of star-formation rates in galaxies.

158.117   Aperture synthesis observations of the neutral hydrogen in the galaxies NGC 4631/NGC 4656.
L. Weliachew, R. Sancisi, M. Guélin.
Astron. Astrophys., Vol. 65, 37 - 45 (1978).
The authors report neutral hydrogen observations of NGC 4631 and NGC 4656. The optical picture of these galaxies is given in a figure where the distribution of neutral hydrogen is also shown.

158.118   A new southern Seyfert 1 galaxy.
R. M. West, A. C. Danks, G. Alcaino.
Astron. Astrophys., Vol. 65, 151 - 152 (1978).
ESO 140-G 43 (Fairall-51) is confirmed as a 14^m type 1 Seyfert galaxy at $V_0 = 4150$ km s^{-1}. $M_V = -20.8$; largest diameter 40 kpc ($H_0 = 55$ km s^{-1} Mpc^{-1}). It has two open spiral arms.

158.119   Surface photometry of elliptical galaxies.
I. R. King.
Astrophys. J., Vol. 222, 1 - 13 (1978).
Photographic surface photometry is presented for 16 giant elliptical galaxies and one S0. The results are presented as radial profiles along the direction halfway between the major and minor axes, and as ellipticities as a function of radius. In four ellipticals the orientation of the major axis shifts significantly with radius. The radial profiles are strikingly similar, but they also differ significantly. Ellipticities decrease outward in some galaxies and increase outward in others.

158.120   Evolutionary synthesis of the stellar population in elliptical galaxies. II. Late M giants and composition effects.   B. M. Tinsley.
Astrophys. J., Vol. 222, 14 - 22 (1978).
Earlier models for elliptical galaxies are revised to agree with recent infrared data, and to incorporate new theoretical isochrones for stars evolving from the main sequence to the base of the giant branch. Infrared colors and spectral features are very sensitive to the luminosity function of late M giants, and they can be reproduced using a function that is consistent with theory and with statistics of old-disk giants. Models with different chemical compositions show that the metal abundance strongly affects all colors, and that its uncertainty seriously limits the accuracy of age determinations. The predicted effects of metallicity on $B - V$ and $V - K$ lead, via empirical color-magnitude relations, to an approximate metallicity-mass relation $Z \propto M^{0.25}$. This relation can be explained in terms of supernova-driven winds, if the initial mass-radius relation is $M_i \propto R_i^{1.5}$, which also leads to qualitative agreement with the present mass-radius relation. The wind model predicts an intrinsic flattening of color-magnitude relations at the bright end.

158.121   Rotation of 10 early-type galaxies.
C. J. Peterson.
Astrophys. J., Vol. 222, 84 - 94 (1978).
Spectra of nine elliptical galaxies and one lenticular system have been measured to determine apparent rotation curves. With two exceptions, only the solid-body part of the rotation curves is observed. The rotation curves of NGC 4459 (S0) and NGC 4621 (E5) are found to flatten beyond the nuclear region. The data show no strong correlation between the central angular velocity and the line-of-sight velocity dispersion. Comparison with the data of this and other studies suggests that the theoretical models do not yet adequately reproduce the rotation curves observed in early-type systems.

158.122   Infrared photometry and polarimetry of NGC 1068.
M. J. Lebofsky, G. H. Rieke, J. C. Kemp.
Astrophys. J., Vol. 222, 95 - 99 (1978).
New observations of NGC 1068 confirm thermal models for the mid-infrared source in the nucleus. Narrow-band 20 μm data show the presence of a silicate emission feature, in detailed agreement with models calculated by Jones et al. (1977). High-spatial-resolution photometry at 1.6 and 2.2 μm and polarimetry from 1.25 to 3.45 μm indicate that the nuclear source is heavily obscured and strongly polarized, and that it may be directly observable only at wavelengths between 1 and 5 μm.

158.123   Detection of neutral hydrogen absorption at $z = 0.045$ in the peculiar galaxy 1506 + 34.
W. A. Baan, A. D. Haschick, P. E. Greenfield.
Astrophys. J., Lett., Vol. 222, L7 - L8 (1978).
Neutral hydrogen absorption at 21 cm has been detected in the radio galaxy 1506 + 34. The absorption feature has a half-width of 110 km s^{-1} and midpoint radial velocity ($c\Delta\lambda/\lambda_0$) of 13,500 km s^{-1}, which is coincident with the optical emission-line velocity of the galaxy of 13,550 km s^{-1}. The high optical depth of the absorbing H I cloud of $\tau = 0.4$ implies a neutral hydrogen column density of $N_H = 8.5 \times 10^{19}$ T_s H I atoms cm^{-2}.

158.124   Detection of a long H I plume emerging from NGC 3628.   A. H. Rots.
Astron. J., Vol. 83, 219 - 223 (1978).
Zwicky (1956) and Kormendy and Bahcall (1974) have described a faint optical plume emerging from the eastern tip of the spiral galaxy NGC 3628. Observations in the 21-cm line of neutral hydrogen reveal the presence of H I in that plume. Although the H I plume extends to a distance of at least 100 kpc from the center of the galaxy, there is a smooth variation in radial velocity over a rather small range and the profile width is very narrow. It seems plausible that this feature is the result of interaction between NGC 3627 and NGC 3628. A reasonably satisfactory two-galaxy-interaction model is presented.

158.125   Photographic photometry of bright galaxies. V. NGC 3898 and NGC 4036.
R. Barbon, L. Benacchio, M. Capaccioli.
Astron. Astrophys., Vol. 65, 165 - 172 (1978).
Surface photographic photometry in the B system is given for NGC 3898 and NGC 4036 up to a threshold brightness level $\mu = 27.0$ mag/arc s^2. Faint features are detected in the luminosity profiles of the two axes in both galaxies. Assuming an Hubble constant $H = 100$ km/s Mpc^{-1}, the overall dimensions of the two objects, the sizes of the nuclear regions and the absolute luminosities appear comparable.

158.126   Polarization of light in the Seyfert galaxy NGC 1068.   A. Elvius.
Astron. Astrophys., Vol. 65, 233 - 238 (1978).
Scans were made through the nucleus of NGC 1068 in four different position angles. Strong polarization (4%) was found in a cloud north-east of the nucleus. Weaker polarization is indicated south-east and north-west of the centre. The north-east cloud complex of dust particles scattering and polarizing the light from the bright nucleus is discussed in relation to models by Jones et al. (1977) for the source of far-infrared radiation in NGC 1068.

158.127   Radioastronomical investigations of galaxies.
R. Wielebinski, translated from the English edition by N. D. Morozova.
Priroda, 1978, No. 4, p. 101 - 111. In Russian.

158.128   Does a hidden mass exist in galaxies?
G. S. Bisnovatyj-Kogan.
Priroda, 1978, No. 4, p. 135 - 136.   In Russian.

158.129   Galaxies with ultraviolet continuum. XI.
B. E. Markarian, V. A. Lipovetskij, Dzh. A. Stepanyan.

Astrofizika, Vol. 13, 397 - 404 (1977). In Russian. English translation in Astrophysics, Vol. 13, No. 3.

The eleventh list of galaxies having intense ultraviolet continuum is presented. The list contains data for 100 objects. The presence of emission lines is either established or suspected among 55 of them. The presence of Seyfert characteristics can be certainly expected of the following objects: No. 1018, 1040, 1044, 1048 and 1095. The object No. 1014 is probably a QSO. It is probable that the objects No. 1015, 1037 and 1055 are variable.

158.130 Spectrophotometry and morphology of the galaxy NGC 6306. I.
M. A. Kazaryan, Eh. E. Khachikyan.
Astrofizika, Vol. 13, 415 - 426 (1977). In Russian. English translation in Astrophysics, Vol. 13, No. 3.

The results of spectrophotometry and morphology of NGC 6306 with UV-continuum in the spectrum are presented. The galaxy consists of many bright knots with strong emission lines. But the highest members of Balmer lines are in absorption. Profiles and equivalent widths of lines are also presented. The electron density n_e is estimated. It is shown that the radial velocities of separate parts of the galaxy differ. It is concluded that NGC 6306 is a very active galaxy with very active star formation. NGC 6306 is physically connected with NGC 6307, which shows no activity.

158.131 Emission line intensity dependence of Seyfert galaxies upon colour index. M. A. Arakelyan.
Astrofizika, Vol. 13, 427 - 436 (1977). In Russian. English translation in Astrophysics, Vol. 13, No. 3.

It is shown that the equivalent width of [O III] λ5007 and the intensity ratio of this line and Hβ in the spectra of Seyfert-type objects decrease with decreasing colour index U−B. Meanwhile both these quantities in the spectra of non-Seyfert galaxies increase with decreasing colour index. The behavior of quantities considered in the spectra of non-Seyfert galaxies can be interpreted as the result of low-electron density of the gas. As to the behavior of the same quantities in the spectra of Seyfert-type objects, it can be ascribed to a higher mean value of electron density and the variation of this value in the colour interval considered.

158.132 On the existence of groups of pairs of galaxies having different physical and kinematical characteristics. L. P. Metik, I. I. Pronik.
Astron. Zh. Akad. Nauk SSSR, Vol. 55, 249 - 261 (1978). In Russian. English translation in Soviet Astron., Vol. 22, No. 2.

47 pairs of galaxies, most of which are tight or interacting, have been considered. All these pairs may be divided into three separate groups according to the definite dependences between the $(U-B)$ colours of their members. Comparison of spectral types, absolute magnitudes, radial velocities and other characteristics of the pair members support the reality of the existence of such groups. Evidence is given that the groups considered differ in origin or evolution.

158.133 The nature of the apparent distribution of faint galaxies. B. I. Fesenko, L. M. Fesenko.
Astron. Zh. Akad. Nauk SSSR, Vol. 55, 262 - 271 (1978). In Russian. English translation in Soviet Astron., Vol. 22, No. 2.

The distribution of galaxies up to the limiting magnitude of $\bar{m} = 19^m - 21^m$ is discussed. If one excludes various distorting factors, the distribution considered is defined completely by the random deviations of the number of galaxy systems with mean multiplicity of visible members of about three or four. That conclusion results from the analysis of Lick counts, of counts in the Jagellonian Field, and possibly from the extra-galactic radio sources distribution. Simple expressions for the ratio of the dispersion of the number of galaxies

observed in an elementary region (e.r.) to the mean number of galaxies in an e.r. are given. Thereby, four models of apparent distribution of galaxies are considered, including the case of superclusters.

158.134 New emission peculiar and interacting galaxies of the southern sky.
T. M. Borchkhadze, R. M. West.
Astrofizika, Vol. 13, 605 - 615 (1977). In Russian. English translation in Astrophysics, Vol. 13, No. 4.

The paper continues the spectroscopic investigation of morphologically peculiar galaxies from the ESO-Uppsala lists and contains a list of southern galaxies with [O III], Hα and Hβ emission lines. Details are given for 37 galaxies in 28 systems; two galaxies, ESO 325−IG41 and 273−IG04, may be classified as Seyfert class 2. The morphological characteristics are discussed. Relative emission line strengths are estimated visually for all galaxies.

158.135 The Andromeda nebula radio emission at 102.5 MHz. Yu. V. Volodin, R. D. Dagkesamanskij.
Astrofizika, Vol. 13, 617 - 626 (1977). In Russian. English translation in Astrophysics, Vol. 13, No. 4.

Observations of the Andromeda nebula (M31) have been made at 102.5 MHz. The observation technique and the reduction of observations are described. Angular sizes, flux densities, spectral indices and emissivity of disk and halo-type components of M31 are estimated. These values are compared with the corresponding physical parameters of our Galaxy.

158.136 On the nature of galaxies with ultraviolet continuum. III. Surface brightness, morphology and activity.
B. E. Markarian, Dzh. A. Stepanyan.
Astrofizika, Vol. 13, 627 - 638 (1977). In Russian. English translation in Astrophysics, Vol. 13, No. 4.

The existence of a dependence between surface brightness and morphology of galaxies is established. The surface brightness of galaxies continuously decreases along Hubble's morphological sequence. This decrease of brightness amounts to $1^m.5$. The comparison of the distribution of the average surface brightness of galaxies up to 16^m with the mentioned dependence shows that half of the galaxies with UV continuum has elliptical-lenticular structure and the other half spiral structure. Calculations show that in color B the average surface brightness of galaxies with UV continuum is higher than the mean surface brightness of all normal galaxies taken together, but is considerably lower than the average surface brightness of elliptical and lenticular galaxies taken separately. In the color U, however, the average surface brightness of galaxies with UV continuum exceeds considerably this parameter of all types of normal galaxies. Thus to the two basic properties of galaxies with UV continuum − very strong ultraviolet radiation and strong emission spectral lines − a third, i.e. very high surface ultraviolet brightness $\gtrsim 21^m.5$, should be added. All these three properties may be used for selection of active galaxies.

158.137 Some notes on the diameters and magnitudes of probable compact galaxies. N. G. Kogoshvili.
Astrofizika, Vol. 13, 639 - 650 (1977). In Russian. English translation in Astrophysics, Vol. 13, No. 4.

Comparison of diameters of probable compact galaxies according to MCG and UGC testifies a possible excess by 40% of the number of compact galaxies found on the basis of the MCG analysis. The values of systematic differences for various systems of stellar magnitudes when transforming them into the system of magnitudes of CGCG are presented.

158.138 Spectrophotometry of three radio galaxies.
M. V. Penston, R. A. E. Fosbury.
Mon. Not. R. Astron. Soc., Vol. 183, 479 - 490 (1978).

Scans of the radio galaxies 3C 33, 3C 327 and PKS

1934—63 are presented. All show strong emission lines but have red continua with stellar absorption features. The two 3C objects, which are classical double radio sources, have He II λ4686 emission stronger with respect to Hβ than any other galaxy yet reported.

158.139   Optical emission-line spectra of Seyfert galaxies and radio galaxies.   D. E. Osterbrock.
Phys. Scr., Vol. 17, (see 012.013), 137 - 143 (1978) = Lick Obs. Bull. No. 774.

Many radio galaxies have strong emission lines in their optical spectra, similar to the emission lines in the spectra of Seyfert galaxies. The range of ionization extends from [O I] and [N I] through [Ne V] and [Fe VII] to [Fe X]. The emission-line spectra of radio galaxies divide into two types, narrow-line radio galaxies whose spectra are indistinguishable from Seyfert 2 galaxies, and broad-line radio galaxies whose spectra are similar to Seyfert 1 galaxies.

158.140   The continuous optical spectra of active nuclei.   P. A. Strittmatter.
Phys. Scr., Vol. 17, (see 012.013), 145 - 147 (1978).

The optical continuum properties of the nuclei of radio galaxies, Seyfert galaxies, N systems, BL Lac objects, and QSOs are reviewed. The observations are in the main consistent with the hypothesis of contributions in varying degrees for a stellar component and a small non-thermal source. The presence of dust strongly influences the continuum spectrum and polarization in the case of Seyferts. The observations have not as yet led to an understanding of the nature of the nonthermal continuum source but offer prospects of so doing.

158.141   Infrared observations of Seyfert galaxies and quasars.   G. Neugebauer.
Phys. Scr., Vol. 17, (see 012.013), 149 - 157 (1978).

The infrared energy distributions of the Seyfert galaxies apparently contain three components: a galactic stellar component, a thermal component from heated dust, plus a nonthermal component. Preliminary data on bright quasars are given. The infrared energy distributions generally increase into the infrared with a power law slope of ~ 1. In detail they differ from power laws with a significant fraction emitting most of their energy near 3 μm.

158.142   Radio observations of active and normal nuclei.   R. D. Ekers.
Phys. Scr., Vol. 17, (see 012.013), 171 - 173 (1978).

At the centre of the Galaxy and in most other spiral galaxies there is a region of enhanced non-thermal radio emission of a few hundred parsecs in extent. The Seyfert galaxies have similar but somewhat more intense radio emission. Elliptical galaxies, radio galaxies and quasars have smaller and variable nuclear radio sources with complex spectral characteristics attributed to self-absorption effects. Using the clear differences between the nuclei of spiral and Seyfert galaxies on the one hand and elliptical and radio galaxies on the other, it can be argued that the quasars are extremely active manifestations of the elliptical and radio galaxies rather than of the Seyfert galaxies.

158.143   Physical state of the emission-line region.   D. E. Osterbrock.
Phys. Scr., Vol. 17, (see 012.013), 285 - 292 (1978) = Lick Obs. Bull. No. 775.

The physical properties of the ionized gas in the active nuclei of Seyfert and radio galaxies derived from their emission-line spectra, are reviewed.

158.144   Future prospects of infrared observations of active nuclei.   G. Neugebauer.
Phys. Scr., Vol. 17, (see 012.013), 325 - 326 (1978).

Future developments in infrared observations are briefly discussed. The prospects for large advances in ground-based broad-band observations probably lie in increased availability of dedicated telescopes and observing time. Several specific areas of potential advancement are described. The plans for the utilization of space for infrared astronomy call for an unbiased all-sky survey in the wavelengths 8—120 μm followed by complementary missions using a cooled moderate sized telescope and a large uncooled telescope.

158.145   Future prospects of radio observations of active nuclei.   H. van der Laan.
Phys. Scr., Vol. 17, (see 012.013), 327 - 328 (1978).

The prospects for future developments in radio observations pertaining to active nuclei are briefly discussed and areas of potential advancement are described.

158.146   Remark on activity in nuclei of spiral galaxies different from that in Seyfert galaxies.
J. H. Oort.
Phys. Scr., Vol. 17, (see 012.013), 369 (1978).

Attention is drawn to some types of large-scale eruptive phenomena which differ from those observed around the nuclei of Seyfert galaxies.

158.147   First results of the Las Campanas survey to classify southern galaxies photographed with the du Pont 2.5-meter reflector.   A. Dressler, A. Sandage.
Publ. Astron. Soc. Pacific, Vol. 90, 5 - 9 (1978).

Hubble types are listed for 30 southern galaxies, newly determined from the first plates of the Las Campanas photographic survey of Shapley-Ames galaxies south of δ = −15°. The photographs were obtained with the 2.5-m du Pont reflector in the course of its commissioning tests. Photographs of twelve of the galaxies are reproduced.

158.148   Radial velocities of 39 galaxies.   C. J. Peterson.
Publ. Astron. Soc. Pacific, Vol. 90, 10 - 13 (1978).

Systematic velocities (heliocentric) are presented for 39 galaxies, of which only six have had velocities published previously. The majority of the spectrograms were obtained with the 1-m Yale telescope at CTIO; intercomparison of duplicate plates suggests that the accuracy of emission-line velocities for image-tube spectrograms of faint galaxies from this instrument is of the order of 33 km s^{-1}.

158.149   Spectra and radial velocities of some southern peculiar galaxies and pairs of galaxies.
M. Pedreros.
Publ. Astron. Soc. Pacific, Vol. 90, 14 - 19 (1978).

Spectroscopic observations of six peculiar galaxies and six pairs of galaxies south of declination −43° were made by using a Carnegie image-tube system mounted on the Cassegrain spectrograph of the 1.5-m telescope at CTIO. The spectra are described and radial velocities are determined from them. Mass estimations are made for the double galaxies and for the nucleus of one individual galaxy.

158.150   Identification of the nucleus in the spiral galaxy NGC 4631.   M. Aaronson.
Publ. Astron. Soc. Pacific, Vol. 90, 28 - 31 (1978).

Broad-band infrared photometry is presented for the edge-on spiral galaxy NGC 4631. The position of peak infrared intensity is found to lie not on the most optically prominent part of the galaxy, but in the nearby area of strongest dust absorption. This position is discussed in relation to the rotation curve measured by G. and A. de Vaucouleurs, and the radio continuum maps of Pooley. It is concluded that the location of the infrared peak is probably the true position of maximum stellar density, i.e., the nucleus.

158.151   Photometry of galaxies in the field of the galaxy cluster Abell 1781.   K. Fritze, M. Lange,

G. M. Richter, D. Stoll.
Astron. Nachr., Band 299, 61 - 68 (1978).

A method for determining the brightness and diameter of galaxies in a large field on Schmidt plates by automatic scanning with a microphotometer is described and applied to the Abell cluster A 1781. The accuracy of this method is tested. The overall errors (r.m.s.) of brightness and diameter are 0.16 mag and 0″.44. The cluster A 1781 has been proved to be a very poor cluster of about 10 members up to $m_B \approx 19^m5$.

158.152 **Variable galactic nuclei.**
D. Hamilton, W. Keel, J. F. Nixon.
Sky Telesc., Vol. 55, 372 - 374 (1978).

158.153 **Observaciones de galaxias en la linea de 21 cm del hidrogeno neutro con el instrumento de sintesis de Westerbork (Holanda).** E. Bajaja.
Astron. extragalactica, (see 012.015), 9 - 36 (1976).

A short description is given of the aperture synthesis and of the Westerbork instrument which applies this technique. The method of reduction and the defects of the system are also described, as well as the results obtained from the observation of some galaxies in the 21 cm line of the H I. The gas distributions found prove definitely its concentration along the optical arms which reproduces the spiral structure. The velocity fields show perturbations consistent with the predictions of the density wave theory.

158.154 **Progressos nas investigações de objetos extragalaticos austrais em ondas milimetricas.**
P. Kaufmann, P. Marques dos Santos, M. A. Bräz.
Astron. extragalactica, (see 012.015), 37 - 40 (1976).

Results are given of the observed extragalactic objects at Itapetinga with the precision radiotelescope, in 22 GHz.

158.155 **Diagrama M-C de Sculptor y Fornax.**
S. Demers, W. E. Kunkel.
Astron. extragalactica, (see 012.015), 83 - 87 (1976).

Photometric observations of the Sculptor and Fornax systems are presented. The characteristics of the respective H-R diagrams lead to the conclusion that, like the globular clusters of our Galaxy, elliptic dwarf galaxies are not all alike.

158.156 **Upper limits to the H I content of the dwarf spheroidal galaxies.** G. R. Knapp, F. J. Kerr, P. F. Bowers.
Astron. J., Vol. 83, 360 - 362 (1978).

A search was made for H I emission from the six dwarf spheroidal satellites of the Galaxy. None was detected, setting limits as low as a few parts in 10^4 by mass.

158.157 **Doppler-shift distributions. II. The relativistic gas.** P. G. Gross.
Astrophys. J., Suppl. Ser., Vol. 36, 305 - 315 (1978) = Warner and Swasey Obs., Case Western Reserve Univ., Cleveland, Ohio, Repr. No. 298.

Earlier theoretical investigations of photon Doppler-shift distributions from absorbing or emitting particles of a Maxwellian gas have extended into the regime of relativistic velocities. The mean, and the next three moments about the mean, of the Doppler-shift distribution $(1 + z)/(1 + z_0)$ of photons are computed as functions of three parameters describing a relativistic gas with an equilibrium velocity distribution. (Here z is the observed Doppler shift, and z_0 is the Hubble redshift of the substratum at the gas relative to the fundamental observer). The four moments are given in four sets of tables for all combinations of the gas-parameter values.

158.158 **Structural investigations in the central part of the Coma cluster on the base of equidensity diagrams.**

II.   N. Richter, W. Högner.
Astron. Nachr., Band 299, 109 - 115 (1978).

On the base of Tautenburg *UBV*-Schmidt plates equidensitometric diameters of galaxies of Rood and Baum's list are given and the ellipticities in the three colours are determined. It was found that there is no difference in the ellipticities depending on the colours, on the average. The percentage of spherical objects (ϵ = 0.00) was found to be 21.5 per cent. The surface brightness for 213 objects in the system V is given. The mean surface brightness is 22.75 mag/□ ″with a dispersion of 2.5 mag/□ ″. It shows only a very slight dependence on the integrated magnitude.

158.159 **Surface brightness of compact galaxies.**
N. Richter, W. Högner.
Astron. Nachr., Band 299, 121 - 123 (1978).

In four fields near the north galactic pole diagrams of the relation between mean surface brightness and integral brightness of compact galaxies are constructed.

158.160 **Optical emission in the radio lobes of radio galaxies.**
W. C. Saslaw, J. A. Tyson, P. Crane.
Astrophys. J., Vol. 222, 435 - 449 (1978).

The authors have found photographic and photometric evidence for optical emission in the radio lobes of 3C285, 3C265, and 3C390.3. They discuss possible mechanisms of optical emission and their implications for models of radio galaxies.

158.161 **Dynamics of the flattened elliptical galaxy NGC 4473.** P. Young, W. L. W. Sargent,
A. Boksenberg, C. R. Lynds, F. D. A. Hartwick.
Astrophys. J., Vol. 222, 450 - 455 (1978).

Spectroscopic observations of the E5 galaxy NGC 4473 to a distance of 45″(= 3300 pc) from the center along the major axis are used to make the first estimates of the radial variation of rotation and velocity dispersion in a galaxy away from the regions of the core.

158.162 **The 2−2.5 micron spectrum of NGC 1068: a detection of extragalactic molecular hydrogen.**
R. I. Thompson, M. J. Lebofsky, G. H. Rieke.
Astrophys. J., Lett., Vol. 222, L49 - L53 (1978).

The infrared spectrum of the Seyfert galaxy NGC 1068 at 2.2 μm is presented. The first extragalactic emission line of H_2 as well as emission lines of H and He are present in the spectrum. Approximately 6000 $M_\odot$ of shocked molecular hydrogen in the nucleus of NGC 1068 are needed to produce the observed emission feature. The observed strength of $B\gamma$ is somewhat lower than predicted, which may be due to a decreased scattering efficiency of the dust grains at 2.1 μm.

158.163 **The Arecibo 2380 MHz survey of bright galaxies.**
L. L. Dressel, J. J. Condon.
Astrophys. J., Suppl. Ser., Vol. 36, 53 - 75 (1978).

A 2380 MHz survey of bright galaxies in the declination range $0° < \delta < +37°$ has been made at Arecibo. It is complete to m_p = +14.5 in all right ascensions except $11^h 30^m < \alpha < 13^h 00^m$, where the limit is m_p = +14.0. Four hundred and fifty-six of the 2095 galaxies observed were detected above 15mJy.

158.164 **The radio halo of NGC 891.**   A. W. Strong.
Astron. Astrophys., Vol. 66, 205 - 209 (1978).

Recent high resolution observations of NGC 891 made with the Westerbork array at 6, 21 and 49 cm provide information on the propagation of relativistic electrons and the distribution of magnetic fields away from the disc. Simple models of diffusive and convective transport of electrons are constructed to account for the spatial distribution of intensity and spectral index. The magnetic field must fall off rather slowly with height above the plane, with most of the intensity decrease resulting from electron propagation and energy losses.

158.165   The radiation of the nucleus of the Seyfert galaxy NGC 4151 in the lines λ 4959 and 5007 Å [O III].
V. I. Pronik.
Izv. Krymskoj Astrofiz. Obs., Vol. 58, 104 - 108 (1978).
In Russian.

The results of a detailed photometry of the λ 4959 Å [O III] line image on unbroadened spectrograms of the nucleus of NGC 4151 are presented. The absence of a clearly defined nucleus of the galaxy in N_1, N_2 [O III] radiations is noted opposed to what observed in integral light.

158.166   The active elliptical galaxy NGC 1052.
R. A. E. Fosbury, U. Mebold, W. M. Goss, M. A. Dopita.
Mon. Not. R. Astron. Soc., Vol. 183, 549 - 568 (1978).

A radio and optical study is presented of the active elliptical galaxy NGC 1052. Spectrophotometric scans and spatially resolved optical spectra of the nuclear regions have been obtained. The diameter of the emission-line region is measured to be 20 arcsec which is the same as the diameter of the radio halo observed with short-baseline interferometers. The total luminosity of the compact nuclear radio source is ample to drive both the radio halo and the optical emission lines. The mass of ionized gas in the galaxy is shown to be about 5 × $10^5 M_\odot$. New 21-cm observations of the galaxy show the presence of an emission line with a width of 470 ± 50 km/s and an intensity corresponding to a neutral hydrogen mass of $(2.2^{+0.6}_{-1.3}) \times 10^9 M_\odot$.

158.167   Clusters of galaxies and the statistics of emission-line galaxies.   G. R. Gisler.
Mon. Not. R. Astron. Soc., Vol. 183, 633 - 643 (1978).

A sample of optical spectra of 1316 galaxies collected from the literature is used to study the relative frequencies of emission-line galaxies inside and outside dense clusters. It is concluded that emission-line galaxies are much less common in dense clusters than in less prominent associations or in the field.

158.168   NGC 5506: an almost Seyfert galaxy.
I. S. Glass.
Mon. Not. R. Astron. Soc., Vol. 183, 85P - 87P (1978).

Infrared observations of the X-ray galaxy NGC 5506 reveal it to have an infrared luminosity of order 10^{37} W, comparable to the brightest Seyferts.

158.169   Two new dwarf galaxies in the Local Group: UKS 1927 − 177 and UKS 2323 − 326.
A. J. Longmore, T. G. Hawarden, B. L. Webster, W. M. Goss, U. Mebold.
Mon. Not. R. Astron. Soc., Vol. 183, 97P - 100P (1978).

Two new dwarf irregular galaxies have been observed in the 21-cm neutral hydrogen line. Distance estimates using the brightest stars place UKS 1927 − 177 at 1.1 Mpc and UKS 2323 − 326 at 1.3 Mpc, indicating that both may be members of the Local Group on the conventional ($\Delta \lesssim 1$ Mpc) definition. Both galaxies have an unusually low surface brightness for their type. The authors tentatively interpret them as very low-mass galaxies which have recently experienced a burst of star formation.

158.170   Curtis Schmidt-thin prism survey for extragalactic emission-line objects. University of Michigan list IV.
G. M. MacAlpine, D. W. Lewis.
Astrophys. J., Suppl. Ser., Vol. 36, 587 - 593 (1978).

Descriptions, positions, and finding charts are presented for 62 apparent emission-line galaxies and 58 QSO candidates. This list contains sources to about continuum magnitude 18.5,

which are located in the roughly 105 square degree region of sky with 1950 equatorial coordinates $1^h.0 < \alpha < 2^h.4$ and $-2°.5 \lesssim \delta \lesssim +2°.5$. Information given for each object includes 1950 coordinates, image dimensions from the Palomar Observatory Sky Survey, an estimated continuum magnitude, a redshift estimate when feasible, alternate designations (if they exist) resulting from membership in other compilations, and (for galaxies) a spectral classification which involves compactness, color, and line-strength parameters.

158.171   H I observations of elliptical galaxies.
G. R. Knapp, F. J. Kerr, B. A. Williams.
Astrophys. J., Vol. 222, 800 - 814 (1978).

This paper describes a very sensitive observational search for H I emission from 38 early-type galaxies, mostly ellipticals, using the Arecibo 305 m telescope. Thirty-two of the galaxies were not detected in H I: with estimates of the velocity width for each galaxy scaled to its luminosity, very low upper limits on the H I content of each galaxy were set. For about half of the galaxies, these limits are inconsistent with the amounts predicted from stellar mass loss. They are also inconsistent with continuing star formation in the galaxies and support the suggestion that gas is continuously removed from these galaxies. Gas was detected in six of the observed galaxies.

158.172   Illusionary warps in H I disks of nearly edge-on spiral galaxies.   G. G. Byrd.
Astrophys. J., Vol. 222, 815 - 820 (1978).

Recently Sancisi (1976) reported on 21 cm observations of five nearly edge-on spiral galaxies. Three of these which show strongly warped 21 cm disks also have neighboring galaxies all so far away that the tidal effects of any close encounter could not have survived to the present. Sancisi mentioned that intergalactic gas might be a cause but concluded that the origin and survival of the warps are puzzling. This paper proposes that the warps in these otherwise normal galaxies are illusions created by their nearly, but not exactly, edge-on orientation, the presence of H I spiral arms in the outer parts of their disks, and the limited resolution of the radio observations.

158.173   The composition gradient across M101.
G. A. Shields, L. Searle.
Astrophys. J., Vol. 222, 821 - 832 (1978).

New observations of the spectra of three H II regions in the disk of M101 are analyzed with the aid of detailed photoionization models. The oxygen abundance decreases from [O/H] = +0.3 at $R = 6$ kpc to −0.7 at $R = 23$ kpc. The form of the oxygen gradient, as defined by these three regions, is consistent with a power law. The He/H ratio decreases slightly with increasing R, and S/O shows little variation. The spectra of the low-excitation H II regions in the inner arms can be understood in terms of a gradient in T_*, the temperature of the ionizing stars.

158.174   X-ray and radio emission from the compact galaxy III Zw 2.   H. W. Schnopper, J. P. Delvaille, A. Epstein, W. Cash, P. Charles, S. Bowyer, R. M. Hjellming, F. N. Owen, W. D. Cotton.
Astrophys. J., Lett., Vol. 222, L91 - L94 (1978).

X-rays from the compact galaxy III Zw 2 have been detected with the rotating modulation collimator on board SAS 3. The (2−10 keV) flux is $(4.3 \pm 0.5) \times 10^{-11}$ ergs cm^{-2} s^{-1}, which corresponds to a luminosity of 1.4×10^{45} ergs s^{-1} at 530 Mpc. Subsequent radio observations between 1.46 and 90 GHz were made at the NRAO Very Large Array and 36 foot (11 m) installations. The radio spectrum rises sharply toward higher frequencies and, below about 100 GHz, is indicative of a selfabsorbed synchrotron spectrum from an extremely compact, rapidly evolving object.

158.175   Enhancement of the jets in NGC 1097.

J. J. Lorre.
Astrophys. J., Lett., Vol. 222, L99 - L103 (1978).

Plates of NGC 1097 taken in four spectral bands by H. Arp have been digitized and processed at JPL's Image Processing Laboratory. The processing has revealed (1) that there is a fourth jet, R4, counter to the L-shaped jet, R1; (2) that the area density of point sources in the jets is no different from that found on the sky; (3) that two of the jets, R1 and R2, are bluer than the night sky, while one of the jets, R3, is redder than the night sky; (4) that the nonstellar objects situated on the jets are noticeably bluer than the night sky; and (5) that color-enhanced images of the spiral structure illustrate the distribution of both gaseous and spiral-arm stellar population disrupted from symmetry by an apparent companion elliptical.

158.176 Radio observations of NGC 5296/7. G. G. Pooley.
Observatory, Vol. 98, 135 - 136 (1978).

158.177 The distribution of light in galaxies.
K. C. Freeman.
IAU Symp. No. 77, (see 012.026), p. 3 - 14 (1978).

The author reviews some recent work on ellipticals and disk galaxies. In summary, the luminosity distributions for both these classes have several complexities, the dynamical significance of which is not yet clear. Both classes turn out to have roughly constant mean surface brightness. The author discusses this topic at the end.

158.178 The past history of star formation in galaxies.
B. M. Tinsley.
IAU Symp. No. 77, (see 012.026), p. 15 - 21 (1978).

158.179 Rotation curves in the outer parts of galaxies from H I observations. E. E. Salpeter.
IAU Symp. No. 77, (see 012.026), p. 23 - 31 (1978).

21 cm observations at the Arecibo Observatory for 9 edge-on spiral galaxies are described. Flat rotation curves are found in most cases.

158.180 The large-scale radio continuum structure of spiral galaxies. P. C. van der Kruit.
IAU Symp. No. 77, (see 012.026), p. 33 - 48 (1978).

This review concerns the large-scale structure of radio continuum emission in spiral galaxies ("the smooth background"), by which is meant the distribution of radio surface brightness at scales larger than, say, 1 kpc.

158.181 The large scale distribution of radio continuum in E and S0 galaxies. R. D. Ekers.
IAU Symp. No. 77, (see 012.026), p. 49 - 55 (1978).

The continuum emission from elliptical and S0 galaxies is completely different from that for the spiral galaxies.

158.182 Some comments on radio observations of spiral arms.
W. W. Shane, J. Bystedt.
IAU Symp. No. 77, (see 012.026), p. 97 - 104 (1978).

158.183 Radio observations of molecules in nearby galaxies.
J. B. Whiteoak.
IAU Symp. No. 77, (see 012.026), p. 131 - 137 (1978).
OH, H_2CO, CO, HCN, H_2O.

158.184 Radio continuum observations of M31 and M33.
E. M. Berkhuijsen.
IAU Symp. No. 77, (see 012.026), p. 149 - 157 (1978).

The two nearest spiral galaxies, M31 and M33, have been extensively studied both at optical as well as at radio wavelengths. New radio observations in various stages of progress are listed, all of which show − or are expected to show − spiral structure. In this paper the Effelsberg maps at 11 cm of M31 and at 6 cm of M33 are discussed and compared with optical data. Some attention is given to the separation of

thermal and nonthermal emission which is essential to any discussion of the origin of the radiation.

158.185 Velocity dispersion in the bulge of M31; dynamical model. G. Monnet, A. Pellet, F. Simien.
IAU Symp. No. 77, (see 012.026), p. 159 - 161 (1978).

158.186 The spiral structure of M31.
E. Athanassoula.
IAU Symp. No. 77, (see 012.026), p. 163 - 167 (1978).

158.187 The kinematics within M31.
M. S. Roberts, R. N. Whitehurst, T. R. Cram.
IAU Symp. No. 77, (see 012.026), p. 169 - 173 (1978).

A new hydrogen line survey of M31 is described. A rotation curve of the galaxy is derived from these data. The northern half of the rotation curve can be traced to 27 kpc, the southern to 30 kpc. They are in general agreement; both sides show an extensive region of essentially constant rotational velocity.

158.188 The three-dimensional distribution of neutral hydrogen in M31.
R. N. Whitehurst, M. S. Roberts, T. R. Cram.
IAU Symp. No. 77, (see 012.026), p. 175 - 181 (1978).

158.189 A warp in the H I distribution at the extreme NE and SW of M31. D. T. Emerson, K. Newton.
IAU Symp. No. 77, (see 012.026), p. 183 - 189 (1978).

158.190 The large-scale distribution of H I in M33 and IC342. J. E. Baldwin.
IAU Symp. No. 77, (see 012.026), p. 191 - 195 (1978).

158.191 A sensitive single-dish HI-survey of the galaxy M33.
W. K. Huchtmeier.
IAU Symp. No. 77, (see 012.026), p. 197 - 201 (1978).

The integrated HI-distribution (over the velocity range −350 to −50 km/s heliocentric) is given. The most striking features of this HI-distribution are its large extent (∼2.2 Holmberg radii) and the different orientation of the lower contours compared to the optical image of the galaxy.

158.192 Radio properties of active nearby spiral galaxies.
A. G. de Bruyn.
IAU Symp. No. 77, (see 012.026), p. 205 - 219 (1978).

158.193 Radio properties of the nuclei in elliptical, S0 and spiral galaxies. R. D. Ekers.
IAU Symp. No. 77, (see 012.026), p. 221 - 225 (1978).

158.194 Infrared, optical, and X-ray properties of the nuclei of nearby galaxies. G. Burbidge.
IAU Symp. No. 77, (see 012.026), p. 227 - 236 (1978).

158.195 Emission from the nuclei of nearby galaxies: evidence for massive black holes? M. J. Rees.
IAU Symp. No. 77, (see 012.026), p. 237 - 244 (1978).

The author ventures some conjectures on scenarios whereby a massive black hole could form in a galactic nucleus, and the characteristic features of accretion onto it. He then suggests how some observed phenomena − particularly the radio and X-ray continuum − can be interpreted in terms of such processes. And in conclusion he mentions some other tests for (or limits on) massive black holes in nearby galaxies.

158.196 Galaxy haloes and the missing mass problem.
S. van den Bergh.
IAU Symp. No. 77, (see 012.026), p. 247 - 265 (1978).

Available evidence on the mass-to-light ratios in binary galaxies, small clusters and in rich clusters is reviewed.

158.197 **On peripheral dynamics.**   A. Toomre.
IAU Symp. No. 77, (see 012.026), p. 267 - 268
(1978).
A review of the observed facts and the observed theories
of distortions of the outer parts of galaxies. − Discussion only.

158.198 **Interacting galaxies: the kinematics of NGC 4038/39
and the H I bridge between M 81 and NGC 3077.**
J. M. van der Hulst.
IAU Symp. No. 77, (see 012.026), p. 269 - 277 (1978).

158.199 **Galaxies with long tails.**   F. Schweizer.
IAU Symp. No. 77, (see 012.026), p. 279 - 285
(1978).
The author would like to show some animals from the
zoo of Alar Toomre. These animals are all characterized by
long tails and belong to two subspecies: The first consists of
animals which always huddle in pairs, each individual having
one long tail. The second, rarer subspecies comprises those
strange animals which live alone but have two tails!

158.200 **H I in the elliptical galaxy NGC 1052.**
K. Reif, U. Mebold, W. M. Goss.
Astron. Astrophys., Vol. 67, L1 - L3 (1978).
The E4 galaxy NGC 1052 has been mapped in the H I
21-cm emission line using the 100-m telescope. The H I associated with this galaxy is $\lesssim 4'$ in size. A systematic change
in radial velocity is measured which suggests that the H I is a
rotating system. The total amplitude of the observed rotation
is ~110 km/s with an orientation of the dynamical major axis
which is $54° \pm 10°$ from the major axis of the light distribution.

158.201 **Carbon monoxide in the Sc galaxy NGC 5236.**
F. Combes, P. J. Encrenaz, R. Lucas, L. Weliachew.
Astron. Astrophys., Vol. 67, L13 - L15 (1978).
Carbon monoxide emission ($J = 1 \rightarrow 0$) has been found
and partially mapped in the disk of the Sc galaxy NGC 5236
(M83). The radial distribution and velocities of CO emission
are compared to the distribution of H I surface density and the
rotation curve.

158.202 **NGC 4472: a very weak radio galaxy?**
R. D. Ekers, C. G. Kotanyi.
Astron. Astrophys., Vol. 67, 47 - 50 (1978).
The authors report radio observations of NGC 4472,
which shows a structure typical of radio galaxies but at a radio-
power level usually associated with "normal" galaxies. It is
concluded that the double structure is related to factors other
than the radio power.

158.203 **The spiral structure of M31: a morphological
approach.**   F. Simien, E. Athanassoula,
A. Pellet, G. Monnet, A. Maucherat, G. Courtès.
Astron. Astrophys., Vol. 67, 73 - 79 (1978).
The problem of the spiral structure of M31 has been
reconsidered, based on a new list of 981 H II regions and the
OB associations of van den Bergh. The distribution of these
arm tracers, as well as arguments concerning the H I density
and an asymmetry in the rotation curve leads to the conclusion that the most probable solution for the unusual structure
of M31 is a one-armed leading pattern (Kalnajs' solution)
forced by the close companion M32, and that the classical
two-armed trailing spiral is far less satisfactory.

158.204 **High frequency radio observations of Seyfert
galaxies.**   W. H. McCutcheon, P. C. Gregory.
Astron. J., Vol. 83, 566 - 573 (1978).
Observations have been made of 42 galaxies (40
Seyferts, 1 BL Lac type galaxy, and 1 emission line galaxy) at
wavelengths of 2.8 and 1.3 cm. Twenty were detected at one
or both frequencies and spectra are presented for 14 of these
using the present and previously published data. Seven, and

possibly nine, have power law spectra. The remaining spectra
are more complex. Four galaxies show variability at one of the
frequencies. Two of them, MRK 348 and MRK 231 vary on
a time scale of a few days at $\lambda = 1.30$ cm. The radio luminosities, calculated for eighteen galaxies, cover a range extending
from the normal spirals at the low end to the quasars at the
high end. The Seyferts with radio luminosities greater than
10^{42} ergs/s are predominantly Class 1 whereas those with
values less than 10^{42} ergs/s belong to both Class 1 and Class 2.

158.205 **Photometry of M87 at faint light levels.**
D. Carter, K. L. Dixon.
Astron. J., Vol. 83, 574 - 582 (1978).
Photographic surface photometry of the giant elliptical
galaxy M87 in B and V wavebands is presented. Problems of
background subtraction are discussed, and two independent
methods of background determination are used. The form
of the ellipticity profile is determined, the ellipticity is shown
to increase with radius in a manner characteristic of most
elliptical galaxies. No variation of position angle of the major
axis with isophote radius is found, although some asymmetrical structure at large radii can be seen on a contour map.

158.206 **Gas envelope parameters and masses of active
nuclei of galaxies.**   Eh. A. Dibaj.
Astron. Zh. Akad. Nauk SSSR, Vol. 55, 456 - 464 (1978).
In Russian. English translation in Soviet Astron., Vol. 22,
No. 3.
For 16 active nuclei of galaxies UBV-magnitudes, redshifts, equivalent widths, H_β luminosities, and Doppler widths
of hydrogen lines are presented. The gas parameters in the
region of formation of broad emission lines are estimated.
The low-velocity (200 − 300 km/sec) motions of gas in the
forbidden-line zone seem to be equilibrated by the gravitational potential of the stellar component of the galaxy. It is
suggested that the high-velocity gas motions are equilibrated
by the gravitational potential of the active nucleus. The
masses of central bodies are found to be ~ $10^8 - 10^9$ solar
masses. Bolometric luminosities of nuclei do not exceed the
Eddington limit corresponding to their masses.

158.207 **Light absorption in spiral galaxies.**   A. V. Zasov.
Astron. Zh. Akad. Nauk SSSR, Vol. 55, 465 - 471
(1978). In Russian. English translation in Soviet Astron.,
Vol. 22, No. 3.
An attempt is made to estimate mean values of optical
depth, internal absorption, and colour excesses of spiral
galaxies on the basis of known statistical dependences of the
brightness and colour indices of the galaxies on their angle
of inclination to the line of sight. The dependences observed
may be explained on the assumption that a considerable
part of the interstellar dust in galaxies is distributed over the
entire thickness of the stellar disk. Best agreement with observations is achieved by a mean value of absorption $A_V = 0^m.3$
for Sa and Sb galaxies and $\approx 0^m.5$ for Sc and Sd galaxies. Corresponding values of colour excesses E_{B-V} and E_{U-V} as
well as of the optical depths of the galaxies are given.

158.208 **X-ray observations of galaxies.**
IAU Circ., No. 3161 (1978).

158.209 **NGC 4151.**
IAU Circ., No. 3173 (1978).

158.210 **MGC−5−23−16.**
IAU Circ., No. 3190 with a corrigendum 3194
(1978).

158.211 **Variable galaxies.**
IAU Circ., No. 3197 (1978).

158.212 **MCG −5−23−16.**

IAU Circ., No. 3202 (1978).

158.213 **MCG −6−30−15.**
IAU Circ., No. 3202 (1978).

158.214 **Spectrophotometry of Seyfert 2 galaxies and narrow-line radio galaxies.** A. T. Koski.
Astrophys. J., Vol. 223, 56 - 73 (1978) = Lick Obs. Bull., No. 786.

Spectrophotometric scans of 20 Seyfert 2 galaxies, three intermediate Seyfert galaxies, and five narrow-line radio galaxies are presented. The author wants to know if the Seyfert 2 galaxies form a homogeneous class, or if they separate into subclasses. Exactly how are the Seyfert 2 galaxies related to the intermediate class? Finally, he determines some of the physical conditions in the emitting gas in these galaxies.

158.215 **Uhuru observations of X-ray emission from Seyfert galaxies.** H. Tananbaum, G. Peters, W. Forman, R. Giacconi, C. Jones, Y. Avni.
Astrophys. J., Vol. 223, 74 - 81 (1978).

Using a point summation technique, the authors have systematically analyzed Uhuru data for X-ray emission from the 88 Seyfert galaxies listed by Weedman, plus MCG 8-11-11 reported by the Ariel 5 group. In addition to measuring the average X-ray intensity for 15 sources reported in the 4U and 2A catalogs, the authors find three new candidate sources. The authors have measured the local X-ray volume emissivity of Seyfert galaxies and find, with standard assumptions, that from 6% to 25% of the diffuse 2−10 keV X-ray background can be attributed to emission from Seyfert galaxies.

158.216 **Planetary nebulae in local group galaxies. VI. An observational determination that M32 is in front of M31.** H. C. Ford, G. H. Jacoby, D. C. Jenner.
Astrophys. J., Vol. 223, 94 - 97 (1978).

Twenty-one centimeter integrated brightness maps of M31 are used to predict a color excess $E(B − V) \approx 0.5$ if M32 is behind M31. A galactic cosecant law applied to M31 predicts a color excess between 0.5 and 0.6. The authors conclude that M32 will be heavily reddened if it is behind M31. The predicted value of $I(H\alpha)/I(H\beta)$ is 5.1 for a nebula in M32 if M32 is behind M31, whereas the predicted ratio is 3.2 if M32 is in front of M31. The observed ratio (2.96) in the planetary nebula M32−1 agrees with the latter value within the observational uncertainty. The authors conclude this to be strong evidence that M32 is in front of M31. Additional arguments in favor of this conclusion are given.

158.217 **Forbidden lines in quasars and active galaxies.** J. M. Shull, R. McCray.
Astrophys. J.,Lett., Vol. 223, L5 - L8 (1978).

The authors consider the effects of photoexcitation and photoionization from the metastable levels of ions in the high radiation environments expected in the nuclei of Seyfert galaxies, broad-line radio galaxies, and quasars. Anomalous line ratios in the narrow-line components and the absence of forbidden lines in the broad-line components of these objects may result from proximity of the emitting region to the central UV source, and do not necessarily imply high electron densities. Detailed calculations are presented for the case of [O III].

158.218 **Detection of an optical halo surrounding the spiral galaxy NGC 4565.** D. J. Hegyi.
IAU Symp. No. 79, (see 012.027), p. 95 - 96 (1978). − Short communication.

158.219 **Morphological investigation of pairs containing Markarian galaxies.** C. Casini, J. Heidmann.
IAU Symp. No. 79, (see 012.027), p. 100 - 101 (1978). Short communication.

158.220 **The results of observations of double galaxies in the UBV system.** A. Tomov.
IAU Symp. No. 79, (see 012.027), p. 102 - 103 (1978). Short communication.

158.221 **On the origin and evolution of pairs of galaxies that have different physical and kinematic characteristics.** I. Pronik, L. Metik.
IAU Symp. No. 79, (see 012.027), p. 103 - 104 (1978). Short communication.

158.222 **Markarian galaxies with double and multiple nuclei.** E. Khachikian (*Eh. E. Khachikyan*).
IAU Symp. No. 79, (see 012.027), p. 105 (1978). − Short communication.

158.223 **Equidensitometric determination of angular diameters and mean surface magnitudes of compact galaxies.** W. Högner, N. Richter.
IAU Symp. No. 79, (see 012.027), p. 105 - 107 (1978). Short communication.

158.224 **Radio haloes around galaxies and in clusters.** V. L. Ginzburg.
IAU Symp. No. 79, (see 012.027), p. 161 - 163 (1978).

158.225 **Markarian galaxies in the vicinity of the Coma cluster.** M. A. Arakelian (*Arakelyan*).
IAU Symp. No. 79, (see 012.027), p. 274 - 275 (1978). Short communication.

158.226 **Remarks on the angular distribution of Markarian galaxies.** P. Flin, M. Urbanik.
IAU Symp. No. 79, (see 012.027), p. 275 - 276 (1978). Short communication.

158.227 **The autocorrelation analysis of deep galaxy samples.** S. Phillipps.
IAU Symp. No. 79, (see 012.027), p. 280 (1978). − Short communication.

158.228 **Counts of faint galaxies.** I. D. Karachentsev, A. I. Kopylov.
IAU Symp. No. 79, (see 012.027), p. 339 - 342 (1978).

158.229 **New evidence for an explosion in galaxy M82.** G. A. Cottrell.
Nature, Vol. 274, 13 - 14 (1978).

158.230 **M82: the exploding galaxy?** D. J. Axon, K. Taylor.
Nature, Vol. 274, 37 - 38 (1978).

The authors draw attention to some spectacular new spectroscopic data, which they believe to be irreconcilable with the more recent scattering models and which naturally lead one to resurrect an expulsion hypothesis.

158.231 **Neutron beams in active galactic nuclei.** D. Eichler, P. J. Wiita.
Nature, Vol. 274, 38 - 39 (1978).

The authors propose an idea in which high energy beams of neutrons can be used to transport energy to relatively large distances from an active central powerhouse assuming that electromagnetic acceleration mechanisms are important. Decay back into charged particles allows radiation to recommence further away from such a generator than in other models.

158.232 **Radio recombination line studies of M82 and other galaxies.** M. B. Bell, E. R. Seaquist.
Astrophys. J., Vol. 223, 378 - 385 (1978).

The authors have searched for recombination lines in a number of nearby galaxies, including M82. They report confirmation of an earlier detection of the H102α line and a new detection of H85α in M82. They have investigated the nature of the line emission mechanism in M82 by comparing the results with published line data at other frequencies and conclude that the lines are produced primarily by stimulated emission in a foreground H II region by nonthermal background radiation. The authors have derived a simple model with $T_e = 5000$ K and $n_e = 150$ cm^{-3} which fits the data and agrees well with other, independent results. In none of the five other nearby galaxies observed was H102α line emission detected.

158.233   **Integrated masses of galaxies.**   J. R. Dickel, H. J. Rood.
Astrophys. J., Vol. 223, 391 - 409 (1978).

Twenty-one centimeter hydrogen line velocities, half-widths, and fluxes are presented for 112 spiral and irregular galaxies, the majority in groups. Integrated Brandt masses (M_T) are derived from the half-widths, taking into account the increased knowledge of the properties of galactic rotation curves obtained by Huchtmeier. Holmberg luminosities corrected for internal and Galactic extinction (L_{pg}) are determined. The sample of galaxies has an average integrated mass-to-light ratio $\langle M_T/L_{pg} \rangle = 11 \pm 1$ solar units (when $H = 50$ km s^{-1} Mpc^{-1}).

158.234   **Binary galaxy statistics. II. Observed axis ratios and position angles for galaxies in pairs.**
R. Arigo, K. Czuhai, E. Hubbard, P. Noerdlinger, K. Wisner.
Astrophys. J., Vol. 223, 410 - 420 (1978).

The authors have measured the major and minor axes of 188 galaxies, comprising all those with known velocities that they could properly identify on the Palomar Sky Survey in Page's and Turner's samples, but excluding known optical pairs. For these 94 pairs, they also measured the angle between the major axis of each galaxy and the line joining their centers, and, when possible, the sense of spiraling. The data were averaged with a novel technique, and the resulting axis ratios agree very well with those of de Vaucouleurs and de Vaucouleurs, and of Nilson, where available. All these data, including standard deviations and corrections for underestimation of the minor axis, are presented with a summary of separations, magnitudes, velocities, and types (on a two-valued scale).

158.235   **Polarization in elliptical galaxies?**   M. Jura.
Astrophys. J., Vol. 223, 421 - 425 (1978).

The author has computed the polarization produced by a cloud of Rayleigh scatterers immersed within a spherically symmetric distribution of light. This calculation might be appropriate for an optically thin interstellar cloud within an elliptical galaxy. Outside the cores of ellipticals, it seems quite possible that the total observed light might be as much as 10% polarized. Consequently, in many cases, polarization is a more sensitive probe of the presence of interstellar matter than direct photometry.

158.236   **Angular momentum in the Local Group.**
J. R. Gott III, T. X. Thuan.
Astrophys. J., Vol. 223, 426 - 436 (1978).

The angular momentum of galaxies in the Local Group is analyzed in terms of the tidal interaction picture. The authors calculate the orbits of M31 and the Galaxy and deduce their masses. They set limits on the collapse times from binding energy considerations. They give a detailed treatment of the tidal torques considering the Galaxy and M31 as an isolated system. Two possible solutions are obtained: one solution where the orbits of M31 and the Galaxy are predominantly radial, and the other with a large motion of the line of sight between the two galaxies. The authors discuss the two solutions in the context of the orbital angular momentum of the Large Magellanic Cloud about the Galaxy and

the observed motion of the local standard of rest relative to the center of mass of the Local Group. They show that other galaxies beyond the Local Group are likely to have had appreciable effects and summarize the main conclusions.

158.237   **The minor-axis brightness profile of the spiral galaxy NGC 4565 and the problem of massive halos.**
J. Kormendy, G. Bruzual A.
Astrophys. J., Lett., Vol. 223, L63 - L66 (1978).

The minor-axis profile of NGC 4565 has been measured between 20 and 25.7 V mag arcsec^{-2} to complement recent faint photometry by Hegyi and Gerber and by Spinrad et al. The composite profile consists of an inner segment only moderately well-fitted by a de Vaucouleurs law, and a shallower outer power, or $r^{1/4}$, law. Both segments are much steeper than profiles of elliptical galaxies. This provides early evidence that disk-galaxy spheroids differ significantly from elliptical galaxies. However, the authors suggest that there is still no evidence that any of the observations has optically detected a "massive halo".

158.238   **Radio continuum emission from the barred spiral galaxy NGC 5383.**   R. Sancisi, R. D. Ekers.
Astron. Astrophys., Vol. 67, L21 - L22 (1978).

Aperture synthesis observations of the radio continuum at 1415 MHz from the barred spiral NGC 5383 (Mark 281) show a bright, slightly extended central source and weaker emission from the region of the bar.

158.239   **Magnetic fields in galaxies as revealed by the polarization of light.**   A. Elvius.
Astrophys. Space Sci., Vol. 55, 49 - 57 (1978).

Some new results on optical polarization in galaxies are reported. These results as well as some other available data indicate the presence of large-scale magnetic fields in galaxies.

158.240   **Surabondance d'azote dans le gaz interstellaire du noyau de la galaxie NGC 6946.**
D. Alloin, A. Chelli.
C. R. Acad. Sci. Paris, Tome 286, Sér. B, 339 - 342 (1978).

The authors present spectrophotometric results of the nuclear regions of the spiral galaxy NGC 6946. From the relative intensities of the Hα, [N II] and [S II] lines, they estimate the electron density and a lower limit of the nitrogen abundance in the emissive gas.

158.241   **The Sculptor dwarf spheroidal galaxy.**
S. van Agt.
Messenger, No. 12, p. 3 - 4 (1978).

158.242   **Whatever happened to NGC 5291?**
H. Pedersen, P. Gammelgård, S. Laustsen.
Messenger, No. 13, p. 11 - 12 (1978).

158.243   **Seyfert galaxies and quasars.**   J. B. Oke.
J. R. Astron. Soc. Canada, Vol. 72, 121 - 137 (1978).

Seyfert galaxies and quasars are looked at first from the standpoint of the continuous spectrum and second in terms of the emission-line spectrum. Seyferts of type II are distinct from those of type I whereas quasars appear to be very luminous type I Seyferts. The nonthermal continuum components come from a region with an extent on the order of one parsec. The broad-line emission region is probably a few parsecs in extent while the sharper lines come from a much larger volume. Narrow emission lines may occur up to a few hundred parsecs from the centre of a Seyfert and as far away as 10,000 parsecs from a quasar nucleus.

158.244   **Les galaxies elliptiques et l'évolution des galaxies.**
L. Gouguenheim.
Astronomie, Vol. 92, 313 - 323 (1978).

158.245 **Ionized gas and dust in active nuclei of galaxies.**
D. E. Osterbrock.
Bull. American Astron. Soc., Vol. 10, 387 (1978). – Abstract.

158.246 **Infrared spectra of active galaxies.**
R. I. Thompson, G. H. Rieke, M. J. Lebofsky,
A. T. Tokunaga.
Bull. American Astron. Soc., Vol. 10, 388 (1978). – Abstract.

158.247 **Discrete cloud models for the broad line emitting**
regions of active galactic nuclei.
E. R. Capriotti, C. B. Foltz, P. L. Byard.
Bull. American Astron. Soc., Vol. 10, 388 (1978). – Abstract.

158.248 **OI λ8446 emission in Seyfert galaxies.**
S. A. Grandi.
Bull. American Astron. Soc., Vol. 10, 388 (1978). – Abstract.

158.249 **A new, bright Seyfert galaxy – NGC 4593.**
D. W. Lewis, G. M. MacAlpine, A. T. Koski.
Bull. American Astron. Soc., Vol. 10, 388 - 389 (1978).
Abstract.

158.250 **Broadband infrared observations of Seyfert galaxies.**
C. W. McAlary, R. A. McLaren, D. R. Crabtree.
Bull. American Astron. Soc., Vol. 10, 389 (1978). – Abstract.

158.251 **Optical polarization of the Seyfert galaxy NGC**
4151. I. Thompson, J. D. L. Landstreet, J. R. P.
Angel, H. S. Stockman, N. J. Woolf, P. G. Martin, J. Maza,
E. A. Beaver.
Bull. American Astron. Soc., Vol. 10, 389 (1978). – Abstract.

158.252 **Far ultraviolet spectrum of the Seyfert galaxy NGC**
4151. G. F. Hartig, A. F. Davidsen.
Bull. American Astron. Soc., Vol. 10, 402 (1978). – Abstract.

158.253 **X-rays from Markarian 541.** W. Cash, P. Charles,
S. Bowyer, H. W. Schnopper, J. P. Delvaille,
A. Epstein.
Bull. American Astron. Soc., Vol. 10, 403 (1978). – Abstract.

158.254 **Systematic errors in galaxy count surveys.**
A. N. Witt, P. S. Tilles.
Bull. American Astron. Soc., Vol. 10, 421 (1978). – Abstract.

158.255 **Physical conditions in dwarf blue galaxies.**
M. Peimbert, S. Torres-Peimbert, J. F. Rayo.
Bull. American Astron. Soc., Vol. 10, 422 (1978). – Abstract.

158.256 **UBVRI photometry of NGC 3623/3627/3628.**
D. J. Hutter, M. S. Burkhead.
Bull. American Astron. Soc., Vol. 10, 422 (1978). – Abstract.

158.257 **Far infrared observations of galaxies with extended**
molecular emission.
K. Sellgren, M. Werner, I. Gatley.
Bull. American Astron. Soc., Vol. 10, 422 (1978). – Abstract.

158.258 **Photometry of spiral galaxy nuclei at 10 μm and**
20 μm.
H. M. Dyck, E. E. Becklin, R. W. Capps.
Bull. American Astron. Soc., Vol. 10, 422 (1978). – Abstract.

158.259 **Surface photometry of two elliptical galaxies.**
T. B. Williams, M. Schwarzschild.
Bull. American Astron. Soc., Vol. 10, 422 (1978). – Abstract.

158.260 **Another galaxy silhouetted in front of NGC 1199?**
G. F. Benedict, G. W. van Citters, J. T. McGraw,
P. M. Rybski.
Bull. American Astron. Soc., Vol. 10, 423 (1978). – Abstract.

158.261 **A H I survey of Nilson dwarf galaxies.**
P. O. Seitzer, T. X. Thuan.
Bull. American Astron. Soc., Vol. 10, 423 (1978). – Abstract.

158.262 **The structure and origin of S0 galaxies.**
D. Burstein.
Bull. American Astron. Soc., Vol. 10, 423 - 424 (1978).
Abstract.

158.263 **X-ray characteristics of normal galaxies.**
D. M. Worrall, E. A. Boldt, S. S. Holt, R. F.
Mushotzky, P. J. Serlemitsos.
Bull. American Astron. Soc., Vol. 10, 424 (1978). – Abstract.

158.264 **NGC 2110: an X-ray emitting, narrow emission line**
galaxy with elliptical morphology.
J. E. McClintock, H. V. Bradt, C. R. Canizares, R. L. Kelley,
J. A. van Paradijs, A. T. Koski.
Bull. American Astron. Soc., Vol. 10, 424 (1978). – Abstract.

158.265 **The HI bridge between M81 and NGC 3077.**
J. M. van der Hulst.
Bull. American Astron. Soc., Vol. 10, 428 (1978). – Abstract.

158.266 **Automated faint object detection.**
J. A. Tyson, J. F. Jarvis.
Bull. American Astron. Soc., Vol. 10, 429 (1978). – Abstract.

158.267 **Extragalactic observations by IUE (*International***
***Ultraviolet Explorer*).** T. R. Gull.
Bull. American Astron. Soc., Vol. 10, 445 (1978). – Abstract.

158.268 **The unusual absorption spectrum of P0237-23.**
M. Barnothy.
Bull. American Astron. Soc., Vol. 10, 448 - 449 (1978).
Abstract.

158.269 **Gravitational effects of M32 on disk stars and spiral**
arms of the Andromeda galaxy. G. G. Byrd.
Bull. American Astron. Soc., Vol. 10, 484 (1978). – Abstract.

158.270 **Ultra-low dispersion spectroscopy of stars and**
galaxies. M. K. V. Bappu, M. Parthasarathy.
Kodaikanal Obs. Bull., Ser. A, Vol. 2, 1 - 5 (1977).
Application of ultra-low dispersion spectroscopy
$(10,000 \text{ Å mm}^{-1})$ is described to study the nuclei of elliptical
galaxies, the quasi-stellar objects and for the discovery of faint
OB stars, reddened stars and red stars. Ultra-low dispersion
spectra (microspectra) were obtained for fifteen elliptical and
three S0 galaxies from the list of Ekers and Ekers (1973) who
classified them as compact and extended sources from the
observations of radio emission at 6 cm. From an analysis of
microspectra and from direct photographs with graded exposure times, the authors find that all compact radio galaxies
in the Ekers list also have optically compact nuclei.

158.271 **Observaciones de galaxias en la línea de 21 cm del**
hidrógeno neutro con el instrumento de síntesis de
Westerbork (Holanda). E. Bajaja.
Bol. Acad. Nac. Cienc. Cordoba, Argentina, Tomo 52, 9 - 36
(1976) = Contrib. Inst. Argentino Radioastron., No. 71.
A short description is given of the aperture synthesis and
of the Westerbork instrument which applies this technique.
The method of reduction and the defects of the system are
also described, as well as the results obtained from the observation of some galaxies in the 21 cm line of the H I. The gas
distributions found prove definitely its concentration along
the optical arms which reproduces the spiral structure.

158.272 **The active region of nuclei of galaxies as a magneto-**
plasma boiler with external heating.
V. N. Kuril'chik.

Astron. Tsirk., No. 979, p. 2 - 4 (1978). In Russian.

158.273   What is 3C120?   V. M. Lyutyj.
         Astron. Tsirk., No. 984, p. 1 - 3 (1978). In Russian.

158.274   Hα-photometry of the core of NGC 4151.
         Yu. S. Kosorukov, A. M. Cherepashchuk.
Astron. Tsirk., No. 990, p. 1 - 3 (1978). In Russian.

158.275   Photometric classification of galaxies and deter-
         mination of their redshifts.
K. Nandy, Z. Sviderskiene, V. Straižys.
Astron. Tsirk., No. 992, p. 2 - 5 (1978). In Russian.

158.276   Invisible mass in spiral galaxies.   E. E. Salpeter.
         Ann. New York Acad. Sci., Vol. 302, (see 012.033),
681 - 684 (1977).

158.277   Nuclei galattici e radio-sorgenti.   G. Setti.
         Mem. Soc. Astron. Italiana, Vol. 47, 159 - 165
(1976).

158.278   On active nuclei of galaxies and QSOs.
         G. C. Perola.
Mem. Soc. Astron. Italiana, Vol. 47, (see 012.036), 387 - 405
(1976).
   The author concentrates on NGC 1068, a typical example
of the Seyfert galaxies, which are characterized by a very
luminous and active nucleus. He briefly describes the properties of QSOs. He presents the most important non-thermal
emission processes which take place in galaxies and QSOs.

158.279   A newly discovered magnitude and color variable
         compact object in Perseus — a bright blue N galaxy
bN?   Y.-h. Chu.
Astron. Circ., No. 1, 5 pp. (1978). In Chinese and English.

         The ESO/Uppsala survey of the ESO (B) Atlas of
the Southern Sky. V.   See Abstr. 002.001.

         The ESO/Uppsala survey of the ESO (B) Atlas of
the Southern Sky — VI.   See Abstr. 002.051.

         Non-stationary phenomena in galaxies.
See Abstr. 003.040.

         Correlation analysis of deep galaxy samples — I.
Techniques with applications to a two-colour sample.
See Abstr. 031.214.

         Numerical mapping technique applied to the photo-
graphic photometry of bright galaxies.
See Abstr. 031.311.

         Express processing of spectrograms of extragalactic
objects. I. Preliminary processing.   See Abstr. 031.419.

         X-ray and gamma-ray line production by nonthermal
ions.   See Abstr. 061.005.

         Cloud confinement by momentum transfer from
suprathermal protons.   See Abstr. 062.034.

         Sustenance of a black hole in a galactic nucleus.
See Abstr. 066.112.

         Spectra of giant stars in distant satellites of the
Galaxy.   See Abstr. 114.010.

         Variable stars in the Leo I dwarf galaxy.
See Abstr. 122.097.

         Supernova remnants in M 33.
See Abstr. 125.008

         A maximum likelihood estimate of the supernova
rate in Sc galaxies.   See Abstr. 125.016.

         Observational consequences of scattering clouds near
galaxies.   See Abstr. 131.230.

         A survey of HII regions in M 31.
See Abstr. 132.016.

         Far infrared observations of extragalactic H II
regions.   See Abstr. 132.027.

         Westerbork 21-cm observations of the low and high
velocity H I in Stephan's Quintet.   See Abstr. 132.032.

         Far-infrared observations of H II regions in M33.
See Abstr. 132.057.

         Reddening of H II regions in M31.
See Abstr. 132.071.

         Infrared detection of luminous stars in M31 and
M33.   See Abstr. 133.019.

         The planetary nebula in the Fornax dwarf galaxy.
See Abstr. 135.014.

         Planetary nebulae in the Andromeda galaxy and its
companions.   See Abstr. 135.030.

         Planetary nebulae in the nuclei of M31 and M32.
See Abstr. 135.104.

         Stellar debris clouds in quasars and related objects.
See Abstr. 141.011.

         M33: radiocontinuum emission at 4850 MHz.
See Abstr. 141.013.

         Multifrequency observations of very large radio gal-
axies: I. 3C326.   See Abstr. 141.015.

         Radio continuum observations of the edge-on disc
galaxy NGC 891.   See Abstr. 141.016.

         Two quasars near a peculiar pair of galaxies.
See Abstr. 141.017.

         High-velocity gas in the jet in the radio galaxy
DA 240.   See Abstr. 141.020.

         He I lines in the spectra of QSOs and Seyfert
galaxies.   See Abstr. 141.022.

         1 millimeter continuum observations of extragalac-
tic objects.   See Abstr. 141.034.

         The existence of Ca II absorption lines in the spec-
trum of the quasar 3C 232 due to the galaxy NGC 3067.
See Abstr. 141.035.

         On the nature of the extragalactic radio recombina-
tion lines.   See Abstr. 141.040.

         Imaging polarimetry of the jets of M87 and 3C 273.
See Abstr. 141.044.

         Radio observations of the head-tail radio galaxy
NGC 1265.   See Abstr. 141.071.

Orbital dynamics of the radio galaxy 3C 129.
See Abstr. 141.083.

The radio structure of 3C 123 at 2.7 and 15 GHz.
See Abstr. 141.088.

Thermal continuum from accretion disks in quasars.
See Abstr. 141.102.

The local space density of quasars and active nuclei.
See Abstr. 141.120.

The line-locking hypothesis, absorption by intervening galaxies, and the $z = 1.95$ peak in redshifts.
See Abstr. 141.130.

The large scale structure of extra-galactic radio
sources (> 1 arcsecond). See Abstr. 141.131.

Self consistent models for the emissive regions of
quasars and Seyfert nuclei. See Abstr. 141.135.

The black flash model of QSOs.
See Abstr. 141.146.

A radio continuum survey of isolated pairs of
galaxies. See Abstr. 141.149.

Excess radio emission from close pairs of galaxies.
See Abstr. 141.150.

Optical emission in the SW radio lobe of 3C 33.
See Abstr. 141.155.

Polarization in the very large radio galaxy NGC 6251
at 610 MHz. See Abstr. 141.158.

A determination of the bivariate size-luminosity
function of radio galaxies. See Abstr. 141.169.

Radio and optical observations of the galaxy 3C293.
See Abstr. 141.170.

Gigantic energy source is discovered in a galaxy by
improved interferometry. See Abstr. 141.171.

Short-lived clouds in atmospheres of QSO's and
active galaxies. See Abstr. 141.200.

Particle reacceleration in radio galaxies.
See Abstr. 141.202.

Surface brightness and shape of radio jets.
See Abstr. 141.203.

A galaxy associated with 3C 273.
See Abstr. 141.220.

The physics of radio galaxies – constraints and
hints provided by observational data.
See Abstr. 141.230.

X-ray spectra and variability of some Seyfert galaxies and other high latitude sources.
See Abstr. 142.008.

A model for compact X-ray sources in galactic
nuclei: application to the giant elliptical galaxy M87.
See Abstr. 142.015.

On the contribution of active galactic nuclei to the
diffuse X-ray background. See Abstr. 142.031.

Hard X-ray spectrum of NGC 4151.
See Abstr. 142.051.

Extragalactic X-ray sources. See Abstr. 142.135.

NGC 5506 and NGC 7582: counterparts to X-ray
sources. See Abstr. 142.203.

Radio observations of the X-ray galaxy NGC 2110.
See Abstr. 142.211.

Diffuse γ-ray background from Seyfert galaxies.
See Abstr. 142.716.

Negative mass instability of flat galaxies.
See Abstr. 151.016.

A model for winds from galactic disks.
See Abstr. 151.018.

Possibilities of observing effects due to Lin's density-wave theory. See Abstr. 151.038.

On the rotation of elliptical galaxies.
See Abstr. 151.048.

The dynamics of the spiral galaxy M81.
See Abstr. 151.068.

On the criterion of applicability of numerical
models of interacting galaxies. See Abstr. 151.071.

Interacting systems. See Abstr. 151.079.

The warping of the galactic plane and changes of
ellipticity. See Abstr. 151.088.

A model of the formation of elliptical galaxies.
See Abstr. 151.089.

Surface brightness distributions for galaxies with
periodic waves of star formation.
See Abstr. 151.096.

Dynamics of the gas in spiral galaxies.
See Abstr. 151.108.

Gravitational separation of elements during certain
stages of galactic evolution. See Abstr. 151.110.

The metallicity of globular clusters in the halo of
M87. See Abstr. 154.033.

The space distribution of globular clusters in M49.
See Abstr. 154.034.

Voie Lactée et galaxies. See Abstr. 155.009.

Multiaperture photometry of peculiar extragalactic
sources. See Abstr. 158.501.

A Westerbork survey of rich clusters of galaxies.
V. Multi-frequency observations of the radio tail galaxy
NGC 6034 in the Hercules cluster.
See Abstr. 160.011

Spectral variations in brightest cluster galaxies.
See Abstr. 160.017.

D and cD galaxies in poor clusters.
See Abstr. 160.021.

Statistical analysis of catalogs of extragalactic objects. IX. The four-point galaxy correlation function. See Abstr. 160.022.

A model of tidal interactions within the NGC 4631 group of galaxies.   See Abstr. 160.040.

Contributions to galaxy photometry. VII. Standard total magnitudes of 139 bright galaxies in the Virgo cluster area.   See Abstr. 160.053.

Statistical analysis of discrepant redshift associations, quintets of galaxies.   See Abstr. 160.062.

Velocity dispersion in small systems of galaxies. See Abstr. 160.068.

Possible optical evidence for ram-pressure-sweeping in the Hydra I cluster of galaxies.   See Abstr. 160.094.

The extragalactic background light and slow star formation in galaxies.   See Abstr. 162.022.

The distance scale within the Local Group. See Abstr. 162.039.

Surface brightness, standard candles and q_0. See Abstr. 162.045.

Time delay effects for measuring cosmological distances.   See Abstr. 162.047.

High redshift 21-cm lines.   See Abstr. 162.052.

The redshift dependence of galaxy surface brightness.   See Abstr. 162.053.

The Hubble modulus of Markarian galaxies. See Abstr. 162.058.

Orientation of spiral galaxies as a test of theories of galaxy formation.   See Abstr. 162.147.

Evolutionary processes in the universe. See Abstr. 162.183.

Peculiar Objects

158.501 Multiaperture photometry of peculiar extragalactic sources.
B. Q. McGimsey, H. R. Miller.
Astrophys. J., Vol. 219, 387 - 391 (1978).
  The surface brightness distributions of several extragalactic sources have been measured. Results for AP Librae and B2 1652+39 are consistent with a model of a nonthermal central source superposed on a giant elliptical galaxy at redshifts of 0.0486 and 0.0337, respectively. The galaxy 0246+18 appears to have a galactic star superposed. Photoelectric UBV observations of five Markarian galaxies indicate that three are similar to BL Lacertae objects in color and spectral index.

158.502 The redshift and other properties of I Zw 1727+5015.
  J. B. Oke.
Astrophys. J., Lett., Vol. 219, L97 - L100 (1978).
  A pair of spectra of the BL Lacertae type of object I Zw 187 (1727+5015), taken 2″ north and 2″ south of the nucleus, show absorption lines typical of a normal elliptical galaxy. The redshift is $z = 0.0554 \pm 0.0003$. This object is very similar to 3C 371 and BL Lac in several respects: (1) all objects have at best very weak emission lines; (2) the underlying galaxies are of similar luminosity within a factor 2.5; (3) the nucleus in all objects is characterized by a power-law spectrum with $\alpha = 1.6$, where $f_\nu \propto \nu^{-\alpha}$; and (4) the luminosity of the power-law component is comparable in all objects; BL Lac at its brightest is much more luminous than I Zw 187 and 3C 371.

158.503 The spectrum and magnitude of the galaxy associated with BL Lacertae.
J. S. Miller, H. B. French, S. A. Hawley.
Astrophys. J., Lett., Vol. 219, L85 - L87 (1978) = Lick Obs. Bull., No. 788.
  The authors present new spectroscopic observations of BL Lacertae, made through an annular aperture, which show convincingly that the diffuse light surrounding the central source has a spectrum which is normal for a luminous elliptical galaxy with a redshift near 0.07.

158.504 Photoelectric intraday observations of BL Lacertae,

3C 66 A, B2 1652+39, and 3C 371.
H. R. Miller, B. Q. McGimsey.
Astrophys. J., Vol. 220, 19 - 24 (1978).
  A search for optical intraday variability in the sources BL Lac, 3C 66 A, B2 1652+39, and 3C 371 has failed to detect significant changes on a time scale ranging from 20 s to a few hours. Day-to-day changes in brightness are present in BL Lac, 3C 66 A, and 3C 371. The data suggest that as 3C 66 A and 3C 371 brighten, the $U - B$ color index becomes more negative, while fairly large changes in $U - B$ occur in BL Lac which may or may not be accompanied by significant changes in V.

158.505 NGC 1275: a BL Lacertae object? P. Veron.
  Nature, Vol. 272, 430 - 431 (1978).
  The author suggests, on the basis of its weak emission line spectrum, its lack of the characteristic broad hydrogen lines and the variability and polarisation of its nucleus, that NGC 1275 should be called a BL Lacertae object rather than a Seyfert galaxy.

158.506 Spectra of the stellar population in three objects related to BL Lacertae.
M.-H. Ulrich.
Astrophys. J., Lett., Vol. 222, L3 - L6 (1978).
  The author presents results on two BL Lacertae objects, B2 1101 + 38 and Markarian 180. For B2 1101 + 38 a new measurement of the redshift based on spectral scans of the nebulosity confirms the value $z = 0.03$ obtained earlier by Ulrich et al. (1975). This BL Lacertae object belongs to a cluster of galaxies. In the case of Markarian 180 the spectral scans of the nebulosity show several absorption lines usually found in the spectrum of elliptical galaxies, at the redshift $z = 0.046$. Markarian 11 has also been observed. Scans of the central region show absorption lines, including Balmer lines, at $z = 0.014$. This galaxy does not seem to be a BL Lacertae object at the present time.

158.507 The nature of the nebulosity around BL Lac objects.
  J. E. Gunn.
Phys. Scr., Vol. 17, (see 012.013), 277 - 279 (1978).

158.508 Physical problems associated with BL Lac objects and QSOs. G. Burbidge.
Phys. Scr., Vol. 17, (see 012.013), 281 - 283 (1978).

Physical problems associated with Compton effect, apparent super-light velocities and the energetics of BL Lac objects and QSOs are discussed.

158.509 A comprehensive radio study of the $z = 0.524$ absorption system in AO 0235 + 164.
A. M. Wolfe, J. J. Broderick, J. J. Condon, K. J. Johnston.
Astrophys. J., Vol. 222, 752 - 760 (1978).

The authors present pencil-beam and VLB observations of the 932 MHz absorption line and continuum of the BL Lacertae object AO 0235 + 164, and pencil-beam observations of its 318 and 430 MHz continuum. All the data available indicate that the $z = 0.524$ absorption system is extremely difficult to understand in terms of ejection from AO 0235 + 164. The data are consistent with absorption by an intervening galaxy with a cosmological redshift of $z = 0.524$.

158.510 On the extragalactic nature of Lacertids. A. N. Deutsch.
Pis'ma Astron. Zh., Vol. 4, 118 - 120 (1978). In Russian.

Measurements of proper motions on three pairs of plates reveal the similarity of parallactic motion of reference stars relative to 8 galaxies and to the radio source ON 325. This indicates the extragalactic nature of ON 325. The coordinates and photographic magnitudes for this object are estimated as well as for the object ON 231, identified on the edge of the plates.

158.511 Peculiar southern galaxies. B. F. Madore.
New Scientist, Vol. 76, 293 - 294 (1977). — Abstr. in Phys. Abstr., Vol. 81, Abstr. 20896 (1978).

158.512 New light on quasars: unraveling the mystery of BL Lacertae. W. D. Metz.
Science, Vol. 200, 1031 - 1033 (1978).

158.513 1418+54: a new, violently variable BL Lacertae object. H. R. Miller.
Astrophys. J., Lett., Vol. 223, L67 - L70 (1978).

The purpose of this Letter is to report on the long-term optical variability of 1418+54 as derived from archival plates.

158.514 Recent photometry of OJ287.
J. Zink, A. J. Pica, J. T. Pollock, D. Kolpanen, P. D. Usher.
Inf. Bull. Variable Stars, No. 1441, 3 pp. (1978).

158.515 Quasars and BL Lac objects as active nuclei of giant galaxies. J. Bergeron.
Messenger, No. 12, p. 5 - 6 (1978).

158.516 Multifrequency observations of the variable quasi-stellar radio source 0735+178.
T. J. Balonek, W. A. Dent.
Bull. American Astron. Soc., Vol. 10, 449 (1978). — Abstract.

158.517 Nonthermal properties of BL Lac objects and the relation to normal QSOs. W. A. Stein.
Bull. American Astron. Soc., Vol. 10, 450 (1978). — Abstract.

158.518 Far red area photometry of PKS 0548-322.
D. Weistrop, B. A. Smith, H. J. Reitsema.
Bull. American Astron. Soc., Vol. 10, 451 (1978). — Abstract.

A handbook of quasistellar and BL Lacertae objects. See Abstr. 003.037.

A photometric sequence for OI 090.4 and additional information on CSV 1180. See Abstr. 113.069.

1 millimeter continuum observations of extragalactic objects. See Abstr. 141.034.

Direct imaging of quasars and BL Lac objects. See Abstr. 141.077.

Size limits for expanding light sources. See Abstr. 141.094.

The 1977 outburst of the quasar 3C 446. See Abstr. 141.198.

Mk 421 = 2A 1102+384. See Abstr. 142.174.

The continuous optical spectra of active nuclei. See Abstr. 158.140.

Extragalactic observations by IUE (*International Ultraviolet Explorer*). See Abstr. 158.267.

Errata

158.901 Erratum: "Seyfert galaxies with large z: an electronographic survey" [Mon. Not. R. Astron. Soc., Vol. 181, 211 - 231 (1977)]. P. A. Wehinger, S. Wyckoff.
Mon. Not. R. Astron. Soc., Vol. 183, 275 (1978).

158.902 Erratum: "Two new faint stellar systems discovered on ESO Schmidt plates" [Astron. Astrophys., Vol. 61, L31 - L33 (1977)]. D. A. Cesarsky, S. Laustsen, J. Lequeux, H.-E. Schuster, R. M. West.
Astron. Astrophys., Vol. 65, 153 (1978).

159 Magellanic Clouds

159.001   Three luminous carbon stars in the Large Magellanic Cloud.
H. B. Richer, B. E. Westerlund, N. Olander.
Astrophys. J., Vol. 219, 452 - 453 (1978).

The properties of three very luminous carbon stars in the Large Magellanic Cloud are discussed. Two of the three are rather blue [$(R - I) \sim 0.88$] and exhibit significant amounts of ^{13}C in their spectra. The third is the first carbon Mira to be found in the LMC.

159.002   Studies of the Large Magellanic Cloud stellar content. III. Spectral types and V magnitudes of 1822 members.   J. Rousseau, N. Martin, L. Prévot, E. Rebeirot, A. Robin, J. P. Brunet.
Astron. Astrophys., Suppl. Ser., Vol. 31, 243 - 260 (1978).

A spectral survey has been made during the period 1971 to 1975 with the ESO 40 cm astrograph at La Silla, equipped with its normal prism, giving an intermediate dispersion of 95 Å mm^{-1} at λ 4026 Å. Long exposure plates taken directly or with an interference filter enabled the authors to obtain spectral types for nearly 1600 stars. In parallel, a V photographic survey has been carried out during the same period with the same astrograph and has led to the determination of V magnitudes for more than 700 stars having no previous photometric data. New spectroscopic and photometric results as well as previous photoelectric UBV values are given in the catalogue together with additional remarks concerning peculiarities of spectra, V magnitudes, and details on double or multiple systems.

159.003   Carbon and M-type giant stars in the Magellanic Clouds.
B. M. Blanco, V. M. Blanco, M. F. McCarthy.
Nature, Vol. 271, 638 - 639 (1978).

Three surveys described here indicate that unsuspected differences in the mixture of C and late M giant stars exist in the nuclear bulge of the Galaxy and in various regions of the Magellanic Clouds. Such studies also yield new information concerning the intrinsic luminosities of C and M giant stars in these three galactic systems.

159.004   On the anomaly of the far UV extinction in the 30 Doradus region.   J. Koornneef.
Astron. Astrophys., Vol. 64, 179 - 193 (1978).

Area-integrated ultraviolet observations made with the Netherlands Astronomical Satellite (ANS), of about 800 fields (2.5×2.5) in a 0.7 square degree region around the giant H II region 30 Doradus in the Large Magellanic Cloud are discussed. The author presents a colour-brightness diagram for about 600 fields with good quality measurements and concludes that most of the spread in colour is caused by differential extinction. By various independent methods he finds that the 2200 Å-feature is deficient by a factor of three, on a logarithmic scale, relative to the average galactic extinction law.

159.005   Very-low-excitation compact nebulae in the Magellanic Clouds.
N. Sanduleak, A. G. D. Philip.
Publ. Astron. Soc. Pacific, Vol. 89, 792 - 794 (1977/78).

A list is given of stellar-like and compact nebular objects in the Magellanic Clouds which were noted on objective-prism plates to have very-low-excitation spectra similar to those characterizing a small group of nebulae in the Milky Way.

159.006   Apollo 16 far-ultraviolet imagery and spectra of the Large Magellanic Cloud.
T. L. Page, G. R. Carruthers.
Space Research, Vol. XVII, (see 012.010), 749 - 755 (1977).

The Large Magellanic Cloud (LMC) was observed by the Naval Research Laboratory's Far-Ultraviolet Camera/Spectrograph (Experiment S-201) from the lunar surface on 22 April 1972 during the Apollo 16 mission. Images were obtained with about 3 arc min resolution, in the 1050–1600 and 1250–1600 Å wavelength ranges, of nearly the entire LMC. Spectra were also obtained in the 1050–1600 and 900–1600 Å ranges, with 30–40 Å resolution, along a strip 0.25° wide passing across the LMC. The images and spectra have been scanned with a microdensitometer, and analyses of the data to date are discussed.

159.007   The Magellanic Stream: the turbulent wake of the Magellanic Clouds in the halo of the Galaxy.
D. S. Mathewson, M. P. Schwarz, J. D. Murray.
Proc. Astron. Soc. Australia, Vol. 3, 133 - 136 (1977).

159.008   Red variable stars in the Magellanic Clouds – II. The field of NGC 371 in the SMC.
T. Lloyd Evans.
Mon. Not. R. Astron. Soc., Vol. 183, 305 - 317 (1978).

A field of 40 × 30 arcmin centred on NGC 371 in the Small Magellanic Cloud was photographed repeatedly over a period of five years. Periods have been found for 29 red variables: most are of small amplitude with variations which are often cyclic rather than strictly periodic. Red variables with properties similar to those of the halo and metal rich globular cluster populations of the Galaxy are absent, except for one or two doubtful cases.

159.009   Red variable stars in the Magellanic Clouds – III. Carbon stars in the field of NGC 419 (SMC).
T. Lloyd Evans.
Mon. Not. R. Astron. Soc., Vol. 183, 319 - 327 (1978).

Photographic photometry on the BVI_K system is presented for a sample of red stars in and around the SMC globular cluster NGC 419. Many of the field stars have $B-V > 2.0$: the colours suggest that they are carbon stars (N stars). The majority of the stars with $B-V > 2.1$, $V-I_k > 2.1$, are variable. These have $M_V \sim -2$, similar to the absolute magnitude of N-type variables of the old disk population in the Galaxy.

159.010   Late type giants in the Large Magellanic Cloud.
V. M. Blanco, M. F. McCarthy.
Astron. extragalactica, (see 012.015), 79 - 82 (1976).

Using an objective prism of very low dispersion together with a Schmidt camera, it was possible to classify late M stars up to about 16.5 apparent magnitude in the Magellanic Cloud. The distribution of these stars is discussed and it is found that the results suggest that the LMC probably has a component of disk population composed of giant M stars.

159.011   Large Magellanic Cloud sources at 3.4-cm wavelength.
R. X. McGee, L. M. Newton, P. W. Butler.
Mon. Not. R. Astron. Soc., Vol. 183, 799 - 804 (1978).

Selected regions of the Large Magellanic Cloud have been surveyed at wavelength 3.4 cm with a 2.5 arcmin half-power beamwidth. Improved spectral indices are given for 35 sources in the LMC area. Contour maps of the 30 Doradus nebula and the four important sources to the south of it are presented.

159.012   The population structure of the Large Magellanic Cloud. II. Count-brightness ratios for the central regions.   E. Hardy.
Astrophys. J., Vol. 223, 98 - 108 (1978).

A combination of luminosity functions and surface photometry, the count-brightness ratio, make it possible to test assumptions on the stellar population content of the crowded central regions of the LMC. By comparison with the van Rhijn function it is found that the LMC central disk is not a scaled version of the galactic disk. The author concludes that intermediate-age stars must contribute to the bright resolved giant branch in agreement with known results on peripheral regions.

159.013 **On the Magellanic Stream, the mass of the Galaxy and the age of the Universe.**   D. Lynden-Bell.
IAU Symp. No. 79, (see 012.027), p. 123 - 130 (1978).

159.014 **The stellar component of the Magellanic Stream.**
W. E. Kunkel.
IAU Symp. No. 79, (see 012.027), p. 130 - 131 (1978).
Short communication.

159.015 **Fine structure in the Magellanic Stream.**
A. G. D. Philip.
IAU Symp. No. 79, (see 012.027), p. 131 - 132 (1978).
Short communication.

159.016 **Emission objects in the Large Magellanic Cloud.**
C. Fehrenbach, M. Duflot, A. Acker.
Astron. Astrophys., Suppl. Ser., Vol. 33, 115 - 123 (1978).
In French.
  Some new information is given on the emission-line objects in the Large Magellanic Cloud: stars, planetary nebulae, compact H II regions and nebulosities with small angular diameter. For the latter objects a relation is found between the angular diameter and the degree of excitation.

159.017 **Some integrated photometric properties of the Large Magellanic Cloud.**   E. Hardy.
Publ. Astron. Soc. Pacific, Vol. 90, 132 - 138 (1978).
  Photoelectric photometry in the UBV and DDO systems of areas in the central regions of the LMC which are free of supergiants is presented. For these smooth areas it is found that $\langle B-V \rangle = 0\overset{m}{.}67 \pm 0\overset{m}{.}06$ and $\langle U-B \rangle = 0\overset{m}{.}03 \pm 0\overset{m}{.}10$. An exponential disk is shown to exist with a length scale $\alpha^{-1} = 1.3$ kpc in all three UBV filters. There is no radial dependence on the colors but a color asymmetry exists if the bar is taken as an axis of symmetry. From all the data available on colors, masses, and length scales for the LMC and M33 it is shown that the underlying disks of both galaxies are similar.

159.018 **Star clusters in the Small Magellanic Cloud.**
M. T. Brück.
Occas. Rep. R. Obs. Edinburgh, No. 1, 7 pp. (1976).
  A paper on the cluster system in the Small Magellanic Cloud (Brück 1975) describes the distribution of star clusters observed on UK 48-inch Schmidt plates but does not list or identify the objects individually. The present list and charts are intended to identify the clusters by coordinate and also on small scale charts.

159.019 **On the origin of the morphological complexity of the Small Magellanic Cloud.**   A. Ardeberg,
E. Maurice.
C. R. Acad. Sci. Paris, Tome 286, Sér. B. 375 - 378 (1978).
In French.
  Recent observations have been used to analyse various encounter models concerning their ability to explain the combined structural and kinematic behaviour of the Small Magellanic Cloud. The apparent morphological complexity may be reasonably described in terms of large-scale displacements, which may in turn be explained as due to cataclysmic disruption caused by gravitational interaction with the Galaxy and/or the Large Magellanic Cloud.

159.020 **Morphological studies of the Large Magellanic Cloud on ESO Schmidt plates.**   E. H. Geyer.

Messenger, No. 12, p. 7 - 9 (1978).

159.021 **The N119 complex in the Large Magellanic Cloud.**
J. Melnick.
Messenger, No. 12, p. 13 - 16 (1978).

159.022 **An upper limit to the absorption of 1 keV diffuse X-rays by the Small Magellanic Cloud.**
A. N. Bunner, W. T. Sanders, W. L. Kraushaar, J. A. Nousek.
Bull. American Astron. Soc., Vol. 10, 389 - 390 (1978).
Abstract.

159.023 **Identifications of faint planetary nebulae in the Magellanic Clouds.**   G. H. Jacoby.
Bull. American Astron. Soc., Vol. 10, 397 (1978). – Abstract.

159.024 **Results of deep objective-prism surveys of the Magellanic Clouds.**
N. Sanduleak, A. G. D. Philip.
Bull. American Astron. Soc., Vol. 10, 414 (1978). – Abstract.

159.025 **UV observations of the Large Magellanic Cloud.**
J. Koornneef.
Mem. Soc. Astron. Italiana, Vol. 47, (see 012.036), 632 (1976).
Summary.

    Remarks on the catalogue of LMC supergiants published by Stock et al. (1976).   See Abstr. 002.037.

    On the numbers of yellow stars in the Large Magellanic Cloud. (Paper II).   See Abstr. 113.059.

    A photoelectric UBV sequence in the region of the wing of the Small Magellanic Cloud.   See Abstr. 113.067.

    uvbyR surface photometry of the 30 Doradus region II.   See Abstr. 113.074.

    A catalogue of carbon stars in the Large Magellanic Cloud.   See Abstr. 114.001.

    Wolf-Rayet stars in the Small Magellanic Cloud.   See Abstr. 114.065.

    A survey of red stars in the direction of the Large Magellanic Cloud. 1. The 30 Doradus region.   See Abstr. 114.087.

    Carbon stars possibly associated with clusters in the Large Magellanic Cloud.   See Abstr. 114.525.

    Spectrophotometric study of the Large Magellanic Cloud emission-line star Hen S 22.   See Abstr. 114.569.

    Wolf-Rayet stars in the Small Magellanic Cloud.   See Abstr. 114.570.

    The eclipsing Small Magellanic Cloud Wolf-Rayet binary HD 5980.   See Abstr. 121.026.

    Long-term light variations of four supergiants in the Large Magellanic Cloud.   See Abstr. 122.025.

    Photometry of cepheid variables in the Large Magellanic Cloud.   See Abstr. 122.042.

    An explanation of the anomalies of the Magellanic Cloud Cepheids.   See Abstr. 122.057.

    The pre-maximum spectrum of a Magellanic Cloud nova.   See Abstr. 124.022.

Nova in Large Magellanic Cloud.
See Abstr. 124.023.

Studies of N132D, a supernova remnant similar to Cassiopeia A in the Large Magellanic Cloud.
See Abstr. 125.050.

A new study of N132D, a Magellanic Cloud supernova remnant similar to Cas A.   See Abstr. 125.064.

Infrared studies of 30 Doradus. I. The 2-μ sources of the inner region.   See Abstr. 132.001.

A giant H II ring and energetic stellar winds in the Large Magellanic Cloud.   See Abstr. 132.065.

Planetary nebulae in the Magellanic Clouds.
See Abstr. 135.029.

Lists of confirmed planetary nebulae in the Magellanic Clouds.   See Abstr. 135.034.

Observation of X-ray eclipses from LMC X-4.
See Abstr. 142.001.

Discovery of eclipsing nature of LMC X-4.
See Abstr. 142.002.

A new measurement of the spin-up rate of SMC X-1.
See Abstr. 142.045.

On two new X-ray sources in the SMC and the high luminosities of the Magellanic X-ray sources.
See Abstr. 142.054.

Observation of new X-ray sources in the Small Magellanic Cloud.   See Abstr. 142.074.

Erste Röntgenquelle in der Großen Magellanschen Wolke identifiziert.   See Abstr. 142.145.

The optical counterparts of SMC X-2 and SMC X-3.
See Abstr. 142.191.

Observation of soft X-ray emission from SMC X-1.
See Abstr. 142.196.

Precise positions of LMC X-1, X-2, and X-3 with the HEAO-1 Modulation Collimator.   See Abstr. 142.197.

The LMC cluster Hodge 11.   See Abstr. 154.020.

Interferometer measurements of H I absorption in the direction of 16 extragalactic radio sources.
See Abstr. 155.002.

160 Groups, Clusters of Galaxies, Superclusters

160.001 **A radio survey of clusters of galaxies. I. 11.1 cm observations of A591, A754, A1066, A1314, A1517, A2094, A2142, A2255, A2256, A2319 and A2462.**
C. G. T. Haslam, P. P. Kronberg, H. Waldthausen, R. Wielebinski, D. Schallwich.
Astron. Astrophys., Suppl. Ser., Vol. 31, 99 - 119 (1978) = M.P.I. Radioastronomie, Bonn, Sonderdruck Ser. A, Nr. 226.

As a part of a programme of studies of clusters of galaxies at centimeter wavelengths, the authors have observed eleven Abell clusters at 11.1 cm using the 100 m radiotelescope. The angular resolution at this wavelength is 4.6 arcmin and an r.m.s. noise limit of 5 mK was reached in most of the observations. This enables the authors to investigate the distribution of the low level emission near point sources in the clusters.

160.002 **A mean density and a correlation function of rich clusters: theory and observations.**
A. G. Doroshkevich, S. F. Shandarin.
Mon. Not. R. Astron. Soc., Vol. 182, 27 - 33 (1978).

The authors investigate a large-scale distribution of rich clusters of galaxies within the framework of the adiabatic theory of galaxy formation. It is shown that a mean density of 'pancakes' is close to the density of Abell's clusters. Angular correlation functions calculated for spatial correlation functions under different assumptions about the spectrum of initial perturbations are compared with Hauser & Peebles points derived from Abell's catalogue.

160.003 **A finding list of southern clusters of galaxies – I.**
M. K. Braid, H. T. MacGillivray.
Mon. Not. R. Astron. Soc., Vol. 182, 241 - 248 (1978).

A finding list is presented for 474 rich clusters of galaxies in 99 fields of the southern sky from reject survey plates taken with the UK 1.2-m Schmidt Telescope in Australia.

160.004 **The evolution of galaxies in clusters. I. ISIT photometry of Cl 0024 + 1654 and 3C 295.**
H. Butcher, A. Oemler, Jr.
Astrophys. J., Vol. 219, 18 - 30 (1978).

This paper presents first results of a program of two-color photometry of very distant ($z \geqslant 0.4$) clusters of galaxies, using the KPNO ISIT vidicon. Photometry is presented for two clusters, Cl 0024 + 1654 ($z = 0.39$) and the cluster around 3C 295 ($z = 0.46$). Both clusters are rich, centrally condensed systems similar to the Coma cluster. However, in contrast to nearby clusters of this type, which contain only elliptical and S0 galaxies, between one-third and one-half of the galaxies in these two distant clusters have the colors of spiral galaxies. The authors discuss the possible implications of this result for the evolutionary history of S0 galaxies.

160.005 **An X-ray and optical study of seven clusters of galaxies.**
D. Maccagni, M. Tarenghi, B. A. Cooke, T. Maccacaro, J. P. Pye, M. J. Ricketts, G. Chincarini.
Astron. Astrophys., Vol. 62, 127 - 133 (1978).

The authors present observations of seven clusters of galaxies associated with X-ray sources detected by the Sky Survey Instrument (SSI) of the University of Leicester on the satellite Ariel V. Three are new X-ray sources and there are four new identifications with clusters of galaxies. All error boxes have an area less than about 1/3 square degree. All clusters have been classified according to the Rood and Sastry and the Bautz and Morgan systems. The new optical material obtained at the 4 m telescope of the Cerro Tololo Inter-American Observatory has been used to give the morphological description of some of the clusters.

160.006 **Ellipticity profiles of elliptical galaxies in the Virgo and Coma clusters.**
G. di Tullio.
Astron. Astrophys., Vol. 62, L17 - L19 (1978).

The run of ellipticity with radius is presented for 22 elliptical galaxies belonging to the Coma and Virgo clusters. The isodensitometer tracings of these galaxies reveal three types of variation of ellipticity with radius here referred to as class A, B, C. None of the theoretical models of ellipticals predicts the existence of all three of the classes. Each class is predicted, however, by one or another of the competing models.

160.007 **The virial mass discrepancy in groups and clusters of galaxies.**
F. D. A. Hartwick.
Astrophys. J., Vol. 219, 345 - 351 (1978).

Evidence is presented to show that the virial mass discrepancy depends on the density of the group. Application of the virial theorem to galaxies in 17 high-density groups yields $M/L \sim 7$ out to distances ~ 100 kpc. Among the low-density groups, statistically significant correlations are found between the group velocity dispersion and group radius. These correlations are interpreted either within the expansion hypothesis or as due to the groups being bound in the presence of a negative cosmological constant. There appears to be little support for the massive halo hypothesis in its usual form.

160.008 **The relation between cluster density and Bautz-Morgan type.**
S. van den Bergh, J. deRoux.
Astrophys. J., Vol. 219, 352 - 353 (1978).

It is shown that galaxy clusters of BM type I, which are dominated by a single cD galaxy, are characterized by a higher number density of galaxies than are clusters in which the first-ranked galaxy is less dominant. This result is consistent with the notion that BM type I clusters are dynamically more evolved than are those of type III.

160.009 **The origin and distribution of gas within rich clusters of galaxies: the evolution of cluster X-ray sources over cosmological time scales.**
L. L. Cowie, S. C. Perrenod.
Astrophys. J., Vol. 219, 354 - 366 (1978).

The authors describe the evolution of the X-ray properties of rich clusters as a function of redshift z for models in which the radiation is thermal bremsstrahlung from hot gas which has accumulated in the cluster either by infall of an intergalactic medium or by mass injection from the member galaxies. The principal approximation in the time-dependent numerical simulations is the assumption that the cluster gravitational potential is static; within this context the authors find that the cluster atmosphere evolves to configurations which are quasi-steady and in approximate hydrostatic equilibrium over Hubble time scales. Scaling laws are given which may be used to generalize the results to a wide variety of cluster and gas parameters.

160.010 **Soft X-ray spectra of the Coma and Perseus clusters of galaxies: constraints on the models.**
R. F. Malina, S. M. Lea, M. Lampton, C. S. Bowyer.
Astrophys. J., Vol. 219, 795 - 802 (1978).

The authors present soft X-ray spectra for the Perseus and Coma clusters obtained with wide-field-of-view instruments. These spectra are particularly sensitive to the amount of photoelectric absorption and to the amount of lower-temperature X-ray emitting gas. The spectra are analyzed with particular reference to the spatial data (in a similar energy band) obtained during the same observing period. The authors discuss in detail the implications of the detection of iron line emission for the interpretation of cluster spectra and for the origin of the X-rays.

160.011 A Westerbork survey of rich clusters of galaxies. V. Multi-frequency observations of the radio tail galaxy NGC 6034 in the Hercules cluster.
E. A. Valentijn, G. C. Perola.
Astron. Astrophys., Vol. 63, 29 - 35 (1978).

Maps of the radio-brightness distribution at 5, 1.4 and 0.6 GHz of the wide-angle head tail radio source 1601+17W1 associated with the elliptical galaxy N 6034 in the Hercules cluster are presented. At 5 and 1.4 GHz the Westerbork observations also provided polarization data. The spectral index distributions $\alpha_{0.6}^{1.4}$ along the trails are given. The remarkable asymmetric source structure is discussed and interpreted as the result of an ejection not normal to the velocity of the galaxy.

160.012 The structure of nearby clusters of galaxies. Hierarchical clustering and an application to the Leo region.
J. Materne.
Astron. Astrophys., Vol. 63, 401 - 409 (1978).

A new method of classifying groups of galaxies, called hierarchical clustering, is presented as a tool for the investigation of nearby groups of galaxies. The method is free from model assumptions about the groups. The scaling of the different coordinates is necessary, and the level from which one accepts the groups as real has to be determined. Hierarchical clustering is applied to an unbiased sample of galaxies in the Leo region. Five distinct groups result which have reasonable physical properties, such as low crossing times and conservative mass-to-light ratios, and which follow a radial velocity-luminosity relation. Only 4 out of 39 galaxies were adopted as field galaxies.

160.013 A group of galaxies in Cetus with a redshift discrepancy.   I. D. Karachentsev, W. G. Tifft.
Astron. Astrophys., Vol. 63, 411 - 413 (1978).

Spectral observations of 6 galaxies in an isolated Cetus group were made. Two galaxies have redshifts appreciably different from the others. This could be due to an effect of projection of the pair on a group of four members.

160.014 The characteristic size of clusters of galaxies: a metric rod used for a determination of q_0.
G. Bruzual A., H. Spinrad.
Astrophys. J., Vol. 220, 1 - 7, with a correction in Vol. 222, 1119 (1978).

Two definitions of the characteristic size of clusters of galaxies are considered. The harmonic mean separation between the brightest members is found to be very sensitive to the criterion used to select cluster galaxies. The mean distance between the brightest cluster galaxies is used to define a convenient annular spacing, allowing construction of surface density profiles for clusters at every z. Cumulative profiles are also considered. Analytical expressions fitted to the profile provide a measurement of a cluster characteristic size. A sample of 54 clusters with redshifts in the range $0.0207 \leq z \leq 0.9470$ are considered. The best solution favors an apparent $q_0 = 0.25 \pm 0.5$. Effects of cluster evolution on cluster sizes and on the determination of q_0 are discussed. No quantitative evaluation of this correction is made.

160.015 The X-ray luminosity function of Abell clusters.
D. A. Schwartz.
Astrophys. J., Vol. 220, 8 - 13 (1978).

X-ray sources identified with Abell clusters of galaxies in distance classes less than or equal to 3 give a quantitative estimate of the volume luminosity function for those clusters emitting between 10^{44} and 10^{45} ergs s^{-1} in X-rays. At higher luminosities, tentative identifications with more distant clusters can be interpreted at least as an upper limit. This limit allows a smooth extension of the luminosity function, but with a fairly steep decrease in the range 10^{45} to 10^{46} ergs s^{-1}. No

direct information is available for luminosities less than 10^{44} ergs s^{-1} because current X-ray surveys are limited to nearby distances with very few Abell clusters.

160.016 Velocity dispersion and global parameters of the southern cluster of galaxies CA 0340−538.
($\equiv$ 3U 0328−52?).   R. J. Havlen, H. Quintana.
Astrophys. J., Vol. 220, 14 - 18 (1978).

The authors report preliminary results of observations of the suspected X-ray cluster of galaxies CA 0340−538, that had tentatively been identified with the X-ray source 4U 0339−54 ($\equiv$ 3U 0328−52) and Ariel 2A 0343−536. A redshift of 0.0576 and a velocity dispersion of 1850 km s^{-1} are derived. The cluster center, size, shape, core radius, and population are found by means of galaxy counts and model fitting. Photometric observations give a total luminosity of 5.4×10^{12} $L_\odot$. This, coupled with a derived virial mass of 2.4×10^{15} $M_\odot$, gives a global mass-to-light ratio of 450 M/L.

160.017 Spectral variations in brightest cluster galaxies.
A. Wilkinson, J. B. Oke.
Astrophys. J., Vol. 220, 376 - 389 (1978).

Multichannel spectrometer scans of 56 of the brightest galaxies in faint clusters have been examined for evidence of color evolution. The dispersion introduced into the sample by some 25% of the galaxies which have unusual spectra is shown to be sufficient to obscure evolutionary variations; but on omitting these galaxies it is found that, over a time equivalent to a redshift of 0.46, $B - V$ decreases by no more than 0.04 mag. A comparative analysis of the spectral energy distribution suggests that their differences may be due to a combination of the dispersion in metallicity among the members of the sample and to anomalous reddening of individual galaxies.

160.018 Protogalaxy interactions in newly formed clusters: galaxy luminosities, colors, and intergalactic gas.
J. Silk.
Astrophys. J., Vol. 220, 390 - 400 (1978).

The role of protogalaxy interactions in galactic evolution is studied during the formation of galaxy clusters. In the early stages of the collapse, coalescent encounters of protogalaxies lead to the development of a galactic luminosity function. Once galaxies acquire appreciable random motions, mutual collisions between galaxies in rich clusters will trigger the collapse of interstellar clouds to form stars. This provides both a source for enriched intracluster gas and an interpretation of the correlation between luminosity and color for cluster elliptical galaxies. Other observational consequences that are considered include optical, X-ray, and diffuse nonthermal radio emission from newly formed clusters of galaxies.

160.019 The elliptical shape of the Coma cluster.
L. Schipper, I. R. King.
Astrophys. J., Vol. 220, 798 - 808 (1978).

The elliptical shape of the Coma cluster is examined quantitatively. The degree of ellipticity is high and depends to some extent on the radial distance of the sample from the Coma center as well as on the brightness of the sample. The elliptical shape does not appear to be caused by rotation; other possible causes are briefly discussed.

160.020 Is the Coma cluster a Zel'dovich disk?
L. A. Thompson, S. A. Gregory.
Astrophys. J., Vol. 220, 809 - 815 (1978).

The two-dimensional structure of the Coma cluster is analyzed with the use of galaxies from a wide-area redshift survey. The authors find that the Coma cluster has an elliptical shape and that the surface density of bright galaxies decreases rapidly beyond a radius of $\sim 3^\circ.1$. They suggest that the observed structure of Coma can be consistently explained using the model of Doroshkevich, Sunyaev, and Zel'dovich which involves the formation of massive protoclusters prior to the epoch of galaxy formation.

160.021   D and cD galaxies in poor clusters.
S. van den Bergh.
Publ. Astron. Soc. Pacific, Vol. 89, 746 - 749 (1977/78).

Inspection of CTIO 4-m plates of Morgan's cD galaxies in poor clusters suggests that they fall into two classes: (A) objects with faint symmetrical envelopes and (B) objects with asymmetrical envelopes that might have been produced by recent tidal encounters. The envelopes of objects belonging to class A are fainter than are those of first-ranked cD galaxies in rich clusters and might have been produced by tidal encounters that took place long ago.

160.022   Statistical analysis of catalogs of extragalactic objects. IX. The four-point galaxy correlation function.   J. N. Fry, P. J. E. Peebles.
Astrophys. J., Vol. 221, 19 - 33 (1978).

The galaxy four-point correlation function is defined and estimated for the Lick and Zwicky catalogs. The fact that the authors find the expected scaling with survey depth gives them reason to believe they have useful measures of the spatial four-point function for galaxy separations in the range of $\sim 0.5 h^{-1}$ Mpc to $\sim 4 h^{-1}$ Mpc ($H = 100h$ km s^{-1} Mpc^{-1}). The results are fairly well described by a simple generalization of the observed behavior of the three-point function. This lends further support to the continuous clustering hierarchy picture for the galaxy distribution, and more generally for the idea that there is a remarkably simple statistical pattern underlying the galaxy distribution.

160.023   The X-ray cluster of galaxies Klemola 44.
G. Chincarini, M. Tarenghi, C. Bettis.
Astrophys. J., Vol. 221, 34 - 40 (1978).

The authors give redshifts for 24 galaxies in the region of the X-ray cluster of galaxies K44. The unweighted velocity dispersion, based on 12 probable members within a circle of 50 arcmin diameter, is $V_D = 351$ km s^{-1}, and the systemic velocity $\langle V_0 \rangle = 8288$ km s^{-1}. The observed X-ray flux is of the order of magnitude the authors would expect from the inverse Compton scattering of the synchrotron electrons (radio source) by the microwave background.

160.024   A search for atomic hydrogen in clusters of galaxies.
M. P. Haynes, R. L. Brown, M. S. Roberts.
Astrophys. J., Vol. 221, 414 - 421 (1978).

The authors have investigated the possibility that clusters of galaxies are bound by discrete, optically thick clouds of atomic hydrogen. This hypothesis is tested in a search for the 21 cm hydrogen line in absorption in the spectra of radio sources located in the direction of galaxy clusters. No absorption lines were seen in any of the 28 sources searched at the appropriate cluster velocity. From these observations the authors infer that the 21 cm optical depth through the clusters is typically less than 0.1. The search for the missing "binding mass" should be focused on cluster material of another form.

160.025   Velocities of head-tail radio galaxies with respect to the intracluster medium.   M.-H. Ulrich.
Astrophys. J., Vol. 221, 422 - 435 (1978).

In four clusters of galaxies having a head-tail radio galaxy, the author has measured σ, the dispersion of the velocities of the galaxies, and ΔV, the difference between the velocity of the head-tail radio galaxy and the mean velocity of the cluster. In these four clusters the velocity of the head-tail radio galaxy is not exceptionally high. The generality of this result is examined by analyzing the distribution of $\Delta V/\sigma$ among the 16 head-tail sources for which ΔV and σ are known. It is found that the distribution of $\Delta V/\sigma$ is consistent with a normal distribution and is therefore not significantly different from the distribution of $\Delta V/\sigma$ which is expected for a set of 16 galaxies selected at random in their respective clusters.

160.026   The evolution of cluster X-ray sources.
S. C. Perrenod.
Bull. American Astron. Soc., Vol. 9, 559 (1978). – Abstract.

160.027   X-ray clusters of galaxies – radio limits to the ionized gas.   L. Rudnick.
Bull. American Astron. Soc., Vol. 9, 559 - 560 (1978). Abstract.

160.028   Comparison of X-ray structure of clusters of galaxies with thermal models of the intracluster gas.
D. Fabricant, K. Topka, F. R. Harnden, Jr., P. Gorenstein.
Bull. American Astron. Soc., Vol. 9, 560 (1978). – Abstract.

160.029   Groups of galaxies.   J. Huchra, M. J. Geller.
Bull. American Astron. Soc., Vol. 9, 583 (1978). Abstract.

160.030   A dynamical and photometric study of the cluster of galaxies CA 0340-538.
R. J. Havlen, H. Quintana.
Bull. American Astron. Soc., Vol. 9, 583 - 584 (1978). Abstract.

160.031   A statistical test for relaxation effects in systems of galaxies.   M. J. Geller, M. Davis.
Bull. American Astron. Soc., Vol. 9, 641 (1978). – Abstract.

160.032   Galaxy counts in the regions surrounding head–tail radio sources.   B. Guindon.
Mon. Not. R. Astron. Soc., Vol. 183, 195 - 200 (1978).

From a study of the galaxy distribution around head–tail radio sources, it is concluded that there is no preferred orientation of the tail and either the galaxy distribution of the cluster as a whole or the location of nearest neighbours.

160.033   Absence of velocity anisotropy in the direction of the Virgo cluster.   J. Kormendy.
IAU Colloq. No. 37, (see 012.009), p. 155 - 158 (1977).

L'auteur présente une méthode de mesure des distances des galaxies elliptiques basée sur une relation entre la brillance superficielle et un rayon déduit de la distribution spatiale de l'intensité de lumière émise. Il confronte ensuite les distances calculées à partir de la vitesse pour 11 galaxies de l'amas de la Vierge, avec celles de 8 objets de plus grand décalage spectral pour la plupart. Ceci implique que l'amas de la Vierge est à une distance plus faible que celle donnée par sa vitesse, résultat négatif confirmant la conclusion de Sandage et Tammann (1974, 1975), à savoir la non-existence d'une vitesse particulière de l'amas de la Vierge.

160.034   Redshift-magnitude bands in clusters of galaxies.
W. G. Tifft.
IAU Colloq. No. 37, (see 012.009), p. 159 - 165 (1977).

This paper presents new material and analysis for three clusters of galaxies. The author assumes familiarity with the initial Coma work (Tifft, 1972, 73). Redshifts and nuclear magnitudes are now available for 108 Perseus galaxies. The sample is nearly complete to $0°.75$ radius and V(6) = 16.8 (6″ scaled from 4″.8 in Coma). No field galaxies are seen, not surprising in view of field studies in the Coma region (Tifft and Gregory, 1976).

160.035   Dynamical friction in clusters of galaxies.
S. D. M. White.
IAU Colloq. No. 37, (see 012.009), p. 243 - 250 (1977).

La formule classique pour la friction dynamique prédit que l'évolution de l'orbite d'une galaxie qui est membre d'un amas sera trop rapide pour être en concordance avec les observations, si toute la masse de l'amas est concentrée dans les galaxies elles-mêmes. Quoique cette formule ait été prouvée sous des conditions très différentes, elle semble bien décrire les effets des rencontres gravitationelles dans un amas qui est stabilisé par sa propre gravitation. La friction dynamique mène à une concentration de galaxies massives au centre d'un amas, laquelle concentration doit avoir des effets importants sur la structure de la galaxie centrale.

160.036 **A search for neutral hydrogen in primordial proto-clusters.**   R. D. Davies, A. Pedlar.
IAU Colloq. No. 37, (see 012.009), p. 283 - 288 (1977).

La prévision de Sunyaev et Zel'dovich que les proto-amas peuvent passer par une phase d'hydrogène neutre pendant leur formation a été testée par la recherche de leur émission de la raie d'hydrogène décalée vers le rouge. Des limites supérieures comprises entre 0.07 et 0.17°K ont été obtenues dans une bande de 2.5 MHz à 327 MHz ce qui correspond à un décalage vers le rouge de z = 3.33. Ces résultats indiquent qu'il ne peut pas exister plus de proto-amas d'hydrogène que la quantité calculée par ces auteurs.

160.037 **Distances of nearby groups and clusters, and the local value to the Hubble ratio.**
G. de Vaucouleurs.
IAU Colloq. No. 37, (see 012.009), p. 301 - 307 (1977).

The correct approach to build up the extragalactic distance scale is to use all available primary (novae, cepheids, RR Lyrae) and secondary indicators (brightest stars, globular clusters, largest H II rings) to calibrate without arbitrary extrapolation all reliable tertiary indicators (magnitudes and diameters of galaxies), precisely corrected for all known effects of type, luminosity class, orientation, internal and galactic extinction and redshift. Such data are now available for over 1000 galaxies in the Second Reference Catalogue. Revised distances to members of the local group from primary indicators and new estimates of distances to the nearest groups from primary and secondary indicators are used to calibrate the tertiary indicators via a new, composite luminosity index.

160.038 **The Hercules cluster of galaxies.**   M. Tarenghi.
        IAU Colloq. No. 37, (see 012.009), p. 313 - 319 (1977).

This communication presents preliminary results of a morphological and spectroscopic study of galaxies in a large area including the Hercules cluster of galaxies. The combination of both the morphological and spectroscopic material strongly suggest the existence of a Hercules supercluster.

160.039 **Anomalous redshifts in groups of galaxies.**
        J. Heidmann.
CNRS Colloq. No. 263, (see 012.009), p. 427 - 444 (1977).

The author reviews the observational data on anomalous redshifts (z or cz) in groups of galaxies. Theoretical aspects are left out, being reviewed by G. Burbidge, Pecker and Rees, as well as cluster data, treated by de Vaucouleurs. In the first part the author treats individual studies and in the second, statistical ones.

160.040 **A model of tidal interactions within the NGC 4631 group of galaxies.**   F. Combes.
Astron. Astrophys., Vol. 65, 47 - 55 (1978).

A model of tidal interactions between the galaxies NGC 4631, 4656 and 4627 is proposed. The 21 cm observations of Weliachew et al. (1977) display four features of neutral hydrogen around NGC 4631; two of them can be explained in terms of a parabolic encounter between the two main galaxies NGC 4631 and 4656. The other two features may be interpreted with the help of the neighbouring galaxy NGC 4627: its mass is too small to perturb significantly the two other galaxies, but it undergoes itself a large damage.

160.041 **Formation times for rich clusters of galaxies.**
        B. M. Lewis.

Proc. Astron. Soc. Australia, Vol. 3, 140 - 142 (1977).

160.042 **Bautz-Morgan classes and the luminosity function for clusters of galaxies.**   A. Dressler.
Astrophys. J., Vol. 222, 23 - 28 (1978).

The distribution of Bautz-Morgan types in the Leir and van den Bergh catalog of 1889 Abell clusters of galaxies is used to discriminate between the "statistical" and "special process" hypotheses for the brightest cluster members. A quantitative measure of Bautz-Morgan types based on m_1, m_2, and m_3 is employed in a Monte Carlo simulation of cluster luminosity functions to predict the distribution of Bautz-Morgan types. It is concluded that the statistical model, regardless of the form of the luminosity function, cannot fulfill all requirements; hence a "special process" model seems favored.

160.043 **X-ray and radio emission from clusters of galaxies: the heating of intracluster gas by relativistic electrons.**   S. M. Lea, G. D. Holman.
Astrophys. J., Vol. 222, 29 - 39 (1978).

The authors discuss the observed correlation between X-ray emission from clusters of galaxies and the presence of steep spectrum radio sources in the context of a thermal model for the X-ray emission. They calculate the rate at which energy is transferred from the relativistic electrons to the thermal gas, and find that under reasonable assumptions about the low-energy cutoff of the electron spectrum this rate is ample to power the X-ray source. The authors show that in general the temperature of the gas is independent of the heating mechanism. They also discuss some constraints on inverse Compton models of X-ray production.

160.044 **Redshift-magnitude bands and the evolution of galaxies. I. New observations.**   W. G. Tifft.
Astrophys. J., Vol. 222, 54 - 60 (1978).

New observations are combined with existing ones to provide well-defined samples of galaxy redshifts and magnitudes for the Perseus and A1367 clusters.

160.045 **The luminosity function of clusters of galaxies.**
        S. van den Bergh.
Astrophys. Space Sci., Vol. 53, 415 - 419 (1978).

Observations by Dressler are used to derive integral luminosity functions for ten rich clusters. All luminosity functions are found to consist of a linear portion and a high-luminosity toe. The size of this toe does not appear to correlate with cluster BM type. A parameter $M_F(0)$, which has a dispersion of only 0.2 mag for the present data sample, is defined.

160.046 **The luminosity function of the cluster of galaxies A2634.**   A. G. Egikyan, A. T. Kalloglyan.
Astrofizika, Vol. 13, 405 - 413 (1977). In Russian. English translation in Astrophysics, Vol. 13, No. 3.

The integral magnitudes of 260 galaxies up to 18^m3 in the B system have been measured in the Abell cluster No. 2634 and the luminosity function of the galaxies has been constructed. The distribution chart of the measured galaxies is given. The differential luminosity function grows not monotonously but has a local maximum at about 15^m8. A segregation of bright and faint galaxies in the cluster is observed.

160.047 **The distribution of mean surface brightness of galaxies in the Coma cluster.**   M. A. Arakelyan.
Astrofizika, Vol. 13, 651 - 655 (1977). In Russian. English translation in Astrophysics, Vol. 13, No. 4.

The dependence of the mean surface brightnesses B of galaxies in the cluster region and their dispersions $\sigma^2(B)$ are considered as functions of the distance from the center of the cluster.

160.048 **A search for X-ray emission from Abell clusters and superclusters.**   M. J. Ricketts.

Mon. Not. Astron. Soc., Vol. 183, 51P - 58P (1978).

Data from the Ariel V Sky Survey Instrument (SSI) has been searched for further X-ray sources associated with Abell clusters. Five new X-ray sources have been found and for four of these the error box includes a single Abell cluster. The fifth coincides with one of the superclusters listed by Murray et al. but not detected by Uhuru.

160.049 **Redshift controversy in the Virgo cluster – a suggestion from Centaurus.** A. P. Fairall.
Mon. Not. R. Astron. Soc., Vol. 183, 59P - 62P (1978).

The spiral galaxies of the Centaurus cluster are concentrated about a centre separated from that of the elliptical and S0 galaxies. This indicates a fundamental dynamical difference between the spiral and elliptical components of that cluster. It therefore suggests that the well-known redshift difference between spirals and ellipticals in the Virgo cluster is a similar dynamical difference, rather than intrinsic redshift in the spirals.

160.050 **Origin of X-ray emission from clusters.**
R. Giacconi.
Phys. Scr., Vol. 17, (see 012.013), 377 - 385 (1978).

The results on the X-ray emission from clusters are reviewed. Correlations with morphological properties of the clusters are discussed. Recent findings on this topic, as well as on the possible detection of X-ray emission from superclusters, are high-lighted.

160.051 **Redshifts for galaxies in the poor cluster AWM-4.**
J. Stauffer, H. Spinrad.
Publ. Astron. Soc. Pacific, Vol. 90, 20 - 21 (1978).

Redshifts for five galaxies in the poor cluster AWM-4 have been obtained. This compact group of galaxies appears not to be a physical association.

160.052 **Further investigations of clusters of galaxies.**
F. W. Baier, W. Mai.
Astron. Nachr., Band 299, 69 - 80 (1978).

The number-density distributions of ten further clusters of galaxies were derived by counting galaxies on the red Palomar Sky Survey prints. For eight isolated clusters the radial number-density distributions and the radial cumulative galaxy distributions were calculated.

160.053 **Contributions to galaxy photometry. VII. Standard total magnitudes of 139 bright galaxies in the Virgo cluster area.** G. de Vaucouleurs, C. Head.
Astrophys. J., Suppl. Ser., Vol. 36, 439 - 449 (1978).

Standard total magnitudes in the B_T system are derived from six sets of data for 139 galaxies in an $18° \times 18°$ area centered on the Virgo cluster. The magnitude range is $9.3 \leq B_T \leq 14.7$ and the mean errors $0.03 \leq \sigma \leq 0.11$ mag. The average mean error is 0.063 mag for 106 primary standards derived from two to six sets and 0.084 mag for 33 secondary standards from the best single sources.

160.054 **The distribution of galaxies in the field of the Hercules supercluster.** F. W. Baier, W. Mai.
Astron. Nachr., Band 299, 125 - 129 (1978).

The number-density distribution of four Abell-clusters in the region of the Hercules supercluster was derived by counting galaxies on the red Palomar Sky Survey prints. For one isolated cluster the radial number-density distribution and the radial cumulative galaxy distribution were calculated.

160.055 **Rich clusters of galaxies.** W. H. Tucker.
Astronomy, Vol. 5, No. 10, p. 28 - 32 (1977).
Abstr. in Phys. Abstr., Vol. 81, Abstr. 8136 (1978).

160.056 **Inhomogeneities in the universe: an observed power law relating mass fraction to density enhancement.**

S. P. Bhavsar.
Astrophys. J., Vol. 222, 412 - 420 (1978).

The spectrum of galaxy cluster sizes is an important cosmological datum. Turner and Gott (1976) have generated a catalog of groups of galaxies by identifying regions of the sky in which the surface number density of galaxies is enhanced. As part of a larger study, group catalogs generated in a similar fashion at different magnitude limits and with a wide range of density enhancements have been investigated. Some preliminary results of this study are presented.

160.057 **Redshift-magnitude bands and the evolution of galaxies. II. Data analysis.** W. G. Tifft.
Astrophys. J., Vol. 222, 421 - 434 (1978).

Using redshift-magnitude band information from the Coma cluster and QSS, the author makes specific predictions on slope and location of redshift-magnitude bands in any cluster or galaxy sample for m_p magnitudes. New data on the A1367 and Perseus clusters are then analyzed. The band pattern in A1367 is readily seen in total magnitudes and is identical in slope and location to the Coma pattern. Using the radial brightening trend, the author shows that the outer part of the Coma cluster contains the same band pattern as the core. Field galaxies and the Virgo cluster are utilized to extend the redshift-magnitude band analysis to low redshifts. An absolute band system which is a scaled version of the QSS system is consistently confirmed. Using the band properties and associated correlations, the author discusses an empirical model for band and galaxy evolution.

160.058 **The evolution of mass and tidal radius of cluster galaxies.** E. Knobloch.
Astrophys. J., Vol. 222, 779 - 783 (1978).

Recent numerical simulations show that in tidally interacting galaxies $dM_T/M_T \approx dR_T/R_T$, where M_T and R_T are respectively the mass and tidal radius of a test galaxy. A simple theory is presented to explain this result and is used to calculate M_T and R_T as functions of elapsed time. An observational test to distinguish between first and second order effects is suggested.

160.059 **The Coma/A1367 supercluster and its environs.**
S. A. Gregory, L. A. Thompson.
Astrophys. J., Vol. 222, 784 - 799 (1978).

The three-dimensional galaxy distribution in the region of space surrounding the two rich clusters Coma and A1367 is analyzed by using a nearly complete redshift sample of 238 galaxies with $m_p < 15.0$ in a 260 degree² region of the sky; 44 of these redshifts are reported here for the first time. The authors find that the two clusters are enveloped in a common supercluster which also contains four groups and a population of isolated galaxies. A number of related topics with more general significance are also discussed.

160.060 **An energy-dependent map of the X-ray emission of the Perseus cluster.** M. P. Ulmer, J. G. Jernigan.
Astrophys. J., Lett., Vol. 222, L85 - L89 (1978).

The authors present SAS 3 observations of the inner 3° diameter region of the Perseus cluster and conclude that the angular extent of the X-ray emission increases significantly with increasing energy. This result is model-independent.

160.061 **Observations of clusters containing radio tail galaxies.** P. Hintzen, W. R. Oegerle, J. S. Scott.
Astron. J., Vol. 83, 478 - 481 (1978).

Spectroscopic observations are reported for galaxies in A347, A779, and Zw 1809+50. These clusters contain at least six radio sources, two of which are well observed radio-tail sources. For A347, the mean cluster redshift (0.0187) implies physical association with A426 (Z = 0.0183), in accord with Zwicky's suggestion that both clusters are members of a single large group. The observed cluster velocity dispersion of A347 and the peculiar radial velocity of 3C66 suggest that this galaxy's radio source may be an intermediate angle tail (angle between lobes $\sim 100°$) rather than a projected narrow angle tail. Finally, the clusters' observed velocity dispersion and the velocity dispersion-X-ray luminosity relation derived by Faber and Dressler are used to estimate X-ray fluxes from the clusters.

160.062 Statistical analysis of discrepant redshift associations, quintets of galaxies.   L. Nottale, M. Moles.
Astron. Astrophys., Vol. 66, 355 - 358 (1978).
The authors study the statistical significance of discrepant redshift quintets of galaxies. Taking into account the small number of known redshifts, they find that the probability that they might be chance projections of field galaxies is only 2×10^{-3} to 3×10^{-4}. It is concluded that the problem of anomalous redshifts in compact groups of galaxies still remains.

160.063 Cosmology II: metrical connection and clusters of galaxies.   A. Maeder.
Astron. Astrophys., Vol. 67, 81 - 86 (1978).
An attempt is made to use the related works on Weyl's geometry for examining the problem of the dynamical mass excess in clusters of galaxies. With a Newtonian approximation in Weyl's geometry, the author shows that at very large scales (i.e. Mpc) a new term may become significant in the equation of motion and energy relation. The expressions obtained lead to much smaller values of the total mass of the systems than in current theory. The data by Rood et al. (1972) and by Abell (1977) for the Coma cluster are examined within this framework.

160.064 Pairs and triplets of clusters of galaxies.
I. D. Karachentsev, A. L. Shcherbanovskij.
Astron. Zh. Akad. Nauk SSSR, Vol. 55, 449 - 455 (1978).
In Russian. English translation in Soviet Astron., Vol. 22, No. 3.
A list is given containing 68 pairs and 24 triplets of clusters of galaxies, obtained by a computer analysis of Abell's catalogue. The ratio of distances between members of a system to the distance from the neighbouring cluster has been taken as a criterion of isolateness of a system. The expected fraction of chance (optical) systems in the list is about 1/4. Indications which show the physical connection between clusters of galaxies in multiple systems are noted.

160.065 Ionized gas in X-ray clusters of galaxies: radio limits.
L. Rudnick.
Astrophys. J., Vol. 223, 37 - 46 (1978).
Radio observations at 2 cm with the 42 m NRAO telescope at Green Bank were made of four X-ray clusters of galaxies (A401, A1656, A2199, and A2319) and one unconfirmed, possible X-ray cluster (A2255). Limits are set on the diminution of the microwave background due to inverse Compton scattering by hot gas in the clusters and are typically on the order of 1 mK at the 90% confidence level. Other contributions to the sky brightness at cm wavelengths are roughly estimated. Limits are derived for small-scale anisotropies in the microwave background, and confirm previous work at a 90% confidence level of $|\delta T/T| \lesssim 2.3 \times 10^{-4}$.

160.066 On the origin and evolution of iron-enriched gas in clusters of galaxies.   D. S. De Young.
Astrophys. J., Vol. 223, 47 - 55 (1978).
Consideration is given to the recent discovery of iron-enriched gas in clusters of galaxies. It is shown that this material may be ejected from the component galaxies as a result of the early evolution of massive stars without the need for an unusual initial mass function. The interaction of the ejected gas clouds is calculated, and it is found that this material is heated and distributed throughout the cluster in a time that is short compared with the Hubble time.

160.067 The photography of groups of galaxies.
B. A. Vorontsov-Velyaminov (*Vorontsov-Vel'yaminov*).
IAU Symp. No. 79, (see 012.027), p. 3 - 10 (1978).
The author shows some nests of galaxies (extradense groups), chains (a particular case of nests) and cases where the structure of the objects is enigmatic, let alone their origin. Some of the VV objects, discovered in 1958 - 1964, and recently photographed with the 6-m telescope are shown.

160.068 Velocity dispersion in small systems of galaxies.
I. D. Karachentsev.
IAU Symp. No. 79, (see 012.027), p. 11 - 20 (1978).
The data presented show that close interacting systems, and also pairs of galaxies both tight and wide have on the average a normal virial mass-to-luminosity ratio $\langle f_c \rangle \simeq 8 \, f_\odot$, which can be explained without the hypothesis of the existence of massive hidden coronas around galaxies.

160.069 Optical studies of small groups of galaxies.
E. L. Turner.
IAU Symp. No. 79, (see 012.027), p. 21 - 29 (1978).
The main results of a statistical study of small groups (Turner and Gott 1976, 1977) are reviewed and compared to N-body simulations of galaxy clustering.

160.070 Nearby small groups of galaxies.
R. B. Tully, J. R. Fisher.
IAU Symp. No. 79, (see 012.027), p. 31 - 47 (1978).
The authors present examples covering the gamut of galaxy associations to be found near by. Characteristic crossing times and virial masses are calculated.

160.071 Hypergalaxies.   J. Einasto.
IAU Symp. No. 79, (see 012.027), p. 51 - 61 (1978).
Contents: Spatial structure of hypergalaxies. Dynamics of hypergalaxies. Morphology and luminosity of galaxies in hypergalaxies. Luminosity function of hypergalaxies. The masses and mass-to-luminosity ratios of hypergalaxies. Interaction between hypergalactic gas and galaxies. Galactic and hypergalactic populations. Hypergalaxies as galaxy communities.

160.072 The galactic neighbourhood.
G. A. Tammann, R. Kraan.
IAU Symp. No. 79, (see 012.027), p. 71 - 91 (1978). – See 20.160.063.

160.073 Three dimensional analysis of groups of galaxies.
J. Materne.
IAU Symp. No. 79, (see 012.027), p. 93 - 94 (1978).

160.074 The scatter in mass-to-luminosity ratios.
J. Einasto.
IAU Symp. No. 79, (see 012.027), p. 96 - 98 (1978). – Short communication.

160.075 Cluster membership of Seyfert galaxies.
K.-H. Schmidt.
IAU Symp. No. 79, (see 012.027), p. 101 - 102 (1978). Short communication.

160.076 Photoelectric surface photometry of the Coma cluster.   J. Melnick, S. White, J. Hoessel.
IAU Symp. No. 79, (see 012.027), p. 135 - 136 (1978).
A photoelectric search is presented for light from a smooth intergalactic background in the Coma cluster of galaxies.

160.077 Radio properties of Abell clusters.
C. Lari, G. C. Perola.
IAU Symp. No. 79, (see 012.027), p. 137 - 147 (1978).

Contents: 1. Radio luminosity function of galaxies in Abell clusters. 2. Radial distribution of radiogalaxies in Abell clusters. 3. Spectral indexes of radio sources in Abell clusters. 4. Radio source structures in Abell clusters. 5. Correlation between radio and X-ray emission in clusters.

160.078   **Radio observations of clusters of galaxies: the tail sources.**   R. D. Ekers.
IAU Symp. No. 79, (see 012.027), p. 149 - 151 (1978).
The author discusses two new observational results concerning the velocities and environment of the head-tail radio galaxies.

160.079   **Extended radio emission in clusters of galaxies: recent Westerbork observations.**
E. A. Valentijn, H. van der Laan.
IAU Symp. No. 79, (see 012.027), p. 153 - 155 (1978).
The authors present results for five clusters studied with the WSRT: the Coma cluster, the Perseus cluster, the Hercules cluster, A2256, and A1314.

160.080   **Clusters with extended radio emission at high frequencies.**   R. Wielebinski.
IAU Symp. No. 79, (see 012.027), p. 157 - 159 (1978).

160.081   **X-ray observations of clusters of galaxies.**   J. L. Culhane.
IAU Symp. No. 79, (see 012.027), p. 165 - 177 (1978).
Contents: X-ray cluster identifications and correlations. The relation between X-ray sources and superclusters. The X-ray structure of clusters. The detection of iron line emission in cluster X-ray spectra.

160.082   **An analogy between simulated and actual clusters of different kinds.**   G. Paál.
IAU Symp. No. 79, (see 012.027), p. 197 - 198 (1978).
Short communication.

160.083   **Structures and number-density distributions in clusters of galaxies.**   F. W. Baier.
IAU Symp. No. 79, (see 012.027), p. 198 - 199 (1978).
Short communication.

160.084   **Radio emission of Abell clusters in the GB and GB2 regions.**   A. Michalec, J. Machalski.
IAU Symp. No. 79, (see 012.027), p. 200 - 202 (1978).
Short communication.

160.085   **The Local Supercluster.**   G. de Vaucouleurs.
IAU Symp. No. 79, (see 012.027), p. 205 - 213 (1978).

160.086   **A tour of the Local Supercluster.**
R. B. Tully, J. R. Fisher.
IAU Symp. No. 79, (see 012.027), p. 214 - 216 (1978).
Short communication.

160.087   **The physical properties of large scale systems from optical observations.**   G. O. Abell.
IAU Symp. No. 79, (see 012.027), p. 253 - 262 (1978).
Contents: The local Supercluster. The Coma Supercluster. Groups of rich clusters. Is there an end to the hierarchy?

160.088   **The structure of the Hercules Supercluster.**
M. Tarenghi, W. G. Tifft, G. Chincarini, H. J. Rood, L. A. Thompson.
IAU Symp. No. 79, (see 012.027), p. 263 - 265 (1978).
The redshifts obtained in the Hercules Region confirm the previous findings in the Coma Supercluster, indicating that galaxies are grouped in large asymmetric structures, groups and superclusters. Clusters are, generally, bound condensations embedded in superclusters. The gaps in the velocity field are a very important feature and can be used to estimate an upper limit for the density of "field" galaxies.

160.089   **Observations of the large scale distribution of galaxies.**   W. G. Tifft, S. A. Gregory.
IAU Symp. No. 79, (see 012.027), p. 267 - 269 (1978).

160.090   **The field luminosity function and nearby groups of galaxies.**   J. Huchra.
IAU Symp. No. 79, (see 012.027), p. 271 - 273 (1978).
The author describes a new determination of the field luminosity function and density plus initial experiments with the use of a redshift catalog to select groups of galaxies.

160.091   **Superclustering of galaxies.**
M. Kalinkov, V. Dermendjiev, B. Staikov, I. Kaneva, B. Tomov, K. Stavrev.
IAU Symp. No. 79, (see 012.027), p. 276 - 279 (1978).
Short communication.
A new type of catalogue of extragalactic objects is nearing completion. The catalogue is a compilation of data for galaxies and clusters of galaxies, together with references, and for counts of galaxies. At present this Metacatalogue contains about 3×10^5 entries – Abell and Zwicky clusters, Zwicky galaxies, Lick counts, Jagellonian counts and some others.

160.092   **On the methods of discovering groups and clusters of galaxies.**   B. I. Fessenko (*Fesenko*).
IAU Symp. No. 79, (see 012.027), p. 279 - 280 (1978).
Short communication.

160.093   **Correlation between supercluster membership and richness for Abell clusters.**   S. C. Perrenod.
Nature, Vol. 274, 39 - 40 (1978).
This letter shows (1) that clusters of galaxies which are located in superclusters (second-order clusters of ~30 Mpc radius) are systematically richer than those in the 'field', and (2) that the expected higher X-ray luminosities of these richer clusters may, therefore, account for much, if not all, of the possibly large X-ray luminosities from superclusters.

160.094   **Possible optical evidence for ram-pressure-sweeping in the Hydra I cluster of galaxies.**
J. S. Gallagher.
Astrophys. J., Vol. 223, 386 - 390 (1978).
An asymmetrical distribution of very faint, filamentary features has been detected in blue light extending from the disk of NGC 3312, a peculiar spiral in the Hydra I cluster. Several interpretations for this type of phenomenon are possible (e.g., collisions, internal activity, an M82 type of event), but the location of NGC 3312 near the cluster core suggests the disturbance is related to the cluster environment. Hydra I is an X-ray source and therefore probably contains an intracluster medium. Thus the interesting possibility exists that NGC 3312 is being stripped of its interstellar medium by ram-pressure, and is an optical analog to a radio head-tail source.

160.095   **Distance to Virgo I cluster via CM data of M 31 group and Virgo I cluster.**   N. Visvanathan.
Astron. Astrophys., Vol. 67, L17 - L19 (1978).
Comparison of the CM data of M31 and its companions NGC 147, NGC 185, and M32 and the Fornax system with that of the E and S0 galaxies and early type spirals of the Virgo I cluster, leads to a distance modulus of the Virgo I cluster as 30.60 ± 0.39.

160.096   **Further X-ray spectra of clusters of galaxies.**
B. W. Smith, R. F. Mushotzky, E. A. Boldt, S. S. Holt, R. E. Rothschild, P. J. Serlemitsos.
Bull. American Astron. Soc., Vol. 10, 424 (1978). – Abstract.

160.097   A systematic survey for faint galaxy clusters.
          J. G. Hoessel, J. E. Gunn.
Bull. American Astron. Soc., Vol. 10, 428 - 429 (1978).
Abstract.

160.098   Virial properties of systems of galaxies.
          H. J. Rood, J. R. Dickel.
Bull. American Astron. Soc., Vol. 10, 429 (1978). – Abstract.

160.099   Generalized χ^2-method for investigation of the
          problem of existence of second-order clusters of
galaxies. I. Theory.   M. Kalinkov.
Astrofiz. issled. (NRB), Vol. 2, 70 - 79 (1977). In Russian.
Abstr. in Ref. zh., 51. Astron., 6.51.909 (1978).

160.100   Generalized χ^2-method for investigation of the
          problem of the existence of second-order clusters
of galaxies. II. Clusters from Abell's and Zwicky's catalogues.
M. Kalinkov, B. Tomov.
Astrofiz. issled. (NRB), Vol. 2, 80 - 114 (1977). In Russian.
Abstr. in Ref. zh., 51. Astron., 6.51.910 (1978).

160.101   Hypergalaxies.   O. Obůrka.
          Říše hvězd, Vol. 59, 137 - 139 (1978). In Czech.

Compton scattering of microwave background radiation by gas in galaxy clusters.   See Abstr. 066.003.

Some problems with the interpretation of recent
microwave background observations in the direction of galaxy
clusters, or, beware of negative antenna temperatures.
See Abstr. 066.025.

Inversion of Limber's relativistic formula.
See Abstr. 066.199.

A new stellar standard sequence in the Coma cluster
of galaxies. (Preliminary report).   See Abstr. 113.049.

Supernovae in the Coma cluster of galaxies.
See Abstr. 125.001.

Upper limits for the microwave pulsed emission
from supernova explosions in clusters of galaxies.
See Abstr. 125.053.

H I observations of interacting groups of galaxies.
See Abstr. 132.033.

Observations of high velocity H I clouds in the
Local Group.   See Abstr. 132.063.

VLA observations of extended sources in rich
clusters of galaxies.   See Abstr. 141.068.

Radio-quiet QSOs in the region of Hercules cluster
of galaxies.   See Abstr. 141.095.

On the distribution of compact groups of compact
galaxies and radio sources.   See Abstr. 141.114.

Classical double sources in the directions of rich
clusters of galaxies.   See Abstr. 141.148.

The radio source 0915+320: a "wide-angle-tail"
source in a group of galaxies.   See Abstr. 141.185.

Radio sources around 3C 465, in the cluster of
galaxies Abell 2634.   See Abstr. 141.194.

Gas stripping from spirals within X-ray clusters of
galaxies.   See Abstr. 142.009.

X-ray line spectroscopy for clusters of galaxies. I.
See Abstr. 142.021.

Evidence for X-ray emission from superclusters of
galaxies determined from Uhuru.   See Abstr. 142.023.

X-ray sources in nearby superclusters.
See Abstr. 142.041.

Compact and diffuse X-ray sources in the core of
the Perseus cluster (Abell 426).   See Abstr. 142.055.

High spatial resolution X-ray observations of the
Perseus cluster of galaxies with the scanning modulation
collimator on HEAO-1.   See Abstr. 142.064.

X-ray spectra of clusters of galaxies from OSO-8.
See Abstr. 142.065.

Theoretical X-ray maps of rich clusters of galaxies.
See Abstr. 142.066.

Soft X-ray structure of the Coma cluster of galaxies.
See Abstr. 142.067.

HEAO-1 observations of soft X-ray emission from
clusters of galaxies.   See Abstr. 142.086.

High-galactic-latitude X-ray sources and southern
clusters of galaxies.   See Abstr. 142.119.

The structure of the Virgo cluster X-ray source.
See Abstr. 142.134.

Fuentes de rayos X extragalácticas.
See Abstr. 142.136.

Entweichgeschwindigkeiten und mittlere Geschwindigkeiten in Stern- und Galaxienhaufen.   See Abstr. 151.044.

On a possibility of stabilization of chains of galaxies
by a continuous background.   See Abstr. 151.057.

Stabilization of systems of galaxies by subclustering.
See Abstr. 151.078.

Mathematical approach to the problem of clustering.
See Abstr. 151.084.

On the dynamical evolution of clusters of galaxies.
See Abstr. 151.086.

Surface brightness and color distributions of
elliptical and S0 galaxies. I. The Coma cluster elliptical
galaxies.   See Abstr. 158.025.

Surface brightness, ellipticity and color distributions
for elliptical galaxies. I. Spiral-rich clusters.
See Abstr. 158.072.

Surface brightness, ellipticity and color distributions
for elliptical galaxies. II. cD and spiral-poor clusters.
See Abstr. 158.073.

On the existence of groups of pairs of galaxies
having different physical and kinematical characteristics.
See Abstr. 158.132.

Photometry of galaxies in the field of the galaxy
cluster Abell 1781.   See Abstr. 158.151.

Structural investigations in the central part of the

Coma cluster on the base of equidensity diagrams. II. See Abstr. 158.158.

Clusters of galaxies and the statistics of emission-line galaxies. See Abstr. 158.167.

Two new dwarf galaxies in the Local Group: UKS 1927 – 177 and UKS 2323 – 326. See Abstr. 158.169.

Galaxy haloes and the missing mass problem. See Abstr. 158.196.

Radio haloes around galaxies and in clusters. See Abstr. 158.224.

Markarian galaxies in the vicinity of the Coma cluster. See Abstr. 158.225.

On the abundances of the chemical elements in intergalactic matter. See Abstr. 161.001.

Are there intergalactic clouds in the Sculptor group? See Abstr. 161.002.

A high sensitivity survey of hydrogen at very high velocities (preliminary report). See Abstr. 161.005.

Gas in galaxy clusters. See Abstr. 161.007.

Interaction between intergalactic medium and galaxies. See Abstr. 161.009.

Possible role of collective relaxation in galaxy correlations. See Abstr. 162.002.

A search for neutral hydrogen in primordial protoclusters at $z = 3.33$ and 4.92. See Abstr. 162.024.

The extension of the Hubble diagram. II. New redshifts and photometry of very distant galaxy clusters: first indication of a deviation of the Hubble diagram from a straight line. See Abstr. 162.031.

The distance scale within the Local Group. See Abstr. 162.039.

21-cm observations of redshifts and their implications for the distance scale. See Abstr. 162.042.

The Hubble diagram for red magnitudes of bright cluster galaxies. See Abstr. 162.046.

On the compatibility of paradoxical redshifts, observed in galaxies and quasars, with the theory of gravitation. See Abstr. 162.057.

Core condensation in heavy halos: a two-stage theory for galaxy formation and clustering. See Abstr. 162.070.

The correlation function for density perturbations in an expanding universe. III. The three-point and predictions of the four-point and higher order correlation functions. See Abstr. 162.072.

Gravitation, primordial stars and the dark mass. See Abstr. 162.086.

On a possible explanation for the 9 Mpc break of the covariance function. See Abstr. 162.114.

The development of structure in the expanding universe. See Abstr. 162.151.

Errata

160.901 Erratum: "Doppler-shift distributions. I. Extragalactic peculiar motions" [Astrophys. J., Vol. 215, 417 - 426 (1977)]. P. G. Gross. Astrophys. J., Vol. 222, 403 (1978).

161 Intergalactic Matter

161.001   On the abundances of the chemical elements in
           intergalactic matter.   P. Biermann.
Astron. Astrophys., Vol. 62, 255 - 258 (1978).
    Recent X-ray observations indicate that the abundance
of iron in the intergalactic medium in clusters of galaxies is of
the order of one-third the cosmic abundance. Since many rich
clusters are dominated by S0 galaxies, their evolution and gas
loss can be expected to provide the key to understand the
intergalactic material and its composition. The author deduces
here for the matter lost from the galaxies (a) the enrichment
in heavy elements and helium, (b) the depletion of deuterium,
and (c) the secondary nucleosynthesis astration. Scaling the
iron abundance from the enrichment in heavy elements he
finds that the observations of iron in intergalactic space can be
accounted for by normal evolution of galaxies with mass loss,
provided that there is only negligible primordial intergalactic
material.

161.002   Are there intergalactic clouds in the Sculptor group?
           M. P. Haynes, M. S. Roberts.
Bull. American Astron. Soc., Vol. 9, 584 (1978). – Abstract.

161.003   Emission spectrum of a hot intergalactic medium.
           R. D. Sherman.
Bull. American Astron. Soc., Vol. 9, 630 (1978). – Abstract.

161.004   The Zel'dovich effect and the intergalactic dust in
           galaxy clusters.   S. Aiello, F. Melchiorri,
F. Mencaraglia.
Astrophys. Space Sci., Vol. 53, 403 - 409 (1978).
    The observations of the Zel'dovich effect in galaxy clus-
ters are reviewed. The failure to detect the effect at short
wavelengths is interpreted as proof of the existence of intergalactic dust. Small and fast spinning grains can emit the power
needed to compensate the decrease in temperature expected
from the Zel'dovich effect.

161.005   A high sensitivity survey of hydrogen at very high
           velocities (preliminary report).
A. N. M. Hulsbosch.
Astron. Astrophys., Vol. 66, L5 - L8 (1978).
    The regions investigated are a $1° \times 1°$ grid around M31,
extending from $l = 100°$ to $140°$ and from $b = -45°$ to $0°$, and
a $2° \times 1°$ grid for $l = 140°$ to $180°$ and $b = -60°$ to $-30°$. The
velocity interval was -1000 to $+1000$ km/s, the detection
limit about 0.05 K. Several objects with large negative velocities were found, in the first region around $V_{lsr} \approx -400$ km/s,
with the extremes HVC 111–7–465 and HVC 114–10–446,
and in the second region around $V_{lsr} \approx -275$ km/s. It is
tentatively concluded that these objects form two hydrogen
streams in the Local Group.

161.006   Material in the vicinity of galaxies.
           B. F. Burke.
IAU Symp. No. 77, (see 012.026), p. 287 - 292 (1978).

161.007   Gas in galaxy clusters.   J. Silk.
           IAU Symp. No. 79, (see 012.027), p. 179 - 188
(1978).
    Gaseous matter almost certainly cannot account for a
significant amount of the binding mass in the cores of rich
clusters. Implications of the variety of upper limits are well
known (Tarter and Silk 1974), and will not be described here.
However intracluster gas provides important clues to the
evolution of galaxy clusters, and the present review is devoted
to elucidating its role.

161.008   Temperature fluctuations and infrared emission

from dust in hot gas.   B. T. Draine.
Bull. American Astron. Soc., Vol. 10, 463 (1978). – Abstract.

161.009   Interaction between intergalactic medium and
           galaxies.   E. Saar, J. Einasto.
IAU Colloq. No. 45, (see 012.032), p. 247 - 269 (1977/78).
    The authors review the properties of the gaseous com-
ponent of the intergalactic medium (both the environment and
the close environs of galaxies) that galaxies are interacting
with. They discuss the interaction of galaxies with the environmental gas both in clusters and in small groups of galaxies.
The interaction of galaxies with their environs is reviewed as
well as collisions between galaxies and giant gas clouds.

161.010   On the origin and evolution of the intracluster gas:
           interaction with galaxies.   F. Takahara, S.
Ikeuchi.
Prog. Theor. Phys., Vol. 58, 1728 - 1741 (1977). – Abstr. in
Phys. Abstr., Vol. 81, Abstr. 45407 (1978).

161.011   Formation of hot intergalactic gas by gas ejection
           from a galaxy in an early explosive era.
S. Ikeuchi.
Prog. Theor. Phys., Vol. 58, 1742 - 1758 (1977). – Abstr. in
Phys. Abstr., Vol. 81, Abstr. 45408 (1978).

          H I observations of Shostak's high-velocity cloud.
See Abstr. 131.012.

          QSO absorption lines and intergalactic hydrogen
clouds.   See Abstr. 141.190.

          The interaction of radio galaxies with the inter-
galactic medium.   See Abstr. 141.232.

          Radiogalaxies and their interaction with the external
medium.   See Abstr. 141.233.

          X-ray line spectroscopy for clusters of galaxies. I.
See Abstr. 142.021.

          High velocity clouds: galactic or extragalactic?
See Abstr. 155.029.

          Far-ultraviolet studies. IV. Spectroscopy of north
and south galactic pole regions observed from Apollo 17.
See Abstr. 157.007.

          A 2.7 GHz continuum survey of the M 81 - M 82
system.   See Abstr. 158.018

          The HI bridge between M81 and NGC 3077.
See Abstr. 158.265.

          The Magellanic Stream: the turbulent wake of the
Magellanic Clouds in the halo of the Galaxy.
See Abstr. 159.007.

          The origin and distribution of gas within rich clus-
ters of galaxies: the evolution of cluster X-ray sources over
cosmological time scales.   See Abstr. 160.009.

          A Westerbork survey of rich clusters of galaxies.
V. Multi-frequency observations of the radio tail galaxy
NGC 6034 in the Hercules cluster.
See Abstr. 160.011.

          A search for atomic hydrogen in clusters of galaxies.

See Abstr. 160.024.

A search for neutral hydrogen in primordial proto-clusters.   See Abstr. 160.036.

X-ray and radio emission from clusters of galaxies: the heating of intracluster gas by relativistic electrons. See Abstr. 160.043.

Ionized gas in X-ray clusters of galaxies: radio limits. See Abstr. 160.065.

On the origin and evolution of iron-enriched gas in clusters of galaxies.   See Abstr. 160.066.

The electrically polarized universe. See Abstr. 162.020.

Rotation measures as a cosmological test. See Abstr. 162.037.

Shock waves in the metagalaxy at large redshifts. See Abstr. 162.067.

162 Structure and Evolution of the Universe, Cosmology

162.001 Primordial black hole formation in an anisotropic Universe.   J. D. Barrow, B. J. Carr.
Mon. Not. R. Astron. Soc., Vol. 182, 537 - 558 (1978).

If the Universe started off with any anisotropy, the shear density would have dominated the matter-radiation density at sufficiently early times. At such times the anisotropy would have inhibited primordial black hole formation. However, observations indicate that the Universe could not have been anisotropy-dominated at times later than 1s, so black hole formation could not have been suppressed on scales larger than about $10^5 M_\odot$. Thus one can reject the possibility that the early Universe was completely chaotic. If the Universe was anisotropy-dominated until only about 10^{-23} s, there would never have been any primordial black holes small enough to evaporate within the lifetime of the Universe through the Hawking process, so the strong observational limits on the number of these small holes would no longer place a restriction on the number of large ones. In this case, large primordial black holes could conceivably have a critical density and there could be enough of them for galaxies to form through black hole clustering.

162.002 Possible role of collective relaxation in galaxy correlations.   W. H. Press, A. P. Lightman.
Astrophys. J., Lett., Vol. 219, L73 - L75 (1978).

Motivated by some recent numerical results, the authors suppose that collective relaxation effects may be important in establishing the internal structure of cosmological condensations. A crude model for this relaxation predicts a power-law correlation function with exponent γ near to the observed value, independent of the cosmological initial conditions, and insensitive to the single parameter of the model.

162.003 Helium production and limits on the anisotropy of the universe.   D. W. Olson.
Astrophys. J., Vol. 219, 777 - 780 (1978).

If most of the presently observed ^{4}He was synthesized in a primordial big-bang stage of the universe, then an indirect method due to Barrow can be used to set upper limits on the global anisotropy of the universe. Numerical results from an improved version of this method are compared with the more direct limits on anisotropy from observations of the microwave background radiation. Limitations in applicability of the method are also discussed.

162.004 Properties of the stars of early generations in the scale covariant cosmology.   A. Maeder.
Astron. Astrophys., Vol. 63, 175 - 181 (1978).

The author examines here some consequences of cosmological propositions which accept that further invariances exist in the theory of gravitation. Within this cosmological framework, stars being very different from the present ones would have existed in the past with higher rate of nuclear transformation and higher luminosity per unit of mass. The basic properties of these stars at an early epoch ($t = 1/5 \, t_0$) are examined with numerical models: the zero-age sequence, the central conditions in stars, the various critical stellar masses, the mass- and gravity-luminosity relation, the paths in the HR diagram, the changes of lifetimes, the evolution in the $\log T_c$ vs $\log \rho_c$ diagram. Expressions are also given for deriving by scaling the above properties at epochs other than the one considered in numerical applications.

162.005 Dynamic measurement of matter creation.
R. C. Ritter, G. T. Gillies, R. T. Rood, J. W. Beams.
Nature, Vol. 271, 228 - 229 (1978).

If a rotor could be spun so that it had an inertial decay time $\sim 10^{11}$ yr, 10^{18} s, it could test matter creation cosmologies

at a significant level. The advantage of this method over previous ones is that the result is independent of the form in which matter is created, providing it stays in place. The authors have designed an experiment to test this process; it is under construction and is described here.

162.006 Smoothing primaeval chaos.   P. Davies.
Nature, Vol. 271, 506 - 507 (1978).

162.007 On falling through a black hole into another universe.   N. D. Birrell, P. C. W. Davies.
Nature, Vol. 272, 35 - 37 (1978).

162.008 Der Urknall.   J. Schmid-Burgk, M. Scholz.
Sterne Weltraum, 17. Jahrg., 91 - 94 (1978).

162.009 Necessary conditions of the relativity of finiteness and infinity of non-empty space.
A. L. Zel'manov, L. I. Kharbediya.
Astron. Zh. Akad. Nauk SSSR, Vol. 55, 186 - 187 (1978). In Russian. English translation in Soviet Astron., Vol. 22, No. 1.

In the case of non-empty homogeneous isotropic cosmological models some necessary conditions of the relativity of finiteness and infinity of space are deduced.

162.010 Baryon number of the universe and anti-proton decay.   K. Sato.
Prog. Theor. Phys., Vol. 58, 385 - 386 (1977). — Abstr. in Phys. Abstr., Vol. 80, Abstr. 91422 (1977).

162.011 Les bases de la cosmologie moderne.
J. Merleau-Ponty.
La recherche en astrophysique, (see 003.001), p. 204 - 214 (1977).

162.012 La renaissance de la cosmologie d'observation.
D. W. Sciama.
La recherche en astrophysique, (see 003.001), p. 215 - 227 (1977).

162.013 L'univers est-il ouvert ou fermé?   J. Audouze.
La recherche en astrophysique, (see 003.001), p. 228 - 234 (1977).

162.014 Coincidences of large numbers in cosmology and microphysics.   Z. Klimek.
Postępy Astron., Tom 24, 223 - 233 (1976). In Polish.

The article gives a review of modern hypotheses concerning the existence of coincidences of large numbers in microphysics and cosmology.

162.015 Preventing singularities in the Einstein-Cartan cosmology.   B. Kuchowicz.
Postępy Astron., Tom 25, 27 - 33 (1977). In Polish.

The singularity in expanding cosmological models is an undesirable consequence of general relativity. It may be removed in the Einstein-Cartan theory of gravitation which is an extension of general relativity ("general relativity with spin"), if only the spin density does increase at a sufficiently fast rate with the contraction of matter.

162.016 Quiescent cosmology.   J. D. Barrow.
Nature, Vol. 272, 211 - 215 (1978).

The entropy level of the Universe makes it likely that its initial state was isotropic and quiescent rather than chaotic. Only if the equation of state for high density matter tends to the stiff, $p = \rho$, form might such an initially isotropic and homogeneous universe be stable and probable. Of those asymptotic states suggested by strong interaction physics, only for such stiff matter is isotropy compatible with large amplitude gravitational waves, random velocities, vorticity and magnetic fields. An explanation for the microwave background is also possible in such a model.

162.017 **Hot hadronic matter in the early universe.**
        R. L. Bowers, P. G. Dykema, A. M. Gleeson.
Astron. Astrophys., Vol. 64, 235 - 242 (1978).

   A fully relativistic equation of state for hot baryonic matter has been used to investigate the strong interaction contribution to the equation of motion of the Friedmann universe. A pronounced softening of the equation of state is observed

162.018 **Light element abundances in a matter—antimatter
        model of the universe.**   J. J. Aly.
Astron. Astrophys., Vol. 64, 273 - 279 (1978).

   This paper is devoted to the problem of light element synthesis in a baryon symmetric big-bang cosmology, in which the universe is constituted at the end of the leptonic era by a nucleon—antinucleon emulsion.

162.019 **Geodesic deviation and absolute motion in cosmology.**   P. D. Noerdlinger.
Astrophys. J., Vol. 220, 373 - 375 (1978).

   Alternate tests for absolute motion in cosmology merit renewed attention because of the Rubin-Ford anisotropy. One such test, "geodesic deviation" (effects of the Riemann tensor), is reexamined. A recent calculation which would suggest a large effect on ultrarelativistic particles is shown to yield a large result only in their own rest frame, and so be essentially inaccessible to experimental test. For slower test particles, the effect of inhomogeneities is estimated and is also shown to dominate in realistic situations. Thus, the test remains only of theoretical interest.

162.020 **The electrically polarized universe.**
        J. Bally, E. R. Harrison.
Astrophys. J., Vol. 220, 743 - 744 (1978).

   It is shown that all gravitationally bound systems — stars, galaxies, and clusters of galaxies — are positively charged and have a charge-to-mass ratio of ~ 100 coulombs per solar mass. The freely expanding intergalactic medium has a compensating negative charge. The immediate physical consequences of an electrically polarized universe are found to be extremely small.

162.021 **Cosmology with another theory of gravity.**
        P. Rastall.
Astrophys. J., Vol. 220, 745 - 748 (1978).

   The homogeneous and isotropic cosmological solutions of a recently developed theory of gravity are discussed. There are no solutions corresponding to closed universes. The solutions corresponding to open universes predict an acceleration constant in the range $-0.03 > q_0 > -0.45$. The open universes with Euclidean space sections have no big bang, and seem to be incompatible with observation.

162.022 **The extragalactic background light and slow star
        formation in galaxies.**   B. M. Tinsley.
Astrophys. J., Vol. 220, 816 - 821 (1978).

   If the past star formation rates in galaxies are constrained to be consistent with their present colors, and if it is assumed that the stellar initial mass function was not weighted toward massive stars at early times, then only early-type galaxies were conspicuously brighter as primeval objects and a recently reported faint value for the visual background light can be readily understood.

162.023 **The evolution of the universe.**   P. Goldreich.
        Focus on the stars, (see 003.002), p. 215 - 256
(1977).

162.024 **A search for neutral hydrogen in primordial
        protoclusters at $z = 3.33$ and 4.92.**
R. D. Davies, A. Pedlar, I. F. Mirabel.
Mon. Not. R. Astron. Soc., Vol. 182, 727 - 733 (1978).

   A search has been made for the redshift neutral hydrogen emission from primordial protoclusters at early epochs in the history of the Universe. A system with 2.5-MHz bandwidth was used to investigate 20 intermediate-latitude fields at frequencies of 328 and 240 MHz which correspond to neutral hydrogen redshifts of $z = 3.33$ and 4.92. The results are used to derive a limit to the properties of protoclusters in the early Universe on the model proposed by Sunyaev & Zeldovich; either their masses are $\lesssim 3 \times 10^{15}\ M_\odot$ or the number of such objects in the early Universe $\lesssim 10^6$.

162.025 **From the origin of the universe to the earliest geo-
        logical times.**   J. E. Jones.
J. British Interplanet. Soc., Vol. 31, 123 - 128, 139 (1978).

162.026 **Clumpy structure of the Universe and general field.**
        G. Chincarini.
Nature, Vol. 272, 515 - 516 (1978).

   This note stresses that the observations set a strong limit on the density of the so called 'field galaxies' and questions the very existence of such objects.

162.027 **The local mean mass density of the universe: new
        methods for studying galaxy clustering.**
M. Davis, M. J. Geller, J. Huchra.
Astrophys. J., Vol. 221, 1 - 18 (1978).

   The authors analyze a catalog of galaxy redshifts nearly complete to a limiting blue magnitude of 13.0 and covering the entire sky. They obtain estimates of the galaxy luminosity function and of the local mean mass density. The redshift catalog is also the basis for three estimators of the spatial correlation function for the galaxy distribution. The authors reexamine the well-studied angular correlation function and they define and explore two new redshift-weighted measures of the distribution. These estimators lead to a revised value for the amplitude of the spatial correlation function. The authors also measure the one-dimensional rms relative peculiar velocity dispersion. Cosmic virial theorems which are cast in terms of the correlation functions and the peculiar velocity dispersion are used to estimate the contribution of matter associated with galaxies to the critical density.

162.028 **Cosmological implications of massive, unstable
        neutrinos.**
D. A. Dicus, E. W. Kolb, V. L. Teplitz.
Astrophys. J., Vol. 221, 327 - 341 (1978).

   The authors consider the cosmological consequences of the existence of massive, neutral, weakly interacting leptons (ν_H). Bounds are calculated on the ν_H mass, lifetime, and allowed interaction strength by assuming minimal modification of the standard big-bang model.

162.029 **The size of the universe.**   G. Gilmore.
        South. Stars, Vol. 27, 65 - 78 (1977).

   This paper is concerned with the large scale structure of the universe and, principally, how to measure it. The author will work through an extended inductive chain and considers uncertainties and justifications at each step.

162.030 **Redshifts as a measure of distance.**   N. V. Zotov.
        South. Stars, Vol. 27, 79 - 81 (1977).

   The distances of far-off objects implied by their redshifts are tabulated, and the time in years that it has taken their light to reach us is also calculated.

162.031 **The extension of the Hubble diagram. II. New red-
        shifts and photometry of very distant galaxy
clusters: first indication of a deviation of the Hubble diagram**

from a straight line.
J. Kristian, A. Sandage, J. A. Westphal.
Astrophys. J., Vol. 221, 383 - 394 (1978).

Redshifts are given for 50 brightest cluster galaxies, extending as far as $z = 0.75$; BVR photometry is given for 33 clusters. These data are combined with earlier data of a similar kind in order to investigate several effects. The measured $B-V$ and $V-R$ colors as a function of redshift are well represented by Whitford's standard-galaxy K corrections, as far as these are defined. Other standard corrections to the measurements are discussed, and formal least-squares values of q_0 are computed. The present sample, cut off at $z = 0.4$ to avoid selection effects and uncertainties in the data, shows the first significant evidence for curvature of the Hubble diagram, with V and R magnitudes giving similar results. The formal value of q_0 (with galaxy evolution ignored) is +1.6 ± 0.4.

162.032 **On the decay of cosmic turbulence.**
        E. Knobloch.
Astrophys. J., Vol. 221, 395 - 398 (1978).

The consideration of homogeneous, incompressible, isotropic, nonrelativistic turbulence in a uniformly expanding universe shows that the derivation of the decay of rms velocity from the Loitsyanskii invariant is in general incorrect. It is shown that this derivation is valid only in the terminal stages of the decay. It is concluded that the Loitsyanskii invariant is not an invariant when the inertial transfer of energy dominates viscous dissipation.

162.033 **The development of density perturbations after the recombination.**
A. Cavaliere, L. Danese, G. De Zotti.
Astrophys. J., Vol. 221, 399 - 406 (1978).

In the framework of the gravitational instability picture for the postrecombination evolution of density enhancements, the authors derive simple expressions relating the virialization time and the dynamical properties of massive ($M \gtrsim 10^{12} M_\odot$) perturbations to their initial density contrast or their total energy. The ensuing predictions are shown to be quite consistent with observational data taken from recent catalogs.

162.034 **A crucial test of the Dirac cosmologies.**
        G. Steigman.
Astrophys. J., Vol. 221, 407 - 411 (1978).

In a cosmology consistent with the Cosmological Principle (large scale, statistical isotropy and homogeneity of the universe), a Planck spectrum is not preserved as the universe evolves unless the number of photons in a comoving volume is conserved. It is shown that a large class of cosmological models based on Dirac's Large Numbers Hypothesis (LNH) violate this constraint. The observed isotropy and spectral distribution of the microwave background radiation thus provide a crucial test of such cosmologies.

162.035 **Quantum cosmological models.**
        M. A. H. MacCallum.
Quantum gravity, (see 012.007), p. 174 - 218 (1975).

162.036 **Positive curvature of the Hubble diagram for galaxy clusters.**
J. Kristian, A. R. Sandage, J. A. Westphal.
Bull. American Astron. Soc., Vol. 9, 586 - 587 (1978).
Abstract.

162.037 **Rotation measures as a cosmological test.**
        P. P. Kronberg, M. Simard-Normandin, M. Reinhardt.
Bull. American Astron. Soc., Vol. 9, 588 (1978). − Abstract.

162.038 **The cosmological principle and quantization.**
        P. G. Gross.
Bull. American Astron. Soc., Vol. 9, 588 (1978). − Abstract.

162.039 **The distance scale within the Local Group.**
        S. van den Bergh.
IAU Colloq. No. 37, (see 012.009), p. 13 - 41 (1977).

The first modern discussion of the extra-galactic distance scale was by Hubble and Humason (1931). We now know that this first bold attempt to measure the size of the Universe failed and that the Universe is, in fact, five or possibly even ten times larger than Hubble and Humason believed. It is perhaps not generally realised that a major portion of this change was due to an increase in distance estimates for the members of the Local Group, which were used as stepping stones to the realm of the nebulae. During the last decade it has become increasingly apparent that individual Local Group galaxies have different chemical compositions. As a result some distance indicators might have characteristic luminosities that differ slightly from galaxy to galaxy. To avoid systematic errors arising from such differences it is important to use as many distance indicators as possible to determine the distances to nearby galaxies.

162.040 **The Hubble constant and the local expansion field.**
        G. A. Tammann.
IAU Colloq. No. 37, (see 012.009), p. 43 - 74 (1977).

On présente une brève revue des méthodes de détermination des distances extragalactiques actuellement disponibles. Plusieurs voies distinctes conduisent à $H_0 = 50 - 55$ km s^{-1} Mpc^{-1}. L'expansion de Hubble dans le superamas local est extrêmement régulière. Il n'y a aucun signe d'un changement de H_0 avec la distance.

162.041 **The distance scale beyond the Local Group – a progress report on work in the southern hemisphere.**
M. G. Smith.
IAU Colloq. No. 37, (see 012.009), p. 75 - 93 (1977).

The persistent discrepancy between recent estimates of the Hubble expansion parameter (ranging from about 50 km s^{-1} Mpc^{-1} to about twice that value) illustrates our poor understanding of possible systematic errors in the application of distance indicators particularly in the critical range of distances beyond the Local Group but closer than the nearer dense clusters of galaxies. It seems appropriate here to concentrate on recent southern work with distance indicators even though much of it is only in a very preliminary stage.

162.042 **21-cm observations of redshifts and their implications for the distance scale.**
R. B. Tully, J. R. Fisher.
IAU Colloq. No. 37, (see 012.009), p. 95 - 117 (1977).

Il existe une corrélation étroite entre la largeur de la raie d'hydrogène neutre à 21 cm et la luminosité des galaxies spirales ainsi qu'entre cette largeur et le diamètre des galaxies spirales. Elle existe pour un échantillon de spirales proches dont la distance est calibrée par les céphéides, ainsi que pour des spirales dans quelques amas. La comparaison entre ces deux relations nous donne la distance de l'amas et une valeur pour la constante de Hubble.

162.043 **Is there evidence of anisotropy in the expansion of the universe?   V. C. Rubin.**
IAU Colloq. No. 37, (see 012.009), p. 119 - 140 (1977).

In the 50 years since Hubble (1929) published the first redshift-magnitude diagram, observational cosmology has been concerned with the evaluation of two numbers, H and q_0. Observational data are not sufficiently precise to detect anisotropies of a few percent, but anisotropies of order 50% would surely have been noted in prior studies of magnitudes and velocities. Modern observations could detect anisotropies of perhaps 10% in H across the sky. It is an anisotropy of this order across the sky which the author here discusses.

162.044 **Observational tests in cosmology.   J. E. Gunn.**
IAU Colloq. No. 37, (see 012.009), p. 183 - 212

(1977).

The author reviews the available data for the determination of the three dimensionless cosmological parameters, q_0, τ_0, and Ω_0, any two of which suffice in principle to determine the structure and time development of the universe up to a scale which is provided by the Hubble constant, assuming that the correct theory of gravity is general relativity with a cosmological constant.

162.045 **Surface brightness, standard candles and q_0.**
V. Petrosian.
IAU Colloq. No. 37, (see 012.009), p. 213 - 218 (1977).

The most direct method of determination of the deceleration parameter (q_0) of the universe is through the study of the redshift-magnitude relation of extragalactic sources. The progress here has been slow because the necessary source for this study must be standard candles; i.e., must have identical absolute total luminosity (bolometric or monochromatic). The author first shows that this although necessary is not a sufficient condition for non-point like (or resolved) sources. He then describes modification of the redshift-magnitude relation for a certain class of non-standard candles using measurement of isophotal surface brightness.

162.046 **The Hubble diagram for red magnitudes of bright cluster galaxies.**
J. Kristian, A. R. Sandage, J. A. Westphal.
IAU Colloq. No. 37, (see 012.009), p. 309 - 311 (1977).

The authors show the current status of a program to extend the Hubble diagram for faint galaxy clusters to larger redshifts. The clusters include (a) at the bright end, previously measured clusters and groups, and (b) at the faint end, a combination of faint Abell clusters, 3CR clusters, and clusters selected randomly from deep 48-inch surveys.

162.047 **Time delay effects for measuring cosmological distances.**   K. Chang, S. Refsdal.
CNRS Colloq. No. 263, (see 012.009), p. 369 - 375 (1977).

Lorsqu'une galaxie compacte et massive se situe entre un observateur O et une source de lumière S, il se peut que par déflection gravitationnelle, la lumière suive différents trajets de durées différentes pour parvenir à l'observateur. Si la source est variable il est possible de mesurer la différence de temps Δt entre les durées des trajets. Cela nous donne la possibilité de déterminer les distances cosmologiques d'une façon purement géométrique. Les problèmes liés à la distribution inconnue des masses et les possibilités observationnelles sont discutés.

162.048 **Possible explanations of non cosmological redshifts.**
J.-C. Pecker.
CNRS Colloq. No. 263, (see 012.009), p. 451 - 479 (1977).

La section I étudie successivement différentes suggestions: perturbations de cosmologies relativistes, chronogéométrie de Ségal, cosmologie à explosions successives d'Ambartsumian, variation des constantes physiques, enfin mécanismes de fatigue de la lumière. Une forme particulière de ces derniers mécanismes est discutée dans la section II, où il est montré qu'on peut, grâce à une interaction, décrite en détails entre les photons de la source et les particules φ (particules scalaires, neutres, de très faible masse et de spin zéro), expliquer les décalages anormaux vers le rouge, mais aussi peut-être (section III) la loi normale de Hubble et le rayonnement à 3 K. Une conclusion montre à quel point le débat reste ouvert.

162.049 **The H_0 value from luminosity classes and 21-cm line observations of galaxies.**
L. Bottinelli, L. Gouguenheim.
CNRS Colloq. No. 263, (see 012.009), p. 481 - 486 (1977).

162.050 **The Hubble constant from Tully and Fisher's rela-**

tion.   J. Heidmann.
CNRS Colloq. No. 263, (see 012.009), p. 487 - 488 (1977).

162.051 **Non-cosmological redshifts.**   J. V. Narlikar.
CNRS Colloq. No. 263, (see 012.009), p. 497 - 499 (1977).

L'introduction de lignes d'univers brisées dans la théorie de la gravitation de Hoyle et Narlikar permet de rendre compte de la création de matière. Cette matière qui se compose de masses inertielles plus petites se manifeste par des décalages spectraux vers le rouge anormaux.

162.052 **High redshift 21-cm lines.**   M. S. Roberts.
CNRS Colloq. No. 263, (see 012.009), p. 501 - 509 (1977).

L'auteur résume l'ensemble des décalages vers le rouge dont on dispose à partir de la raie 21 cm. La plus grande valeur obtenue pour la raie H I en émission est z = 0.0369, et z = 0.692 en absorption. Il y a un excellent accord avec les décalages vers le rouge mesurés optiquement (lorsqu'ils existent). Les deux techniques donnent la même valeur (en moyenne), et montrent que les décalages vers le rouge sont indépendants des longueurs d'onde dans un intervalle de longueur d'onde de $\sim 5 \times 10^5$, et un intervalle en z de -0.0012 à $+0.692$. C'est une condition nécessaire, mais non suffisante pour interpréter les décalages vers le rouge par l'effet Doppler.

162.053 **The redshift dependence of galaxy surface brightness.**   P. Crane, A. W. Hoffman.
CNRS Colloq. No. 263, (see 012.009), p. 531 - 539 (1977).

Les auteurs ont déterminé la brillance superficielle moyenne d'amas à l'intérieur d'un diamètre linéaire projeté de 20.3 kpc en faisant la photométrie de galaxies dans six amas ayant un redshift de z = 0.05 à 0.20. Pour toutes les cosmologies dans lesquelles le décalage vers le rouge est uniquement dû à l'expansion, la brillance superficielle devrait avoir la forme m (SB) = 2.5 a log (1+z) + b où a = 4 et b est une constante d'échelle. Les auteurs trouvent a = 4.1 ± 1.6 et b = 22.12 ± 0.21. Ces résultats supportent fortement l'interprétation des décalages vers le rouge comme dus à l'expansion de l'univers.

162.054 **Anomalous Hubble expansion and inhomogeneous cosmological models.**   S. Mavrides.
CNRS Colloq. No. 263, (see 012.009), p. 549 - 554 (1977).

162.055 **Arguments and evidence concerning non-cosmological redshifts — a partial summary.**
G. Burbidge.
CNRS Colloq. No. 263, (see 012.009), p. 555 - 561 (1977).

Les arguments et les preuves en faveur des décalages non cosmologiques vers le rouge sont résumés.

162.056 **The "conventional" view of redshifts.**   M. J. Rees.
CNRS Colloq. No. 263, (see 012.009), p. 563 - 568 (1977).

Exposé du point de vue "conventionnel" sur les décalages vers le rouge.

162.057 **On the compatibility of paradoxical redshifts, observed in galaxies and quasars, with the theory of gravitation.**   Z. Horák.
Bull. Astron. Inst. Czechoslovakia, Vol. 29, 126 - 128 (1978).

Both Newtonian and Einsteinian theories of gravitation lead to the known velocity-distance relation for "free" galaxies. Yet, within a group of galaxies with a common velocity of recession a companion galaxy may have a greater transverse velocity and thus a greater relativistic redshift than the one of the main galaxy. For a binary system of galaxies the actual distance and the opposite transverse velocities of the members can be computed from their observed redshifts and estimated ratio of masses. The redshift excess of the companion galaxies may thus be explained and the apparent superrelativistic speeds in expanding radio sources avoided, in full agreement with the velocity-distance law.

162.058 **The Hubble modulus of Markarian galaxies.**
        C. J. Krieger.
Astrophys. J., Lett., Vol. 221, L101 - L102 (1978).
        The difference of the Hubble moduli of Markarian galaxies in two regions is the result of observational selection.

162.059 **Is the universe expanding?**   G. F. R. Ellis.
        Gen. Relativ. Gravitation, Vol. 9, 87 - 94 (1978).
        It is shown that spherically symmetric static general relativistic cosmological space-times can reproduce the same cosmological observations as the currently favored Friedmann–Robertson–Walker universes, if the usual assumptions are made about the local physical laws determining the behavior of matter, provided that the universe is inhomogeneous and our galaxy is situated close to one of its centers. Only (1) unverifiable a priori assumptions, (2) detailed physical and astrophysical arguments, or (3) observation of the time variation of cosmological quantities can lead to the conclusion that the universe is not such a static space-time.

162.060 **The problem of scalar field theory in curved space-**
        time.   M. Castagnino.
Gen. Relativ. Gravitation, Vol. 9, 101 - 121 (1978).
        It is demonstrated that (1) there exist infinite G_1 that satisfy Lichnerowicz's conditions (L conditions) in a globally hyperbolic manifold; and, (2) there is no G_1 in an expanding universe that would satisfy those conditions and that would behave as the ordinary Δ_1 of flat space when $x \to x'$. The author thinks that these results present a serious problem for finding a semiclassical theory of scalar field in curved space-time.

162.061 **Newtonian cosmology with a varying gravitational**
        constant.   J. P. Vinti.
Celestial Mech., Vol. 16, 391 - 406 (1977).
        Newtonian cosmology is developed with the assumption that the gravitational constant G diminishes with time. The functional form adoped for $G(t)$, a modification of a suggestion of Dirac, is $G = A(k + t)^{-1}$, where t is the age of the Universe and a small constant k is inserted to avoid a singularity in the two-body problem. If R is the scale factor, normalized to unity at an epoch time τ, the differential equation is then $R^2 \ddot{R} = -(4\pi/3)G(t)\rho_0$. Here ρ_0 is the mean density at the epoch time. With the above form for $G(t)$, the solution is reducible to quadratures. The conditions for a maximum lead to a boundary curve of ρ_{0c} versus H_0 and the numbers indicate strongly that this G-variable Newtonian model corresponds to an open universe.

162.062 **On Newtonian cosmology with a varying gravitation-**
        al constant.   D. G. Saari.
Celestial Mech., Vol. 16, 407 - 409 (1977).
        J. P. Vinti studies the effects a varying gravitational constant has upon Newtonian cosmology. In this paper an alternative approach for the analysis is given which obtains and improves upon some of Vinti's major results.

162.063 **The present revolution in cosmology.** Total sur-
        mounting of the cosmological egocentrism.
G. M. Idlis.
Priroda, 1978, No. 4, p. 74 - 81. In Russian.

162.064 **Long-wavelength perturbations of a Friedmann**
        universe and anisotropy of the relict radiation.
L. P. Grishchuk, Ya. B. Zel'dovich.
Astron. Zh. Akad. Nauk SSSR, Vol. 55, 209 - 215 (1978).
In Russian. English translation in Soviet Astron., Vol. 22, No. 2.

In principle, considerable deviations of properties of the real universe from the parameters of the idealized cosmological Friedmann model could have place on a scale exceeding the size of the horizon. There have been found restrictions on the amplitude of such long-wavelength perturbations, following from the observational fact of the significant isotropy of the relict background radiation ($\delta T/T < 10^{-4}$). It is shown that the data on $\delta T/T$ together with the natural hypothesis on the statistical independence of the perturbations lead to the conclusion that there are no significant (i.e. with amplitude exceeding $\delta T/T$) density perturbations on any spatial scale exceeding the horizon. As far as the metric perturbations are concerned, the amplitude of some types of such perturbations might be significant without being in contradiction with the observational restrictions on $\delta T/T$.

162.065 **Hydrodynamics of primordial black hole forma-**
        tion.
D. K. Nadezhin, I. D. Novikov, A. G. Polnarev.
Astron. Zh. Akad. Nauk SSSR, Vol. 55, 216 - 230 (1978).
In Russian. English translation in Soviet Astron., Vol. 22, No. 2.
        A hydrodynamical picture of primordial black hole (PBH) formation at the early stages of the expanding universe is discussed under the assumption that the expansion of the universe near the singularity is quasi-isotropic. It is clarified of what kind perturbations must be to form PBHs. The role of pressure gradients is evaluated in detail. It is shown that at the moment of PBH formation its mass is considerably smaller than the mass within the cosmological horizon, so catastrophic accretion seems to be unlikely.

162.066 **Possible cosmological consequences of the process**
        of evaporation of primary black holes.
B. V. Vajner, P. D. Nasel'skij.
Astron. Zh. Akad. Nauk SSSR, Vol. 55, 231 - 235 (1978).
In Russian. English translation in Soviet Astron., Vol. 22, No. 2.
        Limitations on the density of distribution of primary black holes with $M < 10^{15}g$ are obtained. On the assumption of the existence of initial density perturbations in the framework of a standard cosmological model, limitations on the fluctuation amplitude are derived.

162.067 **Shock waves in the metagalaxy at large redshifts.**
        L. M. Ozernoj, V. V. Chernomordik.
Astron. Zh. Akad. Nauk SSSR, Vol. 55, 236 - 248 (1978).
In Russian. English translation in Soviet Astron., Vol. 22, No. 2.
        Shock waves in the metagalactic medium produced by young galaxies, quasars and active nuclei of galaxies at redshifts $0 \leq z \lesssim 10$–30 are considered. The authors analyze (1) propagation of shock waves, taking into account cosmological expansion, (2) energy losses due to radiation and (3) the influence of shock waves on the metagalactic gas. As a result of energy losses due to radiation, rarefied hot cavities surrounded by dense cold "walls" may be formed in the intergalactic medium. Due to crossing of shock wave fronts from individual sources, intergalactic gas (if its density is not too low) acquires an inhomogeneous structure in a form of "caves", i.e. cavities, penetrating one another.

162.068 **Charge in cosmological models.**   M. L. Bohra,
        A. L. Mehra.
Gen. Relativ. Gravitation, Vol. 9, 289 - 297 (1978).
        The balanced field equations due to Penney are used to find solutions for cosmological models in the presence of charge. Herein, it is found that the introduction of charge adds additional terms to the Einstein conservation equation and distribution expressions. The curvature parameter is affected and it is concluded that whereas matter affects it positively, the charge does so negatively. There then arises also the pos-

sibility of an evolution of local systems against the background of a global expansion.

162.069 A plan to study the "standard hot big-bang model" of the universe. R. Chandra.
Gen. Relativ. Gravitation, Vol. 9, 329 - 338 (1978).

The dynamics of the "standard hot big-bang model" of the universe according to general relativity with modified cosmical constant are discussed.

162.070 Core condensation in heavy halos: a two-stage theory for galaxy formation and clustering.
S. D. M. White, M. J. Rees.
Mon. Not. R. Astron. Soc., Vol. 183, 341 - 358 (1978).

The authors suggest that most of the material in the Universe condensed at an early epoch into small 'dark' objects. Irrespective of their nature, these objects must subsequently have undergone hierarchical clustering, whose present scale the authors infer from the large-scale distribution of galaxies. The observed sizes of galaxies and their survival through later stages of the hierarchy seem inexplicable without invoking substantial dissipation; this dissipation allows the galaxies to become sufficiently concentrated to survive the disruption of their halos in groups and clusters of galaxies. The authors propose a specific model in which $\Omega \cong 0.2$, the dark matter makes up 80 per cent of the total mass, and half the residual gas has been converted into luminous galaxies by the present time.

162.071 Accelerating Universe revisited. B. M. Tinsley.
Nature, Vol. 273, 208 - 211 (1978).

There is growing evidence that the Hubble constant is of the order of 100 km s^{-1} Mpc^{-1}, rather than only half as great. The purpose of this paper is to discuss the Friedmann models that are consistent with such a large expansion rate, as well as with current estimates of stellar ages and the density of the Universe. The only possible Friedmann models of the Universe are those with a positive cosmological constant and $q_0 < -1$. Various independent tests of this conclusion are suggested.

162.072 The correlation function for density perturbations in an expanding universe. III. The three-point and predictions of the four-point and higher order correlation functions. J. McClelland, J. Silk.
Astrophys. J., Suppl. Ser., Vol. 36, 389 - 404 (1978).

The authors show that the observed two-point and three-point correlation functions for the large-scale distribution of galaxies cannot be explained by having simple perturbations which presently are randomly distributed in space. They develop analytical models of clustered and hierarchical perturbations which can explain these two correlation functions and which provide various predictions for the higher order correlation functions. A possible explanation of these models is to make the single assumption that density fluctuations containing a mass $10^{15} [(\delta\rho_0/\rho_0)z_0]^{-2} \ M_\odot$ are randomly placed at some initial redshift z_0 in the early universe after decoupling, at which time they begin to move freely under their mutual self-gravity.

162.073 Scale covariant cosmology and the temperature of the Earth. V. Canuto, S.-H. Hsieh.
Astron. Astrophys., Vol. 65, 389 - 391 (1978)..

The changes in the temperature of the Earth as predicted by scale covariant cosmological models, that allow for a variation of G and M, are investigated. Recent determination of the early temperature of the Earth seems to favor the case without matter creation.

162.074 Element production in the big bang – how can we make such a bold prediction? P. J. E. Peebles.
AIP Conf. Proc., No. 37 (see 012.016), p. 77 - 84 (1977).

Abstr. in Phys. Abstr., Vol. 81, Abstr. 8177 (1978).

162.075 Viscous dissipation and entropy production in Bianchi type-I universes. N. Caderni, R. Fabbri.
Nuovo Cimento, Lett., Ser. 2, Vol. 20, 185 - 189 (1977).
From Phys. Abstr., Vol. 81, Abstr. 8179 (1978).

162.076 Electric universe. P. C. W. Davies.
Nature, Vol. 273, 268 - 269 (1978).

162.077 Tilting and viscous models in a class of type VI_0 cosmologies. K. A. Dunn, B. O. J. Tupper.
Astrophys. J., Vol. 222, 405 - 411 (1978).

The class of type VI_0 cosmological models with perfect fluid and electromagnetic field which were investigated in a previous article (1976) is shown to contain tilting models which have zero pressure and a null electromagnetic field It is also shown that the class contains viscous fluid models which are necessarily nontilting. Properties of these models are discussed.

162.078 Cosmic heresy? P. C. W. Davies.
Nature, Vol. 273, 336 - 337 (1978).

162.079 Self-similar motions of photon gas and the Friedmann model. N. R. Sibgatullin, O. Yu. Dinariev.
Zh. ehksp. i teor. fiz., Vol. 73, 1599 - 1610 (1977). In Russian. – Abstr. in Ref. zh., 51. Astron., 3.51.939 (1978).

162.080 Gravity, charges, cosmology and coherence. Ya. B. Zel'dovich.
Usp. fiz. nauk, Vol. 123, 487 - 503 (1977). In Russian. – Abstr. in Ref. zh., 51. Astron., 3.51.941 (1978).

162.081 The cosmic background radiation and the new aether drift. R. A. Muller.
Sci. American, Vol. 238, No. 5, p. 64 - 74 (1978).

Sensitive instruments have found slight departures from uniformity in the radiation left by the primordial "big bang". The experiment reveals the earth's motion with respect to the universe as a whole.

162.082 Large-scale structure of the universe. A. G. Doroshkevich.
Zemlya i Vselennaya, 1978, No. 3, p. 62 - 66. In Russian.

162.083 Correlation of Newtonian and relativistic cosmologies. A. B. Evans.
Mon. Not. R. Astron. Soc., Vol. 183, 727 - 748 (1978).

Comoving metrics of relativistic dust universes are interpreted locally in terms of conventional Newtonian cosmology. Quasi-classical equations for these universes show how anisotropies in the motion and the spatial curvature affect the dynamics. Though generally indeterminate, the equations have heuristic and possibly predictive value, especially as regards matter singularities and indefinite expansion, for which they offer simple classifications. Examples are provided by homogeneous world models, including some with an apparently new kind of singularity behaviour.

162.084 Newtonian cosmology with a time-varying constant of gravitation. G. C. McVittie.
Mon. Not. R. Astron. Soc., Vol. 183, 749 - 764 (1978).

Newtonian cosmology is based on the Eulerian equations of fluid mechanics combined with Poisson's equation modified by the introduction of a time-varying G. Spherically symmetric model universes are worked out with instantaneously uniform densities. They are indeterminate unless instantaneous uniformity of the pressure is imposed. When G varies as an inverse power of the time, the models can in some cases be shown to depend on the solution of a second-order differential equation which also occurs in the Friedmann models of general relativity.

162.085   Baryon symmetric big bang cosmology.
   F. W. Stecker.
Nature, Vol. 273, 493 - 497 (1978).

162.086   Gravitation, primordial stars and the dark mass.
   G. A. Shields.
Nature, Vol. 273, 519 - 520 (1978).
   The possibility that the gravitational constant G decreases as the universe expands is a subject of continuing theoretical and experimental interest. Such a cosmology is used here to explain the low luminosity of the 'missing mass' in clusters of galaxies, and the 'heavy haloes' of spiral galaxies, in terms of the rapid initial evolution of a generation of primordial stars.

162.087   Singularities in viscous universes.   M. Heller.
   Acta Cosmologica, Zesz. 7, 7 - 15 (1978).
   Introduction of bulk viscosity into cosmological models may cause different pathological situations connected with the transport of the energy produced in dissipative processes. Conditions to be imposed on the energy-momentum tensor to avoid such pathologies are, in general, sufficient to prove the existence of the singularity. Bulk viscosity introduces quantitatively new features to the model: evolution becomes irreversible, expansion essentially differs from contraction, the thermodynamic "arrow of time" is well defined.

162.088   The redshift phenomenon in systems of different
   scales.   T. Jaakkola.
Acta Cosmologica, Zesz. 7, 17 - 44 (1978).
   The reshift phenomenon as it appears in systems of various scales beginning with the metagalactic redshift field down to that of the solar vicinity is considered. A relation between the size or the mean density of a system on one hand and the strength of the redshift on the other hand is obtained. A generalized distance law of redshift is deduced, applicable e.g. for objects in the Local Supergalaxy and for mapping the redshift fields in external systems. For cosmological distances this law reduces to the Hubble law.

162.089   Uniform relativistic universe models with pressure.
   III. Models with the cosmological constant.
J. Krempeć, B. Krygier.
Acta Cosmologica, Zesz. 7, 45 - 51 (1978).
   The uniform universe models containing interacting matter and radiation are considered for the non-vanishing cosmological constant. The insertion of the cosmological constant into the field equations is treated as the inclusion of the density and pressure of vacuum into the energy-momentum tensor.

162.090   Uniform relativistic universe models with pressure.
   IV. The quadratic state equation.
B. Krygier, J. Krempeć.
Acta Cosmologica, Zesz. 7, 53 - 55 (1978).
   The homogeneous and isotropic universe models with pressure are discussed for the quadratic equation of state. The Einstein field equations have only numerical integrals in this case.

162.091   Fundamental equations for two component world
   models with interaction.   E. Wojciulewitsch.
Acta Cosmologica, Zesz. 7, 75 - 80 (1978).
   The fundamental equations of cosmology have been generalised for world models containing matter in a relativistic as well as in a non relativistic state. The possibility for matter for going from one state to the other will not be excluded and will be called interaction. Without interaction, or when interaction is in equilibrium, a non linear differential equation with constant coefficients appears. The effect of interaction exists in making these coefficients time dependent. Some general relations are derived.

162.092   Local – large scale – global. On certain methodolog-

ical questions of cosmology.   M. Heller.
Acta Cosmologica, Zesz. 7, (see 012.022), 83 - 99 (1978).
   One of the most striking features of cosmology is that we are sentenced to perform only local observations, and we want to draw from them global consequences. Relations between "local" and "global" are discussed.

162.093   A survey of cosmological models.   A. Krasiński.
   Acta Cosmologica, Zesz. 7, (see 012.022), 101 -
118 (1978).
   The paper presents a very concise account of the history of cosmology. The difficulties of the Newtonian theory of gravitation which led to the formulation of general relativity are presented. Then, the basic notions of the general relativity theory are introduced at an elementary level. The cosmological models which were once (or are presently) believed to be the true models of the observed universe, are shortly presented.

162.094   Cylindrical rotating universe.   A. Krasiński.
   Acta Cosmologica, Zesz. 7, (see 012.022), 133 -
138 (1978).

162.095   A simple model of the universe at the "hadron era".
   M. Ostrowski.
Acta Cosmologica, Zesz. 7, (see 012.022), 139 - 145 (1978).

162.096   Cosmological models with bulk viscosity.
   L. Suszycki.
Acta Cosmologica, Zesz. 7, (see 012.022), 147 - 159 (1978).
   The question to what extent the introduction of bulk viscosity into the Friedmann-Lemaître cosmology changes the evolution of world models is discussed. Previous results concerning the flat, dust-filled models with constant viscosity are briefly reviewed.

162.097   Significance of neutrinos in cosmology.
   M. Ostrowski.
Postępy Astron., Tom 26, 51 - 58 (1978). In Polish.
   Cosmic background neutrinos cannot be detected at present. The possible influence of neutrinos for the evolution of a Friedmannian cosmological model is presented. The methods of their detection are discussed.

162.098   Density constraint on local inhomogeneities of a
   Robertson-Walker cosmological universe.
J. Eisenstaedt.
Phys. Rev. D, Vol. 16, 927 - 928 (1977). – Abstr. in Phys. Abstr., Vol. 81, Abstr. 8281 (1978).

162.099   Spatially homogeneous universe.   A. H. Taub.
   Proceedings of the 1st Marcel Grossman meeting on general relativity (see 012.025), p. 231 - 242 (1977). – Abstr. in Phys. Abstr., Vol. 81, Abstr. 8318 (1978).

162.100   Photon condensation in an Einstein universe.
   J. D. Becker, L. Castell.
Contacts between high energy physics and other fields of physics. Schladming, Austria, 24 February - 5 March 1977 = Acta Phys. Austriaca, Suppl. 18 (1977). p. 885 - 897.– Abstr. in Phys. Abstr., Vol. 81, Abstr. 12435 (1978).

162.101   Path-integral quantisation and cosmological particle
   production: an example.   D. M. Chitre, J. B.
Hartle.
Phys. Rev. D, Vol. 16, 251 - 260 (1977). – Abstr. in Phys. Abstr., Vol. 81, Abstr. 12440 (1978).

162.102   Quantum effects in cosmology and black-and-
   white hole physics.
A. A. Starobinsky (Starobinskij).
Proceedings of the 1st Marcel Grossman meeting on general relativity (see 012.025), p. 499 - 501 (1977). – Abstr. in Phys.

Abstr., Vol. 81, Abstr. 12441 (1978).

162.103 **Observational cosmology.**   R. B. Partridge.
Proceedings of the 1st Marcel Grossman meeting on general relativity (see 012.025), p. 617 - 648 (1977). − Abstr. in Phys. Abstr., Vol. 81, Abstr. 12444 (1978).

162.104 **Jeans instability in Lemaître universes.**
F. Occhionero.
Proceedings of the 1st Marcel Grossman meeting on general relativity (see 012.025), p. 653 - 659 (1977). − Abstr. in Phys. Abstr., Vol. 81, Abstr. 12445 (1978).

162.105 **On a quantized scalar field in some Bianchi-type I universe. II. DeWitt's two vacuum states connected causally.**   H. Nariai.
Prog. Theor. Phys., Vol. 58, 842 - 844 (1977). − Abstr. in Phys. Abstr., Vol. 81, Abstr. 12608 (1978).

162.106 **Homogeneous Newtonian cosmologies and their perturbations.**   P. Szekeres, J. R. Rankin.
J. Australian Math. Soc., Ser. B, Vol. 20, 114 - 128 (1977). Abstr. in Phys. Abstr., Vol. 81, Abstr. 16584 (1978).

162.107 **Effect of graviton creation in isotropically expanding universes.**   B. L. Hu, L. Parker.
Phys. Lett. A, Vol. 63A, No. 3, p. 217 - 220 (1977). − Abstr. in Phys. Abstr., Vol. 81, Abstr. 16588 (1978).

162.108 **Is the conventional big bang compatible with the discovery of the quarks?**
T. Nakamura, S. Miyama, K. Sato.
Prog. Theor. Phys., Vol. 58, 1052 - 1054 (1977). − Abstr. in Phys. Abstr., Vol. 81, Abstr. 16589 (1978).

162.109 **Eigenvalue treatment of cosmological models.**
M. Novello, I. Damiao Soares.
Phys. Lett. A, Vol. 64A, 153 - 154 (1977). − Abstr. in Phys. Abstr., Vol. 81, Abstr. 16704 (1978).

162.110 **The kinetic theory of the perturbations in the isotropic universe.**   A. V. Zakharov.
Phys. Lett. A, Vol. 64A, 167 - 168 (1977). − Abstr. in Phys. Abstr., Vol. 81, Abstr. 20920 (1978).

162.111 **The relic radiation spectrum and the thermal history of the Universe.**
L. Danese, G. De Zotti.
Nuovo Cimento Riv., Ser. 2, Vol. 7, 277 - 362 (1977). − Abstr. in Phys. Abstr., Vol. 81, Abstr. 20921 (1978).

162.112 **Galaksernes udvikling i det ekspanderende univers.**
B. Strömgren.
Astron. Tidsskr., Årg. 11, 69 - 81 (1978).

162.113 **Solutions du type Robertson-Walker en théorie unitaire.**   P. Pigeaud.
C. R. Acad. Sci. Paris, Tome 286, Sér. A, 1027 - 1030 (1978).
In the framework of interactions of the gravitational field with a scalar meson field, one finds all solutions of the Robertson-Walker type. In certain particular cases one can completely perform the calculations. In these cases the author is led to coherent cosmological models, and it is possible to evaluate the exact relation between the red-shift and the apparent magnitude of a galactic source.

162.114 **On a possible explanation for the 9 Mpc break of the covariance function.**
S. A. Bonometto, F. Lucchin.
Astron. Astrophys., Vol. 67, L7 - L9 (1978).
The existence of a correlation between the value of the exponent of the covariance function $\xi(r)$ and its sign is stressed. This is put in relation to the possible break in $\xi(r)$ at $\sim 9/H$ Mpc, while the possibility that relevant information on the primaeval fluctuation spectrum may be directly inferred from the sign of $\xi(r)$ at r $\gg 9/H$ Mpc is outlined.

162.115 **Hubble's law and the universal law of gravitation.**
D. Galletto.
Atti Accad. Sci. Torino I, Vol. 110, 335 - 341 (1976). In Italian. − Abstr. in Phys. Abstr., Vol. 81, Abstr. 24577 (1978).

162.116 **Hubble's law and the universal gravitational constant.**
D. Galletto.
Atti Accad. Sci. Torino I, Vol. 110, 401 - 404 (1976). In Italian. − Abstr. in Phys. Abstr., Vol. 81, Abstr. 24578 (1978).

162.117 **A tachyon dust universe.**   S. K. Srivastava.
J. Math. Phys., Vol. 18, 2092 - 2096 (1977). − Abstr. in Phys. Abstr., Vol. 81, Abstr. 24583 (1978).

162.118 **Global structure of the 'Kantowski-Sachs' cosmological models.**   C. B. Collins.
J. Math. Phys., Vol. 18, 2116 - 2124 (1977). −Abstr. in Phys. Abstr., Vol. 81, Abstr. 24584 (1978).

162.119 **Particular exact inhomogeneous solution with matter and pressure.**   N. Tomimura.
Nuovo Cimento B, Ser. 11, Vol. 42B, 1 - 8 (1977). − Abstr. in Phys. Abstr., Vol. 81, Abstr. 24585 (1978).

162.120 **General-relativistic kinetic theory of the Robertson-Walker and Petrov G_4 VIII cosmologies.**
J. R. Ray, J. C. Zimmermann.
Nuovo Cimento B, Ser. 11, Vol. 42B, 183 - 197 (1977). Abstr. in Phys. Abstr., Vol. 81, Abstr. 24586 (1978).

162.121 **Views of the Universe over cosmological time spans.**   J. K. Lawrence.
American J. Phys., Vol. 45, 1164 - 1167 (1977). − Abstr. in Phys. Abstr., Vol. 81, Abstr. 24598 (1978).

162.122 **Evolution of the black-body spectrum in the Friedmann universe.**   J. H. Cooke.
American J. Phys., Vol. 45, 1168 - 1172 (1977). − Abstr. in Phys. Abstr., Vol. 81, Abstr. 24599 (1978).

162.123 **On a nonsingular isotropic cosmological model.**
A. A. Starobinskij.
Pis'ma Astron. Zh., Vol. 4, 155 - 159 (1978). In Russian.
It is shown that if the main contribution to the matter energy density in an isotropic universe is made by a homogeneous massive scalar field (the Parker-Fulling model) then the possibility of the realization of a nonsingular solution, though it exists, is highly improbable at reasonable parameters of the model. Physical reality of such a kind of models is discussed.

162.124 **Relict radiation as a possible product of the quantum era in the expansion of the universe.**
G. E. Gorelik, L. M. Ozernoj.
Pis'ma Astron. Zh., Vol. 4, 160 - 164 (1978). In Russian.
The authors suppose that at the earliest stages of cosmological expansion the universe has created thermal radiation at the same rate as a black hole with mass equal to that contained within the cosmological horizon. This mechanism leads in a natural way to an initial radiation temperature of the order of the Planckian temperature ($\sim 10^{32}$ K) and provides the necessary heating up of the early universe.

162.125 **On spectral lines of prestellar origin.**
Ya. B. Zel'dovich.
Pis'ma Astron. Zh., Vol. 4, 165 - 166 (1978). In Russian.
Weak but narrow spectral features superimposed on the

relict radiation equilibrium spectrum are predicted, due to regions of prestellar gas where medium strong hydrodynamic perturbations are compensating the Hubble expansion. This situation must occur somewhat before the formation of shock waves compressing gas clouds – future protoclusters of galaxies.

162.126   Particle creation by a gravitational field and cosmological singularity.
S. G. Mamaev, V. M. Mostepanenko.
Pis'ma Astron. Zh., Vol. 4, 203 - 206 (1978). In Russian.
  The creation of fermion pairs in Friedmann cosmological models near the singularity is investigated. The expansion law $a(t) \propto t^q$ with $q \leqslant \frac{1}{2}$ is shown to be prohibited since it leads to creation of fermion pairs with infinite energy density and pressure.

162.127   Le problème cosmologique et ses hypothèses VI.
J. Dubois.
Orion, 36. Jahrg., 110 - 114 (1978).

162.128   N-body simulations and the value of Ω.
J. R. Gott III.
IAU Symp. No. 79, (see 012.027), p. 63 - 70 (1978).
  The different suitably corrected statistical virial theorem methods yield values of Ω in the range $0.05 \leqslant \Omega \leqslant 0.18$. These results are consistent with those found by the group catalogue methods and are inconsistent with $\Omega > 1$ due to matter associated with galaxies. The results are consistent with the value of $\Omega = 0.1$ implied from cosmological production of deuterium with $H_0 = 50$ km s^{-1} Mpc^{-1}.

162.129   Large scale clustering in the universe.
P. J. E. Peebles.
IAU Symp. No. 79, (see 012.027), p. 217 - 226 (1978).

162.130   The isotropy of the Universe on scales exceeding the horizon.   L. Grishchuk.
IAU Symp. No. 79, (see 012.027), p. 226 - 227 (1978).
Short communication.

162.131   Results on the large scale distribution of extragalactic objects obtained by the method of statistical reduction.   K. Rudnicki, S. Zieba.
IAU Symp. No. 79, (see 012.027), p. 229 - 239 (1978).

162.132   Has the Universe the cell structure?
M. Jôeveer, J. Einasto.
IAU Symp. No. 79, (see 012.027), p. 241 - 251 (1978).
  To get a clearer picture of the distribution of galaxies in space, the authors have studied the three-dimensional distribution of galaxies and clusters of galaxies. They have used recession velocities of galaxies and mean velocities of clusters as distance indicators supposing, following Sandage and Tammann (1975), that the expansion of the Universe is uniform.

162.133   Discussion of methods on determining the mean matter density of the Universe.
IAU Symp. No. 79, (see 012.027), p. 281 - 286 (1978).

162.134   Search for primordial perturbations of the Universe: observations with RATAN-600 Radio Telescope.
Y. N. (*Yu. N.*) Parijskij.
IAU Symp. No. 79, (see 012.027), p. 315 - 316 (1978).

162.135   The quest for fine-scale anisotropy in the relict radiation.   P. E. Boynton.
IAU Symp. No. 79, (see 012.027), p. 317 - 326 (1978).
  The author presents a summary of those elements of gravitational instability theory which are necessary to establish a meaningful comparison between theory and fine-scale anisotropy data. Currently available upper limits on temperature fluctuations are then tabulated. Finally, a possible "next-generation" fine-scale observing technique is suggested; one with adequate sensitivity to force a confrontation between theory and fact.

162.136   The epoch of galaxy formation.
B. J. T. Jones, M. J. Rees.
IAU Symp. No. 79, (see 012.027), p. 377 - 388 (1978).
  Contents: Observational evidence. The scenario of Doroshkevich et al. Non-dissipative gravitational clustering. Cosmic turbulence. Consequences of early ($z > 100$) star formation. Clusters at $z \gtrsim 2$. A "synthetic compromise" model for the formation of galaxies and their halos.

162.137   Fluctuations of the microwave background radiation.   R. A. Sunyaev (*Syunyaev*).
IAU Symp. No. 79, (see 012.027), p. 393 - 404 (1978).
  Observations of the relic radiation impose stronger and stronger limits on the amplitude of the fluctuations. The theoretical analysis presented shows that background fluctuations at centimetre and millimetre wavelength must exist at a level $\Delta T/T \approx 10^{-5}$ independent of whether there was early secondary reheating or not. At this level, fluctuations must be observed which are associated with known objects – radiogalaxies etc.

162.138   Constraints on the mean density of the Universe which follow from the theories of adiabatic and whirl perturbations.   A. A. Kurskov, L. M. Ozernoy (*Ozernoj*).
IAU Symp. No. 79, (see 012.027), p. 404 - 405 (1978).
Short communication.

162.139   Recent advances in microwave cosmology.
P. Boynton.
IAU Symp. No. 79, (see 012.027), p. 405 - 406 (1978).
Short communication.

162.140   The theory of the large scale structure of the Universe.   Ya. B. Zeldovich (*Zel'dovich*).
IAU Symp. No. 79, (see 012.027), p. 409 - 421 (1978).
  The author investigates adiabatic perturbations. He considers several phases in the development of the perturbations: 1) acoustic oscillations of the radiation-dominated plasma and their attenuation before and during recombination; 2) the growth of small perturbations in the neutral gas; 3) the non-linear growth of perturbations leading to the formation of compressed gas layers – pancakes; 4) the further fate of pancakes, the interaction of pancakes, their decay into galaxies and protoclusters of galaxies. The adiabatic perturbation spectrum possesses a definite cut-off wavelength as already pointed out by Silk. It is shown that this critical wavelength is also reflected in the cell structure of the Universe.

162.141   The internal structure of protoclusters and the formation of galaxies.
A. G. Doroshkevich, E. M. Saar, S. F. Shandarin.
IAU Symp. No. 79, (see 012.027), p. 423 - 425 (1978).
  The authors give a short review of the general picture and main features of the formation of galaxies and clusters of galaxies on the basis of the adiabatic theory.

162.142   The whirl theory of the origin of structure in the Universe.   L. M. Ozernoy (*Ozernoj*).
IAU Symp. No. 79, (see 012.027), p. 427 - 438 (1978).
  The whirl theory of the formation of structure in the Universe is an example of a consistent theory which introduces – by means of initial conditions – non-potential vortex perturbations. This whirl theory explains the rotation of galaxies and enables the author to find the spectrum of cosmological turbulence and the picture of its decay, to explain the origin and magnitude of rotational velocities, as well as

the main dynamical parameters of both galaxies and systems of galaxies, give estimates for the redshifts at which the birth of galaxies and systems of galaxies occurred – and all this by means of substantially only one single parameter which characterizes the amplitude of primeval whirls.

162.143 Origin of protogalactic eddies.   A. D. Chernin.
IAU Symp. No. 79, (see 012.027), p. 439 - 440 (1978).

162.144 Evidence for the gravitational instability picture in a dense Universe.   M. Davis.
IAU Symp. No. 79, (see 012.027), p. 441 - 443 (1978).

162.145 Primeval clusters of galaxies and the X-ray background.   E. M. Kellogg.
IAU Symp. No. 79, (see 012.027), p. 445 - 446 (1978).

162.146 Observational limits on neutral hydrogen in primordial protoclusters.   R. D. Davies.
IAU Symp. No. 79, (see 012.027), p. 447 - 448 (1978).

162.147 Orientation of spiral galaxies as a test of theories of galaxy formation.   J. Jaaniste, E. Saar.
IAU Symp. No. 79, (see 012.027), p. 448 - 449 (1978). Short communication.

162.148 Personal view – The large scale structure of the Universe.   M. S. Longair.
IAU Symp. No. 79, (see 012.027), p. 451 - 461 (1978).

162.149 General-relativistic hierarchical cosmology: an exact model.   P. S. Wesson.
Astrophys. Space Sci., Vol. 54, 489 - 495 (1978).

An exact solution of Einstein's equation is stated in which the density (ϱ), pressure (p), scale factor S and metric coefficients are functions of only one dimensionless self-similar variable, ct/R, where t is cosmic time and R is a co-moving radial coordinate. The solution represents a cosmology that begins as a static sphere having $\varrho \propto R^{-2}$ and evolves into an expanding model which is pressure-free and has a hierarchical type of density law ($\varrho \propto R^{-\theta}$, approximately, with $\theta = a$ number, $0 \leqslant \theta \leqslant 2$). It is suggested that this model should supersede the previous models of Wesson and other workers, since it appears to be the simplest cosmology for a hierarchy.

162.150 The extragalactic distance scale. I. A review of distance indicators: zero points and errors of primary indicators.   G. de Vaucouleurs.
Astrophys. J., Vol. 223, 351 - 363 (1978).

The principal primary, secondary, and tertiary extragalactic distance indicators are reviewed. The zero points and errors of the primary distance indicators, (I) novae, (II) Cepheids, and (III) RR Lyrae and HB stars, are determined. Two subsidiary indicators are considered: (IVa) AB supergiants and (IVb) eclipsing binaries.

162.151 The development of structure in the expanding universe.   J. Silk, S. D. White.
Astrophys. J., Lett., Vol. 223, L59 - L62 (1978).

The authors develop a new model for clustering in an expanding universe, based on an application of the coagulation equation to the collision and aggregation of bound condensations. While the growth rate of clustering is determined by the rate at which density fluctuations reach the nonlinear regime, and therefore depends on the initial fluctuation spectrum, the mass spectrum rapidly approaches a self-similar limiting form. This form is determined by the tidal processes which lead to the merging of condensations, and is not dependent on initial conditions.

162.152 Matter-antimatter hydrodynamics: computation of

the annihilation rate.   J. J. Aly.
Astron. Astrophys., Vol. 67, 199 - 208 (1978).

This paper is devoted to the computation of particle-antiparticle annihilation rate at the boundary between two regions of space filled respectively by matter and antimatter. A general analysis of the problem is done and a simple model, useful to compute this important quantity in all situations encountered in cosmological studies, is established. The annihilation rate is computed analytically in three specific situations: (a) in a matter-antimatter emulsion filling the universe in the radiative and plasma eras of the big-bang; (b) in a situation where there is a strong magnetic field; (c) at the boundary between a hot intergalactic medium and an interantigalactic anti-one.

162.153 The cosmogonical separation of matter and antimatter.   W. B. Thompson.
Astrophys. Space Sci., Vol. 55, 15 - 23 (1978).

Alfvén has shown that in the symmetric cosmology of O. Klein, matter and antimatter can be separated as clouds of solar masses. By considering the dynamics of these clouds it is shown that a further separation process driven by the release of rest energy can separate matter on a galactic scale.

162.154 Le paradoxe d'Olbers ou pourquoi le ciel est noir la nuit.   M. Gabriel.
Bull. Soc. Astron. Liège, Vol. 40, 63 - 65 (1978).

162.155 Long-wave perturbations of Friedmann's universe and anisotropy of relict radiation.
L. P. Grishchuk, Ya. B. Zel'dovich.
Inst. prikl. mat. AN SSSR. Prepr. No. 78. Moskva, 1977. 20 pp. Price 7 Kop. In Russian. – Abstr. in Ref. zh., 51. Astron., 5.51.747 (1978).

162.156 Density fluctuations in an expanding viscous and conducting universe (III).   R. Simon.
Bull. Acad. R. Belgique, Cl. Sci., 5e Sér., Tome 62, 707 - 713 (1976) = Inst. Astrophys., Univ. Liège, Cointe-Ougrée (Belgique), Collect. 8°, No. 672.

A numerical study of the evolution of fluctuations in an Einstein-de Sitter universe is presented. The author assumes white-noise spectra when R→O (big bang) and calculates the power spectrum of density fluctuations at decoupling and well after decoupling, both for isentropic and for isothermal initial fluctuations.

162.157 Model of an elastic expanding universe without peculiarities.   V. I. Bashkov, Yu. G. Ignat'ev.
Tr. Kazan. Gorod. Astron. Obs., No. 41, p. 41 - 45 (1976). In Russian.

162.158 Turbulence in cosmology. III. The effect of non-linear interactions on the Universe expansion law. Quantum turbulence near singularity.
A. M. Krymsky (*Krymskij*), L. S. Marochnik, P. D. Naselsky (*Nasel'skij*), N. V. Pelikhov.
Astrophys. Space Sci., Vol. 55, 299 - 324, 325 - 350 (1978). In Russian and English.

The kinetic equations for the spectra of classical and quantum short-wave turbulences have been obtained, taking account of the influence of the latter on the process of cosmological expansion of a homogeneous and, on average, isotropic Universe. It is possible that a situation exists in which the primordial short-wave turbulence, having had a significant influence on the early metric, would not be observable at the present time. Quantum turbulence has been studied.

162.159 The anisotropy in the microwave background radiation in a FIB metric.   J. M. Barnothy.
Bull. American Astron. Soc., Vol. 10, 396 - 397 (1978).

Abstract.

162.160 **The Hubble law and expansion of the universe.**
K. Brecher, J. Grunsfeld.
Bull. American Astron. Soc., Vol. 10, 429 (1978). – Abstract.

162.161 **The redshift distribution of quasi-stellar objects as an indicator of clustering on gigaparsec scale.**
A. Prakash.
Bull. American Astron. Soc., Vol. 10, 429 (1978). – Abstract.

162.162 **Properties of causally continuous closed universes.**
H. Ishikawa.
J. Math. Phys., Vol. 18, 2375 - 2376 (1977). – Abstr. in Phys. Abstr., Vol. 81, Abstr. 29011 (1978).

162.163 **Cosmological universes with spherical symmetry.**
S. Ram, J. N. S. Kashyap.
Acta Phys. Acad. Sci. Hungaricae, Vol. 42, 245 - 250 (1977). Abstr. in Phys. Abstr., Vol. 81, Abstr. 32615 (1978).

162.164 **Assumptions of the singularity theorems and the rejuvenation of universes.**   J. A. Donald.
Ann. Phys., Vol. 110, 251 - 273 (1978). – Abstr. in Phys. Abstr., Vol. 81, Abstr. 32616 (1978).

162.165 **Cosmological models with a spinor field.**
J. Tafel.
Bull. Acad. Polonaise Sci., Ser. Sci. Math. Astron. Phys., Vol. 25, 593 - 596 (1977). – Abstr. in Phys. Abstr., Vol. 81, Abstr. 32618 (1978).

162.166 **Curvature collineations in some cosmological models.**   K. P. Singh, S. Ram.
Indian J. Pure Appl. Math., Vol. 6, 1023 - 1030 (1975). Abstr. in Phys. Abstr., Vol. 81, Abstr. 32620 (1978).

162.167 **Effective potential for equatorial motion in the Tomimatsu-Sato space-times.**   M. Calvani.
Nuovo Cimento, Lett., Ser. 2, Vol. 21, 115 - 118 (1978). Abstr. in Phys. Abstr., Vol. 81, Abstr. 32621 (1978).

162.168 **Dynamics in the Gödel universe.**
J. Lathrop, R. Teglas.
Nuovo Cimento B, Ser. 11, Vol. 43B, 162 - 172 (1978). Abstr. in Phys. Abstr., Vol. 81, Abstr. 32622 (1978).

162.169 **Entropy generation in the very early universe.**
E. P. T. Liang.
Phys. Rev. D, Vol. 16, No. 12, p. 3369 - 3375 (1977). – Abstr. in Phys. Abstr., Vol. 81, Abstr. 32624 (1978).

162.170 **Hubble's constant from Sc-galaxies and quasars.**
S. V. Vereshchagin, V. V. Syrovoj.
Astron. Tsirk., No. 970, p. 5 - 6 (1977). In Russian.

162.171 **Expanding shearfree spatially homogeneous universes with a nonsynchronous time coordinate and anisotropy of the universe.**   A. J. Fennelly.
J. Math. Phys., Vol. 19, 158 - 163 (1978). – Abstr. in Phys. Abstr., Vol. 81, Abstr. 36740 (1978).

162.172 **Note on the Bertotti-Robinson electromagnetic universe.**   N. Tariq, R. G. McLenaghan.
J. Math. Phys., Vol. 19, 349 - 351 (1978). – Abstr. in Phys. Abstr., Vol. 81, Abstr. 36741 (1978).

162.173 **Angular fluctuations of relict radiation produced by cosmological turbulence.**   A. A. Kurskov, L. M. Ozernoj.
Astrophys. Space Sci., Vol. 56, 51 - 65, 67 - 80 (1978). In Russian and English.

On the basis of a theory for the evolution of cosmological turbulence developed earlier by the present authors, the temperature fluctuations of the relict black-body radiation produced by turbulent motions are calculated. A comparison of the calculations with the available measurements (upper limits) for temperature fluctuations of relict radiation makes it possible to obtain important upper and lower bounds for the initial velocity of the vortex motions.

162.174 **Is there evidence for anisotropy in the Hubble expansion?**   V. C. Rubin.
Ann. New York Acad. Sci., Vol. 302, (see 012.033), 408 - 422 (1977).

Following the initial work of Rubin, Ford, and Rubin that suggested a non-isotropic distribution in the velocities and magnitudes of faint Sc I galaxies, there have been numerous papers detecting, supporting, questioning, dismissing, and refuting such an effect for a variety of extragalactic objects. The author discusses her recent results, and then a selection from the 25 or 30 current papers that examine extragalactic samples for anisotropies.

162.175 **The cosmological constant and cosmological change.**   B. M. Tinsley.
Ann. New York Acad. Sci., Vol. 302, (see 012.033), 423 - 438 (1977).

The author sketches here the current status of the Friedmann models in the light of new developments, especially (1) the recognition of a whole new class of evolutionary corrections to properties of distant galaxies, and (2) a proposed reinstatement of Einstein's lamentable and disinherited cosmological constant.

162.176 **Graviton creation in the early universe.**   L. P. Grishchuk.
Ann. New York Acad. Sci., Vol. 302, (see 012.033), 439 - 444 (1977).

162.177 **Unified treatment of Einstein's and Gödel's universes.**   L. K. Patel.
Curr. Sci., Vol. 46, 808 (1977). – From Phys. Abstr., Vol. 81, Abstr. 41101 (1978).

162.178 **Antiproton instability and symmetric big-bang cosmologies.**   J. Demaret, J. Vandermeulen.
Phys. Lett. B, Vol. 73B, 471 - 474 (1978). – Abstr. in Phys. Abstr., Vol. 81, Abstr. 41104 (1978).

162.179 **Finite-temperature effects in a Robertson-Walker universe.**   G. Kennedy.
J. Phys. A, Vol. 11, L77 - L81 (1978). – Abstr. in Phys. Abstr., Vol. 81, Abstr. 45440 (1978).

162.180 **Mass changing cosmology and solar evolution.**   V. N. Mansfield.
Nuovo Cimento, Lett., Ser. 2, Vol. 21, 390 - 394 (1978). Abstr. in Phys. Abstr., Vol. 81, Abstr. 45441 (1978).

162.181 **Cosmological constraints on the mass and the number of heavy lepton neutrinos.**   K. Sato, M. Kobayashi.
Prog. Theor. Phys., Vol. 58, 1775 - 1789 (1977). – Abstr. in Phys. Abstr., Vol. 81, Abstr. 45444 (1978).

162.182 **On the creation of scalar particles in an isotropic universe.**   H. Nariai, A. Tomimatsu.
Prog. Theor. Phys., Vol. 59, 296 - 298 (1978). – Abstr. in Phys. Abstr., Vol. 81, Abstr. 45445 (1978).

162.183 **Evolutionary processes in the universe.**   V. A. Ambartsumyan.
250 let AN SSSR. Dokl. i mater. yubil. torzhestv. Moskva,

1977, p. 229 - 239. In Russian. – Abstr. in Ref. zh., 51. Astron., 6.51.2 (1978).

162.184 **Quantum gravitational effects in an anisotropic universe.** G. M. Vereshkov, Yu. S. Grishkan, S. V. Ivanov, V. A. Nesterenko, A. N. Poltavtsev. Zh. ehksperim. i teor. fiz., Vol. 73, 1985 - 2007 (1977). In Russian. – Abstr. in Ref. zh., 51. Astron., 6.51.945 (1978).

162.185 **The universe and deuterium.** Z. Mikulášek. Říše hvězd, Vol. 59, 3 - 5 (1978). In Czech.

162.186 **Some properties of a cosmological solution in the Brans-Dicke theory.** A. Miyazaki. Phys. Rev. Lett., Vol. 40, 725 - 727, with a correction p. 1055 (1978).
Properties of a closed cosmological solution with the negative coupling parameter of the scalar field compatible with Mach's principle in the Brans-Dicke theory are summarized.

162.187 **Production of entropy and viscous damping of anisotropy in homogeneous cosmological models: Bianchi Type-I spaces.** N. Caderni, R. Fabbri. Nuovo Cimento B, Ser. 11, Vol. 44B, 228 - 240 (1978). Abstr. in Phys. Abstr., Vol. 81, Abstr. 49572 (1978).

162.188 **Evolution of inhomogeneous plane-symmetric cosmological models.** N. Tomimura. Nuovo Cimento B, Ser. 11, Vol. 44B, 372 - 380 (1978). Abstr. in Phys. Abstr., Vol. 81, Abstr. 49733 (1978).

162.189 **Secondary viscosity effects in cosmology.** A. S. Potupa. Izv. VUZ. Fizika, 1978, No. 1, p. 92 - 97. In Russian. – Abstr. in Ref. zh., 51. Astron., 7.51.962 (1978).

The collapsing universe. See Abstr. 003.022.

What is the world made of? See Abstr. 003.045.

The creation of the universe. See Abstr. 003.049.

Mathematical cosmology, an introduction. See Abstr. 003.080.

The universe. Its beginning and end. See Abstr. 003.100.

Cosmology. See Abstr. 003.117.

The luminosity distance equation in Friedmann cosmology. See Abstr. 014.020.

Scalar-particle production near the singularity in an anisotropic universe. II. Mass creation by gravitational field. See Abstr. 061.009.

Limits on the mass of the muon neutrino in the absence of muon-lepton-number conservation. See Abstr. 061.042.

Implications of a nonzero neutrino mass for the process $\gamma\gamma \rightarrow \nu\bar{\nu}$. See Abstr. 061.043.

General scalar-tensor theory of gravity with constant G. See Abstr. 066.002.

Self-similar growth of primordial black holes. I. Stiff equation of state. See Abstr. 066.006.

Is physics legislated by cosmogony? See Abstr. 066.043.

Quantum fluctuations in gravitational collapse and cosmology. See Abstr. 066.051.

The gauge in general relativity. II. See Abstr. 066.054.

The gauge in general relativity. III. See Abstr. 066.055.

Polarized radiation in relativistic cosmology. See Abstr. 066.057.

Self-similar space–times. I: Three solutions. See Abstr. 066.071.

Self-similar space–times. II: Perturbation scheme. See Abstr. 066.072.

Bimetric gravitation theory on a cosmological basis. See Abstr. 066.073.

Quantum mechanics of electromagnetically bounded spin-$^1/_2$ **particles in an expanding universe: I. Influence of the expansion.** See Abstr. 066.077.

On the induced cosmological term in quantum gravity. See Abstr. 066.088.

Metrical connection in space-time, Newton's and Hubble's laws. See Abstr. 066.093.

Poincaré gauge theory with Weyl-type Lagrangian and gravitation. See Abstr. 066.101.

On a quantized scalar field in a Bianchi-type I universe. See Abstr. 066.104.

Strong gravity with a cosmological constant. See Abstr. 066.107.

Distortions of the cosmic radiation spectrum in baryon-symmetric cosmologies. See Abstr. 066.118.

Space-time singularities. See Abstr. 066.129.

Maximal surfaces in closed and open spacetimes. See Abstr. 066.130.

Positive energy in U_4 **theory.** See Abstr. 066.132.

Inertia and gravity in general relativity. See Abstr. 066.136.

Particle creation and geometry. See Abstr. 066.167.

Covariant treatment of particle creation in curved space-time. See Abstr. 066.168.

Cosmological solution of Einstein's equations in general relativity. See Abstr. 066.169.

Inversion of Limber's relativistic formula. See Abstr. 066.199.

Strong gravity and Dirac numbers. See Abstr. 066.211.

Ray aberration and large-scale anisotropy of the cosmic background radiation.   See Abstr. 066.212.

Transmutation of photons, electrons and gravitons and the density of gravitons in the universe. See Abstr. 066.232.

Calculation of the renormalised quantum stress tensor by adiabatic regularisation in two- and four-dimensional Robertson-Walker space-times.   See Abstr. 066.266.

Quantum field theory in curved space-time: an overview.   See Abstr. 066.274.

Homogeneous and isotropic world models in the Yang-Mills dynamics of gravity. The structure of the adiabats. See Abstr. 066.314.

Shock waves on Earth and in space. See Abstr. 082.059.

Deuterium in the Galaxy.   See Abstr. 131.126.

The angular diameter—redshift test for quasi-stellar radio sources with large redshifts.   See Abstr. 141.003.

New observations of the radio absorption line in 3C 286, with potential application to the direct measurement of cosmological deceleration.   See Abstr. 141.007.

Evidence for nonuniform radial distribution of quasars, regardless of the nature of their redshifts. See Abstr. 141.036.

Free-form analysis of the cosmological evolution of radio sources.   See Abstr. 141.050.

Hubble plot and space density of quasars in a FIB metric.   See Abstr. 141.074.

Spectroscopy and the Hubble diagram of faint radio galaxies.   See Abstr. 141.090.

Hubble diagram of quasars.   See Abstr. 141.091.

Super-light motion in radio sources and its implications for the distance scale problem.   See Abstr. 141.092.

21 cm observations of 3C 286. See Abstr. 141.093.

The Hubble diagrams for quasars. See Abstr. 141.140.

Studies of new complete samples of quasi-stellar radio sources from the 4C and Parkes catalogs. See Abstr. 141.153.

On the Hubble diagrams of quasars and the shapes of their optical luminosity functions derived by a new method.   See Abstr. 141.159.

Quasars and cosmological evolution. See Abstr. 141.186.

The large scale distribution of radio sources. See Abstr. 141.189.

Fuentes de rayos X extragalácticas. See Abstr. 142.136.

Cosmological information from X-ray observations. See Abstr. 142.186.

Galactic evolution and the interpretation of cosmological tests.   See Abstr. 151.037.

On the origin of proto-galactic eddies. See Abstr. 151.050.

The evolution of galaxies: evidence from optical observations.   See Abstr. 151.085.

The problem of formation of galaxies and the "photon whirlwind" hypothesis.   See Abstr. 151.093.

A study of galactic absorption as revealed by the Rubin et al. sample of Sc galaxies.   See Abstr. 155.016.

Kron-Shane corrections and the faint-end anomaly in Zwicky galaxy counts.   See Abstr. 158.075.

The Hubble diagram for Seyfert galaxies and related objects.   See Abstr. 158.107.

The absolute solar motion and the discrete redshift. See Abstr. 158.113.

Doppler-shift distributions. II. The relativistic gas.   See Abstr. 158.157.

On the Magellanic Stream, the mass of the Galaxy and the age of the Universe.   See Abstr. 159.013.

A mean density and a correlation function of rich clusters: theory and observations.   See Abstr. 160.002.

The virial mass discrepancy in groups and clusters of galaxies.   See Abstr. 160.007.

The origin and distribution of gas within rich clusters of galaxies: the evolution of cluster X-ray sources over cosmological time scales.   See Abstr. 160.009.

The characteristic size of clusters of galaxies: a metric rod used for a determination of q_0. See Abstr. 160.014.

Redshift-magnitude bands in clusters of galaxies. See Abstr. 160.034.

Distances of nearby groups and clusters, and the local value of the Hubble ratio.   See Abstr. 160.037.

Formation times for rich clusters of galaxies. See Abstr. 160.041.

Inhomogeneities in the universe: an observed power law relating mass fraction to density enhancement. See Abstr. 160.056.

Cosmology II: metrical connection and clusters of galaxies.   See Abstr. 160.063.

The galactic neighbourhood. See Abstr. 160.072.

The Local Supercluster.   See Abstr. 160.085.

The physical properties of large scale systems from optical observations.   See Abstr. 160.087.

Author Index

The authors are listed in alphabetical order

according to the initial letter following the first names.

Epstein, R. I.
065.040
Epstein, S.
081.022
Erdos, G.
143.064 .066
Eremeev, A. G.
091.044
Eremeev, V. F.
081.040
Erickson, E.
099.510
Erickson, E. F.
064.042
093.027
114.534 .559 .575
122.001
Erickson, N. R.
131.228
Erickson, W. C.
141.165
Eriksson, K. B. S.
022.060
Eriksson, K.-E.
066.053
Erkaev, N. V.
062.116
074.010
Erkes, J. W.
031.237
141.519
Erman, P.
022.048
Ermolaev, Yu. F.
032.540
Ernstson, K.
105.085
Eroschenko, E. A.
See Eroshenko, E. A.
Eroshenko, E. A.
094.102
Eroshenko, E. G.
094.416 .447
Erpylev, N. P.
046.030
Eryushev, N. N.
080.036
Erzhanov, Zh. S.
044.020
Eschrich, K. O.
022.076
Esepkina, N. A.
033.012
Eshleman, R.
091.059
Eshleman, V. R.
008.076 .112
Esipov, V. F.
158.031
Espange, H.
078.009
Espenak, F.
079.303
Esposito, F. P.
066.297
Esposito, L. W.
063.004
100.014 .015 .017 .029
Esposito, P. B.
074.064
092.008
Esser, U.
002.059
Essex, E. A.
083.029 .041 .091
Estabrook, F. B.
066.315
141.218
Etcheto, J.
084.240

Evangelidis, E.
066.273
151.090
Evans, A.
014.017
Evans, A. B.
162.083
Evans, D. A.
003.080
Evans, D. S.
007.000
008.009 .042
084.033
113.046
115.009
117.036
118.022
122.211 .212
126.020
Evans, J. C.
031.215
Evans, J. V.
083.084
Evans, K. D.
074.031
Evans, N.
093.021
Evans, R. G.
114.053
Evans, T. L.
154.002
159.008 .009
Evans, W. D.
142.707
Evans II, N. J.
131.021 .081 .104 .105
       .111 .901
134.002
Evans Jr., J. C.
080.071
Evdokimov, I. Yu.
102.032
Evdokimov, Yu. V.
104.032
Evdokimova, L. S.
042.054
Evdokimova, L. V.
106.039
Everhart, E.
102.001
Everitt, C. W. F.
066.089 .141 .320
Eviatar, A.
099.059
Evseev, O. A.
034.054
113.082
142.226
Evsyukov, N. N.
094.437
Evteev, V. P.
042.093
Ewing, M.
094.112
Ewing, M. S.
141.033 .511
Eyerly, H. W.
032.550
Eyni, M.
074.021
Ezema, P. O.
084.231

Fabbiano, G.
032.507
142.056 .098
Fabbri, R.
066.212 .224
162.075 .187

Faber, S. M.
158.004 .027
Fabian, A.
131.031
Fabian, A. C.
064.070
124.106
142.702
Fabrianesi, R.
113.005
Fabricant, D.
142.067 .068
160.028
Fadeev, Yu. A.
122.014
Faelthammar, C.-G.
006.000
084.261 .299
Faenov, A. Ya.
022.001
071.021
Faer, Yu. N.
083.013 .050 .078
Fahr, H. J.
106.026
131.173
Fahrbach, U.
132.042
Fainberg, E. B.
See Fajnberg, Eh. B.
Fairall, A. P.
158.016 .090 .114
160.049
Fairchild Jr., E. E.
066.220
Faix, L. J.
032.030
Fajnberg, Eh. B.
093.034
094.102 .115 .447
Falciani, R.
071.033
Falgarone, E.
131.137
Falin, J. F.
082.039
Falk, S. W.
125.030 .060 .062
Fall, S. M.
031.214
151.022
158.100
Faller, E.
094.022
Fanale, F. P.
097.034 .070 .155 .172
099.526
Fang, L.-Z.
121.096
141.038
Fanti, C.
141.193
Faraggiana, R.
114.527
Farfal, F.
031.025
Faridi, A. M.
066.246
Farinella, P.
107.029
121.049
Farless, D.
097.502
Farley, T. A.
084.410
Fassio-Canuto, L.
061.053
Fastie, W. G.
131.185
142.103

Marcocci, M.
121.001
Marcos, F. A.
082.037 .038
Marcotte, L. P.
106.059
Marcus, P. S.
062.035
Mardanov, A.
042.140
Mardirossian, F.
121.002 .031 .032 .052
Marec, J. P.
052.024
Marek, J.
022.008
Margolis, J. S.
022.020
Margolis, S. H.
061.023 .031
131.232
Margon, B.
114.526 .537
116.015
126.037
141.162
142.019 .029 .120 .128
Margoni, R.
121.022
Mariani, F.
012.031
Marilli, E.
122.148
Mar'in, B. V.
032.529
Marinbakh, A. V.
014.043
Marioge, J. P.
031.025
Marionni, P. A.
135.102
Maris, G.
075.002
Mariska, J. T.
031.263
076.033
Mark, J. W-K.
151.040 .125
Markarian, B. E.
158.129 .136
Markaryan, E. G.
081.077
Markellos, V. V.
022.102
042.011 .077 .106
Markelov, L. A.
103.651
Markelova, T. N.
073.066
Markert, T.
032.574
142.710
Markey, J. K.
082.115
Markov, M. N.
094.462
Markova, E.
072.029
Markova, L. T.
103.273
Marks, A.
004.033
Markson, R.
085.009
Markus, S.
061.054
Marlborough, J. M.
064.019
Marocchi, D.
080.012

Marochnik, L. S.
061.006
162.158
Marov, M. Ya.
051.022
093.016 .053
Marques Dos Santos, P.
158.154
Marsakov, V. A.
114.089
155.033 .056
Marscher, A. P.
125.018 .019
131.068
141.019
Marsden, B. G.
010.002
098.093
101.004
102.001
103.002 .003
Marsh, J. C. D.
003.096
Marsh, J. G.
052.027 .033
081.032
Marsh, K. A.
065.059
073.057 .075 .083
Marshall, F.
142.076 .089 .207
Marshall, F. E.
142.213
Marshall, H.
142.208
Marshall, H. L.
142.028 .110 .111
Martelli, G.
084.240
Martellini, M.
066.087
Marten, A.
099.029
Marten, M.
003.088
Marti, K.
094.472
105.029
Martin, A. R.
015.025 .027 .028 .029
117.027 .028
Martin, B.
126.029
Martin, E.
084.021
Martin, F.
082.104
Martin, L. J.
097.026
098.020
Martin, L. S.
100.029
Martin, N.
159.002
Martin, P. G.
158.251
Martin, P. M.
105.100
Martin, R. N.
131.120 .167
133.017
Martin, S. F.
072.055
073.089
Martin, T. Z.
097.020
Martin, W. L.
074.064
114.513

Martinek, B.
105.029
Martinet, L.
012.032
151.054 .070
155.009 .061 .063
Martinez-Benjamin, J. J.
042.072
Martres, M.-J.
075.015
Martynenko, V. V.
010.037
Martynov, A. I.
052.028
Martynov, D. Ya.
021.016
098.057
Martynov, Yu. A.
082.016
Marvin, U. B.
094.402
Masevich, A. G.
012.001
013.032
046.015 .017 .030
Mashhoon, B.
066.228
Mason, H. E.
074.065
Mason, K.
125.059
Mason, K. O.
113.057
142.040 .077 .079 .082
.149
Mason, M. L.
097.526
Massa, D.
021.012
Massaguer Navarro, J. M.
065.074
Massaro, E.
141.539
142.732
Massevitch, A. G.
See Masevich, A. G.
Massevitsch, A.
See Masevich, A. G.
Massey, J. S.
014.026
Masson, G.
032.569
Masson, P.
097.110
Mastalka, A.
094.413
Masters, A. R.
126.015
Mastronardi, R.
032.556
Masuda, F.
031.041
Masursky, H.
097.169
Matas, V.
042.118
Materne, J.
160.012 .073
Mathelin, J.-P.
022.037
Mather, J. C.
031.319
Matheson, D. N.
131.146 .176
Mathews, J. D.
063.013
Mathews, W. G.
135.075
142.015
158.009

Subject Index

Starting with Volume 18 of Astronomy and Astrophysics Abstracts, some alterations concerning formation, arrangement, and versatility of the key words have been made. In order to provide an adequate description of a paper, specific key words are used as frequently as possible. References to a whole subject category are suppressed now. The user, therefore, has to refer to the contents at the beginning of each volume.

Whenever possible, the key words are formed in such a way that there are two different supplementary terms, e.g. the pair

interstellar matter
molecules.

An effort is made to choose preferably terms which can be inverted in order to increase the usefulness of this index. In the example given there are the two entries

interstellar matter
molecules

and

molecules
interstellar matter.

Exceptions to the rule of inversion of terms are given in all cases where the second key word is either a very specific one (e.g. Urca processes) or a general one (e.g. history). The use of substantives is preferred. In order to obtain the possibility to extend a one-term key word in a two-term one, combinations as

| | Mars | | sun |
| | atmosphere | or | active regions |

are changed into

| Mars atmosphere | and | solar active regions, |

respectively. The number of cross references indicating such slightly different entries is reduced drastically. In previous volumes combinations like

close binaries	and	binaries
		close binaries,
peculiar A stars	and	A stars
		peculiar,
groups of galaxies	and	galaxies
		groups

have been used. Now only the specific key words

close binaries
peculiar A stars
groups of galaxies

have to be considered as a substitute.

The user is requested to look for more specialized entries, as further references to this topic might exist elsewhere in the index under another current astronomical term.

ASTRONOMY AND ASTROPHYSICS ABSTRACTS

A Publication of the Astronomisches Rechen-Institut Heidelberg
Member of the Abstracting Board
of the International Council of Scientific Unions

Editors: S. Böhme, U. Esser, W. Fricke, I. Heinrich, D. Krahn,
L. D. Schmadel, G. Zech

Volume 1	Literature 1969, Part 1, X + 435 pp. (1969)
Volume 2	Literature 1969, Part 2, X + 516 pp. (1970)
Volume 3	Literature 1970, Part 1, X + 490 pp. (1970)
Volume 4	Literature 1970, Part 2, X + 562 pp. (1971)
Volume 5	Literature 1971, Part 1, X + 505 pp. (1971)
Volume 6	Literature 1971, Part 2, X + 560 pp. (1972)
Volume 7	Literature 1972, Part 1, X + 526 pp. (1972)
Volume 8	Literature 1972, Part 2, X + 594 pp. (1973)
Volume 9	Literature 1973, Part 1, X + 610 pp. (1973)
Volume 10	Literature 1973, Part 2, X + 661 pp. (1974)
Volume 11	Literature 1974, Part 1, X + 579 pp. (1974)
Volume 12	Literature 1974, Part 2, X + 699 pp. (1975)
Volume 13	Literature 1975, Part 1, X + 632 pp. (1975)
Volume 14	Literature 1975, Part 2, X + 747 pp. (1976)
Volume 15/16	Author and Subject Indexes to Volumes 1-10 Literature 1969–1973, VII + 655 pp. (1976)
Volume 17	Literature 1976, Part 1, XII + 645 pp. (1976)
Volume 18	Literature 1976, Part 2, X + 859 pp. (1977)
Volume 19	Literature 1977, Part 1, X + 732 pp. (1977)
Volume 20	Literature 1977, Part 2, X + 786 pp. (1978)
Volume 21	Literature 1978, Part 1, X + 834 pp. (1978)

Published for the Astronomisches Rechen-Institut by
Springer-Verlag Berlin Heidelberg New York

Astronomy and Astrophysics

A European Journal

Recognized as a "Europhysics Journal" by the European
Physical Society
ISSN 0004-6361
Title No. 230

Astronomy & Astrophysics occupies an eminent position
among journals in its field. Established in 1969, it is the result
of the merging of six renowned European journals in astronomy
and astrophysics.

Astronomy & Astrophysics presents papers on all aspects of
astronomy and astrophysics – theoretical, observational, and
instrumental – regardless of the techniques employed – optical,
radio, particles, space vehicles, numerical analysis, etc. Letters
to the editor, research notes and occasional review papers are
also included.

Astronomy & Astrophysics is edited by an international staff
of scientists.

Editors-in-Chief: J. Heidmann, Meudon, France and
H.H. Voigt, Göttingen, FRG

Springer-Verlag
Berlin
Heidelberg
New York

**Published by Springer-Verlag Berlin Heidelberg New York
on behalf of the Board of Directors European Southern
Observatory (ESO)**

Subscription information and sample copies upon request